## VECTOR IDENTITIES

If $\mathbf{u} = u_1\mathbf{i} + u_2\mathbf{j} + u_3\mathbf{k}$     then     (dot product)   $\mathbf{u} \bullet \mathbf{v} = u_1v_1 + u_2v_2 + u_3v_3$

$\mathbf{v} = v_1\mathbf{i} + v_2\mathbf{j} + v_3\mathbf{k}$

$\mathbf{u} = w_1\mathbf{i} + w_2\mathbf{j} + w_3\mathbf{k}$     (cross product)   $\mathbf{u} \times \mathbf{v} = \begin{vmatrix} \mathbf{i} & \mathbf{j} & \mathbf{k} \\ u_1 & u_2 & u_3 \\ v_1 & v_2 & v_3 \end{vmatrix} = (u_2v_3 - u_3v_2)\mathbf{i} + (u_3v_1 - u_1v_3)\mathbf{j} + (u_1v_2 - u_2v_1)\mathbf{k}$

length of $\mathbf{u} = |\mathbf{u}| = \sqrt{\mathbf{u} \bullet \mathbf{u}} = \sqrt{u_1^2 + u_2^2 + u_3^2}$        angle between $\mathbf{u}$ and $\mathbf{v} = \cos^{-1}\left(\dfrac{\mathbf{u} \bullet \mathbf{v}}{|\mathbf{u}||\mathbf{v}|}\right)$

triple product identities        $\mathbf{u} \bullet (\mathbf{v} \times \mathbf{w}) = \mathbf{v} \bullet (\mathbf{w} \times \mathbf{u}) = \mathbf{w} \bullet (\mathbf{u} \times \mathbf{v})$        $\mathbf{u} \times (\mathbf{v} \times \mathbf{w}) = (\mathbf{u} \bullet \mathbf{w})\mathbf{v} - (\mathbf{u} \bullet \mathbf{v})\mathbf{w}$

## IDENTITIES INVOLVING GRADIENT, DIVERGENCE, CURL, AND LAPLACIAN

(del operator) $\nabla = \mathbf{i}\dfrac{\partial}{\partial x} + \mathbf{j}\dfrac{\partial}{\partial y} + \mathbf{k}\dfrac{\partial}{\partial z}$        $\text{curl } \mathbf{F}(x, y, z) = \nabla \times \mathbf{F}(x, y, z) = \begin{vmatrix} \mathbf{i} & \mathbf{j} & \mathbf{k} \\ \dfrac{\partial}{\partial x} & \dfrac{\partial}{\partial y} & \dfrac{\partial}{\partial z} \\ F_1 & F_2 & F_3 \end{vmatrix}$

$\text{grad } \phi(x, y, z) = \nabla\phi(x, y, z) = \dfrac{\partial\phi}{\partial x}\mathbf{i} + \dfrac{\partial\phi}{\partial y}\mathbf{j} + \dfrac{\partial\phi}{\partial z}\mathbf{k}$

$\text{div } \mathbf{F}(x, y, z) = \nabla \bullet \mathbf{F}(x, y, z) = \dfrac{\partial F_1}{\partial x} + \dfrac{\partial F_2}{\partial y} + \dfrac{\partial F_3}{\partial z},$        $= \left(\dfrac{\partial F_3}{\partial y} - \dfrac{\partial F_2}{\partial z}\right)\mathbf{i} + \left(\dfrac{\partial F_1}{\partial z} - \dfrac{\partial F_3}{\partial x}\right)\mathbf{j} + \left(\dfrac{\partial F_2}{\partial x} - \dfrac{\partial F_1}{\partial y}\right)\mathbf{k}$

$\nabla^2\phi(x, y, z) = \nabla \bullet \nabla\phi(x, y, z) = \text{div grad } \phi = \dfrac{\partial^2\phi}{\partial x^2} + \dfrac{\partial^2\phi}{\partial y^2} + \dfrac{\partial^2\phi}{\partial z^2}$

$\nabla(\phi) = \phi\nabla + \nabla\phi$        $\nabla \bullet (\phi\mathbf{F}) = (\nabla\phi) \bullet \mathbf{F} + \phi(\nabla \bullet \mathbf{F})$

$\nabla \times (\phi\mathbf{F}) = (\nabla\phi) \times \mathbf{F} + \phi(\nabla \times \mathbf{F})$        $\nabla \bullet (\mathbf{F} \times \mathbf{G}) = (\nabla \times \mathbf{F}) \bullet \mathbf{G} - \mathbf{F} \bullet (\nabla \times \mathbf{G})$

$\nabla \times (\mathbf{F} \times \mathbf{G}) = (\nabla \bullet \mathbf{G})\mathbf{F} + (\mathbf{G} \bullet \nabla)\mathbf{F} - (\nabla \bullet \mathbf{F})\mathbf{G} - (\mathbf{F} \bullet \nabla)\mathbf{G}$        $\nabla(\mathbf{F} \bullet \mathbf{G}) = \mathbf{F} \times (\nabla \times \mathbf{G}) + \mathbf{G} \times (\nabla \times \mathbf{F}) + (\mathbf{F} \bullet \nabla)\mathbf{G} + (\mathbf{G} \bullet \nabla)\mathbf{F}$

$\nabla \bullet (\nabla \times \mathbf{F}) = 0$    **(div curl** $= 0$)        $\nabla \times (\nabla\phi) = 0$    **(curl grad** $= 0$)

$\nabla \times (\nabla \times \mathbf{F}) = \nabla(\nabla \bullet \mathbf{F}) - \nabla^2\mathbf{F}$    **(curl curl** $=$ **grad div** $-$ Laplacian)

## VERSIONS OF THE FUNDAMENTAL THEOREM OF CALCULUS

$\displaystyle\int_a^b f'(t)\,dt = f(b) - f(a)$     (the one-dimensional **Fundamental Theorem**)

$\displaystyle\int_C \text{grad } \phi \bullet d\mathbf{r} = \phi\big(\mathbf{r}(b)\big) - \phi\big(\mathbf{r}(a)\big)$ if $C$ is the curve $\mathbf{r} = \mathbf{r}(t)$, $(a \le t \le b)$.

$\displaystyle\oint_C \mathbf{F} \bullet d\mathbf{r} = \oint_C F_1(x, y)\,dx + F_2(x, y)\,dy = \iint_R \left(\dfrac{\partial F_2}{\partial x} - \dfrac{\partial F_1}{\partial y}\right) dA$ where $C$ is the positively oriented boundary of $R$   **(Green's Theorem)**

$\displaystyle\oint_C \mathbf{F} \bullet d\mathbf{r} = \oint_C F_1(x, y, z)\,dx + F_2(x, y, z)\,dy + F_3(x, y, z)\,dz = \iint_S \text{curl } \mathbf{F} \bullet \hat{\mathbf{N}}\,dS$ where $C$ is the oriented boundary of $S$.   **(Stokes's Theorem)**

Three-dimensional versions: $S$ is the closed boundary of $D$, with outward normal $\hat{\mathbf{N}}$

$\displaystyle\iiint_D \text{div } \mathbf{F}\,dV = \oiint_S \mathbf{F} \bullet \hat{\mathbf{N}}\,dS$    **Divergence Theorem**        $\displaystyle\iiint_D \text{curl } \mathbf{F}\,dV = -\oiint_S \mathbf{F} \times \hat{\mathbf{N}}\,dS$

$\displaystyle\iiint_D \text{grad } \phi\,dV = \oiint_S \phi\hat{\mathbf{N}}\,dS$

## FORMULAS RELATING TO CURVES IN 3-SPACE

Curve: $\mathbf{r} = \mathbf{r}(t) = x(t)\mathbf{i} + y(t)\mathbf{j} + z(t)\mathbf{k}$    Velocity: $\mathbf{v} = \dfrac{d\mathbf{r}}{dt} = v\hat{\mathbf{T}}$    Speed: $v = |\mathbf{v}| = \dfrac{ds}{dt}$

Arc length: $s = \displaystyle\int_{t_0}^t v\,dt$    Acceleration: $\mathbf{a} = \dfrac{d\mathbf{v}}{dt} = \dfrac{d^2\mathbf{r}}{dt^2}$    Tangential and normal components: $\mathbf{a} = \dfrac{dv}{dt}\hat{\mathbf{T}} + v^2\kappa\hat{\mathbf{N}}$

Unit tangent: $\hat{\mathbf{T}} = \dfrac{\mathbf{v}}{v}$    Binormal: $\hat{\mathbf{B}} = \dfrac{\mathbf{v} \times \mathbf{a}}{|\mathbf{v} \times \mathbf{a}|}$    Normal: $\hat{\mathbf{N}} = \hat{\mathbf{B}} \times \hat{\mathbf{T}} = \dfrac{d\hat{\mathbf{T}}/dt}{|d\hat{\mathbf{T}}/dt|}$

Curvature: $\kappa = \dfrac{|\mathbf{v} \times \mathbf{a}|}{v^3}$    Radius of curvature: $\rho = \dfrac{1}{\kappa}$    Torsion: $\tau = \dfrac{(\mathbf{v} \times \mathbf{a}) \bullet (d\mathbf{a}/dt)}{|\mathbf{v} \times \mathbf{a}|^2}$

The Frenet-Serret formulas:   $\dfrac{d\hat{\mathbf{T}}}{ds} = \kappa\hat{\mathbf{N}},$    $\dfrac{d\hat{\mathbf{N}}}{ds} = -\kappa\hat{\mathbf{T}} + \tau\hat{\mathbf{B}},$    $\dfrac{d\hat{\mathbf{B}}}{ds} = -\tau\hat{\mathbf{N}}$

## ORTHOGONAL CURVILINEAR COORDINATES

transformation: $x = x(u, v, w), \quad y = y(u, v, w), \quad z = z(u, v, w)$

position vector: $\mathbf{r} = x(u, v, w)\mathbf{i} + y(u, v, w)\mathbf{j} + z(u, v, w)\mathbf{k}$

scale factors: $h_u = \left|\dfrac{\partial \mathbf{r}}{\partial u}\right|, \quad h_v = \left|\dfrac{\partial \mathbf{r}}{\partial v}\right|, \quad h_w = \left|\dfrac{\partial \mathbf{r}}{\partial w}\right|$

local basis: $\hat{\mathbf{u}} = \dfrac{1}{h_u}\dfrac{\partial \mathbf{r}}{\partial u}, \quad \hat{\mathbf{v}} = \dfrac{1}{h_v}\dfrac{\partial \mathbf{r}}{\partial v}, \quad \hat{\mathbf{w}} = \dfrac{1}{h_w}\dfrac{\partial \mathbf{r}}{\partial w}$

volume element: $dV = h_u h_v h_w \, du \, dv \, dw$

scalar field: $f(u, v, w)$

vector field: $\mathbf{F}(u, v, w) = F_u(u, v, w)\hat{\mathbf{u}} + F_v(u, v, w)\hat{\mathbf{v}} + F_w(u, v, w)\hat{\mathbf{w}}$

gradient: $\nabla f = \dfrac{1}{h_u}\dfrac{\partial f}{\partial u}\hat{\mathbf{u}} + \dfrac{1}{h_v}\dfrac{\partial f}{\partial v}\hat{\mathbf{v}} + \dfrac{1}{h_w}\dfrac{\partial f}{\partial w}\hat{\mathbf{w}}$

curl: $\nabla \times \mathbf{F} = \dfrac{1}{h_u h_v h_w}\begin{vmatrix} h_u\hat{\mathbf{u}} & h_v\hat{\mathbf{v}} & h_w\hat{\mathbf{w}} \\ \dfrac{\partial}{\partial u} & \dfrac{\partial}{\partial v} & \dfrac{\partial}{\partial w} \\ F_u h_u & F_v h_v & F_w h_w \end{vmatrix}$

divergence: $\nabla \bullet \mathbf{F} = \dfrac{1}{h_u h_v h_w}\left[\dfrac{\partial}{\partial u}\left(h_v h_w F_u\right) + \dfrac{\partial}{\partial v}\left(h_u h_w F_v\right) + \dfrac{\partial}{\partial w}\left(h_u h_v F_w\right)\right]$

## PLANE POLAR COORDINATES

transformation: $x = r\cos\theta, \quad y = r\sin\theta$

position vector: $\mathbf{r} = r\cos\theta\,\mathbf{i} + r\sin\theta\,\mathbf{j}$

scale factors: $h_r = \left|\dfrac{\partial \mathbf{r}}{\partial r}\right| = 1, \quad h_\theta = \left|\dfrac{\partial \mathbf{r}}{\partial \theta}\right| = r$

local basis: $\hat{\mathbf{r}} = \cos\theta\mathbf{i} + \sin\theta\mathbf{j}, \quad \hat{\boldsymbol{\theta}} = -\sin\theta\mathbf{i} + \cos\theta\mathbf{j}$

area element: $dV = r \, dr \, d\theta$

scalar field: $f(r, \theta)$

vector field: $\mathbf{F}(r, \theta) = F_r(r, \theta)\hat{\mathbf{r}} + F_\theta(r, \theta)\hat{\boldsymbol{\theta}}$

gradient: $\nabla f = \dfrac{\partial f}{\partial r}\hat{\mathbf{r}} + \dfrac{1}{r}\dfrac{\partial f}{\partial \theta}\hat{\boldsymbol{\theta}}$

curl: $\nabla \times \mathbf{F} = \left[\dfrac{\partial F_\theta}{\partial r} + \dfrac{F_\theta}{r} - \dfrac{1}{r}\dfrac{\partial F_r}{\partial \theta}\right]\mathbf{k}$

divergence: $\nabla \bullet \mathbf{F} = \dfrac{\partial F_r}{\partial r} + \dfrac{1}{r}F_r + \dfrac{1}{r}\dfrac{\partial F_\theta}{\partial \theta}$

## CYLINDRICAL COORDINATES

transformation: $x = r\cos\theta, \quad y = r\sin\theta, \quad z = z$

position vector: $\mathbf{r} = r\cos\theta\,\mathbf{i} + r\sin\theta\,\mathbf{j} + z\mathbf{k}$

scale factors: $h_r = \left|\dfrac{\partial \mathbf{r}}{\partial r}\right| = 1, \quad h_\theta = \left|\dfrac{\partial \mathbf{r}}{\partial \theta}\right| = r, \quad h_z = \left|\dfrac{\partial \mathbf{r}}{\partial z}\right| = 1$

local basis: $\hat{\mathbf{r}} = \cos\theta\mathbf{i} + \sin\theta\mathbf{j}, \quad \hat{\boldsymbol{\theta}} = -\sin\theta\mathbf{i} + \cos\theta\mathbf{j}, \quad \hat{\mathbf{z}} = \mathbf{k}$

volume element: $dV = r \, dr \, d\theta \, dz$

scalar field: $f(r, \theta, z)$

vector field: $\mathbf{F}(r, \theta, z) = F_r(r, \theta, z)\hat{\mathbf{r}} + F_\theta(r, \theta, z)\hat{\boldsymbol{\theta}} + F_z(r, \theta, z)\mathbf{k}$

gradient: $\nabla f = \dfrac{\partial f}{\partial r}\hat{\mathbf{r}} + \dfrac{1}{r}\dfrac{\partial f}{\partial \theta}\hat{\boldsymbol{\theta}} + \dfrac{\partial f}{\partial z}\mathbf{k}$

curl: $\nabla \times \mathbf{F} = \dfrac{1}{r}\begin{vmatrix} \hat{\mathbf{r}} & r\hat{\boldsymbol{\theta}} & \mathbf{k} \\ \dfrac{\partial}{\partial r} & \dfrac{\partial}{\partial \theta} & \dfrac{\partial}{\partial z} \\ F_r & rF_\theta & F_z \end{vmatrix}$

divergence: $\nabla \bullet \mathbf{F} = \dfrac{\partial F_r}{\partial r} + \dfrac{1}{r}F_r + \dfrac{1}{r}\dfrac{\partial F_\theta}{\partial \theta} + \dfrac{\partial F_z}{\partial z}$

## SPHERICAL COORDINATES

transformation: $x = \rho\sin\phi\cos\theta, \quad y = \rho\sin\phi\sin\theta, \quad z = \rho\cos\phi$

position vector: $\mathbf{r} = \rho\sin\phi\cos\theta\,\mathbf{i} + \rho\sin\phi\sin\theta\,\mathbf{j} + \rho\cos\phi\,\mathbf{k}$

scale factors: $h_\rho = \left|\dfrac{\partial \mathbf{r}}{\partial \rho}\right| = 1, \quad h_\phi = \left|\dfrac{\partial \mathbf{r}}{\partial \phi}\right| = \rho, \quad h_\theta = \left|\dfrac{\partial \mathbf{r}}{\partial \theta}\right| = \rho\sin\phi$

local basis: $\hat{\mathbf{r}} = \sin\phi\cos\theta\,\mathbf{i} + \sin\phi\sin\theta\,\mathbf{j} + \cos\phi\,\mathbf{k}, \quad \hat{\boldsymbol{\phi}} = \cos\phi\cos\theta\,\mathbf{i} + \cos\phi\sin\theta\,\mathbf{j} - \sin\phi\,\mathbf{k}, \quad \hat{\boldsymbol{\theta}} = -\sin\theta\,\mathbf{i} + \cos\theta\,\mathbf{j}$

volume element: $dV = \rho^2\sin\phi \, d\rho \, d\phi \, d\theta$

scalar field: $f(\rho, \phi, \theta)$

vector field: $\mathbf{F}(\rho, \phi, \theta) = F_\rho(\rho, \phi, \theta)\hat{\mathbf{r}} + F_\phi(\rho, \phi, \theta)\hat{\boldsymbol{\phi}} + F_\theta(\rho, \phi, \theta)\hat{\boldsymbol{\theta}}$

gradient: $\nabla f = \dfrac{\partial f}{\partial \rho}\hat{\mathbf{r}} + \dfrac{1}{\rho}\dfrac{\partial f}{\partial \phi}\hat{\boldsymbol{\phi}} + \dfrac{1}{\rho\sin\phi}\dfrac{\partial f}{\partial \theta}\hat{\boldsymbol{\theta}}$

curl: $\nabla \times \mathbf{F} = \dfrac{1}{\rho^2\sin\phi}\begin{vmatrix} \hat{\mathbf{r}} & \rho\hat{\boldsymbol{\phi}} & \rho\sin\phi\hat{\boldsymbol{\theta}} \\ \dfrac{\partial}{\partial \rho} & \dfrac{\partial}{\partial \phi} & \dfrac{\partial}{\partial \theta} \\ F_\rho & \rho F_\phi & \rho\sin\phi F_\theta \end{vmatrix}$

divergence: $\nabla \bullet \mathbf{F} = \dfrac{\partial F_\rho}{\partial \rho} + \dfrac{2}{\rho}F_\rho + \dfrac{1}{\rho}\dfrac{\partial F_\phi}{\partial \phi} + \dfrac{\cot\phi}{\rho}F_\phi + \dfrac{1}{\rho\sin\phi}\dfrac{\partial F_\theta}{\partial \theta}$

A COMPLETE COURSE

# Calculus

## FOURTH EDITION

# ROBERT A. ADAMS

DEPARTMENT OF MATHEMATICS
UNIVERSITY OF BRITISH COLUMBIA

**Addison-Wesley**

An imprint of Addison Wesley Longman Ltd.

Don Mills, Ontario • Reading, Massachusetts • Harlow, England
Melbourne, Australia • Amsterdam, The Netherlands • Bonn, Germany

Publisher: Ron Doleman
Managing Editor: Linda Scott
Editor: First Folio Resource Group Incorporated/Robert Templeton
Cover Design: Anthony Leung
Design: Pronk & Associates
Printing and Binding: Transcontinental

Canadian Cataloguing in Publication Data

Adams, Robert A. (Robert Alexander), 1940-
    Calculus: a complete course

4th ed.
Includes index.
ISBN 0-201-39607-6

1. Calculus.  I. Title.

QA303.A28 1999        515        C98-932922-4

ISBN 0-201-39607-6

Printed and bound in Canada.

A B C D E -TC- 02 01 00 99 98

*To Anne*

# Preface

Most of the changes in this 4th edition have been made to achieve greater efficiency and a more orderly flow of topics. There are two fewer chapters in this edition, although no topics have been removed.

- The Mean-Value Theorem has been moved from Chapter 5 (in the 3rd edition) to the middle of Chapter 2. This results in rather less repetition as the MVT is often needed for such things as determining the relationship between the sign of a derivative and the increasing/decreasing behaviour of the function. Single-variable differential calculus now occupies only three chapters instead of four.

- There is now early emphasis on Taylor Polynomials and Taylor's formula with Lagrange remainder. This topic follows immediately the treatment of linear approximation and precedes the discussion of indeterminate forms, where it can be used in place of l'Hôpital's rules.

- The two chapters on numerical series and power series in the 3rd edition have been combined into a single chapter here. The material of the old section on estimating the sum of a series has been spread among other sections where it is more closely tied to the techniques involved in determining convergence. A new section on Fourier Series has been added.

- The treatment of vectors is now unified in the chapter on 3-dimensional coordinate geometry, rather than being split into plane and 3-dimensional versions as previously.

- The chapter on curves in 3-space (vector-valued functions of one real variable) has been moved to precede the chapters on multivariable calculus. It may be easier for students to deal with functions whose values are vectors before they have to study functions of a vector variable. However, this chapter (Chapter 11) can be delayed as long as it is covered before Chapter 15 on vector fields.

- Some reorganization has been done in the part of Chapter 16 (vector calculus) dealing with the integration theorems of Green, Gauss, and Stokes. Green's Theorem in the plane now precedes Gauss's Divergence Theorem and Stokes's Theorem. Also, some improvements have been made in the treatment of electrostatics and magnetostatics in the applications section, and discussion of the nonstatic electromagnetic case (Maxwell's equations) has been added.

In addition to the above, numerous small local changes have been made here and there to improve the text. Some awkward examples have been removed or replaced with more appropriate ones. Some exercises have been removed and others added. (In some of the longer exercise sets more have been removed than added.) The Explore features in the 3rd edition have been removed, but in some cases the material from them remains as subsections, and may even have been amplified. The remaining Explore features can be found in the Instructor's Manual. The Express Yourself exercises from the Chapter Reviews have also been removed.

# To the Student

When I took my first course in calculus there was no prescribed textbook but a book called *Calculus Made Easy* was recommended to those of us who felt we needed something beyond the notes we could take during lectures. I bought the book, hoping that it would live up to its title and I would have an easy time learning calculus. It didn't and I didn't.

Is calculus a very difficult subject? No, not really, but it sometimes seems that way to students, especially at the beginning, because of the new ideas and techniques involved, and because success in mastering calculus depends on having a very solid basis in precalculus mathematics (algebra, geometry, and trigonometry) to build upon. Learning calculus will provide you with many useful tools for analyzing problems in numerous fields of interest, especially those regarded as "scientific." It is worth your while to acquire those tools, but, like any other worthwhile task, this one requires much effort on your part. No book or instructor can remove that requirement.

In writing this book I have tried to organize material in such a way as to make it as easy as possible, but not at the expense of "sweeping the real difficulties under the rug." You may find some of the concepts difficult to understand when they are first introduced. If so, *reread* the material slowly, if necessary several times; *think about it*; formulate questions to ask fellow students, your TA, or your instructor. Don't delay. It is important to resolve your problems as soon as possible. If you don't understand today's topic, you may not understand how it applies to tomorrow's either. Mathematics is a "linear discipline"; it builds from one idea to the next.

Doing exercises is the best way to deepen your understanding of calculus and to convince yourself that you understand it. There are numerous exercises in this text — too many for you to try them all. Some are "drill" exercises to help you develop your skills in calculation. More important, however, are the problems that develop reasoning skills and your ability to apply the techniques you have learned to concrete situations. In some cases you will have to plan your way through a problem that requires several diffferent "steps" before you can get to the answer. Other exercises are designed to extend the theory developed in the text and therefore enhance your understanding of the concepts of calculus.

The exercises vary greatly in difficulty. Usually, the more difficult ones occur towards the end of exercise sets, but these sets are not strictly graded in this way because exercises on a specific topic tend to be grouped together. Some exercises are marked with an asterisk "∗." This symbol indicates that the exercise is *either* somewhat more theoretical, *or* somewhat more difficult than most. The theoretical ones need not be difficult; sometimes they are quite easy. Most of the problems in the *Challenging Problems* section forming part of the *Chapter Review* at the end of most chapters are also on the difficult side.

Do not be discouraged if you can't do *all* the exercises. Some are very difficult indeed, and only a few very gifted students will be able to do them. However, you should be able to do the vast majority of the exercises. Some will take much more effort than others. When you encounter difficulties with problems proceed as follows:

1. Read and reread the problem until you understand exactly what information it gives you and what you are expected to find or do.

2. If appropriate, draw a diagram illustrating the relationships between the quantities involved in the problem.

3. If necessary, introduce symbols to represent the quantities in the problem. Use appropriate letters (e.g., $V$ for volume, $t$ for time). Don't feel you have to use $x$ or $y$ for everything.

4. Develop a plan of attack. This is usually the hardest part. Look for known relationships; try to recognize patterns; look back at the worked examples in the current or relevant previous sections; try to think of possible connections between the problem at hand and others you have seen before. Can the problem be simplified by making some extra assumptions? If you can solve a simplified version it may help you decide what to do with the given problem. Can the problem be broken down into several cases, each of which is a simpler problem? When reading the examples in the text, be alert for methods that may turn out to be useful in other contexts later.

5. Try to carry out the steps in your plan. If you have trouble with one or more of them you may have to alter the plan.

6. When you arrive at an answer for a problem, always ask yourself whether it is *reasonable*. If it isn't, look back to determine places where you may have made an error.

Answers for most of the odd-numbered exercises are provided at the back of the book. Exceptions are exercises that don't have short answers, for example "Prove that ..." or "Show that ..." problems where the answer is the whole solution. A Student Solutions Manual that contains detailed solutions to even-numbered exercises is available.

Besides ∗ used to mark more difficult or theoretical problems, the following symbols are used to mark exercises of special types:

⬦   exercises pertaining to differential equations or initial-value problems. (It is not used in Sections 4.7 and 8.9, or in Chapter 19, which are wholly concerned with DEs.)

▦   problems requiring the use of a calculator. Often a scientific calculator is needed. Some such problems may require a programmable calculator.

∿   problems requiring the use of either a graphing calculator or mathematical graphing software on a personal computer.

▭   problems requiring the use of a computer. Typically, these will require either computer algebra software (e.g., Maple, Mathematica, Derive), or a spreadsheet program (e.g., Lotus 123, Microsoft Excel, Quattro Pro).

# To the Instructor

Real-variable calculus is usually covered in three or four semesters of study, the first two of which are typically devoted to the differentiation and integration of functions of a single real variable, and the remainder to functions of several real variables, and to vector-valued functions. This book is intended to cover all the material usually encountered in such a four-semester program. There is too much material, and some selection will have to be made on what to include and what to omit. This selection should take into account the background and needs of your students.

Any division of material into "core" and "optional" is necessarily somewhat arbitrary. In the single-variable part of the book, I regard most of the material of Chapters 1–5 as core, with the exception of Sections 1.5, 3.6, and 3.7. I also consider Sections 6.1–6.3 (techniques of integration), 6.5 (improper integrals), 7.1–7.3 (basic geometric applications of integration), 8.2–8.6 (parametric and polar curves), 9.1, 9.2, and 9.5–9.7 (the basics of infinite sequences, series, and power series) as core. These days, scientists and engineers require more training in the proper and effective use of numerical techniques, so Sections 6.6–6.7 (numerical integration) are assuming more importance and might be regarded as "core" for some classes.

For functions of several variables, the background discussion of vectors and geometry in Sections 10.1–10.4, the basic development of partial differentiation in Sections 12.1–12.8, the application to extreme value problems in Sections 13.1–13.3, and most of Chapter 14 on multiple integration should be considered as core.

For vector calculus, Sections 11.1–11.3, all of Chapter 15, and Sections 16.1–16.4 might be considered core.

The remaining sections and even some of the material and examples in core sections is optional in the sense that (with minor exceptions) its prior coverage is not necessary for any of what follows. It is up to individual instructors to decide what is most appropriate for their classes.

The text is designed for general calculus courses, especially those for science and engineering students. Most of the material requires only a reasonable background in high school algebra and analytic geometry. However, some optional material is more subtle and/or theoretical, and is intended mainly for stronger students.

Several supplements are available for use with *Calculus: A Complete Course*. Among these are:

- an **Instructor's Solutions Manual** with detailed solutions to all the exercises, prepared by the author, and a collection of projects.
- a **Student Solutions Manual** with detailed solutions to all the even-numbered exercises, prepared by the author.
- a Maple lab manual, **Calculus: The Maple Way**, by Robert Israel (University of British Columbia) with instruction, examples, and problems dealing with the effective use of the computer algebra system Maple as a tool for calculus.

# Acknowledgments

The first three editions of this material have been used for classes of general science, engineering, and mathematics majors and honours students at the University of British Columbia. I am grateful to many colleagues and students, at UBC and at many other institutions where these books have been used, for their encouragement and useful comments and suggestions.

In preparing the revision of this text I have had guidance from several dedicated reviewers who provided new insight and direction to my writing. I am especially grateful to the following people:

| | | |
|---|---|---|
| Reynir Axelsson | University of Iceland | Iceland |
| James Carrell | University of British Columbia | Canada |
| Kari Eloranta | Helskini University of Technology | Finland |
| Lennart Frennemo | Gothenburg University | Sweden |
| | Chalmers University of Technology | |
| Soren Hector | University College, Gavle-Sandviken | Sweden |
| Robert Israel | University of British Columbia | Canada |
| Turgay Kaptanoglu | Middle East Technical University | Turkey |
| Mark MacLean | University of British Columbia | Canada |
| Henrik Pedersen | University of Odense | Denmark |
| Anders Vretblad | Uppsala University | Sweden |

Thanks also go to Ken MacKenzie (McGill University) who checked the answers for all the new exercises.

I typeset this volume using TEX and PostScript on a PC running Linux, and also generated all of the figures in PostScript using the mathematical graphics software package **MG** developed by my colleague Robert Israel and myself.

Finally, I wish thank the sales and marketing staff of all Addison-Wesley divisions around the world for making the previous editions so successful, and the editorial and production staff at Addison-Wesley for their assistance and encouragement. In particular, I am grateful to Managing Editor, Linda Scott, and especially to my friend, College Publisher Ron Doleman, who has supervised the publication of all of my Addison-Wesley books for the past sixteen years, and who first introduced me to TEX and PostScript.

The expunging of errors and obscurities in a text is an ongoing and asymptotic process; hopefully each edition is better than the previous one. Nevertheless, some such imperfections always remain, and I will be grateful to any readers who call them to my attention, or give me any other suggestions for future improvements.

*R.A.A.*
*Vancouver, Canada*
*August, 1998*

# Contents

# What Is Calculus?

Much of our understanding of the world in which we live depends on our ability to describe how things change. Whether we are concerned with the motion of a pitched baseball or the path of a planet, the temperatures and currents of the oceans or the fluctuations of the stock market, the spread of an epidemic or the concentration of blood in the bloodstream, the growth of an economy or the decay of a radioactive substance, we are constantly forced to analyze relationships among quantities that change with time.

Algebra and geometry are useful tools for describing relationships among *static* quantities, but they do not involve concepts appropriate for describing how a quantity *changes*. For this we need new mathematical operations which go beyond the algebraic operations of addition, subtraction, multiplication, division and the taking of powers and roots. We require operations which measure the way related quantities change.

Calculus provides the tools for describing motion quantitatively. It introduces two new operations called *differentiation* and *integration* which, like addition and subtraction, are opposites of one another; what differentiation does, integration undoes.

For example, consider the motion of a falling rock. The height (in metres) of the rock $t$ seconds after it is dropped from a height of $h_0$ metres is a function $h(t)$ given by

$$h(t) = h_0 - 4.9t^2.$$

The graph of $y = h(t)$ is shown in the figure below:

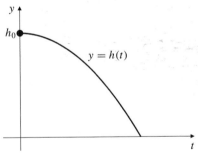

The process of differentiation enables us to find a new function, which we denote $h'(t)$ and call *the derivative* of $h$ with respect to $t$, that represents the *rate of change* of the height of the rock, that is, its *velocity* in metres/second:

$$h'(t) = -9.8t.$$

Conversely, if we know the velocity of the falling rock as a function of time, integration enables us to find the height function $h(t)$.

Calculus was invented independently and in somewhat different ways by two seventeenth-century mathematicians, Sir Isaac Newton and Gottfried Wilhelm Leibniz. Newton's motivation was a desire to analyze the motion of moving objects. Using his calculus he was able to formulate his laws of motion and gravitation, and to *conclude from them* that the planets must move around the sun in elliptical orbits, a fact that had been discovered half a century earlier by Johannes Kepler. Kepler's discovery was empirical, made from years of study of numerical data on the positions of planets.

Many of the most fundamental and important "laws of nature" are conveniently expressed as equations involving rates of change of quantities. Such equations are called *differential equations* and techniques for their study and solution are at the heart of calculus. In the falling rock example the appropriate law is Newton's second law of motion:

Force = mass × acceleration.

The *acceleration*, $-9.8$ m/s$^2$, is the rate of change (the *derivative*) of the velocity, which is in turn the rate of change (the *derivative*) of the height function.

Much of mathematics is related indirectly to the study of motion. We regard *lines* or *curves* as geometric objects, but the ancient Greeks thought of them as paths traced out by moving points. Nevertheless, the study of curves also involves geometric concepts such as tangency and area. The process of differentiation (Chapters 2–4) is closely tied to the geometric problem of finding tangent lines; similarly, integration (Chapters 5–7) is related to the geometric problem of finding areas of regions with curved boundaries.

Both differentiation and integration are defined in terms of a new mathematical operation called a **limit.** The concept of the limit of a function will be developed in Chapter 1. That will be the real beginning of our study of calculus. In the chapter called Preliminaries, we will review some of the background from algebra and geometry needed for the development of calculus that follows.

# CHAPTER P
# Preliminaries

***Introduction*** This preliminary chapter reviews the most important things you should know before beginning calculus. Topics include the real number system, Cartesian coordinates in the plane, equations representing straight lines, circles, and parabolas, functions and their graphs, and, in particular, the trigonometric functions.

Depending on your pre-calculus background, you may or may not be familiar with these topics. If you are, you may want to skim over this material to refresh your understanding of the terms used; if not, you should study this chapter in detail.

## P.1 Real Numbers and the Real Line

Calculus depends on properties of the real number system. **Real numbers** are numbers that can be expressed as decimals, for example

$$5 = 5.00000\ldots$$
$$-\tfrac{3}{4} = -0.750000\ldots$$
$$\tfrac{1}{3} = 0.3333\ldots$$
$$\sqrt{2} = 1.4142\ldots$$
$$\pi = 3.14159\ldots$$

In each case the three dots ... indicate that the sequence of decimal digits goes on forever. For the first three numbers above the patterns of the digits are obvious; we know what all the subsequent digits are. For $\sqrt{2}$ and $\pi$ there are no obvious patterns.

The real numbers can be represented geometrically as points on a number line, which we call the **real line**, shown in Figure P.1. The symbol $\mathbb{R}$ is used to denote either the real number system or, equivalently, the real line.

**Figure P.1**    The real line

The properties of the real number system fall into three categories: algebraic properties, order properties, and completeness. You are already familiar with the algebraic properties; roughly speaking, they assert that real numbers can be added, subtracted, multiplied, and divided (except by zero) to produce more real numbers, and that the usual rules of arithmetic are valid.

The *order properties* of the real numbers refer to the order in which the numbers appear on the real line. If $x$ lies to the left of $y$, then we say that "$x$ is less than $y$" or "$y$ is greater than $x$." These statements are written symbolically as $x < y$ and $y > x$, respectively. The inequality $x \le y$ means that either $x < y$ or $x = y$. The order properties of the real numbers are summarized in the following *rules for inequalities*:

The symbol $\implies$ means "implies."

### Rules for inequalities

If $a$, $b$, and $c$ are real numbers, then:

1. $a < b \qquad \implies \qquad a + c < b + c$
2. $a < b \qquad \implies \qquad a - c < b - c$
3. $a < b$ and $c > 0 \qquad \implies \qquad ac < bc$
4. $a < b$ and $c < 0 \qquad \implies \qquad ac > bc$; in particular, $-a > -b$
5. $a > 0 \qquad \implies \qquad \dfrac{1}{a} > 0$
6. $0 < a < b \qquad \implies \qquad \dfrac{1}{b} < \dfrac{1}{a}$

Rules 1–4 and 6 (for $a > 0$) also hold if $<$ and $>$ are replaced by $\leq$ and $\geq$.

Note especially the rules for multiplying (or dividing) an inequality by a number. If the number is positive, the inequality is preserved; if the number is negative, the inequality is reversed.

The *completeness* property of the real number system is more subtle and difficult to understand. One way to state it is as follows: If $A$ is any set of real numbers having at least one number in it, and if there exists a real number $y$ with the property that $x \leq y$ for every $x$ in $A$, then there exists a *smallest* number $y$ with the same property. Roughly speaking, this says that there can be no holes or gaps on the real line — every point corresponds to a real number. We will not need to deal much with completeness in our study of calculus. It is typically used to prove certain important results, in particular Theorems 8 and 9 in Chapter 1. (These proofs are given in Appendix 3, but are not usually included in elementary calculus courses; they are studied in more advanced courses in mathematical analysis.) Only when we study infinite sequences and series in Chapter 9 will we encounter any direct use of completeness.

The set of real numbers has some important special subsets:

(i) the **natural numbers** or **positive integers**, namely, the numbers 1, 2, 3, 4, ...

(ii) the **integers**, namely, the numbers 0, $\pm 1$, $\pm 2$, $\pm 3$, ...

(iii) the **rational numbers**, that is, numbers that can be expressed in the form of a fraction $m/n$, where $m$ and $n$ are integers, and $n \neq 0$.

The rational numbers are precisely those real numbers with decimal expansions that are either:

(a) terminating (that is, ending with an infinite string of zeros), for example, $3/4 = 0.750000\ldots$, or

(b) repeating (that is, ending with a string of digits that repeats over and over), for example, $23/11 = 2.090909\ldots = 2.\overline{09}$. (The bar indicates the pattern of repeating digits.)

Real numbers that are not rational are called *irrational numbers*.

**Example 1**    Show that each of the numbers    (a) $1.323232\cdots = 1.\overline{32}$, and (b) $0.3405405405\ldots = 0.3\overline{405}$ is a rational number by expressing it as a quotient of two integers.

**Solution**

(a) Let $x = 1.323232\ldots$   Then $x - 1 = 0.323232\ldots$ and

$$100x = 132.323232\ldots = 132 + 0.323232\ldots = 132 + x - 1.$$

Therefore, $99x = 131$ and $x = 131/99$.

(b) Let $y = 0.3405405405\ldots$   Then $10y = 3.405405405\ldots$   and $10y - 3 = 0.405405405\ldots$   Also,

$$10000y = 3405.405405405\ldots = 3405 + 10y - 3.$$

Therefore, $9990y = 3402$ and $y = 3402/9990 = 63/185$.

The set of rational numbers possesses all the algebraic and order properties of the real numbers, but not the completeness property. There is, for example, no rational number whose square is 2. Hence, there is a "hole" on the "rational line" where $\sqrt{2}$ should be.[1] Because the real line has no such "holes," it is the appropriate setting for studying limits and therefore calculus.

## Intervals

A subset of the real line is called an **interval** if it contains at least two numbers and also contains all real numbers between any two of its elements. For example, the set of real numbers $x$ such that $x > 6$ is an interval, but the set of real numbers $y$ such that $y \neq 0$ is not an interval. (Why?) It consists of two intervals.

If $a$ and $b$ are real numbers and $a < b$ we often refer to

(i) the **open interval** from $a$ to $b$, denoted by $(a, b)$, consisting of all real numbers $x$ satisfying $a < x < b$.

(ii) the **closed interval** from $a$ to $b$, denoted by $[a, b]$, consisting of all real numbers $x$ satisfying $a \leq x \leq b$.

(iii) the **half-open interval** $[a, b)$, consisting of all real numbers $x$ satisfying the inequalities $a \leq x < b$.

(iv) the **half-open interval** $(a, b]$, consisting of all real numbers $x$ satisfying the inequalities $a < x \leq b$.

**Figure P.2**    Finite intervals

open interval $(a, b)$
closed interval $[a, b]$
half-open interval $[a, b)$
half-open interval $(a, b]$

These are illustrated in Figure P.2. Note the use of parentheses ( and ) and hollow dots to indicate endpoints of intervals that are not included in the intervals, and square brackets [ and ] and solid dots to indicate endpoints that are included. The

[1]  How do we know that $\sqrt{2}$ is irrational? Suppose, to the contrary, that $\sqrt{2}$ is rational. Then $\sqrt{2} = m/n$, where $m$ and $n$ are integers and $n \neq 0$. We can assume that the fraction $m/n$ has been "reduced to lowest terms"; any common factors have been cancelled out. Now $m^2/n^2 = 2$, so $m^2 = 2n^2$, which is an even integer. Hence $m$ must also be even. (The square of an odd integer is always odd.) Since $m$ is even, we can write $m = 2k$, where $k$ is an integer. Thus $4k^2 = 2n^2$ and $n^2 = 2k^2$, which is even. Thus $n$ is also even. This contradicts the assumption that $\sqrt{2}$ could be written as a fraction $m/n$ in lowest terms; $m$ and $n$ cannot both be even. Accordingly, there can be no rational number whose square is 2.

the interval $(a, \infty)$

the interval $(-\infty, a]$

interval $(-\infty, \infty)$ is the real line

**Figure P.3**   Infinite intervals

endpoints of an interval are also called **boundary points**.

The intervals in Figure P.2 are **finite intervals**; each of them has finite length $b - a$. Intervals can also have infinite length, in which case they are called **infinite intervals**. Figure P.3 shows some examples of infinite intervals. Note that the whole real line $\mathbb{R}$ is an interval, denoted by $(-\infty, \infty)$. The symbol $\infty$ ("infinity") does *not* denote a real number, so we never use a square bracket at an infinite end of an interval.

**Example 2**   Solve the following inequalities. Express the solution sets in terms of intervals and graph them.

(a)   $2x - 1 > x + 3$     (b)   $-\dfrac{x}{3} \geq 2x - 1$     (c)   $\dfrac{2}{x - 1} \geq 5$

**Solution**

(a)   $2x - 1 > x + 3$     Add 1 to both sides.

$\quad\quad 2x > x + 4$     Subtract $x$ from both sides.

$\quad\quad\quad x > 4$     The solution set is the interval $(4, \infty)$.

$(4, \infty)$

$(-\infty, 3/7]$

$(1, 7/5]$

**Figure P.4**   The intervals for Example 2

(b)   $-\dfrac{x}{3} \geq 2x - 1$     Multiply both sides by $-3$.

$\quad\quad x \leq -6x + 3$     Add $6x$ to both sides.

$\quad\quad 7x \leq 3$     Divide both sides by 7.

$\quad\quad x \leq \dfrac{3}{7}$     The solution set is the interval $(-\infty, 3/7]$.

(c)   We would like to multiply both sides of the inequality $2/(x - 1) \geq 5$ by $x - 1$, but we will have to reverse the inequality if $x - 1 < 0$. Therefore we break the problem into two cases:

**CASE I**   $x - 1 > 0$, that is, $x > 1$.

$\quad\dfrac{2}{x - 1} \geq 5$     Multiply both sides by $x - 1$.

$\quad\quad 2 \geq 5x - 5$     Add 5 to both sides.

$\quad\quad 7 \geq 5x$     Divide both sides by 5.

$\quad\quad \dfrac{7}{5} \geq x$     The solutions for this case lie in the interval $(1, 7/5]$.

**CASE II**   $x - 1 < 0$, that is, $x < 1$.

$\quad\dfrac{2}{x - 1} \geq 5$     Multiply both sides by $x - 1$.

$\quad\quad 2 \leq 5x - 5$     Add 5 to both sides.

$\quad\quad 7 \leq 5x$     Divide both sides by 5.

$\quad\quad \dfrac{7}{5} \leq x$     There are no numbers satisfying $x < 1$ and $x \geq 7/5$.

The solution set is the interval $(1, 7/5]$. ∎

Sometimes we will need to solve systems of two or more inequalities that must be satisfied simultaneously. We still solve the inequalities individually, and look for numbers in the intersection of the solution sets.

**Example 3**   Solve the systems of inequalities:
(a) $3 \le 2x + 1 \le 5$, and    (b) $3x - 1 < 5x + 3 \le 2x + 15$.

**Solution**

(a) Using the technique of Example 1, we can solve the inequality $3 \le 2x + 1$ to get $2 \le 2x$, so $x \ge 1$. Similarly, the inequality $2x + 1 \le 5$ leads to $2x \le 4$, so $x \le 2$. The solution set of system (a) is therefore the closed interval $[1, 2]$.

(b) We solve both inequalities as follows:

$$
\left.
\begin{aligned}
3x - 1 &< 5x + 3 \\
-1 - 3 &< 5x - 3x \\
-4 &< 2x \\
-2 &< x
\end{aligned}
\right\}
\quad \text{and} \quad
\left\{
\begin{aligned}
5x + 3 &\le 2x + 15 \\
5x - 2x &\le 15 - 3 \\
3x &\le 12 \\
x &\le 4
\end{aligned}
\right.
$$

The solution set is the interval $(-2, 4]$.

Solving quadratic inequalities depends on solving the corresponding quadratic equations.

**Example 4**   **Quadratic inequalities**
Solve:    (a) $x^2 - 5x + 6 < 0$, and    (b) $2x^2 + 1 > 4x$.

**Solution**

(a) The trinomial $x^2 - 5x + 6$ factors into the product $(x - 2)(x - 3)$, which is negative if and only if exactly one of the factors is negative. Since $x - 3 < x - 2$, this happens when $x - 3 < 0$ and $x - 2 > 0$. Thus we need $x < 3$ and $x > 2$; the solution set is the open interval $(2, 3)$.

(b) The inequality $2x^2 + 1 > 4x$ is equivalent to $2x^2 - 4x + 1 > 0$. The corresponding quadratic equation $2x^2 - 4x + 1 = 0$, which is of the form $Ax^2 + Bx + C = 0$, can be solved by the quadratic formula:

$$
x = \frac{-B \pm \sqrt{B^2 - 4AC}}{2A} = \frac{4 \pm \sqrt{16 - 8}}{4} = 1 \pm \frac{\sqrt{2}}{2},
$$

so the given inequality can be expressed in the form

$$
\left( x - 1 + \tfrac{1}{2}\sqrt{2} \right)\left( x - 1 - \tfrac{1}{2}\sqrt{2} \right) > 0.
$$

This is satisfied if both factors on the left side are positive, or if both are negative. Therefore we require that either $x < 1 - \tfrac{1}{2}\sqrt{2}$ or $x > 1 + \tfrac{1}{2}\sqrt{2}$. The solution set is the *union* of intervals $\left( -\infty, 1 - \tfrac{1}{2}\sqrt{2} \right) \cup \left( 1 + \tfrac{1}{2}\sqrt{2}, \infty \right)$.

Note the use of the symbol $\cup$ to denote the **union** of intervals. A real number is in the union of intervals if it is in at least one of the intervals. We will also need to consider the **intersection** of intervals from time to time. A real number belongs to the intersection of intervals if it belongs to *every one* of the intervals. We will use

$\cap$ to denote intersection. For example

$$[1, 3) \cap [2, 4] = [2, 3) \quad \text{while} \quad [1, 3) \cup [2, 4] = [1, 4].$$

**Example 5** Solve the inequality $\dfrac{3}{x-1} < -\dfrac{2}{x}$ and graph the solution set.

**Solution** We would like to multiply by $x(x-1)$ to clear the inequality of fractions, but this would require considering three cases separately. (What are they?) Instead, we will transpose and combine the two fractions into a single one:

$$\frac{3}{x-1} < -\frac{2}{x}$$

$$\frac{3}{x-1} + \frac{2}{x} < 0$$

$$\frac{5x-2}{x(x-1)} < 0$$

The solution will consist of those values of $x$ for which the numerator and denominator of the latter fraction have opposite signs. This causes us to consider two cases:

**CASE I** $5x - 2 > 0$ and $x(x-1) < 0$. In this case we need $x > 2/5$ and $0 < x < 1$, so the solution set is the interval $(2/5, 1)$.

**CASE II** $5x - 2 < 0$ and $x(x-1) > 0$. In this case we need $x < 2/5$ and either $x < 0$ or $x > 1$, so the solution set is $(-\infty, 0)$.

The solution set of the given inequality is the union of these two intervals, namely $(-\infty, 0) \cup (2/5, 1)$.

the union $(-\infty, 0) \cup (2/5, 1)$

**Figure P.5**   The solution set for Example 5

## The Absolute Value

The **absolute value** or **magnitude** of a number $x$, denoted $|x|$ (read "the absolute value of $x$"), is defined by the formula

$$|x| = \begin{cases} x, & \text{if } x \geq 0 \\ -x, & \text{if } x < 0 \end{cases}$$

The vertical lines in the symbol $|x|$ are called **absolute value bars**.

**Example 6**   $|3| = 3, \quad |0| = 0, \quad |-5| = 5.$

Note that $|x| \geq 0$ for every real number $x$, and $|x| = 0$ only if $x = 0$. People sometimes find it confusing to say that $|x| = -x$ when $x$ is negative, but this is correct since $-x$ is positive in that case. The symbol $\sqrt{a}$ always denotes the *nonnegative* square root of $a$, so an alternative definition of $|x|$ is $|x| = \sqrt{x^2}$.

It is important to remember that $\sqrt{a^2} = |a|$. Do not write $\sqrt{a^2} = a$ unless you already know that $a \geq 0$.

Geometrically, $|x|$ represents the (nonnegative) distance from $x$ to 0 on the real line. More generally, $|x - y|$ represents the (nonnegative) distance between the points $x$ and $y$ on the real line, since this distance is the same as that from the point $x - y$ to 0 (see Figure P.6):

$$|x - y| = \begin{cases} x - y, & \text{if } x \geq y \\ y - x, & \text{if } x < y. \end{cases}$$

**Figure P.6**
$|x - y|$ = distance from $x$ to $y$

The absolute value function has the following properties:

---

**Properties of absolute values**

1. $|-a| = |a|$. A number and its negative have the same absolute value.

2. $|ab| = |a||b|$ and $\left|\dfrac{a}{b}\right| = \dfrac{|a|}{|b|}$. The absolute value of a product (or quotient) of two numbers is the product (or quotient) of their absolute values.

3. $|a \pm b| \le |a| + |b|$ (**the triangle inequality**). The absolute value of a sum of or difference between numbers is less than or equal to the sum of their absolute values.

---

The first two of these properties can be checked by considering the cases where either of $a$ or $b$ is either positive or negative. The third property follows from the first two because $\pm 2ab \le |2ab| = 2|a||b|$. Therefore we have

$$|a \pm b|^2 = (a \pm b)^2 = a^2 \pm 2ab + b^2$$
$$\le |a|^2 + 2|a||b| + |b|^2 = (|a| + |b|)^2,$$

and taking the (positive) square roots of both sides we obtain $|a \pm b| \le |a| + |b|$. This result is called the "triangle inequality" because it follows from the geometric fact that the length of any side of a triangle cannot exceed the sum of the lengths of the other two sides. For instance, if we regard the points $0$, $a$ and $b$ on the number line as the vertices of a degenerate "triangle," then the sides of the triangle have lengths $|a|$, $|b|$, and $|a - b|$. The triangle is degenerate since all three of its vertices lie on a straight line.

## Equations and Inequalities Involving Absolute Values

The equation $|x| = D$ (where $D > 0$) has two solutions, $x = D$ and $x = -D$: the two points on the real line that lie at distance $D$ from the origin. Equations and inequalities involving absolute values can be solved algebraically by breaking them into cases according to the definition of absolute value, but often they can also be solved geometrically by interpreting absolute values as distances. For example, the inequality $|x - a| < D$ says that the distance from $x$ to $a$ is less than $D$, so $x$ must lie between $a - D$ and $a + D$. (Or, equivalently, $a$ must lie between $x - D$ and $x + D$.) If $D$ is a positive number then

---

The symbol $\Longleftrightarrow$ means "if and only if," or "is equivalent to."

---

| | | |
|---|---|---|
| $\|x\| = D$ | $\Longleftrightarrow$ | either $x = -D$ or $x = D$ |
| $\|x\| < D$ | $\Longleftrightarrow$ | $-D < x < D$ |
| $\|x\| \le D$ | $\Longleftrightarrow$ | $-D \le x \le D$ |
| $\|x\| > D$ | $\Longleftrightarrow$ | either $x < -D$ or $x > D$ |

More generally,

$$
\begin{aligned}
|x - a| = D \quad &\Longleftrightarrow \quad \text{either } x = a - D \text{ or } x = a + D \\
|x - a| < D \quad &\Longleftrightarrow \quad a - D < x < a + D \\
|x - a| \le D \quad &\Longleftrightarrow \quad a - D \le x \le a + D \\
|x - a| > D \quad &\Longleftrightarrow \quad \text{either } x < a - D \text{ or } x > a + D
\end{aligned}
$$

**Example 7**   Solve: (a) $|2x + 5| = 3$,   (b) $|3x - 2| \le 1$.

**Solution**

(a) $|2x + 5| = 3 \Longleftrightarrow 2x + 5 = \pm 3$. Thus, either $2x = -3 - 5 = -8$ or $2x = 3 - 5 = -2$. The solutions are $x = -4$ and $x = -1$.

(b) $|3x - 2| \le 1 \Longleftrightarrow -1 \le 3x - 2 \le 1$. We solve this pair of inequalities:

$$
\left.\begin{aligned}
-1 &\le 3x - 2 \\
-1 + 2 &\le 3x \\
1/3 &\le x
\end{aligned}\right\}
\quad \text{and} \quad
\left\{\begin{aligned}
3x - 2 &\le 1 \\
3x &\le 1 + 2 \\
x &\le 1
\end{aligned}\right.
$$

Thus the solutions lie in the interval $[1/3, 1]$.

$\blacksquare$

**Remark**   Here is how part (b) of the Example 7 could have been solved geometrically, by interpreting the absolute value as a distance:

$$
|3x - 2| = \left| 3\left(x - \frac{2}{3}\right) \right| = 3\left| x - \frac{2}{3} \right|.
$$

Thus the given inequality says that

$$
3\left| x - \frac{2}{3} \right| \le 1 \quad \text{or} \quad \left| x - \frac{2}{3} \right| \le \frac{1}{3}.
$$

**Figure P.7**   The solution set for Example 7(b)

This says that distance from $x$ to $2/3$ does not exceed $1/3$. The solutions $x$ therefore lie between $1/3$ and 1, including both of these endpoints. (See Figure P.7.)

**Example 8**   Solve the equation $|x + 1| = |x - 3|$.

**Solution**   The equation says that $x$ is equidistant from $-1$ and 3. Therefore $x$ is the point halfway between $-1$ and 3; $x = (-1 + 3)/2 = 1$. Alternatively, the given equation says that either $x + 1 = x - 3$ or $x + 1 = -(x - 3)$. The first of these equations has no solutions; the second has the solution $x = 1$.

$\blacksquare$

**Example 9**   What values of $x$ satisfy the inequality $\left| 5 - \dfrac{2}{x} \right| < 3$?

**Solution**   We have

$$\left| 5 - \frac{2}{x} \right| < 3 \quad \Longleftrightarrow \quad -3 < 5 - \frac{2}{x} < 3 \qquad \text{Subtract 5 from each member.}$$

$$-8 < -\frac{2}{x} < -2 \qquad \text{Divide each member by } -2.$$

$$4 > \frac{1}{x} > 1 \qquad \text{Take reciprocals.}$$

$$\frac{1}{4} < x < 1$$

In this calculation we manipulated a system of two inequalities simultaneously, rather than split it up into separate inequalities as we have done in previous examples. Note how the various rules for inequalities were used here. Multiplying an inequality by a negative number reverses the inequality. So does taking reciprocals of an inequality in which both sides are positive. The given inequality holds for all $x$ in the open interval $(1/4, 1)$. ∎

# Exercises P.1

In Exercises 1–2, express the given rational number as a repeating decimal. Use a bar to indicate the repeating digits.

**1.** $\dfrac{2}{9}$

**2.** $\dfrac{1}{11}$

In Exercises 3–4, express the given repeating decimal as a quotient of integers in lowest terms.

**3.** $0.\overline{12}$

**4.** $3.2\overline{7}$

**5.** Express the rational numbers $1/7$, $2/7$, $3/7$, and $4/7$ as repeating decimals. (Use a calculator to give as many decimal digits as possible.) Do you see a pattern? Guess the decimal expansions of $5/7$ and $6/7$ and check your guesses.

**6.** Can two different decimals represent the same number? What number is represented by $0.999\ldots = 0.\overline{9}$?

In Exercises 7–12, express the set of all real numbers $x$ satisfying the given conditions as an interval or a union of intervals.

**7.** $x \geq 0$   and   $x \leq 5$

**8.** $x < 2$   and   $x \geq -3$

**9.** $x > -5$   or   $x < -6$

**10.** $x \leq -1$

**11.** $x > -2$

**12.** $x < 4$   or   $x \geq 2$

In Exercises 13–26, solve the given inequality, giving the solution set as an interval or union of intervals.

**13.** $-2x > 4$

**14.** $3x + 5 \leq 8$

**15.** $5x - 3 \leq 7 - 3x$

**16.** $\dfrac{6 - x}{4} \geq \dfrac{3x - 4}{2}$

**17.** $3(2 - x) < 2(3 + x)$

**18.** $x^2 < 9$

**19.** $\dfrac{1}{2 - x} < 3$

**20.** $\dfrac{x + 1}{x} \geq 2$

**21.** $x^2 - 2x \leq 0$

**22.** $6x^2 - 5x \leq -1$

**23.** $x^3 > 4x$

**24.** $x^2 - x \leq 2$

**25.** $\dfrac{x}{2} \geq 1 + \dfrac{4}{x}$

**26.** $\dfrac{3}{x - 1} < \dfrac{2}{x + 1}$

Solve the equations in Exercises 27–32.

**27.** $|x| = 3$

**28.** $|x - 3| = 7$

**29.** $|2t + 5| = 4$

**30.** $|1 - t| = 1$

**31.** $|8 - 3s| = 9$

**32.** $\left| \dfrac{s}{2} - 1 \right| = 1$

In Exercises 33–40, write the interval defined by the given inequality.

**33.** $|x| < 2$

**34.** $|x| \leq 2$

**35.** $|s - 1| \leq 2$

**36.** $|t + 2| < 1$

**37.** $|3x - 7| < 2$

**38.** $|2x + 5| < 1$

**39.** $\left| \dfrac{x}{2} - 1 \right| \leq 1$

**40.** $\left| 2 - \dfrac{x}{2} \right| < \dfrac{1}{2}$

In Exercises 41–42, solve the given inequality by interpreting it as a statement about distances on the real line.

**41.** $|x + 1| > |x - 3|$

**42.** $|x - 3| < 2|x|$

**43.** Do not fall into the trap $|-a| = a$. For what real numbers $a$ is this equation true? For what numbers is it false?

**44.** Solve the equation $|x - 1| = 1 - x$.

**45.** Show that the inequality

$$|a - b| \geq \big| |a| - |b| \big|$$

holds for all real numbers $a$ and $b$.

# P.2   Cartesian Coordinates in the Plane

The positions of all points in a plane can be measured with respect to two perpendicular real lines in the plane intersecting at the 0-point of each. These lines are called **coordinate axes** in the plane. Usually (but not always) we call one of these axes the $x$-axis, and draw it horizontally with numbers $x$ on it increasing to the right, and we call the other the $y$-axis, and draw it vertically with numbers $y$ on it increasing upward. The point of intersection of the coordinate axes (the point where $x$ and $y$ are both zero) is called **the origin**, and is often denoted by the letter $O$.

If $P$ is any point in the plane we can draw a line through $P$ perpendicular to the $x$-axis. If $a$ is the value of $x$ where that line intersects the $x$-axis we call $a$ the **$x$-coordinate** of $P$. Similarly, the **$y$-coordinate** of $P$ is the value of $y$ where a line through $P$ perpendicular to the $y$-axis meets the $y$-axis. The **ordered pair** $(a, b)$ is called the **coordinate pair** or the **Cartesian coordinates** of the point $P$. We refer to the point as $P(a, b)$ to indicate both the name $P$ of the point, and its coordinates $(a, b)$. Note that the $x$-coordinate appears first in a coordinate pair. Coordinate pairs are in one-to-one correspondence with points in the plane; each point has a unique coordinate pair, and each coordinate pair determines a unique point. We call such a set of coordinate axes and the coordinate pairs they determine a **Cartesian coordinate system** in the plane, after the seventeenth-century philosopher René Descartes who created analytic (coordinate) geometry. When equipped with such a coordinate system, a plane is called a **Cartesian plane**.

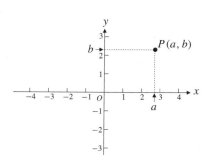

**Figure P.8**    The coordinate axes and the point $P$ with coordinates $(a, b)$

Figure P.9 shows the coordinates of some points in the plane. Note that all points on the $x$-axis have $y$-coordinate 0. We usually just write the $x$-coordinates to label such points. Similarly, points on the $y$-axis have $x = 0$, and we can label such points using their $y$-coordinates only.

The coordinate axes divide the plane into four regions called **quadrants**. These quadrants are numbered I to IV, as shown in Figure P.10. The **first quadrant** is the upper right one; both coordinates of any point in that quadrant are positive numbers. Both coordinates are negative in quadrant III; only $y$ is positive in quadrant II; only $x$ is positive in quadrant IV.

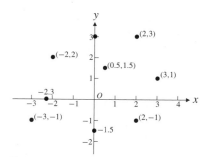

**Figure P.9**    Some points with their coordinates

## Axis Scales

When we plot data in the coordinate plane or graph formulas whose variables have different units of measure, we do not need to use the same scale on the two axes. If, for example, we plot height versus time for a falling rock, there is no reason to place the mark that shows 1 metre on the height axis the same distance from the origin as the mark that shows 1 second on the time axis.

When we graph functions whose variables do not represent physical measurements and when we draw figures in the coordinate plane to study their geometry or trigonometry, we make the scales identical. A vertical unit of distance then looks the same as a horizontal unit. As on a surveyor's map or a scale drawing, line segments that are supposed to have the same length will look as if they do, and angles that are supposed to be equal will look equal. Some of the geometric results we obtain later, such as the relationship between the slopes of perpendicular lines, are valid only if equal scales are used on the two axes.

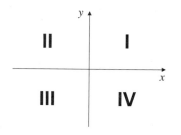

**Figure P.10**    The four quadrants

Computer and calculator displays are another matter. The vertical and horizontal scales on machine-generated graphs usually differ, with resulting distortions in distances, slopes, and angles. Circles may appear elliptical and squares may appear rectangular or even as parallelograms. Right angles may appear as acute or obtuse. Circumstances like these require us to take extra care in interpreting what we see. High-quality computer software for drawing Cartesian graphs usually allows the user to compensate for such scale problems by adjusting the *aspect ratio* (the ratio of vertical to horizontal scale). Some computer screens also allow adjustment within a narrow range. When using graphing software, try to adjust your particular software/hardware configuration so that the horizontal and vertical diameters of a drawn circle appear to be equal.

## Increments and Distances

When a particle moves from one point to another, the net changes in its coordinates are called increments. They are calculated by subtracting the coordinates of the starting point from the coordinates of the ending point. An **increment** in a variable is the net change in the value of the variable. If $x$ changes from $x_1$ to $x_2$, then the increment in $x$ is $\Delta x = x_2 - x_1$.

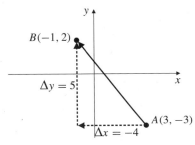

**Figure P.11**   Increments in $x$ and $y$

**Example 1**   Find the increments in the coordinates of a particle that moves from $A(3, -3)$ to $B(-1, 2)$.

**Solution**   The increments (see Figure P.11) are:

$$\Delta x = -1 - 3 = -4 \quad \text{and} \quad \Delta y = 2 - (-3) = 5.$$

If $P(x_1, y_1)$ and $Q(x_2, y_2)$ are two points in the plane, the straight line segment $PQ$ is the hypotenuse of a right triangle $PCQ$, as shown in Figure P.12. The legs $PC$ and $CQ$ of the triangle have lengths

$$|\Delta x| = |x_2 - x_1| \quad \text{and} \quad |\Delta y| = |y_2 - y_1|.$$

These are the *horizontal distance* and *vertical distance* between $P$ and $Q$. By the Pythagorean theorem, the length of $PQ$ is the square root of the sum of the squares of these leg lengths.

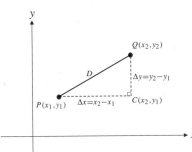

**Figure P.12**   The distance from $P$ to $Q$ is $D = \sqrt{(x_2 - x_1)^2 + (y_2 - y_1)^2}$

> **Distance formula for points in the plane**
>
> The distance $D$ between $P(x_1, y_1)$ and $Q(x_2, y_2)$ is
>
> $$D = \sqrt{(\Delta x)^2 + (\Delta y)^2} = \sqrt{(x_2 - x_1)^2 + (y_2 - y_1)^2}.$$

**Example 2**   The  distance between $A(3, -3)$ and $B(-1, 2)$ in Figure P.11 is

$$\sqrt{(-1 - 3)^2 + (2 - (-3))^2} = \sqrt{(-4)^2 + 5^2} = \sqrt{41} \text{ units.}$$

**Example 3** The distance from the origin $O(0, 0)$ to a point $P(x, y)$ is

$$\sqrt{(x-0)^2 + (y-0)^2} = \sqrt{x^2 + y^2}.$$

## Graphs

The **graph** of an equation (or inequality) involving the variables $x$ and $y$ is the set of all points $P(x, y)$ whose coordinates satisfy the equation (or inequality).

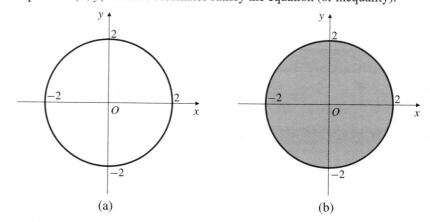

(a) (b)

**Figure P.13**

(a) The circle $x^2 + y^2 = 4$

(b) The disk $x^2 + y^2 \le 4$

**Example 4** The equation $x^2 + y^2 = 4$ represents all points $P(x, y)$ whose distance from the origin is $\sqrt{x^2 + y^2} = \sqrt{4} = 2$. These points lie on the **circle** of radius 2 centred at the origin. This circle is the graph of the equation $x^2 + y^2 = 4$. (See Figure P.13(a).)

**Example 5** Points $(x, y)$ whose coordinates satisfy the inequality $x^2 + y^2 \le 4$ all have distance $\le 2$ from the origin. The graph of the inequality is therefore the disk of radius 2 centred at the origin. (See Figure P.13(b).)

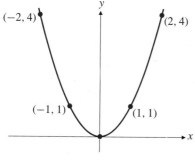

**Figure P.14** The parabola $y = x^2$

**Example 6** Consider the equation $y = x^2$. Some points whose coordinates satisfy this equation are $(0, 0)$, $(1, 1)$, $(-1, 1)$, $(2, 4)$, and $(-2, 4)$. These points (and all others satisfying the equation) lie on a smooth curve called a **parabola**. (See Figure P.14.)

## Straight Lines

Given two points $P_1(x_1, y_1)$ and $P_2(x_2, y_2)$ in the plane, we call the increments $\Delta x = x_2 - x_1$ and $\Delta y = y_2 - y_1$ respectively the **run** and the **rise** between $P_1$ and $P_2$. Two such points always determine a unique **straight line** (usually called simply a **line**) passing through them both. We call the line $P_1 P_2$.

Any non-vertical line in the plane has the property that the ratio

$$m = \frac{\text{rise}}{\text{run}} = \frac{\Delta y}{\Delta x} = \frac{y_2 - y_1}{x_2 - x_1}$$

has the *same value* for every choice of two distinct points $P_1(x_1, y_1)$ and $P_2(x_2, y_2)$ on the line. (See Figure P.15.) The constant $m = \Delta y / \Delta x$ is called the **slope** of the non-vertical line.

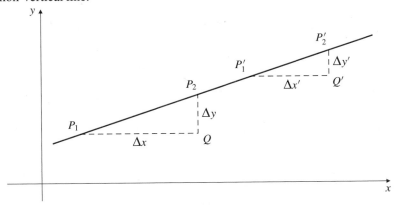

**Figure P.15** $\quad \Delta y / \Delta x = \Delta y' / \Delta x'$ because triangles $P_1 Q P_2$ and $P_1' Q' P_2'$ are similar

**Example 7** The slope of the line joining $A\,(3, -3)$ and $B\,(-1, 2)$ is

$$m = \frac{\Delta y}{\Delta x} = \frac{2 - (-3)}{-1 - 3} = \frac{5}{-4} = -\frac{5}{4}.$$

The slope tells us the direction and steepness of a line. A line with positive slope rises uphill to the right; one with negative slope falls downhill to the right. The greater the absolute value of the slope, the steeper the rise or fall. Since the run $\Delta x$ is zero for a vertical line, we cannot form the ratio $m$; the slope of a vertical line is *undefined*.

The direction of a line can also be measured by an angle. The **inclination** of a line is the smallest counterclockwise angle from the positive direction of the $x$-axis to the line. In Figure P.16 the angle $\phi$ (the Greek letter "phi") is the inclination of the line $L$. The inclination $\phi$ of any line satisfies $0° \leq \phi < 180°$. The inclination of a horizontal line is $0°$ and that of a vertical line is $90°$.

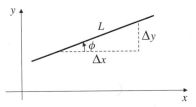

**Figure P.16** Line $L$ has inclination $\phi$

Provided equal scales are used on the coordinate axes, the relationship between the slope $m$ of a non-vertical line and its inclination $\phi$ is shown in Figure P.16:

$$m = \frac{\Delta y}{\Delta x} = \tan \phi.$$

(The trigonometric function tan is defined in Section P.6 below.)

Parallel lines have the same inclination. If they are not vertical, they must therefore have the same slope. Conversely, lines with equal slopes have the same inclination and so are parallel.

If two non-vertical lines, $L_1$ and $L_2$, are perpendicular, their slopes $m_1$ and $m_2$ satisfy $m_1 m_2 = -1$, so each slope is the *negative reciprocal* of the other:

$$m_1 = -\frac{1}{m_2} \quad \text{and} \quad m_2 = -\frac{1}{m_1}.$$

(This result also assumes equal scales on the two coordinate axes.) To see this, observe in Figure P.17 that

$$m_1 = \frac{AD}{BD} \quad \text{and} \quad m_2 = -\frac{AD}{DC}.$$

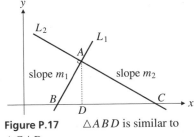

**Figure P.17** $\quad \triangle ABD$ is similar to $\triangle CAD$

Since $\triangle ABD$ is similar to $\triangle CAD$, we have $\dfrac{AD}{BD} = \dfrac{DC}{AD}$, and so

$$m_1 m_2 = \left(\frac{DC}{AD}\right)\left(-\frac{AD}{DC}\right) = -1.$$

## Equations of Lines

Straight lines are particularly simple graphs, and their corresponding equations are also simple. All points on the vertical line through the point $a$ on the $x$-axis have their $x$-coordinates equal to $a$. Thus $x = a$ is the equation of the line. Similarly, $y = b$ is the equation of the horizontal line meeting the $y$-axis at $b$.

**Example 8**   The horizontal and vertical lines passing through the point $(3, 1)$ (Figure P.18) have equations $y = 1$ and $x = 3$, respectively.

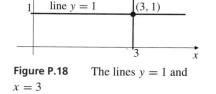

**Figure P.18**   The lines $y = 1$ and $x = 3$

To write an equation for a non-vertical straight line $L$, it is enough to know its slope $m$ and the coordinates of one point $P_1(x_1, y_1)$ on it. If $P(x, y)$ is any other point on $L$, then

$$\frac{y - y_1}{x - x_1} = m,$$

so that

$$y - y_1 = m(x - x_1) \qquad \text{or} \qquad y = m(x - x_1) + y_1.$$

The equation

$$y = m(x - x_1) + y_1$$

is the **point-slope equation** of the line which passes through the point $(x_1, y_1)$ and has slope $m$.

**Example 9**   Find an equation of the line of slope $-2$ through the point $(1, 4)$.

**Solution**   We substitute $x_1 = 1$, $y_1 = 4$, and $m = -2$ into the point-slope form of the equation and obtain

$$y = -2(x - 1) + 4 \qquad \text{or} \qquad y = -2x + 6.$$

**Example 10**   Find an equation of the line through the points $(1, -1)$ and $(3, 5)$.

**Solution**   The slope of the line is $m = \dfrac{5 - (-1)}{3 - 1} = 3$. We can use this slope with either of the two points to write an equation of the line. If we use $(1, -1)$ we get

$$y = 3(x - 1) - 1, \qquad \text{which simplifies to} \quad y = 3x - 4.$$

If we use $(3, 5)$ we get

$$y = 3(x - 3) + 4, \qquad \text{which also simplifies to} \quad y = 3x - 4.$$

Either way, $y = 3x - 4$ is an equation of the line. ∎

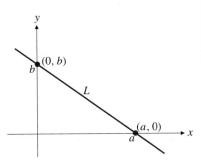

**Figure P.19**  Line $L$ has $x$-intercept $a$ and $y$-intercept $b$

The $y$-coordinate of the point where a non-vertical line intersects the $y$-axis is called the **$y$-intercept** of the line. Similarly, the **$x$-intercept** of a non-horizontal line is the $x$-coordinate of the point where it crosses the $x$-axis. A line with slope $m$ and $y$-intercept $b$ passes through the point $(0, b)$, so its equation is

$$y = m(x - 0) + b \qquad \text{or, more simply,} \quad y = mx + b.$$

A line with slope $m$ and $x$-intercept $a$ passes through $(a, 0)$, and so its equation is

$$y = m(x - a).$$

> The equation $y = mx + b$ is called the **slope–$y$-intercept equation** of the line with slope $m$ and $y$-intercept $b$.
>
> The equation $y = m(x - a)$ is called the **slope–$x$-intercept equation** of the line with slope $m$ and $x$-intercept $a$.

**Example 11**  Find the slope and the two intercepts of the line with equation $8x + 5y = 20$.

**Solution**  Solving the equation for $y$ we get

$$y = \frac{20 - 8x}{5} = -\frac{8}{5}x + 4.$$

Comparing this with the general form $y = mx + b$ of the slope–$y$-intercept equation, we see that the slope of the line is $m = -8/5$, and the $y$-intercept is $b = 4$.

To find the $x$-intercept put $y = 0$ and solve for $x$, obtaining $8x = 20$, or $x = 5/2$. The $x$-intercept is $a = 5/2$. ∎

The equation $Ax + By = C$, (where $A$ and $B$ are not both zero), is called the **general linear equation** in $x$ and $y$ because its graph always represents a straight line, and every line has an equation in this form.

Many important quantities are related by linear equations. Once we know that a relationship between two variables is linear, we can find it from any two pairs of corresponding values, just as we find the equation of a line from the coordinates of two points.

**Example 12**  The relationship between Fahrenheit temperature ($F$) and Celsius temperature ($C$) is given by a linear equation of the form $F = mC + b$. The freezing point of water is $F = 32°$ or $C = 0°$, while the boiling point is $F = 212°$ or $C = 100°$. Thus

$$32 = 0m + b, \qquad \text{and} \qquad 212 = 100m + b,$$

so $b = 32$ and $m = (212 - 32)/100 = 9/5$. The relationship is given by the linear equation

$$F = \frac{9}{5}C + 32 \qquad \text{or} \qquad C = \frac{5}{9}(F - 32).$$

# Exercises P.2

In Exercises 1–4, a particle moves from $A$ to $B$. Find the net increments $\Delta x$ and $\Delta y$ in the particle's coordinates. Also find the distance from $A$ to $B$.

1. $A(0, 3)$, $B(4, 0)$
2. $A(-1, 2)$, $B(4, -10)$
3. $A(3, 2)$, $B(-1, -2)$
4. $A(0.5, 3)$, $B(2, 3)$

5. A particle starts at $A(-2, 3)$ and its coordinates change by $\Delta x = 4$ and $\Delta y = -7$. Find its new position.

6. A particle arrives at the point $(-2, -2)$ after its coordinates experience increments $\Delta x = -5$ and $\Delta y = 1$. From where did it start?

Describe the graphs of the equations and inequalities in Exercises 7–12.

7. $x^2 + y^2 = 1$
8. $x^2 + y^2 = 2$
9. $x^2 + y^2 \leq 1$
10. $x^2 + y^2 = 0$
11. $y \geq x^2$
12. $y < x^2$

In Exercises 13–14, find an equation for (a) the vertical line, and (b) the horizontal line through the given point.

13. $(-2, 5/3)$
14. $(\sqrt{2}, -1.3)$

In Exercises 15–18, write an equation for the line through $P$ with slope $m$.

15. $P(-1, 1)$, $m = 1$
16. $P(-2, 2)$, $m = 1/2$
17. $P(0, b)$, $m = 2$
18. $P(a, 0)$, $m = -2$

In Exercises 19–20, does the given point $P$ lie on, above, or below the given line?

19. $P(2, 1)$, $2x + 3y = 6$
20. $P(3, -1)$, $x - 4y = 7$

In Exercises 21–24, write an equation for the line through the two points.

21. $(0, 0)$, $(2, 3)$
22. $(-2, 1)$, $(2, -2)$
23. $(4, 1)$, $(-2, 3)$
24. $(-2, 0)$, $(0, 2)$

In Exercises 25–26, write an equation for the line with slope $m$ and $y$-intercept $b$.

25. $m = -2$, $b = \sqrt{2}$
26. $m = -1/2$, $b = -3$

In Exercises 27–30, determine the $x$- and $y$-intercepts and the slope of the given lines, and sketch their graphs.

27. $3x + 4y = 12$
28. $x + 2y = -4$

29. $\sqrt{2}x - \sqrt{3}y = 2$
30. $1.5x - 2y = -3$

In Exercises 31–32, find equations for the lines through $P$ that are (a) parallel to, and (b) perpendicular to the given line.

31. $P(2, 1)$, $y = x + 2$
32. $P(-2, 2)$, $2x + y = 4$

33. Find the point of intersection of the lines $3x + 4y = -6$ and $2x - 3y = 13$.

34. Find the point of intersection of the lines $2x + y = 8$ and $5x - 7y = 1$.

35. **(Two-intercept equations)** If a line is neither horizontal nor vertical, and does not pass through the origin, show that its equation can be written in the form $\frac{x}{a} + \frac{y}{b} = 1$, where $a$ is its $x$-intercept and $b$ is its $y$-intercept.

36. Determine the intercepts and sketch the graph of the line $\frac{x}{2} - \frac{y}{3} = 1$.

37. Find the $y$-intercept of the line through the points $(2, 1)$ and $(3, -1)$.

38. A line passes through $(-2, 5)$ and $(k, 1)$ and has $x$-intercept 3. Find $k$.

39. The cost of printing $x$ copies of a pamphlet is $\$C$, where $C = Ax + B$ for certain constants $A$ and $B$. If it costs $\$5,000$ to print 10,000 copies and $\$6,000$ to print 15,000 copies, how much will it cost to print 100,000 copies?

40. **(Fahrenheit versus Celsius)** In the $FC$-plane, sketch the graph of the equation $C = \frac{5}{9}(F - 32)$ linking Fahrenheit and Celsius temperatures found in Example 12. On the same graph sketch the line with equation $C = F$. Is there a temperature at which a Celsius thermometer gives the same numerical reading as a Fahrenheit thermometer? If so, find it.

## Geometry

41. By calculating the lengths of its three sides, show that the triangle with vertices at the points $A(2, 1)$, $B(6, 4)$, and $C(5, -3)$ is isosceles.

42. Show that the triangle with vertices $A(0, 0)$, $B(1, \sqrt{3})$, and $C(2, 0)$ is equilateral.

43. Show that the points $A(2, -1)$, $B(1, 3)$, and $C(-3, 2)$ are three vertices of a square and find the fourth vertex.

**44.** Find the coordinates of the midpoint on the line segment $P_1 P_2$ joining the points $P_1(x_1, y_1)$ and $P_2(x_2, y_2)$.

**45.** Find the coordinates of the point of the line segment joining the points $P_1(x_1, y_1)$ and $P_2(x_2, y_2)$ that is two-thirds of the way from $P_1$ to $P_2$.

**46.** The point $P$ lies on the $x$-axis and the point $Q$ lies on the line $y = -2x$. The point $(2, 1)$ is the midpoint of $PQ$. Find the coordinates of $P$.

In Exercises 47–48, interpret the equation as a statement about distances, and hence determine the graph of the equation.

**47.** $\sqrt{(x-2)^2 + y^2} = 4$

**48.** $\sqrt{(x-2)^2 + y^2} = \sqrt{x^2 + (y-2)^2}$

**49.** For what value of $k$ is the line $2x + ky = 3$ perpendicular to the line $4x + y = 1$? For what value of $k$ are the lines parallel?

**50.** Find the line that passes through the point $(1, 2)$ and through the point of intersection of the two lines $x + 2y = 3$ and $2x - 3y = -1$.

# P.3 Graphs of Quadratic Equations

This section reviews circles, parabolas, ellipses, and hyperbolas, the graphs that are represented by quadratic equations in two variables.

## Circles and Disks

The **circle** having **centre** $C$ and **radius** $a$ is the set of all points in the plane that are at distance $a$ from the point $C$.

The distance from $P(x, y)$ to the point $C(h, k)$ is $\sqrt{(x-h)^2 + (y-k)^2}$, so that the equation of the circle of radius $a > 0$ with centre at $C(h, k)$ is

$$\sqrt{(x-h)^2 + (y-k)^2} = a.$$

A simpler form of this equation is obtained by squaring both sides.

**Figure P.20**    Circle
$(x-1)^2 + (y-3)^2 = 4$

---

**Standard equation of a circle**

The circle with centre $(h, k)$ and radius $a > 0$ has equation

$$(x-h)^2 + (y-k)^2 = a^2.$$

In particular, the circle with centre at the origin $(0, 0)$ and radius $a$ has equation

$$x^2 + y^2 = a^2.$$

---

**Example 1**    The circle with radius 2 and centre $(1, 3)$ (Figure P.20) has equation

$$(x-1)^2 + (y-3)^2 = 4.$$

**Example 2**    The circle having equation $(x+2)^2 + (y-1)^2 = 7$ has centre at the point $(-2, 1)$ and radius $\sqrt{7}$. (See Figure P.21.)

**Figure P.21**    The circle
$(x+2)^2 + (y-1)^2 = 7$

If the squares in the standard equation $(x-h)^2 + (y-k)^2 = a^2$ are multiplied out, and all constant terms collected on the right-hand side, the equation becomes

$$x^2 - 2hx + y^2 - 2ky = a^2 - h^2 - k^2.$$

A quadratic equation of the form

$$x^2 + y^2 + 2ax + 2by = c$$

must represent a circle, a single point, or no points at all. To identify the graph, we complete the squares on the left side of the equation. Since $x^2 + 2ax$ are the first two terms of the square $(x + a)^2 = x^2 + 2ax + a^2$, we add $a^2$ to both sides to complete the square of the $x$ terms. (Note that $a^2$ is *the square of half the coefficient of x*.) Similarly, add $b^2$ to both sides to complete the square of the $y$ terms. The equation then becomes

$$(x + a)^2 + (y + b)^2 = c + a^2 + b^2.$$

If $c + a^2 + b^2 > 0$, the graph is a circle with centre $(-a, -b)$ and radius $\sqrt{c + a^2 + b^2}$. If $c + a^2 + b^2 = 0$, the graph consists of the single point $(-a, -b)$. If $c + a^2 + b^2 < 0$, no points lie on the graph.

**Example 3**   Find the centre and radius of the circle $x^2 + y^2 - 4x + 6y = 3$.

**Solution**   Observe that $x^2 - 4x$ are the first two terms of the binomial square $(x - 2)^2 = x^2 - 4x + 4$, and $y^2 + 6y$ are the first two terms of the square $(y + 3)^2 = y^2 + 6y + 9$. Hence we add $4 + 9$ to both sides of the given equation and obtain

$$x^2 - 4x + 4 + y^2 + 6y + 9 = 3 + 4 + 9 \quad \text{or} \quad (x - 2)^2 + (y + 3)^2 = 16.$$

This is the equation of a circle with centre $(2, -3)$ and radius 4.

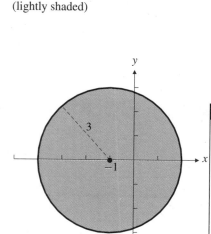

**Figure P.22**   The interior of a circle (darkly shaded), and the exterior (lightly shaded)

The set of all points *inside* a circle is called the **interior** of the circle. The set of all points *outside* the circle is called the **exterior** of the circle. (See Figure P.22.) The interior of a circle is sometimes called an **open disk**. The interior of a circle together with the circle itself is called a **closed disk**, or simply a **disk**. The inequality

$$(x - h)^2 + (y - k)^2 \le a^2$$

represents the disk of radius $|a|$ centred at $(h, k)$.

**Example 4**   Identify the graphs of:
(a) $x^2 + 2x + y^2 \le 8$,   (b) $x^2 + 2x + y^2 < 8$, and   (c) $x^2 + 2x + y^2 > 8$.

**Solution**   We can complete the square in the equation $x^2 + y^2 + 2x = 8$ as follows

$$x^2 + 2x + 1 + y^2 = 8 + 1$$
$$(x + 1)^2 + y^2 = 9.$$

Thus the equation represents the circle of radius 3 with centre at $(-1, 0)$. Inequality (a) represents the (closed) disk with the same radius and centre. (See Figure P.23.) Inequality (b) represents the interior of the circle (or the open disk). Inequality (c) represents the exterior of the circle.

**Figure P.23**   The disk $x^2 + y^2 + 2x \le 8$

## Equations of Parabolas

A **parabola** is a plane curve whose points are equidistant from a fixed point $F$ and a fixed straight line $L$ that does not pass through $F$. The point $F$ is the **focus** of the parabola; the line $L$ is the parabola's **directrix**. The line through $F$ perpendicular to $L$ is the parabola's **axis**. The point $V$ where the axis meets the parabola is the parabola's **vertex**.

Observe that the vertex $V$ of a parabola is halfway between the focus $F$ and the point on the directrix $L$ that is closest to $F$. If the directrix is either horizontal or vertical, and the vertex is at the origin, then the parabola will have a particularly simple equation.

**Example 5**  Find an equation of the parabola having the point $F(0, p)$ as focus and the line $L$ with equation $y = -p$ as directrix.

**Solution**  If $P(x, y)$ is any point on the parabola, then (see Figure P.24) the distances from $P$ to $F$ and $L$ are given by

$$PF = \sqrt{(x - 0)^2 + (y - p)^2} = \sqrt{x^2 + y^2 - 2py + p^2}$$
$$PQ = \sqrt{(x - x)^2 + (y - (-p))^2} = \sqrt{y^2 + 2py + p^2}.$$

For $P$ on the parabola we have $PF = PQ$, so their squares are also equal:

$$x^2 + y^2 - 2py + p^2 = y^2 + 2py + p^2,$$

or, after simplifying,

$$x^2 = 4py \qquad \text{or} \qquad y = \frac{x^2}{4p} \qquad \text{(called **standard forms**)}.$$

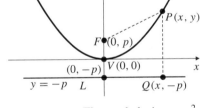

**Figure P.24**   The parabola $4py = x^2$ with focus $F(0, p)$ and directrix $y = -p$

Figure P.24 shows the situation for $p > 0$; the parabola opens upward, and is symmetric about its axis, the $y$-axis. If $p < 0$ the focus $(0, p)$ will lie below the origin and the directrix $y = -p$ will lie above the origin. In this case the parabola will open downward instead of upward. ∎

Figure P.25 shows several parabolas with equations of the form $y = ax^2$ for positive and negative values of $a$.

**Example 6**  An equation for the parabola with focus $(0, 1)$ and directrix $y = -1$, is $y = x^2/4$ or $x^2 = 4y$. (We took $p = 1$ in the standard equation.) ∎

**Example 7**  Find the focus and directrix of the parabola $y = -x^2$.

**Solution**  The given equation matches the standard form $y = x^2/(4p)$ provided $4p = -1$. Thus $p = -1/4$. The focus is $(0, -1/4)$ and the directrix is the line $y = 1/4$. ∎

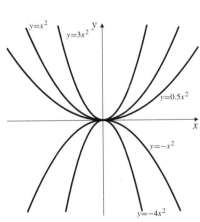

**Figure P.25**   Some parabolas $y = ax^2$

Interchanging the roles of $x$ and $y$ in the derivation of the standard equation above shows that the equation

$$y^2 = 4px \qquad \text{or} \qquad x = \frac{y^2}{4p} \qquad \text{(standard equation)}$$

represents a parabola with focus at $(p, 0)$ and vertical directrix $x = -p$. The axis is the $x$-axis.

## Reflective Properties of Parabolas

One of the chief applications of parabolas involves their use as reflectors of light and radio waves. Rays originating from the focus of a parabola will be reflected in a beam parallel to the axis, as shown in Figure P.26. Similarly, all the rays in a beam striking a parabola parallel to its axis will reflect through the focus. This property is the reason that telescopes and spotlights use parabolic mirrors, and radio telescopes and microwave antennas are parabolic in shape. We will examine this property of parabolas more carefully in Section 8.1.

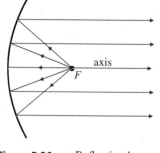

**Figure P.26**   Reflection by a parabola

**Figure P.27**   Horizontal scaling:
(a) the graph $y = 1 - x^2$
(b) graph of (a) compressed horizontally
(c) graph of (a) expanded horizontally

## Scaling a Graph

The graph of an equation can be compressed or expanded horizontally by replacing $x$ with a multiple of $x$. If $a$ is a positive number, replacing $x$ with $ax$ in an equation multiplies horizontal distances in the graph of the equation by a factor $1/a$. (See Figure P.27.) Replacing $y$ with $ay$ will multiply vertical distances in a similar way.

You may find it surprising that, like circles, all parabolas are *similar* geometric figures; they may have different sizes but they all have the same shape. We can change the *size* while preserving the shape of a curve represented by an equation in $x$ and $y$ by scaling both the coordinates by the same amount. If we scale the equation $4py = x^2$ by replacing $x$ and $y$ with $4px$ and $4py$ respectively, we get $4p(4py) = (4px)^2$, or $y = x^2$. Thus the general parabola $4py = x^2$ has the same shape as the specific parabola $y = x^2$, as shown in Figure P.28.

**Figure P.28**   The two parabolas are similar. Compare the parts inside the rectangles.

## Shifting a Graph

The graph of an equation (or inequality) can be shifted $c$ units horizontally by replacing $x$ with $x - c$, or vertically by replacing $y$ with $y - c$.

**Shifts**

To shift a graph $c$ units to the right, replace $x$ in its equation or inequality with $x - c$. (If $c < 0$, the shift will be to the left.)

To shift a graph $c$ units upward, replace $y$ in its equation or inequality with $y - c$. (If $c < 0$, the shift will be downward.)

**Example 8**    The graph of $y = (x - 3)^2$ is the parabola $y = x^2$ shifted 3 units to the right. The graph of $y = (x + 1)^2$ is the parabola $y = x^2$ shifted 1 unit to the left. (See Figure P.29(a).)

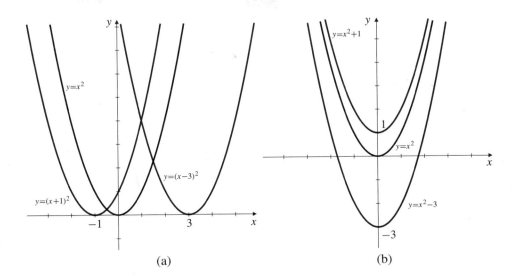

**Figure P.29**

(a) Horizontal shifts of $y = x^2$

(b) Vertical shifts of $y = x^2$

(a)                    (b)

**Example 9**    The graph of $y = x^2 + 1$, (that is, $y - 1 = x^2$), is the parabola $y = x^2$ shifted upward 1 unit. The graph of $y = x^2 - 3$, (that is, $y - (-3) = x^2$), is the parabola $y = x^2$ shifted downward 3 units. (See Figure P.29(b).)

**Example 10**    The circle with equation $(x - h)^2 + (y - k)^2 = a^2$ having centre $(h, k)$ and radius $a$ can be obtained by shifting the circle $x^2 + y^2 = a^2$ of radius $a$ centred at the origin $h$ units to the right and $k$ units upward. These shifts correspond to replacing $x$ with $x - h$ and $y$ with $y - k$.

The graph of $y = ax^2 + bx + c$ is a parabola whose axis is parallel to the $y$-axis. The parabola opens upward if $a > 0$ and downward if $a < 0$. We can complete the square and write the equation in the form $y = a(x - h)^2 + k$ to find the vertex $(h, k)$.

**Example 11**    Describe the graph of $y = x^2 - 4x + 3$.

**Solution**    The equation $y = x^2 - 4x + 3$ represents a parabola, opening upward. To find its vertex and axis we can complete the square:

$$y = x^2 - 4x + 4 - 1 = (x - 2)^2 - 1, \quad \text{so} \quad y - (-1) = (x - 2)^2.$$

This curve is the parabola $y = x^2$ shifted to the right 2 units and down 1 unit. Therefore its vertex is $(2, -1)$ and its axis is the line $x = 2$. Since $y = x^2$ has focus $(0, 1/4)$, the focus of this parabola is $(0 + 2, (1/4) - 1)$, or $(2, -3/4)$. (See Figure P.30.)

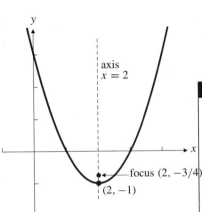

**Figure P.30**    The parabola $y = x^2 - 4x + 3$

## Ellipses and Hyperbolas

If $a$ and $b$ are positive numbers, the equation

$$\frac{x^2}{a^2} + \frac{y^2}{b^2} = 1$$

represents a curve called an **ellipse** that lies wholly within the rectangle $-a \le x \le a$, $-b \le y \le b$. (Why?). If $a = b$ the ellipse is just the circle of radius $a$ centred at the origin. If $a \ne b$ the ellipse is a circle that has been squashed by scaling it by different amounts in the two coordinate directions.

The ellipse has centre at the origin, and it passes through the four points $(a, 0)$, $(0, b)$, $(-a, 0)$, and $(0, -b)$. (See Figure P.31.) The line segments from $(-a, 0)$ to $(a, 0)$ and from $(0, -b)$ to $(0, b)$ are called the **principal axes** of the ellipse; the longer of the two is the **major axis** and the shorter is the **minor axis**.

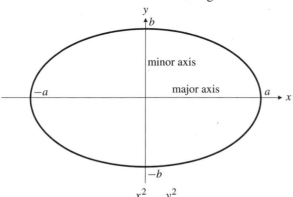

**Figure P.31**    The ellipse $\dfrac{x^2}{a^2} + \dfrac{y^2}{b^2} = 1$

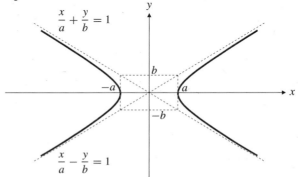

**Figure P.32**    The hyperbola $\dfrac{x^2}{a^2} - \dfrac{y^2}{b^2} = 1$ and its asymptotes

---

**Example 12**    The equation $\dfrac{x^2}{9} + \dfrac{y^2}{4} = 1$ represents an ellipse with major axis from $(-3, 0)$ to $(3, 0)$, and minor axis from $(0, -2)$ to $(0, 2)$. ∎

---

The equation

$$\frac{x^2}{a^2} - \frac{y^2}{b^2} = 1$$

represents a curve called a **hyperbola** that has centre at the origin and passes through the points $(-a, 0)$ and $(a, 0)$. (See Figure P.32.) The curve is in two parts (called branches). Each branch approaches two straight lines (called **asymptotes**) as it recedes far away from the origin. The asymptotes have equations

$$\frac{x}{a} - \frac{y}{b} = 0 \qquad \text{and} \qquad \frac{x}{a} + \frac{y}{b} = 0.$$

The equation $xy = 1$ also represents a hyperbola. This one passes through the points $(-1, -1)$ and $(1, 1)$ and has the coordinate axes as its asymptotes. It is, in fact, the hyperbola $x^2 - y^2 = 2$ rotated $45°$ counterclockwise about the origin. (See Figure P.33.) These hyperbolas are called **rectangular hyperbolas** since their asymptotes intersect at right angles.

We will study ellipses and hyperbolas in more detail in Chapter 8.

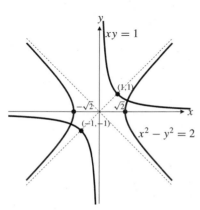

**Figure P.33**    Two rectangular hyperbolas

# Exercises P.3

In Exercises 1–4, write an equation for the circle with centre $C$ and radius $r$.

**1.** $C(0, 0)$,   $r = 4$

**2.** $C(0, 2)$,   $r = 2$

**3.** $C(-2, 0)$,   $r = 3$

**4.** $C(3, -4)$,   $r = 5$

In Exercises 5–8, find the centre and radius of the circle having the given equation.

**5.** $x^2 + y^2 - 2x = 3$

**6.** $x^2 + y^2 + 4y = 0$

**7.** $x^2 + y^2 - 2x + 4y = 4$

**8.** $x^2 + y^2 - 2x - y + 1 = 0$

Describe the regions defined by the inequalities and pairs of inequalities in Exercises 9–16.

**9.** $x^2 + y^2 > 1$

**10.** $x^2 + y^2 < 4$

**11.** $(x + 1)^2 + y^2 \le 4$

**12.** $x^2 + (y - 2)^2 \le 4$

**13.** $x^2 + y^2 > 1$,   $x^2 + y^2 < 4$

**14.** $x^2 + y^2 \le 4$,   $(x + 2)^2 + y^2 \le 4$

**15.** $x^2 + y^2 < 2x$,   $x^2 + y^2 < 2y$

**16.** $x^2 + y^2 - 4x + 2y > 4$,   $x + y > 1$

**17.** Write an inequality that describes the interior of the circle with centre $(-1, 2)$ and radius $\sqrt{6}$.

**18.** Write an inequality that describes the exterior of the circle with centre $(2, -3)$ and radius 4.

**19.** Write a pair of inequalities that describe that part of the interior of the circle with centre $(0, 0)$ and radius $\sqrt{2}$ lying on or to the right of the vertical line through $(1, 0)$.

**20.** Write a pair of inequalities that describe the points that lie outside the circle with centre $(0, 0)$ and radius 2, and inside the circle with centre $(1, 3)$ that passes through the origin.

In Exercises 21–24, write an equation of the parabola having the given focus and directrix.

**21.** Focus: $(0, 4)$     Directrix: $y = -4$

**22.** Focus: $(0, -1/2)$     Directrix: $y = 1/2$

**23.** Focus: $(2, 0)$     Directrix: $x = -2$

**24.** Focus: $(-1, 0)$     Directrix: $x = 1$

In Exercises 25–28, find the parabola's focus and directrix, and make a sketch showing the parabola, focus, and directrix.

**25.** $y = x^2/2$

**26.** $y = -x^2$

**27.** $x = -y^2/4$

**28.** $x = y^2/16$

**29.** Figure P.34 shows the graph $y = x^2$ and four shifted versions of it. Write equations for the shifted versions.

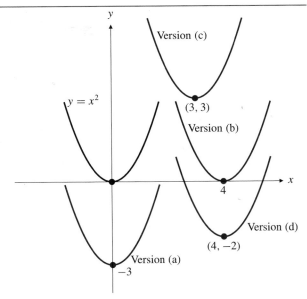

**Figure P.34**

**30.** What equations result from shifting the line $y = mx$
 (a) horizontally to make it pass through the point $(a, b)$,
 (b) vertically to make it pass through $(a, b)$?

In Exercises 31–34, the graph of $y = \sqrt{x + 1}$ is to be scaled in the indicated way. Give the equation of the graph that results from the scaling.

**31.** horizontal distances multiplied by 3

**32.** vertical distances divided by 4

**33.** horizontal distances multiplied by 2/3

**34.** horizontal distances divided by 4 and vertical distances multiplied by 2

In Exercises 35–38, write an equation for the graph obtained by shifting the graph of the given equation as indicated.

**35.** $y = 1 - x^2$          down 1, left 1

**36.** $x^2 + y^2 = 5$          up 2, left 4

**37.** $y = (x - 1)^2 - 1$          down 1, right 1

**38.** $y = \sqrt{x}$          down 2, left 4

Find the points of intersection of the pairs of curves in Exercises 39–42.

**39.** $y = x^2 + 3$,   $y = 3x + 1$

**40.** $y = x^2 - 6$,   $y = 4x - x^2$

**41.** $x^2 + y^2 = 25$,   $3x + 4y = 0$

**42.** $2x^2 + 2y^2 = 5$,   $xy = 1$

In Exercises 43–50, identify and sketch the curve represented by the given equation.

**43.** $\dfrac{x^2}{4} + y^2 = 1$

**44.** $9x^2 + 16y^2 = 144$

**45.** $\dfrac{(x-3)^2}{9} + \dfrac{(y+2)^2}{4} = 1$   **46.** $(x-1)^2 + \dfrac{(y+1)^2}{4} = 4$

**47.** $\dfrac{x^2}{4} - y^2 = 1$   **48.** $x^2 - y^2 = -1$

**49.** $xy = -4$   **50.** $(x-1)(y+2) = 1$

**51.** What is the effect on the graph of an equation in $x$ and $y$ of:

(a) replacing $x$ with $-x$?

(b) replacing $y$ with $-y$?

**52.** What is the effect on the graph of an equation in $x$ and $y$ of replacing $x$ with $-x$ and $y$ with $-y$ simultaneously?

**53.** Sketch the graph of $|x| + |y| = 1$.

# P.4 Functions and Their Graphs

The area of a circle depends on its radius. The temperature at which water boils depends on the altitude above sea level. The interest paid on a cash investment depends on the length of time for which the investment is made.

Whenever one quantity depends on another quantity we say that the former quantity is a function of the latter. For instance, the area $A$ of a circle depends on the radius $r$ according to the formula

$$A = \pi r^2,$$

so we say that the area is a function of the radius. The formula is a *rule* that tells us how to calculate a *unique* (single) output value of the area $A$ for each possible input value of the radius $r$.

The set of all possible input values for the radius is called the **domain** of the function. The set of all output values of the area is the **range** of the function. Since circles cannot have negative radii or areas, the domain and range of the circular area function are both the interval $[0, \infty)$ consisting of all nonnegative real numbers.

The domain and range of a mathematical function can be any sets of objects; they do not have to consist of numbers. Throughout much of this book, however, the domains and ranges of functions we consider will be sets of real numbers.

In calculus we often want to refer to a generic function without having any particular formula in mind. To denote that $y$ is a function of $x$ we write

$$y = f(x),$$

which we read as "$y$ equals $f$ of $x$." In this notation, due to eighteenth-century mathematician Leonhard Euler, the function is represented by the symbol $f$. Also, $x$, called the **independent variable**, represents an input value from the domain of $f$, and $y$, the **dependent variable**, represents the corresponding output value $f(x)$ in the range of $f$.

DEFINITION  **1**

> A **function** $f$ on a set $D$ into a set $S$ is a rule that assigns a *unique* element $f(x)$ in $S$ to each element $x$ in $D$.

In this definition $D = D(f)$ (read "$D$ of $f$") is the domain of the function $f$. The range $R(f)$ of $f$ is the subset of $S$ consisting of all *values* $f(x)$ of the function. Think of a function $f$ as a kind of machine that produces an output value $f(x)$ in its range whenever we feed it an input value $x$ from its domain.

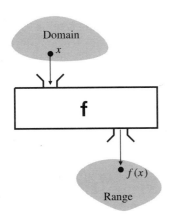

Domain

$x$

**f**

$f(x)$

Range

**Figure P.35**    A function machine

There are several ways to represent a function symbolically. The squaring function which converts any input real number $x$ into its square $x^2$ can be denoted:

(a) by a formula such as $y = x^2$ which uses a dependent variable $y$ to denote the value of the function;

(b) by a formula such as $f(x) = x^2$ which defines a function symbol $f$ to name the function; or

(c) by a mapping rule such as $x \longrightarrow x^2$. (Read this as "$x$ goes to $x^2$.")

In this book we will usually use either (a) or (b) to define functions. Strictly speaking, we should call a function $f$ and not $f(x)$, since the latter denotes the value of the function at the point $x$. However, as is common usage, we will often refer to the function as $f(x)$ in order to name the variable on which $f$ depends. Sometimes it is convenient to use the same letter to denote both a dependent variable and as a function symbol; the circular area function can be written $A = f(r) = \pi r^2$, or as $A = A(r) = \pi r^2$. In the latter case we are using $A$ to denote both the dependent variable and the name of the function.

**Example 1**    The volume of a ball of radius $r$ is given by the function

$$V(r) = \frac{4}{3} \pi r^3$$

for $r \geq 0$. Thus the volume of a ball of radius 3 ft is

$$V(3) = \frac{4}{3} \pi (3)^3 = 36\pi \text{ ft}^3.$$

Note how the variable $r$ is replaced by the special value 3 in the formula defining the function to obtain the value of the function at $r = 3$.

**Example 2**    If a function $F$ is defined for all real numbers $t$ by

$$F(t) = 2t + 3,$$

find the output values of $F$ that correspond to the input values 0, 2, $x + 2$, and $F(2)$.

***Solution***    In each case we substitute the given input for $t$ in the definition of $F$:

$$F(0) = 2(0) + 3 = 0 + 3 = 3$$
$$F(2) = 2(2) + 3 = 4 + 3 = 7$$
$$F(x + 2) = 2(x + 2) + 3 = 2x + 7$$
$$F(F(2)) = F(7) = 2(7) + 3 = 17.$$

## The Domain Convention

A function is not properly defined until its domain is specified. For instance, the function $f(x) = x^2$ defined for all real numbers $x \geq 0$ is different from the function $g(x) = x^2$ defined for all real $x$ because they have different domains, even though they have the same values at every point where both of them are defined. In Chapters

1–9 we will be dealing with real functions (functions whose input and output values are real numbers). When the domain of such a function is not specified explicitly, we will assume that the domain is the largest set of real numbers to which the function assigns real values. Thus, if we talk about the function $x^2$ without specifying a domain, we mean the function $g(x)$ above.

> **The domain convention**
>
> When a function $f$ is defined without specifying its domain, we assume that the domain consists of all real numbers $x$ for which the value $f(x)$ of the function is a real number.

In practice, it is often easy to determine the domain of a function $f(x)$ given by an explicit formula. We just have to exclude those values of $x$ that would result in dividing by 0 or taking even roots of negative numbers.

**Example 3**   **The square root function.** The domain of $f(x) = \sqrt{x}$ is the interval $[0, \infty)$, since negative numbers do not have real square roots. We have $f(0) = 0$, $f(4) = 2$, $f(10) \approx 3.16228$. Note that, although there are *two* numbers whose square is 4, namely $-2$ and 2, only *one* of these numbers, 2, is the square root of 4. (Remember that a function assigns a *unique* value to each element in its domain; it cannot assign two different values to the same input.) The **square root function** $\sqrt{x}$ always denotes the *nonnegative* square root of $x$. The two solutions of the equation $x^2 = 4$ are $x = \sqrt{4} = 2$ and $x = -\sqrt{4} = -2$.

**Example 4**   The domain of the function $h(x) = \dfrac{x}{x^2 - 4}$ consists of all real numbers except $x = -2$ and $x = 2$. Expressed in terms of intervals,

$$D(f) = (-\infty, -2) \cup (-2, 2) \cup (2, \infty).$$

Most of the functions we encounter will have domains that are either intervals or unions of intervals.

**Example 5**   The domain of $S(t) = \sqrt{1 - t^2}$ consists of all real numbers $t$ for which $1 - t^2 \geq 0$. Thus we require that $t^2 \leq 1$, or $-1 \leq t \leq 1$. The domain is the closed interval $[-1, 1]$.

## Graphs of Functions

An old maxim states that "a picture is worth a thousand words." This is certainly true in mathematics; the behaviour of a function is best described by drawing its graph.

The **graph of a function** $f$ is just the graph of the *equation* $y = f(x)$. It consists of those points in the Cartesian plane whose coordinates $(x, y)$ are pairs of input-output values for $f$. Thus $(x, y)$ lies on the graph of $f$ provided $x$ is in the domain of $f$ and $y = f(x)$.

Drawing the graph of a function $f$ sometimes involves making a table of coordinate pairs $(x, f(x))$ for various values of $x$ in the domain of $f$, then plotting these points and connecting them with a "smooth curve."

**Table 1.**

| $x$ | $y = f(x)$ |
|:---:|:---:|
| $-2$ | 4 |
| $-1$ | 1 |
| 0 | 0 |
| 1 | 1 |
| 2 | 4 |

**Example 6**   Graph the function $f(x) = x^2$.

**Solution**   Make a table of $(x, y)$ pairs that satisfy $y = x^2$. (See Table 1.)  Now plot the points and join them with a smooth curve. (See Figure P.36(a).)    ∎

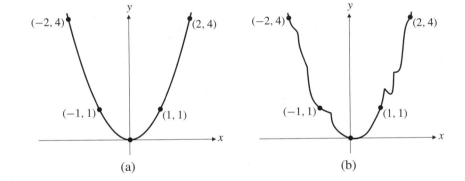

**Figure P.36**

(a)  Correct graph of $f(x) = x^2$

(b)  Incorrect graph of $f(x) = x^2$

How do we know the graph is smooth and doesn't do weird things between the points we have calculated (for example, as shown in Figure P.36(b))? We could, of course, plot more points, spaced more closely together, but how do we know how the graph behaves between the points we have plotted? In Chapter 4, calculus will provide useful tools for answering these questions.

Some functions occur often enough in applications that you should be familiar with their graphs. Some of these are shown in Figures P.37–P.46. Study them for

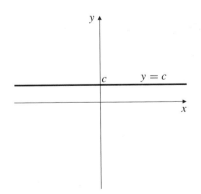

**Figure P.37**    The graph of a constant function $f(x) = c$

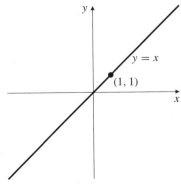

**Figure P.38**    The graph of $f(x) = x$

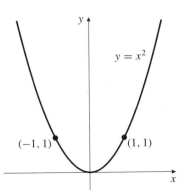

**Figure P.39**    The graph of $f(x) = x^2$

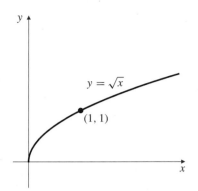

**Figure P.40** The graph of $f(x) = \sqrt{x}$

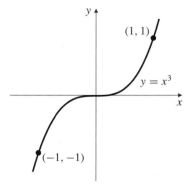

**Figure P.41** The graph of $f(x) = x^3$

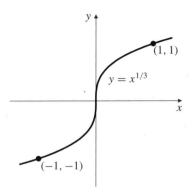

**Figure P.42** The graph of $f(x) = x^{1/3}$

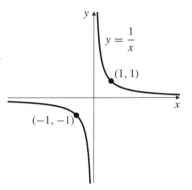

**Figure P.43** The graph of $f(x) = 1/x$

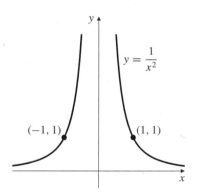

**Figure P.44** The graph of $f(x) = 1/x^2$

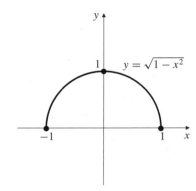

**Figure P.45** The graph of $f(x) = \sqrt{1 - x^2}$

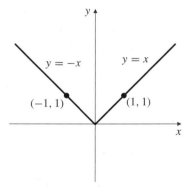

**Figure P.46** The graph of $f(x) = |x|$

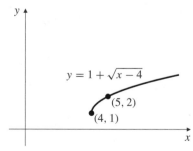

**Figure P.47** The graph of $y = \sqrt{x}$ shifted right 4 units and up 1 unit

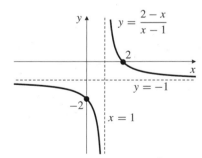

**Figure P.48** The graph of $\dfrac{2 - x}{x - 1}$

a while; they are worth remembering. Note, in particular, the graph of the **absolute value function**, $f(x) = |x|$, shown in Figure P.46. It is made up of the two half-lines $y = -x$ for $x < 0$ and $y = x$ for $x \geq 0$.

If you know the effects of vertical and horizontal shifts on the equations representing graphs (see Section P.3) you can easily sketch some graphs that are shifted versions of ones in Figures P.37–P.46.

**Example 7**  Sketch the graph of $y = 1 + \sqrt{x - 4}$.

**Solution**  This is just the graph of $y = \sqrt{x}$ in Figure P.40 shifted to the right 4 units (because $x$ is replaced by $x - 4$) and up 1 unit. See Figure P.47.  ∎

**Example 8**  Sketch the graph of the function $f(x) = \dfrac{2 - x}{x - 1}$.

**Solution**  It is not immediately obvious that this graph is a shifted version of a known graph. To see that it is, we can divide $x - 1$ into $2 - x$ to get a quotient of $-1$ and a remainder of 1:

$$\frac{2 - x}{x - 1} = \frac{-x + 1 + 1}{x - 1} = \frac{-(x - 1) + 1}{x - 1} = -1 + \frac{1}{x - 1}.$$

Thus the graph is that of $1/x$ from Figure P.43 shifted to the right 1 unit and down 1 unit. See Figure P.48.  ∎

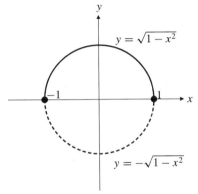

Not every curve you can draw is the graph of a function. A function $f$ can have only one value $f(x)$ for each $x$ in its domain, so no *vertical line* can intersect the graph of a function at more than one point. If $a$ is in the domain of function $f$ then the vertical line $x = a$ will intersect the graph of $f$ at the single point $(a, f(a))$. The circle $x^2 + y^2 = 1$ in Figure P.49 cannot be the graph of a function since some vertical lines intersect it twice. It is, however, the union of the graphs of two functions, namely

$$y = \sqrt{1 - x^2} \qquad \text{and} \qquad y = -\sqrt{1 - x^2},$$

which are, respectively, the upper and lower halves (semicircles) of the given circle.

**Figure P.49**    The circle $x^2 + y^2 = 1$ is not the graph of a function

## Even and Odd Functions; Symmetry and Reflections

It often happens that the graph of a function will have certain kinds of symmetry. The simplest kinds of symmetry relate the values of a function at $x$ and $-x$.

**DEFINITION** ❷

> **Even and odd functions**
>
> Suppose that $-x$ belongs to the domain of $f$ whenever $x$ does. We say that $f$ is an **even function** if
>
> $$f(-x) = f(x) \qquad \text{for every } x \text{ in the domain of } f.$$
>
> We say that $f$ is an **odd function** if
>
> $$f(-x) = -f(x) \qquad \text{for every } x \text{ in the domain of } f.$$

The names *even* and *odd* come from the fact that even powers such as $x^0 = 1$, $x^2$, $x^4, \ldots, x^{-2}, x^{-4}, \ldots$, are even functions, and odd powers such as $x^1 = x$, $x^3$, $\ldots, x^{-1}, x^{-3}, \ldots$, are odd functions. Observe, for example, that $(-x)^4 = x^4$ and $(-x)^{-3} = -x^{-3}$.

Since $(-x)^2 = x^2$, any function that depends only on $x^2$ is even. For instance, the absolute value function $y = |x| = \sqrt{x^2}$ is even.

The graph of an even function is *symmetric about the y-axis*. A horizontal straight line drawn from a point on the graph to the $y$-axis will, if continued an equal distance on the other side of the $y$-axis, come to another point on the graph. (See Figure P.50(a).)

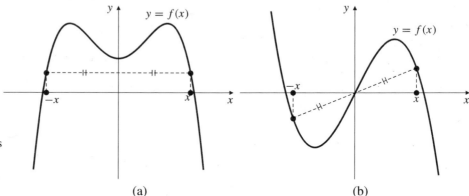

**Figure P.50**

(a) The graph of an even function is symmetric about the $y$-axis.

(b) The graph of an odd function is symmetric about the origin.

(a)    (b)

The graph of an odd function is *symmetric about the origin*. A straight line drawn from a point on the graph to the origin will, if continued an equal distance on the other side of the origin, come to another point on the graph. If an odd function $f$ is defined at $x = 0$ then its value must be zero there: $f(0) = 0$. (See Figure P.50(b).)

If $f(x)$ is even (or odd), then so is any constant multiple of $f(x)$ such as $2f(x)$ or $-5f(x)$. Sums (and differences) of even functions are even; sums (and differences) of odd functions are odd. For example, $f(x) = 3x^4 - 5x^2 - 1$ is even, since it is the sum of three even functions: $3x^4$, $-5x^2$, and $-1 = -x^0$. Similarly, $4x^3 - (2/x)$ is an odd function. The function $g(x) = x^2 - 2x$ is the sum of an even function and an odd function, and is itself neither even nor odd.

Other kinds of symmetry are also possible. For example, the function $g(x) = x^2 - 2x$ can be written in the form $g(x) = (x-1)^2 - 1$. This shows that the values of $g(1 \pm u)$ are equal, so the graph (Figure P.51(a)) is symmetric about the vertical line $x = 1$; it is the parabola $y = x^2$ shifted 1 unit to the right and 1 unit down. Similarly, the graph of $h(x) = x^3 + 1$ is symmetric about the point $(0, 1)$ (Figure P.51(b)).

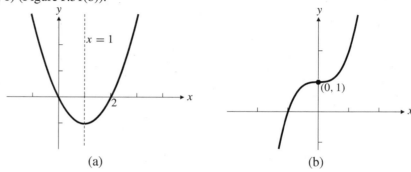

**Figure P.51**

(a) The graph of $g(x) = x^2 - 2x$ is symmetric about $x = 1$.

(b) The graph of $y = h(x) = x^3 + 1$ is symmetric about $(0, 1)$.

(a)    (b)

## Reflections in Straight Lines

The image of an object reflected in a plane mirror appears to be as far behind the mirror as the object is in front of it. Thus the mirror bisects the line from the object to the image at right angles. Given a line $L$ and a point $P$ not on $L$, we call a point $Q$ the **reflection** or the **mirror image** of $P$ in $L$ if $L$ is the right bisector of the line segment $PQ$. The reflection of any graph $G$ in $L$ is the graph consisting of the reflections of all the points of $G$.

Certain reflections of graphs are easily described in terms of the equations of the graphs:

### Reflections in special lines

1. Substituting $-x$ in place of $x$ in an equation in $x$ and $y$ corresponds to reflecting the graph of the equation in the $y$-axis.

2. Substituting $-y$ in place of $y$ in an equation in $x$ and $y$ corresponds to reflecting the graph of the equation in the $x$-axis.

3. Substituting $a - x$ in place of $x$ in an equation in $x$ and $y$ corresponds to reflecting the graph of the equation in the line $x = a/2$.

4. Substituting $b - y$ in place of $y$ in an equation in $x$ and $y$ corresponds to reflecting the graph of the equation in the line $y = b/2$.

5. Interchanging $x$ and $y$ in an equation in $x$ and $y$ corresponds to reflecting the graph of the equation in the line $y = x$.

**Example 9**   Describe and sketch the graph of $y = \sqrt{2-x} - 3$.

**Solution**   The graph of $y = \sqrt{2-x}$ is the reflection of the graph of $y = \sqrt{x}$ (Figure P.40) in the vertical line $x = 1$. The graph of $y = \sqrt{2-x} - 3$ is the result of lowering this reflection by 3 units. See Figure P.52(a). ∎

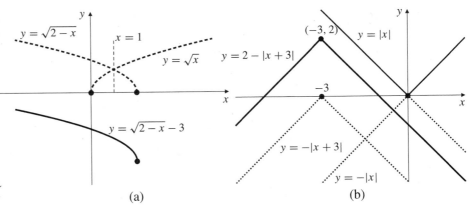

**Figure P.52**

(a) Constructing the graph of $y = \sqrt{2-x} - 3$

(b) Transforming $y = |x|$ to produce the coloured graph

(a)                    (b)

**Example 10**   Express the equation of the coloured graph in Figure P.52(b) in terms of the absolute value function $|x|$.

**Solution**   We can get the coloured graph by first reflecting the graph of $|x|$ (Figure P.46) in the $x$-axis, and then shifting the reflection left 3 units and up 2 units. The reflection of $y = |x|$ in the $x$-axis has equation $-y = |x|$, or $y = -|x|$. Shifting this left 3 units gives $y = -|x + 3|$. Finally, shifting up 2 units gives $y = 2 - |x + 3|$, which is the desired equation. ∎

# Exercises P.4

In Exercises 1–6, find the domain and range of each function.

**1.** $f(x) = 1 + x^2$

**2.** $f(x) = 1 - \sqrt{x}$

**3.** $G(x) = \sqrt{8 - 2x}$

**4.** $F(x) = 1/(x - 1)$

**5.** $h(t) = \dfrac{t}{\sqrt{2 - t}}$

**6.** $g(x) = \dfrac{1}{1 - \sqrt{x - 2}}$

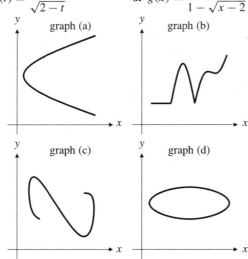

graph (a)

graph (b)

graph (c)

graph (d)

**Figure P.53**

**7.** Which of the graphs in Figure P.53 are graphs of functions $y = f(x)$? Why?

graph (a)

graph (b)

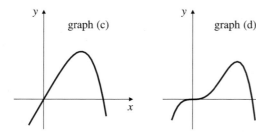

graph (c)

graph (d)

**Figure P.54**

**8.** Figure P.54 shows the graphs of the functions: (i) $x - x^4$, (ii) $x^3 - x^4$, (iii) $x(1 - x)^2$, and (iv) $x^2 - x^3$. Which graph corresponds to each function?

In Exercises 9–10, sketch the graph of the function $f$ by first making a table of values of $f(x)$ at $x = 0$, $x = \pm 1/2$, $x = \pm 1$, $x = \pm 3/2$, and $x = \pm 2$.

**9.** $f(x) = x^4$

**10.** $f(x) = x^{2/3}$

In Exercises 11–22, what (if any) symmetry does the graph of $f$ possess? In particular, is $f$ either even or odd?

**11.** $f(x) = x^2 + 1$

**12.** $f(x) = x^3 + x$

**13.** $f(x) = \dfrac{x}{x^2 - 1}$

**14.** $f(x) = \dfrac{1}{x^2 - 1}$

**15.** $f(x) = \dfrac{1}{x - 2}$

**16.** $f(x) = \dfrac{1}{x + 4}$

**17.** $f(x) = x^2 - 6x$

**18.** $f(x) = x^3 - 2$

**19.** $f(x) = |x^3|$

**20.** $f(x) = |x + 1|$

**21.** $f(x) = \sqrt{2x}$

**22.** $f(x) = \sqrt{(x - 1)^2}$

Sketch the graphs of the functions in Exercises 23–38.

**23.** $f(x) = -x^2$

**24.** $f(x) = 1 - x^2$

**25.** $f(x) = (x - 1)^2$

**26.** $f(x) = (x - 1)^2 + 1$

**27.** $f(x) = 1 - x^3$

**28.** $f(x) = (x + 2)^3$

**29.** $f(x) = \sqrt{x} + 1$

**30.** $f(x) = \sqrt{x + 1}$

**31.** $f(x) = -|x|$

**32.** $f(x) = |x| - 1$

**33.** $f(x) = |x - 2|$

**34.** $f(x) = 1 + |x - 2|$

**35.** $f(x) = \dfrac{2}{x + 2}$

**36.** $f(x) = \dfrac{1}{2 - x}$

**37.** $f(x) = \dfrac{x}{x + 1}$

**38.** $f(x) = \dfrac{x}{1 - x}$

In Exercises 39–46, $f$ refers to the function with domain $[0, 2]$ and range $[0, 1]$ whose graph is shown in Figure P.55. Sketch the graphs of the indicated functions and specify their domains and ranges.

**39.** $f(x) + 2$

**40.** $f(x) - 1$

**41.** $f(x + 2)$

**42.** $f(x - 1)$

**43.** $-f(x)$

**44.** $f(-x)$

**45.** $f(4 - x)$

**46.** $1 - f(1 - x)$

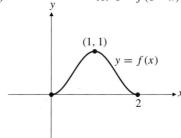

$(1, 1)$

$y = f(x)$

**Figure P.55**

It is often quite difficult to determine the range of a function exactly. In Exercises 47–48, use a graphing utility (calculator or computer) to graph the function $f$, and by zooming in on the graph determine the range of $f$ with accuracy of 2 decimal places.

**47.** $f(x) = \dfrac{x+2}{x^2+2x+3}$    **48.** $f(x) = \dfrac{x-1}{x^2+x}$    **50.** $f(x) = \dfrac{3-2x+x^2}{2-2x+x^2}$

In Exercises 49–52, use a graphing utility to plot the graph of the given function. Examine the graph (zooming in or out as necessary) for symmetries. About what lines and/or points are the graphs symmetric? Try to verify your conclusions algebraically.

**49.** $f(x) = x^4 - 6x^3 + 9x^2 - 1$

**51.** $f(x) = \dfrac{x-1}{x-2}$    **52.** $f(x) = \dfrac{2x^2+3x}{x^2+4x+5}$

**53.** What function $f(x)$, defined on the real line $\mathbb{R}$, is both even and odd?

## P.5    Combining Functions to Make New Functions

Functions can be combined in a variety of ways to make new functions. In this section we examine combinations obtained

(a) algebraically, by adding, subtracting, multiplying, and dividing functions;

(b) by composition, that is, taking functions of functions; and

(c) by piecing together functions defined on separate domains.

### Sums, Differences, Products, Quotients, and Multiples

Like numbers, functions can be added, subtracted, multiplied, and divided (except where the denominator is zero) to produce new functions.

**DEFINITION 3**

**Combining functions**

If $f$ and $g$ are functions, then for every $x$ that belongs to the domains of both $f$ and $g$ we define functions $f+g$, $f-g$, $fg$, and $f/g$ by the formulas:

$$(f+g)(x) = f(x) + g(x)$$
$$(f-g)(x) = f(x) - g(x)$$
$$(fg)(x) = f(x)g(x)$$
$$\left(\frac{f}{g}\right)(x) = \frac{f(x)}{g(x)}, \qquad \text{where } g(x) \neq 0.$$

A special case of the rule for multiplying functions shows how functions can be multiplied by constants. If $c$ is a real number, then the function $cf$ is defined for all $x$ in the domain of $f$ by

$$(cf)(x) = c\,f(x).$$

**Example 1**    Figure P.56(a) shows the graphs of $f(x) = x^2$, $g(x) = x-1$, and their sum $(f+g)(x) = x^2 + x - 1$. Observe that the height of the graph of $f+g$ at any point $x$ is the sum of the heights of the graphs of $f$ and $g$ at that point.

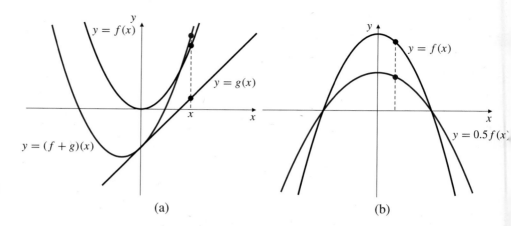

**Figure P.56**

(a) $(f + g)(x) = f(x) + g(x)$

(b) $g(x) = (0.5)f(x)$

(a)

(b)

---

**Example 2**    Figure P.56(b) shows the graphs of $f(x) = 2 - x^2$ and the multiple $g(x) = (0.5)f(x)$. Note how the height of the graph of $g$ at any point $x$ is half the height of the graph of $f$ there. ∎

---

**Example 3**    The functions $f$ and $g$ are defined by the formulas

$$f(x) = \sqrt{x} \quad \text{and} \quad g(x) = \sqrt{1 - x}.$$

Find formulas for the values of $3f, f + g, f - g, fg, f/g$, and $g/f$ at $x$, and specify the domains of each of these functions.

*Solution*    The information is collected in Table 2:

**Table 2.**    Combinations of $f$ and $g$ and their domains

| Function | Formula | Domain |
|:---:|:---:|:---:|
| $f$ | $f(x) = \sqrt{x}$ | $[0, \infty)$ |
| $g$ | $g(x) = \sqrt{1 - x}$ | $(-\infty, 1]$ |
| $3f$ | $(3f)(x) = 3\sqrt{x}$ | $[0, \infty)$ |
| $f + g$ | $(f + g)(x) = f(x) + g(x) = \sqrt{x} + \sqrt{1 - x}$ | $[0, 1]$ |
| $f - g$ | $(f - g)(x) = f(x) - g(x) = \sqrt{x} - \sqrt{1 - x}$ | $[0, 1]$ |
| $fg$ | $(fg)(x) = f(x)g(x) = \sqrt{x(1 - x)}$ | $[0, 1]$ |
| $f/g$ | $\dfrac{f}{g}(x) = \dfrac{f(x)}{g(x)} = \sqrt{\dfrac{x}{1 - x}}$ | $[0, 1)$ |
| $g/f$ | $\dfrac{g}{f}(x) = \dfrac{g(x)}{f(x)} = \sqrt{\dfrac{1 - x}{x}}$ | $(0, 1]$ |

Note that most of the combinations of $f$ and $g$ have domains

$$[0, \infty) \cap (-\infty, 1] = [0, 1],$$

the intersection of the domains of $f$ and $g$. However, the domains of the two quotients $f/g$ and $g/f$ had to be restricted further to remove points where the denominator was zero.

## Composite Functions

There is another method, called **composition**, by which two functions can be combined to form a new function.

**DEFINITION 4**

> **Composite functions**
>
> If $f$ and $g$ are two functions, the **composite** function $f \circ g$ is defined by
>
> $$f \circ g(x) = f(g(x)).$$
>
> The domain of $f \circ g$ consists of those numbers $x$ in the domain of $g$ for which $g(x)$ is in the domain of $f$. In particular, if the range of $g$ is contained in the domain of $f$, then the domain of $f \circ g$ is just the domain of $g$.

As shown in Figure P.57, forming $f \circ g$ is equivalent to arranging the "function machines" $g$ and $f$ in an "assembly line" so that the output of $g$ becomes the input of $f$.

In calculating $f \circ g(x) = f(g(x))$ we first calculate $g(x)$, and then calculate $f$ of the result. We call $g$ the *inner* function and $f$ the *outer* function of the composition. We can, of course, also calculate the composition $g \circ f(x) = g(f(x))$ where $f$ is the inner function, the one that gets calculated first, and $g$ is the outer function that gets calculated last. The functions $f \circ g$ and $g \circ f$ are usually quite different, as the following example shows.

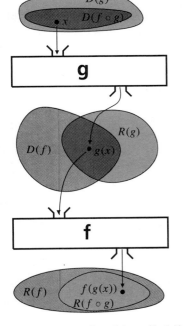

**Figure P.57**    $f \circ g(x) = f(g(x))$

**Example 4**    Given $f(x) = \sqrt{x}$ and $g(x) = x + 1$, calculate the four composite functions $f \circ g(x)$, $g \circ f(x)$, $f \circ f(x)$, and $g \circ g(x)$, and specify the domain of each.

**Solution**    Again, we collect the results in a table:

**Table 3.**    Composites of $f$ and $g$ and their domains

| Function | Formula | Domain |
|:---:|:---:|:---:|
| $f$ | $f(x) = \sqrt{x}$ | $[0, \infty)$ |
| $g$ | $g(x) = x + 1$ | $\mathbb{R}$ |
| $f \circ g$ | $f \circ g(x) = f(g(x)) = f(x + 1) = \sqrt{x + 1}$ | $[-1, \infty)$ |
| $g \circ f$ | $g \circ f(x) = g(f(x)) = g(\sqrt{x}) = \sqrt{x} + 1$ | $[0, \infty)$ |
| $f \circ f$ | $f \circ f(x) = f(f(x)) = f(\sqrt{x}) = \sqrt{\sqrt{x}} = x^{1/4}$ | $[0, \infty)$ |
| $g \circ g$ | $g \circ g(x) = g(g(x)) = g(x + 1) = (x + 1) + 1 = x + 2$ | $\mathbb{R}$ |

To see why, for example, the domain of $f \circ g$ is $[-1, \infty)$, observe that $g(x) = x + 1$ is defined for all real $x$, but belongs to the domain of $f$ only if $x + 1 \geq 0$, that is to say, if $x \geq -1$. ∎

**Example 5**    If $G(x) = \dfrac{1 - x}{1 + x}$, calculate $G \circ G(x)$ and specify its domain.

**Solution**    We calculate

$$G \circ G(x) = G(G(x)) = G\left(\frac{1-x}{1+x}\right) = \frac{1 - \dfrac{1-x}{1+x}}{1 + \dfrac{1-x}{1+x}} = \frac{1+x-1+x}{1+x+1-x} = x.$$

Because the resulting function, $x$, is defined for all real $x$, we might be tempted to say that the domain of $G \circ G$ is $\mathbb{R}$. This is wrong! To belong to the domain of $G \circ G$, $x$ must satisfy two conditions:

(i) $x$ must belong to the domain of $G$, and

(ii) $G(x)$ must belong to the domain of $G$.

The domain of $G$ consists of all real numbers *except* $x = -1$. If we exclude $x = -1$ from the domain of $G \circ G$ condition (i) will be satisfied. Now observe that the equation $G(x) = -1$ has no solution $x$, since it is equivalent to $1 - x = -(1 + x)$ or $1 = -1$. Therefore all numbers $G(x)$ belong to the domain of $G$ and condition (ii) is satisfied with no further restrictions on $x$. The domain of $G \circ G$ is $(-\infty, -1) \cup (-1, \infty)$, that is, all real numbers except $-1$. ∎

## Piecewise Defined Functions

Sometimes it is necessary to define a function by using different formulas on different parts of its domain. One example is the absolute value function

$$|x| = \begin{cases} x & \text{if } x \geq 0 \\ -x & \text{if } x < 0. \end{cases}$$

Here are some other examples:

**Example 6** **The Heaviside function.** The Heaviside function (Figure P.58) (or unit step function) is defined by

$$H(x) = \begin{cases} 1 & \text{if } x \geq 0 \\ 0 & \text{if } x < 0. \end{cases}$$

The function $H(t)$ can be used, for example, to model the voltage applied to an electric circuit by a one volt battery if a switch in the circuit is closed at time $t = 0$. ∎

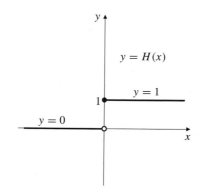

**Figure P.58**   The Heaviside function

**Example 7** **The signum function.** The signum function (Figure P.59) is defined by

$$\text{sgn}(x) = \frac{x}{|x|} = \begin{cases} 1 & \text{if } x > 0, \\ -1 & \text{if } x < 0, \\ \text{undefined} & \text{if } x = 0. \end{cases}$$

The name *signum* is the Latin word meaning "sign." The value of the $\text{sgn}(x)$ tells whether $x$ is positive or negative. Since 0 is neither positive nor negative, $\text{sgn}(0)$ is not defined. The signum function is an odd function.

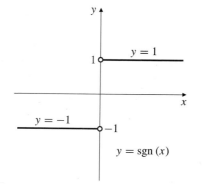

**Figure P.59**   The signum function

**Example 8**    The function

$$f(x) = \begin{cases} (x+1)^2 & \text{if } x < -1, \\ -x & \text{if } -1 \le x < 1, \\ \sqrt{x-1} & \text{if } x \ge 1, \end{cases}$$

is defined on the whole real line but has values given by three different formulas depending on the position of $x$. Its graph is shown in Figure P.60(a). Note how solid and hollow dots are used to indicate, respectively, which endpoints lie on the

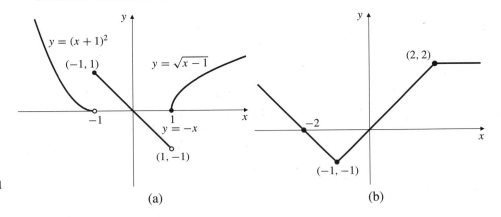

**Figure P.60**    Piecewise defined functions

(a)                                                    (b)

**Example 9**    Find a formula for function $g(x)$ graphed in Figure P.60(b).

***Solution***    The graph consists of parts of three lines. For the part $x < -1$, the line has slope $-1$ and $x$-intercept $-2$, so its equation is $y = -(x+2)$. The middle section is the line $y = x$ for $-1 \le x \le 2$. The right section is $y = 2$ for $x > 2$. Combining these formulas, we write

$$g(x) = \begin{cases} -(x+2) & \text{if } x < -1 \\ -x & \text{if } -1 \le x \le 2 \\ 2 & \text{if } x > 2. \end{cases}$$

Unlike the previous example, it does not matter here which of the two possible formulas we use to define $g(-1)$, since both give the same value. The same is true for $g(2)$.

The following two functions could be defined by different formulas on every interval between consecutive integers, but we will use an easier way to define them.

**Example 10**    **The greatest integer function.** The function whose value at any number $x$ is the *greatest integer less than or equal to $x$* is called the **greatest integer function** or the **integer floor function**. It is denoted $\lfloor x \rfloor$, or, in some books, $[x]$ or $[[x]]$. The graph of $y = \lfloor x \rfloor$ is given in Figure P.61(a). Observe that

$$\lfloor 2.4 \rfloor = 2, \qquad \lfloor 1.9 \rfloor = 1, \qquad \lfloor 0 \rfloor = 0, \qquad \lfloor -1.2 \rfloor = -2,$$
$$\lfloor 2 \rfloor = 2, \qquad \lfloor 0.2 \rfloor = 0, \qquad \lfloor -0.3 \rfloor = -1, \qquad \lfloor -2 \rfloor = -2.$$

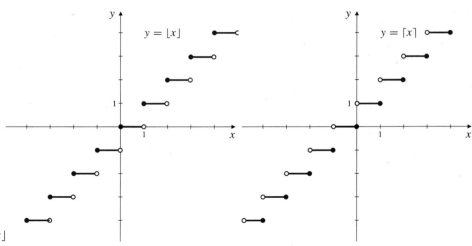

**Figure P.61**

(a) The greatest integer function $\lfloor x \rfloor$

(b) The least integer function $\lceil x \rceil$

(a)

(b)

---

**Example 11** **The least integer function.** The function whose value at any number $x$ is the *smallest integer greater than or equal to $x$* is called the **least integer function** or the **integer ceiling function**. It is denoted $\lceil x \rceil$. Its graph is given in Figure P.61(b). For positive values of $x$, this function might represent, for example, the cost of parking $x$ hours in a parking lot that charges \$1 for each hour or part of an hour. ∎

---

## Exercises P.5

In Exercises 1–2, find the domains of the functions $f + g$, $f - g$, $fg$, $f/g$, and $g/f$, and give formulas for their values.

**1.** $f(x) = x$, $\qquad g(x) = \sqrt{x - 1}$

**2.** $f(x) = \sqrt{1 - x}$, $\qquad g(x) = \sqrt{1 + x}$

Sketch the graphs of the functions in Exercises 3–6 by combining the graphs of simpler functions from which they are built up.

**3.** $x - x^2$ $\qquad\qquad$ **4.** $x^3 - x$

**5.** $x + |x|$ $\qquad\qquad$ **6.** $|x| + |x - 2|$

**7.** If $f(x) = x + 5$ and $g(x) = x^2 - 3$, find the following:

(a) $f \circ g(0)$ $\qquad$ (b) $g(f(0))$

(c) $f(g(x))$ $\qquad$ (d) $g \circ f(x)$

(e) $f \circ f(-5)$ $\qquad$ (f) $g(g(2))$

(g) $f(f(x))$ $\qquad$ (h) $g \circ g(x)$

In Exercises 8–10, construct the following composite functions and specify the domain of each.

(a) $f \circ f(x)$ $\qquad$ (b) $f \circ g(x)$

(c) $g \circ f(x)$ $\qquad$ (d) $g \circ g(x)$

**8.** $f(x) = 2/x$, $\qquad g(x) = x/(1 - x)$

**9.** $f(x) = 1/(1 - x)$, $\qquad g(x) = \sqrt{x - 1}$

**10.** $f(x) = (x + 1)/(x - 1)$, $\qquad g(x) = \text{sgn}(x)$

Find the missing entries in Table 4 (Exercises 11–16).

**Table 4.**

| | $f(x)$ | $g(x)$ | $f \circ g(x)$ |
|---|---|---|---|
| **11.** | $x^2$ | $x + 1$ | |
| **12.** | | $x + 4$ | $x$ |
| **13.** | $\sqrt{x}$ | | $|x|$ |
| **14.** | | $x^{1/3}$ | $2x + 3$ |
| **15.** | $(x + 1)/x$ | | $x$ |
| **16.** | | $x - 1$ | $1/x^2$ |

**17.** Use a graphing utility to examine in order the graphs of the functions

$$y = \sqrt{x}, \qquad\qquad y = 2 + \sqrt{x},$$
$$y = 2 + \sqrt{3 + x}, \qquad y = 1/(2 + \sqrt{3 + x}).$$

Describe the effect on the graph of the change made in the function at each stage.

**18.** Repeat the previous exercise for the functions

$$y = 2x, \qquad y = 2x - 1, \qquad y = 1 - 2x,$$
$$y = \sqrt{1 - 2x}, \qquad y = \frac{1}{\sqrt{1 - 2x}}, \qquad y = \frac{1}{\sqrt{1 - 2x}} - 1.$$

In Exercises 19–24, $f$ refers to the function with domain $[0, 2]$ and range $[0, 1]$ whose graph is shown in Figure P.62. Sketch the graphs of the indicated functions and specify their domains and ranges.

**19.** $2f(x)$ 

**20.** $-(1/2)f(x)$

**21.** $f(2x)$ 

**22.** $f(x/3)$

**23.** $1 + f(-x/2)$ 

**24.** $2f((x - 1)/2)$

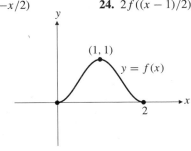

**Figure P.62**

In Exercises 25–26, sketch the graphs of the given functions.

**25.** $f(x) = \begin{cases} x & \text{if } 0 \le x \le 1 \\ 2 - x & \text{if } 1 < x \le 2 \end{cases}$

**26.** $g(x) = \begin{cases} \sqrt{x} & \text{if } 0 \le x \le 1 \\ 2 - x & \text{if } 1 < x \le 2 \end{cases}$

**27.** Find all real values of the constants $A$ and $B$ for which the function $F(x) = Ax + B$ satisfies:
(a) $F \circ F(x) = F(x)$ for all $x$.
(b) $F \circ F(x) = x$ for all $x$.

**Greatest and least integer functions**

**28.** For what values of $x$ is (a) $\lfloor x \rfloor = 0$? (b) $\lceil x \rceil = 0$?

**29.** What real numbers $x$ satisfy the equation $\lfloor x \rfloor = \lceil x \rceil$?

**30.** True or false: $\lceil -x \rceil = -\lfloor x \rfloor$ for all real $x$?

**31.** Sketch the graph of $y = x - \lfloor x \rfloor$.

**32.** Sketch the graph of the function

$$f(x) = \begin{cases} \lfloor x \rfloor & \text{if } x \ge 0 \\ \lceil x \rceil & \text{if } x < 0. \end{cases}$$

Why is $f(x)$ called *the integer part of $x$*?

**Even and odd functions**

**33.** Assume that $f$ is an even function, $g$ is an odd function, and both $f$ and $g$ are defined on the whole real line $\mathbb{R}$. Is each of the following functions even, odd, or neither?

$$f + g, \quad fg, \quad f/g, \quad g/f, \quad f^2 = ff, \quad g^2 = gg$$

$$f \circ g, \quad g \circ f, \quad f \circ f, \quad g \circ g$$

**34.** If $f$ is both an even and an odd function, show that $f(x) = 0$ at every point of its domain.

**35.** Let $f$ be a function whose domain is symmetric about the origin, that is, $-x$ belongs to the domain whenever $x$ does.

(a) Show that $f$ is the sum of an even function and an odd function:

$$f(x) = E(x) + O(x),$$

where $E$ is an even function and $O$ is an odd function. *Hint:* let $E(x) = (f(x) + f(-x))/2$. Show that $E(-x) = E(x)$, so that $E$ is even. Then show that $O(x) = f(x) - E(x)$ is odd.

(b) Show that there is only one way to write $f$ as the sum of an even and an odd function. *Hint:* one way is given in part (a). If also $f(x) = E_1(x) + O_1(x)$, where $E_1$ is even and $O_1$ is odd, show that $E - E_1 = O_1 - O$ and then use Exercise 34 to show that $E = E_1$ and $O = O_1$.

## P.6   The Trigonometric Functions

Most people first encounter the quantities $\cos t$ and $\sin t$ as ratios of sides in a right-angled triangle having $t$ as one of the acute angles. If the sides of the triangle are labelled "hyp" for hypotenuse, "adj" for the side adjacent to angle $t$, and "opp" for the side opposite angle $t$ (see Figure P.63), then

$$\cos t = \frac{\text{adj}}{\text{hyp}} \quad \text{and} \quad \sin t = \frac{\text{opp}}{\text{hyp}}.$$

These ratios depend only on the angle $t$, and not on the particular triangle, since all right-angled triangles having an acute angle $t$ are similar.

In calculus we need more general definitions of $\cos t$ and $\sin t$ as functions defined for *all real numbers $t$*, not just acute angles. Such definitions are phrased in terms of a circle rather than a triangle.

**Figure P.63**   $\cos t = \text{adj}/\text{hyp}$
$\sin t = \text{opp}/\text{hyp}$

Let $C$ be the circle with centre at the origin $O$ and radius 1; its equation is $x^2 + y^2 = 1$. Let $A$ be the point $(1, 0)$ on $C$. For any real number $t$, let $P_t$ be the point on $C$ at distance $|t|$ from $A$, measured along $C$ in the counterclockwise direction if $t > 0$, and the clockwise direction if $t < 0$. For example, since $C$ has circumference $2\pi$, the point $P_{\pi/2}$ is one-quarter of the way counterclockwise around $C$ from $A$; it is the point $(0, 1)$.

We will use the arc length $t$ as a measure of the size of the angle $AOP_t$. See Figure P.64.

**DEFINITION 5**

The **radian measure** of angle $AOP_t$ is $t$ radians:

$$\angle AOP_t = t \text{ radians.}$$

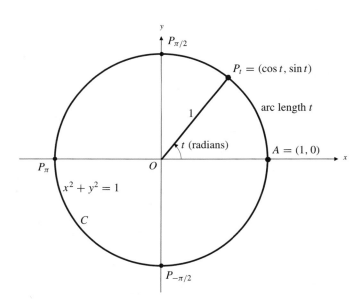

**Figure P.64**    Angle $AOP_t = t$ radians, where $t$ is the length of arc $AP_t$

We are more used to measuring angles in **degrees**. Since $P_\pi$ is the point $(-1, 0)$, halfway ($\pi$ units of distance) around $C$ from $A$, we have

$$\pi \text{ radians} = 180°.$$

To convert degrees to radians, multiply by $\pi/180$; to convert radians to degrees, multiply by $180/\pi$.

**Angle convention**

In calculus, it is assumed that all angles are measured in radians unless degrees or other units are stated explicitly. When we talk about the angle $\pi/3$, we mean $\pi/3$ radians (which is $60°$), not $\pi/3$ degrees.

**Example 1**    **Arc length and sector area.** An arc of a circle of radius $r$ subtends an angle $t$ at the centre of the circle. Find the length $s$ of the arc, and the area $A$ of the sector lying between the arc and the centre of the circle.

**Solution**    The length $s$ of the arc is the same fraction of the circumference $2\pi r$ of the circle that the angle $t$ is of a complete revolution $2\pi$ radians (or $360°$). Thus

$$s = \frac{t}{2\pi}\,(2\pi r) = rt \text{ units.}$$

Similarly, the area $A$ of the circular sector (Figure P.65) is the same fraction of the area $\pi r^2$ of the whole circle:

$$A = \frac{t}{2\pi}\,(\pi r^2) = \frac{r^2 t}{2} \text{ units}^2.$$

(We will show that the area of a circle of radius $r$ is $\pi r^2$ in Section 1.1.)

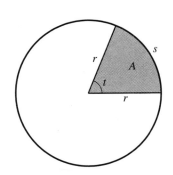

**Figure P.65**

Arc length $s = rt$

Sector area $A = r^2 t/2$

Using the procedure described above we can find the point $P_t$ corresponding to any real number $t$, positive or negative. We define $\cos t$ and $\sin t$ to be the coordinates of $P_t$. (See Figure P.66.)

**DEFINITION 6**

**Cosine and sine**

For any real $t$, the **cosine** of $t$ (abbreviated $\cos t$), and the **sine** of $t$ (abbreviated $\sin t$) are the $x$- and $y$-coordinates of the point $P_t$.

$$\cos t = \text{the } x\text{-coordinate of } P_t$$
$$\sin t = \text{the } y\text{-coordinate of } P_t$$

Because they are defined this way, cosine and sine are often called **the circular functions**.

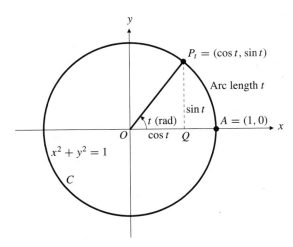

**Figure P.66**    The coordinates of $P_t$ are $(\cos t, \sin t)$

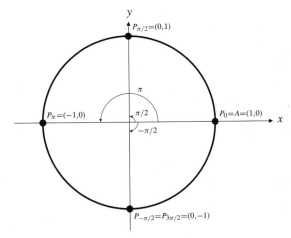

**Figure P.67**    Some special angles

> **Example 2** Examining the coordinates of $P_0 = A$, $P_{\pi/2}$, $P_\pi$, and $P_{-\pi/2} = P_{3\pi/2}$ in Figure P.67, we obtain the following values:
>
> $$\cos 0 = 1 \quad \cos\frac{\pi}{2} = 0 \quad \cos\pi = -1 \quad \cos\left(-\frac{\pi}{2}\right) = \cos\frac{3\pi}{2} = 0$$
>
> $$\sin 0 = 0 \quad \sin\frac{\pi}{2} = 1 \quad \sin\pi = 0 \quad \sin\left(-\frac{\pi}{2}\right) = \sin\frac{3\pi}{2} = -1$$

## Some Useful Identities

Many important properties of $\cos t$ and $\sin t$ follow from the fact that they are coordinates of the point $P_t$ on the circle $C$ with equation $x^2 + y^2 = 1$.

**The range of cosine and sine.** For every real number $t$,

$$-1 \le \cos t \le 1 \qquad \text{and} \qquad -1 \le \sin t \le 1.$$

**The Pythagorean identity.** The coordinates $x = \cos t$ and $y = \sin t$ of $P_t$ must satisfy the equation of the circle. Therefore, for every real number $t$,

$$\cos^2 t + \sin^2 t = 1.$$

(Note that $\cos^2 t$ means $(\cos t)^2$, not $\cos(\cos t)$. This is an unfortunate notation, but it is used everywhere in technical literature, so you have to get used to it!)

**Periodicity.** Since $C$ has circumference $2\pi$, adding $2\pi$ to $t$ causes the point $P_t$ to go one extra complete revolution around $C$ and end up in the same place: $P_{t+2\pi} = P_t$. Thus, for every $t$,

$$\cos(t + 2\pi) = \cos t \qquad \text{and} \qquad \sin(t + 2\pi) = \sin t.$$

This says that cosine and sine are **periodic** with period $2\pi$.

**Cosine is an even function. Sine is an odd function.** Since the circle $x^2 + y^2 = 1$ is symmetric about the $x$-axis, the points $P_{-t}$ and $P_t$ have the same $x$-coordinates and opposite $y$-coordinates (Figure P.68).

$$\cos(-t) = \cos t \qquad \text{and} \qquad \sin(-t) = -\sin t.$$

**Complementary angle identities.** Two angles are complementary if their sum is $\pi/2$ (or, $90°$). The points $P_{(\pi/2)-t}$ and $P_t$ are reflections of each other in the line $y = x$ (Figure P.69), so the $x$-coordinate of one is the $y$-coordinate of the other, and vice versa. Thus

$$\cos\left(\frac{\pi}{2} - t\right) = \sin t \qquad \text{and} \qquad \sin\left(\frac{\pi}{2} - t\right) = \cos t.$$

**Supplementary angle identities.** Two angles are supplementary if their sum is $\pi$ (or $180°$). Since the circle is symmetric about the $y$-axis, $P_{\pi-t}$ and $P_t$ have the same $y$-coordinates and opposite $x$-coordinates. (See Figure P.70.) Thus

$$\cos(\pi - t) = -\cos t \qquad \text{and} \qquad \sin(\pi - t) = \sin t.$$

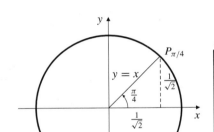

**Figure P.68**    $\cos(-t) = \cos t$

$\sin(-t) = -\sin t$

**Figure P.69**    $\cos((\pi/2) - t) = \sin t$

$\sin((\pi/2) - t) = \cos t$

**Figure P.70**

$\cos(\pi - t) = -\cos t$

$\sin(\pi - t) = \sin t$

## Some Special Angles

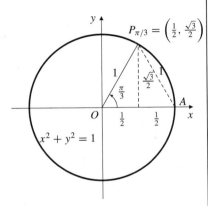

**Figure P.71**    $\sin \dfrac{\pi}{4} = \cos \dfrac{\pi}{4} = \dfrac{1}{\sqrt{2}}$

**Example 3**    Find the sine and cosine of $\pi/4$ (that is, $45°$).

**Solution**    The point $P_{\pi/4}$ lies in the first quadrant on the line $x = y$. To find its coordinates, substitute $y = x$ into the equation $x^2 + y^2 = 1$ of the circle, obtaining $2x^2 = 1$. Thus $x = y = 1/\sqrt{2}$, (see Figure P.71), and

$$\cos(45°) = \cos \frac{\pi}{4} = \frac{1}{\sqrt{2}}, \qquad \sin(45°) = \sin \frac{\pi}{4} = \frac{1}{\sqrt{2}}.$$

**Example 4**    Find the values of sine and cosine of the angles $\pi/3$ (or $60°$) and $\pi/6$ (or $30°$).

**Solution**    The point $P_{\pi/3}$ and the points $O(0, 0)$ and $A(1, 0)$ are the vertices of an equilateral triangle with edge length 1 (see Figure P.72). Thus $P_{\pi/3}$ has $x$-coordinate $1/2$ and $y$-coordinate $\sqrt{1 - (1/2)^2} = \sqrt{3}/2$, and

$$\cos(60°) = \cos \frac{\pi}{3} = \frac{1}{2}, \qquad \sin(60°) = \sin \frac{\pi}{3} = \frac{\sqrt{3}}{2}.$$

Since $\dfrac{\pi}{6} = \dfrac{\pi}{2} - \dfrac{\pi}{3}$, the complementary angle identities now tell us that

$$\cos(30°) = \cos \frac{\pi}{6} = \sin \frac{\pi}{3} = \frac{\sqrt{3}}{2}, \qquad \sin(30°) = \sin \frac{\pi}{6} = \cos \frac{\pi}{3} = \frac{1}{2}.$$

**Figure P.72**    $\cos \pi/3 = 1/2$

$\sin \pi/3 = \sqrt{3}/2$

Table 5 summarizes the values of cosine and sine at multiples of $30°$ and $45°$

between $0°$ and $180°$. The values for $120°$, $135°$, and $150°$ were determined by using the supplementary angle identities; for example,

$$\cos(120°) = \cos\left(\frac{2\pi}{3}\right) = \cos\left(\pi - \frac{\pi}{3}\right) = -\cos\left(\frac{\pi}{3}\right) = -\cos(60°) = -\frac{1}{2}.$$

**Table 5.**   Cosines and sines of special angles

| Degrees | $0°$ | $30°$ | $45°$ | $60°$ | $90°$ | $120°$ | $135°$ | $150°$ | $180°$ |
|---|---|---|---|---|---|---|---|---|---|
| Radians | $0$ | $\dfrac{\pi}{6}$ | $\dfrac{\pi}{4}$ | $\dfrac{\pi}{3}$ | $\dfrac{\pi}{2}$ | $\dfrac{2\pi}{3}$ | $\dfrac{3\pi}{4}$ | $\dfrac{5\pi}{6}$ | $\pi$ |
| Cosine | $1$ | $\dfrac{\sqrt{3}}{2}$ | $\dfrac{1}{\sqrt{2}}$ | $\dfrac{1}{2}$ | $0$ | $-\dfrac{1}{2}$ | $-\dfrac{1}{\sqrt{2}}$ | $-\dfrac{\sqrt{3}}{2}$ | $-1$ |
| Sine | $0$ | $\dfrac{1}{2}$ | $\dfrac{1}{\sqrt{2}}$ | $\dfrac{\sqrt{3}}{2}$ | $1$ | $\dfrac{\sqrt{3}}{2}$ | $\dfrac{1}{\sqrt{2}}$ | $\dfrac{1}{2}$ | $0$ |

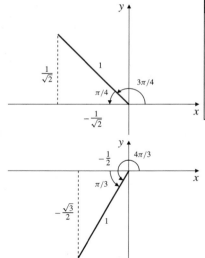

**Example 5**   Find:   (a) $\sin(3\pi/4)$ and   (b) $\cos(4\pi/3)$.

**Solution**   We can draw appropriate triangles in the quadrants where the angles lie to determine the required values. See Figure P.73.

(a) $\sin(3\pi/4) = \sin(\pi - (\pi/4)) = 1/\sqrt{2}$.

(b) $\cos(4\pi/3) = \cos(\pi + (\pi/3)) = -\dfrac{1}{2}$. ∎

While decimal approximations to the values of sine and cosine can be found using a scientific calculator or mathematical tables, it is useful to remember the exact values in the table for angles $0$, $\pi/6$, $\pi/4$, $\pi/3$, and $\pi/2$. They occur frequently in applications.

When we treat sine and cosine as functions we can call the variable they depend on $x$ (as we do with other functions), rather than $t$. The graphs of $\cos x$ and $\sin x$ are shown in Figures P.74 and P.75. In both graphs the pattern between $x = 0$ and $x = 2\pi$ repeats over and over to the left and right. Observe that the graph of $\sin x$ is the graph of $\cos x$ shifted to the right a distance $\pi/2$.

**Figure P.73**   Using suitably placed triangles to find trig functions of special angles

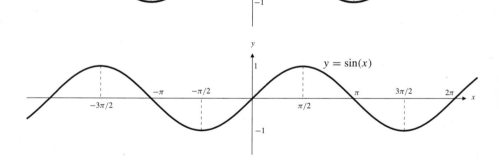

**Figure P.74**   The graph of $\cos x$

**Figure P.75**   The graph of $\sin x$

> **Remember this!**
>
> When using a scientific calculator to calculate any trigonometric functions, be sure you have selected the proper angular mode: degrees or radians.

## The Addition Formulas

The following formulas enable us to determine the cosine and sine of a sum or difference of two angles in terms of the cosines and sines of those angles.

**THEOREM** ❶  **Addition Formulas for Cosine and Sine**

$$\cos(s + t) = \cos s \cos t - \sin s \sin t$$
$$\sin(s + t) = \sin s \cos t + \cos s \sin t$$
$$\cos(s - t) = \cos s \cos t + \sin s \sin t$$
$$\sin(s - t) = \sin s \cos t - \cos s \sin t$$

**PROOF**   We prove the third of these formulas as follows: Let $s$ and $t$ be real numbers and consider the points

$$P_t = (\cos t, \sin t) \qquad P_{s-t} = (\cos(s - t), \sin(s - t))$$
$$P_s = (\cos s, \sin s) \qquad A = (1, 0),$$

as shown in Figure P.76.

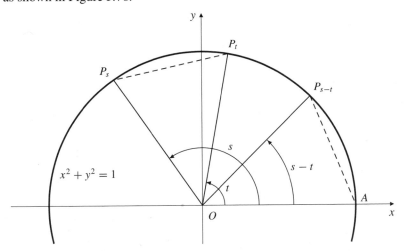

**Figure P.76**     $P_s P_t = P_{s-t} A$

The angle $P_t O P_s = s - t$ radians = angle $A O P_{s-t}$, so the distance $P_s P_t$ is equal to the distance $P_{s-t} A$. Therefore, $(P_s P_t)^2 = (P_{s-t} A)^2$. We express these squared distances in terms of coordinates and expand the resulting squares of binomials:

$$(\cos s - \cos t)^2 + (\sin s - \sin t)^2 = (\cos(s - t) - 1)^2 + \sin^2(s - t),$$

$$\cos^2 s - 2\cos s \cos t + \cos^2 t + \sin^2 s - 2\sin s \sin t + \sin^2 t$$
$$= \cos^2(s - t) - 2\cos(s - t) + 1 + \sin^2(s - t).$$

Since $\cos^2 x + \sin^2 x = 1$ for every $x$, this reduces to

$$\cos(s - t) = \cos s \cos t + \sin s \sin t.$$

Replacing $t$ with $-t$ in the formula above, and recalling that $\cos(-t) = \cos t$ and $\sin(-t) = -\sin t$, we have

$$\cos(s + t) = \cos s \cos t - \sin s \sin t.$$

The complementary angle formulas can be used to obtain either of the addition formulas for sine:

$$\sin(s + t) = \cos\left(\frac{\pi}{2} - (s + t)\right)$$
$$= \cos\left(\left(\frac{\pi}{2} - s\right) - t\right)$$
$$= \cos\left(\frac{\pi}{2} - s\right)\cos t + \sin\left(\frac{\pi}{2} - s\right)\sin t$$
$$= \sin s \cos t + \cos s \sin t,$$

and the other formula again follows if we replace $t$ with $-t$.

---

**Example 6**  Find the value of $\cos(\pi/12) = \cos 15°$.

*Solution*

$$\cos\frac{\pi}{12} = \cos\left(\frac{\pi}{3} - \frac{\pi}{4}\right) = \cos\frac{\pi}{3}\cos\frac{\pi}{4} + \sin\frac{\pi}{3}\sin\frac{\pi}{4}$$
$$= \left(\frac{1}{2}\right)\left(\frac{1}{\sqrt{2}}\right) + \left(\frac{\sqrt{3}}{2}\right)\left(\frac{1}{\sqrt{2}}\right) = \frac{1 + \sqrt{3}}{2\sqrt{2}}$$

---

From the addition formulas, we obtain as special cases certain useful formulas called **double-angle formulas**. Put $s = t$ in the addition formulas for $\sin(s + t)$ and $\cos(s + t)$ to get

$$\sin 2t = 2\sin t \cos t \qquad \text{and}$$
$$\cos 2t = \cos^2 t - \sin^2 t$$
$$= 2\cos^2 t - 1 \qquad (\text{using } \sin^2 t + \cos^2 t = 1)$$
$$= 1 - 2\sin^2 t$$

Solving the last two formulas for $\cos^2 t$ and $\sin^2 t$, we obtain

$$\cos^2 t = \frac{1 + \cos 2t}{2} \qquad \text{and} \qquad \sin^2 t = \frac{1 - \cos 2t}{2},$$

which are sometimes called **half-angle formulas** because they are used to express trig functions of half of the angle $2t$. Later we will find these formulas useful when we have to integrate powers of $\cos x$ and $\sin x$.

## Other Trigonometric Functions

There are four other trigonometric functions, tangent (tan), cotangent (cot), secant (sec), and cosecant (csc), defined in terms of cosine and sine. Their graphs are shown in Figures P.77–P.80.

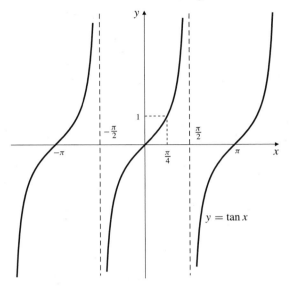

**Figure P.77**    The graph of tan $x$

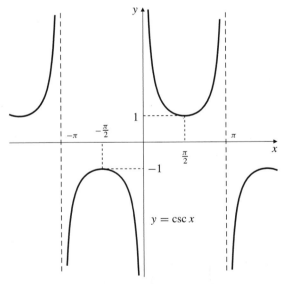

**Figure P.78**    The graph of cot $x$

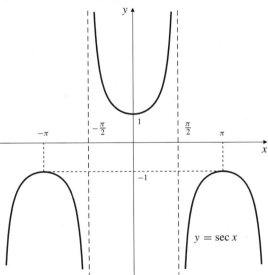

**Figure P.79**    The graph of sec $x$

**Figure P.80**    The graph of csc $x$

**DEFINITION** **7**

**Tangent, cotangent, secant, and cosecant**

$$\tan t = \frac{\sin t}{\cos t} \qquad\qquad \sec t = \frac{1}{\cos t}$$

$$\cot t = \frac{\cos t}{\sin t} = \frac{1}{\tan t} \qquad \csc t = \frac{1}{\sin t}$$

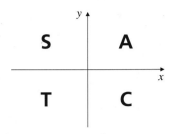

**Figure P.81**    The CAST rule

Observe that each of these functions is undefined (and its graph approaches vertical lines) at points where the function in the denominator of its defining fraction has value 0. Observe also that tangent, cotangent, and cosecant are odd functions and the secant is an even function. Since $|\sin x| \leq 1$ and $|\cos x| \leq 1$ for all $x$, $|\csc x| \geq 1$ and $|\sec x| \geq 1$ for all $x$ where they are defined.

The three functions sine, cosine, and tangent are called the **primary trigonometric functions** while their reciprocals, cosecant, secant, and cotangent are called the **secondary trigonometric functions**. Scientific calculators usually just implement the primary functions; you can use the reciprocal key to find values of the corresponding secondary functions. Figure P.81 shows a useful pattern called the "CAST rule" to help you remember where the primary functions are positive. All three are positive in the first quadrant, marked A. Of the three, only sine is positive in the second quadrant S, only tangent in the third quadrant T, and only cosine in the fourth quadrant C.

---

**Example 7**  Find the sine and tangent of the angle $\theta$ in $\left[\pi, \dfrac{3\pi}{2}\right]$ for which $\cos\theta = -\dfrac{1}{3}$.

**Solution**  From the Pythagorean identity $\sin^2\theta + \cos^2\theta = 1$ we get

$$\sin^2\theta = 1 - \frac{1}{9} = \frac{8}{9}, \quad \text{so} \quad \sin\theta = \pm\sqrt{\frac{8}{9}} = \pm\frac{2\sqrt{2}}{3}.$$

The requirement that $\theta$ should lie in $[\pi, 3\pi/2]$ makes $\theta$ a third quadrant angle. Its sine is therefore negative. We have

$$\sin\theta = -\frac{2\sqrt{2}}{3} \quad \text{and} \quad \tan\theta = \frac{\sin\theta}{\cos\theta} = \frac{-2\sqrt{2}/3}{-1/3} = 2\sqrt{2}.$$

---

Like their reciprocals cosine and sine, the functions secant and cosecant are periodic with period $2\pi$. Tangent and cotangent, however, have period $\pi$, because

$$\tan(x+\pi) = \frac{\sin(x+\pi)}{\cos(x+\pi)} = \frac{\sin x \cos\pi + \cos x \sin\pi}{\cos x \cos\pi - \sin x \sin\pi} = \frac{-\sin x}{-\cos x} = \tan x.$$

Dividing the Pythagorean identity $\sin^2 x + \cos^2 x = 1$ by $\cos^2 x$ and $\sin^2 x$, respectively, leads to two useful alternative versions of that identity:

$$1 + \tan^2 x = \sec^2 x, \qquad 1 + \cot^2 x = \csc^2 x.$$

Addition formulas for tangent and cotangent can be obtained from those for sine and cosine. For example,

$$\tan(s+t) = \frac{\sin(s+t)}{\cos(s+t)} = \frac{\sin s \cos t + \cos s \sin t}{\cos s \cos t - \sin s \sin t}.$$

Now divide the numerator and denominator of the fraction on the right by $\cos s \cos t$ to get

$$\tan(s+t) = \frac{\tan s + \tan t}{1 - \tan s \tan t}.$$

Replacing $t$ by $-t$ leads to

$$\tan(s - t) = \frac{\tan s - \tan t}{1 + \tan s \tan t}.$$

## Trigonometry Review

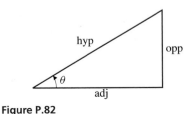

**Figure P.82**

The trigonometric functions are so called because they are often used to express the relationships between the sides and angles of a triangle. As we observed at the beginning of this section, if $\theta$ is one of the acute angles in a right-angled triangle we can refer to the three sides of the triangle as adj (side adjacent $\theta$), opp (side opposite $\theta$), and hyp (hypotenuse). The trigonometric functions of $\theta$ can then be expressed as ratios of these sides, in particular:

$$\sin \theta = \frac{\text{opp}}{\text{hyp}}, \qquad \cos \theta = \frac{\text{adj}}{\text{hyp}}, \qquad \tan \theta = \frac{\text{opp}}{\text{adj}}.$$

**Example 8**   Find the unknown sides $x$ and $y$ of the triangle in Figure P.83.

**Solution**   Here $x$ is the side opposite and $y$ is the side adjacent the $30°$ angle. The hypotenuse of the triangle is 5 units. Thus

$$\frac{x}{5} = \sin 30° = \frac{1}{2} \qquad \text{and} \qquad \frac{y}{5} = \cos 30° = \frac{\sqrt{3}}{2},$$

**Figure P.83**

so $x = \dfrac{5}{2}$ units and $y = \dfrac{5\sqrt{3}}{2}$ units.

**Example 9**   For the triangle in Figure P.84, express sides $x$ and $y$ in terms of side $a$ and angle $\theta$.

**Solution**   The side $x$ is opposite the angle $\theta$ and $y$ is the hypotenuse. The side adjacent $\theta$ is $a$. Thus

$$\frac{x}{a} = \tan \theta \qquad \text{and} \qquad \frac{a}{y} = \cos \theta.$$

**Figure P.84**

Hence $x = a \tan \theta$ and $y = \dfrac{a}{\cos \theta} = a \sec \theta$.

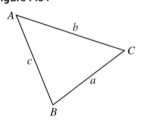

**Figure P.85**   In this triangle the sides are named to correspond to the opposite angles.

When dealing with general (not necessarily right-angled) triangles, it is often convenient to label the vertices with capital letters, which also denote the angles at those vertices, and refer to the sides opposite those vertices by the corresponding lower-case letters. See Figure P.85. Relationships among the sides $a$, $b$, and $c$ and opposite angles $A$, $B$, and $C$ of an arbitrary triangle $ABC$ are given by the following formulas, called the **Sine Law** and the **Cosine Law**.

**THEOREM** **2**

**Sine Law:** $\dfrac{\sin A}{a} = \dfrac{\sin B}{b} = \dfrac{\sin C}{c}$

**Cosine Law:** $a^2 = b^2 + c^2 - 2bc \cos A$

$b^2 = a^2 + c^2 - 2ac \cos B$

$c^2 = a^2 + b^2 - 2ab \cos C$

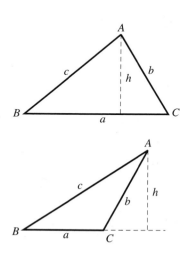

**Figure P.86**

**PROOF** See Figure P.86. Let $h$ be the length of the perpendicular from $A$ to the side $BC$. From right-angled triangles (and using $\sin(\pi - t) = \sin t$ if required), we get $c \sin B = h = b \sin C$. Thus $(\sin B)/b = (\sin C)/c$. By symmetry of the formulas (or by dropping a perpendicular to another side), both of these fractions must be equal to $(\sin A)/a$, so the Sine Law is proved. For the cosine law, observe that

$$c^2 = \begin{cases} h^2 + (a - b\cos C)^2 & \text{if } C \leq \dfrac{\pi}{2} \\[2ex] h^2 + (a + b\cos(\pi - C))^2 & \text{if } C > \dfrac{\pi}{2} \end{cases}$$

$$= h^2 + (a - b\cos C)^2 \qquad (\text{since } \cos(\pi - C) = -\cos C)$$

$$= b^2 \sin^2 C + a^2 - 2ab\cos C + b^2 \cos^2 C$$

$$= a^2 + b^2 - 2ab\cos C.$$

The other versions of the Cosine Law can be proved in a similar way.

---

**Example 10** A triangle has sides $a = 2$ and $b = 3$, and angle $C = 40°$. Find side $c$ and the sine of angle $B$.

**Solution** From the third version of the Cosine Law:

$$c^2 = a^2 + b^2 - 2ab\cos C = 4 + 9 - 12\cos 40° \approx 13 - 12 \times 0.766 = 3.808.$$

Side $c$ is about $\sqrt{3.808} = 1.951$ units in length. Now using Sine Law we get

$$\sin B = b\,\frac{\sin C}{c} \approx 3 \times \frac{\sin 40°}{1.951} \approx \frac{3 \times 0.6428}{1.951} \approx 0.988.$$

A triangle is uniquely determined by any one of the the following sets of data (which correspond to the known cases of congruency of triangles in classical geometry):

1. two sides and the angle contained between them (e.g. Example 10);

2. three sides, no one of which exceeds the sum of the other two in length;

3. two angles and one side; or

4. the hypotenuse and one other side of a right-angled triangle.

In such cases you can always find the unknown sides and angles by using the Pythagorean theorem or the Sine and Cosine Laws, and the fact that the sum of the three angles of a triangle is 180° (or $\pi$ radians).

A triangle is not determined uniquely by two sides and a non-contained angle; there may exist no triangle, one right-angled triangle, or two triangles having such data.

**Example 11**  In triangle $ABC$, angle $B = 30°$, $b = 2$, and $c = 3$. Find $a$.

**Solution**  This is one of the ambiguous cases. By the Cosine Law

$$b^2 = a^2 + c^2 - 2ac \cos B$$
$$4 = a^2 + 9 - 6a(\sqrt{3}/2).$$

Therefore, $a$ must satisfy the equation $a^2 - 3\sqrt{3}a + 5 = 0$. Solving this equation using the quadratic formula, we obtain

$$a = \frac{3\sqrt{3} \pm \sqrt{27 - 20}}{2}$$
$$\approx 1.275 \quad \text{or} \quad 3.921$$

There are two triangles with the given data, as shown in Figure P.87.

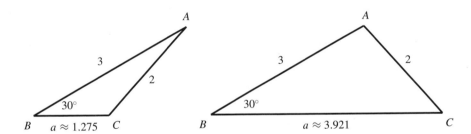

**Figure P.87**  Two triangles with $b = 2$, $c = 3$, $B = 30°$.

# Exercises P.6

Find the values of the quantities in Exercises 1–6 using various formulas presented in this section. Do not use tables or a calculator.

**1.** $\cos \dfrac{3\pi}{4}$   **2.** $\tan -\dfrac{3\pi}{4}$   **3.** $\sin \dfrac{2\pi}{3}$

**4.** $\sin \dfrac{7\pi}{12}$   **5.** $\cos \dfrac{5\pi}{12}$   **6.** $\sin \dfrac{11\pi}{12}$

In Exercises 7–12, express the given quantity in terms of $\sin x$ and $\cos x$.

**7.** $\cos(\pi + x)$   **8.** $\sin(2\pi - x)$   **9.** $\sin\left(\dfrac{3\pi}{2} - x\right)$

**10.** $\cos\left(\dfrac{3\pi}{2} + x\right)$   **11.** $\tan x + \cot x$   **12.** $\dfrac{\tan x - \cot x}{\tan x + \cot x}$

In Exercises 13–16, prove the given identities.

**13.** $\cos^4 x - \sin^4 x = \cos(2x)$

**14.** $\dfrac{1 - \cos x}{\sin x} = \dfrac{\sin x}{1 + \cos x} = \tan \dfrac{x}{2}$

**15.** $\dfrac{1 - \cos x}{1 + \cos x} = \tan^2 \dfrac{x}{2}$

**16.** $\dfrac{\cos x - \sin x}{\cos x + \sin x} = \sec 2x - \tan 2x$

**17.** Express $\sin 3x$ in terms of $\sin x$ and $\cos x$.

**18.** Express $\cos 3x$ in terms of $\sin x$ and $\cos x$.

In Exercises 19–22, sketch the graph of the given function. What is the period of the function?

**19.** $f(x) = \cos 2x$   **20.** $f(x) = \sin \dfrac{x}{2}$

**21.** $f(x) = \sin \pi x$   **22.** $f(x) = \cos \dfrac{\pi x}{2}$

**23.** Sketch the graph of $y = 2 \cos\left(x - \dfrac{\pi}{3}\right)$.

**24.** Sketch the graph of $y = 1 + \sin\left(x + \dfrac{\pi}{4}\right)$.

In Exercises 25–30, one of $\sin\theta$, $\cos\theta$, and $\tan\theta$ is given. Find the other two if $\theta$ lies in the specified interval.

**25.** $\sin\theta = \dfrac{3}{5}$, $\quad\theta$ in $\left[\dfrac{\pi}{2}, \pi\right]$

**26.** $\tan\theta = 2$, $\quad\theta$ in $\left[0, \dfrac{\pi}{2}\right]$

**27.** $\cos\theta = \dfrac{1}{3}$, $\quad\theta$ in $\left[-\dfrac{\pi}{2}, 0\right]$

**28.** $\cos\theta = -\dfrac{5}{13}$, $\quad\theta$ in $\left[\dfrac{\pi}{2}, \pi\right]$

**29.** $\sin\theta = \dfrac{-1}{2}$, $\quad\theta$ in $\left[\pi, \dfrac{3\pi}{2}\right]$

**30.** $\tan\theta = \dfrac{1}{2}$, $\quad\theta$ in $\left[\pi, \dfrac{3\pi}{2}\right]$

**Trigonometry Review**

In Exercises 31–42, $ABC$ is a triangle with a right angle at $C$. The sides opposite angles $A$, $B$, and $C$ are $a$, $b$, and $c$, respectively.

**Figure P.88**

**31.** Find $a$ and $b$ if $c = 2$, $B = \dfrac{\pi}{3}$.

**32.** Find $a$ and $c$ if $b = 2$, $B = \dfrac{\pi}{3}$.

**33.** Find $b$ and $c$ if $a = 5$, $B = \dfrac{\pi}{6}$.

**34.** Express $a$ in terms of $A$ and $c$.

**35.** Express $a$ in terms of $A$ and $b$.

**36.** Express $a$ in terms of $B$ and $c$.

**37.** Express $a$ in terms of $B$ and $b$.

**38.** Express $c$ in terms of $A$ and $a$.

**39.** Express $c$ in terms of $A$ and $b$.

**40.** Express $\sin A$ in terms of $a$ and $c$.

**41.** Express $\sin A$ in terms of $b$ and $c$.

**42.** Express $\sin A$ in terms of $a$ and $b$.

In Exercises 43–52, $ABC$ is an arbitrary triangle with sides $a$, $b$, and $c$, opposite to angles $A$, $B$, and $C$, respectively. Find the indicated quantities. Use tables or a scientific calculator if necessary.

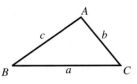

**Figure P.89**

**43.** Find $\sin B$ if $a = 4$, $b = 3$, $A = \dfrac{\pi}{4}$.

**44.** Find $\cos A$ if $a = 2$, $b = 2$, $c = 3$.

**45.** Find $\sin B$ if $a = 2$, $b = 3$, $c = 4$.

**46.** Find $c$ if $a = 2$, $b = 3$, $C = \dfrac{\pi}{4}$.

**47.** Find $a$ if $c = 3$, $A = \dfrac{\pi}{4}$, $B = \dfrac{\pi}{3}$.

**48.** Find $c$ if $a = 2$, $b = 3$, $C = 35°$.

**49.** Find $b$ if $a = 4$, $B = 40°$, $C = 70°$.

**50.** Find $c$ if $a = 1$, $b = \sqrt{2}$, $A = 30°$. (There are two possible answers.)

**51.** Two guy wires stretch from the top $T$ of a vertical pole to points $B$ and $C$ on the ground, where $C$ is 10 m closer to the base of the pole than is $B$. If wire $BT$ makes angle of $35°$ with the horizontal, and wire $CT$ makes an angle of $50°$ with the horizontal, how high is the pole?

**52.** Observers at positions $A$ and $B$ 2 km apart simultaneously measure the angle of elevation of a weather balloon to be $40°$ and $70°$, respectively. If the balloon is directly above a point on the line segment between $A$ and $B$, find the height of the balloon.

**53.** Show that the area of triangle $ABC$ is given by $(1/2)ab\sin C = (1/2)bc\sin A = (1/2)ca\sin B$.

**⁕ 54.** Show that the area of triangle $ABC$ is given by $\sqrt{s(s-a)(s-b)(s-c)}$, where $s = (a+b+c)/2$ is the semi-perimeter of the triangle.

⁕ This symbol is used throughout the book to indicate an exercise that is somewhat more difficult and/or theoretical than most exercises.

**CHAPTER 1**

# Limits and Continuity

**Introduction**   Calculus was created to describe how quantities change. It has two basic procedures that are opposites of one another:

- *differentiation,* for finding the rate of change of a given function, and
- *integration,* for finding a function having a given rate of change.

Both of these procedures are based on the fundamental concept of the *limit* of a function. It is this idea of limit that distinguishes calculus from algebra, geometry, and trigonometry, which are useful for describing static situations.

In this chapter we will introduce the limit concept and develop some of its properties. We begin by considering how limits arise in some basic problems.

## 1.1 Examples of Velocity, Growth Rate, and Area

In this section we consider some examples of phenomena where limits arise in a natural way.

### Average Velocity and Instantaneous Velocity

The position of a moving object is a function of time. The average velocity of the object over a time interval is found by dividing the change in the object's position by the length of the time interval.

**Example 1**   **(The average velocity of a falling rock)**   Physical experiments show that if a rock is dropped from rest near the surface of the earth, in the first $t$ seconds it will fall a distance

$$y = 4.9t^2 \text{ m.}$$

(a) What is the average velocity of the falling rock during the first two seconds?

(b) What is its average velocity from $t = 1$ to $t = 2$?

**Solution**   The *average velocity* of the falling rock over any time interval $[t_1, t_2]$ is the change $\Delta y$ in the distance fallen divided by the length $\Delta t$ of the time interval:

$$\text{average velocity over } [t_1, t_2] = \frac{\Delta y}{\Delta t} = \frac{4.9t_2^2 - 4.9t_1^2}{t_2 - t_1}.$$

(a) In the first two seconds (time interval $[0, 2]$), the average velocity is

$$\frac{\Delta y}{\Delta t} = \frac{4.9(2^2) - 4.9(0^2)}{2 - 0} = 9.8 \text{ m/s}.$$

(b) In the time interval [1, 2], the average velocity is

$$\frac{\Delta y}{\Delta t} = \frac{4.9(2^2) - 4.9(1^2)}{2 - 1} = 14.7 \text{ m/s}.$$

∎

**Example 2**   How fast is the rock in Example 1 falling (a) at time $t = 1$? (b) at time $t = 2$?

**Solution**   We can calculate the average velocity over any time interval, but this question asks for the *instantaneous velocity* at a given time. If the falling rock had a speedometer, what would it show at time $t = 1$? To answer this, we first write the average velocity over the time interval $[1, 1 + h]$ starting at $t = 1$, and having length $h$ seconds:

$$\text{Average velocity over } [1, 1 + h] = \frac{\Delta y}{\Delta t} = \frac{4.9(1 + h)^2 - 4.9(1^2)}{h}.$$

We can't calculate the instantaneous velocity at $t = 1$ by substituting $h = 0$ in this expression, because we can't divide by zero. But we can calculate the average velocities over shorter and shorter time intervals and see whether they seem to get close to a particular number. Table 1 shows the value of $\Delta y / \Delta t$ for some values of $h$ approaching zero. Indeed, it appears that these average velocities get closer and closer to 9.8 m/s as the length of the time interval gets closer and closer to zero. This suggests that the rock is falling at a rate of 9.8 m/s one second after it is dropped.

Similarly, Table 2 shows values of the average velocities over shorter and shorter time intervals $[2, 2 + h]$ starting at $t = 2$. The values suggest that the rock is falling at 19.6 m/s two seconds after it is dropped.

∎

**Table 1.**   Average velocity over $[1, 1 + h]$

| $h$ | $\Delta y / \Delta t$ |
| --- | --- |
| 1 | 14.7000 |
| 0.1 | 10.2900 |
| 0.01 | 9.8490 |
| 0.001 | 9.8049 |
| 0.0001 | 9.8005 |

**Table 2.**   Average velocity over $[2, 2 + h]$

| $h$ | $\Delta y / \Delta t$ |
| --- | --- |
| 1 | 24.5000 |
| 0.1 | 20.0900 |
| 0.01 | 19.6490 |
| 0.001 | 19.6049 |
| 0.0001 | 19.6005 |

In Example 2 the average velocity of the falling rock over the time interval $[t, t + h]$ is

$$\frac{\Delta y}{\Delta t} = \frac{4.9(t + h)^2 - 4.9t^2}{h}.$$

To find the instantaneous velocity (usually just called *the velocity*) at the instants $t = 1$ and $t = 2$, we examined the values of this average velocity for time intervals whose lengths $h$ became smaller and smaller. We were, in fact, finding the *limit of the average velocity as h approaches zero*. This is expressed symbolically in the form

$$\text{velocity at time } t = \lim_{h \to 0} \frac{4.9(t + h)^2 - 4.9t^2}{h}.$$

Read "$\lim_{h \to 0} \ldots$" as "the limit as $h$ approaches zero of $\ldots$" We can't find the limit of the fraction by just substituting $h = 0$ because that would involve dividing by zero. However, we can calculate the limit by first performing some algebraic simplifications on the expression for the average velocity.

**Example 3**    Expand $(t + h)^2$ in the numerator of the average velocity of the rock over $[t, t + h]$ and simplify the result. Hence find the velocity of the falling rock at time $t$ directly, without making a table of values.

**Solution**    The average velocity over $[t, t + h]$ is

$$\frac{4.9(t + h)^2 - 4.9t^2}{h} = \frac{4.9(t^2 + 2th + h^2 - t^2)}{h}$$
$$= \frac{4.9(2th + h^2)}{h}$$
$$= 9.8t + 4.9h.$$

The final form of the expression no longer involves division by $h$. It approaches $9.8t + 4.9(0) = 9.8t$ as $h$ approaches 0. Thus the velocity of the falling rock is $9.8t$ m/s $t$ seconds after it is dropped. In particular, at $t = 1$ and $t = 2$ the velocities are 9.8 m/s and 19.6 m/s, respectively.    ∎

## The Growth of an Algal Culture

In a laboratory experiment, the biomass of an algal culture was measured over a 74 day period by measuring the area in square millimetres occupied by the culture on a microscope slide. These measurements $m$ were plotted against the time $t$ in days, and the points joined by a smooth curve $m = f(t)$, as shown in Figure 1.1.

**Figure 1.1**    The biomass $m$ of an algal culture after $t$ days

Observe that the biomass was about 0.1 on day 10 and had grown to about 1.7 on day 40, an increase of $1.7 - 0.1 = 1.6$ mm$^2$ in a time interval of $40 - 10 = 30$ days. The average rate of growth over the time interval from day 10 to day 40 was therefore

$$\frac{1.7 - 0.1}{40 - 10} = \frac{1.6}{30} \approx 0.053 \text{ mm}^2/\text{d}.$$

This average rate is just the slope of the line joining the points on the graph of $m = f(t)$ corresponding to $t = 10$ and $t = 40$. Similarly, the average rate of growth of the algal biomass over any time interval can be determined by measuring the slope of the line joining the points on the curve corresponding to that time interval. Such lines are called **secant lines** to the curve.

---

**Example 4** How fast is the biomass growing on day 60?

**Solution** To answer this question we could measure the average rates of change over shorter and shorter times around day 60. The corresponding secant lines become shorter and shorter, but their slopes approach a *limit,* namely the slope of the **tangent line** to the graph of $m = f(t)$ at the point where $t = 60$. This tangent line is sketched in Figure 1.1; it seems to go through the points $(2, 0)$ and $(69, 5)$, so that its slope is

$$\frac{5 - 0}{69 - 2} \approx 0.0746 \text{ mm}^2/\text{d}.$$

This is the rate at which the biomass was growing on day 60.

---

## The Area of a Circle

All circles are similar geometric figures; they all have the same shape and differ only in size. The ratio of the circumference $C$ to the diameter $2r$ (twice the radius) has the same value for all circles. The number $\pi$ is defined to be this common ratio:

$$\frac{C}{2r} = \pi \qquad \text{or} \qquad C = 2\pi r.$$

In school we are taught that the ratio of the area $A$ of a circle is this same number $\pi$ times the square of the radius:

$$A = \pi r^2.$$

How can we deduce this area formula from the formula for the circumference that is the definition of $\pi$?

The answer to this question lies in regarding the circle as a "limit" of regular polygons, which are in turn made up of triangles, figures about whose geometry we know a great deal.

Suppose a regular polygon having $n$ sides is inscribed in a circle of radius $r$. (See Figure 1.2.) The perimeter $P_n$ and the area $A_n$ of the polygon are respectively less than the circumference $C$ and the area $A$ of the circle, but if $n$ is large, $P_n$ is *close to* $C$ and $A_n$ is *close to* $A$. (In fact the "circle" in Figure 1.2 was drawn by a computer as a regular polygon having 180 sides, each subtending a $2°$ angle at the centre of the circle. It is very hard to distinguish this 180-sided polygon from a real circle.) We would expect $P_n$ to approach the limit $C$ and $A_n$ to approach the limit $A$ as $n$ grows larger and larger and approaches infinity.

A regular polygon of $n$ sides is the union of $n$ non-overlapping, congruent, isosceles triangles having a common vertex at $O$, the centre of the polygon. One of these triangles, $\triangle OAB$, is shown in Figure 1.2. Since the total angle around the point $O$ is $2\pi$ radians (we are assuming that a circle of radius 1 has circumference $2\pi$), the angle $AOB$ is $2\pi/n$ radians. If $M$ is the midpoint of $AB$, then $OM$ bisects angle $AOB$. Using elementary trigonometry, we can write the length of $AB$ and the area of triangle $OAB$ in terms of the radius $r$ of the circle:

$$|AB| = 2|AM| = 2r \sin \frac{\pi}{n}$$

$$\text{area } OAB = \frac{1}{2}|AB||OM| = \frac{1}{2}\left(2r \sin \frac{\pi}{n}\right)\left(r \cos \frac{\pi}{n}\right)$$

$$= r^2 \sin \frac{\pi}{n} \cos \frac{\pi}{n}.$$

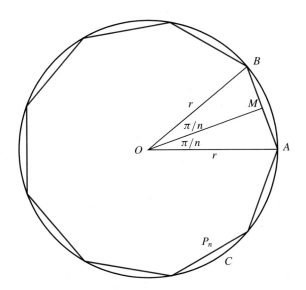

**Figure 1.2**    A regular polygon of $n$
sides inscribed in a circle

The perimeter $P_n$ and area $A_n$ of the polygon are $n$ times these expressions:

$$P_n = 2rn \sin \frac{\pi}{n}$$

$$A_n = r^2 n \sin \frac{\pi}{n} \cos \frac{\pi}{n}.$$

Solving the first equation for $rn \sin(\pi/n) = P_n/2$ and substituting into the second equation, we get

$$A_n = \left(\frac{P_n}{2}\right) r \cos \frac{\pi}{n}.$$

Now the angle $AOM = \pi/n$ approaches 0 as $n$ grows large, so its cosine, $\cos(\pi/n) = |OM|/|OA|$, approaches 1. Since $P_n$ approaches $C = 2\pi r$ as $n$ grows large, the expression for $A_n$ approaches $(2\pi r/2)r(1) = \pi r^2$, which must therefore be the area of the circle.

# Exercises 1.1

Exercises 1–4 refer to an object moving along the $x$-axis in such a way that its position at time $t$ (seconds) is $x = t^2$ ft to the right of the origin.

**1.** Find the average velocity of the object over the time interval $[t, t + h]$.

**2.** Make a table giving the average velocities of the object over time intervals $[2, 2 + h]$, for $h = 1, 0.1, 0.01, 0.001$, and $0.0001$ seconds.

**3.** Use the results from Exercise 2 to guess the instantaneous velocity of the object at $t = 2$ seconds.

**4.** Confirm your guess in Exercise 3 by calculating the limit of the average velocity over $[2, 2 + h]$ as $h$ approaches zero, using the method of Example 3.

Exercises 5–8 refer to the motion of a particle moving along the $x$-axis so that at time $t$ s it is at position $x = 3t^2 - 12t + 1$ m.

**5.** Find the average velocity of the particle over the time intervals $[1, 2]$, $[2, 3]$, and $[1, 3]$.

**6.** Use the method of Example 3 to find the velocity of the particle at $t = 1$, $t = 2$, and $t = 3$.

**7.** In what direction is the particle moving at $t = 1$? $t = 2$? $t = 3$?

**8.** Show that for any positive number $k$, the average velocity of the particle over the time interval $[t - k, t + k]$ is equal to its velocity at time $t$.

In Exercises 9–11, a weight that is suspended by a spring bobs up and down so that its height above the floor at time $t$ seconds is $y$ feet, where

$$y = 2 + \frac{1}{\pi} \sin(\pi t).$$

**9.** Sketch the graph of $y$ as a function of $t$. How high is the weight at $t = 1$ second? In what direction is it moving at that time?

**10.** What is the average velocity of the weight over the time intervals $[1, 2]$, $[1, 1.1]$, $[1, 1.01]$, and $[1, 1.001]$?

**11.** Using the results of Exercise 10, estimate the velocity of the weight at time $t = 1$. What is the significance of the sign of your answer.

Exercises 12–13 refer to the algal biomass graphed in Figure 1.1.

**12.** Approximately how fast is the biomass growing on day 20?

**13.** On about what day is the biomass growing fastest?

**14.** The profits of a small company for each of the first five years of its operation are given in Table 3:

**Table 3.**

| Year | Profit ($1,000s) |
|------|------------------|
| 1990 | 6 |
| 1991 | 27 |
| 1992 | 62 |
| 1993 | 111 |
| 1994 | 174 |

(a) Plot points representing the profit as a function of year on graph paper, and join them by a smooth curve.

(b) What is the average rate of increase in the profits between 1992 and 1994?

(c) Use your graph to estimate the rate of increase in the profits in 1992.

# 1.2    Limits of Functions

In order to speak meaningfully about rates of change, tangent lines, and areas bounded by curves, we have to investigate the process of finding limits. Indeed, the concept of *limit* is the cornerstone on which the development of calculus rests. Before we try to give a definition of a limit, let us look at more examples.

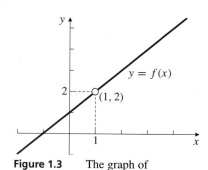

**Figure 1.3**    The graph of $f(x) = \dfrac{x^2 - 1}{x - 1}$

**Example 1**    Describe the behaviour of the function $f(x) = \dfrac{x^2 - 1}{x - 1}$ near $x = 1$.

**Solution**    Note that $f(x)$ is defined for all real numbers $x$ except $x = 1$. (We can't divide by zero.) For any $x \neq 1$ we can simplify the expression for $f(x)$ by factoring the numerator and cancelling common factors:

$$f(x) = \frac{(x - 1)(x + 1)}{x - 1} = x + 1 \qquad \text{for} \quad x \neq 1.$$

The graph of $f$ is the line $y = x + 1$ with one point removed, namely the point $(1, 2)$. This removed point is shown as a "hole" in the graph in Figure 1.3. Even though $f(1)$ is not defined, it is clear that we can make the value of $f(x)$ *as close as we want* to 2 by choosing $x$ *close enough* to 1. Therefore, we say that $f(x)$ approaches arbitrarily close to 2 as $x$ approaches 1, or, more simply, $f(x)$ approaches *the limit* 2 as $x$ approaches 1. We write this as

$$\lim_{x \to 1} f(x) = 2 \qquad \text{or} \qquad \lim_{x \to 1} \frac{x^2 - 1}{x - 1} = 2.$$

■

**Example 2**    What happens to the function $g(x) = (1 + x^2)^{1/x^2}$ as $x$ approaches zero?

**Table 4.**

| $x$ | $g(x)$ |
|---|---|
| $\pm 1.0$ | 2.0000 00000 |
| $\pm 0.1$ | 2.7048 13829 |
| $\pm 0.01$ | 2.7181 45927 |
| $\pm 0.001$ | 2.7182 80469 |
| $\pm 0.0001$ | 2.7182 81815 |
| $\pm 0.00001$ | 1.0000 00000 |

**Solution**    The function $g(x)$ is not defined at $x = 0$. In fact, for the moment it does not appear to be defined for any $x$ whose square $x^2$ is not a rational number. (Recall that if $r = m/n$, where $m$ and $n$ are integers and $n > 0$, then $x^r$ means the $n$th root of $x^m$.) Let us ignore for now the problem of deciding what $g(x)$ means if $x^2$ is irrational, and consider only rational values of $x$. There is no obvious way to simplify the expression for $g(x)$ as we did in Example 1. However, we can use a scientific calculator to obtain approximate values of $g(x)$ for some rational values of $x$ approaching 0. (The values in Table 4 were obtained with a TI-85.)

Except for the last value in the table, the values of $g(x)$ seem to be approaching a certain number $2.71828\ldots$ as $x$ approaches closer and closer to 0. We will show in Section 3.4 that

$$\lim_{x \to 0} g(x) = \lim_{x \to 0} (1 + x^2)^{1/x^2} = e = 2.7\,1828\,1828\,45\,90\,45\ldots.$$

The number $e$ turns out to be very important in mathematics.

Observe that the last entry in the table appears to be wrong. This is because the calculator was unable to distinguish $1 + (0.00001)^2 = 1.0000000001$ from 1, and it therefore calculated $1^{10,000,000,000} = 1$. When you use a calculator or computer to evaluate expressions, always be aware of the effects of round-off errors. In cases like this such effects can be catastrophic.

■

The examples above and those in Section 1.1 suggest the following *informal* definition of limit.

**DEFINITION 1**

> **An informal definition of limit**
>
> If $f(x)$ is defined for all $x$ near $a$, except possibly at $a$ itself, and if we can ensure that $f(x)$ is as close as we want to $L$ by taking $x$ close enough to $a$, we say that the function $f$ approaches the **limit** $L$ as $x$ approaches $a$, and we write
>
> $$\lim_{x \to a} f(x) = L.$$

This definition is *informal* because phrases such as *close as we want* and *close enough* are imprecise; their meaning depends on the context. To a machinist manufacturing a piston *close enough* may mean *within a few thousandths of an inch*. To an astronomer studying distant galaxies *close enough* may mean *within a few thousand light years*. The definition should be clear enough, however, to enable us to recognize and evaluate limits of specific functions. A more precise "formal" definition, given in Section 1.5, is needed if we want to *prove* theorems about limits like Theorems 2–4, stated later in this section.

**Example 3**    Find    (a) $\lim\limits_{x \to a} x$, and    (b) $\lim\limits_{x \to a} c$ (where $c$ is a constant).

**Solution**    In words, part (a) asks: "What does $x$ approach as $x$ approaches $a$?" The answer is surely $a$.

$$\lim_{x \to a} x = a.$$

Similarly, part (b) asks: "What does $c$ approach as $x$ approaches $a$?" The answer here is that $c$ approaches $c$; you can't get any closer to $c$ than by *being c*.

$$\lim_{x \to a} c = c.$$

Example 3 shows that $\lim_{x \to a} f(x)$ can *sometimes* be evaluated by just calculating $f(a)$. This will be the case if $f(x)$ is defined in an open interval containing $x = a$ and the graph of $f$ passes unbroken through the point $(a, f(a))$. The next example shows various ways algebraic manipulations can be used to evaluate $\lim_{x \to a} f(x)$ in situations where $f(a)$ is undefined. This usually happens when $f(x)$ is a fraction with denominator equal to 0 at $x = a$.

**Example 4**    Evaluate:

(a) $\lim\limits_{x \to -2} \dfrac{x^2 + x - 2}{x^2 + 5x + 6}$,    (b) $\lim\limits_{x \to a} \dfrac{\dfrac{1}{x} - \dfrac{1}{a}}{x - a}$, and    (c) $\lim\limits_{x \to 4} \dfrac{\sqrt{x} - 2}{x^2 - 16}$.

**Solution**    Each of these limits involves a fraction whose numerator and denominator are both 0 at the point where the limit is taken.

(a)    $\lim\limits_{x \to -2} \dfrac{x^2 + x - 2}{x^2 + 5x + 6}$          undefined at $x = -2$

Factor numerator and denominator.

$\quad = \lim\limits_{x \to -2} \dfrac{(x + 2)(x - 1)}{(x + 2)(x + 3)}$          Cancel common factors.

$\quad = \lim\limits_{x \to -2} \dfrac{x - 1}{x + 3}$          Evaluate this limit by substituting $x = -2$.

$\quad = \dfrac{-2 - 1}{-2 + 3} = -3.$

(b) $\displaystyle\lim_{x\to a}\frac{\dfrac{1}{x}-\dfrac{1}{a}}{x-a}$ 

undefined at $x = a$

Simplify the numerator.

$$= \lim_{x\to a}\frac{\dfrac{a-x}{ax}}{x-a}$$

$$= \lim_{x\to a}\frac{-(x-a)}{ax(x-a)} \qquad \text{Cancel the common factor.}$$

$$= \lim_{x\to a}\frac{-1}{ax} = -\frac{1}{a^2}.$$

(c) $\displaystyle\lim_{x\to 4}\frac{\sqrt{x}-2}{x^2-16}$

undefined at $x = 4$

Multiply numerator and denominator by the conjugate of the expression in the numerator.

$$= \lim_{x\to 4}\frac{(\sqrt{x}-2)(\sqrt{x}+2)}{(x^2-16)(\sqrt{x}+2)}$$

$$= \lim_{x\to 4}\frac{x-4}{(x-4)(x+4)(\sqrt{x}+2)}$$

$$= \lim_{x\to 4}\frac{1}{(x+4)(\sqrt{x}+2)} = \frac{1}{(4+4)(2+2)} = \frac{1}{32}.$$

A function $f$ may be defined on both sides of $x = a$, but still not have a limit at $x = a$. For example, the function $f(x) = 1/x$ has no limit as $x$ approaches 0. As can be seen in Figure 1.4(a), the values $1/x$ grow ever larger in absolute value as $x$ approaches 0; there is no single number $L$ that they approach.

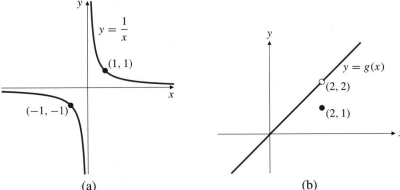

**Figure 1.4**

(a) $\displaystyle\lim_{x\to 0}\frac{1}{x}$ does not exist.

(b) $\displaystyle\lim_{x\to 2} g(x) = 2$, but $g(2) = 1$.

(a)

(b)

The following example shows that even if $f(x)$ is defined at $x = a$, the limit of $f(x)$ as $x$ approaches $a$ may not be equal to $f(a)$.

Always be aware that the existence of $\lim_{x\to a} f(x)$ does not require that $f(a)$ exist, and does not depend on $f(a)$ even if $f(a)$ does exist. It depends only on the values of $f(x)$ for $x$ *near but not equal to a*.

**Example 5** Let $g(x) = \begin{cases} x & \text{if } x \neq 2 \\ 1 & \text{if } x = 2. \end{cases}$ (See Figure 1.4(b).) Then

$$\lim_{x\to 2} g(x) = \lim_{x\to 2} x = 2, \qquad \text{although} \quad g(2) = 1.$$

## One-Sided Limits

Limits are *unique*; if $\lim_{x \to a} f(x) = L$ and $\lim_{x \to a} f(x) = M$, then $L = M$. (See Exercise 31 in Section 1.5.) Although a function $f$ can only have one limit at any particular point, nevertheless, it is useful to be able to describe the behaviour of functions that approach different numbers as $x$ approaches $a$ from one side or the other. (See Figure 1.5.)

**DEFINITION 2**

> ### Left and right limits
>
> If $f(x)$ is defined on some interval $(a, b)$ extending to the right of $x = a$, and if we can ensure that $f(x)$ is as close as we want to $L$ by taking $x$ to the right of $a$ and close enough to $a$, then we say $f(x)$ has **right limit** $L$ at $x = a$, and we write
>
> $$\lim_{x \to a+} f(x) = L.$$
>
> If $f(x)$ is defined on some interval $(c, a)$ extending to the left of $x = a$, and if we can ensure that $f(x)$ is as close as we want to $L$ by taking $x$ to the left of $a$ and close enough to $a$, then we say $f(x)$ has **left limit** $L$ at $x = a$, and we write
>
> $$\lim_{x \to a-} f(x) = L.$$

negative side of $a$
= left-hand side of $a$

$x \to a-$ means $x$ approaches $a$ from the left

positive side of $a$
= right-hand side of $a$

$x \to a+$ means $x$ approaches $a$ from the right

**Figure 1.5**    One-sided approach

Note the use of the suffix $+$ to denote approach from the right (the *positive* side) and the suffix $-$ to denote approach from the left (the *negative* side).

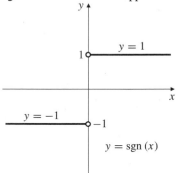

**Figure 1.6**

$\lim_{x \to 0}$ sgn $(x)$ does not exist, because
$\lim_{x \to 0-}$ sgn $(x) = -1$,  $\lim_{x \to 0+}$ sgn $(x) = 1$

**Example 6**    The signum function sgn $(x) = x/|x|$ (see Figure 1.6) has left limit $-1$ and right limit 1 at $x = 0$:

$$\lim_{x \to 0-} \text{sgn}(x) = -1 \quad \text{and} \quad \lim_{x \to 0+} \text{sgn}(x) = 1,$$

because the values of sgn $(x)$ approach $-1$ (they *are* $-1$) if $x$ is negative and approaches 0, and they approach 1 if $x$ is positive and approaches 0. Since these left and right limits are not equal, $\lim_{x \to 0}$ sgn $(x)$ *does not exist*. ∎

As suggested in Example 6, the relationship between ordinary (two-sided) limits and one-sided limits can be stated as follows:

**THEOREM 1**

### Relationship between one-sided and two-sided limits

A function $f(x)$ has limit $L$ at $x = a$ if and only if it has both left and right limits there and these one-sided limits are both equal to $L$:

$$\lim_{x \to a} f(x) = L \quad \Longleftrightarrow \quad \lim_{x \to a-} f(x) = \lim_{x \to a+} f(x) = L.$$

**Example 7**   If $f(x) = \dfrac{|x-2|}{x^2+x-6}$, find: $\lim\limits_{x\to 2+} f(x)$,   $\lim\limits_{x\to 2-} f(x)$,   and   $\lim\limits_{x\to 2} f(x)$.

***Solution***   Observe that $|x-2| = x-2$ if $x > 2$, and $|x-2| = -(x-2)$ if $x < 2$. Therefore

$$\lim_{x\to 2+} f(x) = \lim_{x\to 2+} \frac{x-2}{x^2+x-6} \qquad\qquad \lim_{x\to 2-} f(x) = \lim_{x\to 2-} \frac{-(x-2)}{x^2+x-6}$$

$$= \lim_{x\to 2+} \frac{x-2}{(x-2)(x+3)} \qquad\qquad\qquad = \lim_{x\to 2-} \frac{-(x-2)}{(x-2)(x+3)}$$

$$= \lim_{x\to 2+} \frac{1}{x+3} = \frac{1}{5}, \qquad\qquad\qquad = \lim_{x\to 2-} \frac{-1}{x+3} = -\frac{1}{5}.$$

Since $\lim_{x\to 2-} f(x) \neq \lim_{x\to 2+} f(x)$, the limit $\lim_{x\to 2} f(x)$ does not exist.

$y = \sqrt{1-x^2}$

**Figure 1.7**   $\sqrt{1-x^2}$ has right limit 0 at $-1$ and left limit 0 at 1

**Example 8**   What limits does $g(x) = \sqrt{1-x^2}$ have at $x = -1$ and $x = 1$?

***Solution***   The domain of $g$ is $[-1, 1]$, so $g(x)$ is defined only to the right of $x = -1$ and only to the left of $x = 1$. As can be seen in Figure 1.7,

$$\lim_{x\to -1+} g(x) = 0 \qquad \text{and} \qquad \lim_{x\to 1-} g(x) = 0.$$

$g(x)$ has no left limit or limit at $x = -1$, and no right limit or limit at $x = 1$.

## Rules for Calculating Limits

The following theorems make it easy to calculate limits and one-sided limits of many kinds of functions when we know some elementary limits. We will not prove the theorems here. (See Section 1.5.)

**THEOREM 2**

**Limit Rules**

If $\lim_{x\to a} f(x) = L$, $\lim_{x\to a} g(x) = M$, and $k$ is a constant, then

1. **Limit of a sum:** $\quad\lim\limits_{x\to a}[f(x) + g(x)] = L + M$

2. **Limit of a difference:** $\quad\lim\limits_{x\to a}[f(x) - g(x)] = L - M$

3. **Limit of a product:** $\quad\lim\limits_{x\to a} f(x)g(x) = LM$

4. **Limit of a multiple:** $\quad\lim\limits_{x\to a} kf(x) = kL$

5. **Limit of a quotient:** $\quad\lim\limits_{x\to a} \dfrac{f(x)}{g(x)} = \dfrac{L}{M}$,   if $M \neq 0$.

If $m$ is an integer and $n$ is a positive integer, then

6. **Limit of a power:** $\quad\lim\limits_{x\to a}\left[f(x)\right]^{m/n} = L^{m/n}$, provided $L > 0$ if $n$ is even, and $L \neq 0$ if $m < 0$.

If $f(x) \leq g(x)$ on an interval containing $a$ in its interior, then

7. **Order is preserved:** $\quad L \leq M$

Rules 1–6 are also valid for right limits and left limits. So is Rule 7, under the assumption that $f(x) \leq g(x)$ on an open interval extending in the appropriate direction from $a$.

In words, part 1 of Theorem 2 says that the limit of a sum of functions is the sum of their limits. Similarly, part 5 says that the limit of a quotient of two functions is the quotient of their limits, provided that the limit of the denominator is not zero. Try to state the other parts in words.

We can make use of the limits (a) $\lim_{x \to a} c = c$, (where $c$ is a constant), and (b) $\lim_{x \to a} x = a$, from Example 3, together with parts of Theorem 2 to calculate limits of many combinations of functions.

**Example 9** Find: (a) $\lim_{x \to a} \dfrac{x^2 + x + 4}{x^3 - 2x^2 + 7}$ and (b) $\lim_{x \to 2} \sqrt{2x + 1}$.

*Solution*

(a) The expression $\dfrac{x^2 + x + 4}{x^3 - 2x^2 + 7}$ is formed by combining the basic functions $x$ and $c$ (constant) using addition, subtraction, multiplication, and division. Theorem 2 assures us that the limit of such a combination is the same combination of the limits $a$ and $c$ of the basic functions, provided the denominator does not have limit zero. Thus

$$\lim_{x \to a} \frac{x^2 + x + 4}{x^3 - 2x^2 + 7} = \frac{a^2 + a + 4}{a^3 - 2a^2 + 7} \qquad \text{provided } a^3 - 2a^2 + 7 \neq 0.$$

(b) The same argument as in (a) shows that $\lim_{x \to 2}(2x + 1) = 2(2) + 1 = 5$. Then the Power Rule (part 6 of Theorem 2) assures us that

$$\lim_{x \to 2} \sqrt{2x + 1} = \sqrt{5}.$$

---

**DEFINITION 3**

A **polynomial** in the variable $x$ is a function of the form

$$P(x) = a_n x^n + a_{n-1} x^{n-1} + \cdots + a_1 x + a_0,$$

where $a_n, a_{n-1}, \ldots, a_1$, and $a_0$ are constants and $a_n \neq 0$. The number $n$ is the **degree** of the polynomial $P(x)$. A quotient $P(x)/Q(x)$ of two polynomials is called a **rational function**.

For example,

| | |
|---:|---|
| 3 | is a polynomial of degree 0. |
| $2 - x$ | is a polynomial of degree 1. |
| $2x^3 - 17x + 1$ | is a polynomial of degree 3. |
| $\dfrac{x}{x^2 + 1}$ | is a rational function. |

The following result is an immediate corollary of Theorem 2.

**THEOREM 3**

**Limits of Polynomials and Rational Functions**

1. If $P(x)$ is a polynomial and $a$ is any real number, then

$$\lim_{x \to a} P(x) = P(a).$$

2.  If $P(x)$ and $Q(x)$ are polynomials and $Q(a) \neq 0$, then

$$\lim_{x \to a} \frac{P(x)}{Q(x)} = \frac{P(a)}{Q(a)}.$$

## The Squeeze Theorem

The following theorem will enable us to calculate some very important limits in subsequent chapters. It is called the *Squeeze Theorem* because it refers to a function $g$ whose values are squeezed between the values of two other functions $f$ and $h$ that have the same limit $L$ at a point $a$. Being trapped between the values of two functions which approach $L$, the values of $g$ must also approach $L$. (See Figure 1.8.)

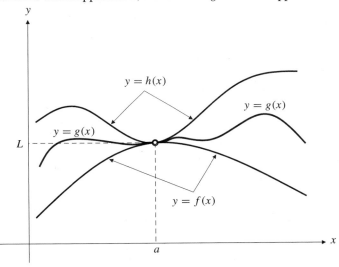

**Figure 1.8**   The graph of $g$ is squeezed between those of $f$ and $h$.

**THEOREM  4**

### The Squeeze Theorem

Suppose that $f(x) \leq g(x) \leq h(x)$ hold for all $x$ in some open interval containing $a$, except possibly at $x = a$ itself. Suppose also that

$$\lim_{x \to a} f(x) = \lim_{x \to a} h(x) = L.$$

Then $\lim_{x \to a} g(x) = L$ also. Similar statements hold for left and right limits.

---

**Example 10**   Given that $3 - x^2 \leq u(x) \leq 3 + x^2$ for all $x \neq 0$, find $\lim_{x \to 0} u(x)$.

***Solution***   Since $\lim_{x \to 0}(3 - x^2) = 3$ and $\lim_{x \to 0}(3 + x^2) = 3$, the Squeeze Theorem implies that $\lim_{x \to 0} u(x) = 3$.

---

**Example 11**   Show that if $\lim_{x \to a} |f(x)| = 0$, then $\lim_{x \to a} f(x) = 0$.

***Solution***   Since $-|f(x)| \leq f(x) \leq |f(x)|$, and $-|f(x)|$ and $|f(x)|$ both have limit 0 as $x$ approaches $a$, so does $f(x)$ by the Squeeze Theorem.

## Exercises 1.2

**1.** Find: (a) $\lim\limits_{x \to -1} f(x)$,  (b) $\lim\limits_{x \to 0} f(x)$,  and (c) $\lim\limits_{x \to 1} f(x)$, for the function $f$ whose graph is shown in Figure 1.9.

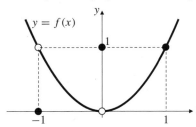

**Figure 1.9**

**2.** For the function $y = g(x)$ graphed in Figure 1.10, find each of the following limits or explain why it does not exist.
(a) $\lim\limits_{x \to 1} g(x)$,  (b) $\lim\limits_{x \to 2} g(x)$,  (c) $\lim\limits_{x \to 3} g(x)$

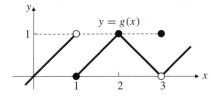

**Figure 1.10**

In Exercises 3–6, find the indicated one-sided limit of the function $g$ whose graph is given in Figure 1.10.

**3.** $\lim\limits_{x \to 1-} g(x)$

**4.** $\lim\limits_{x \to 1+} g(x)$

**5.** $\lim\limits_{x \to 3+} g(x)$

**6.** $\lim\limits_{x \to 3-} g(x)$

In Exercises 7–36, evaluate the limit or explain why it does not exist.

**7.** $\lim\limits_{x \to 4} (x^2 - 4x + 1)$

**8.** $\lim\limits_{x \to 2} 3(1 - x)(2 - x)$

**9.** $\lim\limits_{x \to 3} \dfrac{x + 3}{x + 6}$

**10.** $\lim\limits_{t \to -4} \dfrac{t^2}{4 - t}$

**11.** $\lim\limits_{x \to 1} \dfrac{x^2 - 1}{x + 1}$

**12.** $\lim\limits_{x \to -1} \dfrac{x^2 - 1}{x + 1}$

**13.** $\lim\limits_{x \to 3} \dfrac{x^2 - 6x + 9}{x^2 - 9}$

**14.** $\lim\limits_{x \to -2} \dfrac{x^2 + 2x}{x^2 - 4}$

**15.** $\lim\limits_{h \to 2} \dfrac{1}{4 - h^2}$

**16.** $\lim\limits_{h \to 0} \dfrac{3h + 4h^2}{h^2 - h^3}$

**17.** $\lim\limits_{x \to 9} \dfrac{\sqrt{x} - 3}{x - 9}$

**18.** $\lim\limits_{h \to 0} \dfrac{\sqrt{4 + h} - 2}{h}$

**19.** $\lim\limits_{x \to \pi} \dfrac{(x - \pi)^2}{\pi x}$

**20.** $\lim\limits_{x \to -2} |x - 2|$

**21.** $\lim\limits_{x \to 0} \dfrac{|x - 2|}{x - 2}$

**22.** $\lim\limits_{x \to 2} \dfrac{|x - 2|}{x - 2}$

**23.** $\lim\limits_{t \to 1} \dfrac{t^2 - 1}{t^2 - 2t + 1}$

**24.** $\lim\limits_{x \to 2} \dfrac{\sqrt{4 - 4x + x^2}}{x - 2}$

**25.** $\lim\limits_{t \to 0} \dfrac{t}{\sqrt{4 + t} - \sqrt{4 - t}}$

**26.** $\lim\limits_{x \to 1} \dfrac{x^2 - 1}{\sqrt{x + 3} - 2}$

**27.** $\lim\limits_{t \to 0} \dfrac{t^2 + 3t}{(t + 2)^2 - (t - 2)^2}$

**28.** $\lim\limits_{s \to 0} \dfrac{(s + 1)^2 - (s - 1)^2}{s}$

**29.** $\lim\limits_{y \to 1} \dfrac{y - 4\sqrt{y} + 3}{y^2 - 1}$

**30.** $\lim\limits_{x \to -1} \dfrac{x^3 + 1}{x + 1}$

**31.** $\lim\limits_{x \to 2} \dfrac{x^4 - 16}{x^3 - 8}$

**32.** $\lim\limits_{x \to 8} \dfrac{x^{2/3} - 4}{x^{1/3} - 2}$

**33.** $\lim\limits_{x \to 2} \left( \dfrac{1}{x - 2} - \dfrac{4}{x^2 - 4} \right)$

**34.** $\lim\limits_{x \to 2} \left( \dfrac{1}{x - 2} - \dfrac{1}{x^2 - 4} \right)$

**35.** $\lim\limits_{x \to 0} \dfrac{\sqrt{2 + x^2} - \sqrt{2 - x^2}}{x^2}$

**36.** $\lim\limits_{x \to 0} \dfrac{|3x - 1| - |3x + 1|}{x}$

The limit $\lim\limits_{h \to 0} \dfrac{f(x + h) - f(x)}{h}$ occurs frequently in the study of calculus. (Can you think why?) Evaluate this limit for the functions $f$ in Exercises 37–42.

**37.** $f(x) = x^2$

**38.** $f(x) = x^3$

**39.** $f(x) = \dfrac{1}{x}$

**40.** $f(x) = \dfrac{1}{x^2}$

**41.** $f(x) = \sqrt{x}$

**42.** $f(x) = 1/\sqrt{x}$

Examine the graphs of $\sin(x)$ and $\cos(x)$ in Section P.6 to determine the limits in Exercises 43–46.

**43.** $\lim\limits_{x \to \pi/2} \sin x$

**44.** $\lim\limits_{x \to \pi/4} \cos x$

**45.** $\lim\limits_{x \to \pi/3} \cos x$

**46.** $\lim\limits_{x \to 2\pi/3} \sin x$

**47.** Make a table of values of $f(x) = (\sin x)/x$ for a sequence of values of $x$ approaching 0, say $\pm 1.0$, $\pm 0.1$, $\pm 0.01$, $\pm 0.001$, $\pm 0.0001$, and $\pm 0.00001$. Make sure your calculator is set in *radian mode* rather than degree mode. Guess the value of $\lim\limits_{x \to 0} f(x)$.

**48.** Repeat the Exercise 47 for $f(x) = \dfrac{1 - \cos x}{x^2}$.

In Exercises 49–60, find the indicated one-sided limit or explain why it does not exist.

**49.** $\lim\limits_{x \to 2-} \sqrt{2 - x}$

**50.** $\lim\limits_{x \to 2+} \sqrt{2 - x}$

**51.** $\lim\limits_{x \to -2-} \sqrt{2 - x}$

**52.** $\lim\limits_{x \to -2+} \sqrt{2 - x}$

**53.** $\lim\limits_{x \to 0} \sqrt{x^3 - x}$

**54.** $\lim\limits_{x \to 0-} \sqrt{x^3 - x}$

**55.** $\lim\limits_{x \to 0+} \sqrt{x^3 - x}$

**56.** $\lim\limits_{x \to 0+} \sqrt{x^2 - x^4}$

**57.** $\lim\limits_{x \to a-} \dfrac{|x-a|}{x^2-a^2}$

**58.** $\lim\limits_{x \to a+} \dfrac{|x-a|}{x^2-a^2}$

**59.** $\lim\limits_{x \to 2-} \dfrac{x^2-4}{|x+2|}$

**60.** $\lim\limits_{x \to 2+} \dfrac{x^2-4}{|x+2|}$

Exercises 61–64 refer to the function

$$f(x) = \begin{cases} x-1 & \text{if } x \le -1 \\ x^2+1 & \text{if } -1 < x \le 0 \\ (x+\pi)^2 & \text{if } x > 0. \end{cases}$$

Find the indicated limits.

**61.** $\lim\limits_{x \to -1-} f(x)$

**62.** $\lim\limits_{x \to -1+} f(x)$

**63.** $\lim\limits_{x \to 0+} f(x)$

**64.** $\lim\limits_{x \to 0-} f(x)$

**65.** Suppose $\lim_{x \to 4} f(x) = 2$ and $\lim_{x \to 4} g(x) = -3$. Find:

(a) $\lim\limits_{x \to 4} \big( g(x) + 3 \big)$

(b) $\lim\limits_{x \to 4} x f(x)$

(c) $\lim\limits_{x \to 4} \big( g(x) \big)^2$

(d) $\lim\limits_{x \to 4} \dfrac{g(x)}{f(x)-1}$.

**66.** Suppose $\lim_{x \to a} f(x) = 4$ and $\lim_{x \to a} g(x) = -2$. Find:

(a) $\lim\limits_{x \to a} \big( f(x) + g(x) \big)$

(b) $\lim\limits_{x \to a} f(x) \cdot g(x)$

(c) $\lim\limits_{x \to a} 4g(x)$

(d) $\lim\limits_{x \to a} f(x)/g(x)$.

**67.** If $\lim\limits_{x \to 2} \dfrac{f(x)-5}{x-2} = 3$, find $\lim\limits_{x \to 2} f(x)$.

**68.** If $\lim\limits_{x \to 0} \dfrac{f(x)}{x^2} = -2$, find $\lim\limits_{x \to 0} f(x)$ and $\lim\limits_{x \to 0} \dfrac{f(x)}{x}$.

**Using Graphing Utilities to Find Limits**

Graphing calculators or computer software can be used to evaluate limits at least approximately. Simply "zoom" the plot window to show smaller and smaller parts of the graph near the point where the limit is to be found. Find the following limits by graphical techniques. Where you think it justified, give an exact answer. Otherwise, give the answer correct to four decimal places. Remember to ensure that your calculator or software is set for radian mode when using trigonometric functions.

**69.** $\lim\limits_{x \to 0} \dfrac{\sin x}{x}$

**70.** $\lim\limits_{x \to 0} \dfrac{\sin(2\pi x)}{\sin(3\pi x)}$

**71.** $\lim\limits_{x \to 1-} \dfrac{\sin\sqrt{1-x}}{\sqrt{1-x^2}}$

**72.** $\lim\limits_{x \to 0+} \dfrac{x-\sqrt{x}}{\sqrt{\sin x}}$

**73.** On the same graph plot the three function $y = x \sin(1/x)$, $y = x$, and $y = -x$ for $-0.2 \le x \le 0.2$, $-0.2 \le y \le 0.2$. Describe the behaviour of $f(x) = x \sin(1/x)$ near $x = 0$. Does $\lim_{x \to 0} f(x)$ exist, and if so, what is its value? Could you have predicted this before drawing the graph? Why?

**Using the Squeeze Theorem**

**74.** If $\sqrt{5-2x^2} \le f(x) \le \sqrt{5-x^2}$ for $-1 \le x \le 1$, find $\lim\limits_{x \to 0} f(x)$.

**75.** If $2 - x^2 \le g(x) \le 2\cos x$ for all $x$, find $\lim\limits_{x \to 0} g(x)$.

**76.** (a) Sketch the curves $y = x^2$ and $y = x^4$ on the same graph. Where do they intersect?

(b) The function $f(x)$ satisfies:

$$\begin{cases} x^2 \le f(x) \le x^4 & \text{if } x < -1 \text{ or } x > 1 \\ x^4 \le f(x) \le x^2 & \text{if } -1 \le x \le 1 \end{cases}$$

Find (i) $\lim\limits_{x \to -1} f(x)$, (ii) $\lim\limits_{x \to 0} f(x)$, (iii) $\lim\limits_{x \to 1} f(x)$.

**77.** On what intervals is $x^{1/3} < x^3$? On what intervals is $x^{1/3} > x^3$? If the graph of $y = h(x)$ always lies between the graphs of $y = x^{1/3}$ and $y = x^3$, for what real numbers $a$ can you determine the value of $\lim_{x \to a} h(x)$? Find the limit for each of these values of $a$.

**\* 78.** What is the domain of $x \sin\dfrac{1}{x}$? Evaluate $\lim\limits_{x \to 0} x \sin\dfrac{1}{x}$.

**\* 79.** Suppose $|f(x)| \le g(x)$ for all $x$. What can you conclude about $\lim_{x \to a} f(x)$ if $\lim_{x \to a} g(x) = 0$? What if $\lim_{x \to a} g(x) = 3$?

---

## 1.3 Limits at Infinity and Infinite Limits

**Figure 1.11** The graph of $x/\sqrt{x^2+1}$

In this section we will extend the concept of limit to allow for two situations not covered by the definition of limit or one-sided limit in the previous section:

i) limits at infinity, where $x$ becomes arbitrarily large, positive or negative;

ii) infinite limits, which are not really limits at all but provide useful symbolism for describing the behaviour of functions whose values become arbitrarily large, positive or negative.

## Limits at Infinity

Consider the function

$$f(x) = \frac{x}{\sqrt{x^2 + 1}}$$

whose graph is shown in Figure 1.11, and for which some values (rounded to seven decimal places) are given in Table 5. The values of $f(x)$ approach 1 as $x$ takes on larger and larger positive values, and $-1$ as $x$ takes on negative values that get larger and larger in absolute value. We express this behaviour by writing

**Table 5.**

| $x$ | $f(x) = x/\sqrt{x^2 + 1}$ |
|---|---|
| $-1,000$ | $-0.9999995$ |
| $-100$ | $-0.9999500$ |
| $-10$ | $-0.9950372$ |
| $-1$ | $-0.7071068$ |
| $0$ | $0.0000000$ |
| $1$ | $0.7071068$ |
| $10$ | $0.9950372$ |
| $100$ | $0.9999500$ |
| $1,000$ | $0.9999995$ |

$$\lim_{x \to \infty} f(x) = 1 \qquad \text{“}f(x) \text{ approaches 1 as } x \text{ approaches infinity.”}$$

$$\lim_{x \to -\infty} f(x) = -1 \qquad \text{“}f(x) \text{ approaches } -1 \text{ as } x \text{ approaches negative infinity.”}$$

The graph of $f$ conveys this limiting behaviour by approaching the horizontal lines $y = 1$ as $x$ moves far to the right, and $y = -1$ as $x$ moves far to the left. These lines are called **horizontal asymptotes** of the graph. In general, if a curve approaches a straight line as it recedes very far away from the origin, that line is called an **asymptote** of the curve.

**DEFINITION 4**

> **Limits at infinity and negative infinity (informal definition)**
>
> If the function $f$ is defined on an interval $(a, \infty)$ and if we can ensure that $f(x)$ is as close as we want to the number $L$ by taking $x$ large enough, then we say that $f(x)$ **approaches the limit $L$ as $x$ approaches infinity** and we write
>
> $$\lim_{x \to \infty} f(x) = L.$$
>
> If $f$ is defined on an interval $(-\infty, b)$ and if we can ensure that $f(x)$ is as close as we want to the number $M$ by taking $x$ negative and large enough in absolute value, then we say that $f(x)$ **approaches the limit $M$ as $x$ approaches negative infinity** and we write
>
> $$\lim_{x \to -\infty} f(x) = M.$$

Recall that the symbol $\infty$, called **infinity**, does *not* represent a real number. We cannot use $\infty$ in arithmetic in the usual way, but we can use the phrase "approaches $\infty$" to mean "becomes arbitrarily large positive," and the phrase "approaches $-\infty$" to mean "becomes arbitrarily large negative."

**Example 1** In Figure 1.12, we can see that $\lim_{x \to \infty} 1/x = \lim_{x \to -\infty} 1/x = 0$. The $x$-axis is a horizontal asymptote of the graph $y = 1/x$. ∎

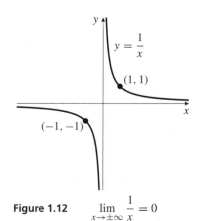

**Figure 1.12** $\quad \displaystyle\lim_{x \to \pm\infty} \frac{1}{x} = 0$

The theorems of Section 1.2 have suitable counterparts for limits at infinity or negative infinity. In particular, it follows from the example above and from the Product Rule for limits that $\lim_{x \to \pm\infty} 1/x^n = 0$ for any positive integer $n$. We will use this fact in the following examples. Example 2 shows how to obtain the limits at $\pm\infty$ for the function $x/\sqrt{x^2 + 1}$ by algebraic means, without resorting to making a table of values or drawing a graph, as we did above.

**Example 2** Evaluate $\lim\limits_{x\to\infty} f(x)$ and $\lim\limits_{x\to-\infty} f(x)$, for $f(x) = \dfrac{x}{\sqrt{x^2+1}}$.

**Solution** Rewrite the expression for $f(x)$ as follows:

$$f(x) = \frac{x}{\sqrt{x^2\left(1+\dfrac{1}{x^2}\right)}} = \frac{x}{\sqrt{x^2}\sqrt{1+\dfrac{1}{x^2}}} \qquad \text{Remember } \sqrt{x^2} = |x|.$$

$$= \frac{x}{|x|\sqrt{1+\dfrac{1}{x^2}}}$$

$$= \frac{\operatorname{sgn} x}{\sqrt{1+\dfrac{1}{x^2}}}, \qquad \text{where } \operatorname{sgn} x = \frac{x}{|x|} = \begin{cases} 1 & \text{if } x > 0 \\ -1 & \text{if } x < 0. \end{cases}$$

The factor $\sqrt{1+(1/x^2)}$ approaches 1 as $x$ approaches $\infty$ or $-\infty$, so $f(x)$ must have the same limits as $x \to \pm\infty$ as does $\operatorname{sgn}(x)$. Therefore

$$\lim_{x\to\infty} f(x) = 1 \qquad \text{and} \qquad \lim_{x\to-\infty} f(x) = -1.$$

## Limits at Infinity for Rational Functions

The only polynomials that have limits at $\pm\infty$ are constant ones, $P(x) = c$. The situation is more interesting for rational functions. Recall that a rational function is a quotient of two polynomials. The following examples show how to render such a function in a form where its limits at infinity and negative infinity (if they exist) are apparent. The way to do this is to *divide the numerator and denominator by the highest power of $x$ appearing in the denominator.* The limits of a rational function at infinity and negative infinity either both fail to exist, or both exist and are equal.

**Example 3** (**Numerator and denominator of the same degree**) Evaluate $\lim_{x\to\pm\infty} \dfrac{2x^2 - x + 3}{3x^2 + 5}$.

**Solution** Divide the numerator and the denominator by $x^2$, the highest power of $x$ appearing in the denominator:

$$\lim_{x\to\pm\infty} \frac{2x^2 - x + 3}{3x^2 + 5} = \lim_{x\to\pm\infty} \frac{2 - (1/x) + (3/x^2)}{3 + (5/x^2)} = \frac{2 - 0 + 0}{3 + 0} = \frac{2}{3}.$$

**Example 4** (**Degree of numerator less than degree of denominator**) Evaluate $\lim_{x\to\pm\infty} \dfrac{5x + 2}{2x^3 - 1}$.

**Solution** Divide the numerator and the denominator by the largest power of $x$ in the denominator, namely $x^3$.

$$\lim_{x\to\pm\infty} \frac{5x+2}{2x^3-1} = \lim_{x\to\pm\infty} \frac{(5/x^2)+(2/x^3)}{2-(1/x^3)} = \frac{0+0}{2-0} = 0.$$

∎

**(Summary of limits at $\pm\infty$ for rational functions)**

Let $P_m(x) = a_m x^m + \cdots + a_0$ and $Q_n(x) = b_n x^n + \cdots + b_0$ be polynomials of degree $m$ and $n$, respectively. Then

$$\lim_{x\to\pm\infty} \frac{P_m(x)}{Q_n(x)}$$

(a) equals zero if $m < n$,

(b) equals $\dfrac{a_m}{b_n}$ if $m = n$,

(c) does not exist if $m > n$.

The limiting behaviour of rational functions at infinity and negative infinity is summarized at the left.

The technique used in the previous examples can also be applied to more general kinds of functions. The function in the following example is not rational, and the limit seems to produce a meaningless $\infty - \infty$ until we resolve matters by rationalizing the numerator.

**Example 5**   Find $\lim_{x\to\infty}(\sqrt{x^2+x}-x)$.

**Solution**   We are trying to find the limit of the difference of two functions, each of which becomes arbitrarily large as $x$ increases to infinity. We rationalize the expression by multiplying the numerator and the denominator (which is 1) by the conjugate expression $\sqrt{x^2+x}+x$:

$$\lim_{x\to\infty}\left(\sqrt{x^2+x}-x\right) = \lim_{x\to\infty} \frac{\left(\sqrt{x^2+x}-x\right)\left(\sqrt{x^2+x}+x\right)}{\sqrt{x^2+x}+x}$$

$$= \lim_{x\to\infty} \frac{x^2+x-x^2}{\sqrt{x^2\left(1+\dfrac{1}{x}\right)}+x}$$

$$= \lim_{x\to\infty} \frac{x}{x\sqrt{1+\dfrac{1}{x}}+x} = \lim_{x\to\infty} \frac{1}{\sqrt{1+\dfrac{1}{x}}+1} = \frac{1}{2}.$$

(Here $\sqrt{x^2} = x$ because $x > 0$ as $x \to \infty$.)

∎

**Remark**   The limit $\lim_{x\to-\infty}(\sqrt{x^2+x}-x)$ is not nearly so subtle. Since $-x > 0$ as $x \to -\infty$, we have $\sqrt{x^2+x}-x > \sqrt{x^2+x}$, which grows arbitrarily large as $x \to -\infty$. The limit does not exist.

## Infinite Limits

A function whose values grow arbitrarily large can sometimes be said to have an infinite limit. Since infinity is not a number, infinite limits are not really limits at all, but they provide a way of describing the behaviour of functions which grow arbitrarily large positive or negative. A few examples will make the terminology clear.

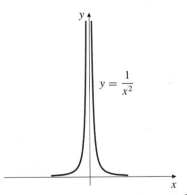

**Figure 1.13**    The graph of $y = 1/x^2$
(not to scale)

**Example 6**    **(A two-sided infinite limit)** Describe the behaviour of the function $f(x) = 1/x^2$ near $x = 0$.

***Solution***    As $x$ approaches 0 from either side the values of $f(x)$ are positive and grow larger and larger (see Figure 1.13), so the limit of $f(x)$ as $x$ approaches 0 *does not exist*. It is nevertheless convenient to describe the behaviour of $f$ near 0 by saying that $f(x)$ *approaches* $\infty$ as $x$ approaches zero. We write

$$\lim_{x \to 0} f(x) = \lim_{x \to 0} \frac{1}{x^2} = \infty.$$

Note that in writing this we are *not* saying that $\lim_{x \to 0} 1/x^2$ exists. Rather, we are saying that that limit *does not exist because $1/x^2$ becomes arbitrarily large near* $x = 0$. Observe how the graph of $f$ approaches the $y$-axis as $x$ approaches 0. The $y$-axis is a **vertical asymptote** of the graph.

**Example 7**    **(One-sided infinite limits)** Describe the behaviour of the function $f(x) = 1/x$ near $x = 0$.

***Solution***    As $x$ approaches 0 from the right the values of $f(x)$ become larger and larger positive numbers, and we say that $f$ has right-hand limit infinity at $x = 0$:

$$\lim_{x \to 0+} f(x) = \infty.$$

Similarly, the values of $f(x)$ become larger and larger negative numbers as $x$ approaches 0 from the left, so $f$ has left-hand limit $-\infty$ at $x = 0$:

$$\lim_{x \to 0-} f(x) = -\infty.$$

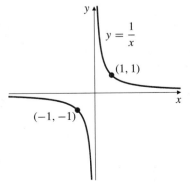

**Figure 1.14**    $\lim_{x \to 0-} 1/x = -\infty$, $\lim_{x \to 0+} 1/x = \infty$

These statements do not say that the one-sided limits *exist*; they do not exist because $\infty$ and $-\infty$ are not numbers. Since the one-sided limits are not equal even as infinite symbols, all we can say about the two-sided $\lim_{x \to 0} f(x)$ is that it does not exist.

**Example 8**    **(Polynomial behaviour at infinity)**
(a) $\lim_{x \to \infty} (3x^3 - x^2 + 2) = \infty$      (b) $\lim_{x \to -\infty} (3x^3 - x^2 + 2) = -\infty$
(c) $\lim_{x \to \infty} (x^4 - 5x^3 - x) = \infty$      (d) $\lim_{x \to -\infty} (x^4 - 5x^3 - x) = \infty$

The highest-degree term of a polynomial dominates the other terms as $|x|$ grows large, so the limits of this term at $\infty$ and $-\infty$ determine the limits of the whole polynomial. For the polynomial in parts (a) and (b) we have

$$3x^3 - x^2 + 2 = 3x^3 \left( 1 - \frac{1}{3x} + \frac{2}{3x^2} \right).$$

The factor in the large parentheses approaches 1 as $x$ approaches $\pm\infty$, so the behaviour of the polynomial is just that of its highest-degree term $3x^3$.

We can now say a bit more about the limits at infinity and negative infinity of a rational function whose numerator has higher degree than the denominator. Earlier in this section we said that such a limit *does not exist*. This is true, but we can assign $\infty$ or $-\infty$ to such limits, as the following example shows.

**Example 9** **(Rational functions with numerator of higher degree)** Evaluate $\displaystyle\lim_{x\to\infty}\frac{x^3+1}{x^2+1}$.

**Solution** Divide the numerator and the denominator by the $x^2$, the largest power of $x$ in the denominator:

$$\lim_{x\to\infty}\frac{x^3+1}{x^2+1} = \lim_{x\to\infty}\frac{x+\dfrac{1}{x^2}}{1+\dfrac{1}{x^2}} = \frac{\lim_{x\to\infty}\left(x+\dfrac{1}{x^2}\right)}{1} = \infty.$$

A polynomial $Q(x)$ of degree $n > 0$ can have at most $n$ *zeros*. That is, there are at most $n$ different real numbers $r$ for which $Q(r) = 0$. If $Q(x)$ is the denominator of a rational function $R(x) = P(x)/Q(x)$, that function will be defined for all $x$ except those finitely many zeros of $Q$. At each of those zeros $R(x)$ may have limits, infinite limits, or one-sided infinite limits. Here are some examples.

**Example 10**

(a) $\displaystyle\lim_{x\to2}\frac{(x-2)^2}{x^2-4} = \lim_{x\to2}\frac{(x-2)^2}{(x-2)(x+2)} = \lim_{x\to2}\frac{x-2}{x+2} = 0.$

(b) $\displaystyle\lim_{x\to2}\frac{x-2}{x^2-4} = \lim_{x\to2}\frac{x-2}{(x-2)(x+2)} = \lim_{x\to2}\frac{1}{x+2} = \frac{1}{4}.$

(c) $\displaystyle\lim_{x\to2+}\frac{x-3}{x^2-4} = \lim_{x\to2+}\frac{x-3}{(x-2)(x+2)} = -\infty.$  (The values are negative for $x > 2$, $x$ near 2.)

(d) $\displaystyle\lim_{x\to2-}\frac{x-3}{x^2-4} = \lim_{x\to2-}\frac{x-3}{(x-2)(x+2)} = \infty.$  (The values are positive for $x < 2$, $x$ near 2.)

(e) $\displaystyle\lim_{x\to2}\frac{x-3}{x^2-4} = \lim_{x\to2}\frac{x-3}{(x-2)(x+2)}$  does not exist.

(f) $\displaystyle\lim_{x\to2}\frac{2-x}{(x-2)^3} = \lim_{x\to2}\frac{-(x-2)}{(x-2)^3} = \lim_{x\to2}\frac{-1}{(x-2)^2} = -\infty.$

In parts (a) and (b) the effect of the zero in the denominator at $x = 2$ is cancelled because the numerator is zero there also. Thus a finite limit exists. This is not true in part (f) because the numerator only vanishes once at $x = 2$, while the denominator vanishes three times there.

## Exercises 1.3

Find the limits in Exercises 1–10.

1. $\displaystyle\lim_{x\to\infty}\frac{x}{2x-3}$

2. $\displaystyle\lim_{x\to\infty}\frac{x}{x^2-4}$

3. $\displaystyle\lim_{x\to\infty}\frac{3x^3-5x^2+7}{8+2x-5x^3}$

4. $\displaystyle\lim_{x\to-\infty}\frac{x^2-2}{x-x^2}$

5. $\displaystyle\lim_{x\to-\infty}\frac{x^2+3}{x^3+2}$

6. $\displaystyle\lim_{x\to\infty}\frac{x^2+\sin x}{x^2+\cos x}$

**7.** $\lim\limits_{x\to\infty} \dfrac{3x + 2\sqrt{x}}{1 - x}$

**8.** $\lim\limits_{x\to\infty} \dfrac{2x - 1}{\sqrt{3x^2 + x + 1}}$

**9.** $\lim\limits_{x\to-\infty} \dfrac{2x - 1}{\sqrt{3x^2 + x + 1}}$

**10.** $\lim\limits_{x\to-\infty} \dfrac{2x - 5}{|3x + 2|}$

In Exercises 11–34 evaluate the indicated limit. If it does not exist, is the limit $\infty$, or $-\infty$, or neither?

**11.** $\lim\limits_{x\to3} \dfrac{1}{3 - x}$

**12.** $\lim\limits_{x\to3} \dfrac{1}{(3 - x)^2}$

**13.** $\lim\limits_{x\to3-} \dfrac{1}{3 - x}$

**14.** $\lim\limits_{x\to3+} \dfrac{1}{3 - x}$

**15.** $\lim\limits_{x\to-5/2} \dfrac{2x + 5}{5x + 2}$

**16.** $\lim\limits_{x\to-2/5} \dfrac{2x + 5}{5x + 2}$

**17.** $\lim\limits_{x\to-(2/5)-} \dfrac{2x + 5}{5x + 2}$

**18.** $\lim\limits_{x\to-(2/5)+} \dfrac{2x + 5}{5x + 2}$

**19.** $\lim\limits_{x\to1-} \dfrac{2x + 3}{x^2 + x - 2}$

**20.** $\lim\limits_{x\to1+} \dfrac{2x + 3}{x^2 + x - 2}$

**21.** $\lim\limits_{x\to2+} \dfrac{x}{(2 - x)^3}$

**22.** $\lim\limits_{x\to1-} \dfrac{x}{\sqrt{1 - x^2}}$

**23.** $\lim\limits_{x\to1+} \dfrac{1}{|x - 1|}$

**24.** $\lim\limits_{x\to1-} \dfrac{1}{|x - 1|}$

**25.** $\lim\limits_{x\to2} \dfrac{x - 3}{x^2 - 4x + 4}$

**26.** $\lim\limits_{x\to1+} \dfrac{\sqrt{x^2 - x}}{x - x^2}$

**27.** $\lim\limits_{x\to\infty} \dfrac{x + x^3 + x^5}{1 + x^2 + x^3}$

**28.** $\lim\limits_{x\to\infty} \dfrac{x^3 + 3}{x^2 + 2}$

* **29.** $\lim\limits_{x\to\infty} \dfrac{x\sqrt{x + 1}\left(1 - \sqrt{2x + 3}\right)}{7 - 6x + 4x^2}$

**30.** $\lim\limits_{x\to\infty} \left(\dfrac{x^2}{x + 1} - \dfrac{x^2}{x - 1}\right)$

* **31.** $\lim\limits_{x\to-\infty} \left(\sqrt{x^2 + 2x} - \sqrt{x^2 - 2x}\right)$

* **32.** $\lim\limits_{x\to\infty} (\sqrt{x^2 + 2x} - \sqrt{x^2 - 2x})$

**33.** $\lim\limits_{x\to\infty} \dfrac{1}{\sqrt{x^2 - 2x} - x}$

**34.** $\lim\limits_{x\to-\infty} \dfrac{1}{\sqrt{x^2 + 2x} - x}$

**35.** What are the horizontal asymptotes of $y = \dfrac{1}{\sqrt{x^2 - 2x} - x}$? What are its vertical asymptotes?

**36.** What are the horizontal and vertical asymptotes of $y = \dfrac{2x - 5}{|3x + 2|}$?

The function $f$ whose graph is shown in Figure 1.15 has domain $[0, \infty)$. Find the following limits of $f$ indicated in Exercises 37–47.

**37.** $\lim\limits_{x\to0+} f(x)$

**38.** $\lim\limits_{x\to1} f(x)$

**39.** $\lim\limits_{x\to2+} f(x)$

**40.** $\lim\limits_{x\to2-} f(x)$

**41.** $\lim\limits_{x\to3-} f(x)$

**42.** $\lim\limits_{x\to3+} f(x)$

**43.** $\lim\limits_{x\to4+} f(x)$

**44.** $\lim\limits_{x\to4-} f(x)$

**45.** $\lim\limits_{x\to5-} f(x)$

**46.** $\lim\limits_{x\to5+} f(x)$

**47.** $\lim\limits_{x\to\infty} f(x)$

**48.** What asymptotes does the graph in Figure 1.15 have?

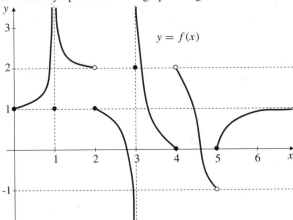

**Figure 1.15**

Exercises 49–54 refer to the greatest integer function $\lfloor x \rfloor$ graphed in Figure 1.16. Find the indicated limit or explain why it does not exist.

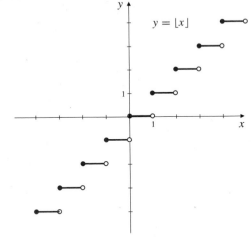

**Figure 1.16**

**49.** $\lim\limits_{x\to3+} \lfloor x \rfloor$

**50.** $\lim\limits_{x\to3-} \lfloor x \rfloor$

**51.** $\lim\limits_{x\to3} \lfloor x \rfloor$

**52.** $\lim\limits_{x\to2.5} \lfloor x \rfloor$

**53.** $\lim\limits_{x\to0+} \lfloor 2 - x \rfloor$

**54.** $\lim\limits_{x\to-3-} \lfloor x \rfloor$

**55.** Parking in a certain parking lot costs \$1.50 for each hour or part of an hour. Sketch the graph of the function $C(t)$ representing the cost of parking for $t$ hours. At what values of $t$ does $C(t)$ have a limit? Evaluate $\lim_{t\to t_0-} C(t)$ and $\lim_{t\to t_0+} C(t)$ for an arbitrary number $t_0 > 0$.

**56.** If $\lim_{x \to 0+} f(x) = L$ find $\lim_{x \to 0-} f(x)$ if (a) $f$ is even, (b) $f$ is odd.

**57.** If $\lim_{x \to 0+} f(x) = A$ and $\lim_{x \to 0-} f(x) = B$, find

(a) $\displaystyle\lim_{x \to 0+} f(x^3 - x)$

(b) $\displaystyle\lim_{x \to 0-} f(x^3 - x)$

(c) $\displaystyle\lim_{x \to 0-} f(x^2 - x^4)$

(d) $\displaystyle\lim_{x \to 0+} f(x^2 - x^4)$.

# 1.4 Continuity

When a car is driven along a highway, its distance from its starting point depends on time in a *continuous* way, changing by small amounts over short intervals of time. But not all quantities change in this way. When the car is parked in a parking lot where the rate is quoted as "$2.00 per hour or portion," the parking charges remain at $2.00 for the first hour and then suddenly jump to $4.00 as soon as the first hour has passed. The function relating parking charges to parking time will be called *discontinuous* at each hour. In this section we will define continuity, and show how to tell whether a function is continuous. We will also examine some important properties possessed by continuous functions.

## Continuity at a Point

Most functions that we encounter have domains that are intervals, or unions of separate intervals. A point $P$ in the domain of such a function is called an **interior point** of the domain if it belongs to some *open* interval contained in the domain. If it is not an interior point, then $P$ is called an **endpoint** of the domain. For example, the domain of the function $f(x) = \sqrt{4 - x^2}$ is the closed interval $[-2, 2]$, which consists of interior points in the interval $(-2, 2)$, a left endpoint $-2$, and a right endpoint $2$. The domain of the function $g(x) = 1/x$ is the union of open intervals $(-\infty, 0) \cup (0, \infty)$, and consists entirely of interior points. Note that, although $0$ is an endpoint of each of those intervals, it does not belong to the domain of $g$.

**DEFINITION** **5**

> **Continuity at an interior point**
>
> We say that a function $f$ is **continuous** at an interior point $c$ of its domain if
>
> $$\lim_{x \to c} f(x) = f(c).$$
>
> If either $\lim_{x \to c} f(x)$ fails to exist, or it exists but is not equal to $f(c)$, then we will say that $f$ is discontinuous at $c$.

In graphical terms, $f$ is continuous at an interior point $c$ of its domain if its graph has no break in it at the point $(c, f(c))$; in other words, if you can draw the graph through that point without lifting your pen from the paper. Consider Figure 1.17. In (a), $f$ is continuous at $c$. In (b), $f$ is discontinuous at $c$ because $\lim_{x \to c} f(x) \neq f(c)$. In (c), $f$ is discontinuous at $c$ because $\lim_{x \to c} f(x)$ does not exist. In both (b) and (c) the graph of $f$ has a break at $x = c$.

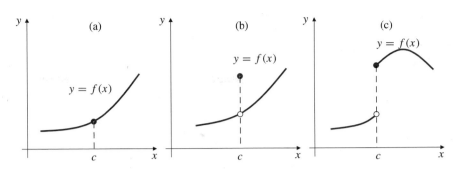

**Figure 1.17**

(a) $f$ is continuous at $c$.

(b) $\lim\limits_{x \to c} f(x) \neq f(c)$.

(c) $\lim\limits_{x \to c} f(x)$ does not exist.

Although a function cannot have a limit at an endpoint of its domain that is not an interior point, it can still have a one-sided limit there. We extend the definition of continuity to provide for such situations.

**Right and left continuity**

We say that $f$ is **right continuous** at $c$ if $\lim\limits_{x \to c+} f(x) = f(c)$.

We say that $f$ is **left continuous** at $c$ if $\lim\limits_{x \to c-} f(x) = f(c)$.

**Figure 1.18** The Heaviside function

**Example 1**  The Heaviside function $H(x)$ whose graph is shown in Figure 1.18 is continuous at every number $x$ except 0. It is right continuous at 0, but is not left continuous or continuous there. ∎

The relationship between continuity and one-sided continuity is summarized in the following theorem.

**THEOREM 5**  Function $f$ is continuous at $c$ if and only if it is both right continuous and left continuous at $c$.

**DEFINITION 7**

**Continuity at an endpoint**

We say that $f$ is continuous at a left endpoint $c$ of its domain if it is right continuous there.

We say that $f$ is continuous at a right endpoint $c$ of its domain if it is left continuous there.

**Example 2**  The function $f(x) = \sqrt{4 - x^2}$ has domain $[-2, 2]$. It is continuous at the right endpoint 2 because it is left continuous there, that is, because $\lim_{x \to 2-} f(x) = 0 = f(2)$. It is continuous at the left endpoint $-2$ because it is right continuous there: $\lim_{x \to -2+} f(x) = 0 = f(-2)$. Of course, $f$ is also continuous at every interior point of its domain. If $-2 < c < 2$ then $\lim_{x \to c} f(x) = \sqrt{4 - c^2} = f(c)$. (See Figure 1.19.) ∎

**Figure 1.19** $f(x) = \sqrt{4 - x^2}$ is continuous at every point of its domain

## Continuity on an Interval

We have defined the concept of continuity at a point. Of greater importance is the concept of continuity on an interval.

**DEFINITION** **8**

**Continuity on an interval**

We say that function $f$ is **continuous on the interval** $I$ if it is continuous at each point of $I$. In particular, we will say that $f$ is **a continuous function** if $f$ is continuous at every point of its domain.

**Example 3**   The function $f(x) = \sqrt{x}$ is a continuous function. Its domain is $[0, \infty)$. It is continuous at the left endpoint 0 because it is right continuous there. Also, $f$ is continuous at every number $c > 0$ since $\lim_{x \to c} \sqrt{x} = \sqrt{c}$.

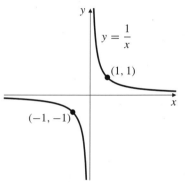

**Figure 1.20**   $1/x$ is continuous on its domain

**Example 4**   The function $g(x) = 1/x$ is also a continuous function. This may seem wrong to you at first glance, because its graph is broken at $x = 0$. (See Figure 1.20.) However, the number 0 is not in the domain of $g$, so it doesn't make sense to talk about $g$ being discontinuous there. A function can only be continuous or discontinuous at a point of its domain. If we were to define $g(0)$ to be some number, say 0, then we would say that $g(x)$ is discontinuous at 0. There is no way of defining $g(0)$ so that $g$ becomes continuous at 0.

**Example 5**   The greatest integer function $\lfloor x \rfloor$ (see Figure 1.16) is continuous on every interval $[n, n + 1)$, where $n$ is an integer. It is right continuous at each integer $n$, but is not left continuous there, so it is discontinuous at the integers.

$$\lim_{x \to n+} \lfloor x \rfloor = n = \lfloor n \rfloor, \qquad \lim_{x \to n-} \lfloor x \rfloor = n - 1 \neq n = \lfloor n \rfloor.$$

## There Are Lots of Continuous Functions

The following functions are continuous wherever they are defined:

(a)  all polynomials;

(b)  all rational functions;

(c)  all rational powers $x^{m/n} = \sqrt[n]{x^m}$;

(d)  the sine, cosine, tangent, secant, cosecant, and cotangent functions defined in Section P.6; and

(e)  the absolute value function $|x|$;

Theorem 3 of Section 1.2 assures us that every polynomial is continuous everywhere on the real line, and every rational function is continuous everywhere on its domain (which consists of all real numbers except the finitely many where its denominator is zero). If $m$ and $n$ are integers and $n \neq 0$, the rational power function $x^{m/n}$ is defined for all positive numbers $x$, and also for all negative numbers $x$ if $n$ is odd. The domain includes 0 if and only if $m/n \geq 0$.

The following theorems show that if we combine continuous functions in various ways, the results will be continuous.

**THEOREM** **6**

**Combining continuous functions**

If the functions $f$ and $g$ are both defined on an interval containing $c$, and are both continuous at $c$, then the following functions are also continuous at $c$:

1. the sum $f + g$ and the difference $f - g$;
2. the product $fg$;
3. the constant multiple $kf$, where $k$ is any number;
4. the quotient $f/g$, (provided $g(c) \neq 0$); and
5. the $n$th root $(f(x))^{1/n}$, provided $f(c) > 0$ if $n$ is even.

The proof involves using the various limit rules in Theorem 2 of Section 1.2. For example,

$$\lim_{x \to c} \big( f(x) + g(x) \big) = \lim_{x \to c} f(x) + \lim_{x \to c} g(x) = f(c) + g(c),$$

so $f + g$ is continuous.

**THEOREM** **7**

**Composites of continuous functions are continuous**

If $f(g(x))$ is defined on an interval containing $c$, and if $f$ is continuous at $L$ and $\lim_{x \to c} g(x) = L$, then

$$\lim_{x \to c} f(g(x)) = f(L) = f\left( \lim_{x \to c} g(x) \right).$$

In particular, if $g$ is continuous at $c$ (so $L = g(c)$), then the composition $f \circ g$ is continuous at $c$:

$$\lim_{x \to c} f(g(x)) = f(g(c)).$$

(See Exercise 37 in Section 1.5.)

**Example 6**   The following functions are continuous everywhere on their respective domains:

(a) $3x^2 - 2x$

(b) $\dfrac{x - 2}{x^2 - 4}$

(c) $|x^2 - 1|$

(d) $\sqrt{x}$

(e) $\sqrt{x^2 - 2x - 5}$

(f) $\dfrac{|x|}{\sqrt{|x + 2|}}$

## Continuous Extensions and Removable Discontinuities

As we have seen in Section 1.2, a rational function may have a limit even at a point where its denominator is zero. If $f(c)$ is not defined, but $\lim_{x \to c} f(x) = L$ exists, we can define a new function $F(x)$ by

$$F(x) = \begin{cases} f(x) & \text{if } x \text{ is in the domain of } f \\ L & \text{if } x = c. \end{cases}$$

$F(x)$ is continuous at $x = c$. It is called the **continuous extension** of $f(x)$ to $x = c$. For rational functions $f$, continuous extensions are usually found by cancelling common factors.

$$y = \frac{x^2 - x}{x^2 - 1}$$

**Figure 1.21**

**Example 7**    Show that $f(x) = \dfrac{x^2 - x}{x^2 - 1}$ has a continuous extension to $x = 1$, and find that extension.

**Solution**    Although $f(1)$ is not defined, if $x \neq 1$ we have

$$f(x) = \frac{x^2 - x}{x^2 - 1} = \frac{x(x-1)}{(x+1)(x-1)} = \frac{x}{x+1}.$$

The function

$$F(x) = \frac{x}{x+1}$$

is equal to $f(x)$ for $x \neq 1$, but is also continuous at $x = 1$, having there the value $1/2$. The graph of $f$ is shown in Figure 1.21. The continuous extension of $f(x)$ to $x = 1$ is $F(x)$. It has the same graph as $f(x)$ except with no hole at $(1, 1/2)$. ∎

If a function $f$ that is discontinuous at a point $a$ can be redefined at that *single point* so that it becomes continuous there, then we say that $f$ has a **removable discontinuity** at $a$.

**Example 8**    The function $g(x) = \begin{cases} x & \text{if } x \neq 2 \\ 1 & \text{if } x = 2 \end{cases}$ has a removable discontinuity at $x = 2$. To remove it, redefine $g(2) = 2$. (See Figure 1.22.) ∎

**Figure 1.22**    $g$ has a removable discontinuity at 2

## Continuous Functions on Closed, Finite Intervals

Continuous functions that are defined on *closed, finite intervals* have special properties that make them particularly useful in mathematics and its applications. We will discuss two of these properties here. Although they may appear obvious, these properties are much more subtle than the results about limits stated earlier in this chapter; their proofs (see Appendix III) require a careful study of the implications of the completeness property of the real numbers.

The first of the properties states that a function $f(x)$ that is continuous on a closed, finite interval $[a, b]$ must have an **absolute maximum value** and an **absolute minimum value.** This means that the values of $f(x)$ at all points of the interval lie between the values of $f(x)$ at two particular points in the interval; the graph of $f$ has a highest point and a lowest point.

**THEOREM 8**

### The Max-Min Theorem

If $f(x)$ is continuous on the closed, finite interval $[a, b]$, then there exist numbers $x_1$ and $x_2$ in $[a, b]$ such that for all $x$ in $[a, b]$,

$$f(x_1) \leq f(x) \leq f(x_2).$$

Thus $f$ has the absolute minimum value $m = f(x_1)$, taken on at the point $x_1$, and the absolute maximum value $M = f(x_2)$, taken on at the point $x_2$.

Many important problems in mathematics and its applications come down to having to find maximum and minimum values of functions. Calculus provides some very useful tools for solving such problems. Observe, however, that the theorem above merely asserts that minimum and maximum values *exist;* it doesn't tell us how to find them. In Chapter 4 we will develop techniques for calculating maximum and minimum values of functions. For now, we can solve some simple maximum or minimum value problems involving quadratic functions by completing the square without using any calculus.

**Example 9**  What is the largest possible area of a rectangular field that can be enclosed by 200 m of fencing?

***Solution***  If the sides of the field are $x$ m and $y$ m (Figure 1.23), then its perimeter is $P = 2x + 2y$ m, and its area is $A = xy$ m$^2$. We are given that $P = 200$, so $x + y = 100$, and $y = 100 - x$. Neither side can be negative, so $x$ must belong to the closed interval [0, 100]. The area of the field can be expressed as a function of $x$ by substituting $100 - x$ for $y$:

$$A = x(100 - x) = 100x - x^2.$$

We want to find the maximum value of the quadratic function $A(x) = 100x - x^2$ on the interval [0, 100]. Theorem 8 assures us that such a maximum exists.

To find the maximum, we complete the square of the function $A(x)$. Note that $x^2 - 100x$ are the first two terms of the square $(x - 50)^2 = x^2 - 100x + 2{,}500$. Thus

$$A(x) = 2{,}500 - (x - 50)^2.$$

Observe that $A(50) = 2{,}500$, and $A(x) < 2{,}500$ if $x \neq 50$, because we are subtracting a positive number $(x - 50)^2$ from 2,500 in this case. Therefore, the maximum value of $A(x)$ is 2,500. The largest field has area 2,500 m$^2$, and is actually a square with dimensions $x = y = 50$ m.

**Figure 1.23**  Rectangular field: perimeter $= 2x + 2y$, area $= xy$

Theorem 8 implies that a function that is continuous on a closed, finite interval is **bounded**. This means that it cannot take on arbitrarily large positive or negative values; there must exist a number $K$ such that

$$|f(x)| \leq K, \qquad \text{that is,} \qquad -K \leq f(x) \leq K.$$

In fact, for $K$ we can use the larger of the numbers $|f(x_1)|$ and $|f(x_2)|$.

The conclusions of Theorem 8 may fail if the function $f$ is not continuous, or if the interval is not closed. See Figures 1.24–1.27 for examples of how such failure can occur.

## Finding Maxima and Minima Graphically

***Remark***  Graphing utilities can be used to find maximum and minimum values of functions on intervals where they are continuous. In particular, the "zoom box" and "trace" facilities of graphing calculators are helpful. Figure 1.28(a) shows the

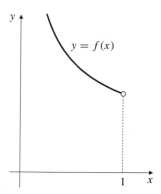

**Figure 1.24** $f(x) = 1/x$ is continuous on the open interval $(0, 1)$. It is not bounded, and has neither a maximum nor a minimum value

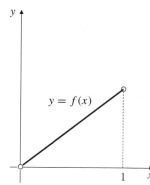

**Figure 1.25** $f(x) = x$ is continuous on the open interval $(0, 1)$. It is bounded but has neither a maximum nor a minimum value

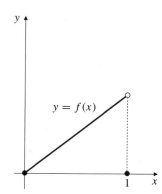

**Figure 1.26** This function is defined on the closed interval $[0, 1]$, but is discontinuous at the endpoint $x = 1$. It has a minimum value but no maximum value

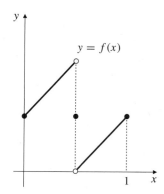

**Figure 1.27** This function is discontinuous at the interior point of its domain, the closed interval $[0, 1]$. It is bounded, but has neither maximum nor minimum values

graph of

$$y = f(x) = \frac{x + 1}{x^2 + 1}$$

on the window $-5 \le x \le 5$, $-2 \le y \le 2$. Observe that $f$ appears to have a maximum value near $x = 0.5$ and a minimum value near $x = -2.5$. Figure 1.28(b) shows the result of expanding the part of the graph in (a) enclosed in the small rectangle (zoom box) to fill the whole screen. Tracing the curve to its highest point gives a more accurate estimate of the maximum value. Further zooming enables us to get any desired degree of accuracy.

**Figure 1.28** Using a "zoom box" to zoom part of a curve (a) near a maximum value to fill the screen (b) without allowing the curve to become flattened

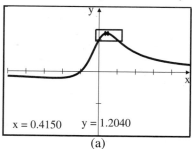

x = 0.4150    y = 1.2040

(a)

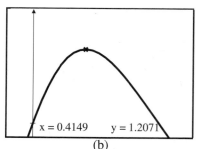

x = 0.4149    y = 1.2071

(b)

Continuing in this way, show that $f(x)$ has maximum value 1.2071 at $x = 0.4142$, each to four significant figures.

The second property of a continuous function defined on a closed, finite interval is that the function takes on all real values between any two of its values. This property is called the **intermediate-value property**.

**THEOREM 9**

**The Intermediate-Value Theorem**

If $f(x)$ is continuous on the interval $[a, b]$ and if $s$ is a number between $f(a)$ and $f(b)$, then there exists a number $c$ in $[a, b]$ such that $f(c) = s$.

In particular, a continuous function defined on a closed interval takes on all values between its minimum value $m$ and its maximum value $M$, so its range is also a closed interval, $[m, M]$.

Figure 1.29 shows a typical situation. The points $(a, f(a))$ and $(b, f(b))$ are on opposite sides of the horizontal line $y = s$. Being unbroken, the graph $y = f(x)$ must cross this line in order to go from one point to the other. In the figure, it crosses the line only once, at $x = c$. If the line $y = s$ were somewhat higher, there might have been three crossings, and three possible values for $c$.

Theorem 9 is the reason why the graph of a function that is continuous on an interval $I$ cannot have any breaks. It must be **connected**, a single, unbroken curve with no jumps.

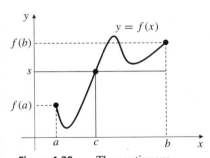

**Figure 1.29**      The continuous function $f$ takes on the value $s$ at some point $c$ between $a$ and $b$

**Example 10**   Determine the intervals on which $f(x) = x^3 - 4x$ is positive and negative.

**Solution**   Since $f(x) = x(x^2 - 4) = x(x - 2)(x + 2)$, $f(x) = 0$ only at $x = 0, 2,$ and $-2$. Because $f$ is continuous on the whole real line, it must have constant sign on each of the intervals $(-\infty, -2)$, $(-2, 0)$, $(0, 2)$, and $(2, \infty)$. (If there were points $a$ and $b$ in one of those intervals, say in $(0, 2)$, such that $f(a) < 0$ and $f(b) > 0$, then by the Intermediate-Value Theorem there would exist $c$ between $a$ and $b$, and therefore between 0 and 2, such that $f(c) = 0$. But we know $f$ has no such zero in $(0, 2)$.)

To find whether $f(x)$ is positive or negative throughout each interval, pick a point in the interval and evaluate $f$ at that point.
Since $f(-3) = -15 < 0$, $f(x)$ is negative on $(-\infty, -2)$.
Since $f(-1) = 3 > 0$, $f(x)$ is positive on $(-2, 0)$.
Since $f(1) = -3 < 0$, $f(x)$ is negative on $(0, 2)$.
Since $f(3) = 15 > 0$, $f(x)$ is positive on $(2, \infty)$. ■

## Finding Roots of Equations

Among the many useful tools that calculus will provide are ones that enable us to calculate solutions to equations of the form $f(x) = 0$ to any desired degree of accuracy. Such a solution is called a **root** of the equation, or a **zero** of the function $f$. Using these tools usually requires previous knowledge that the equation has a solution in some interval. The Intermediate-Value Theorem can provide this information.

**Example 11**   Show that the equation $x^3 - x - 1 = 0$ has a solution in the interval $[1, 2]$.

**Solution**   The function $f(x) = x^3 - x - 1$ is a polynomial, and is therefore continuous everywhere. Now $f(1) = -1$ and $f(2) = 5$. Since 0 lies between $-1$ and 5, the Intermediate-Value Theorem assures us that there must be a number $c$ in $[1, 2]$ such that $f(c) = 0$. ■

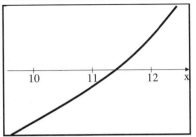

**Figure 1.30**      The function graphed in the upper window has a root between 11 and 12. The small rectangle (zoom box) is zoomed to fill the screen in the lower window, enabling us to estimate that the root is about 11.4. Successive zooms can provide greater precision

You can use a graphing utility to solve an equation $f(x) = 0$. Just graph the function $f(x)$ over a large enough interval so that you can see roughly where its zeros are. Then select one zero at a time and zoom in on it by successively expanding the part of the viewing window near the zero to fill the whole viewing window. Keep zooming until you can estimate the zero to as many decimal places as you want, (or as the calculator or computer will allow).

Another method for finding a root of a function that is continuous and changes sign on an interval involves bisecting the interval many times, each time determining which half of the previous interval must contain the root, because the function has opposite signs at the two ends of that half. This method is slow. For example, if the original interval has length 1, it will take eleven bisections to cut down to an interval of length less than 0.0005 (because $2^{11} > 2,000 = 1/(0.0005)$), and thus to ensure that we have found the root correct to three decimal places. But this method requires no graphics hardware, and is easily implemented with a calculator, preferably one into which the formula for the function can be programmed.

**Example 12**     **(The bisection method)**  Solve the equation $f(x) = x^3 - x - 1 = 0$ of Example 11 correct to three decimal places by successive bisections.

**Solution**     We start out knowing that there is a root in $[1, 2]$. Table 6 shows the results of the bisections.

**Table 6.**   The bisection method for $f(x) = x^3 - x - 1 = 0$

| Bisection Number | $x$ | $f(x)$ | Root in Interval | Midpoint |
|---|---|---|---|---|
|  | 1 | $-1$ |  |  |
|  | 2 | 5 | [1, 2] | 1.5 |
| 1 | 1.5 | 0.8750 | [1, 1.5] | 1.25 |
| 2 | 1.25 | $-0.2969$ | [1.25, 1.5] | 1.375 |
| 3 | 1.375 | 0.2246 | [1.25, 1.375] | 1.3125 |
| 4 | 1.3125 | $-0.0515$ | [1.3125, 1.375] | 1.3438 |
| 5 | 1.3438 | 0.0826 | [1.3125, 1.3438] | 1.3282 |
| 6 | 1.3282 | 0.0147 | [1.3125, 1.3282] | 1.3204 |
| 7 | 1.3204 | $-0.0186$ | [1.3204, 1.3282] | 1.3243 |
| 8 | 1.3243 | $-0.0018$ | [1.3243, 1.3282] | 1.3263 |
| 9 | 1.3263 | 0.0065 | [1.3243, 1.3263] | 1.3253 |
| 10 | 1.3253 | 0.0025 | [1.3243, 1.3253] | 1.3248 |
| 11 | 1.3248 | 0.0003 | [1.3243, 1.3248] | 1.3246 |
| 12 | 1.3246 | $-0.0007$ | [1.3246, 1.3248] |  |

The root is 1.325, rounded to three decimal places. In Section 3.6, calculus will provide us with much faster methods of solving equations such as the one above.

**Remark**     The Max-Min Theorem and the Intermediate-Value Theorem are examples of what mathematicians call **existence theorems**. Such theorems assert that something exists without telling you how to find it. Students sometimes complain that mathematicians worry too much about proving that a problem has a solution and not enough about how to find that solution. They argue: "If I can calculate a solution to a problem, then surely I do not need to worry about whether a solution exists." This is, however, false logic. Suppose we pose the problem, "Find the largest positive integer." Of course this problem has no solution; there is no largest positive integer because we can add 1 to any integer and get a larger integer. Suppose, however, that we forget this and try to calculate a solution. We could proceed as follows:

Let $N$ be the largest positive integer.
Since 1 is a positive integer, therefore $N \geq 1$.
Since $N^2$ is a positive integer, it cannot exceed the largest positive integer.
Therefore $N^2 \leq N$, and so $N^2 - N \leq 0$.

Thus $N(N - 1) \leq 0$ and we must have $N - 1 \leq 0$.

Therefore $N \leq 1$.

Since we also know $N \geq 1$, therefore $N = 1$.

Therefore 1 is the largest positive integer.

The only error we have made here is in the assumption (in the first line) that the problem has a solution.  It is partly to avoid logical pitfalls like this that mathematicians prove existence theorems.

# Exercises 1.4

Exercises 1–3 refer to the function $g$ defined on $[-2, 2]$ whose graph is shown in Figure 1.31.

**1.** State whether $g$ is (i) continuous, (ii) left continuous, (iii) right continuous, and (iv) discontinuous at each of the points $-2, -1, 0, 1$, and 2.

**Figure 1.31**

**2.** At what points in its domain does $g$ have a removable discontinuity, and how should $g$ be redefined at each of those points so as to be continuous there?

**3.** Does $g$ have an absolute maximum value on $[-2, 2]$? an absolute minimum value?

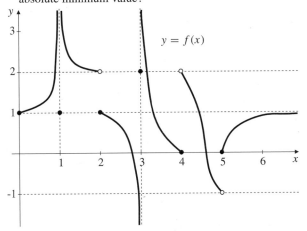

**Figure 1.32**

**4.** At what points is the function $f$ whose graph is shown in Figure 1.32 discontinuous? At which of those points is it left continuous? right continuous?

**5.** Can the function $f$ graphed in Figure 1.32 be redefined at the single point $x = 1$ so that it becomes continuous there?

**6.** The function $\operatorname{sgn}(x) = x/|x|$ is neither continuous nor discontinuous at $x = 0$. How is this possible?

In Exercises 7–12, state where in its domain the given function is continuous, where it is left or right continuous, and where it is just discontinuous.

**7.** $f(x) = \begin{cases} x & \text{if } x < 0 \\ x^2 & \text{if } x \geq 0 \end{cases}$   **8.** $f(x) = \begin{cases} x & \text{if } x < -1 \\ x^2 & \text{if } x \geq -1 \end{cases}$

**9.** $f(x) = \begin{cases} 1/x^2 & \text{if } x \neq 0 \\ 0 & \text{if } x = 0 \end{cases}$   **10.** $f(x) = \begin{cases} x^2 & \text{if } x \leq 1 \\ 0.987 & \text{if } x > 1 \end{cases}$

**11.** The least integer function $\lceil x \rceil$ of Example 11 in Section P.5.

**12.** The parking lot cost function $C(t)$ of Exercise 55 in Section 1.3.

In Exercises 13–16, how should the given function be defined at the given point to be continuous there? Give a formula for the continuous extension to that point.

**13.** $\dfrac{x^2 - 4}{x - 2}$ at $x = 2$   **14.** $\dfrac{1 + t^3}{1 - t^2}$ at $t = -1$

**15.** $\dfrac{t^2 - 5t + 6}{t^2 - t - 6}$ at 3   **16.** $\dfrac{x^2 - 2}{x^4 - 4}$ at $\sqrt{2}$

**17.** Find $k$ so that $f(x) = \begin{cases} x^2 & \text{if } x \leq 2 \\ k - x^2 & \text{if } x > 2 \end{cases}$ is a continuous function.

**18.** Find $m$ so that $g(x) = \begin{cases} x - m & \text{if } x < 3 \\ 1 - mx & \text{if } x \geq 3 \end{cases}$ is continuous for all $x$.

**19.** Does the function $x^2$ have a maximum value on the open interval $-1 < x < 1$? a minimum value? Explain.

**20.** The Heaviside function of Example 1 has both absolute maximum and minimum values on the interval $[-1, 1]$, but it is not continuous on that interval. Does this violate the Max-Min Theorem? Why?

Exercises 21–24 involve finding maximum and minimum values of functions. They can all be done by the method of Example 9.

**21.** The sum of two nonnegative numbers is 8. What is the largest possible value of their product?

**22.** The sum of two nonnegative numbers is 8. What is (a) the smallest, and (b) the largest possible value for the sum of their squares?

23. A software company estimates that if it assigns $x$ programmers to work on the project, it can develop a new product in $T$ days, where

$$T = 100 - 30x + 3x^2.$$

How many programmers should the company assign in order to complete the development as quickly as possible?

24. It costs a desk manufacturer $\$(245x - 30x^2 + x^3)$ to send a shipment of $x$ desks to its warehouse. How many desks should it include in each shipment to minimize the average shipping cost per desk?

Find the intervals on which the functions $f(x)$ in Exercises 25–28 are positive and negative.

25. $f(x) = \dfrac{x^2 - 1}{x}$

26. $f(x) = x^2 + 4x + 3$

27. $f(x) = \dfrac{x^2 - 1}{x^2 - 4}$

28. $f(x) = \dfrac{x^2 + x - 2}{x^3}$

29. Show that $f(x) = x^3 + x - 1$ has a zero between $x = 0$ and $x = 1$.

30. Show that the equation $x^3 - 15x + 1 = 0$ has three solutions in the interval $[-4, 4]$.

31. Show that the function $F(x) = (x - a)^2(x - b)^2 + x$ has the value $(a + b)/2$ at some point $x$.

32. **(A fixed-point theorem)** Suppose that $f$ is continuous on the closed interval $[0, 1]$ and that $0 \le f(x) \le 1$ for every $x$

in $[0, 1]$. Show that there must exist a number $c$ in $[0, 1]$ such that $f(c) = c$. ($c$ is called a fixed point of the function $f$.) *Hint:* if $f(0) = 0$ or $f(1) = 1$ you are done. If not, apply the Intermediate-Value Theorem to $g(x) = f(x) - x$.

33. If an even function $f$ is right continuous at $x = 0$, show that it is continuous at $x = 0$.

34. If an odd function $f$ is right continuous at $x = 0$, show that it is continuous at $x = 0$ and that it satisfies $f(0) = 0$.

Use a graphing utility to find the maximum and minimum values of the functions in Exercises 35–38, and the points $x$ where they occur. Obtain three decimal place accuracy for all answers.

35. $f(x) = \dfrac{x^2 - 2x}{x^4 + 1}$    on    $[-5, 5]$

36. $f(x) = \dfrac{\sin x}{6 + x}$    on    $[-\pi, \pi]$

37. $f(x) = x^2 + \dfrac{4}{x}$    on    $[1, 3]$

38. $f(x) = \sin(\pi x) + x(\cos(\pi x) + 1)$    on    $[0, 1]$

Use a graphing utility, or a programmable calculator and the Bisection Method to solve the equations in Exercises 39–42 to three decimal places. As a first step, try to guess a small interval that you can be sure contains a root.

39. $x^3 + x - 1 = 0$

40. $\cos x - x = 0$

41. $\sin x + 1 - x^2 = 0$    (two roots)

42. $x^4 - x - 1 = 0$    (two roots)

## 1.5 The Formal Definition of Limit

The *informal* definition of limit given in Section 1.2 is not precise enough to enable us to prove results about limits such as those given in Theorems 2–4 of Section 1.2. A more precise *formal* definition is based on the idea of controlling the input $x$ of a function $f$ so that the output $f(x)$ will lie in a specific interval.

The material of this section is optional.

**Example 1**    The area of a circular disk of radius $r$ cm is $A = \pi r^2$ cm$^2$. A machinist is required to manufacture a circular metal disk having area $400\pi$ cm$^2$ within an error tolerance of $\pm 5$ cm$^2$. How close to 20 cm must the machinist control the radius of the disk to achieve this?

**Solution**    The machinist wants $|\pi r^2 - 400\pi| < 5$, that is,

$$400\pi - 5 < \pi r^2 < 400\pi + 5,$$

or, equivalently,

$$\sqrt{400 - (5/\pi)} < r < \sqrt{400 + (5/\pi)}$$
$$19.96017 < r < 20.03975.$$

Thus the machinist needs $|r - 20| < 0.03975$; she must ensure that the radius of the disk differs from 20 cm by less than 0.4 mm so that the area of the disk will lie within the required error tolerance. ∎

When we say that $f(x)$ has limit $L$ as $x$ approaches $a$, we are really saying that we can ensure that the *error* $|f(x) - L|$ will be less than *any* allowed tolerance, no matter how small, by taking $x$ *close enough* to $a$ (but not equal to $a$). It is traditional to use $\epsilon$, the Greek letter epsilon, for the size of the allowable *error,* and $\delta$, the Greek letter delta, for the *difference* $x - a$ that measures how close $x$ must be to $a$ to ensure that the error is within that tolerance. These are the letters that Cauchy and Weierstrass used in their pioneering work on limits and continuity in the nineteenth century.

If $\epsilon$ is any positive number, *no matter how small*, we must be able to ensure that $|f(x) - L| < \epsilon$ by restricting $x$ to be *close enough to* (but not equal to) $a$. How close is close enough? It is sufficient that the distance $|x - a|$ from $x$ to $a$ be less than a positive number $\delta$ that depends on $\epsilon$. (See Figure 1.33.) If we can find such a $\delta$ for any positive $\epsilon$ we are entitled to conclude that $\lim_{x \to a} f(x) = L$.

**DEFINITION 9**

**Figure 1.33** If $x \neq a$ and $|x - a| < \delta$, then $|f(x) - L| < \epsilon$

---

**A formal definition of limit**

We say that $f(x)$ approaches the limit $L$ as $x$ approaches $a$, and we write

$$\lim_{x \to a} f(x) = L,$$

if the following condition is satisfied:
for every number $\epsilon > 0$ there exists a number $\delta > 0$, depending on $\epsilon$, such that

$$0 < |x - a| < \delta \qquad \text{implies} \qquad |f(x) - L| < \epsilon.$$

---

The formal definition of limit does not tell you how to find the limit of a function, but it does enable you to verify that a suspected limit is correct. The following examples show how it can be used to verify limit statements for specific functions. The first gives a formal verification of the two limits found in Example 3 of Section 1.2.

**Example 2** (**Two important limits**) Verify: (a) $\lim_{x \to a} x = a$, and (b) $\lim_{x \to a} k = k$ ($k =$ constant).

**Solution**

(a) Let $\epsilon > 0$ be given. We must find $\delta > 0$ so that

$$0 < |x - a| < \delta \qquad \text{implies} \qquad |x - a| < \epsilon.$$

Clearly we can take $\delta = \epsilon$ and the implication above will be true. This proves that $\lim_{x \to a} x = a$.

(b) Let $\epsilon > 0$ be given. We must find $\delta > 0$ so that

$$0 < |x - a| < \delta \qquad \text{implies} \qquad |k - k| < \epsilon.$$

Since $k - k = 0$ we can use any positive number for $\delta$ and the implication above will be true. This proves that $\lim_{x \to a} k = k$.   ∎

---

**Example 3**   Verify that $\lim_{x \to 2} x^2 = 4$.

**Solution**   Here $a = 2$ and $L = 4$. Let $\epsilon$ be a given positive number. We want to find $\delta > 0$ so that if $0 < |x - 2| < \delta$, then $|f(x) - 4| < \epsilon$. Now

$$|f(x) - 4| = |x^2 - 4| = |(x + 2)(x - 2)| = |x + 2||x - 2|.$$

We want the expression above to be less than $\epsilon$. We can make the factor $|x - 2|$ as small as we wish by choosing $\delta$ properly, but we need to control the factor $|x + 2|$ so that it does not become too large. If we first assume $\delta \leq 1$, and require that $|x - 2| < \delta$, then we have

$$|x - 2| < 1 \quad \Rightarrow \quad 1 < x < 3 \quad \Rightarrow \quad 3 < x + 2 < 5$$
$$\Rightarrow \quad |x + 2| < 5.$$

Hence

$$|f(x) - 4| < 5|x - 2| \quad \text{if} \quad |x - 2| < \delta \leq 1.$$

But $5|x - 2| < \epsilon$ if $|x - 2| < \epsilon/5$. Therefore, if we take $\delta = \min\{1, \epsilon/5\}$, the *minimum* (the smaller) of the two numbers 1 and $\epsilon/5$, then

$$|f(x) - 4| < 5|x - 2| < 5 \times \frac{\epsilon}{5} = \epsilon \quad \text{if} \quad |x - 2| < \delta.$$

This proves that $\lim_{x \to 2} f(x) = 4$.   ∎

## Using the Definition of Limit to Prove Theorems

We do not usually rely on the formal definition of limit to verify specific limits such as those in the two examples above. Rather, we appeal to general theorems about limits, and in particular Theorems 2–4 of Section 1.2. The definition is used to prove these theorems. As an example, we prove part 1 of Theorem 2, the *Sum Rule*.

---

**Example 4**   **(Proving the rule for the limit of a sum)**   If $\lim_{x \to a} f(x) = L$ and $\lim_{x \to a} g(x) = M$, prove that $\lim_{x \to a} \big(f(x) + g(x)\big) = L + M$.

*Solution* Let $\epsilon > 0$ be given. We want to find a positive number $\delta$ such that

$$0 < |x - a| < \delta \quad \Rightarrow \quad \big|(f(x) + g(x)) - (L + M)\big| < \epsilon.$$

Observe that

$$\begin{aligned}
\big|(f(x)+g(x)) - (L+M)\big| & \qquad \text{Regroup terms.}\\
= \big|(f(x) - L) + (g(x) - M)\big| & \qquad \text{(Use the triangle inequality:}\\
& \qquad |a + b| \le |a| + |b|).\\
\le |f(x) - L| + |g(x) - M|. &
\end{aligned}$$

Since $\lim_{x \to a} f(x) = L$ and $\epsilon/2$ is a positive number, there exists a number $\delta_1 > 0$ such that

$$0 < |x - a| < \delta_1 \quad \Rightarrow \quad |f(x) - L| < \epsilon/2.$$

Similarly, since $\lim_{x \to a} g(x) = M$, there exists a number $\delta_2 > 0$ such that

$$0 < |x - a| < \delta_2 \quad \Rightarrow \quad |g(x) - M| < \epsilon/2.$$

Let $\delta = \min\{\delta_1, \delta_2\}$, the smaller of $\delta_1$ and $\delta_2$. If $0 < |x - a| < \delta$, then $|x - a| < \delta_1$, so $|f(x) - L| < \epsilon/2$, and $|x - a| < \delta_2$, so $|g(x) - M| < \epsilon/2$. Therefore

$$\big|(f(x) + g(x)) - (L + M)\big| < \frac{\epsilon}{2} + \frac{\epsilon}{2} = \epsilon.$$

This shows that $\lim_{x \to a}(f(x) + g(x)) = L + M$.

## Other Kinds of Limits

The formal definition of limit can be modified to give precise definitions of one-sided limits, limits at infinity, and infinite limits. We give some of the definitions here and leave you to supply the others.

**DEFINITION 10**

> **Right limits**
> We say that $f(x)$ has **right limit** $L$ at $a$, and we write
>
> $$\lim_{x \to a+} f(x) = L,$$
>
> if the following condition is satisfied:
> for every number $\epsilon > 0$ there exists a number $\delta > 0$, depending on $\epsilon$, such that
>
> $$a < x < a + \delta \qquad \text{implies} \qquad |f(x) - L| < \epsilon.$$

**Figure 1.34** If $a < x < a + \delta$, then $|f(x) - L| < \epsilon$

Notice how the condition $0 < |x - a| < \delta$ in the definition of limit becomes $a < x < a + \delta$ in the right limit case (Figure 1.34). The definition for a left limit is formulated in a similar way.

**Example 5** Show that $\lim\limits_{x\to 0+} \sqrt{x} = 0$.

**Solution** Let $\epsilon > 0$ be given. If $x > 0$, then $|\sqrt{x} - 0| = \sqrt{x}$. We can ensure that $\sqrt{x} < \epsilon$ by requiring $x < \epsilon^2$. Thus we can take $\delta = \epsilon^2$ and the condition of the definition will be satisfied:

$$0 < x < \delta = \epsilon^2 \qquad \text{implies} \qquad |\sqrt{x} - 0| < \epsilon.$$

Therefore $\lim_{x\to 0+} \sqrt{x} = 0$.

To claim that a function $f$ has a limit $L$ at infinity, we must be able to ensure that the error $|f(x) - L|$ is less than any given positive number $\epsilon$ by restricting $x$ to be *sufficiently large*, that is, by requiring $x > R$ for some positive number $R$ depending on $\epsilon$.

**DEFINITION 11**

**Limit at infinity**

We say that $f(x)$ **approaches the limit $L$ as $x$ approaches infinity**, and we write

$$\lim_{x\to\infty} f(x) = L,$$

if the following condition is satisfied:
for every number $\epsilon > 0$ there exists a number $R$, depending on $\epsilon$, such that

$$x > R \qquad \text{implies} \qquad |f(x) - L| < \epsilon.$$

You are invited to formulate a version of the definition of a limit at negative infinity.

**Example 6** Show that $\lim\limits_{x\to\infty} \dfrac{1}{x} = 0$.

**Solution** Let $\epsilon$ be a given positive number. For $x > 0$ we have

$$\left| \frac{1}{x} - 0 \right| = \frac{1}{|x|} = \frac{1}{x} < \epsilon \qquad \text{provided} \qquad x > \frac{1}{\epsilon}.$$

Therefore the condition of the definition is satisfied with $R = 1/\epsilon$. We have shown that $\lim_{x\to\infty} 1/x = 0$.

To show that $f(x)$ has an infinite limit at $a$ we must ensure that $f(x)$ is larger than any given positive number ($f(x) > B$) by restricting $x$ to a sufficiently small interval centred at $a$, and requiring that $x \neq a$.

**DEFINITION 12**

**Infinite limits**

We say that $f(x)$ approaches infinity as $x$ approaches $a$, and write

$$\lim_{x \to a} f(x) = \infty,$$

if for every positive real number $B$ we can find a positive number $\delta$ (depending on $B$), such that

$$0 < |x - a| < \delta \quad \text{implies} \quad f(x) > B.$$

Try to formulate the corresponding definition for the concept $\lim_{x \to a} f(x) = -\infty$. Then try to modify both definitions to cover the case of infinite one-sided limits and infinite limits at infinity.

**Example 7** Verify that $\displaystyle\lim_{x \to 0} \frac{1}{x^2} = \infty$.

**Solution** Let $B$ be any positive number. We have

$$\frac{1}{x^2} > B \quad \text{provided that} \quad x^2 < \frac{1}{B}.$$

If $\delta = 1/\sqrt{B}$, then

$$0 < |x| < \delta \quad \Rightarrow \quad x^2 < \delta^2 = \frac{1}{B} \quad \Rightarrow \quad \frac{1}{x^2} > B.$$

Therefore $\lim_{x \to 0} 1/x^2 = \infty$. ∎

# Exercises 1.5

1. The length $L$ of a metal rod is given in terms of the temperature $T$ (°C) by $L = 39.6 + 0.025T$ cm. Within what range of temperature must the rod be kept if its length must be maintained within ±1 mm of 40 cm?

2. What is the largest tolerable error in the 20 cm edge length of a cubical cardboard box if the volume of the box must be within ±1.2% of 8,000 cm³?

In Exercises 3–6, in what interval must $x$ be confined if $f(x)$ must be within the given distance $\epsilon$ of the number $L$?

3. $f(x) = 2x - 1$, $\quad L = 3$, $\quad \epsilon = 0.02$

4. $f(x) = x^2$, $\quad L = 4$, $\quad \epsilon = 0.1$

5. $f(x) = \sqrt{x}$, $\quad L = 1$, $\quad \epsilon = 0.1$

6. $f(x) = 1/x$, $\quad L = -2$, $\epsilon = 0.01$

In Exercises 7–10, find a number $\delta > 0$ such that if $|x - a| < \delta$, then $|f(x) - L|$ will be less than the given number $\epsilon$.

7. $f(x) = 3x + 1$, $\quad a = 2$, $\quad L = 7$, $\quad \epsilon = 0.03$

8. $f(x) = \sqrt{2x + 3}$, $\quad a = 3$, $\quad L = 3$, $\quad \epsilon = 0.01$

9. $f(x) = x^3$, $\quad a = 2$, $\quad L = 8$, $\quad \epsilon = 0.2$

10. $f(x) = 1/(x + 1)$, $a = 0$, $\quad L = 1$, $\quad \epsilon = 0.05$

In Exercises 11–20, use the formal definition of limit to verify the indicated limit.

11. $\displaystyle\lim_{x \to 1} (3x + 1) = 4$

12. $\displaystyle\lim_{x \to 2} (5 - 2x) = 1$

13. $\displaystyle\lim_{x \to 0} x^2 = 0$

14. $\displaystyle\lim_{x \to 2} \frac{x - 2}{1 + x^2} = 0$

15. $\displaystyle\lim_{x \to 1/2} \frac{1 - 4x^2}{1 - 2x} = 2$

16. $\displaystyle\lim_{x \to -2} \frac{x^2 + 2x}{x + 2} = -2$

**17.** $\lim_{x \to 1} \dfrac{1}{x+1} = \dfrac{1}{2}$

**18.** $\lim_{x \to -1} \dfrac{x+1}{x^2-1} = -\dfrac{1}{2}$

**19.** $\lim_{x \to 1} \sqrt{x} = 1$

**20.** $\lim_{x \to 2} x^3 = 8$

Give formal definitions of the limit statements in Exercises 21–26.

**21.** $\lim_{x \to a-} f(x) = L$

**22.** $\lim_{x \to -\infty} f(x) = L$

**23.** $\lim_{x \to a} f(x) = -\infty$

**24.** $\lim_{x \to \infty} f(x) = \infty$

**25.** $\lim_{x \to a+} f(x) = -\infty$

**26.** $\lim_{x \to a-} f(x) = \infty$

Use formal definitions of the various kinds of limits to prove the statements in Exercises 27–30.

**27.** $\lim_{x \to 1+} \dfrac{1}{x-1} = \infty$

**28.** $\lim_{x \to 1-} \dfrac{1}{x-1} = -\infty$

**29.** $\lim_{x \to \infty} \dfrac{1}{\sqrt{x^2+1}} = 0$

**30.** $\lim_{x \to \infty} \sqrt{x} = \infty$

**Proving Theorems with the Definition of Limit**

* **31.** Prove that limits are unique; that is, if $\lim_{x \to a} f(x) = L$ and $\lim_{x \to a} f(x) = M$, prove that $L = M$. *Hint:* suppose $L \neq M$ and let $\epsilon = |L - M|/3$.

* **32.** If $\lim_{x \to a} g(x) = M$, show that there exists a number $\delta > 0$ such that

$$0 < |x - a| < \delta \quad \Rightarrow \quad |g(x)| < 1 + |M|.$$

(*Hint:* Take $\epsilon = 1$ in the definition of limit.) This says that the values of $g(x)$ are **bounded** near a point where $g$ has a limit.

* **33.** If $\lim_{x \to a} f(x) = L$ and $\lim_{x \to a} g(x) = M$, prove that $\lim_{x \to a} f(x)g(x) = LM$ (the Product Rule part of Theorem 2). *Hint:* reread Example 4. Let $\epsilon > 0$ and write

$$
\begin{aligned}
|f(x)g(x) - LM| &= |f(x)g(x) - Lg(x) + Lg(x) - LM| \\
&= |(f(x) - L)g(x) + L(g(x) - M)| \\
&\leq |(f(x) - L)g(x)| + |L(g(x) - M)| \\
&= |g(x)||f(x) - L| + |L||g(x) - M|
\end{aligned}
$$

Now try to make each term in the last line less than $\epsilon/2$ by taking $x$ close enough to $a$. You will need the result of Exercise 32.

* **34.** If $\lim_{x \to a} g(x) = M$, where $M \neq 0$, show that there exists a number $\delta > 0$ such that

$$0 < |x - a| < \delta \quad \Rightarrow \quad |g(x)| > |M|/2.$$

* **35.** If $\lim_{x \to a} g(x) = M$, where $M \neq 0$, show that

$$\lim_{x \to a} \frac{1}{g(x)} = \frac{1}{M}.$$

*Hint:* you will need the result of Exercise 34.

**36.** Use the facts proved in Exercises 33 and 35 to prove the Quotient Rule (part 5 of Theorem 2): If $\lim_{x \to a} f(x) = L$ and $\lim_{x \to a} g(x) = M$, where $M \neq 0$, then

$$\lim_{x \to a} \frac{f(x)}{g(x)} = \frac{L}{M}.$$

* **37.** Use the definition of limit twice to prove Theorem 7 of Section 1.4; that is, if $f$ is continuous at $L$ and if $\lim_{x \to c} g(x) = L$, then

$$\lim_{x \to c} f(g(x)) = f(L) = f\left(\lim_{x \to c} g(x)\right).$$

* **38.** Prove the Squeeze Theorem (Theorem 4 in Section 1.2). *Hint:* if $f(x) \leq g(x) \leq h(x)$, then

$$
\begin{aligned}
|g(x) - L| &= |g(x) - f(x) + f(x) - L| \\
&\leq |g(x) - f(x)| + |f(x) - L| \\
&\leq |h(x) - f(x)| + |f(x) - L| \\
&= |h(x) - L - (f(x) - L)| + |f(x) - L| \\
&\leq |h(x) - L| + |f(x) - L| + |f(x) - L|
\end{aligned}
$$

Now you can make each term in the last expression less than $\epsilon/3$.

# Chapter Review

## Key Ideas

- **What do the following statements and phrases mean?**
  ◇ the average rate of change of $f(x)$ on $[a, b]$
  ◇ the instantaneous rate of change of $f(x)$ at $x = a$
  ◇ $\lim_{x \to a} f(x) = L$
  ◇ $\lim_{x \to a+} f(x) = L$,    $\lim_{x \to a-} f(x) = L$
  ◇ $\lim_{x \to \infty} f(x) = L$,    $\lim_{x \to -\infty} f(x) = L$
  ◇ $\lim_{x \to a} f(x) = \infty$,    $\lim_{x \to a+} f(x) = -\infty$
  ◇ $f$ is continuous at $c$.
  ◇ $f$ is left continuous at $c$.
  ◇ $f$ is right continuous at $c$.
  ◇ $f$ has a continuous extension to $c$.

◇ $f$ takes on maximum and minimum values on interval $I$.

◇ $f$ is bounded on interval $I$.

◇ $f$ has the intermediate-value property on interval $I$.

- **State as many "laws of limits" as you can.**
- **What properties must a function have if it is continuous and its domain is a closed, finite interval?**
- **How can you find zeros (roots) of a continuous function?**

## Review Exercises

**1.** Find the average rate of change of $x^3$ over $[1, 3]$.

**2.** Find the average rate of change of $1/x$ over $[-2, -1]$.

**3.** Find the rate of change of $x^3$ at $x = 2$.

**4.** Find the rate of change of $1/x$ at $x = -3/2$.

Evaluate the limits in Exercises 5–30 or explain why they do not exist.

**5.** $\displaystyle\lim_{x \to 1} (x^2 - 4x + 7)$

**6.** $\displaystyle\lim_{x \to 2} \frac{x^2}{1 - x^2}$

**7.** $\displaystyle\lim_{x \to 1} \frac{x^2}{1 - x^2}$

**8.** $\displaystyle\lim_{x \to 2} \frac{x^2 - 4}{x^2 - 5x + 6}$

**9.** $\displaystyle\lim_{x \to 2} \frac{x^2 - 4}{x^2 - 4x + 4}$

**10.** $\displaystyle\lim_{x \to 2-} \frac{x^2 - 4}{x^2 - 4x + 4}$

**11.** $\displaystyle\lim_{x \to -2+} \frac{x^2 - 4}{x^2 + 4x + 4}$

**12.** $\displaystyle\lim_{x \to 4} \frac{2 - \sqrt{x}}{x - 4}$

**13.** $\displaystyle\lim_{x \to 3} \frac{x^2 - 9}{\sqrt{x} - \sqrt{3}}$

**14.** $\displaystyle\lim_{h \to 0} \frac{h}{\sqrt{x + 3h} - \sqrt{x}}$

**15.** $\displaystyle\lim_{x \to 0+} \sqrt{x - x^2}$

**16.** $\displaystyle\lim_{x \to 0} \sqrt{x - x^2}$

**17.** $\displaystyle\lim_{x \to 1} \sqrt{x - x^2}$

**18.** $\displaystyle\lim_{x \to 1-} \sqrt{x - x^2}$

**19.** $\displaystyle\lim_{x \to \infty} \frac{1 - x^2}{3x^2 - x - 1}$

**20.** $\displaystyle\lim_{x \to -\infty} \frac{2x + 100}{x^2 + 3}$

**21.** $\displaystyle\lim_{x \to -\infty} \frac{x^3 - 1}{x^2 + 4}$

**22.** $\displaystyle\lim_{x \to \infty} \frac{x^4}{x^2 - 4}$

**23.** $\displaystyle\lim_{x \to 0+} \frac{1}{\sqrt{x - x^2}}$

**24.** $\displaystyle\lim_{x \to 1/2} \frac{1}{\sqrt{x - x^2}}$

**25.** $\displaystyle\lim_{x \to \infty} \sin x$

**26.** $\displaystyle\lim_{x \to \infty} \frac{\cos x}{x}$

**27.** $\displaystyle\lim_{x \to 0} x \sin \frac{1}{x}$

**28.** $\displaystyle\lim_{x \to 0} \sin \frac{1}{x^2}$

**29.** $\displaystyle\lim_{x \to -\infty} [x + \sqrt{x^2 - 4x + 1}]$

**30.** $\displaystyle\lim_{x \to \infty} [x + \sqrt{x^2 - 4x + 1}]$

At what, if any, points in its domain is the function $f$ in Exercises 31–38 discontinuous? Is $f$ either left or right continuous at these points? In Exercises 35 and 36, $H$ refers to the Heaviside function: $H(x) = 1$ if $x \geq 0$, and $H(x) = 0$ if $x < 0$.

**31.** $f(x) = x^3 - 4x^2 + 1$

**32.** $f(x) = \dfrac{x}{x + 1}$

**33.** $f(x) = \begin{cases} x^2 & \text{if } x > 2 \\ x & \text{if } x \leq 2 \end{cases}$

**34.** $f(x) = \begin{cases} x^2 & \text{if } x > 1 \\ x & \text{if } x \leq 1 \end{cases}$

**35.** $f(x) = H(x - 1)$

**36.** $f(x) = H(9 - x^2)$

**37.** $f(x) = |x| + |x + 1|$

**38.** $f(x) = \begin{cases} |x|/|x + 1| & \text{if } x \neq -1 \\ 1 & \text{if } x = -1 \end{cases}$

## Challenging Problems

**1.** Show that the average rate of change of the function $x^3$ over the interval $[a, b]$, where $0 < a < b$, is equal to the instantaneous rate of change of $x^3$ at $x = \sqrt{(a^2 + ab + b^2)/3}$. Is this point to the left or to the right of the midpoint $(a + b)/2$ of the interval $[a, b]$?

**2.** Evaluate $\displaystyle\lim_{x \to 0} \frac{x}{|x - 1| - |x + 1|}$.

**3.** Evaluate $\displaystyle\lim_{x \to 3} \frac{|5 - 2x| - |x - 2|}{|x - 5| - |3x - 7|}$.

**4.** Evaluate $\displaystyle\lim_{x \to 64} \frac{x^{1/3} - 4}{x^{1/2} - 8}$.

**5.** Evaluate $\displaystyle\lim_{x \to 1} \frac{\sqrt{3 + x} - 2}{\sqrt[3]{7 + x} - 2}$.

**6.** The equation $ax^2 + 2x - 1 = 0$, where $a$ is a constant, has two roots if $a > -1$ and $a \neq 0$:

$$r_+(a) = \frac{-1 + \sqrt{1 + a}}{a} \text{ and } r_-(a) = \frac{-1 - \sqrt{1 + a}}{a}.$$

(a) What happens to the root $r_-(a)$ when $a \to 0$ ?

(b) Investigate numerically what happens to the root $r_+(a)$ when $a \to 0$ by trying the values $a = 1, \pm0.1, \pm0.01, \ldots$. For values such as $a = 10^{-8}$, the limited precision of your calculator may produce some interesting results. What happens, and why?

(c) Evaluate $\lim_{a \to 0} r_+(a)$ mathematically by using the identity

$$\sqrt{A} - \sqrt{B} = \frac{A - B}{\sqrt{A} + \sqrt{B}}.$$

**7.** TRUE or FALSE? If TRUE, give reasons; if FALSE, give a counterexample.

(a) If $\lim_{x \to a} f(x)$ exists but $\lim_{x \to a} g(x)$ does not exist, then $\lim_{x \to a} (f(x) + g(x))$ does not exist.

(b) If neither $\lim_{x \to a} f(x)$ nor $\lim_{x \to a} g(x)$ exists, then $\lim_{x \to a} (f(x) + g(x))$ does not exist.

(c) If $f$ is continuous at $a$, then so is $|f|$.

(d) If $|f|$ is continuous at $a$, then so is $f$.

(e) If $f(x) < g(x)$ for all $x$ in an interval around $a$, and if $\lim_{x \to a} f(x)$ and $\lim_{x \to a} g(x)$ both exist, then $\lim_{x \to a} f(x) < \lim_{x \to a} g(x)$.

**8.** (a) If $f$ is a continuous function defined on a closed interval $[a, b]$, show that $R(f)$ is a closed interval.

   (b) What are the possibilities for $R(f)$ if $D(f)$ is an open interval $(a, b)$?

**9.** Consider the function $f(x) = \dfrac{x^2 - 1}{|x^2 - 1|}$. Find all points where $f$ is not continuous. Does $f$ have one-sided limits at those points, and if so, what are they?

**10.** Find the minimum value of $f(x) = 1/(x - x^2)$ on the interval $(0, 1)$. Explain how you know such a minimum value must exist.

**∗ 11.** (a) Suppose $f$ is a continuous function on the interval $[0, 1]$, and $f(0) = f(1)$. Show that $f(a) = f\left(a + \dfrac{1}{2}\right)$ for some $a \in \left[0, \dfrac{1}{2}\right]$.

   *Hint:* let $g(x) = f\left(x + \dfrac{1}{2}\right) - f(x)$, and use the Intermediate-Value Theorem.

   (b) If $n$ is an integer larger than 2, show that
   $$f(a) = f\left(a + \dfrac{1}{n}\right) \text{ for some } a \in \left[0, 1 - \dfrac{1}{n}\right].$$

# CHAPTER 2
# Differentiation

***Introduction*** Two fundamental problems are considered in calculus. The **problem of slopes** is concerned with finding the slope of (the tangent line to) a given curve at a given point on the curve. The **problem of areas** is concerned with finding the area of a plane region bounded by curves and straight lines. The solution of the problem of slopes is the subject of **differential calculus**. As we will see, it has many applications in mathematics and other disciplines. The problem of areas is the subject of **integral calculus**, which we begin in Chapter 5.

## 2.1 Tangent Lines and Their Slopes

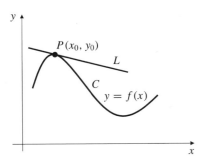

**Figure 2.1**    $L$ is tangent to $C$ at $P$

This section deals with the problem of finding a straight line $L$ that is tangent to a curve $C$ at a point $P$. As is often the case in mathematics, the most important step in the solution of such a fundamental problem is making a suitable definition.

For simplicity, and to avoid certain problems best postponed until later, we will not deal with the most general kinds of curves now, but only with those that are the *graphs of continuous functions*. Let $C$ be the graph of $y = f(x)$ and let $P$ be the point $(x_0, y_0)$ on $C$, so that $y_0 = f(x_0)$. We assume that $P$ is not an endpoint of $C$. Therefore $C$ extends some distance on both sides of $P$. (See Figure 2.1.)

What do we mean when we say that the line $L$ is tangent to $C$ at $P$? Past experience with tangent lines to circles does not help us to define tangency for more general curves. A tangent line to a circle (Figure 2.2) has the following properties:

(i)  It meets the circle at only one point.

(ii)  The circle lies on only one side of the line.

(iii)  The tangent is perpendicular to the line joining the centre of the circle to the point of contact.

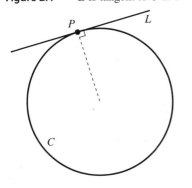

**Figure 2.2**    $L$ is tangent to $C$ at $P$

Most curves do not have obvious *centres*, so (iii) is useless for characterizing tangents to them. The curves in Figure 2.3 show that (i) and (ii) cannot be used to define tangency either. In particular, Figure 2.3(d) is not "smooth" at $P$ so that curve should not have any tangent line there. A tangent line should have the "same direction" as the curve does at the point of tangency.

A reasonable definition of tangency can be stated in terms of limits. If $Q$ is a point on $C$ different from $P$, then the line through $P$ and $Q$ is called a **secant line** to the curve. This line rotates around $P$ as $Q$ moves along the curve. If $L$ is a line through $P$ whose slope is the limit of the slopes of these secant lines $PQ$ as $Q$ approaches $P$ along $C$ (Figure 2.4), then $L$ is tangent to $C$ at $P$.

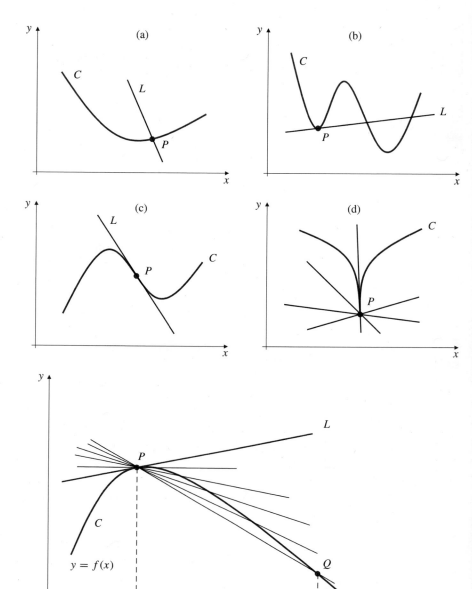

**Figure 2.3**

(a) $L$ meets $C$ only at $P$, but is not tangent to $C$

(b) $L$ meets $C$ at several points but is tangent to $C$ at $P$

(c) $L$ is tangent to $C$ at $P$ but crosses $C$ at $P$

(d) Many lines meet $C$ only at $P$ but none of them is tangent to $C$ at $P$

**Figure 2.4** Secant lines $PQ$ approach tangent line $L$ as $Q$ approaches $P$ along the curve $C$

Since $C$ is the graph of the *function* $y = f(x)$, then vertical lines can meet $C$ only once. Since $P = (x_0, f(x_0))$, a different point $Q$ on the graph must have a different $x$-coordinate, say $x_0 + h$, where $h \neq 0$. Thus $Q = (x_0 + h, f(x_0 + h))$, and the slope of the line $PQ$ is

$$\frac{f(x_0 + h) - f(x_0)}{h}.$$

This expression is called the **Newton quotient** or **difference quotient** for $f$ at $x_0$. Note that $h$ can be positive or negative, depending on whether $Q$ is to the right or left of $P$.

**Non-vertical tangent lines**

Suppose that the function $f$ is continuous at $x = x_0$ and that

$$\lim_{h \to 0} \frac{f(x_0 + h) - f(x_0)}{h} = m$$

exists. Then the straight line having slope $m$ and passing through the point $P = (x_0, f(x_0))$ is called the **tangent line** (or simply the **tangent**) to the graph of $y = f(x)$ at $P$. An equation of this tangent is

$$y = m(x - x_0) + y_0.$$

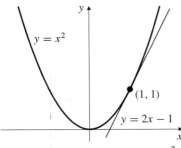

**Figure 2.5**    The tangent to $y = x^2$ at $(1, 1)$

**Example 1**    Find an equation of the tangent line to the curve $y = x^2$ at the point $(1, 1)$.

**Solution**    Here $f(x) = x^2$, $x_0 = 1$, and $y_0 = f(1) = 1$. The slope of the required tangent is:

$$m = \lim_{h \to 0} \frac{f(1 + h) - f(1)}{h} = \lim_{h \to 0} \frac{(1 + h)^2 - 1}{h}$$

$$= \lim_{h \to 0} \frac{1 + 2h + h^2 - 1}{h}$$

$$= \lim_{h \to 0} \frac{2h + h^2}{h} = \lim_{h \to 0} 2 + h = 2.$$

Accordingly, the equation of the tangent line at $(1, 1)$ is $y = 2(x - 1) + 1$, or $y = 2x - 1$. See Figure 2.5.

Definition 1 deals only with tangents that have finite slopes and are, therefore, not vertical. It is also possible for the graph of a continuous function to have a *vertical* tangent line.

**Figure 2.6**    The $y$-axis is tangent to $y = x^{1/3}$ at the origin

**Example 2**    Consider the graph of the function $f(x) = \sqrt[3]{x} = x^{1/3}$, which is shown in Figure 2.6. The graph is a smooth curve and it seems evident that the $y$-axis is tangent to this curve at the origin. Let us try to calculate the limit of the Newton quotient for $f$ at $x = 0$:

$$\lim_{h \to 0} \frac{f(0 + h) - f(0)}{h} = \lim_{h \to 0} \frac{h^{1/3}}{h} = \lim_{h \to 0} \frac{1}{h^{2/3}} = \infty.$$

Although the limit does not exist, the slope of the secant line joining the origin to another point $Q$ on the curve approaches infinity as $Q$ approaches the origin from either side.

**Example 3** On the other hand, the function $f(x) = x^{2/3}$ whose graph is shown in Figure 2.7, does not have a tangent line at the origin because it is not "smooth" there. In this case the Newton quotient is

$$\frac{f(0 + h) - f(0)}{h} = \frac{h^{2/3}}{h} = \frac{1}{h^{1/3}},$$

which has no limit as $h$ approaches zero. (The right limit is $\infty$; the left limit is $-\infty$.) We say this curve has a **cusp** at the origin. A cusp is an infinitely sharp point; if you were travelling along the curve you would have to stop and turn $180°$ at the origin. ∎

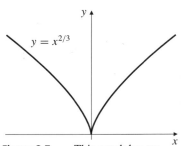

**Figure 2.7** This graph has no tangent at the origin

In the light of the two examples above, we extend the definition of tangent line to allow for vertical tangents as follows:

**DEFINITION 2**

**Vertical tangents**

If $f$ is continuous at $P = (x_0, y_0)$, where $y_0 = f(x_0)$, and if either

$$\lim_{h \to 0} \frac{f(x_0 + h) - f(x_0)}{h} = \infty \quad \text{or} \quad \lim_{h \to 0} \frac{f(x_0 + h) - f(x_0)}{h} = -\infty$$

then the vertical line $x = x_0$ is tangent to the graph $y = f(x)$ at $P$. If the limit of the Newton quotient fails to exist in any other way than by being $\infty$ or $-\infty$ the graph $y = f(x)$ has no tangent line at $P$.

**Example 4** Does the graph of $y = |x|$ have a tangent line at $x = 0$?

**Solution** The Newton quotient here is

$$\frac{|0 + h| - |0|}{h} = \frac{|h|}{h} = \text{sgn}\, h = \begin{cases} 1, & \text{if } h > 0 \\ -1, & \text{if } h < 0. \end{cases}$$

Since $\text{sgn}\, h$ has different right and left limits at 0 (namely 1 and $-1$) the Newton quotient has no limit as $h \to 0$, so $y = |x|$ has no tangent line at $(0, 0)$. (See Figure 2.8.) The graph does not have a cusp at the origin, but it is kinked at that point; *it suddenly changes direction and is not smooth.* Curves have tangents only at points where they are smooth. The graphs of $y = x^{2/3}$ and $y = |x|$ have tangent lines everywhere except at the origin, where they are not smooth. ∎

**Figure 2.8** $y = |x|$ has no tangent at the origin

**DEFINITION 3**

**The slope of a curve**

The **slope** of a curve $C$ at a point $P$ is the slope of the tangent line to $C$ at $P$ if such a tangent line exists. In particular, the slope of the graph of $y = f(x)$ at the point $x_0$ is

$$\lim_{h \to 0} \frac{f(x_0 + h) - f(x_0)}{h}.$$

**Example 5** Find the slope of the curve $y = x/(3x + 2)$ at the point $x = -2$.

**Solution** If $x = -2$, then $y = 1/2$, so the required slope is

$$
\begin{aligned}
m &= \lim_{h \to 0} \frac{\dfrac{-2+h}{3(-2+h)+2} - \dfrac{1}{2}}{h} \\
&= \lim_{h \to 0} \frac{-4 + 2h - (-6 + 3h + 2)}{2(-6 + 3h + 2)h} \\
&= \lim_{h \to 0} \frac{-1}{2(-4 + 3h)} = \frac{1}{8}.
\end{aligned}
$$

## Normals

If a curve $C$ has a tangent line $L$ at point $P$, then the straight line $N$ through $P$ perpendicular to $L$ is called the **normal** to $C$ at $P$. If $L$ is horizontal, then $N$ is vertical; if $L$ is vertical, then $N$ is horizontal. If $L$ is neither horizontal nor vertical, then, as shown in Section P.2, the slope of $N$ is the negative reciprocal of the slope of $L$:

$$
\text{slope of the normal} \; = \; \frac{-1}{\text{slope of the tangent}}.
$$

**Example 6** Find an equation of the normal to $y = x^2$ at $(1, 1)$.

**Solution** By Example 1, the tangent to $y = x^2$ at $(1, 1)$ has slope 2. Hence the normal has slope $-1/2$, and its equation is

$$
y = -\frac{1}{2}(x - 1) + 1 \qquad \text{or} \qquad y = -\frac{x}{2} + \frac{3}{2}.
$$

**Example 7** Find equations of the straight lines that are tangent and normal to the curve $y = \sqrt{x}$ at the point $(4, 2)$.

**Solution** The slope of the tangent at $(4, 2)$ (Figure 2.9) is

$$
\begin{aligned}
m &= \lim_{h \to 0} \frac{\sqrt{4+h} - 2}{h} = \lim_{h \to 0} \frac{(\sqrt{4+h} - 2)(\sqrt{4+h} + 2)}{h(\sqrt{4+h} + 2)} \\
&= \lim_{h \to 0} \frac{4 + h - 4}{h(\sqrt{4+h} + 2)} \\
&= \lim_{h \to 0} \frac{1}{\sqrt{4+h} + 2} = \frac{1}{4}.
\end{aligned}
$$

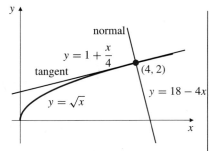

The tangent line has equation

$$y = \frac{1}{4}(x-4) + 2 \quad \text{or} \quad x - 4y + 4 = 0,$$

and the normal has equation

$$y = -4(x-4) + 2 \quad \text{or} \quad y = -4x + 18.$$

**Figure 2.9**    The tangent and normal to $y = \sqrt{x}$ at $(4, 2)$

## Exercises 2.1

In Exercises 1–14 find an equation of the straight line tangent to the given curve at the point indicated.

**1.** $y = 3x - 1$ at $(1, 2)$       **2.** $y = x/2$ at $(a, a/2)$

**3.** $y = 2x^2 - 5$ at $(2, 3)$       **4.** $y = 6 - x - x^2$ at $x = -2$

**5.** $y = x^3 + 8$ at $x = -2$       **6.** $y = \dfrac{1}{x^2 + 1}$ at $(0, 1)$

**7.** $y = \dfrac{1}{x^2}$ at $x = 3$       **8.** $y = \dfrac{1}{x^2 + 1}$ at $x = -1$

**9.** $y = \sqrt{x + 1}$ at $x = 3$       **10.** $y = \dfrac{1}{\sqrt{x}}$ at $x = 9$

**11.** $y = \dfrac{2x}{x + 2}$ at $x = 2$       **12.** $y = \sqrt{5 - x^2}$ at $x = 1$

**13.** $y = x^2$ at $x = x_0$       **14.** $y = \dfrac{1}{x}$ at $\left(a, \dfrac{1}{a}\right)$

Do the graphs of the functions $f$ in Exercises 15–19 have tangent lines at the given points or not? If yes, what is the tangent line?

**15.** $f(x) = \sqrt{|x|}$ at $x = 0$       **16.** $f(x) = (x - 1)^{4/3}$ at $x = 1$

**17.** $f(x) = (x + 2)^{3/5}$ at $x = -2$

**18.** $f(x) = |x^2 - 1|$ at $x = 1$

**19.** $f(x) = \begin{cases} \sqrt{x} & \text{if } x \geq 0 \\ -\sqrt{-x} & \text{if } x < 0 \end{cases}$ at $x = 0$

**20.** Find the slope of the curve $y = x^2 - 1$ at the point $x = x_0$. What is the equation of the tangent line to $y = x^2 - 1$ that has slope $-3$?

**21.** (a) Find the slope of $y = x^3$ at the point $x = a$.

　　(b) Find the equations of the straight lines having slope 3 that are tangent to $y = x^3$.

**22.** Find all points on the curve $y = x^3 - 3x$ where the tangent line is parallel to the $x$-axis.

**23.** Find all points on the curve $y = x^3 - x + 1$ where the tangent line is parallel to the line $y = 2x + 5$.

**24.** Find all points on the curve $y = 1/x$ where the tangent line is perpendicular to the line $y = 4x - 3$.

**25.** For what value of the constant $k$ is the line $x + y = k$ normal to the curve $y = x^2$?

**26.** For what value of the constant $k$ do the curves $y = kx^2$ and $y = k(x - 2)^2$ intersect at right angles? *Hint:* where do the curves intersect? What are their slopes there?

Use a graphics utility to plot the following curves. Where does the curve have a horizontal tangent? Does the curve fail to have a tangent line anywhere?

**27.** $y = x^3(5 - x)^2$       **28.** $y = 2x^3 - 3x^2 - 12x + 1$

**29.** $y = |x^2 - 1| - x$       **30.** $y = |x + 1| - |x - 1|$

**31.** $y = (x^2 - 1)^{1/3}$       **32.** $y = ((x^2 - 1)^2)^{1/3}$

**\* 33.** If line $L$ is tangent to curve $C$ at point $P$, then the smaller angle between $L$ and the secant line $PQ$ joining $P$ to another point $Q$ on $C$ approaches 0 as $Q$ approaches $P$ along $C$. Is the converse true; if the angle between $PQ$ and line $L$ (which passes through $P$) approaches 0, must $L$ be tangent to $C$?

**\* 34.** Let $P(x)$ be a polynomial. If $a$ is a real number, then $P(x)$ can be expressed in the form

$$P(x) = a_0 + a_1(x - a) + a_2(x - a)^2 + \cdots + a_n(x - a)^n$$

for some $n \geq 0$. If $\ell(x) = m(x - a) + b$, show that the straight line $y = \ell(x)$ is tangent to the graph of $y = P(x)$ at $x = a$ provided $P(x) - \ell(x) = (x - a)^2 Q(x)$, where $Q(x)$ is a polynomial.

## 2.2 The Derivative

A straight line has the property that its slope is the same at all points. For any other graph, however, the slope may vary from point to point. Thus the slope of the graph of $y = f(x)$ at the point $x$ is itself a function of $x$. At any point $x$ where the graph has a finite slope we say that $f$ is differentiable, and we call the slope the derivative of $f$. The derivative is therefore the limit of the Newton quotient.

**DEFINITION 4**

> The **derivative** of a function $f$ is another function $f'$ defined by
> $$f'(x) = \lim_{h \to 0} \frac{f(x+h) - f(x)}{h}$$
> at all points $x$ for which the limit exists (i.e., is a finite real number). If $f'(x)$ exists, we say that $f$ is **differentiable** at $x$.

The domain of the derivative $f'$ (read "$f$ prime") is the set of numbers $x$ in the domain of $f$ where the graph of $f$ has a *non-vertical* tangent line, and the value $f'(x_0)$ of $f'$ at such a point $x_0$ is the slope of the tangent line to $y = f(x)$ there. Thus the equation of the tangent line to $y = f(x)$ at $(x_0, f(x_0))$ is

$$y = f(x_0) + f'(x_0)(x - x_0).$$

The domain $\mathcal{D}(f')$ of $f'$ may be smaller than the domain $\mathcal{D}(f)$ of $f$ because it contains only those points in $\mathcal{D}(f)$ at which $f$ is differentiable. Values of $x$ in $\mathcal{D}(f)$ where $f$ is not differentiable are called **singular points** of $f$.

**Remark**   The value of the derivative of $f$ at a particular point $x_0$ can be expressed as a limit in either of two ways:

$$f'(x_0) = \lim_{h \to 0} \frac{f(x_0 + h) - f(x_0)}{h} = \lim_{x \to x_0} \frac{f(x) - f(x_0)}{x - x_0}.$$

In the second limit $x_0 + h$ is replaced by $x$, so that $h = x - x_0$ and $h \to 0$ is equivalent to $x \to x_0$.

The process of calculating the derivative $f'$ of a given function $f$ is called **differentiation**. The graph of $f'$ can often be sketched directly from that of $f$ by visualizing slopes, a procedure called **graphical differentiation**. In Figure 2.10 the graphs of $f'$ and $g'$ were obtained by measuring the slopes at the corresponding points in the graphs of $f$ and $g$ lying above them. The height of the graph $y = f'(x)$ at $x$ is the slope of the graph of $y = f(x)$ at $x$. Note that $-1$ and $1$ are singular points of $f$. $f(-1)$ and $f(1)$ are defined, but $f'(-1)$ and $f'(1)$ are not defined; the graph of $f$ has no tangent at $-1$ or at $1$.

A function is differentiable on a set $S$ if it is differentiable at every point $x$ in $S$. Typically, the functions we encounter are defined on intervals or unions of intervals. If $f$ is defined on a closed interval $[a, b]$, Definition 4 does not allow for the existence of a derivative at the endpoints $x = a$ or $x = b$. (Why?)   As

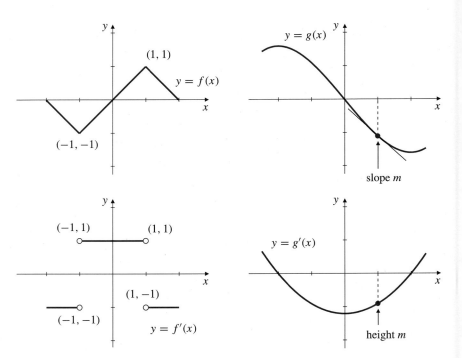

**Figure 2.10** Graphical differentiation

we did for continuity in Section 1.4, we extend the definition to allow for a **right derivative** at $x = a$ and a **left derivative** at $x = b$:

$$f'_+(a) = \lim_{h \to 0+} \frac{f(a+h) - f(a)}{h}, \qquad f'_-(b) = \lim_{h \to 0-} \frac{f(b+h) - f(b)}{h}.$$

We now say that $f$ is **differentiable** on $[a, b]$ if $f'(x)$ exists for all $x$ in $(a, b)$ and $f'_+(a)$ and $f'_-(b)$ both exist.

## Some Important Derivatives

We now give several examples of the calculation of derivatives algebraically from the definition of derivative. Some of these are the basic building blocks from which more complicated derivatives can be calculated later. They are collected in Table 1 later in this section, and should be memorized.

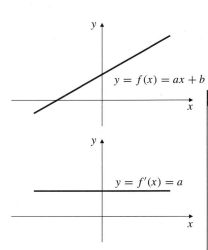

**Figure 2.11** The derivative of the linear function $f(x) = ax + b$ is the constant function $f'(x) = a$

**Example 1** (**The derivative of a linear function**) Show that if $f(x) = ax + b$, then $f'(x) = a$.

**Solution** The result is apparent from the graph of $f$ (Figure 2.11), but we will do the calculation using the definition:

$$
\begin{aligned}
f'(x) &= \lim_{h \to 0} \frac{f(x+h) - f(x)}{h} \\
&= \lim_{h \to 0} \frac{a(x+h) + b - (ax + b)}{h} \\
&= \lim_{h \to 0} \frac{ah}{h} = a.
\end{aligned}
$$

An important special case of Example 1 says that the derivative of a constant function is the zero function:

If $g(x) = c$ (constant),    then    $g'(x) = 0.$

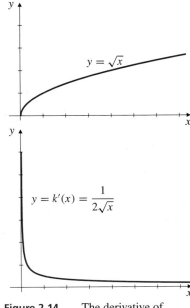

**Figure 2.12**    The derivative of $f(x) = x^2$ is $f'(x) = 2x$

**Figure 2.13**    The derivative of $g(x) = 1/x$ is $g'(x) = -1/x^2$

**Figure 2.14**    The derivative of $k(x) = \sqrt{x}$ is $k'(x) = 1/(2\sqrt{x})$

**Example 2**    Use the definition of the derivative to calculate the derivatives of the functions:

$$\text{(a) } f(x) = x^2, \quad \text{(b) } g(x) = \frac{1}{x}, \quad \text{and} \quad \text{(c) } k(x) = \sqrt{x}.$$

**Solution**    See Figures 2.12–2.14 for the graphs of these functions and of their derivatives.

$$\text{(a) } f'(x) = \lim_{h \to 0} \frac{f(x+h) - f(x)}{h}$$

$$= \lim_{h \to 0} \frac{(x+h)^2 - x^2}{h}$$

$$= \lim_{h \to 0} \frac{2hx + h^2}{h} = \lim_{h \to 0} (2x + h) = 2x.$$

$$\text{(b) } g'(x) = \lim_{h \to 0} \frac{g(x+h) - g(x)}{h}$$

$$= \lim_{h \to 0} \frac{\dfrac{1}{x+h} - \dfrac{1}{x}}{h}$$

$$= \lim_{h \to 0} \frac{x - (x+h)}{h(x+h)x} = \lim_{h \to 0} -\frac{1}{(x+h)x} = -\frac{1}{x^2}.$$

(c)  $k'(x) = \lim\limits_{h \to 0} \dfrac{k(x+h) - k(x)}{h}$

$= \lim\limits_{h \to 0} \dfrac{\sqrt{x+h} - \sqrt{x}}{h}$

$= \lim\limits_{h \to 0} \dfrac{\sqrt{x+h} - \sqrt{x}}{h} \times \dfrac{\sqrt{x+h} + \sqrt{x}}{\sqrt{x+h} + \sqrt{x}}$

$= \lim\limits_{h \to 0} \dfrac{x+h-x}{h(\sqrt{x+h} + \sqrt{x})} = \lim\limits_{h \to 0} \dfrac{1}{\sqrt{x+h} + \sqrt{x}} = \dfrac{1}{2\sqrt{x}}.$

Note that $k$ is not differentiable at $x = 0$. (See Figure 2.14.)  Since 0 is in the domain of $k$, it is a singular point of $k$.

The three derivative formulas calculated in Example 2 are special cases of the following **General Power Rule**:

If $f(x) = x^r$,    then    $f'(x) = r\, x^{r-1}.$

This formula, which we will verify in Section 3.3, is valid for *all values of r and x for which $x^{r-1}$ makes sense as a real number.*

**Example 3**    (**Differentiating powers**)

If $f(x) = x^{5/3}$, then $f'(x) = \dfrac{5}{3}x^{(5/3)-1} = \dfrac{5}{3}x^{2/3}$ for all real $x$.

If $g(t) = \dfrac{1}{\sqrt{t}} = t^{-1/2}$, then $g'(t) = -\dfrac{1}{2}t^{-(1/2)-1} = -\dfrac{1}{2}t^{-3/2}$ for $t > 0$.

Eventually we will prove all appropriate cases of the General Power Rule. For the time being, here is a proof of the case $r = n$, a positive integer, based on the *factoring of a difference of nth powers*:

$$a^n - b^n = (a - b)(a^{n-1} + a^{n-2}b + a^{n-3}b^2 + \cdots + ab^{n-2} + b^{n-1}).$$

(Check this formula by multiplying the two factors on the right-hand side.)  If $f(x) = x^n$, $a = x + h$ and $b = x$, then $a - b = h$ and,

$$f'(x) = \lim_{h \to 0} \frac{(x+h)^n - x^n}{h}$$

$$= \lim_{h \to 0} \frac{h \overbrace{[(x+h)^{n-1} + (x+h)^{n-2}x + (x+h)^{n-3}x^2 + \cdots + x^{n-1}]}^{n \text{ terms}}}{h}$$

$$= nx^{n-1}.$$

An alternative proof based on the product rule and mathematical induction, will be given in Section 2.3. The factorization method used above can also be used to demonstrate the General Power Rule for negative integers, $r = -n$, and reciprocals of integers, $r = 1/n$. (See Exercises 52 and 54 at the end of this section.)

**Example 4**    **(Differentiating the absolute value function)**  Verify that:

$$\text{If } f(x) = |x|, \quad \text{then} \quad f'(x) = \frac{x}{|x|} = \text{sgn}\, x.$$

**Solution**   We have

$$f(x) = \begin{cases} x, & \text{if } x \geq 0 \\ -x, & \text{if } x < 0 \end{cases}.$$

Thus, from Example 1 above, $f'(x) = 1$ if $x > 0$ and $f'(x) = -1$ if $x < 0$. Also, Example 4 of Section 2.1 shows that $f$ is not differentiable at $x = 0$; which is a singular point of $f$. Therefore (see Figure 2.15),

$$f'(x) = \begin{cases} 1, & \text{if } x > 0 \\ -1, & \text{if } x < 0 \end{cases} = \frac{x}{|x|} = \text{sgn}\, x.$$

Table 1 collects the elementary derivatives calculated above. Beginning in Section 2.3 we will develop general rules for calculating the derivatives of functions obtained by combining simpler functions. Thereafter we will seldom have to revert to the definition of the derivative and to the calculation of limits to evaluate derivatives. It is important, therefore, to remember the derivatives of some elementary functions. Memorize those in Table 1.

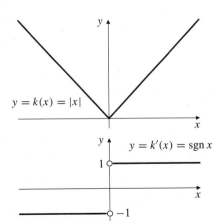

$y = k(x) = |x|$

$y = k'(x) = \text{sgn}\, x$

**Figure 2.15**   The derivative of $|x|$ is $\text{sgn}\, x = x/|x|$

**Table 1.**   Some elementary functions and their derivatives

| $f(x)$ | $f'(x)$ |
|--------|---------|
| $c$ (constant) | $0$ |
| $x$ | $1$ |
| $x^2$ | $2x$ |
| $\dfrac{1}{x}$ | $-\dfrac{1}{x^2}$  $(x \neq 0)$ |
| $\sqrt{x}$ | $\dfrac{1}{2\sqrt{x}}$  $(x > 0)$ |
| $x^r$ | $r\, x^{r-1}$  $(x^{r-1}$ real$)$ |
| $|x|$ | $\dfrac{x}{|x|} = \text{sgn}\, x$ |

## Leibniz Notation

Because functions can be written in different ways, it is useful to have more than one notation for derivatives. If $y = f(x)$ we can use the dependent variable $y$ to represent the function, and we can denote the derivative of the function with respect to $x$ in any of the following ways:

$$D_x y = y' = \frac{dy}{dx} = \frac{d}{dx} f(x) = f'(x) = Df(x).$$

Often the most convenient way of referring to the derivative of a function given explicitly as an expression in the variable $x$ is to write $\frac{d}{dx}$ in front of that expression. The symbol $\frac{d}{dx}$ is a *differential operator*, and should be read "the derivative with respect to $x$ of ..."   For example:

$$\frac{d}{dx}x^2 = 2x \quad \text{(the derivative with respect to } x \text{ of } x^2 \text{ is } 2x)$$

$$\frac{d}{dx}\sqrt{x} = \frac{1}{2\sqrt{x}}$$

$$\frac{d}{dt}t^{100} = 100t^{99}$$

if $y = u^3$, then $\dfrac{dy}{du} = 3u^2$.

The value of the derivative of a function at a particular number $x_0$ in its domain can also be expressed in several ways:

$$D_x y \bigg|_{x=x_0} = y' \bigg|_{x=x_0} = \frac{dy}{dx}\bigg|_{x=x_0} = \frac{d}{dx}f(x)\bigg|_{x=x_0} = f'(x_0) = Df(x_0).$$

Do not confuse the expressions

$$\frac{d}{dx}f(x) \text{ and } \frac{d}{dx}f(x)\bigg|_{x=x_0}.$$

The first expression represents a *function*, $f'(x)$. The second represents a *number*, $f'(x_0)$.

The symbol $\bigg|_{x=x_0}$ is called an **evaluation symbol**. It signifies that the expression preceding it should be evaluated at $x = x_0$. Thus

$$\frac{d}{dx}x^4\bigg|_{x=-1} = 4x^3\bigg|_{x=-1} = 4(-1)^3 = -4.$$

Here is another example in which a derivative is computed from the definition, this time for a somewhat more complicated function.

**Example 5**   Use the definition of derivative to calculate $\dfrac{d}{dx}\left(\dfrac{x}{x^2+1}\right)\bigg|_{x=2}.$

**Solution**   We could calculate $\dfrac{d}{dx}\left(\dfrac{x}{x^2+1}\right)$ and then substitute $x = 2$, but it is easier to put $x = 2$ in the expression for the Newton quotient before taking the limit:

$$\frac{d}{dx}\left(\frac{x}{x^2+1}\right)\bigg|_{x=2} = \lim_{h\to 0}\frac{\dfrac{2+h}{(2+h)^2+1} - \dfrac{2}{2^2+1}}{h}$$

$$= \lim_{h\to 0}\frac{\dfrac{2+h}{5+4h+h^2} - \dfrac{2}{5}}{h}$$

$$= \lim_{h\to 0}\frac{5(2+h) - 2(5+4h+h^2)}{5(5+4h+h^2)h}$$

$$= \lim_{h\to 0}\frac{-3h - 2h^2}{5(5+4h+h^2)h}$$

$$= \lim_{h\to 0}\frac{-3 - 2h}{5(5+4h+h^2)} = -\frac{3}{25}.$$

The notations $dy/dx$ and $\frac{d}{dx}f(x)$ are called **Leibniz notations** for the derivative, after Gottfried Wilhelm Leibniz (1646–1716), one of the creators of calculus, who used such notations. The main ideas of calculus were developed independently by Leibniz and Isaac Newton (1642–1727); the latter used notations similar to the prime ($y'$) notations we use here.

The Leibniz notation is suggested by the definition of derivative. The Newton quotient $[f(x+h) - f(x)]/h$, whose limit we take to find the derivative $dy/dx$, can be written in the form $\Delta y/\Delta x$, where $\Delta y = f(x+h) - f(x)$ is the increment in $y$, and $\Delta x = (x+h) - x = h$ is the corresponding increment in $x$ as we pass from the point $(x, f(x))$ to the point $(x+h, f(x+h))$ on the graph of $f$. (See Figure 2.16.) Using symbols:

$$\frac{dy}{dx} = \lim_{\Delta x \to 0} \frac{\Delta y}{\Delta x}.$$

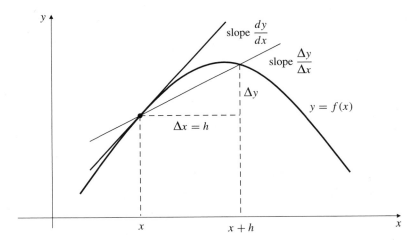

**Figure 2.16** $\quad \dfrac{dy}{dx} = \lim\limits_{\Delta x \to 0} \dfrac{\Delta y}{\Delta x}$

## Differentials

The Newton quotient $\Delta y/\Delta x$ is actually the quotient of two quantities, $\Delta y$ and $\Delta x$. It is not at all clear, however, that the derivative $dy/dx$, the limit of $\Delta y/\Delta x$ as $\Delta x$ approaches zero, can be regarded as a quotient. If $y$ is a continuous function of $x$, then $\Delta y$ approaches zero when $\Delta x$ approaches zero, so $dy/dx$ appears to be the meaningless quantity $0/0$. Nevertheless, it is sometimes useful to be able to refer to quantities $dy$ and $dx$ in such a way that their quotient is the derivative $dy/dx$. We can justify this by regarding $dx$ as a new *independent* variable (called **the differential of $x$**) and defining a new *dependent* variable $dy$ (**the differential of $y$**) as a function of $x$ and $dx$ by

$$dy = \frac{dy}{dx}dx = f'(x)\,dx.$$

For example, if $y = x^2$, we can write $dy = 2x\,dx$ to mean the same thing as $dy/dx = 2x$. This *differential notation* will be used for the interpretation and manipulation of integrals beginning in Chapter 5.

Note that, defined as above, differentials are merely variables that may or may not be small in absolute value. The differentials $dy$ and $dx$ were originally regarded (by Leibniz and his successors) as "infinitesimals" (infinitely small but nonzero) quantities whose quotient $dy/dx$ gave the slope of the tangent line (a secant line meeting the graph of $y = f(x)$ at two points infinitely close together). It can be shown that such "infinitesimal" quantities cannot exist (as real numbers). It is possible to extend the number system to contain infinitesimals and use these to develop calculus, but we will not consider this approach here.

## Derivatives Have the Intermediate-Value Property

Is a function $f$ defined on an interval $I$ necessarily the derivative of some other function defined on $I$? The answer is no; some functions are derivatives and some are not. Although a derivative need not be a continuous function, it must, like a continuous function, have the intermediate-value property: on an interval $[a, b]$, a derivative $f'(x)$ takes on every value between $f'(a)$ and $f'(b)$. An everywhere-defined step function such as the Heaviside $H(x)$ function of Example 1 in Section 1.4 does not have this property on, say, the interval $[-1, 1]$, so cannot be the derivative of a function on that interval. This argument does not apply to the signum function, which is the derivative of the absolute value function on any interval (see Example 4) even though it does not have the intermediate-value property on an interval containing the origin. Note, however, that the signum function is not defined at the origin.

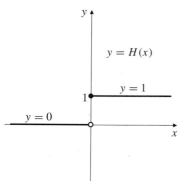

If $g(x)$ is continuous on an interval $I$, then $g(x) = f'(x)$ for some function $f$ that is differentiable on $I$. We will discuss this further in Chapter 5, and prove it in Appendix IV.

**Figure 2.17**   This function is not a derivative on $[-1, 1]$; it does not have the intermediate-value property.

## Exercises 2.2

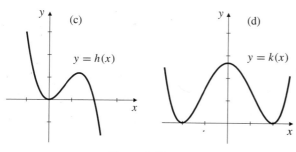

**Figure 2.18**

Make rough sketches of the graphs of the derivatives of the functions in Exercises 1–4.

**1.** The function $f$ graphed in Figure 2.18(a).

**2.** The function $g$ graphed in Figure 2.18(b).

**3.** The function $h$ graphed in Figure 2.18(c).

**4.** The function $k$ graphed in Figure 2.18(d).

**5.** Where is the function $f$ graphed in Figure 2.18(a) differentiable?

**6.** Where is the function $g$ graphed in Figure 2.18(b) differentiable?

Use a graphics utility with differentiation capabilities to plot the graphs of the following functions and their derivatives. Observe the relationships between the graph of $y$ and that of $y'$ in each case. What features of the graph of $y$ can you infer from the graph of $y'$?

**7.** $y = 3x - x^2 - 1$    **8.** $y = x^3 - 3x^2 + 2x + 1$

**9.** $y = |x^3 - x|$    **10.** $y = |x^2 - 1| - |x^2 - 4|$

In Exercises 11–24, calculate the derivative of the given function directly from the definition of derivative.

**11.** $y = x^2 - 3x$    **12.** $f(x) = 1 + 4x - 5x^2$

**13.** $f(x) = x^3$    **14.** $y = \frac{1}{3}x^3 - x$

**15.** $g(x) = \frac{2 - x}{2 + x}$    **16.** $s = \frac{1}{3 + 4t}$

**17.** $F(t) = \sqrt{2t + 1}$

**18.** $f(x) = \dfrac{3}{4}\sqrt{2 - x}$

**19.** $y = x + \dfrac{1}{x}$

**20.** $z = \dfrac{s}{1 + s}$

**21.** $F(x) = \dfrac{1}{\sqrt{1 + x^2}}$

**22.** $y = \dfrac{1}{x^2}$

**23.** $y = \dfrac{1}{\sqrt{1 + x}}$

**24.** $f(t) = \dfrac{t^2 - 3}{t^2 + 3}$

**25.** How should the function $f(x) = x \operatorname{sgn} x$ be defined at $x = 0$ so that it is continuous there? Is it then differentiable there?

**26.** How should the function $g(x) = x^2 \operatorname{sgn} x$ be defined at $x = 0$ so that it is continuous there? Is it then differentiable there?

**27.** Where does $h(x) = |x^2 + 3x + 2|$ fail to be differentiable?

**28.** Using a calculator, find the slope of the secant line to $y = x^3 - 2x$ passing through the points corresponding to $x = 1$ and $x = 1 + \Delta x$, for several values of $\Delta x$ of decreasing size, say $\Delta x = \pm 0.1, \pm 0.01, \pm 0.001, \pm 0.0001$. (Make a table.) Also, calculate $\dfrac{d}{dx}\left(x^3 - 2x\right)\Big|_{x=1}$ using the definition of derivative.

**29.** Repeat the Exercise 28 for the function $f(x) = \dfrac{1}{x}$ and the points $x = 2$ and $x = 2 + \Delta x$.

Using the definition of derivative, find equations for the tangent lines to the curves in Exercises 30–33 at the points indicated.

**30.** $y = 5 + 4x - x^2$ at the point where $x = 2$

**31.** $y = \sqrt{x + 6}$ at the point $(3, 3)$

**32.** $y = \dfrac{t}{t^2 - 2}$ at the point where $t = -2$

**33.** $y = \dfrac{2}{t^2 + t}$ at the point where $t = a$

Calculate the derivatives of the functions in Exercises 34–39 using the General Power Rule. Where is each derivative valid?

**34.** $f(x) = x^{-17}$

**35.** $g(t) = t^{22}$

**36.** $y = x^{1/3}$

**37.** $y = x^{-1/3}$

**38.** $t^{-2.25}$

**39.** $s^{119/4}$

In Exercises 40–50, you may use the formulas for derivatives established in this section.

**40.** Calculate $\dfrac{d}{ds}\sqrt{s}\,\Big|_{s=9}$.

**41.** Find $F'(\tfrac{1}{4})$ if $F(x) = \dfrac{1}{x}$.

**42.** Find $f'(8)$ if $f(x) = x^{-2/3}$.

**43.** Find $dy/dt\big|_{t=4}$ if $y = t^{1/4}$.

**44.** Find an equation of the straight line tangent to the curve $y = \sqrt{x}$ at $x = x_0$.

**45.** Find an equation of the straight line normal to the curve $y = 1/x$ at the point where $x = a$.

**46.** Show that the curve $y = x^2$ and the straight line $x + 4y = 18$ intersect at right angles at one of their two intersection points. *Hint:* find the product of their slopes at their intersection points.

**47.** There are two distinct straight lines that pass through the point $(1, -3)$ and are tangent to the curve $y = x^2$. Find their equations. *Hint:* draw a sketch. The points of tangency are not given; let them be denoted $(a, a^2)$.

**48.** Find equations of two straight lines that have slope $-2$ and are tangent to the graph of $y = 1/x$.

**49.** Find the slope of a straight line that passes through the point $(-2, 0)$ and is tangent to the curve $y = \sqrt{x}$.

**∗ 50.** Show that there are two distinct tangent lines to the curve $y = x^2$ passing through the point $(a, b)$ provided $b < a^2$. How many tangent lines to $y = x^2$ pass through $(a, b)$ if $b = a^2$? If $b > a^2$?

**∗ 51.** Show that the derivative of an odd differentiable function is even, and that the derivative of an even differentiable function is odd.

**∗ 52.** Prove the case $r = -n$, ($n$ is a positive integer), of the General Power Rule; that is, prove that

$$\frac{d}{dx} x^{-n} = -n\, x^{-n-1}.$$

Use the factorization of a difference of $n$th powers given in this section.

**∗ 53.** Use the factoring of a difference of cubes:

$$a^3 - b^3 = (a - b)(a^2 + ab + b^2),$$

to help you calculate the derivative of $f(x) = x^{1/3}$ directly from the definition of derivative.

**∗ 54.** Prove the General Power Rule for $\frac{d}{dx} x^r$, where $r = 1/n$, $n$ being a positive integer. (*Hint:*

$$\frac{d}{dx} x^{1/n} = \lim_{h \to 0} \frac{(x + h)^{1/n} - x^{1/n}}{h}$$

$$= \lim_{h \to 0} \frac{(x + h)^{1/n} - x^{1/n}}{((x + h)^{1/n})^n - (x^{1/n})^n}.$$

Apply the factorization of the difference of $n$th powers to the denominator of the latter quotient.)

**55.** Give a proof of the power rule $\frac{d}{dx} x^n = nx^{n-1}$ for positive integers $n$ using the Binomial Theorem:

$$(x + h)^n = x^n + \frac{n}{1} x^{n-1} h + \frac{n(n - 1)}{1 \times 2} x^{n-2} h^2$$

$$+ \frac{n(n - 1)(n - 2)}{1 \times 2 \times 3} x^{n-3} h^3 + \cdots + h^n.$$

**∗ 56.** Use right and left derivatives, $f'_+(a)$ and $f'_-(a)$, to define the concept of a half-line starting at $(a, f(a))$ being a right or left tangent to the graph of $f$ at $x = a$. Show that the graph has a tangent line at $x = a$ if and only if it has right and left tangents that are opposite halves of the same straight line. What are the left and right tangents to the graphs of $y = x^{1/3}$, $y = x^{2/3}$, and $y = |x|$ at $x = 0$?

# 2.3   Differentiation Rules

If every derivative had to be calculated directly from the definition of derivative as in the examples of Section 2.2, calculus would indeed be a painful subject. Fortunately there is an easier way. We will develop several general *differentiation rules* that enable us to calculate the derivatives of complicated combinations of functions easily if we already know the derivatives of the elementary functions from which they are constructed. For instance, we will be able to find the derivative of $\dfrac{x^2}{\sqrt{x^2+1}}$ if we know the derivatives of $x^2$ and $\sqrt{x}$. The rules we develop in this section tell us how to differentiate sums, constant multiples, products, and quotients of functions whose derivatives we already know. In Section 2.4 we will learn how to differentiate composite functions.

Before developing these differentiation rules we need to establish one obvious but very important theorem which states, roughly, that the graph of a function cannot possibly have a break at a point where it is smooth.

**THEOREM 1**

### Differentiability implies continuity

If $f$ is differentiable at $x$, then $f$ is continuous at $x$.

**PROOF**   Since $f$ is differentiable at $x$, we know that

$$\lim_{h \to 0} \frac{f(x+h) - f(x)}{h} = f'(x)$$

exists. In order to prove that $f$ is continuous at $x$, we need to show that

$$\lim_{h \to 0} f(x+h) = f(x).$$

Using the limit rules (Theorem 2 of Section 1.2), we have

$$\lim_{h \to 0} f(x+h) = \lim_{h \to 0} \left( f(x) + \frac{f(x+h) - f(x)}{h} \times h \right)$$
$$= f(x) + f'(x) \times 0 = f(x).$$

## Sums and Constant Multiples

The derivative of a sum (or difference) of functions is the sum (or difference) of the derivatives of those functions. The derivative of a constant multiple of a function is the same constant multiple of the derivative of the function.

**THEOREM 2**

### Differentiation rules for sums, differences, and constant multiples

If functions $f$ and $g$ are differentiable at $x$, and if $C$ is a constant, then the functions $f + g$, $f - g$, and $Cf$ are all differentiable at $x$ and

$$(f + g)'(x) = f'(x) + g'(x),$$
$$(f - g)'(x) = f'(x) - g'(x),$$
$$(Cf)'(x) = Cf'(x).$$

**PROOF**  The proofs of all three assertions are straightforward, using the corresponding limit rules from Theorem 2 of Section 1.2. For the sum, we have

$$
\begin{aligned}
(f+g)'(x) &= \lim_{h \to 0} \frac{(f+g)(x+h) - (f+g)(x)}{h} \\
&= \lim_{h \to 0} \frac{(f(x+h) + g(x+h)) - (f(x) + g(x))}{h} \\
&= \lim_{h \to 0} \left( \frac{f(x+h) - f(x)}{h} + \frac{g(x+h) - g(x)}{h} \right) \\
&= f'(x) + g'(x),
\end{aligned}
$$

because the limit of a sum is the sum of the limits. The proof for the difference $f - g$ is similar. For the constant multiple, we have

$$
\begin{aligned}
(Cf)'(x) &= \lim_{h \to 0} \frac{Cf(x+h) - Cf(x)}{h} \\
&= C \lim_{h \to 0} \frac{f(x+h) - f(x)}{h} = Cf'(x).
\end{aligned}
$$

The rule for differentiating sums extends to sums of any finite number of terms:

$$(f_1 + f_2 + \cdots + f_n)' = f_1' + f_2' + \cdots + f_n'. \qquad (*)$$

## Mathematical Induction

Mathematical induction is a technique for proving that a statement about an integer $n$ is true for every integer $n$ greater than or equal to some lowest integer $n_0$. The proof requires us to carry out two steps:

STEP 1. Prove that the statement is true for $n = n_0$.

STEP 2. Prove that if the statement is true for some integer $n = k$, where $k \geq n_0$, then it is also true for the next larger integer, $n = k + 1$.

Step 2 prevents there from being a smallest integer greater than $n_0$ for which the statement is false. Being true for $n_0$, the statement must therefore be true for all larger integers.

To see this we can use a technique called **mathematical induction.** (See the note in the margin.) Theorem 2 shows that the case $n = 2$ is true; this is STEP 1. For STEP 2, we must show that if the formula $(*)$ holds for some integer $n = k \geq 2$, then it must also hold for $n = k + 1$. Therefore, assume that

$$(f_1 + f_2 + \cdots + f_k)' = f_1' + f_2' + \cdots + f_k'.$$

Then we have

$$
\begin{aligned}
(f_1 + f_2 &+ \cdots + f_k + f_{k+1})' \\
&= \Big( \underbrace{(f_1 + f_2 + \cdots + f_k)}_{\text{Let this function be } f} + f_{k+1} \Big) \\
&= (f + f_{k+1})' \qquad \text{(Now use the known case } n = 2.) \\
&= f' + f_{k+1}' \\
&= f_1' + f_2' + \cdots + f_k' + f_{k+1}'.
\end{aligned}
$$

With both steps verified, we can claim that $(*)$ holds for any $n \geq 2$ *by induction.* In particular, therefore, the derivative of any polynomial is the sum of the derivatives of its terms.

**Example 1**   Calculate the derivatives of the functions:

(a) $2x^3 - 5x^2 + 4x + 7$,   (b) $f(x) = 5\sqrt{x} + \dfrac{3}{x} - 18$, and   (c) $y = \dfrac{1}{7}t^4 - 3t^{7/3}$.

**Solution**   Each of these functions is a sum of constant multiples of functions that we already know how to differentiate.

(a) $\dfrac{d}{dx}(2x^3 - 5x^2 + 4x + 7) = 2(3x^2) - 5(2x) + 4(1) + 0 = 6x^2 - 10x + 4.$

(b) $f'(x) = 5\left(\dfrac{1}{2\sqrt{x}}\right) + 3\left(-\dfrac{1}{x^2}\right) - 18(0) = \dfrac{5}{2\sqrt{x}} - \dfrac{3}{x^2}.$

(c) $\dfrac{dy}{dt} = \dfrac{1}{7}(4t^3) - 3\left(\dfrac{7}{3}t^{4/3}\right) = \dfrac{4}{7}t^3 - 7t^{4/3}.$

**Example 2**   Find an equation of the tangent to the curve $y = \dfrac{3x^3 - 4}{x}$ at the point on the curve where $x = -2$.

**Solution**   If $x = -2$, then $y = 14$. The slope of the curve at $(-2, 14)$ is

$$\left.\dfrac{dy}{dx}\right|_{x=-2} = \left.\dfrac{d}{dx}\left(3x^2 - \dfrac{4}{x}\right)\right|_{x=-2} = \left.\left(6x + \dfrac{4}{x^2}\right)\right|_{x=-2} = -11.$$

An equation of the tangent line is $y = 14 - 11(x + 2)$, or $y = -11x - 8$.

## The Product Rule

The rule for differentiating a product of functions is a little more complicated than that for sums. It is *not* true that the derivative of a product is the product of the derivatives.

**THEOREM 3**

**The Product Rule**

If functions $f$ and $g$ are differentiable at $x$, then their product $fg$ is also differentiable at $x$, and

$$(fg)'(x) = f'(x)g(x) + f(x)g'(x).$$

**PROOF**   We set up the Newton quotient for $fg$ and then add 0 to the numerator in a way that enables us to involve the Newton quotients for $f$ and $g$ separately:

$$(fg)'(x) = \lim_{h\to 0}\dfrac{f(x+h)g(x+h) - f(x)g(x)}{h}$$
$$= \lim_{h\to 0}\dfrac{f(x+h)g(x+h) - f(x)g(x+h) + f(x)g(x+h) - f(x)g(x)}{h}$$
$$= \lim_{h\to 0}\left(\dfrac{f(x+h) - f(x)}{h}g(x+h) + f(x)\dfrac{g(x+h) - g(x)}{h}\right)$$
$$= f'(x)g(x) + f(x)g'(x).$$

To get the last line we have used the fact that $f$ and $g$ are differentiable and the fact that $g$ is therefore continuous (Theorem 1), as well as limit rules from Theorem 2 of Section 1.2. A graphical proof of the Product Rule is suggested by Figure 2.19.

**Example 3**    Find the derivative of $(x^2 + 1)(x^3 + 4)$ using and without using the Product Rule.

**Solution**    Using the Product Rule with $f(x) = x^2 + 1$ and $g(x) = x^3 + 4$, we calculate

$$\frac{d}{dx}\left((x^2 + 1)(x^3 + 4)\right) = 2x(x^3 + 4) + (x^2 + 1)(3x^2) = 5x^4 + 3x^2 + 8x.$$

On the other hand, we can calculate the derivative by first multiplying the two binomials and then differentiating the resulting polynomial:

$$\frac{d}{dx}\left((x^2 + 1)(x^3 + 4)\right) = \frac{d}{dx}(x^5 + x^3 + 4x^2 + 4) = 5x^4 + 3x^2 + 8x.$$

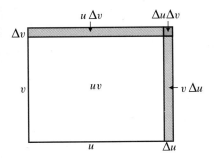

**Figure 2.19**

**A graphical proof of the Product Rule**

Here $u = f(x)$ and $v = g(x)$, so that the rectangular area $uv$ represents $f(x)g(x)$. If $x$ changes by an amount $\Delta x$, the corresponding increments in $u$ and $v$ are $\Delta u$ and $\Delta v$. The change in the area of the rectangle is

$$\Delta(uv)$$
$$= (u + \Delta u)(v + \Delta v) - uv$$
$$= (\Delta u)v + u(\Delta v) + (\Delta u)(\Delta v),$$

the sum of the three shaded areas. Dividing by $\Delta x$ and taking the limit as $\Delta x \to 0$, we get

$$\frac{d}{dx}(uv) = \left(\frac{du}{dx}\right)v + u\left(\frac{dv}{dx}\right),$$

since

$$\lim_{\Delta x \to 0} \frac{\Delta u}{\Delta x}\Delta v = \frac{du}{dx} \times 0 = 0.$$

**Example 4**    Find $\dfrac{dy}{dx}$ if $y = \left(2\sqrt{x} + \dfrac{3}{x}\right)\left(3\sqrt{x} - \dfrac{2}{x}\right)$.

**Solution**    Applying the Product Rule with $f$ and $g$ being the two functions enclosed in the large parentheses, we obtain

$$\frac{dy}{dx} = \left(\frac{1}{\sqrt{x}} - \frac{3}{x^2}\right)\left(3\sqrt{x} - \frac{2}{x}\right) + \left(2\sqrt{x} + \frac{3}{x}\right)\left(\frac{3}{2\sqrt{x}} + \frac{2}{x^2}\right)$$
$$= 6 - \frac{5}{2x^{3/2}} + \frac{12}{x^3}.$$

**Example 5**    Let $y = uv$ be the product of the functions $u$ and $v$. Find $y'(2)$ if $u(2) = 2$, $u'(2) = -5$, $v(2) = 1$, and $v'(2) = 3$.

**Solution**    From the Product Rule we have

$$y' = (uv)' = u'v + uv'.$$

Therefore

$$y'(2) = u'(2)v(2) + u(2)v'(2) = (-5)(1) + (2)(3) = -5 + 6 = 1.$$

---

**Example 6**   Use mathematical induction to verify the formula $\dfrac{d}{dx}x^n = n\,x^{n-1}$ for all positive integers $n$.

**Solution**   For $n = 1$ the formula says that $\frac{d}{dx}x^1 = 1 = 1x^0$, so the formula is true in this case. We must show that if the formula is true for $n = k \geq 1$, then it is also true for $n = k + 1$. Therefore, assume that

$$\frac{d}{dx}x^k = kx^{k-1}.$$

Using the Product Rule we calculate

$$\frac{d}{dx}x^{k+1} = \frac{d}{dx}\left(x^k x\right) = (kx^{k-1})(x) + (x^k)(1) = (k+1)x^k = (k+1)x^{(k+1)-1}.$$

Thus the formula is true for $n = k + 1$ also. The formula is true for all integers $n \geq 1$ *by induction.*

---

The Product Rule can be extended to products of any number of factors, for instance:

$$(fgh)'(x) = f'(x)(gh)(x) + f(x)(gh)'(x)$$
$$= f'(x)g(x)h(x) + f(x)g'(x)h(x) + f(x)g(x)h'(x).$$

In general, the derivative of a product of $n$ functions will have $n$ terms; each term will be the same product but with one of the factors replaced by its derivative:

$$(f_1 f_2 f_3 \cdots f_n)' = f_1' f_2 f_3 \cdots f_n + f_1 f_2' f_3 \cdots f_n + \cdots + f_1 f_2 f_3 \cdots f_n'.$$

This can be proved by mathematical induction. See Exercise 58 at the end of this section.

## The Reciprocal Rule

**THEOREM 4**

**The Reciprocal Rule**

If $f$ is differentiable at $x$ and $f(x) \neq 0$, then $1/f$ is differentiable at $x$, and

$$\left(\frac{1}{f}\right)'(x) = \frac{-f'(x)}{(f(x))^2}.$$

**PROOF**   Using the definition of the derivative, we calculate

$$\frac{d}{dx}\frac{1}{f(x)} = \lim_{h \to 0} \frac{\dfrac{1}{f(x+h)} - \dfrac{1}{f(x)}}{h}$$

$$= \lim_{h \to 0} \frac{f(x) - f(x+h)}{hf(x+h)f(x)}$$

$$= \lim_{h \to 0} \left(\frac{-1}{f(x+h)f(x)}\right)\frac{f(x+h) - f(x)}{h}$$

$$= \frac{-1}{(f(x))^2}f'(x)$$

Again we have to use the continuity of $f$ (Theorem 1) and the limit rules from Section 1.2.

**Example 7**   Differentiate the functions

(a) $\dfrac{1}{x^2 + 1}$, and   (b) $f(t) = \dfrac{1}{t + \dfrac{1}{t}}$.

**Solution**   Using the Reciprocal Rule:

(a) $\dfrac{d}{dx}\left(\dfrac{1}{x^2 + 1}\right) = \dfrac{-2x}{(x^2 + 1)^2}$.

(b) $f'(t) = \dfrac{-1}{\left(t + \dfrac{1}{t}\right)^2}\left(1 - \dfrac{1}{t^2}\right) = \dfrac{-t^2}{(t^2 + 1)^2}\dfrac{t^2 - 1}{t^2} = \dfrac{1 - t^2}{(t^2 + 1)^2}$.

We can use the Reciprocal Rule to confirm the General Power Rule for negative integers:

$$\frac{d}{dx}x^{-n} = -n\,x^{-n-1},$$

since we have already proved the rule for positive integers. We have

$$\frac{d}{dx}x^{-n} = \frac{d}{dx}\frac{1}{x^n} = \frac{-n\,x^{n-1}}{(x^n)^2} = -n\,x^{-n-1}.$$

**Example 8**   (**Differentiating sums of reciprocals**)

$$\frac{d}{dx}\left(\frac{x^2 + x + 1}{x^3}\right) = \frac{d}{dx}\left(\frac{1}{x} + \frac{1}{x^2} + \frac{1}{x^3}\right)$$
$$= \frac{d}{dx}(x^{-1} + x^{-2} + x^{-3})$$
$$= -x^{-2} - 2x^{-3} - 3x^{-4} = -\frac{1}{x^2} - \frac{2}{x^3} - \frac{3}{x^4}.$$

## The Quotient Rule

The Product Rule and the Reciprocal Rule can be combined to provide a rule for differentiating a quotient of two functions. Observe that

$$\frac{d}{dx}\left(\frac{f(x)}{g(x)}\right) = \frac{d}{dx}\left(f(x)\frac{1}{g(x)}\right) = f'(x)\frac{1}{g(x)} + f(x)\left(-\frac{g'(x)}{(g(x))^2}\right)$$
$$= \frac{g(x)f'(x) - f(x)g'(x)}{(g(x))^2}.$$

Thus we have proved the following Quotient Rule.

**THEOREM  5**   **The Quotient Rule**

If $f$ and $g$ are differentiable at $x$, and if $g(x) \neq 0$, then the quotient $f/g$ is differentiable at $x$ and

$$\left(\frac{f}{g}\right)'(x) = \frac{g(x)f'(x) - f(x)g'(x)}{(g(x))^2}.$$

Sometimes students have trouble remembering this rule. (Getting the order of the terms in the numerator wrong will reverse the sign.) Try to remember (and use) the Quotient Rule in the following form:

$$(\text{quotient})'$$
$$= \frac{(\text{denominator}) \times (\text{numerator})' - (\text{numerator}) \times (\text{denominator})'}{(\text{denominator})^2}$$

**Example 9**   Find the derivatives of

(a) $y = \dfrac{1 - x^2}{1 + x^2}$,   (b) $\dfrac{\sqrt{t}}{3 - 5t}$,   and   (c) $f(\theta) = \dfrac{a + b\theta}{m + n\theta}$.

**Solution**   We use the Quotient Rule in each case.

(a) $\dfrac{dy}{dx} = \dfrac{(1 + x^2)(-2x) - (1 - x^2)(2x)}{(1 + x^2)^2} = -\dfrac{4x}{(1 + x^2)^2}.$

(b) $\dfrac{d}{dt}\left(\dfrac{\sqrt{t}}{3 - 5t}\right) = \dfrac{(3 - 5t)\dfrac{1}{2\sqrt{t}} - \sqrt{t}(-5)}{(3 - 5t)^2} = \dfrac{3 + 5t}{2\sqrt{t}(3 - 5t)^2}.$

(c) $f'(\theta) = \dfrac{(m + n\theta)(b) - (a + b\theta)(n)}{(m + n\theta)^2} = \dfrac{mb - na}{(m + n\theta)^2}.$

In all three parts of Example 9 the Quotient Rule yielded fractions with numerators that were complicated but could be simplified algebraically. It is advisable to attempt such simplifications when calculating derivatives; the usefulness of derivatives in applications of calculus often depends on such simplifications.

**Example 10**   Find equations of any lines that pass through the point $(-1, 0)$ and are tangent to the curve $y = (x - 1)/(x + 1)$.

**Solution**   The point $(-1, 0)$ does not lie on the curve, so it is not the point of tangency. Suppose a line is tangent to the curve at $x = a$, so the point of tangency is $(a, (a - 1)/(a + 1))$. Note that $a$ can not be $-1$. The slope of the line must be

$$\left.\frac{dy}{dx}\right|_{x=a} = \left.\frac{(x + 1)(1) - (x - 1)(1)}{(x + 1)^2}\right|_{x=a} = \frac{2}{(a + 1)^2}.$$

If the line also passes through $(-1, 0)$, its slope must also be given by

$$\frac{\dfrac{a-1}{a+1} - 0}{a - (-1)} = \frac{a-1}{(a+1)^2}.$$

Equating these two expressions for the slope, we get an equation to solve for $a$:

$$\frac{a-1}{(a+1)^2} = \frac{2}{(a+1)^2} \qquad \Longrightarrow \qquad a - 1 = 2.$$

Thus $a = 3$, and the slope of the line is $2/4^2 = 1/8$. There is only one line through $(-1, 0)$ tangent to the given curve, and its equation is

$$y = 0 + \frac{1}{8}(x+1) \qquad \text{or} \qquad x - 8y + 1 = 0.$$ ∎

**Remark** Derivatives of quotients of functions where the denominator is a monomial, such as in Example 8, are usually easier to do by breaking the quotient into several fractions (as was done in that example) rather than by using the Quotient Rule.

# Exercises 2.3

In Exercises 1–36, calculate the derivatives of the given functions. Simplify your answers whenever possible.

**1.** $y = 3x^2 - 5x - 7$

**2.** $y = x^8 - x^4$

**3.** $y = \dfrac{x^3}{3} - \dfrac{x^2}{2} + x$

**4.** $y = 4x^{1/2} - \dfrac{5}{x}$

**5.** $f(x) = Ax^2 + Bx + C$

**6.** $f(x) = \dfrac{6}{x^3} + \dfrac{2}{x^2} - 2$

**7.** $z = \dfrac{s^5 - s^3}{15}$

**8.** $y = x^{45} - x^{-45}$

**9.** $g(t) = t^{1/3} + 2t^{1/4} + 3t^{1/5}$

**10.** $y = 3\sqrt[3]{t^2} - \dfrac{2}{\sqrt{t^3}}$

**11.** $u = \dfrac{3}{5}x^{5/3} - \dfrac{5}{3}x^{-3/5}$

**12.** $F(x) = (3x - 2)(1 - 5x)$

**13.** $y = \sqrt{x}\left(5 - x - \dfrac{x^2}{3}\right)$

**14.** $g(t) = \dfrac{1}{2t - 3}$

**15.** $y = \dfrac{1}{x^2 + 5x}$

**16.** $y = \dfrac{4}{3 - x}$

**17.** $f(t) = \dfrac{\pi}{2 - \pi t}$

**18.** $g(y) = \dfrac{2}{1 - y^2}$

**19.** $f(x) = \dfrac{1 - 4x^2}{x^3}$

**20.** $g(u) = \dfrac{u\sqrt{u} - 3}{u^2}$

**21.** $y = \dfrac{2 + t + t^2}{\sqrt{t}}$

**22.** $z = \dfrac{x - 1}{x^{2/3}}$

**23.** $y = \dfrac{3}{x + \sqrt{x}}$

**24.** $y = \dfrac{x - 2}{x + 2}$

**25.** $f(x) = \dfrac{3 - 4x}{3 + 4x}$

**26.** $z = \dfrac{t^2 + 2t}{t^2 - 1}$

**27.** $s = \dfrac{1 + \sqrt{t}}{1 - \sqrt{t}}$

**28.** $f(x) = \dfrac{x^3 - 4}{x + 1}$

**29.** $f(x) = \dfrac{ax + b}{cx + d}$

**30.** $F(t) = \dfrac{t^2 + 7t - 8}{t^2 - t + 1}$

**31.** $f(x) = (1 + x)(1 + 2x)(1 + 3x)(1 + 4x)$

**32.** $f(r) = (r^{-2} + r^{-3} - 4)(r^2 + r^3 + 1)$

**33.** $y = (x^2 + 4)(\sqrt{x} + 1)(5x^{2/3} - 2)$

**34.** $y = \dfrac{(x^2 + 1)(x^3 + 2)}{(x^2 + 2)(x^3 + 1)}$

**∗ 35.** $y = \dfrac{x}{2x + \dfrac{1}{3x + 1}}$

**∗ 36.** $f(x) = \dfrac{(\sqrt{x} - 1)(2 - x)(1 - x^2)}{\sqrt{x}(3 + 2x)}$

Calculate the derivatives in Exercises 37–40, given that $f(2) = 2$ and $f'(2) = 3$.

**37.** $\dfrac{d}{dx}\left(\dfrac{x^2}{f(x)}\right)\Big|_{x=2}$

**38.** $\dfrac{d}{dx}\left(\dfrac{f(x)}{x^2}\right)\Big|_{x=2}$

**39.** $\dfrac{d}{dx}\left(x^2 f(x)\right)\Big|_{x=2}$

**40.** $\dfrac{d}{dx}\left(\dfrac{f(x)}{x^2 + f(x)}\right)\Big|_{x=2}$

**41.** Find $\dfrac{d}{dx}\left(\dfrac{x^2 - 4}{x^2 + 4}\right)\Big|_{x=-2}$.

**42.** Find $\dfrac{d}{dt}\left(\dfrac{t(1 + \sqrt{t})}{5 - t}\right)\Big|_{t=4}$.

**43.** If $f(x) = \dfrac{\sqrt{x}}{x + 1}$, find $f'(2)$.

**44.** Find $\dfrac{d}{dt}\left((1 + t)(1 + 2t)(1 + 3t)(1 + 4t)\right)\Big|_{t=0}$.

**45.** Find an equation of the tangent line to $y = \dfrac{2}{3 - 4\sqrt{x}}$ at the point $(1, -2)$.

**46.** Find equations of the tangent and normal to $y = \dfrac{x + 1}{x - 1}$ at $x = 2$.

**47.** Find the points on the curve $y = x + 1/x$ where the tangent line is horizontal.

**48.** Find the equations of all horizontal lines that are tangent to the curve $y = x^2(4 - x^2)$.

**49.** Find the coordinates of all points where the curve $y = \dfrac{1}{x^2 + x + 1}$ has a horizontal tangent line.

**50.** Find the coordinates of points on the curve $y = \dfrac{x + 1}{x + 2}$ where the tangent line is parallel to the line $y = 4x$.

**51.** Find the equation of the straight line that passes through the point $(0, b)$ and is tangent to the curve $y = 1/x$. Assume $b \neq 0$.

**★ 52.** Show that the curve $y = x^2$ intersects the curve $y = 1/\sqrt{x}$ at right angles.

**53.** Find two straight lines that are tangent to $y = x^3$ and pass through the point $(2, 8)$.

**54.** Find two straight lines that are tangent to $y = x^2/(x - 1)$ and pass through the point $(2, 0)$.

**55.** (A Square Root Rule) Show that if $f$ is differentiable at $x$ and $f(x) > 0$, then

$$\frac{d}{dx}\sqrt{f(x)} = \frac{f'(x)}{2\sqrt{f(x)}}.$$

Use this Square Root Rule to find the derivative of $\sqrt{x^2 + 1}$.

**56.** Show that $f(x) = |x^3|$ is differentiable at every real number $x$, and find its derivative.

**Mathematical Induction**

**57.** Use mathematical induction to prove that $\dfrac{d}{dx}x^{n/2} = \dfrac{n}{2}x^{(n/2)-1}$ for every positive integer $n$. Then use the Reciprocal Rule to get the same result for negative integers $n$.

**58.** Use mathematical induction to prove the formula for the derivative of a product of $n$ functions given earlier in this section.

---

## 2.4 The Chain Rule

Although we can differentiate $\sqrt{x}$ and $x^2 + 1$, we can not yet differentiate $\sqrt{x^2 + 1}$. To do this, we need a rule that tells us how to differentiate *composites* of functions whose derivatives we already know. This rule is known as the Chain Rule, and is the most often used of all the differentiation rules.

**Example 1** The function $\dfrac{1}{x^2 - 4}$ is the composite $f(g(x))$ of $f(u) = \dfrac{1}{u}$ and $g(x) = x^2 - 4$, which have derivatives

$$f'(u) = \frac{-1}{u^2} \qquad \text{and} \qquad g'(x) = 2x.$$

According to the Reciprocal Rule (which is a special case of the Chain Rule),

$$\frac{d}{dx}f(g(x)) = \frac{d}{dx}\left(\frac{1}{x^2 - 4}\right) = \frac{-2x}{(x^2 - 4)^2} = \frac{-1}{(x^2 - 4)^2}(2x)$$
$$= f'(g(x))g'(x).$$

This example suggests that the derivative of a composite function $f(g(x))$ is the the derivative of $f$ evaluated at $g(x)$ multiplied by the derivative of $g$ evaluated at $x$. This is the Chain Rule:

$$\frac{d}{dx} f(g(x)) = f'(g(x))\, g'(x).$$

**THEOREM 6**

**The Chain Rule**

If $f(u)$ is differentiable at $u = g(x)$, and $g(x)$ is differentiable at $x$, then the composite function $f \circ g(x) = f(g(x))$ is differentiable at $x$, and

$$(f \circ g)'(x) = f'(g(x))g'(x).$$

In terms of Leibniz notation, if $y = f(u)$ where $u = g(x)$, then $y = f(g(x))$ and:

at $u$, $y$ is changing $\dfrac{dy}{du}$ times as fast as $u$ is changing;

at $x$, $u$ is changing $\dfrac{du}{dx}$ times as fast as $x$ is changing.

Therefore, at $x$, $y = f(u) = f(g(x))$ is changing $\dfrac{dy}{du} \times \dfrac{du}{dx}$ times as fast as $x$ is changing. That is:

$$\frac{dy}{dx} = \frac{dy}{du}\frac{du}{dx}, \qquad \text{where } \frac{dy}{du} \text{ is evaluated at } u = g(x).$$

It appears as though the symbol $du$ cancels from the numerator and denominator, but this is not meaningful because $dy/du$ was not defined as the quotient of two quantities, but rather as a single quantity, the derivative of $y$ with respect to $u$.

We would like to prove Theorem 6 by writing

$$\frac{\Delta y}{\Delta x} = \frac{\Delta y}{\Delta u}\frac{\Delta u}{\Delta x}$$

and taking the limit as $\Delta x \to 0$. Such a proof is valid for most composite functions, but not all. (See Exercise 46 at the end of this section.) A correct proof will be given later in this section, but first we do more examples to give a better idea of how the Chain Rule works.

**Example 2**  Find the derivative of $y = \sqrt{x^2 + 1}$.

**Solution**  Here $y = f(g(x))$, where $f(u) = \sqrt{u}$ and $g(x) = x^2 + 1$. Since the derivatives of $f$ and $g$ are

$$f'(u) = \frac{1}{2\sqrt{u}} \qquad \text{and} \qquad g'(x) = 2x,$$

the Chain Rule gives

$$\frac{dy}{dx} = \frac{d}{dx}f(g(x)) = f'(g(x)) \cdot g'(x)$$

$$= \frac{1}{2\sqrt{g(x)}} \cdot g'(x) = \frac{1}{2\sqrt{x^2+1}} \cdot (2x) = \frac{x}{\sqrt{x^2+1}}.$$

**Outside and Inside Functions**

In the composite $f(g(x))$, the function $f$ is "outside," and the function $g$ is "inside." The Chain Rule says that the derivative of the composite is the derivative $f'$ of the outside function evaluated at the inside function $g(x)$, multiplied by the derivative $g'(x)$ of the inside function:
$$\frac{d}{dx}f(g(x)) = f'(g(x)) \times g'(x).$$

Usually, when applying the Chain Rule, we do not introduce symbols to represent the functions being composed, but rather just proceed to calculate the derivative of the "outside" function, and then multiply by the derivative of whatever is "inside." You can say to yourself: "the derivative of $f$ of something is $f'$ of that thing, multiplied by the derivative of that thing."

**Example 3**    Find derivatives of the following functions:

(a) $(7x-3)^{10}$,    (b) $f(t) = |t^2 - 1|$,    and    (c) $\left(3x + \dfrac{1}{(2x+1)^3}\right)^{1/4}$.

**Solution**

(a) Here the outside function is the tenth power; it must be differentiated first, and the result multiplied by the derivative of the expression $7x - 3$:

$$\frac{d}{dx}(7x-3)^{10} = 10(7x-3)^9(7) = 70(7x-3)^9.$$

(b) Here we are differentiating the absolute value of something. The derivative is signum of that thing, multiplied by the derivative of that thing:

$$f'(t) = \left(\operatorname{sgn}(t^2-1)\right)(2t) = \frac{2t(t^2-1)}{|t^2-1|} = \begin{cases} 2t & \text{if } t < -1 \text{ or } t > 1 \\ -2t & \text{if } -1 < t < 1 \\ \text{undefined} & \text{if } t = \pm 1. \end{cases}$$

(c) Here we will need to use the Chain Rule twice. We begin by differentiating the $1/4$ power of something, but the something involves the $-3$rd power of $2x + 1$, and the derivative of that will also require the Chain Rule:

$$\frac{d}{dx}\left(3x + \frac{1}{(2x+1)^3}\right)^{1/4} = \frac{1}{4}\left(3x + \frac{1}{(2x+1)^3}\right)^{-3/4}\frac{d}{dx}\left(3x + \frac{1}{(2x+1)^3}\right)$$

$$= \frac{1}{4}\left(3x + \frac{1}{(2x+1)^3}\right)^{-3/4}\left(3 - \frac{3}{(2x+1)^4}\frac{d}{dx}(2x+1)\right)$$

$$= \frac{3}{4}\left(1 - \frac{2}{(2x+1)^4}\right)\left(3x + \frac{1}{(2x+1)^3}\right)^{-3/4}.$$

When you start to feel comfortable with the Chain Rule, you may want to "save a line or two" by carrying out the whole differentiation in one step:

$$\frac{d}{dx}\left(3x + \frac{1}{(2x+1)^3}\right)^{1/4} = \frac{1}{4}\left(3x + \frac{1}{(2x+1)^3}\right)^{-3/4}\left(3 - \frac{3}{(2x+1)^4}(2)\right)$$

$$= \frac{3}{4}\left(1 - \frac{2}{(2x+1)^4}\right)\left(3x + \frac{1}{(2x+1)^3}\right)^{-3/4}.$$

Use of the Chain Rule produces products of factors that do not usually come out in the order you would naturally write them. Often you will want to rewrite the result with the factors in a different order. This is obvious in parts (a) and (c) of the example above. In monomials (expressions that are products of factors), it is common to write the factors in order of increasing complexity from left to right, with numerical factors coming first. One time when you would *not* waste time doing this, or trying to make any other simplification, is when you are going to evaluate the derivative at a particular number. In this case, substitute the number as soon as you have calculated the derivative, before doing any simplification:

$$\frac{d}{dx}(x^2-3)^{10}\Big|_{x=2} = 10(x^2-3)^9(2x)\Big|_{x=2} = (10)(1^9)(4) = 40.$$

**Example 4**   Suppose that $f$ is a differentiable function on the real line. In terms of the derivative $f'$ of $f$, express the derivatives of:

(a) $f(3x)$,   (b) $f(x^2)$,   (c) $f(\pi f(x))$, and   (d) $[f(3-2f(x))]^4$.

**Solution**

(a) $\dfrac{d}{dx}f(3x) = \big(f'(3x)\big)(3) = 3f'(3x)$.

(b) $\dfrac{d}{dx}f(x^2) = \big(f'(x^2)\big)(2x) = 2xf'(x^2)$.

(c) $\dfrac{d}{dx}f(\pi f(x)) = \big(f'(\pi f(x))\big)(\pi f'(x)) = \pi f'(x)f'(\pi f(x))$.

(d) $\dfrac{d}{dx}[f(3-2f(x))]^4 = 4[f(3-2f(x))]^3 f'(3-2f(x))(-2f'(x))$

$$= -8f'(x)f'(3-2f(x))[f(3-2f(x))]^3.$$

## Finding Derivatives by Computer Algebra

Computer algebra systems such as Derive, Maple, and Mathematica know the derivatives of elementary functions and can calculate the derivatives of combinations of these functions symbolically, using differentiation rules. For example, the commands for calculating the derivative of $\sqrt{1+2x^2}$ and then evaluating the result at $x=2$ are:

```
for Derive:    dif(sqrt(1+2x^2),x) <Enter>
               S(implify)<Enter>
               M(anage)S(ubstitute)<Enter> 2<Enter>
               S(implify)<Enter>
for Maple V:   f := diff(sqrt(1+2*x^2),x);<Enter>
               simplify(subs(x=2,f));<Enter>
for Mathematica:  D[Sqrt[1+2x^2],x]<Enter>
               %/.x->2 <Enter>
```

## Building the Chain Rule into Differentiation Formulas

If $u$ is a differentiable function of $x$ and $y = u^n$, then the Chain Rule gives

$$\frac{d}{dx}u^n = \frac{dy}{dx} = \frac{dy}{du}\frac{du}{dx} = nu^{n-1}\frac{du}{dx}.$$

The formula

$$\frac{d}{dx}u^n = nu^{n-1}\frac{du}{dx}$$

is just the formula $\frac{d}{dx}x^n = nx^{n-1}$ with an application of the Chain Rule built in, so that it applies to functions of $x$ rather than just to $x$. Some other differentiation rules with built-in Chain Rule applications are:

$$\frac{d}{dx}\left(\frac{1}{u}\right) = \frac{-1}{u^2}\frac{du}{dx} \qquad \text{(the Reciprocal Rule)}$$

$$\frac{d}{dx}\sqrt{u} = \frac{1}{2\sqrt{u}}\frac{du}{dx} \qquad \text{(the Square Root Rule)}$$

$$\frac{d}{dx}x^r = r\,u^{r-1}\frac{du}{dx} \qquad \text{(the General Power Rule)}$$

$$\frac{d}{dx}|u| = \text{sgn}\,u\,\frac{du}{dx} = \frac{u}{|u|}\frac{du}{dx} \qquad \text{(the Absolute Value Rule)}$$

## Proof of the Chain Rule (Theorem 6)

Suppose that $f$ is differentiable at the point $u = g(x)$ and that $g$ is differentiable at $x$. Let the function $E(k)$ be defined by

$$E(0) = 0,$$

$$E(k) = \frac{f(u+k) - f(u)}{k} - f'(u), \qquad \text{if } k \neq 0.$$

By the definition of derivative, $\lim_{k\to 0} E(k) = f'(u) - f'(u) = 0 = E(0)$, so $E(k)$ is continuous at $k = 0$. Also, whether $k = 0$ or not, we have

$$f(u+k) - f(u) = \big(f'(u) + E(k)\big)k.$$

Now put $u = g(x)$ and $k = g(x+h) - g(x)$, so that $u + k = g(x+h)$, and obtain

$$f(g(x+h)) - f(g(x)) = \big(f'(g(x)) + E(k)\big)(g(x+h) - g(x)).$$

Since $g$ is differentiable at $x$, $\lim_{h\to 0}[g(x+h) - g(x)]/h = g'(x)$. Also, $g$ is continuous at $x$ by Theorem 1, so $\lim_{h\to 0} k = \lim_{h\to 0}(g(x+h) - g(x)) = 0$. Since $E$ is continuous at 0, $\lim_{h\to 0} E(k) = \lim_{k\to 0} E(k) = E(0) = 0$. Hence

$$\frac{d}{dx}f(g(x)) = \lim_{h\to 0}\frac{f(g(x+h)) - f(g(x))}{h}$$

$$= \lim_{h\to 0}\big(f'(g(x)) + E(k)\big)\frac{g(x+h) - g(x)}{h}$$

$$= \big(f'(g(x)) + 0\big)g'(x) = f'(g(x))g'(x),$$

which was to be proved.

# Exercises 2.4

Find the derivatives of the functions in Exercises 1–16.

**1.** $y = (2x + 3)^6$

**2.** $y = \left(1 - \dfrac{x}{3}\right)^{99}$

**3.** $f(x) = (4 - x^2)^{10}$

**4.** $y = \sqrt{1 - 3x^2}$

**5.** $F(t) = \left(2 + \dfrac{3}{t}\right)^{-10}$

**6.** $(1 + x^{2/3})^{3/2}$

**7.** $\dfrac{3}{5 - 4x}$

**8.** $(1 - 2t^2)^{-3/2}$

**9.** $y = |1 - x^2|$

**10.** $f(t) = |2 + t^3|$

**11.** $y = 4x + |4x - 1|$

**12.** $y = (2 + |x|^3)^{1/3}$

**13.** $y = \dfrac{1}{2 + \sqrt{3x + 4}}$

**14.** $f(x) = \left(1 + \sqrt{\dfrac{x - 2}{3}}\right)^4$

**15.** $z = \left(u + \dfrac{1}{u - 1}\right)^{-5/3}$

**16.** $y = \dfrac{x^5\sqrt{3 + x^6}}{(4 + x^2)^3}$

**17.** Sketch the graph of the function in Exercise 10.

**18.** Sketch the graph of the function in Exercise 11.

Verify that the General Power Rule holds for the functions in Exercises 19–21.

**19.** $x^{1/4} = \sqrt{\sqrt{x}}$

**20.** $x^{3/4} = \sqrt{x\sqrt{x}}$

**21.** $x^{3/2} = \sqrt{(x^3)}$

In Exercises 22–29, express the derivative of the given function in terms of the derivative $f'$ of the differentiable function $f$.

**22.** $f(2t + 3)$

**23.** $f(5x - x^2)$

**24.** $\left[f\left(\dfrac{2}{x}\right)\right]^3$

**25.** $\sqrt{3 + 2f(x)}$

**26.** $f\left(\sqrt{3 + 2t}\right)$

**27.** $f\left(3 + 2\sqrt{x}\right)$

**28.** $f\left(2f(3f(x))\right)$

**29.** $f\left(2 - 3f(4 - 5t)\right)$

**30.** Find $\dfrac{d}{dx}\left(\dfrac{\sqrt{x^2 - 1}}{x^2 + 1}\right)\bigg|_{x=-2}$.

**31.** Find $\dfrac{d}{dt}\sqrt{3t - 7}\,\bigg|_{t=3}$.

**32.** If $f(x) = \dfrac{1}{\sqrt{2x + 1}}$, find $f'(4)$.

**33.** If $y = (x^3 + 9)^{17/2}$, find $y'\big|_{x=-2}$.

**34.** Find $F'(0)$ if $F(x) = (1 + x)(2 + x)^2(3 + x)^3(4 + x)^4$.

**\* 35.** Calculate $y'$ if $y = (x + ((3x)^5 - 2)^{-1/2})^{-6}$. Try to do it all in one step.

In Exercises 36–39, find an equation of the tangent line to the given curve at the given point.

**36.** $y = \sqrt{1 + 2x^2}$ at $x = 2$

**37.** $y = (1 + x^{2/3})^{3/2}$ at $x = -1$

**38.** $y = (ax + b)^8$ at $x = b/a$

**39.** $y = 1/(x^2 - x + 3)^{3/2}$ at $x = -2$

**40.** Show that the derivative of $f(x) = (x - a)^m(x - b)^n$ vanishes at some point between $a$ and $b$ if $m$ and $n$ are positive integers.

Use a computer algebra system to evaluate and simplify the derivatives of the functions in Exercises 41–44.

**41.** $y = \sqrt{x^2 + 1} + \dfrac{1}{(x^2 + 1)^{3/2}}$

**42.** $y = \dfrac{(x^2 - 1)(x^2 - 4)(x^2 - 9)}{x^6}$

**43.** $\dfrac{dy}{dt}\bigg|_{t=2}$ if $y = (t + 1)(t^2 + 2)(t^3 + 3)(t^4 + 4)(t^5 + 5)$

**44.** $f'(1)$ if $f(x) = \dfrac{(x^2 + 3)^{1/2}(x^3 + 7)^{1/3}}{(x^4 + 15)^{1/4}}$

**45.** Does the Chain Rule enable you to calculate the derivatives of $|x|^2$ and $|x^2|$ at $x = 0$? Do these functions have derivatives at $x = 0$? Why?

**\* 46.** What is wrong with the following "proof" of the Chain Rule? Let $k = g(x + h) - g(x)$. Then $\lim_{h\to 0} k = 0$. Thus

$$\lim_{h\to 0} \frac{f(g(x + h)) - f(g(x))}{h}$$

$$= \lim_{h\to 0} \frac{f(g(x + h)) - f(g(x))}{g(x + h) - g(x)} \cdot \frac{g(x + h) - g(x)}{h}$$

$$= \lim_{h\to 0} \frac{f(g(x) + k) - f(g(x))}{k} \cdot \frac{g(x + h) - g(x)}{h}$$

$$= f'(g(x))\, g'(x).$$

## 2.5 Derivatives of Trigonometric Functions

The trigonometric functions, especially sine and cosine, play a very important role in the mathematical modelling of real-world phenomena. In particular, they arise whenever quantities fluctuate in a periodic way. Elastic motions, vibrations, and waves of all kinds naturally involve the trigonometric functions, and many physical and mechanical laws are formulated as differential equations having these functions as solutions.

In this section we will calculate the derivatives of the six trigonometric functions. We only have to work hard for one of them, sine; the others then follow from known identities and the differentiation rules of Section 2.3.

## Some Special Limits

First we have to establish some trigonometric limits that we will need to calculate the derivatives of sine. It is assumed throughout that the arguments of the trigonometric functions are measured in radians.

**THEOREM 7**

The functions $\sin\theta$ and $\cos\theta$ are continuous at every value of $\theta$. In particular, at $\theta = 0$ we have:

$$\lim_{\theta\to 0}\sin\theta = \sin 0 = 0 \qquad \text{and} \qquad \lim_{\theta\to 0}\cos\theta = \cos 0 = 1.$$

**Figure 2.20**    It appears that $\lim_{\theta\to 0}(\sin\theta)/\theta = 1$

This result is obvious from the graphs of sine and cosine, so we will not prove it here. A proof can be based on the Squeeze Theorem (Theorem 4 of Section 1.2). The method is suggested in Exercise 64 at the end of this section.

The graph of the function $y = (\sin\theta)/\theta$ is shown in Figure 2.20. Although it is not defined at $\theta = 0$, this function appears to have limit 1 as $\theta$ approaches 0.

**THEOREM 8**

**An important trigonometric limit**

$$\lim_{\theta\to 0}\frac{\sin\theta}{\theta} = 1 \qquad \text{(where } \theta \text{ is in radians).}$$

**PROOF** Let $0 < \theta < \pi/2$, and represent $\theta$ as shown in Figure 2.21. Points $A(1, 0)$ and $P(\cos\theta, \sin\theta)$ lie on the unit circle $x^2 + y^2 = 1$. The area of the circular sector $OAP$ lies between the areas of triangles $OAP$ and $OAT$:

$$\text{Area } \triangle OAP < \text{Area sector } OAP < \text{Area } \triangle OAT.$$

As shown in Section P.5, the area of a circular sector having central angle $\theta$ (radians) and radius 1 is is $\theta/2$. The area of a triangle is $(1/2) \times$ base $\times$ height, so

$$\text{Area } \triangle OAP = \frac{1}{2}(1)(\sin\theta) = \frac{\sin\theta}{2},$$

$$\text{Area } \triangle OAT = \frac{1}{2}(1)(\tan\theta) = \frac{\sin\theta}{2\cos\theta}.$$

**Figure 2.21**    Area $\triangle OAP$
$<$ Area sector $OAP$
$<$ Area $\triangle OAT$

Thus

$$\frac{\sin\theta}{2} < \frac{\theta}{2} < \frac{\sin\theta}{2\cos\theta},$$

or, upon multiplication by the positive number $2/\sin\theta$,

$$1 < \frac{\theta}{\sin\theta} < \frac{1}{\cos\theta}.$$

Now take reciprocals, thereby reversing the inequalities:

$$1 > \frac{\sin\theta}{\theta} > \cos\theta.$$

Since $\lim_{\theta\to 0+}\cos\theta = 1$ by Theorem 7, the Squeeze Theorem gives

$$\lim_{\theta\to 0+}\frac{\sin\theta}{\theta} = 1.$$

Finally, note that $\sin\theta$ and $\theta$ are *odd functions*. Therefore $f(\theta) = (\sin\theta)/\theta$ is an *even function*: $f(-\theta) = f(\theta)$, as shown in Figure 2.20. This symmetry implies that the left limit at 0 must have the same value as the right limit:

$$\lim_{\theta\to 0-}\frac{\sin\theta}{\theta} = 1 = \lim_{\theta\to 0+}\frac{\sin\theta}{\theta},$$

so $\lim_{\theta\to 0}(\sin\theta)/\theta = 1$ by Theorem 1 of Section 1.2.

Theorem 8 can be combined with limit rules and known trigonometric identities to yield other trigonometric limits.

**Example 1**   Show that $\lim_{h\to 0}\dfrac{\cos h - 1}{h} = 0.$

**Solution**   Using the half-angle formula $\cos h = 1 - 2\sin^2(h/2)$, we calculate

$$\lim_{h\to 0}\frac{\cos h - 1}{h} = \lim_{h\to 0}-\frac{2\sin^2(h/2)}{h} \qquad \text{Let } \theta = h/2.$$

$$= -\lim_{\theta\to 0}\frac{\sin\theta}{\theta}\,\sin\theta = -(1)(0) = 0.$$

## The Derivatives of Sine and Cosine

To calculate the derivative of $\sin x$ we need the addition formula for sine (see Section P.6):

$$\sin(x + h) = \sin x \, \cos h + \cos x \, \sin h.$$

**THEOREM 9**   **The derivative of the sine function is the cosine function.**

$$\frac{d}{dx}\sin x = \cos x$$

**PROOF**  We use the definition of derivative, the addition formula for sine, the rules for combining limits, Theorem 8, and the result of Example 1:

$$\frac{d}{dx}\sin x = \lim_{h\to 0}\frac{\sin(x+h)-\sin x}{h}$$

$$= \lim_{h\to 0}\frac{\sin x \cos h + \cos x \sin h - \sin x}{h}$$

$$= \lim_{h\to 0}\frac{\sin x(\cos h - 1) + \cos x \sin h}{h}$$

$$= \lim_{h\to 0}\sin x \cdot \lim_{h\to 0}\frac{\cos h - 1}{h} + \lim_{h\to 0}\cos x \cdot \lim_{h\to 0}\frac{\sin h}{h}$$

$$= (\sin x)\cdot(0) + (\cos x)\cdot(1) = \cos x.$$

**THEOREM 10**

**The derivative of the cosine function is minus the sine function.**

$$\frac{d}{dx}\cos x = -\sin x$$

**PROOF**  We could mimic the proof for sine above, using the addition rule for cosine, $\cos(x+h) = \cos x \cos h - \sin x \sin h$. An easier way is to make use of the complementary angle identities, $\sin((\pi/2)-x) = \cos x$ and $\cos((\pi/2)-x) = \sin x$, and the special case of the Chain Rule from Section 2.3:

$$\frac{d}{dx}\cos x = \frac{d}{dx}\sin\left(\frac{\pi}{2}-x\right) = (-1)\cos\left(\frac{\pi}{2}-x\right) = -\sin x.$$

Notice the minus sign in the derivative of cosine. The derivative of the sine is the cosine, but the derivative of the cosine is *minus* the sine. This is shown graphically in Figure 2.22.

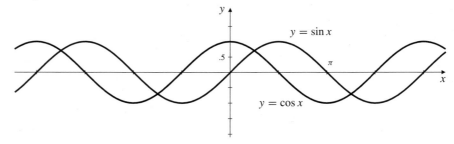

**Figure 2.22**  The sine and cosine plotted together. The slope of the sine curve at $x$ is $\cos x$; the slope of the cosine curve at $x$ is $-\sin x$.

**Example 2**  Evaluate the derivatives of the following functions:

(a) $\sin(\pi x) + \cos(3x)$,  (b) $x^2 \sin\sqrt{x}$,  and  (c) $\dfrac{\cos x}{1 - \sin x}$.

**Solution**

(a) By the Sum Rule and the Chain Rule:

$$\frac{d}{dx}(\sin(\pi x)+\cos(3x)) = \cos(\pi x)(\pi) - \sin(3x)(3) = \pi\cos(\pi x) - 3\sin(3x).$$

(b) By the Product and Chain Rules:

$$\frac{d}{dx}(x^2 \sin \sqrt{x}) = 2x \sin \sqrt{x} + x^2 \cos \sqrt{x}\,\frac{1}{2\sqrt{x}} = 2x \sin \sqrt{x} + \frac{1}{2}x^{3/2} \cos \sqrt{x}.$$

(c) By the Quotient Rule:

$$\begin{aligned}
\frac{d}{dx}\frac{\cos x}{1 - \sin x} &= \frac{(1 - \sin x)(-\sin x) - (\cos x)(0 - \cos x)}{(1 - \sin x)^2} \\
&= \frac{-\sin x + \sin^2 x + \cos^2 x}{(1 - \sin x)^2} \\
&= \frac{1 - \sin x}{(1 - \sin x)^2} = \frac{1}{1 - \sin x}
\end{aligned}$$

We used the identity $\sin^2 x + \cos^2 x = 1$ to simplify the middle line.

Using trigonometric identities can sometimes change the way a derivative is calculated. Carrying out a differentiation in different ways can lead to different-looking answers, but they should be equal if no errors have been made.

**Example 3**   Use two different methods to find the derivative of the function $f(t) = \sin t \cos t$.

**Solution**   By the Product Rule:

$$f'(t) = (\cos t)(\cos t) + (\sin t)(-\sin t) = \cos^2 t - \sin^2 t.$$

On the other hand, since $\sin(2t) = 2 \sin t \cos t$, we have

$$f'(t) = \frac{d}{dt}\left(\frac{1}{2}\sin(2t)\right) = (2)\left(\frac{1}{2}\right)\cos(2t) = \cos(2t).$$

The two answers are really the same, since $\cos(2t) = \cos^2 t - \sin^2 t$. (See Section P.5.)

It is very important to remember that the formulas for the derivatives of $\sin x$ and $\cos x$ were obtained under the assumption that $x$ is measured in *radians*. Since $180° = \pi$ radians, $x° = \pi x/180$ radians. By the Chain Rule,

$$\frac{d}{dx}\sin(x°) = \frac{d}{dx}\sin\left(\frac{\pi x}{180}\right) = \frac{\pi}{180}\cos\left(\frac{\pi x}{180}\right) = \frac{\pi}{180}\cos(x°).$$

Similarly, the derivative of $\cos(x°)$ is $-(\pi/180)\sin(x°)$.

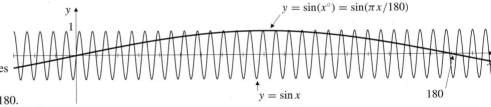

**Figure 2.23**   $\sin(x°)$ oscillates much more slowly than does $\sin x$. Its maximum slope is $\pi/180$.

## Continuity

The six trigonometric functions are differentiable, and therefore continuous (by Theorem 1), everywhere on their domains. This means that we can calculate the limits of most trigonometric functions as $x \to a$ by evaluating them at $x = a$.

The three "co-" functions (cosine, cotangent, and cosecant) have explicit minus signs in their derivatives.

# The Derivatives of the Other Trigonometric Functions

Because $\sin x$ and $\cos x$ are differentiable everywhere, the functions

$$\tan x = \frac{\sin x}{\cos x} \qquad \sec x = \frac{1}{\cos x}$$
$$\cot x = \frac{\cos x}{\sin x} \qquad \csc x = \frac{1}{\sin x}$$

are differentiable at every value of $x$ at which they are defined. Their derivatives can be calculated by the Quotient or Reciprocal Rules, and are as follows:

$$\frac{d}{dx}\tan x = \sec^2 x \qquad \frac{d}{dx}\sec x = \sec x \tan x$$
$$\frac{d}{dx}\cot x = -\csc^2 x \qquad \frac{d}{dx}\csc x = -\csc x \cot x.$$

**Example 4**  Verify the derivative formulas for $\tan x$ and $\sec x$.

**Solution**  We use the Quotient Rule for tangent and the Reciprocal Rule for secant:

$$\frac{d}{dx}\tan x = \frac{d}{dx}\left(\frac{\sin x}{\cos x}\right) = \frac{\cos x \dfrac{d}{dx}(\sin x) - \sin x \dfrac{d}{dx}(\cos x)}{\cos^2 x}$$
$$= \frac{\cos x \cos x - \sin x(-\sin x)}{\cos^2 x} = \frac{\cos^2 x + \sin^2 x}{\cos^2 x}$$
$$= \frac{1}{\cos^2 x} = \sec^2 x.$$

$$\frac{d}{dx}\sec x = \frac{d}{dx}\left(\frac{1}{\cos x}\right) = \frac{-1}{\cos^2 x}\frac{d}{dx}(\cos x)$$
$$= \frac{-1}{\cos^2 x}(-\sin x) = \frac{1}{\cos x}\cdot\frac{\sin x}{\cos x}$$
$$= \sec x \tan x.$$

**Example 5**

(a) $\dfrac{d}{dx}\left[3x + \cot\left(\dfrac{x}{2}\right)\right] = 3 + \dfrac{1}{2}\left[-\csc^2\left(\dfrac{x}{2}\right)\right] = 3 - \dfrac{1}{2}\csc^2\left(\dfrac{x}{2}\right)$

(b) $\dfrac{d}{dx}\left(\dfrac{3}{\sin(2x)}\right) = \dfrac{d}{dx}(3\csc(2x))$
$$= (2)3(-\csc(2x)\cot(2x)) = -6\csc(2x)\cot(2x).$$

**Example 6**  Find the tangent and normal lines to the curve $y = \tan(\pi x/4)$ at the point $(1, 1)$.

**Solution**  The slope of the tangent to $y = \tan(\pi x/4)$ at $(1, 1)$ is:

$$\left.\frac{dy}{dx}\right|_{x=1} = \frac{\pi}{4}\sec^2(\pi x/4)\bigg|_{x=1} = \frac{\pi}{4}\sec^2\left(\frac{\pi}{4}\right) = \frac{\pi}{4}\left(\sqrt{2}\right)^2 = \frac{\pi}{2}.$$

The tangent is the line

$$y = 1 + \frac{\pi}{2}(x-1), \qquad \text{or} \qquad y = \frac{\pi x}{2} - \frac{\pi}{2} + 1.$$

The normal has slope $m = -2/\pi$, so its point-slope equation is

$$y = 1 - \frac{2}{\pi}(x-1) \qquad \text{or} \qquad y = -\frac{2x}{\pi} + \frac{2}{\pi} + 1.$$

# Exercises 2.5

**1.** Verify the formula for the derivative of $\csc x = 1/(\sin x)$.

**2.** Verify the formula for the derivative of $\cot x = (\cos x)/(\sin x)$.

Find the derivatives of the functions in Exercises 3–38. Simplify your answers whenever possible. Also be on the lookout for ways you might simplify the given expression before differentiating it.

**3.** $y = \cos 3x$

**4.** $y = \sin 2x$

**5.** $y = \tan \pi x$

**6.** $y = \sin \dfrac{x}{5}$

**7.** $y = \cos \dfrac{2x}{\pi}$

**8.** $y = \sec ax$

**9.** $y = \cot(4 - 3x)$

**10.** $y = \sin((\pi - x)/3)$

**11.** $f(x) = \cos(s - rx)$

**12.** $y = \sin(Ax + B)$

**13.** $\sin(\pi x^2)$

**14.** $\cos(\sqrt{x})$

**15.** $y = \sqrt{1 + \cos x}$

**16.** $\sin(2\cos x)$

**17.** $f(x) = \cos(x + \sin x)$

**18.** $g(\theta) = \tan(\theta \sin \theta)$

**19.** $u = \sin^3(\pi x/2)$

**20.** $y = \sec(1/x)$

**21.** $F(t) = \sin at \cos at$

**22.** $G(\theta) = \dfrac{\sin a\theta}{\cos b\theta}$

**23.** $\sin(2x) - \cos(2x)$

**24.** $\cos^2 x - \sin^2 x$

**25.** $\tan x + \cot x$

**26.** $\sec x - \csc x$

**27.** $\tan x - x$

**28.** $\tan(3x)\cot(3x)$

**29.** $t \cos t - \sin t$

**30.** $t \sin t + \cos t$

**31.** $\dfrac{\sin x}{1 + \cos x}$

**32.** $\dfrac{\cos x}{1 + \sin x}$

**33.** $x^2 \cos(3x)$

**34.** $g(t) = \sqrt{(\sin t)/t}$

**35.** $v = \sec(x^2)\tan(x^2)$

**36.** $z = \dfrac{\sin \sqrt{x}}{1 + \cos \sqrt{x}}$

**37.** $\sin(\cos(\tan t))$

**38.** $f(s) = \cos(s + \cos(s + \cos s))$

**39.** Given that $\sin 2x = 2 \sin x \cos x$, deduce that $\cos 2x = \cos^2 x - \sin^2 x$.

**40.** Given that $\cos 2x = \cos^2 x - \sin^2 x$, deduce that $\sin 2x = 2 \sin x \cos x$.

In Exercises 41–44, find equations for the lines that are tangent and normal to the curve $y = f(x)$ at the given point.

**41.** $y = \sin x$, $(\pi, 0)$

**42.** $y = \tan(2x)$, $(0, 0)$

**43.** $y = \sqrt{2}\cos(x/4)$, $(\pi, 1)$  **44.** $y = \cos^2 x$, $\left(\dfrac{\pi}{3}, \dfrac{1}{4}\right)$

**45.** Find an equation of the line tangent to the curve $y = \sin(x^\circ)$ at the point where $x = 45$.

**46.** Find an equation of the straight line normal to $y = \sec(x^\circ)$ at the point where $x = 60$.

**47.** Find the points on the curve $y = \tan x$, $-\pi/2 < x < \pi/2$, where the tangent is parallel to the line $y = 2x$.

**48.** Find the points on the curve $y = \tan(2x)$, $-\pi/4 < x < \pi/4$, where the normal is parallel to the line $y = -x/8$.

**49.** Show that the graphs of $y = \sin x$, $y = \cos x$, $y = \sec x$, and $y = \csc x$ have horizontal tangents.

**50.** Show that the graphs of $y = \tan x$ and $y = \cot x$ never have horizontal tangents.

Do the graphs of the functions in Exercises 51–54 have any horizontal tangents in the interval $0 \leq x \leq 2\pi$? If so, where? If not, why not?

**51.** $y = x + \sin x$        **52.** $y = 2x + \sin x$

**53.** $y = x + 2\sin x$        **54.** $y = x + 2\cos x$

Find the limits in Exercises 55–58.

**55.** $\lim\limits_{x \to 0} \dfrac{\tan(2x)}{x}$        **56.** $\lim\limits_{x \to \pi} \sec(1 + \cos x)$

**57.** $\lim\limits_{x \to 0} (x^2 \csc x \cot x)$        **58.** $\lim\limits_{x \to 0} \cos\left(\dfrac{\pi - \pi \cos^2 x}{x^2}\right)$

**59.** Use the method of Example 1 to evaluate $\lim\limits_{h \to 0} \dfrac{1 - \cos h}{h^2}$.

**60.** Find values of $a$ and $b$ that make

$$f(x) = \begin{cases} ax + b, & x < 0 \\ 2\sin x + 3\cos x, & x \geq 0 \end{cases}$$

differentiable at $x = 0$.

**61.** How many straight lines that pass through the origin are tangent to $y = \cos x$? Find (to six decimal places) the slopes of the two such lines that have the largest positive slopes.

Use a computer algebra system to evaluate and simplify the derivatives of the functions in Exercises 62–63.

**62.** $\dfrac{d}{dx} \dfrac{x \cos(x \sin x)}{x + \cos(x \cos x)}\bigg|_{x=0}$

**63.** $\dfrac{d}{dx} \left( \sqrt{2x^2 + 3} \sin(x^2) - \dfrac{(2x^2 + 3)^{3/2} \cos(x^2)}{x} \right)\bigg|_{x=\sqrt{\pi}}$

**\* 64.** (**The continuity of sine and cosine**)

(a) Prove that

$$\lim\limits_{\theta \to 0} \sin \theta = 0 \quad \text{and} \quad \lim\limits_{\theta \to 0} \cos \theta = 1$$

as follows: use the fact that the length of chord $AP$ is less than the length of arc $AP$ in Figure 2.24 to show that

$$\sin^2 \theta + (1 - \cos \theta)^2 < \theta^2.$$

Then deduce that $0 \leq |\sin \theta| < |\theta|$ and $0 \leq |1 - \cos \theta| < |\theta|$. Then use the Squeeze Theorem from Section 1.2.

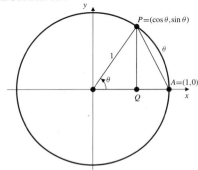

**Figure 2.24**

(b) Part (a) says that $\sin \theta$ and $\cos \theta$ are continuous at $\theta = 0$. Use the addition formulas to prove that they are therefore continuous at every $\theta$.

# 2.6 The Mean-Value Theorem

If you set out in a car at 1:00 p.m. and arrive in a town 150 km away from your starting point at 3:00 p.m., then you have travelled at an average speed of $150/2 = 75$ km/h. Although you may not have travelled at constant speed, you must have been going 75 km/h at *at least one instant* during your journey, for if your speed was always less than 75 km/h you would have gone less than 150 km in 2 hours, and if your speed was always more than 75 km/h you would have gone more than 150 km in 2 hours. In order to get from a value less than 75 km/h to a value greater than 75 km/h your speed, which is a continuous function of time, must pass through the value 75 km/h at some intermediate time.

The conclusion that the average speed over a time interval must be equal to the instantaneous speed at some time in that interval is an instance of an important mathematical principle. In geometric terms it says that if $A$ and $B$ are two points on a smooth curve, then there is at least one point $C$ on the curve between $A$ and $B$ where the tangent line is parallel to the chord line $AB$. See Figure 2.25.

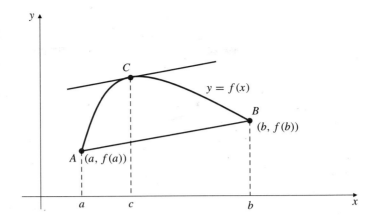

**Figure 2.25** There is a point $C$ on the curve where the tangent is parallel to the chord $AB$.

This principle is stated more precisely in the following theorem.

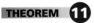 **11** **The Mean-Value Theorem**

Suppose that the function $f$ is continuous on the closed, finite interval $[a, b]$ and that it is differentiable on the open interval $(a, b)$. Then there exists a point $c$ in the open interval $(a, b)$ such that

$$\frac{f(b) - f(a)}{b - a} = f'(c).$$

This says that the slope of the chord line joining the points $(a, f(a))$ and $(b, f(b))$ is equal to the slope of the tangent line to the curve $y = f(x)$ at the point $(c, f(c))$, so the two lines are parallel.

We will prove the Mean-Value Theorem later in this section. For now we make several observations.

1. The hypotheses of the Mean-Value Theorem are all necessary for the conclusion; if $f$ fails to be continuous at even one point of $[a, b]$, or fails to be differentiable at even one point of $(a, b)$, then there may be no point where the tangent line is parallel to the secant line $AB$. See Figure 2.26.

**Figure 2.26** Functions that fail to satisfy the hypotheses of the Mean-Value Theorem, and for which the conclusion is false:

(a) $f$ is discontinuous at endpoint $b$

(b) $f$ is discontinuous at $p$

(c) $f$ is not differentiable at $p$

2. The Mean-Value Theorem gives no indication of how many points $C$ there may be on the curve between $A$ and $B$ where the tangent is parallel to $AB$. If the curve is itself the straight line $AB$, then every point on the line between $A$ and $B$ has the required property. In general there may be more than one point (see Figure 2.27); the Mean-Value Theorem asserts only that there must be at least one.

**Figure 2.27** For this curve there are three points $C$ where the tangent is parallel to the chord $AB$

3. The Mean-Value Theorem gives us no information on how to find the point $c$, that it says must exist. For some simple functions it is possible to calculate $c$ (see the following example), but doing so is usually of no practical value. As we shall see, the importance of the Mean-Value Theorem lies in its use as a theoretical tool. It belongs to a class of theorems called *existence theorems*, as do the Max-Min Theorem and the Intermediate-Value Theorem (Theorems 8 and 9 of Section 1.4).

**Example 1** Verify the conclusion of the Mean-Value Theorem for $f(x) = \sqrt{x}$ on the interval $[a, b]$, where $0 \le a < b$.

**Solution** The theorem says that there must be a number $c$ in the interval $(a, b)$ such that

$$f'(c) = \frac{f(b) - f(a)}{b - a}$$

$$\frac{1}{2\sqrt{c}} = \frac{\sqrt{b} - \sqrt{a}}{b - a} = \frac{\sqrt{b} - \sqrt{a}}{(\sqrt{b} - \sqrt{a})(\sqrt{b} + \sqrt{a})} = \frac{1}{\sqrt{b} + \sqrt{a}}.$$

Thus $2\sqrt{c} = \sqrt{a} + \sqrt{b}$, and $c = \left(\dfrac{\sqrt{b} + \sqrt{a}}{2}\right)^2$. Since $a < b$ we have

$$a = \left(\frac{\sqrt{a} + \sqrt{a}}{2}\right)^2 < c < \left(\frac{\sqrt{b} + \sqrt{b}}{2}\right)^2 = b,$$

so $c$ lies in the interval $(a, b)$. ∎

The following two examples are more representative of how the Mean-Value Theorem is actually used.

**Example 2** Show that $\sin x < x$ for all $x > 0$.

**Solution** If $x > 2\pi$ then $\sin x \le 1 < 2\pi < x$. If $0 < x \le 2\pi$, then, by the Mean-Value Theorem, there exists $c$ in the open interval $(0, 2\pi)$ such that

$$\frac{\sin x}{x} = \frac{\sin x - \sin 0}{x - 0} = \frac{d}{dx} \sin x \bigg|_{x=c} = \cos c < 1.$$

Thus $\sin x < x$ in this case too. ∎

**Example 3** Show that $\sqrt{1+x} < 1 + \dfrac{x}{2}$ for $x > 0$ and for $-1 \le x < 0$.

**Solution** If $x > 0$, apply the Mean-Value Theorem to $f(x) = \sqrt{1+x}$ on the interval $[0, x]$. There exists $c$ in $(0, x)$ such that

$$\frac{\sqrt{1+x} - 1}{x} = \frac{f(x) - f(0)}{x - 0} = f'(c) = \frac{1}{2\sqrt{1+c}} < \frac{1}{2}.$$

The last inequality holds because $c > 0$. Multiplying by the positive number $x$ and transposing the $-1$ gives $\sqrt{1+x} < 1 + \dfrac{x}{2}$.

If $-1 \le x < 0$ we apply the Mean-Value Theorem to $f(x) = \sqrt{1+x}$ on the interval $[x, 0]$. There exists $c$ in $(x, 0)$ such that

$$\frac{\sqrt{1+x} - 1}{x} \frac{1 - \sqrt{1+x}}{-x} = \frac{f(0) - f(x)}{0 - x} = f'(c) = \frac{1}{2\sqrt{1+c}} > \frac{1}{2},$$

(because $0 < 1 + c < 1$). Now we must multiply by the negative number $x$, which reverses the inequality, $\sqrt{1+x} - 1 < \dfrac{x}{2}$, and the required inequality again follows by transposing the $-1$. ∎

## Increasing and Decreasing Functions

Intervals on which the graph of a function $f$ has positive or negative slope provide useful information about the behaviour of $f$. The Mean-Value Theorem enables us to determine such intervals by considering the sign of the derivative $f'$.

**DEFINITION 5**

**Increasing and decreasing functions**

Suppose that the function $f$ is defined on an interval $I$, and that $x_1$ and $x_2$ are two points of $I$.
(a) If $f(x_2) > f(x_1)$ whenever $x_2 > x_1$ we say $f$ is **increasing** on $I$.
(b) If $f(x_2) < f(x_1)$ whenever $x_2 > x_1$ we say $f$ is **decreasing** on $I$.
(c) If $f(x_2) \ge f(x_1)$ whenever $x_2 > x_1$ we say $f$ is **nondecreasing** on $I$.
(d) If $f(x_2) \le f(x_1)$ whenever $x_2 > x_1$ we say $f$ is **nonincreasing** on $I$.

Figure 2.28 illustrates these terms. Note the distinction between *increasing* and *nondecreasing*. If a function is increasing (or decreasing) on an interval it must take different values at different points. (Such a function is called **one-to-one**.) A nondecreasing function (or a nonincreasing function) may be constant on all or part of an interval, and may therefore not be one-to-one.

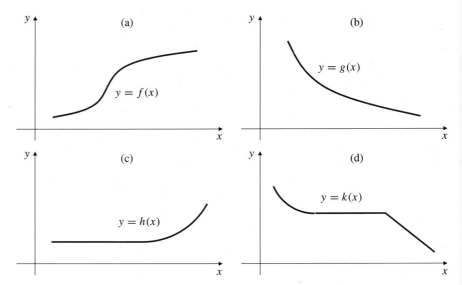

**Figure 2.28**

(a) Function $f$ is increasing

(b) Function $g$ is decreasing

(c) Function $h$ is nondecreasing

(d) Function $k$ is nonincreasing

**THEOREM 12**

Let $J$ be an open interval, and let $I$ be an interval consisting of all the points in $J$ and possibly one or both of the endpoints of $J$. Suppose that $f$ is continuous on $I$ and differentiable on $J$.

(a) If $f'(x) > 0$ for all $x$ in $J$ then $f$ is increasing on $I$.

(b) If $f'(x) < 0$ for all $x$ in $J$ then $f$ is decreasing on $I$.

(c) If $f'(x) \geq 0$ for all $x$ in $J$ then $f$ is nondecreasing on $I$.

(d) If $f'(x) \leq 0$ for all $x$ in $J$ then $f$ is nonincreasing on $I$.

**PROOF**    Let $x_1$ and $x_2$ be points in $I$ with $x_2 > x_1$. By the Mean-Value Theorem there exists a point $c$ in $(x_1, x_2)$ (and therefore in $J$) such that

$$\frac{f(x_2) - f(x_1)}{x_2 - x_1} = f'(c),$$

and hence $f(x_2) - f(x_1) = (x_2 - x_1) f'(c)$. Since $x_2 - x_1 > 0$, the difference $f(x_2) - f(x_1)$ has the same sign as $f'(c)$, and may be zero if $f'(c)$ is zero. Thus all four conclusions follow from the corresponding parts of Definition 5.

**Remark**    Despite Theorem 12, $f'(x_0) > 0$ does *not* imply that $f$ is increasing on *any* interval containing $x_0$. For a counterexample, see Exercise 21 at the end of this section.

**Example 4**    On what intervals is the function $f(x) = x^3 - 12x + 1$ increasing? On what intervals is it decreasing?

**Solution**    We have $f'(x) = 3x^2 - 12 = 3(x - 2)(x + 2)$. Observe that $f'(x) > 0$ if $x < -2$ or $x > 2$, and $f'(x) < 0$ if $-2 < x < 2$. Therefore, $f$ is increasing on the intervals $(-\infty, -2)$ and $(2, \infty)$, and is decreasing on the interval $(-2, 2)$. See Figure 2.29.

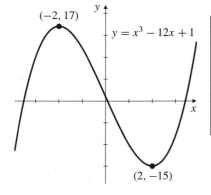

**Figure 2.29**

A function $f$ whose derivative satisfies $f'(x) \geq 0$ on an interval can still be increasing there, rather than just nondecreasing as assured by Theorem 12(c). This will happen if $f'(x) = 0$ only at isolated points, so that $f$ is assured to be increasing on intervals to the left and right of these points.

**Example 5** Show that $f(x) = x^3$ is increasing on any interval.

**Solution** Let $x_1$ and $x_2$ be any two real numbers satsifying $x_1 < x_2$. Since $f'(x) = 3x^2 > 0$ except at $x = 0$, Theorem 12(a) tells us that $f(x_1) < f(x_2)$ if either $x_1 < x_2 \leq 0$ or $0 \leq x_1 < x_2$. If $x_1 < 0 < x_2$ then $f(x_1) < 0 < f(x_2)$. Thus $f$ is increasing on every interval.  ∎

If a function is constant on an interval, then its derivative is zero on that interval. The Mean-Value Theorem provides a converse of this fact.

**THEOREM 13**

If $f$ is continuous on an interval $I$, and $f'(x) = 0$ at every interior point of $I$ (that is, at every point of $I$ that is not an endpoint of $I$), then $f(x) = C$, a constant, on $I$.

**PROOF** Pick a point $x_0$ in $I$ and let $C = f(x_0)$. If $x$ is any other point of $I$, then the Mean-Value Theorem says that there exists a point $c$ between $x_0$ and $x$ such that

$$\frac{f(x) - f(x_0)}{x - x_0} = f'(c).$$

The point $c$ must belong to $I$ because an interval contains all points between any two of its points, and $c$ cannot be an endpoint of $I$ since $c \neq x_0$ and $c \neq x$. Since $f'(c) = 0$ for all such points $c$, we have $f(x) - f(x_0) = 0$ for all $x$ in $I$, and $f(x) = f(x_0) = C$ as claimed.

We will see how Theorem 13 can be used to establish identities for new functions encountered in later chapters. We will also use it when finding antiderivatives in Section 2.10.

## Proof of the Mean-Value Theorem

The Mean-Value Theorem is one of those deep results that is based on the completeness of the real number system via the fact that a continuous function on a closed, finite interval takes on a maximum and minimum value (Theorem 8 of Section 1.4). Before giving the proof we establish two preliminary results.

**THEOREM 14**

If $f$ is defined on an open interval $(a, b)$ and achieves a maximum (or minimum) value at the point $c$ in $(a, b)$, and if $f'(c)$ exists, then $f'(c) = 0$. (Values of $x$ where $f'(x) = 0$ are called **critical points** of the function $f$.)

**PROOF** Suppose that $f$ has a maximum value at $c$. Then $f(x) - f(c) \leq 0$ whenever $x$ is in $(a, b)$. If $c < x < b$, then

$$\frac{f(x) - f(c)}{x - c} \leq 0, \qquad \text{so} \quad f'(c) = \lim_{x \to c+} \frac{f(x) - f(c)}{x - c} \leq 0.$$

Similarly, if $a < x < c$, then

$$\frac{f(x) - f(c)}{x - c} \geq 0, \qquad \text{so} \quad f'(c) = \lim_{x \to c-} \frac{f(x) - f(c)}{x - c} \geq 0.$$

Thus $f'(c) = 0$. The proof for a minimum value at $c$ is similar.

THEOREM **15**

**Rolle's theorem**

Suppose that the function $g$ is continuous on the closed, finite interval $[a, b]$ and that it is differentiable on the open interval $(a, b)$. If $g(a) = g(b)$, then there exists a point $c$ in the open interval $(a, b)$ such that $g'(c) = 0$.

**PROOF**   If $g(x) = g(a)$ for every $x$ in $[a, b]$, then $g$ is a constant function, so $g'(c) = 0$ for every $c$ in $(a, b)$. Therefore, suppose there exists $x_0$ in $(a, b)$ such that $g(x_0) \neq g(a)$. Let us assume that $g(x_0) > g(a)$. (The proof if $g(x_0) < g(a)$ is similar.) By the max-min theorem (Theorem 8 of Section 1.4), being continuous on $[a, b]$, $g$ must have a maximum value at some point $c$ in $[a, b]$. Since

$$g(c) \geq g(x_0) > g(a) = g(b),$$

$c$ cannot be either $a$ or $b$. Therefore $c$ is in the open interval $(a, b)$, and so $g$ is differentiable at $c$. By Theorem 14, $c$ must be a critical point of $g$: $g'(c) = 0$.

**Remark**   Rolle's theorem is a special case of the Mean-Value Theorem in which the chord line has slope 0, and so the corresponding parallel tangent line must also have slope 0. We can deduce the Mean-Value Theorem from this special case.

**Proof of the Mean-Value Theorem**   Suppose $f$ satisfies the conditions of the Mean-Value Theorem. Let

$$g(x) = f(x) - \left( f(a) + \frac{f(b) - f(a)}{b - a} (x - a) \right).$$

(For $a \leq x \leq b$, $g(x)$ is the vertical displacement between the curve $y = f(x)$ and the chord line

$$y = f(a) + \frac{f(b) - f(a)}{b - a} (x - a)$$

joining $(a, f(a))$ and $(b, f(b))$. See Figure 2.30.)

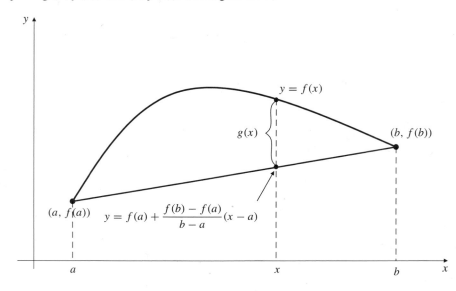

**Figure 2.30**   $g(x)$ is the vertical distance between the graph of $f$ and the chord line

The function $g$ is also continuous on $[a, b]$ and differentiable on $(a, b)$ because $f$ has these properties. In addition, $g(a) = g(b) = 0$. By Rolle's theorem, there is some point $c$ in $(a, b)$ such that $g'(c) = 0$. Since

$$g'(x) = f'(x) - \frac{f(b) - f(a)}{b - a},$$

it follows that

$$f'(c) = \frac{f(b) - f(a)}{b - a}.$$

Many of the applications we will make of the Mean-Value Theorem in later chapters will actually use the following generalized version of it.

**THEOREM 16**

**The Generalized Mean-Value Theorem**

If functions $f$ and $g$ are both continuous on $[a, b]$ and differentiable on $(a, b)$, and if $g'(x) \neq 0$ for every $x$ in $(a, b)$, then there exists a number $c$ in $(a, b)$ such that

$$\frac{f(b) - f(a)}{g(b) - g(a)} = \frac{f'(c)}{g'(c)}.$$

**PROOF** Note that $g(b) \neq g(a)$ because otherwise there would be some number in $(a, b)$ where $g' = 0$. Hence neither denominator above can be zero. Apply the Mean-Value Theorem to

$$h(x) = \big(f(b) - f(a)\big)\big(g(x) - g(a)\big) - \big(g(b) - g(a)\big)\big(f(x) - f(a)\big).$$

Since $h(a) = h(b) = 0$, there exists $c$ in $(a, b)$ such that $h'(c) = 0$. Thus

$$\big(f(b) - f(a)\big)g'(c) - \big(g(b) - g(a)\big)f'(c) = 0,$$

and the result follows on division by the $g$ factors.

## Exercises 2.6

In Exercises 1–3, illustrate the Mean-Value Theorem by finding any points in the open interval $(a, b)$ where the tangent line to $y = f(x)$ is parallel to the chord line joining $(a, f(a))$ and $(b, f(b))$.

**1.** $f(x) = x^2$ on $[a, b]$   **2.** $f(x) = \dfrac{1}{x}$ on $[1, 2]$

**3.** $f(x) = x^3 - 3x + 1$ on $[-2, 2]$

$*$ **4.** By applying the Mean-Value Theorem to
$f(x) = \cos x + \dfrac{x^2}{2}$ on the interval $[0, x]$, and using the

result of Example 2, show that

$$\cos x > 1 - \frac{x^2}{2}$$

for $x > 0$. This inequality is also true for $x < 0$. Why?

**5.** Show that $\tan x > x$ for $0 < x < \pi/2$.

**6.** Let $r > 1$. If $x > 0$ or $-1 \leq x < 0$, show that $(1 + x)^r > 1 + rx$.

**7.** Let $0 < r < 1$. If $x > 0$ or $-1 \leq x < 0$, show that $(1 + x)^r < 1 + rx$.

Find the intervals of increase and decrease of the functions in Exercises 8–15.

**8.** $f(x) = x^2 + 2x + 2$

**9.** $f(x) = x^3 - 4x + 1$

**10.** $f(x) = x^3 + 4x + 1$

**11.** $f(x) = (x^2 - 4)^2$

**12.** $f(x) = \dfrac{1}{x^2 + 1}$

**13.** $f(x) = x^3(5 - x)^2$

**14.** $f(x) = x - 2\sin x$

**15.** $f(x) = x + \sin x$

**∗ 16.** If $f(x)$ is differentiable on an interval $I$ and vanishes at $n \geq 2$ distinct points of $I$, prove that $f'(x)$ must vanish at least $n - 1$ points in $I$.

**∗ 17.** If $f''(x)$ exists on an interval $I$ and if $f$ vanishes at at least three distinct points of $I$, prove that $f''$ must vanish at some point in $I$.

**∗ 18.** Generalize Exercise 17 to a function for which $f^{(n)}$ exists on $I$ and for which $f$ vanishes at at least $n + 1$ distinct points in $I$.

**∗ 19.** Suppose $f$ is twice differentiable on an interval $I$ (that is, $f''$ exists on $I$). Suppose that the points 0 and 2 belong to $I$

and that $f(0) = f(1) = 0$ and $f(2) = 1$. Prove that:

(a) $f'(a) = \dfrac{1}{2}$ for some point $a$ in $I$.

(b) $f''(b) > \dfrac{1}{2}$ for some point $b$ in $I$.

(c) $f'(c) = \dfrac{1}{7}$ for some point $c$ in $I$.

**20.** What is wrong with the following "proof" of the Generalized Mean-Value Theorem? By the Mean-Value Theorem, $f(b) - f(a) = (b - a)f'(c)$ for some $c$ between $a$ and $b$, and similarly $g(b) - g(a) = (b - a)g'(c)$ for some such $c$. Hence $(f(b) - f(a))/(g(b) - g(a)) = f'(c)/g'(c)$, as required.

**∗ 21.** Let $f(x) = \begin{cases} x + 2x^2 \sin(1/x) & \text{if } x \neq 0, \\ 0 & \text{if } x = 0. \end{cases}$

(a) Show that $f'(0) = 1$. (*Hint:* use the definition of derivative.)

(b) Show that any interval containing $x = 0$ also contains points where $f'(x) < 0$, so $f$ cannot be increasing on such an interval.

## 2.7 Using Derivatives

In this section we will look at some examples of ways in which derivatives are used to represent and interpret changes and rates of change in the world around us. It is natural to think of change in terms of dependence on time, such as the velocity of a moving object, but there is no need to be so restrictive. Change with respect to variables other than time can be treated in the same way. For example, a physician may want to know how small changes in dosage can affect the body's response to a drug. An economist may want to study how foreign investment changes with respect to variations in a country's interest rates. These questions can all be formulated in terms of rate of change of a function with respect to a variable.

### Approximating Small Changes

If one quantity, say $y$, is a function of another quantity $x$, that is

$$y = f(x),$$

we sometimes want to know how a change in the value of $x$ by an amount $\Delta x$ will affect the value of $y$. The exact change $\Delta y$ in $y$ is given by

$$\Delta y = f(x + \Delta x) - f(x),$$

but if the change $\Delta x$ is small, then we can get a good approximation to $\Delta y$ by using the fact that $\Delta y/\Delta x$ is approximately the derivative $dy/dx$. Thus

$$\Delta y = \frac{\Delta y}{\Delta x}\Delta x \approx \frac{dy}{dx}\Delta x = f'(x)\,\Delta x.$$

Sometimes changes in a quantity are measured with respect to the size of the quantity. The **relative change** in $x$ is the ratio $\Delta x/x$; the **percentage change** in $x$ is the relative change expressed as a percentage:

$$\text{relative change in } x = \frac{\Delta x}{x}$$

$$\text{percentage change in } x = 100\,\frac{\Delta x}{x}.$$

**Example 1** By approximately what percentage does the area of a circle increase if the radius increases by 2%?

**Solution** The area $A$ of a circle is given in terms of the radius $r$ by $A = \pi r^2$. Thus

$$\Delta A \approx \frac{dA}{dr}\,\Delta r = 2\pi r\,\Delta r.$$

We divide this approximation by $A = \pi r^2$ to get an approximation that links the relative changes in $A$ and $r$:

$$\frac{\Delta A}{A} \approx \frac{2\pi r\,\Delta r}{\pi r^2} = 2\,\frac{\Delta r}{r}.$$

If $r$ increases by 2%, then $\Delta r = \frac{2}{100}\,r$, so

$$\frac{\Delta A}{A} \approx 2 \times \frac{2}{100} = \frac{4}{100}.$$

Thus $A$ increases by approximately 4%.

## Average and Instantaneous Rates of Change

Recall the concept of average rate of change of a function over an interval, introduced in Section 1.1. The derivative of the function is the limit of this average rate as the length of the interval goes to zero, and so represents the rate of change of the function at a given value of its variable.

**DEFINITION 6**

> The **average rate of change** of a function $f(x)$ with respect to $x$ over the interval from $a$ to $a + h$ is
>
> $$\frac{f(a + h) - f(a)}{h}.$$
>
> The **(instantaneous) rate of change** of $f$ with respect to $x$ at $x = a$ is the derivative
>
> $$f'(a) = \lim_{h \to 0} \frac{f(a + h) - f(a)}{h},$$
>
> provided the limit exists.

It is conventional to use the word *instantaneous* even when $x$ does not represent time, although the word is frequently omitted. When we say *rate of change* we mean *instantaneous rate of change*.

**Example 2** How fast is area $A$ of a circle increasing with respect to its radius when the radius is 5 m?

**Solution** The rate of change of the area with respect to the radius is

$$\frac{dA}{dr} = \frac{d}{dr}(\pi r^2) = 2\pi r.$$

When $r = 5$ m, the area is changing at rate $2\pi \times 5 = 10\pi$ m$^2$/m. This means that a small change $\Delta r$ m in the radius when the radius is 5 m would result in a change of about $10\pi \Delta r$ m$^2$ in the area of the circle.

The above example suggests that the appropriate units for the rate of change of a quantity $y$ with respect to another quantity $x$ are units of $y$ per unit of $x$.

If $f'(x_0) = 0$, we say that $f$ is **stationary** at $x_0$, and call $x_0$ a **critical point** of $f$. The corresponding point $(x_0, f(x_0))$ on the graph of $f$ is also called a **critical point** of the graph. The graph has a horizontal tangent at a critical point, and $f$ may or may not have a maximum or minimum value there. (See Figure 2.31.) It is still possible for $f$ to be increasing or decreasing on an open interval containing a critical point. (See point $a$ in Figure 2.31.)

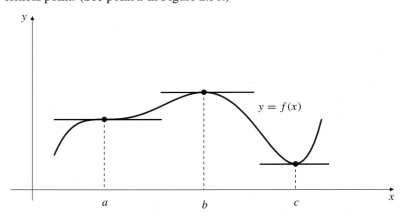

**Figure 2.31**    Critical points of $f$

**Example 3** Suppose the temperature at a certain location $t$ hours after noon on a certain day is $T°$C ($T$ degrees Celsius), where

$$T = \tfrac{1}{3}t^3 - 3t^2 + 8t + 10 \qquad (\text{for } 0 \le t \le 5).$$

How fast is the temperature rising or falling at 1:00 p.m.? at at 3:00 p.m.? At what instants is the temperature stationary?

**Solution** The rate of change of the temperature is given by

$$\frac{dT}{dt} = t^2 - 6t + 8 = (t - 2)(t - 4).$$

If $t = 1$, then $\dfrac{dT}{dt} = 3$, so the temperature is rising at rate $3°$C/hour at 1:00 p.m.

If $t = 3$, then $\dfrac{dT}{dt} = -1$, so the temperature is falling at a rate of $1°$C/hour at 3:00 p.m.

The temperature is stationary when $\dfrac{dT}{dt} = 0$, that is at 2:00 p.m. and 4:00 p.m. ∎

## Rate of a Chemical Reaction

Suppose that one molecule of each of two reagents A and B combine in solution to form a single molecule of the product C, and, in reverse, a molecule of C may decompose to produce molecules of A and B. Such a bidirectional reaction is denoted

$$A + B \rightleftharpoons C.$$

The direction and rate at which such a reaction proceeds depends on the concentrations of A, B, and C in the solution. The concentration of A is denoted $[A]$, and is measured in moles per litre. (A mole is about $6.023 \times 10^{23}$ molecules.)

According to the Law of Mass Action, the rate at which C is produced from A and B is proportional to the product of the concentrations of A and B:

rate of production of C:     $\dfrac{d[C]}{dt} = k_1[A][B].$

Similarly, the rate at which C decomposes is the negative of the rate at which A (or B) is produced from C, and so is proportional to the concentration of C:

rate of decomposition of C:     $\dfrac{d[C]}{dt} = -\dfrac{d[A]}{dt} = -k_2[C].$

The rate at which the concentration of C in the solution changes is the rate at which it increases in the reaction producing C, minus the rate at which it is lost in the decomposition reaction:

$$\dfrac{d[C]}{dt} = k_1[A][B] - k_2[C].$$

The reaction will be in *equilibrium* if $d[C]/dt = 0$, that is, if

$$\dfrac{[A][B]}{[C]} = \dfrac{k_2}{k_1} \quad \text{(constant)}.$$

At equilibrium, C is being produced from A and B at the same rate it is decomposing into A and B.

If $k_2$ is very small or zero (or if C is removed from the solution by precipitation or escapes as a gas), the reaction will continue to produce C until one of A or B is used up. We describe such a reaction as going to completion. In Section 7.8, we will learn how to solve some rate of reaction equations.

## Sensitivity to Change

When a small change in $x$ produces a large change in the value of a function $f(x)$, we say that the function is very **sensitive** to changes in $x$. The derivative $f'(x)$ is a measure of the sensitivity of the dependence of $f$ on $x$.

**Example 4**    (**Dosage of a medicine**) A pharmacologist studying a drug that has been developed to lower blood pressure determines experimentally that the average reduction $R$ in blood pressure resulting from a daily dosage of $x$ mg of the drug is given by

$$R = 24.2 \left(1 + \frac{x - 13}{\sqrt{x^2 - 26x + 529}}\right) \quad \text{mm Hg.}$$

(The units are millimetres of mercury (Hg).) Determine the sensitivity of $R$ to dosage $x$ at dosage levels of 5 mg, 15 mg, and 35 mg. At which of these dosage levels would an increase in the dosage have the greatest effect?

**Solution**    The sensitivity of $R$ to $x$ is $dR/dx$. We have

$$\frac{dR}{dx} = 24.2 \left(\frac{\sqrt{x^2 - 26x + 529}(1) - (x - 13)\dfrac{x - 13}{\sqrt{x^2 - 26x + 529}}}{x^2 - 26x + 529}\right)$$

$$= 24.2 \left(\frac{x^2 - 26x + 529 - (x^2 - 26x + 169)}{(x^2 - 26x + 529)^{3/2}}\right)$$

$$= \frac{8712}{(x^2 - 26x + 529)^{3/2}}.$$

At dosages $x = 5$ mg, 15 mg, and 35 mg we have sensitivities of

$$\left.\frac{dR}{dx}\right|_{x=5} = 0.998 \text{ mm Hg/mg}, \qquad \left.\frac{dR}{dx}\right|_{x=15} = 1.254 \text{ mm Hg/mg},$$

$$\left.\frac{dR}{dx}\right|_{x=35} = 0.355 \text{ mm Hg/mg}.$$

Among these three levels, the greatest sensitivity is at 15 mg. Increasing the dosage from 15 to 16 mg/day could be expected to further reduce average blood pressure by about 1.25 mm Hg.

## Derivatives in Economics

Just as physicists use terms such as *velocity* and *acceleration* to refer to derivatives of certain quantities, so also economists have their own specialized vocabulary for derivatives. They call them marginals. In economics the term **marginal** denotes the rate of change of a quantity with respect to a variable on which it depends. For example, the **cost of production** $C(x)$ in a manufacturing operation is a function of $x$, the number of units of product produced. The **marginal cost of production** is the rate of change of $C$ with respect to $x$, so it is $dC/dx$. Sometimes the marginal cost of production is loosely defined to be the extra cost of producing one more unit, that is

$$\Delta C = C(x + 1) - C(x).$$

To see why this is approximately correct, observe from Figure 2.32 that if the slope of $C = f(x)$ does not change quickly near $x$, then the difference quotient $\Delta C/\Delta x$ will be close to its limit, the derivative $dC/dx$, even if $\Delta x = 1$.

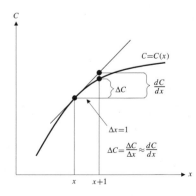

**Figure 2.32**    The marginal cost $dC/dx$ is approximately the extra cost $\Delta C$ of producing $\Delta x = 1$ more unit

**Example 5** (**Marginal Tax Rates**) If your marginal income tax rate is 35% and your income increases by $1,000, you can expect to have to pay an extra $350 in income taxes. This does not mean that you pay 35% of your entire income in taxes. It just means that at your current income level $I$, the rate of increase of taxes $T$ with respect to income is $dT/dI = 0.35$. You will pay $0.35 out of every extra dollar you earn in taxes. Of course, if your income increases greatly, you may land in a higher tax bracket and your marginal rate will increase. ∎

**Example 6** (**Marginal Cost of Production**) The cost of producing $x$ tons of coal per day in a mine is $\$C(x)$ where

$$C(x) = 4{,}200 + 5.40x - 0.001x^2 + 0.000002x^3.$$

(a) What is the average cost of producing each ton if the daily production level is 1,000 tons? 2,000 tons?

(b) Find the marginal cost of production if the daily production level is 1,000 tons. 2,000 tons.

(c) If the production level increases slightly from 1,000 tons or from 2,000 tons, what will happen to the average cost per ton?

**Solution**

(a) The average cost per ton of coal is

$$\frac{C(x)}{x} = \frac{4{,}200}{x} + 5.40 - 0.001x + 0.000002x^2.$$

If $x = 1{,}000$, the average cost per ton is $C(1{,}000)/1{,}000 = \$10.6$ /ton. If $x = 2{,}000$, the average cost per ton is $C(2{,}000)/2{,}000 = \$13.5$ /ton.

(b) The marginal cost of production is

$$C'(x) = 5.40 - 0.002x + 0.000006x^2.$$

If $x = 1{,}000$, the marginal cost is $C'(1{,}000) = \$9.4$ /ton. If $x = 2{,}000$, the marginal cost is $C'(2{,}000) = \$25.4$ /ton.

(c) If the production level $x$ is increased slightly from $x = 1{,}000$, then the average cost per ton will drop because the cost is increasing at a rate lower than the average cost. At $x = 2{,}000$ the opposite is true; an increase in production will increase the average cost per ton. ∎

Economists sometimes prefer to measure relative rates of change that do not depend on the units used to measure the quantities involved. They use the term **elasticity** for such relative rates.

**Example 7** (**Elasticity of demand**) The demand $y$ for a certain product (that is the amount that can be sold) typically depends on the price $p$ charged for the product: $y = f(p)$. The marginal demand $dy/dp = f'(p)$ (which is typically negative) depends on the units used to measure $y$ and $p$. The *elasticity of the demand* is the quantity

$$-\frac{p}{y}\frac{dy}{dp}, \qquad \text{(the} - \text{sign ensures elasticity is positive)}$$

which is independent of units and provides a good measure of the sensitivity of demand to changes in price. To see this, suppose that new units of demand and price are introduced, that are multiples of the old units. In terms of the new units the demand and price are now $Y$ and $P$, where

$$Y = k_1 y \qquad \text{and} \qquad P = k_2 p.$$

Thus $Y = k_1 f(P/k_2)$ and so $dY/dP = (k_1/k_2) f'(P/k_2) = (k_1/k_2) f'(p)$ by the Chain Rule. It follows that the elasticity has the same value:

$$-\frac{P}{Y}\frac{dY}{dP} = -\frac{k_2 p}{k_1 y}\frac{k_1}{k_2} f'(p) = -\frac{p}{y}\frac{dy}{dp}.$$

---

# Exercises 2.7

In Exercises 1–6, find the approximate percentage changes in the given function $y = f(x)$ resulting from an increase of 2% in the value of $x$.

**1.** $y = x^2$

**2.** $y = 1/x$

**3.** $y = 1/x^2$

**4.** $y = x^3$

**5.** $y = \sqrt{x}$

**6.** $y = x^{-2/3}$

**7.** By approximately what percentage will the volume ($V = \frac{4}{3}\pi r^3$) of a ball of radius $r$ increase if the radius increases by 2%?

**8.** By about what percentage will the edge length of an ice cube decrease if the cube loses 6% of its volume by melting?

**9.** Find the rate of change of the area of a square with respect to the length of its side when the side is 4 ft.

**10.** Find the rate of change of the side of a square with respect to the area of the square when the area is 16 m².

**11.** Find the rate of change of the diameter of a circle with respect to its area.

**12.** Find the rate of change of the area of a circle with respect to its diameter.

**13.** Find the rate of change of the volume of a sphere ($V = \frac{4}{3}\pi r^3$) with respect to its radius $r$ when the radius is 2 metres.

**14.** What is the rate of change of the area $A$ of a square with respect to the length $L$ of the diagonal of the square?

**15.** What is the rate of change of the circumference $C$ of a circle with respect to the area $A$ of the circle?

**16.** Find the rate of change of the side $s$ of a cube with respect to the volume $V$ of the cube.

What are the critical points of the functions in Exercises 17–20? On what intervals is each function increasing and decreasing?

**17.** $f(x) = x^2 - 4$

**18.** $f(x) = x^3 - 12x + 1$

**19.** $y = x^3 + 6x^2$

**20.** $y = 1 - x - x^5$

**21.** Show that $f(x) = x^3$ is increasing on the whole real line even though $f'(x)$ is not positive at every point.

**22.** On what intervals is $f(x) = x + 2\sin x$ increasing?

Use a graphing utility or a computer algebra system to find the critical points of the functions in Exercises 23–26 correct to 6 decimal places.

**23.** $f(x) = \dfrac{x^2 - x}{x^2 - 4}$

**24.** $f(x) = \dfrac{2x + 1}{x^2 + x + 1}$

**25.** $f(x) = x - \sin\left(\dfrac{x}{x^2 + x + 1}\right)$

**26.** $f(x) = \dfrac{\sqrt{1 - x^2}}{\cos(x + 0.1)}$

**27.** The volume of water in a tank $t$ minutes after it starts draining is

$$V(t) = 350(20 - t)^2 \text{ litres.}$$

(a) How fast is the water draining out after 5 minutes? after 15 minutes?

(b) What is the average rate at which water is draining out during the time interval from 5 to 15 minutes?

**28.** **(Poiseuille's Law)** The flow rate $F$ (in litres per minute) of a liquid through a pipe is proportional to the fourth power of the radius of the pipe:

$$F = kr^4.$$

Approximately what percentage increase is needed in the radius of the pipe to increase the flow rate by 10%?

**29. (Gravitational force)** The gravitational force $F$ with which the earth attracts an object in space is given by $F = k/r^2$, where $k$ is a constant and $r$ is the distance from the object to the centre of the earth. If $F$ decreases with respect to $r$ at rate $1\,pound/mile$ when $r = 4,000$ mi, how fast does $F$ change with respect to $r$ when $r = 8,000$ mi?

**30. (Sensitivity of revenue to price)** The sales revenue $\$R$ from a software product depends on the price $\$p$ charged by the distributor according to the formula

$$R = 4,000p - 10p^2.$$

(a) How sensitive is $R$ to $p$ when $p = \$100$? $p = \$200$? $p = \$300$?

(b) Which of these three is the most reasonable price for the distributor to charge? Why?

**31. (Marginal cost)** The cost of manufacturing $x$ refrigerators is $\$C(x)$, where

$$C(x) = 8,000 + 400x - 0.5x^2.$$

(a) Find the marginal cost if 100 refrigerators are manufactured.

(b) Show that the marginal cost is approximately the difference in cost of manufacturing 101 refrigerators instead of 100.

**32. (Marginal profit)** If a plywood factory produces $x$ sheets of plywood per day its profit that day will be $\$P(x)$, where

$$P(x) = 8x - 0.005x^2 - 1,000.$$

(a) Find the marginal profit. For what values of $x$ is the marginal profit positive? negative?

(b) How many sheets should be produced each day to generate maximum profits?

**33.** The cost $C$ (in dollars) of producing $n$ widgets per month in a widget factory is given by

$$C = \frac{80,000}{n} + 4n + \frac{n^2}{100}.$$

Find the marginal cost of production if the number of widgets manufactured each month is (a) 100 and (b) 300.

**∗ 34.** In a mining operation the cost $C$ (in dollars) of extracting each tonne of ore is given by

$$C = 10 + \frac{20}{x} + \frac{x}{1,000},$$

where $x$ is the number of tonnes extracted each day. (For small $x$, $C$ decreases as $x$ increases because of economies of scale, but for large $x$, $C$ increases with $x$ because of overloaded equipment and labour overtime.) If each tonne of ore can be sold for $13, how many tonnes should be extracted each day to maximize the daily profit of the mine?

**∗ 35. (Average cost and marginal cost)** If it costs a manufacturer $C(x)$ dollars to produce $x$ items then his average cost of production is $C(x)/x$ dollars per item. Typically the average cost is a decreasing function of $x$ for small $x$ and an increasing function of $x$ for large $x$. (Why?) Show that the value of $x$ that minimizes the average cost makes the average cost equal to the marginal cost.

**36. (Constant elasticity)** Show that if demand $y$ is related to price $p$ by the equation $y = Cp^{-r}$, where $C$ and $r$ are positive constants, then the elasticity of demand (see Example 7) is the constant $r$.

# 2.8 Higher-Order Derivatives

If the derivative $y' = f'(x)$ of a function $y = f(x)$ is itself differentiable at $x$, we can calculate *its* derivative, which we call the **second derivative** of $f$ and denote by $y'' = f''(x)$. As is the case for first derivatives, second derivatives can be denoted by various notations depending on the context. Some of the more common ones are

$$y'' = f''(x) = \frac{d^2y}{dx^2} = \frac{d}{dx}\frac{d}{dx}f(x) = \frac{d^2}{dx^2}f(x) = D_x^2 y = D_x^2 f(x).$$

Similarly, you can consider third-, fourth-, and in general $n$th-order derivatives. The prime notation is inconvenient for derivatives of high order, so we denote the order by a superscript in parentheses (to distinguish it from an exponent): the $n$th derivative of $y = f(x)$ is

$$y^{(n)} = f^{(n)}(x) = \frac{d^n y}{dx^n} = \frac{d^n}{dx^n} f(x) = D_x^n y = D_x^n f(x),$$

and it is defined to be the derivative of the $(n-1)$st derivative. For $n = 1$, $2$, and $3$, primes are still normally used: $f^{(2)}(x) = f''(x)$, $f^{(3)}(x) = f'''(x)$. It is sometimes convenient to denote $f^{(0)}(x) = f(x)$, that is, to regard a function as its own zeroth-order derivative.

**Example 1**   The **velocity** of a moving object is the (instantaneous) rate of change of the position of the object with respect to time; if the object moves along the $x$-axis and is at position $x = f(t)$ at time $t$, then its velocity at that time is

$$v = \frac{dx}{dt} = f'(t).$$

Similarly, the **acceleration** of the object is the rate of change of the velocity. Thus the acceleration is the *second derivative* of the position:

$$a = \frac{dv}{dt} = \frac{d^2 x}{dt^2} = f''(t).$$

We will investigate the relationships among position, velocity, and acceleration further in Section 2.11. ∎

**Example 2**   If $y = x^3$, then $y' = 3x^2$, $y'' = 6x$, $y''' = 6$, $y^{(4)} = 0$, and all higher derivatives are zero. ∎

In general, if $f(x) = x^n$ (where $n$ is a positive integer), then

$$f^{(k)}(x) = n(n-1)(n-2)...(n-(k-1))\, x^{n-k}$$

$$= \begin{cases} \dfrac{n!}{(n-k)!}\, x^{n-k} & \text{if } 0 \le k \le n \\ 0 & \text{if } k > n, \end{cases}$$

where $n!$ (called $n$ **factorial**) is defined by:

$$0! = 1$$
$$1! = 0! \times 1 = 1 \times 1 = 1$$
$$2! = 1! \times 2 = 1 \times 2 = 2$$
$$3! = 2! \times 3 = 1 \times 2 \times 3 = 6$$
$$4! = 3! \times 4 = 1 \times 2 \times 3 \times 4 = 24$$
$$\vdots$$
$$n! = (n-1)! \times n = 1 \times 2 \times 3 \times \cdots \times (n-1) \times n.$$

It follows that if $P$ is a polynomial of degree $n$,

$$P(x) = a_n x^n + a_{n-1} x^{n-1} + \cdots + a_1 x + a_0,$$

where $a_n$, $a_{n-1}$, ... , $a_1$, $a_0$ are constants, then $P^{(k)}(x) = 0$ for $k > n$. For $k \leq n$, $P^{(k)}$ is a polynomial of degree $n - k$; in particular, $P^{(n)}(x) = n!\, a_n$, a constant function.

---

**Example 3**  Show that if $A$, $B$, and $k$ are constants, then the function

$$y = A\cos(kt) + B\sin(kt)$$

is a solution of the *second-order* **differential equation of simple harmonic motion** (see Section 3.7):

$$\frac{d^2 y}{dt^2} + k^2 y = 0.$$

**Solution**  To be a solution, the function $y(t)$ must satisfy the differential equation *identically*; that is,

$$\frac{d^2}{dt^2}\, y(t) = k^2 y(t) = 0$$

must hold for every real number $t$.  We verify this by calculating the first two derivatives of the given function $y(t) = A\cos(kt) + B\sin(kt)$ and observing that the second derivative plus $k^2 y(t)$ is, in fact, zero everywhere:

$$\frac{dy}{dt} = -Ak\sin(kt) + Bk\cos(kt)$$

$$\frac{d^2 y}{dt^2} = -Ak^2\cos(kt) - Bk^2\sin(kt) = -k^2 y(t),$$

$$\frac{d^2 y}{dt^2} + k^2 y(t) = 0.$$

---

**Example 4**  Find the $n$th derivative, $y^{(n)}$, of $y = \dfrac{1}{1+x} = (1+x)^{-1}$.

**Solution**  Begin by calculating the first few derivatives:

$$y' = -(1+x)^{-2}$$
$$y'' = -(-2)(1+x)^{-3} = 2(1+x)^{-3}$$
$$y''' = 2(-3)(1+x)^{-4} = -3!(1+x)^{-4}$$
$$y^{(4)} = -3!(-4)(1+x)^{-5} = 4!(1+x)^{-5}$$

The pattern here is becoming obvious.  It seems that

$$y^{(n)} = (-1)^n n!(1+x)^{-n-1}.$$

Note the use of $(-1)^n$ to denote a positive sign if $n$ is even and a negative sign if $n$ is odd.

We have not yet actually proved that the above formula is correct for every $n$, although it is clearly correct for $n = 1, 2, 3$, and 4. To complete the proof we use mathematical induction (Section 2.2). Suppose that the formula is valid for $n = k$, where $k$ is some positive integer. Consider $y^{(k+1)}$:

$$y^{(k+1)} = \frac{d}{dx}\, y^{(k)} = \frac{d}{dx}\left((-1)^k k!(1+x)^{-k-1}\right)$$
$$= (-1)^k k!(-k-1)(1+x)^{-k-2} = (-1)^{k+1}(k+1)!(1+x)^{-(k+1)-1}.$$

This is what the formula predicts for the $(k+1)$st derivative. Therefore, if the formula for $y^{(n)}$ is correct for $n = k$, then it is also correct for $n = k+1$. Since the formula is known to be true for $n = 1$, it must therefore be true for every integer $n \geq 1$ *by induction*. ∎

**Example 5**   Find a formula for $f^{(n)}(x)$, given that $f(x) = \sin(ax + b)$.

**Solution**   Begin by calculating several derivatives:

$$f'(x) = a\, \cos(ax+b)$$
$$f''(x) = -a^2 \sin(ax+b) = -a^2 f(x)$$
$$f'''(x) = -a^3 \cos(ax+b) = -a^2 f'(x)$$
$$f^{(4)}(x) = a^4 \sin(ax+b) = a^4 f(x)$$
$$f^{(5)}(x) = a^5 \cos(ax+b) = a^4 f'(x)$$
$$\vdots$$

The pattern is pretty obvious here. Each new derivative is $-a^2$ times the second previous one. A formula that gives all the derivatives is

$$f^{(n)}(x) = \begin{cases} (-1)^k\, a^n \sin(ax+b) & \text{if } n = 2k \\ (-1)^k\, a^n \cos(ax+b) & \text{if } n = 2k+1 \end{cases} \quad (k = 0, 1, 2, \ldots),$$

which can also be verified by induction. ∎

Our final example shows that it is not always easy to obtain a formula for the $n$th derivative of a function.

**Example 6**   Calculate $f'$, $f''$, and $f'''$ for $f(x) = \sqrt{x^2+1}$. Can you see enough of a pattern to predict $f^{(4)}$?

**Solution**   Since $f(x) = (x^2+1)^{1/2}$ we have

$$f'(x) = \tfrac{1}{2}(x^2+1)^{-1/2}(2x) = x(x^2+1)^{-1/2},$$
$$f''(x) = (x^2+1)^{-1/2} + x\left(-\tfrac{1}{2}\right)(x^2+1)^{-3/2}(2x)$$
$$= (x^2+1)^{-3/2}(x^2+1-x^2) = (x^2+1)^{-3/2},$$
$$f'''(x) = -\tfrac{3}{2}(x^2+1)^{-5/2}(2x) = -3x(x^2+1)^{-5/2}.$$

Although the expression obtained from each differentiation simplified somewhat, the pattern of these derivatives is not (yet) obvious enough to enable us to predict the formula for $f^{(4)}(x)$ without having to calculate it. In fact,

$$f^{(4)}(x) = 3(4x^2 - 1)(x^2 + 1)^{-7/2},$$

so the pattern (if there is one) doesn't become any clearer at this stage. ∎

## Exercises 2.8

Find $y'$, $y''$, and $y'''$ for the functions in Exercises 1–12.

**1.** $y = (3 - 2x)^7$

**2.** $y = x^2 - \dfrac{1}{x}$

**3.** $y = \dfrac{6}{(x-1)^2}$

**4.** $y = \sqrt{ax+b}$

**5.** $y = x^{1/3} - x^{-1/3}$

**6.** $y = x^{10} + 2x^8$

**7.** $y = (x^2 + 3)\sqrt{x}$

**8.** $y = \dfrac{x-1}{x+1}$

**9.** $y = \tan x$

**10.** $y = \sec x$

**11.** $y = \cos(x^2)$

**12.** $y = \dfrac{\sin x}{x}$

In Exercises 13–23, calculate enough derivatives of the given function to enable you to guess the general formula for $f^{(n)}(x)$. Then verify your guess using mathematical induction.

**13.** $f(x) = \dfrac{1}{x}$

**14.** $f(x) = \dfrac{1}{x^2}$

**15.** $f(x) = \dfrac{1}{2-x}$

**16.** $f(x) = \sqrt{x}$

**17.** $f(x) = \dfrac{1}{a+bx}$

**18.** $f(x) = x^{2/3}$

**19.** $f(x) = \cos(ax)$

**20.** $f(x) = x\cos x$

**21.** $f(x) = x\sin(ax)$

**\* 22.** $f(x) = \dfrac{1}{|x|}$

**\* 23.** $f(x) = \sqrt{1-3x}$

**24.** If $y = \tan kx$, show that $y'' = 2k^2 y(1 + y^2)$.

**25.** If $y = \sec kx$, show that $y'' = k^2 y(2y^2 - 1)$.

**26.** Use mathematical induction to prove that the $n$th derivative of $y = \sin(ax + b)$ is given by the formula asserted at the end of Example 5.

**27.** Use mathematical induction to prove that the $n$th derivative of $y = \tan x$ is of the form $P_{n+1}(\tan x)$, where $P_{n+1}$ is a polynomial of degree $n + 1$.

**28.** If $f$ and $g$ are twice-differentiable functions, show that $(fg)'' = f''g + 2f'g' + fg''$.

**\* 29.** State and prove the results analogous to that of Exercise 28 but for $(fg)^{(3)}$ and $(fg)^{(4)}$. Can you guess the formula for $(fg)^{(n)}$?

## 2.9 Implicit Differentiation

We know how to find the slope of a curve which is the graph of a function $y = f(x)$ by calculating the derivative of $f$. But not all curves are the graphs of such functions. To be the graph of a function $f(x)$, the curve must not intersect any vertical lines at more than one point.

Curves are generally the graphs of *equations* in two variables. Such equations can be written in the form

$$F(x, y) = 0,$$

where $F(x, y)$ denotes an expression involving the two variables $x$ and $y$. For example, a circle with centre at the origin and radius 5 has equation

$$x^2 + y^2 - 25 = 0,$$

so $F(x, y) = x^2 + y^2 - 25$ for that circle.

Sometimes we can solve an equation $F(x, y) = 0$ for $y$, and so find explicit formulas for one or more functions $y = f(x)$ defined by the equation. Usually, however, we are not able to solve the equation. However, we can still regard it as defining $y$ as one or more functions of $x$ *implicitly*, even it we cannot solve for these functions *explicitly*. Moreover, we still find the derivative $dy/dx$ of these implicit solutions by a technique called **implicit differentiation**. The idea is to differentiate the given equation with respect to $x$, regarding $y$ as a function of $x$ having derivative $dy/dx$, or $y'$.

**Example 1**    Find $dy/dx$ if $y^2 = x$.

**Solution**    The equation $y^2 = x$ defines two differentiable functions of $x$; in this case we know them explicitly. They are $y_1 = \sqrt{x}$ and $y_2 = -\sqrt{x}$, (Figure 2.33) having derivatives defined for $x > 0$ by

$$\frac{dy_1}{dx} = \frac{1}{2\sqrt{x}} \quad \text{and} \quad \frac{dy_2}{dx} = -\frac{1}{2\sqrt{x}}.$$

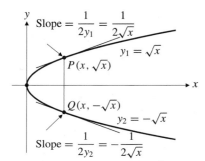

Slope $= \dfrac{1}{2y_1} = \dfrac{1}{2\sqrt{x}}$

$y_1 = \sqrt{x}$

$P(x, \sqrt{x})$

$Q(x, -\sqrt{x})$

$y_2 = -\sqrt{x}$

Slope $= \dfrac{1}{2y_2} = -\dfrac{1}{2\sqrt{x}}$

**Figure 2.33**    The equation $y^2 = x$ defines two differentiable functions of $x$ on the interval $x \geq 0$

However, we can find the slope of the curve $y^2 = x$ at any point $(x, y)$ satisfying that equation without first solving the equation for $y$. To find $dy/dx$ we simply differentiate both sides of the equation $y^2 = x$ with respect to $x$, treating $y$ as a differentiable function of $x$, and using the Chain Rule to differentiate $y^2$:

$$\frac{d}{dx}(y^2) = \frac{d}{dx}(x) \qquad \left( \text{The Chain Rule gives } \frac{d}{dx}\,y^2 = 2y\frac{dy}{dx}. \right)$$

$$2y\frac{dy}{dx} = 1$$

$$\frac{dy}{dx} = \frac{1}{2y}.$$

Observe that this agrees with the derivatives we calculated above for *both* of the explicit solutions $y_1 = \sqrt{x}$ and $y_2 = -\sqrt{x}$:

$$\frac{dy_1}{dx} = \frac{1}{2y_1} = \frac{1}{2\sqrt{x}} \quad \text{and} \quad \frac{dy_2}{dx} = \frac{1}{2y_2} = \frac{1}{2(-\sqrt{x})} = -\frac{1}{2\sqrt{x}}.$$

**Example 2**    Find the slope of circle $x^2 + y^2 = 25$ at the point $(3, -4)$.

**Solution**    The circle is not the graph of a single function of $x$. Again, it combines the graphs of two functions, $y_1 = \sqrt{25 - x^2}$ and $y_2 = -\sqrt{25 - x^2}$ (Figure 2.34). The point $(3, -4)$ lies on the graph of $y_2$, so we can find the slope by calculating explicitly:

$$\left.\frac{dy_2}{dx}\right|_{x=3} = -\left.\frac{-2x}{2\sqrt{25 - x^2}}\right|_{x=3} = -\frac{-6}{2\sqrt{25 - 9}} = \frac{3}{4}.$$

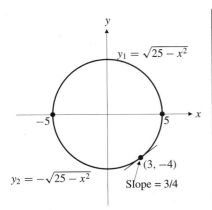

**Figure 2.34**    The circle combines the graphs of two functions. The graph of $y_2$ is the lower semicircle, and passes through $(3, -4)$

But we can also solve the problem more easily by differentiating the given equation of the circle implicitly with respect to $x$:

$$\frac{d}{dx}(x^2) + \frac{d}{dx}(y^2) = \frac{d}{dx}(25)$$

$$2x + 2y\frac{dy}{dx} = 0$$

$$\frac{dy}{dx} = -\frac{x}{y}.$$

The slope at $(3, -4)$ is $-x/y\big|_{(3,-4)} = -3/(-4) = 3/4.$

**Example 3**    Find $\dfrac{dy}{dx}$ if $y\sin x = x^3 + \cos y$.

**Solution**    This time we cannot solve the equation for $y$ as an explicit function of $x$, so we *must* use implicit differentiation.

$$\frac{d}{dx}(y\sin x) = \frac{d}{dx}(x^3) + \frac{d}{dx}(\cos y) \qquad \left(\begin{array}{l}\text{Use the Product Rule}\\ \text{on the left side.}\end{array}\right)$$

$$(\sin x)\frac{dy}{dx} + y\cos x = 3x^2 - (\sin y)\frac{dy}{dx}$$

$$(\sin x + \sin y)\frac{dy}{dx} = 3x^2 - y\cos x$$

$$\frac{dy}{dx} = \frac{3x^2 - y\cos x}{\sin x + \sin y}$$

To find $dy/dx$ by implicit differentiation:

1. Differentiate both sides of the equation with respect to $x$, regarding $y$ as a function of $x$ and using the Chain Rule to differentiate functions of $y$.

2. Collect terms with $dy/dx$ on one side of the equation and solve for $dy/dx$ by dividing by its coefficient.

In the examples above the derivatives $dy/dx$ calculated by implicit differentiation depend on $y$, or on both $y$ and $x$, rather than just on $x$. This is to be expected because an equation in $x$ and $y$ can define more than one function of $x$, and the implicitly calculated derivative must apply to each of the solutions. For example, in Example 2, the derivative $dy/dx = -x/y$ also gives the slope $-3/4$ at the point $(3, 4)$ on the circle. When you use implicit differentiation to find the slope of a curve at a point, you will usually have to know both coordinates of the point.

There are subtle dangers involved in calculating derivatives implicitly. When you use the Chain Rule to differentiate an equation involving $y$ with respect to $x$ you are automatically assuming that the equation defines $y$ as a differentiable function of $x$. This need not be the case. To see what can happen, consider the problem of finding $y' = dy/dx$ from the equation

$$x^2 + y^2 = K, \tag{$*$}$$

where $K$ is a constant. Just as in Example 2 (where $K = 25$), implicit differentiation gives

$$2x + 2yy' = 0 \qquad \text{or} \qquad y' = -\frac{x}{y}.$$

This formula will give the slope of the curve (∗) at any point on the curve where $y \neq 0$. For $K > 0$, (∗) represents a circle centred at the origin having radius $\sqrt{K}$. This circle has a finite slope, except at the two points where it crosses the $x$-axis (where $y = 0$). If $K = 0$, the equation represents only a single point, the origin. The concept of slope of a point is meaningless. For $K < 0$, there are no real points whose coordinates satisf y equation (∗), so $y'$ is meaningless here too. T he point of this is that being able to calculate $y'$ from a given equation by implicit differentiation does not guarantee that $y'$ actually represents the slope of anything.

If $(x_0, y_0)$ is a point on the graph of the equation $F(x, y) = 0$, there is a theorem that can justify our use of implicit differentiation to find the slope of the graph there. We cannot give a careful statement or proof of this **implicit function theorem** here, but roughly speaking, it says that part of  the graph of $F(x, y) = 0$ near $(x_0, y_0)$ is the graph of a function of $x$ that is differentiable at $x_0$, provided that $F(x, y)$ is a "smooth" function, and that the derivative

$$\frac{d}{dy} F(x_0, y) \Big|_{y=y_0} \neq 0.$$

For the circle $x^2 + y^2 - K = 0$ (where $K > 0$) this condition says that $2y_0 \neq 0$, which is the condition that the derivative $y' = -x/y$ should exist at $(x_0, y_0)$.

**A useful strategy**

When you use implicit differentiation to find the value of a derivative at a particular point, it is best to substitute the coordinates of the point immediately after you carry out the differentiation, and before you solve for the derivative $dy/dx$. It is easier to solve an equation involving numbers than one with algebraic expressions.

**Example 4**    Find an equation of the tangent to the curve $x^2 + xy + 2y^3 = 4$ at the point $(-2, 1)$.

**Solution**    To find the slope of the tangent we differentiate the given equation implicitly with respect to $x$. You must use the Product Rule to differentiate the $xy$ term:

$$2x + y + xy' + 6y^2 y' = 0.$$

Now substitute the coordinates $x = -2$, $y = 1$, and solve the resulting equation for $y'$.

$$-4 + 1 - 2y' + 6y' = 0 \qquad \Rightarrow \qquad y' = \frac{3}{4}.$$

The slope of the tangent at $(-2, 1)$ is $3/4$, and its equation is

$$y = \frac{3}{4}(x + 2) + 1 \qquad \text{or} \quad 3x - 4y = -10.$$

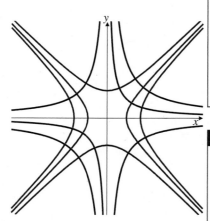

**Figure 2.35**    Some hyperbolas in the family $x^2 - y^2 = a$ (colour) intersecting some hyperbolas in the family $xy = b$ (black) at right angles

**Example 5**    Show that for any constants $a$ and $b$, the curves $x^2 - y^2 = a$ and $xy = b$ intersect at right angles, that is, at any point where they intersect their tangents are perpendicular.

**Solution**    The slope at any point on $x^2 - y^2 = a$ is given by $2x - 2yy' = 0$, or $y' = x/y$. The slope at any point on $xy = b$ is given by $y + xy' = 0$, or $y' = -y/x$. If the two curves (they are both hyperbolas if $a \neq 0$ and $b \neq 0$) intersect at $(x_0, y_0)$, then their slopes at that point are $x_0/y_0$ and $-y_0/x_0$, respectively. Clearly these slopes are negative reciprocals, so the tangent line to one curve is the normal line to the other at that point. Hence, the curves intersect at right angles. (See Figure 2.35.)

## Higher-Order Derivatives

**Example 6** Find $y'' = \dfrac{d^2y}{dx^2}$ if $xy + y^2 = 2x$.

**Solution** To start, differentiate both sides of the equation with respect to $x$:

$$y + xy' + 2yy' = 2 \quad \Longrightarrow \quad y' = \frac{2-y}{x+2y}.$$

Now differentiate $y'$ using the Quotient Rule:

$$y'' = \frac{(x+2y)(-y') - (2-y)(1+2y')}{(x+2y)^2} = \frac{y-2-(x+4)y'}{(x+2y)^2}.$$

Observe that this expression for $y''$ depends on $y'$ as well as on $x$ and $y$. We can eliminate $y'$ by substituting the expression obtained above:

$$
\begin{aligned}
y'' &= \frac{y-2-(x+4)\dfrac{2-y}{x+2y}}{(x+2y)^2} \\[2mm]
&= \frac{(y-2)(x+2y) - (x+4)(2-y)}{(x+2y)^3} \\[2mm]
&= \frac{2xy + 2y^2 - 4x - 8}{(x+2y)^3} = \frac{-8}{(x+2y)^3}.
\end{aligned}
$$

We used the original equation to simplify the numerator in the last line. ∎

## The General Power Rule

Until now, we have only proven the General Power Rule

$$\frac{d}{dx}x^r = r\,x^{r-1}$$

for integer exponents $r$ and a few special rational exponents such as $r = 1/2$. Using implicit differentiation, we can give the proof for any rational exponent $r = m/n$, where $m$ and $n$ are integers, and $n \neq 0$.

If $y = x^{m/n}$, then $y^n = x^m$. Differentiating implicitly with respect to $x$, we obtain

$$n\,y^{n-1}\frac{dy}{dx} = m\,x^{m-1}, \qquad \text{and so}$$

$$\frac{dy}{dx} = \frac{m}{n}x^{m-1}y^{1-n} = \frac{m}{n}x^{m-1}x^{(m/n)(1-n)} = \frac{m}{n}x^{m-1+(m/n)-m} = \frac{m}{n}x^{(m/n)-1}.$$

## Exercises 2.9

In Exercises 1–8, find $dy/dx$ in terms of $x$ and $y$.

**1.** $xy - x + 2y = 1$

**2.** $x^3 + y^3 = 1$

**3.** $x^2 + xy = y^3$

**4.** $x^3 y + xy^5 = 2$

**5.** $x^2 y^3 = 2x - y$

**6.** $x^2 + 4(y - 1)^2 = 4$

**7.** $\dfrac{x - y}{x + y} = \dfrac{x^2}{y} + 1$

**8.** $x\sqrt{x + y} = 8 - xy$

In Exercises 9–16, find an equation of the tangent to the given curve at the given point.

**9.** $2x^2 + 3y^2 = 5$ at $(1, 1)$

**10.** $x^2 y^3 - x^3 y^2 = 12$ at $(-1, 2)$

**11.** $\dfrac{x}{y} + \left(\dfrac{y}{x}\right)^3 = 2$ at $(-1, -1)$

**12.** $x + 2y + 1 = \dfrac{y^2}{x - 1}$ at $(2, -1)$

**13.** $2x + y - \sqrt{2}\sin(xy) = \pi/2$ at $\left(\dfrac{\pi}{4}, 1\right)$

**14.** $\tan(xy^2) = \dfrac{2xy}{\pi}$ at $\left(-\pi, \dfrac{1}{2}\right)$

**15.** $x\sin(xy - y^2) = x^2 - 1$ at $(1, 1)$

**16.** $\cos\left(\dfrac{\pi y}{x}\right) = \dfrac{x^2}{y} - \dfrac{17}{2}$ at $(3, 1)$

In Exercises 17–20, find $y''$ in terms of $x$ and $y$.

**17.** $xy = x + y$

**18.** $x^2 + 4y^2 = 4$

**∗ 19.** $x^3 - y^2 + y^3 = x$

**∗ 20.** $x^3 - 3xy + y^3 = 1$

**21.** For $x^2 + y^2 = a^2$ show that $y'' = -\dfrac{a^2}{y^3}$.

**22.** For $Ax^2 + By^2 = C$ show that $y'' = -\dfrac{AC}{B^2 y^3}$.

**∗ 23.** Show that the ellipse $x^2 + 2y^2 = 2$ and the hyperbola $2x^2 - 2y^2 = 1$ intersect at right angles.

**∗ 24.** Show that the ellipse $x^2/a^2 + y^2/b^2 = 1$ and the hyperbola $x^2/A^2 - y^2/B^2 = 1$ intersect at right angles if $A^2 \leq a^2$ and $a^2 - b^2 = A^2 + B^2$. (This says that the ellipse and the hyperbola have the same foci.)

**25.** If $z = \tan\dfrac{x}{2}$ show that

$$\dfrac{dx}{dz} = \dfrac{2}{1 + z^2}, \quad \sin x = \dfrac{2z}{1 + z^2}, \quad \text{and } \cos x = \dfrac{1 - z^2}{1 + z^2}.$$

**∗ 26.** Use implicit differentiation to find $y'$ if $(x - y)/(x + y) = x/y + 1$. Now show that there are, in fact, no points on that curve, so the derivative you calculated is meaningless. This is another example that demonstrates the dangers of calculating something when you don't know whether or not it exists.

---

## 2.10   Antiderivatives and Initial-Value Problems

Throughout this chapter we have been concerned with the problem of finding the derivative $f'$ of a given function $f$. The reverse problem: given the derivative $f'$, find $f$, is also interesting and important. It is the problem studied in *integral calculus* and is generally harder to solve than the problem of finding a derivative. We will take a preliminary look at this problem in this section, and will return to it in more detail in Chapter 5.

### Antiderivatives

We begin by defining an antiderivative of a function $f$ to be a function $F$ whose derivative is $f$. It is appropriate to require that $F'(x) = f(x)$ on an *interval*.

**DEFINITION**  **7**

An **antiderivative** of a function $f$ on an interval $I$ is another function $F$ satisfying

$$F'(x) = f(x) \quad \text{for } x \text{ in } I.$$

**Example 1**

(a) $F(x) = x$ is an antiderivative of the function $f(x) = 1$ on any interval, because $F'(x) = 1 = f(x)$ everywhere.

(b) $G(x) = \frac{1}{2}x^2$ is an antiderivative of the function $g(x) = x$ on any interval, because $G'(x) = \frac{1}{2}(2x) = x = g(x)$ everywhere.

(c) $R(x) = -\frac{1}{3}\cos(3x)$ is an antiderivative of $r(x) = \sin(3x)$ on any interval, because $R'(x) = -\frac{1}{3}(-3\sin(3x)) = \sin(3x) = r(x)$ everywhere.

(d) $F(x) = -1/x$ is an antiderivative of $f(x) = 1/x^2$ on any interval not containing $x = 0$, because $F'(x) = 1/x^2 = f(x)$ everywhere except at $x = 0$. ∎

Antiderivatives are not unique; indeed, if $C$ is any constant, then $F(x) = x + C$ is an antiderivative of $f(x) = 1$ on any interval. You can always add a constant to an antiderivative $F$ of a function $f$ on an interval and get another antiderivative of $f$. More importantly, *all* antiderivatives of $f$ on an interval can be obtained by adding constants to any particular one. If $F$ and $G$ are both antiderivatives of $f$ on an interval $I$, then

$$\frac{d}{dx}\big(G(x) - F(x)\big) = f(x) - f(x) = 0$$

on $I$, so $G(x) - F(x) = C$ (a constant) on $I$ by Theorem 13 of Section 2.6. Thus $G(x) = F(x) + C$ on $I$.

Note that neither this conclusion, nor Theorem 13, are valid over a set that is not an interval. For example, the derivative of

$$\operatorname{sgn} x = \begin{cases} -1 & \text{if } x < 0 \\ 1 & \text{if } x > 0 \end{cases}$$

is 0 for all $x \neq 0$, but $\operatorname{sgn} x$ is not constant for all $x \neq 0$. $\operatorname{sgn} x$ has *different* constant values on the two intervals $(-\infty, 0)$ and $(0, \infty)$ comprising its domain.

## The Indefinite Integral

The *general antiderivative* of a function $f(x)$ on an interval $I$ is $F(x) + C$, where $F(x)$ is any particular antiderivative of $f(x)$ on $I$ and $C$ is a constant. This general antiderivative is called the indefinite integral of $f(x)$ on $I$, and is denoted

$$\int f(x)\,dx.$$

**DEFINITION 8**

The **indefinite integral** of $f(x)$ on interval $I$ is

$$\int f(x)\,dx = F(x) + C \qquad \text{on } I,$$

where $F'(x) = f(x)$ for all $x$ in $I$.

The symbol $\int$ is called an **integral sign**. It is shaped like an elongated "S" for reasons that will only become apparent when we study the *definite integral* in Chapter 5. Just as you regard $dy/dx$ as a single symbol representing the derivative of $y$ with respect to $x$, so you should regard $\int f(x)\,dx$ as a single symbol representing the indefinite integral (general antiderivative) of $f$ with respect to $x$. The constant $C$ is called a **constant of integration**.

**Example 2**

(a) $\displaystyle\int x\,dx = \frac{1}{2}x^2 + C$ on any interval.

(b) $\displaystyle\int (x^3 - 5x^2 + 7)\,dx = \frac{1}{4}x^4 - \frac{5}{3}x^3 + 7x + C$ on any interval.

(c) $\displaystyle\int \left(\frac{1}{x^2} + \frac{2}{\sqrt{x}}\right)\,dx = -\frac{1}{x} + 4\sqrt{x} + C$ on any interval not containing $x = 0$.

All three formulas above can be checked by differentiating the right-hand sides.

Finding antiderivatives is generally more difficult than finding derivatives; many functions do not have antiderivatives that can be expressed as combinations of finitely many elementary functions. However, *every formula for a derivative can be rephrased as a formula for an antiderivative*. For instance,

$$\text{Since } \frac{d}{dx}\sin x = \cos x, \quad \text{therefore } \int \cos x\,dx = \sin x + C.$$

We will develop several techniques for finding antiderivatives in Chapter 5. Until then, we must content ourselves with being able to write a few simple antiderivatives based on the known derivatives of elementary functions:

(a) $\displaystyle\int dx = \int 1\,dx = x + C$  (b) $\displaystyle\int x\,dx = \frac{x^2}{2} + C$

(c) $\displaystyle\int x^2\,dx = \frac{x^3}{3} + C$  (d) $\displaystyle\int \frac{1}{x^2}\,dx = \int \frac{dx}{x^2} = -\frac{1}{x} + C$

(e) $\displaystyle\int \frac{1}{\sqrt{x}}\,dx = 2\sqrt{x} + C$  (f) $\displaystyle\int x^r\,dx = \frac{x^{r+1}}{r+1} + C \;\;(r \neq -1)$

(g) $\displaystyle\int \sin x\,dx = -\cos x + C$  (h) $\displaystyle\int \cos x\,dx = \sin x + C$

(i) $\displaystyle\int \sec^2 x\,dx = \tan x + C$  (j) $\displaystyle\int \csc^2 x\,dx = -\cot x + C$

(k) $\displaystyle\int \sec x \tan x\,dx = \sec x + C$  (l) $\displaystyle\int \csc x \cot x\,dx = -\csc x + C$

Observe that formulas (a)–(e) are special cases of formula (f). For the moment, $r$ must be rational in (f), but this restriction will be removed later.

The rule for differentiating sums and constant multiples of functions translates into a similar rule for antiderivatives, as reflected in parts (b) and (c) of Example 2 above.

The graphs of the different antiderivatives of the same function on the same interval are vertically displaced versions of the same curve, as shown in Figure 2.36. In general, only one of these curves will pass through any given point, so we can obtain a unique antiderivative of a given function on an interval by requiring the antiderivative to take a prescribed value at a particular point $x$.

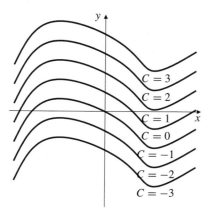

**Figure 2.36**   Graphs of various antiderivatives of the same function

**Example 3**   Find the function $f(x)$ whose derivative is $f'(x) = 6x^2 - 1$ for all real $x$, and for which $f(2) = 10$.

**Solution**   Since $f'(x) = 6x^2 - 1$, we have

$$f(x) = \int (6x^2 - 1)\, dx = 2x^3 - x + C$$

for some constant $C$. Since $f(2) = 10$, we have

$$10 = f(2) = 16 - 2 + C.$$

Thus $C = -4$ and $f(x) = 2x^3 - x - 4$. (By direct calculation we can verify that $f'(x) = 6x^2 - 1$ and $f(2) = 10$.)

**Example 4**   Find the function $g(t)$ whose derivative is $\dfrac{t+5}{t^{3/2}}$ and whose graph passes through the point $(4, 1)$.

**Solution**   We have

$$\begin{aligned}
g(t) &= \int \frac{t+5}{t^{3/2}}\, dt \\
&= \int (t^{-1/2} + 5t^{-3/2})\, dt \\
&= 2t^{1/2} - 10t^{-1/2} + C
\end{aligned}$$

Since the graph of $y = g(t)$ must pass through $(4, 1)$, we require that

$$1 = g(4) = 4 - 5 + C,$$

Hence, $C = 2$ and

$$g(t) = 2t^{1/2} - 10t^{-1/2} + 2 \qquad \text{for } t > 0.$$

## Differential Equations and Initial-Value Problems

A **differential equation** (abbreviated DE) is an equation involving one or more derivatives of an unknown function. Any function whose derivatives satisfy the differential equation *identically on an interval* is called a **solution** of the equation on that interval. For instance, the function $y = x^3 - x$ is a solution of the differential equation

$$\frac{dy}{dx} = 3x^2 - 1$$

on the whole real line. This differential equation has more than one solution; in fact, $y = x^3 - x + C$ is a solution for any value of the constant $C$.

**Example 5** Show that for any constants $A$ and $B$, the function $y = Ax^3 + B/x$ is a solution of the differential equation $x^2y'' - xy' - 3y = 0$ on any interval not containing 0.

**Solution** If $y = Ax^3 + B/x$, then for $x \neq 0$ we have

$$y' = 3Ax^2 - B/x^2 \quad \text{and} \quad y'' = 6Ax + 2B/x^3.$$

Therefore,

$$x^2y'' - xy' - 3y = 6Ax^3 + \frac{2B}{x} - 3Ax^3 + \frac{B}{x} - 3Ax^3 - \frac{3B}{x} = 0,$$

provided $x \neq 0$. This is what had to be proved.

The **order** of a differential equation is the order of the highest-order derivative appearing in the equation. The DE in Example 5 is a *second-order* DE since it involves $y''$ and no higher derivatives of $y$. Note that the solution verified in Example 5 involves two arbitrary constants, $A$ and $B$. This solution is called a **general solution** to the equation since it can be shown that every solution is of this form for some choice of the constants $A$ and $B$. A **particular solution** of the equation is obtained by assigning specific values to these constants. The general solution of an $n$th-order differential equation typically involves $n$ arbitrary constants.

An **initial-value problem** (abbreviated IVP) is a problem that consists of:

(i) a differential equation (to be solved for an unknown function), and

(ii) prescribed values for the solution and enough of its derivatives at a particular point (the initial point) to determine values for all the arbitrary constants in the general solution of the DE and so yield a particular solution.

**Remark** It is common to use the same symbol, say $y$, to denote both the dependent variable and the function which is the solution to a DE or an IVP; that is, we call the solution function $y = y(x)$ rather than $y = f(x)$.

**Example 6** Use the result of Example 5 to solve the following initial-value problem.

$$\begin{cases} x^2y'' - xy' - 3y = 0 & (x > 0) \\ y(1) = 2 \\ y'(1) = -6 \end{cases}$$

**Solution** As shown in Example 5, the DE $x^2y'' - xy' - 3y = 0$ has solution $y = Ax^3 + B/x$, which has derivative $y' = 3Ax^2 - B/x^2$. At $x = 1$ we must have $y = 2$ and $y' = -6$. Therefore

$$\begin{aligned} A+B &= 2 \\ 3A-B &= -6. \end{aligned}$$

Solving these two linear equations for $A$ and $B$, we get $A = -1$ and $B = 3$. Hence $y = -x^3 + 3/x$ for $x > 0$ is the solution of the IVP.

One of the simplest kinds of differential equation is the equation

$$\frac{dy}{dx} = f(x),$$

which is to be solved for $y$ as a function of $x$. Evidently the solution is

$$y = \int f(x)\, dx.$$

Our ability to find the unknown function $y(x)$ depends on our ability to find an antiderivative of $f$.

**Example 7**   Solve the initial-value problem

$$\begin{cases} y' = \dfrac{3 + 2x^2}{x^2} \\ y(-2) = 1. \end{cases}$$

Where is the solution valid?

**Solution**

$$y = \int \left( \frac{3}{x^2} + 2 \right) dx = -\frac{3}{x} + 2x + C$$

$$1 = y(-2) = \frac{3}{2} - 4 + C$$

Therefore $C = \frac{7}{2}$ and

$$y = -\frac{3}{x} + 2x + \frac{7}{2}.$$

Although the solution function appears to be defined for all $x$ except 0, it is only a solution of the given initial-value problem for $x < 0$. This is because $(-\infty, 0)$ is the largest interval which contains the initial point $-2$ but not the point $x = 0$, where the DE is undefined. ∎

**Example 8**   Solve the second-order IVP

$$\begin{cases} y'' = \sin x \\ y(\pi) = 2 \\ y'(\pi) = -1. \end{cases}$$

**Solution**   Since $(y')' = y'' = \sin x$, we have

$$y'(x) = \int \sin x\, dx = -\cos x + C_1.$$

The initial condition for $y'$ gives

$$-1 = y'(\pi) = -\cos \pi + C_1 = 1 + C_1,$$

so that $C_1 = -2$, and $y'(x) = -(\cos x + 2)$. Thus

$$y(x) = -\int (\cos x + 2)\,dx$$
$$= -\sin x - 2x + C_2.$$

The initial condition for $y$ now gives

$$2 = y(\pi) = -\sin \pi - 2\pi + C_2 = -2\pi + C_2,$$

so that $C_2 = 2 + 2\pi$. The solution to the given IVP is

$$y = 2 + 2\pi - \sin x - 2x,$$

and is valid for all $x$.  ∎

Differential equations and initial-value problems are of great importance in applications of calculus, especially for expressing in mathematical form certain laws of nature that involve rates of change of quantities. A large portion of the total mathematical endeavour of the last two hundred years has been devoted to their study. They are usually treated in separate courses on differential equations, but we will discuss them from time to time in this book when appropriate. Throughout this book, exercises about differential equations and initial-value problems are designated with the symbol ⬥.

## Exercises 2.10

In Exercises 1–14, find the given indefinite integrals.

**1.** $\displaystyle\int 5\,dx$

**2.** $\displaystyle\int x^2\,dx$

**3.** $\displaystyle\int \sqrt{x}\,dx$

**4.** $\displaystyle\int x^{12}\,dx$

**5.** $\displaystyle\int x^3\,dx$

**6.** $\displaystyle\int (x + \cos x)\,dx$

**7.** $\displaystyle\int \tan x \cos x\,dx$

**8.** $\displaystyle\int \frac{1 + \cos^3 x}{\cos^2 x}\,dx$

**9.** $\displaystyle\int (a^2 - x^2)\,dx$

**10.** $\displaystyle\int (A + Bx + Cx^2)\,dx$

**11.** $\displaystyle\int (2x^{1/2} + 3x^{1/3})\,dx$

**12.** $\displaystyle\int \frac{6(x-1)}{x^{4/3}}\,dx$

**13.** $\displaystyle\int \left( \frac{x^3}{3} - \frac{x^2}{2} + x - 1 \right) dx$

**14.** $\displaystyle 105 \int (1 + t^2 + t^4 + t^6)\,dt$

In Exercises 15–22, find the given indefinite integrals. This may require guessing the form of an antiderivative, and then checking by differentiation. For instance, you might suspect that

$\int \cos(5x - 2)\,dx = k \sin(5x - 2) + C$ for some $k$. Differentiating the answer shows that $k$ must be $1/5$.

**15.** $\displaystyle\int \cos(2x)\,dx$

**16.** $\displaystyle\int \sin\left(\frac{x}{2}\right) dx$

**⬥ 17.** $\displaystyle\int \frac{dx}{(1 + x)^2}$

**⬥ 18.** $\displaystyle\int \sec(1 - x)\tan(1 - x)\,dx$

**⬥ 19.** $\displaystyle\int \sqrt{2x + 3}\,dx$

**⬥ 20.** $\displaystyle\int \frac{4}{\sqrt{x + 1}}\,dx$

**21.** $\displaystyle\int 2x \sin(x^2)\,dx$

**⬥ 22.** $\displaystyle\int \frac{2x}{\sqrt{x^2 + 1}}\,dx$

Use trigonometric identities such as $\sec^2 x = 1 + \tan^2 x$, $\sin(2x) = 2 \sin x \cos x$, and $\cos(2x) = 2\cos^2 x - 1 = 1 - 2\sin^2 x$ to help you evaluate the indefinite integrals in Exercises 23–26.

**⬥ 23.** $\displaystyle\int \tan^2 x\,dx$

**⬥ 24.** $\displaystyle\int \sin x \cos x\,dx$

**⬥ 25.** $\displaystyle\int \cos^2 x\,dx$

**⬥ 26.** $\displaystyle\int \sin^2 x\,dx$

**Differential equations**

In Exercises 27–42, find the solution $y = y(x)$ to the given initial-value problem. On what interval is the solution valid? (Note that exercises involving differential equations are prefixed with the symbol ❖.)

❖ **27.** $\begin{cases} y' = x - 2 \\ y(0) = 3 \end{cases}$

❖ **28.** $\begin{cases} y' = x^{-2} - x^{-3} \\ y(-1) = 0 \end{cases}$

❖ **29.** $\begin{cases} y' = 3\sqrt{x} \\ y(4) = 1 \end{cases}$

❖ **30.** $\begin{cases} y' = x^{1/3} \\ y(0) = 5 \end{cases}$

❖ **31.** $\begin{cases} y' = Ax^2 + Bx + C \\ y(1) = 1 \end{cases}$

❖ **32.** $\begin{cases} y' = x^{-9/7} \\ y(1) = -4 \end{cases}$

❖ **33.** $\begin{cases} y' = \cos x \\ y(\pi/6) = 2 \end{cases}$

❖ **34.** $\begin{cases} y' = \sin(2x) \\ y(\pi/2) = 1 \end{cases}$

❖ **35.** $\begin{cases} y' = \sec^2 x \\ y(0) = 1 \end{cases}$

❖ **36.** $\begin{cases} y' = \sec^2 x \\ y(\pi) = 1 \end{cases}$

❖ **37.** $\begin{cases} y'' = 2 \\ y'(0) = 5 \\ y(0) = -3 \end{cases}$

❖ **38.** $\begin{cases} y'' = x^{-4} \\ y'(1) = 2 \\ y(1) = 1 \end{cases}$

❖ **39.** $\begin{cases} y'' = x^3 - 1 \\ y'(0) = 0 \\ y(0) = 8 \end{cases}$

❖ **40.** $\begin{cases} y'' = 5x^2 - 3x^{-1/2} \\ y'(1) = 2 \\ y(1) = 0 \end{cases}$

❖ **41.** $\begin{cases} y'' = \cos x \\ y(0) = 0 \\ y'(0) = 1 \end{cases}$

❖ **42.** $\begin{cases} y'' = x + \sin x \\ y(0) = 2 \\ y'(0) = 0 \end{cases}$

❖ **43.** Show that for any constants $A$ and $B$ the function $y = y(x) = Ax + B/x$ satisfies the *second-order differential equation* $x^2 y'' + xy' - y = 0$ for $x \neq 0$. Find a function $y$ satisfying the initial-value problem:

$$\begin{cases} x^2 y'' + xy' - y = 0 \qquad (x > 0) \\ y(1) = 2 \\ y'(1) = 4. \end{cases}$$

❖ **44.** Show that for any constants $A$ and $B$ the function $y = Ax^{r_1} + Bx^{r_2}$ satisfies, for $x > 0$, the differential equation $ax^2 y'' + bxy' + cy = 0$, provided that $r_1$ and $r_2$ are two distinct rational roots of the quadratic equation $ar(r - 1) + br + c = 0$.

Use the result of Exercise 44 to solve the initial-value problems in Exercises 45–46 on the interval $x > 0$.

❖ **45.** $\begin{cases} 4x^2 y'' + 4xy' - y = 0 \\ y(4) = 2 \\ y'(4) = -2 \end{cases}$

❖ **46.** $\begin{cases} x^2 y'' - 6y = 0 \\ y(1) = 1 \\ y'(1) = 1 \end{cases}$

## 2.11  Velocity and Acceleration

### Velocity and Speed

Suppose that an object is moving along a straight line (say the $x$-axis) so that its position $x$ is a function of time $t$, say $x = x(t)$. (We are using $x$ to represent both the dependent variable and the function.) Suppose we are measuring $x$ in metres and $t$ in seconds. The **average velocity** of the object over the time interval $[t, t + h]$ is the change in position divided by the change in time, that is, the Newton quotient

$$v_{\text{average}} = \frac{\Delta x}{\Delta t} = \frac{x(t + h) - x(t)}{h} \text{ m/s}.$$

The **velocity** $v(t)$ of the object at time $t$ is the limit of this average velocity as $h \to 0$. Thus it is the rate of change (the derivative) of position with respect to time;

$$\text{Velocity: } \quad v(t) = \frac{dx}{dt} = x'(t).$$

Besides telling us how fast the object is moving, the velocity also tells us in which direction it is moving. If $v(t) > 0$, then $x$ is increasing, so the object is moving to the right; if $v(t) < 0$, then $x$ is decreasing, so the object is moving to the left. At a critical point of $x$, that is, a time $t$ when $v(t) = 0$, the object is instantaneously at rest—at that instant it is not moving in either direction.

We distinguish between the term *velocity* (which involves direction of motion as well as the rate) and **speed**, which only involves the rate, and not the direction. The speed is the absolute value of the velocity:

$$\text{Speed:} \quad s(t) = |v(t)| = \left|\frac{dx}{dt}\right|.$$

A speedometer gives us the speed an automobile is moving; it does not give the velocity. The speedometer does not start to show negative values if the automobile turns around and heads in the opposite direction.

## Example 1

(a) Determine the velocity $v(t)$ at time $t$ of an object moving along the $x$-axis so that at time $t$ its position is given by

$$x = v_0 t + \frac{1}{2}at^2,$$

where $v_0$ and $a$ are constants.

(b) Draw the graph of the function $v(t)$, and show that the area under the graph and above the $t$-axis, over the interval $[t_1, t_2]$ is equal to the distance the object travels in that time interval.

**Solution** The velocity is given by

$$v(t) = \frac{dx}{dt} = v_0 + at.$$

Its graph is a straight line with slope $a$ and intercept $v_0$ on the vertical (velocity) axis. The area under the graph (shaded in Figure 2.37) is the sum of the areas of a rectangle and a triangle. Each has base $t_2 - t_1$. The rectangle has height $v(t_1) = v_0 + at_1$, and the triangle has height $a(t_2 - t_1)$. (Why?) Thus the shaded area is equal to

$$\begin{aligned}
\text{Area} &= (t_2 - t_1)(v_0 + at_1) + \frac{1}{2}(t_2 - t_1)[a(t_2 - t_1)] \\
&= (t_2 - t_1)\left[v_0 + at_1 + \frac{a}{2}(t_2 - t_1)\right] \\
&= (t_2 - t_1)\left[v_0 + \frac{a}{2}(t_2 + t_1)\right] \\
&= v_0(t_2 - t_1) + \frac{a}{2}(t_2^2 - t_1^2) \\
&= x(t_2) - x(t_1),
\end{aligned}$$

which is the distance travelled by the object between times $t_1$ and $t_2$.

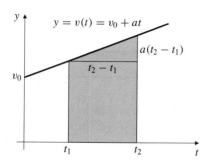

**Figure 2.37** The shaded area equals the distance travelled between $t_1$ and $t_2$

**Remark** In Example 1 we differentiated the position $x$ to get the velocity $v$, and then used the area under the velocity graph to recover information about the position.

It appears that there is a connection between finding areas and finding functions that have given derivatives (i.e. finding antiderivatives). This connection, which we will explore in Chapter 5, is, perhaps, the most important idea in calculus!

## Acceleration

The derivative of the velocity also has a useful interpretation. The rate of change of the velocity with respect to time is the **acceleration** of the moving object. It is measured in units of distance/time$^2$. The value of the acceleration at time $t$ is

$$\text{Acceleration:} \quad a(t) = v'(t) = \frac{dv}{dt} = \frac{d^2x}{dt^2}.$$

The acceleration is the *second derivative* of the position. If $a(t) > 0$, the velocity is increasing. This does not necessarily mean that the speed is increasing; if the object is moving to the left ($v(t) < 0$) and accelerating to the right ($a(t) > 0$), then it is actually slowing down. The object is speeding up only when the velocity and acceleration have the same sign.

**Table 2.**   Velocity, acceleration, and speed

| If velocity is | and acceleration is | then object is | and speed is |
|---|---|---|---|
| positive | positive | moving right | increasing |
| positive | negative | moving right | decreasing |
| negative | positive | moving left | decreasing |
| negative | negative | moving left | increasing |

If $a(t_0) = 0$, then the velocity and the speed are stationary at $t_0$. If $a(t) = 0$ during an interval of time, then the velocity is unchanging and, therefore, constant over that interval.

**Example 2**   A point $P$ moves along the $x$-axis in such a way that its position at time $t$ s (seconds) is given by

$$x = 2t^3 - 15t^2 + 24t \text{ ft.}$$

(a) Find the velocity and acceleration of $P$ at time $t$.

(b) In which direction and how fast is $P$ moving at $t = 2$ s? Is it speeding up or slowing down at that time?

(c) When is $P$ instantaneously at rest? When is its speed instantaneously not changing?

(d) When is $P$ moving to the left? to the right?

(e) When is $P$ speeding up? slowing down?

*Solution*

(a) The velocity and acceleration of $P$ at time $t$ are

$$v = \frac{dx}{dt} = 6t^2 - 30t + 24 = 6(t-1)(t-4) \text{ ft/s}, \quad \text{and}$$

$$a = \frac{dv}{dt} = 12t - 30 = 6(2t - 5) \text{ ft/s}^2.$$

(b) At $t = 2$ we have $v = -12$ and $a = -6$. Thus $P$ is moving to the left with speed 12 ft/s and, since the velocity and acceleration are both negative, its speed is increasing.

(c) $P$ is at rest when $v = 0$, that is, when $t = 1$ s, or $t = 4$ s. Its speed is unchanging when $a = 0$, that is at $t = 5/2$ s.

(d) The velocity is continuous for all $t$, so, by the Intermediate-Value Theorem, has a constant sign on the intervals between the points where it is 0. By examining the values of $v(t)$ at $t = 0$, 2, and 5 (or by analyzing the signs of the factors $(t-1)$ and $(t-4)$ in the expression for $v(t)$), we conclude that $v(t) < 0$ (and $P$ is moving to the left) on time interval $(1, 4)$. $v(t) > 0$ (and $P$ is moving to the right) on time intervals $(-\infty, 1)$ and $(4, \infty)$.

(e) The acceleration $a$ is negative for $t < 5/2$ and is positive for $t > 5/2$. Table 3 combines this information with information about $v$ to show where $P$ is speeding up and slowing down.

**Table 3.**    Data for Example 2

| Interval | $v(t)$ is | $a(t)$ is | $P$ is |
|---|---|---|---|
| $(-\infty, 1)$ | positive | negative | slowing down |
| $(1, 5/2)$ | negative | negative | speeding up |
| $(5/2, 4)$ | negative | positive | slowing down |
| $(4, \infty)$ | positive | positive | speeding up |

The motion of $P$ is shown in Figure 2.38.

$\blacksquare$

**Figure 2.38**    The motion of $P$ in Example 2

**Example 3**    An object is hurled upward from the roof of a building 10 m high. It rises and then falls back; its height above the ground $t$ seconds after it is thrown is

$$y = -4.9\,t^2 + 8t + 10 \text{ m,}$$

until it strikes the ground. What is the greatest height above the ground that the object attains? With what speed does the object strike the ground?

**Solution**    Refer to Figure 2.39. The vertical velocity at time $t$ during flight is

$$v(t) = -2(4.9)\,t + 8 = -9.8\,t + 8 \text{ m/s.}$$

The object is rising when $v > 0$, that is when $0 < t < 8/9.8$, and is falling for $t > 8/9.8$. Thus the object is at its maximum height at time $t = 8/9.8 \approx 0.8163$ s, and this maximum height is

$$y_{max} = -4.9\left(\frac{8}{9.8}\right)^2 + 8\left(\frac{8}{9.8}\right) + 10 \approx 13.27 \text{ m.}$$

The time $t$ at which the object strikes the ground is the positive root of the quadratic equation obtained by setting $y = 0$,

$$-4.9t^2 + 8t + 10 = 0,$$

namely,

$$t = \frac{-8 - \sqrt{64 + 196}}{-9.8} \approx 2.462 \text{ s.}$$

The velocity at this time is $v = -(9.8)(2.462) + 8 \approx -16.12$. Thus the object strikes the ground with a speed of about 16.12 m/s. ■

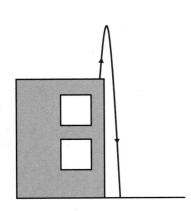

**Figure 2.39**

## Falling Under Gravity

According to Newton's Second Law of Motion, a rock of mass $m$ acted on by an unbalanced force $F$ will experience an acceleration $a$ proportional to and in the same direction as $F$; with appropriate units of force, $F = ma$. If the rock is sitting on the ground it is acted on by two forces, the force of gravity acting downward and reaction of the ground acting upward. These forces balance, so there is no resulting acceleration. On the other hand, if the rock is up in the air and is unsupported, the gravitational force on it will be unbalanced and the rock will experience downward acceleration. It will fall.

According to Newton's Universal Law of Gravitation, the force by which the earth attracts the rock is proportional to the mass $m$ of the rock, and inversely proportional to the square of its distance $r$ from the centre of the earth: $F = km/r^2$. If the relative change $\Delta r/r$ is small, as will be the case if the rock remains near the surface of the earth, then $F = mg$, where $g = k/r^2$ is approximately constant. It follows that $ma = F = mg$, and the rock experiences *constant* downward acceleration $g$. Since $g$ does not depend on $m$, all objects experience the same acceleration when falling near the surface of the earth, provided we ignore air resistance, and any other forces that may be acting them. Newton's laws therefore imply that if the height of such an object at time $t$ is $y(t)$, then

$$\frac{d^2y}{dt^2} = -g.$$

The negative sign is needed because the gravitational acceleration is downward, the opposite direction to that of increasing $y$. Physical experiments give the following approximate values for $g$ at the surface of the earth:

$$g = 32 \text{ ft/s}^2 \quad \text{or} \quad g = 9.8 \text{ m/s}^2.$$

**Example 4** A rock falling freely near the surface of the earth is subject to a constant downward acceleration $g = 9.8$, if the effect of air resistance is neglected. If the height and velocity of the rock are $y_0$ and $v_0$ at time $t = 0$, find the height $y(t)$ of the rock at any later time $t$ until the rock strikes the ground.

**Solution** This example asks for a solution $y(t)$ to the second-order initial-value problem:

$$\begin{cases} y''(t) = -g \\ y(0) = y_0 \\ y'(0) = v_0. \end{cases}$$

We have

$$y'(t) = -\int g\,dt = -gt + C_1$$
$$v_0 = y'(0) = 0 + C_1.$$

Thus $C_1 = v_0$.

$$y'(t) = -gt + v_0$$
$$y(t) = \int(-gt + v_0)dt = -\frac{1}{2}gt^2 + v_0 t + C_2$$
$$y_0 = y(0) = 0 + 0 + C_2.$$

Thus $C_2 = y_0$. Finally, therefore,

$$y(t) = -\frac{1}{2}gt^2 + v_0 t + y_0.$$

**Example 5** A ball is thrown down with an initial speed of 20 ft/s from the top of a cliff, and it strikes the ground at the bottom of the cliff after 5 s. How high is the cliff?

**Solution** We will apply the result of Example 4. Here we have $g = 32$ ft/s$^2$, $v_0 = -20$ ft/s, and $y_0$ is the unknown height of the cliff. The height of the ball $t$ seconds after it is thrown down is

$$y(t) = -16t^2 - 20t + y_0 \text{ ft.}$$

At $t = 5$ the ball reaches the ground, so $y(5) = 0$:

$$0 = -16(25) - 20(5) + y_0 \qquad \Rightarrow \qquad y_0 = 500.$$

The cliff is 500 ft high.

**Example 6**   **(Stopping distance)**   A car is travelling at 72 km/h. At a certain instant its brakes are applied to produce a constant deceleration of 0.8 m/s$^2$. How far does the car travel before coming to a stop?

**Solution**   Let $s(t)$ be the distance the car travels in the $t$ seconds after the brakes are applied. Then $s''(t) = -0.8$ (m/s$^2$), so the velocity at time $t$ is given by

$$s'(t) = \int -0.8 \, dt = -0.8t + C_1.$$

Since $s'(0) = 72$ km/h $= 72 \times 1000/3600 = 20$ m/s, we have $C_1 = 20$. Thus,

$$s'(t) = 20 - 0.8t$$

and

$$s(t) = \int (20 - 0.8t) \, dt = 20t - 0.4t^2 + C_2.$$

Since $s(0) = 0$, we have $C_2 = 0$ and $s(t) = 20t - 0.4t^2$. When the car has stopped, its velocity will be 0. Hence the stopping time is the solution $t$ of the equation

$$0 = s'(t) = 20 - 0.8t,$$

that is, $t = 25$ s. The distance travelled during deceleration is $s(25) = 250$ m. ∎

# Exercises 2.11

In Exercises 1–4, a point moves along the $x$-axis so that its position $x$ at time $t$ is specified by the given function. In each case determine the following:

(a) the time intervals on which the point is moving to the right, and,   (b) to the left;

(c) the time intervals on which the point is accelerating to the right, and,   (d) to the left;

(e) the time intervals when the particle is speeding up, and, (f) slowing down;

(g) the acceleration at times when the velocity is zero;

(h) the average velocity over the time interval [0, 4].

1. $x = t^2 - 4t + 3$
2. $x = 4 + 5t - t^2$
3. $x = t^3 - 4t + 1$
4. $x = \dfrac{t}{t^2 + 1}$

5. A ball is thrown upward from ground level with an initial speed of 9.8 m/s so that its height in metres after $t$ seconds is given by $y = 9.8t - 4.9t^2$. What is the acceleration of the ball at any time $t$? How high does the ball go? How fast is it moving when it strikes the ground?

6. A ball is thrown downward from the top of a 100-metre-high tower with an initial speed of 2 m/s. Its height in metres

above the ground $t$ seconds later is $y = 100 - 2t - 4.9t^2$. How long does it take to reach the ground? What is its average velocity during the fall? At what instant is its velocity equal to its average velocity?

* 7. **(Take-off distance)** The distance an aircraft travels along a runway before takeoff is given by $D = t^2$, where $D$ is measured in metres from the starting point and $t$ is measured in seconds from the time the brake is released. If the aircraft will become airborne when its speed reaches 200 km/h, how long will it take to become airborne, and what distance will it travel in that time?

8. **(Projectiles on Mars)** A projectile fired upward from the surface of the earth falls back to the ground after 10 seconds. How long would it take to fall back to the surface if it is fired upward on Mars with the same initial velocity? $g_{\text{Mars}} = 3.72$ m/s$^2$.

9. A ball is thrown upward with initial velocity $v_0$ m/s and reaches a maximum height of $h$ metres. How high would it have gone if its initial velocity was $2v_0$? How fast must it be thrown upward to achieve a maximum height of $2h$ metres?

10. How fast would the ball in the previous exercise have to be thrown upward on Mars in order to achieve a maximum

height of $3h$ metres?

11. A rock falls from the top of a cliff and hits the ground at the base of the cliff at a speed of 160 ft/s. How high is the cliff?

12. A rock is thrown down from the top of a cliff with the initial speed of 32 ft/s and hits the ground at the base of the cliff at a speed of 160 ft/s. How high is the cliff?

13. **(Distance travelled while braking)** With full brakes applied, a freight train can decelerate at a constant rate of $1/6$ m/s$^2$. How far will the train travel while braking to a full stop from an initial speed of 60 km/h?

* 14. Show that if the position $x$ of a moving point is given by a quadratic function of $t$, $x = At^2 + Bt + C$, then the average velocity over any time interval $[t_1, t_2]$ is equal to the instantaneous velocity at the midpoint of that time interval.

* 15. **(Piecewise motion)** The position of an object moving along the $s$-axis is given at time $t$ by

$$s = \begin{cases} t^2 & \text{if } 0 \le t \le 2 \\ 4t - 4 & \text{if } 2 < t < 8 \\ -68 + 20t - t^2 & \text{if } 8 \le t \le 10. \end{cases}$$

Determine the velocity and acceleration at any time $t$. Is the velocity continuous? Is the acceleration continuous? What is the maximum velocity and when is it attained?

**(Rocket flight with limited fuel)** Figure 2.40 shows the velocity $v$ in ft/s of a small rocket that was fired from the top of a tower at time $t = 0$ ($t$ in seconds), accelerated with constant upward

acceleration until its fuel was used up, then fell back to the ground at the foot of the tower. The whole flight lasted 14 seconds. Exercises 16–19 refer to this rocket.

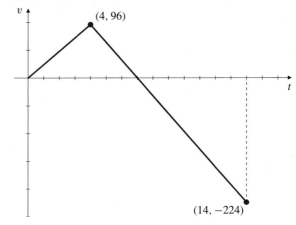

**Figure 2.40**

16. What was the acceleration of the rocket while its fuel lasted?

17. How long was the rocket rising?

* 18. What is the maximum height above ground that the rocket reached?

* 19. How high was the tower from which the rocket was fired?

20. Redo Example 6 using instead a non-constant deceleration, $s''(t) = -t$ m/s$^2$.

# Chapter Review

## Key Ideas

- **What do the following statements and phrases mean?**
  ◇ Line $L$ is tangent to curve $C$ at point $P$.
  ◇ the Newton quotient of $f(x)$ at $x = a$
  ◇ the derivative $f'(x)$ of the function $f(x)$
  ◇ $f$ is differentiable at $x = a$.
  ◇ the slope of the graph $y = f(x)$ at $x = a$
  ◇ $f$ is increasing (or decreasing) on interval $I$.
  ◇ $f$ is nondecreasing (or nonincreasing) on interval $I$.
  ◇ the average rate of change of $f(x)$ on $[a, b]$
  ◇ the rate of change of $f(x)$ at $x = a$
  ◇ $c$ is a critical point of $f(x)$.
  ◇ the second derivative of $f(x)$ at $x = a$
  ◇ an antiderivative of $f$ on interval $I$
  ◇ the indefinite integral of $f$ on interval $I$
  ◇ differential equation       ◇ initial-value problem
  ◇ velocity       ◇ speed       ◇ acceleration

- **State the following differentiation rules:**
  ◇ the rule for differentiating a sum of functions
  ◇ the rule for differentiating a constant multiple of a function
  ◇ the Product Rule       ◇ the Reciprocal Rule
  ◇ the Quotient Rule       ◇ the Chain Rule

- **State the Mean-Value Theorem.**

- **State the Generalized Mean-Value Theorem.**

- **State the derivatives of the following functions:**
  ◇ $x$       ◇ $x^2$       ◇ $1/x$       ◇ $\sqrt{x}$
  ◇ $x^n$       ◇ $|x|$       ◇ $\sin x$       ◇ $\cos x$
  ◇ $\tan x$       ◇ $\cot x$       ◇ $\sec x$       ◇ $\csc x$

- **What is a proof by mathematical induction?**

## Review Exercises

Use the definition of derivative to calculate the derivatives in Exercises 1–4.

1. $\dfrac{dy}{dx}$ if $y = (3x + 1)^2$       2. $\dfrac{d}{dx}\sqrt{1 - x^2}$

**3.** $f'(2)$ if $f(x) = \dfrac{4}{x^2}$

**4.** $g'(9)$ if $g(t) = \dfrac{t-5}{1+\sqrt{t}}$

**5.** Find the tangent to $y = \cos(\pi x)$ at $x = 1/6$.

**6.** Find the normal to $y = \tan(x/4)$ at $x = \pi$.

Calculate the derivatives of the functions in Exercises 7–12.

**7.** $\dfrac{1}{x - \sin x}$

**8.** $\dfrac{1 + x + x^2 + x^3}{x^4}$

**9.** $(4 - x^{2/5})^{-5/2}$

**10.** $\sqrt{2 + \cos^2 x}$

**11.** $\tan\theta - \theta\sec^2\theta$

**12.** $\dfrac{\sqrt{1+t^2}-1}{\sqrt{1+t^2}+1}$

Evaluate the limits in Exercises 13–16 by interpreting each as a derivative.

**13.** $\displaystyle\lim_{h\to 0}\dfrac{(x+h)^{20}-x^{20}}{h}$

**14.** $\displaystyle\lim_{x\to 2}\dfrac{\sqrt{4x+1}-3}{x-2}$

**15.** $\displaystyle\lim_{x\to\pi/6}\dfrac{\cos(2x)-(1/2)}{x-\pi/6}$

**16.** $\displaystyle\lim_{x\to -a}\dfrac{(1/x^2)-(1/a^2)}{x+a}$

In Exercises 17–24, express the derivatives of the given functions in terms of the derivatives $f'$ and $g'$ of the differentiable functions $f$ and $g$.

**17.** $f(3-x^2)$

**18.** $[f(\sqrt{x})]^2$

**19.** $f(2x)\sqrt{g(x/2)}$

**20.** $\dfrac{f(x)-g(x)}{f(x)+g(x)}$

**21.** $f(x+(g(x))^2)$

**22.** $f\left(\dfrac{g(x^2)}{x}\right)$

**23.** $f(\sin x)\,g(\cos x)$

**24.** $\sqrt{\dfrac{\cos f(x)}{\sin g(x)}}$

**25.** Find the tangent to the curve $x^3 y + 2xy^3 = 12$ at the point $(2, 1)$.

**26.** Find the slope of the curve $3\sqrt{2}x\sin(\pi y) + 8y\cos(\pi x) = 2$ at the point $\left(\frac{1}{3}, \frac{1}{4}\right)$.

Find the indefinite integrals in Exercises 27–30.

**27.** $\displaystyle\int \dfrac{1+x^4}{x^2}\,dx$

**28.** $\displaystyle\int \dfrac{1+x}{\sqrt{x}}\,dx$

**29.** $\displaystyle\int \dfrac{2+3\sin x}{\cos^2 x}\,dx$

**30.** $\displaystyle\int (2x+1)^4\,dx$

**31.** Find $f(x)$ given that $f'(x) = 12x^2 + 12x^3$, and $f(1) = 0$.

**32.** Find $g(x)$ if $g'(x) = \sin(x/3) + \cos(x/6)$ and the graph of $g$ passes through the point $(\pi, 2)$.

**33.** Differentiate $x\sin x + \cos x$ and $x\cos x - \sin x$, and use the results to find the indefinite integrals

$$I_1 = \int x\cos x\,dx \quad\text{and}\quad I_2 = \int x\sin x\,dx.$$

**34.** Suppose that $f'(x) = f(x)$ for every $x$, and let $g(x) = x f(x)$. Calculate the first several derivatives of $g$ and guess a formula for the $n$th order derivative $g^{(n)}(x)$. Verify your guess by induction.

**35.** Find an equation of the straight line that passes through the origin and is tangent to the curve $y = x^3 + 2$.

**36.** Find an equation of the straight lines that pass through the point $(0, 1)$ and are tangent to the curve $y = \sqrt{2 + x^2}$.

**37.** Show that $\dfrac{d}{dx}\left(\sin^n x\,\sin(nx)\right) = n\,\sin^{n-1} x\,\sin((n+1)x)$. At what points $x$ in $[0, \pi]$ does the graph of $y = \sin^n x\,\sin(nx)$ have a horizontal tangent. Assume that $n \ge 2$.

**38.** Find differentiation formulas for $y = \sin^n x\,\cos(nx)$, $y = \cos^n x\,\sin(nx)$, and $y = \cos^n x\,\cos(nx)$ analogous to the one given for $y = \sin^n x\,\sin(nx)$ in the previous exercise.

**39.** Let $Q$ be the point $(0, 1)$. Find all points $P$ on the curve $y = x^2$ such that the line $PQ$ is normal to $y = x^2$ at $P$. What is the shortest distance from $Q$ to the curve $y = x^2$?

**40.** **(Average and marginal profit)** Figure 2.41 shows the the graph of the profit $\$P(x)$ realized by a grain exporter from its sale of $x$ tonnes of wheat. Thus, the average profit per tonne is $\$P(x)/x$. Show that the maximum average profit occurs when the average profit equals the marginal profit. What is the geometric significance of this fact in the figure?

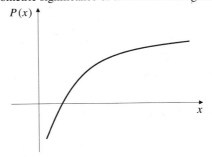

**Figure 2.41**

**41.** **(Gravitational attraction)** The gravitational attraction of the earth on a mass $m$ at distance $r$ from the centre of the earth is a continuous function $F(r)$ given for $r \ge 0$ by

$$F(r) = \begin{cases} \dfrac{mgR^2}{r^2} & \text{if } r \ge R \\ mkr & \text{if } 0 \le r < R \end{cases}$$

where $R$ is the radius of the earth, and $g$ is the acceleration due to gravity at the surface of the earth.

(a) Find the constant $k$ in terms of $g$ and $R$.

(b) $F$ decreases as $m$ moves away from the surface of the earth, either upward or downward. Show that $F$ decreases as $r$ increases from $R$ at twice the rate at which $F$ decreases as $r$ decreases from $R$.

**42.** **(Compressibility of a gas)** The isothermal compressibility of a gas is the relative rate of change of the volume $V$ with respect to the pressure $P$, at a constant temperature $T$, that is,

$$\frac{1}{V}\frac{dV}{dP}.$$

For a sample of an ideal gas, the temperature, pressure, and volume satisfy the equation $PV = kT$, where $k$ is a constant related to the number of molecules of gas present in the sample. Show that the isothermal compressibility of such a gas is the negative reciprocal of the pressure:

$$\frac{1}{V}\frac{dV}{dP} = -\frac{1}{P}.$$

43. A ball is thrown upward with an initial speed of 10 m/s from the top of a building. A second ball is thrown upward with an initial speed of 20 m/s from the ground. Both balls achieve the same maximum height above the ground. How tall is the building?

44. A ball is dropped from the top of a 60 m high tower at the same instant that a second ball is thrown upward from the ground at the base of the tower. The balls collide at a height of 30 m above the ground. With what initial velocity was the second ball thrown? How fast is each ball moving when they collide?

45. **(Braking distance)** A car's brakes can decelerate the car at 20 ft/s$^2$. How fast can the car travel if it must be able to stop in a distance of 160 ft?

46. **(Measuring variations in $g$)** The period $P$ of a pendulum of length $L$ is given by

$$P = 2\pi\sqrt{L/g},$$

where $g$ is the acceleration of gravity.

(a) Assuming that $L$ remains fixed, show that a 1% increase in $g$ results in approximately a 1/2% decrease in the period $P$. (Variations in the period of a pendulum can be used to detect small variations in $g$ from place to place on the earth's surface.)

(b) For fixed $g$, what percentage change in $L$ will produce a 1% increase in $P$?

## Challenging Problems

1. René Descartes, the inventor of analytic geometry, calculated the tangent to a parabola, (or a circle or other quadratic curve) at a given point $(x_0, y_0)$ on the curve by looking for a straight line through $(x_0, y_0)$ having only one intersection with the given curve. Illustrate his method by writing the equation of a line through $(a, a^2)$, having arbitrary slope $m$, and then finding the value of $m$ for which the line has only one intersection with the parabola $y = x^2$. Why does the method not work for more general curves?

2. Given that $f'(x) = 1/x$ and $f(2) = 9$, find:

(a) $\displaystyle\lim_{x\to 2}\frac{f(x^2+5) - f(9)}{x-2}$    (b) $\displaystyle\lim_{x\to 2}\frac{\sqrt{f(x)} - 3}{x - 2}$

3. Suppose that $f'(4) = 3$, $g'(4) = 7$, $g(4) = 4$, and $g(x) \neq 4$ for $x \neq 4$. Find:

(a) $\displaystyle\lim_{x\to 4}\big(f(x) - f(4)\big)$    (b) $\displaystyle\lim_{x\to 4}\frac{f(x) - f(4)}{x^2 - 16}$

(c) $\displaystyle\lim_{x\to 4}\frac{f(x) - f(4)}{\sqrt{x} - 2}$    (d) $\displaystyle\lim_{x\to 4}\frac{f(x) - f(4)}{(1/x) - (1/4)}$

(e) $\displaystyle\lim_{x\to 4}\frac{f(x) - f(4)}{g(x) - 4}$    (f) $\displaystyle\lim_{x\to 4}\frac{f(g(x)) - f(4)}{x - 4}$

4. Let $f(x) = \begin{cases} x & \text{if } x = 1,\ 1/2,\ 1/3,\ 1/4,\ \dots \\ x^2 & \text{otherwise.} \end{cases}$

(a) Find all points at which $f$ is continuous. In particular, is it continuous at $x = 0$?

(b) Is the following statement true or false? Justify your answer. For any two real numbers $a$ and $b$, there is some $x$ between $a$ and $b$ such that $f(x) = (f(a) + f(b))/2$.

(c) Find all points at which $f$ is differentiable. In particular, is it differentiable at $x = 0$?

5. Suppose $f(0) = 0$ and $|f(x)| > \sqrt{|x|}$ for all $x$. Show that $f'(0)$ does not exist.

6. Suppose that $f$ is a function satisfying the following conditions: $f'(0) = k$, $f(0) \neq 0$, and $f(x + y) = f(x)f(y)$ for all $x$ and $y$. Show that $f(0) = 1$ and that $f'(x) = k\,f(x)$ for every $x$. (We will study functions with these properties in Chapter 3.)

7. Suppose the function $g$ satisfies the conditions: $g'(0) = k$, and $g(x + y) = g(x) + g(y)$ for all $x$ and $y$. Show that:

(a) $g(0) = 0$,    (b) $g'(x) = k$ for all $x$, and

(c) $g(x) = kx$ for all $x$. *Hint:* let $h(x) = g(x) - g'(0)x$.

8. (a) If $f$ is differentiable at $x$, show that

(i) $\displaystyle\lim_{h\to 0}\frac{f(x) - f(x - h)}{h} = f'(x)$

(ii) $\displaystyle\lim_{h\to 0}\frac{f(x + h) - f(x - h)}{2h} = f'(x)$

(b) Show that the existence of the limit in (i) guarantees that $f$ is differentiable at $x$.

(c) Show that the existence of the limit in (ii) does *not* guarantee that $f$ is differentiable at $x$. *Hint:* consider the function $f(x) = |x|$ at $x = 0$.

9. Show that there is a line through $(a, 0)$ that is tangent to the curve $y = x^3$ at $x = 3a/2$. If $a \neq 0$, is there any other line through $(a, 0)$ that is tangent to the curve? If $(x_0, y_0)$ is an arbitrary point, what is the maximum number of lines through $(x_0, y_0)$ that can be tangent to $y = x^3$? the minimum number?

10. Make a sketch showing that there are two straight lines, each of which is tangent to both of the parabolas $y = x^2 + 4x + 1$ and $y = -x^2 + 4x - 1$. Find equations of the two lines.

11. Show that if $b > 1/2$, there are three straight lines through $(0, b)$, each of which is normal to the curve $y = x^2$. How many such lines are there if $b = 1/2$? if $b < 1/2$?

12. **(Distance from a point to a curve)** Find the point on the curve $y = x^2$ that is closest to the point $(3, 0)$. *Hint:* the line from $(3, 0)$ to the closest point $Q$ on the parabola is normal to the parabola at $Q$.

* 13. **(Envelope of a family of lines)** Show that for each value of the parameter $m$, the line $y = mx - (m^2/4)$ is tangent to the parabola $y = x^2$. (The parabola is called the *envelope* of the family of lines $y = mx - (m^2/4)$.) Find $f(m)$ such that the family of lines $y = mx + f(m)$ has envelope the parabola $y = Ax^2 + Bx + C$.

* 14. **(Common tangents)** Consider the two parabolas with equations $y = x^2$ and $y = Ax^2 + Bx + C$. We assume that $A \neq 0$, and if $A = 1$ then either $B \neq 0$ or $C \neq 0$, so that the two equations do represent different parabolas. Show that:

    (a) the two parabolas are tangent to each other if $B^2 = 4C(A - 1)$.

    (b) the parabolas have two common tangent lines if and only if $A \neq 1$ and $A(B^2 - 4C(A - 1)) > 0$.

    (c) the parabolas have exactly one common tangent line if either $A = 1$ and $B \neq 0$, or $A \neq 1$ and $B^2 = 4C(A - 1)$.

    (d) the parabolas have no common tangent lines if either $A = 1$ and $B = 0$, or $A \neq 1$ and $A(B^2 - 4C(A - 1)) < 0$.

Make sketches illustrating each of these possibilities.

15. Let $C$ be the graph of $y = x^3$.

    (a) Show that if $a \neq 0$ then the tangent to $C$ at $x = a$ also intersects $C$ at a second point $x = b$.

    (b) Show that the slope of $C$ at $x = b$ is four times its slope at $x = a$.

    (c) Can any line be tangent to $C$ at more than one point?

    (d) Can any line be tangent to the graph of $y = Ax^3 + Bx^2 + Cx + D$ at more than one point?

* 16. Let $C$ be the graph of $y = x^4 - 2x^2$.

    (a) Find all horizontal lines that are tangent to $C$.

    (b) One of the lines found in (a) is tangent to $C$ at two different points. Show that there are no other lines that have this property.

    (c) Find an equation of a straight line that is tangent to the graph of $y = x^4 - 2x^2 + x$ at two different points. Can there exist more than one such line? Why?

17. **(Double tangents)** A line tangent to the quartic (fourth-degree polynomial) curve $C$ with equation $y = ax^4 + bx^3 + cx^2 + dx + e$ at $x = p$ may intersect $C$ at zero, one, or two other points. If it meets $C$ at only one other point $x = q$ it must be tangent to $C$ at that point also, and it is thus a "double tangent."

    (a) Find the condition that must be satisfied by the coefficients of the quartic to ensure that there does exist such a

double tangent, and show that there cannot be more than one such double tangent. Illustrate this by applying your results to $y = x^4 - 2x^2 + x - 1$.

    (b) If the line $PQ$ is tangent to $C$ at two distinct points $x = p$ and $x = q$, show that $PQ$ is parallel to the line tangent to $C$ at $x = (p + q)/2$.

    (c) If the line $PQ$ is tangent to $C$ at two distinct points $x = p$ and $x = q$, show that $C$ has two distinct inflection points $R$ and $S$, and that $RS$ is parallel to $PQ$.

18. **(Rocket with a parachute)** A rocket is fired from the top of a tower at time $t = 0$. It experiences constant upward acceleration until its fuel is used up. Thereafter its acceleration is the constant downward acceleration of gravity until, during its fall, it deploys a parachute that gives it a constant upward acceleration again to slow it down. The rocket hits the ground near the base of the tower. The upward velocity $v$ (in metres per second) is graphed against time in Figure 2.42. From information in the figure determine:

    (a) How long did the fuel last?

    (b) When was the rocket's height maximum?

    (c) When was the parachute deployed?

    (d) What was the rocket's upward acceleration while its motor was firing?

    (e) What was the maximum height achieved by the rocket?

    (f) How high was the tower from which the rocket was fired?

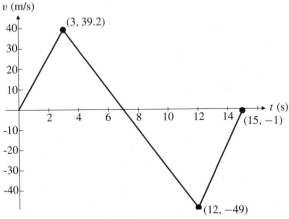

**Figure 2.42**

19. Verify the following formulas for every positive integer $n$:

    (a) $\dfrac{d^n}{dx^n} \cos(ax) = a^n \cos\left(ax + \dfrac{n\pi}{2}\right)$

    (b) $\dfrac{d^n}{dx^n} \sin(ax) = a^n \sin\left(ax + \dfrac{n\pi}{2}\right)$

    (c) $\dfrac{d^n}{dx^n}\left(\cos^4 x + \sin^4 x\right) = 4^{n-1} \cos\left(4x + \dfrac{n\pi}{2}\right)$

# CHAPTER 3

# Transcendental Functions

***Introduction*** With the exception of the trigonometric functions, all the functions we have encountered so far have been of three main types, *polynomials, rational functions* (quotients of polynomials), and *algebraic functions* (fractional powers of rational functions). On an interval in its domain, each of these functions can be constructed from real numbers and a single real variable $x$ by using finitely many arithmetic operations (addition, subtraction, multiplication, and division), and by taking finitely many roots (fractional powers). Functions that cannot be so constructed are called **transcendental functions.** The only examples of these that we have seen so far are the trigonometric functions.

Much of the importance of calculus, and many of its most useful applications, result from its ability to illuminate the behaviour of transcendental functions that arise naturally when we try to model concrete problems in mathematical terms. This chapter is devoted to developing other transcendental functions, including exponential and logarithmic functions, and the inverse trigonometric functions.

Some of these functions "undo" what other ones "do," and vice versa. When a pair of functions behaves this way, we call each the inverse of the other. We begin the chapter by studying inverse functions in general.

## 3.1 Inverse Functions

Consider the function

$$f(x) = x^3$$

whose graph is shown in Figure 3.1. Like any function, $f(x)$ has only one value for each $x$ in its domain (the whole real line $\mathbb{R}$). In geometric terms, any *vertical* line meets the graph of $f$ at only one point. For this function $f$, any *horizontal* line also meets the graph at only one point. This means that different values of $x$ always give different values to $f(x)$. Such a function is said to be *one-to-one*.

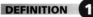 **DEFINITION 1**

> A function $f$ is **one-to-one** if $f(x_1) \neq f(x_2)$ whenever $x_1$ and $x_2$ belong to the domain of $f$ and $x_1 \neq x_2$, or, equivalently, if
>
> $$f(x_1) = f(x_2) \quad \Longrightarrow \quad x_1 = x_2.$$

A function defined on a single interval is one-to-one there if it is either increasing or decreasing.

Because $f(x) = x^3$ is one-to-one, the equation

$$y = x^3$$

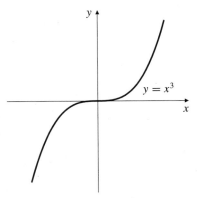

**Figure 3.1**　The graph of $f(x) = x^3$

Do not confuse the $-1$ in $f^{-1}$ with an exponent. The inverse $f^{-1}$ is *not* the reciprocal $1/f$. If we want to denote the reciprocal $1/f(x)$ with an exponent we can write it as $\left(f(x)\right)^{-1}$.

**Figure 3.2**

(a) $f$ is one-to-one and has an inverse. $y = f(x)$ means the same thing as $x = f^{-1}(y)$

(b) $g$ is not one-to-one

has a unique solution $x$ for every given value of $y$ in the range of $f$. Specifically,

$$x = y^{1/3}.$$

This equation defines $x$ as a function of $y$. We call this new function the *inverse of* $f$ and denote it $f^{-1}$. Thus

$$f^{-1}(y) = y^{1/3}.$$

Whenever a function $f$ is one-to-one, for any number $y$ in its range there will always exist a single number $x$ in its domain such that $y = f(x)$. Since $x$ is determined uniquely by $y$, it is a function of $y$. We write $x = f^{-1}(y)$ and call $f^{-1}$ the inverse of $f$. The function $f$ whose graph is shown in Figure 3.2(a) is one-to-one and has an inverse. The function $g$ whose graph is shown in Figure 3.2(b) is not one-to-one and does not have an inverse.

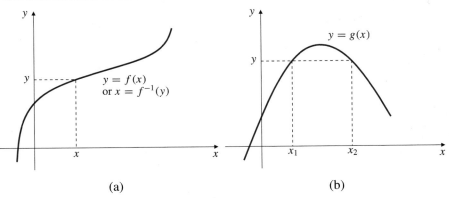

(a)                                  (b)

We usually like to write functions with the domain variable called $x$ rather than $y$, so we reverse the roles of $x$ and $y$ and reformulate the above definition as follows.

**DEFINITION 2**

If $f$ is one to one, then it has an **inverse function** $f^{-1}$. The value of $f^{-1}(x)$ is the unique number $y$ in the domain of $f$ for which $f(y) = x$. Thus

$$y = f^{-1}(x) \quad \Longleftrightarrow \quad x = f(y).$$

As observed above, $y = x^3$ is equivalent to $x = y^{1/3}$, or, reversing the roles of $x$ and $y$,

$$y = x^{1/3} \quad \Longleftrightarrow \quad x = y^3.$$

**Example 1**　Show that $f(x) = 2x - 1$ is one-to-one, and find its inverse $f^{-1}(x)$.

**Solution**　Since $f'(x) = 2 > 0$ on $\mathbb{R}$, $f$ is increasing and therefore one-to-one there. Let $y = f^{-1}(x)$. Then

$$x = f(y) = 2y - 1.$$

Solving this equation for $y$ gives $y = \dfrac{x+1}{2}$. Thus $f^{-1}(x) = \dfrac{x+1}{2}$.

There are several things you should remember about the relationship between a function $f$ and its inverse $f^{-1}$. The most important one is that the two equations

$$y = f^{-1}(x) \qquad \text{and} \qquad x = f(y)$$

*say the same thing.* They are equivalent just as, for example $y = x+1$ and $x = y-1$ are equivalent. Either of the equations can be replaced by the other. This implies that the domain of $f^{-1}$ is the range of $f$, and vice versa.

The inverse of a one-to-one function is itself one-to-one and so also has an inverse. Not surprisingly, the inverse of $f^{-1}$ is $f$:

$$y = (f^{-1})^{-1}(x) \quad \Longleftrightarrow \quad x = f^{-1}(y) \quad \Longleftrightarrow \quad y = f(x).$$

We can substitute either of the equations $y = f^{-1}(x)$ or $x = f(y)$ into the other and obtain the **cancellation identities**:

$$f\big(f^{-1}(x)\big) = x, \qquad f^{-1}\big(f(y)\big) = y.$$

The first of these identities holds for all $x$ in the domain of $f^{-1}$, and the second for all $y$ in the domain of $f$. If $S$ is any set of real numbers and $I_S$ denotes the **identity function** on $S$, defined by

$$I_S(x) = x \quad \text{for all } x \text{ in } S,$$

then the cancellation identities say that if $D(f)$ is the domain of $f$, then

$$f \circ f^{-1} = I_{D(f^{-1})}, \qquad \text{and} \qquad f^{-1} \circ f = I_{D(f)},$$

where $f \circ g(x)$ denotes the composition $f\big(g(x)\big)$.

If the coordinates of a point $P = (a, b)$ are exchanged to give those of a new point $Q = (b, a)$, then each point is the reflection of the other in the line $x = y$. (To see this, note that the line $PQ$ has slope $-1$, and so is perpendicular to $y = x$. Also, the midpoint of $PQ$ is $\left(\frac{a+b}{2}, \frac{b+a}{2}\right)$ which lies on $y = x$.) It follows that the graphs of the equations $x = f(y)$ and $y = f(x)$ are reflections of each other in the line $x = y$. Since the equation $x = f(y)$ is equivalent to $y = f^{-1}(x)$, the graphs of the functions $f^{-1}$ and $f$ are reflections of each other in $y = x$. See Figure 3.3.

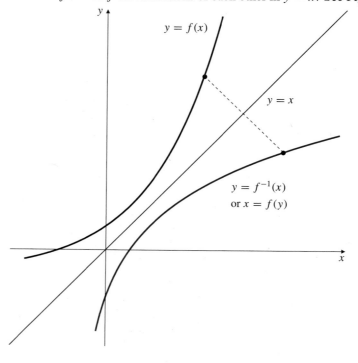

**Figure 3.3**    The graph of $y = f^{-1}(x)$ is the reflection of the graph of $y = f(x)$ in the line $y = x$

Here is a summary of the properties of inverse functions discussed above:

**Properties of inverse functions**

1. $y = f^{-1}(x) \iff x = f(y)$.
2. The domain of $f^{-1}$ is the range of $f$.
3. The range of $f^{-1}$ is the domain of $f$.
4. $f^{-1}(f(x)) = x$ for all $x$ in the domain of $f$.
5. $f(f^{-1}(x)) = x$ for all $x$ in the domain of $f^{-1}$.
6. $(f^{-1})^{-1}(x) = f(x)$ for all $x$ in the domain of $f$.
7. The graph of $f^{-1}$ is the reflection of the graph of $f$ in the line $x = y$.

**Example 2**    Show that $g(x) = \sqrt{2x+1}$ is invertible and find its inverse.

**Solution**    If $g(x_1) = g(x_2)$, then $\sqrt{2x_1 + 1} = \sqrt{2x_2 + 1}$. Squaring both sides we get $2x_1 + 1 = 2x_2 + 1$, which implies that $x_1 = x_2$. Thus $g$ is one-to-one and invertible. Let $y = g^{-1}(x)$ then

$$x = g(y) = \sqrt{2y+1}.$$

It follows that $x \geq 0$ and $x^2 = 2y + 1$. Therefore $y = \dfrac{x^2 - 1}{2}$, and

$$g^{-1}(x) = \frac{x^2 - 1}{2} \qquad \text{for } x \geq 0.$$

(The restriction $x \geq 0$ applies since the range of $g$ is $[0, \infty)$.) See Figure 3.4(a) for the graphs of $g$ and $g^{-1}$. ∎

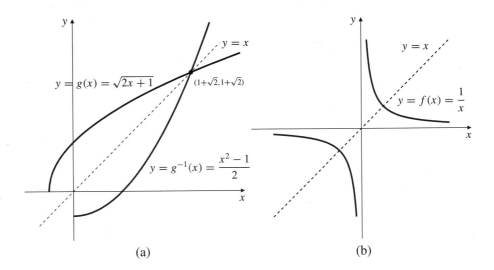

**Figure 3.4**

(a) The graphs of $g(x) = \sqrt{2x+1}$ and its inverse

(b) The graph of the self-inverse function $f(x) = 1/x$

**DEFINITION 3**

A function $f$ is **self-inverse** if $f^{-1} = f$, that is, if $f(f(x)) = x$ for every $x$ in the domain of $f$.

**Example 3**    The function $f(x) = 1/x$ is self-inverse. If $y = f^{-1}(x)$, then $x = f(y) = 1/y$. Therefore $y = 1/x$, so $f^{-1}(x) = \dfrac{1}{x} = f(x)$. See Figure 3.4(b). The graph of any self-inverse function must be its own reflection in the line $x = y$, and must therefore be symmetric about that line. ∎

## Inverting Non-One-to-One Functions

Many important functions such as the trigonometric functions are not one-to-one on their whole domains. It is still possible to define an inverse for such a function, but we have to restrict the domain of the function artificially so that the restricted function is one-to-one.

As an example, consider the function $f(x) = x^2$. Unrestricted, its domain is the whole real line and it is not one-to-one since $f(-a) = f(a)$ for any $a$. Let us define a new function $F(x)$ equal to $f(x)$, but having a smaller domain, so that it is one-to-one. We can use the interval $[0, \infty)$ as the domain of $F$:

$$F(x) = x^2 \qquad \text{for} \quad 0 \le x < \infty.$$

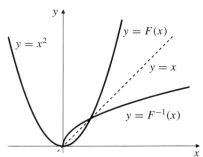

**Figure 3.5**    The restriction $F$ of $x^2$ to $[0, \infty)$ and its inverse $F^{-1}$

The graph of $F$ is shown in Figure 3.5; it is the right half of the parabola $y = x^2$, the graph of $f$. Evidently $F$ is one-to-one, and so has an inverse $F^{-1}$ which we calculate as follows:

Let $y = F^{-1}(x)$ then $x = F(y) = y^2$ and $y \ge 0$. Thus $y = \sqrt{x}$. Hence $F^{-1}(x) = \sqrt{x}$.

This method of restricting the domain of a non one-to-one function to make it invertible will be used extensively when we invert the trigonometric functions in Section 3.5.

## Derivatives of Inverse Functions

Suppose that the function $f$ is differentiable on an interval $(a, b)$ and that either $f'(x) > 0$ for $a < x < b$ (so that $f$ is increasing on $(a, b)$), or $f'(x) < 0$ for $a < x < b$ (so that $f$ is decreasing on $(a, b)$). In either case $f$ is one-to-one on $(a, b)$, and has an inverse, $f^{-1}$, defined by

$$y = f^{-1}(x) \quad \Longleftrightarrow \quad x = f(y), \qquad (a < y < b).$$

Since we are assuming that the graph $y = f(x)$ has a *non-horizontal* tangent line at any $x$ in $(a, b)$, its reflection, the graph $y = f^{-1}(x)$, has a *non-vertical* tangent line at any $x$ in the interval between $f(a)$ and $f(b)$. Therefore $f^{-1}$ is differentiable at any such $x$. (See Figure 3.6.)

Let $y = f^{-1}(x)$. We want to find $dy/dx$. Solve the equation $y = f^{-1}(x)$ for $x = f(y)$ and differentiate implicitly with respect to $x$ to obtain

$$1 = f'(y)\,\frac{dy}{dx}, \qquad \text{so} \qquad \frac{dy}{dx} = \frac{1}{f'(y)} = \frac{1}{f'\left(f^{-1}(x)\right)}.$$

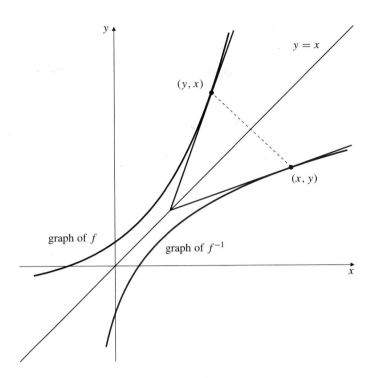

**Figure 3.6**   Tangents to the graphs of $f$ and $f^{-1}$

Therefore the slope of the graph of $f^{-1}$ at $(x, y)$ is the reciprocal of the slope of the graph of $f$ at $(y, x)$ (Figure 3.6), and

$$\frac{d}{dx} f^{-1}(x) = \frac{1}{f'\left(f^{-1}(x)\right)}.$$

In terms of Leibniz notation we have $\left.\dfrac{dy}{dx}\right|_x = \dfrac{1}{\left.\dfrac{dx}{dy}\right|_{y=f^{-1}(x)}}.$

**Example 4**   Show that $f(x) = x^3 + x$ is one-to-one on the whole real line, and, noting that $f(2) = 10$, find $\left(f^{-1}\right)'(10)$.

**Solution**   Since $f'(x) = 3x^2 + 1 > 0$ for all real numbers $x$, $f$ is increasing, and therefore one-to-one and invertible. If $y = f^{-1}(x)$, then

$$x = f(y) = y^3 + y \quad \Longrightarrow \quad 1 = (3y^2 + 1)y'$$
$$\Longrightarrow \quad y' = \frac{1}{3y^2 + 1}.$$

Now $x = f(2) = 10$ implies $y = f^{-1}(10) = 2$. Thus

$$\left(f^{-1}\right)'(10) = \left.\frac{1}{3y^2 + 1}\right|_{y=2} = \frac{1}{13}.$$

# Exercises 3.1

Show that the functions $f$ in Exercises 1–12 are one-to-one and calculate the inverse functions $f^{-1}$. Specify the domains and ranges of $f$ and $f^{-1}$.

**1.** $f(x) = x - 1$

**2.** $f(x) = 2x - 1$

**3.** $f(x) = \sqrt{x - 1}$

**4.** $f(x) = -\sqrt{x - 1}$

**5.** $f(x) = x^3$

**6.** $f(x) = 1 + \sqrt[3]{x}$

**7.** $f(x) = x^2, \quad x \le 0$

**8.** $f(x) = (1 - 2x)^3$

**9.** $f(x) = \dfrac{1}{x + 1}$

**10.** $f(x) = \dfrac{x}{1 + x}$

**11.** $f(x) = \dfrac{1 - 2x}{1 + x}$

**12.** $f(x) = \dfrac{x}{\sqrt{x^2 + 1}}$

In Exercises 13–20, $f$ is a one-to-one function with inverse $f^{-1}$. Calculate the inverses of the given functions in terms of $f^{-1}$.

**13.** $g(x) = f(x) - 2$

**14.** $h(x) = f(2x)$

**15.** $k(x) = -3f(x)$

**16.** $m(x) = f(x - 2)$

**17.** $p(x) = \dfrac{1}{1 + f(x)}$

**18.** $q(x) = \dfrac{f(x) - 3}{2}$

**19.** $r(x) = 1 - 2f(3 - 4x)$

**20.** $s(x) = \dfrac{1 + f(x)}{1 - f(x)}$

**21.** Show that $f(x) = \begin{cases} x^2 + 1 & \text{if } x \ge 0 \\ x + 1 & \text{if } x < 0 \end{cases}$ is one-to-one. What is its inverse?

**22.** Find $f^{-1}(2)$ if $f(x) = x^3 + x$.

**23.** Assume that the function $f(x)$ satisfies $f'(x) = \dfrac{1}{x}$, and that $f$ is one-to-one. If $y = f^{-1}(x)$, show that $dy/dx = y$.

**24.** Show that $f(x) = \dfrac{4x^3}{x^2 + 1}$ has an inverse and find $\left(f^{-1}\right)'(2)$.

**25.** Find $\left(f^{-1}\right)'(x)$ if $f(x) = 1 + 2x^3$.

**∗ 26.** Find $\left(f^{-1}\right)'(-2)$ if $f(x) = x\sqrt{3 + x^2}$.

**▦ 27.** If $f(x) = x^2/(1 + \sqrt{x})$, find $f^{-1}(2)$ correct to five decimal places.

**▦ 28.** If $g(x) = 2x + \sin x$, show that $g$ is invertible, and find $g^{-1}(2)$ and $(g^{-1})'(2)$ correct to five decimal places.

**29.** Show that $f(x) = x \sec x$ is one-to-one on $(-\pi/2, \pi/2)$. What is the domain of $f^{-1}(x)$? Find $(f^{-1})'(0)$.

**30.** If $f$ and $g$ have respective inverses $f^{-1}$ and $g^{-1}$, show that the composite function $f \circ g$ has inverse $(f \circ g)^{-1} = g^{-1} \circ f^{-1}$.

**∗ 31.** For what values of the constants $a$, $b$, and $c$ is the function $f(x) = (x - a)/(bx - c)$ self-inverse?

**32.** Can an even function be self-inverse? An odd function?

# 3.2 Exponential and Logarithmic Functions

This section reviews exponential and logarithmic functions in a form that you are likely to have encountered in your previous mathematical studies. In the following sections we will approach these functions from a different point of view, and will learn how to find their derivatives.

## Exponentials

An **exponential function** is a function of the form $f(x) = a^x$, where the **base** $a$ is a positive constant and the **exponent** $x$ is the variable. Do not confuse such functions with **power** functions like $f(x) = x^a$ where the base is variable and the exponent is constant. The exponential function $a^x$ can be defined for integer and rational exponents $x$ as follows:

**DEFINITION** **4**

> **Exponential Functions**
>
> If $a > 0$, then
>
> $$a^0 = 1$$
> $$a^n = \underbrace{a \cdot a \cdot a \cdots a}_{n \text{ factors}} \quad \text{if } n = 1, 2, 3, \ldots$$
> $$a^{-n} = \frac{1}{a^n} \quad \text{if } n = 1, 2, 3, \ldots$$
> $$a^{m/n} = \sqrt[n]{a^m} \quad \text{if } n = 1, 2, 3, \ldots \quad \text{and } m = \pm 1, \pm 2, \pm 3, \ldots.$$

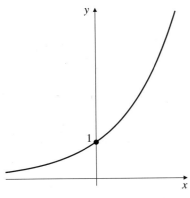

**Figure 3.7**  $y = 2^r$ for rational $r$

In this definition, $\sqrt[n]{a}$ is the number $b > 0$ that satisfies $b^n = a$.

How should we define $a^x$ if $x$ is not rational? For example, what does $2^\pi$ mean? In order to calculate a derivative of $a^x$ we will want the function to be defined for all real numbers $x$, not just rational ones.

In Figure 3.7 we plot points with coordinates $(x, 2^x)$ for many, closely spaced rational values of $x$. They appear to lie on a smooth curve. The definition of $a^x$ can be extended to irrational $x$ in such a way that $a^x$ becomes a differentiable function of $x$ on the whole real line. We will do so in the next section. For the moment, if $x$ is irrational we can regard $a^x$ as being the limit of values $a^r$ for rational numbers $r$ approaching $x$:

$$a^x = \lim_{\substack{r \to x \\ r \text{ rational}}} a^r.$$

**Example 1**  Since the irrational number $\pi = 3.14159265359\ldots$ is the limit of the sequence of rational numbers

$$r_1 = 3, \quad r_2 = 3.1, \quad r_3 = 3.14, \quad r_4 = 3.141, \quad r_5 = 3.1415, \quad \ldots$$

we can calculate $2^\pi$ as the limit of the corresponding sequence

$$2^3 = 8, \quad 2^{3.1} = 8.5741877\ldots, \quad 2^{3.14} = 8.8152409\ldots.$$

This gives $2^\pi = \lim_{n \to \infty} 2^{r_n} = 8.824977827\ldots$. ∎

Exponential functions satisfy several identities called *laws of exponents*:

> **Laws of exponents**
>
> If $a > 0$ and $b > 0$, and $x$ and $y$ are any real numbers, then
>
> (i)  $a^0 = 1$  (ii)  $a^{x+y} = a^x a^y$
>
> (iii)  $a^{-x} = \dfrac{1}{a^x}$  (iv)  $a^{x-y} = \dfrac{a^x}{a^y}$
>
> (v)  $(a^x)^y = a^{xy}$  (vi)  $(ab)^x = a^x b^x$

These identities can be proved for rational exponents using the definitions above. They remain true for irrational exponents, but we can't show that until the next section.

If $a = 1$ then $a^x = 1^x = 1$ for every $x$. If $a > 1$ then $a^x$ is an increasing function of $x$; if $0 < a < 1$, then $a^x$ is decreasing. The graphs of some typical exponential functions are shown in Figure 3.8(a). They all pass through the point $(0,1)$ since $a^0 = 1$ for every $a > 0$. Observe that $a^x > 0$ for all $a > 0$ and all real $x$, and that

$$\text{If } a > 1, \qquad \text{then} \quad \lim_{x \to -\infty} a^x = 0 \quad \text{and} \quad \lim_{x \to \infty} a^x = \infty.$$

$$\text{If } 0 < a < 1, \text{ then} \quad \lim_{x \to -\infty} a^x = \infty \quad \text{and} \quad \lim_{x \to \infty} a^x = 0.$$

The graph of $y = a^x$ has the $x$-axis as a horizontal asymptote if $a \neq 1$. It is asymptotic on the left (as $x \to -\infty$) if $a > 1$ and on the right (as $x \to \infty$) if $0 < a < 1$.

**Figure 3.8**

(a) Graphs of some exponential
    functions

(b) Graphs of some logarithmic
    functions

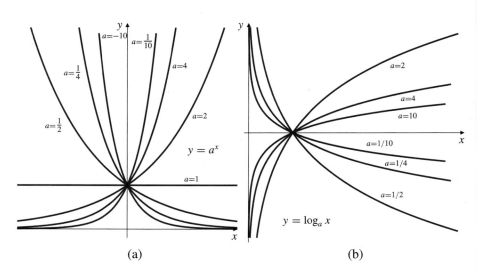

(a)

(b)

## Logarithms

The function $f(x) = a^x$ is a one-to-one function provided that $a > 0$ and $a \neq 1$. Therefore $f$ has an inverse which we call a logarithmic function.

**DEFINITION 5**

> **Logarithms**
>
> If $a > 0$ and $a \neq 1$, the function $\log_a x$, called **the logarithm of $x$ to the base $a$**, is the inverse of the one-to-one function $a^x$:
>
> $$y = \log_a x \quad \Longleftrightarrow \quad x = a^y, \qquad (a > 0, \quad a \neq 1).$$

Since $x = a^y$ takes only positive values, therefore the domain of $\log_a x$ is the interval $(0, \infty)$. Since $a^y$ is defined for all real $y$, therefore $y = \log_a x$ takes on all real values. Since $a^x$ and $\log_a x$ are inverse functions, the following **cancellation identities** hold:

$$\log_a \left( a^x \right) = x \qquad \text{for all real } x$$
$$a^{\log_a x} = x \qquad \text{for all} \quad x > 0.$$

The graphs of some typical logarithmic functions are shown in Figure 3.8(b). They all pass through the point $(1, 0)$. Each graph is the reflection in the line $y = x$ of the corresponding exponential graph in Figure 3.8(a).

Corresponding to the laws of exponents we have several identities involving logarithms:

---

**Laws of logarithms**

If $a > 0$, $b > 0$, $a \neq 1$, and $b \neq 1$, then

(i) $\log_a 1 = 0$            (ii) $\log_a(xy) = \log_a x + \log_a y$

(iii) $\log_a\left(\dfrac{1}{x}\right) = -\log_a x$      (iv) $\log_a\left(\dfrac{x}{y}\right) = \log_a x - \log_a y$

(v) $\log_a\left(x^y\right) = y\,\log_a x$       (vi) $\log_a x = \dfrac{\log_b x}{\log_b a}$

---

These identities can be derived from the laws of exponents.

**Example 2**   If $a > 0$, $x > 0$, and $y > 0$, verify that $\log_a(xy) = \log_a x + \log_a y$, using laws of exponents.

**Solution**   Let $u = \log_a x$ and $v = \log_a y$. By the defining property of inverse functions, $x = a^u$ and $y = a^v$. Thus $xy = a^u a^v = a^{u+v}$. Inverting again, we get $\log_a(xy) = u + v = \log_a x + \log_a y$. ∎

Logarithm law (vi) presented above shows that if you know logarithms to a particular base $b$, you can calculate logarithms to any other base $a$. Scientific calculators usually have built-in programs for calculating logarithms to base 10, and to base $e$, a special number that we will discover in Section 3.3. Logarithms to any base can be calculated using either of these functions. For example, computer scientists sometimes need to use logarithms to base 2. Using a scientific calculator, you can readily calculate

$$\log_2 13 = \frac{\log_{10} 13}{\log_{10} 2} = \frac{1.11394335231\ldots}{0.301029995664\ldots} = 3.70043971814\ldots$$

The laws of logarithms can sometimes be used to simplify complicated expressions.

**Example 3**   Simplify
(a) $\log_2 10 + \log_2 12 - \log_2 15$,    (b) $\log_{a^2} a^3$, and    (c) $3^{\log_9 4}$.

**Solution**

(a) $\log_2 10 + \log_2 12 - \log_2 15 = \log_2 \dfrac{10 \times 12}{15}$      (laws (ii) and (iv))

$\qquad\qquad\qquad\qquad\qquad\qquad = \log_2 8$

$\qquad\qquad\qquad\qquad\qquad\qquad = \log_2 2^3 = 3.$     (cancellation identity)

(b)  $\log_{a^2} a^3 = 3 \log_{a^2} a$                    (law (v))

$\qquad\qquad = \dfrac{3}{2} \log_{a^2} a^2$                    (law (v) again)

$\qquad\qquad = \dfrac{3}{2}.$                    (cancellation identity)

(c)  $3^{\log_9 4} = 3^{(\log_3 4)/(\log_3 9)}$                    (law (vi))

$\qquad\qquad = 4^{1/\log_3 3^2} = 4^{1/2} = 2.$        (cancellation identity)

**Example 4**    Solve the equation $3^{x-1} = 2^x$.

**Solution**    We can take logarithms of both sides of the equation to any base $a$ and get

$$(x - 1) \log_a 3 = x \log_a 2$$
$$(\log_a 3 - \log_a 2)x = \log_a 3$$
$$x = \frac{\log_a 3}{\log_a 3 - \log_a 2} = \frac{\log_a 3}{\log_a (3/2)}.$$

The numerical value of $x$ can be found using the "log" function on a scientific calculator. (This function is $\log_{10}$.) The value is $x = 2.7095\ldots$.

Corresponding to the asymptotic behaviour of the exponential functions, the logarithmic functions also exhibit asymptotic behaviour. Their graphs are all asymptotic to the $y$-axis as $x \to 0$ from the right:

If  $a > 1$,    then  $\displaystyle\lim_{x \to 0+} \log_a x = -\infty$  and  $\displaystyle\lim_{x \to \infty} \log_a x = \infty.$

If  $0 < a < 1$, then  $\displaystyle\lim_{x \to 0+} \log_a x = \infty$  and  $\displaystyle\lim_{x \to \infty} \log_a x = -\infty.$

# Exercises 3.2

Simplify the expressions in Exercises 1–18.

**1.** $\dfrac{3^3}{\sqrt{3^5}}$

**2.** $2^{1/2} 8^{1/2}$

**3.** $\left(x^{-3}\right)^{-2}$

**4.** $\left(\dfrac{1}{2}\right)^x 4^{x/2}$

**5.** $\log_5 125$

**6.** $\log_4 \left(\dfrac{1}{8}\right)$

**7.** $\log_{1/3} 3^{2x}$

**8.** $2^{\log_4 8}$

**9.** $10^{-\log_{10}(1/x)}$

**10.** $x^{1/(\log_a x)}$

**11.** $(\log_a b)(\log_b a)$

**12.** $\log_x \left(x(\log_y y^2)\right)$

**13.** $(\log_4 16)(\log_4 2)$

**14.** $\log_{15} 75 + \log_{15} 3$

**15.** $\log_6 9 + \log_6 4$

**16.** $2 \log_3 12 - 4 \log_3 6$

**17.** $\log_a(x^4 + 3x^2 + 2) + \log_a(x^4 + 5x^2 + 6)$
$\quad - 4 \log_a \sqrt{x^2 + 2}$

**18.** $\log_\pi (1 - \cos x) + \log_\pi (1 + \cos x) - 2 \log_\pi \sin x$

Use the base 10 exponential and logarithm functions $10^x$ and $\log x$ ($= \log_{10} x$) on a scientific calculator to evaluate the expressions or solve the equations in Exercises 19–24.

**19.** $3^{\sqrt{2}}$

**20.** $\log_3 5$

**21.** $2^{2x} = 5^{x+1}$

**22.** $x^{\sqrt{2}} = 3$

**23.** $\log_x 3 = 5$

**24.** $\log_3 x = 5$

Use the laws of exponents to prove the laws of logarithms in Exercises 25–28.

**25.** $\log_a \left( \dfrac{1}{x} \right) = -\log_a x$

**26.** $\log_a \left( \dfrac{x}{y} \right) = \log_a x - \log_a y$

**27.** $\log_a (x^y) = y \log_a x$

**28.** $\log_a x = (\log_b x)/(\log_b a)$

**29.** Solve $\log_4(x + 4) - 2\log_4(x + 1) = \dfrac{1}{2}$ for $x$.

**30.** Solve $2\log_3 x + \log_9 x = 10$ for $x$.

**\* 31.** Suppose that $f(x) = a^x$ is differentiable at $x = 0$, and that

$f'(0) = k$, where $k \neq 0$. Prove that $f$ is differentiable at any real number $x$, and that

$$f'(x) = k\, a^x = k\, f(x).$$

**\* 32.** Continuing Exercise 31, prove that $f^{-1}(x) = \log_a x$ is differentiable at any $x > 0$, and that

$$(f^{-1})'(x) = \dfrac{1}{kx}.$$

## 3.3 The Natural Logarithm and Exponential

In this section we are going to define a function $\ln x$, called the *natural* logarithm of $x$, in a way that does not at first seem to have anything to do with the logarithms considered in Section 3.2. We will show, however, that it has the same properties as those logarithms, and in the end we will see that $\ln x = \log_e x$, the logarithm of $x$ to a certain specific base $e$. We will show that $\ln x$ is a one-to-one function, defined for all positive real numbers. It must therefore have an inverse, $e^x$, that we will call *the* exponential function. Our final goal is to arrive at a definition of the exponential functions $a^x$ (for any $a > 0$) that is valid for any real number $x$ instead of just rational numbers, and that is known to be continuous and even differentiable without our having to assume those properties as we did in Section 3.2.

### The Natural Logarithm

**Table 1.** Derivatives of integer powers

| $f(x)$ | $f'(x)$ |
|---|---|
| $\vdots$ | $\vdots$ |
| $x^3$ | $3x^2$ |
| $x^2$ | $2x$ |
| $x^1$ | $x^0 = 1$ |
| $x^0$ | $0$ |
| $x^{-1}$ | $-x^{-2}$ |
| $x^{-2}$ | $-2x^{-3}$ |
| $x^{-3}$ | $-3x^{-4}$ |
| $\vdots$ | $\vdots$ |

Table 1 lists the derivatives of integer powers of $x$. Those derivatives are multiples of integer powers of $x$, but one integer power, $x^{-1}$, is conspicuously absent from the list of derivatives; we do not yet know a function whose derivative is $x^{-1} = 1/x$. We are going to remedy this situation by defining a function $\ln x$ in such a way that it will have derivative $1/x$.

To get a hint as to how this can be done, review Example 1 of Section 2.11. In that example we showed that the area under the graph of the velocity of a moving object in a time interval was equal to the distance travelled by the object in that time interval. Since the derivative of distance is velocity, measuring the area provided a way of finding a function (the distance) that had a given derivative (the velocity). This relationship between area and derivatives is one of the most important ideas in calculus. It is called the **Fundamental Theorem of Calculus**. We will explore it fully in Chapter 5, but we will make use of the idea now to define $\ln x$, which we want to have derivative $1/x$.

**DEFINITION** **6**

**The natural logarithm**

For $x > 0$, let $A_x$ be the area of the plane region bounded by the curve $y = 1/t$, the $t$-axis, and the vertical lines $t = 1$ and $t = x$. The function $\ln x$ is defined by

$$\ln x = \begin{cases} A_x & \text{if } x \geq 1, \\ -A_x & \text{if } 0 < x < 1, \end{cases}$$

as shown in Figure 3.9.

The definition implies that $\ln 1 = 0$, that $\ln x > 0$ if $x > 1$, that $\ln x < 0$ if $0 < x < 1$, and that $\ln$ is a one-to-one function. We now show that if $y = \ln x$, then $y' = 1/x$. The proof of this result is similar to the proof we will give for the Fundamental Theorem of Calculus in Section 5.5.

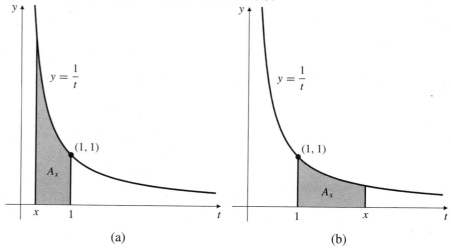

**Figure 3.9**

(a) $\ln x = -$area $A_x$ if $0 < x < 1$

(b) $\ln x =$ area $A_x$ if $x \geq 1$

**THEOREM** **1**

If $x > 0$, then

$$\frac{d}{dx} \ln x = \frac{1}{x}.$$

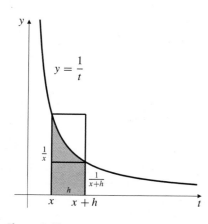

**Figure 3.10**

**PROOF** For $x > 0$ and $h > 0$, $\ln(x + h) - \ln x$ is the area of the plane region bounded by $y = 1/t$, $y = 0$, and the vertical lines $t = x$ and $t = x + h$; it is the shaded area in Figure 3.10. Comparing this area with that of two rectangles, we see that

$$\frac{h}{x + h} < \text{shaded area} = \ln(x + h) - \ln x < \frac{h}{x}.$$

Hence the Newton quotient for $\ln x$ satisfies

$$\frac{1}{x + h} < \frac{\ln(x + h) - \ln x}{h} < \frac{1}{x}.$$

Letting $h$ approach 0 from the right, we obtain (by the Squeeze Theorem applied to one-sided limits)

$$\lim_{h \to 0+} \frac{\ln(x + h) - \ln x}{h} = \frac{1}{x}.$$

A similar argument shows that if $0 < x + h < x$, then

$$\frac{1}{x} < \frac{\ln(x+h) - \ln x}{h} < \frac{1}{x+h},$$

so that

$$\lim_{h \to 0-} \frac{\ln(x+h) - \ln x}{h} = \frac{1}{x}.$$

Combining these two one-sided limits we get the desired result:

$$\frac{d}{dx} \ln x = \lim_{h \to 0} \frac{\ln(x+h) - \ln x}{h} = \frac{1}{x}.$$

The two properties $(d/dx) \ln x = 1/x$ and $\ln 1 = 0$ are sufficient to determine the function $\ln x$ completely. We can deduce from these two properties that $\ln x$ satisfies the appropriate laws of logarithms:

**THEOREM** **2**

**Properties of the natural logarithm**

(i)   $\ln(xy) = \ln x + \ln y$       (ii)   $\ln\left(\dfrac{1}{x}\right) = -\ln x$

(iii)   $\ln\left(\dfrac{x}{y}\right) = \ln x - \ln y$       (iv)   $\ln\left(x^r\right) = r \ln x$

Because we do not want to *assume* that exponentials are continuous, (as we did in Section 3.2), we should regard (iv) for the moment as only valid for exponents $r$ that are rational numbers.

**PROOF**   We will only prove part (i), as the other parts are proved by the same method. If $y > 0$ is a constant, then by the Chain Rule,

$$\frac{d}{dx}\left(\ln(xy) - \ln x\right) = \frac{y}{xy} - \frac{1}{x} = 0 \quad \text{for all } x > 0.$$

Theorem 13 of Section 2.6 now tells us that $(\ln(xy) - \ln x) = C$ (a constant) for $x > 0$. Putting $x = 1$ we get $C = \ln y$ and identity (i) follows.

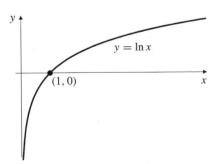

**Figure 3.11**     The graph of $\ln x$

Part (iv) of Theorem 2 shows that $\ln(2^n) = n \ln 2 \to \infty$. Therefore we also have $\ln(1/2)^n = -n \ln 2 \to -\infty$ as $n \to \infty$. Since $(d/dx) \ln x = 1/x > 0$ for $x > 0$, it follows that $\ln x$ is increasing, so we must have (see Figure 3.11)

$$\lim_{x \to \infty} \ln x = \infty, \qquad \lim_{x \to 0+} \ln x = -\infty.$$

**Example 1** Show that $\dfrac{d}{dx} \ln |x| = \dfrac{1}{x}$ for any $x \neq 0$. Hence find $\displaystyle\int \dfrac{1}{x}\, dx$.

**Solution** If $x > 0$, then

$$\frac{d}{dx} \ln |x| = \frac{d}{dx} \ln x = \frac{1}{x}$$

by Theorem 1. If $x < 0$, then, using the Chain Rule,

$$\frac{d}{dx} \ln |x| = \frac{d}{dx} \ln(-x) = \frac{1}{-x}(-1) = \frac{1}{x}.$$

Therefore $\dfrac{d}{dx} \ln |x| = \dfrac{1}{x}$, and on any interval not containing $x = 0$,

$$\int \frac{1}{x}\, dx = \ln |x| + C.$$

**Example 2** Find the derivatives of (a) $\ln |\cos x|$, and (b) $\ln(x + \sqrt{x^2 + 1})$. Simplify your answers as much as possible.

**Solution**

(a) Using the result of the Example 1 and the Chain Rule, we have

$$\frac{d}{dx} \ln |\cos x| = \frac{1}{\cos x}(-\sin x) = -\tan x.$$

(b) $\dfrac{d}{dx} \ln(x + \sqrt{x^2 + 1}) = \dfrac{1}{x + \sqrt{x^2 + 1}}\left(1 + \dfrac{2x}{2\sqrt{x^2 + 1}}\right)$

$$= \frac{1}{x + \sqrt{x^2 + 1}} \cdot \frac{\sqrt{x^2 + 1} + x}{\sqrt{x^2 + 1}}$$

$$= \frac{1}{\sqrt{x^2 + 1}}.$$

## The Exponential Function

The function $\ln x$ is one-to-one on its domain, the interval $(0, \infty)$, and so it has an inverse there. For the moment, let us call this inverse $\exp x$. Thus

$$y = \exp x \iff x = \ln y \quad (y > 0).$$

Since $\ln 1 = 0$, therefore $\exp 0 = 1$. The domain of $\exp$ is $(-\infty, \infty)$, the range of $\ln$. The range of $\exp$ is $(0, \infty)$, the domain of $\ln$. We have cancellation identities

$$\ln(\exp x) = x \quad \text{for all real } x, \quad \text{and} \quad \exp(\ln x) = x \quad \text{for } x > 0.$$

We can deduce various properties of $\exp$ from corresponding properties of $\ln$. Not surprisingly, they are properties we would expect an exponential function to have.

**THEOREM 3**

### Properties of the exponential function

(i)   $(\exp x)^r = \exp(rx)$     (ii)   $\exp(x+y) = (\exp x)(\exp y)$

(iii)   $\exp(-x) = \dfrac{1}{\exp(x)}$     (iv)   $\exp(x - y) = \dfrac{\exp x}{\exp y}$

For the moment, identity (i) is asserted only for rational numbers $r$.

**PROOF**   We prove only identity (i); the rest are done similarly. If $u = (\exp x)^r$, then, by Theorem 2(iv), $\ln u = r \ln(\exp x) = rx$. Therefore, $u = \exp(rx)$.

Now we make an important definition!

Let   $e = \exp(1)$.

The number $e$ satisfies $\ln e = 1$, so the area bounded by the curve $y = 1/t$, the $t$-axis, and the vertical lines $t = 1$ and $t = e$ must be equal to 1 square unit. See Figure 3.12.   The number $e$ is one of the most important numbers in mathematics. Like $\pi$, it is irrational and not a zero of any polynomial with rational coefficients. (Such numbers are called **transcendental**.) Its value is between 2 and 3, and begins

$$e = 2.7\ 1828\ 1828\ 45\ 90\ 45 \ldots.$$

Later on we will learn that
$$e = 1 + \frac{1}{1!} + \frac{1}{2!} + \frac{1}{3!} + \frac{1}{4!} + \cdots,$$
a formula from which the value of $e$ can be calculated to any desired precision.

Theorem 3(i) shows that $\exp r = \exp(1r) = (\exp 1)^r = e^r$ holds for any rational number $r$. Now here is a crucial observation. We only know what $e^r$ means if $r$ is a rational number (if $r = m/n$ then $e^r = \sqrt[n]{e^m}$). But $\exp x$ is defined for all *real* $x$, rational or not. Since $e^r = \exp r$ when $r$ is rational, we can use $\exp x$ as a *definition* of what $e^x$ means for any real number $x$, and there will be no contradiction if $x$ happens to be rational.

$$e^x = \exp x \qquad \text{for all real } x.$$

Theorem 3 can now be restated in terms of $e^x$:

(i)   $(e^x)^y = e^{xy}$     (ii)   $e^{x+y} = e^x\,e^y$

(iii)   $e^{-x} = \dfrac{1}{e^x}$     (iv)   $e^{x-y} = \dfrac{e^x}{e^y}$

The graph of $e^x$ is the reflection of the graph of its inverse, $\ln x$, in the line $x = y$. Both graphs are shown for comparison in Figure 3.13. Observe that the $x$-axis is a horizontal asymptote of the graph of $y = e^x$ as $x \to -\infty$. We have

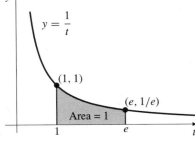

**Figure 3.12**    The definition of $e$

**Figure 3.13**    The graphs of $e^x$ and $\ln x$

$$\lim_{x \to -\infty} e^x = 0, \qquad \lim_{x \to \infty} e^x = \infty.$$

Since $\exp x = e^x$ actually *is* an exponential function, its inverse must actually *be* a logarithm:

$$\ln x = \log_e x.$$

The derivative of $y = e^x$ is calculated by implicit differentiation:

$$y = e^x \quad \Longrightarrow \quad x = \ln y$$

$$\Longrightarrow \quad 1 = \frac{1}{y}\frac{dy}{dx}$$

$$\Longrightarrow \quad \frac{dy}{dx} = y = e^x.$$

Thus the exponential function has the remarkable property that it is its own derivative, and therefore also its own antiderivative:

$$\frac{d}{dx}e^x = e^x, \qquad \int e^x\, dx = e^x + C.$$

**Example 3**    Find the derivatives of

(a) $e^{x^2 - 3x}$,   (b) $\sqrt{1 + e^{2x}}$, and   (c) $\dfrac{e^x - e^{-x}}{e^x + e^{-x}}$.

**Solution**

(a) $\dfrac{d}{dx}e^{x^2 - 3x} = e^{x^2 - 3x}(2x - 3) = (2x - 3)e^{x^2 - 3x}$.

(b) $\dfrac{d}{dx}\sqrt{1 + e^{2x}} = \dfrac{1}{2\sqrt{1 + e^{2x}}}\left(e^{2x}(2)\right) = \dfrac{e^{2x}}{\sqrt{1 + e^{2x}}}$.

(c) $\dfrac{d}{dx}\dfrac{e^x - e^{-x}}{e^x + e^{-x}} = \dfrac{(e^x + e^{-x})(e^x - (-e^{-x})) - (e^x - e^{-x})(e^x + (-e^{-x}))}{(e^x + e^{-x})^2}$

$$= \dfrac{(e^x)^2 + 2e^x e^{-x} + (e^{-x})^2 - [(e^x)^2 - 2e^x e^{-x} + (e^{-x})^2]}{(e^x + e^{-x})^2}$$

$$= \dfrac{4e^{x-x}}{(e^x + e^{-x})^2} = \dfrac{4}{(e^x + e^{-x})^2}.$$

**Example 4**    Let $f(t) = e^{at}$. Find:   (a) $f^{(n)}(t)$,   and   (b) $\int f(t)\, dt$.

**Solution**

(a) We have   $f'(t) = a\,e^{at}$

$$f''(t) = a^2\,e^{at}$$

$$f'''(t) = a^3\,e^{at}$$

$$\vdots$$

$$f^{(n)}(t) = a^n\,e^{at}.$$

(b) Also, $\int f(t)\, dt = \int e^{at}\, dt = \dfrac{1}{a} e^{at} + C$, since $\dfrac{d}{dt} \dfrac{1}{a} e^{at} = e^{at}$. ∎

## General Exponentials and Logarithms

We can use the fact that $e^x$ is now defined for *all* real $x$ to define the arbitrary exponential $a^x$ (where $a > 0$) for all real $x$. If $r$ is rational, then $\ln(a^r) = r \ln a$; therefore $a^r = e^{r \ln a}$. However, $e^{x \ln a}$ is defined for all real $x$, so we can use it as a definition of $a^x$ with no possibility of contradiction arising if $x$ is rational.

**DEFINITION 7**

> **The general exponential $a^x$**
> $$a^x = e^{x \ln a}, \qquad (a > 0, \quad x \text{ real.})$$

**Example 5**  Evaluate $2^\pi$, using the natural logarithm (ln) and exponential (exp or $e^x$) keys on a scientific calculator, but not using the $y^x$ or ^ keys.

**Solution**  $2^\pi = e^{\pi \ln 2} = 8.8249778\ldots$. If your calculator has a ^ key, or an $x^y$ or $y^x$ key, the chances are that it is implemented in terms of the exp and ln functions. ∎

The laws of exponents for $a^x$ as presented in Section 3.2 can now be obtained from those for $e^x$, as can the derivative

$$\frac{d}{dx} a^x = \frac{d}{dx} e^{x \ln a} = e^{x \ln a} \ln a = a^x \ln a.$$

We can also verify the General Power Rule for $x^a$, where $a$ is any real number, provided $x > 0$:

$$\frac{d}{dx} x^a = \frac{d}{dx} e^{a \ln x} = e^{a \ln x} \frac{a}{x} = \frac{a\, x^a}{x} = a\, x^{a-1}.$$

**Example 6**  Show that the graph of $f(x) = x^\pi - \pi^x$ has a negative slope at $x = \pi$.

**Solution**  
$$f'(x) = \pi\, x^{\pi-1} - \pi^x\, \ln \pi$$
$$f'(\pi) = \pi\, \pi^{\pi-1} - \pi^\pi\, \ln \pi = \pi^\pi (1 - \ln \pi).$$

Since $\pi > 3 > e$, we have $\ln \pi > \ln e = 1$, so $1 - \ln \pi < 0$. Since $\pi^\pi = e^{\pi \ln \pi} > 0$, we have $f'(\pi) < 0$. Thus the graph $y = f(x)$ has negative slope at $x = \pi$. ∎

Do not confuse $x^\pi$, which is a power function of $x$, and $\pi^x$, which is an exponential function of $x$.

**Example 7**  Find the critical point of $y = x^x$.

**Solution**  We can't differentiate $x^x$ by treating it as a power (like $x^a$), because the exponent varies. We can't treat it as an exponential (like $a^x$), because the base varies. We can differentiate it if we first write it in terms of the exponential function, $x^x = e^{x \ln x}$, and then use the Chain Rule and the Product Rule:

$$\frac{dy}{dx} = \frac{d}{dx} e^{x \ln x} = e^{x \ln x} \left( \ln x + x \left( \frac{1}{x} \right) \right) = x^x (1 + \ln x).$$

Now $x^x$ is defined only for $x > 0$, and is itself never 0. (Why?) Therefore the critical point occurs where $1 + \ln x = 0$, that is, $\ln x = -1$, or $x = 1/e$.

Finally, observe that $(d/dx)a^x = a^x \ln a$ is negative for all $x$ if $0 < a < 1$, and is positive for all $x$ if $a > 1$. Thus $a^x$ is one-to-one and has an inverse function, $\log_a x$, provided $a > 0$ and $a \neq 1$. Its properties follow in the same way as in Section 3.2. If $y = \log_a x$, then $x = a^y$ and, differentiating implicitly with respect to $x$, we get

$$1 = a^y \ln a \frac{dy}{dx} = x \ln a \frac{dy}{dx}.$$

Thus the derivative of $\log_a x$ is given by

$$\frac{d}{dx} \log_a x = \frac{1}{x \ln a}.$$

Since $\log_a x$ can be expressed in terms of logarithms to any other base, say $e$,

$$\log_a x = \frac{\ln x}{\ln a},$$

we normally use only natural logarithms. Exceptions are found in chemistry, acoustics, and other sciences where "logarithmic scales" are used to measure quantities for which a one unit increase in the measure corresponds to a tenfold increase in the quantity. Logarithms to base 10 are used in defining such scales. In computer science, where powers of 2 play a central role, logarithms to base 2 are often encountered.

## Logarithmic Differentiation

Suppose we want to differentiate a function of the form

$$y = (f(x))^{g(x)}, \qquad \text{(for } f(x) > 0\text{)}.$$

Since the variable appears in both the base and the exponent, neither the general power rule, $(d/dx)x^a = ax^{a-1}$, nor the exponential rule, $(d/dx)a^x = a^x \ln a$, can be directly applied. One method for finding the derivative of such a function is to express it in the form

$$y = e^{g(x) \ln f(x)},$$

and then differentiate, using the Product Rule to handle the exponent. This is the method used in Example 7.

The derivative in Example 7 can also be obtained by taking natural logarithms of both sides of the equation $y = x^x$ and differentiating implicitly:

$$\ln y = x \ln x$$

$$\frac{1}{y} \frac{dy}{dx} = \ln x + \frac{x}{x} = 1 + \ln x$$

$$\frac{dy}{dx} = y(1 + \ln x) = x^x (1 + \ln x).$$

This latter technique is called **logarithmic differentiation**.

**Example 8**   Find $f'(t)$ if $f(t) = (\sin t)^{\ln t}$, where $0 < t < \pi$.

**Solution**   We have $\ln y = \ln t \, \ln \sin t$. Thus

$$\frac{1}{y}\frac{dy}{dt} = \frac{1}{t}\ln \sin t + \ln t \frac{\cos t}{\sin t}$$

$$\frac{dy}{dt} = y\left(\frac{\ln \sin t}{t} + \ln t \, \cot t\right)$$

$$= (\sin t)^{\ln t}\left(\frac{\ln \sin t}{t} + \ln t \, \cot t\right).$$

Logarithmic differentiation is also useful for finding the derivatives of functions expressed as products and quotients of many factors. Taking logarithms reduces these products and quotients to sums and differences. This usually makes the calculation easier than it would be using the Product and Quotient Rules, especially if the derivative is to be evaluated at a specific point.

**Example 9**   Differentiate $y = [(x+1)(x+2)(x+3)]/(x+4)$.

**Solution**   $\ln|y| = \ln|x+1| + \ln|x+2| + \ln|x+3| - \ln|x+4|$. Thus

$$\frac{1}{y}y' = \frac{1}{x+1} + \frac{1}{x+2} + \frac{1}{x+3} - \frac{1}{x+4}$$

$$y' = \frac{(x+1)(x+2)(x+3)}{x+4}\left(\frac{1}{x+1} + \frac{1}{x+2} + \frac{1}{x+3} - \frac{1}{x+4}\right)$$

$$= \frac{(x+2)(x+3)}{x+4} + \frac{(x+1)(x+3)}{x+4} + \frac{(x+1)(x+2)}{x+4}$$

$$- \frac{(x+1)(x+2)(x+3)}{(x+4)^2}.$$

**Example 10**   Find $\left.\dfrac{du}{dx}\right|_{x=1}$ if $u = \sqrt{(x+1)(x^2+1)(x^3+1)}$.

**Solution**

$$\ln u = \frac{1}{2}\left(\ln(x+1) + \ln(x^2+1) + \ln(x^3+1)\right)$$

$$\frac{1}{u}\frac{du}{dx} = \frac{1}{2}\left(\frac{1}{x+1} + \frac{2x}{x^2+1} + \frac{3x^2}{x^3+1}\right)$$

At $x = 1$ we have $u = \sqrt{8} = 2\sqrt{2}$. Hence

$$\left.\frac{du}{dx}\right|_{x=1} = \sqrt{2}\left(\frac{1}{2} + 1 + \frac{3}{2}\right) = 3\sqrt{2}.$$

# Exercises 3.3

Simplify the expressions given in Exercises 1–10.

**1.** $e^3/\sqrt{e^5}$

**2.** $\ln\left(e^{1/2}e^{2/3}\right)$

**3.** $e^{5\ln x}$

**4.** $e^{(3\ln 9)/2}$

**5.** $\ln\dfrac{1}{e^{3x}}$

**6.** $e^{2\ln\cos x} + \left(\ln e^{\sin x}\right)^2$

**7.** $3\ln 4 - 4\ln 3$

**8.** $4\ln\sqrt{x} + 6\ln(x^{1/3})$

**9.** $2\ln x + 5\ln(x - 2)$

**10.** $\ln(x^2 + 6x + 9)$

Solve the equations in Exercises 11–14 for $x$.

**11.** $2^{x+1} = 3^x$

**12.** $3^x = 9^{1-x}$

**13.** $\dfrac{1}{2^x} = \dfrac{5}{8^{x+3}}$

**14.** $2^{x^2-3} = 4^x$

Find the domains of the functions in Exercises 15–16.

**15.** $\ln\dfrac{x}{2 - x}$

**16.** $\ln(x^2 - x - 2)$

Solve the inequalities in Exercises 17–18.

**17.** $\ln(2x - 5) > \ln(7 - 2x)$   **18.** $\ln(x^2 - 2) \le \ln x$

In Exercises 19–50, differentiate the given functions. If possible, simplify your answers.

**19.** $y = e^{5x}$

**20.** $y = xe^x - x$

**21.** $y = \dfrac{x}{e^{2x}}$

**22.** $y = x^2 e^{x/2}$

**23.** $y = \ln(3x - 2)$

**24.** $y = \ln|3x - 2|$

**25.** $y = \ln(1 + e^x)$

**26.** $y = 2\ln\sqrt{x^2 + 2}$

**27.** $y = \dfrac{e^x + e^{-x}}{2}$

**28.** $f(x) = e^{(x^2)}$

**29.** $y = e^{(e^x)}$

**30.** $x = e^{3t}\ln t$

**31.** $y = \dfrac{e^x}{1 + e^x}$

**32.** $f(x) = \dfrac{e^x - e^{-x}}{e^x + e^{-x}}$

**33.** $y = e^x\sin x$

**34.** $y = e^{-x}\cos x$

**35.** $y = \ln\ln x$

**36.** $y = x\ln x - x$

**37.** $y = x^2\ln x - \dfrac{x^2}{2}$

**38.** $y = \ln|\sin x|$

**39.** $y = 5^{2x+1}$

**40.** $y = 2^{(x^2-3x+8)}$

**41.** $g(x) = t^x x^t$

**42.** $h(t) = t^x - x^t$

**43.** $f(s) = \log_a(bs + c)$

**44.** $g(x) = \log_x(2x + 3)$

**45.** $y = x^{\sqrt{x}}$

**46.** $y = (1/x)^{\ln x}$

**47.** $y = \ln|\sec x + \tan x|$

**48.** $y = \ln|x + \sqrt{x^2 - a^2}|$

**49.** $y = \ln(\sqrt{x^2 + a^2} - x)$

**50.** $y = (\cos x)^x - x^{\cos x}$

**51.** Find the $n$th derivative of $f(x) = xe^{ax}$.

**52.** Show that the $n$th derivative of $(ax^2 + bx + c)e^x$ is a function of the same form but with different constants.

**53.** Find the first four derivatives of $e^{x^2}$.

**54.** Find the $n$th derivative of $\ln(2x + 1)$.

**55.** Differentiate (a) $f(x) = (x^x)^x$ and (b) $g(x) = x^{(x^x)}$. Which function grows more rapidly as $x$ grows large?

* **56.** Solve the equation $x^{x^{x^{\cdot^{\cdot^{\cdot}}}}} = a$, where $a > 0$. The exponent tower goes on forever.

Use logarithmic differentiation to find the required derivatives in Exercises 57–59.

**57.** $f(x) = (x - 1)(x - 2)(x - 3)(x - 4)$. Find $f'(x)$.

**58.** $F(x) = \dfrac{\sqrt{1 + x}(1 - x)^{1/3}}{(1 + 5x)^{4/5}}$. Find $F'(0)$.

**59.** $f(x) = \dfrac{(x^2 - 1)(x^2 - 2)(x^2 - 3)}{(x^2 + 1)(x^2 + 2)(x^2 + 3)}$. Find $f'(2)$. Also find $f'(1)$.

**60.** At what points does the graph $y = x^2 e^{-x^2}$ have a horizontal tangent line?

**61.** Let $f(x) = xe^{-x}$. Determine where $f$ is increasing and where it is decreasing. Sketch the graph of $f$.

**62.** Find the equation of a straight line of slope 4 that is tangent to the graph of $y = \ln x$.

**63.** Find an equation of the straight line tangent to the curve $y = e^x$ and passing through the origin.

**64.** Find an equation of the straight line tangent to the curve $y = \ln x$ and passing through the origin.

**65.** Find an equation of the straight line that is tangent to $y = 2^x$ and that passes through the point $(1, 0)$.

**66.** For what values of $a > 0$ does the curve $y = a^x$ intersect the straight line $y = x$?

**67.** Find the slope of the curve $e^{xy}\ln\dfrac{x}{y} = x + \dfrac{1}{y}$ at $(e, 1/e)$.

**68.** Find an equation of the straight line tangent to the curve $xe^y + y - 2x = \ln 2$ at the point $(1, \ln 2)$.

**69.** Find the derivative of $f(x) = Ax\cos\ln x + Bx\sin\ln x$. Use the result to help you find the indefinite integrals
$$\int\cos\ln x\,dx \text{ and } \int\sin\ln x\,dx.$$

* **70.** Let $F_{A,B}(x) = Ae^x\cos x + Be^x\sin x$. Show that $(d/dx)F_{A,B}(x) = F_{A+B,B-A}(x)$.

* **71.** Using the results of Exercise 70, find
(a) $(d^2/dx^2)F_{A,B}(x)$ and (b) $(d^3/dx^3)e^x\cos x$.

* **72.** Find $\dfrac{d}{dx}(Ae^{ax}\cos bx + Be^{ax}\sin bx)$ and use the answer to help you evaluate
(a) $\displaystyle\int e^{ax}\cos bx\,dx$ and (b) $\displaystyle\int e^{ax}\sin bx\,dx$.

**73.** Prove identity (ii) of Theorem 2 by examining the derivative of the left side minus the right side as was done in the proof of identity (i).

**74.** Deduce identity (iii) of Theorem 2 from identities (i) and (ii).

**75.** Prove identity (iv) of Theorem 2 for rational exponents $r$ by the same method as used for Exercise 73.

**∗ 76.** Let $x > 0$ and let $F(x)$ be the area bounded by the curve $y = t^2$, the $t$-axis, and the vertical lines $t = 0$ and $t = x$. Using the method of the proof of Theorem 1, show that $F'(x) = x^2$. Hence find an explicit formula for $F(x)$. What is the area of the region bounded by $y = t^2$, $y = 0$, $t = 0$ and $t = 2$?

**∗ 77.** Carry out the following steps to show that $2 < e < 3$. Let

$f(t) = 1/t$ for $t > 0$.

(a) Show that the area under $y = f(t)$, above $y = 0$, and between $t = 1$ and $t = 2$ is less than 1 square unit. Deduce that $e > 2$.

(b) Show that all tangent lines to the graph of $f$ lie below the graph. *Hint:* $f''(t) = 2/t^3 > 0$.

(c) Find the lines $T_2$ and $T_3$ that are tangent to $y = f(t)$ at $t = 2$ and $t = 3$, respectively.

(d) Find the area $A_2$ under $T_2$, above $y = 0$, and between $t = 1$ and $t = 2$. Also find the area $A_3$ under $T_3$, above $y = 0$, and between $t = 2$ and $t = 3$.

(e) Show that $A_2 + A_3 > 1$ square unit. Deduce that $e < 3$.

# 3.4 Growth and Decay

In this section we will study the use of exponential functions to model the growth rates of quantities whose rate of growth is directly related to their size. The growth of such quantities is typically governed by differential equations whose solutions involve exponential functions. Before delving into this topic, we prepare the way by examining the growth behaviour of exponential and logarithmic functions.

## The Growth of Exponentials and Logarithms

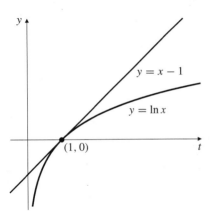

**Figure 3.14**   $\ln x \le x - 1$ for $x > 0$

In Section 3.3 we showed that both $e^x$ and $\ln x$ grow large (approach infinity) as $x$ grows large. However, $e^x$ increases very rapidly as $x$ increases, and $\ln x$ increases very slowly. In fact, $e^x$ increases, for large $x$, faster than any positive power of $x$ (no matter how large the power), while $\ln x$ increases more slowly than any positive power of $x$ (no matter how small the power). In order to verify this behaviour we start with an inequality satisfied by $\ln x$. The straight line $y = x - 1$ is tangent to the curve $y = \ln x$ at the point $(1, 0)$. The following theorem asserts that the curve lies below that line. (See Figure 3.14.)

**THEOREM 4**

If $x > 0$, then $\ln x \le x - 1$.

**PROOF**   Let $g(x) = \ln x - (x - 1)$ for $x > 0$. Then $g(1) = 0$ and

$$g'(x) = \frac{1}{x} - 1 \quad \begin{cases} > 0 & \text{if } 0 < x < 1 \\ < 0 & \text{if } x > 1. \end{cases}$$

As observed in Section 2.6, these inequalities imply that $g$ is increasing on $(0, 1)$ and decreasing on $(1, \infty)$. Thus $g(x) \le g(1) = 0$ for all $x > 0$, and so $\ln x \le x - 1$ for all such $x$.

**THEOREM 5**

**The growth properties of exp and ln**

If $a > 0$ then

(a) $\displaystyle \lim_{x \to \infty} \frac{x^a}{e^x} = 0,$

(b) $\displaystyle \lim_{x \to \infty} \frac{\ln x}{x^a} = 0,$

(c) $\displaystyle \lim_{x \to -\infty} |x|^a\, e^x = 0,$

(d) $\displaystyle \lim_{x \to 0+} x^a \ln x = 0.$

Each of these limits makes a statement about who "wins" in a contest between an exponential or logarithm and a power. For example, in part (a), the denominator $e^x$ grows large as $x \to \infty$ and so tries to make the fraction $x^a/e^x$ approach 0. On the other hand, if $a$ is a large positive number, the numerator $x^a$ also grows large, and so tries to make the fraction approach infinity. The assertion of (a) is that in this contest between the exponential and the power, the exponential is stronger and wins; the fraction approaches 0. The content of Theorem 5 can be paraphrased as follows:

> In a struggle between a power and an exponential, the exponential wins.
> In a struggle between a power and a logarithm the power wins.

**PROOF**   First we prove part (b). Let $x > 1$, $a > 0$, and let $s = a/2$. Since $\ln(x^s) = s \ln x$, we have, using Theorem 4,

$$0 < s \ln x = \ln(x^s) \le x^s - 1 < x^s.$$

Thus $0 < \ln x < \dfrac{1}{s} x^s$, and, dividing by $x^a = x^{2s}$,

$$0 < \frac{\ln x}{x^a} < \frac{1}{s} \frac{x^s}{x^{2s}} = \frac{1}{s\, x^s}.$$

Now $1/(s\, x^s) \to 0$ as $x \to \infty$ (since $s > 0$), and therefore by the Squeeze Theorem,

$$\lim_{x \to \infty} \frac{\ln x}{x^a} = 0.$$

Next we deduce part (d) from part (b) by substituting $x = 1/t$. As $x \to 0+$, we have $t \to \infty$, so

$$\lim_{x \to 0+} x^a \ln x = \lim_{t \to \infty} \frac{\ln(1/t)}{t^a} = \lim_{t \to \infty} \frac{-\ln t}{t^a} = -0 = 0.$$

Now we deduce (a) from (b). If $x = \ln t$, then $t \to \infty$ as $x \to \infty$, so

$$\lim_{x \to \infty} \frac{x^a}{e^x} = \lim_{t \to \infty} \frac{(\ln t)^a}{t} = \lim_{t \to \infty} \left( \frac{\ln t}{t^{1/a}} \right)^a = 0^a = 0.$$

Finally, (c) follows from (a) via the substitution $x = -t$:

$$\lim_{x \to -\infty} |x|^a e^x = \lim_{t \to \infty} |-t|^a e^{-t} = \lim_{t \to \infty} \frac{t^a}{e^t} = 0.$$

## Exponential Growth and Decay Models

Many natural processes involve quantities that increase or decrease at a rate proportional to their size. For example, the mass of a culture of bacteria growing in a medium supplying adequate nourishment will increase at a rate proportional to that mass. The value of an investment bearing interest that is continuously compounding increases at a rate proportional to that value. The mass of undecayed radioactive material in a sample decreases at a rate proportional to that mass.

All of these phenomena, and others exhibiting similar behaviour, can be modelled mathematically in the same way. If $y = y(t)$ denotes the value of a quantity $y$ at time $t$, and if $y$ changes at a rate proportional to its size, then

$$\frac{dy}{dt} = ky,$$

where $k$ is the constant of proportionality. The above equation is called the **differential equation of exponential growth or decay**, because, for any value of the constant $C$, the function $y = Ce^{kt}$ satisfies the equation. In fact, if $y(t)$ is any solution of the differential equation $y' = ky$ then

$$\frac{d}{dt}\left(\frac{y(t)}{e^{kt}}\right) = \frac{e^{kt}y'(t) - ke^{kt}y(t)}{e^{2kt}} = \frac{y'(t) - ky(t)}{e^{kt}} = 0 \quad \text{for all } t.$$

Thus $y(t)/e^{kt} = C$, a constant, and $y(t) = Ce^{kt}$. Since $y(0) = Ce^0 = C$,

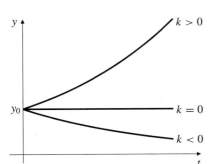

The initial-value problem $\begin{cases} \dfrac{dy}{dt} = ky \\ y(0) = y_0 \end{cases}$ has unique solution $y = y_0 e^{kt}$.

**Figure 3.15**  Solutions of the initial-value problem $dy/dt = ky$, $y(0) = y_0$, for $k > 0$, $k = 0$ and $k < 0$

If $y_0 > 0$, then $y(t)$ is an increasing function of $t$ if $k > 0$ and a decreasing function of $t$ if $k < 0$. We say that the quantity $y$ exhibits **exponential growth** if $k > 0$, and **exponential decay** if $k < 0$. (See Figure 3.15.)

**Example 1**  (**Growth of a cell culture**) A certain cell culture grows at a rate proportional to the number of cells present. If the culture contains 500 cells initially, and 800 after 24 hours, how many cells will be present after a further 12 hours?

**Solution**  Let $y(t)$ be the number of cells present $t$ hours after there were 500 cells. Thus $y(0) = 500$ and $y(24) = 800$. Because $dy/dt = ky$, we have

$$y(t) = y(0)e^{kt} = 500e^{kt}.$$

Therefore $800 = y(24) = 500e^{24k}$, and so $24k = \ln\frac{800}{500} = \ln(1.6)$. It follows that $k = (1/24)\ln(1.6)$, and

$$y(t) = 500e^{(t/24)\ln(1.6)} = 500(1.6)^{t/24}.$$

We want to know $y$ when $t = 36$: $y(36) = 500e^{(36/24)\ln(1.6)} = 500(1.6)^{3/2} \approx 1012$. The cell count grew to about 1012 in the 12 hours after it was 800.  ∎

Exponential growth is characterized by a **fixed doubling time**. If $T$ is the time at which $y$ has doubled from its size at $t = 0$, then $2y(0) = y(T) = y(0)e^{kT}$. Therefore $e^{kT} = 2$. Since $y(t) = y(0)e^{kt}$, we have

$$y(t + T) = y(0)e^{k(t+T)} = e^{kT}y(0)e^{kt} = 2y(t),$$

that is, $T$ units of time are required for $y$ to double from any value. Similarly, exponential decay involves a fixed halving time (usually called the **half-life**). If $y(T) = \frac{1}{2}y(0)$, then $e^{kT} = \frac{1}{2}$ and

$$y(t + T) = y(0)e^{k(t+T)} = \frac{1}{2}y(t).$$

**Example 2** (**Radioactive decay**) A radioactive material has a half-life of 1,200 years. What percentage of the original radioactivity of a sample is left after 10 years? How many years are required to reduce the radioactivity by 10 percent?

**Solution** Let $p(t)$ be the percentage of the original radioactivity left after $t$ years. Thus $p(0) = 100$ and $p(1,200) = 50$. Since the radioactivity decreases at a rate proportional to itself, $dp/dt = kp$ and

$$p(t) = 100e^{kt}.$$

Now $50 = p(1,200) = 100e^{1,200k}$, and so

$$k = \frac{1}{1,200} \ln \frac{50}{100} = -\frac{\ln 2}{1,200}.$$

The percentage left after 10 years is

$$p(10) = 100e^{10k} = 100e^{-10(\ln 2)/1,200} \approx 99.424.$$

If after $t$ years 90 percent of the radioactivity is left, then

$$90 = 100e^{kt},$$
$$kt = \ln \frac{90}{100},$$
$$t = \frac{1}{k} \ln(0.9) = -\frac{1,200}{\ln 2} \ln(0.9) \approx 182.4,$$

so it will take a little over 182 years to reduce the radioactivity by 10 percent. ∎

Sometimes an exponential growth or decay problem will involve a quantity that changes at a rate proportional to the difference between itself and a fixed value:

$$\frac{dy}{dt} = k(y - a).$$

In this case the change of dependent variable $u(t) = y(t) - a$ should be used to convert the differential equation to the standard form. Observe that $u(t)$ changes at the same rate as $y(t)$ (that is, $du/dt = dy/dt$), and so it satisfies

$$\frac{du}{dt} = ku.$$

**Example 3** (**Newton's law of cooling**) A hot object introduced into a cooler environment will cool at a rate proportional to the excess of its temperature above that of its environment. If a cup of coffee sitting in a room maintained at a temperature of 20°C cools from 80°C to 50°C in five minutes, how much longer will it take to cool to 40°C?

**Solution** Let $y(t)$ be the temperature of the coffee $t$ minutes after it was 80°C. Thus $y(0) = 80$ and $y(5) = 50$. Newton's law says that $dy/dt = k(y - 20)$ in this case, so let $u(t) = y(t) - 20$. Thus $u(0) = 60$ and $u(5) = 30$. We have

$$\frac{du}{dt} = \frac{dy}{dt} = k(y - 20) = ku.$$

Thus

$$u(t) = 60e^{kt},$$
$$30 = u(5) = 60e^{5k},$$
$$5k = \ln \tfrac{1}{2} = -\ln 2.$$

We want to know $t$ such that $y(t) = 40$, that is, $u(t) = 20$.

$$20 = u(t) = 60e^{-(t/5)\ln 2}$$
$$-\frac{t}{5}\ln 2 = \ln\frac{20}{60} = -\ln 3,$$
$$t = 5\frac{\ln 3}{\ln 2} \approx 7.92.$$

The coffee will take about $7.92 - 5 = 2.92$ minutes to cool from $50°C$ to $40°C$. ∎

## Interest on Investments

Suppose that \$10,000 is invested at an annual rate of interest of 8%. Thus the value of the investment at the end of 1 year will be \$10,000(1.08) = \$10,800. If this amount remains invested for a second year at the same rate it will grow to \$10,000(1.08)$^2$ = \$11,664, and in general, $n$ years after the original investment was made, it will be worth \$10,000(1.08)$^n$.

Now suppose that the 8% rate is *compounded semiannually* so that the interest is actually paid at a rate of 4% per six-month period. After 1 year (2 interest periods) the \$10,000 will grow to \$10,000(1.04)$^2$ = \$10,816. This is \$16 more than was obtained when the 8% was compounded only once per year. The extra \$16 is the interest paid in the second six-month period on the \$400 interest earned in the first six-month period. Continuing in this way, if the 8% interest is compounded *monthly* (12 periods per year and $\frac{8}{12}$% paid per period) or *daily* (365 periods per year and $\frac{8}{365}$% paid per period), then the original \$10,000 would grow in 1 year to $\$10,000\left(1 + \frac{8}{1,200}\right)^{12} = \$10,830$ or $\$10,000\left(1 + \frac{8}{36,500}\right)^{365} = \$10,832.78$, respectively.

For any given *nominal* interest rate, the investment grows more if the compounding period is shorter. In general, an original investment of \$$A$ invested at $r$% per annum compounded $n$ times per year grows in one year to

$$\$A\left(1 + \frac{r}{100n}\right)^n.$$

It is natural to ask how well we can do with our investment if we let the number of periods in a year approach infinity, that is, we compound the interest *continuously*. The answer is that in 1 year the \$$A$ will grow to

$$\$A \lim_{n\to\infty}\left(1 + \frac{r}{100n}\right)^n = \$Ae^{r/100}.$$

For example, at 8% per annum compounded continuously, our \$10,000 will grow in 1 year to $\$10,000e^{0.08} \approx \$10,832.87$. (Note that this is just a few cents more than we get compounding daily.) To justify this result we need the following theorem.

**THEOREM** **6** For every real number $x$,

$$e^x = \lim_{n \to \infty} \left(1 + \frac{x}{n}\right)^n$$

**PROOF** If $x = 0$, there is nothing to prove; both sides of the identity are 1. If $x \neq 0$ let $h = x/n$. As $n$ tends to infinity, $h$ approaches 0. Thus

$$\lim_{n \to \infty} \ln\left(1 + \frac{x}{n}\right)^n = \lim_{n \to \infty} n \ln\left(1 + \frac{x}{n}\right)$$

$$= \lim_{n \to \infty} x \frac{\ln\left(1 + \frac{x}{n}\right)}{\dfrac{x}{n}}$$

$$= x \lim_{h \to 0} \frac{\ln(1 + h)}{h} \qquad \text{(where } h = x/n)$$

$$= x \lim_{h \to 0} \frac{\ln(1 + h) - \ln 1}{h} \qquad \text{(since } \ln 1 = 0)$$

$$= x \left(\frac{d}{dt} \ln t\right)\Big|_{t=1} \qquad \text{(by the definition of derivative)}$$

$$= x \frac{1}{t}\Big|_{t=1} = x.$$

Since ln is differentiable, it is continuous. Hence, by Theorem 7 of Section 1.4,

$$\ln\left(\lim_{n \to \infty} \left(1 + \frac{x}{n}\right)^n\right) = \lim_{n \to \infty} \ln\left(1 + \frac{x}{n}\right)^n = x.$$

Since the exponential function is the inverse of the natural logarithm, we conclude

$$\lim_{n \to \infty} \left(1 + \frac{x}{n}\right)^n = e^x.$$

In the case $x = 1$ the formula given in Theorem 6 takes the following form:

$$e = \lim_{n \to \infty} \left(1 + \frac{1}{n}\right)^n.$$

We can use this formula to compute approximations to $e$, as shown in Table 2. In a sense we have cheated in obtaining the numbers in this table; they were produced using the $y^x$ function on a scientific calculator. However, this function is actually computed as $e^{x \ln y}$. In any event, the formula in this table is not a very efficient way to calculate $e$ to any great accuracy. Only four decimal places are correct for $n = 100,000$. A much better way is to use the series

$$e = 1 + \frac{1}{1!} + \frac{1}{2!} + \frac{1}{3!} + \frac{1}{4!} + \cdots = 1 + 1 + \frac{1}{2} + \frac{1}{6} + \frac{1}{24} + \cdots,$$

which we will establish in Section 4.7.

A final word about interest rates. Financial institutions sometimes quote *effective* rates of interest rather than *nominal* rates. The effective rate tells you what the actual effect of the interest rate will be after one year. Thus $10,000 invested at an effective rate of 8% will grow to $10,800.00 in one year regardless of the compounding period. A nominal rate of 8% per annum compounded daily is equivalent to an effective rate of about 8.3278%.

**Table 2.**

| $n$ | $\left(1 + \dfrac{1}{n}\right)^n$ |
|---|---|
| 1 | 2 |
| 10 | $2.59374\cdots$ |
| 100 | $2.70481\cdots$ |
| 1,000 | $2.71692\cdots$ |
| 10,000 | $2.71815\cdots$ |
| 100,000 | $2.71827\cdots$ |

## Logistic Growth

Few quantities in nature can sustain exponential growth over extended periods of time; the growth is usually limited by external constraints. For example, suppose a small number of rabbits (of both sexes) is introduced to a small island where there were no rabbits previously, and where there are no predators who eat rabbits. By virtue of natural fertility, the number of rabbits might be expected to grow exponentially, but this growth will eventually be limited by the food supply available to the rabbits. Suppose the island can grow enough food to supply a population of $L$ rabbits indefinitely. If there are $y(t)$ rabbits in the population at time $t$, we would expect $y(t)$ to grow at a rate proportional to $y(t)$ provided $y(t)$ is quite small (much less than $L$). But as the numbers increase, it will be harder for the rabbits to find enough food, and we would expect the rate of increase to approach 0 as $y(t)$ gets closer and closer to $L$. One possible model for such behaviour is the differential equation

$$\frac{dy}{dt} = ky\left(1 - \frac{y}{L}\right),$$

which is called the **logistic equation** since it models growth that is limited by the *supply* of necessary resources. Observe that $dy/dt > 0$ if $0 < y < L$ and that this rate is small if $y$ is small (there are few rabbits to reproduce) or if $y$ is close to $L$ (there are almost as many rabbits as the available resources can feed). Observe also that $dy/dt < 0$ if $y > L$; there being more animals than the resources can feed, the rabbits die at a greater rate than they are born. Of course the steady-state populations $y = 0$ and $y = L$ are solutions of the logistic equation; for both of these $dy/dt = 0$. We will examine techniques for solving differential equations like the logistic equation in Section 7.8. For now, we invite the reader to verify by differentiation that the solution satisfying $y(0) = y_0$, is

$$y = \frac{Ly_0}{y_0 + (L - y_0)e^{-kt}}.$$

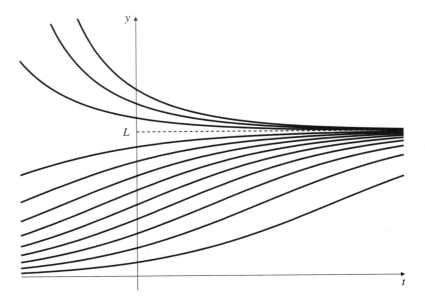

**Figure 3.16**  Some logistic curves

Observe that, as expected, if $0 < y_0 < L$ then

$$\lim_{t \to \infty} y(t) = L, \qquad \lim_{t \to -\infty} y(t) = 0.$$

The solution given above also holds for $y_0 > L$. However, the solution does not approach 0 as $t$ approaches $-\infty$ in this case. It has a vertical asymptote at a certain negative value of $t$. (See Exercise 28 below.) The graphs of solutions of the logistic equation for various positive values of $y_0$ are given Figure 3.16.

## Exercises 3.4

Evaluate the limits in Exercises 1–8.

**1.** $\lim_{x \to \infty} x^3 e^{-x}$

**2.** $\lim_{x \to \infty} x^{-3} e^x$

**3.** $\lim_{x \to \infty} \dfrac{2e^x - 3}{e^x + 5}$

**4.** $\lim_{x \to \infty} \dfrac{x - 2e^{-x}}{x + 3e^{-x}}$

**5.** $\lim_{x \to 0+} x \ln x$

**6.** $\lim_{x \to 0+} \dfrac{\ln x}{x}$

**7.** $\lim_{x \to 0} x \left( \ln |x| \right)^2$

**8.** $\lim_{x \to \infty} \dfrac{(\ln x)^3}{\sqrt{x}}$

**9. (Bacterial growth)** Bacteria grow in a certain culture at a rate proportional to the amount present. If there are 100 bacteria present initially and the amount doubles in 1 hour, how many will there be after a further $1\frac{1}{2}$ hours?

**10. (Dissolving sugar)** Sugar dissolves in water at a rate proportional to the amount still undissolved. If there were 50 kg of sugar present initially, and at the end of 5 hours only 20 kg is left, how much longer will it take until 90 percent of the sugar is disssolved?

**11. (Radioactive decay)** A radioactive substance decays at a rate proportional to the amount present. If 30 percent of such a substance decays in 15 years, what is the half-life of the substance?

**12. (Half-life of radium)** If the half-life of radium is 1,690 years, what percentage of the amount present now will be remaining after (a) 100 years, (b) 1,000 years?

**13.** Find the half-life of a radioactive substance if after 1 year 99.57 percent of an initial amount still remains.

**14. (Bacterial growth)** In a certain culture where the rate of growth of bacteria is proportional to the number present, the number triples in 3 days. If at the end of 7 days there are 10 million bacteria present in the culture, how many were present initially?

**15. (Weight of a newborn)** In the first few weeks after birth, a baby gains weight at a rate proportional to its weight. A baby weighing 4 kg at birth weighs 4.4 kg after two weeks. How much did it weigh 5 days after birth?

**16. (Electric current)** When a simple electrical circuit containing inductance and resistance but no capacitance has

the electromotive force removed, the rate of decrease of the current is proportional to the current. If the current is $I(t)$ amperes $t$ seconds after cutoff, and if $I = 40$ when $t = 0$, and $I = 15$ when $t = 0.01$, find a formula for $I(t)$.

**17. (Continuously compounding interest)** How much money needs to be invested today at a nominal rate of 4% compounded continuously, in order that it should grow to $10,000 in 7 years?

**18. (Continuously compounding interest)** Money invested at compound interest (with instantaneous compounding) accumulates at a rate proportional to the amount present. If an initial investment of $1,000 grows to $1,500 in exactly 5 years, find (a) the doubling time for the investment, and (b) the effective annual rate of interest being paid.

**19. (Purchasing power)** If the purchasing power of the dollar is decreasing at an effective rate of 9 percent annually, how long will it take for the purchasing power to be reduced to 25 cents?

**∗ 20. (Effective interest rate)** A bank claims to pay interest at an effective rate of 9.5% on an investment account. If the interest is actually being compounded monthly, what is the nominal rate of interest being paid on the account?

**Differential equations of the form** $y' = a + by$

**21.** Suppose that $f(x)$ satisfies the differential equation

$$f'(x) = a + bf(x), \qquad (*)$$

where $a$ and $b$ are constants.

(a) Solve equation $(*)$ by substituting $u(x) = a + bf(x)$ and solving the simpler differential equation that results for $u(x)$.

(b) Solve the initial-value problem:

$$\begin{cases} \dfrac{dy}{dx} = a + by \\ y(0) = y_0 \end{cases}$$

22. **(Drug concentrations in the blood)** A drug is introduced into the bloodstream intravenously at a constant rate, and breaks down and is eliminated from the body at a rate proportional to its concentration in the blood. The concentration $x(t)$ of the drug in the blood satisfies the differential equation

$$\frac{dx}{dt} = a - bx,$$

where $a$ and $b$ are positive constants.

(a) What is the limiting concentration $\lim_{t \to \infty} x(t)$ of the drug in the blood?

(b) Find the concentration of the drug in the blood at time $t$, given that the concentration was zero at $t = 0$.

(c) How long after $t = 0$ will it take for the concentration to rise to half its limiting value?

23. **(Cooling)** Use Newton's law of cooling to determine the reading on a thermometer 5 minutes after it is taken from an oven at $72°C$ to the outdoors where the temperature is $20°C$, if the reading dropped to $48°C$ after one minute.

24. **(Cooling)** An object is placed in a freezer maintained at a temperature of $-5°C$. If the object cools from $45°C$ to $20°C$ in 40 minutes, how many more minutes will it take to cool to $0°C$?

25. **(Warming)** If an object in a room warms up from $5°C$ to $10°C$ in 4 minutes, and if the room is being maintained at $20°C$, how much longer will the object take to warm up to $15°C$? Assume the object warms at a rate proportional to the difference between its temperature and room temperature.

**The logistic equation**

* 26. Suppose the quantity $y(t)$ exhibits logistic growth. If the values of $y(t)$ at times $t = 0$, $t = 1$, and $t = 2$ are $y_0$, $y_1$, and $y_2$, respectively, find an equation satisfied by the limiting value $L$ of $y(t)$, and solve it for $L$. If $y_0 = 3$, $y_1 = 5$, and $y_2 = 6$, find $L$.

* 27. Show that a solution $y(t)$ of the logistic equation having $0 < y(0) < L$ is increasing most rapidly when its value is $L/2$. (*Hint:* You do not need to use the formula for the solution to see this.)

* 28. If $y_0 > L$, find the interval on which the given solution of the logistic equation is valid. What happens to the solution as $t$ approaches the left endpoint of this interval?

* 29. If $y_0 < 0$, find the interval on which the given solution of the logistic equation is valid. What happens to the solution as $t$ approaches the right endpoint of this interval?

30. **(Modelling an epidemic)** The number $y$ of persons infected by a highly contagious virus is modelled by a logistic curve

$$y = \frac{L}{1 + Me^{-kt}},$$

where $t$ is measured in months from the time the outbreak was discovered. At that time there were 200 infected persons, and the number grew to 1,000 after 1 month. Eventually, the number levelled out at 10,000. Find the values of the parameters $L$, $M$, and $k$ of the model.

31. Continuing the previous exercise, how many people were infected 3 months after the outbreak was discovered, and how fast was the number growing at that time?

# 3.5 The Inverse Trigonometric Functions

The six trigonometric functions are periodic, and hence not one-to-one. However, as we did with the function $x^2$ in Section 3.1, we can restrict their domains in such a way that the restricted functions are one-to-one and invertible.

## The Inverse Sine (or Arcsine) Function

Let us define a function $\operatorname{Sin} x$ (note the capital letter) to be $\sin x$, restricted so that its domain is the interval $-\frac{\pi}{2} \le x \le \frac{\pi}{2}$:

**DEFINITION 8**

> **The restricted function $\operatorname{Sin} x$**
>
> $$\operatorname{Sin} x = \sin x \qquad \text{if } -\frac{\pi}{2} \le x \le \frac{\pi}{2}.$$

Since its derivative $\cos x$ is positive on the interval $\left(-\frac{\pi}{2}, \frac{\pi}{2}\right)$, the function $\operatorname{Sin} x$ is increasing on its domain, and so is a one-to-one function. It has domain $\left[-\frac{\pi}{2}, \frac{\pi}{2}\right]$ and range $[-1, 1]$. (See Figure 3.17.)

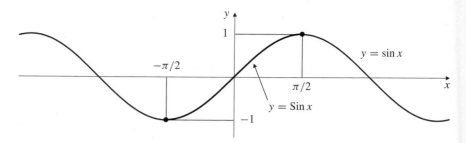

**Figure 3.17** The graph of $\operatorname{Sin} x$ forms part of the graph of $\sin x$

Being one-to-one, Sin has an inverse function which is denoted $\sin^{-1}$ (or, in some books and computer programs, by arcsin, Arcsin, or asin), and which is called the **inverse sine** or **arcsine** function.

**DEFINITION** **9**

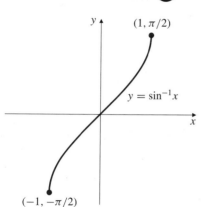

**Figure 3.18** The arcsine function

> **The inverse sine function $\sin^{-1}x$ or $\arcsin x$**
> $$y = \sin^{-1} x \quad \Longleftrightarrow \quad x = \operatorname{Sin} y$$
> $$\Longleftrightarrow \quad x = \sin y \quad \text{and} \quad -\frac{\pi}{2} \le y \le \frac{\pi}{2}$$

The graph of $\sin^{-1}$ is shown in Figure 3.18; it is the reflection of the graph of Sin in the line $y = x$. The domain of $\sin^{-1}$ is $[-1, 1]$ (the range of Sin), and the range of $\sin^{-1}$ is $\left[-\frac{\pi}{2}, \frac{\pi}{2}\right]$ (the domain of Sin). The cancellation identities for Sin and $\sin^{-1}$ are

$$\sin^{-1}(\sin x) = \arcsin(\sin x) = x \qquad \text{for } -\frac{\pi}{2} \le x \le \frac{\pi}{2}$$

$$\sin(\sin^{-1} x) = \sin(\arcsin x) = x \qquad \text{for } -1 \le x \le 1$$

**Remark** As for the general inverse function $f^{-1}$, be aware that $\sin^{-1} x$ does *not* represent the *reciprocal* $1/\sin x$. (We already have a perfectly good name for the reciprocal of $\sin x$; we call it $\csc x$.) We should think of $\sin^{-1} x$ as "the angle between $-\frac{\pi}{2}$ and $\frac{\pi}{2}$ whose sine is $x$."

**Example 1**

(a) $\sin^{-1}\frac{1}{2} = \frac{\pi}{6}$ (because $\sin\frac{\pi}{6} = \frac{1}{2}$ and $-\frac{\pi}{2} < \frac{\pi}{6} < \frac{\pi}{2}$).

(b) $\sin^{-1}\left(-\frac{1}{\sqrt{2}}\right) = -\frac{\pi}{4}$ (because $\sin\left(-\frac{\pi}{4}\right) = -\frac{1}{\sqrt{2}}$ and $-\frac{\pi}{2} < -\frac{\pi}{4} < \frac{\pi}{2}$).

(c) $\sin^{-1}(-1) = -\frac{\pi}{2}$ (because $\sin\left(-\frac{\pi}{2}\right) = -1$.)

(d) $\sin^{-1} 2$ is not defined. (2 is not in the range of sine.)

**Example 2** Find (a) $\sin\left(\sin^{-1} 0.7\right)$, (b) $\sin^{-1}(\sin 0.3)$, (c) $\sin^{-1}\left(\sin\frac{4\pi}{5}\right)$, (d) $\cos\left(\sin^{-1} 0.6\right)$.

**Solution**

(a) $\sin\left(\sin^{-1} 0.7\right) = 0.7$ (cancellation identity).

(b) $\sin^{-1}(\sin 0.3) = 0.3$ (cancellation identity).

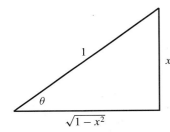

**Figure 3.19**

(c) The number $\frac{4\pi}{5}$ does not lie in $\left[-\frac{\pi}{2}, \frac{\pi}{2}\right]$ so we can't apply the cancellation identity directly. However, $\sin \frac{4\pi}{5} = \sin\left(\pi - \frac{\pi}{5}\right) = \sin \frac{\pi}{5}$ by the supplementary angle identity. Therefore, $\sin^{-1}\left(\sin \frac{4\pi}{5}\right) = \sin^{-1}\left(\sin \frac{\pi}{5}\right) = \frac{\pi}{5}$ (by cancellation).

(d) Let $\theta = \sin^{-1} 0.6$, as shown in the right triangle in Figure 3.19, which has hypotenuse 1 and side opposite $\theta$ equal to 0.6. By the Pythagorean Theorem, the side adjacent $\theta$ is $\sqrt{1 - (0.6)^2} = 0.8$. Thus $\cos\left(\sin^{-1} 0.6\right) = \cos\theta = 0.8$. ∎

**Example 3**    Simplify the expression $\tan(\sin^{-1} x)$.

**Solution**    We want the tangent of an angle whose sine is $x$. Suppose first that $0 \le x < 1$. As in Example 2, we draw a right triangle (Figure 3.20) with one angle $\theta$, and label the sides so that $\theta = \sin^{-1} x$. The side opposite $\theta$ is $x$, and hypotenuse 1. The remaining side is $\sqrt{1 - x^2}$, and we have

$$\tan(\sin^{-1} x) = \tan\theta = \frac{x}{\sqrt{1 - x^2}}.$$

Because both sides of the above equation are odd functions of $x$, the same result holds for $-1 < x < 0$. ∎

**Figure 3.20**

Now let us use implicit differentiation to find the derivative of the inverse sine function. If $y = \sin^{-1} x$ then $x = \sin y$ and $-\frac{\pi}{2} \le y \le \frac{\pi}{2}$. Differentiating with respect to $x$, we obtain

$$1 = (\cos y)\frac{dy}{dx}.$$

Since $-\frac{\pi}{2} \le y \le \frac{\pi}{2}$ we know that $\cos y \ge 0$. Therefore

$$\cos y = \sqrt{1 - \sin^2 y} = \sqrt{1 - x^2},$$

and $dy/dx = 1/\cos y = 1/\sqrt{1 - x^2}$.

$$\frac{d}{dx}\sin^{-1} x = \frac{d}{dx}\arcsin x = \frac{1}{\sqrt{1 - x^2}}.$$

Note that the inverse sine function is differentiable only on the *open* interval $(-1, 1)$; the slope of its graph approaches infinity as $x \to -1+$ or $x \to 1-$. (See Figure 3.18.)

**Example 4**    Find the derivative of $\sin^{-1}\left(\frac{x}{a}\right)$ and hence evaluate $\int \frac{dx}{\sqrt{a^2 - x^2}}$, where $a > 0$.

**Solution**    By the Chain Rule,

$$\frac{d}{dx}\sin^{-1}\frac{x}{a} = \frac{1}{\sqrt{1-\dfrac{x^2}{a^2}}}\frac{1}{a} = \frac{1}{\sqrt{\dfrac{a^2-x^2}{a^2}}}\frac{1}{a} = \frac{1}{\sqrt{a^2-x^2}} \qquad \text{if } a > 0.$$

Hence,

$$\int \frac{1}{\sqrt{a^2-x^2}}\,dx = \sin^{-1}\frac{x}{a} + C \qquad (a > 0).$$

---

**Example 5**    Find the solution $y$ of the following initial-value problem:

$$\begin{cases} y' = \dfrac{4}{\sqrt{2-x^2}} & (-\sqrt{2} < x < \sqrt{2}) \\ y(1) = 2\pi. \end{cases}$$

**Solution**    Using the integral from the previous example, we have

$$y = \int \frac{4}{\sqrt{2-x^2}}\,dx = 4\sin^{-1}\left(\frac{x}{\sqrt{2}}\right) + C$$

for some constant $C$. Also $2\pi = y(1) = 4\sin^{-1}(1/\sqrt{2}) + C = 4\left(\frac{\pi}{4}\right) + C = \pi + C$. Thus $C = \pi$ and $y = 4\sin^{-1}(x/\sqrt{2}) + \pi$.

---

**Example 6**    (**A sawtooth curve**) Let $f(x) = \sin^{-1}(\sin x)$ for all real $x$.

(a) Calculate and simplify $f'(x)$.

(b) Where is $f$ differentiable? Where is $f$ continuous?

(c) Use your results from (a) and (b) to sketch the graph of $f$.

**Solution**    (a) Using the Chain Rule and the Pythagorean identity we calculate

$$f'(x) = \frac{1}{\sqrt{1-(\sin x)^2}}(\cos x)$$

$$= \frac{\cos x}{\sqrt{\cos^2 x}} = \frac{\cos x}{|\cos x|} = \begin{cases} 1 & \text{if } \cos x > 0 \\ -1 & \text{if } \cos x < 0. \end{cases}$$

(b) $f$ is differentiable at all points where $\cos x \neq 0$, that is, everywhere except at odd multiples of $\pi/2$, namely $\pm\frac{\pi}{2}, \pm\frac{3\pi}{2}, \pm\frac{5\pi}{2}, \ldots$.
Since sin is continuous everywhere and has values in $[-1, 1]$, and since $\sin^{-1}$ is continuous on $[-1, 1]$, therefore $f$ is continuous on the whole real line.

(c) Since $f$ is continuous, its graph has no breaks. The graph consists of straight line segments of slopes alternating between 1 and $-1$ on intervals between consecutive odd multiples of $\pi/2$. Since $f'(x) = 1$ on the interval $\left[-\frac{\pi}{2}, \frac{\pi}{2}\right]$ (where $\cos x \geq 0$), the graph must be as shown in Figure 3.21.

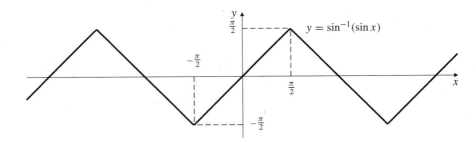

**Figure 3.21**    A sawtooth graph

## The Inverse Tangent (or Arctangent) Function

The inverse tangent function is defined in a manner similar to the inverse sine. We begin by restricting tangent to an interval where it is one-to-one; in this case we use the open interval $\left(-\frac{\pi}{2}, \frac{\pi}{2}\right)$. See Figure 3.22(a).

> **The restricted function Tan $x$**
>
> $$\text{Tan}\, x = \tan x \qquad \text{if} \ -\frac{\pi}{2} < x < \frac{\pi}{2}.$$

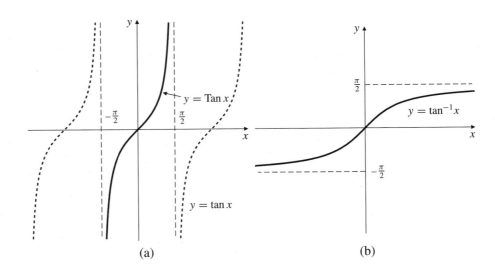

**Figure 3.22**

(a) The graph of Tan $x$

(b) The graph of $\tan^{-1} x$

(a)                (b)

The inverse of the function Tan is called the **inverse tangent** function and is denoted $\tan^{-1}$ (or arctan, Arctan, or atan). The function $\tan^{-1}$ has domain the whole real line (the range of Tan). Its range is the open interval $\left(-\frac{\pi}{2}, \frac{\pi}{2}\right)$.

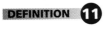

> **The inverse tangent function $\tan^{-1} x$ or $\arctan x$**
>
> $$y = \tan^{-1} x \iff x = \text{Tan}\, y$$
> $$\iff x = \tan y \quad \text{and} \quad -\frac{\pi}{2} < y < \frac{\pi}{2}$$

The graph of $\tan^{-1}$ is shown in Figure 3.22(b); it is the reflection of the graph of Tan in the line $y = x$.

The cancellation identities for Tan and $\tan^{-1}$ are

$$\tan^{-1}(\tan x) = \arctan(\tan x) = x \qquad \text{for} \quad -\frac{\pi}{2} < x < \frac{\pi}{2}$$

$$\tan(\tan^{-1} x) = \tan(\arctan x) = x \qquad \text{for} \quad -\infty < x < \infty$$

**Example 7**    Evaluate:    (a) $\tan(\tan^{-1} 3)$,    (b) $\tan^{-1}\left(\tan \dfrac{3\pi}{4}\right)$, and    (c) $\cos(\tan^{-1} 2)$.

**Solution**

(a) $\tan(\tan^{-1} 3) = 3$ by cancellation.

(b) $\tan^{-1}\left(\tan \frac{3\pi}{4}\right) = \tan^{-1}(-1) = -\frac{\pi}{4}$.

(c) $\cos(\tan^{-1} 2) = \cos\theta = \frac{1}{\sqrt{5}}$ via the triangle in Figure 3.23. Alternatively, we have $\tan(\tan^{-1} 2) = 2$, so $\sec^2(\tan^{-1} 2) = 1 + 2^2 = 5$. Thus $\cos^2(\tan^{-1} 2) = \dfrac{1}{5}$. Since cosine is positive on the range of $\tan^{-1}$, we have $\cos(\tan^{-1} 2) = \frac{1}{\sqrt{5}}$. ∎

The derivative of the inverse tangent function is also found by implicit differentiation: if $y = \tan^{-1} x$ then $x = \tan y$ and

$$1 = (\sec^2 y)\frac{dy}{dx} = (1 + \tan^2 y)\frac{dy}{dx} = (1 + x^2)\frac{dy}{dx}.$$

Thus

$$\frac{d}{dx}\tan^{-1} x = \frac{1}{1 + x^2}.$$

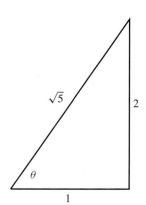

**Figure 3.23**

**Example 8**    Find $\dfrac{d}{dx}\tan^{-1}\left(\dfrac{x}{a}\right)$, and hence evaluate $\displaystyle\int \frac{1}{x^2 + a^2}\,dx$.

**Solution**    We have

$$\frac{d}{dx}\tan^{-1}\left(\frac{x}{a}\right) = \frac{1}{1 + \dfrac{x^2}{a^2}}\frac{1}{a} = \frac{a}{a^2 + x^2},$$

and hence

$$\int \frac{dx}{a^2 + x^2} = \frac{1}{a}\tan^{-1}\left(\frac{x}{a}\right) + C.$$

> **Example 9** Prove that $\tan^{-1}\left(\dfrac{x-1}{x+1}\right) = \tan^{-1}x - \dfrac{\pi}{4}$ for $x > -1$.
>
> **Solution** Let $f(x) = \tan^{-1}\left(\dfrac{x-1}{x+1}\right) - \tan^{-1}x$. On the interval $(-1, \infty)$ we have, by the Chain Rule and the Quotient Rule,
>
> $$f'(x) = \frac{1}{1+\left(\dfrac{x-1}{x+1}\right)^2} \frac{(x+1)-(x-1)}{(x+1)^2} - \frac{1}{1+x^2}$$
>
> $$= \frac{(x+1)^2}{(x^2+2x+1)+(x^2-2x+1)}\frac{2}{(x+1)^2} - \frac{1}{1+x^2}$$
>
> $$= \frac{2}{2+2x^2} - \frac{1}{1+x^2} = 0.$$
>
> Hence $f(x) = C$ (constant) on that interval. We can find $C$ by finding $f(0)$:
>
> $$C = f(0) = \tan^{-1}(-1) - \tan^{-1}0 = -\frac{\pi}{4}.$$
>
> Hence the given identity holds on $(-1, \infty)$. ∎

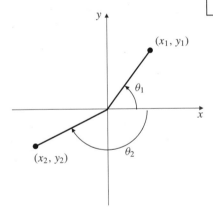

**Figure 3.24**

$\theta_1 = \tan^{-1}(y_1/x_1)$
$= \text{atan}(y_1/x_1)$
$= \text{atan2}(x_1, y_1)$
$\theta_2 = \text{atan2}(x_2, y_2)$

**Remark** Some computer programs, especially spreadsheets, implement two versions of the arctangent function, usually called "atan" and "atan2." The function atan is just the function $\tan^{-1}$ that we have defined; $\text{atan}(y/x)$ gives the angle in radians, between the line from the origin to the point $(x, y)$ and the positive $x$-axis, provided $(x, y)$ lies in quadrants I or IV of the plane. The function atan2 is a function of two variables: $\text{atan2}(x, y)$ gives that angle for any point $(x, y)$ not on the $y$-axis. See Figure 3.24.

## Other Inverse Trigonometric Functions

The function $\cos x$ is one-to-one on the interval $[0, \pi]$, so we could define the **inverse cosine function**, $\cos^{-1}x$ (or $\arccos x$, or $\text{Arccos}\,x$, or $\text{acos}\,x$), so that

$$y = \cos^{-1}x \iff x = \cos y \quad \text{and} \quad 0 \le y \le \pi.$$

However, $\cos y = \sin\left(\frac{\pi}{2} - y\right)$, (the complementary angle identity), and $\frac{\pi}{2} - y$ is in the interval $\left[-\frac{\pi}{2}, \frac{\pi}{2}\right]$ when $0 \le y \le \pi$. Thus the definition above would lead to

$$y = \cos^{-1}x \iff x = \sin\left(\frac{\pi}{2} - y\right) \iff \sin^{-1}x = \frac{\pi}{2} - y = \frac{\pi}{2} - \cos^{-1}x.$$

It is easier to use this result to define $\cos^{-1}x$ directly:

**DEFINITION 12**

> **The inverse cosine function $\cos^{-1}x$ or $\arccos x$**
> $$\cos^{-1}x = \frac{\pi}{2} - \sin^{-1}x \qquad \text{for} \quad -1 \le x \le 1.$$

The cancellation identities for $\cos^{-1}x$ are

$$\cos^{-1}(\cos x) = \arccos(\cos x) = x \qquad \text{for } 0 \le x \le \pi$$
$$\cos(\cos^{-1} x) = \cos(\arccos x) = x \qquad \text{for } -1 \le x \le 1$$

The derivative of $\cos^{-1} x$ is the negative of that of $\sin^{-1} x$ (why?):

$$\frac{d}{dx}\cos^{-1} x = -\frac{1}{\sqrt{1-x^2}}.$$

The graph of $\cos^{-1}$ is shown in Figure 3.25(a).

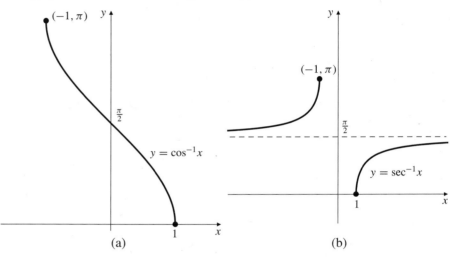

**Figure 3.25** The graphs of $\cos^{-1}$ and $\sec^{-1}$

Scientific calculators usually implement only the primary trigonometric functions, sine, cosine, and tangent, and the inverses of these three. The secondary functions, secant, cosecant, and cotangent are calculated using the reciprocal key; to calculate $\sec x$ you calculate $\cos x$ and take the reciprocal of the answer. The inverses of the secondary trigonometric functions are also easily expressed in terms of those of their reciprocal functions. For example, we define:

**DEFINITION 13**

> **The inverse secant function $\sec^{-1} x$ (or arcsec $x$)**
>
> $$\sec^{-1} x = \cos^{-1}\left(\frac{1}{x}\right) \qquad \text{for } |x| \ge 1.$$

The domain of $\sec^{-1}$ is the union of intervals $(-\infty, -1] \cup [1, \infty)$, and its range is $\left[0, \frac{\pi}{2}\right) \cup \left(\frac{\pi}{2}, \pi\right]$. The graph of $y = \sec^{-1} x$ is shown in Figure 3.25(b). It is the reflection in the line $y = x$ of that part of the graph of $\sec x$ for $x$ between 0 and $\pi$. Observe that

$$\sec(\sec^{-1} x) = \sec\left(\cos^{-1}\left(\frac{1}{x}\right)\right)$$

$$= \frac{1}{\cos\left(\cos^{-1}\left(\frac{1}{x}\right)\right)} = \frac{1}{\frac{1}{x}} = x \qquad \text{for } |x| \ge 1,$$

$$\sec^{-1}(\sec x) = \cos^{-1}\left(\frac{1}{\sec x}\right)$$

$$= \cos^{-1}(\cos x) = x \qquad \text{for } x \text{ in } [0, \pi], \; x \neq \frac{\pi}{2}.$$

We calculate the derivative of $\sec^{-1}$ from that of $\cos^{-1}$:

$$\frac{d}{dx}\sec^{-1}x = \frac{d}{dx}\cos^{-1}\left(\frac{1}{x}\right) = \frac{-1}{\sqrt{1 - \frac{1}{x^2}}}\left(-\frac{1}{x^2}\right)$$

$$= \frac{1}{x^2}\sqrt{\frac{x^2}{x^2-1}} = \frac{1}{x^2}\frac{|x|}{\sqrt{x^2-1}} = \frac{1}{|x|\sqrt{x^2-1}}.$$

Note that we had to use $\sqrt{x^2} = |x|$ in the last line. There are negative values of $x$ in the domain of $\sec^{-1}$. Observe in Figure 3.25(b) that the slope of $y = \sec^{-1}(x)$ is always positive.

$$\frac{d}{dx}\sec^{-1}x = \frac{1}{|x|\sqrt{x^2-1}}.$$

The corresponding integration formula takes different forms on intervals where $x \geq 1$ or $x \leq -1$:

$$\int \frac{1}{x\sqrt{x^2-1}}\,dx = \begin{cases} \sec^{-1}x + C & \text{on intervals where } x \geq 1 \\ -\sec^{-1}x + C & \text{on intervals where } x \leq -1 \end{cases}$$

Finally, note that $\csc^{-1}$ and $\cot^{-1}$ are defined similarly to $\sec^{-1}$. They are seldom encountered.

> **DEFINITION 14**
>
> **The inverse cosecant and inverse cotangent functions**
>
> $$\csc^{-1}x = \sin^{-1}\left(\frac{1}{x}\right), \qquad (|x| \geq 1), \qquad \cot^{-1}x = \tan^{-1}\left(\frac{1}{x}\right), \qquad (x \neq 0)$$

Some authors prefer to define $\sec^{-1}$ as the inverse of the restriction of $\sec x$ to the separated intervals $[0, \pi/2)$ and $[\pi, 3\pi/2)$ because this prevents the absolute value from appearing in the formula for the derivative. However, it is much harder to calculate values with that definition. Our definition makes it easy to obtain a value such as $\sec^{-1}(-3)$ from a calculator. Scientific calculators usually have just the inverses of sine, cosine, and tangent built in.

# Exercises 3.5

In Exercises 1–12, evaluate the given expression.

**1.** $\sin^{-1}\frac{\sqrt{3}}{2}$

**2.** $\cos^{-1}\left(\frac{-1}{2}\right)$

**3.** $\tan^{-1}(-1)$

**4.** $\sec^{-1}\sqrt{2}$

**5.** $\sin(\sin^{-1}0.7)$

**6.** $\cos(\sin^{-1}0.7)$

**7.** $\tan^{-1}\left(\tan\frac{2\pi}{3}\right)$

**8.** $\sin^{-1}(\cos 40°)$

**9.** $\cos^{-1}(\sin(-0.2))$

**10.** $\sin\left(\cos^{-1}\left(\frac{-1}{3}\right)\right)$

**11.** $\cos\left(\tan^{-1}\frac{1}{2}\right)$

**12.** $\tan(\tan^{-1}200)$

In Exercises 13–18, simplify the given expression.

**13.** $\sin(\cos^{-1}x)$

**14.** $\cos(\sin^{-1}x)$

**15.** $\cos(\tan^{-1}x)$

**16.** $\sin(\tan^{-1}x)$

**17.** $\tan(\cos^{-1}x)$

**18.** $\tan(\sec^{-1}x)$

In Exercises 19–32, differentiate the given function and simplify the answer whenever possible.

**19.** $y = \sin^{-1}\frac{2x-1}{3}$

**20.** $y = \tan^{-1}(ax + b)$

**21.** $y = \cos^{-1} \dfrac{x-b}{a}$      **22.** $f(x) = x \sin^{-1} x$

**23.** $f(t) = t \tan^{-1} t$      **24.** $u = z^2 \sec^{-1}(1 + z^2)$

**25.** $F(x) = (1 + x^2) \tan^{-1} x$    **26.** $y = \sin^{-1} \dfrac{a}{x}$

**27.** $G(x) = \dfrac{\sin^{-1} x}{\sin^{-1} 2x}$      **28.** $H(t) = \dfrac{\sin^{-1} t}{\sin t}$

**29.** $f(x) = (\sin^{-1} x^2)^{1/2}$    **30.** $y = \cos^{-1} \dfrac{a}{\sqrt{a^2 + x^2}}$

**31.** $y = \sqrt{a^2 - x^2} + a \sin^{-1} \dfrac{x}{a}$    $(a > 0)$

**32.** $y = a \cos^{-1}\left(1 - \dfrac{x}{a}\right) - \sqrt{2ax - x^2}$    $(a > 0)$

**33.** Find the slope of the curve $\tan^{-1}\left(\dfrac{2x}{y}\right) = \dfrac{\pi x}{y^2}$ at the point $(1, 2)$.

**34.** Find equations of two straight lines tangent to the graph of $y = \sin^{-1} x$ and having slope 2.

**35.** Show that, on their respective domains, $\sin^{-1}$ and $\tan^{-1}$ are increasing functions and $\cos^{-1}$ is a decreasing function.

**36.** The derivative of $\sec^{-1} x$ is positive for every $x$ in the domain of $\sec^{-1}$. Does this imply that $\sec^{-1}$ is increasing on its domain? Why?

**37.** Sketch the graph of $\csc^{-1} x$ and find its derivative.

**38.** Sketch the graph of $\cot^{-1} x$ and find its derivative.

**39.** Show that $\tan^{-1} x + \cot^{-1} x = \frac{\pi}{2}$ for $x > 0$. What is the sum if $x < 0$?

**40.** Find the derivative of $g(x) = \tan(\tan^{-1} x)$ and sketch the graph of $g$.

In Exercises 41–44, plot the graphs of the given functions by first calculating and simplifying the derivative of the function. Where is each function continuous? Where is it differentiable?

**∗ 41.** $\cos^{-1}(\cos x)$      **∗ 42.** $\sin^{-1}(\cos x)$

**∗ 43.** $\tan^{-1}(\tan x)$      **∗ 44.** $\tan^{-1}(\cot x)$

**45.** Show that $\sin^{-1} x = \tan^{-1}\left(\dfrac{x}{\sqrt{1 - x^2}}\right)$ if $|x| < 1$.

**46.** Show that $\sec^{-1} x = \begin{cases} \tan^{-1} \sqrt{x^2 - 1} & \text{if } x \geq 1 \\ \pi - \tan^{-1} \sqrt{x^2 - 1} & \text{if } x \leq -1 \end{cases}$

**47.** Show that $\tan^{-1} x = \sin^{-1}\left(\dfrac{x}{\sqrt{1 + x^2}}\right)$ for all $x$.

**48.** Show that $\sec^{-1} x = \begin{cases} \sin^{-1} \dfrac{\sqrt{x^2 - 1}}{x} & \text{if } x \geq 1 \\ \pi - \sin^{-1} \dfrac{\sqrt{x^2 - 1}}{x} & \text{if } x \leq -1 \end{cases}$

**∗ 49.** Show that the function $f(x)$ of Example 9 is also constant on the interval $(-\infty, -1)$. Find the value of the constant. *Hint:* find $\lim_{x \to -\infty} f(x)$.

**∗ 50.** Find the derivative of $f(x) = x - \tan^{-1}(\tan x)$. What does your answer imply about $f(x)$? Calculate $f(0)$ and $f(\pi)$. Is there a contradiction here?

**∗ 51.** Find the derivative of $f(x) = x - \sin^{-1}(\sin x)$ for $-\pi \leq x \leq \pi$ and sketch the graph of $f$ on that interval.

In Exercises 52–55, solve the initial-value problems.

**❖ 52.** $\begin{cases} y' = \dfrac{1}{1 + x^2} \\ y(0) = 1 \end{cases}$    **❖ 53.** $\begin{cases} y' = \dfrac{1}{9 + x^2} \\ y(3) = 2 \end{cases}$

**❖ 54.** $\begin{cases} y' = \dfrac{1}{\sqrt{1 - x^2}} \\ y(1/2) = 1 \end{cases}$    **❖ 55.** $\begin{cases} y' = \dfrac{4}{\sqrt{25 - x^2}} \\ y(0) = 0 \end{cases}$

# 3.6 Hyperbolic Functions

Any function defined on the real line can be expressed (in a unique way) as the sum of an even function and an odd function. (See Exercise 35 of Section P.5.) The **hyperbolic functions** $\cosh x$ and $\sinh x$ are respectively the even and odd functions whose sum is the exponential function $e^x$.

**DEFINITION 15**

> **The hyperbolic cosine and hyperbolic sine functions**
>
> For any real $x$ the **hyperbolic cosine**, $\cosh x$, and the **hyperbolic sine**, $\sinh x$ are defined by
>
> $$\cosh x = \frac{e^x + e^{-x}}{2}, \qquad \sinh x = \frac{e^x - e^{-x}}{2}.$$

(The symbols *sinh* is somewhat hard to pronounce as written. Some people say "shine," and others say "sinch.") Recall that cosine and sine are called *circular functions* because, for any $t$, the point $(\cos t, \sin t)$ lies on the circle with equation $x^2 + y^2 = 1$. Similarly, cosh and sinh are called *hyperbolic functions* because the point $(\cosh t, \sinh t)$ lies on the rectangular hyperbola with equation $x^2 - y^2 = 1$, that is,

$$\cosh^2 t - \sinh^2 t = 1 \quad \text{for any real } t.$$

To see this, observe that

$$\cosh^2 t - \sinh^2 t = \left(\frac{e^t + e^{-t}}{2}\right)^2 - \left(\frac{e^t - e^{-t}}{2}\right)^2$$

$$= \frac{1}{4}\left(e^{2t} + 2 + e^{-2t} - (e^{2t} - 2 + e^{-2t})\right)$$

$$= \frac{1}{4}(2 + 2) = 1.$$

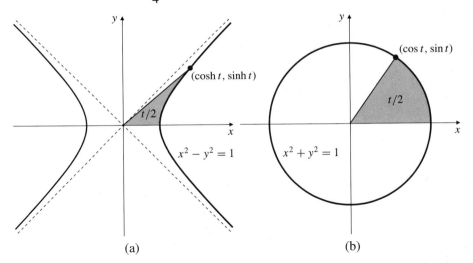

**Figure 3.26** Both shaded areas are $t/2$ square units

(a)  (b)

There is no interpretation of $t$ as an arc length or angle as there was in the circular case; however, the *area* of the *hyperbolic sector* bounded by $y = 0$, the hyperbola $x^2 - y^2 = 1$ and the ray from the origin to $(\cosh t, \sinh t)$ is $t/2$ square units (see Exercise 21 of Section 8.4), just as is the area of the circular sector bounded by $y = 0$, the circle $x^2 + y^2 = 1$ and the ray from the origin to $(\cos t, \sin t)$. (See Figure 3.26.)

Observe that, similar to the corresponding values of $\cos x$ and $\sin x$, we have

$$\cosh 0 = 1, \quad \text{and} \quad \sinh 0 = 0,$$

and $\cosh x$, like $\cos x$, is an even function, and $\sinh x$, like $\sin x$, is an odd function:

$$\cosh(-x) = \cosh x, \qquad \sinh(-x) = -\sinh x$$

Many other properties of the hyperbolic functions resemble those of the corresponding circular functions, sometimes with signs changed.

**Example 1** Show that

$$\frac{d}{dx}\cosh x = \sinh x \quad \text{and} \quad \frac{d}{dx}\sinh x = \cosh x.$$

**Solution** We have

$$\frac{d}{dx}\cosh x = \frac{d}{dx}\frac{e^x + e^{-x}}{2} = \frac{e^x + e^{-x}(-1)}{2} = \sinh x$$

$$\frac{d}{dx}\sinh x = \frac{d}{dx}\frac{e^x - e^{-x}}{2} = \frac{e^x - e^{-x}(-1)}{2} = \cosh x.$$

The following addition formulas and double angle formulas can be checked algebraically by using the definition of cosh and sinh and the laws of exponents:

$$\cosh(x + y) = \cosh x \cosh y + \sinh x \sinh y$$

$$\sinh(x + y) = \sinh x \cosh y + \cosh x \sinh y$$

$$\cosh(2x) = \cosh^2 x + \sinh^2 x = 1 + 2\sinh^2 x = 2\cosh^2 x - 1$$

$$\sinh(2x) = 2\sinh x \cosh x.$$

The graphs of cosh and sinh are shown in Figure 3.27. The graph $y = \cosh x$ is called a **catenary**. A chain hanging by its ends will assume the shape of a catenary.

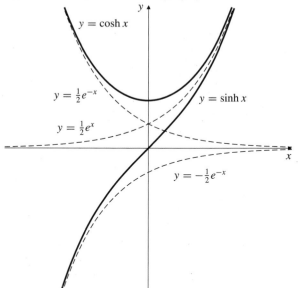

**Figure 3.27**    The graphs of cosh and sinh, and some exponential graphs to which they are asymptotic

By analogy with the trigonometric functions, four other hyperbolic functions can be defined in terms of cosh and sinh.

**DEFINITION 16**

**Other hyperbolic functions**

$$\tanh x = \frac{\sinh x}{\cosh x} = \frac{e^x - e^{-x}}{e^x + e^{-x}} \qquad \text{sech } x = \frac{1}{\cosh x} = \frac{2}{e^x + e^{-x}}$$

$$\coth x = \frac{\cosh x}{\sinh x} = \frac{e^x + e^{-x}}{e^x - e^{-x}} \qquad \text{csch } x = \frac{1}{\sinh x} = \frac{2}{e^x - e^{-x}}$$

Multiplying the numerator and denominator of the fraction defining $\tanh x$ by $e^{-x}$ and $e^x$, respectively, we obtain

$$\lim_{x\to\infty} \tanh x = \lim_{x\to\infty} \frac{1 - e^{-2x}}{1 + e^{-2x}} = 1, \quad \text{and}$$

$$\lim_{x\to-\infty} \tanh x = \lim_{x\to-\infty} \frac{e^{2x} - 1}{e^{2x} + 1} = -1,$$

so that the graph of $y = \tanh x$ has two horizontal asymptotes. The graph resembles that of the inverse tangent function in shape, as you can see in Figure 3.28.

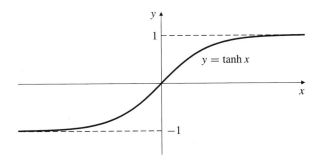

**Figure 3.28**     The graph of $\tanh x$

The derivatives of the remaining hyperbolic functions

$$\frac{d}{dx} \tanh x = \operatorname{sech}^2 x \qquad \frac{d}{dx} \operatorname{sech} x = -\operatorname{sech} x \tanh x$$

$$\frac{d}{dx} \coth x = -\operatorname{csch}^2 x \qquad \frac{d}{dx} \operatorname{csch} x = -\operatorname{csch} x \coth x$$

are easily calculated from those of $\cosh x$ and $\sinh x$ using the Reciprocal and Quotient Rules. For example,

$$\frac{d}{dx} \tanh x = \frac{d}{dx} \frac{\sinh x}{\cosh x} = \frac{(\cosh x)(\cosh x) - (\sinh x)(\sinh x)}{\cosh^2 x}$$

$$= \frac{1}{\cosh^2 x} = \operatorname{sech}^2 x.$$

## Inverse Hyperbolic Functions

The functions sinh and tanh are increasing and therefore one-to-one and invertible on the whole real line. Their inverses are denoted $\sinh^{-1}$ and $\tanh^{-1}$ respectively:

$$y = \sinh^{-1} x \quad \Longleftrightarrow \quad x = \sinh y,$$
$$y = \tanh^{-1} x \quad \Longleftrightarrow \quad x = \tanh y.$$

Since the hyperbolic functions are defined in terms of exponentials, it is not surprising that their inverses can be expressed in terms of logarithms.

**Example 2** Express $\sinh^{-1} x$ and $\tanh^{-1} x$ in terms of logarithms.

**Solution** Let $y = \sinh^{-1} x$. Then

$$x = \sinh y = \frac{e^y - e^{-y}}{2} = \frac{(e^y)^2 - 1}{2e^y}.$$

(We multiplied the numerator and denominator of the first fraction by $e^y$ to get the second fraction.) Therefore

$$(e^y)^2 - 2xe^y - 1 = 0.$$

This is a quadratic equation in $e^y$, and it can be solved by the quadratic formula:

$$e^y = \frac{2x \pm \sqrt{4x^2 + 4}}{2} = x \pm \sqrt{x^2 + 1}.$$

Note that $\sqrt{x^2 + 1} > x$. Since $e^y$ cannot be negative, we need to use the positive square root:

$$e^y = x + \sqrt{x^2 + 1}.$$

Hence $y = \ln\left(x + \sqrt{x^2 + 1}\right)$, and we have

$$\sinh^{-1} x = \ln\left(x + \sqrt{x^2 + 1}\right).$$

Now let $y = \tanh^{-1} x$. Then

$$x = \tanh y = \frac{e^y - e^{-y}}{e^y + e^{-y}} = \frac{e^{2y} - 1}{e^{2y} + 1} \qquad (-1 < x < 1),$$

$$xe^{2y} + x = e^{2y} - 1,$$

$$e^{2y} = \frac{1 + x}{1 - x}, \qquad y = \frac{1}{2} \ln\left(\frac{1 + x}{1 - x}\right).$$

Thus

$$\tanh^{-1} x = \frac{1}{2} \ln\left(\frac{1 + x}{1 - x}\right), \qquad (-1 < x < 1).$$

Since cosh is not one-to-one its domain must be restricted before an inverse can be defined. Let us define the principal value of cosh to be

$$\text{Cosh } x = \cosh x \qquad (x \geq 0).$$

The inverse, $\cosh^{-1}$, is then defined by

$$y = \cosh^{-1} x \quad \Longleftrightarrow \quad x = \text{Cosh } y$$
$$\Longleftrightarrow \quad x = \cosh y \qquad (y \geq 0).$$

As we did for $\sinh^{-1}$, we can obtain the formula

$$\cosh^{-1}x = \ln\left(x + \sqrt{x^2 - 1}\right), \qquad (x \geq 1).$$

# Exercises 3.6

**1.** Verify the formulas for the derivatives of $\operatorname{sech} x$, $\operatorname{csch} x$, and $\coth x$ given in this section.

**2.** Verify the addition formulas

$$\cosh(x + y) = \cosh x \cosh y + \sinh x \sinh y$$
$$\sinh(x + y) = \sinh x \cosh y + \cosh x \sinh y.$$

Proceed by expanding the right-hand side of each identity in terms of exponentials. Find similar formulas for $\cosh(x - y)$ and $\sinh(x - y)$.

**3.** Obtain addition formulas for $\tanh(x + y)$ and $\tanh(x - y)$ from those for sinh and cosh.

**4.** Sketch the graphs of $y = \coth x$, $y = \operatorname{sech} x$, and $y = \operatorname{csch} x$, showing any asymptotes.

**5.** Calculate the derivatives of $\sinh^{-1} x$, $\cosh^{-1}x$, and $\tanh^{-1} x$. Hence express each of the indefinite integrals

$$\int \frac{dx}{\sqrt{x^2 + 1}}, \qquad \int \frac{dx}{\sqrt{x^2 - 1}}, \qquad \int \frac{dx}{1 - x^2}$$

in terms of inverse hyperbolic functions.

**6.** Calculate the derivatives of the functions $\sinh^{-1}(x/a)$, $\cosh^{-1}(x/a)$, and $\tanh^{-1}(x/a)$ (where $a > 0$) and use your answers to provide formulas for certain indefinite integrals.

**7.** Simplify the following expressions: (a) $\sinh \ln x$, (b) $\cosh \ln x$, (c) $\tanh \ln x$, (d) $\dfrac{\cosh \ln x + \sinh \ln x}{\cosh \ln x - \sinh \ln x}$

**8.** Let $\operatorname{csch}^{-1}x = \sinh^{-1}(1/x)$. Find the domain, range, and derivative of $\operatorname{csch}^{-1}x$, and sketch its graph. Express $\operatorname{csch}^{-1}x$ in terms of logarithms.

**9.** Do an analogous version of Exercise 8 for the function $\coth^{-1} x$.

**\* 10.** Define $\operatorname{Sech} x$ at a suitably restricted version of $\operatorname{sech} x$ and repeat Exercise 8 for the function $\operatorname{Sech}^{-1}x$.

**❖ 11.** Show that the functions $f_{A,B}(x) = Ae^{kx} + Be^{-kx}$ and $g_{C,D}(x) = C \cosh kx + D \sinh kx$ are both solutions of the differential equation $y'' - k^2 y = 0$. (They are both general solutions.) Express $f_{A,B}$ in terms of $g_{C,D}$, and express $g_{C,D}$ in terms of $f_{A,B}$.

**❖ 12.** Show that $h_{L,M}(x) = L \cosh k(x - a) + M \sinh k(x - a)$ is also a solution of the differential equation in the previous exercise. Express $h_{L,M}$ in terms of the function $f_{A,B}$ above.

**❖ 13.** Solve the initial-value problem $y'' - k^2 y = 0$, $\quad y(a) = y_0$, $y'(a) = v_0$. Express the solution in terms of the function $h_{L,M}$ of Exercise 12.

# 3.7 Second-Order Linear DEs with Constant Coefficients

In this section we develop a recipe for solving differential equations of the form

$$a y'' + b y' + cy = 0, \qquad (*)$$

where $a$, $b$, and $c$ are constants and $a \neq 0$, and illustrate its application to the study of simple and damped harmonic motion. Equation $(*)$ is a **second-order, linear, homogeneous** differential equation. The *second-order* refers to the presence of a second derivative, the terms *linear* and *homogeneous* refer to the fact that if $y_1(t)$ and $y_2(t)$ are two solutions of the equation, then so is $y(t) = Ay_1(t) + By_2(t)$ for any constants $A$ and $B$.

> If $ay_1''(t) + by_1'(t) + cy_1(t) = 0$ and $ay_2''(t) + by_2'(t) + cy_2(t) = 0$,
> and if $y(t) = Ay_1(t) + By_2(t)$, then $ay''(t) + by'(t) + cy(t) = 0$.

(Throughout this section we will assume that the independent variable in our functions is $t$ rather than $x$, so the prime ($'$) refers to the derivative $d/dt$. This is because in most applications of such equations the independent variable is time.)

Equations of type ($*$) arise in many applications of mathematics. Later in this section we will examine the equation of simple harmonic motion,

$$y'' + \omega^2 y = 0,$$

(which is of type ($*$) with $a = 1$, $b = 0$ and $c = \omega^2$) governing the vibration of a mass suspended by an elastic spring. If the motion of the mass is impeded by a resistance proportional to its velocity (supplied, say, by friction with the air) then the equation of motion would look like ($*$) with a positive constant $b$. The mass will still vibrate if the resistance is not too large ($b^2 < 4ac$) but the amplitude of the vibration will decay exponentially with time.

Equation ($*$) also arises in the study of current flowing in an electric circuit having resistors, capacitors and inductors connected in series. In this case $a$ measures the inductance, $b$ the resistance and $c$ the reciprocal of the capacitance of the circuit. In both mechanical (vibrating mass) and electrical applications the constants $a$, $b$, and $c$ are positive.

Let us try to find a solution of equation ($*$) having the form $y = e^{rt}$. Substituting this expression into equation ($*$), we obtain

$$ar^2 e^{rt} + bre^{rt} + ce^{rt} = 0.$$

Since $e^{rt}$ is never zero, $y = e^{rt}$ will be a solution of the differential equation ($*$) if and only if $r$ satisfies the quadratic **auxiliary equation**

$$ar^2 + br + c = 0, \qquad (**)$$

which has roots given by the quadratic formula:

$$r = \frac{-b \pm \sqrt{b^2 - 4ac}}{2a} = -\frac{b}{2a} \pm \frac{\sqrt{D}}{2a},$$

where $D = b^2 - 4ac$ is called the **discriminant** of the auxiliary equation ($**$).

There are three cases to consider, depending on whether the discriminant $D$ is positive, zero, or negative.

**CASE I**  Suppose $D = b^2 - 4ac > 0$. Then the auxiliary equation has two different real roots, $r_1$ and $r_2$, given by

$$r_1 = \frac{-b - \sqrt{D}}{2a}, \qquad r_2 = \frac{-b + \sqrt{D}}{2a}.$$

(Sometimes these roots can be found easily by factoring the left side of the auxiliary equation.) In this case both $y = y_1(t) = e^{r_1 t}$ and $y = y_2(t) = e^{r_2 t}$ are solutions of the differential equation ($*$), and neither is a multiple of the other. As noted above, the function

$$y = A e^{r_1 t} + B e^{r_2 t},$$

is also a solution for any choice of the constants $A$ and $B$. Since the differential equation is of second order and this solution involves two arbitrary constants we suspect it is the **general solution**, that is, that every solution of the differential equation can be written in this form. Exercise 17 at the end of this section outlines a way to prove this.

**CASE II**   Suppose $D = b^2 - 4ac = 0$. Then the auxiliary equation has two equal roots $r_1 = r_2 = -b/(2a) = r$, say. Certainly $y = e^{rt}$ is a solution of ($*$). We can find the general solution by letting $y = e^{rt}u(t)$, and calculating the first two derivatives of $y$.

$$y' = e^{rt}\big(u'(t) + ru(t)\big)$$
$$y'' = e^{rt}\big(u''(t) + 2ru'(t) + r^2 u(t)\big)$$

Substituting these expressions into ($*$), we obtain

$$e^{rt}\big(au''(t) + (2ar + b)u'(t) + (ar^2 + br + c)u(t)\big) = 0.$$

Since $e^{rt} \neq 0$, $2ar + b = 0$, and $r$ satisfies ($**$), this equation reduces to $u''(t) = 0$, which has general solution $u(t) = A + Bt$ for arbitrary constants $A$ and $B$. Thus the general solution of ($*$) in this case is

$$y = A e^{rt} + Bt\, e^{rt}.$$

**CASE III**   Suppose $D = b^2 - 4ac < 0$. Then the auxiliary equation ($**$) has complex conjugate roots given by

$$r = \frac{-b \pm \sqrt{b^2 - 4ac}}{2a} = k \pm i\omega,$$

where $k = -b/(2a)$, $\omega = \sqrt{4ac - b^2}/(2a)$, and $i$ is the imaginary unit ($i^2 = -1$; see Appendix I). As in CASE II, we perform the substitution $y = e^{kt}u(t)$. The details are left to the reader, but the resulting differential equation for $u(t)$ is

$$u''(t) + \omega^2 u(t) = 0,$$

for which a general solution is

$$u(t) = A \cos(\omega t) + B \sin(\omega t),$$

where $A$ and $B$ are any constants. (See Exercises 18–21 below.) Therefore the general solution of ($*$) in this case is

$$y = A e^{kt} \cos(\omega t) + B e^{kt} \sin(\omega t).$$

The following examples illustrate the solution of ($*$) in each of these three cases.

**Example 1** Find the general solution of $y'' + y' - 2y = 0$.

**Solution** The auxiliary equation is $r^2 + r - 2 = 0$, or $(r + 2)(r - 1) = 0$. The auxiliary roots are $r_1 = -2$ and $r_2 = 1$, which are real and unequal. According to Case I, the general solution of the differential equation is

$$y = A e^{-2t} + B e^t.$$

**Example 2** Find the general solution of $y'' + 6y' + 9y = 0$.

**Solution** The auxiliary equation is $r^2 + 6r + 9 = 0$, or $(r + 3)^2 = 0$, which has equal roots $r = -3$. According Case II, the general solution of the differential equation is

$$y = A e^{-3t} + B t e^{-3t}.$$

**Example 3** Find the general solution of $y'' + 4y' + 13y = 0$.

**Solution** The auxiliary equation is $r^2 + 4r + 13 = 0$, which has solutions

$$r = \frac{-4 \pm \sqrt{16 - 52}}{2} = \frac{-4 \pm \sqrt{-36}}{2} = -2 \pm 3i.$$

Thus $k = -2$ and $\omega = 3$. According to Case III, the general solution of the given differential equation is

$$y = A e^{-2t} \cos(3t) + B e^{-2t} \sin(3t).$$

Initial-value problems for $ay'' + by' + cy = 0$ specify values for $y$ and $y'$ at an initial point. These values can be used to determine the values of the constants $A$ and $B$ in the general solution, so the initial-value problem has a unique solution.

**Example 4** Solve the initial-value problem

$$\begin{cases} y'' + 2y' + 2y = 0 \\ y(0) = 2 \\ y'(0) = -3. \end{cases}$$

**Solution** The auxiliary equation is $r^2 + 2r + 2$, which has roots

$$r = \frac{-2 \pm \sqrt{4 - 8}}{2} = -1 \pm i.$$

Thus Case III applies, $k = -1$ and $\omega = 1$. Thus the differential equation has the general solution

$$y = A\,e^{-t}\cos t + B\,e^{-t}\sin t.$$

Also

$$\begin{aligned}
y' &= e^{-t}\left(-A\cos t - B\sin t - A\sin t + B\cos t\right)\\
&= (B - A)\,e^{-t}\cos t - (A + B)\,e^{-t}\sin t.
\end{aligned}$$

Applying the initial conditions $y(0) = 2$ and $y'(0) = -3$, we obtain $A = 2$ and $B - A = -3$. Hence $B = -1$ and the initial-value problem has the solution

$$y = 2\,e^{-t}\cos t - e^{-t}\sin t.$$

## Simple Harmonic Motion

Many natural phenomena exhibit periodic behaviour. The swinging of a clock pendulum, the vibrating of a plucked guitar string or drum membrane, the altitude of a rider on a rotating ferris wheel, the motion of an object floating in wavy seas, and the voltage produced by an alternating current generator are but a few examples where quantities depend on time in a periodic way. Being periodic, the circular functions sine and cosine provide a useful model for such behaviour.

It often happens that a quantity displaced from an equilibrium value experiences a restoring force that tends to move it back in the direction of its equilibrium. Besides the obvious examples of elastic motions in physics, one can imagine such a model applying, say, to a biological population in equilibrium with its food supply, or the price of a commodity in an elastic economy where increasing price causes decreasing demand and hence decreasing price. In the simplest models, the restoring force is proportional to the amount of displacement from equilibrium. Such a force causes the quantity to oscillate sinusoidally; we say that it executes *simple harmonic motion*.

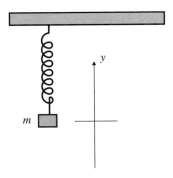

As a specific example, suppose a mass $m$ is suspended by an elastic spring so that it hangs unmoving in its equilibrium position. If it is displaced vertically by an amount $y$ from this position, a force is exerted by the spring, directed to restore the mass to its equilibrium position. (See Figure 3.29.) This force is proportional to the displacement (Hooke's Law); its magnitude is $-ky$, where $k$ is a positive constant called the **spring constant**. Assuming the spring is weightless, this force imparts to the mass $m$ an acceleration $d^2y/dt^2$ that satisfies, by Newton's Second Law, $m(d^2y/dt^2) = -ky$, (mass $\times$ acceleration = force). Dividing this equation by $m$, we obtain the equation

**Figure 3.29**

$$\frac{d^2y}{dt^2} + \omega^2 y = 0, \qquad \text{where} \quad \omega^2 = \frac{k}{m}.$$

The second-order differential equation

$$\frac{d^2y}{dt^2} + \omega^2 y = 0$$

is called the **equation of simple harmonic motion**. Its auxiliary equation, $r^2 + \omega^2 = 0$, has complex roots $r = \pm i\omega$, so it has general solution

$$y = A \cos \omega t + B \sin \omega t,$$

where $A$ and $B$ are arbitrary constants.

For any values of the constants $R$ and $t_0$, the function

$$y = R \cos\big(\omega(t - t_0)\big)$$

is also a general solution of the differential equation of simple harmonic motion. If we expand this formula using the addition formula for cosine, we get

$$y = R \cos \omega t_0 \cos \omega t + R \sin \omega t_0 \sin \omega t$$
$$= A \cos \omega t + B \sin \omega t,$$

where

$$A = R \cos(\omega t_0) \qquad B = R \sin(\omega t_0)$$
$$R^2 = A^2 + B^2 \qquad \tan(\omega t_0) = B/A.$$

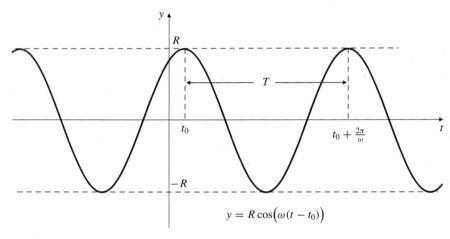

$$y = R \cos\big(\omega(t - t_0)\big)$$

**Figure 3.30**    Simple harmonic motion

The constants $A$ and $B$ are related to the position $y_0$ and the velocity $v_0$ of the mass $m$ at time $t = 0$:

$$y_0 = y(0) = A \cos 0 + B \sin 0 = A,$$
$$v_0 = y'(0) = -A\omega \sin 0 + B\omega \cos 0 = B\omega.$$

The constant $R = \sqrt{A^2 + B^2}$ is called the **amplitude** of the motion. Because $\cos x$ oscillates between $-1$ and $1$, the displacement $y$ varies between $-R$ and $R$. Note in Figure 3.30 that the graph of the displacement as a function of time is the curve $y = R \cos \omega t$ shifted $t_0$ units to the right. The number $t_0$ is called the **time-shift**. (The related quantity $\omega t_0$ is called a **phase-shift**.) The **period** of this curve is $T = 2\pi/\omega$; it is the time interval between consecutive instants when the mass is at the same height moving in the same direction. The reciprocal $1/T$ of the period is called the **frequency** of the motion. It is usually measured in Hertz (Hz), that is, cycles per second. The quantity $\omega = 2\pi/T$ is called the **circular frequency**. It is measured in radians per second since 1 cycle = 1 revolution = $2\pi$ radians.

**Example 5** Solve the initial-value problem

$$\begin{cases} y'' + 16y = 0 \\ y(0) = -6 \\ y'(0) = 32. \end{cases}$$

Find the amplitude, frequency and period of the solution.

**Solution** Here $\omega^2 = 16$ so $\omega = 4$. The solution is of the form

$$y = A\cos(4t) + B\sin(4t).$$

Since $y(0) = -6$, we have $A = -6$. Also, $y'(t) = -4A\sin(4t) + 4B\cos(4t)$. Since $y'(0) = 32$, we have $4B = 32$, or $B = 8$. Thus the solution is

$$y = -6\cos(4t) + 8\sin(4t).$$

The amplitude is $\sqrt{(-6)^2 + 8^2} = 10$, the frequency is $\omega/(2\pi) = 4/(2\pi) \approx 0.637$ Hz, and the period is $2\pi/\omega \approx 1.57$ rad/s.

**Example 6** (**Spring-mass problem**) Suppose that a 100 gram mass is suspended from a spring and that a force of $3 \times 10^4$ dynes ($3 \times 10^4$ g cm/s$^2$) is required to produce a displacement from equilibrium of 1/3 cm. At time $t = 0$ the mass is pulled down 2 cm below equilibrium and flicked upwards with a velocity of 60 cm/s. Find its subsequent displacement at any time $t > 0$. Find the frequency, period, amplitude, and time-shift of the motion. Express the position of the mass at time $t$ in terms of the amplitude and the time-shift.

**Solution** The spring constant $k$ is determined from Hooke's Law, $F = -ky$. Here $F = -3 \times 10^4$ g cm/s$^2$ is the force of the spring on the mass displaced 1/3 cm:

$$-3 \times 10^4 = -\frac{1}{3}k,$$

so $k = 9 \times 10^4$ g/s$^2$. Hence the circular frequency is $\omega = \sqrt{k/m} = 30$ radians per second, the frequency is $\omega/2\pi = 15/\pi \approx 4.77$ Hz, and the period is $2\pi/\omega \approx 0.209$ seconds.

Since the displacement at time $t = 0$ is $y_0 = -2$ and the velocity at that time is $v_0 = 60$, the subsequent displacement is $y = A\cos(30t) + B\sin(30t)$, where $A = y_0 = -2$ and $B = v_0/\omega = 60/30 = 2$. Thus

$$y = -2\cos(30t) + 2\sin(30t), \qquad (y \text{ in cm}, t \text{ in seconds}).$$

The amplitude of the motion is $R = \sqrt{(-2)^2 + 2^2} = 2\sqrt{2} \approx 2.83$ cm. The time-shift $t_0$ must satisfy

$$-2 = A = R\cos(\omega t_0) = 2\sqrt{2}\cos(30t_0),$$
$$2 = B = R\sin(\omega t_0) = 2\sqrt{2}\sin(30t_0),$$

so $\sin(30t_0) = 1/\sqrt{2} = -\cos(30t_0)$. Hence the phase-shift is $30t_0 = 3\pi/4$ radians, and the time-shift is $t_0 = \pi/40 \approx 0.0785$ seconds. The position of the mass at time $t > 0$ is also given by

$$y = 2\sqrt{2}\cos\left(30(t - \frac{\pi}{40})\right).$$

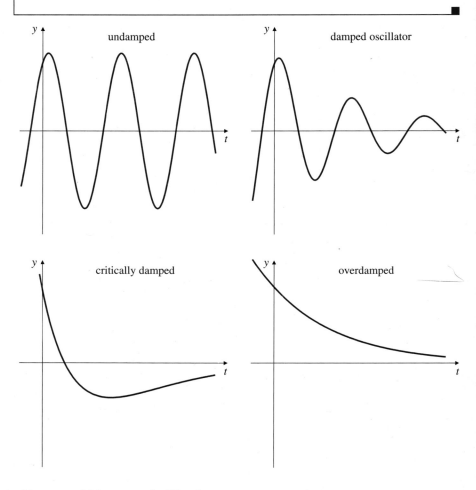

**Figure 3.31**

Undamped oscillator ($b = 0$)
Damped oscillator ($b > 0$, $b^2 < 4ac$)
Critically damped case ($b^2 = 4ac$)
Overdamped case ($b^2 > 4ac$)

## Damped Harmonic Motion

If $a$ and $c$ are positive, then equation

$$ay'' + by' + cy = 0$$

is the differential equation of simple harmonic motion and has oscillatory solutions of fixed amplitude as shown above. If $a > 0$, $b > 0$, and $c > 0$, then the roots of the auxiliary equation are either negative real numbers or, if $b^2 < 4ac$, complex numbers $k \pm i\omega$ with negative real parts $k = -b/(2a)$ (Case III). In this latter case the solutions still oscillate, but the amplitude diminishes exponentially as $t \to \infty$ because of the factor $e^{kt} = e^{-(b/2a)t}$. (See Exercise 16 below.) A system whose behaviour is modelled by such an equation is said to exhibit **damped harmonic motion**. If $b^2 = 4ac$ (Case II) the system is said to be **critically damped**, and if $b^2 > 4ac$ (Case I) it is **overdamped**. In these cases the behaviour is no longer oscillatory. (See Figure 3.31. Imagine a mass suspended by a spring in a jar of oil.)

## Exercises 3.7

In Exercises 1–12, find the general solutions for the given equations.

1. $y'' + 7y' + 10y = 0$
2. $y'' - 2y' - 3y = 0$
3. $y'' + 2y' = 0$
4. $4y'' - 4y' - 3y = 0$
5. $y'' + 8y' + 16y = 0$
6. $y'' - 2y' + y = 0$
7. $y'' - 6y' + 10y = 0$
8. $9y'' + 6y' + y = 0$
9. $y'' + 2y' + 5y = 0$
10. $y'' - 4y' + 5y = 0$
11. $y'' + 2y' + 3y = 0$
12. $y'' + y' + y = 0$

In Exercises 13–15, solve the given initial-value problems.

13. $\begin{cases} 2y'' + 5y' - 3y = 0 \\ y(0) = 1 \\ y'(0) = 0. \end{cases}$

14. $\begin{cases} y'' + 10y' + 25y = 0 \\ y(1) = 0 \\ y'(1) = 2. \end{cases}$

15. $\begin{cases} y'' + 4y' + 5y = 0 \\ y(0) = 2 \\ y'(0) = 2. \end{cases}$

16. If $a > 0$, $b > 0$, and $c > 0$, prove that all solutions of the differential equation $ay'' + by' + cy = 0$ satisfy $\lim_{t \to \infty} y(t) = 0$.

* 17. Prove that the solution given in the discussion of Case I, namely $y = A e^{r_1 t} + B e^{r_2 t}$, is the general solution for that case as follows: first let $y = e^{r_1 t} u$ and show that $u$ satisfies the equation

$$u'' - (r_2 - r_1)u' = 0.$$

Then let $v = u'$, so that $v$ must satisfy $v' = (r_2 - r_1)v$. The general solution of this equation is $v = C e^{(r_2 - r_1)t}$, as shown in the discussion of the equation $y' = ky$ at the beginning of Section 4.4. Hence find $u$ and $y$.

### Simple harmonic motion

Exercises 18–21 all refer to the differential equation of simple harmonic motion:

$$\frac{d^2 y}{dt^2} + \omega^2 y = 0, \qquad (\omega \neq 0). \qquad (\dagger)$$

Together they show that $y = A \cos \omega t + B \sin \omega t$ is a *general solution* of this equation, that is, that every solution is of this form for some choice of the constants $A$ and $B$.

◆ 18. Show that $y = A \cos \omega t + B \sin \omega t$ is a solution of $(*)$.
◆ 19. If $f(t)$ is any solution of $(\dagger)$, show that $\omega^2 (f(t))^2 + (f'(t))^2$ is constant.
◆ 20. If $g(t)$ is a solution of $(\dagger)$ satisfying $g(0) = g'(0) = 0$, show that $g(t) = 0$ for all $t$.
◆ 21. Suppose that $f(t)$ is any solution of the differential equation $(\dagger)$. Show that $f(t) = A \cos \omega t + B \sin \omega t$, where $A = f(0)$ and $B\omega = f'(0)$.
   (*Hint:* Let $g(t) = f(t) - A \cos \omega t - B \sin \omega t$.)

In Exercises 22–24, solve the given initial-value problems. For each problem determine the circular frequency, the frequency, the period, and the amplitude of the solution.

◆ 22. $\begin{cases} y'' + 4y = 0 \\ y(0) = 2 \\ y'(0) = -5 \end{cases}$

◆ 23. $\begin{cases} y'' + 100y = 0 \\ y(0) = 0 \\ y'(0) = 3 \end{cases}$

◆ 24. $\begin{cases} y'' + \dfrac{4\pi^2}{T^2} y = 0 \\ y(0) = R \\ y'(0) = 0 \end{cases}$

◆ 25. Show that $y = \alpha \cos(\omega(t - c)) + \beta \sin(\omega(t - c))$ is a solution of the differential equation $y'' + \omega^2 y = 0$, and that it satisfies $y(c) = \alpha$ and $y'(c) = \beta\omega$. Express the solution in the form $y = A \cos(\omega t) + B \sin(\omega t)$ for certain values of the constants $A$ and $B$ depending on $\alpha$, $\beta$, $c$, and $\omega$.

◆ 26. Solve $\begin{cases} y'' + y = 0 \\ y(2) = 3 \\ y'(2) = -4 \end{cases}$

◆ 27. Solve $\begin{cases} y'' + \omega^2 y = 0 \\ y(a) = A \\ y'(a) = B \end{cases}$

◆ 28. What mass should be suspended from the spring in Example 6 to provide a system whose natural frequency of oscillation is 10 Hz? Find the displacement of such a mass from its equilibrium position $t$ s after it is pulled down 1 cm from equilibrium and flicked upward with a speed of 2 cm/s. What is the amplitude of this motion?

◆ 29. A mass of 400 g suspended from a certain elastic spring will oscillate with a frequency of 24 Hz. What would be the frequency if the 400 g mass were replaced with a 900 g mass? A 100 g mass?

## Chapter Review

### Key Ideas

- State the laws of exponents.
- State the laws of logarithms.
- What is the significance of the number $e$?

- **What do the following statements and phrases mean?**
  ◇ $f$ is one-to-one.
  ◇ $f$ is invertible.
  ◇ Function $f^{-1}$ is the inverse of function $f$.
  ◇ $a^b = c$
  ◇ $\log_a b = c$

⋄ the natural logarithm of $x$

⋄ logarithmic differentiation

⋄ the half-life of a varying quantity

⋄ The quantity $y$ exhibits exponential growth.

⋄ The quantity $y$ exhibits logistic growth.

⋄ $y = \sin^{-1} x$    ⋄ $y = \tan^{-1} x$

⋄ The quantity $y$ exhibits simple harmonic motion.

⋄ The quantity $y$ exhibits damped harmonic motion.

● **Define the functions** $\sinh x$, $\cosh x$, **and** $\tanh x$.

● **What kinds of functions satisfy second-order differential equations with constant coefficients?**

## Review Exercises

**1.** If $f(x) = 3x + x^3$, show that $f$ has an inverse and find the slope of $y = f^{-1}(x)$ at $x = 0$.

**2.** Let $f(x) = \sec^2 x \tan x$. Show that $f$ is increasing on the interval $(-\pi/2, \pi/2)$, and hence one-to-one and invertible there. What is the domain of $f^{-1}$? Find $(f^{-1})'(2)$. *Hint:* $f(\pi/4) = 2$.

Exercises 3–5 refer to the function $f(x) = x\, e^{-x^2}$.

**3.** Find $\lim_{x \to \infty} f(x)$ and $\lim_{x \to -\infty} f(x)$.

**4.** On what intervals is $f$ increasing? decreasing?

**5.** What are the maximum and minimum values of $f(x)$?

**6.** Find the points on the graph of $y = e^{-x} \sin x$, $(0 \le x \le 2\pi)$, where the graph has a horizontal tangent line.

**7.** Suppose that a function $f(x)$ satisfies $f'(x) = x\, f(x)$ for all real $x$, and $f(2) = 3$. Calculate the derivative of $f(x)/e^{x^2/2}$, and use the result to help you find $f(x)$ explicitly.

**8.** A lump of modelling clay is being rolled out so that it maintains the shape of a circular cylinder. If the length is increasing at a rate proportional to itself, show that the radius is decreasing at a rate proportional to itself.

**9.** (a) What nominal interest rate, compounded continuously, will cause an investment to double in 5 years?

(b) By about how many days will the doubling time in part (a) increase if the nominal interest rate drops by 0.5%?

**10.** (**A poor man's natural logarithm**)

(a) Show that $\lim\limits_{h \to 0} \dfrac{a^h - 1}{h} = \ln a$ if $a > 0$, and hence that
$$\lim_{n \to \infty} n(a^{1/n} - 1) = \ln a.$$

(b) Most calculators, even non-scientific ones, have a square root key. If $n$ is a power of 2, say $n = 2^k$, then $a^{1/n}$ can be calculated by entering $a$ and hitting the square root key $k$ times:
$$a^{1/2^k} = \sqrt{\sqrt{\cdots \sqrt{a}}} \qquad (k \text{ square roots}).$$

Then you can subtract 1 and multiply by $n$ to get an approximation for $\ln a$. Use $n = 2^{10} = 1024$ and

$n = 2^{11} = 2048$ to find approximations for $\ln 2$. Based on the agreement of these two approximations, quote a value of $\ln 2$ to as many decimal places as you feel justified.

**11.** A nonconstant function $f$ satisfies $\dfrac{d}{dx}\big(f(x)\big)^2 = \big(f'(x)\big)^2$ for all $x$. If $f(0) = 1$, find $f(x)$.

**12.** If $f(x) = (\ln x)/x$, show that $f'(x) > 0$ for $0 < x < e$ and $f'(x) < 0$ for $x > e$, so that $f(x)$ has a maximum value at $x = e$. Use this to show that $e^{\pi} > \pi^e$.

**13.** Find an equation of a straight line that passes through the origin and is tangent to the curve $y = x^x$.

**14.** (a) Find $x \ne 2$ such that $\dfrac{\ln x}{x} = \dfrac{\ln 2}{2}$.

(b) Find $b > 1$ such that there is **no** $x \ne b$ with $\dfrac{\ln x}{x} = \dfrac{\ln b}{b}$.

**15.** Investment account A bears simple interest at a certain rate. Investment account B bears interest at the same nominal rate, but compounded instantaneously. If $1,000 is invested in each account, B produces $10 more in interest after one year than does A. Find the nominal rate both accounts use.

**16.** Express $\cos^{-1} x$, $\cot^{-1} x$, and $\csc^{-1} x$ in terms of $\tan^{-1}$.

**17.** Express $\cos^{-1} x$, $\cot^{-1} x$, and $\csc^{-1} x$ in terms of $\sin^{-1}$.

**18.** (**A warming problem**) A bottle of milk at $5°C$ is removed from a refrigerator into a room maintained at $20°C$. After 12 minutes the temperature of the milk is $12°C$. How much longer will it take for the milk to warm up to $18°C$?

**19.** (**A cooling problem**) A kettle of hot water at $96°C$ is allowed to sit in an air-conditioned room. The water cools to $60°C$ in 10 minutes and then to $40°C$ in another 10 minutes. What is the temperature of the room?

**20.** Show that $e^x > 1 + x$ if $x \ne 0$.

**21.** Use mathematical induction to show that
$$e^x > 1 + x + \frac{x^2}{2!} + \cdots + \frac{x^n}{n!}$$
if $x > 0$ and $n$ is any positive integer.

## Challenging Problems

**1.** (a) Show that the function $f(x) = x^x$ is strictly increasing on $[e^{-1}, \infty)$.

(b) If $g$ is the inverse function to $f$ of part (a), show that
$$\lim_{y \to \infty} \frac{g(y) \ln(\ln y)}{\ln y} = 1.$$

*Hint:* start with the equation $y = x^x$ and take the ln of both sides twice.

**Two models for incorporating air resistance into the analysis of the motion of a falling body**

2. **(Air resistance proportional to speed)** An object falls under gravity near the surface of the earth, and its motion is impeded by air resistance proportional to its speed. Its velocity $v$ therefore satisfies the equation

$$\frac{dv}{dt} = -g - kv, \qquad (*)$$

where $k$ is a positive constant depending on such factors as the shape and density of the object and the density of the air.

(a) Find the velocity of the object as a function of time $t$, given that it was $v_0$ at $t = 0$.

(b) Find the limiting velocity $\lim_{t \to \infty} v(t)$. Observe that this can be done either directly from $(*)$, or from the solution found in (a).

c) If the object was at height $y_0$ at time $t = 0$, find its height $y(t)$ at any time during its fall.

* 3. **(Air resistance proportional to the square of speed)** Under certain conditions a better model for the effect of air resistance on a moving object is one where the resistance is proportional to the square of the speed. For an object falling under constant gravitational acceleration $g$, the appropriate equation of motion is

$$\frac{dv}{dt} = -g - kv|v|,$$

where $k > 0$. Note that $v|v|$ is used instead of $v^2$ to ensure that the resistance is always in the opposite direction to the velocity. For an object falling from rest at time $t = 0$, we have $v(0) = 0$ and $v(t) < 0$ for $t > 0$, so the equation of motion becomes

$$\frac{dv}{dt} = -g + kv^2.$$

We are not (yet) in a position to solve this equation. However we can verify its solution.

(a) Verify that the velocity is given for $t \geq 0$ by

$$v(t) = \sqrt{\frac{g}{k}} \, \frac{1 - e^{2t\sqrt{gk}}}{1 + e^{2t\sqrt{gk}}}.$$

(b) What is the limiting velocity $\lim_{t \to \infty} v(t)$?

(c) Also verify that if the falling object was at height $y_0$ at time $t = 0$, then its height at subsequent times during its fall is given by

$$y(t) = y_0 + \sqrt{\frac{g}{k}} \, t - \frac{1}{k} \ln\left(\frac{1 + e^{2t\sqrt{gk}}}{2}\right).$$

4. **(A model for the spread of a new techology)** When a new and superior technology is introduced, the percentage $p$ of potential clients that adopt it might be expected to increase logistically with time. However, even newer technologies are continually being introduced, so adoption of a particular one will fall off exponentially over time. The following model exhibits this behaviour:

$$\frac{dp}{dt} = kp\left(1 - \frac{p}{e^{-bt}M}\right).$$

This DE suggests that the growth in $p$ is logistic but that the asymptotic limit is not a constant but rather $e^{-bt}M$, which decreases exponentially with time.

(a) Show that the change of variable $p = e^{-bt}y(t)$ transforms the equation above into a standard logistic equation, and hence find an explicit formula for $p(t)$ given that $p(0) = p_0$. It will be necessary to assume that $M \ll 100k/(b+k)$ to ensure that $p(t) < 100$.

(b) If $k = 10$, $b = 1$, $M = 90$, and $p_0 = 1$, how large will $p(t)$ become before it starts to decrease?

# CHAPTER 4

# Some Applications of Derivatives

***Introduction***  Differential calculus can be used to analyze many kinds of problems and situations that arise in applied disciplines. Calculus has made and will continue to make significant contributions to every field of human endeavour that uses quantitative measurement to further its aims. From economics to physics and from biology to sociology, problems can be found whose solutions can be aided by the use of some calculus.

In this chapter we will examine several kinds of problems to which the techniques we have already learned can be applied. These problems arise both outside and within mathematics. We will deal with the following kinds of problems:

1. Related rates problems – where the rates of change of related quantities are analyzed.
2. Graphing problems – where derivatives are used to illuminate the behaviour of functions.
3. Optimization problems – where a quantity is to be maximized or minimized.
4. Root finding methods – where we try to find numerical solutions of equations.
5. Approximation problems – where complicated functions are approximated by polynomials,
6. Evaluation of limits.

Do not assume that most of the problems we present here are "real-world" problems. Such problems are usually too complex to be treated in a general calculus course. However, the problems we consider, while sometimes artificial, do show how calculus can be applied in concrete situations.

## 4.1 Related Rates

When two or more quantities that change with time are linked by an equation, that equation can be differentiated with respect to time to produce an equation linking the rates of change of the quantities. Any one of these rates may then be determined when the others, and the values of the quantities themselves, are known. We will consider a couple of examples before formulating a list of procedures for dealing with such problems.

**Figure 4.1**

**Example 1**  An aircraft is flying horizontally at a speed of 600 km/h. How fast is the distance between the aircraft and a radio beacon increasing 1 minute after the aircraft passes 5 km directly above the beacon?

***Solution***  A diagram is useful here; see Figure 4.1. Let $C$ be the point on the aircraft's path directly above the beacon $B$. Let $A$ be the position of the aircraft $t$ minutes after it is at $C$, and let $x$ and $s$ be the distances $CA$ and $BA$. From the right triangle $BCA$ we have

$$s^2 = x^2 + 5^2.$$

We differentiate this equation implicitly with respect to $t$ to obtain

$$2s\frac{ds}{dt} = 2x\frac{dx}{dt}.$$

We are given that $dx/dt = 600$ km/h $= 10$ km/min. Therefore, $x = 10$ km at time $t = 1$ min. At that time $s = \sqrt{10^2 + 5^2} = 5\sqrt{5}$ km, and is increasing at rate

$$\frac{ds}{dt} = \frac{x}{s}\frac{dx}{dt} = \frac{10}{5\sqrt{5}}(600) = \frac{1,200}{\sqrt{5}} \approx 536.7 \text{ km/h}.$$

One minute after the aircraft passes over the beacon, its distance from the beacon is increasing at about 537 km/h. ∎

---

**Example 2**   How fast is the area of a rectangle changing if one side is 10 cm long and is increasing at a rate of 2 cm/s and the other side is 8 cm long and is decreasing at a rate of 3 cm/s?

**Solution**   Let the lengths of the sides of the rectangle at time $t$ be $x$ cm and $y$ cm, respectively. Thus the area at time $t$ is $A = xy$ cm$^2$. (See Figure 4.2.) We want to know the value of $dA/dt$ when $x = 10$ and $y = 8$, given that $dx/dt = 2$ and $dy/dt = -3$. (Note the negative sign to indicate that $y$ is decreasing.) Since all the quantities in the equation $A = xy$ are functions of time, we can differentiate that equation implicitly with respect to time and obtain

$$\left.\frac{dA}{dt}\right|_{\substack{x=10,\\y=8}} = \left.\left(\frac{dx}{dt}y + x\frac{dy}{dt}\right)\right|_{\substack{x=10,\\y=8}} = 2(8) + 10(-3) = -14.$$

At the time in question, the area of the rectangle is decreasing at a rate of 14 cm$^2$/s. ∎

**Figure 4.2**   Rectangle with sides changing

$A = xy$

$y$

$x$

## Procedures for Related-Rates Problems

In view of these examples we can formulate a few general procedures for dealing with related-rates problems.

> **How to Solve Related-Rates Problems**
>
> 1. Read the problem very carefully. Try to understand the relationships among the variable quantities. What is given? What is to be found?
>
> 2. Make a sketch if appropriate.
>
> 3. Define any symbols you want to use that are not defined in the statement of the problem. Express given and required quantities, and rates in terms of these symbols.
>
> 4. Discover from a careful reading of the problem or consideration of the sketch one or more equations linking the variable quantities. (You will need as many equations as quantities or rates to be found in the problem.)

5. Differentiate the equation(s) implicitly with respect to time, regarding all variable quantities as functions of time. You can manipulate the equation(s) algebraically before the differentiation is performed (for instance, you could solve for the quantities whose rates are to be found), but it is usually easier to differentiate the equations as they are originally obtained and solve for the desired items later.

6. Substitute any given values for the quantities and their rates, then solve the resulting equation(s) for the unknown quantities and rates.

7. Make a concluding statement answering the question asked. Is your answer "reasonable?" If not, check back through your solution to see what went wrong.

**Example 3**   Air is being pumped into a spherical balloon. The volume of the balloon is increasing at a rate of 20 cm$^3$/s when the radius is 30 cm. How fast is the radius increasing at that time?

**Solution**   Let $r$ and $V$ denote the radius and volume of the balloon, respectively, at time $t$. We are given that $dV/dt = 20$ cm$^3$/s when $r = 30$ cm. We want to find $dr/dt$ at that time. A diagram is not of much help in this problem, but we do have to know the formula for the volume of a sphere:

$$V = \frac{4}{3} \pi r^3.$$

Differentiating this equation implicitly with respect to time $t$, we obtain

$$\frac{dV}{dt} = 4\pi r^2 \frac{dr}{dt}.$$

We are given that $dV/dt = 20$ when $r = 30$. At this time,

$$20 = 4\pi \times 900 \frac{dr}{dt}, \qquad \text{so} \qquad \frac{dr}{dt} = \frac{1}{180\pi}.$$

The radius is increasing at $1/180\pi \approx 0.00177$ cm/s when it is 30 cm. ∎

Observe that while we wanted $dr/dt$ ultimately, we did not solve the initial equation $V = \frac{4}{3}\pi r^3$ for $r$ before differentiating. It is usually easier to differentiate the relationship between the variables in its original form and then solve for the desired rate.

**Example 4**   A lighthouse $L$ is located on a small island 2 km from the nearest point $A$ on a long, straight shoreline. If the lighthouse lamp rotates at 3 revolutions per minute, how fast is the illuminated spot $P$ on the shoreline moving along the shoreline when it is 4 km from $A$?

**Solution**   Referring to Figure 4.3, let $x$ be the distance $AP$ and let $\theta$ be the angle $\angle PLA$. Then $x = 2 \tan \theta$ and

$$\frac{dx}{dt} = 2 \sec^2 \theta \frac{d\theta}{dt}.$$

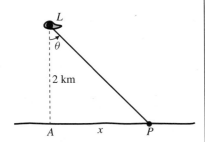

**Figure 4.3**

Now

$$\frac{d\theta}{dt} = 3 \text{ rev/min} \times 2\pi \text{ radians/rev} = 6\pi \text{ radians/min}.$$

When $x = 4$ we have $\tan\theta = 2$ and $\sec^2\theta = 1 + \tan^2\theta = 5$. Thus

$$\frac{dx}{dt} = 2 \times 5 \times 6\pi = 60\pi \approx 188.5.$$

The spot of light is moving along the shoreline at a rate of about 188.5 km/min when it is 4 km from $A$.

(Note that it was essential to convert the rate of change of $\theta$ from revolutions per minute to radians per minute. If $\theta$ were not measured in radians we could not assert that $(d/d\theta)\tan\theta = \sec^2\theta$.) ∎

**Example 5**   A water tank is in the shape of an inverted right circular cone with depth 5 m and top radius 2 m. Water leaks out of the tank at a rate proportional to the depth of the water in the tank. When the water in the tank is 4 m deep it is leaking out at a rate $1/12$ m³/min; how fast is the water level in the tank dropping at that time?

**Solution**   Let $r$ and $h$ denote the surface radius and depth of water in the tank at time $t$ (both measured in metres). Thus the volume $V$ (in m³) of water in the tank at time $t$ is

$$V = \frac{1}{3}\pi r^2 h.$$

Using similar triangles in Figure 4.4, we can find a relationship between $r$ and $h$:

$$\frac{r}{h} = \frac{2}{5}, \qquad \text{so} \qquad r = \frac{2h}{5}.$$

Thus

$$V = \frac{1}{3}\pi \left(\frac{2h}{5}\right)^2 h = \frac{4\pi}{75}h^3.$$

Differentiating this equation with respect to $t$ we obtain

$$\frac{dV}{dt} = \frac{4\pi}{25}h^2\frac{dh}{dt}.$$

We are told that the volume of water in the tank is decreasing at a rate proportional to $h$, that is, $dV/dt = kh$, and that the rate is $1/12$ m³/min when $h = 4$ m. Thus $-(1/12) = 4k$ and $k = -(1/48)$. Hence

$$-\frac{1}{48}h = \frac{dV}{dt} = \frac{4\pi}{25}h^2\frac{dh}{dt}.$$

It follows that

$$\frac{dh}{dt} = -\frac{25}{192\pi}\frac{1}{h},$$

and

$$\frac{dh}{dt}\bigg|_{h=4} = -\frac{25}{768\pi}.$$

When the water in the tank is 4 m deep its level is dropping at a rate of $25/(768\pi)$ m/min, or about 1.036 cm/min.  ∎

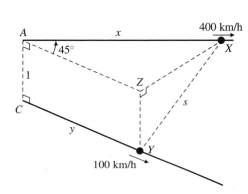

**Figure 4.4**  The conical tank of Example 5        **Figure 4.5**  Aircraft paths in Example 6

---

**Example 6**  At a certain instant an aircraft flying due east at 400 km/h passes directly over a car traveling due southeast at 100 km/h on a straight, level road. If the aircraft is flying at an altitude of 1 km, how fast is the distance between the aircraft and the car increasing 36 s after the aircraft passes directly over the car?

**Solution**  A good diagram is essential here. See Figure 4.5. Let time $t$ be measured in hours from the time the aircraft was at position $A$ directly above the car at position $C$. Let $X$ and $Y$ be the positions of the aircraft and the car, respectively, at time $t$. Let $x$ be the distance $AX$, $y$ be the distance $CY$, and let $s$ be the distance $XY$, all measured in kilometres. Let $Z$ be the point 1 km above $Y$. Since angle $XAZ = 45°$, the Pythagorean theorem and Cosine Law yield

$$s^2 = 1 + (ZX)^2 = 1 + x^2 + y^2 - 2xy\cos 45°$$
$$= 1 + x^2 + y^2 - \sqrt{2}xy.$$

Thus

$$2s\frac{ds}{dt} = 2x\frac{dx}{dt} + 2y\frac{dy}{dt} - \sqrt{2}\frac{dx}{dt}y - \sqrt{2}x\frac{dy}{dt}$$
$$= 400(2x - \sqrt{2}y) + 100(2y - \sqrt{2}x),$$

since $dx/dt = 400$ and $dy/dt = 100$. When $t = 1/100$ (that is, 36 s after $t = 0$) we have $x = 4$ and $y = 1$. Hence

$$s^2 = 1 + 16 + 1 - 4\sqrt{2} = 18 - 4\sqrt{2}$$
$$s \approx 3.5133.$$
$$\frac{ds}{dt} = \frac{1}{2s}\left(400(8 - \sqrt{2}) + 100(2 - 4\sqrt{2})\right) \approx 322.86.$$

The aircraft and the car are separating at a rate of about 323 km/h after 36 s.

(Note that it was necessary to convert 36 s to hours in the solution. In general all measurements should be in compatible units.)

# Exercises 4.1

1. Find the rate of change of the area of a square whose side is 8 cm long, if the side length is increasing at 2 cm/min.

2. The area of a square is decreasing at 2 ft$^2$/s. How fast is the side length changing when it is 8 ft?

3. A pebble dropped into a pond causes a circular ripple to expand outward from the point of impact. How fast is the area enclosed by the ripple increasing when the radius is 20 cm and is increasing at a rate of 4 cm/s?

4. The area of a circle is decreasing at a rate of 2 cm$^2$/min. How fast is the radius of the circle changing when the area is 100 cm$^2$?

5. The area of a circle is increasing at 1/3 km$^2$/h. Express the rate of change of the radius of the circle as a function of (a) the radius $r$, and (b) the area $A$ of the circle.

6. At a certain instant the length of a rectangle is 16 m and the width is 12 m. The width is increasing at 3 m/s. How fast is the length changing if the area of the rectangle is not changing?

7. How fast is the surface area of a cube changing when the volume of the cube is 64 cm$^3$ and is increasing at 2 cm$^3$/s?

8. The volume of a right circular cylinder is 60 cm$^3$ and is increasing at 2 cm$^3$/min at a time when the radius is 5 cm and is increasing at 1 cm/min. How fast is the height of the cylinder changing at that time?

9. How fast is the volume of a rectangular box changing when the length is 6 cm, the width is 5 cm, and the depth is 4 cm, if the length and depth are both increasing at a rate of 1 cm/s and the width is decreasing at a rate of 2 cm/s?

10. The area of a rectangle is increasing at a rate of 5 m$^2$/s while the length is increasing at a rate of 10 m/s. If the length is 20 m and the width is 16 m, how fast is the width changing?

11. A point moves on the curve $y = x^2$. How fast is $y$ changing when $x = -2$ and $x$ is decreasing at a rate 3?

12. A point is moving to the right along the first-quadrant portion of the curve $x^2 y^3 = 72$. When the point has coordinates $(3, 2)$ its horizontal velocity is 2 units/s. What is its vertical velocity?

13. The point $P$ moves so that at time $t$ it is at the intersection of the curves $xy = t$ and $y = tx^2$. How fast is the distance of $P$ from the origin changing at time $t = 2$?

14. **(Radar guns)** A policeman is standing near a highway using a radar gun to catch speeders. (See Figure 4.6.) He aims the gun at a car that has just passed his position and, when the gun is pointing at an angle of 45° to the direction of the highway, notes that the distance between the car and the gun is increasing at a rate of 100 km/h. How fast is the car travelling?

**Figure 4.6**

15. If the radar gun of Exercise 14 is aimed at a car travelling at 90 km/h along a straight road, what will its reading be at an instant when it is aimed in a direction making an angle of 30° with the road?

16. The top of a ladder 5 m long rests against a vertical wall. If the base of the ladder is being pulled away from the base of the wall at a rate of 1/3 m/s, how fast is the top of the ladder slipping down the wall when it is 3 m above the base of the wall?

17. A man 2 m tall walks toward a lamp post on level ground at a rate of 0.5 m/s. If the lamp is 5 m high on the post, how

fast is the length of the man's shadow decreasing when he is 3 m from the post? How fast is the shadow of his head moving at that time?

18. A woman 6 ft tall is walking at 2 ft/s along a straight path on level ground. There is a lamp post 5 ft to the side of the path. A light 15 ft high on the lamp post casts the woman's shadow on the ground. How fast is the length of her shadow changing when the woman is 12 feet from the point on the path closest to the lamp post?

19. **(Cost of production)** It costs a coal mine owner $C each day to maintain a production of $x$ tons of coal, where $C = 10,000 + 3x + x^2/8,000$. At what rate is the production increasing when it is 12,000 tons and the daily cost is increasing at $600 per day?

20. **(Distance between ships)** At 1:00 p.m. ship $A$ is 25 km due north of ship $B$. If ship $A$ is sailing west at a rate of 16 km/h and ship $B$ is sailing south at 20 km/h, find the rate at which the distance between the two ships is changing at 1:30 p.m.

21. What is the first time after 3:00 p.m. that the hands of the clock are together?

22. **(Tracking a balloon)** A balloon released at point $A$ rises vertically with a constant speed of 5 m/s. Point $B$ is level with and 100 m distant from point $A$. How fast is the angle of elevation of the balloon at $B$ changing when the balloon is 200 m above $A$?

23. Sawdust is falling onto a pile at a rate of 1/2 m³/min. If the pile maintains the shape of a right circular cone with height equal to half the diameter of its base, how fast is the height of the pile increasing when the pile is 3 m high?

24. **(Conical tank)** A water tank is in the shape of an inverted right circular cone with top radius 10 m and depth 8 m. Water is flowing in at a rate of 1/10 m³/min. How fast is the depth of water in the tank increasing when the water is 4 m deep?

25. **(Leaky tank)** Repeat Exercise 24 with the added assumption that water is leaking out of the bottom of the tank at a rate of $h^3/1,000$ m³/min when the depth of water in the tank is $h$ m. How full can the tank get in this case?

26. **(Another leaky tank)** Water is pouring into a leaky tank at a rate of 10 m³/h. The tank is a cone with vertex down, 9 m in depth and 6 m in diameter at the top. The surface of water in the tank is rising at a rate of 20 cm/h when the depth is 6 m. How fast is the water leaking out at that time?

27. **(Kite flying)** How fast must you let out line if the kite you are flying is 30 m high, 40 m horizontally away from you, and moving horizontally away from you at a rate of 10 m/min?

28. **(Ferris wheel)** You are riding on a Ferris wheel of diameter 20 m. The wheel is rotating at 1 revolution per minute. How fast are you rising or falling when you are 6 m horizontally away from the vertical line passing through the centre of the wheel?

29. **(Distance between aircraft)** An aircraft is 144 km east of an airport and is travelling west at 200 km/h. At the same time, a second aircraft at the same altitude is 60 km north of the airport and travelling north at 150 km/h. How fast is the distance between the two aircraft changing?

30. **(Production rate)** If a truck factory employs $x$ workers and has daily operating expenses of $\$y$, it can produce $P = (1/3)x^{0.6}y^{0.4}$ trucks per year. How fast are the daily expenses decreasing when they are $10,000 and the number of workers is 40, if the number of workers is increasing at 1 per day and production is remaining constant?

31. A lamp is located at point $(3, 0)$ in the $xy$-plane. An ant is crawling in the first quadrant of the plane and the lamp casts its shadow onto the $y$-axis. How fast is the ant's shadow moving along the $y$-axis when the ant is at position $(1, 2)$ and moving so that its $x$-coordinate is increasing at rate 1/3 units/s and its $y$-coordinate is decreasing at 1/4 units/s?

32. A straight highway and a straight canal intersect at right angles, the highway crossing over the canal on a bridge 20 m above the water. A boat travelling at 20 km/h passes under the bridge just as a car travelling at 80 km/h passes over it. How fast are the boat and car separating after one minute?

33. **(Filling a trough)** The cross section of a water trough is an equilateral triangle with top edge horizontal. If the trough is 10 m long and 30 cm deep, and if water is flowing in at a rate of 1/4 m³/min, how fast is the water level rising when the water is 20 cm deep at the deepest?

34. **(Draining a pool)** A rectangular swimming pool is 8 m wide and 20 m long. (See Figure 4.7.) Its bottom is a sloping plane, the depth increasing from 1 m at the shallow end to 3 m at the deep end. Water is draining out of the pool at a rate of 1 m³/min. How fast is the surface of the water falling when the depth of water at the deep end is (a) 2.5 m? (b) 1 m?

8 m    20 m

3 m    1 m

**Figure 4.7**

* 35. One end of a 10 m long ladder is on the ground and the ladder is supported part way along its length by resting on top of a 3 m high fence. (See Figure 4.8.) If the bottom of the ladder is 4 m from the base of the fence and is being dragged along the ground away from the fence at a rate of 1/5 m/s, how fast is the free top end of the ladder moving (a) vertically, and (b) horizontally?

**Figure 4.8**

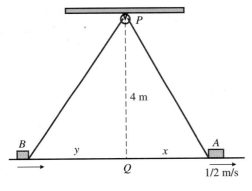

**Figure 4.9**

* **36.** Two crates, $A$ and $B$, are on the floor of a warehouse. The crates are joined by a rope 15 m long, each crate being hooked at floor level to an end of the rope. The rope is stretched tight and pulled over a pulley $P$ that is attached to a rafter 4 m above a point $Q$ on the floor directly between the two crates. (See Figure 4.9.) If crate $A$ is 3 m from $Q$ and is being pulled directly away from $Q$ at a rate of 1/2 m/s, how fast is crate $B$ moving toward $Q$?

**37. (Tracking a rocket)** Shortly after launch, a rocket is 100 km high and 50 km down range. If it is travelling at 4 km/s at an angle of 30° above the horizontal, how fast is its angle of elevation, as measured at the launch site, changing?

**38. (Shadow of a falling ball)** A lamp is 20 m high on a pole. At time $t = 0$ a ball is dropped from a point level with the lamp and 10 m away from it. The ball falls under gravity (acceleration 9.8 m/s²) until it hits the ground. How fast is the shadow of the ball moving along the ground (a) one second after it is dropped? (b) just as the ball hits the ground?

**39. (Tracking a rocket)** A rocket blasts off at time $t = 0$ and climbs vertically with acceleration 10 m/s². The progress of the rocket is monitored by a tracking station located 2 km horizontally away from the launch pad. How fast is the tracking station antenna rotating upward 10 s after launch?

## 4.2 Extreme Values

The first derivative of a function is a source of much useful information about the behaviour of the function. As we have already seen, the sign of $f'$ tells us whether $f$ is increasing or decreasing. In this section we use this information to find maximum and minimum values of functions. In Section 4.5 we will put the techniques developed here to use in solving problems requiring the finding of maximum and minimum values.

### Maximum and Minimum Values

Recall (from Section 1.4) that a function has a maximum value at $x_0$ if $f(x) \le f(x_0)$ for all $x$ in the domain of $f$. The maximum value is $f(x_0)$. To be more precise, we should call such a maximum value an *absolute* or *global* maximum because it is the largest value that $f$ attains anywhere on its entire domain.

**DEFINITION 1**

> **Absolute extreme values**
>
> Function $f$ has an **absolute maximum value** $f(x_0)$ at the point $x_0$ in its domain if $f(x) \le f(x_0)$ holds for every $x$ in the domain of $f$.
> Similarly, $f$ has an **absolute minimum value** $f(x_1)$ at the point $x_1$ in its domain if $f(x) \ge f(x_1)$ holds for every $x$ in the domain of $f$.

A function can have at most one absolute maximum or minimum value, though this value can be assumed at many points. For example, $f(x) = \sin x$ has absolute maximum value 1 occurring at every point of the form $x = (\pi/2) + 2n\pi$ where $n$ is an integer. Of course a function need not have any absolute extreme values. The function $f(x) = 1/x$ becomes arbitrarily large as $x$ approaches 0 from the right, so has no finite absolute maximum. (Remember, $\infty$ is not a number, and so is not a value of $f$.) Even a bounded function may not have an absolute maximum or minimum value. The function $g(x) = x$ with domain specified to be the *open* interval $(0, 1)$ has neither; the range of $g$ is also the interval $(0, 1)$ and there is no largest or smallest number in this interval. Of course, if the domain of $g$ were extended to be the *closed* interval $[0, 1]$ then $g$ would have both a maximum value, 1, and a minimum value, 0.

Maximum and minimum values of a function are collectively referred to as **extreme values**. The following theorem is a restatement (and slight generalization) of Theorem 8 of Section 1.4. It will prove very useful in some circumstances when we want to find extreme values.

**THEOREM 1**

**Existence of extreme values**

If the domain of the function $f$ is a *closed, finite interval*, or a union of finitely many such intervals, and if $f$ is *continuous* on that domain, then $f$ must have an absolute maximum value and an absolute minimum value.

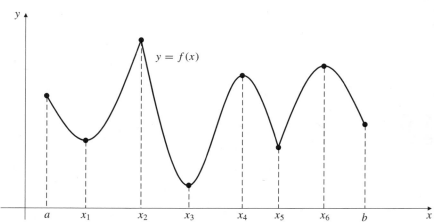

**Figure 4.10**    Local extreme values

Consider the graph $y = f(x)$ shown in Figure 4.10. Evidently the absolute maximum value of $f$ is $f(x_2)$ and the absolute minimum value is $f(x_3)$. In addition to these extreme values, $f$ has several other "local" maximum and minimum values corresponding to points on the graph that are higher or lower than neighbouring points. Observe that $f$ has *local maximum values* at $a$, $x_2$, $x_4$ and $x_6$, and local minimum values at $x_1$, $x_3$, $x_5$ and $b$. The absolute maximum is the highest of the local maxima; the absolute minimum is the lowest of the local minima.

**DEFINITION 2**

> **Local extreme values**
>
> Function $f$ has a **local maximum value (loc max)** $f(x_0)$ at the point $x_0$ in its domain provided there exists a number $h > 0$ such that $f(x) \leq f(x_0)$ whenever $x$ is in the domain of $f$ and $|x - x_0| < h$.
>
> Similarly, $f$ has a **local minimum value (loc min)** $f(x_1)$ at the point $x_1$ in its domain provided there exists a number $h > 0$ such that $f(x) \geq f(x_1)$ whenever $x$ is in the domain of $f$ and $|x - x_1| < h$.

Thus $f$ has a local maximum (or minimum) value at $x$ if it has an absolute maximum (or minimum) value at $x$ when its domain is restricted to points sufficiently near $x$. Geometrically, the graph of $f$ is at least as high (or low) at $x$ as it is at nearby points.

## Critical Points, Singular Points, and Endpoints

Figure 4.10 suggests that a function $f(x)$ can have local extreme values only at points $x$ that belong to one of three classes:

(i) **critical points** of $f$ (points $x$ in $D(f)$ where $f'(x) = 0$),

(ii) **singular points** of $f$ (points $x$ in $D(f)$ where $f'(x)$ is not defined), and

(iii) **endpoints** of the domain of $f$ (points in $D(f)$ that do not belong to any open interval contained in $D(f)$).

In Figure 4.10, $x_1$, $x_3$, $x_4$ and $x_6$ are critical points, $x_2$ and $x_5$ are singular points, and $a$ and $b$ are endpoints.

**THEOREM 2**

**Locating extreme values**

If the function $f$ is defined on an interval $I$ and has a local maximum (or local minimum) value at point $x = x_0$ in $I$, then $x_0$ must be either a critical point of $f$, a singular point of $f$, or an endpoint of $I$.

**PROOF**  Suppose that $f$ has a local maximum value at $x_0$, and that $x_0$ is neither an endpoint of the domain of $f$ nor a singular point of $f$. Then for some $h > 0$, $f(x)$ is defined on the open interval $(x_0 - h, x_0 + h)$ and has an absolute maximum (for that interval) at $x_0$. Also, $f'(x_0)$ exists. By Theorem 14 of Section 2.6, $f'(x_0) = 0$. The proof for the case where $f$ has a local minimum value at $x_0$ is similar.

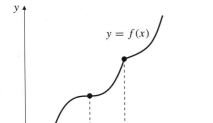

**Figure 4.11**    A function need not have extreme values at a critical point or a singular point

Although a function cannot have extreme values anywhere other than at endpoints, critical points, and singular points, it need not have extreme values at such points. Figure 4.11 shows the graph of a function with a critical point $x_0$ and a singular point $x_1$ at neither of which it has an extreme value. It is more difficult to draw the graph of a function whose domain has an endpoint at which the function fails to have an extreme value. See Exercise 53 at the end of this section for an example of such a function.

## Finding Absolute Extreme Values

If a function $f$ is defined on a closed interval, or a union of finitely many closed intervals, Theorem 1 assures us that $f$ must have an absolute maximum value and an absolute minimum value. Theorem 2 tells us how to find them. We need only check the values of $f$ at any critical points, singular points, and endpoints.

**Example 1**    Find the maximum and minimum values of the function

$$g(x) = x^3 - 3x^2 - 9x + 2$$

on the interval $-2 \leq x \leq 2$.

**Solution**    Since $g$ is a polynomial, it can have no singular points. For critical points, we calculate

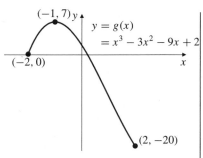

**Figure 4.12**    $g$ has maximum and minimum values 7 and $-20$ respectively

$$g'(x) = 3x^2 - 6x - 9 = 3(x^2 - 2x - 3)$$
$$= 3(x + 1)(x - 3)$$
$$= 0 \quad \text{if} \quad x = -1 \text{ or } x = 3.$$

However, $x = 3$ is not in the domain of $g$, so we can ignore it. We need to consider only the values of $g$ at the critical point $x = -1$, and at the endpoints $x = -2$ and $x = 2$:

$$g(-2) = 0, \qquad g(-1) = 7, \qquad g(2) = -20.$$

The maximum value of $g(x)$ on $-2 \le x \le 2$ is 7, at the critical point $x = -1$, and the minimum value is $-20$, at the endpoint $x = 2$. See Figure 4.12.  ∎

**Example 2**    Find the maximum and minimum values of $h(x) = 3x^{2/3} - 2x$ on the interval $[-1, 1]$.

**Solution**   The derivative of $h$ is

$$h'(x) = 3\left(\frac{2}{3}\right) x^{-1/3} - 2 = 2(x^{-1/3} - 1).$$

Note that $x^{-1/3}$ is not defined at the point $x = 0$ in $D(h)$, so $x = 0$ is a singular point of $h$. Also, $h$ has a critical point where $x^{-1/3} = 1$, that is, at $x = 1$, which also happens to be an endpoint of the domain of $h$. We must therefore examine the values of $h$ at the points $x = 0$ and $x = 1$, as well as at the other endpoint $x = -1$. We have

$$h(-1) = 5, \qquad h(0) = 0, \qquad h(1) = 1.$$

The function $h$ has maximum value 5 at the endpoint $-1$, and minimum value 0 at the singular point $x = 0$. See Figure 4.13.  ∎

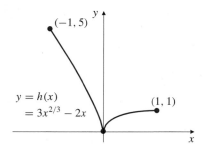

**Figure 4.13**    $h$ has absolute minimum value 0 at a singular point

## The First Derivative Test

Most functions you will encounter in elementary calculus have nonzero derivatives everywhere on their domains except possibly at a finite number of critical points, singular points, and endpoints of their domains. On intervals between these points the derivative exists and is not zero, so the function is either increasing or decreasing there. If $f$ is continuous and increases to the left of $x_0$ and decreases to the right, then it must have a local maximum value at $x_0$. The following theorem collects several results of this type together.

**THEOREM** ❸    **The first derivative test**

PART I. Testing interior critical points and singular points.

Suppose that $f$ is continuous at $x_0$, and $x_0$ is not an endpoint of the domain of $f$.

(a) If there exists an open interval $(a, b)$ containing $x_0$ such that $f'(x) > 0$ on $(a, x_0)$ and $f'(x) < 0$ on $(x_0, b)$, then $f$ has a local maximum value at $x_0$.

(b) If there exists an open interval $(a, b)$ containing $x_0$ such that $f'(x) < 0$ on

$(a, x_0)$ and $f'(x) > 0$ on $(x_0, b)$ then $f$ has a local minimum value at $x_0$.

PART II. Testing endpoints of the domain.

Suppose $x_0$ is a left endpoint of the domain of $f$ and $f$ is right continuous at $x_0$.

(c) If $f'(x) > 0$ on some interval $(x_0, b)$, then $f$ has a local minimum value at $x_0$.

(d) If $f'(x) < 0$ on some interval $(x_0, b)$, then $f$ has a local maximum value at $x_0$.

Suppose $x_0$ is a right endpoint of the domain of $f$ and $f$ is left continuous at $x_0$.

(e) If $f'(x) > 0$ on some interval $(a, x_0)$, then $f$ has a local maximum value at $x_0$.

(f) If $f'(x) < 0$ on some interval $(a, x_0)$, then $f$ has a local minimum value at $x_0$.

**Remark** If $f'$ is positive (or negative) on *both* sides of a critical or singular point, then $f$ has neither a maximum nor a minimum value at that point.

**Example 3** Find the local and absolute extreme values of $f(x) = x^4 - 2x^2 - 3$ on the interval $[-2, 2]$. Sketch the graph of $f$.

**Solution** We begin by calculating and factoring the derivative $f'(x)$:

$$f'(x) = 4x^3 - 4x = 4x(x^2 - 1) = 4x(x - 1)(x + 1).$$

The critical points are $0$, $-1$, and $1$. The corresponding values are $f(0) = -3$, $f(-1) = f(1) = -4$. There are no singular points. The values of $f$ at the endpoints $-2$ and $2$ are $f(-2) = f(2) = 5$. The factored form of $f'(x)$ is also convenient for determining the sign of $f'(x)$ on intervals between these endpoints and critical points. On $(1, 2)$ all the factors are positive so $f'(x) > 0$ there; on $(0, 1)$ the factor $(x - 1)$ is negative while the other factors are still positive, so $f'(x) < 0$ there. Similarly, $f'(x) > 0$ on $(-1, 0)$ because exactly two of its factors are negative, and $f'(x) < 0$ on $(-2, -1)$ because three of its factors are negative on that interval. This information about the sign of $f'(x)$ on intervals between points where $f'(x)$ is $0$ can also be obtained (provided $f'(x)$ is continuous, as it is here) by using the intermediate-value theorem (Theorem 9 of Section 1.4). We summarize the positive/negative properties of $f'(x)$ and the implied increasing/decreasing behaviour of $f(x)$ in chart form:

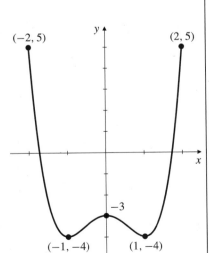

(−2, 5)   (2, 5)

(−1, −4)   (1, −4)

−3

**Figure 4.14** The graph $y = x^4 - 2x^2 - 3$

| | EP | | CP | | CP | | CP | | EP |
|---|---|---|---|---|---|---|---|---|---|
| $f'$ | $-2$ | $-$ | $-1$ | $+$ | $0$ | $-$ | $1$ | $+$ | $2$ |
| $f$ | loc max | ↘ | loc min | ↗ | loc max | ↘ | loc min | ↗ | loc max |

Note how the sloping arrows indicate visually the appropriate classification of the endpoints (EP) and critical points (CP) as determined by the First Derivative Test. We will make extensive use of such charts in future sections. The graph of $f$ is shown in Figure 4.14. Since the domain is a closed, finite interval, $f$ must have absolute maximum and minimum values. These are $5$ (at $\pm2$) and $-4$ (at $\pm1$).mfig[cpone]f-cponeThe graph $y = x^4 - 2x^2 - 3$

**Example 4** Find and classify the local and absolute extreme values of the function $f(x) = x - x^{2/3}$ with domain $[-1, 2]$. Sketch the graph of $f$.

**Solution** $f'(x) = 1 - \frac{2}{3}x^{-1/3} = \left(x^{1/3} - \frac{2}{3}\right)/x^{1/3}$. There is a singular point, $x = 0$, and a critical point, $x = 8/27$. The endpoints are $x = -1$ and $x = 2$. The values of $f$ at these points are $f(-1) = -2$, $f(0) = 0$, $f(8/27) = -4/27$, and $f(2) = 2 - 2^{2/3} \approx 0.4126$ (see Figure 4.15). Another interesting point on the graph is the $x$-intercept at $x = 1$. Information from $f'$ is summarized in the chart:

|   | EP |   | SP |   | CP |   | EP |
|---|----|----|----|----|----|----|----|
| $f'$ | $-1$ | $+$ | $0$ | $-$ | $\frac{8}{27}$ | $+$ | $2$ |
| $f$ | loc min | ↗ | loc max | ↘ | loc min | ↗ | loc max |

There are two local minima and two local maxima. The absolute maximum of $f$ is $2 - 2^{2/3}$ at $x = 2$; the absolute minimum is $-2$ at $x = -1$.  ∎

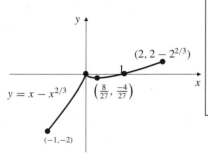

$y = x - x^{2/3}$

$(2, 2 - 2^{2/3})$

$\left(\frac{8}{27}, \frac{-4}{27}\right)$

$(-1, -2)$

**Figure 4.15** The graph for Example 4

## Functions Not Defined on Closed, Finite Intervals

If the function $f$ is not defined on a closed, finite interval, then Theorem 1 can not be used to guarantee the existence of maximum and minimum values for $f$. Of course, $f$ may still have such extreme values. In many applied situations we will want to find extreme values of functions defined on infinite and/or open intervals. The following theorem adapts Theorem 1 to cover some such situations.

**THEOREM 4**

**Existence of extreme values on open intervals**

If $f$ is continuous on the open interval $(a, b)$, and if

$$\lim_{x \to a+} f(x) = L \quad \text{and} \quad \lim_{x \to b-} f(x) = M,$$

then the following conclusions hold.

(i) If $f(u) > L$ and $f(u) > M$ for some $u$ in $(a, b)$, then $f$ has an absolute maximum value on $(a, b)$.

(ii) If $f(v) < L$ and $f(v) < M$ for some $v$ in $(a, b)$, then $f$ has an absolute minimum value on $(a, b)$.

In this theorem $a$ may be $-\infty$, in which case $\lim_{x \to a+}$ should be replaced with $\lim_{x \to -\infty}$, and $b$ may be $\infty$, in which case $\lim_{x \to b-}$ should be replaced with $\lim_{x \to \infty}$. Also, either or both of $L$ and $M$ may be either $\infty$ or $-\infty$.

**PROOF** We prove part (i); the proof of (ii) is similar. We are given that there is a number $u$ in $(a, b)$ such that $f(u) > L$ and $f(u) > M$. Here $L$ and $M$ may be finite numbers or $-\infty$. Since $\lim_{x \to a+} f(x) = L$, there must exist a number $x_1$ in $(a, u)$ such that

$$f(x) < f(u) \quad \text{whenever} \quad a < x < x_1.$$

Similarly, there must exist a number $x_2$ in $(u, b)$ such that

$$f(x) < f(u) \quad \text{whenever} \quad x_2 < x < b.$$

(See Figure 4.16.)  Thus $f(x) < f(u)$ at all points of $(a, b)$ that are not in the closed, finite subinterval $[x_1, x_2]$.  By Theorem 1, the function $f$, being continuous on $[x_1, x_2]$, must have an absolute maximum value on that interval, say at the point $w$.  Since $u$ belongs to $[x_1, x_2]$, we must have $f(w) \geq f(u)$, so $f(w)$ is the maximum value of $f(x)$ for all of $(a, b)$.

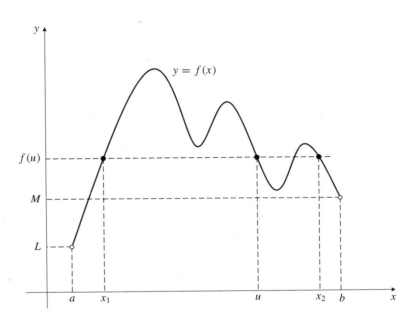

**Figure 4.16**

Theorem 2 still tells us where to look for extreme values.  There are no endpoints to consider in an open interval, but we must still look at the values of the function at any critical points or singular points in the interval.

**Figure 4.17**    $f$ has minimum value 4 at $x = 2$

**Example 5**   Show that $f(x) = x + (4/x)$ has an absolute minimum value on the interval $(0, \infty)$, and find that minimum value.

**Solution**   We have

$$\lim_{x \to 0+} f(x) = \infty \quad \text{and} \quad \lim_{x \to \infty} f(x) = \infty.$$

Since $f(1) = 2 < \infty$, Theorem 4 guarantees that $f$ must have an absolute minimum value at some point in $(0, \infty)$.  To find the minimum value we must check the values of $f$ at any critical points or singular points in the interval.  We have

$$f'(x) = 1 - \frac{4}{x^2} = \frac{x^2 - 4}{x^2} = \frac{(x-2)(x+2)}{x^2},$$

which equals 0 only at $x = 2$ and $x = -2$.  Since $f$ has domain $(0, \infty)$, it has no singular points and only one critical point, namely $x = 2$, where $f$ has the value $f(2) = 4$.  This must be the minimum value of $f$ on $(0, \infty)$.  (See Figure 4.17.)

---

**Example 6**   Let $f(x) = x e^{-x^2}$. Find and classify the critical points of $f$, evaluate $\lim_{x \to \pm\infty} f(x)$, and use these results to help you sketch the graph of $f$.

**Solution**   $f'(x) = e^{-x^2}(1 - 2x^2) = 0$ only if $1 - 2x^2 = 0$ since the exponential is always positive. Thus the critical points are $\pm\frac{1}{\sqrt{2}}$. We have $f\left(\pm\frac{1}{\sqrt{2}}\right) = \pm\frac{1}{\sqrt{2e}}$. $f'$ is positive (or negative) when $1 - 2x^2$ is positive (or negative). We summarize the intervals where $f$ is increasing and decreasing in chart form:

|  |  | CP |  | CP |  |
|---|---|---|---|---|---|
| $f'(x)$ | $-$ | $-\frac{1}{\sqrt{2}}$ | $+$ | $\frac{1}{\sqrt{2}}$ | $-$ |
| $f(x)$ | $\searrow$ | loc min | $\nearrow$ | loc max | $\searrow$ |

Note that $f(0) = 0$ and that $f$ is an odd function ($f(-x) = -f(x)$) so the graph is symmetric about the origin. Also,

$$\lim_{x \to \pm\infty} x e^{-x^2} = \left(\lim_{x \to \pm\infty} \frac{1}{x}\right)\left(\lim_{x \to \pm\infty} \frac{x^2}{e^{x^2}}\right) = 0 \times 0 = 0,$$

because $\lim_{x \to \pm\infty} x^2 e^{-x^2} = \lim_{u \to \infty} u e^{-u} = 0$ by Theorem 5 of Section 3.4. Since $f(x)$ is positive at $x = 1/\sqrt{2}$ and is negative at $x = -1/\sqrt{2}$, $f$ must have absolute maximum and minimum values by Theorem 4. These values can only be the values $\pm 1/\sqrt{2e}$ at the two critical points. The graph is shown in Figure 4.18. The $x$-axis is an asymptote as $x \to \pm\infty$. ∎

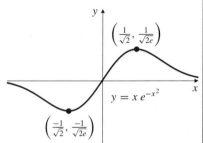

**Figure 4.18**   The graph for Example 6

# Exercises 4.2

In Exercises 1–17, determine whether the given function has any local or absolute extreme values, and find those values if possible.

**1.** $f(x) = x + 2$ on $[-1, 1]$   **2.** $f(x) = x + 2$ on $(-\infty, 0]$

**3.** $f(x) = x + 2$ on $[-1, 1)$   **4.** $f(x) = x^2 - 1$

**5.** $f(x) = x^2 - 1$ on $[-2, 3]$   **6.** $f(x) = x^2 - 1$ on $(2, 3)$

**7.** $f(x) = x^3 + x - 4$ on $[a, b]$

**8.** $f(x) = x^3 + x - 4$ on $(a, b)$

**9.** $f(x) = x^5 + x^3 + 2x$ on $(a, b]$

**10.** $f(x) = \dfrac{1}{x - 1}$   **11.** $f(x) = \dfrac{1}{x - 1}$ on $(0, 1)$

**12.** $f(x) = \dfrac{1}{x - 1}$ on $[2, 3]$   **13.** $|x - 1|$ on $[-2, 2]$

**14.** $|x^2 - x - 2|$ on $[-3, 3]$   **15.** $f(x) = \dfrac{1}{x^2 + 1}$

**16.** $f(x) = (x + 2)^{2/3}$   **17.** $f(x) = (x - 2)^{1/3}$

In Exercises 18–45, locate and classify all local extreme values of the given function. Determine whether any of these extreme values are absolute. Sketch the graph of the function.

**18.** $f(x) = x^2 + 2x$

**19.** $f(x) = x^3 - 3x - 2$

**20.** $f(x) = (x^2 - 4)^2$

**21.** $f(x) = x(x - 1)^2$

**22.** $f(x) = x^4 + 4x$

**23.** $f(x) = x^3(x - 1)^2$

**24.** $f(x) = x^2(x - 1)^2$

**25.** $f(x) = x(x^2 - 1)^2$

**26.** $f(x) = \dfrac{x}{x^2 + 1}$

**27.** $f(x) = \dfrac{x^2}{x^2 + 1}$

**28.** $f(x) = \dfrac{x}{\sqrt{x^4 + 1}}$

**29.** $f(x) = x\sqrt{2 - x^2}$

**30.** $f(x) = x + \sin x$

**31.** $f(x) = x - 2\sin x$

**32.** $f(x) = x - 2\tan^{-1} x$

**33.** $f(x) = 2x - \sin^{-1} x$

**34.** $f(x) = e^{-x^2/2}$

**35.** $f(x) = x \, 2^{-x}$

**36.** $f(x) = x^2 e^{-x^2}$

**37.** $f(x) = \dfrac{\ln x}{x}$

**38.** $f(x) = |x + 1|$

**39.** $f(x) = |x^2 - 1|$

**40.** $f(x) = \sin |x|$

**41.** $f(x) = |\sin x|$

**\* 42.** $f(x) = (x - 1)^{2/3} - (x + 1)^{2/3}$

**\* 43.** $f(x) = (x - 1)^{1/3} + (x + 1)^{1/3}$

**∗ 44.** $f(x) = x - x^{1/3}$

In Exercises 45–50 determine whether or not the given function has absolute maximum or absolute minimum values. Justify your answers. Find the extreme values if you can.

**45.** $\dfrac{x}{\sqrt{x^2 + 1}}$

**46.** $\dfrac{x}{\sqrt{x^4 + 1}}$

**47.** $x\sqrt{4 - x^2}$

**48.** $\dfrac{x^2}{\sqrt{4 - x^2}}$

**∗ 49.** $\dfrac{1}{x \sin x}$ on $(0, \pi)$

**∗ 50.** $\dfrac{\sin x}{x}$

**51.** If a function has an absolute maximum value must it have any local maximum values? If a function has a local maximum value must it have an absolute maximum value? Give reasons for your answers.

**52.** If the function $f$ has an absolute maximum value and $g(x) = |f(x)|$, must $g$ have an absolute maximum value? Justify your answer.

**∗ 53. (A function with no max or min at an endpoint)** Let

$$f(x) = \begin{cases} x \sin \dfrac{1}{x} & \text{if } x > 0, \\ 0 & \text{if } x = 0. \end{cases}$$

Show that $f$ is continuous on $[0, \infty)$ and differentiable on $(0, \infty)$ but that it has neither a local maximum nor a local minimum value at the endpoint $x = 0$.

# 4.3 Concavity and Inflections

Like the first derivative, the second derivative of a function also provides useful information about the behaviour of the function and the shape of its graph; it determines whether the graph is *bending upward* (that is, has increasing slope) or *bending downward* (has decreasing slope) as we move along the graph toward the right.

**DEFINITION 3**

We say that the function $f$ is **concave up** on an open interval $I$ if it is differentiable there and the derivative $f'$ is an increasing function on $I$. Similarly, $f$ is **concave down** on $I$ if $f'$ exists and is decreasing on $I$.

The terms "concave up" and "concave down" are used to describe the graph of the function as well as the function itself.

Note that concavity is defined only for differentiable functions, and even for those, only on intervals on which their derivatives are not constant. According to the above definition, a function is neither concave up nor concave down on an interval where its graph is a straight line segment. We say the function has no concavity on such an interval. We also say a function has opposite concavity on two intervals if it is concave up on one interval and concave down on the other.

**Example 1** The function $f$ whose graph is shown in Figure 4.19 is concave up on the interval $(a, b)$ and concave down on the interval $(b, c)$.

Some geometric observations can be made about concavity:

(i) If $f$ is concave up on an interval, then, on that interval, the graph of $f$ lies above its tangents, and chords joining points on the graph lie above the graph.

(ii) If $f$ is concave down on an interval, then, on that interval, the graph of $f$ lies below its tangents, and chords to the graph lie below the graph.

(iii) If the graph of $f$ has a tangent at a point, and if the concavity of $f$ is opposite on opposite sides of that point, then the graph crosses its tangent at that point.

(This occurs at the point $(b, f(b))$ in Figure 4.19. Such a point is called an *inflection point* of the graph of $f$.)

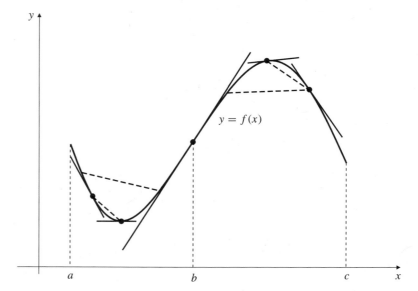

**Figure 4.19**    $f$ is concave up on $(a, b)$ and concave down on $(b, c)$

**DEFINITION 4**

**Inflection points**

We say that the point $(x_0, f(x_0))$ *is* an inflection point of the curve $y = f(x)$ (or that the function $f(x)$ has an **inflection point** at $x = x_0$) if the following two conditions are satisfied:

(a)  the graph $y = f(x)$ has a tangent line at $x_0$, and

(b)  the concavity of $f$ is opposite on opposite sides of $x_0$.

Note that (a) implies that either $f$ is differentiable at $x_0$ or its graph has a vertical tangent line there, and that (b) implies that the graph crosses its tangent line at $x_0$. An inflection point of a function $f$ is a point on the graph of a function, rather than a point in its domain like a critical point or a singular point. A function may or may not have an inflection point at a critical point or singular point. In general, a point $P$ is an inflection point (or simply *an inflection*) of a curve $C$ (which is not necessarily the graph of a function) if $C$ has a tangent at $P$ and arcs of $C$ extending in opposite directions from $P$ are on opposite sides of that tangent line.

**Example 2**    In Figure 4.20, the curve $y = x^3$ has an inflection point at $x = 0$; it has a horizontal tangent there and the function $x^3$ is concave down on $(-\infty, 0)$ and concave up on $(0, \infty)$.

The function $g$ in Figure 4.21 is concave down to the left of $x = a$ and concave up to the right of $x = a$, but $g$ has no inflection point at $x = a$ because its graph has no tangent line there. ($x = a$ is a singular point of $g$.)

The function $h(x) = x^{1/3}$ in Figure 4.22 has an inflection at $x = 0$ even though $x = 0$ is a singular point of $h$. The graph $y = h(x)$ has a vertical tangent at $x = 0$, and has opposite concavity on opposite sides of that point.

If a function $f$ has a second derivative $f''$, the sign of that second derivative tells us whether the first derivative $f'$ is increasing or decreasing, and hence determines the concavity of $f$.

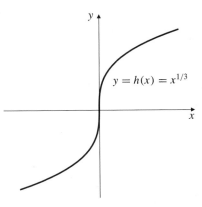

**Figure 4.20**  $x = 0$ is a critical point of $f(x) = x^3$, and $f$ has an inflection point there

**Figure 4.21**  The concavity of $g$ is opposite on opposite sides of $a$, but $g$ does not have an inflection point at $x = a$

**Figure 4.22**  This graph of $h$ has an inflection point at the origin even though $x = 0$ is a singular point of $h$

**THEOREM** **5**

**Concavity and the second derivative**

   (a) If $f''(x) > 0$ on interval $I$, then $f$ is concave up on $I$.

   (b) If $f''(x) < 0$ on interval $I$, then $f$ is concave down on $I$.

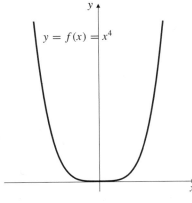

**Figure 4.23**  $f''(0) = 0$, but $f$ does not have an inflection point at 0

   (c) If $f$ has an inflection point at $x_0$ and $f''(x_0)$ exists, then $f''(x_0) = 0$.

**PROOF**  Parts (a) and (b) follow from applying Theorem 12 of Section 2.6 to the derivative $f'$ of $f$. If $f$ has an inflection point at $x_0$ and $f''(x_0)$ exists, then $f$ must be differentiable in an open interval containing $x_0$. Since $f'$ is increasing on one side of $x_0$ and decreasing on the other side, it must have a local maximum or minimum value at $x_0$. By Theorem 2, $f''(x_0) = 0$.

Theorem 5 tells us that to find (the $x$-coordinates of) inflection points of a twice differentiable function $f$ we need only look at points where $f''(x) = 0$. However, not every such point has to be an inflection point. For example, $f(x) = x^4$, whose graph is shown in Figure 4.23, does not have an inflection point at $x = 0$ even though $f''(0) = 12x^2|_{x=0} = 0$. In fact, $x^4$ is concave up on every interval.

**Example 3**  Determine the intervals of concavity of $f(x) = x^6 - 10x^4$, and the inflection points of its graph.

**Solution**  We have

$$f'(x) = 6x^5 - 40x^3,$$
$$f''(x) = 30x^4 - 120x^2 = 30x^2(x-2)(x+2).$$

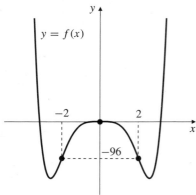

**Figure 4.24**    The graph
$y = f(x) = x^6 - 10x^4$

Having factored $f''(x)$ in this manner, we can see that it vanishes only at $x = -2$, $x = 0$ and $x = 2$. On the intervals $(-\infty, -2)$ and $(2, \infty)$, $f''(x) > 0$ so $f$ is concave up. On $(-2, 0)$ and $(0, 2)$, $f''(x) < 0$ so $f$ is concave down. $f''(x)$ changes sign as we pass through $-2$ and $2$. Since $f(\pm 2) = -96$, the graph of $f$ has inflection points at $(\pm 2, -96)$. However, $f''(x)$ does not change sign at $x = 0$, since $x^2 > 0$ for both positive and negative $x$. Thus there is no inflection point at $0$. As was the case for the first derivative, information about the sign of $f''(x)$ and the consequent concavity of $f$ can be conveniently conveyed in a chart:

The graph of $f$ is sketched in Figure 4.24.

---

**Example 4**    Determine the intervals of increase and decrease, the local extreme values, and the concavity of $f(x) = x^4 - 2x^3 + 1$. Use the information to sketch the graph of $f$.

**Solution**

$$f'(x) = 4x^3 - 6x^2 = 2x^2(2x - 3) = 0 \quad \text{at } x = 0 \text{ and } x = 3/2,$$
$$f''(x) = 12x^2 - 12x = 12x(x - 1) = 0 \quad \text{at } x = 0 \text{ and } x = 1.$$

The behaviour of $f$ is summarized in the following charts:

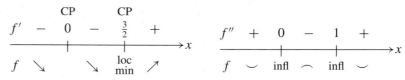

Note that $f$ has an inflection at the critical point $x = 0$. We calculate the values of $f$ at the "interesting values of $x$" in the charts:

$$f(0) = 1, \qquad f(1) = 0, \qquad f\left(\tfrac{3}{2}\right) = -\tfrac{11}{16}.$$

The graph of $f$ is sketched in Figure 4.25.

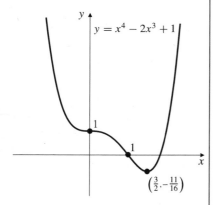

**Figure 4.25**    The function of Example 4

## The Second Derivative Test

A function $f$ will have a local maximum (or minimum) value at a critical point if its graph is concave downward (or upward) in an interval containing that point. In fact, we can often use the value of the second derivative at the critical point to determine whether the function has a local maximum or a local minimum value there.

**THEOREM 6**

### The second derivative test

(a) If $f'(x_0) = 0$ and $f''(x_0) < 0$, then $f$ has a local maximum value at $x_0$.

(b) If $f'(x_0) = 0$ and $f''(x_0) > 0$, then $f$ has a local minimum value at $x_0$.

(c) If $f'(x_0) = 0$ and $f''(x_0) = 0$, no conclusion can be drawn; $f$ may have a local maximum at $x_0$, or a local minimum, or it may have an inflection point instead.

***PROOF***   Suppose that $f'(x_0) = 0$ and $f''(x_0) < 0$. Since

$$\lim_{h \to 0} \frac{f'(x_0 + h)}{h} = \lim_{h \to 0} \frac{f'(x_0 + h) - f'(x_0)}{h} = f''(x_0) < 0,$$

it follows that $f'(x_0 + h) < 0$ for all sufficiently small positive $h$, and $f'(x_0 + h) > 0$ for all sufficiently small negative $h$. By the first derivative test (Theorem 3), $f$ must have a local maximum value at $x_0$. The proof of the local minimum case is similar.

The functions $f(x) = x^4$ (Figure 4.23), $f(x) = -x^4$, and $f(x) = x^3$ (Figure 4.20) all satisfy $f'(0) = 0$ and $f''(0) = 0$. But $x^4$ has a minimum value at $x = 0$, $-x^4$ has a maximum value at $x = 0$, and $x^3$ has neither a maximum nor a minimum value at $x = 0$, but has an inflection there. Therefore, we can not make any conclusion about the nature of a critical point based on knowing that $f''(x) = 0$ there.

---

**Example 5**   Find and classify the critical points of $f(x) = x^2 e^{-x}$.

***Solution***

$$f'(x) = (2x - x^2)e^{-x} = x(2 - x)e^{-x} = 0 \quad \text{at } x = 0 \text{ and } x = 2,$$
$$f''(x) = (2 - 4x + x^2)e^{-x}$$
$$f''(0) = 2 > 0, \qquad f''(2) = -2e^{-2} < 0.$$

Thus $f$ has a local minimum value at $x = 0$ and a local maximum value at $x = 2$. See Figure 4.26.

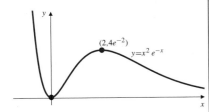

**Figure 4.26**   The critical points of $f(x) = x^2 e^{-x}$

For many functions the second derivative is more complicated to calculate than the first derivative, so the First Derivative Test is likely to be of more use in classifying critical points than is the Second Derivative Test. Also note that the First Derivative Test can classify local extreme values that occur at endpoints and singular points as well as at critical points.

It is possible to generalize the Second Derivative Test to obtain a higher derivative test to deal with some situations where the second derivative is zero at a critical point. (See Exercise 40 at the end of this section.)

## Exercises 4.3

In Exercises 1–22, determine the intervals of constant concavity of the given function and locate any inflection points.

**1.** $f(x) = \sqrt{x}$

**2.** $f(x) = 2x - x^2$

**3.** $f(x) = x^2 + 2x + 3$

**4.** $f(x) = x - x^3$

**5.** $f(x) = 10x^3 - 3x^5$

**6.** $f(x) = 10x^3 + 3x^5$

**7.** $f(x) = (3 - x^2)^2$

**8.** $f(x) = (2 + 2x - x^2)^2$

**9.** $f(x) = (x^2 - 4)^3$

**10.** $f(x) = \dfrac{x}{x^2 + 3}$

**11.** $f(x) = \sin x$

**12.** $f(x) = \cos 3x$

**13.** $f(x) = x + \sin 2x$

**14.** $f(x) = x - 2\sin x$

**15.** $f(x) = \tan^{-1} x$

**16.** $f(x) = x\, e^x$

**17.** $f(x) = e^{-x^2}$

**18.** $f(x) = \dfrac{\ln(x^2)}{x}$

**19.** $f(x) = \ln(1 + x^2)$

**20.** $f(x) = (\ln x)^2$

**21.** $f(x) = \dfrac{x^3}{3} - 4x^2 + 12x - \dfrac{25}{3}$

**22.** $f(x) = (x - 1)^{1/3} + (x + 1)^{1/3}$

**23.** Discuss the concavity of the linear function
$f(x) = ax + b$. Does it have any inflections?

Classify the critical points of the functions in Exercises 24–35 using the Second Derivative Test whenever possible.

**24.** $f(x) = 3x^3 - 36x - 3$

**25.** $f(x) = x(x-2)^2 + 1$

**26.** $f(x) = x + \dfrac{4}{x}$

**27.** $f(x) = x^3 + \dfrac{1}{x}$

**28.** $f(x) = \dfrac{x}{2^x}$

**29.** $f(x) = \dfrac{x}{1 + x^2}$

**30.** $f(x) = xe^x$

**31.** $f(x) = x \ln x$

**32.** $f(x) = (x^2 - 4)^2$

**33.** $f(x) = (x^2 - 4)^3$

**34.** $f(x) = (x^2 - 3)e^x$

**35.** $f(x) = x^2 e^{-2x^2}$

**36.** Let $f(x) = x^2$ if $x \geq 0$ and $f(x) = -x^2$ if $x < 0$. Is 0 a critical point of $f$? Does $f$ have an inflection point there? Is $f''(0) = 0$? If a function has a nonvertical tangent line at an inflection point, does the second derivative of the function necessarily vanish at that point?

**\* 37.** Verify that if $f$ is concave up on an interval, then its graph lies above its tangent lines on that interval. *Hint:* suppose $f$ is concave up on an open interval containing $x_0$. Let $h(x) = f(x) - f(x_0) - f'(x_0)(x - x_0)$. Show that $h$ has a local minimum value at $x_0$ and hence that $h(x) \geq 0$ on the interval. Show that $h(x) > 0$ if $x \neq x_0$.

**\* 38.** Verify that the graph $y = f(x)$ crosses its tangent line at an inflection point. *Hint:* consider separately the cases where the tangent line is vertical and nonvertical.

**39.** Let $f_n(x) = x^n$ and $g_n(x) = -x^n$, $(n = 2, 3, 4, \ldots)$. Determine whether each function has a local maximum, a local minimum, or an inflection point at $x = 0$.

**\* 40.** **(Higher derivative test)** Use your conclusions from the previous Exercise to suggest a generalization of the second derivative test that applies when
$$f'(x_0) = f''(x_0) = \ldots = f^{(k-1)}(x_0) = 0, \ f^{(k)}(x_0) \neq 0,$$

for some $k \geq 2$.

**\* 41.** This problem shows that no test based solely on the signs of derivatives at $x_0$ can determine whether every function with a critical point at $x_0$ has a local maximum or minimum, or an inflection point there. Let

$$f(x) = \begin{cases} e^{-1/x^2} & \text{if } x \neq 0, \\ 0 & \text{if } x = 0. \end{cases}$$

Prove the following

(a) $\lim_{x \to 0} x^{-n} f(x) = 0$ for $n = 0, 1, 2, 3, \ldots$

(b) $\lim_{x \to 0} P(1/x) f(x) = 0$ for every polynomial $P$.

(c) For $x \neq 0$, $f^{(k)}(x) = P_k(1/x) f(x) (k = 1, 2, 3, \ldots)$ where $P_k$ is a polynomial.

(d) $f^{(k)}(0)$ exists and equals 0 for $k = 1, 2, 3, \ldots$

(e) $f$ has a local minimum at $x = 0$; $-f$ has a local maximum at $x = 0$.

(f) If $g(x) = xf(x)$, then $g^{(k)}(0) = 0$ for every positive integer $k$ and $g$ has an inflection point at $x = 0$.

**\* 42.** A function may have neither a local maximum nor a local minimum nor an inflection at a critical point. Show this by considering the following function.

$$f(x) = \begin{cases} x^2 \sin \dfrac{1}{x} & \text{if } x \neq 0 \\ 0 & \text{if } x = 0 \end{cases}$$

Show that $f'(0) = f(0) = 0$, so the $x$-axis is tangent to the graph of $f$ at $x = 0$; but $f'(x)$ is not continuous at $x = 0$, so $f''(0)$ does not exist. Show that the concavity of $f$ is not constant on any interval with endpoint 0.

## 4.4  Sketching the Graph of a Function

When sketching the graph $y = f(x)$ of a function $f$ we have three sources of useful information:

(i) **the function $f$ itself**, from which we determine the coordinates of some points on the graph, the symmetry of the graph, and any asymptotes,

(ii) **the first derivative**, $f'$, from which we determine the intervals of increase and decrease and the location of any local extreme values, and

(iii) **the second derivative**, $f''$, from which we determine the concavity and inflection points, and sometimes extreme values.

Items (ii) and (iii) have been explored in the previous two sections. In this section we consider what we can learn from the function itself about the shape of its graph, and then we illustrate the entire sketching procedure with several examples using all three sources of information.

We could sketch a graph by plotting the coordinates of many points on it and joining them by a suitably smooth curve. This is what graphics calculators and computer software do. When carried out by hand (without a computer or calculator), this simplistic approach is at best tedious, and at worst can fail to reveal the most interesting aspects of the graph (singular points, extreme values, and so on). We could also compute the slope at each of the plotted points and, by drawing short line segments through these points with the appropriate slopes, ensure that the sketched graph passes through each plotted point with the correct slope. A more efficient procedure is to obtain the coordinates of only a few points and use qualitative information from the function and its first and second derivatives to determine the *shape* of the graph between these points.

Besides critical and singular points and inflections, a graph may have other "interesting" points. The **intercepts** (points at which the graph intersects the coordinate axes) are usually among these. When sketching any graph it is wise to try to find all such intercepts, that is, all points with coordinates $(x, 0)$ and $(0, y)$ that lie on the graph. Of course, not every graph will have such points, and even when they do exist it may not always be possible to compute them exactly. Whenever a graph is made up of several disconnected pieces (called **components**), the coordinates of *at least one point on each component* must be obtained. It can sometimes be useful to determine the slopes at those points too. Vertical asymptotes (discussed below) usually break the graph of a function into components.

Realizing that a given function possesses some symmetry can aid greatly in obtaining a good sketch of its graph. In Section P.4 we discussed odd and even functions and observed that odd functions have graphs that are symmetric about the origin, while even functions have graphs that are symmetric about the $y$-axis, as shown in Figure 4.27. These are the symmetries you are most likely to notice, but functions can have other symmetries. For example, the graph of $f(x) = 2 + (x - 1)^2$ will certainly be symmetric about the line $x = 1$, and the graph of $2 + (x - 3)^3$ is symmetric about the point $(3, 2)$.

## Asymptotes

Some of the curves we have sketched in previous sections have had **asymptotes**, that is, straight lines to which the curve draws arbitrarily near as it recedes to infinite distance from the origin. Asymptotes are of three types: vertical, horizontal, and oblique.

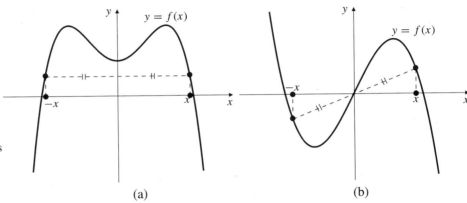

**Figure 4.27**

(a) The graph of an even function is symmetric about the $y$-axis

(b) The graph of an odd function is symmetric about the origin

**DEFINITION  5**

> The graph of $y = f(x)$ has a **vertical asymptote** at $x = a$ if
>
> $$\text{either} \quad \lim_{x \to a-} f(x) = \pm\infty, \quad \text{or} \quad \lim_{x \to a+} f(x) = \pm\infty, \quad \text{or both.}$$

This situation tends to arise when $f(x)$ is a quotient of two expressions and the denominator is zero at $x = a$.

**Example 1**   Find the vertical asymptotes of $f(x) = \dfrac{1}{x^2 - x}$. How does the graph approach these asymptotes?

**Solution**   The denominator $x^2 - x = x(x-1)$ approaches 0 as $x$ approaches 0 or 1, so $f$ has vertical asymptotes at $x = 0$ and $x = 1$ (Figure 4.28). Since $x(x-1)$ is positive on $(-\infty, 0)$ and on $(1, \infty)$, and is negative on $(0, 1)$, we have

$$\lim_{x \to 0-} \frac{1}{x^2 - x} = \infty \qquad \lim_{x \to 1-} \frac{1}{x^2 - x} = -\infty$$

$$\lim_{x \to 0+} \frac{1}{x^2 - x} = -\infty \qquad \lim_{x \to 1+} \frac{1}{x^2 - x} = \infty.$$

**DEFINITION  6**

> The graph of $y = f(x)$ has a **horizontal asymptote** $y = L$ if
>
> $$\text{either} \quad \lim_{x \to \infty} f(x) = L, \quad \text{or} \quad \lim_{x \to -\infty} f(x) = L, \quad \text{or both.}$$

**Example 2**   Find the horizontal asymptotes of

(a) $f(x) = \dfrac{1}{x^2 - x}$   and   (b) $g(x) = \dfrac{x^4 + x^2}{x^4 + 1}$.

**Solution**

(a) The function $f$ has horizontal asymptote $y = 0$ (Figure 4.28) since

$$\lim_{x \to \pm\infty} \frac{1}{x^2 - x} = \lim_{x \to \pm\infty} \frac{1/x^2}{1 - (1/x)} = \frac{0}{1} = 0.$$

(b) The function $g(x)$ has horizontal asymptote $y = 1$ (Figure 4.29) since

$$\lim_{x \to \pm\infty} \frac{x^4 + x^2}{x^4 + 1} = \lim_{x \to \pm\infty} \frac{1 + (1/x^2)}{1 + (1/x^4)} = \frac{1}{1} = 1.$$

Observe that the graph of $g$ crosses its asymptote twice. (There is a popular misconception among students that curves cannot cross their asymptotes. Exercise 43 below gives an example of a curve which crosses its asymptote infinitely often.)

**Figure 4.28**  **Figure 4.29**  **Figure 4.30**

The horizontal asymptotes of both functions $f$ and $g$ in Example 2 are **two-sided**, which means that the graphs approach the asymptotes as $x$ approaches both infinity and negative infinity. The function $\tan^{-1} x$ has two **one-sided** asymptotes, $y = \pi/2$ (as $x \to \infty$) and $y = -(\pi/2)$ (as $x \to -\infty$). See Figure 4.30.

It can also happen that the graph of a function $f(x)$ approaches a nonhorizontal straight line as $x$ approaches $\infty$ or $-\infty$ (or both). Such a line is called an *oblique asymptote* of the graph.

**DEFINITION 7**

The straight line $y = ax + b$, (where $a \neq 0$), is an **oblique asymptote** of the graph of $y = f(x)$ if

either $\quad \lim_{x \to -\infty} \big(f(x) - (ax+b)\big) = 0, \quad$ or $\quad \lim_{x \to \infty} \big(f(x) - (ax+b)\big) = 0,$

or both.

**Example 3**  Consider the function

$$f(x) = \frac{x^2 + 1}{x} = x + \frac{1}{x},$$

whose graph is shown in Figure 4.31(a). The straight line $y = x$ is a *two-sided* oblique asymptote of the graph of $f$ because

$$\lim_{x \to \pm\infty} \big(f(x) - x\big) = \lim_{x \to \pm\infty} \frac{1}{x} = 0.$$

**Example 4**  The graph of $y = \dfrac{x\, e^x}{1 + e^x}$ is shown in Figure 4.31(b). It has a horizontal asymptote $y = 0$ at the left and an oblique asymptote $y = x$ at the right:

$$\lim_{x \to -\infty} \frac{x\, e^x}{1 + e^x} = \frac{0}{1} = 0 \quad \text{and}$$

$$\lim_{x \to \infty} \left( \frac{x\, e^x}{1 + e^x} - x \right) = \lim_{x \to \infty} \frac{x(e^x - 1 - e^x)}{1 + e^x} = \lim_{x \to \infty} \frac{-x}{1 + e^x} = 0.$$

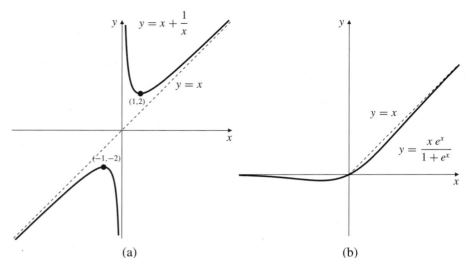

**Figure 4.31**

(a) $y = f(x)$ has a two-sided oblique asymptote, $y = x$

(b) This graph has a horizontal asymptote at the left and an oblique asymptote at the right

Recall that a **rational function** is a function of the form $f(x) = P(x)/Q(x)$, where $P$ and $Q$ are polynomials. It is possible to be quite specific about the asymptotes of a rational function.

---

**Asymptotes of a rational function**

Suppose that $f(x) = \dfrac{P_m(x)}{Q_n(x)}$, where $P_m$ and $Q_n$ are polynomials of degree $m$ and $n$, respectively. Suppose also that $P_m$ and $Q_n$ have no common linear factors. Then

(a) The graph of $f$ has a vertical asymptote at every position $x$ such that $Q_n(x) = 0$.

(b) The graph of $f$ has a two-sided horizontal asymptote $y = 0$ if $m < n$.

(c) The graph of $f$ has a two-sided horizontal asymptote $y = L$, $(L \neq 0)$ if $m = n$. $L$ is the quotient of the coefficients of the highest degree terms in $P_m$ and $Q_n$.

(d) The graph of $f$ has a two-sided oblique asymptote if $m = n + 1$. This asymptote can be found by dividing $Q_n$ into $P_m$ to obtain a linear quotient, $ax + b$, and remainder, $R$, a polynomial of degree at most $n - 1$. That is,

$$f(x) = ax + b + \frac{R(x)}{Q_n(x)}.$$

The oblique asymptote is $y = ax + b$.

(e) The graph of $f$ has no horizontal or oblique asymptotes if $m > n + 1$.

---

**Example 5** Find the oblique asymptote of $y = \dfrac{x^3}{x^2 + x + 1}$.

**Solution** We can either obtain the quotient by long division:

$$\frac{x^3}{x^2+x+1} = x-1+\frac{1}{x^2+x+1}$$

$$\begin{array}{r} x-1 \\ x^2+x+1\overline{)x^3\phantom{+x^2+x}} \\ x^3+x^2+x \\ \hline -x^2-x \\ -x^2-x-1 \\ \hline 1 \end{array}$$

or we can obtain the same result by "short division":

$$\frac{x^3}{x^2+x+1}=\frac{x^3+x^2+x-x^2-x-1+1}{x^2+x+1}=x-1+\frac{1}{x^2+x+1}.$$

In any event we see that the oblique asymptote has equation $y=x-1$.

## Examples of Formal Curve Sketching

Here is a checklist of things to consider when you are asked to make a careful sketch of the graph $y=f(x)$. It will, of course, not always be possible to obtain every item of information mentioned in the list.

**Checklist for curve sketching**

1. Calculate $f'(x)$ and $f''(x)$, and express the results in factored form.
2. Examine $f(x)$ to determine its domain and the following items:
   (a) any vertical asymptotes. (Look for zeros of denominators.)
   (b) any horizontal or oblique asymptotes. (Consider $\lim_{x\to\pm\infty} f(x)$.)
   (c) any obvious symmetry. (Is $f$ even or odd?)
   (d) any easily calculated intercepts (points with coordinates $(x,0)$ or $(0,y)$), or endpoints or other "obvious" points. You will add to this list when you know any critical points, singular points, and inflection points. Eventually you should make sure you know the coordinates of at least one point on every component of the graph.
3. Examine $f'(x)$ for the following:
   (a) any critical points.
   (b) any points where $f'$ is not defined. (These will include singular points, endpoints of the domain of $f$, and vertical asymptotes.)
   (c) intervals on which $f'$ is positive or negative. It's a good idea to convey this information in the form of a chart such as those used in the examples. Conclusions about where $f$ is increasing and decreasing and classification of some critical and singular points as local maxima and minima can also be indicated on the chart.
4. Examine $f''(x)$ for the following:
   (a) points where $f''(x)=0$.
   (b) points where $f''(x)$ is undefined. (These will include singular points, endpoints, vertical asymptotes, and possibly other points as well, where $f'$ is defined but $f''$ isn't.)
   (c) intervals where $f''$ is positive and negative, and where $f$ is therefore concave up or down. Use a chart.
   (d) any inflection points.

When you have obtained as much of this information as possible, make a careful sketch that reflects *everything* you have learned about the function. Consider where best to place the axes and what scale to use on each so the "interesting features" of the graph show up most clearly. Be alert for seeming inconsistencies in the information — that is a strong suggestion you may have made an error somewhere. For example, if you have determined that $f(x) \to \infty$ as $x$ approaches the vertical asymptote $x = a$ from the right, and also that $f$ is decreasing and concave down on the interval $(a, b)$ then you have very likely made an error. (Try to sketch such a situation to see why.)

**Example 6** Sketch the graph of $y = \dfrac{x^2 + 2x + 4}{2x}$.

**Solution** It is useful to rewrite the function $y$ in the form

$$y = \frac{x}{2} + 1 + \frac{2}{x},$$

since this form not only shows clearly that $y = (x/2) + 1$ is an oblique asymptote, but also makes it easier to calculate the derivatives

$$y' = \frac{1}{2} - \frac{2}{x^2} = \frac{x^2 - 4}{2x^2}, \qquad y'' = \frac{4}{x^3}$$

From $y$:  Domain: all $x$ except 0.   Vertical asymptote: $x = 0$,

Oblique asymptote: $y = \dfrac{x}{2} + 1$, $\quad y - \left(\dfrac{x}{2} + 1\right) = \dfrac{2}{x} \to 0$ as $x \to \pm\infty$.

Symmetry: none obvious ($y$ is neither odd nor even).

Intercepts: none. $x^2 + 2x + 4 = (x + 1)^2 + 3 \geq 3$ for all $x$, and $y$ is not defined at $x = 0$.

From $y'$:  Critical points: $x = \pm 2$; points $(-2, -1)$ and $(2, 3)$.
$y'$ not defined at $x = 0$ (vertical asymptote).

From $y''$:  $y'' = 0$ nowhere; $y''$ undefined at $x = 0$.

The graph is shown in Figure 4.32.

■

**Example 7** Sketch the graph of $f(x) = \dfrac{x^2 - 1}{x^2 - 4}$.

**Solution** We have

$$f'(x) = \frac{-6x}{(x^2 - 4)^2}, \qquad f''(x) = \frac{6(3x^2 + 4)}{(x^2 - 4)^3}.$$

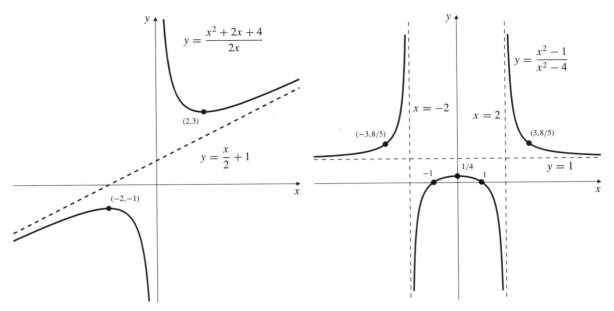

**Figure 4.32**                                    **Figure 4.33**

From $f$:   Domain: all $x$ except $\pm 2$.   Vertical asymptotes: $x = -2$ and $x = 2$.
Horizontal asymptote: $y = 1$ (as $x \to \pm\infty$).
Symmetry: about the $y$-axis ($y$ is even).
Intercepts: $(0, 1/4)$, $(-1, 0)$ and $(1, 0)$.
Other points: $(-3, 8/5)$, $(3, 8/5)$. (The two vertical asymptotes divide the graph into three components; we need points on each. The outer components require points with $|x| > 2$.)

From $f'$:   Critical point: $x = 0$; $f'$ not defined at $x = 2$ or $x = -2$.

$$
\begin{array}{ccccccccc}
 & & \text{ASY} & & \text{CP} & & \text{ASY} & & \\
f' & + & -2 & + & 0 & - & 2 & - & \\
\hline
 & & & & & & & & \to x \\
f & \nearrow & & \nearrow & \begin{array}{c}\text{loc}\\\text{max}\end{array} & \searrow & & \searrow &
\end{array}
$$

From $f''$:   $f''(x) = 0$ nowhere; $f''$ not defined at $x = 2$ or $x = -2$.

$$
\begin{array}{ccccccc}
f'' & + & -2 & - & 2 & + & \\
\hline
 & & & & & & \to x \\
f & \smile & & \frown & & \smile &
\end{array}
$$

The graph is shown in Figure 4.33.  ■

---

**Example 8**   Sketch the graph of $y = xe^{-x^2/2}$.

**Solution**   We have $y' = (1 - x^2)e^{-x^2/2}$,   $y'' = x(x^2 - 3)e^{-x^2/2}$.

From $y$:  Domain: all $x$.
Horizontal asymptote: $y = 0$. Note that if $t = x^2/2$, then $|xe^{-x^2/2}| = \sqrt{2t}\,e^{-t} \to 0$ as $t \to \infty$ (hence as $x \to \pm\infty$).
Symmetry: about the origin ($y$ is odd). Intercepts: $(0, 0)$.

From $y'$:  Critical points: $x = \pm 1$; points $(\pm 1, \pm 1/\sqrt{e}) \approx (\pm 1, \pm 0.61)$.

From $y''$:  $y'' = 0$ at $x = 0$ and $x = \pm\sqrt{3}$; points $(0, 0)$, $(\pm\sqrt{3}, \pm\sqrt{3}e^{-3/2}) \approx (\pm 1.73, \pm 0.39)$.

The graph is shown in Figure 4.34.

---

**Example 9**  Sketch the graph of $f(x) = (x^2 - 1)^{2/3}$.

**Solution**  $f'(x) = \dfrac{4}{3}\dfrac{x}{(x^2-1)^{1/3}}$,  $f''(x) = \dfrac{4}{9}\dfrac{x^2-3}{(x^2-1)^{4/3}}$.

From $f$:  Domain: all $x$.
Asymptotes: none. ($f(x)$ grows like $x^{4/3}$ as $x \to \pm\infty$).
Symmetry: about the $y$-axis ($f$ is an even function).
Intercepts: $(\pm 1, 0)$, $(0, 1)$.

From $f'$:  Critical points: $x = 0$; singular points: $x = \pm 1$.

From $f''$:  $f''(x) = 0$ at $x = \pm\sqrt{3}$; points $(\pm\sqrt{3}, 2^{2/3}) \approx (\pm 1.73, 1.59)$; $f''(x)$ not defined at $x = \pm 1$.

The graph is sketched in Figure 4.35.

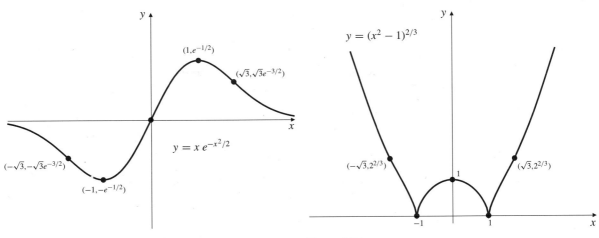

Figure 4.34                    Figure 4.35

***Remark*** **Using a Graphing Utility** The techniques for curve sketching developed above are useful only for graphs of functions that are simple enough to allow you to calculate and analyze their derivatives. In practice, if you want to study the behaviour of a function using its graph, you will likely want to use a graphing calculator or a computer graphing program to produce the graph quickly and painlessly. However, to make effective use of such a utility, you have to decide on a viewing window; what part of the graph do you want to see, and what horizontal and vertical scales should be used? An inappropriate choice of viewing window can cause you to miss significant features of the graph.

# Exercises 4.4

**1.** Figure 4.36 shows the graphs of a function $f$, its two derivatives $f'$ and $f''$, and another function $g$. Which graph corresponds to each function?

inspecting the graph. (E.g., symmetry, asymptotes, intercepts, intervals of increase and decrease, critical and singular points, local maxima and minima, intervals of constant concavity, inflection points.)

Figure 4.36

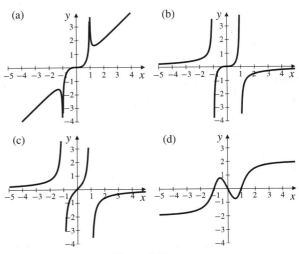

Figure 4.37

**2.** List, for each function graphed in Figure 4.36, such information that you can determine (approximately) by

**3.** Figure 4.37 shows the graphs of four functions:

$$f(x) = \frac{x}{1 - x^2} \qquad g(x) = \frac{x^3}{1 - x^4}$$

$$h(x) = \frac{x^3 - x}{\sqrt{x^6 + 1}} \qquad k(x) = \frac{x^3}{\sqrt{|x^4 - 1|}}$$

Which graph corresponds to each function?

**4.** Repeat Exercise 2 for the graphs in Figure 4.37.

In Exercises 5–6, sketch the graph of a function which has the given properties. Identify any critical points, singular points, local maxima and minima, and inflection points. Assume that $f$ is continuous and its derivatives exist everywhere unless the contrary is implied or explicitly stated.

**5.** $f(0) = 1$, $f(\pm 1) = 0$, $f(2) = 1$, $\lim_{x \to \infty} f(x) = 2$, $\lim_{x \to -\infty} f(x) = -1$, $f'(x) > 0$ on $(-\infty, 0)$ and on $(1, \infty)$, $f'(x) < 0$ on $(0, 1)$, $f''(x) > 0$ on $(-\infty, 0)$ and on $(0, 2)$, and $f''(x) < 0$ on $(2, \infty)$.

**6.** $f(-1) = 0$, $f(0) = 2$, $f(1) = 1$, $f(2) = 0$, $f(3) = 1$, $\lim_{x \to \pm\infty}(f(x) + 1 - x) = 0$, $f'(x) > 0$ on $(-\infty, -1)$, $(-1, 0)$ and $(2, \infty)$, $f'(x) < 0$ on $(0, 2)$, $\lim_{x \to -1} f'(x) = \infty$, $f''(x) > 0$ on $(-\infty, -1)$ and on $(1, 3)$, and $f''(x) < 0$ on $(-1, 1)$ and on $(3, \infty)$.

In Exercises 7–41, sketch the graphs of the given functions, making use of any suitable information you can obtain from the function and its first and second derivatives.

**7.** $y = (x^2 - 1)^3$

**8.** $y = x(x^2 - 1)^2$

**9.** $y = \dfrac{2 - x}{x}$

**10.** $y = \dfrac{x - 1}{x + 1}$

**11.** $y = \dfrac{x^3}{1 + x}$

**12.** $y = \dfrac{1}{4 + x^2}$

**13.** $y = \dfrac{1}{2 - x^2}$

**14.** $y = \dfrac{x}{x^2 - 1}$

**15.** $y = \dfrac{x^2}{x^2 - 1}$

**16.** $y = \dfrac{x^3}{x^2 - 1}$

**17.** $y = \dfrac{x^3}{x^2 + 1}$

**18.** $y = \dfrac{x^2}{x^2 + 1}$

**19.** $y = \dfrac{x^2 - 4}{x + 1}$

**20.** $y = \dfrac{x^2 - 2}{x^2 - 1}$

**21.** $y = \dfrac{x^3 - 4x}{x^2 - 1}$

**22.** $y = \dfrac{x^2 - 1}{x^2}$

**23.** $y = \dfrac{x^5}{(x^2 - 1)^2}$

**24.** $y = \dfrac{(2 - x)^2}{x^3}$

**25.** $y = \dfrac{1}{x^3 - 4x}$

**26.** $y = \dfrac{x}{x^2 + x - 2}$

**27.** $y = \dfrac{x^3 - 3x^2 + 1}{x^3}$

**28.** $y = x + \sin x$

**29.** $y = x + 2\sin x$

**30.** $y = e^{-x^2}$

**31.** $y = xe^x$

**32.** $y = e^{-x} \sin x$, $(x \ge 0)$

**33.** $y = x^2 e^{-x^2}$

**34.** $y = x^2 e^x$

**35.** $y = \dfrac{\ln x}{x}$, $(x > 0)$

**36.** $y = \dfrac{\ln x}{x^2}$, $(x > 0)$

**37.** $y = \dfrac{1}{\sqrt{4 - x^2}}$

**38.** $y = \dfrac{x}{\sqrt{x^2 + 1}}$

**39.** $y = (x^2 - 1)^{1/3}$

**∗ 40.** $y = x^{1/3} + (x - 1)^{2/3}$

**∗ 41.** $y = (x - 1)^{2/3} - (x + 1)^{2/3}$

**∗ 42.** What is $\lim_{x \to 0+} x \ln x$? $\lim_{x \to 0} x \ln |x|$? If $f(x) = x \ln |x|$ for $x \ne 0$ is it possible to define $f(0)$ in such a way that $f$ is continuous on the whole real line? Sketch the graph of $f$.

**43.** What straight line is an asymptote of the curve $y = \dfrac{\sin x}{1 + x^2}$? At what points does the curve cross this asymptote?

# 4.5 Extreme-Value Problems

In this section we solve various word problems that, when translated into mathematical terms, require the finding of a maximum or minimum value of a function of one variable. Such problems can range from simple to very complex and difficult; they can be phrased in terminology appropriate to some other discipline or they can be already partially translated into a more mathematical context. We have already encountered a few such problems in earlier chapters.

Let us consider a few examples before attempting to formulate any general principles for dealing with such problems.

**Example 1**   Find the largest possible product of two nonnegative numbers whose sum is 10.

**Solution**   Let the numbers be $x$ and $10 - x$, so that their sum is evidently 10. Since neither number can be negative, we require $0 \leq x \leq 10$. The product of the numbers,

$$P(x) = x(10 - x) = 10x - x^2,$$

$P$ is differentiable everywhere. Since $P(0) = 0$, $P(10) = 0$, and $P(1) = 9 > 0$, the maximum value of $P(x)$ must occur at a critical point in $(0, 10)$ by Theorem 4. To find critical points we solve

$$0 = P'(x) = 10 - 2x.$$

The only critical point is $x = 5$, so this must maximize the product $P(x)$. The largest possible product is $P(5) = 25$.

$$y \qquad A = xy \qquad y$$

$$x$$

**Figure 4.38**

**Example 2**   A rectangular animal enclosure is to be constructed having one side along an existing long wall and the other three sides fenced. If 100 m of fence are available, what is the largest possible area for the enclosure?

**Solution**   This problem, like many others, is essentially a geometric one. A sketch should be made at the outset, as we have done in Figure 4.38. Let the length and width of the enclosure be $x$ and $y$ m, respectively, and let its area be $A$ m$^2$. Thus $A = xy$. Since the total length of the fence is 100 m, we must have $x + 2y = 100$. $A$ appears to be a function of two variables, $x$ and $y$, but these variables are not independent; they are related by the *constraint* $x + 2y = 100$. This constraint equation can be solved for one variable in terms of the other, and $A$ can therefore be written as a function of only one variable:

$$x = 100 - 2y,$$
$$A = A(y) = (100 - 2y)y = 100y - 2y^2.$$

Evidently we require $y \geq 0$ and $y \leq 50$, (that is, $x \geq 0$), in order that the area make sense. (It would otherwise be negative.) Thus we must maximize the function $A(y)$ on the interval $[0, 50]$. Being continuous on this closed, finite interval, $A$ must have a maximum value, by Theorem 1. Clearly $A(0) = A(50) = 0$ and $A(y) > 0$ for $0 < y < 50$. Hence the maximum cannot occur at an endpoint. Since $A$ has no singular points, the maximum must occur at a critical point. To find any critical points, we set

$$0 = A'(y) = 100 - 4y.$$

Therefore $y = 25$. Since $A$ must have a maximum value and there is only one possible point where it can be, the maximum must occur at $y = 25$. The greatest possible area for the enclosure is therefore $A(25) = 1,250$ m$^2$.

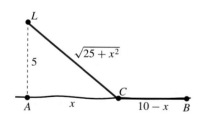

**Figure 4.39**

**Example 3**    A lighthouse $L$ is located on a small island 5 km north of a point $A$ on a straight east-west shoreline. A cable is to be laid from $L$ to point $B$ on the shoreline 10 km east of $A$. The cable will be laid through the water in a straight line from $L$ to a point $C$ on the shoreline between $A$ and $B$, and from there to $B$ along the shoreline. The part of the cable lying in the water costs \$5,000/km and the part along the shoreline costs \$3,000/km.

(a)  Where should $C$ be chosen to minimize the total cost of the cable?

(b)  Where should $C$ be chosen if $B$ is only 3 km from $A$?

*Solution*

(a)  Let $C$ be $x$ km from $A$ towards $B$. Thus $0 \le x \le 10$. The length of $LC$ is $\sqrt{25 + x^2}$ km and the length of $CB$ is $10 - x$ km, as illustrated in Figure 4.39. Hence the total cost of the cable is \$$T$ where

$$T = T(x) = 5{,}000\sqrt{25 + x^2} + 3{,}000(10 - x), \qquad (0 \le x \le 10).$$

$T$ is continuous on the closed, finite interval $[0, 10]$, and so has a minimum value that may occur at one of the endpoints $x = 0$ or $x = 10$, or at a critical point in the interval $(0, 10)$. ($T$ has no singular points.) To find any critical points, we set

$$0 = \frac{dT}{dx} = \frac{5{,}000x}{\sqrt{25 + x^2}} - 3{,}000.$$

Thus
$$5{,}000x = 3{,}000\sqrt{25 + x^2}$$
$$25x^2 = 9(25 + x^2)$$
$$16x^2 = 225$$
$$x^2 = \frac{225}{16} = \frac{15^2}{4^2}.$$

The critical points are $x = \pm 15/4$. Only one critical point, $x = 15/4 = 3.75$, lies in the interval $(0, 10)$. Since $T(0) = 55{,}000$, $T(15/4) = 50{,}000$, and $T(10) \approx 55{,}902$, the critical point evidently provides the minimum value for $T(x)$. For minimal cost, $C$ should be 3.75 km from $A$.

(b)  If $B$ is 3 km from $A$, the corresponding total cost function is

$$T(x) = 5{,}000\sqrt{25 + x^2} + 3{,}000(3 - x), \qquad (0 \le x \le 3),$$

which differs from the total cost function $T(x)$ of part (a) only in the added constant (9,000 rather than 30,000). It therefore has the same critical points, $x = \pm 15/4$, neither of which lie in the interval $(0, 3)$. Since $T(0) = 34{,}000$ and $T(3) \approx 29{,}155$, in this case we should choose $x = 3$. To minimize the total cost, the cable should go straight from $L$ to $B$.  ∎

## Procedure for Solving Extreme-Value Problems

Based on our experience with the examples above we can formulate a checklist of steps involved in solving optimization problems.

**Solving extreme-value problems**

1. Read the problem very carefully, perhaps more than once. You must understand clearly what is given and what must be found.

2. Make a diagram if appropriate. Many problems have a geometric component and a good diagram can often be an essential part of the solution process.

3. Define any symbols you wish to use that are not already specified in the statement of the problem.

4. Express the quantity $Q$ to be maximized or minimized as a function of one or more variables.

5. If $Q$ depends on $n$ variables, where $n > 1$, find $n - 1$ equations (constraints) linking these variables. (If this cannot be done, the problem cannot be solved by single-variable techniques.)

6. Use the constraints to eliminate variables and hence express $Q$ as a function of only one variable. Determine the interval(s) in which this variable must lie for the problem to make sense. Alternatively, regard the constraints as implicitly defining $n - 1$ of the variables, and hence $Q$, as functions of the remaining variable. (It is usually better to avoid this implicit method in an extreme-value problem if you can.)

7. Find the required extreme value of $Q$ using the techniques of Section 4.2. Remember to consider any critical points, singular points, and endpoints. Make sure to give a convincing argument that your extreme value is the one being sought; e.g if you are looking for a maximum, the value you have found should not be a minimum.

8. Make a concluding statement answering the question asked. Is your answer for the question "reasonable?" If not, check back through the solution to see what went wrong.

**Example 4**  Find the length of the shortest ladder that can extend from a vertical wall, over a fence 2 m high located 1 m away from the wall, to a point on the ground outside the fence.

**Solution**  Let $\theta$ be the angle of inclination of the ladder, as shown in Figure 4.40. Using the two right-angled triangles in the figure, we obtain the length $L$ of the ladder as a function of $\theta$:

$$L = L(\theta) = \frac{1}{\cos \theta} + \frac{2}{\sin \theta},$$

where $0 < \theta < \pi/2$. Since

$$\lim_{\theta \to (\pi/2)-} L(\theta) = \infty \quad \text{and} \quad \lim_{\theta \to 0+} L(\theta) = \infty,$$

$L(\theta)$ must have a minimum value on $(0, \pi/2)$, occurring at a critical point. ($L$ has no singular points in $(0, \pi/2)$.) To find any critical points, we set

$$0 = L'(\theta) = \frac{\sin \theta}{\cos^2 \theta} - \frac{2 \cos \theta}{\sin^2 \theta} = \frac{\sin^3 \theta - 2 \cos^3 \theta}{\cos^2 \theta \sin^2 \theta}.$$

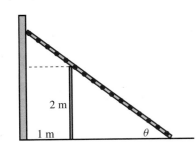

**Figure 4.40**

Any critical point satisfies $\sin^3 \theta = 2 \cos^3 \theta$, or, equivalently, $\tan^3 \theta = 2$. We don't need to solve this equation for $\theta = \tan^{-1}(2^{1/3})$ since it is really the corresponding value of $L(\theta)$ that we want. Observe that

$$\sec^2 \theta = 1 + \tan^2 \theta = 1 + 2^{2/3}.$$

It follows that

$$\cos \theta = \frac{1}{(1 + 2^{2/3})^{1/2}} \quad \text{and} \quad \sin \theta = \tan \theta \cos \theta = \frac{2^{1/3}}{(1 + 2^{2/3})^{1/2}}.$$

Therefore the minimal value of $L(\theta)$ is

$$\frac{1}{\cos \theta} + \frac{2}{\sin \theta} = (1 + 2^{2/3})^{1/2} + 2 \frac{(1 + 2^{2/3})^{1/2}}{2^{1/3}} = \left(1 + 2^{2/3}\right)^{3/2} \approx 4.16.$$

The shortest ladder which can extend from the wall over the fence to the ground outside is about 4.16 m long.

∎

**Example 5**    Find the most economical shape of a cylindrical tin can.

**Solution**    This problem is stated in a rather vague way. We must consider what is meant by "most economical" and even "shape." Without further information, we can take one of two points of view:

(i) the volume of the tin can is to be regarded as given and we must choose the dimensions to minimize the total surface area, or

(ii) the total surface area is given (we can use just so much metal) and we must choose the dimensions to maximize the volume.

We will discuss other possible interpretations later. Since a cylinder is determined by its radius and height, its shape is determined by the ratio radius/height (Figure 4.41). Let $r$, $h$, $S$, and $V$ denote, respectively, the radius, height, total surface area, and volume of the can. The volume of a cylinder is the base area times the height, so

$$V = \pi r^2 h.$$

The surface of the can is made up of the cylindrical wall and circular disks for the top and bottom. The disks each have area $\pi r^2$ and the cylindrical wall is really just a rolled-up rectangle with base $2\pi r$ (the circumference of the can) and height $h$. Therefore the total surface area of the can is

$$S = 2\pi r^2 + 2\pi r h.$$

Let us use interpretation (i); $V$ is a given constant, and $S$ is to be minimized. We can use the equation for $V$ to eliminate one of the two variables $r$ and $h$ on which $S$ depends. Say we solve for $h = V/(\pi r^2)$ and substitute into the equation for $S$ to obtain $S$ as a function of $r$ alone:

$$S = S(r) = 2\pi r \frac{V}{\pi r^2} + 2\pi r^2 = \frac{2V}{r} + 2\pi r^2 \quad (0 < r < \infty).$$

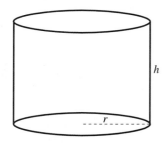

**Figure 4.41**

Evidently, $\lim_{r \to 0+} S(r) = \infty$ and $\lim_{r \to \infty} S(r) = \infty$. Being differentiable and therefore continuous on $(0, \infty)$, $S(r)$ must have a minimum value, and it must occur at a critical point. To find any critical points,

$$0 = S'(r) = -\frac{2V}{r^2} + 4\pi r$$

$$r^3 = \frac{2V}{4\pi} = \frac{1}{2\pi}\pi r^2 h = \frac{1}{2}r^2 h.$$

Thus $h = 2r$ at the critical point of $S$. Under interpretation (i), the most economical can is shaped so that its height equals the diameter of its base. You are encouraged to show that interpretation (ii) leads to the same conclusion.   ∎

**Remark**   There is another way to solve Example 5 that shows directly that interpretations (i) and (ii) must give the same solution. Again we start from the two equations

$$V = \pi r^2 h \qquad \text{and} \qquad S = 2\pi rh + 2\pi r^2.$$

If we regard $h$ as a function of $r$ and differentiate implicitly, we obtain

$$\frac{dV}{dr} = 2\pi rh + \pi r^2 \frac{dh}{dr},$$

$$\frac{dS}{dr} = 2\pi h + 2\pi r \frac{dh}{dr} + 4\pi r.$$

Under interpretation (i) $V$ is constant and we want a critical point of $S$; under interpretation (ii), $S$ is constant and we want a critical point of $V$. In *either* case, $dV/dr = 0$ and $dS/dr = 0$. Hence both interpretations yield

$$2\pi rh + \pi r^2 \frac{dh}{dr} = 0, \qquad \text{and} \qquad 2\pi h + 4\pi r + 2\pi r \frac{dh}{dr} = 0.$$

If we divide the first equation by $\pi r^2$ and the second equation by $2\pi r$ and subtract to eliminate $dh/dr$ we again get $h = 2r$.

**Remark**   **Modifying Example 5**  Given the sparse information provided in the statement of the problem in Example 5, interpretations (i) and (ii) are the best we can do. The problem could be made more meaningful economically (from the point of view, say, of a tin can manufacturer) if more elements were brought into it. For example:

(a) Most cans use thicker material for the cylindrical wall than for the top and bottom disks. If the cylindrical wall material costs $A$ per unit area and the material for the top and bottom costs $B$ per unit area, we might prefer to minimize the total cost for materials for a can of given volume. What is the optimal shape if $A = 2B$?

(b) Large numbers of cans are to be manufactured. The material is probably being cut out of sheets of metal. The cylindrical walls are made by bending up rectangles, and rectangles can be cut from the sheet with little or no waste. There will, however, always be a proportion of material wasted when the disks are cut out. The exact proportion will depend on how the disks are arranged;

two possible arrangements are shown in Figure 4.42. What is the optimal shape of the can if a square packing of disks is used? a hexagonal packing? Any such modification of the original problem will alter the optimal shape to some extent. In "real-world" problems, many factors may have to be taken into account to come up with a "best" strategy.

(c) The problem makes no provision for costs of manufacturing the can other than the cost of sheet metal. There may also be costs for joining the opposite edges of the rectangle to make the cylinder, and for joining the top and bottom disks to the cylinder. These costs may be proportional to the lengths of the joins.

In most of the examples above the maximum or minimum value being sought occurred at a critical point. Our final example is one where this is not the case.

Square Packing:
each disk uses up a square

Hexagonal Packing:
each disk uses up a hexagon

**Figure 4.42**  Square and hexagonal packing of disks in a plane

**Example 6**  A man can run twice as fast as he can swim. He is standing at point $A$ on the edge of a circular swimming pool 40 m in diameter, and he wishes to get to the diametrically opposite point $B$ as quickly as possible. He can run around the edge to point $C$, then swim directly from $C$ to $B$. Where should $C$ be chosen to minimize the total time taken to get from $A$ to $B$?

**Solution**  It is convenient to describe the position of $C$ in terms of the angle $AOC$, where $O$ is the centre of the pool. (See Figure 4.43.) Let $\theta$ denote this angle. Clearly $0 \le \theta \le \pi$. (If $\theta = 0$ the man swims the whole way; if $\theta = \pi$ he runs the whole way.) The radius of the pool is 20 m, so arc $AC = 20\theta$. Since angle $BOC = \pi - \theta$, we have angle $BOL = (\pi - \theta)/2$, and chord $BC = 2BL = 40 \sin((\pi - \theta)/2)$.

Suppose the man swims at a rate $k$ m/s and therefore runs at a rate $2k$ m/s. If $t$ is the total time he takes to get from $A$ to $B$, then

$$t = t(\theta) = \text{time running} + \text{time swimming}$$
$$= \frac{20\theta}{2k} + \frac{40}{k} \sin \frac{\pi - \theta}{2}.$$

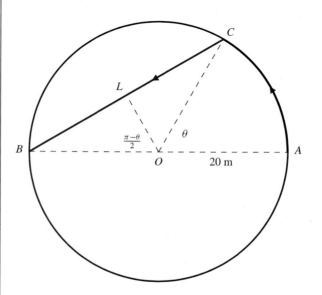

**Figure 4.43**  Running and swimming to get from $A$ to $B$

(We are assuming that no time is wasted in jumping into the water at $C$.) The domain of $t$ is $[0, \pi]$ and $t$ has no singular points. Since $t$ is continuous on a closed, finite interval it must have a minimum value, and that value must occur at a critical point or an endpoint. For critical points,

$$0 = t'(\theta) = \frac{10}{k} - \frac{20}{k} \cos \frac{\pi - \theta}{2}.$$

Thus

$$\cos \frac{\pi - \theta}{2} = \frac{1}{2}, \qquad \frac{\pi - \theta}{2} = \frac{\pi}{3}, \qquad \theta = \frac{\pi}{3}.$$

This is the only critical value of $\theta$ lying in the interval $[0, \pi]$. We have

$$t\left(\frac{\pi}{3}\right) = \frac{10\pi}{3k} + \frac{40}{k} \sin \frac{\pi}{3} = \frac{10}{k}\left(\frac{\pi}{3} + \frac{4\sqrt{3}}{2}\right) \approx \frac{45.11}{k}.$$

We must also look at the endpoints $\theta = 0$ and $\theta = \pi$:

$$t(0) = \frac{40}{k}, \qquad t(\pi) = \frac{10\pi}{k} \approx \frac{31.4}{k}.$$

Evidently $t(\pi)$ is the least of these three times. To get from $A$ to $B$ as quickly as possible, the man should run the entire distance. ∎

**Remark**   This problem shows how important it is to check every candidate point to see whether it gives a maximum or minimum. Here the critical point $\theta = \pi/3$ yielded the *worst* possible strategy; running one-third of the way around and then swimming the remainder would take the greatest time, not the least.

## Exercises 4.5

1. Two positive numbers have sum 7. What is the largest possible value for their product?

2. Two positive numbers have product 8. What is the smallest possible value for their sum?

3. Two nonnegative numbers have sum 60. What are the numbers if the product of one of them and the square of the other is maximal?

4. Two numbers have sum 16. What are the numbers if the product of the cube of one and the fifth power of the other is as large as possible?

5. The sum of two nonnegative numbers is 10. What is the smallest value of the sum of the cube of one number and the square of the other?

6. Two nonnegative numbers have sum $n$. What is the smallest possible value for the sum of their squares?

7. Among all rectangles of given area, show that the square has the least perimeter.

8. Among all rectangles of given perimeter, show that the square has the greatest area.

9. Among all isosceles triangles of given perimeter, show that the equilateral triangle has the greatest area.

10. Find the largest possible area for an isosceles triangle if the length of each of its two equal sides is 10 m.

11. Find the area of the largest rectangle that can be inscribed in a semicircle of radius $R$ if one side of the rectangle lies along the diameter of the semicircle.

12. Find the largest possible perimeter of a rectangle inscribed in a semicircle of radius $R$ if one side of the rectangle lies along the diameter of the semicircle. (It is interesting to note that the rectangle with the largest perimeter has a different shape than the one with the largest area, obtained in Exercise 11.)

13. A rectangle with sides parallel to the coordinate axes is

inscribed in the ellipse

$$\frac{x^2}{a^2} + \frac{y^2}{b^2} = 1.$$

Find the largest possible area for this rectangle.

14. Let $ABC$ be a triangle right-angled at $C$ and having area $S$. Find the maximum area of a rectangle inscribed in the triangle if (a) one corner of the rectangle lies at $C$, or (b) one side of the rectangle lies along the hypotenuse, $AB$.

15. **(Designing a billboard)** A billboard is to be made with 100 m² of printed area, and with margins of 2 m at the top and bottom and 4 m on each side. Find the outside dimensions of the billboard if its total area is to be a minimum.

16. **(Designing a box)** A box is to be made from a rectangular sheet of cardboard 70 cm by 150 cm by cutting equal squares out of the four corners and bending up the resulting four flaps to make the sides of the box. (The box has no top.) What is the largest possible volume of the box?

17. **(Using rebates to maximize profit)** An automobile manufacturer sells 2,000 cars per month, at an average profit of $1,000 per car. Market research indicates that for each $50 of factory rebate the manufacturer offers to buyers it can expect to sell 200 more cars each month. How much of a rebate should it offer to maximize its monthly profit?

18. **(Maximizing rental profit)** All 80 rooms in a motel will be rented each night if the manager charges $40 or less per room. If he charges $(40 + x)$ per room, then $2x$ rooms will remain vacant. If each rented room costs the manager $10 per day and each unrented room $2 per day in overhead, how much should the manager charge per room to maximize his daily profit?

19. **(Minimizing travel time)** You are in a dune buggy in the desert 12 km due south of the nearest point $A$ on a straight east-west road. You wish to get to point $B$ on the road 10 km east of $A$. If your dune buggy can average 15 km/h travelling over the desert, and 39 km/h travelling on the road, toward what point on the road should you head in order to minimize your travel time to $B$?

20. Repeat Exercise 19, but assume that $B$ is only 4 km from $A$.

21. A one-metre length of stiff wire is cut into two pieces. One piece is bent into a circle, the other piece into a square. Find the length of the part used for the square if the sum of the areas of the circle and the square is (a) maximum, and (b) minimum.

22. Find the area of the largest rectangle which can be drawn so that each of its sides passes through a different vertex of a rectangle having sides $a$ and $b$.

23. What is the length of the shortest line segment having one end on the $x$-axis, the other end on the $y$-axis, and passing through the point $(9, \sqrt{3})$?

24. **(Getting around a corner)** Find the length of the longest beam that can be carried horizontally around the corner from a hallway of width $a$ m to a hallway of width $b$ m. (See Figure 4.44; assume the beam has no width.)

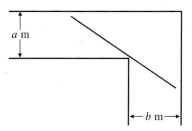

**Figure 4.44**

25. If the height of both hallways in Exercise 24 is $c$ m, and if the beam need not be carried horizontally, how long can it be and still get around the corner? *Hint:* you can use the result of the previous exercise to do this one easily.

26. The fence in Example 4 is demolished and a new fence is built 2 m away from the wall. How high can the fence be if a 6 m ladder must be able to extend from the wall, over the fence, to the ground outside?

27. Find the shortest distance from the origin to the curve $x^2 y^4 = 1$.

28. Find the shortest distance from the point $(8, 1)$ to the curve $y = 1 + x^{3/2}$.

29. Find the dimensions of the largest right-circular cylinder that can be inscribed in a sphere of radius $R$.

30. Find the dimensions of the circular cylinder of greatest volume that can be inscribed in a cone of base radius $R$ and height $H$ if the base of the cylinder lies in the base of the cone.

31. A box with square base and no top has a volume of 4 m³. Find the dimensions of the most economical box.

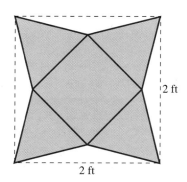

**Figure 4.45**

32. **(Folding a pyramid)** A pyramid with a square base and four faces, each in the shape of an isosceles triangle, is made by cutting away four triangles from a 2 ft square piece of cardboard as shown in Figure 4.45 and bending up the resulting triangles to form the walls of the pyramid. What is the largest volume the pyramid can have? *Hint:* the volume

of a pyramid having base area $A$ and height $h$ measured perpendicular to the base is $V = \frac{1}{3}Ah$.

33. **(Getting the most light)** A window has perimeter 10 m and is in the shape of a rectangle with top edge replaced by a semicircle. Find the dimensions of the rectangle if the window admits the greatest amount of light.

34. **(Fuel tank design)** A fuel tank is made of a cylindrical part capped by hemispheres at each end. If the hemispheres are twice as expensive per unit area as the cylindrical wall, and if the volume of the tank is $V$, find the radius and height of the cylindrical part to minimize the total cost. The surface area of a sphere of radius $r$ is $4\pi r^2$; its volume is $\frac{4}{3}\pi r^3$.

35. **(Reflection of light)** Light travels in such a way that it requires the minimum possible time to get from one point to another. A ray of light from $C$ reflects off a plane mirror $AB$ at $X$ and then passes through $D$. (See Figure 4.46.) Show that the rays $CX$ and $XD$ make equal angles with the normal to $AB$ at $X$. (*Remark:* You may wish to give a proof based on elementary geometry without using any calculus, or you can minimize the travel time on $CXD$.)

**Figure 4.46**

* 36. **(Snell's law)** If light travels with speed $v_1$ in one medium and speed $v_2$ in a second medium, and if the two media are separated by a plane interface, show that a ray of light passing from point $A$ in one medium to point $B$ in the other is bent at the interface in such a way that

$$\frac{\sin i}{\sin r} = \frac{v_1}{v_2},$$

where $i$ and $r$ are the angles of incidence and refraction, as is shown in Figure 4.47. This is known as Snell's Law. Deduce it from the least time principle stated in Exercise 35.

**Figure 4.47**

37. **(Cutting the stiffest beam)** The stiffness of a wooden beam of rectangular cross section is proportional to the product of the width and the cube of the depth of the cross section.

Find the width and depth of the stiffest beam that can be cut out of a circular log of radius $R$.

38. Find the equation of the straight line of maximum slope tangent to the curve $y = 1 + 2x - x^3$.

39. A quantity $Q$ grows according to the differential equation

$$\frac{dQ}{dt} = kQ^3(L - Q)^5,$$

where $k$ and $L$ are positive constants. How large is $Q$ when it is growing most rapidly?

* 40. Find the smallest possible volume of a right-circular cone that can contain a sphere of radius $R$. (The volume of a cone of base radius $r$ and height $h$ is $\frac{1}{3}\pi r^2 h$.)

* 41. **(The best view of a mural)** How far back from a mural should one stand to view it best if the mural is 10 ft high and the bottom of it is 2 ft above eye level? (See Figure 4.48.)

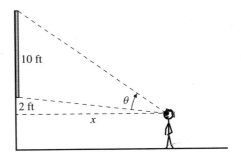

**Figure 4.48**

* 42. **(Improving the enclosure of Example 2)** An enclosure is to be constructed having part of its boundary along an existing straight wall. The other part of the boundary is to be fenced in the shape of an arc of a circle. If 100 m of fencing are available, what is the area of the largest possible enclosure? Into what fraction of a circle is the fence bent?

* 43. **(Designing a dixie cup)** A sector is cut out of a circular disk of radius $R$ and the remaining part of the disk is bent up so that the two edges join and a cone is formed (Figure 4.49). What is the largest possible volume for the cone?

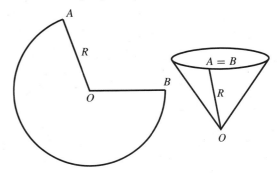

**Figure 4.49**

\* **44. (Minimize the fold)** One corner of a strip of paper $a$ cm wide is folded up so that it lies along the opposite edge (Figure 4.50). Find the least possible length for the fold line.

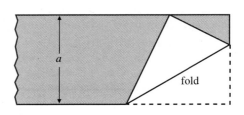

**Figure 4.50**

# 4.6 Finding Roots of Equations

Finding solutions (roots) of equations is an important mathematical problem to which calculus can make significant contributions. There are only a few general classes of equations of the form $f(x) = 0$ that we can solve exactly. These include **linear equations**:

$$ax + b = 0 \quad \Rightarrow \quad x = -\frac{b}{a},$$

and **quadratic equations**:

$$ax^2 + bx + c = 0 \quad \Rightarrow \quad x = \frac{-b \pm \sqrt{b^2 - 4ac}}{2a}.$$

Cubic and quartic (3rd and 4th degree polynomial) equations can also be solved, but the formulas are very complicated. We usually solve these and most other equations approximately by using numerical methods, often with the aid of a calculator or computer.

In Section 1.4 we discussed the Bisection Method for approximating a root of an equation $f(x) = 0$. That method uses the Intermediate-Value Theorem and depends only on the continuity of $f$ and our ability to find an interval $[x_1, x_2]$ that must contain the root because $f(x_1)$ and $f(x_2)$ have opposite signs. The method is rather slow; it requires between three and four iterations to gain one significant figure of precision in the root being approximated.

If we know that $f$ is more than just continuous, we can devise better (that is, faster) methods for finding roots of $f(x) = 0$. We study two such methods in this section:

(a) **Newton's Method**, which requires that $f$ be differentiable, and which is usually very efficient, and

(b) **Fixed-Point Iteration**, which is concerned with equations of a different form: $f(x) = x$.

Like the Bisection Method, both of these methods require that we have at the outset a rough idea of where a root can be found, and they generate sequences of approximations that get closer and closer to the root.

## Newton's Method

We want to find a **root** of the equation $f(x) = 0$, that is, a number $r$ such that $f(r) = 0$. Such a number is also called a **zero** of the function $f$. If $f$ is differentiable near the root, then tangent lines can be used to produce a sequence of approximations to the root that approaches the root quite quickly. The idea is as follows. (See Figure 4.51.) Make an initial guess at the root, say $x = x_0$. Draw the tangent line to $y = f(x)$ at $(x_0, f(x_0))$, and find $x_1$, the $x$-intercept of this tangent line. Under certain circumstances $x_1$ will be closer to the root than $x_0$ was. The process can be repeated over and over to get numbers $x_2, x_3, \ldots$, getting closer and closer to the root $r$. The number $x_{n+1}$ is the $x$-intercept of the tangent line to $y = f(x)$ at $(x_n, f(x_n))$.

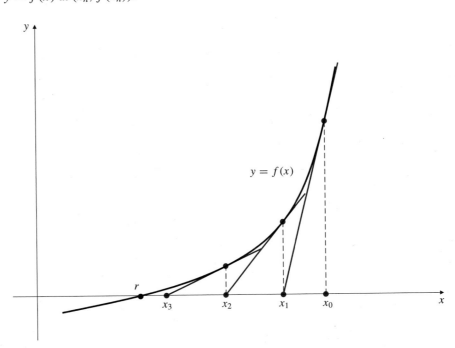

**Figure 4.51**

The tangent line to $y = f(x)$ at $x = x_0$ has equation

$$y = f(x_0) + f'(x_0)(x - x_0).$$

Since the point $(x_1, 0)$ lies on this line, we have $0 = f(x_0) + f'(x_0)(x_1 - x_0)$. Hence

$$x_1 = x_0 - \frac{f(x_0)}{f'(x_0)}.$$

Similar formulas produce $x_2$ from $x_1$, then $x_3$ from $x_2$, and so on. The formula producing $x_{n+1}$ from $x_n$ is

$$x_{n+1} = x_n - \frac{f(x_n)}{f'(x_n)},$$

and is known as the **Newton's method formula**. We usually use a calculator or computer to calculate the successive approximations $x_1$, $x_2$, $x_3$, ..., and observe whether these numbers appear to converge to a limit. If $\lim_{n\to\infty} x_n = r$ exists, and if $f/f'$ is continuous near $r$, then $r$ must be a root of $f$ because

$$\lim_{n\to\infty} x_{n+1} = \lim_{n\to\infty} x_n - \lim_{n\to\infty} \frac{f(x_n)}{f'(x_n)}$$

$$r = r - \frac{f(r)}{f'(r)},$$

from which it follows that $f(r) = 0$. This method is known as **Newton's method** .

**Example 1**    Use Newton's method to find the only real root of the equation $x^3 - x - 1 = 0$ correct to 10 decimal places.

**Solution**    We have $f(x) = x^3 - x - 1$ and $f'(x) = 3x^2 - 1$. Since $f$ is continuous and since $f(1) = -1$ and $f(2) = 5$, the equation has a root in the interval $[1, 2]$. Let us make the initial guess $x_0 = 1.5$. Newton's formula here is

$$x_{n+1} = x_n - \frac{x_n^3 - x_n - 1}{3x_n^2 - 1} = \frac{2x_n^3 + 1}{3x_n^2 - 1},$$

so that, for example, the approximation $x_1$ is given by

$$x_1 = \frac{2(1.5)^3 + 1}{3(1.5)^2 - 1} \approx 1.347826\ldots.$$

Using a scientific calculator, we calculated the values Table 1:

**Table 1.**

| $n$ | $x_n$ | $f(x_n)$ |
|---|---|---|
| 0 | 1.5 | 0.875 000 000 000 $\cdots$ |
| 1 | 1.347 826 086 96 $\cdots$ | 0.100 682 173 091 $\cdots$ |
| 2 | 1.325 200 398 95 $\cdots$ | 0.002 058 361 917 $\cdots$ |
| 3 | 1.324 718 174 00 $\cdots$ | 0.000 000 924 378 $\cdots$ |
| 4 | 1.324 717 957 24 $\cdots$ | 0.000 000 000 000 $\cdots$ |
| 5 | 1.324 717 957 24 $\cdots$ | |

Evidently $r = 1.3247179572$ correctly rounded to 10 decimal places.

Observe the behaviour of the numbers $x_n$. By the third iteration, $x_3$, we have apparently achieved a precision of six decimal places, and by $x_4$ over 10 decimal places. It is characteristic of Newton's method that when you begin to get close to the root the convergence can be very rapid. Compare these results with those obtained for the same equation by the Bisection Method in Example 12 of Section 1.4; there we achieved only 3 decimal place precision after 11 iterations.

**Example 2**    Solve the equation $x^3 = \cos x$ to 11 decimal places.

**Solution**    We are looking for the $x$-coordinate $r$ of the intersection of the curves $y = x^3$ and $y = \cos x$. From Figure 4.52 it appears that the curves intersect slightly to the left of $x = 1$. Let us start with the guess $x_0 = 0.8$. If $f(x) = x^3 - \cos x$, then $f'(x) = 3x^2 + \sin x$. Newton's formula for this function is

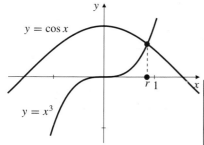

**Figure 4.52**    Solving $x^3 = \cos x$

$$x_{n+1} = x_n - \frac{x_n^3 - \cos x_n}{3x_n^2 + \sin x_n} = \frac{2x_n^3 + x_n \sin x_n + \cos x_n}{3x_n^2 + \sin x_n}.$$

The approximations $x_1, x_2, \ldots$ are given in Table 2:

**Table 2.**

| $n$ | $x_n$ | $f(x_n)$ |
|---|---|---|
| 0 | 0.8 | $-0.184\,706\,709\,347\cdots$ |
| 1 | $0.870\,034\,801\,135\cdots$ | $0.013\,782\,078\,762\cdots$ |
| 2 | $0.865\,494\,102\,425\cdots$ | $0.000\,006\,038\,051\cdots$ |
| 3 | $0.865\,474\,033\,493\cdots$ | $0.000\,000\,001\,176\cdots$ |
| 4 | $0.865\,474\,033\,102\cdots$ | $0.000\,000\,000\,000\cdots$ |
| 5 | $0.865\,474\,033\,102\cdots$ | |

The two curves intersect at $x = 0.86547403310$, rounded to 11 decimal places.  ∎

**Remark**   Example 2 above shows how useful a sketch can be for determining an initial guess $x_0$. Even a rough sketch of the graph of $y = f(x)$ can show you how many roots the equation $f(x) = 0$ has, and approximately where they are. Usually, the closer the initial approximation is to the actual root, the smaller the number of iterations needed to achieve the desired precision. Similarly, for an equation of the form $g(x) = h(x)$, making a sketch of the graphs of $g$ and $h$ (on the same set of axes) can suggest starting approximations for any intersection points. In either case, you can then apply Newton's method to improve the approximations.

**Remark**   When using Newton's method to solve an equation that is of the form $g(x) = h(x)$, (such as the one in Example 2), we must rewrite the equation in the form $f(x) = 0$, and apply Newton's method to $f$. Usually we just use $f(x) = g(x) - h(x)$, although $f(x) = (g(x)/h(x)) - 1$ is also a possibility.

**Remark**   If your calculator is programmable you should learn to program the Newton's method formula for a given equation so that generating new iterations requires pressing only a few buttons. If your calculator has graphing capabilities, you can use them to locate a good initial guess.

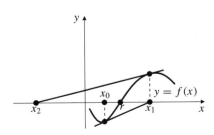

**Figure 4.53**   Here the Newton's method iterations do not converge to the root

Newton's method does not always work as well as it does in the preceding examples. If the first derivative $f'$ is very small near the root, or if the second derivative $f''$ is very large near the root, a single iteration of the formula can take us from quite close to the root to quite far away. Figure 4.53 illustrates this possibility. (Also see Exercises 15 and 16 at the end of this section.)

The following theorem gives sufficient conditions for the Newton approximations to converge to a root $r$ of the equation $f(x) = 0$ if the initial guess $x_0$ is sufficiently close to that root.

**THEOREM**  **7**

**Error bounds for Newton's method**

Suppose that $f$, $f'$, and $f''$ are continuous on an interval containing $x_n$, $x_{n+1}$, and a root $x = r$ of $f(x) = 0$. Suppose also that there exist constants $K$ and $L > 0$ such that for all $x$ in $I$ we have

    (i)  $|f''(x)| \le K$, and

    (ii)  $|f'(x)| \ge L$.

Then

(a) $|x_{n+1} - r| \leq \dfrac{K}{2L}|x_{n+1} - x_n|^2$, and

(b) $|x_{n+1} - r| \leq \dfrac{K}{2L}|x_n - r|^2.$

Conditions (i) and (ii) assert that near $r$ the slope of $y = f(x)$ is not too small in size and does not change too rapidly. If $K/(2L) < 1$, the theorem shows that $x_n$ converges quickly to $r$ once $n$ becomes large enough that $|x_n - r| < 1$.

The proof of Theorem 7 depends on the Mean-Value Theorem. We will not give it since the theorem is of little practical use. In practice, we calculate successive approximations using Newton's formula and observe whether they seem to converge to a limit. If they do, and if the values of $f$ at these approximations approach 0, we can be confident that we have located a root.

## Fixed-Point Iteration

A number $r$ satisfying the equation $f(r) = r$ is called a **fixed point** of the function $f$ because $f$ leaves that number unchanged. For certain kinds of functions, fixed points can be found by starting with an initial "guess" $x_0$ and calculating successive approximations $x_1 = f(x_0), x_2 = f(x_1), \ldots$. In general,

$$x_{n+1} = f(x_n), \qquad \text{for } n = 0, 1, 2, \ldots.$$

Let us begin by investigating a simple example:

**Example 3**    Find a root of the equation $\cos x = 5x$.

**Solution**    This equation is of the form $f(x) = x$, where $f(x) = \frac{1}{5}\cos x$. Since $\cos x$ is close to 1 for $x$ near 0, we see that $\frac{1}{5}\cos x$ will be close to $\frac{1}{5}$ when $x = \frac{1}{5}$. This suggests that a reasonable first guess at the fixed point is $x_0 = \frac{1}{5} = 0.2$. The values of subsequent approximations

$$x_1 = \frac{1}{5}\cos(x_0), \qquad x_2 = \frac{1}{5}\cos(x_1), \qquad x_3 = \frac{1}{5}\cos(x_2), \ldots$$

are presented in Table 3. The root is $0.19616428$ to eight decimal places.   ∎

**Table 3.**

| $n$ | $x_n$ |
| --- | --- |
| 0 | 0.2 |
| 1 | 0.196 013 32 |
| 2 | 0.196 170 16 |
| 3 | 0.196 164 05 |
| 4 | 0.196 164 29 |
| 5 | 0.196 164 28 |
| 6 | 0.196 164 28 |

Why did the method used in Example 3 work? Will it work for any function $f$? In order to answer these questions, examine the polygonal line in Figure 4.54. Starting at $x_0$ it goes vertically to the curve $y = f(x)$, the height there being $x_1$. Then it goes horizontally to the line $y = x$, meeting that line at a point whose $x$-coordinate must therefore also be $x_1$. Then the process repeats; the line goes vertically to the curve $y = f(x)$ and horizontally to $y = x$, arriving at $x = x_2$. The line continues in this way, "spiralling" closer and closer to the intersection of $y = f(x)$ and $y = x$. Each value of $x_n$ is closer to the fixed point $r$ than the previous value.

Now consider the function $f$ whose graph appears in Figure 4.55(a). If we try the same method there, starting with $x_0$, the polygonal line spirals outward, away from the root, and the resulting values $x_n$ will not "converge" to the root as they did in

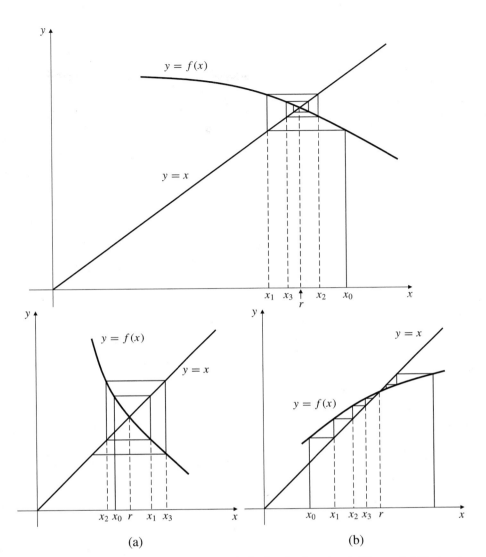

**Figure 4.54**     Iterations of $x_{n+1} = f(x_n)$ "spiral" towards the fixed point

**Figure 4.55**

(a) A function $f$ for which the iterations $x_{n+1} = f(x_n)$ do not converge

(b) "Staircase" convergence to the fixed point

(a)

(b)

Example 3. To see why the method works in for the function in Figure 4.54 but not for the function in Figure 4.55(a), observe the slopes of the two graphs $y = f(x)$, near the fixed point $r$. Both slopes are negative, but in Figure 4.54 the absolute value of the slope is less than 1 while the absolute value of the slope of $f$ in Figure 4.55(a) is greater than 1. Close consideration of the graphs should convince you that it is this fact that caused the points $x_n$ to get closer to $r$ in Figure 4.54 and farther from $r$ in Figure 4.55(a).

A third example, Figure 4.55(b), shows that the method can be expected to work for functions whose graphs have positive slope near the fixed point $r$, provided that the slope is less than 1. In this case the polygonal line forms a "staircase" rather than a "spiral" and the successive approximations $x_n$ increase towards the root if $x_0 < r$ and decrease towards it if $x_0 > r$.

The following theorem guarantees that the method of fixed-point iteration will work for a particular class of functions.

**THEOREM  8**     **A fixed-point theorem**

Suppose that $f$ is defined on an interval $I = [a, b]$ and satisfies the following two conditions:

(i) $f(x)$ belongs to $I$ whenever $x$ belongs to $I$, and

(ii) there exists a constant $K$ with $0 < K < 1$ such that for every $u$ and $v$ in $I$,

$$|f(u) - f(v)| \le K|u - v|.$$

Then $f$ has a fixed point $r$ in $I$,

$$f(r) = r,$$

and starting with any number $x_0$ in $I$, the iterates

$$x_1 = f(x_0), \quad x_2 = f(x_1), \quad \dots$$

converge to $r$.

You are invited to prove this theorem by a method outlined in Exercises 24 and 25 at the end of this section.

## "Solve" Routines

Many of the more advanced models of scientific calculators, and most computer based mathematics software have built-in routines for solving general equations numerically, or, in a few cases, symbolically. These "Solve" routines assume continuity of the left and right sides of the given equations and often require the user to specify an interval in which to search for the root, or an initial guess at the value of the root, or both. Typically the calculator or computer software also has graphing capabilities and you are expected to use them to get an idea of how many roots the equation has and roughly where they are located before invoking the solving routines. It may also be possible to specify a *tolerance* on the value of the left side — the right side of the equation. For instance, if we want a solution to the equation $f(x) = 0$, it may be more important to us to be sure that an approximate solution $\hat{x}$ satisfies $|f(\hat{x})| < 0.0001$ than it is to be sure that $\hat{x}$ is within any particular distance of the actual root.

The methods used by the solve routines vary from one calculator or software package to another and are frequently very sophisticated, making use of numerical differentiation and other techniques to find roots very quickly, even when the search interval is large.

If you have an advanced scientific calculator and/or computer software with similar capabilities, it is well worth your while to read the manuals that describe how to make effective use of your hardware/software for solving equations. Applications of mathematics to solving "real-world" problems frequently require finding approximate solutions of equations that are intractable by exact methods.

## Exercises 4.6

In Exercises 1–10, use Newton's method to solve the given equations to the precision permitted by your calculator.

**1.** Find $\sqrt{2}$ by solving $x^2 - 2 = 0$.

**2.** Find $\sqrt{3}$ by solving $x^2 - 3 = 0$.

**3.** Find the root of $x^3 + 2x - 1 = 0$ between 0 and 1.

**4.** Find the root of $x^3 + 2x^2 - 2 = 0$ between 0 and 1.

**5.** Find the two roots of $x^4 - 8x^2 - x + 16 = 0$ between 1 and 3.

**6.** Find the three roots of $x^3 + 3x^2 - 1 = 0$ between $-3$ and $1$.

**7.** Solve $\sin x = 1 - x$. Make a sketch to help you make a first guess $x_0$.

**8.** Solve $\cos x = x^2$. How many roots are there?

**9.** How many roots does the equation $\tan x = x$ have? Find the one between $\pi/2$ and $3\pi/2$.

**10.** Solve $\dfrac{1}{1 + x^2} = \sqrt{x}$, by rewriting it in the form $(1 + x^2)\sqrt{x} - 1 = 0$

**11.** If your calculator has a built-in Solve routine, or if you use computer software with such a routine, use it to solve the equations in the previous 10 exercises.

Find the maximum and minimum values of the functions in Exercises 12–13.

**12.** $\dfrac{\sin x}{1 + x^2}$

**13.** $\dfrac{\cos x}{1 + x^2}$

**14.** Let $f(x) = x^2$. The equation $f(x) = 0$ clearly has solution $x = 0$. Find the Newton's method iterations $x_1$, $x_2$, and $x_3$, starting with $x_0 = 1$.

(a) What is $x_n$?

(b) How many iterations are needed to find the root with error less than $0.0001$ in absolute value?

(c) How many iterations are needed to get an approximation $x_n$ for which $|f(x_n)| < 0.0001$?

(d) Why do the Newton's method iterations converge more slowly here than in the examples done in this section?

**15. (Oscillation)** Apply Newton's method to

$$f(x) = \begin{cases} \sqrt{x}, & x \ge 0, \\ \sqrt{-x}, & x < 0, \end{cases}$$

starting with the initial guess $x_0 = a > 0$. Calculate $x_1$ and $x_2$. What happens? (Make a sketch.) If you ever observed this behaviour when you were using Newton's method to find a root of an equation, what would you do next?

**16. (Divergent oscillations)** Apply Newton's method to $f(x) = x^{1/3}$ with $x_0 = 1$. Calculate $x_1$, $x_2$, $x_3$, and $x_4$. What is happening? Find a formula for $x_n$.

**17. (Convergent oscillations)** Apply Newton's method to find $f(x) = x^{2/3}$ with $x_0 = 1$. Calculate $x_1$, $x_2$, $x_3$, and $x_4$. What is happening? Find a formula for $x_n$.

Use fixed-point iteration to solve the equations in Exercises 18–22. Obtain 5 decimal place precision.

**18.** $1 + \frac{1}{4}\sin x = x$

**19.** $\cos \dfrac{x}{3} = x$

**20.** $(x + 9)^{1/3} = x$

**21.** $\dfrac{1}{2 + x^2} = x$

**22.** Solve $x^3 + 10x - 10 = 0$ by rewriting it in the form $1 - \frac{1}{10}x^3 = x$.

**23.** Let $f(x)$ be a differentiable function whose derivative $f'(x)$ is never zero. Let

$$N(x) = x - \frac{f(x)}{f'(x)}.$$

Show that $r$ is a root of $f(x) = 0$ if and only if $r$ is a fixed point of $N(x)$. What are the successive approximations $x_{n+1} = N(x_n)$ starting from $x_0$ in this case?

Exercises 24–25 constitute a proof of Theorem 8.

* **24.** Condition (ii) of Theorem 8 implies that $f$ is continuous on $I = [a, b]$. Use condition (i) to show that $f$ has a fixed point $r$ on $I$. *Hint:* apply the Intermediate-Value Theorem to $g(x) = f(x) - x$ on $[a, b]$.

* **25.** Use condition (ii) of Theorem 8 and mathematical induction to show that

$$|x_n - r| \le K^n |x_0 - r|.$$

Since $0 < K < 1$, we know that $K^n \to 0$ as $n \to \infty$. This shows that $\lim_{n \to \infty} x_n = r$.

## 4.7  Linear Approximations

Many problems in applied mathematics are too difficult to be solved exactly — all we can hope to do is find approximate solutions that are correct to within some acceptably small tolerance. In this section we will examine how knowledge of the values of a function and its first derivative at a point can help us find approximate values for the function at nearby points.

The tangent to the graph $y = f(x)$ at $x = a$ describes the behaviour of that graph near the point $P = (a, f(a))$ better than any other straight line through $P$, because it goes through $P$ in the same direction as the curve $y = f(x)$. (See Figure 4.56.) We exploit this fact by using the height to the tangent line to calculate approximate values of $f(x)$ for values of $x$ near $a$. The tangent line has equation

$y = f(a) + f'(a)(x - a)$. We call the right side of this equation the linearization of $f$ about $x = a$.

---

**DEFINITION 8**

The **linearization**, or **linear approximation** of the function $f$ about $x = a$ is the function $L(x)$ defined by

$$L(x) = f(a) + f'(a)(x - a).$$

---

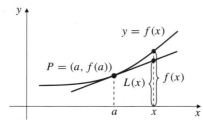

**Figure 4.56**    The linearization of $f(x)$ about $x = a$

**Example 1**    Find the linearizations for (a) $f(x) = \sqrt{1 + x}$ about $x = 0$, and (b) $g(x) = 1/x$ about $x = 1/2$.

**Solution**

(a) Since $f'(x) = 1/(2\sqrt{1 + x})$, we have $f(0) = 1$ and $f'(0) = 1/2$. The linearization of $f(x)$ about $x = 0$ is

$$L(x) = 1 + \frac{1}{2}(x - 0) = 1 + \frac{x}{2}.$$

(b) Since $g'(x) = -1/x^2$, we have $g(1/2) = 2$ and $g'(1/2) = -4$. The linearization of $g(x)$ about $x = 1/2$ is

$$L(x) = 2 - 4\left(x - \frac{1}{2}\right) = 4 - 4x.$$

■

## Approximating Values of Functions

We have already made use of linearization in Section 2.7, where it was disguised as the formula

$$\Delta y \approx \frac{dy}{dx} \Delta x$$

and used to approximate a small change $\Delta y = f(a + \Delta x) - f(a)$ in the values of function $f$ corresponding to the small change in the argument of the function from $a$ to $a + \Delta x$. This is just the linear approximation

$$f(a + \Delta x) \approx f(a) + f'(a)\Delta x.$$

**Example 2**    A ball of ice melts so that its radius decreases from 5 cm to 4.92 cm. By approximately how much does the volume of the ball decrease?

**Solution**    The volume $V$ of a ball of radius $r$ is given by $V = \frac{4}{3}\pi r^3$, so

$$\Delta V \approx \frac{4}{3}\pi\,(3r^2)\,\Delta r = 4\pi r^2\,\Delta r.$$

For $r = 5$ and $\Delta r = -0.08$, we have

$$\Delta V \approx 4\pi\,(5^2)(-0.08) = -8\pi \approx -25.13.$$

The volume of the ball decreases by about 25 cm$^3$.

■

The following example illustrates the use of linearization to find an approximate value of a function near a point where the values of the function and its derivative is known.

---

**Example 3**   Use the linearization for $\sqrt{x}$ about $x = 25$ to find an approximate value for $\sqrt{26}$.

**Solution**   If $f(x) = \sqrt{x}$, then $f'(x) = 1/(2\sqrt{x})$. Since we know that $f(25) = 5$ and $f'(25) = 1/10$, the linearization of $f(x)$ about $x = 25$ is

$$L(x) = 5 + \frac{1}{10}(x - 25).$$

Putting $x = 26$, we get

$$\sqrt{26} = f(26) \approx L(26) = 5 + \frac{1}{10}(26 - 25) = 5.1.$$

---

If we use the square root function on a calculator we can obtain the "true value" of $\sqrt{26}$ (actually, just another approximation, though presumably a rather better one): $\sqrt{26} = 5.0990195\cdots$, but if we have such a calculator we don't need the approximation in the first place. Approximations are useful when there is no easy way to obtain the true value. However, if we don't know the true value, we would at least like to have some way of determining how good the approximation must be; that is, we want an *estimate for the error*. After all, *any number* is an approximation to $\sqrt{26}$; but the error may be unacceptably large. For instance, the size of the error in the approximation $\sqrt{26} \approx 1{,}000{,}000$ is greater than 999,994.

## Error Analysis

In any approximation, the **error** is defined by

> error = true value − approximate value.

If the linearization of $f(x)$ about $x = a$ is used to approximate $f(x)$ near $a$, that is,

$$f(x) \approx L(x) = f(a) + f'(a)(x - a),$$

then the error $E(x)$ in this approximation is

$$E(x) = f(x) - L(x) = f(x) - f(a) - f'(a)(x - a).$$

It is the vertical distance at $x$ between the graph of $f$ and the tangent line to that graph at $x = a$, as shown in Figure 4.57. Observe that if $x$ is "near" $a$, then $E(x)$ is small compared to the horizontal distance between $x$ and $a$.

The following theorem and its corollaries gives us a way to estimate this error if we know bounds for the *second derivative* of $f$.

**THEOREM 9**

**An error formula for linearization**

If $f''(t)$ exists for all $t$ in an interval containing $a$ and $x$, then there exists some point $X$ between $a$ and $x$ such that the error $E(x) = f(x) - L(x)$ in the linear approximation $f(x) \approx L(x) = f(a) + f'(a)(x - a)$ satisfies

$$E(x) = \frac{f''(X)}{2}(x-a)^2.$$

**PROOF** Let us assume that $x > a$. (The proof for $x < a$ is similar.) Since

$$E(t) = f(t) - f(a) - f'(a)(t-a),$$

we have $E'(t) = f'(t) - f'(a)$. We apply the Generalized Mean-Value Theorem (Theorem 16 of Section 2.6) to the two functions $E(t)$ and $(t-a)^2$ on $[a, x]$. Noting that $E(a) = 0$, we obtain a number $c$ in $(a, x)$ such that

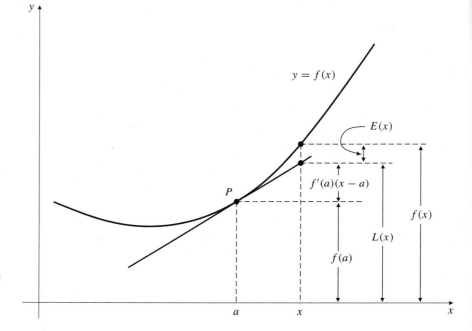

**Figure 4.57** $f(x)$ and its linearization $L(x)$ about $x = a$. $E(x)$ is the error in the approximation $f(x) \approx L(x)$.

$$\frac{E(x)}{(x-a)^2} = \frac{E(x) - E(a)}{(x-a)^2 - (a-a)^2} = \frac{E'(c)}{2(c-a)} = \frac{f'(c) - f'(a)}{2(c-a)} = \frac{1}{2}f''(X)$$

for some $X$ in $(a, c)$; the latter expression is a consequence of applying the Mean-Value Theorem again, this time to $f'(t)$ on $[a, c]$. Thus

$$E(x) = \frac{f''(X)}{2}(x-a)^2$$

as claimed.

The following three corollaries are immediate consequences of Theorem 9.

**Corollary A.** If $f''(t)$ has constant sign (i.e. is always positive or always negative) between $a$ and $x$, then the error $E(x)$ in the linear approximation $f(x) \approx L(x)$ has that same sign. If $f''(t) > 0$ between $a$ and $x$, then $f(x) > L(x)$; if $f''(t) < 0$ between $a$ and $x$, then $f(x) < L(x)$.

**Corollary B.** If $|f''(t)| < K$ for all $t$ between $a$ and $x$, then $|E(x)| < (K/2)(x-a)^2$.

**Corollary C.** If $f''(t)$ satisfies $M < f''(t) < N$ for all $t$ between $a$ and $x$ (where $M$ and $N$ are constants), then

$$L(x) + \frac{M}{2}(x-a)^2 < f(x) < L(x) + \frac{N}{2}(x-a)^2.$$

If $M$ and $N$ have the same sign, a better approximation to $f(x)$ is given by the midpoint of this interval containing $f(x)$:

$$f(x) \approx L(x) + \frac{M+N}{4}(x-a)^2.$$

For this approximation the error is less than half the length of the interval:

$$|\text{Error}| < \frac{N-M}{4}(x-a)^2.$$

---

**Example 4**    Determine the sign and estimate the size of the error in the approximation $\sqrt{26} \approx 5.1$ obtained in Example 3. Use these to give an interval that you can be sure contains $\sqrt{26}$.

**Solution**    For $f(t) = t^{1/2}$, we have

$$f'(t) = \frac{1}{2}t^{-1/2}, \qquad \text{and} \qquad f''(t) = -\frac{1}{4}t^{-3/2}.$$

For $25 < t < 26$, we have $f''(t) < 0$, so $\sqrt{26} = f(26) < L(26) = 5.1$. Also, $t^{3/2} > 25^{3/2} = 125$, so $|f''(t)| < (1/4)(1/125) = 1/500$ and

$$|E(26)| < \frac{1}{2} \times \frac{1}{500} \times (26-25)^2 = \frac{1}{1,000} = 0.001.$$

Therefore, $f(26) > L(26) - 0.001 = 5.099$, and $\sqrt{26}$ is in the interval $(5.099, 5.1)$. ∎

---

**Remark**    We can use Corollary C of Theorem 9 and the fact that $\sqrt{26} < 5.1$ to find a better (i.e. smaller) interval containing $\sqrt{26}$ as follows. If $25 < t < 26$, then $t^{3/2} < 26^{3/2} < 5.1^3$. Thus $t^{-3/2} > (1/5.1)^3$ and $f''(t) < -(1/4)(1/5.1^3)$. By Corollary C,

$$E(26) < -\frac{1}{8} \times \frac{1}{5.1^3} \times (26-25)^2 < -0.00094.$$

Thus $\sqrt{26} < 5.1 - 0.00094$, and $\sqrt{26}$ lies in $(5.099, 5.09906)$. The midpoint of this interval is a better approximation to $\sqrt{26}$ than 5.1 is:

$$\sqrt{26} \approx \frac{5.099 + 5.09906}{2} = 5.09903,$$

with error less than 0.00003 in absolute value.

**Example 5**    Use a suitable linearization to find an approximate value for $\cos(36°) = \cos(\pi/5)$. Is the true value greater than or less than your approximation? Estimate the size of the error and give an interval that you can be sure contains $\cos(36°)$.

**Solution**    Let $f(t) = \cos t$, so that $f'(t) = -\sin t$ and $f''(t) = -\cos t$. The value of $a$ nearest to $36°$ for which we know $\cos a$ is $a = 30° = \pi/6$, so we use the linearization at that point:

$$L(x) = \cos\frac{\pi}{6} - \sin\frac{\pi}{6}\left(x - \frac{\pi}{6}\right) = \frac{\sqrt{3}}{2} - \frac{1}{2}\left(x - \frac{\pi}{6}\right).$$

Since $(\pi/5) - (\pi/6) = \pi/30$, our approximation is

$$\cos(36°) = \cos\frac{\pi}{5} \approx L\left(\frac{\pi}{5}\right) = \frac{\sqrt{3}}{2} - \frac{1}{2}\left(\frac{\pi}{30}\right) \approx 0.81367.$$

If $(\pi/6) < t < (\pi/5)$, then $f''(t) < 0$ and $|f''(t)| < \cos(\pi/6) = \sqrt{3}/2$. Therefore $\cos(36°) < 0.81367$ and

$$|E(36°)| < \frac{\sqrt{3}}{4}\left(\frac{\pi}{30}\right)^2 < 0.00475.$$

Thus $0.81367 - 0.00475 < \cos(36°) < 0.81367$, so $\cos(36°)$ lies in the interval $(0.80892, 0.81367)$.    ∎

**Remark**    Again it is possible to get a smaller interval and a better approximation to $\cos(36°)$ by using Corollary C of Theorem 9. Using the above result we conclude that if $(\pi/6) < t < (\pi/5)$, then $\cos t > \cos(36°) > 0.80892$ and therefore $f''(t) < -0.80892$. Thus

$$E(36°) < \frac{-0.80892}{2}\left(\frac{\pi}{30}\right)^2 < -0.00443,$$

and $\cos(36°) < 0.81367 - 0.00443 = 0.80924$. So $\cos(36°)$ lies in the interval $(0.80892, 0.80924)$. A better approximation to $\cos(36°)$ is the midpoint of this interval:

$$\cos(36°) \approx \frac{0.80892 + 0.80924}{2} = 0.80908,$$

with error less than 0.00016 in absolute value. (A scientific calculator tells us that $\cos(36°) \approx 0.80902$.)

**Remark**    The error in the linearization of $f(x)$ about $x = a$ can be interpreted in terms of differentials (see Section 2.2) as follows. If $x - a = \Delta x = dx$, then the change in $f(x)$ as we pass from $x = a$ to $x = a + \Delta x$ is $f(a + \Delta x) - f(a) = \Delta y$, and the corresponding change in the linearization $L(x)$ is $f'(a)(x - a) = f'(a)\,dx$, which is just the value at $x = a$ of the differential $dy = f'(x)\,dx$. Thus

$$E(x) = \Delta y - dy.$$

The fact that the approximating line is tangent to the graph of $f$ at $x = a$ implies that the error is small compared with $\Delta x$ as $\Delta x$ approaches 0. In fact,

$$\lim_{\Delta x \to 0} \frac{\Delta y - dy}{\Delta x} = \lim_{\Delta x \to 0} \left( \frac{\Delta y}{\Delta x} - \frac{dy}{dx} \right) = \frac{dy}{dx} - \frac{dy}{dx} = 0.$$

If $|f''(t)| \le K$ near $t = a$, we can assert a stronger result than this, namely,

$$\left| \frac{\Delta y - dy}{(\Delta x)^2} \right| = \left| \frac{E(x)}{(\Delta x)^2} \right| \le \frac{K}{2} = \text{constant},$$

so $|\Delta y - dy| \le \dfrac{K}{2} (\Delta x)^2$.

## An Application to Special Relativity

The following example uses a linear approximation to deduce the equivalence $E = mc^2$ between mass and energy from the relativistic expression for the mass of a moving particle.

> **Example 6**  (**The equivalence of mass and energy**) In Einstein's special theory of relativity, the momentum of a particle having rest mass $m_0$ moving at speed $v$ is given by
>
> $$p = \frac{m_0 v}{\sqrt{1 - \dfrac{v^2}{c^2}}},$$
>
> where $c$ is the speed of light. This is consistent with the particle's having an apparent mass of $m = m_0/\sqrt{1 - (v^2/c^2)}$ if we use the Newtonian interpretation of momentum as mass $\times$ velocity. Note that $m \to m_0$ as $v \to 0+$ and $m \to \infty$ as $v \to c-$.
>
> Using the linearization of $(1 - t)^{-1/2}$ about $t = 0$, namely $L(t) = 1 + (t/2)$, which is a good approximation to $(1 - t)^{-1/2}$ for small values of $|t|$, we conclude, taking $t = v^2/c^2$, that if $v$ is small compared to $c$, then
>
> $$m \approx m_0 \left( 1 + \frac{v^2}{2c^2} \right) = m_0 + \frac{1}{c^2} \left( \frac{1}{2} m_0 v^2 \right).$$
>
> In going from speed 0 to speed $v$, the apparent mass of the particle has increased by
>
> $$\Delta m = m - m_0 \approx \frac{1}{c^2} \Delta KE,$$
>
> where
>
> $$\Delta KE = \frac{1}{2} m_0 (v^2 - 0^2) = \frac{1}{2} m_0 v^2$$
>
> is the change in the (classical) kinetic energy of the particle. Thus the increase in kinetic energy is about $c^2$ times the increase in mass.

## Exercises 4.7

In Exercises 1–10, find the linearization of the given function at the given point.

1. $x^2$ at $x = 3$

2. $x^{-3}$ at $x = 2$

3. $\sqrt{4 - x}$ at $x = 0$

4. $\sqrt{3 + x^2}$ at $x = 1$

**5.** $1/(1+x)^2$ at $x = 2$

**6.** $1/\sqrt{x}$ at $x = 4$

**7.** $\sin x$ at $x = \pi$

**8.** $\cos(2x)$ at $x = \pi/3$

**9.** $\sin^2 x$ at $x = \pi/6$

**10.** $\tan x$ at $x = \pi/4$

**11.** By approximately how much does the area of a square increase if its side length increases from 10 cm to 10.4 cm?

**12.** By about how much must the edge length of a cube decrease from 20 cm to reduce the volume of the cube by 12 cm$^3$?

**13.** A spacecraft orbits the earth at an distance of 4,100 miles from the centre of the earth. By about how much will the circumference of its orbit decrease if the radius decreases by 10 miles?

**14.** **(Acceleration of gravity)** The acceleration $a$ of gravity at an altitude of $h$ miles above the surface of the earth is given by

$$a = g\left(\frac{R}{R+h}\right)^2,$$

where $g \approx 32$ ft/s$^2$ is the acceleration at the surface of the earth, and $R \approx 3960$ miles is the radius of the earth. By about what percentage will $a$ decrease if $h$ increases from 0 to 10 miles?

In Exercises 15–22, use a suitable linearization to approximate the indicated value. Determine the sign of the error and estimate its size. Use this information to specify an interval you can be sure contains the value.

**15.** $\sqrt{50}$

**16.** $\sqrt{47}$

**17.** $\sqrt[4]{85}$

**18.** $\dfrac{1}{2.003}$

**19.** $\cos 46°$

**20.** $\sin\dfrac{\pi}{5}$

**21.** $\sin(3.14)$

**22.** $\sin 33°$

Use Corollary C of Theorem 9 in the manner described in the remarks following Examples 4 and 5 to find better intervals and better approximations to the values in Exercises 23–26.

**23.** $\sqrt{50}$ as first approximated in Exercise 15.

**24.** $\sqrt{47}$ as first approximated in Exercise 16.

**25.** $\sqrt[4]{85}$ as first approximated in Exercise 17.

**26.** $\sin 33°$ as first approximated in Exercise 22.

**27.** If $f(2) = 4$, $f'(2) = -1$ and $0 \le f''(x) \le 1/x$ for all $x > 0$, find the smallest interval you can that contains $f(3)$.

**28.** If $f(2) = 4$, $f'(2) = -1$ and $\dfrac{1}{2x} \le f''(x) \le \dfrac{1}{x}$ for $2 \le x \le 3$, find the best approximation you can for $f(3)$.

**29.** If $g(2) = 1$, $g'(2) = 2$, and $|g''(x)| < 1 + (x-2)^2$ for all $x > 0$, find the best approximation you can for $g(1.8)$. How large can the error be?

**30.** Show that the linearization of $\sin\theta$ about $\theta = 0$ is $L(\theta) = \theta$. What is the percentage error in the approximation $\sin\theta \approx \theta$ if $|\theta|$ is less than $17°$?

**31.** A spherical balloon is inflated so that its radius increases from 20.00 cm to 20.20 cm in 1 minute. By approximately how much has its volume increased in that minute?

# 4.8 Taylor Polynomials

The linearization of a function $f(x)$ about $x = a$, namely the linear function

$$P_1(x) = L(x) = f(a) + f'(a)(x - a),$$

describes the behaviour of $f(x)$ near $x = a$ better than any other polynomial of degree 1 because both $P_1$ and $f$ have the same value and the same derivative at $x = a$:

$$P_1(a) = f(a), \qquad \text{and} \qquad P_1'(a) = f'(a).$$

(We are now using the symbol $P_1$ instead of $L$ to stress the fact that the linearization is a polynomial of degree 1.)

We can obtain even better approximations to $f(x)$ by using quadratic or higher-degree polynomials and matching more derivatives at $x = a$. For example, if $f$ is twice differentiable near $x = a$, then the quadratic polynomial

$$P_2(x) = f(a) + f'(a)(x - a) + \frac{f''(a)}{2}(x - a)^2$$

satisfies $P_2(a) = f(a)$, $P_2'(a) = f'(a)$ and $p_2''(a) = f''(a)$, and describes the behaviour of $f(x)$ near $x = a$ better than any other polynomial of degree 2.

In general, if $f^{(n)}(x)$ exists in an open interval containing $x = a$, then the polynomial

$$P_n(x) = f(a) + \frac{f'(a)}{1!}(x-a) + \frac{f''(a)}{2!}(x-a)^2 + \frac{f'''(a)}{3!}(x-a)^3 + \cdots + \frac{f^{(n)}(a)}{n!}(x-a)^n$$

matches $f$ and its first $n$ derivatives at $x = a$,

$$P_n(a) = f(a), \quad P_n'(a) = f'(a), \quad \ldots, \quad P_n^{(n)}(a) = f^{(n)}(a),$$

and so describes $f(x)$ near $x = a$ better than any other polynomial of degree $n$. $P_n$ is called **the Taylor polynomial of degree $n$ for $f(x)$ about $x = a$.** (If $a = 0$, the Taylor polynomials are sometimes called Maclaurin polynomials.) The Taylor polynomial of degree 0 is just the constant function $P_0(x) = f(a)$.

**Example 1** Find the following Taylor polynomials:

(a) $P_2(x)$ for $f(x) = \sqrt{x}$ about $x = 1$.

(b) $P_3(x)$ for $g(x) = \sin x$ about $x = 0$.

(c) $P_n(x)$ for $h(x) = e^x$ about $x = a$.

**Solution** (a) $f'(x) = (1/2)x^{-1/2}$, $f''(x) = -(1/4)x^{-3/2}$. Thus

$$P_2(x) = f(1) + f'(1)(x - 1) + \frac{f''(1)}{2!}(x-1)^2$$

$$= 1 + \frac{1}{2}(x - 1) - \frac{1}{8}(x-1)^2.$$

(b) $g'(x) = \cos x$, $g''(x) = -\sin x$, $g'''(x) = -\cos x$. Thus

$$P_3(x) = g(0) + g'(0)x + \frac{g''(0)}{2!}x^2 + \frac{g'''(0)}{3!}x^3$$

$$= x - \frac{1}{6}x^3.$$

(c) Evidently $h^{(n)}(x) = e^x$ for every positive integer $n$, so

$$P_n(x) = e^a + \frac{e^a}{1!}(x - a) + \frac{e^a}{2!}(x-a)^2 + \cdots + \frac{e^a}{n!}(x-a)^n.$$

**Example 2** Use Taylor polynomials for $e^x$ about $x = 0$ to find successive approximations to $e = e^1$. Stop when you think you have three decimal places correct.

**Solution** Since every derivative of $e^x$ is $e^x$ and so is 1 at $x = 0$, the Taylor polynomials for $e^x$ about $x = 0$ are

$$P_n(x) = 1 + \frac{x}{1!} + \frac{x^2}{2!} + \frac{x^3}{3!} + \cdots + \frac{x^n}{n!}.$$

Thus we have for $x = 1$, adding one more term at each step,

$$P_0(1) = 1$$

$$P_1(1) = 1 + \frac{1}{1!} = 2$$

$$P_2(1) = P_1(1) + \frac{1}{2!} = P_1(1) + \frac{1}{2} = 2.5$$

$$P_3(1) = P_2(1) + \frac{1}{3!} = P_2(1) + \frac{1}{6} = 2.6666$$

$$P_4(1) = P_3(1) + \frac{1}{4!} = P_3(1) + \frac{1}{24} = 2.7083$$

$$P_5(1) = P_4(1) + \frac{1}{5!} = P_4(1) + \frac{1}{120} = 2.7166$$

$$P_6(1) = P_5(1) + \frac{1}{6!} = P_5(1) + \frac{1}{720} = 2.7180$$

$$P_7(1) = P_6(1) + \frac{1}{7!} = P_6(1) + \frac{1}{5040} = 2.7182$$

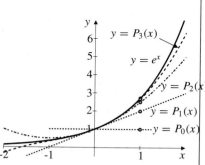

**Figure 4.58**    Taylor polynomials for $e^x$ about $x = 0$.

It appears that $e \approx 2.718$ to 3 decimal places. We will verify in Example 4 below that $P_7(1)$ does indeed give this much precision. The graphs of $e^x$ and its first four Taylor polynomials are shown in Figure 4.58. ∎

## Taylor's Formula

The following theorem provides a formula for the error in a Taylor approximation $f(x) \approx P_n(x)$ similar to that provided for linear approximation by Theorem 9.

**THEOREM 10**

**Taylor's theorem**

If the $(n + 1)$st-order derivative, $f^{(n+1)}(t)$, exists for all $t$ in an interval containing $a$ and $x$, and if $P_n(x)$ is the Taylor polynomial of degree $n$ for $f(x)$ about $x = a$, that is,

$$P_n(x) = f(a) + f'(a)(x - a) + \frac{f''(a)}{2!}(x - a)^2 + \cdots + \frac{f^{(n)}(a)}{n!}(x - a)^n,$$

then the formula $f(x) = P_n(x) + E_n(x)$ (called **Taylor's Formula**) holds where the error term $E_n(x)$ (also called "**the Lagrange remainder**") is given by

$$E_n(x) = \frac{f^{(n+1)}(X)}{(n + 1)!}(x - a)^{n+1},$$

where $X$ is some number between $a$ and $x$.

**PROOF**    $E_n(x) = f(x) - P_n(x)$ is the error in the approximation $f(x) \approx P_n(x)$. Observe that the case $n = 0$ of this theorem, namely

$$f(x) = P_0(x) + E_0(x) = f(a) + \frac{f'(X)}{1!}(x - a)$$

is just the Mean-Value Theorem

$$\frac{f(x) - f(a)}{x - a} = f'(X) \quad \text{for some } X \text{ between } a \text{ and } x.$$

Also note that the case $n = 1$ is just the error formula for linearization given in Theorem 9.

We will complete the proof for higher $n$ using mathematical induction. Suppose, therefore, that we have proved the case $n = k - 1$ where $k \geq 2$ is an integer. Thus we are assuming that if $f$ is *any* function whose $k$th derivative exists on an interval containing $a$ and $x$, then

$$E_{k-1}(x) = \frac{f^{(k)}(X)}{k!} (x - a)^k,$$

where $X$ is some number between $a$ and $x$. Let us consider the next higher case: $n = k$. As in the proof of Theorem 9, we assume $x > a$ (the case $x < a$ is similar) and apply the Generalized Mean-Value Theorem to the functions $E_k(t)$ and $(t - a)^{k+1}$ on $[a, x]$. Since $E_k(a) = 0$, we obtain a number $c$ in $(a, x)$ such that

$$\frac{E_k(x)}{(x - a)^{k+1}} = \frac{E_k(x) - E_k(a)}{(x - a)^{k+1} - (a - a)^{k+1}} = \frac{E_k'(c)}{(k + 1)(c - a)^k}.$$

Now

$$E_k'(c) = \frac{d}{dt} \left( f(t) - f(a) - f'(a)(t - a) - \frac{f''(a)}{2!}(t - a)^2 \right.$$
$$\left. - \cdots - \frac{f^{(k)}(a)}{k!}(t - a)^k \right)\Bigg|_{t=c}$$
$$= f'(c) - f'(a) - f''(a)(c - a) - \cdots - \frac{f^{(k)}(a)}{(k - 1)!}(c - a)^{k-1}.$$

This last expression is just $E_{k-1}(c)$ for the function $f'(t)$ instead of $f(t)$. By the induction assumption it is equal to

$$\frac{(f')^{(k)}(X)}{k!}(c - a)^k = \frac{f^{(k+1)}(X)}{k!}(c - a)^k,$$

for some $X$ between $a$ and $c$. Therefore

$$E_k(x) = \frac{f^{(k+1)}(X)}{(k + 1)!}(x - a)^{k+1}.$$

We have shown that the case $n = k$ of Taylor's theorem is true if the case $n = k - 1$ is true, and the inductive proof is complete.

**Example 3** Find the degree 2 Taylor approximation to $\sqrt{26}$ based on values of $f(x) = \sqrt{x}$ and its derivatives at 25. Estimate the size of the error, and specify an interval that you can be sure contains $\sqrt{26}$.

**Solution** The first three derivatives of $f$ are:

$$f'(x) = \frac{1}{2}x^{-1/2}, \quad f''(x) = -\frac{1}{4}x^{-3/2}, \quad \text{and} \quad f'''(x) = \tfrac{3}{8}x^{-5/2}.$$

The required approximation is

$$\sqrt{26} = f(26) \approx P_2(26) = f(25) + f'(25)(26 - 25) + \frac{f''(25)}{2}(26 - 25)^2$$

$$= 5 + \frac{1}{10} - \frac{1}{2 \times 4 \times 125} = 5.09900.$$

For $25 < t < 26$, we have

$$|f'''(t)| \leq \frac{3}{8}\frac{1}{25^{5/2}} = \frac{3}{8 \times 3125} = \frac{3}{25000}.$$

Thus the error in the approximation satisfies

$$|E_2(26)| \leq \frac{3}{25000 \times 6}(26 - 25)^3 = \frac{1}{50000} = 0.00002.$$

Therefore $\sqrt{26}$ lies in the interval $(5.09898, 5.09902)$. ∎

**Example 4** Use Taylor's Theorem to confirm that the Taylor polynomial $P_7(x)$ for $e^x$ about $x = 0$ is sufficient to give $e$ correct to three decimal places as claimed in Example 2.

**Solution** The error in the approximation $e^x \approx P_n(x)$ satisfies

$$E_n(x) = \frac{e^X}{(n + 1)!}x^{n+1}, \quad \text{for some } X \text{ between } 0 \text{ and } x.$$

If $x = 1$, then $0 < X < 1$, so $e^X < e < 3$ and $0 < E_n(1) < 3/(n + 1)!$. To get an approximation for $e = e^1$ correct to three decimal places, we need to have $E_n(1) < 0.0005$. Since $3/(8!) = 3/40320 \approx 0.000074$, but $3/(7!) = 3/5040 \approx 0.00059$, we can be sure $n = 7$ will do, but we cannot be sure $n = 6$ will do:

$$e \approx 1 + 1 + \frac{1}{2!} + \frac{1}{3!} + \frac{1}{4!} + \frac{1}{5!} + \frac{1}{6!} + \frac{1}{7!} \approx 2.7182 \approx 2.718$$

to three decimal places. ∎

## Big-O Notation

**DEFINITION 9**

We say that $f(x) = O\big(u(x)\big)$ as $x \to a$ (read this "$f(x)$ is big-oh of $u(x)$ as $x$ approaches $a$") provided that

$$|f(x)| \le K|u(x)|$$

holds for some constant $K$ on some open interval containing $x = a$.

Similarly, $f(x) = g(x) + O\big(u(x)\big)$ as $x \to a$ if $f(x) - g(x) = O\big(u(x)\big)$ as $x \to a$, that is, if

$$|f(x) - g(x)| \le K|u(x)| \quad \text{near } a.$$

For example, $\sin x = O(x)$ as $x \to 0$ because $|\sin x| \le |x|$ near 0.

The following properties of big-O notation follow from the definition:

(i) If $f(x) = O\big(u(x)\big)$ as $x \to a$, then $Cf(x) = O\big(u(x)\big)$ as $x \to a$ for any constant $C$.

(ii) If $f(x) = O\big(u(x)\big)$ as $x \to a$ and $g(x) = O\big(u(x)\big)$ as $x \to a$, then $f(x) \pm g(x) = O\big(u(x)\big)$ as $x \to a$.

(iii) If $f(x) = O\big((x - a)^k u(x)\big)$ as $x \to a$, then $f(x)/(x - a)^k = O\big(u(x)\big)$ as $x \to a$ for any constant $k$.

Taylor's Theorem says that if $f^{(n+1)}(t)$ exists on an interval containing $a$ and $x$, and if $P_n(x)$ is the Taylor polynomial for $f(x)$ about $x = a$, then, as $x \to a$,

$$f(x) = P_n(x) + O\big((x - a)^{n+1}\big).$$

This is a statement about how closely the Taylor polynomial $P_n(x)$ approximates $f(x)$ near $x = a$. The following theorem shows that *only* the Taylor polynomial $P_n(x)$ approximates $f(x)$ this closely.

**THEOREM 11**

If $f(x) = Q_n(x) + O\big((x - a)^{n+1}\big)$ as $x \to a$, where $Q_n$ is a polynomial of degree at most $n$, then $Q_n(x) = P_n(x)$, the Taylor polynomial for $f(x)$ about $x = a$.

**PROOF** If $P_n$ is the Taylor polynomial, then properties (i) and (ii) of big-O imply that $R_n(x) = Q_n(x) - P_n(x) = O\big((x - a)^{n+1}\big)$ as $x \to a$. By replacing $x$ with $a + (x - a)$ and expanding powers, we can write $R_n(x)$ in the form

$$R_n(x) = c_0 + c_1(x - a) + c_2(x - a)^2 + \cdots + c_n(x - a)^n.$$

If $R_n(x)$ is not identically zero, then there is a smallest coefficient $c_k$ ($k \le n$), such that $c_k \ne 0$, but $c_j = 0$ for $0 \le j \le k - 1$. Thus

$$R_n(x) = (x - a)^k \big(c_k + c_{k+1}(x - a) + \cdots + c_n(x - a)^{n-k}\big).$$

Therefore, $\lim_{x \to a} R_n(x)/(x - a)^k = c_k \ne 0$. However, by property (iii) above we have $R_n(x)/(x - a)^k = O\big((x - a)^{n+1-k}\big)$. Thus $R_n(x) \to 0$ as $x \to a$. This contradiction shows that $R_n(x)$ must be identically zero. Therefore $Q_n(x) = P_n(x)$ for all $x$.

Here is a list of Taylor Formulas about $x = 0$ for some elementary functions, with error terms expressed using big-O notation. It is worthwhile remembering these.

$$\frac{1}{1-x} = 1 + x + x^2 + x^3 + \cdots + x^n + O(x^{n+1})$$

$$\ln(1+x) = x - \frac{x^2}{2} + \frac{x^3}{3} - \cdots + (-1)^{n-1}\frac{x^n}{n} + O(x^{n+1})$$

$$e^x = 1 + x + \frac{x^2}{2!} + \frac{x^3}{3!} + \cdots + \frac{x^n}{n!} + O(x^{n+1})$$

$$\cos x = 1 - \frac{x^2}{2!} + \frac{x^4}{4!} - \cdots + (-1)^n\frac{x^{2n}}{(2n)!} + O(x^{2n+2})$$

$$\sin x = x - \frac{x^3}{3!} + \frac{x^5}{5!} - \cdots + (-1)^n\frac{x^{2n+1}}{(2n+1)!} + O(x^{2n+3})$$

$$\tan^{-1} x = x - \frac{x^3}{3} + \frac{x^5}{5} - \cdots + (-1)^n\frac{x^{2n+1}}{2n+1} + O(x^{2n+3})$$

We can obtain Taylor polynomials for new functions from others already known. As long as the error term is of higher degree than the polynomial obtained, the polynomial must be the Taylor polynomial by Theorem 11. We illustrate this with a few examples.

**Example 5**  Find the Taylor polynomial of degree $2n$ for $\cosh x$ about $x = 0$.

**Solution**  Write the Taylor formula for $e^x$ with $n$ replaced by $2n + 1$, and then rewrite that with $x$ replaced by $-x$. We get

$$e^x = 1 + x + \frac{x^2}{2!} + \frac{x^3}{3!} + \cdots + \frac{x^{2n}}{(2n)!} + \frac{x^{2n+1}}{(2n+1)!} + O(x^{2n+2})$$

$$e^{-x} = 1 - x + \frac{x^2}{2!} - \frac{x^3}{3!} + \cdots + \frac{x^{2n}}{(2n)!} - \frac{x^{2n+1}}{(2n+1)!} + O(x^{2n+2}).$$

Now average these two to get

$$\cosh x = \frac{e^x + e^{-x}}{2} = 1 + \frac{x^2}{2!} + \frac{x^4}{4!} + \cdots + \frac{x^{2n}}{(2n)!} + O(x^{2n+2}).$$

Thus the Taylor polynomial for $\cosh x$ about $x = 0$ is

$$P_{2n}(x) = 1 + \frac{x^2}{2!} + \frac{x^4}{4!} + \cdots + \frac{x^{2n}}{(2n)!}.$$

**Example 6**  Obtain the Taylor polynomial of degree 3 for $e^{2x}$ about $x = 1$ from the Taylor polynomial for $e^x$ about $x = 0$.

**Solution**  Writing $x = 1 + (x - 1)$, we have
$$e^{2x} = e^{2+2(x-1)} = e^2 e^{2(x-1)}$$

$$= e^2\left[1 + 2(x-1) + \frac{2^2(x-1)^2}{2!} + \frac{2^3(x-1)^3}{3!} + O((x-1)^4)\right].$$

The Taylor polynomial of degree 3 for $e^{2x}$ about $x = 1$ must be

$$P_3(x) = e^2 + 2e^2(x - 1) + 2e^2(x - 1)^2 + \frac{4e^2}{3}(x - 1)^3.$$

■

**Example 7** Find the Taylor polynomial $P_2(x)$ for $\ln x$ about $x = 2$.

*Solution* We replace $x$ with $2 + (x - 2)$.

$$\ln x = \ln\big(2 + (x - 2)\big) = \ln\left[2\left(1 + \frac{x - 2}{2}\right)\right] = \ln 2 + \ln\left(1 + \frac{x - 2}{2}\right)$$

$$= \ln 2 + \frac{x - 2}{2} - \frac{1}{2}\left(\frac{x - 2}{2}\right)^2 + O\big((x - 2)^3\big).$$

Therefore, $P_2(x) = \ln 2 + \frac{1}{2}(x - 2) - \frac{1}{8}(x - 2)^2$.

■

# Exercises 4.8

Calculate the indicated Taylor polynomials for the functions in Exercises 1–6 by using the definition of Taylor polynomial directly.

**1.** for $e^{-x}$ about $x = 0$, degree 4.

**2.** for $\cos x$ about $x = \pi/4$, degree 3.

**3.** for $\ln x$ about $x = e$, degree 4.

**4.** for $\sec x$ about $x = 0$, degree 3.

**5.** for $\sqrt{x}$ about $x = 4$, degree 3.

**6.** for $1/(2 + x)$ about $x = 1$, degree $n$.

In Exercises 7–12 use degree 2 Taylor polynomials for the given function near the point specified to approximate the indicated value. Estimate the error and write the smallest interval you can be sure contains the value.

**7.** $f(x) = x^{1/3}$ near 8; approximate $9^{1/3}$.

**8.** $f(x) = \sqrt{x}$ near 64; approximate $\sqrt{61}$.

**9.** $f(x) = \dfrac{1}{x}$ near 1; approximate $\dfrac{1}{1.02}$.

**10.** $f(x) = \tan^{-1} x$ near 1; approximate $\tan^{-1}(0.97)$.

**11.** $f(x) = e^x$ near 0; approximate $e^{-0.5}$.

**12.** $f(x) = \sin x$ near $\pi/4$; approximate $\sin(47°)$.

In Exercises 13–18, write the indicated case of Taylor's Formula for the given function. What is the Lagrange remainder in each case?

**13.** $f(x) = \sin x$, $a = 0$, $n = 7$

**14.** $f(x) = \cos x$, $a = 0$, $n = 6$

**15.** $f(x) = \sin x$, $a = \pi/4$, $n = 4$

**16.** $f(x) = \dfrac{1}{1 - x}$, $a = 0$, $n = 6$

**17.** $f(x) = \ln x$, $a = 1$, $n = 6$

**18.** $f(x) = \tan x$, $a = 0$, $n = 3$

Find the requested Taylor polynomials in Exercises 19–24 by using known Taylor polynomials and changing variables as in Examples 6 and 7.

**19.** $P_3(x)$ for $e^{3x}$ about $x = -1$.

**20.** $P_8(x)$ for $e^{-x^2}$ about $x = 0$.

**21.** $P_4(x)$ for $\sin^2 x$ about $x = 0$. *Hint:* $\sin^2 x = \dfrac{1 - \cos 2x}{2}$.

**22.** $P_5(x)$ for $\sin x$ about $x = \pi$.

**23.** $P_6(x)$ for $1/(1 + 2x^2)$ about $x = 0$

**24.** $P_8(x)$ for $\cos(3x - \pi)$ about $x = 0$.

**25.** Find the Taylor polynomial $P_{2n+1}(x)$ for $\sinh x$ about $x = 0$ by suitably combining polynomials for $e^x$ and $e^{-x}$.

**26.** By suitably combining Taylor polynomials for $\ln(1 + x)$ and $\ln(1 - x)$ about $x = 0$, find the Taylor polynomial of degree $2n + 1$ about $x = 0$ for $\tanh^{-1}(x) = \dfrac{1}{2}\ln\left(\dfrac{1 + x}{1 - x}\right)$.

**27.** Write Taylor's Formula for $f(x) = e^{-x}$ with $a = 0$ and use it to calculate $1/e$ to 5 decimal places. (You may use a calculator, but not the $e^x$ function on it.)

**\* 28.** Write the general form of Taylor's Formula for $f(x) = \sin x$ about $x = 0$ with Lagrange remainder. How large need $n$ be taken to ensure that the corresponding Taylor polynomial approximation will give the sine of 1 radian correct to 5 decimal places?

29. What is the best degree 2 approximation to the function $f(x) = (x-1)^2$ near $x = 0$? What is the error in this approximation? Now answer the same questions for $g(x) = x^3 + 2x^2 + 3x + 4$. Can the constant $1/6 = 1/3!$, in the error formula for the degree 2 approximation, be improved (i.e., made smaller)?

## 4.9 Indeterminate Forms

In Section 2.5 we showed that

$$\lim_{x \to 0} \frac{\sin x}{x} = 1.$$

We could not readily see this by substituting $x = 0$ into the function $(\sin x)/x$ because both $\sin x$ and $x$ are zero at $x = 0$. We call $(\sin x)/x$ an **indeterminate form** of type $[0/0]$ at $x = 0$. The limit of such an indeterminate form can be any number. For instance, each of the quotients $kx/x$, $x/x^3$, and $x^3/x^2$ is an indeterminate form of type $[0/0]$ at $x = 0$, but

$$\lim_{x \to 0} \frac{kx}{x} = k, \qquad \lim_{x \to 0} \frac{x}{x^3} = \infty, \qquad \lim_{x \to 0} \frac{x^3}{x^2} = 0.$$

There are other types of indeterminate forms. Table 4 lists them together with an example of each type.

**Table 4.**   Types of indeterminate forms

| Type | Example |
|------|---------|
| $[0/0]$ | $\displaystyle\lim_{x \to 0} \frac{\sin x}{x}$ |
| $[\infty/\infty]$ | $\displaystyle\lim_{x \to 0} \frac{\ln(1/x^2)}{\cot(x^2)}$ |
| $[0 \cdot \infty]$ | $\displaystyle\lim_{x \to 0+} x \ln \frac{1}{x}$ |
| $[\infty - \infty]$ | $\displaystyle\lim_{x \to (\pi/2)-} \left( \tan x - \frac{1}{\pi - 2x} \right)$ |
| $[0^0]$ | $\displaystyle\lim_{x \to 0+} x^x$ |
| $[\infty^0]$ | $\displaystyle\lim_{x \to (\pi/2)-} (\tan x)^{\cos x}$ |
| $[1^\infty]$ | $\displaystyle\lim_{x \to \infty} \left( 1 + \frac{1}{x} \right)^x$ |

Indeterminate forms of type $[0/0]$ are the most common. They can often be evaluated quite easily by using known Taylor formulas.

**Example 1** Evaluate $\lim\limits_{x\to 0}\dfrac{2\sin x - \sin(2x)}{2e^x - 2 - 2x - x^2}$.

**Solution** Both the numerator and denominator approach 0 as $x \to 0$. Let us repace the trig and exponential functions with their degree 3 Maclaurin polynomials plus error terms written in big-O notation:

$$\lim_{x\to 0}\frac{2\sin x - \sin(2x)}{2e^x - 2 - 2x - x^2}$$

$$= \lim_{x\to 0}\frac{2\left(x - \dfrac{x^3}{3!} + O(x^5)\right) - \left(2x - \dfrac{2^3 x^3}{3!} + O(x^5)\right)}{2\left(1 + x + \dfrac{x^2}{2!} + \dfrac{x^3}{3!} + O(x^4)\right) - 2 - 2x - x^2}$$

$$= \lim_{x\to 0}\frac{-\dfrac{x^3}{3} + \dfrac{4x^3}{3} + O(x^5)}{\dfrac{x^3}{3} + O(x^4)}$$

$$= \lim_{x\to 0}\frac{1 + O(x^2)}{\dfrac{1}{3} + O(x)} = \frac{1}{\dfrac{1}{3}} = 3.$$

Observe how we used the properties of big-O as listed in the previous section. We needed to use Maclaurin polynomials of degree at least 3 because all lower degree terms cancelled out in the numerator and the denominator.

**Example 2** Evaluate $\lim\limits_{x\to 1}\dfrac{\ln x}{x^2 - 1}$.

**Solution** This is also of type [0/0]. We begin by substituting $x = 1 + t$. Note that $x \to 1$ corresponds to $t \to 0$. We can use a known Maclaurin polynomial for $\ln(1 + t)$. For this limit even the degree 1 polynomial $P_1(t) = t$ with error $O(t^2)$ will do.

$$\lim_{x\to 1}\frac{\ln x}{x^2 - 1} = \lim_{t\to 0}\frac{\ln(1 + t)}{(1 + t)^2 - 1} = \lim_{t\to 0}\frac{\ln(1 + t)}{2t + t^2}$$

$$= \lim_{t\to 0}\frac{t + O(t^2)}{2t + t^2} = \lim_{t\to 0}\frac{1 + O(t)}{2 + t} = \frac{1}{2}.$$

## l'Hôpital's Rules

You can evaluate many indeterminate forms of type [0/0] with simple algebra, typically by cancelling common factors. Examples can be found in Sections 1.2 and 1.3. Otherwise you can use the method of Taylor polynomials, if the appropriate polynomials are known or can be calculated easily. We will now develop a third method called l'Hôpital's Rule* for evaluating limits of indeterminate forms of the

---

\* The Marquis de l'Hôpital (1661–1704), for whom these rules are named, published the first textbook on calculus. The circumflex (ˆ) did not come into use in the French language until after the French Revolution. The Marquis would have written his name "l'Hospital."

types $[0/0]$ and $[\infty/\infty]$. The other types of indeterminate forms can usually be reduced to one of these two by algebraic manipulation and the taking of logarithms.

**THEOREM 12**

### The first l'Hôpital Rule

Suppose the functions $f$ and $g$ are differentiable on the interval $(a, b)$, and $g'(x) \neq 0$ there. Suppose also that

(i) $\lim\limits_{x \to a+} f(x) = \lim\limits_{x \to a+} g(x) = 0$, and

(ii) $\lim\limits_{x \to a+} \dfrac{f'(x)}{g'(x)} = L$ (where $L$ is finite or $\infty$ or $-\infty$).

Then

$$\lim_{x \to a+} \frac{f(x)}{g(x)} = L.$$

Similar results hold if every occurrence of $\lim_{x \to a+}$ is replaced by $\lim_{x \to b-}$ or even $\lim_{x \to c}$ where $a < c < b$. The cases $a = -\infty$ and $b = \infty$ are also allowed.

***PROOF***   We prove the case involving $\lim_{x \to a+}$ for finite $a$. Define

$$F(x) = \begin{cases} f(x) & \text{if } a < x < b \\ 0 & \text{if } x = a \end{cases} \qquad \text{and} \qquad G(x) = \begin{cases} g(x) & \text{if } a < x < b \\ 0 & \text{if } x = a \end{cases}$$

Then $F$ and $G$ are continuous on the interval $[a, x]$ and differentiable on the interval $(a, x)$ for every $x$ in $(a, b)$. By the Generalized Mean-Value Theorem (Theorem 16 of Section 2.6) there exists a number $c$ in $(a, x)$ such that

$$\frac{f(x)}{g(x)} = \frac{F(x)}{G(x)} = \frac{F(x) - F(a)}{G(x) - G(a)} = \frac{F'(c)}{G'(c)} = \frac{f'(c)}{g'(c)}.$$

Since $a < c < x$, if $x \to a+$, then necessarily $c \to a+$, so we have

$$\lim_{x \to a+} \frac{f(x)}{g(x)} = \lim_{c \to a+} \frac{f'(c)}{g'(c)} = L.$$

The case involving $\lim_{x \to b-}$ for finite $b$ is proved similarly. The cases where $a = -\infty$ or $b = \infty$ follow from the cases already considered via the change of variable $x = 1/t$:

$$\lim_{x \to \infty} \frac{f(x)}{g(x)} = \lim_{t \to 0+} \frac{f\left(\dfrac{1}{t}\right)}{g\left(\dfrac{1}{t}\right)} = \lim_{t \to 0+} \frac{f'\left(\dfrac{1}{t}\right)\left(\dfrac{-1}{t^2}\right)}{g'\left(\dfrac{1}{t}\right)\left(\dfrac{-1}{t^2}\right)} = \lim_{x \to \infty} \frac{f'(x)}{g'(x)} = L.$$

---

Note that in applying l'Hôpital's Rule we calculate the quotient of the derivatives, *not* the derivative of the quotient.

---

**Example 3**   Reevaluate $\lim\limits_{x \to 1} \dfrac{\ln x}{x^2 - 1}$. (See Example 2.)

***Solution***   We have

$$\lim_{x \to 1} \frac{\ln x}{x^2 - 1} \qquad \begin{bmatrix} 0 \\ 0 \end{bmatrix}$$

$$= \lim_{x \to 1} \frac{1/x}{2x} = \lim_{x \to 1} \frac{1}{2x^2} = \frac{1}{2}.$$

This example illustrates how calculations based on l'Hôpital's Rule are carried out. Having identified the limit as that of a [0/0] indeterminate form, we replace it by the limit of the quotient of derivatives; the existence of this latter limit will justify the equality. It is possible that the limit of the quotient of derivatives may still be indeterminate, in which case a second application of l'Hôpital's Rule can be made. Such applications may be strung out until a limit can finally be extracted, which then justifies all the previous applications of the rule.

**Remark**   The solution above seems easier than that of Example 2, and we might be tempted to think that l'Hôpital's are easier to use than Taylor polynomials. It was easier here because we only had to apply l'Hôpital's Rule once. If we try to redo Example 1 by l'Hôpital's Rule we will have to use the rule three times (which corresponds to the fact that degree 3 polynomials were needed in Example 1).

**Example 4**   Evaluate $\displaystyle\lim_{x\to 0}\frac{2\sin x - \sin(2x)}{2e^x - 2 - 2x - x^2}$.

**Solution**   We have

$$\lim_{x\to 0}\frac{2\sin x - \sin(2x)}{2e^x - 2 - 2x - x^2} \qquad \left[\frac{0}{0}\right]$$

$$= \lim_{x\to 0}\frac{2\cos x - 2\cos(2x)}{2e^x - 2 - 2x} \qquad \text{cancel the 2's}$$

$$= \lim_{x\to 0}\frac{\cos x - \cos(2x)}{e^x - 1 - x} \qquad \text{still} \quad \left[\frac{0}{0}\right]$$

$$= \lim_{x\to 0}\frac{-\sin x + 2\sin(2x)}{e^x - 1} \qquad \text{still} \quad \left[\frac{0}{0}\right]$$

$$= \lim_{x\to 0}\frac{-\cos x + 4\cos(2x)}{e^x} = \frac{-1+4}{1} = 3.$$

Note that l'Hôpital's Rule was used three times here.

**Example 5**   Evaluate   (a) $\displaystyle\lim_{x\to(\pi/2)-}\frac{2x-\pi}{\cos^2 x}$   and   (b) $\displaystyle\lim_{x\to 1+}\frac{x}{\ln x}$.

**Solution**

(a)   $\displaystyle\lim_{x\to(\pi/2)-}\frac{2x-\pi}{\cos^2 x} \qquad \left[\frac{0}{0}\right]$

$$= \lim_{x\to(\pi/2)-}\frac{2}{-2\sin x\cos x} = -\infty$$

(b)   l'Hôpital's Rule can not be used to evaluate $\lim_{x\to 1+} x/(\ln x)$, because this is not an indeterminate form. The denominator approaches 0 as $x \to 1+$, but the numerator does not approach 0. Since $\ln x > 0$ for $x > 1$, we have, directly,

$$\lim_{x\to 1+}\frac{x}{\ln x} = \infty.$$

(Had we tried to apply l'Hôpital's Rule, we would have been led to the erroneous answer $\lim_{x\to 1+}(1/(1/x)) = 1$.)

**Example 6**    Evaluate $\lim\limits_{x \to 0+} \left( \dfrac{1}{x} - \dfrac{1}{\sin x} \right)$.

**Solution**    The indeterminate form here is of type $[\infty - \infty]$ to which l'Hôpital's Rule can not be applied. However it becomes $[0/0]$ after we combine the fractions into one fraction.

$$\lim_{x \to 0+} \left( \frac{1}{x} - \frac{1}{\sin x} \right) \qquad [\infty - \infty]$$

$$= \lim_{x \to 0+} \frac{\sin x - x}{x \sin x} \qquad \left[ \frac{0}{0} \right]$$

$$= \lim_{x \to 0+} \frac{\cos x - 1}{\sin x + x \cos x} \qquad \left[ \frac{0}{0} \right]$$

$$= \lim_{x \to 0+} \frac{-\sin x}{2 \cos x - x \sin x} = \frac{-0}{2} = 0.$$

A version of l'Hôpital's Rule also holds for indeterminate forms of the type $[\infty/\infty]$.

**THEOREM 13**

**The second l'Hôpital Rule**

Suppose that $f$ and $g$ are differentiable on the interval $(a, b)$ and that $g'(x) \neq 0$ there. Suppose also that

(i) $\lim\limits_{x \to a+} g(x) = \pm\infty$, and

(ii) $\lim\limits_{x \to a+} \dfrac{f'(x)}{g'(x)} = L$   (where $L$ is finite, or $\infty$ or $-\infty$).

Then

$$\lim_{x \to a+} \frac{f(x)}{g(x)} = L.$$

Again similar results hold for $\lim_{x \to b-}$ and for $\lim_{x \to c}$, and the cases $a = -\infty$ and $b = \infty$ are allowed.

The proof of the second l'Hôpital Rule is technically rather more difficult than that of the first Rule and we will not give it here. A sketch of the proof is outlined in Exercise 34 at the end of this section.

**Remark**    Do **not** try to use l'Hôpital's Rules to evaluate limits which are not indeterminate of type $[0/0]$ or $[\infty/\infty]$; such attempts will almost always lead to false conclusions as observed in Example 5(b) above. (Strictly speaking, the second l'Hôpital Rule can be applied to the form $[a/\infty]$, but there is no point to doing so if $a$ is not infinite, since the limit is obviously 0 in that case.)

**Example 7**    Evaluate    (a) $\lim\limits_{x \to \infty} \dfrac{x^2}{e^x}$,    and    (b) $\lim\limits_{x \to 0+} x^a \ln x$, where $a > 0$.

**Solution**    Both of these limits are covered by Theorem 5 in Section 3.4. We do them here by l'Hôpital's Rule.

(a) $\quad \lim_{x \to \infty} \dfrac{x^2}{e^x} \qquad \left[\dfrac{\infty}{\infty}\right]$

$\quad = \lim_{x \to \infty} \dfrac{2x}{e^x} \qquad \text{still } \left[\dfrac{\infty}{\infty}\right]$

$\quad = \lim_{x \to \infty} \dfrac{2}{e^x} = 0.$

Similarly, one can show that $\lim_{x \to \infty} x^n/e^x = 0$ for any positive integer $n$ by repeated applications of l'Hôpital's Rule.

(b) $\quad \lim_{x \to 0+} x^a \ln x \quad (a > 0) \qquad [0 \cdot (-\infty)]$

$\quad = \lim_{x \to 0+} \dfrac{\ln x}{x^{-a}} \qquad \left[\dfrac{-\infty}{\infty}\right]$

$\quad = \lim_{x \to 0+} \dfrac{1/x}{-ax^{-a-1}} = \lim_{x \to 0+} \dfrac{x^a}{-a} = 0.$ ∎

The easiest way to deal with indeterminate forms of types $[0^0]$, $[\infty^0]$, and $[1^\infty]$ is to take logarithms of the expressions involved. The next three examples illustrate the technique.

**Example 8** Evaluate $\lim_{x \to 0+} x^x$.

**Solution** This indeterminate form is of type $[0^0]$. Let $y = x^x$. Then

$$\lim_{x \to 0+} \ln y = \lim_{x \to 0+} x \ln x = 0,$$

by Example 7(b). Hence $\lim_{x \to 0} x^x = \lim_{x \to 0+} y = e^0 = 1.$ ∎

**Example 9** Evaluate $\lim_{x \to (\pi/2)-} (\tan x)^{\cos x}$.

**Solution** This indeterminate form is of type $[\infty^0]$. Again it is appropriate to take logarithms. Let $y = (\tan x)^{\cos x}$. Then

$$\lim_{x \to (\pi/2)-} \ln y = \lim_{x \to (\pi/2)-} \cos x \ln \tan x \qquad [0 \cdot \infty]$$

$$= \lim_{x \to (\pi/2)-} \dfrac{\ln \tan x}{\sec x} \qquad \left[\dfrac{\infty}{\infty}\right]$$

$$= \lim_{x \to (\pi/2)-} \dfrac{\dfrac{1}{\tan x} \sec^2 x}{\sec x \tan x}$$

$$= \lim_{x \to (\pi/2)-} \dfrac{\sec x}{\tan^2 x}$$

$$= \lim_{x \to (\pi/2)-} \dfrac{\cos x}{\sin^2 x} = 0.$$

Thus $\lim_{x \to (\pi/2)-} (\tan x)^{\cos x} = e^0 = 1.$ ∎

**Example 10**   Evaluate $\lim\limits_{x\to\infty}\left(1+\sin\dfrac{3}{x}\right)^x$.

**Solution**   This indeterminate form is of type $1^\infty$. Let $y=\left(1+\sin\dfrac{3}{x}\right)^x$. Then, taking ln of both sides,

$$\lim_{x\to\infty}\ln y=\lim_{x\to\infty}x\ln\left(1+\sin\frac{3}{x}\right)\qquad[\infty\cdot 0]$$

$$=\lim_{x\to\infty}\frac{\ln\left(1+\sin\dfrac{3}{x}\right)}{\dfrac{1}{x}}\qquad\left[\frac{0}{0}\right]$$

$$=\lim_{x\to\infty}\frac{\dfrac{1}{1+\sin\dfrac{3}{x}}\left(\cos\dfrac{3}{x}\right)\left(-\dfrac{3}{x^2}\right)}{-\dfrac{1}{x^2}}=\lim_{x\to\infty}\frac{3\cos\dfrac{3}{x}}{1+\sin\dfrac{3}{x}}=3$$

Hence $\lim\limits_{x\to\infty}\left(1+\sin\dfrac{3}{x}\right)^x=e^3$.   ∎

# Exercises 4.9

Evaluate the limits in Exercises 1–32.

**1.** $\lim\limits_{x\to 0}\dfrac{3x}{\tan 4x}$

**2.** $\lim\limits_{x\to 2}\dfrac{\ln(2x-3)}{x^2-4}$

**3.** $\lim\limits_{x\to 0}\dfrac{\sin ax}{\sin bx}$

**4.** $\lim\limits_{x\to 0}\dfrac{1-\cos ax}{1-\cos bx}$

**5.** $\lim\limits_{x\to 0}\dfrac{\sin^{-1}x}{\tan^{-1}x}$

**6.** $\lim\limits_{x\to 1}\dfrac{x^{1/3}-1}{x^{2/3}-1}$

**7.** $\lim\limits_{x\to 0}x\cot x$

**8.** $\lim\limits_{x\to 0}\dfrac{1-\cos x}{\ln(1+x^2)}$

**9.** $\lim\limits_{t\to\pi}\dfrac{\sin^2 t}{t-\pi}$

**10.** $\lim\limits_{x\to 0}\dfrac{10^x-e^x}{x}$

**11.** $\lim\limits_{x\to\pi/2}\dfrac{\cos 3x}{\pi-2x}$

**12.** $\lim\limits_{x\to 1}\dfrac{\ln(ex)-1}{\sin\pi x}$

**13.** $\lim\limits_{x\to\infty}x\sin\dfrac{1}{x}$

**14.** $\lim\limits_{x\to 0}\dfrac{x-\sin x}{x^3}$

**15.** $\lim\limits_{x\to 0}\dfrac{x-\sin x}{x-\tan x}$

**16.** $\lim\limits_{x\to 0}\dfrac{2-x^2-2\cos x}{x^4}$

**17.** $\lim\limits_{x\to 0+}\dfrac{\sin^2 x}{\tan x-x}$

**18.** $\lim\limits_{r\to\pi/2}\dfrac{\ln\sin r}{\cos r}$

**19.** $\lim\limits_{t\to\pi/2}\dfrac{\sin t}{t}$

**20.** $\lim\limits_{x\to 1-}\dfrac{\arccos x}{x-1}$

**21.** $\lim\limits_{x\to\infty}x(2\tan^{-1}x-\pi)$

**22.** $\lim\limits_{t\to(\pi/2)-}(\sec t-\tan t)$

**23.** $\lim\limits_{t\to 0}\left(\dfrac{1}{t}-\dfrac{1}{te^{at}}\right)$

**24.** $\lim\limits_{x\to 0+}x^{\sqrt{x}}$

**\* 25.** $\lim\limits_{x\to 0+}(\csc x)^{\sin^2 x}$

**\* 26.** $\lim\limits_{x\to 1+}\left(\dfrac{x}{x-1}-\dfrac{1}{\ln x}\right)$

**\* 27.** $\lim\limits_{t\to 0}\dfrac{3\sin t-\sin 3t}{3\tan t-\tan 3t}$

**\* 28.** $\lim\limits_{x\to 0}\left(\dfrac{\sin x}{x}\right)^{1/x^2}$

**\* 29.** $\lim\limits_{t\to 0}(\cos 2t)^{1/t^2}$

**\* 30.** $\lim\limits_{x\to 0+}\dfrac{\csc x}{\ln x}$

**\* 31.** $\lim\limits_{x\to 1-}\dfrac{\ln\sin\pi x}{\csc\pi x}$

**\* 32.** $\lim\limits_{x\to 0}(1+\tan x)^{1/x}$

**33.** **(A Newton quotient for the second derivative)** Evaluate $\lim_{h\to 0}\dfrac{f(x+h)-2f(x)+f(x-h)}{h^2}$ if $f$ is a twice differentiable function.

**\* 34.** **(Proof of the second l'Hôpital Rule)** Fill in the details of the following outline of a proof of the second l'Hôpital Rule (Theorem 13) for the case where $a$ and $L$ are both finite. Let $a<x<t<b$ and show that there exists $c$ in $(x,t)$ such that

$$\frac{f(x)-f(t)}{g(x)-g(t)}=\frac{f'(c)}{g'(c)}.$$

Now juggle the above equation algebraically into the form

$$\frac{f(x)}{g(x)} - L = \frac{f'(c)}{g'(c)} - L + \frac{1}{g(x)}\left(f(t) - g(t)\frac{f'(c)}{g'(c)}\right).$$

It follows that

$$\left|\frac{f(x)}{g(x)} - L\right|$$

$$\leq \left|\frac{f'(c)}{g'(c)} - L\right| + \frac{1}{|g(x)|}\left(|f(t)| + |g(t)|\left|\frac{f'(c)}{g'(c)}\right|\right).$$

Now show that the right side of the above inequality can be made as small as you wish (say less than a positive number $\epsilon$) by choosing first $t$ and then $x$ close enough to $a$. Remember, you are given that $\lim_{c\to a+}\left(f'(c)/g'(c)\right) = L$ and $\lim_{x\to a+}|g(x)| = \infty$.

# Chapter Review

## Key Ideas

- **What do the following words, phrases, and statements mean?**

  ◇ velocity   ◇ speed   ◇ acceleration

  ◇ critical point of $f$   ◇ singular point of $f$

  ◇ inflection point of $f$

  ◇ $f$ has absolute maximum value $M$.

  ◇ $f$ has a local minimum value at $x = c$.

  ◇ vertical asymptote   ◇ horizontal asymptote

  ◇ oblique asymptote

  ◇ the linearization of $f(x)$ about $x = a$

  ◇ the Taylor polynomial of degree $n$ of $f(x)$ about $x = a$

  ◇ Taylor's formula with Lagrange remainder

  ◇ $f(x) = O\left((x - a)^n\right)$ as $x \to a$.

  ◇ a root of $f(x) = 0$   ◇ a fixed point of $f(x)$

  ◇ an indeterminate form   ◇ l'Hôpital's Rules

- **Describe how to estimate the error in a linear (tangent line) approximation to the value of a function.**

- **Describe how to find a root of an equation $f(x) = 0$ by using Newton's Method. When will this method work well?**

## Review Exercises

1. A ball is thrown upward with an initial speed of 10 m/s from the top of a building. A second ball is thrown upward with an initial speed of 20 m/s from the ground. Both balls achieve the same maximum height above the ground. How tall is the building?

2. A ball is dropped from the top of a 60 m high tower at the same instant that a second ball is thrown upward from the ground at the base of the tower. The balls collide at a height of 30 m above the ground. With what initial velocity was the second ball thrown? How fast is each ball moving when they collide?

3. **(Braking distance)** A car's brakes can decelerate the car at 20 ft/s². How fast can the car travel if it must be able to stop in a distance of 160 ft?

4. **(Gravitational attraction)** The gravitational attraction of the earth on a mass $m$ at distance $r$ from the centre of the earth is a continuous function of $r$ for $r \geq 0$, given by

$$F = \begin{cases} \dfrac{mgR^2}{r^2} & \text{if } r \geq R \\ mkr & \text{if } 0 \leq r < R, \end{cases}$$

where $R$ is the radius of the earth, and $g$ is the acceleration due to gravity at the surface of the earth.

   (a) Find the constant $k$ in terms of $g$ and $R$.

   (b) $F$ decreases as $m$ moves away from the surface of the earth, either upward or downward. Show that $F$ decreases as $r$ increases from $R$ at twice the rate at which $F$ decreases as $r$ decreases from $R$.

5. **(Resistors in parallel)** Two variable resistors $R_1$ and $R_2$ are connected in parallel so that their combined resistance $R$ is given by

$$\frac{1}{R} = \frac{1}{R_1} + \frac{1}{R_2}.$$

At an instant when $R_1 = 250$ ohms and $R_2 = 1000$ ohms, $R_1$ is increasing at a rate of 100 ohms/minute. How fast must $R_2$ be changing at that moment (a) to keep $R$ constant? and (b) to enable $R$ to increase at a rate of 10 ohms/minute?

6. **(Gas law)** The volume $V$ (in m³), pressure $P$ (in kilopascals, kPa) and temperature $T$ (in kelvins) for a sample of a certain gas satisfy the equation $pV = 5.0T$.

   (a) How rapidly does the pressure increase if the temperature is 400 K and increasing at 4 K/min, while the gas is kept confined in a volume of 2.0 m³?

   (b) How rapidly does the pressure decrease if the volume is 2 m³ and increases at 0.05 m³/min while the temperature is kept constant at 400 K?

7. **(The size of a print run)** It costs a publisher $10,000 to set up the presses for a print run of a book, and $8 to cover the material costs for each book printed. In addition, machinery servicing, labour, and warehousing add another $6.25 \times 10^{-7}x^2$ to the cost of each book if $x$ copies are manufactured during the printing. How many copies should the publisher print in order to minimize the average cost per book?

8. **(Maximizing profit)** A bicycle wholesaler must pay the manufacturer $75 for each bicycle. Market research tells the wholesaler that if she charges her customers $x$ per bicycle, she can expect to sell $N(x) = 4.5 \times 10^6/x^2$ of them. What price should she charge to maximize her profit, and how many bicycles should she order from the manufacturer?

9. Find the largest possible volume of a right-circular cone that can be inscribed in a sphere of radius $R$.

10. **(Minimizing production costs)** The cost $C(x)$ of production in a factory varies with the amount $x$ of product manufactured. The cost may rise sharply with $x$ for $x$ small, and more slowly for larger values of $x$ because of economies of scale. However, if $x$ becomes too large, the resources of the factory can be overtaxed, and the cost can begin to rise quickly again. Figure 4.59 shows the graph of a typical such cost function $C(x)$.

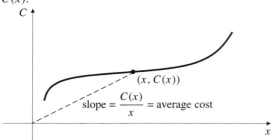

**Figure 4.59**

If $x$ units are manufactured, the average cost per unit is $C(x)/x$, which is the slope of the line from the origin to the point $(x, C(x))$ on the graph.

(a) If it is desired to choose $x$ to minimize this average cost per unit (as would be the case if all units produced could be sold for the same price), show that $x$ should be chosen to make the average cost equal to the marginal cost:

$$\frac{C(x)}{x} = C'(x).$$

(b) Interpret the conclusion of (a) geometrically in the figure.

(c) If the average cost equals the marginal cost for some $x$, does $x$ necessarily minimize the average cost?

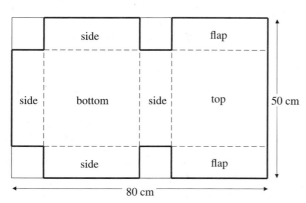

**Figure 4.60**

11. **(Box design)** Four squares are cut out of a rectangle of cardboard 50 cm by 80 cm, as shown Figure 4.60, and the remaining piece is folded into a closed, rectangular box, with two extra flaps tucked in. What is the largest possible volume for such a box?

12. **(Yield from an orchard)** A certain orchard has 60 trees, and produces an average of 800 apples per tree per year. If the density of trees is increased, the yield per tree drops; for each additional tree planted the average yield per tree is reduced by 10 apples per year. How many more trees should be planted to maximize the total annual yield of apples from the orchard?

13. **(Rotation of a tracking antenna)** What is the maximum rate at which the antenna in Exercise 39 of Section 4.1 must be able to turn in order to track the rocket during its entire vertical ascent?

14. An oval table has outer edge in the shape of the curve $x^2 + y^4 = 1/8$, where $x$ and $y$ are measured in metres. What is the width of the narrowest hallway in which the table can be turned horizontally through 180 degrees?

15. A hollow iron ball whose shell is 2 cm thick weighs half as much as it would if it were solid iron throughout. What is the radius of the ball?

16. **(Range of a cannon fired from a hill)** A cannon ball is fired with a speed of 200 ft/s at an angle of 45° above the horizontal from the top of a hill whose height at a horizontal distance $x$ ft from the top is $y = 1{,}000/(1 + (x/500)^2)$ ft above sea level. How far does the cannon ball travel horizontally before striking the ground?

17. **(Linear approximation for a pendulum)** Because $\sin\theta \approx \theta$ for small values of $|\theta|$, the nonlinear equation of motion of a simple pendulum

$$\frac{d^2\theta}{dt^2} = -\frac{g}{L}\sin\theta,$$

which determines the displacement angle $\theta(t)$ away from vertical at time $t$ for a simple pendulum, is frequently approximated by the simpler linear equation

$$\frac{d^2\theta}{dt^2} = -\frac{g}{L}\theta,$$

when the maximum displacement of the pendulum is not large. What is the percentage error in the right side of the equation if $|\theta|$ does not exceed 20 degrees?

18. Find the Taylor polynomial of degree 6 for $\sin^2 x$ about $x = 0$ and use it to help you evaluate

$$\lim_{x \to 0} \frac{3\sin^2 x - 3x^2 + x^4}{x^6}.$$

19. Use a degree 2 Taylor polynomial for $\tan^{-1} x$ about $x = 1$ to find an approximate value for $\tan^{-1}(1.1)$. Estimate the size of the error by using Taylor's formula.

**20.** The line $2y = 10x - 19$ is tangent to $y = f(x)$ at $x = 2$. If an initial approximation $x_0 = 2$ is made for a root of $f(x) = 0$ and Newton's method is applied once, what will be the new approximation that results?

**21.** Find all solutions of the equation $\cos x = (x - 1)^2$ to 10 decimal places.

**22.** Find the shortest distance from the point $(2, 0)$ to the curve $y = \ln x$.

**23.** A car is travelling at night along a level, curved road whose equation is $y = e^x$. At a certain instant its headlights illuminate a signpost located at the point $(1, 1)$. Where is the car at that instant?

**24.** If $f'''(t)$ is continuous on an interval containing $t = x$, find

$$\lim_{h \to 0} \frac{f(x + 2h) - 2f(x + h) + 2f(x - h) - f(x - 2h)}{2h^3}.$$

## Challenging Problems

**1. (Growth of a crystal)** A single cubical salt crystal is growing in a beaker of salt solution. The crystal's volume $V$ increases at a rate proportional to its surface area and to the amount by which its volume is less than a limiting volume $V_0$:

$$\frac{dV}{dt} = kx^2(V_0 - V),$$

where $x$ is the edge length of the crystal at time $t$.

(a) Using $V = x^3$, transform the equation above to one giving the rate of change $dx/dt$ of the edge length $x$ in terms of $x$.

(b) Show that the growth rate of the edge of the crystal decreases with time, but remains positive as long as $x < x_0 = V_0^{1/3}$.

(c) Find the volume of the crystal when its edge length is growing at half the rate it was initially.

**2. (A Review of Calculus!)** You are in a tank (the military variety) moving down the $y$-axis toward the origin. At time $t = 0$ you are 4 km from the origin, and 10 min later you are 2 km from the origin. Your speed is decreasing; it is proportional to your distance from the origin. You know that an enemy tank is waiting somewhere on the positive $x$-axis, but there is a high wall along the curve $xy = 1$ (all distances in kilometres) preventing you from seeing just where it is. How fast must your gun turret be capable of turning to maximize your chances of surviving the encounter?

**3. (The economics of blood testing)** Suppose that it is necessary to perform a blood test on a large number $N$ of individuals to detect the presence of a virus. If each test costs $\$C$, then the total cost of the testing program is $\$NC$. If the proportion of people in the population who have the virus is not large, this cost can be greatly reduced by adopting the following strategy. Divide the $N$ samples of blood into $N/x$ groups of $x$ samples each. Pool the blood in each group to make a single sample for that group and test it. If it tests negative, no further testing is necessary for individuals in that group. If the group sample tests positive, test all the individuals in that group.

Suppose that the fraction of individuals in the population infected with the virus is $p$, so the fraction uninfected is $q = 1 - p$. The probability that a given individual is unaffected is $q$, so the probability that all $x$ individuals in a group are unaffected is $q^x$. Therefore, the probability that a pooled sample is infected is $1 - q^x$. Each group requires one test, and the infected groups require an extra $x$ tests. Therefore the expected total number of tests to be performed is

$$T = \frac{N}{x} + \frac{N}{x}(1 - q^x)x = N\left(\frac{1}{x} + 1 - q^x\right).$$

For example, if $p = 0.01$, so that $q = 0.99$, and $x = 20$, then the expected number of tests required is $T = 0.23N$, a reduction of over 75%. But maybe we can do better by making a different choice for $x$.

(a) For $q = 0.99$, find the number $x$ of samples in a group that minimizes $T$. (That is, solve $dT/dx = 0$.) Show that the minimizing value of $x$ satisfies

$$x = \frac{(0.99)^{-x/2}}{\sqrt{-\ln(0.99)}}.$$

(b) Use the technique of fixed-point iteration (see Section 4.6) to solve the equation in (a) for $x$. Start with $x = 20$, say.

**4. (Measuring variations in $g$)** The period $P$ of a pendulum of length $L$ is given by

$$P = 2\pi\sqrt{L/g},$$

where $g$ is the acceleration of gravity.

(a) Assuming that $L$ remains fixed, show that a 1% increase in $g$ results in approximately a 1/2% decrease in the period $P$. (Variations in the period of a pendulum can be used to detect small variations in $g$ from place to place on the earth's surface.)

(b) For fixed $g$, what percentage change in $L$ will produce a 1% increase in $P$?

**5. (Torricelli's Law)** The rate at which a tank drains is proportional to the square root of the depth of liquid in the tank above the level of the drain: If $V(t)$ is the volume of liquid in the tank at time $t$ and $y(t)$ is the height of the surface of the liquid above the drain, then $dV/dt = -k\sqrt{y}$, where $k$ is a constant depending on the size of the drain. For a cylindrical tank with constant cross-sectional area $A$ with drain at the bottom:

(a) Verify that the depth $y(t)$ of liquid in the tank at time $t$ satisfies $dy/dt = -(k/A)\sqrt{y}$.

(b) Verify that if the depth of liquid in the tank at $t = 0$ is $y_0$, then the depth at subsequent times during the draining process is $y = \left(\sqrt{y_0} - \dfrac{kt}{2A}\right)^2$.

(c) If the tank drains completely in time $T$, express the depth $y(t)$ at time $t$ in terms of $y_0$ and $T$.

(d) In terms of $T$, how long does it take for half the liquid in the tank to drain out?

**6.** If a conical tank with top radius $R$ and depth $H$, drains according to Torricelli's Law and empties in time $T$, show that the depth of liquid in the tank at time $t$ $(0 < t < T)$ is

$$y = y_0 \left(1 - \frac{t}{T}\right)^{2/5},$$

where $y_0$ is the depth at $t = 0$.

**7.** Find the largest possible area of a right-angled triangle whose perimeter is $P$.

**8.** Find a tangent to the graph of $y = x^3 + ax^2 + bx + c$ that is not parallel to any other tangent.

**9.** **(Branching angles for electric wires and pipes)**

(a) The resistance offered by a wire to the flow of electric current through it is proportional to its length and inversely proportional to its cross-sectional area. Thus the resistance $R$ of a wire of length $L$ and radius $r$ is $R = kL/r^2$, where $k$ is a positive constant.

A long straight wire of length $L$ and radius $r_1$ extends from $A$ to $B$. A second straight wire of smaller radius $r_2$ is to be connected between a point $P$ on $AB$ and a point $C$ at distance $h$ from $B$ such that $CB$ is perpendicular to $AB$. (See Figure 4.61.) Find the value of the angle $\theta = \angle BPC$ that minimizes the total resistance of the path $APC$, that is, the resistance of $AP$ plus the resistance of $PC$.

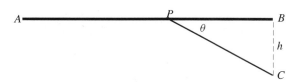

**Figure 4.61**

(b) The resistance of a pipe (for example a blood vessel) to the flow of liquid through it is, by Poiseuille's Law, proportional to its length and inversely proportional to the *fourth power* of its radius: $R = kL/r^4$. If the situation in part (a) represents pipes instead of wires, find the value of $\theta$ that minimizes the total resistance of the path $APC$. How does your answer relate to the answer for part (a)? Could you have predicted this relationship?

**Figure 4.62**

**∗ 10.** **(The range of a spurt)** A cylindrical water tank sitting on a horizontal table has a small hole located on its vertical wall at height $h$ above the bottom of the tank. Water escapes from the tank horizontally through the hole, and then curves down under the influence of gravity to strike the table at a distance $R$ from the base of the tank, as shown in Figure 4.62. (We ignore air resistance.) Torricelli's Law implies that the speed $v$ at which water escapes through the hole is proportional to the square root of the depth of the hole below the surface of the water: if the depth of water in the tank at time $t$ is $y(t) > h$, then $v = k\sqrt{y - h}$, where the constant $k$ depends on the size of the hole.

(a) Find the range $R$ in terms of $v$ and $h$.

(b) For a given depth $y$ of water in the tank, how high should the hole be to maximize $R$?

(c) Suppose that the depth of water in the tank at time $t = 0$ is $y_0$, that the range $R$ of the spurt is $R_0$ at that time, and that the water level drops to the height $h$ of the hole in $T$ minutes. Find, as a function of $t$, the range $R$ of the water that escaped through the hole at time $t$.

**∗ 11.** **(Designing a dustpan)** Equal squares are cut out of two adjacent corners of a square of sheet metal having sides of length 25 cm. The three resulting flaps are bent up, as shown in Figure 4.63, to form the sides of a dustpan. Find the maximum volume of a dustpan made in this way.

**Figure 4.63**

# CHAPTER 5

# Integration

*Introduction*   The second fundamental problem addressed by calculus is the problem of areas, that is, the problem of determining the area of a region of the plane bounded by various curves. Like the problem of tangents considered in Chapter 2, many practical problems in various disciplines require the evaluation of areas for their solution, and the solution of the problem of areas necessarily involves the notion of limits. On the surface the problem of areas appears unrelated to the problem of tangents. However, we will see that the two problems are very closely related; one is the inverse of the other. Finding an area is equivalent to finding an antiderivative or, as we prefer to say, finding an integral. The relationship between areas and antiderivatives is called the Fundamental Theorem of Calculus. When we have proved it we will be able to find areas at will, provided only that we can integrate (that is, antidifferentiate) the various functions we encounter.

We would like to have at our disposal a set of integration rules similar to the differentiation rules developed in Chapter 2. We can find the derivative of any differentiable function using those differentiation rules. Unfortunately, integration is generally more difficult; indeed, some fairly simple functions are not themselves derivatives of simple functions. For example, $e^{x^2}$ is not the derivative of any finite combination of elementary functions. Nevertheless, we will expend some effort in Section 5.6, and Sections 6.1–6.4 to develop techniques for integrating as many functions as possible. Later in Chapter 6 we will examine how to approximate areas bounded by graphs of functions that we cannot antidifferentiate.

## 5.1 Sums and Sigma Notation

When we begin calculating areas in the next section we will often encounter sums of values of functions. We need to have a convenient notation for representing sums of arbitrary (possibly large) numbers of terms, and we need to develop techniques for evaluating some such sums.

We use the symbol $\sum$ to represent a sum; it is an enlarged Greek capital letter "S" called *sigma*.

**DEFINITION 1**

**Sigma notation**

If $m$ and $n$ are integers with $m \leq n$, and if $f$ is a function defined at the integers $m$, $m + 1$, $m + 2$, ..., $n$, the symbol $\sum_{i=m}^{n} f(i)$ represents the sum of the values of $f$ at those integers:

$$\sum_{i=m}^{n} f(i) = f(m) + f(m + 1) + f(m + 2) + \cdots + f(n).$$

The explicit sum appearing on the right side of this equation is the **expansion** of the sum represented in sigma notation on the left side.

**Example 1** $\displaystyle\sum_{i=1}^{5} i^2 = 1^2 + 2^2 + 3^2 + 4^2 + 5^2 = 55.$

The $i$ that appears in the symbol $\sum_{i=m}^{n} f(i)$ is called an **index of summation**. To evaluate $\sum_{i=m}^{n} f(i)$, replace the index $i$ with the integers $m$, $m + 1$, ..., $n$, successively, and sum the results. Observe that the value of the sum does not depend on what we call the index; the index does not appear on the right side of the definition. If we use another letter in place of $i$ in the sum in Example 1 we still get the same value for the sum:

$$\sum_{k=1}^{5} k^2 = 1^2 + 2^2 + 3^2 + 4^2 + 5^2 = 55.$$

The index of summation is a *dummy variable* used to represent an arbitrary point where the function is evaluated to produce a term to be included in the sum. On the other hand, the sum $\sum_{i=m}^{n} f(i)$ does depend on the two numbers $m$ and $n$, called the **limits of summation**; $m$ is the **lower limit** and $n$ is the **upper limit**.

**Example 2** (Examples of sums using sigma notation)

$$\sum_{j=1}^{20} j = 1 + 2 + 3 + \cdots + 18 + 19 + 20$$

$$\sum_{i=0}^{n} x^i = x^0 + x^1 + x^2 + \cdots + x^{n-1} + x^n$$

$$\sum_{m=1}^{n} 1 = \underbrace{1 + 1 + 1 + \cdots + 1}_{n \text{ terms}}$$

$$\sum_{k=-2}^{3} \frac{1}{k+7} = \frac{1}{5} + \frac{1}{6} + \frac{1}{7} + \frac{1}{8} + \frac{1}{9} + \frac{1}{10}$$

Sometimes we use a subscripted variable $a_i$ to denote the $i$th term of a general sum instead of using the functional notation $f(i)$:

$$\sum_{i=m}^{n} a_i = a_m + a_{m+1} + a_{m+2} + \cdots + a_n.$$

In particular, an **infinite series** is such a sum with infinitely many terms:

$$\sum_{n=1}^{\infty} a_n = a_1 + a_2 + a_3 + \cdots$$

When no final term follows the $\cdots$, it is understood that the terms go on forever. We will study infinite series in Chapter 9.

When adding finitely many numbers, the order in which they are added is unimportant; any order will give the same sum. If all the numbers have a common factor, then that factor can be removed from each term and multiplied after the sum is evaluated: $ca + cb = c(a + b)$. These laws of arithmetic translate into the following *linearity* rule for finite sums; if $A$ and $B$ are constants, then

$$\sum_{i=m}^{n} \left(A f(i) + B g(i)\right) = A \sum_{i=m}^{n} f(i) + B \sum_{i=m}^{n} g(i).$$

Both of the sums $\sum_{j=m}^{m+n} f(j)$ and $\sum_{i=0}^{n} f(i+m)$ have the same expansion, namely $f(m) + f(m+1) + \cdots + f(m+n)$. Therefore the two sums are equal.

$$\sum_{j=m}^{m+n} f(j) = \sum_{i=0}^{n} f(i+m).$$

This equality can also derived by substituting $i+m$ for $j$ everywhere $j$ appears on the left side, noting that $i+m = m$ reduces to $i = 0$ and $i + m = m + n$ reduces to $i = n$. It is often convenient to make such a **change of index** in a summation.

**Example 3**   Express $\sum_{j=3}^{17} \sqrt{1+j^2}$ in the form $\sum_{i=1}^{n} f(i)$.

**Solution**   Let $j = i+2$, Then $j = 3$ corresponds to $i = 1$ and $j = 17$ corresponds to $i = 15$. Thus

$$\sum_{j=3}^{17} \sqrt{1+j^2} = \sum_{i=1}^{15} \sqrt{1+(i+2)^2}.$$

■

## Evaluating Sums

When a sum like

$$S = \sum_{i=1}^{n} i = 1 + 2 + 3 + \cdots + n$$

involves a variable or large number of terms we would like to have a formula giving the value of the sum in **closed form**, that is, not using the ellipsis $(\cdots)$. For the above sum, the formula is

$$S = \sum_{i=1}^{n} i = \frac{n(n+1)}{2}.$$

This can be verified by writing the sum forwards and backwards and adding the two to get

$$
\begin{array}{rcccccccc}
S = & 1 & + & 2 & + & 3 & + \cdots + (n-1) + & n \\
S = & n & + (n-1) & + (n-2) & + \cdots + & 2 & + & 1 \\
\hline
2S = & (n+1) + (n+1) + (n+1) + \cdots + (n+1) + (n+1) = n(n+1)
\end{array}
$$

The formula for $S$ follows when we divide by 2.

It is not usually this easy to evaluate a general sum in closed form. We can only simplify $\sum_{i=m}^{n} f(i)$ for a small class of functions $f$. The only such formulas we will need in the next sections are collected in Theorem 1.

**THEOREM 1**

**Summation formulas**

(a) $\displaystyle\sum_{i=1}^{n} 1 = \underbrace{1 + 1 + 1 + \cdots + 1}_{n \text{ terms}} = n$

(b) $\displaystyle\sum_{i=1}^{n} i = 1 + 2 + 3 + \cdots + n = \frac{n(n+1)}{2}$

(c) $\displaystyle\sum_{i=1}^{n} i^2 = 1^2 + 2^2 + 3^2 + \cdots + n^2 = \frac{n(n+1)(2n+1)}{6}$

(d) $\displaystyle\sum_{i=1}^{n} r^{i-1} = 1 + r + r^2 + r^3 + \cdots + r^{n-1} = \frac{r^n - 1}{r - 1}$   if $r \neq 1$.

**PROOF**  Formula (a) is trivial; the sum of $n$ ones is $n$. One proof of formula (b) was given above. Three others are suggested in Exercises 42–44.

To prove (c) we write $n$ copies of the identity

$$(k+1)^3 - k^3 = 3k^2 + 3k + 1,$$

one for each value of $k$ from 1 to $n$, and add them up:

$$
\begin{array}{rclcrclclcl}
2^3 & - & 1^3 & = & 3 \times 1^2 & + & 3 \times 1 & + & 1 \\
3^3 & - & 2^3 & = & 3 \times 2^2 & + & 3 \times 2 & + & 1 \\
4^3 & - & 3^3 & = & 3 \times 3^2 & + & 3 \times 3 & + & 1 \\
\vdots & & \vdots & & \vdots & & \vdots & & \vdots \\
n^3 & - & (n-1)^3 & = & 3(n-1)^2 & + & 3(n-1) & + & 1 \\
(n+1)^3 & - & n^3 & = & 3\,n^2 & + & 3n & + & 1 \\
\hline
(n+1)^3 & - & 1^3 & = & 3\left(\sum_{i=1}^{n} i^2\right) & + & 3\left(\sum_{i=1}^{n} i\right) & + & n \\
& & & = & 3\left(\sum_{i=1}^{n} i^2\right) & + & \dfrac{3n(n+1)}{2} & + & n.
\end{array}
$$

We used formula (b) in the last line. The final equation can be solved for the desired sum to give formula (c). Note the cancellations that occurred when we added up the left sides of the $n$ equations. The term $2^3$ in the first line cancelled the $-2^3$ in the second line, and so on, leaving us with only two terms, the $(n+1)^3$ from the $n$th line and the $-1^3$ from the first line:

$$\sum_{k=1}^{n} \left((k+1)^3 - k^3\right) = (n+1)^3 - 1^3.$$

This is an example of what we call a **telescoping sum**. In general, a sum of the form $\sum_{i=m}^{n}\left(f(i+1) - f(i)\right)$ telescopes to the closed form $f(n+1) - f(m)$ because all but the first and last terms cancel out.

To prove formula (d), let $s = \sum_{i=1}^{n} r^{i-1}$ and subtract $s$ from $rs$:

$$(r-1)s = rs - s = (r + r^2 + r^3 + \cdots + r^n) - (1 + r + r^2 + \cdots + r^{n-1})$$
$$= r^n - 1.$$

The result follows on division by $r - 1$.

**Example 4**   Evaluate $\displaystyle\sum_{k=m+1}^{n} (6k^2 - 4k + 3)$, where $1 \le m < n$.

**Solution**   Using the rules of summation and various summation formulas from Theorem 1 we calculate

$$\sum_{k=1}^{n}(6k^2 - 4k + 3) = 6\sum_{k=1}^{n} k^2 - 4\sum_{k=1}^{n} k + 3\sum_{k=1}^{n} 1$$

$$= 6\,\frac{n(n+1)(2n+1)}{6} - 4\,\frac{n(n+1)}{2} + 3n$$

$$= 2n^3 + n^2 + 2n$$

Thus

$$\sum_{k=m+1}^{n}(6k^2 - 4k + 3) = \sum_{k=1}^{n}(6k^2 - 4k + 3) - \sum_{k=1}^{m}(6k^2 - 4k + 3)$$

$$= 2n^3 + n^2 + 2n - 2m^3 - m^2 - 2m.$$

# Exercises 5.1

Expand the sums in Exercises 1–8.

**1.** $\displaystyle\sum_{i=1}^{4} i^3$

**2.** $\displaystyle\sum_{j=1}^{100} \frac{j}{j+1}$

**3.** $\displaystyle\sum_{i=1}^{n} 3^i$

**4.** $\displaystyle\sum_{i=0}^{n-1} \frac{(-1)^i}{i+1}$

**5.** $\displaystyle\sum_{j=3}^{n} \frac{(-2)^j}{(j-2)^2}$

**6.** $\displaystyle\sum_{j=1}^{n} \frac{j^2}{n^3}$

**7.** $\displaystyle\sum_{n=1}^{k} \sin\frac{n\pi}{3k}$

**8.** $\displaystyle\sum_{k=2}^{n} \frac{e^{-k}}{n+k}$

Write the sums in Exercises 9–18 using sigma notation. (Note that the answers are not unique.)

**9.** $5 + 6 + 7 + 8 + 9$

**10.** $5^3 + 6^3 + 7^3 + 8^3 + 9^3$

**11.** $2 + 2 + 2 + \cdots + 2$   (200 terms)

**12.** $\dfrac{1}{7} + \dfrac{1}{8} + \dfrac{1}{9} + \cdots + \dfrac{1}{99}$

**13.** $2^2 - 3^2 + 4^2 - 5^2 + \cdots - 99^2$

**14.** $1 + 2x + 3x^2 + 4x^3 + \cdots + 100x^{99}$

**15.** $1 + x + x^2 + x^3 + \cdots + x^n$

**16.** $1 - x + x^2 - x^3 + \cdots + x^{2n}$

**17.** $1 - \dfrac{1}{4} + \dfrac{1}{9} - \dfrac{1}{16} + \cdots + \dfrac{(-1)^{n-1}}{n^2}$

**18.** $\dfrac{1}{2} + \dfrac{2}{4} + \dfrac{3}{8} + \dfrac{4}{16} + \cdots + \dfrac{n}{2^n}$

Express the sums in Exercises 19–20 in the form $\sum_{i=1}^{n} f(i)$.

**19.** $\displaystyle\sum_{j=0}^{99} \sin(j)$

**20.** $\displaystyle\sum_{k=-5}^{m} \frac{1}{k^2 + 1}$

Find closed form values for the sums in Exercises 21–36.

**21.** $\displaystyle\sum_{i=2}^{6} (i - 1)$

**22.** $\displaystyle\sum_{j=1}^{1,000} (2j + 3)$

**23.** $\displaystyle\sum_{i=1}^{n} i^2 + 2i$

**24.** $\displaystyle\sum_{k=1}^{n-1} (3k^2 - 4)$

**25.** $\displaystyle\sum_{i=m}^{n} (2i - 1)$   $(m < n)$

**26.** $\displaystyle\sum_{j=0}^{n-1} (4j + 4^j)$

**27.** $\displaystyle\sum_{k=1}^{n} (\pi^k - 3)$

**28.** $\displaystyle\sum_{i=1}^{n} (2^i - i^2)$

**29.** $\displaystyle\sum_{m=1}^{n} \ln m$

**30.** $\displaystyle\sum_{i=0}^{n} e^{i/n}$

**31.** The sum in Exercise 11.     **32.** The sum in Exercise 15.

**33.** The sum in Exercise 16.

*  **34.** The sum in Exercise 14. *Hint:* differentiate the sum $\sum_{i=0}^{100} x^i$.

* **35.** The sum in Exercise 13. *Hint:* the sum is

$$\sum_{k=1}^{49} \left( (2k)^2 - (2k+1)^2 \right) = \sum_{k=1}^{49} (-4k-1).$$

* **36.** The sum in Exercise 18. *Hint:* apply the method of proof of Theorem 1(d) to this sum.

**37.** Verify the formula for the value of a telescoping sum:

$$\sum_{i=m}^{n} \left( f(i+1) - f(i) \right) = f(n+1) - f(m).$$

Why is the word "telescoping" used to describe this sum?

In Exercises 38–40, evaluate the given telescoping sums.

**38.** $\displaystyle\sum_{n=1}^{10} \left( n^4 - (n-1)^4 \right)$     **39.** $\displaystyle\sum_{j=1}^{m} (2^j - 2^{j-1})$

**40.** $\displaystyle\sum_{i=m}^{2m} \left( \frac{1}{i} - \frac{1}{i+1} \right)$

**41.** Show that $\dfrac{1}{j(j+1)} = \dfrac{1}{j} - \dfrac{1}{j+1}$, and hence evaluate

$$\sum_{j=1}^{n} \frac{1}{j(j+1)}.$$

**42.** Figure 5.1 shows a square of side $n$ subdivided into $n^2$ smaller squares of side 1. How many small squares are shaded? Obtain the closed form expression for $\sum_{i=1}^{n} i$ by considering the sum of the areas of the shaded squares.

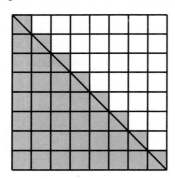

**Figure 5.1**

**43.** Write $n$ copies of the identity $(k+1)^2 - k^2 = 2k+1$, one for each integer $k$ from 1 to $n$, and add them up to obtain the formula

$$\sum_{i=1}^{n} i = \frac{n(n+1)}{2}$$

in a manner similar to the proof of Theorem 1(c).

**44.** Use mathematical induction to prove Theorem 1(b).

**45.** Use mathematical induction to prove Theorem 1(c).

**46.** Use mathematical induction to prove Theorem 1(d).

**47.** Figure 5.2 shows a square of side $\sum_{i=1}^{n} i = n(n+1)/2$ subdivided into a small square of side 1 and $n-1$ $L$-shaped regions whose short edges are 2, 3, ..., $n$. Show that the area of the $L$-shaped region with short side $i$ is $i^3$, and hence verify that

$$\sum_{i=1}^{n} i^3 = \frac{n^2(n+1)^2}{4}.$$

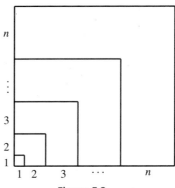

**Figure 5.2**

* **48.** Write $n$ copies of the identity

$$(k+1)^4 - k^4 = 4k^3 + 6k^2 + 4k + 1,$$

one for each integer $k$ from 1 to $n$, and add them up to obtain the formula

$$\sum_{i=1}^{n} i^3 = \frac{n^2(n+1)^2}{4},$$

in a manner similar to the proof of Theorem 1(c).

**49.** Use mathematical induction to verify the formula for the sum of cubes given in Exercise 48.

**50.** Extend the method of Exercises 47 and 48 to find a closed form expression for $\sum_{i=1}^{n} i^4$. You will probably want to use computer algebra software to do all the algebra.

**51.** Most computer algebra systems will find closed form expressions for many kinds of sums. Use such a system to find $\sum_{i=1}^{n} i^k$ for $k = 5, 6, 7, 8$. Observe the term involving the highest power of $n$ in each case. Predict the highest power term in $\sum_{i=1}^{n} i^{10}$ and verify your prediction.

# 5.2 Areas as Limits of Sums

We began the study of derivatives in Chapter 2 by defining what is meant by a tangent line to a curve at a particular point. We would like to begin the study of integrals by defining what is meant by the **area** of a plane region, but a definition of area is much harder to give than a definition of tangency. Let us assume, therefore, that we know intuitively what area means and list some of its properties. (See Figure 5.3.)

(i) The area of a plane region is a nonnegative real number of *square units*.

(ii) The area of a rectangle with width $w$ and height $h$ is $A = wh$.

(iii) The areas of congruent plane regions are equal.

(iv) If region $S$ is contained in region $R$, then the area of $S$ is less than or equal to that of $R$.

(v) If region $R$ is a union of (finitely many) non-overlapping regions, then the area of $R$ is the sum of the areas of those regions.

Using these five properties we can calculate the area of any **polygon** (a region bounded by straight line segments). First we note that properties (iii) and (v) show that the area of a parallelogram is the same of that of a rectangle having the same base width and height. Any triangle can be butted against a congruent copy of itself to form a parallelogram, so a triangle has area half the base width times the height. Finally, any polygon can be subdivided into finitely many non-overlapping triangles so its area is the sum of the areas of those triangles.

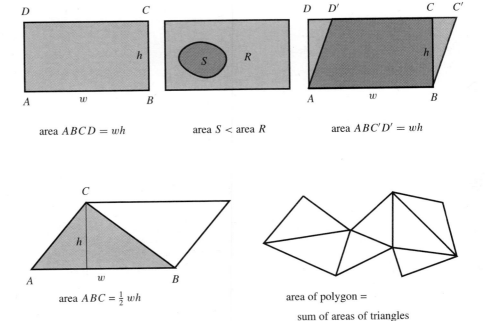

area $ABCD = wh$      area $S <$ area $R$      area $ABC'D' = wh$

area $ABC = \frac{1}{2} wh$      area of polygon = sum of areas of triangles

**Figure 5.3**    Properties of area

We can't go beyond polygons without taking limits. If a region has a curved boundary its area can only be approximated by using rectangles or triangles; calculating the exact area requires the evaluation of a limit. We showed how this could be done for a circle in Section 1.1.

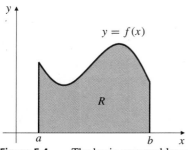

**Figure 5.4** The basic area problem: find the area of region $R$

## The Basic Area Problem

In this section we are going to consider how to find the area of a region $R$ lying under the graph $y = f(x)$ of a nonnegative-valued, continuous function $f$, above the $x$-axis and between the vertical lines $x = a$ and $x = b$, where $a < b$. (See Figure 5.4.) To accomplish this we proceed as follows. Divide the interval $[a, b]$ into $n$ subintervals by using division points:

$$a = x_0 < x_1 < x_2 < x_3 < \cdots < x_{n-1} < x_n = b.$$

Denote by $\Delta x_i$ the length of the $i$th subinterval $[x_{i-1}, x_i]$:

$$\Delta x_i = x_i - x_{i-1}, \qquad (i = 1, 2, 3, \ldots, n).$$

Vertically above each subinterval $[x_{i-1}, x_i]$, build a rectangle whose base has length $\Delta x_i$ and whose height is $f(x_i)$. The area of this rectangle is $f(x_i) \Delta x_i$. Form the sum of these areas:

$$S_n = f(x_1)\,\Delta x_1 + f(x_2)\,\Delta x_2 + f(x_3)\,\Delta x_3 + \cdots + f(x_n)\,\Delta x_n = \sum_{i=1}^{n} f(x_i)\,\Delta x_i.$$

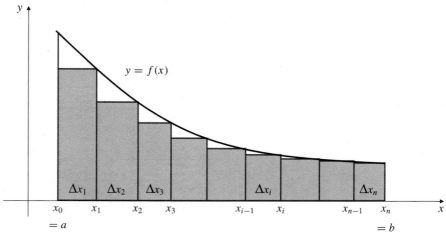

**Figure 5.5** Approximating the area under the graph of a decreasing function using rectangles

The rectangles are shown shaded in Figure 5.5 for a decreasing function $f$. For an increasing function, the tops of the rectangles would lie above the graph of $f$ rather than below it. Evidently $S_n$ is an approximation to the area of the region $R$, and the approximation gets better as $n$ increases, provided we choose the points $a = x_0 < x_1 < \cdots < x_n = b$ in such a way that the width $\Delta x_i$ of the widest rectangle approaches zero.

Observe in Figure 5.6, for example, that subdividing a subinterval into two smaller subintervals reduces the error in the approximation by reducing that part of the area under the curve that is not contained in the rectangles. It is reasonable, therefore, to calculate the area of $R$ by finding the limit of $S_n$ as $n \to \infty$ with the restriction that the largest of the subinterval widths $\Delta x_i$ must approach zero:

$$\text{Area of } R = \lim_{\substack{n \to \infty \\ \max \Delta x_i \to 0}} S_n.$$

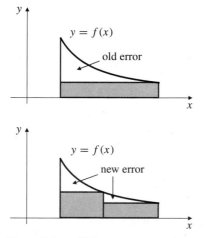

**Figure 5.6** Using more rectangles makes the error smaller

Sometimes, but not always, it is useful to choose the points $x_i$ ($0 \le i \le n$) in $[a, b]$ in such a way that all the subinterval lengths $\Delta x_i$ are all equal. In this case we have

$$\Delta x_i = \Delta x = \frac{b - a}{n}, \qquad x_i = a + i\Delta x = a + \frac{i}{n}(b - a).$$

## Some Area Calculations

We devote the rest of this section to some examples in which we apply the technique described above for finding areas under graphs of functions by approximating with rectangles. Let us begin with a region for which we already know the area so we can satisfy ourselves that the method does give the correct value.

**Example 1** Find the area $A$ of the region lying under the straight line $y = x + 1$, above the $x$-axis and between the lines $x = 0$ and $x = 2$.

*Solution* The region is shaded in Figure 5.7(a). It is a *trapezoid* (a four-sided polygon with one pair of parallel sides) and has area 4 square units. (It can be divided into a rectangle and a triangle, each of area 2 square units.) We will calculate the area as a limit of sums of areas of rectangles constructed as described above. Divide the interval $[0, 2]$ into $n$ subintervals *of equal length* by points

$$x_0 = 0, \; x_1 = \frac{2}{n}, \; x_2 = \frac{4}{n}, \; x_3 = \frac{6}{n}, \; \ldots \; x_n = \frac{2n}{n} = 2.$$

The value of $y = x + 1$ at $x = x_i$ is $x_i + 1 = \dfrac{2i}{n} + 1$ and the $i$th subinterval, $\left[\dfrac{2(i-1)}{n}, \dfrac{2i}{n}\right]$, has length $\Delta x_i = \dfrac{2}{n}$. Observe that $\Delta x_i \to 0$ as $n \to \infty$. The sum of the areas of the approximating rectangles shown in Figure 5.7(a) is

$$
\begin{aligned}
S_n &= \sum_{i=1}^{n} \left(\frac{2i}{n} + 1\right) \frac{2}{n} \\
&= \left(\frac{2}{n}\right) \left[\frac{2}{n} \sum_{i=1}^{n} i + \sum_{i=1}^{n} 1\right] \qquad \text{(Use parts (b) and (a) of Theorem 1.)} \\
&= \left(\frac{2}{n}\right) \left[\frac{2}{n} \frac{n(n+1)}{2} + n\right] \\
&= 2\frac{n+1}{n} + 2.
\end{aligned}
$$

Therefore the required area $A$ is given by

$$A = \lim_{n \to \infty} S_n = \lim_{n \to \infty} \left(2\frac{n+1}{n} + 2\right) = 2 + 2 = 4 \text{ square units.}$$

**Example 2** Find the area of the region bounded by the parabola $y = x^2$ and the straight lines $y = 0$, $x = 0$, and $x = b > 0$.

*Solution* The area $A$ of the region is the limit of the sum $S_n$ of areas of the rectangles shown in Figure 5.7(b). Again we have used

equal subintervals, each of length $b/n$. The height of the $i$th rectangle is $(ib/n)^2$. Thus,

**Figure 5.7**

(a) The region of Example 1

(b) The region of Example 2

$$y = x + 1$$

$$\frac{2}{n} \quad \frac{4}{n} \quad \frac{6}{n} \qquad \frac{2n}{n}$$

(a)

$$y = x^2$$

$$\frac{b}{n} \quad \frac{2b}{n} \quad \frac{3b}{n} \qquad \frac{(n-1)b}{n} \qquad \frac{nb}{n} = b$$

(b)

$$S_n = \sum_{i=1}^{n} \left(\frac{ib}{n}\right)^2 \frac{b}{n} = \frac{b^3}{n^3} \sum_{i=1}^{n} i^2 = \frac{b^3}{n^3} \frac{n(n+1)(2n+1)}{6},$$

by formula (c) of Theorem 1. Hence the required area is

$$A = \lim_{n \to \infty} S_n = \lim_{n \to \infty} b^3 \frac{(n+1)(2n+1)}{6n^2} = \frac{b^3}{3} \text{ square units.} \qquad \blacksquare$$

Finding an area under the graph of $y = x^k$ over an interval $I$ becomes more and more difficult as $k$ increases if we continue to try to subdivide $I$ into subintervals of equal length. (See Exercise 14 at the end of this section for the case $k = 3$.) It is, however, possible to find the area for arbitrary $k$ if we subdivide the interval $I$ into subintervals whose lengths increase in geometric progression. Our final example illustrates this.

**Example 3** Let $b > a > 0$ and let $k$ be any real number except $-1$. Show that the area $A$ of the region bounded by $y = x^k$, $y = 0$, $x = a$ and $x = b$ is

$$A = \frac{b^{k+1} - a^{k+1}}{k+1} \text{ square units.}$$

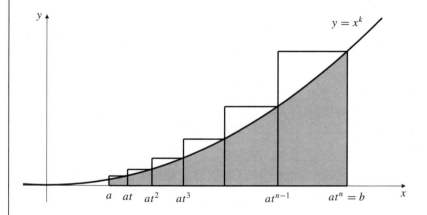

**Figure 5.8** For this partition the subinterval lengths increase exponentially

**Solution**  Let $t = (b/a)^{1/n}$ and let

$$x_0 = a, \ x_1 = at, \ x_2 = at^2, \ x_3 = at^3, \ \ldots \ x_n = at^n = b.$$

These points subdivide the interval $[a, b]$ into $n$ subintervals of which the $i$th, $[x_{i-1}, x_i]$, has length $\Delta x_i = at^{i-1}(t - 1)$. If $f(x) = x^k$, then $f(x_i) = a^k t^{ki}$. The sum of the areas of the rectangles shown in Figure 5.8 is:

**DANGER**

This is a long and rather difficult example. Either skip over it or take your time and check each step carefully.

$$S_n = \sum_{i=1}^{n} f(x_i)\,\Delta x_i$$

$$= \sum_{i=1}^{n} a^k\, t^{ki}\, at^{i-1}(t - 1)$$

$$= a^{k+1}\,(t - 1)\,t^k \sum_{i=1}^{n} t^{(k+1)(i-1)}$$

$$= a^{k+1}\,(t - 1)\,t^k \sum_{i=1}^{n} r^{(i-1)} \qquad \text{where } r = t^{k+1}$$

$$= a^{k+1}\,(t - 1)\,t^k\, \frac{r^n - 1}{r - 1} \qquad \text{(by Theorem 1(d))}$$

$$= a^{k+1}\,(t - 1)\,t^k\, \frac{t^{(k+1)n} - 1}{t^{k+1} - 1}.$$

Now replace $t$ with its value $(b/a)^{1/n}$ and rearrange factors to obtain

$$S_n = a^{k+1} \left( \left(\frac{b}{a}\right)^{1/n} - 1 \right) \left(\frac{b}{a}\right)^{k/n} \frac{\left(\dfrac{b}{a}\right)^{k+1} - 1}{\left(\dfrac{b}{a}\right)^{(k+1)/n} - 1}$$

$$= \left(b^{k+1} - a^{k+1}\right) c^{k/n}\, \frac{c^{1/n} - 1}{c^{(k+1)/n} - 1}, \qquad \text{where } c = \frac{b}{a}.$$

Of the three factors in the final line above, the first does not depend on $n$, and the second, $c^{k/n}$, approaches $c^0 = 1$ as $n \to \infty$. The third factor is an indeterminate form of type $[0/0]$, which we evaluate using l'Hôpital's Rule. First let $u = 1/n$. Then

$$\lim_{n \to \infty} \frac{c^{1/n} - 1}{c^{(k+1)/n} - 1} = \lim_{u \to 0+} \frac{c^u - 1}{c^{(k+1)u} - 1} \qquad \left[\frac{0}{0}\right]$$

$$= \lim_{u \to 0+} \frac{c^u \ln c}{(k + 1)\, c^{(k+1)u} \ln c} = \frac{1}{k + 1}.$$

Therefore the required area is

$$A = \lim_{n \to \infty} S_n = \left(b^{k+1} - a^{k+1}\right) \times 1 \times \frac{1}{k + 1} = \frac{b^{k+1} - a^{k+1}}{k + 1} \quad \text{square units.}$$

It is, as you can see, rather difficult to calculate areas bounded by curves by the methods developed above. Fortunately, there is an easier way, as we will discover in Section 6.5.

***Remark***    For technical reasons it was necessary to assume $a > 0$ in Example 3. The result is also valid for $a = 0$ provided $k > -1$. In this case we have $\lim_{a \to 0+} a^{k+1} = 0$, so the area under $y = x^k$, above $y = 0$ between $x = 0$ and $x = b > 0$ is $A = b^{k+1}/(k + 1)$ square units. For $k = 2$ this agrees with the result of Example 2.

---

**Example 4**    Identify the limit $L = \lim\limits_{n \to \infty} \sum\limits_{i=1}^{n} \dfrac{n - i}{n^2}$ as an area, and hence evaluate it.

***Solution***    We can rewrite the $i$th term of the sum so that it depends on $i/n$:

$$L = \lim_{n \to \infty} \sum_{i=1}^{n} \left(1 - \frac{i}{n}\right) \frac{1}{n}.$$

The terms now appear to be the areas of rectangles of base $1/n$ and heights $1 - x_i$, $(1 \le i \le n)$, where

$$x_1 = \frac{1}{n}, \quad x_2 = \frac{2}{n}, \quad x_3 = \frac{3}{n}, \quad \ldots, \quad x_n = \frac{n}{n}.$$

Thus the limit $L$ is the area under the curve $y = 1 - x$ from $x = 0$ to $x = 1$. (See Figure 5.9.) This region is a triangle having area $1/2$ square unit, so $L = 1/2$. ∎

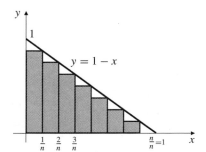

**Figure 5.9**    Recognizing a sum of areas

---

## Exercises 5.2

Use the techniques of Examples 1 and 2 (with subintervals of equal length) to find the areas of the regions specified in Exercises 1–13.

1. Below $y = 3x$, above $y = 0$, from $x = 0$ to $x = 1$.

2. Below $y = 2x + 1$, above $y = 0$, from $x = 0$ to $x = 3$.

3. Below $y = 2x - 1$, above $y = 0$, from $x = 1$ to $x = 3$.

4. Below $y = 3x + 4$, above $y = 0$, from $x = -1$ to $x = 2$.

5. Below $y = x^2$, above $y = 0$, from $x = 1$ to $x = 3$.

6. Below $y = x^2 + 1$, above $y = 0$, from $x = 0$ to $x = a > 0$.

7. Below $y = x^2 + 2x + 3$, above $y = 0$, from $x = -1$ to $x = 2$.

8. Above $y = x^2 - 1$, below $y = 0$.

9. Above $y = 1 - x$, below $y = 0$, from $x = 2$ to $x = 4$.

10. Above $y = x^2 - 2x$, below $y = 0$.

11. Below $y = 4x - x^2 + 1$, above $y = 1$.

* 12. Below $y = e^x$, above $y = 0$, from $x = 0$ to $x = b > 0$.

* 13. Below $y = 2^x$, above $y = 0$, from $x = -1$ to $x = 1$.

14. Use the formula $\sum_{i=1}^{n} i^3 = n^2(n + 1)^2/4$, from Exercises 47–49 of Section 5.1, to find the area of the region lying under $y = x^3$, above the $x$-axis, and between the vertical lines at $x = 0$ and $x = b > 0$.

15. Use the subdivision of $[a, b]$ given in Example 3 to find the area under $y = 1/x$, above $y = 0$ from $x = a > 0$ to $x = b > a$. Why should your answer not be surprising?

In Exercises 16–19, interpret the given sum $S_n$ as a sum of areas of rectangles approximating the area of a certain region in the plane, and hence evaluate $\lim_{n \to \infty} S_n$.

16. $S_n = \sum\limits_{i=1}^{n} \dfrac{2}{n}\left(1 - \dfrac{i}{n}\right)$

17. $S_n = \sum\limits_{i=1}^{n} \dfrac{2}{n}\left(1 - \dfrac{2i}{n}\right)$

18. $S_n = \sum\limits_{i=1}^{n} \dfrac{2n + 3i}{n^2}$

* 19. $S_n = \sum\limits_{j=1}^{n} \dfrac{1}{n}\sqrt{1 - (j/n)^2}$

## 5.3 The Definite Integral

In this section we generalize and make more precise the procedure used for finding areas developed in Section 5.2, and we use it to define the *definite integral* of a function $f$ on an interval $I$. Let us assume, for the time being, that $f(x)$ is defined and continuous on the closed, finite interval $[a, b]$. We no longer assume that the values of $f$ are not negative.

### Partitions and Riemann Sums

Let $P$ be a finite set of points arranged in order between $a$ and $b$ on the real line, say

$$P = \{x_0, \, x_1, \, x_2, \, x_3, \, \ldots, \, x_{n-1}, \, x_n\},$$

where $a = x_0 < x_1 < x_2 < x_3 < \cdots < x_{n-1} < x_n = b$. Such a set $P$ is called a **partition** of $[a, b]$; it divides $[a, b]$ into $n$ subintervals of which the $i$th is $[x_{i-1}, x_i]$. We call these the subintervals of the partition $P$. The number $n$ depends on the particular partition, so we write $n = n(P)$. The length of the $i$th subinterval of $P$ is

$$\Delta x_i = x_i - x_{i-1},$$

and we call the greatest of these numbers $\Delta x_i$ the **norm** of the partition $P$ and denote it $\|P\|$:

$$\|P\| = \max_{1 \le i \le n} \Delta x_i.$$

Since $f$ is continuous on each subinterval $[x_{i-1}, x_i]$ of $P$, it takes on maximum and minimum values at points of that interval (by Theorem 8 of Section 1.4). Thus there are numbers $l_i$ and $u_i$ in $[x_{i-1}, x_i]$ such that

$$f(l_i) \le f(x) \le f(u_i) \qquad \text{whenever } x_{i-1} \le x \le x_i.$$

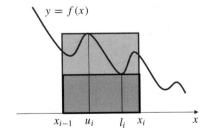

$y = f(x)$

$x_{i-1}$   $u_i$   $l_i$   $x_i$     $x$

**Figure 5.10**

If $f(x) \ge 0$ on $[a, b]$, then $f(l_i)\,\Delta x_i$ and $f(u_i)\,\Delta x_i$ represent the areas of rectangles having the interval $[x_{i-1}, x_i]$ on the $x$-axis as base, and having tops passing through the lowest and highest points, respectively, on the graph of $f$ on that interval. (See Figure 5.10.) If $A_i$ is that part of the area under $y = f(x)$ and above the $x$-axis that lies in the vertical strip between $x = x_{i-1}$ and $x = x_i$, then

$$f(l_i)\,\Delta x_i \le A_i \le f(u_i)\,\Delta x_i.$$

If $f$ can have negative values, then one or both of $f(l_i)\,\Delta x_i$ and $f(u_i)\,\Delta x_i$ can be negative, and will then represent the negative of the area of a rectangle lying below the $x$-axis. In any event we always have $f(l_i)\,\Delta x_i \le f(u_i)\,\Delta x_i$.

**DEFINITION 2**

**Upper and lower Riemann sums**

The **lower (Riemann) sum** $L(f, P)$ and the **upper (Riemann) sum** $U(f, P)$ for the function $f$ and the partition $P$ are defined by:

$$L(f, P) = f(l_1)\,\Delta x_1 + f(l_2)\,\Delta x_2 + \cdots + f(l_n)\,\Delta x_n$$

$$= \sum_{i=1}^{n} f(l_i)\,\Delta x_i,$$

$$U(f, P) = f(u_1)\,\Delta x_1 + f(u_2)\,\Delta x_2 + \cdots + f(u_n)\,\Delta x_n$$

$$= \sum_{i=1}^{n} f(u_i)\,\Delta x_i.$$

Figure 5.11 illustrates these Riemann sums as sums of *signed* areas of rectangles; any such areas that lie below the *x*-axis are counted as negative.

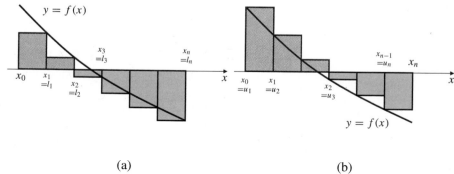

**Figure 5.11** (a) A lower Riemann sum, and (b) an upper Riemann sum for a decreasing function $f$. The areas of rectangles shaded in colour are counted as positive, those shaded in grey are counted as negative

(a)         (b)

**Example 1** Calculate lower and upper Riemann sums for the function $f(x) = 1/x$ on the interval $[1, 2]$, corresponding to the partition $P$ of $[1, 2]$ into 4 subintervals of equal length.

**Solution** The partition $P$ consists of the points $x_0 = 1$, $x_1 = 5/4$, $x_2 = 3/2$, $x_3 = 7/4$, and $x_4 = 2$. Since $1/x$ is decreasing on $[1, 2]$, its minimum and maximum values on the $i$th subinterval $[x_{i-1}, x_i]$ are $1/x_i$ and $1/x_{i-1}$, respectively. Thus, the lower and upper Riemann sums are

$$L(f, P) = \frac{1}{4}\left(\frac{4}{5} + \frac{2}{3} + \frac{4}{7} + \frac{1}{2}\right) = \frac{533}{840} \approx 0.6345$$

$$U(f, P) = \frac{1}{4}\left(1 + \frac{4}{5} + \frac{2}{3} + \frac{4}{7}\right) = \frac{319}{420} \approx 0.7595.$$

**Example 2** Calculate the lower and upper Riemann sums for the function $f(x) = x^2$ on the interval $[0, a]$ (where $a > 0$), corresponding to the partition $P_n$ of $[0, a]$ into $n$ subintervals of equal length.

**Solution** Each subinterval of $P_n$ has length $\Delta x = a/n$, and the division points are given by $x_i = ia/n$ for $i = 0, 1, 2, \ldots, n$. Since $x^2$ is increasing on $[0, a]$, its minimum and maximum values over the $i$th subinterval $[x_{i-1}, x_i]$ occur at $l_i = x_{i-1}$ and $u_i = x_i$, respectively. Thus the lower Riemann sum of $f$ for $P_n$ is

$$L(f, P_n) = \sum_{i=1}^{n}(x_{i-1})^2\Delta x = \frac{a^3}{n^3}\sum_{i=1}^{n}(i-1)^2$$

$$= \frac{a^3}{n^3}\sum_{j=0}^{n-1}j^2 = \frac{a^3}{n^3}\frac{(n-1)n(2(n-1)+1)}{6} = \frac{(n-1)(2n-1)a^3}{6n^2},$$

where we have used Theorem 1(c) of Section 5.1 to evaluate the sum of squares.

Similarly, the upper Riemann sum is

$$U(f, P_n) = \sum_{i=1}^{n} (x_i)^2 \Delta x$$

$$= \frac{a^3}{n^3} \sum_{i=1}^{n} i^2 = \frac{a^3}{n^3} \frac{n(n+1)(2n+1)}{6} = \frac{(n+1)(2n+1)a^3}{6n^2}.$$

## The Definite Integral

If we calculate $L(f, P)$ and $U(f, P)$ for partitions $P$ having more and more points spaced closer and closer together, we expect that, in the limit, these Riemann sums will converge to a common value that will be the area bounded by $y = f(x)$, $y = 0$, $x = a$, and $x = b$ if $f(x) \geq 0$ on $[a, b]$. This is indeed the case, but we cannot fully prove it yet.

If $P_1$ and $P_2$ are two partitions of $[a, b]$ such that every point of $P_1$ also belongs to $P_2$, then we say that $P_2$ is a **refinement** of $P_1$. It is not difficult to show that in this case

$$L(f, P_1) \leq L(f, P_2) \leq U(f, P_2) \leq U(f, P_1);$$

adding more points to a partition increases the lower sum and decreases the upper sum. (See Exercise 18 at the end of this section.) Given any two partitions, $P_1$ and $P_2$, we can form their **common refinement** $P$, which consists of all of the points of $P_1$ and $P_2$. Thus

$$L(f, P_1) \leq L(f, P) \leq U(f, P) \leq U(f, P_2).$$

Hence every lower sum is less than or equal to every upper sum. Since the real numbers are complete, there must exist *at least one* real number $I$ such that

$$L(f, P) \leq I \leq U(f, P), \qquad \text{for every partition } P.$$

If there is *only one* such number, we will call it the definite integral of $f$ on $[a, b]$.

---

**DEFINITION 3**

**The definite integral**

Suppose there is exactly one number $I$ such that for every partition $P$ of $[a, b]$ we have

$$L(f, P) \leq I \leq U(f, P).$$

Then we say that the function $f$ is **integrable** on $[a, b]$, and we call $I$ the **definite integral** of $f$ on $[a, b]$. The definite integral is denoted by the symbol

$$I = \int_a^b f(x)\, dx.$$

We stress at once that the definite integral of $f(x)$ over $[a, b]$ is a *number*; it is not a function of $x$. It depends on the numbers $a$ and $b$ and on the particular function $f$, but not on the variable $x$ (which is a **dummy variable** like the variable $i$ in the sum $\sum_{i=1}^{n} f(i)$). You can replace $x$ with any other variable without changing the value of the integral:

$$\int_a^b f(x)\,dx = \int_a^b f(t)\,dt.$$

The various parts of the symbol $\int_a^b f(x)\,dx$ have their own names:

(i) $\int$ is called the **integral sign**; it resembles the letter S since it represents the limit of a sum.

(ii) $a$ and $b$ are called the **limits of integration**; $a$ is the **lower limit**, $b$ is the **upper limit**.

(iii) The function $f$ is the **integrand**; $x$ is the **variable of integration**.

(iv) $dx$ is the **differential of** $x$. It replaces $\Delta x$ in the Riemann sums. If an integrand depends on more than one variable, the differential tells you which one is the variable of integration.

**Example 3**  Show that $f(x) = x^2$ is integrable over the interval $[0, a]$, where $a > 0$, and evaluate $\int_0^a x^2\,dx$.

**Solution**  We evaluate the limits as $n \to \infty$ of the lower and upper sums of $f$ over $[0, a]$ obtained in Example 2 above.

$$\lim_{n\to\infty} L(f, P_n) = \lim_{n\to\infty} \frac{(n-1)(2n-1)a^3}{6n^2} = \frac{a^3}{3}$$

$$\lim_{n\to\infty} U(f, P_n) = \lim_{n\to\infty} \frac{(n+1)(2n+1)a^3}{6n^2} = \frac{a^3}{3}$$

If $L(f, P_n) \le I \le U(f, P_n)$, we must have $I = a^3/3$. Thus $f(x) = x^2$ is integrable over $[0, a]$, and

$$\int_0^a f(x)\,dx = \int_0^a x^2\,dx = \frac{a^3}{3}.$$

For all partitions $P$ of $[a, b]$ we have

$$L(f, P) \le \int_a^b f(x)\,dx \le U(f, P).$$

If $f(x) \ge 0$ on $[a, b]$, then the area of the region $R$ bounded by the graph of $y = f(x)$, the $x$-axis, and the lines $x = a$ and $x = b$ is $A$ square units, where $A = \int_a^b f(x)\,dx$. If $f(x) \le 0$ on $[a, b]$, the area of $R$ is $-\int_a^b f(x)\,dx$ square units. For general $f$, $\int_a^b f(x)\,dx$ is the area of that part of $R$ lying above the $x$-axis minus the area of that part lying below the $x$-axis. (See Figure 5.12.) You can think of $\int_a^b f(x)\,dx$ as a "sum" of "areas" of infinitely many rectangles with heights $f(x)$ and "infinitesimally small widths" $dx$; it is a limit of the upper and lower Riemann sums.

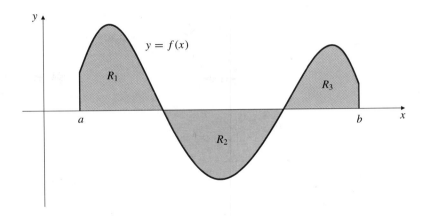

**Figure 5.12**    $\displaystyle\int_a^b f(x)\,dx$ equals
area $R_1$ − area $R_2$ + area $R_3$

## General Riemann Sums

Let $P = \{x_0, x_1, x_2, \ldots, x_n\}$, where $a = x_0 < x_1 < x_2 < \cdots < x_n = b$, be a partition of $[a, b]$ having norm $\|P\| = \max_{1\le i\le n} \Delta x_i$. In each subinterval $[x_{i-1}, x_i]$ of $P$ pick a point $c_i$ (called a *tag*). Let $c = (c_1, c_2, \ldots, c_n)$ denote the set of these tags. The sum

$$R(f, P, c) = \sum_{i=1}^{n} f(c_i)\,\Delta x_i$$
$$= f(c_1)\,\Delta x_1 + f(c_2)\,\Delta x_2 + f(c_3)\,\Delta x_3 + \cdots + f(c_n)\,\Delta x_n$$

is called the **Riemann sum** of $f$ on $[a, b]$ corresponding to partition $P$ and tags $c$.

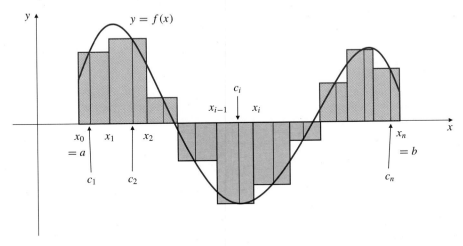

**Figure 5.13**    The Riemann sum $R(f, P)$ is the sum of areas of the rectangles shaded in colour minus the sum of the areas of the rectangles shaded in grey

Note in Figure 5.13 that $R(f, P, c)$ is a sum of *signed* areas of rectangles between the $x$-axis and the curve $y = f(x)$. For any choice of the tags $c$, the Riemann sum $R(f, P, c)$ satisfies

$$L(f, P) \le R(f, P, c) \le U(f, P).$$

Therefore, if $f$ is integrable on $[a, b]$, then its integral is the limit of such Riemann sums, where the limit is taken as the number $n(P)$ of subintervals of $P$ increases to infinity in such a way the lengths of all the subintervals approach zero. That is,

$$\lim_{\substack{n(P)\to\infty \\ \|P\|\to 0}} R(f, P, c) = \int_a^b f(x)\, dx.$$

As we will see in Chapter 7, many applications of integration depend on recognizing that a limit of Riemann sums is a definite integral.

**THEOREM 2**

If $f$ is continuous on $[a, b]$, then $f$ is integrable on $[a, b]$.

As noted above, we cannot prove this theorem in its full generality now. The proof makes use of the completeness property of the real numbers, and is given in Appendix 4. We can, however, make the following observation. In order to prove that $f$ is integrable on $[a, b]$, it is sufficient that, for any given positive number $\epsilon$, we should be able to find a partition $P$ of $[a, b]$ for which $U(f, P) - L(f, P) < \epsilon$. This condition prevents there being more than one number $I$ that is both greater than every lower sum and less than every upper sum. It is not difficult to find such a partition if the function $f$ is nondecreasing (or if it is nonincreasing) on $[a, b]$. (See Exercise 17 at the end of this section.) Therefore nondecreasing and nonincreasing continuous functions are integrable; so, therefore, is any continuous function that is the sum of a nondecreasing and a nonincreasing function. This class of functions includes any continuous functions we are likely to encounter in concrete applications of calculus but, unfortunately, does not include all continuous functions.

In Section 5.4 we will extend the definition of the definite integral to certain kinds of functions that are not continuous.

**Example 4**   Express the limit $\displaystyle\lim_{n\to\infty} \sum_{i=1}^{n} \frac{2}{n}\left(1 + \frac{2i - 1}{n}\right)^{1/3}$ as a definite integral.

**Solution**   We want to interpret the sum as a Riemann sum for $f(x) = (1 + x)^{1/3}$. The factor $2/n$ suggests that the interval of integration has length 2 and is partitioned into $n$ equal subintervals, each of length $2/n$. Let $c_i = (2i - 1)/n$ for $i = 1, 2, 3, \ldots, n$. As $n \to \infty$, $c_1 = 1/n \to 0$ and $c_n = (2n - 1)/n \to 2$. Thus the interval is $[0, 2]$, and the points of the partition are $x_i = 2i/n$. Observe that $x_{i-1} = (2i - 2)/n < c_i < 2i/n = x_i$ for each $i$, so that the sum is indeed a Riemann sum for $f(x)$ over $[0, 2]$. Since $f$ is continuous on that interval, it is integrable there, and

$$\lim_{n\to\infty} \sum_{i=1}^{n} \frac{2}{n}\left(1 + \frac{2i - 1}{n}\right)^{1/3} = \int_0^2 (1 + x)^{1/3}\, dx.$$

## Exercises 5.3

In Exercises 1–6, let $P_n$ denote the partition of the given interval $[a, b]$ into $n$ subintervals of equal length $\Delta x_i = (b - a)/n$. Evaluate $L(f, P_n)$ and $U(f, P_n)$ for the given functions $f$ and the given values of $n$.

1. $f(x) = x$ on $[0, 2]$, with $n = 8$
2. $f(x) = x^2$ on $[0, 4]$, with $n = 4$
3. $f(x) = e^x$ on $[-2, 2]$, with $n = 4$

**4.** $f(x) = \ln x$ on $[1, 2]$, with $n = 5$

**5.** $f(x) = \sin x$ on $[0, \pi]$, with $n = 6$

**6.** $f(x) = \cos x$ on $[0, 2\pi]$, with $n = 4$

In Exercises 7–10, calculate $L(f, P_n)$ and $U(f, P_n)$ for the given function $f$ over the given interval $[a, b]$, where $P_n$ is the partition of the interval into $n$ subintervals of equal length $\Delta x = (b - a)/n$. Show that

$$\lim_{n \to \infty} L(f, P_n) = \lim_{n \to \infty} U(f, P_n).$$

Hence $f$ is integrable on $[a, b]$. (Why?) What is $\int_a^b f(x)\,dx$?

**7.** $f(x) = x$, $[a, b] = [0, 1]$

**8.** $f(x) = 1 - x$, $[a, b] = [0, 2]$

**9.** $f(x) = x^3$, $[a, b] = [0, 1]$

**10.** $f(x) = e^x$, $[a, b] = [0, 3]$

In Exercises 11–16, express the given limit as a definite integral.

**11.** $\lim\limits_{n \to \infty} \sum\limits_{i=1}^{n} \frac{1}{n} \sqrt{\frac{i}{n}}$

**12.** $\lim\limits_{n \to \infty} \sum\limits_{i=1}^{n} \frac{1}{n} \sqrt{\frac{i - 1}{n}}$

**13.** $\lim\limits_{n \to \infty} \sum\limits_{i=1}^{n} \frac{\pi}{n} \sin\left(\frac{\pi i}{n}\right)$

**14.** $\lim\limits_{n \to \infty} \sum\limits_{i=1}^{n} \frac{2}{n} \ln\left(1 + \frac{2i}{n}\right)$

**\* 15.** $\lim\limits_{n \to \infty} \sum\limits_{i=1}^{n} \frac{1}{n} \tan^{-1}\left(\frac{2i - 1}{2n}\right)$

**\* 16.** $\lim\limits_{n \to \infty} \sum\limits_{i=1}^{n} \frac{n}{n^2 + i^2}$

**\* 17.** If $f$ is continuous and nondecreasing on $[a, b]$, and $P_n$ is the partition of $[a, b]$ into $n$ subintervals of equal length $(\Delta x_i = (b - a)/n$ for $1 \le i \le n)$, show that

$$U(f, P_n) - L(f, P_n) = \frac{(b - a)\big(f(b) - f(a)\big)}{n}.$$

Since we can make the right side as small as we please by choosing $n$ large enough, $f$ must be integrable on $[a, b]$.

**\* 18.** Let $P = \{a = x_0 < x_1 < x_2 < \cdots < x_n = b\}$ be a partition of $[a, b]$ and let $P'$ be a refinement of $P$ having one more point, $x'$, satisfying, say, $x_{i-1} < x' < x_i$ for some $i$ between 1 and $n$. Show that $L(f, P) \le L(f, P') \le U(f, P') \le U(f, P)$ for any continuous function $f$. (*Hint:* consider the maximum and minimum values of $f$ on the intervals $[x_{i-1}, x_i]$, $[x_{i-1}, x']$ and $[x', x_i]$.) Hence deduce that $L(f, P) \le L(f, P'') \le U(f, P'') \le U(f, P)$ if $P''$ is *any* refinement of $P$.

# 5.4 Properties of the Definite Integral

It is convenient to extend the definition of the definite integral $\int_a^b f(x)\,dx$ to allow $a = b$ and $a > b$ as well as $a < b$. The extension still involves partitions $P$ having $x_0 = a$ and $x_n = b$ with intermediate points occurring in order between these end points, so that if $a = b$, then we must have $\Delta x_i = 0$ for every $i$, and hence the integral is zero. If $a > b$ we have $\Delta x_i < 0$ for each $i$, so the integral will be negative for positive functions $f$ and vice versa.

Some of the most important properties of the definite integral are summarized in the following theorem.

**THEOREM 3**

Let $f$ and $g$ be integrable on an interval containing the points $a$, $b$, and $c$. Then

(a) An integral over an interval of zero length is zero.

$$\int_a^a f(x)\,dx = 0.$$

(b) Reversing the limits of integration changes the sign of the integral.

$$\int_b^a f(x)\,dx = -\int_a^b f(x)\,dx.$$

(c) An integral depends linearly on the integrand. If $A$ and $B$ are constants, then

$$\int_a^b \big(Af(x) + Bg(x)\big)\,dx = A\int_a^b f(x)\,dx + B\int_a^b g(x)\,dx.$$

(d) An integral depends additively on the interval of integration.

$$\int_a^b f(x)\,dx + \int_b^c f(x)\,dx = \int_a^c f(x)\,dx.$$

(e) If $a \le b$ and $f(x) \le g(x)$ for $a \le x \le b$, then

$$\int_a^b f(x)\,dx \le \int_a^b g(x)\,dx.$$

(f) The **triangle inequality** for definite integrals. If $a \le b$, then

$$\left| \int_a^b f(x)\,dx \right| \le \int_a^b |f(x)|\,dx.$$

(g) The integral of an odd function over an interval symmetric about zero is zero. If $f$ is an odd function $(f(-x) = -f(x))$, then

$$\int_{-a}^a f(x)\,dx = 0.$$

(h) The integral of an even function over an interval symmetric about zero is twice the integral over the positive half of the interval. If $f$ is an even function $(f(-x) = f(x))$, then

$$\int_{-a}^a f(x)\,dx = 2\int_0^a f(x)\,dx.$$

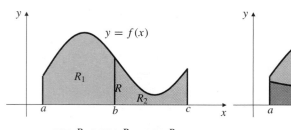

**Figure 5.14**

(a) Property (d) of Theorem 3

(b) Property (e) of Theorem 3

area $R_1$ + area $R_2$ = area $R$

$$\int_a^b f(x)\,dx + \int_b^c f(x)\,dx = \int_a^c f(x)\,dx$$

(a)

area $S \le$ area $R$

$$\int_a^b f(x)\,dx \le \int_a^b g(x)\,dx$$

(b)

All these properties can be deduced from the definition of the definite integral. Most of them should appear intuitively reasonable if you regard the integrals as representing (signed) areas. For instance, properties (d) and (e) are, respectively, properties (v) and (iv) of areas mentioned in the first paragraph of Section 5.2. (See Figure 5.14.) Property (f) is a generalization of the triangle inequality for numbers:

$$|x + y| \le |x| + |y|, \quad \text{or more generally}, \quad \left| \sum_{i=1}^{n} x_i \right| \le \sum_{i=1}^{n} |x_i|.$$

It follows from property (e), (assuming that $|f|$ is integrable on $[a, b]$), since $-|f(x)| \le f(x) \le |f(x)|$. The symmetry properties (g) and (h), which are illustrated in Figure 5.15, are particularly useful and should always be kept in mind when evaluating definite integrals as they can save much unnecessary work.

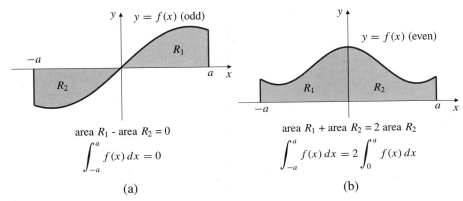

**Figure 5.15**

(a) Property (g) of Theorem 3

(b) Property (h) of Theorem 3

area $R_1$ - area $R_2 = 0$

$$\int_{-a}^{a} f(x)\, dx = 0$$

(a)

area $R_1$ + area $R_2 = 2$ area $R_2$

$$\int_{-a}^{a} f(x)\, dx = 2 \int_{0}^{a} f(x)\, dx$$

(b)

As yet we have no easy method for evaluating definite integrals. However, some such integrals can be simplified by using various properties in Theorem 3, and others can be interpreted as known areas.

**Example 1**   Evaluate

(a) $\displaystyle\int_{-2}^{2} (2 + 5x)\, dx$,   (b) $\displaystyle\int_{0}^{3} (2 + x)\, dx$,   and   (c) $\displaystyle\int_{-3}^{3} \sqrt{9 - x^2}\, dx$.

**Solution**   See Figures 5.16–5.18.

(a) By the linearity property (c), $\int_{-2}^{2}(2+5x)\, dx = \int_{-2}^{2} 2\, dx + 5 \int_{-2}^{2} x\, dx$. The first integral on the right represents the area of a rectangle of width 4 and height 2 (Figure 5.16), so it has value 8. The second integral on the right is 0 because its integrand is odd and the interval is symmetric about 0. Thus

$$\int_{-2}^{2} (2 + 5x)\, dx = 8 + 0 = 8.$$

(b) $\int_{0}^{3}(2 + x)\, dx$ represents the area of the trapezoid in Figure 5.17. Adding the areas of the rectangle and triangle comprising this trapezoid, we get

$$\int_{0}^{3} (2 + x)\, dx = (3)(2) + \frac{1}{2}(3)(3) = \frac{21}{2}.$$

While areas are measured in squared units of length, definite integrals are numbers, and have no units. Even when you use an area to find an integral, do not quote units for the integral.

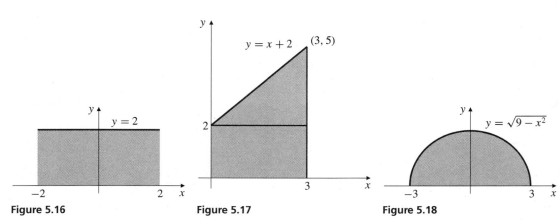

**Figure 5.16**  **Figure 5.17**  **Figure 5.18**

(c) $\int_{-3}^{3} \sqrt{9 - x^2}\,dx$ represents the area of a semicircle of radius 3 (Figure 5.18), so

$$\int_{-3}^{3} \sqrt{9 - x^2}\,dx = \frac{1}{2}\,\pi (3^2) = \frac{9\pi}{2}.$$

## A Mean-Value Theorem for Integrals

Let $f$ be a function continuous on the interval $[a, b]$. Then $f$ assumes a minimum value $m$ and a maximum value $M$ on the interval, say at points $x = l$ and $x = u$, respectively:

$$m = f(l) \le f(x) \le f(u) = M \qquad \text{for all } x \text{ in } [a, b].$$

For the 2-point partition $P$ of $[a, b]$ having $x_0 = a$ and $x_1 = b$, we have

$$m(b - a) = L(f, P) \le \int_{a}^{b} f(x)\,dx \le U(f, P) = M(b - a).$$

Therefore

$$f(l) = m \le \frac{1}{b - a} \int_{a}^{b} f(x)\,dx \le M = f(u).$$

By the Intermediate-Value Theorem, $f(x)$ must take on every value between the two values $f(l)$ and $f(u)$ at some point between $l$ and $u$ (Figure 5.19). Hence there is a number $c$ between $l$ and $u$ such that

$$f(c) = \frac{1}{b - a} \int_{a}^{b} f(x)\,dx,$$

that is, $\int_{a}^{b} f(x)\,dx$ is equal to the area $(b - a)f(c)$ of a rectangle with base width $b - a$ and height $f(c)$ for some $c$ between $a$ and $b$. This is the Mean-Value Theorem for integrals.

**THEOREM 4**

### The Mean-Value Theorem for integrals

If $f$ is continuous on $[a, b]$, then there exists a point $c$ in $[a, b]$ such that

$$\int_{a}^{b} f(x)\,dx = (b - a)f(c).$$

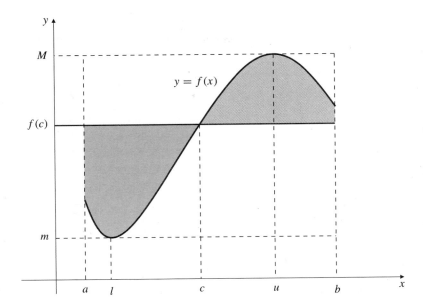

**Figure 5.19**  Half of the area between $y = f(x)$ and the horizontal line $y = f(c)$ lies above the line and the other half lies below the line

Observe in Figure 5.19 that the area below the curve $y = f(x)$ and above the line $y = f(c)$ is equal to the area above $y = f(x)$ and below $y = f(c)$. In this sense, $f(c)$ is the average value of the function $f(x)$ on the interval $[a, b]$.

**DEFINITION** ❹

> **Average value of a function**
>
> If $f$ is integrable on $[a, b]$, then the **average value** or **mean value** of $f$ on $[a, b]$, denoted by $\bar{f}$, is
>
> $$\bar{f} = \frac{1}{b - a} \int_a^b f(x)\, dx.$$

**Example 2**  Find the average value of $f(x) = 2x$ on the interval $[1, 5]$.

**Solution**  The average value (see Figure 5.20) is

$$\bar{f} = \frac{1}{5 - 1} \int_1^5 2x\, dx = \frac{1}{4}\left(4 \times 2 + \frac{1}{2}(4 \times 8)\right) = 6.$$

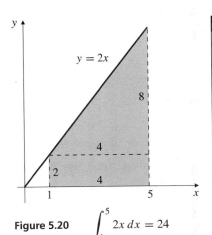

**Figure 5.20**  $\displaystyle\int_1^5 2x\, dx = 24$

## Definite Integrals of Piecewise Continuous Functions

The definition of integrability and the definite integral given above can be extended to a wider class than just continuous functions. One simple but very important extension is to the class of *piecewise continuous functions*.

Consider the graph $y = f(x)$ shown in Figure 5.21(a). Although $f$ is not continuous at all points in $[a, b]$ (it is discontinuous at $c_1$ and $c_2$), clearly the region lying under the graph and above the $x$-axis between $x = a$ and $x = b$ does have an area. We would like to represent this area as

$$\int_a^{c_1} f(x)\,dx + \int_{c_1}^{c_2} f(x)\,dx + \int_{c_2}^b f(x)\,dx.$$

This is reasonable because there are continuous functions on $[a, c_1]$, $[c_1, c_2]$ and $[c_2, b]$ equal to $f(x)$ on the corresponding open intervals, $(a, c_1)$, $(c_1, c_2)$ and $(c_2, b)$.

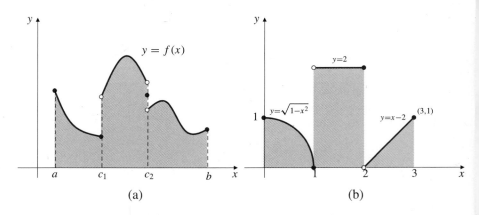

**Figure 5.21**    Two piecewise continuous functions

(a)                    (b)

**DEFINITION 5**

**Piecewise continuous functions**

Let $c_0 < c_1 < c_2 < \cdots < c_n$ be a finite set of points on the real line. A function $f$ defined on $[c_0, c_n]$ except possibly at some of the points $c_i$, $(0 \le i \le n)$, is called **piecewise continuous** on that interval if for each $i$, $(1 \le i \le n)$, there exists a function $F_i$ continuous on the *closed* interval $[c_{i-1}, c_i]$ such that

$$f(x) = F_i(x) \qquad \text{on the } \textit{open} \text{ interval} \quad (c_{i-1}, c_i).$$

In this case, we define the definite integral of $f$ from $c_0$ to $c_n$ to be

$$\int_{c_0}^{c_n} f(x)\,dx = \sum_{i=1}^n \int_{c_{i-1}}^{c_i} F_i(x)\,dx.$$

**Example 3**   Find $\displaystyle\int_0^3 f(x)\,dx$, where $f(x) = \begin{cases} \sqrt{1 - x^2} & \text{if } 0 \le x \le 1 \\ 2 & \text{if } 1 < x \le 2 \\ x - 2 & \text{if } 2 < x \le 3. \end{cases}$

**Solution**   The value of the integral is the sum of the shaded areas in Figure 5.21(b):

$$\int_0^3 f(x)\,dx = \int_0^1 \sqrt{1 - x^2}\,dx + \int_1^2 2\,dx + \int_2^3 (x - 2)\,dx$$

$$= \left(\tfrac{1}{4} \times \pi \times 1^2\right) + (2 \times 1) + \left(\tfrac{1}{2} \times 1 \times 1\right) = \frac{\pi + 10}{4}.$$

# Exercises 5.4

**1.** Simplify $\int_a^b f(x)\,dx + \int_b^c f(x)\,dx + \int_c^a f(x)\,dx$.

**2.** Simplify $\int_0^2 3f(x)\,dx + \int_1^3 3f(x)\,dx - \int_0^3 2f(x)\,dx - \int_1^2 3f(x)\,dx$.

Evaluate the integrals in Exercises 3–16 by using the properties of the definite integral and interpreting integrals as areas.

**3.** $\int_{-2}^2 (x+2)\,dx$

**4.** $\int_0^2 (3x+1)\,dx$

**5.** $\int_a^b x\,dx$

**6.** $\int_{-1}^2 (1-2x)\,dx$

**7.** $\int_{-\sqrt{2}}^{\sqrt{2}} \sqrt{2-t^2}\,dt$

**8.** $\int_{-\sqrt{2}}^0 \sqrt{2-x^2}\,dx$

**9.** $\int_{-\pi}^{\pi} \sin(x^3)\,dx$

**10.** $\int_{-a}^a (a-|s|)\,ds$

**11.** $\int_{-1}^1 (u^5 - 3u^3 + \pi)\,du$

**12.** $\int_0^2 \sqrt{2x-x^2}\,dx$

**13.** $\int_{-4}^4 (e^x - e^{-x})\,dx$

**14.** $\int_{-3}^3 (2+t)\sqrt{9-t^2}\,dt$

**\* 15.** $\int_0^1 \sqrt{4-x^2}\,dx$

**\* 16.** $\int_1^2 \sqrt{4-x^2}\,dx$

Given that $\int_0^a x^2\,dx = \dfrac{a^3}{3}$, evaluate the integrals in Exercises 17–24.

**17.** $\int_0^2 6x^2\,dx$

**18.** $\int_1^2 6x^2\,dx$

**19.** $\int_0^3 (1+x^2)\,dx$

**20.** $\int_2^3 (x^2-4)\,dx$

**21.** $\int_{-2}^2 (4-t^2)\,dt$

**22.** $\int_0^2 (v^2-v)\,dv$

**23.** $\int_0^1 (x^2+\sqrt{1-x^2})\,dx$

**24.** $\int_{-6}^6 x^2(2+\sin x)\,dx$

The definition of $\ln x$ as an area in Section 3.3 implies that

$$\int_1^x \frac{1}{t}\,dt = \ln x$$

for $x > 0$. Use this to evaluate the integrals in Exercises 25–28.

**25.** $\int_1^2 \frac{1}{x}\,dx$

**26.** $\int_2^4 \frac{1}{t}\,dt$

**27.** $\int_{1/3}^1 \frac{1}{t}\,dt$

**28.** $\int_{1/4}^3 \frac{1}{s}\,ds$

Find the average values of the functions in Exercises 29–34 over the given intervals.

**29.** $f(x) = x + 2$ over $[0, 4]$

**30.** $g(x) = x + 2$ over $[a, b]$

**31.** $f(t) = 1 + \sin t$ over $[-\pi, \pi]$

**32.** $k(x) = x^2$ over $[0, 3]$

**33.** $f(x) = \sqrt{4 - x^2}$ over $[0, 2]$

**34.** $g(s) = 1/s$ over $[1/2, 2]$

**Piecewise continuous functions**

**35.** Evaluate $\int_{-1}^2 \operatorname{sgn} x\,dx$. Recall that $\operatorname{sgn} x$ is 1 if $x > 0$ and $-1$ if $x < 0$.

**36.** Find $\int_{-3}^2 f(x)\,dx$ where $f(x) = \begin{cases} 1+x & \text{if } x < 0 \\ 2 & \text{if } x \geq 0. \end{cases}$

**37.** Find $\int_0^2 g(x)\,dx$ where $g(x) = \begin{cases} x^2 & \text{if } 0 \leq x \leq 1 \\ x & \text{if } 1 < x \leq 2. \end{cases}$

**38.** Evaluate $\int_0^3 (2-|x|)\,dx$.

**\* 39.** Evaluate $\int_0^2 \sqrt{4-x^2}\,\operatorname{sgn}(x-1)\,dx$.

**40.** Evaluate $\int_0^{3.5} \lfloor x \rfloor\,dx$, where $\lfloor x \rfloor$ is the greatest integer less than or equal to $x$. (See Example 10 of section P.5.)

Evaluate the integrals in Exercises 41–42 by inspecting the graphs of the integrands.

**41.** $\int_{-3}^4 \left( |x+1| - |x-1| + |x+2| \right) dx$

**42.** $\int_0^3 \dfrac{x^2 - x}{|x-1|}\,dx$

**43.** Find the average value of the function $f(x) = |x+1|\operatorname{sgn} x$ on the interval $[-2, 2]$.

**44.** If $a < b$ and $f$ is continuous on $[a, b]$, show that

$$\int_a^b \left( f(x) - \bar{f} \right) dx = 0.$$

**45.** Suppose that $a < b$ and $f$ is continuous on $[a, b]$. Find the constant $k$ that minimizes the integral $\int_a^b \left( f(x) - k \right)^2 dx$.

## 5.5 The Fundamental Theorem of Calculus

In this section we demonstrate the relationship between the definite integral defined in Section 5.3 and the indefinite integral (or general antiderivative) introduced in Section 2.10. A consequence of this relationship is that we will be able to calculate definite integrals of functions whose antiderivatives we can find.

In Section 3.3 we wanted to find a function whose derivative was $1/x$. We solved this problem by defining the desired function, ($\ln x$), in terms of the area under the graph of $y = 1/x$. This idea motivates, and is a special case of, the following theorem.

**THEOREM 5**

**The Fundamental Theorem of Calculus**

Suppose that the function $f$ is continuous on an interval $I$ containing the point $a$.

PART I. Let the function $F$ be defined on $I$ by

$$F(x) = \int_a^x f(t)\,dt.$$

Then $F$ is differentiable on $I$ and $F'(x) = f(x)$ there. Thus $F$ is an antiderivative of $f$ on $I$:

$$\frac{d}{dx}\int_a^x f(t)\,dt = f(x).$$

PART II. If $G(x)$ is *any* antiderivative of $f(x)$ on $I$, so that $G'(x) = f(x)$ on $I$, then for any $b$ in $I$ we have

$$\int_a^b f(x)\,dx = G(b) - G(a).$$

**PROOF** Using the definition of the derivative, we calculate

$$
\begin{aligned}
F'(x) &= \lim_{h\to 0}\frac{F(x+h) - F(x)}{h} \\
&= \lim_{h\to 0}\frac{1}{h}\left(\int_a^{x+h} f(t)\,dt - \int_a^x f(t)\,dt\right) \\
&= \lim_{h\to 0}\frac{1}{h}\int_x^{x+h} f(t)\,dt \qquad\text{by Theorem 3(d)} \\
&= \lim_{h\to 0}\frac{1}{h}hf(c) \qquad\qquad\text{for some } c = c(h) \text{ (depending on } h) \\
&\qquad\qquad\qquad\qquad\qquad\qquad\text{between } x \text{ and } x + h \text{ (Theorem 4)} \\
&= \lim_{c\to x} f(c) \qquad\qquad\qquad\text{since } c \to x \text{ as } h \to 0 \\
&= f(x) \qquad\qquad\qquad\qquad\text{since } f \text{ is continuous.}
\end{aligned}
$$

Also, if $G'(x) = f(x)$, then $F(x) = G(x) + C$ on $I$ for some constant $C$ (by Theorem 13 of Section 2.6). Hence

$$\int_a^x f(t)\,dt = F(x) = G(x) + C.$$

Let $x = a$ and obtain $0 = G(a) + C$ via Theorem 3(a), so $C = -G(a)$. Now let $x = b$ to get

$$\int_a^b f(t)\, dt = G(b) + C = G(b) - G(a).$$

Of course, we can replace $t$ with $x$ (or any other variable) as the variable of integration on the left-hand side.

**Remark**   You should remember *both* conclusions of the Fundamental Theorem; they are both useful. Part I concerns the derivative of an integral; it tells you how to differentiate a definite integral with respect to its upper limit. Part II concerns the integral of a derivative; it tells you how to evaluate a definite integral if you can find an antiderivative of the integrand.

**DEFINITION  6**

To facilitate the evaluation of definite integrals using the Fundamental Theorem, we define the **evaluation symbol**

$$F(x)\Big|_a^b = F(b) - F(a).$$

Thus

$$\int_a^b f(x)\, dx = \left( \int f(x)\, dx \right)\Big|_a^b,$$

where $\int f(x)\, dx$ denotes the indefinite integral or general antiderivative of $f$. (See Section 2.10.)  When evaluating a definite integral this way, we will omit the constant of integration $(+C)$ from the indefinite integral because it cancels out in the subtraction:

$$(F(x) + C)\Big|_a^b = F(b) + C - (F(a) + C) = F(b) - F(a) = F(x)\Big|_a^b.$$

*Any* antiderivative of $f$ can be used to calculate the definite integral.

**Example 1**   Evaluate   (a) $\displaystyle\int_0^a x^2\, dx$   and   (b) $\displaystyle\int_{-1}^2 (x^2 - 3x + 2)\, dx$.

**Solution**

(a) $\displaystyle\int_0^a x^2\, dx = \frac{1}{3}x^3\Big|_0^a = \frac{1}{3}a^3 - \frac{1}{3}0^3 = \frac{a^3}{3}$   (because $\dfrac{d}{dx}\dfrac{x^3}{3} = x^2$).

(b) $\displaystyle\int_{-1}^2 (x^2 - 3x + 2)\, dx = \left( \frac{1}{3}x^3 - \frac{3}{2}x^2 + 2x \right)\Big|_{-1}^2$

$$= \frac{1}{3}(8) - \frac{3}{2}(4) + 4 - \left( \frac{1}{3}(-1) - \frac{3}{2}(1) + (-2) \right) = \frac{9}{2}.$$

**Example 2**    Find the area $A$ of the plane region lying above the $x$-axis and under the curve $y = 3x - x^2$.

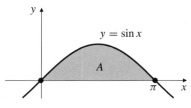

**Figure 5.22**

**Solution**    We need to find the points where the curve $y = 3x - x^2$ meets the $x$-axis. These are solutions of the equation

$$0 = 3x - x^2 = x(3 - x).$$

The only roots are $x = 0$ and $x = 3$. (See Figure 5.22.) Hence the area of the region is given by

$$A = \int_0^3 (3x - x^2)\, dx = \left( \frac{3}{2} x^2 - \frac{1}{3} x^3 \right) \Big|_0^3$$

$$= \frac{27}{2} - \frac{27}{3} - (0 - 0) = \frac{27}{6} = \frac{9}{2} \text{ square units.}$$

**Example 3**    Find the area under the curve $y = \sin x$, above $y = 0$ from $x = 0$ to $x = \pi$.

**Solution**    The required area, illustrated in Figure 5.23, is

$$A = \int_0^\pi \sin x\, dx = -\cos x \Big|_0^\pi = -(-1 - (1)) = 2 \text{ square units.}$$

**Figure 5.23**

Note that while the definite integral is a pure number, an area is a geometric quantity that implicitly involves units. If the units along the $x$- and $y$-axes are, for example, metres, the area should be quoted in square metres ($m^2$). If units of length along the $x$-axis and $y$-axis are not specified, areas should be quoted in square units.

**Example 4**    Find the area of the region $R$ lying above the line $y = 1$ and below the curve $y = 5/(x^2 + 1)$.

**Solution**    The region $R$ is shaded in Figure 5.24. To find the intersections of $y = 1$ and $y = 5/(x^2 + 1)$, we must solve these equations simultaneously:

$$1 = \frac{5}{x^2 + 1}$$

so $x^2 + 1 = 5$, $x^2 = 4$, and $x = \pm 2$.

The area $A$ of the region $R$ is the area under the curve $y = 5/(x^2 + 1)$ and above the $x$-axis between $x = -2$ and $x = 2$, minus the area of a rectangle of width 4 and height 1.

$$A = \int_{-2}^2 \frac{5}{x^2 + 1}\, dx - 4 = 2 \int_0^2 \frac{5}{x^2 + 1}\, dx - 4$$

$$= 10 \tan^{-1} x \Big|_0^2 - 4 = 10 \tan^{-1} 2 - 4 \text{ square units.}$$

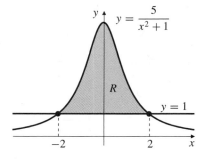

**Figure 5.24**

Observe the use of symmetry to replace the lower limit of integration by 0. It is easier to substitute 0 into the antiderivative than $-2$.

**Example 5** Find the average value of $f(x) = e^{-x} + \cos x$ on the interval $[-\pi/2, 0]$.

**Solution** The average value is

$$\bar{f} = \frac{1}{0 - \left(-\dfrac{\pi}{2}\right)} \int_{-(\pi/2)}^{0} (e^{-x} + \cos x)\, dx$$

$$= \frac{2}{\pi} \left(-e^{-x} + \sin x\right)\Big|_{-(\pi/2)}^{0}$$

$$= \frac{2}{\pi} \left(-1 + 0 + e^{\pi/2} - (-1)\right) = \frac{2}{\pi}\, e^{\pi/2}.$$

Beware of integrals of the form $\int_a^b f(x)\, dx$ where $f$ is not continuous at *all* points in the interval $[a, b]$. The Fundamental Theorem does not apply in such cases.

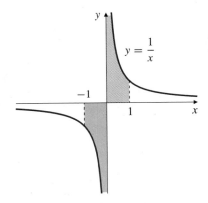

**Figure 5.25**

**Example 6** We know that $\dfrac{d}{dx} \ln|x| = \dfrac{1}{x}$ if $x \neq 0$. It is *incorrect*, however, to state that

$$\int_{-1}^{1} \frac{dx}{x} = \ln|x|\,\Big|_{-1}^{1} = 0 - 0 = 0,$$

even though $1/x$ is an odd function. In fact, $1/x$ is discontinuous and unbounded on $[-1, 1]$, and it is not integrable on $[-1, 0]$ or $[0, 1]$ (Figure 5.25). Observe that

$$\lim_{c \to 0+} \int_c^1 \frac{1}{x}\, dx = \lim_{c \to 0+} -\ln c = \infty,$$

so both shaded regions in Figure 5.25 have infinite area. Integrals of this type are called **improper integrals**. We deal with them in Section 6.5.

We now give some examples illustrating the first conclusion of the Fundamental Theorem.

**Example 7** Find the derivatives of the following functions:

(a) $F(x) = \displaystyle\int_x^3 e^{-t^2}\, dt,$ (b) $G(x) = x^2 \displaystyle\int_{-4}^{5x} e^{-t^2}\, dt,$ (c) $H(x) = \displaystyle\int_{x^2}^{x^3} e^{-t^2}\, dt.$

**Solution** The solutions involve applying the first conclusion of the Fundamental Theorem together with other differentiation rules.

(a) Observe that $F(x) = -\int_3^x e^{-t^2}\, dt$ (by Theorem 3(b)). Therefore, by the Fundamental Theorem, $F'(x) = -e^{-x^2}$.

(b) By the Product Rule and the Chain Rule,

$$G'(x) = 2x \int_{-4}^{5x} e^{-t^2}\, dt + x^2 \frac{d}{dx} \int_{-4}^{5x} e^{-t^2}\, dt$$

$$= 2x \int_{-4}^{5x} e^{-t^2}\, dt + x^2 e^{-(5x)^2}(5)$$

$$= 2x \int_{-4}^{5x} e^{-t^2}\, dt + 5x^2 e^{-25x^2}.$$

(c) Split the integral into a difference of integrals in each of which the variable $x$ appears only in the upper limit.

$$H(x) = \int_0^{x^3} e^{-t^2}\, dt - \int_0^{x^2} e^{-t^2}\, dt$$

$$H'(x) = e^{-(x^3)^2}(3x^2) - e^{-(x^2)^2}(2x)$$

$$= 3x^2 e^{-x^6} - 2x e^{-x^4}.$$

Parts (b) and (c) of Example 7 are examples of the following formulas that build the Chain Rule into the first conclusion of the Fundamental Theorem.

$$\frac{d}{dx} \int_a^{g(x)} f(t)\, dt = f\big(g(x)\big)\, g'(x)$$

$$\frac{d}{dx} \int_{h(x)}^{g(x)} f(t)\, dt = f\big(g(x)\big)\, g'(x) - f\big(h(x)\big)\, h'(x)$$

**Example 8**  Solve the **integral equation** $f(x) = 2 + 3 \int_4^x f(t)\, dt$.

**Solution**  Differentiate the integral equation to get $f'(x) = 3f(x)$, the DE for exponential growth, having solution $f(x) = Ce^{3x}$. Now put $x = 4$ into the integral equation to get $f(4) = 2$. Hence $2 = Ce^{12}$, so $C = 2e^{-12}$. Therefore the integral equation has solution $f(x) = 2e^{3x-12}$.

We conclude with an example showing how the Fundamental Theorem can be used to evaluate limits of Riemann sums.

**Figure 5.26**

**Example 9**  Evaluate $\displaystyle\lim_{n\to\infty} \frac{1}{n} \sum_{j=1}^n \cos\left(\frac{j\pi}{2n}\right)$.

**Solution**  The sum involves values of $\cos x$ at the right endpoints of the $n$ subintervals of the partition

$$0, \quad \frac{\pi}{2n}, \quad \frac{2\pi}{2n}, \quad \frac{3\pi}{2n}, \quad \ldots, \quad \frac{n\pi}{2n}$$

of the interval $[0, \pi/2]$. Since each of the subintervals of this partition has length $\pi/(2n)$, and since $\cos x$ is continuous on $[0, \pi/2]$, we have, expressing the limit of a Riemann sum as an integral (see Figure 5.26),

$$\lim_{n\to\infty} \frac{\pi}{2n} \sum_{j=1}^n \cos\left(\frac{j\pi}{2n}\right) = \int_0^{\pi/2} \cos x\, dx = \sin x \Big|_0^{\pi/2} = 1 - 0 = 1.$$

The given sum differs from the Riemann sum above only in that is missing the factor $\pi/2$. Thus

$$\lim_{n\to\infty} \frac{1}{n} \sum_{j=1}^n \cos\left(\frac{j\pi}{2n}\right) = \frac{2}{\pi}.$$

## Exercises 5.5

Evaluate the definite integrals in Exercises 1–22.

**1.** $\displaystyle\int_0^2 x^3\,dx$

**2.** $\displaystyle\int_0^4 \sqrt{x}\,dx$

**3.** $\displaystyle\int_{1/2}^1 \frac{1}{x^2}\,dx$

**4.** $\displaystyle\int_4^9 s^{-1/2}\,ds$

**5.** $\displaystyle\int_0^1 (x^5 + x^3)\,dx$

**6.** $\displaystyle\int_{-2}^{-1} \left(\frac{1}{x^2} - \frac{1}{x^3}\right)dx$

**7.** $\displaystyle\int_{-1}^2 (3x^2 - 4x + 2)\,dx$

**8.** $\displaystyle\int_1^2 \left(\frac{2}{x^3} - \frac{x^3}{2}\right)dx$

**9.** $\displaystyle\int_{-2}^2 (x^2 + 3)^2\,dx$

**10.** $\displaystyle\int_4^9 \left(\sqrt{x} - \frac{1}{\sqrt{x}}\right)dx$

**11.** $\displaystyle\int_{-\pi/4}^{-\pi/6} \cos x\,dx$

**12.** $\displaystyle\int_0^{\pi/3} \sec^2\theta\,d\theta$

**13.** $\displaystyle\int_{\pi/4}^{\pi/3} \sin\theta\,d\theta$

**14.** $\displaystyle\int_0^{2\pi} (1 + \sin u)\,du$

**15.** $\displaystyle\int_{-\pi}^{\pi} e^x\,dx$

**16.** $\displaystyle\int_{-2}^2 (e^x - e^{-x})\,dx$

**17.** $\displaystyle\int_0^e a^x\,dx \quad (a > 0)$

**18.** $\displaystyle\int_{-1}^1 2^x\,dx$

**19.** $\displaystyle\int_{-1}^1 \frac{dx}{1 + x^2}$

**20.** $\displaystyle\int_0^{1/2} \frac{dx}{\sqrt{1 - x^2}}$

**∗ 21.** $\displaystyle\int_{-1}^1 \frac{dx}{\sqrt{4 - x^2}}$

**∗ 22.** $\displaystyle\int_{-2}^0 \frac{dx}{4 + x^2}$

Find the area of the region $R$ specified in Exercises 23–34. It is helpful to make a sketch of the region.

**23.** Bounded by $y = x^4$, $y = 0$, $x = 0$, and $x = 1$

**24.** Bounded by $y = 1/x$, $y = 0$, $x = e$, and $x = e^2$

**25.** Above $y = x^2 - 4x$ and below the $x$-axis

**26.** Bounded by $y = 5 - 2x - 3x^2$, $y = 0$, $x = -1$, and $x = 1$

**27.** Bounded by $y = x^2 - 3x + 3$ and $y = 1$

**28.** Below $y = \sqrt{x}$ and above $y = \dfrac{x}{2}$

**29.** Above $y = x^2$ and to the right of $x = y^2$

**30.** Above $y = |x|$ and below $y = 12 - x^2$

**31.** Bounded by $y = x^{1/3} - x^{1/2}$, $y = 0$, $x = 0$, and $x = 1$

**32.** Under $y = e^{-x}$ and above $y = 0$ from $x = -a$ to $x = 0$

**33.** Below $y = 1 - \cos x$ and above $y = 0$ between two consecutive intersections of these graphs

**34.** Below $y = x^{-1/3}$ and above $y = 0$ from $x = 1$ to $x = 27$

Find the integrals of the piecewise continuous functions in Exercises 35–36.

**35.** $\displaystyle\int_0^{3\pi/2} |\cos x|\,dx$

**36.** $\displaystyle\int_1^3 \frac{\text{sgn}\,(x - 2)}{x^2}\,dx$

In Exercises 37–40, find the average values of the given functions over the intervals specified.

**37.** $f(x) = 1 + x + x^2 + x^3$ over $[0, 2]$

**38.** $f(x) = e^{3x}$ over $[-2, 2]$

**39.** $f(x) = 2^x$ over $[0, 1/\ln 2]$

**40.** $g(t) = \begin{cases} 0 & \text{if } 0 \le t \le 1 \\ 1 & \text{if } 1 < t \le 3 \end{cases}$ over $[0, 3]$

Find the indicated derivatives in Exercises 41–48.

**41.** $\displaystyle\frac{d}{dx}\int_2^x \frac{\sin t}{t}\,dt$

**42.** $\displaystyle\frac{d}{dt}\int_t^3 \frac{\sin x}{x}\,dx$

**43.** $\displaystyle\frac{d}{dx}\int_{x^2}^0 \frac{\sin t}{t}\,dt$

**44.** $\displaystyle\frac{d}{dx}x^2 \int_0^{x^2} \frac{\sin u}{u}\,du$

**45.** $\displaystyle\frac{d}{dt}\int_{-\pi}^t \frac{\cos y}{1 + y^2}\,dy$

**46.** $\displaystyle\frac{d}{d\theta}\int_{\sin\theta}^{\cos\theta} \frac{1}{1 - x^2}\,dx$

**47.** $\displaystyle\frac{d}{dx}F(\sqrt{x})$, if $F(t) = \int_0^t \cos(x^2)\,dx$ $\quad F'(t) = \cos(t^2)$

**48.** $H'(2)$, if $H(x) = 3x \displaystyle\int_4^{x^2} e^{-\sqrt{t}}\,dt$

**49.** Solve the integral equation $f(x) = \pi\left(1 + \displaystyle\int_1^x f(t)\,dt\right)$.

**50.** Solve the integral equation $f(x) = 1 - \displaystyle\int_0^x f(t)\,dt$.

**∗ 51.** Criticize the following erroneous calculation:

$$\int_{-1}^1 \frac{dx}{x^2} = -\frac{1}{x}\bigg|_{-1}^1 = -1 + \frac{1}{-1} = -2.$$

Exactly where did the error occur? Why is $-2$ an unreasonable value for the integral?

**∗ 52.** Use a definite integral to define a function $F(x)$ having derivative $\dfrac{\sin x}{1 + x^2}$ for all $x$ and satisfying $F(17) = 0$.

**∗ 53.** Does the function $F(x) = \displaystyle\int_0^{2x - x^2} \cos\left(\frac{1}{1 + t^2}\right)dt$ have a maximum or a minimum value? Justify your answer.

Evaluate the limits in Exercises 54–56.

**∗ 54.** $\displaystyle\lim_{n\to\infty} \frac{1}{n}\left(\left(1 + \frac{1}{n}\right)^5 + \left(1 + \frac{2}{n}\right)^5 + \cdots + \left(1 + \frac{n}{n}\right)^5\right)$.

**∗ 55.** $\displaystyle\lim_{n\to\infty} \frac{\pi}{n}\left(\sin\frac{\pi}{n} + \sin\frac{2\pi}{n} + \sin\frac{3\pi}{n} + \cdots + \sin\frac{n\pi}{n}\right)$.

**∗ 56.** $\displaystyle\lim_{n\to\infty} \left(\frac{n}{n^2 + 1} + \frac{n}{n^2 + 4} + \frac{n}{n^2 + 9} + \cdots + \frac{n}{2n^2}\right)$.

## 5.6 The Method of Substitution

As we have seen, the evaluation of definite integrals is most easily carried out if we can antidifferentiate the integrand. In this section and Sections 6.1–6.4 we develop some *techniques of integration*, that is, methods for finding antiderivatives of functions. Although the techniques we develop can be used for a large class of functions, they will not work for all functions we might want to integrate. If a definite integral involves an integrand whose antiderivative is either impossible or very difficult to find, we may wish, instead, to approximate the definite integral by numerical means. Techniques for doing that will be presented in Sections 6.6–6.9.

Let us begin by assembling a table of some known indefinite integrals. These results have all emerged during our development of differentiation formulas for elementary functions. You should *memorize* them.

**Some elementary integrals**

1. $\displaystyle\int 1\,dx = x + C$

2. $\displaystyle\int x\,dx = \frac{1}{2}x^2 + C$

3. $\displaystyle\int x^2\,dx = \frac{1}{3}x^3 + C$

4. $\displaystyle\int \frac{1}{x^2}\,dx = -\frac{1}{x} + C$

5. $\displaystyle\int \sqrt{x}\,dx = \frac{2}{3}x^{3/2} + C$

6. $\displaystyle\int \frac{1}{\sqrt{x}}\,dx = 2\sqrt{x} + C$

7. $\displaystyle\int x^r\,dx = \frac{1}{r+1}x^{r+1} + C \quad (r \neq -1)$

8. $\displaystyle\int \frac{1}{x}\,dx = \ln|x| + C$

9. $\displaystyle\int \sin ax\,dx = -\frac{1}{a}\cos ax + C$

10. $\displaystyle\int \cos ax\,dx = \frac{1}{a}\sin ax + C$

11. $\displaystyle\int \sec^2 ax\,dx = \frac{1}{a}\tan ax + C$

12. $\displaystyle\int \csc^2 ax\,dx = -\frac{1}{a}\cot ax + C$

13. $\displaystyle\int \sec ax \tan ax\,dx = \frac{1}{a}\sec ax + C$

14. $\displaystyle\int \csc ax \cot ax\,dx = -\frac{1}{a}\csc ax + C$

15. $\displaystyle\int \frac{1}{\sqrt{a^2 - x^2}}\,dx = \sin^{-1}\frac{x}{a} + C \quad (a > 0)$

16. $\displaystyle\int \frac{1}{a^2 + x^2}\,dx = \frac{1}{a}\tan^{-1}\frac{x}{a} + C$

17. $\displaystyle\int e^{ax}\,dx = \frac{1}{a}e^{ax} + C$

18. $\displaystyle\int b^{ax}\,dx = \frac{1}{a\ln b}b^{ax} + C$

19. $\displaystyle\int \cosh ax\,dx = \frac{1}{a}\sinh ax + C$

20. $\displaystyle\int \sinh ax\,dx = \frac{1}{a}\cosh ax + C$

Note that formulas 1–6 are special cases of formula 7, which holds on any interval where $x^r$ makes sense. The linearity formula

$$\int (A\,f(x) + B\,g(x))\,dx = A\int f(x)\,dx + B\int g(x)\,dx$$

makes it possible to integrate sums and constant multiples of functions.

**Example 1** (Combining elementary integrals)

(a) $\displaystyle\int (x^4 - 3x^3 + 8x^2 - 6x - 7)\,dx = \frac{x^5}{5} - \frac{3x^4}{4} + \frac{8x^3}{3} - 3x^2 - 7x + C$

(b) $\displaystyle\int \left(5x^{3/5} - \frac{3}{2+x^2}\right) dx = \frac{25}{8}x^{8/5} - \frac{3}{\sqrt{2}} \tan^{-1} \frac{x}{\sqrt{2}} + C$

(c) $\displaystyle\int (4\cos 5x - 5\sin 3x)\, dx = \frac{4}{5}\sin 5x + \frac{5}{3}\cos 3x + C$

(d) $\displaystyle\int \left(\frac{1}{\pi x} + a^{\pi x}\right) dx = \frac{1}{\pi}\ln |x| + \frac{1}{\pi \ln a}a^{\pi x} + C, \quad (a > 0).$

∎

Sometimes it is necessary to manipulate an integrand so that the method can be applied.

**Example 2**
$$\int \frac{(x+1)^3}{x}\, dx = \int \frac{x^3 + 3x^2 + 3x + 1}{x}\, dx$$
$$= \int \left(x^2 + 3x + 3 + \frac{1}{x}\right) dx$$
$$= \frac{1}{3}x^3 + \frac{3}{2}x^2 + 3x + \ln |x| + C.$$

When an integral cannot be evaluated by inspection, as those in Examples 1–2 can, we require one or more special techniques. The most important of these techniques is the **method of substitution**, the integral version of the Chain Rule. If we rewrite the Chain Rule, $\frac{d}{dx} f(g(x)) = f'(g(x))\, g'(x)$, in integral form, we obtain

$$\int f'(g(x))\, g'(x)\, dx = f(g(x)) + C.$$

Observe that the following formalism would produce this latter formula even if we did not already know it was true:

Let $u = g(x)$. Then $du/dx = g'(x)$, or in differential form, $du = g'(x)\, dx$. Thus

$$\int f'(g(x))\, g'(x)\, dx = \int f'(u)\, du = f(u) + C = f(g(x)) + C.$$

**Example 3** (**Examples of substitution**) Find the indefinite integrals:

(a) $\displaystyle\int \frac{x}{x^2+1}\, dx,$ (b) $\displaystyle\int \frac{\sin(3\ln x)}{x}\, dx,$ and (c) $\displaystyle\int e^x\sqrt{1+e^x}\, dx.$

**Solution**

(a) $\displaystyle\int \frac{x}{x^2+1}\, dx$ $\qquad$ Let $u = x^2 + 1$.

$\qquad\qquad\qquad\qquad\qquad\quad$ Then $du = 2x\, dx$, and

$\qquad\qquad\qquad\qquad\qquad\quad$ $x\, dx = \frac{1}{2}\, du.$

$\qquad = \dfrac{1}{2}\displaystyle\int \dfrac{du}{u} = \dfrac{1}{2}\ln |u| + C = \dfrac{1}{2}\ln(x^2 + 1) + C = \ln\sqrt{x^2+1} + C.$

(Both versions of the final answer are equally acceptable.)

(b) $\displaystyle\int \frac{\sin(3\ln x)}{x}\, dx$ $\qquad$ Let $u = 3\ln x$.

$\qquad\qquad\qquad\qquad\qquad\quad$ Then $du = \dfrac{3}{x}\, dx.$

$\qquad = \dfrac{1}{3}\displaystyle\int \sin u\, du = -\dfrac{1}{3}\cos u + C = -\dfrac{1}{3}\cos(3\ln x) + C.$

(c) $\displaystyle\int e^x \sqrt{1 + e^x}\, dx$        Let $v = 1 + e^x$.

Then $dv = e^x\, dx$.

$$= \int v^{1/2}\, dv = \frac{2}{3} v^{3/2} + C = \frac{2}{3}(1 + e^x)^{3/2} + C.$$

∎

Sometimes the appropriate substitutions are not as obvious as they were in Example 3, and it may be necessary to manipulate the integrand algebraically to put it into a better form for substitution.

**Example 4**   Evaluate   (a) $\displaystyle\int \frac{1}{x^2 + 4x + 5}\, dx$   and   (b) $\displaystyle\int \frac{dx}{\sqrt{e^{2x} - 1}}$.

**Solution**

(a) $\displaystyle\int \frac{dx}{x^2 + 4x + 5} = \int \frac{dx}{(x+2)^2 + 1}$        Let $t = x + 2$.

Then $dt = dx$.

$$= \int \frac{dt}{t^2 + 1}$$

$$= \tan^{-1} t + C = \tan^{-1}(x + 2) + C.$$

(b) $\displaystyle\int \frac{dx}{\sqrt{e^{2x} - 1}} = \int \frac{dx}{e^x \sqrt{1 - e^{-2x}}}$

$$= \int \frac{e^{-x}\, dx}{\sqrt{1 - (e^{-x})^2}}$$        Let $u = e^{-x}$.

Then $du = -e^{-x}\, dx$.

$$= -\int \frac{du}{\sqrt{1 - u^2}}$$

$$= -\sin^{-1} u + C = -\sin^{-1}\left(e^{-x}\right) + C.$$

∎

The method of substitution cannot be *forced* to work. There is no substitution that will do much good with the integral $\int x(2 + x^7)^{1/5}\, dx$, for instance. However, the integral $\int x^6(2 + x^7)^{1/5}\, dx$ will yield to the substitution $u = 2 + x^7$. The substitution $u = g(x)$ is more likely to work if $g'(x)$ is a factor of the integrand.

The following theorem simplifies the use of the method of substitution in definite integrals.

**THEOREM**   **6**

**Substitution in a definite integral**

Suppose that $g$ is a differentiable function on $[a, b]$, and satisfies $g(a) = A$ and $g(b) = B$. Also suppose that $f$ is continuous on the range of $g$. Then

$$\int_a^b f\big(g(x)\big)\, g'(x)\, dx = \int_A^B f(u)\, du.$$

**PROOF**   Let $F$ be an antiderivative of $f$; $F'(u) = f(u)$. Then

$$\frac{d}{dx} F\big(g(x)\big) = F'\big(g(x)\big)\, g'(x) = f\big(g(x)\big)\, g'(x).$$

Thus

$$\int_a^b f\big(g(x)\big)\, g'(x)\, dx = F\big(g(x)\big)\Big|_a^b = F\big(g(b)\big) - F\big(g(a)\big)$$

$$= F(B) - F(A) = F(u)\Big|_A^B = \int_A^B f(u)\, du.$$

**Example 5**  Evaluate the integral $I = \displaystyle\int_0^8 \frac{\cos\sqrt{x+1}}{\sqrt{x+1}}\, dx$.

**Solution**  **Method I.** Let $u = \sqrt{x+1}$. Then $du = \dfrac{dx}{2\sqrt{x+1}}$. If $x = 0$, then $u = 1$; if $x = 8$, then $u = 3$. Thus

$$I = 2\int_1^3 \cos u\, du = 2\sin u\Big|_1^3 = 2\sin 3 - 2\sin 1.$$

**Method II.** We use the same substitution as in Method I, but we do not transform the limits of integration from $x$ values to $u$ values. Hence we must return to the variable $x$ before substituting in the limits:

$$I = 2\int_{x=0}^{x=8} \cos u\, du = 2\sin u\,\Big|_{x=0}^{x=8} = 2\sin\sqrt{x+1}\,\Big|_0^8 = 2\sin 3 - 2\sin 1.$$

Note that the limits *must* be written $x = 0$ and $x = 8$ at any stage where the variable is not $x$. It would have been *wrong* to write

$$I = 2\int_0^8 \cos u\, du$$

because this would imply that $u$, rather than $x$, goes from 0 to 8. Method I gives the shorter solution, and is therefore preferable. However, in cases where the transformed limits (the $u$-limits) are very complicated, you might prefer to use Method II.

**Example 6**  Find the area of the region bounded by $y = \left(2 + \sin\dfrac{x}{2}\right)^2 \cos\dfrac{x}{2}$, $y = 0$, $x = 0$, and $x = \pi$.

**Solution**  Because $y \geq 0$ when $0 \leq x \leq \pi$, the required area is

$$A = \int_0^\pi \left(2 + \sin\frac{x}{2}\right)^2 \cos\frac{x}{2}\, dx \qquad\qquad \text{Let } v = 2 + \sin\frac{x}{2}.$$

$$\text{Then } dv = \frac{1}{2}\,\cos\frac{x}{2}\, dx.$$

$$= 2\int_2^3 v^2\, dv = \frac{2}{3}\, v^3\,\Big|_2^3 = \frac{2}{3}\,(27 - 8) = \frac{38}{3}\ \text{square units.}$$

**Remark** The condition that $f$ be continuous on the range of the function $u = g(x)$ (for $a \le x \le b$) is essential in Theorem 6. Using the substitution $u = x^2$ in the integral $\int_{-1}^{1} x \csc(x^2)\,dx$ leads to the erroneous conclusion

$$\int_{-1}^{1} x\,\csc(x^2)\,dx = \frac{1}{2}\int_{1}^{1}\csc u\,du = 0.$$

Although $x\,\csc(x^2)$ is an odd function, it is not continuous at 0, and it happens that the given integral represents the difference of *infinite* areas. If we assume that $f$ is continuous on an interval containing $A$ and $B$, then it suffices to know that $u = g(x)$ is one-to-one as well as differentiable. In this case the range of $g$ will lie between $A$ and $B$, so the condition of Theorem 6 will be satisfied.

## Trigonometric Integrals

The method of substitution is often useful for evaluating trigonometric integrals. We begin by listing the integrals of the four trigonometric functions whose integrals we have not yet seen. They arise often in applications and should be memorized.

---

**Integrals of tangent, cotangent, secant, and cosecant**

$$\int \tan x\,dx = \ln|\sec x| + C,$$

$$\int \cot x\,dx = \ln|\sin x| + C = -\ln|\csc x| + C,$$

$$\int \sec x\,dx = \ln|\sec x + \tan x| + C,$$

$$\int \csc x\,dx = -\ln|\csc x + \cot x| + C = \ln|\csc x - \cot x| + C.$$

---

All of these can, of course, be checked by differentiating the right-hand sides. The first two can be evaluated directly by rewriting $\tan x$ or $\cot x$ in terms of $\sin x$ and $\cos x$ and using an appropriate substitution. For example,

$$\int \tan x\,dx = \int \frac{\sin x}{\cos x}\,dx \qquad \text{Let } u = \cos x.$$
$$\hspace{4cm} \text{Then } du = -\sin x\,dx.$$

$$= -\int \frac{du}{u} = -\ln|u| + C$$

$$= -\ln|\cos x| + C = \ln\left|\frac{1}{\cos x}\right| + C = \ln|\sec x| + C.$$

The integral of $\sec x$ can be evaluated by rewriting it in the form

$$\int \sec x\,dx = \int \frac{\sec x(\sec x + \tan x)}{\sec x + \tan x}$$

and using the substitution $u = \sec x + \tan x$. The integral of $\csc x$ can be evaluated similarly. (Show that the two versions given for that integral are equivalent!)

We now consider integrals of the form

$$\int \sin^m x\,\cos^n x\,dx.$$

If either $m$ or $n$ is an odd, positive integer, the integral can be done easily by substitution. If, say, $n = 2k + 1$ where $k$ is an integer, then we can use the identity $\sin^2 x + \cos^2 x = 1$ to rewrite the integral in the form

$$\int \sin^m x \, (1 - \sin^2 x)^k \cos x \, dx,$$

which can be integrated using the substitution $u = \sin x$. Similarly, $u = \cos x$ can be used if $m$ is an odd integer.

---

**Example 7**    Evaluate:    (a) $\displaystyle\int \sin^3 x \, \cos^8 x \, dx$, and    (b) $\displaystyle\int \cos^5 ax \, dx$.

**Solution**

(a) $\displaystyle\int \sin^3 x \, \cos^8 x \, dx = \int (1 - \cos^2 x) \cos^8 x \sin x \, dx$    Let $u = \cos x$,
$$\hspace{9cm} du = -\sin x \, dx.$$

$$= -\int (1 - u^2) \, u^8 \, du = \int (u^{10} - u^8) \, du$$

$$= \frac{u^{11}}{11} - \frac{u^9}{9} + C = \frac{1}{11} \cos^{11} x - \frac{1}{9} \cos^9 x + C$$

(b) $\displaystyle\int \cos^5 ax \, dx = \int (1 - \sin^2 ax)^2 \cos ax \, dx$    Let $u = \sin ax$,
$$\hspace{8cm} du = a \cos ax \, dx.$$

$$= \frac{1}{a} \int (1 - u^2)^2 \, du = \frac{1}{a} \int (1 - 2u^2 + u^4) \, du$$

$$= \frac{1}{a} \left( u - \frac{2}{3} u^3 + \frac{1}{5} u^5 \right) + C$$

$$= \frac{1}{a} \left( \sin ax - \frac{2}{3} \sin^3 ax + \frac{1}{5} \sin^5 ax \right) + C$$

---

If the powers of $\sin x$ and $\cos x$ are both even, then we can make use of the *double-angle formulas* (see Section P.6):

$$\cos^2 x = \frac{1}{2}(1 + \cos 2x) \qquad \text{and} \qquad \sin^2 x = \frac{1}{2}(1 - \cos 2x).$$

---

**Example 8**    **(Integrating even powers of sine and cosine)**  Verify the formulas:

$$\int \cos^2 x \, dx = \frac{1}{2}(x + \sin x \cos x) + C,$$

$$\int \sin^2 x \, dx = \frac{1}{2}(x - \sin x \cos x) + C.$$

These integrals are encountered frequently, and are worth remembering.

**Solution**    Each of the integrals follows from the corresponding double-angle identity. We do the first; the second is similar.

$$\int \cos^2 x \, dx = \frac{1}{2} \int (1 + \cos 2x) \, dx$$

$$= \frac{x}{2} + \frac{1}{4} \sin 2x + C$$

$$= \frac{1}{2}(x + \sin x \cos x) + C \quad \text{(since } \sin 2x = 2 \sin x \cos x\text{).}$$

(Note that there is no point in inserting the constant of integration $C$ until the last integral has been evaluated.)

**Example 9**   Evaluate $\int \sin^4 x \, dx$.

**Solution**   We will have to apply the double-angle formula twice.

$$\int \sin^4 x \, dx = \frac{1}{4} \int (1 - \cos 2x)^2 \, dx$$

$$= \frac{1}{4} \int (1 - 2\cos 2x + \cos^2 2x) \, dx$$

$$= \frac{x}{4} - \frac{1}{4} \sin 2x + \frac{1}{8} \int (1 + \cos 4x) \, dx$$

$$= \frac{x}{4} - \frac{1}{4} \sin 2x + \frac{x}{8} + \frac{1}{32} \sin 4x + C$$

$$= \frac{3}{8} x - \frac{1}{4} \sin 2x + \frac{1}{32} \sin 4x + C$$

Using the identities $\sec^2 x = 1 + \tan^2 x$ and $\csc^2 x = 1 + \cot^2 x$ and one of the substitutions $u = \sec x$, $u = \tan x$, $u = \csc x$, or $u = \cot x$, we can evaluate integrals of the form

$$\int \sec^m x \, \tan^n x \, dx \qquad \text{or} \qquad \int \csc^m x \, \cot^n x \, dx,$$

unless $m$ is odd and $n$ is even. (If this is the case, these integrals can be handled by integration by parts; see Section 6.1.)

**Example 10**   **(Integrals involving secants and tangents)**   Evaluate the integrals:

(a) $\int \tan^2 x \, dx$,    (b) $\int \sec^4 t \, dt$,    and    (c) $\int \sec^3 x \, \tan^3 x \, dx$.

**Solution**

(a) $\int \tan^2 x \, dx = \int (\sec^2 x - 1) \, dx = \tan x - x + C.$

(b) $\int \sec^4 t \, dt = \int (1 + \tan^2 t) \sec^2 t \, dt \qquad$ Let $u = \tan t$,

$$du = \sec^2 t \, dt.$$

$$= \int (1 + u^2) \, du = u + \frac{1}{3} u^3 + C = \tan t + \frac{1}{3} \tan^3 t + C.$$

(c) $\displaystyle\int \sec^3 x\ \tan^3 x\ dx$

$\displaystyle= \int \sec^2 x\ (\sec^2 x - 1)\ \sec x\ \tan x\ dx$    Let $u = \sec x$,
$\qquad\qquad\qquad\qquad\qquad\qquad\qquad\qquad du = \sec x\ \tan x\ dx.$

$\displaystyle= \int (u^4 - u^2)\ du = \frac{u^5}{5} - \frac{u^3}{3} + C = \frac{1}{5}\sec^5 x - \frac{1}{3}\sec^3 x + C.$

∎

## Exercises 5.6

Evaluate the integrals in Exercises 1–48. Remember to include a constant of integration with the indefinite integrals. Your answers may appear different from those in the Answers section, but still be correct. For example, evaluating $I = \int \sin x \cos x\ dx$ using the substitution $u = \sin x$ leads to the answer $I = \frac{1}{2}\sin^2 x + C$; using $u = \cos x$ leads to $I = -\frac{1}{2}\cos^2 x + C$; and rewriting $I = \frac{1}{2}\int \sin(2x)\ dx$ leads to $I = -\frac{1}{4}\cos(2x) + C$. These answers are all equivalent up to different choices for the constant of integration: $\frac{1}{2}\sin^2 x = -\frac{1}{2}\cos^2 x + \frac{1}{2} = -\frac{1}{4}\cos(2x) + \frac{1}{4}$.

You can always check your own answer to an indefinite integral by differentiating it to get back to the integrand. This is often easier than comparing your answer with the answer in the back of the book. You may find integrals that you can't do, but you should not make mistakes in those you can do, because the answer is so easily checked. (This is a good thing to remember during tests and exams.)

**1.** $\displaystyle\int e^{5-2x}\ dx$

**2.** $\displaystyle\int \cos(ax + b)\ dx$

**3.** $\displaystyle\int \sqrt{3x + 4}\ dx$

**4.** $\displaystyle\int e^{2x} \sin(e^{2x})\ dx$

**5.** $\displaystyle\int \frac{x^2}{(x^3 + 2)^{5/2}}\ dx$

**6.** $\displaystyle\int (x + 2)(x^2 + 4x + 9)^{1/3}\ dx$

**7.** $\displaystyle\int \frac{x\ dx}{(4x^2 + 1)^5}$

**8.** $\displaystyle\int \frac{\sin\sqrt{x}}{\sqrt{x}}\ dx$

**9.** $\displaystyle\int x\ e^{x^2}\ dx$

**10.** $\displaystyle\int x^2 2^{x^3+1}\ dx$

**11.** $\displaystyle\int \frac{\cos x}{4 + \sin^2 x}\ dx$

**12.** $\displaystyle\int \frac{\sec^2 x}{\sqrt{1 - \tan^2 x}}\ dx$

**\* 13.** $\displaystyle\int \frac{e^x + 1}{e^x - 1}\ dx$

**14.** $\displaystyle\int \frac{\ln t}{t}\ dt$

**15.** $\displaystyle\int \frac{ds}{\sqrt{4 - 5s}}$

**16.** $\displaystyle\int \frac{x + 1}{\sqrt{x^2 + 2x + 3}}\ dx$

**17.** $\displaystyle\int \frac{t\ dt}{\sqrt{4 - t^4}}$

**18.** $\displaystyle\int \frac{x^2\ dx}{2 + x^6}$

**\* 19.** $\displaystyle\int \frac{dx}{e^x + 1}$

**\* 20.** $\displaystyle\int \frac{dx}{e^x + e^{-x}}$

**21.** $\displaystyle\int \tan x\ \ln \cos x\ dx$

**22.** $\displaystyle\int \frac{x + 1}{\sqrt{1 - x^2}}\ dx$

**23.** $\displaystyle\int \frac{2t + 3}{t^2 + 9}\ dt$

**24.** $\displaystyle\int \frac{ax + b}{\sqrt{A^2 - B^2 x^2}}\ dx$

**25.** $\displaystyle\int \frac{dx}{x^2 + 6x + 13}$

**26.** $\displaystyle\int \frac{dx}{\sqrt{4 + 2x - x^2}}$

**27.** $\displaystyle\int \sin^3 x \cos^5 x\ dx$

**28.** $\displaystyle\int \sin^4 t \cos^5 t\ dt$

**29.** $\displaystyle\int \sin ax \cos^2 ax\ dx$

**30.** $\displaystyle\int \sin^2 x \cos^2 x\ dx$

**31.** $\displaystyle\int \sin^6 x\ dx$

**32.** $\displaystyle\int \cos^4 x\ dx$

**33.** $\displaystyle\int \sec^5 x \tan x\ dx$

**34.** $\displaystyle\int \sec^6 x \tan^2 x\ dx$

**35.** $\displaystyle\int \sqrt{\tan x}\ \sec^4 x\ dx$

**36.** $\displaystyle\int \sin^{-2/3} x \cos^3 x\ dx$

**37.** $\displaystyle\int \cos x \sin^4(\sin x)\ dx$

**38.** $\displaystyle\int \frac{\sin^3 \ln x \cos^3 \ln x}{x}\ dx$

**39.** $\displaystyle\int \frac{\sin^2 x}{\cos^4 x}\ dx$

**40.** $\displaystyle\int \frac{\sin^3 x}{\cos^4 x}\ dx$

**41.** $\displaystyle\int \csc^5 x \cot^5 x\ dx$

**42.** $\displaystyle\int \frac{\cos^4 x}{\sin^8 x}\ dx$

**43.** $\displaystyle\int_0^4 x^3 (x^2 + 1)^{-\frac{1}{2}}\ dx$

**44.** $\displaystyle\int_1^{\sqrt{e}} \frac{\sin(\pi \ln x)}{x}\ dx$

**45.** $\displaystyle\int_0^{\pi/2} \sin^4 x\ dx$

**46.** $\displaystyle\int_{\pi/4}^{\pi} \sin^5 x\ dx$

**47.** $\displaystyle\int_e^{e^2} \frac{dt}{t\ \ln t}$

**48.** $\displaystyle\int_{\frac{\pi^2}{16}}^{\frac{\pi^2}{9}} \frac{2^{\sin\sqrt{x}} \cos\sqrt{x}}{\sqrt{x}}\ dx$

* **49.** Use the identities $\cos 2\theta = 2\cos^2\theta - 1 = 1 - 2\sin^2\theta$ and $\sin\theta = \cos\left(\dfrac{\pi}{2} - \theta\right)$ to help you evaluate the following:

$$\int_0^{\pi/2} \sqrt{1 + \cos x}\, dx \quad \text{and} \quad \int_0^{\pi/2} \sqrt{1 - \sin x}\, dx$$

**50.** Find the area of the region bounded by $y = x/(x^2 + 16)$, $y = 0$, $x = 0$, and $x = 2$.

**51.** Find the area of the region bounded by $y = x/(x^4 + 16)$, $y = 0$, $x = 0$, and $x = 2$.

**52.** Express the area bounded by the ellipse $(x^2/a^2) + (y^2/b^2) = 1$ as a definite integral. Make a substitution that converts this integral into one representing the area of a circle, and hence evaluate it.

* **53.** Use the addition formulas for $\sin(x \pm y)$ and $\cos(x \pm y)$ from Section P.6 to establish the following identities:

$$\cos x \, \cos y = \frac{1}{2}\Big(\cos(x - y) + \cos(x + y)\Big),$$

$$\sin x \, \sin y = \frac{1}{2}\Big(\cos(x - y) - \cos(x + y)\Big),$$

$$\sin x \, \cos y = \frac{1}{2}\Big(\sin(x + y) + \sin(x - y)\Big).$$

* **54.** Use the identities established in the previous exercise to calculate the following integrals:

$$\int \cos ax \, \cos bx \, dx, \quad \int \sin ax \, \sin bx \, dx,$$

$$\text{and} \quad \int \sin ax \, \cos bx \, dx.$$

* **55.** If $m$ and $n$ are integers, show that:

(i) $\displaystyle\int_{-\pi}^{\pi} \cos mx \, \cos nx \, dx = 0$ if $m \neq n$,

(ii) $\displaystyle\int_{-\pi}^{\pi} \sin mx \, \sin nx \, dx = 0$ if $m \neq n$,

(iii) $\displaystyle\int_{-\pi}^{\pi} \sin mx \, \cos nx \, dx = 0$.

# 5.7 Areas of Plane Regions

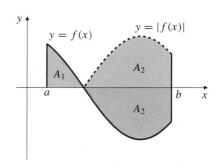

**Figure 5.27**

In this section we review and extend the use of definite integrals to represent plane areas. Recall that the integral $\int_a^b f(x)\, dx$ measures the area between the graph of $f$ and the $x$-axis from $x = a$ to $x = b$, but it counts as *negative* any part of this area that lies below the $x$-axis. (We are assuming that $a < b$.) In order to express the total area bounded by $y = f(x)$, $y = 0$, $x = a$, and $x = b$, counting all of the area positively, we should integrate the *absolute value* of $f$ (see Figure 5.27):

$$\int_a^b f(x)\, dx = A_1 - A_2 \quad \text{and} \quad \int_a^b |f(x)|\, dx = A_1 + A_2.$$

There is no "rule" for integrating $\int_a^b |f(x)|\, dx$; one must break the integral into a sum of integrals over intervals where $f(x) > 0$ (so $|f(x)| = f(x)$), and intervals where $f(x) < 0$ (so $|f(x)| = -f(x)$).

**Example 1**   The area bounded by $y = \cos x$, $y = 0$, $x = 0$, and $x = 3\pi/2$ (see Figure 5.28) is

$$A = \int_0^{3\pi/2} |\cos x|\, dx$$

$$= \int_0^{\pi/2} \cos x \, dx + \int_{\pi/2}^{3\pi/2} (-\cos x)\, dx$$

$$= \sin x \Big|_0^{\pi/2} - \sin x \Big|_{\pi/2}^{3\pi/2}$$

$$= (1 - 0) - (-1 - 1) = 3 \text{ square units.}$$

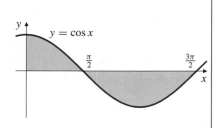

**Figure 5.28**

## Areas Between Two Curves

Suppose that a plane region $R$ is bounded by the graphs of two continuous functions, $y = f(x)$ and $y = g(x)$, and the vertical straight lines $x = a$ and $x = b$, as shown in Figure 5.29(a). Assume that $a < b$ and that $f(x) \leq g(x)$ on $[a, b]$, so the graph of $f$ lies below that of $g$. If $f(x) \geq 0$ on $[a, b]$, then the area $A$ of $R$ is the area above the $x$-axis and under the graph of $g$ minus the area above the $x$-axis and under the graph of $f$:

$$A = \int_a^b g(x)\, dx - \int_a^b f(x)\, dx = \int_a^b \left(g(x) - f(x)\right) dx.$$

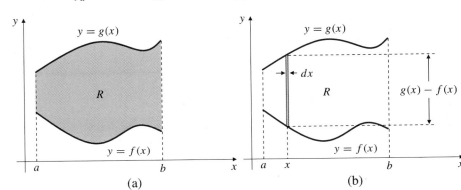

**Figure 5.29**

(a)  The region $R$ lying between two graphs

(b)  An area element of the region $R$

It is useful to regard this formula as expressing $A$ as the "sum" (that is, the integral) of *infinitely many* **area elements**

$$dA = (g(x) - f(x))\, dx,$$

corresponding to values of $x$ between $a$ and $b$. Each such area element is the area of an infinitely thin vertical rectangle of width $dx$ and height $g(x) - f(x)$ located at position $x$ (see Figure 5.29(b)). Even if $f$ and $g$ can take on negative values on $[a, b]$, this interpretation, and the resulting area formula

$$A = \int_a^b \left(g(x) - f(x)\right) dx$$

remain valid, provided that $f(x) \leq g(x)$ on $[a, b]$ so that all the area elements $dA$ have positive area. Using integrals to represent a quantity as a *sum* of *differential elements* (i.e., a sum of little bits of the quantity) is a very helpful approach. We will do this often in Chapter 7. Of course, what we are really doing is identifying the integral as a *limit* of a suitable Riemann sum.

More generally, if the restriction $f(x) \leq g(x)$ is removed, then the vertical rectangle of width $dx$ at position $x$ extending between the graphs of $f$ and $g$ has height $|f(x) - g(x)|$ and hence area

$$dA = |f(x) - g(x)|\, dx.$$

(See Figure 5.30.) Hence the total area lying between the graphs $y = f(x)$ and $y = g(x)$ and between the vertical lines $x = a$ and $x = b > a$ is given by

$$A = \int_a^b |f(x) - g(x)|\, dx.$$

In order to evaluate this integral, we have to determine the intervals on which $f(x) > g(x)$ or $f(x) < g(x)$, and break the integral into a sum of integrals over each of these intervals.

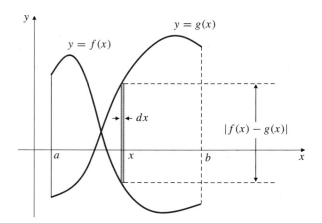

**Figure 5.30**    An area element for the region between $y = f(x)$ and $y = g(x)$

**Example 2**    Find the area of the bounded, plane region $R$ lying between the curves $y = x^2 - 2x$ and $y = 4 - x^2$.

**Solution**    We must find the intersections of the curves, so we solve the equations simultaneously:

$$x^2 - 2x = y = 4 - x^2$$
$$2x^2 - 2x - 4 = 0$$
$$2(x - 2)(x + 1) = 0 \quad \text{so } x = 2 \text{ or } x = -1.$$

The curves are sketched in Figure 5.31, and the bounded (finite) region between them is shaded. (A sketch should always be made in problems of this sort.) Since $4 - x^2 \geq x^2 - 2x$ for $-1 \leq x \leq 2$, the area $A$ of $R$ is given by

$$A = \int_{-1}^{2} \left( (4 - x^2) - (x^2 - 2x) \right) dx$$

$$= \int_{-1}^{2} (4 - 2x^2 + 2x) \, dx$$

$$= \left( 4x - \frac{2}{3}x^3 + x^2 \right) \Big|_{-1}^{2}$$

$$= 4(2) - \frac{2}{3}(8) + 4 - \left( -4 + \frac{2}{3} + 1 \right) = 9 \text{ square units.}$$

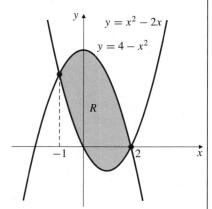

**Figure 5.31**

Note that in representing the area as an integral we *must subtract the height $y$ to the lower curve from the height $y$ to the upper curve* to get a positive area element $dA$. Subtracting the wrong way would have produced a negative value for the area.    ∎

**Example 3** Find the total area $A$ lying between the curves $y = \sin x$ and $y = \cos x$ from $x = 0$ to $x = 2\pi$.

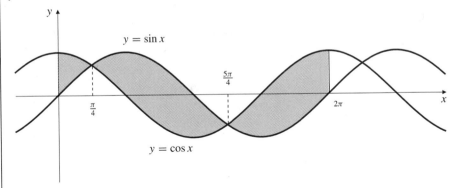

**Figure 5.32**

**Solution** The region is shaded in Figure 5.32. The graphs of sine and cosine cross at $x = \pi/4$ and $x = 5\pi/4$. The required area is

$$A = \int_0^{\pi/4} (\cos x - \sin x)\, dx + \int_{\pi/4}^{5\pi/4} (\sin x - \cos x)\, dx$$

$$+ \int_{5\pi/4}^{2\pi} (\cos x - \sin x)\, dx$$

$$= (\sin x + \cos x)\Big|_0^{\pi/4} - (\cos x + \sin x)\Big|_{\pi/4}^{5\pi/4} + (\sin x + \cos x)\Big|_{5\pi/4}^{2\pi}$$

$$= (\sqrt{2} - 1) + (\sqrt{2} + \sqrt{2}) + (1 + \sqrt{2}) = 4\sqrt{2} \text{ square units.}$$

It is sometimes more convenient to use horizontal area elements instead of vertical ones, and integrate over an interval of the $y$-axis instead of the $x$-axis. This is usually the case if the region whose area we want to find is bounded by curves whose equations are written in terms of functions of $y$. The region $R$ lying to the right of $x = f(y)$ and to the left of $x = g(y)$, and between the horizontal lines $y = c$ and $y = d > c$ (see Figure 5.33(a)) has area element $dA = \big(g(y) - f(y)\big)\, dy$ and its area is

$$A = \int_c^d \big(g(y) - f(y)\big)\, dy.$$

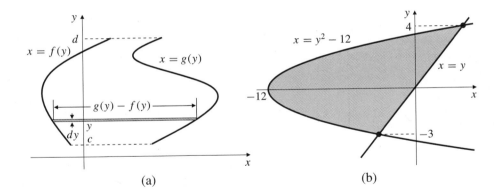

**Figure 5.33**

(a) A horizontal area element

(b) The finite region bounded by $x = y^2 - 12$ and $x = y$

(a)

(b)

**Example 4**    Find the area of the plane region lying to the right of the parabola $x = y^2 - 12$ and to the left of the straight line $y = x$, as illustrated in Figure 5.33(b).

**Solution**    For the intersections of the curves:

$$y^2 - 12 = x = y$$
$$y^2 - y - 12 = 0$$
$$(y - 4)(y + 3) = 0 \quad \text{so } y = 4 \text{ or } y = -3.$$

Observe that $y^2 - 12 \le y$ for $-3 \le y \le 4$. Thus the area is

$$A = \int_{-3}^{4} \left(y - (y^2 - 12)\right) dy = \left(\frac{y^2}{2} - \frac{y^3}{3} + 12y\right)\Big|_{-3}^{4} = \frac{343}{6} \text{ square units.}$$

Of course, the same result could have been obtained by integrating in the $x$ direction, but the integral would have been more complicated:

$$A = \int_{-12}^{-3} \left(\sqrt{12 + x} - (-\sqrt{12 + x})\right) dx + \int_{-3}^{4} \left(\sqrt{12 + x} - x\right) dx;$$

different integrals are required over the intervals where the region is bounded below by the parabola and by the straight line. ∎

## Exercises 5.7

In Exercises 1–16, find the area of the plane region bounded by the given curves. In each case you should make a sketch of the region.

1. $y = x, \quad y = x^2$

2. $y = \sqrt{x}, \quad y = x^2$

3. $y = x^2 - 5, \quad y = 3 - x^2$

4. $y = x^2 - 2x, \quad y = 6x - x^2$

5. $2y = 4x - x^2, \quad 2y + 3x = 6$

6. $x - y = 7, \quad x = 2y^2 - y + 3$

7. $y = x^3, \quad y = x$

8. $y = x^3, \quad y = x^2$

9. $y = x^3, \quad x = y^2$

10. $x = y^2, \quad x = 2y^2 - y - 2$

11. $y = \dfrac{1}{x}, \quad 2x + 2y = 5$

12. $y = (x^2 - 1)^2, \quad y = 1 - x^2$

13. $y = \dfrac{1}{2}x^2, \quad y = \dfrac{1}{x^2 + 1}$

14. $y = \dfrac{4x}{3 + x^2}, \quad y = 1$

15. $y = \dfrac{4}{x^2}, \quad y = 5 - x^2$

16. $x = y^2 - \pi^2, \quad x = \sin y$

Find the areas of the regions described in Exercises 17–28. It is helpful to sketch the regions before writing an integral to represent the area.

17. Bounded by $y = \sin x$ and $y = \cos x$, and between two consecutive intersections of these curves

18. Bounded by $y = \sin^2 x$ and $y = 1$, and between two consecutive intersections of these curves

19. Bounded by $y = \sin x$ and $y = \sin^2 x$, between $x = 0$ and $x = \pi/2$

20. Bounded by $y = \sin^2 x$ and $y = \cos^2 x$, and between two consecutive intersections of these curves

21. Under $y = 4x/\pi$ and above $y = \tan x$, between $x = 0$ and the first intersection of the curves to the right of $x = 0$

22. Bounded by $y = x^{1/3}$ and the component of $y = \tan(\pi x/4)$ that passes through the origin

23. Bounded by $y = 2$ and the component of $y = \sec x$ that passes through the point $(0, 1)$

24. Bounded by $y = \sqrt{2}\cos(\pi x/4)$ and $y = |x|$

25. Bounded by $y = \sin(\pi x/2)$ and $y = x$

26. Bounded by $y = \sin(\pi x/2)$ and $y = x^3$

27. Bounded by $y = (x - 1)^2$ and $y = 1/(1 + x^2)$

28. Bounded by $y = e^x$ and $y = x + 2$

29. Find the total area enclosed by the curve $y^2 = x^2 - x^4$.

30. Find the area of the closed loop of the curve $y^2 = x^4(2 + x)$ that lies to the left of the origin.

**31.** Find the area of the finite plane region that is bounded by the curve $y = e^x$, the line $x = 0$, and the tangent line to $y = e^x$ at $x = 1$.

**\* 32.** Find the area of the finite plane region bounded by the curve $y = x^3$ and the tangent line to that curve at the point $(1, 1)$. *Hint:* find the other point at which that tangent line meets the curve.

## Chapter Review

### Key Ideas

- **What do the following phrases mean?**
  - ◇ sigma notation
  - ◇ a partition of an interval
  - ◇ a Riemann sum
  - ◇ a definite integral
  - ◇ an indefinite integral
  - ◇ an integrable function
  - ◇ an area element
  - ◇ an evaluation symbol
  - ◇ the triangle inequality for integrals
  - ◇ a piecewise continuous function
  - ◇ the average value of function $f$ on $[a, b]$
  - ◇ the method of substitution

- **State the Mean-Value Theorem for integrals.**
- **State the Fundamental Theorem of Calculus.**
- **List as many properties of the definite integral as you can.**
- **What is the relationship between the definite integral and the indefinite integral of a function $f$ on an interval $[a, b]$?**
- **What is the derivative of $\int_{f(x)}^{g(x)} h(t)\, dt$ with respect to $x$?**
- **How can the area between the graphs of two functions be calculated?**

### Review Exercises

**1.** Show that $\dfrac{2j + 1}{j^2(j + 1)^2} = \dfrac{1}{j^2} - \dfrac{1}{(j + 1)^2}$, and hence evaluate

$$\sum_{j=1}^{n} \frac{2j + 1}{j^2(j + 1)^2}.$$

**2.** **(Stacking balls)** A display of golf balls in a sporting goods store is built in the shape or a pyramid with a rectangular base measuring 40 balls long and 30 balls wide. The next layer up is 39 balls by 29 balls, etc. How many balls are in the pyramid?

**3.** Let $P_n = \{x_0 = 1, x_1, x_2, \ldots, x_n = 3\}$ be a partition of $[1, 3]$ into $n$ subintervals of equal length, and let $f(x) = x^2 - 2x + 3$. Evaluate $\int_1^3 f(x)\, dx$ by finding $\lim_{n \to \infty} \sum_{i=1}^{n} f(x_i)\, \Delta x_i$.

**4.** Interpret $R_n = \displaystyle\sum_{i=1}^{n} \frac{1}{n}\sqrt{1 + \frac{i}{n}}$ as a Riemann sum for a certain function $f$ on the interval $[0, 1]$, and hence evaluate $\lim_{n \to \infty} R_n$.

Evaluate the integrals in Exercises 5–8 without using the Fundamental Theorem of Calculus.

**5.** $\displaystyle\int_{-\pi}^{\pi} (2 - \sin x)\, dx$

**6.** $\displaystyle\int_0^{\sqrt{5}} \sqrt{5 - x^2}\, dx$

**7.** $\displaystyle\int_1^3 \left(1 - \frac{x}{2}\right) dx$

**8.** $\displaystyle\int_0^{\pi} \cos x\, dx$

Find the average values of the functions in Exercises 9–10 over the indicated intervals.

**9.** $f(x) = 2 - \sin x^3$ on $[-\pi, \pi]$

**10.** $h(x) = |x - 2|$ on $[0, 3]$

Find the derivatives of the functions in Exercises 11–14.

**11.** $f(t) = \displaystyle\int_{13}^{t} \sin(x^2)\, dx$

**12.** $f(x) = \displaystyle\int_{-13}^{\sin x} \sqrt{1 + t^2}\, dt$

**13.** $g(s) = \displaystyle\int_{4s}^{1} e^{\sin u}\, du$

**14.** $g(\theta) = \displaystyle\int_{e^{\sin \theta}}^{e^{\cos \theta}} \ln x\, dx$

**15.** Solve the integral equation $2 f(x) + 1 = 3 \displaystyle\int_x^1 f(t)\, dt$.

**16.** Use the substitution $x = \pi - u$ to show that

$$\int_0^{\pi} x\, f(\sin x)\, dx = \frac{\pi}{2} \int_0^{\pi} f(\sin x)\, dx$$

for any function $f$ continuous on $[0, 1]$.

Find the areas of the finite plane regions bounded by the indicated graphs in Exercises 17–22.

**17.** $y = 2 + x - x^2$, and $y = 0$

**18.** $y = (x - 1)^2$, $y = 0$, and $x = 0$

**19.** $x = y - y^4$, and $x = 0$

**20.** $y = 4x - x^2$, and $y = 3$

**21.** $y = \sin x$, $y = \cos 2x$, $x = 0$, and $x = \pi/6$

**22.** $y = 5 - x^2$, and $y = 4/x^2$

Evaluate the integrals in Exercises 23–30.

**23.** $\displaystyle\int x^2 \cos(2x^3 + 1)\, dx$

**24.** $\displaystyle\int_1^e \frac{\ln x}{x}\, dx$

**25.** $\displaystyle\int_0^4 \sqrt{9t^2 + t^4}\, dt$

**26.** $\displaystyle\int \sin^3(\pi x)\, dx$

**27.** $\displaystyle\int_0^{\ln 2} \frac{e^u}{4 + e^{2u}}\, du$

**28.** $\displaystyle\int_1^{\sqrt[4]{e}} \frac{\tan^2 \pi \ln x}{x}\, dx$

**29.** $\displaystyle\int \frac{\sin\sqrt{2s+1}}{\sqrt{2s+1}}\,ds$   **30.** $\displaystyle\int \cos^2\frac{t}{5}\,\sin^2\frac{t}{5}\,dt$

**31.** Find the minimum value of $F(x) = \displaystyle\int_0^{x^2-2x} \frac{1}{1+t^2}\,dt$. Does $F$ have a maximum value? Why?

**32.** Find the maximum value of $\int_a^b (4x - x^2)\,dx$ for intervals $[a, b]$ where $a < b$. How do you know such a maximum value exists?

**33.** An object moves along the x-axis so that its position at time $t$ is given by the function $x(t)$. In Section 2.11 we defined the average velocity of the object over the time interval $[t_0, t_1]$ to be $v_{av} = \big(x(t_1) - x(t_0)\big)/(t_1 - t_0)$. Show that $v_{av}$ is, in fact, the average value of the velocity function $v(t) = dx/dt$ over the interval $[t_0, t_1]$.

**34.** If an object falls from rest under constant gravitational acceleration, show that its average height during the time $T$ of its fall is its height at time $T/\sqrt{3}$.

## Challenging Problems

**1.** Evaluate the upper and lower Riemann sums, $U(f, P_n)$ and $L(f, P_n)$, for $f(x) = 1/x$ on the interval $[1, 2]$ for the partition $P_n$ with division points $x_i = 2^{i/n}$ for $0 \le i \le n$. Verify that $\lim_{n\to\infty} U(f, P_n) = \ln 2 = \lim_{n\to\infty} L(f, P_n)$.

**2.** (a) Use the addition formulas for $\cos(a + b)$ and $\cos(a - b)$ to show that

$$\cos\big((j+\tfrac{1}{2})t\big) - \cos\big((j-\tfrac{1}{2})t\big) = -2\sin(\tfrac{1}{2}t)\,\sin(jt),$$

and hence deduce that if $t/(2\pi)$ is not an integer, then

$$\sum_{j=1}^n \sin(jt) = \frac{\cos\frac{t}{2} - \cos\big((n+\tfrac{1}{2})t\big)}{2\sin\frac{t}{2}}$$

(b) Use the result to part (a) to evaluate $\int_0^{\pi/2} \sin x\,dx$ as a limit of a Riemann sum.

**3.** (a) Use the method of the previous exercise to show that if $t/(2\pi)$ is not an integer, then

$$\sum_{j=1}^n \cos(jt) = \frac{\sin\big((n+\tfrac{1}{2})t\big) - \sin\frac{t}{2}}{2\sin\frac{t}{2}}$$

(b) Use the result to part (a) to evaluate $\int_0^{\pi/3} \cos x\,dx$ as a limit of a Riemann sum.

**4.** Let $f(x) = 1/x^2$ and let $1 = x_0 < x_1 < x_2 < \cdots < x_n = 2$, so that $\{x_0, x_1, x_2, \ldots, x_n\}$ is a partition of $[1, 2]$ into $n$ subintervals. Show that $c_i = \sqrt{x_{i-1}x_i}$ is in the $i$th subinterval $[x_{i-1}, x_i]$ of the partition, and evaluate the Riemann sum $\sum_{i=1}^n f(c_i)\,\Delta x_i$. What does this imply about $\int_1^2 (1/x^2)\,dx$?

**5.** (a) Use mathematical induction to verify that for every positive integer $k$, $\sum_{j=1}^n j^k = \dfrac{n^{k+1}}{k+1} + \dfrac{n^k}{2} + P_{k-1}(n)$, where $P_{k-1}$ is a polynomial of degree at most $k - 1$. *Hint:* start by iterating the identity

$$(j + 1)^{k+1} - j^{k+1} = (k + 1)j^k + \frac{(k + 1)k}{2}j^{k-1}$$
$$+ \text{ lower powers of } j$$

for $j = 1, 2, 3, \ldots, k$ and adding.

(b) Deduce from (a) that $\displaystyle\int_0^a x^k\,dx = \frac{a^{k+1}}{k+1}$.

**6.** Let $C$ be the cubic curve $y = ax^3 + bx^2 + cx + d$, and let $P$ be any point on $C$. The tangent to $C$ at $P$ meets $C$ again at point $Q$. The tangent to $C$ at $Q$ meets $C$ again at $R$. Show that the area between $C$ and the tangent at $Q$ is 16 times the area between $C$ and the tangent at $P$.

**7.** Let $C$ be the cubic curve $y = ax^3 + bx^2 + cx + d$, and let $P$ be any point on $C$. The tangent to $C$ at $P$ meets $C$ again at point $Q$. Let $R$ be the inflection point of $C$. Show that $R$ lies between $P$ and $Q$ on $C$, and that $QR$ divides the area between $C$ and its tangent at $P$ in the ratio 16/11.

**8.** **(Double tangents)** Let line $PQ$ be tangent to the graph $C$ of the quartic polynomial $f(x) = ax^4 + bx^3 + cx^2 + dx + e$ at two distinct points $P = (p, f(p))$ and $Q = (q, f(q))$. Let $U = (u, f(u))$ and $V = (v, f(v))$ be the other two points where the line tangent to $C$ at $T = ((p+q)/2, f((p+q)/2))$ meets $C$. If $A$ and $B$ are the two inflection points of $C$ let $R$ and $S$ be the other two points where $AB$ meets $C$. (See Figure 5.34. Also see Challenging Problem 17 in Chapter 2 for more background.)

(a) Find the ratio of the area bounded by $UV$ and $C$ to the area bounded by $PQ$ and $C$.

(b) Show that the area bounded by $RS$ and $C$ is divided at $A$ and $B$ into three parts in the ratio $1 : 2 : 1$.

**Figure 5.34**

# CHAPTER 6
# Techniques of Integration

***Introduction***  This chapter is completely concerned with how to evaluate integrals. The first four sections continue our search, begun in Section 5.6, for ways to find antiderivatives, and therefore also definite integrals by the Fundamental Theorem of Calculus. Section 6.5 deals with the problem of finding definite integrals of functions over infinite intervals, or over intervals where the functions are not bounded. The remaining four sections deal with techniques of *numerical integration* that can be used to find approximate values of definite integrals when an antiderivative cannot be found.

It is not necessary to cover the material of this chapter before proceeding to the various applications of integration discussed in Chapter 7, but some of the examples and exercises in that chapter do depend on techniques presented here.

## 6.1  Integration by Parts

Our next general method for antidifferentiation is called **integration by parts**. Just as the method of substitution can be regarded as inverse to the Chain Rule for differentiation, so the method for integration by parts is inverse to the Product Rule for differentiation.

Suppose that $U(x)$ and $V(x)$ are two differentiable functions. According to the Product Rule,

$$\frac{d}{dx}\big(U(x)V(x)\big) = U(x)\frac{dV}{dx} + V(x)\frac{dU}{dx}.$$

Integrating both sides of this equation and transposing terms, we obtain

$$\int U(x)\frac{dV}{dx}\,dx = U(x)V(x) - \int V(x)\frac{dU}{dx}\,dx$$

or, more simply,

$$\int U\,dV = UV - \int V\,dU.$$

The above formula serves as a *pattern* for carrying out integration by parts, as we will see in the examples below. In each application of the method, we break up the given integrand into a product of two pieces, $U$ and $V'$, where $V'$ is readily integrated and where $\int VU'\,dx$ is usually (but not always) a *simpler* integral than $\int UV'\,dx$. The technique is called integration by parts because it replaces one integral with the sum of an integrated term and another integral that remains to be evaluated. That is, it accomplishes only *part* of the original integration.

**Example 1**    $\displaystyle\int xe^x\,dx$                Let    $U = x$        $dV = e^x\,dx$

Then  $dU = dx$        $V = e^x$

$$= xe^x - \int e^x\,dx \qquad \text{(this is, } UV - \int V\,dU\text{)}$$

$$= xe^x - e^x + C.$$

■

Note the form in which the integration by parts is carried out. We indicate at the side what choices we are making for $U$ and $dV$ and then calculate $dU$ and $V$ from these. However, we do not actually substitute $U$ and $V$ into the integral; instead we use the formula $\int U\,dV = UV - \int V\,dU$ as a pattern or mnemonic device to replace the given integral by the equivalent partially integrated form on the second line.

Note also that had we included a constant of integration with $V$, (for example, $V = e^x + K$), that constant would cancel out in the next step:

$$\int xe^x\,dx = x(e^x + K) - \int (e^x + K)\,dx$$

$$= xe^x + Kx - e^x - Kx + C = xe^x - e^x + C.$$

In general, do not include a constant of integration with $V$ or on the right-hand side until the last integral has been evaluated.

Study the various parts of the following example carefully; they show the various ways in which integration by parts is used, and they give some insights into what choices should be made for $U$ and $dV$ in various situations. An improper choice can result in making an integral harder rather than easier. Look for a factor of the integrand that is easily integrated, and include $dx$ with that factor to make up $dV$. Then $U$ is the remaining factor of the integrand. Sometimes it is necessary to take $dV = dx$ only. When breaking up an integrand using integration by parts, choose $U$ and $dV$ so that, if possible, $V\,dU$ is "simpler" (easier to integrate) than $U\,dV$.

**Example 2**    Use integration by parts to evaluate:

(a) $\displaystyle\int \ln x\,dx$,    (b) $\displaystyle\int x^2 \sin x\,dx$,    (c) $\displaystyle\int x \tan^{-1} x\,dx$,    (d) $\displaystyle\int \sin^{-1} x\,dx$.

**Solution**

(a)  $\displaystyle\int \ln x\,dx$                Let    $U = \ln x$        $dV = dx$

Then  $dU = dx/x$        $V = x$

$$= x \ln x - \int x \frac{1}{x}\,dx$$

$$= x \ln x - x + C.$$

(b) We have to integrate by parts twice this time:

$$\int x^2 \sin x \, dx \qquad\qquad \text{Let} \quad U = x^2 \qquad dV = \sin x \, dx$$
$$\text{Then } dU = 2x \, dx \quad V = -\cos x$$

$$= -x^2 \cos x + 2 \int x \cos x \, dx \qquad \text{Let} \quad U = x \qquad dV = \cos x \, dx$$
$$\text{Then } dU = dx \quad V = \sin x$$

$$= -x^2 \cos x + 2 \left( x \sin x - \int \sin x \, dx \right)$$

$$= -x^2 \cos x + 2x \sin x + 2 \cos x + C.$$

(c) $$\int x \tan^{-1} x \, dx \qquad\qquad \text{Let} \quad U = \tan^{-1} x \qquad dV = x \, dx$$
$$\text{Then } dU = dx/(1+x^2) \quad V = \tfrac{1}{2} x^2$$

$$= \frac{1}{2} x^2 \tan^{-1} x - \frac{1}{2} \int \frac{x^2}{1+x^2} \, dx$$

$$= \frac{1}{2} x^2 \tan^{-1} x - \frac{1}{2} \int \left( 1 - \frac{1}{1+x^2} \right) dx$$

$$= \frac{1}{2} x^2 \tan^{-1} x - \frac{1}{2} x + \frac{1}{2} \tan^{-1} x + C.$$

(d) $$\int \sin^{-1} x \, dx \qquad\qquad \text{Let} \quad U = \sin^{-1} x \qquad dV = dx$$
$$\text{Then } dU = dx/\sqrt{1-x^2} \quad V = x$$

$$= x \sin^{-1} x - \int \frac{x}{\sqrt{1-x^2}} \, dx \qquad \text{Let } u = 1 - x^2,$$
$$du = -2x \, dx$$

$$= x \sin^{-1} x + \frac{1}{2} \int u^{-1/2} \, du$$

$$= x \sin^{-1} x + u^{1/2} + C = x \sin^{-1} x + \sqrt{1-x^2} + C.$$

∎

The following are two useful rules of thumb for choosing $U$ and $dV$:

(i) If the integrand involves a polynomial multiplied by an exponential, a sine or a cosine, or some other readily integrable function, try $U =$ the polynomial, and $dV =$ the rest.

(ii) If the integrand involves a logarithm, an inverse trigonometric function, or some other function that is not readily integrable but whose derivative is readily calculated, try that function for $U$ and let $dV$ equal the rest.

(Of course, these "rules" come with no guarantee. They may fail to be helpful if "the rest" is not of a suitable form. There remain many integrals that cannot be evaluated by any of the standard techniques presented in this chapter.)

The following two examples illustrate a frequently occurring and very useful phenomenon. It may happen after one or two integrations by parts, with the possible application of some known identity, that the original integral reappears on the right-hand side. Unless its coefficient there is 1, we have an equation that can be solved for that integral.

**Example 3**    Evaluate $I = \displaystyle\int \sec^3 x \, dx$.

**Solution**    Start by integrating by parts:

$$I = \int \sec^3 x \, dx \qquad \text{Let} \quad U = \sec x \qquad\qquad dV = \sec^2 x \, dx$$
$$\qquad\qquad\qquad\qquad\qquad \text{Then } dU = \sec x \, \tan x \, dx \qquad V = \tan x$$

$$= \sec x \, \tan x - \int \sec x \, \tan^2 x \, dx$$

$$= \sec x \, \tan x - \int \sec x (\sec^2 x - 1) \, dx$$

$$= \sec x \, \tan x - \int \sec^3 x \, dx + \int \sec x \, dx$$

$$= \sec x \, \tan x - I + \ln|\sec x + \tan x|.$$

This is an equation that can be solved for the desired integral $I$: Since $2I = \sec x \, \tan x + \ln|\sec x + \tan x|$, we have

$$\int \sec^3 x \, dx = I = \frac{1}{2} \sec x \, \tan x + \frac{1}{2} \ln|\sec x + \tan x| + C.$$

This integral occurs frequently in applications and is worth remembering. ∎

**Example 4**    Find $I = \displaystyle\int e^{ax} \cos bx \, dx$.

**Solution**    If either $a = 0$ or $b = 0$ the integral is easy to do, so let us assume $a \neq 0$ and $b \neq 0$. We have

$$I = \int e^{ax} \cos bx \, dx, \qquad \text{Let} \quad U = e^{ax} \qquad\qquad dV = \cos bx \, dx$$
$$\qquad\qquad\qquad\qquad\qquad\qquad \text{Then } dU = a \, e^{ax} \, dx \qquad V = (1/b) \sin bx$$

$$= \frac{1}{b} e^{ax} \sin bx - \frac{a}{b} \int e^{ax} \sin bx \, dx$$

$$\qquad\qquad\qquad\qquad\qquad \text{Let} \quad U = e^{ax} \qquad\qquad dV = \sin bx \, dx$$
$$\qquad\qquad\qquad\qquad\qquad \text{Then } dU = ae^{ax} dx \qquad V = -(\cos bx)/b$$

$$= \frac{1}{b} e^{ax} \sin bx - \frac{a}{b} \left( -\frac{1}{b} e^{ax} \cos bx + \frac{a}{b} \int e^{ax} \cos bx \, dx \right)$$

$$= \frac{1}{b} e^{ax} \sin bx + \frac{a}{b^2} e^{ax} \cos bx - \frac{a^2}{b^2} I.$$

Thus

$$\left( 1 + \frac{a^2}{b^2} \right) I = \frac{1}{b} e^{ax} \sin bx + \frac{a}{b^2} e^{ax} \cos bx + C_1$$

and

$$\int e^{ax} \cos bx \, dx = I = \frac{b \, e^{ax} \, \sin bx + a \, e^{ax} \, \cos bx}{b^2 + a^2} + C.$$



Observe that after the first integration by parts we had an integral that was different from, but no simpler than, the original integral. At this point we might have become discouraged and given up on this method. However, perseverance proved worthwhile; a second integration by parts returned the original integral $I$ in an equation that could be solved for $I$. Having chosen to let $U$ be the exponential in the first integration by parts (we could have let it be the cosine), we made the same choice for $U$ in the second integration by parts. Had we switched horses in midstream and decided to let $U$ be the trigonometric function the second time, we would have obtained

$$I = \frac{1}{b} e^{ax} \sin bx - \frac{1}{b} e^{ax} \sin bx + I;$$

we would have *undone* what we accomplished in the first step.

If we want to evaluate a definite integral by the method of integration by parts, we must remember to include the appropriate evaluation symbol with the integrated term.

**Example 5**  **(A definite integral)**

$$\int_1^e x^3 (\ln x)^2 \, dx \qquad \text{Let} \quad U = (\ln x)^2 \qquad dV = x^3 \, dx$$
$$\text{Then } dU = (2 \ln x \, dx)/x \qquad V = x^4/4$$

$$= \frac{x^4}{4} (\ln x)^2 \Big|_1^e - \frac{1}{2} \int_1^e x^3 \ln x \, dx \qquad \text{Let} \quad U = \ln x \qquad dV = x^3 \, dx$$
$$\text{Then} \quad dU = dx/x \qquad V = x^4/4$$

$$= \frac{e^4}{4} (1^2) - 0 - \frac{1}{2} \left( \frac{x^4}{4} \ln x \Big|_1^e - \frac{1}{4} \int_1^e x^3 \, dx \right)$$

$$= \frac{e^4}{4} - \frac{e^4}{8} + \frac{1}{8} \frac{x^4}{4} \Big|_1^e = \frac{e^4}{8} + \frac{e^4}{32} - \frac{1}{32} = \frac{5}{32} e^4 - \frac{1}{32}.$$

## Reduction Formulas

Consider the problem of finding $\int x^4 e^{-x} \, dx$. We can, as in Example 1, proceed by using integration by parts four times. Each time will reduce the power of $x$ by 1. Since this is repetitive and tedious, we prefer the following approach. For $n \geq 0$ let

$$I_n = \int x^n e^{-x} \, dx.$$

We want to find $I_4$. If we integrate by parts, we obtain a formula for $I_n$ in terms of $I_{n-1}$:

$$I_n = \int x^n e^{-x} \, dx \qquad \text{Let} \quad U = x^n \qquad dV = e^{-x} \, dx$$
$$\text{Then } dU = nx^{n-1} \, dx \qquad V = -e^{-x}$$

$$= -x^n e^{-x} + n \int x^{n-1} e^{-x} \, dx = -x^n e^{-x} + n I_{n-1}.$$

The formula

$$I_n = -x^n e^{-x} + n I_{n-1}$$

is called a **reduction formula** because it gives the value of the integral $I_n$ in terms of $I_{n-1}$, an integral corresponding to a reduced value of the exponent $n$. Starting with

$$I_0 = \int x^0 e^{-x}\, dx = \int e^{-x}\, dx = -e^{-x} + C,$$

we can apply the reduction formula four times to get

$$I_1 = -xe^{-x} + I_0 = -e^{-x}(x+1) + C_1$$
$$I_2 = -x^2 e^{-x} + 2I_1 = -e^{-x}(x^2 + 2x + 2) + C_2$$
$$I_3 = -x^3 e^{-x} + 3I_2 = -e^{-x}(x^3 + 3x^2 + 6x + 6) + C_3$$
$$I_4 = -x^4 e^{-x} + 4I_3 = -e^{-x}(x^4 + 4x^3 + 12x^2 + 24x + 24) + C_4.$$

**Example 6**    Obtain and use a reduction formula to evaluate

$$I_n = \int_0^{\pi/2} \cos^n x\, dx \qquad (n = 0,\ 1,\ 2,\ 3,\ \dots).$$

**Solution**    Observe first that

$$I_0 = \int_0^{\pi/2} dx = \frac{\pi}{2} \quad \text{and} \quad I_1 = \int_0^{\pi/2} \cos x\, dx = \sin x \Big|_0^{\pi/2} = 1.$$

Now let $n \geq 2$:

$$I_n = \int_0^{\pi/2} \cos^n x\, dx = \int_0^{\pi/2} \cos^{n-1} x \cos x\, dx$$
$$U = \cos^{n-1} x, \qquad dV = \cos x\, dx$$
$$dU = -(n-1)\cos^{n-2} x \sin x\, dx, \qquad V = \sin x$$
$$= \sin x \cos^{n-1} x \Big|_0^{\pi/2} + (n-1)\int_0^{\pi/2} \cos^{n-2} x \sin^2 x\, dx$$
$$= 0 - 0 + (n-1)\int_0^{\pi/2} \cos^{n-2} x\, (1 - \cos^2 x)\, dx$$
$$= (n-1)I_{n-2} - (n-1)I_n.$$

Transposing the term $-(n-1)I_n$, we obtain $nI_n = (n-1)I_{n-2}$, or

$$I_n = \frac{n-1}{n} I_{n-2},$$

which is the required reduction formula. It is valid for $n \geq 2$, which was needed to ensure that $\cos^{n-1}(\pi/2) = 0$. If $n \geq 2$ is an *even integer*, we have

$$I_n = \frac{n-1}{n} I_{n-2} = \frac{n-1}{n} \cdot \frac{n-3}{n-2} I_{n-4} = \cdots$$
$$= \frac{n-1}{n} \cdot \frac{n-3}{n-2} \cdot \frac{n-5}{n-4} \cdots \frac{5}{6} \cdot \frac{3}{4} \cdot \frac{1}{2} \cdot I_0$$
$$= \frac{n-1}{n} \cdot \frac{n-3}{n-2} \cdot \frac{n-5}{n-4} \cdots \frac{5}{6} \cdot \frac{3}{4} \cdot \frac{1}{2} \cdot \frac{\pi}{2}.$$

If $n \geq 3$ is an *odd* integer, we have

$$
\begin{aligned}
I_n &= \frac{n-1}{n} \cdot \frac{n-3}{n-2} \cdot \frac{n-5}{n-4} \cdots \frac{6}{7} \cdot \frac{4}{5} \cdot \frac{2}{3} \cdot I_1 \\
&= \frac{n-1}{n} \cdot \frac{n-3}{n-2} \cdot \frac{n-5}{n-4} \cdots \frac{6}{7} \cdot \frac{4}{5} \cdot \frac{2}{3}.
\end{aligned}
$$

See Exercise 38 for an interesting consequence of these formulas.

## Exercises 6.1

Evaluate the integrals in Exercises 1–28.

**1.** $\displaystyle\int x \cos x \, dx$

**2.** $\displaystyle\int (x+3)e^{2x} \, dx$

**3.** $\displaystyle\int x^2 \cos \pi x \, dx$

**4.** $\displaystyle\int (x^2 - 2x)e^{kx} \, dx$

**5.** $\displaystyle\int x^3 \ln x \, dx$

**6.** $\displaystyle\int x(\ln x)^3 \, dx$

**7.** $\displaystyle\int \tan^{-1} x \, dx$

**8.** $\displaystyle\int x^2 \tan^{-1} x \, dx$

**9.** $\displaystyle\int x \sin^{-1} x \, dx$

**10.** $\displaystyle\int x^5 e^{-x^2} \, dx$

**11.** $\displaystyle\int_0^{\pi/4} \sec^5 x \, dx$

**12.** $\displaystyle\int \tan^2 x \sec x \, dx$

**13.** $\displaystyle\int e^{2x} \sin 3x \, dx$

**14.** $\displaystyle\int x e^{\sqrt{x}} \, dx$

**\* 15.** $\displaystyle\int_{1/2}^1 \frac{\sin^{-1} x}{x^2} \, dx$

**16.** $\displaystyle\int_0^1 \sqrt{x} \sin(\pi \sqrt{x}) \, dx$

**17.** $\displaystyle\int x \sec^2 x \, dx$

**18.** $\displaystyle\int x \sin^2 x \, dx$

**19.** $\displaystyle\int \cos(\ln x) \, dx$

**20.** $\displaystyle\int_1^e \sin(\ln x) \, dx$

**21.** $\displaystyle\int \frac{\ln(\ln x)}{x} \, dx$

**22.** $\displaystyle\int_0^4 \sqrt{x} e^{\sqrt{x}} \, dx$

**23.** $\displaystyle\int \arccos x \, dx$

**24.** $\displaystyle\int x \sec^{-1} x \, dx$

**25.** $\displaystyle\int_1^2 \sec^{-1} x \, dx$

**\* 26.** $\displaystyle\int (\sin^{-1} x)^2 \, dx$

**\* 27.** $\displaystyle\int x(\tan^{-1} x)^2 \, dx$

**\* 28.** $\displaystyle\int x e^x \cos x \, dx$

**29.** Find the area below $y = e^{-x} \sin x$ and above $y = 0$ from $x = 0$ to $x = \pi$.

**30.** Find the area of the finite plane region bounded by the curve $y = \ln x$, the line $y = 1$, and the tangent line to $y = \ln x$ at $x = 1$.

**Reduction formulas**

**31.** Obtain a reduction formula for $I_n = \int (\ln x)^n \, dx$ and use it to evaluate $I_4$.

**32.** Obtain a reduction formula for $I_n = \int_0^{\pi/2} x^n \sin x \, dx$ and use it to evaluate $I_6$.

**33.** Obtain a reduction formula for $I_n = \int \sin^n x \, dx$ (where $n \geq 2$), and use it to find $I_6$ and $I_7$.

**34.** Obtain a reduction formula for $I_n = \int \sec^n x \, dx$ (where $n \geq 3$), and use it to find $I_6$ and $I_7$.

**\* 35.** By writing

$$
\begin{aligned}
I_n &= \int \frac{dx}{(x^2 + a^2)^n} \\
&= \frac{1}{a^2} \int \frac{dx}{(x^2 + a^2)^{n-1}} - \frac{1}{a^2} \int x \frac{x}{(x^2 + a^2)^n} \, dx
\end{aligned}
$$

and integrating the last integral by parts, using $U = x$, obtain a reduction formula for $I_n$. Use this formula to find $I_3$.

**\* 36.** If $f$ is twice differentiable on $[a, b]$ and $f(a) = f(b) = 0$, show that

$$
\int_a^b (x - a)(b - x) f''(x) \, dx = -2 \int_a^b f(x) \, dx.
$$

(*Hint:* Use integration by parts on the left-hand side twice.) This formula will be used in the Section 6.6 to construct an error estimate for the Trapezoid Rule approximation formula.

**\* 37.** If $f$ and $g$ are two functions having continuous second derivatives on the interval $[a, b]$, and if $f(a) = g(a) = f(b) = g(b) = 0$, show that

$$
\int_a^b f(x) g''(x) \, dx = \int_a^b f''(x) g(x) \, dx.
$$

What other assumptions about the values of $f$ and $g$ at $a$ and $b$ would give the same result?

* **38. (The Wallis Product)** Let $I_n = \int_0^{\pi/2} \cos^n x \, dx$.

(a) Use the fact that $0 \le \cos x \le 1$ for $0 \le x \le \pi/2$ to show that $I_{2n+2} \le I_{2n+1} \le I_{2n}$, for $n = 0, 1, 2, \dots$.

(b) Use the reduction formula $I_n = ((n-1)/n)I_{n-2}$ obtained in Example 6, together with the result of (a) to show that
$$\lim_{n \to \infty} \frac{I_{2n+1}}{I_{2n}} = 1.$$

(c) Combine the result of (b) with the explicit formulas

obtained for $I_n$ (for even and odd $n$) in Example 6 to show that

$$\lim_{n \to \infty} \frac{2}{1} \cdot \frac{2}{3} \cdot \frac{4}{3} \cdot \frac{4}{5} \cdot \frac{6}{5} \cdot \frac{6}{7} \cdots \frac{2n}{2n-1} \cdot \frac{2n}{2n+1} = \frac{\pi}{2}.$$

This interesting product formula for $\pi$ is due to the seventeenth century English mathematician John Wallis, and is referred to as the Wallis Product.

## 6.2  Inverse Substitutions

The substitutions considered in the Section 5.6 were direct substitutions in the sense that we simplified an integrand by replacing an expression appearing in it with a single variable. In this section we consider the reverse approach; we replace the variable of integration with a function of a new variable. Such substitutions, called *inverse substitutions*, would appear on the surface to make the integral more complicated. That is, substituting $x = g(u)$ in the integral

$$\int_a^b f(x) \, dx$$

leads to the more "complicated" integral

$$\int_{x=a}^{x=b} f\big(g(u)\big) g'(u) \, du.$$

As we will see, however, sometimes such substitutions can actually simplify an integrand, transforming the integral into one that can be evaluated by inspection or to which other techniques can readily be applied.

### The Inverse Trigonometric Substitutions

Three very useful inverse substitutions are:

$$x = a \sin \theta, \qquad x = a \tan \theta, \quad \text{and} \quad x = a \sec \theta.$$

These correspond to the direct substitutions:

$$\theta = \sin^{-1} \frac{x}{a}, \qquad \theta = \tan^{-1} \frac{x}{a}, \quad \text{and} \quad \theta = \sec^{-1} \frac{x}{a} = \cos^{-1} \frac{a}{x}.$$

**The inverse sine substitution**

Integrals involving $\sqrt{a^2 - x^2}$ (where $a > 0$) can frequently be reduced to a simpler form by means of the substitution

$$x = a \sin \theta \quad \text{or, equivalently,} \quad \theta = \sin^{-1} \frac{x}{a}.$$

Observe that $\sqrt{a^2 - x^2}$ makes sense only if $-a \le x \le a$, which corresponds to $-\pi/2 \le \theta \le \pi/2$. Since $\cos\theta \ge 0$ for such $\theta$, we have

$$\sqrt{a^2 - x^2} = \sqrt{a^2(1 - \sin^2\theta)} = \sqrt{a^2 \cos^2\theta} = a\cos\theta.$$

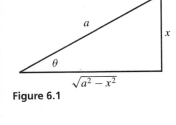

**Figure 6.1**

(If $\cos\theta$ were not nonnegative, we would have obtained $a|\cos\theta|$ instead.) If needed, the other trigonometric functions of $\theta$ can be recovered in terms of $x$ by examining a right-angled triangle labelled to correspond to the substitution. The expression $\sqrt{a^2 - x^2}$ suggests that $a$ should be the hypotenuse (see Figure 6.1):

$$\cos\theta = \frac{\sqrt{a^2 - x^2}}{a} \qquad \text{and} \qquad \tan\theta = \frac{x}{\sqrt{a^2 - x^2}}.$$

---

**Example 1**   Evaluate $\displaystyle\int \frac{1}{(5 - x^2)^{3/2}}\, dx$.

**Solution**   Refer to Figure 6.2.

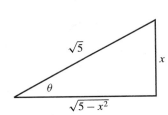

**Figure 6.2**

$$\int \frac{1}{(5 - x^2)^{3/2}}\, dx \qquad\qquad \text{Let } x = \sqrt{5}\sin\theta,$$
$$\qquad\qquad\qquad\qquad dx = \sqrt{5}\cos\theta\, d\theta$$

$$= \int \frac{\sqrt{5}\cos\theta\, d\theta}{5^{3/2}\cos^3\theta}$$

$$= \frac{1}{5}\int \sec^2\theta\, d\theta = \frac{1}{5}\tan\theta + C = \frac{1}{5}\frac{x}{\sqrt{5 - x^2}} + C$$

---

**Example 2**   Find the area of the circular segment shaded in Figure 6.3.

**Solution**   The area is

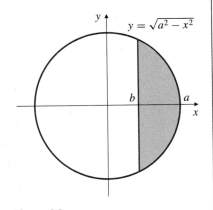

**Figure 6.3**

$$A = 2\int_b^a \sqrt{a^2 - x^2}\, dx \qquad\qquad \text{Let } x = a\sin\theta,$$
$$\qquad\qquad\qquad\qquad dx = a\cos\theta\, d\theta$$

$$= 2\int_{x=b}^{x=a} a^2 \cos^2\theta\, d\theta$$

$$= a^2 \left(\theta + \sin\theta\cos\theta\right)\Big|_{x=b}^{x=a} \qquad \text{(as in Example 8 of Section 5.6)}$$

$$= a^2 \left(\sin^{-1}\frac{x}{a} + \frac{x\sqrt{a^2 - x^2}}{a^2}\right)\Big|_b^a \qquad \text{(See Figure 6.1.)}$$

$$= \frac{\pi}{2}a^2 - a^2 \sin^{-1}\frac{b}{a} - b\sqrt{a^2 - b^2} \text{ square units.}$$

> **The inverse tangent substitution**
>
> Integrals involving $\sqrt{a^2 + x^2}$ or $\dfrac{1}{x^2 + a^2}$ (where $a > 0$) are often simplified by the substitution
>
> $$x = a \tan\theta \quad \text{or, equivalently,} \quad \theta = \tan^{-1}\frac{x}{a}.$$

Since $x$ can take any real value, we have $-\pi/2 < \theta < \pi/2$, so $\sec\theta > 0$ and

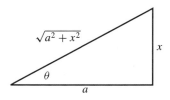

**Figure 6.4**

$$\sqrt{a^2 + x^2} = a\sqrt{1 + \tan^2\theta} = a\sec\theta.$$

Other trigonometric functions of $\theta$ can be expressed in terms of $x$ by referring to a right-angled triangle with legs $a$ and $x$ and hypotenuse $\sqrt{a^2 + x^2}$ (see Figure 6.4):

$$\sin\theta = \frac{x}{\sqrt{a^2 + x^2}} \quad \text{and} \quad \cos\theta = \frac{a}{\sqrt{a^2 + x^2}}.$$

**Example 3**   Evaluate   (a) $\displaystyle\int \frac{1}{\sqrt{4 + x^2}}\,dx$   and   (b) $\displaystyle\int \frac{1}{(1 + 9x^2)^2}\,dx$.

***Solution***   Figures 6.5 and 6.6 illustrate parts (a) and (b), respectively.

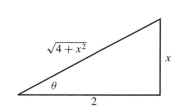

**Figure 6.5**

(a) $\displaystyle\int \frac{1}{\sqrt{4 + x^2}}\,dx$   Let $x = 2\tan\theta$,
        $dx = 2\sec^2\theta\,d\theta$

$$= \int \frac{2\sec^2\theta}{2\sec\theta}\,d\theta$$

$$= \int \sec\theta\,d\theta$$

$$= \ln|\sec\theta + \tan\theta| + C = \ln\left|\frac{\sqrt{4 + x^2}}{2} + \frac{x}{2}\right| + C$$

$$= \ln\left(\sqrt{4 + x^2} + x\right) + C_1, \qquad \text{where } C_1 = C - \ln 2.$$

(Note that $\sqrt{4 + x^2} + x > 0$ for all $x$, so we do not need an absolute value on it.)

(b) $\displaystyle\int \frac{1}{(1 + 9x^2)^2}\,dx$   Let $3x = \tan\theta$,
        $3\,dx = \sec^2\theta\,d\theta$,
        $1 + 9x^2 = \sec^2\theta$

$$= \frac{1}{3}\int \frac{\sec^2\theta\,d\theta}{\sec^4\theta}$$

$$= \frac{1}{3}\int \cos^2\theta\,d\theta = \frac{1}{6}(\theta + \sin\theta\,\cos\theta) + C$$

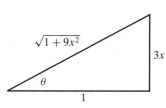

**Figure 6.6**

$$= \frac{1}{6}\tan^{-1}(3x) + \frac{1}{6}\frac{3x}{\sqrt{1 + 9x^2}}\frac{1}{\sqrt{1 + 9x^2}} + C$$

$$= \frac{1}{6}\tan^{-1}(3x) + \frac{1}{2}\frac{x}{1 + 9x^2} + C$$

### The inverse secant substitution

Integrals involving $\sqrt{x^2 - a^2}$ (where $a > 0$) can frequently be simplified by using the substitution

$$x = a \sec \theta \quad \text{or, equivalently,} \quad \theta = \sec^{-1}\frac{x}{a}.$$

We must be more careful with this substitution. Although

$$\sqrt{x^2 - a^2} = a\sqrt{\sec^2 \theta - 1} = a\sqrt{\tan^2 \theta} = a|\tan \theta|,$$

we cannot always drop the absolute value from the tangent. Observe that $\sqrt{x^2 - a^2}$ makes sense for $x \geq a$ and for $x \leq -a$.

If $x \geq a$, then $0 \leq \theta = \sec^{-1}\frac{x}{a} = \arccos\frac{a}{x} < \frac{\pi}{2}$, and $\tan \theta \geq 0$.

If $x \leq -a$, then $\frac{\pi}{2} < \theta = \sec^{-1}\frac{x}{a} = \arccos\frac{a}{x} \leq \pi$, and $\tan \theta \leq 0$.

In the first case $\sqrt{x^2 - a^2} = a \tan \theta$; in the second case $\sqrt{x^2 - a^2} = -a \tan \theta$.

---

**Example 4**  Find $I = \displaystyle\int \frac{dx}{\sqrt{x^2 - a^2}}$, where $a > 0$.

**Solution**  For the moment assume that $x \geq a$. If $x = a \sec \theta$, then $dx = a \sec \theta \tan \theta \, d\theta$ and $\sqrt{x^2 - a^2} = a \tan \theta$ (Figure 6.7). Thus

$$I = \int \sec \theta \, d\theta = \ln|\sec \theta + \tan \theta| + C$$

$$= \ln\left|\frac{x}{a} + \frac{\sqrt{x^2 - a^2}}{a}\right| + C = \ln|x + \sqrt{x^2 - a^2}| + C_1,$$

where $C_1 = C - \ln a$. If $x \leq -a$, let $u = -x$ so that $u \geq a$ and $du = -dx$. We have

$$I = -\int \frac{du}{\sqrt{u^2 - a^2}} = -\ln|u + \sqrt{u^2 - a^2}| + C_1$$

$$= \ln\left|\frac{1}{-x + \sqrt{x^2 - a^2}} \frac{x + \sqrt{x^2 - a^2}}{x + \sqrt{x^2 - a^2}}\right| + C_1$$

$$= \ln\left|\frac{x + \sqrt{x^2 - a^2}}{-a^2}\right| + C_1 = \ln|x + \sqrt{x^2 - a^2}| + C_2,$$

where $C_2 = C_1 - 2\ln a$. Thus, in either case, we have

$$I = \ln|x + \sqrt{x^2 - a^2}| + C.$$

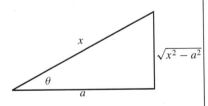

**Figure 6.7**

## Completing the Square

Quadratic expressions of the form $Ax^2 + Bx + C$ are often found in integrands. These can be written as sums or differences of squares using the procedure of completing the square. First factor out $A$ so that the remaining expression begins with $x^2 + 2bx$, where $2b = B/A$. These are the first two terms of $(x + b)^2 = x^2 + 2bx + b^2$. Add the third term $b^2 = B^2/4A^2$ and then subtract it again.

$$
\begin{aligned}
Ax^2 + Bx + C &= A\left(x^2 + \frac{B}{A}x + \frac{C}{A}\right) \\
&= A\left(x^2 + \frac{B}{A}x + \frac{B^2}{4A^2} + \frac{C}{A} - \frac{B^2}{4A^2}\right) \\
&= A\left(x + \frac{B}{2A}\right)^2 + \frac{4AC - B^2}{4A}
\end{aligned}
$$

The substitution $u = x + \dfrac{B}{2A}$ should then be made.

**Example 5**   Evaluate   (a) $\displaystyle\int \frac{1}{\sqrt{2x - x^2}}\,dx$ and   (b) $\displaystyle\int \frac{x}{4x^2 + 12x + 13}\,dx$.

**Solution**

(a) $\displaystyle\int \frac{1}{\sqrt{2x - x^2}}\,dx = \int \frac{dx}{\sqrt{1 - (1 - 2x + x^2)}}$

$$
\begin{aligned}
&= \int \frac{dx}{\sqrt{1 - (x-1)^2}} &&\text{Let } u = x - 1, \\
& &&du = dx \\
&= \int \frac{du}{\sqrt{1 - u^2}} = \sin^{-1} u + C = \sin^{-1}(x - 1) + C.
\end{aligned}
$$

(b) $\displaystyle\int \frac{x}{4x^2 + 12x + 13}\,dx = \int \frac{x\,dx}{4\left(x^2 + 3x + \dfrac{9}{4} + 1\right)}$

$$
\begin{aligned}
&= \frac{1}{4}\int \frac{x\,dx}{\left(x + \dfrac{3}{2}\right)^2 + 1} &&\text{Let } u = x + (3/2), \\
& &&du = dx, \\
& &&x = u - (3/2) \\
&= \frac{1}{4}\int \frac{u\,du}{u^2 + 1} - \frac{3}{8}\int \frac{du}{u^2 + 1} &&\text{In the first integral} \\
& &&\text{let } v = u^2 + 1, \\
& &&dv = 2u\,du \\
&= \frac{1}{8}\int \frac{dv}{v} - \frac{3}{8}\tan^{-1} u \\
&= \frac{1}{8}\ln|v| - \frac{3}{8}\tan^{-1} u + C \\
&= \frac{1}{8}\ln(4x^2 + 12x + 13) - \frac{3}{8}\tan^{-1}\left(x + \frac{3}{2}\right) + C_1,
\end{aligned}
$$

where $C_1 = C - (\ln 4)/8$.

## Other Inverse Substitutions

Integrals involving $\sqrt{ax + b}$ can sometimes be made simpler with the substitution $ax + b = u^2$.

**Example 6**

$$\int \frac{1}{1 + \sqrt{2x}}\, dx \qquad \text{Let } 2x = u^2,$$
$$\qquad\qquad\qquad\qquad\qquad 2\, dx = 2u\, du$$

$$= \int \frac{u}{1 + u}\, du = \int \frac{1 + u - 1}{1 + u}\, du$$

$$= \int \left(1 - \frac{1}{1 + u}\right) du \qquad \text{Let } v = 1 + u,$$
$$\qquad\qquad\qquad\qquad\qquad dv = du$$

$$= u - \int \frac{dv}{v} = u - \ln|v| + C$$

$$= \sqrt{2x} - \ln\left(1 + \sqrt{2x}\right) + C$$

Sometimes integrals involving $\sqrt[n]{ax + b}$ will be much simplified by the hybrid substitution $ax + b = u^n$, $a\, dx = n\, u^{n-1}\, du$.

**Example 7**

$$\int_{-1/3}^{2} \frac{x}{\sqrt[3]{3x + 2}}\, dx \qquad \text{Let } 3x + 2 = u^3,$$
$$\qquad\qquad\qquad\qquad\qquad\qquad 3\, dx = 3u^2\, du$$

$$= \int_{1}^{2} \frac{u^3 - 2}{3u}\, u^2\, du$$

$$= \frac{1}{3} \int_{1}^{2} (u^4 - 2u)\, du = \frac{1}{3} \left(\frac{u^5}{5} - u^2\right)\Bigg|_{1}^{2} = \frac{16}{15}$$

Note that the limits were changed in this definite integral. $u = 1$ when $x = -1/3$, and, coincidentally, $u = 2$ when $x = 2$.

If more than one fractional power is present, it may be possible to eliminate all of them at once.

**Example 8** Evaluate $\displaystyle\int \frac{1}{x^{1/2}(1 + x^{1/3})}\, dx$.

**Solution** We can eliminate both the square root and the cube root by using the inverse substitution $x = u^6$. (The power 6 is chosen because 6 is the least common multiple of 2 and 3.)

$$\int \frac{dx}{x^{1/2}(1 + x^{1/3})} \qquad \text{Let } x = u^6,$$
$$\qquad\qquad\qquad\qquad\qquad dx = 6u^5\, du$$

$$= 6 \int \frac{u^5\, du}{u^3(1 + u^2)} = 6 \int \frac{u^2}{1 + u^2}\, du = 6 \int \left(1 - \frac{1}{1 + u^2}\right) du$$

$$= 6\,(u - \tan^{-1} u) + C = 6\,(x^{1/6} - \tan^{-1} x^{1/6}) + C.$$

## The tan(θ/2) Substitution

There is a certain special substitution that can transform an integral whose integrand is a rational function of $\sin\theta$ and $\cos\theta$ (that is, a quotient of polynomials in $\sin\theta$ and $\cos\theta$) into a rational function of $x$. The substitution is

$$x = \tan\frac{\theta}{2} \quad \text{or, equivalently,} \quad \theta = 2\tan^{-1}x.$$

Observe that

$$\cos^2\frac{\theta}{2} = \frac{1}{\sec^2\dfrac{\theta}{2}} = \frac{1}{1+\tan^2\dfrac{\theta}{2}} = \frac{1}{1+x^2},$$

so

$$\cos\theta = 2\cos^2\frac{\theta}{2} - 1 = \frac{2}{1+x^2} - 1 = \frac{1-x^2}{1+x^2}$$

$$\sin\theta = 2\sin\frac{\theta}{2}\cos\frac{\theta}{2} = 2\tan\frac{\theta}{2}\cos^2\frac{\theta}{2} = \frac{2x}{1+x^2}.$$

Also, $dx = \dfrac{1}{2}\sec^2\dfrac{\theta}{2}\,d\theta$, so

$$d\theta = 2\cos^2\frac{\theta}{2}\,dx = \frac{2\,dx}{1+x^2}.$$

In summary:

**The tan(θ/2) substitution**

If $x = \tan(\theta/2)$ then

$$\cos\theta = \frac{1-x^2}{1+x^2}, \qquad \sin\theta = \frac{2x}{1+x^2}, \qquad \text{and} \qquad d\theta = \frac{2\,dx}{1+x^2}.$$

Note that $\cos\theta$, $\sin\theta$, and $d\theta$ all involve only rational functions of $x$. We will be investigating general techniques for integrating rational functions of $x$ in Section 6.3.

**Example 9**
$$\int \frac{1}{2+\cos\theta}\,d\theta \qquad \text{Let } x = \tan(\theta/2), \text{ so}$$

$$\cos\theta = \frac{1-x^2}{1+x^2},$$

$$d\theta = \frac{2\,dx}{1+x^2}$$

$$= \int \frac{\dfrac{2\,dx}{1+x^2}}{2+\dfrac{1-x^2}{1+x^2}} = 2\int \frac{1}{3+x^2}\,dx$$

$$= \frac{2}{\sqrt3}\tan^{-1}\frac{x}{\sqrt3} + C$$

$$= \frac{2}{\sqrt3}\tan^{-1}\left(\frac{1}{\sqrt3}\tan\frac{\theta}{2}\right) + C.$$

# Exercises 6.2

Evaluate the integrals in Exercises 1–36.

**1.** $\displaystyle\int \frac{dx}{\sqrt{1-4x^2}}$    **2.** $\displaystyle\int \frac{x^2\,dx}{\sqrt{1-4x^2}}$

**3.** $\displaystyle\int \frac{x^2\,dx}{\sqrt{9-x^2}}$    **4.** $\displaystyle\int \frac{dx}{x\sqrt{1-4x^2}}$

**5.** $\displaystyle\int \frac{dx}{x^2\sqrt{9-x^2}}$    **6.** $\displaystyle\int \frac{dx}{x\sqrt{9-x^2}}$

**7.** $\displaystyle\int \frac{x+1}{\sqrt{9-x^2}}\,dx$    **8.** $\displaystyle\int \frac{dx}{\sqrt{9+x^2}}$

**9.** $\displaystyle\int \frac{x^3\,dx}{\sqrt{9+x^2}}$    **10.** $\displaystyle\int \frac{\sqrt{9+x^2}}{x^4}\,dx$

**11.** $\displaystyle\int \frac{dx}{(a^2-x^2)^{3/2}}$    **12.** $\displaystyle\int \frac{dx}{(a^2+x^2)^{3/2}}$

**13.** $\displaystyle\int \frac{x^2\,dx}{(a^2-x^2)^{3/2}}$    **14.** $\displaystyle\int \frac{dx}{(1+2x^2)^{5/2}}$

**15.** $\displaystyle\int \frac{dx}{x\sqrt{x^2-a^2}}$    **16.** $\displaystyle\int \frac{dx}{x^2\sqrt{x^2-a^2}}$

**17.** $\displaystyle\int \frac{dx}{x^2+2x+10}$    **18.** $\displaystyle\int \frac{dx}{x^2+x+1}$

**19.** $\displaystyle\int \frac{dx}{(4x^2+4x+5)^2}$    **20.** $\displaystyle\int \frac{x\,dx}{x^2-2x+3}$

**21.** $\displaystyle\int \frac{x\,dx}{\sqrt{2ax-x^2}}$    **22.** $\displaystyle\int \frac{dx}{(4x-x^2)^{3/2}}$

**23.** $\displaystyle\int \frac{x\,dx}{(3-2x-x^2)^{3/2}}$    **24.** $\displaystyle\int \frac{dx}{(x^2+2x+2)^2}$

**25.** $\displaystyle\int \frac{dx}{(1+x^2)^3}$    **26.** $\displaystyle\int \frac{x^2\,dx}{(1+x^2)^2}$

**\*27.** $\displaystyle\int \frac{\sqrt{1-x^2}}{x^3}\,dx$    **28.** $\displaystyle\int \sqrt{9+x^2}\,dx$

**29.** $\displaystyle\int \frac{dx}{2+\sqrt{x}}$    **30.** $\displaystyle\int \frac{dx}{1+x^{1/3}}$

**\*31.** $\displaystyle\int \frac{1+x^{1/2}}{1+x^{1/3}}\,dx$    **\*32.** $\displaystyle\int \frac{x\sqrt{2-x^2}}{\sqrt{x^2+1}}\,dx$

**33.** $\displaystyle\int_{-\ln 2}^{0} e^x\sqrt{1-e^{2x}}\,dx$    **34.** $\displaystyle\int_{0}^{\pi/2} \frac{\cos x}{\sqrt{1+\sin^2 x}}\,dx$

**35.** $\displaystyle\int_{-1}^{\sqrt{3}-1} \frac{dx}{x^2+2x+2}$    **36.** $\displaystyle\int_{1}^{2} \frac{dx}{x^2\sqrt{9-x^2}}$

In Exercises 37–39, evaluate the integral using the special substitution $x=\tan(\theta/2)$ as in Example 9.

**\*37.** $\displaystyle\int \frac{d\theta}{2+\sin\theta}$    **\*38.** $\displaystyle\int_{0}^{\pi/2} \frac{d\theta}{1+\cos\theta+\sin\theta}$

**\*39.** $\displaystyle\int \frac{d\theta}{3+2\cos\theta}$

**40.** Find the area of the region bounded by $y=(2x-x^2)^{-1/2}$, $y=0$, $x=1/2$, and $x=1$.

**41.** Find the area of the region lying below $y=9/(x^4+4x^2+4)$ and above $y=1$.

**42.** Find the average value of the function $f(x)=(x^2-4x+8)^{-3/2}$ over the interval $[0,4]$.

**43.** Find the area inside the circle $x^2+y^2=a^2$ and above the line $y=b$, $(-a\le b\le a)$.

**44.** Find the area inside both of the circles $x^2+y^2=1$ and $(x-2)^2+y^2=4$.

**45.** Find the area in the first quadrant, above the hyperbola $xy=12$ and inside the circle $x^2+y^2=25$.

**46.** Find the area to the left of $\dfrac{x^2}{a^2}+\dfrac{y^2}{b^2}=1$ and to the right of the line $x=c$, where $-a\le c\le a$.

**\*47.** Find the area of the region bounded by the $x$-axis, the hyperbola $x^2-y^2=1$, and the straight line from the origin to the point $(\sqrt{1+Y^2},Y)$ on that hyperbola. (Assume $Y>0$.) In particular, show that the area is $t/2$ square units if $Y=\sinh t$.

**\*48.** Evaluate the integrals

$$\int \frac{dx}{\sqrt{x^2-a^2}} \quad\text{and}\quad \int \frac{dx}{x^2\sqrt{x^2-a^2}},$$

using the substitution $x=a\cosh u$. (*Hint:* Review the properties of the hyperbolic functions in Section 3.6.) This substitution is an alternative to $x=a\sec\theta$ when dealing with $\sqrt{x^2-a^2}$.

## 6.3 Integrals of Rational Functions

In this section we are concerned with integrals of the form

$$\int \frac{P(x)}{Q(x)}\, dx,$$

where $P$ and $Q$ are polynomials. Recall that a **polynomial** is a function $P$ of the form

$$P(x) = a_n x^n + a_{n-1} x^{n-1} + \cdots + a_2 x^2 + a_1 x + a_0$$

where $n$ is a nonnegative integer, $a_0$, $a_1$, $a_2$, $\ldots$, $a_n$ are constants, and $a_n \neq 0$. We call $n$ the **degree** of $P$. A quotient $P(x)/Q(x)$ of two polynomials is called a **rational function**. We need normally concern ourselves only with rational functions $P(x)/Q(x)$ where the degree of $P$ is less than that of $Q$. If the degree of $P$ equals or exceeds the degree of $Q$, then we can use long division or some equivalent procedure to express the fraction $P(x)/Q(x)$ as a polynomial plus another fraction $R(x)/Q(x)$ where $R$, the remainder in the division, has degree less than that of $Q$.

**Example 1**  Evaluate $\displaystyle\int \frac{x^3 + 3x^2}{x^2 + 1}\, dx$.

**Solution**  The numerator has degree 3 and the denominator has degree 2 so we need to divide. We use long division:

$$
\begin{array}{r}
x + 3 \phantom{xxxxx} \\
x^2 + 1 \,\overline{\smash{)}\, x^3 + 3x^2 \phantom{xxxx}} \\
\underline{x^3 \phantom{xxxx} + x} \\
3x^2 - x \phantom{xx}\\
\underline{3x^2 \phantom{xxx} + 3} \\
- x - 3
\end{array}
$$

$$\frac{x^3 + 3x^2}{x^2 + 1} = x + 3 - \frac{x + 3}{x^2 + 1}$$

Thus

$$\int \frac{x^3 + 3x^2}{x^2 + 1}\, dx = \int (x + 3)\, dx - \int \frac{x}{x^2 + 1}\, dx - 3 \int \frac{dx}{x^2 + 1}$$

$$= \frac{1}{2} x^2 + 3x - \frac{1}{2} \ln(x^2 + 1) - 3 \tan^{-1} x + C.$$

**Example 2**  Evaluate $\displaystyle\int \frac{x}{2x - 1}\, dx$.

**Solution**  The numerator and denominator have the same degree, 1, so division is again required. In this case the division can be carried out by manipulation of the integrand:

$$\frac{x}{2x - 1} = \frac{1}{2}\frac{2x}{2x - 1} = \frac{1}{2}\frac{2x - 1 + 1}{2x - 1} = \frac{1}{2}\left(1 + \frac{1}{2x - 1}\right),$$

a process that we can call *short division*. We have

$$\int \frac{x}{2x - 1}\, dx = \frac{1}{2} \int \left(1 + \frac{1}{2x - 1}\right) dx = \frac{x}{2} + \frac{1}{4} \ln|2x - 1| + C.$$

In the discussion that follows, we always assume that any necessary division has been performed and the quotient polynomial has been integrated. The remaining basic problem with which we will deal in this section is the following:

> **The basic problem**
>
> Evaluate $\displaystyle\int \frac{P(x)}{Q(x)}\,dx$ where the degree of $P <$ the degree of $Q$.

The complexity of this problem depends on the degree of $Q$.

## Linear and Quadratic Denominators

Suppose that $Q(x)$ has degree 1. Thus $Q(x) = ax + b$, where $a \neq 0$. Then $P(x)$ must have degree 0, and be a constant $c$. We have $P(x)/Q(x) = c/(ax + b)$. The substitution $u = ax + b$ leads to

$$\int \frac{c}{ax + b}\,dx = \frac{c}{a} \int \frac{du}{u} = \frac{c}{a} \ln|u| + C,$$

so that, for $c = 1$,

> **The case of a linear denominator**
>
> $$\int \frac{1}{ax + b}\,dx = \frac{1}{a} \ln|ax + b| + C.$$

Now suppose that $Q(x)$ is quadratic, that is, has degree 2. For purposes of this discussion we can assume that $Q(x)$ is either of the form $x^2 + a^2$ or of the form $x^2 - a^2$, since completing the square and making the appropriate change of variable can always reduce a quadratic denominator to this form, as shown in Section 6.2. Since $P(x)$ can be at most a linear function, $P(x) = Ax + B$, we are led to consider the following four integrals:

$$\int \frac{x\,dx}{x^2 + a^2}, \qquad \int \frac{x\,dx}{x^2 - a^2}, \qquad \int \frac{dx}{x^2 + a^2}, \quad \text{and} \quad \int \frac{dx}{x^2 - a^2}.$$

(If $a = 0$, there are only two integrals; each is easily evaluated.) The first two integrals yield to the substitution $u = x^2 \pm a^2$; the third is a known integral. The fourth integral can be done with the substitution $x = a \sin\theta$ if $|x| < |a|$, and with the substitution $x = a \sec\theta$ if $|x| > |a|$, but we will evaluate it by a different method below. The values of all four integrals are given in the following box:

> **The case of a quadratic denominator**
>
> $$\int \frac{x\,dx}{x^2 + a^2} = \frac{1}{2} \ln(x^2 + a^2) + C,$$
>
> $$\int \frac{x\,dx}{x^2 - a^2} = \frac{1}{2} \ln|x^2 - a^2| + C,$$
>
> $$\int \frac{dx}{x^2 + a^2} = \frac{1}{a} \tan^{-1} \frac{x}{a} + C,$$
>
> $$\int \frac{dx}{x^2 - a^2} = \frac{1}{2a} \ln\left|\frac{x - a}{x + a}\right| + C.$$

To obtain the last formula in the box let us try to write the integrand as a sum of two fractions with linear denominators:

$$\frac{1}{x^2 - a^2} = \frac{1}{(x-a)(x+a)} = \frac{A}{x-a} + \frac{B}{x+a} = \frac{Ax + Aa + Bx - Ba}{x^2 - a^2},$$

where we have added the two fractions together again in the last step. If this equation is to hold identically for all $x$ (except $x = \pm a$), then the numerators on the left and right sides must be identical as polynomials in $x$. The equation $(A + B)x + (Aa - Ba) = 1 = 0x + 1$ can hold for all $x$ only if

$$A + B = 0 \qquad \text{(the coefficient of } x)$$
$$Aa - Ba = 1 \qquad \text{(the constant term)}$$

Solving this pair of linear equations for the unknowns $A$ and $B$, we get $A = 1/(2a)$ and $B = -1/(2a)$. Therefore,

$$\begin{aligned}
\int \frac{dx}{x^2 - a^2} &= \frac{1}{2a} \int \frac{dx}{x-a} - \frac{1}{2a} \int \frac{dx}{x+a} \\
&= \frac{1}{2a} \ln|x - a| - \frac{1}{2a} \ln|x + a| + C \\
&= \frac{1}{2a} \ln\left|\frac{x-a}{x+a}\right| + C.
\end{aligned}$$

## Partial Fractions

The technique used above, involving the writing of a complicated fraction as a sum of simpler fractions, is called the **method of partial fractions**. Suppose that a polynomial $Q(x)$ is of degree $n$ and that its highest degree term is $x^n$ (with coefficient 1). Suppose also that $Q$ factors into a product of $n$ *distinct* linear (degree 1) factors, say

$$Q(x) = (x - a_1)(x - a_2) \cdots (x - a_n),$$

where $a_i \neq a_j$ if $i \neq j$, $1 \leq i, j \leq n$. If $P(x)$ is a polynomial of degree smaller than $n$, then $P(x)/Q(x)$ has a **partial fraction decomposition** of the form

$$\frac{P(x)}{Q(x)} = \frac{A_1}{x - a_1} + \frac{A_2}{x - a_2} + \cdots + \frac{A_n}{x - a_n}$$

for certain values of the constants $A_1, A_2, \ldots, A_n$. We do not attempt to give any formal proof of this assertion here; such a proof belongs in an algebra course. (See Theorem 1 below for the statement of a more general result.)

Given that $P(x)/Q(x)$ has a partial fraction decomposition as claimed above, there are two methods for determining the constants $A_1, A_2, \ldots, A_n$. The first of these methods, and one that generalizes most easily to the more complicated decompositions considered below, is to add up the fractions in the decomposition, obtaining a new fraction $S(x)/Q(x)$ with numerator $S(x)$, a polynomial of degree one less than that of $Q(x)$. This new fraction will be identical to the original fraction $P(x)/Q(x)$ if $S$ and $P$ are identical polynomials. The constants $A_1, A_2, \ldots, A_n$ are determined by solving the $n$ linear equations resulting from equating the coefficients of like powers of $x$ in the two polynomials $S$ and $P$.

The second method depends on the following observation: If we multiply the partial fraction decomposition by $x - a_j$, we get

$$(x - a_j) \frac{P(x)}{Q(x)}$$

$$= A_1 \frac{x - a_j}{x - a_1} + \cdots + A_{j-1} \frac{x - a_j}{x - a_{j-1}} + A_j + A_{j+1} \frac{x - a_j}{x - a_{j+1}} + \cdots + A_n \frac{x - a_j}{x - a_n}.$$

All terms on the right side are 0 at $x = a_j$ except the $j$th term, $A_j$. Hence

$$A_j = \lim_{x \to a_j} (x - a_j) \frac{P(x)}{Q(x)}$$

$$= \frac{P(a_j)}{(a_j - a_1) \cdots (a_j - a_{j-1})(a_j - a_{j+1}) \cdots (a_j - a_n)},$$

for $1 \le j \le n$. In practice, you can use this method to find each number $A_j$ by cancelling the factor $x - a_j$ from the denominator of $P(x)/Q(x)$ and evaluating the resulting expression at $x = a_j$.

**Example 3**   Evaluate $\displaystyle\int \frac{(x+4)}{x^2 - 5x + 6} \, dx$.

**Solution**   The partial fraction decomposition takes the form

$$\frac{x+4}{x^2 - 5x + 6} = \frac{x+4}{(x-2)(x-3)} = \frac{A}{x-2} + \frac{B}{x-3}.$$

We calculate $A$ and $B$ by both of the methods suggested above.

**Method I.** Add the partial fractions

$$\frac{x+4}{x^2 - 5x + 6} = \frac{Ax - 3A + Bx - 2B}{(x-2)(x-3)},$$

and equate the coefficient of $x$ and the constant terms in the numerators on both sides to obtain

$$A + B = 1 \qquad \text{and} \qquad -3A - 2B = 4.$$

Solve these equations to get $A = -6$ and $B = 7$.

**Method II.** To find $A$, cancel $x - 2$ from the denominator of the expression $P(x)/Q(x)$ and evaluate the result at $x = 2$. Obtain $B$ similarly.

$$A = \left. \frac{x+4}{x-3} \right|_{x=2} = -6, \qquad \text{and} \qquad B = \left. \frac{x+4}{x-2} \right|_{x=3} = 7.$$

In either case we have

$$\int \frac{(x+4)}{x^2 - 5x + 6} \, dx = -6 \int \frac{1}{x-2} \, dx + 7 \int \frac{1}{x-3} \, dx$$

$$= -6 \ln |x - 2| + 7 \ln |x - 3| + C.$$

**Example 4** Evaluate $I = \displaystyle\int \frac{x^3 + 2}{x^3 - x}\, dx$.

**Solution** Since the numerator does not have degree smaller than the denominator, we must divide

$$I = \int \frac{x^3 - x + x + 2}{x^3 - x}\, dx = \int \left( 1 + \frac{x + 2}{x^3 - x} \right) dx = x + \int \frac{x + 2}{x^3 - x}\, dx.$$

Now we can use the method of partial fractions.

$$\frac{x + 2}{x^3 - x} = \frac{x + 2}{x(x - 1)(x + 1)} = \frac{A}{x} + \frac{B}{x - 1} + \frac{C}{x + 1}$$
$$= \frac{A(x^2 - 1) + B(x^2 + x) + C(x^2 - x)}{x(x - 1)(x + 1)}$$

We have

$$\begin{aligned} A \ + \ B \ + \ C &= 0 \qquad \text{(coefficient of } x^2\text{)} \\ B \ - \ C &= 1 \qquad \text{(coefficient of } x\text{)} \\ -A \qquad\qquad &= 2 \qquad \text{(constant term).} \end{aligned}$$

It follows that $A = -2$, $B = 3/2$, and $C = 1/2$. We can also find these values using the Method II of the previous example:

$$A = \left.\frac{x + 2}{(x - 1)(x + 1)}\right|_{x=0} = -2, \qquad B = \left.\frac{x + 2}{x(x + 1)}\right|_{x=1} = \frac{3}{2}, \quad \text{and}$$
$$C = \left.\frac{x + 2}{x(x - 1)}\right|_{x=-1} = \frac{1}{2}.$$

Finally, we have

$$I = x - 2\int \frac{1}{x}\, dx + \frac{3}{2}\int \frac{1}{x - 1}\, dx + \frac{1}{2}\int \frac{1}{x + 1}\, dx$$
$$= x - 2\ln|x| + \frac{3}{2}\ln|x - 1| + \frac{1}{2}\ln|x + 1| + C.$$

Next we consider a rational function whose denominator has a quadratic factor that is equivalent to a sum of squares and that cannot, therefore, be further factored into a product of real linear factors.

**Example 5** Evaluate $\displaystyle\int \frac{2 + 3x + x^2}{x(x^2 + 1)}\, dx$.

**Solution** Note that the numerator has degree 2 and the denominator degree 3, so no division is necessary. If we decompose the integrand as a sum of two simpler fractions, we want one with denominator $x$ and one with denominator $x^2 + 1$. The appropriate form of the decomposition turns out to be

$$\frac{2 + 3x + x^2}{x(x^2 + 1)} = \frac{A}{x} + \frac{Bx + C}{x^2 + 1} = \frac{A(x^2 + 1) + Bx^2 + Cx}{x(x^2 + 1)}.$$

Note that corresponding to the quadratic (degree 2) denominator we use a linear (degree 1) numerator. Equating coefficients in the two numerators, we obtain

$$
\begin{aligned}
A + B &= 1 && \text{(coefficient of } x^2) \\
C &= 3 && \text{(coefficient of } x) \\
A &= 2 && \text{(constant term).}
\end{aligned}
$$

Hence $A = 2$, $B = -1$, and $C = 3$. We have, therefore,

$$
\int \frac{2 + 3x + x^2}{x(x^2 + 1)}\, dx = 2 \int \frac{1}{x}\, dx - \int \frac{x}{x^2 + 1}\, dx + 3 \int \frac{1}{x^2 + 1}\, dx
$$

$$
= 2 \ln |x| - \frac{1}{2} \ln(x^2 + 1) + 3 \tan^{-1} x + C.
$$

We remark that addition of the fractions is the only reasonable real-variable method for determining the constants $A$, $B$, and $C$ here. We could determine $A$ by Method II of Example 3, but there is no simple equivalent way of finding $B$ or $C$ without using complex numbers.

---

**Example 6**  Evaluate $I = \displaystyle\int \frac{1}{x^3 + 1}\, dx$.

**Solution**  Here $Q(x) = x^3 + 1 = (x + 1)(x^2 - x + 1)$. The latter factor has no real roots, and so has no real linear subfactors. We have

$$
\frac{1}{x^3 + 1} = \frac{1}{(x + 1)(x^2 - x + 1)} = \frac{A}{x + 1} + \frac{Bx + C}{x^2 - x + 1}
$$

$$
= \frac{A(x^2 - x + 1) + B(x^2 + x) + C(x + 1)}{(x + 1)(x^2 - x + 1)}
$$

$$
\begin{aligned}
A + B &= 0 && \text{(coefficient of } x^2) \\
- A + B + C &= 0 && \text{(coefficient of } x) \\
A + C &= 1 && \text{(constant term).}
\end{aligned}
$$

Hence $A = 1/3$, $B = -1/3$, and $C = 2/3$. We have

$$
I = \frac{1}{3} \int \frac{1}{x + 1} - \frac{1}{3} \int \frac{x - 2}{x^2 - x + 1}\, dx
$$

$$
= \frac{1}{3} \ln |x + 1| - \frac{1}{3} \int \frac{x - \dfrac{1}{2} - \dfrac{3}{2}}{\left(x - \dfrac{1}{2}\right)^2 + \dfrac{3}{4}}\, dx \qquad \text{Let } u = x - 1/2,
$$
$$
\qquad\qquad\qquad\qquad\qquad\qquad\qquad\qquad du = dx
$$

$$
= \frac{1}{3} \ln |x + 1| - \frac{1}{3} \int \frac{u}{u^2 + \dfrac{3}{4}}\, du + \frac{1}{2} \int \frac{1}{u^2 + \dfrac{3}{4}}\, du
$$

$$
= \frac{1}{3} \ln |x + 1| - \frac{1}{6} \ln\left(u^2 + \frac{3}{4}\right) + \frac{1}{2} \frac{2}{\sqrt{3}} \tan^{-1}\left(\frac{2u}{\sqrt{3}}\right) + C
$$

$$
= \frac{1}{3} \ln |x + 1| - \frac{1}{6} \ln(x^2 - x + 1) + \frac{1}{\sqrt{3}} \tan^{-1}\left(\frac{2x - 1}{\sqrt{3}}\right) + C.
$$

We require one final refinement of the method of partial fractions. If any of the linear or quadratic factors of $Q(x)$ is *repeated* (say $m$ times), then the partial fraction decomposition of $P(x)/Q(x)$ requires $m$ distinct fractions corresponding to that factor. The denominators of these fractions have exponents increasing from 1 to $m$, and the numerators are all constants where the repeated factor is linear, or linear where the repeated factor is quadratic. (See Theorem 1 below.)

**Example 7**   Evaluate $\displaystyle\int \frac{1}{x(x-1)^2}\, dx$.

**Solution**   The appropriate partial fraction decomposition here is

$$\frac{1}{x(x-1)^2} = \frac{A}{x} + \frac{B}{x-1} + \frac{C}{(x-1)^2}$$
$$= \frac{A(x^2 - 2x + 1) + B(x^2 - x) + Cx}{x(x-1)^2}.$$

Equating coefficients of $x^2$, $x$, and 1 in the numerators of both sides, we get

$$\begin{array}{rcll} A + B & = 0 & \text{(coefficient of } x^2) \\ -2A - B + C & = 0 & \text{(coefficient of } x) \\ A & = 1 & \text{(constant term).} \end{array}$$

Hence $A = 1$, $B = -1$, $C = 1$, and

$$\int \frac{1}{x(x-1)^2}\, dx = \int \frac{1}{x}\, dx - \int \frac{1}{x-1}\, dx + \int \frac{1}{(x-1)^2}\, dx$$
$$= \ln|x| - \ln|x-1| - \frac{1}{x-1} + C$$
$$= \ln\left|\frac{x}{x-1}\right| - \frac{1}{x-1} + C.$$

∎

**Example 8**   Evaluate $\displaystyle I = \int \frac{x^2 + 2}{4x^5 + 4x^3 + x}\, dx$.

**Solution**   The denominator factors to $x(2x^2 + 1)^2$, so the appropriate partial fraction decomposition is

$$\frac{x^2 + 2}{x(2x^2 + 1)^2} = \frac{A}{x} + \frac{Bx + C}{2x^2 + 1} + \frac{Dx + E}{(2x^2 + 1)^2}$$
$$= \frac{A(4x^4 + 4x^2 + 1) + B(2x^4 + x^2) + C(2x^3 + x) + Dx^2 + Ex}{x(2x^2 + 1)^2}$$

Thus

$$\begin{array}{rcll} 4A + 2B & = 0 & \text{(coefficient of } x^4) \\ 2C & = 0 & \text{(coefficient of } x^3) \\ 4A + B + D & = 1 & \text{(coefficient of } x^2) \\ C + E & = 0 & \text{(coefficient of } x) \\ A & = 2 & \text{(constant term).} \end{array}$$

Solving these equations, we get $A = 2$, $B = -4$, $C = 0$, $D = -3$, and $E = 0$.

$$I = 2 \int \frac{dx}{x} - 4 \int \frac{x\,dx}{2x^2 + 1} - 3 \int \frac{x\,dx}{(2x^2 + 1)^2} \qquad \text{Let } u = 2x^2 + 1,$$
$$du = 4x\,dx$$

$$= 2\ln|x| - \int \frac{du}{u} - \frac{3}{4} \int \frac{du}{u^2}$$

$$= 2\ln|x| - \ln|u| + \frac{3}{4u} + C$$

$$= \ln\left(\frac{x^2}{2x^2 + 1}\right) + \frac{3}{4}\frac{1}{2x^2 + 1} + C.$$

The following theorem summarizes the various aspects of the method of partial fractions.

**THEOREM** ❶ **Partial fraction decompositions of rational functions**

Let $P$ and $Q$ be real polynomials with real coefficients, and suppose that the degree of $P$ is less than the degree of $Q$. Then

(a) $Q(x)$ can be factored into the product of a constant $K$, real linear factors of the form $x - a_i$, and real quadratic factors of the form $x^2 + b_i x + c_i$ having no real roots. The linear and quadratic factors may be repeated.

$$Q(x) = K(x - a_1)^{m_1}(x - a_2)^{m_2} \cdots (x - a_j)^{m_j}(x^2 + b_1 x + c_1)^{n_1}$$
$$\cdots (x^2 + b_k x + c_k)^{n_k}.$$

The degree of $Q$ is $m_1 + m_2 + \cdots + m_j + 2n_1 + 2n_2 + \cdots + 2n_k$.

(b) The rational function $P(x)/Q(x)$ can be expressed as a sum of partial fractions as follows:

(i) corresponding to each factor $(x - a)^m$ of $Q(x)$ the decomposition contains a sum of fractions of the form

$$\frac{A_1}{x - a} + \frac{A_2}{(x - a)^2} + \cdots + \frac{A_m}{(x - a)^m};$$

(ii) corresponding to each factor $(x^2 + bx + c)^n$ of $Q(x)$ the decomposition contains a sum of fractions of the form

$$\frac{B_1 x + C_1}{x^2 + bx + c} + \frac{B_2 x + C_2}{(x^2 + bx + c)^2} + \cdots + \frac{B_n x + C_n}{(x^2 + bx + c)^n}.$$

The constants $A_1$, $A_2$, ..., $A_m$, $B_1$, $B_2$, ..., $B_n$, $C_1$, $C_2$, ..., $C_n$ can be determined by adding up the fractions in the decomposition and equating the coefficients of like powers of $x$ in the numerator of the sum with those in $P(x)$.

We will not attempt to prove this theorem here; it is related to the Fundamental Theorem of Algebra given in Appendix II.

Note that part (a) does not tell us how to find the factors of $Q(x)$; it tells us only what form they have. We must know the factors of $Q$ before we can make use of partial fractions to integrate the rational function $P(x)/Q(x)$. Partial fraction decompositions are also used in other mathematical situations, in particular to solve certain problems involving differential equations.

## Exercises 6.3

Evaluate the integrals in Exercises 1–34.

**1.** $\displaystyle\int \frac{2\,dx}{2x-3}$

**2.** $\displaystyle\int \frac{dx}{5-4x}$

**3.** $\displaystyle\int \frac{x\,dx}{\pi x+2}$

**4.** $\displaystyle\int \frac{x^2}{x-4}\,dx$

**5.** $\displaystyle\int \frac{1}{x^2-9}\,dx$

**6.** $\displaystyle\int \frac{dx}{5-x^2}$

**7.** $\displaystyle\int \frac{dx}{a^2-x^2}$

**8.** $\displaystyle\int \frac{dx}{b^2-a^2x^2}$

**9.** $\displaystyle\int \frac{x^2\,dx}{x^2+x-2}$

**10.** $\displaystyle\int \frac{x\,dx}{3x^2+8x-3}$

**11.** $\displaystyle\int \frac{x-2}{x^2+x}\,dx$

**12.** $\displaystyle\int \frac{dx}{x^3+9x}$

**13.** $\displaystyle\int \frac{dx}{1-6x+9x^2}$

**14.** $\displaystyle\int \frac{x\,dx}{2+6x+9x^2}$

**15.** $\displaystyle\int \frac{x^2+1}{6x-9x^2}\,dx$

**16.** $\displaystyle\int \frac{x^3+1}{12+7x+x^2}\,dx$

**17.** $\displaystyle\int \frac{dx}{x(x^2-a^2)}$

**18.** $\displaystyle\int \frac{dx}{x^4-a^4}$

**\* 19.** $\displaystyle\int \frac{x^3\,dx}{x^3-a^3}$

**20.** $\displaystyle\int \frac{dx}{x^3+2x^2+2x}$

**21.** $\displaystyle\int \frac{dx}{x^3-4x^2+3x}$

**22.** $\displaystyle\int \frac{x^2+1}{x^3+8}\,dx$

**23.** $\displaystyle\int \frac{dx}{(x^2-1)^2}$

**24.** $\displaystyle\int \frac{x^2\,dx}{(x^2-1)(x^2-4)}$

**25.** $\displaystyle\int \frac{dx}{x^4-3x^3}$

**26.** $\displaystyle\int \frac{x\,dx}{(x^2-x+1)^2}$

**\* 27.** $\displaystyle\int \frac{t\,dt}{(t+1)(t^2+1)^2}$

**\* 28.** $\displaystyle\int \frac{dt}{(t-1)(t^2-1)^2}$

**\* 29.** $\displaystyle\int \frac{dx}{x(3+x^2)\sqrt{1-x^2}}$

**\* 30.** $\displaystyle\int \frac{dx}{e^{2x}-4e^x+4}$

**\* 31.** $\displaystyle\int \frac{dx}{x(1+x^2)^{3/2}}$

**\* 32.** $\displaystyle\int \frac{dx}{x(1-x^2)^{3/2}}$

**\* 33.** $\displaystyle\int \frac{dx}{x^2(x^2-1)^{3/2}}$

**\* 34.** $\displaystyle\int \frac{d\theta}{\cos\theta(1+\sin\theta)}$

**\* 35.** Suppose that $P$ and $Q$ are polynomials such that the degree of $P$ is smaller than that of $Q$. If

$$Q(x) = (x-a_1)(x-a_2)\cdots(x-a_n),$$

where $a_i \neq a_j$ if $i \neq j$ $(1 \leq i, j \leq n)$, so that $P(x)/Q(x)$ has partial fraction decomposition

$$\frac{P(x)}{Q(x)} = \frac{A_1}{x-a_1} + \frac{A_2}{x-a_2} + \cdots + \frac{A_n}{x-a_n},$$

show that

$$A_j = \frac{P(a_j)}{Q'(a_j)} \qquad (1 \leq j \leq n).$$

This gives yet another method for computing the constants in a partial fraction decomposition if the denominator factors completely into distinct linear factors.

## 6.4   Integration Using Computer Algebra or Tables

Although anyone who uses calculus should be familiar with the basic techniques of integration, just as anyone who uses arithmetic should be familiar with the techniques of multiplication and division, technology is steadily eroding the necessity for being able to do long, complicated integrals by such methods. In fact, today there are several computer programs that can manipulate mathematical expressions symbolically (rather than just numerically) and that can carry out, with little or no assistance from us, the various algebraic steps and limit calculations that are required to calculate and simplify both derivatives and integrals. Much pain can be avoided and time saved by having the computer evaluate a complicated integral such as

$$\int \frac{1+x+x^2}{(x^4-1)(x^4-16)^2}\,dx$$

rather than doing it by hand using partial fractions.  Even without the aid of a computer, we can use tables of standard integrals such as the ones in the back endpapers of this book to help us evaluate complicated integrals. Using computers or tables can nevertheless require that we perform some simplifications beforehand, and can make demands on our ability to interpret the answers we get. We give a few examples below.

## Using Computer Algebra Programs for Integration

All computer algebra systems are capable of evaluating both indefinite and definite integrals symbolically, as well as giving numerical approximations for those definite integrals that have numerical values.  The following examples show how to get three such software packages, Derive (version 2.57), Maple V (release 4), and Mathematica (version 2.1) to evaluate the integrals

$$\text{(a)} \quad I = \int 2^x \sqrt{1 + (2^x)^2}\, dx \quad \text{and} \quad \text{(b)} \quad J = \int_0^\pi e^{ax} \sin bx\, dx.$$

**Derive**

On being given the input

```
1:      int(2^x sqrt(1+(2^x)^2),x)
```

Derive responds with

```
              x - 1       2 x                    2 x             x
             2         SQRT(2    + 1)    LN(SQRT(2    + 1) + 2 )
    2:       ----------------------- + ----------------------------
                      LN(2)                      2 LN(2)
```

To the input

```
3:      int(exp(a x) sin(b x), x, 0, pi)
```

Derive responds with

```
           pi a
          #e     (a SIN(pi b) - b COS(pi b))              b
    4:    ----------------------------------- + ---------
                           2    2                  2    2
                          a  + b                  a  + b
```

**Maple V**

On being given the input

```
    > int(2^x*sqrt(1+(2^x)^2),x);
```

Maple responds with

$$\frac{1}{2} \frac{2^x \sqrt{1 + (2^x)^2}}{\ln(2)} + \frac{1}{2} \frac{\operatorname{arcsinh}(2^x)}{\ln(2)}$$

Now the command

```
    > convert(",ln);
```

will force Maple to convert the inverse hyperbolic sine to logarithm:

$$\frac{1}{2} \frac{2^x \sqrt{1 + (2^x)^2}}{\ln(2)} + \frac{1}{2} \frac{\ln(2^x + \sqrt{1 + (2^x)^2})}{\ln(2)}$$

To the input

```
> int(exp(a*x)*sin(b*x),x=0..Pi);
```

Maple responds with

$$\frac{(-b\cos(\pi b) + a\sin(\pi b))e^{(\pi a)}}{a^2 + b^2} + \frac{b}{a^2 + b^2}$$

**Mathematica**

```
In[1]:= Integrate[2^x*Sqrt[1+(2^x)^2],x]

             x              2 x             x
          2   Sqrt[1 + 2     ]    ArcSinh[2  ]
Out[1]=  ----------------------- + ------------
                  2 Log[2]           2 Log[2]

In[2]:= Integrate[Exp[a x] Sin[b x], {x,0,Pi}]
                            a Pi                    a Pi
            b         b E       Cos[b Pi]    a E       Sin[b Pi]
Out[2]=  ------- -  ------------------- + -------------------
           2    2       2    2                  2    2
          a  + b        a  + b                  a  + b
```

Note that Mathematica uses Log instead of ln for the natural logarithm. None of the programs includes a constant of integration.

***Remark***   While computer algebra programs can do very complicated integrals that yield to standard methods of integration, it is sometimes necessary to give them a little help. Neither Maple nor Mathematica (in the versions mentioned above) were able to evaluate

$$\int 2^x \sqrt{1 + 4^x}\, dx,$$

which is clearly equivalent to the integral evaluated in part (a) of the three examples above. Apparently, neither program recognizes that the identity $4^x = (2^x)^2$ is useful in the context of this integral. Mathematica carried out an integration by parts, giving

```
                                             x
                                            2
                               Integrate[------------, x] Log[4]
              x              x                     x
          2 2   Sqrt[1 + 4  ]             Sqrt[1 + 4  ]
Out[5]=  ------------------- + -------------------------------
          2 Log[2] + Log[4]         2 Log[2] + Log[4]
```

Both programs were able to do the integral only after $4^x$ was replaced by $(2^x)^2$ or $2^{2x}$. Although much less powerful a program than either Maple or Mathematica, Derive evaluated the integral without requiring any help.

***Remark***   Another problem that can arise is getting an answer expressed in terms of functions that you don't understand. For example, all three programs will tell you that

$$\int e^{-x^2}\, dx = \frac{\sqrt{\pi}}{2}\, \mathrm{Erf}(x).$$

Here Erf refers to the **error function** that is, in fact, defined by

$$\text{Erf}(x) = \frac{2}{\sqrt{\pi}} \int_0^x e^{-t^2} \, dt.$$

Similarly, Maple and Mathematica (but not Derive) will cheerfully tell you that

$$\int \sin(x^2) \, dx = \sqrt{\frac{\pi}{2}} \, \text{FresnelS} \left( \sqrt{\frac{2}{\pi}} x \right),$$

but you are expected to know about the **Fresnel sine integral** function

$$\text{FresnelS}(x) = \int_0^x \sin \frac{\pi t^2}{2} \, dt.$$

In summary, computer algebra programs can be very useful for calculating integrals symbolically for large classes of functions, but you may get some surprises when you use these programs, and you may have to do some of the work to get an answer useful in the context of the problem on which you are working. All of these programs, and many of the more sophisticated scientific calculators, are able to evaluate definite integrals numerically to any desired degree of accuracy even if symbolic antiderivatives cannot be found. We will discuss techniques of numerical integration in Sections 6.6–6.9.

## Using Integral Tables

When a computer algebra program is not available you can still get some help evaluating integrals by using an Integral Table such as the one in the back endpapers of this book. Besides giving the values of the common elementary integrals that you likely remember while you are studying calculus, they also give many more complicated integrals, especially ones representing standard types that often arise in applications. Familiarize yourself with the main headings under which the integrals are classified. Using the tables usually means massaging your integral using simple substitutions until you get it into the form of one of the integrals in the table.

**Example 1** Use the table to evaluate $I = \displaystyle\int \frac{t^5}{\sqrt{3 - 2t^4}} \, dt.$

**Solution** This integral doesn't resemble any in the tables, but there are numerous integrals in the tables involving $\sqrt{a^2 - x^2}$. We can begin to put the integral into this form with the substitution $t^2 = u$, so that $2t \, dt = du$. Thus

$$I = \frac{1}{2} \int \frac{u^2}{\sqrt{3 - 2u^2}} \, du.$$

This is not quite what we want yet; let us get rid of the 2 multiplying the $u^2$ under the square root. One way to do this is with the change of variable $\sqrt{2}u = x$, so that $du = dx/\sqrt{2}$:

$$I = \frac{1}{4\sqrt{2}} \int \frac{x^2}{\sqrt{3 - x^2}} \, dx.$$

Now the denominator is of the form $\sqrt{a^2 - x^2}$ for $a = \sqrt{3}$. Looking through the part of the table (in the back endpapers) dealing with integrals involving $\sqrt{a^2 - x^2}$ we find the third one, which says that

$$\int \frac{x^2}{\sqrt{a^2 - x^2}} \, dx = -\frac{x}{2}\sqrt{a^2 - x^2} + \frac{a^2}{2}\sin^{-1}\frac{x}{a} + C.$$

Thus

$$I = \frac{1}{4\sqrt{2}} \left( -\frac{x}{2}\sqrt{3 - x^2} + \frac{3}{2}\sin^{-1}\frac{x}{\sqrt{3}} \right) + C_1$$

$$= -\frac{t^2}{8}\sqrt{3 - 2t^4} + \frac{3}{8\sqrt{2}}\sin^{-1}\frac{\sqrt{2}t^2}{\sqrt{3}} + C_1.$$

Many of the integrals in the table are reduction formulas. (An integral appears on both sides of the equation.) These can be iterated to simplify integrals as in some of the examples and exercises of Section 6.1.

**Example 2** Evaluate $I = \displaystyle\int_0^1 \frac{1}{(x^2 + 1)^3} \, dx$.

**Solution** The fourth integral in the table of Miscellaneous Algebraic Integrals says that if $n \neq 1$, then

$$\int \frac{dx}{(a^2 \pm x^2)^n} = \frac{1}{2a^2(n - 1)} \left( \frac{x}{(a^2 \pm x^2)^{n-1}} + (2n - 3) \int \frac{dx}{(a^2 \pm x^2)^{n-1}} \right)$$

Using $a = 1$ and the "+" signs we have

$$\int_0^1 \frac{dx}{(1 + x^2)^n} = \frac{1}{2(n - 1)} \left( \frac{x}{(1 + x^2)^{n-1}} \Big|_0^1 + (2n - 3) \int_0^1 \frac{dx}{(1 + x^2)^{n-1}} \right)$$

$$= \frac{1}{2^n(n - 1)} + \frac{2n - 3}{2(n - 1)} \int_0^1 \frac{dx}{(1 + x^2)^{n-1}}.$$

Thus we have

$$I = \frac{1}{16} + \frac{3}{4} \int_0^1 \frac{dx}{(1 + x^2)^2}$$

$$= \frac{1}{16} + \frac{3}{4} \left( \frac{1}{4} + \frac{1}{2} \int_0^1 \frac{dx}{1 + x^2} \right)$$

$$= \frac{1}{16} + \frac{3}{16} + \frac{3}{8}\tan^{-1}x \Big|_0^1 = \frac{1}{4} + \frac{3\pi}{32}.$$

# Exercises 6.4

1. Use a computer algebra program to check any of the integrals you have done in the exercises from Sections 5.6 and 6.1–6.3, as well as any of the integrals you have been unable to do.

2. Use a computer algebra program to evaluate the integral in the opening paragraph of this section.

3. Use a computer algebra program to reevaluate the integrals in Examples 1 and 2.

4. Use a computer algebra program to evaluate the integrals in Exercises 5–14.

Use the integral tables to help you find the integrals in Exercises 5–14.

5. $\displaystyle\int \frac{x^2}{\sqrt{x^2 - 2}}\, dx$

6. $\displaystyle\int \sqrt{(x^2 + 4)^3}\, dx$

7. $\displaystyle\int \frac{dt}{t^2\sqrt{3t^2 + 5}}$

8. $\displaystyle\int \frac{dt}{t\sqrt{3t - 5}}$

9. $\displaystyle\int x^4 (\ln x)^4\, dx$

10. $\displaystyle\int x^7 e^{x^2}\, dx$

11. $\displaystyle\int x\sqrt{2x - x^2}\, dx$

12. $\displaystyle\int \frac{\sqrt{2x - x^2}}{x^2}\, dx$

13. $\displaystyle\int \frac{dx}{(\sqrt{4x - x^2})^3}$

14. $\displaystyle\int \frac{dx}{(\sqrt{4x - x^2})^4}$

# 6.5 Improper Integrals

Up to this point we have considered definite integrals of the form

$$I = \int_a^b f(x)\, dx$$

where the integrand $f$ is *continuous* on the *closed, finite* interval $[a, b]$. Since such a function is necessarily *bounded*, the integral $I$ is necessarily a finite number; for positive $f$ it corresponds to the area of a **bounded region** of the plane, a region contained inside some disc of finite radius with centre at the origin. Such integrals are also called **proper integrals**. We are now going to generalize the definite integral to allow for two possibilities excluded in the situation described above:

(i) We may have $a = -\infty$ or $b = \infty$ or both.

(ii) $f$ may be unbounded as $x$ approaches $a$ or $b$ or both.

Integrals satisfying (i) are called **improper integrals of type I**; integrals satisfying (ii) are called **improper integrals of type II**. Either type of improper integral corresponds (for positive $f$) to the area of a region in the plane that "extends to infinity" in some direction and therefore is *unbounded*. As we will see, such integrals may or may not have finite values. The ideas involved are best introduced by examples.

## Improper Integrals of Type I

**Example 1** Find the area of $A$ the region lying under the curve $y = 1/x^2$ and above the $x$-axis to the right of $x = 1$. (See Figure 6.8(a).)

**Solution** We would like to calculate the area with an integral

$$A = \int_1^\infty \frac{dx}{x^2},$$

**Figure 6.8**

(a) $A = \displaystyle\int_1^\infty \dfrac{1}{x^2}\, dx$

(b) $A = \displaystyle\lim_{R\to\infty} \int_1^R \dfrac{1}{x^2}\, dx$

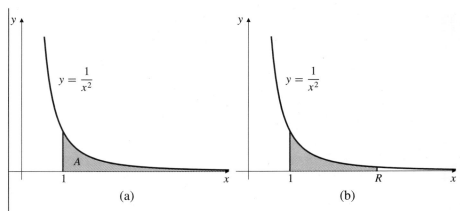

(a)        (b)

which is improper of type I since its interval of integration is infinite. It is not immediately obvious whether the area is finite or not; the region has an infinitely long "spike" along the $x$-axis, but this spike becomes infinitely thin as $x$ approaches $\infty$. In order to evaluate this improper integral, we interpret it as a limit of proper integrals over intervals $[1, R]$ as $R \to \infty$. (See Figure 6.8(b).)

$$A = \int_1^\infty \frac{dx}{x^2} = \lim_{R\to\infty} \int_1^R \frac{dx}{x^2} = \lim_{R\to\infty} \left(-\frac{1}{x}\right)\bigg|_1^R$$

$$= \lim_{R\to\infty} \left(-\frac{1}{R} + 1\right) = 1$$

Since the limit exists (is finite), we say that the improper integral *converges*. The region has finite area, $A = 1$ square unit. ∎

**Example 2**   Find the area of the region under $y = 1/x$, above $y = 0$, and to the right of $x = 1$. (See Figure 6.9.)

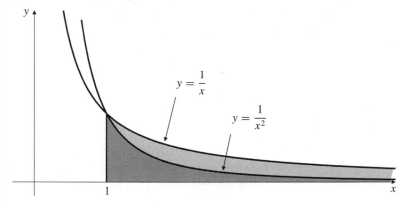

**Figure 6.9**   The area shaded in colour is infinite

**Solution**   This area is given by the improper integral

$$A = \int_1^\infty \frac{dx}{x} = \lim_{R\to\infty} \int_1^R \frac{dx}{x} = \lim_{R\to\infty} \ln x \bigg|_1^R = \lim_{R\to\infty} \ln R = \infty.$$

We say that this improper integral *diverges to infinity*. Observe that the region has a similar shape to the region under $y = 1/x^2$ considered in the above example, but its "spike" is somewhat thicker at each value of $x > 1$. Evidently the extra thickness makes a big difference; this region has *infinite* area. ∎

**DEFINITION** **1**

**Improper integrals of type I**

If $f$ is continuous on $[a, \infty)$, we define the improper integral of $f$ over $[a, \infty)$ as a limit of proper integrals:

$$\int_a^\infty f(x)\,dx = \lim_{R \to \infty} \int_a^R f(x)\,dx.$$

Similarly, if $f$ is continuous on $(-\infty, b]$ then we define

$$\int_{-\infty}^b f(x)\,dx = \lim_{R \to -\infty} \int_R^b f(x)\,dx.$$

In either case, if the limit exists (is a finite number), we say that the improper integral **converges**; if the limit does not exist, we say that the improper integral **diverges**. If the limit is $\infty$ (or $-\infty$), we say the improper integral **diverges to infinity** (or **diverges to negative infinity**).

The integral $\int_{-\infty}^\infty f(x)\,dx$ is, for $f$ continuous on the real line, improper of type I at both endpoints. We break it into two separate integrals:

$$\int_{-\infty}^\infty f(x)\,dx = \int_{-\infty}^0 f(x)\,dx + \int_0^\infty f(x)\,dx.$$

The integral on the left converges if and only if *both* integrals on the right converge.

**Example 3**  Evaluate $\displaystyle\int_{-\infty}^\infty \frac{1}{1+x^2}\,dx$.

**Solution**  By the (even) symmetry of the integrand (see Figure 6.10), we have

$$\int_{-\infty}^\infty \frac{dx}{1+x^2} = \int_{-\infty}^0 \frac{dx}{1+x^2} + \int_0^\infty \frac{dx}{1+x^2}$$

$$= 2 \lim_{R \to \infty} \int_0^R \frac{dx}{1+x^2}$$

$$= 2 \lim_{R \to \infty} \tan^{-1} R = 2\left(\frac{\pi}{2}\right) = \pi.$$

The use of symmetry here requires some justification. At the time we used it we did not know whether each of the half-line integrals was finite or infinite. However, since both are positive, even if they are infinite, their sum would still be twice one of them. If one had been positive and the other negative, we would not have been justified in cancelling them to get 0 until we knew that they were finite. ($\infty + \infty = \infty$, but $\infty - \infty$ is not defined.) In any event, the given integral converges to $\pi$. ∎

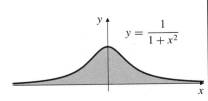

$y = \dfrac{1}{1+x^2}$

**Figure 6.10**

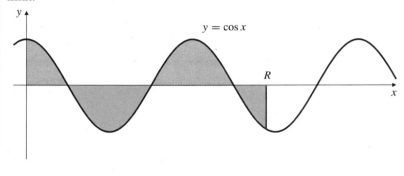

**Example 4**    $\displaystyle\int_0^\infty \cos x\, dx = \lim_{R\to\infty} \int_0^R \cos x\, dx = \lim_{R\to\infty} \sin R.$

This limit does not exist (and it is not $\infty$ or $-\infty$), so all we can say is that the given integral diverges. (See Figure 6.11.) As $R$ increases the integral alternately adds and subtracts the areas of the hills and valleys but does not approach any unique limit.

$y = \cos x$

$R$

**Figure 6.11**    Not every divergent improper integral diverges to $\infty$ or $-\infty$

## Improper Integrals of Type II

**DEFINITION 2**

**Improper integrals of type II**

If $f$ is continuous on the interval $(a, b]$, and is possibly unbounded near $a$, we define the improper integral

$$\int_a^b f(x)\, dx = \lim_{c\to a+} \int_c^b f(x)\, dx.$$

Similarly, if $f$ is continuous on $[a, b)$, and is possibly unbounded near $b$, we define

$$\int_a^b f(x)\, dx = \lim_{c\to b-} \int_a^c f(x)\, dx.$$

These improper integrals may converge, diverge, diverge to infinity, or diverge to negative infinity.

**Example 5**    Find the area of the region $S$ lying under $y = 1/\sqrt{x}$, above the $x$-axis, between $x = 0$ and $x = 1$.

**Solution**    The area $A$ is given by

$$A = \int_0^1 \frac{1}{\sqrt{x}}\, dx,$$

which is an improper integral of type II since the integrand is unbounded near $x = 0$. The region $S$ has a "spike" extending to infinity along the $y$-axis, a vertical asymptote of the integrand, as shown in Figure 6.12. As we did for improper integrals of type I, we express such integrals as limits of proper integrals.

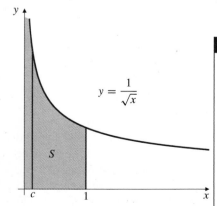

$y = \dfrac{1}{\sqrt{x}}$

$S$

$c$    $1$

**Figure 6.12**    The shaded area is finite

$$A = \lim_{c \to 0+} \int_c^1 x^{-1/2}\, dx = \lim_{c \to 0+} 2x^{1/2}\Big|_c^1 = \lim_{c \to 0+} (2 - 2\sqrt{c}) = 2.$$

This integral converges, and $S$ has a finite area of 2 square units.   ∎

While improper integrals of type I are always easily recognized because of the infinite limits of integration, improper integrals of type II can be somewhat harder to spot. You should be alert for singularities of integrands and especially points where they have vertical asymptotes. It may be necessary to break an improper integral into several improper integrals if it is improper at both endpoints or at points inside the interval of integration. For example,

$$\int_{-1}^1 \frac{\ln|x|\,dx}{\sqrt{1-x}} = \int_{-1}^0 \frac{\ln|x|\,dx}{\sqrt{1-x}} + \int_0^{1/2} \frac{\ln|x|\,dx}{\sqrt{1-x}} + \int_{1/2}^1 \frac{\ln|x|\,dx}{\sqrt{1-x}}.$$

Each integral on the right is improper because of a singularity of its integrand at one endpoint.

**Example 6**   Evaluate each of the following integrals or show that it diverges:

$$\text{(a)} \quad \int_0^1 \frac{1}{x}\, dx, \qquad \text{(b)} \quad \int_0^2 \frac{1}{\sqrt{2x - x^2}}\, dx, \text{ and} \qquad \text{(c)} \quad \int_0^1 \ln x\, dx.$$

**Solution**

(a) $\displaystyle \int_0^1 \frac{1}{x}\, dx = \lim_{c \to 0+} \int_c^1 \frac{1}{x}\, dx = \lim_{c \to 0+} (\ln 1 - \ln c) = \infty.$
This integral diverges to infinity.

(b) $\displaystyle \int_0^2 \frac{1}{\sqrt{2x - x^2}}\, dx = \int_0^2 \frac{1}{\sqrt{1 - (x-1)^2}}\, dx \qquad$ Let $u = x - 1,$
$$\hspace{10cm} du = dx$$

$$= \int_{-1}^1 \frac{1}{\sqrt{1 - u^2}}\, du$$

$$= 2\int_0^1 \frac{1}{\sqrt{1 - u^2}}\, du \qquad \text{(by symmetry)}$$

$$= 2 \lim_{c \to 1-} \int_0^c \frac{1}{\sqrt{1 - u^2}}\, du$$

$$= 2 \lim_{c \to 1-} \sin^{-1} u \Big|_0^c = 2 \lim_{c \to 1-} \sin^{-1} c = \pi.$$

This integral converges to $\pi$. Observe how a change of variable can be made even before an improper integral is expressed as a limit of proper integrals.

(c) $\displaystyle\int_0^1 \ln x\, dx = \lim_{c\to 0+}\int_c^1 \ln x\, dx$   (See Example 2(a) of Section 6.1 for the evaluation of the indefinite integral.)

$$= \lim_{c\to 0+} (x\ln x - x)\Big|_c^1$$

$$= \lim_{c\to 0+} (0 - 1 - c\ln c + c)$$

$$= -1 + 0 - \lim_{c\to 0+}\frac{\ln c}{1/c} \qquad \left[\frac{-\infty}{\infty}\right]$$

$$= -1 - \lim_{c\to 0+}\frac{1/c}{-(1/c^2)} \qquad \text{(by l'Hôpital's Rule)}$$

$$= -1 - \lim_{c\to 0+}(-c) = -1 + 0 = -1.$$

The integral converges to $-1$.

The following theorem summarizes the behaviour of improper integrals of types I and II for powers of $x$.

**THEOREM** **2**  **$p$-integrals**

If $0 < a < \infty$, then

(a) $\displaystyle\int_a^\infty x^{-p}\, dx$ $\begin{cases} \text{converges to } \dfrac{a^{1-p}}{p-1} & \text{if } p > 1 \\ \text{diverges to } \infty & \text{if } p \le 1 \end{cases}$

(b) $\displaystyle\int_0^a x^{-p}\, dx$ $\begin{cases} \text{converges to } \dfrac{a^{1-p}}{1-p} & \text{if } p < 1 \\ \text{diverges to } \infty & \text{if } p \ge 1. \end{cases}$

**PROOF**  We prove part (b) only. The proof of part (a) is similar, and is left as an exercise. Also, the case $p = 1$ of part (b) is similar to Example 6(a) above, so we need consider only the cases $p < 1$ and $p > 1$. If $p < 1$, then we have

$$\int_0^a x^{-p}\, dx = \lim_{c\to 0+}\int_c^a x^{-p}\, dx$$

$$= \lim_{c\to 0+}\frac{x^{-p+1}}{-p+1}\Big|_c^a$$

$$= \lim_{c\to 0+}\frac{a^{1-p} - c^{1-p}}{1-p} = \frac{a^{1-p}}{1-p}$$

because $1 - p > 0$. If $p > 1$, then

$$\int_0^a x^{-p}\, dx = \lim_{c\to 0+}\int_c^a x^{-p}\, dx$$

$$= \lim_{c\to 0+}\frac{x^{-p+1}}{-p+1}\Big|_c^a$$

$$= \lim_{c\to 0+}\frac{c^{-(p-1)} - a^{-(p-1)}}{p-1} = \infty.$$

The integrals in Theorem 2 are called **$p$-integrals**. It is very useful to know when they converge and diverge when you have to decide whether certain other improper integrals converge or not and you can't find the appropriate antiderivatives. (See the discussion of estimating convergence below.) Note that $\int_0^\infty x^{-p}\, dx$ does not converge for any value of $p$.

**Remark**   If $f$ is continuous on the interval $[a, b]$ so that $\int_a^b f(x)\, dx$ is a proper definite integral, then treating the integral as improper will lead to the same value:

$$\lim_{c \to a+} \int_c^b f(x)\, dx = \int_a^b f(x)\, dx = \lim_{c \to b-} \int_a^c f(x)\, dx.$$

This justifies the definition of the definite integral of a piecewise continuous function given in Section 5.4. To integrate a function defined to be different continuous functions on different intervals, we merely add the integrals of the various component functions over their respective intervals. Any of these integrals may be proper or improper; if any are improper all must converge or the given integral will diverge.

---

**Example 7**   Evaluate $\int_0^2 f(x)\, dx$, where $f(x) = \begin{cases} 1/\sqrt{x} & \text{if } 0 < x \le 1 \\ x - 1 & \text{if } 1 < x \le 2. \end{cases}$

**Solution**   The graph of $f$ is shown in Figure 6.13. We have

$$\int_0^2 f(x)\, dx = \int_0^1 \frac{dx}{\sqrt{x}} + \int_1^2 (x - 1)\, dx$$

$$= \lim_{c \to 0+} \int_c^1 \frac{dx}{\sqrt{x}} + \left( \frac{x^2}{2} - x \right) \Big|_1^2 = 2 + \left( 2 - 2 - \frac{1}{2} + 1 \right) = \frac{5}{2};$$

the first integral on the right is improper but convergent (see Example 5 above) and the second is proper.

■

---

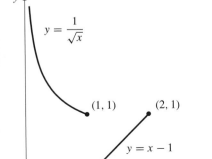

**Figure 6.13**

## Estimating Convergence and Divergence

When an improper integral cannot be evaluated by the Fundamental Theorem of Calculus because an antiderivative can't be found, we may still be able to determine whether the integral converges by comparing it with simpler integrals. The following theorem is central to this approach.

**THEOREM 3**

**A comparison theorem for integrals**

Let $-\infty \le a < b \le \infty$ and suppose that functions $f$ and $g$ are continuous on the interval $(a, b)$ and satisfy $0 \le f(x) \le g(x)$. If $\int_a^b g(x)\, dx$ converges, then so does $\int_a^b f(x)\, dx$, and

$$\int_a^b f(x)\, dx \le \int_a^b g(x)\, dx.$$

Equivalently, if $\int_a^b f(x)\, dx$ diverges to $\infty$, then so does $\int_a^b g(x)\, dx$.

**PROOF** Since both integrands are nonnegative, there are only two possibilities for each integral: it can either converge to a nonnegative number or diverge to $\infty$. Since $f(x) \le g(x)$ on $(a, b)$, it follows by Theorem 3(e) of Section 5.4 that if $a < r < s < b$, then

$$\int_r^s f(x)\, dx \le \int_r^s g(x)\, dx.$$

This theorem now follows by taking limits as $r \to a+$ and $s \to b-$.

**Example 8** Show that $\displaystyle\int_0^\infty e^{-x^2}\, dx$ converges and find an upper bound for its value.

**Solution** We can't integrate $e^{-x^2}$, but we can integrate $e^{-x}$. We would like to use the inequality $e^{-x^2} \le e^{-x}$, but this is only valid for $x \ge 1$. (See Figure 6.14.) Therefore we break the integral into two parts.

On $[0, 1]$ we have $0 < e^{-x^2} \le 1$, so

$$0 < \int_0^1 e^{-x^2}\, dx \le \int_0^1 dx = 1.$$

On $[1, \infty)$ we have $x^2 \ge x$, so $-x^2 \le -x$ and $0 < e^{-x^2} \le e^{-x}$. Thus

$$0 < \int_1^\infty e^{-x^2}\, dx \le \int_1^\infty e^{-x}\, dx = \lim_{R \to \infty} \left.\frac{e^{-x}}{-1}\right|_1^R$$

$$= \lim_{R \to \infty} \left(\frac{1}{e} - \frac{1}{e^R}\right) = \frac{1}{e}.$$

Hence $\int_0^\infty e^{-x^2}\, dx$ converges and its value is not greater than $1 + (1/e)$.

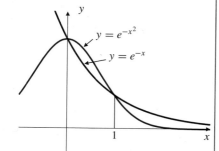

**Figure 6.14** Comparing $e^{-x^2}$ and $e^{-x}$

We remark that the above integral is in fact equal to $\frac{1}{2}\sqrt{\pi}$, although we cannot prove this now. It is normally proved using techniques of multivariable calculus.

For large or small values of $x$ many integrands behave like powers of $x$. If so, they can be compared with $p$-integrals.

**Example 9** Determine whether $\displaystyle\int_0^\infty \frac{dx}{\sqrt{x + x^3}}$ converges.

**Solution** The integral is improper of both types, so we write

$$\int_0^\infty \frac{dx}{\sqrt{x + x^3}} = \int_0^1 \frac{dx}{\sqrt{x + x^3}} + \int_1^\infty \frac{dx}{\sqrt{x + x^3}} = I_1 + I_2.$$

On $(0, 1]$ we have $\sqrt{x + x^3} > \sqrt{x}$, so

$$I_1 < \int_0^1 \frac{dx}{\sqrt{x}} = 2 \qquad \text{(by Theorem 2).}$$

On $[1, \infty)$ we have $\sqrt{x + x^3} > \sqrt{x^3}$, so

$$I_2 < \int_1^\infty x^{-3/2}\, dx = 2 \qquad \text{(by Theorem 2)}.$$

Hence the given integral converges and its value is less than 4.  ∎

**Example 10**   Determine whether $\displaystyle\int_2^\infty \frac{dx}{\ln x}$ converges.

**Solution**   Recall that $\ln x \le x - 1$ for all $x > 0$. In particular, $\ln x < x$ for $x \ge 2$, so

$$\int_2^\infty \frac{dx}{\ln x} > \int_2^\infty \frac{dx}{x},$$

which diverges to infinity by Theorem 2.   Hence the given integral diverges to infinity by Theorem 3.  ∎

## Exercises 6.5

In Exercises 1–22, evaluate the given integral or show that it diverges.

**1.** $\displaystyle\int_2^\infty \frac{1}{(x-1)^3}\, dx$

**2.** $\displaystyle\int_3^\infty \frac{1}{(2x-1)^{2/3}}\, dx$

**3.** $\displaystyle\int_0^\infty e^{-2x}\, dx$

**4.** $\displaystyle\int_{-\infty}^{-1} \frac{dx}{x^2+1}$

**5.** $\displaystyle\int_{-1}^{1} \frac{dx}{(x+1)^{2/3}}$

**6.** $\displaystyle\int_0^a \frac{dx}{a^2-x^2}$

**7.** $\displaystyle\int_0^1 \frac{1}{(1-x)^{1/3}}\, dx$

**8.** $\displaystyle\int_0^1 \frac{1}{x\sqrt{1-x}}\, dx$

**9.** $\displaystyle\int_0^{\pi/2} \frac{\cos x\, dx}{(1-\sin x)^{2/3}}$

**10.** $\displaystyle\int_0^\infty x\, e^{-x}\, dx$

**11.** $\displaystyle\int_0^1 \frac{dx}{\sqrt{x(1-x)}}$

**12.** $\displaystyle\int_0^\infty \frac{x}{1+2x^2}\, dx$

**13.** $\displaystyle\int_0^\infty \frac{x\, dx}{(1+2x^2)^{3/2}}$

**14.** $\displaystyle\int_0^{\pi/2} \sec x\, dx$

**15.** $\displaystyle\int_0^{\pi/2} \tan x\, dx$

**16.** $\displaystyle\int_e^\infty \frac{dx}{x\ln x}$

**17.** $\displaystyle\int_1^e \frac{dx}{x\sqrt{\ln x}}$

**18.** $\displaystyle\int_e^\infty \frac{dx}{x(\ln x)^2}$

**19.** $\displaystyle\int_{-\infty}^\infty \frac{x}{1+x^2}\, dx$

**20.** $\displaystyle\int_{-\infty}^\infty \frac{x}{1+x^4}\, dx$

**21.** $\displaystyle\int_{-\infty}^\infty x\, e^{-x^2}\, dx$

**22.** $\displaystyle\int_{-\infty}^\infty e^{-|x|}\, dx$

**23.** Find the area below $y = 0$, above $y = \ln x$, and to the right of $x = 0$.

**24.** Find the area below $y = e^{-x}$, above $y = e^{-2x}$, and to the right of $x = 0$.

**25.** Find the area of a region that lies above $y = 0$, to the right of $x = 1$, and under the curve $y = \dfrac{4}{2x+1} - \dfrac{2}{x+2}$.

**26.** Find the area of the plane region that lies under the graph of $y = x^{-2} e^{-1/x}$, above the $x$-axis, and to the right of the $y$-axis.

**27.** Prove Theorem 2(a) by directly evaluating the integrals involved.

**28.** Evaluate $\int_{-1}^1 (x\,\operatorname{sgn} x)/(x+2)\, dx$. Recall that $\operatorname{sgn} x = x/|x|$.

**29.** Evaluate $\int_0^2 x^2 \operatorname{sgn}(x-1)\, dx$.

In Exercises 30–41, state whether the given integral converges or diverges, and justify your claim.

**30.** $\displaystyle\int_0^\infty \frac{x^2}{x^5+1}\, dx$

**31.** $\displaystyle\int_0^\infty \frac{dx}{1+\sqrt{x}}$

**32.** $\displaystyle\int_2^\infty \frac{x\sqrt{x}\, dx}{x^2-1}$

**33.** $\displaystyle\int_0^\infty e^{-x^3}\, dx$

**34.** $\displaystyle\int_0^\infty \frac{dx}{\sqrt{x}+x^2}$

**35.** $\displaystyle\int_{-1}^1 \frac{e^x}{x+1}\, dx$

**36.** $\displaystyle\int_0^\pi \frac{\sin x}{x}\,dx$

**\* 37.** $\displaystyle\int_0^\infty \frac{|\sin x|}{x^2}\,dx$

**\* 38.** $\displaystyle\int_0^{\pi^2} \frac{dx}{1-\cos\sqrt{x}}$

**\* 39.** $\displaystyle\int_{-\pi/2}^{\pi/2} \csc x\,dx$

**\* 40.** $\displaystyle\int_2^\infty \frac{dx}{\sqrt{x}\,\ln x}$

**\* 41.** $\displaystyle\int_0^\infty \frac{dx}{xe^x}$

**\* 42.** Given that $\int_0^\infty e^{-x^2}\,dx = \dfrac{1}{2}\sqrt{\pi}$, evaluate

(a) $\displaystyle\int_0^\infty x^2 e^{-x^2}\,dx$ and (b) $\displaystyle\int_0^\infty x^4 e^{-x^2}\,dx$.

**\* 43.** If $f$ is continuous on $[a, b]$, show that

$$\lim_{c\to a+} \int_c^b f(x)\,dx = \int_a^b f(x)\,dx.$$

*Hint:* a continuous function on a closed, finite interval is *bounded*: there exists a positive constant $K$ such that $|f(x)| \le K$ for all $x$ in $[a, b]$. Use this fact, together with parts (d) and (f) of Theorem 3 of Section 5.4, to show that

$$\lim_{c\to a+}\left(\int_a^b f(x)\,dx - \int_c^b f(x)\,dx\right) = 0.$$

Similarly, show that

$$\lim_{c\to b-} \int_a^c f(x)\,dx = \int_a^b f(x)\,dx.$$

**\* 44. (The gamma function)** The gamma function $\Gamma(x)$ is defined by the improper integral

$$\Gamma(x) = \int_0^\infty t^{x-1} e^{-t}\,dt.$$

($\Gamma$ is the Greek capital letter gamma.)

(a) Show that the integral converges for $x > 0$.

(b) Use integration by parts to show that $\Gamma(x+1) = x\Gamma(x)$ for $x > 0$.

(c) Show that $\Gamma(n+1) = n!$ for $n = 0, 1, 2, \ldots$.

(d) Given that $\int_0^\infty e^{-x^2}\,dx = \frac{1}{2}\sqrt{\pi}$, show that $\Gamma(\frac{1}{2}) = \sqrt{\pi}$ and $\Gamma(\frac{3}{2}) = \frac{1}{2}\sqrt{\pi}$.

In view of (c), $\Gamma(x+1)$ is often written $x!$ and regarded as a real-valued extension of the factorial function. Some scientific calculators (in particular, HP calculators) with the factorial function $n!$ built in actually calculate the Gamma function rather than just the integral factorial. Check whether your calculator does this by asking it for 0.5!. If you get an error message it's not using the gamma function.

## 6.6 The Trapezoid and Midpoint Rules

Most of the applications of integration, within and outside of mathematics, involve the definite integral

$$I = \int_a^b f(x)\,dx.$$

Thanks to the Fundamental Theorem of Calculus, we can evaluate such definite integrals by first finding an antiderivative of $f$. This is why we have spent considerable time on developing techniques of integration. There are, however, two obstacles that can prevent our calculating $I$ in this way:

(i) Finding an antiderivative of $f$ in terms of familiar functions may be impossible, or at least very difficult.

(ii) We may not be given a formula for $f(x)$ as a function of $x$; for instance, $f(x)$ may be an unknown function whose values at certain points of the interval $[a, b]$ have been determined by experimental measurement.

In the next two sections we investigate the problem of approximating the value of the definite integral $I$ using only the values of $f(x)$ at finitely many points of $[a, b]$. Obtaining such an approximation is called **numerical integration**. Upper and lower sums (or, indeed, any Riemann sum) can be used for this purpose, but these usually require much more calculation to yield a desired precision than the methods we will develop here. We will develop three methods for evaluating definite integrals numerically: the Trapezoid Rule, the Midpoint Rule, and Simpson's Rule. All of these methods can be easily implemented on a small computer or using a scientific calculator. The wide availability of these devices makes numerical integration a steadily more important tool for the user of mathematics. Some of the more advanced calculators have built-in routines for numerical integration.

All the techniques we consider require us to calculate the values of $f(x)$ at a set of equally spaced points in $[a, b]$. The computational "expense" involved in determining an approximate value for the integral $I$ will be roughly proportional to the number of function values required, so that the fewer function evaluations needed to achieve a desired degree of accuracy for the integral, the better we will regard the technique. Time is money, even in the world of computers.

## The Trapezoid Rule

We assume that $f(x)$ is continuous on $[a, b]$ and subdivide $[a, b]$ into $n$ subintervals of equal length $h = (b - a)/n$ using the $n + 1$ points

$$x_0 = a, \quad x_1 = a + h, \quad x_2 = a + 2h, \quad \dots, \quad x_n = a + nh = b.$$

We assume that the value of $f(x)$ at each of these points is known:

$$y_0 = f(x_0), \quad y_1 = f(x_1), \quad y_2 = f(x_2), \quad \dots, \quad y_n = f(x_n).$$

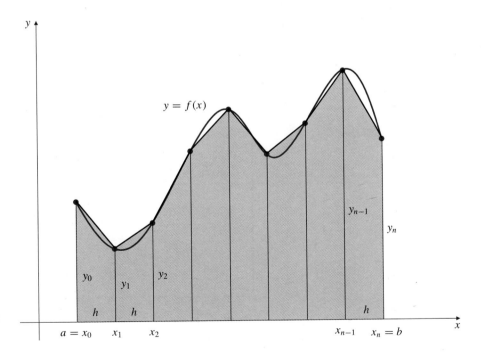

**Figure 6.15**    The area under $y = f(x)$ is approximated by the sum of the areas of $n$ trapezoids

The Trapezoid Rule approximates $\int_a^b f(x)\,dx$ by using straight line segments between the points $(x_{j-1}, y_{j-1})$ and $(x_j, y_j)$, $(1 \le j \le n)$, to approximate the graph of $f$, as shown in Figure 6.15, and summing the areas of the resulting $n$ *trapezoids*. A **trapezoid** is a four-sided polygon with one pair of parallel sides. (For our discussion we assume $f$ is positive so we can talk about "areas," but the resulting formulas apply to any continuous function $f$.)

The first trapezoid has vertices $(x_0, 0)$, $(x_0, y_0)$, $(x_1, y_1)$, and $(x_1, 0)$. The two parallel sides are vertical and have lengths $y_0$ and $y_1$. The perpendicular distance between them is $h = x_1 - x_0$. The area of this trapezoid is $h$ times the average of the parallel sides:

$$h\,\frac{y_0 + y_1}{2} \text{ square units.}$$

This can be seen geometrically by considering the trapezoid as the non-overlapping union of a rectangle and a triangle; see Figure 6.16. We use this trapezoidal area to approximate the integral of $f$ over the first subinterval $[x_0, x_1]$:

$$\int_{x_0}^{x_1} f(x)\,dx \approx h\,\frac{y_0 + y_1}{2}.$$

We can approximate the integral of $f$ over any subinterval in the same way:

$$\int_{x_{j-1}}^{x_j} f(x)\,dx \approx h\,\frac{y_{j-1} + y_j}{2}, \qquad (1 \le j \le n).$$

It follows that the original integral $I$ can be approximated by the sum of these trapezoidal areas:

$$\int_a^b f(x)\,dx \approx h\left(\frac{y_0 + y_1}{2} + \frac{y_1 + y_2}{2} + \frac{y_2 + y_3}{2} + \cdots + \frac{y_{n-1} + y_n}{2}\right)$$
$$= h\left(\frac{1}{2}y_0 + y_1 + y_2 + y_3 + \cdots + y_{n-1} + \frac{1}{2}y_n\right).$$

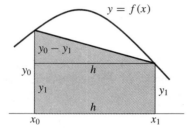

**Figure 6.16** The trapezoid has area
$y_1 h + \frac{1}{2}(y_0 - y_1)h = \frac{1}{2}h(y_0 + y_1)$

**DEFINITION 3**

**The Trapezoid Rule**

The $n$-subinterval **Trapezoid Rule** approximation to $\int_a^b f(x)\,dx$, denoted $T_n$, is given by

$$T_n = h\left(\frac{1}{2}y_0 + y_1 + y_2 + y_3 + \cdots + y_{n-1} + \frac{1}{2}y_n\right)$$
$$= \frac{h}{2}\left(y_0 + 2\sum_{j=1}^{n-1} y_j + y_n\right).$$

We now illustrate the Trapezoid Rule by using it to approximate an integral whose value we already know:

$$I = \int_1^2 \frac{1}{x}\,dx = \ln 2 = 0.69314718\ldots.$$

(This value, and those of all the approximations quoted in these sections, were calculated using a scientific calculator.) We will use the same integral to illustrate other methods for approximating definite integrals later.

**Example 1** Calculate the Trapezoid Rule approximations $T_4$, $T_8$, and $T_{16}$ for

$$I = \int_1^2 \frac{1}{x}\, dx.$$

**Solution**  For $n = 4$ we have $h = (2 - 1)/4 = 1/4$; for $n = 8$ we have $h = 1/8$; for $n = 16$ we have $h = 1/16$. Therefore

$$T_4 = \frac{1}{4}\left[\frac{1}{2}(1) + \frac{4}{5} + \frac{2}{3} + \frac{4}{7} + \frac{1}{2}\left(\frac{1}{2}\right)\right] = 0.69702381\dots$$

$$T_8 = \frac{1}{8}\left[\frac{1}{2}(1) + \frac{8}{9} + \frac{4}{5} + \frac{8}{11} + \frac{2}{3} + \frac{8}{13} + \frac{4}{7} + \frac{8}{15} + \frac{1}{2}\left(\frac{1}{2}\right)\right]$$

$$= \frac{1}{8}\left[4\,T_4 + \frac{8}{9} + \frac{8}{11} + \frac{8}{13} + \frac{8}{15}\right] = 0.69412185\dots$$

$$T_{16} = \frac{1}{16}\left[8\,T_8 + \frac{16}{17} + \frac{16}{19} + \frac{16}{21} + \frac{16}{23} + \frac{16}{25} + \frac{16}{27} + \frac{16}{29} + \frac{16}{31}\right]$$

$$= 0.69339120\dots.$$

Note how the function values used to calculate $T_4$ were reused in the calculation of $T_8$, and similarly how those in $T_8$ were reused for $T_{16}$. When several approximations are needed it is very useful to double the number of subintervals for each new calculation, so that previously calculated values of $f$ can be reused. ∎

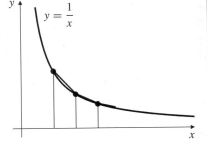

**Figure 6.17**  The trapezoid areas are greater than the area under the curve if the curve is concave upward

All Trapezoid Rule approximations to $I = \int_1^2 (1/x)\, dx$ are greater than the true value of $I$. This is because the graph of $y = 1/x$ is concave up on $[1, 2]$, and therefore the tops of the approximating trapezoids lie above the curve. (See Figure 6.17.)

We can calculate the exact errors in the three approximations since we know that $I = \ln 2 = 0.69314718\dots$ (Remember that the error in an approximation is always taken to be the true value minus the approximate value.)

$$I - T_4 = 0.69314718\dots - 0.69702381\dots = -0.00387663\dots$$

$$I - T_8 = 0.69314718\dots - 0.69412185\dots = -0.00097467\dots$$

$$I - T_{16} = 0.69314718\dots - 0.69339120\dots = -0.00024402\dots.$$

Observe that the size of the error decreases to about a quarter of its previous value each time we double $n$. We will show below that this is to be expected for a "well-behaved" function like $1/x$.

Example 1 is somewhat artificial in the sense that we know the actual value of the integral so we really don't need an approximation. In practical applications of numerical integration we do not know the actual value. It is tempting to calculate several approximations for increasing values of $n$ until the two most recent ones agree to within a prescribed error tolerance. For example, we might be inclined to claim that $\ln 2 \approx 0.69\dots$ from a comparison of $T_4$ and $T_8$, and further comparison of $T_{16}$ and $T_8$ suggests that the third decimal place is probably 3: $I \approx 0.693\dots$. Although this approach cannot be justified in general, it is frequently used in practice.

## The Midpoint Rule

A somewhat simpler approximation to $\int_a^b f(x)\,dx$, based on the partition of $[a, b]$ into $n$ equal subintervals, involves forming a Riemann sum of the areas of rectangles whose heights are taken at the midpoints of the $n$ subintervals. (See Figure 6.18.)

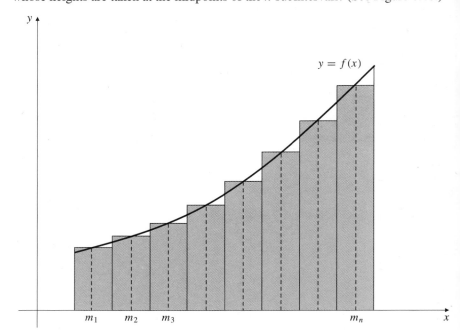

**Figure 6.18**    The Midpoint Rule approximation $M_n$ to $\int_a^b f(x)\,dx$ is the Riemann sum based on the heights to the graph of $f$ at the midpoints of the subintervals of the partition

**DEFINITION 4**

**The Midpoint Rule**

If $h = (b - a)/n$, let $m_j = a + \left(j - \frac{1}{2}\right)h$ for $1 \le j \le n$. The **Midpoint Rule** approximation to $\int_a^b f(x)\,dx$, denoted $M_n$, is given by

$$M_n = h\big(f(m_1) + f(m_2) + \cdots + f(m_n)\big) = h\sum_{j=1}^{n} f(m_j).$$

**Example 2**    Find the Midpoint Rule approximations $M_4$ and $M_8$ for the integral $I = \int_1^2 \frac{1}{x}\,dx$ and compare their actual errors with those obtained for the Trapezoid Rule approximations above.

**Solution**    To find $M_4$, the interval $[1, 2]$ is divided into four equal subintervals,

$$\left[1, \frac{5}{4}\right], \quad \left[\frac{5}{4}, \frac{3}{2}\right], \quad \left[\frac{3}{2}, \frac{7}{4}\right], \quad \text{and} \quad \left[\frac{7}{4}, 2\right].$$

The midpoints of these intervals are 9/8, 11/8, 13/8 and 15/8, respectively. The midpoints of the subintervals for $M_8$ are obtained in a similar way. The required Midpoint Rule approximations are

$$M_4 = \frac{1}{4}\left[\frac{8}{9} + \frac{8}{11} + \frac{8}{13} + \frac{8}{15}\right] = 0.69121989\ldots$$

$$M_8 = \frac{1}{8}\left[\frac{16}{17} + \frac{16}{19} + \frac{16}{21} + \frac{16}{23} + \frac{16}{25} + \frac{16}{27} + \frac{16}{29} + \frac{16}{31}\right] = 0.69266055\ldots$$

The errors in these approximations are

$$I - M_4 = 0.69314718\ldots - 0.69121989\ldots = 0.00192729\ldots$$
$$I - M_8 = 0.69314718\ldots - 0.69266055\ldots = 0.00048663\ldots$$

These errors are of opposite sign and about *half the size* of the corresponding Trapezoid Rule errors $I - T_4$ and $I - T_8$. Figure 6.19 suggests the reason for this. The rectangular area $hf(m_j)$ is equal to the area of the trapezoid formed by the tangent line to $y = f(x)$ at $(m_j, f(m_j))$. The shaded region above the curve is the part of the Trapezoid Rule error due to the $j$th subinterval. The shaded area below the curve is the corresponding Midpoint Rule error.

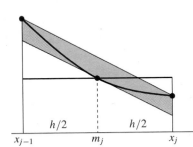

**Figure 6.19**   The Midpoint Rule error, the area shaded in colour, is opposite in sign and about half the size of the Trapezoid Ruler error, the area shaded in grey

One drawback of the Midpoint Rule is that we cannot reuse values of $f$ calculated for $M_n$ when we calculate $M_{2n}$. However, to calculate $T_{2n}$ we can use the data values already calculated for $T_n$ and $M_n$. Specifically,

$$T_{2n} = \tfrac{1}{2}(T_n + M_n).$$

A good strategy for using these methods to obtain a value for an integral $I$ to a desired degree of accuracy is to calculate successively:

$$T_n, \quad M_n, \quad T_{2n} = \frac{T_n + M_n}{2}, \quad M_{2n}, \quad T_{4n} = \frac{T_{2n} + M_{2n}}{2}, \quad M_{4n}, \quad \cdots$$

until two consecutive terms agree sufficiently closely. If a single quick approximation is needed, $M_n$ is a better choice than $T_n$.

## Error Estimates

The following theorem provides a bound for the error in the Trapezoid and Midpoint Rule approximations in terms of the second derivative of the integrand.

**THEOREM 4**

**Error estimates for the Trapezoid and Midpoint Rules**

If $f$ has a continuous second derivative on $[a, b]$ and satisfies $|f''(x)| \le K$ there, then

$$\left|\int_a^b f(x)\,dx - T_n\right| \le \frac{K(b-a)}{12}h^2 = \frac{K(b-a)^3}{12n^2},$$

$$\left|\int_a^b f(x)\,dx - M_n\right| \le \frac{K(b-a)}{24}h^2 = \frac{K(b-a)^3}{24n^2},$$

where $h = (b - a)/n$. Note that these error bounds decrease like the square of the subinterval length as $n$ increases.

**PROOF**   We will prove only the Trapezoid Rule error estimate here. (The one for the Midpoint Rule is a little easier to prove; the method is suggested in Exercise

14 below.) The straight line approximating $y = f(x)$ in the first subinterval $[x_0, x_1] = [a, a + h]$ passes through the two points $(x_0, y_0)$ and $(x_1, y_1)$. Its equation is $y = A + B(x - x_0)$, where

$$A = y_0 \qquad \text{and} \qquad B = \frac{y_1 - y_0}{x_1 - x_0} = \frac{y_1 - y_0}{h}.$$

Let the function $g(x)$ be the vertical distance between the graph of $f$ and this line:

$$g(x) = f(x) - A - B(x - x_0).$$

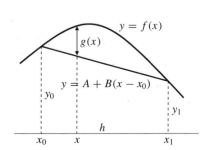

**Figure 6.20**

Since the integral of $A + B(x - x_0)$ over $[x_0, x_1]$ is the area of the first trapezoid, which is $h(y_0 + y_1)/2$ (see Figure 6.20), the integral of $g(x)$ over $[x_0, x_1]$ is the error in the approximation of $\int_{x_0}^{x_1} f(x)\, dx$ by the area of the trapezoid:

$$\int_{x_0}^{x_1} f(x)\, dx - h \frac{y_0 + y_1}{2} = \int_{x_0}^{x_1} g(x)\, dx.$$

Now $g$ is twice differentiable, and $g''(x) = f''(x)$. Also $g(x_0) = g(x_1) = 0$. Two integrations by parts (see Exercise 36 of Section 6.1) show that

$$\int_{x_0}^{x_1} (x - x_0)(x_1 - x)\, f''(x)\, dx = \int_{x_0}^{x_1} (x - x_0)(x_1 - x)\, g''(x)\, dx$$
$$= -2 \int_{x_0}^{x_1} g(x)\, dx.$$

By the triangle inequality for definite integrals (Theorem 3(f) of Section 5.4)

$$\left| \int_{x_0}^{x_1} f(x)\, dx - h \frac{y_0 + y_1}{2} \right| \le \frac{1}{2} \int_{x_0}^{x_1} (x - x_0)(x_1 - x)\, |f''(x)|\, dx.$$
$$\le \frac{K}{2} \int_{x_0}^{x_1} \left( -x^2 + (x_0 + x_1)x - x_0 x_1 \right) dx$$
$$= \frac{K}{12} (x_1 - x_0)^3 = \frac{K}{12} h^3.$$

A similar estimate holds on each subinterval $[x_{j-1}, x_j]$ $(1 \le j \le n)$. Therefore,

$$\left| \int_a^b f(x)\, dx - T_n \right| = \left| \sum_{j=1}^{n} \left( \int_{x_{j-1}}^{x_j} f(x)\, dx - h \frac{y_{j-1} + y_j}{2} \right) \right|$$
$$\le \sum_{j=1}^{n} \left| \int_{x_{j-1}}^{x_j} f(x)\, dx - h \frac{y_{j-1} + y_j}{2} \right|$$
$$= \sum_{j=1}^{n} \frac{K}{12} h^3 = \frac{K}{12} nh^3 = \frac{K(b - a)}{12} h^2,$$

since $nh = b - a$.

We illustrate this error estimate for the approximations of Examples 1 and 2 above.

**Example 3**  Obtain bounds for the errors for $T_4$, $T_8$, $T_{16}$, $M_4$, and $M_8$ for

$$I = \int_1^2 \frac{1}{x}\, dx.$$

**Solution**  If $f(x) = 1/x$, then $f'(x) = -1/x^2$ and $f''(x) = 2/x^3$. On $[1, 2]$ we have $|f''(x)| \le 2$, so we may take $K = 2$ in the estimate. Thus

$$|I - T_4| \le \frac{2(2-1)}{12}\left(\frac{1}{4}\right)^2 = 0.0104\ldots,$$

$$|I - M_4| \le \frac{2(2-1)}{24}\left(\frac{1}{4}\right)^2 = 0.0052\ldots,$$

$$|I - T_8| \le \frac{2(2-1)}{12}\left(\frac{1}{8}\right)^2 = 0.0026\ldots,$$

$$|I - M_8| \le \frac{2(2-1)}{24}\left(\frac{1}{8}\right)^2 = 0.0013\ldots,$$

$$|I - T_{16}| \le \frac{2(2-1)}{12}\left(\frac{1}{16}\right)^2 = 0.00065\ldots.$$

The actual errors calculated earlier are considerably smaller than these bounds, because $|f''(x)|$ is rather smaller than $K = 2$ over most of the interval $[0, 2]$.

**Remark**  Error bounds are not usually as easily obtained as they are in Example 3. In particular, if an exact formula for $f(x)$ is not known (as is usually the case if the values of $f$ are obtained from experimental data) then we have no method of calculating $f''(x)$, and so we can't determine $K$. Theorem 4 is of more theoretical than practical importance. It shows us that, for a "well-behaved" function $f$, the Midpoint Rule error is typically about half as large as the Trapezoid Rule error, and that both the Trapezoid Rule and Midpoint Rule errors can be expected to decrease like $1/n^2$ as $n$ increases; in terms of big-O notation,

$$I = T_n + O\left(\frac{1}{n^2}\right), \quad \text{and} \quad I = M_n + O\left(\frac{1}{n^2}\right) \qquad \text{as } n \to \infty.$$

Of course, actual errors are not equal to the error bounds so they won't always be cut to exactly a quarter of their size when we double $n$.

# Exercises 6.6

In Exercises 1–4, calculate the approximations $T_4$, $M_4$, $T_8$, $M_8$, and $T_{16}$ for the given integrals. (Use a scientific calculator or computer spreadsheet program.) Also calculate the exact value of each integral and so determine the exact error in each approximation. Compare these exact errors with the bounds for the size of the error supplied by Theorem 4.

**1.** $I = \int_0^2 (1 + x^2)\, dx$

**2.** $I = \int_0^1 e^{-x}\, dx$

**3.** $I = \int_0^{\pi/2} \sin x\, dx$

**4.** $I = \int_0^1 \frac{dx}{1 + x^2}$

**5.** Figure 6.21 shows the graph of a function $f$ over the interval $[1, 9]$. Using values from the graph, find the Trapezoid Rule estimates $T_4$ and $T_8$ for $\int_1^9 f(x)\,dx$.

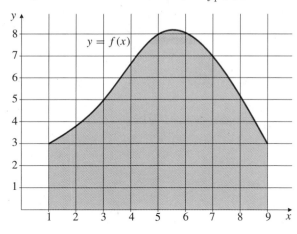

**Figure 6.21**

**6.** Obtain the best Midpoint Rule approximation that you can for $\int_1^9 f(x)\,dx$ from the data in Figure 6.21.

**7.** The map of a region is traced on the grid in Figure 6.22 where 1 unit in both the vertical and horizontal directions represents 10 km. Use the Trapezoid Rule to obtain two estimates for the area of the region.

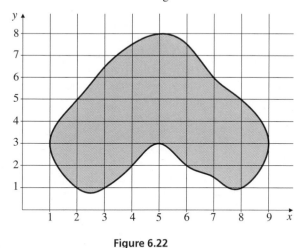

**Figure 6.22**

**8.** Find a Midpoint Rule estimate for the area of the region in the previous exercise.

**Table 1.**

| $x$ | $f(x)$ | $x$ | $f(x)$ |
|------|--------|------|--------|
| 0.0 | 1.4142 | 0.1 | 1.4124 |
| 0.2 | 1.4071 | 0.3 | 1.3983 |
| 0.4 | 1.3860 | 0.5 | 1.3702 |
| 0.6 | 1.3510 | 0.7 | 1.3285 |
| 0.8 | 1.3026 | 0.9 | 1.2734 |
| 1.0 | 1.2411 | 1.1 | 1.2057 |
| 1.2 | 1.1772 | 1.3 | 1.1258 |
| 1.4 | 1.0817 | 1.5 | 1.0348 |
| 1.6 | 0.9853 |     |        |

**9.** Find $T_4$, $M_4$, $T_8$, $M_8$, and $T_{16}$ for $\int_0^{1.6} f(x)\,dx$ for the function $f$ whose values are given in Table 1.

**10.** Find the approximations $M_8$ and $T_{16}$ for $\int_0^1 e^{-x^2}\,dx$. Quote a value for the integral to as many decimal places as you feel are justified.

**11.** Repeat Exercise 10 for $\int_0^{\pi/2} \frac{\sin x}{x}\,dx$. (Assume the integrand is 1 at $x = 0$.)

**12.** Compute the actual error in the approximation $\int_0^1 x^2\,dx \approx T_1$ and use it to show that the constant 12 in the estimate of Theorem 4 cannot be improved. That is, show that the absolute value of the actual error is as large as allowed by that estimate.

**13.** Repeat the previous exercise for $M_1$.

**∗ 14.** Prove the error estimate for the Midpoint Rule in Theorem 4 as follows: If $x_1 - x_0 = h$ and $m_1$ is the midpoint of $[x_0, x_1]$ use the error estimate for the tangent line approximation (Theorem 4 of Section 3.5) to show that

$$|f(x) - f(m_1) - f'(m_1)(x - m_1)| \le \frac{K}{2}(x - m_1)^2.$$

Use this inequality to show that

$$\left| \int_{x_0}^{x_1} f(x)\,dx - f(m_1)h \right|$$
$$= \left| \int_{x_0}^{x_1} \left( f(x) - f(m_1) - f'(m_1)(x - m_1) \right) dx \right|$$
$$\le \frac{K}{24} h^3.$$

Complete the proof the same way used for the Trapezoid Rule estimate in Theorem 4.

## 6.7  Simpson's Rule

The Trapezoid Rule approximation to $\int_a^b f(x)\,dx$ results from approximating the graph of $f$ by straight line segments through adjacent pairs of data points on the graph. Intuitively, we would expect to do better if we approximate the graph by more general curves. Since straight lines are the graphs of linear functions, the simplest obvious generalization is to use the class of quadratic functions, that is, to approximate the graph of $f$ by segments of parabolas. This is the basis of Simpson's Rule.

Suppose that we are given three points in the plane, one on each of three equally spaced vertical lines, spaced, say, $h$ units apart. If we choose the middle of these lines as the $y$-axis, then the coordinates of the three points will be, say, $(-h, y_L)$, $(0, y_M)$, and $(h, y_R)$, as illustrated in Figure 6.23.

Constants $A$, $B$, and $C$ can be chosen so that the parabola $y = A + Bx + Cx^2$ passes through these points; substituting the coordinates of the three points into the equation of the parabola, we get

$$\left. \begin{array}{l} y_L = A - Bh + Ch^2 \\ y_M = A \\ y_R = A + Bh + Ch^2 \end{array} \right\} \quad \Rightarrow \quad A = y_M \quad \text{and} \quad 2Ch^2 = y_L - 2y_M + y_R.$$

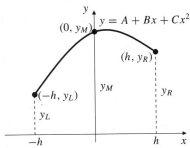

**Figure 6.23**   Fitting a quadratic graph through three points with equal horizontal spacing

Now we have

$$\int_{-h}^{h} (A + Bx + Cx^2)\, dx = \left( Ax + \frac{B}{2} x^2 + \frac{C}{3} x^3 \right)\Big|_{-h}^{h} = 2Ah + \frac{2}{3} Ch^3$$
$$= h \left( 2y_M + \frac{1}{3} (y_L - 2y_M + y_R) \right)$$
$$= \frac{h}{3} (y_L + 4y_M + y_R).$$

Thus the area of the plane region bounded by the parabolic arc, the interval of length $2h$ on the $x$-axis, and the left and right vertical lines is equal to $(h/3)$ times the sum of the heights of the region at the left and right edges and four times the height at the middle. (It is independent of the position of the $y$-axis.)

Now suppose that we are given the same data for $f$ as we were given for the Trapezoid Rule, that is, we know the values $y_j = f(x_j)$ $(0 \le j \le n)$ at $n + 1$ equally spaced points

$$x_0 = a, \quad x_1 = a + h, \quad x_2 = a + 2h, \quad \dots, \quad x_n = a + nh = b,$$

where $h = (b-a)/n$. We can approximate the graph of $f$ over *pairs* of the subintervals $[x_{j-1}, x_j]$ using parabolic segments, and use the integrals of the corresponding quadratic functions to approximate the integrals of $f$ over these subintervals. Since we need to use the subintervals two at a time we must assume that $n$ is *even*. Using the integral computed for the parabolic segment above, we have

$$\int_{x_0}^{x_2} f(x)\, dx \approx \frac{h}{3} (y_0 + 4y_1 + y_2)$$
$$\int_{x_2}^{x_4} f(x)\, dx \approx \frac{h}{3} (y_2 + 4y_3 + y_4)$$
$$\vdots$$
$$\int_{x_{n-2}}^{x_n} f(x)\, dx \approx \frac{h}{3} (y_{n-2} + 4y_{n-1} + y_n).$$

Adding these $n/2$ individual approximations we get the Simpson's Rule approximation to the integral $\int_a^b f(x)\,dx$.

**DEFINITION 5**

> **Simpson's Rule**
>
> The **Simpson's Rule** approximation to $\int_a^b f(x)\,dx$ based on a subdivision of $[a, b]$ into an even number $n$ of subintervals of equal length $h = (b-a)/n$ is denoted $S_n$, and is given by:
>
> $$\int_a^b f(x)\,dx \approx S_n$$
> $$= \frac{h}{3}\left(y_0 + 4y_1 + 2y_2 + 4y_3 + 2y_4 + \cdots + 2y_{n-2} + 4y_{n-1} + y_n\right)$$
> $$= \frac{h}{3}\left(y_{\text{"ends"}} + 4y_{\text{"odds"}} + 2y_{\text{"evens"}}\right)$$

Note that the Simpson's Rule approximation $S_n$ requires no more data than does the Trapezoid Rule approximation $T_n$; both require the values of $f(x)$ at $n+1$ equally spaced points. However, Simpson's Rule treats the data differently, weighting successive values either $1/3$, $2/3$, or $4/3$. As we will see, this can produce a much better approximation to the integral of $f$.

**Example 1**    Calculate the approximations $S_4$, $S_8$, and $S_{16}$ for $I = \int_1^2 \frac{1}{x}\,dx$ and compare them with the actual value $I = \ln 2 = 0.69314718\ldots$, and with the values of $T_4$, $T_8$, and $T_{16}$ obtained in Example 1 of Section 6.6.

**Solution**    We calculate

$$S_4 = \frac{1}{12}\left[1 + 4\left(\frac{4}{5}\right) + 2\left(\frac{2}{3}\right) + 4\left(\frac{4}{7}\right) + \frac{1}{2}\right] = 0.69325397\ldots,$$

$$S_8 = \frac{1}{24}\left[1 + \frac{1}{2} + 4\left(\frac{8}{9} + \frac{8}{11} + \frac{8}{13} + \frac{8}{15}\right)\right.$$
$$\left. + 2\left(\frac{4}{5} + \frac{2}{3} + \frac{4}{7}\right)\right] = 0.69315453\ldots,$$

$$S_{16} = \frac{1}{48}\left[1 + \frac{1}{2}\right.$$
$$+ 4\left(\frac{16}{17} + \frac{16}{19} + \frac{16}{21} + \frac{16}{23} + \frac{16}{25} + \frac{16}{27} + \frac{16}{29} + \frac{16}{31}\right)$$
$$\left. + 2\left(\frac{8}{9} + \frac{4}{5} + \frac{8}{11} + \frac{2}{3} + \frac{8}{13} + \frac{4}{7} + \frac{8}{15}\right)\right] = 0.69314765\ldots.$$

The errors are

$$I - S_4 = 0.69314718\ldots - 0.69325397\ldots = -0.00010679,$$
$$I - S_8 = 0.69314718\ldots - 0.69315453\ldots = -0.00000735,$$
$$I - S_{16} = 0.69314718\ldots - 0.69314765\ldots = -0.00000047.$$

These errors are evidently much smaller than the corresponding errors for the Trapezoid or Midpoint Rule approximations.

**Remark**   Simpson's Rule $S_{2n}$ make use of the same $2n + 1$ data values that $T_n$ and $M_n$ together use. It is not difficult to verify that

$$S_{2n} = \frac{T_n + 2M_n}{3}, \qquad S_{2n} = \frac{2T_{2n} + M_n}{3}, \qquad \text{and} \qquad S_{2n} = \frac{4T_{2n} - T_n}{3}.$$

Figure 6.19 and Theorem 4 in Section 6.6 suggest why the first of these formulas ought to yield a particularly good approximation to $I$.

Obtaining an error estimate for Simpson's Rule is more difficult than it was for the Trapezoid Rule. We state the appropriate estimate in the following theorem, but we do not attempt any proof. Proofs can be found in textbooks on numerical analysis.

**THEOREM  5**

**Error estimate for Simpson's Rule**

If $f$ has a continuous fourth derivative on the interval $[a, b]$, satisfying $|f^{(4)}(x)| \leq K$ there, then

$$\left| \int_a^b f(x)\, dx - S_n \right| \leq \frac{K(b-a)}{180} h^4 = \frac{K(b-a)^5}{180 n^4},$$

where $h = (b - a)/n$.

Observe that, as $n$ increases, the error decreases as the fourth power of $h$, and hence as $1/n^4$. Using the big-O notation we have

$$\int_a^b f(x)\, dx = S_n + O\left(\frac{1}{n^4}\right) \qquad \text{as } n \to \infty.$$

This accounts for the fact that $S_n$ is a much better approximation than is $T_n$, provided that $h$ is small and $|f^{(4)}(x)|$ is not unduly large compared with $|f''(x)|$. Note also that for any (even) $n$, $S_n$ gives the exact value of the integral of any *cubic* function $f(x) = A + Bx + Cx^2 + Dx^3$; $f^{(4)}(x) = 0$ identically for such $f$, so we can take $K = 0$ in the error estimate.

**Example 2**   Obtain bounds for the absolute values of the errors in the approximations of Example 1.

**Solution**   If $f(x) = 1/x$, then

$$f'(x) = -\frac{1}{x^2}, \qquad f''(x) = \frac{2}{x^3}, \qquad f^{(3)}(x) = -\frac{6}{x^4}, \qquad f^{(4)}(x) = \frac{24}{x^5}.$$

Clearly, $|f^{(4)}(x)| \leq 24$ on $[1, 2]$, so we can take $K = 24$ in the estimate of Theorem 5. We have

$$|I - S_4| \leq \frac{24(2-1)}{180}\left(\frac{1}{4}\right)^4 \approx 0.00052083,$$

$$|I - S_8| \leq \frac{24(2-1)}{180}\left(\frac{1}{8}\right)^4 \approx 0.00003255,$$

$$|I - S_{16}| \leq \frac{24(2-1)}{180}\left(\frac{1}{16}\right)^4 \approx 0.00000203.$$

Again we observe that the actual errors are well within these bounds.

**Example 3** A function $f$ satisfies $|f^{(4)}(x)| \le 7$ on the interval $[1, 3]$, and the values $f(1.0) = 0.1860$, $f(1.5) = 0.9411$, $f(2.0) = 1.1550$, $f(2.5) = 1.4511$, and $f(3.0) = 1.2144$. Find the best possible Simpson's rule approximation to $I = \int_1^3 f(x)\,dx$, based on this data. Give a bound for the size of the error, and specify the smallest interval you can that must contain the value of $I$.

**Solution** We take $n = 4$, so that $h = (3 - 1)/4 = 0.5$, and we obtain

$$I = \int_1^3 f(x)\,dx$$

$$\approx S_4 = \frac{0.5}{3}\Big(0.1860 + 4(0.9411 + 1.4511) + 2(1.1550) + 1.2144\Big)$$

$$= 2.2132.$$

Since $|f^{(4)}(x)| \le 7$ on $[1, 3]$ we have

$$|I - S_4| \le \frac{7(3 - 1)}{180}(0.5)^4 < 0.0049.$$

$I$ must therefore satisfy

$$2.2132 - 0.0049 < I < 2.2132 + 0.0049 \quad \text{or} \quad 2.2083 < I < 2.2181.$$

---

# Exercises 6.7

In Exercises 1–4, find Simpson's Rule approximations $S_4$ and $S_8$ for the given functions. Compare your results with the actual values of the integrals and with the corresponding Trapezoid Rule approximations obtained in Exercises 1–4 of Section 6.6.

**1.** $I = \int_0^2 (1 + x^2)\,dx$     **2.** $I = \int_0^1 e^{-x}\,dx$

**3.** $I = \int_0^{\pi/2} \sin x\,dx$     **4.** $I = \int_0^1 \dfrac{dx}{1 + x^2}$

**5.** Find the Simpson's Rule approximation $S_8$ for the integral in Exercise 5 of Section 6.6.

**6.** Find the best Simpson's Rule approximation that you can for the area of the region in Exercise 7 of Section 6.6.

**7.** Use Theorem 5 to obtain bounds for the errors in the approximations obtained in Exercises 2 and 3 above.

**8.** Verify that $S_{2n} = \dfrac{T_n + 2M_n}{3} = \dfrac{2T_{2n} + M_n}{3}$, where $T_n$ and

$M_n$ refer to the appropriate Trapezoid and Midpoint Rule approximations. Deduce that $S_{2n} = \dfrac{4T_{2n} - T_n}{3}$.

**9.** Find $S_4$, $S_8$, and $S_{16}$ for $\int_0^{1.6} f(x)\,dx$ for the function $f$ whose values are tabulated in Exercise 9 of Section 6.6.

**10.** Find the Simpson's Rule approximations $S_8$ and $S_{16}$ for $\int_0^1 e^{-x^2}\,dx$. Quote a value for the integral to the number of decimal places you feel is justified based on comparing the two approximations.

**\* 11.** Compute the actual error in the approximation $\int_0^1 x^4\,dx \approx S_2$ and use it to show that the constant 180 in the estimate of Theorem 5 cannot be improved.

**\* 12.** Since Simpson's Rule is based on quadratic approximation, it is not surprising that it should give an exact value for an integral of $A + Bx + Cx^2$. It is more surprising that it is exact for a cubic function as well. Verify by direct calculation that $\int_0^1 x^3\,dx = S_2$.

The numerical methods described in Sections 6.6 and 6.7 are suitable for finding approximate values for integrals of the form

$$I = \int_a^b f(x)\, dx,$$

where $[a, b]$ is a finite interval and the integrand $f$ is "well-behaved" on $[a, b]$. In particular, $I$ must be a *proper* integral. There are many other methods for dealing with such integrals, some of which we mention later in this section. First, however, we consider what can be done if the function $f$ isn't "well-behaved" on $[a, b]$. We mean by this that either the integral is improper, or $f$ doesn't have sufficiently many continuous derivatives on $[a, b]$ to justify whatever numerical methods we want to use.

The ideas of this section are best presented by means of concrete examples.

**Example 1** How can you evaluate the integral $I = \int_0^1 \sqrt{x}\, e^x\, dx$ numerically?

**Solution** Although $I$ is a proper integral, with integrand $f(x) = \sqrt{x}\, e^x$ satisfying $f(x) \to 0$ as $x \to 0+$, nevertheless the standard numerical methods can be expected to perform poorly for $I$ because the derivatives of $f$ are not bounded near 0. This problem is easily remedied; just make the change of variable $x = t^2$ and rewrite $I$ in the form

$$I = 2 \int_0^1 t^2 e^{t^2}\, dt,$$

whose integrand $g(t) = t^2 e^{t^2}$ has bounded derivatives near 0. The latter integral can be efficiently approximated by the methods of Sections 6.6 and 6.7.

## Approximating Improper Integrals

**Example 2** Describe how to evaluate $I = \int_0^1 \dfrac{\cos x}{\sqrt{x}}\, dx$ numerically.

**Solution** The integral is improper, but convergent because, on $[0, 1]$,

$$0 < \frac{\cos x}{\sqrt{x}} \le \frac{1}{\sqrt{x}} \qquad \text{and} \qquad \int_0^1 \frac{dx}{\sqrt{x}} = 2.$$

However, since $\lim_{x \to 0+} \dfrac{\cos x}{\sqrt{x}} = \infty$, we cannot directly apply any of the techniques developed in Sections 6.6 and 6.7. ($y_0$ is infinite.) The substitution $x = t^2$ removes this difficulty:

$$I = \int_0^1 \frac{\cos t^2}{t}\, 2t\, dt = 2 \int_0^1 \cos t^2\, dt.$$

The latter integral is not improper, and is well-behaved. Numerical techniques can be applied to evaluate it.

**Example 3**    Describe how to evaluate $I = \displaystyle\int_0^2 f(x)\,dx$ numerically, where

$$f(x) = \frac{1}{\sqrt{x(4-x^2)}}.$$

**Solution**    This integral is improper because the integrand $f$ approaches infinity at both ends of the interval $(0, 2)$. First break the integral into two integrals, each with only one singularity:

$$I = \int_0^1 f(x)\,dx + \int_1^2 f(x)\,dx = I_1 + I_2.$$

In $I_1$ make the substitution $x = t^2$:

$$I_1 = \int_0^1 \frac{dx}{\sqrt{x(4-x^2)}} = \int_0^1 \frac{2t\,dt}{\sqrt{t^2(4-t^4)}} = 2\int_0^1 \frac{dt}{\sqrt{4-t^4}}.$$

$I_2$ is improper at $x = 2$ since the denominator involves the factor $\sqrt{2-x}$. In this integral the appropriate substitution is $2 - x = t^2$, or $x = 2 - t^2$. In this case,

$$x(4-x^2) = (2-t^2)\big(4 - (2-t^2)^2\big) = t^2(2-t^2)(4-t^2).$$

Thus we have

$$I_2 = \int_1^2 \frac{dx}{\sqrt{x(4-x^2)}} = \int_0^1 \frac{2t\,dt}{t\sqrt{(2-t^2)(4-t^2)}} = 2\int_0^1 \frac{dt}{\sqrt{(2-t^2)(4-t^2)}}.$$

Recombining these two integrals, we obtain

$$I = 2\int_0^1 \left( \frac{1}{\sqrt{4-t^4}} + \frac{1}{\sqrt{(2-t^2)(4-t^2)}} \right) dt.$$

The integrand is now well-behaved on $[0, 1]$, and the integral can be evaluated by numerical methods.    ∎

**Example 4**    Show how to evaluate $I = \displaystyle\int_0^\infty \frac{dx}{\sqrt{2+x^2+x^4}}$ by numerical means.

**Solution**    Here the integral is improper of type I; the interval of integration is infinite. Although there is no singularity at $x = 0$, it is still useful to break the integral into two parts:

$$I = \int_0^1 \frac{dx}{\sqrt{2+x^2+x^4}} + \int_1^\infty \frac{dx}{\sqrt{2+x^2+x^4}} = I_1 + I_2.$$

$I_1$ is proper. In $I_2$ make the change of variable $x = 1/t$:

$$I_2 = \int_0^1 \frac{dt}{t^2\sqrt{2 + \dfrac{1}{t^2} + \dfrac{1}{t^4}}} = \int_0^1 \frac{dt}{\sqrt{2t^4 + t^2 + 1}}.$$

This is also a proper integral. If desired, $I_1$ and $I_2$ can be recombined into a single integral before numerical methods are applied:

$$I = \int_0^1 \left( \frac{1}{\sqrt{2 + x^2 + x^4}} + \frac{1}{\sqrt{2x^4 + x^2 + 1}} \right) dx.$$

Example 4 suggests that when an integral is taken over an infinite interval, a change of variable should be made to convert the integral to a finite interval.

## Using Taylor's Formula

Taylor's Formula (see Section 4.8) can sometimes be useful for evaluating integrals. Here is an example.

**Example 5**   Use Taylor's Formula for $f(x) = e^x$, obtained in Section 4.8, to evaluate the integral $\int_0^1 e^{x^2} dx$ to within an error of less than $10^{-4}$.

**Solution**   In Example 4 of Section 4.7 we showed that

$$f(x) = e^x = 1 + x + \frac{x^2}{2!} + \frac{x^3}{3!} + \cdots + \frac{x^n}{n!} + E_n(x),$$

where

$$E_n(x) = \frac{e^X}{(n+1)!} x^{n+1}$$

for some $X$ between 0 and $x$. If $0 \le x \le 1$, then $0 \le X \le 1$, so $e^X \le e < 3$. Therefore

$$|E_n(x)| \le \frac{3}{(n+1)!} x^{n+1}.$$

Now replace $x$ by $x^2$ in the formula for $e^x$ above and integrate from 0 to 1:

$$\int_0^1 e^{x^2} dx = \int_0^1 \left( 1 + x^2 + \frac{x^4}{2!} + \cdots + \frac{x^{2n}}{n!} \right) dx + \int_0^1 E_n(x^2) dx$$

$$= 1 + \frac{1}{3} + \frac{1}{5 \times 2!} + \cdots + \frac{1}{(2n+1)n!} + \int_0^1 E_n(x^2) dx.$$

We want the error to be less than $10^{-4}$, so we estimate the remainder term:

$$\left| \int_0^1 E_n(x^2) dx \right| \le \frac{3}{(n+1)!} \int_0^1 x^{2(n+1)} dx = \frac{3}{(n+1)!(2n+3)} < 10^{-4},$$

provided $(2n+3)(n+1)! > 30,000$. Since $13 \times 6! = 9,360$ and $15 \times 7! = 75,600$ we need $n = 6$. Thus

$$\int_0^1 e^{x^2} dx = 1 + \frac{1}{3} + \frac{1}{5 \times 2!} + \frac{1}{7 \times 3!} + \frac{1}{9 \times 4!} + \frac{1}{11 \times 5!} + \frac{1}{13 \times 6!}$$

$$\approx 1.46264,$$

with error less than $10^{-4}$.

## Romberg Integration

Using Taylor's Formula, it is possible to verify that for a function $f$ having continuous derivatives up to order $2m + 2$ on $[a, b]$ the error $E_n = I - T_n$ in the Trapezoid Rule approximation $T_n$ to $I = \int_a^b f(x)\, dx$ satisfies

$$E_n = I - T_n = \frac{C_1}{n^2} + \frac{C_2}{n^4} + \frac{C_3}{n^6} + \cdots + \frac{C_m}{n^{2m}} + O\left(\frac{1}{n^{2m+2}}\right),$$

where the constants $C_j$ depend on the $2j$th derivative of $f$. It is possible to use this formula to obtain higher order approximations to $I$, starting with Trapezoid Rule approximations. The technique is known as **Romberg integration** or **Richardson Extrapolation**.

To begin, suppose we have constructed Trapezoid Rule approximations for values of $n$ that are powers of 2: $n = 1, 2, 4, 8, \ldots$. Accordingly, let us define

$$T_k^0 = T_{2^k}. \quad \text{Thus} \quad T_0^0 = T_1, \quad T_1^0 = T_2, \quad T_2^0 = T_4, \quad \ldots$$

Using the formula for $T_{2^k} = I - E_{2^k}$ given above, we write

$$T_k^0 = I - \frac{C_1}{4^k} - \frac{C_2}{4^{2k}} - \cdots - \frac{C_m}{4^{mk}} + O\left(\frac{1}{4^{(m+1)k}}\right) \quad \text{(as } k \to \infty\text{)}.$$

Similarly, replacing $k$ by $k + 1$, we get

$$T_{k+1}^0 = I - \frac{C_1}{4^{k+1}} - \frac{C_2}{4^{2(k+1)}} - \cdots - \frac{C_m}{4^{m(k+1)}} + O\left(\frac{1}{4^{(m+1)(k+1)}}\right).$$

If we multiply the formula for $T_{k+1}^0$ by 4 and subtract the formula for $T_k^0$, the terms involving $C_1$ will cancel out. The first term on the right will be $4I - I = 3I$, so let us also divide by 3 and define $T_{k+1}^1$ to be the result. Then as $k \to \infty$, we have

$$T_{k+1}^1 = \frac{4T_{k+1}^0 - T_k^0}{3} = I - \frac{C_2^1}{4^{2k}} - \frac{C_3^1}{4^{3k}} - \cdots - \frac{C_m^1}{4^{mk}} + O\left(\frac{1}{4^{(m+1)k}}\right).$$

(The $C_i^1$ are new constants.) Unless these constants are much larger than the previous ones, $T_{k+1}^1$ ought to be a better approximation to $I$ than $T_{k+1}^0$ since we have eliminated the lowest order (and therefore the largest) of the error terms, $C_1/4^{k+1}$. In fact, Exercise 8 in Section 6.7 shows that $T_{k+1}^1 = S_{2^{k+1}}$, the Simpson's Rule approximation based on $2^{k+1}$ subintervals.

We can continue the process of eliminating error terms begun above. Replacing $k + 1$ by $k + 2$ in the expression for $T_{k+1}^1$ we obtain

$$T_{k+2}^1 = I - \frac{C_2^1}{4^{2(k+1)}} - \frac{C_3^1}{4^{3(k+1)}} - \cdots - \frac{C_m^1}{4^{m(k+1)}} + O\left(\frac{1}{4^{(m+1)(k+1)}}\right).$$

To eliminate $C_2^1$ we can multiply the second formula by 16, subtract the first formula, and divide by 15. Denoting the result $T_{k+2}^2$, we have as $k \to \infty$,

$$T_{k+2}^2 = \frac{16T_{k+2}^1 - T_{k+1}^1}{15} = I - \frac{C_3^2}{4^{3k}} - \cdots - \cdots - \frac{C_m^2}{4^{mk}} + O\left(\frac{1}{4^{(m+1)k}}\right).$$

We can proceed in this way, eliminating one error term after another. In general, for $j < m$ and $k \geq 0$,

$$T_{k+j}^{j} = \frac{4^j T_{k+j}^{j-1} - T_{k+j-1}^{j-1}}{4^j - 1} = I - \frac{C_{j+1}^j}{4^{(j+1)k}} - \cdots - \frac{C_m^j}{4^{mk}} + O\left(\frac{1}{4^{(m+1)k}}\right).$$

The big-O term refers to $k \to \infty$ for fixed $j$. All this looks very complicated, but it is not difficult to carry out in practice, especially with the aid of a computer spreadsheet. Let $R_j = T_j^{j}$, called a **Romberg approximation** to $I$, and calculate the entries in the following scheme in order from left to right and down each column when you come to it:

**Scheme for calculating Romberg approximations**

$$T_0^0 = T_1 = R_0 \longrightarrow \quad T_1^0 = T_2 \quad \longrightarrow \quad T_2^0 = T_4 \quad \longrightarrow \quad T_3^0 = T_8 \quad \longrightarrow$$
$$\downarrow \qquad\qquad \downarrow \qquad\qquad \downarrow$$
$$T_1^1 = S_2 = R_1 \qquad T_2^1 = S_4 \qquad T_3^1 = S_8$$
$$\downarrow \qquad\qquad \downarrow$$
$$T_2^2 = R_2 \qquad T_3^2$$
$$\downarrow$$
$$T_3^3 = R_3$$

Stop when $T_j^{j-1}$ and $R_j$ differ by less than the acceptable error, and quote $R_j$ as the Romberg approximation to $\int_a^b f(x)\,dx$.

The top line in the scheme is made up of the Trapezoid Rule approximations $T_1, T_2, T_4, T_8, \ldots$. Elements in subsequent rows are calculated by the formulas

**Formulas for calculating Romberg approximations**

$$T_1^1 = \frac{4T_1^0 - T_0^0}{3} \qquad T_2^1 = \frac{4T_2^0 - T_1^0}{3} \qquad T_3^1 = \frac{4T_3^0 - T_2^0}{3} \quad \cdots$$

$$T_2^2 = \frac{16T_2^1 - T_1^1}{15} \qquad T_3^2 = \frac{16T_3^1 - T_2^1}{15} \quad \cdots$$

$$T_3^3 = \frac{64T_3^2 - T_2^2}{63} \quad \cdots$$

In general, if $1 \leq j \leq k$, then $\quad T_k^j = \dfrac{4^j T_k^{j-1} - T_{k-1}^{j-1}}{4^j - 1}$.

Each new entry is calculated from the one above and the one to the left of that one.

**Example 6**   Calculate the Romberg approximations $R_0, R_1, R_2, R_3$, and $R_4$ for the integral $I = \displaystyle\int_1^2 \frac{1}{x}\,dx$.

**Solution**   We will carry all calculations to eight decimal places. Since we must obtain $R_4$, we will need to find all the entries in the first five columns of the scheme. First we calculate the first two Trapezoid Rule approximations:

$$R_0 = T_0^0 = T_1 = \frac{1}{2} + \frac{1}{4} = 0.75000000,$$

$$T_1^0 = T_2 = \frac{1}{2}\left[\frac{1}{2}(1) + \frac{2}{3} + \frac{1}{2}\left(\frac{1}{2}\right)\right] = 0.70833333.$$

The remaining required Trapezoid Rule approximations were calculated in Example 1 of Section 6.6, so we will just record them here:

$$T_2^0 = T_4 = 0.69702381,$$
$$T_3^0 = T_8 = 0.69412185,$$
$$T_4^0 = T_{16} = 0.69339120.$$

Now we calculate down the columns from left to right. For the second column:

$$R_1 = S_2 = T_1^1 = \frac{4T_1^0 - T_0^0}{3} = 0.69444444;$$

the third column:

$$S_4 = T_2^1 = \frac{4T_2^0 - T_1^0}{3} = 0.69325397,$$

$$R_2 = T_2^2 = \frac{16T_2^1 - T_1^1}{15} = 0.69317460;$$

the fourth column:

$$S_8 = T_3^1 = \frac{4T_3^0 - T_2^0}{3} = 0.69315453,$$

$$T_3^2 = \frac{16T_3^1 - T_2^1}{15} = 0.69314790,$$

$$R_3 = T_3^3 = \frac{64T_3^2 - T_2^2}{63} = 0.69314748;$$

and the fifth column:

$$S_{16} = T_4^1 = \frac{4T_4^0 - T_3^0}{3} = 0.69314765,$$

$$T_4^2 = \frac{16T_4^1 - T_3^1}{15} = 0.69314719,$$

$$T_4^3 = \frac{64T_4^2 - T_3^2}{63} = 0.69314718,$$

$$R_4 = T_4^4 = \frac{256T_4^3 - T_3^3}{255} = 0.69314718.$$

Since $T_4^3$ and $R_4$ agree to the eight decimal places we are calculating, we conclude that

$$I = \int_1^2 \frac{dx}{x} = \ln 2 \approx 0.69314718\ldots,$$

correct to eight decimal places. ∎

The various approximations calculated above suggest that for any given value of $n = 2^k$, the Romberg approximation $R_n$ should give the best value obtainable for the integral based on the $n + 1$ data values $y_0, y_1, \ldots, y_n$. This is so only if the derivatives $f^{(n)}(x)$ do not grow too rapidly as $n$ increases.

## Other Methods

As developed above, the Trapezoid, Midpoint, Simpson, and Romberg methods all involved using equal subdivisions of the interval $[a, b]$. There are other methods that avoid this restriction. In particular, **Gaussian approximations** involve selecting evaluation points and weights in an optimal way so as to give the most accurate results for "well-behaved" functions. See Exercises 11–13 below. You can consult a text on numerical analysis to learn more about this method.

Finally, we note that even when you apply one of the methods of Sections 6.6 and 6.7, it may be advisable for you to break up the integral into two or more integrals over smaller intervals and then use different subinterval lengths $h$ for each of the different integrals. You will want to evaluate the integrand at more points in an interval where its graph is changing direction erratically than in one where the graph is better behaved.

## Exercises 6.8

Rewrite the integrals in Exercises 1–6 in a form to which numerical methods can be readily applied.

**1.** $\displaystyle\int_0^1 \frac{dx}{x^{1/3}(1 + x)}$

**2.** $\displaystyle\int_0^1 \frac{e^x}{\sqrt{1 - x}}\, dx$

**3.** $\displaystyle\int_{-1}^1 \frac{e^x}{\sqrt{1 - x^2}}\, dx$

**4.** $\displaystyle\int_1^\infty \frac{dx}{x^2 + \sqrt{x} + 1}$

**\* 5.** $\displaystyle\int_0^{\pi/2} \frac{dx}{\sqrt{\sin x}}$

**6.** $\displaystyle\int_0^\infty \frac{dx}{x^4 + 1}$

**7.** Find $T_2$, $T_4$, $T_8$, and $T_{16}$ for $\int_0^1 \sqrt{x}\, dx$ and find the actual errors in these approximations. Do the errors decrease like $1/n^2$ as $n$ increases? Why?

**8.** Transform the integral $I = \int_1^\infty e^{-x^2}\, dx$ using the substitution $x = 1/t$ and calculate the Simpson's Rule approximations $S_2$, $S_4$, and $S_8$ for the resulting integral (whose integrand has limit 0 as $t \to 0+$). Quote the value of $I$ to the accuracy you feel is justified. Do the approximations converge as quickly as you might expect? Can you think of a reason why they might not?

**9.** Evaluate $I = \int_0^1 e^{-x^2}\, dx$, by the Taylor's Formula method of Example 5, to within an error of $10^{-4}$.

**10.** Recall that $\int_0^\infty e^{-x^2}\, dx = \frac{1}{2}\sqrt{\pi}$. Combine this fact with the result of the previous exercise to evaluate

$$I = \int_1^\infty e^{-x^2}\, dx$$ to three decimal places.

**11. (Gaussian approximation)** Find constants $A$ and $u$, with $u$

between 0 and 1, such that

$$\int_{-1}^1 f(x)\, dx = Af(-u) + Af(u)$$

holds for every cubic polynomial
$f(x) = ax^3 + bx^2 + cx + d$. For a general function $f(x)$ defined on $[-1, 1]$, the approximation

$$\int_{-1}^1 f(x)\, dx \approx Af(-u) + Af(u)$$

is called a *Gaussian* approximation.

**12.** Use the method of Exercise 11 to approximate the integrals of (a) $x^4$, (b) $\cos x$, and (c) $e^x$, over the interval $[-1, 1]$, and find the error in each approximation.

**13. (Another Gaussian approximation)** Find constants $A$ and $B$, and $u$ between 0 and 1, such that

$$\int_{-1}^1 f(x)\, dx = Af(-u) + Bf(0) + Af(u)$$

holds for every quintic polynomial
$f(x) = ax^5 + bx^4 + cx^3 + dx^2 + ex + f$.

**14.** Use the Gaussian approximation

$$\int_{-1}^1 f(x)\, dx \approx Af(-u) + Bf(0) + Af(u),$$

where $A$, $B$, and $u$ are as determined in Exercise 13, to find approximations for the integrals of (a) $x^6$, (b) $\cos x$, and (c) $e^x$, over the interval $[-1, 1]$, and find the error in each approximation.

**15.** Calculate sufficiently many Romberg approximations $R_1, R_2, R_3, \ldots$ for the integral

$$\int_0^1 e^{-x^2}\, dx$$

to be confident you have evaluated the integral correctly to six decimal places.

**16.** Use the values of $f(x)$ given in the table accompanying Exercise 9 in Section 6.6 to calculate the Romberg approximations $R_1, R_2$ and $R_3$ for the integral

$$\int_0^{1.6} f(x)\, dx$$

in that exercise.

**∗ 17.** The Romberg approximation $R_2$ for $\int_a^b f(x)\, dx$ requires five values of $f$, $y_0 = f(a)$, $y_1 = f(a+h), \ldots,$ $y_4 = f(x+4h) = f(b)$, where $h = (b-a)/4$. Write the formula for $R_2$ explicitly in terms of these five values.

**∗ 18.** Explain why the change of variable $x = 1/t$ is not suitable for transforming the integral $\int_\pi^\infty \dfrac{\sin x}{1+x^2}\, dx$ into a form to which numerical methods can be applied. Try to devise a method whereby this integral could be approximated to any desired degree of accuracy.

**∗ 19.** If $f(x) = \dfrac{\sin x}{x}$ for $x \neq 0$ and $f(0) = 1$, show that $f''(x)$ has a finite limit as $x \to 0$. Hence $f''$ is bounded on finite intervals $[0, a]$ and Trapezoid Rule approximations $T_n$ to $\int_0^a \dfrac{\sin x}{x}\, dx$ converge suitably quickly as $n$ increases. Higher derivatives are also bounded (Taylor's Formula is useful for showing this) so Simpson's Rule and higher-order approximations can also be used effectively.

# Chapter Review

## Key Ideas

- **What do the following phrases mean?**

  ◇ integration by parts          ◇ a reduction formula

  ◇ an inverse substitution      ◇ a rational function

  ◇ the method of partial fractions

  ◇ a computer algebra system

  ◇ an improper integral of type I

  ◇ an improper integral of type II

  ◇ a $p$-integral                ◇ the Trapezoid Rule

  ◇ the Midpoint Rule            ◇ Simpson's Rule

- **Describe the inverse sine and inverse tangent substitutions.**
- **What is the significance of the comparison theorem for improper integrals?**
- **When is numerical integration necessary?**

## Summary of Techniques of Integration

Students sometimes have difficulty deciding which method to use to evaluate a given integral. Often no one method will suffice to produce the whole solution, but one method may lead to a different, possibly simpler integral that can then be dealt with on its own merits. Here are a few guidelines:

1. First, and always, be alert for simplifying substitutions. Even when these don't accomplish the whole integration, they can lead to integrals to which some other method can be applied.

2. If the integral involves a quadratic expression $Ax^2 + Bx + C$ with $A \neq 0$ and $B \neq 0$, complete the square. A simple substitution then reduces the quadratic expression to a sum or difference of squares.

3. Integrals of products of trigonometric functions can sometimes be evaluated or rendered simpler by the use of appropriate trigonometric identities such as:

$$\sin^2 x + \cos^2 x = 1$$
$$\sec^2 x = 1 + \tan^2 x$$
$$\csc^2 x = 1 + \cot^2 x$$
$$\sin x \, \cos x = \tfrac{1}{2} \sin 2x$$
$$\sin^2 x = \tfrac{1}{2}(1 - \cos 2x)$$
$$\cos^2 x = \tfrac{1}{2}(1 + \cos 2x).$$

4. Integrals involving $(a^2 - x^2)^{1/2}$ can be transformed using $x = a \sin\theta$. Integrals involving $(a^2 + x^2)^{1/2}$ or $1/(a^2 + x^2)$ may yield to $x = a \tan\theta$. Integrals involving $(x^2 - a^2)^{1/2}$ can be transformed using $x = a \sec\theta$ or $x = a \cosh\theta$.

5. Use integration by parts for integrals of functions such as products of polynomials and transcendental functions, and for inverse trigonometric functions and logarithms. Be alert for ways of using integration by parts to obtain formulas representing complicated integrals in terms of simpler ones.

6. Use partial fractions to integrate rational functions whose denominators can be factored into real linear and quadratic factors. Remember to divide the polynomials first, if necessary, to reduce the fraction to one whose numerator has degree smaller than that of its denominator.

7. There is a table of integrals at the back of this book. If you can't do an integral directly, try to use the methods above to convert it to the form of one of the integrals in the table.

8. If you can't find any way to evaluate a definite integral for which you need a numerical value, consider using a computer or calculator and one of the numerical methods presented in Sections 6.6–6.8.

# Review Exercises on Techniques of Integration

Here is an opportunity to get more practice evaluating integrals. Unlike the exercises in Sections 5.6 and 6.1–6.3, which used only the technique of the particular section, these exercises are grouped randomly so you will have to decide which techniques to use.

1. $\displaystyle\int \frac{x\,dx}{2x^2 + 5x + 2}$

2. $\displaystyle\int \frac{x\,dx}{(x-1)^3}$

3. $\displaystyle\int \sin^3 x \cos^3 x \, dx$

4. $\displaystyle\int \frac{(1+\sqrt{x})^{1/3}}{\sqrt{x}} \, dx$

5. $\displaystyle\int \frac{3\,dx}{4x^2 - 1}$

6. $\displaystyle\int (x^2 + x - 2) \sin 3x \, dx$

7. $\displaystyle\int \frac{\sqrt{1-x^2}}{x^4} \, dx$

8. $\displaystyle\int x^3 \cos(x^2) \, dx$

9. $\displaystyle\int \frac{x^2\,dx}{(5x^3 - 2)^{2/3}}$

10. $\displaystyle\int \frac{dx}{x^2 + 2x - 15}$

11. $\displaystyle\int \frac{dx}{(4+x^2)^2}$

12. $\displaystyle\int (\sin x + \cos x)^2 \, dx$

13. $\displaystyle\int 2^x \sqrt{1 + 4^x} \, dx$

14. $\displaystyle\int \frac{\cos x}{1 + \sin^2 x} \, dx$

15. $\displaystyle\int \frac{\sin^3 x}{\cos^7 x} \, dx$

16. $\displaystyle\int \frac{x^2\,dx}{(3 + 5x^2)^{3/2}}$

17. $\displaystyle\int e^{-x} \sin(2x) \, dx$

18. $\displaystyle\int \frac{2x^2 + 4x - 3}{x^2 + 5x} \, dx$

19. $\displaystyle\int \cos(3 \ln x) \, dx$

20. $\displaystyle\int \frac{dx}{4x^3 + x}$

21. $\displaystyle\int \frac{x \ln(1+x^2)}{1 + x^2} \, dx$

22. $\displaystyle\int \sin^2 x \cos^4 x \, dx$

23. $\displaystyle\int \frac{x^2}{\sqrt{2 - x^2}} \, dx$

24. $\displaystyle\int \tan^4 x \sec x \, dx$

25. $\displaystyle\int \frac{x^2\,dx}{(4x + 1)^{10}}$

26. $\displaystyle\int x \sin^{-1} \frac{x}{2} \, dx$

27. $\displaystyle\int \sin^5(4x) \, dx$

28. $\displaystyle\int \frac{dx}{x^5 - 2x^3 + x}$

29. $\displaystyle\int \frac{dx}{2 + e^x}$

30. $\displaystyle\int x^3 3^x \, dx$

31. $\displaystyle\int \frac{\sin^2 x \cos x}{2 - \sin x} \, dx$

32. $\displaystyle\int \frac{x^2 + 1}{x^2 + 2x + 2} \, dx$

33. $\displaystyle\int \frac{dx}{x^2\sqrt{1 - x^2}}$

34. $\displaystyle\int x^3 (\ln x)^2 \, dx$

35. $\displaystyle\int \frac{x^3}{\sqrt{1 - 4x^2}} \, dx$

36. $\displaystyle\int \frac{e^{1/x}\,dx}{x^2}$

37. $\displaystyle\int \frac{x+1}{\sqrt{x^2 + 1}} \, dx$

38. $\displaystyle\int e^{(x^{1/3})} \, dx$

39. $\displaystyle\int \frac{x^3 - 3}{x^3 - 9x} \, dx$

40. $\displaystyle\int \frac{10^{\sqrt{x+2}}}{\sqrt{x+2}} \, dx$

41. $\displaystyle\int \sin^5 x \cos^9 x \, dx$

42. $\displaystyle\int \frac{x^2\,dx}{\sqrt{x^2 - 1}}$

43. $\displaystyle\int \frac{x\,dx}{x^2 + 2x - 1}$

44. $\displaystyle\int \frac{2x - 3}{\sqrt{4 - 3x + x^2}} \, dx$

45. $\displaystyle\int x^2 \sin^{-1}(2x) \, dx$

46. $\displaystyle\int \frac{\sqrt{3x^2 - 1}}{x} \, dx$

47. $\displaystyle\int \cos^4 x \sin^4 x \, dx$

48. $\displaystyle\int \sqrt{x - x^2} \, dx$

49. $\displaystyle\int \frac{dx}{(4+x)\sqrt{x}}$

50. $\displaystyle\int x \tan^{-1} \frac{x}{3} \, dx$

51. $\displaystyle\int \frac{x^4 - 1}{x^3 + 2x^2} \, dx$

52. $\displaystyle\int \frac{dx}{x(x^2 + 4)^2}$

53. $\displaystyle\int \frac{\sin(2 \ln x)}{x} \, dx$

54. $\displaystyle\int \frac{\sin(\ln x)}{x^2} \, dx$

55. $\displaystyle\int \frac{e^{2 \tan^{-1} x}}{1 + x^2} \, dx$

56. $\displaystyle\int \frac{x^3 + x - 2}{x^2 - 7} \, dx$

57. $\displaystyle\int \frac{\ln(3 + x^2)}{3 + x^2} x \, dx$

58. $\displaystyle\int \cos^7 x \, dx$

59. $\displaystyle\int \frac{\sin^{-1}(x/2)}{(4 - x^2)^{1/2}} \, dx$

60. $\displaystyle\int \tan^4(\pi x) \, dx$

61. $\displaystyle\int \frac{(x+1)\,dx}{\sqrt{x^2 + 6x + 10}}$

62. $\displaystyle\int e^x (1 - e^{2x})^{5/2} \, dx$

63. $\displaystyle\int \frac{x^3\,dx}{(x^2 + 2)^{7/2}}$

64. $\displaystyle\int \frac{x^2}{2x^2 - 3} \, dx$

65. $\displaystyle\int \frac{x^{1/2}}{1 + x^{1/3}} \, dx$

66. $\displaystyle\int \frac{dx}{x(x^2 + x + 1)^{1/2}}$

67. $\displaystyle\int \frac{1+x}{1 + \sqrt{x}} \, dx$

68. $\displaystyle\int \frac{x\,dx}{4x^4 + 4x^2 + 5}$

69. $\displaystyle\int \frac{x\,dx}{(x^2 - 4)^2}$

70. $\displaystyle\int \frac{dx}{x^3 + x^2 + x}$

71. $\displaystyle\int x^2 \tan^{-1} x \, dx$

72. $\displaystyle\int e^x \sec(e^x) \, dx$

**73.** $\displaystyle\int \frac{dx}{4\sin x - 3\cos x}$

**74.** $\displaystyle\int \frac{dx}{x^{1/3} - 1}$

**75.** $\displaystyle\int \frac{dx}{\tan x + \sin x}$

**76.** $\displaystyle\int \frac{x\,dx}{\sqrt{3 - 4x - 4x^2}}$

**77.** $\displaystyle\int \frac{\sqrt{x}}{1 + x}\,dx$

**78.** $\displaystyle\int \sqrt{1 + e^x}\,dx$

**79.** $\displaystyle\int \frac{x^4\,dx}{x^3 - 8}$

**80.** $\displaystyle\int x e^x \cos x\,dx$

## Other Review Exercises

**1.** Evaluate $I = \int x e^x \cos x\,dx$ and $J = \int x e^x \sin x\,dx$ by differentiating $e^x\big((ax+b)\cos x + (cx+d)\sin x\big)$ and examining coefficients.

**2.** For which real numbers $r$ is the following reduction formula (obtained using integration by parts) valid?

$$\int_0^\infty x^r e^{-x}\,dx = r \int_0^\infty x^{r-1} e^{-x}\,dx$$

Evaluate the integrals in Exercises 3–6 or show that they diverge.

**3.** $\displaystyle\int_0^{\pi/2} \csc x\,dx$

**4.** $\displaystyle\int_1^\infty \frac{1}{x + x^3}\,dx$

**5.** $\displaystyle\int_0^1 \sqrt{x}\,\ln x\,dx$

**6.** $\displaystyle\int_{-1}^1 \frac{dx}{x\sqrt{1 - x^2}}$

**7.** Show that the integral $I = \int_0^\infty (1/(\sqrt{x}\,e^x))\,dx$ converges and that its value satisfies $I < (2e + 1)/e$.

**8.** By measuring the areas enclosed by contours on a topographic map, a geologist determines the cross-sectional areas $A$ (m$^2$) through a 60 m high hill at various heights $h$ (m) given in Table 2.

**Table 2.**

| $h$ | 0 | 10 | 20 | 30 | 40 | 50 | 60 |
|---|---|---|---|---|---|---|---|
| $A$ | 10,200 | 9,200 | 8,000 | 7,100 | 4,500 | 2,400 | 100 |

If she uses the Trapezoid Rule to estimate the volume of the hill (which is $V = \int_0^{60} A(h)\,dh$), what will be her estimate, to the nearest 1,000 m$^3$?

**9.** What will be the geologist's estimate of the volume of the hill in Exercise 8 if she uses Simpson's Rule instead of the Trapezoid Rule?

**10.** Find the Trapezoid Rule and Midpoint Rule approximations $T_4$ and $M_4$ for the integral $I = \int_0^1 \sqrt{2 + \sin(\pi x)}\,dx$. Quote the results to five decimal places. Quote a value of $I$ to as many decimal places as you feel are justified by these approximations.

**11.** Use the results of Exercise 10 to calculate the Trapezoid Rule approximation $T_8$ and the Simpson's Rule approximation $S_8$ for the integral $I$ in that exercise. Quote a value of $I$ to as many decimal places as you feel are justified by these approximations.

**12.** Devise a way to evaluate $I = \int_{1/2}^\infty x^2/(x^5 + x^3 + 1)\,dx$ numerically, and use it to find $I$ correct to three decimal places.

**13.** You want to approximate the integral $I = \int_0^4 f(x)\,dx$ of an unknown function $f(x)$, and you measure the following values of $f$:

**Table 3.**

| $x$ | 0 | 1 | 2 | 3 | 4 |
|---|---|---|---|---|---|
| $f(x)$ | 0.730 | 1.001 | 1.332 | 1.729 | 2.198 |

(a) What are the approximations $T_4$ and $S_4$ to $I$ that you calculate with these data.

(b) You then decide to make more measurements in order to calculate $T_8$ and $S_8$. You obtain $T_8 = 5.5095$. What do you obtain for $S_8$?

(c) You have theoretical reasons to believe that $f(x)$ is, in fact, a polynomial of degree 3. Do your calculations support this theory or not? Why?

## Challenging Problems

**1.** (a) Some people think that $\pi = 22/7$. Prove that this is not so by showing that

$$\int_0^1 \frac{x^4(1 - x)^4}{x^2 + 1}\,dx = \frac{22}{7} - \pi.$$

(b) If $I = \int_0^1 x^4(1 - x)^4\,dx$, show that

$$\frac{22}{7} - I < \pi < \frac{22}{7} - \frac{I}{2}.$$

(c) Evaluate $I$ and hence determine an explicit small interval containing $\pi$.

**2.** (a) Find a reduction formula for $\int (1 - x^2)^n\,dx$.

(b) Show that if $n$ is a positive integer, then

$$\int_0^1 (1 - x^2)^n\,dx = \frac{2^{2n}(n!)^2}{(2n + 1)!}.$$

(c) Use your reduction formula to evaluate $\int (1 - x^2)^{-3/2}\,dx$.

**3.** (a) Show that $x^4 + x^2 + 1$ factors into a product of two real quadratics, and evaluate $\int (x^2 + 1)/(x^4 + x^2 + 1)\,dx$. *Hint:* $x^4 + x^2 + 1 = (x^2 + 1)^2 - x^2$.

(b) Use the same method to find $\int (x^2 + 1)/(x^4 + 1)\,dx$.

**4.** Let $I_{m,n} = \int_0^1 x^m (\ln x)^n\,dx$.

(a) Show that $I_{m,n} = (-1)^n \int_0^\infty x^n e^{-(m+1)x}\,dx$.

(b) Show that $I_{m,n} = \dfrac{(-1)^n n!}{(m + 1)^{n+1}}$.

**5.** Let $I_n = \int_0^1 x^n e^{-x}\,dx$.

(a) Show that $0 < I_n < \dfrac{1}{n+1}$, and hence that $\lim_{n \to \infty} I_n = 0$.

(b) Show that $I_n = nI_{n-1} - \dfrac{1}{e}$ for $n \geq 1$, and $I_0 = 1 - \dfrac{1}{e}$.

(c) Verify by induction that $I_n = n!\left(1 - \dfrac{1}{e}\sum_{j=0}^{n}\dfrac{1}{j!}\right)$.

(d) Deduce from (a) and (c) that $\displaystyle\lim_{n \to \infty} \sum_{j=0}^{n} \dfrac{1}{j!} = e$.

* **6.** If $K$ is very large, which of the approximations $T_{100}$ (Trapezoidal Rule), $M_{100}$ (Midpoint Rule), $S_{100}$ (Simpson's Rule) will be closest to the true value for $\int_0^1 e^{-Kx}\,dx$? Which will be farthest? Justify your answers. (*Caution:* this is trickier than it sounds!)

**7.** Simpson's Rule gives the exact definite integral for a cubic $f$. Suppose you want a numerical integration rule that gives the exact answer for a polynomial of degree 5. You might approximate the integral over the subinterval $[m-h, m+h]$ by something of the form

$$2h\left(af(m-h) + bf\left(m - \frac{h}{2}\right) + cf(m)\right.$$
$$\left. + bf\left(m + \frac{h}{2}\right) + af(m+h)\right),$$

for some constants $a$, $b$, and $c$.

(a) Determine $a$, $b$, and $c$ for which this will work. (*Hint:* take $m = 0$ to make things simple.)

(b) Use this method to approximate $\int_0^1 e^{-x}\,dx$ using first one and then two of these intervals (thus evaluating the integrand at nine points).

* **8.** The convergence of improper integrals can be a more delicate matter when the integrand changes sign. Here is one method that can be used to prove convergence in some cases where the comparison theorem fails.

(a) Suppose that $f(x)$ is differentiable on $[1, \infty)$, $f'(x)$ is continuous there, $f'(x) < 0$, and $\lim_{x \to \infty} f(x) = 0$. Show that $\int_1^\infty f'(x)\cos(x)\,dx$ converges. *Hint:* what is $\int_1^\infty |f'(x)|\,dx$ ?

(b) Under the same hypotheses, show that $\int_1^\infty f(x)\sin x\,dx$ converges. *Hint:* integrate by parts, and use (a).

(c) Show that $\int_1^\infty \dfrac{\sin x}{x}\,dx$ converges but $\int_1^\infty \dfrac{|\sin x|}{x}\,dx$ diverges. *Hint:* $|\sin x| \geq \sin^2 x = \dfrac{1 - \cos(2x)}{2}$. Note that (b) would work just as well with $\sin x$ replaced by $\cos(2x)$.

# CHAPTER 7

# Applications of Integration

**Introduction**  Numerous quantities in mathematics, physics, economics, biology, and indeed any quantitative science, can be conveniently represented by definite integrals, both proper and improper.  In addition to measuring plane areas, the problem that motivated the definition of the definite integral, we can use these integrals to express volumes of solids, lengths of curves, areas of surfaces, forces, work, energy, pressure, probabilities, dollar values of a stream of payments, and a variety of other quantities that are in one sense or another equivalent to areas under graphs.

In addition, as we have seen previously, many of the basic principles that govern the behaviour of our world are expressed in terms of differential equations and initial-value problems.  Indefinite integration is a key tool in the solution of such problems.

In this chapter we examine some of these applications. For the most part they are independent of one another, and for that reason some of the later sections in this chapter can be regarded as optional material. The material of Sections 7.1–7.3, however, should be regarded as core, because these ideas will arise again in the study of multivariable calculus.

## 7.1 Volumes of Solids of Revolution

In this section we show how volumes of certain three-dimensional regions (or *solids*) can be expressed as definite integrals and thereby determined. We will not attempt to give a definition of *volume*, but will rely on our intuition and experience with solid objects to provide enough insight for us to specify the volumes of certain simple solids. For example, if the base of a rectangular box is a rectangle of length $l$ and width $w$ (and therefore area $A = lw$), and if the box has height $h$, then its volume is $V = Ah = lwh$. If $l$, $w$, and $h$ are measured in *units* (for example, inches), then the volume is expressed in *cubic units* (cubic inches).

A rectangular box is a special case of a solid called a **prism** or **cylinder**. (See Figure 7.1.) Such a solid has a flat base, occupying a plane region having area $A$. Every cross-section of the solid in a plane parallel to the base is congruent to the base, and so has the same area. If the solid has height $h$ (so that its top is in a plane parallel to the base and $h$ units above it), then the volume of the solid is $V = Ah$. Such solids are usually called *prisms* if the base is bounded by straight lines, and *cylinders* if the base is bounded by curves.  In particular, if the base is a circular disk of radius $r$ and the top of the solid is directly above the base, then the solid is a **right-circular cylinder**. If it has height $h$ then its volume is $V = \pi r^2 h$ cubic units. Cylinders and prisms are said to be **right** if their side walls are perpendicular to their bases; otherwise they are **oblique**. Obliqueness has no effect on the volume $V = Ah$, for which $h$ is always measured in a direction perpendicular to the base.

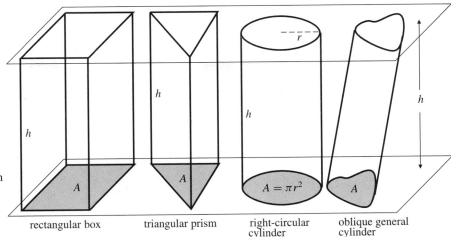

**Figure 7.1** The volume of any prism or cylinder is the base area times the height (measured perpendicular to the base) $V = Ah$

rectangular box     triangular prism     right-circular cylinder     oblique general cylinder

## Volumes by Slicing

Knowing the volume of a cylinder enables us to determine the volumes of some more general solids. We can divide solids into thin "slices" by parallel planes. (Think of a loaf of sliced bread.) Each slice is approximately a cylinder of very small "height"; the height is the thickness of the slice. See Figure 7.2, where the height is measured horizontally in the direction of the $x$-axis. If we know the cross-sectional area of each slice we can determine its volume and sum these volumes to find the volume of the solid.

**Figure 7.2** Slicing a solid perpendicular to an axis

**Figure 7.3** The volume of a slice

To be specific, suppose that the solid $S$ lies between planes perpendicular to the $x$-axis at positions $x = a$ and $x = b$ and that the cross-sectional area of $S$ in the plane perpendicular to the $x$-axis at $x$ is a known function $A(x)$, for $a \leq x \leq b$. We assume that $A(x)$ is continuous on $[a, b]$. If $a = x_0 < x_1 < x_2 < \cdots < x_{n-1} < x_n = b$, then $P = \{x_0, x_1, x_2, \ldots, x_{n-1}, x_n\}$ is a partition of $[a, b]$ into $n$ subintervals, and the planes perpendicular to the $x$-axis at the $x_1, x_2, \ldots, x_{n-1}$ divides the solid into $n$ slices of which the $i$th has thickness $\Delta x_i = x_i - x_{i-1}$. The volume $\Delta V_i$ of that slice lies between the maximum and minimum values of $A(x) \Delta x_i$ for values of $x$ in $[x_{i-1}, x_i]$ (Figure 7.3), and so

$$\Delta V_i = A(c_i) \Delta x_i$$

for some $c_i$ in $[x_{i-1}, x_i]$, by the Intermediate-Value Theorem. The volume of the solid is therefore given by the Riemann sum

$$V = \sum_{i=1}^{n} \Delta V_i = \sum_{i=1}^{n} A(c_i)\, \Delta x_i.$$

Letting $n$ approach infinity in such a way that $\max \Delta x_i$ approaches 0, we obtain the definite integral of $A(x)$ over $[a, b]$ as the limit of this Riemann sum. Therefore:

> The volume $V$ of a solid between $x = a$ and $x = b$ having cross-sectional area $A(x)$ at position $x$ is
>
> $$V = \int_a^b A(x)\, dx.$$

There is another way to obtain this formula, and others of a similar nature. Consider a slice of the solid between the planes perpendicular to the $x$-axis at positions $x$ and $x + \Delta x$. Since $A(x)$ is continuous it doesn't change much in a short interval, so if $\Delta x$ is small, then the slice has volume $\Delta V$ approximately equal to the volume of a cylinder of base area $A(x)$ and height $\Delta x$:

$$\Delta V \approx A(x)\, \Delta x.$$

The error in this approximation is small compared to the size of $\Delta V$. This suggests, correctly, that the **volume element**, that is, the volume of an infinitely thin slice of thickness $dx$ is $dV = A(x)\, dx$, and that the volume of the solid is the "sum" (that is, the integral) of these volume elements between the two ends of the solid, $x = a$ and $x = b$ (see Figure 7.4):

$$V = \int_{x=a}^{x=b} dV \qquad \text{where} \qquad dV = A(x)\, dx.$$

We will use this *differential element* approach to model other applications that result in integrals rather than setting up explicit Riemann sums each time. Even though this argument does not constitute a proof of the formula, you are strongly encouraged to think of the formula this way; the volume is the integral of the volume elements.

**Figure 7.4**    The volume element

## Solids of Revolution

Many common solids have circular cross-sections in planes perpendicular to some axis. Such solids are called **solids of revolution** because they can be generated by rotating a plane region about an axis in that plane so that it sweeps out the solid. For example, a solid ball is generated by rotating a half-disk about the diameter of that half-disk (Figure 7.5(a)). Similarly, a solid right-circular cone is generated by rotating a right-angled triangle about one of its legs (Figure 7.5(b)).

If the region $R$ bounded by $y = f(x)$, $y = 0$, $x = a$, and $x = b$ is rotated about the $x$-axis, then the cross-section of the solid generated in the plane perpendicular to the $x$-axis at $x$ is a circular disk of radius $|f(x)|$. The area of this cross-section is $A(x) = \pi\big(f(x)\big)^2$, so the volume of the solid of revolution is

$$V = \pi \int_a^b (f(x))^2\, dx.$$

**Example 1** **(The volume of a ball)** Find the volume of a solid ball of radius $a$.

**Solution** The ball can be generated by rotating the half-disk, $0 \leq y \leq \sqrt{a^2 - x^2}$, $-a \leq x \leq a$ about the $x$-axis. See the cutaway view in Figure 7.5(a). Therefore its volume is

$$V = \pi \int_{-a}^{a} (\sqrt{a^2 - x^2})^2 \, dx = 2\pi \int_{0}^{a} (a^2 - x^2) \, dx$$

$$= 2\pi \left( a^2 x - \frac{x^3}{3} \right) \Big|_{0}^{a} = 2\pi \left( a^3 - \frac{1}{3}a^3 \right) = \frac{4}{3}\pi a^3 \text{ cubic units.}$$

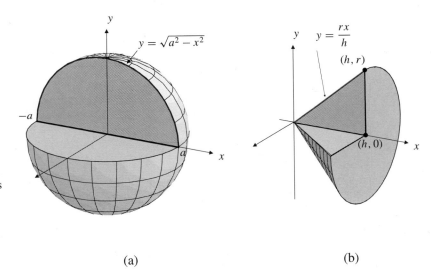

**Figure 7.5**

(a) The ball is generated by rotating the half-disk $0 \leq y \leq \sqrt{a^2 - x^2}$ (shown in colour) about the $x$-axis

(b) The cone of base radius $r$ and height $h$ is generated by rotating the triangle $0 \leq x \leq h$, $0 \leq y \leq rx/h$ (in colour) about the $x$-axis

(a)                    (b)

**Example 2** **(The volume of a right-circular cone)** Find the volume of the right-circular cone of base radius $r$ and height $h$ that is generated by rotating the triangle with vertices $(0, 0)$, $(h, 0)$, and $(h, r)$ about the $x$-axis.

**Solution** The line from $(0, 0)$ to $(h, r)$ has equation $y = rx/h$. Thus the volume of the cone (see the cutaway view in Figure 7.5(b)) is

$$V = \pi \int_{0}^{h} \left( \frac{rx}{h} \right)^2 \, dx = \pi \left( \frac{r}{h} \right)^2 \frac{x^3}{3} \Big|_{0}^{h} = \frac{1}{3}\pi r^2 h \text{ cubic units.}$$

Improper integrals can represent volumes of unbounded solids. If the improper integral converges, the unbounded solid has a finite volume.

**Example 3** Find the volume of the infinitely long horn that is generated by rotating the region bounded by $y = 1/x$ and $y = 0$ and lying to the right of $x = 1$ about the $x$-axis. The horn is illustrated in Figure 7.6.

**Solution** The volume of the horn is

$$V = \pi \int_1^\infty \left(\frac{1}{x}\right)^2 dx = \pi \lim_{R \to \infty} \int_1^R \frac{1}{x^2} dx$$

$$= -\pi \lim_{R \to \infty} \frac{1}{x}\Big|_1^R = -\pi \lim_{R \to \infty} \left(\frac{1}{R} - 1\right) = \pi \text{ cubic units.}$$

It is interesting to note that this finite volume arises from rotating a region that itself has infinite area: $\int_1^\infty dx/x = \infty$. We have a paradox: it takes an infinite amount of paint to paint the region, but only a finite amount to fill the horn obtained by rotating the region. (How can you resolve this paradox?)

**Figure 7.6** Cutaway view of an infinitely long horn

The following example shows how to deal with a problem where the axis of rotation is not the $x$-axis. Just rotate a suitable area element about the axis to form a volume element.

**Example 4** A ring-shaped solid is generated by rotating the finite plane region $R$ bounded by the curve $y = x^2$ and the line $y = 1$ about the line $y = 2$. Find its volume.

**Solution** First we solve the pair of equations $y = x^2$ and $y = 1$ to obtain the intersections at $x = -1$ and $x = 1$. The solid lies between these two values of $x$. The area element of $R$ at position $x$ is a vertical strip of width $dx$ extending upward from $y = x^2$ to $y = 1$. When $R$ is rotated about the line $y = 2$, this area element sweeps out a thin, washer-shaped volume element of thickness $dx$ and radius $2 - x^2$, having a hole of radius 1 through the middle. (See Figure 7.7.) The cross-sectional area of this element is the area of a circle of radius $2 - x^2$ minus the area of the hole, a circle of radius 1. Thus

$$dV = \left(\pi(2 - x^2)^2 - \pi(1)^2\right) dx = \pi(3 - 4x^2 + x^4) dx.$$

Since the solid extends from $x = -1$ to $x = 1$, its volume is

$$V = \pi \int_{-1}^1 (3 - 4x^2 + x^4) dx = 2\pi \int_0^1 (3 - 4x^2 + x^4) dx$$

$$= 2\pi \left(3x - \frac{4x^3}{3} + \frac{x^5}{5}\right)\Big|_0^1 = 2\pi \left(3 - \frac{4}{3} + \frac{1}{5}\right) = \frac{56\pi}{15} \text{ cubic units.}$$

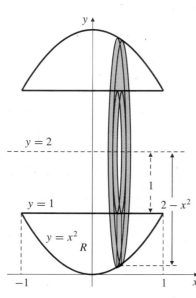

**Figure 7.7**

Sometimes we want to rotate a region bounded by curves with equations of the form $x = g(y)$ about the $y$-axis. In this case, the roles of $x$ and $y$ are reversed, and we use horizontal slices instead of vertical ones.

**Example 5**   Find the volume of the solid generated by rotating the region to the right of the $y$-axis and to the left of the curve $x = 2y - y^2$ about the $y$-axis.

**Solution**   For intersections of $x = 2y - y^2$ and $x = 0$, we have

$$2y - y^2 = 0 \qquad \Longrightarrow \qquad y = 0 \quad \text{or} \quad y = 2.$$

The solid lies between the horizontal planes at $y = 0$ and $y = 2$. A horizontal area element at height $y$ and having thickness $dy$ rotates about the $y$-axis to generate a thin disk-shaped volume element of radius $2y - y^2$ and thickness $dy$. (See Figure 7.8.) Its volume is

$$dV = \pi(2y - y^2)^2 \, dy = \pi(4y^2 - 4y^3 + y^4) \, dy.$$

Thus the volume of the solid is

$$V = \pi \int_0^2 (4y^2 - 4y^3 + y^4) \, dy$$

$$= \pi \left( \frac{4y^3}{3} - y^4 + \frac{y^5}{5} \right) \Bigg|_0^2$$

$$= \pi \left( \frac{32}{3} - 16 + \frac{32}{5} \right) = \frac{16\pi}{15} \text{ cubic units.}$$

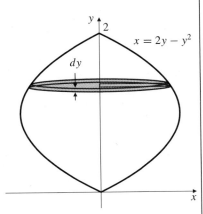

**Figure 7.8**

## Cylindrical Shells

Suppose that the region $R$ bounded by $y = f(x)$, $y = 0$, $x = a \geq 0$, and $x = b > a$ is rotated about the $y$-axis to generate a solid of revolution. In order to find the volume of the solid using (plane) slices, we would need to know the cross-sectional area $A(y)$ in each plane of height $y$, and this would entail solving the equation $y = f(x)$ for one or more solutions of the form $x = g(y)$. In practice this can be inconvenient or impossible.

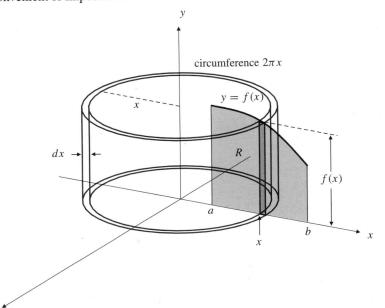

**Figure 7.9**   When rotated around the $y$-axis, the area element of width $dx$ under $y = f(x)$ at $x$ generates a cylindrical shell of height $f(x)$, circumference $2\pi x$, and hence volume $dV = 2\pi x \, f(x) \, dx$

The standard area element of $R$ at position $x$ is a vertical strip of width $dx$, height $f(x)$, and area $dA = f(x)\,dx$. When $R$ is rotated about the $y$-axis this strip sweeps out a volume element in the shape of a circular **cylindrical shell** having radius $x$, height $f(x)$, and thickness $dx$. (See Figure 7.9.) Regard this shell as a rolled-up rectangular slab with dimensions $2\pi x$, $f(x)$, and $dx$; evidently it has volume

$$dV = 2\pi x\, f(x)\,dx.$$

The volume of the solid of revolution is the sum *(integral)* of the volumes of such shells with radii ranging from $a$ to $b$:

> The volume of the solid obtained by rotating the plane region
> $0 \le y \le f(x),\, 0 \le a < x < b$ about the $y$-axis is
>
> $$V = 2\pi \int_a^b x\, f(x)\,dx.$$

**Example 6**    **(The volume of a torus)**  A disk of radius $a$ has centre at the point $(b, 0)$ where $b > a > 0$. The disk is rotated about the $y$-axis to generate a **torus**, (a doughnut-shaped solid), illustrated in Figure 7.10.  Find its volume.

**Solution**    The circle with centre at $(b, 0)$ and having radius $a$ has equation $(x - b)^2 + y^2 = a^2$, so its upper semicircle is the graph of the function

$$f(x) = \sqrt{a^2 - (x - b)^2}.$$

We will double the volume of the upper half of the torus, which is generated by rotating the half-disk $0 \le y \le \sqrt{a^2 - (x - b)^2}$, $b - a \le x \le b + a$ about the $y$-axis. The volume of the complete torus is

$$V = 2 \times 2\pi \int_{b-a}^{b+a} x\sqrt{a^2 - (x-b)^2}\,dx \qquad \text{Let } u = x - b,$$
$$\phantom{V} \qquad\qquad\qquad\qquad\qquad\qquad\qquad du = dx.$$
$$= 4\pi \int_{-a}^{a} (u + b)\sqrt{a^2 - u^2}\,du$$
$$= 4\pi \int_{-a}^{a} u\sqrt{a^2 - u^2}\,du + 4\pi b \int_{-a}^{a} \sqrt{a^2 - u^2}\,du$$
$$= 0 + 4\pi b \frac{\pi a^2}{2} = 2\pi^2 a^2 b \text{ cubic units.}$$

(The first of the final two integrals is 0 because the integrand is odd and the interval is symmetric about 0; the second is the area of a semicircle of radius $a$.) Note that the volume of the torus is $(\pi a^2)(2\pi b)$, that is, the area of the disk being rotated times the distance travelled by the centre of that disk as it rotates about the $y$-axis. This result will be generalized by Pappus's Theorem in Section 7.5.

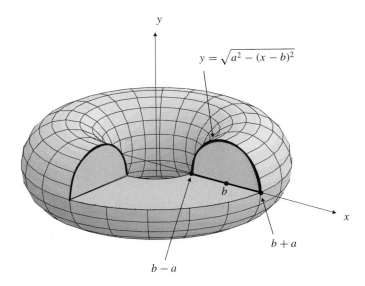

$$y = \sqrt{a^2 - (x - b)^2}$$

**Figure 7.10**    Cutaway view of a torus

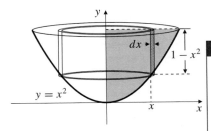

**Figure 7.11**    A parabolic bowl

**Example 7**    Find the volume of a bowl obtained by revolving the parabolic arc $y = x^2$, $0 \le x \le 1$ about the $y$-axis.

**Solution**    The interior of the bowl corresponds to revolving the region given by $x^2 \le y \le 1$, $0 \le x \le 1$ about the $y$-axis. The area element at position $x$ has height $1 - x^2$, and generates a cylindrical shell of volume $dV = 2\pi x(1 - x^2)\, dx$. (See Figure 7.11.) Thus the volume of the bowl is

$$V = 2\pi \int_0^1 x(1 - x^2)\, dx$$

$$= 2\pi \left( \frac{x^2}{2} - \frac{x^4}{4} \right) \Big|_0^1 = \frac{\pi}{2} \text{ cubic units.}$$

We have described two methods for determining the volume of a solid of revolution, slicing and cylindrical shells. The choice of method for a particular solid is usually dictated by the form of the equations defining the region being rotated, and by the axis of rotation. The volume element $dV$ can always be determined by rotating a suitable area element $dA$ about the axis of rotation. If the region is bounded by vertical lines and one or more graphs of the form $y = f(x)$, the appropriate area element is a vertical strip of width $dx$. If the rotation is about the $x$-axis or any other horizontal line, this strip generates a disk- or washer-shaped slice of thickness $dx$. If the rotation is about the $y$-axis or any other vertical line, the strip generates a cylindrical shell of thickness $dx$. On the other hand, if the region being rotated is bounded by horizontal lines and one or more graphs of the form $x = g(y)$, it is easier to use a horizontal strip of width $dy$ as the area element, and this generates a slice if the rotation is about a vertical line and a cylindrical shell if the rotation is about a horizontal line. For very simple regions either method can be made to work easily.

**Table 1.**   Volumes of solids of revolution

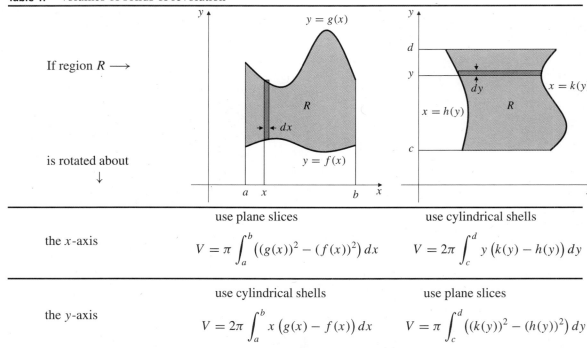

|  | use plane slices | use cylindrical shells |
|---|---|---|
| the $x$-axis | $V = \pi \displaystyle\int_a^b \left((g(x))^2 - (f(x))^2\right) dx$ | $V = 2\pi \displaystyle\int_c^d y\left(k(y) - h(y)\right) dy$ |
|  | use cylindrical shells | use plane slices |
| the $y$-axis | $V = 2\pi \displaystyle\int_a^b x\left(g(x) - f(x)\right) dx$ | $V = \pi \displaystyle\int_c^d \left((k(y))^2 - (h(y))^2\right) dy$ |

Our final example involves rotation about a vertical line other than the $y$-axis.

**Example 8**   The triangular region bounded by $y = x$, $y = 0$, and $x = a > 0$ is rotated about the line $x = b > a$. (See Figure 7.12.) Find the volume of the solid so generated.

**Figure 7.12**

**Solution**   Here the vertical area element at $x$ generates a cylindrical shell of radius $b - x$, height $x$, and thickness $dx$. Its volume is $dV = 2\pi(b - x)x\,dx$, and the volume of the solid is

$$V = 2\pi \int_0^a (b-x)\,x\,dx = 2\pi \left(\frac{bx^2}{2} - \frac{x^3}{3}\right)\Bigg|_0^a = \pi\left(a^2 b - \frac{2a^3}{3}\right) \text{ cubic units.}$$

# Exercises 7.1

Find the volume of each solid $S$ in Exercises 1–4 in two ways, using the method of slicing and the method of cylindrical shells.

1. $S$ is generated by rotating about the $x$-axis the region bounded by $y = x^2$, $y = 0$, and $x = 1$.

2. $S$ is generated by rotating the region of Exercise 1 about the $y$-axis.

3. $S$ is generated by rotating about the $x$-axis the region bounded by $y = x^2$ and $y = \sqrt{x}$ between $x = 0$ and $x = 1$.

4. $S$ is generated by rotating the region of Exercise 3 about the $y$-axis.

Find the volumes of the solids obtained if the plane regions $R$ described in Exercises 5–10 are rotated about (a) the $x$-axis and (b) the $y$-axis.

5. $R$ is bounded by $y = x(2 - x)$ and $y = 0$ between $x = 0$ and $x = 2$.

6. $R$ is the finite region bounded by $y = x$ and $y = x^2$.

7. $R$ is the finite region bounded by $y = x$ and $x = 4y - y^2$.

8. $R$ is bounded by $y = 1 + \sin x$ and $y = 1$ from $x = 0$ to $x = \pi$.

9. $R$ is bounded by $y = 1/(1 + x^2)$, $y = 2$, $x = 0$, and $x = 1$.

10. $R$ is the finite region bounded by $y = 1/x$ and $3x + 3y = 10$.

11. The triangular region with vertices $(0, -1)$, $(1, 0)$, and $(0, 1)$ is rotated about the line $x = 2$. Find the volume of the solid so generated.

12. Find the volume of the solid generated by rotating the region $0 \le y \le 1 - x^2$ about the line $y = 1$.

13. What percentage of the volume of a ball of radius 2 is removed if a hole of radius 1 is drilled through the centre of the ball?

14. A cylindrical hole is bored through the centre of a ball of radius $R$. If the length of the hole is $L$, show that the volume of the remaining part of the ball depends only on $L$ and not on $R$.

15. A cylindrical hole of radius $a$ is bored through a solid right-circular cone of height $h$ and base radius $b > a$. If the axis of the hole lies along that of the cone, find the volume of the remaining part of the cone.

16. Find the volume of the solid obtained by rotating a circular disk about one of its tangent lines.

17. A plane slices a ball of radius $a$ into two pieces. If the plane passes $b$ units away from the centre of the ball (where $b < a$), find the volume of the smaller piece.

18. Water partially fills a hemispherical bowl of radius 30 cm so that the maximum depth of the water is 20 cm. What volume of water is in the bowl?

19. Find the volume of the ellipsoid of revolution obtained by rotating the ellipse $(x^2/a^2) + (y^2/b^2) = 1$ about the $x$-axis.

20. Recalculate the volume of the torus of Example 6 by slicing perpendicular to the $y$-axis rather than using cylindrical shells.

21. The region $R$ bounded by $y = e^{-x}$ and $y = 0$ and lying to the right of $x = 0$ is rotated (a) about the $x$-axis and (b) about the $y$-axis. Find the volume of the solid of revolution generated in each case.

22. The region $R$ bounded by $y = x^{-k}$ and $y = 0$ and lying to the right of $x = 1$ is rotated about the $x$-axis. Find all real values of $k$ for which the solid so generated has finite volume.

23. Repeat Exercise 22 with rotation about the $y$-axis.

24. The region shaded in Figure 7.13 is rotated about the $x$-axis. Use Simpson's Rule to find the volume of the resulting solid.

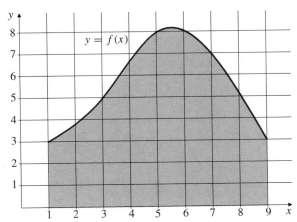

**Figure 7.13**

25. The region shaded in Figure 7.13 is rotated about the $y$-axis. Use Simpson's Rule to find the volume of the resulting solid.

26. The region shaded in Figure 7.13 is rotated about the line $x = -1$. Use Simpson's Rule to find the volume of the resulting solid.

\* 27. Find the volume of the solid generated by rotating the finite region in the first quadrant bounded by the coordinate axes and the curve $x^{2/3} + y^{2/3} = 4$ about either of the coordinate axes. (Both volumes are the same. Why?)

\* 28. Given that the surface area of a sphere of radius $r$ is $kr^2$ for some constant $k$, express the volume of a ball of radius $R$ as an integral of volume elements that are the volumes of spherical shells of varying radii and thickness $dr$. Hence find $k$.

The following problems are *very difficult*. You will need some ingenuity and a lot of hard work to solve them by the techniques available to you now.

\* 29. A wine glass in the shape of a right-circular cone of height $h$ and semi-vertical angle $\alpha$ (see Figure 7.14) is filled with

wine. Slowly a ball is lowered into the glass, displacing
wine and causing it to overflow. Find the radius $R$ of the ball
that causes the greatest volume of wine to overflow out of
the glass.

* **30.** The finite plane region bounded by the curve $xy = 1$ and the
straight line $2x + 2y = 5$ is rotated about that line to
generate a solid of revolution. Find the volume of that solid.

**Figure 7.14**

# 7.2   Other Volumes by Slicing

The method of slicing introduced in Section 7.1 can be used to determine volumes
of solids that are not solids of revolution. All we need to know is the area of
cross-section of the solid in every plane perpendicular to some fixed axis. If that
axis is the $x$-axis, if the solid lies between the planes at $x = a$ and $x = b > a$,
and if the cross-sectional area in the plane at $x$ is the continuous (or even piecewise
continuous) function $A(x)$, then the volume of the solid is

$$V = \int_a^b A(x)\, dx.$$

In this section we consider some examples that are not solids of revolution.

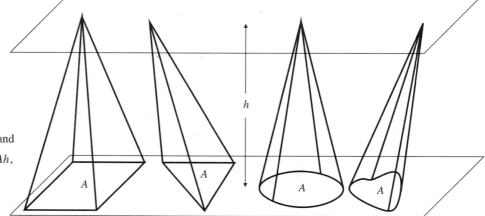

**Figure 7.15**   Some pyramids and
cones. Each has volume $V = \dfrac{1}{3}Ah$,
where $A$ is the area of the base
and $h$ is the height measured
perpendicular to the base

**Pyramids** and **cones** are solids consisting of all points on line segments which join
a fixed point, the **vertex**, to all the points in a region lying in a plane not containing
the vertex. The region is called the **base** of the pyramid or cone. Some pyramids
and cones are shown in Figure 7.15. If the base is bounded by straight lines, the
solid is called a pyramid; if the base has a curved boundary the solid is called a
cone. All pyramids and cones have volume

$$V = \frac{1}{3} Ah,$$

where $A$ is the area of the base region $R$ and $h$ is the height measured perpendicularly from the vertex to the plane of the base.

**Example 1**  Verify the formula for the volume of a pyramid with rectangular base of area $A$ and height $h$.

**Solution**  Cross-sections of the pyramid in planes parallel to the base are similar rectangles. If the origin is at the vertex of the pyramid and the $x$-axis is perpendicular to the base, then the cross-section at position $x$ is a rectangle whose dimensions are $x/h$ times the corresponding dimensions of the base. For example, in Figure 7.16(a), the length $LM$ is $x/h$ times the length $PQ$, as can be seen from the similar triangles $OLM$ and $OPQ$. Thus the area of the rectangular cross-section at $x$ is

$$A(x) = \left(\frac{x}{h}\right)^2 A.$$

The volume of the pyramid is therefore

$$V = \int_0^h \left(\frac{x}{h}\right)^2 A \, dx = \frac{A}{h^2} \frac{x^3}{3} \Big|_0^h = \frac{1}{3} Ah \text{ cubic units.}$$

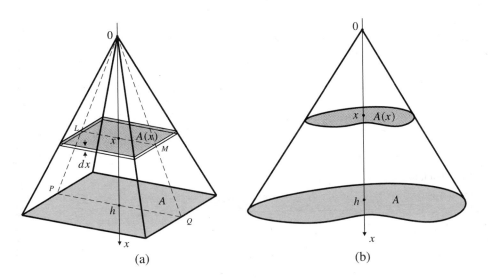

**Figure 7.16**

(a) A rectangular pyramid

(b) A general cone

(a)

(b)

A similar argument, resulting in the same formula for the volume, holds for a cone, that is, a pyramid with a more general (curved) shape to its base, such as that in Figure 7.16(b). Although it is not as obvious as in the case of the pyramid, the cross-section at $x$ still has area $(x/h)^2$ times that of the base.

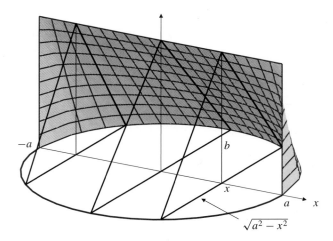

**Figure 7.17**   The tent of Example 2 with the front covering removed to show the shape more clearly

---

**Example 2**   A tent has a circular base of radius $a$ metres, and is supported by a horizontal ridge bar held at height $b$ metres above a diameter of the base by vertical supports at each end of the diameter. The material of the tent is stretched tight so that each cross section perpendicular to the ridge bar is an isosceles triangle. (See Figure 7.17.) Find the volume of the tent.

**Solution**   Let the $x$-axis be the diameter of the base under the ridge bar. The cross-section at position $x$ has base length $2\sqrt{a^2 - x^2}$, and so has area

$$A(x) = \frac{1}{2}(2\sqrt{a^2 - x^2})b = b\sqrt{a^2 - x^2}.$$

Thus the volume of the solid is

$$V = \int_{-a}^{a} b\sqrt{a^2 - x^2}\, dx = b\int_{-a}^{a} \sqrt{a^2 - x^2}\, dx = b\frac{\pi a^2}{2} = \frac{\pi}{2}a^2 b \text{ m}^3.$$

Note that we evaluated the last integral by inspection. It is the area of a half-disk of radius $a$.

---

**Example 3**   Two circular cylinders, each having radius $a$, intersect so that their axes meet at right angles. Find the volume of the region lying inside both cylinders.

**Solution**   We represent the cylinders in a three-dimensional Cartesian coordinate system where the plane containing the $x$- and $y$-axes is horizontal and the $z$-axis is vertical. One-eighth of the solid is represented in Figure 7.18, that part corresponding to all three coordinates being positive. The two cylinders have axes along the $x$- and $y$-axes, respectively. The cylinder with axis along the $x$-axis intersects the plane of the $y$- and $z$-axes in a circle of radius $a$.

Similarly, the other cylinder meets the plane of the $x$- and $z$-axes in a circle of radius $a$. It follows that if the region lying inside both cylinders (and having $x \geq 0$, $y \geq 0$, and $z \geq 0$) is sliced horizontally, then the slice at height $z$ above the $xy$-plane is a square of side $\sqrt{a^2 - z^2}$ and has area $A(z) = a^2 - z^2$. The volume $V$ of the whole region, being eight times that of the part shown, is

$$V = 8 \int_0^a (a^2 - z^2)\, dz = 8 \left( a^2 z - \frac{z^3}{3} \right) \Big|_0^a = \frac{16}{3} a^3 \text{ cubic units.}$$

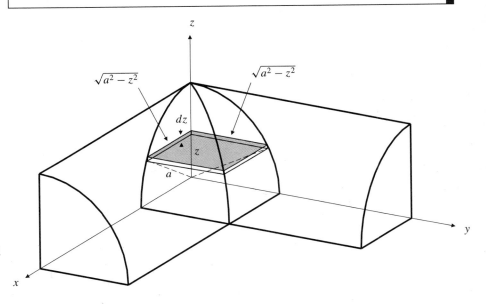

**Figure 7.18** One-eighth of the solid lying inside two perpendicular cylindrical pipes. The horizontal slice shown is square

# Exercises 7.2

1. A solid is 2 m high. The cross-section of the solid at height $x$ above its base has area $3x$ square metres. Find the volume of the solid.

2. A solid extends from $x = 1$ to $x = 3$. The cross-section of the solid in the plane perpendicular to the $x$-axis at $x$ is a square of side $x$. Find the volume of the solid.

3. A solid is 6 ft high. Its horizontal cross-section at height $z$ ft above its base is a rectangle with length $2 + z$ ft and width $8 - z$ ft. Find the volume of the solid.

4. A solid extends along the $x$-axis from $x = 1$ to $x = 4$. Its cross-section at position $x$ is an equilateral triangle with edge length $\sqrt{x}$. Find the volume of the solid.

5. A solid is $h$ cm high. Its horizontal cross-section at any height $y$ above its base is a circular sector having radius $a$ cm and angle $2\pi\left(1 - (y/h)\right)$ radians. Find the volume of the solid.

6. The opposite ends of a solid are at $x = 0$ and $x = 2$. The area of cross-section of the solid in a plane perpendicular to the $x$-axis at $x$ is $kx^3$ square units. The volume of the solid is 4 cubic units. Find $k$.

7. Find the cross-sectional area of a solid in any horizontal plane at height $z$ above its base if the volume of that part of the solid lying below any such plane is $z^3$ cubic units.

8. All the cross-sections of a solid in horizontal planes are squares. The volume of the part of the solid lying below any plane of height $z$ is $4z$ cubic units, where $0 < z < h$, the height of the solid. Find the edge length of the square cross-section at height $z$ for $0 < z < h$.

9. A solid has a circular base of radius $r$. All sections of the solid perpendicular to a particular diameter of the base are squares. Find the volume of the solid.

10. Repeat Exercise 9 but with sections that are equilateral

triangles instead of squares.

11. The base of a solid is an isosceles right-angled triangle with equal legs measuring 12 cm. Each cross-section perpenducular to one of these legs is half of a circular disk. Find the volume of the solid.

12. **(Cavalieri's Principle)** Two solids have equal cross-sectional areas at equal heights above their bases. If both solids have the same height, show that they both have the same volume.

13. The top of a circular cylinder of radius $r$ is a plane inclined at an angle to the horizontal. (See Figure 7.19.) If the lowest and highest points on the top are at heights $a$ and $b$, respectively, above the base, find the volume of the cylinder. (Note that there is an easy geometric way to get the answer, but you should also try to do it by slicing. You can use either rectangular or trapezoidal slices.)

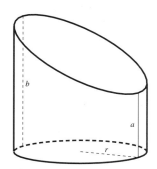

**Figure 7.19**

∗ 14. **(Volume of an ellipsoid)** Find the volume enclosed by the ellipsoid

$$\frac{x^2}{a^2} + \frac{y^2}{b^2} + \frac{z^2}{c^2} = 1.$$

*Hint:* this is not a solid of revolution. As in Example 3, the $z$-axis is perpendicular to the plane of the $x$- and $y$-axes. Each horizontal plane $z = k$ $(-c \le k \le c)$ intersects the ellipsoid in an ellipse $(x/a)^2 + (y/b)^2 = 1 - (k/c)^2$. Thus $dV = dz \times$ the area of this ellipse. The area of the ellipse $(x/a)^2 + (y/b)^2 = 1$ is $\pi ab$.

**Figure 7.20**

∗ 15. **(Notching a log)** A 45° notch is cut to the centre of a cylindrical log having radius 20 cm, as shown in Figure 7.20. One plane face of the notch is perpendicular to the axis of the log. What volume of wood was removed from the log by cutting the notch?

16. **(A smaller notch)** Repeat Exercise 15, but assume that the notch penetrates only one quarter way (10 cm) into the log.

17. What volume of wood is removed from a 3 in thick board if a circular hole of radius 2 in is drilled through it with the axis of the hole tilted at an angle of 45° to board?

∗ 18. **(More intersecting cylinders)** The axes of two circular cylinders intersect at right angles. If the radii of the cylinders are $a$ and $b$ $(a > b > 0)$, show that the region lying inside both cylinders has volume

$$V = 8 \int_0^b \sqrt{b^2 - z^2} \sqrt{a^2 - z^2}\, dz.$$

*Hint:* review Example 3. Try to make a similar diagram, showing only one-eighth of the region. The integral is not easily evaluated.

19. A circular hole of radius 2 cm is drilled through the middle of a circular log of radius 4 cm, with the axis of the hole perpendicular to the axis of the log. Find the volume of wood removed from the log. *Hint:* this is very similar to Exercise 18. You will need to use numerical methods or a calculator with a numerical integration function to get the answer.

# 7.3  Arc Length and Surface Area

In this section we consider how integrals can be used to find the lengths of curves and the areas of the surfaces of solids of revolution.

## Arc Length

If $A$ and $B$ are two points in the plane, let $|AB|$ denote the distance between $A$ and $B$, that is, the length of the straight line segment $AB$.

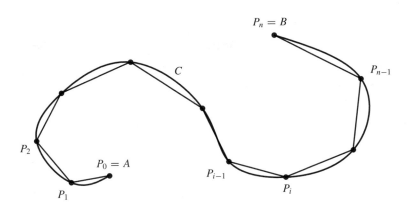

**Figure 7.21** A polygonal approximation to a curve $C$

Given a curve $C$ joining the two points $A$ and $B$, we would like to define what is meant by the *length* of the curve $C$ from $A$ to $B$. Suppose we choose points $A = P_0$, $P_1$, $P_2$, ..., $P_{n-1}$, and $P_n = B$ in order along the curve, as shown in Figure 7.21. The polygonal line $P_0 P_1 P_2 \ldots P_{n-1} P_n$ constructed by joining adjacent pairs of these points with straight line segments forms a *polygonal approximation* to $C$, having length

$$L_n = |P_0 P_1| + |P_1 P_2| + \cdots + |P_{n-1} P_n| = \sum_{i=1}^{n} |P_{i-1} P_i|.$$

Intuition tells us that the shortest curve joining two points is a straight line segment, so the length $L_n$ of any such polygonal approximation to $C$ can not exceed the length of $C$. If we increase $n$ by adding more vertices to the polygonal line between existing vertices, $L_n$ cannot get smaller and may increase. If there exists a finite number $K$ such that $L_n \leq K$ for every polygonal approximation to $C$, then there will be a smallest such number $K$ (by the completeness of the real numbers), and we call this smallest $K$ the arc length of $C$.

**DEFINITION 1**

> The **arc length** of the curve $C$ from $A$ to $B$ is the smallest real number $s$ such that the length $L_n$ of every polygonal approximation to $C$ satisfies $L_n \leq s$.

A curve with a finite arc length is said to be **rectifiable**. Its arc length $s$ is the limit of the lengths $L_n$ of polygonal approximations as $n \to \infty$ in such a way that the maximum segment length $|P_{i-1} P_i| \to 0$.

It is possible to construct continuous curves that are bounded (they do not go off to infinity anywhere) but are not rectifiable; they have infinite length. To avoid such pathological examples, we will assume that our curves are **smooth**; they will be defined by functions having continuous derivatives.

## The Arc Length of the Graph of a Function

Let $f$ be a function defined on a closed, finite interval $[a, b]$, and having a continuous derivative $f'$ there. If $C$ is the graph of $f$, that is, the graph of the equation $y = f(x)$, then any partition of $[a, b]$ provides a polygonal approximation to $C$. For the partition

$$\{a = x_0 < x_1 < x_2 < \cdots < x_n = b\},$$

let $P_i$ be the point $(x_i, f(x_i))$, $(0 \le i \le n)$. The length of the polygonal line $P_0 P_1 P_2 \ldots P_{n-1} P_n$ is

$$L_n = \sum_{i=1}^{n} |P_{i-1} P_i| = \sum_{i=1}^{n} \sqrt{(x_i - x_{i-1})^2 + (f(x_i) - f(x_{i-1}))^2}$$

$$= \sum_{i=1}^{n} \sqrt{1 + \left( \frac{f(x_i) - f(x_{i-1})}{x_i - x_{i-1}} \right)^2} \Delta x_i,$$

where $\Delta x_i = x_i - x_{i-1}$. By the Mean-Value Theorem there exists a number $c_i$ in the interval $[x_{i-1}, x_i]$ such that

$$\frac{f(x_i) - f(x_{i-1})}{x_i - x_{i-1}} = f'(c_i),$$

so we have $L_n = \sum_{i=1}^{n} \sqrt{1 + \left( f'(c_i) \right)^2} \Delta x_i$.

Thus $L_n$ is a Riemann sum for $\int_a^b \sqrt{1 + (f'(x))^2} \, dx$. Being the limit of such Riemann sums as $n \to \infty$ in such a way that $\max(\Delta x_i) \to 0$, that integral is the length of the curve $C$.

The arc length $s$ of the curve $y = f(x)$ from $x = a$ to $x = b$ is given by

$$s = \int_a^b \sqrt{1 + \left( f'(x) \right)^2} \, dx = \int_a^b \sqrt{1 + \left( \frac{dy}{dx} \right)^2} \, dx.$$

You can regard the integral formula above as giving the arc length $s$ of $C$ as a "sum" of **arc length elements**

$$s = \int_{x=a}^{x=b} ds \qquad \text{where} \qquad ds = \sqrt{1 + \left( f'(x) \right)^2} \, dx.$$

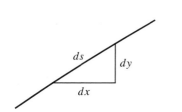

**Figure 7.22**    A differential triangle

Figure 7.22 provides a convenient way to remember this, and also suggests how we can arrive at similar formulas for arc length elements of other kinds of curves. The *differential triangle* in the figure suggests that

$$(ds)^2 = (dx)^2 + (dy)^2.$$

Dividing this equation by $(dx)^2$, taking the square root, and then multiplying by $dx$, we get

$$\left( \frac{ds}{dx} \right)^2 = 1 + \left( \frac{dy}{dx} \right)^2$$

$$\frac{ds}{dx} = \sqrt{1 + \left( \frac{dy}{dx} \right)^2}$$

$$ds = \sqrt{1 + \left( \frac{dy}{dx} \right)^2} \, dx = \sqrt{1 + \left( f'(x) \right)^2} \, dx.$$

A similar argument shows that for a curve specified by an equation of the form $x = g(y)$, $(c \le y \le d)$, the arc length element is

$$ds = \sqrt{1 + \left( \frac{dx}{dy} \right)^2} \, dy = \sqrt{1 + \left( g'(y) \right)^2} \, dy.$$

**Example 1** Find the length of the curve $y = x^{2/3}$ from $x = 1$ to $x = 8$.

**Solution** Since $dy/dx = \frac{2}{3}x^{-1/3}$ is continuous between $x = 1$ and $x = 8$, the length of the curve is given by

$$
\begin{aligned}
s &= \int_1^8 \sqrt{1 + \frac{4}{9}x^{-2/3}}\,dx \\
&= \int_1^8 \frac{\sqrt{9x^{2/3}+4}}{3x^{1/3}}\,dx \qquad \text{Let } u = 9x^{2/3}+4, \\
&\qquad\qquad\qquad\qquad\qquad\qquad du = 6x^{-1/3}\,dx \\
&= \frac{1}{18}\int_{13}^{40} u^{1/2}\,du = \frac{1}{27}u^{3/2}\Big|_{13}^{40} = \frac{40\sqrt{40}-13\sqrt{13}}{27} \text{ units.}
\end{aligned}
$$

■

**Example 2** Find the length of the curve $y = x^4 + \dfrac{1}{32x^2}$ from $x = 1$ to $x = 2$.

**Solution** Here $\dfrac{dy}{dx} = 4x^3 - \dfrac{1}{16x^3}$ and

$$
\begin{aligned}
1 + \left(\frac{dy}{dx}\right)^2 &= 1 + \left(4x^3 - \frac{1}{16x^3}\right)^2 \\
&= 1 + (4x^3)^2 - \frac{1}{2} + \left(\frac{1}{16x^3}\right)^2 \\
&= (4x^3)^2 + \frac{1}{2} + \left(\frac{1}{16x^3}\right)^2 = \left(4x^3 + \frac{1}{16x^3}\right)^2 .
\end{aligned}
$$

The expression in the last set of parentheses is positive for $1 \le x \le 2$, so the length of the curve is

$$
\begin{aligned}
s &= \int_1^2 \left(4x^3 + \frac{1}{16x^3}\right)dx = \left(x^4 - \frac{1}{32x^2}\right)\Big|_1^2 \\
&= 16 - \frac{1}{128} - \left(1 - \frac{1}{32}\right) = 15 + \frac{3}{128} \text{ units.}
\end{aligned}
$$

■

The examples above are deceptively simple; the curves were chosen so that the arc length integrals could be easily evaluated. For instance, the number 32 in the curve in Example 2 was chosen so the expression $1 + (dy/dx)^2$ would turn out to be a perfect square and its square root would cause no problems. Because of the square root in the formula, arc length problems for most curves lead to integrals that are difficult or impossible to evaluate without using numerical techniques.

**Example 3** (**Manufacturing corrugated panels**) Flat rectangular sheets of metal 2 m wide are to be formed into corrugated roofing panels 2 m wide by bending them into the sinusoidal shape shown in Figure 7.23. The period of the cross-sectional sine curve is 20 cm. Its amplitude is 5 cm, so the panel is 10 cm thick. How long should the flat sheets be cut if the resulting panels must be 5 m long?

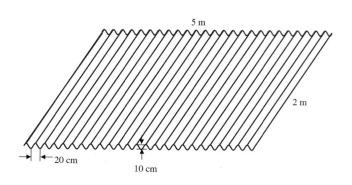

**Figure 7.23** A corrugated roofing panel

**Solution** One period of the sinusoidal cross-section is shown in Figure 7.24. The distances are all in metres; the 5 cm amplitude is shown as 1/20 m, and the 20 cm period corresponds to 2/10 m. The curve has equation

$$y = \frac{1}{20} \sin(10\pi x).$$

Note that 25 periods are required to produce a 5 m long panel. The length of the flat sheet required is 25 times the length of one period of the sine curve:

$$s = 25 \int_0^{2/10} \sqrt{1 + \left(\frac{\pi}{2} \cos(10\pi x)\right)^2}\, dx \qquad \text{Let } t = 10\pi x,$$

$$dt = 10\pi\, dx$$

$$= \frac{5}{2\pi} \int_0^{2\pi} \sqrt{1 + \frac{\pi^2}{4} \cos^2 t}\, dt = \frac{10}{\pi} \int_0^{\pi/2} \sqrt{1 + \frac{\pi^2}{4} \cos^2 t}\, dt.$$

The integral can be evaluated numerically using the techniques of the previous chapter or by using the definite integral function on an advanced scientific calculator. The value is $s \approx 7.32$. The flat metal sheet should be about 7.32 m long to a yield 5 m long finished panel. ∎

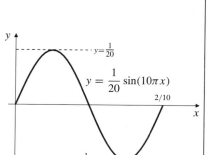

**Figure 7.24**

If integrals needed for standard problems such as arc lengths of simple curves cannot be evaluated exactly, they are sometimes used to define new functions whose values are tabulated or built in to computer programs. An example of this is the complete elliptic integral function that arises in the next example.

**Example 4** (**The circumference of an ellipse**) Find the circumference of the ellipse

$$\frac{x^2}{a^2} + \frac{y^2}{b^2} = 1,$$

where $a \geq b > 0$. See Figure 7.25.

***Solution***  The upper half of the ellipse has equation $y = b\sqrt{1 - \dfrac{x^2}{a^2}} = \dfrac{b}{a}\sqrt{a^2 - x^2}$.

Hence

$$\frac{dy}{dx} = -\frac{b}{a}\frac{x}{\sqrt{a^2 - x^2}}$$

and so

$$1 + \left(\frac{dy}{dx}\right)^2 = 1 + \frac{b^2}{a^2}\frac{x^2}{a^2 - x^2}$$

$$= \frac{a^4 - (a^2 - b^2)x^2}{a^2(a^2 - x^2)}.$$

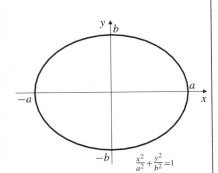

Figure 7.25

The circumference of the ellipse is four times the arc length of the part lying in the first quadrant, so

$$s = 4\int_0^a \frac{\sqrt{a^4 - (a^2 - b^2)x^2}}{a\sqrt{a^2 - x^2}}\,dx \qquad \text{Let } x = a\sin t,$$
$$\phantom{s} \qquad\qquad\qquad\qquad\qquad\qquad dx = a\cos t\,dt$$

$$= 4\int_0^{\pi/2} \frac{\sqrt{a^4 - (a^2 - b^2)a^2\sin^2 t}}{a(a\cos t)}\,a\cos t\,dt$$

$$= 4\int_0^{\pi/2} \sqrt{a^2 - (a^2 - b^2)\sin^2 t}\,dt$$

$$= 4a\int_0^{\pi/2} \sqrt{1 - \frac{a^2 - b^2}{a^2}\sin^2 t}\,dt$$

$$= 4a\int_0^{\pi/2} \sqrt{1 - \epsilon^2\sin^2 t}\,dt \text{ units,}$$

where $\epsilon = (\sqrt{a^2 - b^2})/a$ is the *eccentricity* of the ellipse. (See Section 9.1 for a discussion of ellipses.) Note that $0 < \epsilon < 1$. The function $E(\epsilon)$, defined by

$$E(\epsilon) = \int_0^{\pi/2} \sqrt{1 - \epsilon^2\sin^2 t}\,dt,$$

is called the **complete elliptic integral of the second kind**. The integral cannot be evaluated by elementary techniques for general $\epsilon$, though numerical methods can be applied to find approximate values for any given value of $\epsilon$. Tables of values of $E(\epsilon)$ for various values of $\epsilon$ can be found in collections of mathematical tables. As shown above, the circumference of the ellipse is given by $4a\,E(\epsilon)$. Note that for $a = b$ we have $\epsilon = 0$, and the formula returns the circumference of a circle; $s = 4a(\pi/2) = 2\pi a$ units.

## Areas of Surfaces of Revolution

When a plane curve is rotated (in three dimensions) about a line in the plane of the curve, it sweeps out a **surface of revolution**. For instance, a sphere of radius $a$ is generated by rotating a semicircle of radius $a$ about the diameter of that semicircle. The area of a surface of revolution can be found by integrating an area element $dS$ constructed by rotating the arc length element $ds$ of the curve about the given line. If the radius of rotation of an arc length element $ds$ is $r$, then it generates, on rotation, a circular band of width $ds$ and length (circumference) $2\pi r$. The area of this band is, therefore,

$$dS = 2\pi r\, ds,$$

as shown in Figure 7.26. The areas of surfaces of revolution around various lines can be obtained by integrating $dS$ with appropriate choices of $r$. Here are some important special cases.

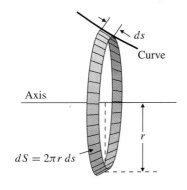

**Figure 7.26**    The circular band generated by rotating arc length element $ds$ about the axis

---

**Area of a surface of revolution**

If $f'(x)$ is continuous on $[a, b]$ and the curve $y = f(x)$ is rotated about the $x$-axis, then area of the surface of revolution so generated is

$$S = 2\pi \int_{x=a}^{x=b} |y|\, ds = 2\pi \int_{a}^{b} |f(x)|\sqrt{1 + (f'(x))^2}\, dx.$$

If the rotation is about the $y$-axis, the surface area is

$$S = 2\pi \int_{x=a}^{x=b} |x|\, ds = 2\pi \int_{a}^{b} |x|\sqrt{1 + (f'(x))^2}\, dx.$$

If $g'(y)$ is continuous on $[c, d]$ and the curve $x = g(y)$ is rotated about the $x$-axis, then area of the surface of revolution so generated is

$$S = 2\pi \int_{y=c}^{y=d} |y|\, ds = 2\pi \int_{c}^{d} |y|\sqrt{1 + (g'(y))^2}\, dy.$$

If the rotation is about the $y$-axis, the surface area is

$$S = 2\pi \int_{y=c}^{y=d} |x|\, ds = 2\pi \int_{c}^{d} |g(y)|\sqrt{1 + (g'(y))^2}\, dy.$$

---

***Remark***    Students sometimes wonder whether such complicated formulas are actually necessary. Why not just use $dS = 2\pi y\, dx$ for the area element when $y = f(x)$ is rotated about the $x$-axis instead of the more complicated area element $dS = 2\pi y\, ds$? After all, we are regarding $dx$ and $ds$ as both being infinitely small, and we certainly used $dx$ for the width of the disk-shaped volume element when we rotated the region under $y = f(x)$ about the $x$-axis to generate a solid of revolution. The reason is somewhat subtle. For small thickness $\Delta x$, the volume of a slice of the solid of revolution is only approximately $\pi y^2 \Delta x$, but the error is *small compared to the volume of this slice.* On the other hand, if we use $2\pi y \Delta x$ as an approximation to the area of a thin band of the surface of revolution corresponding to an $x$ interval of width $\Delta x$, the error is *not small compared to the area of that band.* If, for instance, the curve $y = f(x)$ has slope 1 at $x$, then the width of the band is really $\Delta s = \sqrt{2}\, \Delta x$, so that the area of the band is $\Delta S = \sqrt{2} y\, \Delta x$, not just $y\, \Delta x$. Always use the appropriate arc length element along the curve when you rotate a curve to find the area of a surface of revolution.

**Example 5**   (**Surface area of a sphere**) Find the area of the surface of a sphere of radius $a$.

**Solution**   Such a sphere can be generated by rotating the semicircle with equation $y = \sqrt{a^2 - x^2}$, $(-a \le x \le a)$, about the $x$-axis. (See Figure 7.27.) Since

$$\frac{dy}{dx} = -\frac{x}{\sqrt{a^2 - x^2}} = -\frac{x}{y},$$

the area of the sphere is given by

$$S = 2\pi \int_{-a}^{a} y\sqrt{1 + \left(\frac{x}{y}\right)^2}\, dx$$

$$= 4\pi \int_{0}^{a} \sqrt{y^2 + x^2}\, dx$$

$$= 4\pi \int_{0}^{a} \sqrt{a^2}\, dx = 4\pi ax \Big|_{0}^{a} = 4\pi a^2 \text{ square units.}$$

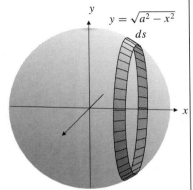

**Figure 7.27**   An area element on a sphere

**Example 6**   (**Surface area of a parabolic dish**) Find the surface area of a parabolic reflector whose shape is obtained by rotating the parabolic arc $y = x^2$, $(0 \le x \le 1)$, about the $y$-axis, as illustrated in Figure 7.28.

**Solution**   The arc length element for the parabola $y = x^2$ is $ds = \sqrt{1 + 4x^2}\, dx$, so the required surface area is

$$S = 2\pi \int_{0}^{1} x\sqrt{1 + 4x^2}\, dx \qquad \text{Let } u = 1 + 4x^2,$$
$$\hspace{6cm} du = 8x\, dx$$

$$= \frac{\pi}{4} \int_{1}^{5} u^{1/2}\, du$$

$$= \frac{\pi}{6} u^{3/2} \Big|_{1}^{5} = \frac{\pi}{6}(5\sqrt{5} - 1) \text{ square units.}$$

**Figure 7.28**   The area element is a horizontal band here

## Exercises 7.3

In Exercises 1–16, find the lengths of the given curves.

**1.** $y = 2x - 1$ from $x = 1$ to $x = 3$

**2.** $y = ax + b$ from $x = A$ to $x = B$

**3.** $y = \frac{2}{3}x^{3/2}$ from $x = 0$ to $x = 8$

**4.** $y^2 = (x - 1)^3$ from $(1, 0)$ to $(2, 1)$

**5.** $y^3 = x^2$ from $(-1, 1)$ to $(1, 1)$

**6.** $2(x + 1)^3 = 3(y - 1)^2$ from $(-1, 1)$ to $(0, 1 + \sqrt{2/3})$

**7.** $y = \frac{x^3}{12} + \frac{1}{x}$ from $x = 1$ to $x = 4$

**8.** $y = \frac{x^3}{3} + \frac{1}{4x}$ from $x = 1$ to $x = 2$

**9.** $4y = 2\ln x - x^2$ from $x = 1$ to $x = e$

**10.** $y = x^2 - \frac{\ln x}{8}$ from $x = 1$ to $x = 2$

**11.** $y = \ln(1 - x^2)$ from $x = -(1/2)$ to $x = 1/2$

**12.** $y = \dfrac{e^x + e^{-x}}{2}$ ($= \cosh x$) from $x = 0$ to $x = a$

**13.** $y = \ln \cos x$ from $x = \pi/6$ to $x = \pi/4$

**∗ 14.** $y = x^2$ from $x = 0$ to $x = 2$

**∗ 15.** $y = \ln \dfrac{e^x - 1}{e^x + 1}$ from $x = 2$ to $x = 4$

**∗ 16.** $y = \ln x$ from $x = 1$ to $x = e$

**17.** Find the circumference of the closed curve
$x^{2/3} + y^{2/3} = a^{2/3}$. *Hint:* the curve is symmetric about both
coordinate axes (why?), so one-quarter of it lies in the first
quadrant.

Use numerical methods (or a calculator with integration
function) to find the lengths of the curves in Exercises 18–21 to
four decimal places.

**18.** $y = x^4$ from $x = 0$ to $x = 1$

**19.** $y = x^{1/3}$ from $x = 1$ to $x = 2$

**20.** the circumference of the ellipse $3x^2 + y^2 = 3$

**21.** the shorter arc of of the ellipse $x^2 + 2y^2 = 2$ between $(0, 1)$
and $(1, 1/\sqrt{2})$

In Exercises 22–29, find the areas of the surfaces obtained by
rotating the given curve about the indicated lines.

**22.** $y = x^2$, $(0 \le x \le 2)$, about the $y$-axis

**23.** $y = x^3$, $(0 \le x \le 1)$, about the $x$-axis

**24.** $y = x^{3/2}$, $(0 \le x \le 1)$, about the $x$-axis

**25.** $y = x^{3/2}$, $(0 \le x \le 1)$, about the $y$-axis

**26.** $y = e^x$, $(0 \le x \le 1)$, about the $x$-axis

**27.** $y = \sin x$, $(0 \le x \le \pi)$, about the $x$-axis

**28.** $y = \dfrac{x^3}{12} + \dfrac{1}{x}$, $(1 \le x \le 4)$, about the $x$-axis

**29.** $y = \dfrac{x^3}{12} + \dfrac{1}{x}$, $(1 \le x \le 4)$, about the $y$-axis

**30.** **(Surface area of a cone)** Find the area of the curved surface
of a right-circular cone of base radius $r$ and height $h$ by
rotating the straight line segment from $(0, 0)$ to $(r, h)$ about
the $y$-axis.

**31.** **(How much icing on a doughnut?)** Find the surface area of
the torus (doughnut) obtained by rotating the circle
$(x - b)^2 + y^2 = a^2$ about the $y$-axis.

**32.** **(Area of a prolate spheroid)** Find the area of the surface
obtained by rotating the ellipse $x^2 + 4y^2 = 4$ about the
$x$-axis.

**33.** **(Area of an oblate spheroid)** Find the area of the surface
obtained by rotating the ellipse $x^2 + 4y^2 = 4$ about the
$y$-axis.

**∗ 34.** The ellipse of Example 4 is rotated about the line $y = c > b$
to generate a doughnut with elliptical cross-sections.
Express the surface area of this doughnut in terms of the
complete elliptic integral function $E(\epsilon)$ introduced in that
example.

**∗ 35.** Express the integral formula obtained for the length of the
metal sheet in Example 3 in terms of the complete elliptic
integral function $E(\epsilon)$ introduced in Example 4.

**36.** **(An interesting property of spheres)** If two parallel planes
intersect a sphere, show that the surface area of that part of
the sphere lying between the two planes depends only on the
radius of the sphere and the distance between the planes, and
not on the position of the planes.

**37.** For what real values of $k$ does the surface generated by
rotating the curve $y = x^k$, $(0 < x \le 1)$, about the $y$-axis
have a finite surface area?

**∗ 38.** The curve $y = \ln x$, $(0 < x \le 1)$, is rotated about the $y$-axis.
Find the area of the area of the horn-shaped surface so
generated.

**39.** A hollow container in the shape of an infinitely long horn is
generated by rotating the curve $y = 1/x$, $(1 \le x < \infty)$,
about the $x$-axis.

　(a) Find the volume of the container.

　(b) Show that the container has infinite surface area.

　(c) How do you explain the "paradox" that the container
　　　can be filled with a finite volume of paint but requires
　　　infinitely much paint to cover its surface?

# 7.4　Mass, Moments, and Centre of Mass

Many quantities of interest in physics, mechanics, ecology, finance and other disciplines are described in terms of densities over regions of space, the plane, or even
the real line. To determine the total value of such a quantity we must add up (integrate) the contributions from the various places where the quantity is distributed.
Examples include the mass of an object, the charge on an electrical conductor, the
accumulation of an investment over a period of time, the force on a dam, and the
probability of a complex event, to name but a few. We will examine some of these
in the remaining sections of this chapter.

## Mass and Density

If a solid object is made of a homogeneous material, we would expect different parts of the solid that have the same volume to have the same mass as well. We express this homogeneity by saying that the object has constant density, that density being the mass divided by the volume for the whole object or for any part of it. Thus a rectangular brick with dimensions 20 cm, 10 cm, and 8 cm would have volume $V = 20 \times 10 \times 8 = 1{,}600$ cm$^3$, and if it was made of material having constant density $\delta = 3$ g/cm$^3$, it would have mass $m = \delta V = 3 \times 1{,}600 = 4{,}800$ g. (We will use the lower case Greek letter delta ($\delta$) to represent density.)

If the density of the material constituting a solid object is not constant, but varies from point to point in the object, no such simple relationship exists between mass and volume. If the density $\delta = \delta(P)$ is a *continuous* function of position $P$ we could subdivide the solid into many small volume elements and, by regarding $\delta$ as approximately constant over each such element, determine the masses of all the elements and add them up to get the mass of the solid. The mass $\Delta m$ of a volume element $\Delta V$ containing the point $P$ would satisfy

$$\Delta m \approx \delta(P)\,\Delta V,$$

so the mass $m$ of the solid can be approximated:

$$m = \sum \Delta m \approx \sum \delta(P)\,\Delta V.$$

Such approximations become exact as we pass to the limit of differential mass and volume elements, $dm = \delta(P)\,dV$, so we expect to be able to calculate masses as integrals, that is, as the limits of such sums:

$$m = \int dm = \int \delta(P)\,dV.$$

**Example 1**  The density of a solid vertical cylinder of height $H$ cm and base area $A$ cm$^2$ is $\delta = \delta_0(1 + h)$ g/cm$^3$, where $h$ is the height in centimetres above the base and $\delta_0$ is a constant. Find the mass of the cylinder.

**Solution**  See Figure 7.29(a). A slice of the solid at height $h$ above the base and having thickness $dh$ is a circular disk of volume $dV = A\,dh$. Since the density is constant over this disk, the mass of the volume element is

$$dm = \delta\,dV = \delta_0(1 + h)\,A\,dh.$$

Therefore the mass of the whole cylinder is

$$m = \int_0^H \delta_0 A(1 + h)\,dh = \delta_0 A\left(H + \frac{H^2}{2}\right)\ \text{g}.$$

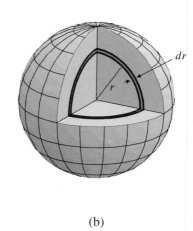

**Figure 7.29**

(a) A solid cylinder whose density varies with height

(b) Cutaway view of a planet whose density depends on distance from the centre

(a)

(b)

**Example 2** (**Using spherical shells**) The density of a spherical planet of radius $R$ m varies with distance $r$ from the centre according to the formula

$$\delta = \frac{\delta_0}{1 + r^2} \ \text{kg/m}^3.$$

Find the mass of the planet.

**Solution** Recall that the surface area of a sphere of radius $r$ is $4\pi r^2$. The planet can be regarded as being composed of concentric spherical shells having radii between 0 and $R$. The volume of a shell of radius $r$ and thickness $dr$ (see Figure 7.29(b)) is equal to its surface area times its thickness:

$$dV = 4\pi r^2 \, dr.$$

Therefore, the mass of this shell is

$$dm = \delta \, dV = 4\pi \delta_0 \frac{r^2}{1 + r^2} \, dr.$$

We add the masses of these shells to find the mass of the whole planet:

$$m = 4\pi \delta_0 \int_0^R \frac{r^2}{1 + r^2} \, dr = 4\pi \delta_0 \int_0^R \left(1 - \frac{1}{1 + r^2}\right) dr$$

$$= 4\pi \delta_0 (r - \tan^{-1} r)\Big|_0^R = 4\pi \delta_0 (R - \tan^{-1} R) \ \text{kg}.$$

Similar techniques can be applied to find masses of one- and two-dimensional objects such as wires and thin plates that have variable densities of the forms mass/unit length (**line density**) and mass/unit area (**areal density**).

**Example 3**   A wire of variable composition is stretched along the $x$-axis from $x = 0$ to $x = L$ cm. Find the mass of the wire if the line density at position $x$ is $\delta(x) = kx$ g/cm, where $k$ is a positive constant.

**Solution**   The mass of a length element $dx$ of the wire located at position $x$ is given by $dm = \delta(x)\,dx = kx\,dx$. Thus the mass of the wire is

$$m = \int_0^L kx\,dx = \left(\frac{kx^2}{2}\right)\Big|_0^L = \frac{kL^2}{2}\ \text{g.}$$

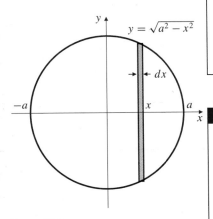

**Figure 7.30**

**Example 4**   Find the mass of a disk of radius $a$ cm having centre at the origin in the $xy$-plane if the areal density at position $(x, y)$ is $\delta = k(2a + x)$ g/cm$^2$. Here $k$ is a constant.

**Solution**   The density depends only on the horizontal coordinate $x$, and so is constant along vertical lines on the disk. This suggests that thin vertical strips should be used as area elements. A vertical strip of thickness $dx$ at $x$ has area $dA = 2\sqrt{a^2 - x^2}\,dx$ (see Figure 7.30), and so has mass

$$dm = \delta\,dA = 2k(2a + x)\sqrt{a^2 - x^2}\,dx.$$

Hence the mass of the disk is

$$m = \int_{x=-a}^{x=a} dm = 2k \int_{-a}^{a} (2a + x)\sqrt{a^2 - x^2}\,dx$$

$$= 4ak \int_{-a}^{a} \sqrt{a^2 - x^2}\,dx + 2k \int_{-a}^{a} x\sqrt{a^2 - x^2}\,dx$$

$$= 4ak\frac{\pi a^2}{2} + 0 = 2\pi ka^3\ \text{g.}$$

We used the area of a semicircle to evaluate the first integral. The second integral is zero because the integrand is odd and the interval is symmetric about $x = 0$.

Distributions of mass along one-dimensional structures (lines or curves) necessarily lead to integrals of functions of one variable, but distributions of mass on a surface or in space can lead to integrals involving functions of more than one variable. Such integrals are studied in multivariable calculus. In the examples above, the given densities were functions of only one variable, so these problems, though higher dimensional in nature, led to integrals of functions of only one variable and could be solved by the methods at hand.

## Moments and Centres of Mass

A mass $m$ located at position $x$ on the $x$-axis is said to have **moment** $xm$ about the point $x = 0$ or, more generally, moment $(x - x_0)m$ about the point $x = x_0$. If the $x$-axis is a horizontal arm hinged at $x_0$, the moment about $x_0$ measures the tendency of the weight of the mass $m$ to cause the arm to rotate. If several masses $m_1$, $m_2$, $m_3$, ..., $m_n$ are located at the points $x_1$, $x_2$, $x_3$, ..., $x_n$, respectively, then the total moment of the system of masses about the point $x = x_0$ is the sum of the individual moments (see Figure 7.31):

$$M_{x=x_0} = (x_1 - x_0)m_1 + (x_2 - x_0)m_2 + \cdots + (x_n - x_0)m_n = \sum_{j=1}^{n}(x_j - x_0)m_j.$$

**Figure 7.31**    A system of discrete masses on a line

The **centre of mass** of the system of masses is the point $\bar{x}$ about which the total moment of the system is zero. Thus

$$0 = \sum_{j=1}^{n}(x_j - \bar{x})m_j = \sum_{j=1}^{n}x_j m_j - \bar{x}\sum_{j=1}^{n}m_j.$$

The centre of mass of the system is therefore given by

$$\bar{x} = \frac{\displaystyle\sum_{j=1}^{n}x_j m_j}{\displaystyle\sum_{j=1}^{n}m_j} = \frac{M_{x=0}}{m},$$

where $m$ is the total mass of the system and $M_{x=0}$ is the total moment about $x = 0$. If you think of the $x$-axis as being a weightless wire supporting the masses, then $\bar{x}$ is the point at which the wire could be supported and remain in perfect balance (equilibrium), not tipping either way. Even if the axis represents a nonweightless support, say a seesaw, supported at $x = \bar{x}$, it will remain balanced after the masses are added, provided it was balanced beforehand. For many purposes a system of masses behaves as though its total mass were concentrated at its centre of mass.

Now suppose that a one-dimensional distribution of mass with continuously variable line density $\delta(x)$ lies along the interval $[a, b]$ of the $x$-axis. An element of length $dx$ at position $x$ contains mass $dm = \delta(x)\,dx$, and so its moment is $dM_{x=0} = x\,dm = x\delta(x)\,dx$ about $x = 0$. The total moment about $x = 0$ is the *sum* (integral) of these moment elements:

$$M_{x=0} = \int_a^b x\delta(x)\,dx.$$

Since the total mass is

$$m = \int_a^b \delta(x)\,dx,$$

we obtain the following formula for the centre of mass.

The centre of mass of a distribution of mass with line density $\delta(x)$ on the interval $[a, b]$ is given by

$$\bar{x} = \frac{M_{x=0}}{m} = \frac{\displaystyle\int_a^b x\delta(x)\,dx}{\displaystyle\int_a^b \delta(x)\,dx}.$$

**Example 5**   At what point can the wire of Example 3 be suspended so that it will balance?

**Solution**   In Example 3 we evaluated the mass of the wire to be $kL^2/2$ g. Its moment about $x = 0$ is

$$
\begin{aligned}
M_{x=0} &= \int_0^L x\delta(x)\,dx \\
&= \int_0^L kx^2\,dx = \left(\frac{kx^3}{3}\right)\Big|_0^L = \frac{kL^3}{3} \text{ g-cm.}
\end{aligned}
$$

(Note that the appropriate units for the moment are units of mass times units of distance: in this case gram-centimetres.) The centre of mass of the wire is

$$
\bar{x} = \frac{kL^3/3}{kL^2/2} = \frac{2L}{3}.
$$

The wire will be balanced if suspended at position $x = 2L/3$ cm. ∎

## Two- and Three-Dimensional Examples

The system of mass considered in Example 5 is one-dimensional and lies along a straight line. If mass is distributed in a plane or in space, similar considerations prevail. For a system of masses $m_1$ at $(x_1, y_1)$, $m_2$ at $(x_2, y_2)$, ..., $m_n$ at $(x_n, y_n)$, the **moment about $x = 0$** is

$$
M_{x=0} = x_1 m_1 + x_2 m_2 + \cdots + x_n m_n = \sum_{j=1}^n x_j m_j,
$$

and the **moment about $y = 0$** is

$$
M_{y=0} = y_1 m_1 + y_2 m_2 + \cdots + y_n m_n = \sum_{j=1}^n y_j m_j.
$$

The **centre of mass** is the point $(\bar{x}, \bar{y})$ where

$$
\bar{x} = \frac{M_{x=0}}{m} = \frac{\displaystyle\sum_{j=1}^n x_j m_j}{\displaystyle\sum_{j=1}^n m_j} \quad \text{and} \quad \bar{y} = \frac{M_{y=0}}{m} = \frac{\displaystyle\sum_{j=1}^n y_j m_j}{\displaystyle\sum_{j=1}^n m_j}.
$$

For continuous distributions of mass, the sums become appropriate integrals.

**Example 6**   Find the centre of mass of a rectangular plate that occupies the region $0 \le x \le a$, $0 \le y \le b$ if the areal density of the material in the plate at position $(x, y)$ is $ky$.

**Solution**   Since the density is independent of $x$ and the rectangle is symmetric about the line $x = a/2$, the $x$-coordinate of the centre of mass must be $\bar{x} = a/2$.

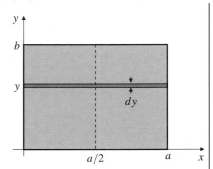

**Figure 7.32**

A thin horizontal strip of width $dy$ at height $y$ (see Figure 7.32) has mass $dm = aky\,dy$. The moment of this strip about $y = 0$ is $dM_{y=0} = y\,dm = kay^2\,dy$. Hence the mass and moment about $y = 0$ of the whole plate are

$$m = ka \int_0^b y\,dy = \frac{kab^2}{2}$$

$$M_{y=0} = ka \int_0^b y^2\,dy = \frac{kab^3}{3}.$$

Therefore $\bar{y} = M_{y=0}/m = 2b/3$, and the centre of mass of the plate is $(a/2, 2b/3)$. The plate would be balanced if supported at this point.

For distributions of mass in three-dimensional space one defines, analogously, the moments $M_{x=0}$, $M_{y=0}$, and $M_{z=0}$ of the system of mass about the planes $x = 0$, $y = 0$, and $z = 0$, respectively. For a discrete system of masses $m_1$ at $(x_1, y_1, z_1)$, $\ldots$, $m_n$ at $(x_n, y_n, z_n)$, we have

$$M_{x=0} = x_1 m_1 + x_2 m_2 + \cdots + x_n m_n,$$

with similar formulas for $M_{y=0}$ and $M_{z=0}$. The centre of mass is $(\bar{x}, \bar{y}, \bar{z})$ where

$$\bar{x} = \frac{M_{x=0}}{m}, \qquad \bar{y} = \frac{M_{y=0}}{m}, \qquad \text{and} \qquad \bar{z} = \frac{M_{z=0}}{m},$$

$m$ being the total mass: $m = m_1 + m_2 + \cdots + m_n$. Again the sums are replaced with integrals for continuous distributions of mass.

**Example 7** Find the centre of mass of a solid hemisphere of radius $R$ feet if its density at height $z$ ft above the base plane of the hemisphere is $\delta_0 z$ lb/ft$^3$.

**Solution** The solid is symmetric about the vertical axis (let us call it the $z$-axis), and the density is constant in planes perpendicular to this axis. Therefore the centre of mass must lie somewhere on this axis. A slice of the solid at height $z$ above the base, and having thickness $dz$, is a disk of radius $\sqrt{R^2 - z^2}$. (See Figure 7.33.) Its volume is $dV = \pi(R^2 - z^2)\,dz$, and its mass is $dm = \delta_0 z\,dV = \delta_0 \pi(R^2 z - z^3)\,dz$. Its moment about the base plane $z = 0$ is $dM_{z=0} = z\,dm = \delta_0 \pi(R^2 z^2 - z^4)\,dz$. The mass of the solid is

**Figure 7.33** Mass element of a solid hemisphere with density depending on height

$$m = \delta_0 \pi \int_0^R (R^2 z - z^3)\,dz = \delta_0 \pi \left( \frac{R^2 z^2}{2} - \frac{z^4}{4} \right)\Bigg|_0^R = \frac{\pi}{4} \delta_0 R^4 \text{ lb.}$$

The moment of the hemisphere about the plane $z = 0$ is

$$M_{z=0} = \delta_0 \pi \int_0^R (R^2 z^2 - z^4)\,dz = \delta_0 \pi \left( \frac{R^2 z^3}{3} - \frac{z^5}{5} \right)\Bigg|_0^R = \frac{2\pi}{15} \delta_0 R^5 \text{ lb-ft.}$$

The centre of mass therefore lies along the axis of symmetry of the hemisphere at height $\bar{z} = M_{z=0}/m = 8R/15$ ft above the base.

**Example 8** Find the centre of mass of a plate that occupies the region $a \le x \le b, 0 \le y \le f(x)$, if the density at any point $(x, y)$ is $\delta(x)$.

**Solution** The appropriate area element is shown in Figure 7.34. It has area $f(x)\,dx$, mass

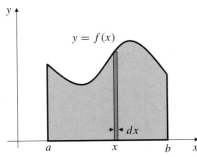

Figure 7.34    Mass element of a plate

$$dm = \delta(x)f(x)\,dx,$$

and moment about $x = 0$

$$dM_{x=0} = x\delta(x)f(x)\,dx.$$

Since the density depends only on $x$, the mass element $dm$ has constant density, and so the $y$-coordinate of *its* centre of mass is at its midpoint: $\bar{y}_{dm} = \frac{1}{2}f(x)$. Therefore the moment of the mass element $dm$ about $y = 0$ is

$$dM_{y=0} = \bar{y}_{dm}\, dm = \frac{1}{2}\delta(x)\big(f(x)\big)^2 dx.$$

Thus the coordinates of the centre of mass of the plate are $\bar{x} = \dfrac{M_{x=0}}{m}$ and $\bar{y} = \dfrac{M_{y=0}}{m}$ where

$$m = \int_a^b \delta(x)f(x)\,dx,$$

$$M_{x=0} = \int_a^b x\delta(x)f(x)\,dx,$$

$$M_{y=0} = \frac{1}{2}\int_a^b \delta(x)\big(f(x)\big)^2 dx.$$

**Remark** Similar formulas can be obtained if the density depends on $y$ instead of $x$, provided that the region admits a suitable horizontal area element. (For example, the region might be specified by $c \le y \le d, 0 \le x \le g(y)$.) Finding centres of mass for plates that occupy regions specified by functions of $x$, but where the density depends on $y$, generally requires the use of "double integrals." Such problems are therefore studied in multi-variable calculus.

## Exercises 7.4

Find the mass and centre of mass for the systems in Exercises 1–16. Be alert for symmetries.

1. A straight wire of length $L$ cm, where the density at distance $s$ cm from one end is $\delta(s) = \sin \pi s/L$ g/cm

2. A straight wire along the $x$-axis from $x = 0$ to $x = L$ if the density is constant $\delta_0$, but the cross-sectional radius of the wire varies so that its value at $x$ is $a + bx$

3. A quarter-circular plate having radius $a$, constant areal density $\delta_0$, and occupying the region $x^2 + y^2 \le a^2, x \ge 0$,

$y \ge 0$

4. A quarter-circular plate of radius $a$ occupying the region $x^2 + y^2 \le a^2, x \ge 0, y \ge 0$, having areal density $\delta(x) = \delta_0 x$

5. A plate occupying the region $0 \le y \le 4 - x^2$ if the areal density at $(x, y)$ is $ky$

6. A right-triangular plate with legs 2 m and 3 m if the areal density at any point $P$ is $5h$ kg/m$^2$, $h$ being the distance of $P$ from the shorter leg

7. A square plate of edge $a$ cm if the areal density at $P$ is $kx$ g/cm$^2$, where $x$ is the distance from $P$ to one edge of the square

8. The plate in Exercise 7, but with areal density $kr$ g/cm$^2$, where $r$ is the distance (in cm) from $P$ to one of the diagonals of the square

9. A plate of density $\delta(x)$ occupying the region $a \le x \le b$, $f(x) \le y \le g(x)$

10. A rectangular brick with dimensions 20 cm, 10 cm, and 5 cm, if the density at $P$ is $kx$ g/cm$^3$ where $x$ is the distance from $P$ to one of the $10 \times 5$ faces

11. A solid ball of radius $R$ m if the density at $P$ is $z$ kg/m$^3$, where $z$ is the distance from $P$ to a plane at distance $2R$ m from the centre of the ball

12. A right-circular cone of base radius $a$ cm and height $b$ cm if the density at point $P$ is $kz$ g/cm$^3$, where $z$ is the distance of $P$ from the base of the cone

\* 13. The solid occupying the quarter of a ball of radius $a$ centred at the origin having as base the region $x^2 + y^2 \le a^2, x \ge 0$ in the $xy$-plane, if the density at height $z$ above the base is $\delta_0 z$

\* 14. The cone of Exercise 12, but with density at $P$ equal to $kx$ g/cm$^3$, where $x$ is the distance of $P$ from the axis of symmetry of the cone. *Hint:* use a cylindrical shell centred on the axis of symmetry as volume element. This element has constant density, so its centre of mass is known, and its moment can be determined from its mass.

\* 15. A semicircular plate occupying the region $x^2 + y^2 \le a^2$, $y \ge 0$, if the density at distance $s$ from the origin is $ks$ g/cm$^2$

\* 16. The wire in Exercise 1 if it is bent in a semicircle

17. It is estimated that the density of matter in the neighbourhood of a gas giant star is given by $\delta(r) = Ce^{-kr^2}$, where $C$ and $k$ are positive constants and $r$ is the distance from the centre of the star. The radius of the star is indeterminate, but can be taken to be infinite since $\delta(r)$ decreases very rapidly for large $r$. Find the approximate mass of the star in terms of $C$ and $k$.

18. Find the average distance $\bar{r}$ of matter in the star of Exercise 17 from the centre of the star. $\bar{r}$ is given by $\int_0^\infty r \, dm / \int_0^\infty dm$, where $dm$ is the mass element at distance $r$ from the centre of the star.

## 7.5 Centroids

If matter is distributed uniformly in a system so that the density $\delta$ is constant, then that density cancels out of the numerator and denominator in sum or integral expressions for coordinates of the centre of mass. In such cases the centre of mass depends only on the *shape* of the object, that is, on geometric properties of the region occupied by the object, and we call it the **centroid** of the region.

Centroids are calculated using the same formulas as those used for centres of mass, except that the density (being constant) is taken to be unity, so the mass is just the length, area, or volume of the region, and the moments are referred to as **moments of the region** rather than of any mass occupying the region. If we set $\delta(x) = 1$ in the formulas obtained in Example 8 of Section 7.4, we obtain the following result:

---

**The centroid of a standard plane region**

The centroid of the plane region $a \le x \le b, 0 \le y \le f(x)$, is $(\bar{x}, \bar{y})$, where $\bar{x} = \dfrac{M_{x=0}}{A}$, $\bar{y} = \dfrac{M_{y=0}}{A}$, and

$$A = \int_a^b f(x) \, dx, \quad M_{x=0} = \int_a^b x f(x) \, dx, \quad M_{y=0} = \frac{1}{2} \int_a^b \left(f(x)\right)^2 \, dx.$$

---

Thus, for example, $\bar{x}$ is the *average value* of the function $x$ over the region.

The centroids of some regions are obvious by symmetry. The centroid of a circular disk or an elliptical disk is at the centre of the disk. The centroid of a rectangle is at the centre also; the centre is the point of intersection of the diagonals. The centroid of any region lies on any axes of symmetry of the region.

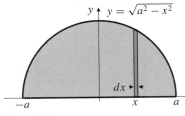

**Figure 7.35**

**Example 1**   What is the average value of $y$ over the half-disk $-a \le x \le a$, $0 \le y \le \sqrt{a^2 - x^2}$? Find the centroid of the half-disk.

**Solution**   By symmetry, the centroid lies on the $y$-axis, so its $x$-coordinate is $\bar{x} = 0$. (See Figure 7.35.) Since the area of the half-disk is $A = \frac{1}{2}\pi a^2$, the average value of $y$ over the half-disk is

$$\bar{y} = \frac{M_{y=0}}{A} = \frac{2}{\pi a^2}\frac{1}{2}\int_{-a}^{a}(a^2 - x^2)\,dx = \frac{2}{\pi a^2}\frac{2a^3}{3} = \frac{4a}{3\pi}.$$

The centroid of the half-disk is $\left(0, \dfrac{4a}{3\pi}\right)$.

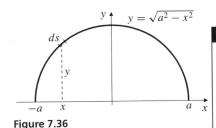

**Figure 7.36**

**Example 2**   Find the centroid of the semicircle $y = \sqrt{a^2 - x^2}$.

**Solution**   Here the "region" is a one-dimensional curve, having length rather than area. Again $\bar{x} = 0$ by symmetry. A short arc of length $ds$ at height $y$ on the semicircle has moment $dM_{y=0} = y\,ds$ about $y = 0$. (See Figure 7.36.) Since

$$ds = \sqrt{1 + \left(\frac{dy}{dx}\right)^2}\,dx = \sqrt{1 + \frac{x^2}{a^2 - x^2}}\,dx = \frac{a\,dx}{\sqrt{a^2 - x^2}},$$

and since $y = \sqrt{a^2 - x^2}$ on the semicircle, we have

$$M_{y=0} = \int_{-a}^{a}\sqrt{a^2 - x^2}\,\frac{a\,dx}{\sqrt{a^2 - x^2}} = a\int_{-a}^{a}dx = 2a^2.$$

Since the length of the semicircle is $\pi a$, we have $\bar{y} = \dfrac{M_{y=0}}{\pi a} = \dfrac{2a}{\pi}$, and the centroid of the semicircle is $\left(0, \dfrac{2a}{\pi}\right)$. Note that the centroid of a semicircle of radius $a$ is not the same as that of half-disk of radius $a$. Note also that the centroid of the semicircle does not lie on the semicircle itself.

**THEOREM 1**   **The centroid of a triangle**

The centroid of a triangle is the point at which all three medians of the triangle intersect.

**PROOF**   Recall that a median of a triangle is a straight line joining one vertex of the triangle to the midpoint of the opposite side. Given any median of a triangle we will show that the centroid lies on that median. Thus the centroid must lie on all three medians.

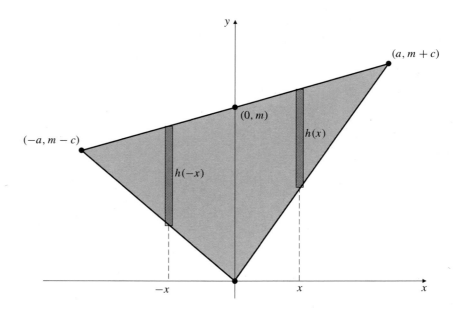

**Figure 7.37**

Adopt a coordinate system where the median in question lies along the $y$-axis, and such that a vertex of the triangle is at the origin. (See Figure 7.37.) Let the midpoint of the opposite side be $(0, m)$. Then the other two vertices of the triangle must have coordinates of the form $(-a, m - c)$ and $(a, m + c)$ so that $(0, m)$ will be the midpoint between them. The two vertical area elements shown in the figure are at the same distance on opposite sides of the $y$-axis, so they have the same heights $h(-x) = h(x)$ (by similar triangles), and the same area. The sum of the moments about $x = 0$ of these area elements is

$$dM_{x=0} = -xh(-x)\,dx + xh(x)\,dx = 0,$$

and so the moment of the whole triangle about $x = 0$ is

$$M_{x=0} = \int_{x=-a}^{x=a} dM_{x=0} = 0.$$

Therefore the centroid of the triangle lies on the $y$-axis.

**Remark**  By solving simultaneously the equations of any two medians of a triangle, we can verify the following formula:

**Coordinates of the centroid of a triangle**

The coordinates of the centroid of a triangle are the averages of the corresponding coordinates of the three vertices of the triangle. The triangle with vertices $(x_1, y_1)$, $(x_2, y_2)$, and $(x_3, y_3)$ has centroid

$$(\bar{x}, \bar{y}) = \left( \frac{x_1 + x_2 + x_3}{3}, \frac{y_1 + y_2 + y_3}{3} \right).$$

If a region is a union of non-overlapping subregions, then any moment of the region is the sum of the corresponding moments of the subregions. This fact enables us to calculate the centroid of the region if we know the centroids and areas of all the subregions.

**Example 3**  Find the centroid of the trapezoid with vertices $(0, 0)$, $(1, 0)$, $(1, 2)$, and $(0, 1)$.

***Solution***  The trapezoid is the union of a square and a (non-overlapping) triangle, as shown in Figure 7.38. By symmetry, the square has centroid $(\bar{x}_S, \bar{y}_S) = \left(\frac{1}{2}, \frac{1}{2}\right)$, and its area is $A_S = 1$. The triangle has area $A_T = \frac{1}{2}$, and its centroid is $(\bar{x}_T, \bar{y}_T)$, where

$$\bar{x}_T = \frac{0+1+1}{3} = \frac{2}{3} \quad \text{and} \quad \bar{y}_T = \frac{1+1+2}{3} = \frac{4}{3}.$$

Continuing to use subscripts $S$ and $T$ to denote the square and triangle, respectively, we calculate

$$M_{x=0} = M_{S;x=0} + M_{T;x=0} = A_S\bar{x}_S + A_T\bar{x}_T = \frac{1}{2} \times 1 + \frac{1}{2} \times \frac{2}{3} = \frac{5}{6}$$

$$M_{y=0} = M_{S;y=0} + M_{T;y=0} = A_S\bar{y}_S + A_T\bar{y}_T = \frac{1}{2} \times 1 + \frac{1}{2} \times \frac{4}{3} = \frac{7}{6}$$

Since the area of the trapezoid is $A = A_S + A_T = \frac{3}{2}$, its centroid is

$$(\bar{x}, \bar{y}) = \left(\frac{5}{6} \bigg/ \frac{3}{2}, \frac{7}{6} \bigg/ \frac{3}{2}\right) = \left(\frac{5}{9}, \frac{7}{9}\right).$$

**Figure 7.38**

---

**Example 4**  Find the centroid of the solid region obtained by rotating about the $y$-axis the first quadrant region lying between the $x$-axis and the parabola $y = 4-x^2$.

***Solution***  By symmetry, the centroid of the parabolic solid will lie on its axis of symmetry, the $y$-axis. A thin disk-shaped slice of the solid at height $y$ and having thickness $dy$ (see Figure 7.39) has volume

$$dV = \pi x^2\, dy = \pi(4 - y)\, dy,$$

and has moment about the base plane

$$dM_{y=0} = y\, dV = \pi(4y - y^2)\, dy.$$

Hence the volume of the solid is

$$V = \pi \int_0^4 (4 - y)\, dy = \pi \left(4y - \frac{y^2}{2}\right)\bigg|_0^4 = \pi(16 - 8) = 8\pi,$$

and its moment about $y = 0$ is

$$M_{y=0} = \pi \int_0^4 (4y - y^2)\, dy = \pi \left(2y^2 - \frac{y^3}{3}\right)\bigg|_0^4 = \pi \left(32 - \frac{64}{3}\right) = \frac{32}{3}\pi.$$

**Figure 7.39**

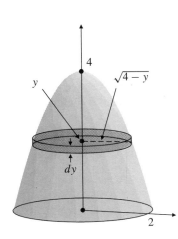

Hence the centroid is located at $\bar{y} = \dfrac{32\pi}{3} \times \dfrac{1}{8\pi} = \dfrac{4}{3}.$

## Pappus's Theorem

The following theorem relates volumes or surface areas of revolution to the centroid of the region or curve being rotated.

**THEOREM 2**

**Pappus's Theorem**

(a) If a plane region $R$ lies on one side of a line $L$ in that plane and is rotated about $L$ to generate a solid of revolution, then the volume $V$ of that solid is the product of the area of $R$ and the distance travelled by the centroid of $R$ under the rotation; that is,

$$V = 2\pi \bar{r} A,$$

where $A$ is the area of $R$ and $\bar{r}$ is the perpendicular distance from the centroid of $R$ to $L$.

(b) If a plane curve $C$ lies on one side of a line $L$ in that plane and is rotated about that line to generate a surface of revolution, then the area $S$ of that surface is the length of $C$ times the distance travelled by the centroid of $C$;

$$S = 2\pi \bar{r} s,$$

where $s$ is the length of the curve $C$ and $\bar{r}$ is the perpendicular distance from the centroid of $C$ to the line $L$.

**PROOF** We prove part (a). The proof of (b) is similar, and is left as an exercise.

Let us take $L$ to be the $y$-axis and suppose that $R$ lies between $x = a$ and $x = b$ where $0 \le a < b$. Thus $\bar{r} = \bar{x}$, the $x$-coordinate of the centroid of $R$. Let $dA$ denote the area of a thin strip of $R$ at position $x$ and having width $dx$. This strip generates, on rotation about $L$, a cylindrical shell of volume $dV = 2\pi x \, dA$, so the volume of the solid of revolution is

$$V = 2\pi \int_{x=a}^{x=b} x \, dA = 2\pi M_{x=0} = 2\pi \bar{x} A = 2\pi \bar{r} A.$$

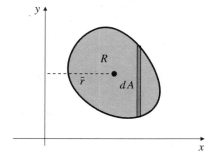

**Figure 7.40**

As the following examples illustrate, Pappus's theorem can be used in two ways: either the centroid can be determined when the appropriate volume or surface area is known, or the volume or surface area can be determined if the centroid of the rotating region or curve is known.

**Example 5** Use Pappus's theorem to find the centroid of the semicircle $y = \sqrt{a^2 - x^2}$.

**Solution** The centroid of the semicircle lies on its axis of symmetry, the $y$-axis, so it is located at a point with coordinates $(0, \bar{y})$. Since the semicircle has length $\pi a$ units and generates, on rotation about the $x$-axis, a sphere having area $4\pi a^2$ square units, we obtain, using part (b) of Pappus's theorem,

$$4\pi a^2 = 2\pi (\pi a) \bar{y}.$$

Thus $\bar{y} = 2a/\pi$, as shown previously in Example 2.

**Example 6**   Use Pappus's theorem to find the volume and surface area of the torus (doughnut) obtained by rotating the disk $(x - b)^2 + y^2 \leq a^2$ about the $y$-axis. Here $0 < a < b$. (See Figure 7.10 in Section 7.1.)

**Solution**   The centroid of the disk is at $(b, 0)$, which is at distance $\bar{r} = b$ units from the axis of rotation. Since the disk has area $\pi a^2$ square units, the volume of the torus is

$$V = 2\pi b(\pi a^2) = 2\pi^2 a^2 b \text{ cubic units.}$$

The surface area of the torus (in case you want to have icing on the doughnut) is

$$S = 2\pi b(2\pi a) = 4\pi^2 ab \text{ square units,}$$

the surface being obtained by rotating a circle of radius $a$, which has length (circumference) $2\pi a$.

## Exercises 7.5

Find the centroids of the geometric structures in Exercises 1–21. Be alert for symmetries and opportunities to use Pappus's theorem.

1. The quarter-disk $x^2 + y^2 \leq r^2$, $x \geq 0$, $y \geq 0$

2. The region $0 \leq y \leq 9 - x^2$

3. The region $0 \leq x \leq 1$, $0 \leq y \leq \dfrac{1}{\sqrt{1 + x^2}}$

4. The circular disk sector $x^2 + y^2 \leq r^2$, $0 \leq y \leq x$

5. The circular disk segment $0 \leq y \leq \sqrt{4 - x^2} - 1$

6. The semi-elliptic disk $0 \leq y \leq b\sqrt{1 - (x/a)^2}$

7. The quadrilateral with vertices (in clockwise order) $(0, 0)$, $(3, 1)$, $(4, 0)$, and $(2, -2)$

8. The region bounded by the semicircle $y = \sqrt{1 - (x - 1)^2}$, the $y$-axis, and the line $y = x - 2$.

9. A hemispherical surface of radius $r$

10. A solid half ball of radius $r$

11. A solid cone of base radius $r$ and height $h$

12. A conical surface of base radius $r$ and height $h$

13. The region in Figure 7.41(a)

14. The region in Figure 7.41(b)

15. The region in Figure 7.41(c)

16. The region in Figure 7.41(d)

17. The plane region $0 \leq y \leq \sin x$, $0 \leq x \leq \pi$

18. The plane region $0 \leq y \leq \cos x$, $0 \leq x \leq \pi/2$

19. The quarter-circle arc $x^2 + y^2 = r^2$, $x \geq 0$, $y \geq 0$

20. The solid obtained by rotating the region in Figure 7.41(a) about the $y$-axis

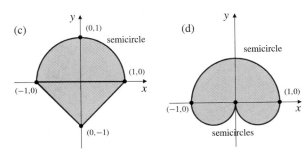

**Figure 7.41**

21. The solid obtained by rotating the plane region $0 \leq y \leq 2x - x^2$ about the line $y = -2$.

22. The line segment from $(1, 0)$ to $(0, 1)$ is rotated about the line $x = 2$ to generate part of a conical surface. Find the area of that surface.

23. The triangle with vertices $(0, 0)$, $(1, 0)$, and $(0, 1)$ is rotated about the line $x = 2$ to generate a certain solid. Find the volume of that solid.

24. An equilateral triangle of edge $s$ cm is rotated about one of its edges to generate a solid. Find the volume and surface area of that solid.

**25.** Find to five decimal places the coordinates of the centroid of the region $0 \le x \le \pi/2$, $0 \le y \le \sqrt{x} \cos x$.

**26.** Find to five decimal places the coordinates of the centroid of the region $0 < x \le \pi/2$, $\ln(\sin x) \le y \le 0$.

**27.** Find the centroid of the infinitely long spike-shaped region lying between the $x$-axis and the curve $y = (x+1)^{-3}$ and to the right of the $y$-axis.

**28.** Show that the curve $y = e^{-x^2}$ $(-\infty < x < \infty)$ generates a surface of finite area when rotated about the $x$-axis. What does this imply about the location of the centroid of this infinitely long curve?

**29.** Obtain formulas for the coordinates of the centroid of the plane region $c \le y \le d$, $0 < f(y) \le x \le g(y)$.

**30.** Prove part (b) of Pappus's theorem (Theorem 2).

**31.** **(Stability of a floating object)** Determining the orientation that a floating object will assume is a problem of critical importance to ship designers. Boats must be designed to float stably in an upright position; if the boat tilts somewhat from upright the forces on it must be such as to right it again. The two forces on a floating object that need to be taken into account are its weight **W**, and the balancing buoyant force **B** $= -$**W**. The weight **W** must be treated for mechanical purposes as being applied at the centre of mass CM of the object. The buoyant force, however, acts at the *centre of buoyancy* CB, which is the centre of mass of the water displaced by the object, and is therefore the centroid of the "hole in the water" made by the object. For example, consider a channel marker buoy consisting of a hemispherical hull surmounted by a conical tower supporting a navigation light. The buoy has a vertical axis of symmetry. If it is upright, both the centre of mass and the centre of buoyancy lie on this line, as shown in Figure 7.42(left).

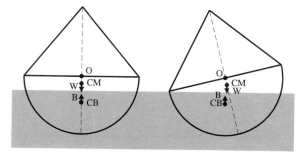

**Figure 7.42**

Is this upright flotation of the buoy stable? It is if the centre of mass CM lies below the centre O of the hemispherical hull, as shown in the figure. To see why, imagine the buoy tilted slightly from the vertical as shown in the right figure. Observe that the centre of mass still lies on the axis of symmetry of the buoy, but the centre of buoyancy lies on the vertical line through O. The forces **W** and **B** no longer act along the same line, but their torques are such as to rotate the buoy back to a vertical upright position. If CM had been above O in the left figure, the torques would have been such as to tip the buoy over once it was displaced even slightly from the vertical.

A wooden beam has a square cross-section and has specific gravity 0.5, so that it will float with half of its volume submerged. (See Figure 7.43.) Assuming it will float horizontally in the water, what is the stable orientation of the square cross section with respect to the surface of the water? In particular, will the beam float with a flat face upward, or an edge upward? Prove your assertions. You may find a symbolic algebra computer program useful.

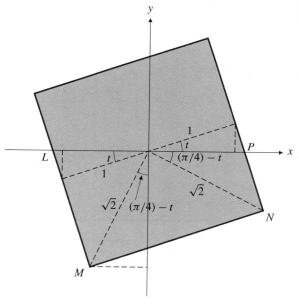

**Figure 7.43**

# 7.6  Other Physical Applications

In this section we present some examples of the use of integration to calculate quantities arising in physics and mechanics.

## Hydrostatic Pressure

The **pressure** $p$ at depth $h$ beneath the surface of a liquid is the *force per unit area* exerted on a horizontal plane surface at that depth due to the weight of the liquid above it. Hence $p$ is given by

$$p = \delta g h,$$

where $\delta$ is the density of the liquid, and $g$ is the acceleration produced by gravity where the fluid is located. (See Figure 7.44.) For water at the surface of the earth we have, approximately, $\delta = 1000 \text{ kg/m}^3$ and $g = 9.8 \text{ m/s}^2$, so the pressure at depth $h$ metres is

$$p = 9800h \text{ N/m}^2.$$

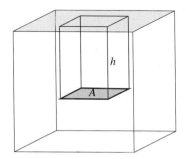

**Figure 7.44**    The volume of liquid above the area $A$ is $V = Ah$. The weight of this liquid is $\delta V g = \delta g h A$, so the pressure (force per unit area) at depth $h$ is $p = \delta g h$

The unit of force used here is the newton (N); $1 \text{ N} = 1 \text{ kg·m/s}^2$, the force that imparts an acceleration of $1 \text{ m/s}^2$ to a mass of 1 kg.

The molecules in a liquid interact in such a way that the pressure at any depth acts equally in all directions; the pressure against a vertical surface is the same as that against a horizontal surface at the same depth. This is **Pascal's principle**.

The total force exerted by a liquid on a horizontal surface (say the bottom of a tank holding the liquid) is found by multiplying the area of that surface by the pressure at the depth of the surface below the top of the liquid. For nonhorizontal surfaces, however, the pressure is not constant over the whole surface, and the total force cannot be determined so easily. In this case we divide the surface into area elements $dA$, each at some particular depth $h$, and we then sum (that is, integrate) the corresponding force elements $dF = \delta g h \, dA$ to find the total force.

**Example 1**    One vertical wall of a water trough is a semicircular plate of radius $R$ metres with curved edge downward. If the trough is full, so that the water comes up to the top of the plate, find the total force of the water on the plate.

*Solution*    A horizontal strip of the surface of the plate at depth $h$ m and having width $dh$ m, (see Figure 7.45), has length $2\sqrt{R^2 - h^2}$ m and hence its area is $dA = 2\sqrt{R^2 - h^2} \, dh \text{ m}^2$. The force of the water on this strip is

$$dF = \delta g h \, dA = 2\delta g h \sqrt{R^2 - h^2} \, dh.$$

Thus the total force on the plate is

**Figure 7.45**

$$F = \int_{h=0}^{h=R} dF = 2\delta g \int_0^R h\sqrt{R^2 - h^2}\, dh \qquad \text{Let } u = R^2 - h^2,$$
$$du = -2h\, dh$$

$$= \delta g \int_0^{R^2} u^{1/2}\, du = \delta g\, \frac{2}{3} u^{3/2} \Big|_0^{R^2}$$

$$\approx \frac{2}{3} \times 9{,}800 R^3 \approx 6{,}533 R^3 \text{ N}.$$

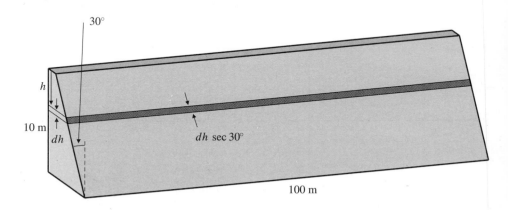

**Figure 7.46**

---

**Example 2**    (**Force on a dam**)  Find the total force on a section of a dam 100 m long and having a vertical height of 10 m, if the surface holding back the water is inclined at an angle of 30° to the vertical and the water comes up to the top of the dam.

**Solution**    The water in a horizontal layer of thickness $dh$ m at depth $h$ m makes contact with the dam along a slanted strip of width $dh \sec 30° = (2/\sqrt{3})\, dh$ m. (See Figure 7.46.) The area of this strip is $dA = (200/\sqrt{3})\, dh$ m², and the force of water against the strip is

$$dF = \delta g h\, dA = \frac{200}{\sqrt{3}} \times 1{,}000 \times 9.8 h\, dh \approx 1{,}131{,}600 h\, dh \text{ N}.$$

The total force on the dam section is therefore

$$F \approx 1{,}131{,}600 \int_0^{10} h\, dh = 1{,}131{,}600 \times \frac{10^2}{2} \approx 5.658 \times 10^7 \text{ N}.$$

## Work

When a force acts on an object to move that object, it is said to have done **work** on the object. The amount of work done by a constant force is measured by the product of the force and the distance through which it moves the object. This assumes that the force is in the direction of the motion.

Work = Force × Distance

Work is always related to a particular force. If other forces acting on an object cause it to move in a direction opposite to the force $F$, then work is said to have been done *against* the force $F$.

Suppose that a force in the direction of the $x$-axis moves an object from $x = a$ to $x = b$ on that axis, and that the force varies continuously with the position $x$ of the object; that is, $F = F(x)$ is a continuous function. The element of work done by the force in moving the object through a very short distance from $x$ to $x + dx$ is $dW = F(x)\, dx$, so the total work done by the force is

$$W = \int_{x=a}^{x=b} dF = \int_a^b F(x)\, dx.$$

**Example 3**   (**Stretching or compressing a spring**) By **Hooke's Law**, the force $F(x)$ required to extend (or compress) an elastic spring to $x$ units longer (or shorter) than its natural length is proportional to $x$:

$$F(x) = kx,$$

where $k$ is the **spring constant** for the particular spring. If a force of 2,000 N is required to extend a certain spring to 4 cm longer than its natural length, how much work must be done to extend it that far?

**Solution**   Since $F(x) = kx = 2{,}000$ N when $x = 4$ cm, we must have $k = 2{,}000/4 = 500$ N/cm. The work done in extending the spring 4 cm is

$$W = \int_0^4 kx\, dx = k\frac{x^2}{2}\Big|_0^4 = 500\,\frac{\text{N}}{\text{cm}} \times \frac{4^2\ \text{cm}}{2} = 4{,}000\ \text{N·cm} = 40\ \text{N·m}.$$

40 newton-metres (joules) of work must be done to stretch the spring 4 cm.  ∎

**Example 4**   (**Work done to pump out a tank**)  Water fills a tank in the shape of a right-circular cone with top radius 3 m and depth 4 m. How much work must be done (against gravity) to pump all the water out of the tank over the top edge of the tank?

**Solution**   A thin disk-shaped slice of water at height $h$ above the vertex of the tank has radius $r$ (see Figure 7.47), where $r = \frac{3}{4}h$ by similar triangles. The volume of this slice is

$$dV = \pi r^2\, dh = \frac{9}{16}\pi h^2\, dh,$$

and its *weight* (the force of gravity on the mass of water in the slice) is

$$dF = \delta g\, dV = \frac{9}{16}\delta g\, \pi h^2\, dh.$$

The water in this disk must be raised (against gravity) a distance $(4 - h)$ m by the pump. The work required to do this is

**Figure 7.47**

$$dW = \frac{9}{16}\delta g\,\pi(4-h)h^2\,dh.$$

The total work that must be done to empty the tank is the sum (integral) of all these elements of work for disks at depths between 0 and 4 metres:

$$W = \int_0^4 \frac{9}{16}\delta g\,\pi(4h^2 - h^3)\,dh$$

$$= \frac{9}{16}\delta g\,\pi\left(\frac{4h^3}{3} - \frac{h^4}{4}\right)\Bigg|_0^4$$

$$= \frac{9\pi}{16}\times 1{,}000 \times 9.8 \times \frac{64}{3} \approx 3.69 \times 10^5\ \text{N·m.}$$

**Example 5**   (**Work to raise material into orbit**)   The gravitational force of the earth on a mass $m$ located at height $h$ above the surface of the earth is given by

$$F(h) = \frac{Km}{(R+h)^2},$$

where $R$ is the radius of the earth and $K$ is a constant that is independent of $m$ and $h$. Determine, in terms of $K$ and $R$ the work that must be done against gravity to raise an object from the surface of the earth to:

(a) a height $H$ above the surface of the earth, and

(b) an infinite height above the surface of the earth.

**Solution**   The work done to raise the mass $m$ from height $h$ to height $h + dh$ is

$$dW = \frac{Km}{(R+h)^2}\,dh.$$

(a) The total work to raise it from height $h = 0$ to height $h = H$ is

$$W = \int_0^H \frac{Km}{(R+h)^2}\, dh = \frac{-Km}{R+h}\Big|_0^H = Km\left(\frac{1}{R} - \frac{1}{R+H}\right).$$

If $R$ and $H$ are measured in metres and $F$ is measured in newtons, then $W$ is measured in newton-metres (N·m) or joules.

(b) The total work necessary to raise the mass $m$ to an infinite height is

$$W = \int_0^\infty \frac{Km}{(R+h)^2}\, dh = \lim_{H\to\infty} Km\left(\frac{1}{R} - \frac{1}{R+H}\right) = \frac{Km}{R}.$$

## Potential and Kinetic Energy

The units of work (force $\times$ distance) are the same as those of energy. Work done against a force may be regarded as storing up energy for future use or for conversion to other forms. Such stored energy is called **potential energy** (P.E.). For instance, in extending or compressing an elastic spring, we are doing work against the tension in the spring and hence storing energy in the spring. When work is done against a (variable) force $F(x)$ to move an object from $x = a$ to $x = b$, the potential energy stored is

$$\text{P.E.} = -\int_a^b F(x)\, dx.$$

Since the work is being done against $F$, the signs of $F(x)$ and $b - a$ are opposite, so the integral is negative; the explicit negative sign is included so that the calculated potential energy will be positive.

One of the forms of energy into which potential energy can be converted is **kinetic energy** (K.E.), the energy of motion. If an object of mass $m$ is moving with velocity $v$, it has kinetic energy

$$\text{K.E.} = \frac{1}{2}m v^2.$$

For example, if an object is raised and then dropped, it accelerates downward under gravity as more and more of the potential energy stored in it when it was raised is converted to kinetic energy.

Consider the change in potential energy stored in a mass $m$ as it moves along the $x$-axis from $a$ to $b$ under the influence of a force $F(x)$ depending only on $x$:

$$\text{P.E.}(b) - \text{P.E.}(a) = -\int_a^b F(x)\, dx.$$

(The change in P.E. is negative if $m$ is moving in the direction of $F$.) According to Newton's second law of motion, the force $F(x)$ causes the mass $m$ to accelerate, with acceleration $dv/dt$ given by

$$F(x) = m\frac{dv}{dt} \qquad (\text{force} = \text{mass} \times \text{acceleration}).$$

By the Chain Rule we can rewrite $dv/dt$ in the form

$$\frac{dv}{dt} = \frac{dv}{dx}\frac{dx}{dt} = v\frac{dv}{dx},$$

so $F(x) = mv\dfrac{dv}{dx}$. Hence

$$P.E.(b) - P.E.(a) = -\int_a^b mv\frac{dv}{dx}\,dx$$
$$= -m\int_{x=a}^{x=b} v\,dv$$
$$= -\frac{1}{2}mv^2\Big|_{x=a}^{x=b}$$
$$= K.E.(a) - K.E.(b).$$

It follows that

$$P.E.(b) + K.E.(b) = P.E.(a) + K.E.(a).$$

This shows that the total energy (potential + kinetic) remains constant as the mass $m$ moves under the influence of a force $F$, *depending only on position.* Such a force is said to be **conservative** and the above result is called the **law of conservation of energy**.

**Example 6** (Escape velocity) Use the result of Example 5 together with the following known values,

(a) the radius $R$ of the earth is about 6,400 km, or $6.4 \times 10^6$ m,

(b) the acceleration of gravity $g$ at the surface of the earth is about 9.8 m/s$^2$,

to determine the constant $K$ in the gravitational force formula of Example 5, and hence to determine the escape velocity for a projectile fired vertically from the surface of the earth. The **escape velocity** is the (minimum) speed that such a projectile must have at firing to ensure that it will continue to move farther and farther away from the earth and not fall back.

**Solution** According to the formula of Example 5, the force of gravity on a mass $m$ kg at the surface of the earth ($h = 0$) is

$$F = \frac{Km}{(R+0)^2} = \frac{Km}{R^2}.$$

According to Newton's second law of motion, this force is related to the acceleration of gravity $(g)$ there by the equation $F = mg$. Thus

$$\frac{Km}{R^2} = mg$$

and $K = gR^2$.

According to the law of conservation of energy, the projectile must have sufficient kinetic energy at firing to do the work necessary to raise the mass $m$ to infinite height. By the result of Example 5, this required energy is $Km/R$. If the initial velocity of the projectile is $v$, we want

$$\frac{1}{2}mv^2 \geq \frac{Km}{R}.$$

Thus $v$ must satisfy

$$v \geq \sqrt{\frac{2K}{R}} = \sqrt{2gR} \approx \sqrt{2 \times 9.8 \times 6.4 \times 10^6} \approx 1.12 \times 10^4 \text{ m/s}.$$

Thus the escape velocity is approximately 11.2 km/s, and is independent of the mass $m$. In this calculation we have neglected any air resistance near the surface of the earth. Such resistance depends on velocity rather than on position, so it is not a conservative force. The effect of such resistance would be to use up (convert to heat) some of the initial kinetic energy and so raise the escape velocity.

## Exercises 7.6

1. A tank has a square base 2 m on each side and vertical sides 6 m high. If the tank is filled with water, find the total force exerted by the water (a) on the bottom of the tank, and (b) on one of the four vertical walls of the tank.

2. A swimming pool 20 m long and 8 m wide has a sloping plane bottom so that the depth of the pool is 1 m at one end and 3 m at the other end. Find the total force exerted on the bottom if the pool is full of water.

3. A dam 200 m long and 24 m high presents a sloping face of 26 m slant height to the water in a reservoir behind the dam (Figure 7.48). If the surface of the water is level with the top of the dam, what is the total force of the water on the dam?

4. A pyramid with a square base 4 m on each side and four equilateral triangular faces, sits on the level bottom of a lake at a place where the lake is 10 m deep. Find the total force of the water on each of the triangular faces.

5. A lock on a canal has a gate in the shape of a vertical rectangle 5 m wide and 20 m high. If the water on one side of the gate comes up to the top of the gate and the water on the other side comes only 6 m up the gate, find the total force that must be exerted to hold the gate in place.

6. If 100 N·cm of work must be done to compress an elastic spring to 3 cm shorter than its natural length, how much work must be done to compress it 1 cm further?

7. Find the total work that must be done to pump all the water in the tank of Exercise 1 out over the top of the tank.

8. Find the total work that must be done to pump all the water in the swimming pool of Exercise 2 out over the top edge of the pool.

9. Find the work that must be done to pump all the water in a full hemispherical bowl of radius $a$ m to a height $h$ m above the top of the bowl.

* 10. A bucket is raised vertically from ground level at a constant speed of 2 m/min by a winch. If the bucket weighs 1 kg and contains 15 kg of water when it starts up, but loses water by leakage at a rate of 1 kg/min thereafter, how much work must be done by the winch to raise the bucket to a height of 10 m?

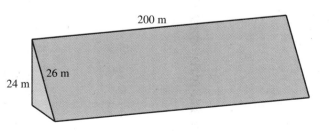

200 m

26 m

24 m

**Figure 7.48**

## 7.7 Applications in Business, Finance, and Ecology

If the rate of change $f'(x)$ of a function $f(x)$ is known, the change in value of the function over an interval from $x = a$ to $x = b$ is just the integral of $f'$ over $[a, b]$:

$$f(b) - f(a) = \int_a^b f'(x)\,dx.$$

For example, if the speed of a moving car at time $t$ is $s(t)$ km/h, then the distance travelled by the car during the time interval $[0, T]$ (hours) is $\int_0^T s(t)\,dt$ km.

Similar situations arise naturally in business and economics, where the rates of change are often called marginals.

**Example 1** **(Finding total revenue from marginal revenue)** A supplier of calculators realizes a marginal revenue of $\$15 - 5e^{-x/50}$ per calculator when $x$ calculators have been sold. What will be her total revenue from the sale of 100 calculators?

**Solution** The marginal revenue is the rate of change of revenue with respect to the number of calculators sold. Thus the revenue from the sale of $dx$ calculators after $x$ have already been sold is

$$dR = (15 - 5e^{-x/50})\,dx$$

dollars. The total revenue from the sale of the first 100 calculators is $\$R$, where

$$R = \int_{x=0}^{x=100} dR = \int_0^{100} (15 - 5e^{-x/50})\,dx$$

$$= \left(15x + 250e^{-x/50}\right)\Big|_0^{100}$$

$$= 1,500 + 250e^{-2} - 250 \approx 1,283.83,$$

that is, about $\$1,284$.

### The Present Value of a Stream of Payments

Suppose that you have a business that generates income continuously at a variable rate $P(t)$ dollars per year at time $t$, and that you expect this income to continue for the next $T$ years. How much is the business worth today?

The answer surely depends on interest rates. One dollar to be received $t$ years from now is worth less than one dollar received today, which could be invested at interest to yield more than one dollar $t$ years from now. The higher the interest rate, the lower the value today of a payment that is not due until some time in the future.

To analyze this situation, suppose that the interest rate is a nominal $100\delta\%$, compounded continuously. As shown in Section 3.4, an investment of $\$1$ today will grow to

$$\lim_{n \to \infty} \left(1 + \frac{\delta}{n}\right)^{nt} = e^{\delta t}$$

after $t$ years. Therefore, a payment of \$1 after $t$ years must be worth only \$$e^{-\delta t}$ today. This is called the *present value* of the future payment. When viewed this way, the interest rate $\delta$ is frequently called a *discount rate*; it represents the amount by which future payments are discounted.

Returning to the business income problem, in the short time interval from $t$ to $t + dt$, the business produces income \$$P(t)\, dt$, of which the present value is \$$e^{-\delta t} P(t)\, dt$. Therefore, the present value \$$V$ of the income stream over the time interval $[0, T]$ is the "sum" of these contributions:

$$V = \int_0^T e^{-\delta t} P(t)\, dt.$$

**Example 2**   What is the present value of a constant, continual stream of payments at a rate of \$10,000 per year, to continue forever, starting now? Assume an interest rate of 6% per annum, compounded continuously.

**Solution**   The required present value is

$$V = \int_0^\infty e^{-0.06t} 10,000\, dt = 10,000 \frac{e^{-0.06t}}{-0.06}\Big|_0^\infty \approx \$166,667.$$

## The Economics of Exploiting Renewable Resources

As noted in Section 3.4, the rate of increase of a biological population sometimes conforms to a logistic model[1]

$$\frac{dx}{dt} = kx\left(1 - \frac{x}{L}\right).$$

Here $x = x(t)$ is the size (or biomass) of the population at time $t$, $k$ is the natural rate at which the population would grow if its food supply were unlimited, and $L$ is the natural limiting size of the population — the carrying capacity of its environment. Such models are thought to apply, for example, to the Antarctic blue whale, and to several species of fish and trees. If the resource is harvested (say, the fish are caught) at a rate $h(t)$ units per year at time $t$, then the population grows at a slower rate

$$\frac{dx}{dt} = kx\left(1 - \frac{x}{L}\right) - h(t). \tag{$*$}$$

In particular, if we harvest the population at its current rate of growth,

$$h(t) = kx\left(1 - \frac{x}{L}\right),$$

then $dx/dt = 0$ and the population will maintain a constant size. Assume that each unit of harvest produces an income of \$$p$ for the fishing industry. The total annual income from harvesting the resource at its current rate of growth will be

$$T = ph(t) = pkx\left(1 - \frac{x}{L}\right).$$

---

[1]   This example was suggested by Professor C. W. Clark, of the University of British Columbia.

Considered as a function of $x$, this total annual income is quadratic and has a maximum value when $x = L/2$, the value that ensures $dT/dx = 0$. The industry can maintain a stable maximum annual income by ensuring that the population level remains at half the maximal size of the population with no harvesting.

The analysis above, however, does not take into account the discounted value of future harvests. If the discount rate is $\delta$, compounded continuously, then the present value of the income $\$ph(t)\,dt$ due between $t$ and $t + dt$ years from now is $e^{-\delta t}\,ph(t)\,dt$. The total present value of all income from the fishery in future years is

$$T = \int_0^\infty e^{-\delta t}\,ph(t)\,dt.$$

What fishing strategy will maximize $T$? If we substitute for $h(t)$ from equation $(*)$ governing the growth rate of the population, we get

$$T = \int_0^\infty pe^{-\delta t}\left[kx\left(1 - \frac{x}{L}\right) - \frac{dx}{dt}\right] dt$$

$$= \int_0^\infty kpe^{-\delta t}x\left(1 - \frac{x}{L}\right) dt - \int_0^\infty pe^{-\delta t}\frac{dx}{dt}\,dt.$$

Integrate by parts in the last integral above, taking $U = pe^{-\delta t}$ and $dV = \dfrac{dx}{dt}\,dt$:

$$T = \int_0^\infty kpe^{-\delta t}x\left(1 - \frac{x}{L}\right) dt - \left[pe^{-\delta t}x\Big|_0^\infty + \int_0^\infty p\delta e^{-\delta t}x\,dt\right]$$

$$= px(0) + \int_0^\infty pe^{-\delta t}\left[kx\left(1 - \frac{x}{L}\right) - \delta x\right] dt.$$

To make this expression as large as possible we should choose the population size $x$ to maximize the quadratic expression

$$Q(x) = kx\left(1 - \frac{x}{L}\right) - \delta x$$

at as early a time $t$ as possible, and keep the population size constant at that level thereafter. The maximum occurs where $Q'(x) = k - (2kx/L) - \delta = 0$, that is, where

$$x = \frac{L}{2} - \frac{\delta L}{2k} = (k - \delta)\frac{L}{2k}.$$

The maximum present value of the fishery is realized if the population level $x$ is held at this value. Note that this population level is smaller than the optimal level $L/2$ we obtained by ignoring the discount rate. The higher the discount rate $\delta$, the smaller will be the income-maximizing population level. More unfortunately, if $\delta \geq k$, the model predicts greatest income from fishing the species to *extinction* immediately! (See Figure 7.49.)

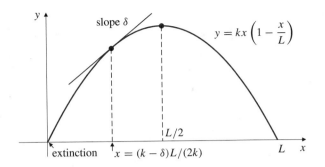

**Figure 7.49**    The greater the discount rate $\delta$, the smaller the population size $x$ that will maximize the present value of future income from harvesting. If $\delta \geq k$ the model predicts fishing the species to extinction

Of course this model fails to take into consideration other factors that may affect the fishing strategy, such as the increased cost of harvesting when the population level is small, and the effect of competition among various parts of the fishing industry. Nevertheless, it does explain the regrettable fact that, under some circumstances, an industry based on a renewable resource can find it in its best interest to destroy the resource. This is especially likely to happen when the natural growth rate $k$ of the resource is low, as it is for the case of whales and most trees. There is good reason not to allow economics alone to dictate the management of the resource.

# Exercises 7.7

1. **(Cost of production)** The marginal cost of production in a coal mine is $6 - 2 \times 10^{-3}x + 6 \times 10^{-6}x^2$ per ton after the first $x$ tons are produced each day. In addition, there is a fixed cost of $4,000 per day to open the mine. Find the total cost of production on a day when 1,000 tons are produced.

2. **(Total sales)** The sales of a new computer chip are modelled by $s(t) = te^{-t/10}$, where $s(t)$ is the number of thousands of chips sold per week, $t$ weeks after the chip was introduced to the market. How many chips were sold in the first year.

3. **(Internet connection rates)** An internet service provider charges clients at a continuously decreasing marginal rate of $4/(1 + \sqrt{t})$ per hour when the client has already used $t$ hours during a month. How much will be billed to a client who uses $x$ hours in a month? ($x$ need not be an integer.)

4. **(Total revenue from declining sales)** The price per kilogram of maple syrup in a store rises at a constant rate from $10 at the beginning of the year to $15 at the end of the year. As the price rises, the quantity sold decreases; the sales rate is $400/(1 + 0.1t)$ kg/year at time $t$ years, $(0 \leq t \leq 1)$. What total revenue does the store obtain from sales of the syrup during the year?

**(Stream of payment problems)** Find the present value of a continuous stream of payments of $1,000 per year for the periods and discount rates given in Exercises 5–10. In each case the discount rate is compounded continuously.

5. 10 years at a discount rate of 2%

6. 10 years at a discount rate of 5%

7. 10 years beginning two years from now at a discount rate of 8%

8. 25 years beginning 10 years from now at a discount rate of 5%

9. For all future time at a discount rate of 2%

10. Beginning in 10 years and continuing forever after at a discount rate of 5%

11. Find the present value of a continuous stream of payments over a 10 year period beginning at a rate of $1,000 per year now and increasing steadily at $100 per year. The discount rate is 5%.

12. Find the present value of a continuous stream of payments over a 10 year period beginning at a rate of $1,000 per year now and increasing steadily at 10% per year. The discount rate is 5%.

13. Money flows continuously into an account at a rate of $5,000 per year. If the account earns interest at a rate of 5% compounded continuously, how much will be in the account after 10 years?

14. Money flows continuously into an account beginning at a rate of $5,000 per year and increasing at 10% per year. Interest causes the account to grow at a real rate of 6% (so that $1 grows to $1.06^t$ in $t$ years). How long will it take for the balance in the account to reach $1,000,000?

15. If the discount rate $\delta$ varies with time, say $\delta = \delta(t)$, show that the present value of a payment of $P$ due $t$ years from now is $Pe^{-\lambda(t)}$, where

$$\lambda(t) = \int_0^t \delta(\tau)\,d\tau.$$

What is the value of a stream of payments due at a rate $P(t)$ at time $t$, from $t = 0$ to $t = T$?

16. **(Discount rates and population models)** Suppose that the growth rate of a population is a function of the population size: $dx/dt = F(x)$. (For the logistic model, $F(x) = kx(1 - (x/L))$.) If the population is harvested at rate $h(t)$ at time $t$, then $x(t)$ satisfies

    $$\frac{dx}{dt} = F(x) - h(t).$$

    Show that the value of $x$ that maximizes the present value of all future harvests satisfies $F'(x) = \delta$, where $\delta$ is the (continuously compounded) discount rate. *Hint:* mimic the argument used above for the logistic case.

17. **(Managing a fishery)** The carrying capacity of a certain lake is $L = 80,000$ of a certain species of fish. The natural growth rate of this species is 12% per year ($k = 0.12$). Each fish is worth \$6. The discount rate is 5%. What population of fish should be maintained in the lake to maximize the present value of all future revenue from harvesting the fish? What is the annual revenue resulting from maintaining this population level?

18. **(Blue whales)** It is speculated that the natural growth rate of the Antarctic blue whale population is about 2% per year ($k = 0.02$), and that the carrying capacity of its habitat is about $L = 150,000$. One blue whale is worth, on average, \$10,000. Assuming that the blue whale population satisfies a logistic model, and using the data above, find the following:

    (a) the maximum sustainable annual harvest of blue whales.

    (b) the annual revenue resulting from the maximum annual sustainable harvest.

    (c) the annual interest generated if the whale population (assumed to be at the level $L/2$ supporting the maximum sustainable harvest) is exterminated and the proceeds invested at 2%. (d) at 5%

    (e) the total present value of all future revenue if the population is maintained at the level $L/2$ and the discount rate is 5%

* 19. The model developed above does not allow for the costs of harvesting. Try to devise a way to alter the model to take this into account. Typically, the cost of catching a fish goes up as the number of fish goes down.

# 7.8 Probability

Probability theory is a very important field of application of the definite integral. This subject cannot, of course, be developed thoroughly here—an adequate presentation requires one or more whole courses—but we can give a brief introduction that suggests some of the ways integrals are used in probability theory.

The **probability** of an event occurring is a real number between 0 and 1 that measures the proportion of times the event can be expected to occur in a large number of trials. If the occurrence of an event is certain, its probability is 1; if the event cannot possibly occur, its probability is 0. For example, the probability that a tossed coin will land heads is 1/2 because we would expect it to land heads about half the time if it were tossed a great many times. In such a tossing of a coin there are only two possible outcomes, heads or tails, each equally likely, that is, each having probability 1/2. (We are assuming the coin won't ever land standing on its edge.) For any toss, let $X = 0$ if the outcome is heads, and let $X = 1$ if the outcome is tails. $X$ is called a **discrete random variable**. The probability that $X = 0$ is 1/2 and the probability that $X = 1$ is 1/2, so we write

$$\Pr(X = 0) = \frac{1}{2} \quad \text{and} \quad \Pr(X = 1) = \frac{1}{2}.$$

Note that $\Pr(X = 0) + \Pr(X = 1) = 1$, since it is certain that the coin will land either heads or tails.

**Example 1**    A single die is rolled so that it will show one of the numbers 1 to 6 on top when it stops. If $X$ denotes the number showing on any roll, then $X$ is a discrete random variable. Assuming no one value of $X$ is any more likely than any other, the probability that the number showing is $x$ must be 1/6 for each possible value of $x$; that is,

$$\Pr(X = x) = \frac{1}{6} \qquad \text{for each } x \text{ in } \{1, 2, 3, 4, 5, 6\}.$$

The discrete random variable $X$ is therefore said to be distributed **uniformly**. Again we note that

$$\sum_{n=1}^{6} \Pr(X = n) = 1,$$

reflecting the fact that the rolled die must certainly give one of the six possible outcomes. The probability that a roll will produce a value from 1 to 4 is

$$\Pr(1 \le X \le 4) = \sum_{n=1}^{4} \Pr(X = n) = \frac{1}{6} + \frac{1}{6} + \frac{1}{6} + \frac{1}{6} = \frac{2}{3}.$$

■

Now we consider an example with a continuous range of possible outcomes.

**Figure 7.50**

(a) $X$ is the number of degrees in the acute angle the needle makes with the line

(b) The probability density function $f$ of the random variable $X$

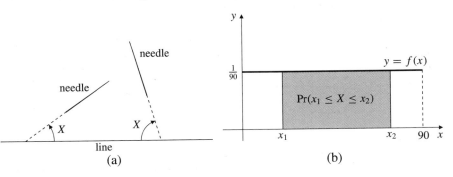

(a)

(b)

**Example 2**    Suppose that a needle is dropped at random on a flat table with a straight line drawn on it. For each drop, let $X$ be the number of degrees in the (acute) angle that the needle makes with the line. (See Figure 7.50(a).) Evidently $X$ can take any real value in the interval $[0, 90]$; therefore $X$ is called a **continuous random variable**. The probability that $X$ takes on any particular real value is 0. (There are infinitely many real numbers in $[0, 90]$ and none is more likely than any other.) However, the probability that $X$ lies in some interval, say $[10, 20]$, is the same as the probability that it lies in any other interval of the same length. Since the interval has length 10 and the interval of all possible values of $X$ has length 90, this probability is

$$\Pr(10 \le X \le 20) = \frac{10}{90} = \frac{1}{9}.$$

More generally, if $0 \le x_1 \le x_2 \le 90$, then

$$\Pr(x_1 \leq X \leq x_2) = \frac{1}{90}(x_2 - x_1).$$

This situation can be conveniently represented as follows: Let $f(x)$ be defined on the interval $[0, 90]$, taking at each point the constant value $1/90$:

$$f(x) = \frac{1}{90}, \qquad 0 \leq x \leq 90.$$

The area under the graph of $f$ is 1, and $\Pr(x_1 \leq X \leq x_2)$ is equal to the area under that part of the graph lying over the interval $[x_1, x_2]$. (See Figure 7.50(b).) The function $f(x)$ is called the **probability density function** for the random variable $X$. Since $f(x)$ is constant on its domain, $X$ is said to be **uniformly distributed**.

---

**DEFINITION 2**

**Probability density functions**

A function defined on an interval $[a, b]$ is a probability density function for a continuous random variable $X$ distributed on $[a, b]$ if, whenever $x_1$ and $x_2$ satisfy $a \leq x_1 \leq x_2 \leq b$, we have

$$\Pr(x_1 \leq X \leq x_2) = \int_{x_1}^{x_2} f(x)\, dx.$$

In order to be such a probability density function, $f$ must satisfy two conditions:

(a) $f(x) \geq 0$ on $[a, b]$,     (probability cannot be negative), and

(b) $\int_a^b f(x)\, dx = 1$,     ($\Pr(a \leq X \leq b) = 1$).

---

These ideas extend to random variables distributed on semi-infinite or infinite intervals, but the integrals appearing will be improper in those cases.

In the example of the dropping needle, the probability density function has a horizontal straight line graph, and we termed such a probability distribution uniform. The uniform probability density function on the interval $[a, b]$ is

$$f(x) = \begin{cases} \dfrac{1}{b-a} & \text{if } a \leq x \leq b \\ 0 & \text{otherwise.} \end{cases}$$

Many other functions are commonly encountered as density functions for continuous random variables.

**Example 3** (**The exponential distribution**) The length of time $T$ that any particular atom in a radioactive sample survives before decaying is a random variable taking values in $[0, \infty)$. It has been observed that the proportion of atoms that survive to time $t$ becomes small exponentially as $t$ increases; thus

$$\Pr(T \geq t) = Ce^{-kt}.$$

Let $f$ be the probability density function for the random variable $T$. Then

$$\int_t^\infty f(x)\,dx = \Pr(T \geq t) = Ce^{-kt}.$$

Differentiating this equation with respect to $t$ (using the Fundamental Theorem of Calculus), we obtain $-f(t) = -Cke^{-kt}$, so $f(t) = Cke^{-kt}$. $C$ is determined by the requirement that $\int_0^\infty f(t)\,dt = 1$. We have

$$1 = Ck \int_0^\infty e^{-kt}\,dt = \lim_{R\to\infty} Ck \int_0^R e^{-kt}\,dt = -C \lim_{R\to\infty} (e^{-kR} - 1) = C.$$

Thus $C = 1$ and $f(t) = ke^{-kt}$. Note that $\Pr(T \geq (\ln 2)/k) = e^{-k(\ln 2)/k} = 1/2$, reflecting the fact that the half-life of such a radioactive sample is $(\ln 2)/k$.

---

**Example 4**  For what value of $C$ is $f(x) = C(1 - x^2)$ a probability density function on $[-1, 1]$? If $X$ is a random variable with this density what is the probability that $X \leq 1/2$?

**Solution**  Observe that $f(x) \geq 0$ on $[-1, 1]$ if $C \geq 0$. Since

$$\int_{-1}^1 f(x)\,dx = C \int_{-1}^1 (1 - x^2)\,dx = 2C\left(x - \frac{x^3}{3}\right)\Big|_0^1 = \frac{4C}{3},$$

$f(x)$ will be a probability density function if $C = 3/4$. In this case

$$\Pr\left(X \leq \frac{1}{2}\right) = \frac{3}{4} \int_{-1}^{1/2} (1 - x^2)\,dx = \frac{3}{4}\left(x - \frac{x^3}{3}\right)\Big|_{-1}^{1/2}$$

$$= \frac{3}{4}\left(\frac{1}{2} - \frac{1}{24} - (-1) + \frac{-1}{3}\right) = \frac{27}{32}.$$

## Expectation, Mean, Variance, and Standard Deviation

Consider a simple gambling game in which the player pays the house $C$ dollars for the privilege of rolling a single die and in which he wins $X$ dollars where $X$ is the number showing on top of the rolled die. In each game the possible winnings are 1, 2, 3, 4, 5, or 6 dollars, each with probability 1/6. In $n$ games the player can expect to win about $n/6 + 2n/6 + 3n/6 + 4n/6 + 5n/6 + 6n/6 = 21n/6 = 7n/2$ dollars, so that his expected *average winnings per game* are 7/2 dollars, $3.50. If $C > 3.5$, the player can expect, on average, to lose money. The amount 3.5 is called the **expectation**, or **mean**, of the discrete random variable $X$. The mean is usually denoted by $\mu$, the Greek letter *mu* (pronounced "mew").

In general, if a random variable can take on values $x_1$ with probability $p_1$, $x_2$ with probability $p_2$, ..., and $x_n$ with probability $p_n$ (where $p_1 + p_2 + \cdots + p_n = 1$), the mean $\mu$ or expectation $E(X)$ of that random variable $X$ is given by

$$\mu = E(X) = \sum_{i=1}^n x_i p_i.$$

We formulate an analogous definition for the mean or expectation of a continuous random variable as follows:

**DEFINITION 3**

> **Mean or expectation**
>
> If $X$ is a continuous random variable on $[a, b]$ with probability density function $f(x)$, the **mean** (denoted $\mu$), or **expectation** of $X$ (denoted $E(X)$), is
>
> $$\mu = E(X) = \int_a^b x f(x)\, dx.$$

Note that in this usage $E(X)$ does not define a function of $X$ but a constant (parameter) associated with the probability distribution of $X$. Note also that if $f(x)$ were a mass density such as that studied in Section 7.4, then $\mu$ would be the moment of the mass about 0 and, since the total mass would be $\int_a^b f(x)\, dx = 1$, $\mu$ would in fact be the centre of mass.

More generally, it can be shown that the **expectation** of any function $g(X)$ of the random variable $X$ is

$$E(g(X)) = \int_a^b g(x) f(x)\, dx.$$

**DEFINITION 4**

> **Variance**
>
> The **variance** of a random variable $X$ with density $f(x)$ on $[a, b]$ is the expectation of the square of the distance of $X$ from its mean $\mu$. The variance is denoted $\sigma^2$ or $\mathrm{Var}(X)$.
>
> $$\sigma^2 = \mathrm{Var}(X) = E((X - \mu)^2) = \int_a^b (x - \mu)^2 f(x)\, dx.$$

The symbol $\sigma$ is the lower case Greek letter *sigma*. (The symbol $\Sigma$ used for summation is an upper case sigma.) Since $\int_a^b f(x)\, dx = 1$, the expression above for the variance can be rewritten as follows:

$$\sigma^2 = \mathrm{Var}(X) = \int_a^b x^2 f(x)\, dx - 2\mu \int_a^b x f(x)\, dx + \mu^2 \int_a^b f(x)\, dx$$

$$= \int_a^b x^2 f(x)\, dx - 2\mu^2 + \mu^2 = E(X^2) - \mu^2,$$

that is,

$$\sigma^2 = \mathrm{Var}(X) = E(X^2) - \mu^2 = E(X^2) - (E(X))^2.$$

**DEFINITION 5**

**Standard deviation**

The **standard deviation** of the random variable $X$ is $\sigma$, the square root of the variance. Thus it is the square root of the mean square deviation of $X$ from its mean:

$$\sigma = \left( \int_a^b (x - \mu)^2 f(x) \, dx \right)^{1/2} = \sqrt{E(X^2) - \mu^2}.$$

The standard deviation gives a measure of how spread out the probability distribution of $X$ is. The smaller the standard deviation, the more concentrated is the area under the density curve around the mean, and so the smaller is the probability that a value of $X$ will be far away from the mean. (See Figure 7.51.)

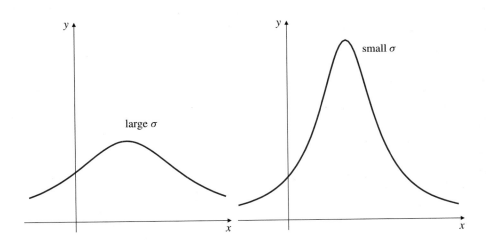

**Figure 7.51**    Densities with large and small standard deviations

**Example 5**    Find the mean $\mu$ and the standard deviation $\sigma$ of a random variable $X$ distributed uniformly on the interval $[a, b]$. Find $\Pr(\mu - \sigma \leq X \leq \mu + \sigma)$.

**Solution**    The probability density function is $f(x) = 1/(b - a)$ on $[a, b]$, so the mean is given by

$$\mu = E(X) = \int_a^b \frac{x}{b - a} \, dx = \frac{1}{b - a} \frac{x^2}{2} \Big|_a^b = \frac{1}{2} \frac{b^2 - a^2}{b - a} = \frac{b + a}{2}.$$

Hence the mean is, as might have been anticipated, the midpoint of $[a, b]$. The expectation of $X^2$ is given by

$$E(X^2) = \int_a^b \frac{x^2}{b - a} \, dx = \frac{1}{b - a} \frac{x^3}{3} \Big|_a^b = \frac{1}{3} \frac{b^3 - a^3}{b - a} = \frac{b^2 + ab + a^2}{3}.$$

Hence the variance is

$$\sigma^2 = E(X^2) - \mu^2 = \frac{b^2 + ab + a^2}{3} - \frac{b^2 + 2ab + a^2}{4} = \frac{(b-a)^2}{12},$$

and the standard deviation is

$$\sigma = \frac{b-a}{2\sqrt{3}} \approx 0.29(b-a).$$

Finally,

$$\Pr(\mu - \sigma \le X \le \mu + \sigma) = \int_{\mu-\sigma}^{\mu+\sigma} \frac{dx}{b-a} = \frac{1}{b-a}\frac{2(b-a)}{2\sqrt{3}} = \frac{1}{\sqrt{3}} \approx 0.577.$$

**Example 6**    Find the mean $\mu$ and the standard deviation $\sigma$ of a random variable $X$ distributed exponentially with density function $f(x) = ke^{-kx}$ on the interval $[0, \infty)$. Find $\Pr(\mu - \sigma \le X \le \mu + \sigma)$.

**Solution**    We use integration by parts to find the mean:

$$\mu = E(X) = k \int_0^\infty xe^{-kx}\, dx$$

$$= \lim_{R\to\infty} k \int_0^R xe^{-kx}\, dx \qquad \text{Let} \qquad U = x \qquad dV = e^{-kx}\, dx$$
$$\text{Then} \quad dU = dx \qquad V = -e^{-kx}/k$$

$$= \lim_{R\to\infty} \left( -xe^{-kx}\Big|_0^R + \int_0^R e^{-kx}\, dx \right)$$

$$= \lim_{R\to\infty} \left( -Re^{-kR} - \frac{1}{k}\left(e^{-kR} - 1\right) \right) = \frac{1}{k}, \qquad \text{since } k > 0.$$

Thus the mean of the exponential distribution is $1/k$. This fact can be quite useful in determining the value of $k$ for an exponentially distributed random variable. A similar integration by parts enables us to evaluate

$$E(X^2) = k \int_0^\infty x^2 e^{-kx}\, dx = 2 \int_0^\infty xe^{-kx}\, dx = \frac{2}{k^2},$$

so that the variance of the exponential distribution is

$$\sigma^2 = E(X^2) - \mu^2 = \frac{1}{k^2}$$

and the standard deviation is equal to the mean

$$\sigma = \mu = \frac{1}{k}.$$

Now we have

$$\Pr(\mu - \sigma \le X \le \mu + \sigma) = \Pr(0 \le X \le 2/k)$$
$$= k \int_0^{2/k} e^{-kx}\, dx$$
$$= -e^{-kx}\Big|_0^{2/k}$$
$$= 1 - e^{-2} \approx 0.86,$$

which is independent of the value of $k$. Exponential densities for small and large values of $k$ are graphed in Figure 7.52.

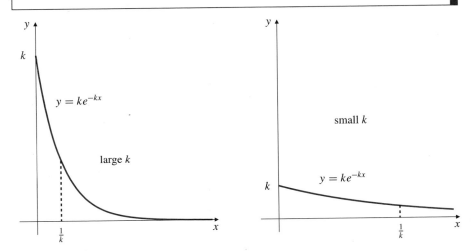

**Figure 7.52**    Exponential density functions

## The Normal Distribution

The most important probability distributions are the so-called **normal** or **Gaussian** distributions. Such distributions govern the behaviour of many interesting random variables, in particular those associated with random errors in measurements. There is a family of normal distributions, all related to the particular normal distribution called the **standard normal distribution**, which has the following probability density function:

**DEFINITION 6**

> **The standard normal probability density**
> $$f(z) = \frac{1}{\sqrt{2\pi}} e^{-z^2/2}, \qquad -\infty < z < \infty.$$

It is common to use $z$ to denote the random variable in the standard normal distribution; the other normal distributions are obtained from this one by a change of variable. The graph of the standard normal density has a pleasant bell shape, as shown in Figure 7.53.

As we have noted previously, the function $e^{-z^2}$ has no elementary antiderivative, so the improper integral

$$I = \int_{-\infty}^{\infty} e^{-z^2/2}\, dz$$

cannot be evaluated using the Fundamental Theorem of Calculus, though it is a convergent improper integral. The integral can be evaluated using techniques of

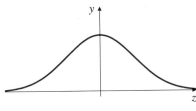

**Figure 7.53**    The standard normal density function $f(z) = \dfrac{1}{\sqrt{2\pi}} e^{-z^2/2}$

multivariable calculus involving double integrals of functions of two variables. (We do so in Chapter 15.) The value is $I = \sqrt{2\pi}$, which ensures that the above-defined standard normal density $f(z)$ is indeed a probability density function:

$$\int_{-\infty}^{\infty} f(z)\, dz = \frac{1}{\sqrt{2\pi}} \int_{-\infty}^{\infty} e^{-z^2/2}\, dz = 1.$$

Since $ze^{-z^2/2}$ is an odd function of $z$ and its integral on $(-\infty, \infty)$ converges, the mean of the standard normal distribution is 0:

$$\mu = E(Z) = \frac{1}{\sqrt{2\pi}} \int_{-\infty}^{\infty} ze^{-z^2/2}\, dz = 0.$$

We calculate the variance of the standard normal distribution using integration by parts as follows:

$$
\begin{aligned}
\sigma^2 &= E(Z^2) \\
&= \frac{1}{\sqrt{2\pi}} \int_{-\infty}^{\infty} z^2 e^{-z^2/2}\, dz \\
&= \frac{1}{\sqrt{2\pi}} \lim_{R \to \infty} \int_{-R}^{R} z^2 e^{-z^2/2}\, dz \qquad \text{Let} \quad U = z \qquad dV = ze^{-z^2/2}\, dz \\
&\qquad\qquad\qquad\qquad\qquad\qquad\qquad\qquad\qquad\quad \text{Then } dU = dz \qquad V = -e^{-z^2/2} \\
&= \frac{1}{\sqrt{2\pi}} \lim_{R \to \infty} \left( -ze^{-z^2/2}\Big|_{-R}^{R} + \int_{-R}^{R} e^{-z^2/2}\, dz \right) \\
&= \frac{1}{\sqrt{2\pi}} \lim_{R \to \infty} (-2Re^{-R^2/2}) + \frac{1}{\sqrt{2\pi}} \int_{-\infty}^{\infty} e^{-z^2/2}\, dz \\
&= 0 + 1 = 1.
\end{aligned}
$$

Hence the standard deviation of the standard normal distribution is 1.

Other normal distributions are obtained from the standard normal distribution by a change of variable.

---

**DEFINITION 7**

**The general normal distribution**

A random variable $X$ on $(-\infty, \infty)$ is said to be *normally distributed with mean $\mu$ and standard deviation $\sigma$* (where $\mu$ is any real number and $\sigma > 0$) if its probability density function $f_{\mu,\sigma}$ is given in terms of the standard normal density $f$ by

$$f_{\mu,\sigma}(x) = \frac{1}{\sigma} f\left( \frac{x - \mu}{\sigma} \right) = \frac{1}{\sigma\sqrt{2\pi}}\, e^{-(x-\mu)^2/(2\sigma^2)}.$$

---

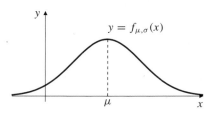

**Figure 7.54**    A general normal density with mean $\mu$

(See Figure 7.54.) Using the change of variable $z = (x - \mu)/\sigma$, $dz = dx/\sigma$, we can verify that

$$\int_{-\infty}^{\infty} f_{\mu,\sigma}(x)\, dx = \int_{-\infty}^{\infty} f(z)\, dz = 1,$$

so $f_{\mu,\sigma}(x)$ is indeed a probability density function. Using the same change of variable, we can show that

$$E(X) = \mu \qquad \text{and} \qquad E((X - \mu)^2) = \sigma^2.$$

Hence $f_{\mu,\sigma}(x)$ does indeed have a mean $\mu$ and a standard deviation $\sigma$.

Because $e^{-z^2/2}$ cannot be easily antidifferentiated, we cannot determine normal probabilities (that is, areas) by using the Fundamental Theorem of Calculus. Numerical integrations can be performed, or one can consult a book of statistical tables for computed areas under the standard normal curve. Specifically, these tables usually provide values for what is called the **cumulative distribution function** of a random variable with standard normal distribution. This is the function

$$F(z) = \frac{1}{\sqrt{2\pi}} \int_{-\infty}^{z} e^{-x^2/2}\, dx = \Pr(Z \le z),$$

which represents the area under the standard normal density function from $-\infty$ up to $z$, as shown in Figure 7.55.

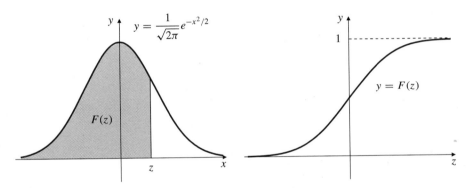

**Figure 7.55**   The cumulative distribution function $F(z)$ for the standard normal distribution is the area under the standard normal density function from $-\infty$ to $z$

For use in the following examples and exercises, we include here an abbreviated version of such a table.

**Table 2.**   Values of the standard normal distribution function $F(z)$ (rounded to three decimal places)

| z | 0.0 | 0.1 | 0.2 | 0.3 | 0.4 | 0.5 | 0.6 | 0.7 | 0.8 | 0.9 |
|---|-----|-----|-----|-----|-----|-----|-----|-----|-----|-----|
| −3.0 | 0.001 | 0.001 | 0.001 | 0.001 | 0.000 | 0.000 | 0.000 | 0.000 | 0.000 | 0.000 |
| −2.0 | 0.023 | 0.018 | 0.014 | 0.011 | 0.008 | 0.006 | 0.005 | 0.004 | 0.003 | 0.002 |
| −1.0 | 0.159 | 0.136 | 0.115 | 0.097 | 0.081 | 0.067 | 0.055 | 0.045 | 0.036 | 0.029 |
| −0.0 | 0.500 | 0.460 | 0.421 | 0.382 | 0.345 | 0.309 | 0.274 | 0.242 | 0.212 | 0.184 |
| 0.0 | 0.500 | 0.540 | 0.579 | 0.618 | 0.655 | 0.692 | 0.726 | 0.758 | 0.788 | 0.816 |
| 1.0 | 0.841 | 0.864 | 0.885 | 0.903 | 0.919 | 0.933 | 0.945 | 0.955 | 0.964 | 0.971 |
| 2.0 | 0.977 | 0.982 | 0.986 | 0.989 | 0.992 | 0.994 | 0.995 | 0.997 | 0.997 | 0.998 |
| 3.0 | 0.999 | 0.999 | 0.999 | 1.000 | 1.000 | 1.000 | 1.000 | 1.000 | 1.000 | 1.000 |

**Example 7**   If $Z$ is a standard normal random variable, find

(a) $\Pr(-1.2 \le Z \le 2.0)$   and   (b) $\Pr(Z \ge 1.5)$.

**Solution**   Using values from the table we obtain

$$\Pr(-1.2 \le Z \le 2.0) = \Pr(Z \le 2.0) - \Pr(Z < -1.2)$$
$$= F(2.0) - F(-1.2) \approx 0.977 - 0.115$$
$$= 0.862$$

$$\Pr(Z \ge 1.5) = 1 - \Pr(Z < 1.5)$$
$$= 1 - F(1.5) \approx 1 - 0.933 = 0.067.$$

**Example 8** A random variable $X$ is distributed normally with mean 2 and standard deviation 0.4. Find

(a) $\Pr(1.8 \le X \le 2.4)$  and  (b) $\Pr(X > 2.4)$.

**Solution** Since $X$ is distributed normally with mean 2 and standard deviation 0.4, $Z = (X - 2)/0.4$ is distributed according to the standard normal distribution (with mean 0 and standard deviation 1). Accordingly,

$$\Pr(1.8 \le X \le 2.4) = \Pr(-0.5 \le Z \le 1)$$
$$= F(1) - F(-0.5) \approx 0.841 - 0.309 = 0.532,$$

$$\Pr(X > 2.4) = \Pr(Z > 1) = 1 - \Pr(Z \le 1)$$
$$= 1 - F(1) \approx 1 - 0.841 = 0.159.$$

## Exercises 7.8

For each function $f(x)$ in Exercises 1–7, find the following:

(a) the value of $C$ for which $f$ is a probability density on the given interval,

(b) the mean $\mu$, variance $\sigma^2$, and standard deviation $\sigma$ of the probability density $f$, and

(c) $\Pr(\mu - \sigma \le X \le \mu + \sigma)$, that is, the probability that the random variable $X$ is no further than one standard deviation away from its mean.

1. $f(x) = Cx$ on $[0, 3]$   2. $f(x) = Cx$ on $[1, 2]$

3. $f(x) = Cx^2$ on $[0, 1]$   4. $f(x) = C \sin x$ on $[0, \pi]$

5. $f(x) = C(x - x^2)$ on $[0, 1]$

6. $f(x) = C x e^{-kx}$ on $[0, \infty)$, $(k > 0)$

7. $f(x) = C e^{-x^2}$ on $[0, \infty)$. *Hint:* use properties of the standard normal density to show that $\int_0^\infty e^{-x^2} \, dx = \sqrt{\pi}/2$.

8. Is it possible for a random variable to be uniformly distributed on the whole real line? Explain why.

9. Carry out the calculations to show that the normal density $f_{\mu,\sigma}(x)$ defined in the text is a probability density function and has mean $\mu$ and standard deviation $\sigma$.

* 10. Show that $f(x) = \dfrac{2}{\pi(1 + x^2)}$ is a probability density on

$[0, \infty)$. Find the expectation of $X$ for this density. If a machine generates values of a random variable $X$ distributed with density $f(x)$, how much would you be willing to pay, per game, to play a game in which you operate the machine to produce a value of $X$ and win $X$ dollars? Explain.

11. Calculate $\Pr(|X - \mu| \ge 2\sigma)$ for

(a) the uniform distribution on $[a, b]$,

(b) the exponential distribution with density $f(x) = ke^{-kx}$ on $[0, \infty)$, and

(c) the normal distribution with density $f_{\mu,\sigma}(x)$.

12. The length of time $T$ (in hours) between malfunctions of a computer system is an exponentially distributed random variable. If the average length of time between successive malfunctions is 20 hours, find the probability that the system, having just had a malfunction corrected, will operate without malfunction for at least 12 hours.

13. The number $X$ of metres of cable produced any day by a cable-making company is a normally distributed random variable with mean 5,000 and standard deviation 200. On what fraction of the days the company operates will the number of metres of cable produced exceed 5,500?

## 7.9 First-Order Differential Equations

This final section on applications of integration concentrates on application of the indefinite integral rather than of the definite integral. We can use the techniques of integration developed in Chapters 5 and 6 to solve certain kinds of first-order differential equations that arise in a variety of modelling situations. We have already seen some examples of applications of differential equations to modelling growth and decay phenomena in Section 3.4.

### Separable Equations

Consider the logistic equation introduced in Section 3.4 to model the growth of an animal population with a limited food supply:

$$\frac{dy}{dt} = ky\left(1 - \frac{y}{L}\right),$$

where $y(t)$ is the size of the population at time $t$, $k$ is a positive constant related to the fertility of the population, and $L$ is the steady-state population size that can be sustained by the available food supply. This equation is an example of a class of first-order differential equations called **separable equations** because when they are written in terms of differentials they can be separated with only the dependent variable on one side of the equation and only the independent variable on the other. The logistic equation can be written in the form

$$\frac{L\,dy}{y(L-y)} = k\,dt,$$

and solved by integrating both sides. Expanding the left side in partial fractions and integrating, we get

$$\int\left(\frac{1}{y} + \frac{1}{L-y}\right)dy = kt + C.$$

Assuming that $0 < y < L$, we therefore obtain

$$\ln y - \ln(L-y) = kt + C,$$

$$\ln\left(\frac{y}{L-y}\right) = kt + C.$$

We can solve this equation for $y$ by taking exponentials of both sides:

$$\frac{y}{L-y} = e^{kt+C} = C_1 e^{kt}$$

$$y = (L-y)C_1 e^{kt}$$

$$y = \frac{C_1 L e^{kt}}{1 + C_1 e^{kt}}.$$

where $C_1 = e^C$.

Generally, separable equations are of the form

$$\frac{dy}{dx} = f(x)g(y).$$

We solve them by rewriting them in the form

$$\frac{dy}{g(y)} = f(x)\,dx$$

and integrating both sides.

**Example 1**   Solve the equation $\dfrac{dy}{dx} = \dfrac{x}{y}$.

**Solution**   We rewrite the equation in the form $y\,dy = x\,dx$ and integrate both sides to get

$$\frac{1}{2}y^2 = \frac{1}{2}x^2 + C,$$

or $y^2 - x^2 = C_1$ where $C_1 = 2C$ is an arbitrary constant. The solution curves are rectangular hyperbolas with asymptotes $y = x$ and $y = -x$. ∎

**Example 2**   Solve the initial-value problem

$$\begin{cases} \dfrac{dy}{dx} = x^2 y^3 \\ y(1) = 3. \end{cases}$$

**Solution**   Separating the differential equation gives $\dfrac{dy}{y^3} = x^2\,dx$. Thus

$$\int \frac{dy}{y^3} = \int x^2\,dx \qquad \text{and so} \qquad \frac{-1}{2y^2} = \frac{x^3}{3} + C.$$

Since $y = 3$ when $x = 1$, we have $-\frac{1}{18} = \frac{1}{3} + C$ and $C = -\frac{7}{18}$. Substituting this value into the above solution and solving for $y$, we obtain

$$y(x) = \frac{3}{\sqrt{7 - 6x^3}}.$$

This solution is valid for $x \le \left(\frac{7}{6}\right)^{1/3}$. ∎

**Example 3**   Solve the **integral equation** $y(x) = 3 + 2\displaystyle\int_1^x t\,y(t)\,dt$.

**Solution**   Differentiating the integral equation with respect to $x$ gives

$$\frac{dy}{dx} = 2x\,y(x) \qquad \text{or} \qquad \frac{dy}{y} = 2x\,dx.$$

Thus $\ln|y(x)| = x^2 + C$, and solving for $y$, $y(x) = C_1 e^{x^2}$. Putting $x = 1$ in the integral equation provides an initial value: $y(1) = 3 + 0 = 3$, so $C_1 = 3/e$ and

$$y(x) = 3e^{x^2-1}.$$ ∎

**Example 4**   **(A solution concentration problem)**   Initially a tank contains 1,000 litres of brine with 50 kilograms of dissolved salt. If brine containing 10 g of salt per litre is flowing into the tank at a constant rate of 10 L/min, if the contents of the tank are kept thoroughly mixed at all times, and if the solution also flows out at 10 L/min, how much salt remains in the tank at the end of 40 min?

**Solution**   Let $x(t)$ be the number of kilograms of salt in solution in the tank after $t$ minutes. Thus $x(0) = 50$. Salt is coming into the tank at a rate of 10 g/L $\times$ 10 L/min = 100 g/min = 1/10 kg/min. At all times the tank contains 1,000 L of liquid, so the concentration of salt in the tank at time $t$ is $x/1,000$ kg/L. Since the contents flow out at 10 L/min, salt is being removed at a rate of $10x/1,000 = x/100$ kg/min. Therefore,

$$\frac{dx}{dt} = \text{rate in} - \text{rate out} = \frac{1}{10} - \frac{x}{100} = \frac{10 - x}{100}$$

or

$$\frac{dx}{10 - x} = \frac{dt}{100}.$$

Integrating both sides of this equation, we obtain

$$-\ln|10 - x| = \frac{t}{100} + C.$$

Observe that $x(t) \neq 10$ for any finite time $t$ (since $\ln 0$ is not defined). Since $x(0) = 50 > 10$, it follows that $x(t) > 10$ for all $t > 0$. ($x(t)$ is necessarily continuous so it cannot take any value less than 10 without somewhere taking the value 10 by the Intermediate-Value Theorem.) Hence we can drop the absolute value from the solution above and obtain

$$\ln(x - 10) = -\frac{t}{100} - C.$$

Since $x(0) = 50$, we have $-C = \ln 40$ and

$$x = x(t) = 10 + 40e^{-t/100}.$$

After 40 min there will be $10 + 40e^{-0.4} \approx 36.8$ kg of salt in the tank.  ∎

**Example 5**   **(A rate of reaction problem)**   In a chemical reaction that goes to completion in solution, one molecule of each of two reactants, $A$ and $B$, combine to form each molecule of the product $C$. According to the law of mass action, the reaction proceeds at a rate proportional to the product of the concentrations of $A$ and $B$ in the solution (see Section 2.7). Thus if there were initially present $a$ molecules/cm$^3$ of $A$ and $b$ molecules/cm$^3$ of $B$, then the number $x(t)$ of molecules/cm$^3$ of $C$ present at time $t$ thereafter is determined by the differential equation

$$\frac{dx}{dt} = k(a - x)(b - x).$$

We solve this equation by the technique of partial fraction decomposition under the assumption that $b \neq a$.

$$\int \frac{dx}{(a-x)(b-x)} = k \int dt = kt + C.$$

Since

$$\frac{1}{(a-x)(b-x)} = \frac{1}{b-a}\left(\frac{1}{a-x} - \frac{1}{b-x}\right),$$

and since necessarily $x \leq a$ and $x \leq b$, we have

$$\frac{1}{b-a}\left(-\ln(a-x) + \ln(b-x)\right) = kt + C,$$

or

$$\ln\left(\frac{b-x}{a-x}\right) = (b-a)\,kt + C_1, \quad \text{where } C_1 = (b-a)C.$$

By assumption, $x(0) = 0$, so $C_1 = \ln(b/a)$ and

$$\ln\frac{a}{b}\left(\frac{b-x}{a-x}\right) = (b-a)\,kt.$$

This equation can be solved for $x$ to yield $x = x(t) = \dfrac{ab(e^{(b-a)kt} - 1)}{be^{(b-a)kt} - a}$. $\blacksquare$

**Example 6**   Find a family of curves, each of which intersects every parabola with equation of the form $y = Cx^2$ at right angles.

**Solution**   The family of parabolas $y = Cx^2$ satisfies the differential equation

$$\frac{d}{dx}\left(\frac{y}{x^2}\right) = \frac{d}{dx}C = 0,$$

that is,

$$x^2\frac{dy}{dx} - 2xy = 0 \quad \text{or} \quad \frac{dy}{dx} = \frac{2y}{x}.$$

Any curve that meets the parabolas $y = Cx^2$ at right angles must, at any point $(x, y)$ on it, have slope equal to the negative reciprocal of the slope of the particular parabola passing through that point. Thus such a curve must satisfy

$$\frac{dy}{dx} = -\frac{x}{2y}.$$

**Figure 7.56**    The parabolas $y = C_1x^2$ and the ellipses $x^2 + 2y^2 = C_2$ intersect at right angles

Separation of the variables leads to $2y\,dy = -x\,dx$, and integration of both sides then yields $y^2 = -\frac{1}{2}x^2 + C_1$ or $x^2 + 2y^2 = C_2$, $(C_2 = 2C_1)$. This equation represents a family of ellipses centred at the origin. Each ellipse meets each parabola at right angles, as shown in Figure 7.56. When the curves of one family intersect the curves of a second family at right angles, each family is called the family of **orthogonal trajectories** of the other family.

## First-Order Linear Equations

A first-order **linear** differential equation is one of the type

$$\frac{dy}{dx} + p(x)y = q(x),$$

where $p(x)$ and $q(x)$ are given functions, which we assume to be continuous. We can solve such equations (that is, find $y$ as a function of $x$) by the following procedure.

Let $\mu(x)$ be any antiderivative of $p(x)$:

$$\mu(x) = \int p(x)\,dx \quad \text{and} \quad \frac{d\mu}{dx} = p(x).$$

If $y = y(x)$ satisfies the given equation, then we calculate, using the Product Rule,

$$\frac{d}{dx}\left(e^{\mu(x)}y(x)\right) = e^{\mu(x)}\frac{dy}{dx} + e^{\mu(x)}\frac{d\mu}{dx}\,y(x)$$

$$= e^{\mu(x)}\left(\frac{dy}{dx} + p(x)y\right) = e^{\mu(x)}q(x).$$

Therefore $e^{\mu(x)}\,y(x) = \displaystyle\int e^{\mu(x)}\,q(x)\,dx$, or

$$y(x) = e^{-\mu(x)} \int e^{\mu(x)} q(x)\, dx.$$

We reuse this method, rather than the final formula, in the examples below. $e^{\mu(x)}$ is called an **integrating factor** of the given differential equation because, if we multiply the equation by $e^{\mu(x)}$, the left side becomes the derivative of $e^{\mu(x)} y(x)$.

**Example 7**   Solve $\dfrac{dy}{dx} + \dfrac{y}{x} = 1$ for $x > 0$.

**Solution**   Here we want $d\mu/dx = 1/x$, so $\mu(x) = \ln x$ (for $x > 0$). Thus $e^{\mu(x)} = x$, and we calculate

$$\frac{d}{dx}(xy) = x\frac{dy}{dx} + y = x\left(\frac{dy}{dx} + \frac{y}{x}\right) = x,$$

and

$$xy = \int x\, dx = \frac{1}{2}x^2 + C.$$

Finally,

$$y = \frac{1}{x}\left(\frac{1}{2}x^2 + C\right) = \frac{x}{2} + \frac{C}{x}.$$

This is a solution of the given equation for any value of the constant $C$.

**Example 8**   Solve $\dfrac{dy}{dx} + xy = x^3$.

**Solution**   Here $p(x) = x$ so $\mu(x) = x^2/2$. We calculate

$$\frac{d}{dx}\left(e^{x^2/2}y\right) = e^{x^2/2}\frac{dy}{dx} + e^{x^2/2}xy = x^3 e^{x^2/2}.$$

Thus,

$$e^{x^2/2}\, y = \int x^3\, e^{x^2/2}\, dx \qquad \text{Let} \qquad U = x^2 \qquad dV = x\, e^{x^2/2}\, dx$$
$$\qquad\qquad\qquad\qquad\qquad \text{Then } dU = 2x\, dx \qquad V = e^{x^2/2}$$
$$= x^2\, e^{x^2/2} - 2\int x\, e^{x^2/2}\, dx$$
$$= x^2\, e^{x^2/2} - 2\, e^{x^2/2} + C,$$

and, finally, $y = x^2 - 2 + Ce^{-x^2/2}$.

Our final example reviews a typical *stream of payments* problem of the sort considered in Section 7.7. This time we treat the problem as an initial-value problem for a differential equation.

**Example 9**   A savings account is opened with a deposit of $A$ dollars. At any time $t$ years thereafter, money is being continually deposited into the account at a rate of $(C + Dt)$ dollars per year. If interest is also being paid into the account at a nominal rate of $100R\%$ per year, compounded continuously, find the balance $B(t)$ dollars in the account after $t$ years. Illustrate the solution for the data $A = 5,000$, $C = 1,000$, $D = 200$, $R = 0.13$, and $t = 5$.

**Solution**   As noted in Section 3.4, continuous compounding of interest at a nominal rate of $100R\%$ causes \$1.00 to grow to $\$e^{Rt}$ in $t$ years. Without subsequent deposits, the balance in the account would grow according to the differential equation of exponential growth:

$$\frac{dB}{dt} = RB.$$

Allowing for additional growth due to the continual deposits, we observe that $B$ must satisfy the differential equation

$$\frac{dB}{dt} = RB + (C + Dt)$$

or, equivalently, $dB/dt - RB = C + Dt$. This is a linear equation for $B$ having $p(t) = -R$. Hence we may take $\mu(t) = -Rt$ and $e^{\mu(t)} = e^{-Rt}$. We now calculate

$$\frac{d}{dt}\left(e^{-Rt}B(t)\right) = e^{-Rt}\frac{dB}{dt} - Re^{-Rt}B(t) = (C + Dt)e^{-Rt},$$

and

$$e^{-Rt}B(t) = \int (C + Dt)e^{-Rt}\,dt \qquad \text{Let} \quad U = C + Dt \quad dV = e^{-Rt}\,dt$$
$$\text{Then } dU = D\,dt \qquad V = -e^{-Rt}/R$$
$$= -\frac{C + Dt}{R}e^{-Rt} + \frac{D}{R}\int e^{-Rt}\,dt$$
$$= -\frac{C + Dt}{R}e^{-Rt} - \frac{D}{R^2}e^{-Rt} + K, \qquad (K = \text{constant}).$$

Hence

$$B(t) = -\frac{C + Dt}{R} - \frac{D}{R^2} + Ke^{Rt}.$$

Since $A = B(0) = -\dfrac{C}{R} - \dfrac{D}{R^2} + K$, we have $K = A + \dfrac{C}{R} + \dfrac{D}{R^2}$ and

$$B(t) = \left(A + \frac{C}{R} + \frac{D}{R^2}\right)e^{Rt} - \frac{C + Dt}{R} - \frac{D}{R^2}.$$

For the illustration $A = 5,000$, $C = 1,000$, $D = 200$, $R = 0.13$, and $t = 5$, we obtain, using a calculator, $B(5) = 19,762.82$. The account will contain \$19,762.82, after 5 years, under these circumstances. ∎

# Exercises 7.9

Solve the differential equations in Exercises 1–16.

**1.** $\dfrac{dy}{dx} = \dfrac{y}{2x}$

**2.** $\dfrac{dy}{dx} = \dfrac{3y-1}{x}$

**3.** $\dfrac{dy}{dx} = \dfrac{x^2}{y^2}$

**4.** $\dfrac{dy}{dx} = x^2 y^2$

**5.** $\dfrac{dY}{dt} = tY$

**6.** $\dfrac{dx}{dt} = e^x \sin t$

**7.** $\dfrac{dy}{dx} = 1 - y^2$

**8.** $\dfrac{dy}{dx} = 1 + y^2$

**9.** $\dfrac{dy}{dt} = 2 + e^y$

**10.** $\dfrac{dy}{dx} = y^2(1 - y)$

**11.** $\dfrac{dy}{dx} - \dfrac{2y}{x} = x^2$

**12.** $\dfrac{dy}{dx} + \dfrac{2y}{x} = \dfrac{1}{x^2}$

**13.** $\dfrac{dy}{dx} + 2y = 3$

**14.** $\dfrac{dy}{dx} + y = e^x$

**15.** $\dfrac{dy}{dx} + y = x$

**16.** $\dfrac{dy}{dx} + 2e^x y = e^x$

Solve the integral equations in Exercises 17–20.

**17.** $y(x) = 2 + \displaystyle\int_0^x \dfrac{t}{y(t)}\, dt$

**18.** $y(x) = 1 + \displaystyle\int_0^x \dfrac{\big(y(t)\big)^2}{1+t^2}\, dt$

**19.** $y(x) = 1 + \displaystyle\int_1^x \dfrac{y(t)\, dt}{t(t+1)}$

**20.** $y(x) = 3 + \displaystyle\int_0^x e^{-y(t)}\, dt$

**21.** Why is the solution given for the chemical reaction rate problem in Example 5 not valid for $a = b$? Find the solution for the case $a = b$.

**22.** An object of mass $m$ falling near the surface of the earth is retarded by air resistance proportional to its velocity so that, according to Newton's second law of motion,

$$m\dfrac{dv}{dt} = mg - kv,$$

where $v = v(t)$ is the velocity of the object at time $t$, and $g$ is the acceleration of gravity near the surface of the earth. Assuming that the object falls from rest at time $t = 0$, that is, $v(0) = 0$, find the velocity $v(t)$ for any $t > 0$ (up until the object strikes the ground). Show $v(t)$ approaches a limit as $t \to \infty$. Do you need the explicit formula for $v(t)$ to determine this limiting velocity?

**23.** Repeat Exercise 22 except assuming that the air resistance is proportional to the square of the velocity so that the equation of motion is

$$m\dfrac{dv}{dt} = mg - kv^2.$$

**24.** Find the amount in a savings account after one year if the initial balance in the account was $1,000, if the interest is paid continuously into the account at a nominal rate of 10% per annum, compounded continuously, and if the account is being continuously depleted (by taxes, say) at a rate of $y^2/1,000,000$ dollars per year, where $y = y(t)$ is the balance in the account after $t$ years. How large can the account grow? How long will it take the account to grow to half this balance?

**25.** Find the family of curves each of which intersects all of the hyperbolas $xy = C$ at right angles.

**26.** Repeat the solution concentration problem in Example 4, changing the rate of inflow of brine into the tank to 12 L/min but leaving all the other data as they were in that example. Note that the volume of liquid in the tank is no longer constant as time increases.

# Chapter Review

## Key Ideas

- **What do the following phrases mean?**
  - ◇ a solid of revolution
  - ◇ a volume element
  - ◇ the arc length of a curve
  - ◇ the moment of a point mass $m$ about $x = 0$
  - ◇ the centre of mass of a distribution of mass
  - ◇ the centroid of a plane region
  - ◇ a first-order separable differential equation
  - ◇ a first-order linear differential equation

- Let $D$ be the plane region $0 \le y \le f(x)$, $a \le x \le b$. Use integrals to represent the following:
  - ◇ the volume generated by revolving $D$ about the $x$-axis
  - ◇ the volume generated by revolving $D$ about the $y$-axis
  - ◇ the moment of $D$ about the $y$-axis
  - ◇ the moment of $D$ about the $x$-axis
  - ◇ the centroid of $D$

- Let $C$ be the curve $y = f(x)$, $a \le x \le b$. Use integrals to represent the following:
  - ◇ the length of $C$
  - ◇ the area of the surface generated by revolving $C$ about the

x-axis

◇ the area of the surface generated by revolving C about the y-axis

## Review Exercises

**1.** The figure shows cross-sections along the axes of two circular spools. The left spool will hold 1,000 metres of thread if wound full with no bulging. How many metres of thread of the same size will the right spool hold?

**Figure 7.57**

**2.** Water sitting in a bowl evaporates at a rate proportional to its surface area. Show that the depth of water in the bowl decreases at a constant rate, regardless of the shape of the bowl.

**3.** A barrel is 4 ft high and its volume is 16 cubic feet. Its top and bottom are circular disks of radius 1 ft, and its side wall is obtained by rotating the parabola $x = a - by^2$, $-2 \le y \le 2$, about the y-axis. Find, approximately, the values of the positive constants $a$ and $b$.

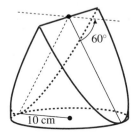

**Figure 7.58**

**4.** The solid in Figure 7.58 is cut from a vertical cylinder of radius 10 cm by two planes making angles of 60° with the horizontal. Find its volume.

**5.** Find to four decimal places the value of the positive constant $a$ for which the curve $y = (1/a) \cosh ax$ has arc length 2 units between $x = 0$ and $x = 1$.

**6.** Find the area of the surface obtained by rotating the curve $y = \sqrt{x}$, $(0 \le x \le 6)$, about the x-axis.

**7.** Find the centroid of the plane region $x \ge 0$, $y \ge 0$, $x^2 + 4y^2 \le 4$.

**8.** A thin plate in the shape of a circular disk has radius 3 ft and constant areal density. A circular hole of radius 1 ft is cut out

of the disk, centred 1 ft from the centre of the disk. Find the centre of mass of the remaining part of the disk.

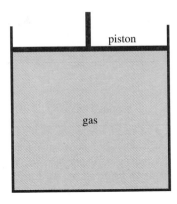

**Figure 7.59**

**9.** According to Boyle's Law, the product of the pressure and volume of a gas remains constant if the gas expands or is compressed isothermally. The cylinder in Figure 7.59 is filled with a gas that exerts a force of 1,000 N on the piston when the piston is 20 cm above the base of the cylinder. How much work is done by the piston if it compresses the gas isothermally by descending to a height of 5 cm above the base?

**10.** Suppose two functions $f$ and $g$ have the following property: for any $a > 0$, the solid produced by revolving the region of the $xy$ plane bounded by $y = f(x)$, $y = g(x)$, $x = 0$, and $x = a$ about the $x$ axis has the same volume as the solid produced by revolving the same region about the y-axis. What can you say about $f$ and $g$?

**11.** Find the equation of a curve that passes through the point $(2, 4)$ and has slope $3y/(x - 1)$ at any point $(x, y)$ on it.

**12.** Find a family of curves that intersect every ellipse of the form $3x^2 + 4y^2 = C$ at right angles.

**13.** The income and expenses of a seasonal business result in deposits and withdrawals from its bank account that correspond to a flow rate into the account of $\$P(t)$/year at time $t$ years, where $P(t) = 10,000 \sin(2\pi t)$. If the account earns interest at an instantaneous rate of 4% per year, and has \$8,000 in it at time $t = 0$, how much is in the account two years later?

## Challenging Problems

**1.** The curve $y = e^{-kx} \sin x$, $(x \ge 0)$, is revolved about the x-axis to generate a string of "beads" whose volumes decrease to the right if $k > 0$.

(a) Show that the ratio of the volume of the $(n + 1)$st bead to that of the $n$th bead depends on $k$, but not on $n$.

(b) For what value of $k$ is the ratio in part (a) equal to 1/2?

(c) Find the total volume of all the beads as a function of $k > 0$.

**Figure 7.60**

2. **(Conservation of earth)** A landscaper wants to create on level ground a ring-shaped pool having an outside radius of 10 m and a maximum depth of 1 m surrounding a hill that will be built up using all the earth excavated from the pool. (See Figure 7.60.) She decides to use a fourth-degree polynomial to determine the cross-sectional shape of the hill and pool bottom: at distance $r$ metres from the centre of the development the height above or below normal ground level will be

$$h(r) = a(r^2 - 100)(r^2 - k^2) \text{ metres},$$

for some $a > 0$, where $k$ is the inner radius of the pool. Find $k$ and $a$ so that the requirements given above are all satisfied. How much earth must be moved from the pool to build the hill?

3. **(Rocket design)** The nose of a rocket is a solid of revolution of base radius $r$ and height $h$ that must join smoothly to the cylindrical body of the rocket. (See Figure 7.61.) Taking the origin at the tip of the nose and the $x$-axis along the central axis of the rocket, various nose shapes can be obtained by revolving the cubic curve $y = f(x) = ax + bx^2 + cx^3$ about the $x$-axis. The cubic must have slope 0 at $x = h$, and its slope must be positive for $0 < x < h$. Find the particular cubic that maximizes the volume of the nose. Also show that this choice of the conic makes the slope $dy/dx$ at the origin as large as possible, and hence corresponds to the bluntest nose.

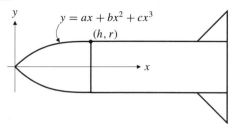

**Figure 7.61**

4. **(Quadratic splines)** Let $A = (x_1, y_1)$, $B = (x_2, y_2)$, and $C = (x_3, y_3)$ be three points with $x_1 < x_2 < x_3$. A function $f(x)$ whose graph passes through the three points is a *quadratic spline* if $f(x)$ is a quadratic function on $[x_1, x_2]$ and a possibly different quadratic function on $[x_2, x_3]$, and the two quadratics have the same slope at $x_2$. For this problem, take $A = (0, 1)$, $B = (1, 2)$, and $C = (3, 0)$.

(a) Find a one parameter family $f(x, m)$ of quadratic splines through $A$, $B$, and $C$, having slope $m$ at $B$.

(b) Find the value of $m$ for which the length of the graph $y = f(x, m)$ between $x = 0$ and $x = 3$ is minimum. What is this minimum length? Compare it with the length of the polygonal line $ABC$.

5. **(The volume of an $n$-dimensional ball)** Euclidean $n$-dimensional space consists of *points* $(x_1, x_2, \ldots, x_n)$ with $n$ real coordinates. By analogy with the 3-dimensional case, we call the set of such points that satisfy the inequality $x_1^2 + x_2^2 + \cdots + x_n^2 \le r^2$ the $n$-dimensional *ball* centred at the origin. For example, the 1-dimensional ball is the interval $-r \le x_1 \le r$, which has *volume* (that is, *length*), $V_1(r) = 2r$. The 2-dimensional ball is the disk $x_1^2 + x_2^2 \le r^2$, which has *volume* (that is, *area*),

$$V_2(r) = \pi r^2 = \int_{-r}^{r} 2\sqrt{r^2 - x^2} \, dx$$

$$= \int_{-r}^{r} V_1\left(\sqrt{r^2 - x^2}\right) dx.$$

The 3-dimensional ball $x_1^2 + x_2^2 + x_3^2 \le r^2$ has volume

$$V_3(r) = \frac{4}{3}\pi r^3 = \int_{-r}^{r} \pi \left(\sqrt{r^2 - x^2}\right)^2 dx$$

$$= \int_{-r}^{r} V_2\left(\sqrt{r^2 - x^2}\right) dx.$$

By analogy with these formulas, the volume $V_n(r)$ of the $n$-dimensional ball of radius $r$ is the integral of the volume of the $(n - 1)$-dimensional ball of radius $\sqrt{r^2 - x^2}$ from $x = -r$ to $x = r$:

$$V_n(r) = \int_{-r}^{r} V_{n-1}\left(\sqrt{r^2 - x^2}\right) dx.$$

Using a computer algebra program, calculate $V_4(r)$, $V_5(r)$, ..., $V_{10}(r)$, and guess formulas for $V_{2n}(r)$ (the even-dimensional balls) and $V_{2n+1}(r)$ (the odd-dimensional balls). If your computer algebra software is sufficiently powerful, you may be able to verify your guesses by induction. Otherwise, use them to predict $V_{11}(r)$ and $V_{12}(r)$, then check your predictions by starting from $V_{10}(r)$.

* 6. **(Buffon's needle problem)** A horizontal flat surface is ruled with parallel lines 10 cm apart, as shown in Figure 7.62. A needle 5 cm long is dropped at random onto the surface. Find the probability that the needle intersects one of the lines. *Hint:* Let the "lower" end of the needle (the end further down the page in the figure) be considered to be the reference point. (If both ends are the same height, use the left end.) Let $y$ be the distance from the reference point to the nearest line above it, and let $\theta$ be the angle between the needle and the

line extending to the right of the reference point in the figure. What are the possible values of $y$ and $\theta$? In a plane with Cartesian coordinates $\theta$ and $y$ sketch the region consisting of all points $(\theta, y)$ corresponding to possible positions of the needle. Also sketch the region corresponding to those positions for which the needle crosses one of the parallel lines. The required probability is the area of the second region divided by the area of the first.

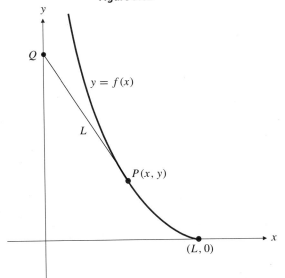

**Figure 7.62**

**Figure 7.63**

* **7.** (**The path of a trailer**) Find the equation $y = f(x)$ of a curve in the first quadrant of the $xy$-plane, starting from the point $(L, 0)$, and having the property that if the tangent line to the curve at $P$ meets the $y$-axis at $Q$, then the length of $PQ$ is the constant $L$. (See Figure 7.63. This curve is called a **tractrix** after the Latin participle *tractus* meaning *dragged*. It is the path of the rear end $P$ of a trailer of length $L$, originally lying along the $x$-axis, as the trailer is pulled (dragged) by a tractor $Q$ moving along the $y$-axis away from the origin.)

* **8.** (**Approximating the surface area of an ellipsoid**) A physical geographer studying the flow of streams around oval stones needed to calculate the surface areas of many such stones which he modelled as ellipsoids:

$$\frac{x^2}{a^2} + \frac{y^2}{b^2} + \frac{z^2}{c^2} = 1.$$

He wanted a simple formula for the surface area so that he could implement it in a spreadsheet containing the measurements $a$, $b$, and $c$ of the stones. Unfortunately, there is no exact formula for the area of a general ellipsoid in terms of elementary functions. However, there are such formulas for ellipsoids of revolution, where two of the three semi-axes are equal. These ellipsoids are called spheroids; an *oblate spheroid* (like the earth) has its two longer semi-axes equal, a *prolate spheroid* (like an American football) has its two shorter semi-axes equal. A reasonable approximation to the area of a general ellipsoid can be obtained by linear interpolation between these two.

To be specific, assume the semi-axes are arranged in decreasing order $a \geq b \geq c$, and let the surface area be $S(a, b, c)$.

(a) Calculate $S(a, a, c)$, the area of an oblate spheroid.

(b) Calculate $S(a, c, c)$, the area of a prolate spheroid.

(c) Construct an approximation for $S(a, b, c)$ that divides the interval from $S(a, a, c)$ to $S(a, c, c)$ in the same ratio that $b$ divides the interval from $a$ to $c$.

(d) Approximate the area of the ellipsoid

$$\frac{x^2}{9} + \frac{y^2}{4} + z^2 = 1$$

using the above method.

# CHAPTER 8

# Conics, Parametric Curves, and Polar Curves

***Introduction*** Up to this point we have been concerned with curves in the plane mostly as the graphs of functions. The curves provided useful visual information about the behaviour of the functions. In this chapter we begin to look at curves as interesting objects in their own right. First, we examine conic sections, curves with quadratic equations obtained by intersecting a plane with a right circular cone. Then we consider curves that can be described by two parametric equations that give the coordinates of points on the curve as functions of a parameter. If this parameter is time, the equations describe the path of a moving point in the plane. Finally, we consider curves described by equations in a new coordinate system called polar coordinates, in which a point is located by giving its distance and direction from the origin.

## 8.1 Conics

Circles, ellipses, parabolas, and hyperbolas are called **conic sections**, (or, more simply, just **conics**), because they are curves in which planes intersect right-circular cones.

To be specific, suppose that a line $A$ is fixed in space, and $V$ is a point fixed on $A$. The **right-circular cone** having **axis** $A$, **vertex** $V$, and **semi-vertical angle** $\alpha$ is the surface consisting of all points on straight lines through $V$ that make angle $\alpha$ with the line $A$. (See Figure 8.1.) The cone has two halves (called **nappes**) lying on opposite sides of the vertex $V$. Any plane $P$ that does not pass through $V$ will intersect the cone (one or both nappes) in a curve $C$. (See Figure 8.2.) If a line normal (that is, perpendicular) to $P$ makes angle $\theta$ with the axis $A$ of the cone, where $0 \le \theta \le \pi/2$, then

$$
\begin{aligned}
&C \text{ is a circle if} && \theta = 0 \\
&C \text{ is an ellipse if} && 0 < \theta < \frac{\pi}{2} - \alpha \\
&C \text{ is a parabola if} && \theta = \frac{\pi}{2} - \alpha \\
&C \text{ is a hyperbola if} && \theta > \frac{\pi}{2} - \alpha.
\end{aligned}
$$

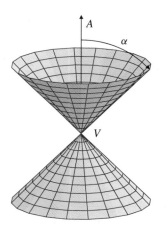

**Figure 8.1** A cone with vertex $V$, axis $A$, and semi-vertical angle $\alpha$

Since planes are represented by first-degree equations and cones by second-degree equations, all conics can be represented analytically (in terms of Cartesian coordinates $x$ and $y$ in the plane of the conic) by a second-degree equation of the general form

$$
Ax^2 + Bxy + Cy^2 + Dx + Ey + F = 0,
$$

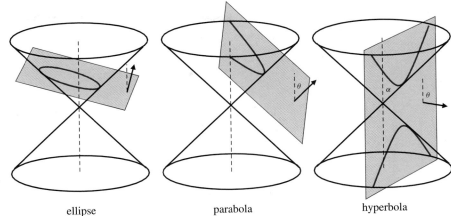

**Figure 8.2** Planes intersecting cones in an ellipse, a parabola, and a hyperbola

ellipse          parabola          hyperbola

where $A$, $B$, ..., $F$ are constants. However, such an equation can represent a conic, the empty set, a single point, or, if the left-hand side factors into linear factors,

$$(A_1 x + B_1 y + C_1)(A_2 x + B_2 y + C_2) = 0,$$

one or two straight lines.

After straight lines the conic sections are the simplest of plane curves. They have many properties that make them useful in applications of mathematics; that is why we include a discussion of them here. Much of this material is optional from the point of view of a calculus course, but familiarity with the properties of conics can be very important in some applications. Most of the properties of conics were discovered by the Greek geometer, Apollonius of Perga, about 200 BC. It is remarkable that he was able to obtain these properties using only the techniques of classical Euclidean geometry; today most of these properties are expressed more conveniently using analytic geometry and specific coordinate systems.

## Parabolas

**Parabolas**

A **parabola** consists of points in the plane that are equidistant from a given point (the **focus** of the parabola) and a given straight line (the **directrix** of the parabola). The line through the focus perpendicular to the directrix is called the **principal axis** (or simply the **axis**) of the parabola. The **vertex** of the parabola is the point where the parabola crosses its principal axis. It is on the axis halfway between the focus and the directrix.

**Example 1**   Find an equation of the parabola with focus at the point $F = (a, 0)$ and directrix the line $L$ with equation $x = -a$.

**Solution**   The parabola has axis along the $x$-axis and vertex at the origin. (See Figure 8.3.) If $P = (x, y)$ is any point on the parabola, then the distance from $P$ to $F$ is equal to the distance from $P$ to the nearest point $Q$ on $L$. Thus

$$\sqrt{(x - a)^2 + y^2} = x + a$$
$$\text{or } x^2 - 2ax + a^2 + y^2 = x^2 + 2ax + a^2,$$

or, upon simplification, $y^2 = 4ax$.

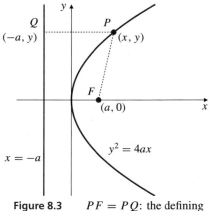

**Figure 8.3**   $PF = PQ$: the defining property of a parabola

Similarly, we can obtain standard equations for parabolas with vertices at the origin and foci at $(-a, 0)$, $(0, a)$ and $(0, -a)$:

**Table 1.**   Standard equations of parabolas

| Focus | Directrix | Equation |
|:-----:|:---------:|:--------:|
| $(a, 0)$ | $x = -a$ | $y^2 = 4ax$ |
| $(-a, 0)$ | $x = a$ | $y^2 = -4ax$ |
| $(0, a)$ | $y = -a$ | $x^2 = 4ay$ |
| $(0, -a)$ | $y = a$ | $x^2 = -4ay$ |

## The Focal Property of a Parabola

All of the conic sections have interesting and useful focal properties relating to the way in which surfaces of revolution they generate reflect light if the surfaces are mirrors. For instance, a circle will clearly reflect back along the same path any ray of light incident along a line passing through its centre. The focal properties of parabolas, ellipses, and hyperbolas can be derived from the reflecting property of a straight line (that is, a plane mirror) by elementary geometrical arguments.

Light travels in straight lines in a medium of constant optical density (one where the speed of light is constant). This is a consequence of the physical Principle of Least Action that asserts that in travelling between two points, light takes the path requiring the minimum travel time. Given a straight line $L$ in a plane and two points $A$ and $B$ in the plane on the same side of $L$, the point $P$ on $L$ for which the sum of the distances $AP + PB$ is minimum is such that $AP$ and $PB$ make equal angles with $L$, or equivalently, with the normal to $L$ at $P$. (See Figure 8.4.) If $B'$ is the point such that $L$ is the right bisector of the line segment $BB'$, then $P$ is the intersection of $L$ and $AB'$. Since one side of a triangle cannot exceed the sum of the other two sides,

$$AP + PB = AP + PB' = AB' \leq AQ + QB' = AQ + QB.$$

**Reflection by a straight line**

The point $P$ on $L$ at which a ray from $A$ reflects so as to pass through $B$ is the point that minimizes the sum of the distances $AP + PB$.

**Figure 8.4**   Reflection by a straight line

**Figure 8.5**   Reflection by a parabola

Now consider a parabola with focus $F$ and directrix $D$. Let $P$ be on the parabola and let $T$ be the line tangent to the parabola at $P$. (See Figure 8.5.) Let $Q$ be any point on $T$. Then $FQ$ meets the parabola at a point $X$ between $F$ and $Q$. Let $M$ and $N$ be points on $D$ such that $MX$ and $NP$ are perpendicular to $D$, and let $A$ be a point on the line through $N$ and $P$ that lies on the same side of the parabola as $F$. We have

$$FP + PA = NP + PA = NA \le MX + XA = FX + XA$$
$$\le FX + XQ + QA = FQ + QA.$$

Thus, among all points $Q$ on the line $T$, $Q = P$ is the one that minimizes the sum of distances $FQ + QA$. By the observation made for straight lines above, $FP$ and $PA$ make equal angles with $T$ and so also with the normal to the parabola at $P$. (The parabola and the tangent line have the same normal at $P$.)

> **DANGER**
>
> Consider the equalities and inequalities in this chain one at a time. Why is each one true?

> **Reflection by a parabola**
>
> Any ray from the focus will be reflected parallel to the axis of the parabola. Equivalently, any incident ray parallel to the axis of the parabola will be reflected through the focus.

## Ellipses

> **DEFINITION 2**
>
> **Ellipses**
> An **ellipse** consists of all points in the plane, the sum of whose distances from two fixed points (the **foci**) is constant.

**Example 2**   Find the ellipse with foci at the points $(-c, 0)$ and $(c, 0)$ if the sum of the distances from any point $P$ on the ellipse to these two foci is $2a$.

**Solution**   The ellipse passes through the four points $(a, 0)$, $(-a, 0)$, $(0, b)$ and $(0, -b)$, where $b^2 = a^2 - c^2$. (See Figure 8.6.) Also, if $P = (x, y)$ is on the ellipse, then

$$\sqrt{(x-c)^2 + y^2} + \sqrt{(x+c)^2 + y^2} = 2a.$$

Transposing one term from the left side to the right side and squaring, we get

$$(x-c)^2 + y^2 = 4a^2 - 4a\sqrt{(x+c)^2 + y^2} + (x+c)^2 + y^2$$

Now we expand the squares, cancel terms, transpose, and square again:

$$a\sqrt{(x+c)^2 + y^2} = a^2 + cx$$
$$a^2(x^2 + 2cx + c^2 + y^2) = a^4 + 2a^2cx + c^2x^2$$
$$(a^2 - c^2)x^2 + a^2y^2 = a^2(a^2 - c^2)$$

Finally, replace $a^2 - c^2$ with $b^2$ and divide by $a^2b^2$ to get the standard equation of the ellipse:

$$\frac{x^2}{a^2} + \frac{y^2}{b^2} = 1.$$

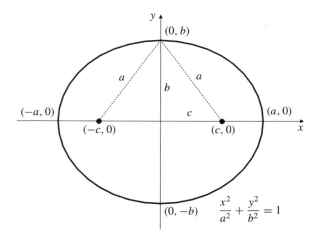

**Figure 8.6**    An ellipse

The following quantities describe this ellipse:

$a$    is the **semi-major axis**

$b$    is the **semi-minor axis**

$c = \sqrt{a^2 - b^2}$    is the **semi-focal separation**.

The point halfway between the foci is called the **centre** of the ellipse. In the example above it is the origin. Note that $a > b$ in this example. If $a < b$ then the ellipse has its foci at $(0, c)$ and $(0, -c)$, where $c = \sqrt{b^2 - a^2}$. The line containing the foci (the **major axis**), and the line through the centre perpendicular to that line (the **minor axis**) are called the **principal axes** of the ellipse.

The **eccentricity** of an ellipse is the ratio of the semi-focal separation to the semi-major axis. We denote the eccentricity $\varepsilon$. For the ellipse $\dfrac{x^2}{a^2} + \dfrac{y^2}{b^2} = 1$ with

$a > b$

$$\varepsilon = \frac{c}{a} = \frac{\sqrt{a^2 - b^2}}{a}.$$

Note that $\varepsilon < 1$ for any ellipse; the greater the value of $\varepsilon$ the more elongated (less circular) is the ellipse. If $\varepsilon = 0$ so that $a = b$ and $c = 0$, the two foci coincide and the ellipse is a circle.

## The Focal Property of an Ellipse

Let $P$ be any point on an ellipse having foci $F_1$ and $F_2$. The normal to the ellipse at $P$ bisects the angle between the lines $F_1 P$ and $F_2 P$.

> **Reflection by an ellipse**
>
> Any ray coming from one focus of an ellipse will be reflected through the other focus.

To see this, observe that if $Q$ is any point on the line $T$ tangent to the ellipse at $P$ then $F_1 Q$ meets the ellipse at a point $X$ between $F_1$ and $Q$ (see Figure 8.7) and so

$$F_1 P + P F_2 = F_1 X + X F_2 \leq F_1 X + X Q + Q F_2 = F_1 Q + Q F_2.$$

Among all points on $T$, $P$ is the one that minimizes the sum of the distances to $F_1$ and $F_2$. This implies that the normal to the ellipse at $P$ bisects the angle $F_1 P F_2$.

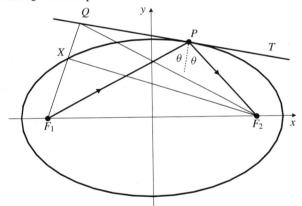

**Figure 8.7**    A ray from one focus of an ellipse is reflected to the other focus

## The Directrices of an Ellipse

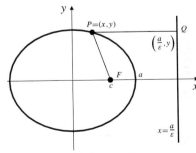

**Figure 8.8**    A focus and corresponding directrix of an ellipse

If $a > b > 0$, each of the lines $x = a/\varepsilon$ and $x = -a/\varepsilon$ is called a **directrix** of the ellipse $\dfrac{x^2}{a^2} + \dfrac{y^2}{b^2} = 1$. If $P$ is on the ellipse, then the ratio of the distance from $P$ to a focus to its distance from the corresponding directrix is equal to the eccentricity $\varepsilon$. If $P = (x, y)$, $F$ is the focus $(c, 0)$, $Q$ is on the corresponding directrix $x = a/\varepsilon$ and $PQ$ is perpendicular to the directrix, then (see Figure 8.8)

$$PF^2 = (x - c)^2 + y^2$$
$$= x^2 - 2cx + c^2 + b^2 \left( 1 - \frac{x^2}{a^2} \right)$$
$$= x^2 \left( \frac{a^2 - b^2}{a^2} \right) - 2cx + a^2 - b^2 + b^2$$
$$= \varepsilon^2 x^2 - 2\varepsilon a x + a^2 \qquad \text{(because } c = \varepsilon a\text{)}$$
$$= (a - \varepsilon x)^2.$$

Thus $PF = a - \varepsilon x$. Also, $QP = (a/\varepsilon) - x = (a - \varepsilon x)/\varepsilon$. Therefore $PF/QP = \varepsilon$, as asserted.

A parabola may be considered as the limiting case of an ellipse whose eccentricity has increased to 1. The distance between the foci is infinite, so the centre, one focus and its corresponding directrix have moved off to infinity leaving only one focus and its directrix in the finite plane.

## Hyperbolas

**DEFINITION 3**

**Hyperbolas**

A **hyperbola** consists of all points in the plane, the difference of whose distances from two fixed points (the **foci**) is constant.

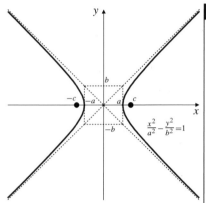

**Figure 8.9**

**Example 3** If the foci of a hyperbola are $F_1 = (c, 0)$ and $F_2 = (-c, 0)$, and the difference of the distances from a point $P = (x, y)$ on the hyperbola to these foci is $2a$ (where $a < c$), then

$$PF_2 - PF_1 = \sqrt{(x+c)^2 + y^2} - \sqrt{(x-c)^2 + y^2} = \begin{cases} 2a & \text{(right branch)} \\ -2a & \text{(left branch)}. \end{cases}$$

(See Figure 8.9.) Simplifying this equation by squaring and transposing as was done for the ellipse in Example 2, we obtain the standard equation for the hyperbola:

$$\frac{x^2}{a^2} - \frac{y^2}{b^2} = 1,$$

where $b^2 = c^2 - a^2$. ∎

The points $(a, 0)$ and $(-a, 0)$ (called the **vertices**) lie on the hyperbola, one on each branch. (The two branches correspond to the intersections of the plane of the hyperbola with the two nappes of a cone.) Some parameters used to describe the hyperbola are

$a$    the **semi-transverse axis**

$b$    the **semi-conjugate axis**

$c = \sqrt{a^2 + b^2}$    the **semi-focal separation**.

The midpoint of the line segment $F_1 F_2$ (in this case the origin) is called the **centre** of the hyperbola. The line through the centre, the vertices, and the foci is the **transverse axis**. The line through the centre perpendicular to the transverse axis is the **conjugate axis**. The conjugate axis does not intersect the hyperbola. If a rectangle with sides $2a$ and $2b$ is drawn centred at the centre of the hyperbola and with two sides tangent to the hyperbola at the vertices, the two diagonal lines of the rectangle are **asymptotes** of the hyperbola. They have equations $(x/a) \pm (y/b) = 0$; that is, they are solutions of the degenerate equation

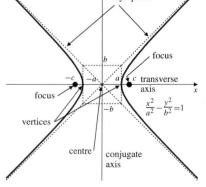

**Figure 8.10** Terms associated with a hyperbola

$$\frac{x^2}{a^2} - \frac{y^2}{b^2} = 0.$$

The hyperbola approaches arbitrarily close to these lines as it recedes from the origin. (See Figure 8.10.) A **rectangular** hyperbola is one whose asymptotes are perpendicular lines.

The eccentricity of the hyperbola is

$$\varepsilon = \frac{c}{a} = \frac{\sqrt{a^2 + b^2}}{a}.$$

**Figure 8.11** The directrices of a hyperbola

Note that $\varepsilon > 1$. The lines $x = \pm(a/\varepsilon)$ are called the **directrices** of the hyperbola $(x^2/a^2) - (y^2/b^2) = 1$. (See Figure 8.11.) In a manner similar to that used for the ellipse, you can show that if $P$ is on the hyperbola, then

$$\frac{\text{distance from } P \text{ to a focus}}{\text{distance from } P \text{ to the corresponding directrix}} = \varepsilon.$$

The eccentricity of a rectangular hyperbola is $\sqrt{2}$.

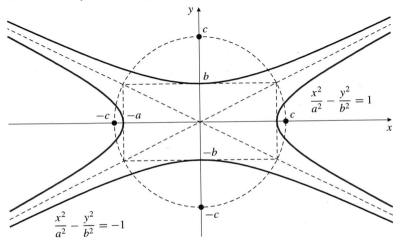

**Figure 8.12** Two conjugate hyperbolas and their common asymptotes

A hyperbola with the same asymptotes as $x^2/a^2 - y^2/b^2 = 1$, but with transverse axis along the $y$-axis, vertices at $(0, b)$ and $(0, -b)$ and foci at $(0, c)$ and $(0, -c)$ is represented by the equation

$$\frac{x^2}{a^2} - \frac{y^2}{b^2} = -1.$$

The two hyperbolas are said to be **conjugate** to one another. (See Figure 8.12.) The **conjugate axis** of a hyperbola is the transverse axis of the conjugate hyperbola. Together, the transverse and conjugate axes of a hyperbola are called its **principal axes**.

## The Focal Property of a Hyperbola

Let $P$ be any point on a hyperbola with foci $F_1$ and $F_2$. Then the tangent line to the hyperbola at $P$ bisects the angle between the lines $F_1 P$ and $F_2 P$.

> **Reflection by a hyperbola**
>
> A ray from one focus of a hyperbola is reflected by the hyperbola so that it appears to have come from the other focus.

To see this, let $P$ be on the right branch, let $T$ be the line tangent to the hyperbola at $P$, and let $C$ be a circle of large radius centred at $F_2$. (See Figure 8.13.) Let $F_2P$ intersect this circle at $D$. Let $Q$ be any point on $T$. Then $QF_1$ meets the hyperbola at $X$ between $Q$ and $F_1$, and $F_2X$ meets $C$ at $E$. Since $X$ is on the radial line $F_2E$, it is closer to $E$ than it is to other points on $C$. That is, $XE \leq XD$. Thus

> **■ DANGER**
>
> Check the equalities and inequalities in this chain one at a time to make sure you understand why it is true.

$$
\begin{aligned}
F_1P + PD &= F_1P + F_2D - F_2P \\
&= F_2D - (F_2P - F_1P) \\
&= F_2E - (F_2X - F_1X) \\
&= F_1X + F_2E - F_2X \\
&= F_1X + XE \\
&\leq F_1X + XD \\
&\leq F_1X + XQ + QD = F_1Q + QD.
\end{aligned}
$$

$P$ is the point on $T$ which minimizes the sum of distances to $F_1$ and $D$, and therefore the normal to the hyperbola at $P$ bisects the angle $F_1PD$. Therefore $T$ bisects the angle $F_1PF_2$.

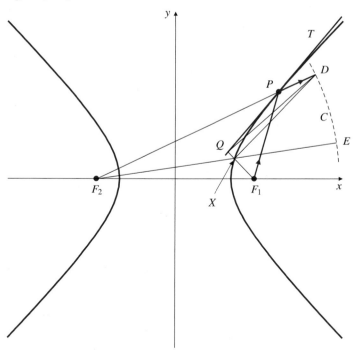

**Figure 8.13**   A ray from one focus is reflected along a line from the other focus

## Classifying General Conics

A second-degree equation in two variables,

$$
Ax^2 + Bxy + Cy^2 + Dx + Ey + F = 0, \qquad (A^2 + B^2 + C^2 > 0)
$$

generally represents a conic curve, but in certain degenerate cases it may represent two straight lines ($x^2 - y^2 = 0$ represents the lines $x = y$ and $x = -y$), one straight line ($x^2 = 0$ represents the line $x = 0$), a single point ($x^2 + y^2 = 0$ represents the origin), or no points at all ($x^2 + y^2 = -1$ is not satisfied by any point in the plane).

The nature of the set of points represented by a given second-degree equation can be determined by rewriting the equation in a form that can be recognized as one of the standard types. If $B = 0$, this rewriting can be accomplished by completing the squares in the $x$ and $y$ terms.

**Example 4**   Describe the curve with equation $x^2 + 2y^2 + 6x - 4y + 7 = 0$.

**Solution**   We complete the squares in the $x$ and $y$ terms, and rewrite the equation in the form

$$x^2 + 6x + 9 + 2(y^2 - 2y + 1) = 9 + 2 - 7 = 4$$

$$\frac{(x+3)^2}{4} + \frac{(y-1)^2}{2} = 1.$$

Therefore it represents an ellipse with centre at $(-3, 1)$, semi-major axis $a = 2$ and semi-minor axis $b = \sqrt{2}$. Since $c = \sqrt{a^2 - b^2} = \sqrt{2}$, the foci are $(-3 \pm \sqrt{2}, 1)$. See Figure 8.14.

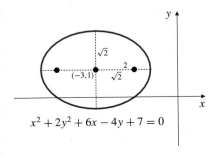

$x^2 + 2y^2 + 6x - 4y + 7 = 0$

**Figure 8.14**

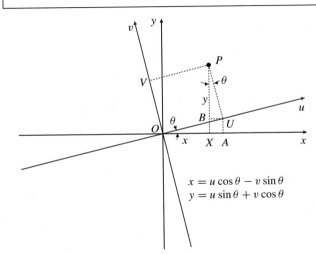

$$x = u \cos\theta - v \sin\theta$$
$$y = u \sin\theta + v \cos\theta$$

**Figure 8.15**   Rotation of axes

If $B \neq 0$, the equation has an $xy$ term, and it cannot represent a circle. To see what it does represent, we can rotate the coordinate axes to produce an equation with no $xy$ term. Let new coordinate axes (a $u$-axis and an $v$-axis) have the same origin but be rotated an angle $\theta$ from the $x$- and $y$-axes respectively. (See Figure 8.15.) If point $P$ has coordinates $(x, y)$ with respect to the old axes and coordinates $(u, v)$ with respect to the new axes, then an analysis of triangles in the figure shows that

$$x = OA - XA = OU \cos\theta - OV \sin\theta = u \cos\theta - v \sin\theta$$
$$y = XB + BP = OU \sin\theta + OV \cos\theta = u \sin\theta + v \cos\theta.$$

Substituting these expressions into the equation

$$Ax^2 + Bxy + Cy^2 + Dx + Ey + F = 0, \qquad (A^2 + B^2 + C^2 > 0)$$

**DANGER**

A lengthy calculation is needed here. The details have been omitted.

leads to a new equation

$$A'u^2 + B'uv + C'v^2 + D'u + E'v + F = 0,$$

where

$$A' = A \cos^2 \theta + B \cos \theta \sin \theta + C \sin^2 \theta$$
$$B' = (2C - 2A) \cos \theta \sin \theta + B(\cos^2 \theta - \sin^2 \theta)$$
$$C' = A \sin^2 \theta - B \cos \theta \sin \theta + C \cos^2 \theta$$
$$D' = D \cos \theta + E \sin \theta$$
$$E' = -D \sin \theta + E \cos \theta.$$

Note that $F$ remains unchanged. If we choose $\theta$ so that

$$\tan 2\theta = \frac{B}{A - C}, \qquad (\theta = \frac{\pi}{4} \text{ if } A = C, \ B \neq 0),$$

then $B' = 0$, and the new equation can then be analyzed as described previously.

**Example 5**    What curve is represented by the equation $xy = 1$?

**Solution**    The reader is likely well aware that the given equation represents a rectangular hyperbola with the coordinate axes as asymptotes. Since the given equation involves $A = C = D = E = 0$ and $B = 1$, it is appropriate to rotate the axes through angle $\pi/4$ so that

$$x = \frac{1}{\sqrt{2}}(u - v), \qquad y = \frac{1}{\sqrt{2}}(u + v).$$

The transformed equation is

$$\frac{u^2}{2} - \frac{v^2}{2} = 1$$

which is, as suspected, a rectangular hyperbola with vertices at $u = \pm\sqrt{2}$, $v = 0$, foci at $u = \pm 2$, $v = 0$, and asymptotes $u = \pm v$. Hence $xy = 1$ represents a rectangular hyperbola with coordinate axes as asymptotes, vertices at $(1, 1)$ and $(-1, -1)$, and foci at $(\sqrt{2}, \sqrt{2})$ and $(-\sqrt{2}, -\sqrt{2})$.    ∎

**Example 6**    Show that $x^2 + 2\sqrt{3}xy - y^2 + 2\sqrt{3}x + 2y = 6$ represents a hyperbola. What are its principal axes? its vertices? its foci? its asymptotes?

**Solution**    Here $A = 1$, $B = 2\sqrt{3}$, $C = -1$, $D = 2\sqrt{3}$, $E = 2$ and $F = -6$. We rotate the axes through angle $\theta$ satisfying $\tan 2\theta = B/(A - C) = \sqrt{3}$, that is, $\theta = \pi/6$, and obtain $A' = 2$, $B' = 0$, $C' = -2$, $D' = 4$, and $E' = 0$. The transformed equation is

$$2u^2 - 2v^2 + 4u = 6,$$

or, with completed square,

$$(u + 1)^2 - v^2 = 4.$$

This is a rectangular hyperbola with centre at $u = -1$, $v = 0$, vertices at $u = -1 \pm 2$, $v = 0$, foci at $u = -1 \pm 2\sqrt{2}$, $v = 0$, and asymptotes $u + 1 \pm v = 0$. Since the rotation of axes is specified by

$$x = \left(\cos\frac{\pi}{6}\right)u - \left(\sin\frac{\pi}{6}\right)v = \frac{\sqrt{3}}{2}u - \frac{1}{2}v$$

$$y = \left(\sin\frac{\pi}{6}\right)u + \left(\cos\frac{\pi}{6}\right)v = \frac{u}{2} + \frac{\sqrt{3}}{2}v,$$

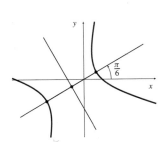

**Figure 8.16**

the given equation represents a rectangular hyperbola with centre $\left(-\frac{\sqrt{3}}{2}, -\frac{1}{2}\right)$ and transverse and conjugate axes through that point inclined at angle $\frac{\pi}{6}$ to the $x$- and $y$-axes respectively. (See Figure 8.16.)  Its vertices are at the points $\left(\frac{\sqrt{3}}{2}, \frac{1}{2}\right)$ and $\left(-\frac{3\sqrt{3}}{2}, -\frac{3}{2}\right)$, its foci at the points $\left(-\frac{\sqrt{3}(1+2\sqrt{2})}{2}, -\frac{1+2\sqrt{2}}{2}\right)$ and $\left(\frac{\sqrt{3}(-1+2\sqrt{2})}{2}, \frac{-1+2\sqrt{2}}{2}\right)$, and its asymptotes are $\sqrt{3}x + y + 2 \pm (\sqrt{3}y - x) = 0$. ∎

## Exercises 8.1

Find equations of the conics specified in Exercises 1–6.

**1.** ellipse with foci at $(0, \pm2)$ and semi-major axis 3.

**2.** ellipse with foci at $(0, 1)$ and $(4, 1)$ and eccentricity $1/2$.

**3.** parabola with focus at $(2, 3)$ and vertex at $(2, 4)$.

**4.** parabola passing through the origin and having focus at $(0, -1)$ and axis along $y = -1$.

**5.** hyperbola with foci at $(0, \pm2)$ and semi-transverse axis 1.

**6.** hyperbola with foci at $(\pm5, 1)$ and asymptotes $x = \pm(y - 1)$.

In Exercises 7–15, identify and sketch the set of points in the plane satisfying the given equation. Specify the asymptotes of any hyperbolas.

**7.** $x^2 + y^2 + 2x = -1$   **8.** $x^2 + 4y^2 - 4y = 0$

**9.** $4x^2 + y^2 - 4y = 0$   **10.** $4x^2 - y^2 - 4y = 0$

**11.** $x^2 + 2x - y = 3$   **12.** $x + 2y + 2y^2 = 1$

**13.** $x^2 - 2y^2 + 3x + 4y = 2$

**14.** $9x^2 + 4y^2 - 18x + 8y = -13$

**15.** $9x^2 + 4y^2 - 18x + 8y = 23$

**16.** Identify and sketch the curve which is the graph of the equation $(x - y)^2 - (x + y)^2 = 1$.

**\* 17.**  Light rays in the $xy$-plane coming from the point $(3, 4)$ reflect in a parabola so that they form a beam parallel to the $x$-axis. The parabola passes through the origin. Find its equation. (There are two possible answers.)

**18.** Light rays in the $xy$-plane coming from the origin are reflected by an ellipse so that they converge at the point $(3, 0)$. Find all possible equations for the ellipse.

In Exercises 19–24, identify the conic and find its centre, principal axes, foci, and eccentricity. Specify the asymptotes of any hyperbolas.

**19.** $xy + x - y = 2$

**\* 20.** $x^2 + 4xy + y^2 - 2\sqrt{2}(x - y) = 2$

**\* 21.** $x^2 + xy = \dfrac{1}{2\sqrt{2}}$

**\* 22.** $x^2 + 2xy + y^2 = 4x - 4y + 4$

**\* 23.** $8x^2 + 12xy + 17y^2 = 20$

**\* 24.** $x^2 - 4xy + 4y^2 + 2x + y = 0$

**25.** The *focus-directrix definition of a conic* defines a conic as a set of points $P$ in the plane which satisfy the condition

$$\frac{\text{distance from } P \text{ to } F}{\text{distance from } P \text{ to } D} = \varepsilon$$

where $F$ is a fixed point, $D$ a fixed straight line, and $\varepsilon$ a fixed positive number. The conic is an ellipse, a parabola, or a hyperbola according to whether $\varepsilon < 1$, $\varepsilon = 1$, or $\varepsilon > 1$. Find the equation of the conic if $F$ is the origin and $D$ is the line $x = -p$.

Another parameter associated with conics is the **semi-latus rectum**, usually denoted $\ell$. For a circle it is equal to the radius. For other conics it is half the length of the chord through a focus and perpendicular to the axis (for a parabola), the major axis (for an ellipse), or the transverse axis (for a hyperbola). That chord is called the **latus rectum** of the conic.

**26.** Show that the semi-latus rectum of the parabola is twice the distance from the vertex to the focus.

**27.** Show that the semi-latus rectum for an ellipse with semi-major axis $a$ and semi-minor axis $b$ is $\ell = b^2/a$.

**28.** Show that the formula in the above exercise also gives the semi-latus rectum of a hyperbola with semi-transverse axis $a$ and semi-conjugate axis $b$.

* **29.** Suppose a plane intersects a right-circular cone in an ellipse, and that two spheres (one on each side of the plane) are inscribed between the cone and the plane so that each is tangent to the cone around a circle and is also tangent to the plane at a point. Show that the points where these two spheres touch the plane are the foci of the ellipse. *Hint:* all tangent lines drawn to a sphere from a given point outside the sphere are equal in length. The distance between the two circles in which the spheres intersect the cone, measured

along generators of the cone (i.e. straight lines lying on the cone), is the same for all generators.

* **30.** State and prove a result analogous to that in the above exercise but pertaining to a hyperbola.

* **31.** Suppose a plane intersects a right-circular cone in a parabola with vertex at $V$. Suppose that a sphere is inscribed between the cone and the plane as in the previous exercises, and is tangent to the plane of the parabola at point $F$. Show that the chord to the parabola through $F$ which is perpendicular to $FV$ has length equal to that of the latus rectum of the parabola. Therefore $F$ is the focus of the parabola.

# 8.2 Parametric Curves

Suppose that an object moves around in the $xy$-plane so that the coordinates of its position at any time $t$ are continuous functions of the variable $t$:

$$x = f(t), \qquad y = g(t).$$

The path followed by the moving object is a curve $C$ in the plane which is specified by the two equations above. We call these equations, parametric equations of $C$. A curve specified by a particular pair of parametric equations is called a parametric curve.

**DEFINITION 4**

> **Parametric curves**
>
> A **parametric curve** $C$ in the plane consists of an ordered pair $(f, g)$ of continuous functions each defined on the same interval $I$. The equations
>
> $$x = f(t), \qquad y = g(t), \qquad \text{for } t \text{ in } I$$
>
> are called **the parametric equations** of the curve $C$. The independent variable $t$ is called the **parameter**.

Note that the parametric curve $C$ was *not* defined as a set of points in the plane, but rather as the ordered pair of functions whose range is that set of points. Different pairs of functions can give the same set of points in the plane, but we may still want to regard them as different parametric curves. Nevertheless, we will often refer to the set of points (the path traced out by $(x, y)$ as $t$ traverses $I$) as the curve $C$. The axis (real line) of the parameter $t$ is distinct from the coordinate axes of the plane of the curve. (See Figure 8.17.) We will usually denote the parameter by $t$; in many applications the parameter represents time, but this need not always be the case. Because $f$ and $g$ are assumed to be continuous, the curve $x = f(t)$, $y = g(t)$ has no breaks in it. A parametric curve has a *direction* (indicated, say, by arrowheads), namely the direction corresponding to increasing values of the parameter $t$, as shown in Figure 8.17.

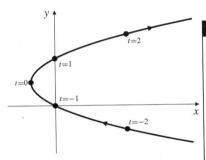

**Figure 8.17** A parametric curve

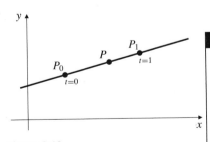

**Figure 8.18** The parabola defined parametrically by $x = t^2 - 1$, $y = t + 1$, $(-\infty < t < \infty)$

**Example 1** Sketch and identify the parametric curve

$$x = t^2 - 1, \qquad y = t + 1 \qquad (-\infty < t < \infty).$$

**Solution** We could construct a table of values of $x$ and $y$ for various values of $t$, thus getting the coordinates of a number of points on a curve. However, for this example it is easier to *eliminate the parameter* from the pair of parametric equations, thus producing a single equation in $x$ and $y$ whose graph is the desired curve:

$$t = y - 1, \qquad x = t^2 - 1 = (y - 1)^2 - 1 = y^2 - 2y.$$

All points on the curve lie on the parabola $x = y^2 - 2y$. Since $y \to \pm\infty$ as $t \to \pm\infty$, the parametric curve is the whole parabola. (See Figure 8.18.) ∎

Although the curve in Example 1 is more easily identified when the parameter is eliminated, there is a loss of information in going to the nonparametric form. Specifically, we lose the sense of the curve as the path of a moving point and hence also the direction of the curve. If the $t$ in the parametric form denotes the time at which an object is at the point $(x, y)$, the nonparametric equation $x = y^2 - 2y$ no longer tells us where the object is at any particular time $t$.

**Figure 8.19**

**Example 2** (**Parametric equations of a straight line**) The straight line passing through the two points $P_0 = (x_0, y_0)$ and $P_1 = (x_1, y_1)$ (see Figure 8.19) has parametric equations

$$\begin{cases} x = x_0 + t(x_1 - x_0) \\ y = y_0 + t(y_1 - y_0) \end{cases} \qquad (-\infty < t < \infty).$$

To see that these equations represent a straight line, note that

$$\frac{y - y_0}{x - x_0} = \frac{y_1 - y_0}{x_1 - x_0} = \text{constant} \qquad (\text{assuming } x_1 \neq x_0).$$

The point $P = (x, y)$ is at position $P_0$ when $t = 0$ and at $P_1$ when $t = 1$. If $t = 1/2$, then $P$ is the midpoint between $P_0$ and $P_1$. Note that the line segment from $P_0$ to $P_1$ corresponds to values of $t$ between 0 and 1. ∎

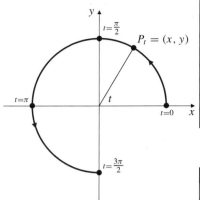

**Figure 8.20**

**Example 3**    (**An arc of a circle**)  Sketch and identify the curve $x = 3\cos t$, $y = 3\sin t$, $(0 \leq t \leq 3\pi/2)$.

**Solution**    Since $x^2 + y^2 = 9\cos^2 t + 9\sin^2 t = 9$, all points on the curve lie on the circle $x^2 + y^2 = 9$. As $t$ increases from 0 through $\pi/2$ and $\pi$ to $3\pi/2$, the point $(x, y)$ moves from $(3, 0)$ through $(0, 3)$ and $(-3, 0)$ to $(0, -3)$. The parametric curve is three-quarters of the circle. See Figure 8.20. The parameter $t$ has geometric significance in this example. If $P_t$ is the point on the curve corresponding to parameter value $t$, then $t$ is the angle at the centre of the circle corresponding to the arc from the initial point to $P_t$.

■

**Example 4**    (**Parametric equations of an ellipse**)  Sketch and identify the curve $x = a\cos t$, $y = b\sin t$, $(0 \leq t \leq 2\pi)$, where $a > b > 0$.

**Solution**    Observe that

$$\frac{x^2}{a^2} + \frac{y^2}{b^2} = \cos^2 t + \sin^2 t = 1.$$

Therefore the curve is all or part of an ellipse with major axis from $(-a, 0)$ to $(a, 0)$ and minor axis from $(0, -b)$ to $(0, b)$. As $t$ increases from 0 to $2\pi$, the point $(x, y)$ moves counterclockwise around the ellipse starting from $(a, 0)$ and returning to the same point. Thus the curve is the whole ellipse.

Figure 8.21(a) shows how the parameter $t$ can be interpreted as an angle and how the points on the ellipse can be obtained using circles of radii $a$ and $b$. Since the curve starts and ends at the same point, it is called a **closed curve**.

■

**Figure 8.21**

(a)  An ellipse parametrized in terms of an angle, and constructed with the help of two circles

(b)  A self-intersecting parametric curve

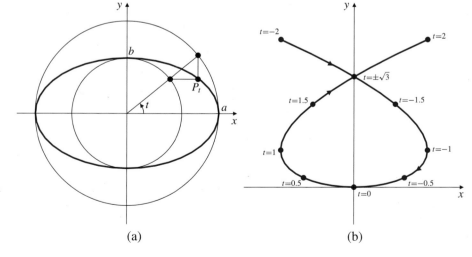

**Example 5**    Sketch the parametric curve

$$x = t^3 - 3t, \qquad y = t^2 \qquad (-2 \leq t \leq 2).$$

*Solution*  We could eliminate the parameter and obtain

$$x^2 = t^2(t^2 - 3)^2 = y(y - 3)^2,$$

but this doesn't help much since we do not recognize this curve from its Cartesian equation. Instead let us calculate the coordinates of some points:

**Table 2.**  Coordinates of some points on the curve of Example 5

| $t$ | $-2$ | $-\dfrac{3}{2}$ | $-1$ | $-\dfrac{1}{2}$ | $0$ | $\dfrac{1}{2}$ | $1$ | $\dfrac{3}{2}$ | $2$ |
|---|---|---|---|---|---|---|---|---|---|
| $x$ | $-2$ | $\dfrac{9}{8}$ | $2$ | $\dfrac{11}{8}$ | $0$ | $-\dfrac{11}{8}$ | $-2$ | $-\dfrac{9}{8}$ | $2$ |
| $y$ | $4$ | $\dfrac{9}{4}$ | $1$ | $\dfrac{1}{4}$ | $0$ | $\dfrac{1}{4}$ | $1$ | $\dfrac{9}{4}$ | $4$ |

Note that the curve is symmetric about the $y$-axis because $x$ is an odd function of $t$ and $y$ is an even function of $t$. (At $t$ and $-t$, $x$ has opposite values but $y$ has the same value.)

The curve intersects itself on the $y$-axis. (See Figure 8.21(b).) To find this self-intersection set $x = 0$:

$$0 = x = t^3 - 3t = t(t - \sqrt{3})(t + \sqrt{3}).$$

For $t = 0$ the curve is at $(0, 0)$, but for $t = \pm\sqrt{3}$ the curve is at $(0, 3)$. The self-intersection occurs because the curve passes through the same point for two different values of the parameter. ∎

## General Plane Curves and Parametrizations

According to Definition 4, a parametric curve always involves a particular set of parametric equations; it is not just a set of points in the plane. When we are interested in considering a curve solely as a set of points (a *geometric object*), we need not be concerned with any particular pair of parametric equations representing that curve. In this case we call the curve simply a *plane curve*.

**DEFINITION 5**

**Plane curves**

A **plane curve** is a set of points $(x, y)$ in the plane such that $x = f(t)$ and $y = g(t)$ for some $t$ in an interval $I$, where $f$ and $g$ are continuous functions defined on $I$. Any such interval $I$ and function pair $(f, g)$ that generate the points of $C$ is called a **parametrization** of $C$.

Since a plane curve does not involve any specific parametrization, it has no specific direction.

**Example 6**    The circle $x^2 + y^2 = 1$ is a plane curve. Each of the following is a possible parametrization of $C$:

(i)  $x = \cos t,\ y = \sin t,\qquad (0 \le t \le 2\pi)$

(ii)  $x = \sin s^2,\ y = \cos s^2,\qquad (0 \le s \le \sqrt{2\pi})$

(iii)  $x = \cos(\pi u + 1),\ y = \sin(\pi u + 1),\qquad (-1 \le u \le 1)$

(iv)  $x = 1 - t^2,\ y = t\sqrt{2 - t^2},\qquad (-\sqrt{2} \le t \le \sqrt{2})$

To verify that any of these represents the circle, substitute the appropriate functions for $x$ and $y$ in the expression $x^2 + y^2$, and show that the result simplifies to the value 1. This shows that the parametric curve lies on the circle. Then examine the ranges of $x$ and $y$ as the parameter varies over its domain. For example, for (iv) we have

$$x^2 + y^2 = (1 - t^2)^2 + (t\sqrt{2 - t^2})^2 = 1 - 2t^2 + t^4 + 2t^2 - t^4 = 1,$$

and $(x, y)$ moves from $(-1, 0)$ through $(0, -1)$ to $(1, 0)$ as $t$ increases from $-\sqrt{2}$ through $-1$ to 0, and then continues on through $(0, 1)$ back to $(-1, 0)$ as $t$ continues to increase from 0 through 1 to $\sqrt{2}$.

There are, of course, infinitely many other possible parametrizations of this curve.    ∎

**Example 7**    If $f$ is a continuous function on an interval $I$, then the graph of $f$ is a plane curve. One obvious parametrization of this curve is

$$x = t,\qquad y = f(t),\qquad (t \text{ in } I).$$

∎

**The brachistochrone and tautochrone problems**
Suppose a wire is bent into a curve from point $A$ to a lower point $B$ and a bead can slide without friction along the wire. If the bead is released at $A$ it will fall towards $B$. What curve should be used to minimize the time it takes to fall from $A$ to $B$? This problem, known as the *brachistochrone* (Greek for "shortest time") problem, has as its solution part of an upside down arch of a cycloid. Moreover, it takes the same amount of time for the bead to slide from any point on the curve to the lowest point $B$, making the cycloid the solution of the *tautochrone* ("equal time") problem as well. We will examine these matters further in the Challenging Exercises at the end of Chapter 11.

## Some Interesting Plane Curves

We complete this section by parametrizing two curves that arise in the physical world.

**Example 8**    **(A cycloid)** If a circle rolls without slipping along a straight line, find the path followed by a point fixed on the circle. This path is called a **cycloid**.

**Solution**    Suppose that the line on which the circle rolls is the $x$-axis, that the circle has radius $a$ and lies above the line, and that the point whose motion we follow is originally at the origin $O$. See Figure 8.22. After the circle has rolled through an angle $t$ it is tangent to the line at $T$, and the point whose path we are trying to find has moved to position $P$ as shown in the figure. Since no slipping occurs,

segment $OT = \text{arc } PT = at.$

Let $PQ$ be perpendicular to $TC$, as shown in the figure. If $P$ has coordinates $(x, y)$, then

$$x = OT - PQ = at - a\sin(\pi - t) = at - a\sin t,$$
$$y = TC + CQ = a + a\cos(\pi - t) = a - a\cos t.$$

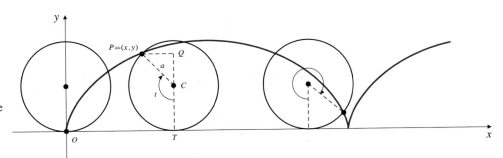

**Figure 8.22**     Each arch of the cycloid is traced out by $P$ as the wheel rolls through one complete revolution

The parametric equations of the cycloid are therefore

$$x = a(t - \sin t), \qquad y = a(1 - \cos t).$$

Observe that the cycloid has a cusp at the points where it returns to the $x$-axis, i.e., at points corresponding to $t = 2n\pi$, where $n$ is an integer. Even though the functions $x$ and $y$ are everywhere differentiable functions of $t$, the curve is not smooth everywhere. We shall consider such matters in the next section.  ∎

**Example 9**   (**An involute of a circle**)  A string is wound around a fixed circle. One end is unwound in such a way that the part of the string not lying on the circle is extended in a straight line. The curve followed by this free end of the string is called an **involute** of the circle. (The involute of any curve is the path traced out by the end of the curve as the curve is straightened out beginning at that end.)

Suppose the circle has equation $x^2 + y^2 = a^2$ and suppose the end of the string being unwound starts at the point $A = (a, 0)$. At some subsequent time during the unwinding let $P$ be the position of the end of the string and let $T$ be the point where the string leaves the circle. The line $PT$ must be tangent to the circle at $T$. We parametrize the path of $P$ in terms of the angle $AOT$, which we denote by $t$. Let points $R$ on $OA$ and $S$ on $TR$ be as shown in Figure 8.23. $TR$ is perpendicular to $OA$ and to $PS$. Note that

$$OR = OT \cos t = a \cos t, \qquad RT = OT \sin t = a \sin t.$$

Since angle $OTP$ is 90°, we have angle $ST = t$, and so, since $PT = \text{arc } AT = at$ because the string does not stretch or slip on the circle,

$$SP = TP \sin t = at \sin t, \qquad ST = TP \cos t = at \cos t.$$

If $P$ has coordinates $(x, y)$, then $x = OR + SP$, and $y = RT - ST$:

$$x = a \cos t + at \sin t, \qquad y = a \sin t - at \cos t, \qquad (t \geq 0).$$

These are parametric equations of the involute.  ∎

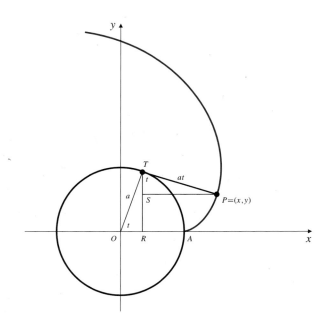

**Figure 8.23**    An involute of a circle

# Exercises 8.2

In Exercises 1–10, sketch the given parametric curve, showing its direction with an arrow. Eliminate the parameter to give a Cartesian equation in $x$ and $y$ whose graph contains the parametric curve.

1. $x = 1 + 2t$, $y = t^2$, $(-\infty < t < \infty)$

2. $x = 2 - t$, $y = t + 1$, $(0 \le t < \infty)$

3. $x = \dfrac{1}{t}$, $y = t - 1$, $(0 < t < 4)$

4. $x = \dfrac{1}{1+t^2}$, $y = \dfrac{t}{1+t^2}$, $(-\infty < t < \infty)$

5. $x = 3 \sin 2t$, $y = 3 \cos 2t$, $\left(0 \le t \le \dfrac{\pi}{3}\right)$

6. $x = a \sec t$, $y = b \tan t$, $\left(-\dfrac{\pi}{2} < t < \dfrac{\pi}{2}\right)$

7. $x = 3 \sin \pi t$, $y = 4 \cos \pi t$, $(-1 \le t \le 1)$

8. $x = \cos \sin s$, $y = \sin \sin s$, $(-\infty < s < \infty)$

9. $x = \cos^3 t$, $y = \sin^3 t$, $(0 \le t \le 2\pi)$

10. $x = 1 - \sqrt{4 - t^2}$, $y = 2 + t$, $(-2 \le t \le 2)$

11. Describe the parametric curve $x = \cosh t$, $y = \sinh t$, and find its Cartesian equation.

12. Describe the parametric curve $x = 2 - 3 \cosh t$, $y = -1 + 2 \sinh t$.

13. Describe the curve $x = t \cos t$, $y = t \sin t$, $(0 \le t \le 4\pi)$.

14. Show that each of the following sets of parametric equations represents a different arc of the parabola with equation $2(x + y) = 1 + (x - y)^2$.

    (i) $x = \cos^4 t$, $y = \sin^4 t$,

    (ii) $x = \sec^4 t$, $y = \tan^4 t$,

    (iii) $x = \tan^4 t$, $y = \sec^4 t$

15. Find a parametrization of the parabola $y = x^2$ using as parameter the slope of the tangent line at the general point.

16. Find a parametrization of the circle $x^2 + y^2 = R^2$ using as parameter the slope $m$ of the line joining the general point to the point $(R, 0)$. Does the parametrization fail to give any point on the circle?

17. A circle of radius $a$ is centred at the origin $O$. $T$ is a point on the circle such that $OT$ makes angle $t$ with the positive $x$-axis. The tangent to the circle at $T$ meets the $x$-axis at $X$. The point $P = (x, y)$ is at the intersection of the vertical line through $X$ and the horizontal line through $T$. Find, in terms of the parameter $t$, parametric equations for the curve $C$ traced out by $P$ as $T$ moves around the circle. Also, eliminate $t$ and find an equation for $C$ in $x$ and $y$. Sketch $C$.

18. Repeat Exercise 17 with the following modification. $OT$ meets a second circle of radius $b$ centred at $O$ at the point $Y$. $P = (x, y)$ is at the intersection of the vertical line through $X$ and the horizontal line through $Y$.

* 19. (**The folium of Descartes**) Eliminate the parameter from the parametric equations

$$x = \frac{3t}{1+t^3}, \qquad y = \frac{3t^2}{1+t^3} \qquad (t \ne -1),$$

and hence find an ordinary equation in $x$ and $y$ for this curve. The parameter $t$ can be interpreted as the slope of the line joining the general point $(x, y)$ to the origin. Sketch the curve and show that the line $x + y = -1$ is an asymptote.

* **20. (A prolate cycloid)** A railroad wheel has a flange extending below the level of the track on which the wheel rolls. If the radius of the wheel is $a$, and that of the flange is $b > a$, find parametric equations of the path of a point $P$ at the circumference of the flange as the wheel rolls along the track. (Note that for a portion of each revolution of the wheel, $P$ is moving backward.) Try to sketch the graph of this prolate cycloid.

* **21. (Hypocycloids)** If a circle of radius $b$ rolls, without slipping, around the inside of a fixed circle of radius $a > b$, a point on the circumference of the rolling circle traces a curve called a hypocycloid. If the fixed circle is centred at the origin and the point tracing the curve starts at $(a, 0)$, show that the hypocycloid has parametric equations

$$x = (a - b)\cos t + b\cos\left(\frac{a - b}{b}t\right),$$

$$y = (a - b)\sin t - b\sin\left(\frac{a - b}{b}t\right),$$

where $t$ is the angle between the positive $x$-axis and the line from the origin to the point at which the rolling circle touches the fixed circle.

If $a = 2$ and $b = 1$, show that the hypocycloid becomes a straight line segment.

If $a = 4$ and $b = 1$, show that the parametric equations of the hypocycloid simplify to $x = 4\cos^3 t$, $y = 4\sin^3 t$. This curve is called a hypocycloid of four cusps or an **astroid**. (See Figure 8.24.) It has Cartesian equation $x^{2/3} + y^{2/3} = 4^{2/3}$.

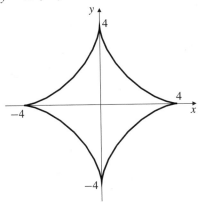

**Figure 8.24**   The astroid   $x^{2/3} + y^{2/3} = 4^{2/3}$

Hypocycloids resemble the curves produced by a popular children's toy called Spirograph, but Spirograph curves result from following a point inside the disc of the rolling circle rather than on its circumference, and they therefore do not have sharp cusps.

* **22. (The witch of Agnesi)**

(a) Show that the curve traced out by the point $P$ constructed from a circle as shown in Figure 8.25 has parametric equations $x = \tan t$,   $y = \cos^2 t$ in terms of the angle $t$ shown. (*Hint:* you will need to make extensive use of similar triangles.)

(b) Use a trigonometric identity to eliminate $t$ from the parametric equations and hence find an ordinary Cartesian equation for the curve.

This curve is named for the Italian mathematician Maria Agnesi (1718-1799), one of the foremost women scholars of her century, and author of an important calculus text. The term *witch* is due to a mistranslation of the Italian word *versiera* ("turning curve") that she used to describe the curve. The word is similar to *avversiera* ("wife of the devil" or "witch").

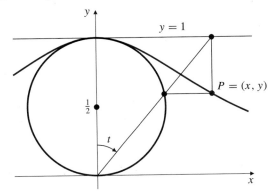

**Figure 8.25**   The witch of Agnesi

In Exercises 23–26, obtain a graph of the curve $x = \sin(mt)$, $y = \sin(nt)$ for the given values of $m$ and $n$. Such curves are called **Lissajous figures.** They arise in the analysis of electrical signals using an oscilloscope. A signal of fixed but unknown frequency is applied to the vertical input, and a control signal is applied to the horizontal input. The horizontal frequency is varied until a stable Lissajous figure is observed. The (known) frequency of the control signal and the shape of the figure then determine the unknown frequency.

**23.** $m = 1$,   $n = 2$         **24.** $m = 1$,   $n = 3$

**25.** $m = 2$,   $n = 3$         **26.** $m = 2$,   $n = 5$

**27. (Epicycloids)** Use a graphing calculator or computer graphing program to investigate the behaviour of curves with equations of the form

$$x = \left(1 + \frac{1}{n}\right)\cos t - \frac{1}{n}\cos(nt)$$

$$y = \left(1 + \frac{1}{n}\right)\sin t - \frac{1}{n}\sin(nt)$$

for various integer and fractional values of $n \geq 3$. Can you formulate any principles governing the behaviour of such curves?

**28. (More hypocycloids)** Use a graphing calculator or computer graphing program to investigate the behaviour of curves with equations of the form

$$x = \left(1 + \frac{1}{n}\right)\cos t + \frac{1}{n}\cos((n-1)t)$$

$$y = \left(1 + \frac{1}{n}\right)\sin t - \frac{1}{n}\sin((n-1)t)$$

for various integer and fractional values of $n \geq 3$. Can you formulate any principles governing the behaviour of these curves?

# 8.3 Smooth Parametric Curves and Their Slopes

We say that a plane curve is *smooth* if it has a tangent line at each point $P$ and this tangent turns in a continuous way as $P$ moves along the curve. (That is, the angle between the tangent line at $P$ and some fixed line, the $x$-axis say, is a continuous function of the position of $P$.)

If the curve $C$ is the graph of function $f$, then $C$ is certainly smooth on any interval where the derivative $f'(x)$ exists and is a continuous function of $x$. It may also be smooth on intervals containing isolated singular points; for example, the curve $y = x^{1/3}$ is smooth everywhere even though $dy/dx$ does not exist at $x = 0$.

For parametric curves $x = f(t)$, $y = g(t)$, the situation is more complicated. Even if $f$ and $g$ have continuous derivatives everywhere, such curves may fail to be smooth at certain points, specifically points where $f'(t) = g'(t) = 0$.

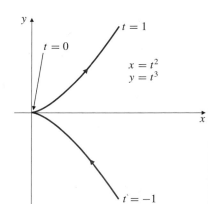

**Figure 8.26**    This curve is not smooth at the origin, but has a cusp there

**Example 1**    Consider the parametric curve $x = f(t) = t^2$, $y = g(t) = t^3$. Even though $f'(t) = 2t$ and $g'(t) = 3t^2$ are continuous for all $t$, the curve is not smooth at $t = 0$. (See Figure 8.26.) Observe that both $f'$ and $g'$ vanish at $t = 0$: $f'(0) = g'(0) = 0$. If we regard the parametric equations as specifying the position at time $t$ of a moving point $P$, then the horizontal velocity is $f'(t)$ and the vertical velocity is $g'(t)$. Both velocities are 0 at $t = 0$, so $P$ has come to a stop at that instant. When it starts moving again it need not move in the direction it was going before it stopped. The cycloid of Example 8 of Section 8.2 is another example where a parametric curve is not smooth at points where $dx/dt$ and $dy/dt$ both vanish.

## The Slope of a Parametric Curve

The following theorem confirms that a parametric curve is smooth at points where the derivatives of its coordinate functions are continuous and not both zero.

**THEOREM    1**

Let $C$ be the parametric curve $x = f(t)$, $y = g(t)$, where $f'(t)$ and $g'(t)$ are continuous on an interval $I$. If $f'(t) \neq 0$ on $I$, then $C$ is smooth and has at each $t$ a tangent line with slope

$$\frac{dy}{dx} = \frac{g'(t)}{f'(t)}.$$

If $g'(t) \neq 0$ on $I$, then $C$ is smooth and has at each $t$ a normal line with slope

$$-\frac{dx}{dy} = -\frac{f'(t)}{g'(t)}.$$

Thus $C$ is smooth except possibly at points where $f'(t)$ and $g'(t)$ are both 0.

**PROOF** If $f'(t) \neq 0$ on $I$, then $f$ is either increasing or decreasing on $I$ and so is one-to-one and invertible. The part of $C$ corresponding to values of $t$ in $I$ has ordinary equation $y = g(f^{-1}(x))$ and hence slope

$$\frac{dy}{dx} = g'(f^{-1}(x)) \frac{d}{dx} f^{-1}(x) = \frac{g'(f^{-1}(x))}{f'(f^{-1}(x))} = \frac{g'(t)}{f'(t)}.$$

We have used here the formula

$$\frac{d}{dx} f^{-1}(x) = \frac{1}{f'(f^{-1}(x))}$$

for the derivative of an inverse function obtained in Section 3.1. This slope is a continuous function of $t$, so the tangent to $C$ turns continuously for $t$ in $I$. The proof for $g'(t) \neq 0$ is similar. In this case the slope of the normal is a continuous function of $t$, so the normal turns continuously. Therefore, so does the tangent.

If $f'$ and $g'$ are continuous, and both vanish at some point $t_0$, then the curve $x = f(t)$, $y = g(t)$ *may or may not* be smooth around $t_0$. Example 1 was an example of a curve that was not smooth at such a point.

**Example 2** The curve with parametrization $x = t^3$, $y = t^6$ is just the parabola $y = x^2$, so it is smooth everywhere, although $dx/dt = 3t^2$ and $dy/dt = 6t^5$ both vanish at $t = 0$. ∎

**Tangents and normals to parametric curves**

If $f'$ and $g'$ are continuous and not both 0 at $t_0$, then the parametric equations

$$\begin{cases} x = f(t_0) + f'(t_0)(t - t_0) \\ y = g(t_0) + g'(t_0)(t - t_0) \end{cases} \qquad (-\infty < t < \infty),$$

represent the tangent line to the parametric curve $x = f(t)$, $y = g(t)$ at the point $(f(t_0), g(t_0))$. The normal line there has parametric equations

$$\begin{cases} x = f(t_0) + g'(t_0)(t - t_0) \\ y = g(t_0) - f'(t_0)(t - t_0) \end{cases} \qquad (-\infty < t < \infty).$$

Both lines pass through $(f(t_0), g(t_0))$ when $t = t_0$.

**Example 3** Find equations of the tangent and normal lines to the parametric curve $x = t^2 - t$, $y = t^2 + t$ at the point where $t = 2$.

**Solution** At $t = 2$ we have $x = 2$, $y = 6$ and

$$\frac{dx}{dt} = 2t - 1 = 3, \qquad \frac{dy}{dx} = 2t + 1 = 5.$$

Hence the tangent and the normal lines have parametric equations

$$\text{Tangent: } \begin{cases} x = 2 + 3(t-2) = 3t - 4 \\ y = 6 + 5(t-2) = 5t - 4 \end{cases}$$

$$\text{Normal: } \begin{cases} x = 2 + 5(t-2) = 5t - 8 \\ y = 6 - 3(t-2) = -3t + 12 \end{cases}$$

The concavity of a parametric curve can be determined using the second derivatives of the parametric equations. The procedure is just to calculate $d^2y/dx^2$ using the chain rule:

$$\frac{d^2y}{dx^2} = \frac{d}{dx}\frac{dy}{dx} = \frac{d}{dx}\frac{g'(t)}{f'(t)} = \frac{d}{dt}\left(\frac{g'(t)}{f'(t)}\right)\frac{dt}{dx}$$

$$= \frac{f'(t)g''(t) - g'(t)f''(t)}{(f'(t))^2}\frac{1}{f'(t)}.$$

**Concavity of a parametric curve**

On an interval where $f'(t) \neq 0$, the parametric curve $x = f(t)$, $y = g(t)$ has concavity determined by

$$\frac{d^2y}{dx^2} = \frac{f'(t)g''(t) - g'(t)f''(t)}{(f'(t))^3}.$$

## Sketching Parametric Curves

As in the case of graphs of functions, derivatives provide useful information about the shape of a parametric curve. At points where $dy/dt = 0$ but $dx/dt \neq 0$, the tangent is horizontal; at points where $dx/dt = 0$ but $dy/dt \neq 0$, the tangent is vertical. For points where $dx/dt = dy/dt = 0$, anything can happen; it is wise to calculate left- and right-hand limits of the slope $dy/dx$ as the parameter $t$ approaches one of these points. Concavity can be determined using the formula obtained above. We illustrate these ideas by reconsidering a parametric curve encountered in the previous section.

**Example 4** Use slope and concavity information to sketch the graph of the parametric curve

$$x = f(t) = t^3 - 3t, \qquad y = g(t) = t^2, \qquad (-2 \leq t \leq 2)$$

previously encountered in Example 5 of Section 8.2.

**Solution** We have

$$f'(t) = 3(t^2 - 1) = 3(t-1)(t+1), \qquad g'(t) = 2t.$$

The curve has a horizontal tangent at $t = 0$, that is, at $(0, 0)$, and vertical tangents at $t = \pm 1$, that is, at $(2, 1)$ and $(-2, 1)$. Directional information for the curve between these points is summarized in the following chart.

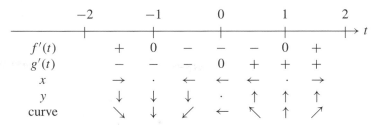

| | | $-2$ | | $-1$ | | $0$ | | $1$ | | $2$ | |
|---|---|---|---|---|---|---|---|---|---|---|---|---|
| $f'(t)$ | | $+$ | | $0$ | $-$ | $-$ | $-$ | $0$ | | $+$ | | |
| $g'(t)$ | | $-$ | | $-$ | $-$ | $0$ | $+$ | $+$ | | $+$ | | |
| $x$ | | $\rightarrow$ | $\cdot$ | | $\leftarrow$ | $\leftarrow$ | $\leftarrow$ | | $\cdot$ | | $\rightarrow$ | |
| $y$ | | $\downarrow$ | | $\downarrow$ | | $\downarrow$ | $\cdot$ | $\uparrow$ | | $\uparrow$ | $\uparrow$ | |
| curve | | $\searrow$ | | $\downarrow$ | $\swarrow$ | $\leftarrow$ | $\nwarrow$ | | $\uparrow$ | | $\nearrow$ | |

For concavity we calculate the second derivative $d^2y/dx^2$ by the formula obtained above. Since $f''(t) = 6t$ and $g''(t) = 2$, we have

$$\frac{d^2y}{dx^2} = \frac{f'(t)g''(t) - g'(t)f''(t)}{(f'(t))^3}$$

$$= \frac{3(t^2 - 1)(2) - 2t(6t)}{[3(t^2 - 1)]^3} = -\frac{2}{9}\frac{t^2 + 1}{(t^2 - 1)^3},$$

which is never zero but which fails to be defined at $t = \pm 1$. The concavity is shown in the following chart:

| $d^2y/dx^2$ | | $-$ | $-1$ | $+$ | $1$ | $-$ | |
|---|---|---|---|---|---|---|---|
| curve | | $\frown$ | | $\smile$ | | $\frown$ | $t$ |

The curve is sketched in Figure 8.27.  ∎

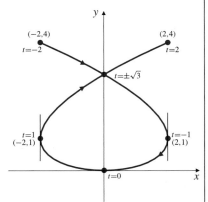

**Figure 8.27**

# Exercises 8.3

In Exercises 1–8, find the coordinates of the points at which the given parametric curve has (a) a horizontal tangent, (b) a vertical tangent.

**1.** $x = t^2 + 1$, $y = 2t - 4$     **2.** $x = t^2 - 2t$, $y = t^2 + 2t$

**3.** $x = t^2 - 2t$, $y = t^3 - 12t$

**4.** $x = t^3 - 3t$, $y = 2t^3 + 3t^2$

**5.** $x = te^{-t^2/2}$, $y = e^{-t^2}$

**6.** $x = \sin t$, $y = \sin t - t\cos t$

**7.** $x = \sin 2t$, $y = \sin t$     **8.** $x = \dfrac{3t}{1 + t^3}$, $y = \dfrac{3t^2}{1 + t^3}$

Find the slopes of the curves in Exercises 9–12 at the points indicated.

**9.** $x = t^3 + t$, $y = 1 - t^3$, at $t = 1$

**10.** $x = t^4 - t^2$, $y = t^3 + 2t$, at $t = -1$

**11.** $x = \cos 2t$, $y = \sin t$, at $t = \pi/6$

**12.** $x = e^{2t}$, $y = te^{2t}$, at $t = -2$

Find parametric equations of the tangents to the curves in Exercises 13–14 at the indicated points.

**13.** $x = t^3 - 2t$, $y = t + t^3$, at $t = 1$

**14.** $x = t - \cos t$, $y = 1 - \sin t$, at $t = \pi/4$

**15.** Show that the curve $x = t^3 - t$, $y = t^2$ has two different tangent lines at the point $(0, 1)$ and find their slopes.

**16.** Find the slopes of two lines that are tangent to $x = \sin t$, $y = \sin 2t$ at the origin.

Where, if anywhere, do the curves in Exercises 17–20 fail to be smooth?

**17.** $x = t^3$, $y = t^2$

**18.** $x = (t - 1)^4$, $y = (t - 1)^3$

**19.** $x = t\sin t$, $y = t^3$     **20.** $x = t^3$, $y = t - \sin t$

In Exercises 21–25, sketch the graphs of the given parametric curves, making use of information from the first two derivatives. Unless otherwise stated, the parameter interval for each curve is the whole real line.

**21.** $x = t^2 - 2t$, $y = t^2 - 4t$     **22.** $x = t^3$, $y = 3t^2 - 1$

**23.** $x = t^3 - 3t$, $y = \dfrac{2}{1 + t^2}$

**24.** $x = t^3 - 3t - 2$, $y = t^2 - t - 2$

**25.** $x = \cos t + t\sin t$, $y = \sin t - t\cos t$, $(t \geq 0)$. (See Example 9 of Section 8.2.)

## 8.4 Arc Lengths and Areas for Parametric Curves

In this section we look at the problems of finding lengths of curves defined parametrically, areas of surfaces of revolution obtained by rotating parametric curves, and areas of plane regions bounded by parametric curves.

### Arc Lengths and Surface Areas

Let $C$ be a smooth parametric curve with equations

$$x = f(t), \qquad y = g(t), \qquad a \le t \le b.$$

(We assume that $f'(t)$ and $g'(t)$ are continuous on the interval $[a, b]$ and are never both zero.) From the differential triangle with legs $dx$ and $dy$ and hypotenuse $ds$ (see Figure 8.28), we obtain $(ds)^2 = (dx)^2 + (dy)^2$, so we have

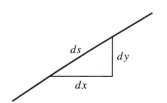

**Figure 8.28**     A differential triangle

> **The arc length element for a parametric curve**
>
> $$ds = \frac{ds}{dt}\, dt = \sqrt{\left(\frac{ds}{dt}\right)^2}\, dt = \sqrt{\left(\frac{dx}{dt}\right)^2 + \left(\frac{dy}{dt}\right)^2}\, dt$$

The length of the curve $C$ is given by

$$s = \int_{t=a}^{t=b} ds = \int_a^b \sqrt{\left(\frac{dx}{dt}\right)^2 + \left(\frac{dy}{dt}\right)^2}\, dt.$$

**Example 1**     Find the length of the parametric curve

$$x = e^t \cos t, \qquad y = e^t \sin t, \qquad (0 \le t \le 2).$$

**Solution**     We have

$$\frac{dx}{dt} = e^t(\cos t - \sin t), \qquad \frac{dy}{dt} = e^t(\sin t + \cos t).$$

Squaring these formulas, adding and simplifying, we get

$$\begin{aligned}
\left(\frac{ds}{dt}\right)^2 &= e^{2t}(\cos t - \sin t)^2 + e^{2t}(\sin t + \cos t)^2 \\
&= e^{2t}\left(\cos^2 t - 2\cos t \sin t + \sin^2 t + \sin^2 t + 2\sin t \cos t + \cos^2 t\right) \\
&= 2e^{2t}.
\end{aligned}$$

The length of the curve is therefore

$$s = \int_0^2 \sqrt{2e^{2t}}\, dt = \sqrt{2} \int_0^2 e^t\, dt = \sqrt{2}\,(e^2 - 1) \text{ units.}$$

Parametric curves can be rotated around various axes to generate surfaces of revolution. The areas of these surfaces can be found by the same procedure used for graphs of functions, with the appropriate version of $ds$.

If the smooth parametric curve with equations

$$x = f(t), \qquad y = g(t), \qquad a \le t \le b,$$

is rotated about the $x$-axis, the area $S$ of the surface so generated is given by

$$S = 2\pi \int_{t=a}^{t=b} |y| \, ds = 2\pi \int_a^b |g(t)| \sqrt{(f'(t))^2 + (g'(t))^2} \, dt.$$

If the rotation is about the $y$-axis, then the area is

$$S = 2\pi \int_{t=a}^{t=b} |x| \, ds = 2\pi \int_a^b |f(t)| \sqrt{(f'(t))^2 + (g'(t))^2} \, dt.$$

**Example 2**  Find the area of the surface of revolution obtained by rotating the astroid curve

$$x = a \cos^3 t, \qquad y = a \sin^3 t$$

(where $a > 0$) about the $x$-axis.

**Solution**  The curve is symmetric about both coordinate axes. (See Figure 8.29.) The entire surface will be generated by rotating the upper half of the curve, and, in fact, we need only rotate the first quadrant part and multiply by two. The first quadrant part of the curve corresponds to $0 \le t \le \pi/2$. We have

$$\frac{dx}{dt} = -3a \cos^2 t \sin t, \qquad \frac{dy}{dt} = 3a \sin^2 t \cos t.$$

Accordingly, the arc length element is

$$ds = \sqrt{9a^2 \cos^4 t \sin^2 t + 9a^2 \sin^4 t \cos^2 t} \, dt$$

$$= 3a \cos t \sin t \sqrt{\cos^2 t + \sin^2 t} \, dt$$

$$= 3a \cos t \sin t \, dt.$$

Therefore the required surface area is

$$S = 2 \times 2\pi \int_0^{\pi/2} a \sin^3 t \, 3a \cos t \sin t \, dt$$

$$= 12\pi a^2 \int_0^{\pi/2} \sin^4 t \cos t \, dt \qquad \text{Let } u = \sin t,$$
$$\qquad\qquad\qquad\qquad\qquad\qquad\qquad\qquad du = \cos t \, dt$$

$$= 12\pi a^2 \int_0^1 u^4 \, du = \frac{12\pi a^2}{5} \text{ square units.}$$

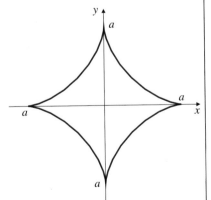

**Figure 8.29**

## Areas Bounded by Parametric Curves

Consider the parametric curve $C$ with equations $x = f(t)$, $y = g(t)$, ($a \le t \le b$), where $f$ is differentiable and $g$ is continuous on $[a, b]$. For the moment let us also assume that $f'(t) \geq 0$ and $g(t) \geq 0$ on $[a, b]$, so $\mathcal{C}$ has no points below the $x$-axis and is traversed from left to right as $t$ increases from $a$ to $b$.

The region under $\mathcal{C}$ and above the $x$-axis has area element given by $dA = y\, dx = g(t) f'(t)\, dt$, and so its area (see Figure 8.30) is

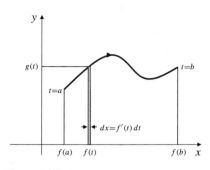

$$A = \int_a^b g(t) f'(t)\, dt.$$

**Figure 8.30**

Similar arguments can be given for three other cases:

If $f'(t) \geq 0$ and $g(t) \leq 0$ on $[a, b]$, then $A = -\int_a^b g(t) f'(t)\, dt$,

If $f'(t) \leq 0$ and $g(t) \geq 0$ on $[a, b]$, then $A = -\int_a^b g(t) f'(t)\, dt$,

If $f'(t) \leq 0$ and $g(t) \leq 0$ on $[a, b]$, then $A = \int_a^b g(t) f'(t)\, dt$,

where $A$ is the (positive) area bounded by $\mathcal{C}$, the $x$-axis, and the vertical lines $x = f(a)$ and $x = f(b)$. Combining these results we can see that

$$\int_a^b g(t) f'(t)\, dt = A_1 - A_2,$$

where $A_1$ is the area lying vertically between $\mathcal{C}$ and that part of the $x$-axis consisting of points $x = f(t)$ such that $g(t) f'(t) \geq 0$, and $A_2$ is a similar area corresponding to points where $g(t) f'(t) < 0$. This formula is valid for arbitrary continuous $g$ and differentiable $f$. See Figure 8.31 for generic examples. In particular, if $\mathcal{C}$ is a non-self-intersecting closed curve, then the area of the region bounded by $\mathcal{C}$ is given by

$$A = \int_a^b g(t) f'(t)\, dt \qquad \text{if } \mathcal{C} \text{ is traversed clockwise as } t \text{ increases,}$$

$$A = -\int_a^b g(t) f'(t)\, dt \qquad \text{if } \mathcal{C} \text{ is traversed counterclockwise,}$$

both of which are illustrated in Figure 8.32.

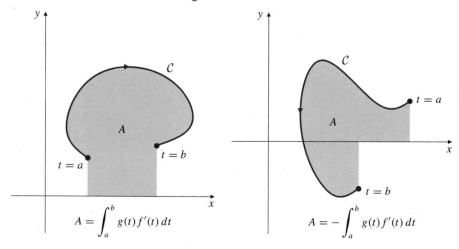

**Figure 8.31**     Areas defined by parametric curves

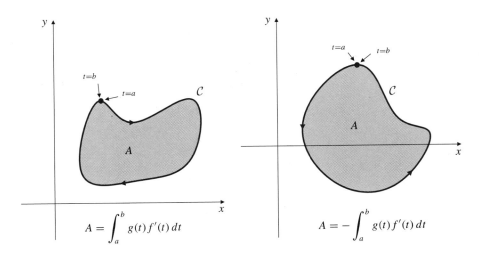

**Figure 8.32**    Areas bounded by closed parametric curves

$$A = \int_a^b g(t) f'(t) \, dt$$

$$A = -\int_a^b g(t) f'(t) \, dt$$

**Example 3**    Find the area bounded by the ellipse $x = a \cos s$, $y = b \sin s$, $(0 \le s \le 2\pi)$.

**Solution**    This ellipse is traversed counterclockwise. (See Example 4 in Section 8.2.) The area enclosed is

$$A = -\int_0^{2\pi} b \sin s \, (-a \sin s) \, ds$$

$$= \frac{ab}{2} \int_0^{2\pi} (1 - \cos 2s) \, ds$$

$$= \frac{ab}{2} s \Big|_0^{2\pi} - \frac{ab}{4} \sin 2s \Big|_0^{2\pi} = \pi a b \text{ square units.}$$

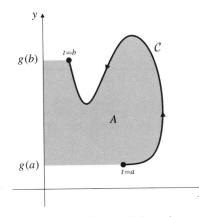

**Figure 8.33**    The shaded area is

$$A = \int_a^b f(t) g'(t) \, dt$$

**Example 4**    Find the area above the $x$-axis and under one arch of the cycloid $x = at - a \sin t$, $y = a - a \cos t$.

**Solution**    Part of the cycloid is shown in Figure 8.22 in Section 8.2. One arch corresponds to the parameter interval $0 \le t \le 2\pi$. Since $y = a(1 - \cos t) \ge 0$ and $dx/dt = a(1 - \cos t) \ge 0$, the area under one arch is

$$A = \int_0^{2\pi} a^2 (1 - \cos t)^2 \, dt = a^2 \int_0^{2\pi} \left( 1 - 2 \cos t + \frac{1 + \cos 2t}{2} \right) dt$$

$$= a^2 \left( t - 2 \sin t + \frac{t}{2} + \frac{\sin 2t}{4} \right) \Big|_0^{2\pi} = 3 \pi a^2 \text{ square units.}$$

Similar arguments to those used above show that if $f$ is continuous and $g$ is differentiable, then we can also interpret

$$\int_a^b f(t) g'(t) \, dt = \int_{t=a}^{t=b} x \, dy = A_1 - A_2,$$

where $A_1$ is the area of the region lying *horizontally* between the parametric curve $x = f(t)$, $y = g(t)$, $(a \leq t \leq b)$ and that part of the $y$-axis consisting of points $y = g(t)$ such that $f(t)g'(t) \geq 0$, and $A_2$ is the area of a similar region corresponding to $f(t)g'(t) < 0$. For example, the region shaded in Figure 8.33 has area $\int_a^b f(t)g'(t)\,dt$. Green's Theorem in Section 16.3 provides a more coherent approach to finding such areas.

## Exercises 8.4

Find the lengths of the curves in Exercises 1–8.

**1.** $x = 3t^2$, $y = 2t^3$, $(0 \leq t \leq 1)$

**2.** $x = 1 + t^3$, $y = 1 - t^2$, $(-1 \leq t \leq 2)$

**3.** $x = a\cos^3 t$, $y = a\sin^3 t$, $(0 \leq t \leq 2\pi)$

**4.** $x = \ln(1 + t^2)$, $y = 2\tan^{-1} t$, $(0 \leq t \leq 1)$

**5.** $x = t^2 \sin t$, $y = t^2 \cos t$, $(0 \leq t \leq 2\pi)$

**6.** $x = \cos t + t \sin t$, $y = \sin t - t \cos t$, $(0 \leq t \leq 2\pi)$

**7.** $x = t + \sin t$, $y = \cos t$, $(0 \leq t \leq \pi)$

**8.** $x = \sin^2 t$, $y = 2 \cos t$, $(0 \leq t \leq \pi/2)$

**9.** Find the length of one arch of the cycloid $x = at - a \sin t$, $y = a - a \cos t$. (One arch corresponds to $0 \leq t \leq 2\pi$.)

**10.** Find the area of the surfaces obtained by rotating one arch of the cycloid in Exercise 9 about (a) the $x$-axis, (b) the $y$-axis.

**11.** Find the area of the surface generated by rotating the curve $x = e^t \cos t$, $y = e^t \sin t$, $(0 \leq t \leq \pi/2)$ about the $x$-axis.

**12.** Find the area of the surface generated by rotating the curve of Exercise 11 about the $y$-axis.

**13.** Find the area of the surface generated by rotating the curve $x = 3t^2$, $y = 2t^3$, $(0 \leq t \leq 1)$ about the $y$-axis.

**14.** Find the area of the surface generated by rotating the curve $x = 3t^2$, $y = 2t^3$, $(0 \leq t \leq 1)$ about the $x$-axis.

In Exercises 15–20, sketch and find the area of the region $R$ described in terms of the given parametric curves.

**15.** $R$ is the closed loop bounded by $x = t^3 - 4t$, $y = t^2$, $(-2 \leq t \leq 2)$.

**16.** $R$ is bounded by the astroid $x = a\cos^3 t$, $y = a\sin^3 t$, $(0 \leq t \leq 2\pi)$.

**17.** $R$ is bounded by the coordinate axes and the parabolic arc $x = \sin^4 t$, $y = \cos^4 t$.

**18.** $R$ is bounded by $x = \cos s \sin s$, $y = \sin^2 s$, $(0 \leq s \leq \pi/2)$, and the $y$-axis.

**19.** $R$ is bounded by the oval $x = (2 + \sin t)\cos t$, $y = (2 + \sin t)\sin t$.

**∗ 20.** $R$ is bounded by the $x$-axis, the hyperbola $x = \sec t$, $y = \tan t$ and the ray joining the origin to the point $(\sec t_0, \tan t_0)$.

**21.** Show that the region bounded by the $x$-axis and the hyperbola $x = \cosh t$, $y = \sinh t$ (where $t > 0$), and the ray from the origin to the point $(\cosh t_0, \sinh t_0)$ has area $t_0/2$ square units. This proves a claim made at the beginning of Section 3.6.

**22.** Find the volume of the solid obtained by rotating about the $x$-axis the region bounded by that axis and one arch of the cycloid $x = at - a \sin t$, $y = a - a \cos t$. (See Example 8 in Section 8.2.)

**23.** Find the volume generated by rotating about the $x$-axis the region lying under the astroid $x = a\cos^3 t$, $y = a\sin^3 t$ and above the $x$-axis.

## 8.5   Polar Coordinates and Polar Curves

The **polar coordinate system** is an alternative to the rectangular (Cartesian) coordinate system for describing the location of points in a plane. Sometimes it is more important to know how far, and in what direction, a point is from the origin than it is to know its Cartesian coordinates. In the polar coordinate system there is an origin (or **pole**), $O$, and a **polar axis**, a ray (that is, a half-line) extending from $O$ horizontally to the right. The position of any point $P$ in the plane is then determined by its polar coordinates $[r, \theta]$, where

(i) $r$ is the distance from $O$ to $P$, and

(ii) $\theta$ is the angle that the ray $OP$ makes with the polar axis (counterclockwise angles being considered positive).

We will use square brackets [, ] for polar coordinates of a point to distinguish them from rectangular (Cartesian) coordinates. Figure 8.34 shows some points with their polar coordinates. The rectangular coordinate axes $x$ and $y$ are usually shown on a polar graph. The polar axis coincides with the positive $x$-axis.

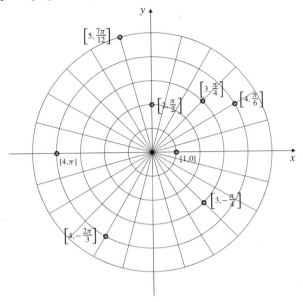

**Figure 8.34**    Polar coordinates of some points in the plane

Unlike rectangular coordinates, the polar coordinates of a point are not unique. The polar coordinates $[r, \theta_1]$ and $[r, \theta_2]$ represent the same point provided $\theta_1$ and $\theta_2$ differ by an integer multiple of $2\pi$:

$$\theta_2 = \theta_1 + 2n\pi, \qquad \text{where } n = 0, \pm 1, \pm 2, \ldots .$$

For instance, the polar coordinates

$$\left[1, \frac{\pi}{4}\right], \quad \left[1, \frac{9\pi}{4}\right], \quad \text{and} \quad \left[1, -\frac{7\pi}{4}\right]$$

all represent the same point with Cartesian coordinates $\left(\frac{1}{\sqrt{2}}, \frac{1}{\sqrt{2}}\right)$. Similarly, $[2, \pi]$ and $[2, -\pi]$ both represent the point with Cartesian coordinates $(-2, 0)$, and $[4, 0]$ and $[4, 2\pi]$ both represent the point with Cartesian coordinates $(4, 0)$. In addition, the origin $O$ has polar coordinates $[0, \theta]$ for any value of $\theta$. (If we go zero distance from $O$, it doesn't matter in what direction we go.)

Sometimes we need to interpret polar coordinates $[r, \theta]$ where $r < 0$. The appropriate interpretation for this "negative distance" $r$ is that it represents a positive distance $-r$ measured in the *opposite direction* (that is, in the direction $\theta + \pi$):

$$[r, \theta] = [-r, \theta + \pi].$$

For example, $[-1, \pi/4] = [1, 5\pi/4]$. Allowing $r < 0$ increases the number of different sets of polar coordinates that represent the same point.

If we want to consider both rectangular and polar coordinate systems in the same plane, and we choose the positive $x$-axis as the polar axis, then the relationships between the rectangular coordinates of a point and its polar coordinates are as shown in Figure 8.35.

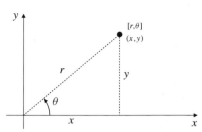

**Figure 8.35**    Relating Cartesian and polar coordinates of a point

### Polar–rectangular conversion

$$x = r \cos \theta \qquad x^2 + y^2 = r^2$$

$$y = r \sin \theta \qquad \tan \theta = \frac{y}{x}$$

A single equation in $x$ and $y$ generally represents a curve in the plane with respect to the Cartesian coordinate system. Similarly, a single equation in $r$ and $\theta$ generally represents a curve with respect to the polar coordinate system. The conversion formulas above can be used to convert one representation of a curve into the other.

**Example 1**    The straight line $2x - 3y = 5$ has polar equation $r(2 \cos \theta - 3 \sin \theta) = 5$, or

$$r = \frac{5}{2 \cos \theta - 3 \sin \theta}.$$

**Example 2**    Find the Cartesian equation of the curve represented by the polar equation $r = 2a \cos \theta$ and hence identify the curve.

**Solution**    The polar equation can be transformed to Cartesian coordinates if we first multiply it by $r$:

$$r^2 = 2ar \cos \theta$$
$$x^2 + y^2 = 2ax$$
$$(x - a)^2 + y^2 = a^2$$

The given polar equation $r = 2a \cos \theta$ thus represents a circle with centre $(a, 0)$ and radius $a$ as shown in Figure 8.36. Observe from the equation that $r \to 0$ as $\theta \to \pm\pi/2$. This corresponds in the figure to the fact that the circle approaches the origin in the vertical direction.

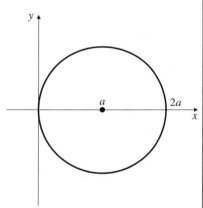

**Figure 8.36**    The circle $r = 2a \cos \theta$

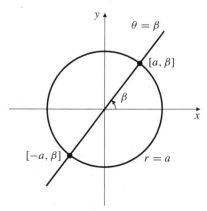

**Figure 8.37**    Coordinate curves for the polar coordinate system

## Some Polar Curves

Figure 8.37 shows the graphs of the polar equations $r = a$ and $\theta = \beta$, where $a$ and $\beta$ (Greek *beta*) are constants. These are, respectively, the circle with radius $|a|$ centred at the origin, and a line through the origin making angle $\beta$ with the polar axis. Note that the line and the circle meet at two points, with polar coordinates $[a, \beta]$ and $[-a, \beta]$. The "coordinate curves" for polar coordinates, that is, the curves with equations $r =$ constant and $\theta =$ constant, are circles and lines through the origin, respectively. The "coordinate curves" for Cartesian coordinates, $x =$ constant and $y =$ constant, are vertical and horizontal straight lines. Cartesian graph paper is ruled with vertical and horizontal lines; polar graph paper is ruled with concentric circles and radial lines emanating from the origin, as shown in Figure 8.34.

The graph of an equation of the form $r = f(\theta)$ is called the **polar graph** of the function $f$. Some polar graphs can be recognized easily if the polar equation is transformed to rectangular form. For others, this transformation does not help;

the rectangular equation may be too complicated to be recognizable. In these cases one must resort to constructing a table of values and plotting points.

---

**Example 3**    Sketch and identify the curve $r = 2a \cos(\theta - \theta_0)$.

***Solution***    We proceed as in Example 2.

$$r^2 = 2ar \cos(\theta - \theta_0) = 2ar \cos\theta_0 \cos\theta + 2ar \sin\theta_0 \sin\theta$$
$$x^2 + y^2 = 2a \cos\theta_0 x + 2a \sin\theta_0 y$$
$$x^2 - 2a \cos\theta_0 x + a^2 \cos^2\theta_0 + y^2 - 2a \sin\theta_0 y + a^2 \sin^2\theta_0 = a^2$$
$$(x - a \cos\theta_0)^2 + (y - a \sin\theta_0)^2 = a^2.$$

This is a circle of radius $a$ that passes through the origin in the directions, $\theta = \theta_0 \pm \frac{\pi}{2}$, that make $r = 0$. (See Figure 8.38.) Its centre has Cartesian coordinates $(a \cos\theta_0, a \sin\theta_0)$ and hence polar coordinates $[a, \theta_0]$. For $\theta_0 = \pi/2$ we have $r = 2a \sin\theta$ as the equation of a circle of radius $a$ centred on the $y$-axis.

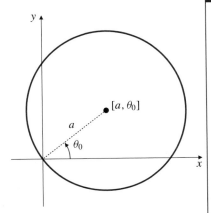

**Figure 8.38**    The circle $r = 2a \cos(\theta - \theta_0)$

Comparing Examples 2 and 3, we are led to formulate the following principle.

---

**Rotating a polar graph**

The polar graph with equation $r = f(\theta - \theta_0)$ is the polar graph with equation $r = f(\theta)$ rotated through angle $\theta_0$ about the origin.

---

**Example 4**    Sketch the polar curve $r = a(1 - \cos\theta)$, where $a > 0$.

***Solution***    Transformation to rectangular coordinates is not much help here; the resulting equation is $(x^2 + y^2 + ax)^2 = a^2(x^2 + y^2)$ (verify this) which we do not recognize. Therefore we will make a table of values and plot some points.

**Table 3.**

| $\theta$ | 0 | $\pm\dfrac{\pi}{6}$ | $\pm\dfrac{\pi}{4}$ | $\pm\dfrac{\pi}{3}$ | $\pm\dfrac{\pi}{2}$ | $\pm\dfrac{2\pi}{3}$ | $\pm\dfrac{3\pi}{4}$ | $\pm\dfrac{5\pi}{6}$ | $\pi$ |
|---|---|---|---|---|---|---|---|---|---|
| $r$ | 0 | $0.13a$ | $0.29a$ | $0.5a$ | $a$ | $1.5a$ | $1.71a$ | $1.87a$ | $2a$ |

Because it is shaped like a heart, this curve is called a **cardioid**. Observe the cusp at the origin in Figure 8.39. As in the previous example, the curve enters the origin in the directions $\theta$ that make $r = f(\theta) = 0$. In this case, the only such direction is $\theta = 0$. It is important, when sketching polar graphs, to show clearly any directions of approach to the origin.

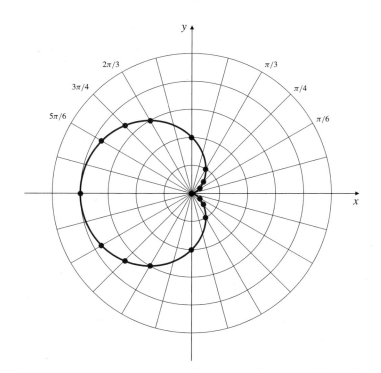

**Figure 8.39**    The cardioid
$r = a(1 - \cos\theta)$

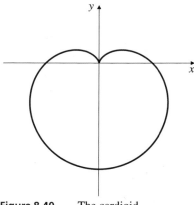

**Figure 8.40**    The cardioid
$r = a(1 - \sin\theta)$

**Direction of a polar graph at the origin**

A polar graph $r = f(\theta)$ approaches the origin from the direction $\theta$ for which $f(\theta) = 0$.

The equation $r = a(1 - \cos(\theta - \theta_0))$ represents a cardioid of the same size and shape as that in Figure 8.39, but rotated through an angle $\theta_0$ counterclockwise about the origin. Its cusp is in the direction $\theta = \theta_0$. In particular, $r = a(1 - \sin\theta)$ has a vertical cusp, as shown in Figure 8.40.

It is not usually necessary to make a detailed table of values to sketch a polar curve with a simple equation of the form $r = f(\theta)$. It is essential to determine those values of $\theta$ for which $r = 0$ and indicate them on the graph with rays. It is also useful to determine points where the curve is farthest from the origin. (Where is $f(\theta)$ maximum or minimum?) Except possibly at the origin, polar curves will be smooth wherever $f(\theta)$ is a differentiable function of $\theta$.

**Example 5**    Sketch the polar graphs    (a) $r = \cos(2\theta)$,    (b) $r = \sin(3\theta)$, and (c) $r^2 = \cos(2\theta)$.

**Solution**    The graphs are shown in Figures 8.41–8.43. Observe how the curves (a) and (c) approach the origin in the directions $\theta = \pm\frac{\pi}{4}$ and $\theta = \pm\frac{3\pi}{4}$, and curve (b) approaches in the directions $\theta = 0$, $\pi$, $\pm\frac{\pi}{3}$ and $\pm\frac{2\pi}{3}$. This curve is traced out twice as $\theta$ increases from $-\pi$ to $\pi$. So is curve (c) if we allow both square roots $r = \pm\sqrt{\cos(2\theta)}$. Note that there are no points on curve (c) between $\theta = \pm\frac{\pi}{4}$ and $\theta = \pm\frac{3\pi}{4}$ because $r^2$ cannot be negative.

Curve (c) is called a **lemniscate**. Lemniscates are curves consisting of points $P$ such that the product of the distances from $P$ to certain fixed points is constant. For the curve (c) these fixed points are $\left(\pm\frac{1}{\sqrt{2}}, 0\right)$.

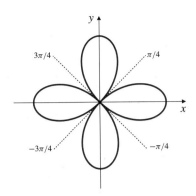

**Figure 8.41**    Curve (a): the polar curve $r = \cos(2\theta)$

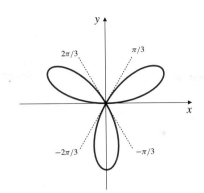

**Figure 8.42**    Curve (b): the polar curve $r = \sin(3\theta)$

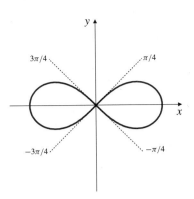

**Figure 8.43**    Curve (c): the lemniscate $r^2 = \cos(2\theta)$

In all of the examples above the functions $f(\theta)$ are periodic and $2\pi$ is a period of each of them, so each line through the origin could meet the polar graph at most twice. ($\theta$ and $\theta + \pi$ determine the same line.) If $f(\theta)$ does not have period $2\pi$, then the curve can wind around the origin many times. Two such *spirals* are shown in Figure 8.44, the **equiangular spiral** $r = \theta$ and the **exponential spiral** $r = e^{-\theta/3}$, each sketched for positive values of $\theta$.

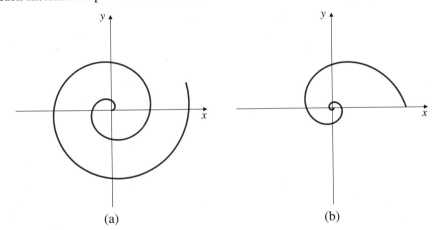

(a)                                          (b)

**Figure 8.44**

(a) The equiangular spiral $r = \theta$

(b) The exponential spiral $r = e^{-\theta/3}$

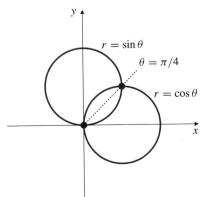

**Figure 8.45**

## Intersections of Polar Curves

Because the polar coordinates of points are not unique, finding the intersection points of two polar curves can be more complicated than the similar problem for Cartesian graphs. Of course, the polar curves $r = f(\theta)$ and $r = g(\theta)$ will intersect at any points $[r_0, \theta_0]$ for which

$$f(\theta_0) = g(\theta_0), \qquad \text{and} \qquad r_0 = f(\theta_0),$$

but there may be other intersections as well. In particular, if both curves pass through the origin, then the origin will be an intersection point even though it may not show up in solving $f(\theta) = g(\theta)$, because the curves may be at the origin for different values of $\theta$. For example, the two circles $r = \cos\theta$ and $r = \sin\theta$ intersect at the origin and also at the point $[1/\sqrt{2}, \pi/4]$, even though only the latter point is obtained by solving the equation $\cos\theta = \sin\theta$. (See Figure 8.45.)

**Example 6**    Find the intersections of the curves $r = \sin\theta$ and $r = 1 - \sin\theta$.

**Solution**    Since both functions of $\theta$ are periodic with period $2\pi$, we need only look for solutions satisfying $0 \le \theta \le 2\pi$. Solving the equation

$$\sin\theta = 1 - \sin\theta,$$

we get $\sin\theta = 1/2$, so that $\theta = \pi/6$ or $\theta = 5\pi/6$. Both curves have $r = 1/2$ at these points, so the two curves intersect at $[1/2, \pi/6]$ and $[1/2, 5\pi/6]$. Also, the origin lies on the curve $r = \sin\theta$ (for $\theta = 0$ and $\theta = 2\pi$), and on the curve $r = 1 - \sin\theta$ (for $\theta = \pi/2$). Therefore, the origin is also an intersection point of the curves. (See Figure 8.46.)

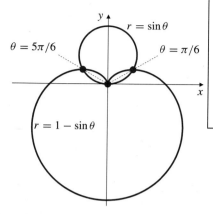

**Figure 8.46**

Finally, if negative values of $r$ are allowed, then the curves $r = f(\theta)$, $y = g(\theta)$ will also intersect at $[r_1, \theta_1] = [r_2, \theta_2]$ if, for some integer $k$,

$$\theta_1 = \theta_2 + (2k+1)\pi \qquad \text{and} \qquad r_1 = f(\theta_1) = -g(\theta_2) = -r_2.$$

See Exercise 28 for an example.

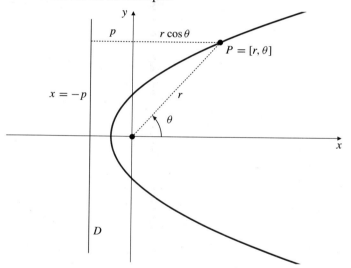

**Figure 8.47**    A conic curve with eccentricity $\varepsilon$, focus at the origin, and directrix $x = -p$

## Polar Conics

Let $D$ be the vertical straight line $x = -p$ and let $\varepsilon$ be a positive real number. The set of points $P$ in the plane which satisfy the condition

$$\frac{\text{distance of } P \text{ from the origin}}{\text{perpendicular distance from } P \text{ to } D} = \varepsilon$$

is a conic section with eccentricity $\varepsilon$, focus at the origin, and corresponding directrix $D$, as observed in Section 8.1. (It is an ellipse if $\varepsilon < 1$, a parabola if $\varepsilon = 1$ and a hyperbola if $\varepsilon > 1$.) If $P$ has polar coordinates $(r, \theta)$, then the condition above becomes (see Figure 8.47)

$$\frac{r}{p + r\cos\theta} = \varepsilon,$$

or, solving for $r$,

$$r = \frac{\varepsilon p}{1 - \varepsilon \cos \theta}.$$

Examples of the three possibilities (ellipse, parabola and hyperbola) are shown in Figures 8.48–8.50. Note that for the hyperbola, the directions of the asymptotes are the angles that make the denominator $1 - \varepsilon \cos \theta = 0$.

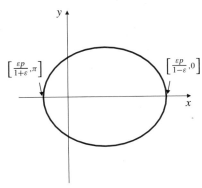

**Figure 8.48**   Ellipse: $\varepsilon < 1$

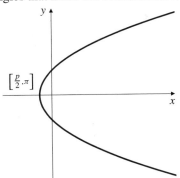

**Figure 8.49**   Parabola: $\varepsilon = 1$

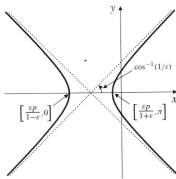

**Figure 8.50**   Hyperbola: $\varepsilon > 1$

## Exercises 8.5

In Exercises 1–12, transform the given polar equation to rectangular coordinates and hence identify the curve represented.

**1.** $r = 3 \sec \theta$

**2.** $r = -2 \csc \theta$

**3.** $r = \dfrac{5}{3 \sin \theta - 4 \cos \theta}$

**4.** $r = \sin \theta + \cos \theta$

**5.** $r^2 = \csc 2\theta$

**6.** $r = \sec \theta \tan \theta$

**7.** $r = \sec \theta (1 + \tan \theta)$

**8.** $r = \dfrac{2}{\sqrt{\cos^2 \theta + 4 \sin^2 \theta}}$

**9.** $r = \dfrac{1}{1 - \cos \theta}$

**10.** $r = \dfrac{2}{2 - \cos \theta}$

**11.** $r = \dfrac{2}{1 - 2 \sin \theta}$

**12.** $r = \dfrac{2}{1 + \sin \theta}$

In Exercises 13–24, sketch the polar graphs of the given equations.

**13.** $r = 1 + \sin \theta$

**14.** $r = 1 - \cos(\theta + \frac{\pi}{4})$

**15.** $r = 1 + 2 \cos \theta$

**16.** $r = 1 - 2 \sin \theta$

**17.** $r = 2 + \cos \theta$

**18.** $r = 2 \sin 2\theta$

**19.** $r = \cos 3\theta$

**20.** $r = 2 \cos 4\theta$

**21.** $r^2 = 4 \sin 2\theta$

**22.** $r^2 = 4 \cos 3\theta$

**23.** $r^2 = \sin 3\theta$

**24.** $r = \ln \theta$

Find all intersections of the pairs of curves in Exercises 25–28.

**25.** $r = \sqrt{3} \cos \theta$,    $r = \sin \theta$

**26.** $r^2 = 2 \cos(2\theta)$,    $r = 1$

**27.** $r = 1 + \cos \theta$,    $r = 3 \cos \theta$

**∗ 28.** $r = \theta$,    $r = \theta + \pi$

**∗ 29.** Sketch the graph of the equation $r = 1/\theta$, $\theta > 0$. Show that this curve has a horizontal asymptote. Does $r = 1/(\theta - \alpha)$ have an asymptote?

**30.** How many leaves does the curve $r = \cos n\theta$ have? The curve $r^2 = \cos n\theta$? Distinguish the cases where $n$ is odd and even.

**31.** Show that the polar graph $r = f(\theta)$ (where $f$ is continuous) can be written as a parametric curve with parameter $\theta$.

In Exercises 32–37, use computer graphing software or a graphing calculator to plot various members of the given families of polar curves, and try to observe patterns that would enable you to predict behaviour of other memebers of the families.

**32.** $r = \cos t \cos(mt)$,    $m = 1, 2, 3, \ldots$

**33.** $r = 1 + \cos t \cos(mt)$,    $m = 1, 2, 3, \ldots$

**34.** $r = \sin(2t) \sin(mt)$,    $m = 2, 3, 4, 5, \ldots$

**35.** $r = 1 + \sin(2t) \sin(mt)$,    $m = 2, 3, 4, 5, \ldots$

**36.** $r = C + \cos t \cos(2t)$ for $C = 0$, $C = 1$, values of $C$ between 0 and 1, and values of $C$ greater than 1

**37.** $r = C + \cos t \sin(3t)$ for $C = 0$, $C = 1$, values of $C$ between 0 and 1, values of $C$ less than 0, and values of $C$ greater than 1

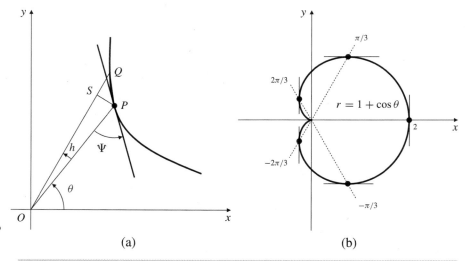

**38.** Plot the curve $r = \ln\theta$ for $0 < \theta \le 2\pi$. It intersects itself at point $P$. Thus there are two values $\theta_1$ and $\theta_2$ between 0 and $2\pi$ for which $[f(\theta_1), \theta_1] = [f(\theta_2), \theta_2]$. What equations must be satisfied by $\theta_1$ and $\theta_2$? Find $\theta_1$ and $\theta_2$, and the Cartesian coordinates of $P$ correct to six decimal places.

**39.** Simultaneously plot the two curves $r = \ln\theta$ and $r = 1/\theta$, for $0 < \theta \le 2\pi$. The two curves intersect at two points. What equations must be satisfied by the $\theta$ values of these points? What are their Cartesian coordinates to six decimal places?

## 8.6   Slopes, Areas, and Arc Lengths for Polar Curves

There is a simple formula that can be used to determine the direction of the tangent line to a polar curve $r = f(\theta)$ at a point $P = [r, \theta]$ other than the origin. Let $Q$ be a point on the curve nearby $P$ corresponding to polar angle $\theta + h$. Draw $PS$ perpendicular to $OQ$. Observe that $PS = f(\theta)\sin h$ and $SQ = OQ - OS = f(\theta + h) - f(\theta)\cos h$. If the tangent line to $r = f(\theta)$ at $P$ makes angle $\Psi$ (Greek *psi*) with the radial line $OP$ as shown in Figure 8.51(a), then $\Psi$ is the limit of the angle $SQP$ as $h \to 0$. Thus

$$\tan\Psi = \lim_{h\to 0}\frac{PS}{SQ} = \lim_{h\to 0}\frac{f(\theta)\sin h}{f(\theta + h) - f(\theta)\cos h} \qquad \begin{bmatrix}0\\0\end{bmatrix}$$

$$= \lim_{h\to 0}\frac{f(\theta)\cos h}{f'(\theta + h) + f(\theta)\sin h} \qquad \text{(by l'Hôpital's Rule)}$$

$$= \frac{f(\theta)}{f'(\theta)} = \frac{r}{dr/d\theta}.$$

**Figure 8.51**

(a) The angle $\Psi$ is the limit of angle $SQP$ as $h \to 0$

(b) Horizontal and vertical tangents to a cardioid

(a)                              (b)

---

**Tangent direction for a polar curve**

At any point $P$ other than the origin on the polar curve $r = f(\theta)$, the angle between the radial line from the origin to $P$ and the tangent to the curve is given by

$$\tan\Psi = \frac{f(\theta)}{f'(\theta)}.$$

In particular, $\Psi = \pi/2$ if $f'(\theta) = 0$.
If $f(\theta_0) = 0$ and the curve has a tangent line at $\theta_0$ then that tangent line has equation $\theta = \theta_0$.

The formula above can be used to find points where a polar graph has horizontal or vertical tangents:

$$\Psi + \theta = \pi, \qquad \text{so } \tan \Psi = -\tan \theta \qquad \text{for a horizontal tangent,}$$

$$\Psi + \theta = \frac{\pi}{2}, \qquad \text{so } \tan \Psi = \cot \theta \qquad \text{for a vertical tangent.}$$

**Remark**   Since for parametric curves horizontal and vertical tangents correspond to $dy/dt = 0$ and $dx/dt = 0$, respectively, it is usually easier to find the critical points of $y = f(\theta) \sin \theta$ for horizontal tangents, and of $x = f(\theta) \cos \theta$ for vertical tangents.

**Example 1**   Find the points on the cardioid $r = 1 + \cos \theta$ where the tangent lines are vertical or horizontal.

**Solution**   We have $y = (1+\cos \theta) \sin \theta$ and $x = (1+\cos \theta) \cos \theta$. For horizontal tangents

$$0 = \frac{dy}{d\theta} = -\sin^2 \theta + \cos^2 \theta + \cos \theta$$

$$= 2\cos^2 \theta + \cos \theta - 1$$

$$= (2\cos \theta - 1)(\cos \theta + 1).$$

The solutions are $\cos \theta = \frac{1}{2}$ and $\cos \theta = -1$, that is, $\theta = \pm \pi/3$ and $\theta = \pi$. There are horizontal tangents at $\left[ \frac{3}{2}, \pm \frac{\pi}{3} \right]$. At $\theta = \pi$, we have $r = 0$. The curve does not have a tangent line at the origin (it has a cusp). See Figure 8.51(b).

For vertical tangents

$$0 = \frac{dx}{d\theta} = -\sin \theta - 2\cos \theta \sin \theta = -\sin \theta (1 + 2\cos \theta).$$

The solutions are $\sin \theta = 0$ and $\cos \theta = -\frac{1}{2}$, that is, $\theta = 0, \pi, \pm 2\pi/3$. There are vertical tangent lines at $[2, 0]$ and $\left[ \frac{1}{2}, \pm \frac{2\pi}{3} \right]$. ∎

## Areas Bounded by Polar Curves

The basic area problem in polar coordinates is that of finding the area $A$ of the region $R$ bounded by the polar graph $r = f(\theta)$ and the two rays $\theta = \alpha$ and $\theta = \beta$. We assume that $\beta > \alpha$ and that $f$ is continuous for $\alpha \leq \theta \leq \beta$. See Figure 8.52.

A suitable area element in this case is a sector of angular width $d\theta$, as shown in Figure 8.52. For infinitesimal $d\theta$ this is just a sector of a circle of radius $r = f(\theta)$:

$$dA = \frac{d\theta}{2\pi} \pi r^2 = \frac{1}{2} r^2 \, d\theta = \frac{1}{2} \left( f(\theta) \right)^2 d\theta.$$

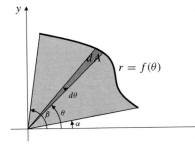

**Figure 8.52**   An area element in polar coordinates

> **Area in polar coordinates**
>
> The region bounded by $r = f(\theta)$ and the rays $\theta = \alpha$ and $\theta = \beta$, $(\alpha < \beta)$, has area
>
> $$A = \frac{1}{2} \int_\alpha^\beta \left( f(\theta) \right)^2 d\theta.$$

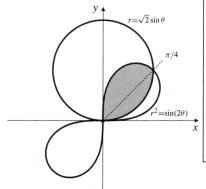

**Figure 8.53**

**Example 2**    Find the area bounded by the cardioid $r = a(1+\cos\theta)$, as illustrated in Figure 8.53.

**Solution**    By symmetry, the area is twice that of the top half:

$$A = 2 \times \frac{1}{2} \int_0^\pi a^2 (1 + \cos\theta)^2 \, d\theta$$

$$= a^2 \int_0^\pi (1 + 2\cos\theta + \cos^2\theta) \, d\theta$$

$$= a^2 \int_0^\pi \left( 1 + 2\cos\theta + \frac{1 + \cos 2\theta}{2} \right) d\theta$$

$$= a^2 \left( \frac{3}{2}\theta + 2\sin\theta + \frac{1}{4}\sin 2\theta \right) \Big|_0^\pi = \frac{3}{2}\pi a^2 \text{ square units.}$$

■

**Example 3**    Find the area of the region that lies inside the circle $r = \sqrt{2}\sin\theta$ and inside the lemniscate $r^2 = \sin 2\theta$.

**Solution**    The region is shaded in Figure 8.54. Besides intersecting at the origin, the curves intersect at the first quadrant point satisfying

$$2\sin^2\theta = \sin 2\theta = 2\sin\theta\cos\theta.$$

Thus $\sin\theta = \cos\theta$ and $\theta = \pi/4$. The required area is

$$A = \frac{1}{2}\int_0^{\pi/4} 2\sin^2\theta \, d\theta + \frac{1}{2}\int_{\pi/4}^{\pi/2} \sin 2\theta \, d\theta$$

$$= \int_0^{\pi/4} \frac{1 - \cos 2\theta}{2} \, d\theta - \frac{1}{4}\cos 2\theta \Big|_{\pi/4}^{\pi/2}$$

$$= \frac{\pi}{8} - \frac{1}{4}\sin 2\theta \Big|_0^{\pi/4} + \frac{1}{4} = \frac{\pi}{8} - \frac{1}{4} + \frac{1}{4} = \frac{\pi}{8} \text{ square units.}$$

■

**Figure 8.54**

## Arc Lengths for Polar Curves

The arc length element for the polar curve $r = f(\theta)$ can be determined from the differential triangle shown in Figure 8.55. The leg $r \, d\theta$ of the triangle is obtained as the arc length of a circular arc of radius $r$ subtending angle $d\theta$ at the origin. We have

$$(ds)^2 = (dr)^2 + r^2(d\theta)^2 = \left[ \left( \frac{dr}{d\theta} \right)^2 + r^2 \right] (d\theta)^2,$$

and so we obtain the following formula:

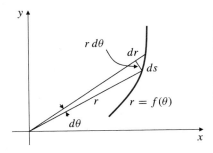

**Figure 8.55**    The arc length element for a polar curve

**Arc length element for a polar curve**

The arc length element for the polar curve $r = f(\theta)$ is

$$ds = \sqrt{\left(\frac{dr}{d\theta}\right)^2 + r^2}\, d\theta = \sqrt{\left(f'(\theta)\right)^2 + \left(f(\theta)\right)^2}\, d\theta.$$

This arc length element can also be derived from that for a parametric curve. See Exercise 26 at the end of this section.

**Example 4**    Find the total length of the cardioid $r = a(1 + \cos\theta)$.

**Solution**    The total length is twice the length from $\theta = 0$ to $\theta = \pi$. (Review Figure 8.53.) Since $dr/d\theta = -a\sin\theta$ for the cardioid, the arc length is

$$
\begin{aligned}
s &= 2\int_0^\pi \sqrt{a^2\sin^2\theta + a^2(1+\cos\theta)^2}\, d\theta \\
&= 2\int_0^\pi \sqrt{2a^2 + 2a^2\cos\theta}\, d\theta \qquad \text{(but } 1 + \cos\theta = 2\cos^2(\theta/2)\text{)} \\
&= 2\sqrt{2}a\int_0^\pi \sqrt{2\cos^2\frac{\theta}{2}}\, d\theta \\
&= 4a\int_0^\pi \cos\frac{\theta}{2}\, d\theta = 8a\sin\frac{\theta}{2}\Big|_0^\pi = 8a \text{ units.}
\end{aligned}
$$

---

# Exercises 8.6

In Exercises 1–11, sketch and find the areas of the given polar regions $R$.

1. $R$ lies between the origin and the spiral $r = \sqrt{\theta}$, $0 \le \theta \le 2\pi$.

2. $R$ lies between the origin and the spiral $r = \theta$, $0 \le \theta \le 2\pi$.

3. $R$ is bounded by the curve $r^2 = a^2\cos 2\theta$.

4. $R$ is one leaf of the curve $r = \sin 3\theta$.

5. $R$ is bounded by the curve $r = \cos 4\theta$.

6. $R$ lies inside both of the circles $r = a$ and $r = 2a\cos\theta$.

7. $R$ lies inside the cardioid $r = 1 - \cos\theta$ and outside the circle $r = 1$.

8. $R$ lies inside the cardioid $r = a(1 - \sin\theta)$ and inside the circle $r = a$.

9. $R$ lies inside the cardioid $r = 1 + \cos\theta$ and outside the circle $r = 3\cos\theta$.

10. $R$ is bounded by the lemniscate $r^2 = 2\cos 2\theta$ and is outside the circle $r = 1$.

11. $R$ is bounded by the smaller loop of the curve

$r = 1 + 2\cos\theta$.

Find the lengths of the polar curves in Exercises 12–14

12. $r = \theta^2$, $0 \le \theta \le \pi$       13. $r = e^{a\theta}$, $-\pi \le \theta \le \pi$

14. $r = a\theta$, $0 \le \theta \le 2\pi$

15. Show that the total arc length of the lemniscate $r^2 = \cos 2\theta$
is $4\displaystyle\int_0^{\pi/4} \sqrt{\sec 2\theta}\, d\theta$.

16. One leaf of the lemniscate $r^2 = \cos 2\theta$ is rotated (a) about the $x$-axis, and (b) about the $y$-axis. Find the area of the surface generated in each case.

* 17. Determine the angles at which the straight line $\theta = \pi/4$ intersects the cardioid $r = 1 + \sin\theta$.

* 18. At what points do the curves $r^2 = 2\sin 2\theta$ and $r = 2\cos\theta$ intersect? At what angle do the curves intersect at each of these points?

* 19. At what points do the curves $r = 1 - \cos\theta$ and $r = 1 - \sin\theta$ intersect? At what angle do the curves intersect at each of these points?

In Exercises 20–25, find all points on the given curve where the tangent line is horizontal, vertical, or does not exist.

* **20.** $r = \cos\theta + \sin\theta$

* **21.** $r = 2\cos\theta$

* **22.** $r^2 = \cos 2\theta$

* **23.** $r = \sin 2\theta$

* **24.** $r = e^\theta$

* **25.** $r = 2(1 - \sin\theta)$

**26.** The polar curve $r = f(\theta)$, $(\alpha \le \theta \le \beta)$ can be parametrized

$$x = r\cos\theta = f(\theta)\cos\theta, \qquad y = r\sin\theta = f(\theta)\sin\theta.$$

Derive the formula for the arc length element for the polar curve from that for a parametric curve.

# Chapter Review

## Key Ideas

- **What do the following phrases mean?**

  ◇ a conic section

  ◇ an ellipse

  ◇ a parabola

  ◇ a hyperbola

  ◇ a parametric curve

  ◇ a parametrization of a curve

  ◇ a smooth curve

  ◇ a polar curve

- **What is the focus-directrix definition of a conic?**

- **How can you find the slope of a parametric curve?**

- **How can you find the length of a parametric curve?**

- **How can you find the length of a polar curve?**

- **How can you find the area bounded by a polar curve?**

## Review Exercises

In Exercises 1–4, describe the conic having the given equation. Give its foci and principal axes, and, if it is a hyperbola, its asymptotes.

**1.** $x^2 + 2y^2 = 2$

**2.** $9x^2 - 4y^2 = 36$

**3.** $x + y^2 = 2y + 3$

**4.** $2x^2 + 8y^2 = 4x - 48y$

Identify the parametric curves in Exercises 5–10.

**5.** $x = t$, $y = 2 - t$, $(0 \le t \le 2)$

**6.** $x = 2\sin 3t$, $y = 2\cos 3t$, $(0 \le t \le 1/2)$

**7.** $x = \cosh t$, $y = \sinh^2 t$,

**8.** $x = e^t$, $y = e^{-2t}$, $(-1 \le t \le 1)$

**9.** $x = \cos(t/2)$, $y = 4\sin(t/2)$, $(0 \le t \le \pi)$

**10.** $x = \cos t + \sin t$, $y = \cos t - \sin t$, $(0 \le t \le 2\pi)$

In Exercises 11–14, determine the points where the given parametric curves have horizontal and vertical tangents, and sketch the curves.

**11.** $x = \dfrac{4}{1 + t^2}$, $y = t^3 - 3t$

**12.** $x = t^3 - 3t$, $y = t^3 + 3t$

**13.** $x = t^3 - 3t$, $y = t^3$

**14.** $x = t^3 - 3t$, $y = t^3 - 12t$

**15.** Find the area bounded by the part of the curve $x = t^3 - t$, $y = |t^3|$ that forms a closed loop.

**16.** Find the volume of the solid of revolution generated by rotating the closed loop in the previous exercise about the $y$-axis.

**17.** Find the length of the curve $x = e^t - t$, $y = 4e^{t/2}$ from $t = 0$ to $t = 2$.

**18.** Find the area of the surface obtained by rotating the arc in the previous exercise about the $x$-axis.

Sketch the polar graphs of the equations in Exercises 19–24.

**19.** $r = \theta$, $\left(-\dfrac{3\pi}{2} \le \theta \le \dfrac{3\pi}{2}\right)$

**20.** $r = |\theta|$, $(-2\pi \le \theta \le 2\pi)$

**21.** $r = 1 + \cos 2\theta$

**22.** $r = 2 + \cos 2\theta$

**23.** $r = 1 + 2\cos 2\theta$

**24.** $r = 1 - \sin 3\theta$

**25.** Find the area of one of the two larger loops of the curve in Exercise 23.

**26.** Find the area of one of the two smaller loops of the curve in Exercise 23.

**27.** Find the area of the smaller of the two loops enclosed by the curve $r = 1 + \sqrt{2}\sin\theta$.

**28.** Find the area of the region inside the cardioid $r = 1 + \cos\theta$ and to the left of the line $x = 1/4$.

## Challenging Problems

**1.** A glass in the shape of a circular cylinder of radius 4 cm is more than half filled with water. If the glass is tilted by an angle $\theta$ from the vertical, where $\theta$ is small enough that no water spills out, find the surface area of the water.

**2.** Show that a plane that is not parallel to the axis of a circular cylinder intersects the cylinder in an ellipse. *Hint:* you can do this by the same method used in Exercise 29 of Section 8.1.

**3.** Given two points $F_1$ and $F_2$ that are foci of an ellipse, and a third point $P$ on the ellipse, describe a geometric method (using straight edge and compass) for constructing the tangent line to the ellipse at $P$. *Hint:* think about the reflection property of ellipses.

**4.** Let $C$ be a parabola with vertex $V$, and let $P$ be any point on the parabola. Let $R$ be the point where the tangent to the parabola at $P$ intersects the axis of the parabola. (Thus the axis is the line $RV$.) Let $Q$ be the point on $RV$ such that $PQ$ is perpendicular to $RV$. Show that $V$ bisects the line segment

*RQ.* How does this result suggest a geometric method for constructing a tangent to a parabola at a point on it, given the axis and vertex of the parabola?

5. A barrel has the shape of a solid of revolution obtained by rotating about its major axis the part of an ellipse lying between lines through its foci perpendicular to that axis. The barrel is 4 ft high and 2 ft in radius at its middle. What is its volume?

6. (a) Show that any straight line not passing through the origin can be written in polar form as $r = \dfrac{a}{\cos(\theta - \theta_0)}$, where $a$ and $\theta_0$ are constants. What is the geometric significance of these constants?

   (b) Let $r = g(\theta)$ be the polar equation of a straight line that does not pass through the origin. Show that
   $$g^2 + 2(g')^2 - gg'' = 0.$$

   (c) Let $r = f(\theta)$ be the polar equation of a curve, where $f''$ is continuous and $r \neq 0$ in some interval of values of $\theta$. Show that the curve is turning towards the origin if $f^2 + 2(f')^2 - ff'' > 0$, and away from the origin if $< 0$. *Hint:* let $r = g(\theta)$ be the polar equation of a straight line tangent to the curve, and use part (b). How do $f$, $f'$, and $f''$ relate to $g$, $g'$, and $g''$ at the point of tangency?

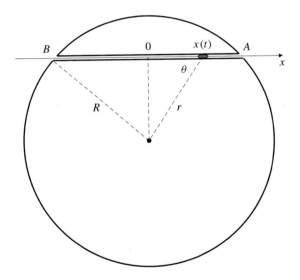

**Figure 8.56**

7. **(Fast trip, but it might get hot)** If we assume that the density of the earth is uniform throughout, then it can be shown that the acceleration of gravity at a distance $r \leq R$ from the centre of the earth is directed towards the centre of the earth and has magnitude $a(r) = rg/R$, where $g$ is the usual acceleration of

gravity at the surface ($g \approx 32$ ft/s$^2$), and $R$ is the radius of the earth ($R \approx 3960$ mi). Suppose that a straight tunnel $AB$ is drilled through the earth between any two points $A$ and $B$ on the surface, say Atlanta and Baghdad. (See Figure 8.56.) Suppose that a vehicle is constructed that can slide without friction or air resistance through this tunnel. Show that such a vehicle will, if released at one end of the tunnel, fall back and forth between $A$ and $B$, executing simple harmonic motion with period $2\pi\sqrt{R/g}$. How many minutes will the round trip take? What is surprising here is that this period does not depend on where $A$ and $B$ are, or on the distance between them. *Hint:* Let the $x$-axis lie along the tunnel, with origin at the point closest to the centre of the earth. When the vehicle is at position with $x$-coordinate $x(t)$, its acceleration along the tunnel is the component of the gravitational acceleration along the tunnel, that is, $-a(r)\cos\theta$, where $\theta$ is the angle between the line of the tunnel and the line from the vehicle to the centre of the earth.

* 8. **(Search and Rescue)** Two coast guard stations pick up a distress signal from a ship and use radio direction finders to locate it. Station O observes that the distress signal is coming from the northeast (45 degrees east of north), while station $P$, which is 100 miles north of station O, observes that the signal is coming from due east. Each station's direction finder is accurate to within $\pm 3$ degrees.

   (a) How large an area of the ocean must a rescue aircraft search to ensure that it finds the foundering ship?

   (b) If the accuracy of the direction finders is within $\pm\varepsilon$, how sensitive is the search area to changes in $\varepsilon$ when $\varepsilon = 3$ degrees? (Express your answer in square miles per degree.)

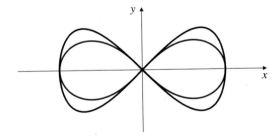

**Figure 8.57**

9. Figure 8.57 shows the graphs of the parametric curve $x = \sin t$, $y = \frac{1}{2}\sin(2t)$, $0 \leq t \leq 2\pi$ and the polar curve $r^2 = \cos(2\theta)$. Each has the shape of an "$\infty$." Which curve is which? Find the area inside the outer curve and outside the inner curve.

# CHAPTER 9

# Sequences, Series, and Power Series

**Introduction**  An infinite series is a sum that involves infinitely many terms. Since addition is carried out on two numbers at a time, the evaluation of the sum of an infinite series necessarily involves finding a limit. Complicated functions $f(x)$ can frequently be expressed as series of simpler functions. For example, many of the transcendental functions we have encountered can be expressed as series of powers of $x$ so that they resemble polynomials of infinite degree. Such series can be differentiated and integrated term by term, and they play a very important role in the study of calculus.

## 9.1 Sequences and Convergence

By a **sequence** (or an **infinite sequence**) we mean an ordered list having a first element but no last element. For our purposes the elements (called **terms**) of a sequence will always be real numbers, though much of our discussion could be applied to complex numbers as well. Examples of sequences are:

$\{1,\ 2,\ 3,\ 4,\ 5,\ \ldots\}$ the sequence of positive integers,

$\left\{-\dfrac{1}{2},\ \dfrac{1}{4},\ -\dfrac{1}{8},\ \dfrac{1}{16},\ \ldots\right\}$ the sequence of positive integer powers of $-\dfrac{1}{2}$.

The terms of a sequence are usually listed in braces { } as shown. The ellipsis (...) should be read "and so on."

An infinite sequence is a special kind of function, one whose domain is a set of integers extending from some starting integer to infinity. The starting integer is usually 1, so the domain is the set of positive integers. The sequence $\{a_1,\ a_2,\ a_3,\ a_4,\ \ldots\}$ is the function $f$ that takes the value $f(n) = a_n$ at each positive integer $n$. A sequence can be specified in three ways:

(i) We can list the first few terms followed by ... *if the pattern is obvious.*

(ii) We can provide a formula for the **general term** $a_n$ as a function of $n$.

(iii) We can provide a formula for calculating the term $a_n$ as a function of earlier terms $a_1, a_2, \ldots, a_{n-1}$ and specify enough of the beginning terms so the process of computing higher terms can begin.

In each case it must be possible to determine any term of the sequence, though it may be necessary to calculate all the preceding terms first.

### Example 1  (Some examples of sequences)

(a) $\{n\} = \{1,\ 2,\ 3,\ 4,\ 5,\ \ldots\}$

(b) $\left\{\left(-\dfrac{1}{2}\right)^n\right\} = \left\{-\dfrac{1}{2},\ \dfrac{1}{4},\ -\dfrac{1}{8},\ \dfrac{1}{16},\ \ldots\right\}$

(c) $\left\{\dfrac{n-1}{n}\right\} = \left\{0, \dfrac{1}{2}, \dfrac{2}{3}, \dfrac{3}{4}, \dfrac{4}{5}, \cdots\right\}$

(d) $\{(-1)^{n-1}\} = \{\cos((n-1)\pi)\} = \{1, -1, 1, -1, 1, \ldots\}$

(e) $\left\{\dfrac{n^2}{2^n}\right\} = \left\{\dfrac{1}{2}, 1, \dfrac{9}{8}, 1, \dfrac{25}{32}, \dfrac{36}{64}, \dfrac{49}{128}, \cdots\right\}$

(f) $\left\{\left(1+\dfrac{1}{n}\right)^n\right\} = \left\{2, \left(\dfrac{3}{2}\right)^2, \left(\dfrac{4}{3}\right)^3, \left(\dfrac{5}{4}\right)^4, \cdots\right\}$

(g) $\left\{\dfrac{\cos(n\pi/2)}{n}\right\} = \left\{0, -\dfrac{1}{2}, 0, \dfrac{1}{4}, 0, -\dfrac{1}{6}, 0, \dfrac{1}{8}, 0, \ldots\right\}$

(h) $a_1 = 1, \quad a_{n+1} = \sqrt{6 + a_n}, \quad (n = 1, 2, 3, \ldots).$

In this case $\{a_n\} = \{1, \sqrt{7}, \sqrt{6 + \sqrt{7}}, \ldots\}.$

Note that there is no *obvious* formula for $a_n$ as an explicit function of $n$ here, but we can still calculate $a_n$ for any desired value of $n$ provided we first calculate all the earlier values $a_2, a_3, \ldots, a_{n-1}$.

(i) $a_1 = 1, a_2 = 1, a_{n+2} = a_n + a_{n+1}, \quad (n = 1, 2, 3, \ldots)$

Here $\{a_n\} = \{1, 1, 2, 3, 5, 8, 13, 21, \ldots\}.$

This is called the **Fibonacci sequence**. Each term after the second is the sum of the previous two terms.

In parts (a)–(g) of Example 1, the formulas on the left sides define the general term of each sequence $\{a_n\}$ as an explicit function of $n$. In parts (h) and (i) we say the sequence $\{a_n\}$ is defined **recursively** or **inductively**; each term must be calculated from previous ones rather than directly as a function of $n$.

The following definitions introduce terminology used to describe various properties of sequences.

---

**DEFINITION 1**

**Terms for describing sequences**

(a) The sequence $\{a_n\}$ is **bounded below** by $L$, and $L$ is a **lower bound** for $\{a_n\}$, if $a_n \geq L$ for every $n = 1, 2, 3, \ldots$. The sequence is **bounded above** by $M$, and $M$ is an **upper bound**, if $a_n \leq M$ for every such $n$.

The sequence $\{a_n\}$ is **bounded** if it is both bounded above and bounded below. In this case there is a constant $K$ such that $|a_n| \leq K$ for every $n = 1, 2, 3, \ldots$. (We can take $K$ to be the larger of $-L$ and $M$.)

(b) The sequence $\{a_n\}$ is **positive** if it is bounded below by zero, that is, if $a_n \geq 0$ for every $n = 1, 2, 3, \ldots$; it is **negative** if $a_n \leq 0$ for every $n$.

(c) The sequence $\{a_n\}$ is **increasing** if $a_{n+1} \geq a_n$ for every $n = 1, 2, 3, \ldots$; it is **decreasing** if $a_{n+1} \leq a_n$ for every such $n$. The sequence is said to be **monotonic** if it is either increasing or decreasing. (Note that the terminology here is looser than that we have used for functions, where we would have used *nondecreasing* and *nonincreasing* to describe this behaviour.)

(d) The sequence $\{a_n\}$ is **alternating** if $a_n a_{n+1} < 0$ for every $n = 1, 2, \ldots$, that is, if any two consecutive terms have opposite sign.

**Example 2**    (Describing some sequences)

(a) The sequence $\{n\} = \{1, 2, 3, \dots\}$ is positive, increasing, and bounded below. A lower bound for the sequence is 1, or any smaller number. The sequence is not bounded above.

(b) $\left\{\dfrac{n-1}{n}\right\} = \left\{0, \dfrac{1}{2}, \dfrac{2}{3}, \dfrac{3}{4}, \dots\right\}$ is positive, bounded, and increasing. Here 0 is a lower bound and 1 is an upper bound.

(c) $\left\{\left(-\dfrac{1}{2}\right)^n\right\} = \left\{-\dfrac{1}{2}, \dfrac{1}{4}, -\dfrac{1}{8}, \dfrac{1}{16}, \dots\right\}$ is bounded and alternating.    Here $-1/2$ is a lower bound and $1/4$ is an upper bound.

(d) $\{(-1)^n n\} = \{-1, 2, -3, 4, -5, \dots\}$ is alternating but not bounded either above or below.

When you want to show that a sequence is increasing you can try to show that the inequality $a_{n+1} - a_n \geq 0$ holds for $n \geq 1$. Alternatively, if $a_n = f(n)$ for a differentiable function $f(x)$, you can show that $f$ is an increasing function on $[1, \infty)$ by showing that $f'(x) \geq 0$ there. Similar approaches are useful for showing that a sequence is decreasing.

**Example 3**    If $a_n = \dfrac{n}{n^2 + 1}$, show that the sequence $\{a_n\}$ is decreasing.

**Solution**    Since $a_n = f(n)$, where $f(x) = \dfrac{x}{x^2 + 1}$ and

$$f'(x) = \frac{(x^2 + 1)(1) - x(2x)}{(x^2 + 1)^2} = \frac{1 - x^2}{(x^2 + 1)^2} \leq 0 \quad \text{for } x \geq 1,$$

the function $f(x)$ is decreasing on $[1, \infty)$ and therefore $\{a_n\}$ is a decreasing sequence.

The sequence $\left\{\dfrac{n^2}{2^n}\right\} = \left\{\dfrac{1}{2}, 1, \dfrac{9}{8}, 1, \dfrac{25}{32}, \dfrac{36}{64}, \dfrac{49}{128}, \dots\right\}$ is obviously positive therefore bounded below. It seems clear that from the fourth term on, all the terms are getting smaller. However, $a_2 > a_1$ and $a_3 > a_2$. Since $a_{n+1} \leq a_n$ only if $n \geq 3$, we say that this sequence is **ultimately decreasing**. The adverb *ultimately* is used to describe any termwise property of a sequence that the terms have from some point on, but not necessarily at the beginning of the sequence. Thus the sequence

$$\{n - 100\} = \{-99, -98, \dots, -2, -1, 0, 1, 2, 3, \dots\}$$

is *ultimately positive* even though the first 99 terms are negative, and the sequence

$$\left\{(-1)^n + \frac{4}{n}\right\} = \left\{3, 3, \frac{1}{3}, 2, -\frac{1}{5}, \frac{5}{3}, -\frac{3}{7}, \frac{3}{2}, \dots\right\}$$

is *ultimately alternating* even though the first few terms do not alternate.

## Convergence of Sequences

Central to the study of sequences is the notion of convergence. The concept of the limit of a sequence is a special case of the concept of the limit of a function $f(x)$

as $x \to \infty$. We say that the sequence $\{a_n\}$ **converges to the limit** $L$, and we write $\lim_{n \to \infty} a_n = L$, provided the distance from $a_n$ to $L$ on the real line approaches 0 as $n$ increases toward $\infty$. We state this definition more formally as follows:

**DEFINITION 2**

> **Limit of a sequence**
> We say that sequence $\{a_n\}$ converges to the limit $L$, and we write $\lim_{n \to \infty} a_n = L$, if for every positive real number $\epsilon$ there exists an integer $N$ (which may depend on $\epsilon$) such that if $n > N$, then $|a_n - L| < \epsilon$.

This definition is illustrated in Figure 9.1.

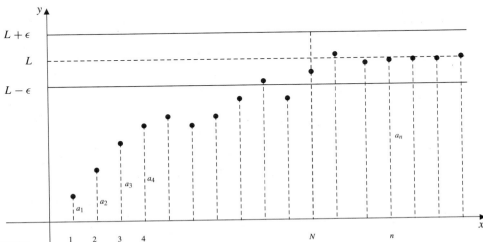

**Figure 9.1**    A convergent sequence

**Example 4**    Show that $\lim_{n \to \infty} \dfrac{c}{n^p} = 0$ for any real number $c$ and any $p > 0$.

**Solution**    Let $\epsilon > 0$ be given. Then

$$\left| \frac{c}{n^p} \right| < \epsilon \quad \text{if} \quad n^p > \frac{|c|}{\epsilon},$$

that is, if $n > N = \left( \dfrac{|c|}{\epsilon} \right)^{1/p}$. By Definition 2, $\lim_{n \to \infty} \dfrac{c}{n^p} = 0$. ∎

Every sequence $\{a_n\}$ must either **converge** to a finite limit $L$ or **diverge**. That is, either $\lim_{n \to \infty} a_n = L$ exists (is a real number), or $\lim_{n \to \infty} a_n$ does not exist. If $\lim_{n \to \infty} a_n = \infty$, we can say that the sequence diverges to $\infty$; if $\lim_{n \to \infty} a_n = -\infty$, we can say that it diverges to $-\infty$. If $\lim_{n \to \infty} a_n$ simply does not exist (but is not $\infty$ or $-\infty$), we can only say that the sequence diverges.

**Example 5**    (**Examples of convergent and divergent sequences**)

(a) $\{(n - 1)/n\}$ converges to 1; $\lim_{n \to \infty} (n - 1)/n = \lim_{n \to \infty} \left( 1 - (1/n) \right) = 1$.

(b) $\{n\} = \{1, 2, 3, 4, \ldots\}$ diverges to $\infty$.

(c) $\{-n\} = \{-1, -2, -3, -4, \ldots\}$ diverges to $-\infty$.

(d) $\{(-1)^n\} = \{-1, 1, -1, 1, -1, \ldots\}$ simply diverges.

(e) $\{(-1)^n n\} = \{-1, 2, -3, 4, -5, \ldots\}$ diverges (but not to $\infty$ or $-\infty$ even though $\lim_{n \to \infty} |a_n| = \infty$).

The limit of a sequence is equivalent to the limit of a function as its argument approaches infinity:

> If $\lim_{x \to \infty} f(x) = L$ and $a_n = f(n)$, then $\lim_{n \to \infty} a_n = L$.

Because of this, the standard rules for limits of functions (Theorems 2 and 4 of Section 1.2) also hold for limits of sequences, with the appropriate changes of notation. Thus, if $\{a_n\}$ and $\{b_n\}$ converge, then

$$\lim_{n \to \infty} (a_n \pm b_n) = \lim_{n \to \infty} a_n \pm \lim_{n \to \infty} b_n,$$

$$\lim_{n \to \infty} c a_n = c \lim_{n \to \infty} a_n,$$

$$\lim_{n \to \infty} a_n b_n = (\lim_{n \to \infty} a_n)(\lim_{n \to \infty} b_n),$$

$$\lim_{n \to \infty} \frac{a_n}{b_n} = \frac{\lim_{n \to \infty} a_n}{\lim_{n \to \infty} b_n} \qquad \text{assuming } \lim_{n \to \infty} b_n \neq 0.$$

If $a_n \leq b_n$ ultimately, then $\lim_{n \to \infty} a_n \leq \lim_{n \to \infty} b_n$.

If $a_n \leq b_n \leq c_n$ ultimately, and $\lim_{n \to \infty} a_n = L = \lim_{n \to \infty} c_n$, then $\lim_{n \to \infty} b_n = L$.

The limits of many explicitly defined sequences can be evaluated using these properties in a manner similar to the methods used for limits of the form $\lim_{x \to \infty} f(x)$ in Section 1.3.

**Example 6**    Calculate the limits of the sequences

(a) $\left\{\dfrac{2n^2 - n - 1}{5n^2 + n - 3}\right\}$,    (b) $\left\{\dfrac{\cos n}{n}\right\}$,    (c) $\{\sqrt{n^2 + 2n} - n\}$.

**Solution**

(a) We divide the numerator and denominator of the expression for $a_n$ by the highest power of $n$ in the denominator, that is, by $n^2$:

$$\lim_{n \to \infty} \frac{2n^2 - n - 1}{5n^2 + n - 3} = \lim_{n \to \infty} \frac{2 - (1/n) - (1/n^2)}{5 + (1/n) - (3/n^2)} = \frac{2 - 0 - 0}{5 + 0 - 0} = \frac{2}{5},$$

since $\lim_{n \to \infty} 1/n = 0$ and $\lim_{n \to \infty} 1/n^2 = 0$. The sequence converges and its limit is 2/5.

(b) Since $|\cos n| \leq 1$ for every $n$, we have

$$-\frac{1}{n} \leq \frac{\cos n}{n} \leq \frac{1}{n} \qquad \text{for} \quad n \geq 1.$$

Now $\lim_{n \to \infty} -1/n = 0$ and $\lim_{n \to \infty} 1/n = 0$, so by the sequence version of the Squeeze Theorem, $\lim_{n \to \infty} (\cos n)/n = 0$ also. The sequence converges to 0.

(c) For this sequence we multiply the numerator and the denominator (which is 1) by the conjugate of the expression in the numerator:

$$\lim_{n\to\infty} (\sqrt{n^2 + 2n} - n) = \lim_{n\to\infty} \frac{(\sqrt{n^2 + 2n} - n)(\sqrt{n^2 + 2n} + n)}{\sqrt{n^2 + 2n} + n}$$

$$= \lim_{n\to\infty} \frac{2n}{\sqrt{n^2 + 2n} + n} = \lim_{n\to\infty} \frac{2}{\sqrt{1 + (2/n)} + 1} = 1.$$

The sequence converges to 1.

**Example 7**  Evaluate $\lim_{n\to\infty} n \tan^{-1}\left(\dfrac{1}{n}\right)$.

**Solution**  For this example it is best to replace the $n$th term of the sequence by the corresponding function of a real variable $x$ and take the limit as $x \to \infty$. We use l'Hôpital's rule:

$$\lim_{n\to\infty} n \tan^{-1}\left(\frac{1}{n}\right) = \lim_{x\to\infty} x \tan^{-1}\left(\frac{1}{x}\right)$$

$$= \lim_{x\to\infty} \frac{\tan^{-1}\left(\dfrac{1}{x}\right)}{\dfrac{1}{x}} \qquad \left[\frac{0}{0}\right]$$

$$= \lim_{x\to\infty} \frac{\dfrac{1}{1 + (1/x^2)}\left(-\dfrac{1}{x^2}\right)}{-\left(\dfrac{1}{x^2}\right)} = \lim_{x\to\infty} \frac{1}{1 + \dfrac{1}{x^2}} = 1.$$

**THEOREM 1**

If $\{a_n\}$ converges, then $\{a_n\}$ is bounded.

**PROOF**  Suppose $\lim_{n\to\infty} a_n = L$. According to Definition 2, for $\epsilon = 1$ there exists a number $N$ such that if $n \geq N$ then $|a_n - L| < 1$; therefore $|a_n| < 1 + |L|$ for such $n$. (Why is this true?) If $K$ denotes the largest of the numbers $|a_1|$, $|a_2|$, $\ldots$, $|a_{N-1}|$, and $1 + |L|$, then $|a_n| \leq K$ for every $n = 1, 2, 3, \ldots$. Hence $\{a_n\}$ is bounded.

The converse of Theorem 1 is false; the sequence $\{(-1)^n\}$ is bounded but does not converge.

The *completeness property* of the real number system (see Section P.1) can be reformulated in terms of sequences to read as follows:

**Bounded monotonic sequences converge**

If the sequence $\{a_n\}$ is bounded above and is (ultimately) increasing, then it converges. The same conclusion holds if $\{a_n\}$ is bounded below and is (ultimately) decreasing.

Thus, a bounded, ultimately monotonic sequence is convergent. (See Figure 9.2.)

**Figure 9.2**    An ultimately increasing sequence that is bounded above

**Example 8**    Let $a_n$ be defined recursively by

$$a_1 = 1, \qquad a_{n+1} = \sqrt{6 + a_n} \qquad (n = 1, 2, 3, \ldots).$$

Show that $\lim_{n\to\infty} a_n$ exists and find its value.

**Solution**    Observe that $a_2 = \sqrt{6+1} = \sqrt{7} > a_1$. If $a_{k+1} > a_k$, then $a_{k+2} = \sqrt{6 + a_{k+1}} > \sqrt{6 + a_k} = a_{k+1}$, so $\{a_n\}$ is increasing, by induction. Now observe that $a_1 = 1 < 3$. If $a_k < 3$, then $a_{k+1} = \sqrt{6 + a_k} < \sqrt{6 + 3} = 3$, so $a_n < 3$ for every $n$ by induction. Since $\{a_n\}$ is increasing and bounded above, $\lim_{n\to\infty} a_n = a$ exists, by completeness. Since $\sqrt{6 + x}$ is a continuous function of $x$, we have

$$a = \lim_{n\to\infty} a_{n+1} = \lim_{n\to\infty} \sqrt{6 + a_n} = \sqrt{6 + \lim_{n\to\infty} a_n} = \sqrt{6 + a}.$$

Thus $a^2 = 6 + a$, or $a^2 - a - 6 = 0$, or $(a - 3)(a + 2) = 0$. This quadratic has roots $a = 3$ and $a = -2$. Since $a_n \geq 1$ for every $n$, we must have $a \geq 1$. Therefore, $a = 3$ and $\lim_{n\to\infty} a_n = 3$. ∎

**Example 9**    Does $\left\{ \left(1 + \dfrac{1}{n}\right)^n \right\}$ converge or not?

**Solution**    We could make an effort to show that the given sequence is, in fact, increasing and bounded above. (See Exercise 32 at the end of this section.) However, we already know the answer. By Theorem 6 of Section 3.4,

$$\lim_{n\to\infty} \left(1 + \frac{1}{n}\right)^n = e^1 = e.$$

∎

**THEOREM**    **2**    If $\{a_n\}$ is (ultimately) increasing, then either it is bounded above, and therefore convergent, or else it is not bounded above and diverges to infinity.

The proof of this theorem is left as an exercise. A corresponding result holds for (ultimately) decreasing sequences.

The following theorem evaluates two important limits that find frequent application in the study of series.

**THEOREM 3**

(a)   If $|x| < 1$, then $\lim_{n \to \infty} x^n = 0$.

(b)   If $x$ is any real number, then $\lim_{n \to \infty} \dfrac{x^n}{n!} = 0$.

**PROOF**   For part (a) observe that

$$\lim_{n \to \infty} \ln |x|^n = \lim_{n \to \infty} n \ln |x| = -\infty,$$

since $\ln |x| < 0$ when $|x| < 1$. Accordingly, since $e^x$ is continuous,

$$\lim_{n \to \infty} |x|^n = \lim_{n \to \infty} e^{\ln |x|^n} = e^{\lim_{n \to \infty} \ln |x|^n} = 0.$$

Since $-|x|^n \le x^n \le |x|^n$, we have $\lim_{n \to \infty} x^n = 0$ by the Squeeze Theorem.

For part (b), pick any $x$ and let $N$ be an integer such that $N > |x|$. If $n > N$ we have

$$\left| \frac{x^n}{n!} \right| = \frac{|x|}{1} \frac{|x|}{2} \frac{|x|}{3} \cdots \frac{|x|}{N-1} \frac{|x|}{N} \frac{|x|}{N+1} \cdots \frac{|x|}{n}$$

$$< \frac{|x|^{N-1}}{(N-1)!} \frac{|x|}{N} \frac{|x|}{N} \frac{|x|}{N} \cdots \frac{|x|}{N}$$

$$= \frac{|x|^{N-1}}{(N-1)!} \left( \frac{|x|}{N} \right)^{n-N+1} = K \left( \frac{|x|}{N} \right)^n,$$

where $K = \dfrac{|x|^{N-1}}{(N-1)!} \left( \dfrac{|x|}{N} \right)^{1-N}$ is a constant that is independent of $n$. Since $|x|/N < 1$, we have $\lim_{n \to \infty} (|x|/N)^n = 0$ by part (a). Thus $\lim_{n \to \infty} |x^n/n!| = 0$, and so $\lim_{n \to \infty} x^n/n! = 0$.

---

**Example 10**   Find $\lim_{n \to \infty} \dfrac{3^n + 4^n + 5^n}{5^n}$.

**Solution**   $\lim_{n \to \infty} \dfrac{3^n + 4^n + 5^n}{5^n} = \lim_{n \to \infty} \left[ \left( \dfrac{3}{5} \right)^n + \left( \dfrac{4}{5} \right)^n + 1 \right] = 0 + 0 + 1 = 1$, by Theorem 3(a).

# Exercises 9.1

In Exercises 1–13, determine whether the given sequence is (a) bounded (above or below), (b) positive or negative (ultimately), (c) increasing, decreasing, or alternating, and (d) convergent, divergent, divergent to $\infty$ or $-\infty$.

**1.** $\left\{\dfrac{2n^2}{n^2+1}\right\}$   **2.** $\left\{\dfrac{2n}{n^2+1}\right\}$

**3.** $\left\{4-\dfrac{(-1)^n}{n}\right\}$   **4.** $\left\{\sin\dfrac{1}{n}\right\}$

**5.** $\left\{\dfrac{n^2-1}{n}\right\}$   **6.** $\left\{\dfrac{e^n}{\pi^n}\right\}$

**7.** $\left\{\dfrac{e^n}{\pi^{n/2}}\right\}$   **8.** $\left\{\dfrac{(-1)^n n}{e^n}\right\}$

**9.** $\left\{\dfrac{2^n}{n^n}\right\}$   **10.** $\left\{\dfrac{(n!)^2}{(2n)!}\right\}$

**11.** $\left\{n\cos\left(\dfrac{n\pi}{2}\right)\right\}$   **12.** $\left\{\dfrac{\sin n}{n}\right\}$

**13.** $\{1,\ 1,\ -2,\ 3,\ 3,\ -4,\ 5,\ 5,\ -6,\ \ldots\}$

In Exercises 14–29, evaluate, wherever possible, the limit of the sequence $\{a_n\}$.

**14.** $a_n = \dfrac{5-2n}{3n-7}$   **15.** $a_n = \dfrac{n^2-4}{n+5}$

**16.** $a_n = \dfrac{n^2}{n^3+1}$   **17.** $a_n = (-1)^n\dfrac{n}{n^3+1}$

**18.** $a_n = \dfrac{n^2-2\sqrt{n}+1}{1-n-3n^2}$   **19.** $a_n = \dfrac{e^n-e^{-n}}{e^n+e^{-n}}$

**20.** $a_n = n\sin\dfrac{1}{n}$   **21.** $a_n = \left(\dfrac{n-3}{n}\right)^n$

**22.** $a_n = \dfrac{n}{\ln(n+1)}$   **23.** $a_n = \sqrt{n+1}-\sqrt{n}$

**24.** $a_n = n - \sqrt{n^2-4n}$

**25.** $a_n = \sqrt{n^2+n} - \sqrt{n^2-1}$

**26.** $a_n = \left(\dfrac{n-1}{n+1}\right)^n$   **27.** $a_n = \dfrac{(n!)^2}{(2n)!}$

**28.** $a_n = \dfrac{n^2 2^n}{n!}$   **29.** $a_n = \dfrac{\pi^n}{1+2^{2n}}$

**30.** Let $a_1 = 1$ and $a_{n+1} = \sqrt{1+2a_n}$ ($n = 1, 2, 3, \ldots$). Show that $\{a_n\}$ is increasing and bounded above. (*Hint:* show that 3 is an upper bound.) Hence conclude that the sequence converges, and find its limit.

* **31.** Repeat Exercise 30 for the sequence defined by $a_1 = 3$, $a_{n+1} = \sqrt{15+2a_n}$, $n = 1, 2, 3, \ldots$. This time you will have to guess an upper bound.

* **32.** Let $a_n = \left(1+\dfrac{1}{n}\right)^n$ so that $\ln a_n = n\ln\left(1+\dfrac{1}{n}\right)$. Use properties of the logarithm function to show that (a) $\{a_n\}$ is increasing and (b) $e$ is an upper bound for $\{a_n\}$.

* **33.** Prove Theorem 2. Also, state an analogous theorem pertaining to ultimately decreasing sequences.

**34.** If $\{|a_n|\}$ is bounded, prove that $\{a_n\}$ is bounded.

**35.** If $\lim_{n\to\infty}|a_n| = 0$, prove that $\lim_{n\to\infty} a_n = 0$.

* **36.** Which of the following statements are TRUE and which are FALSE? Justify your answers.

   (a) If $\lim_{n\to\infty} a_n = \infty$ and $\lim_{n\to\infty} b_n = L > 0$, then $\lim_{n\to\infty} a_n b_n = \infty$.

   (b) If $\lim_{n\to\infty} a_n = \infty$ and $\lim_{n\to\infty} b_n = -\infty$, then $\lim_{n\to\infty}(a_n + b_n) = 0$.

   (c) If $\lim_{n\to\infty} a_n = \infty$ and $\lim_{n\to\infty} b_n = -\infty$, then $\lim_{n\to\infty} a_n b_n = -\infty$.

   (d) If neither $\{a_n\}$ nor $\{b_n\}$ converges, then $\{a_n b_n\}$ does not converge.

   (e) If $\{|a_n|\}$ converges, then $\{a_n\}$ converges.

## 9.2 Infinite Series

An **infinite series**, usually just called a **series**, is a formal sum of infinitely many terms; for instance,

$$a_1 + a_2 + a_3 + a_4 + \cdots$$

is a series formed by adding the terms of the sequence $\{a_n\}$. This series is also denoted $\sum_{n=1}^{\infty} a_n$:

$$\sum_{n=1}^{\infty} a_n = a_1 + a_2 + a_3 + a_4 + \cdots$$

For example,

$$\sum_{n=1}^{\infty} \frac{1}{n} = 1 + \frac{1}{2} + \frac{1}{3} + \frac{1}{4} + \cdots$$

$$\sum_{n=1}^{\infty} \frac{(-1)^{n-1}}{2^{n-1}} = 1 - \frac{1}{2} + \frac{1}{4} - \frac{1}{8} + \frac{1}{16} - \cdots.$$

It is sometimes necessary or useful to start the sum from some index other than 1:

$$\sum_{n=0}^{\infty} a^n = 1 + a + a^2 + a^3 + \cdots$$

$$\sum_{n=2}^{\infty} \frac{1}{\ln n} = \frac{1}{\ln 2} + \frac{1}{\ln 3} + \frac{1}{\ln 4} + \cdots.$$

Note that the latter series would make no sense if we had started the sum from $n = 1$; the first term would have been undefined.

When necessary, we can change the index of summation to start at a different value. This is accomplished by a substitution as illustrated in Example 3 of Section 5.1. For instance, using the substitution $n = m - 2$, we can rewrite $\sum_{n=1}^{\infty} a_n$ in the form $\sum_{m=3}^{\infty} a_{m-2}$. Both sums give rise to the same expansion

$$\sum_{n=1}^{\infty} a_n = a_1 + a_2 + a_3 + \cdots = \sum_{m=3}^{\infty} a_{m-2}.$$

Addition is an operation that is carried out on two numbers at a time. If we want to calculate the finite sum

$$a_1 + a_2 + a_3,$$

we could proceed by adding $a_1 + a_2$ and then adding $a_3$ to this sum, or else we might first add $a_2 + a_3$ and then add $a_1$ to the sum. Of course the associative law for addition assures us we will get the same answer both ways. This is the reason the symbol $a_1 + a_2 + a_3$ makes sense; we would otherwise have to write $(a_1 + a_2) + a_3$ or $a_1 + (a_2 + a_3)$. This reasoning extends to any sum $a_1 + a_2 + \cdots + a_n$ of finitely many terms, but it is not obvious what should be meant by a sum with infinitely many terms:

$$a_1 + a_2 + a_3 + a_4 + \cdots.$$

We no longer have any assurance that the terms can be added up in any order to yield the same sum. In fact, we will see in Section 9.4 that in certain circumstances, changing the order of terms in a series can actually change the sum of the series. The interpretation we place on the infinite sum is that of adding from left to right, as suggested by the grouping

$$\cdots ((((a_1 + a_2) + a_3) + a_4) + a_5) + \cdots.$$

We accomplish this by defining a new sequence $\{s_n\}$, called the **sequence of partial sums** of the series $\sum_{n=1}^{\infty} a_n$ so that $s_n$ is the sum of the first $n$ terms of the series:

$$s_1 = a_1$$
$$s_2 = s_1 + a_2 = a_1 + a_2$$
$$s_3 = s_2 + a_3 = a_1 + a_2 + a_3$$
$$\vdots$$
$$s_n = s_{n-1} + a_n = a_1 + a_2 + a_3 + \cdots + a_n = \sum_{j=1}^{n} a_j.$$
$$\vdots$$

We then define the sum of the infinite series to be the limit of this sequence of partial sums.

**DEFINITION 3**

> **Convergence of a series**
>
> We say that the series $\sum_{n=1}^{\infty} a_n$ **converges to the sum** $s$, and we write
>
> $$\sum_{n=1}^{\infty} a_n = s,$$
>
> if $\lim_{n \to \infty} s_n = s$, where $s_n$ is the $n$th partial sum of $\sum_{n=1}^{\infty} a_n$:
>
> $$s_n = a_1 + a_2 + a_3 + \cdots + a_n = \sum_{j=1}^{n} a_j.$$
>
> Thus, the *series* $\sum_{n=1}^{\infty} a_n$ converges if and only if the *sequence* $\{s_n\}$ of its partial sums converges.

Similarly, a series is said to diverge to infinity, diverge to negative infinity, or simply diverge, if its sequence of partial sums does so. It must be stressed that the convergence of the series $\sum_{n=1}^{\infty} a_n$ depends on the convergence of the sequence $\{s_n\} = \{\sum_{j=1}^{n} a_j\}$, *not* the sequence $\{a_n\}$.

## Geometric Series

**DEFINITION 4**

> **Geometric series**
>
> A series of the form $\sum_{n=1}^{\infty} a\, r^{n-1} = a + ar + ar^2 + ar^3 + \cdots$, whose $n$th term is $a_n = a\, r^{n-1}$ is called a **geometric series**. The number $a$ is the first term. The number $r$ is called the **common ratio of** the series since it is the value of the ratio of the $(n+1)$st term to the $n$th term for any $n \geq 1$:
>
> $$\frac{a_{n+1}}{a_n} = \frac{ar^n}{ar^{n-1}} = r \qquad n = 1, 2, 3, \ldots.$$

The $n$th partial sum $s_n$ of a geometric series is calculated as follows:

$$s_n = a + ar + ar^2 + ar^3 + \cdots + ar^{n-1}$$
$$r s_n = \quad ar + ar^2 + ar^3 + \cdots + ar^{n-1} + ar^n.$$

The second equation is obtained by multiplying the first by $r$. Subtracting these two equations (note the cancellations), we get $(1 - r)s_n = a - ar^n$. If $r \neq 1$, we can divide by $1 - r$ and get a formula for $s_n$.

---

**Partial sums of geometric series**

If $r \neq 1$, then the $n$th partial sum of a geometric series $\sum_{n=1}^{\infty} ar^{n-1}$ is

$$s_n = a + ar + ar^2 + \cdots + ar^{n-1} = \frac{a(1 - r^n)}{1 - r}.$$

If $r = 1$, we have, instead, $s_n = na$.

---

If $a = 0$ then $s_n = 0$ for every $n$, and $\lim_{n \to \infty} s_n = 0$. Now suppose $a \neq 0$. If $|r| < 1$, then $\lim_{n \to \infty} r^n = 0$, so $\lim_{n \to \infty} s_n = a/(1 - r)$. If $r > 1$, then $\lim_{n \to \infty} r^n = \infty$, and $\lim_{n \to \infty} s_n = \infty$ if $a > 0$, or $\lim_{n \to \infty} s_n = -\infty$ if $a < 0$. The same conclusion holds if $r = 1$, since $s_n = na$ in this case. If $r \leq -1$, $\lim_{n \to \infty} r^n$ does not exist and neither does $\lim_{n \to \infty} s_n$. Hence we conclude that

$$\sum_{n=1}^{\infty} ar^{n-1} \begin{cases} \text{converges to } 0 & \text{if } a = 0 \\ \text{converges to } \dfrac{a}{1 - r} & \text{if } |r| < 1 \\ \text{diverges to } \infty & \text{if } r \geq 1 \text{ and } a > 0 \\ \text{diverges to } -\infty & \text{if } r \geq 1 \text{ and } a < 0 \\ \text{diverges} & \text{if } r \leq -1. \end{cases}$$

The representation of the function $1/(1 - x)$ as the sum of a geometric series,

$$\frac{1}{1 - x} = \sum_{n=0}^{\infty} x^n = 1 + x + x^2 + x^3 + \cdots \quad \text{for } -1 < x < 1,$$

will be important in our discussion of power series later in this chapter.

---

**Example 1**  (Examples of geometric series and their sums)

(a) $1 + \dfrac{1}{2} + \dfrac{1}{4} + \dfrac{1}{8} + \cdots = \sum_{n=1}^{\infty} \left(\dfrac{1}{2}\right)^{n-1} = \dfrac{1}{1 - \dfrac{1}{2}} = 2$. Here $a = 1$ and $r = \dfrac{1}{2}$.

Since $|r| < 1$, the series converges.

(b) $\pi - e + \dfrac{e^2}{\pi} - \dfrac{e^3}{\pi^2} + \cdots = \sum_{n=1}^{\infty} \pi \left(-\dfrac{e}{\pi}\right)^{n-1}$  Here $a = \pi$ and $r = -\dfrac{e}{\pi}$.

$$= \frac{\pi}{1 - \left(-\dfrac{(e)}{\pi}\right)} = \frac{\pi^2}{\pi + e}.$$

The series converges since $\left|-\dfrac{e}{\pi}\right| < 1$.

(c) $1 + 2^{1/2} + 2 + 2^{3/2} + \cdots = \sum_{n=1}^{\infty} (\sqrt{2})^{n-1}$. This series diverges to $\infty$ since $a = 1 > 0$ and $r = \sqrt{2} > 1$.

(d) $1 - 1 + 1 - 1 + 1 - \cdots = \sum_{n=1}^{\infty} (-1)^{n-1}$. This series diverges since $r = -1$.

(e) Let $x = 0.323232 \cdots = 0.\overline{32}$; then

$$x = \frac{32}{100} + \frac{32}{100^2} + \frac{32}{100^3} + \cdots = \sum_{n=1}^{\infty} \frac{32}{100} \left( \frac{1}{100} \right)^{n-1} = \frac{32}{100} \frac{1}{1 - \frac{1}{100}} = \frac{32}{99}.$$

This is an alternative to the method of Example 1 of Section P.1 for representing repeating decimals as quotients of integers. ∎

**Example 2** If money earns interest at a constant effective rate of 8% per year, how much should you pay today for an annuity that will pay you (a) $1,000 at the end of each of the next 10 years and (b) $1,000 at the end of every year forever?

**Solution** A payment of $1,000 that is due to be received $n$ years from now has present value $\$1{,}000 \times \left( \dfrac{1}{1.08} \right)^n$ (since $\$A$ would grow to $\$A(1.08)^n$ in $n$ years). Thus $1,000 payments at the end of each of the next $n$ years are worth $\$s_n$ at the present time, where

$$s_n = 1{,}000 \left[ \frac{1}{1.08} + \left( \frac{1}{1.08} \right)^2 + \cdots + \left( \frac{1}{1.08} \right)^n \right]$$

$$= \frac{1{,}000}{1.08} \left[ 1 + \frac{1}{1.08} + \left( \frac{1}{1.08} \right)^2 + \cdots + \left( \frac{1}{1.08} \right)^{n-1} \right].$$

$$= \frac{1{,}000}{1.08} \frac{1 - \left( \frac{1}{1.08} \right)^n}{1 - \frac{1}{1.08}} = \frac{1{,}000}{0.08} \left[ 1 - \left( \frac{1}{1.08} \right)^n \right]$$

(a) The present value of 10 future payments is $\$s_{10} = \$6{,}710.08$.

(b) The present value of future payments continuing forever is

$$\$ \lim_{n \to \infty} s_n = \frac{\$1{,}000}{0.08} = \$12{,}500.$$

∎

## Telescoping Series and Harmonic Series

**Example 3** Show that the series

$$\sum_{n=1}^{\infty} \frac{1}{n(n+1)} = \frac{1}{1 \times 2} + \frac{1}{2 \times 3} + \frac{1}{3 \times 4} + \frac{1}{4 \times 5} + \cdots$$

converges and find its sum.

**Solution**  Since $\dfrac{1}{n(n+1)} = \dfrac{1}{n} - \dfrac{1}{n+1}$ we can write the partial sum $s_n$ in the form

$$s_n = \frac{1}{1 \times 2} + \frac{1}{2 \times 3} + \frac{1}{3 \times 4} + \cdots + \frac{1}{(n-1)n} + \frac{1}{n(n+1)}$$

$$= \left(1 - \frac{1}{2}\right) + \left(\frac{1}{2} - \frac{1}{3}\right) + \left(\frac{1}{3} - \frac{1}{4}\right) + \cdots$$

$$\cdots + \left(\frac{1}{n-1} - \frac{1}{n}\right) + \left(\frac{1}{n} - \frac{1}{n+1}\right)$$

$$= 1 - \frac{1}{2} + \frac{1}{2} - \frac{1}{3} + \frac{1}{3} - \cdots - \frac{1}{n} + \frac{1}{n} - \frac{1}{n+1}$$

$$= 1 - \frac{1}{n+1}.$$

Therefore $\lim_{n \to \infty} s_n = 1$ and the series converges to 1:

$$\sum_{n=1}^{\infty} \frac{1}{n(n+1)} = 1.$$

This is an example of a **telescoping series**, so called because the partial sums *fold up* into a simple form when the terms are expanded in partial fractions. Other examples can be found in the exercises at the end of this section. As these examples show, the method of partial fractions can be a useful tool for series as well as for integrals.

**Example 4**  Show that the **harmonic series**

$$\sum_{n=1}^{\infty} \frac{1}{n} = 1 + \frac{1}{2} + \frac{1}{3} + \frac{1}{4} + \cdots$$

diverges to infinity.

**Solution**  If $s_n$ is its $n$th partial sum of the harmonic series, then

$$s_n = 1 + \frac{1}{2} + \frac{1}{3} + \cdots + \frac{1}{n}$$

$$= \text{sum of areas of rectangles shaded in Figure 9.3}$$

$$> \text{area under } y = \frac{1}{x} \text{ from 1 to } n+1$$

$$= \int_1^{n+1} \frac{dx}{x} = \ln(n+1).$$

Now $\lim_{n \to \infty} \ln(n+1) = \infty$. Therefore $\lim_{n \to \infty} s_n = \infty$ and

$$\sum_{n=1}^{\infty} \frac{1}{n} = 1 + \frac{1}{2} + \frac{1}{3} + \cdots \qquad \text{diverges to infinity.}$$

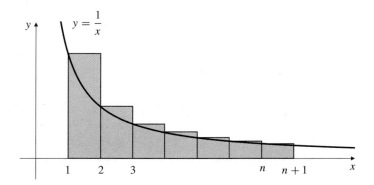

**Figure 9.3**   A partial sum of the harmonic series

Like geometric series, the harmonic series will often be encountered in subsequent sections.

## Some Theorems About Series

**THEOREM** **4**   If $\sum_{n=1}^{\infty} a_n$ converges, then $\lim_{n \to \infty} a_n = 0$.

**PROOF**   If $s_n = a_1 + a_2 + \cdots + a_n$, then $s_n - s_{n-1} = a_n$. If $\sum_{n=1}^{\infty} a_n$ converges, then $\lim_{n \to \infty} s_n = s$ exists, and $\lim_{n \to \infty} s_{n-1} = s$. Hence $\lim_{n \to \infty} a_n = s - s = 0$.

**Remark**   Theorem 4 is *very important* for the understanding of infinite series. Students often err either in forgetting that *a series cannot converge if its terms do not approach zero* or in confusing this result with its *converse*, which is false. The converse would say that if $\lim_{n \to \infty} a_n = 0$, then $\sum_{n=1}^{\infty} a_n$ must converge. The harmonic series is a counterexample showing the falsehood of this assertion:

$$\lim_{n \to \infty} \frac{1}{n} = 0 \quad \text{but} \quad \sum_{n=1}^{\infty} \frac{1}{n} \text{ diverges to infinity.}$$

When considering whether a given series converges, the first question you should ask yourself is "Does the $n$th term approach 0 as $n$ approaches $\infty$?" If the answer is *no*, then the series does *not* converge. If the answer is *yes*, then the series *may or may not* converge. If the sequence of terms $\{a_n\}$ tends to a nonzero limit $L$, then $\sum_{n=1}^{\infty} a_n$ diverges to infinity if $L > 0$ and diverges to negative infinity if $L < 0$.

### Example 5

(a) $\sum_{n=1}^{\infty} \frac{n}{2n-1}$ diverges to infinity since $\lim_{n \to \infty} \frac{n}{2n+1} = 1/2 > 0$.

(b) $\sum_{n=1}^{\infty} (-1)^n n \sin(1/n)$ diverges since

$$\lim_{n \to \infty} \left| (-1)^n n \sin \frac{1}{n} \right| = \lim_{n \to \infty} \frac{\sin(1/n)}{1/n} = \lim_{x \to 0+} \frac{\sin x}{x} = 1 \neq 0.$$

The following theorem asserts that it is only the *ultimate* behaviour of $\{a_n\}$ that determines whether $\sum_{n=1}^{\infty} a_n$ converges. Any finite number of terms can be dropped from the beginning of a series without affecting the convergence; the convergence depends only on the *tail* of the series. Of course, the actual sum of the series depends on *all* the terms.

**THEOREM 5** $\sum_{n=1}^{\infty} a_n$ converges if and only if $\sum_{n=N}^{\infty} a_n$ converges for any integer $N \geq 1$.

**THEOREM 6** If $\{a_n\}$ is ultimately positive, then the series $\sum_{n=1}^{\infty} a_n$ must either converge (if its partial sums are bounded above) or diverge to infinity (if its partial sums are not bounded above).

The proofs of these two theorems are posed as exercises at the end of this section. The following theorem is just a reformulation of standard laws of limits.

**THEOREM 7** If $\sum_{n=1}^{\infty} a_n$ and $\sum_{n=1}^{\infty} b_n$ converge to $A$ and $B$, respectively, then

(a) $\sum_{n=1}^{\infty} c a_n$ converges to $cA$ (where $c$ is any constant);

(b) $\sum_{n=1}^{\infty} (a_n \pm b_n)$ converges to $A \pm B$;

(c) If $a_n \leq b_n$ for all $n = 1, 2, 3, \ldots$, then $A \leq B$.

---

**Example 6** Find the sum of the series $\displaystyle\sum_{n=1}^{\infty} \frac{1 + 2^{n+1}}{3^n}$.

**Solution** The given series is the sum of two geometric series,

$$\sum_{n=1}^{\infty} \frac{1}{3^n} = \sum_{n=1}^{\infty} \frac{1}{3}\left(\frac{1}{3}\right)^{n-1} = \frac{1/3}{1 - (1/3)} = \frac{1}{2} \quad \text{and}$$

$$\sum_{n=1}^{\infty} \frac{2^{n+1}}{3^n} = \sum_{n=1}^{\infty} \frac{4}{3}\left(\frac{2}{3}\right)^{n-1} = \frac{4/3}{1 - (2/3)} = 4.$$

Thus its sum is $\dfrac{1}{2} + 4 = \dfrac{9}{2}$ by Theorem 7(b).

---

## Exercises 9.2

In Exercises 1–18, find the sum of the given series, or show that the series diverges (possibly to infinity or negative infinity). Exercises 11–14 are telescoping series and should be done by partial fractions as suggested in Example 3 in this section.

**1.** $\dfrac{1}{3} + \dfrac{1}{9} + \dfrac{1}{27} + \cdots = \displaystyle\sum_{n=1}^{\infty} \dfrac{1}{3^n}$

**2.** $3 - \dfrac{3}{4} + \dfrac{3}{16} - \dfrac{3}{64} + \cdots = \displaystyle\sum_{n=1}^{\infty} 3\left(-\dfrac{1}{4}\right)^{n-1}$

**3.** $\displaystyle\sum_{n=5}^{\infty} \dfrac{1}{(2 + \pi)^{2n}}$

**4.** $\displaystyle\sum_{n=0}^{\infty} \dfrac{5}{10^{3n}}$

**5.** $\displaystyle\sum_{n=2}^{\infty} \dfrac{(-5)^n}{8^{2n}}$

**6.** $\displaystyle\sum_{n=0}^{\infty} \dfrac{1}{e^n}$

**7.** $\displaystyle\sum_{k=0}^{\infty} \dfrac{2^{k+3}}{e^{k-3}}$

**8.** $\displaystyle\sum_{j=1}^{\infty} \pi^{j/2} \cos(j\pi)$

**9.** $\displaystyle\sum_{n=1}^{\infty} \dfrac{3 + 2^n}{2^{n+2}}$

**10.** $\displaystyle\sum_{n=0}^{\infty} \dfrac{3 + 2^n}{3^{n+2}}$

**11.** $\displaystyle\sum_{n=1}^{\infty} \dfrac{1}{n(n + 2)} = \dfrac{1}{1 \times 3} + \dfrac{1}{2 \times 4} + \dfrac{1}{3 \times 5} + \cdots$

**12.** $\displaystyle\sum_{n=1}^{\infty} \dfrac{1}{(2n - 1)(2n + 1)} = \dfrac{1}{1 \times 3} + \dfrac{1}{3 \times 5} + \dfrac{1}{5 \times 7} + \cdots$

**13.** $\displaystyle\sum_{n=1}^{\infty} \frac{1}{(3n-2)(3n+1)} = \frac{1}{1 \times 4} + \frac{1}{4 \times 7} + \frac{1}{7 \times 10} + \cdots$

* **14.** $\displaystyle\sum_{n=1}^{\infty} \frac{1}{n(n+1)(n+2)}$

$= \dfrac{1}{1 \times 2 \times 3} + \dfrac{1}{2 \times 3 \times 4} + \dfrac{1}{3 \times 4 \times 5} + \cdots$

**15.** $\displaystyle\sum_{n=1}^{\infty} \frac{1}{2n-1}$      **16.** $\displaystyle\sum_{n=1}^{\infty} \frac{n}{n+2}$

**17.** $\displaystyle\sum_{n=1}^{\infty} n^{-1/2}$      **18.** $\displaystyle\sum_{n=1}^{\infty} \frac{2}{n+1}$

**19.** Obtain a simple expression for the partial sum $s_n$ of the series $\sum_{n=1}^{\infty} (-1)^n$ and use it to show that this series diverges.

**20.** Find the sum of the series

$$\frac{1}{1} + \frac{1}{1+2} + \frac{1}{1+2+3} + \frac{1}{1+2+3+4} + \cdots.$$

**21.** When dropped, an elastic ball bounces back up to a height three-quarters of that from which it fell. If the ball is

dropped from a height of 2 m and allowed to bounce up and down indefinitely, what is the total distance it travels before coming to rest?

**22.** If a bank account pays 10% simple interest into an account once a year, what is the balance in the account at the end of 8 years if $1,000 is deposited into the account at the beginning of each of the eight years? (Assume there was no balance in the account initially.)

* **23.** Prove Theorem 5.     * **24.** Prove Theorem 6.

* **25.** State a theorem analogous to Theorem 6 but for a negative sequence.

In Exercises 26–31, decide whether the given statement is TRUE or FALSE. If it is true, prove it. If it is false, give a counterexample showing the falsehood.

* **26.** If $a_n = 0$ for every $n$, then $\sum a_n$ converges.

* **27.** If $\sum a_n$ converges, then $\sum (1/a_n)$ diverges to infinity.

* **28.** If $\sum a_n$ and $\sum b_n$ both diverge, then so does $\sum (a_n + b_n)$.

* **29.** If $a_n \geq c > 0$ for every $n$, then $\sum a_n$ diverges to infinity.

* **30.** If $\sum a_n$ diverges and $\{b_n\}$ is bounded, then $\sum a_n b_n$ diverges.

* **31.** If $a_n > 0$ and $\sum a_n$ converges, then $\sum (a_n)^2$ converges.

# 9.3 Convergence Tests for Positive Series

In the previous section we saw a few examples of convergent series (geometric and telescoping series) whose sums could be determined exactly because the partial sums $s_n$ could be expressed in closed form as explicit functions of $n$ whose limits as $n \to \infty$ could be evaluated. It is not usually possible to do this with a given series, and therefore it is not usually possible to determine the sum of the series exactly. However, there are many techniques for determining whether a given series converges and, if it does, for approximating the sum to any desired degree of accuracy.

In this section we deal exclusively with *positive series*, that is, series of the form

$$\sum_{n=1}^{\infty} a_n = a_1 + a_2 + a_3 + \cdots$$

where $a_n \geq 0$ for all $n \geq 1$. As noted in Theorem 6, such a series will converge if its partial sums are bounded above and will diverge to infinity otherwise. All our results apply equally well to *ultimately* positive series since convergence or divergence depends only on the *tail* of a series.

## The Integral Test

The integral test provides a means for determining whether an ultimately positive series converges or diverges by comparing it with an improper integral that behaves similarly. Example 4 in Section 9.2 is an example of the use of this technique. We formalize the method in the following theorem.

**THEOREM  8**

### The integral test

Suppose that $a_n = f(n)$, where $f(x)$ is positive, continuous, and nonincreasing on an interval $[N, \infty)$ for some positive integer $N$. Then

$$\sum_{n=1}^{\infty} a_n \qquad \text{and} \qquad \int_{N}^{\infty} f(t)\, dt$$

either both converge or both diverge to infinity.

**PROOF**   Let $s_n = a_1 + a_2 + \cdots + a_n$. If $n > N$, we have

$$
\begin{aligned}
s_n &= s_N + a_{N+1} + a_{N+2} + \cdots + a_n \\
&= s_N + f(N+1) + f(N+2) + \cdots + f(n) \\
&= s_N + \text{ sum of areas of rectangles shaded in Figure 9.4(a)} \\
&\leq s_N + \int_{N}^{\infty} f(t)\, dt.
\end{aligned}
$$

If the improper integral $\int_{N}^{\infty} f(t)\, dt$ converges, then the sequence $\{s_n\}$ is bounded above and $\sum_{n=1}^{\infty} a_n$ converges.

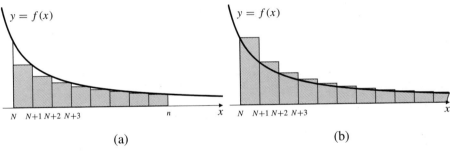

**Figure 9.4**

(a)                                          (b)

Conversely, suppose that $\sum_{n=1}^{\infty} a_n$ converges to the sum $s$. Then

$$
\begin{aligned}
\int_{N}^{\infty} f(t)\, dt &= \text{ area under } y = f(t) \text{ above } y = 0 \text{ from } t = N \text{ to } t = \infty \\
&\leq \text{ sum of areas of shaded rectangles in Figure 9.4(b)} \\
&= a_N + a_{N+1} + a_{N+2} + \cdots \\
&= s - s_{N-1} < \infty,
\end{aligned}
$$

so the improper integral represents a finite area and is thus convergent. (We omit the remaining details showing that $\lim_{R \to \infty} \int_{N}^{R} f(t)\, dt$ exists; like the series case, the argument depends on the completeness of the real numbers.)

**Remark**   If $a_n = f(n)$, where $f$ is positive, continuous, and nonincreasing on $[1, \infty)$, then Theorem 8 assures us that $\sum_{n=1}^{\infty} a_n$ and $\int_{1}^{\infty} f(x)\, dx$ both converge or both diverge to infinity. It does *not* tell us that the sum of the series is equal to the value of the integral. The two are not likely to be equal in the case of convergence. However, as we see below, integrals can help us approximate the sum of a series.

The principal use of the integral test is to establish the result of the following example concerning the series $\sum_{n=0}^{\infty} n^{-p}$, which is called a ***p*-series**. This result should be memorized; we will frequently compare the behaviour of other series with *p*-series later in this and subsequent sections.

**Example 1**    (*p*-series)  Show that

$$\sum_{n=1}^{\infty} n^{-p} = \sum_{n=1}^{\infty} \frac{1}{n^p} \begin{cases} \text{converges if } p > 1 \\ \text{diverges to infinity if } p \le 1. \end{cases}$$

**Solution**   Observe that if $p > 0$, then $f(x) = x^{-p}$ is positive, continuous, and decreasing on $[1, \infty)$. By the integral test, the *p*-series converges for $p > 1$ and diverges for $0 < p \le 1$ by comparison with $\int_1^{\infty} x^{-p} \, dx$. (See Theorem 2(a) of Section 6.5.) If $p \le 0$, then $\lim_{n \to \infty}(1/n^p) \ne 0$, so the series cannot converge in this case. Being a positive series, it must diverge to infinity.    ∎

**Remark**   The harmonic series $\sum_{n=1}^{\infty} n^{-1}$ (the case $p = 1$ of the *p*-series) is on the borderline between convergence and divergence, although it diverges. While its terms decrease towards 0 as $n$ increases, they do not decrease *fast enough* to allow the sum of the series to be finite. If $p > 1$, the terms of $\sum_{n=1}^{\infty} n^{-p}$ decrease towards zero fast enough that their sum is finite. We can refine the distinction between convergence and divergence at $p = 1$ by using terms that decrease faster than $1/n$, but not as fast as $1/n^q$ for any $q > 1$. Recall that $\ln n$ grows more slowly than any positive power of $n$ as $n$ increases. Accordingly, if $p > 0$ and $q > 1$, then

$$\frac{1}{n^q} < \frac{1}{n(\ln n)^p} < \frac{1}{n}.$$

The question now arises whether $\sum_{n=2}^{\infty} 1/(n(\ln n)^p)$ converges or not.

**Example 2**    For what values of $p$ does the series $\displaystyle\sum_{n=2}^{\infty} \frac{1}{n(\ln n)^p}$ converge?

**Solution**   We had to start the series at $n = 2$ because $\ln 1 = 0$. Let $p > 0$. Since $f(x) = 1/(x(\ln x)^p))$ is positive, continuous, and decreasing on $[2, \infty)$, the given series should be compared with

$$\int_2^{\infty} \frac{dx}{x(\ln x)^p}.$$

The substitution $u = \ln x$ transforms this integral to the equivalent one

$$\int_{\ln 2}^{\infty} \frac{du}{u^p},$$

which converges if $p > 1$ and diverges if $0 < p < 1$, again by Theorem 2(a) of Section 6.5. Therefore the given series converges if $p > 1$ and diverges if $0 < p < 1$. For $p = 0$ the given series is the harmonic series, so it diverges to infinity. For $p < 0$ the terms are ultimately greater than those of the harmonic series, so the partial sums of the series are not bounded above, and the series diverges to infinity in this case also.    ∎

This process of fine-tuning Example 1 exhibited in Example 2 can be extended even further. (See Exercise 36 below.)

## Using Integral Bounds to Estimate the Sum of a Series

Suppose that $a_k = f(k)$ for $k = n + 1$, $n + 2$, $n + 3$, $\ldots$, where $f$ is a positive, continuous function, decreasing at least on the interval $[n, \infty)$. We have:

$$s - s_n = \sum_{k=n+1}^{\infty} f(k)$$

$$= \text{sum of areas of rectangles shaded in Figure 9.5(a)}$$

$$\leq \int_{n}^{\infty} f(x)\, dx.$$

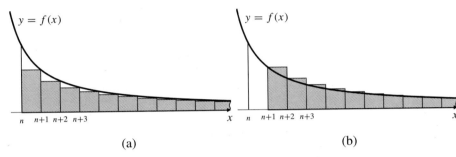

**Figure 9.5**

(a)

(b)

Similarly,

$$s - s_n = \text{sum of areas of rectangles in Figure 9.5(b)}$$

$$\geq \int_{n+1}^{\infty} f(x)\, dx.$$

If we define

$$A_n = \int_{n}^{\infty} f(x)\, dx$$

then we can combine the above inequalities to obtain

$$A_{n+1} \leq s - s_n \leq A_n,$$

or, equivalently:

$$s_n + A_{n+1} \leq s \leq s_n + A_n.$$

The error in the approximation $s \approx s_n$ satisfies $0 \leq s - s_n \leq A_n$. However, since $s$ must lie in the interval $[s_n + A_{n+1}, s_n + A_n]$, we can do better by using the midpoint $s_n^*$ of this interval as an approximation for $s$. The error is then less than half the length $A_n - A_{n+1}$ of the interval:

---

**A better integral approximation**

The error $|s - s_n^*|$ in the approximation

$$s \approx s_n^* = s_n + \frac{A_{n+1} + A_n}{2}, \qquad \text{where} \quad A_n = \int_{n}^{\infty} f(x)\, dx,$$

satisfies $\quad |s - s_n^*| \leq \dfrac{A_n - A_{n+1}}{2}.$

---

(Whenever a quantity is known to lie in a certain interval, the midpoint of that interval can be used to approximate the quantity, and the absolute value of the error in that approximation does not exceed half the length of the interval.)

**Example 3**    Find the best approximation $s_n^*$ to the sum $s$ of the series $\sum_{n=1}^{\infty} 1/n^2$, making use of the partial sum $s_n$ of the first $n$ terms. How large would $n$ have to be to ensure that the approximation $s \approx s_n^*$ has error less than 0.001 in absolute value? How large would $n$ have to be to ensure that the approximation $s \approx s_n$ has error less than 0.001 in absolute value?

**Solution**    Since $f(x) = 1/x^2$ is positive, continuous, and decreasing on $[1, \infty)$ for any $n = 1, 2, 3, \ldots$, we have

$$s_n + A_{n+1} \le s \le s_n + A_n$$

where

$$A_n = \int_n^{\infty} \frac{dx}{x^2} = \lim_{R \to \infty} \left( -\frac{1}{x} \right) \Big|_n^{\infty} = \frac{1}{n}.$$

The best approximation to $s$ using $s_n$ is

$$s_n^* = s_n + \frac{1}{2} \left( \frac{1}{n+1} + \frac{1}{n} \right) = s_n + \frac{2n+1}{2n(n+1)}$$

$$= 1 + \frac{1}{4} + \frac{1}{9} + \cdots + \frac{1}{n^2} + \frac{2n+1}{2n(n+1)}.$$

The error in this approximation satisfies

$$|s - s_n^*| \le \frac{1}{2} \left( \frac{1}{n} - \frac{1}{n+1} \right) = \frac{1}{2n(n+1)} \le 0.001$$

provided $2n(n+1) \ge 1/0.001 = 1,000$. It is easily checked that this condition is satisfied if $n \ge 22$; the approximation

$$s \approx s_{22}^* = 1 + \frac{1}{4} + \frac{1}{9} + \cdots + \frac{1}{22^2} + \frac{45}{44 \times 23}$$

will have error with absolute value not exceeding 0.001.

Had we used the approximation $s \approx s_n$ we could only have concluded that

$$0 \le s - s_n \le A_n = \frac{1}{n} < 0.001,$$

provided $n > 1000$; we would need 1000 terms of the series to get the desired accuracy.

## Comparison Tests

The next test we consider for positive series is analogous to the comparison theorem for improper integrals. (See Theorem 3 of Section 6.5.) It enables us to determine the convergence or divergence of one series by comparing it with another series that is known to converge or diverge.

**THEOREM** **9**

**A comparison test**

Let $\{a_n\}$ and $\{b_n\}$ be sequences for which there exists a positive constant $K$ such that, ultimately, $0 \leq a_n \leq K b_n$.

(a) If the series $\sum_{n=1}^{\infty} b_n$ converges, then so does the series $\sum_{n=1}^{\infty} a_n$.

(b) If the series $\sum_{n=1}^{\infty} a_n$ diverges to infinity, then so does the series $\sum_{n=1}^{\infty} b_n$.

> **DANGER**
>
> Theorem 9 does *not* say that if $\sum a_n$ converges, then $\sum b_n$ converges. It is possible that the *smaller* sum may be finite while the *larger* one is infinite. (Do not confuse a theorem with its converse.)

**PROOF** Since a series converges if and only if its tail converges (Theorem 5), we can assume, without loss of generality, that the condition $0 \leq a_n \leq K b_n$ holds for all $n \geq 1$. Let $s_n = a_1 + a_2 + \cdots + a_n$ and $S_n = b_1 + b_2 + \cdots + b_n$. Then $s_n \leq K S_n$. If $\sum b_n$ converges, then $\{S_n\}$ is convergent and hence is bounded by Theorem 1. Hence $\{s_n\}$ is bounded above. By Theorem 6, $\sum a_n$ converges. Since the convergence of $\sum b_n$ guarantees that of $\sum a_n$, if the latter series diverges to infinity, then the former cannot converge either, so it must diverge to infinity too.

---

**Example 4** Do the following series converge or not? Give reasons for your answer.

(a) $\displaystyle\sum_{n=1}^{\infty} \frac{1}{2^n + 1}$,     (b) $\displaystyle\sum_{n=1}^{\infty} \frac{3n + 1}{n^3 + 1}$,     (c) $\displaystyle\sum_{n=2}^{\infty} \frac{1}{\ln n}$.

**Solution** In each case we must find a suitable comparison series that we already know converges or diverges.

(a) Since $0 < \dfrac{1}{2^n + 1} < \dfrac{1}{2^n}$ for $n = 1, 2, 3, \ldots$, and since $\sum_{n=1}^{\infty} (1/2^n)$ is a convergent geometric series, the series $\sum_{n=1}^{\infty} (1/(2^n + 1))$ also converges by comparison.

(b) Observe that $\dfrac{3n + 1}{n^3 + 1}$ behaves like $\dfrac{3}{n^2}$ for large $n$, so we would expect to compare the series with the convergent $p$-series $\sum_{n=1}^{\infty} n^{-2}$. We have, for $n \geq 1$,

$$\frac{3n + 1}{n^3 + 1} = \frac{3n}{n^3 + 1} + \frac{1}{n^3 + 1} < \frac{3n}{n^3} + \frac{1}{n^3} < \frac{3}{n^2} + \frac{1}{n^2} = \frac{4}{n^2}.$$

Thus the given series converges by Theorem 9.

(c) For $n = 2, 3, 4, \ldots$, we have $0 < \ln n < n$. Thus $\dfrac{1}{\ln n} > \dfrac{1}{n}$. Since $\sum_{n=2}^{\infty} \dfrac{1}{n}$ diverges to infinity (it is a harmonic series), so does $\sum_{n=2}^{\infty} \dfrac{1}{\ln n}$ by comparison.

---

The following theorem provides a version of the comparison test that is not quite as general as Theorem 9, but is often easier to apply in specific cases.

**THEOREM 10**

**A limit comparison test**

Suppose that $\{a_n\}$ and $\{b_n\}$ are positive sequences and that

$$\lim_{n \to \infty} \frac{a_n}{b_n} = L,$$

where $L$ is either a nonnegative finite number or $+\infty$.

(a) If $L < \infty$ and $\sum_{n=1}^{\infty} b_n$ converges, then $\sum_{n=1}^{\infty} a_n$ also converges.

(b) If $L > 0$ and $\sum_{n=1}^{\infty} b_n$ diverges to infinity, then so does $\sum_{n=1}^{\infty} a_n$.

**PROOF** If $L < \infty$, then for $n$ sufficiently large, we have $b_n > 0$ and

$$0 \leq \frac{a_n}{b_n} \leq L + 1,$$

so $0 \le a_n \le (L+1)b_n$. Hence $\sum_{n=1}^{\infty} a_n$ converges if $\sum_{n=1}^{\infty} b_n$ converges, by Theorem 9(a).

If $L > 0$, then for $n$ sufficiently large

$$\frac{a_n}{b_n} \ge \frac{L}{2}.$$

Therefore $0 < b_n \le (2/L)a_n$, and $\sum_{n=1}^{\infty} a_n$ diverges to infinity if $\sum_{n=1}^{\infty} b_n$ does, by Theorem 9(b).

---

**Example 5**   Do the following series converge or not?  Give reasons for your answers.

(a) $\displaystyle\sum_{n=1}^{\infty} \frac{1}{1+\sqrt{n}}$,   (b) $\displaystyle\sum_{n=1}^{\infty} \frac{n+5}{n^3 - 2n + 3}$,   (c) $\displaystyle\sum_{n=1}^{\infty} \frac{2^n + 1}{3^n - 1}$.

**Solution**   Again we must make appropriate choices for comparison series.

(a) The terms of this series decrease like $1/\sqrt{n}$. Observe that

$$L = \lim_{n\to\infty} \frac{\dfrac{1}{1+\sqrt{n}}}{\dfrac{1}{\sqrt{n}}} = \lim_{n\to\infty} \frac{\sqrt{n}}{1+\sqrt{n}} = \lim_{n\to\infty} \frac{1}{(1/\sqrt{n})+1} = 1.$$

Since the $p$-series $\sum_{n=1}^{\infty} 1/\sqrt{n}$ diverges to infinity ($p = 1/2$), so does the series $\sum_{n=1}^{\infty} 1/(1+\sqrt{n})$, by the limit comparison test.

(b) For large $n$, the terms behave like $n/n^3$ so let us compare the series with the $p$-series $\sum_{n=1}^{\infty} 1/n^2$, which we know converges.

$$L = \lim_{n\to\infty} \frac{\dfrac{n+5}{n^3 - 2n + 3}}{\dfrac{1}{n^2}} = \lim_{n\to\infty} \frac{n^3 + 5n^2}{n^3 - 2n + 3} = 1.$$

Since $L < \infty$, the series $\displaystyle\sum_{n=1}^{\infty} \frac{n+5}{n^3 - 2n + 3}$ also converges by the limit comparison test.

(c) In this case the terms behave like $(2/3)^n$ for large $n$. We have

$$L = \lim_{n\to\infty} \frac{\dfrac{2^n + 1}{3^n - 1}}{\left(\dfrac{2}{3}\right)^n} = \lim_{n\to\infty} \frac{2^n + 1}{2^n} \frac{3^n}{3^n - 1} = \lim_{n\to\infty} \frac{1 + \dfrac{1}{2^n}}{1 - \dfrac{1}{3^n}} = 1.$$

Again $L < \infty$ and the series $\displaystyle\sum_{n=1}^{\infty} \frac{2^n + 1}{3^n - 1}$ converges by comparison with the convergent geometric series $\displaystyle\sum_{n=1}^{\infty} \left(\frac{2}{3}\right)^n$.

In order to apply the original version of the comparison test (Theorem 9) successfully, it is important to have an intuitive feeling for whether the given series converges or diverges. The form of the comparison will depend on whether you are trying to prove convergence or divergence. For instance, if you did not know intuitively that

$$\sum_{n=1}^{\infty} \frac{1}{100n + 20,000}$$

would have to diverge to infinity, you might try to argue that

$$\frac{1}{100n + 20,000} < \frac{1}{n} \qquad \text{for } n = 1, 2, 3, \ldots.$$

While true, this doesn't help at all. $\sum_{n=1}^{\infty} 1/n$ diverges to infinity, and therefore Theorem 9 yields no information from this comparison. We could, of course, argue instead that

$$\frac{1}{100n + 20,000} \geq \frac{1}{20,100n} \qquad \text{if } n \geq 1,$$

and conclude by Theorem 9 that $\sum_{n=1}^{\infty} (1/(100n + 20,000))$ diverges to infinity by comparison with the divergent series $\sum_{n=1}^{\infty} 1/n$. An easier way is to use Theorem 10 and the fact that

$$L = \lim_{n \to \infty} \frac{\dfrac{1}{100n + 20,000}}{\dfrac{1}{n}} = \lim_{n \to \infty} \frac{n}{100n + 20,000} = \frac{1}{100} > 0.$$

However, the limit comparison test Theorem 10 has a disadvantage when compared to the ordinary comparison test Theorem 9. It can fail in certain cases because the limit $L$ does not exist. In such cases it is possible that the ordinary comparison test may still work.

**Example 6** Test the series $\displaystyle\sum_{n=1}^{\infty} \frac{1 + \sin n}{n^2}$ for convergence.

*Solution* Since

$$\lim_{n \to \infty} \frac{\dfrac{1 + \sin n}{n^2}}{\dfrac{1}{n^2}} = \lim_{n \to \infty} (1 + \sin n)$$

does not exist, the limit comparison test gives us no information. However, since $\sin n \leq 1$, we have

$$0 \leq \frac{1 + \sin n}{n^2} \leq \frac{2}{n^2} \qquad \text{for } n = 1, 2, 3, \ldots.$$

The given series does, in fact, converge by comparison with $\sum_{n=1}^{\infty} 1/n^2$, using the ordinary comparison test.

## The Ratio and Root Tests

**THEOREM  11**

**The ratio test**

Suppose that $a_n > 0$ (ultimately) and that $\rho = \lim\limits_{n \to \infty} \dfrac{a_{n+1}}{a_n}$ exists or is $+\infty$.

(a) If $0 \le \rho < 1$, then $\sum_{n=1}^{\infty} a_n$ converges.

(b) If $1 < \rho \le \infty$, then $\lim_{n \to \infty} a_n = \infty$ and $\sum_{n=1}^{\infty} a_n$ diverges to infinity.

(c) If $\rho = 1$, this test gives no information; the series may either converge or diverge to infinity.

**PROOF**   Here $\rho$ is the lower case Greek letter *rho*, (pronounced *row*).

(a) Suppose $\rho < 1$. Pick a number $r$ such that $\rho < r < 1$. Since we are given that $\lim_{n \to \infty} a_{n+1}/a_n = \rho$, therefore $a_{n+1}/a_n \le r$ for $n$ sufficiently large; that is, $a_{n+1} \le r a_n$ for $n \ge N$, say. In particular,

$$a_{N+1} \le r a_N$$
$$a_{N+2} \le r a_{N+1} \le r^2 a_N$$
$$a_{N+3} \le r a_{N+2} \le r^3 a_N$$
$$\vdots$$
$$a_{N+k} \le r^k a_N \qquad (k = 0, 1, 2, 3, \ldots)$$

Hence $\sum_{n=N}^{\infty} a_n$ converges by comparison with the convergent geometric series $\sum_{k=0}^{\infty} r^k$. It follows that $\sum_{n=1}^{\infty} a_n = \sum_{n=1}^{N-1} a_n + \sum_{n=N}^{\infty} a_n$ must also converge.

(b) Now suppose that $\rho > 1$. Pick a number $r$ such that $1 < r < \rho$. Since $\lim_{n \to \infty} a_{n+1}/a_n = \rho$, therefore $a_{n+1}/a_n \ge r$ for $n$ sufficiently large, say for $n \ge N$. We assume $N$ is chosen large enough that $a_N > 0$. It follows by an argument similar to that used in part (a) that $a_{N+k} \ge r^k a_N$ for $k = 0, 1, 2, \ldots$, and since $r > 1$, $\lim_{n \to \infty} a_n = \infty$. Therefore $\sum_{n=1}^{\infty} a_n$ diverges to infinity.

(c) If $\rho$ is computed for the series $\sum_{n=1}^{\infty} 1/n$ and $\sum_{n=1}^{\infty} 1/n^2$, we get $\rho = 1$ in each case. Since the first series diverges to infinity and the second converges, the ratio test cannot distinguish between convergence and divergence if $\rho = 1$.

All $p$-series fall into the indecisive category where $\rho = 1$, as does $\sum_{n=1}^{\infty} a_n$, where $a_n$ is any rational function of $n$. The ratio test is most useful for series whose terms decrease at least exponentially fast. The presence of factorials in a term also suggest that the ratio test might be useful.

**Example 7**   Test for convergence:

(a) $\displaystyle\sum_{n=1}^{\infty} \frac{99^n}{n!}$,   (b) $\displaystyle\sum_{n=1}^{\infty} \frac{n^5}{2^n}$,   (c) $\displaystyle\sum_{n=1}^{\infty} \frac{n!}{n^n}$,   (d) $\displaystyle\sum_{n=1}^{\infty} \frac{(2n)!}{(n!)^2}$

**Solution**   We use the ratio test for each of these series.

(a) $\rho = \lim\limits_{n \to \infty} \dfrac{99^{n+1}}{(n+1)!} \left/ \dfrac{99^n}{n!} \right. = \lim\limits_{n \to \infty} \dfrac{99}{n+1} = 0 < 1.$

Thus $\sum_{n=1}^{\infty} (99^n/n!)$ converges.

(b) $\rho = \lim_{n\to\infty} \dfrac{(n+1)^5}{2^{n+1}} \Big/ \dfrac{n^5}{2^n} = \lim_{n\to\infty} \dfrac{1}{2}\left(\dfrac{n+1}{n}\right)^5 = \dfrac{1}{2} < 1.$

Hence $\sum_{n=1}^{\infty}(n^5/2^n)$ converges.

(c) $\rho = \lim_{n\to\infty} \dfrac{(n+1)!}{(n+1)^{n+1}} \Big/ \dfrac{n!}{n^n} = \lim_{n\to\infty} \dfrac{(n+1)!n^n}{(n+1)^{n+1}n!} = \lim_{n\to\infty} \left(\dfrac{n}{n+1}\right)^n$

$\qquad\qquad = \lim_{n\to\infty} \dfrac{1}{\left(1+\dfrac{1}{n}\right)^n} = \dfrac{1}{e} < 1.$

Thus $\sum_{n=1}^{\infty}(n!/n^n)$ converges.

(d) $\rho = \lim_{n\to\infty} \dfrac{(2(n+1))!}{((n+1)!)^2} \Big/ \dfrac{(2n)!}{(n!)^2} = \lim_{n\to\infty} \dfrac{(2n+2)(2n+1)}{(n+1)^2} = 4 > 1.$

Thus $\sum_{n=1}^{\infty}(2n)!/(n!)^2$ diverges to infinity. ∎

The following theorem is very similar to the ratio test, but is less frequently used. Its proof is left as an exercise. (See Exercise 37.) For examples of series to which it can be applied, see Exercises 38 and 39.

**THEOREM 12**

**The root test**

Suppose that $a_n > 0$ (ultimately) and that $\sigma = \lim_{n\to\infty}(a_n)^{1/n}$ exists or is $+\infty$.

(a) If $0 \le \sigma < 1$, then $\sum_{n=1}^{\infty} a_n$ converges.

(b) If $1 < \sigma \le \infty$, then $\lim_{n\to\infty} a_n = \infty$ and $\sum_{n=1}^{\infty} a_n$ diverges to infinity.

(c) If $\sigma = 1$, this test gives no information; the series may either converge or diverge to infinity.

## Using Geometric Bounds to Estimate the Sum of a Series

Suppose that an inequality of the form

$$0 \le a_k \le Kr^k$$

holds for $k = n+1, n+2, n+3, \ldots$, where $K$ and $r$ are constants and $r < 1$. We can then use a geometric series to bound the tail of $\sum_{n=1}^{\infty} a_n$.

$$0 \le s - s_n = \sum_{k=n+1}^{\infty} a_k \le \sum_{k=n+1}^{\infty} Kr^k$$
$$= Kr^{n+1}(1 + r + r^2 + \cdots)$$
$$= \dfrac{Kr^{n+1}}{1-r}.$$

Since $r < 1$ the series converges and the error approaches $0$ at an exponential rate as $n$ increases.

**Example 8** In Section 9.6 we will show that

$$e = \frac{1}{0!} + \frac{1}{1!} + \frac{1}{2!} + \frac{1}{3!} + \cdots = \sum_{n=0}^{\infty} \frac{1}{n!}$$

(Recall that $0! = 1$.) Estimate the error if the sum $s_n$ of the first $n$ terms of the series is used to approximate $e$. Find $e$ to three-decimal-place accuracy using the series.

**Solution** We have

$$s_n = \frac{1}{0!} + \frac{1}{1!} + \frac{1}{2!} + \frac{1}{3!} + \cdots + \frac{1}{(n-1)!}$$

$$= 1 + 1 + \frac{1}{2} + \frac{1}{6} + \frac{1}{24} + \cdots + \frac{1}{(n-1)!}.$$

(Since the series starts with the term for $n = 0$, the $n$th term is $1/(n-1)!$.) We can estimate the error in the approximation $s \approx s_n$ as follows:

$$0 < s - s_n = \frac{1}{n!} + \frac{1}{(n+1)!} + \frac{1}{(n+2)!} + \frac{1}{(n+3)!} + \cdots$$

$$= \frac{1}{n!}\left(1 + \frac{1}{n+1} + \frac{1}{(n+1)(n+2)} + \frac{1}{(n+1)(n+2)(n+3)} + \cdots\right)$$

$$< \frac{1}{n!}\left(1 + \frac{1}{n+1} + \frac{1}{(n+1)^2} + \frac{1}{(n+1)^3} + \cdots\right)$$

since $n + 2 > n + 1$, $n + 3 > n + 1$, and so on. The latter series is geometric, so

$$0 < s - s_n < \frac{1}{n!} \frac{1}{1 - \dfrac{1}{n+1}} = \frac{n+1}{n!n}.$$

If we want to evaluate $e$ accurately to three decimal places, then we must ensure that the error is less than 5 in the fourth decimal place, that is, that the error is less than 0.0005. Hence we want

$$\frac{n+1}{n} \frac{1}{n!} < 0.0005 = \frac{1}{2000}.$$

Since $7! = 5040$, but $6! = 720$, we can use $n = 7$ but no smaller. We have

$$e \approx 1 + 1 + \frac{1}{2!} + \frac{1}{3!} + \frac{1}{4!} + \frac{1}{5!} + \frac{1}{6!}$$

$$= 2 + \frac{1}{2} + \frac{1}{6} + \frac{1}{24} + \frac{1}{120} + \frac{1}{720} \approx 2.718 \quad \text{to three decimal places.}$$

It is appropriate to use geometric series to bound the tails of positive series whose convergence would be demonstrated by the ratio test. Such series converge ultimately faster than any $p$-series $\sum_{n=1}^{\infty} n^{-p}$, for which the limit ratio is $\rho = 1$.

# Exercises 9.3

In Exercises 1–26, determine whether the given series converges or diverges by using any appropriate test. The $p$-series can be used for comparison, as can geometric series. Be alert for series whose terms do not approach 0.

**1.** $\displaystyle\sum_{n=1}^{\infty} \frac{1}{n^2+1}$

**2.** $\displaystyle\sum_{n=1}^{\infty} \frac{n}{n^4-2}$

**3.** $\displaystyle\sum_{n=1}^{\infty} \frac{n^2+1}{n^3+1}$

**4.** $\displaystyle\sum_{n=1}^{\infty} \frac{\sqrt{n}}{n^2+n+1}$

**5.** $\displaystyle\sum_{n=1}^{\infty} \left| \sin \frac{1}{n^2} \right|$

**6.** $\displaystyle\sum_{n=8}^{\infty} \frac{1}{\pi^n+5}$

**7.** $\displaystyle\sum_{n=2}^{\infty} \frac{1}{(\ln n)^3}$

**8.** $\displaystyle\sum_{n=1}^{\infty} \frac{1}{\ln(3n)}$

**9.** $\displaystyle\sum_{n=1}^{\infty} \frac{1}{\pi^n-n^\pi}$

**10.** $\displaystyle\sum_{n=0}^{\infty} \frac{1+n}{2+n}$

**11.** $\displaystyle\sum_{n=1}^{\infty} \frac{1+n^{4/3}}{2+n^{5/3}}$

**12.** $\displaystyle\sum_{n=1}^{\infty} \frac{n^2}{1+n\sqrt{n}}$

**13.** $\displaystyle\sum_{n=3}^{\infty} \frac{1}{n\ln n\sqrt{\ln\ln n}}$

**14.** $\displaystyle\sum_{n=2}^{\infty} \frac{1}{n\ln n(\ln\ln n)^2}$

**15.** $\displaystyle\sum_{n=1}^{\infty} \frac{1-(-1)^n}{n^4}$

**16.** $\displaystyle\sum_{n=1}^{\infty} \frac{1+(-1)^n}{\sqrt{n}}$

**17.** $\displaystyle\sum_{n=1}^{\infty} \frac{1}{2^n(n+1)}$

**18.** $\displaystyle\sum_{n=1}^{\infty} \frac{n^4}{n!}$

**19.** $\displaystyle\sum_{n=1}^{\infty} \frac{n!}{n^2e^n}$

**20.** $\displaystyle\sum_{n=1}^{\infty} \frac{(2n)!6^n}{(3n)!}$

**21.** $\displaystyle\sum_{n=2}^{\infty} \frac{\sqrt{n}}{3^n\ln n}$

**22.** $\displaystyle\sum_{n=0}^{\infty} \frac{n^{100}2^n}{\sqrt{n!}}$

**23.** $\displaystyle\sum_{n=1}^{\infty} \frac{(2n)!}{(n!)^3}$

**24.** $\displaystyle\sum_{n=1}^{\infty} \frac{1+n!}{(1+n)!}$

**25.** $\displaystyle\sum_{n=4}^{\infty} \frac{2^n}{3^n-n^3}$

**26.** $\displaystyle\sum_{n=1}^{\infty} \frac{n^n}{\pi^n n!}$

In Exercises 27–30, use $s_n$ and integral bounds to find the smallest interval that you can be sure contains the sum $s$ of the series. If the midpoint $s_n^*$ of this interval is used to approximate $s$, how large should $n$ be chosen to ensure that the error is less than 0.001?

**27.** $\displaystyle\sum_{k=1}^{\infty} \frac{1}{k^4}$

**28.** $\displaystyle\sum_{k=1}^{\infty} \frac{1}{k^3}$

**29.** $\displaystyle\sum_{k=1}^{\infty} \frac{1}{k^{3/2}}$

**30.** $\displaystyle\sum_{k=1}^{\infty} \frac{1}{k^2+4}$

For each positive series in Exercises 31–34, find the best upper bound you can for the error $s - s_n$ encountered if the partial sum $s_n$ is used to approximate the sum $s$ of the series. How many terms of each series do you need to be sure that the approximation has error less than 0.001?

**31.** $\displaystyle\sum_{k=1}^{\infty} \frac{1}{2^k k!}$

**32.** $\displaystyle\sum_{n=1}^{\infty} \frac{1}{(2n-1)!}$

**33.** $\displaystyle\sum_{n=0}^{\infty} \frac{2^n}{(2n)!}$

**34.** $\displaystyle\sum_{n=1}^{\infty} \frac{1}{n^n}$

**35.** Use the integral test to show that $\displaystyle\sum_{n=1}^{\infty} \frac{1}{1+n^2}$ converges. Show that the sum $s$ of the series is less than $\pi/2$.

**∗ 36.** Show that $\sum_{n=3}^{\infty}(1/(n\ln n(\ln\ln n)^p))$ converges if and only if $p>1$. Generalize this result to series of the form

$$\sum_{n=N}^{\infty} \frac{1}{n(\ln n)(\ln\ln n)\cdots(\ln_j n)(\ln_{j+1} n)^p}$$

where $\ln_j n = \underbrace{\ln\ln\ln\ln\cdots\ln}_{j\ \ln's} n$.

**∗ 37.** Prove the root test. *Hint:* mimic the proof of the ratio test.

**38.** Use the root test to show that $\displaystyle\sum_{n=1}^{\infty} \frac{2^{n+1}}{n^n}$ converges.

**∗ 39.** Use the root test from Exercise 38 to test for convergence

$$\sum_{n=1}^{\infty} \left( \frac{n}{n+1} \right)^{n^2}.$$

**40.** Repeat Exercise 38, but use the ratio test instead of the root test.

**∗ 41.** Try to use the ratio test to determine whether $\displaystyle\sum_{n=1}^{\infty} \frac{2^{2n}(n!)^2}{(2n)!}$ converges. What happens? Now observe that

$$\frac{2^{2n}(n!)^2}{(2n)!} = \frac{[2n(2n-2)(2n-4)\cdots 6\times 4\times 2]^2}{2n(2n-1)(2n-2)\cdots 4\times 3\times 2\times 1}$$

$$= \frac{2n}{2n-1} \times \frac{2n-2}{2n-3} \times \cdots \times \frac{4}{3} \times \frac{2}{1}.$$

Does the given series converge or not? Why?

**∗ 42.** Determine whether the series $\displaystyle\sum_{n=1}^{\infty} \frac{(2n)!}{2^{2n}(n!)^2}$ converges.
*Hint:* proceed as in Exercise 41. Show that $a_n \geq 1/(2n)$.

**∗ 43.** (a) Show that if $k>0$ and $n$ is a positive integer, then

$$n < \frac{1}{k}(1+k)^n.$$

(b) Use the estimate in (a) with $0 < k < 1$ to obtain an upper bound for the sum of the series $\sum_{n=0}^{\infty} n/2^n$. For what value of $k$ is this bound least?

(c) If we use the sum $s_n$ of the first $n$ terms to approximate the sum $s$ of the series in (b), obtain an upper bound for the error $s - s_n$ using the inequality from (a). For given $n$, find $k$ to minimize this upper bound.

**\* 44. (Improving the convergence of a series)** We know that $\sum_{n=1}^{\infty} 1/(n(n + 1)) = 1$. (See Example 3 of Section 9.2.) Since

$$\frac{1}{n^2} = \frac{1}{n(n + 1)} + c_n, \quad \text{where} \quad c_n = \frac{1}{n^2(n + 1)},$$

we have $\displaystyle\sum_{n=1}^{\infty} \frac{1}{n^2} = 1 + \sum_{n=1}^{\infty} c_n.$

The series, $\sum_{n=1}^{\infty} c_n$ converges more rapidly than does $\sum_{n=1}^{\infty} 1/n^2$ because its terms decrease like $1/n^3$. Hence fewer terms of that series will be needed to compute $\sum_{n=1}^{\infty} 1/n^2$ to any desired degree of accuracy than would be needed if we calculated with $\sum_{n=1}^{\infty} 1/n^2$ directly. Using integral upper and lower bounds, determine a value of $n$ for which the modified partial sum $s_n^*$ for the series $\sum_{n=1}^{\infty} c_n$ approximates the sum of that series with error less than 0.001 in absolute value. Hence determine $\sum_{n=1}^{\infty} 1/n^2$ to within 0.001 of its true value.
(The technique exibited in this exercise is known as *improving the convergence* of a series. It can be applied to estimating the sum $\sum a_n$ if we know the sum $\sum b_n$ and if $a_n - b_n = c_n$, where $|c_n|$ decreases faster than $|a_n|$ as $n$ tends to infinity.)

# 9.4  Absolute and Conditional Convergence

All of the series $\sum_{n=1}^{\infty} a_n$ considered in the previous section were ultimately positive; that is, $a_n \geq 0$ for $n$ sufficiently large. We now drop this restriction and allow arbitrary real terms $a_n$. We can, however, always obtain a positive series from any given series by replacing all the terms with their absolute values.

**DEFINITION 5**

> **Absolute convergence**
>
> The series $\sum_{n=1}^{\infty} a_n$ is said to be **absolutely convergent** if $\sum_{n=1}^{\infty} |a_n|$ converges.

The series

$$s = \sum_{n=1}^{\infty} \frac{(-1)^n}{n^2} = -1 + \frac{1}{4} - \frac{1}{9} + \frac{1}{16} - \cdots$$

converges absolutely since

$$S = \sum_{n=1}^{\infty} \left| \frac{(-1)^n}{n^2} \right| = \sum_{n=1}^{\infty} \frac{1}{n^2} = 1 + \frac{1}{4} + \frac{1}{9} + \frac{1}{16} + \cdots$$

converges. It seems reasonable that the first series must converge, and its sum $s$ should satisfy $-S \leq s \leq S$. In general, the cancellation that occurs because some terms are negative and others positive makes it *easier* for a series to converge than if all the terms are of one sign. We verify this insight in the following theorem.

**THEOREM 13**    If a series converges absolutely, then it converges.

**PROOF**    Let $\sum_{n=1}^{\infty} a_n$ be absolutely convergent, and let $b_n = a_n + |a_n|$ for each $n$. Since $-|a_n| \leq a_n \leq |a_n|$, we have $0 \leq b_n \leq 2|a_n|$ for each $n$. Thus $\sum_{n=1}^{\infty} b_n$ converges by the comparison test. Therefore $\sum_{n=1}^{\infty} a_n = \sum_{n=1}^{\infty} b_n - \sum_{n=1}^{\infty} |a_n|$ also converges.

Again you are cautioned not to confuse the statement of Theorem 13 with the converse statement, which is false. We will show later in this section that the **alternating harmonic series**

$$\sum_{n=1}^{\infty} \frac{(-1)^{n-1}}{n} = 1 - \frac{1}{2} + \frac{1}{3} - \frac{1}{4} + \frac{1}{5} - \cdots$$

converges, although it does not converge absolutely. If we replace all the terms by their absolute values we get the divergent harmonic series:

$$\sum_{n=1}^{\infty} \frac{1}{n} = 1 + \frac{1}{2} + \frac{1}{3} + \frac{1}{4} + \cdots = \infty.$$

**DANGER**

Although absolute convergence implies convergence, convergence does *not* imply absolute convergence.

**DEFINITION** **6**

> **Conditional convergence**
>
> If $\sum_{n=1}^{\infty} a_n$ is convergent, but not absolutely convergent, then we say that it is **conditionally convergent** or that it **converges conditionally**.

The alternating harmonic series is an example of a conditionally convergent series.

The comparison tests, the integral test and the ratio test, can each be used to test for absolute convergence. They should be applied to the series $\sum_{n=1}^{\infty} |a_n|$. For the ratio test we calculate $\rho = \lim_{n\to\infty} |a_{n+1}/a_n|$. If $\rho < 1$, then $\sum_{n=1}^{\infty} a_n$ converges absolutely. If $\rho > 1$, then $\lim_{n\to\infty} |a_n| = \infty$, so both $\sum_{n=1}^{\infty} |a_n|$ and $\sum_{n=1}^{\infty} a_n$ must diverge. If $\rho = 1$ we get no information; the series $\sum_{n=1}^{\infty} a_n$ may converge absolutely, it may converge conditionally, or it may diverge.

**Example 1**  Test for absolute convergence:

(a) $\displaystyle\sum_{n=1}^{\infty} \frac{(-1)^{n-1}}{2n-1}$,     (b) $\displaystyle\sum_{n=1}^{\infty} \frac{n\cos(n\pi)}{2^n}$

**Solution**

(a) $\displaystyle\lim_{n\to\infty} \left| \frac{(-1)^{n-1}}{2n-1} \right| \Big/ \frac{1}{n} = \lim_{n\to\infty} \frac{n}{2n-1} = \frac{1}{2} > 0.$ Since the harmonic series $\sum_{n=1}^{\infty}(1/n)$ diverges to infinity, therefore $\sum_{n=1}^{\infty}((-1)^{n-1}/(2n-1))$ does not converge absolutely (comparison test).

(b) $\rho = \displaystyle\lim_{n\to\infty} \left| \frac{(n+1)\cos((n+1)\pi)}{2^{n+1}} \Big/ \frac{n\cos(n\pi)}{2^n} \right| = \lim_{n\to\infty} \frac{n+1}{2n} = \frac{1}{2} < 1.$ (Note that $\cos(n\pi)$ is just a fancy way of writing $(-1)^n$.) Therefore (ratio test) $\sum_{n=1}^{\infty}((n\cos(n\pi))/2^n)$ converges absolutely. ∎

## The Alternating Series Test

We cannot use any of the previously developed tests to show that the alternating harmonic series converges; all of those tests apply only to (ultimately) positive series, and so they can test only for absolute convergence. Demonstrating convergence that is not absolute is generally harder to do. We present only one test that can establish such convergence; this test can only be used on a very special kind of series.

**THEOREM  14**

**The alternating series test**

Suppose that the sequence $\{a_n\}$ is positive, decreasing, and converges to 0, that is, suppose that

(i) $a_n \geq 0$ for $n = 1, 2, 3, \ldots$, and

(ii) $a_{n+1} \leq a_n$ for $n = 1, 2, 3, \ldots$, and

(iii) $\lim_{n \to \infty} a_n = 0$.

Then the alternating series

$$\sum_{n=1}^{\infty} (-1)^{n-1} a_n = a_1 - a_2 + a_3 - a_4 + a_5 - \cdots$$

converges.

**! DANGER**

Read this proof slowly and think about why each statement is true.

**PROOF**   Since the sequence $\{a_n\}$ is decreasing, we have $a_{2n+1} \geq a_{2n+2}$. Therefore $s_{2n+2} = s_{2n} + a_{2n+1} - a_{2n+2} \geq s_{2n}$ for $n = 1, 2, 3, \ldots$; the even partial sums $\{s_n\}$ form an increasing sequence. Similarly $s_{2n+1} = s_{2n-1} - a_{2n} + a_{2n+1} \leq s_{2n-1}$, so the odd partial sums $\{s_{2n-1}\}$ form a decreasing sequence. Since $s_{2n} = s_{2n-1} - a_{2n} \leq s_{2n-1}$, we can say, for any $n \geq 1$, that

$$s_2 \leq s_4 \leq s_6 \leq \cdots \leq s_{2n} \leq s_{2n-1} \leq s_{2n-3} \leq \cdots \leq s_5 \leq s_3 \leq s_1.$$

Hence $s_2$ is a lower bound for the decreasing sequence $\{s_{2n-1}\}$ and $s_1$ is an upper bound for the increasing sequence $\{s_{2n}\}$. Both of these sequences therefore converge by the completeness of the real numbers:

$$\lim_{n \to \infty} s_{2n-1} = s_{\text{odd}}, \qquad \lim_{n \to \infty} s_{2n} = s_{\text{even}}.$$

Now $a_{2n} = s_{2n-1} - s_{2n}$, so $0 = \lim_{n \to \infty} a_{2n} = \lim_{n \to \infty} (s_{2n-1} - s_{2n}) = s_{\text{odd}} - s_{\text{even}}$. Therefore $s_{\text{odd}} = s_{\text{even}} = s$, say. Every partial sum $s_n$ is either of the form $s_{2n-1}$ or of the form $s_{2n}$. Thus $\lim_{n \to \infty} s_n = s$ exists and the series $\sum (-1)^{n-1} a_n$ converges to this sum $s$.

**Remark**   Note that the series $\sum_{n=1}^{\infty} (-1)^{n-1} a_n$ begins with a positive term, $a_1$. The conclusion of Theorem 14 also holds for the series $\sum_{n=1}^{\infty} (-1)^n a_n$, which starts with a negative term, $-a_1$. (It is just the negative of the first series.) The theorem also remains valid if the conditions that $\{a_n\}$ is positive and decreasing are replaced by the corresponding *ultimate* versions:

(i)   $a_n \geq 0$   and   (ii) $a_{n+1} \leq a_n$   for $n = N, N+1, N+2, \ldots$.

**Remark**   The proof of Theorem 14 shows that every even partial sum is less than or equal to $s$ and every odd partial sum is greater than or equal to $s$. (The reverse is true if $(-1)^n$ is used instead of $(-1)^{n-1}$.) That is, $s$ lies between $s_{2n}$ and either $s_{2n-1}$ or $s_{2n+1}$. It follows that, for odd or even $n$,

$$|s - s_n| \leq |s_{n+1} - s_n| = a_{n+1}.$$

This proves the following theorem.

THEOREM **15**

**Error estimate for alternating series**

If the sequence $\{a_n\}$ satisfies the conditions of the alternating series test (Theorem 14), so that the series $\sum_{n=1}^{\infty}(-1)^{n-1}a_n$ (or $\sum_{n=1}^{\infty}(-1)^{n}a_n$) converges to the sum $s$, then for any $n \geq 1$, the $n$th partial sum $s_n$ of the series satisfies

$$|s - s_n| \leq |a_{n+1}|.$$

That is, the size of the error involved in using $s_n$ as an approximation to $s$ is less than the size of the first omitted term.

**Example 2** How many terms of the series $\sum_{n=1}^{\infty}\dfrac{(-1)^n}{1 + 2^n}$ are needed to compute the sum of the series with error less that 0.001?

**Solution** This series satisfies the hypotheses for Theorem 15. If we use the partial sum of the first $n$ terms of the series to approximate the sum of the series, the error will satisfy

$$|\text{error}| \leq |\text{first omitted term}| = \frac{1}{1 + 2^{n+1}}.$$

This error is less than 0.001 if $1 + 2^{n+1} > 1,000$. Since $2^{10} = 1,024$, $n + 1 = 10$ will do; we need 9 terms of the series to compute the sum to within 0.001 of its actual value.

When determining the convergence of a given series, it is best to consider first whether the series converges absolutely. If it does not, then there remains the possibility of conditional convergence.

**Example 3** Test the following series for absolute and conditional convergence:

(a) $\displaystyle\sum_{n=1}^{\infty}\frac{(-1)^{n-1}}{n}$, (b) $\displaystyle\sum_{n=2}^{\infty}\frac{\cos(n\pi)}{\ln n}$, (c) $\displaystyle\sum_{n=1}^{\infty}\frac{(-1)^{n-1}}{n^4}$

**Solution** The absolute values of the terms in series (a) and (b) are $1/n$ and $1/(\ln n)$, respectively. Since $1/(\ln n) > 1/n$, and $\sum_{n=k}^{\infty}1/n$ diverges to infinity, neither series (a) nor (b) converges absolutely. However, both $\{1/n\}$ and $\{1/(\ln n)\}$ are positive, decreasing sequences that converge to 0. Therefore both (a) and (b) converge by Theorem 14. Each of these series is conditionally convergent.

Series (c) is absolutely convergent because $|(-1)^{n-1}/n^4| = 1/n^4$, and $\sum_{n=1}^{\infty}1/n^4$ is a convergent $p$-series ($p = 4 > 1$). We could establish its convergence using Theorem 14, but there is no need to do that since every absolutely convergent series is convergent (Theorem 13).

**Example 4**   For what values of $x$ does the series $\displaystyle\sum_{n=1}^{\infty} \frac{(x-5)^n}{n\,2^n}$ converge absolutely? converge conditionally? diverge?

**Solution**   For such series whose terms involve functions of a variable $x$, it is usually wisest to begin testing for absolute convergence with the ratio test. We have

$$\rho = \lim_{n \to \infty} \left| \frac{(x-5)^{n+1}}{(n+1)2^{n+1}} \middle/ \frac{(x-5)^n}{n\,2^n} \right| = \lim_{n \to \infty} \frac{n}{n+1} \left| \frac{x-5}{2} \right| = \left| \frac{x-5}{2} \right|.$$

The series converges absolutely if $|(x-5)/2| < 1$. This inequality is equivalent to $|x-5| < 2$, (the distance from $x$ to 5 is less than 2), that is, $3 < x < 7$. If $x < 3$ or $x > 7$, then $|(x-5)/2| > 1$. The series diverges; its terms do not approach zero.

If $x = 3$, the series is $\sum_{n=1}^{\infty}((-1)^n/n)$, which converges conditionally (it is an alternating harmonic series); if $x = 7$, the series is the harmonic series $\sum_{n=1}^{\infty} 1/n$, which diverges to infinity. Hence the given series converges absolutely on the open interval $(3, 7)$, converges conditionally at $x = 3$, and diverges everywhere else. ∎

**Example 5**   For what values of $x$ does the series $\displaystyle\sum_{n=0}^{\infty}(n+1)^2 \left( \frac{x}{x+2} \right)^n$ converge absolutely? converge conditionally? diverge?

**Solution**   Again we begin with the ratio test.

$$\rho = \lim_{n \to \infty} \left| (n+2)^2 \left( \frac{x}{x+2} \right)^{n+1} \middle/ (n+1)^2 \left( \frac{x}{x+2} \right)^n \right|$$

$$= \lim_{n \to \infty} \left( \frac{n+2}{n+1} \right)^2 \left| \frac{x}{x+2} \right| = \left| \frac{x}{x+2} \right| = \frac{|x|}{|x+2|}.$$

The series converges absolutely if $|x|/|x+2| < 1$. This condition says that the distance from $x$ to 0 is less than the distance from $x$ to $-2$. Hence $x > -1$. The series diverges if $|x|/|x+2| > 1$, that is, if $x < -1$. If $x = -1$, the series is $\sum_{n=0}^{\infty}(-1)^n(n+1)^2$, which diverges. We conclude that the series converges absolutely for $x > -1$, converges conditionally nowhere, and diverges for $x \le -1$. ∎

When using the alternating series test, it is important to verify (at least mentally) that *all three conditions* (i)–(iii) are satisfied. (As mentioned above, conditions (i) and (ii) need only be satisfied ultimately.)

**Example 6**   Test for convergence:

(a)   $\displaystyle\sum_{n=1}^{\infty}(-1)^{n-1}\frac{n+1}{n}$,

(b)   $\displaystyle 1 - \frac{1}{4} + \frac{1}{3} - \frac{1}{16} + \frac{1}{5} - \cdots = \sum_{n=1}^{\infty}(-1)^{n-1}a_n$, where

$$a_n = \begin{cases} 1/n & \text{if } n \text{ is odd}, \\ 1/n^2 & \text{if } n \text{ is even}. \end{cases}$$

### Solution

(a) Here $a_n = (n + 1)/n$ is positive and decreases as $n$ increases. However, $\lim_{n\to\infty} a_n = 1 \neq 0$. The alternating series test does not apply. In fact, the given series diverges, because its terms do not approach $0$.

(b) This series alternates, $a_n$ is positive, and $\lim_{n\to\infty} a_n = 0$. However, $\{a_n\}$ is not decreasing (even ultimately). Once again the alternating series test cannot be applied. In fact, since

$$-\frac{1}{4} - \frac{1}{16} - \cdots - \frac{1}{(2n)^2} - \cdots \qquad \text{converges, and}$$

$$1 + \frac{1}{3} + \frac{1}{5} + \cdots + \frac{1}{2n - 1} + \cdots \qquad \text{diverges to infinity,}$$

it is readily seen that the given series diverges to infinity. ∎

## Rearranging the Terms in a Series

The basic difference between absolute and conditional convergence is that when a series $\sum_{n=1}^{\infty} a_n$ converges absolutely it does so because its terms $\{a_n\}$ decrease in size fast enough that their sum can be finite even if no cancellation occurs due to terms of opposite sign. If cancellation is required to make the series converge (because the terms decrease slowly), then the series can only converge conditionally.

Consider the alternating harmonic series

$$1 - \frac{1}{2} + \frac{1}{3} - \frac{1}{4} + \frac{1}{5} - \frac{1}{6} + \cdots.$$

This series converges, but only conditionally. If we take the subseries containing only the positive terms, we get the series

$$1 + \frac{1}{3} + \frac{1}{5} + \frac{1}{7} + \cdots,$$

which diverges to infinity. Similarly, the subseries of negative terms

$$-\frac{1}{2} - \frac{1}{4} - \frac{1}{6} - \frac{1}{8} - \cdots$$

diverges to negative infinity.

If a series converges absolutely, the subseries consisting of positive terms and the subseries consisting of negative terms must each converge to a finite sum. If a series converges conditionally, the positive and negative subseries will both diverge, to $\infty$ and $-\infty$, respectively.

Using these facts we can answer a question raised at the beginning of Section 9.2. If we rearrange the terms of a convergent series so that they are added in a different order, must the rearranged series converge or not, and if it does will it converge to the same sum? The answer depends on whether the original series was absolutely convergent or merely conditionally convergent.

**THEOREM 16**   **Convergence of rearrangements of a series**

(a) If the terms of an absolutely convergent series are rearranged so that addition occurs in a different order, the rearranged series still converges to the same sum as the original series.

(b) If a series is conditionally convergent, and $L$ is any real number, then the terms of the series can be rearranged so as to make the series converge (conditionally) to the sum $L$. It can also be rearranged so as to diverge to $\infty$ or to $-\infty$, or just to diverge.

Part (b) shows that conditional convergence is a rather suspect kind of convergence, being dependent on the order in which the terms are added. We will not present a formal proof of the theorem, but will give an example suggesting what is involved. (See also Exercise 30 below.)

**Example 7**   In Section 9.5 we will show that the alternating harmonic series

$$\sum_{n=1}^{\infty} \frac{(-1)^{n-1}}{n} = 1 - \frac{1}{2} + \frac{1}{3} - \frac{1}{4} + \frac{1}{5} - \frac{1}{6} + \frac{1}{7} - \cdots$$

converges (conditionally) to the sum $\ln 2$. Describe how to rearrange its terms so that it converges to 8 instead.

**Solution**   Start adding terms of the positive subseries

$$1 + \frac{1}{3} + \frac{1}{5} + \cdots$$

and keep going until the partial sum exceeds 8. (It will, eventually, because the positive subseries diverges to infinity.) Then add the first term $-1/2$ of the negative subseries

$$-\frac{1}{2} - \frac{1}{4} - \frac{1}{6} - \cdots.$$

This will reduce the partial sum below 8 again. Now resume adding terms of the positive subseries until the partial sum climbs above 8 once more. Then add the second term of the negative subseries and the partial sum will drop below 8. Keep repeating this procedure, alternately adding terms of the positive subseries to force the sum above 8 and then terms of the negative subseries to force it below 8. Since both subseries have infinitely many terms and diverge to $\infty$ and $-\infty$, respectively, eventually every term of the original series will be included, and the partial sums of the new series will oscillate back and forth around 8, converging to that number. Of course, any number other than 8 could also be used in place of 8.

# Exercises 9.4

Determine whether the series in Exercises 1–12 converge absolutely, converge conditionally, or diverge.

**1.** $\displaystyle\sum_{n=1}^{\infty} \frac{(-1)^{n-1}}{\sqrt{n}}$

**2.** $\displaystyle\sum_{n=1}^{\infty} \frac{(-1)^{n}}{n^2 + \ln n}$

**3.** $\displaystyle\sum_{n=1}^{\infty} \frac{\cos(n\pi)}{(n+1)\ln(n+1)}$

**4.** $\displaystyle\sum_{n=1}^{\infty} \frac{(-1)^{2n}}{2^n}$

**5.** $\displaystyle\sum_{n=0}^{\infty} \frac{(-1)^n(n^2-1)}{n^2+1}$

**6.** $\displaystyle\sum_{n=1}^{\infty} \frac{(-2)^n}{n!}$

**7.** $\displaystyle\sum_{n=1}^{\infty} \frac{(-1)^n}{n\pi^n}$

**8.** $\displaystyle\sum_{n=0}^{\infty} \frac{-n}{n^2+1}$

**9.** $\displaystyle\sum_{n=1}^{\infty} (-1)^n \frac{20n^2-n-1}{n^3+n^2+33}$

**10.** $\displaystyle\sum_{n=1}^{\infty} \frac{100\cos(n\pi)}{2n+3}$

**11.** $\displaystyle\sum_{n=1}^{\infty} \frac{n!}{(-100)^n}$

**12.** $\displaystyle\sum_{n=10}^{\infty} \frac{\sin(n+1/2)\pi}{\ln\ln n}$

For the series in Exercises 13–16, find the smallest integer $n$ that ensures that the partial sum $s_n$ approximates the sum $s$ of the series with error less than 0.001 in absolute value.

**13.** $\displaystyle\sum_{n=1}^{\infty} (-1)^{n-1} \frac{n}{n^2+1}$

**14.** $\displaystyle\sum_{n=0}^{\infty} \frac{(-1)^n}{(2n)!}$

**15.** $\displaystyle\sum_{n=1}^{\infty} (-1)^{n-1} \frac{n}{2^n}$

**16.** $\displaystyle\sum_{n=0}^{\infty} (-1)^n \frac{3^n}{n!}$

Determine the values of $x$ for which the series in Exercises 17–24 converge absolutely, converge conditionally, or diverge.

**17.** $\displaystyle\sum_{n=0}^{\infty} \frac{x^n}{\sqrt{n+1}}$

**18.** $\displaystyle\sum_{n=1}^{\infty} \frac{(x-2)^n}{n^2 2^{2n}}$

**19.** $\displaystyle\sum_{n=0}^{\infty} (-1)^n \frac{(x-1)^n}{2n+3}$

**20.** $\displaystyle\sum_{n=1}^{\infty} \frac{1}{2n-1} \left(\frac{3x+2}{-5}\right)^n$

**21.** $\displaystyle\sum_{n=2}^{\infty} \frac{x^n}{2^n \ln n}$

**22.** $\displaystyle\sum_{n=1}^{\infty} \frac{(4x+1)^n}{n^3}$

**23.** $\displaystyle\sum_{n=1}^{\infty} \frac{(2x+3)^n}{n^{1/3} 4^n}$

**24.** $\displaystyle\sum_{n=1}^{\infty} \frac{1}{n} \left(1+\frac{1}{x}\right)^n$

**\* 25.** Does the alternating series test apply directly to the series $\sum_{n=1}^{\infty}(1/n)\sin(n\pi/2)$? Determine whether the series converges.

**\* 26.** Show that the series $\sum_{n=1}^{\infty} a_n$ converges absolutely if $a_n = 10/n^2$ for even $n$ and $a_n = -1/10n^3$ for odd $n$.

**\* 27.** Which of the following statements are TRUE and which are FALSE? Justify your assertion of truth, or give a counterexample to show falsehood.

(a) If $\sum_{n=1}^{\infty} a_n$ converges, then $\sum_{n=1}^{\infty} (-1)^n a_n$ converges.

(b) If $\sum_{n=1}^{\infty} a_n$ converges and $\sum_{n=1}^{\infty} (-1)^n a_n$ converges, then $\sum_{n=1}^{\infty} a_n$ converges absolutely.

(c) If $\sum_{n=1}^{\infty} a_n$ converges absolutely, then

$$\sum_{n=1}^{\infty} (-1)^n a_n \text{ converges absolutely.}$$

**\* 28.** (a) Use a Riemann sum argument to show that

$$\ln n! \geq \int_1^n \ln t\, dt = n\ln n - n + 1.$$

(b) For what values of $x$ does the series $\sum_{n=1}^{\infty} \frac{n! x^n}{n^n}$ converge absolutely? converge conditionally? diverge? (*Hint:* first use the ratio test. To test the cases where $\rho = 1$, you may find the inequality in part (a) useful.)

**\* 29.** For what values of $x$ does the series $\sum_{n=1}^{\infty} \frac{(2n)! x^n}{2^{2n}(n!)^2}$ converge absolutely? converge conditionally? diverge? *Hint:* see Exercise 42 of Section 9.3.

**\* 30.** Devise procedures for rearranging the terms of the alternating harmonic series so that the rearranged series (a) diverges to $\infty$, (b) converges to $-2$.

# 9.5 Power Series

This section is concerned with a special kind of infinite series called a *power series* that may be thought of as a polynomial of infinite degree.

---

**DEFINITION 7**

**Power series**

A series of the form

$$\sum_{n=0}^{\infty} a_n(x-c)^n = a_0 + a_1(x-c) + a_2(x-c)^2 + a_3(x-c)^3 + \cdots$$

is called a **power series in powers of** $x-c$ or a **power series about the point** $x = c$. The constants $a_0, a_1, a_2, \ldots$ are called the **coefficients** of the power series.

Since the terms of a power series are functions of a variable $x$, the series may or may not converge for each value of $x$. For those values of $x$ for which the series does converge, the sum defines a function of $x$. For example, if $-1 < x < 1$, then

$$1 + x + x^2 + x^3 + \cdots = \frac{1}{1-x}.$$

The geometric series on the left side is a power series *representation* of the function $1/(1-x)$ in powers of $x$ (or about the point $x = 0$). Note that the representation is valid only in the open interval $(-1, 1)$ even though $1/(1-x)$ is defined for all real $x$ except $x = 1$. For $x = -1$ and for $|x| > 1$ the series does not converge, so it cannot represent $1/(1-x)$ at these points.

The point $c$ is the **centre of convergence** of the power series $\sum_{n=0}^{\infty} a_n(x-c)^n$. The series certainly converges (to $a_0$) at $x = c$. (All the terms except possibly the first are 0.) Theorem 17 below shows that if the series converges anywhere else, then it converges on an interval (possibly infinite) centred at $x = c$, and it converges absolutely everywhere on that interval except possibly at one or both of the endpoints if the interval is finite. The geometric series

$$1 + x + x^2 + x^3 + \cdots$$

is an example of this behaviour. It has centre of convergence $c = 0$, and converges only on the interval $(-1, 1)$, centred at 0. The convergence is absolute at every point of the interval. Another example is the series

$$\sum_{n=1}^{\infty} \frac{1}{n\, 2^n} (x-5)^n = \frac{x-5}{2} + \frac{(x-5)^2}{2 \times 2^2} + \frac{(x-5)^3}{3 \times 2^3} + \cdots,$$

which we discussed in Example 4 of Section 9.4. We showed that this series converges on the interval $[3, 7)$, an interval with centre $x = 5$, and that the convergence is absolute on the open interval $(3, 7)$, but is only conditional at the endpoint $x = 3$.

**THEOREM 17**

For any power series $\sum_{n=0}^{\infty} a_n(x-c)^n$ one of the following alternatives must hold:

(i) the series may converge only at $x = c$, or

(ii) the series may converge at every real number $x$, or

(iii) there may exist a positive real number $R$ such that the series converges at every $x$ satisfying $|x - c| < R$ and diverges at every $x$ satisfying $|x - c| > R$. In this case the series may or may not converge at either of the two *endpoints* $x = c - R$ and $x = c + R$.

In each of these cases the convergence is absolute except possibly at the endpoints $x = c - R$ and $x = c + R$ in case (iii).

**PROOF**    We observed above that every power series converges at its centre of convergence; only the first term can be nonzero so the convergence is absolute. To prove the rest of this theorem, it suffices to show that if the series converges at any number $x_0 \neq c$, then it converges absolutely at every number $x$ closer to $c$ than $x_0$ is, that is, at every $x$ satisfying $|x - c| < |x_0 - c|$. This means that convergence at any $x_0 \neq c$ implies absolute convergence on $(c - x_0, c + x_0)$, so the set of points $x$ where the series converges must be an interval centred at $c$.

Suppose, therefore, that $\sum_{n=0}^{\infty} a_n(x_0-c)^n$ converges. Then $\lim a_n(x_0-c)^n = 0$, and so $|a_n(x_0-c)^n| \leq K$ for all $n$, where $K$ is some constant (Theorem 1 of Section 9.1). If $r = |x - c|/|x_0 - c| < 1$, then

$$\sum_{n=0}^{\infty} |a_n(x-c)^n| = \sum_{n=0}^{\infty} |a_n(x_0-c)^n| \left|\frac{x-c}{x_0-c}\right|^n \leq K \sum_{n=0}^{\infty} r^n = \frac{K}{1-r} < \infty.$$

Thus $\sum_{n=0}^{\infty} a_n(x - c)^n$ converges absolutely.

By Theorem 17, the set of values $x$ for which the power series $\sum_{n=0}^{\infty} a_n(x - c)^n$ converges is an interval centred at $x = c$. We call this interval the **interval of convergence** of the power series. It must be of one of the following forms:

    (i) the isolated point $x = c$ (a degenerate closed interval $[c, c]$),

    (ii) the entire line $(-\infty, \infty)$,

    (iii) a finite interval centred at $c$:

$$[c - R, c + R], \text{ or } [c - R, c + R), \text{ or } (c - R, c + R], \text{ or } (c - R, c + R).$$

The number $R$ in (iii) is called the **radius of convergence** of the power series. In case (i) we say the radius of convergence is $R = 0$; in case (ii) it is $R = \infty$.

The radius of convergence, $R$, can often be found by using the ratio test on the power series: if

$$\rho = \lim_{n \to \infty} \left| \frac{a_{n+1}(x - c)^{n+1}}{a_n(x - c)^n} \right| = \left( \lim_{n \to \infty} \left| \frac{a_{n+1}}{a_n} \right| \right) |x - c|$$

exists, then the series $\sum_{n=0}^{\infty} a_n(x - c)^n$ converges absolutely where $\rho < 1$, that is, where

$$|x - c| < R = \frac{1}{L} = 1 \Bigg/ \lim_{n \to \infty} \left| \frac{a_{n+1}}{a_n} \right|.$$

The series diverges if $|x - c| > R$.

---

**Radius of convergence**

Suppose that $L = \lim_{n \to \infty} \left| \frac{a_{n+1}}{a_n} \right|$ exists or is $\infty$. Then the power series $\sum_{n=0}^{\infty} a_n(x - c)^n$ has radius of convergence $R = 1/L$ (If $L = 0$ then $R = \infty$; if $L = \infty$ then $R = 0$.)

---

**Example 1**    Determine the centre, radius, and interval of convergence of

$$\sum_{n=0}^{\infty} \frac{(2x + 5)^n}{(n^2 + 1)3^n}.$$

**Solution**    The series can be rewritten

$$\sum_{n=0}^{\infty} \left( \frac{2}{3} \right)^n \frac{1}{n^2 + 1} \left( x + \frac{5}{2} \right)^n.$$

The centre of convergence is $x = -5/2$. The radius of convergence, $R$, is given by

$$\frac{1}{R} = L = \lim \left| \frac{\left(\frac{2}{3}\right)^{n+1} \frac{1}{(n+1)^2+1}}{\left(\frac{2}{3}\right)^n \frac{1}{n^2+1}} \right| = \lim \frac{2}{3} \frac{n^2+1}{(n+1)^2+1} = \frac{2}{3}.$$

Thus $R = 3/2$. The series converges absolutely on $(-5/2 - 3/2, -5/2 + 3/2) = (-4, -1)$, and it diverges on $(-\infty, -4)$ and on $(-1, \infty)$. At $x = -1$ the series is $\sum_{n=0}^{\infty} 1/(n^2+1)$; at $x = -4$ it is $\sum_{n=0}^{\infty}(-1)^n/(n^2+1)$. Both series converge (absolutely). The interval of convergence of the given power series is therefore $[-4, -1]$.

---

**Example 2**    Determine the radii of convergence of the series

(a) $\displaystyle\sum_{n=0}^{\infty} \frac{x^n}{n!}$    and    (b) $\displaystyle\sum_{n=0}^{\infty} n! x^n$.

**Solution**

(a) $L = \left| \lim \frac{1}{(n+1)!} \bigg/ \frac{1}{n!} \right| = \lim \frac{n!}{(n+1)!} = \lim \frac{1}{n+1} = 0$. Thus $R = \infty$.

This series converges (absolutely) for all $x$. The sum is $e^x$ as will be shown in Example 1 in the next section.

(b) $L = \left| \lim \frac{(n+1)!}{n!} \right| = \lim(n+1) = \infty$. Thus $R = 0$

This series converges only at its centre of convergence, $x = 0$.

---

## Algebraic Operations on Power Series

To simplify the following discussion, we will consider only power series with $x = 0$ as centre of convergence, that is, series of the form

$$\sum_{n=0}^{\infty} a_n x^n = a_0 + a_1 x + a_2 x^2 + a_3 x^3 + \cdots.$$

Any properties we demonstrate for such series extend automatically to power series of the form $\sum_{n=0}^{\infty} a_n(y-c)^n$ via the change of variable $x = y - c$.

First we observe that series having the same centre of convergence can be added or subtracted on whatever interval is common to their intervals of convergence. The following theorem is a simple consequence of Theorem 7 of Section 9.2 and does not require a proof.

**THEOREM 18**    Let $\sum_{n=0}^{\infty} a_n x^n$ and $\sum_{n=0}^{\infty} b_n x^n$ be two power series with radii of convergence $R_a$ and $R_b$ respectively, and let $c$ be a constant. Then

(i) $\sum_{n=0}^{\infty}(c a_n) x^n$ has radius of convergence $R_a$, and

$$\sum_{n=0}^{\infty}(c a_n) x^n = c \sum_{n=0}^{\infty} a_n x^n$$

wherever the series on the right converges.

(ii) $\sum_{n=0}^{\infty}(a_n + b_n)\,x^n$ has radius of convergence $R$ at least as large as the smaller of $R_a$ and $R_b$ ($R \geq \min\{R_a, R_b\}$), and

$$\sum_{n=0}^{\infty}(a_n + b_n)\,x^n = \sum_{n=0}^{\infty} a_n\,x^n + \sum_{n=0}^{\infty} b_n\,x^n$$

wherever both series on the right converge.

The situation regarding multiplication and division of power series is more complicated. We will mention only the results and will not attempt any proofs of our assertions. A textbook in mathematical analysis will provide more details.

Long multiplication of the form

$$(a_0 + a_1 x + a_2 x^2 + \cdots)(b_0 + b_1 x + b_2 x^2 + \cdots)$$
$$= a_0 b_0 + (a_0 b_1 + a_1 b_0)x + (a_0 b_2 + a_1 b_1 + a_2 b_0)x^2 + \cdots$$

leads us to conjecture the formula

$$\left(\sum_{n=0}^{\infty} a_n x^n\right)\left(\sum_{n=0}^{\infty} b_n x^n\right) = \sum_{n=0}^{\infty} c_n x^n,$$

where

$$c_n = a_0 b_n + a_1 b_{n-1} + \cdots + a_n b_0 = \sum_{j=0}^{n} a_j b_{n-j}.$$

The series $\sum_{n=0}^{\infty} c_n x^n$ is called the **Cauchy product** of the series $\sum_{n=0}^{\infty} a_n x^n$ and $\sum_{n=0}^{\infty} b_n x^n$. Like the sum, the Cauchy product also has radius of convergence at least equal to the lesser of those of the factor series.

**Example 3**  Since

$$\frac{1}{1 - x} = 1 + x + x^2 + x^3 + \cdots = \sum_{n=0}^{\infty} x^n$$

holds for $-1 < x < 1$, we can determine a power series representation for $1/(1-x)^2$ by taking the Cauchy product of this series with itself. Since $a_n = b_n = 1$ for $n = 0, 1, 2, \ldots$ we have

$$c_n = \sum_{j=0}^{n} 1 = n + 1$$

and

$$\frac{1}{(1-x)^2} = 1 + 2x + 3x^2 + 4x^3 + \cdots \sum_{n=0}^{\infty} (n+1)x^n,$$

which must also hold for $-1 < x < 1$. The same series can be obtained by direct long multiplication of the series:

$$
\begin{array}{rcccccccccc}
 & 1 & + & x & + & x^2 & + & x^3 & + & \cdots \\
\times & 1 & + & x & + & x^2 & + & x^3 & + & \cdots \\
\hline
 & 1 & + & x & + & x^2 & + & x^3 & + & \cdots \\
 & & & x & + & x^2 & + & x^3 & + & \cdots \\
 & & & & & x^2 & + & x^3 & + & \cdots \\
 & & & & & & & x^3 & + & \cdots \\
 & & & & & & & & & \cdots \\
\hline
 & 1 & + & 2x & + & 3x^2 & + & 4x^3 & + & \cdots
\end{array}
$$

Long division can also be performed on power series, but there is no simple rule for determining the coefficients of the quotient series. The radius of convergence of the quotient series is not less than the least of the three numbers $R_1$, $R_2$, and $R_3$, where $R_1$ and $R_2$ are the radii of convergence of the divisor and dividend series and $R_3$ is the distance from the centre of convergence to the nearest complex number where the divisor series has sum equal to 0.

**Example 4**    To illustrate this point, observe that 1 and $1 - x$ are both power series with infinite radii of convergence:

$$1 = 1 + 0x + 0x^2 + 0x^3 + \cdots \qquad \text{for all } x,$$
$$1 - x = 1 - x + 0x^2 + 0x^3 + \cdots \qquad \text{for all } x.$$

Their quotient, $1/(1 - x)$, however, only has radius of convergence 1, the distance from the centre of convergence $x = 0$ to the point $x = 1$ where the denominator vanishes:

$$\frac{1}{1-x} = 1 + x + x^2 + x^3 + \cdots \qquad \text{for} \quad |x| < 1.$$

## Differentiation and Integration of Power Series

If a power series has a positive radius of convergence, it can be differentiated or integrated term by term. The resulting series will converge to the appropriate derivative or integral of the sum of the original series everywhere except possibly at the endpoints of the interval of convergence of the original series. This very important fact ensures that, for purposes of calculation, power series behave just like polynomials, the easiest functions to differentiate and integrate. We formalize the differentiation and integration properties of power series in the following theorem.

**THEOREM  19**

**Term by term differentiation and integration of power series**

If the series $\sum_{n=0}^{\infty} a_n x^n$ converges to the sum $f(x)$ on an interval $(-R, R)$, where

$R > 0$; that is,

$$f(x) = \sum_{n=0}^{\infty} a_n x^n = a_0 + a_1 x + a_2 x^2 + a_3 x^3 + \cdots, \qquad (-R < x < R),$$

then $f$ is differentiable on $(-R, R)$ and

$$f'(x) = \sum_{n=1}^{\infty} n a_n x^{n-1} = a_1 + 2a_2 x + 3a_3 x^2 + \cdots, \qquad (-R < x < R).$$

Also, $f$ is integrable over any closed subinterval of $(-R, R)$, and if $|x| \leq R$, then

$$\int_0^x f(t)\, dt = \sum_{n=0}^{\infty} \frac{a_n}{n+1} x^{n+1} = a_0 x + \frac{a_1}{2} x^2 + \frac{a_2}{3} x^3 + \cdots.$$

**PROOF**  Let $x$ satisfy $-R < x < R$ and choose $H > 0$ such that $|x| + H < R$. By Theorem 17, we then have[1]

$$\sum_{n=1}^{\infty} |a_n| (|x| + H)^n = K < \infty.$$

The Binomial Theorem (see Section 9.9) shows that if $n \geq 1$, then

$$(x + h)^n = x^n + n x^{n-1} h + \sum_{k=2}^{n} \binom{n}{k} x^{n-k} h^k.$$

Therefore, if $|h| \leq H$ we have

$$\begin{aligned}
|(x+h)^n - x^n - n x^{n-1} h| &= \left| \sum_{k=2}^{n} \binom{n}{k} x^{n-k} h^k \right| \\
&\leq \sum_{k=2}^{n} \binom{n}{k} |x|^{n-k} \frac{|h|^k}{H^k} H^k \\
&\leq \frac{|h|^2}{H^2} \sum_{k=0}^{n} \binom{n}{k} |x|^{n-k} H^k \\
&= \frac{|h|^2}{H^2} (|x| + H)^n.
\end{aligned}$$

Also

$$|n x^{n-1}| = \frac{n |x|^{n-1} H}{H} \leq \frac{1}{H} (|x| + H)^n.$$

Thus

$$\sum_{n=1}^{\infty} |n a_n x^{n-1}| \leq \frac{1}{H} \sum_{n=1}^{\infty} |a_n| (|x| + H)^n = \frac{K}{H} < \infty$$

---

[1] This proof is due to R. Výborný, *American Mathematical Monthly*, April 1987.

so the series $\sum_{n=1}^{\infty} n a_n x^{n-1}$ converges (absolutely) to $g(x)$, say. Now

$$\left| \frac{f(x+h) - f(x)}{h} - g(x) \right| = \left| \sum_{n=1}^{\infty} \frac{a_n(x+h)^n - a_n x^n - n a_n x^{n-1} h}{h} \right|$$

$$\leq \frac{1}{|h|} \sum_{n=1}^{\infty} |a_n| |(x+h)^n - x^n - n x^{n-1} h|$$

$$\leq \frac{|h|}{H^2} \sum_{n=1}^{\infty} |a_n| (|x| + H)^n \leq \frac{K|h|}{H^2}.$$

Letting $h$ approach zero, we obtain $|f'(x) - g(x)| \leq 0$, so $f'(x) = g(x)$, as required.

Now observe that since $|a_n/(n+1)| \leq |a_n|$, the series

$$h(x) = \sum_{n=0}^{\infty} \frac{a_n}{n+1} x^{n+1}$$

converges (absolutely) at least on the interval $(-R, R)$. Using the differentiation result proved above, we obtain

$$h'(x) = \sum_{n=0}^{\infty} a_n x^n = f(x).$$

Since $h(0) = 0$, we have

$$\int_0^x f(t) \, dt = \int_0^x h'(t) \, dt = h(t) \Big|_0^x = h(x),$$

as required.

Together, these results imply that the termwise differentiated or integrated series have the same radius of convergence as the given series. In fact, as the following examples illustrate, the interval of convergence of the differentiated series is the same as that of the original series except for the *possible* loss of one or both endpoints if the original series converges at endpoints of its interval of convergence. Similarly, the integrated series will converge everywhere on the interval of convergence of the original series and possibly at one or both endpoints of that interval, even if the original series does not converge at the endpoints.

Being differentiable on $(-R, R)$, where $R$ is the radius of convergence, the sum $f(x)$ of a power series is necessarily continuous on that open interval. If the series happens to converge at either or both of the endpoints $-R$ and $R$, then $f$ is also continuous (on one side) up to these endpoints. This result is stated formally in the following theorem. We will not prove it here; the interested reader is referred to textbooks on mathematical analysis for a proof.

**THEOREM 20**

### Abel's Theorem

The sum of a power series is a continuous function everywhere on the interval of convergence of the series. In particular, if $\sum_{n=0}^{\infty} a_n R^n$ converges for some $R > 0$, then

$$\lim_{x \to R-} \sum_{n=0}^{\infty} a_n x^n = \sum_{n=0}^{\infty} a_n R^n,$$

and if $\sum_{n=0}^{\infty} a_n(-R)^n$ converges, then

$$\lim_{x \to -R+} \sum_{n=0}^{\infty} a_n x^n = \sum_{n=0}^{\infty} a_n(-R)^n.$$

The following examples show how the above theorems are applied to obtain power series representations for functions.

**Example 5**   Find power series representations for the functions

(a) $\dfrac{1}{(1-x)^2}$,   (b) $\dfrac{1}{(1-x)^3}$,   (c) $\ln(1+x)$

by starting with the geometric series

$$\frac{1}{1-x} = \sum_{n=0}^{\infty} x^n = 1 + x + x^2 + x^3 + \cdots \qquad (-1 < x < 1)$$

and using differentiation, integration, and substitution. Where is each series valid?

**Solution**

(a) Differentiate the geometric series term by term according to Theorem 19 to obtain

$$\frac{1}{(1-x)^2} = \sum_{n=1}^{\infty} n x^{n-1} = 1 + 2x + 3x^2 + 4x^3 + \cdots \qquad (-1 < x < 1).$$

This is the same result obtained by multiplication of series in Example 3 above.

(b) Differentiate again to get, for $-1 < x < 1$,

$$\frac{2}{(1-x)^3} = \sum_{n=2}^{\infty} n(n-1) x^{n-2} = (1 \times 2) + (2 \times 3)x + (3 \times 4)x^2 + \cdots$$

Now divide by 2:

$$\frac{1}{(1-x)^3} = \sum_{n=2}^{\infty} \frac{n(n-1)}{2} x^{n-2} = 1 + 3x + 6x^2 + 10x^3 + \cdots \qquad (-1 < x < 1).$$

(c) Substitute $-t$ in place of $x$ in the original geometric series:

$$\frac{1}{1+t} = \sum_{n=0}^{\infty} (-1)^n t^n = 1 - t + t^2 - t^3 + t^4 - \cdots \qquad (-1 < t < 1).$$

Integrate from 0 to $x$, where $|x| < 1$, to get

$$\ln(1+x) = \int_0^x \frac{dt}{1+t} = \sum_{n=0}^{\infty} (-1)^n \int_0^x t^n \, dt$$

$$= \sum_{n=0}^{\infty} (-1)^n \frac{x^{n+1}}{n+1} = x - \frac{x^2}{2} + \frac{x^3}{3} - \frac{x^4}{4} + \cdots \quad (-1 < x \le 1).$$

Note that the latter series converges (conditionally) at the endpoint $x = 1$ as well as on the interval $-1 < x < 1$. Since $\ln(1+x)$ is continuous at $x = 1$, Theorem 20 assures us the series must converge to that function at $x = 1$ also. In particular, therefore, the alternating harmonic series converges to $\ln 2$:

$$\ln 2 = 1 - \frac{1}{2} + \frac{1}{3} - \frac{1}{4} + \frac{1}{5} - \cdots = \sum_{n=0}^{\infty} \frac{(-1)^n}{n+1}.$$

This would not, however, be a very useful formula for calculating the value of $\ln 2$. (Why not?)

---

**Example 6**   Use the geometric series of the previous example to find a power series representation for $\tan^{-1} x$.

**Solution**   Substitute $-t^2$ for $x$ in the geometric series. Since $0 \le t^2 < 1$ whenever $-1 < t < 1$, we obtain

$$\frac{1}{1+t^2} = 1 - t^2 + t^4 - t^6 + t^8 - \cdots \quad (-1 < t < 1).$$

Now integrate from 0 to $x$, where $|x| < 1$:

$$\tan^{-1} x = \int_0^x \frac{dt}{1+t^2} = \int_0^x (1 - t^2 + t^4 - t^6 + t^8 - \cdots) \, dt$$

$$= x - \frac{x^3}{3} + \frac{x^5}{5} - \frac{x^7}{7} + \frac{x^9}{9} - \cdots$$

$$= \sum_{n=0}^{\infty} (-1)^n \frac{x^{2n+1}}{2n+1} \quad (-1 < x < 1).$$

However, note that the series also converges (conditionally) at $x = -1$ and 1. Since $\tan^{-1}$ is continuous at $\pm 1$, the above series representation for $\tan^{-1} x$ also holds for these values, by Theorem 20. Letting $x = 1$ we get another interesting result:

$$\frac{\pi}{4} = 1 - \frac{1}{3} + \frac{1}{5} - \frac{1}{7} + \frac{1}{9} - \cdots.$$

Again, however, this would not be a good formula with which to calculate a numerical value of $\pi$. (Why not?)

---

**Example 7**   Find the sum of the series $\displaystyle\sum_{n=1}^{\infty} \frac{n^2}{2^n}$ by first finding the sum of the power series

$$\sum_{n=1}^{\infty} n^2 x^n = x + 4x^2 + 9x^3 + 16x^4 + \cdots.$$

**Solution**  Observe (in Example 5(a)) how the process of differentiating the geometric series produces a series with coefficients 1, 2, 3, .... Start with the series obtained for $1/(1-x)^2$ and multiply it by $x$ to obtain

$$\sum_{n=1}^{\infty} nx^n = x + 2x^2 + 3x^3 + 4x^4 + \cdots = \frac{x}{(1-x)^2}.$$

Now differentiate again to get a series with coefficients $1^2$, $2^2$, $3^2$, ...

$$\sum_{n=1}^{\infty} n^2 x^{n-1} = 1 + 4x + 9x^2 + 16x^3 + \cdots = \frac{d}{dx} \frac{x}{(x-1)^2} = \frac{1+x}{(1-x)^3}.$$

Multiplication by $x$ again gives the desired power series

$$\sum_{n=1}^{\infty} n^2 x^n = x + 4x^2 + 9x^3 + 16x^4 + \cdots = \frac{x(1+x)}{(1-x)^3}.$$

Differentiation and multiplication by $x$ do not change the radius of convergence, so this series converges to the indicated function for $-1 < x < 1$. Putting $x = 1/2$, we get

$$\sum_{n=1}^{\infty} \frac{n^2}{2^n} = \frac{\frac{1}{2} \times \frac{3}{2}}{\frac{1}{8}} = 6.$$

The following example illustrates how substitution can be used to obtain power series representations of functions with centres of convergence different from 0.

**Example 8**  Find a series representation of $f(x) = 1/(2 + x)$ in powers of $x - 1$. What is the interval of convergence of this series?

**Solution**  Let $t = x - 1$ so that $x = t + 1$. We have

$$\frac{1}{2+x} = \frac{1}{3+t} = \frac{1}{3} \frac{1}{1 + \dfrac{t}{3}}$$

$$= \frac{1}{3} \left( 1 - \frac{t}{3} + \frac{t^2}{3^2} - \frac{t^3}{3^3} + \cdots \right) \qquad (-1 < t/3 < 1)$$

$$= \sum_{n=0}^{\infty} (-1)^n \frac{t^n}{3^{n+1}} \qquad (-3 < t < 3)$$

$$= \sum_{n=0}^{\infty} (-1)^n \frac{(x-1)^n}{3^{n+1}} \qquad (-2 < x < 4).$$

Note that the radius of convergence of this series is 3, the distance from the centre of convergence, 1, to the point $-2$ where the denominator is 0. We could have predicted this in advance.

## Exercises 9.5

Determine the centre, radius, and interval of convergence of each of the power series in Exercises 1–8.

**1.** $\displaystyle\sum_{n=0}^{\infty} \frac{x^{2n}}{\sqrt{n+1}}$

**2.** $\displaystyle\sum_{n=0}^{\infty} 3n\,(x+1)^n$

**3.** $\displaystyle\sum_{n=1}^{\infty} \frac{1}{n}\left(\frac{x+2}{2}\right)^n$

**4.** $\displaystyle\sum_{n=1}^{\infty} \frac{(-1)^n}{n^4 2^{2n}}\,x^n$

**5.** $\displaystyle\sum_{n=0}^{\infty} n^3 (2x-3)^n$

**6.** $\displaystyle\sum_{n=1}^{\infty} \frac{e^n}{n^3}\,(4-x)^n$

**7.** $\displaystyle\sum_{n=0}^{\infty} \frac{(1+5^n)}{n!}\,x^n$

**8.** $\displaystyle\sum_{n=1}^{\infty} \frac{(4x-1)^n}{n^n}$

**9.** Use multiplication of series to find a power series representation of $1/(1-x)^3$ valid in the interval $(-1, 1)$.

**10.** Determine the Cauchy product of the series $1 + x + x^2 + x^3 + \cdots$ and $1 - x + x^2 - x^3 + \cdots$. On what interval and to what function does the product series converge?

**11.** Determine the power series expansion of $1/(1-x)^2$ by formally dividing $1 - 2x + x^2$ into 1.

Starting with the power series representation

$$\frac{1}{1-x} = 1 + x + x^2 + x^3 + \cdots \qquad (-1 < x < 1)$$

determine power series representations for the functions indicated in Exercises 12–20. On what interval is each representation valid?

**12.** $\dfrac{1}{2-x}$ in powers of $x$

**13.** $\dfrac{1}{(2-x)^2}$ in powers of $x$

**14.** $\dfrac{1}{1+2x}$ in powers of $x$

**15.** $\ln(2-x)$ in powers of $x$

**16.** $\dfrac{1}{x}$ in powers of $x-1$

**17.** $\dfrac{1}{x^2}$ in powers of $x+2$

**18.** $\dfrac{1-x}{1+x}$ in powers of $x$

**19.** $\dfrac{x^3}{1-2x^2}$ in powers of $x$

**20.** $\ln x$ in powers of $x-4$

Determine the interval of convergence and the sum of each of the series in Exercises 21–26.

**21.** $1 - 4x + 16x^2 - 64x^3 + \cdots = \displaystyle\sum_{n=0}^{\infty}(-1)^n (4x)^n$

**\* 22.** $3 + 4x + 5x^2 + 6x^3 + \cdots = \displaystyle\sum_{n=0}^{\infty}(n+3)x^n$

**\* 23.** $\dfrac{1}{3} + \dfrac{x}{4} + \dfrac{x^2}{5} + \dfrac{x^3}{6} + \cdots = \displaystyle\sum_{n=0}^{\infty} \frac{x^n}{n+3}$

**\* 24.** $1 \times 3 - 2 \times 4x + 3 \times 5x^2 - 4 \times 6x^3 + \cdots$
$$= \sum_{n=0}^{\infty}(-1)^n (n+1)(n+3)\,x^n$$

**\* 25.** $2 + 4x^2 + 6x^4 + 8x^6 + 10x^8 + \cdots = \displaystyle\sum_{n=0}^{\infty} 2(n+1)\,x^{2n}$

**\* 26.** $1 - \dfrac{x^2}{2} + \dfrac{x^4}{3} - \dfrac{x^6}{4} + \dfrac{x^8}{5} - \cdots = \displaystyle\sum_{n=0}^{\infty} \frac{(-1)^n x^{2n}}{n+1}$

Use the technique (or the result) of Example 7 to find the sums of the numerical series in Exercises 27–32.

**27.** $\displaystyle\sum_{n=1}^{\infty} \frac{n}{3^n}$

**28.** $\displaystyle\sum_{n=0}^{\infty} \frac{n+1}{2^n}$

**\* 29.** $\displaystyle\sum_{n=0}^{\infty} \frac{(n+1)^2}{\pi^n}$

**\* 30.** $\displaystyle\sum_{n=1}^{\infty} \frac{(-1)^n n(n+1)}{2^n}$

**31.** $\displaystyle\sum_{n=1}^{\infty} \frac{(-1)^{n-1}}{n2^n}$

**32.** $\displaystyle\sum_{n=3}^{\infty} \frac{1}{n2^n}$

# 9.6 Taylor and Maclaurin Series

If a power series $\sum_{n=0}^{\infty} a_n (x-c)^n$ has a positive radius of convergence $R$, then the sum of the series defines a function $f(x)$ on the interval $(c-R, c+R)$. We say that the power series is a **representation** of $f(x)$ on that interval. What relationship exists between the function $f(x)$ and the coefficients $a_0,\ a_1,\ a_2,\ \ldots$ of the power series? The following theorem answers this question.

**THEOREM 21**  Suppose the series

$$f(x) = \sum_{n=0}^{\infty} a_n (x - c)^n = a_0 + a_1(x - c) + a_2(x - c)^2 + a_3(x - c)^3 + \cdots$$

converges to $f(x)$ for $c - R < x < c + R$, where $R > 0$. Then

$$a_k = \frac{f^{(k)}(c)}{k!} \qquad \text{for } k = 0, 1, 2, 3, \ldots.$$

**PROOF**  This proof requires that we differentiate the series for $f(x)$ term by term several times, a process justified by Theorem 19 (suitably reformulated for powers of $x - c$):

$$f'(x) = \sum_{n=1}^{\infty} n a_n (x - c)^{n-1} = a_1 + 2a_2(x - c) + 3a_3(x - c)^2 + \cdots$$

$$f''(x) = \sum_{n=2}^{\infty} n(n - 1) a_n (x - c)^{n-2} = 2a_2 + 6a_3(x - c) + 12a_4(x - c)^2 + \cdots$$

$$\vdots$$

$$f^{(k)}(x) = \sum_{n=k}^{\infty} n(n - 1)(n - 2) \cdots (n - k + 1) a_n (x - c)^{n-k}$$

$$= k! a_k + \frac{(k + 1)!}{1!} a_{k+1}(x - c) + \frac{(k + 2)!}{2!} a_{k+2}(x - c)^2 + \cdots.$$

Each series converges for $c - R < x < c + R$. Setting $x = c$, we obtain $f^{(k)}(c) = k! a_k$, which proves the theorem.

Theorem 21 shows that a function $f(x)$ that has a power series representation with centre at $c$ and positive radius of convergence must have derivatives of all orders in an interval around $x = c$, and it can have only one representation as a power series in powers of $x - c$, namely

$$f(x) = \sum_{n=0}^{\infty} \frac{f^{(n)}(c)}{n!} (x - c)^n = f(c) + f'(c)(x - c) + \frac{f''(c)}{2!}(x - c)^2 + \cdots.$$

Such a series is called a Taylor series or, if $c = 0$, a Maclaurin series.

**DEFINITION 8**

**Taylor and Maclaurin series**

If $f(x)$ has derivatives of all orders at $x = c$ (that is, if $f^{(k)}(c)$ exists for $k = 0, 1, 2, 3, \ldots$), then the series

$$\sum_{k=0}^{\infty} \frac{f^{(k)}(c)}{k!} (x - c)^k$$

$$= f(c) + f'(c)(x - c) + \frac{f''(c)}{2!} (x - c)^2 + \frac{f^{(3)}(c)}{3!} (x - c)^3 + \cdots$$

is called the **Taylor series of $f$ about $x = c$** (or the **Taylor series of $f$ in powers of $x - c$**). If $c = 0$, the term **Maclaurin series** is usually used in place of Taylor series.

Note that the partial sums of such Taylor (or Maclaurin) series are just the Taylor (or Maclaurin) polynomials studied in Section 4.8.

The Taylor series is a power series as defined in the previous section. Theorem 17 implies that $c$ must be the centre of any interval on which such a series converges, but the definition of Taylor series makes no requirement that the series should converge anywhere except at the point $x = c$ where the series is just $f(0) + 0 + 0 + \cdots$. The series exists provided all the derivatives of $f$ exist at $x = c$; in practice this means that each derivative must exist in an open interval containing $x = c$. (Why?) However, the series may converge nowhere except at $x = c$, and if it does converge elsewhere, it may converge to something other than $f(x)$. (See Exercise 40 at the end of this section for an example where this happens.) If the Taylor series does converge to $f(x)$ in an open interval containing $c$, then we will say that $f$ is analytic at $x = c$.

**DEFINITION 9**

**Analytic functions**

A function $f(x)$ is **analytic** at $x = c$ if $f(x)$ is the sum of a power series in powers of $x - c$ having positive radius of convergence. (The series is its Taylor series.) If $f$ is analytic at each point of an open interval $I$, then we say it is analytic on the interval $I$.

Most, but not all, of the elementary functions encountered in calculus are analytic wherever they have derivatives of all orders. On the other hand, whenever a power series converges on an open interval containing $c$, then its sum $f(x)$ is analytic at $c$, and the given series is the Taylor series of $f(x)$ about $x = c$.

## Maclaurin Series for Some Elementary Functions

Calculating Taylor and Maclaurin series for a function $f$ directly from Definition 8 is practical only when we can find a formula for the $n$th derivative of $f$. Examples of such functions include $(ax + b)^r$, $e^{ax+b}$, $\ln(ax + b)$, $\sin(ax + b)$, $\cos(ax + b)$, and sums of such functions.

**Example 1** Find the Taylor series for $e^x$ about $x = c$. Where does the series converge to $e^x$? Where is $e^x$ analytic? What is the Maclaurin series for $e^x$?

**Solution** Since all the derivatives of $f(x) = e^x$ are $e^x$, we have $f^{(n)}(c) = e^c$ for every integer $n \geq 0$. Thus the Taylor series for $e^x$ about $x = c$ is

$$\sum_{n=0}^{\infty} \frac{e^c}{n!}(x - c)^n = e^c + e^c(x - c) + \frac{e^c}{2!}(x - c)^2 + \frac{e^c}{3!}(x - c)^3 + \cdots.$$

The radius of convergence $R$ of this series is given by

$$\frac{1}{R} = \lim_{n \to \infty} \left| \frac{e^c/(n + 1)!}{e^c/n!} \right| = \lim_{n \to \infty} \frac{n!}{(n + 1)!} = \lim_{n \to \infty} \frac{1}{n + 1} = 0.$$

Thus the radius of convergence is $R = \infty$ and the series converges for all $x$. Suppose the sum is $g(x)$:

$$g(x) = e^c + e^c(x - c) + \frac{e^c}{2!}(x - c)^2 + \frac{e^c}{3!}(x - c)^3 + \cdots.$$

By Theorem 19, we have

$$g'(x) = 0 + e^c + \frac{e^c}{2!}2(x - c) + \frac{e^c}{3!}3(x - c)^2 + \cdots$$

$$= e^c + e^c(x - c) + \frac{e^c}{2!}(x - c)^2 + \cdots = g(x).$$

Also, $g(c) = e^c + 0 + 0 + \cdots = e^c$. Since $g(x)$ satisfies the differential equation $g'(x) = g(x)$ of exponential growth, we have $g(x) = Ce^x$. Substituting $x = c$ gives $e^c = g(c) = Ce^c$, so $C = 1$. Thus the Taylor series for $e^x$ in powers of $x - c$ converges to $e^x$ for every real number $x$:

$$e^x = \sum_{n=0}^{\infty} \frac{e^c}{n!}(x - c)^n$$

$$= e^c + e^c(x - c) + \frac{e^c}{2!}(x - c)^2 + \frac{e^c}{3!}(x - c)^3 + \cdots \qquad \text{(for all } x\text{)}.$$

In particular, setting $c = 0$ we obtain the Maclaurin series for $e^x$:

$$e^x = \sum_{n=0}^{\infty} \frac{x^n}{n!} = 1 + x + \frac{x^2}{2!} + \frac{x^3}{3!} + \cdots \qquad \text{(for all } x\text{)}.$$

**Example 2** Find the Maclaurin series for (a) $\sin x$, and (b) $\cos x$. Where does each series converge?

**Solution** Let $f(x) = \sin x$. Then we have $f(0) = 0$ and:

$$f'(x) = \cos x \qquad\qquad f'(0) = 1$$
$$f''(x) = -\sin x \qquad\qquad f''(0) = 0$$
$$f^{(3)}(x) = -\cos x \qquad\qquad f^{(3)}(0) = -1$$
$$f^{(4)}(x) = \sin x \qquad\qquad f^{(4)}(0) = 0$$
$$f^{(5)}(x) = \cos x \qquad\qquad f^{(5)}(0) = 1$$

$$\vdots \qquad\qquad\qquad \vdots$$

Thus, the Maclaurin series for $\sin x$ is

$$g(x) = 0 + x + 0 - \frac{x^3}{3!} + 0 + \frac{x^5}{5!} + 0 - \cdots$$

$$= x - \frac{x^3}{3!} + \frac{x^5}{5!} - \frac{x^7}{7!} + \cdots = \sum_{n=0}^{\infty} \frac{(-1)^n}{(2n+1)!} x^{2n+1}.$$

We have denoted the sum by $g(x)$ since we don't yet know whether the series converges to $\sin x$. The series does converge for all $x$ by the ratio test:

$$\lim_{n\to\infty} \left| \frac{\dfrac{(-1)^{n+1}}{(2(n+1)+1)!} x^{2(n+1)+1}}{\dfrac{(-1)^n}{(2n+1)!} x^{2n+1}} \right| = \lim_{n\to\infty} \frac{(2n+1)!}{(2n+3)!} |x|^2$$

$$= \lim_{n\to\infty} \frac{|x|^2}{(2n+3)(2n+2)} = 0.$$

Now we can differentiate the function $g(x)$ twice to get

$$g'(x) = 1 - \frac{x^2}{2!} + \frac{x^4}{4!} - \frac{x^6}{6!} + \cdots$$

$$g''(x) = -x + \frac{x^3}{3!} - \frac{x^5}{5!} + \frac{x^7}{7!} - \cdots = -g(x).$$

Thus $g(x)$ satisfies the differential equation $g''(x) + g(x) = 0$ of simple harmonic motion. The general solution of this equation, as observed in Section 3.1, is

$$g(x) = A \cos x + B \sin x.$$

Observe, from the series, that $g(0) = 0$ and $g'(0) = 1$. These values determine that $A = 0$ and $B = 1$. Thus $g(x) = \sin x$ and $g'(x) = \cos x$ for all $x$.

We have therefore demonstrated that

$$\sin x = \sum_{n=0}^{\infty} \frac{(-1)^n}{(2n+1)!} x^{2n+1} = x - \frac{x^3}{3!} + \frac{x^5}{5!} - \frac{x^7}{7!} + \cdots \qquad \text{(for all } x\text{)}$$

$$\cos x = \sum_{n=0}^{\infty} \frac{(-1)^n}{(2n)!} x^{2n} = 1 - \frac{x^2}{2!} + \frac{x^4}{4!} - \frac{x^6}{6!} + \cdots \qquad \text{(for all } x\text{)}$$

Theorem 21 shows that we can use any available means to find a power series converging to a given function on an interval, and the series obtained will turn out to be the Taylor series. In Section 9.5 several series were constructed by manipulating a geometric series. These include:

**Some Maclaurin series**

$$\frac{1}{1-x} = \sum_{n=0}^{\infty} x^n = 1 + x + x^2 + x^3 + \cdots \qquad (-1 < x < 1)$$

$$\frac{1}{(1-x)^2} = \sum_{n=1}^{\infty} n x^{n-1} = 1 + 2x + 3x^2 + 4x^3 + \cdots \qquad (-1 < x < 1)$$

$$\ln(1+x) = \sum_{n=1}^{\infty} \frac{(-1)^{n-1}}{n} x^n = x - \frac{x^2}{2} + \frac{x^3}{3} - \frac{x^4}{4} + \cdots \quad (-1 < x \le 1)$$

$$\tan^{-1} x = \sum_{n=0}^{\infty} \frac{(-1)^n}{2n+1} x^{2n+1} = x - \frac{x^3}{3} + \frac{x^5}{5} - \frac{x^7}{7} + \cdots \quad (-1 \le x \le 1)$$

These series, together with the intervals on which they converge, are frequently used hereafter and should be memorized.

## Other Maclaurin and Taylor Series

Series can be combined in various ways to generate new series. For example, we can find the Maclaurin series for $e^{-x}$ by replacing $x$ with $-x$ in the Maclaurin series for $x$:

$$e^{-x} = \sum_{n=0}^{\infty} \frac{(-1)^n}{n!} x^n = 1 - x + \frac{x^2}{2!} - \frac{x^3}{3!} + \cdots \quad \text{(for all } x.)$$

The series for $e^x$ and $e^{-x}$ can then be subtracted or added and the results divided by 2 to obtain Maclaurin series for the hyperbolic functions $\sinh x$ and $\cosh x$:

$$\sinh x = \frac{e^x - e^{-x}}{2} = \sum_{n=0}^{\infty} \frac{x^{2n+1}}{(2n+1)!} = x + \frac{x^3}{3!} + \frac{x^5}{5!} + \cdots \text{ (for all real } x\text{)}$$

$$\cosh x = \frac{e^x + e^{-x}}{2} = \sum_{n=0}^{\infty} \frac{x^{2n}}{(2n)!} = 1 + \frac{x^2}{2!} + \frac{x^4}{4!} + \cdots \qquad \text{(for all real } x\text{)}.$$

**Remark** Observe the similarity between the series for $\sin x$ and $\sinh x$ and between those for $\cos x$ and $\cosh x$. If we were to allow complex numbers (numbers of the form $z = x + iy$, where $i^2 = -1$ and $x$ and $y$ are real — see Appendix I) as arguments for our functions, and if we were to demonstrate that our operations on series could be extended to series of complex numbers, we would see that $\cos x = \cosh(ix)$ and $\sin x = -i \sinh(ix)$. In fact,

$$e^{ix} = \cos x + i \sin x \qquad \text{and} \qquad e^{-ix} = \cos x - i \sin x,$$

so

$$\cos x = \frac{e^{ix} + e^{-ix}}{2} \qquad \text{and} \qquad \sin x = \frac{e^{ix} - e^{-ix}}{2i}.$$

Such formulas are encountered in the study of functions of a complex variable; from the complex point of view the trigonometric and exponential functions are just different manifestations of the same basic function, a complex exponential $e^z = e^{x+iy}$. We content ourselves here with having mentioned the interesting relationships above and invite the reader to verify them formally by calculating with series. (Such formal calculations do not, of course, constitute a proof since we have not established the various rules covering series of complex numbers.)

**Example 3**  Obtain Maclaurin series for the following functions:

(a)  $e^{-x^2/3}$,     (b)  $\dfrac{\sin(x^2)}{x}$,     (c)  $\sin^2 x$.

*Solution*

(a)  We substitute $-x^2/3$ for $x$ in the Maclaurin series for $e^x$:

$$e^{-x^2/3} = 1 - \frac{x^2}{3} + \frac{1}{2!}\left(\frac{x^2}{3}\right)^2 - \frac{1}{3!}\left(\frac{x^2}{3}\right)^3 + \cdots$$

$$= \sum_{n=0}^{\infty}(-1)^n \frac{1}{3^n n!}\, x^{2n} \qquad \text{(for all real } x\text{).}$$

(b)  For all $x \neq 0$ we have

$$\frac{\sin x^2}{x} = \frac{1}{x}\left(x^2 - \frac{(x^2)^3}{3!} + \frac{(x^2)^5}{5!} - \cdots\right)$$

$$= x - \frac{x^5}{3!} + \frac{x^9}{5!} - \cdots = \sum_{n=0}^{\infty}(-1)^n \frac{x^{4n+1}}{(2n+1)!}.$$

Note that $f(x) = (\sin(x^2))/x$ is not defined at $x = 0$ but does have a limit (namely 0) as $x$ approaches 0. If we define $f(0) = 0$ (the continuous extension of $f(x)$ to $x = 0$), then the series converges to $f(x)$ for all $x$.

(c)  We use a trigonometric identity to express $\sin^2 x$ in terms of $\cos 2x$ and then use the Maclaurin series for $\cos x$ with $x$ replaced by $2x$.

$$\sin^2 x = \frac{1 - \cos 2x}{2} = \frac{1}{2} - \frac{1}{2}\left(1 - \frac{(2x)^2}{2!} + \frac{(2x)^4}{4!} - \cdots\right)$$

$$= \frac{1}{2}\left(\frac{(2x)^2}{2!} - \frac{(2x)^4}{4!} + \frac{(2x)^6}{6!} - \cdots\right)$$

$$= \sum_{n=0}^{\infty}(-1)^n \frac{2^{2n+1}}{(2n+2)!}\, x^{2n+2} \qquad \text{(for all real } x\text{).}$$

Taylor series about points other than 0 can often be obtained from known Maclaurin series by a change of variable.

**Example 4** Find the Taylor series for $\ln x$ in powers of $x - 2$. Where does the series converge to $\ln x$?

**Solution** Note that

$$\ln x = \ln(2 + (x - 2)) = \ln\left[2\left(1 + \frac{x - 2}{2}\right)\right] = \ln 2 + \ln(1 + t),$$

where $t = (x - 2)/2$. We use the known Maclaurin series for $\ln(1 + t)$:

$$\begin{aligned}
\ln x &= \ln 2 + \ln(1 + t) \\
&= \ln 2 + t - \frac{t^2}{2} + \frac{t^3}{3} - \frac{t^4}{4} - \cdots \\
&= \ln 2 + \frac{x - 2}{2} - \frac{(x - 2)^2}{2 \times 2^2} + \frac{(x - 2)^3}{3 \times 2^3} - \frac{(x - 2)^4}{4 \times 2^4} + \cdots \\
&= \ln 2 + \sum_{n=1}^{\infty} \frac{(-1)^{n-1}}{n\,2^n} (x - 2)^n.
\end{aligned}$$

Since the series for $\ln(1 + t)$ is valid for $-1 < t \le 1$, this series for $\ln x$ is valid for $-1 < (x - 2)/2 \le 1$, that is, for $0 < x \le 4$.

**Example 5** Find the Taylor series for $\cos x$ about the point $x = \pi/3$. Where is the series valid?

**Solution** We use the addition formula for cosine:

$$\begin{aligned}
\cos x &= \cos\left(x - \frac{\pi}{3} + \frac{\pi}{3}\right) = \cos\left(x - \frac{\pi}{3}\right)\cos\frac{\pi}{3} - \sin\left(x - \frac{\pi}{3}\right)\sin\frac{\pi}{3} \\
&= \frac{1}{2}\left[1 - \frac{1}{2!}\left(x - \frac{\pi}{3}\right)^2 + \frac{1}{4!}\left(x - \frac{\pi}{3}\right)^4 - \cdots\right] \\
&\quad - \frac{\sqrt{3}}{2}\left[\left(x - \frac{\pi}{3}\right) - \frac{1}{3!}\left(x - \frac{\pi}{3}\right)^3 + \cdots\right] \\
&= \frac{1}{2} - \frac{\sqrt{3}}{2}\left(x - \frac{\pi}{3}\right) - \frac{1}{2}\frac{1}{2!}\left(x - \frac{\pi}{3}\right)^2 + \frac{\sqrt{3}}{2}\frac{1}{3!}\left(x - \frac{\pi}{3}\right)^3 \\
&\quad + \frac{1}{2}\frac{1}{4!}\left(x - \frac{\pi}{3}\right)^4 - \cdots.
\end{aligned}$$

This series representation is valid for all $x$. A similar calculation would enable us to expand $\cos x$ or $\sin x$ in powers of $x - c$ for any real $c$; both functions are analytic at every point of the real line.

Sometimes it is quite difficult, if not impossible, to find a formula for the general term of a Maclaurin or Taylor series. In such cases it is usually possible to obtain the first few terms before the calculations get too cumbersome. Had we attempted to solve Example 3(c) by multiplying the series for $\sin x$ by itself we might have found ourselves in this bind. Other examples occur when it is necessary to substitute one series into another or to divide one by another.

**Example 6**    Obtain the first three nonzero terms of the Maclaurin series for    (a) $\ln \cos x$, and    (b) $\tan x$.

**Solution**

(a) $\ln \cos x = \ln \left( 1 + \left( -\frac{x^2}{2!} + \frac{x^4}{4!} - \frac{x^6}{6!} + \cdots \right) \right)$

$$= \left( -\frac{x^2}{2!} + \frac{x^4}{4!} - \frac{x^6}{6!} + \cdots \right) - \frac{1}{2} \left( -\frac{x^2}{2!} + \frac{x^4}{4!} - \frac{x^6}{6!} + \cdots \right)^2$$

$$+ \frac{1}{3} \left( -\frac{x^2}{2!} + \frac{x^4}{4!} - \frac{x^6}{6!} + \cdots \right)^3 - \cdots$$

$$= -\frac{x^2}{2} + \frac{x^4}{24} - \frac{x^6}{720} + \cdots - \frac{1}{2} \left( \frac{x^4}{4} - \frac{x^6}{24} + \cdots \right)$$

$$+ \frac{1}{3} \left( -\frac{x^6}{8} + \cdots \right) - \cdots$$

$$= -\frac{x^2}{2} - \frac{x^4}{12} - \frac{x^6}{45} - \cdots$$

Note that at each stage of the calculation we kept only enough terms to ensure that we could get all the terms with powers up to $x^6$. Being an even function, $\ln \cos x$ has only even powers in its Maclaurin series. We cannot find the general term of this series, and only with considerable computational effort can we find many more terms than we have already found. We could also try to calculate terms by using the formula $a_k = f^{(k)}(0)/k!$ but even this becomes difficult after the first few values of $k$.

(b) $\tan x = (\sin x)/(\cos x)$. We can obtain the first three terms of the Maclaurin series for $\tan x$ by long division of the series for $\cos x$ into that for $\sin x$:

$$
\begin{array}{r}
x + \dfrac{x^3}{3} + \dfrac{2}{15}x^5 + \cdots \\[1em]
1 - \dfrac{x^2}{2} + \dfrac{x^4}{24} \overline{\smash{\big)}\ x - \dfrac{x^3}{6} + \dfrac{x^5}{120} - \cdots } \\[1em]
x - \dfrac{x^3}{2} + \dfrac{x^5}{24} - \cdots \\[0.5em]
\hline
\dfrac{x^3}{3} - \dfrac{x^5}{30} + \cdots \\[1em]
\dfrac{x^3}{3} - \dfrac{x^5}{6} + \cdots \\[0.5em]
\hline
\dfrac{2x^5}{15} - \cdots \\[1em]
\dfrac{2x^5}{15} - \cdots \\
\hline
\end{array}
$$

Thus $\tan x = x + \frac{1}{3}x^3 + \frac{2}{15}x^5 + \cdots$.

Again we cannot easily find all the terms of the series. This Maclaurin series for $\tan x$ converges for $|x| < \pi/2$, but we cannot demonstrate this fact by the techniques we have at our disposal now. Note that the series for $\tan x$ could also have been derived from that of $\ln \cos x$ obtained in part (a) because we have $\tan x = -\dfrac{d}{dx} \ln \cos x$. ∎

# Exercises 9.6

Find Maclaurin series representations for the functions in Exercises 1–14. For what values of $x$ is each representation valid?

**1.** $e^{3x+1}$

**2.** $\cos(2x^3)$

**3.** $\sin(x - \pi/4)$

**4.** $\cos(2x - \pi)$

**5.** $x^2 \sin(x/3)$

**6.** $\cos^2(x/2)$

**7.** $\sin x \cos x$

**8.** $\tan^{-1}(5x^2)$

**9.** $\dfrac{1 + x^3}{1 + x^2}$

**10.** $\ln(2 + x^2)$

**11.** $\ln \dfrac{1 - x}{1 + x}$

**12.** $(e^{2x^2} - 1)/x^2$

**13.** $\cosh x - \cos x$

**14.** $\sinh x - \sin x$

Find the required Taylor series representations of the functions in Exercises 15–26. Where is each series representation valid?

**15.** $f(x) = e^{-2x}$ about the point $x = -1$

**16.** $f(x) = \sin x$ about the point $x = \pi/2$

**17.** $f(x) = \cos x$ in powers of $x - \pi$

**18.** $f(x) = \ln x$ in powers of $x - 3$

**19.** $f(x) = \ln(2 + x)$ in powers of $x - 2$

**20.** $f(x) = e^{2x+3}$ in powers of $x + 1$

**21.** $f(x) = \sin x - \cos x$ about $x = \dfrac{\pi}{4}$

**22.** $f(x) = \cos^2 x$ about $x = \dfrac{\pi}{8}$

**23.** $f(x) = 1/x^2$ in powers of $x + 2$

**24.** $f(x) = \dfrac{x}{1 + x}$ in powers of $x - 1$

**25.** $f(x) = x \ln x$ in powers of $x - 1$

**26.** $f(x) = xe^x$ in powers of $x + 2$

Find the first three nonzero terms in the Maclaurin series for the functions in Exercises 27–30.

**27.** $\sec x$

**28.** $\sec x \tan x$

**29.** $\tan^{-1}(e^x - 1)$

**30.** $e^{\tan^{-1} x} - 1$

**∗ 31.** Use the fact that $(\sqrt{1 + x})^2 = 1 + x$ to find the first three nonzero terms of the Maclaurin series for $\sqrt{1 + x}$.

**32.** Does $\csc x$ have a Maclaurin series? Why? Find the first three nonzero terms of the Taylor series for $\csc x$ about the point $x = \pi/2$.

Find the sums of the series in Exercises 33–36.

**33.** $1 + x^2 + \dfrac{x^4}{2!} + \dfrac{x^6}{3!} + \dfrac{x^8}{4!} + \cdots$

**∗ 34.** $x^3 - \dfrac{x^9}{3! \times 4} + \dfrac{x^{15}}{5! \times 16} - \dfrac{x^{21}}{7! \times 64} + \dfrac{x^{27}}{9! \times 256} - \cdots$

**35.** $1 + \dfrac{x^2}{3!} + \dfrac{x^4}{5!} + \dfrac{x^6}{7!} + \dfrac{x^8}{9!} + \cdots$

**∗ 36.** $1 + \dfrac{1}{2 \times 2!} + \dfrac{1}{4 \times 3!} + \dfrac{1}{8 \times 4!} + \cdots$

**37.** Let $P(x) = 1 + x + x^2$. Find (a) the Maclaurin series for $P(x)$ and  (b) the Taylor series for $P(x)$ about $x = 1$.

**∗ 38.** Verify by direct calculation that $f(x) = 1/x$ is analytic at $x = a$ for every $a \neq 0$.

**∗ 39.** Verify by direct calculation that $\ln x$ is analytic at $x = a$ for every $a > 0$.

**∗ 40.** Review Exercise 41 of Section 4.3. It shows that the function

$$f(x) = \begin{cases} e^{-1/x^2} & \text{if } x \neq 0 \\ 0 & \text{if } x = 0 \end{cases}$$

has derivatives of all orders at every point of the real line, and $f^{(k)}(0) = 0$ for every positive integer $k$. What is the Maclaurin series for $f(x)$? What is the interval of convergence of this Maclaurin series? On what interval does the series converge to $f(x)$? Is $f$ analytic at $x = 0$?

**∗ 41.** By direct multiplication of the Maclaurin series for $e^x$ and $e^y$ show that $e^x e^y = e^{x+y}$.

### Approximating the Values of Functions

We have seen in Section 4.8 how Taylor and Maclaurin polynomials (the partial sums of Taylor and Maclaurin series) can be used as polynomial approximations to more complicated functions. Example 4 of Section 4.8 is an example of this. In that example the Lagrange remainder in Taylor's Formula is used to determine how many terms of the Maclaurin series for $e^x$ are needed to calculate $e^1 = e$ correct to three decimal places. We will reconsider Taylor's Formula in the next section. In the following examples we use other methods to estimate the error when Taylor polynomials are used to approximate function values.

**Example 1**   Use Maclaurin series to find $\sqrt{e}$ correct to four decimal places, that is, with error less than 0.00005 in absolute value.

**Solution**   For the approximation we use the sum of the first $n$ terms of the Maclaurin series:

$$\sqrt{e} = e^{1/2} = 1 + \frac{1}{2} + \frac{1}{2!}\frac{1}{2^2} + \frac{1}{3!}\frac{1}{2^3} + \cdots$$

$$\approx 1 + \frac{1}{2} + \frac{1}{2!}\frac{1}{2^2} + \frac{1}{3!}\frac{1}{2^3} + \cdots + \frac{1}{(n-1)!}\frac{1}{2^{n-1}} = s_n$$

The error in this approximation can be bounded by a geometric series if we use the inequalities $n+1 < n+2 < n+3 < \cdots$.

$$0 < \sqrt{e} - s_n = \frac{1}{n!}\frac{1}{2^n} + \frac{1}{(n+1)!}\frac{1}{2^{n+1}} + \frac{1}{(n+2)!}\frac{1}{2^{n+2}} + \cdots$$

$$= \frac{1}{n!}\frac{1}{2^n}\left[1 + \frac{1}{2(n+1)} + \frac{1}{2^2(n+1)(n+2)} + \frac{1}{2^3(n+1)(n+2)(n+3)} + \cdots\right]$$

$$< \frac{1}{n!}\frac{1}{2^n}\left[1 + \frac{1}{2(n+1)} + \left(\frac{1}{2(n+1)}\right)^2 + \left(\frac{1}{2(n+1)}\right)^3 + \cdots\right]$$

$$= \frac{1}{n!}\frac{1}{2^n}\frac{1}{1 - \dfrac{1}{2(n+1)}} = \frac{2(n+1)}{(2n+1)n!\,2^n}.$$

We want this error to be less than $0.00005 = 1/20,000$. Thus, we want

$$\frac{2^n n!(2n+1)}{2n+2} > 20,000.$$

This is true if $n = 6$; $2^6 6!(13)/14 \approx 42,789 > 20,000$. Thus

$$\sqrt{e} \approx 1 + \frac{1}{2} + \frac{1}{8} + \frac{1}{48} + \frac{1}{384} + \frac{1}{3840} \approx 1.6487,$$

rounded to four decimal places.

When the terms $a_n$ of a series (i) alternate in sign, (ii) decrease steadily in size, and (iii) approach zero as $n \to \infty$, then the error involved in using a partial sum of the series as an approximation to the sum of the series has the same sign as, and is smaller in absolute value than the first omitted term. This was proved in Section 9.4 and provides an easy estimate for the error when the three conditions above are satisfied.

**Example 2**  Find $\cos 43°$ with error less than $1/10,000$.

**Solution**  We give two alternative solutions:

METHOD I. We can use the Maclaurin series:

$$\cos 43° = \cos \frac{43\pi}{180} = 1 - \frac{1}{2!}\left(\frac{43\pi}{180}\right)^2 + \frac{1}{4!}\left(\frac{43\pi}{180}\right)^4 - \cdots.$$

Now $43\pi/180 \approx 0.75049 \cdots < 1$, so the series above must satisfy the conditions (i)–(iii) mentioned above. If we truncate the series after the $n$th term

$$(-1)^{n-1}\frac{1}{(2n-2)!}\left(\frac{43\pi}{180}\right)^{2n-2},$$

then the error $E$ will satisfy

$$|E| \le \frac{1}{(2n)!}\left(\frac{43\pi}{180}\right)^{2n} < \frac{1}{(2n)!}.$$

The error will not exceed $1/10,000$ if $(2n)! > 10,000$, so $n = 4$ will do; ($8! = 40,320$).

$$\cos 43° \approx 1 - \frac{1}{2!}\left(\frac{43\pi}{180}\right)^2 + \frac{1}{4!}\left(\frac{43\pi}{180}\right)^4 - \frac{1}{6!}\left(\frac{43\pi}{180}\right)^6 \approx 0.73135\cdots$$

METHOD II. Since $43°$ is close to $45°$ we can do a bit better by using the Taylor series about $x = \pi/4$ instead of the Maclaurin series.

$$\begin{aligned}
\cos 43° &= \cos\left(\frac{\pi}{4} - \frac{\pi}{90}\right) \\
&= \cos\frac{\pi}{4}\cos\frac{\pi}{90} + \sin\frac{\pi}{4}\sin\frac{\pi}{90} \\
&= \frac{1}{\sqrt{2}}\left[\left(1 - \frac{1}{2!}\left(\frac{\pi}{90}\right)^2 + \frac{1}{4!}\left(\frac{\pi}{90}\right)^4 - \cdots\right) \right. \\
&\qquad\qquad \left. + \left(\frac{\pi}{90} - \frac{1}{3!}\left(\frac{\pi}{90}\right)^3 + \cdots\right)\right].
\end{aligned}$$

Since

$$\frac{1}{4!}\left(\frac{\pi}{90}\right)^4 < \frac{1}{3!}\left(\frac{\pi}{90}\right)^3 < \frac{1}{20,000},$$

we need only the first two terms of the first series and the first term of the second series:

$$\cos 43° \approx \frac{1}{\sqrt{2}}\left(1 + \frac{\pi}{90} - \frac{1}{2}\left(\frac{\pi}{90}\right)^2\right) \approx 0.731358\cdots.$$

(In fact, $\cos 43° = 0.7313537\cdots$.)

When finding approximate values of functions, it is best, whenever possible, to use a power series about a point as close as possible to the point where the approximation is desired.

## Functions Defined by Integrals

Many functions that can be expressed as simple combinations of elementary functions cannot be antidifferentiated by elementary techniques; their antiderivatives are not simple combinations of elementary functions. We can, however, often find the Taylor series for the antiderivatives of such functions and hence approximate their definite integrals.

**Example 3**   Find the Maclaurin series for

$$E(x) = \int_0^x e^{-t^2}\,dt$$

and use it to evaluate $E(1)$ correct to three decimal places.

**Solution**   The Maclaurin series for $E(x)$ is given by

$$E(x) = \int_0^x \left(1 - t^2 + \frac{t^4}{2!} - \frac{t^6}{3!} + \frac{t^8}{4!} - \cdots\right)dt$$

$$= \left(t - \frac{t^3}{3} + \frac{t^5}{5 \times 2!} - \frac{t^7}{7 \times 3!} + \frac{t^9}{9 \times 4!} - \cdots\right)\Bigg|_0^x$$

$$= x - \frac{x^3}{3} + \frac{x^5}{5 \times 2!} - \frac{x^7}{7 \times 3!} + \frac{x^9}{9 \times 4!} - \cdots = \sum_{n=0}^{n}(-1)^n \frac{x^{2n+1}}{(2n+1)n!},$$

and is valid for all real $x$ because the series for $e^{-t^2}$ is valid for all real $t$. Therefore

$$E(1) = 1 - \frac{1}{3} + \frac{1}{5 \times 2!} - \frac{1}{7 \times 3!} + \cdots$$

$$\approx 1 - \frac{1}{3} + \frac{1}{5 \times 2!} - \frac{1}{7 \times 3!} + \cdots + \frac{(-1)^{n-1}}{(2n-1)(n-1)!}.$$

We stopped with the $n$th term. The error in this approximation does not exceed the first omitted term, so it will be less than 0.0005, provided $(2n+1)n! > 2,000$. Since $13 \times 6! = 9360$, $n = 6$ will do. Thus

$$E(1) \approx 1 - \frac{1}{3} + \frac{1}{10} - \frac{1}{42} + \frac{1}{216} - \frac{1}{1320} \approx 0.747,$$

rounded to three decimal places.

## Indeterminate Forms

Examples 1 and 2 of Section 4.9 showed how Maclaurin polynomials could be used for evaluating the limits of indeterminate forms. Here are two more examples, this time using the series directly and keeping enough terms to allow cancellation of the [0/0] factors.

**Example 4**   Evaluate   (a) $\lim\limits_{x \to 0} \dfrac{x - \sin x}{x^3}$   and   (b) $\lim\limits_{x \to 0} \dfrac{(e^{2x} - 1)\ln(1 + x^3)}{(1 - \cos 3x)^2}$

**Solution**

(a) $\lim\limits_{x \to 0} \dfrac{x - \sin x}{x^3}$   $\begin{bmatrix} 0 \\ 0 \end{bmatrix}$

$$= \lim_{x \to 0} \frac{x - \left(x - \dfrac{x^3}{3!} + \dfrac{x^5}{5!} - \cdots\right)}{x^3}$$

$$= \lim_{x \to 0} \frac{\dfrac{x^3}{3!} - \dfrac{x^5}{5!} + \cdots}{x^3}$$

$$= \lim_{x \to 0} \left(\frac{1}{3!} - \frac{x^2}{5!} + \cdots\right) = \frac{1}{3!} = \frac{1}{6}.$$

(b) $\lim\limits_{x \to 0} \dfrac{(e^{2x} - 1)\ln(1 + x^3)}{(1 - \cos 3x)^2}$   $\begin{bmatrix} 0 \\ 0 \end{bmatrix}$

$$= \lim_{x \to 0} \frac{\left(1 + (2x) + \dfrac{(2x)^2}{2!} + \dfrac{(2x)^3}{3!} + \cdots - 1\right)\left(x^3 - \dfrac{x^6}{2} + \cdots\right)}{\left(1 - \left(1 - \dfrac{(3x)^2}{2!} + \dfrac{(3x)^4}{4!} - \cdots\right)\right)^2}$$

$$= \lim_{x \to 0} \frac{2x^4 + 2x^5 + \cdots}{\left(\dfrac{9}{2}x^2 - \dfrac{3^4}{4!}x^4 + \cdots\right)^2}$$

$$= \lim_{x \to 0} \frac{2 + 2x + \cdots}{\left(\dfrac{9}{2} - \dfrac{3^4}{4!}x^2 + \cdots\right)^2} = \frac{2}{\left(\dfrac{9}{2}\right)^2} = \frac{8}{81}.$$

You can check that the second of these examples is much more difficult if attempted using l'Hôpital's Rule.

## Exercises 9.7

Use Maclaurin or Taylor series to calculate the function values indicated in Exercises 1–12, with error less than $5 \times 10^{-5}$ in absolute value.

1. $e^{0.2}$

2. $1/e$

3. $e^{1.2}$

4. $\sin(0.1)$

5. $\cos 5°$

6. $\ln(6/5)$

7. $\ln(0.9)$

8. $\sin 80°$

9. $\cos 65°$

10. $\tan^{-1} 0.2$

11. $\cosh(1)$

12. $\ln(3/2)$

Find Maclaurin series for the functions in Exercises 13–17.

13. $I(x) = \displaystyle\int_0^x \frac{\sin t}{t}\, dt$

14. $J(x) = \displaystyle\int_0^x \frac{e^t - 1}{t}\, dt$

**15.** $K(x) = \int_{1}^{1+x} \frac{\ln t}{t-1}\, dt$    **16.** $L(x) = \int_{0}^{x} \cos(t^2)\, dt$

**17.** $M(x) = \int_{0}^{x} \frac{\tan^{-1} t^2}{t^2}\, dt$

**18.** Find $L(0.5)$ correct to three decimal places, with $L$ defined as in Exercise 16.

**19.** Find $I(1)$ correct to three decimal places, with $I$ defined as in Exercise 13.

Evaluate the limits in Exercises 20–25.

**20.** $\lim_{x\to 0} \dfrac{\sin(x^2)}{\sinh x}$

**21.** $\lim_{x\to 0} \dfrac{1-\cos(x^2)}{(1-\cos x)^2}$

**22.** $\lim_{x\to 0} \dfrac{(e^x - 1 - x)^2}{x^2 - \ln(1 + x^2)}$

**23.** $\lim_{x\to 0} \dfrac{2\sin 3x - 3\sin 2x}{5x - \tan^{-1} 5x}$

**24.** $\lim_{x\to 0} \dfrac{\sin(\sin x) - x}{x(\cos(\sin x) - 1)}$

**25.** $\lim_{x\to 0} \dfrac{\sinh x - \sin x}{\cosh x - \cos x}$

**⬧ 26.** Series can provide a useful tool for solving differential equations. The method is developed in Section 17.9, but for now you can experiment as follows: calculate the first two derivatives of the series $y = \sum_{n=0}^{\infty} a_n x^n$ and substitute the series for $y$, $y'$ and $y''$ into the differential equation

$$y'' + xy' + y = 0.$$

Thereby determine a relationship between $a_{n+2}$ and $a_n$ that enables all the coefficients in the series to be determined in terms of $a_0$ and $a_1$. Solve the initial-value problem

$$\begin{cases} y'' + xy' + y = 0 \\ y(0) = 1 \\ y'(0) = 0. \end{cases}$$

**⬧ 27.** Solve the initial-value problem

$$\begin{cases} y'' + xy' + 2y = 0 \\ y(0) = 1 \\ y'(0) = 2 \end{cases}$$

by the method suggested in Exercise 26.

## 9.8   Taylor's Formula Revisited

Theorem 10 of Section 4.8 (Taylor's Theorem with Lagrange Remainder) provides a formula for the error involved when the Taylor polynomial

$$P_n(x) = \sum_{k=0}^{n} \frac{f^{(k)}(c)}{k!} (x - c)^k$$

of a function $f(x)$ about $x = c$ is used to approximate $f(x)$ for values of $x \neq c$. Specifically, it states the following:

**THEOREM 22**

**Taylor's Theorem with Lagrange remainder**

If the $(n+1)$st derivative of $f$ exists on an interval containing $c$ and $x$, and if $P_n(x)$ is the Taylor polynomial of degree $n$ for $f$ about the point $x = c$, then Taylor's Formula

$$f(x) = P_n(x) + E_n(x)$$

holds, where the error term $E_n(x)$ is given by

$$E_n(x) = \frac{f^{(n+1)}(X)}{(n+1)!} (x - c)^{n+1}$$

for some $X$ between $c$ and $x$. ($E_n(x)$ is called the Lagrange Remainder in Taylor's Formula.)

Observe that the Lagrange form of the remainder, $E_n(x)$, looks just like the $(n + 1)$st degree term in $P_{n+1}(x)$, except that $c$ in $f^{(n+1)}(c)$ has been replaced by an unknown number $X$ between $c$ and $x$. The cases $n = 0$ and $n = 1$ of Taylor's Formula with Lagrange remainder are just the Mean-Value Theorem (theorem 11 of Section 2.6) and the error formula for linear approximation (theorem 9 of Section 4.7) respectively.

**Example 1** Use Taylor's Theorem to determine how many terms of the Maclaurin series for $\cos x$ are needed to calculate $\cos 10°$ correctly to five decimal places.

**Solution** Being an even function, $f(x) = \cos x$ has only even degree terms in its Maclaurin series. The Maclaurin polynomials $P_{2n}$ and $P_{2n+1}$ for $f(x)$ are therefore equal:

$$P_{2n}(x) = P_{2n+1}(x) = \sum_{j=0}^{n} \frac{(-1)^j}{(2j)!} x^{2j}$$

$$1 - \frac{x^2}{2!} + \frac{x^4}{4!} - \cdots + \frac{(-1)^n}{(2n)!} x^{2n}.$$

It makes good sense to use the remainder $E_{2n+1}$ rather than the remainder $E_{2n}$; it is likely to be smaller, and therefore assure us of more accuracy for any given value of $n$. Since $f^{(2n+2)}(x) = (-1)^n \cos x$, we have, for some $X$ between 0 and $x$,

$$|E_{2n+1}(x)| = \left| \frac{(-1)^n \cos X}{(2n + 2)!} x^{2n+2} \right| \leq \frac{|x|^{2n+2}}{(2n + 2)!}.$$

For $x = 10° = \pi/18 \approx 0.174533 < 0.2$ radians, we will have five decimal place accuracy if

$$\frac{0.2^{2n+2}}{(2n + 2)!} < 0.000005.$$

This is satisfied if $n = 2$ ($0.2^6/6! < 9 \times 10^{-8}$), but not $n = 1$. Thus

$$\cos 10° = \cos \frac{\pi}{18} \approx 1 - \frac{1}{2} \left( \frac{\pi}{18} \right)^2 + \frac{1}{24} \left( \frac{\pi}{18} \right)^4 \approx 0.98481$$

to five decimal places.

## Using Taylor's Theorem to Find Taylor and Maclaurin Series

If a function $f$ has derivatives of all orders, we can write Taylor's Formula for any $n$:

$$f(x) = P_n(x) + E_n(x).$$

If we can show that $\lim_{n \to \infty} E_n(x) = 0$ for all $x$ in some interval $I$, then we are entitled to conclude, for $x$ in $I$, that

$$f(x) = \lim_{n \to \infty} P_n(x) = \sum_{k=0}^{\infty} \frac{f^{(k)}(c)}{k!} (x - c)^k,$$

that is, we will have expressed $f(x)$ as the sum of an infinite series of terms which are multiples of positive integer powers of $x - c$, and the series converges for all $x$ in $I$. This series is the **Taylor series** representation of $f$ in powers of $x - c$ (or the **Maclaurin series** if $c = 0$).

---

**Example 2**    Use Taylor's Theorem to find the Maclaurin series for $f(x) = e^x$. Where does the series converge to $f(x)$?

**Solution**    Since $e^x$ is positive and increasing, $e^X \le e^{|x|}$ for any $X \le |x|$. Since $f^{(k)}(x) = e^x$ for any $k$ we have, taking $c = 0$ in the Lagrange remainder in Taylor's Formula,

$$|E_n(x)| = \left| \frac{f^{(n+1)}(X)}{(n+1)!} x^{n+1} \right|$$

$$\le \frac{e^X}{(n+1)!} |x|^{n+1} \le e^{|x|} \frac{|x|^{n+1}}{(n+1)!} \to 0 \text{ as } n \to \infty$$

for any real $x$, as shown in Theorem 3(b) of Section 9.1. Therefore

$$e^x = \sum_{k=0}^{\infty} \frac{x^k}{k!} = 1 + x + \frac{x^2}{2!} + \frac{x^3}{3!} + \cdots,$$

and the series converges to $e^x$ for all real numbers $x$. ∎

## Taylor's Theorem with Integral Remainder

The following theorem is another version of Taylor's Theorem, where the remainder in Taylor's Formula is expressed as an integral.

**THEOREM 23**

**Taylor's Theorem with integral remainder**

If the $(n+1)$st derivative of $f$ exists on an interval containing $c$ and $x$, and if $P_n(x)$ is the Taylor polynomial of degree $n$ for $f$ about the point $x = c$, then the remainder $E_n(x) = f(x) - P_n(x)$ in Taylor's Formula is given by

$$E_n(x) = \frac{1}{n!} \int_c^x (x - t)^n f^{(n+1)}(t)\, dt.$$

**PROOF**    We start with the Fundamental Theorem of Calculus written in the form

$$f(x) = f(c) + \int_c^x f'(t)\, dt = P_0(x) + E_0(x).$$

(Note that the Fundamental Theorem is just the special case $n = 0$ of Taylor's Formula with integral remainder.) We now apply integration by parts to the integral, setting

$$U = f'(t), \qquad dV = dt$$
$$dU = f''(t)\, dt, \qquad V = -(x - t).$$

(We have broken our usual rule about not including a constant of integration with $V$. In this case we have included the constant $-x$ in $V$ in order to have $V$ vanish when $t = x$.) We have

$$f(x) = f(c) - f'(t)(x - t)\Big|_{t=c}^{t=x} + \int_c^x (x - t)f''(t)\,dt$$

$$= f(c) + f'(c)(x - c) + \int_c^x (x - t)f''(t)\,dt$$

$$= P_1(x) + E_1(x).$$

We have thus proved the case $n = 1$ of Taylor's Formula with integral remainder.

Let us complete the proof for general $n$ by mathematical induction. Suppose that Taylor's Formula holds with Integral Remainder for some $n = k$:

$$f(x) = P_k(x) + E_k(x) = P_k(x) + \frac{1}{k!} \int_c^x (x - t)^k f^{(k+1)}(t)\,dt.$$

Again we integrate by parts. Let

$$U = f^{(k+1)}(t), \qquad dV = (x - t)^k\,dt,$$
$$dU = f^{(k+2)}(t)\,dt, \qquad V = \frac{-1}{k+1}(x - t)^{k+1}.$$

We have

$$f(x) = P_k(x) + \frac{1}{k!}\left( -\frac{f^{(k+1)}(t)(x - t)^{k+1}}{k+1}\Big|_{t=c}^{t=x} + \int_c^x \frac{(x - t)^{k+1} f^{(k+2)}(t)}{k+1}\,dt \right)$$

$$= P_k(x) + \frac{f^{(k+1)}(c)}{(k+1)!}(x - c)^{k+1} + \frac{1}{(k+1)!} \int_c^x (x - t)^{k+1} f^{(k+2)}(t)\,dt$$

$$= P_{k+1}(x) + E_{k+1}(x).$$

Thus Taylor's Formula with integral remainder is valid for $n = k + 1$ if it is valid for $n = k$. Having been shown to be valid for $n = 0$ (and $n = 1$), it must therefore be valid for every positive integer $n$.

**Remark**   Using one or the other of the versions of Taylor's Theorem given in this section, all the basic Maclaurin and Taylor series given in Section 9.6 can be verified without having to use the theory of power series.

## Exercises 9.8

1. Estimate the error if the Maclaurin polynomial of degree 5 for $\sin x$ is used to approximate $\sin(0.2)$.

2. Estimate the error if the Maclaurin polynomial of degree 6 for $\cos x$ is used to approximate $\cos(1)$.

3. Estimate the error if the Maclaurin polynomial of degree 4 for $e^{-x}$ is used to approximate $e^{-0.5}$.

4. Estimate the error if the Maclaurin polynomial of degree 2 for $\sec x$ is used to approximate $\sec(0.2)$.

5. Estimate the error if the Maclaurin polynomial of degree 3 for $\ln(\cos x)$ is used to approximate $\ln(\cos 0.1)$.

6. Estimate the error if the Taylor polynomial of degree 3 for $\tan^{-1} x$ in powers of $x - 1$ is used to approximate $\tan^{-1} 0.99$.

**7.** Estimate the error if the Taylor polynomial of degree 4 for $\ln x$ in powers of $x - 2$ is used to approximate $\ln(1.95)$.

Use Taylor's Formula to establish the Maclaurin series for the functions in Exercises 8–15.

**8.** $e^{-x}$

**9.** $2^x$

**10.** $\cos x$

**11.** $\sin x$

**12.** $\sin^2 x$

**13.** $\dfrac{1}{1 - x}$

**\* 14.** $\ln(1 + x)$ (Use the integral remainder.)

**\* 15.** $\dfrac{x}{2 + 3x}$ (Use Exercise 13.)

Use Taylor's Formula to obtain the Taylor series indicated in Exercises 16–21.

**16.** for $e^x$ in powers of $x - a$

**17.** for $\sin x$ in powers of $x - (\pi/6)$

**18.** for $\cos x$ in powers of $x - (\pi/4)$

**\* 19.** for $\ln x$ in powers of $x - 1$ (Use the integral remainder.)

**\* 20.** for $\ln x$ in powers of $x - 2$

**21.** for $1/x$ in powers of $x + 2$ (Use Exercise 13.)

# 9.9  The Binomial Theorem and Binomial Series

**Example 1**   Use Taylor's Formula to prove the Binomial Theorem: if $n$ is a positive integer then

$$(a + x)^n = a^n + n\,a^{n-1}x + \frac{n(n-1)}{2!}a^{n-2}x^2 + \cdots + n\,ax^{n-1} + x^n$$

$$= \sum_{k=0}^{n} \binom{n}{k} a^{n-k}x^k,$$

where $\dbinom{n}{k} = \dfrac{n!}{(n-k)!\,k!}$.

**Solution**   Let $f(x) = (a + x)^n$. Then

$$f'(x) = n(a + x)^{n-1} = \frac{n!}{(n-1)!}(a + x)^{n-1}$$

$$f''(x) = \frac{n!}{(n-1)!}(n-1)(a + x)^{n-2} = \frac{n!}{(n-2)!}(a + x)^{n-2}$$

$$\vdots$$

$$f^{(k)}(x) = \frac{n!}{(n-k)!}(a + x)^{n-k} \qquad (0 \le k \le n).$$

In particular, $f^{(n)}(x) = \dfrac{n!}{0!}(a + x)^{n-n} = n!$, a constant, and

$$f^{(k)}(x) = 0 \qquad \text{for all } x, \text{ if } k > n.$$

For $0 \le k \le n$ we have $f^{(k)}(0) = \dfrac{n!}{(n-k)!}a^{n-k}$. Thus, by Taylor's Theorem with Lagrange remainder,

$$(a + x)^n = f(x) = \sum_{k=0}^{n} \frac{f^{(k)}(0)}{k!} x^k + \frac{f^{(n+1)}(X)}{(n+1)!} x^{n+1}$$

$$= \sum_{k=0}^{n} \frac{n!}{(n-k)!k!} a^{n-k}x^k + 0 = \sum_{k=0}^{n} \binom{n}{k} a^{n-k}x^k.$$

This is, in fact, the Maclaurin *series* for $(a + x)^n$, not just the Maclaurin polynomial of degree $n$. Since all higher-degree terms are zero, the series has only finitely many nonzero terms and so converges for all $x$. ∎

***Remark***  If $f(x) = (a + x)^r$ where $a > 0$ and $r$ is any real number, then calculations similar to those above show that the Maclaurin polynomial of degree $n$ for $f$ is

$$P_n(x) = a^r + \sum_{k=1}^{n} \frac{r(r-1)(r-2)\cdots(r-k+1)}{k!} a^{r-k}x^k.$$

However, if $r$ is not a positive integer, then there will be no positive integer $n$ for which the remainder $E_n(x) = f(x) - P_n(x)$ vanishes identically, and the corresponding Maclaurin series will not be a polynomial.

## The Binomial Series

To simplify the discussion of the function $(a + x)^r$ when $r$ is not a positive integer, we take $a = 1$ and consider the function $(1 + x)^r$. Results for the general case follow via the identity

$$(a + x)^r = a^r \left(1 + \frac{x}{a}\right)^r,$$

valid for any $a > 0$.

If $r$ is any real number and $x > -1$, then the $k$th derivative of $(1 + x)^r$ is

$$r(r-1)(r-2)\cdots(r-k+1)(1+x)^{r-k}, \qquad (k = 1, 2, \ldots).$$

Thus the Maclaurin series for $(1 + x)^r$ is

$$1 + \sum_{k=1}^{n} \frac{r(r-1)(r-2)\cdots(r-k+1)}{k!} x^k,$$

which is called the **binomial series**. The following theorem shows that the binomial series does, in fact, converge to $(1 + x)^r$ if $|x| < 1$. We could accomplish this by writing Taylor's Formula for $(1 + x)^r$ with $c = 0$ and showing that the remainder $E_n(x) \to 0$ as $n \to \infty$. (We would need to use the integral form of the remainder to prove this for all $|x| < 1$.) However, we will use an easier method, similar to the one used for the exponential and trigonometric functions in Section 9.6.

**THEOREM 24**

**The binomial series**

If $|x| < 1$, then

$$(1 + x)^r = 1 + rx + \frac{r(r-1)}{2!} x^2 + \frac{r(r-1)(r-2)}{3!} x^3 + \cdots$$

$$= 1 + \sum_{n=1}^{\infty} \frac{r(r-1)(r-2)\cdots(r-n+1)}{n!} x^n \qquad (-1 < x < 1).$$

**PROOF**   If $|x| < 1$, then the series

$$f(x) = 1 + \sum_{n=1}^{\infty} \frac{r(r-1)(r-2)\cdots(r-n+1)}{n!} x^n$$

converges by the ratio test, since

$$\rho = \lim_{n\to\infty} \left| \frac{\dfrac{r(r-1)(r-2)\cdots(r-n+1)(r-n)}{(n+1)!} x^{n+1}}{\dfrac{r(r-1)(r-2)\cdots(r-n+1)}{n!} x^n} \right|$$

$$= \lim_{n\to\infty} \left| \frac{r-n}{n+1} \right| |x| = |x| < 1.$$

Note that $f(0) = 1$. We need to show that $f(x) = (1+x)^r$ for $|x| < 1$.

By Theorem 19, we can differentiate the series for $f(x)$ termwise on $|x| < 1$ to obtain

$$f'(x) = \sum_{n=1}^{\infty} \frac{r(r-1)(r-2)\cdots(r-n+1)}{(n-1)!} x^{n-1}$$

$$= \sum_{n=0}^{\infty} \frac{r(r-1)(r-2)\cdots(r-n)}{n!} x^n.$$

We have replaced $n$ with $n+1$ to get the second version of the sum from the first version. Adding the second version to $x$ times the first version, we get

$$(1+x)f'(x) = \sum_{n=0}^{\infty} \frac{r(r-1)(r-2)\cdots(r-n)}{n!} x^n$$

$$+ \sum_{n=1}^{\infty} \frac{r(r-1)(r-2)\cdots(r-n+1)}{(n-1)!} x^n$$

$$= r + \sum_{n=1}^{\infty} \frac{r(r-1)(r-2)\cdots(r-n+1)}{n!} x^n \Big[ (r-n) + n \Big]$$

$$= r\, f(x).$$

The differential equation $(1+x)f'(x) = rf(x)$ implies that

$$\frac{d}{dx} \frac{f(x)}{(1+x)^r} = \frac{(1+x)^r f'(x) - r(1+x)^{r-1} f(x)}{(1+x)^{2r}} = 0$$

for all $x$ satisfying $|x| < 1$. Thus $f(x)/(1+x)^r$ is constant on that interval, and since $f(0) = 1$, the constant must be 1. Thus $f(x) = (1+x)^r$.

**Remark**   For some values of $r$ the binomial series may converge at the endpoints $x = 1$ or $x = -1$. As observed above, if $r$ is a positive integer, the series has only finitely many nonzero terms, and so converges for all $x$.

**Example 2** Find the Maclaurin series for $\dfrac{1}{\sqrt{1+x}}$.

**Solution** Here $r = -(1/2)$:

$$\frac{1}{\sqrt{1+x}} = (1+x)^{-1/2}$$

$$= 1 - \frac{1}{2}x + \frac{1}{2!}\left(-\frac{1}{2}\right)\left(-\frac{3}{2}\right)x^2 + \frac{1}{3!}\left(-\frac{1}{2}\right)\left(-\frac{3}{2}\right)\left(-\frac{5}{2}\right)x^3 + \cdots$$

$$= 1 - \frac{1}{2}x + \frac{1 \times 3}{2^2 2!}x^2 - \frac{1 \times 3 \times 5}{2^3 3!}x^3 + \cdots$$

$$= 1 + \sum_{n=1}^{\infty}(-1)^n \frac{1 \times 3 \times 5 \times \cdots \times (2n-1)}{2^n n!}x^n.$$

This series converges for $-1 < x \leq 1$. (Use the alternating series test to get the endpoint $x = 1$.)

**Example 3** Find the Maclaurin series for $\sin^{-1} x$.

**Solution** Replace $x$ with $-t^2$ in the series obtained in the previous example to get

$$\frac{1}{\sqrt{1-t^2}} = 1 + \sum_{n=1}^{\infty} \frac{1 \times 3 \times 5 \times \cdots \times (2n-1)}{2^n n!} t^{2n} \qquad (-1 < x < 1).$$

Now integrate $t$ from 0 to $x$:

$$\sin^{-1} x = \int_0^x \frac{dt}{\sqrt{1-t^2}} = \int_0^x \left(1 + \sum_{n=1}^{\infty} \frac{1 \times 3 \times 5 \times \cdots \times (2n-1)}{2^n n!} t^{2n}\right) dt$$

$$= x + \sum_{n=1}^{\infty} \frac{1 \times 3 \times 5 \times \cdots \times (2n-1)}{2^n n!(2n+1)} x^{2n+1}$$

$$= x + \frac{x^3}{6} + \frac{3}{40}x^5 + \cdots \qquad (-1 < x < 1).$$

# Exercises 9.9

Find Maclaurin series representations for the functions in Exercises 1–6. Use the binomial series to calculate the answers.

**1.** $\sqrt{1+x}$

**2.** $x\sqrt{1-x}$

**3.** $\sqrt{4+x}$

**4.** $\dfrac{1}{\sqrt{4+x^2}}$

**5.** $(1-x)^{-2}$

**6.** $(1+x)^{-3}$

\* **7. (Binomial coefficients)** Show that the binomial coefficients

$$\binom{n}{k} = \frac{n!}{k!\,(n-k)!}$$

satisfy

i) $\dbinom{n}{0} = \dbinom{n}{n} = 1$ for every $n$, and

ii) if $0 \leq k \leq n$, then $\dbinom{n}{k-1} + \dbinom{n}{k} = \dbinom{n+1}{k}$.

It follows that, for fixed $n \geq 1$, the binomial coefficients $\binom{n}{0}$, $\binom{n}{1}$, $\binom{n}{2}$, ..., $\binom{n}{n}$ are the elements of the $n$th row of **Pascal's triangle**

$$
\begin{array}{ccccccccccc}
 & & & & & 1 & & 1 & & & \\
 & & & & 1 & & 2 & & 1 & & \\
 & & & 1 & & 3 & & 3 & & 1 & \\
 & & 1 & & 4 & & 6 & & 4 & & 1 \\
 & 1 & & 5 & & 10 & & 10 & & 5 & & 1
\end{array}
$$

where each element with value $> 1$ is the sum of the two diagonally above it.

* **8. (An inductive proof of the Binomial Theorem)** Use mathematical induction and the results of Exercise 7 to prove the Binomial Theorem:

$$(a+b)^n = \sum_{k=0}^{n} \binom{n}{k} a^{n-k} b^k$$

$$= a^n + na^{n-1}b + \binom{n}{2}a^{n-2}b^2 + \binom{n}{3}a^{n-3}b^3 + \cdots + b^n.$$

* **9. (The Leibniz Rule)** Use mathematical induction, the Product Rule, and Exercise 7 to verify the Leibniz Rule for the $n$th derivative of a product of two functions:

$$(fg)^{(n)} = \sum_{k=0}^{n} \binom{n}{k} f^{(n-k)} g^{(k)}$$

$$= f^{(n)}g + nf^{(n-1)}g' + \binom{n}{2}f^{(n-2)}g''$$

$$+ \binom{n}{3}f^{(n-3)}g^{(3)} + \cdots + fg^{(n)}.$$

## 9.10    Fourier Series

As we have seen, power series representations of functions make it possible to approximate those functions as closely as we want in intervals near a particular point of interest by using partial sums of the series, that is, polynomials. However, in many important applications of mathematics, the functions involved are required to be periodic. For example, much of electronic engineering is concerned with the analysis and manipulation of *waveforms*, which are periodic functions of time. Polynomials are not periodic functions, and for this reason power series are not well suited to representing such functions.

Much more appropriate for the representations of periodic functions over extended intervals are certain infinite series of periodic functions called Fourier series.

### Periodic Functions

Recall that a function $f$ defined on the real line is **periodic** with period $T$ if

$$f(t+T) = f(t) \quad \text{for all real } t. \tag{$*$}$$

This implies that $f(t+mT) = f(t)$ for any integer $m$, so that if $T$ is a period of $f$, then so is any multiple $mT$ of $T$. The smallest positive number $T$ for which $(*)$ holds is called **the fundamental period**, or simply **the period** of $f$.

The entire graph of a function with period $T$ can be obtained by shifting the part of the graph in any half-open interval of length $T$ (for example, the interval $[0, T)$) to the left or right by integer multiples of the period $T$. Figure 9.6 shows the graph of a function of period 2.

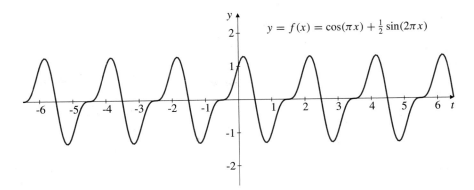

**Figure 9.6**    This function has period 2. Observe how the graph repeats the part in the interval [0, 2) over and over to the left and right

---

**Example 1**    The functions $g(t) = \cos(\pi t)$ and $h(t) = \sin(\pi t)$ are both periodic with period 2:

$$g(t + 2) = \cos(\pi t + 2\pi) = \cos(\pi t) = g(t).$$

The function $k(t) = \sin(2\pi t)$ also has period 2, but this is not its fundamental period. The fundamental period is 1:

$$k(t + 1) = \sin(2\pi t + 2\pi) = \sin(2\pi t) = k(t).$$

The sum $f(t) = g(t) + \frac{1}{2}k(t) = \cos(\pi t) + \frac{1}{2}\sin(2\pi t)$, graphed in Figure 9.6, has period 2, the least common multiple of the periods of its two terms.

---

**Example 2**    For any positive integer $n$, the functions

$$f_n(t) = \cos(n\omega t) \quad \text{and} \quad g_n(t) = \sin(n\omega t)$$

all have period $T = 2\pi/\omega$; the fundamental periods of $f_n$ and $g_n$ are $T/n$. The subject of Fourier series is concerned with expressing general functions with period $T$ as series whose terms are real multiples of these functions.

## Fourier Series

It can be shown (but we won't do it here) that if $f(t)$ is periodic with fundamental period $T$, is continuous, and has a piecewise continuous derivative on the real line, then $f(t)$ is everywhere the sum of a series of the form

$$f(t) = \frac{a_0}{2} + \sum_{n=1}^{\infty}\left(a_n \cos(n\omega t) + b_n \sin(n\omega t)\right), \qquad (**)$$

called the **Fourier series** of $f$, where $\omega = 2\pi/T$ and the sequences $\{a_n\}_{n=0}^{\infty}$ and $\{b_n\}_{n=1}^{\infty}$ are the **Fourier coefficients** of $f$. Determining the values of these coefficients for a given such function $f$ is made possible by the following identities, valid for integers $m$ and $n$, which are easily proved by using the addition formulas for sine and cosine. (See Exercises 53–55 in Section 5.6.)

$$\int_0^T \cos(n\omega t)\, dt = \begin{cases} 0 & \text{if } n \neq 0 \\ T & \text{if } n = 0 \end{cases}$$

$$\int_0^T \sin(n\omega t)\, dt = 0$$

$$\int_0^T \cos(m\omega t) \cos(n\omega t)\, dt = \begin{cases} 0 & \text{if } m \neq n \\ T/2 & \text{if } m = n \end{cases}$$

$$\int_0^T \sin(m\omega t) \sin(n\omega t)\, dt = \begin{cases} 0 & \text{if } m \neq n \\ T/2 & \text{if } m = n \end{cases}$$

$$\int_0^T \cos(m\omega t) \sin(n\omega t)\, dt = 0.$$

If we multiply equation (∗∗) by $\cos(m\omega t)$ (or by $\sin(m\omega t)$) and integrate the resulting equation over $[0, T]$ term by term, all the terms on the right except the one involving $a_m$ (or $b_m$) will be 0. (The term-by-term integration requires justification, but we won't try to do that here either.) The integration results in

$$\int_0^T f(t) \cos(m\omega t)\, dt = \frac{1}{2} T a_m$$

$$\int_0^T f(t) \sin(m\omega t)\, dt = \frac{1}{2} T b_m.$$

(Note that the first of these formulas is even valid for $m = 0$ because we chose to call the constant term in the Fourier series $a_0/2$ instead of $a_0$.) Since the integrands are all periodic with period $T$, the integrals can be taken over any interval of length $T$; it is often convenient to use $[-T/2, T/2]$ instead of $[0, T]$. The Fourier coefficients of $f$ are therefore given by

$$a_n = \frac{2}{T} \int_{-T/2}^{T/2} f(t) \cos(n\omega t)\, dt \quad (n = 0,\ 1,\ 2,\ \ldots)$$

$$b_n = \frac{2}{T} \int_{-T/2}^{T/2} f(t) \sin(n\omega t)\, dt \quad (n = 1,\ 2,\ 3,\ \ldots),$$

where $\omega = 2\pi/T$.

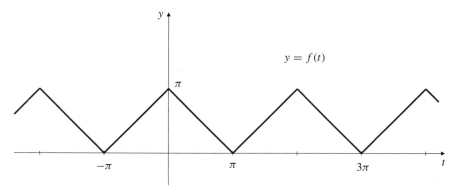

**Figure 9.7**    A sawtooth function of period $2\pi$

**Example 3**    Find the Fourier series of the sawtooth function $f(t)$ of period $2\pi$ whose values in the interval $[-\pi, \pi]$ are given by $f(t) = \pi - |t|$. (See Figure 9.7.)

**Solution**    Here $T = 2\pi$ and $\omega = 2\pi/(2\pi) = 1$. Since $f(t)$ is an even function, therefore $f(t)\sin(nt)$ is odd, and so all the Fourier sine coefficients $b_n$ are zero:

$$b_n = \frac{2}{2\pi} \int_{-\pi}^{\pi} f(t)\sin(nt)\,dt = 0.$$

Also, $f(t)\cos(nt)$ is an even function, so

$$
\begin{aligned}
a_n &= \frac{2}{2\pi} \int_{-\pi}^{\pi} f(t)\cos(nt)\,dt \\
&= \frac{2}{\pi} \int_{0}^{\pi} (\pi - t)\cos(nt)\,dt \\
&= \begin{cases} \pi & \text{if } n = 0 \\ 0 & \text{if } n \neq 0 \text{ and } n \text{ is even} \\ 4/(\pi n^2) & \text{if } n \text{ is odd} \end{cases}
\end{aligned}
$$

Since odd positive integers $n$ are of the form $n = 2k - 1$, where $k$ is a positive integer, the Fourier series of $f$ is given by

$$f(t) = \frac{\pi}{2} + \sum_{k=1}^{\infty} \frac{4}{\pi(2k-1)^2} \cos\big((2k-1)t\big).$$

## Convergence of Fourier Series

The partial sums of a Fourier series are called Fourier polynomials because they can be expressed as polynomials in $\sin(\omega t)$ and $\cos(\omega t)$, although we will not actually try to write them that way.  The Fourier polynomial of order $m$ of the periodic function $f$ having period $T$ is

$$f_m(t) = \frac{a_0}{2} + \sum_{n=1}^{m} \big(a_n \cos(n\omega t) + b_n \sin(n\omega t)\big),$$

where $\omega = 2\pi/T$.

**Example 4**    The Fourier polynomial of order 3 of the sawtooth function of Example 3 is

$$f_3(t) = \frac{\pi}{2} + \frac{4}{\pi}\cos t + \frac{4}{9\pi}\cos(3t).$$

The graph of this function is shown in Figure 9.8.  Observe that it appears to be a reasonable approximation to the graph of $f$ in Figure 9.7, but, being a finite sum of differentiable functions, $f_3(t)$ is itself differentiable everywhere, even at the integer multiples of $\pi$ where $f$ is not differentiable.

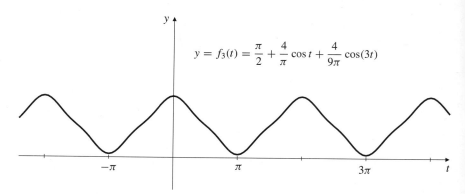

$$y = f_3(t) = \frac{\pi}{2} + \frac{4}{\pi}\cos t + \frac{4}{9\pi}\cos(3t)$$

**Figure 9.8** The Fourier polynomial approximation $f_3(t)$ to the sawtooth function of Example 3

As noted earlier, the Fourier series of a function $f(t)$ that is periodic, continuous, and has a piecewise continuous derivative on the real line converges to $f(t)$ at each real number $t$. However, the Fourier coefficients (and hence the Fourier series) can be calculated (by the formulas given above) for periodic functions with piecewise continuous derivative even if the functions are not themselves continuous, but only piecewise continuous.

Recall that $f(t)$ is piecewise continuous on the interval $[a, b]$ if there exists a partition $\{a = x_0 < x_1 < x_2 < \cdots < x_k = b\}$ of $[a, b]$ and functions $F_1, F_2, \ldots, F_k$, such that

(i) $F_i$ is continuous on $[x_{i-1}, x_i]$, and

(ii) $f(t) = F_i(t)$ on $(x_{i-1}, x_i)$.

The integral of such a function $f$ is the sum of integrals of the functions $F_i$:

$$\int_a^b f(t)\,dt = \sum_{i=1}^{k} \int_{x_{i-1}}^{x_i} F_i(t)\,dt.$$

Since $f(t)\cos(n\omega t)$ and $f(t)\sin(n\omega t)$ are piecewise continuous if $f$ is, the Fourier coefficients of a piecewise continuous, periodic function can be calculated by the same formulas given for a continuous periodic function. The question of where and to what the Fourier series converges in this case is answered by the following theorem, proved in textbooks on Fourier analysis.

**THEOREM 25**

The Fourier series of a piecewise continuous, periodic function $f$ with piecewise continuous derivative converges to that function at every point $t$ where $f$ is continuous. Moreover, if $f$ is discontinuous at $t = c$, then $f$ has different, but finite, left and right limits at $c$:

$$\lim_{t \to c-} f(t) = f(c-), \quad \text{and} \quad \lim_{t \to c+} f(t) = f(c+).$$

The Fourier series of $f$ converges at $t = c$ to the average of these left and right limits:

$$\frac{a_0}{2} + \sum_{n=1}^{\infty} \left(a_n \cos(n\omega c) + b_n \sin(n\omega c)\right) = \frac{f(c-) + f(c+)}{2},$$

where $\omega = 2\pi/T$.

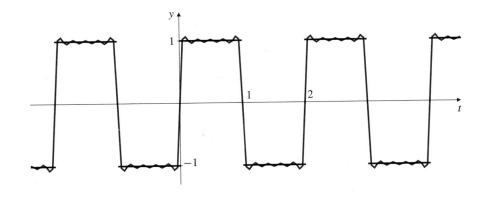

**Figure 9.9** The piecewise continuous function $f$ (black) of Example 4 and its Fourier polynomial $f_{15}$ (colour)

$$f_{15}(t) = \sum_{k=1}^{8} \frac{4\sin\big((2k-1)\pi t\big)}{(2k-1)\pi}$$

**Example 5**  Calculate the Fourier series for the periodic function $f$ with period 2 satisfying

$$f(t) = \begin{cases} -1 & \text{if } -1 < x < 0 \\ 1 & \text{if } 0 < x < 1. \end{cases}$$

Where does $f$ fail to be continuous? To what does the Fourier series of $f$ converge at these points?

**Solution**  Here $T = 2$ and $\omega = 2\pi/2 = \pi$. Since $f$ is an odd function, its cosine coefficients are all zero:

$$a_n = \int_{-1}^{1} f(t)\cos(n\pi t)\, dt$$

$$= -\int_{-1}^{0} \cos(n\pi t)\, dt + \int_{0}^{1} \cos(n\pi t)\, dt = 0.$$

The same symmetry implies that

$$b_n = \int_{-1}^{1} f(t)\sin(n\pi t)\, dt$$

$$= 2\int_{0}^{1} \sin(n\pi t)\, dt = -\frac{2\cos(n\pi t)}{n\pi}\bigg|_{0}^{1}$$

$$= -\frac{2}{n\pi}\big((-1)^n - 1\big) = \begin{cases} 4/(n\pi) & \text{if } n \text{ is odd} \\ 0 & \text{if } n \text{ is even.} \end{cases}$$

Odd integers $n$ are of the form $n = 2k - 1$ for $k = 1, 2, 3, \ldots$. Therefore, the Fourier series of $f$ is

$$\frac{4}{\pi}\sum_{k=1}^{\infty} \frac{1}{2k-1}\sin\big((2k-1)\pi t\big)$$

$$= \frac{4}{\pi}\left( \sin(\pi t) + \frac{1}{3}\sin(3\pi t) + \frac{1}{5}\sin(5\pi t) + \cdots \right).$$

Note that $f$ is continuous except at the points where $t$ is an integer. At each of these points $f$ jumps from $-1$ to $1$ or from $1$ to $-1$, so the average of the left and right limits of $f$ at these points is $0$. Observe that the sum of the Fourier series is $0$ at integer values of $t$, in accordance with Theorem 25.

The graphs of $f$ (black) and its Fourier polynomial $f_{15}$ (colour) are shown in Figure 9.9.

## Fourier Cosine and Sine Series

As observed in Example 3 and Example 5, even functions have no sine terms in their Fourier series, and odd functions have no cosine terms (including the constant term $a_0/2$). It is often necessary in applications to find a Fourier series representation of a given function defined on a finite interval $[0, a]$ having either no sine terms (a **Fourier cosine series**) or no cosine terms (a **Fourier sine series**). This is accomplished by extending the domain of $f$ to $[-a, 0)$ so as to make $f$ either even or odd on $[-a, a]$,

$$f(-t) = f(t) \text{ if } -a \leq t < 0 \text{ for the even extension}$$
$$f(-t) = -f(t) \text{ if } -a \leq t < 0 \text{ for the odd extension,}$$

and then calculating its Fourier series considering the extended $f$ to have period $2a$. (If we want the odd extension, we may have to redefine $f(0)$ to be $0$.)

---

**Example 6**    Find the Fourier cosine series of $g(t) = \pi - t$ defined on $[0, \pi]$.

**Solution**    The even extension of $g(t)$ to $[-\pi, \pi]$ is the function $f$ of Example 3. Thus the Fourier cosine series of $g$ is

$$\frac{\pi}{2} + \sum_{k=1}^{\infty} \frac{4}{\pi(2k-1)^2} \cos\big((2k-1)t\big).$$

---

**Example 7**    Find the Fourier sine series of $h(t) = 1$ defined on $[0, 1]$.

**Solution**    If we redefine $h(0) = 0$, then the odd extension of $h$ to $[-1, 1]$ coincides with the function $f(t)$ of Example 5 except that the latter function is undefined at $t = 0$. The Fourier sine series of $h$ is the series obtained in Example 5, namely

$$\frac{4}{\pi} \sum_{k=1}^{\infty} \frac{1}{2k-1} \sin\big((2k-1)\pi t\big).$$

---

**Remark**    Fourier cosine and sine series are treated from a different perspective in Section 13.4.

## Exercises 9.10

In Exercises 1–4, what is the fundamental period of the given function?

**1.** $f(t) = \sin(3t)$

**2.** $g(t) = \cos(3 + \pi t)$

**3.** $h(t) = \cos^2 t$

**4.** $k(t) = \sin(2t) + \cos(3t)$

In Exercises 5–8, find the Fourier series of the given function.

**5.** $f(t) = t$, $-\pi < t \leq \pi$, $f$ has period $2\pi$.

**6.** $f(t) = \begin{cases} 0 & \text{if } 0 \le t < 1 \\ 1 & \text{if } 1 \le t < 2, \end{cases}$ $f$ has period 2.

**7.** $f(t) = \begin{cases} 0 & \text{if } -1 \le t < 0 \\ t & \text{if } 0 \le t < 1, \end{cases}$ $f$ has period 2.

**8.** $f(t) = \begin{cases} t & \text{if } 0 \le t < 1 \\ 1 & \text{if } 1 \le t < 2 \\ 3 - t & \text{if } 2 \le t < 3, \end{cases}$ $f$ has period 3.

**9.** What is the Fourier cosine series of the function $h(t)$ of Example 7?

**10.** Calculate the Fourier sine series of the function $g(t)$ of Example 6.

**11.** Find the Fourier sine series of $f(t) = t$ on $[0, 1]$.

**12.** Find the Fourier cosine series of $f(t) = t$ on $[0, 1]$.

**13.** Use the result of Example 3 to evaluate

$$\sum_{n=1}^{\infty} \frac{1}{(2n-1)^2} = 1 + \frac{1}{3^2} + \frac{1}{5^2} + \cdots.$$

**14.** Verify that if $f$ is an even function of period $T$, then the Fourier sine coefficients $b_n$ of $f$ are all zero and the Fourier cosine coefficients $a_n$ of $f$ are given by

$$a_n = \frac{4}{T} \int_0^{T/2} f(t) \cos(n\omega t)\, dt, \qquad n = 0, 1, 2, \dots,$$

where $\omega = 2\pi/T$. State and verify the corresponding result for odd functions $f$.

# Chapter Review

## Key Ideas

- **What does it mean to say that the sequence $\{a_n\}$**
  - ◇ is bounded above?
  - ◇ is alternating?
  - ◇ converges?
  - ◇ is ultimately positive?
  - ◇ is increasing?
  - ◇ diverges to infinity?
- **What does it mean to say that the series $\sum_{n=1}^{\infty} a_n$**
  - ◇ converges?
  - ◇ is geometric?
  - ◇ is a $p$-series?
  - ◇ converges absolutely?
  - ◇ diverges?
  - ◇ is telescoping?
  - ◇ is positive?
  - ◇ converges conditionally?
- **State the following convergence tests for series.**
  - ◇ the integral test
  - ◇ the limit comparison test
  - ◇ the alternating series test
  - ◇ the comparsion test
  - ◇ the ratio test
- **How can you find bounds for the tail of a series?**
- **What is a bound for the tail of an alternating series?**
- **What do the following phrases mean?**
  - ◇ a power series
  - ◇ radius of convergence
  - ◇ a Taylor series
  - ◇ a Taylor polynomial
  - ◇ an analytic function
  - ◇ interval of convergence
  - ◇ centre of convergence
  - ◇ a Maclaurin series
  - ◇ a binomial series
- **Where is the sum of a power series differentiable?**
- **Where does the integral of a power series converge?**
- **Where is the sum of a power series continuous?**
- **State Taylor's Theorem with Lagrange remainder.**
- **State Taylor's Theorem with integral remainder.**
- **What is the binomial theorem?**
- **What is a Fourier series?**
- **What is a Fourier cosine series? a Fourier sine series?**

## Review Exercises

In Exercises 1–4, determine whether the given sequence converges or not, and find its limit if it does converge.

**1.** $\left\{ \dfrac{(-1)^n e^n}{n!} \right\}$

**2.** $\left\{ \dfrac{n^{100} + 2^n \pi}{2^n} \right\}$

**3.** $\left\{ \dfrac{\ln n}{\tan^{-1} n} \right\}$

**4.** $\left\{ \dfrac{(-1)^n n^2}{\pi n(n - \pi)} \right\}$

**5.** Let $a_1 > \sqrt{2}$, and let

$$a_{n+1} = \frac{a_n}{2} + \frac{1}{a_n} \quad \text{for} \quad n = 1. 2, 3, \dots$$

Show that $\{a_n\}$ is decreasing and that $a_n > \sqrt{2}$ for $n \ge 1$. Why must $\{a_n\}$ converge? Find $\lim_{n\to\infty} a_n$.

**6.** Find the limit of the sequence $\{\ln \ln(n + 1) - \ln \ln n\}$.

Evaluate the sums of the series in Exercises 7–10.

**7.** $\displaystyle\sum_{n=1}^{\infty} 2^{-(n-5)/2}$

**8.** $\displaystyle\sum_{n=0}^{\infty} \frac{4^{n-1}}{(\pi - 1)^{2n}}$

**9.** $\displaystyle\sum_{n=1}^{\infty} \frac{1}{n^2 - \frac{1}{4}}$

**10.** $\displaystyle\sum_{n=1}^{\infty} \frac{1}{n^2 - \frac{9}{4}}$

Determine whether the series in Exercises 11–16 converge or not. Give reasons for your answers.

**11.** $\displaystyle\sum_{n=1}^{\infty} \frac{n-1}{n^3}$

**12.** $\displaystyle\sum_{n=1}^{\infty} \frac{n + 2^n}{1 + 3^n}$

**13.** $\displaystyle\sum_{n=1}^{\infty} \frac{n}{(1+n)(1+n\sqrt{n})}$

**14.** $\displaystyle\sum_{n=1}^{\infty} \frac{n^2}{(1+2^n)(1+n\sqrt{n})}$

**15.** $\displaystyle\sum_{n=1}^{\infty} \frac{3^{2n+1}}{n!}$

**16.** $\displaystyle\sum_{n=1}^{\infty} \frac{n!}{(n+2)!+1}$

Do the series in Exercises 17–20 converge absolutely, converge conditionally, or diverge?

**17.** $\displaystyle\sum_{n=1}^{\infty} \frac{(-1)^{n-1}}{1+n^3}$

**18.** $\displaystyle\sum_{n=1}^{\infty} \frac{(-1)^n}{2^n - n}$

**19.** $\displaystyle\sum_{n=10}^{\infty} \frac{(-1)^{n-1}}{\ln \ln n}$

**20.** $\displaystyle\sum_{n=1}^{\infty} \frac{n^2 \cos(n\pi)}{1+n^3}$

For what values of $x$ do the series in Exercises 21–22 converge absolutely? converge conditionally? diverge?

**21.** $\displaystyle\sum_{n=1}^{\infty} \frac{(x-2)^n}{3^n \sqrt{n}}$

**22.** $\displaystyle\sum_{n=1}^{\infty} \frac{(5-2x)^n}{n}$

Determine the sums of the series in Exercises 23–24 to within 0.001.

**23.** $\displaystyle\sum_{n=1}^{\infty} \frac{1}{n^3}$

**24.** $\displaystyle\sum_{n=1}^{\infty} \frac{1}{4+n^2}$

In Exercises 25–32, find Maclaurin series for the given functions. State where each series converges to the function.

**25.** $\dfrac{1}{3-x}$

**26.** $\dfrac{x}{3-x^2}$

**27.** $\ln(e+x^2)$

**28.** $\dfrac{1-e^{-2x}}{x}$

**29.** $x \cos^2 x$

**30.** $\sin(x+(\pi/3))$

**31.** $(8+x)^{-1/3}$

**32.** $(1+x)^{1/3}$

Find Taylor series for the functions in Exercises 33–34 about the indicated points $x = c$.

**33.** $1/x, \quad c = \pi$

**34.** $\sin x + \cos x, \quad c = \pi/4$

Find the Maclaurin polynomial of the indicated degree for the functions in Exercises 35–38.

**35.** $e^{x^2+2x}$, degree 3

**36.** $\sin(1+x)$, degree 3

**37.** $\cos(\sin x)$, degree 4

**38.** $\sqrt{1+\sin x}$, degree 4

**39.** What function has Maclaurin series

$$1 - \frac{x}{2!} + \frac{x^2}{4!} - \cdots = \sum_{n=0}^{\infty} \frac{(-1)^n x^n}{(2n)!}?$$

**40.** A function $f(x)$ has Maclaurin series

$$1 + x^2 + \frac{x^4}{2^2} + \frac{x^6}{3^2} + \cdots = 1 + \sum_{n=1}^{\infty} \frac{x^{2n}}{n^2}.$$

Find $f^{(k)}(0)$ for all positive integers $k$.

Find the sums of the series in 41–44.

**41.** $\displaystyle\sum_{n=0}^{\infty} \frac{n+1}{\pi^n}$

**\* 42.** $\displaystyle\sum_{n=0}^{\infty} \frac{n^2}{\pi^n}$

**43.** $\displaystyle\sum_{n=1}^{\infty} \frac{1}{ne^n}$

**\* 44.** $\displaystyle\sum_{n=2}^{\infty} \frac{(-1)^n \pi^{2n-4}}{(2n-1)!}$

**45.** If $S(x) = \displaystyle\int_0^x \sin(t^2)\,dt$, find $\displaystyle\lim_{x\to 0} \frac{x^3 - 3S(x)}{x^7}$.

**46.** Use series to evaluate $\displaystyle\lim_{x\to 0} \frac{(x - \tan^{-1} x)(e^{2x} - 1)}{2x^2 - 1 + \cos(2x)}$.

**47.** How many nonzero terms in the Maclaurin series for $e^{-x^4}$ are needed to evaluate $\int_0^{1/2} e^{-x^4}\,dx$ correct to five decimal places? Evaluate the integral to that accuracy.

**48.** Estimate the size of the error if the Taylor polynomial of degree 4 about $x = \pi/2$ for $f(x) = \ln \sin x$ is used to approximate $\ln \sin(1.5)$.

**49.** Find the Fourier sine series of $f(t) = \pi - t$ on $[0, \pi]$.

**50.** Find the Fourier series of $f(t) = \begin{cases} 1 & \text{if } -\pi < t \le 0 \\ t & \text{if } 0 < t \le \pi. \end{cases}$

## Challenging Problems

**1.** (**A refinement of the ratio test**) Suppose $a_n > 0$ and $a_{n+1}/a_n \ge n/n+1$ for all $n$. Show that $\sum_{n=1}^{\infty} a_n$ diverges. *Hint:* $a_n \ge K/n$ for some constant $K$.

**\* 2.** (**Summation by parts**) Let $\{u_n\}$ and $\{v_n\}$ be two sequences and let $s_n = \sum_{k=1}^{n} v_k$.

(a) Show that $\sum_{k=1}^{n} u_k v_k = u_{n+1}s_n + \sum_{k=1}^{n}(u_k - u_{k+1})s_n$. (*Hint:* write $v_n = s_n - s_{n-1}$, with $s_0 = 0$, and rearrange the sum.)

(b) If $\{u_n\}$ is positive, decreasing, and convergent to 0, and if $\{v_n\}$ has bounded partial sums, $|s_n| \le K$ for all $n$, where $K$ is a constant, show that $\sum_{n=1}^{\infty} u_n v_n$ converges. (*Hint:* show that the series $\sum_{n=1}^{\infty}(u_n - u_{n+1})s_n$ converges by comparing it to the telescoping series $\sum_{n=1}^{\infty}(u_n - u_{n+1})$.)

**\* 3.** Show that $\sum_{n=1}^{\infty}(1/n)\sin(nx)$ converges for every $x$. *Hint:* if $x$ is an integer multiple of $\pi$ all the terms in the series are 0 so there is nothing to prove. Otherwise, $\sin(x/2) \ne 0$. In this case show that

$$\sum_{n=1}^{N} \sin(nx) = \frac{\cos(x/2) - \cos((N+1/2)x)}{2\sin(x/2)}$$

using the identity

$$\sin a \sin b = \frac{\cos(a-b) - \cos(a+b)}{2}$$

to make the sum telescope. Then apply the result of Exercise 2(b) with $u_n = 1/n$ and $v_n = \sin(nx)$.

**4.** Let $a_1, a_2, a_3, \ldots$ be those positive integers that do not contain the digit 0 in their decimal representations. Thus $a_1 = 1$, $a_2 = 2, \ldots a_9 = 9$, $a_{10} = 11, \ldots a_{18} = 19$, $a_{19} = 21, \ldots$ $a_{90} = 99$, $a_{91} = 111$, etc. Show that the series $\sum\limits_{n=1}^{\infty} 1/a_n$ converges, and that the sum is less than 90. (*Hint:* how many of these integers have $m$ digits? Each term $1/a_n$, where $a_n$ has $m$ digits, is less than $10^{-m+1}$.)

* **5. (Using an integral to improve convergence)** Recall the error formula for the Midpoint Rule, according to which

$$\int_{k-1/2}^{k+1/2} f(x)\, dx - f(k) = \frac{f''(c)}{24},$$

where $k - (1/2) \leq c \leq k + (1/2)$.

(a) If $f''(x)$ is a decreasing function of $x$, show that

$$f'(k + \tfrac{3}{2}) - f'(k + \tfrac{1}{2}) \leq f''(c) \leq f'(k - \tfrac{1}{2}) - f'(k - \tfrac{3}{2}).$$

(b) If (i) $f''(x)$ is a decreasing function of $x$, (ii) $\int_{N+1/2}^{\infty} f(x)\, dx$ converges, and (iii) $f'(x) \to 0$ as $x \to \infty$, show that

$$\frac{f'(N - \tfrac{1}{2})}{24} \leq \sum_{n=N+1}^{\infty} f(n) - \int_{N+1/2}^{\infty} f(x)\, dx \leq \frac{f'(N + \tfrac{3}{2})}{24}.$$

(c) Use the result of part (b) to approximate $\sum_{n=1}^{\infty} 1/n^2$ to within 0.001.

* **6. (The number $e$ is irrational.)** Start with $e = \sum_{n=0}^{\infty} 1/n!$.

(a) Use the technique of Example 8 in Section 9.3 to show that for any $n > 0$,

$$0 < e - \sum_{j=0}^{n} \frac{1}{j!} < \frac{1}{n!\, n}.$$

(Note that the sum here has $n + 1$ terms rather than $n$ terms.)

(b) Suppose that $e$ is a rational number, say $e = M/N$ for certain positive integers $M$ and $N$. Show that $N!\left(e - \sum_{j=0}^{N}(1/j!)\right)$ is an integer.

(c) Combine parts (a) and (b) to show that there is an integer between 0 and $1/N$. Why is this not possible? Conclude that $e$ cannot be a rational number.

**7.** Let

$$f(x) = \sum_{k=0}^{\infty} \frac{2^{2k} k!}{(2k+1)!} x^{2k+1}$$

$$= x + \frac{2}{3} x^3 + \frac{4}{3 \times 5} x^5 + \frac{8}{3 \times 5 \times 7} x^7 + \ldots.$$

(a) Find the radius of convergence of this power series.

(b) Show that $f'(x) = 1 + 2x f(x)$.

(c) What is $\dfrac{d}{dx}\left(e^{-x^2} f(x)\right)$?

(d) Express $f(x)$ in terms of an integral.

* **8. (The number $\pi$ is irrational)** The last problem in the Chapter Review at the end of Chapter 10 shows how to prove that $e$ is irrational by assuming the contrary and deducing a contradiction. In this problem you will show that $\pi$ is also irrational. The proof for $\pi$ is also by contradiction, but is rather more complicated and will be broken down into several parts.

(a) Let $f(x)$ be a polynomial, and let

$$g(x) = f(x) - f''(x) + f^{(4)}(x) - f^{(6)}(x) + \cdots$$

$$= \sum_{j=0}^{\infty} (-1)^j f^{(2j)}(x).$$

(Since $f$ is a polynomial all but a finite number of terms in the above sum are identically zero, so there are no convergence problems.) Verify that

$$\frac{d}{dx}\left(g'(x) \sin x - g(x) \cos x\right) = f(x) \sin x,$$

and hence that

$$\int_0^{\pi} f(x) \sin x\, dx = g(\pi) + g(0).$$

(b) Suppose that $\pi$ is rational, say $\pi = m/n$, where $m$ and $n$ are positive integers. You will show that this leads to a contradiction, and thus cannot be true. Choose a positive integer $k$ such that $(\pi m)^k/k! < 1/2$. (Why is this possible?) Consider the polynomial

$$f(x) = \frac{x^k(m - nx)^k}{k!} = \frac{1}{k!} \sum_{j=0}^{k} \binom{k}{j} m^{k-j}(-n)^j x^{j+k}.$$

Show that $0 < f(x) < 1/2$ for $0 < x < \pi$, and hence that

$$0 < \int_0^{\pi} f(x) \sin x\, dx < 1.$$

Thus $0 < g(\pi) + g(0) < 1$, where $g(x)$ is defined as in part (a).

(c) Show that the $i$th derivative of $f(x)$ is given by

$$f^{(i)}(x) = \frac{1}{k!} \sum_{j=0}^{k} \binom{k}{j} m^{k-j}(-n)^j \frac{(j+k)!}{(j+k-i)!} x^{j+k-i}.$$

(d) Show that $f^{(i)}(0)$ is an integer for $i = 0, 1, 2, \ldots$. (*Hint:* observe for $i < k$ that $f^{(i)}(0) = 0$, and for $i > 2k$ that $f^{(i)}(x) = 0$ for all $x$. For $k \leq i \leq 2k$, show that only one term in the sum for $f^{(i)}(0)$ is not 0, and that this term is an integer. You will need the fact that the binomial coefficients $\binom{k}{j}$ are integers.)

(e) Show that $f(\pi - x) = f(x)$ for all $x$, and hence that $f^{(i)}(\pi)$ is also an integer for each $i = 0, 1, 2, \ldots$. Therefore, if $g(x)$ is defined as in (a), then $g(\pi) + g(0)$ is an integer. This contradicts the conclusion of part (b), and so shows that $\pi$ cannot be rational.

*   **9. (An asymptotic series)** Use integration by parts to show that

$$\int_0^x e^{-1/t}\,dt = e^{-1/x} \sum_{n=2}^{N} (-1)^n (n-1)!\,x^n$$
$$+ (-1)^{N+1} N! \int_0^x t^{N-1} e^{-1/t}\,dt.$$

Why can't you just use a Maclaurin series to approximate this integral? Using $N = 5$, find an approximate value for $\int_0^{0.1} e^{-1/t}\,dt$, and estimate the error. Estimate the error for $N = 10$ and $N = 20$.

Note that the series $\sum_{n=2}^{\infty} (-1)^n (n-1)!\,x^n$ **diverges** for any $x \neq 0$. This is an example of what is called an **asymptotic series**. Even though it diverges, a properly chosen partial sum gives a good approximation to our function when $x$ is small.

# CHAPTER 10

# Vectors and Coordinate Geometry in 3-Space

**Introduction**  A complete real-variable calculus program involves the study of

(i) real-valued functions of a single real variable,

(ii) vector-valued functions of a single real variable,

(iii) real-valued functions of a real vector variable,

(iv) vector-valued functions of a real vector variable.

Chapters 1–9 were concerned with item (i). The remaining chapters deal with items (ii), (iii), and (iv). Specifically, Chapter 11 deals with vector-valued functions of a single real variable. Chapters 12–14 are concerned with the differentiation and integration of real-valued functions of several real variables, that is, of a real vector variable. Chapters 15 and 16 present aspects of the calculus of functions whose domains and ranges both have dimension greater than one, that is, vector-valued functions of a vector variable. Most of the time we will limit our attention to vector functions with domains and ranges in the plane, or in 3-dimensional space.

In this chapter we will lay the foundation for multi-variable and vector calculus by extending the concepts of analytic geometry to three or more dimensions, and by introducing vectors as a convenient way of dealing with several variables as a single entity. We also introduce matrices, as these will prove useful for formulating some of the concepts of calculus. This chapter is not intended to be a course in linear algebra. We develop only those aspects that we will use in later chapters, and omit most proofs.

## 10.1  Analytic Geometry in 3 Dimensions

We say that the physical world in which we live is three dimensional because through any point there can pass three, and no more, straight lines that are **mutually perpendicular**, that is to say, each of them is perpendicular to the other two. This is equivalent to the fact that we require three numbers to locate a point in space with respect to some reference point (the **origin**). One way to use three numbers to locate a point is by having them represent (signed) distances from the origin, measured in the directions of three mutually perpendicular lines passing through the origin. We call such a set of lines a Cartesian coordinate system, and each of the lines is called a coordinate axis. We usually call these axes the $x$-axis, the $y$-axis, and the $z$-axis, regarding the $x$- and $y$-axes as lying in a horizontal plane and the $z$-axis as vertical. Moreover, the coordinate system should have a **right-handed orientation**. This means that the thumb, forefinger, and middle finger of the right hand can be extended so as to point respectively in the directions of the positive $x$-axis, the positive $y$-axis, and the positive $z$-axis. For the more mechanically minded, a right-handed screw will advance in the positive $z$ direction if twisted in the direction of rotation from the positive $x$-axis towards the positive $y$-axis. (See Figure 10.1(a).)

**Figure 10.1**

(a) The screw moves upward when twisted counterclockwise as seen from above

(b) The three coordinates of a point in 3-space

(a)

(b)

With respect to such a Cartesian coordinate system, the **coordinates** of a point $P$ in 3-space constitute an ordered triple of real numbers, $(x, y, z)$. The numbers $x$, $y$, and $z$ are, respectively, the signed distances of $P$ from the origin, measured in the directions of the $x$-axis, the $y$-axis, and the $z$-axis. (See Figure 10.1(b).)

Let $Q$ be the point with coordinates $(x, y, 0)$. Then $Q$ lies in the $xy$-plane (the plane containing the $x$- and $y$-axes) directly under (or over) $P$. We say that $Q$ is the vertical projection of $P$ onto the $xy$-plane. If $r$ is the distance from the origin $O$ to $P$ and $s$ is the distance from $O$ to $Q$, then, using two right-angled triangles, we have

$$s^2 = x^2 + y^2 \quad \text{and} \quad r^2 = s^2 + z^2 = x^2 + y^2 + z^2.$$

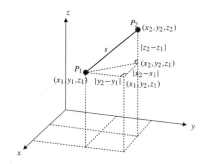

**Figure 10.2**

Thus the distance from $P$ to the origin is given by

$$r = \sqrt{x^2 + y^2 + z^2}.$$

Similarly, the distance $s$ between points $P_1 = (x_1, y_1, z_1)$ and $P_2 = (x_2, y_2, z_2)$ (see Figure 10.2) is

$$s = \sqrt{(x_2 - x_1)^2 + (y_2 - y_1)^2 + (z_2 - z_1)^2}.$$

**Example 1**    Show that the triangle with vertices $A = (1, -1, 2)$, $B = (3, 3, 8)$, and $C = (2, 0, 1)$ has a right angle.

**Solution**    We calculate the lengths of the three sides of the triangle:

$$a = |BC| = \sqrt{(2-3)^2 + (0-3)^2 + (1-8)^2} = \sqrt{59}$$
$$b = |AC| = \sqrt{(2-1)^2 + (0+1)^2 + (1-2)^2} = \sqrt{3}$$
$$c = |AB| = \sqrt{(3-1)^2 + (3+1)^2 + (8-2)^2} = \sqrt{56}$$

By the cosine law, $a^2 = b^2 + c^2 - 2bc \cos A$. In this case $a^2 = 59 = 3 + 56 = b^2 + c^2$, so that $2bc \cos A$ must be 0. Therefore $\cos A = 0$ and $A = 90°$. ∎

**Figure 10.3**    The first octant

Just as the $x$- and $y$-axes divide the $xy$-plane into four quadrants, so also the three **coordinate planes** in 3-space (the $xy$-plane, the $xz$-plane, and the $yz$-plane) divide 3-space into eight **octants**. We call the octant in which $x \geq 0$, $y \geq 0$, and $z \geq 0$ the **first octant**. When drawing graphs in 3-space it is sometimes easier to draw only the part lying in the first octant (Figure 10.3).

An equation or inequality involving the three variables $x$, $y$, and $z$ defines a subset of points in 3-space whose coordinates satisfy the equation or inequality. A single equation usually represents a surface (a two-dimensional object) in 3-space.

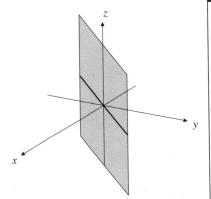

**Figure 10.4**    Equation $x = y$ defines a vertical plane

---

**Example 2**    **(Some equations and the surfaces they represent)**

(a) The equation $z = 0$ represents all points with coordinates $(x, y, 0)$, that is, the $xy$-plane. The equation $z = -2$ represents all points with coordinates $(x, y, -2)$, that is, the horizontal plane passing through the point $(0, 0, -2)$ on the $z$-axis.

(b) The equation $x = y$ represents all points with coordinates $(x, x, z)$. This is a vertical plane containing the straight line with equation $x = y$ in the $xy$-plane. The plane also contains the $z$-axis. (See Figure 10.4.)

(c) The equation $x + y + z = 1$ represents all points the sum of whose coordinates is 1. This set is a plane that passes through the three points $(1, 0, 0)$, $(0, 1, 0)$, and $(0, 0, 1)$. These points are not collinear (they do not lie on a straight line) so there is only one plane passing through all three. (See Figure 10.5.) The equation $x + y + z = 0$ represents a plane parallel to the one with equation $x + y + z = 1$, but passing through the origin.

(d) The equation $x^2 + y^2 = 4$ represents all points on the vertical circular cylinder containing the circle with equation $x^2 + y^2 = 4$ in the $xy$-plane. This cylinder has radius 2 and axis along the $z$-axis. (See Figure 10.6.)

(e) The equation $z = x^2$ represents all points with coordinates $(x, y, x^2)$. This surface is a parabolic cylinder tangent to the $xy$-plane along the $y$-axis. (See Figure 10.7.)

(f) The equation $x^2 + y^2 + z^2 = 25$ represents all points $(x, y, z)$ at distance 5 from the origin. This set of points is a *sphere* of radius 5 centred at the origin.

---

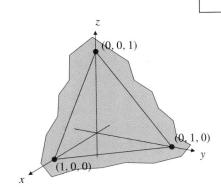

**Figure 10.5**    The plane with equation $x + y + z = 1$

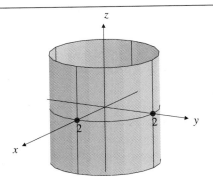

**Figure 10.6**    The circular cylinder with equation $x^2 + y^2 = 4$

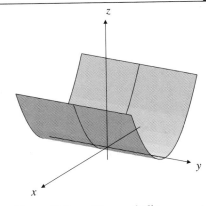

**Figure 10.7**    The parabolic cylinder with equation $z = x^2$

Observe that equations in $x$, $y$, and $z$ need not involve each variable explicitly. When one of the variables is missing from the equation, the equation represents a surface *parallel to* the axis of the missing variable. Such a surface may be a plane or a cylinder. For example, if $z$ is absent from the equation, the equation represents in 3-space a vertical (that is, parallel to the $z$-axis) surface containing the curve with the same equation in the $xy$-plane.

Occasionally a single equation may not represent a two-dimensional object (a surface). It can represent a one-dimensional object (a line or curve), a zero-dimensional object (one or more points), or even nothing at all.

**Example 3**   Identify the graphs of the equations:   (a) $y^2 + (z - 1)^2 = 4$, (b) $y^2 + (z - 1)^2 = 0$,   (c) $x^2 + y^2 + z^2 = 0$, and   (d) $x^2 + y^2 + z^2 = -1$.

*Solution*

(a) Since $x$ is absent, the equation $y^2 + (z - 1)^2 = 4$ represents an object parallel to the $x$-axis. In the $yz$-plane the equation represents a circle of radius 2 centred at $(y, z) = (0, 1)$. In 3-space it represents a horizontal circular cylinder, parallel to the $x$-axis, with axis one unit above the $x$-axis.

(b) Since squares cannot be negative, the equation $y^2 + (z - 1)^2 = 0$ implies that $y = 0$ and $z = 1$, so it represents points $(x, 0, 1)$. All these points lie on the line parallel to the $x$-axis and one unit above it.

(c) As in part (b), $x^2 + y^2 + z^2 = 0$ implies that $x = 0$, $y = 0$, and $z = 0$. The equation represents only one point, the origin.

(d) The equation $x^2 + y^2 + z^2 = -1$ is not satisfied by any real numbers $x$, $y$, and $z$, so represents no points at all.

A single inequality in $x$, $y$, and $z$ typically represents points lying on one side of the surface represented by the corresponding equation (together with points on the surface if the inequality is not strict).

**Example 4**

(a) The inequality $z > 0$ represents all points above the $xy$-plane.

(b) The inequality $x^2 + y^2 \geq 4$ says that the square of the distance from $(x, y, z)$ to the nearest point $(x, y, 0)$ on the $z$-axis is at least 4. This inequality represents all points lying on or outside the cylinder of Example 2(d).

(c) The inequality $x^2 + y^2 + z^2 \leq 25$ says that the square of the distance from $(x, y, z)$ to the origin is no greater than 25. It represents the solid ball of radius 5 centred at the origin, which consists of all points lying inside or on the sphere of Example 2(f).

Two equations in $x$, $y$, and $z$ normally represent a one-dimensional object, the line or curve along which the two surfaces represented by the two equations intersect. Any point whose coordinates satisfy both equations must lie on both the surfaces, so must lie on their intersection.

**Example 5**   What sets of points in 3-space are represented by the pairs of equations?

(a) $\begin{cases} x + y + z = 1 \\ y - 2x = 0 \end{cases}$   (b) $\begin{cases} x^2 + y^2 + z^2 = 1 \\ x + y = 1 \end{cases}$

### Solution

(a) The equation $x + y + z = 1$ represents the oblique plane of Example 2(c), and the equation $y - 2x = 0$ represents a vertical plane through the origin and the point $(1, 2, 0)$. Together these two equations represent the line of intersection of the two planes. This line passes through, for example, the points $(0, 0, 1)$ and $(\frac{1}{3}, \frac{2}{3}, 0)$. (See Figure 10.8(a).)

(b) The equation $x^2 + y^2 + z^2 = 1$ represents a sphere of radius 1 with centre at the origin, and $x + y = 1$ represents a vertical plane through the points $(1, 0, 0)$ and $(0, 1, 0)$. The two surfaces intersect in a circle, as shown in Figure 10.8(b). The line from $(1, 0, 0)$ to $(0, 1, 0)$ is a diameter of the circle, so the centre of the circle is $(\frac{1}{2}, \frac{1}{2}, 0)$, and its radius is $\sqrt{2}/2$.

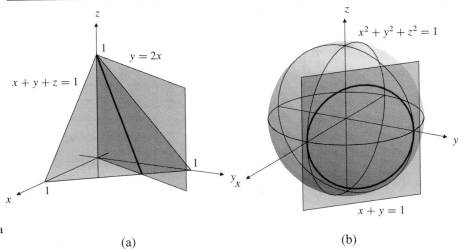

**Figure 10.8**

(a)  The two planes intersect in a straight line

(b)  The plane intersects the sphere in a circle

(a)                              (b)

In Sections 10.4 and 10.5 we will see many more examples of geometric objects in 3-space represented by simple equations.

## Euclidean *n*-Space

Mathematicians and users of mathematics frequently need to consider *n*-**dimensional space** where $n$ is greater than three, and may even be infinite. Students sometimes have difficulty in visualizing a space of dimension four or higher. The secret to dealing with these spaces is to regard the points in *n*-space as *being* ordered *n*-tuples of real numbers; that is, $(x_1, x_2, \ldots, x_n)$ is a point in *n*-space instead of just being the coordinates of such a point. We stop thinking of points as existing in physical space and start thinking of them as algebraic objects. We usually denote *n*-space by the symbol $\mathbb{R}^n$ to show that its points are *n*-tuples of *real* numbers. Thus $\mathbb{R}^2$ and $\mathbb{R}^3$ denote the plane and 3-space, respectively. Note that in passing from $\mathbb{R}^3$ to $\mathbb{R}^n$ we have altered the notation a bit — in $\mathbb{R}^3$ we called the coordinates $x$, $y$, and $z$ while in $\mathbb{R}^n$ we called them $x_1, x_2, \ldots$ and $x_n$ so as not to run out of letters. We could, of course, talk about coordinates $(x_1, x_2, x_3)$ in $\mathbb{R}^3$, and $(x_1, x_2)$ in the plane $\mathbb{R}^2$, but $(x, y, z)$ and $(x, y)$ are traditionally used there.

Although we think of points in $\mathbb{R}^n$ as *n*-tuples rather than geometric objects, we do not want to lose all sight of the underlying geometry. By analogy with the two- and three-dimensional cases, we still consider the quantity

$$\sqrt{(y_1 - x_1)^2 + (y_2 - x_2)^2 + \cdots + (y_n - x_n)^2}$$

as representing the *distance* between the points with coordinates $(x_1, x_2, \ldots, x_n)$ and $(y_1, y_2, \ldots, y_n)$. Also, we call the $(n-1)$-dimensional set of points in $\mathbb{R}^n$ which satisfy the equation $x_n = 0$ a **hyperplane**, by analogy with the plane $z = 0$ in $\mathbb{R}^3$.

## Describing Sets in the Plane, 3-Space, and *n*-Space

We conclude this section by collecting together some definitions of terms used to describe sets of points in $\mathbb{R}^n$ for $n \geq 2$. These terms generalize the notions of open and closed intervals and endpoints used to describe sets on the real line $\mathbb{R}$. We state the definitions for $\mathbb{R}^n$, but we are most interested in the cases where $n = 2$ or $n = 3$.

A **neighbourhood** of a point $P$ in $\mathbb{R}^n$ is a set of the form

$$B_r(P) = \{Q \in \mathbb{R}^n : \text{distance from } Q \text{ to } P < r\}$$

for some $r > 0$.

For $n = 1$, if $p \in \mathbb{R}$, then $B_r(p)$ is the **open interval** $(p - r, p + r)$ centred at $p$.

For $n = 2$, $B_r(P)$ is the **open disk** of radius $r$ centred at point $P$.

For $n = 3$, $B_r(P)$ is the **open ball** of radius $r$ centred at point $P$.

A set $S$ is **open** in $\mathbb{R}^n$ if every point of $S$ has a neighbourhood contained in $S$. Every neighbourhood is itself an open set. Other examples of open sets in $\mathbb{R}^2$ include the sets of points $(x, y)$ such that $x > 0$, or such that $y > x^2$, or even such that $y \neq x^2$. Typically sets defined by strict inequalities (using ">" and "<") are open. Examples in $\mathbb{R}^3$ include the sets of points $(x, y, z)$ satisfying $x + y + z > 2$, or $1 < x < 3$.

The whole space $\mathbb{R}^n$ is an open set in itself. For technical reasons the empty set (containing no points) is also considered to be open. (No point in the empty set fails to have a neighbourhood contained in the empty set.)

The **complement**, $S^c$, of a set $S$ in $\mathbb{R}^n$ is the set of all points in $\mathbb{R}^n$ that do not belong to $S$. For example, the complement of the set of points $(x, y)$ in $\mathbb{R}^2$ such that $x > 0$ is the set of points for which $x \leq 0$. A set is said to be **closed** if its complement is open. Typically sets defined by nonstrict inequalities (using "$\geq$" and "$\leq$") are closed. Closed intervals are closed sets in $\mathbb{R}$. Since the whole space and the empty set are both open in $\mathbb{R}^n$, and are complements of each other, they are also both closed. They are the only sets that are both open and closed.

A point $P$ is called a **boundary point** of a set $S$ if every neighbourhood of $P$ contains both points in $S$ and points in $S^c$. The **boundary**, bdry$(S)$, of a set $S$ is the set of all boundary points of $S$. For example the boundary of the closed disk $x^2 + y^2 \leq 1$ in $\mathbb{R}^2$ is the circle $x^2 + y^2 = 1$. A closed set contains all its boundary points. An open set contains none of its boundary points.

A point $P$ is an **interior point** of a set $S$ if it belongs to $S$ but not to the boundary of $S$. $P$ is an **exterior point** of $S$ if it belongs to the complement of $S$ but not to the boundary of $S$. The **interior**, int$(S)$, and **exterior**, ext$(S)$, of $S$ consist of all the interior points and exterior points of $S$ respectively. Both int$(S)$ and ext$(S)$ are open sets. If $S$ is open, then int$(S) = S$. If $S$ is closed, then ext$(S) = S^c$. See Figure 10.9.

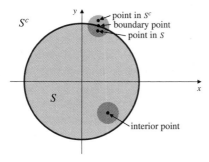

**Figure 10.9**    The closed disk $S$ consisting of points $(x, y) \in \mathbb{R}^2$ satisfying $x^2 + y^2 \leq 1$. Note the shaded neighbourhoods of the boundary point and the interior point.

bdry$(S)$ is the circle $x^2 + y^2 = 1$.

int$(S)$ is the open disk $x^2 + y^2 < 1$.

ext$(S)$ is the open set $x^2 + y^2 > 1$.

# Exercises 10.1

Find the distance between the pairs of points in Exercises 1–4.

**1.** $(0, 0, 0)$ and $(2, -1, -2)$    **2.** $(-1, -1, -1)$ and $(1, 1, 1)$

**3.** $(1, 1, 0)$ and $(0, 2, -2)$    **4.** $(3, 8, -1)$ and $(-2, 3, -6)$

**5.** What is the shortest distance from the point $(x, y, z)$ to (a) the $xy$-plane? (b) the $x$-axis?

**6.** Show that the triangle with vertices $(1, 2, 3)$, $(4, 0, 5)$, and $(3, 6, 4)$ has a right angle.

**7.** Find the angle $A$ in the triangle with vertices $A = (2, -1, -1)$, $B = (0, 1, -2)$, and $C = (1, -3, 1)$.

**8.** Show that the triangle with vertices $(1, 2, 3)$, $(1, 3, 4)$, and $(0, 3, 3)$ is equilateral.

**9.** Find the area of the triangle with vertices $(1, 1, 0)$, $(1, 0, 1)$, and $(0, 1, 1)$.

**10.** What is the distance from the origin to the point $(1, 1, \ldots, 1)$ in $\mathbb{R}^n$?

**11.** What is the distance from the point $(1, 1, \ldots, 1)$ in $n$-space to the closest point on the $x_1$-axis?

In Exercises 12–23, describe (and sketch if possible) the set of points in $\mathbb{R}^3$ which satisfy the given equation or inequality.

**12.** $z = 2$

**13.** $y \geq -1$

**14.** $z = x$

**15.** $x + y = 1$

**16.** $x^2 + y^2 + z^2 = 4$

**17.** $(x - 1)^2 + (y + 2)^2 + (z - 3)^2 = 4$

**18.** $x^2 + y^2 + z^2 = 2z$

**19.** $y^2 + z^2 \leq 4$

**20.** $x^2 + z^2 = 4$

**21.** $z = y^2$

**22.** $z \geq \sqrt{x^2 + y^2}$

**23.** $x + 2y + 3z = 6$

In Exercises 24–32, describe (and sketch if possible) the set of points in $\mathbb{R}^3$ which satisfy the given pair of equations or inequalities.

**24.** $\begin{cases} x = 1 \\ y = 2 \end{cases}$

**25.** $\begin{cases} x = 1 \\ y = z \end{cases}$

**26.** $\begin{cases} x^2 + y^2 + z^2 = 4 \\ z = 1 \end{cases}$

**27.** $\begin{cases} x^2 + y^2 + z^2 = 4 \\ x^2 + y^2 + z^2 = 4x \end{cases}$

**28.** $\begin{cases} x^2 + y^2 + z^2 = 4 \\ x^2 + z^2 = 1 \end{cases}$

**29.** $\begin{cases} x^2 + y^2 = 1 \\ z = x \end{cases}$

**30.** $\begin{cases} y \geq x \\ z \leq y \end{cases}$

**31.** $\begin{cases} x^2 + y^2 \leq 1 \\ z \geq y \end{cases}$

**32.** $\begin{cases} x^2 + y^2 + z^2 \leq 1 \\ \sqrt{x^2 + y^2} \leq z \end{cases}$

In Exercises 33–36, specify the boundary and the interior of the plane sets $S$ whose points $(x, y)$ satisfy the given conditions. Is $S$ open, closed, or neither?

**33.** $0 < x^2 + y^2 < 1$    **34.** $x \geq 0, \quad y < 0$

**35.** $x + y = 1$    **36.** $|x| + |y| \leq 1$

In Exercises 37–40, specify the boundary and the interior of the sets $S$ in 3-space whose points $(x, y, z)$ satisfy the given conditions. Is $S$ open, closed, or neither?

**37.** $1 \leq x^2 + y^2 + z^2 \leq 4$    **38.** $x \geq 0, \quad y > 1, \quad z < 2$

**39.** $(x - z)^2 + (y - z)^2 = 0$    **40.** $x^2 + y^2 < 1, \ y + z > 2$

---

## 10.2 Vectors

A **vector** is a quantity that involves both **magnitude** (size or length) and **direction**. Such quantities are represented geometrically by arrows (directed line segments) and are often actually identified with these arrows. For instance, the vector $\overrightarrow{AB}$ is an arrow with tail at the point $A$ and head at the point $B$. In print, such a vector is usually denoted by a single letter in boldface type,

**Figure 10.10**    The vector $\mathbf{v} = \overrightarrow{AB}$

$$\mathbf{v} = \overrightarrow{AB}.$$

(See Figure 10.10.) In handwriting, an arrow over a letter ($\vec{v} = \overrightarrow{AB}$) can be used to denote a vector. The *magnitude* of the vector $\mathbf{v}$ is the length of the arrow and is denoted $|\mathbf{v}|$ or $|\overrightarrow{AB}|$.

While vectors have magnitude and direction, they do not generally have *position*, that is, they are not regarded as being in a particular place. Two vectors, $\mathbf{u}$ and $\mathbf{v}$, are considered *equal* if they have *the same length and the same direction*, even

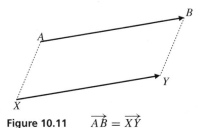

**Figure 10.11**   $\overrightarrow{AB} = \overrightarrow{XY}$

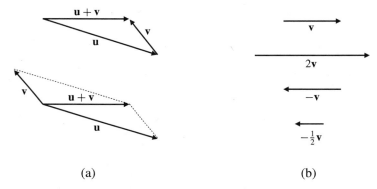

**Figure 10.12**   Components of a vector

if their representative arrows do not coincide. The arrows must be parallel, have the same length, and point in the same direction. In Figure 10.11, for example, if $ABYX$ is a parallelogram, then $\overrightarrow{AB} = \overrightarrow{XY}$.

For the moment, we consider plane vectors, that is, vectors whose representative arrows lie in a plane. If we introduce a Cartesian coordinate system into the plane, we can talk about the $x$ and $y$ components of any vector. If $A = (a, b)$ and $P = (p, q)$, as shown in Figure 10.12, then the $x$ and $y$ components of $\overrightarrow{AP}$ are, respectively, $p - a$ and $q - b$. Note that if $O$ is the origin and $X$ is the point $(p - a, q - b)$, then

$$|\overrightarrow{AP}| = \sqrt{(p-a)^2 + (q-b)^2} = |\overrightarrow{OX}|$$

$$\text{slope of } \overrightarrow{AP} = \frac{q-b}{p-a} = \text{slope of } \overrightarrow{OX}.$$

Hence $\overrightarrow{AP} = \overrightarrow{OX}$. In general, two vectors are equal if and only if they have the same $x$ components and $y$ components.

**Figure 10.13**

(a) Vector addition

(b) Scalar multiplication

There are two important algebraic operations defined for vectors: addition and scalar multiplication.

---

**DEFINITION 1**

**Vector addition**

Given two vectors $\mathbf{u}$ and $\mathbf{v}$, their **sum $\mathbf{u} + \mathbf{v}$** is defined as follows. If an arrow representing $\mathbf{v}$ is placed with its tail at the head of an arrow representing $\mathbf{u}$, then an arrow from the tail of $\mathbf{u}$ to the head of $\mathbf{v}$ represents $\mathbf{u} + \mathbf{v}$. Equivalently, if $\mathbf{u}$ and $\mathbf{v}$ have tails at the same point, then $\mathbf{u} + \mathbf{v}$ is represented by an arrow with its tail at that point and its head at the opposite vertex of the parallelogram spanned by $\mathbf{u}$ and $\mathbf{v}$. This is shown in Figure 10.13(a).

---

**DEFINITION 2**

**Scalar multiplication**

If $\mathbf{v}$ is a vector and $t$ is a real number (also called a **scalar**), then the **scalar multiple $t\mathbf{v}$** is a vector with magnitude $|t|$ times that of $\mathbf{v}$ and direction the same as $\mathbf{v}$ if $t > 0$, or opposite to that of $\mathbf{v}$ if $t < 0$. See Figure 10.13(b). If $t = 0$, then $t\mathbf{v}$ has zero length and therefore no particular direction. It is the **zero vector**, denoted $\mathbf{0}$.

Suppose that **u** has components $a$ and $b$ and that **v** has components $x$ and $y$. Then the components of $\mathbf{u} + \mathbf{v}$ are $a + x$ and $b + y$, and those of $t\mathbf{v}$ are $tx$ and $ty$. See Figure 10.14.

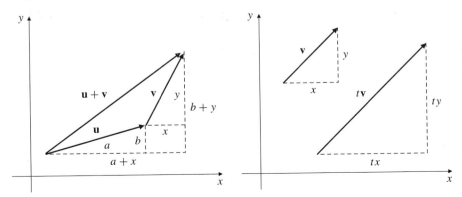

**Figure 10.14**     The components of a sum of vectors or a scalar multiple of a vector is the same sum or multiple of the corresponding components of the vectors

In $\mathbb{R}^2$ we single out two particular vectors for special attention. They are

(i)  the vector **i** from the origin to the point $(1, 0)$, and

(ii)  the vector **j** from the origin to the point $(0, 1)$.

Thus **i** has components 1 and 0, and **j** has components 0 and 1. These vectors are called the **standard basis vectors** in the plane. The vector **r** from the origin to the point $(x, y)$ has components $x$ and $y$ and can be expressed in the form

$$\mathbf{r} = x\mathbf{i} + y\mathbf{j}.$$

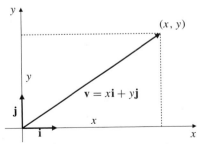

**Figure 10.15**     Any vector is a linear combination of the basis vectors

We say that we have written **r** as a **linear combination of the standard basis vectors**. (See Figure 10.15.) The vector **r** is called the **position vector** of the point $(x, y)$. A position vector has its tail at the origin and its head at the point whose position it is specifying. The length of **r** is $|\mathbf{r}| = \sqrt{x^2 + y^2}$.

More generally, the vector $\overrightarrow{AP}$ from $A = (a, b)$ to $P = (p, q)$ in Figure 10.12 can be written as a linear combination of the standard basis vectors:

$$\overrightarrow{AP} = (p - a)\mathbf{i} + (q - b)\mathbf{j}.$$

Sums and scalar multiples of vectors are easily expressed in terms of components. If $\mathbf{u} = u_1\mathbf{i} + u_2\mathbf{j}$ and $\mathbf{v} = v_1\mathbf{i} + v_2\mathbf{j}$, and if $t$ is a scalar (i.e. a real number) then

$$\mathbf{u} + \mathbf{v} = (u_1 + v_1)\mathbf{i} + (u_2 + v_2)\mathbf{j},$$
$$t\mathbf{u} = (tu_1)\mathbf{i} + (tu_2)\mathbf{j}.$$

The zero vector is $\mathbf{0} = 0\mathbf{i} + 0\mathbf{j}$. It has length zero and no specific direction. For any vector **u** we have $0\mathbf{u} = \mathbf{0}$. A **unit vector** is a vector of length 1. The standard basis vectors **i** and **j** are unit vectors. Given any nonzero vector **v**, we can form a unit vector $\hat{\mathbf{v}}$ in the same direction as **v** by multiplying **v** by the reciprocal of its length (a scalar):

$$\hat{\mathbf{v}} = \left(\frac{1}{|\mathbf{v}|}\right)\mathbf{v}.$$

**Example 1**   If $A = (2, -1)$, $B = (-1, 3)$, and $C = (0, 1)$, express each of the following vectors as a linear combination of the standard basis vectors:

(a) $\overrightarrow{AB}$    (b) $\overrightarrow{BC}$    (c) $\overrightarrow{AC}$    (d) $\overrightarrow{AB} + \overrightarrow{BC}$    (e) $2\overrightarrow{AC} - 3\overrightarrow{CB}$

(f) a unit vector in the direction of $\overrightarrow{AB}$.

**Solution**   We have

(a) $\overrightarrow{AB} = (-1 - 2)\mathbf{i} + (3 - (-1))\mathbf{j} = -3\mathbf{i} + 4\mathbf{j}$

(b) $\overrightarrow{BC} = (0 - (-1))\mathbf{i} + (1 - 3)\mathbf{j} = \mathbf{i} - 2\mathbf{j}$

(c) $\overrightarrow{AC} = (0 - 2)\mathbf{i} + (1 - (-1))\mathbf{j} = -2\mathbf{i} + 2\mathbf{j}$

(d) $\overrightarrow{AB} + \overrightarrow{BC} = \overrightarrow{AC} = -2\mathbf{i} + 2\mathbf{j}$

(e) $2\overrightarrow{AC} - 3\overrightarrow{CB} = 2(-2\mathbf{i} + 2\mathbf{j}) - 3(-\mathbf{i} + 2\mathbf{j}) = -\mathbf{i} - 2\mathbf{j}$

(f) A unit vector in the direction of $\overrightarrow{AB}$ is $\dfrac{\overrightarrow{AB}}{|\overrightarrow{AB}|} = -\dfrac{3}{5}\mathbf{i} + \dfrac{4}{5}\mathbf{j}$. ∎

Implicit in the above example is the fact that the operations of addition and scalar multiplication obey appropriate algebraic rules, such as

$$\mathbf{u} + \mathbf{v} = \mathbf{v} + \mathbf{u}$$
$$(\mathbf{u} + \mathbf{v}) + \mathbf{w} = \mathbf{u} + (\mathbf{v} + \mathbf{w})$$
$$\mathbf{u} - \mathbf{v} = \mathbf{u} + (-1)\mathbf{v}$$
$$t(\mathbf{u} + \mathbf{v}) = t\mathbf{u} + t\mathbf{v}$$

## Vectors in 3-Space

The algebra and geometry of vectors described there extends to spaces of any number of dimensions; we can still think of vectors as represented by arrows, and sums and scalar multiples are formed just as for plane vectors.

Given a Cartesian coordinate system in 3-space, we define three **standard basis vectors**, $\mathbf{i}$, $\mathbf{j}$, and $\mathbf{k}$, represented by arrows from the origin to the points $(1, 0, 0)$, $(0, 1, 0)$, and $(0, 0, 1)$, respectively. (See Figure 10.16.) Any vector in 3-space can be written as a *linear combination* of these basis vectors; for instance, the position vector of the point $(x, y, z)$ is given by

$$\mathbf{r} = x\mathbf{i} + y\mathbf{j} + z\mathbf{k}.$$

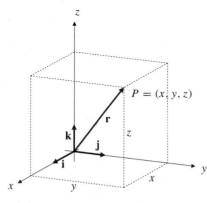

**Figure 10.16**   The standard basis vectors $\mathbf{i}$, $\mathbf{j}$, and $\mathbf{k}$

We say that $\mathbf{r}$ has **components** $x$, $y$, and $z$. The length of $\mathbf{r}$ is

$$|\mathbf{r}| = \sqrt{x^2 + y^2 + z^2}.$$

If $P_1 = (x_1, y_1, z_1)$ and $P_2 = (x_2, y_2, z_2)$ are two points in 3-space, then the vector $\mathbf{v} = \overrightarrow{P_1 P_2}$ from $P_1$ to $P_2$ has components $x_2 - x_1$, $y_2 - y_1$, and $z_2 - z_1$, and is therefore represented in terms of the standard basis vectors by

$$\mathbf{v} = \overrightarrow{P_1 P_2} = (x_2 - x_1)\mathbf{i} + (y_2 - y_1)\mathbf{j} + (z_2 - z_1)\mathbf{k}.$$

**Example 2**   If $\mathbf{u} = 2\mathbf{i} + \mathbf{j} - 2\mathbf{k}$ and $\mathbf{v} = 3\mathbf{i} - 2\mathbf{j} - \mathbf{k}$, find $\mathbf{u} + \mathbf{v}$, $\mathbf{u} - \mathbf{v}$, $3\mathbf{u} - 2\mathbf{v}$, $|\mathbf{u}|$, $|\mathbf{v}|$, and a unit vector $\hat{\mathbf{u}}$ in the direction of $\mathbf{u}$.

**Solution**   We have

$$\mathbf{u} + \mathbf{v} = (2 + 3)\mathbf{i} + (1 - 2)\mathbf{j} + (-2 - 1)\mathbf{k} = 5\mathbf{i} - \mathbf{j} - 3\mathbf{k}$$
$$\mathbf{u} - \mathbf{v} = (2 - 3)\mathbf{i} + (1 + 2)\mathbf{j} + (-2 + 1)\mathbf{k} = -\mathbf{i} + 3\mathbf{j} - \mathbf{k}$$
$$3\mathbf{u} - 2\mathbf{v} = (6 - 6)\mathbf{i} + (3 + 4)\mathbf{j} + (-6 + 2)\mathbf{k} = 7\mathbf{j} - 4\mathbf{k}$$
$$|\mathbf{u}| = \sqrt{4 + 1 + 4} = 3, \qquad |\mathbf{v}| = \sqrt{9 + 4 + 1} = \sqrt{14}$$
$$\hat{\mathbf{u}} = \left(\frac{1}{|\mathbf{u}|}\right)\mathbf{u} = \frac{2}{3}\mathbf{i} + \frac{1}{3}\mathbf{j} - \frac{2}{3}\mathbf{k}.$$

The following example illustrates the way vectors can be used to solve problems involving relative velocities. If $A$ moves with velocity $\mathbf{v}_{A\,\mathrm{rel}\,B}$ relative to $B$, and $B$ moves with velocity $\mathbf{v}_{B\,\mathrm{rel}\,C}$ relative to $C$, then $A$ moves with velocity $\mathbf{v}_{A\,\mathrm{rel}\,C}$ relative to $C$, where

$$\mathbf{v}_{A\,\mathrm{rel}\,C} = \mathbf{v}_{A\,\mathrm{rel}\,B} + \mathbf{v}_{B\,\mathrm{rel}\,C}.$$

**Example 3**   An aircraft cruises at a speed of 300 km/h in still air. If the wind is blowing from the east at 100 km/h, in what direction should the aircraft head in order to fly in a straight line from city $P$ to city $Q$, 400 km north northeast of $P$? How long will the trip take?

**Solution**   The problem is two-dimensional, so we use plane vectors. Let us choose our coordinate system so that the $x$- and $y$-axes point east and north, respectively. Figure 10.17 illustrates the three velocities that must be considered. The velocity of the air relative to the ground is

$$\mathbf{v}_{\mathrm{air\ rel\ ground}} = -100\,\mathbf{i}.$$

If the aircraft heads in a direction making angle $\theta$ with the positive direction of the $x$-axis then the velocity of the aircraft relative to the air is

$$\mathbf{v}_{\mathrm{aircraft\ rel\ air}} = 300\cos\theta\,\mathbf{i} + 300\sin\theta\,\mathbf{j}.$$

Thus the velocity of the aircraft relative to the ground is

$$\mathbf{v}_{\mathrm{aircraft\ rel\ ground}} = \mathbf{v}_{\mathrm{aircraft\ rel\ air}} + \mathbf{v}_{\mathrm{air\ rel\ ground}}$$
$$= (300\cos\theta - 100)\,\mathbf{i} + 300\sin\theta\,\mathbf{j}.$$

We want this latter velocity to be in a north-northeasterly direction, that is, in the direction making angle $3\pi/8 = 67.5°$ with the positive direction of the $x$-axis. Thus we will have

$$\mathbf{v}_{\mathrm{aircraft\ rel\ ground}} = v\,[(\cos 67.5°)\,\mathbf{i} + (\sin 67.5°)\,\mathbf{j}],$$

where $v$ is the actual groundspeed of the aircraft. Comparing the two expressions for $\mathbf{v}_{\mathrm{aircraft\ rel\ ground}}$ we obtain

**Figure 10.17**   Velocity diagram for the aircraft in Example 3

$300(\cos\theta\mathbf{i} + \sin\theta\mathbf{j})$

$67.5°$

$\theta$

$-100\mathbf{i}$

$$300 \cos\theta - 100 = v \cos 67.5°$$
$$300 \sin\theta = v \sin 67.5°.$$

Eliminating $v$ between these two equations we get

$$300 \cos\theta \sin 67.5° - 300 \sin\theta \cos 67.5° = 100 \sin 67.5°,$$

or,

$$3 \sin(67.5° - \theta) = \sin 67.5°.$$

Therefore the aircraft should head in direction $\theta$ given by

$$\theta = 67.5° - \text{Arcsin}\left(\frac{1}{3} \sin 67.5°\right) \approx 49.56°,$$

that is, 49.56° north of east. The groundspeed is now seen to be

$$v = 300 \sin\theta / \sin 67.5° \approx 247.15 \text{ km/h}.$$

Thus the 400 km trip will take about $400/247.15 \approx 1.616$ hours, or, about 1 hour and 37 minutes.  ∎

## Hanging Cables and Chains

What shape does a heavy cable or chain assume when it is suspended from both ends and allowed to hang under gravity? In Section 3.6 we claimed that the shape would be a **catenary** curve having the shape of the graph of the hyperbolic cosine function. We will demonstrate this now, using vectors to keep track of the various forces acting on the cable.

Suppose that the cable has line density $\delta$ (units of mass per unit length), and hangs as shown in Figure 10.18. Let us choose a coordinate system so that the lowest point $L$ on the cable is at $(0, y_0)$; we will specify the value of $y_0$ later. If $P = (x, y)$ is another point on the cable, there are three forces acting on the arc $LP$ of the cable between $L$ and $P$. These are all forces that we can represent using horizontal and vertical components.

(i) The horizontal tension $\mathbf{H} = -H\mathbf{i}$ at $L$. This is the force that the part of the cable to the left of $L$ exerts on the arc $LP$ at $L$.

(ii) The tangential tension $\mathbf{T} = T_h\mathbf{i} + T_v\mathbf{j}$. This is the force the part of the cable to the right of $P$ exerts on arc $LP$ at $P$.

(iii) The weight $\mathbf{W} = -\delta g s\mathbf{j}$ of arc $LP$, where $g$ is the acceleration of gravity and $s$ is the length of the arc $LP$.

Since the cable is not moving, these three forces must balance; their vector sum must be zero.

$$\mathbf{T} + \mathbf{H} + \mathbf{W} = \mathbf{0}$$
$$(T_h - H)\mathbf{i} + (T_v - \delta g s)\mathbf{j} = \mathbf{0}$$

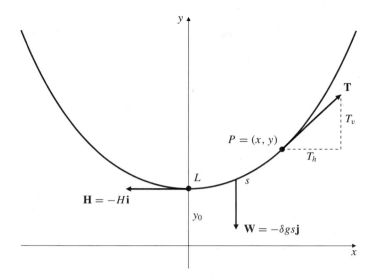

**Figure 10.18**    A hanging cable and the forces acting on arc $LP$

Thus $T_h = H$ and $T_v = \delta gs$. Since $\mathbf{T}$ is tangent to the cable at $P$, the slope of the cable there is

$$\frac{dy}{dx} = \frac{T_v}{T_h} = \frac{\delta gs}{H} = as,$$

where $a = \delta g/H$ is a constant for the given cable. Differentiating with respect to $x$, and using the fact, from our study of arc length, that

$$\frac{ds}{dx} = \sqrt{1 + \left(\frac{dy}{dx}\right)^2},$$

we obtain a second-order differential equation

$$\frac{d^2y}{dx^2} = a\frac{ds}{dx} = a\sqrt{1 + \left(\frac{dy}{dx}\right)^2}$$

to be solved for the equation of the curve along which the hanging cable lies. The appropriate initial conditions are $y = y_0$ and $dy/dx = 0$ at $x = 0$.

Since the differential equation depends on $dy/dx$, rather than $y$, we substitute $m(x) = dy/dx$ and obtain a first-order equation for $m$:

$$\frac{dm}{dx} = a\sqrt{1 + m^2}.$$

This equation is separable; we integrate it using the substitution $m = \sinh u$:

$$\int \frac{1}{\sqrt{1 + m^2}}\,dm = \int a\,dx$$

$$\int du = \int \frac{\cosh u}{\sqrt{1 + \sinh^2 u}}\,du = ax + C_1$$

$$\sinh^{-1} m = u = ax + C_1$$

$$m = \sinh(ax + C_1).$$

Since $m = dy/dx = 0$ at $x = 0$, we have $0 = \sinh C_1$, so $C_1 = 0$, and

$$\frac{dy}{dx} = m = \sinh(ax).$$

This equation is easily integrated to find $y$. (Had we used a tangent substitution instead of the hyperbolic sine substitution for $m$ we would have had more trouble here.)

$$y = \frac{1}{a}\cosh(ax) + C_2.$$

If we choose $y_0 = y(0) = 1/a$, then, substituting $x = 0$ we will get $C_2 = 0$. With this choice of $y_0$, we therefore find that the equation of the curve along which the hanging cable lies is the catenary

$$y = \frac{1}{a}\cosh(ax).$$

**Remark**   If a hanging cable bears loads other than its own weight, it will assume a different shape. For example, a cable supporting a level suspension bridge whose weight per unit length is much greater than that of the cable will assume the shape of a parabola. See Exercise 34 below.

## The Dot Product and Projections

There is another operation on vectors in any dimension by which two vectors are combined to produce a number called their dot product.

**DEFINITION 3**

> **The dot product of two vectors**
> Given two vectors, $\mathbf{u} = u_1\mathbf{i} + u_2\mathbf{j}$ and $\mathbf{v} = v_1\mathbf{i} + v_2\mathbf{j}$ in $\mathbb{R}^2$, we define their **dot product $\mathbf{u} \bullet \mathbf{v}$** to be the sum of the products of their corresponding components:
>
> $$\mathbf{u} \bullet \mathbf{v} = u_1v_1 + u_2v_2.$$
>
> Similarly, for vectors $\mathbf{u} = u_1\mathbf{i} + u_2\mathbf{j} + u_3\mathbf{k}$ and $\mathbf{v} = v_1\mathbf{i} + v_2\mathbf{j} + v_3\mathbf{k}$ in $\mathbb{R}^3$,
>
> $$\mathbf{u} \bullet \mathbf{v} = u_1v_1 + u_2v_2 + u_3v_3.$$

The dot product has the following algebraic properties, easily checked using the definition above:

$$\mathbf{u} \bullet \mathbf{v} = \mathbf{v} \bullet \mathbf{u} \qquad \text{(commutative law)}$$
$$\mathbf{u} \bullet (\mathbf{v} + \mathbf{w}) = \mathbf{u} \bullet \mathbf{v} + \mathbf{u} \bullet \mathbf{w} \qquad \text{(distributive law)}$$
$$(t\mathbf{u}) \bullet \mathbf{v} = \mathbf{u} \bullet (t\mathbf{v}) = t(\mathbf{u} \bullet \mathbf{v}) \qquad \text{(for real } t)$$
$$\mathbf{u} \bullet \mathbf{u} = |\mathbf{u}|^2.$$

where $\theta$ is the angle between $\mathbf{u}$ and $\mathbf{v}$. The real significance of the dot product is shown by the following result, which could have been used as the definition of dot product:

**THEOREM  1**   If $\theta$ is the angle between the directions of $\mathbf{u}$ and $\mathbf{v}$ ($0 \le \theta \le \pi$), then

$$\mathbf{u} \bullet \mathbf{v} = |\mathbf{u}||\mathbf{v}| \cos\theta.$$

In particular, $\mathbf{u} \bullet \mathbf{v} = 0$ if and only if $\mathbf{u}$ and $\mathbf{v}$ are perpendicular. (Of course the zero vector is perpendicular to every vector.)

**PROOF**   Refer to Figure 10.19 and apply the Cosine Law (Theorem 3 of Section P.6) to the triangle with the arrows $\mathbf{u}$, $\mathbf{v}$, and $\mathbf{u} - \mathbf{v}$ as sides.

$$
\begin{aligned}
|\mathbf{u}|^2 + |\mathbf{v}|^2 - 2|\mathbf{u}|\,|\mathbf{v}| \cos\theta &= |\mathbf{u} - \mathbf{v}|^2 = (\mathbf{u} - \mathbf{v}) \bullet (\mathbf{u} - \mathbf{v}) \\
&= \mathbf{u} \bullet (\mathbf{u} - \mathbf{v}) - \mathbf{v} \bullet (\mathbf{u} - \mathbf{v}) \\
&= \mathbf{u} \bullet \mathbf{u} - \mathbf{u} \bullet \mathbf{v} - \mathbf{v} \bullet \mathbf{u} + \mathbf{v} \bullet \mathbf{v} \\
&= |\mathbf{u}|^2 + |\mathbf{v}|^2 - 2\mathbf{u} \bullet \mathbf{v}
\end{aligned}
$$

Hence $|\mathbf{u}||\mathbf{v}| \cos\theta = \mathbf{u} \bullet \mathbf{v}$, as claimed.

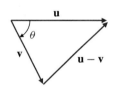

**Figure 10.19**

**Example 4**   Find the angle $\theta$ between the vectors $\mathbf{u} = 2\mathbf{i} + \mathbf{j} - 2\mathbf{k}$ and $\mathbf{v} = 3\mathbf{i} - 2\mathbf{j} - \mathbf{k}$.

**Solution**   Solving the formula $\mathbf{u} \bullet \mathbf{v} = |\mathbf{u}||\mathbf{v}| \cos\theta$ for $\theta$, we obtain

$$
\begin{aligned}
\theta &= \cos^{-1} \frac{\mathbf{u} \bullet \mathbf{v}}{|\mathbf{u}||\mathbf{v}|} = \cos^{-1}\left( \frac{(2)(3) + (1)(-2) + (-2)(-1)}{3\sqrt{14}} \right) \\
&= \cos^{-1} \frac{2}{\sqrt{14}} \approx 57.69°.
\end{aligned}
$$

It is sometimes useful to project one vector along another. We define both scalar and vector projections of $\mathbf{u}$ in the direction of $\mathbf{v}$:

**DEFINITION  4**

**Scalar and vector projections**

The **scalar projection** $s$ of any vector $\mathbf{u}$ in the direction of a nonzero vector $\mathbf{v}$ is the dot product of $\mathbf{u}$ with a unit vector in the direction of $\mathbf{v}$. Thus it is the *number*

$$s = \frac{\mathbf{u} \bullet \mathbf{v}}{|\mathbf{v}|} = |\mathbf{u}| \cos\theta$$

where $\theta$ is the angle between $\mathbf{u}$ and $\mathbf{v}$.

The **vector projection**, $\mathbf{u_v}$, of $\mathbf{u}$ in the direction of $\mathbf{v}$ (see Figure 10.20) is the scalar multiple of a unit vector $\hat{\mathbf{v}}$ in the direction of $\mathbf{v}$, by the scalar projection of $\mathbf{u}$ in the direction of $\mathbf{v}$, that is,

$$\text{vector projection of } \mathbf{u} \text{ along } \mathbf{v} = \mathbf{u_v} = \frac{\mathbf{u} \bullet \mathbf{v}}{|\mathbf{v}|} \hat{\mathbf{v}} = \frac{\mathbf{u} \bullet \mathbf{v}}{|\mathbf{v}|^2} \mathbf{v}.$$

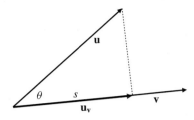

**Figure 10.20**    The scalar projection $s$ and the vector projection $\mathbf{u_v}$ of vector $\mathbf{u}$ along vector $\mathbf{v}$

Note that $|s|$ is the length of the line segment along the line of $\mathbf{v}$ obtained by dropping perpendiculars to that line from the tail and head of $\mathbf{u}$. (See Figure 10.20.) Also, $s$ is negative if $\theta > 90°$.

It is often necessary to express a vector as a sum of two other vectors parallel and perpendicular to a given direction.

**Example 5**    Express the vector $3\mathbf{i} + \mathbf{j}$ as a sum of vectors $\mathbf{u} + \mathbf{v}$, where $\mathbf{u}$ is parallel to the vector $\mathbf{i} + \mathbf{j}$ and $\mathbf{v}$ is perpendicular to $\mathbf{u}$.

**Solution**

METHOD I (Using vector projection): Note that $\mathbf{u}$ must be the vector projection of $3\mathbf{i} + \mathbf{j}$ in the direction of $\mathbf{i} + \mathbf{j}$. Thus

$$\mathbf{u} = \frac{(3\mathbf{i} + \mathbf{j}) \bullet (\mathbf{i} + \mathbf{j})}{|\mathbf{i} + \mathbf{j}|^2}(\mathbf{i} + \mathbf{j}) = \frac{4}{2}(\mathbf{i} + \mathbf{j}) = 2\mathbf{i} + 2\mathbf{j}$$

$$\mathbf{v} = 3\mathbf{i} + \mathbf{j} - \mathbf{u} = \mathbf{i} - \mathbf{j}.$$

METHOD II (From basic principles): Since $\mathbf{u}$ is parallel to $\mathbf{i} + \mathbf{j}$ and $\mathbf{v}$ is perpendicular to $\mathbf{u}$, we have

$$\mathbf{u} = t(\mathbf{i} + \mathbf{j}) \qquad \text{and} \qquad \mathbf{v} \bullet (\mathbf{i} + \mathbf{j}) = 0,$$

for some scalar $t$. We want $\mathbf{u} + \mathbf{v} = 3\mathbf{i} + \mathbf{j}$. Take the dot product of this equation with $\mathbf{i} + \mathbf{j}$:

$$\mathbf{u} \bullet (\mathbf{i} + \mathbf{j}) + \mathbf{v} \bullet (\mathbf{i} + \mathbf{j}) = (3\mathbf{i} + \mathbf{j}) \bullet (\mathbf{i} + \mathbf{j})$$
$$t(\mathbf{i} + \mathbf{j}) \bullet (\mathbf{i} + \mathbf{j}) + 0 = 4.$$

Thus $2t = 4$, and so $t = 2$. Therefore,

$$\mathbf{u} = 2\mathbf{i} + 2\mathbf{j} \qquad \text{and} \qquad \mathbf{v} = 3\mathbf{i} + \mathbf{j} - \mathbf{u} = \mathbf{i} - \mathbf{j}.$$

## Vectors in *n*-Space

All the above ideas make sense for vectors in spaces of any dimension. Vectors in $\mathbb{R}^n$ can expressed as linear combinations of the $n$ unit vectors

$\mathbf{e_1}$    from the origin to the point    $(1, 0, 0, \ldots, 0)$

$\mathbf{e_2}$    from the origin to the point    $(0, 1, 0, \ldots, 0)$

$\vdots$

$\mathbf{e_n}$    from the origin to the point    $(0, 0, 0, \ldots, 1)$.

These vectors constitute a *standard basis* in $\mathbb{R}^n$. The $n$-vector $\mathbf{x}$ with components $x_1, x_2, \ldots, x_n$ is expressed in the form

$$\mathbf{x} = x_1\mathbf{e_1} + x_2\mathbf{e_2} + \cdots + x_n\mathbf{e_n}.$$

The length of $\mathbf{x}$ is $|\mathbf{x}| = \sqrt{x_1{}^2 + x_2{}^2 + \cdots + x_n{}^2}$. The angle between two vectors $\mathbf{x}$ and $\mathbf{y}$ is defined to be

$$\theta = \cos^{-1} \frac{\mathbf{x} \bullet \mathbf{y}}{|\mathbf{x}||\mathbf{y}|},$$

where

$$\mathbf{x} \bullet \mathbf{y} = x_1 y_1 + x_2 y_2 + \cdots + x_n y_n.$$

We will not make much use of $n$-vectors for $n > 3$ but you should be aware that everything said up until now for 2-vectors or 3-vectors extends to $n$-vectors.

## Exercises 10.2

**1.** Let $A = (-1, 2)$, $B = (2, 0)$, $C = (1, -3)$, $D = (0, 4)$.
Express each of the following vectors as a linear combination of the standard basis vectors $\mathbf{i}$ and $\mathbf{j}$ in $\mathbb{R}^2$.

(a) $\overrightarrow{AB}$,   (b) $\overrightarrow{BA}$,   (c) $\overrightarrow{AC}$,   (d) $\overrightarrow{BD}$,   (e) $\overrightarrow{DA}$,

(f) $\overrightarrow{AB} - \overrightarrow{BC}$,   (g) $\overrightarrow{AC} - 2\overrightarrow{AB} + 3\overrightarrow{CD}$,

(h) $\dfrac{\overrightarrow{AB} + \overrightarrow{AC} + \overrightarrow{AD}}{3}$.

In Exercises 2–5, calculate the following for the given vectors $\mathbf{u}$ and $\mathbf{v}$:

(a) $\mathbf{u} + \mathbf{v}$,    $\mathbf{u} - \mathbf{v}$,    $2\mathbf{u} - 3\mathbf{v}$,

(b) the lengths $|\mathbf{u}|$ and $|\mathbf{v}|$,

(c) unit vectors $\hat{\mathbf{u}}$ and $\hat{\mathbf{v}}$ in the directions of $\mathbf{u}$ and $\mathbf{v}$, respectively,

(d) the dot product $\mathbf{u} \bullet \mathbf{v}$,

(e) the angle between $\mathbf{u}$ and $\mathbf{v}$,

(f) the scalar projection of $\mathbf{u}$ in the direction of $\mathbf{v}$,

g) the vector projection of $\mathbf{v}$ along $\mathbf{u}$.

**2.** $\mathbf{u} = \mathbf{i} - \mathbf{j}$ and $\mathbf{v} = \mathbf{j} + 2\mathbf{k}$

**3.** $\mathbf{u} = 3\mathbf{i} + 4\mathbf{j} - 5\mathbf{k}$ and $\mathbf{v} = 3\mathbf{i} - 4\mathbf{j} - 5\mathbf{k}$

**4.** Use vectors to show that the triangle with vertices $(-1, 1)$, $(2, 5)$, and $(10, -1)$ is right-angled.

In Exercises 5–8, prove the stated geometric result using vectors.

**5.** The line segment joining the midpoints of two sides of a triangle is parallel to and half as long as the third side.

**6.** If $P$, $Q$, $R$, and $S$ are midpoints of sides $AB$, $BC$, $CD$, and $DA$, respectively, of quadrilateral $ABCD$, then $PQRS$ is a parallelogram.

* **7.** The diagonals of any parallelogram bisect each other.

* **8.** The medians of any triangle meet in a common point. (A median is a line joining one vertex to the midpoint of the opposite side. The common point is the *centroid* of the triangle.)

**9.** A weather vane mounted on the top of a car moving due north at 50 km/h indicates that the wind is coming from the west. When the car doubles its speed, the weather vane indicates that the wind is coming from the northwest. From what direction is the wind coming, and what is its speed?

**10.** A straight river 500 m wide flows due east at a constant speed of 3 km/h. If you can row your boat at a speed of 5 km/h in still water, in what direction should you head if you wish to row from point $A$ on the south shore to point $B$ on the north shore directly north of $A$? How long will the trip take?

* **11.** In what direction should you head to cross the river in Exercise 10 if you can only row at 2 km/h, and you wish to row from $A$ to point $C$ on the north shore, $k$ km downstream from $B$. For what values of $k$ is the trip not possible?

**12.** A certain aircraft flies with an airspeed of 750 km/h. In what direction should it head in order to make progress in a true easterly direction if the wind is from the northeast at 100 km/h? How long will it take to complete a trip to a city 1500 km from its starting point?

**13.** For what value of $t$ is the vector $2t\mathbf{i} + 4\mathbf{j} - (10 + t)\mathbf{k}$ perpendicular to the vector $\mathbf{i} + t\mathbf{j} + \mathbf{k}$?

**14.** Find the angle between a diagonal of a cube and one of the edges of the cube.

**15.** Find the angle between a diagonal of a cube and a diagonal of one of the faces of the cube. Give all possible answers.

**16.** **(Direction cosines)** If a vector $\mathbf{u}$ in $\mathbb{R}^3$ makes angles $\alpha$, $\beta$, and $\gamma$ with the coordinate axes, show that

$$\hat{\mathbf{u}} = \cos \alpha \mathbf{i} + \cos \beta \mathbf{j} + \cos \gamma \mathbf{k}$$

is a unit vector in the direction of $\mathbf{u}$, so $\cos^2 \alpha + \cos^2 \beta + \cos^2 \gamma = 1$. The numbers $\cos \alpha$, $\cos \beta$, and $\cos \gamma$ are called the direction cosines of $\mathbf{u}$.

**17.** Find a unit vector that makes equal angles with the three coordinate axes.

**18.** Find the three angles of the triangle with vertices $(1, 0, 0)$, $(0, 2, 0)$, and $(0, 0, 3)$.

**19.** If $\mathbf{r}_1$ and $\mathbf{r}_2$ are the position vectors of two points, $P_1$ and $P_2$, and $\lambda$ is a real number, show that

$$\mathbf{r} = (1 - \lambda)\mathbf{r}_1 + \lambda\mathbf{r}_2$$

is the position vector of a point $P$ on the straight line joining $P_1$ and $P_2$. Where is $P$ if $\lambda = 1/2$? if $\lambda = 2/3$? if $\lambda = -1$? if $\lambda = 2$?

**20.** Let $\mathbf{a}$ be a nonzero vector. Descibe the set of all points in 3-space whose position vectors $\mathbf{r}$ satisfy $\mathbf{a} \bullet \mathbf{r} = 0$.

**21.** Let $\mathbf{a}$ be a nonzero vector and let $b$ be any real number. Descibe the set of all points in 3-space whose position vectors $\mathbf{r}$ satisfy $\mathbf{a} \bullet \mathbf{r} = b$.

In Exercises 22–24, $\mathbf{u} = 2\mathbf{i} + \mathbf{j} - 2\mathbf{k}$, $\mathbf{v} = \mathbf{i} + 2\mathbf{j} - 2\mathbf{k}$, and $\mathbf{w} = 2\mathbf{i} - 2\mathbf{j} + \mathbf{k}$.

**22.** Find two unit vectors each of which is perpendicular to $\mathbf{u}$ and $\mathbf{v}$.

**23.** Find a vector $\mathbf{x}$ satisfying the system of equations $\mathbf{x} \bullet \mathbf{u} = 9$, $\mathbf{x} \bullet \mathbf{v} = 4$, $\mathbf{x} \bullet \mathbf{w} = 6$.

**24.** Find two unit vectors each of which makes equal angles with $\mathbf{u}$, $\mathbf{v}$, and $\mathbf{w}$.

**25.** Find a unit vector that bisects the angle between any two nonzero vectors $\mathbf{u}$ and $\mathbf{v}$.

**26.** Given two nonparallel vectors $\mathbf{u}$ and $\mathbf{v}$, describe the set of all points whose position vectors $\mathbf{r}$ are of the form $\mathbf{r} = \lambda\mathbf{u} + \mu\mathbf{v}$, where $\lambda$ and $\mu$ are arbitrary real numbers.

**27. (The triangle inequality)** Let $\mathbf{u}$ and $\mathbf{v}$ be two vectors.

(a) Show that $|\mathbf{u} + \mathbf{v}|^2 = |\mathbf{u}|^2 + 2\mathbf{u} \bullet \mathbf{v} + |\mathbf{v}|^2$.

(b) Show that $\mathbf{u} \bullet \mathbf{v} \leq |\mathbf{u}||\mathbf{v}|$.

(c) Deduce from (a) and (b) that $|\mathbf{u} + \mathbf{v}| \leq |\mathbf{u}| + |\mathbf{v}|$.

**28.** (a) Why is the inequality in Exercise 27(c) called a triangle inequality?

(b) What conditions on $\mathbf{u}$ and $\mathbf{v}$ imply that $|\mathbf{u} + \mathbf{v}| = |\mathbf{u}| + |\mathbf{v}|$?

**29. (Orthonormal bases)** Let $\mathbf{u} = \frac{3}{5}\mathbf{i} + \frac{4}{5}\mathbf{j}$, $\mathbf{v} = \frac{4}{5}\mathbf{i} - \frac{3}{5}\mathbf{j}$, and $\mathbf{w} = \mathbf{k}$.

(a) Show that $|\mathbf{u}| = |\mathbf{v}| = |\mathbf{w}| = 1$ and $\mathbf{u} \bullet \mathbf{v} = \mathbf{u} \bullet \mathbf{w} = \mathbf{v} \bullet \mathbf{w} = 0$. The vectors $\mathbf{u}$, $\mathbf{v}$, and $\mathbf{w}$ are mutually perpendicular unit vectors, and as such are said to constitute an **orthonormal basis** for $\mathbb{R}^3$.

(b) If $\mathbf{r} = x\mathbf{i} + y\mathbf{j} + z\mathbf{k}$, show by direct calculation that

$$\mathbf{r} = (\mathbf{r} \bullet \mathbf{u})\mathbf{u} + (\mathbf{r} \bullet \mathbf{v})\mathbf{v} + (\mathbf{r} \bullet \mathbf{w})\mathbf{w}.$$

**30.** Show that if $\mathbf{u}$, $\mathbf{v}$, and $\mathbf{w}$ are any three mutually perpendicular unit vectors in $\mathbb{R}^3$ and $\mathbf{r} = a\mathbf{u} + b\mathbf{v} + c\mathbf{w}$, then $a = \mathbf{r} \bullet \mathbf{u}$, $b = \mathbf{r} \bullet \mathbf{v}$, and $c = \mathbf{r} \bullet \mathbf{w}$.

**31. (Resolving a vector in perpendicular directions)** If $\mathbf{a}$ is a nonzero vector and $\mathbf{w}$ is any vector, find vectors $\mathbf{u}$ and $\mathbf{v}$ such that $\mathbf{w} = \mathbf{u} + \mathbf{v}$, $\mathbf{u}$ is parallel to $\mathbf{a}$, and $\mathbf{v}$ is perpendicular to $\mathbf{a}$.

**32. (Expressing a vector as a linear combination of two other vectors with which it is coplanar)** Suppose that $\mathbf{u}$, $\mathbf{v}$, and $\mathbf{r}$ are position vectors of points $U$, $V$, and $P$ respectively, that $\mathbf{u}$ is not parallel to $\mathbf{v}$, and that $P$ lies in the plane containing the origin, $U$ and $V$. Show that there exist numbers $\lambda$ and $\mu$ such that $\mathbf{r} = \lambda\mathbf{u} + \mu\mathbf{v}$. *Hint:* resolve both $\mathbf{v}$ and $\mathbf{r}$ as sums of vectors parallel and perpendicular to $\mathbf{u}$ as suggested in Exercise 31.

**\* 33.** Given constants $r$, $s$ and $t$, with $r \neq 0$ and $s \neq 0$, and given a vector $\mathbf{a}$ satisfying $|\mathbf{a}|^2 > 4rst$, solve the system of equations

$$\begin{cases} r\mathbf{x} + s\mathbf{y} = \mathbf{a} \\ \mathbf{x} \bullet \mathbf{y} = t \end{cases}$$

for the unknown vectors $\mathbf{x}$ and $\mathbf{y}$.

**Hanging cables**

**34. (A suspension bridge)** If a hanging cable is supporting weight with constant horizontal line density (so that the weight supported by the arc $LP$ in Figure 10.18 is $\delta gx$ rather than $\delta gs$, show that the cable assumes the shape of a parabola rather than a catenary. Such is likely to be the case for the cables of a suspension bridge.

**35.** At a point $P$, 10 m away horizontally from its lowest point $L$, a cable makes an angle $55°$ with the horizontal. Find the length of the cable between $L$ and $P$.

**36.** Calculate the length $s$ of the arc $LP$ of the hanging cable in Figure 10.18 using the equation $y = (1/a) \cosh(ax)$ obtained for the cable. Hence verify that the magnitude $T = |\mathbf{T}|$ of the tension in the cable at any point $P = (x, y)$ is $T = \delta gy$.

**37.** A cable 100 m long hangs between two towers 90 m apart so that its ends are attached at the same height on the two towers. How far below that height is the lowest point on the cable?

# 10.3 The Cross Product in 3-Space

There is defined, *in 3-space only*, another kind of product of two vectors called a cross product or vector product, and denoted $\mathbf{u} \times \mathbf{v}$.

**DEFINITION 5**

For any vectors $\mathbf{u}$ and $\mathbf{v}$ in $\mathbb{R}^3$, the **cross product** $\mathbf{u} \times \mathbf{v}$ is the unique vector satisfying the three conditions:

(i) $(\mathbf{u} \times \mathbf{v}) \bullet \mathbf{u} = 0$ and $(\mathbf{u} \times \mathbf{v}) \bullet \mathbf{v} = 0$,

(ii) $|\mathbf{u} \times \mathbf{v}| = |\mathbf{u}||\mathbf{v}| \sin \theta$, where $\theta$ is the angle between $\mathbf{u}$ and $\mathbf{v}$, and

(iii) $\mathbf{u}$, $\mathbf{v}$, and $\mathbf{u} \times \mathbf{v}$ form a right-handed triad.

If $\mathbf{u}$ and $\mathbf{v}$ are parallel, condition (ii) says that $\mathbf{u} \times \mathbf{v} = \mathbf{0}$, the zero vector. Otherwise, through any point in $\mathbb{R}^3$ there is a unique straight line that is perpendicular to both $\mathbf{u}$ and $\mathbf{v}$. Condition (i) says that $\mathbf{u} \times \mathbf{v}$ is parallel to this line. Condition (iii) determines which of the two directions along this line is the direction of $\mathbf{u} \times \mathbf{v}$; a right-handed screw advances in the direction of $\mathbf{u} \times \mathbf{v}$ if rotated in the direction from $\mathbf{u}$ towards $\mathbf{v}$. (This is equivalent to saying that the thumb, forefinger, and middle finger of the right hand can be made to point in the directions of $\mathbf{u}$, $\mathbf{v}$, and $\mathbf{u} \times \mathbf{v}$, respectively.)

If $\mathbf{u}$ and $\mathbf{v}$ have their tails at the point $P$, then $\mathbf{u} \times \mathbf{v}$ is normal (i.e., perpendicular) to the plane through $P$ in which $\mathbf{u}$ and $\mathbf{v}$ lie, and, by condition (ii), $\mathbf{u} \times \mathbf{v}$ has length equal to the area of the parallelogram spanned by $\mathbf{u}$ and $\mathbf{v}$. (See Figure 10.21.) These properties make the cross product very useful for the description of tangent planes and normal lines to surfaces in $\mathbb{R}^3$.

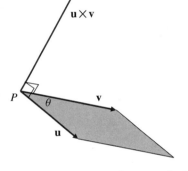

**Figure 10.21** $\mathbf{u} \times \mathbf{v}$ is perpendicular to both $\mathbf{u}$ and $\mathbf{v}$ and has length equal to the area of the shaded parallelogram

The definition of cross product given above does not involve any coordinate system and therefore does not directly show the components of the cross product with respect to the standard basis. These components are provided by the following theorem:

**THEOREM 2**

**Components of the cross product**

If $\mathbf{u} = u_1 \mathbf{i} + u_2 \mathbf{j} + u_3 \mathbf{k}$ and $\mathbf{v} = v_1 \mathbf{i} + v_2 \mathbf{j} + v_3 \mathbf{k}$ then

$$\mathbf{u} \times \mathbf{v} = (u_2 v_3 - u_3 v_2)\mathbf{i} + (u_3 v_1 - u_1 v_3)\mathbf{j} + (u_1 v_2 - u_2 v_1)\mathbf{k}.$$

***PROOF*** First we observe that the vector

$$\mathbf{w} = (u_2 v_3 - u_3 v_2)\mathbf{i} + (u_3 v_1 - u_1 v_3)\mathbf{j} + (u_1 v_2 - u_2 v_1)\mathbf{k}$$

is perpendicular to both $\mathbf{u}$ and $\mathbf{v}$ since

$$\mathbf{u} \bullet \mathbf{w} = u_1(u_2 v_3 - u_3 v_2) + u_2(u_3 v_1 - u_1 v_3) + u_3(u_1 v_2 - u_2 v_1) = 0,$$

and similarly $\mathbf{v} \bullet \mathbf{w} = 0$. Thus $\mathbf{u} \times \mathbf{v}$ is parallel to $\mathbf{w}$. Next we show that $\mathbf{w}$ and $\mathbf{u} \times \mathbf{v}$ have the same length. In fact

$$
\begin{aligned}
|\mathbf{w}|^2 &= (u_2 v_3 - u_3 v_2)^2 + (u_3 v_1 - u_1 v_3)^2 + (u_1 v_2 - u_2 v_1)^2 \\
&= u_2^2 v_3^2 + u_3^2 v_2^2 - 2 u_2 v_3 u_3 v_2 + u_3^2 v_1^2 + u_1^2 v_3^2 \\
&\quad - 2 u_3 v_1 u_1 v_3 + u_1^2 v_2^2 + u_2^2 v_1^2 - 2 u_1 v_2 u_2 v_1,
\end{aligned}
$$

while

$$|\mathbf{u} \times \mathbf{v}|^2 = |\mathbf{u}|^2 |\mathbf{v}|^2 \sin^2 \theta$$
$$= |\mathbf{u}|^2 |\mathbf{v}|^2 (1 - \cos^2 \theta)$$
$$= |\mathbf{u}|^2 |\mathbf{v}|^2 - (\mathbf{u} \bullet \mathbf{v})^2$$
$$= (u_1^2 + u_2^2 + u_3^2)(v_1^2 + v_2^2 + v_3^2) - (u_1 v_1 + u_2 v_2 + u_3 v_3)^2$$
$$= u_1^2 v_1^2 + u_1^2 v_2^2 + u_1^2 v_3^2 + u_2^2 v_1^2 + u_2^2 v_2^2 + u_2^2 v_3^2 + u_3^2 v_1^2 + u_3^2 v_2^2 + u_3^2 v_3^2$$
$$- u_1^2 v_1^2 - u_2^2 v_2^2 - u_3^2 v_3^2 - 2 u_1 v_1 u_2 v_2 - 2 u_1 v_1 u_3 v_3 - 2 u_2 v_2 u_3 v_3$$
$$= |\mathbf{w}|^2.$$

Since $\mathbf{w}$ is parallel to, and has the same length as $\mathbf{u} \times \mathbf{v}$, we must have either $\mathbf{u} \times \mathbf{v} = \mathbf{w}$ or $\mathbf{u} \times \mathbf{v} = -\mathbf{w}$. It remains to be shown that the first of these is the correct choice. To see this, suppose that the triad of vectors $\mathbf{u}$, $\mathbf{v}$, and $\mathbf{w}$ are rigidly rotated in 3-space so that $\mathbf{u}$ points in the direction of the positive $x$-axis and $\mathbf{v}$ lies in the upper half of the $xy$-plane. Then $\mathbf{u} = u_1 \mathbf{i}$, and $\mathbf{v} = v_1 \mathbf{i} + v_2 \mathbf{j}$, where $u_1 > 0$ and $v_2 > 0$. By the "right-hand rule" $\mathbf{u} \times \mathbf{v}$ must point in the direction of the positive $z$-axis. But $\mathbf{w} = u_1 v_2 \mathbf{k}$ does point in that direction, so $\mathbf{u} \times \mathbf{v} = \mathbf{w}$, as asserted.

The formula for the cross product in terms of components may seem awkward and asymmetric. As we shall see, however, it can be written more easily in terms of a determinant. We introduce determinants later in this section.

---

**Example 1**    (Calculating cross products)

(a)    $\mathbf{i} \times \mathbf{i} = \mathbf{0}$,    $\mathbf{i} \times \mathbf{j} = \mathbf{k}$,    $\mathbf{j} \times \mathbf{i} = -\mathbf{k}$,
   $\mathbf{j} \times \mathbf{j} = \mathbf{0}$,    $\mathbf{j} \times \mathbf{k} = \mathbf{i}$,    $\mathbf{k} \times \mathbf{j} = -\mathbf{i}$,
   $\mathbf{k} \times \mathbf{k} = \mathbf{0}$,    $\mathbf{k} \times \mathbf{i} = \mathbf{j}$,    $\mathbf{i} \times \mathbf{k} = -\mathbf{j}$.

(b)    $(2\mathbf{i} + \mathbf{j} - 3\mathbf{k}) \times (-2\mathbf{j} + 5\mathbf{k})$
$$= \big((1)(5) - (-2)(-3)\big)\mathbf{i} + \big((-3)(0) - (2)(5)\big)\mathbf{j} + \big((2)(-2) - (1)(0)\big)\mathbf{k}$$
$$= -\mathbf{i} - 10\mathbf{j} - 4\mathbf{k}.$$

---

The cross product has some, but not all of the properties we usually ascribe to products. We summarize its algebraic properties as follows:

---

**Properties of the cross product**

If $\mathbf{u}$, $\mathbf{v}$, and $\mathbf{w}$ are any vectors in $\mathbb{R}^3$, and $t$ is a real number (a scalar), then

(i) $\mathbf{u} \times \mathbf{u} = \mathbf{0}$,

(ii) $\mathbf{u} \times \mathbf{v} = -\mathbf{v} \times \mathbf{u}$,    (The cross product is **anticommutative**.)

(iii) $(\mathbf{u} + \mathbf{v}) \times \mathbf{w} = \mathbf{u} \times \mathbf{w} + \mathbf{v} \times \mathbf{w}$,

(iv) $\mathbf{u} \times (\mathbf{v} + \mathbf{w}) = \mathbf{u} \times \mathbf{v} + \mathbf{u} \times \mathbf{w}$,

(v) $(t\mathbf{u}) \times \mathbf{v} = \mathbf{u} \times (t\mathbf{v}) = t(\mathbf{u} \times \mathbf{v})$,

(vi) $\mathbf{u} \bullet (\mathbf{u} \times \mathbf{v}) = \mathbf{v} \bullet (\mathbf{u} \times \mathbf{v}) = \mathbf{0}$.

---

These identities are all easily verified using the components or the definition of the cross product or by using properties of determinants discussed below. They are left as exercises for the reader. Note the absence of an associative law. The cross product is not associative. (See Exercise 21 at the end of this section.) In general,

$$\mathbf{u} \times (\mathbf{v} \times \mathbf{w}) \neq (\mathbf{u} \times \mathbf{v}) \times \mathbf{w}.$$

## Determinants

In order to simplify certain formulas such as the component representation of the cross product we introduce $2 \times 2$ and $3 \times 3$ **determinants**. General $n \times n$ determinants are normally studied in courses on linear algebra; we will encounter them in Section 10.6. In this section we will outline enough of the properties of determinants to enable us to use them as shorthand in some otherwise complicated formulas.

A determinant is an expression that involves the elements of a square array (matrix) of numbers. The determinant of the $2 \times 2$ array of numbers

$$\begin{matrix} a & b \\ c & d \end{matrix}$$

is denoted by enclosing the array between vertical bars, and its value is the number $ad - bc$:

$$\begin{vmatrix} a & b \\ c & d \end{vmatrix} = ad - bc.$$

**Figure 10.22**

This is the product of elements in the *downward diagonal* of the array, minus the product of elements in the *upward diagonal* as shown in Figure 10.22. For example,

$$\begin{vmatrix} 1 & 2 \\ 3 & 4 \end{vmatrix} = (1)(4) - (2)(3) = -2.$$

Similarly, the determinant of a $3 \times 3$ array of numbers is defined by

$$\begin{vmatrix} a & b & c \\ d & e & f \\ g & h & i \end{vmatrix} = aei + bfg + cdh - gec - hfa - idb.$$

**Figure 10.23**

Observe that each of the six products in the value of the determinant involves exactly one element from each row and exactly one from each column of the array. As such, each term is the product of elements in a *diagonal* of an *extended* array obtained by repeating the first two columns of the array to the right of the third column, as shown in Figure 10.23. The value of the determinant is the sum of products corresponding to the three complete *downward* diagonals minus the sum corresponding of the three *upward* diagonals. With practice you will be able to form these diagonal products without having to write the extended array.

If we group the terms in the expansion of the determinant to factor out the elements of the first row we obtain

$$\begin{vmatrix} a & b & c \\ d & e & f \\ g & h & i \end{vmatrix} = a(ei - fh) - b(di - fg) + c(dh - eg)$$

$$= a \begin{vmatrix} e & f \\ h & i \end{vmatrix} - b \begin{vmatrix} d & f \\ g & i \end{vmatrix} + c \begin{vmatrix} d & e \\ g & h \end{vmatrix}.$$

The $2 \times 2$ determinants appearing here (called *minors* of the given $3 \times 3$ determinant) are obtained by deleting the row and column containing the corresponding element from the original $3 \times 3$ determinant. This process is called *expanding* the $3 \times 3$ determinant *in minors* about the first row.

Such expansions in minors can be carried out about any row or column. Note that a minus sign appears in any term whose minor is obtained by deleting the $i$th row and $j$th column, where $i + j$ is an *odd* number. For example, we can expand the above determinant in minors about the second column as follows:

$$\begin{vmatrix} a & b & c \\ d & e & f \\ g & h & i \end{vmatrix} = -b \begin{vmatrix} d & f \\ g & i \end{vmatrix} + e \begin{vmatrix} a & c \\ g & i \end{vmatrix} - h \begin{vmatrix} a & c \\ d & f \end{vmatrix}$$

$$= -bdi + bfg + eai - ecg - haf + hcd.$$

(Of course this is the same value as the one obtained previously.)

**Example 2**

$$\begin{vmatrix} 1 & 4 & -2 \\ -3 & 1 & 0 \\ 2 & 2 & -3 \end{vmatrix} = 3 \begin{vmatrix} 4 & -2 \\ 2 & -3 \end{vmatrix} + 1 \begin{vmatrix} 1 & -2 \\ 2 & -3 \end{vmatrix}$$

$$= 3(-8) + 1 = -23.$$

We expanded about the second row; the third column would also have been a good choice. (Why?)

Any row (or column) of a determinant may be regarded as the components of a vector. Then the determinant is a *linear function* of that vector. For example

$$\begin{vmatrix} a & b & c \\ d & e & f \\ sx + tl & sy + tm & sz + tn \end{vmatrix} = s \begin{vmatrix} a & b & c \\ d & e & f \\ x & y & z \end{vmatrix} + t \begin{vmatrix} a & b & c \\ d & e & f \\ l & m & n \end{vmatrix}$$

because the determinant is a linear function of its third row. This and other properties of determinants follow directly from the definition. Some other properties are summarized below. These are stated for rows and for $3 \times 3$ determinants but similar statements can be made for columns and for determinants of any order.

**Properties of determinants**

(i) If two rows of a determinant are interchanged, then the determinant changes sign:

$$\begin{vmatrix} d & e & f \\ a & b & c \\ g & h & i \end{vmatrix} = - \begin{vmatrix} a & b & c \\ d & e & f \\ g & h & i \end{vmatrix}.$$

(ii) If two rows of a determinant are equal, the determinant has value 0:

$$\begin{vmatrix} a & b & c \\ a & b & c \\ g & h & i \end{vmatrix} = 0.$$

(iii) If a multiple of one row of a determinant is added to another row, the value of the determinant remains unchanged:

$$\begin{vmatrix} a & b & c \\ d+ta & e+tb & f+tc \\ g & h & i \end{vmatrix} = \begin{vmatrix} a & b & c \\ d & e & f \\ g & h & i \end{vmatrix}.$$

## The Cross Product as a Determinant

The elements of a determinant are usually numbers, because they have to be multiplied to get the value of the determinant. However, it is possible to use vectors as the elements of *one row* (or column) of a determinant. When expanding in minors about that row (or column) the minor for each vector element is a number that determines the scalar multiple of the vector. The formula for the cross product of

$$\mathbf{u} = u_1\mathbf{i} + u_2\mathbf{j} + u_3\mathbf{k} \quad \text{and} \quad \mathbf{v} = v_1\mathbf{i} + v_2\mathbf{j} + v_3\mathbf{k}$$

presented in Theorem 2 can be expressed symbolically as a determinant with the standard basis vectors as the elements of the first row:

$$\mathbf{u} \times \mathbf{v} = \begin{vmatrix} \mathbf{i} & \mathbf{j} & \mathbf{k} \\ u_1 & u_2 & u_3 \\ v_1 & v_2 & v_3 \end{vmatrix} = \begin{vmatrix} u_2 & u_3 \\ v_2 & v_3 \end{vmatrix}\mathbf{i} - \begin{vmatrix} u_1 & u_3 \\ v_1 & v_3 \end{vmatrix}\mathbf{j} + \begin{vmatrix} u_1 & u_2 \\ v_1 & v_2 \end{vmatrix}\mathbf{k}.$$

The formula for the cross product given in that theorem is just the expansion of this determinant in minors about the first row.

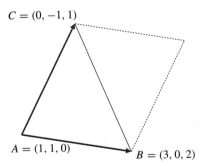

$C = (0, -1, 1)$

$A = (1, 1, 0)$

$B = (3, 0, 2)$

**Figure 10.24**

**Example 3** Find the area of the triangle with vertices at the three points $A = (1, 1, 0)$, $B = (3, 0, 2)$, and $C = (0, -1, 1)$.

**Solution** Two sides of the triangle (Figure 10.24) are given by the vectors:

$$\overrightarrow{AB} = 2\mathbf{i} - \mathbf{j} + 2\mathbf{k} \quad \text{and} \quad \overrightarrow{AC} = -\mathbf{i} - 2\mathbf{j} + \mathbf{k}.$$

The area of the triangle is half the area of the parallelogram spanned by $\overrightarrow{AB}$ and $\overrightarrow{AC}$. By the definition of cross product, the area of the triangle must therefore be

$$\frac{1}{2}|\overrightarrow{AB} \times \overrightarrow{AC}| = \frac{1}{2}|\begin{vmatrix} \mathbf{i} & \mathbf{j} & \mathbf{k} \\ 2 & -1 & 2 \\ -1 & -2 & 1 \end{vmatrix}|$$

$$= \frac{1}{2}|3\mathbf{i} - 4\mathbf{j} - 5\mathbf{k}| = \frac{1}{2}\sqrt{9 + 16 + 25} = \frac{5}{2}\sqrt{2} \text{ square units.}$$

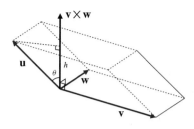

**Figure 10.25**

A **parallelepiped** is the three-dimensional analogue of a parallelogram. It is a solid with three pairs of parallel planar faces. Each face is in the shape of a parallelogram. A rectangular brick is a special case of a parallelepiped in which nonparallel faces intersect at right angles. We say that a parallelepiped is **spanned** by three vectors coinciding with three of its edges that meet at one vertex. (See Figure 10.25.)

**Example 4** Find the volume of the parallelepiped spanned by **u**, **v** and **w**.

**Solution** The volume of the parallelepiped is equal to the area of one of its faces, say the face spanned by **v** and **w**, multiplied by the height of the parallelepiped measured in a direction perpendicular to that face. The area of the face is $|\mathbf{v} \times \mathbf{w}|$. Since $\mathbf{v} \times \mathbf{w}$ is perpendicular to the face, the height $h$ of the parallelepiped will be the absolute value of the scalar projection of **u** along $\mathbf{v} \times \mathbf{w}$. If $\theta$ is the angle between **u** and $\mathbf{v} \times \mathbf{w}$, then the volume of the parallelepiped is given by

$$\text{Volume} = |\mathbf{u}||\mathbf{v} \times \mathbf{w}||\cos\theta| = |\mathbf{u} \bullet (\mathbf{v} \times \mathbf{w})| \text{ cubic units.}$$

**DEFINITION 6**

The quantity $\mathbf{u} \bullet (\mathbf{v} \times \mathbf{w})$ is called the **scalar triple product** of the vectors **u**, **v**, and **w**.

The scalar triple product is easily expressed in terms of a determinant. If $\mathbf{u} = u_1\mathbf{i} + u_2\mathbf{j} + u_3\mathbf{k}$, and similar representations hold for **v** and **w**, then

$$\mathbf{u} \bullet (\mathbf{v} \times \mathbf{w}) = u_1 \begin{vmatrix} v_2 & v_3 \\ w_2 & w_3 \end{vmatrix} - u_2 \begin{vmatrix} v_1 & v_3 \\ w_1 & w_3 \end{vmatrix} + u_3 \begin{vmatrix} v_1 & v_2 \\ w_1 & w_2 \end{vmatrix}$$

$$= \begin{vmatrix} u_1 & u_2 & u_3 \\ v_1 & v_2 & v_3 \\ w_1 & w_2 & w_3 \end{vmatrix}.$$

The volume of the parallelepiped spanned by **u**, **v**, and **w** is the absolute value of this determinant.

Using the properties of the determinant, it is easily verified that

$$\mathbf{u} \bullet (\mathbf{v} \times \mathbf{w}) = \mathbf{v} \bullet (\mathbf{w} \times \mathbf{u}) = \mathbf{w} \bullet (\mathbf{u} \times \mathbf{v}).$$

(See Exercise 18 below.) Note that **u**, **v**, and **w** remain in the same *cyclic order* in these three expressions. Reversing the order would introduce a factor $-1$:

$$\mathbf{u} \bullet (\mathbf{v} \times \mathbf{w}) = -\mathbf{u} \bullet (\mathbf{w} \times \mathbf{v}).$$

Three vectors in 3-space are said to be **coplanar** if the parallelepiped they span has zero volume; if their tails coincide, three such vectors must line in the same plane.

$$\mathbf{u}, \mathbf{v}, \text{ and } \mathbf{w} \text{ are coplanar} \iff \mathbf{u} \bullet (\mathbf{v} \times \mathbf{w}) = 0$$

$$\iff \begin{vmatrix} u_1 & u_2 & u_3 \\ v_1 & v_2 & v_3 \\ w_1 & w_2 & w_3 \end{vmatrix} = 0.$$

Three vectors are certainly coplanar if any of them is **0**, or if any pair of them is parallel. If neither of these degenerate conditions apply, they are only coplanar if any one of them can be expressed as a linear combination of the other two. (See Exercise 20 below.)

## Applications of Cross Products

Cross products are of considerable importance in mechanics and electromagnetic theory, as well as in the study of motion in general. For example:

(a) The linear velocity **v** of a particle located at position **r** in a body rotating with angular velocity $\Omega$ about the origin is given by $\mathbf{v} = \Omega \times \mathbf{r}$. (See Section 11.2 for more details.)

(b) The angular momentum of a planet of mass $m$ moving with velocity **v** in its orbit around the sun is given by $\mathbf{h} = \mathbf{r} \times m\mathbf{v}$, where **r** is the position vector of the planet relative to the sun as origin. (See Section 11.6.)

(c) If a particle of electric charge $q$ is travelling with velocity **v** through a magnetic field whose strength and direction are given by vector **B**, then the force that the field exerts on the particle is given by $\mathbf{F} = q\mathbf{v} \times \mathbf{B}$. The electron beam in a television tube is controlled by magnetic fields using this principle.

(d) The torque **T** of a force **F** applied at the point $P$ with position vector **r** about another point $P_0$ with position vector $\mathbf{r}_0$ is defined to be

$$\mathbf{T} = \overrightarrow{P_0 P} \times \mathbf{F} = (\mathbf{r} - \mathbf{r}_0) \times \mathbf{F}.$$

This torque measures the effectiveness of the force **F** in causing rotation about $P_0$. The direction of **T** is along the axis through $P_0$ about which **F** acts to rotate $P$.

**Figure 10.26**    The force on the handle is 500 N in a direction directly towards you

**Example 5**    An automobile wheel has centre at the origin and axle along the $y$-axis. One of the retaining nuts holding the wheel is at position $P_0 = (0, 0, 10)$. (Distances are measured in centimetres.) A bent tire wrench with arm 25 cm long and inclined at an angle of 60° to the direction of its handle is fitted to the nut in an upright direction, as shown in Figure 10.26. If a horizontal force $\mathbf{F} = 500\mathbf{i}$ newtons is applied to the handle of the wrench, what is its torque on the nut? What part (component) of this torque is effective in trying to rotate the nut about its horizontal axis? What is the effective torque trying to rotate the wheel?

**Solution**    The nut is at position $\mathbf{r}_0 = 10\mathbf{k}$, and the handle of the wrench is at position

$$\mathbf{r} = 25\cos 60°\mathbf{j} + (10 + 25\sin 60°)\mathbf{k} \approx 12.5\mathbf{j} + 31.65\mathbf{k}.$$

The torque of the force **F** on the nut is

$$\mathbf{T} = (\mathbf{r} - \mathbf{r}_0) \times \mathbf{F}$$
$$\approx (12.5\mathbf{j} + 21.65\mathbf{k}) \times 500\mathbf{i} \approx 10{,}825\mathbf{j} - 6{,}250\mathbf{k},$$

which is at right angles to **F** and to the arm of the wrench. Only the horizontal component of this torque is effective in turning the nut. This component is 10,825 N·cm or 108.24 N·m in magnitude. For the effective torque on the wheel itself, we have to replace $\mathbf{r}_0$ by **0**, the position of the centre of the wheel. In this case the horizontal torque is

$$31.65\mathbf{k}\times500\mathbf{i} \approx 15,825\mathbf{j},$$

that is, about 158.25 N·m.                                                    ∎

## Exercises 10.3

**1.** Calculate $\mathbf{u}\times\mathbf{v}$ if $\mathbf{u} = \mathbf{i} - 2\mathbf{j} + 3\mathbf{k}$ and $\mathbf{v} = 3\mathbf{i} + \mathbf{j} - 4\mathbf{k}$.

**2.** Calculate $\mathbf{u}\times\mathbf{v}$ if $\mathbf{u} = \mathbf{j} + 2\mathbf{k}$ and $\mathbf{v} = -\mathbf{i} - \mathbf{j} + \mathbf{k}$.

**3.** Find the area of the triangle with vertices $(1, 2, 0)$, $(1, 0, 2)$, and $(0, 3, 1)$.

**4.** Find a unit vector perpendicular to the plane containing the points $(a, 0, 0)$, $(0, b, 0)$, and $(0, 0, c)$. What is the area of the triangle with these vertices?

**5.** Find a unit vector perpendicular to the vectors $\mathbf{i} + \mathbf{j}$ and $\mathbf{j} + 2\mathbf{k}$.

**6.** Find a unit vector with positive **k** component that is perpendicular to both $2\mathbf{i} - \mathbf{j} - 2\mathbf{k}$ and $2\mathbf{i} - 3\mathbf{j} + \mathbf{k}$.

Verify the identities in Exercises 7–11, either by using the definition of cross product or by using properties of determinants.

**7.** $\mathbf{u}\times\mathbf{u} = \mathbf{0}$

**8.** $\mathbf{u}\times\mathbf{v} = -\mathbf{v}\times\mathbf{u}$

**9.** $(\mathbf{u} + \mathbf{v})\times\mathbf{w} = \mathbf{u}\times\mathbf{w} + \mathbf{v}\times\mathbf{w}$

**10.** $(t\mathbf{u})\times\mathbf{v} = \mathbf{u}\times(t\mathbf{v}) = t(\mathbf{u}\times\mathbf{v})$

**11.** $\mathbf{u}\bullet(\mathbf{u}\times\mathbf{v}) = \mathbf{v}\bullet(\mathbf{u}\times\mathbf{v}) = 0$

**12.** Obtain the addition formula

$$\sin(\alpha - \beta) = \sin\alpha\cos\beta - \cos\alpha\sin\beta$$

by examining the cross product of the two unit vectors $\mathbf{u} = \cos\beta\mathbf{i} + \sin\beta\mathbf{j}$ and $\mathbf{v} = \cos\alpha\mathbf{i} + \sin\alpha\mathbf{j}$. Assume $0 \le \alpha - \beta \le \pi$. *Hint:* regard **u** and **v** as position vectors. What is the area of the parallelogram they span?

**13.** If $\mathbf{u} + \mathbf{v} + \mathbf{w} = \mathbf{0}$ show that $\mathbf{u}\times\mathbf{v} = \mathbf{v}\times\mathbf{w} = \mathbf{w}\times\mathbf{u}$.

**14.** **(Volume of a tetrahedron)** A **tetrahedron** is a pyramid with a triangular base and three other triangular faces. It has four vertices and six edges. Like any pyramid or cone, its volume is equal to $\frac{1}{3}Ah$, where $A$ is the area of the base and $h$ is the height measured perpendicular to the base. If **u**, **v**, and **w** are vectors coinciding with the three edges of a tetrahedron that meet at one vertex, show that the tetrahedron has volume given by

$$\text{Volume} = \frac{1}{6}|\mathbf{u}\bullet(\mathbf{v}\times\mathbf{w})| = \frac{1}{6}\left|\begin{matrix} u_1 & u_2 & u_3 \\ v_1 & v_2 & v_3 \\ w_1 & w_2 & w_3 \end{matrix}\right|.$$

Thus the volume of a tetrahedron spanned by three vectors is one-sixth of the volume of the parallelepiped spanned by the same vectors.

**15.** Find the volume of the tetrahedron with vertices $(1, 0, 0)$, $(1, 2, 0)$, $(2, 2, 2)$, and $(0, 3, 2)$.

**16.** Find the volume of the parallelepiped spanned by the diagonals of the three faces of a cube of side $a$ that meet at one vertex of the cube.

**17.** For what value of $k$ do the four points $(1, 1, -1)$, $(0, 3, -2)$, $(-2, 1, 0)$, and $(k, 0, 2)$ all lie on a plane?

**18.** **(The scalar triple product)** Verify the identities

$$\mathbf{u}\bullet(\mathbf{v}\times\mathbf{w}) = \mathbf{v}\bullet(\mathbf{w}\times\mathbf{u}) = \mathbf{w}\bullet(\mathbf{u}\times\mathbf{v}).$$

**19.** If $\mathbf{u}\bullet(\mathbf{v}\times\mathbf{w}) \neq 0$ and **x** is an arbitrary 3-vector, find the numbers $\lambda$, $\mu$, and $\nu$ such that

$$\mathbf{x} = \lambda\mathbf{u} + \mu\mathbf{v} + \nu\mathbf{w}.$$

**20.** If $\mathbf{u}\bullet(\mathbf{v}\times\mathbf{w}) = 0$ but $\mathbf{v}\times\mathbf{w} \neq \mathbf{0}$, show that there are constants $\lambda$ and $\mu$ such that

$$\mathbf{u} = \lambda\mathbf{v} + \mu\mathbf{w}.$$

*Hint:* use the result of Exercise 19 with **u** in place of **x** and $\mathbf{v}\times\mathbf{w}$ in place of **u**.

**21.** Calculate $\mathbf{u}\times(\mathbf{v}\times\mathbf{w})$ and $(\mathbf{u}\times\mathbf{v})\times\mathbf{w}$, given that $\mathbf{u} = \mathbf{i} + 2\mathbf{j} + 3\mathbf{k}$, $\mathbf{v} = 2\mathbf{i} - 3\mathbf{j}$, and $\mathbf{w} = \mathbf{j} - \mathbf{k}$. Why would you not expect these to be equal?

**22.** Does the notation $\mathbf{u}\bullet\mathbf{v}\times\mathbf{w}$ make sense? Why? How about the notation $\mathbf{u}\times\mathbf{v}\times\mathbf{w}$?

**23.** **(The vector triple product)** The product $\mathbf{u}\times(\mathbf{v}\times\mathbf{w})$ is called a **vector triple product**. Since it is perpendicular to $\mathbf{v}\times\mathbf{w}$ it must lie in the plane of **v** and **w**. Show that

$$\mathbf{u}\times(\mathbf{v}\times\mathbf{w}) = (\mathbf{u}\bullet\mathbf{w})\mathbf{v} - (\mathbf{u}\bullet\mathbf{v})\mathbf{w}.$$

*Hint:* this can be done by direct calculation of the components of both sides of the equation, but the job is much easier if you choose coordinate axes so that **v** lies along the *x*-axis and **w** lies in the *xy*-plane.

24. If **u**, **v**, and **w** are mutually perpendicular vectors, show that $\mathbf{u}\times(\mathbf{v}\times\mathbf{w}) = \mathbf{0}$. What is $\mathbf{u}\bullet(\mathbf{v}\times\mathbf{w})$ in this case?

25. Show that $\mathbf{u}\times(\mathbf{v}\times\mathbf{w}) + \mathbf{v}\times(\mathbf{w}\times\mathbf{u}) + \mathbf{w}\times(\mathbf{u}\times\mathbf{v}) = \mathbf{0}$.

26. Find all vectors **x** that satisfy the equation

$$(-\mathbf{i}+2\mathbf{j}+3\mathbf{k})\times\mathbf{x} = \mathbf{i}+5\mathbf{j}-3\mathbf{k}.$$

27. Show that the equation

$$(-\mathbf{i}+2\mathbf{j}+3\mathbf{k})\times\mathbf{x} = \mathbf{i}+5\mathbf{j}$$

has no solutions for the unknown vector **x**.

28. What condition must be satisfied by the nonzero vectors **a** and **b** to guarantee that the equation $\mathbf{a}\times\mathbf{x} = \mathbf{b}$ has a solution for **x**? Is the solution unique?

# 10.4 Planes and Lines

A single equation in the three variables, *x*, *y*, and *z*, constitutes a single constraint on the freedom of the point $P = (x, y, z)$ to lie anywhere in 3-space. Such a constraint usually results in the loss of exactly one *degree of freedom* and so forces *P* to lie on a two-dimensional surface. For example, the equation

$$x^2 + y^2 + z^2 = 4$$

states that the point $(x, y, z)$ is at distance 2 from the origin. All points satisfying this condition lie on a **sphere** (that is, the surface of a ball) of radius 2 centred at the origin. The equation above therefore represents that sphere, and the sphere is the graph of the equation. In this section we will investigate the graphs of linear equations in three variables.

## Planes in 3-Space

Let $P_0 = (x_0, y_0, z_0)$ be a point in $\mathbb{R}^3$ with position vector

$$\mathbf{r}_0 = x_0\mathbf{i} + y_0\mathbf{j} + z_0\mathbf{k}.$$

If $\mathbf{n} = A\mathbf{i} + B\mathbf{j} + C\mathbf{k}$ is any given *nonzero* vector, then there exists exactly one **plane** (flat surface) passing through $P_0$ and perpendicular to **n**. We say that **n** is a **normal vector** to the plane. The plane is the set of all points *P* for which $\overrightarrow{P_0P}$ is perpendicular to **n**. (See Figure 10.27.)

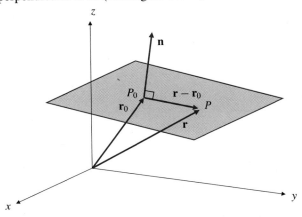

**Figure 10.27** The plane through $P_0$ with normal **n** contains all points *P* for which $\overrightarrow{P_0P}$ is perpendicular to **n**

If $P = (x, y, z)$ has position vector $\mathbf{r}$, then $\overrightarrow{P_0P} = \mathbf{r} - \mathbf{r}_0$. This vector is perpendicular to $\mathbf{n}$ if and only if $\mathbf{n} \bullet (\mathbf{r} - \mathbf{r}_0) = 0$. This is the equation of the plane in vector form. We can rewrite it in terms of coordinates to obtain the corresponding scalar equation.

---

**The equations of a plane**

The plane having nonzero normal vector $\mathbf{n} = A\mathbf{i} + B\mathbf{j} + C\mathbf{k}$, and passing through the point $P_0 = (x_0, y_0, z_0)$ with position vector $\mathbf{r}_0$ has equation

$$\mathbf{n} \bullet (\mathbf{r} - \mathbf{r}_0) = 0$$

in vector form, or, equivalently,

$$A(x - x_0) + B(y - y_0) + C(z - z_0) = 0$$

in scalar form

---

The scalar form can be written more simply as $Ax + By + Cz = D$, where $D = Ax_0 + By_0 + Cz_0$.

If at least one of the constants $A$, $B$, and $C$ is not zero, then the *linear equation* $Ax + By + Cz = D$ always represents a plane in $\mathbb{R}^3$. For example, if $A \neq 0$, it represents the plane through $(D/A, 0, 0)$ with normal vector $\mathbf{n} = A\mathbf{i} + B\mathbf{j} + C\mathbf{k}$. A vector normal to a plane can always be determined from the coefficients of $x$, $y$, and $z$. If the constant term $D = 0$, then the plane must pass through the origin.

**Example 1** (**Recognizing and writing the equations of planes**)

(a) The equation $2x - 3y - 4z = 0$ represents a plane that passes through the origin and is normal (perpendicular) to the vector $\mathbf{n} = 2\mathbf{i} - 3\mathbf{j} - 4\mathbf{k}$.

(b) The plane that passes through the point $(2, 0, 1)$ and is perpendicular to the straight line passing through the points $(1, 1, 0)$ and $(4, -1, -2)$ has normal vector $\mathbf{n} = (4 - 1)\mathbf{i} + (-1 - 1)\mathbf{j} + (-2 - 0)\mathbf{k} = 3\mathbf{i} - 2\mathbf{j} - 2\mathbf{k}$. Therefore its equation is $3(x-2)-2(y-0)-2(z-1) = 0$, or, more simply, $3x-2y-2z = 4$.

(c) The plane with equation $2x - y = 1$ has a normal $2\mathbf{i} - \mathbf{j}$ that is perpendicular to the $z$-axis. The plane is therefore parallel to the $z$-axis. Note that the equation is independent of $z$. In the $xy$-plane the equation $2x - y = 1$ represents a straight line; in 3-space it represents a plane containing that line and parallel to the $z$-axis. What does the equation $y = z$ represent in $\mathbb{R}^3$? The equation $y = -2$?

(d) The equation $2x + y + 3z = 6$ represents a plane with normal $\mathbf{n} = 2\mathbf{i} + \mathbf{j} + 3\mathbf{k}$. In this case we cannot directly read from the equation the coordinates of a particular point on the plane, but it is not difficult to discover some points. For instance, if we put $y = z = 0$ in the equation we get $x = 3$, and so $(3, 0, 0)$ is a point on the plane. We say that the $x$-**intercept** of the plane is 3 since $(3, 0, 0)$ is the point where the plane intersects the $x$-axis. Similarly, the $y$-intercept is 6 and the $z$-intercept is 2 because the plane intersects the $y$- and $z$-axes at $(0, 6, 0)$ and $(0, 0, 2)$, respectively.

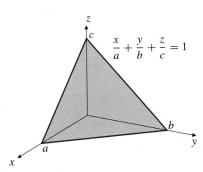

**Figure 10.28** The plane with intercepts $a$, $b$, and $c$ on the coordinate axes

(e) In general, if $a$, $b$, and $c$ are all nonzero, the plane with equation

$$\frac{x}{a} + \frac{y}{b} + \frac{z}{c} = 1$$

has intercepts $a$, $b$, and $c$ on the coordinate axes. (See Figure 10.28.) ∎

**Example 2** Find an equation of the plane that passes through the three points $P = (1, 1, 0)$, $Q = (0, 2, 1)$, and $R = (3, 2, -1)$.

**Solution** We need to find a vector, $\mathbf{n}$, normal to the plane. Such a vector will be perpendicular to the vectors $\overrightarrow{PQ} = -\mathbf{i} + \mathbf{j} + \mathbf{k}$ and $\overrightarrow{PR} = 2\mathbf{i} + \mathbf{j} - \mathbf{k}$. Therefore we can use

$$\mathbf{n} = \overrightarrow{PQ} \times \overrightarrow{PR} = \begin{vmatrix} \mathbf{i} & \mathbf{j} & \mathbf{k} \\ -1 & 1 & 1 \\ 2 & 1 & -1 \end{vmatrix} = -2\mathbf{i} + \mathbf{j} - 3\mathbf{k}.$$

We can use this normal vector together with the coordinates of any one of the three given points to write the equation of the plane. Using point $P$ leads to the equation $-2(x - 1) + 1(y - 1) - 3(z - 0) = 0$, or

$$2x - y + 3z = 1.$$

You can check that using either $Q$ or $R$ leads to the same equation. (If the cross product $\overrightarrow{PQ} \times \overrightarrow{PR}$ had been the zero vector what would have been true about the three points $P$, $Q$, and $R$? Would they have determined a unique plane?) ∎

**Example 3** Show that the two planes

$$x - y = 3 \qquad \text{and} \qquad x + y + z = 0$$

intersect and find a vector, $\mathbf{v}$, parallel to their line of intersection.

**Solution** The two planes have normal vectors

$$\mathbf{n}_1 = \mathbf{i} - \mathbf{j} \qquad \text{and} \qquad \mathbf{n}_2 = \mathbf{i} + \mathbf{j} + \mathbf{k},$$

respectively. Since these vectors are not parallel, the planes are not parallel, and so they intersect in a straight line perpendicular to both $\mathbf{n}_1$ and $\mathbf{n}_2$. This line must therefore be parallel to

$$\mathbf{v} = \mathbf{n}_1 \times \mathbf{n}_2 = \begin{vmatrix} \mathbf{i} & \mathbf{j} & \mathbf{k} \\ 1 & -1 & 0 \\ 1 & 1 & 1 \end{vmatrix} = -\mathbf{i} - \mathbf{j} + 2\mathbf{k}.$$

∎

A family of planes intersecting in a straight line is called a **pencil of planes**. (See Figure 10.29.) Such a pencil of planes is determined by any two nonparallel planes in it, since these have a unique line of intersection. If the two nonparallel planes

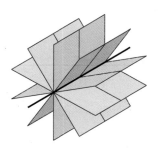

**Figure 10.29**    A pencil of planes

have equations

$$A_1x + B_1y + C_1z = D_1 \quad \text{and} \quad A_2x + B_2y + C_2z = D_2$$

then, for any value of the real number $\lambda$, the equation

$$A_1x + B_1y + C_1z - D_1 + \lambda(A_2x + B_2y + C_2z - D_2) = 0$$

represents a plane in the pencil. To see this, observe that the equation is linear, and so represents a plane, and that any point $(x, y, z)$ satisfying the equations of both given planes also satisfies this equation for any value of $\lambda$. Any plane in the pencil except the second defining plane, $A_2x + B_2y + C_2z = D_2$, can be obtained by suitably choosing $\lambda$.

---

**Example 4**    Find an equation of the plane passing through the line of intersection of the two planes

$$x + y - 2z = 6 \quad \text{and} \quad 2x - y + z = 2$$

and also passing through the point $(-2, 0, 1)$.

**Solution**    For any constant $\lambda$, the equation

$$x + y - 2z - 6 + \lambda(2x - y + z - 2) = 0$$

represents a plane, and is satisfied by the coordinates of all points on the line of intersection of the given planes. This plane passes through the point $(-2, 0, 1)$ if $-2 - 2 - 6 + \lambda(-4 + 1 - 2) = 0$, that is, if $\lambda = -2$. The equation of the required plane therefore simplifies to $3x - 3y + 4z + 2 = 0$. (This solution would not have worked if the given point had been on the second plane, $2x - y + z = 2$. Why?)

---

## Lines in 3-Space

As we observed above, any two nonparallel planes in $\mathbb{R}^3$ determine a unique (straight) line of intersection, and a vector parallel to this line can be obtained by taking the cross product of normal vectors to the two planes.

Suppose that $\mathbf{r}_0 = x_0\mathbf{i} + y_0\mathbf{j} + z_0\mathbf{k}$ is the position vector of point $P_0$ and $\mathbf{v} = a\mathbf{i} + b\mathbf{j} + c\mathbf{k}$ is a nonzero vector. There is a unique line passing through $P_0$ parallel to $\mathbf{v}$. If $\mathbf{r} = x\mathbf{i} + y\mathbf{j} + z\mathbf{k}$ is the position vector of any other point, $P$, on the line then $\mathbf{r} - \mathbf{r}_0$ lies along the line and so is parallel to $\mathbf{v}$. (See Figure 10.30.) Thus $\mathbf{r} - \mathbf{r}_0 = t\mathbf{v}$ for some real number $t$. This equation, usually rewritten in the form

$$\mathbf{r} = \mathbf{r}_0 + t\mathbf{v},$$

is called the **vector parametric equation of the straight line**. All points on the line can be obtained as $t$ ranges from $-\infty$ to $\infty$. The vector $\mathbf{v}$ is called a **direction vector** of the line.

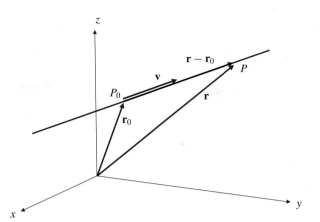

**Figure 10.30** The line through $P_0$ parallel to **v**

Breaking the vector parametric equation down into its components yields the **scalar parametric** equations of the line:

$$\begin{cases} x = x_0 + at \\ y = y_0 + bt \\ z = z_0 + ct. \end{cases} \qquad (-\infty < t < \infty)$$

These appear to be *three* linear equations, but the parameter $t$ can be eliminated to give *two* linear equations in $x$, $y$, and $z$. If $a \neq 0$, $b \neq 0$, and $c \neq 0$, then we can solve each of the scalar equations for $t$, and so obtain

$$\frac{x - x_0}{a} = \frac{y - y_0}{b} = \frac{z - z_0}{c},$$

which is called the **standard form** for the equations of the straight line through $(x_0, y_0, z_0)$ parallel to **v**. The standard form must be modified if any component of **v** vanishes. For example, if $c = 0$ the equations are

$$\frac{x - x_0}{a} = \frac{y - y_0}{b}, \quad z = z_0.$$

Note that none of above equations for straight lines is unique; each depends on the particular choice of the point $(x_0, y_0, z_0)$ on the line. In general, you can always use the equations of two nonparallel planes to represent their line of intersection.

**Example 5** **(Equations of straight lines)**

(a) The equations

$$\begin{cases} x = 2 + t \\ y = 3 \\ z = -4t \end{cases}$$

represent the straight line through $(2, 3, 0)$ parallel to the vector $\mathbf{i} - 4\mathbf{k}$.

(b) The straight line through $(1, -2, 3)$ perpendicular to the plane $x - 2y + 4z = 5$ is parallel to the normal vector $\mathbf{i} - 2\mathbf{j} + 4\mathbf{k}$ of the plane. Therefore the line has vector parametric equation

$$\mathbf{r} = \mathbf{i} - 2\mathbf{j} + 3\mathbf{k} + t(\mathbf{i} - 2\mathbf{j} + 4\mathbf{k}),$$

or scalar parametric equations

$$\begin{cases} x = 1 + t \\ y = -2 - 2t \\ z = 3 + 4t \end{cases}$$

Its standard form equations are

$$\frac{x-1}{1} = \frac{y+2}{-2} = \frac{z-3}{4}.$$

---

**Example 6**   Find a direction vector for the line of intersection of the two planes

$$x + y - z = 0 \qquad \text{and} \qquad y + 2z = 6,$$

and find a set of equations for the line in standard form.

**Solution**   The two planes have respective normals $\mathbf{n}_1 = \mathbf{i} + \mathbf{j} - \mathbf{k}$ and $\mathbf{n}_2 = \mathbf{j} + 2\mathbf{k}$. Thus a direction vector of their line of intersection is

$$\mathbf{v} = \mathbf{n}_1 \times \mathbf{n}_2 = 3\mathbf{i} - 2\mathbf{j} + \mathbf{k}.$$

We need to know one point on the line in order to write equations in standard form. We can find a point by assigning a value to one coordinate and calculating the other two from the given equations. For instance, taking $z = 0$ in the two equations we are led to $y = 6$ and $x = -6$, so $(-6, 6, 0)$ is one point on the line. Thus the line has standard form equations

$$\frac{x+6}{3} = \frac{y-6}{-2} = z.$$

This answer is not unique; the coordinates of any other point on the line could be used in place of $(-6, 6, 0)$. You could even find a direction vector $\mathbf{v}$ by subtracting the position vectors of two different points on the line.

## Distances

The **distance** between two geometric objects always means the minimum distance between two points, one in each object. In the case of *flat* objects like lines or planes defined by linear equations, such minimum distances can usually be determined by geometric arguments without having to use calculus.

---

**Example 7**   **(Distance from a point to a plane)**

(a) Find the distance from the point $P_0 = (x_0, y_0, z_0)$ to the plane $\mathcal{P}$ having equation $Ax + By + Cz = D$.

(b) What is the distance from $(2, -1, 3)$ to the plane $2x - 2y - z = 9$?

**Figure 10.31** The distance from $P_0$ to the plane $\mathcal{P}$ is the length of the vector projection of $PP_0$ along the normal $\mathbf{n}$ to $\mathcal{P}$, where $P$ is any point on $\mathcal{P}$

### Solution

(a) Let $\mathbf{r}_0$ be the position vector of $P_0$ and let $\mathbf{n} = A\mathbf{i} + B\mathbf{j} + C\mathbf{k}$, be the normal to $\mathcal{P}$. Let $P_1$ be the point on $\mathcal{P}$ that is closest to $P_0$. Then $PP_0$ is perpendicular to $\mathcal{P}$, and so is parallel to $\mathbf{n}$. The distance from $P_0$ to $\mathcal{P}$ is $s = |\overrightarrow{P_1 P_0}|$. If $P$, having position vector $\mathbf{r}$, is any point on $\mathcal{P}$ then $s$ is the length of the projection of $\overrightarrow{PP_0} = \mathbf{r}_0 - \mathbf{r}$ in the direction of $\mathbf{n}$. (See Figure 10.31.) Thus

$$s = \left| \frac{\overrightarrow{PP_0} \bullet \mathbf{n}}{|\mathbf{n}|} \right| = \frac{|(\mathbf{r}_0 - \mathbf{r}) \bullet \mathbf{n}|}{|\mathbf{n}|} = \frac{|\mathbf{r}_0 \bullet \mathbf{n} - \mathbf{r} \bullet \mathbf{n}|}{|\mathbf{n}|}.$$

Since $P = (x, y, z)$ lies on $\mathcal{P}$ we have $\mathbf{r} \bullet \mathbf{n} = Ax + By + Cz = D$. In terms of the coordinates $(x_0, y_0, z_0)$ of $P_0$, we can therefore represent the distance from $P_0$ to $\mathcal{P}$ as

$$s = \frac{|Ax_0 + By_0 + Cz_0 - D|}{\sqrt{A^2 + B^2 + C^2}}.$$

(b) The distance from $(2, -1, 3)$ to the plane $2x - 2y - z = 9$ is

$$s = \frac{|2(2) - 2(-1) - 1(3) - 9|}{\sqrt{2^2 + (-2)^2 + (-1)^2}} = \frac{|-6|}{3} = 2 \text{ units}.$$

■

**Example 8** (Distance from a point to a line)

(a) Find the distance from the point $P_0$ to the straight line $\mathcal{L}$ through $P_1$ parallel to the nonzero vector $\mathbf{v}$.

(b) What is the distance from $(2, 0, -3)$ to the line $\mathbf{r} = \mathbf{i} + (1 + 3t)\mathbf{j} - (3 - 4t)\mathbf{k}$?

### Solution

(a) Let $\mathbf{r}_0$ and $\mathbf{r}_1$ be the position vectors of $P_0$ and $P_1$ respectively. The point $P_2$ on $\mathcal{L}$ that is closest to $P_0$ is such that $P_2 P_0$ is perpendicular to $\mathcal{L}$. The distance from $P_0$ to $\mathcal{L}$ is

$$s = |P_2P_0| = |P_1P_0|\sin\theta = |\mathbf{r}_0 - \mathbf{r}_1|\sin\theta,$$

where $\theta$ is the angle between $\mathbf{r}_0 - \mathbf{r}_1$ and $\mathbf{v}$. (See Figure 10.32(a).) Since

$$|(\mathbf{r}_0 - \mathbf{r}_1)\times\mathbf{v}| = |\mathbf{r}_0 - \mathbf{r}_1|\,|\mathbf{v}|\,\sin\theta,$$

we have

$$s = \frac{|(\mathbf{r}_0 - \mathbf{r}_1)\times\mathbf{v}|}{|\mathbf{v}|}.$$

(b) The line $\mathbf{r} = \mathbf{i} + (1 + 3t)\mathbf{j} - (3 - 4t)\mathbf{k}$ passes through $P_1 = (1, 1, -3)$ and is parallel to $\mathbf{v} = 3\mathbf{j} + 4\mathbf{k}$. The distance from $P_0 = (2, 0, -3)$ to this line is

$$s = \frac{\left|\big((2-1)\mathbf{i} + (0-1)\mathbf{j} + (-3+3)\mathbf{k}\big)\times(3\mathbf{j}+4\mathbf{k})\right|}{\sqrt{3^2+4^2}}$$

$$= \frac{|(\mathbf{i}-\mathbf{j})\times(3\mathbf{j}+4\mathbf{k})|}{5} = \frac{|-4\mathbf{i}-4\mathbf{j}+3\mathbf{k}|}{5} = \frac{\sqrt{41}}{5}\ \text{units.}$$

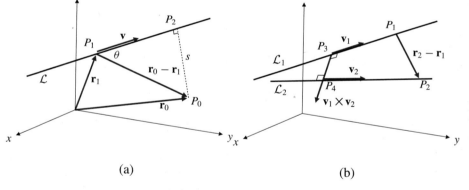

**Figure 10.32**

(a) The distance from $P_0$ to the line $\mathcal{L}$ is $s = |P_0P_1|\sin\theta$

(b) The distance between lines is the length of the projection of $P_1P_2$ along the vector $\mathbf{v}_1\times\mathbf{v}_2$

(a)                         (b)

**Example 9**   (**The distance between two lines**)   Find the distance between the two lines $\mathcal{L}_1$ through point $P_1$ parallel to vector $\mathbf{v}_1$ and $\mathcal{L}_2$ through point $P_2$ parallel to vector $\mathbf{v}_2$.

**Solution**   Let $\mathbf{r}_1$ and $\mathbf{r}_2$ be the position vectors of points $P_1$ and $P_2$, respectively. If $P_3$ and $P_4$ (with position vectors $\mathbf{r}_3$ and $\mathbf{r}_4$) are the points on $\mathcal{L}_1$ and $\mathcal{L}_2$, respectively, that are closest to one another, then $\overrightarrow{P_3P_4}$ is perpendicular to both lines, and is therefore parallel to $\mathbf{v}_1\times\mathbf{v}_2$. (See Figure 10.32(b).) $\overrightarrow{P_3P_4}$ is the vector projection of $\overrightarrow{P_1P_2} = \mathbf{r}_2 - \mathbf{r}_1$ along $\mathbf{v}_1\times\mathbf{v}_2$. Therefore, the distance $s = |\overrightarrow{P_3P_4}|$ between the lines is given by

$$s = |\mathbf{r}_4 - \mathbf{r}_3| = \frac{|(\mathbf{r}_2 - \mathbf{r}_1)\bullet(\mathbf{v}_1\times\mathbf{v}_2)|}{|\mathbf{v}_1\times\mathbf{v}_2|}.$$

# Exercises 10.4

**1.** A single equation involving the coordinates $(x, y, z)$ need not always represent a two-dimensional "surface" in $\mathbb{R}^3$. For example, $x^2 + y^2 + z^2 = 0$ represents the single point $(0, 0, 0)$, which has dimension zero. Give examples of single equations in $x$, $y$, and $z$ that represent

(a) a (one-dimensional) straight line,

(b) the whole of $\mathbb{R}^3$,

(c) no points at all (that is, the empty set).

In Exercises 2–9, find equations of the planes satisfying the given conditions.

**2.** Passing through $(0, 2, -3)$ and normal to the vector $4\mathbf{i} - \mathbf{j} - 2\mathbf{k}$

**3.** Passing through the origin and having normal $\mathbf{i} - \mathbf{j} + 2\mathbf{k}$

**4.** Passing through $(1, 2, 3)$ and parallel to the plane $3x + y - 2z = 15$

**5.** Passing through the three points $(1, 1, 0)$, $(2, 0, 2)$, and $(0, 3, 3)$

**6.** Passing through the three points $(-2, 0, 0)$, $(0, 3, 0)$, and $(0, 0, 4)$

**7.** Passing through $(1, 1, 1)$ and $(2, 0, 3)$ and perpendicular to the plane $x + 2y - 3z = 0$

**8.** Passing through the line of intersection of the planes $2x + 3y - z = 0$ and $x - 4y + 2z = -5$, and passing through the point $(-2, 0, -1)$

**9.** Passing through the line $x + y = 2$, $y - z = 3$, and perpendicular to the plane $2x + 3y + 4z = 5$

**10.** Under what geometric condition will three distinct points in $\mathbb{R}^3$ not determine a unique plane passing through them? How can this condition be expressed algebraically in terms of the position vectors, $\mathbf{r}_1$, $\mathbf{r}_2$, and $\mathbf{r}_3$, of the three points?

**11.** Give a condition on the position vectors of four points that guarantees that the four points are *coplanar*, that is, all lie on one plane.

Describe geometrically the one-parameter families of planes in Exercises 12–14. ($\lambda$ is a real parameter.)

**12.** $x + y + z = \lambda$.

**13.** $x + \lambda y + \lambda z = \lambda$.

**14.** $\lambda x + \sqrt{1 - \lambda^2}\, y = 1$.

In Exercises 15–19, find equations of the line specified in vector and scalar parametric forms and in standard form.

**15.** Through the point $(1, 2, 3)$ and parallel to $2\mathbf{i} - 3\mathbf{j} - 4\mathbf{k}$

**16.** Through $(-1, 0, 1)$ and perpendicular to the plane $2x - y + 7z = 12$

**17.** Through the origin and parallel to the line

$$x + 2y - z = 2, \qquad 2x - y + 4z = 5$$

**18.** Through $(2, -1, -1)$ and parallel to each of the two planes $x + y = 0$ and $x - y + 2z = 0$

**19.** Through $(1, 2, -1)$ and making equal angles with the positive directions of the coordinate axes

In Exercises 20–22, find the equations of the given line in standard form.

**20.** $\mathbf{r} = (1 - 2t)\mathbf{i} + (4 + 3t)\mathbf{j} + (9 - 4t)\mathbf{k}$.

**21.** $\begin{cases} x = 4 - 5t \\ y = 3t \\ z = 7 \end{cases}$

**22.** $\begin{cases} x - 2y + 3z = 0 \\ 2x + 3y - 4z = 4 \end{cases}$

**23.** Under what conditions on the position vectors of four distinct points $P_1$, $P_2$, $P_3$, and $P_4$ will the straight line through $P_1$ and $P_2$ intersect the straight line through $P_3$ and $P_4$ at a unique point?

Find the required distances in Exercises 24–27.

**24.** From the origin to the plane $x + 2y + 3z = 4$

**25.** From $(1, 2, 0)$ to the plane $3x - 4y - 5z = 2$

**26.** From the origin to the line $x + y + z = 0$, $2x - y - 5z = 1$

**27.** Between the lines

$$\begin{cases} x + 2y = 3 \\ y + 2z = 3 \end{cases} \quad \text{and} \quad \begin{cases} x + y + z = 6 \\ x - 2z = -5 \end{cases}$$

**28.** Show that the line

$$x - 2 = \frac{y + 3}{2} = \frac{z - 1}{4}$$

is parallel to the plane $2y - z = 1$. What is the distance between the line and the plane?

In Exercises 29–30, describe the one-parameter families of straight lines represented by the given equations. ($\lambda$ is a real parameter.)

**\* 29.** $(1 - \lambda)(x - x_0) = \lambda(y - y_0)$, $z = z_0$.

**\* 30.** $\dfrac{x - x_0}{\sqrt{1 - \lambda^2}} = \dfrac{y - y_0}{\lambda} = z - z_0$.

**31.** Why does the factored second-degree equation

$$(A_1 x + B_1 y + C_1 z - D_1)(A_2 x + B_2 y + C_2 z - D_2) = 0$$

represent a pair of planes rather than a single straight line?

## 10.5 Quadric Surfaces

The most general second-degree equation in three variables is

$$Ax^2 + By^2 + Cz^2 + Dxy + Exz + Fyz + Gx + Hy + Iz = J.$$

We will not attempt the (rather difficult) task of classifying all the surfaces that can be represented by such an equation, but will examine some interesting special cases. Let us observe at the outset that if the above equation can be factored in the form

$$(A_1 x + B_1 y + C_1 z - D_1)(A_2 x + B_2 y + C_2 z - D_2) = 0$$

then the graph is, in fact, a pair of planes,

$$A_1 x + B_1 y + C_1 z = D_1 \quad \text{and} \quad A_2 x + B_2 y + C_2 z = D_2,$$

or one plane if the two linear equations represent the same plane. This is considered a degenerate case. Where such factorization is not possible, the surface (called a **quadric surface**) will not be flat, although there may still be straight lines that lie on the surface. Nondegenerate quadric surfaces fall into the following six categories.

**Spheres.** The equation $x^2 + y^2 + z^2 = a^2$ represents a sphere of radius $a$ centred at the origin. More generally,

$$(x - x_0)^2 + (y - y_0)^2 + (z - z_0)^2 = a^2$$

represents a sphere of radius $a$ centred at the point $(x_0, y_0, z_0)$. If a quadratic equation in $x$, $y$, and $z$ has equal coefficients for the $x^2$, $y^2$, and $z^2$ terms and has no other second-degree terms, then it will represent, if any surface at all, a sphere. The centre can be found by completing the squares as for circles in the plane.

**Cylinders.** The equation $x^2 + y^2 = a^2$, being independent of $z$, represents a **right-circular cylinder** of radius $a$ and axis along the $z$-axis. (See Figure 10.33(a).) The intersection of the cylinder with the horizontal plane $z = k$ is the circle with equations

$$\begin{cases} x^2 + y^2 = a^2 \\ z = k. \end{cases}$$

Quadric cylinders also come in other shapes — elliptic, parabolic and hyperbolic. For instance, $z = x^2$ represents a parabolic cylinder with vertex line along the $y$-axis. (See Figure 10.33(b).) In general, an equation in two variables only will represent a cylinder in 3-space.

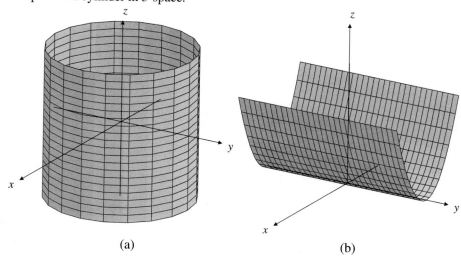

**Figure 10.33**

(a) The circular cylinder $x^2 + y^2 = a^2$

(b) The parabolic cylinder $z = x^2$

(a)     (b)

**Cones.** The equation $z^2 = x^2 + y^2$ represents a **right-circular cone** with axis along the $z$-axis. The surface is generated by rotating about the $z$-axis the line $z = y$ in the $yz$-plane. This *generator* makes an angle of $45°$ with the axis of the cone. Cross-sections of the cone in planes parallel to the $xy$-plane are circles. (See Figure 10.34(a).) The equation $x^2 + y^2 = a^2 z^2$ also represents a right circular cone with vertex at the origin and axis along the $z$-axis, but having semi-vertical angle $\alpha = \tan^{-1} a$. A circular cone has plane cross-sections that are elliptical, parabolic, and hyperbolic. Conversely, any nondegenerate quadric cone has a direction perpendicular to which the cross-sections of the cone are circular. In that sense, every quadric cone is a circular cone, though it may be *oblique* rather than right-circular in that the line joining the centres of the circular cross-sections need not be perpendicular to those cross-sections. (See Exercise 24 at the end of this section.)

**Ellipsoids.** The equation

$$\frac{x^2}{a^2} + \frac{y^2}{b^2} + \frac{z^2}{c^2} = 1$$

represents an **ellipsoid** with *semi-axes* $a$, $b$, and $c$. (See Figure 10.34(b).) The surface is oval, and it is enclosed inside the rectangular parallelepiped $-a \le x \le a$, $-b \le y \le b$, $-c \le z \le c$. If $a = b = c$ the ellipsoid is a sphere. In general, all plane cross-sections of ellipsoids are ellipses. This is easy to see for cross-sections parallel to coordinate planes, but somewhat harder to see for other planes.

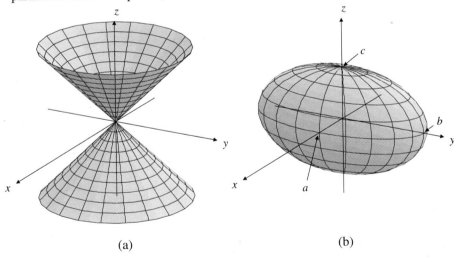

**Figure 10.34**

(a) The circular cone $a^2 z^2 = x^2 + y^2$

(b) The ellipsoid $\dfrac{x^2}{a^2} + \dfrac{y^2}{b^2} + \dfrac{z^2}{c^2} = 1$

(a)

(b)

**Paraboloids.** The equations

$$z = \frac{x^2}{a^2} + \frac{y^2}{b^2} \qquad \text{and} \qquad z = \frac{x^2}{a^2} - \frac{y^2}{b^2}$$

represent, respectively, an **elliptic paraboloid** and a **hyperbolic paraboloid**. (See Figures 10.35(a) and (b).) Cross-sections in planes $z = k$ ($k$ a positive constant) are ellipses (circles if $a = b$) and hyperbolas, respectively. Parabolic reflective mirrors have the shape of circular paraboloids. The hyperbolic paraboloid is a **ruled surface**. (A ruled surface is one through every point of which there passes a straight line lying wholly on the surface. Cones and cylinders are also examples of ruled surfaces.) There are two one-parameter families of straight lines that lie on the hyperbolic paraboloid, namely

$$\begin{cases} \lambda z = \dfrac{x}{a} - \dfrac{y}{b} \\ \dfrac{1}{\lambda} = \dfrac{x}{a} + \dfrac{y}{b} \end{cases} \qquad \text{and} \qquad \begin{cases} \mu z = \dfrac{x}{a} + \dfrac{y}{b} \\ \dfrac{1}{\mu} = \dfrac{x}{a} - \dfrac{y}{b} \end{cases}$$

where $\lambda$ and $\mu$ are real parameters. Every point on the hyperbolic paraboloid lies on one line of each family.

**Figure 10.35**

(a) The elliptic paraboloid
$$z = \frac{x^2}{a^2} + \frac{y^2}{b^2}$$

(b) The hyperbolic paraboloid
$$z = \frac{x^2}{a^2} - \frac{y^2}{b^2}$$

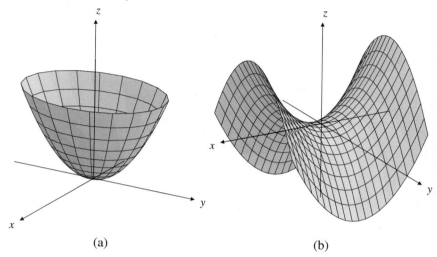

(a)                              (b)

**Hyperboloids.** The equation

$$\frac{x^2}{a^2} + \frac{y^2}{b^2} - \frac{z^2}{c^2} = 1$$

**Figure 10.36**

(a) The hyperboloid of one sheet
$$\frac{x^2}{a^2} + \frac{y^2}{b^2} - \frac{z^2}{c^2} = 1$$

(b) The hyperboloid of two sheets
$$\frac{x^2}{a^2} + \frac{y^2}{b^2} - \frac{z^2}{c^2} = -1$$

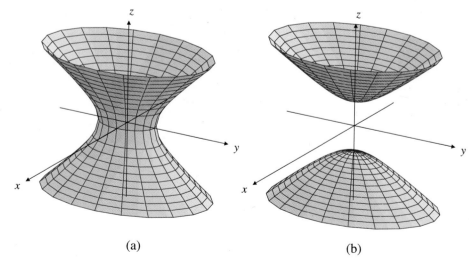

(a)                              (b)

represents a surface called a **hyperboloid of one sheet**. (See Figure 10.36(a).) The equation

$$\frac{x^2}{a^2} + \frac{y^2}{b^2} - \frac{z^2}{c^2} = -1$$

represents a **hyperboloid of two sheets**. (See Figure 10.36(b).) Both surfaces have elliptical cross-sections in horizontal planes and hyperbolic cross-sections in vertical planes. Both are *asymptotic* to the elliptic cone with equation

$$\frac{x^2}{a^2} + \frac{y^2}{b^2} = \frac{z^2}{c^2};$$

they approach arbitrarily close to the cone as they recede arbitrarily far away from the origin. Like the hyperbolic paraboloid, the hyperboloid of one sheet is a ruled surface.

## Exercises 10.5

Identify the surfaces represented by the equations in Exercises 1–16 and sketch their graphs.

**1.** $x^2 + 4y^2 + 9z^2 = 36$    **2.** $x^2 + y^2 + 4z^2 = 4$

**3.** $2x^2 + 2y^2 + 2z^2 - 4x + 8y - 12z + 27 = 0$

**4.** $x^2 + 4y^2 + 9z^2 + 4x - 8y = 8$

**5.** $z = x^2 + 2y^2$    **6.** $z = x^2 - 2y^2$

**7.** $x^2 - y^2 - z^2 = 4$    **8.** $-x^2 + y^2 + z^2 = 4$

**9.** $z = xy$    **10.** $x^2 + 4z^2 = 4$

**11.** $x^2 - 4z^2 = 4$    **12.** $y = z^2$

**13.** $x = z^2 + z$    **14.** $x^2 = y^2 + 2z^2$

**15.** $(z - 1)^2 = (x - 2)^2 + (y - 3)^2$

**16.** $(z - 1)^2 = (x - 2)^2 + (y - 3)^2 + 4$

Describe and sketch the geometric objects represented by the systems of equations in Exercises 17–20.

**17.** $\begin{cases} x^2 + y^2 + z^2 = 4 \\ x + y + z = 1 \end{cases}$    **18.** $\begin{cases} x^2 + y^2 = 1 \\ z = x + y \end{cases}$

**19.** $\begin{cases} z^2 = x^2 + y^2 \\ z = 1 + x \end{cases}$    **20.** $\begin{cases} x^2 + 2y^2 + 3z^2 = 6 \\ y = 1 \end{cases}$

**21.** Find two one-parameter families of straight lines that lie on the hyperboloid of one sheet

$$\frac{x^2}{a^2} + \frac{y^2}{b^2} - \frac{z^2}{c^2} = 1.$$

**22.** Find two one-parameter families of straight lines that lie on the hyperbolic paraboloid $z = xy$.

**23.** The equation $2x^2 + y^2 = 1$ represents a cylinder with elliptical cross-sections in planes perpendicular to the $z$-axis. Find a vector **a** perpendicular to which the cylinder has circular cross-sections.

**\* 24.** The equation $z^2 = 2x^2 + y^2$ represents a cone with elliptical cross-sections in planes perpendicular to the $z$-axis. Find a vector **a** perpendicular to which the cone has circular cross-sections. *Hint:* do Exercise 23 first, and use its result.

# 10.6    A Little Matrix Algebra

Differential calculus is essentially the study of linear approximations to functions. The tangent line to the graph $y = f(x)$ at $x = x_0$ provides the "best linear approximation" to $f(x)$ near $x_0$. Differentiation of functions of several variables can also be viewed as a process of finding *best linear approximations*. Therefore the language of linear algebra can be very useful for expressing certain concepts in the calculus of several variables.

Linear algebra is a vast subject, and is usually studied independently of calculus. This is unfortunate, because understanding the relationship between the two subjects can greatly enhance your understanding and appreciation of each of them. Knowledge of linear algebra, and therefore familiarity with the material covered in this section, is *not essential* for fruitful study of the rest of this book. However, we shall from time to time comment on the significance of the subject at hand from the point of view of linear algebra. To this end we need only a little of the terminology and content of linear algebra, especially that part pertaining to matrix manipulation and systems of linear equations. In the rest of this section we present an outline of this material. Some students will already be familiar with it; others will encounter it later. We make no attempt at completeness here, and refer interested students to standard linear algebra texts for proofs of some assertions. Students proceeding beyond this book to further study of advanced calculus and differential equations will certainly need a much more extensive background in linear algebra.

## Matrices

An $m \times n$ **matrix** $\mathcal{A}$ is a rectangular array of $mn$ numbers arranged in $m$ rows and $n$ columns. If $a_{ij}$ is the element in the $i$th row and the $j$th column, then

$$\mathcal{A} = \begin{pmatrix} a_{11} & a_{12} & \cdots & a_{1n} \\ a_{21} & a_{22} & \cdots & a_{2n} \\ \vdots & \vdots & & \vdots \\ a_{m1} & a_{m2} & \cdots & a_{mn} \end{pmatrix}.$$

Sometimes, as a shorthand notation, we write $\mathcal{A} = (a_{ij})$. In this case $i$ is assumed to range from 1 to $m$ and $j$ from 1 to $n$. If $m = n$ we say that $\mathcal{A}$ is a square matrix. The elements $a_{ij}$ of the matrices we use in this book will always be real numbers.

The **transpose** of an $m \times n$ matrix $\mathcal{A}$ is the $n \times m$ matrix $\mathcal{A}^T$ whose rows are the columns of $\mathcal{A}$:

$$\mathcal{A}^T = \begin{pmatrix} a_{11} & a_{21} & \cdots & a_{m1} \\ a_{12} & a_{22} & \cdots & a_{m2} \\ \vdots & \vdots & & \vdots \\ a_{1n} & a_{2n} & \cdots & a_{mn} \end{pmatrix}.$$

Matrix $\mathcal{A}$ is called **symmetric** if $\mathcal{A}^T = \mathcal{A}$. Symmetric matrices are necessarily square. Observe that $(\mathcal{A}^T)^T = \mathcal{A}$ for *every* matrix $\mathcal{A}$. Frequently we want to consider an $n$-vector $\mathbf{x}$ as an $n \times 1$ matrix having $n$ rows and one column:

$$\mathbf{x} = \begin{pmatrix} x_1 \\ x_2 \\ \vdots \\ x_n \end{pmatrix}.$$

As such, $\mathbf{x}$ is called a **column vector**. $\mathbf{x}^T$ has one row and $n$ columns, and so is called a **row vector**:

$$\mathbf{x}^T = (x_1 \ x_2 \ \cdots \ x_n).$$

Most of the usefulness of matrices depends on the following definition of matrix multiplication, which enables two arrays to be combined into a single one in a manner that preserves linear relationships.

**DEFINITION** **7**

---

**Multiplying matrices**

If $A = (a_{ij})$ is an $m \times n$ matrix and $B = (b_{ij})$ is an $n \times p$ matrix, then the product $AB$ is the $m \times p$ matrix $C = (c_{ij})$ with elements given by

$$c_{ij} = \sum_{k=1}^{n} a_{ik}b_{kj}, \qquad i = 1, \ldots, m, \quad j = 1, \ldots, p.$$

That is, $c_{ij}$ is the *dot product* of the $i$th row of $A$ and the $j$th column of $B$ (both of which are $n$-vectors).

---

Note that only *some* pairs of matrices can be multiplied. The product $AB$ is only defined if the number of columns of $A$ is equal to the number of rows of $B$.

**Example 1**

$$\begin{pmatrix} 1 & 0 & 3 \\ 2 & 1 & -1 \end{pmatrix} \begin{pmatrix} 2 & 1 & 1 & 0 \\ 0 & -1 & 3 & 1 \\ 1 & 0 & 4 & 5 \end{pmatrix} = \begin{pmatrix} 5 & 1 & 13 & 15 \\ 3 & 1 & 1 & -4 \end{pmatrix}$$

The left factor has 2 rows and 3 columns, and the right factor has 3 rows and 4 columns. Therefore the product has 2 rows and 4 columns. The element in the first row and third column of the product, 13, is the dot product of the first row, $(1, 0, 3)$, of the left factor and the third column, $(1, 3, 4)$, of the second factor:

$$1 \times 1 + 0 \times 3 + 3 \times 4 = 13.$$

With a little practice you can easily calculate the elements of a matrix product by simultaneously running your left index finger across rows of the left factor and your right index finger down columns of the right factor while taking the dot products. ∎

**Example 2**

$$\begin{pmatrix} 1 & 2 & 3 \\ 0 & 1 & -1 \\ -2 & 3 & 0 \end{pmatrix} \begin{pmatrix} x \\ y \\ z \end{pmatrix} = \begin{pmatrix} x + 2y + 3z \\ y - z \\ -2x + 3y \end{pmatrix}$$

The product of a $3 \times 3$ matrix with a column 3-vector is a column 3-vector. ∎

Matrix multiplication is *associative*. This means that

$$A(BC) = (AB)C$$

(provided $A$, $B$, and $C$ have dimensions compatible with the formation of the various products) and therefore it makes sense to write $ABC$. However, matrix multiplication is *not commutative*. Indeed, if $A$ is an $m \times n$ matrix and $B$ is an $n \times p$ matrix then the product $AB$ is defined, but the product $BA$ is not defined unless $m = p$. Even if $A$ and $B$ are square matrices of the same size it is not necessarily true that $AB = BA$.

**Example 3**

$$\begin{pmatrix} 1 & 2 \\ 3 & 0 \end{pmatrix}\begin{pmatrix} 1 & -1 \\ 1 & 1 \end{pmatrix} = \begin{pmatrix} 3 & 1 \\ 3 & -3 \end{pmatrix} \quad \text{but} \quad \begin{pmatrix} 1 & -1 \\ 1 & 1 \end{pmatrix}\begin{pmatrix} 1 & 2 \\ 3 & 0 \end{pmatrix} = \begin{pmatrix} -2 & 2 \\ 4 & 2 \end{pmatrix}$$

The reader should verify that if the product $\mathcal{AB}$ is defined then the transpose of the product is the product of the transposes *in the reverse order*:

$$(\mathcal{AB})^T = \mathcal{B}^T \mathcal{A}^T.$$

## Determinants and Matrix Inverses

In Section 10.3 we introduced $2 \times 2$ and $3 \times 3$ determinants as certain algebraic expressions associated with $2 \times 2$ and $3 \times 3$ square arrays of numbers. In general, it is possible to define the determinant $\det(\mathcal{A})$ for any square matrix. For an $n \times n$ matrix $\mathcal{A}$ we continue to denote

$$\det(\mathcal{A}) = \begin{vmatrix} a_{11} & a_{12} & \cdots & a_{1n} \\ a_{21} & a_{22} & \cdots & a_{2n} \\ \vdots & \vdots & \ddots & \vdots \\ a_{n1} & a_{n2} & \cdots & a_{nn} \end{vmatrix}.$$

We will not attempt to give a formal definition of the determinant here, but will note that the properties of determinants stated for the $3 \times 3$ case in Section 10.3 continue to be true. In particular, an $n \times n$ determinant can be expanded in minors about any row or column and so expressed as a sum of multiples of $(n - 1) \times (n - 1)$ determinants. Continuing this process, we can eventually reduce the evaluation of any $n \times n$ determinant to the evaluation of (perhaps many) $2 \times 2$ or $3 \times 3$ determinants. It is important to realize that the "diagonal" method for evaluating $2 \times 2$ or $3 \times 3$ determinants does not extend to $4 \times 4$ or higher-order determinants.

**Example 4**

$$\begin{vmatrix} 2 & 1 & 0 & 1 \\ 1 & 0 & 1 & 1 \\ 3 & 0 & 0 & 2 \\ -1 & 1 & 1 & 0 \end{vmatrix} = - \begin{vmatrix} 2 & 1 & 1 \\ 3 & 0 & 2 \\ -1 & 1 & 0 \end{vmatrix} - \begin{vmatrix} 2 & 1 & 1 \\ 1 & 0 & 1 \\ 3 & 0 & 2 \end{vmatrix}$$

$$= - \left( -3 \begin{vmatrix} 1 & 1 \\ 1 & 0 \end{vmatrix} - 2 \begin{vmatrix} 2 & 1 \\ -1 & 1 \end{vmatrix} \right) - \left( -1 \begin{vmatrix} 1 & 1 \\ 3 & 2 \end{vmatrix} \right)$$

$$= 3(0 - 1) + 2(2 + 1) + 1(2 - 3) = 2.$$

We expanded the $4 \times 4$ determinant in minors about the third column to obtain the two $3 \times 3$ determinants. The first of these was expanded about the second row, the other about the second column.

In addition to the properties stated in Section 10.3, determinants have two other very important properties, which are stated in the following theorem.

**THEOREM 3**

If $\mathcal{A}$ and $\mathcal{B}$ are $n \times n$ matrices then

(a) $\det(\mathcal{A}^T) = \det(\mathcal{A})$.

(b)    $\det(\mathcal{A}\mathcal{B}) = \det(\mathcal{A})\det(\mathcal{B})$.

We will not attempt any proof of this or other theorems in this section. The reader is referred to texts on linear algebra. Part (a) is not very difficult to prove, even in the case of general $n$. Part (b) cannot really be proved in general without a formal definition of determinant. However, the reader should verify (b) for $2 \times 2$ matrices by direct calculation.

We say that the square matrix $\mathcal{A}$ is **singular** if $\det(\mathcal{A}) = 0$. If $\det(\mathcal{A}) \neq 0$ we say that $\mathcal{A}$ is **nonsingular** or **invertible**.

**Remark**   If $\mathcal{A}$ is a $3 \times 3$ matrix then $\det(\mathcal{A})$ is the scalar triple product of the rows of $\mathcal{A}$, and its absolute value is the volume of the parallelepiped spanned by those rows. Therefore $\mathcal{A}$ is nonsingular if and only if its rows span a parallelepiped of positive volume; the row vectors cannot all lie in the same plane. The same may be said of the columns of $\mathcal{A}$.

In general, an $n \times n$ matrix is singular if its rows (or columns), considered as vectors, satisfy one or more linear equations of the form

$$c_1 \mathbf{x}_1 + c_2 \mathbf{x}_2 + \cdots + c_n \mathbf{x}_n = \mathbf{0},$$

with at least one nonzero coefficient $c_i$. A set of vectors satisfying such a linear equations is called **linearly dependent** because one of the vectors can always be expressed as a linear combination of the others; if $c_1 \neq 0$ then

$$\mathbf{x}_1 = -\frac{c_2}{c_1} \mathbf{x}_2 - \frac{c_3}{c_1} \mathbf{x}_3 - \cdots - \frac{c_n}{c_1} \mathbf{x}_n.$$

All linear combinations of the vectors in a linearly dependent set of $n$ vectors in $\mathbb{R}^n$ must lie in a **subspace** of dimension lower than $n$.

The $n \times n$ **identity matrix** is the matrix

$$\mathcal{I} = \begin{pmatrix} 1 & 0 & \cdots & 0 \\ 0 & 1 & \cdots & 0 \\ \vdots & \vdots & \ddots & \vdots \\ 0 & 0 & \cdots & 1 \end{pmatrix}$$

with "1" in every position on the **main diagonal** and "0" in every other position. Evidently $\mathcal{I}$ commutes with every $n \times n$ matrix: $\mathcal{I}\mathcal{A} = \mathcal{A}\mathcal{I} = \mathcal{A}$. Also $\det(\mathcal{I}) = 1$. The identity matrix plays the same role in matrix algebra that the number 1 plays in arithmetic.

Any nonzero number $x$ has a reciprocal $x^{-1}$ such that $xx^{-1} = x^{-1}x = 1$. A similar situation holds for square matrices. The **inverse** of a *nonsingular* square matrix $\mathcal{A}$ is a nonsingular square matrix $\mathcal{A}^{-1}$ satisfying

$$\mathcal{A}\mathcal{A}^{-1} = \mathcal{A}^{-1}\mathcal{A} = \mathcal{I}.$$

**THEOREM 4**

Every nonsingular square matrix $\mathcal{A}$ has a *unique* inverse $\mathcal{A}^{-1}$. Moreover, the inverse satisfies

(a)    $\det(\mathcal{A}^{-1}) = \frac{1}{\det(\mathcal{A})}$,

(b)    $(\mathcal{A}^{-1})^T = (\mathcal{A}^T)^{-1}$.

We will not have much cause to calculate inverses, but we note that it can be done by solving systems of linear equations, as the following simple example illustrates.

**Example 5** Show that the matrix $\mathcal{A} = \begin{pmatrix} 1 & -1 \\ 1 & 1 \end{pmatrix}$ is nonsingular and find its inverse.

**Solution** $\det(\mathcal{A}) = \begin{vmatrix} 1 & -1 \\ 1 & 1 \end{vmatrix} = 1 + 1 = 2.$ Therefore $\mathcal{A}$ is nonsingular and invertible. Let $\mathcal{A}^{-1} = \begin{pmatrix} a & b \\ c & d \end{pmatrix}$. Then

$$\begin{pmatrix} 1 & 0 \\ 0 & 1 \end{pmatrix} = \begin{pmatrix} 1 & -1 \\ 1 & 1 \end{pmatrix} \begin{pmatrix} a & b \\ c & d \end{pmatrix} = \begin{pmatrix} a-c & b-d \\ a+c & b+d \end{pmatrix},$$

so $a$, $b$, $c$, and $d$ must satisfy the systems of equations

$$\begin{cases} a - c = 1 \\ a + c = 0 \end{cases} \qquad \begin{cases} b - d = 0 \\ b + d = 1. \end{cases}$$

Evidently $a = b = d = 1/2$, $c = -1/2$, and

$$\mathcal{A}^{-1} = \begin{pmatrix} \frac{1}{2} & \frac{1}{2} \\ -\frac{1}{2} & \frac{1}{2} \end{pmatrix}.$$

Generally, matrix inversion is not carried out by the method of the above example, but rather by an orderly process of performing operations on the rows of the matrix to transform it into the identity. When the same operations are performed on the rows of the identity matrix, the inverse of the original matrix results. See a text on linear algebra for a description of the method.

## Linear Transformations

A function $F$ whose domain is the $m$-dimensional space $\mathbb{R}^m$ and whose range is contained in the $n$-dimensional space $\mathbb{R}^n$ is called a **linear transformation from $\mathbb{R}^m$ to $\mathbb{R}^n$** if it satisfies

$$F(\lambda \mathbf{x} + \mu \mathbf{y}) = \lambda F(\mathbf{x}) + \mu F(\mathbf{y})$$

for all points $\mathbf{x}$ and $\mathbf{y}$ in $\mathbb{R}^m$ and all real numbers $\lambda$ and $\mu$. To such a linear transformation $F$ there corresponds an $n \times m$ matrix $\mathcal{F}$ such that for all $\mathbf{x}$ in $\mathbb{R}^m$,

$$F(\mathbf{x}) = \mathcal{F}\mathbf{x},$$

or expressed in terms of the components of $\mathbf{x}$,

$$F(x_1, x_2, \cdots, x_m) = \mathcal{F} \begin{pmatrix} x_1 \\ x_2 \\ \vdots \\ x_m \end{pmatrix}.$$

We say that $\mathcal{F}$ is a **matrix representation** of the linear transformation $F$. If $m = n$ so that $F$ maps $\mathbb{R}^m$ into itself, then $\mathcal{F}$ is a square matrix. In this case $\mathcal{F}$ is nonsingular if and only if $F$ is one-to-one and has the whole of $\mathbb{R}^m$ as range.

A composition of linear transformations is still a linear transformation and so will have a matrix representation. The real motivation lying behind the definition of matrix multiplication is that the matrix representation of a *composition* of linear transformations is the *product* of the individual matrix representations of the transformations being composed.

**THEOREM 5**

If $F$ is a linear transformation from $\mathbb{R}^m$ to $\mathbb{R}^n$ represented by the $n \times m$ matrix $\mathcal{F}$, and if $G$ is a linear transformation from $\mathbb{R}^n$ to $\mathbb{R}^p$ represented by the $p \times n$ matrix $\mathcal{G}$, then the composition $G \circ F$ defined by

$$G \circ F(x_1, x_2, \ldots, x_m) = G\Big(F(x_1, x_2, \ldots, x_m)\Big)$$

is itself a linear transformation from $\mathbb{R}^m$ to $\mathbb{R}^p$ represented by the $p \times m$ matrix $\mathcal{G}\mathcal{F}$. That is,

$$G\Big(F(\mathbf{x})\Big) = \mathcal{G}\mathcal{F}\mathbf{x}.$$

## Linear Equations

A system of $n$ linear equations in $n$ unknowns:

$$a_{11}x_1 + a_{12}x_2 + \cdots + a_{1n}x_n = b_1$$
$$a_{21}x_1 + a_{22}x_2 + \cdots + a_{2n}x_n = b_2$$
$$\vdots$$
$$a_{n1}x_1 + a_{n2}x_2 + \cdots + a_{nn}x_n = b_n$$

can be written compactly as a single matrix equation,

$$\mathcal{A}\mathbf{x} = \mathbf{b},$$

where

$$\mathcal{A} = \begin{pmatrix} a_{11} & a_{12} & \cdots & a_{1n} \\ a_{21} & a_{22} & \cdots & a_{2n} \\ \vdots & \vdots & \ddots & \vdots \\ a_{n1} & a_{n2} & \cdots & a_{nn} \end{pmatrix}, \qquad \mathbf{x} = \begin{pmatrix} x_1 \\ x_2 \\ \vdots \\ x_n \end{pmatrix}, \quad \text{and} \quad \mathbf{b} = \begin{pmatrix} b_1 \\ b_2 \\ \vdots \\ b_n \end{pmatrix}.$$

Compare the equation $\mathcal{A}\mathbf{x} = \mathbf{b}$ with the equation $ax = b$ for a single unknown $x$. The equation $ax = b$ has the unique solution $x = a^{-1}b$ provided $a \neq 0$. By analogy, the linear system $\mathcal{A}\mathbf{x} = \mathbf{b}$ has a unique solution given by

$$\mathbf{x} = \mathcal{A}^{-1}\mathbf{b},$$

provided $\mathcal{A}$ is nonsingular. To see this, just multiply both sides of the equation $\mathcal{A}\mathbf{x} = \mathbf{b}$ on the left by $\mathcal{A}^{-1}$.

If $\mathcal{A}$ is singular, then the system $\mathcal{A}\mathbf{x} = \mathbf{b}$ may or may not have a solution, and if a solution exists it will not be unique. Consider the case $\mathbf{b} = \mathbf{0}$ (the zero vector). Then any vector $\mathbf{x}$ perpendicular to all the rows of $\mathcal{A}$ will satisfy the system. Since the rows of $\mathcal{A}$ lie in a space of dimension less than $n$ (because $\det(\mathcal{A}) = 0$) there will be at least a line of such vectors $\mathbf{x}$. Thus solutions of $\mathcal{A}\mathbf{x} = \mathbf{0}$ are not unique if $\mathcal{A}$ is singular. The same must be true of the system $\mathcal{A}^T \mathbf{y} = \mathbf{0}$; there will be nonzero vectors $\mathbf{y}$ satisfying it if $\mathcal{A}$ is singular. But then, if the system $\mathcal{A}\mathbf{x} = \mathbf{b}$ has any solution $\mathbf{x}$ we must have

$$(\mathbf{y} \bullet \mathbf{b}) = \mathbf{y}^T \mathbf{b} = \mathbf{y}^T \mathcal{A}\mathbf{x} = (\mathbf{x}^T \mathcal{A}^T \mathbf{y})^T = (\mathbf{x}^T \mathbf{0})^T = (0).$$

Hence $A\mathbf{x} = \mathbf{b}$ can only have solutions for those vectors $\mathbf{b}$ that are perpendicular to *every* solution $\mathbf{y}$ of $A^T\mathbf{y} = \mathbf{0}$.

A system of $m$ linear equations in $n$ unknowns may or may not have any solutions if $n < m$. It will have solutions if some $m - n$ of the equations are *linear combinations* (sums of multiples) of the other $n$ equations. If $n > m$, then we can try to solve the $m$ equations for $m$ of the variables, allowing the solutions to depend on the other $n - m$ variables. Such a solution exists if the determinant of the coefficients of the $m$ variables for which we want to solve is not zero. This is a special case of the **implicit function theorem** that we will examine in Section 12.8.

**Example 6**   Solve $\begin{cases} 2x + y - 3z = 4 \\ x + 2y + 6z = 5 \end{cases}$ for $x$ and $y$ in terms of $z$.

*Solution*   The system can be expressed in the form

$$A\begin{pmatrix} x \\ y \end{pmatrix} = \begin{pmatrix} 4 + 3z \\ 5 - 6z \end{pmatrix} \qquad \text{where} \quad A = \begin{pmatrix} 2 & 1 \\ 1 & 2 \end{pmatrix}$$

$A$ has determinant 3 and inverse $A^{-1} = \begin{pmatrix} 2/3 & -1/3 \\ -1/3 & 2/3 \end{pmatrix}$. Thus

$$\begin{pmatrix} x \\ y \end{pmatrix} = A^{-1}\begin{pmatrix} 4 + 3z \\ 5 - 6z \end{pmatrix} = \begin{pmatrix} 2/3 & -1/3 \\ -1/3 & 2/3 \end{pmatrix}\begin{pmatrix} 4 + 3z \\ 5 - 6z \end{pmatrix} = \begin{pmatrix} 1 + 4z \\ 2 - 5z \end{pmatrix}$$

The solution is $x = 1 + 4z$, $y = 2 - 5z$. (Of course, this solution could have been found by elimination of $x$ or $y$ from the given equations without using matrix methods.)

We conclude this section by stating a result of some theoretical importance expressing the solution of the system $A\mathbf{x} = \mathbf{b}$ for nonsingular $A$ in terms of determinants.

**THEOREM 6**

**Cramer's Rule**

Let $A$ be a nonsingular $n \times n$ matrix. Then the solution $\mathbf{x}$ of the system

$$A\mathbf{x} = \mathbf{b}$$

has components given by

$$x_1 = \frac{\det(A_1)}{\det(A)}, \quad x_2 = \frac{\det(A_2)}{\det(A)}, \quad \cdots, \quad x_n = \frac{\det(A_n)}{\det(A)},$$

where $A_j$ is the matrix $A$ with its $j$th column replaced by the column vector $\mathbf{b}$. That is

$$\det(A_j) = \begin{vmatrix} a_{11} & \cdots & a_{1(j-1)} & b_1 & a_{1(j+1)} & \cdots & a_{1n} \\ a_{21} & \cdots & a_{2(j-1)} & b_2 & a_{2(j+1)} & \cdots & a_{2n} \\ \vdots & & \vdots & \vdots & \vdots & & \vdots \\ a_{n1} & \cdots & a_{n(j-1)} & b_n & a_{n(j+1)} & \cdots & a_{nn} \end{vmatrix}.$$

The following example provides a concrete illustration of use of Cramer's Rule to solve a specific linear system. However, Cramer's Rule is primarily used in a more general (theoretical) context; it is not efficient to use determinants to calculate solutions of linear systems.

**Example 7** Find the point of intersection of the three planes

$$x + y + 2z = 1$$
$$3x + 6y - z = 0$$
$$x - y - 4z = 3.$$

**Solution** The solution of the linear system above provides the coordinates of the intersection point. The determinant of the coefficient matrix of this system is

$$\det(\mathcal{A}) = \begin{vmatrix} 1 & 1 & 2 \\ 3 & 6 & -1 \\ 1 & -1 & -4 \end{vmatrix} = -32,$$

so the system does have a unique solution. We have

$$x = \frac{1}{-32} \begin{vmatrix} 1 & 1 & 2 \\ 0 & 6 & -1 \\ 3 & -1 & -4 \end{vmatrix} = \frac{-64}{-32} = 2,$$

$$y = \frac{1}{-32} \begin{vmatrix} 1 & 1 & 2 \\ 3 & 0 & -1 \\ 1 & 3 & -4 \end{vmatrix} = \frac{32}{-32} = -1,$$

$$z = \frac{1}{-32} \begin{vmatrix} 1 & 1 & 1 \\ 3 & 6 & 0 \\ 1 & -1 & 3 \end{vmatrix} = \frac{0}{-32} = 0.$$

The intersection point is $(2, -1, 0)$.

# Exercises 10.6

Evaluate the matrix products in Exercises 1–4.

**1.** $\begin{pmatrix} 3 & 0 & -2 \\ 1 & 1 & 2 \\ -1 & 1 & -1 \end{pmatrix} \begin{pmatrix} 2 & 1 \\ 3 & 0 \\ 0 & -2 \end{pmatrix}$

**2.** $\begin{pmatrix} 1 & 1 & 1 \\ 0 & 1 & 1 \\ 0 & 0 & 1 \end{pmatrix} \begin{pmatrix} 1 & 1 & 1 \\ 0 & 1 & 1 \\ 0 & 0 & 1 \end{pmatrix}$

**3.** $\begin{pmatrix} a & b \\ c & d \end{pmatrix} \begin{pmatrix} w & x \\ y & z \end{pmatrix}$  **4.** $\begin{pmatrix} w & x \\ y & z \end{pmatrix} \begin{pmatrix} a & b \\ c & d \end{pmatrix}$

**5.** Evaluate $\mathcal{A}\mathcal{A}^T$ and $\mathcal{A}^2 = \mathcal{A}\mathcal{A}$ where

$$\mathcal{A} = \begin{pmatrix} 1 & 1 & 1 & 1 \\ 0 & 1 & 1 & 1 \\ 0 & 0 & 1 & 1 \\ 0 & 0 & 0 & 1 \end{pmatrix}$$

**6.** Evaluate $\mathbf{x}\mathbf{x}^T$, $\mathbf{x}^T\mathbf{x}$, and $\mathbf{x}^T\mathcal{A}\mathbf{x}$ where

$$\mathbf{x} = \begin{pmatrix} x \\ y \\ z \end{pmatrix} \quad \text{and} \quad \mathcal{A} = \begin{pmatrix} a & p & q \\ p & b & r \\ q & r & c \end{pmatrix}.$$

Evaluate the determinants in Exercises 7–8.

**7.** $\begin{vmatrix} 2 & 3 & -1 & 0 \\ 4 & 0 & 2 & 1 \\ 1 & 0 & -1 & 1 \\ -2 & 0 & 0 & 1 \end{vmatrix}$  **8.** $\begin{vmatrix} 1 & 1 & 1 & 1 \\ 1 & 2 & 3 & 4 \\ -2 & 0 & 2 & 4 \\ 3 & -3 & 2 & -2 \end{vmatrix}$

**9.** Show that if $\mathcal{A} = (a_{ij})$ is an $n \times n$ matrix for which $a_{ij} = 0$ whenever $i > j$, then $\det(\mathcal{A}) = \prod_{k=1}^{n} a_{kk}$, the product of the elements on the main diagonal of $\mathcal{A}$.

**10.** Show that $\begin{vmatrix} 1 & 1 \\ x & y \end{vmatrix} = y - x$, and

$$\begin{vmatrix} 1 & 1 & 1 \\ x & y & z \\ x^2 & y^2 & z^2 \end{vmatrix} = (y - x)(z - x)(z - y).$$

Try to generalize this result to the $n \times n$ case.

**11.** Verify the associative law $(\mathcal{A}\mathcal{B})\mathcal{C} = \mathcal{A}(\mathcal{B}\mathcal{C})$ by direct calculation for three arbitrary $2 \times 2$ matrices.

**12.** Show that $\det(\mathcal{A}^T) = \det(\mathcal{A})$ for $n \times n$ matrices by induction on $n$. Start with the $2 \times 2$ case.

**13.** Verify by direct calculation that $\det(\mathcal{AB}) = \det(\mathcal{A})\det(\mathcal{B})$ holds for two arbitrary $2 \times 2$ matrices.

**14.** Let $\mathcal{A}_\theta = \begin{pmatrix} \cos\theta & \sin\theta \\ -\sin\theta & \cos\theta \end{pmatrix}$. Show that $(\mathcal{A}_\theta)^T = (\mathcal{A}_\theta)^{-1} = \mathcal{A}_{-\theta}$.

Find the inverses of the matrices in Exercises 15–16.

**15.** $\begin{pmatrix} 1 & 1 & 1 \\ 0 & 1 & 1 \\ 0 & 0 & 1 \end{pmatrix}$  **16.** $\begin{pmatrix} 1 & 0 & -1 \\ -1 & 1 & 0 \\ 2 & 1 & 3 \end{pmatrix}$

**17.** Use your result from Exercise 16 to solve the linear system
$$\begin{cases} x - z = -2 \\ -x + y = 1 \\ 2x + y + 3z = 13. \end{cases}$$

**18.** Solve the system of Exercise 17 by using Cramer's Rule.

**19.** Solve the system $\begin{cases} x_1 + x_2 + x_3 + x_4 = 0 \\ x_1 + x_2 + x_3 - x_4 = 4 \\ x_1 + x_2 - x_3 - x_4 = 6 \\ x_1 - x_2 - x_3 - x_4 = 2. \end{cases}$

**20.** Verify Theorem 5 for the special case where $F$ and $G$ are linear transformations from $\mathbb{R}^2$ to $\mathbb{R}^2$.

# Chapter Review

## Key Ideas

• **What is each of the following?**

◇ a neighbourhood  ◇ an open set  ◇ a closed set

◇ the boundary of a set  ◇ the interior of a set

◇ a vector in 3-space  ◇ the dot product of vectors

◇ the cross product of two vectors in $\mathbb{R}^3$

◇ a scalar triple product  ◇ a vector triple product

◇ a matrix  ◇ a determinant

◇ a plane  ◇ a straight line  ◇ a cone

◇ a cylinder  ◇ an ellipsoid  ◇ a paraboloid

◇ a hyperboloid of 1 sheet  ◇ a hyperboloid of 2 sheets

◇ the transpose of a matrix  ◇ the inverse of a matrix

◇ a linear transformation

• **What is the angle between the vectors u and v?**
• **How do you calculate u × v, given the components of u and v?**
• **What is an equation of the plane through $P_0$ having normal vector N?**
• **What is an equation of the straight line through $P_0$ parallel to a?**
• **Given two $3 \times 3$ matrices A and B, how do you calculate AB?**
• **What is the distance from $P_0$ to the plane $Ax + By + Cz + D = 0$?**
• **What is Cramer's Rule and how is it used?**

## Review Exercises

Describe the sets of points in 3-space that satisfy the given equations or inequalities in Exercises 1–18.

**1.** $x + 3z = 3$

**2.** $y - z \geq 1$

**3.** $x + y + z \geq 0$

**4.** $x - 2y - 4z = 8$

**5.** $y = 1 + x^2 + z^2$

**6.** $y = z^2$

**7.** $x = y^2 - z^2$

**8.** $z = xy$

**9.** $x^2 + y^2 + 4z^2 < 4$

**10.** $x^2 + y^2 - 4z^2 = 4$

**11.** $x^2 - y^2 - 4z^2 = 0$

**12.** $x^2 - y^2 - 4z^2 = 4$

**\*13.** $(x - z)^2 + y^2 = 1$

**\*14.** $(x - z)^2 + y^2 = z^2$

**15.** $\begin{cases} x + 2y = 0 \\ z = 3 \end{cases}$

**16.** $\begin{cases} x + y + 2z = 1 \\ x + y + z = 0 \end{cases}$

**17.** $\begin{cases} x^2 + y^2 + z^2 = 4 \\ x + y + z = 3 \end{cases}$

**18.** $\begin{cases} x^2 + z^2 \leq 1 \\ x - y \geq 0 \end{cases}$

Find equations of the planes and lines specified in Exercises 19–28.

**19.** The plane through the origin perpendicular to the line
$$\frac{x - 1}{2} = \frac{y + 3}{-1} = \frac{z + 2}{3}$$

**20.** The plane through $(2, -1, 1)$ and $(1, 0, -1)$ parallel to the line in Exercise 19

**21.** The plane through $(2, -1, 1)$ perpendicular to the planes $x - y + z = 0$ and $2x + y - 3z = 2$

**22.** The plane through $(-1, 1, 0)$, $(0, 4, -1)$, and $(2, 0, 0)$

**23.** The plane containing the line of intersection of the planes $x + y + z = 0$ and $2x + y - 3z = 2$, and passing through the point $(2, 0, 1)$

**24.** The plane containing the line of intersection of the planes $x + y + z = 0$ and $2x + y - 3z = 2$, and perpendicular to the plane $x - 2y - 5z = 17$

**25.** The vector parametric equation of the line through $(2, 1, -1)$ and $(-1, 0, 1)$

**26.** Standard form equations of the line through $(1, 0, -1)$ parallel to each of the planes $x - y = 3$ and $x + 2y + z = 1$

**27.** Scalar parametric equations of the line through the origin perpendicular to the plane $3x - 2y + 4z = 5$

**28.** The vector parametric equation of the line that joins points on the two lines

$$\mathbf{r} = (1+t)\mathbf{i} - t\mathbf{j} - (2+2t)\mathbf{k}$$
$$\mathbf{r} = 2t\mathbf{i} + (t-2)\mathbf{j} - (1+3t)\mathbf{k}$$

and is perpendicular to both those lines

Express the given conditions or quantities in Exercises 29–30 in terms of dot and cross products.

**29.** The three points with position vectors $\mathbf{r}_1$, $\mathbf{r}_2$, and $\mathbf{r}_3$ all lie on a straight line

**30.** The four points with position vectors $\mathbf{r}_1$, $\mathbf{r}_2$, $\mathbf{r}_3$, and $\mathbf{r}_4$ do not all lie on a plane

**31.** Find the area of the triangle with vertices $(1, 2, 1)$, $(4, -1, 1)$, and $(3, 4, -2)$.

**32.** Find the volume of the tetrahedron with vertices $(1, 2, 1)$, $(4, -1, 1)$, $(3, 4, -2)$, and $(2, 2, 2)$.

**33.** Show that the matrix

$$\mathcal{A} = \begin{pmatrix} 1 & 0 & 0 & 0 \\ 2 & 1 & 0 & 0 \\ 3 & 2 & 1 & 0 \\ 4 & 3 & 2 & 1 \end{pmatrix}$$

has an inverse, and find the inverse $\mathcal{A}^{-1}$.

**34.** Let $\mathcal{A} = \begin{pmatrix} 1 & 1 & 1 \\ 2 & 1 & 0 \\ 1 & 0 & -1 \end{pmatrix}$. What condition must the vector $\mathbf{b}$ satsify in order that the equation $\mathcal{A}\mathbf{x} = \mathbf{b}$ has solutions $\mathbf{x}$? What are the solutions $\mathbf{x}$ if $\mathbf{b}$ satisfies the condition?

## Challenging Problems

**1.** Show that the distance $d$ from point $P$ to the line $AB$ can be expressed in terms of the position vectors of $P$, $A$, and $B$ by

$$d = \frac{|(\mathbf{r}_A - \mathbf{r}_P) \times (\mathbf{r}_B - \mathbf{r}_P)|}{|\mathbf{r}_A - \mathbf{r}_B|}$$

**2.** For any vectors $\mathbf{u}$, $\mathbf{v}$, $\mathbf{w}$, and $\mathbf{x}$, show that

$$(\mathbf{u} \times \mathbf{v}) \times (\mathbf{w} \times \mathbf{x}) = \big((\mathbf{u} \times \mathbf{v}) \bullet \mathbf{x}\big)\mathbf{w} - \big((\mathbf{u} \times \mathbf{v}) \bullet \mathbf{w}\big)\mathbf{x}$$
$$= \big((\mathbf{w} \times \mathbf{x}) \bullet \mathbf{u}\big)\mathbf{v} - \big((\mathbf{w} \times \mathbf{x}) \bullet \mathbf{v}\big)\mathbf{u}.$$

In particular, show that

$$(\mathbf{u} \times \mathbf{v}) \times (\mathbf{u} \times \mathbf{w}) = \big((\mathbf{u} \times \mathbf{v}) \bullet \mathbf{w}\big)\mathbf{u}.$$

**3.** Show that the area $A$ of a triangle with vertices $(x_1, y_1, 0)$, $(x_2, y_2, 0)$, and $(x_3, y_3, 0)$, in the $xy$-plane is given by

$$A = \frac{1}{2} \left| \begin{vmatrix} x_1 & y_1 & 1 \\ x_2 & y_2 & 1 \\ x_3 & y_3 & 1 \end{vmatrix} \right|.$$

**4.** (a) If $L_1$ and $L_2$ are two **skew** (that is, non-parallel and non-intersecting) lines, show that there is a pair of parallel planes $P_1$ and $P_2$ such that $L_1$ lies in $P_1$ and $L_2$ lies in $P_2$.

 (b) Find parallel planes containing the lines $L_1$ through $(1, 1, 0)$ and $(2, 0, 1)$ and $L_2$ through $(0, 1, 1)$ and $(1, 2, 2)$.

**5.** What condition must the vectors $\mathbf{a}$ and $\mathbf{b}$ satisfy to ensure that the equation $\mathbf{a} \times \mathbf{x} = \mathbf{b}$ has solutions? If this condition is satisfied, find all solutions of the equation. Describe the set of solutions.

# CHAPTER 11

# Vector Functions and Curves

***Introduction*** This chapter is concerned with functions of a single real variable that have *vector* values. Such functions can be thought of as parametric representations of curves, and we will examine them from both a *kinematic* point of view (involving position, velocity and acceleration of a moving particle) and a *geometric* point of view (involving tangents, normals, curvature, and torsion). Finally, we will work through a simple derivation of Kepler's laws of planetary motion.

## 11.1 Vector Functions of One Variable

In this section we will examine several aspects of differential and integral calculus as applied to **vector-valued functions** of a single real variable. Such functions can be used to represent curves parametrically. It is natural to interpret a vector-valued function of the real variable $t$ as giving the position, at time $t$, of a point or "particle" moving around in space. Derivatives of this *position vector* are then other vector-valued functions giving the velocity and acceleration of the particle. To motivate the study of vector functions we will consider such a vectorial description of motion in 3-space. Some of our examples will involve motion in the plane; in this case the third components of the vectors will be 0 and will be omitted.

If a particle moves around in 3-space, its motion can be described by giving the three coordinates of its position as functions of time $t$:

$$x = x(t), \qquad y = y(t), \qquad \text{and} \qquad z = z(t).$$

It is more convenient, however, to replace these three equations by a single vector equation,

$$\mathbf{r} = \mathbf{r}(t),$$

giving the position vector of the moving particle as a function of $t$. (Recall that the position vector of a point is the vector from the origin to that point.) In terms of the standard basis vectors $\mathbf{i}$, $\mathbf{j}$, and $\mathbf{k}$, the position of the particle at time $t$ is

position: $\quad \mathbf{r} = \mathbf{r}(t) = x(t)\,\mathbf{i} + y(t)\,\mathbf{j} + z(t)\,\mathbf{k}.$

As $t$ increases, the particle moves along a *path*, a curve $C$ in 3-space. We assume that $C$ is a *continuous curve*; the particle cannot instantaneously jump from one point to a distant point. This is equivalent to requiring that the component functions $x(t)$, $y(t)$, and $z(t)$ are continuous functions of $t$, and we therefore say that $\mathbf{r}(t)$ is a continuous vector function of $t$.

In the time interval from $t$ to $t + \Delta t$ the particle moves from position $\mathbf{r}(t)$ to position $\mathbf{r}(t + \Delta t)$. Therefore its **average velocity** is

$$\frac{\mathbf{r}(t + \Delta t) - \mathbf{r}(t)}{\Delta t},$$

which is a vector parallel to the secant vector from $\mathbf{r}(t)$ to $\mathbf{r}(t + \Delta t)$. If the average velocity has a limit as $\Delta t \to 0$, then we say that $\mathbf{r}$ is **differentiable** at $t$, and we call the limit the (instantaneous) **velocity** of the particle at time $t$. We denote the velocity vector by $\mathbf{v}(t)$.

$$\text{velocity:} \quad \mathbf{v}(t) = \lim_{\Delta t \to 0} \frac{\mathbf{r}(t + \Delta t) - \mathbf{r}(t)}{\Delta t} = \frac{d}{dt}\mathbf{r}(t).$$

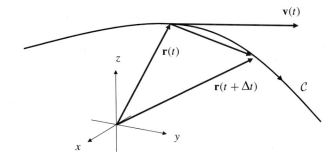

**Figure 11.1**    The velocity $\mathbf{v}(t)$ is the derivative of the position $\mathbf{r}(t)$, and is tangent to the path of motion at the point with position vector $\mathbf{r}(t)$

This velocity vector has direction tangent to the path $\mathcal{C}$ at the point $\mathbf{r}(t)$, (see Figure 11.1), and points in the direction of motion. The length of the velocity vector, $v(t) = |\mathbf{v}(t)|$, is called the **speed** of the particle:

$$\text{speed:} \quad v(t) = |\mathbf{v}(t)|.$$

Wherever the velocity vector exists, is continuous, and does not vanish, the path $\mathcal{C}$ is a **smooth** curve, that is, it has a continuously turning tangent line. The path may not be smooth at points where the velocity is zero even if the components of the velocity vector are smooth functions of $t$.

**Example 1**    Consider the curve $\mathbf{r} = t^3\mathbf{i} + t^2\mathbf{j}$. Its component functions $t^3$ and $t^2$ have continuous derivatives of all orders. However, the curve is not smooth at the origin ($t = 0$) where its velocity $\mathbf{v} = 3t^2\mathbf{i} + 2t\mathbf{j} = \mathbf{0}$. (See Figure 11.2.) The curve is smooth at all other points where $\mathbf{v}(t) \neq \mathbf{0}$.

**Figure 11.2**    The components of $\mathbf{r}(t)$ are smooth functions of $t$, but the curve fails to be smooth at the origin where $\mathbf{v} = \mathbf{0}$.

The rules for addition and scalar multiplication of vectors imply that

$$
\begin{aligned}
\mathbf{v} &= \frac{d\mathbf{r}}{dt} \\
&= \lim_{\Delta t \to 0} \left( \frac{x(t + \Delta t) - x(t)}{\Delta t}\mathbf{i} + \frac{y(t + \Delta t) - y(t)}{\Delta t}\mathbf{j} + \frac{z(t + \Delta t) - z(t)}{\Delta t}\mathbf{k} \right) \\
&= \frac{dx}{dt}\mathbf{i} + \frac{dy}{dt}\mathbf{j} + \frac{dz}{dt}\mathbf{k}.
\end{aligned}
$$

Thus the vector function $\mathbf{r}$ is differentiable at $t$ if and only if its three scalar components, $x$, $y$, and $z$, are differentiable at $t$. In general, vector functions can be differentiated (or integrated) by differentiating (or integrating) their component functions, provided that the basis vectors with respect to which the components are taken are fixed in space and not changing with time.

Continuing our analysis of the moving particle, we define the **acceleration** of the particle to be the time derivative of the velocity:

$$\text{acceleration:} \quad \mathbf{a}(t) = \frac{d\mathbf{v}}{dt} = \frac{d^2\mathbf{r}}{dt^2}.$$

Newton's Second Law of Motion asserts that this acceleration is proportional to, and *in the same direction as*, the force $\mathbf{F}$ causing the motion: if the particle has mass $m$ then the law is expressed by the *vector equation* $\mathbf{F} = m\mathbf{a}$.

**Example 2**   Describe the curve $\mathbf{r} = t\mathbf{i} + t^2\mathbf{j} + t^3\mathbf{k}$. Find the velocity and acceleratioJn vectors for this curve at $(1, 1, 1)$.

**Solution**   Since the scalar parametric equations for the curve are

$$x = t, \quad y = t^2, \quad \text{and} \quad z = t^3,$$

which satisfy $y = x^2$ and $z = x^3$, the curve is the curve of intersection of the two cylinders $y = x^2$ and $z = x^3$. At any time $t$ the velocity and acceleration vectors are given by

$$\mathbf{v} = \frac{d\mathbf{r}}{dt} = \mathbf{i} + 2t\mathbf{j} + 3t^2\mathbf{k},$$

$$\mathbf{a} = \frac{d\mathbf{v}}{dt} = 2\mathbf{j} + 6t\mathbf{k}.$$

The point $(1, 1, 1)$ on the curve corresponds to $t = 1$, so the velocity and acceleration at that point are $\mathbf{v} = \mathbf{i} + 2\mathbf{j} + 3\mathbf{k}$ and $\mathbf{a} = 2\mathbf{j} + 6\mathbf{k}$.

**Example 3**   Find the velocity, speed, and acceleration, and describe the motion of a particle whose position at time $t$ is

$$\mathbf{r} = 3\cos\omega t\, \mathbf{i} + 4\cos\omega t\, \mathbf{j} + 5\sin\omega t\, \mathbf{k}.$$

**Solution**   The velocity, speed, and acceleration are readily calculated:

$$\mathbf{v} = \frac{d\mathbf{r}}{dt} = -3\omega\sin\omega t\, \mathbf{i} - 4\omega\sin\omega t\, \mathbf{j} + 5\omega\cos\omega t\, \mathbf{k}$$

$$v = |\mathbf{v}| = 5\omega$$

$$\mathbf{a} = \frac{d\mathbf{v}}{dt} = -3\omega^2\cos\omega t\, \mathbf{i} - 4\omega^2\cos\omega t\, \mathbf{j} - 5\omega^2\sin\omega t\, \mathbf{k} = -\omega^2\mathbf{r}.$$

Observe that $|\mathbf{r}| = 5$. Therefore the path of the particle lies on the sphere with equation $x^2 + y^2 + z^2 = 25$. Since $x = 3\cos\omega t$ and $y = 4\cos\omega t$, the path also lies on the vertical plane $4x = 3y$. Hence the particle moves around a circle of radius 5 centred at the origin and lying in the plane $4x = 3y$. Observe also that $\mathbf{r}$ is periodic with period $2\pi/\omega$. Therefore the particle makes one revolution around the circle in time $2\pi/\omega$. The acceleration is always in the direction of $-\mathbf{r}$, that is, towards the origin. The term **centripetal acceleration** is used to describe such a "centre-seeking" acceleration.

**Example 4**   (**The projectile problem**) Describe the path followed by a particle experiencing a constant downward acceleration, $-g\mathbf{k}$, caused by gravity. Assume that at time $t = 0$ the position of the particle is $\mathbf{r}_0$ and its velocity is $\mathbf{v}_0$.

**Solution**   If the position of the particle at time $t$ is $\mathbf{r}(t)$, then its acceleration is $d^2\mathbf{r}/dt^2$. The position of the particle can be found by solving the *initial-value problem*

$$\frac{d^2\mathbf{r}}{dt^2} = -g\mathbf{k}, \qquad \frac{d\mathbf{r}}{dt}\bigg|_{t=0} = \mathbf{v}_0, \qquad \mathbf{r}(0) = \mathbf{r}_0.$$

We integrate the differential equation twice. Each integration introduces a *vector* constant of integration that we can determine from the given data by evaluating at $t = 0$:

$$\frac{d\mathbf{r}}{dt} = -gt\mathbf{k} + \mathbf{v}_0$$

$$\mathbf{r} = -\frac{gt^2}{2}\mathbf{k} + \mathbf{v}_0 t + \mathbf{r}_0.$$

The latter equation represents a parabola in the vertical plane passing through the point with position vector $\mathbf{r}_0$ and parallel to the vector $\mathbf{v}_0$. (See Figure 11.3.) The parabola has scalar parametric equations

$$x = u_0 t + x_0$$
$$y = v_0 t + y_0$$
$$z = -\frac{gt^2}{2} + w_0 t + z_0,$$

where $r_0 = x_0\mathbf{i} + y_0\mathbf{j} + z_0\mathbf{k}$ and $\mathbf{v}_0 = u_0\mathbf{i} + v_0\mathbf{j} + w_0\mathbf{k}$.  ∎

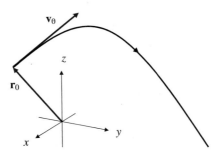

**Figure 11.3**   The path of a projectile fired from position $\mathbf{r}_0$ with velocity $\mathbf{v}_0$

**Example 5**   An object moves to the right along the curve $y = x^2$ with constant speed $v = 5$. Find the velocity and acceleration of the object when it is at the point $(1, 1)$.

**Solution**   The position of the object at time $t$ is

$$\mathbf{r} = x\mathbf{i} + x^2\mathbf{j},$$

where $x$, the $x$-coordinate of the object's position, is a function of $t$. The object's velocity, speed, and acceleration at time $t$ are given by

$$\mathbf{v} = \frac{d\mathbf{r}}{dt} = \frac{dx}{dt}\mathbf{i} + 2x\frac{dx}{dt}\mathbf{j} = \frac{dx}{dt}(\mathbf{i} + 2x\mathbf{j})$$

$$v = |\mathbf{v}| = \left|\frac{dx}{dt}\right|\sqrt{1 + (2x)^2} = \frac{dx}{dt}\sqrt{1 + 4x^2}$$

$$\mathbf{a} = \frac{d\mathbf{v}}{dt} = \frac{d^2x}{dt^2}(\mathbf{i} + 2x\mathbf{j}) + 2\left(\frac{dx}{dt}\right)^2\mathbf{j}.$$

(In the speed calculation we used $|dx/dt| = dx/dt$ because the object is moving to the right.) We are given that the speed is constant; $v = 5$. Therefore

$$\frac{dx}{dt} = \frac{5}{\sqrt{1 + 4x^2}}.$$

When $x = 1$, we have $dx/dt = 5/\sqrt{1+4} = \sqrt{5}$, so the velocity of the object at that point is $\mathbf{v} = \sqrt{5}\mathbf{i} + 2\sqrt{5}\mathbf{j}$. Now we can calculate

$$\frac{d^2x}{dt^2} = \frac{d}{dt}\frac{5}{\sqrt{1 + 4x^2}} = \left(\frac{d}{dx}\frac{5}{\sqrt{1 + 4x^2}}\right)\frac{dx}{dt}$$

$$= -\frac{5}{2(1 + 4x^2)^{3/2}}(8x)\frac{5}{\sqrt{1 + 4x^2}} = -\frac{100x}{(1 + 4x^2)^2}.$$

At $x = 1$, we have $d^2x/dt^2 = -4$. Thus the acceleration at that point is $\mathbf{a} = -4(\mathbf{i} + 2\mathbf{j}) + 10\mathbf{j} = -4\mathbf{i} + 2\mathbf{j}$. ∎

**Remark**   Note that we used $x$ as the parameter for the curve in the above example, so we could use $t$ for time. If you want to analyze motion along a curve $\mathbf{r} = \mathbf{r}(t)$ where $t$ is just a parameter, not necessarily time, then you will have to use a different symbol, say $u$, for time. The *physical velocity and acceleration* of a particle moving along the curve are then

$$\mathbf{v} = \frac{d\mathbf{r}}{du} = \frac{dt}{du}\frac{d\mathbf{r}}{dt} \quad \text{and} \quad \mathbf{a} = \frac{d\mathbf{v}}{du} = \frac{d^2t}{du^2}\frac{d\mathbf{r}}{dt} + \left(\frac{dt}{du}\right)^2\frac{d^2\mathbf{r}}{dt^2}.$$

Be careful how you interpret $t$ in a problem where time is meaningful.

## Differentiating Combinations of Vectors

Vectors and scalars can be combined in a variety of ways to form other vectors or scalars. Vectors can be added and multiplied by scalars, and can be factors in dot and cross products. Appropriate differentiation rules apply to all such combinations of vector and scalar functions; we summarize them in the following theorem.

**THEOREM ❶**   **Differentiation rules for vector functions**

Let $\mathbf{u}(t)$ and $\mathbf{v}(t)$ be differentiable vector-valued functions, and let $\lambda(t)$ be a differentiable scalar-valued functions. Then $\mathbf{u}(t) + \mathbf{v}(t)$, $\lambda(t)\mathbf{u}(t)$, $\mathbf{u}(t) \bullet \mathbf{v}(t)$, $\mathbf{u}(t) \times \mathbf{v}(t)$, and $\mathbf{u}(\lambda(t))$ are differentiable, and

(a) $\quad \dfrac{d}{dt}\Big(\mathbf{u}(t) + \mathbf{v}(t)\Big) = \mathbf{u}'(t) + \mathbf{v}'(t)$

(b) $\quad \dfrac{d}{dt}\Big(\lambda(t)\mathbf{u}(t)\Big) = \lambda'(t)\mathbf{u}(t) + \lambda(t)\mathbf{u}'(t)$

(c) $\quad \dfrac{d}{dt}\Big(\mathbf{u}(t) \bullet \mathbf{v}(t)\Big) = \mathbf{u}'(t) \bullet \mathbf{v}(t) + \mathbf{u}(t) \bullet \mathbf{v}'(t)$

(d) $\quad \dfrac{d}{dt}\Big(\mathbf{u}(t) \times \mathbf{v}(t)\Big) = \mathbf{u}'(t) \times \mathbf{v}(t) + \mathbf{u}(t) \times \mathbf{v}'(t)$

(e) $\quad \dfrac{d}{dt}\Big(\mathbf{u}\big(\lambda(t)\big)\Big) = \lambda'(t)\mathbf{u}'\big(\lambda(t)\big).$

Also, at any point where $\mathbf{u}(t) \neq 0$,

(f) $\qquad\qquad\qquad \dfrac{d}{dt}|\mathbf{u}(t)| = \dfrac{\mathbf{u}(t) \bullet \mathbf{u}'(t)}{|\mathbf{u}(t)|}.$

**Remark** Formulas (b), (c), and (d) are versions of the Product Rule. Formula (e) is a version of the Chain Rule. Formula (f) is also a case of the Chain Rule applied to $|\mathbf{u}| = \sqrt{\mathbf{u} \bullet \mathbf{u}}$. All have the obvious form. Note that the order of the factors is the same in the terms on both sides of the cross product formula (d). It is essential that the order be preserved because, unlike the dot product or the product of a vector with a scalar, the cross product is *not commutative*.

**Remark** The formula for the derivative of a cross product is a special case of that for the derivative of a $3 \times 3$ determinant. Since every term in the expansion of a determinant of any order is a product involving one element from each row (or column), the general Product Rule implies that the derivative of an $n \times n$ determinant whose elements are functions will be the sum of $n$ such $n \times n$ determinants, each with the elements of one of the rows (or columns) differentiated. For the $3 \times 3$ case we have

$$\frac{d}{dt}\begin{vmatrix} a_{11}(t) & a_{12}(t) & a_{13}(t) \\ a_{21}(t) & a_{22}(t) & a_{23}(t) \\ a_{31}(t) & a_{32}(t) & a_{33}(t) \end{vmatrix} = \begin{vmatrix} a_{11}'(t) & a_{12}'(t) & a_{13}'(t) \\ a_{21}(t) & a_{22}(t) & a_{23}(t) \\ a_{31}(t) & a_{32}(t) & a_{33}(t) \end{vmatrix}$$

$$+ \begin{vmatrix} a_{11}(t) & a_{12}(t) & a_{13}(t) \\ a_{21}'(t) & a_{22}'(t) & a_{23}'(t) \\ a_{31}(t) & a_{32}(t) & a_{33}(t) \end{vmatrix} + \begin{vmatrix} a_{11}(t) & a_{12}(t) & a_{13}(t) \\ a_{21}(t) & a_{22}(t) & a_{23}(t) \\ a_{31}'(t) & a_{32}'(t) & a_{33}'(t) \end{vmatrix}.$$

**Example 6** Show that the speed of a moving particle remains constant over an interval of time if and only if the acceleration is perpendicular to the velocity throughout that interval.

**Solution** Since $\big(v(t)\big)^2 = \mathbf{v}(t) \bullet \mathbf{v}(t)$ we have

$$2v(t)\frac{dv}{dt} = \frac{d}{dt}\big(v(t)\big)^2 = \frac{d}{dt}\mathbf{v}(t) \bullet \mathbf{v}(t)$$
$$= \mathbf{a}(t) \bullet \mathbf{v}(t) + \mathbf{v}(t) \bullet \mathbf{a}(t) = 2\mathbf{v}(t) \bullet \mathbf{a}(t).$$

If we assume that $v(t) \neq 0$, it follows that $dv/dt = 0$ if and only if $\mathbf{v} \bullet \mathbf{a} = 0$. The speed is constant if and only if the velocity is perpendicular to the acceleration.

---

**Example 7**    If $\mathbf{u}$ is three times differentiable, calculate and simplify the triple product derivative

$$\frac{d}{dt}\left(\mathbf{u} \bullet \left(\frac{d\mathbf{u}}{dt} \times \frac{d^2\mathbf{u}}{dt^2}\right)\right).$$

**Solution**    Using various versions of the Product Rule, we calculate

$$\frac{d}{dt}\left(\mathbf{u}\bullet\left(\frac{d\mathbf{u}}{dt} \times \frac{d^2\mathbf{u}}{dt^2}\right)\right)$$
$$= \frac{d\mathbf{u}}{dt} \bullet \left(\frac{d\mathbf{u}}{dt} \times \frac{d^2\mathbf{u}}{dt^2}\right) + \mathbf{u} \bullet \left(\frac{d^2\mathbf{u}}{dt^2} \times \frac{d^2\mathbf{u}}{dt^2}\right) + \mathbf{u} \bullet \left(\frac{d\mathbf{u}}{dt} \times \frac{d^3\mathbf{u}}{dt^3}\right)$$
$$= 0 + 0 + \mathbf{u} \bullet \left(\frac{d\mathbf{u}}{dt} \times \frac{d^3\mathbf{u}}{dt^3}\right) = \mathbf{u} \bullet \left(\frac{d\mathbf{u}}{dt} \times \frac{d^3\mathbf{u}}{dt^3}\right).$$

The first term vanishes because $d\mathbf{u}/dt$ is perpendicular to its cross product with another vector; the second term vanishes because of the cross product of identical vectors.

---

## Exercises 11.1

In Exercises 1–14, find the velocity, speed and acceleration at time $t$ of the particle whose position is $\mathbf{r}(t)$. Describe the path of the particle.

**1.** $\mathbf{r} = \mathbf{i} + t\mathbf{j}$

**2.** $\mathbf{r} = t^2\mathbf{i} + \mathbf{k}$

**3.** $\mathbf{r} = t^2\mathbf{j} + t\mathbf{k}$

**4.** $\mathbf{r} = \mathbf{i} + t\mathbf{j} + t\mathbf{k}$

**5.** $\mathbf{r} = t^2\mathbf{i} - t^2\mathbf{j} + \mathbf{k}$

**6.** $\mathbf{r} = t\mathbf{i} + t^2\mathbf{j} + t^2\mathbf{k}$

**7.** $\mathbf{r} = a\cos t\,\mathbf{i} + a\sin t\,\mathbf{j} + ct\mathbf{k}$

**8.** $\mathbf{r} = a\cos\omega t\,\mathbf{i} + b\mathbf{j} + a\sin\omega t\,\mathbf{k}$

**9.** $\mathbf{r} = 3\cos t\,\mathbf{i} + 4\cos t\,\mathbf{j} + 5\sin t\,\mathbf{k}$

**10.** $\mathbf{r} = 3\cos t\,\mathbf{i} + 4\sin t\,\mathbf{j} + t\mathbf{k}$

**11.** $\mathbf{r} = ae^t\mathbf{i} + be^t\mathbf{j} + ce^t\mathbf{k}$

**12.** $\mathbf{r} = at\cos\omega t\,\mathbf{i} + at\sin\omega t\,\mathbf{j} + b\ln t\,\mathbf{k}$

**13.** $\mathbf{r} = e^{-t}\cos(e^t)\mathbf{i} + e^{-t}\sin(e^t)\mathbf{j} - e^t\mathbf{k}$

**14.** $\mathbf{r} = a\cos t\sin t\,\mathbf{i} + a\sin^2 t\,\mathbf{j} + a\cos t\,\mathbf{k}$

**15.** A particle moves around the circle $x^2 + y^2 = 25$ at constant speed, making one revolution in 2 seconds. Find its acceleration when it is at $(3, 4)$.

**16.** A particle moves to the right along the curve $y = 3/x$. If its speed is 10 when it passes through the point $\left(2, \frac{3}{2}\right)$, what is its velocity at that time?

**17.** A point $P$ moves along the curve of intersection of the cylinder $z = x^2$ and the plane $x + y = 2$ in the direction of increasing $y$ with constant speed $v = 3$. Find the velocity of $P$ when it is at $(1, 1, 1)$.

**18.** An object moves along the curve $y = x^2$, $z = x^3$, with constant vertical speed $dz/dt = 3$. Find the velocity and acceleration of the object when it is at the point $(2, 4, 8)$.

**19.** A particle moves along the curve with parametrization $\mathbf{r} = 3u\mathbf{i} + 3u^2\mathbf{j} + 2u^3\mathbf{k}$ in the direction corresponding to increasing $u$, and with a constant speed of 6. Find the velocity and acceleration of the particle when it is at the point $(3, 3, 2)$.

**20.** A particle moves along the curve of intersection of the cylinders $y = -x^2$ and $z = x^2$ in the direction in which $x$ increases. (All distances are in cm.) At the instant when the particle is at the point $(1, -1, 1)$, its speed is 9 cm/s, and that speed is increasing at a rate of 3 cm/s$^2$. Find the velocity.

and acceleration of the particle at that instant.

**21.** Show that if the dot product of the velocity and acceleration of a moving particle is positive (or negative) then the speed of the particle is increasing (or decreasing).

**22.** Verify the formula for the derivative of a dot product given in Theorem 1(c).

**23.** Verify the formula for the derivative of a $3 \times 3$ determinant in the second remark following Theorem 1. Use this formula to verify the formula for the derivative of the cross product in Theorem 1.

**24.** If the position and velocity vectors of a moving particle are always perpendicular show that the path of the particle lies on a sphere.

**25.** Generalize the previous problem to the case where the velocity of the particle is always perpendicular to the line joining the particle to a fixed point $P_0$.

**26.** What can be said about the motion of a particle at a time when its position and velocity satisfy $\mathbf{r} \bullet \mathbf{v} > 0$? What can be said when $\mathbf{r} \bullet \mathbf{v} < 0$?

In Exercises 27–32, assume that the vector functions encountered have continuous derivatives of all required orders.

**27.** Show that $\dfrac{d}{dt}\left(\dfrac{d\mathbf{u}}{dt} \times \dfrac{d^2\mathbf{u}}{dt^2}\right) = \dfrac{d\mathbf{u}}{dt} \times \dfrac{d^3\mathbf{u}}{dt^3}$.

**28.** Write the Product Rule for $\dfrac{d}{dt}\Big(\mathbf{u} \bullet (\mathbf{v} \times \mathbf{w})\Big)$.

**29.** Write the Product Rule for $\dfrac{d}{dt}\Big(\mathbf{u} \times (\mathbf{v} \times \mathbf{w})\Big)$.

**30.** Expand and simplify: $\dfrac{d}{dt}\left(\mathbf{u} \times \left(\dfrac{d\mathbf{u}}{dt} \times \dfrac{d^2\mathbf{u}}{dt^2}\right)\right)$.

**31.** Expand and simplify: $\dfrac{d}{dt}\Big((\mathbf{u} + \mathbf{u}'') \bullet (\mathbf{u} \times \mathbf{u}')\Big)$.

**32.** Expand and simplify: $\dfrac{d}{dt}\Big((\mathbf{u} \times \mathbf{u}') \bullet (\mathbf{u}' \times \mathbf{u}'')\Big)$.

**33.** If at all times $t$ the position and velocity vectors of a moving particle satisfy $\mathbf{v}(t) = 2\mathbf{r}(t)$, and if $\mathbf{r}(0) = \mathbf{r}_0$, find $\mathbf{r}(t)$ and the acceleration $\mathbf{a}(t)$. What is the path of motion?

**◈ 34.** Verify that $\mathbf{r} = \mathbf{r}_0 \cos(\omega t) + (\mathbf{v}_0/\omega) \sin(\omega t)$ satisfies the initial-value problem

$$\frac{d^2\mathbf{r}}{dt^2} = -\omega^2 \mathbf{r}, \qquad \mathbf{r}'(0) = \mathbf{v}_0, \qquad \mathbf{r}(0) = \mathbf{r}_0.$$

(It is the unique solution.) Describe the path $\mathbf{r}(t)$. What is the path if $\mathbf{r}_0$ is perpendicular to $\mathbf{v}_0$?

**◈ 35.** **(Free fall with air resistance)** A projectile falling under gravity and slowed by air resistance proportional to its speed has position satisfying

$$\frac{d^2\mathbf{r}}{dt^2} = -g\mathbf{k} - c\frac{d\mathbf{r}}{dt},$$

where $c$ is a positive constant. If $\mathbf{r} = \mathbf{r}_0$ and $d\mathbf{r}/dt = \mathbf{v}_0$ at time $t = 0$ find $\mathbf{r}(t)$. (*Hint:* let $\mathbf{w} = e^{ct}(d\mathbf{r}/dt)$.) Show that the solution approaches that of the projectile problem given in this section as $c \to 0$.

# 11.2 Some Applications of Vector Differentiation

Many interesting problems in mechanics involve the differentiation of vector functions. This section is devoted to a brief discussion of a few of these.

## Motion Involving Varying Mass

The **momentum p** of a moving object is the product of its (scalar) mass $m$ and its (vector) velocity $\mathbf{v}$; $\mathbf{p} = m\mathbf{v}$. Newton's second law of motion states that the rate of change of *momentum* is equal to the external force acting on the object:

$$\mathbf{F} = \frac{d\mathbf{p}}{dt} = \frac{d}{dt}(m\mathbf{v}).$$

It is only when the mass of the object remains constant that this law reduces to the more familiar $\mathbf{F} = m\mathbf{a}$. When mass is changing you must deal with momentum rather than acceleration.

**Example 1** (**The changing velocity of a rocket**) A rocket accelerates by burning its onboard fuel. If the exhaust gases are ejected with constant velocity $\mathbf{v}_e$ *relative to the rocket*, and if the rocket ejects $p\%$ of its initial mass while its engines are firing, by what amount will the velocity of the rocket change? Assume the rocket is in deep space so that gravitational and other external forces acting on it can be neglected.

**Solution** Since the rocket is not acted on by any external forces (i.e., $\mathbf{F} = \mathbf{0}$), Newton's law implies that the total momentum of the rocket and its exhaust gases will remain constant. At time $t$ the rocket has mass $m(t)$ and velocity $\mathbf{v}(t)$. At time $t + \Delta t$ the rocket's mass is $m + \Delta m$ (where $\Delta m < 0$), its velocity is $\mathbf{v} + \Delta \mathbf{v}$, and the mass $-\Delta m$ of exhaust gases has escaped with velocity $\mathbf{v} + \mathbf{v}_e$ (relative to a coordinate system fixed in space). Equating total momenta at $t$ and $t + \Delta t$ we obtain

$$(m + \Delta m)(\mathbf{v} + \Delta \mathbf{v}) + (-\Delta m)(\mathbf{v} + \mathbf{v}_e) = m\mathbf{v}.$$

Simplifying this equation and dividing by $\Delta t$ gives

$$(m + \Delta m)\frac{\Delta \mathbf{v}}{\Delta t} = \frac{\Delta m}{\Delta t}\mathbf{v}_e,$$

and, on taking the limit as $\Delta t \to 0$,

$$m\frac{d\mathbf{v}}{dt} = \frac{dm}{dt}\mathbf{v}_e.$$

Suppose that the engine fires from $t = 0$ to $t = T$. By the Fundamental Theorem of Calculus, the velocity of the rocket will change by

$$\mathbf{v}(T) - \mathbf{v}(0) = \int_0^T \frac{d\mathbf{v}}{dt}\,dt = \left(\int_0^T \frac{1}{m}\frac{dm}{dt}\,dt\right)\mathbf{v}_e$$

$$= \Big(\ln m(T) - \ln m(0)\Big)\mathbf{v}_e = -\ln\left(\frac{m(0)}{m(T)}\right)\mathbf{v}_e.$$

Since $m(0) > m(T)$, we have $\ln\big(m(0)/m(T)\big) > 0$ and, as was to be expected, the change in velocity of the rocket is in the opposite direction to the exhaust velocity $\mathbf{v}_e$. If $p\%$ of the mass of the rocket is ejected during the burn then the velocity of the rocket will change by the amount $-\mathbf{v}_e \ln(100/(100 - p))$. ∎

**Remark** It is interesting that this model places no restriction on how great a velocity the rocket can achieve, provided that a sufficiently large percentage of its initial mass is fuel. See Exercise 1 at the end of the section.

## Circular Motion

The angular speed $\Omega$ of a rotating body is its rate of rotation measured in radians per unit time. For instance, a lighthouse lamp rotating at a rate of three revolutions per minute has an angular speed of $\Omega = 6\pi$ radians per minute. It is useful to represent the rate of rotation of a rigid body about an axis in terms of an **angular velocity** vector rather than just the scalar angular speed. The angular velocity vector, $\mathbf{\Omega}$, has magnitude equal to the angular speed, $\Omega$, and direction along the axis of rotation such that if the extended right thumb points in the direction of $\Omega$ then the fingers surround the axis in the direction of rotation.

If the origin of the coordinate system is on the axis of rotation, and $\mathbf{r} = \mathbf{r}(t)$ is the position vector at time $t$ of a point $P$ in the rotating body, then $P$ moves around a circle of radius $D = |\mathbf{r}(t)| \sin\theta$, where $\theta$ is the (constant) angle between $\Omega$ and $\mathbf{r}(t)$. (See Figure 11.4.) Thus $P$ travels a distance $2\pi D$ in time $2\pi/\Omega$, and its linear speed is

$$\frac{\text{distance}}{\text{time}} = \frac{2\pi D}{2\pi/\Omega} = \Omega D = |\Omega||\mathbf{r}(t)| \sin\theta = |\Omega \times \mathbf{r}(t)|.$$

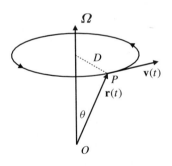

**Figure 11.4**    Rotation with angular velocity $\Omega$: $\mathbf{v} = \Omega \times \mathbf{r}$

Since the direction of $\Omega$ was defined so that $\Omega \times \mathbf{r}(t)$ would point in the direction of motion of $P$, the linear velocity of $P$ at time $t$ is given by

$$\frac{d\mathbf{r}}{dt} = \mathbf{v}(t) = \Omega \times \mathbf{r}(t).$$

**Example 2**    The position vector $\mathbf{r}(t)$ of a moving particle $P$ satisfies the initial-value problem

$$\begin{cases} \dfrac{d\mathbf{r}}{dt} = 2\mathbf{i} \times \mathbf{r}(t) \\ \mathbf{r}(0) = \mathbf{i} + 3\mathbf{j}. \end{cases}$$

Find $\mathbf{r}(t)$ and describe the motion of $P$.

**Solution**    There are two ways to solve this problem. We will do it both ways.

METHOD I:    By the discussion above, the given differential equation is consistent with rotation about the $x$-axis with angular velocity $2\mathbf{i}$, so that the angular speed is 2 and the motion is counterclockwise as seen from far out on the positive $x$-axis. Therefore the particle $P$ moves on a circle in a plane $x = $ constant and centred on the $x$-axis. Since $P$ is at $(1, 3, 0)$ at time $t = 0$, the plane of motion is $x = 1$ and the radius of the circle is 3. Therefore the circle has a parametric equation of the form

$$\mathbf{r} = \mathbf{i} + 3\cos(\lambda t)\mathbf{j} + 3\sin(\lambda t)\mathbf{k}.$$

$P$ travels once around this circle ($2\pi$ radians) in time $t = 2\pi/\lambda$, so the angular speed is $\lambda$. Therefore $\lambda = 2$ and the motion of the particle is given by

$$\mathbf{r} = \mathbf{i} + 3\cos(2t)\mathbf{j} + 3\sin(2t)\mathbf{k}.$$

METHOD II:    Break the given vector differential equation into components:

$$\frac{dx}{dt}\mathbf{i} + \frac{dy}{dt}\mathbf{j} + \frac{dz}{dt}\mathbf{k} = 2\mathbf{i} \times (x\mathbf{i} + y\mathbf{j} + z\mathbf{k}) = -2z\mathbf{j} + 2y\mathbf{k}$$

$$\frac{dx}{dt} = 0, \qquad \frac{dy}{dt} = -2z, \qquad \frac{dz}{dt} = 2y.$$

The first equation implies that $x =$ constant. Since $x(0) = 1$, we have $x(t) = 1$ for all $t$. Differentiate the second equation with respect to $t$ and substitute the third equation. This leads to the equation of simple harmonic motion for $y$,

$$\frac{d^2y}{dt^2} = -2\frac{dz}{dt} = -4y,$$

for which a general solution is

$$y = A\cos(2t) + B\sin(2t).$$

Thus $z = -\frac{1}{2}(dy/dt) = A\sin(2t) - B\cos(2t)$. Since $y(0) = 3$ and $z(0) = 0$, we have $A = 3$ and $B = 0$. Thus the particle $P$ travels counterclockwise around the circular path

$$\mathbf{r} = \mathbf{i} + 3\cos(2t)\mathbf{j} + 3\sin(2t)\mathbf{k}$$

in the plane $x = 1$ with angular speed 2.

∎

**Remark** Newton's second law states that $\mathbf{F} = (d/dt)m\mathbf{v} = d\mathbf{p}/dt$, where $\mathbf{p} = m\mathbf{v}$ is the (linear) momentum of a particle of mass $m$ moving under the influence of a force $\mathbf{F}$. This law may be reformulated in a manner appropriate for describing rotational motion as follows. If $\mathbf{r}(t)$ is the position of the particle at time $t$, then, since $\mathbf{v} \times \mathbf{v} = \mathbf{0}$,

$$\frac{d}{dt}(\mathbf{r} \times \mathbf{p}) = \frac{d}{dt}\left(\mathbf{r} \times (m\mathbf{v})\right) = \mathbf{v} \times (m\mathbf{v}) + \mathbf{r} \times \frac{d}{dt}(m\mathbf{v}) = \mathbf{r} \times \mathbf{F}.$$

The quantities $\mathbf{H} = \mathbf{r} \times (m\mathbf{v})$ and $\mathbf{T} = \mathbf{r} \times \mathbf{F}$ are respectively the **angular momentum** of the particle about the origin and the **torque** of $\mathbf{F}$ about the origin. We have shown that

$$\mathbf{T} = \frac{d\mathbf{H}}{dt};$$

the torque of the external forces is equal to the rate of change of the angular momentum of the particle. This is the analogue for rotational motion of $\mathbf{F} = d\mathbf{p}/dt$.

## Rotating Frames and the Coriolis Effect

The procedure of differentiating a vector function by differentiating its components is valid only if the basis vectors themselves do not depend on the variable of differentiation. In some situations in mechanics this is not the case. For instance, in modelling large scale weather phenomena the analysis is affected by the fact that a coordinate system fixed with respect to the earth is, in fact, rotating (along with the earth) relative to directions fixed in space.

In order to understand the effect that the rotation of the coordinate system has on representations of velocity and acceleration, let us consider two Cartesian coordinate frames (that is, systems of axes), sharing a common origin, one "fixed" in space and the other rotating with angular velocity $\Omega$ about an axis through that origin. Let $\hat{\mathbf{I}}$, $\hat{\mathbf{J}}$, and $\hat{\mathbf{K}}$ be the standard basis vectors of the fixed coordinate frame, and let $\mathbf{i}$, $\mathbf{j}$, and $\mathbf{k}$ be the basis vectors of the rotating frame. The latter three are constant vectors in their own frame but are functions of time in the fixed frame. We showed above that the velocity of a point with position $\mathbf{r}$ rotating about an axis with angular velocity $\Omega$ is $\Omega \times \mathbf{r}$. Therefore

$$\frac{d\,\mathbf{i}}{dt} = \Omega \times \mathbf{i}, \qquad \frac{d\,\mathbf{j}}{dt} = \Omega \times \mathbf{j}, \qquad \text{and} \qquad \frac{d\,\mathbf{k}}{dt} = \Omega \times \mathbf{k}.$$

Any vector function can be expressed in terms of either basis. If $\mathbf{r}(t)$ is the position vector of a moving particle it can be expressed in the fixed frame as

$$\mathbf{r} = \hat{x}(t)\hat{\mathbf{I}} + \hat{y}(t)\hat{\mathbf{J}} + \hat{z}(t)\hat{\mathbf{K}},$$

or in the rotating frame as

$$\mathbf{r} = x(t)\mathbf{i} + y(t)\mathbf{j} + z(t)\mathbf{k}.$$

The velocity of the moving particle can be expressed in the fixed frame, $\mathcal{F}$, as

$$\mathbf{v}_{\mathcal{F}} = \frac{d\hat{x}}{dt}\hat{\mathbf{I}} + \frac{d\hat{y}}{dt}\hat{\mathbf{J}} + \frac{d\hat{z}}{dt}\hat{\mathbf{K}}.$$

Similarly, the velocity of the particle, measured with respect to the rotating frame, $\mathcal{R}$, is

$$\mathbf{v}_{\mathcal{R}} = \frac{dx}{dt}\mathbf{i} + \frac{dy}{dt}\mathbf{j} + \frac{dz}{dt}\mathbf{k}.$$

These vectors are, however, not equal. (For instance, a point at rest with respect to the rotating frame would be seen as moving with respect to the fixed frame.) If we want to express the fixed-frame velocity of the particle, $\mathbf{v}_{\mathcal{F}}$, in terms of the rotating basis, we must also account for the rate of change of the basis vectors in the rotating frame; using three applications of the Product Rule we have

$$\begin{aligned} \mathbf{v}_{\mathcal{F}} &= \frac{dx}{dt}\mathbf{i} + x\frac{d\mathbf{i}}{dt} + \frac{dy}{dt}\mathbf{j} + y\frac{d\mathbf{j}}{dt} + \frac{dz}{dt}\mathbf{k} + z\frac{d\mathbf{k}}{dt} \\ &= \mathbf{v}_{\mathcal{R}} + x\Omega \times \mathbf{i} + y\Omega \times \mathbf{j} + z\Omega \times \mathbf{k} \\ &= \mathbf{v}_{\mathcal{R}} + \Omega \times \mathbf{r}. \end{aligned}$$

This formula shows that differentiation with respect to time has different effects in the two coordinate systems. If we denote by $(d/dt)_{\mathcal{F}}$ and $(d/dt)_{\mathcal{R}}$ the operators of time differentiation in the two frames then the formula above can be written

$$\left(\frac{d}{dt}\right)_{\mathcal{F}} \mathbf{r}(t) = \left(\left(\frac{d}{dt}\right)_{\mathcal{R}} + \Omega \times\right)\mathbf{r}(t),$$

or, in terms of operators,

$$\left(\frac{d}{dt}\right)_{\mathcal{F}} = \left(\left(\frac{d}{dt}\right)_{\mathcal{R}} + \Omega \times\right).$$

These operators can be applied to *any* vector function, and in particular to the function $\mathbf{v}_{\mathcal{F}} = \mathbf{v}_{\mathcal{R}} + \Omega \times \mathbf{r}$. Since $(d/dt)_{\mathcal{F}}\mathbf{v}_{\mathcal{F}}$ is the acceleration $\mathbf{a}_{\mathcal{F}}$ of the particle in the fixed frame, and $(d/dt)_{\mathcal{R}}\mathbf{v}_{\mathcal{R}}$ is the acceleration $\mathbf{a}_{\mathcal{R}}$ in the rotating frame, we calculate, assuming that the angular velocity $\Omega$ of the rotating frame is constant,

$$\begin{aligned} \mathbf{a}_{\mathcal{F}} &= \left(\frac{d}{dt}\right)_{\mathcal{F}}(\mathbf{v}_{\mathcal{R}} + \Omega \times \mathbf{r}) \\ &= \left(\left(\frac{d}{dt}\right)_{\mathcal{R}} + \Omega \times\right)(\mathbf{v}_{\mathcal{R}} + \Omega \times \mathbf{r}) \\ &= \mathbf{a}_{\mathcal{R}} + 2\Omega \times \mathbf{v}_{\mathcal{R}} + \Omega \times (\Omega \times \mathbf{r}). \end{aligned}$$

The term $2\boldsymbol{\Omega}\times\mathbf{v}_{\mathcal{R}}$ is called the **Coriolis acceleration** and the term $\boldsymbol{\Omega}\times(\boldsymbol{\Omega}\times\mathbf{r})$ is called the **centripetal acceleration**.

In order to interpret these accelerations, suppose that the origin is at the centre of the earth, the fixed frame is fixed with respect to the stars, and the rotating frame is attached to the rotating earth. Suppose that the particle has mass $m$ and is acted upon by an external force $\mathbf{F}$. If we view the moving particle from a vantage point fixed in space (so that we see the fixed frame as unmoving) then, in accordance with Newton's second law, we observe that $\mathbf{F} = m\mathbf{a}_{\mathcal{F}}$. If, instead, we are sitting on the earth, and regarding the rotating frame as stationary, we would still want to regard $\mathbf{a}_{\mathcal{R}}$ as the force per unit mass acting on the particle. Thus we would write

$$\mathbf{a}_{\mathcal{R}} = \frac{\mathbf{F}}{m} - 2\boldsymbol{\Omega}\times\mathbf{v}_{\mathcal{R}} - \boldsymbol{\Omega}\times(\boldsymbol{\Omega}\times\mathbf{r}),$$

and would interpret the $\boldsymbol{\Omega}$ terms as other *forces* per unit mass acting on the particle. The term $-\boldsymbol{\Omega}\times(\boldsymbol{\Omega}\times\mathbf{r})$ is a vector pointing directly away from the axis of rotation of the earth and we would interpret it as a **centrifugal force**. (If the earth were spinning fast enough, we would expect objects on the surface to be flung into space.) The term $-2\boldsymbol{\Omega}\times\mathbf{v}_{\mathcal{R}}$ is at right angles to both the velocity and the polar axis of the earth. We would call this a **Coriolis force**. The centrifugal and Coriolis forces are not "real" forces acting on the particle. They are fictitious forces which compensate for the fact that we are measuring acceleration with respect to a frame which we are regarding as fixed although it is really rotating and hence accelerating.

**Example 3** (**Winds around the eye of a storm**) The circulation of winds around a storm centre is an example of the Coriolis effect. The eye of a storm is an area of low pressure sucking air towards it. The direction of rotation of the earth is such that the angular velocity $\boldsymbol{\Omega}$ points north and is parallel to the earth's axis of rotation. At any point $P$ on the surface of the earth we can express $\boldsymbol{\Omega}$ as a sum of tangential (to the earth's surface) and normal components (see Figure 11.5(a)),

$$\boldsymbol{\Omega}(P) = \boldsymbol{\Omega}_T(P) + \boldsymbol{\Omega}_N(P).$$

**Figure 11.5**

(a) Tangential and normal components of the angular velocity of the earth in the northern and southern hemispheres

(b) In the northern hemisphere the tangential Coriolis force deflects winds to the right of the path towards the low pressure area $L$ so the winds move counterclockwise around the low

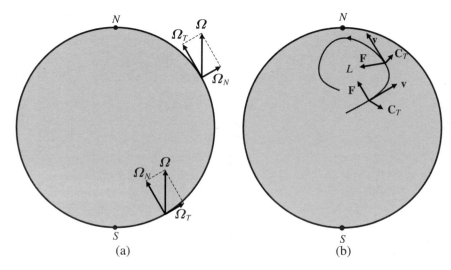

(a)

(b)

If $P$ is in the northern hemisphere $\Omega_N(P)$ points upwards (away from the centre of the earth). At such a point the Coriolis "force" $\mathbf{C} = -2\Omega(P) \times \mathbf{v}$ on a particle of air moving with horizontal velocity $\mathbf{v}$ would itself have horizontal and normal components

$$\mathbf{C} = -2\Omega_T \times \mathbf{v} - 2\Omega_N \times \mathbf{v} = \mathbf{C}_N + \mathbf{C}_T.$$

The tangential component of the Coriolis force, $\mathbf{C}_T = -2\Omega_N \times \mathbf{v}$ is $90°$ to the right of $\mathbf{v}$ (that is, clockwise from $\mathbf{v}$). Therefore, particles of air which are being sucked towards the eye of the storm experience Coriolis deflection to the right and so actually spiral into the eye in a counterclockwise direction. The opposite is true in the southern hemisphere where the normal component of $\Omega_N$ is downward (into the earth). The suction force $\mathbf{F}$, the velocity $\mathbf{v}$, and the component of the Coriolis force tangential to the earth's surface, $\mathbf{C}_T$, are shown at two positions on the path of an air particle spiralling around a low pressure area in the northern hemisphere in Figure 11.5(b).

**Remark** Strong winds spiralling inwards around low pressure areas are called **cyclones**. Strong winds spiralling outwards around high pressure areas are called **anticyclones**. These latter spiral counterclockwise in the southern hemisphere and clockwise in the northern hemisphere.

# Exercises 11.2

1. What fraction of its total initial mass would the rocket considered in Example 1 have to burn as fuel in order to accelerate in a straight line from rest to the speed of its own exhaust gases? to twice that speed?

* 2. When run at maximum power output the motor in a self-propelled tank car can accelerate the full car (mass $M$ kg) along a horizontal track at $a$ m/sec². The tank is full at time zero but the contents pour out of a hole in the bottom at rate $k$ kg/sec thereafter. If the car is at rest at time zero and full forward power is turned on at that time, how fast will it be moving at any time $t$ before the tank is empty?

❖ 3. Solve the initial-value problem

$$\frac{d\mathbf{r}}{dt} = \mathbf{k} \times \mathbf{r}, \qquad \mathbf{r}(0) = \mathbf{i} + \mathbf{k}.$$

Describe the curve $\mathbf{r} = \mathbf{r}(t)$.

❖ 4. An object moves so that its position vector $\mathbf{r}(t)$ satisfies

$$\frac{d\mathbf{r}}{dt} = \mathbf{a} \times (\mathbf{r}(t) - \mathbf{b})$$

and $\mathbf{r}(0) = \mathbf{r}_0$. Here $\mathbf{a}$, $\mathbf{b}$, and $\mathbf{r}_0$ are given constant vectors with $\mathbf{a} \neq \mathbf{0}$. Describe the path along which the object moves.

**The Coriolis effect**

* 5. A satellite is in a low, circular orbit around the earth, passing over the north and south poles. It makes one revolution every two hours. From the point of view of an observer standing on the earth at the equator, what is the approximate value of the Coriolis force acting on the satellite as it crosses over his head heading south? (Assume the radius of the satellite's orbit is approximately the radius of the earth.) In what direction would the satellite seem to the observer to be moving as it crosses over his head?

* 6. Describe the tangential and normal components of the Coriolis force on a particle moving with horizontal velocity $\mathbf{v}$ at (a) the north pole, (b) the south pole, and (c) the equator. In general, what is the effect of the normal component of the Coriolis force near the eye of a storm?

# 11.3 Curves and Parametrizations

In this section we will consider curves as geometric objects rather than as paths of moving particles. Everyone has an intuitive idea of what a curve is, but it is difficult to give a formal definition of a curve as a geometric object (that is, as a certain kind of set of points) without involving the concept of parametric representation. We will avoid this difficulty by continuing to regard a curve in 3-space as the set of points whose positions are given by the position vector function

$$\mathbf{r} = \mathbf{r}(t) = x(t)\mathbf{i} + y(t)\mathbf{j} + z(t)\mathbf{k}, \qquad a \le t \le b.$$

However, the parameter $t$ need no longer represent time, or any other specific physical quantity.

Curves can be very pathological. For instance, there exist continuous curves that pass through every point in a cube. It is difficult to think of such a curve as a one-dimensional object. In order to avoid such strange objects we *assume* hereafter that the defining function $\mathbf{r}(t)$ has a *continuous* first derivative, $d\mathbf{r}/dt$, which we will continue to call "velocity" and denote by $\mathbf{v}(t)$ by analogy with the physical case where $t$ is time. (We also continue to call $v(t) = |\mathbf{v}(t)|$ the "speed.") As we will see later, this implies that the curve has an *arc length* between any two points corresponding to parameter values $t_1$ and $t_2$; if $t_1 < t_2$ this arc length is

$$\int_{t_1}^{t_2} v(t)\,dt = \int_{t_1}^{t_2} |\mathbf{v}(t)|\,dt = \int_{t_1}^{t_2} \left|\frac{d\mathbf{r}}{dt}\right|\,dt.$$

Frequently we will want $\mathbf{r}(t)$ to have continuous derivatives of higher order. Whenever needed, we will assume that the "acceleration", $\mathbf{a}(t) = d^2\mathbf{r}/dt^2$, and even the third derivative, $d^3\mathbf{r}/dt^3$, are continuous. Of course, most of the curves we encounter in practice have parametrizations with continuous derivatives of all orders.

It must be recalled, however, that no assumptions on the continuity of derivatives of the function $\mathbf{r}(t)$ are sufficient to guarantee that the curve $\mathbf{r} = \mathbf{r}(t)$ is a "smooth" curve. It may fail to be smooth at a point where $\mathbf{v} = \mathbf{0}$. (See Example 1 in Section 11.1.) We will show in the next section that if, besides being continuous, the velocity vector $\mathbf{v}(t)$ is *never the zero vector*, then the curve $\mathbf{r} = \mathbf{r}(t)$ is **smooth** in the sense that it has a continuously turning tangent line.

Although we have said that a curve is a set of points given by a parametric equation $\mathbf{r} = \mathbf{r}(t)$, there is no *unique* way of representing a given curve parametrically. Just as two cars can travel the same highway at different speeds and stopping and starting at different places, so also different parametrizations can define the same curve. Typically, a given curve can have infinitely many different parametrizations.

**Example 1** Show that each of the vector functions

$$\mathbf{r}_1(t) = \sin t\,\mathbf{i} + \cos t\,\mathbf{j}, \qquad (-\pi/2 \le t \le \pi/2),$$
$$\mathbf{r}_2(t) = (t-1)\mathbf{i} + \sqrt{2t - t^2}\,\mathbf{j}, \qquad (0 \le t \le 2), \quad \text{and}$$
$$\mathbf{r}_3(t) = t\sqrt{2 - t^2}\,\mathbf{i} + (1 - t^2)\,\mathbf{j}, \qquad (-1 \le t \le 1)$$

all represent the same curve. Describe the curve.

**Solution** All three functions represent points in the $xy$-plane. The function $\mathbf{r}_1(t)$ starts at the point $(-1, 0)$ with position vector $\mathbf{r}_1(-\pi/2) = -\mathbf{i}$, and ends at the point $(1, 0)$ with position vector $\mathbf{i}$. It lies in the half of the $xy$-plane where $y \geq 0$ (because $\cos t \geq 0$ for $(-\pi/2 \leq t \leq \pi/2)$). Finally, all points on the curve are at distance 1 from the origin:

$$|\mathbf{r}_1(t)| = (\sin t)^2 + (\cos t)^2 = 1.$$

Therefore, $\mathbf{r}_1(t)$ represents the semicircle $y = \sqrt{1 - x^2}$ in the $xy$-plane traversed from left to right.

The other two functions have the same properties: both graphs lie in $y \geq 0$,

$$\mathbf{r}_2(0) = -\mathbf{i}, \qquad \mathbf{r}_2(2) = \mathbf{i}, \qquad |\mathbf{r}_2(t)| = (t-1)^2 + 2t - t^2 = 1,$$
$$\mathbf{r}_3(-1) = -\mathbf{i}, \qquad \mathbf{r}_3(1) = \mathbf{i}, \qquad |\mathbf{r}_3(t)| = t^2(2 - t^2) + (1 - t^2)^2 = 1.$$

**Figure 11.6** Three parametrizations of the semicircle $\mathcal{C}$ are given in Example 1

Thus all three functions represent the same semicircle. Of course, the three parametrizations trace out the curve with different velocities. ∎

The curve $\mathbf{r} = \mathbf{r}(t)$, $(a \leq t \leq b)$ is called a **closed curve** if $\mathbf{r}(a) = \mathbf{r}(b)$ that is, if the curve begins and ends at the same point. The curve $\mathcal{C}$ is **non-self-intersecting** if there exists some parametrization $\mathbf{r} = \mathbf{r}(t)$, $(a \leq t \leq b)$, of $\mathcal{C}$ that is one-to-one except that the endpoints could be the same;

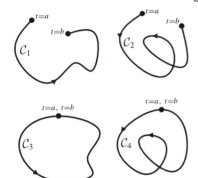

$$\mathbf{r}(t_1) = \mathbf{r}(t_2) \quad a \leq t_1 < t_2 \leq b \quad \Longrightarrow \quad t_1 = a \quad \text{and} \quad t_2 = b.$$

Such a curve can be closed, but otherwise does not intersect itself; it is then called a **simple closed curve.** Circles and ellipses are examples of simple closed curves. Every parametrization of a particular curve determines one of two possible **orientations** corresponding to the direction along the curve in which the parameter is increasing. All three parametrizations of the semicircle in the Example 1 orient the semicircle clockwise as viewed from a point above the $xy$-plane. This orientation is shown by the arrowheads on the curve in Figure 11.6. The same semicircle could be given the opposite orientation by, for example, the parametrization

**Figure 11.7** Curves $\mathcal{C}_1$ and $\mathcal{C}_3$ are non-self-intersecting
Curves $\mathcal{C}_2$ and $\mathcal{C}_4$ intersect themselves
Curves $\mathcal{C}_1$ and $\mathcal{C}_2$ are not closed
Curves $\mathcal{C}_3$ and $\mathcal{C}_4$ are closed
Curve $\mathcal{C}_3$ is a simple closed curve

$$\mathbf{r}(t) = \cos t\, \mathbf{i} + \sin t\, \mathbf{j}, \qquad 0 \leq t \leq \pi.$$

## Parametrizing the Curve of Intersection of Two Surfaces

Frequently a curve is specified as the intersection of two surfaces with given Cartesian equations. We may want to represent the curve by parametric equations. There is no unique way to do do this, but if one of the given surfaces is a cylinder parallel to a coordinate axis (so its equation is independent of one of the variables) we can begin by parametrizing that surface. The following examples clarify the method.

**Example 2** Parametrize the curve of intersection of the plane $3x + 2y + z = 4$ and the elliptic cylinder $x^2 + 4y^2 = 4$.

**Solution** We begin with the equation $x^2 + 4y^2 = 4$, which is independent of $z$. It can be parametrized in many ways; one convenient way is

$$x = 2\cos t, \qquad y = \sin t, \qquad (0 \le t \le 2\pi).$$

The equation of the plane can then be solved for $z$, so that $z$ can be expressed in terms of $t$:

$$z = 4 - 3x - 2y = 4 - 6\cos t - 2\sin t.$$

Thus the given surfaces intersect in the curve

$$\mathbf{r} = 2\cos t\,\mathbf{i} + \sin t\,\mathbf{j} + (4 - 6\cos t - 2\sin t)\mathbf{k} \qquad (0 \le t \le 2\pi).$$

**Example 3** Find a parametric representation of the curve of intersection of the two surfaces

$$x^2 + y + z = 2 \qquad \text{and} \qquad xy + z = 1.$$

**Solution** Here, neither given equation is independent of a variable, but we can obtain a third equation representing a surface containing the curve of intersection of the two given surfaces by subtracting the two given equations to eliminate $z$:

$$x^2 + y - xy = 1.$$

This equation is readily parametrized. If, for example, we let $x = t$, then

$$t^2 + y(1 - t) = 1, \qquad \text{so} \quad y = \frac{1 - t^2}{1 - t} = 1 + t.$$

Either of the given equations can then be used to express $z$ in terms of $t$:

$$z = 1 - xy = 1 - t(1 + t) = 1 - t - t^2.$$

Thus a possible parametrization of the curve is

$$\mathbf{r} = t\,\mathbf{i} + (1 + t)\mathbf{j} + (1 - t - t^2)\mathbf{k}.$$

Of course, this answer is not unique. Many other parametrizations can be found for the curve, providing orientations in either direction.

## Arc Length

We now consider how to define and calculate the length of a curve. Let $\mathcal{C}$ be a bounded, continuous curve specified by

$$\mathbf{r} = \mathbf{r}(t), \qquad a \le t \le b.$$

Subdivide the closed interval $[a, b]$ into $n$ subintervals by points

$$a = t_0 < t_1 < t_2 < \cdots < t_{n-1} < t_n = b.$$

The points $\mathbf{r}_i = \mathbf{r}(t_i)$, $(0 \le i \le n)$, subdivide $C$ into $n$ arcs. If we use the chord length $|\mathbf{r}_i - \mathbf{r}_{i-1}|$ as an approximation to the arc length between $\mathbf{r}_{i-1}$ and $\mathbf{r}_i$, then the sum

$$s_n = \sum_{i=1}^{n} |\mathbf{r}_i - \mathbf{r}_{i-1}|$$

approximates the length of $C$ by the length of a polygonal line. (See Figure 11.8.) Evidently any such approximation is less than or equal to the actual length of $C$. We say that $C$ is **rectifiable** if there exists a constant $K$ such that $s_n \le K$ for every $n$ and every choice of the points $t_i$. In this case, the completeness axiom of the real number system assures us that there will be a smallest such number $K$. We call this smallest $K$ the **length** of $C$ and denote it by $s$.

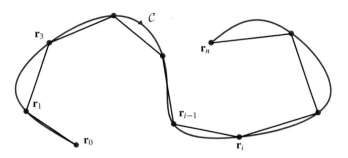

**Figure 11.8**    A polygonal approximation to a curve $C$. The length of the polygonal line cannot exceed the length of the curve

Let $\Delta t_i = t_i - t_{i-1}$ and $\Delta \mathbf{r}_i = \mathbf{r}_i - \mathbf{r}_{i-1}$. Then $s_n$ can be written in the form

$$s_n = \sum_{i=1}^{n} \left| \frac{\Delta \mathbf{r}_i}{\Delta t_i} \right| \Delta t_i.$$

If $\mathbf{r}(t)$ has a continuous derivative $\mathbf{v}(t)$, then

$$s = \lim_{\max \Delta t_i \to 0} s_n = \int_a^b \left| \frac{d\mathbf{r}}{dt} \right| dt = \int_a^b |\mathbf{v}(t)| \, dt = \int_a^b v(t) \, dt.$$

In kinematic terms, this formula states that the distance travelled by a moving particle is the integral of the speed.

**Remark**    Although the above formula is expressed in terms of the parameter $t$, the arc length, as defined above, is a strictly geometric property of the curve $C$. It is independent of the particular parametrization used to represent $C$. See Exercise 27 at the end of this section.

If $s(t)$ denotes the arc length of that part of $C$ corresponding to parameter values in $[a, t]$ then

$$\frac{ds}{dt} = \frac{d}{dt} \int_a^t v(\tau) \, d\tau = v(t)$$

so that the **arc length element** for $C$ is given by

$$ds = v(t) \, dt = \left| \frac{d}{dt} \mathbf{r}(t) \right| dt.$$

The length of $C$ is the integral of these arc length elements: we write

$$\int_C ds = \text{length of } C = \int_a^b v(t)\, dt.$$

Several familiar formulas for arc length follow from the above formula by using specific parametrizations of curves. For instance, the arc length element $ds$ for the Cartesian plane curve $y = f(x)$ on $[a, b]$ is obtained by using $x$ as parameter: here $\mathbf{r} = x\mathbf{i} + f(x)\mathbf{j}$ so $\mathbf{v} = \mathbf{i} + f'(x)\mathbf{j}$ and

$$ds = \sqrt{1 + \left(f'(x)\right)^2}\, dx.$$

Similarly, the arc length element $ds$ for a plane polar curve $r = g(\theta)$ can be calculated from the parametrization

$$\mathbf{r}(\theta) = g(\theta) \cos\theta\, \mathbf{i} + g(\theta) \sin\theta\, \mathbf{j}.$$

It is

$$ds = \sqrt{\left(g(\theta)\right)^2 + \left(g'(\theta)\right)^2}\, d\theta.$$

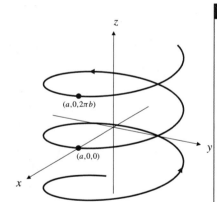

**Figure 11.9**    The helix
$x = a \cos t$
$y = a \sin t$
$z = bt$

**Example 4**    Find the length $s$ of that part of the **circular helix**

$$\mathbf{r} = a \cos t\, \mathbf{i} + a \sin t\, \mathbf{j} + bt\, \mathbf{k}$$

between the points $(a, 0, 0)$ and $(a, 0, 2\pi b)$.

**Solution**    This curve spirals around the $z$-axis, rising as it turns. (See Figure 11.9.) It lies on the surface of the circular cylinder $x^2 + y^2 = a^2$. We have

$$\mathbf{v} = \frac{d\mathbf{r}}{dt} = -a \sin t\, \mathbf{i} + a \cos t\, \mathbf{j} + b\mathbf{k}$$
$$v = \sqrt{a^2 + b^2},$$

so that in terms of the parameter $t$ the helix is traced out at constant speed. The required length $s$ corresponds to parameter interval $[0, 2\pi]$. Thus

$$s = \int_0^{2\pi} v(t)\, dt = \int_0^{2\pi} \sqrt{a^2 + b^2}\, dt = 2\pi \sqrt{a^2 + b^2}.$$

## Piecewise Smooth Curves

As observed earlier, a parametric curve $C$ given by $\mathbf{r} = \mathbf{r}(t)$ can fail to be smooth at points where $d/vecr/dt = \mathbf{0}$. If there are finitely many such points we will say that the curve is piecewise smooth.

In general, a **piecewise smooth curve** $C$ consists of a finite number of smooth arcs, $C_1, C_2, \ldots, C_k$, as shown in Figure 11.10.

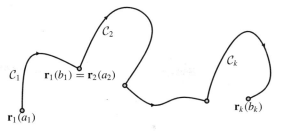

**Figure 11.10**   A piecewise smooth curve

In this case we express $C$ as the sum of the individual arcs:

$$C = C_1 + C_2 + \cdots + C_k.$$

Each arc $C_i$ can have its own parametrization

$$\mathbf{r} = \mathbf{r}_i(t), \qquad (a_i \le t \le b_i),$$

where $\mathbf{v}_i = d\mathbf{r}_i/dt \ne \mathbf{0}$ for $a_i < t < b_i$. The fact that $C_{i+1}$ must begin at the point where $C_i$ ends requires the conditions

$$\mathbf{r}_{i+1}(a_{i+1}) = \mathbf{r}_i(b_i) \qquad \text{for } 1 \le i \le k - 1.$$

If also $\mathbf{r}_k(b_k) = \mathbf{r}_1(a_1)$, then $C$ is a closed piecewise smooth curve.

The length of a piecewise smooth curve $C = C_1 + C_2 + \cdots + C_k$ is the sum of the lengths of its component arcs:

$$\text{length of } C = \sum_{i=1}^{k} \int_{a_i}^{b_i} \left| \frac{d\mathbf{r}_i}{dt} \right| dt.$$

## The Arc-Length Parametrization

The selection of a particular parameter in terms of which to specify a given curve will usually depend on the problem in which the curve arises; there is no one "right way" to parametrize a curve. However, there is one parameter that is "natural" in that it arises from the geometry (shape and size) of the curve itself, and not from any particular coordinate system in which the equation of the curve is to be expressed. This parameter is the *arc length* measured from some particular point (the *initial point*) on the curve. The position vector of an arbitrary point $P$ on the curve can be specified as a function of the arc length $s$ along the curve from the initial point $P_0$ to $P$,

$$\mathbf{r} = \mathbf{r}(s).$$

This equation is called an **arc-length parametrization** or **intrinsic parametrization** of the curve. Since $ds = v(t)\,dt$ for any parametrization $\mathbf{r} = \mathbf{r}(t)$, for the arc-length parametrization we have $ds = v(s)\,ds$. Thus $v(s) = 1$, identically; a *curve parametrized in terms of arc length is traced at unit speed.* Although it is seldom easy (and usually not possible), to find $\mathbf{r}(s)$ explicitly when the curve is given in terms of some other parameter, smooth curves always have such parametrizations (see Exercise 28 at the end of this section), and they will prove useful when we develop the fundamentals of the *differential geometry* for 3-space curves in the next section.

Suppose that a curve is specified in terms of an arbitrary parameter $t$. If the arc length over a parameter interval $[t_0, t]$,

$$s = s(t) = \int_{t_0}^{t} \left| \frac{d}{d\tau} \mathbf{r}(\tau) \right| d\tau,$$

can be evaluated explicitly, and if the equation $s = s(t)$ can be explicitly solved for $t$ as a function of $s$ ($t = t(s)$) then the curve can be reparametrized in terms of arc length by substituting for $t$ in the original parametrization:

$$\mathbf{r} = \mathbf{r}(t(s)).$$

**Example 5**   Parametrize the circular helix

$$\mathbf{r} = a \cos t \, \mathbf{i} + a \sin t \, \mathbf{j} + bt \, \mathbf{k}$$

in terms of the arc length measured from the point $(a, 0, 0)$ in the direction of increasing $t$.

**Solution**   The initial point corresponds to $t = 0$. As shown in Example 4, we have $ds/dt = \sqrt{a^2 + b^2}$ so

$$s = s(t) = \int_{0}^{t} \sqrt{a^2 + b^2} \, d\tau = \sqrt{a^2 + b^2} \, t.$$

Therefore $t = s/\sqrt{a^2 + b^2}$ and the arc-length parametrization is

$$\mathbf{r}(s) = a \cos\left( \frac{s}{\sqrt{a^2 + b^2}} \right) \mathbf{i} + a \sin\left( \frac{s}{\sqrt{a^2 + b^2}} \right) \mathbf{j} + \frac{bs}{\sqrt{a^2 + b^2}} \mathbf{k}.$$

# Exercises 11.3

In Exercises 1–4, find the required parametrization of the first quadrant part of the circular arc $x^2 + y^2 = a^2$.

**1.** In terms of the $y$-coordinate, oriented counterclockwise

**2.** In terms of the $x$-coordinate, oriented clockwise

**3.** In terms of the angle between the tangent line and the positive $x$-axis, oriented counterclockwise

**4.** In terms of arc length measured from $(0, a)$, oriented clockwise

**5.** The cylinders $z = x^2$ and $z = 4y^2$ intersect in two curves, one of which passes through the point $(2, -1, 4)$. Find a parametrization of that curve using $t = y$ as parameter.

**6.** The plane $x + y + z = 1$ intersects the cylinder $z = x^2$ in a parabola. Parametrize the parabola using $t = x$ as parameter.

In Exercises 7–10, parametrize the curve of intersection of the given surfaces. *Note*: the answers are not unique.

**7.** $x^2 + y^2 = 9$ and $z = x + y$

**8.** $z = \sqrt{1 - x^2 - y^2}$ and $x + y = 1$

**9.** $z = x^2 + y^2$ and $2x - 4y - z - 1 = 0$

**10.** $yz + x = 1$ and $xz - x = 1$

**11.** The plane $z = 1 + x$ intersects the cone $z^2 = x^2 + y^2$ in a parabola. Try to parametrize the parabola using as parameter: (a) $t = x$, (b) $t = y$, and (c) $t = z$. Which of these choices for $t$ leads to a single parametrization that represents the whole parabola? What is that parametrization? What happens with the other two choices?

**∗ 12.** The plane $x + y + z = 1$ intersects the sphere $x^2 + y^2 + z^2 = 1$ in a circle $C$. Find the centre $\mathbf{r}_0$ and radius $r$ of $C$. Also find two perpendicular unit vectors $\hat{\mathbf{v}}_1$ and $\hat{\mathbf{v}}_2$ parallel to the plane of $C$. (*Hint:* to be specific, show that $\hat{\mathbf{v}}_1 = (\mathbf{i} - \mathbf{j})/\sqrt{2}$ is one such vector, and then find a second that is perpendicular to $\hat{\mathbf{v}}_1$.) Use your results to construct a parametrization of $\mathcal{C}$.

13. Find the length of the curve $\mathbf{r} = t^2\mathbf{i} + t^2\mathbf{j} + t^3\mathbf{k}$ from $t = 0$ to $t = 1$.

14. For what values of the parameter $\lambda$ is the length $s(T)$ of the curve $\mathbf{r} = t\mathbf{i} + \lambda t^2\mathbf{j} + t^3\mathbf{k}$, $(0 \le t \le T)$ given by $s(T) = T + T^3$?

15. Express the length of the curve $\mathbf{r} = at^2\mathbf{i} + bt\mathbf{j} + c\ln t\mathbf{k}$, $(1 \le t \le T)$ as a definite integral. Evaluate the integral if $b^2 = 4ac$.

16. Describe the parametric curve $\mathcal{C}$ given by

$$x = a\cos t\sin t, \qquad y = a\sin^2 t, \qquad z = bt.$$

What is the length of $\mathcal{C}$ between $t = 0$ and $t = T > 0$?

17. Find the length of the conical helix $\mathbf{r} = t\cos t\,\mathbf{i} + t\sin t\,\mathbf{j} + t\,\mathbf{k}$, $(0 \le t \le 2\pi)$. Why is the curve called a conical helix?

18. Describe the intersection of the sphere $x^2 + y^2 + z^2 = 1$ and the elliptic cylinder $x^2 + 2z^2 = 1$. Find the total length of this intersection curve.

19. Let $\mathcal{C}$ be the curve $x = e^t\cos t$, $y = e^t\sin t$, $z = t$ between $t = 0$ and $t = 2\pi$. Find the length of $\mathcal{C}$.

20. Find the length of the piecewise smooth curve $\mathbf{r} = t^3\mathbf{i} + t^2\mathbf{j}$, $(-1 \le t \le 2)$.

21. Describe the piecewise smooth curve $\mathcal{C} = \mathcal{C}_1 + \mathcal{C}_2$ where $\mathbf{r}_1(t) = t\mathbf{i} + t\mathbf{j}$, $(0 \le t \le 1)$, and $\mathbf{r}_2(t) = (1 - t)\mathbf{i} + (1 + t)\mathbf{j}$, $(0 \le t \le 1)$.

* 22. A cable of length $L$ and circular cross-section of radius $a$ is wound around a cylindrical spool of radius $b$ with no overlapping and so that adjacent windings touch one another. What length of the spool is covered by the cable?

In Exercises 23–26, reparametrize the given curve in the same orientation in terms of arc length measured from the point where $t = 0$.

23. $\mathbf{r} = At\mathbf{i} + Bt\mathbf{j} + Ct\mathbf{k}$,     $(A^2 + B^2 + C^2 > 0)$

24. $\mathbf{r} = e^t\mathbf{i} + \sqrt{2}t\mathbf{j} - e^{-t}\mathbf{k}$

* 25. $\mathbf{r} = a\cos^3 t\,\mathbf{i} + a\sin^3 t\,\mathbf{j} + b\cos 2t\,\mathbf{k}$,     $(0 \le t \le \dfrac{\pi}{2})$

* 26. $\mathbf{r} = 3t\cos t\,\mathbf{i} + 3t\sin t\,\mathbf{j} + 2\sqrt{2}t^{3/2}\mathbf{k}$

* 27. Let $\mathbf{r} = \mathbf{r}_1(t)$, $(a \le t \le b)$, and $\mathbf{r} = \mathbf{r}_2(u)$, $(c \le u \le d)$, be two parametrizations of the same curve $\mathcal{C}$, each one-to-one on its domain, and each giving $\mathcal{C}$ the same orientation (so that $\mathbf{r}_1(a) = \mathbf{r}_2(c)$ and $\mathbf{r}_1(b) = \mathbf{r}_2(d)$). Then for each $t$ in $[a, b]$ there is a unique $u = u(t)$ such that $\mathbf{r}_2(u(t)) = \mathbf{r}_1(t)$. Show that

$$\int_a^b \left|\frac{d}{dt}\mathbf{r}_1(t)\right| dt = \int_c^d \left|\frac{d}{du}\mathbf{r}_2(u)\right| du,$$

and thus that the length of $\mathcal{C}$ is independent of parametrization.

* 28. If the curve $\mathbf{r} = \mathbf{r}(t)$ has continuous, nonvanishing velocity $\mathbf{v}(t)$ on the interval $[a, b]$, and if $t_0$ is some point in $[a, b]$, show that the function

$$s = g(t) = \int_{t_0}^t |\mathbf{v}(u)|\, du$$

is an increasing function on $[a, b]$, and so has an inverse:

$$t = g^{-1}(s) \iff s = g(t).$$

Hence show that the curve can be parametrized in terms of arc length measured from $\mathbf{r}(t_0)$.

---

# 11.4 Curvature, Torsion, and the Frenet Frame

In this section and the next we develop the fundamentals of differential geometry of curves in 3-space. We will introduce several new scalar and vector functions associated with a curve $\mathcal{C}$. The most important of these are the curvature and torsion of the curve, and a right-handed triad of mutually perpendicular unit vectors forming a basis at any point on the curve, and called the Frenet frame. The curvature and torsion measure the rate at which a curve is turning (away from its tangent line) and twisting (out of the plane in which it is turning) at any point.

## The Unit Tangent Vector

The velocity vector $\mathbf{v}(t) = d\mathbf{r}/dt$ is tangent to the parametric curve $\mathbf{r} = \mathbf{r}(t)$ at the point $\mathbf{r}(t)$, and points in the direction of the orientation of the curve there. Since we are assuming that $\mathbf{v}(t) \neq \mathbf{0}$ we can find a **unit tangent vector**, $\hat{\mathbf{T}}(t)$, at $\mathbf{r}(t)$ by dividing $\mathbf{v}(t)$ by its length:

$$\hat{\mathbf{T}}(t) = \frac{\mathbf{v}(t)}{v(t)} = \frac{d\mathbf{r}}{dt} \bigg/ \left|\frac{d\mathbf{r}}{dt}\right|.$$

Recall that a curve parametrized in terms of arc length, $\mathbf{r} = \mathbf{r}(s)$, is traced at unit speed; $v(s) = 1$. In terms of arc-length parametrization, the unit tangent vector is

$$\hat{\mathbf{T}}(s) = \frac{d\mathbf{r}}{ds}.$$

**Example 1** Find the unit tangent vector, $\hat{\mathbf{T}}$, for the circular helix of Example 4 of Section 11.3, both in terms of $t$ and in terms of the arc-length parameter $s$.

**Solution** In terms of $t$ we have

$$\mathbf{r} = a\cos t\,\mathbf{i} + a\sin t\,\mathbf{j} + bt\mathbf{k}$$
$$\mathbf{v}(t) = -a\sin t\,\mathbf{i} + a\cos t\,\mathbf{j} + b\mathbf{k}$$
$$v(t) = \sqrt{a^2\sin^2 t + a^2\cos^2 t + b^2} = \sqrt{a^2 + b^2}$$
$$\hat{\mathbf{T}}(t) = -\frac{a}{\sqrt{a^2 + b^2}}\sin t\,\mathbf{i} + \frac{a}{\sqrt{a^2 + b^2}}\cos t\,\mathbf{j} + \frac{b}{\sqrt{a^2 + b^2}}\mathbf{k}.$$

In terms of the arc-length parameter (see Example 5 of Section 11.3)

$$\mathbf{r}(s) = a\cos\left(\frac{s}{\sqrt{a^2 + b^2}}\right)\mathbf{i} + a\sin\left(\frac{s}{\sqrt{a^2 + b^2}}\right)\mathbf{j} + \frac{bs}{\sqrt{a^2 + b^2}}\mathbf{k}$$
$$\hat{\mathbf{T}}(s) = \frac{d\mathbf{r}}{ds} = -\frac{a}{\sqrt{a^2 + b^2}}\sin\left(\frac{s}{\sqrt{a^2 + b^2}}\right)\mathbf{i} + \frac{a}{\sqrt{a^2 + b^2}}\cos\left(\frac{s}{\sqrt{a^2 + b^2}}\right)\mathbf{j}$$
$$+ \frac{b}{\sqrt{a^2 + b^2}}\mathbf{k}.$$

**Remark** If the curve $\mathbf{r} = \mathbf{r}(t)$ has a continuous, nonvanishing velocity $\mathbf{v}(t)$ then the unit tangent vector $\hat{\mathbf{T}}(t)$ is a continuous function of $t$. The angle $\theta(t)$ between $\hat{\mathbf{T}}(t)$ and any fixed unit vector $\hat{\mathbf{u}}$ is also continuous in $t$:

$$\theta(t) = \cos^{-1}(\hat{\mathbf{T}}(t) \bullet \hat{\mathbf{u}}).$$

Thus, as asserted previously, the curve is *smooth* in the sense that it has a continuously turning tangent line. The rate of this turning is quantified by the curvature, which we introduce now.

## Curvature and the Unit Normal

In the rest of this section we will deal abstractly with a curve $\mathcal{C}$ parametrized in terms of arc length measured from some point on it:

$$\mathbf{r} = \mathbf{r}(s).$$

In the next section we return to curves with arbitrary parametrizations and apply the principles developed in this section to specific problems. Throughout we assume that the parametric equations of curves have continuous derivatives up to third order on the intervals where they are defined.

Having unit length, the tangent vector $\hat{\mathbf{T}}(s) = d\mathbf{r}/ds$ satisfies $\hat{\mathbf{T}}(s) \bullet \hat{\mathbf{T}}(s) = 1$. Differentiating this equation with respect to $s$ we get

$$2\hat{\mathbf{T}}(s) \bullet \frac{d\hat{\mathbf{T}}}{ds} = 0,$$

so that $d\hat{\mathbf{T}}/ds$ is perpendicular to $\hat{\mathbf{T}}(s)$.

**DEFINITION 1**

> **Curvature and radius of curvature**
>
> The **curvature** of $\mathcal{C}$ at the point $\mathbf{r}(s)$ is the length of $d\hat{\mathbf{T}}/ds$ there. It is denoted by the Greek letter $\kappa$ (*kappa*):
>
> $$\kappa(s) = \left| \frac{d\hat{\mathbf{T}}}{ds} \right|.$$
>
> The **radius of curvature**, denoted $\rho$ (the Greek letter *rho*) is the reciprocal of the curvature:
>
> $$\rho(s) = \frac{1}{\kappa(s)}.$$

As we will see below, the curvature of $\mathcal{C}$ at $\mathbf{r}(s)$ measures the rate of turning of the tangent line to the curve there. The radius of curvature is the radius of the circle through $\mathbf{r}(s)$ that most closely approximates the curve $\mathcal{C}$ near that point.

According to its definition, $\kappa(s) \geq 0$ everywhere on $\mathcal{C}$. If $\kappa(s) \neq 0$ we can divide $d\hat{\mathbf{T}}/ds$ by its length, $\kappa(s)$, and obtain a unit vector $\hat{\mathbf{N}}(s)$ in the same direction. This unit vector is called the **unit principal normal** to $\mathcal{C}$ at $\mathbf{r}(s)$, or, more commonly, just the **unit normal**.

$$\hat{\mathbf{N}}(s) = \frac{1}{\kappa(s)} \frac{d\hat{\mathbf{T}}}{ds} = \frac{d\hat{\mathbf{T}}}{ds} \bigg/ \left| \frac{d\hat{\mathbf{T}}}{ds} \right|$$

Note that $\hat{\mathbf{N}}(s)$ is perpendicular to $\mathcal{C}$ at $\mathbf{r}(s)$, and points in the direction that $\hat{\mathbf{T}}$ and therefore $\mathcal{C}$ is turning. The principal normal is not defined at points where the curvature $\kappa(s)$ is zero. For instance, a straight line has no principal normal. Figure 11.11(a) shows $\hat{\mathbf{T}}$ and $\hat{\mathbf{N}}$ at a point on a typical curve.

**Example 2**   Let $a > 0$. Show that the curve $\mathcal{C}$ given by

$$\mathbf{r} = a \cos\left(\frac{s}{a}\right)\mathbf{i} + a \sin\left(\frac{s}{a}\right)\mathbf{j}$$

is a circle in the $xy$-plane having radius $a$ and centre at the origin, and that it is parametrized in terms of arc length. Find the curvature, the radius of curvature, and the unit tangent and principal normal vectors at any point on $\mathcal{C}$.

***Solution***   Since

$$|\mathbf{r}(s)| = a\sqrt{\left(\cos\left(\frac{s}{a}\right)\right)^2 + \left(\sin\left(\frac{s}{a}\right)\right)^2} = a,$$

$\mathcal{C}$ is indeed a circle of radius $a$ centred at the origin in the $xy$-plane. Since the speed

**Figure 11.11**

(a) The unit tangent and principal normal vectors for a curve

(b) The unit tangent and principal normal vectors for a circle

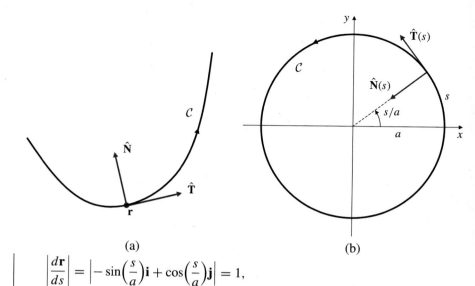

(a)                                          (b)

$$\left|\frac{d\mathbf{r}}{ds}\right| = \left|-\sin\left(\frac{s}{a}\right)\mathbf{i} + \cos\left(\frac{s}{a}\right)\mathbf{j}\right| = 1,$$

the parameter $s$ must represent arc length, and hence the unit tangent vector is

$$\hat{\mathbf{T}}(s) = -\sin\left(\frac{s}{a}\right)\mathbf{i} + \cos\left(\frac{s}{a}\right)\mathbf{j}.$$

Therefore

$$\frac{d\hat{\mathbf{T}}}{ds} = -\frac{1}{a}\cos\left(\frac{s}{a}\right)\mathbf{i} - \frac{1}{a}\sin\left(\frac{s}{a}\right)\mathbf{j}$$

and the curvature and radius of curvature at $\mathbf{r}(s)$ are

$$\kappa(s) = \left|\frac{d\hat{\mathbf{T}}}{ds}\right| = \frac{1}{a}, \qquad \rho(s) = \frac{1}{\kappa(s)} = a.$$

Finally, the unit principal normal is

$$\hat{\mathbf{N}}(s) = -\cos\left(\frac{s}{a}\right)\mathbf{i} - \sin\left(\frac{s}{a}\right)\mathbf{j} = -\frac{1}{a}\mathbf{r}(s).$$

Note that the curvature and radius of curvature are constant; the latter is in fact the radius of the circle. The circle and its unit tangent and normal vectors at a typical point are sketched in Figure 11.11(b). Note that $\hat{\mathbf{N}}$ points towards the centre of the circle.

**Remark**  Another observation can be made about the above example. The position vector $\mathbf{r}(s)$ makes angle $\theta = s/a$ with the positive $x$-axis, and therefore $\hat{\mathbf{T}}(s)$ makes the same angle with the positive $y$-axis. Therefore the rate of rotation of $\hat{\mathbf{T}}$ with respect to $s$ is

$$\frac{d\theta}{ds} = \frac{1}{a} = \kappa.$$

That is, $\kappa$ is the rate at which $\hat{\mathbf{T}}$ is turning (measured with respect to arc length). This observation extends to a general smooth curve.

| THEOREM | **2** |
|---|---|

**Curvature is the rate of turning of the unit tangent**

Let $\kappa > 0$ on an interval containing $s$, and let $\Delta\theta$ be the angle between $\hat{\mathbf{T}}(s + \Delta s)$ and $\hat{\mathbf{T}}(s)$, the unit tangent vectors at neighbouring points on the curve. Then

$$\kappa(s) = \lim_{\Delta s \to 0} \frac{\Delta\theta}{\Delta s}.$$

**PROOF** Let $\Delta\hat{\mathbf{T}} = \hat{\mathbf{T}}(s + \Delta s) - \hat{\mathbf{T}}(s)$. Because both $\hat{\mathbf{T}}(s)$ and $\hat{\mathbf{T}}(s + \Delta s)$ are unit vectors, $|\Delta\hat{\mathbf{T}}/\Delta\theta|$ is the ratio of the length of a chord to the length of the corresponding arc on a circle of radius 1. (See Figure 11.12.) Thus

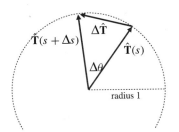

$$\lim_{\Delta s \to 0} \left| \frac{\Delta\hat{\mathbf{T}}}{\Delta\theta} \right| = 1, \qquad \text{and}$$

$$\kappa(s) = \lim_{\Delta s \to 0} \left| \frac{\Delta\hat{\mathbf{T}}}{\Delta s} \right| = \lim_{\Delta s \to 0} \left| \frac{\Delta\hat{\mathbf{T}}}{\Delta\theta} \right| \left| \frac{\Delta\theta}{\Delta s} \right| = \lim_{\Delta s \to 0} \left| \frac{\Delta\theta}{\Delta s} \right|.$$

**Figure 11.12**   $\Delta\hat{\mathbf{T}} \approx \Delta\theta$ for small $\Delta s$

The unit tangent $\hat{\mathbf{T}}$ and unit normal $\hat{\mathbf{N}}$ at a point $\mathbf{r}(s)$ on a curve $\mathcal{C}$ are regarded as having their tails at that point. They are perpendicular, and $\hat{\mathbf{N}}$ points in the direction towards which $\hat{\mathbf{T}}(s)$ turns as $s$ increases. The plane passing through $\mathbf{r}(s)$ and containing the vectors $\hat{\mathbf{T}}(s)$ and $\hat{\mathbf{N}}(s)$ is called the **osculating plane** of $\mathcal{C}$ at $\mathbf{r}(s)$ (from the Latin *osculum*, meaning *kiss*). For a *plane curve*, such as the circle in Example 2, the osculating plane is just the plane containing the curve. For more general three-dimensional curves the osculating plane varies from point to point; at any point it is that plane which comes closest to containing the part of the curve near that point. The osculating plane is not properly defined at a point where $\kappa(s) = 0$, though if such points are isolated, it can sometimes be defined as a limit of osculating planes for neighbouring points.

Still assuming that $\kappa(s) \neq 0$, let

$$\mathbf{r}_c(s) = \mathbf{r}(s) + \rho(s)\hat{\mathbf{N}}(s).$$

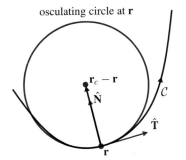

osculating circle at **r**

**Figure 11.13**   An osculating circle

For each $s$ the point with position vector $\mathbf{r}_c(s)$ lies in the osculating plane of $\mathcal{C}$ at $\mathbf{r}(s)$, on the concave side of $\mathcal{C}$ and at distance $\rho(s)$ from $\mathbf{r}(s)$. It is called the **centre of curvature** of $\mathcal{C}$ for the point $\mathbf{r}(s)$. The circle in the osculating plane having centre at the centre of curvature and radius equal to the radius of curvature $\rho(s)$ is called the **osculating circle** for $\mathcal{C}$ at $\mathbf{r}(s)$. Among all circles that pass through the point $\mathbf{r}(s)$, the osculating circle is the one that best describes the behaviour of $\mathcal{C}$ near that point. Of course, the osculating circle of a circle at any point is the same circle. A typical example of an osculating circle is shown in Figure 11.13.

## Torsion and Binormal, The Frenet–Serret Formulas

At any point $\mathbf{r}(s)$ on the curve $\mathcal{C}$ where $\hat{\mathbf{T}}$ and $\hat{\mathbf{N}}$ are defined, a third unit vector, the **unit binormal** $\hat{\mathbf{B}}$, is defined by the formula

$$\hat{\mathbf{B}} = \hat{\mathbf{T}} \times \hat{\mathbf{N}}.$$

Note that $\hat{\mathbf{B}}(s)$ is normal to the osculating plane of $\mathcal{C}$ at $\mathbf{r}(s)$; if $\mathcal{C}$ is a plane curve, then $\hat{\mathbf{B}}$ is a constant vector, independent of $s$ on any interval where $\kappa(s) \neq 0$. At each point on $\mathcal{C}$, the three vectors $\{\hat{\mathbf{T}}, \hat{\mathbf{N}}, \hat{\mathbf{B}}\}$ constitute a right-handed basis of mutually perpendicular unit vectors like the standard basis $\{\mathbf{i}, \mathbf{j}, \mathbf{k}\}$. (See Figure 11.14.) This basis is called the **Frenet frame** at the point. Note that

$$\hat{\mathbf{B}} \times \hat{\mathbf{T}} = \hat{\mathbf{N}} \text{ and } \hat{\mathbf{N}} \times \hat{\mathbf{B}} = \hat{\mathbf{T}}.$$

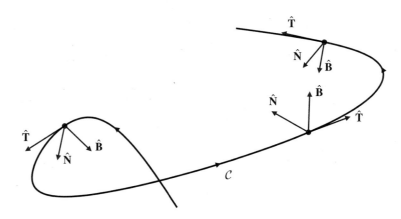

**Figure 11.14** The Frenet frame $\{\hat{\mathbf{T}}, \hat{\mathbf{N}}, \hat{\mathbf{B}}\}$ at some points on $\mathcal{C}$

Since $1 = \hat{\mathbf{B}}(s) \bullet \hat{\mathbf{B}}(s)$, therefore $\hat{\mathbf{B}}(s) \bullet (d\hat{\mathbf{B}}/ds) = 0$, and $d\hat{\mathbf{B}}/ds$ is perpendicular to $\hat{\mathbf{B}}(s)$. Also, differentiating $\hat{\mathbf{B}} = \hat{\mathbf{T}} \times \hat{\mathbf{N}}$ we obtain

$$\frac{d\hat{\mathbf{B}}}{ds} = \frac{d\hat{\mathbf{T}}}{ds} \times \hat{\mathbf{N}} + \hat{\mathbf{T}} \times \frac{d\hat{\mathbf{N}}}{ds} = \kappa\hat{\mathbf{N}} \times \hat{\mathbf{N}} + \hat{\mathbf{T}} \times \frac{d\hat{\mathbf{N}}}{ds} = \hat{\mathbf{T}} \times \frac{d\hat{\mathbf{N}}}{ds}.$$

Therefore $d\hat{\mathbf{B}}/ds$ is also perpendicular to $\hat{\mathbf{T}}$. Being perpendicular to both $\hat{\mathbf{T}}$ and $\hat{\mathbf{B}}$, $d\hat{\mathbf{B}}/ds$ must be parallel to $\hat{\mathbf{N}}$. This fact is the basis for our definition of torsion.

**DEFINITION 2**

> **Torsion**
>
> There exists a function $\tau(s)$ such that
>
> $$\frac{d\hat{\mathbf{B}}}{ds} = -\tau(s)\hat{\mathbf{N}}(s).$$
>
> The number $\tau(s)$ is called the **torsion** of $\mathcal{C}$ at $\mathbf{r}(s)$.

The torsion measures the degree of twisting that the curve exhibits near a point, that is, the extent to which the curve fails to be planar. It may be positive or negative, depending on the right-handedness or left-handedness of the twisting. We will present an example later in this section.

Theorem 2 has an analogue for torsion, for which the proof is similar. It states that the absolute value of the torsion, $|\tau(s)|$, at point $\mathbf{r}(s)$ on the curve $\mathcal{C}$ is the rate of turning of the unit binormal:

$$\lim_{\Delta s \to 0} \frac{\Delta\Psi}{\Delta s} = |\tau(s)|,$$

where $\Delta\Psi$ is the angle between $\hat{\mathbf{B}}(s + \Delta s)$ and $\hat{\mathbf{B}}(s)$.

**Example 3** (**The circular helix**) As observed in Example 5 of Section 11.3, the parametric equation

$$\mathbf{r}(s) = a\cos(cs)\mathbf{i} + a\sin(cs)\mathbf{j} + bcs\mathbf{k}, \qquad \text{where } c = \frac{1}{\sqrt{a^2 + b^2}},$$

represents a circular helix wound on the cylinder $x^2 + y^2 = a^2$, and parametrized in terms of arc length. Assume $a > 0$. Find the curvature and torsion functions $\kappa(s)$ and $\tau(s)$ for this helix, and also the unit vectors comprising the Frenet frame at any point $\mathbf{r}(s)$ on the helix.

**Solution** In Example 1 we calculated the unit tangent vector to be

$$\hat{\mathbf{T}}(s) = -ac\sin(cs)\mathbf{i} + ac\cos(cs)\mathbf{j} + bc\mathbf{k}.$$

Differentiating again leads to

$$\frac{d\hat{\mathbf{T}}}{ds} = -ac^2\cos(cs)\mathbf{i} - ac^2\sin(cs)\mathbf{j},$$

so that the curvature of the helix is

$$\kappa(s) = \left|\frac{d\hat{\mathbf{T}}}{ds}\right| = ac^2 = \frac{a}{a^2 + b^2},$$

and the unit normal vector is

$$\hat{\mathbf{N}}(s) = \frac{1}{\kappa(s)}\frac{d\hat{\mathbf{T}}}{ds} = -\cos(cs)\mathbf{i} - \sin(cs)\mathbf{j}.$$

Now we have

$$\hat{\mathbf{B}}(s) = \hat{\mathbf{T}}(s)\times\hat{\mathbf{N}}(s) = \begin{vmatrix} \mathbf{i} & \mathbf{j} & \mathbf{k} \\ -ac\sin(cs) & ac\cos(cs) & bc \\ -\cos(cs) & -\sin(cs) & 0 \end{vmatrix}$$

$$= bc\sin(cs)\mathbf{i} - bc\cos(cs)\mathbf{j} + ac\mathbf{k}.$$

Differentiating this formula leads to

$$\frac{d\hat{\mathbf{B}}}{ds} = bc^2\cos(cs)\mathbf{i} + bc^2\sin(cs)\mathbf{j} = -bc^2\hat{\mathbf{N}}(s).$$

Therefore, the torsion is given by

$$\tau(s) = -(-bc^2) = \frac{b}{a^2 + b^2}.$$

**Remark** Observe that the curvature $\kappa(s)$ and the torsion $\tau(s)$ are both constant (i.e. independent of $s$) for a circular helix. In the above example, $\tau > 0$ (assuming that $b > 0$). This corresponds to the fact that the helix is *right-handed*. (See Figure 11.9

in the previous section.) If you grasp the helix with your right hand so your fingers surround it in the direction of increasing $s$ (counterclockwise, looking down from the positive $z$-axis), then your thumb also points in the axial direction corresponding to increasing $s$ (the upward direction). Had we started with a left-handed helix, such as

$$\mathbf{r} = a \sin t\, \mathbf{i} + a \cos t\, \mathbf{j} + bt\mathbf{k}, \qquad (a, b > 0),$$

we would have obtained $\tau = -b/(a^2 + b^2)$.

Making use of the formulas $d\hat{\mathbf{T}}/ds = \kappa \hat{\mathbf{N}}$ and $d\hat{\mathbf{B}}/ds = -\tau \hat{\mathbf{N}}$ we can calculate $d\hat{\mathbf{N}}/ds$ as well:

$$\frac{d\hat{\mathbf{N}}}{ds} = \frac{d}{ds}(\hat{\mathbf{B}} \times \hat{\mathbf{T}}) = \frac{d\hat{\mathbf{B}}}{ds} \times \hat{\mathbf{T}} + \hat{\mathbf{B}} \times \frac{d\hat{\mathbf{T}}}{ds}$$
$$= -\tau \hat{\mathbf{N}} \times \hat{\mathbf{T}} + \kappa \hat{\mathbf{B}} \times \hat{\mathbf{N}} = -\kappa \hat{\mathbf{T}} + \tau \hat{\mathbf{B}}.$$

Together, the three formulas

$$\frac{d\hat{\mathbf{T}}}{ds} = \kappa \hat{\mathbf{N}}$$

$$\frac{d\hat{\mathbf{N}}}{ds} = -\kappa \hat{\mathbf{T}} + \tau \hat{\mathbf{B}}$$

$$\frac{d\hat{\mathbf{B}}}{ds} = -\tau \hat{\mathbf{N}}$$

**Figure 11.15** $\hat{\mathbf{T}}$, $\hat{\mathbf{N}}$, and $\hat{\mathbf{B}}$, and their directions of change

are known as **The Frenet–Serret formulas**. They are of fundamental importance in the theory of curves in 3-space. The Frenet–Serret formulas can be written in matrix form as follows:

$$\frac{d}{ds}\begin{pmatrix} \hat{\mathbf{T}} \\ \hat{\mathbf{N}} \\ \hat{\mathbf{B}} \end{pmatrix} = \begin{pmatrix} 0 & \kappa & 0 \\ -\kappa & 0 & \tau \\ 0 & -\tau & 0 \end{pmatrix} \begin{pmatrix} \hat{\mathbf{T}} \\ \hat{\mathbf{N}} \\ \hat{\mathbf{B}} \end{pmatrix}.$$

Using the Frenet–Serret formulas we can show that the shape of a curve with non-vanishing curvature is completely determined by the curvature and torsion functions $\kappa(s)$ and $\tau(s)$.

**THEOREM 3**

**The Fundamental Theorem of Space Curves**

Let $\mathcal{C}_1$ and $\mathcal{C}_2$ be two curves, both of which have the same non-vanishing curvature function $\kappa(s)$ and the same torsion function $\tau(s)$. Then the curves are congruent. That is, one can be moved rigidly (translated and rotated) so as to coincide exactly with the other.

**PROOF** We require $\kappa \neq 0$ because $\hat{\mathbf{N}}$ and $\hat{\mathbf{B}}$ are not defined where $\kappa = 0$. Move $\mathcal{C}_2$ rigidly so that its initial point coincides with the initial point of $\mathcal{C}_1$ and so that the Frenet frames of both curves coincide at that point. Let $\hat{\mathbf{T}}_1$, $\hat{\mathbf{T}}_2$, $\hat{\mathbf{N}}_1$, $\hat{\mathbf{N}}_2$, $\hat{\mathbf{B}}_1$, and $\hat{\mathbf{B}}_2$ be the unit tangents, normals and binormals for the two curves. Let

$$f(s) = \hat{\mathbf{T}}_1(s) \bullet \hat{\mathbf{T}}_2(s) + \hat{\mathbf{N}}_1(s) \bullet \hat{\mathbf{N}}_2(s) + \hat{\mathbf{B}}_1(s) \bullet \hat{\mathbf{B}}_2(s).$$

We calculate the derivative of $f(s)$ using the Product Rule and the Frenet–Serret formulas:

$$f'(s) = \hat{\mathbf{T}}'_1 \bullet \hat{\mathbf{T}}_2 + \hat{\mathbf{T}}_1 \bullet \hat{\mathbf{T}}'_2 + \hat{\mathbf{N}}'_1 \bullet \hat{\mathbf{N}}_2 + \hat{\mathbf{N}}_1 \bullet \hat{\mathbf{N}}'_2 + \hat{\mathbf{B}}'_1 \bullet \hat{\mathbf{B}}_2 + \hat{\mathbf{B}}_1 \bullet \hat{\mathbf{B}}'_2$$

$$= \kappa\hat{\mathbf{N}}_1 \bullet \hat{\mathbf{T}}_2 + \kappa\hat{\mathbf{T}}_1 \bullet \hat{\mathbf{N}}_2 - \kappa\hat{\mathbf{T}}_1 \bullet \hat{\mathbf{N}}_2 + \tau\hat{\mathbf{B}}_1 \bullet \hat{\mathbf{N}}_2 - \kappa\hat{\mathbf{N}}_1 \bullet \hat{\mathbf{T}}_2$$

$$\qquad + \tau\hat{\mathbf{N}}_1 \bullet \hat{\mathbf{B}}_2 - \tau\hat{\mathbf{N}}_1 \bullet \hat{\mathbf{B}}_2 - \tau\hat{\mathbf{B}}_1 \bullet \hat{\mathbf{N}}_2$$

$$= 0.$$

Therefore $f(s)$ is constant. Since the frames coincide at $s = 0$, the constant must be 3:

$$\hat{\mathbf{T}}_1(s) \bullet \hat{\mathbf{T}}_2(s) + \hat{\mathbf{N}}_1(s) \bullet \hat{\mathbf{N}}_2(s) + \hat{\mathbf{B}}_1(s) \bullet \hat{\mathbf{B}}_2(s) = 3.$$

However, each dot product cannot exceed 1 since the factors are unit vectors. Therefore each dot product must be equal to 1. In particular, $\hat{\mathbf{T}}_1(s) \bullet \hat{\mathbf{T}}_2(s) = 1$ for all $s$, and hence

$$\frac{d\mathbf{r}_1}{ds} = \hat{\mathbf{T}}_1(s) = \hat{\mathbf{T}}_2(s) = \frac{d\mathbf{r}_2}{ds}.$$

Integrating with respect to $s$ and using the fact that both curves start from the same point when $s = 0$, we obtain $\mathbf{r}_1(s) = \mathbf{r}_2(s)$ for all $s$, which is what we wanted to show.

**Remark** It is a consequence of the above theorem that any curve having nonzero constant curvature and constant torsion must, in fact, be a circle (if the torsion is zero) or a circular helix (if the torsion is nonzero). See Exercises 7 and 8 below.

## Exercises 11.4

Find the unit tangent vector $\hat{\mathbf{T}}(t)$ for the curves in Exercises 1–4.

1. $\mathbf{r} = t\mathbf{i} - 2t^2\mathbf{j} + 3t^3\mathbf{k}$

2. $\mathbf{r} = a\sin\omega t\,\mathbf{i} + a\cos\omega t\,\mathbf{k}$

3. $\mathbf{r} = \cos t\sin t\,\mathbf{i} + \sin^2 t\,\mathbf{j} + \cos t\,\mathbf{k}$

4. $\mathbf{r} = a\cos t\,\mathbf{i} + b\sin t\,\mathbf{j} + t\mathbf{k}$

5. Show that if $\kappa(s) = 0$ for all $s$, then the curve $\mathbf{r} = \mathbf{r}(s)$ is a straight line.

6. Show that if $\tau(s) = 0$ for all $s$, then the curve $\mathbf{r} = \mathbf{r}(s)$ is a plane curve. *Hint:* show that $\mathbf{r}(s)$ lies in the plane through $\mathbf{r}(0)$ with normal $\hat{\mathbf{B}}(0)$.

7. Show that if $\kappa(s) = C$ is a positive constant, and $\tau(s) = 0$, for all $s$, then the curve $\mathbf{r} = \mathbf{r}(s)$ is a circle. *Hint:* find a circle having the given constant curvature. Then use Theorem 3.

8. Show that if the curvature $\kappa(s)$ and the torsion $\tau(s)$ are both nonzero constants, then the curve $\mathbf{r} = \mathbf{r}(s)$ is a circular helix. *Hint:* find a helix having the given curvature and torsion.

## 11.5 Curvature and Torsion for General Parametrizations

The formulas developed above for curvature and torsion as well as for the unit normal and binormal vectors are not very useful if the curve we want to analyze is not expressed in terms of the arc length parameter. We will now consider how to find these quantities in terms of a general parametrization $\mathbf{r} = \mathbf{r}(t)$. We will express them all in terms of the velocity, $\mathbf{v}(t)$, the speed, $v(t) = |\mathbf{v}(t)|$, and the acceleration,

$\mathbf{a}(t)$. First observe that

$$\mathbf{v} = \frac{d\mathbf{r}}{dt} = \frac{d\mathbf{r}}{ds}\frac{ds}{dt} = v\hat{\mathbf{T}}$$

$$\mathbf{a} = \frac{d\mathbf{v}}{dt} = \frac{dv}{dt}\hat{\mathbf{T}} + v\frac{d\hat{\mathbf{T}}}{dt}$$

$$= \frac{dv}{dt}\hat{\mathbf{T}} + v\frac{d\hat{\mathbf{T}}}{ds}\frac{ds}{dt} = \frac{dv}{dt}\hat{\mathbf{T}} + v^2\kappa\hat{\mathbf{N}}$$

$$\mathbf{v}\times\mathbf{a} = v\frac{dv}{dt}\hat{\mathbf{T}}\times\hat{\mathbf{T}} + v^3\kappa\hat{\mathbf{T}}\times\hat{\mathbf{N}} = v^3\kappa\hat{\mathbf{B}}$$

Note that $\hat{\mathbf{B}}$ is in the direction of $\mathbf{v}\times\mathbf{a}$. From these formulas we obtain useful formulas for $\hat{\mathbf{T}}$, $\hat{\mathbf{B}}$, and $\kappa$:

$$\hat{\mathbf{T}} = \frac{\mathbf{v}}{v}, \qquad \hat{\mathbf{B}} = \frac{\mathbf{v}\times\mathbf{a}}{|\mathbf{v}\times\mathbf{a}|}, \qquad \kappa = \frac{|\mathbf{v}\times\mathbf{a}|}{v^3}.$$

There are several ways to calculate $\hat{\mathbf{N}}$. Perhaps the easiest is

$$\hat{\mathbf{N}} = \hat{\mathbf{B}}\times\hat{\mathbf{T}}.$$

Sometimes it may be easier to use $\dfrac{d\hat{\mathbf{T}}}{dt} = \dfrac{d\hat{\mathbf{T}}}{ds}\dfrac{ds}{dt} = v\dfrac{d\hat{\mathbf{T}}}{ds} = v\kappa\hat{\mathbf{N}}$ to calculate

$$\hat{\mathbf{N}} = \frac{1}{v\kappa}\frac{d\hat{\mathbf{T}}}{dt} = \frac{\rho}{v}\frac{d\hat{\mathbf{T}}}{dt} = \frac{d\hat{\mathbf{T}}}{dt}\bigg/\left|\frac{d\hat{\mathbf{T}}}{dt}\right|.$$

The torsion remains to be calculated. Observe that

$$\frac{d\mathbf{a}}{dt} = \frac{d}{dt}\left(\frac{dv}{dt}\hat{\mathbf{T}} + v^2\kappa\hat{\mathbf{N}}\right).$$

This differentiation will produce several terms. The only one that involves $\hat{\mathbf{B}}$ is the one that comes from evaluating $v^2\kappa(d\hat{\mathbf{N}}/dt) = v^3\kappa(d\hat{\mathbf{N}}/ds) = v^3\kappa(\tau\hat{\mathbf{B}} - \kappa\hat{\mathbf{T}})$. Therefore

$$\frac{d\mathbf{a}}{dt} = \lambda\hat{\mathbf{T}} + \mu\hat{\mathbf{N}} + v^3\kappa\tau\hat{\mathbf{B}},$$

for certain scalars $\lambda$ and $\mu$. Since $\mathbf{v}\times\mathbf{a} = v^3\kappa\hat{\mathbf{B}}$, it follows that

$$(\mathbf{v}\times\mathbf{a})\bullet\frac{d\mathbf{a}}{dt} = (v^3\kappa)^2\tau = |\mathbf{v}\times\mathbf{a}|^2\tau.$$

Hence

$$\tau = \frac{(\mathbf{v}\times\mathbf{a})\bullet(d\mathbf{a}/dt)}{|\mathbf{v}\times\mathbf{a}|^2}.$$

**Example 1**   Find the curvature, the torsion, and the Frenet frame at a general point on the curve

$$\mathbf{r} = (t + \cos t)\mathbf{i} + (t - \cos t)\mathbf{j} + \sqrt{2}\sin t\,\mathbf{k}.$$

Describe this curve.

**Solution**   We calculate the various quantities using the recipe given above. First the preliminaries:

$$\mathbf{v} = (1 - \sin t)\mathbf{i} + (1 + \sin t)\mathbf{j} + \sqrt{2}\cos t\,\mathbf{k}$$

$$\mathbf{a} = -\cos t\,\mathbf{i} + \cos t\,\mathbf{j} - \sqrt{2}\sin t\,\mathbf{k}$$

$$\frac{d\mathbf{a}}{dt} = \sin t\,\mathbf{i} - \sin t\,\mathbf{j} - \sqrt{2}\cos t\,\mathbf{k}$$

$$\mathbf{v}\times\mathbf{a} = \begin{vmatrix} \mathbf{i} & \mathbf{j} & \mathbf{k} \\ 1-\sin t & 1+\sin t & \sqrt{2}\cos t \\ -\cos t & \cos t & -\sqrt{2}\sin t \end{vmatrix}$$

$$= -\sqrt{2}(1+\sin t)\mathbf{i} - \sqrt{2}(1-\sin t)\mathbf{j} + 2\cos t\,\mathbf{k}$$

$$(\mathbf{v}\times\mathbf{a}) \bullet \frac{d\mathbf{a}}{dt} = -\sqrt{2}\sin t\,(1+\sin t) + \sqrt{2}\sin t\,(1-\sin t) - 2\sqrt{2}\cos^2 t$$

$$= -2\sqrt{2}$$

$$v = |\mathbf{v}| = \sqrt{2 + 2\sin^2 t + 2\cos^2 t} = 2$$

$$|\mathbf{v}\times\mathbf{a}| = \sqrt{2(2 + 2\sin^2 t) + 4\cos^2 t} = \sqrt{8} = 2\sqrt{2}.$$

Thus we have

$$\kappa = \frac{|\mathbf{v}\times\mathbf{a}|}{v^3} = \frac{2\sqrt{2}}{8} = \frac{1}{2\sqrt{2}}$$

$$\tau = \frac{(\mathbf{v}\times\mathbf{a}) \bullet (d\mathbf{a}/dt)}{|\mathbf{v}\times\mathbf{a}|^2} = \frac{-2\sqrt{2}}{(2\sqrt{2})^2} = -\frac{1}{2\sqrt{2}}$$

$$\hat{\mathbf{T}} = \frac{\mathbf{v}}{v} = \frac{1-\sin t}{2}\mathbf{i} + \frac{1+\sin t}{2}\mathbf{j} + \frac{1}{\sqrt{2}}\cos t\,\mathbf{k}$$

$$\hat{\mathbf{B}} = \frac{\mathbf{v}\times\mathbf{a}}{|\mathbf{v}\times\mathbf{a}|} = -\frac{1+\sin t}{2}\mathbf{i} - \frac{1-\sin t}{2}\mathbf{j} + \frac{1}{\sqrt{2}}\cos t\,\mathbf{k}$$

$$\hat{\mathbf{N}} = \hat{\mathbf{B}}\times\hat{\mathbf{T}} = -\frac{1}{\sqrt{2}}\cos t\,\mathbf{i} + \frac{1}{\sqrt{2}}\cos t\,\mathbf{j} - \sin t\,\mathbf{k}.$$

Since the curvature and torsion are both constant (they are therefore constant when expressed in terms of any parametrization) the curve must be a circular helix by Theorem 3. It is left-handed, since $\tau < 0$. By Example 3 in Section 11.4, it is congruent to the helix

$$\mathbf{r} = a\cos t\,\mathbf{i} + a\sin t\,\mathbf{j} + bt\,\mathbf{k},$$

provided that $a/(a^2 + b^2) = 1/(2\sqrt{2}) = -b/(a^2 + b^2)$. Solving these equations gives $a = \sqrt{2}$ and $b = -\sqrt{2}$, so the helix is wound on a cylinder of radius $\sqrt{2}$. The axis of this cylinder is the line $x = y, z = 0$, as can be seen by inspecting the components of $\mathbf{r}(t)$.

**Example 2**    (**Curvature of the graph of a function of one variable**)  Find the curvature of the plane curve with equation $y = f(x)$ at an arbitrary point $(x, f(x))$ on the curve.

**Solution**    The graph can be parametrized $\mathbf{r} = x\mathbf{i} + f(x)\mathbf{j}$. Thus

$$\mathbf{v} = \mathbf{i} + f'(x)\mathbf{j},$$
$$\mathbf{a} = f''(x)\mathbf{j}$$
$$\mathbf{v} \times \mathbf{a} = f''(x)\mathbf{k}.$$

Therefore the curvature is

$$\kappa(x) = \frac{|f''(x)|}{\left(1 + (f'(x))^2\right)^{3/2}}.$$

## Tangential and Normal Acceleration

In the formula obtained above for the acceleration in terms of the unit tangent and normal,

$$\mathbf{a} = \frac{dv}{dt}\hat{\mathbf{T}} + v^2\kappa\hat{\mathbf{N}},$$

the term $(dv/dt)\hat{\mathbf{T}}$ is called the **tangential** acceleration and the term $v^2\kappa\hat{\mathbf{N}}$ is called the **normal** or **centripetal** acceleration. This latter component is directed towards the centre of curvature and its magnitude is $v^2\kappa = v^2/\rho$. Highway and railway designers attempt to bank curves in such a way that the resultant of the corresponding "centrifugal force," $-m(v^2/\rho)\hat{\mathbf{N}}$, and the weight, $-mg\mathbf{k}$, of the vehicle will be normal to the surface at a desired speed. (See Exercise 14 below.)

**Example 3**    A bead of mass $m$ slides without friction down a wire bent in the shape of the curve $y = x^2$, under the influence of the gravitational force $-mg\mathbf{j}$. The speed of the bead is $v$ as it passes through the point $(1, 1)$. Find, at that instant, the magnitude of the normal acceleration of the bead and the rate of change of its speed.

**Solution**    By Example 2, the curvature of $y = x^2$ at $(1, 1)$ is

$$\kappa = \frac{2}{(1 + 4x^2)^{3/2}}\bigg|_{x=1} = \frac{2}{5\sqrt{5}}.$$

Thus the magnitude of the normal acceleration of the bead at that point is $v^2\kappa = 2v^2/(5\sqrt{5})$.

The rate of change of the speed, $dv/dt$, is the tangential component of the acceleration, and is due entirely to the tangential component of the gravitational force since there is no friction:

$$\frac{dv}{dt} = g\cos\theta = g(-\mathbf{j}) \bullet \hat{\mathbf{T}},$$

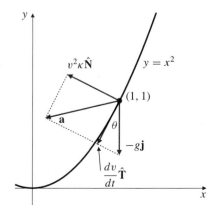

**Figure 11.16**    The acceleration of a sliding bead

where $\theta$ is the angle between $\hat{\mathbf{T}}$ and $-\mathbf{j}$. (See Figure 11.16.) Since the slope of $y = x^2$ at $(1, 1)$ is 2, we have $\hat{\mathbf{T}} = -(\mathbf{i} + 2\mathbf{j})/\sqrt{5}$, and therefore $dv/dt = 2g/\sqrt{5}$.

**Remark** The definition of centripetal acceleration given above is consistent with the one that arose in the discussion of rotating frames in Section 11.2. If $\mathbf{r}(t)$ is the position of a moving particle at time $t$ we can regard the motion at any instant as being a rotation about the centre of curvature, so that the angular velocity must be $\mathbf{\Omega} = \Omega \hat{\mathbf{B}}$. The linear velocity is $\mathbf{v} = \mathbf{\Omega} \times (\mathbf{r} - \mathbf{r}_c) = v\hat{\mathbf{T}}$, so the speed is $v = \Omega \rho$, and $\mathbf{\Omega} = (v/\rho)\hat{\mathbf{B}}$. As developed in Section 11.2, the centripetal acceleration is

$$\mathbf{\Omega} \times (\mathbf{\Omega} \times (\mathbf{r} - \mathbf{r}_c)) = \mathbf{\Omega} \times \mathbf{v} = \frac{v^2}{\rho}\hat{\mathbf{B}} \times \hat{\mathbf{T}} = \frac{v^2}{\rho}\hat{\mathbf{N}}.$$

## Evolutes

The centre of curvature $\mathbf{r}_c(t)$ of a given curve can itself trace out another curve as $t$ varies. This curve is called the **evolute** of the given curve $\mathbf{r}(t)$.

**Example 4** Find the evolute of the exponential spiral

$$\mathbf{r} = ae^{-t}\cos t\,\mathbf{i} + ae^{-t}\sin t\,\mathbf{j}.$$

**Solution** The curve is a plane curve so $\tau = 0$. We will take a shortcut to the curvature and the unit normal without calculating $\mathbf{v} \times \mathbf{a}$. First we calculate

$$\mathbf{v} = ae^{-t}\Big(-(\cos t + \sin t)\mathbf{i} - (\sin t - \cos t)\mathbf{j}\Big)$$

$$\frac{ds}{dt} = v = \sqrt{2}ae^{-t}$$

$$\hat{\mathbf{T}}(t) = \frac{1}{\sqrt{2}}\Big(-(\cos t + \sin t)\mathbf{i} - (\sin t - \cos t)\mathbf{j}\Big)$$

$$\frac{d\hat{\mathbf{T}}}{ds} = \frac{1}{(ds/dt)}\frac{d\hat{\mathbf{T}}}{dt} = \frac{1}{2ae^{-t}}\Big((\sin t - \cos t)\mathbf{i} - (\cos t + \sin t)\mathbf{j}\Big)$$

$$\kappa(t) = \left|\frac{d\hat{\mathbf{T}}}{ds}\right| = \frac{1}{\sqrt{2}ae^{-t}}.$$

It follows that the radius of curvature is $\rho(t) = \sqrt{2}ae^{-t}$. Since $d\hat{\mathbf{T}}/ds = \kappa\hat{\mathbf{N}}$, therefore $\hat{\mathbf{N}} = \rho(d\hat{\mathbf{T}}/ds)$. The centre of curvature is

$$\mathbf{r}_c(t) = \mathbf{r}(t) + \rho(t)\hat{\mathbf{N}}(t)$$

$$= \mathbf{r}(t) + \rho^2\frac{d\hat{\mathbf{T}}}{ds}$$

$$= ae^{-t}\Big(\cos t\,\mathbf{i} + \sin t\,\mathbf{j}\Big)$$

$$\quad + 2a^2e^{-2t}\frac{1}{2ae^{-t}}\Big((\sin t - \cos t)\mathbf{i} - (\cos t + \sin t)\mathbf{j}\Big)$$

$$= ae^{-t}\Big(\sin t\,\mathbf{i} - \cos t\,\mathbf{j}\Big)$$

$$= ae^{-t}\Big(\cos(t - \frac{\pi}{2})\mathbf{i} + \sin(t - \frac{\pi}{2})\mathbf{j}\Big).$$

Thus, interestingly, the evolute of the exponential spiral is the same exponential spiral rotated 90° clockwise in the plane. (See Figure 11.17(a).)

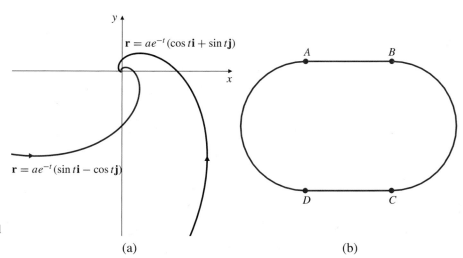

**Figure 11.17**

(a) The evolute of an exponential spiral is another exponential spiral

(b) The shape of a model train track

## An Application to Track Design

Model trains frequently come with two kinds of track sections, straight and curved. The curved sections are arcs of a circle of radius $R$, and the track is intended to be laid out in the shape shown in Figure 11.17(b); $AB$ and $CD$ are straight, and $BC$ and $DA$ are semicircles. The track looks smooth, but is it smooth enough?

The track is held together by friction and occasionally it can come apart as the train is racing around. It is especially likely to come apart at the points $A$, $B$, $C$, and $D$. To see why, assume that the train is travelling at constant speed $v$. Then the tangential acceleration, $(dv/dt)\hat{\mathbf{T}}$, is zero and the total acceleration is just the centripetal acceleration, $\mathbf{a} = (v^2/\rho)\hat{\mathbf{N}}$. Therefore, $|\mathbf{a}| = 0$ along the straight sections, and $|\mathbf{a}| = v^2\kappa = v^2/R$ on the semicircular sections. The acceleration is *discontinuous* at the points $A$, $B$, $C$, and $D$, and the reactive force exerted by the train on the track is also discontinuous at these points. There is a "shock" or "jolt" as the train enters or leaves a curved part of the track. In order to avoid such stress points, tracks should be designed so that the curvature varies continuously from point to point.

**Example 5**    Existing track along the negative $x$-axis, and along the ray $y = x - 1$, $x \geq 2$, is to be joined smoothly by track along the transition curve $y = f(x)$, $0 \leq x \leq 2$, where $f(x)$ is a polynomial of degree as small as possible. Find $f(x)$ so that a train moving along the track will not experience discontinuous acceleration at the joins.

**Solution**    The situation is shown in Figure 11.18. The polynomial $f(x)$ must be chosen so that the track is continuous, has continuous slope, and has continuous curvature at $x = 0$ and $x = 1$. Since the curvature of $y = f(x)$ is

$$\kappa = f''(x)\left(1 + (f'(x))^2\right)^{-3/2},$$

we need only arrange that $f$, $f'$, and $f''$ take the same values at $x = 0$ and $x = 2$ that the straight sections do there:

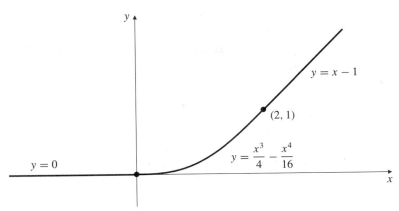

**Figure 11.18**    Joining two straight tracks with a curved track

$$f(0) = 0, \qquad f'(0) = 0, \qquad f''(0) = 0,$$
$$f(2) = 1, \qquad f'(2) = 1, \qquad f''(2) = 0.$$

These six independent conditions suggest we should try a polynomial of degree five involving six arbitrary coefficients:

$$f(x) = A + Bx + Cx^2 + Dx^3 + Ex^4 + Fx^5$$
$$f'(x) = B + 2Cx + 3Dx^2 + 4Ex^3 + 5Fx^4$$
$$f''(x) = 2C + 6Dx + 12Ex^2 + 20Fx^3.$$

The three conditions at $x = 0$ imply that $A = B = C = 0$. Those at $x = 2$ imply that

$$8D + 16E + \phantom{0}32F = f(2) = 1$$
$$12D + 32E + \phantom{0}80F = f'(2) = 1$$
$$12D + 48E + 160F = f''(2) = 0.$$

This system has solution $D = 1/4$, $E = -1/16$, and $F = 0$, so we should use $f(x) = (x^3/4) - (x^4/16)$. ∎

*Remark*  Road and railroad builders do not usually use polynomial graphs as transition curves. Other kinds of curves called **clothoids** and **lemniscates** are usually used. (See Exercise 7 in the Review Exercises at the end of this chapter.)

## Exercises 11.5

Find the radius of curvature of the curves in Exercises 1–4 at the points indicated.

**1.** $y = x^2$ at $x = 0$ and at $x = \sqrt{2}$

**2.** $y = \cos x$ at $x = 0$ and at $x = \pi/2$

**3.** $\mathbf{r} = 2t\mathbf{i} + (1/t)\mathbf{j} - 2t\mathbf{k}$ at $(2, 1, -2)$

**4.** $\mathbf{r} = t^3\mathbf{i} + t^2\mathbf{j} + t\mathbf{k}$ at the point where $t = 1$

Find the Frenet frames $\{\hat{\mathbf{T}}, \hat{\mathbf{N}}, \hat{\mathbf{B}}\}$ for the curves in Exercises 5–6 at the points indicated.

**5.** $\mathbf{r} = t\mathbf{i} + t^2\mathbf{j} + 2\mathbf{k}$ at $(1, 1, 2)$

**6.** $\mathbf{r} = t\mathbf{i} + t^2\mathbf{j} + t\mathbf{k}$ at $(1, 1, 1)$

In Exercises 7–8, find the unit tangent, normal, and binormal vectors, and the curvature and torsion at a general point on the given curve.

**7.** $\mathbf{r} = t\mathbf{i} + \dfrac{t^2}{2}\mathbf{j} + \dfrac{t^3}{3}\mathbf{k}$ **8.** $\mathbf{r} = e^t(\cos t\,\mathbf{i} + \sin t\,\mathbf{j} + \mathbf{k})$

**9.** Find the curvature and torsion of the parametric curve

$$x = 2 + \sqrt{2}\cos t, \quad y = 1 - \sin t, \quad z = 3 + \sin t$$

at an arbitrary point $t$. What is the curve?

**10.** A particle moves along the plane curve $y = \sin x$ in the direction of increasing $x$ with constant horizontal speed $dx/dt = k$. Find the tangential and normal components of the acceleration of the particle when it is at position $x$.

**11.** Find the unit tangent, normal and binormal, and the curvature and torsion for the curve

$$\mathbf{r} = \sin t \cos t\, \mathbf{i} + \sin^2 t\, \mathbf{j} + \cos t\, \mathbf{k}$$

at the points (i) $t = 0$ and (ii) $t = \pi/4$.

**12.** A particle moves on an elliptical path in the $xy$-plane so that its position at time $t$ is $\mathbf{r} = a \cos t\, \mathbf{i} + b \sin t\, \mathbf{j}$. Find the tangential and normal components of its acceleration at time $t$. At what points is the tangential acceleration zero?

**13.** Find the maximum and minimum values for the curvature of the ellipse $x = a \cos t$, $y = b \sin t$, where $a > b > 0$.

**14.** A level, curved road lies along the curve $y = x^2$ in the horizontal $xy$-plane. Find the normal acceleration at position $(x, x^2)$ of a vehicle travelling along the road with constant speed $v_0$. Find, as a function of $x$, the angle at which the road should be banked (i.e. the angle between the vertical and the normal to the surface of the road) so that the result of the centrifugal and gravitational ($-mg\mathbf{k}$) forces acting on the vehicle is always normal to the surface of the road.

**15.** Find the curvature of the plane curve $y = e^x$ at $x$. Find the equation of the evolute of this curve.

**16.** Show that the curvature of the plane polar graph $r = f(\theta)$ at a general point $\theta$ is

$$\kappa(\theta) = \frac{\left|2\big(f'(\theta)\big)^2 + \big(f(\theta)\big)^2 - f(\theta)f''(\theta)\right|}{\left[\big(f'(\theta)\big)^2 + \big(f(\theta)\big)^2\right]}.$$

**17.** Find the curvature of the cardioid $r = a(1 - \cos\theta)$.

**∗ 18.** Find the curve $\mathbf{r} = \mathbf{r}(t)$ for which $\kappa(t) = 1$ and $\tau(t) = 1$ for all $t$, and $\mathbf{r}(0) = \hat{\mathbf{T}}(0) = \mathbf{i}$, $\hat{\mathbf{N}}(0) = \mathbf{j}$, and $\hat{\mathbf{B}}(0) = \mathbf{k}$.

**19.** Suppose the curve $\mathbf{r} = \mathbf{r}(t)$ satisfies $\dfrac{d\mathbf{r}}{dt} = \mathbf{c} \times \mathbf{r}(t)$, where $\mathbf{c}$ is a constant vector. Use curvature and torsion to show that the curve is the circle in which the plane through $\mathbf{r}(0)$ normal to $\mathbf{c}$ intersects a sphere with radius $|\mathbf{r}(0)|$ centred at the origin.

**20.** Find the evolute of the circular helix $\mathbf{r} = a \cos t\, \mathbf{i} + a \sin t\, \mathbf{j} + bt\, \mathbf{k}$.

**21.** Find the evolute of the parabola $y = x^2$.

**22.** Find the evolute of the ellipse $x = 2 \cos t$, $y = \sin t$.

**23.** Find the polynomial $f(x)$ of lowest degree so that track along $y = f(x)$ from $x = -1$ to $x = 1$ joins with existing straight tracks $y = -1$, $x \le -1$ and $y = 1$, $x \ge 1$ sufficiently smoothly that a train moving at constant speed will not experience discontinuous acceleration at the joins.

**∗ 24.** Help out model train manufacturers. Design a track segment $y = f(x)$, $-1 \le x \le 0$, to provide a jolt-free link between a straight track section $y = 1$, $x \le -1$, and a semicircular arc section $x^2 + y^2 = 1$, $x \ge 0$.

**∗ 25.** If the position $\mathbf{r}$, velocity $\mathbf{v}$, and acceleration $\mathbf{a}$ of a moving particle satisfy $\mathbf{a}(t) = \lambda(t)\mathbf{r}(t) + \mu(t)\mathbf{v}(t)$, where $\lambda(t)$ and $\mu(t)$ are scalar functions of time $t$, and if $\mathbf{v} \times \mathbf{a} \ne \mathbf{0}$, show that the path of the particle lies in a plane.

# 11.6   Kepler's Laws of Planetary Motion

The German mathematician and astronomer, Johannes Kepler (1571–1630), was a student and colleague of Danish astronomer Tycho Brahe (1546–1601). Over a lifetime of observing the positions of planets without the aid of a telescope, Brahe compiled a vast amount of data, which Kepler analyzed. Although Polish astronomer Nicolaus Copernicus (1473–1543) had postulated that the earth and other planets moved around the sun, the religious and philosophical climate in Europe at the end of the sixteenth century still favoured explaining the motion of heavenly bodies in terms of circular orbits around the earth. It was known that planets such as Mars could not move on circular orbits centred at the earth, but models were proposed in which they moved on other circles (epicycles) whose centres moved on circles centred at the earth.

Brahe's observations of Mars were sufficiently detailed that Kepler realized that no simple model based on circles could be made to conform very closely with the actual orbit. He was, however, able to fit a more general ellipse with one focus at the sun, and, based on this success and on Brahe's data on other planets, he formulated the following three laws of planetary motion.

> **Kepler's Laws**
>
> 1. The planets move on elliptical orbits with the sun at one focus.
> 2. The radial line from the sun to a planet sweeps out equal areas in equal times.
> 3. The squares of the periods of revolution of the planets around the sun are proportional to the cubes of the major axes of their orbits.

Kepler's statement of the third law actually says that the squares of the periods of revolution of the planets are proportional to the cubes of their mean distances from the sun. The mean distance of points on an ellipse from a focus of the ellipse is equal to the semi-major axis. (See Exercise 17 at the end of this section.) Therefore the two statements are equivalent.

The choice of ellipses was reasonable once it became clear that circles would not work. The properties of the conic sections were well understood, having been developed by the Greek mathematician Apollonius of Perga around 200 BC. Nevertheless, based, as it was, on observations rather than theory, Kepler's formulation of his laws without any causal explanation was a truly remarkable feat. The theoretical underpinnings came later when Newton, with the aid of his newly created calculus, showed that Kepler's laws implied an inverse square gravitational force. (See Review Exercises 14–16 at the end of this chapter.) Newton believed that his universal gravitational law also implied Kepler's laws, but his writings fail to provide a proof that is convincing by today's standards.[1]

Later in this section we will derive Kepler's laws from the gravitational law by an elegant method that exploits vector differentiation to the fullest. First, however, we need to attend to some preliminaries.

## Ellipses in Polar Coordinates

The polar coordinates $[r, \theta]$ of a point in the plane whose distance from the origin is $\varepsilon$ times its distance from the line $x = p$ (see Figure 11.19) satisfy $r = \varepsilon(p - r\cos\theta)$, or, solving for $r$,

$$r = \frac{\ell}{1 + \varepsilon\cos\theta},$$

where $\ell = \varepsilon p$. For $0 \le \varepsilon < 1$ this equation represents an ellipse having **eccentricity** $\varepsilon$. (It is a circle if $\varepsilon = 0$.) To see this, let us transform the equation to Cartesian coordinates:

$$x^2 + y^2 = r^2 = \varepsilon^2(p - r\cos\theta)^2 = \varepsilon^2(p - x)^2 = \varepsilon^2(p^2 - 2px + x^2).$$

With some algebraic manipulation this equation can be juggled into the form

$$\frac{\left(x + \dfrac{\varepsilon\ell}{1 - \varepsilon^2}\right)^2}{\left(\dfrac{\ell}{1 - \varepsilon^2}\right)^2} + \frac{y^2}{\left(\dfrac{\ell}{\sqrt{1 - \varepsilon^2}}\right)^2} = 1,$$

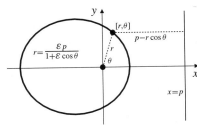

**Figure 11.19**    An ellipse with focus at the origin, directrix $x = p$, and eccentricity $\varepsilon$

---

[1]   There are interesting articles debating the historical significance of Newton's work by Robert Weinstock, Curtis Wilson, and others in *The College Mathematics Journal,* volume 25 #3, 1994.

which can be recognized as an ellipse with **centre** at the point $C = (-c, 0)$, where $c = \varepsilon\ell/(1 - \varepsilon^2)$, and semi-axes $a$ and $b$ given by

$$a = \frac{\ell}{1 - \varepsilon^2}, \qquad \textbf{(semi-major axis)},$$

$$b = \frac{\ell}{\sqrt{1 - \varepsilon^2}}, \qquad \textbf{(semi-minor axis)}.$$

(See Section 8.1 for a discussion of ellipses and other conic sections in Cartesian coordinates.)

The Cartesian equation of the ellipse shows that the curve is symmetric about the lines $x = -c$ and $y = 0$, and so has a second focus at $F = (-2c, 0)$ and a second directrix with equation $x = -2c - p$. (See Figure 11.20.) The ends of the major axis are $A = (a - c, 0)$ and $A' = (-a - c, 0)$, and the ends of the minor axis are $B = (-c, b)$ and $B' = (-c, -b)$.

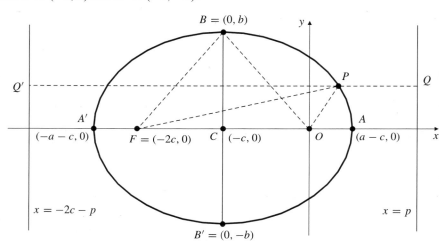

**Figure 11.20** The sum of the distances from any point $P$ on the ellipse to the two foci $O$ and $F$ is constant, $\varepsilon$ times the distance between the directrices

If $P$ is any point on the ellipse then the distance $OP$ is $\varepsilon$ times the distance $PQ$ from $P$ to the right directrix. Similarly, the distance $FQ$ is $\varepsilon$ times the distance $Q'P$ from $P$ to the left directrix. Thus the sum of the focal radii $OP + FP$ is the constant $\varepsilon Q'Q = \varepsilon(2c + 2p)$, regardless of where $P$ is on the ellipse. Taking for $P$ the two points $A$ and $B$ we get for this sum,

$$2a = (a - c) + (a + c) = OA + FA = OB + FB = 2\sqrt{b^2 + c^2}.$$

It follows that

$$a^2 = b^2 + c^2, \qquad c = \sqrt{a^2 - b^2} = \frac{\ell\varepsilon}{1 - \varepsilon^2} = \varepsilon a.$$

The number $\ell$ is called the **semi-latus rectum** of the ellipse; the latus rectum is the width measured along the line through a focus, perpendicular to the major axis. (See Figure 11.21.)

**Remark** The polar equation $r = \ell/(1 + \varepsilon \cos \theta)$ represents a *bounded* curve only if $\varepsilon < 1$: in this case we have $\ell/(1 + \varepsilon) \le r \le \ell/(1 - \varepsilon)$ for all directions $\theta$. If $\varepsilon = 1$ the equation represents a parabola, and if $\varepsilon > 1$ a hyperbola. It is possible

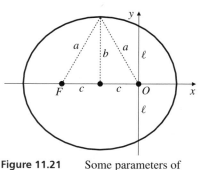

**Figure 11.21** Some parameters of an ellipse

for objects to travel on parabolic or hyperbolic orbits, but they will approach the sun only once, rather than continue to loop around it. Some comets have hyperbolic orbits.

## Polar Components of Velocity and Acceleration

Let $\mathbf{r}(t)$ be the position vector at time $t$ of a particle $P$ moving in the $xy$-plane. We construct two unit vectors at $P$, the vector $\hat{\mathbf{r}}$ points in the direction of the position vector $\mathbf{r}$, and the vector $\hat{\boldsymbol{\theta}}$ is rotated 90° counterclockwise from $\hat{\mathbf{r}}$. (See Figure 11.22.) If $P$ has polar coordinates $[r, \theta]$ then $\hat{\mathbf{r}}$ points in the direction of increasing $r$ at $P$, and $\hat{\boldsymbol{\theta}}$ points in the direction of increasing $\theta$. Evidently

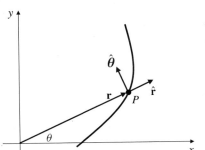

**Figure 11.22** Basis vectors in the direction of increasing $r$ and $\theta$

$$\hat{\mathbf{r}} = \cos\theta\,\mathbf{i} + \sin\theta\,\mathbf{j}$$
$$\hat{\boldsymbol{\theta}} = -\sin\theta\,\mathbf{i} + \cos\theta\,\mathbf{j}.$$

Note that $\hat{\mathbf{r}}$ and $\hat{\boldsymbol{\theta}}$ do not depend on $r$ but only on $\theta$:

$$\frac{d\hat{\mathbf{r}}}{d\theta} = \hat{\boldsymbol{\theta}} \qquad \text{and} \qquad \frac{d\hat{\boldsymbol{\theta}}}{d\theta} = -\hat{\mathbf{r}}.$$

The pair $\{\hat{\mathbf{r}}, \hat{\boldsymbol{\theta}}\}$ form a reference frame (a basis) at $P$ so that vectors in the plane can be expressed in terms of these two unit vectors. The $\hat{\mathbf{r}}$ component of a vector is called the **radial component**, and the $\hat{\boldsymbol{\theta}}$ component is called the **transverse component**. The frame varies from point to point, so we must remember that $\hat{\mathbf{r}}$ and $\hat{\boldsymbol{\theta}}$ are both functions of $t$. In terms of this moving frame, the position $\mathbf{r}(t)$ of $P$ can be expressed very simply:

$$\mathbf{r} = r\hat{\mathbf{r}},$$

where $r = r(t) = |\mathbf{r}(t)|$ is the distance from $P$ to the origin at time $t$.

We are going to differentiate this equation with respect to $t$ in order to express the velocity and acceleration of $P$ in terms of the moving frame. Along the path of motion, $\mathbf{r}$ can be regarded as a function of either $\theta$ or $t$; $\theta$ is itself a function of $t$. So as to avoid confusion, let us adopt a notation that is used extensively in mechanics and that resembles the notation originally used by Newton in his calculus.

A dot over a quantity denotes the time derivative of that quantity. Two dots denotes the second derivative with respect to time. Thus:

$$\dot{u} = du/dt \qquad \text{and} \qquad \ddot{u} = d^2u/dt^2.$$

First let us record the time derivatives of the vectors $\hat{\mathbf{r}}$ and $\hat{\boldsymbol{\theta}}$. By the Chain Rule, we have

$$\dot{\hat{\mathbf{r}}} = \frac{d\hat{\mathbf{r}}}{d\theta}\frac{d\theta}{dt} = \dot{\theta}\hat{\boldsymbol{\theta}}$$

$$\dot{\hat{\boldsymbol{\theta}}} = \frac{d\hat{\boldsymbol{\theta}}}{d\theta}\frac{d\theta}{dt} = -\dot{\theta}\hat{\mathbf{r}}.$$

Now the velocity of $P$ is

$$\mathbf{v} = \dot{\mathbf{r}} = \frac{d}{dt}(r\hat{\mathbf{r}}) = \dot{r}\hat{\mathbf{r}} + r\dot{\theta}\hat{\boldsymbol{\theta}}.$$

> **Polar components of velocity**
>
> The **radial component of velocity** is $\dot{r}$.
> The **transverse component of velocity** is $r\dot{\theta}$.

Since $\hat{\mathbf{r}}$ and $\hat{\boldsymbol{\theta}}$ are perpendicular unit vectors, the speed of $P$ is given by

$$v = |\mathbf{v}| = \sqrt{\dot{r}^2 + r^2\dot{\theta}^2}.$$

Similarly, the acceleration of $P$ can be expressed in terms of radial and transverse components:

$$\mathbf{a} = \dot{\mathbf{v}} = \ddot{\mathbf{r}} = \frac{d}{dt}(\dot{r}\hat{\mathbf{r}} + r\dot{\theta}\hat{\boldsymbol{\theta}})$$
$$= \ddot{r}\hat{\mathbf{r}} + \dot{r}\dot{\theta}\hat{\boldsymbol{\theta}} + \dot{r}\dot{\theta}\hat{\boldsymbol{\theta}} + r\ddot{\theta}\hat{\boldsymbol{\theta}} - r\dot{\theta}^2\hat{\mathbf{r}}$$
$$= (\ddot{r} - r\dot{\theta}^2)\hat{\mathbf{r}} + (r\ddot{\theta} + 2\dot{r}\dot{\theta})\hat{\boldsymbol{\theta}}.$$

> **Polar components of acceleration**
>
> The **radial component of acceleration** is $\ddot{r} - r\dot{\theta}^2$.
> The **transverse component of acceleration** is $r\ddot{\theta} + 2\dot{r}\dot{\theta}$.

## Central Forces and Kepler's Second Law

Polar coordinates are most appropriate for analyzing motion due to a **central force** that is always directed towards (or away from) a single point, the origin: $\mathbf{F} = \lambda\mathbf{r}$, where the scalar $\lambda$ depends on the position $\mathbf{r}$ of the object. If the velocity and acceleration of the object are $\mathbf{v} = \dot{\mathbf{r}}$ and $\mathbf{a} = \dot{\mathbf{v}}$, then Newton's second law of motion ($\mathbf{F} = m\mathbf{a}$) says that $\mathbf{a}$ is parallel to $\mathbf{r}$. Therefore

$$\frac{d}{dt}(\mathbf{r}\times\mathbf{v}) = \dot{\mathbf{r}}\times\mathbf{v} + \mathbf{r}\times\dot{\mathbf{v}} = \mathbf{v}\times\mathbf{v} + \mathbf{r}\times\mathbf{a} = \mathbf{0} + \mathbf{0} = \mathbf{0},$$

and $\mathbf{r} \times \mathbf{v} = \mathbf{h}$, a constant vector representing the object's angular momentum per unit mass about the origin. This says that $\mathbf{r}$ is always perpendicular to $\mathbf{h}$, so motion due to a central force always takes place in a *plane* through the origin.

If we choose the $z$-axis to be in the direction of $\mathbf{h}$ and let $|\mathbf{h}| = h$, then $\mathbf{h} = h\mathbf{k}$, and the path of the object is in the $xy$-plane. In this case the position and velocity of the object satisfy

$$\mathbf{r} = r\hat{\mathbf{r}} \qquad \text{and} \qquad \mathbf{v} = \dot{r}\hat{\mathbf{r}} + r\dot{\theta}\hat{\boldsymbol{\theta}}.$$

Since $\hat{\mathbf{r}}\times\hat{\boldsymbol{\theta}} = \mathbf{k}$, we have

$$h\mathbf{k} = \mathbf{r}\times\mathbf{v} = r\dot{r}\hat{\mathbf{r}}\times\hat{\mathbf{r}} + r^2\dot{\theta}\hat{\mathbf{r}}\times\hat{\boldsymbol{\theta}} = r^2\dot{\theta}\mathbf{k}.$$

Hence, for any motion under a central force,

$$r^2\dot{\theta} = h, \qquad \text{(a constant for the path of motion)}.$$

This formula is equivalent to Kepler's second law; if $A(t)$ is the area in the plane of motion bounded by the orbit and radial lines $\theta = \theta_0$ and $\theta = \theta(t)$ then

$$A(t) = \frac{1}{2} \int_{\theta_0}^{\theta(t)} r^2 \, d\theta,$$

so that

$$\frac{dA}{dt} = \frac{dA}{d\theta} \frac{d\theta}{dt} = \frac{1}{2} r^2 \dot{\theta} = \frac{h}{2}.$$

Thus area is being swept out at the constant rate $h/2$, and equal areas are swept out in equal times. Note that this law does not depend on the magnitude or direction of the force on the moving object other than the fact that it is *central*. You can also derive the equation $r^2 \dot{\theta} = h$ (constant) directly from the fact that the transverse accleration is zero:

$$(d/dt)(r^2 \dot{\theta}) = 2r\dot{r}\dot{\theta} + r^2\ddot{\theta} = r(2\dot{r}\dot{\theta} + r\ddot{\theta}) = 0.$$

**Example 1**    An object moves along the polar curve $r = \theta^{-2}$ under the influence of a force attracting it towards the origin. If the speed of the object is $v_0$ at the instant when $\theta = 1$, find the magnitude of the acceleration of the object at any point on its path as a function of its distance $r$ from the origin.

**Solution**    Since the force is central, we know that the transverse acceleration is zero and that $r^2 \dot{\theta} = h$ is constant. Differentiating the equation of the path with respect to time twice, and expressing the results in terms of $r$, we obtain

$$\dot{r} = -2\theta^{-3}\dot{\theta} = -2r^{3/2}\dot{\theta} = -2r^{-1/2}h$$
$$\ddot{r} = r^{-3/2}\dot{r}h = -2r^{-2}h^2.$$

Hence the radial component of acceleration is

$$a_r = \ddot{r} - r(\dot{\theta})^2 = -2r^{-2}h^2 - rh^2r^{-4} = -h^2\left(\frac{2}{r^2} + \frac{1}{r^3}\right).$$

At $\theta = 1$ we have $r = 1$ and the square of the speed is

$$v_0^2 = (\dot{r})^2 + (r\dot{\theta})^2 = 4h^2 + h^2 = 5h^2.$$

Thus $h^2 = v_0^2/5$, and the magnitude of the acceleration of the object is

$$|a_r| = \frac{v_0^2}{5}\left(\frac{2}{r^2} + \frac{1}{r^3}\right).$$

## Derivation of Kepler's First and Third Laws

The planets and the sun move around their common centre of mass. Since the sun is vastly more massive than the planets that centre of mass is quite close to the sun. For example, the joint centre of mass of the sun and the earth lies inside the sun. For the following derivation we will take the sun and a planet as *point masses* and consider the sun to be fixed at the origin. We will specify the directions of the coordinate axes later, when the need arises.

According to Newton's law of gravitation, the force that the sun exerts on a planet of mass $m$ whose position vector is $\mathbf{r}$ is

$$\mathbf{F} = -\frac{km}{r^2}\hat{\mathbf{r}} = -\frac{km}{r^3}\mathbf{r},$$

where $k$ is a positive constant depending on mass of the sun, and $\hat{\mathbf{r}} = \mathbf{r}/r$.

As observed above, the fact that the force on the planet is always directed towards the origin implies that $\mathbf{r} \times \mathbf{v}$ is constant. We choose the direction of the $z$-axis so that $\mathbf{r} \times \mathbf{v} = h\mathbf{k}$, so the motion will be in the $xy$-plane and $r^2\dot{\theta} = h$. We have not yet specified the directions of the $x$- and $y$-axes, but will do so shortly. Using polar coordinates in the $xy$-plane, we calculate

$$\frac{d\mathbf{v}}{d\theta} = \frac{\dot{\mathbf{v}}}{\dot{\theta}} = \frac{-\dfrac{k}{r^2}\hat{\mathbf{r}}}{\dfrac{h}{r^2}} = -\frac{k}{h}\hat{\mathbf{r}}.$$

Since $d\hat{\boldsymbol{\theta}}/d\theta = -\hat{\mathbf{r}}$, we can integrate the differential equation above to find $\mathbf{v}$:

$$\mathbf{v} = -\frac{k}{h}\int \hat{\mathbf{r}}\,d\theta = \frac{k}{h}\hat{\boldsymbol{\theta}} + \mathbf{C},$$

where $\mathbf{C}$ is a vector constant of integration. Therefore, we have shown that

$$|\mathbf{v} - \mathbf{C}| = \frac{k}{h}.$$

This result, known as **Hamilton's Theorem**, says that as a planet moves around its orbit, its velocity vector (when positioned with its tail at the origin) traces out a *circle*. It is perhaps surprising that there is a circle associated with the orbit of a planet after all. Only it is not the *position* vector that moves on a circle, but the *velocity* vector. (See Figure 11.23.)

Recall that so far we have specified only the position of the origin and the direction of the $z$-axis. Therefore the $xy$-plane is determined, but not the directions of the $x$-axis or the $y$-axis. Let us choose these axes in the $xy$-plane so that $\mathbf{C}$ is in the direction of the $y$-axis; say $\mathbf{C} = (\varepsilon k/h)\mathbf{j}$, where $\varepsilon$ is a positive constant. We therefore have

$$\mathbf{v} = \frac{k}{h}(\hat{\boldsymbol{\theta}} + \varepsilon\mathbf{j}).$$

The position of the $x$-axis is now determined by the fact that the three vectors $\mathbf{i}$, $\mathbf{j}$, and $\mathbf{k}$ are mutually perpendicular and form a right-handed basis. We calculate $\mathbf{r} \times \mathbf{v}$ again. Remember that $\mathbf{r} = r\cos\theta\,\mathbf{i} + r\sin\theta\,\mathbf{j}$, and also $\mathbf{r} = r\hat{\mathbf{r}}$:

$$hk = \mathbf{r} \times \mathbf{v} = \frac{k}{h}(r\hat{\mathbf{r}} \times \hat{\boldsymbol{\theta}} + r\varepsilon\cos\theta\,\mathbf{i}\times\mathbf{j} + r\varepsilon\sin\theta\,\mathbf{j}\times\mathbf{j})$$

$$= \frac{k}{h}r(1 + \varepsilon\cos\theta)\mathbf{k}.$$

Thus $h = \dfrac{kr}{h}(1 + \varepsilon\cos\theta)$, or, solving for $r$,

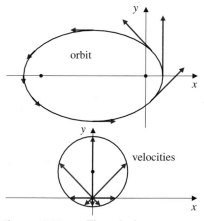

**Figure 11.23** The velocity vectors define a circle

$$r = \frac{h^2/k}{1 + \varepsilon \cos \theta}.$$

This is the polar equation of the orbit. If $\varepsilon < 1$ it is an ellipse with one focus at the origin (the sun) and with parameters given by

**Semi-latus rectum:**        $\ell = \dfrac{h^2}{k}$

**Semi-major axis:**        $a = \dfrac{h^2}{k(1 - \varepsilon^2)} = \dfrac{\ell}{1 - \varepsilon^2}$

**Semi-minor axis:**        $b = \dfrac{h^2}{k\sqrt{1 - \varepsilon^2}} = \dfrac{\ell}{\sqrt{1 - \varepsilon^2}}$

**Semi-focal separation:**        $c = \sqrt{a^2 - b^2} = \dfrac{\varepsilon\ell}{1 - \varepsilon^2}.$

We have deduced Kepler's first law! The choices we made for the coordinate axes result in **perihelion** (the point on the orbit which is closest to the sun) being on the positive $x$-axis ($\theta = 0$).

**Example 2**    A planet's orbit has eccentricity $\varepsilon$ (where $0 < \varepsilon < 1$) and its speed at perihelion is $v_P$. Find its speed $v_A$ at **aphelion** (the point on its orbit farthest from the sun).

**Solution**    At perihelion and aphelion the planet's radial velocity $\dot{r}$ is zero (since $r$ is minimum or maximum), so the velocity is entirely transverse. Thus $v_P = r_P \dot{\theta}_P$ and $v_A = r_A \dot{\theta}_A$. Since $r^2 \dot{\theta} = h$ has the same value at all points of the orbit, we have

$$r_P v_P = r_P^2 \dot{\theta}_P = h = r_A^2 \dot{\theta}_A = r_A v_A.$$

The planet's orbit has equation

$$r = \frac{\ell}{1 + \varepsilon \cos \theta},$$

so perihelion corresponds to $\theta = 0$ and aphelion to $\theta = \pi$:

$$r_P = \frac{\ell}{1 + \varepsilon} \qquad \text{and} \qquad r_A = \frac{\ell}{1 - \varepsilon}.$$

Therefore, $v_A = \dfrac{r_P}{r_A} v_P = \dfrac{1 - \varepsilon}{1 + \varepsilon} v_P.$

We can obtain Kepler's third law from the other two as follows. Since the radial line from the sun to a planet sweeps out area at a constant rate $h/2$, the total area $A$ enclosed by the orbit is $A = (h/2)T$ where $T$ is the period of revolution. The area of an ellipse with semi-axes $a$ and $b$ is $A = \pi ab$. Since $b^2 = \ell a = h^2 a/k$, we have

$$T^2 = \frac{4}{h^2} A^2 = \frac{4}{h^2} \pi^2 a^2 b^2 = \frac{4\pi^2}{k} a^3.$$

Note how the final expression for $T^2$ does not depend on $h$, which is a constant for the orbit of any one planet, but varies from planet to planet. The constant $4\pi^2/k$ does not depend on the particular planet. ($k$ depends on the mass of the sun and a universal gravitational constant.) Thus

$$T^2 = \frac{4\pi^2}{k}a^3$$

says that the square of the period of a planet is proportional to the length, $2a$, of the major axis of the orbit, the proportionality extending over all the planets. This is Kepler's third law. Modern astronomical data show that $T^2/a^3$ varies by only about three-tenths of one percent over the nine known planets.

## Conservation of Energy

Solving the second-order differential equation of motion $\mathbf{F} = m\ddot{\mathbf{r}}$ to find the orbit of a planet requires two integrations. In the above derivation we exploited properties of the cross product to make these integrations easy. More traditional derivations of Kepler's laws usually begin with separating the radial and transverse components in the equation of motion:

$$\ddot{r} - r\dot{\theta}^2 = -\frac{k}{r^2}, \qquad r\ddot{\theta} + 2\dot{r}\dot{\theta} = 0.$$

As observed earlier, the second equation above implies that $r^2\dot{\theta} = h = $ constant, which is Kepler's second law. This can be used to eliminate $\theta$ from the first equation to give

$$\ddot{r} - \frac{h^2}{r^3} = -\frac{k}{r^2}.$$

Therefore

$$\frac{d}{dt}\left(\frac{\dot{r}^2}{2} + \frac{h^2}{2r^2}\right) = \dot{r}\left(\ddot{r} - \frac{h^2}{r^3}\right) = -\frac{k}{r^2}\dot{r}.$$

If we integrate this equation we obtain

$$\frac{1}{2}\left(\dot{r}^2 + \frac{h^2}{r^2}\right) - \frac{k}{r} = E.$$

This is a **conservation of energy** law. The first term on the left is $v^2/2$, the kinetic energy (per unit mass) of the planet. The term $-k/r$ is the potential energy per unit mass. It is difficult to integrate this equation and to find $r$ as a function of $t$. In any event, we really want $r$ as a function of $\theta$ so that we can recognize that we have an ellipse. A way to obtain this is suggested in Exercise 18 below.

*Remark*    The procedure used above to demonstrate Kepler's laws in fact shows that if any object moves under the influence of a force that attracts it towards the origin (or repels it away from the origin) and has magnitude proportional to the reciprocal of the square of distance from the origin, then the object must move in a plane orbit whose shape is a conic section. If the total energy $E$ defined above is *negative*, then the orbit is *bounded* and must therefore be an ellipse. If $E = 0$, the orbit is a parabola. If $E > 0$, then the orbit is a hyperbola. Hyperbolic orbits are typical for repulsive forces, but may also occur for attractions if the object has high enough velocity (exceeding the *escape velocity*). See Exercise 22 below for an example.

# Exercises 11.6

1. **(Polar ellipses)** Fill in the details of the calculation suggested in the text to transform the polar equation of an ellipse, $r = \ell/(1 + \varepsilon \cos\theta)$, where $0 < \varepsilon < 1$, to Cartesian coordinates in a form showing the centre and semi-axes explicitly.

### Polar components of velocity and acceleration

2. A particle moves on the circle with polar equation $r = k$. $(k > 0)$. What are the radial and transverse components of its velocity and acceleration? Show that the transverse component of the acceleration is equal to the rate of change of the speed of the particle.

3. Find the radial and transverse components of velocity and acceleration of a particle moving at unit speed along the exponential spiral $r = e^\theta$. Express your answers in terms of the angle $\theta$.

4. If a particle moves along the polar curve $r = \theta$ under the influence of a central force attracting it to the origin, find the magnitude of the acceleration as a function of $r$ and the speed of the particle.

5. A particle of mass $m$ moves along a curve in the $xy$-plane having polar equation $r = 1/\theta$ under the influence of a central force directed towards the origin. Show that the magnitude of this force is inversely proportional to the cube of the particle's distance from the origin.

### Deductions from Kepler's Laws

6. The mean distance from the earth to the sun is approximately 150 million km. Halley's Comet approaches perihelion (comes closest to the sun) in its elliptical orbit approximately every 76 years. Estimate the major axis of the orbit of Halley's Comet.

7. The mean distance from the moon to the earth is about 385,000 km, and its period of revolution around the earth is about 27 days (the siderial month). At approximately what distance from the centre of the earth, and in what plane, should a communications satellite be inserted into circular orbit if it must remain directly above the the same position on the earth at all times?

8. An asteroid is in a circular orbit around the sun. If its period of revolution is $T$ find the radius of its orbit.

* 9. If the asteroid in Exercise 8 is instantaneously stopped in its orbit it will fall towards the sun. How long will it take to get there? *Hint:* you can do this question easily if instead you regard the asteroid as *almost* stopped, so that it goes into a highly eccentric elliptical orbit whose major axis is a bit greater than the radius of the original circular orbit.

10. Find the eccentricity of an asteroid's orbit if the asteroid's speed at perihelion is twice its speed at aphelion.

11. Show that the orbital speed of a planet is constant if and only if the orbit is circular. *Hint:* use the conservation of energy identity.

12. A planet's distance from the sun at perihelion is 80% of its distance at aphelion. Find the ratio of its speeds at perihelion and aphelion, and the eccentricity of its orbit.

* 13. As a result of a collision, an asteroid originally in a circular orbit about the sun suddenly has its velocity cut in half, so that it falls into an elliptical orbit with maximum distance from the sun equal to the radius of the original circular orbit. Find the eccentricity of its new orbit.

14. If the speeds of a planet at perihelion and aphelion are $v_P$ and $v_A$, what is its speed when it is at the ends of the minor axis of its orbit?

15. What fraction of its "year" (i.e. the period of its orbit) does a planet spend traversing the half of its orbit that is closest to the sun? Give your answer in terms of the eccentricity $\varepsilon$ of the planet's orbit.

* 16. Suppose that a planet is travelling at speed $v_0$ at an instant when it is at distance $r_0$ from the sun. Show that the period of the planet's orbit is

$$T = \frac{2\pi}{\sqrt{k}} \left( \frac{2}{r_0} - \frac{v_0^2}{k} \right)^{-3/2}.$$

*Hint:* the quantity $\dfrac{k}{r} - \dfrac{1}{2}v^2$ is constant at all points of the orbit, as shown in the discussion of conservation of energy. Find the value of this expression at perihelion in terms of the semi-major axis, $a$.

* 17. The sum of the distances from a point $P$ on an ellipse $\mathcal{E}$ to the foci of $\mathcal{E}$ is the constant $2a$, the length of the major axis of the ellipse. Use this fact in a *geometric* argument to show that the mean distance from points $P$ to one focus of $\mathcal{E}$ is $a$. That is, show that

$$\frac{1}{c(\mathcal{E})} \int_{\mathcal{E}} r \, ds = a,$$

where $c(\mathcal{E})$ is the circumference of $\mathcal{E}$ and $r$ is the distance from a point on $\mathcal{E}$ to one focus.

* 18. **(A direct approach to Kepler's 1st Law)** The result of eliminating $\theta$ between the equations for the radial and transverse components of acceleration for a planet is

$$\ddot{r} - \frac{h^2}{r^3} = -\frac{k}{r^2}.$$

Show that the change of dependent and independent variables:

$$r(t) = \frac{1}{u(\theta)}, \qquad \theta = \theta(t),$$

transforms this equation to the simpler equation

$$\frac{d^2u}{d\theta^2} + u = \frac{k}{h^2}.$$

Show that the solution of this equation is

$$u = \frac{k}{h^2}\left(1 + \varepsilon\cos(\theta - \theta_0)\right),$$

where $\varepsilon$ and $\theta_0$ are constants. Hence show that the orbit is elliptical if $|\varepsilon| < 1$.

* **19. (What if gravitation were an inverse cube law?)** Use the technique of Exercise 18 to find the trajectory of an object of unit mass attracted to the origin by a force of magnitude $f(r) = k/r^3$. Are there any orbits that do not approach infinity or the origin as $t \to \infty$?

**20.** Use the conservation of energy formula to show that if $E < 0$, the orbit must be bounded; i.e. it cannot get arbitrarily far away from the origin.

* **21. (Polar hyperbolas)** If $\varepsilon > 1$ then the equation

$$r = \frac{\ell}{1 + \varepsilon\cos\theta}$$

represents a hyperbola rather than an ellipse. Sketch the hyperbola, find its centre and the directions of its asymptotes, and determine its semi-transverse axis, its semi-conjugate axis, and semi-focal separation in terms of $\ell$ and $\varepsilon$. (See Section 8.1 for definitions of terms.)

* **22. (Hyperbolic orbits)** A meteor travels from infinity on a hyperbolic orbit passing near the sun. At a very large distance from the sun it has speed $v_\infty$. The asymptotes of its orbit pass at perpendicular distance $D$ from the sun. (See

Figure 11.24.) Show that the angle $\delta$ through which the meteor's path is deflected by the gravitational attraction of the sun is given by

$$\cot\left(\frac{\delta}{2}\right) = \frac{Dv_\infty^2}{k}.$$

(*Hint:* you will need the result of Exercise 21.) The same analysis, and result, holds for electrostatic attraction or repulsion; $f(r) = \pm k/r^2$ in that case also. The constant $k$ depends on the charges of two particles and $r$ is the distance between them.

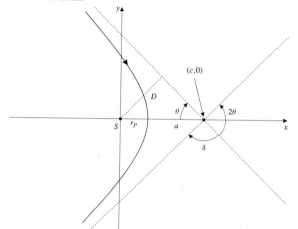

**Figure 11.24**    Path of a meteor

---

# Chapter Review

## Key Ideas

- **What is a vector function of a real variable, and why does it represent a curve?**
- **State the Product Rule for the derivative of $u(t) \bullet \left(v(t) \times w(t)\right)$.**
- **What do the following phrases mean?**
  - ◇ angular velocity
  - ◇ angular momentum
  - ◇ centripetal acceleration
  - ◇ Coriolis acceleration
  - ◇ arc-length parametrization
  - ◇ central force
- **Find the following quantities associated with a parametric curve $C$ with parametrization $r = r(t)$, $(a \le t \le b)$.**
  - ◇ the velocity $v(t)$
  - ◇ the speed $v(t)$
  - ◇ the arc length
  - ◇ the acceleration $a(t)$
  - ◇ the unit tangent $\hat{T}(t)$
  - ◇ the unit normal $\hat{N}(t)$
  - ◇ the curvature $\kappa(t)$
  - ◇ the radius of curvature $\rho(t)$
  - ◇ the osculating plane
  - ◇ the osculating circle
  - ◇ the unit binormal $\hat{B}(t)$
  - ◇ the torsion $\tau(t)$
  - ◇ the tangential acceleration
  - ◇ the normal acceleration
  - ◇ the evolute
- **State the Frenet–Serret formulas.**
- **State Kepler's laws of planetary motion.**
- **What are the radial and transverse components of velocity and acceleration?**

## Review Exercises

**1.** If $r(t)$, $v(t)$, and $a(t)$ represent the position, velocity, and acceleration at time $t$ of a particle moving in 3-space, and if, at every time $t$, the $a$ is perpendicular to both $r$ and $v$, show that the vector $r(t) - tv(t)$ has constant length.

**2.** Describe the parametric curve

$$r = t\cos t\,i + t\sin t\,j + (2\pi - t)k,$$

$(0 \le t \le 2\pi)$ and find its length.

3. A particle moves along the curve of intersection of the surfaces $y = x^2$ and $z = 2x^3/3$ with constant speed $v = 6$. It is moving in the direction of increasing $x$. Find its velocity and acceleration when it is at the point $(1, 1, 2/3)$.

4. A particle moves along the curve $y = x^2$ in the $xy$-plane so that at time $t$ its speed is $v = t$. Find its acceleration at time $t = 3$ if it is at the point $(\sqrt{2}, 2)$ at that time.

5. Find the curvature and torsion at a general point of the curve $\mathbf{r} = e^t\mathbf{i} + \sqrt{2}t\mathbf{j} + e^{-t}\mathbf{k}$.

6. A particle moves on the curve of Exercise 5 so that it is at position $\mathbf{r}(t)$ at time $t$. Find its normal acceleration and tangential acceleration at any time $t$. What is its minimum speed?

7. **(A clothoid curve)** The plane curve $\mathcal{C}$ in Figure 11.25 has parametric equations

$$x(s) = \int_0^s \cos\frac{kt^2}{2}\,dt \quad \text{and} \quad y(s) = \int_0^s \sin\frac{kt^2}{2}\,dt.$$

Verify that $s$ is, in fact, arc length along $\mathcal{C}$ measured from $(0, 0)$, and that the curvature of $\mathcal{C}$ is given by $\kappa(s) = ks$. Because the curvature changes linearly with distance along the curve, such curves, called *clothoids*, are useful for joining track sections of different curvatures.

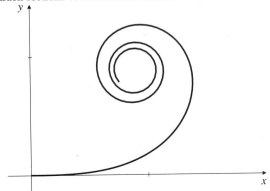

**Figure 11.25**    A clothoid curve

8. A particle moves along the polar curve $r = e^{-\theta}$ with constant angular speed $\dot{\theta} = k$. Express its velocity and acceleration in terms of radial and transverse components depending only on the distance $r$ from the origin.

**Some properties of cycloids**

Exercises 9–12 all deal with the cycloid

$$\mathbf{r} = a(t - \sin t)\mathbf{i} + a(1 - \cos t)\mathbf{j}.$$

Recall that this curve is the path of a point on the circumference of a circle of radius $a$ rolling along the $x$-axis. (See Section 8.2.)

9. Find the arc length $s = s(T)$ of the part of the cycloid from $t = 0$ to $t = T \le 2\pi$.

10. Find the arc-length parametrization $\mathbf{r} = \mathbf{r}(s)$ of the arch $0 \le t \le 2\pi$ of the cycloid, with $s$ measured from the point $(0, 0)$.

11. Find the *evolute* of the cycloid, that is, find parametric equations of the centre of curvature $\mathbf{r} = \mathbf{r}_c(t)$ of the cycloid. Show that the evolute is the same cycloid translated $\pi a$ units to the right and $2a$ units downward.

12. A string of length $4a$ has one end fixed at the origin and is wound along the arch of the cycloid to the right of the origin. Since that arch has total length $8a$, the free end of the string lies at the highest point $A$ of the arch. Find the path followed by the free end $Q$ of the string as it is unwound from the cycloid and is held taught during the unwinding. (See Figure 11.26.) If the string leaves the cycloid at $P$, then

$$(\text{arc } OP) + PQ = 4a.$$

The path of $Q$ is called the *involute* of the cycloid. Show that, like the evolute, the involute is also a translate of the original cycloid. In fact, the cycloid is the evolute of its involute.

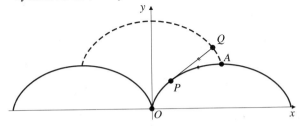

**Figure 11.26**

13. Let $P$ be a point in 3-space with spherical coordinates $(\rho, \phi, \theta)$. Suppose that $P$ is not on the $z$-axis. Find a triad of mutually perpendicular unit vectors, $\{\hat{\boldsymbol{\rho}}, \hat{\boldsymbol{\phi}}, \hat{\boldsymbol{\theta}}\}$, at $P$ in the directions of increasing $\rho$, $\phi$, and $\theta$ respectively. Is the triad right- or left-handed?

**Kepler's laws imply Newton's law of gravitation**

In Exercises 14–16, it is *assumed* that a planet of mass $m$ moves in an elliptical orbit $r = \ell/(1 + \varepsilon\cos\theta)$, with focus at the origin (the sun), under the influence of a force $\mathbf{F} = \mathbf{F}(\mathbf{r})$ that depends only on the position of the planet.

14. Assuming Kepler's second law, show that $\mathbf{r}\times\mathbf{v} = \mathbf{h}$ is constant, and hence that $r^2\dot{\theta} = h$ is constant.

15. Use Newton's second law of motion ($\mathbf{F} = m\ddot{\mathbf{r}}$) to show that $\mathbf{r}\times\mathbf{F}(\mathbf{r}) = \mathbf{0}$. Therefore $\mathbf{F}(\mathbf{r})$ is parallel to $\mathbf{r}$: $\mathbf{F}(\mathbf{r}) = -f(\mathbf{r})\hat{\mathbf{r}}$, for some scalar-valued function $f(\mathbf{r})$, and transverse component of $\mathbf{F}(\mathbf{r})$ is zero.

16. By direct calculation of the radial acceleration of the planet, show that $f(\mathbf{r}) = mh^2/(r^2)$, where $r = |\mathbf{r}|$. Thus $\mathbf{F}$ is an attraction to the origin, proportional to the mass of the planet, and inversely proportional to the square of its distance from the sun.

## Challenging Problems

1. Let $P$ be a point on the surface of the earth at $45°$ north latitude. Use a coordinate system with origin at $P$ and basis vectors $\mathbf{i}$ and $\mathbf{j}$ pointing east and north, respectively, so that $\mathbf{k}$ points vertically upward.

(a) Express the angular velocity $\boldsymbol{\Omega}$ of the earth in terms of the basis vectors at $P$. What is the magnitude $\Omega$ of $\boldsymbol{\Omega}$ in rad/s?

(b) Find the Coriolis acceleration $\mathbf{a}_C = 2\boldsymbol{\Omega} \times \mathbf{v}$ of an object falling vertically with speed $v$ above $P$.

(c) If the object in (b) drops from rest from a height of 100 metres above $P$, approximately where will it strike the ground? Ignore air resistance, but not the Coriolis acceleration. Since the Coriolis acceleration is much smaller than the gravitational acceleration in magnitude, you can use the vertical velocity as a good approximation to the actual velocity of the object at any time during its fall.

* **2. (The spin of a baseball)** When a ball is thrown with spin about an axis that is not parallel to its velocity, it experiences a lateral acceleration due to differences in friction along its sides. This spin acceleration is given by

$$\mathbf{a}_s = k\mathbf{S} \times \mathbf{v},$$

where $\mathbf{v}$ is the velocity of the ball, $\mathbf{S}$ is the angular velocity of its spin, and $k$ is a positive constant depending on the surface of the ball. Suppose that a ball for which $k = 0.001$ is thrown horizontally along the $x$-axis with an initial speed of 70 ft/s and a spin of 1,000 radians/s about a vertical axis. Its velocity $\mathbf{v}$ must satisfy

$$\begin{cases} \dfrac{d\mathbf{v}}{dt} = (0.001)(1,000\mathbf{k}) \times \mathbf{v} - 32\mathbf{k} = \mathbf{k} \times \mathbf{v} - 32\mathbf{k} \\ \mathbf{v}(0) = 70\mathbf{i}, \end{cases}$$

since the acceleration of gravity is 32 ft/s$^2$.

(a) Show that the components of $\mathbf{v} = v_1\mathbf{i} + v_2\mathbf{j} + v_3\mathbf{k}$ satisfy

$$\begin{cases} \dfrac{dv_1}{dt} = -v_2 \\ v_1(0) = 70 \end{cases} \quad \begin{cases} \dfrac{dv_2}{dt} = v_1 \\ v_2(0) = 0 \end{cases} \quad \begin{cases} \dfrac{dv_3}{dt} = -32 \\ v_3(0) = 0. \end{cases}$$

(b) Solve these equations, and find the position of the ball $t$ seconds after it is thrown. Assume that it is thrown from the origin at time $t = 0$.

(c) At $t = 1/5$ s, how far, and in what direction, has the ball deviated from the parabolic path it would have followed if it had been thrown without spin?

* **3. (Charged particles moving in magnetic fields)** Magnetic fields exert forces on moving charged particles. If a particle of mass $m$ and charge $q$ is moving with velocity $\mathbf{v}$ in a magnetic field $\mathbf{B}$, then it experiences a force $\mathbf{F} = q\mathbf{v} \times \mathbf{B}$, and hence its velocity is governed by the equation

$$m\frac{d\mathbf{v}}{dt} = q\mathbf{v} \times \mathbf{B}.$$

For this exercise, suppose that the magnetic field is constant and vertical, say $\mathbf{B} = B\mathbf{k}$ (as, for example, in a cathode-ray tube). If the moving particle has initial velocity $\mathbf{v}_0$, then its velocity at time $t$ is determined by

$$\begin{cases} \dfrac{d\mathbf{v}}{dt} = \omega\mathbf{v} \times \mathbf{k}, & \text{where } \omega = \dfrac{qB}{m} \\ \mathbf{v}(0) = \mathbf{v}_0. \end{cases}$$

(a) Show that $\mathbf{v} \bullet \mathbf{k} = \mathbf{v}_0 \bullet \mathbf{k}$ and $|\mathbf{v}| = |\mathbf{v}_0|$ for all $t$.

(b) Let $\mathbf{w}(t) = \mathbf{v}(t) - (\mathbf{v}_0 \bullet \mathbf{k})\mathbf{k}$, so that $\mathbf{w}$ is perpendicular to $\mathbf{k}$ for all $t$. Show that $\mathbf{w}$ satisfies

$$\begin{cases} \dfrac{d^2\mathbf{w}}{dt^2} = -\omega^2\mathbf{w} \\ \mathbf{w}(0) = \mathbf{v}_0 - (\mathbf{v}_0 \bullet \mathbf{k})\mathbf{k} \\ \mathbf{w}'(0) = \omega\mathbf{v}_0 \times \mathbf{k}. \end{cases}$$

(c) Solve the initial-value problem in (b) for $\mathbf{w}(t)$, and hence find $\mathbf{v}(t)$.

(d) Find the position vector $\mathbf{r}(t)$ of the particle at time $t$ if it is at the origin at time $t = 0$. Verify that the path of the particle is, in general, a circular helix. Under what circumstances is the path a straight line? a circle?

* **4. (The tautochrone)** The parametric equations

$$x = a(\theta - \sin\theta) \quad \text{and} \quad y = a(\cos\theta - 1)$$

(for $0 \le \theta \le 2\pi$) describe an arch of the cycloid followed by a point on a circle of radius $a$ rolling along the underside of the $x$-axis. Suppose the curve is made of wire along which a bead can slide without friction. (See Figure 11.27.) If the bead slides from rest under gravity, starting at a point having parameter value $\theta_0$, show that the time it takes for the bead to fall to the lowest point on the arch (corresponding to $\theta = \pi$) is a *constant*, independent of the starting position $\theta_0$. Thus two such beads released simultaneously from different positions along the wire will always collide at the lowest point. For this reason, the cycloid is sometimes called the *tautochrone*, from the Greek for "constant time." *Hint:* when the bead has fallen from height $y(\theta_0)$ to height $y(\theta)$ its speed is $v = \sqrt{2g\big(y(\theta_0) - y(\theta)\big)}$. (Why?) The time for the bead to fall to the bottom is

$$T = \int_{\theta=\theta_0}^{\theta=\pi} \frac{1}{v}\, ds,$$

where $ds$ is the arc length element along the cycloid.

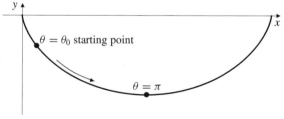

**Figure 11.27**

5. **(The Drop of Doom)** An amusement park ride at the West Edmonton Mall in Alberta, Canada, gives thrill seekers a taste of free-fall. It consists of a car moving along a track consisting of straight vertical and horizontal sections joined by a smooth curve. The car drops from the top and falls vertically under gravity for $10 - 2\sqrt{2} \approx 7.2$ m before entering the curved section at $B$. It falls another $2\sqrt{2} \approx 2.8$ m as it whips around the curve and into the horizontal section $DE$ at ground level, where brakes are applied to stop it. (Thus, the total vertical drop from $A$ to $D$ or $E$ is 10 m, a figure, like the others in this problem, chosen for mathematical convenience rather than engineering precision.) For purposes of this problem it is helpful to take the coordinate axes at a $45°$ angle to the vertical, so that the two straight sections of the track lie along the graph $y = |x|$. The curved section then goes from $(-2, 2)$ to $(2, 2)$, and can be taken to be symmetric about the $y$-axis. With this coordinate system, the gravitational acceleration is in the direction of $\mathbf{i} - \mathbf{j}$. (See Figure 11.28.)

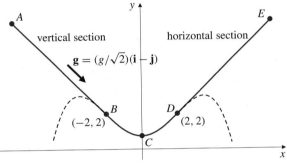

**Figure 11.28**

(a) Find a fourth-degree polynomial whose graph can be used to link the two straight sections of track without producing discontinuous accelerations for the falling car. (Why is fourth degree adequate?)

(b) Ignoring friction and air resistance, how fast is the car moving when it enters the curve at $B$? at the midpoint $C$ of the curve? and when it leaves the curve at $D$?

(c) Find the magnitude of the normal acceleration and of the total acceleration of the car as it passes through $C$.

6. **(A chase problem)** A fox and a hare are running in the $xy$-plane. Both are running at the same speed $v$. The hare is running up the $y$-axis; at time $t = 0$ it is at the origin. The fox is always running straight towards the hare. At time $t = 0$ the fox is at the point $(a, 0)$, where $a > 0$. Let the fox's position at time $t$ be $(x(t), y(t))$.

(a) Verify that the tangent to the fox's path at time $t$ has slope

$$\frac{dy}{dx} = \frac{y(t) - vt}{x(t)}.$$

(b) Show that the equation of the path of the fox satisfies the equation

$$x\frac{d^2y}{dx^2} = \sqrt{1 + \left(\frac{dy}{dx}\right)^2}.$$

*Hint:* differentiate the equation in (a) with respect to $t$. On the left side note that $(d/dt) = (dx/dt)(d/dx)$.

(c) Solve the equation in (b) by substituting $u(x) = dy/dx$ and separating variables. Note that $y = 0$ and $u = 0$ when $x = a$.

7. Suppose the earth is a perfect sphere of radius $a$. You set out from the point on the equator whose spherical coordinates are $(\rho, \phi, \theta) = (a, \pi/2, 0)$ and travel on the surface of the earth at constant speed $v$, always moving towards the northeast ($45°$ east of north).

(a) Will you ever get to the north pole? If so, how long will it take to get there?

(b) Find the functions $\phi(t)$ and $\theta(t)$ that are the angular spherical coordinates of your position at time $t > 0$.

(c) How many times does your path cross the meridian $\theta = 0$?

# CHAPTER 12
# Partial Differentiation

***Introduction*** This chapter is concerned with extending the idea of the derivative to real functions of a vector variable, that is, to functions depending on several real variables.

## 12.1 Functions of Several Variables

The notation $y = f(x)$ is used to indicate that the variable $y$ depends on the single real variable $x$, that is, that $y$ is a function of $x$. The domain of such a function $f$ is a set of real numbers. Many quantities can be regarded as depending on more than one real variable, and thus to be functions of more than one variable. For example, the volume of a circular cylinder of radius $r$ and height $h$ is given by $V = \pi r^2 h$; we say that $V$ is a function of the two variables $r$ and $h$. If we choose to denote this function by $f$, then we would write $V = f(r, h)$ where

$$f(r, h) = \pi r^2 h, \qquad (r > 0, \quad h > 0).$$

Thus $f$ is a function of two variables having as *domain* the set of points in the $rh$-plane with coordinates $(r, h)$ satisfying $r > 0$ and $h > 0$. Similarly, the relationship $w = f(x, y, z) = x + 2y - 3z$ defines $w$ as a function of the three variables $x$, $y$, and $z$, with domain the whole of $\mathbb{R}^3$, or, if we state explicitly, some particular subset of $\mathbb{R}^3$.

By analogy with the corresponding definition for functions of one variable, we define a function of $n$ variables as follows:

**DEFINITION 1**

> A **function** $f$ of $n$ real variables is a rule that assigns a *unique* real number $f(x_1, x_2, \ldots, x_n)$ to each point $(x_1, x_2, \ldots, x_n)$ in some subset $\mathcal{D}(f)$ of $\mathbb{R}^n$. $\mathcal{D}(f)$ is called the **domain** of $f$. The set of real numbers $f(x_1, x_2, \ldots, x_n)$ obtained from points in the domain is called the **range** of $f$.

As for functions of one variable, it is conventional to take the domain of such a function of $n$ variables to be the largest set of points $(x_1, x_2, \ldots, x_n)$ for which $f(x_1, x_2, \ldots, x_n)$ makes sense as a real number, unless the domain is explicitly stated to be a smaller set.

Most of the examples we consider hereafter will be functions of two or three independent variables. When a function $f$ depends on two variables we will usually call these independent variables $x$ and $y$, and use $z$ to denote the dependent variable which represents the value of the function; that is, $z = f(x, y)$. We will normally use $x$, $y$, and $z$ as the independent variables of a function of three variables, and $w$ as the value of the function; $w = f(x, y, z)$. Some definitions will be given, and some theorems will be stated (and proved) only for the two-variable case, but extensions to three or more variables will usually be obvious.

## Graphical Representations

The graph of a function $f$ of one variable (that is, the graph of the equation $y = f(x)$) is the set of points in the $xy$-plane having coordinates $(x, f(x))$, where $x$ is in the domain of $f$. Similarly, the graph of a function $f$ of 2 variables, (the graph of the equation $z = f(x, y)$), is the set of points in 3-space having coordinates $(x, y, f(x, y))$, where $(x, y)$ belongs to the domain of $f$. This graph is a surface in $\mathbb{R}^3$ lying above (if $f(x, y) > 0$) or below (if $f(x, y) < 0$) the domain of $f$ in the $xy$-plane. (See Figure 12.1.) The graph of a function of three variables is a three-dimensional *hypersurface* in 4-space, $\mathbb{R}^4$. In general, the graph of a function of $n$ variables is an $n$-dimensional *surface* in $\mathbb{R}^{n+1}$. We will not attempt to *draw* graphs of functions of more than two variables!

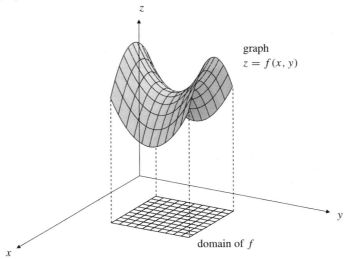

graph
$z = f(x, y)$

domain of $f$

**Figure 12.1**    The graph of $f(x, y)$ is the surface with equation $z = f(x, y)$ defined for points $(x, y)$ in the domain of $f$

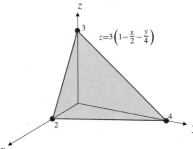

$z = 3\left(1 - \frac{x}{2} - \frac{y}{4}\right)$

**Figure 12.2**

**Example 1**    Consider the function

$$f(x, y) = 3\left(1 - \frac{x}{2} - \frac{y}{4}\right), \qquad (0 \le x \le 2, \quad 0 \le y \le 4 - 2x).$$

The graph of $f$ is the plane triangular surface with vertices at $(2, 0, 0)$, $(0, 4, 0)$, and $(0, 0, 3)$. (See Figure 12.2.) If the domain of $f$ had not been explicitly stated to be a particular set in the $xy$-plane, the graph would have been the whole plane through these three points.

**Example 2**    Consider $f(x, y) = \sqrt{9 - x^2 - y^2}$. The expression under the square root can not be negative, so the domain is the disk $x^2 + y^2 \le 9$ in the $xy$-plane.

If we square the equation $z = \sqrt{9 - x^2 - y^2}$ we can rewrite the result in the form $x^2 + y^2 + z^2 = 9$. This is a sphere of radius 3 centred at the origin. However, the graph of $f$ is only the upper hemisphere where $z \ge 0$. (See Figure 12.3.)

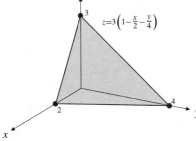

$z = \sqrt{9 - x^2 - y^2}$

**Figure 12.3**

Since it is necessary to project the surface $z = f(x, y)$ onto a two-dimensional page, most such graphs are difficult to sketch without considerable artistic talent and training. Nevertheless, you should always try to visualize such a graph, and sketch it as best you can. Sometimes it is convenient to sketch only part of a graph, for instance, the part lying in the first octant. It is also helpful to determine (and sketch) the intersections of the graph with various planes, especially the coordinate planes, and planes parallel to the coordinate planes. (See Figure 12.1.)

Some mathematical software packages will produce plots of three-dimensional graphs to help you get a feeling for how the corresponding functions behave. Figure 12.1 is an example of such a computer-drawn graph. Along with all the other mathematical graphics in this book, it was produced using the mathematical graphics software package **MG**. Another example is given in Figure 12.4.

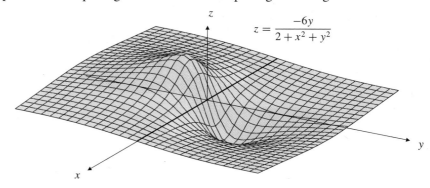

$$z = \frac{-6y}{2 + x^2 + y^2}$$

**Figure 12.4**    The graph of
$$\frac{-6y}{2 + x^2 + y^2}$$

Another way to represent the function $f(x, y)$ graphically is to produce a two-dimensional *topographic map* of the surface $z = f(x, y)$. In the $xy$-plane we sketch the curves $f(x, y) = C$ for various values of the constant $C$. These curves are called **level curves** of $f$ because they are the vertical projections onto the $xy$-plane of the curves in which the graph $z = f(x, y)$ intersects the horizontal (level) planes $z = C$. The graph and some level curves of the function $f(x, y) = x^2 + y^2$ are shown in Figure 12.5. The graph is a circular paraboloid; the level curves are circles centred at the origin.

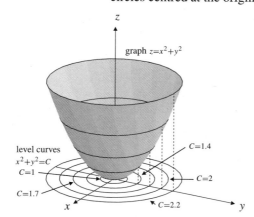

**Figure 12.5**    The graph of $f(x, y) = x^2 + y^2$ and some of level curves of $f$

**Figure 12.6**    Level curves (contours) representing elevation in a topographic map

The contour curves in the topographic map in Figure 12.6 show the elevations, in 100 m increments above sea level, on part of Nelson Island on the British Columbia coast. Since these contours are drawn for equally spaced values of $C$, the spacing of the contours themselves conveys information about the relative steepness at various places on the mountains; the land is steepest where the contour lines are closest together. Observe also that the streams shown cross the contours at right angles. They take the route of steepest descent. Isotherms (curves of constant temperature) and isobars (curves of constant pressure) on weather maps are also examples of level curves.

**Example 3**   The level curves of the function $f(x, y) = 3\left(1 - \dfrac{x}{2} - \dfrac{y}{4}\right)$ of Example 1 are the segments of the straight lines

$$3\left(1 - \frac{x}{2} - \frac{y}{4}\right) = C \qquad \text{or} \qquad \frac{x}{2} + \frac{y}{4} = 1 - \frac{C}{3}, \qquad (0 \le C \le 3)$$

that lie in the first quadrant. Several such level curves are shown in Figure 12.7(a). They correspond to equally spaced values of $C$, and their equal spacing indicates the uniform steepness of the graph of $f$. ∎

**Figure 12.7**

(a) Level curves of $3\left(1 - \dfrac{x}{2} - \dfrac{y}{4}\right)$

(b) Level curves of $\sqrt{9 - x^2 - y^2}$

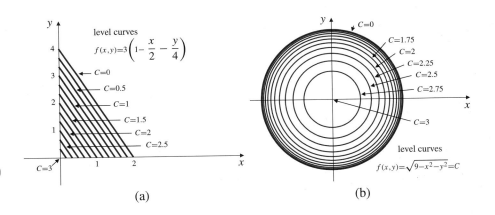

(a)                    (b)

**Example 4**   The level curves of the function $f(x, y) = \sqrt{9 - x^2 - y^2}$ of Example 2 are the concentric circles

$$\sqrt{9 - x^2 - y^2} = C \qquad \text{or} \qquad x^2 + y^2 = 9 - C^2, \qquad (0 \le C \le 3).$$

Observe the spacing of these circles in Figure 12.7(b); they are plotted for several equally spaced values of $C$. The bunching of the circles as $C \to 0+$ indicates the steepness of the hemispherical surface that is the graph of $f$. ∎

A function determines its level curves with any given spacing between consecutive values of $C$. However, level curves only determine the function if *all of them* are known.

**Example 5**   The level curves of the function $f(x, y) = x^2 - y^2$ are the curves $x^2 - y^2 = C$. For $C = 0$ the level "curve" is the pair of straight lines $x = y$ and $x = -y$. For other values of $C$ the level curves are rectangular hyperbolas with these lines as asymptotes. (See Figure 12.8(a).) The graph of $f$ is the saddle-like hyperbolic paraboloid in Figure 12.8(b). ∎

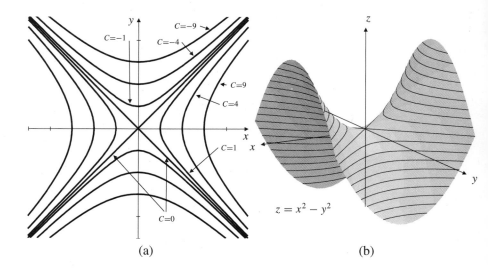

**Figure 12.8**

(a) Level curves of $x^2 - y^2$

(b) The graph of $x^2 - y^2$

---

**Example 6**    Describe and sketch some level curves of the function $z = g(x, y)$ defined by $z \geq 0$, $x^2 + (y - z)^2 = 2z^2$. Also sketch the graph of $g$.

**Solution**    The level curve $z = g(x, y) = C$ (where $C$ is a positive constant) has equation $x^2 + (y - C)^2 = 2C^2$, and is, therefore, a circle of radius $\sqrt{2}C$ centred at $(0, C)$. Level curves for $C$ in increments of 0.1 from 0 to 1 are shown in Figure 12.9(a). These level curves intersect rays from the origin at equal spacing (the spacing is different for different rays) indicating that the surface $z = g(x, y)$ is an oblique circular cone. See Figure 12.9(b). ∎

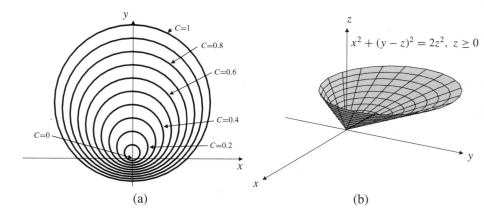

**Figure 12.9**

(a) Level curves of $z = g(x, y)$ for Example 6

(b) The graph of $z = g(x, y)$

Although the *graph* of a function $f(x, y, z)$ of three variables cannot easily be drawn (it is a three-dimensional *hypersurface* in 4-space), such a function has **level surfaces** in 3-space that can, perhaps be drawn. These level surfaces have equations $f(x, y, z) = C$ for various choices of the constant $C$. For instance, the level surfaces of the function $f(x, y, z) = x^2 + y^2 + z^2$ are concentric spheres centred at the origin. Figure 12.10 shows a few level surfaces of the function $f(x, y, z) = x^2 - z$. They are parabolic cylinders.

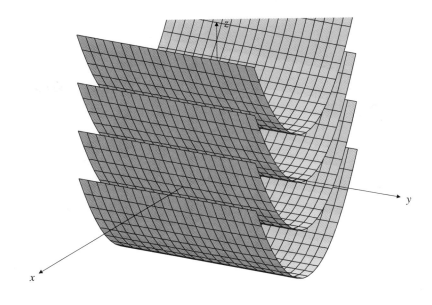

**Figure 12.10** Level surfaces of
$f(x, y, z) = x^2 - z$

# Exercises 12.1

Specify the domains of the functions in Exercises 1–10.

**1.** $f(x, y) = \dfrac{x + y}{x - y}$

**2.** $f(x, y) = \sqrt{xy}$

**3.** $f(x, y) = \dfrac{x}{x^2 + y^2}$

**4.** $f(x, y) = \dfrac{xy}{x^2 - y^2}$

**5.** $f(x, y) = \sqrt{4x^2 + 9y^2 - 36}$

**6.** $f(x, y) = \dfrac{1}{\sqrt{x^2 - y^2}}$

**7.** $f(x, y) = \ln(1 + xy)$

**8.** $f(x, y) = \sin^{-1}(x + y)$

**9.** $f(x, y, z) = \dfrac{xyz}{x^2 + y^2 + z^2}$

**10.** $f(x, y, z) = \dfrac{e^{xyz}}{\sqrt{xyz}}$

Sketch the graphs of the functions in Exercises 11–18.

**11.** $f(x, y) = x$, $\quad (0 \le x \le 2, \quad 0 \le y \le 3)$

**12.** $f(x, y) = \sin x$, $\quad (0 \le x \le 2\pi, \quad 0 \le y \le 1)$

**13.** $f(x, y) = y^2$, $\quad (-1 \le x \le 1, \quad -1 \le y \le 1)$

**14.** $f(x, y) = 4 - x^2 - y^2$, $\quad (x^2 + y^2 \le 4, \ x \ge 0, \ y \ge 0)$

**15.** $f(x, y) = \sqrt{x^2 + y^2}$

**16.** $f(x, y) = 4 - x^2$

**17.** $f(x, y) = |x| + |y|$

**18.** $f(x, y) = 6 - x - 2y$

Sketch some of the level curves of the functions in Exercises 19–26.

**19.** $f(x, y) = x - y$

**20.** $f(x, y) = x^2 + 2y^2$

**21.** $f(x, y) = xy$

**22.** $f(x, y) = \dfrac{x^2}{y}$

**23.** $f(x, y) = \dfrac{x - y}{x + y}$

**24.** $f(x, y) = \dfrac{y}{x^2 + y^2}$

**25.** $f(x, y) = xe^{-y}$

**26.** $f(x, y) = \sqrt{\dfrac{1}{y} - x^2}$

Exercises 27–28 refer to Figure 12.11, which shows contours of a hilly region with heights given in metres.

**27.** At which of the points $A$ and $B$ is the landscape steepest? How do you know?

**28.** Describe the topography of the region near point $C$.

**Figure 12.11**

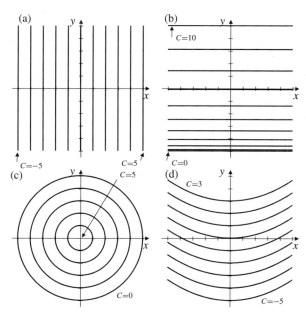

**Figure 12.12**

**34.** If we assume $z \geq 0$, the equation $4z^2 = (x - z)^2 + (y - z)^2$ defines $z$ as a function of $x$ and $y$. Sketch some level curves of this function. Describe its graph.

**35.** Find $f(x, y)$ if each level curve $f(x, y) = C$ is a circle centred at the origin and having radius
(a) $C$,    (b) $C^2$,    (c) $\sqrt{C}$,    (d) $\ln C$.

**36.** Find $f(x, y, z)$ if for each constant $C$ the level surface $f(x, y, z) = C$ is a plane having intercepts $C^3$, $2C^3$, and $3C^3$ on the $x$-axis, the $y$-axis and the $z$-axis respectively.

Describe the level surfaces of the functions specified in Exercises 37–41.

**37.** $f(x, y, z) = x^2 + y^2 + z^2$

**38.** $f(x, y, z) = x + 2y + 3z$

**39.** $f(x, y, z) = x^2 + y^2$    **40.** $f(x, y, z) = \dfrac{x^2 + y^2}{z^2}$

**41.** $f(x, y, z) = |x| + |y| + |z|$

**42.** Describe the "level hypersurfaces" of the function

$$f(x, y, z, t) = x^2 + y^2 + z^2 + t^2.$$

Describe the graphs of the functions $f(x, y)$ for which families of level curves $f(x, y) = C$ are shown in the figures referred to in Exercises 29–32. Assume that each family corresponds to equally spaced values of $C$, and that the behaviour of the family is representative of all such families for the function.

**29.** See Figure 12.12(a).       **30.** See Figure 12.12(b).

**31.** See Figure 12.12(c).       **32.** See Figure 12.12(d).

**33.** Are the curves $y = (x - C)^2$ level curves of a function $f(x, y)$? What property must a family of curves in a region of the $xy$-plane have to be the family of level curves of a function defined in the region?

Use 3-dimensional computer graphing software to examine the graphs and the level curves of the functions in Exercises 43–48.

**43.** $\dfrac{1}{1 + x^2 + y^2}$    **44.** $\dfrac{\cos x}{1 + y^2}$

**45.** $\dfrac{y}{1 + x^2 + y^2}$    **46.** $\dfrac{x}{(x^2 - 1)^2 + y^2}$

**47.** $xy$    **48.** $\dfrac{1}{xy}$

## 12.2    Limits and Continuity

Before reading this section you should review the concepts of neighbourhood, open and closed sets, boundary and interior points introduced in Section 10.1.

The concept of the limit of a function of several variables is similar to that for functions of one variable. For clarity we present the definition for functions of two variables only; the general case is similar.

We say that $f(x, y)$ approaches the limit $L$ as the point $(x, y)$ approaches the point $(a, b)$, and we write

$$\lim_{(x,y)\to(a,b)} f(x, y) = L,$$

if all points of a neighbourhood of $(a, b)$, except possibly the point $(a, b)$ itself, belong to the domain of $f$, and if $f(x, y)$ approaches $L$ as $(x, y)$ approaches $(a, b)$. We make this more formal in the following definition.

**DEFINITION** ❷

> We say that $\lim_{(x,y)\to(a,b)} f(x, y) = L$ if and only if for every positive number $\epsilon$ there exists a positive number $\delta = \delta(\epsilon)$, depending on $\epsilon$, such that
>
> $$|f(x, y) - L| < \epsilon \quad \text{whenever} \quad 0 < \sqrt{(x-a)^2 + (y-b)^2} < \delta.$$

If a limit exists it is unique. For a single-variable function $f$, the existence of $\lim_{x\to a} f(x)$ implies that $f(x)$ approaches the same finite number as $x$ approaches $a$ from either the right or the left. Similarly, for a function of two variables, $\lim_{(x,y)\to(a,b)} f(x, y) = L$ exists only if $f(x, y)$ approaches the same number $L$ no matter how $(x, y)$ approaches $(a, b)$ in the $xy$-plane. In particular, $(x, y)$ can approach $(a, b)$ along any curve through $(a, b)$. It is not necessary that $L = f(a, b)$ even if $f(a, b)$ is defined. The examples below illustrate these assertions.

All the usual laws of limits extend to functions of several variables in the obvious way. For example, if $\lim_{(x,y)\to(a,b)} f(x, y) = L$ and $\lim_{(x,y)\to(a,b)} g(x, y) = M$, then

$$\lim_{(x,y)\to(a,b)} \big(f(x, y) \pm g(x, y)\big) = L \pm M,$$

$$\lim_{(x,y)\to(a,b)} f(x, y)\, g(x, y) = LM,$$

$$\lim_{(x,y)\to(a,b)} \frac{f(x, y)}{g(x, y)} = \frac{L}{M}, \qquad \text{provided } M \neq 0.$$

Also, if $F(t)$ is continuous at $t = L$, then

$$\lim_{(x,y)\to(a,b)} F\big(f(x, y)\big) = F(L).$$

**Example 1**

(a) $\quad \lim_{(x,y)\to(2,3)} 2x - y^2 = 4 - 9 = -5,$

(b) $\quad \lim_{(x,y)\to(a,b)} x^2 y = a^2 b,$

(c) $\quad \lim_{(x,y)\to(\pi/3,2)} y \sin(x/y) = 2 \sin(\pi/6) = 1.$

The following examples show that the requirement that $f(x, y)$ approach the same limit *no matter how* $(x, y)$ approaches $(a, b)$ can be very restrictive.

**Example 2** Investigate the limiting behaviour of $f(x, y) = \dfrac{2xy}{x^2 + y^2}$ as $(x, y)$ approaches $(0, 0)$.

**Solution** Note that $f(x, y)$ is defined at all points of the $xy$-plane except the origin $(0, 0)$. We can still ask whether $\lim_{(x,y)\to(0,0)} f(x, y)$ exists or not. If we let $(x, y)$ approach $(0, 0)$ along the $x$-axis ($y = 0$), then $f(x, y) = f(x, 0) \to 0$ (because $f(x, 0) = 0$ identically). Thus $\lim_{(x,y)\to(0,0)} f(x, y)$ must be 0 if it exists at all. Similarly, at all points of the $y$-axis we have $f(x, y) = f(0, y) = 0$.

However, at points of the line $x = y$, $f$ has a different constant value; $f(x, x) = 1$. Since the limit of $f(x, y)$ is 1 as $(x, y)$ approaches $(0, 0)$ along this line, it follows that $f(x, y)$ cannot have a unique limit at the origin. That is,

$$\lim_{(x,y)\to(0,0)} \frac{2xy}{x^2 + y^2} \qquad \text{does not exist.}$$

Observe that $f(x, y)$ has a constant value on any ray from the origin (on the ray $y = kx$ the value is $2k/(1 + k^2)$), but these values differ on different rays. The level curves of $f$ are the rays from the origin (with the origin itself removed). It is difficult to sketch the graph of $f$ near the origin. The first-octant part of the graph is the "hood-shaped" surface in Figure 12.13(a).

**Figure 12.13**

(a) $f(x, y)$ has different limits as $(x, y) \to (0, 0)$ along different straight lines

(b) $f(x, y)$ has the same limit 0 as $(x, y) \to (0, 0)$ along any straight line, but has limit 1 as $(x, y) \to (0, 0)$ along $y = x^2$

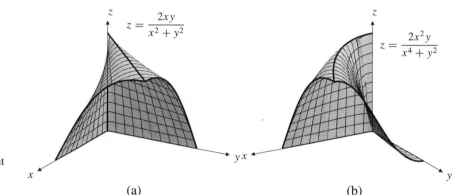

(a)

(b)

$$z = \frac{2xy}{x^2 + y^2}$$

$$z = \frac{2x^2 y}{x^4 + y^2}$$

---

**Example 3**   Investigate the limiting behaviour of $f(x, y) = \dfrac{2x^2 y}{x^4 + y^2}$ as $(x, y)$ approaches $(0, 0)$.

**Solution**   As in Example 2, $f(x, y)$ vanishes identically on the coordinate axes so $\lim_{(x,y)\to(0,0)} f(x, y)$ must be 0 if it exists at all. If we examine $f(x, y)$ at points of the ray $y = kx$, we obtain

$$f(x, kx) = \frac{2kx^3}{x^4 + k^2 x^2} = \frac{2kx}{x^2 + k^2} \to 0, \qquad \text{as} \quad x \to 0 \quad (k \neq 0).$$

Thus $f(x, y) \to 0$ as $(x, y) \to (0, 0)$ along *any* straight line through the origin. We might be tempted to conclude, therefore, that $\lim_{(x,y)\to(0,0)} f(x, y) = 0$, but this is incorrect. Observe the behaviour of $f(x, y)$ along the curve $y = x^2$:

$$f(x, x^2) = \frac{2x^4}{x^4 + x^4} = 1.$$

Thus $f(x, y)$ does not approach 0 as $(x, y)$ approaches the origin along this curve, and so $\lim_{(x,y)\to(0,0)} f(x, y)$ does not exist. The level curves of $f$ are pairs of parabolas of the form $y = kx^2$, $y = x^2/k$ with the origin removed. See Figure 12.13(b) for the first octant part of the graph of $f$.

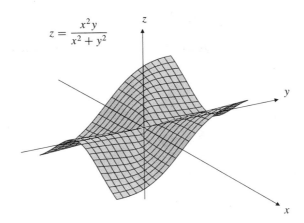
$$z = \frac{x^2 y}{x^2 + y^2}$$

**Figure 12.14**

$$\lim_{(x,y)\to(0,0)} \frac{x^2 y}{x^2 + y^2} = 0$$

**Example 4** Show that the function $f(x, y) = \dfrac{x^2 y}{x^2 + y^2}$ does have a limit at the origin; specifically,

$$\lim_{(x,y)\to(0,0)} \frac{x^2 y}{x^2 + y^2} = 0.$$

**Solution** This function is also defined everywhere except at the origin. Observe that since $x^2 \le x^2 + y^2$, we have

$$|f(x, y) - 0| = \left| \frac{x^2 y}{x^2 + y^2} \right| \le |y| \le \sqrt{x^2 + y^2},$$

which approaches zero as $(x, y) \to (0, 0)$. (See Figure 12.14.) Formally, if $\epsilon > 0$ is given and we take $\delta = \epsilon$, then $|f(x, y) - 0| < \epsilon$ whenever $0 < \sqrt{x^2 + y^2} < \delta$, so $f(x, y)$ has limit 0 as $(x, y) \to (0, 0)$ by Definition 2. ∎

As for functions of one variable, continuity at an interior point of the domain is defined directly in terms of the limit.

**DEFINITION 3**

The function $f(x, y)$ is continuous at the point $(a, b)$ if

$$\lim_{(x,y)\to(a,b)} f(x, y) = f(a, b).$$

It remains true that sums, differences, products, quotients, and compositions of continuous functions are continuous. The functions of Examples 2 and 3 above are continuous wherever they are defined, that is, at all points except the origin. There is no way to define $f(0, 0)$ so that these functions become continuous at the origin. They show that the continuity of the single-variable functions $f(x, b)$ at $x = a$ and $f(a, y)$ at $y = b$ does *not* imply that $f(x, y)$ is continuous at $(a, b)$. In fact, even if $f(x, y)$ is continuous along every straight line through $(a, b)$ it still need not be continuous at $(a, b)$. (See Exercises 16–17 below.) Note, however, that the function $f(x, y)$ of Example 4, though not defined at the origin, has a continuous extension to that point. If we extend the domain of $f$ by defining

$f(0, 0) = \lim_{(x,y)\to(0,0)} f(x, y) = 0$, then $f$ is continuous on the whole $xy$-plane.

As for functions of one variable, the existence of a limit of a function at a point does not imply that the function is continuous at that point. The function

$$f(x, y) = \begin{cases} 0 & \text{if } (x, y) \neq (0, 0) \\ 1 & \text{if } (x, y) = (0, 0) \end{cases}$$

satisfies $\lim_{(x,y)\to(0,0)} f(x, y) = 0$, which is not equal to $f(0, 0)$, so $f$ is not continuous at $(0, 0)$. Of course we can *make* $f$ continuous at $(0, 0)$ by redefining its value at that point to be 0.

Definition 2 requires that $f(x, y)$ be defined at all points of a neighbourhood of $(a, b)$ (except the point $(a, b)$ itself) in order that $\lim_{(x,y)\to(a,b)} f(x, y)$ can exist. This means that continuity is defined by Definition 3 only for *interior* points of the domain of $f$. We can modify the definition of limit to allow for limits at points on the *boundary* of the domain of $f$ in a manner somewhat similar to that used for *one-sided limits* of functions of one variable. Definition 4 below is thus an extension of Definition 2. With this extended definition of limit, Definition 3 can be used to determine the continuity of a function at a point on the boundary of its domain.

---

**DEFINITION 4**

**Extended definition of limit**

We say that $\lim_{(x,y)\to(a,b)} f(x, y) = L$, provided that

(i) every neighbourhood of $(a, b)$ contains points of the domain of $f$ different from $(a, b)$, and

(ii) for every positive number $\epsilon$ there exists a positive number $\delta = \delta(\epsilon)$ such that $|f(x, y) - L| < \epsilon$ holds whenever $(x, y)$ is in the domain of $f$ and satisfies $0 < \sqrt{(x - a)^2 + (y - b)^2} < \delta$.

---

Condition (i) is included in Definition 4 because it is not appropriate to consider limits at *isolated* points of the domain of $f$, that is, points with neighbourhoods that contain no other points of the domain.

**Example 5** The function $f(x, y) = \sqrt{1 - x^2 - y^2}$ is continuous at all points of its domain, the *closed* disk $x^2 + y^2 \leq 1$. Of course $(x, y)$ can approach points of the bounding circle $x^2 + y^2 = 1$ only from within the disk. ∎

---

## Exercises 12.2

In Exercises 1–12, evaluate the indicated limit or explain why it does not exist.

**1.** $\lim_{(x,y)\to(2,-1)} xy + x^2$

**2.** $\lim_{(x,y)\to(0,0)} \sqrt{x^2 + y^2}$

**3.** $\lim_{(x,y)\to(0,0)} \dfrac{x^2 + y^2}{y}$

**4.** $\lim_{(x,y)\to(0,0)} \dfrac{x}{x^2 + y^2}$

**5.** $\lim_{(x,y)\to(1,\pi)} \dfrac{\cos(xy)}{1 - x - \cos y}$

**6.** $\lim_{(x,y)\to(0,1)} \dfrac{x^2(y - 1)^2}{x^2 + (y - 1)^2}$

**7.** $\lim_{(x,y)\to(0,0)} \dfrac{y^3}{x^2 + y^2}$

**8.** $\lim_{(x,y)\to(0,0)} \dfrac{\sin(x - y)}{\cos(x + y)}$

**9.** $\lim_{(x,y)\to(0,0)} \dfrac{\sin(xy)}{x^2 + y^2}$

**10.** $\lim_{(x,y)\to(1,2)} \dfrac{2x^2 - xy}{4x^2 - y^2}$

**11.** $\lim_{(x,y)\to(0,0)} \dfrac{x^2 y^2}{x^2 + y^4}$

**12.** $\lim_{(x,y)\to(0,0)} \dfrac{x^2 y^2}{2x^4 + y^4}$

13. How can the function

$$f(x, y) = \frac{x^2 + y^2 - x^3 y^3}{x^2 + y^2}, \qquad (x, y) \neq (0, 0),$$

be defined at the origin so that it becomes continuous at all points of the $xy$-plane?

14. How can the function

$$f(x, y) = \frac{x^3 - y^3}{x - y}, \qquad (x \neq y),$$

be defined along the line $x = y$ so that the resulting function is continuous on the whole $xy$-plane?

15. What is the domain of

$$f(x, y) = \frac{x - y}{x^2 - y^2}?$$

Does $f(x, y)$ have a limit by Definition 2 as $(x, y) \to (1, 1)$? By Definition 4? Can the domain of $f$ be extended so that the resulting function is continuous at $(1, 1)$? Can the domain be extended so that the resulting function is continuous everywhere in the $xy$-plane?

* 16. Given a function $f(x, y)$ and a point $(a, b)$ in its domain, define single-variable functions $g$ and $h$ as follows:

$$g(x) = f(x, b), \qquad h(y) = f(a, y).$$

If $g$ is continuous at $x = a$ and $h$ is continuous at $y = b$ does it follow that $f$ is continuous at $(a, b)$? Conversely, does the continuity of $f$ at $(a, b)$ guarantee the continuity of $g$ at $a$ and the continuity of $h$ at $b$? Justify your answers.

* 17. Let $\mathbf{u} = u\mathbf{i} + v\mathbf{j}$ be a unit vector, and let

$$f_{\mathbf{u}}(t) = f(a + tu, b + tv)$$

be the single-variable function obtained by restricting the domain of $f(x, y)$ to points of the straight line through $(a, b)$ parallel to $\mathbf{u}$. If $f_{\mathbf{u}}(t)$ is continuous at $t = 0$ for every unit vector $\mathbf{u}$, does it follow that $f$ is continuous at $(a, b)$? Conversely, does the continuity of $f$ at $(a, b)$ guarantee the continuity of $f_{\mathbf{u}}(t)$ at $t = 0$? Justify your answers.

* 18. What condition must the nonnegative integers $m, n$, and $p$ satisfy to guarantee that $\lim_{(x,y) \to (0,0)} x^m y^n / (x^2 + y^2)^p$ exists? Prove your answer.

* 19. What condition must the constants $a, b$, and $c$ satisfy to guarantee that $\lim_{(x,y) \to (0,0)} xy / (ax^2 + bxy + cy^2)$ exists in the sense of Definition 4? Prove your answer.

* 20. Can the function $f(x, y) = \dfrac{\sin x \sin^3 y}{1 - \cos(x^2 + y^2)}$ be defined at $(0, 0)$ in such a way that it becomes continuous there? If so, how?

21. Use 2- and 3-dimensional mathematical graphing software to examine the graph and level curves of the function $f(x, y)$ of Example 2 on the region $-1 \leq x \leq 1$, $-1 \leq y \leq 1$, $(x, y) \neq (0, 0)$ How would you describe the behaviour of the graph near $(x, y) = (0, 0)$?

22. Use 2- and 3-dimensional mathematical graphing software to examine the graph and level curves of the function $f(x, y)$ of Example 3 on the region $-1 \leq x \leq 1$, $-1 \leq y \leq 1$, $(x, y) \neq (0, 0)$ How would you describe the behaviour of the graph near $(x, y) = (0, 0)$?

23. The graph of a single-variable function $f(x)$ that is continuous on an interval is a curve that has no *breaks* in it there and that intersects any vertical line through a point in the interval exactly once. What analogous statement can you make about the graph of a bivariate function $f(x, y)$ that is continuous on a region of the $xy$-plane?

## 12.3 Partial Derivatives

In this section we begin the process of extending the concepts and techniques of single-variable calculus to functions of more than one variable. It is convenient to begin by considering the rate of change of such functions with respect to one variable at a time. Thus a function of $n$ variables has $n$ *first-order partial derivatives,* one with respect to each of its independent variables. For a function of two variables we make this precise in the following definition.

**DEFINITION 5**

The **first partial derivatives** of the function $f(x, y)$ **with respect to the variables** $x$ and $y$ are the functions $f_1(x, y)$ and $f_2(x, y)$ given by

$$f_1(x, y) = \lim_{h \to 0} \frac{f(x + h, y) - f(x, y)}{h}$$

$$f_2(x, y) = \lim_{k \to 0} \frac{f(x, y + k) - f(x, y)}{k}$$

provided these limits exist.

Each of the two partial derivatives is the limit of a Newton quotient in one of the variables. Observe that $f_1(x, y)$ is just the ordinary first derivative of $f(x, y)$ considered as a function of $x$ only, regarding $y$ as a constant parameter. Similarly, $f_2(x, y)$ is the first derivative of $f(x, y)$ considered as a function of $y$ alone, with $x$ held fixed.

**Example 1**  If $f(x, y) = x^2 \sin y$, then

$$f_1(x, y) = 2x \sin y \qquad \text{and} \qquad f_2(x, y) = x^2 \cos y.$$

The subscripts "1" and "2" in the notations for the partial derivatives refer to the "first" and "second" variables of $f$. For functions of one variable we use the notation $f'$ for the derivative; the *prime* ($'$) denotes differentiation with respect to the only variable on which $f$ depends. For functions $f$ of two variables we use $f_1$ or $f_2$ to show the variable of differentiation. Do not confuse these subscripts with subscripts used for other purposes, for example to denote the components of vectors.

The partial derivative $f_1(a, b)$ measures the rate of change of $f(x, y)$ with respect to $x$ at $x = a$ while $y$ is held fixed at $b$. In graphical terms, the surface $z = f(x, y)$ intersects the vertical plane $y = b$ in a curve. If we take horizontal and vertical lines through the point $(0, b, 0)$ as coordinate axes in the plane $y = b$, then the curve has equation $z = f(x, b)$, and its slope at $x = a$ is $f_1(a, b)$. (See Figure 12.15.) Similarly, $f_2(a, b)$ represents the rate of change of $f$ with respect to $y$ at $y = b$ with $x$ held fixed at $a$. The surface $z = f(x, y)$ intersects the vertical plane $x = a$ in a curve $z = f(a, y)$ whose slope at $y = b$ is $f_2(a, b)$. (See Figure 12.16.)

Various notations can be used to denote the partial derivatives of $z = f(x, y)$ considered as functions of $x$ and $y$:

**Notations for first partial derivatives**

$$\frac{\partial z}{\partial x} = \frac{\partial}{\partial x} f(x, y) = f_1(x, y) = D_1 f(x, y)$$

$$\frac{\partial z}{\partial y} = \frac{\partial}{\partial y} f(x, y) = f_2(x, y) = D_2 f(x, y)$$

The symbol "$\partial$" should be read as "partial" so "$\partial z / \partial x$" is "partial $z$ with respect to $x$." The reason for distinguishing "$\partial$" from the "$d$" of ordinary derivatives of single-variable functions will be made clear later.

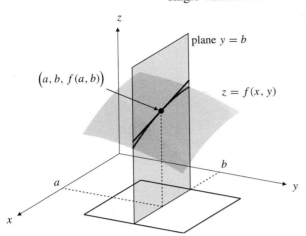

**Figure 12.15**    $f_1(a, b)$ is the slope of the curve of intersection of $z = f(x, y)$ and the vertical plane $y = b$ at $x = a$

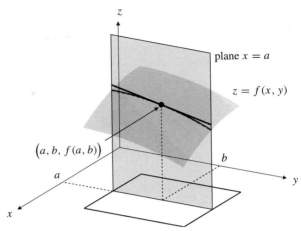

**Figure 12.16**    $f_2(a, b)$ is the slope of the curve of intersection of $z = f(x, y)$ and the vertical plane $x = a$ at $y = b$

Values of partial derivatives at a particular point $(a, b)$ are denoted similarly:

**Values of partial derivatives**

$$\left.\frac{\partial z}{\partial x}\right|_{(a,b)} = \left.\left(\frac{\partial}{\partial x} f(x, y)\right)\right|_{(a,b)} = f_1(a, b) = D_1 f(a, b)$$

$$\left.\frac{\partial z}{\partial y}\right|_{(a,b)} = \left.\left(\frac{\partial}{\partial y} f(x, y)\right)\right|_{(a,b)} = f_2(a, b) = D_2 f(a, b)$$

Some authors prefer to use $f_x$ or $\partial f/\partial x$ and $f_y$ or $\partial f/\partial y$ instead of $f_1$ and $f_2$. However, this can lead to problems of ambiguity when compositions of functions arise. For instance, suppose $f(x, y) = x^2 y$. By $f_1(x^2, xy)$ we clearly mean

$$\left.\left(\frac{\partial}{\partial x} f(x, y)\right)\right|_{(x^2, xy)} = \left. 2xy \right|_{(x^2, xy)} = (2)(x^2)(xy) = 2x^3 y.$$

But does $f_x(x^2, xy)$ mean the same thing? One could argue that $f_x(x^2, xy)$ should mean

$$\frac{\partial}{\partial x}\left(f(x^2, xy)\right) = \frac{\partial}{\partial x}\left((x^2)^2(xy)\right) = \frac{\partial}{\partial x}(x^5 y) = 5x^4 y.$$

In order to avoid such ambiguities we usually prefer to use $f_1$ and $f_2$ instead of $f_x$ and $f_y$. (However, in some situations where no confusion is likely to occur we may still use the notations $f_x$ and $f_y$, and also $\partial f/\partial x$ and $\partial f/\partial y$.)

All the standard differentiation rules for sums, products, reciprocals, and quotients continue to apply to partial derivatives.

**Example 2**    Find $\partial z/\partial x$ and $\partial z/\partial y$ if $z = x^3 y^2 + x^4 y + y^4$.

**Solution**    $\partial z/\partial x = 3x^2 y^2 + 4x^3 y$   and   $\partial z/\partial y = 2x^3 y + x^4 + 4y^3$.

**Example 3**    Find $f_1(0, \pi)$ if $f(x, y) = e^{xy} \cos(x + y)$.

**Solution**

$$f_1(x, y) = y \, e^{xy} \cos(x + y) - e^{xy} \sin(x + y),$$
$$f_1(0, \pi) = \pi \, e^0 \cos(\pi) - e^0 \sin(\pi) = -\pi.$$

The single-variable version of the Chain Rule also continues to apply to, say, $f\big(g(x, y)\big)$:

$$\frac{\partial}{\partial x} f\big(g(x, y)\big) = f'\big(g(x, y)\big) g_1(x, y), \quad \frac{\partial}{\partial y} f\big(g(x, y)\big) = f'\big(g(x, y)\big) g_2(x, y).$$

We will develop versions of the Chain Rule for more complicated compositions of multivariate functions in Section 12.5.

**Example 4**    If $f$ is an everywhere differentiable function of one variable show that $z = f(x/y)$ satisfies the *partial differential equation*

$$x \frac{\partial z}{\partial x} + y \frac{\partial z}{\partial y} = 0.$$

**Solution**    By the (single-variable) Chain Rule,

$$\frac{\partial z}{\partial x} = f'\left(\frac{x}{y}\right)\left(\frac{1}{y}\right) \quad \text{and} \quad \frac{\partial z}{\partial y} = f'\left(\frac{x}{y}\right)\left(\frac{-x}{y^2}\right).$$

Hence

$$x \frac{\partial z}{\partial x} + y \frac{\partial z}{\partial y} = f'\left(\frac{x}{y}\right)\left(x \times \frac{1}{y} + y \times \frac{-x}{y^2}\right) = 0.$$

Definition 5 can be extended in the obvious way to cover functions of more than two variables. If $f$ is a function of $n$ variables $x_1, x_2, \ldots, x_n$ then $f$ has $n$ first partial derivatives, $f_1(x_1, x_2, \ldots, x_n)$, $f_2(x_1, x_2, \ldots, x_n)$, $\ldots$, $f_n(x_1, x_2, \ldots, x_n)$, one with respect to each variable.

**Example 5**

$$\frac{\partial}{\partial z}\left(\frac{2xy}{1 + xz + yz}\right) = -\frac{2xy}{(1 + xz + yz)^2}(x + y).$$

Again all the standard differentiation rules are applied to calculate partial derivatives.

**Remark**    If a single-variable function $f(x)$ has a derivative $f'(a)$ at $x = a$, then $f$ is necessarily continuous at $x = a$. This property does *not* extend to partial derivatives. Even if all the first partial derivatives of a function of several variables exist at a point, the function may still fail to be continuous at that point. See Exercise 36 below.

## Tangent Planes and Normal Lines

If the graph $z = f(x, y)$ is a "smooth" surface near the point $P$ with coordinates $\big(a, b, f(a, b)\big)$, then that graph will have a **tangent plane** and a **normal line** at $P$. The normal line is the line through $P$ that is perpendicular to the surface; for instance, a line joining a point on a sphere to the centre of the sphere is normal to the sphere. Any nonzero vector that is parallel to the normal line at $P$ is called a normal vector to the surface at $P$. The tangent plane to the surface $z = f(x, y)$ at $P$ is the plane through $P$ that is perpendicular to the normal line at $P$.

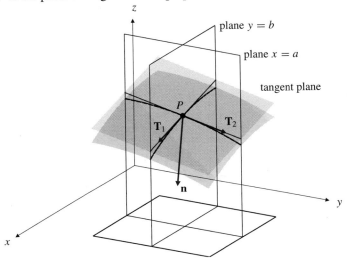

**Figure 12.17**   The tangent plane and a normal vector to $z = f(x, y)$ at $P = \big(a, b, f(a, b)\big)$

Let us assume that the surface $z = f(x, y)$ has a *nonvertical* tangent plane (and therefore a *nonhorizontal* normal line) at point $P$. (Later in this chapter we will state precise conditions that guarantee that the graph of a function has a nonvertical tangent plane at a point.) The tangent plane intersects the vertical plane $y = b$ in a straight line that is tangent at $P$ to the curve of intersection of the surface $z = f(x, y)$ and the plane $y = b$. (See Figures 12.15 and 12.17.) This line has slope $f_1(a, b)$, and so is parallel to the vector $\mathbf{T}_1 = \mathbf{i} + f_1(a, b)\mathbf{k}$. Similarly, the tangent plane intersects the vertical plane $x = a$ in a straight line having slope $f_2(a, b)$. This line is therefore parallel to the vector $\mathbf{T}_2 = \mathbf{j} + f_2(a, b)\mathbf{k}$. It follows that the tangent plane, and therefore the surface $z = f(x, y)$ itself, has normal vector

$$\mathbf{n} = \mathbf{T}_2 \times \mathbf{T}_1 = \begin{vmatrix} \mathbf{i} & \mathbf{j} & \mathbf{k} \\ 0 & 1 & f_2(a, b) \\ 1 & 0 & f_1(a, b) \end{vmatrix} = f_1(a, b)\mathbf{i} + f_2(a, b)\mathbf{j} - \mathbf{k}.$$

**Normal vector to $z = f(x, y)$ at $\big(a, b, f(a, b)\big)$**

$$\mathbf{n} = f_1(a, b)\mathbf{i} + f_2(a, b)\mathbf{j} - \mathbf{k}$$

Since the tangent plane passes through $P = (a, b, f(a, b))$, it has equation

$$f_1(a, b)(x - a) + f_2(a, b)(y - b) - (z - f(a, b)) = 0,$$

or, equivalently,

**Equation of the tangent plane to $z = f(x, y)$ at $\big(a, b, f(a, b)\big)$**

$$z = f(a, b) + f_1(a, b)(x - a) + f_2(a, b)(y - b).$$

We shall obtain this result by a different method in Section 12.7.

The normal line to $z = f(x, y)$ at $(a, b, f(a, b))$ has direction vector $f_1(a, b)\mathbf{i} + f_2(a, b)\mathbf{j} - \mathbf{k}$, and so has equations

$$\frac{x - a}{f_1(a, b)} = \frac{y - b}{f_2(a, b)} = \frac{z - f(a, b)}{-1}$$

with suitable modifications if either $f_1(a, b) = 0$ or $f_2(a, b) = 0$.

**Example 6**    Find a normal vector and equations of the tangent plane and normal line to the graph $z = \sin(xy)$ at the point where $x = \pi/3$ and $y = -1$.

**Solution**    The point on the graph has coordinates $(\pi/3, -1, -\sqrt{3}/2)$ Now

$$\frac{\partial z}{\partial x} = y \, \cos(xy) \qquad \text{and} \qquad \frac{\partial z}{\partial y} = x \, \cos(xy).$$

At $(\pi/3, -1)$ we have $\partial z/\partial x = -1/2$ and $\partial z/\partial y = \pi/6$. Therefore the surface has normal vector $\mathbf{n} = -(1/2)\mathbf{i} + (\pi/6)\mathbf{j} - \mathbf{k}$, and tangent plane

$$z = \frac{-\sqrt{3}}{2} - \frac{1}{2}\left(x - \frac{\pi}{3}\right) + \frac{\pi}{6}(y + 1),$$

or, more simply, $3x - \pi y + 6z = 2\pi - 3\sqrt{3}$. The normal line has equation

$$\frac{x - \dfrac{\pi}{3}}{-1} = \frac{y + 1}{\dfrac{\pi}{6}} = \frac{z + \dfrac{\sqrt{3}}{2}}{-1} \qquad \text{or} \qquad \frac{6x - 2\pi}{-3} = \frac{6y + 6}{\pi} = \frac{6z + 3\sqrt{3}}{-6}.$$

**Example 7**    What horizontal plane is tangent to the surface

$$z = x^2 - 4xy - 2y^2 + 12x - 12y - 1$$

and what is the point of tangency?

**Solution**    A plane is horizontal only if its equation is of the form $z = k$, that is, it is independent of $x$ and $y$. Therefore we must have $\partial z/\partial x = \partial z/\partial y = 0$ at the point of tangency. The equations

$$\frac{\partial z}{\partial x} = 2x - 4y + 12 = 0$$

$$\frac{\partial z}{\partial y} = -4x - 4y - 12 = 0$$

have solution $x = -4$, $y = 1$. For these values we have $z = -31$ so the required tangent plane has equation $z = -31$ and the point of tangency is $(-4, 1, -31)$.

## Distance from a Point to a Surface: A Geometric Example

**Example 8**  Find the distance from the point $(3, 0, 0)$ to the hyperbolic paraboloid with equation $z = x^2 - y^2$.

**Solution**  This is an optimization problem of a sort we will deal with in a more systematic way in the next chapter. However, such problems involving minimizing distances from points to surfaces can frequently be solved using geometric methods. If $Q = (X, Y, Z)$ is the point on the surface $z = x^2 - y^2$ that is closest to $P = (3, 0, 0)$ then the vector $\overrightarrow{PQ} = (X - 3)\mathbf{i} + Y\mathbf{j} + Z\mathbf{k}$ must be normal to the surface at $Q$. (See Figure 12.18(a).) Using the partial derivatives of $z = x^2 - y^2$ we know that the vector $\mathbf{n} = 2X\mathbf{i} - 2Y\mathbf{j} - \mathbf{k}$ is normal to the surface at $Q$. Thus $\overrightarrow{PQ}$ must be parallel to $\mathbf{n}$, and so $\overrightarrow{PQ} = t\mathbf{n}$ for some scalar $t$. Separated into components, this vector equation states that

$$X - 3 = 2Xt, \qquad Y = -2Yt, \qquad \text{and} \qquad Z = -t.$$

The middle equation implies that either $Y = 0$ or $t = -\frac{1}{2}$. We must consider both of these possibilities.

**CASE I**  If $Y = 0$, then

$$X = \frac{3}{1 - 2t} \qquad \text{and} \qquad Z = -t.$$

But $Z = X^2 - Y^2$, so we must have

$$-t = \frac{9}{(1 - 2t)^2}.$$

This is a cubic equation in $t$, so we might expect to have to solve it numerically, for instance by using Newton's method. However, if we try small integer values of $t$, we will quickly discover that $t = -1$ is a solution. The graphs of both sides of the equation are shown in Figure 12.18(b). They show that $t = -1$ is the only real solution. Calculating the corresponding values of $X$ and $Z$, we obtain $(1, 0, 1)$ as a candidate for $Q$. The distance from this point to $P$ is $\sqrt{5}$.

**Figure 12.18**

(a) If $Q$ is the point on $z = x^2 - y^2$ closest to $P$ then $\overrightarrow{PQ}$ is normal to the surface

(b) Equation $-t = \dfrac{9}{(1 - 2t)^2}$ has only one real root, $t = -1$

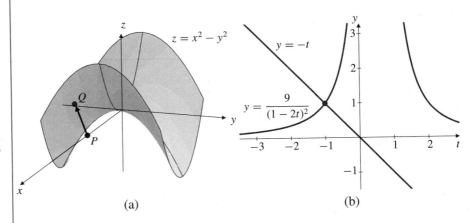

(a)

(b)

**CASE II** If $t = -1/2$, then $X = 3/2$, $Z = 1/2$, and $Y = \pm\sqrt{X^2 - Z} = \pm\sqrt{7}/2$, and the distance from these points to $P$ is $\sqrt{17}/2$.

Since $\frac{17}{4} < 5$, the points $(3/2, \pm\sqrt{7}/2, 1/2)$ are the points on $z = x^2 - y^2$ closest to $(3, 0, 0)$, and the distance from $(3,0,0)$ to the surface is $\sqrt{17}/2$ units. ∎

## Exercises 12.3

In Exercises 1–10, find all the first partial derivatives of the functions specified and evaluate them at the given point.

**1.** $f(x, y) = x - y + 2$, $\quad (3, 2)$

**2.** $f(x, y) = xy + x^2$, $\quad (2, 0)$

**3.** $f(x, y, z) = x^3 y^4 z^5$, $\quad (0, -1, -1)$

**4.** $g(x, y, z) = \dfrac{xz}{y + z}$, $\quad (1, 1, 1)$

**5.** $z = \tan^{-1}\left(\dfrac{y}{x}\right)$, $\quad (-1, 1)$

**6.** $w = \ln(1 + e^{xyz})$, $\quad (2, 0, -1)$

**7.** $f(x, y) = \sin(x\sqrt{y})$, $\quad \left(\dfrac{\pi}{3}, 4\right)$

**8.** $f(x, y) = \dfrac{1}{\sqrt{x^2 + y^2}}$, $\quad (-3, 4)$

**9.** $w = x^{(y \ln z)}$, $\quad (e, 2, e)$

**10.** $g(x_1, x_2, x_3, x_4) = \dfrac{x_1 - x_2^2}{x_3 + x_4^2}$, $\quad (3, 1, -1, -2)$

In Exercises 11–12, calculate the first partial derivatives of the given functions at $(0, 0)$. You will have to use Definition 5.

**11.** $f(x, y) = \begin{cases} \dfrac{2x^3 - y^3}{x^2 + 3y^2} & \text{if } (x, y) \neq (0, 0) \\ 0 & \text{if } (x, y) = (0, 0). \end{cases}$

**12.** $f(x, y) = \begin{cases} \dfrac{x^2 - 2y^2}{x - y} & \text{if } x \neq y \\ 0 & \text{if } x = y. \end{cases}$

In Exercises 13–22, find equations of the tangent plane and normal line to the graph of the given function at the point with specified values of $x$ and $y$.

**13.** $f(x, y) = x^2 - y^2$ at $(-2, 1)$

**14.** $f(x, y) = \dfrac{x - y}{x + y}$ at $(1, 1)$

**15.** $f(x, y) = \cos(x/y)$ at $(\pi, 4)$

**16.** $f(x, y) = e^{xy}$ at $(2, 0)$

**17.** $f(x, y) = \dfrac{x}{x^2 + y^2}$ at $(1, 2)$

**18.** $f(x, y) = y e^{-x^2}$ at $(0, 1)$

**19.** $f(x, y) = \ln(x^2 + y^2)$ at $(1, -2)$

**20.** $f(x, y) = \dfrac{2xy}{x^2 + y^2}$ at $(0, 2)$

**21.** $f(x, y) = \tan^{-1}(y/x)$ at $(1, -1)$

**22.** $f(x, y) = \sqrt{1 + x^3 y^2}$ at $(2, 1)$

**23.** Find the coordinates of all points on the surface with equation $z = x^4 - 4xy^3 + 6y^2 - 2$ where the surface has a horizontal tangent plane.

**24.** Find all horizontal planes that are tangent to the surface with equation $z = xye^{-(x^2+y^2)/2}$. At what points are they tangent?

In Exercises 25–31, show that the given function satisfies the given partial differential equation.

**◆ 25.** $z = x e^y$, $\quad x\dfrac{\partial z}{\partial x} = \dfrac{\partial z}{\partial y}$

**◆ 26.** $z = \dfrac{x + y}{x - y}$, $\quad x\dfrac{\partial z}{\partial x} + y\dfrac{\partial z}{\partial y} = 0$

**◆ 27.** $z = \sqrt{x^2 + y^2}$, $\quad x\dfrac{\partial z}{\partial x} + y\dfrac{\partial z}{\partial y} = z$

**◆ 28.** $w = x^2 + yz$, $\quad x\dfrac{\partial w}{\partial x} + y\dfrac{\partial w}{\partial y} + z\dfrac{\partial w}{\partial z} = 2w$

**◆ 29.** $w = \dfrac{1}{x^2 + y^2 + z^2}$, $\quad x\dfrac{\partial w}{\partial x} + y\dfrac{\partial w}{\partial y} + z\dfrac{\partial w}{\partial z} = -2w$

**◆ 30.** $z = f(x^2 + y^2)$, where $f$ is any differentiable function of one variable,
$$y\dfrac{\partial z}{\partial x} - x\dfrac{\partial z}{\partial y} = 0.$$

**◆ 31.** $z = f(x^2 - y^2)$, where $f$ is any differentiable function of one variable,
$$y\dfrac{\partial z}{\partial x} + x\dfrac{\partial z}{\partial y} = 0.$$

**32.** Give a formal definition of the three first partial derivatives of the function $f(x, y, z)$.

**33.** What is an equation of the "tangent hyperplane" to the graph $w = f(x, y, z)$ at $\big(a, b, c, f(a, b, c)\big)$?

**∗ 34.** Find the distance from the point $(1, 1, 0)$ to the circular paraboloid with equation $z = x^2 + y^2$.

**∗ 35.** Find the distance from the point $(0, 0, 1)$ to the elliptic paraboloid having equation $z = x^2 + 2y^2$.

**∗ 36.** Let $f(x, y) = \begin{cases} \dfrac{2xy}{x^2 + y^2}, & \text{if } (x, y) \neq (0, 0) \\ 0, & \text{if } (x, y) = (0, 0). \end{cases}$

Note that $f$ is not continuous at $(0,0)$. (See Example 2 of Section 12.2.) Therefore its graph is not smooth there. Show, however, that $f_1(0, 0)$ and $f_2(0, 0)$ both exist. Hence the existence of partial derivatives does not imply that a function of several variables is continuous. This is in contrast to the single-variable case.

**37.** Determine $f_1(0, 0)$ and $f_2(0, 0)$ if they exist, where

$$f(x, y) = \begin{cases} (x^3 + y) \sin \dfrac{1}{x^2 + y^2} & \text{if } (x, y) \neq (0, 0) \\ 0 & \text{if } (x, y) = (0, 0). \end{cases}$$

**38.** Calculate $f_1(x, y)$ for the function in Exercise 37. Is $f_1(x, y)$ continuous at $(0, 0)$?

**\* 39.** Let $f(x, y) = \begin{cases} \dfrac{x^3 - y^3}{x^2 + y^2}, & \text{if } (x, y) \neq (0, 0) \\ 0, & \text{if } (x, y) = (0, 0). \end{cases}$

Calculate $f_1(x, y)$ and $f_2(x, y)$ at all points $(x, y)$ in the plane. Is $f$ continuous at $(0, 0)$? Are $f_1$ and $f_2$ continuous at $(0, 0)$?

**\* 40.** Let $f(x, y, z) = \begin{cases} \dfrac{xy^2 z}{x^4 + y^4 + z^4}, & \text{if } (x, y, z) \neq (0, 0, 0) \\ 0, & \text{if } (x, y, z) = (0, 0, 0). \end{cases}$

Find $f_1(0, 0, 0)$, $f_2(0, 0, 0)$, and $f_3(0, 0, 0)$. Is $f$ continuous at $(0, 0, 0)$? Are $f_1$, $f_2$, and $f_3$ continuous at $(0, 0, 0)$?

# 12.4 Higher-Order Derivatives

Partial derivatives of second and higher orders are calculated by taking partial derivatives of already calculated partial derivatives. The order in which the differentiations are performed is indicated in the notations used. If $z = f(x, y)$, we can calculate *four* partial derivatives of second order, namely

two **pure** second partial derivatives with respect to $x$ or $y$,

$$\frac{\partial^2 z}{\partial x^2} = \frac{\partial}{\partial x} \frac{\partial z}{\partial x} = f_{11}(x, y) = f_{xx}(x, y),$$

$$\frac{\partial^2 z}{\partial y^2} = \frac{\partial}{\partial y} \frac{\partial z}{\partial y} = f_{22}(x, y) = f_{yy}(x, y),$$

and two **mixed** second partial derivatives with respect to $x$ and $y$,

$$\frac{\partial^2 z}{\partial x \partial y} = \frac{\partial}{\partial x} \frac{\partial z}{\partial y} = f_{21}(x, y) = f_{yx}(x, y),$$

$$\frac{\partial^2 z}{\partial y \partial x} = \frac{\partial}{\partial y} \frac{\partial z}{\partial x} = f_{12}(x, y) = f_{xy}(x, y).$$

Again we remark that the notations $f_{11}$, $f_{12}$, $f_{21}$, and $f_{22}$ are usually preferable to $f_{xx}$, $f_{xy}$, $f_{yx}$, and $f_{yy}$, although the latter are often used in partial differential equations. Note that $f_{12}$ indicates differentiation of $f$ *first* with respect to its first variable and *then* with respect to its second variable; $f_{21}$ indicates the opposite order of differentiation. The subscript closest to $f$ indicates which differentiation occurs first.

Similarly, if $w = f(x, y, z)$ then

$$\frac{\partial^5 w}{\partial y \partial x \partial y^2 \partial z} = \frac{\partial}{\partial y} \frac{\partial}{\partial x} \frac{\partial}{\partial y} \frac{\partial}{\partial y} \frac{\partial w}{\partial z} = f_{32212}(x, y, z) = f_{zyyxy}(x, y, z).$$

**Example 1** Find the four second partial derivatives of $f(x, y) = x^3 y^4$.

**Solution**

$$f_1(x, y) = 3x^2 y^4, \qquad\qquad f_2(x, y) = 4x^3 y^3,$$

$$f_{11}(x, y) = \frac{\partial}{\partial x}(3x^2 y^4) = 6xy^4, \qquad f_{21}(x, y) = \frac{\partial}{\partial x}(4x^3 y^3) = 12x^2 y^3,$$

$$f_{12}(x, y) = \frac{\partial}{\partial y}(3x^2 y^4) = 12x^2 y^3, \qquad f_{22}(x, y) = \frac{\partial}{\partial y}(4x^3 y^3) = 12x^3 y^2.$$

**Example 2** Calculate $f_{223}(x, y, z)$, $f_{232}(x, y, z)$, and $f_{322}(x, y, z)$ for the function $f(x, y, z) = e^{x-2y+3z}$.

**Solution**
$$\begin{aligned}
f_{223}(x, y, z) &= \frac{\partial}{\partial z}\frac{\partial}{\partial y}\frac{\partial}{\partial y} e^{x-2y+3z} \\
&= \frac{\partial}{\partial z}\frac{\partial}{\partial y}\left(-2e^{x-2y+3z}\right) \\
&= \frac{\partial}{\partial z}\left(4e^{x-2y+3z}\right) = 12\, e^{x-2y+3z}.
\end{aligned}$$

$$\begin{aligned}
f_{232}(x, y, z) &= \frac{\partial}{\partial y}\frac{\partial}{\partial z}\frac{\partial}{\partial y} e^{x-2y+3z} \\
&= \frac{\partial}{\partial y}\frac{\partial}{\partial z}\left(-2e^{x-2y+3z}\right) \\
&= \frac{\partial}{\partial y}\left(-6e^{x-2y+3z}\right) = 12\, e^{x-2y+3z}.
\end{aligned}$$

$$\begin{aligned}
f_{322}(x, y, z) &= \frac{\partial}{\partial y}\frac{\partial}{\partial y}\frac{\partial}{\partial z} e^{x-2y+3z} \\
&= \frac{\partial}{\partial y}\frac{\partial}{\partial y}\left(3e^{x-2y+3z}\right) \\
&= \frac{\partial}{\partial y}\left(-6e^{x-2y+3z}\right) = 12\, e^{x-2y+3z}.
\end{aligned}$$

In both of the examples above observe that the mixed partial derivatives taken with respect to the same variables but in different orders turned out to be equal. This is not a coincidence. It will always occur for sufficiently smooth functions. In particular the mixed partial derivatives involved are required to be *continuous*. The following theorem presents a more precise statement of this important phenomenon.

**THEOREM** 1

**Equality of mixed partials**

Suppose that two mixed $n$th order partial derivatives of a function $f$ involve the same differentiations but in different orders. If those partials are continuous at a point $P$, and if $f$ and all partials of $f$ of order less than $n$ are continuous in a neighbourhood of $P$, then the two mixed partials are equal at the point $P$.

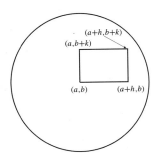

**Figure 12.19**   A rectangle contained in the disk where $f$ and certain partials are continuous

**PROOF**   We shall prove only a representative special case, showing the equality of $f_{12}(a, b)$ and $f_{21}(a, b)$ for a function $f$ of two variables, provided $f_{12}$ and $f_{21}$ are defined and $f_1$, $f_2$, and $f$ are continuous throughout a disk of positive radius centred at $(a, b)$, and $f_{12}$ and $f_{21}$ are continuous at $(a, b)$. Let $h$ and $k$ have sufficiently small absolute values that the point $(a + h, b + k)$ lies in this disk. Then so do all points of the rectangle with sides parallel to the coordinate axes and diagonally opposite corners at $(a, b)$ and $(a + h, b + k)$. (See Figure 12.19.)

Let $Q = f(a + h, b + k) - f(a + h, b) - f(a, b + k) + f(a, b)$ and define single-variable functions $u(x)$ and $v(y)$ by

$$u(x) = f(x, b + k) - f(x, b) \quad \text{and} \quad v(y) = f(a + h, y) - f(a, y).$$

Evidently $Q = u(a + h) - u(a)$ and also $Q = v(b + k) - v(b)$. By the (single-variable) Mean-Value Theorem, there exists a number $\theta_1$ satisfying $0 < \theta_1 < 1$ (so that $a + \theta_1 h$ lies between $a$ and $a + h$) such that

$$Q = u(a+h) - u(a) = h\, u'(a+\theta_1 h) = h\left[f_1(a+\theta_1 h, b+k) - f_1(a+\theta_1 h, b)\right].$$

Now we apply the Mean-Value Theorem again, this time to $f_1$ considered as a function of its second variable, and obtain another number $\theta_2$ satisfying $0 < \theta_2 < 1$ such that

$$f_1(a + \theta_1 h, b + k) - f_1(a + \theta_1 h, b) = k\, f_{12}(a + \theta_1 h, b + \theta_2 k).$$

Thus $Q = hk\, f_{12}(a + \theta_1 h, b + \theta_2 k)$. Two similar applications of the Mean-Value Theorem to $Q = v(b + k) - v(b)$ lead to $Q = hk\, f_{21}(a + \theta_3 h, b + \theta_4 k)$, where $\theta_3$ and $\theta_4$ are two numbers each between 0 and 1. Equating these two expressions for $Q$ and cancelling the common factor $hk$, we obtain

$$f_{12}(a + \theta_1 h, b + \theta_2 k) = f_{21}(a + \theta_3 h, b + \theta_4 k).$$

Since $f_{12}$ and $f_{21}$ are continuous at $(a, b)$, we can let $h$ and $k$ approach zero to obtain $f_{12}(a, b) = f_{21}(a, b)$, as required.

**⚠ DANGER**

The Mean-Value Theorem is used four times in this proof, each time to write a difference of the form $g(p + m) - g(p)$ in the form $g'(c)m$, where $c$ is some number between $p$ and $p + m$. It is convenient to write $c$ in the form $p + \theta m$, where $\theta$ is some number between 0 and 1.

Exercise 16 at the end of this section develops an example of a function for which $f_{12}$ and $f_{21}$ exist but are not continuous at $(0, 0)$, and for which $f_{12}(0, 0) \neq f_{21}(0, 0)$.

## The Laplace and Wave Equations

Many important and interesting phenomena are modelled by functions of several variables that satisfy certain *partial differential equations*. In the following examples we encounter two particular partial differential equations that arise frequently in mathematics and the physical sciences. Exercises 17–19 at the end of this section introduce another such equation with important applications.

**Example 3**   Show that for any real number $k$ the functions

$$z = e^{kx} \cos(ky) \qquad \text{and} \qquad z = e^{kx} \sin(ky)$$

satisfy the partial differential equation

$$\frac{\partial^2 z}{\partial x^2} + \frac{\partial^2 z}{\partial y^2} = 0$$

at every point in the $xy$-plane.

**Solution**   For $z = e^{kx} \cos(ky)$ we have

$$\frac{\partial z}{\partial x} = k\, e^{kx} \cos(ky), \qquad \frac{\partial z}{\partial y} = -k\, e^{kx} \sin(ky),$$

$$\frac{\partial^2 z}{\partial x^2} = k^2\, e^{kx} \cos(ky), \qquad \frac{\partial^2 z}{\partial y^2} = -k^2\, e^{kx} \cos(ky).$$

Thus

$$\frac{\partial^2 z}{\partial x^2} + \frac{\partial^2 z}{\partial y^2} = k^2\, e^{kx} \cos(ky) - k^2\, e^{kx} \cos(ky) = 0.$$

The calculation for $z = e^{kx} \sin(ky)$ is similar.

**Remark**   The partial differential equation in the above example is called the (two-dimensional) **Laplace equation**. A function of two variables having continuous second partial derivatives in a region of the plane is said to be **harmonic** there if it satisfies Laplace's equation. Such functions play a critical role in the theory of differentiable functions of a *complex variable,* and are used to model various physical quantities such as steady-state temperature distributions, fluid flows, and electric and magnetic potential fields. Harmonic functions have many interesting properties. They have derivatives of all orders, and are *analytic,* that is, they are the sums of their (multivariable) Taylor series. Moreover, a harmonic function can achieve maximum and minimum values only on the boundary of its domain. Laplace's equation, and therefore harmonic functions, can be considered in any number of dimensions. (See Exercises 13 and 14 below.)

**Example 4**   If $f$ and $g$ are any twice differentiable functions of one variable, show that

$$w = f(x - ct) + g(x + ct)$$

satisfies the partial differential equation

$$\frac{\partial^2 w}{\partial t^2} = c^2 \frac{\partial^2 w}{\partial x^2}.$$

**Solution**  Using the Chain Rule for functions of one variable we obtain

$$\frac{\partial w}{\partial t} = -c\, f'(x - ct) + c\, g'(x + ct), \qquad \frac{\partial w}{\partial x} = f'(x - ct) + g'(x + ct),$$

$$\frac{\partial^2 w}{\partial t^2} = c^2 f''(x - ct) + c^2 g''(x + ct), \qquad \frac{\partial^2 w}{\partial x^2} = f''(x - ct) + g''(x + ct).$$

Thus $w$ satisfies the given differential equation.  ∎

**Remark**  The partial differential equation in the above example is called the (one-dimensional) **wave equation**. If $t$ measures time, then $f(x - ct)$ represents a waveform travelling to the right along the $x$-axis with speed $c$. (See Figure 12.20.) Similarly, $g(x + ct)$ represents a waveform travelling to the left with speed $c$. Unlike the solutions of Laplace's equation that must be infinitely differentiable, solutions of the wave equation need only have enough derivatives to satisfy the differential equation. The functions $f$ and $g$ are otherwise arbitrary.

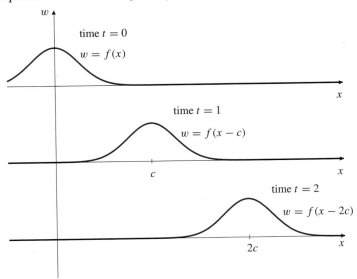

**Figure 12.20**   $w = f(x - ct)$ represents a waveform moving to the right with speed $c$

# Exercises 12.4

In Exercises 1–6, find all the second partial derivatives of the given function.

**1.** $z = x^2(1 + y^2)$

**2.** $f(x, y) = x^2 + y^2$

**3.** $w = x^3 y^3 z^3$

**4.** $z = \sqrt{3x^2 + y^2}$

**5.** $z = x\, e^y - y\, e^x$

**6.** $f(x, y) = \ln\big(1 + \sin(xy)\big)$

* **7.** How many mixed partial derivatives of order 3 can a function of three variables have? If they are all continuous, how many different values can they have at one point? Find the mixed partials of order 3 for $f(x, y, z) = x\, e^{xy} \cos(xz)$

that involve two differentiations with respect to $z$ and one with respect to $x$.

Show that the functions in Exercises 8–12 are harmonic in the plane regions indicated.

**8.** $f(x, y) = A(x^2 - y^2) + Bxy$ in the whole plane ($A$ and $B$ are constants.)

**9.** $f(x, y) = 3x^2 y - y^3$ in the whole plane (Can you think of another polynomial of degree 3 in $x$ and $y$ that is also harmonic?)

**10.** $f(x, y) = \dfrac{x}{x^2 + y^2}$ everywhere except at the origin

**11.** $f(x, y) = \ln(x^2 + y^2)$ everywhere except at the origin

**12.** $\tan^{-1}(y/x)$ except at points on the $y$-axis

**13.** Show that $w = e^{3x+4y} \sin(5z)$ is harmonic in all of $\mathbb{R}^3$, that is, it satisfies everywhere the 3-dimensional Laplace equation

$$\frac{\partial^2 w}{\partial x^2} + \frac{\partial^2 w}{\partial y^2} + \frac{\partial^2 w}{\partial z^2} = 0.$$

**14.** Assume that $f(x, y)$ is harmonic in the $xy$-plane. Show that each of the functions $z\, f(x, y)$, $x\, f(y, z)$, and $y\, f(z, x)$ is harmonic in the whole of $\mathbb{R}^3$. What condition should the constants $a$, $b$, and $c$ satisfy to ensure that $f(ax + by, cz)$ is harmonic in $\mathbb{R}^3$?

**15.** Suppose the functions $u(x, y)$ and $v(x, y)$ have continuous second partial derivatives and satisfy the **Cauchy–Riemann equations**

$$\frac{\partial u}{\partial x} = \frac{\partial v}{\partial y} \qquad \text{and} \qquad \frac{\partial v}{\partial x} = -\frac{\partial u}{\partial y}.$$

Show that $u$ and $v$ are both harmonic.

**\* 16.** Let $F(x, y) = \begin{cases} \dfrac{2xy(x^2 - y^2)}{x^2 + y^2} & \text{if } (x, y) \neq (0, 0) \\ 0 & \text{if } (x, y) = (0, 0) \end{cases}$

Calculate $F_1(x, y)$, $F_2(x, y)$, $F_{12}(x, y)$, and $F_{21}(x, y)$ at points $(x, y) \neq (0, 0)$. Also calculate these derivatives at $(0, 0)$. Observe that $F_{21}(0, 0) = 2$ and $F_{12}(0, 0) = -2$. Does this result contradict Theorem 1? Explain why.

**The heat (diffusion) equation**

**◈ 17.** Show that the function $u(x, t) = t^{-1/2} e^{-x^2/4t}$ satisfies the partial differential equation

$$\frac{\partial u}{\partial t} = \frac{\partial^2 u}{\partial x^2}.$$

This equation is called the **one-dimensional heat equation** as it models heat diffusion in an insulated rod (with $u(x, t)$ representing the temperature at position $x$ at time $t$) and other similar phenomena.

**◈ 18.** Show that the function $u(x, y, t) = t^{-1} e^{-(x^2+y^2)/4t}$ satisfies the **two-dimensional heat equation**

$$\frac{\partial u}{\partial t} = \frac{\partial^2 u}{\partial x^2} + \frac{\partial^2 u}{\partial y^2}.$$

**◈ 19.** By comparing the results of Exercises 17 and 18, guess a solution to the **three-dimensional heat equation**

$$\frac{\partial u}{\partial t} = \frac{\partial^2 u}{\partial x^2} + \frac{\partial^2 u}{\partial y^2} + \frac{\partial^2 u}{\partial z^2}.$$

Verify your guess.

**Biharmonic functions**

A function $u(x, y)$ with continuous partials of fourth order is **biharmonic** if $\dfrac{\partial^2 u}{\partial x^2} + \dfrac{\partial^2 u}{\partial y^2}$ is a harmonic function.

**◈ 20.** Show that $u(x, y)$ is biharmonic if and only if it satisfies the biharmonic equation

$$\frac{\partial^4 u}{\partial x^4} + 2\frac{\partial^4 u}{\partial x^2 \partial y^2} + \frac{\partial^4 u}{\partial y^4} = 0$$

**21.** Verify that $u(x, y) = x^4 - 3x^2 y^2$ is biharmonic.

**22.** Show that if $u(x, y)$ is harmonic then $v(x, y) = xu(x, y)$ and $w(x, y) = yu(x, y)$ are biharmonic.

Use the result of Exercise 22 to show that the functions in Exercises 23–25 are biharmonic.

**23.** $x\, e^x \sin y$

**24.** $y \ln(x^2 + y^2)$

**25.** $\dfrac{xy}{x^2 + y^2}$

**◈ 26.** Propose a definition of a biharmonic function of three variables, and prove results analogous to those of Exercises 20 and 22 for biharmonic functions $u(x, y, z)$.

## 12.5 The Chain Rule

The Chain Rule for functions of one variable is a formula that gives the derivative of a composition $f\big(g(x)\big)$ of two functions $f$ and $g$:

$$\frac{d}{dx} f\big(g(x)\big) = f'\big(g(x)\big)g'(x).$$

The situation for several variables is more complicated. If $f$ depends on more than one variable, and any of those variables can be functions of one or more other variables, we cannot expect a simple formula for partial derivatives of the composition to cover all possible cases. We must come to think of the Chain Rule as a *procedure for differentiating compositions* rather than as a formula for their derivatives. In order to motivate a formulation of the Chain Rule for functions of two variables we begin with a concrete example.

**Example 1** Suppose you are hiking in a mountainous region for which you have a map. Let $(x, y)$ be the coordinates of your position on the map (that is, the horizontal coordinates of your actual position in the region). Let $z = f(x, y)$ denote the height of land (above sea level, say), at position $(x, y)$. Suppose you are walking along a trail so that your position at time $t$ is given by $x = u(t)$ and $y = v(t)$. (These are parametric equations of the trail on the map.) At time $t$ your altitude above sea level is given by the composite function

$$z = f\big(u(t), v(t)\big) = g(t),$$

a function of only one variable. How fast is your altitude changing with respect to time at time $t$?

**Solution** The answer is the derivative of $g(t)$:

$$g'(t) = \lim_{h \to 0} \frac{g(t + h) - g(t)}{h} = \lim_{h \to 0} \frac{f\big(u(t + h), v(t + h)\big) - f\big(u(t), v(t)\big)}{h}$$

$$= \lim_{h \to 0} \frac{f\big(u(t + h), v(t + h)\big) - f\big(u(t), v(t + h)\big)}{h}$$

$$+ \lim_{h \to 0} \frac{f\big(u(t), v(t + h)\big) - f\big(u(t), v(t)\big)}{h}.$$

We added 0 to the numerator of the Newton quotient in a creative way so as to separate the quotient into the sum of two quotients, in the first of which the difference of values of $f$ involves only the first variable of $f$, and in the second of which the difference involves only the second variable of $f$. The single-variable Chain Rule suggests that the sum of the two limits above is

$$g'(t) = f_1\big(u(t), v(t)\big)u'(t) + f_2\big(u(t), v(t)\big)v'(t).$$

The above formula is the Chain Rule for $\dfrac{d}{dt} f\big(u(t), v(t)\big)$. In terms of Leibniz notation we have

**A version of the Chain Rule**

If $z$ is a function of $x$ and $y$ with continuous first partial derivatives, and if $x$ and $y$ are differentiable functions of $t$, then

$$\frac{dz}{dt} = \frac{\partial z}{\partial x} \frac{dx}{dt} + \frac{\partial z}{\partial y} \frac{dy}{dt}.$$

Note that there are two terms in the expression for $dz/dt$ (or $g'(t)$), one arising from each variable of $f$ that depends on $t$.

Now consider a function $f$ of two variables, $x$ and $y$, each of which is in turn a function of two other variables, $s$ and $t$:

$$z = f(x, y), \qquad \text{where} \qquad x = u(s, t) \quad \text{and} \quad y = v(s, t).$$

We can form the composite function

$$z = f\big(u(s, t), v(s, t)\big) = g(s, t).$$

For instance, if $f(x, y) = x^2 + 3y$, where $u(s, t) = st^2$ and $v(s, t) = s - t$, then $g(s, t) = s^2 t^4 + 3(s - t)$.

Let us assume that $f$, $u$, and $v$ have first partial derivatives with respect to their respective variables, and that those of $f$ are continuous. Then $g$ has first partial derivatives given by

$$g_1(s, t) = f_1\big(u(s, t), v(s, t)\big)u_1(s, t) + f_2\big(u(s, t), v(s, t)\big)v_1(s, t),$$
$$g_2(s, t) = f_1\big(u(s, t), v(s, t)\big)u_2(s, t) + f_2\big(u(s, t), v(s, t)\big)v_2(s, t).$$

These formulas can be expressed more simply using Leibniz notation:

**Another version of the Chain Rule**

If $z$ is a function of $x$ and $y$ with continuous first partial derivatives, and if $x$ and $y$ depend on $s$ and $t$, then

$$\frac{\partial z}{\partial s} = \frac{\partial z}{\partial x}\frac{\partial x}{\partial s} + \frac{\partial z}{\partial y}\frac{\partial y}{\partial s},$$
$$\frac{\partial z}{\partial t} = \frac{\partial z}{\partial x}\frac{\partial x}{\partial t} + \frac{\partial z}{\partial y}\frac{\partial y}{\partial t}.$$

This can be deduced from the version obtained in Example 1 by allowing $u$ and $v$ there to depend on two variables, but holding one of them fixed while we differentiate with respect to the other. A more formal proof of this simple, but representative case of the Chain Rule will be given in the next section.

The two equations in the box above can be combined into a single matrix equation

$$\begin{pmatrix} \dfrac{\partial z}{\partial s} & \dfrac{\partial z}{\partial t} \end{pmatrix} = \begin{pmatrix} \dfrac{\partial z}{\partial x} & \dfrac{\partial z}{\partial y} \end{pmatrix} \begin{pmatrix} \dfrac{\partial x}{\partial s} & \dfrac{\partial x}{\partial t} \\ \dfrac{\partial y}{\partial s} & \dfrac{\partial y}{\partial t} \end{pmatrix}.$$

We will comment on the significance of this matrix form at the end of the next section.

In general, if $z$ is a function of several "primary" variables, and each of these depends on some "secondary" variables then the partial derivative of $z$ with respect to one of the secondary variables will have several terms, one for the contribution to the derivative arising from each of the primary variables on which $z$ depends.

**Remark**   Note the significance of the various subscripts denoting partial derivatives in the functional form of the Chain Rule:

$$g_1(s, t) = f_1\big(u(s, t), v(s, t)\big)u_1(s, t) + f_2\big(u(s, t), v(s, t)\big)v_1(s, t).$$

The "1" in $g_1(s, t)$ refers to differentiation with respect to $s$, the first variable on which $g$ depends. By contrast, the "1" in $f_1(u(s, t), v(s, t))$ refers to differentiation with respect to $x$, the first variable on which $f$ depends. (This derivative is then evaluated at $x = u(s, t), \; y = v(s, t)$.)

**Example 2**  If $z = \sin(x^2 y)$ where $x = st^2$ and $y = s^2 + \dfrac{1}{t}$, find $\partial z/\partial s$ and $\partial z/\partial t$

(a) by direct substitution and the single-variable form of the Chain Rule, and

(b) by using the (two-variable) Chain Rule.

**Solution**    (a) By direct substitution:

$$z = \sin\left((st^2)^2\left(s^2 + \frac{1}{t}\right)\right) = \sin(s^4 t^4 + s^2 t^3),$$

$$\frac{\partial z}{\partial s} = (4s^3 t^4 + 2st^3)\cos(s^4 t^4 + s^2 t^3),$$

$$\frac{\partial z}{\partial t} = (4s^4 t^3 + 3s^2 t^2)\cos(s^4 t^4 + s^2 t^3).$$

(b) Using the Chain Rule:

$$\frac{\partial z}{\partial s} = \frac{\partial z}{\partial x}\frac{\partial x}{\partial s} + \frac{\partial z}{\partial y}\frac{\partial y}{\partial s}$$

$$= \left(2xy\cos(x^2 y)\right)t^2 + \left(x^2\cos(x^2 y)\right)2s$$

$$= \left(2st^2\left(s^2 + \frac{1}{t}\right)t^2 + 2s^3 t^4\right)\cos(s^4 t^4 + s^2 t^3)$$

$$= (4s^3 t^4 + 2st^3)\cos(s^4 t^4 + s^2 t^3),$$

$$\frac{\partial z}{\partial t} = \frac{\partial z}{\partial x}\frac{\partial x}{\partial t} + \frac{\partial z}{\partial y}\frac{\partial y}{\partial t}$$

$$= \left(2xy\cos(x^2 y)\right)2st + \left(x^2\cos(x^2 y)\right)\left(\frac{-1}{t^2}\right)$$

$$= \left(2st^2(s^2 + \frac{1}{t})2st + s^2 t^4\left(\frac{-1}{t^2}\right)\right)\cos(s^4 t^4 + s^2 t^3)$$

$$= (4s^4 t^3 + 3s^2 t^2)\cos(s^4 t^4 + s^2 t^3).$$

Note that we still had to use direct substitution on the derivatives obtained in (b) in order to show that the values were the same as those obtained in (a). ∎

**Example 3**  Find $\dfrac{\partial}{\partial x} f(x^2 y, x + 2y)$ and $\dfrac{\partial}{\partial y} f(x^2 y, x + 2y)$ in terms of the partial derivatives of $f$, assuming that these partial derivatives are continuous.

**Solution**    We have

$$\frac{\partial}{\partial x} f(x^2 y, x + 2y) = f_1(x^2 y, x + 2y)\frac{\partial}{\partial x}(x^2 y) + f_2(x^2 y, x + 2y)\frac{\partial}{\partial x}(x + 2y)$$

$$= 2xy f_1(x^2 y, x + 2y) + f_2(x^2 y, x + 2y),$$

$$\frac{\partial}{\partial y} f(x^2 y, x + 2y) = f_1(x^2 y, x + 2y)\frac{\partial}{\partial y}(x^2 y) + f_2(x^2 y, x + 2y)\frac{\partial}{\partial y}(x + 2y)$$

$$= x^2 f_1(x^2 y, x + 2y) + 2 f_2(x^2 y, x + 2y).$$

**Example 4**   Express the partial derivatives of $z = h(s, t) = f\big(g(s, t)\big)$ in terms of the partial derivatives of $f$ and $g$.

**Solution**   The partial derivatives of $h$ can be calculated using the single-variable version of the Chain Rule:

$$h_1(s, t) = \frac{\partial z}{\partial s} = \frac{dz}{dx}\frac{\partial x}{\partial s} = f'\big(g(s, t)\big)g_1(s, t),$$

$$h_2(s, t) = \frac{\partial z}{\partial t} = \frac{dz}{dx}\frac{\partial x}{\partial t} = f'\big(g(s, t)\big)g_2(s, t).$$

The following example involves a hybrid application of the Chain Rule to a function that depends both directly and indirectly on the variable of differentiation.

**Example 5**   Find $dz/dt$ where $z = f(x, y, t)$, $x = g(t)$, and $y = h(t)$. (Assume that $f$, $g$, and $h$ all have continuous derivatives.)

**Solution**   Since $z$ depends on $t$ through each of the three variables of $f$, there will be three terms in the appropriate Chain Rule:

$$\frac{dz}{dt} = \frac{\partial z}{\partial x}\frac{dx}{dt} + \frac{\partial z}{\partial y}\frac{dy}{dt} + \frac{\partial z}{\partial t}$$
$$= f_1(x, y, t)g'(t) + f_2(x, y, t)h'(t) + f_3(x, y, t).$$

**Remark**   In the above example we can easily distinguish between the meanings of the symbols $dz/dt$ and $\partial z/\partial t$. If, however, we had been dealing with the situation

$$z = f(x, y, s, t), \qquad \text{where} \quad x = g(s, t) \quad \text{and} \quad y = h(s, t),$$

then the meaning of the symbol $\partial z/\partial t$ would be unclear; it could refer to the simple partial derivative of $f$ with respect to its fourth primary variable (i.e., $f_4(x, y, s, t)$), or it could refer to the derivative of the composite function $f(g(s, t), h(s, t), s, t)$. Three of the four primary variables of $f$ depend on $t$, and therefore contribute to the rate of change of $z$ with respect to $t$. The partial derivative $f_4(x, y, s, t)$ denotes the contribution of only one of these three variables. It is conventional to use $\partial z/\partial t$ to denote the derivative of the composite function with respect to the secondary variable $t$:

$$\frac{\partial z}{\partial t} = \frac{\partial}{\partial t}f(g(s, t), h(s, t), s, t)$$
$$= f_1(x, y, s, t)g_2(s, t) + f_2(x, y, s, t)h_2(s, t) + f_4(x, y, s, t).$$

When necessary, we can denote the contribution coming from the primary variable $t$ by

$$\left(\frac{\partial z}{\partial t}\right)_{x,y,s} = \frac{\partial}{\partial t}f(x, y, s, t) = f_4(x, y, s, t).$$

Here the subscripts denote those primary variables of $f$ whose contributions to the rate of change of $z$ with respect to $t$ are being *ignored*. Of course, in the situation described above, $(\partial z/\partial t)_s$ means the same as $\partial z/\partial t$.

In applications, the variables that contribute to a particular partial derivative will usually be clear from the context. The following example contains such an application. This is an example of a procedure called *differentiation following the motion.*

**Example 6** Atmospheric temperature depends on position and time. If we denote position by three spatial coordinates $x$, $y$, and $z$ (measured in km), and time by $t$ (measured in hours), then the temperature $T$ is a function of four variables, $T(x, y, z, t)$.

(a) If a thermometer is attached to a weather balloon that moves through the atmosphere on a path with parametric equations $x = f(t)$, $y = g(t)$, and $z = h(t)$, what is the rate of change at time $T$ of the temperature recorded by the thermometer?

(b) Find the rate of change of the recorded temperature at time $t = 1$ if

$$T(x, y, z, t) = \frac{xy}{1+z}(1+t),$$

and if the balloon moves along the curve

$$x = t, \qquad y = 2t, \qquad z = t - t^2.$$

**Solution**

(a) Here the rate of change of the thermometer reading must take into account the change in position of the thermometer as well as increasing time. Thus none of the four variables of $T$ can be ignored in the differentiation. The rate is given by

$$\frac{dT}{dt} = \frac{\partial T}{\partial x}\frac{dx}{dt} + \frac{\partial T}{\partial y}\frac{dy}{dt} + \frac{\partial T}{\partial z}\frac{dz}{dt} + \frac{\partial T}{\partial t}.$$

The term $\partial T/\partial t$ refers only to the rate of change of the temperature with respect to time at a fixed position in the atmosphere. The other three terms arise from the motion of the balloon.

(b) The values of the three coordinates and their derivatives at $t = 1$ are $x = 1$, $y = 2$, $z = 0$, $dx/dt = 1$, $dy/dt = 2$, and $dz/dt = -1$. Also, at $t = 1$,

$$\frac{\partial T}{\partial x} = \frac{y}{1+z}(1+t) = 4, \qquad \frac{\partial T}{\partial z} = \frac{-xy}{(1+z)^2}(1+t) = -4,$$

$$\frac{\partial T}{\partial y} = \frac{x}{1+z}(1+t) = 2, \qquad \frac{\partial T}{\partial t} = \frac{xy}{1+z} = 2.$$

Thus

$$\left.\frac{dT}{dt}\right|_{t=1} = (4)(1) + (2)(2) + (-4)(-1) + 2 = 14.$$

The recorded temperature is increasing at a rate of 14 °/h at time $t = 1$. ∎

The discussion and examples above show that the Chain Rule for functions of several variables can take different forms depending on the numbers of variables of the various functions being composed. As an aid in determining the correct form of the Chain Rule in a given situation you can construct a chart showing which variables depend on which. Figure 12.21(a) shows such a chart for the temperature function of Example 6. The Chain Rule for $dT/dt$ involves a term for every route from $T$ to $t$ in the chart. The route from $T$ through $x$ to $t$ produces the term $\dfrac{\partial T}{\partial x}\dfrac{dx}{dt}$ and so on.

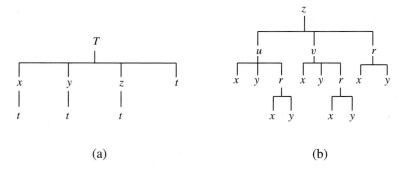

**Figure 12.21**

(a) Chart showing the dependence of $T$ on $t$ in Example 6

(b) Dependence chart for Example 7

(a)

(b)

**Example 7**   Write the appropriate Chain Rule for $\partial z/\partial x$, where $z$ depends on $u$, $v$, and $r$; $u$ and $v$ depend on $x$, $y$, and $r$; and $r$ depends on $x$ and $y$.

**Solution**   The appropriate chart is shown in Figure 12.21(b). There are five routes from $z$ to $x$:

$$\frac{\partial z}{\partial x} = \frac{\partial z}{\partial u}\frac{\partial u}{\partial x} + \frac{\partial z}{\partial u}\frac{\partial u}{\partial r}\frac{\partial r}{\partial x} + \frac{\partial z}{\partial v}\frac{\partial v}{\partial x} + \frac{\partial z}{\partial v}\frac{\partial v}{\partial r}\frac{\partial r}{\partial x} + \frac{\partial z}{\partial r}\frac{\partial r}{\partial x}.$$

## Homogeneous Functions

A function $f(x_1, \ldots, x_n)$ is said to be **positively homogeneous of degree** $k$ if, for every point $(x_1, x_2, \ldots, x_n)$ in its domain and every real number $t > 0$, we have

$$f(tx_1, tx_2, \ldots, tx_n) = t^k f(x_1, \ldots, x_n).$$

For example

$$f(x, y) = x^2 + xy - y^2 \qquad \text{is positively homogeneous of degree 2,}$$
$$f(x, y) = \sqrt{x^2 + y^2} \qquad \text{is positively homogeneous of degree 1,}$$
$$f(x, y) = \frac{2xy}{x^2 + y^2} \qquad \text{is positively homogeneous of degree 0,}$$
$$f(x, y, z) = \frac{x - y + 5z}{yz - z^2} \qquad \text{is positively homogeneous of degree } -1,$$
$$f(x, y) = x^2 + y \qquad \text{is not positively homogeneous.}$$

Observe that a positively homogeneous function of degree 0 remains constant along rays from the origin. More generally, along such rays a positively homogeneous function of degree $k$ grows or decays proportionally to the $k$th power of distance from the origin.

**THEOREM 2**

**Euler's Theorem**

If $f(x_1, \ldots, x_n)$ has continuous first partial derivatives and is positively homogeneous of degree $k$, then

$$\sum_{i=1}^{n} x_i f_i(x_1, \ldots, x_n) = kf(x_1, \ldots, x_n).$$

**PROOF**   Differentiate the equation $f(tx_1, tx_2, \ldots, tx_n) = t^k f(x_1, \ldots, x_n)$ with respect to $t$ to get

$$x_1 f_1(tx_1, \ldots, tx_n) + x_2 f_2(tx_1, \ldots, tx_n) + \ldots + x_n f_n(tx_1, \ldots, tx_n)$$
$$= k\, t^{k-1} f(x_1, \ldots, x_n).$$

Now substitute $t = 1$ to get the desired result.

Note that Exercises 26–29 in Section 12.3 illustrate this theorem.

## Higher-Order Derivatives

Applications of the Chain Rule to higher-order derivatives can become quite complicated. It is important to keep in mind at each stage which variables are independent of one another.

**Example 8**   Calculate $\dfrac{\partial^2}{\partial x \partial y} f(x^2 - y^2, xy)$ in terms of partial derivatives of the function $f$. Assume that the partials of $f$ are continuous.

**Solution**   In this problem symbols for the primary variables on which $f$ depends are not stated explicitly. Let them be $u$ and $v$. The problem therefore asks us to find

$$\frac{\partial^2}{\partial x \partial y} f(u, v), \qquad \text{where} \quad u = x^2 - y^2 \quad \text{and} \quad v = xy.$$

First differentiate with respect to $y$:

$$\frac{\partial}{\partial y} f(u, v) = -2yf_1(u, v) + xf_2(u, v).$$

Now differentiate this result with respect to $x$. Note that the second term on the right is a product to two functions of $x$ so we need to use the Product Rule:

$$\frac{\partial^2}{\partial x \partial y} f(u, v) = -2y\Big(2xf_{11}(u, v) + yf_{12}(u, v)\Big)$$
$$+ f_2(u, v) + x\Big(2xf_{21}(u, v) + yf_{22}(u, v)\Big)$$
$$= f_2(u, v) - 4xyf_{11}(u, v) + 2(x^2 - y^2)f_{12}(u, v) + xyf_{22}(u, v).$$

In the last step we have used the fact that the mixed partials of $f$ are continuous so we could equate $f_{12}$ and $f_{21}$.

Review the above calculation very carefully and make sure you understand what is being done at each step. Note that all the derivatives of $f$ that appear are evaluated at $(u, v) = (x^2 - y^2, xy)$, not at $(x, y)$, because $x$ and $y$ are not themselves the primary variables on which $f$ depends.

**Example 9**    Show that $f(x^2 - y^2, 2xy)$ is a harmonic function if $f(x, y)$ is harmonic.

**Solution**    Let $u = x^2 - y^2$ and $v = 2xy$. If $z = f(u, v)$, then

$$\frac{\partial z}{\partial x} = 2xf_1(u, v) + 2yf_2(u, v),$$

$$\frac{\partial z}{\partial y} = -2yf_1(u, v) + 2xf_2(u, v),$$

$$\frac{\partial^2 z}{\partial x^2} = 2f_1(u, v) + 2x\big(2xf_{11}(u, v) + 2yf_{12}(u, v)\big)$$
$$+ 2y\big(2xf_{21}(u, v) + 2yf_{22}(u, v)\big)$$
$$= 2f_1(u, v) + 4x^2 f_{11}(u, v) + 8xyf_{12}(u, v) + 4y^2 f_{22}(u, v),$$

$$\frac{\partial^2 z}{\partial y^2} = -2f_1(u, v) - 2y\big(-2yf_{11}(u, v) + 2xf_{12}(u, v)\big)$$
$$+ 2x\big(-2yf_{21}(u, v) + 2xf_{22}(u, v)\big)$$
$$= -2f_1(u, v) + 4y^2 f_{11}(u, v) - 8xyf_{12}(u, v) + 4x^2 f_{22}(u, v).$$

Therefore

$$\frac{\partial^2 z}{\partial x^2} + \frac{\partial^2 z}{\partial y^2} = 4(x^2 + y^2)\big(f_{11}(u, v) + f_{22}(u, v)\big) = 0$$

because $f$ is given to be harmonic. Thus $z = f(x^2 - y^2, 2xy)$ is a harmonic function of $x$ and $y$. ∎

In the following example we show that the two-dimensional Laplace differential equation (see Example 3 in Section 12.4) takes the form

$$\frac{\partial^2 z}{\partial r^2} + \frac{1}{r}\frac{\partial z}{\partial r} + \frac{1}{r^2}\frac{\partial^2 z}{\partial \theta^2} = 0$$

when stated for a function $z$ expressed in terms of polar coordinates $r$ and $\theta$.

**Example 10**    (**Laplace's equation in polar coordinates**) If $z = f(x, y)$ has continuous partial derivatives of second order, and if $x = r\cos\theta$, and $y = r\sin\theta$, show that

$$\frac{\partial^2 z}{\partial r^2} + \frac{1}{r}\frac{\partial z}{\partial r} + \frac{1}{r^2}\frac{\partial^2 z}{\partial \theta^2} = \frac{\partial^2 z}{\partial x^2} + \frac{\partial^2 z}{\partial y^2}.$$

***Solution***   It is possible to do this in two different ways; we can start with either side and use the Chain Rule to show that it is equal to the other side. Here we will calculate the partial derivatives with respect to $r$ and $\theta$ that appear on the left side and express them in terms of partial derivatives with respect to $x$ and $y$. The other approach, involving expressing partial derivatives with respect to $x$ and $y$ in terms of partial derivatives with respect to $r$ and $\theta$, is a little harder. (See Exercise 24 below.) However, we would have to do it that way if we were not given the form of the differential equation in polar coordinates and had to find it.

First note that,

$$\frac{\partial x}{\partial r} = \cos\theta, \qquad \frac{\partial x}{\partial \theta} = -r\sin\theta, \qquad \frac{\partial y}{\partial r} = \sin\theta, \qquad \frac{\partial y}{\partial \theta} = r\cos\theta.$$

Thus

$$\frac{\partial z}{\partial r} = \frac{\partial z}{\partial x}\frac{\partial x}{\partial r} + \frac{\partial z}{\partial y}\frac{\partial y}{\partial r} = \cos\theta\,\frac{\partial z}{\partial x} + \sin\theta\,\frac{\partial z}{\partial y}.$$

**DANGER**

This is a difficult, but important example. Examine each step carefully to make sure you understand what is being done.

Now differentiate with respect to $r$ again. Remember that $r$ and $\theta$ are independent variables, so that the factors $\cos\theta$ and $\sin\theta$ can be regarded as constants. However, $\partial z/\partial x$ and $\partial z/\partial y$ depend on $x$ and $y$, and therefore on $r$ and $\theta$.

$$\frac{\partial^2 z}{\partial r^2} = \cos\theta\,\frac{\partial}{\partial r}\frac{\partial z}{\partial x} + \sin\theta\,\frac{\partial}{\partial r}\frac{\partial z}{\partial y}$$

$$= \cos\theta\left(\cos\theta\,\frac{\partial^2 z}{\partial x^2} + \sin\theta\,\frac{\partial^2 z}{\partial y\partial x}\right) + \sin\theta\left(\cos\theta\,\frac{\partial^2 z}{\partial x\partial y} + \sin\theta\,\frac{\partial^2 z}{\partial y^2}\right)$$

$$= \cos^2\theta\,\frac{\partial^2 z}{\partial x^2} + 2\cos\theta\sin\theta\,\frac{\partial^2 z}{\partial x\partial y} + \sin^2\theta\,\frac{\partial^2 z}{\partial y^2}.$$

We have used the equality of mixed partials in the last line. Similarly,

$$\frac{\partial z}{\partial \theta} = -r\,\sin\theta\,\frac{\partial z}{\partial x} + r\,\cos\theta\,\frac{\partial z}{\partial y}.$$

When we differentiate a second time with respect to $\theta$ we can regard $r$ as constant, but each term above is still a product of two functions that depend on $\theta$. Thus

$$\frac{\partial^2 z}{\partial \theta^2} = -r\left(\cos\theta\,\frac{\partial z}{\partial x} + \sin\theta\,\frac{\partial}{\partial \theta}\frac{\partial z}{\partial x}\right) + r\left(-\sin\theta\,\frac{\partial z}{\partial y} + \cos\theta\,\frac{\partial}{\partial \theta}\frac{\partial z}{\partial y}\right)$$

$$= -r\,\frac{\partial z}{\partial r} - r\,\sin\theta\left(-r\,\sin\theta\,\frac{\partial^2 z}{\partial x^2} + r\,\cos\theta\,\frac{\partial^2 z}{\partial y\partial x}\right)$$

$$+ r\,\cos\theta\left(-r\,\sin\theta\,\frac{\partial^2 z}{\partial x\partial y} + r\,\cos\theta\,\frac{\partial^2 z}{\partial y^2}\right)$$

$$= -r\,\frac{\partial z}{\partial r} + r^2\left(\sin^2\theta\,\frac{\partial^2 z}{\partial x^2} - 2\sin\theta\cos\theta\,\frac{\partial^2 z}{\partial x\partial y} + \cos^2\theta\,\frac{\partial^2 z}{\partial y^2}\right).$$

Combining these results, we obtain the desired formula:

$$\frac{\partial^2 z}{\partial r^2} + \frac{1}{r}\frac{\partial z}{\partial r} + \frac{1}{r^2}\frac{\partial^2 z}{\partial \theta^2} = \frac{\partial^2 z}{\partial x^2} + \frac{\partial^2 z}{\partial y^2}.$$

# Exercises 12.5

In Exercises 1–4, write appropriate versions of the Chain Rule for the indicated derivatives.

1. $\partial w/\partial t$ if $w = f(x, y, z)$, where $x = g(s, t)$, $y = h(s, t)$, and $z = k(s, t)$

2. $\partial w/\partial t$ if $w = f(x, y, z)$, where $x = g(s)$, $y = h(s, t)$, and $z = k(t)$

3. $\partial z/\partial u$ if $z = g(x, y)$, where $y = f(x)$ and $x = h(u, v)$

4. $dw/dt$ if $w = f(x, y)$, $x = g(r, s)$, $y = h(r, t)$, $r = k(s, t)$, and $s = m(t)$

5. If $w = f(x, y, z)$, where $x = g(y, z)$ and $y = h(z)$, state appropriate versions of the Chain Rule for $\dfrac{dw}{dz}$, $\left(\dfrac{\partial w}{\partial z}\right)_x$, and $\left(\dfrac{\partial w}{\partial z}\right)_{x,y}$.

6. Use two different methods to calculate $\partial u/\partial t$ if $u = \sqrt{x^2 + y^2}$, $x = e^{st}$, and $y = 1 + s^2 \cos t$.

7. Use two different methods to calculate $\partial z/\partial x$ if $z = \tan^{-1}(u/v)$, $u = 2x + y$ and $v = 3x - y$.

8. Use two methods to calculate $dz/dt$ given that $z = txy^2$, $x = t + \ln(y + t^2)$ and $y = e^t$.

In Exercises 9–12, find the indicated derivatives, assuming that the function $f(x, y)$ has continuous first partial derivatives.

9. $\dfrac{\partial}{\partial x} f(2x, 3y)$

10. $\dfrac{\partial}{\partial x} f(2y, 3x)$

11. $\dfrac{\partial}{\partial x} f(y^2, x^2)$

12. $\dfrac{\partial}{\partial y} f\big(yf(x, t), f(y, t)\big)$

13. Suppose that the temperature $T$ in a certain liquid varies with depth $z$ and time $t$ according to the formula $T = e^{-t}z$. Find the rate of change of temperature with respect to time at a point that is moving through the liquid so that at time $t$ its depth is $f(t)$. What is this rate if $f(t) = e^t$? What is happening in this case?

14. Suppose the strength $E$ of an electric field in space varies with position $(x, y, z)$ and time $t$ according to the formula $E = f(x, y, z, t)$. Find the rate of change with respect to time of the electric field strength measured by an instrument moving through space along the helix $x = \sin t$, $y = \cos t$, $z = t$.

In Exercises 15–20, assume that $f$ has continuous partial derivatives of all orders.

15. If $z = f(x, y)$ where $x = 2s + 3t$ and $y = 3s - 2t$ find
    (a) $\dfrac{\partial^2 z}{\partial s^2}$,
    (b) $\dfrac{\partial^2 z}{\partial s \partial t}$, and
    (c) $\dfrac{\partial^2 z}{\partial t^2}$.

16. If $f(x, y)$ is harmonic show that $f\left(\dfrac{x}{x^2 + y^2}, \dfrac{-y}{x^2 + y^2}\right)$ is also harmonic.

17. If $x = t \sin s$ and $y = t \cos s$ find $\dfrac{\partial^2}{\partial s \partial t} f(x, y)$.

18. Find $\dfrac{\partial^3}{\partial x \partial y^2} f(2x + 3y, xy)$ in terms of partial derivatives of $f$.

19. Find $\dfrac{\partial^2}{\partial y \partial x} f(y^2, xy, -x^2)$ in terms of partial derivatives of the function $f$.

20. Find $\dfrac{\partial^3}{\partial t^2 \partial s} f(s^2 - t, s + t^2)$ in terms of partial derivatives of $f$.

21. Suppose that $u(x, y)$ and $v(x, y)$ have continuous second partial derivatives and satisfy the Cauchy–Riemann equations

    $$\frac{\partial u}{\partial x} = \frac{\partial v}{\partial y} \quad \text{and} \quad \frac{\partial v}{\partial x} = -\frac{\partial u}{\partial y}.$$

    Suppose also that $f(u, v)$ is a harmonic function of $u$ and $v$. Show that $f\big(u(x, y), v(x, y)\big)$ is a harmonic function of $x$ and $y$. *Hint: $u$ and $v$ are harmonic functions by Exercise 15 in Section 12.4.*

22. If $r^2 = x^2 + y^2 + z^2$, verify that $u(x, y, z) = 1/r$ is harmonic throughout $\mathbb{R}^3$ except at the origin.

* 23. If $x = e^s \cos t$, $y = e^s \sin t$ and $z = u(x, y) = v(s, t)$ show that

    $$\frac{\partial^2 z}{\partial s^2} + \frac{\partial^2 z}{\partial t^2} = (x^2 + y^2)\left(\frac{\partial^2 z}{\partial x^2} + \frac{\partial^2 z}{\partial y^2}\right).$$

* 24. **(Converting Laplace's equation to polar coordinates)** The transformation to polar coordinates, $x = r \cos \theta$, $y = r \sin \theta$ implies that $r^2 = x^2 + y^2$ and $\tan \theta = y/x$. Use these equations to show that

    $$\frac{\partial r}{\partial x} = \cos \theta \qquad \frac{\partial r}{\partial y} = \sin \theta$$

    $$\frac{\partial \theta}{\partial x} = -\frac{\sin \theta}{r} \qquad \frac{\partial \theta}{\partial y} = \frac{\cos \theta}{r}.$$

    Use these formulas to help you express $\dfrac{\partial^2 u}{\partial x^2} + \dfrac{\partial^2 u}{\partial y^2}$ in terms of partials of $u$ with respect to $r$ and $\theta$, and hence reprove the formula for the Laplace differential equation in polar coordinates given in Example 10.

25. If $u(x, y) = r^2 \ln r$, where $r^2 = x^2 + y^2$, verify that $u$ is a biharmonic function by showing that

    $$\left(\frac{\partial^2}{\partial x^2} + \frac{\partial^2}{\partial y^2}\right)\left(\frac{\partial^2 u}{\partial x^2} + \frac{\partial^2 u}{\partial y^2}\right) = 0.$$

26. If $f(x, y)$ is positively homogeneous of degree $k$ and has continuous partial derivatives of second order, show that

    $$x^2 f_{11}(x, y) + 2xy f_{12}(x, y) + y^2 f_{22}(x, y)$$
    $$= k(k - 1) f(x, y).$$

* **27.** Generalize the result of Exercise 26 to functions of $n$ variables.

* **28.** Generalize the results of Exercises 26 and 27 to expressions involving $m$th-order partial derivatives of the function $f$.

Exercises 29–30 revisit Exercise 16 of Section 12.4. Let

$$F(x, y) = \begin{cases} \dfrac{2xy(x^2 - y^2)}{x^2 + y^2} & \text{if } (x, y) \neq (0, 0) \\ 0 & \text{if } (x, y) = (0, 0). \end{cases}$$

**29.** (a) Show that $F(x, y) = -F(y, x)$ for all $(x, y)$.

(b) Show that $F_1(x, y) = -F_2(y, x)$ and $F_{12}(x, y) = -F_{21}(y, x)$ for $(x, y) \neq (0, 0)$.

(c) Show that $F_1(0, y) = -2y$ for all $y$, and hence that $F_{12}(0, 0) = 2$.

(d) Deduce that $F_2(x, 0) = 2x$ and $F_{21}(0, 0) = 2$.

**30.** (a) Use Exercise 29(b) to find $F_{12}(x, x)$ for $x \neq 0$.

(b) Is $F_{12}(x, y)$ continuous at $(0, 0)$? Why?

◈ **31.** Use the change of variables $\xi = x + ct$, $\eta = x$ to transform the partial differential equation

$$\frac{\partial u}{\partial t} = c\,\frac{\partial u}{\partial x}, \qquad (c = \text{constant}),$$

into the simpler equation $\partial v/\partial \eta = 0$, where $v(\xi, \eta) = v(x + ct, x) = u(x, t)$. This equation says that $v(\xi, \eta)$ does not depend on $\eta$, so $v = f(\xi)$ for some arbitrary differentiable function $f$. What is the corresponding "general solution" $u(x, t)$ of the original partial differential equation?

◈ **32.** Having considered Exercise 31, guess a "general solution" $w(r, s)$ of the second-order partial differential equation

$$\frac{\partial^2}{\partial r \partial s}\, w(r, s) = 0.$$

Your answer should involve two arbitrary functions.

◈ **33.** Use the change of variables $r = x + ct$, $s = x - ct$, $w(r, s) = u(x, t)$ to transform the one-dimensional wave equation

$$\frac{\partial^2 u}{\partial t^2} = c^2\,\frac{\partial^2 u}{\partial x^2}$$

to a simpler form. Now use the result of Exercise 32 to find the *general solution* of this wave equation in the form given in Example 4 in Section 12.4.

◈ **34.** Show that the initial-value problem for the one-dimensional wave equation:

$$\begin{cases} u_{tt}(x, t) = c^2 u_{xx}(x, t), \\ u(x, 0) = p(x), \\ u_t(x, 0) = q(x). \end{cases}$$

has the solution

$$u(x, t) = \frac{1}{2}\Big[p(x - ct) + p(x + ct)\Big] + \frac{1}{2c}\int_{x-ct}^{x+ct} q(s)\,ds.$$

(Note that we have used subscripts "$x$" and "$t$" instead of "1" and "2" to denote the partial derivatives here. This is common usage in dealing with partial differential equations.)

***Remark*** The initial-value problem in Exercise 33 gives the small lateral displacement $u(x, t)$ at position $x$ at time $t$ of a vibrating string held under tension along the $x$-axis. The function $p(x)$ gives the *initial* displacement at position $x$, that is, the displacement at time $t = 0$. Similarly, $q(x)$ gives the initial velocity at position $x$. Observe that the position at time $t$ depends only on values of these initial data at points no further than $ct$ units away. This is consistent with the previous observation that the solutions of the wave equation represent waves travelling with speed $c$.

## 12.6 Linear Approximations, Differentiability, and Differentials

The tangent line to the graph $y = f(x)$ at $x = a$ provides a convenient approximation for values of $f(x)$ for $x$ near $a$:

$$f(x) \approx L(x) = f(a) + f'(a)(x - a).$$

Here $L(x)$ is the **linearization** of $f$ at $a$; its graph is the tangent line to $y = f(x)$ there. The mere existence of $f'(a)$ is sufficient to guarantee that the error in the approximation $f(x) \approx L(x)$ is small compared to the distance $h = x - a$ between

$a$ and $x$, that is to say,

$$\lim_{h \to 0} \frac{f(a+h) - L(a+h)}{h} = \lim_{h \to 0} \frac{f(a+h) - f(a) - f'(a)h}{h}$$

$$= \lim_{h \to 0} \frac{f(a+h) - f(a)}{h} - f'(a)$$

$$= f'(a) - f'(a) = 0.$$

Similarly, we can use the tangent plane

$$z = f(a, b) + f_1(a, b)(x - a) + f_2(a, b)(y - b)$$

to $z = f(x, y)$ at $(a, b)$ to define a linear approximation to $f(x, y)$ near $(a, b)$:

The **linearization** (linear approximation) of $f(x, y)$ at $(a, b)$ is given by

$$f(x, y) \approx L(x, y) = f(a, b) + f_1(a, b)(x - a) + f_2(a, b)(y - b).$$

**Example 1** Find an approximate value for $f(x, y) = \sqrt{2x^2 + e^{2y}}$ at $(2.2, -0.2)$.

**Solution** It is convenient to use the linearization at $(2, 0)$, where the values of $f$ and its partials are easily evaluated:

$$f(2, 0) = 3,$$

$$f_1(x, y) = \frac{2x}{\sqrt{2x^2 + e^{2y}}}, \qquad f_1(2, 0) = \frac{4}{3},$$

$$f_2(x, y) = \frac{e^{2y}}{\sqrt{2x^2 + e^{2y}}}, \qquad f_2(2, 0) = \frac{1}{3}.$$

Thus $L(x, y) = 3 + \dfrac{4}{3}(x - 2) + \dfrac{1}{3}(y - 0)$, and

$$f(2.2, -0.2) \approx L(2.2, -0.2) = 3 + \frac{4}{3}(2.2 - 2) + \frac{1}{3}(-0.2 - 0) = 3.2 .$$

Unlike the single-variable case, the mere existence of the partial derivatives $f_1(a, b)$ and $f_2(a, b)$ does not even imply that $f$ is continuous at $(a, b)$, let alone that the error in the linearization is small compared to the distance $\sqrt{(x - a)^2 + (y - b)^2}$ between $(a, b)$ and $(x, y)$. We adopt this latter condition as our definition of what it means for a function to be *differentiable* at a point.

**DEFINITION 6**

We say that the function $f(x, y)$ is **differentiable** at the point $(a, b)$ if

$$\lim_{(h,k) \to (0,0)} \frac{f(a+h, b+k) - f(a, b) - h\, f_1(a, b) - k f_2(a, b)}{\sqrt{h^2 + k^2}} = 0.$$

This definition and the following theorems can be generalized to functions of any number of variables in the obvious way. For the sake of simplicity, we state them for the two-variable case only.

The function $f(x, y)$ is differentiable at the point $(a, b)$ if and only if the surface $z = f(x, y)$ has a *nonvertical tangent plane* at $(a, b)$. This implies that $f_1(a, b)$ and $f_2(a, b)$ must exist, and that $f$ must be continuous at $(a, b)$. (Recall, however, that the existence of the partial derivatives does *not* even imply that $f$ is continuous let alone differentiable. Compare this with the single-variable situation.) In particular, the function is *continuous* wherever it is differentiable. We will prove a two-variable version of the Mean-Value Theorem and use it to show that functions with *continuous* first partial derivatives are differentiable.

**THEOREM 3**

**A mean-value theorem**

If $f_1(x, y)$ and $f_2(x, y)$ are continuous in a neighbourhood of the point $(a, b)$, and if the absolute values of $h$ and $k$ are sufficiently small, then there exist numbers $\theta_1$ and $\theta_2$, each between 0 and 1, such that

$$f(a + h, b + k) - f(a, b) = hf_1(a + \theta_1 h, b + k) + kf_2(a, b + \theta_2 k).$$

**PROOF**   The proof of this theorem is very similar to that of Theorem 1 in Section 12.4, so we give only a sketch here. The reader can fill in the details. Write

$$f(a+h, b+k) - f(a, b) = \big(f(a+h, b+k) - f(a, b+k)\big) + \big(f(a, b+k) - f(a, b)\big),$$

and then apply the single-variable Mean-Value Theorem separately to $f(x, b + k)$ on the interval between $a$ and $a + h$, and to $f(a, y)$ on the interval between $b$ and $b + k$ to get the desired result.

**THEOREM 4**

If $f_1$ and $f_2$ are continuous in a neighbourhood of the point $(a, b)$, then $f$ is differentiable at $(a, b)$.

**PROOF**   Using Theorem 3 and the facts that

$$\left| \frac{h}{\sqrt{h^2 + k^2}} \right| \leq 1 \qquad \text{and} \qquad \left| \frac{k}{\sqrt{h^2 + k^2}} \right| \leq 1,$$

we estimate

$$\left| \frac{f(a + h, b + k) - f(a, b) - hf_1(a, b) - kf_2(a, b)}{\sqrt{h^2 + k^2}} \right|$$

$$= \left| \frac{h}{\sqrt{h^2 + k^2}} \Big( f_1(a + \theta_1 h, b + k) - f_1(a, b) \Big) \right.$$

$$+ \left. \frac{k}{\sqrt{h^2 + k^2}} \Big( f_2(a, b + \theta_2 k) - f_2(a, b) \Big) \right|$$

$$\leq \left| f_1(a + \theta_1 h, b + k) - f_1(a, b) \right| + \left| f_2(a, b + \theta_2 k) - f_2(a, b) \right|.$$

Since $f_1$ and $f_2$ are continuous at $(a, b)$ each of these latter terms approaches 0 as $h$ and $k$ approach 0. This is what we needed to prove.

We illustrate differentiability with an example where we can calculate directly the error in the tangent plane approximation.

**Example 2**    Calculate $f(x+h, y+k) - f(x, y) - f_1(x, y)h - f_2(x, y)k$ if $f(x, y) = x^3 + xy^2$.

**Solution**    Since $f_1(x, y) = 3x^2 + y^2$ and $f_2(x, y) = 2xy$, we have

$$f(x+h, y+k) - f(x, y) - f_1(x, y)h - f_2(x, y)k$$
$$= (x+h)^3 + (x+h)(y+k)^2 - x^3 - xy^2 - (3x^2 + y^2)h - 2xyk$$
$$= 3xh^2 + h^3 + 2yhk + hk^2 + xk^2.$$

Observe that the result above is a polynomial in $h$ and $k$ with no term of degree less than two in these variables. Therefore this difference approaches zero like the *square* of the distance $\sqrt{h^2 + k^2}$ from $(x, y)$ to $(x+h, y+k)$ as $(h, k) \to (0, 0)$, so the condition for differentiability is certainly satisfied:

$$\lim_{(h,k) \to (0,0)} \frac{3xh^2 + h^3 + 2yhk + hk^2 + xk^2}{\sqrt{h^2 + k^2}} = 0.$$

This quadratic behaviour is the case for any function $f$ with continuous *second* partial derivatives. (See Exercise 19 at the end of this section.)    ∎

## Proof of the Chain Rule

We are now able to give a formal statement and proof of a simple but representative case of the Chain Rule for multivariate functions.

**THEOREM 5**

**A Chain Rule**

Let $z = f(x, y)$, where $x = u(s, t)$ and $y = v(s, t)$. Suppose that

    (i) $u(a, b) = p$ and $v(a, b) = q$,

    (ii) the first partial derivatives of $u$ and $v$ exist at the point $(a, b)$, and

    (iii) $f$ is differentiable at the point $(p, q)$.

Then $z = w(s, t) = f(u(s, t), v(s, t))$ has first partial derivatives with respect to $s$ and $t$ at $(a, b)$, and

$$w_1(a, b) = f_1(p, q)u_1(a, b) + f_2(p, q)v_1(a, b),$$
$$w_2(a, b) = f_1(p, q)u_2(a, b) + f_2(p, q)v_2(a, b),$$

that is,

$$\frac{\partial z}{\partial s} = \frac{\partial z}{\partial x}\frac{\partial x}{\partial s} + \frac{\partial z}{\partial y}\frac{\partial y}{\partial s} \quad \text{and} \quad \frac{\partial z}{\partial t} = \frac{\partial z}{\partial x}\frac{\partial x}{\partial t} + \frac{\partial z}{\partial y}\frac{\partial y}{\partial t}.$$

**PROOF**    Define a function $E$ of two variables as follows: $E(0, 0) = 0$, and if $(h, k) \neq (0, 0)$, then

$$E(h, k) = \frac{f(p+h, q+k) - f(p, q) - hf_1(p, q) - kf_2(p, q)}{\sqrt{h^2 + k^2}}.$$

Observe that $E(h, k)$ is continuous at $(0, 0)$, because $f$ is differentiable at $(p, q)$. Now

$$f(p+h, q+k) - f(p, q) = hf_1(p, q) + kf_2(p, q) + \sqrt{h^2 + k^2}\, E(h, k).$$

In this formula put $h = u(a + \sigma, b) - u(a, b)$ and $k = v(a + \sigma, b) - v(a, b)$ and divide by $\sigma$ to obtain

$$\frac{w(a + \sigma, b) - w(a, b)}{\sigma} = \frac{f(u(a + \sigma, b), v(a + \sigma, b)) - f(u(a, b), v(a, b))}{\sigma}$$

$$= \frac{f(p + h, q + k) - f(p, q)}{\sigma}$$

$$= f_1(p, q)\frac{h}{\sigma} + f_2(p, q)\frac{k}{\sigma} + \sqrt{\left(\frac{h}{\sigma}\right)^2 + \left(\frac{k}{\sigma}\right)^2}\, E(h, k).$$

We want to let $\sigma$ approach $0$ in this formula. Note that

$$\lim_{\sigma \to 0}\frac{h}{\sigma} = \lim_{\sigma \to 0}\frac{u(a + \sigma, b) - u(a, b)}{\sigma} = u_1(a, b),$$

and, similarly, $\lim_{\sigma \to 0}(k/\sigma) = v_1(a, b)$. Since $(h, k) \to (0, 0)$ if $\sigma \to 0$, we have

$$w_1(a, b) = f_1(p, q)u_1(a, b) + f_2(p, q)v_1(a, b).$$

The proof for $w_2$ is similar.

## Differentials

If the first partial derivatives of a function $z = f(x_1, \ldots, x_n)$ exist at a point we may construct a **differential** $dz$ or $df$ of the function at that point in a manner similar to that used for functions of one variable:

$$dz = df = \frac{\partial z}{\partial x_1}\, dx_1 + \frac{\partial z}{\partial x_2}\, dx_2 + \cdots + \frac{\partial z}{\partial x_n}\, dx_n$$

$$= f_1(x_1, \ldots, x_n)\, dx_1 + \cdots + f_n(x_1, \ldots, x_n)\, dx_n.$$

Here the differential $dz$ is considered to be a function of the $2n$ independent variables $x_1, x_2, \ldots, x_n, dx_1, dx_2, \ldots, dx_n$.

For a differentiable function $f$, the differential $df$ is an approximation to the change $\Delta f$ in value of the function given by

$$\Delta f = f(x_1 + dx_1, \ldots, x_n + dx_n) - f(x_1, \ldots, x_n).$$

The error in this approximation is small compared to the distance between the two points in the domain of $f$, that is,

$$\frac{\Delta f - df}{\sqrt{(dx_1)^2 + \cdots + (dx_n)^2}} \to 0 \quad \text{if all } dx_i \to 0, \qquad (1 \le i \le n).$$

In this sense, differentials are just another way of looking at linearization.

**Example 3**    Find the percentage change in the period

$$T = 2\pi \sqrt{\frac{L}{g}}$$

of a simple pendulum if the length $L$ of the pendulum increases by 2% and the acceleration of gravity $g$ decreases by 0.6%.

**Solution**    We calculate the differential of $T$:

$$dT = \frac{\partial T}{\partial L} dL + \frac{\partial T}{\partial g} dg$$

$$= \frac{2\pi}{2\sqrt{Lg}} dL - \frac{2\pi \sqrt{L}}{2g^{3/2}} dg.$$

We are given that $dL = \dfrac{2}{100} L$ and $dg = -\dfrac{6}{1,000} g$. Thus

$$dT = \frac{1}{100} 2\pi \sqrt{\frac{L}{g}} - \left(-\frac{6}{1,000}\right) \frac{2\pi}{2} \sqrt{\frac{L}{g}} = \frac{13}{1,000} T.$$

Therefore the period $T$ of the pendulum increases by 1.3%. ∎

## Functions from *n*-space to *m*-space

This is an optional topic. A vector $\mathbf{f} = (f_1, f_2, \ldots, f_m)$ of $m$ functions, each depending on $n$ variables $(x_2, x_2, \ldots, x_n)$ defines a *transformation* (that is, a function) from $\mathbb{R}^n$ to $\mathbb{R}^m$; specifically, if $\mathbf{x} = (x_2, x_2, \ldots, x_n)$ is a point in $\mathbb{R}^n$, and

$$y_1 = f_1(x_1, x_2, \ldots, x_n)$$
$$y_2 = f_2(x_1, x_2, \ldots, x_n)$$
$$\vdots$$
$$y_m = f_m(x_1, x_2, \ldots, x_n),$$

then $\mathbf{y} = (y_1, y_2, \ldots, y_m)$ is the point in $\mathbb{R}^m$ that corresponds to $\mathbf{x}$ under the transformation $\mathbf{f}$. We can write these equations more compactly as

$$\mathbf{y} = \mathbf{f}(\mathbf{x}).$$

Information about the rate of change of $\mathbf{y}$ with respect to $\mathbf{x}$ is contained in the various partial derivatives $\partial y_i / \partial x_j$ ($1 \le i \le m$, $1 \le j \le n$), and is conveniently organized into an $m \times n$ matrix, $D\mathbf{f}(\mathbf{x})$, called the **Jacobian matrix** of the transformation $f$:

$$D\mathbf{f}(\mathbf{x}) = \begin{pmatrix} \dfrac{\partial y_1}{\partial x_1} & \dfrac{\partial y_1}{\partial x_2} & \cdots & \dfrac{\partial y_1}{\partial x_n} \\ \dfrac{\partial y_2}{\partial x_1} & \dfrac{\partial y_2}{\partial x_2} & \cdots & \dfrac{\partial y_2}{\partial x_n} \\ \vdots & \vdots & & \vdots \\ \dfrac{\partial y_m}{\partial x_1} & \dfrac{\partial y_m}{\partial x_2} & \cdots & \dfrac{\partial y_m}{\partial x_n} \end{pmatrix}$$

The linear transformation (see Section 11.6) represented by the Jacobian matrix is called **the derivative** of the transformation $\mathbf{f}$.

It is not our purpose to enter into a study of such *vector-valued functions of a vector variable* at this point, but we can observe here that the composition of two such transformations corresponds to forming the matrix product of their Jacobian matrices.

To see this, let $\mathbf{y} = \mathbf{f}(\mathbf{x})$ be a transformation from $\mathbb{R}^n$ to $\mathbb{R}^m$ as described above, and let $\mathbf{z} = \mathbf{g}(\mathbf{y})$ be another such transformation from $\mathbb{R}^m$ to $\mathbb{R}^k$ given by

$$z_1 = g_1(y_1, y_2, \ldots, y_m)$$
$$z_2 = g_2(y_1, y_2, \ldots, y_m)$$
$$\vdots$$
$$z_k = g_k(y_1, y_2, \ldots, y_m),$$

which has the $k \times m$ Jacobian matrix

$$D\mathbf{g}(\mathbf{y}) = \begin{pmatrix} \dfrac{\partial z_1}{\partial y_1} & \dfrac{\partial z_1}{\partial y_2} & \cdots & \dfrac{\partial z_1}{\partial y_m} \\ \dfrac{\partial z_2}{\partial y_1} & \dfrac{\partial z_2}{\partial y_2} & \cdots & \dfrac{\partial z_2}{\partial y_m} \\ \vdots & \vdots & & \vdots \\ \dfrac{\partial z_k}{\partial y_1} & \dfrac{\partial z_k}{\partial y_2} & \cdots & \dfrac{\partial z_k}{\partial y_m} \end{pmatrix}$$

Then the composition $\mathbf{z} = \mathbf{g} \circ \mathbf{f}(\mathbf{x}) = \mathbf{g}(\mathbf{f}(\mathbf{x}))$ given by

$$z_1 = g_1\big(f_1(x_1, \ldots, x_n), \ldots, f_m(x_1, \ldots, x_n)\big)$$
$$z_2 = g_2\big(f_1(x_1, \ldots, x_n), \ldots, f_m(x_1, \ldots, x_n)\big)$$
$$\vdots$$
$$z_k = g_k\big(f_1(x_1, \ldots, x_n), \ldots, f_m(x_1, \ldots, x_n)\big)$$

has, according to the Chain Rule, the $k \times n$ Jacobian matrix

$$\begin{pmatrix} \dfrac{\partial z_1}{\partial x_1} & \dfrac{\partial z_1}{\partial x_2} & \cdots & \dfrac{\partial z_1}{\partial x_n} \\ \dfrac{\partial z_2}{\partial x_1} & \dfrac{\partial z_2}{\partial x_2} & \cdots & \dfrac{\partial z_2}{\partial x_n} \\ \vdots & \vdots & & \vdots \\ \dfrac{\partial z_k}{\partial x_1} & \dfrac{\partial z_k}{\partial x_2} & \cdots & \dfrac{\partial z_k}{\partial x_n} \end{pmatrix} = \begin{pmatrix} \dfrac{\partial z_1}{\partial y_1} & \dfrac{\partial z_1}{\partial y_2} & \cdots & \dfrac{\partial z_1}{\partial y_m} \\ \dfrac{\partial z_2}{\partial y_1} & \dfrac{\partial z_2}{\partial y_2} & \cdots & \dfrac{\partial z_2}{\partial y_m} \\ \vdots & \vdots & & \vdots \\ \dfrac{\partial z_k}{\partial y_1} & \dfrac{\partial z_k}{\partial y_2} & \cdots & \dfrac{\partial z_k}{\partial y_m} \end{pmatrix} \begin{pmatrix} \dfrac{\partial y_1}{\partial x_1} & \dfrac{\partial y_1}{\partial x_2} & \cdots & \dfrac{\partial y_1}{\partial x_n} \\ \dfrac{\partial y_2}{\partial x_1} & \dfrac{\partial y_2}{\partial x_2} & \cdots & \dfrac{\partial y_2}{\partial x_n} \\ \vdots & \vdots & & \vdots \\ \dfrac{\partial y_m}{\partial x_1} & \dfrac{\partial y_m}{\partial x_2} & \cdots & \dfrac{\partial y_m}{\partial x_n} \end{pmatrix}.$$

This is, in fact, the Chain Rule for compositions of transformations:

$$D(\mathbf{g} \circ \mathbf{f})(\mathbf{x}) = D\mathbf{g}(\mathbf{f}(\mathbf{x}))D\mathbf{f}(\mathbf{x}).$$

In this context we can regard the function $f(x, y)$, say, as a transformation from $\mathbb{R}^2$ to $\mathbb{R}$. Its derivative is the linear transformation with matrix

$$Df(x, y) = \big(f_1(x, y), f_2(x, y)\big).$$

The transformation $\mathbf{y} = \mathbf{f}(\mathbf{x})$ also defines a vector $d\mathbf{y}$ of differentials of the variables $y_i$ in terms of the vector $d\mathbf{x}$ of differentials of the variables $x_j$. Writing $d\mathbf{y}$ and $d\mathbf{x}$ as column vectors we have

$$d\mathbf{y} = \begin{pmatrix} dy_1 \\ dy_2 \\ \vdots \\ dy_m \end{pmatrix} = \begin{pmatrix} \dfrac{\partial y_1}{\partial x_1} & \dfrac{\partial y_1}{\partial x_2} & \cdots & \dfrac{\partial y_1}{\partial x_n} \\ \dfrac{\partial y_2}{\partial x_1} & \dfrac{\partial y_2}{\partial x_2} & \cdots & \dfrac{\partial y_2}{\partial x_n} \\ \vdots & \vdots & & \vdots \\ \dfrac{\partial y_m}{\partial x_1} & \dfrac{\partial y_m}{\partial x_2} & \cdots & \dfrac{\partial y_m}{\partial x_n} \end{pmatrix} \begin{pmatrix} dx_1 \\ dx_2 \\ \vdots \\ dx_n \end{pmatrix} = D\mathbf{f}(\mathbf{x})d\mathbf{x}.$$

**Example 4** Find the Jacobian matrix $D\mathbf{f}(1, 0)$ for the transformation from $\mathbb{R}^2$ to $\mathbb{R}^3$ given by

$$\mathbf{f}(x, y) = \left(xe^y + \cos(\pi y), x^2, x - e^y\right)$$

and use it to find an approximate value for $\mathbf{f}(1.02, 0.01)$.

**Solution** $D\mathbf{f}(x, y)$ is the $3 \times 2$ matrix whose $j$th row consists of the partial derivatives of the $j$th component of $\mathbf{f}$ with respect to $x$ and $y$. Thus

$$D\mathbf{f}(1, 0) = \begin{pmatrix} e^y & xe^y - \pi\sin(\pi y) \\ 2x & 0 \\ 1 & -e^y \end{pmatrix}\Bigg|_{(1,0)} = \begin{pmatrix} 1 & 1 \\ 2 & 0 \\ 1 & -1 \end{pmatrix}.$$

Since $f(1, 0) = (2, 1, 0)$ and $d\mathbf{x} = \begin{pmatrix} 0.02 \\ 0.01 \end{pmatrix}$, we have

$$d\mathbf{f} = D\mathbf{f}(1, 0)\,d\mathbf{x} = \begin{pmatrix} 1 & 1 \\ 2 & 0 \\ 1 & -1 \end{pmatrix}\begin{pmatrix} 0.02 \\ 0.01 \end{pmatrix} = \begin{pmatrix} 0.03 \\ 0.04 \\ 0.01 \end{pmatrix}.$$

Therefore, $\mathbf{f}(1.02, 0.01) \approx (2.03, 1.04, 0.01)$.

For transformations between spaces of the same dimension (say from $\mathbb{R}^n$ to $\mathbb{R}^n$) the Jacobian matrices are square, and have determinants. These Jacobian determinants will play a role in our consideration of implicit functions and inverse functions in Section 12.8, and in changes of variables in multiple integrals in Chapter 14.

## Exercises 12.6

In Exercises 1–6, use suitable linearizations to find approximate values for the given functions at the points indicated.

1. $f(x, y) = x^2 y^3$ at $(3.1, 0.9)$

2. $f(x, y) = \tan^{-1}\left(\dfrac{y}{x}\right)$ at $(3.01, 2.99)$

3. $f(x, y) = \sin(\pi xy + \ln y)$ at $(0.01, 1.05)$

4. $f(x, y) = \dfrac{24}{x^2 + xy + y^2}$ at $(2.1, 1.8)$

5. $f(x, y, z) = \sqrt{x + 2y + 3z}$ at $(1.9, 1.8, 1.1)$

6. $f(x, y) = x\,e^{y+x^2}$ at $(2.05, -3.92)$

7. The edges of a rectangular box are each measured to within an accuracy of 1% of their values. What is the approximate maximum percentage error in

(a) the calculated volume of the box,

(b) the calculated area of one of the faces of the box, and

(c) the calculated length of a diagonal of the box?

**8.** The radius and height of a right-circular conical tank are measured to be 25 ft and 21 ft, respectively. Each measurement is accurate to within 0.5 in. By about how much can the calculated volume of the tank be in error?

**9.** By approximately how much can the calculated area of the conical surface of the tank in Exercise 8 be in error?

**10.** Two sides and the contained angle of a triangular plot of land are measured to be 224 m, 158 m, and $64°$, respectively. The length measurements were accurate to within 0.4 m and the angle measurement to within $2°$. What is the approximate maximum percentage error if the area of the plot is calculated from these measurements?

**11.** The angle of elevation of the top of a tower is measured at two points $A$ and $B$ on the ground in the same direction from the base of the tower. The angles are $50°$ at $A$ and $35°$ at $B$, each measured to within $1°$. The distance $AB$ is measured to be 100 m with error at most 0.1%. What is the calculated height of the building, and by about how much can it be in error? To which of the three measurements is the calculated height most sensitive?

**12.** By approximately what percentage will the value of

$$w = \frac{x^2 y^3}{z^4}$$ increase or decrease if $x$ increases by 1%, $y$

increases by 2%, and $z$ and increases by 3%?

**13.** Find the Jacobian matrix for the transformation $\mathbf{f}(r, \theta) = (x, y)$, where

$$x = r \cos \theta \quad \text{and} \quad y = r \sin \theta.$$

(Although $(r, \theta)$ can be regarded as *polar coordinates* in the $xy$-plane, they are Cartesian coordinates in their own $r\theta$-plane.)

**14.** Find the Jacobian matrix for the transformation $\mathbf{f}(\rho, \phi, \theta) = (x, y, z)$, where

$$x = \rho \sin \phi \cos \theta, \quad y = \rho \sin \phi \sin \theta, \quad z = \rho \cos \phi.$$

Here $(\rho, \phi, \theta)$ are *spherical coordinates* in $xyz$-space. They will be formally introduced in Section 14.5.

**15.** Find the Jacobian matrix $D\mathbf{f}(x, y, z)$ for the transformation of $\mathbb{R}^3$ to $\mathbb{R}^2$ given by

$$\mathbf{f}(x, y, z) = (x^2 + yz, y^2 - x \ln z).$$

Use $D\mathbf{f}(2, 2, 1)$ to help you find an approximate value for $\mathbf{f}(1.98, 2.01, 1.03)$.

**16.** Find the Jacobian matrix $D\mathbf{g}(1, 3, 3)$ for the transformation of $\mathbb{R}^3$ to $\mathbb{R}^3$ given by

$$\mathbf{g}(r, s, t) = (r^2 s, r^2 t, s^2 - t^2)$$

and use the result to find an approximate value for $\mathbf{g}(0.99, 3.02, 2.97)$.

**17.** Prove that if $f(x, y)$ is differentiable at $(a, b)$, then $f(x, y)$ is continuous at $(a, b)$.

**\* 18.** Prove the following version of the Mean-Value Theorem: if $f(x, y)$ has first partial derivatives continuous near every point of the straight line segment joining the points $(a, b)$ and $(a + h, b + k)$ then there exists a number $\theta$ satisfying $0 < \theta < 1$ such that

$$f(a + h, b + k) = f(a, b) + h f_1(a + \theta h, b + \theta k)$$
$$+ k f_2(a + \theta h, b + \theta k).$$

(*Hint:* apply the single-variable Mean-Value Theorem to $g(t) = f(a + th, b + tk)$.) Why could we not have used this result in place of Theorem 3 to prove Theorem 4 and hence the version of the Chain Rule given in this section?

**\* 19.** Generalize Exercise 18 as follows: show that if $f(x, y)$ has continuous partial derivatives of second order near the point $(a, b)$, then there exists a number $\theta$ satisfying $0 < \theta < 1$ such that for $h$ and $k$ sufficiently small in absolute value,

$$f(a + h, b + k) = f(a, b) + h f_1(a, b) + k f_2(a, b)$$
$$+ h^2 f_{11}(a + \theta h, b + \theta k)$$
$$+ 2hk f_{12}(a + \theta h, b + \theta k)$$
$$+ k^2 f_{22}(a + \theta h, b + \theta k).$$

Hence show that there is a constant $K$ such that for all sufficiently small $h$ and $k$,

$$\left| f(a + h, b + k) - f(a, b) - h f_1(a, b) - k f_2(a, b) \right|$$
$$\le K(h^2 + k^2).$$

## 12.7 Gradients and Directional Derivatives

A first partial derivative of a function of several variables gives the rate of change of that function with respect to distance measured in the direction of one of the coordinate axes. In this section we will develop a method for finding the rate of change of such a function with respect to distance measured in *any direction* in the domain of the function.

To begin, it is useful to combine the first partial derivatives of a function into a single *vector function* called a **gradient**. For simplicity, we will develop and interpret the gradient for functions of two variables. Extension to functions of three or more variables is straightforward and will be discussed later in the section.

**DEFINITION 7**

At any point $(x, y)$ where the first partial derivatives of the function $f(x, y)$ exist, we define the **gradient vector** $\nabla f(x, y) = \mathbf{grad}\ f(x, y)$ by

$$\nabla f(x, y) = \mathbf{grad}\ f(x, y) = f_1(x, y)\mathbf{i} + f_2(x, y)\mathbf{j}.$$

Recall that $\mathbf{i}$ and $\mathbf{j}$ denote the unit basis vectors from the origin to the points $(1,0)$ and $(0,1)$ respectively. The symbol $\nabla$, called *del* or *nabla*, is a *vector differential operator*:

$$\nabla = \mathbf{i}\,\frac{\partial}{\partial x} + \mathbf{j}\,\frac{\partial}{\partial y}.$$

We can *apply* this operator to a function $f(x, y)$ by writing the operator to the left of the function. The result is the gradient of the function

$$\nabla f(x, y) = \left(\mathbf{i}\,\frac{\partial}{\partial x} + \mathbf{j}\,\frac{\partial}{\partial y}\right) f(x, y) = f_1(x, y)\mathbf{i} + f_2(x, y)\mathbf{j}.$$

We will make extensive use of the del operator in Chapter 16.

**Example 1** If $f(x, y) = x^2 + y^2$ then $\nabla f(x, y) = 2x\mathbf{i} + 2y\mathbf{j}$. In particular, $\nabla f(1, 2) = 2\mathbf{i} + 4\mathbf{j}$. Observe that this vector is perpendicular to the tangent line $x + 2y = 5$ to the circle $x^2 + y^2 = 5$ at $(1, 2)$. This circle is the level curve of $f$ that passes through the point $(1, 2)$. (See Figure 12.22(a).) As the following theorem shows, this perpendicularity is not a coincidence. ∎

**Figure 12.22**

(a) The gradient of $f(x, y) = x^2 + y^2$ at $(1, 2)$ is normal to the level curve of $f$ through $(1, 2)$

(b) The angle between $\nabla f(a, b)$ and a chord vector to the level curve of $f$ through $(a, b)$ approaches $90°$ as the length of the chord approaches zero

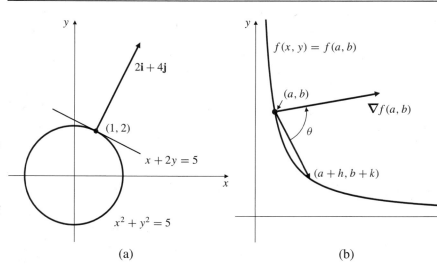

(a)

(b)

**THEOREM  6**

If $f(x, y)$ is differentiable at the point $(a, b)$ and $\nabla f(a, b) \neq \mathbf{0}$, then $\nabla f(a, b)$ is a normal vector to the level curve of $f$ that passes through $(a, b)$.

**PROOF**    The angle $\theta$ between the vector $\nabla f(a, b)$ and the vector $h\mathbf{i} + k\mathbf{j}$ from the point $(a, b)$ to the point $(a + h, b + k)$ (see Figure 12.22(b)) satisfies

$$\cos \theta = \frac{\nabla f(a, b) \bullet (h\mathbf{i} + k\mathbf{j})}{|\nabla f(a, b)|\sqrt{h^2 + k^2}} = \frac{1}{|\nabla f(a, b)|} \frac{hf_1(a, b) + kf_2(a, b)}{\sqrt{h^2 + k^2}}.$$

If $(a + h, b + k)$ lies on the level curve of $f$ that passes through $(a, b)$, then $f(a + h, b + k) = f(a, b)$, and so, since $f$ is differentiable at $(a, b)$, we have

$$\lim_{h, k \to 0} \frac{hf_1(a, b) + kf_2(a, b)}{\sqrt{h^2 + k^2}}$$

$$= -\lim_{h, k \to 0} \frac{f(a + h, b + k) - f(a, b) - hf_1(a, b) - kf_2(a, b)}{\sqrt{h^2 + k^2}} = 0.$$

Hence $\cos \theta \to 0$ and $\theta \to \pi/2$ as $h, k \to 0$. This says that $\nabla f(a, b)$ is perpendicular to the tangent to the level curve $f(x, y) = f(a, b)$.

## Directional Derivatives

The first partial derivatives $f_1(a, b)$ and $f_2(a, b)$ give the rates of change of $f(x, y)$ at $(a, b)$ measured in the directions of the positive $x$- and $y$-axes, respectively. If we want to know how fast $f(x, y)$ changes value as we move through the domain of $f$ at $(a, b)$ in some other direction we require a more general **directional derivative**. We can specify the direction by means of a nonzero vector. It is most convenient to use a *unit vector*.

**DEFINITION  8**

Let $\mathbf{u} = u\mathbf{i} + v\mathbf{j}$ be a unit vector, so that $u^2 + v^2 = 1$. The **directional derivative** of $f(x, y)$ at $(a, b)$ in the direction of $\mathbf{u}$ is the rate of change of $f(x, y)$ with respect to distance measured at $(a, b)$ along a ray in the direction of $\mathbf{u}$ in the $xy$-plane. This directional derivative is given by

$$D_{\mathbf{u}} f(a, b) = \lim_{h \to 0+} \frac{f(a + hu, b + hv) - f(a, b)}{h}.$$

It is also given by

$$D_{\mathbf{u}} f(a, b) = \frac{d}{dt} f(a + tu, b + tv)\Big|_{t=0}$$

if the derivative on the right side exists.

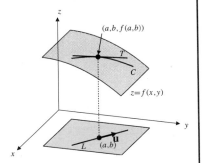

**Figure 12.23**    Unit vector **u** determines a line $L$ through $(a, b)$ in the domain of $f$. The vertical plane containing $L$ intersects the graph of $f$ in a curve $C$ whose tangent $T$ at $(a, b, f(a, b))$ has slope $D_{\mathbf{u}} f(a, b)$

Observe that directional derivatives in directions parallel to the coordinate axes are given directly by first partials: $D_{\mathbf{i}} f(a, b) = f_1(a, b)$, $D_{\mathbf{j}} f(a, b) = f_2(a, b)$, $D_{-\mathbf{i}} f(a, b) = -f_1(a, b)$, and $D_{-\mathbf{j}} f(a, b) = -f_2(a, b)$. The following theorem shows how the gradient can be used to calculate any directional derivative.

**THEOREM 7**

**Using the gradient to find directional derivatives**

If $f$ is differentiable at $(a, b)$ and $\mathbf{u} = u\mathbf{i} + v\mathbf{j}$ is a unit vector, then the directional derivative of $f$ at $(a, b)$ in the direction of $\mathbf{u}$ is given by

$$D_{\mathbf{u}}f(a, b) = \mathbf{u} \bullet \nabla f(a, b).$$

**PROOF** By the Chain Rule:

$$D_{\mathbf{u}}f(a, b) = \frac{d}{dt}f(a + tu, b + tv)\Big|_{t=0}$$
$$= uf_1(a, b) + vf_2(a, b) = \mathbf{u} \bullet \nabla f(a, b).$$

We already know that having partial derivatives at a point does not imply that a function is continuous there, let alone that it is differentiable. The same can be said about directional derivatives. It is possible for a function to have a directional derivative in every direction at a given point and still not be continuous at that point. See Exercise 37 at the end of this section for an example of such a function.

Given any nonzero vector $\mathbf{v}$, we can always obtain a unit vector in the same direction by dividing $\mathbf{v}$ by its length. The directional derivative of $f$ at $(a, b)$ in the direction of $\mathbf{v}$ is therefore given by

$$D_{\mathbf{v}/|\mathbf{v}|}f(a, b) = \frac{\mathbf{v}}{|\mathbf{v}|} \bullet \nabla f(a, b).$$

**Example 2** Find the rate of change of $f(x, y) = y^4 + 2xy^3 + x^2y^2$ at $(0, 1)$ measured in each of the following directions:

(a) $\mathbf{i} + 2\mathbf{j}$,  (b) $\mathbf{j} - 2\mathbf{i}$,  (c) $3\mathbf{i}$, and  (d) $\mathbf{i} + \mathbf{j}$.

**Solution** We calculate

$$\nabla f(x, y) = (2y^3 + 2xy^2)\mathbf{i} + (4y^3 + 6xy^2 + 2x^2y)\mathbf{j},$$
$$\nabla f(0, 1) = 2\mathbf{i} + 4\mathbf{j}.$$

(a) The directional derivative of $f$ at $(0, 1)$ in the direction of $\mathbf{i} + 2\mathbf{j}$ is

$$\frac{\mathbf{i} + 2\mathbf{j}}{|\mathbf{i} + 2\mathbf{j}|} \bullet (2\mathbf{i} + 4\mathbf{j}) = \frac{2 + 8}{\sqrt{5}} = 2\sqrt{5}.$$

Observe that $\mathbf{i} + 2\mathbf{j}$ points in the same direction as $\nabla f(0, 1)$ so the directional derivative is positive and equal to the length of $\nabla f(0, 1)$.

(b) The directional derivative of $f$ at $(0, 1)$ in the direction of $\mathbf{j} - 2\mathbf{i}$ is

$$\frac{-2\mathbf{i} + \mathbf{j}}{|-2\mathbf{i} + \mathbf{j}|} \bullet (2\mathbf{i} + 4\mathbf{j}) = \frac{-4 + 4}{\sqrt{5}} = 0.$$

Since $\mathbf{j} - 2\mathbf{i}$ is perpendicular to $\nabla f(0, 1)$, it is tangent to the level curve of $f$ through $(0, 1)$, so the directional derivative in that direction is zero.

(c) The directional derivative of $f$ at $(0, 1)$ in the direction of $3\mathbf{i}$ is

$$\mathbf{i} \bullet (2\mathbf{i} + 4\mathbf{j}) = 2.$$

As noted previously, the directional derivative of $f$ in the direction of the positive $x$-axis is just $f_1(0, 1)$.

(d) The directional derivative of $f$ at $(0, 1)$ in the direction of $\mathbf{i} + \mathbf{j}$ is

$$\frac{\mathbf{i} + \mathbf{j}}{|\mathbf{i} + \mathbf{j}|} \bullet (2\mathbf{i} + 4\mathbf{j}) = \frac{2 + 4}{\sqrt{2}} = 3\sqrt{2}.$$

If we move along the surface $z = f(x, y)$ through the point $(0, 1, 1)$ in a direction making horizontal angles of 45 degrees with the positive directions of the $x$- and $y$-axes, we would be rising at a rate of $3\sqrt{2}$ vertical units per horizontal unit moved. ∎

**Remark** A direction in the plane can be specified by a polar angle. The direction making angle $\phi$ with the positive direction of the $x$-axis corresponds to the unit vector

$$\mathbf{u}_\phi = \cos\phi\,\mathbf{i} + \sin\phi\,\mathbf{j},$$

so the directional derivative of $f$ at $(x, y)$ in that direction is

$$D_\phi f(x, y) = D_{\mathbf{u}_\phi} f(x, y) = \mathbf{u}_\phi \bullet \nabla f(x, y) = f_1(x, y)\cos\phi + f_2(x, y)\sin\phi.$$

Note the use of the symbol $D_\phi f(x, y)$ to denote a derivative of $f$ with respect to *distance* measured in the direction $\phi$.

As observed in the previous example, Theorem 7 provides a useful interpretation for the gradient vector. For any unit vector $\mathbf{u}$ we have

$$D_{\mathbf{u}} f(a, b) = \mathbf{u} \bullet \nabla f(a, b) = |\nabla f(a, b)|\cos\theta,$$

where $\theta$ is the angle between the vectors $\mathbf{u}$ and $\nabla f(a, b)$. Since $\cos\theta$ only takes on values between $-1$ and $1$, $D_{\mathbf{u}} f(a, b)$ only takes on values between $-|\nabla f(a, b)|$ and $|\nabla f(a, b)|$. Moreover, $D_{\mathbf{u}} f(a, b) = -|\nabla f(a, b)|$ if and only if $\mathbf{u}$ points in the opposite direction to $\nabla f(a, b)$ (so that $\cos\theta = -1$), and $D_{\mathbf{u}} f(a, b) = |\nabla f(a, b)|$ if and only if $\mathbf{u}$ points in the same direction as $\nabla f(a, b)$ (so that $\cos\theta = 1$). The directional derivative is zero in the direction $\theta = \pi/2$; this is the direction of the (tangent line to the) level curve of $f$ through $(a, b)$.

We summarize these properties of the gradient as follows:

**Geometric properties of the gradient vector**

(i) At $(a, b)$, $f(x, y)$ increases most rapidly in the direction of the gradient vector $\nabla f(a, b)$. The maximum rate of increase is $|\nabla f(a, b)|$.

(ii) At $(a, b)$, $f(x, y)$ decreases most rapidly in the direction of $-\nabla f(a, b)$. The maximum rate of decrease is $|\nabla f(a, b)|$.

(iii) The rate of change of $f(x, y)$ at $(a, b)$ is zero in directions tangent to the level curve of $f$ that passes through $(a, b)$.

Look again at the topographic map in Figure 12.6 in Section 12.1. The streams on the map flow in the direction of steepest descent, that is, in the direction of $-\nabla f$, where $f$ measures the elevation of land. The streams therefore cross the contours (the level curves of $f$) at right angles.

**Example 3** The temperature at position $(x, y)$ in a region of the $xy$-plane is $T°C$ where

$$T(x, y) = x^2 e^{-y}.$$

In what direction at the point $(2, 1)$ does the temperature increase most rapidly? What is the rate of increase of $f$ in that direction?

**Solution** We have

$$\nabla T(x, y) = 2x\,e^{-y}\mathbf{i} - x^2\,e^{-y}\mathbf{j},$$

$$\nabla T(2, 1) = \frac{4}{e}\mathbf{i} - \frac{4}{e}\mathbf{j} = \frac{4}{e}(\mathbf{i} - \mathbf{j}).$$

At $(2, 1)$, $T(x, y)$ increases most rapidly in the direction of the vector $\mathbf{i} - \mathbf{j}$. The rate of increase in this direction is $|\nabla T(2, 1)| = 4\sqrt{2}/e$ °C/unit distance.

∎

**Example 4** A hiker is standing beside a stream on the side of a mountain, examining her map of the region. The height of land (in kilometres) at any point $(x, y)$ is given by

$$h(x, y) = \frac{20}{3 + x^2 + 2y^2},$$

where $x$ and $y$ (also in kilometres) denote the coordinates of the point on the hiker's map. The hiker is at the point $(3, 2)$.

(a) What is the direction of flow of the stream at $(3, 2)$ on the hiker's map? How fast is the stream descending at her location?

(b) Find the equation of the path of the stream on the hiker's map.

(c) At what angle to the path of the stream (on the map) should the hiker set out if she wishes to climb at a $15°$ inclination to the horizontal?

(d) Make a sketch of the hiker's map, showing some curves of constant elevation, and showing the stream.

**Solution**

(a) We begin by calculating the gradient of $h$ and its length at $(3, 2)$:

$$\nabla h(x, y) = -\frac{20}{(3 + x^2 + 2y^2)^2}(2x\mathbf{i} + 4y\mathbf{j}),$$

$$\nabla h(3, 2) = -\frac{1}{20}(6\mathbf{i} + 8\mathbf{j}) = -\frac{1}{10}(3\mathbf{i} + 4\mathbf{j}),$$

$$|\nabla h(3, 2)| = 0.5.$$

The stream is flowing in the direction whose horizontal projection at $(3, 2)$ is $-\nabla h(3, 2)$, that is, in the horizontal direction of the vector $3\mathbf{i} + 4\mathbf{j}$. The stream is descending at a rate of 0.5 vertical units per horizontal unit travelled.

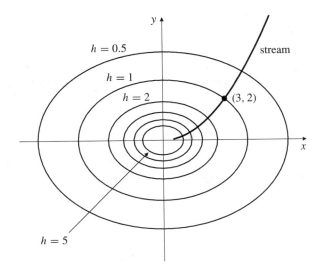

**Figure 12.24** The hiker's map. Unlike most mountains, this one has elliptical cross-sections

(b) Coordinates on the map are the coordinates $(x, y)$ in the domain of the height function $h$. We can find an equation of the path of the stream on a map of the region by setting up a differential equation for a change of position along the path. If the vector $d\mathbf{r} = dx\,\mathbf{i} + dy\,\mathbf{j}$ is tangent to the path of the stream at point $(x, y)$ on the map then $d\mathbf{r}$ is parallel to $\nabla h(x, y)$. Hence the components of these two vectors are proportional:

$$\frac{dx}{2x} = \frac{dy}{4y} \qquad \text{or} \qquad \frac{dy}{y} = \frac{2dx}{x}.$$

Integrating both sides of this equation, we get $\ln y = 2\ln x + \ln C$, or $y = Cx^2$. Since the path of the stream passes through $(3, 2)$, we have $C = 2/9$ and the equation is $9y = 2x^2$.

(c) Suppose the hiker moves away from $(3, 2)$ in the direction of the unit vector $\mathbf{u}$. She will be ascending at an inclination of $15°$ if the directional derivative of $h$ in the direction of $\mathbf{u}$ is $\tan 15° \approx 0.268$. If $\theta$ is the angle between $\mathbf{u}$ and the upstream direction then

$$0.5\cos\theta = |\nabla h(3, 2)|\cos\theta = D_{\mathbf{u}}h(3, 2) \approx 0.268.$$

Hence $\cos\theta \approx 0.536$ and $\theta \approx 57.6°$. She should set out in a direction making a horizontal angle of about $58°$ with the upstream direction.

(d) A suitable sketch of the map is given in Figure 12.24.

---

**Example 5** Find the second directional derivative of $f(x, y)$ in the direction making angle $\phi$ with the positive $x$-axis.

**Solution** As observed earlier, the first directional derivative is

$$D_\phi f(x, y) = (\cos\phi\,\mathbf{i} + \sin\phi\,\mathbf{j}) \bullet \nabla f(x, y) = f_1(x, y)\cos\phi + f_2(x, y)\sin\phi.$$

The second directional derivative is therefore

$$D_\phi^2 f(x, y) = D_\phi \left( D_\phi f(x, y) \right)$$

$$= (\cos\phi \mathbf{i} + \sin\phi \mathbf{j}) \bullet \nabla\left( f_1(x, y)\cos\phi + f_2(x, y)\sin\phi \right)$$

$$= \left( f_{11}(x, y)\cos\phi + f_{21}(x, y)\sin\phi \right)\cos\phi$$

$$+ \left( f_{12}(x, y)\cos\phi + f_{22}(x, y)\sin\phi \right)\sin\phi$$

$$= f_{11}(x, y)\cos^2\phi + 2f_{12}(x, y)\cos\phi\sin\phi + f_{22}(x, y)\sin^2\phi.$$

Note that if $\phi = 0$ or $\phi = \pi$ (so the directional derivative is in a direction parallel to the $x$-axis) then $D_\phi^2 f(x, y) = f_{11}(x, y)$. Similarly, $D_\phi^2 f(x, y) = f_{22}(x, y)$ if $\phi = \pi/2$ or $3\pi/2$.  ∎

## Rates Perceived by a Moving Observer

Suppose that an observer is moving around in the $xy$-plane measuring the value of a function $f(x, y)$ defined in the plane as he passes through each point $(x, y)$. (For instance, $f(x, y)$ might be the temperature at $(x, y)$.) If the observer is moving with velocity $\mathbf{v}$ at the instant when he passes through the point $(a, b)$, how fast would he observe $f(x, y)$ to be changing at that moment?

At the moment in question the observer is moving in the direction of the unit vector $\mathbf{v}/|\mathbf{v}|$. The rate of change of $f(x, y)$ at $(a, b)$ in that direction is

$$D_{\mathbf{v}/|\mathbf{v}|} f(a, b) = \frac{\mathbf{v}}{|\mathbf{v}|} \bullet \nabla f(a, b)$$

measured in units of $f$ per unit distance in the $xy$-plane. To convert this rate to units of $f$ per unit time, we must multiply by the speed of the observer, $|\mathbf{v}|$ units of distance per unit time. Thus the time rate of change of $f(x, y)$ as measured by the observer passing through $(a, b)$ is

$$|\mathbf{v}| \times \frac{\mathbf{v}}{|\mathbf{v}|} \bullet \nabla f(a, b) = \mathbf{v} \bullet \nabla f(a, b).$$

It is natural to extend our use of the symbol $D_{\mathbf{v}} f(a, b)$ to represent this rate even though $\mathbf{v}$ is not (necessarily) a unit vector. Thus

> The rate of change of $f(x, y)$ at $(a, b)$ as measured by an observer moving through $(a, b)$ with velocity $\mathbf{v}$ is
>
> $$D_{\mathbf{v}} f(a, b) = \mathbf{v} \bullet \nabla f(a, b)$$
>
> units of $f$ per unit time.

If the hiker in Example 4 moves away from $(3, 2)$ with horizontal velocity $\mathbf{v} = -\mathbf{i} - \mathbf{j}$ km/h, then she will be rising at a rate

$$\mathbf{v} \bullet \nabla h(3, 2) = (-\mathbf{i} - \mathbf{j}) \bullet \left( -\frac{1}{10}(3\mathbf{i} + 4\mathbf{j}) \right) = \frac{7}{10} \text{ km/h.}$$

As defined here, $D_{\mathbf{v}} f$ is the spatial component of the derivative of $f$ following the motion. See Example 6 in Section 12.5. The rate of change of the reading on the moving thermometer in that example can be expressed as

$$\frac{dT}{dt} = D_{\mathbf{v}}T(x, y, z, t) + \frac{\partial T}{\partial t}$$

where $\mathbf{v}$ is the velocity of the moving thermometer and $D_{\mathbf{v}}T = \mathbf{v} \bullet \nabla T$. The gradient is being taken with respect to the *three spatial variables* only. (See below for the gradient in 3-space.)

## The Gradient in Three and More Dimensions

By analogy with the two-dimensional case, a function $f(x_1, x_2, \ldots, x_n)$ of $n$ variables possessing first partial derivatives has gradient given by

$$\nabla f(x_1, x_2, \ldots, x_n) = \frac{\partial f}{\partial x_1}\mathbf{e_1} + \frac{\partial f}{\partial x_2}\mathbf{e_2} + \cdots + \frac{\partial f}{\partial x_n}\mathbf{e_n},$$

where $\mathbf{e}_j$ is the unit vector from the origin to the unit point on the $j$th coordinate axis. In particular, for a function of three variables,

$$\nabla f(x, y, z) = \frac{\partial f}{\partial x}\mathbf{i} + \frac{\partial f}{\partial y}\mathbf{j} + \frac{\partial f}{\partial z}\mathbf{k}.$$

The level surface of $f(x, y, z)$ passing through $(a, b, c)$ has a tangent plane there if $f$ is differentiable at $(a, b, c)$ and $\nabla f(a, b, c) \neq \mathbf{0}$.

For functions of any number of variables the vector $\nabla f(P_0)$ is normal to the "level surface" of $f$ passing through the point $P_0$ (that is, the (hyper)surface with equation $f(x_1, \ldots, x_n) = f(P_0)$), and, if $f$ is differentiable at $P_0$, the rate of change of $f$ at $P_0$ in the direction of the unit vector $\mathbf{u}$ is given by $\mathbf{u} \bullet \nabla f(P_0)$. Equations of tangent planes to surfaces in 3-space can be found easily with the aid of gradients.

**Example 6** Let $f(x, y, z) = x^2 + y^2 + z^2$.

(a) Find $\nabla f(x, y, z)$ and $\nabla f(1, -1, 2)$.

(b) Find an equation of the tangent plane to the sphere $x^2 + y^2 + z^2 = 6$ at the point $(1, -1, 2)$.

(c) What is the maximum rate of increase of $f$ at $(1, -1, 2)$?

(d) What is the rate of change with respect to distance of $f$ at $(1, -1, 2)$ measured in the direction from that point towards the point $(3, 1, 1)$?

*Solution*

(a) $\nabla f(x, y, z) = 2x\mathbf{i} + 2y\mathbf{j} + 2z\mathbf{k}$, so $\nabla f(1, -1, 2) = 2\mathbf{i} - 2\mathbf{j} + 4\mathbf{k}$.

(b) The required tangent plane has $\nabla f(1, -1, 2)$ as normal. (See Figure 12.25(a).) Therefore its equation is given by $2(x - 1) - 2(y + 1) + 4(z - 2) = 0$ or, more simply, $x - y + 2z = 6$.

(c) The maximum rate of increase of $f$ at $(1, -1, 2)$ is $|\nabla f(1, -1, 2)| = 2\sqrt{6}$, and it occurs in the direction of the vector $\mathbf{i} - \mathbf{j} + 2\mathbf{k}$.

(d) The direction from $(1, -1, 2)$ towards $(3, 1, 1)$ is specified by $2\mathbf{i} + 2\mathbf{j} - \mathbf{k}$. The rate of change of $f$ with respect to distance in this direction is

$$\frac{2\mathbf{i} + 2\mathbf{j} - \mathbf{k}}{\sqrt{4 + 4 + 1}} \bullet (2\mathbf{i} - 2\mathbf{j} + 4\mathbf{k}) = \frac{4 - 4 - 4}{3} = -\frac{4}{3},$$

that is, $f$ decreases at rate $4/3$ of a unit per horizontal unit moved. ∎

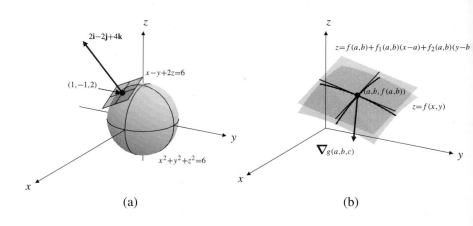

**Figure 12.25**

(a) The tangent plane to
   $x^2 + y^2 + z^2 = 6$ at $(1, -1, 2)$

(b) The gradient of $f(x, y) - z$ at
   $(a, b, f(a, b))$ is normal to the
   tangent plane to $z = f(x, y)$ at
   that point

(a)                                          (b)

**Example 7**   The graph of a *function* $f(x, y)$ of two variables is the graph of the *equation* $z = f(x, y)$ in 3-space. This surface is the level surface $g(x, y, z) = 0$ of the 3-variable function

$$g(x, y, z) = f(x, y) - z.$$

If $f$ is differentiable at $(a, b)$ and $c = f(a, b)$ then $g$ is differentiable at $(a, b, c)$ and

$$\nabla g(a, b, c) = f_1(a, b)\mathbf{i} + f_2(a, b)\mathbf{j} - \mathbf{k}$$

is normal to $g(x, y, z) = 0$ at $(a, b, c)$. (Note that $\nabla g(a, b, c) \neq \mathbf{0}$, since its $z$ component is $-1$.) It follows that the graph of $f$ has nonvertical tangent plane at $(a, b)$ given by

$$f_1(a, b)(x - a) + f_2(a, b)(y - b) - (z - c) = 0,$$

or

$$z = f(a, b) + f_1(a, b)(x - a) + f_2(a, b)(y - b).$$

(See Figure 12.25(b).) This result was obtained by a different argument in Section 12.3.

**DANGER**

Make sure you understand the difference between the graph of a function and a level curve or level surface of that function. (See the discussion following this example.) Here the surface $z = f(x, y)$ is the *graph* of the function $f$, but it is also a *level surface* of a *different* function $g$.

Students sometimes confuse graphs of functions with level curves or surfaces of those functions. In the above example we are talking about a *level surface* of the function $g(x, y, z)$ that happens to coincide with the *graph* of a different function, $f(x, y)$. Do not confuse that surface with the graph of $g$, which is a three-dimensional *hypersurface* in 4-space having equation $w = g(x, y, z)$. Similarly, do not confuse the tangent *plane* to the graph of $f(x, y)$, (that is, the plane obtained in the above example), with the tangent *line* to the level curve of $f(x, y)$ passing through $(a, b)$ and lying in the $xy$-plane. This line has an equation involving only $x$ and $y$, namely $f_1(a, b)(x - a) + f_2(a, b)(y - b) = 0$.

**Example 8**   Find a vector tangent to the curve of intersection of the two surfaces

$$z = x^2 - y^2 \qquad \text{and} \qquad xyz + 30 = 0$$

at the point $(-3, 2, 5)$.

**Solution**    The coordinates of the given point satisfy the equations of both surfaces so the point lies on the curve of intersection of the two surfaces. A vector tangent to this curve at that point will be perpendicular to the normals to both surfaces, that is, to the vectors

$$\mathbf{n}_1 = \nabla(x^2 - y^2 - z)\Big|_{(-3,2,5)} = 2x\mathbf{i} - 2y\mathbf{j} - \mathbf{k}\Big|_{(-3,2,5)} = -6\mathbf{i} - 4\mathbf{j} - \mathbf{k},$$

$$\mathbf{n}_2 = \nabla(xyz + 30)\Big|_{(-3,2,5)} = (yz\mathbf{i} + xz\mathbf{j} + xy\mathbf{k})\Big|_{(-3,2,5)} = 10\mathbf{i} - 15\mathbf{j} - 6\mathbf{k}.$$

For the tangent vector $\mathbf{T}$ we can therefore use the cross product of these normals

$$\mathbf{T} = \mathbf{n}_1 \times \mathbf{n}_2 = \begin{vmatrix} \mathbf{i} & \mathbf{j} & \mathbf{k} \\ -6 & -4 & -1 \\ 10 & -15 & -6 \end{vmatrix} = 9\mathbf{i} - 46\mathbf{j} + 130\mathbf{k}.$$

# Exercises 12.7

In Exercises 1–6, find:

(a) the gradient of the given function at the point indicated,

(b) an equation of the plane tangent to the graph of the given function at the point whose $x$ and $y$ coordinates are given, and

(c) an equation of the straight line tangent, at the given point, to the level curve of the given function passing through that point.

**1.** $f(x, y) = x^2 - y^2$ at $(2, -1)$

**2.** $f(x, y) = \dfrac{x - y}{x + y}$ at $(1, 1)$

**3.** $f(x, y) = \dfrac{x}{x^2 + y^2}$ at $(1, 2)$

**4.** $f(x, y) = e^{xy}$ at $(2, 0)$

**5.** $f(x, y) = \ln(x^2 + y^2)$ at $(1, -2)$

**6.** $f(x, y) = \sqrt{1 + xy^2}$ at $(2, -2)$

In Exercises 7–9, find an equation of the tangent plane to the level surface of the given function that passes through the given point.

**7.** $f(x, y, z) = x^2 y + y^2 z + z^2 x$ at $(1, -1, 1)$

**8.** $f(x, y, z) = \cos(x + 2y + 3z)$ at $(\dfrac{\pi}{2}, \pi, \pi)$

**9.** $f(x, y, z) = y\, e^{-x^2} \sin z$ at $(0, 1, \pi/3)$

In Exercises 10–13, find the rate of change of the given function at the given point in the specified direction.

**10.** $f(x, y) = 3x - 4y$ at $(0, 2)$ in the direction of the vector $-2\mathbf{i}$

**11.** $f(x, y) = x^2 y$ at $(-1, -1)$ in the direction of the vector $\mathbf{i} + 2\mathbf{j}$

**12.** $f(x, y) = \dfrac{x}{1 + y}$ at $(0, 0)$ in the direction of the vector $\mathbf{i} - \mathbf{j}$

**13.** $f(x, y) = x^2 + y^2$ at $(1, -2)$ in the direction making a (positive) angle of $60°$ with the positive $x$-axis

**14.** Let $f(x, y) = \ln |\mathbf{r}|$ where $\mathbf{r} = x\mathbf{i} + y\mathbf{j}$. Show that
$$\nabla f = \frac{\mathbf{r}}{|\mathbf{r}|^2}.$$

**15.** Let $f(x, y, z) = |\mathbf{r}|^{-n}$ where $\mathbf{r} = x\mathbf{i} + y\mathbf{j} + z\mathbf{k}$. Show that
$$\nabla f = \frac{-n\mathbf{r}}{|\mathbf{r}|^{n+2}}.$$

**16.** Show that, in terms of polar coordinates $(r, \theta)$ (where $x = r \cos\theta$, and $y = r \sin\theta$), the gradient of a function $f(r, \theta)$ is given by

$$\nabla f = \frac{\partial f}{\partial r}\hat{\mathbf{r}} + \frac{1}{r}\frac{\partial f}{\partial \theta}\hat{\boldsymbol{\theta}},$$

where $\hat{\mathbf{r}}$ is a unit vector in the direction of the position vector $\mathbf{r} = x\mathbf{i} + y\mathbf{j}$, and $\hat{\boldsymbol{\theta}}$ is a unit vector at right angles to $\hat{\mathbf{r}}$ in the direction of increasing $\theta$.

**17.** In what directions at the point $(2, 0)$ does the function $f(x, y) = xy$ have rate of change $-1$? Are there directions in which the rate is $-3$? How about $-2$?

**18.** In what directions at the point $(a, b, c)$ does the function $f(x, y, z) = x^2 + y^2 - z^2$ increase at half of its maximal rate at that point?

**19.** Find $\nabla f(a, b)$ for the differentiable function $f(x, y)$ given the directional derivatives

$$D_{(\mathbf{i}+\mathbf{j})/\sqrt{2}} f(a, b) = 3\sqrt{2} \text{ and } D_{(3\mathbf{i}-4\mathbf{j})/5} f(a, b) = 5.$$

**20.** If $f(x, y)$ is differentiable at $(a, b)$, what condition should angles $\phi_1$ and $\phi_2$ satisfy in order that the gradient $\nabla f(a, b)$ can be determined from the values of the directional derivatives $D_{\phi_1} f(a, b)$ and $D_{\phi_2} f(a, b)$?

**21.** The temperature $T(x, y)$ at points of the $xy$-plane is given by $T(x, y) = x^2 - 2y^2$.

   (a) Draw a contour diagram for $T$ showing some isotherms (curves of constant temperature).

   (b) In what direction should an ant at position $(2, -1)$ move if it wishes to cool off as quickly as possible?

   (c) If the ant moves in that direction at speed $k$ (units distance per unit time), at what rate does it experience the decrease of temperature?

   (d) At what rate would the ant experience the decrease of temperature if it moved from $(2, -1)$ at speed $k$ in the direction of the vector $-\mathbf{i} - 2\mathbf{j}$?

   (e) Along what curve through $(2, -1)$ should the ant move in order to continue to experience maximum rate of cooling?

**22.** Find an equation of the curve in the $xy$-plane that passes through the point $(1, 1)$ and intersects all level curves of the function $f(x, y) = x^4 + y^2$ at right angles.

**23.** Find an equation of the curve in the $xy$-plane that passes through the point $(2, -1)$ and that intersects every curve with equation of the form $x^2 y^3 = K$ at right angles.

**24.** Find the second directional derivative of $e^{-x^2 - y^2}$ at the point $(a, b) \neq (0, 0)$ in the direction directly away from the origin.

**25.** Find the second directional derivative of $f(x, y, z) = xyz$ at $(2, 3, 1)$ in the direction of the vector $\mathbf{i} - \mathbf{j} - \mathbf{k}$.

**26.** Find a vector tangent to the curve of intersection of the two cylinders $x^2 + y^2 = 2$ and $y^2 + z^2 = 2$ at the point $(1, -1, 1)$

**27.** Repeat Exercise 26 for the surfaces $x + y + z = 6$ and $x^2 + y^2 + z^2 = 14$ and the point $(1, 2, 3)$

**28.** The temperature in 3-space is given by
$$T(x, y, z) = x^2 - y^2 + z^2 + xz^2.$$
At time $t = 0$ a fly passes through the point $(1, 1, 2)$, flying along the curve of intersection of the surfaces $z = 3x^2 - y^2$ and $2x^2 + 2y^2 - z^2 = 0$. If the fly's speed is 7, what rate of temperature change does it experience at $t = 0$?

**29.** State and prove a version of Theorem 6 for a function of three variables.

**30.** What is the level surface of $f(x, y, z) = \cos(x + 2y + 3z)$ that passes through $(\pi, \pi, \pi)$? What is the tangent plane to that level surface at that point? (Compare this exercise with Exercise 8 above.)

**31.** If $\nabla f(x, y) = 0$ throughout the disk $x^2 + y^2 < r^2$ prove that $f(x, y)$ is constant throughout the disk.

**32.** Theorem 6 implies that the level curve of $f(x, y)$ passing through $(a, b)$ is smooth (has a tangent line) at $(a, b)$ provided $f$ is differentiable at $(a, b)$ and satisfies $\nabla f(a, b) \neq \mathbf{0}$. Show that the level curve need not be smooth at $(a, b)$ if $\nabla f(a, b) = \mathbf{0}$. (*Hint:* consider $f(x, y) = y^3 - x^2$ at $(0, 0)$.)

**33.** If $\mathbf{v}$ is a nonzero vector, express $D_{\mathbf{v}}(D_{\mathbf{v}} f)$ in terms of the components of $\mathbf{v}$ and the second partials of $f$. What is the interpretation of this quantity for a moving observer?

**\* 34.** An observer moves so that his position, velocity and acceleration at time $t$ are given by $\mathbf{r}(t) = x(t)\mathbf{i} + y(t)\mathbf{j} + z(t)\mathbf{k}$, $\mathbf{v}(t) = d\mathbf{r}/dt$ and $\mathbf{a}(t) = d\mathbf{v}/dt$. If the temperature in the vicinity of the observer depends only on position, $T = T(x, y, z)$, express the second time derivative of temperature as measured by the observer in terms of $D_{\mathbf{v}}$ and $D_{\mathbf{a}}$.

**\* 35.** Repeat Exercise 34 but with $T$ depending explicitly on time as well as position: $T = T(x, y, z, t)$.

**36.** Let $f(x, y) = \begin{cases} \dfrac{\sin(xy)}{\sqrt{x^2 + y^2}} & \text{if } (x, y) \neq (0, 0) \\ 0 & \text{if } (x, y) = (0, 0). \end{cases}$

   (a) Calculate $\nabla f(0, 0)$.

   (b) Use the definition of directional derivative to calculate $D_{\mathbf{u}} f(0, 0)$, where $\mathbf{u} = (\mathbf{i} + \mathbf{j})/\sqrt{2}$.

   (b) Is $f(x, y)$ differentiable at $(0, 0)$? Why?

**37.** Let $f(x, y) = \begin{cases} 2x^2 y/(x^4 + y^2) & \text{if } (x, y) \neq (0, 0) \\ 0 & \text{if } (x, y) = (0, 0). \end{cases}$
Use the definition of directional derivative as a limit (Definition 8) to show that $D_{\mathbf{u}} f(0, 0)$ exists for every unit vector $\mathbf{u} = u\mathbf{i} + v\mathbf{j}$ in the plane. Specifically, show that $D_{\mathbf{u}} f(0, 0) = 0$ if $v = 0$ and $D_{\mathbf{u}} f(0, 0) = 2u^2/v$ if $v \neq 0$. However, as was shown in Example 3 in Section 12.2, $f(x, y)$ has no limit as $(x, y) \to (0, 0)$, and so is not continuous there. Even if a function has directional derivatives in all directions at a point it may not be continuous at that point.

# 12.8 Implicit Functions

When we study the calculus of functions of one variable we encounter examples of functions that are defined implicitly as solutions of equations in two variables. Suppose, for example, that $F(x, y) = 0$ is such an equation. Suppose that the point $(a, b)$ satisfies the equation, and that $F$ has continuous first partial derivatives (and so is differentiable) at all points near $(a, b)$. Can the equation be solved for $y$ as a function of $x$ near $(a, b)$? That is, does there exist a function $y(x)$ defined in some interval $I = (a - h, a + h)$ (where $h > 0$) satisfying $y(a) = b$ and such that

$$F\left(x, y(x)\right) = 0$$

holds for all $x$ in the interval $I$? If there is such a function $y(x)$ we can try to find its derivative at $x = a$ by differentiating the equation $F(x, y) = 0$ implicitly with respect to $x$, and evaluating the result at $(a, b)$:

$$F_1(x, y) + F_2(x, y) \frac{dy}{dx} = 0,$$

so that

$$\frac{dy}{dx}\bigg|_{x=a} = -\frac{F_1(a, b)}{F_2(a, b)} \qquad \text{provided} \qquad F_2(a, b) \neq 0.$$

Observe, however, that the condition $F_2(a, b) \neq 0$ required for the calculation of $y'(a)$ will itself guarantee that the solution $y(x)$ exists. This condition, together with the differentiability of $F(x, y)$ near $(a, b)$, implies that the level curve $F(x, y) = F(a, b)$ has *nonvertical* tangent lines near $(a, b)$, so some part of the level curve near $(a, b)$ must be the graph of a function of $x$. (See Figure 12.26; the part of the curve $F(x, y) = 0$ in the shaded disk centred at $P_0 = (a, b)$ is the graph of a function $y(x)$ because vertical lines meet that part of the curve only once. The only points on the curve where a disk with that property cannot be drawn are the three points $V_1$, $V_2$, and $V_3$ where the curve has a vertical tangent, that is, where $F_2(x, y) = 0$.) This is a special case of the Implicit Function Theorem, which we will state more generally later in this section.

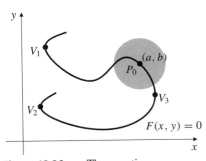

**Figure 12.26** The equation $F(x, y) = 0$ can be solved for $y$ as a function of $x$ near $P_0$, or near any other point except the three points where the curve has a vertical tangent

A similar situation holds for equations involving several variables. We can, for example, ask whether the equation

$$F(x, y, z) = 0$$

defines $z$ as a function of $x$ and $y$ (say $z = z(x, y)$) near some point $P_0 = (x_0, y_0, z_0)$ satisfying the equation. If so, and if $F$ has continuous first partials near $P_0$, then the partial derivatives of $z$ can be found at $(x_0, y_0)$ by implicit differentiation of the equation $F(x, y, z) = 0$ with respect to $x$ and $y$:

$$F_1(x, y, z) + F_3(x, y, z) \frac{\partial z}{\partial x} = 0 \quad \text{and} \quad F_2(x, y, z) + F_3(x, y, z) \frac{\partial z}{\partial y} = 0,$$

so that

$$\frac{\partial z}{\partial x}\bigg|_{(x_0, y_0)} = -\frac{F_1(x_0, y_0, z_0)}{F_3(x_0, y_0, z_0)} \quad \text{and} \quad \frac{\partial z}{\partial y}\bigg|_{(x_0, y_0)} = -\frac{F_2(x_0, y_0, z_0)}{F_3(x_0, y_0, z_0)},$$

provided $F_3(x_0, y_0, z_0) \neq 0$. Since $F_3$ is the $z$ component of the gradient of $F$, this condition implies that the level surface of $F$ through $P_0$ does not have a horizontal normal vector, so it is not vertical (that is, it is not parallel to the $z$-axis). Therefore, part of the surface near $P_0$ must indeed be the graph of a function $z = z(x, y)$. Similarly, $F(x, y, z) = 0$ can be solved for $x$ as a function of $y$ and $z$ near points where $F_1 \neq 0$, and for $y = y(x, z)$ near points where $F_2 \neq 0$.

**Example 1** Near what points on the sphere $x^2 + y^2 + z^2 = 1$ can the equation of the sphere be solved for $z$ as a function of $x$ and $y$? Find $\partial z/\partial x$ and $\partial z/\partial y$ at such points.

**Solution** The sphere is the level surface $F(x, y, z) = 0$ of the function

$$F(x, y, z) = x^2 + y^2 + z^2 - 1.$$

The above equation can be solved for $z = z(x, y)$ near $P_0 = (x_0, y_0, z_0)$ provided that $P_0$ is not on the *equator* of the sphere, that is, the circle $x^2 + y^2 = 1$, $z = 0$. The equator consists of those points that satisfy $F_3(x, y, z) = 0$. If $P_0$ is not on the equator, then it is on either the upper or the lower hemisphere. The upper hemisphere has equation $z = z(x, y) = \sqrt{1 - x^2 - y^2}$ and the lower hemisphere has equation $z = z(x, y) = -\sqrt{1 - x^2 - y^2}$.

If $z \neq 0$, we can calculate the partial derivatives of the solution $z = z(x, y)$ by implicitly differentiating the equation of the sphere: $x^2 + y^2 + z^2 = 1$:

$$2x + 2z\frac{\partial z}{\partial x} = 0, \qquad \text{so} \qquad \frac{\partial z}{\partial x} = -\frac{x}{z},$$

$$2y + 2z\frac{\partial z}{\partial y} = 0, \qquad \text{so} \qquad \frac{\partial z}{\partial y} = -\frac{y}{z}.$$

∎

## Systems of Equations

Experience with linear equations shows us that systems of such equations can generally be solved for as many variables as there are equations in the system. We would expect, therefore, that a pair of equations in several variables might determine two of those variables as functions of the remaining ones. For instance we might expect the equations

$$\begin{cases} F(x, y, z, w) = 0 \\ G(x, y, z, w) = 0 \end{cases}$$

to possess, nearby some point that satisfies them, solutions of one or more of the forms

$$\begin{cases} x = x(z, w) \\ y = y(z, w), \end{cases} \qquad \begin{cases} x = x(y, w) \\ z = z(y, w), \end{cases} \qquad \begin{cases} x = x(y, z) \\ w = w(y, z), \end{cases}$$

$$\begin{cases} y = y(x, w) \\ z = z(x, w), \end{cases} \qquad \begin{cases} y = y(x, z) \\ w = w(x, z), \end{cases} \qquad \begin{cases} z = z(x, y) \\ w = w(x, y). \end{cases}$$

Where such solutions exist, we should be able to differentiate the given system of equations implicitly to find partial derivatives of the solutions.

If you are given a single equation $F(x, y, z) = 0$ and asked to find $\partial x/\partial z$, you would understand that $x$ is intended to be a function of the remaining variables $y$ and $z$, so there would be no chance of misinterpreting which variable is to be held constant in calculating the partial derivative. Suppose, however, that you are asked to calculate $\partial x/\partial z$ given the system $F(x, y, z, w) = 0$, $G(x, y, z, w) = 0$. The question implies that $x$ is one of the dependent variables and $z$ is one of the independent variables, but does not imply which of $y$ and $w$ is the other dependent variable and which is the other independent variable. In short, which of the situations

$$\begin{cases} x = x(z, w) \\ y = y(z, w) \end{cases} \quad \text{and} \quad \begin{cases} x = x(y, z) \\ w = w(y, z) \end{cases}$$

are we dealing with? As it stands, the question is ambiguous. To avoid this ambiguity we can specify *in the notation for the partial derivative* which variable is to be regarded as the other independent variable and therefore *held fixed* during the differentiation. Thus

$$\left(\frac{\partial x}{\partial z}\right)_w \quad \text{implies the interpretation} \quad \begin{cases} x = x(z, w) \\ y = y(z, w), \end{cases}$$

$$\left(\frac{\partial x}{\partial z}\right)_y \quad \text{implies the interpretation} \quad \begin{cases} x = x(y, z) \\ w = w(y, z). \end{cases}$$

**Example 2**    Given the equations $F(x, y, z, w) = 0$ and $G(x, y, z, w) = 0$, where $F$ and $G$ have continuous first partial derivatives, calculate $(\partial x/\partial z)_w$.

**Solution**    We differentiate the two equations with respect to $z$, regarding $x$ and $y$ as functions of $z$ and $w$, and holding $w$ fixed:

$$F_1 \frac{\partial x}{\partial z} + F_2 \frac{\partial y}{\partial z} + F_3 = 0$$

$$G_1 \frac{\partial x}{\partial z} + G_2 \frac{\partial y}{\partial z} + G_3 = 0$$

(Note that the terms $F_4(\partial w/\partial z)$ and $G_4(\partial w/\partial z)$ are not present because $w$ and $z$ are independent variables, and $w$ is being held fixed during the differentiation.) The pair of equations above is linear in $\partial x/\partial z$ and $\partial y/\partial z$. Eliminating $\partial y/\partial z$ (or using Cramer's Rule, Theorem 6 of Section 10.6) we obtain

$$\left(\frac{\partial x}{\partial z}\right)_w = -\frac{F_3 G_2 - F_2 G_3}{F_1 G_2 - F_2 G_1}.$$

In the light of the examples considered above, you should not be too surprised to learn that the nonvanishing of the denominator $F_1 G_2 - F_2 G_1$ at some point $P_0 = (x_0, y_0, z_0, w_0)$ satisfying the system $F = 0$, $G = 0$ is sufficient to guarantee that the system does indeed have a solution of the form $x = x(z, w)$, $y = y(z, w)$ near $P_0$. We will not, however, attempt to prove this fact here.

**Example 3**    Let $x$, $y$, $u$, and $v$ be related by the equations

$$\begin{cases} u = x^2 + xy - y^2 \\ v = 2xy + y^2 \end{cases}$$

Find (a) $(\partial x/\partial u)_v$, and (b) $(\partial x/\partial u)_y$ at the point where $x = 2$ and $y = -1$.

### Solution

(a) To calculate $(\partial x/\partial u)_v$ we regard $x$ and $y$ as functions of $u$ and $v$, and differentiate the given equations with respect to $u$, holding $v$ constant:

$$1 = \frac{\partial u}{\partial u} = (2x + y)\frac{\partial x}{\partial u} + (x - 2y)\frac{\partial y}{\partial u}$$

$$0 = \frac{\partial v}{\partial u} = 2y\frac{\partial x}{\partial u} + (2x + 2y)\frac{\partial y}{\partial u}$$

At $x = 2$, $y = -1$ we have

$$1 = 3\frac{\partial x}{\partial u} + 4\frac{\partial y}{\partial u}$$

$$0 = -2\frac{\partial x}{\partial u} + 2\frac{\partial y}{\partial u}.$$

Eliminating $\partial y/\partial u$ leads to the result $(\partial x/\partial u)_v = 1/7$.

(b) To calculate $(\partial x/\partial u)_y$ we regard $x$ and $v$ as functions of $y$ and $u$, and differentiate the given equations with respect to $u$, holding $y$ constant:

$$1 = \frac{\partial u}{\partial u} = (2x + y)\frac{\partial x}{\partial u}$$

$$\frac{\partial v}{\partial u} = 2y\frac{\partial x}{\partial u}.$$

At $x = 2$, $y = -1$ the first equation immediately gives $(\partial x/\partial u)_y = 1/3$. ∎

It often happens that the independent variables in a problem are either clear from the context or can be chosen at the outset. In either case, ambiguity is not likely to occur and we can safely omit subscripts showing which variables are being held constant. The following example is taken from the thermodynamics of an ideal gas.

**Example 4** (**Reversible changes in an ideal gas**) The state of a closed system containing $n$ moles of an ideal gas is characterized by three state variables, pressure, volume, and temperature ($P$, $V$, and $T$, respectively), which satisfy the equation of state for an ideal gas:

$$PV = nRT,$$

where $R$ is a universal constant. This equation can be regarded as defining one of the three variables as a function of the other two. The internal energy, $E$, and the entropy, $S$, of the system are thermodynamic quantities that depend on the state variables and hence may be expressed as functions of any two of $P$, $V$, and $T$. Let us choose $V$ and $T$ for the independent variables and so write $E = E(V, T)$ and $S = S(V, T)$. For reversible processes in the system, the first and second laws of thermodynamics imply that infinitesimal changes in these quantities satisfy the differential equation

$$T\,dS = dE + P\,dV.$$

Deduce that for such processes, $E$ is independent of $V$ and so depends only on the temperature $T$.

***Solution***   We calculate the differentials $dS$ and $dE$ and substitute them into the differential equation to obtain

$$T\left(\frac{\partial S}{\partial V}\,dV + \frac{\partial S}{\partial T}\,dT\right) = \frac{\partial E}{\partial V}\,dV + \frac{\partial E}{\partial T}\,dT + P\,dV.$$

Divide by $T$, substitute $nR/V$ for $P/T$ (from the equation of state), and collect coefficients of $dV$ and $dT$ on opposite sides of the equation to get

$$\left(\frac{\partial S}{\partial V} - \frac{1}{T}\frac{\partial E}{\partial V} - \frac{nR}{V}\right)dV = \left(\frac{1}{T}\frac{\partial E}{\partial T} - \frac{\partial S}{\partial T}\right)dT.$$

Since $dV$ and $dT$ are independent variables, both coefficients must vanish. Hence

$$\frac{\partial S}{\partial V} = \frac{1}{T}\frac{\partial E}{\partial V} + \frac{nR}{V}$$

$$\frac{\partial S}{\partial T} = \frac{1}{T}\frac{\partial E}{\partial T}.$$

Now differentiate the first of these equations with respect to $T$ and the second with respect to $V$. Using equality of mixed partials for both $S$ and $E$, we obtain the desired result:

$$\frac{\partial}{\partial T}\left(\frac{1}{T}\frac{\partial E}{\partial V} + \frac{nR}{V}\right) = \frac{\partial^2 S}{\partial T\,\partial V} = \frac{\partial^2 S}{\partial V\,\partial T} = \frac{\partial}{\partial V}\left(\frac{1}{T}\frac{\partial E}{\partial T}\right)$$

$$\frac{-1}{T^2}\frac{\partial E}{\partial V} + \frac{1}{T}\frac{\partial^2 E}{\partial T\,\partial V} = \frac{1}{T}\frac{\partial^2 E}{\partial V\,\partial T}$$

$$\frac{-1}{T^2}\frac{\partial E}{\partial V} = 0.$$

It follows that $\partial E/\partial V = 0$, and so $E$ is independent of $V$.   ∎

## Jacobian Determinants

Partial derivatives obtained by implicit differentiation of systems of equations are fractions, the numerators and denominators of which are conveniently expressed in terms of certain determinants called Jacobians.

**DEFINITION 9**

The **Jacobian determinant** (or simply **the Jacobian**) of the two functions, $u = u(x, y)$ and $v = v(x, y)$, with respect to two variables, $x$ and $y$, is the determinant

$$\frac{\partial(u, v)}{\partial(x, y)} = \begin{vmatrix} \dfrac{\partial u}{\partial x} & \dfrac{\partial u}{\partial y} \\[2mm] \dfrac{\partial v}{\partial x} & \dfrac{\partial v}{\partial y} \end{vmatrix}.$$

Similarly, the Jacobian of two functions, $F(x, y, \ldots)$ and $G(x, y, \ldots)$, with respect to the variables, $x$ and $y$, is the determinant

$$\frac{\partial(F, G)}{\partial(x, y)} = \begin{vmatrix} \dfrac{\partial F}{\partial x} & \dfrac{\partial F}{\partial y} \\[2mm] \dfrac{\partial G}{\partial x} & \dfrac{\partial G}{\partial y} \end{vmatrix} = \begin{vmatrix} F_1 & F_2 \\ G_1 & G_2 \end{vmatrix}.$$

The definition above can be extended in the obvious way to give the Jacobian of $n$ functions (or variables) with respect to $n$ variables. For example the Jacobian of three functions, $F$, $G$, and $H$, with respect to three variables, $x$, $y$, and $z$, is the determinant

$$\frac{\partial(F, G, H)}{\partial(x, y, z)} = \begin{vmatrix} F_1 & F_2 & F_3 \\ G_1 & G_2 & G_3 \\ H_1 & H_2 & H_3 \end{vmatrix}.$$

Jacobians are the determinants of the square Jacobian matrices corresponding to transformations of $\mathbb{R}^n$ to $\mathbb{R}^n$ as discussed briefly in Section 12.6.

**Example 5**    In terms of Jacobians, the value of $(\partial x/\partial z)_w$, obtained from the system of equations

$$F(x, y, z, w) = 0, \qquad G(x, y, z, w) = 0$$

in Example 2, can be expressed in the form

$$\left(\frac{\partial x}{\partial z}\right)_w = -\frac{\dfrac{\partial(F, G)}{\partial(z, y)}}{\dfrac{\partial(F, G)}{\partial(x, y)}}.$$

Observe the pattern here. The denominator is the Jacobian of $F$ and $G$ with respect to the two *dependent* variables, $x$ and $y$. The numerator is the same Jacobian except that the dependent variable $x$ is replaced by the independent variable $z$. ∎

The pattern observed above is general. Given $n$ equations in $m + n$ variables,

$$\begin{cases} F_{(1)}(x_1, x_2, \ldots, x_m, y_1, y_2, \ldots, y_n) = 0 \\ F_{(2)}(x_1, x_2, \ldots, x_m, y_1, y_2, \ldots, y_n) = 0 \\ \quad \vdots \\ F_{(n)}(x_1, x_2, \ldots, x_m, y_1, y_2, \ldots, y_n) = 0, \end{cases}$$

we can calculate, for instance,

$$\left(\frac{\partial y_i}{\partial x_j}\right)_{x_1, \ldots, x_{j-1}, x_{j+1}, \ldots, x_m} = -\frac{\dfrac{\partial(F_{(1)}, F_{(2)}, \ldots, F_{(n)})}{\partial(y_1, \ldots, x_j, \ldots, y_n)}}{\dfrac{\partial(F_{(1)}, F_{(2)}, \ldots, F_{(n)})}{\partial(y_1, \ldots, y_i, \ldots, y_n)}}.$$

This is a consequence of Cramer's Rule (Theorem 6 of Section 10.6) applied to the $n$ linear equations in the $n$ unknowns $\partial y_1/\partial x_j, \ldots, \partial y_n/\partial x_j$ obtained by differentiating each of the equations in the given system with respect to $x_j$.

**Example 6** If the equations $x = u^2 + v^2$, and $y = uv$ are solved for $u$ and $v$ in terms of $x$ and $y$, find, where possible,

$$\frac{\partial u}{\partial x}, \qquad \frac{\partial u}{\partial y}, \qquad \frac{\partial v}{\partial x}, \qquad \text{and} \qquad \frac{\partial v}{\partial y}.$$

Hence show that $\dfrac{\partial(u, v)}{\partial(x, y)} = 1 \Big/ \dfrac{\partial(x, y)}{\partial(u, v)}$, provided the denominator is not zero.

**Solution** The given equations can be rewritten in the form

$$F(u, v, x, y) = u^2 + v^2 - x = 0$$
$$G(u, v, x, y) = uv - y = 0.$$

Let

$$J = \frac{\partial(F, G)}{\partial(u, v)} = \begin{vmatrix} 2u & 2v \\ v & u \end{vmatrix} = 2(u^2 - v^2) = \frac{\partial(x, y)}{\partial(u, v)}.$$

If $u^2 \neq v^2$, then $J \neq 0$ and we can calculate the required partial derivatives:

$$\frac{\partial u}{\partial x} = -\frac{1}{J} \frac{\partial(F, G)}{\partial(x, v)} = -\frac{1}{J} \begin{vmatrix} -1 & 2v \\ 0 & u \end{vmatrix} = \frac{u}{2(u^2 - v^2)}$$

$$\frac{\partial u}{\partial y} = -\frac{1}{J} \frac{\partial(F, G)}{\partial(y, v)} = -\frac{1}{J} \begin{vmatrix} 0 & 2v \\ -1 & u \end{vmatrix} = \frac{-2v}{2(u^2 - v^2)}$$

$$\frac{\partial v}{\partial x} = -\frac{1}{J} \frac{\partial(F, G)}{\partial(u, x)} = -\frac{1}{J} \begin{vmatrix} 2u & -1 \\ v & 0 \end{vmatrix} = \frac{-v}{2(u^2 - v^2)}$$

$$\frac{\partial v}{\partial y} = -\frac{1}{J} \frac{\partial(F, G)}{\partial(u, y)} = -\frac{1}{J} \begin{vmatrix} 2u & 0 \\ v & -1 \end{vmatrix} = \frac{2u}{2(u^2 - v^2)}.$$

Thus

$$\frac{\partial(u, v)}{\partial(x, y)} = \frac{1}{J^2} \begin{vmatrix} u & -2v \\ -v & 2u \end{vmatrix} = \frac{J}{J^2} = \frac{1}{J} = \frac{1}{\dfrac{\partial(x, y)}{\partial(u, v)}}.$$

■

**Remark** Note in the above example that $\partial u / \partial x \neq 1/(\partial x / \partial u)$. This should be contrasted with the single-variable situation where, if $y = f(x)$ and $dy/dx \neq 0$ then $x = f^{-1}(y)$ and $dx/dy = 1/(dy/dx)$. This is another reason for distinguishing between "$\partial$" and "$d$". It is the Jacobian rather than any single partial derivative that takes the place of the ordinary derivative in such situations.

**Remark** Let us look briefly at the general case of invertible transformations from $\mathbb{R}^n$ to $\mathbb{R}^n$. Suppose that $\mathbf{y} = \mathbf{f}(\mathbf{x})$ and $\mathbf{z} = \mathbf{g}(\mathbf{y})$ are both functions from $\mathbb{R}^n$ to $\mathbb{R}^n$ whose components have continuous first partial derivatives. As shown in Section

12.6, the Chain Rule implies that

$$
\begin{pmatrix} \dfrac{\partial z_1}{\partial x_1} & \cdots & \dfrac{\partial z_1}{\partial x_n} \\ \vdots & \ddots & \vdots \\ \dfrac{\partial z_n}{\partial x_1} & \cdots & \dfrac{\partial z_n}{\partial x_n} \end{pmatrix} = \begin{pmatrix} \dfrac{\partial z_1}{\partial y_1} & \cdots & \dfrac{\partial z_1}{\partial y_n} \\ \vdots & \ddots & \vdots \\ \dfrac{\partial z_n}{\partial y_1} & \cdots & \dfrac{\partial z_n}{\partial y_n} \end{pmatrix} \begin{pmatrix} \dfrac{\partial y_1}{\partial x_1} & \cdots & \dfrac{\partial y_1}{\partial x_n} \\ \vdots & \ddots & \vdots \\ \dfrac{\partial y_n}{\partial x_1} & \cdots & \dfrac{\partial y_n}{\partial x_n} \end{pmatrix}.
$$

This is just the Chain Rule for the composition $\mathbf{z} = \mathbf{g}(\mathbf{f}(\mathbf{x}))$. It follows from Theorem 3(b) of Section 10.6 that the determinants of these matrices satisfy a similar equation:

$$
\frac{\partial(z_1 \cdots z_n)}{\partial(x_1 \cdots x_n)} = \frac{\partial(z_1 \cdots z_n)}{\partial(y_1 \cdots y_n)} \frac{\partial(y_1 \cdots y_n)}{\partial(x_1 \cdots x_n)}.
$$

If $\mathbf{f}$ is one-to-one and $\mathbf{g}$ is the inverse of $\mathbf{f}$, then $\mathbf{z} = \mathbf{g}(\mathbf{f}(\mathbf{x})) = \mathbf{x}$, and $\partial(z_1 \cdots z_n)/\partial(x_1 \cdots x_n) = 1$, the determinant of the identity matrix. Thus

$$
\frac{\partial(x_1 \cdots x_n)}{\partial(y_1 \cdots y_n)} = \frac{1}{\dfrac{\partial(y_1 \cdots y_n)}{\partial(x_1 \cdots x_n)}}.
$$

In fact, the nonvanishing of either of these determinants is sufficient to guarantee that $\mathbf{f}$ is one-to-one and so has an inverse. This is a special case of the Implicit Function Theorem stated below.

We will encounter Jacobians again when we study transformations of coordinates in multiple integrals in Chapter 14.

## The Implicit Function Theorem

The Implicit Function Theorem guarantees that systems of equations can be solved for certain variables as functions of other variables under certain circumstances. Before stating it we consider a simple illustrative example.

**Example 7**  Consider the system of linear equations

$$
F(x, y, s, t) = a_1 x + b_1 y + c_1 s + d_1 t + e_1 = 0
$$
$$
G(x, y, s, t) = a_2 x + b_2 y + c_2 s + d_2 t + e_2 = 0.
$$

This system can be written in matrix form:

$$
A \begin{pmatrix} x \\ y \end{pmatrix} + C \begin{pmatrix} s \\ t \end{pmatrix} + \mathcal{E} = \begin{pmatrix} 0 \\ 0 \end{pmatrix},
$$

where

$$
A = \begin{pmatrix} a_1 & b_1 \\ a_2 & b_2 \end{pmatrix}, \qquad C = \begin{pmatrix} c_1 & d_1 \\ c_2 & d_2 \end{pmatrix}, \qquad \text{and} \qquad \mathcal{E} = \begin{pmatrix} e_1 \\ e_2 \end{pmatrix}.
$$

The equations can be solved for $x$ and $y$ as functions of $s$ and $t$ provided $\det(\mathcal{A}) \neq 0$, for this implies the existence of the inverse matrix $\mathcal{A}^{-1}$ (Theorem 3 of Section 11.6), so

$$\begin{pmatrix} x \\ y \end{pmatrix} = -\mathcal{A}^{-1} \left( \mathcal{C} \begin{pmatrix} s \\ t \end{pmatrix} + \mathcal{E} \right).$$

Observe that $\det(\mathcal{A}) = \partial(F, G)/\partial(x, y)$, so the nonvanishing of this Jacobian guarantees that the equations can be solved for $x$ and $y$. ∎

We conclude this section with a statement, without proof, of a general form of the Implicit Function Theorem.

**THEOREM 8**

**The Implicit Function Theorem**

Consider a system of $n$ equations in $n + m$ variables,

$$\begin{cases} F_{(1)}(x_1, x_2, \ldots, x_m, y_1, y_2, \ldots, y_n) = 0 \\ F_{(2)}(x_1, x_2, \ldots, x_m, y_1, y_2, \ldots, y_n) = 0 \\ \quad \vdots \\ F_{(n)}(x_1, x_2, \ldots, x_m, y_1, y_2, \ldots, y_n) = 0, \end{cases}$$

and a point $P_0 = (a_1, a_2, \ldots, a_m, b_1, b_2, \ldots, b_n)$ that satisfies the system. Suppose each of the functions $F_{(i)}$ has continuous first partial derivatives with respect to each of the variables $x_j$ and $y_k$, $(i = 1, \ldots, n, \ j = 1, \ldots, m, \ k = 1, \ldots, n)$, near $P_0$. Finally, suppose that

$$\left. \frac{\partial(F_{(1)}, F_{(2)}, \ldots, F_{(n)})}{\partial(y_1, y_2, \ldots, y_n)} \right|_{P_0} \neq 0.$$

Then the system can be solved for $y_1, y_2, \ldots, y_n$ as functions of $x_1, x_2, \ldots, x_m$ near $P_0$. That is, there exist functions

$$\phi_1(x_1, \ldots, x_m), \ldots, \phi_n(x_1, \ldots, x_m)$$

such that

$$\phi_j(a_1, \ldots, a_m) = b_j, \qquad (j = 1, \ldots, n),$$

and such that the equations

$$F_{(1)}\left( x_1, \ldots, x_m, \phi_1(x_1, \ldots, x_m), \ldots, \phi_n(x_1, \ldots, x_m) \right) = 0,$$

$$F_{(2)}\left( x_1, \ldots, x_m, \phi_1(x_1, \ldots, x_m), \ldots, \phi_n(x_1, \ldots, x_m) \right) = 0,$$

$$\vdots$$

$$F_{(n)}\left( x_1, \ldots, x_m, \phi_1(x_1, \ldots, x_m), \ldots, \phi_n(x_1, \ldots, x_m) \right) = 0,$$

hold for all $(x_1, \ldots, x_m)$ sufficiently near $(a_1, \ldots, a_m)$.

**Example 8**   Show that the system

$$\begin{cases} xy^2 + xzu + yv^2 = 3 \\ x^3yz + 2xv - u^2v^2 = 2 \end{cases}$$

can be solved for $(u, v)$ as a (vector) function of $(x, y, z)$ near the point $P_0$ where $(x, y, z, u, v) = (1, 1, 1, 1, 1)$, and find the value of $\partial v / \partial y$ for the solution at $(x, y, z) = (1, 1, 1)$.

**Solution**   Let $\begin{cases} F(x, y, z, u, v) = xy^2 + xzu + yv^2 - 3 \\ G(x, y, z, u, v) = x^3yz + 2xv - u^2v^2 - 2 \end{cases}$. Then

$$\frac{\partial(F, G)}{\partial(u, v)}\bigg|_{P_0} = \begin{vmatrix} xz & 2yv \\ -2uv^2 & 2x - 2u^2v \end{vmatrix}\bigg|_{P_0} = \begin{vmatrix} 1 & 2 \\ -2 & 0 \end{vmatrix} = 4.$$

Since this Jacobian is not zero, the Implicit Function Theorem assures us that the given equations can be solved for $u$ and $v$ as functions of $x$, $y$, and $z$, that is, for $(u, v) = \mathbf{f}(x, y, z)$. Since

$$\frac{\partial(F, G)}{\partial(u, y)}\bigg|_{P_0} = \begin{vmatrix} xz & 2xy + v^2 \\ -2uv^2 & x^3z \end{vmatrix}\bigg|_{P_0} = \begin{vmatrix} 1 & 3 \\ -2 & 1 \end{vmatrix} = 7,$$

we have

$$\left(\frac{\partial v}{\partial y}\right)_{x,z} = -\frac{\dfrac{\partial(F, G)}{\partial(u, y)}}{\dfrac{\partial(F, G)}{\partial(u, v)}}\bigg|_{P_0} = -\frac{7}{4},$$

by the formula preceding Example 6.

# Exercises 12.8

In Exercises 1–12, calculate the indicated derivative from the given equation(s). What condition on the variables will guarantee the existence of a solution that has the indicated derivative? Assume that any general functions $F$, $G$, and $H$ have continuous first partial derivatives.

**1.** $\dfrac{dx}{dy}$ if $xy^3 + x^4y = 2$   **2.** $\dfrac{\partial x}{\partial y}$ if $xy^3 = y - z$

**3.** $\dfrac{\partial z}{\partial y}$ if $z^2 + xy^3 = \dfrac{xz}{y}$   **4.** $\dfrac{\partial y}{\partial z}$ if $e^{yz} - x^2z \ln y = \pi$

**5.** $\dfrac{\partial x}{\partial w}$ if $x^2y^2 + y^2z^2 + z^2t^2 + t^2w^2 - xw = 0$

**6.** $\dfrac{dy}{dx}$ if $F(x, y, x^2 - y^2) = 0$

**7.** $\dfrac{\partial u}{\partial x}$ if $G(x, y, z, u, v) = 0$

**8.** $\dfrac{\partial z}{\partial x}$ if $F(x^2 - z^2, y^2 + xz) = 0$

**9.** $\dfrac{\partial w}{\partial t}$ if $H(u^2w, v^2t, wt) = 0$

**10.** $\left(\dfrac{\partial y}{\partial x}\right)_u$ if $xyuv = 1$ and $x + y + u + v = 0$

**11.** $\left(\dfrac{\partial x}{\partial y}\right)_z$ if $x^2 + y^2 + z^2 + w^2 = 1$, and $x + 2y + 3z + 4w = 2$

**12.** $\dfrac{du}{dx}$ if $x^2y + y^2u - u^3 = 0$ and $x^2 + yu = 1$

**13.** If $x = u^3 + v^3$ and $y = uv - v^2$ are solved for $u$ and $v$ in terms of $x$ and $y$ evaluate

$$\frac{\partial u}{\partial x}, \quad \frac{\partial u}{\partial y}, \quad \frac{\partial v}{\partial x}, \quad \frac{\partial v}{\partial y}, \quad \text{and} \quad \frac{\partial(u, v)}{\partial(x, y)}$$

at the point where $u = 1$ and $v = 1$.

**14.** Near what points $(r, s)$ can the transformation

$$x = r^2 + 2s, \qquad y = s^2 - 2r$$

be solved for $r$ and $s$ as functions of $x$ and $y$? Calculate the values of the first partial derivatives of the solution at the origin.

**15.** Evaluate the Jacobian $\partial(x, y)/\partial(r, \theta)$ for the transformation to polar coordinates: $x = r \cos \theta$, $y = r \sin \theta$. Near what points $(r, \theta)$ is the transformation one-to-one and therefore invertible to give $r$ and $\theta$ as functions of $x$ and $y$?

**16.** Evaluate the Jacobian $\partial(x, y, z)/\partial(\rho, \phi, \theta)$ where

$$x = \rho \sin \phi \cos \theta, \ \ y = \rho \sin \phi \sin \theta, \ \text{and} \ z = \rho \cos \phi.$$

This is the transformation from Cartesian to spherical polar coordinates in 3-space that we will consider in Section 14.6. Near what points is the transformation one-to-one and hence invertible to give $\rho$, $\phi$, and $\theta$ as functions of $x$, $y$, and $z$?

**17.** Show that the equations

$$\begin{cases} xy^2 + zu + v^2 = 3 \\ x^3 z + 2y - uv = 2 \\ xu + yv - xyz = 1 \end{cases}$$

can be solved for $x$, $y$, and $z$ as functions of $u$ and $v$ near the point $P_0$ where $(x, y, z, u, v) = (1, 1, 1, 1, 1)$, and find $(\partial y/\partial u)_v$ at $(u, v) = (1, 1)$.

**18.** Show that the equations $\begin{cases} xe^y + uz - \cos v = 2 \\ u \cos y + x^2 v - yz^2 = 1 \end{cases}$ can be solved for $u$ and $v$ as functions of $x$, $y$, and $z$ near the point $P_0$ where $(x, y, z) = (2, 0, 1)$ and $(u, v) = (1, 0)$, and find $(\partial u/\partial z)_{x,y}$ at $(x, y, z) = (2, 0, 1)$.

**19.** Find $dx/dy$ from the system

$$F(x, y, z, w) = 0, \ \ G(x, y, z, w) = 0, \ \ H(x, y, z, w) = 0.$$

**20.** Given the system

$$F(x, y, z, u, v) = 0$$
$$G(x, y, z, u, v) = 0$$
$$H(x, y, z, u, v) = 0,$$

how many possible interpretations are there for $\partial x/\partial y$? Evaluate them.

**21.** Given the system

$$F(x_1, x_2, \ldots, x_8) = 0$$
$$G(x_1, x_2, \ldots, x_8) = 0$$
$$H(x_1, x_2, \ldots, x_8) = 0,$$

how many possible interpretations are there for the partial $\partial x_1/\partial x_2$? Evaluate $\left( \partial x_1/\partial x_2 \right)_{x_4, x_6, x_7, x_8}$.

**22.** If $F(x, y, z) = 0$ determines $z$ as a function of $x$ and $y$, calculate $\partial^2 z/\partial x^2$, $\partial^2 z/\partial x \partial y$ and $\partial^2 z/\partial y^2$ in terms of the partial derivatives of $F$.

**23.** If $x = u + v$, $y = uv$ and $z = u^2 + v^2$ define $z$ as a function of $x$ and $y$, find $\partial z/\partial x$, $\partial z/\partial y$ and $\partial^2 z/\partial x \partial y$.

**24.** A certain gas satisfies the law $pV = T - \dfrac{4p}{T^2}$,

where $p$ = pressure, $V$ = volume, and $T$ = temperature.
(a) Calculate $\partial T/\partial p$ and $\partial T/\partial V$ at the point where $p = V = 1$ and $T = 2$.
(b) If measurements of $p$ and $V$ yield the values $p = 1 \pm 0.001$ and $V = 1 \pm 0.002$, find the approximate maximum error in the calculated value $T = 2$.

**25.** If $F(x, y, z) = 0$ show that $\left( \dfrac{\partial x}{\partial y} \right)_z \left( \dfrac{\partial y}{\partial z} \right)_x \left( \dfrac{\partial z}{\partial x} \right)_y = -1$.

Derive analogous results for $F(x, y, z, u) = 0$ and for $F(x, y, z, u, v) = 0$. What is the general case?

**\* 26.** If the equations $F(x, y, u, v) = 0$ and $G(x, y, u, v) = 0$ are solved for $x$ and $y$ as functions of $u$ and $v$, show that

$$\frac{\partial(x, y)}{\partial(u, v)} = \frac{\partial(F, G)}{\partial(u, v)} \bigg/ \frac{\partial(F, G)}{\partial(x, y)}.$$

**\* 27.** If the equations $x = f(u, v)$, $y = g(u, v)$ can be solved for $u$ and $v$ in terms of $x$ and $y$, show that

$$\frac{\partial(u, v)}{\partial(x, y)} = 1 \bigg/ \frac{\partial(x, y)}{\partial(u, v)}.$$

*Hint:* use the result of Exercise 26.

**\* 28.** If $x = f(u, v)$, $y = g(u, v)$, $u = h(r, s)$, and $v = k(r, s)$, then $x$ and $y$ can be expressed as functions of $r$ and $s$. Verify by direct calculation that

$$\frac{\partial(x, y)}{\partial(r, s)} = \frac{\partial(x, y)}{\partial(u, v)} \frac{\partial(u, v)}{\partial(r, s)}.$$

This is a special case of the Chain Rule for Jacobians.

**\* 29.** Two functions, $f(x, y)$ and $g(x, y)$, are said to be *functionally dependent* if one is a function of the other; that is, if there exists a single-variable function $k(t)$ such that $f(x, y) = k\big(g(x, y)\big)$ for all $x$ and $y$. Show that in this case $\partial(f, g)/\partial(x, y)$ vanishes identically. Assume that all necessary derivatives exist.

**\* 30.** Prove the converse of the previous exercise as follows: Let $u = f(x, y)$ and $v = g(x, y)$, and suppose that $\partial(u, v)/\partial(x, y) = \partial(f, g)/\partial(x, y)$ is identically zero for all $x$ and $y$. Show that $(\partial u/\partial x)_v$ is identically zero. Hence $u$, considered as a function of $x$ and $v$, is independent of $x$; that is, $u = k(v)$ for some function $k$ of one variable. Why does this imply that $f$ and $g$ are functionally dependent?

## 12.9 Taylor Series and Approximations

As is the case for functions of one variable, power series representations and their partial sums (Taylor polynomials) can provide an efficient method for determining the behavior of a smooth function of several variables near a point in its domain. In this section we will look briefly at the extension of Taylor series to such functions. As usual, we will develop the machinery for functions of two variables, but the extension to more variables should be clear.

As a starting point, recall Taylor's Formula for a function $F(x)$ with continuous partial derivatives of order up to $n + 1$ on the interval $[a, a + h]$:

$$F(a+h) = F(a) + F'(a)h + \frac{F''(a)}{2!}h^2 + \cdots + \frac{F^{(n)}(a)}{n!}h^n + \frac{F^{(n+1)}(X)}{(n+1)!}h^{n+1},$$

where $X$ is some number between $a$ and $a + h$. (The last term in the formula is the *Lagrange* form of the remainder.) In the special case where $a = 0$ and $h = 1$, this formula becomes

$$F(1) = F(0) + F'(0) + \frac{F''(0)}{2!} + \cdots + \frac{F^{(n)}(0)}{n!} + \frac{F^{(n+1)}(\theta)}{(n+1)!}$$

for some $\theta$ between 0 and 1.

Now suppose that $f(x, y)$ has continuous partial derivatives up to order $n + 1$ at all points in an open set containing the line segment joining the points $(a, b)$ and $(a + h, b + k)$ in its domain. We can find the Taylor Formula for $f(a + h, b + k)$ in powers of $h$ and $k$ by applying the one-variable formula above to the function

$$F(t) = f(a + th, b + tk), \qquad 0 \le t \le 1.$$

Clearly $F(0) = f(a, b)$ and $F(1) = f(a + h, b + k)$. Let us calculate some derivatives of $F$:

$$F'(t) = hf_1(a + th, b + tk) + kf_2(a + th, b + tk),$$
$$F''(t) = h^2 f_{11}(a + th, b + tk) + 2hkf_{12}(a + th, b + tk)$$
$$\qquad + k^2 f_{22}(a + th, b + tk),$$
$$F'''(t) = \left( h^3 f_{111} + 3h^2 k f_{112} + 3hk^2 f_{122} + k^3 f_{222} \right)\Big|_{(a+th, b+tk)}.$$

The pattern of binomial coefficients is pretty obvious here, but the notation, involving subscripts to denote partial derivatives of $f$, becomes more and more unwieldy as the order of the derivatives increases. The notation can be simplified greatly by using $D_1 f$ and $D_2 f$ to denote the first partials of $f$ with respect to its first and second variables. Since $h$ and $k$ are constant, and mixed partials commute ($D_1 D_2 f = D_2 D_1 f$), we have

$$(h^2 D_1^2 f + 2hk D_1 D_2 f + k^2 D_2^2 f = (hD_1 + kD_2)^2 f,$$

and so on. Therefore

$$F'(t) = \left( hD_1 + kD_2 \right) f(a + th, b + tk),$$
$$F''(t) = \left( hD_1 + kD_2 \right)^2 f(a + th, b + tk),$$
$$F'''(t) = \left( hD_1 + kD_2 \right)^3 f(a + th, b + tk),$$
$$\vdots$$
$$F^{(m)}(t) = \left( hD_1 + kD_2 \right)^m f(a + th, b + tk).$$

In particular, $F^{(m)}(0) = (hD_1 + kD_2)^m f(a, b)$. Hence the Taylor Formula for $f(a + h, b + k)$ is

$$f(a + h, b + k) = \sum_{m=0}^{n} \frac{1}{m!} (hD_1 + kD_2)^m f(a, b) + R_n(h, k)$$

$$= \sum_{m=0}^{n} \sum_{j=0}^{m} C_{mj} D_1^j D_2^{m-j} f(a, b) h^j k^{m-j} + R_n(h, k),$$

where, using the binomial expansion, we have

$$C_{mj} = \frac{1}{m!} \binom{m}{j} = \frac{1}{m!} \frac{m!}{j!(m-j)!} = \frac{1}{j!(m-j)!},$$

and where the remainder term is given by

$$R_n(h, k) = \frac{1}{(n+1)!} (hD_1 + kD_2)^{n+1} f(a + \theta h, b + \theta k)$$

for some $\theta$ between 0 and 1. If $f$ has partial derivatives of all orders and

$$\lim_{n \to \infty} R_n(h, k) = 0,$$

then $f(a + h, b + k)$ can be expressed as a Taylor series in powers of $h$ and $k$:

$$f(a + h, b + k) = \sum_{m=0}^{\infty} \sum_{j=0}^{m} \frac{1}{j!(m-j)!} D_1^j D_2^{m-j} f(a, b) h^j k^{m-j}.$$

As for functions of one variable, the Taylor polynomial of degree $n$,

$$P_n(x, y) = \sum_{m=0}^{n} \sum_{j=0}^{m} \frac{1}{j!(m-j)!} D_1^j D_2^{m-j} f(a, b) (x - a)^j (y - b)^{m-j},$$

provides the "best" $n$th-degree polynomial approximation to $f(x, y)$ near $(a, b)$. For $n = 1$ this approximation reduces to the tangent plane approximation

$$f(x, y) \approx f(a, b) + f_1(a, b)(x - a) + f_2(a, b)(y - b).$$

**Example 1**   Find a second-degree polynomial approximation to the function $f(x, y) = \sqrt{x^2 + y^3}$ near the point $(1,2)$ and use it to estimate $\sqrt{(1.02)^2 + (1.97)^3}$.

**Solution**   For the second-degree approximation we need the values of the partial derivatives of $f$ up to second order at $(1,2)$. We have

$$f(x, y) = \sqrt{x^2 + y^3} \qquad\qquad f(1, 2) = 3$$

$$f_1(x, y) = \frac{x}{\sqrt{x^2 + y^3}} \qquad\qquad f_1(1, 2) = \frac{1}{3}$$

$$f_2(x, y) = \frac{3y^2}{2\sqrt{x^2 + y^3}} \qquad\qquad f_2(1, 2) = 2$$

$$f_{11}(x, y) = \frac{y^3}{(x^2 + y^3)^{3/2}} \qquad\qquad f_{11}(1, 2) = \frac{8}{27}$$

$$f_{12}(x, y) = \frac{-3xy^2}{2(x^2 + y^3)^{3/2}} \qquad\qquad f_{12}(1, 2) = -\frac{2}{9}$$

$$f_{22}(x, y) = \frac{12x^2 y + 3y^4}{4(x^2 + y^3)^{3/2}} \qquad\qquad f_{22}(1, 2) = \frac{2}{3}.$$

Thus

$$f(1 + h, 2 + k) = 3 + \frac{1}{3}h + 2k + \frac{1}{2!}\left(\frac{8}{27}h^2 + 2(-\frac{2}{9})hk + \frac{2}{3}k^2\right)$$

or, setting $x = 1 + h$ and $y = 2 + k$,

$$f(x, y) = 3 + \frac{1}{3}(x - 1) + 2(y - 2) + \frac{4}{27}(x - 1)^2 - \frac{2}{9}(x - 1)(y - 2) + \frac{1}{3}(y - 2)^2.$$

This is the required second-degree Taylor polynomial for $f$ near (1,2). Therefore

$$\begin{aligned}
\sqrt{(1.02)^2 + (1.97)^3} &= f(1 + 0.02, 2 - 0.03) \\
&\approx 3 + \frac{1}{3}(0.02) + 2(-0.03) + \frac{4}{27}(0.02)^2 \\
&\quad - \frac{2}{9}(0.02)(-0.03) + \frac{1}{3}(-0.03)^2 \\
&\approx 2.9471593.
\end{aligned}$$

(For comparison purposes: the true value is 2.9471636.... The approximation is accurate to six significant figures.) ∎

As observed for functions of one variable, it is not usually necessary to calculate derivatives in order to determine the coefficients in a Taylor series or Taylor polynomial. It is often much easier to perform algebraic manipulations on known series. For instance, the above example could have been done by writing $f$ in the form

$$\begin{aligned}
f(1 + h, 2 + k) &= \sqrt{(1 + h)^2 + (2 + k)^3} \\
&= \sqrt{9 + 2h + h^2 + 12k + 6k^2 + k^3} \\
&= 3\sqrt{1 + \frac{2h + h^2 + 12k + 6k^2 + k^3}{9}}
\end{aligned}$$

and then applying the binomial expansion

$$\sqrt{1 + t} = 1 + \frac{1}{2}t + \frac{1}{2!}\left(\frac{1}{2}\right)\left(-\frac{1}{2}\right)t^2 + \cdots$$

with $t = \dfrac{2h + h^2 + 12k + 6k^2 + k^3}{9}$ to obtain the terms up to second degree in $h$ and $k$.

**Example 2**  Find the Taylor polynomial of degree 3 in powers of $x$ and $y$ for the function $f(x, y) = e^{x-2y}$.

**Solution**  The required Taylor polynomial will be the Taylor polynomial of degree 3 for $e^t$ evaluated at $t = x - 2y$:

$$P_3(x, y) = 1 + (x - 2y) + \frac{1}{2!}(x - 2y)^2 + \frac{1}{3!}(x - 2y)^3$$

$$= 1 + x - 2y + \frac{1}{2}x^2 - 2xy + 2y^2 + \frac{1}{6}x^3 - x^2y + 2xy^2 - \frac{4}{3}y^3.$$

## Approximating Implicit Functions

In the previous section we saw how to determine whether an equation in several variables could be solved for one of those variables as a function of the others. Even when such a solution is known to exist, it is not usually possible to find an exact formula for it. However, if the equation involves only smooth functions, then the solution will have a Taylor series. We can determine at least the first several coefficients in that series, and thus obtain a useful approximation to the solution. The following example shows the technique.

**Example 3**  Show that the equation

$$\sin(x + y) = xy + 2x$$

has a solution of the form $y = f(x)$ near $x = 0$ satisfying $f(0) = 0$, and find the terms up to fourth degree for the Taylor series for $f(x)$ in powers of $x$.

**Solution**  The given equation can be written in the form $F(x, y) = 0$, where

$$F(x, y) = \sin(x + y) - xy - 2x.$$

Since $F(0, 0) = 0$ and $F_2(0, 0) = \cos(0) = 1 \neq 0$, the equation has a solution $y = f(x)$ near $x = 0$ satisfying $f(0) = 0$ by the implicit function theorem. It is not possible to calculate $f(x)$ exactly, but it will have a Maclaurin series of the form

$$y = f(x) = a_1 x + a_2 x^2 + a_3 x^3 + a_4 x^4 + \cdots.$$

(There is no constant term because $f(0) = 0$.) We can substitute this series into the given equation and keep track of terms up to degree 4 in order to calculate the coefficients $a_1$, $a_2$, $a_3$ and $a_4$. For the left side we use the Maclaurin series for sin to obtain

$$\sin(x + y) = \sin\Big((1 + a_1)x + a_2 x^2 + a_3 x^3 + a_4 x^4 + \cdots\Big)$$

$$= (1 + a_1)x + a_2 x^2 + a_3 x^3 + a_4 x^4 + \cdots$$

$$- \frac{1}{3!}\Big((1 + a_1)x + a_2 x^2 + \cdots\Big)^3 + \cdots$$

$$= (1 + a_1)x + a_2 x^2 + \Big(a_3 - \frac{1}{6}(1 + a_1)^3\Big)x^3$$

$$+ \Big(a_4 - \frac{3}{6}(1 + a_1)^2 a_2\Big)x^4 + \cdots.$$

The right side is

$$xy + 2x = 2x + a_1 x^2 + a_2 x^3 + a_3 x^4 + \cdots .$$

Equating coefficients of like powers of $x$, we obtain

$$1 + a_1 = 2 \qquad a_1 = 1$$
$$a_2 = a_1 \qquad a_2 = 1$$
$$a_3 - \frac{1}{6}(1 + a_1)^3 = a_2 \qquad a_3 = \frac{7}{3}$$
$$a_4 - \frac{1}{2}(1 + a_1)^2 a_2 = a_3 \qquad a_4 = \frac{13}{3}.$$

Thus

$$y = f(x) = x + x^2 + \frac{7}{3}x^3 + \frac{13}{3}x^4 + \cdots .$$

(We could have obtained more terms in the series by keeping track of higher powers of $x$ in the substitution process.)

**Remark** From the series for $f(x)$ obtained above we can determine the values of the first four derivatives of $f$ at $x = 0$. Remember that

$$a_k = \frac{f^{(k)}(0)}{k!}.$$

We have, therefore,

$$f'(0) = a_1 = 1 \qquad f''(0) = 2!a_2 = 2$$
$$f'''(0) = 3!a_3 = 14 \qquad f^{(4)}(0) = 4!a_4 = 104.$$

We could have done the example by first calculating these derivatives by implicit differentiation of the given equation and then determined the series coefficients from them. This would have been much the more difficult way to do it. (Try it and see.)

# Exercises 12.9

In Exercises 1–6, find the Taylor series for the given function about the indicated point.

**1.** $f(x, y) = \dfrac{1}{2 + xy^2}; \quad (0, 0)$

**2.** $f(x, y) = \ln(1 + x + y + xy); \quad (0, 0)$

**3.** $f(x, y) = \tan^{-1}(x + xy); \quad (0, -1)$

**4.** $f(x, y) = x^2 + xy + y^3; \quad (1, -1)$

**5.** $f(x, y) = e^{x^2 + y^2}; \quad (0, 0)$

**6.** $f(x, y) = \sin(2x + 3y); \quad (0, 0)$

In Exercises 7–12, find Taylor polynomials of the indicated degree for the given functions near the given point.

**7.** $f(x, y) = \dfrac{1}{2 + x - 2y}$, degree 3, near $(2, 1)$

**8.** $f(x, y) = \ln(x^2 + y^2)$, degree 3, near $(1, 0)$

**9.** $f(x, y) = \displaystyle\int_0^{x+y^2} e^{-t^2} \, dt$, degree 3, near $(0, 0)$

**10.** $f(x, y) = \cos(x + \sin y)$, degree 4, near $(0, 0)$

**11.** $f(x, y) = \dfrac{\sin x}{y}$, degree 2, near $(\frac{\pi}{2}, 1)$

**12.** $f(x, y) = \dfrac{1 + x}{1 + x^2 + y^4}$, degree 2, near $(0, 0)$

In Exercises 13–14, show that the given equation has a solution of the form $y = f(x)$ for $x$ near the indicated point $x = a$, and taking on the indicated value at that point. Find the first three nonzero terms of the Taylor series for $f(x)$ in powers of $x - a$.

* **13.** $x \sin y = y + \sin x$, near $x = 0$, with $f(0) = 0$

* **14.** $\sqrt{1 + xy} = 1 + x + \ln(1 + y)$, near $x = 0$, with $f(0) = 0$

* **15.** Show that the equation $x + 2y + z + e^{2z} = 1$ has a solution of the form $z = f(x, y)$ near $x = 0$, $y = 0$, where

$f(0, 0) = 0$. Find the Taylor polynomial of degree 2 for $f(x, y)$ in powers of $x$ and $y$.

* **16.** Use series methods to find the value of the partial derivative $f_{112}(0, 0)$ given that $f(x, y) = \arctan(x + y)$.

* **17.** Use series methods to evaluate

$$\frac{\partial^{4n}}{\partial x^{2n} \partial y^{2n}} \frac{1}{1 + x^2 + y^2}\bigg|_{(0,0)}.$$

# Chapter Review

## Key Ideas

• **What do the following sentences or phrases mean?**

◇ $\mathcal{S}$ is the graph of $f(x, y)$.

◇ $\mathcal{C}$ is a level curve of $f(x, y)$.

◇ $\lim_{(x, y) \to (a, b)} f(x, y) = L$.

◇ $f(x, y)$ is continuous at $(a, b)$.

◇ the partial derivative $(\partial/\partial x) f(x, y)$

◇ the tangent plane to $z = f(x, y)$ at $(a, b)$

◇ pure second partials      ◇ mixed second partials

◇ $f(x, y)$ is a harmonic function.

◇ $L(x, y)$ is the linearization of $f(x, y)$ at $(a, b)$.

◇ the differential of $z = f(x, y)$

◇ $f(x, y)$ is differentiable at $(a, b)$.

◇ the gradient of $f(x, y)$ at $(a, b)$

◇ the directional derivative of $f(x, y)$ at $(a, b)$ in direction $\mathbf{v}$

◇ the Jacobian determinant $\partial(x, y)/\partial(u, v)$

• **Under what conditions are two mixed partial derivatives equal?**

• **State the Chain Rule for $z = f(x, y)$, where $x = g(u, v)$, and $y = h(u, v)$.**

• **Describe the process of calculating partial derivatives of implicitly defined functions.**

• **What is the Taylor series of $f(x, y)$ about $(a, b)$?**

## Review Exercises

**1.** Sketch some level curves of the function $x + \dfrac{4y^2}{x}$.

**2.** Sketch some isotherms (curves of constant temperature) for the temperature function

$$T = \frac{140 + 30x^2 - 60x + 120y^2}{8 + x^2 - 2x + 4y^2} \quad (°\text{C}).$$

What is the coolest location?

**3.** Sketch some level curves of the polynomial function $f(x, y) = x^3 - 3xy^2$. Why do you think the graph of this function is called a *monkey saddle*?

**4.** Let $f(x, y) = \begin{cases} \dfrac{x^3}{x^2 + y^2} & \text{if } (x, y) \neq (0, 0) \\ 0 & \text{if } (x, y) = (0, 0). \end{cases}$

Calculate each of the following partial derivatives or explain why it does not exist: $f_1(0, 0)$, $f_2(0, 0)$, $f_{21}(0, 0)$, $f_{12}(0, 0)$.

**5.** Let $f(x, y) = \dfrac{x^3 - y^3}{x^2 - y^2}$. Where is $f(x, y)$ continuous? To what additional set of points does $f(x, y)$ have a continuous extension? In particular, can $f$ be extended to be continuous at the origin? Can $f$ be defined at the origin in such a way that its first partial derivatives exist there?

**6.** The surface $\mathcal{S}$ is the graph of the function $z = f(x, y)$, where $f(x, y) = e^{x^2 - 2x - 4y^2 + 5}$.

(a) Find an equation of the tangent plane to $\mathcal{S}$ at the point $(1, -1, 1)$.

(b) Sketch a representative sample of the level curves of the function $f(x, y)$.

**7.** Consider the surface $\mathcal{S}$ with equation $x^2 + y^2 + 4z^2 = 16$.

(a) Find an equation for the tangent plane to $\mathcal{S}$ at the point $(a, b, c)$ on $\mathcal{S}$.

(b) For which points $(a, b, c)$ on $\mathcal{S}$ does the tangent plane to $\mathcal{S}$ at $(a, b, c)$ pass through the point $(0, 0, 4)$? Describe this set of points geometrically.

(c) For which points $(a, b, c)$ on $\mathcal{S}$ is the tangent plane to $\mathcal{S}$ at $(a, b, c)$ parallel to the plane $x + y + 2\sqrt{2}z = 97$?

**8.** Two variable resistors $R_1$ and $R_2$ are connected in parallel so that their combined resistance $R$ is given by

$$\frac{1}{R} = \frac{1}{R_1} + \frac{1}{R_2}.$$

If $R_1 = 100$ ohms $\pm 5\%$ and $R_2 = 25$ ohms $\pm 2\%$, by approximately what percentage can the calculated value of their combined resistance $R = 20$ ohms be in error?

9. You have measured two sides of a triangular field and the angle between them. The side measurements are 150 m and 200 m, each accurate to within $\pm 1$ m. The angle measurement is $30°$, accurate to within $\pm 2°$. What area do you calculate for the field and what is your estimate of the maximum percentage error in this area?

10. Suppose that $T(x, y, z) = x^3 y + y^3 z + z^3 x$ gives the temperature at the point $(x, y, z)$ in 3-space.

   (a) Calculate the directional derivative of $T$ at $(2, -1, 0)$ in the direction towards the point $(1, 1, 2)$.

   (b) A fly is moving through space with constant speed 5. At time $t = 0$ the fly crosses the surface $2x^2 + 3y^2 + z^2 = 11$ at right angles at the point $(2, -1, 0)$, moving in the direction of increasing temperature. Find $dT/dt$ at $t = 0$ as experienced by the fly.

11. Consider the function $f(x, y, z) = x^2 y + yz + z^2$.

   (a) Find the directional derivative of $f$ at $(1, -1, 1)$ in the direction of the vector $\mathbf{i} + \mathbf{k}$.

   (b) An ant is crawling on the plane $x + y + z = 1$ through $(1, -1, 1)$. Suppose it crawls so as to keep $f$ constant. In what direction is it going as it passes through $(1, -1, 1)$?

   (c) Another ant crawls on the plane $x + y + z = 1$, moving in the direction of the greatest rate of increase of $f$. Find its direction as it goes through $(1, -1, 1)$.

12. Let $f(x, y, z) = (x^2 + z^2) \sin \dfrac{\pi xy}{2} + yz^2$. Let $P_0$ be the point $(1, 1, -1)$.

   (a) Find the gradient of $f$ at $P_0$.

   (b) Find the linearization $L(x, y, z)$ of $f$ at $P_0$.

   (c) Find an equation for the tangent plane at $P_0$ to the level surface of $f$ through $P_0$.

   (d) If a bird flies through $P_0$ with speed 5, heading directly towards the point $(2, -1, 1)$, what is the rate of change of $f$ as seen by the bird as it passes through $P_0$?

   (e) In what direction from $P_0$ should the bird fly at speed 5 to experience the greatest rate of increase of $f$?

13. Verify that for any constant $k$ the function $u(x, y) = k\left(\ln \cos(x/k) - \ln \cos(y/k)\right)$ satisfies the *minimal surface equation*

$$(1 + u_x^2)u_{yy} - u_x u_y u_{xy} + (1 + u_y^2)u_{xx} = 0.$$

14. The equations $F(x, y, z) = 0$ and $G(x, y, z) = 0$ can define any two of the variables $x$, $y$, and $z$ as functions of the remaining variable. Show that

$$\frac{dx}{dy} \frac{dy}{dz} \frac{dz}{dx} = 1.$$

15. The equations $\begin{cases} x = u^3 - uv \\ y = 3uv + 2v^2 \end{cases}$ define $u$ and $v$ as functions of $x$ and $y$ near the point $P$ where $(u, v, x, y) = (-1, 2, 1, 2)$.

   (a) Find $\dfrac{\partial u}{\partial x}$ and $\dfrac{\partial u}{\partial y}$ at $P$.

   (b) Find the approximate value of $u$ when $x = 1.02$ and $y = 1.97$.

16. The equations $\begin{cases} u = x^2 + y^2 \\ v = x^2 - 2xy^2 \end{cases}$ define $x$ and $y$ implicitly as functions of $u$ and $v$ for values of $(x, y)$ near $(1, 2)$ and values of $(u, v)$ near $(5, -7)$.

   (a) Find $\dfrac{\partial x}{\partial u}$ and $\dfrac{\partial y}{\partial u}$ at $(u, v) = (5, -7)$.

   (b) If $z = \ln(y^2 - x^2)$, find $\dfrac{\partial z}{\partial u}$ at $(u, v) = (5, -7)$.

## Challenging Problems

1. (a) If the graph of a function $f(x, y)$ that is differentiable at $(a, b)$ contains part of a straight line through $(a, b)$, show that the line lies in the tangent plane to $z = f(x, y)$ at $(a, b)$.

   (b) If $g(t)$ is a differentiable function of $t$, describe the surface $z = yg(x/y)$ and show that all its tangent planes pass through the origin.

2. A particle moves in 3-space in such a way that its direction of motion at any point is perpendicular to the level surface of

$$f(x, y, z) = 4 - x^2 - 2y^2 + 3z^2$$

through that point. If the path of the particle passes through the point $(1, 1, 8)$, show that it also passes through $(2, 4, 1)$. Does it pass through $(3, 7, 0)$?

3. **(The Laplace operator in spherical coordinates)** If $u(x, y, z)$ has continuous second partial derivatives and

$$v(\rho, \phi, \theta) = u(\rho \sin \phi \cos \theta, \rho \sin \phi \sin \theta, \rho \cos \phi),$$

show that

$$\frac{\partial^2 v}{\partial \rho^2} + \frac{2}{\rho} \frac{\partial v}{\partial \rho} + \frac{\cot \phi}{\rho^2} \frac{\partial v}{\partial \phi} + \frac{1}{\rho^2} \frac{\partial^2 v}{\partial \phi^2} + \frac{1}{\rho^2 \sin^2 \phi} \frac{\partial^2 v}{\partial \theta^2}$$
$$= \frac{\partial^2 u}{\partial x^2} + \frac{\partial^2 u}{\partial y^2} + \frac{\partial^2 u}{\partial z^2}.$$

You can do this by hand, but it is a lot easier using computer algebra.

4. **(Spherically expanding waves)** If $f$ is a twice differentiable function of one variable and $\rho = \sqrt{x^2 + y^2 + z^2}$, show that $u(x, y, z, t) = \dfrac{f(\rho - ct)}{\rho}$ satisfies the three-dimensional wave equation

$$\frac{\partial^2 u}{\partial t^2} = c^2 \left( \frac{\partial^2 u}{\partial x^2} + \frac{\partial^2 u}{\partial y^2} + \frac{\partial^2 u}{\partial z^2} \right).$$

What is the geometric significance of this solution as a function of increasing time $t$? *Hint:* you may want to use the result of Exercise 3. In this case $v(\rho, \phi, \theta)$ is independent of $\phi$ and $\theta$.

# CHAPTER 13

# Applications of Partial Derivatives

***Introduction*** In this chapter we will discuss some of the ways partial derivatives contribute to the understanding and solution of problems in applied mathematics. Many such problems can be put in the context of determining maximum or minimum values for functions of several variables, and the first four sections of this chapter deal with that subject. The remaining sections discuss some miscellaneous problems involving the differentiation of functions with respect to parameters, and also Newton's method for approximating solutions of systems of nonlinear equations. Much of the material in this chapter may be considered *optional*. Only Sections 13.1–13.3 contain *core material*, and even parts of those sections can be omitted (for example, the discussion of linear programming in Section 13.2).

## 13.1 Extreme Values

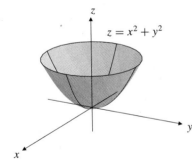

**Figure 13.1** $x^2 + y^2$ has minimum value 0 at the origin

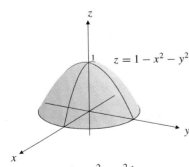

**Figure 13.2** $1 - x^2 - y^2$ has maximum value 1 at the origin

The function $f(x, y) = x^2 + y^2$, part of whose graph is shown in Figure 13.1, has a minimum value of 0; this value occurs at the origin $(0, 0)$. Similarly, the function $g(x, y) = 1 - x^2 - y^2$, part of whose graph appears in Figure 13.2, has a maximum value of 1 at $(0, 0)$. What techniques could be used to discover these facts if they were not evident from a diagram? Finding maximum and minimum values of functions of several variables is, like its single-variable counterpart, the crux of many applications of advanced calculus to problems that arise in other disciplines. Unfortunately, this problem is often much more complicated than in the single-variable case. Our discussion will begin by developing the techniques for functions of two variables. Some of the techniques extend to functions of more variables in obvious ways. The extension of those that do not will be discussed later in the section.

Let us begin by reviewing what we know about the single-variable case. Recall that a function $f(x)$ has a *local maximum value* (or a *local minimum value*) at a point $a$ in its domain if $f(x) \le f(a)$ (or $f(x) \ge f(a)$) for all $x$ in the domain of $f$ that are *sufficiently close* to $a$. If the appropriate inequality holds *for all* $x$ in the domain of $f$, then we say that $f$ has an *absolute maximum* (or *absolute minimum*) value at $a$. Moreover, such local or absolute extreme values can occur only at points of one of the following three types:

(a) critical points — where $f'(x) = 0$, or

(b) singular points — where $f'(x)$ does not exist, or

(c) endpoints of the domain of $f$.

A similar situation exists for functions of several variables. For example, we say that a function of two variables has a **local maximum** or **relative maximum** value at the point $(a, b)$ in its domain if $f(x, y) \le f(a, b)$ for all points $(x, y)$ in the domain of $f$ that are *sufficiently close* to the point $(a, b)$. If the inequality holds *for all* $(x, y)$ in the domain of $f$, then we say that $f$ has a **global maximum** or **absolute maximum**) value at $(a, b)$. Similar definitions obtain for local (relative) and absolute (global) minimum values. In practice, the words *absolute* or *global* are usually omitted, and we refer simply to *the maximum* or *the minimum* value of $f$.

The following theorem shows that there are three possibilities for points where extreme values can occur, analogous to those for the single-variable case.

 **THEOREM 1**

### Necessary conditions for extreme values

A function $f(x, y)$ can have a local or absolute extreme value at a point $(a, b)$ in its domain only if $(a, b)$ is

(a) a **critical point** of $f$, that is, a point satisfying $\nabla f(a, b) = \mathbf{0}$, or

(b) a **singular point** of $f$, that is, a point where $\nabla f(a, b)$ does not exist, or

(c) a **boundary point** of the domain of $f$.

**PROOF** Suppose that $(a, b)$ belongs to the domain of $f$. If $(a, b)$ is not on the boundary of the domain of $f$, then it must belong to the interior of that domain, and if $(a, b)$ is not a singular point of $f$ then $\nabla f(a, b)$ exists. Finally, if $(a, b)$ is not a critical point of $f$, then $\nabla f(a, b) \neq \mathbf{0}$ and so $f$ has a positive directional derivative in the direction of $\nabla f(a, b)$ and a negative directional derivative in the direction of $-\nabla f(a, b)$. That is, $f$ is increasing as we move from $(a, b)$ in one direction and decreasing as we move in the opposite direction. Hence $f$ cannot have either a maximum or a minimum value at $(a, b)$. Therefore, any point where an extreme value occurs must be either a critical point or a singular point of $f$, or a boundary point of the domain of $f$.

Note that Theorem 1 remains valid with unchanged proof for functions of any number of variables. Of course, Theorem 1 does not guarantee that a given function will have any extreme values. It only tells us where to look to find any that may exist. Theorem 2, below, provides conditions that guarantee the existence of absolute maximum and minimum values for a continuous function. It is analogous to the Max-Min Theorem (Section 1.4) for functions of one variable. The proof is beyond the scope of this book; an interested student should consult an elementary text on mathematical analysis. A set in $\mathbb{R}^n$ is **bounded** if it is contained inside some *ball* $x_1^2 + x_2^2 + \cdots + x_n^2 \leq R^2$ of finite radius $R$. A set on the real line is bounded if it is contained in an interval of finite length.

 **THEOREM 2**

### Sufficient conditions for extreme values

If $f$ is a *continuous* function of $n$ variables whose domain is a *closed* and *bounded* set in $\mathbb{R}^n$, then the range of $f$ is a bounded set of real numbers, and there are points in its domain where $f$ takes on absolute maximum and minimum values.

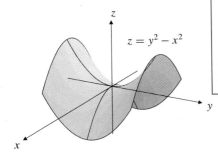

**Example 1**   The function $f(x, y) = x^2 + y^2$ (see Figure 13.1) has a critical point at $(0, 0)$ since $\nabla f = 2x\mathbf{i} + 2y\mathbf{j}$ and both components of $\nabla f$ vanish at $(0, 0)$. Since

$$f(x, y) > 0 = f(0, 0) \quad \text{if} \quad (x, y) \neq (0, 0),$$

$f$ must have (absolute) minimum value 0 at that point. If the domain of $f$ is not restricted $f$ has no maximum value. Similarly, $g(x, y) = 1 - x^2 - y^2$ has (absolute) maximum value 1 at its critical point $(0, 0)$. (See Figure 13.2.) ∎

**Figure 13.3**   $y^2 - x^2$ has a saddle point at $(0, 0)$

**Example 2**   The function $h(x, y) = y^2 - x^2$ also has a critical point at $(0, 0)$ but has neither a local maximum nor a local minimum value at that point. Observe that $h(0, 0) = 0$ but $h(x, 0) < 0$ and $h(0, y) > 0$ for all nonzero values of $x$ and $y$. (See Figure 13.3.) The graph of $h$ is a hyperbolic paraboloid. In view of its shape we call the critical point $(0, 0)$ a **saddle point** of $h$. ∎

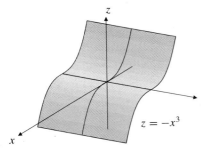

In general we will somewhat loosely call any *interior critical* point of the domain of a function $f$ of several variables a **saddle point** if $f$ does not have a local maximum or minimum value there. Even for functions of two variables, the graph will not always look like a saddle near a saddle point. For instance, the function $f(x, y) = -x^3$ has a whole line of *saddle points* along the $y$-axis (see Figure 13.4), though its graph does not resemble a saddle anywhere. These points resemble inflection points of a function of one variable. Saddle points are higher-dimensional analogues of such horizontal inflection points.

**Figure 13.4**    A line of saddle points

**Example 3**   The function $f(x, y) = \sqrt{x^2 + y^2}$ has no critical points, but does have a singular point at $(0, 0)$ where it has a local (and absolute) minimum value, zero. The graph of $f$ is a circular cone. (See Figure 13.5(a).) ∎

**Figure 13.5**

(a) $\sqrt{x^2 + y^2}$ has a minimum value at the singular point $(0, 0)$

(b) When restricted to the disk $x^2 + y^2 \leq 1$, the function $1 - x$ has maximum and minimum values at boundary points

(a)

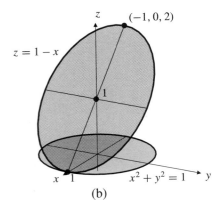

(b)

**Example 4** The function $f(x, y) = 1 - x$ is defined everywhere in the $xy$-plane and has no critical or singular points. ($\nabla f(x, y) = -\mathbf{i}$ at every point $(x, y)$.) Therefore $f$ has no extreme values. However, if we restrict the domain of $f$ to the points in the disk $x^2 + y^2 \le 1$ (a closed bounded set in the $xy$-plane) then $f$ does have absolute maximum and minimum values, as it must by Theorem 2. The maximum value is 2 at the boundary point $(-1, 0)$ and the minimum value is 0 at $(1,0)$. (See Figure 13.5(b).)

∎

## Classifying Critical Points

The above examples were very simple ones; it was immediately obvious in each case whether the function had a local maximum or local minimum or a saddle point at the critical or singular point. For more complicated functions, it may be harder to classify the interior critical points. In theory such a classification can always be made by considering the difference

$$\Delta f = f(a + h, b + k) - f(a, b)$$

for small values of $h$ and $k$, where $(a, b)$ is the critical point in question. If the difference is always nonnegative (or nonpositive) for small $h$ and $k$, then $f$ must have a local minimum (or maximum) at $(a, b)$; if the difference is negative for some points $(h, k)$ arbitrarily near $(0, 0)$, and positive for others, then $f$ must have a saddle point at $(a, b)$.

**Example 5** Find and classify the critical points of $f(x, y) = 2x^3 - 6xy + 3y^2$.

**Solution** The critical points must satisfy the system of equations:

$$0 = f_1(x, y) = 6x^2 - 6y \quad \Longleftrightarrow \quad x^2 = y$$
$$0 = f_2(x, y) = -6x + 6y \quad \Longleftrightarrow \quad x = y.$$

Together, these equations imply that $x^2 = x$ so that $x = 0$ or $x = 1$. Therefore the critical points are $(0, 0)$ and $(1, 1)$.

Consider $(0, 0)$. Here $\Delta f$ is given by

$$\Delta f = f(h, k) - f(0, 0) = 2h^3 - 6hk + 3k^2.$$

Since $f(h, 0) - f(0, 0) = 2h^3$ is positive for small positive $h$ and negative for small negative $h$, $f$ cannot have a maximum or minimum value at $(0, 0)$. Therefore $(0, 0)$ is a saddle point.

Now consider $(1, 1)$. Here $\Delta f$ is given by

$$
\begin{aligned}
\Delta f &= f(1 + h, 1 + k) - f(1, 1) \\
&= 2(1 + h)^3 - 6(1 + h)(1 + k) + 3(1 + k)^2 - (-1) \\
&= 2 + 6h + 6h^2 + 2h^3 - 6 - 6h - 6k - 6hk + 3 + 6k + 3k^2 + 1 \\
&= 6h^2 - 6hk + 3k^2 + 2h^3 \\
&= 3(h - k)^2 + h^2(3 + 2h).
\end{aligned}
$$

Both terms in the latter expression are nonnegative if $|h| < 3/2$, and they are not both zero unless $h = k = 0$. Hence $\Delta f > 0$ for small $h$ and $k$, and $f$ has a local minimum value $-1$ at $(1, 1)$. ∎

The method used to classify critical points in the above example takes on a "brute force" aspect if the function involved is more complicated. However, it provides the basis for developing a *second derivative test* similar to that for functions of one variable. The two-variable version is the subject of the following theorem.

**THEOREM 3**

**A second derivative test**

Suppose that $(a, b)$ is a critical point of $f(x, y)$ interior to the domain of $f$. Suppose also that the second partial derivatives of $f$ are continuous in a neighbourhood of $(a, b)$ and have at that point the values

$$A = f_{11}(a, b), \quad B = f_{12}(a, b) = f_{21}(a, b), \quad \text{and} \quad C = f_{22}(a, b).$$

(a) If $B^2 < AC$ and $A > 0$, then $f$ has a local minimum value at $(a, b)$.

(b) If $B^2 < AC$ and $A < 0$, then $f$ has a local maximum value at $(a, b)$.

(c) If $B^2 > AC$, then $f$ has a saddle point at $(a, b)$.

(d) If $B^2 = AC$, this test provides no information; $f$ may have a local maximum or a local minimum value or a saddle point at $(a, b)$.

**PROOF** Since $f_1(a, b) = f_2(a, b) = 0$, and the second partial derivatives of $f$ are continuous near $(a, b)$, the case $n = 1$ of Taylor's Formula (see Section 12.9) gives

$$f(a + h, b + k) - f(a, b) = \frac{1}{2}\left(\bar{A}h^2 + 2\bar{B}hk + \bar{C}k^2\right) = \frac{1}{2}\bar{Q}(h, k),$$

where $\bar{A}$, $\bar{B}$, and $\bar{C}$ are, respectively, the values of the partial derivatives $f_{11}$, $f_{12}$, and $f_{22}$ at some point $(a + \theta h, b + \theta k)$ on the line segment from $(a, b)$ to $(a + h, b + k)$ (i.e., $\theta$ is between 0 and 1), and where $\bar{Q}(h, k) = \bar{A}h^2 + 2\bar{B}hk + \bar{C}k^2$. In order to prove that $f$ has a local minimum (or a local maximum) at $(a, b)$, we need to show that $\bar{Q}(h, k) \geq 0$ (or $\bar{Q}(h, k) \leq 0$) at *all* points $(h, k)$ sufficiently close to $(0, 0)$. If $\bar{Q}(h, k) > 0$ at some points $(h, k)$ arbitrarily close to $(0, 0)$ and $\bar{Q}(h, k) < 0$ at other such points, then $f$ will have a saddle point at $(a, b)$.

Let $Q(h, k) = Ah^2 + 2Bhk + Ck^2$. Since the second partial derivatives of $f$ are continuous at $(a, b)$, any of the inequalities $A > 0$, $A < 0$, $B^2 < AC$, or $B^2 > AC$ satisfied by the coefficients $A$, $B$, and $C$ of $Q$ will imply the same inequalities for the coefficients $\bar{A}$, $\bar{B}$, and $\bar{C}$ of $\bar{Q}$, provided $(h, k)$ is close enough to $(0, 0)$ so that $(a + \theta h, b + \theta k)$ is sufficiently close to $(a, b)$. Hence it suffices to consider the sign of $Q(h, k)$ under the conditions stated in the theorem.

If $A \neq 0$, then completing the square in the quadratic function $Q$, we obtain

$$Q(h, k) = A\left[\left(h + \frac{B}{A}k\right)^2 + \frac{AC - B^2}{A^2}k^2\right].$$

If $B^2 < AC$, then the expression in the square brackets above is a sum of positive expressions, so, for all $(h, k) \neq (0, 0)$, $Q(h, k)$ will have the same sign as $A$, and $f$ will have a local minimum or maximum at $(a, b)$ according to whether $A > 0$ or $A < 0$.

If $B^2 > AC$, then the expression in the square brackets is a difference of positive quantities. It is positive at points $(h, 0)$ and it is negative at points $(-Bk/A, k)$. Thus $f$ must have a saddle point at $(a, b)$.

If $A = 0$ but $B^2 \neq AC$, then $B \neq 0$ and $Q(h, k) = k(2Bh + Ck)$. This has both positive and negative values in the four angular regions lying between the lines $k = 0$ and $2Bh + Ck = 0$. In this case $f$ must have a saddle point.

Consideration of the functions $f(x, y) = x^4 + y^4$, $g(x, y) = -x^4 - y^4$ and $h(x, y) = x^4 - y^4$ will readily convince you that if $B^2 = AC$, then a function can have a minimum, a maximum or a saddle point.

**Example 6** Reconsider Example 5 and use the second derivative test to classify the two critical points $(0, 0)$ and $(1, 1)$ of $f(x, y) = 2x^3 - 6xy + 3y^2$.

**Solution** We have

$$f_{11}(x, y) = 12x, \quad f_{12}(x, y) = -6, \quad \text{and} \quad f_{22}(x, y) = 6.$$

At $(0, 0)$ we therefore have

$$A = 0, \quad B = -6, \quad C = 6, \quad \text{and} \quad B^2 - AC = 36 > 0$$

so $(0, 0)$ is a saddle point. At $(1, 1)$ we have

$$A = 12 > 0, \quad B = -6, \quad C = 6, \quad \text{and} \quad B^2 - AC = -36 < 0$$

so $f$ must have a local minimum at $(1, 1)$.

**Example 7** Find and classify the critical points of

$$f(x, y) = xy\, e^{-(x^2+y^2)/2}.$$

Does $f$ have absolute maximum and minimum values? Why?

**Solution** We begin by calculating the first- and second-order partial derivatives of $f$:

$$f_1(x, y) = y(1 - x^2)\, e^{-(x^2+y^2)/2},$$
$$f_2(x, y) = x(1 - y^2)\, e^{-(x^2+y^2)/2},$$
$$f_{11}(x, y) = xy(x^2 - 3)\, e^{-(x^2+y^2)/2},$$
$$f_{12}(x, y) = (1 - x^2)(1 - y^2)\, e^{-(x^2+y^2)/2},$$
$$f_{22}(x, y) = xy(y^2 - 3)\, e^{-(x^2+y^2)/2}.$$

At any critical point $f_1 = 0$ and $f_2 = 0$ so the critical points are the solutions of the system of equations

$$y(1 - x^2) = 0$$
$$x(1 - y^2) = 0.$$

The first of these equations says that either $y = 0$ or $x = \pm 1$. The second equation says that either $x = 0$ or $y = \pm 1$. There are five points satisfying both conditions: $(0, 0)$, $(1, 1)$, $(1, -1)$, $(-1, 1)$ and $(-1, -1)$. We classify them using the second derivative test.

At $(0, 0)$ we have $A = C = 0$, $B = 1$, so that $B^2 - AC = 1 > 0$. Thus $f$ has a saddle point at $(0, 0)$.

At $(1, 1)$ and $(-1, -1)$ we have $A = C = -2/e < 0$, $B = 0$. It follows that $B^2 - AC = -4/e^2 < 0$. Thus $f$ has local maximum values at these points. The value of $f$ is $1/e$ at each point.

At $(1, -1)$ and $(-1, 1)$ we have $A = C = 2/e > 0$, $B = 0$. If follows that $B^2 - AC = -4/e^2 < 0$. Thus $f$ has local minimum values at these points. The value of $f$ at each of them is $-1/e$.

Indeed, $f$ has absolute maximum and minimum values—namely, the values obtained above as local extrema. To see why, observe that $f(x, y)$ approaches 0 as the point $(x, y)$ recedes to infinity in any direction, because the negative exponential dominates the power factor $xy$ for large $x^2 + y^2$. Pick a number between 0 and the local maximum value $1/e$ found above, say the number $1/(2e)$. For some $R$, we must have $|f(x, y)| \le 1/(2e)$ whenever $x^2 + y^2 \ge R^2$. On the closed disk $x^2 + y^2 \le R^2$ $f$ must have absolute maximum and minimum values by Theorem 2. These cannot occur on the boundary circle $x^2 + y^2 = R^2$ because $|f|$ is smaller there ($\le 1/(2e)$) than it is at the critical points considered above. Since $f$ has no singular points, the absolute maximum and minimum values for the disk, and therefore for the whole plane, must occur at those critical points. ∎

**Example 8**    Find the dimensions of a rectangular box of given volume $V$ having the least possible surface area.

**Solution**    If the dimensions of the box are $x$, $y$, and $z$, then we want to minimize

$$S = 2xy + 2yz + 2xz$$

subject to the restriction that $xyz = V$, the required volume. We can use this restriction to reduce the number of variables on which $S$ depends, for instance, by substituting

$$z = \frac{V}{xy}.$$

Then $S$ becomes a function of the two variables $x$ and $y$:

$$S = S(x, y) = 2xy + \frac{2V}{x} + \frac{2V}{y}.$$

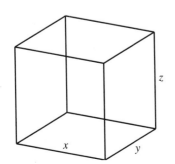

**Figure 13.6**

A real box has positive dimensions, so the domain of $S$ should consist of only those points $(x, y)$ that satisfy $x > 0$ and $y > 0$. If either $x$ or $y$ approach 0 or $\infty$, then $S \to \infty$, so the minimum value of $S$ must occur at a critical point. ($S$ has no singular points.) For critical points we solve the equations

$$0 = \frac{\partial S}{\partial x} = 2y - \frac{2V}{x^2} \qquad \Longleftrightarrow \qquad x^2 y = V,$$

$$0 = \frac{\partial S}{\partial y} = 2x - \frac{2V}{y^2} \qquad \Longleftrightarrow \qquad xy^2 = V.$$

Thus $x^2 y - xy^2 = 0$, or $xy(x - y) = 0$. Since $x > 0$ and $y > 0$, this implies that $x = y$. Therefore $x^3 = V$, and so $x = y = V^{1/3}$. Therefore, $z = V/(xy) = V^{1/3}$. Since there is only one critical point it must minimize $S$. (Why?) The box having minimal surface area is a cube of edge length $V^{1/3}$. ∎

**Remark** The preceding problem is a *constrained* extreme value problem in three variables; the equation $xyz = V$ is a *constraint* limiting the freedom of $x$, $y$, and $z$. We used the constraint to eliminate one variable, $z$, and so to reduce the problem to a *free* (that is, *unconstrained*) problem in two variables. This is the same technique that was used in Section 4.5 to reduce constrained problems in two variables to free problems in one variable. In Section 13.3 we will develop another method for solving constrained extreme value problems.

## Quadratic Forms

It is not immediately apparent how Theorem 3 can be extended to functions of more than two variables. In order to make such an extension we shall rephrase that theorem in terms of *quadratic forms*.

If $\mathbf{u} = u\mathbf{i} + v\mathbf{j}$ is a unit vector, then the *second directional derivative* of $f(x, y)$ at $(a, b)$ in the direction of $\mathbf{u}$ is given by

$$D_{\mathbf{u}}^2 f(a, b) = \mathbf{u} \bullet \nabla \Big( \mathbf{u} \bullet \nabla f \Big)(a, b)$$
$$= Au^2 + 2Buv + Cv^2,$$

where $A$, $B$, and $C$ are constants involving $\mathbf{u}$ and the second partial derivatives of $f$ (see Example 5 of Section 12.7, where $\mathbf{u} = \cos \phi \mathbf{i} + \sin \phi \mathbf{j}$.) Being a homogeneous polynomial of degree two, the expression

$$Q(u, v) = Au^2 + 2Buv + Cv^2$$

is called a **quadratic form** in the variables $u$ and $v$. Such a quadratic form is said to be **positive definite** (or **negative definite**) if $Q(u, v) > 0$ (or $Q(u, v) < 0$) for *all* nonzero vectors $\mathbf{u}$. Otherwise it is said to be **indefinite**.

Theorem 3 says that $f$ has a local minimum or a local maximum at the critical point $(a, b)$ if the second directional derivative $Q(u, v) = D_{\mathbf{u}}^2 f(a, b)$ is a positive definite or negative definite in the components of $\mathbf{u}$. It says also that $f$ has a saddle point at $(a, b)$ if $Q(u, v)$ is positive for some vectors $\mathbf{u}$ and negative for others. This is not the same as saying $Q$ is indefinite. For instance, $Q$ might satisfy $Q(u, v) \geq 0$ (or $Q(u, v) \leq 0$) for all nonzero vectors $\mathbf{u}$, with equality holding for some such vectors, or it can even happen that $Q(u, v)$ vanishes for all vectors $\mathbf{u}$. Such things can happen only in the indeterminate case $B^2 = AC$ of Theorem 3.

An extension of Theorem 3 to functions of any number of variables can be phrased in terms of quadratic forms. For a function $f(x, y, z)$ with continuous second partial derivatives, the second directional derivative at $(a, b, c)$ in the direction of the unit vector $\mathbf{u} = u\mathbf{i} + v\mathbf{j} + w\mathbf{k}$ is the quadratic form

$$Q(u, v, w) = D_{\mathbf{u}}^2 f(a, b, c) = Au^2 + Bv^2 + Cw^2 + 2Duv + 2Euw + 2Fvw,$$

where

$$A = f_{11}(a, b, c), \quad B = f_{22}(a, b, c), \quad C = f_{33}(a, b, c),$$
$$D = f_{12}(a, b, c), \quad E = f_{13}(a, b, c), \quad F = f_{23}(a, b, c).$$

Again, $f$ will have a local minimum, a local maximum, or a saddle point at the critical point $(a, b, c)$ if $Q(u, v, w)$ is positive definite, negative definite, or has positive values for some vectors $\mathbf{u}$ and negative values for others. This can frequently be determined by a process of successive completion of squares, as the example below shows. If $Q$ does not satisfy one of these conditions, then no *second derivative test* can determine the nature of the critical point. One can still resort to "brute force" methods, as in Example 5.

**Example 9** Find and classify the critical points of the function
$f(x, y, z) = x^2y + y^2z + z^2 - 2x.$

**Solution** The equations that determine the critical points are

$$0 = f_1(x, y, z) = 2xy - 2,$$
$$0 = f_2(x, y, z) = x^2 + 2yz,$$
$$0 = f_3(x, y, z) = y^2 + 2z.$$

The third equation implies $z = -y^2/2$ and the second then implies $y^3 = x^2$. From the first equation we get $y^{5/2} = 1$. Thus $y = 1$ and $z = -\frac{1}{2}$. Since $xy = 1$, we must have $x = 1$. The only critical point is $P = (1, 1, -\frac{1}{2})$. Evaluating the second partial derivatives of $f$ at this point we get

$$A = 2, \quad B = -1, \quad C = 2, \quad D = 2, \quad E = 0, \text{ and } \quad F = 2.$$

Hence the second directional derivative of $f$ at P in the direction of the unit vector $\mathbf{u}$ is

$$\begin{aligned}
Q(u, v, w) &= 2u^2 - v^2 + 2w^2 + 4uv + 4vw \\
&= 2(u^2 + 2uv + v^2) + 2(w^2 + 2vw + v^2) - 5v^2 \\
&= 2(u + v)^2 + 2(w + v)^2 - 5v^2.
\end{aligned}$$

Observe that $Q(1, 0, 0) > 0$ and $Q(0, 1, 0) < 0$, so $P$ must be a saddle point of the function $f$.

**Remark** There is an orderly procedure for determining whether a quadratic form in any number of variables is positive definite. We shall describe it for the three-variable case but the extension to higher dimensions is similar. The quadratic form

$$Q(u, v, w) = Au^2 + Bv^2 + Cw^2 + 2Duv + 2Euw + 2Fvw$$

can be written as

$$Q(u, v, w) = \mathbf{u}^T Q \mathbf{u}$$

where

$$\mathbf{u} = \begin{pmatrix} u \\ v \\ w \end{pmatrix} \quad \text{and} \quad Q = \begin{pmatrix} A & D & E \\ D & B & F \\ E & F & C \end{pmatrix}.$$

$Q$ is positive definite if

$$A > 0, \qquad \begin{vmatrix} A & D \\ D & B \end{vmatrix} > 0, \quad \text{and} \quad \begin{vmatrix} A & D & E \\ D & B & F \\ E & F & C \end{vmatrix} > 0.$$

The reader is encouraged to prove this (see Exercise 28 below) by using the square-completion technique of Example 9. We can deduce an equivalent test for negative definiteness. $Q(u, v, w)$ is negative definite if $-Q(u, v, w)$ is positive definite. Thus $Q$ is negative definite if

$$A < 0, \qquad \begin{vmatrix} A & D \\ D & B \end{vmatrix} > 0, \quad \text{and} \quad \begin{vmatrix} A & D & E \\ D & B & F \\ E & F & C \end{vmatrix} < 0.$$

If $\det(Q) \neq 0$ but neither $Q$ nor $-Q$ is positive definite, then $Q$ will be positive for some values of $(u, v, w)$ and negative for others.

## Exercises 13.1

In Exercises 1–15, find and classify the critical points of the given functions.

**1.** $f(x, y) = x^2 + 2y^2 - 4x + 4y$

**2.** $f(x, y) = xy - x + y$         **3.** $f(x, y) = x^3 + y^3 - 3xy$

**4.** $f(x, y) = x^4 + y^4 - 4xy$    **5.** $f(x, y) = \dfrac{x}{y} + \dfrac{8}{x} - y$

**6.** $f(x, y) = \cos(x + y)$        **7.** $f(x, y) = x \sin y$

**8.** $f(x, y) = \cos x + \cos y$    **9.** $f(x, y) = x^2 y\, e^{-(x^2+y^2)}$

**10.** $f(x, y) = \dfrac{xy}{2 + x^4 + y^4}$   **11.** $f(x, y) = x\, e^{-x^3+y^3}$

**12.** $f(x, y) = \dfrac{1}{1 - x + y + x^2 + y^2}$

**13.** $f(x, y) = \left(1 + \dfrac{1}{x}\right)\left(1 + \dfrac{1}{y}\right)\left(\dfrac{1}{x} + \dfrac{1}{y}\right)$

**\* 14.** $f(x, y, z) = xyz - x^2 - y^2 - z^2$

**\* 15.** $f(x, y, z) = xy + x^2 z - x^2 - y - z^2$

**\* 16.** Show that $f(x, y, z) = 4xyz - x^4 - y^4 - z^4$ has a local maximum value at the point $(1,1,1)$.

**17.** Find the maximum and minimum values of $f(x, y) = xy\, e^{-x^2-y^4}$.

**18.** Find the maximum and minimum values of $f(x, y) = x/(1 + x^2 + y^2)$.

**\* 19.** Find the maximum and minimum values of $f(x, y, z) = xyz\, e^{-x^2-y^2-z^2}$. How do you know that such extreme values exist?

**20.** Find the minimum value of $f(x, y) = x + 8y + \dfrac{1}{xy}$ in the first quadrant $x > 0$, $y > 0$. How do you know that a minimum exists?

**21.** Find the dimensions of the rectangular box with no top having given volume $V$ and the least possible total surface area of its five faces.

**22.** The material used to make the bottom of a rectangular box is twice as expensive per unit area as the material used to make the top or side walls. Find the dimensions of the box of given volume $V$ for which the cost of materials is minimum.

**23.** Find the volume of the largest rectangular box (with faces parallel to the coordinate planes) that can be inscribed inside the ellipsoid

$$\frac{x^2}{a^2} + \frac{y^2}{b^2} + \frac{z^2}{c^2} = 1.$$

**24.** Find the three positive numbers $a$, $b$, and $c$, whose sum is 30 and for which the expression $ab^2c^3$ is maximum.

**25.** Find the critical points of the function $z = g(x, y)$ that satisfies the equation $e^{2zx-x^2} - 3e^{2zy+y^2} = 2$.

**\* 26.** Classify the critical points of the function $g$ in the previous exercise.

**\* 27.** Let $f(x, y) = (y - x^2)(y - 3x^2)$. Show that the origin is a critical point of $f$ and that the restriction of $f$ to every straight line through the origin has a local minimum value at the origin. (That is, show that $f(x, kx)$ has a local minimum value at $x = 0$ for every $k$, and that $f(0, y)$ has a local

minimum value at $y = 0$.) Does $f(x, y)$ have a local minimum value at the origin? What happens to $f$ on the curve $y = 2x^2$? What does the second derivative test say about this situation?

**\* 28.** Prove the determinant tests for the positive and negative definiteness of the three-variable quadratic forms given in the remark at the end of this section.

**\* 29.** State conditions on the coefficients $A, B, \ldots, J$ that will guarantee that the quadratic form

$$
\begin{aligned}
Q(w, x, y, z) =\ & Aw^2 + Bx^2 + Cy^2 \\
& + Dz^2 + 2Ewx + 2Fwy \\
& + 2Gwz + 2Hxy + 2Ixz + 2Jyz
\end{aligned}
$$

is positive definite. (Generalize the assertion made for the three-variable forms in the text.)

# 13.2  Extreme Values of Functions Defined on Restricted Domains

Much of the previous section was concerned with techniques for determining whether a critical point of a function provides a local maximum or minimum value, or is a saddle point. In this section we address the problem of determining absolute maximum and minimum values for functions that have them — usually functions whose domains are restricted to subsets of $\mathbb{R}^2$ (or $\mathbb{R}^n$) having nonempty interiors. In Example 7 of Section 13.1 we had to *prove* that the given function had absolute extreme values. If, however, we are dealing with a continuous function on a domain that is closed and bounded, then we can rely on Theorem 2 to guarantee the existence of such extreme values, but we will always have to check boundary points as well as any interior critical or singular points to find them. The following examples illustrate the technique.

**Example 1**    Find the maximum and minimum values of $f(x, y) = 2xy$ on the closed disk $x^2 + y^2 \le 4$.

***Solution***    Since $f$ is continuous and the disk is closed, $f$ must have absolute maximum and minimum values at some points of the disk. The first partial derivatives of $f$ are

$$f_1(x, y) = 2y \qquad \text{and} \qquad f_2(x, y) = 2x,$$

so there are no singular points and the only critical point is $(0, 0)$, where $f$ has the value 0.

We must still consider values of $f$ on the boundary circle $x^2 + y^2 = 4$. We can express $f$ as a function of a single variable on this circle, by using a convenient parametrization of the circle, say

$$x = 2\cos t, \qquad y = 2\sin t, \qquad (-\pi \le t \le \pi).$$

We have

$$f\left(2\cos t, 2\sin t\right) = 8\cos t \sin t = g(t).$$

We must find any extreme values of $g(t)$. We can do this in either of two ways. If we rewrite $g(t) = 4\sin 2t$, it is clear that $g(t)$ has maximum value 4 (at $t = \frac{\pi}{4}$ and $-\frac{3\pi}{4}$), and minimum value $-4$ (at $t = -\frac{\pi}{4}$ and $\frac{3\pi}{4}$). Alternatively, we can differentiate $g$ to find its critical points:

$$0 = g'(t) = -8\sin^2 t + 8\cos^2 t \qquad \Longleftrightarrow \qquad \tan^2 t = 1$$
$$\Longleftrightarrow \qquad t = \pm\frac{\pi}{4} \text{ or } \pm\frac{3\pi}{4},$$

which again yield the maximum value 4 and the minimum value $-4$. (It is not necessary to check the endpoints $t = -\pi$ and $t = \pi$; since $g$ is everywhere differentiable and is periodic with period $\pi$, any absolute maximum or minimum will occur at a critical point.)

In any event, $f$ has maximum value 4 at the boundary points $(\sqrt{2}, \sqrt{2})$ and $(-\sqrt{2}, -\sqrt{2})$, and minimum value $-4$ at the boundary points $(\sqrt{2}, -\sqrt{2})$ and $(-\sqrt{2}, \sqrt{2})$. It is easily shown by the second derivative test (or otherwise) that the interior critical point $(0, 0)$ is a saddle point.

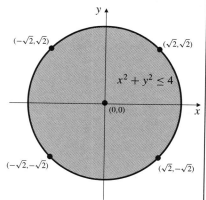

**Figure 13.7**   Points that are candidates for extreme values in Example 1

**Example 2**   Find the extreme values of the function $f(x, y) = x^2 y e^{-(x+y)}$ on the triangular region $T$ given by $x \ge 0$, $y \ge 0$, and $x + y \le 4$.

**Solution**   First we look for critical points:

$$0 = f_1(x, y) = xy(2 - x)e^{-(x+y)} \qquad \Longleftrightarrow \quad x = 0, \ y = 0, \text{ or } x = 2,$$
$$0 = f_2(x, y) = x^2(1 - y)e^{-(x+y)} \qquad \Longleftrightarrow \quad x = 0 \text{ or } y = 1.$$

The critical points are $(0, y)$ for any $y$ and $(2, 1)$. Only $(2, 1)$ is an interior point of $T$. (See Figure 13.8.) $f(2, 1) = 4/e^3 \approx 0.199$. The boundary of $T$ consists of three straight line segments. On two of these, the coordinate axes, $f$ is identically zero. The third segment is given by

$$y = 4 - x, \qquad 0 \le x \le 4,$$

so the values of $f$ on this segment can be expressed as a function of $x$ alone:

$$g(x) = f(x, 4 - x) = x^2(4 - x)e^{-4}, \qquad 0 \le x \le 4.$$

Note that $g(0) = g(4) = 0$ and $g(x) > 0$ if $0 < x < 4$. The critical points of $g$ are given by $0 = g'(x) = (8x - 3x^2)e^{-4}$, so they are $x = 0$ and $x = 8/3$. We have

$$g\left(\frac{8}{3}\right) = f\left(\frac{8}{3}, \frac{4}{3}\right) = \frac{256}{27}e^{-4} \approx 0.174 < f(2, 1).$$

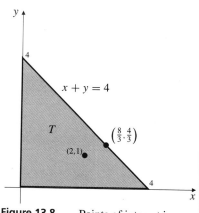

**Figure 13.8**   Points of interest in Example 2

We conclude that the maximum value of $f$ over the region $T$ is $4/e^3$, and that it occurs at the interior critical point $(2, 1)$. The minimum value of $f$ is zero and occurs at all points of the two perpendicular boundary segments. Note that $f$ has neither a local maximum nor a local minimum at the boundary point $(8/3, 4/3)$, although $g$ has a local maximum there. Of course that point is not a saddle point of $f$ either. It is not a critical point of $f$.  ■

**Example 3**   Among all triangles with vertices on the unit circle $x^2 + y^2 = 1$, find those that have largest area.

**Solution**   Intuition tells us that the equilateral triangles must have the largest area. However, proving this can be quite difficult unless a good choice of variables in which to set up the problem analytically is made. Let one vertex of the triangle be the point $P$ with coordinates $(1, 0)$ and let the other two vertices, $Q$ and $R$, be as shown in Figure 13.10. There is no harm in assuming that $Q$ lies on the upper semicircle and $R$ on the lower, and that the origin $O$ is inside triangle $PQR$. Let $PQ$ and $PR$ make angles $\theta$ and $\phi$, respectively, with the negative direction of the $x$-axis. Clearly $0 \le \theta \le \pi/2$ and $0 \le \phi \le \pi/2$. The lines from $O$ to $Q$ and $R$ make equal angles $\Psi$ with the line $QR$, where $2\theta + 2\phi + 2\Psi = \pi$. Dropping perpendiculars from $O$ to the three sides of the triangle $PQR$, we can write the area $A$ of the triangle as the sum of the areas of six small, right-angled triangles:

$$A = 2 \times \frac{1}{2} \sin\theta \cos\theta + 2 \times \frac{1}{2} \sin\phi \cos\phi + 2 \times \frac{1}{2} \sin\Psi \cos\Psi$$

$$= \frac{1}{2}(\sin 2\theta + \sin 2\phi + \sin 2\Psi).$$

Since $2\Psi = \pi - 2(\theta + \phi)$, we can express $A$ as a function of the two variables $\theta$ and $\phi$:

$$A = A(\theta, \phi) = \frac{1}{2}(\sin 2\theta + \sin 2\phi + \sin 2(\theta + \phi)).$$

The domain of $A$ is the triangle $\theta \ge 0$, $\phi \ge 0$, $\theta + \phi \le \pi/2$. $A = 0$ at the vertices of the triangle and is positive elsewhere. (See Figure 13.9.) We show that the maximum value of $A(\theta, \phi)$ on any edge of the triangle is 1 and occurs at the midpoint of that edge. On the edge $\theta = 0$ we have

$$A(0, \phi) = \frac{1}{2}(\sin 2\phi + \sin 2\phi) = \sin 2\phi \le 1 = A(0, \pi/4).$$

Similarly, on $\phi = 0$, $A(\theta, 0) \le 1 = A(\pi/4, 0)$. On the edge $\theta + \phi = \pi/2$ we have

$$A\left(\theta, \frac{\pi}{2} - \theta\right) = \frac{1}{2}(\sin 2\theta + \sin(\pi - 2\theta))$$

$$= \sin 2\theta \le 1 = A\left(\frac{\pi}{4}, \frac{\pi}{4}\right).$$

We must now check for any interior critical points of $A(\theta, \phi)$. (There are no singular points.) For critical points we have

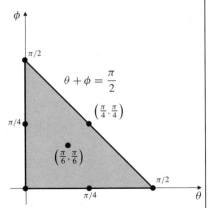

**Figure 13.9**    The domain of $A(\theta, \phi)$

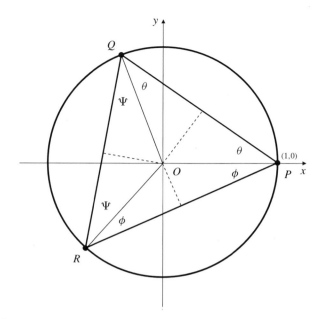

**Figure 13.10**   Where should $Q$ and $R$ be to ensure that triangle $PQR$ has maximum area?

$$0 = \frac{\partial A}{\partial \theta} = \cos 2\theta + \cos(2\theta + 2\phi),$$

$$0 = \frac{\partial A}{\partial \phi} = \cos 2\phi + \cos(2\theta + 2\phi),$$

so the critical points satisfy $\cos 2\theta = \cos 2\phi$, and hence $\theta = \phi$. We now substitute this equation into either of the above equations to determine $\theta$:

$$\cos 2\theta + \cos 4\theta = 0$$
$$2\cos^2 2\theta + \cos 2\theta - 1 = 0$$
$$(2\cos 2\theta - 1)(\cos 2\theta + 1) = 0$$
$$\cos 2\theta = \frac{1}{2} \quad \text{or} \quad \cos 2\theta = -1.$$

The only solution leading to an interior point of the domain of $A$ is $\theta = \phi = \pi/6$. Note that

$$A\left(\frac{\pi}{6}, \frac{\pi}{6}\right) = \frac{1}{2}\left(\frac{\sqrt{3}}{2} + \frac{\sqrt{3}}{2} + \frac{\sqrt{3}}{2}\right) = \frac{3\sqrt{3}}{4} > 1,$$

this interior critical point maximizes the area of the inscribed triangle. Finally, observe that for $\theta = \phi = \pi/6$, we have $\Psi = \pi/6$ also, so the largest triangle is indeed equilateral.  ∎

**Remark**   Since the area $A$ of the inscribed triangle must have a maximum value ($A$ is continuous and its domain is closed and bounded), a strictly geometric argument can be used to show that the largest triangle is equilateral. If an inscribed triangle has two unequal sides, its area can be made larger by moving the common vertex of these two sides along the circle to increase its perpendicular distance from the opposite side of the triangle.

## Linear Programming

Linear programming is a branch of linear algebra that develops systematic techniques for finding maximum or minimum values of a *linear function* subject to several *linear inequality constraints*. Such problems arise frequently in management science and operations research. Because of their linear nature they do not usually involve calculus in their solution; linear programming is frequently presented in courses on *finite mathematics*. We will not attempt any formal study of linear programming here, but will make a few observations for comparison with the more general nonlinear extreme value problems considered above that involve calculus in their solution.

The inequality $ax + by \leq c$ is an example of a linear inequality in two variables. The *solution set* of this inequality consists of a half-plane lying on one side of the straight line $ax + by = c$. The solution set of a system of several two-variable linear inequalities is an intersection of such half-planes, so it is a *convex* region of the plane bounded by a *polygonal line*. If it is a bounded set then it is a convex polygon together with its interior. (A set is called **convex** if it contains the entire line segment between any two of its points. On the real line the convex sets are intervals.)

Let us examine a simple concrete example that involves only two variables and a few constraints.

**Example 4**   Find the maximum value of $F(x, y) = 2x + 7y$ subject to the constraints

$$x + 2y \leq 6, \quad 2x + y \leq 6, \quad x \geq 0, \quad \text{and} \quad y \geq 0.$$

**Solution**   The solution set $S$ of the system of four constraint equations is shown in Figure 13.11. It is the quadrilateral region with vertices $(0, 0)$, $(3, 0)$, $(2, 2)$, and $(0, 3)$. Several level curves of the linear function $F$ are also shown in the figure. They are parallel straight lines with slope $-\frac{2}{7}$. We want the line that gives $F$ the greatest value and that still intersects $S$. Evidently this is the line $F = 21$ that passes through the vertex $(0, 3)$ of $S$. The maximum value of $F$ subject to the constraints is 21. ∎

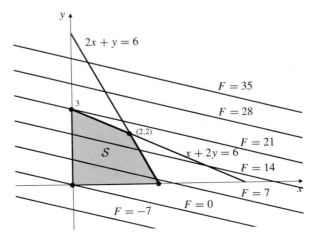

**Figure 13.11**   The shaded region is the solution set for the constraint inequalities in Example 4

As this simple example illustrates, a *linear* function with domain restricted by *linear inequalities* does not achieve maximum or minimum values at points in the interior of its domain (if that domain has an interior). Any such extreme value occurs at a boundary point of the domain, or a set of such boundary points. Where an extreme value occurs at a set of boundary points that set will *always* contain at least one vertex. This phenomenon holds in general for extreme value problems for linear functions in any number of variables with domains restricted by any number of linear inequalities. For problems involving three variables the domain will be a convex region of $\mathbb{R}^3$ bounded by planes. For a problem involving $n$ variables the domain will be a convex region in $\mathbb{R}^n$ bounded by $(n-1)$-dimensional hyperplanes. Such *polyhedral* regions still have vertices (where $n$ hyperplanes intersect), and maximum or minimum values of linear functions subject to the constraints will still occur at subsets of the boundary containing such vertices. These problems can therefore be solved by evaluating the linear function to be extremized (it is called the **objective function**) at all the vertices and selecting the greatest or least value.

In practice, linear programming problems can involve hundreds or even thousands of variables and even more constraints. Such problems need to be solved with computers, but even then it is extremely inefficient, if not impossible, to calculate all the vertices of the constraint solution set and the values of the objective function at them. Much of the study of linear programming therefore centres on devising techniques for getting to (or at least near) the optimizing vertex in as few steps as possible. Usually this involves criteria whereby large numbers of vertices can be rejected on geometric grounds. We will not delve into such techniques here, but will content ourselves with one more example to illustrate, in a very simple case, how the underlying geometry of a problem can be used to reduce the number of vertices that must be considered.

**Example 5**   A tailor has 230 m of a certain fabric and has orders for up to 20 suits, up to 30 jackets, and up to 40 pairs of slacks to be made from the fabric. Each suit requires 6 m, each jacket 3 m, and each pair of slacks 2 m of the fabric. If the tailor's profit is $20 per suit, $14 per jacket, and $12 per pair of slacks, how many of each should he make to realize the maximum profit from his supply of the fabric?

**Solution**   Suppose he makes $x$ suits, $y$ jackets, and $z$ pairs of slacks. Then his profit will be

$$P = 20x + 14y + 12z.$$

The constraints posed in the problem are

$$x \geq 0, \qquad x \leq 20,$$
$$y \geq 0, \qquad y \leq 30,$$
$$z \geq 0, \qquad z \leq 40,$$

$$6x + 3y + 2z \leq 230.$$

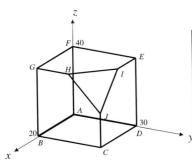

**Figure 13.12**    The convex set of points satisfying the constraints in Example 5

The last inequality is due to the limited supply of fabric. The solution set is shown in Figure 13.12. It has ten vertices, $A, B, \ldots, J$. Since $P$ increases in the direction of the vector $\nabla P = 20\mathbf{i} + 14\mathbf{j} + 12\mathbf{k}$ which points into the first octant, its maximum value cannot occur at any of the vertices $A, B, \ldots, G$. (Think about why.) Thus we need look only at the vertices $H$, $I$, and $J$.

$$H = (20, 10, 40), \qquad P = 1020 \text{ at } H.$$
$$I = (10, 30, 40), \qquad P = 1100 \text{ at } I.$$
$$J = (20, 30, 10), \qquad P = 940 \text{ at } J.$$

Thus the tailor should make 10 suits, 30 jackets and 40 pairs of slacks to realize the maximum profit, $1,100, from the fabric. ∎

# Exercises 13.2

1. Find the maximum and minimum values of
   $f(x, y) = x - x^2 + y^2$ on the rectangle $0 \le x \le 2$, $0 \le y \le 1$.

2. Find the maximum and minimum values of
   $f(x, y) = xy - 2x$ on the rectangle $-1 \le x \le 1, 0 \le y \le 1$.

3. Find the maximum and minimum values of
   $f(x, y) = xy - y^2$ on the disk $x^2 + y^2 \le 1$.

4. Find the maximum and minimum values of
   $f(x, y) = x + 2y$ on the disk $x^2 + y^2 \le 1$.

5. Find the maximum value of $f(x, y) = xy - x^3 y^2$ over the square $0 \le x \le 1, 0 \le y \le 1$.

6. Find the maximum and minimum values of
   $f(x, y) = xy(1 - x - y)$ over the triangle with vertices $(0, 0)$, $(1, 0)$, and $(0, 1)$.

7. Find the maximum and minimum values of
   $f(x, y) = \sin x \cos y$ on the closed triangular region bounded by the coordinate axes and the line $x + y = 2\pi$.

8. Find the maximum value of
   $$f(x, y) = \sin x \sin y \sin(x + y)$$
   over the triangle bounded by the coordinate axes and the line $x + y = \pi$.

9. The temperature at all points in the disk $x^2 + y^2 \le 1$ is given by
   $$T = (x + y)\, e^{-x^2 - y^2}.$$
   Find the maximum and minimum temperatures at points of the disk.

10. Find the maximum and minimum values of
    $$f(x, y) = \frac{x - y}{1 + x^2 + y^2}$$

on the upper half-plane $y \ge 0$.

11. Find the maximum and minimum values of $xy^2 + yz^2$ over the ball $x^2 + y^2 + z^2 \le 1$.

12. Find the maximum and minimum values of $xz + yz$ over the ball $x^2 + y^2 + z^2 \le 1$.

13. Consider the function $f(x, y) = xy\, e^{-xy}$ with domain the first quadrant: $x \ge 0$, $y \ge 0$. Show that $\lim_{x \to \infty} f(x, kx) = 0$. Does $f$ have a limit as $(x, y)$ recedes arbitrarily far from the origin in the first quadrant? Does $f$ have a maximum value in the first quadrant?

14. Repeat Exercise 13 for the function $f(x, y) = xy^2\, e^{-xy}$.

15. In a certain community there are two breweries in competition, so that sales of each negatively affect the profits of the other. If brewery A produces $x$ litres of beer per month and brewery B produces $y$ litres per month, then brewery A's monthly profit $P and brewery B's monthly profit $Q are assumed to be
    $$P = 2x - \frac{2x^2 + y^2}{10^6},$$
    $$Q = 2y - \frac{4y^2 + x^2}{2 \times 10^6}.$$
    Find the sum of the profits of the two breweries if each brewery independently sets its own production level to maximize its own profit, and assumes its competitor does likewise. Find the sum of the profits if the two breweries cooperate to determine their respective productions to maximize that sum.

16. Equal angle bends are made at equal distances from the two ends of a 100 m long straight length of fence so the resulting three segment fence can be placed along an existing wall to make an enclosure of trapezoidal shape. What is the largest possible area for such an enclosure?

**17.** Maximize $Q(x, y) = 2x + 3y$ subject to the constraints $x \geq 0$, $y \geq 0$, $y \leq 5$, $x + 2y \leq 12$, and $4x + y \leq 12$.

**18.** Minimize $F(x, y, z) = 2x + 3y + 4z$ subject to $x \geq 0$, $y \geq 0$, $z \geq 0$, $x + y \geq 2$, $y + z \geq 2$, and $x + z \geq 2$.

**19.** A textile manufacturer produces two grades of wool-cotton-polyester fabric. The deluxe grade has composition (by weight) 20% wool, 50% cotton, and 30% polyester, and it sells for \$3 per kilogram. The standard grade has composition 10% wool, 40% cotton, and 50% polyester, and sells for \$2 per kilogram. If he has in stock 2000 kg of wool and 6000 kg each of cotton and polyester, how many kilograms of fabric of each grade should he manufacture to maximize his revenue?

**20.** A 10 hectare parcel of land is zoned for building densities of 6 detached houses per hectare, 8 duplex units per hectare, or 12 apartments per hectare. The developer who owns the land can make a profit of \$40,000 per house, \$20,000 per duplex unit, and \$16,000 per apartment that he builds. Municipal bylaws require him to build at least as many apartments as houses or duplex units. How many of each type of dwelling should he build to maximize his profit?

## 13.3   Lagrange Multipliers

A constrained extreme-value problem is one in which the variables of the function to be maximized or minimized are not completely independent of one another, but must satisfy one or more constraint equations or inequalities. For instance, the problems

$$\text{maximize} \quad f(x, y) \quad \text{subject to} \quad g(x, y) = C,$$

and

$$\text{minimize} \quad f(x, y, z, w) \quad \text{subject to} \quad g(x, y, z, w) = C_1,$$
$$\text{and} \quad h(x, y, z, w) = C_2$$

have, respectively, one and two constraint equations, while the problem

$$\text{maximize} \quad f(x, y, z) \quad \text{subject to} \quad g(x, y, z) \leq C$$

has a single constraint inequality.

Generally, inequality constraints can be regarded as restricting the domain of the function to be extremized to a smaller set that still has interior points. Section 13.2 was devoted to such problems. In each of the first three examples of that section we looked for *free* (that is, *unconstrained*) extreme values in the interior of the domain, and then examined the boundary of the domain, which was specified by one or more *constraint equations*. In Example 1 we parametrized the boundary and expressed the function to be extremized as a function of the parameter, thus reducing the boundary case to a free problem in one variable instead of a constrained problem in two variables. In Example 2 the boundary consisted of three line segments, on two of which the function was obviously zero. We solved the equation for the third boundary segment for $y$ in terms of $x$, again in order to express the values of $f(x, y)$ on that segment as a function of one free variable. A similar approach was used in Example 3 to deal with the triangular boundary of the domain of the area function $A(\theta, \phi)$.

The reduction of extremization problems with equation constraints to free problems with fewer independent variables is only feasible when the constraint equations can be solved either explicitly for some variables in terms of others, or parametrically for all variables in terms of some parameters. It is often very difficult or impossible to solve the constraint equations, so we need another technique.

## The Method of Lagrange Multipliers

A technique for finding extreme values of $f(x, y)$ subject to the equality constraint $g(x, y) = 0$ can be based on the following theorem.

**THEOREM 4**

Suppose that $f$ and $g$ have continuous first partial derivatives near the point $P_0 = (x_0, y_0)$ on the curve $C$ with equation $g(x, y) = 0$. Suppose also that, when restricted to points on $C$, the function $f(x, y)$ has a local maximum or minimum value at $P_0$. Finally, suppose that

(i) $P_0$ is not an endpoint of $C$, and

(ii) $\nabla g(P_0) \neq \mathbf{0}$.

Then there exists a number $\lambda_0$ such that $(x_0, y_0, \lambda_0)$ is a critical point of the **Lagrangian function**

$$L(x, y, \lambda) = f(x, y) + \lambda g(x, y).$$

**PROOF**  Together, (i) and (ii) imply that $C$ is smooth enough to have a tangent line at $P_0$, and that $\nabla g(P_0)$ is normal to that tangent line. If $\nabla f(P_0)$ is not parallel to $\nabla g(P_0)$, then $\nabla f(P_0)$ has a nonzero vector projection $\mathbf{v}$ along the tangent line to $C$ at $P_0$. (See Figure 13.13.)  Therefore $f$ has a positive directional derivative at $P_0$ in the direction of $\mathbf{v}$, and a negative directional derivative in the opposite direction. Thus $f(x, y)$ increases or decreases as we move away from $P_0$ along $C$ in the direction of $\mathbf{v}$ or $-\mathbf{v}$, and $f$ cannot have a maximum or minimum value at $P_0$. Since we are assuming that $f$ *does* have an extreme value at $P_0$, it must be that $\nabla f(P_0)$ is parallel to $\nabla g(P_0)$. Since $\nabla g(P_0) \neq \mathbf{0}$, there must exist a real number $\lambda_0$ such that $\nabla f(P_0) = -\lambda_0 \nabla g(P_0)$, or

$$\nabla (f + \lambda_0 g)(P_0) = \mathbf{0}.$$

The two components of the above vector equation assert that $\partial L/\partial x = 0$ and $\partial L/\partial y = 0$ at $(x_0, y_0, \lambda_0)$. The third equation that must be satisfied by a critical point of $L$ is $\partial L/\partial \lambda = g(x, y) = 0$. This is satisfied at $(x_0, y_0, \lambda_0)$ because $P_0$ lies on $C$. Thus $(x_0, y_0, \lambda_0)$ is a critical point of $L(x, y, \lambda)$.

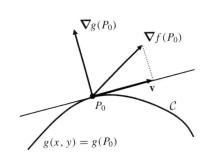

**Figure 13.13**  If $\nabla f(P_0)$ is not a multiple of $\nabla g(P_0)$, then $\nabla f(P_0)$ has a nonzero projection $\mathbf{v}$ tangent to the level curve of $g$ through $P_0$

Theorem 4 suggests that to find candidates for points on the curve $g(x, y) = 0$ at which $f(x, y)$ is maximum or minimum, we should look for critical points of the Lagrangian function

$$L(x, y, \lambda) = f(x, y) + \lambda g(x, y).$$

At any critical point of $L$ we must have

$$0 = \frac{\partial L}{\partial x} = f_1(x, y) + \lambda g_1(x, y),$$

$$0 = \frac{\partial L}{\partial y} = f_2(x, y) + \lambda g_2(x, y),$$

i.e., $\nabla f$ is parallel to $\nabla g$,

and

$$0 = \frac{\partial L}{\partial \lambda} = g(x, y)$$

the constraint equation.

Note, however, that it is *assumed* that the constrained problem *has a solution*. Theorem 4 does not guarantee that a solution exists; it only provides a means for finding a solution already known to exist. It is usually necessary to satisfy yourself that the problem you are trying to solve has a solution before using this method to find the solution.

Let us put the method to a concrete test:

**Example 1**    Find the shortest distance from the origin to the curve $x^2 y = 16$.

**Solution**    The graph of $x^2 y = 16$ is shown in Figure 13.14. There appear to be two points on the curve that are closest to the origin, and no points that are farthest from the origin. (The curve is unbounded.) To find the closest points it is sufficient to minimize the square of the distance from the point $(x, y)$ on the curve to the origin, that is, to solve the problem

$$\text{minimize} \quad f(x, y) = x^2 + y^2 \quad \text{subject to} \quad g(x, y) = x^2 y - 16 = 0.$$

Let $L(x, y, \lambda) = x^2 + y^2 + \lambda(x^2 y - 16)$. For critical points of $L$ we want

$$0 = \frac{\partial L}{\partial x} = 2x + 2\lambda xy = 2x(1 + \lambda y) \tag{A}$$

$$0 = \frac{\partial L}{\partial y} = 2y + \lambda x^2 \tag{B}$$

$$0 = \frac{\partial L}{\partial \lambda} = x^2 y - 16. \tag{C}$$

Equation (A) requires that either $x = 0$ or $\lambda y = -1$. However, $x = 0$ is inconsistent with equation (C). Therefore $\lambda y = -1$. From equation (B) we now have

$$0 = 2y^2 + \lambda y x^2 = 2y^2 - x^2.$$

Thus $x = \pm\sqrt{2} y$, and (C) now gives $2y^3 = 16$, so $y = 2$. There are, therefore, two candidates for points on $x^2 y = 16$ closest to the origin, $(\pm 2\sqrt{2}, 2)$. Both of these points are at distance $\sqrt{8 + 4} = 2\sqrt{3}$ units from the origin, so this must be the minimum distance from the origin to the curve. Some level curves of $x^2 + y^2$ are shown, along with the constraint curve $x^2 y = 16$ in Figure 13.14. Observe how the constraint curve is tangent to the level curve passing through the minimizing points $(\pm 2\sqrt{2}, 2)$, reflecting the fact that the two curves have parallel normals there.

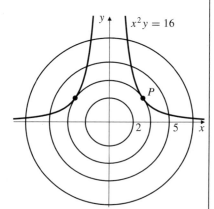

**Figure 13.14**    The level curve of the function representing the square of distance from the origin is tangent to the curve $x^2 y = 16$ at the two points on that curve that are closest to the origin

**Remark**    In the above example we could, of course, have solved the constraint equation for $y = 16/x^2$, substituted into $f$, and thus reduced the problem to one of finding the (unconstrained) minimum value of

$$F(x) = f\left(x, \frac{16}{x^2}\right) = x^2 + \frac{256}{x^4}.$$

The reader is invited to verify that this gives the same result.

The number $\lambda$ that occurs in the Lagrangian function is called a **Lagrange multiplier**. The technique for solving an extreme-value problem with equation constraints by looking for critical points of an unconstrained problem in more variables (the original variables plus a Lagrange multiplier corresponding to each constraint equation) is called **the method of Lagrange multipliers**. It can be expected to give results as long as the function to be maximized or minimized (called the **objective function**) and the constraint equations have *smooth* graphs in a neighbourhood of the points where the extreme values occur, and these points are not on *edges* of those graphs. See Example 3 and Exercise 26 at the end of this section.

**Example 2** Find the points on the curve $17x^2 + 12xy + 8y^2 = 100$ that are closest to and farthest away from the origin.

**Solution** The quadratic form on the left side of the equation above is positive definite as can be seen by completing a square. Hence the curve is bounded and must have points closest to and farthest from the origin. (In fact, the curve is an ellipse with centre at the origin and oblique principal axes. The problem asks us to find the ends of the major and minor axes.)

Again we want to extremize $x^2 + y^2$ subject to an equation constraint. The Lagrangian in this case is

$$L(x, y, \lambda) = x^2 + y^2 + \lambda(17x^2 + 12xy + 8y^2 - 100),$$

and its critical points are given by

$$0 = \frac{\partial L}{\partial x} = 2x + \lambda(34x + 12y) \tag{A}$$

$$0 = \frac{\partial L}{\partial y} = 2y + \lambda(12x + 16y) \tag{B}$$

$$0 = \frac{\partial L}{\partial \lambda} = 17x^2 + 12xy + 8y^2 - 100. \tag{C}$$

Solving each of equations (A) and (B) for $\lambda$ and equating the two expressions for $\lambda$ obtained, we get

$$\frac{-2x}{34x + 12y} = \frac{-2y}{12x + 16y} \qquad \text{or} \qquad 12x^2 + 16xy = 34xy + 12y^2.$$

This equation simplifies to

$$2x^2 - 3xy - 2y^2 = 0. \tag{D}$$

We multiply equation (D) by 4 and add the result to equation (C) to get $25x^2 = 100$, so that $x = \pm 2$. Finally, we substitute each of these values of $x$ into (D) and obtain (for each) two values of $y$ from the resulting quadratics:

For $x = 2$: $\quad y^2 + 3y - 4 = 0$, $\qquad$ For $x = -2$: $\quad y^2 - 3y - 4 = 0$,
$$(y - 1)(y + 4) = 0. \qquad\qquad\qquad (y + 1)(y - 4) = 0.$$

We therefore obtain four candidate points: $(2, 1)$, $(-2, -1)$, $(2, -4)$, and $(-2, 4)$. The first two points are closest to the origin (they are the ends of the minor axis of the ellipse); the second pair are farthest from the origin (the ends of the major axis). (See Figure 13.15.)

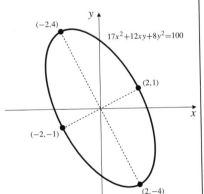

**Figure 13.15** The points on the ellipse that are closest to and farthest from the origin

Considering the geometric underpinnings of the method of Lagrange multipliers, we would not expect the method to work if the level curves of the functions involved are not smooth, or if the maximum or minimum occurs at an endpoint of the constraint curve. One of the pitfalls of the method is that the level curves of functions may not be smooth, even though the functions themselves are differentiable. Problems can occur where a gradient vanishes, as the following example shows.

**Example 3**   Find the minimum value of $f(x, y) = y$ subject to the constraint equation $g(x, y) = y^3 - x^2 = 0$.

**Solution**   The semicubical parabola $y^3 = x^2$ has a cusp at the origin. (See Figure 13.16.) Clearly, $f(x, y) = y$ has minimum value 0 at that point. Suppose, however, that we try to solve the problem using the method of Lagrange multipliers. The Lagrangian here is

$$L(x, y, \lambda) = y + \lambda(y^3 - x^2),$$

which has critical points given by

$$-2\lambda x = 0,$$
$$1 + 3\lambda y^2 = 0,$$
$$y^3 - x^2 = 0.$$

Observe that $y = 0$ cannot satisfy the second equation, and, in fact, the three equations have *no solution* $(x, y, \lambda)$.

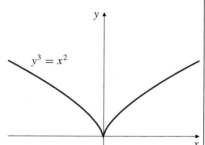

**Figure 13.16**   The minimum of $y$ occurs at a point on the curve where the curve has no tangent line

**Remark**   The method of Lagrange multipliers breaks down in the above example because $\nabla g = \mathbf{0}$ at the solution point, and therefore the curve $g(x, y) = 0$ need not be smooth there. (In this case, it isn't smooth!) The geometric condition that $\nabla f$ should be parallel to $\nabla g$ at the solution point is meaningless in this case. When applying the method of Lagrange multipliers, be aware that an extreme value may occur at

    (i)   a critical point of the Lagrangian,

    (ii)   a point where $\nabla g = \mathbf{0}$,

    (iii)   a point where $\nabla f$ or $\nabla g$ does not exist, or

    (iv)   an "endpoint" of the constraint set.

This situation is similar to that for extreme values of a function $f$ of one variable, which can occur at a critical point of $f$, a singular point of $f$, or an endpoint of the domain of $f$.

## Problems with More than One Constraint

Next consider a three-dimensional problem requiring us to find a maximum or minimum value of a function of three variables subject to two equation constraints:

    **extremize**   $f(x, y, z)$   **subject to**   $g(x, y, z) = 0$ and $h(x, y, z) = 0$.

Again we assume that the problem has a solution, say at the point $P_0 = (x_0, y_0, z_0)$, that the functions $f$, $g$ and $h$ have continuous first partial derivatives near $P_0$. Also, we assume that $\mathbf{T} = \nabla g(P_0) \times \nabla h(P_0) \neq \mathbf{0}$. These conditions imply that the surfaces $g(x, y, z) = 0$ and $h(x, y, z) = 0$ are smooth near $P_0$ and are not tangent to each other there, so they must intersect in a curve $C$ that is smooth near $P_0$. The curve $C$ has tangent vector $\mathbf{T}$ at $P_0$. The same geometric argument used in the proof of Theorem 4 again shows that $\nabla f(P_0)$ must be perpendicular to $\mathbf{T}$. (If not, then it would have a nonzero vector projection along $\mathbf{T}$, and $f$ would have a nonzero directional derivative in the directions $\pm\mathbf{T}$ and would therefore increase and decrease as we moved away from $P_0$ along $C$ in opposite directions.)

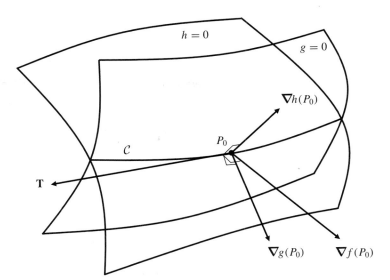

**Figure 13.17**    At $P_0$, $\nabla f$, $\nabla g$, and $\nabla h$ are all perpendicular to $\mathbf{T}$. Thus $\nabla f$ is in the plane spanned by $\nabla g$ and $\nabla h$

Since $\nabla g(P_0)$ and $\nabla h(P_0)$ are nonzero and both are perpendicular to $\mathbf{T}$, $\nabla f(P_0)$ must lie in the plane spanned by these two vectors and hence must be a linear combination of them:

$$\nabla f(x, y, z) = -\lambda_0 \nabla g(x, y, z) - \mu_0 \nabla h(x, y, z)$$

for some constants $\lambda_0$ and $\mu_0$. It follows that $(x_0, y_0, z_0, \lambda_0, \mu_0)$ is a critical point of the Lagrangian function

$$L(x, y, z, \lambda, \mu) = f(x, y, z) + \lambda g(x, y, z) + \mu h(x, y, z).$$

We look for triples $(x, y, z)$ that extremize $f(x, y, z)$ subject to the two constraints $g(x, y, z) = 0$ and $h(x, y, z) = 0$ among the points $(x, y, z, \lambda, \mu)$ that are critical points of the above Lagrangian function, and we therefore solve the system of equations

$$
\begin{aligned}
f_1(x, y, z) + \lambda g_1(x, y, z) + \mu h_1(x, y, z) &= 0, \\
f_2(x, y, z) + \lambda g_2(x, y, z) + \mu h_2(x, y, z) &= 0, \\
f_3(x, y, z) + \lambda g_3(x, y, z) + \mu h_3(x, y, z) &= 0, \\
g(x, y, z) &= 0, \\
h(x, y, z) &= 0.
\end{aligned}
$$

Solving such a system can be very difficult. It should be noted that in using the method of Lagrange multipliers instead of solving the constraint equations, we have traded the problem of having to solve two equations for two variables as *functions* of a third one for a problem of having to solve five equations for *numerical* values of five unknowns.

**Example 4**    If $f(x, y, z) = xy + 2z$, find the maximum and minimum values of $f$ on the circle that is the intersection of the plane $x + y + z = 0$ and the sphere $x^2 + y^2 + z^2 = 24$.

**Solution** The function $f$ is continuous and the circle is a closed bounded set in 3-space. Therefore maximum and minimum values must exist. We look for critical points of the Lagrangian

$$L = xy + 2z + \lambda(x + y + z) + \mu(x^2 + y^2 + z^2 - 24).$$

Setting the first partial derivatives of $L$ equal to zero, we obtain

$$y + \lambda + 2\mu x = 0, \quad (A)$$
$$x + \lambda + 2\mu y = 0, \quad (B)$$
$$2 + \lambda + 2\mu z = 0, \quad (C)$$
$$x + y + z = 0, \quad (D)$$
$$x^2 + y^2 + z^2 - 24 = 0. \quad (E)$$

Subtracting (A) from (B) we get $(x - y)(1 - 2\mu) = 0$. Therefore either $\mu = \frac{1}{2}$ or $x = y$. We analyze both possibilities.

**CASE I** If $\mu = \frac{1}{2}$ we obtain from (B) and (C)

$$x + \lambda + y = 0 \quad \text{and} \quad 2 + \lambda + z = 0.$$

Thus $x + y = 2 + z$. Combining this with (D) we get $z = -1$ and $x + y = 1$. Now by (E), $x^2 + y^2 = 24 - z^2 = 23$. Since $x^2 + y^2 + 2xy = (x + y)^2 = 1$, we have $2xy = 1 - 23 = -22$ and $xy = -11$. Now $(x - y)^2 = x^2 + y^2 - 2xy = 23 + 22 = 45$, so $x - y = \pm 3\sqrt{5}$. Combining this with $x + y = 1$ we obtain two critical points arising from $\mu = \frac{1}{2}$, namely $\left((1 + 3\sqrt{5})/2, (1 - 3\sqrt{5})/2, -1\right)$ and $\left((1 - 3\sqrt{5})/2, (1 + 3\sqrt{5})/2, -1\right)$. At both of these points we find that $f(x, y, z) = xy + 2z = -11 - 2 = -13$.

**CASE II** If $x = y$, then (D) implies that $z = -2x$, and (E) then gives $6x^2 = 24$, so $x = \pm 2$. Therefore points $(2, 2, -4)$ and $(-2, -2, 4)$ must be considered. We have $f(2, 2, -4) = 4 - 8 = -4$ and $f(-2, -2, 4) = 4 + 8 = 12$.

We conclude that the maximum value of $f$ on the circle is 12 and the minimum value is $-13$.

---

When none of the equations factors, try to combine two or more of them to produce an equation that does factor.

---

The method of Lagrange multipliers can be applied to find extreme values of a function of $n$ variables,

$$f(x_1, x_2, \ldots, x_n),$$

subject to $m \leq n - 1$ constraints,

$$g_{(1)}(x_1, x_2, \ldots, x_n) = 0,$$
$$g_{(2)}(x_1, x_2, \ldots, x_n) = 0,$$
$$\vdots$$
$$g_{(m)}(x_1, x_2, \ldots, x_n) = 0.$$

Assuming that the problem has a solution at the point $P_0$, that $f$ and all of the functions $g_{(j)}$ have continuous first partial derivatives at all points near $P_0$, and that the intersection of the constraint (hyper)surfaces is smooth near $P_0$, then we should look for $P_0$ among the critical points of the $(n + m)$-variable Lagrangian function

$$L(x_1, \ldots, x_n, \lambda_1, \ldots, \lambda_m) = f(x_1, \ldots, x_n) + \lambda_1 g_{(1)}(x_1, \ldots, x_n)$$
$$+ \lambda_2 g_{(2)}(x_1, \ldots, x_n) + \cdots + \lambda_m g_{(m)}(x_1, \ldots, x_n).$$

We will not attempt to prove this general assertion. (A proof can be based on the Implicit Function Theorem.)

# Exercises 13.3

**1.** Use the method of Lagrange multipliers to maximize $x^3 y^5$ subject to the constraint $x + y = 8$.

**2.** Find the shortest distance from the point $(3, 0)$ to the parabola $y = x^2$,

  (a) by reducing to an unconstrained problem in one variable, and

  (b) by using the method of Lagrange multipliers.

**3.** Find the distance from the origin to the plane $x + 2y + 2z = 3$,

  (a) using a geometric argument (no calculus),

  (b) by reducing the problem to an unconstrained problem in two variables, and

  (c) using the method of Lagrange multipliers.

**4.** Find the maximum and minimum values of the function $f(x, y, z) = x + y - z$ over the sphere $x^2 + y^2 + z^2 = 1$.

**5.** Use the Lagrange multiplier method to find the greatest and least distances from the point $(2, 1, -2)$ to the sphere with equation $x^2 + y^2 + z^2 = 1$. (Of course, the answer could be obtained more easily using a simple geometric argument.)

**6.** Find the shortest distance from the origin to the surface $xyz^2 = 2$.

**7.** Find $a$, $b$, and $c$ so that the volume $V = 4\pi abc/3$ of an ellipsoid $\dfrac{x^2}{a^2} + \dfrac{y^2}{b^2} + \dfrac{z^2}{c^2} = 1$ passing through the point $(1, 2, 1)$ is as small as possible.

**8.** Find the ends of the major and minor axes of the ellipse $3x^2 + 2xy + 3y^2 = 16$.

**9.** Find the maximum and minimum values of $f(x, y, z) = xyz$ on the sphere $x^2 + y^2 + z^2 = 12$.

**10.** Find the maximum and minimum values of $x + 2y - 3z$ over the ellipsoid $x^2 + 4y^2 + 9z^2 \le 108$.

**11.** Find the maximum and minimum values of the function $f(x, y, z) = x$ over the curve of intersection of the plane $z = x + y$ and the ellipsoid $x^2 + 2y^2 + 2z^2 = 8$.

**12.** Find the maximum and minimum values of $f(x, y, z) = x^2 + y^2 + z^2$ on the ellipse formed by the intersection of the cone $z^2 = x^2 + y^2$ and the plane $x - 2z = 3$.

**13.** Find the maximum and minimum values of $f(x, y, z) = 4 - z$ on the ellipse formed by the intersection of the cylinder $x^2 + y^2 = 8$ and the plane $x + y + z = 1$.

**14.** Find the maximum and minimum values of $f(x, y, z) = x + y^2 z$ subject to the constraints $y^2 + z^2 = 2$ and $z = x$.

**\* 15.** Use the method of Lagrange multipliers to find the shortest distance between the straight lines $x = y = z$ and $x = -y$, $z = 2$. (There are, of course, much easier ways to get the answer. This is an object lesson in the folly of shooting sparrows with cannons.)

**16.** Find the maximum and minimum values of the $n$-variable function $x_1 + x_2 + \cdots + x_n$ subject to the constraint $x_1^2 + x_2^2 + \cdots + x_n^2 = 1$.

**17.** Repeat Exercise 16 for the function $x_1 + 2x_2 + 3x_3 + \cdots + nx_n$ with the same constraint.

**18.** Find the most economical shape of a rectangular box with no top.

**19.** Find the maximum volume of a rectangular box with faces parallel to the coordinate planes if one corner is at the origin and the diagonally opposite corner lies on the plane $4x + 2y + z = 2$.

**20.** Find the maximum volume of a rectangular box with faces parallel to the coordinate planes if one corner is at the origin and the diagonally opposite corner is on the first octant part of the surface $xy + 2yz + 3xz = 18$.

**21.** A rectangular box having no top and having a prescribed volume $V$ m$^3$ is to be constructed using two different materials. The material used for the bottom and front of the box is five times as costly (per square metre) as the material used for the back and the other two sides. What should be the dimensions of the box to minimize the cost of materials?

**\* 22.** Find the maximum and minimum values of $xy + z^2$ on the ball $x^2 + y^2 + z^2 \le 1$. Use Lagrange multipliers to treat the boundary case.

**\* 23.** Repeat Exercise 22 but handle the boundary case by parametrizing the sphere $x^2 + y^2 + z^2 = 1$ using

$$x = \sin\phi \cos\theta, \quad y = \sin\phi \sin\theta, \quad \text{and} \quad z = \cos\phi,$$

where $0 \le \phi \le \pi$ and $0 \le \theta \le 2\pi$.

* **24.** If $\alpha$, $\beta$, and $\gamma$ are the angles of a triangle, show that

$$\sin\frac{\alpha}{2}\sin\frac{\beta}{2}\sin\frac{\gamma}{2} \le \frac{1}{8}.$$

For what triangles does equality occur?

* **25.** Suppose that $f$ and $g$ have continuous first partial derivatives throughout the $xy$-plane, and suppose that $g_2(a, b) \ne 0$. This implies that the equation $g(x, y) = g(a, b)$ defines $y$ implicitly as a function of $x$ nearby the point $(a, b)$. Use the Chain Rule to show that if $f(x, y)$ has a local extreme value at $(a, b)$ subject to the constraint $g(x, y) = g(a, b)$, then for some number $\lambda$ the point $(a, b, \lambda)$ is a critical point of the function

$$L(x, y, \lambda) = f(x, y) + \lambda g(x, y).$$

This constitutes a more formal justification of the method of Lagrange multipliers in this case.

**26.** What is the shortest distance from the point $(0, -1)$ to the curve $y = \sqrt{1 - x^2}$? Can this problem be solved by the Lagrange multiplier method? Why?

**27.** Example 3 showed that the method of Lagrange multipliers might fail to find a point that extremizes $f(x, y)$ subject to the constraint $g(x, y) = 0$ if $\nabla g = \mathbf{0}$ at the extremizing point. Can the method also fail if $\nabla f = \mathbf{0}$ at the extremizing point? Why?

# 13.4 The Method of Least Squares

Important optimization problems arise in the statistical analysis of experimental data. Frequently experiments are designed to measure the values of one or more quantities supposed constant, or to demonstrate a supposed functional relationship among variable quantities. Experimental error is usually present in the measurements, and experiments need to be repeated several times in order to arrive at *mean* or *average* values of the quantities being measured.

Consider a very simple example. An experiment to measure a certain physical constant $c$ is repeated $n$ times, yielding the values $c_1, c_2, \ldots, c_n$. If none of the measurements is suspected of being faulty, intuition tells us that we should use the mean value $\bar{c} = (c_1 + c_2 + \cdots + c_n)/n$ as the value of $c$ determined by the experiments. Let us see how this intuition can be justified.

Various methods for determining $c$ from the data values are possible. We could, for instance, choose $c$ to minimize the sum $T$ of its distances from the data points:

$$T = |c - c_1| + |c - c_2| + \cdots + |c - c_n|.$$

This is unsatisfactory for a number of reasons. Since absolute values have singular points it is difficult to determine the minimizing value of $c$. More importantly, $c$ may not be determined uniquely. If $n = 2$ any point in the interval between $c_1$ and $c_2$ will give the same minimum value to $T$. (See Exercise 24 at the end of this section for a generalization of this phenomenon.)

A more promising approach is to minimize the sum $S$ of *squares* of the distances from $c$ to the data points:

$$S = (c - c_1)^2 + (c - c_2)^2 + \cdots + (c - c_n)^2 = \sum_{i=1}^{n}(c - c_i)^2.$$

This function of $c$ is smooth and its (unconstrained) minimum value will occur at a critical point $\bar{c}$ given by

$$0 = \frac{dS}{dc}\bigg|_{c=\bar{c}} = \sum_{i=1}^{n} 2(\bar{c} - c_i) = 2n\bar{c} - 2\sum_{i=1}^{n}c_i.$$

Thus $\bar{c}$ is the *mean* of the data values:

$$\bar{c} = \frac{1}{n}\sum_{i=1}^{n} c_i = \frac{c_1 + c_2 + \cdots + c_n}{n}.$$

The technique used to obtain $\bar{c}$ above is an example of what is called **the method of least squares.** It has the following geometric interpretation. If the data values $c_1, c_2, \ldots, c_n$ are regarded as components of a vector $\mathbf{c}$ in $\mathbb{R}^n$, and $\mathbf{w}$ is the vector with components $1, 1, \ldots, 1$, then the vector projection of $\mathbf{c}$ in the direction of $\mathbf{w}$,

$$\mathbf{c_w} = \frac{\mathbf{c} \bullet \mathbf{w}}{|\mathbf{w}|^2}\,\mathbf{w} = \frac{c_1 + c_2 + \cdots + c_n}{n}\,\mathbf{w},$$

has all its components equal to the average of the data values. Thus determining $c$ from the data by the method of least squares corresponds to finding the vector projection of the data vector onto the one-dimensional subspace of $\mathbb{R}^n$ spanned by $\mathbf{w}$. Had there been no error in the measurements $c_i$, then $\mathbf{c}$ would have been equal to $c\mathbf{w}$.

## Linear Regression

In scientific investigations it is often believed that the response of a system is a certain kind of function of one or more input variables. An investigator can set up an experiment to measure the response of the system for various values of those variables in order to determine the parameters of the function.

For example, suppose that the response $y$ of a system is suspected to depend on the input $x$ according to the linear relationship

$$y = ax + b,$$

where the values of $a$ and $b$ are unknown. An experiment set up to measure values of $y$ corresponding to several values of $x$ yields $n$ data points, $(x_i, y_i), i = 1, 2, \ldots, n$. If the supposed linear relationship is valid, these data points should lie *approximately* along a straight line, but not exactly on one because of experimental error. Suppose the points are as shown in Figure 13.18. The linear relationship seems reasonable in this case. We want to find values of $a$ and $b$ so that the straight line $y = ax + b$ "best" fits the data.

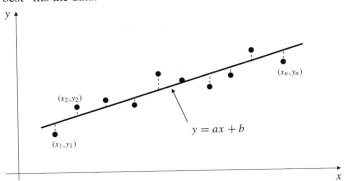

**Figure 13.18** Fitting a straight line through experimental data

In this situation the method of least squares requires that $a$ and $b$ be chosen to minimize the sum $S$ of the squares of the vertical displacements of the data points from the line:

$$S = \sum_{i=1}^{n} (y_i - ax_i - b)^2.$$

This is an unconstrained minimum problem in two variables, $a$ and $b$. The minimum will occur at a critical point of $S$ that satisfies

$$0 = \frac{\partial S}{\partial a} = -2 \sum_{i=1}^{n} x_i(y_i - ax_i - b),$$

$$0 = \frac{\partial S}{\partial b} = -2 \sum_{i=1}^{n} (y_i - ax_i - b).$$

These equations can be rewritten

$$\left( \sum_{i=1}^{n} x_i^2 \right) a + \left( \sum_{i=1}^{n} x_i \right) b = \sum_{i=1}^{n} x_i y_i,$$

$$\left( \sum_{i=1}^{n} x_i \right) a + n b = \sum_{i=1}^{n} y_i.$$

Solving this pair of linear equations we obtain the desired parameters:

$$a = \frac{n \left( \sum_{i=1}^{n} x_i y_i \right) - \left( \sum_{i=1}^{n} x_i \right)\left( \sum_{i=1}^{n} y_i \right)}{n \left( \sum_{i=1}^{n} x_i^2 \right) - \left( \sum_{i=1}^{n} x_i \right)^2} = \frac{\overline{xy} - \bar{x}\bar{y}}{\overline{x^2} - (\bar{x})^2},$$

$$b = \frac{\left( \sum_{i=1}^{n} x_i^2 \right)\left( \sum_{i=1}^{n} y_i \right) - \left( \sum_{i=1}^{n} x_i \right)\left( \sum_{i=1}^{n} x_i y_i \right)}{n \left( \sum_{i=1}^{n} x_i^2 \right) - \left( \sum_{i=1}^{n} x_i \right)^2} = \frac{\overline{x^2}\bar{y} - \bar{x}\overline{xy}}{\overline{x^2} - (\bar{x})^2}.$$

In these formulas we have used a bar to indicate the mean value of a quantity; thus $\overline{xy} = (1/n) \sum_{i=0}^{n} x_i y_i$, and so on.

This procedure for fitting the "best" straight line through data points by the method of least squares is called **linear regression**, and the line $y = ax + b$ obtained in this way is called the **empirical regression line** corresponding to the data. Some scientific calculators with statistical features provide for linear regression by accumulating the sums of $x_i$, $y_i$, $x_i^2$, and $x_i y_i$ in various registers, and keeping track of the number $n$ of data points entered in another register. At any time it has available the information necessary to calculate $a$ and $b$, and the value of $y$ corresponding to any given $x$.

**Example 1**    Find the empirical regression line for the data $(x, y) = (0, 2.10)$, $(1, 1.92)$, $(2, 1.84)$, and $(3, 1.71)$, $(4, 1.64)$. What is the predicted value of $y$ at $x = 5$?

**Solution**    We have

$$\bar{x} = \frac{0 + 1 + 2 + 3 + 4}{5} = 2,$$

$$\bar{y} = \frac{2.10 + 1.92 + 1.84 + 1.71 + 1.64}{5} = 1.842,$$

$$\overline{xy} = \frac{(0)(2.10) + (1)(1.92) + (2)(1.84) + (3)(1.71) + (4)(1.64)}{5} = 3.458,$$

$$\overline{x^2} = \frac{0^2 + 1^2 + 2^2 + 3^2 + 4^2}{5} = 6.$$

Therefore

$$a = \frac{3.458 - (2)(1.842)}{6 - 2^2} = -0.113,$$

$$b = \frac{(6)(1.842) - (2)(3.458)}{6 - 2^2} = 2.068,$$

and the empirical regression line is

$$y = 2.068 - 0.113x.$$

The predicted value of $y$ at $x = 5$ is $2.068 - 0.113 \times 5 = 1.503$. ∎

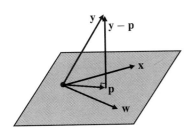

**Figure 13.19**    $\mathbf{p} = a\mathbf{x} + b\mathbf{w}$ is the projection of $\mathbf{y}$ onto the plane spanned by $\mathbf{x}$ and $\mathbf{w}$

Linear regression can also be interpreted in terms of vector projection. The data points define two vectors $\mathbf{x}$ and $\mathbf{y}$ in $\mathbb{R}^n$ with components $x_1, x_2, \dots, x_n$ and $y_1, y_2, \dots, y_n$ respectively. Let $\mathbf{w}$ be the vector with components $1, 1, \dots, 1$. Finding the coefficients $a$ and $b$ for the regression line corresponds to finding the orthogonal projection of $\mathbf{y}$ onto the two-dimensional subspace (plane) in $\mathbb{R}^n$ spanned by $\mathbf{x}$ and $\mathbf{w}$. (See Figure 13.19.) This projection is $\mathbf{p} = a\mathbf{x} + b\mathbf{w}$. In fact, the two equations obtained above by setting the partial derivatives of $S$ equal to zero are just the two conditions

$$(\mathbf{y} - \mathbf{p}) \bullet \mathbf{x} = 0,$$
$$(\mathbf{y} - \mathbf{p}) \bullet \mathbf{w} = 0,$$

stating that $\mathbf{y}$ minus its projection onto the subspace is perpendicular to the subspace. The angle between $\mathbf{y}$ and this $\mathbf{p}$ provides a measure of how well the empirical regression line fits the data; the smaller the angle the better the fit.

Linear regression can be used to find specific functional relationships of types other than linear if suitable transformations are applied to the data.

**Example 2**    Find the values of constants $K$ and $s$ for which the curve

$$y = Kx^s$$

best fits the experimental data points $(x_i, y_i)$, $i = 1, 2, \dots, n$. (Assume all data values are positive.)

**Solution**    Observe that the required functional form corresponds to a linear relationship between $\ln y$ and $\ln x$:

$$\ln y = \ln K + s \ln x.$$

If we determine the parameters $a$ and $b$ of the empirical regression line $\eta = a\xi + b$ corresponding to the transformed data $(\xi_i, \eta_i) = (\ln x_i, \ln y_i)$, then $s = a$ and $K = e^b$ are the required values.    ■

**Remark**    It should be stressed that the constants $K$ and $s$ obtained by the method used in the solution above are not the same as those that would be obtained by direct application of the least squares method to the untransformed problem, that is, by minimizing $\sum_{i=1}^{n}(y_i - Kx_i^s)^2$. This latter problem cannot readily be solved. (Try it!)

Generally, the method of least squares is applied to fit an equation in which the response is expressed as a sum of constants times functions of one or more input variables. The constants are determined as critical points of the sum of squared deviations of the actual response values from the values predicted by the equation.

## Applications of the Least Squares Method to Integrals

The method of least squares can be used to find approximations to reasonably well-behaved (say, piecewise continuous) functions as sums of constants times specified functions. The idea is to choose the constants to minimize the *integral* of the square of the difference.

For example, suppose we want to approximate the continuous function $f(x)$ over the interval $[0, 1]$ by a linear function $g(x) = px + q$. The method of least squares would require that $p$ and $q$ be chosen to minimize the integral

$$I(p, q) = \int_0^1 \Big(f(x) - px - q\Big)^2 dx.$$

Assuming that we can "differentiate through the integral" (we will investigate this issue in Section 13.5) the critical point of $I(p, q)$ can be found from

$$0 = \frac{\partial I}{\partial p} = -2\int_0^1 x\Big(f(x) - px - q\Big)dx,$$

$$0 = \frac{\partial I}{\partial q} = -2\int_0^1 \Big(f(x) - px - q\Big)dx.$$

Thus

$$\frac{p}{3} + \frac{q}{2} = \int_0^1 xf(x)\,dx,$$

$$\frac{p}{2} + q = \int_0^1 f(x)\,dx,$$

and solving this linear system for $p$ and $q$ we get

$$p = \int_0^1 (12x - 6)f(x)\,dx,$$

$$q = \int_0^1 (4 - 6x)f(x)\,dx.$$

The following example concerns the approximation of a function by a **trigonometric polynomial**. Such approximations form the basis for the study of **Fourier series**, which are of fundamental importance in the solution of boundary-value problems for the Laplace, heat, and wave equations, and other partial differential equations that arise in applied mathematics.

**Example 3**   Use a least squares integral to approximate $f(x)$ by the sum

$$\sum_{k=1}^{n} b_k \sin kx$$

on the interval $0 \le x \le \pi$.

**Solution**   We want to choose the constants to minimize

$$I = \int_0^\pi \left( f(x) - \sum_{k=1}^n b_k \sin kx \right)^2 dx.$$

For each $1 \le j \le n$ we have

$$0 = \frac{\partial I}{\partial b_j} = -2 \int_0^\pi \left( f(x) - \sum_{k=1}^n b_k \sin kx \right) \sin jx \, dx.$$

Thus

$$\sum_{k=1}^n b_k \int_0^\pi \sin kx \sin jx \, dx = \int_0^\pi f(x) \sin jx \, dx.$$

However, if $j \ne k$, then $\sin kx \sin jx$ is an even function, so that

$$\int_0^\pi \sin kx \sin jx \, dx = \frac{1}{2} \int_{-\pi}^\pi \sin kx \sin jx \, dx$$

$$= \frac{1}{4} \int_{-\pi}^\pi \left( \cos(k-j)x - \cos(k+j)x \right) dx = 0.$$

If $j = k$, then we have

$$\int_0^\pi \sin^2 jx \, dx = \frac{1}{2} \int_0^\pi (1 - \cos 2jx) \, dx = \frac{\pi}{2},$$

so that

$$b_j = \frac{2}{\pi} \int_0^\pi f(x) \sin jx \, dx.$$

**Remark**   If $f$ is continuous on $0 \le x \le \pi$, it can be shown that

$$\lim_{n \to \infty} \int_0^\pi \left( f(x) - \sum_{k=1}^n b_k \sin kx \right)^2 dx = 0,$$

with the coefficients $b_k$ given as above. It follows that

$$f(x) = \sum_{k=1}^\infty b_k \sin kx, \qquad 0 < x < \pi.$$

This is known as a **Fourier sine series** representation of $f(x)$ on the interval $(0, \pi)$. The function $f$ can also be approximated using a sum of cosines (see Exercise 21 below), and such approximations lead to an exact representation of $f$ as a **Fourier cosine series**:

$$f(x) = \frac{a_0}{2} + \sum_{k=1}^{\infty} a_k \cos kx, \qquad 0 < x < \pi,$$

where

$$a_j = \frac{2}{\pi} \int_0^{\pi} f(x) \cos jx \, dx.$$

**Remark**   Fourier series are treated with more detail in Section 9.10. Representing a function as the sum of a Fourier series is analogous to representing a vector as a sum of basis vectors. If we think of continuous functions on the interval $[0, \pi]$ as "vectors" with addition and scalar multiplication defined pointwise:

$$(f + g)(x) = f(x) + g(x), \qquad (cf)(x) = cf(x),$$

and with the "dot product" defined as

$$f \bullet g = \int_0^{\pi} f(x)g(x) \, dx,$$

then the functions $e_k(x) = \sqrt{2/\pi} \sin kx$ form a "basis." As shown in the example above, $e_j \bullet e_j = 1$, and if $k \neq j$ then $e_k \bullet e_j = 0$. Thus these "basis vectors" are "mutually perpendicular unit vectors." The Fourier coefficients $b_j$ of a function $f$ are the components of $f$ with respect to that basis.

# Exercises 13.4

1. A generator is to be installed in a factory to supply power to $n$ machines located at positions $(x_i, y_i)$, $i = 1, 2, \ldots, n$. Where should the generator be located to minimize the sum of the squares of its distances from the machines?

2. The relationship $y = ax^2$ is known to hold between certain variables. Given the experimental data $(x_i, y_i)$, $i = 1, 2, \ldots, n$, determine a value for $a$ by the method of least squares.

3. Repeat Exercise 2 but with the relationship $y = ae^x$.

4. Use the method of least squares to find the plane $z = ax + by + c$ that best fits the data $(x_i, y_i, z_i)$, $i = 1, 2, \ldots, n$.

5. Repeat Exercise 4 using a vector projection argument instead of the method of least squares.

In Exercises 6–11, show how to adapt linear regression to determine the two parameters $p$ and $q$ so that the given relationship fits the experimental data $(x_i, y_i)$, $i = 1, 2, \ldots, n$.

In which of these situations are the values of $p$ and $q$ obtained identical to those obtained by direct application of the method of least squares with no change of variable?

6. $y = p + qx^2$

7. $y = pe^{qx}$

8. $y = \ln(p + qx)$

9. $y = px + qx^2$

10. $y = \sqrt{px + q}$

11. $y = pe^x + qe^{-x}$

12. Find the parabola of the form $y = p + qx^2$ that best fits the data $(x, y) = (1, 0.11)$, $(2, 1.62)$, $(3, 4.07)$, $(4, 7.55)$, $(6, 17.63)$, and $(7, 24.20)$. No value of $y$ was measured at $x = 5$. What value would you predict at this point?

13. Use the method of least squares to find constants $a$, $b$ and $c$ so that the relationship $y = ax^2 + bx + c$ best describes the experimental data $(x_i, y_i)$, $i = 1, 2, \ldots, n$, $(n \geq 3)$. How is this situation interpreted in terms of vector projection?

14. How can the result of Exercise 13 be used to fit a curve of the form $y = pe^x + q + re^{-x}$ through the same data points?

15. Find the value of the constant $a$ for which the function $f(x) = ax^2$ best approximates the function $g(x) = x^3$ on the interval $[0, 1]$, in the sense that the integral

$$I = \int_0^1 \big(f(x) - g(x)\big)^2 \, dx$$

is minimized. What is the minimum value of $I$?

16. Find $a$ to minimize $I = \int_0^\pi \big(ax(\pi - x) - \sin x\big)^2 \, dx$. What is the minimum value of the integral?

17. Repeat Exercise 15 with the function $f(x) = ax^2 + b$ and the same $g$. Find $a$ and $b$.

* 18. Find $a$, $b$, and $c$ to minimize $\int_0^1 (x^3 - ax^2 - bx - c)^2 \, dx$. What is the minimum value of the integral?

* 19. Find $a$ and $b$ to minimize $\int_0^\pi (\sin x - ax^2 - bx)^2 \, dx$.

20. Find $a$, $b$, and $c$ to minimize the integral

$$J = \int_{-1}^1 \big(x - a \sin \pi x - b \sin 2\pi x - c \sin 3\pi x\big)^2 \, dx.$$

* 21. Find constants $a_j$, $j = 0, 1, \ldots, n$, to minimize

$$\int_0^\pi \left( f(x) - \frac{a_0}{2} - \sum_{k=1}^n a_k \cos kx \right)^2 dx.$$

22. Find the Fourier sine series for the function $f(x) = x$ on $0 < x < \pi$. To what function would you expect the series to converge in the interval $-\pi < x < 0$?

23. Repeat Exercise 22, but obtaining instead a Fourier cosine series.

24. Suppose $x_1, x_2, \ldots, x_n$ satisfy $x_i \le x_j$ whenever $i < j$. Find $x$ that minimizes $\sum_{i=1}^n |x - x_i|$. Treat the cases $n$ odd and $n$ even separately. For what values of $n$ is $x$ unique? *Hint:* use no calculus in this problem.

## 13.5 Parametric Problems

In this section we will briefly examine three unrelated situations in which one wants to differentiate a function with respect to a parameter rather than one of the basic variables of the function. Such situations arise frequently in mathematics and its applications.

### Differentiating Integrals with Parameters

The Fundamental Theorem of Calculus shows how to differentiate a definite integral with respect to the upper limit of integration:

$$\frac{d}{dx} \int_a^x f(t) \, dt = f(x).$$

We are going to look at a different problem about differentiating integrals. If the integrand of a definite integral also depends on variables other than the variable of integration, then the integral will be a function of those other variables. How are we to find the derivative of such a function? For instance, consider the function $F(x)$ defined by

$$F(x) = \int_a^b f(x, t) \, dt.$$

We would like to be able to calculate $F'(x)$ by taking the derivative inside the integral:

$$F'(x) = \frac{d}{dx} \int_a^b f(x, t) \, dt = \int_a^b \frac{\partial}{\partial x} f(x, t) \, dt.$$

Observe that we use "$d/dx$" outside the integral and "$\partial/\partial x$" inside because the integral is a function of $x$ only, but the integrand $f$ is a function of both $x$ and $t$. If the integrand depends on more than one parameter, then partial derivatives would be needed inside and outside the integral:

$$\frac{\partial}{\partial x} \int_a^b f(x, y, t)\, dt = \int_a^b \frac{\partial}{\partial x} f(x, y, t)\, dt.$$

The operation of taking a derivative with respect to a parameter inside the integral, or *differentiating through the integral* as it is usually called, seems plausible. We differentiate sums term by term, and integrals are the limits of sums. However, both the differentiation and integration operations involve the taking of limits, (limits of Newton quotients for derivatives, limits of Riemann sums for integrals). Differentiating through the integral requires changing the order in which the two limits are taken, and therefore requires justification.

We have already seen another example of change of order of limits. When we assert that two mixed partial derivatives with respect to the same variables are equal,

$$\frac{\partial^2 f}{\partial x \partial y} = \frac{\partial^2 f}{\partial y \partial x},$$

we are, in fact, saying that limits corresponding to differentiation with respect to $x$ and $y$ can be taken in either order with the same result. This is not true in general; we proved it under the assumption that both of the mixed partials were *continuous*. (See Theorem 1 and Exercise 16 of Section 12.4.) In general, some assumptions are required to justify the interchange of limits. The following theorem gives one set of conditions that justify the interchange of limits involved in differentiating through the integral.

**THEOREM 5**

**Differentiating through an integral**

Suppose that for every $x$ satisfying $c < x < d$ the following conditions hold:

(i) the integrals

$$\int_a^b f(x, t)\, dt \qquad \text{and} \qquad \int_a^b f_1(x, t)\, dt$$

both exist (either as proper or convergent improper integrals).

(ii) $f_{11}(x, t)$ exists and satisfies

$$|f_{11}(x, t)| \le g(t), \qquad a < t < b,$$

where

$$\int_a^b g(t)\, dt = K < \infty.$$

Then for each $x$ satisfying $c < x < d$ we have

$$\frac{d}{dx} \int_a^b f(x, t)\, dt = \int_a^b \frac{\partial}{\partial x} f(x, t)\, dt.$$

**PROOF**  Let

$$F(x) = \int_a^b f(x, t)\, dt.$$

If $c < x < d$, $h \neq 0$, and $|h|$ is sufficiently small that $c < x + h < d$, then, by Taylor's Formula,

$$f(x + h, t) = f(x, t) + h f_1(x, t) + \frac{h^2}{2} f_{11}(x + \theta h, t)$$

for some $\theta$ between 0 and 1. Therefore

$$\left| \frac{F(x+h) - F(x)}{h} - \int_a^b f_1(x, t)\, dt \right|$$

$$= \left| \int_a^b \frac{f(x+h, t) - f(x, t)}{h}\, dt - \int_a^b f_1(x, t)\, dt \right|$$

$$\leq \int_a^b \left| \frac{f(x+h, t) - f(x, t)}{h} - f_1(x, t) \right| dt$$

$$= \int_a^b \left| \frac{h}{2} f_{11}(x + \theta h, t) \right| dt$$

$$\leq \frac{h}{2} \int_a^b g(t)\, dt = \frac{Kh}{2} \;\to\; 0 \text{ as } h \to 0.$$

Therefore

$$F'(x) = \lim_{h \to 0} \frac{F(x+h) - F(x)}{h} = \int_a^b f_1(x, t)\, dt,$$

which is the desired result.

**Remark**  It can be shown that the conclusion of Theorem 5 also holds under the sole assumption that $f_1(x, t)$ is continuous on the *closed, bounded* rectangle $c \leq x \leq d$, $a \leq t \leq b$. We cannot prove this here—the proof depends on a subtle property called *uniform continuity* possessed by continuous functions on closed bounded sets in $\mathbb{R}^n$. In any event, Theorem 5 is more useful for our purposes because it allows for improper integrals.

**Example 1**  Evaluate $\displaystyle\int_0^\infty t^n e^{-t}\, dt.$

**Solution**  Starting with the convergent improper integral

$$\int_0^\infty e^{-s}\, ds = \lim_{R \to \infty} \frac{e^{-s}}{-1} \Big|_0^R = \lim_{R \to \infty} (1 - e^{-R}) = 1$$

we introduce a parameter by substituting $s = xt$, $ds = x\,dt$ (where $x > 0$) and get

$$\int_0^\infty e^{-xt}\,dt = \frac{1}{x}.$$

Now differentiate $n$ times (each resulting integral converges):

$$\int_0^\infty -t\,e^{-xt}\,dt = -\frac{1}{x^2},$$

$$\int_0^\infty (-t)^2 e^{-xt}\,dt = (-1)^2 \frac{2}{x^3},$$

$$\vdots$$

$$\int_0^\infty (-t)^n e^{-xt}\,dt = (-1)^n \frac{n!}{x^{n+1}}.$$

Putting $x = 1$ we get

$$\int_0^\infty t^n e^{-t}\,dt = n!.$$

Note that this result could be obtained by integration by parts ($n$ times) or a reduction formula. This method is a little easier.

**Remark**   The reader should check that the function $f(x, t) = t^k e^{-xt}$ satisfies the conditions of Theorem 5 for $x > 0$ and $k \geq 0$. We will normally not make a point of this.

**Example 2**   Evaluate $F(x, y) = \displaystyle\int_0^\infty \frac{e^{-xt} - e^{-yt}}{t}\,dt$ for $x > 0$, $y > 0$.

**Solution**   We have

$$\frac{\partial F}{\partial x} = -\int_0^\infty e^{-xt}\,dt = -\frac{1}{x} \quad \text{and} \quad \frac{\partial F}{\partial y} = \int_0^\infty e^{-yt}\,dt = \frac{1}{y}.$$

It follows that

$$F(x, y) = -\ln x + C_1(y) \quad \text{and} \quad F(x, y) = \ln y + C_2(x).$$

Comparing these two formulas for $F$, we are forced to conclude that $C_1(y) = \ln y + C$ for some constant $C$. Therefore

$$F(x, y) = \ln y - \ln x + C = \ln\frac{y}{x} + C.$$

Since $F(1, 1) = 0$, we must have $C = 0$ and $F(x, y) = \ln(y/x)$.

**Remark**   We can combine Theorem 5 and the Fundamental Theorem of Calculus to differentiate an integral with respect to a parameter that appears in the limits of integration as well as in the integrand. If

$$F(x, b, a) = \int_a^b f(x, t) \, dt,$$

then, by the Chain Rule,

$$\frac{d}{dx} F\big(x, b(x), a(x)\big) = \frac{\partial F}{\partial x} + \frac{\partial F}{\partial b} \frac{db}{dx} + \frac{\partial F}{\partial a} \frac{da}{dx}.$$

Accordingly, we have

$$\frac{d}{dx} \int_{a(x)}^{b(x)} f(x, t) \, dt$$

$$= \int_{a(x)}^{b(x)} \frac{\partial}{\partial x} f(x, t) \, dt + f\big(x, b(x)\big) b'(x) - f\big(x, a(x)\big) a'(x).$$

We require that $a(x)$ and $b(x)$ be differentiable at $x$, and for the application of Theorem 5, that $a \le a(x) \le b$ and $a \le b(x) \le b$ for all $x$ satisfying $c < x < d$.

**Example 3**    Solve the *integral equation*

$$f(x) = a - \int_b^x (x - t) f(t) \, dt.$$

***Solution***    Assume, for the moment, that the equation has a sufficiently well behaved solution to allow for differentiation through the integral. Differentiating twice, we get

$$f'(x) = -(x - x) f(x) - \int_b^x f(t) \, dt = - \int_b^x f(t) \, dt,$$

$$f''(x) = -f(x).$$

The latter equation is the differential equation of simple harmonic motion. Observe that the given equation for $f$ and that for $f'$ imply the initial conditions

$$f(b) = a \qquad \text{and} \qquad f'(b) = 0.$$

Accordingly, we write the general solution of $f''(x) = -f(x)$ in the form

$$f(x) = A \cos(x - b) + B \sin(x - b).$$

The initial conditions then imply $A = a$ and $B = 0$, so the required solution is $f(x) = a \cos(x - b)$. Finally, we note that this function is indeed smooth enough to allow the differentiations through the integral, and so is the solution of the given integral equation. (If you wish, verify it in the integral equation.)    ∎

## Envelopes

An equation $f(x, y, c) = 0$ that involves a parameter $c$ as well as the variables $x$ and $y$ represents a family of curves in the $xy$-plane. Consider, for instance, the family

$$f(x, y, c) = \frac{x}{c} + cy - 2 = 0.$$

This family consists of straight lines with intercepts $(2c, 2/c)$ on the coordinate axes. Several of these lines are sketched in Figure 13.20. It appears that there is a curve to which all these lines are tangent. This curve is called the envelope of the family of lines.

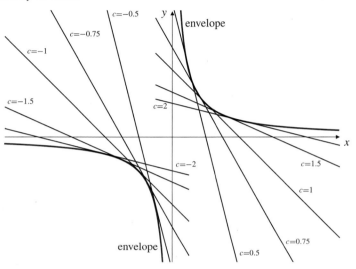

**Figure 13.20**    A family of straight lines and their envelope

In general, a curve $\mathcal{C}$ is called the **envelope** of the family of curves with equations $f(x, y, c) = 0$ if, for each value of $c$, the curve $f(x, y, c) = 0$ is tangent to $\mathcal{C}$ at some point depending on $c$.

For the family of lines in Figure 13.20 it appears that the envelope may be the rectangular hyperbola $xy = 1$. We will verify this after developing a method for determining the equation of the envelope of a family of curves. We assume that the function $f(x, y, c)$ has continuous first partials, and that the envelope is a smooth curve.

For each $c$, the curve $f(x, y, c) = 0$ is tangent to the envelope at a point $(x, y)$ that depends on $c$. Let us express this dependence in the explicit form $x = g(c)$, $y = h(c)$; these equations are parametric equations of the envelope. Since $(x, y)$ lies on the curve $f(x, y, c) = 0$, we have

$$f\big(g(c), h(c), c\big) = 0.$$

Differentiating this equation with respect to $c$, we obtain

$$f_1 g'(c) + f_2 h'(c) + f_3 = 0, \tag{$*$}$$

where the partials of $f$ are evaluated at $\big(g(c), h(c), c\big)$.

The slope of the curve $f(x, y, c) = 0$ at $\big(g(c), h(c), c\big)$ can be obtained by differentiating its equation implicitly with respect to $x$:

$$f_1 + f_2 \frac{dy}{dx} = 0.$$

**■ DANGER**

This is a subtle argument. Take your time and try to understand each step in the development.

On the other hand, the slope of the envelope $x = g(c)$, $y = h(c)$ at that point is $dy/dx = h'(c)/g'(c)$. Since the curve and the envelope are tangent at $f\big(g(c), h(c), c\big)$, these slopes must be equal. Therefore

$$f_1 g'(c) + f_2 h'(c) = 0.$$

Combining this with equation (∗) we get $f_3(x, y, c) = 0$ at all points of the envelope.

> The equation of the envelope can be found by eliminating $c$ between the two equations
>
> $$f(x, y, c) = 0 \qquad \text{and} \qquad \frac{\partial}{\partial c} f(x, y, c) = 0.$$

**Example 4**   Find the envelope of the family of straight lines

$$f(x, y, c) = \frac{x}{c} + cy - 2 = 0.$$

**Solution**   We eliminate $c$ between the equations

$$f(x, y, c) = \frac{x}{c} + cy - 2 = 0 \qquad \text{and} \qquad f_3(x, y, c) = -\frac{x}{c^2} + y = 0.$$

These equations can be solved for $x = c$ and $y = 1/c$, and hence imply that the envelope is $xy = 1$, as we conjectured earlier. ∎

**Example 5**   Find the envelope of the family of circles

$$(x - c)^2 + y^2 = c.$$

**Solution**   Here $f(x, y, c) = (x - c)^2 + y^2 - c$. The equation of the envelope is obtained by eliminating $c$ from the pair of equations

$$f(x, y, c) = (x - c)^2 + y^2 - c = 0,$$

$$\frac{\partial}{\partial c} f(x, y, c) = -2(x - c) - 1 = 0.$$

From the second equation, $x = c - \frac{1}{2}$, and then from the first, $y^2 = c - \frac{1}{4}$. Hence the envelope is the parabola

$$x = y^2 - \frac{1}{4}.$$

This envelope and some of the circles in the family are sketched in Figure 13.21. ∎

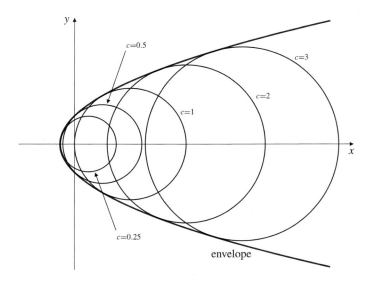

**Figure 13.21**   Circles $(x - c)^2 + y^2 = c$ and their envelope

A similar technique can be used to find the envelope of a family of surfaces. This will be a surface tangent to each member of the family.

**Example 6**   (**The Mach cone**)  Suppose that sound travels at speed $c$ in still air, and that a supersonic aircraft is travelling at speed $v > c$ along the $x$-axis, so that its position at time $t$ is $(vt, 0, 0)$. Find the envelope at time $t$ of the sound waves created by the aircraft at previous times.

**Solution**   The sound created by the aircraft at time $\tau < t$ spreads out as a spherical wave front at speed $c$. The centre of this wave front is $(v\tau, 0, 0)$, the position of the aircraft at time $\tau$. At time $t$ the radius of this wave front is $c(t - \tau)$, so its equation is

$$f(x, y, z, \tau) = (x - v\tau)^2 + y^2 + z^2 - c^2(t - \tau)^2 = 0. \tag{$*$}$$

At time $t$ the envelope of all these wave fronts created at earlier times $\tau$ is obtained by eliminating the parameter $\tau$ from the above equation and the equation

$$\frac{\partial}{\partial \tau} f(x, y, z, \tau) = -2v(x - v\tau) + 2c^2(t - \tau) = 0.$$

Solving this latter equation for $\tau$, we get

$$\tau = \frac{vx - c^2 t}{v^2 - c^2}.$$

Thus

$$x - v\tau = x - \frac{v^2 x - vc^2 t}{v^2 - c^2} = \frac{c^2}{v^2 - c^2}(vt - x)$$

$$t - \tau = t - \frac{vx - c^2 t}{v^2 - c^2} = \frac{v}{v^2 - c^2}(vt - x).$$

We substitute these two expressions into equation $(*)$ to eliminate $\tau$:

$$\frac{c^4}{(v^2 - c^2)^2}(vt - x)^2 + y^2 + z^2 - \frac{c^2 v^2}{(v^2 - c^2)^2}(vt - x)^2 = 0$$

$$y^2 + z^2 = \frac{c^2}{(v^2 - c^2)^2}(v^2 - c^2)(vt - x)^2 = \frac{c^2}{v^2 - c^2}(vt - x)^2.$$

The envelope is the cone

$$x = vt - \frac{\sqrt{v^2 - c^2}}{c}\sqrt{y^2 + z^2},$$

which extends backwards in the $x$ direction from its vertex at $(vt, 0, 0)$, the position of the aircraft at time $t$. This is called the **Mach cone**. The sound of the aircraft can not be heard at any point until the cone reaches that point. ∎

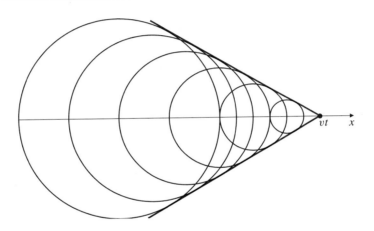

**Figure 13.22**    The Mach cone

## Equations with Perturbations

In applied mathematics one frequently encounters intractable equations for which at least approximate solutions are desired. Sometimes such equations result from adding an extra term to what would otherwise be a simple and easily solved equation. This extra term is called a **perturbation** of the simpler equation. Often the perturbation has coefficient smaller than the other terms in the equation, that is, it is a **small perturbation**. When this is the case you can find approximate solutions to the perturbed equation by replacing the small coefficient by a parameter and calculating Maclaurin polynomials in that parameter. One example should serve to clarify the method.

**Example 7**    Find an approximate solution of the equation

$$y + \frac{1}{50}\ln(1 + y) = x^2.$$

**Solution**    Without the logarithm term the equation would clearly have the solution $y = x^2$. Let us replace the coefficient $1/50$ with the parameter $\epsilon$ and look for a solution $y = y(x, \epsilon)$ to the equation

$$y + \epsilon \ln(1 + y) = x^2$$

in the form

$$y = y(x, \epsilon) = y(x, 0) + \epsilon y_\epsilon(x, 0) + \frac{\epsilon^2}{2!} y_{\epsilon\epsilon}(x, 0) + \cdots,$$

where the subscripts $\epsilon$ denote derivatives with respect to $\epsilon$. We shall calculate the terms up to second order in $\epsilon$. Evidently $y(x, 0) = x^2$. Differentiating the equation twice with respect to $\epsilon$ and evaluating the results at $\epsilon = 0$, we obtain

$$\frac{\partial y}{\partial \epsilon} + \ln(1 + y) + \frac{\epsilon}{1 + y} \frac{\partial y}{\partial \epsilon} = 0,$$

$$\frac{\partial^2 y}{\partial \epsilon^2} + \frac{2}{1 + y} \frac{\partial y}{\partial \epsilon} + \epsilon \frac{\partial}{\partial \epsilon} \left( \frac{1}{1 + y} \frac{\partial y}{\partial \epsilon} \right) = 0,$$

$$y_\epsilon(x, 0) = -\ln(1 + x^2),$$

$$y_{\epsilon\epsilon}(x, 0) = \frac{2}{1 + x^2} \ln(1 + x^2).$$

Hence

$$y(x, \epsilon) = x^2 - \epsilon \ln(1 + x^2) + \frac{\epsilon^2}{1 + x^2} \ln(1 + x^2) + \cdots,$$

and the given equation has the approximate solution

$$y \approx x^2 - \frac{\ln(1 + x^2)}{50} + \frac{\ln(1 + x^2)}{2500(1 + x^2)}.$$

Similar perturbation techniques can be used for systems of equations and for differential equations.

# Exercises 13.5

**1.** Let $F(x) = \displaystyle\int_0^1 t^x \, dt = \frac{1}{x + 1}$ for $x > -1$. By repeated differentiation of $F$ evaluate the integral

$$\int_0^1 t^x (\ln t)^n \, dt.$$

**2.** By replacing $t$ with $xt$ in the well-known integral

$$\int_{-\infty}^{\infty} e^{-t^2} \, dt = \sqrt{\pi},$$

and differentiating with respect to $x$, evaluate

$$\int_{-\infty}^{\infty} t^2 e^{-t^2} \, dt \quad \text{and} \quad \int_{-\infty}^{\infty} t^4 e^{-t^2} \, dt.$$

**3.** Evaluate $\displaystyle\int_{-\infty}^{\infty} \frac{e^{-xt^2} - e^{-yt^2}}{t^2} \, dt$ for $x > 0, \ y > 0$.

**4.** Evaluate $\displaystyle\int_0^1 \frac{t^x - t^y}{\ln t} \, dt$ for $x > -1, \ y > -1$.

**5.** Given that $\displaystyle\int_0^{\infty} e^{-xt} \sin t \, dt = \frac{1}{1 + x^2}$ for $x > 0$ (which can be shown by integration by parts) evaluate

$$\int_0^{\infty} t e^{-xt} \sin t \, dt \quad \text{and} \quad \int_0^{\infty} t^2 e^{-xt} \sin t \, dt.$$

**\* 6.** Referring to Exercise 5, for $x > 0$ evaluate

$$F(x) = \int_0^{\infty} e^{-xt} \frac{\sin t}{t} \, dt.$$

Show that $\lim_{x \to \infty} F(x) = 0$ and hence evaluate the integral

$$\int_0^\infty \frac{\sin t}{t} \, dt = \lim_{x \to 0} F(x).$$

**7.** Evaluate $\displaystyle\int_0^\infty \frac{dt}{x^2 + t^2}$ and use the result to help you evaluate

$$\int_0^\infty \frac{dt}{(x^2 + t^2)^2}, \quad \text{and} \quad \int_0^\infty \frac{dt}{(x^2 + t^2)^3}.$$

**8.** Evaluate $\displaystyle\int_0^x \frac{dt}{x^2 + t^2}$ and use the result to help you evaluate

$$\int_0^x \frac{dt}{(x^2 + t^2)^2}, \quad \text{and} \quad \int_0^x \frac{dt}{(x^2 + t^2)^3}.$$

**9.** Find $f^{(n+1)}(a)$ if $f(x) = 1 + \displaystyle\int_a^x (x - t)^n f(t) \, dt$.

Solve the integral equations in Exercises 10–12.

**10.** $f(x) = Cx + D + \displaystyle\int_0^x (x - t) f(t) \, dt$

**11.** $f(x) = x + \displaystyle\int_0^x (x - 2t) f(t) \, dt$

**12.** $f(x) = 1 + \displaystyle\int_0^1 (x + t) f(t) \, dt$

Find the envelopes of the families of curves in Exercises 13–18.

**13.** $y = 2cx - c^2$

**14.** $y - (x - c) \cos c = \sin c$

**15.** $x \cos c + y \sin c = 1$

**16.** $\dfrac{x}{\cos c} + \dfrac{y}{\sin c} = 1$

**17.** $y = c + (x - c)^2$

**18.** $(x - c)^2 + (y - c)^2 = 1$

**19.** Does every one-parameter family of curves in the plane have an envelope? Try to find the envelope of $y = x^2 + c$.

**20.** For what values of $k$ does the family of curves $x^2 + (y - c)^2 = kc^2$ have an envelope?

**21.** Try to find the envelope of the family $y^3 = (x + c)^2$. Are the curves of the family tangent to the envelope? What have you actually found in this case? Compare with Example 3 of Section 13.3.

**∗ 22.** Show that if a two-parameter family of surfaces $f(x, y, z, \lambda, \mu) = 0$ has an envelope, then the equation of that envelope can be obtained by eliminating $\lambda$ and $\mu$ from the three equations

$$f(x, y, z, \lambda, \mu) = 0,$$

$$\frac{\partial}{\partial \lambda} f(x, y, z, \lambda, \mu) = 0,$$

$$\frac{\partial}{\partial \mu} f(x, y, z, \lambda, \mu) = 0.$$

**23.** Find the envelope of the two-parameter family of planes

$$x \sin \lambda \cos \mu + y \sin \lambda \sin \mu + z \cos \lambda = 1.$$

**24.** Find the envelope of the two-parameter family of spheres

$$(x - \lambda)^2 + (y - \mu)^2 + z^2 = \frac{\lambda^2 + \mu^2}{2}.$$

In Exercises 25–27, find the terms up to second power in $\epsilon$ in the solution $y$ of the given equation.

**25.** $y + \epsilon \sin \pi y = x$

**26.** $y^2 + \epsilon e^{-y^2} = 1 + x^2$

**27.** $2y + \dfrac{\epsilon x}{1 + y^2} = 1$

**28.** Use perturbation methods to evaluate $y$ with error less than $10^{-8}$ given that $y + (y^5/100) = 1/2$.

**∗ 29.** Use perturbation methods to find approximate values for $x$ and $y$ from the system

$$x + 2y + \frac{1}{100} e^{-x} = 3, \quad x - y + \frac{1}{100} e^{-y} = 0.$$

Calculate all terms up to second order in $\epsilon = 1/100$.

# 13.6 Newton's Method

A frequently encountered problem in applied mathematics is to determine, to some desired degree of accuracy, a root (that is, a solution $r$) of an equation of the form

$$f(r) = 0.$$

Such a root is called a **zero** of the function $f$. In Section 4.6 we encountered Newton's method, a simple but powerful method for determining roots of functions that are sufficiently smooth. The method involves *guessing* an approximate value $x_0$ for a root $r$ of the function $f$, and then calculating successive approximations $x_1, x_2, \ldots$, using the formula

$$x_{n+1} = x_n - \frac{f(x_n)}{f'(x_n)}, \qquad n = 0, 1, 2, \cdots .$$

If the initial guess $x_0$ is not too far from $r$, and if $|f'(x)|$ is *not too small* and $|f''(x)|$ is *not too large* near $r$ then the successive approximations $x_1, x_2, \ldots$ will converge very rapidly to $r$. Recall that each new approximation $x_{n+1}$ is obtained as the $x$-intercept of the tangent line drawn to the graph of $f$ at the previous approximation, $x_n$. The tangent line to the graph $y = f(x)$ at $x = x_n$ has equation

$$y - f(x_n) = f'(x_n)(x - x_n).$$

(See Figure 13.23.) The $x$-intercept, $x_{n+1}$, of this line is determined by setting $y = 0$, $x = x_{n+1}$ in this equation, so is given by the formula in the shaded box above.

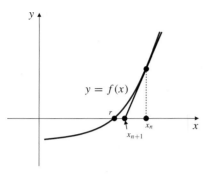

**Figure 13.23**    $x_{n+1}$ is the $x$-intercept of the tangent at $x_n$

Newton's method can be extended to finding solutions of systems of $m$ equations in $m$ variables. We will show here how to adapt the method to find approximations to a solution $(x, y)$ of the pair of equations

$$\begin{cases} f(x, y) = 0 \\ g(x, y) = 0 \end{cases}$$

starting from an initial guess $(x_0, y_0)$. Under auspicious circumstances, we will observe the same rapid convergence of approximations to the root that typifies the single-variable case.

The idea is as follows. The two surfaces $z = f(x, y)$ and $z = g(x, y)$ intersect in a curve which itself intersects the $xy$-plane at the point whose coordinates are the desired solution. If $(x_0, y_0)$ is near that point then the tangent planes to the two surfaces at $(x_0, y_0)$ will intersect in a straight line. This line meets the $xy$-plane at a point $(x_1, y_1)$ that should be even closer to the solution point than was $(x_0, y_0)$. We can easily determine $(x_1, y_1)$. The tangent planes to $z = f(x, y)$ and $z = g(x, y)$ at $(x_0, y_0)$ have equations

$$z = f(x_0, y_0) + f_1(x_0, y_0)(x - x_0) + f_2(x_0, y_0)(y - y_0),$$
$$z = g(x_0, y_0) + g_1(x_0, y_0)(x - x_0) + g_2(x_0, y_0)(y - y_0).$$

The line of intersection of these two planes meets the $xy$-plane at the point $(x_1, y_1)$ satisfying

$$f_1(x_0, y_0)(x_1 - x_0) + f_2(x_0, y_0)(y_1 - y_0) + f(x_0, y_0) = 0,$$
$$g_1(x_0, y_0)(x_1 - x_0) + g_2(x_0, y_0)(y_1 - y_0) + g(x_0, y_0) = 0.$$

Solving these two equations for $x_1$ and $y_1$, we obtain

$$x_1 = x_0 - \frac{fg_2 - f_2g}{f_1g_2 - f_2g_1}\bigg|_{(x_0, y_0)} = x_0 - \frac{\begin{vmatrix} f & f_2 \\ g & g_2 \end{vmatrix}}{\begin{vmatrix} f_1 & f_2 \\ g_1 & g_2 \end{vmatrix}}\Bigg|_{(x_0, y_0)},$$

$$y_1 = y_0 - \frac{f_1g - fg_1}{f_1g_2 - f_2g_1}\bigg|_{(x_0, y_0)} = y_0 - \frac{\begin{vmatrix} f_1 & f \\ g_1 & g \end{vmatrix}}{\begin{vmatrix} f_1 & f_2 \\ g_1 & g_2 \end{vmatrix}}\Bigg|_{(x_0, y_0)}.$$

Observe that the denominator in each of these expressions is the Jacobian determinant $\partial(f, g)/\partial(x, y)\big|_{(x_0, y_0)}$. This is another instance where the Jacobian is the appropriate multivariable analogue of the derivative of a function of one variable.

Continuing in this way we generate successive approximations $(x_n, y_n)$ according to the formulas

$$x_{n+1} = x_n - \frac{\begin{vmatrix} f & f_2 \\ g & g_2 \end{vmatrix}}{\begin{vmatrix} f_1 & f_2 \\ g_1 & g_2 \end{vmatrix}}\Bigg|_{(x_n, y_n)},$$

$$y_{n+1} = y_n - \frac{\begin{vmatrix} f_1 & f \\ g_1 & g \end{vmatrix}}{\begin{vmatrix} f_1 & f_2 \\ g_1 & g_2 \end{vmatrix}}\Bigg|_{(x_n, y_n)}.$$

We stop when the desired accuracy has been achieved.

**Example 1** Find the four roots of the system

$$x^2 - xy + 2y^2 - 10 = 0, \qquad x^3 y^2 - 2 = 0$$

with sufficient accuracy to ensure that the left sides of the equations vanish to the sixth decimal place.

**Solution** A sketch of the graphs of the two equations (see Figure 13.24) in the $xy$-plane indicates that there are roots near the points $(1, 2.5)$, $(3.5, 0.5)$, $(3, -0.5)$, and $(1, -2)$. Application of Newton's method requires successive computations of the quantities

$$f(x, y) = x^2 - xy + 2y^2 - 10, \qquad f_1(x, y) = 2x - y, \qquad f_2(x, y) = -x + 4y,$$
$$g(x, y) = x^3 y^2 - 2, \qquad\qquad g_1(x, y) = 3x^2 y^2, \qquad g_2(x, y) = 2x^3 y.$$

Using a calculator or computer we can calculate successive values of $(x_n, y_n)$ starting from each of the four points $(x_0, y_0)$ mentioned above:

**Table 1.** Root near $(1, 2.5)$

| $n$ | $x_n$ | $y_n$ | $f(x_n, y_n)$ | $g(x_n, y_n)$ |
|---|---|---|---|---|
| 0 | 1.000000 | 2.500000 | 1.000000 | 4.250000 |
| 1 | 0.805839 | 2.378102 | 0.043749 | 0.959420 |
| 2 | 0.721568 | 2.365659 | 0.006363 | 0.102502 |
| 3 | 0.710233 | 2.363735 | 0.000114 | 0.001708 |
| 4 | 0.710038 | 2.363701 | 0.000000 | 0.000001 |
| 5 | 0.710038 | 2.363701 | 0.000000 | 0.000000 |

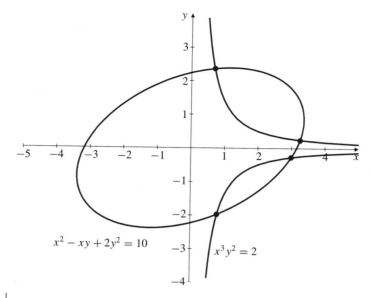

$$x^2 - xy + 2y^2 = 10$$

$$x^3 y^2 = 2$$

**Figure 13.24**    The two graphs intersect at four points

The values in these tables were calculated sequentially in a spreadsheet by the method suggested on the next page. They were rounded for inclusion in the table but the unrounded values were used in subsequent calculations. If you actually use the (rounded) values of $x_n$ and $y_n$ given in the table to calculate $f(x_n, y_n)$ and $g(x_n, y_n)$ your results may vary slightly. For example, the value of $f(x_n, y_n)$ calculated from the values of $x_n$ and $y_n$ given for $n = 5$ in Table 1 is 0.000001.

**Table 2.**    Root near $(3.5, 0.5)$

| $n$ | $x_n$ | $y_n$ | $f(x_n, y_n)$ | $g(x_n, y_n)$ |
|---|---|---|---|---|
| 0 | 3.500000 | 0.500000 | 1.000000 | 8.718750 |
| 1 | 3.308687 | 0.337643 | 0.058259 | 2.129355 |
| 2 | 3.273924 | 0.255909 | 0.011728 | 0.298145 |
| 3 | 3.266437 | 0.240187 | 0.000433 | 0.010584 |
| 4 | 3.266148 | 0.239587 | 0.000001 | 0.000015 |
| 5 | 3.266148 | 0.239586 | 0.000000 | 0.000000 |

**Table 3.**    Root near $(3, -0.5)$

| $n$ | $x_n$ | $y_n$ | $f(x_n, y_n)$ | $g(x_n, y_n)$ |
|---|---|---|---|---|
| 0 | 3.000000 | -0.500000 | 1.000000 | 4.750000 |
| 1 | 2.977072 | -0.329806 | 0.062360 | 0.870022 |
| 2 | 3.004434 | -0.275270 | 0.005205 | 0.054976 |
| 3 | 3.006167 | -0.271350 | 0.000027 | 0.000318 |
| 4 | 3.006178 | -0.271327 | 0.000000 | 0.000000 |

**Table 4.**    Root near $(1, -2)$

| $n$ | $x_n$ | $y_n$ | $f(x_n, y_n)$ | $g(x_n, y_n)$ |
|---|---|---|---|---|
| 0 | 1.000000 | -2.000000 | 1.000000 | 2.000000 |
| 1 | 0.847826 | -1.956522 | 0.033554 | 0.332865 |
| 2 | 0.803190 | -1.971447 | 0.001772 | 0.013843 |
| 3 | 0.801180 | -1.972071 | 0.000004 | 0.000028 |
| 4 | 0.801176 | -1.972072 | 0.000000 | 0.000000 |

The desired approximations to the four roots of the system are the $x_n$ and $y_n$ values in the last lines of the above tables. Note the rapidity of convergence in each case. However, many function evaluations are needed for each iteration of the method. For large systems Newton's method is computationally too inefficient to be practical. Other methods requiring more iterations but many fewer calculations per iteration are used in practice.

# Implementing Newton's Method Using a Spreadsheet

A computer spreadsheet is an ideal environment in which to calculate Newton's method approximations. For a pair of equations in two unknowns such as the system in Example 1, you can proceed as follows:

(i) In the first nine cells of the first row (A1–I1) put the labels n, x, y, f, g, f1, f2, g1, and g2.

(ii) In cells A2–A9 put the numbers 0, 1, 2, ..., 7.

(iii) In cells B2 and C2 put the starting values $x_0$ and $y_0$.

(iv) In cells D2–I2 put formulas for calculating $f(x, y)$, $g(x, y)$, ..., $g_2(x, y)$ in terms of values of $x$ and $y$ assumed to be in B2 and C2.

(v) In cells B3 and C3 store the Newton's method formulas for calculating $x_1$ and $y_1$ in terms of the values $x_0$ and $y_0$, using values calculated in the second row. For instance, cell B3 should contain the formula

$$\texttt{+B2-(D2*I2-G2*E2)/(F2*I2-G2*H2).}$$

(vi) Replicate the formulas in cells D2–I2 to cells D3–I3.

(vii) Replicate the formulas in cells B3–I3 to the cells B4–I9.

You can now inspect the successive approximations $x_n$ and $y_n$ in columns B and C. To use different starting values, just replace the numbers in cells B2 and C2. To solve a different system of (two) equations, replace the contents of cells D2–I2. You may wish to save this spreadsheet for reuse with the exercises below, or other systems you may want to solve later.

**Remark** While a detailed analysis of the convergence of Newton's method approximations is beyond the scope of this book, a few observations can be made. At each step in the approximation process we must divide by $J$, the Jacobian determinant of $f$ and $g$ with respect to $x$ and $y$ evaluated at the most recently obtained approximation. Assuming that the functions and partial derivatives involved in the formulas are continuous, the larger the value of $J$ at the actual solution, the more likely are the approximations to converge to the solution, and to do so rapidly. If $J$ vanishes (or is very small) at the solution, the successive approximations may not converge, even if the initial guess is quite close to the solution. Even if the first partials of $f$ and $g$ are large at the solution, their Jacobian may be small if their gradients are nearly parallel there. Thus we cannot expect convergence to be rapid when the curves $f(x, y) = 0$ and $g(x, y) = 0$ intersect at a very small angle.

Newton's method can be applied to systems of $m$ equations in $m$ variables; the formulas are the obvious generalizations of those for two functions given above.

# Exercises 13.6

Find the solutions of the systems in Exercises 1–6, so that the left-hand sides of the equations vanish up to six decimal places. These can be done with the aid of a scientific calculator, but that approach will be very time consuming. It is much easier to program the Newton's method formulas on a computer to generate the required approximations. In each case try to determine reasonable *initial guesses* by sketching graphs of the equations.

**1.** $y - e^x = 0$, $\quad x - \sin y = 0$

**2.** $x^2 + y^2 - 1 = 0$, $\quad y - e^x = 0$ (two solutions)

**3.** $x^4 + y^2 - 16 = 0$, $\quad xy - 1 = 0$ (four solutions)

**4.** $x(1 + y^2) - 1 = 0$, $\quad y(1 + x^2) - 2 = 0$ (one solution)

**5.** $y - \sin x = 0,$ $\quad x^2 + (y+1)^2 - 2 = 0$ (two solutions)

**6.** $\sin x + \sin y - 1 = 0,$ $\quad y^2 - x^3 = 0$ (two solutions)

**7.** Write formulas for obtaining successive Newton's method approximations to a solution of the system

$$f(x, y, z) = 0, \quad g(x, y, z) = 0, \quad h(x, y, z) = 0,$$

starting from an initial guess $(x_0, y_0, z_0)$.

**8.** Use the formulas from Exercise 7 to find the first octant intersection point of the surfaces $y^2 + z^2 = 3,\ x^2 + z^2 = 2,$ and $x^2 - z = 0$.

**9.** The equations $y - x^2 = 0$ and $y - x^3 = 0$ evidently have the solutions $x = y = 0$ and $x = y = 1$. Try to obtain these solutions using the two-variable form of Newton's method with starting values (a) $x_0 = y_0 = 0.1$, and (b) $x_0 = y_0 = 0.9$. How many iterations are required to obtain six decimal place accuracy for the appropriate solution in each case? How do you account for the difference in the behaviour of Newton's method for these equations near $(0, 0)$ and $(1, 1)$?

# Chapter Review

## Key Ideas

• **What is meant by the following phrases?**

  ◇ a critical point of $f(x, y)$      ◇ a singular point of $f(x, y)$

  ◇ an absolute maximum value of $f(x, y)$

  ◇ a local minimum value of $f(x, y)$

  ◇ a saddle point of $f(x, y)$      ◇ a quadratic form

  ◇ a constraint      ◇ linear programming

  ◇ an envelope of a family of curves

• **State the second derivative test for a critical point of $f(x, y)$.**

• **Describe the method of Lagrange multipliers.**

• **Describe the method of least squares.**

• **Describe Newton's method for two equations.**

## Review Exercises

In Exercises 1–4, find and classify all the critical points of the given functions.

**1.** $xy\, e^{-x+y}$

**2.** $x^2 y - 2xy^2 + 2xy$

**3.** $\dfrac{1}{x} + \dfrac{4}{y} + \dfrac{9}{4 - x - y}$

**4.** $x^2 y(2 - x - y)$

**5.** Let $f(x, y, z) = x^2 + y^2 + z^2 + 1/(x^2 + y^2 + z^2)$. Does $f$ have a minimum value? If so, what is it and where is it assumed?

**6.** Show that $x^2 + y^2 + z^2 - xy - xz - yz$ has a local minimum value at $(0, 0, 0)$. Is the minimum value assumed anywhere else?

**7.** Find the absolute maximum and minimum values of $f(x, y) = xye^{-x^2 - 4y^2}$. Justify your answer.

**8.** Let $f(x, y) = (4x^2 - y^2)e^{-x^2 + y^2}$.

  (a) Find the maximum and minimum values of $f(x, y)$ on the $xy$-plane.

  (b) Find the maximum and minimum values of $f(x, y)$ on the wedge-shaped region $0 \le y \le 3x$.

**9.** A wire of length $L$ cm is cut into at most three pieces and each piece is bent into a square. What is the (a) minimum and (b) maximum of the sum of the areas of the squares?

**10.** A delivery service will accept parcels in the shape of rectangular boxes the sum of whose girth and height is at most 120 inches. (The girth is the perimeter of a horizontal cross-section.) What is the largest possible volume of such a box?

**11.** Find the area of the smallest ellipse $(x/a)^2 + (y/b)^2 = 1$ that contains the rectangle $-1 \le x \le 1, -2 \le y \le 2$.

**12.** Find the volume of the smallest ellipsoid

$$\frac{x^2}{a^2} + \frac{y^2}{b^2} + \frac{z^2}{c^2} = 1$$

that contains the rectangular box $-1 \le x \le 1, -2 \le y \le 2, -3 \le z \le 3$.

**13.** Find the volume of the smallest region of the form

$$0 \le z \le a\left(1 - \frac{x^2}{b^2} - \frac{y^2}{c^2}\right)$$

that contains the box $-1 \le x \le 1, -2 \le y \le 2, 0 \le z \le 2$.

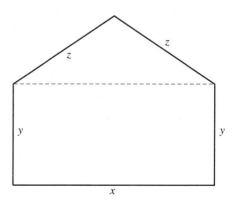

**Figure 13.25**

**14.** A window has the shape of a rectangle surmounted by an isosceles triangle. What are the dimensions $x$, $y$, and $z$ of the window (see Figure 13.25) if its perimeter is $L$ and its area is maximum?

**15.** A widget manufacturer determines that if she manufactures $x$ thousands of widgets per month, and sells the widgets for $y$ dollars each, then her monthly profit (in thousands of dollars) will be $P = xy - \frac{1}{27}x^2y^3 - x$. If her factory is capable of producing at most 3,000 widgets per month, and government regulations prevent her from charging more than $2 per widget, how many should she manufacture, and how much should she charge for each, to maximize her monthly profit?

**16.** Find the envelope of the curves $y = (x - c)^3 + 3c$.

**17.** Find an approximate solution $y(x, \epsilon)$ of the equation $y + \epsilon x e^y = -2x$ having terms up to second degree in $\epsilon$.

## Challenging Problems

**1. (Fourier series)**
Show that the constants $a_k$, $(k = 0, 1, 2, \ldots, n)$, and $b_k$, $(k = 1, 2, \ldots, n)$, that minimize the integral

$$I_n = \int_{-\pi}^{\pi} \left[ f(x) - \frac{a_0}{2} - \sum_{k=0}^{n}\left(a_k \cos kx + b_k \sin kx\right) \right]^2 dx$$

are given by

$$a_k = \frac{1}{\pi}\int_{\pi}^{\pi} f(x)\cos kx\, dx, \quad b_k = \frac{1}{\pi}\int_{\pi}^{\pi} f(x)\sin kx\, dx.$$

Note that these numbers, called the **Fourier coefficients** of $f$ on $[-\pi, \pi]$, do not depend on $n$. If they can be calculated for all positive integers $k$, then the series

$$\frac{a_0}{2} + \sum_{k=0}^{\infty}\left(a_k \cos kx + b_k \sin kx\right)$$

is called the **full range Fourier series** of $f$.

**2.** This is a continuation of Exercise 1. Find the (full range) Fourier coefficients $a_k$ and $b_k$ of

$$f(x) = \begin{cases} 0 & \text{if } -\pi \leq x < 0 \\ x & \text{if } 0 \leq x \leq \pi. \end{cases}$$

What is the minimum value of $I_n$ in this case? How does it behave as $n \to \infty$?

\* **3.** Evaluate $\displaystyle\int_0^x \frac{\ln(tx + 1)}{1 + t^2}\, dt.$

\* **4. (Steiner's problem)** The problem of finding a point in the plane (or a higher-dimensional space) that minimizes the sum of its distances from $n$ given points is very difficult. The case $n = 3$ is known as Steiner's problem. If $P_1 P_2 P_3$ is a triangle whose largest angle is less than $120°$, there is a point $Q$ inside the triangle so that the lines $QP_1$, $QP_2$, and $QP_3$ make equal $120°$ angles with one another. Show that the sum of the distances from the vertices of the triangle to a point $P$ is minimum when $P = Q$. *Hint:* first show that if $P = (x, y)$ and $P_i = (x_i, y_i)$, then

$$\frac{d|PP_i|}{dx} = \cos\theta_i \quad \text{and} \quad \frac{d|PP_i|}{dy} = \sin\theta_i,$$

where $\theta_i$ is the angle between $\overrightarrow{P_iP}$ and the positive direction of the $x$-axis. Hence show that the minimal point $P$ satisfies two trigonometric equations involving $\theta_1$, $\theta_2$, and $\theta_3$. Then try to show that any two of those angles differ by $\pm 2\pi/3$. Where should $P$ be taken if the triangle has an angle of $120°$ or greater?

# CHAPTER 14

# Multiple Integration

***Introduction*** In this chapter we extend the concept of definite integral to functions of several variables. Defined as limits of Riemann sums, like the one-dimensional definite integral, such multiple integrals can be evaluated using successive single definite integrals. They are used to represent and calculate quantities specified in terms of densities in regions of the plane or spaces of higher dimension. In the simplest instance, the volume of a three-dimensional region is given by a *double integral* of its height over the two-dimensional plane region that is its base.

## 14.1 Double Integrals

The definition of the definite integral, $\int_a^b f(x)\,dx$, is motivated by the *standard area problem*, namely the problem of finding the area of the plane region bounded by the curve $y = f(x)$, the $x$-axis, and the lines $x = a$ and $x = b$. Similarly, we can motivate the double integral of a function of two variables over a domain $D$ in the plane by means of the *standard volume problem* of finding the volume of the three-dimensional region $S$ bounded by the surface $z = f(x, y)$, the $xy$-plane, and the cylinder parallel to the $z$-axis passing through the boundary of $D$. (See Figure 14.1. $D$ is called the **domain of integration**.) We will call such a three-dimensional region $S$ a "solid," although we are not implying that it is filled with any particular substance. We will define the double integral of $f(x, y)$ over the domain $D$,

$$\iint_D f(x, y)\,dA,$$

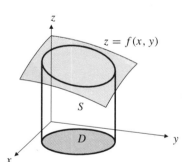

**Figure 14.1** A *solid* region $S$ lying above domain $D$ in the $xy$-plane, and below the surface $z = f(x, y)$.

in such a way that its value will give the volume of the solid $S$ whenever $D$ is a "reasonable" domain and $f$ is a "reasonable" function with positive values.

Let us start with the case where $D$ is a closed rectangle with sides parallel to the coordinate axes in the $xy$-plane, and $f$ is a bounded function on $D$. If $D$ consists of the points $(x, y)$ such that $a \le x \le b$ and $c \le y \le d$, we can form a **partition** $P$ of $D$ into small rectangles by partitioning each of the intervals $[a, b]$ and $[c, d]$, say by points

$$a = x_0 < x_1 < x_2 < \cdots < x_{m-1} < x_m = b,$$
$$c = y_0 < y_1 < y_2 < \cdots < y_{n-1} < y_n = d.$$

The partition $P$ of $D$ then consists of the $mn$ rectangles $R_{ij}$ ($1 \le i \le m$, $1 \le j \le n$), consisting of points $(x, y)$ for which $x_{i-1} \le x \le x_i$ and $y_{j-1} \le y \le y_j$. (See Figure 14.2.)

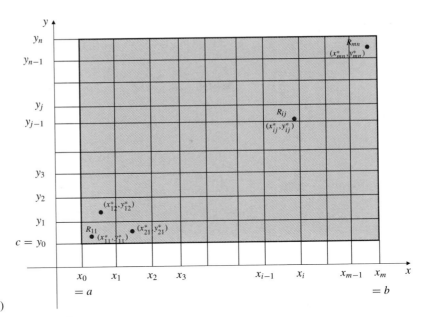

**Figure 14.2** A partition of $D$ (the large shaded rectangle) into smaller rectangles $R_{ij}$ ($1 \le i \le m$, $1 \le j \le n$)

The rectangle $R_{ij}$ has area

$$\Delta A_{ij} = \Delta x_i \Delta y_j = (x_i - x_{i-1})(y_j - y_{j-1})$$

and *diameter* (that is, diagonal length)

$$\mathrm{diam}(R_{ij}) = \sqrt{(\Delta x_i)^2 + (\Delta y_j)^2} = \sqrt{(x_i - x_{i-1})^2 + (y_j - y_{j-1})^2}.$$

The **norm** of the partition $P$ is the largest of these subrectangle diameters:

$$\|P\| = \max_{\substack{1 \le i \le m \\ 1 \le j \le n}} \mathrm{diam}(R_{ij}).$$

Now we pick an arbitrary point $(x_{ij}^*, y_{ij}^*)$ in each of the rectangles $R_{ij}$ and form the **Riemann sum**

$$R(f, P) = \sum_{i=1}^{m} \sum_{j=1}^{n} f(x_{ij}^*, y_{ij}^*) \, \Delta A_{ij},$$

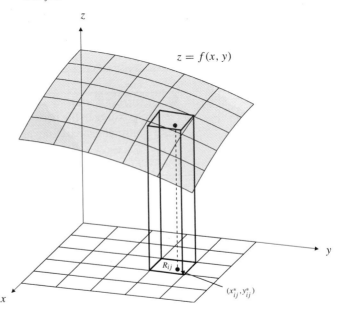

**Figure 14.3** A rectangular box above rectangle $R_{ij}$. The Riemann sum is a sum of volumes of such boxes

which is the sum of $mn$ terms, one for each rectangle in the partition. (Here the *double summation* indicates the sum as $i$ goes from 1 to $m$ of terms, each of which is itself a sum as $j$ goes from 1 to $n$.) The term corresponding to rectangle $R_{ij}$ is, if $f(x_{ij}^*, y_{ij}^*) \geq 0$, the volume of the rectangular box whose base is $R_{ij}$, and whose height is the value of $f$ at $(x_{ij}^*, y_{ij}^*)$. (See Figure 14.3.) Therefore, for positive functions $f$, the Riemann sum $R(f, P)$ approximates the volume above $D$ and under the graph of $f$. The double integral of $f$ over $D$ is defined to be the limit of such Riemann sums, provided the limit exists as $\|P\| \to 0$ independently of how the points $(x_{ij}^*, y_{ij}^*)$ are chosen. We make this precise in the following definition.

**DEFINITION 1**

**The double integral over a rectangle**

We say that $f$ is **integrable** over the rectangle $D$, and has **double integral**

$$I = \iint_D f(x, y)\, dA,$$

if for every positive number $\epsilon$ there exists a number $\delta$ depending on $\epsilon$, such that

$$|R(f, P) - I| < \epsilon$$

holds for every partition $P$ of $D$ satisfying $\|P\| < \delta$ and for all choices of the points $(x_{ij}^*, y_{ij}^*)$ in the subrectangles of $P$.

The "$dA$" that appears in the expression for the double integral is an *area element*. It represents the limit of the $\Delta A = \Delta x \, \Delta y$ in the Riemann sum, and can also be written $dx \, dy$ or $dy \, dx$, the order being unimportant. When we evaluate double integrals by *iteration* in the next section, $dA$ will be replaced with a product of differentials $dx$ and $dy$, and the order will be important.

As is true for functions of one variable, functions that are continuous on $D$ are integrable on $D$. Of course, many bounded but discontinuous functions are also integrable, but an exact description of the class of integrable functions is beyond the scope of this text.

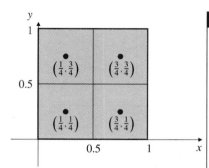

**Figure 14.4** The partitioned square of Example 1

**Example 1** Let $D$ be the square $0 \leq x \leq 1$, $0 \leq y \leq 1$. Use a Riemann sum corresponding to the partition of $D$ into four smaller squares with points selected at the centre of each to find an approximate value for

$$\iint_D (x^2 + y)\, dA.$$

**Solution** The required partition $P$ is formed by the lines $x = 1/2$ and $y = 1/2$, which divide $D$ into four squares, each of area $\Delta A = 1/4$. The centres of these squares are the points $\left(\frac{1}{4}, \frac{1}{4}\right)$, $\left(\frac{1}{4}, \frac{3}{4}\right)$, $\left(\frac{3}{4}, \frac{1}{4}\right)$, and $\left(\frac{3}{4}, \frac{3}{4}\right)$. (See Figure 14.4.) Therefore, the required approximation is

$$\iint_D (x^2 + y) \, dA \approx R(x^2 + y, P) = \left( \frac{1}{16} + \frac{1}{4} \right) \frac{1}{4} + \left( \frac{1}{16} + \frac{3}{4} \right) \frac{1}{4}$$

$$+ \left( \frac{9}{16} + \frac{1}{4} \right) \frac{1}{4} + \left( \frac{9}{16} + \frac{3}{4} \right) \frac{1}{4} = \frac{13}{16} = 0.8125.$$

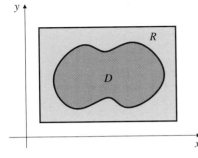

**Figure 14.5**

## Double Integrals over More General Domains

It is often necessary to use double integrals of bounded functions $f(x, y)$ over domains that are not rectangles. If the domain $D$ is *bounded*, we can choose a rectangle $R$ with sides parallel to the coordinate axes such that $D$ is contained inside $R$. (See Figure 14.5.) If $f(x, y)$ is defined on $D$ we can extend its domain to be $R$ by defining $f(x, y) = 0$ for points in $R$ that are outside of $D$. The integral of $f$ over $D$ can then be defined to be the integral of the extended function over the rectangle $R$.

**DEFINITION 2**

If $f(x, y)$ is defined and bounded on domain $D$, let $\hat{f}$ be the extension of $f$ that is zero everywhere outside $D$:

$$\hat{f}(x, y) = \begin{cases} f(x, y) & \text{if } (x, y) \text{ belongs to } D \\ 0 & \text{if } (x, y) \text{ does not belong to } D. \end{cases}$$

If $D$ is a *bounded* domain, then it is contained in some rectangle $R$ with sides parallel to the coordinate axes. We say that $f$ is **integrable** over $D$ and define the **double integral** of $f$ over $D$ to be

$$\iint_D f(x, y) \, dA = \iint_R \hat{f}(x, y) \, dA$$

provided that $\hat{f}$ is integrable over $R$.

This definition makes sense because the values of $\hat{f}$ in the part of $R$ outside of $D$ are all zero, so do not contribute anything to the value of the integral. However, even if $f$ is continuous on $D$, $\hat{f}$ will not be continuous on $R$ unless $f(x, y) \to 0$ as $(x, y)$ approaches the boundary of $D$. Nevertheless, if $f$ and $D$ are "well-behaved," the integral will exist. We cannot delve too deeply into what constitutes *well-behaved*, but assert, without proof, the following theorem that will assure us that most of the double integrals we encounter do, in fact, exist.

**THEOREM 1**

If $f$ is continuous on a *closed, bounded* domain $D$ whose boundary consists of finitely many curves of finite length, then $f$ is integrable on $D$.

According to Theorem 2 of Section 13.1, a continuous function is bounded if its domain is closed and bounded. Generally, however, it is not necessary to restrict our domains to be closed. If $D$ is a bounded domain and $\text{int}(D)$ is its interior (an open set), and if $f$ is integrable on $D$, then

$$\iint_D f(x, y) \, dA = \iint_{\text{int}(D)} f(x, y) \, dA.$$

We will discuss *improper double integrals* of unbounded functions or over unbounded domains in Section 14.3.

## Properties of the Double Integral

Some properties of double integrals are analogous to properties of the one-dimensional definite integral and require little comment: if $f$ and $g$ are integrable over $D$, and if $L$ and $M$ are constants, then

(a) $\displaystyle\iint_D f(x, y)\, dA = 0$ if $D$ has zero area.

(b) **Area of a domain:** $\displaystyle\iint_D 1\, dA = $ area of $D$.

(c) **Integrals representing volumes:**

If $f(x, y) \geq 0$ on $D$, then $\displaystyle\iint_D f(x, y)\, dA = V \geq 0$, where $V$ is the volume of the solid lying vertically above $D$ and below the surface $z = f(x, y)$.

(d) If $f(x, y) \leq 0$ on $D$, then $\displaystyle\iint_D f(x, y)\, dA = -V \leq 0$, where $V$ is the volume of the solid lying vertically below $D$ and above the surface $z = f(x, y)$.

(e) **Linear dependence on the integrand:**
$$\iint_D \Big(Lf(x, y) + Mg(x, y)\Big)\, dA = L\iint_D f(x, y)\, dA + M\iint_D g(x, y)\, dA.$$

(f) **Inequalities are preserved:**

If $f(x, y) \leq g(x, y)$ on $D$, then $\displaystyle\iint_D f(x, y)\, dA \leq \iint_D g(x, y)\, dA$.

(g) **The triangle inequality:** $\displaystyle\left|\iint_D f(x, y)\, dA\right| \leq \iint_D |f(x, y)|\, dA$.

(h) **Additivity of domains:** If $D_1, D_2, \ldots, D_k$ are nonoverlapping domains on each of which $f$ is integrable, then $f$ is integrable over the union $D = D_1 \cup D_2 \cup \cdots \cup D_k$ and
$$\iint_D f(x, y)\, dA = \sum_{j=1}^{k} \iint_{D_j} f(x, y)\, dA.$$

Nonoverlapping domains can share boundary points, but have no interior points in common.

## Double Integrals by Inspection

As yet we have not said anything about how to *evaluate* a double integral. The main technique for doing this, called *iteration*, will be developed in the next section, but it is worth pointing out that double integrals can sometimes be evaluated using symmetry arguments or by interpreting them as volumes that we already know.

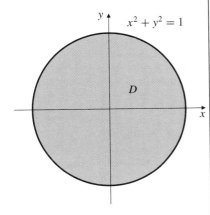

**Figure 14.6**

**Figure 14.7**

**Example 2** If $R$ is the rectangle $a \le x \le b, c \le y \le d$, then

$$\iint_R 3\, dA = 3 \times \text{area of } R = 3(b-a)(d-c).$$

Here the integrand is $f(x, y) = 3$ and the integral is equal to the volume of the solid box of height 3 whose base is the rectangle $R$. (See Figure 14.6.) ∎

**Example 3** Evaluate $I = \iint_{x^2+y^2 \le 1} (\sin x + y^3 + 3)\, dA.$

**Solution** The integral can be expressed as the sum of three integrals by property (e) of double integrals:

$$I = \iint_{x^2+y^2 \le 1} \sin x\, dA + \iint_{x^2+y^2 \le 1} y^3\, dA + \iint_{x^2+y^2 \le 1} 3\, dA$$
$$= I_1 + I_2 + I_3.$$

The domain of integration (Figure 14.7) is a circular disk of radius 1 centred at the origin. Since $f(x, y) = \sin x$ is an *odd* function of $x$, its graph bounds as much volume below the $xy$-plane in the region $x < 0$ as it does above the $xy$-plane in the region $x > 0$. These two contributions to the double integral cancel, and so $I_1 = 0$. Note that symmetry of *both* the domain *and* the integrand is necessary for this argument.

Similarly, $I_2 = 0$ because $y^3$ is an odd function and $D$ is symmetric about the $x$-axis.

Finally,

$$I_3 = \iint_D 3\, dA = 3 \times \text{area of } D = 3\pi.$$

Thus $I = 0 + 0 + 3\pi = 3\pi$. ∎

**Example 4** If $D$ is the disk of Example 3, the integral

$$\iint_D \sqrt{1 - x^2 - y^2}\, dA$$

represents the volume of a hemisphere of radius 1, and so has the value $2\pi/3$. ∎

When evaluating double integrals, always be alert for situations such as those in the above examples. You can save much time by not trying to calculate an integral whose value should be obvious without calculation.

## Exercises 14.1

Exercises 1–6 refer to the double integral

$$I = \iint_D (5 - x - y)\, dA,$$

where $D$ is the rectangle $0 \le x \le 3, 0 \le y \le 2$. $P$ is the partition of $D$ into six squares of side 1 as shown in Figure 14.8.

In Exercises 1–5, calculate the Riemann sums for $I$ corresponding to the given choices of points $(x_{ij}^*, y_{ij}^*)$.

**1.** $(x_{ij}^*, y_{ij}^*)$ is the upper left corner of each square.

**2.** $(x_{ij}^*, y_{ij}^*)$ is the upper right corner of each square.

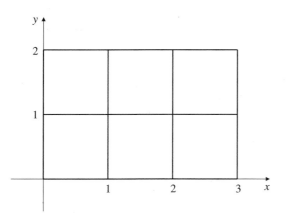

**Figure 14.8**

3. $(x_{ij}^*, y_{ij}^*)$ is the lower left corner of each square.
4. $(x_{ij}^*, y_{ij}^*)$ is the lower right corner of each square.
5. $(x_{ij}^*, y_{ij}^*)$ is the centre of each square.
6. Evaluate $I$ by interpreting it as a volume.

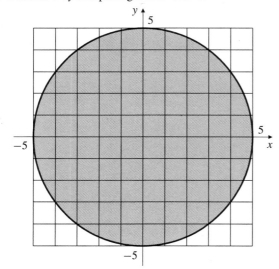

**Figure 14.9**

In Exercises 7–10, $D$ is the disk $x^2 + y^2 \le 25$ and $P$ is the partition of the square $-5 \le x \le 5$, $-5 \le y \le 5$ into one hundred $1 \times 1$ squares, as shown in Figure 14.9. Approximate the double integral

$$J = \iint_D f(x, y)\, dA,$$

where $f(x, y) = 1$ by calculating the Riemann sums $R(f, P)$ corresponding to the indicated choice of points in the small squares. *Hint:* using symmetry will make the job easier.

7. $(x_{ij}^*, y_{ij}^*)$ is the corner of each square closest to the origin.
8. $(x_{ij}^*, y_{ij}^*)$ is the corner of each square farthest from the origin.
9. $(x_{ij}^*, y_{ij}^*)$ is the centre of each square.
10. Evaluate $J$.
11. Repeat Exercise 5 using the integrand $e^x$ instead of $5 - x - y$.
12. Repeat Exercise 9 using $f(x, y) = x^2 + y^2$ instead of $f(x, y) = 1$.

In Exercises 13–22, evaluate the given double integral by inspection.

13. $\displaystyle\iint_R dA$, where $R$ is the rectangle $-1 \le x \le 3$, $-4 \le y \le 1$

14. $\displaystyle\iint_D (x + 3)\, dA$, where $D$ is the half-disk $0 \le y \le \sqrt{4 - x^2}$

15. $\displaystyle\iint_T (x + y)\, dA$, where $T$ is the parallelogram having the points $(2,2)$, $(1, -1)$, $(-2, -2)$, and $(-1, 1)$ as vertices

16. $\displaystyle\iint_{|x|+|y|\le 1} \left(x^3 \cos(y^2) + 3 \sin y - \pi\right) dA$

17. $\displaystyle\iint_{x^2+y^2\le 1} (4x^2 y^3 - x + 5)\, dA$

18. $\displaystyle\iint_{x^2+y^2\le a^2} \sqrt{a^2 - x^2 - y^2}\, dA$

19. $\displaystyle\iint_{x^2+y^2\le a^2} (a - \sqrt{x^2 + y^2})\, dA$

20. $\displaystyle\iint_S (x + y)\, dA$, where $S$ is the square $0 \le x \le a, 0 \le y \le a$

21. $\displaystyle\iint_T (1 - x - y)\, dA$, where $T$ is the triangle with vertices $(0,0)$, $(1,0)$, and $(0,1)$

22. $\displaystyle\iint_R \sqrt{b^2 - y^2}\, dA$, where $R$ is the rectangle $0 \le x \le a, 0 \le y \le b$

# 14.2 Iteration of Double Integrals in Cartesian Coordinates

The existence of the double integral $\iint_D f(x, y)\, dA$ depends on $f$ and the domain $D$. As we shall see, evaluation of double integrals is easiest when the domain of integration is of *simple* type.

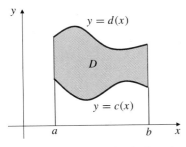

Figure 14.10    A $y$-simple domain

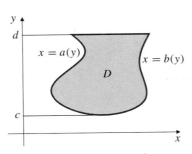

Figure 14.11    An $x$-simple domain

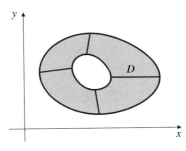

Figure 14.12    A regular domain

We say that the domain $D$ in the $xy$-plane is $y$-**simple** if it is bounded by two vertical lines $x = a$ and $x = b$, and two continuous graphs $y = c(x)$ and $y = d(x)$ between these lines. (See Figure 14.10.) Lines parallel to the $y$-axis intersect a $y$-simple domain in an *interval* (possibly a single point) if at all. Similarly, $D$ is $x$-**simple** if it is bounded by horizontal lines $y = c$ and $y = d$, and two continuous graphs $x = a(y)$ and $x = b(y)$ between these lines. (See Figure 14.11.) Many of the domains over which we will take integrals are $y$-simple, $x$-simple, or both. For example, rectangles, triangles, and disks are both $x$-simple and $y$-simple. Those domains that are neither one or the other will usually be unions of finitely many nonoverlapping subdomains that are both $x$-simple and $y$-simple. We will call such domains **regular**. The shaded region in Figure 14.12 is divided into four subregions, each of which is both $x$-simple and $y$-simple.

It can be shown that a bounded, continuous function $f(x, y)$ is integrable over a bounded $x$-simple or $y$-simple domain, and therefore over any regular domain.

Unlike the examples in the previous section, most double integrals cannot be evaluated by inspection. We need a technique for evaluating double integrals similar to the technique for evaluating single definite integrals in terms of antiderivatives. Since the double integral represents a volume we can evaluate it for simple domains by a slicing technique.

Suppose, for instance, that $D$ is $y$-simple, and is bounded by $x = a$, $x = b$, $y = c(x)$, and $y = d(x)$, as shown in Figure 14.13(a). Then $\iint_D f(x, y)\, dA$ represents the volume of the solid region inside the vertical cylinder through the boundary of $D$ and between the $xy$-plane and the surface $z = f(x, y)$. Consider the cross-section of this solid in the vertical plane perpendicular to the $x$-axis at position $x$. Note that $x$ is constant in that plane. If we use the projections of the $y$- and $z$-axes onto the plane as coordinate axes there, the cross-section is a plane region bounded by vertical lines $y = c(x)$ and $y = d(x)$, by the horizontal line $z = 0$, and by the curve $z = f(x, y)$. The area of the cross-section is therefore given by

$$A(x) = \int_{c(x)}^{d(x)} f(x, y)\, dy.$$

The double integral, $\iint_D f(x, y)\, dA$ is obtained by summing the volumes of "thin" slices of area $A(x)$ and thickness $dx$ between $x = a$ and $x = b$ and is therefore given by

$$\iint_D f(x, y)\, dA = \int_a^b A(x)\, dx = \int_a^b \left( \int_{c(x)}^{d(x)} f(x, y)\, dy \right) dx.$$

Notationally, it is common to omit the large parentheses and write

$$\iint_D f(x, y)\, dA = \int_a^b \int_{c(x)}^{d(x)} f(x, y)\, dy\, dx,$$

or

$$\iint_D f(x, y)\, dA = \int_a^b dx \int_{c(x)}^{d(x)} f(x, y)\, dy.$$

The latter form shows more clearly which variable corresponds to which limits of integration.

**Figure 14.13**

(a) Integrals over $y$-simple domains should be sliced perpendicular to the $x$-axis

(b) Integrals over $x$-simple domains should be sliced perpendicular to the $y$-axis

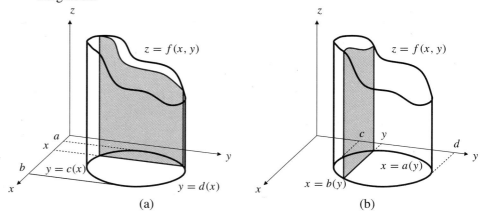

The expressions on the right-hand sides of the above formulas are called **iterated** integrals. **Iteration** is the process of reducing the problem of evaluating a double (or multiple) integral to one of evaluating two (or more) successive single definite integrals. In the above iteration, the integral

$$\int_{c(x)}^{d(x)} f(x, y)\, dy$$

is called the **inner** integral since it must be evaluated first. It is evaluated using standard techniques, treating $x$ as a constant. The result of this evaluation is a function of $x$ alone (note that both the integrand and the limits of the inner integral can depend on $x$), and is the integrand of the **outer** integral in which $x$ is the variable of integration.

For double integrals over $x$-simple domains we can slice perpendicular to the $y$ axis and obtain an iterated integral with the outer integral in the $y$ direction. (See Figure 14.13(b).) We summarize the above discussion in the following theorem.

**THEOREM 2**

### Iteration of Double Integrals

If $f(x, y)$ is continuous on the bounded $y$-simple domain $D$ given by $a \le x \le b$ and $c(x) \le y \le d(x)$, then

$$\iint_D f(x, y)\, dA = \int_a^b dx \int_{c(x)}^{d(x)} f(x, y)\, dy.$$

Similarly, if $f$ is continuous on the $x$-simple domain $D$ given by $c \le y \le d$ and $a(y) \le x \le b(y)$, then

$$\iint_D f(x, y)\, dA = \int_c^d dy \int_{a(y)}^{b(y)} f(x, y)\, dx.$$

**Remark**  The symbol $dA$ in the double integral is replaced in the iterated integrals by the $dx$ and the $dy$. Accordingly, $dA$ is frequently written $dx\, dy$ or $dy\, dx$ even in the double integral. The three expressions

$$\iint_D f(x, y)\, dx\, dy, \qquad \iint_D f(x, y)\, dy\, dx, \qquad \text{and} \qquad \iint_D f(x, y)\, dA$$

all stand for the double integral of $f$ over $D$. Only when the double integral is iterated does the order of $dx$ and $dy$ become important. Later in this chapter we will iterate double integrals in polar coordinates and $dA$ will take the form $r\, dr\, d\theta$.

It is not always necessary to make a three-dimensional sketch of the solid volume represented by a double integral. In order to iterate the integral properly (in one direction or the other) it is usually sufficient to make a sketch of *the domain D* over which the integral is taken. The direction of iteration can be shown by a line along which the inner integral is taken. The following examples illustrate this.

**Example 1**  Find the volume of the solid lying above the square $Q$ defined by $0 \le x \le 1$, $1 \le y \le 2$, and below the plane $z = 4 - x - y$.

**Solution**  The square $Q$ is both $x$-simple and $y$-simple, so the double integral giving the volume can be iterated in either direction. We will do it both ways just for practice. The horizontal line at height $y$ in Figure 14.14 suggests that we first integrate with respect to $x$ along this line (from 0 to 1) and then integrate the result with respect to $y$ from 1 to 2. Iterating the double integral in this direction, we calculate

$$\text{Volume of } Q = \iint_Q (4 - x - y)\, dA$$

$$= \int_1^2 dy \int_0^1 (4 - x - y)\, dx$$

$$= \int_1^2 dy \left(4x - \frac{x^2}{2} - xy\right)\Big|_0^1$$

$$= \int_1^2 \left(\frac{7}{2} - y\right) dy$$

$$= \left(\frac{7y}{2} - \frac{y^2}{2}\right)\Big|_1^2 = 2 \text{ cubic units.}$$

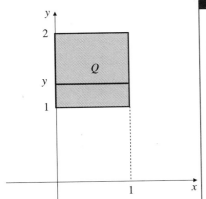

**Figure 14.14**  The horizontal line through $Q$ indicates iteration with the inner integral in the $x$ direction

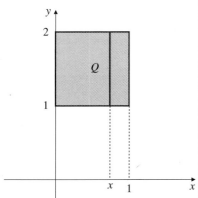

**Figure 14.15** The vertical line through $Q$ indicates iteration with the inner integral in the $y$ direction

Using the opposite iteration, as suggested by Figure 14.15, we calculate

$$\text{Volume of } Q = \iint_Q (4 - x - y)\, dA$$

$$= \int_0^1 dx \int_1^2 (4 - x - y)\, dy$$

$$= \int_0^1 dx \left(4y - xy - \frac{y^2}{2}\right)\Bigg|_1^2$$

$$= \int_0^1 \left(\frac{5}{2} - x\right) dx$$

$$= \left(\frac{5x}{2} - \frac{x^2}{2}\right)\Bigg|_0^1 = 2 \text{ cubic units.}$$

It is comforting to get the same answer both ways! Note that because $Q$ is a rectangle with sides parallel to the coordinate axes, the limits of the inner integrals do not depend on the variables of the outer integrals in either iteration. This cannot be expected to happen with more general domains.

---

**Example 2** Evaluate $\iint_T xy\, dA$ over the triangle $T$ with vertices $(0, 0)$, $(1, 0)$, and $(1, 1)$.

**Solution** The triangle $T$ is shown in Figure 14.16. It is both $x$-simple and $y$-simple. Using the iteration corresponding to slicing in the direction shown in the figure, we obtain:

$$\iint_T xy\, dA = \int_0^1 dx \int_0^x xy\, dy$$

$$= \int_0^1 dx \left(\frac{xy^2}{2}\right)\Bigg|_{y=0}^{y=x}$$

$$= \int_0^1 \frac{x^3}{2}\, dx = \frac{x^4}{8}\Bigg|_0^1 = \frac{1}{8}.$$

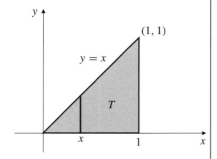

**Figure 14.16** The triangular domain $T$ with vertical line indicating iteration with inner integral in the $y$ direction

Iteration in the other direction (Figure 14.17) leads to the same value:

$$\iint_T xy\, dA = \int_0^1 dy \int_y^1 xy\, dx$$

$$= \int_0^1 dy \left(\frac{yx^2}{2}\right)\Bigg|_{x=y}^{x=1}$$

$$= \int_0^1 \frac{y}{2}(1 - y^2)\, dy$$

$$= \left(\frac{y^2}{4} - \frac{y^4}{8}\right)\Bigg|_0^1 = \frac{1}{8}.$$

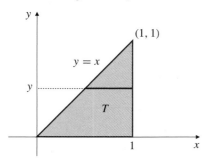

**Figure 14.17** The triangular domain $T$ with horizontal line indicating iteration with inner integral in the $x$ direction

In both of the examples above the double integral could be evaluated easily using either possible iteration. (We did them both ways just to illustrate that fact.) It often occurs, however, that a double integral is easily evaluated if iterated in one direction, and very difficult, or impossible, if iterated in the other direction. Sometimes you will even encounter iterated integrals whose evaluation requires that they be expressed as double integrals and then reiterated in the opposite direction.

**Example 3** Evaluate the iterated integral $I = \int_0^1 dx \int_{\sqrt{x}}^1 e^{y^3} dy$.

**Solution** We cannot antidifferentiate $e^{y^3}$ to evaluate the inner integral in this iteration, so we express $I$ as a double integral and identify the region over which it is taken:

$$I = \iint_D e^{y^3} dA$$

where $D$ is the region shown in Figure 14.18. Reiterating with the $x$ integration on the inside we get

$$I = \int_0^1 dy \int_0^{y^2} e^{y^3} dx$$

$$= \int_0^1 e^{y^3} dy \int_0^{y^2} dx$$

$$= \int_0^1 y^2 e^{y^3} dy = \frac{e^{y^3}}{3}\Big|_0^1 = \frac{e-1}{3}.$$

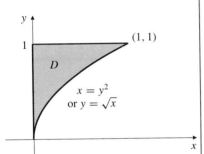

**Figure 14.18** The region corresponding to the iterated integral in Example 3

The following is an example of the calculation of the volume of a somewhat awkward solid. Even though it is not always necessary to sketch solids to find their volumes, you are encouraged to sketch them whenever possible. When we encounter triple integrals over three-dimensional regions later in this chapter it will usually be necessary to sketch the regions. Get as much practice as you can.

**Example 4** Sketch and find the volume of the solid bounded by the planes $y = 0$, $z = 0$, and $z = a - x + y$, and the parabolic cylinder $y = a - (x^2/a)$, where $a$ is a positive constant.

**Solution** The solid is shown in Figure 14.19. Its base is the parabolic segment $D$ in the $xy$-plane bounded by $y = 0$ and $y = a - (x^2/a)$, so the volume of the solid is given by

$$V = \iint_D (a - x + y) dA = \iint_D (a + y) dA.$$

(Note how we used symmetry to drop the $x$ term from the integrand. This term is an odd function of $x$ and $D$ is symmetric about the $y$-axis.) Iterating the double integral in the direction suggested by the slice shown in the figure, we obtain

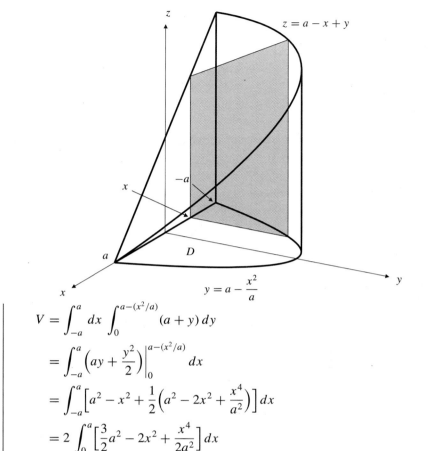

**Figure 14.19**    The solid in Example 4, sliced perpendicular to the $x$-axis

$$V = \int_{-a}^{a} dx \int_{0}^{a-(x^2/a)} (a+y)\, dy$$

$$= \int_{-a}^{a} \left(ay + \frac{y^2}{2}\right)\Bigg|_{0}^{a-(x^2/a)} dx$$

$$= \int_{-a}^{a} \left[a^2 - x^2 + \frac{1}{2}\left(a^2 - 2x^2 + \frac{x^4}{a^2}\right)\right] dx$$

$$= 2 \int_{0}^{a} \left[\frac{3}{2}a^2 - 2x^2 + \frac{x^4}{2a^2}\right] dx$$

$$= \left(3a^2 x - \frac{4x^3}{3} + \frac{x^5}{5a^2}\right)\Bigg|_{0}^{a}$$

$$= 3a^3 - \frac{4}{3}a^3 + \frac{1}{5}a^3 = \frac{28}{15}a^3 \text{ cubic units.}$$

## Exercises 14.2

In Exercises 1–4, calculate the given iterated integrals.

**1.** $\displaystyle\int_{0}^{1} dx \int_{0}^{x} (xy + y^2)\, dy$     **2.** $\displaystyle\int_{0}^{1}\int_{0}^{y} (xy + y^2)\, dx\, dy$

**3.** $\displaystyle\int_{0}^{\pi}\int_{-x}^{x} \cos y\, dy\, dx$     **4.** $\displaystyle\int_{0}^{2} dy \int_{0}^{y} y^2 e^{xy}\, dx$

In Exercises 5–14, evaluate the double integrals by iteration.

**5.** $\displaystyle\iint_{R} (x^2 + y^2)\, dA$, where $R$ is the rectangle $0 \le x \le a$, $0 \le y \le b$

**6.** $\displaystyle\iint_{R} x^2 y^2\, dA$, where $R$ is the rectangle of Exercise 5

**7.** $\displaystyle\iint_{S} (\sin x + \cos y)\, dA$, where $S$ is the square $0 \le x \le \pi/2$, $0 \le y \le \pi/2$

**8.** $\displaystyle\iint_{T} (x - 3y)\, dA$, where $T$ is the triangle with vertices $(0, 0)$, $(a, 0)$, and $(0, b)$

**9.** $\displaystyle\iint_{R} xy^2\, dA$, where $R$ is the finite region in the first quadrant bounded by the curves $y = x^2$ and $x = y^2$

**10.** $\displaystyle\iint_{D} x \cos y\, dA$, where $D$ is the finite region in the first quadrant bounded by the coordinate axes and the curve $y = 1 - x^2$

**11.** $\iint_D \ln x \, dA$, where $D$ is the finite region in the first quadrant bounded by the line $2x + 2y = 5$ and the hyperbola $xy = 1$

**12.** $\iint_T \sqrt{a^2 - y^2} \, dA$, where $T$ is the triangle with vertices $(0, 0)$, $(a, 0)$, and $(a, a)$.

**13.** $\iint_R \dfrac{x}{y} e^y \, dA$, where $R$ is the region

$0 \le x \le 1, x^2 \le y \le x$

**14.** $\iint_T \dfrac{xy}{1 + x^4} \, dA$, where $T$ is the triangle with vertices $(0, 0)$, $(1, 0)$, and $(1, 1)$

In Exercises 15–18, sketch the domain of integration and evaluate the given iterated integrals.

**15.** $\displaystyle\int_0^1 dy \int_y^1 e^{-x^2} \, dx$    **16.** $\displaystyle\int_0^{\pi/2} dy \int_y^{\pi/2} \dfrac{\sin x}{x} \, dx$

**17.** $\displaystyle\int_0^1 dx \int_x^1 \dfrac{y^\lambda}{x^2 + y^2} \, dy \quad (\lambda > 0)$

**18.** $\displaystyle\int_0^1 dx \int_x^{x^{1/3}} \sqrt{1 - y^4} \, dy$

In Exercises 19–28, find the volumes of the indicated solids.

**19.** Under $z = 1 - x^2$ and above the region $0 \le x \le 1$, $0 \le y \le x$

**20.** Under $z = 1 - x^2$ and above the region $0 \le y \le 1$, $0 \le x \le y$

**21.** Under $z = 1 - x^2 - y^2$ and above the region $x \ge 0$, $y \ge 0$, $x + y \le 1$

**22.** Under $z = 1 - y^2$ and above $z = x^2$

**23.** Under the surface $z = 1/(x + y)$ and above the region in the

$xy$-plane bounded by $x = 1$, $x = 2$, $y = 0$, and $y = x$

**24.** Under the surface $z = x^2 \sin(y^4)$ and above the triangle in the $xy$-plane with vertices $(0, 0)$, $(0, \pi^{1/4})$, and $(\pi^{1/4}, \pi^{1/4})$

**25.** Above the $xy$-plane and under the surface $z = 1 - x^2 - 2y^2$

**26.** Above the triangle with vertices $(0, 0)$, $(a, 0)$, and $(0, b)$ and under the plane $z = 2 - (x/a) - (y/b)$

**27.** Inside the two cylinders $x^2 + y^2 = a^2$ and $y^2 + z^2 = a^2$

**28.** Inside the cylinder $x^2 + 2y^2 = 8$, above the plane $z = y - 4$ and below the plane $z = 8 - x$

**∗ 29.** Suppose that $f(x, t)$ and $f_1(x, t)$ are continuous on the rectangle $a \le x \le b$ and $c \le t \le d$. Let

$$g(x) = \int_c^d f(x, t) \, dt \quad \text{and} \quad G(x) = \int_c^d f_1(x, t) \, dt.$$

Show that $g'(x) = G(x)$ for $a < x < b$. *Hint:* evaluate $\int_a^x G(u) \, du$ by reversing the order of iteration. Then differentiate the result. This is a different version of Theorem 5 of Section 13.5.

**∗ 30.** Let $F'(x) = f(x)$ and $G'(x) = g(x)$ on the interval $a \le x \le b$. Let $T$ be the triangle with vertices $(a, a)$, $(b, a)$, and $(b, b)$. By iterating $\iint_T f(x)g(y) \, dA$ in both directions, show that

$$\int_a^b f(x)G(x) \, dx$$

$$= F(b)G(b) - F(a)G(a) - \int_a^b g(y)F(y) \, dy.$$

(This is an alternative derivation of the formula for integration by parts.)

---

## 14.3 Improper Integrals and a Mean-Value Theorem

In order to simplify matters, the definition of the double integral given in Section 14.1 required that the domain $D$ be bounded, and that the integrand $f$ be bounded on $D$. As in the single-variable case, **improper double integrals** can arise if either the domain of integration is unbounded or the integrand is unbounded near any part of the boundary of the domain.

### Improper Integrals of Positive Functions

If $f(x, y) \ge 0$ on the domain $D$ then such an improper integral must either exist (that is, converge to a finite value) or be infinite (diverge to infinity). Convergence or divergence of improper double integrals of such *positive* functions can be determined by iterating them and determining the convergence or divergence of any single improper integrals that result.

**Example 1**    Evaluate $I = \iint_R e^{-x^2}\,dA$. Here $R$ is the region where $x \geq 0$ and $-x \leq y \leq x$. (See Figure 14.20.)

**Solution**    We iterate with the outer integral in the $x$ direction:

$$I = \int_0^\infty dx \int_{-x}^x e^{-x^2}\,dy$$

$$= \int_0^\infty e^{-x^2}\,dx \int_{-x}^x dy$$

$$= 2 \int_0^\infty x e^{-x^2}\,dx$$

This is an improper integral that can be expressed as a limit:

$$I = 2 \lim_{r \to \infty} \int_0^r x e^{-x^2}\,dx$$

$$= 2 \lim_{r \to \infty} \left( -\frac{1}{2} e^{-x^2} \right) \Big|_0^r$$

$$= \lim_{r \to \infty} (1 - e^{-r^2}) = 1.$$

The given integral converges; its value is 1.

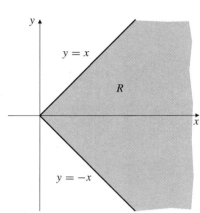

**Figure 14.20**    An unbounded sector of the plane

**Example 2**    If $D$ is the region lying above the $x$-axis, under the curve $y = 1/x$, and to the right of the line $x = 1$, determine whether the double integral

$$\iint_D \frac{dA}{x+y}$$

converges or not.

**Solution**    The region $D$ is sketched in Figure 14.21. We have

$$\iint_D \frac{dA}{x+y} = \int_1^\infty dx \int_0^{1/x} \frac{dy}{x+y}$$

$$= \int_1^\infty \ln(x+y) \Big|_{y=0}^{y=1/x} dx$$

$$= \int_1^\infty \left( \ln\left(x + \frac{1}{x}\right) - \ln x \right) dx$$

$$= \int_1^\infty \ln\left( \frac{x + \frac{1}{x}}{x} \right) dx = \int_1^\infty \ln\left( 1 + \frac{1}{x^2} \right) dx.$$

It happens that this integral can be evaluated exactly (see Exercise 28 below), but we are only asked to determine whether it converges or not, and that is more easily accomplished by estimating it. Since $0 < \ln(1+u) < u$ if $u > 0$ we have

$$0 < \iint_D \frac{dA}{x+y} < \int_1^\infty \frac{1}{x^2}\,dx = 1.$$

Therefore the given integral converges, and its value lies between 0 and 1.

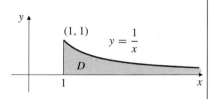

**Figure 14.21**    The domain of the integrand in Example 2

**Example 3**  Evaluate $\int_D \dfrac{1}{(x+y)^2}\,dA$, where $D$ is the region $0 \le x \le 1$, $0 \le y \le x^2$.

**Solution**  The integral is improper because the integrand is unbounded as $(x, y)$ approaches $(0, 0)$, a boundary point of $D$. Nevertheless, iteration leads to a proper integral:

$$\int_D \frac{1}{(x+y)^2}\,dA = \lim_{c \to 0+} \int_c^1 dx \int_0^{x^2} \frac{1}{(x+y)^2}\,dy$$

$$= \lim_{c \to 0+} \int_c^1 dx \left(-\frac{1}{x+y}\right)\Big|_{y=0}^{y=x^2}$$

$$= \lim_{c \to 0+} \int_c^1 dx \left(\frac{1}{x} - \frac{1}{x^2+x}\right) dx$$

$$= \int_0^1 \frac{1}{x+1}\,dx = \ln(x+1)\Big|_0^1 = \ln 2.$$

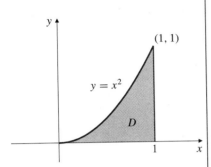

**Figure 14.22**    The function $\dfrac{1}{(x+y)^2}$ is unbounded on $D$

**Example 4**  Determine the convergence or divergence of $I = \displaystyle\iint_D \frac{dA}{xy}$, where $D$ is the bounded region in the first quadrant lying between the line $y = x$ and the parabola $y = x^2$.

**Solution**  The domain $D$ is shown in Figure 14.23. Again the integral is improper because the integrand $1/(xy)$ is unbounded as $(x, y)$ approaches the boundary point $(0, 0)$. We have

$$I = \iint_D \frac{dA}{xy} = \int_0^1 \frac{dx}{x} \int_{x^2}^x \frac{dy}{y}$$

$$= \int_0^1 \frac{1}{x}(\ln x - \ln x^2)\,dx = -\int_0^1 \frac{\ln x}{x}\,dx.$$

If we substitute $\ln x = t$ (so that $x = e^t$ and $dx = e^t\,dt$) in this integral, we obtain

$$I = -\int_{-\infty}^0 \frac{t}{e^t}\,e^t\,dt = -\int_{-\infty}^0 t\,dt,$$

which diverges to infinity.

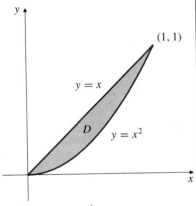

**Figure 14.23**    $\dfrac{1}{xy}$ is unbounded on $D$

**Remark**  In each of the examples above, the integrand was nonnegative on the domain of integration. Nonpositive integrands could have been handled similarly, but we cannot deal here with the convergence of general improper double integrals with integrands $f(x, y)$ that take both positive and negative values on the domain $D$ of the integral. We remark, however, that such an integral cannot converge unless

$$\iint_E f(x, y)\,dA$$

is finite for every bounded, regular subdomain $E$ of $D$. We cannot, in general, determine the convergence of the given integral by looking at the convergence of iterations. The double integral may diverge even if its iterations converge. (See Exercise 21 at the end of this section.) In fact, opposite iterations may even give different values.) This happens because of cancellation of infinite volumes of opposite sign. (Similar behaviour in one dimension is exemplified by the integral $\int_{-1}^{1} dx/x$, which does not exist, although it represents the difference of "equal" but infinite areas.) It can be shown (for a large class of functions containing, for example, continuous functions) that an improper double integral of $f(x, y)$ over $D$ converges if the integral of $|f(x, y)|$ over $D$ converges:

$$\int_D |f(x, y)|\, dA \text{ converges} \quad \Rightarrow \quad \int_D f(x, y)\, dA \text{ converges.}$$

In this case any iterations will converge to the same value. Such double integrals are called **absolutely convergent** by analogy with absolutely convergent infinite series.

## A Mean-Value Theorem for Double Integrals

Let $D$ be a set in the $xy$-plane that is closed and bounded and has positive area $A = \iint_D dA$. Suppose that $f(x, y)$ is continuous on $D$. Then there exist points $(x_1, y_1)$ and $(x_2, y_2)$ in $D$ where $f$ assumes minimum and maximum values (see Theorem 2 of Section 13.1); that is

$$f(x_1, y_1) \le f(x, y) \le f(x_2, y_2)$$

for all points $(x, y)$ in $D$. If we integrate this inequality over $D$ we obtain

$$f(x_1, y_1)A = \iint_D f(x_1, y_1)\, dA$$
$$\le \iint_D f(x, y)\, dA \le \iint_D f(x_2, y_2)\, dA = f(x_2, y_2)A.$$

Therefore, dividing by $A$, we find that the *number*

$$\bar{f} = \frac{1}{A} \iint_D f(x, y)\, dA$$

lies between the minimum and maximum values of $f$ on $D$:

$$f(x_1, y_1) \le \bar{f} \le f(x_2, y_2).$$

A set $D$ in the plane is said to be **connected** if any two points in it can be joined by a continuous parametric curve $x = x(t)$, $y = y(t)$, $(0 \le t \le 1)$, lying in $D$. Suppose this curve joins $(x_1, y_1)$ (where $t = 0$) and $(x_2, y_2)$ (where $t = 1$). Let $g(t)$ be defined by

$$g(t) = f(x(t), y(t)), \quad 0 \le t \le 1.$$

Then $g$ is continuous and takes the values $f(x_1, y_1)$ at $t = 0$ and $f(x_2, y_2)$ at $t = 1$. By the Intermediate-Value Theorem there exists a number $t_0$ between 0 and 1 such that $\bar{f} = g(t_0) = f(x_0, y_0)$ where $x_0 = x(t_0)$ and $y_0 = y(t_0)$. Thus we have found a point $(x_0, y_0)$ in $D$ such that

$$\frac{1}{\text{area of } D} \iint_D f(x, y)\, dA = f(x_0, y_0).$$

We have therefore proved the following version of the Mean-Value Theorem.

**THEOREM** **3**    **A Mean-Value Theorem for Double Integrals**

If the function $f(x, y)$ is continuous on a closed, bounded, connected set $D$ in the $xy$-plane then there exists a point $(x_0, y_0)$ in $D$ such that

$$\iint_D f(x, y)\, dA = f(x_0, y_0) \times (\text{area of } D).$$

By analogy with the definition of average value for one-variable functions, we make the following definition.

**DEFINITION** **3**    The **average value** or **mean value** of an integrable function $f(x, y)$ over the set $D$ is the number

$$\bar{f} = \frac{1}{\text{area of } D} \iint_D f(x, y)\, dA.$$

If $f(x, y) \geq 0$ on $D$, then the cylinder with base $D$ and constant height $\bar{f}$ has volume equal to that of the solid region lying above $D$ and below the surface $z = f(x, y)$. It is often very useful to interpret a double integral in terms of the average value of the function which is its integrand.

**Example 5**    The average value of $x$ over a domain $D$ having area $A$ is

$$\bar{x} = \frac{1}{A} \iint_D x\, dA.$$

Of course, $\bar{x}$ is just the $x$-coordinate of the centroid of the region $D$.

**Example 6**    A large number of points $(x, y)$ are chosen at random in the triangle $T$ with vertices $(0, 0)$, $(1, 0)$, and $(1, 1)$. What is the approximate average value of $x^2 + y^2$ for these points?

**Solution**    The approximate average value of $x^2 + y^2$ for the randomly chosen points will be the average value of that function over the triangle, namely

$$\frac{1}{1/2} \iint_T (x^2 + y^2)\, dA = 2 \int_0^1 dx \int_0^x (x^2 + y^2)\, dy$$

$$= 2 \int_0^1 \left( x^2 y + \frac{1}{3} y^3 \right) \Big|_0^x dx = \frac{8}{3} \int_0^1 x^3\, dx = \frac{2}{3}.$$

**Example 7** Let $(a, b)$ be an interior point of a domain $D$ on which $f(x, y)$ is continuous. For sufficiently small positive $r$, the closed circular disk $D_r$ with centre at $(a, b)$ and radius $r$ is contained in $D$. Show that

$$\lim_{r \to 0} \frac{1}{\pi r^2} \iint_{D_r} f(x, y) \, dA = f(a, b).$$

**Solution** If $D_r$ is contained in $D$, then by Theorem 3

$$\frac{1}{\pi r^2} \iint_{D_r} f(x, y) \, dA = f(x_0, y_0)$$

for some point $(x_0, y_0)$ in $D_r$. As $r \to 0$, the point $(x_0, y_0)$ approaches $(a, b)$. Since $f$ is continuous at $(a, b)$, we have $f(x_0, y_0) \to f(a, b)$. Thus

$$\lim_{r \to 0} \frac{1}{\pi r^2} \iint_{D_r} f(x, y) \, dA = f(a, b).$$

# Exercises 14.3

In Exercises 1–12, determine whether the given integral converges or not. Try to evaluate those that do exist.

**1.** $\displaystyle\iint_Q e^{-x-y} \, dA$, where $Q$ is the first quadrant of the $xy$-plane

**2.** $\displaystyle\iint_Q \frac{dA}{(1+x^2)(1+y^2)}$, where $Q$ is the first quadrant of the $xy$-plane.

**3.** $\displaystyle\iint_S \frac{y}{1+x^2} \, dA$, where $S$ is the strip $0 < y < 1$ in the $xy$-plane

**4.** $\displaystyle\iint_T \frac{1}{x\sqrt{y}} \, dA$ over the triangle $T$ with vertices $(0, 0)$, $(1, 1)$, and $(1, 2)$

**5.** $\displaystyle\iint_Q \frac{x^2 + y^2}{(1+x^2)(1+y^2)} \, dA$, where $Q$ is the first quadrant of the $xy$-plane

**6.** $\displaystyle\iint_H \frac{1}{1+x+y} \, dA$, where $H$ is the half-strip $0 \leq x < \infty$, $0 < y < 1$

**7.** $\displaystyle\iint_{\mathbb{R}^2} e^{-(|x|+|y|)} \, dA$ **8.** $\displaystyle\iint_{\mathbb{R}^2} e^{-|x+y|} \, dA$

**9.** $\displaystyle\iint_T \frac{1}{x^3} e^{-y/x} \, dA$, where $T$ is the region satisfying $x \geq 1$ and $0 \leq y \leq x$

**10.** $\displaystyle\iint_T \frac{dA}{x^2 + y^2}$, where $T$ is the region in Exercise 9

**\* 11.** $\displaystyle\iint_Q e^{-xy} \, dA$, where $Q$ is the first quadrant of the $xy$-plane

**12.** $\displaystyle\iint_R \frac{1}{x} \sin \frac{1}{x} \, dA$, where $R$ is the region $2/\pi \leq x < \infty$, $0 \leq y \leq 1/x$

**13.** Evaluate

$$I = \iint_S \frac{dA}{x + y},$$

where $S$ is the square $0 \leq x \leq 1$, $0 \leq y \leq 1$,

(a) by direct iteration of the double integral,

(b) by using the symmetry of the integrand and the domain to write

$$I = 2 \iint_T \frac{dA}{x + y},$$

where $T$ is the triangle with vertices $(0, 0)$, $(1, 0)$, and $(1, 1)$.

**14.** Find the volume of the solid lying above the square $S$ of Exercise 13, and under the surface $z = 2xy/(x^2 + y^2)$.

In Exercises 15–20, $a$ and $b$ are given real numbers. $D_k$ is the region $0 \leq x \leq 1$, $0 \leq y \leq x^k$, and $R_k$ is the region $1 \leq x < \infty$, $0 \leq y \leq x^k$. Find all real values of $k$ for which the given integral converges.

**15.** $\displaystyle\iint_{D_k} \frac{dA}{x^a}$ **16.** $\displaystyle\iint_{D_k} y^b \, dA$

**17.** $\displaystyle\iint_{R_k} x^a \, dA$ **18.** $\displaystyle\iint_{R_k} \frac{dA}{y^b}$

**19.** $\displaystyle\iint_{D_k} x^a y^b \, dA$       **20.** $\displaystyle\iint_{R_k} x^a y^b \, dA$

**\* 21.** Evaluate both iterations of the improper integral

$$\iint_S \frac{x - y}{(x + y)^3} \, dA,$$

where $S$ is the square $0 < x < 1, 0 < y < 1$. Show that the above improper double integral does not exist by considering

$$\iint_T \frac{x - y}{(x + y)^3} \, dA,$$

where $T$ is that part of the square $S$ lying under the line $x = y$.

In Exercises 22–24, find the average value of the given function over the given region.

**22.** $x^2$ over the rectangle $a \le x \le b, c \le y \le d$

**23.** $x^2 + y^2$ over the triangle $0 \le x \le a, 0 \le y \le a - x$

**24.** $1/x$ over the region $0 \le x \le 1, x^2 \le y \le \sqrt{x}$

**\* 25.** Find the average distance from points in the quarter-disk $x^2 + y^2 \le a^2, x \ge 0, y \ge 0$, to the line $x + y = 0$.

**26.** Does $f(x, y) = x$ have an average value over the region $0 \le x < \infty, 0 \le y \le \dfrac{1}{1 + x^2}$? If so what is it?

**27.** Does $f(x, y) = xy$ have an average value over the region $0 \le x < \infty, 0 \le y \le \dfrac{1}{1 + x^2}$? If so what is it?

**\* 28.** Find the exact value of the integral in Example 2. *Hint:* integrate by parts in $\int_1^\infty \ln\left(1 + (1/x^2)\right) dx$.

**\* 29.** Let $(a, b)$ be an interior point of a domain $D$ on which the function $f(x, y)$ is continuous. For small enough $h^2 + k^2$ the rectangle $R_{hk}$ with vertices $(a, b), (a + h, b), (a, b + k)$, and $(a + h, b + k)$ is contained in $D$. Show that

$$\lim_{(h,k) \to (0,0)} \frac{1}{hk} \iint_{R_{hk}} f(x, y) = f(a, b).$$

*Hint:* see Example 7.

**\* 30. (Another proof of equality of mixed partials)** Suppose that $f_{12}(x, y)$ and $f_{21}(x, y)$ are continuous in a neighbourhood of the point $(a, b)$. Without assuming the equality of these mixed partial derivatives, show that

$$\iint_R f_{12}(x, y) \, dA = \iint_R f_{21}(x, y) \, dA,$$

where $R$ is the rectangle with vertices $(a, b), (a + h, b),$ $(a, b + k)$, and $(a + h, b + k)$ and $h^2 + k^2$ is sufficiently small. Now use the result of Exercise 29 to show that $f_{12}(a, b) = f_{21}(a, b)$. (This reproves Theorem 1 of Section 12.4. However, in that theorem we only assumed continuity of the mixed partials *at* $(a, b)$. Here we assume the continuity *at all points sufficiently near* $(a, b)$.)

---

## 14.4 Double Integrals in Polar Coordinates

For many applications of double integrals either the domain of integration, or the integrand function, or both may be more easily expressed in terms of polar coordinates than in terms of Cartesian coordinates. Recall that a point $P$ with Cartesian coordinates $(x, y)$ can also be located by its polar coordinates $[r, \theta]$, where $r$ is the distance from $P$ to the origin $O$, and $\theta$ is the angle $OP$ makes with the positive direction of the $x$-axis. (Positive angles $\theta$ are measured counterclockwise.) The polar and Cartesian coordinates of $P$ are related by the transformations

$$x = r \cos\theta, \qquad r^2 = x^2 + y^2,$$
$$y = r \sin\theta, \qquad \tan\theta = y/x.$$

Consider the problem of finding the volume $V$ of the solid region lying above the $xy$-plane and beneath the paraboloid $z = 1 - x^2 - y^2$. Since the paraboloid intersects the $xy$-plane in the circle $x^2 + y^2 = 1$, the volume is given in Cartesian coordinates by

$$V = \iint_{x^2 + y^2 \le 1} (1 - x^2 - y^2) \, dA = \int_{-1}^{1} dx \int_{-\sqrt{1-x^2}}^{\sqrt{1-x^2}} (1 - x^2 - y^2) \, dy.$$

Evaluating this iterated integral would require considerable effort. However, we can express the same volume in terms of polar coordinates as

$$V = \iint_{r \le 1} (1 - r^2) \, dA.$$

In order to iterate this integral, we have to know the form that the *area element* $dA$ takes in polar coordinates.

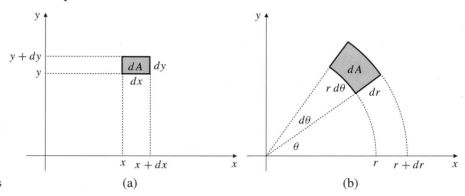

**Figure 14.24**

(a) $dA = dx \, dy$ in Cartesian coordinates

(b) $dA = r \, dr \, d\theta$ in polar coordinates

In the Cartesian formula for $V$, the area element $dA = dx \, dy$ represents the area of the "infinitesimal" region bounded by the coordinate lines at $x$, $x + dx$, $y$, and $y + dy$. (See Figure 14.24(a).) In the polar formula, the area element $dA$ should represent the area of the "infinitesimal" region bounded by the coordinate circles with radii $r$ and $r + dr$, and coordinate rays from the origin at angles $\theta$ and $\theta + d\theta$. (See Figure 14.24(b).) Observe that $dA$ is approximately the area of a rectangle with dimensions $dr$ and $r \, d\theta$. The error in this approximation becomes negligible *compared to the size of $dA$* as $dr$ and $d\theta$ approach zero. Thus, in transforming a double integral between Cartesian and polar coordinates the area element transforms according to the formula

$$dx \, dy = dA = r \, dr \, d\theta.$$

In order to iterate the polar form of the double integral for $V$ considered above, we can regard the domain of integration as a set in a plane having *Cartesian coordinates* $r$ and $\theta$. In the $xy$ Cartesian plane the domain is a disk $r \le 1$ (see Figure 14.25), but in the $r\theta$ Cartesian plane (with perpendicular $r$- and $\theta$-axes) the domain is the rectangle $R$ specified by $0 \le r \le 1$ and $0 \le \theta \le 2\pi$. (See Figure 14.26.) The area element in the $r\theta$-plane is $dA^* = dr \, d\theta$, so area is not preserved under the transformation to polar coordinates ($dA = r \, dA^*$). Thus the polar integral for $V$ is really a Cartesian integral in the $r\theta$-plane, with integrand modified by the inclusion of an extra factor $r$ to compensate for the change of area. It can be evaluated by standard iteration methods:

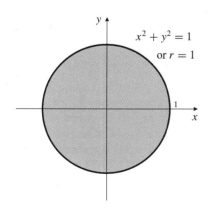

**Figure 14.25** The domain in the $xy$-plane

**Figure 14.26** The domain in the $r\theta$-plane

$$V = \iint_R (1 - r^2) r \, dA^* = \int_0^{2\pi} d\theta \int_0^1 (1 - r^2) r \, dr$$

$$= \int_0^{2\pi} \left( \frac{r^2}{2} - \frac{r^4}{4} \right) \Big|_0^1 \, d\theta = \frac{\pi}{2} \text{ units}^3.$$

**Remark**   It is not necessary to sketch the region $R$ in the $r\theta$-plane. We are used to thinking of polar coordinates in terms of distances and angles in the $xy$-plane, and can easily understand from looking at the disk in Figure 14.25 that the iteration of the integral in polar coordinates corresponds to $0 \le \theta \le 2\pi$ and $0 \le r \le 1$. That is, we should be able to write the iteration

$$V = \int_0^{2\pi} d\theta \int_0^1 (1 - r^2) r \, dr$$

directly from consideration of the domain of integration in the $xy$-plane.

**Example 1**   If $R$ is that part of the annulus $0 < a^2 \le x^2 + y^2 \le b^2$ lying in the first quadrant and below the line $y = x$, evaluate

$$I = \iint_R \frac{y^2}{x^2} \, dA.$$

**Solution**   Figure 14.27 shows the region $R$. It is specified in polar coordinates by $0 \le \theta \le \pi/4$ and $a \le r \le b$. Since

$$\frac{y^2}{x^2} = \frac{r^2 \sin^2 \theta}{r^2 \cos^2 \theta} = \tan^2 \theta,$$

we have

$$I = \int_0^{\pi/4} \tan^2\theta \, d\theta \int_a^b r \, dr$$

$$= \frac{1}{2}(b^2 - a^2) \int_0^{\pi/4} \left( \sec^2\theta - 1 \right) d\theta$$

$$= \frac{1}{2}(b^2 - a^2)(\tan\theta - \theta) \Big|_0^{\pi/4}$$

$$= \frac{1}{2}(b^2 - a^2) \left( 1 - \frac{\pi}{4} \right) = \frac{4 - \pi}{8}(b^2 - a^2).$$

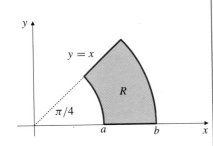

**Figure 14.27**

**Example 2**   (**Area of a polar region**)   Derive the formula for the area of the polar region $R$ bounded by the curve $r = f(\theta)$ and the rays $\theta = \alpha$ and $\theta = \beta$. (See Figure 14.28.)

**Solution**   The area $A$ of $R$ is numerically equal to the volume of a cylinder of height 1 above the region $R$:

$$A = \iint_R dx \, dy = \iint_R r \, dr \, d\theta$$

$$= \int_\alpha^\beta d\theta \int_0^{f(\theta)} r \, dr = \frac{1}{2} \int_\alpha^\beta \left( f(\theta) \right)^2 d\theta.$$

Observe that the inner integral in the iteration involves integrating $r$ along the ray specified by $\theta$ from 0 to $f(\theta)$. The final result is the formula obtained for the area in Section 8.6.

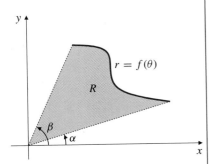

**Figure 14.28**    A standard area problem for polar coordinates

There is no firm rule as to when one should or should not convert a double integral from Cartesian to polar coordinates. In Example 1 above, the conversion was strongly suggested by the shape of the domain, but was also indicated by the fact that the integrand, $y^2/x^2$, becomes a function of $\theta$ alone when converted to polar coordinates. It is usually wise to switch to polar coordinates if the switch simplifies the iteration (that is, if the *domain* is "simpler" when expressed in terms of polar coordinates), even if the form of the integrand is made more complicated.

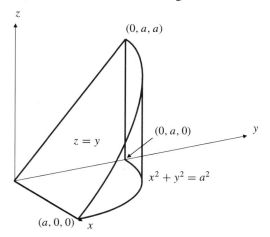

**Figure 14.29**    This volume is easily calculated using iteration in polar coordinates

**Example 3**    Find the volume of the solid lying in the first octant, inside the cylinder $x^2 + y^2 = a^2$, and under the plane $z = y$.

**Solution**    The solid is shown in Figure 14.29. The base is a quarter disk, which is expressed in polar coordinates by the inequalities $0 \le \theta \le \pi/2$ and $0 \le r \le a$. The height is given by $z = y = r \sin\theta$. The solid has volume

$$V = \int_0^{\pi/2} d\theta \int_0^a (r\sin\theta)r\,dr = \int_0^{\pi/2} \sin\theta\,d\theta \int_0^a r^2\,dr = \frac{1}{3}a^3 \text{ units}^3.$$

∎

The following example establishes the value of a definite integral which plays a very important role in probability theory and statistics. (See the discussion of the normal distribution in Section 7.8.) It is interesting that this single-variable integral cannot be evaluated by the techniques of single-variable calculus.

**Example 4**    (**A very important integral**) Show that

$$\int_{-\infty}^{\infty} e^{-x^2}\,dx = \sqrt{\pi}.$$

**Solution**    The improper integral converges, and its value does not depend on what symbol we use for the variable of integration. Therefore we can express the square of the integral as a product of two identical integrals, but with their variables of integration named differently. We then interpret this product as an improper double integral and reiterate it in polar coordinates:

$$\left(\int_{-\infty}^{\infty} e^{-x^2}\,dx\right)^2 = \int_{-\infty}^{\infty} e^{-x^2}\,dx \int_{-\infty}^{\infty} e^{-y^2}\,dy$$

$$= \iint_{\mathbb{R}^2} e^{-(x^2+y^2)}\,dA$$

$$= \int_0^{2\pi} d\theta \int_0^{\infty} e^{-r^2} r\,dr$$

$$= 2\pi \lim_{R\to\infty} \left(-\frac{1}{2} e^{-r^2}\right)\Big|_0^R = \pi.$$

The $r$ integral, a convergent improper integral, was evaluated with the aid of the substitution $u = r^2$.

As our final example of iteration in polar coordinates let us try something a little more demanding.

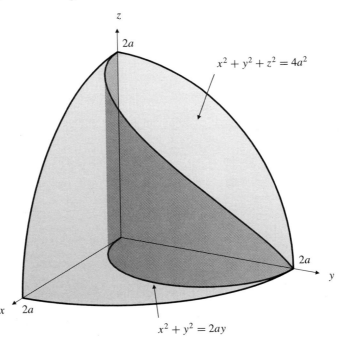

**Figure 14.30**    The first octant part of the intersection of the cylinder $x^2 + y^2 = 2ay$ and the sphere $x^2 + y^2 + z^2 = 4a^2$

**Example 5**    Find the volume of the solid region lying inside both the sphere $x^2 + y^2 + z^2 = 4a^2$ and the cylinder $x^2 + y^2 = 2ay$.

**Solution**    The sphere is centred at the origin and has radius $2a$. The equation of the cylinder becomes

$$x^2 + (y - a)^2 = a^2$$

if we complete the square in the $y$ terms. Thus it is a vertical circular cylinder of radius $a$ having its axis along the vertical line through $(0, a, 0)$. The $z$-axis lies on the cylinder. One-quarter of the required volume lies in the first octant. This part is shown in Figure 14.30.

If we use polar coordinates in the $xy$-plane, then the sphere has equation $r^2 + z^2 = 4a^2$ and the cylinder has equation $r^2 = 2ar \sin \theta$ or, more simply, $r = 2a \sin \theta$. The first octant portion of the volume lies above the region specified by the inequalities $0 \le \theta \le \pi/2$ and $0 \le r \le 2a \sin \theta$. Therefore the total volume is

$$V = 4 \int_0^{\pi/2} d\theta \int_0^{2a \sin \theta} \sqrt{4a^2 - r^2}\, r\, dr \qquad \text{(Let } u = 4a^2 - r^2.)$$

$$= 2 \int_0^{\pi/2} d\theta \int_{4a^2 \cos^2 \theta}^{4a^2} \sqrt{u}\, du$$

$$= \frac{4}{3} \int_0^{\pi/2} (8a^3 - 8a^3 \cos^3 \theta)\, d\theta \qquad \text{(Let } v = \sin \theta.)$$

$$= \frac{16}{3} \pi a^3 - \frac{32}{3} a^3 \int_0^1 (1 - v^2)\, dv$$

$$= \frac{16}{3} \pi a^3 - \frac{64}{9} a^3 = \frac{16}{9} (3\pi - 4) a^3 \text{ cubic units.}$$

## Change of Variables in Double Integrals

The transformation of a double integral to polar coordinates is just a special case of a general change of variables formula for double integrals. Suppose that $x$ and $y$ are expressed as functions of two other variables $u$ and $v$ by the equations

$$x = x(u, v)$$
$$y = y(u, v).$$

We regard these equations as defining a **transformation** (or mapping) from points $(u, v)$ in a $uv$-Cartesian plane to points $(x, y)$ in the $xy$-plane. We say that the transformation is *one-to-one* from the set $S$ in the $uv$-plane *onto* the set $D$ in the $xy$-plane provided:

    (i) every point in $S$ gets mapped to a point in $D$,

    (ii) every point in $D$ is the image of a point in $S$, and

    (iii) different points in $S$ get mapped to different points in $D$.

If the transformation is one-to-one, the defining equations can be solved for $u$ and $v$ as functions of $x$ and $y$, and the resulting **inverse transformation**:

$$u = u(x, y)$$
$$v = v(x, y)$$

is one-to-one from $D$ onto $S$.

    Let us assume that the functions $x(u, v)$ and $y(u, v)$ have continuous first partial derivatives, and that the Jacobian determinant

$$\frac{\partial(x, y)}{\partial(u, v)} \ne 0 \qquad \text{at} \quad (u, v).$$

As noted in Section 12.8, the Implicit Function Theorem implies that the transformation is one-to-one near $(u, v)$ and the inverse transformation also has continuous first partial derivatives and nonzero Jacobian satisfying

$$\frac{\partial(u, v)}{\partial(x, y)} = \frac{1}{\dfrac{\partial(x, y)}{\partial(u, v)}} \qquad \text{on } D.$$

**Example 6** The transformation $x = r \cos\theta$, $y = r \sin\theta$ to polar coordinates has Jacobian

$$\frac{\partial(x, y)}{\partial(r, \theta)} = \begin{vmatrix} \cos\theta & -r\sin\theta \\ \sin\theta & r\cos\theta \end{vmatrix} = r.$$

Near any point except the origin (where $r = 0$), the transformation is one-to-one. (In fact it is one-to-one from any set in the $r\theta$-plane that does not contain the axis $r = 0$ and lies in, say, the strip $0 \le \theta < 2\pi$.) ∎

A one-to-one transformation can be used to transform the double integral

$$\iint_D f(x, y)\, dA$$

to a double integral over the corresponding set $S$ in the $uv$-plane. Under the transformation, the integrand $f(x, y)$ becomes $g(u, v) = f\big(x(u, v), y(u, v)\big)$. We must discover how to express the area element $dA = dx\, dy$ in terms of the area element $du\, dv$ in the $uv$-plane.

For any fixed value of $u$, (say $u = c$), the equations

$$x = x(u, v) \quad \text{and} \quad y = y(u, v)$$

define a parametric curve (with $v$ as parameter) in the $xy$-plane. This curve is called a $u$-curve corresponding to the value $u = c$. Similarly, for fixed $v$ the equations define a parametric curve (with parameter $u$) called a $v$-curve. Consider the differential area element bounded by the $u$-curves corresponding to nearby values $u$ and $u + du$ and the $v$-curves corresponding to nearby values $v$ and $v + dv$. Since these curves are smooth, for small values of $du$ and $dv$ the area element is approximately a parallelogram, and its area is approximately

$$dA = |\overrightarrow{PQ} \times \overrightarrow{PR}|,$$

where $P$, $Q$, and $R$ are the points shown in Figure 14.31. The error in this approximation becomes negligible compared to $dA$ as $du$ and $dv$ approach zero.

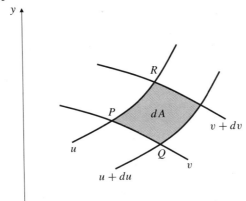

**Figure 14.31** The image in the $xy$ plane of the area element $du\, dv$ in the $uv$-plane

Now $\overrightarrow{PQ} = dx\,\mathbf{i} + dy\,\mathbf{j}$ where

$$dx = \frac{\partial x}{\partial u}du + \frac{\partial x}{\partial v}dv \quad \text{and} \quad dy = \frac{\partial y}{\partial u}du + \frac{\partial y}{\partial v}dv.$$

However, $dv = 0$ along the $v$-curve $PQ$ so

$$\overrightarrow{PQ} = \frac{\partial x}{\partial u}du\,\mathbf{i} + \frac{\partial y}{\partial u}du\,\mathbf{j}.$$

Similarly,

$$\overrightarrow{PR} = \frac{\partial x}{\partial v}dv\,\mathbf{i} + \frac{\partial y}{\partial v}dv\,\mathbf{j}.$$

Hence

$$dA = \left\| \begin{matrix} \mathbf{i} & \mathbf{j} & \mathbf{k} \\ \frac{\partial x}{\partial u}du & \frac{\partial y}{\partial u}du & 0 \\ \frac{\partial x}{\partial v}dv & \frac{\partial y}{\partial v}dv & 0 \end{matrix} \right\| = \left| \frac{\partial(x,\,y)}{\partial(u,\,v)} \right| du\,dv;$$

that is, the absolute value of the Jacobian $\partial(x,\,y)/\partial(u,\,v)$ is the ratio between corresponding area elements in the $xy$-plane and the $uv$-plane:

$$dA = dx\,dy = \left| \frac{\partial(x,\,y)}{\partial(u,\,v)} \right| du\,dv.$$

The following theorem summarizes the change of variables procedure for a double integral.

**THEOREM 4**

**Change of variables formula for double integrals**

Let $x = x(u,\,v)$, $y = y(u,\,v)$ be a one-to-one transformation from a domain $S$ in the $uv$-plane onto a domain $D$ in the $xy$-plane. Suppose that the functions $x$ and $y$, and their first partial derivatives with respect to $u$ and $v$ are continuous in $S$. If $f(x,\,y)$ is integrable on $D$, and if $g(u,\,v) = f(x(u,\,v),\,y(u,\,v))$, then $g$ is integrable on $S$ and

$$\iint_D f(x,\,y)\,dx\,dy = \iint_S g(u,\,v) \left| \frac{\partial(x,\,y)}{\partial(u,\,v)} \right| du\,dv.$$

**Remark** It is not necessary that $S$ or $D$ be closed, or that the transformation be one-to-one on the boundary of $S$. The transformation to polar coordinates maps the rectangle $0 < r < 1$, $0 \le \theta < 2\pi$ one-to-one onto the punctured disk $0 < x^2 + y^2 < 1$ and, as in the first example in this section, we can transform an integral over the closed disk $x^2 + y^2 \le 1$ to one over the closed rectangle $0 \le r \le 1$, $0 \le \theta \le 2\pi$.

**Example 7** Use an appropriate change of variables to find the area of the elliptic disk $E$ given by

$$\frac{x^2}{a^2} + \frac{y^2}{b^2} \le 1.$$

**Solution** Under the transformation $x = au$, $y = bv$, the elliptic disk $E$ is the one-to-one image of the circular disk $D$ given by $u^2 + v^2 \le 1$. Assuming $a > 0$ and $b > 0$, we have

$$dx\,dy = \left| \frac{\partial(x, y)}{\partial(u, v)} \right| du\,dv = \left\| \begin{matrix} a & 0 \\ 0 & b \end{matrix} \right\| du\,dv = ab\,du\,dv.$$

Therefore, the area of $E$ is given by

$$\iint_E 1\,dx\,dy = \iint_D ab\,du\,dv = ab \times (\text{area of } D) = \pi ab \text{ square units.}$$

It is often tempting to try to use the change of variable formula to transform the domain of a double integral into a rectangle so that iteration will be easy. As the following example shows, this usually involves defining the inverse transformation ($u$ and $v$ in terms of $x$ and $y$). Remember that inverse transformations have reciprocal Jacobians.

**Example 8** Find the area of the finite plane region bounded by the four parabolas $y = x^2$, $y = 2x^2$, $x = y^2$, and $x = 3y^2$.

**Solution** The region, call it $D$, is sketched in Figure 14.32. Let

$$u = \frac{y}{x^2} \qquad \text{and} \qquad v = \frac{x}{y^2}.$$

Then the region $D$ corresponds to the rectangle $R$ in the $uv$-plane given by $1 \le u \le 2$ and $1 \le v \le 3$. Since

$$\frac{\partial(u, v)}{\partial(x, y)} = \left| \begin{matrix} -2y/x^3 & 1/x^2 \\ 1/y^2 & -2x/y^3 \end{matrix} \right| = \frac{3}{x^2 y^2} = 3u^2 v^2,$$

it follows that

$$\left| \frac{\partial(x, y)}{\partial(u, v)} \right| = \frac{1}{3u^2 v^2}$$

and so the area of $D$ is given by

$$\iint_D dx\,dy = \iint_R \frac{1}{3u^2 v^2}\,du\,dv$$

$$= \frac{1}{3} \int_1^2 \frac{du}{u^2} \int_1^3 \frac{dv}{v^2} = \frac{1}{3} \times \frac{1}{2} \times \frac{2}{3} = \frac{1}{9} \text{ square units.}$$

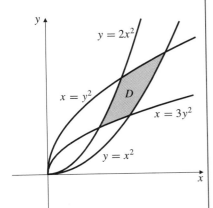

**Figure 14.32**

The following example shows what can happen if the transformation of a double integral is not one-to-one.

**Example 9**   Let $D$ be the square $0 \le x \le 1, 0 \le y \le 1$ in the $xy$-plane, and let $S$ be the square $0 \le u \le 1, 0 \le v \le 1$ in the $uv$-plane. Show that the transformation

$$x = 4u - 4u^2, \qquad y = v$$

maps $S$ onto $D$, and use it to transform the integral $I = \iint_D dx\,dy$. Compare the value of $I$ with that of the transformed integral.

**Solution**   Since $x = 4u - 4u^2 = 1 - (1 - 2u)^2$, the minimum value of $x$ on the interval $0 \le u \le 1$ is 0 (at $u = 0$ and $u = 1$), and the maximum value is 1 (at $u = \frac{1}{2}$). Therefore, $x = 4u - 4u^2$ maps the interval $0 \le u \le 1$ onto the interval $0 \le x \le 1$. Since $y = v$ clearly maps $0 \le v \le 1$ onto $0 \le y \le 1$, the given transformation maps $S$ onto $D$. Since

$$dx\,dy = \left| \frac{\partial(x, y)}{\partial(u, v)} \right| du\,dv = \left| \begin{matrix} 4 - 8u & 0 \\ 0 & 1 \end{matrix} \right| du\,dv = |4 - 8u|\,du\,dv,$$

transforming $I$ leads to the integral

$$J = \iint_S |4 - 8u|\,du\,dv = 4 \int_0^1 dv \int_0^1 |1 - 2u|\,du = 8 \int_0^{1/2} (1 - 2u)\,du = 2.$$

However, $I = \iint_D dx\,dy = $ area of $D = 1$. The reason that $J \ne I$ is that the transformation is not one-to-one from $S$ onto $D$; it actually maps $S$ onto $D$ twice. The rectangle $R$ defined by $0 \le u \le \frac{1}{2}$ and $0 \le v \le 1$ is mapped one-to-one onto $D$ by the transformation, so the appropriate transformed integral is $\iint_R |4 - 8u|\,du\,dv$, which is equal to $I$.

## Exercises 14.4

In Exercises 1–6, evaluate the given double integral over the disk $D$ given by $x^2 + y^2 \le a^2$, where $a > 0$.

**1.** $\displaystyle\iint_D (x^2 + y^2)\,dA$

**2.** $\displaystyle\iint_D \sqrt{x^2 + y^2}\,dA$

**3.** $\displaystyle\iint_D \frac{1}{\sqrt{x^2 + y^2}}\,dA$

**4.** $\displaystyle\iint_D |x|\,dA$

**5.** $\displaystyle\iint_D x^2\,dA$

**6.** $\displaystyle\iint_D x^2 y^2\,dA$

In Exercises 7–10, evaluate the given double integral over the quarter-disk $Q$ given by $x \ge 0, y \ge 0$, and $x^2 + y^2 \le a^2$, where $a > 0$.

**7.** $\displaystyle\iint_Q y\,dA$

**8.** $\displaystyle\iint_Q (x + y)\,dA$

**9.** $\displaystyle\iint_Q e^{x^2 + y^2}\,dA$

**10.** $\displaystyle\iint_Q \frac{2xy}{x^2 + y^2}\,dA$

**11.** Evaluate $\displaystyle\iint_S (x + y)\,dA$, where $S$ is the region in the first quadrant lying inside the disk $x^2 + y^2 \le a^2$ and under the line $y = \sqrt{3}x$.

**12.** Find $\displaystyle\iint_S x\,dA$, where $S$ is the disk segment $x^2 + y^2 \le 2$, $x \ge 1$.

**13.** Evaluate $\displaystyle\iint_T (x^2 + y^2)\,dA$, where $T$ is the triangle with vertices $(0, 0)$, $(1, 0)$, and $(1, 1)$.

**14.** Evaluate $\displaystyle\iint_{x^2 + y^2 \le 1} \ln(x^2 + y^2)\,dA$.

**15.** Find the average distance from the origin to points in the disk $x^2 + y^2 \le a^2$.

**16.** Find the average value of $e^{-(x^2 + y^2)}$ over the annular region $0 < a \le \sqrt{x^2 + y^2} \le b$.

**17.** For what values of $k$, and to what value, does the integral $\displaystyle\iint_{x^2 + y^2 \le 1} \frac{dA}{(x^2 + y^2)^k}$ converge?

**18.** For what values of $k$, and to what value, does the integral $\displaystyle\iint_{\mathbb{R}^2} \frac{dA}{(1 + x^2 + y^2)^k}$ converge?

**19.** Evaluate $\iint_D xy \, dA$, where $D$ is the plane region satisfying $x \geq 0, 0 \leq y \leq x$, and $x^2 + y^2 \leq a^2$.

**20.** Evaluate $\iint_C y \, dA$, where $C$ is the upper half of the cardioid disk $r \leq 1 + \cos\theta$.

**21.** Find the volume lying between the paraboloids $z = x^2 + y^2$ and $3z = 4 - x^2 - y^2$.

**22.** Find the volume lying inside both the sphere $x^2 + y^2 + z^2 = a^2$ and the cylinder $x^2 + y^2 = ax$.

**23.** Find the volume lying inside both the sphere $x^2 + y^2 + z^2 = 2a^2$ and the cylinder $x^2 + y^2 = a^2$.

**24.** Find the volume of the region lying above the $xy$-plane, inside the cylinder $x^2 + y^2 = 4$ and below the plane $z = x + y + 4$.

**\* 25.** Find the volume of the region lying inside all three of the circular cylinders $x^2 + y^2 = a^2$, $x^2 + z^2 = a^2$, and $y^2 + z^2 = a^2$. *Hint:* make a good sketch of the first octant part of the region and use symmetry whenever possible.

**26.** Find the volume of the region lying inside the circular cylinder $x^2 + y^2 = 2y$ and inside the parabolic cylinder $z^2 = y$.

**\* 27.** Many points are chosen at random in the disk $x^2 + y^2 \leq 1$. Find the approximate average value of the distance from these points to the nearest side of the smallest square that contains the disk.

**\* 28.** Find the average value of $x$ over the segment of the disk $x^2 + y^2 \leq 4$ lying to the right of $x = 1$. What is the centroid of the segment?

**29.** Find the volume enclosed by the ellipsoid

$$\frac{x^2}{a^2} + \frac{y^2}{b^2} + \frac{z^2}{c^2} = 1.$$

**30.** Find the volume of the region in the first octant below the paraboloid

$$z = 1 - \frac{x^2}{a^2} - \frac{y^2}{b^2}.$$

*Hint:* use the change of variables $x = au$, $y = bv$.

**\* 31.** Evaluate $\iint_{|x|+|y|\leq a} e^{x+y} \, dA$.

**32.** Find $\iint_P (x^2 + y^2) \, dA$, where $P$ is the parallelogram bounded by the lines $x + y = 1$, $x + y = 2$, $3x + 4y = 5$, and $3x + 4y = 6$.

**33.** Find the area of the region in the first quadrant bounded by the curves $xy = 1$, $xy = 4$, $y = x$, and $y = 2x$.

**34.** Evaluate $\iint_R (x^2 + y^2) \, dA$, where $R$ is the region in the first quadrant bounded by $y = 0$, $y = x$, $xy = 1$, and $x^2 - y^2 = 1$.

**\* 35.** Let $T$ be the triangle with vertices $(0, 0)$, $(1, 0)$, and $(0, 1)$. Evaluate the integral $\iint_T e^{(y-x)/(y+x)} \, dA$,

(a) by transforming to polar coordinates, and

(b) using the change of variables $u = y - x$, $v = y + x$.

**36.** Use the method of Example 7 to find the area of the region inside the ellipse $4x^2 + 9y^2 = 36$ and above the line $2x + 3y = 6$.

**\* 37.** (**The error function**) The error function, $\text{Erf}(x)$, is defined for $x \geq 0$ by

$$\text{Erf}(x) = \frac{2}{\sqrt{\pi}} \int_0^x e^{-t^2} \, dt.$$

Show that $\left(\text{Erf}(x)\right)^2 = \frac{4}{\pi} \int_0^{\pi/4} \left(1 - e^{-x^2/\cos^2\theta}\right) d\theta.$

Hence deduce that $\text{Erf}(x) \geq \sqrt{1 - e^{-x^2}}$.

**\* 38.** (**The gamma and beta functions**) The gamma function $\Gamma(x)$ and the beta function $B(x, y)$ are defined by

$$\Gamma(x) = \int_0^\infty t^{x-1} e^{-t} \, dt, \quad (x > 0),$$

$$B(x, y) = \int_0^1 t^{x-1}(1 - t)^{y-1} \, dt, \quad (x > 0, \ y > 0).$$

The gamma function was introduced in Exercise 44 of Section 6.5, where it was to be shown that

$$\Gamma(x + 1) = x\Gamma(x), \quad \text{and}$$
$$\Gamma(n + 1) = n!, \quad (n = 0, 1, 2, \ldots).$$

Deduce the following further properties of these functions:

(a) $\Gamma(x) = 2 \int_0^\infty s^{2x-1} e^{-s^2} \, ds$, $\quad (x > 0)$,

(b) $\Gamma\left(\frac{1}{2}\right) = \sqrt{\pi}$, $\quad \Gamma\left(\frac{3}{2}\right) = \frac{1}{2}\sqrt{\pi}$,

(c) If $x > 0$ and $y > 0$, then

$$B(x, y) = 2 \int_0^{\pi/2} \cos^{2x-1}\theta \sin^{2y-1}\theta \, d\theta,$$

(d) $B(x, y) = \dfrac{\Gamma(x)\Gamma(y)}{\Gamma(x + y)}.$

# 14.5 Triple Integrals

Now that we have seen how to extend definite integration to two-dimensional domains the extension to three (or more) dimensions is straightforward. For a bounded function $f(x, y, z)$ defined on a rectangular box $B$ ($x_0 \le x \le x_1$, $y_0 \le y \le y_1$, $z_0 \le z \le z_1$), the **triple integral** of $f$ over $B$,

$$\iiint_B f(x, y, z)\, dV \qquad \text{or} \qquad \iiint_B f(x, y, z)\, dx\, dy\, dz,$$

can be defined as a suitable limit of Riemann sums corresponding to partitions of $B$ into subboxes by planes parallel to each of the coordinate planes. We omit the details. Triple integrals over more general domains are defined by extending the function to be zero outside the domain and integrating over a rectangular box containing the domain.

All the properties of double integrals mentioned in Section 14.1 have analogues for triple integrals. In particular, a continuous function is integrable over a closed, bounded domain. If $f(x, y, z) = 1$ on the domain $D$ then the triple integral gives the volume of $D$:

$$\text{Volume of } D = \iiint_D dV.$$

The triple integral of a positive function $f(x, y, z)$ can be interpreted as the "hypervolume" (that is, the four-dimensional volume) of a region in 4-space having the set $D$ as its three-dimensional "base" and having its top on the hypersurface $w = f(x, y, z)$. This is not a particularly useful interpretation; many more useful ones arise in applications. For instance, if $\delta(x, y, z)$ represents the density (mass per unit volume) at position $(x, y, z)$ in a substance occupying the domain $D$ in 3-space then the mass $m$ of the solid is the "sum" of mass elements $dm = \delta(x, y, z)\, dV$ occupying volume elements $dV$:

$$\text{mass} = \iiint_D \delta(x, y, z)\, dV.$$

Some triple integrals can be evaluated by inspection, using symmetry and known volumes.

**Example 1**   Evaluate

$$\iiint_{x^2+y^2+z^2 \le a^2} (2 + x - \sin z)\, dV.$$

**Solution**   The domain of integration is the ball of radius $a$ centred at the origin. The integral of 2 over this ball is twice the ball's volume, that is $8\pi a^3/3$. The integrals of $x$ and $\sin z$ over the ball are both zero since both functions are odd in one of the variables and the domain is symmetric about each coordinate plane. (For instance, for every volume element $dV$ in the half of the ball where $x > 0$, there is a corresponding element in the other half where $x$ has the same size but the opposite sign. The contributions from these two elements cancel one another.) Thus

$$\iiint_{x^2+y^2+z^2 \le a^2} (2 + x - \sin z)\, dV = \frac{8}{3}\pi a^3 + 0 + 0 = \frac{8}{3}\pi a^3.$$

Most triple integrals are evaluated by an iteration procedure similar to that used for double integrals. We slice the domain $D$ with a plane parallel to one of the coordinate planes, double integrate the function with respect to two variables over that slice, and then integrate the result with respect to the remaining variable. Some examples should serve to make the procedure clear.

**Example 2** Let $B$ be the rectangular box $0 \leq x \leq a, 0 \leq y \leq b, 0 \leq z \leq c$. Evaluate

$$I = \iiint_B (xy^2 + z^3)\, dV.$$

**Solution** As indicated in Figure 14.33(a), we will slice with planes perpendicular to the $z$-axis, so the $z$ integral will be outermost in the iteration. The slices are rectangles, so the double integrals over them can be immediately iterated also. We do it with the $y$ integral outer and the $x$ integral inner, as suggested by the line shown in the slice.

$$
\begin{aligned}
I &= \int_0^c dz \int_0^b dy \int_0^a (xy^2 + z^3)\, dx \\
&= \int_0^c dz \int_0^b dy \left( \frac{x^2 y^2}{2} + xz^3 \right) \Bigg|_{x=0}^{x=a} \\
&= \int_0^c dz \int_0^b \left( \frac{a^2 y^2}{2} + az^3 \right) dy \\
&= \int_0^c dz \left( \frac{a^2 y^3}{6} + ayz^3 \right) \Bigg|_{y=0}^{y=b} \\
&= \int_0^c \left( \frac{a^2 b^3}{6} + abz^3 \right) dz \\
&= \left( \frac{a^2 b^3 z}{6} + \frac{abz^4}{4} \right) \Bigg|_{z=0}^{z=c} = \frac{a^2 b^3 c}{6} + \frac{abc^4}{4}.
\end{aligned}
$$

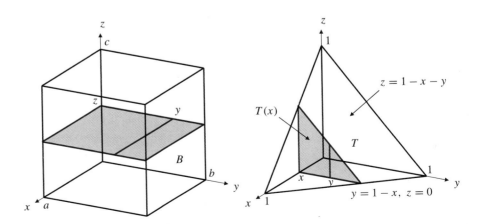

**Figure 14.33**

(a) The iteration in Example 2

(b) The iteration in Example 3

(a)

(b)

**Example 3**    If $T$ is the tetrahedron with vertices $(0, 0, 0)$, $(1, 0, 0)$, $(0, 1, 0)$, and $(0, 0, 1)$, evaluate

$$I = \iiint_T y \, dV.$$

**Solution**    The tetrahedron is shown in Figure 14.33(b). The plane slice in the plane normal to the $x$-axis at position $x$ is the triangle $T(x)$ shown in that figure. $x$ is constant and $y$ and $z$ are variables in the slice. The double integral of $y$ over $T(x)$ is a function of $x$. We evaluate it by integrating first in the $z$ direction and then in the $y$ direction as suggested by the vertical line shown in the slice:

$$\iint_{T(x)} y \, dA = \int_0^{1-x} dy \int_0^{1-x-y} y \, dz$$

$$= \int_0^{1-x} y(1 - x - y) \, dy$$

$$= \left( (1-x)\frac{y^2}{2} - \frac{y^3}{3} \right)\Big|_0^{1-x} = \frac{1}{6}(1-x)^3.$$

The value of the triple integral $I$ is the integral of this expression with respect to the remaining variable $x$, to sum the contributions from all such slices between $x = 0$ and $x = 1$:

$$I = \int_0^1 \frac{1}{6}(1-x)^3 \, dx = -\frac{1}{24}(1-x)^4 \Big|_0^1 = \frac{1}{24}.$$

In the above solution we carried out the iteration in two steps in order to show the procedure clearly. In practice, triple integrals are iterated in one step, with no explicit mention made of the double integral over the slice. Thus, using the iteration suggested by Figure 14.33(b) we would immediately write

$$I = \int_0^1 dx \int_0^{1-x} dy \int_0^{1-x-y} y \, dz.$$

The evaluation proceeds as above, starting with the right (that is, inner) integral, followed by the middle integral and then the left (outer) integral. The triple integral represents the "sum" of elements $y \, dV$ over the three-dimensional region $T$. The above iteration corresponds to "summing" (that is, integrating) first along a vertical line (the $z$ integral), then summing these one-dimensional sums in the $y$ direction to get the double sum of all elements in the plane slice, and finally summing these double sums in the $x$ direction to add up the contributions from all the slices. The iteration can be carried out in other directions; there are six possible iterations corresponding to different orders of doing the $x$, $y$, and $z$ integrals. The other five are

$$I = \int_0^1 dx \int_0^{1-x} dz \int_0^{1-x-z} y \, dy,$$

$$I = \int_0^1 dy \int_0^{1-y} dx \int_0^{1-x-y} y \, dz,$$

$$I = \int_0^1 dy \int_0^{1-y} dz \int_0^{1-y-z} y \, dx,$$

$$I = \int_0^1 dz \int_0^{1-z} dx \int_0^{1-x-z} y \, dy,$$

$$I = \int_0^1 dz \int_0^{1-z} dy \int_0^{1-y-z} y \, dx.$$

You should verify these by drawing diagrams analogous to Figure 14.33(b). Of course, all six iterations give the same result.

It is sometimes difficult to visualize the region of 3-space over which a given triple integral is taken. In such situations try to determine the *projection* of that region on one or other of the coordinate planes. For instance, if a region $R$ is bounded by two surfaces with given equations, combining these equations to eliminate one variable will yield the equation of a cylinder (not necessarily circular) with axis parallel to the axis of the eliminated variable. This cylinder will then determine the projection of $R$ onto the coordinate plane perpendicular to that axis. The following example illustrates the use of this technique to find a volume bounded by two surfaces. The volume is expressed as a triple integral with unit integrand.

**Example 4**    Find the volume of the region $R$ lying below the plane $z = 3 - 2y$ and above the paraboloid $z = x^2 + y^2$.

**Solution**    The region $R$ is shown in Figure 14.34. It is symmetric about the $yz$-plane, and so its volume $V$ may be found by doubling that part lying in the half-space $x > 0$. The two surfaces bounding $R$ intersect on the vertical cylinder $x^2 + y^2 = 3 - 2y$, or $x^2 + (y + 1)^2 = 4$, which lies between $y = -3$ and $y = 1$. Thus

$$V = \iiint_R dV = 2 \int_{-3}^1 dy \int_0^{\sqrt{3-2y-y^2}} dx \int_{x^2+y^2}^{3-2y} dz$$

$$= 2 \int_{-3}^1 dy \int_0^{\sqrt{3-2y-y^2}} (3 - 2y - x^2 - y^2) \, dx$$

$$= 2 \int_{-3}^1 \left( (3 - 2y - y^2)x - \frac{x^3}{3} \right) \Big|_0^{\sqrt{3-2y-y^2}} dy$$

$$= \frac{4}{3} \int_{-3}^1 \left( 3 - 2y - y^2 \right)^{3/2} dy$$

$$= \frac{4}{3} \int_{-3}^1 \left( 4 - (y + 1)^2 \right)^{3/2} dy.$$

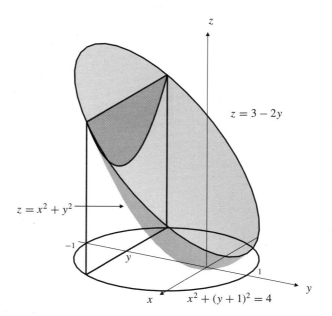

$z = 3 - 2y$

$z = x^2 + y^2$

$x^2 + (y+1)^2 = 4$

**Figure 14.34** The volume above a paraboloid and under a slanting plane

In the last integral we make the change of variable $y + 1 = 2\sin\theta$ and obtain

$$V = \frac{64}{3} \int_{-\pi/2}^{\pi/2} \cos^4\theta \, d\theta$$

$$= \frac{128}{3} \int_{0}^{\pi/2} \left(\frac{1+\cos 2\theta}{2}\right)^2 d\theta$$

$$= \frac{32}{3} \int_{0}^{\pi/2} \left(1 + 2\cos 2\theta + \frac{1+\cos 4\theta}{2}\right) d\theta = 8\pi \text{ cubic units.}$$

As was the case for double integrals, it is sometimes necessary to reiterate a given iterated integral so that the integrations are performed in a different order. This task is most easily accomplished if we can translate the given iteration into a sketch of the region of integration. The ability to deduce the shape of the region from the limits in the iterated integral is a skill that one acquires with practice. You should first determine the projection of the region on a coordinate plane — the plane of the two variables in the outer integrals of the given iteration.

It is also possible to reiterate an iterated integral in a different order by manipulating the limits of integration algebraically. We will illustrate both approaches (graphical and algebraic) in the following examples.

**Example 5** Express the iterated integral

$$I = \int_{0}^{1} dy \int_{y}^{1} dz \int_{0}^{z} f(x, y, z) \, dx$$

as a triple integral and sketch the region over which it is taken. Reiterate the integral in such a way that the integrations are performed in the order: first $y$, then $z$, then $x$ (that is, the opposite order to the given iteration).

**Figure 14.35**

(a) The solid region for the triple integral in Example 5 sliced corresponding to the given iteration

(b) The same solid sliced to conform to the desired iteration

**Solution** We express $I$ as an uniterated triple integral:

$$I = \iiint_R f(x, y, z)\, dV.$$

The outer integral in the given iteration shows that the region $R$ lies between the planes $y = 0$ and $y = 1$. For each such value of $y$, $z$ must lie between $y$ and 1. Therefore $R$ lies below the plane $z = 1$ and above the plane $z = y$, and the projection of $R$ onto the $yz$-plane is the triangle with vertices $(0, 0, 0)$, $(0, 0, 1)$, and $(0, 1, 1)$. Through any point $(0, y, z)$ in this triangle, a line parallel to the $x$-axis intersects $R$ between $x = 0$ and $x = z$. Thus the solid is bounded by the five planes $x = 0$, $y = 0$, $z = 1$, $y = z$, and $z = x$. It is sketched in Figure 14.35(a), with slice and line corresponding to the given iteration.

The required iteration corresponds to the slice and line shown in Figure 14.35(b). Therefore it is

$$I = \int_0^1 dx \int_x^1 dz \int_0^z f(x, y, z)\, dy.$$

$\blacksquare$

**Example 6** Use algebra to write an iteration of the integral

$$I = \int_0^1 dx \int_x^1 dy \int_x^y f(x, y, z)\, dz$$

with the order of integrations reversed.

**Solution** From the given iteration we can write three sets of inequalities satisfied by the outer variable $x$, the middle variable $y$, and the inner variable $z$. We write these in order as follows:

$$0 \le x \le 1 \qquad \text{inequalities for } x$$
$$x \le y \le 1 \qquad \text{inequalities for } y$$
$$x \le z \le y \qquad \text{inequalities for } z.$$

Note that the limits for each variable can be constant or can depend only on variables whose inequalities are on lines above the line for that variable. (In this case, the limits for $x$ must both be constant, those for $y$ can depend on $x$, and those for $z$ can depend on both $x$ and $y$.) This is a requirement for iterated integrals – outer integrals cannot depend on the variables of integration of the inner integrals.

We want to construct an equivalent set of inequalities with those for $z$ on the top line, then those for $y$, then those for $x$ on the bottom line. The limits for $z$ must be constants. From the inequalities above we determine that $0 \le x \le z$ and $z \le y \le 1$. Thus $z$ must satisfy $0 \le z \le 1$. The inequalities for $y$ can depend on $z$. Since $z \le y$ and $y \le 1$, we have $z \le y \le 1$. Finally, the limits for $x$ can depend on both $y$ and $z$. We have $0 \le x, x \le y$, and $x \le z$. Since we have already determined that $z \le y$, we must have $0 \le x \le z$. Thus the revised inequalities are

$$0 \le z \le 1 \qquad \text{inequalities for } z$$
$$z \le y \le 1 \qquad \text{inequalities for } y$$
$$0 \le x \le z \qquad \text{inequalities for } x$$

and the required iteration is

$$I = \int_0^1 dz \int_z^1 dy \int_0^z f(x, y, z)\, dx.$$

## Exercises 14.5

In Exercises 1–12, evaluate the triple integrals over the indicated region $R$. Be alert for simplifications and auspicious orders of iteration.

**1.** $\iiint_R (1 + 2x - 3y)\, dV$, over the box $-a \le x \le a$, $-b \le y \le b$, $-c \le z \le c$

**2.** $\iiint_B xyz\, dV$ over the box $0 \le x \le 1$, $-2 \le y \le 0$, $1 \le z \le 4$

**3.** $\iiint_D (3 + 2xy)\, dV$ over the solid hemispherical dome $D$ given by $x^2 + y^2 + z^2 \le 4$ and $z \ge 0$

**4.** $\iiint_R x\, dV$, over the tetrahedron bounded by the coordinate planes and the plane $\dfrac{x}{a} + \dfrac{y}{b} + \dfrac{z}{c} = 1$

**5.** $\iiint_R (x^2 + y^2)\, dV$ over the cube $0 \le x, y, z \le 1$

**6.** $\iiint_R (x^2 + y^2 + z^2)\, dV$ over the cube of Exercise 5

**7.** $\iiint_R (xy + z^2)\, dV$, over the set $0 \le z \le 1 - |x| - |y|$

**8.** $\iiint_R yz^2 e^{-xyz}\, dV$, over the cube $0 \le x, y, z \le 1$

**9.** $\iiint_R \sin(\pi y^3)\, dV$, over the pyramid with vertices $(0, 0, 0)$, $(0, 1, 0)$, $(1, 1, 0)$, $(1, 1, 1)$, and $(0, 1, 1)$

**10.** $\iiint_R y\, dV$, over that part of the cube $0 \le x, y, z \le 1$ lying above the plane $y + z = 1$ and below the plane $x + y + z = 2$

**11.** $\iiint_R \dfrac{1}{(x + y + z)^3}\, dV$, over the region bounded by the six planes $z = 1$, $z = 2$, $y = 0$, $y = z$, $x = 0$, and $x = y + z$

**12.** $\iiint_R \cos x \cos y \cos z\, dV$, over the tetrahedron defined by $x \ge 0$, $y \ge 0$, $z \ge 0$, and $x + y + z \le \pi$

**13.** Evaluate $\iiint_{\mathbb{R}^3} e^{-x^2 - 2y^2 - 3z^2}\, dV$. *Hint:* use the result of Example 4 of Section 14.4.

**14.** Find the volume of the region lying inside the cylinder $x^2 + 4y^2 = 4$, above the $xy$-plane and below the plane $z = 2 + x$.

**15.** Find $\iiint_T x\, dV$ where $T$ is the tetrahedron bounded by the planes $x = 1$, $y = 1$, $z = 1$ and $x + y + z = 2$.

**16.** Sketch the region $R$ in the first octant of 3-space that has finite volume and is bounded by the surfaces $x = 0$, $z = 0$, $x + y = 1$, and $z = y^2$. Write six different iterations of the triple integral of $f(x, y, z)$ over $R$.

In Exercises 17–20, express the given iterated integral as a triple integral and sketch the region over which it is taken. Reiterate the integral so that the outermost integral is with respect to $x$ and the innermost is with respect to $z$.

**\* 17.** $\displaystyle\int_0^1 dz \int_0^{1-z} dy \int_0^1 f(x, y, z)\, dx$

**\* 18.** $\displaystyle\int_0^1 dz \int_z^1 dy \int_0^y f(x, y, z)\, dx$

**\* 19.** $\displaystyle\int_0^1 dz \int_z^1 dx \int_0^{x-z} f(x, y, z)\, dy$

**\* 20.** $\displaystyle\int_0^1 dy \int_0^{\sqrt{1-y^2}} dz \int_{y^2+z^2}^1 f(x, y, z)\, dx$

**21.** Repeat Exercise 17 using the method of Example 6.

**22.** Repeat Exercise 18 using the method of Example 6.

**23.** Repeat Exercise 19 using the method of Example 6.

**24.** Repeat Exercise 20 using the method of Example 6.

**25.** Rework Example 5 using the method of Example 6.

**26.** Rework Example 6 using the method of Example 5.

In Exercises 27–28, evaluate the given iterated integral by reiterating it in a different order. (You will need to make a good sketch of the region.)

**\* 27.** $\displaystyle\int_0^1 dz \int_z^1 dx \int_0^x e^{x^3}\, dy$

**\* 28.** $\displaystyle\int_0^1 dx \int_0^{1-x} dy \int_y^1 \frac{\sin(\pi z)}{z(2-z)}\, dz$

**29.** Define the average value of an integrable function $f(x, y, z)$ over a region $R$ of 3-space. Find the average value of $x^2 + y^2 + z^2$ over the cube $0 \le x \le 1$, $0 \le y \le 1$, $0 \le z \le 1$.

**\* 30.** State a Mean-Value Theorem for triple integrals analogous to Theorem 3 of Section 14.3. Use it to prove that if $f(x, y, z)$ is continuous near the point $(a, b, c)$ and if $B_\epsilon(a, b, c)$ is the ball of radius $\epsilon$ centred at $(a, b, c)$ then

$$\lim_{\epsilon \to 0} \frac{3}{4\pi\epsilon^3} \iiint_{B_\epsilon(a,b,c)} f(x, y, z)\, dV = f(a, b, c).$$

## 14.6 Change of Variables in Triple Integrals

The change of variables formula for a double integral extends to triple (or higher-order) integrals. Consider the transformation:

$$x = x(u, v, w),$$
$$y = y(u, v, w),$$
$$z = z(u, v, w),$$

where $x$, $y$, and $z$ have continuous first partial derivatives with respect to $u$, $v$, and $w$. Near any point where the Jacobian $\partial(x, y, z)/\partial(u, v, w)$ is nonzero, the transformation scales volume elements according to the formula

$$dV = dx\, dy\, dz = \left| \frac{\partial(x, y, z)}{\partial(u, v, w)} \right| du\, dv\, dw.$$

Thus, if the transformation is one-to-one from a domain $S$ in $uvw$-space onto a domain $D$ in $xyz$-space, and if

$$g(u, v, w) = f\big(x(u, v, w),\, y(u, v, w),\, z(u, v, w)\big),$$

then

$$\iiint_D f(x, y, z)\, dx\, dy\, dz = \iiint_S g(u, v, w) \left| \frac{\partial(x, y, z)}{\partial(u, v, w)} \right| du\, dv\, dw.$$

The proof is similar to that of the two-dimensional case given in Section 14.4. (See Exercise 35 at the end of this section.)

---

**Example 1**    Under the change of variables $x = au$, $y = bv$, $z = cw$, the solid ellipsoid $E$ given by

$$\frac{x^2}{a^2} + \frac{y^2}{b^2} + \frac{z^2}{c^2} \le 1$$

becomes the ball $B$ given by $u^2 + v^2 + w^2 \le 1$. Since the Jacobian of this transformation is

$$\frac{\partial(x, y, z)}{\partial(u, v, w)} = \begin{vmatrix} a & 0 & 0 \\ 0 & b & 0 \\ 0 & 0 & c \end{vmatrix} = abc$$

the volume of the ellipsoid is given by

$$\begin{aligned} \text{Volume of } E &= \iiint_E dx\, dy\, dz \\ &= \iiint_B abc\, du\, dv\, dw = abc \times (\text{Volume of } B) \\ &= \frac{4}{3}\pi abc \text{ cubic units.} \end{aligned}$$

---

## Cylindrical Coordinates

Among the most useful alternatives to Cartesian coordinates in 3-space are two coordinate systems that generalize plane polar coordinates. The simpler of these is the system of **cylindrical coordinates**. It uses ordinary plane polar coordinates in the $xy$-plane while retaining the Cartesian $z$ coordinate for measuring vertical distances. Thus each point in 3-space has cylindrical coordinates $[r, \theta, z]$ related to its Cartesian coordinates $(x, y, z)$ by the transformation

$$x = r\cos\theta, \quad y = r\sin\theta, \quad z = z.$$

Figure 14.36 shows how a point $P$ is located by its cylindrical coordinates $[r, \theta, z]$ as well as by its Cartesian coordinates $(x, y, z)$. Note that the distance from the origin to $P$ is

$$d = \sqrt{r^2 + z^2} = \sqrt{x^2 + y^2 + z^2}.$$

---

**Example 2**    The point with Cartesian coordinates $(1, 1, 1)$ has cylindrical coordinates $[\sqrt{2}, \pi/4, 1]$. The point with Cartesian coordinates $(0, 2, -3)$ has cylindrical coordinates $[2, \pi/2, -3]$. The point with cylindrical coordinates $[4, -\pi/3, 5]$ has Cartesian coordinates $(2, -2\sqrt{3}, 5)$.

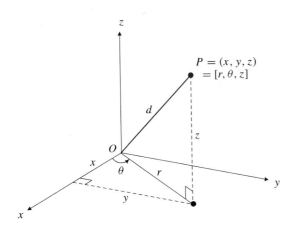

**Figure 14.36**   The cylindrical coordinates of a point

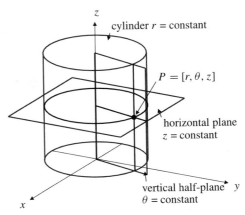

**Figure 14.37**   The coordinate surfaces for cylindrical coordinates

The coordinate surfaces in cylindrical coordinates are the $r$-surfaces (vertical circular cylinders centred on the $z$-axis), the $\theta$-surfaces (vertical planes containing the $z$-axis), and the $z$-surfaces (horizontal planes). (See Figure 14.37.) Cylindrical coordinates lend themselves to representing domains that are bounded by such surfaces, and in general to problems with axial symmetry (around the $z$-axis).

The **volume element in cylindrical coordinates** is

$$dV = r\, dr\, d\theta\, dz,$$

which is easily seen by examining the infinitesimal "box" bounded by the coordinate surfaces corresponding to values $r, r+dr, \theta, \theta+d\theta, z$, and $z+dz$ (see Figure 14.38), or by calculating the Jacobian

$$\frac{\partial(x, y, z)}{\partial(r, \theta, z)} = \begin{vmatrix} \cos\theta & -r\sin\theta & 0 \\ \sin\theta & r\cos\theta & 0 \\ 0 & 0 & 1 \end{vmatrix} = r.$$

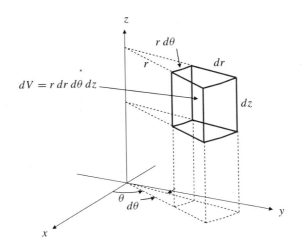

**Figure 14.38**   The volume element in cylindrical coordinates

**Example 3**    Evaluate

$$\iiint_D (x^2 + y^2)\,dV$$

over the first octant region bounded by the cylinders $x^2 + y^2 = 1$ and $x^2 + y^2 = 4$ and the planes $z = 0$, $z = 1$, $x = 0$, and $x = y$.

**Solution**    In terms of cylindrical coordinates the region is bounded by $r = 1$, $r = 2$, $\theta = \pi/4$, $\theta = \pi/2$, $z = 0$, and $z = 1$. (See Figure 14.39. It is a rectangular coordinate box in $r\theta z$-space.) Since the integrand is $x^2 + y^2 = r^2$ the integral is

$$\iiint_D (x^2 + y^2)\,dV = \int_0^1 dz \int_{\pi/4}^{\pi/2} d\theta \int_1^2 r^2\, r\, dr$$

$$= (1 - 0)\left(\frac{\pi}{2} - \frac{\pi}{4}\right)\left(\frac{2^4}{4} - \frac{1^4}{4}\right) = \frac{15}{16}\pi.$$

This integral would have been much more difficult to evaluate using Cartesian coordinates.

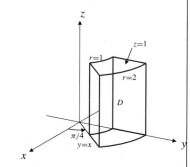

**Figure 14.39**

**Example 4**    Use a triple integral to find the volume of the solid region $R$ inside the sphere $x^2 + y^2 + z^2 = 6$ and above the paraboloid $z = x^2 + y^2$.

**Solution**    One-quarter of the required volume lies in the first octant. It is shown in Figure 14.40. The two surfaces intersect on the vertical cylinder

$$6 - x^2 - y^2 = z^2 = (x^2 + y^2)^2,$$

or, in terms of cylindrical coordinates,

$$r^4 + r^2 - 6 = 0$$
$$(r^2 + 3)(r^2 - 2) = 0.$$

The only relevant solution to this equation is $r = \sqrt{2}$. Thus the required volume lies above the disk of radius $\sqrt{2}$ centred at the origin in the $xy$-plane. The total volume $V$ is

$$V = \iiint_R dV = \int_0^{2\pi} d\theta \int_0^{\sqrt{2}} r\, dr \int_{r^2}^{\sqrt{6-r^2}} dz$$

$$= 2\pi \int_0^{\sqrt{2}} \left(r\sqrt{6 - r^2} - r^3\right) dr$$

$$= 2\pi \left[-\frac{1}{3}(6 - r^2)^{3/2} - \frac{r^4}{4}\right]\Big|_0^{\sqrt{2}}$$

$$= 2\pi \left[\frac{6\sqrt{6}}{3} - \frac{8}{3} - 1\right] = \frac{2\pi}{3}(6\sqrt{6} - 11) \text{ cubic units.}$$

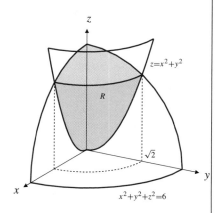

**Figure 14.40**

## Spherical Coordinates

In the system of **spherical coordinates** a point $P$ in 3-space is represented by the ordered triple $[\rho, \phi, \theta]$, where $\rho$ is the distance from $P$ to the origin $O$, $\phi$ is the angle the radial line $OP$ makes with the positive direction of the $z$-axis, and $\theta$ is the angle between the plane containing $P$ and the $z$-axis and the $xz$-plane. (See Figure 14.41.) It is conventional to consider spherical coordinates restricted in such a way that $\rho \geq 0$, $0 \leq \phi \leq \pi$, and $0 \leq \theta < 2\pi$ (or $-\pi < \theta \leq \pi$). Every point not on the $z$-axis then has exactly one spherical coordinate representation, and the transformation from Cartesian coordinates $(x, y, z)$ to spherical coordinates $[\rho, \phi, \theta]$ is one-to-one off the $z$-axis. Using the right-angled triangles in the figure, we can see that this transformation is given by:

$$x = \rho \sin \phi \cos \theta$$
$$y = \rho \sin \phi \sin \theta$$
$$z = \rho \cos \phi.$$

Observe that

$$\rho^2 = x^2 + y^2 + z^2 = r^2 + z^2,$$

and that the $r$ coordinate in cylindrical coordinates is related to $\rho$ and $\phi$ by

$$r = \sqrt{x^2 + y^2} = \rho \sin \phi.$$

Thus, also

$$\tan \phi = \frac{r}{z} = \frac{\sqrt{x^2 + y^2}}{z} \qquad \text{and} \qquad \tan \theta = \frac{y}{x}.$$

If $\phi = 0$ or $\phi = \pi$, then $r = 0$, so the $\theta$ coordinate is irrelevant at points on the $z$-axis.

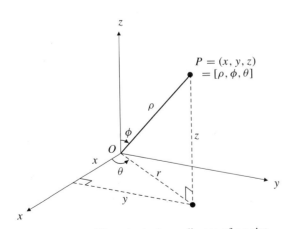

**Figure 14.41**    The spherical coordinates of a point

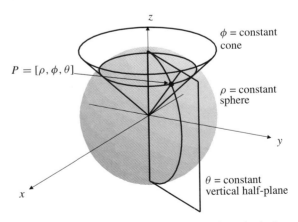

**Figure 14.42**    The coordinate surfaces for spherical coordinates

Some coordinate surfaces for spherical coordinates are shown in Figure 14.42. The $\rho$-surfaces ($\rho$ = constant) are spheres centred at the origin; the $\phi$-surfaces ($\phi$ =

constant) are circular cones with the $z$-axis as axis; the $\theta$-surfaces ($\theta = $ constant) are vertical half-planes with edge along the $z$-axis. If we take a coordinate system with origin at the centre of the earth, $z$-axis through the north pole, and $x$-axis through the intersection of the Greenwich meridian and the equator, then the intersection of the surface of the earth with the $\phi$-surfaces are the *parallels of latitude*, and the intersection with the $\theta$-surfaces are the *meridians of longitude*. Since latitude is measured from $90°$ at the north pole to $-90°$ at the south pole, while $\phi$ is measured from 0 at the north pole to $\pi$ (= $180°$) at the south pole, the coordinate $\phi$ is frequently referred to as the **colatitude** coordinate. $\theta$ is the **longitude** coordinate. Observe that $\theta$ has the same significance in spherical coordinates as it does in cylindrical coordinates.

**Example 5**  Find:

(a) the Cartesian coordinates of the point $P$ with spherical coordinates $[2, \pi/3, \pi/2]$, and

(b) the spherical coordinates of the point $Q$ with Cartesian coordinates $(1, 1, \sqrt{2})$.

***Solution***

(a) If $\rho = 2$, $\phi = \pi/3$, and $\theta = \pi/2$, then

$$x = 2\sin(\pi/3)\cos(\pi/2) = 0$$
$$y = 2\sin(\pi/3)\sin(\pi/2) = \sqrt{3}$$
$$z = 2\cos(\pi/3) = 1.$$

The Cartesian coordinates of $P$ are $(0, \sqrt{3}, 1)$.

(b) Given that

$$\rho \sin\phi \cos\theta = x = 1$$
$$\rho \sin\phi \sin\theta = y = 1$$
$$\rho \cos\phi = z = \sqrt{2}$$

we calculate that $\rho^2 = 1 + 1 + 2 = 4$, so $\rho = 2$. Also $r^2 = 1 + 1 = 2$, so $r = \sqrt{2}$. Thus $\tan\phi = r/z = 1$, so $\phi = \pi/4$. Also, $\tan\theta = y/x = 1$, so $\theta = \pi/4$ or $5\pi/4$. Since $x > 0$, we must have $\theta = \pi/4$. The spherical coordinates of $Q$ are $[2, \pi/4, \pi/4]$.

■

***Remark***  You may wonder why we write spherical coordinates in the order $\rho, \phi, \theta$ rather than $\rho, \theta, \phi$. The reason, which will not become apparent until Chapter 16, is so that the triad of unit vectors at any point $P$ pointing in the directions of increasing $\rho$, increasing $\phi$, and increasing $\theta$ form a *right-hand basis* rather than a left-hand one.

The **volume element in spherical coordinates** is

$$dV = \rho^2 \sin\phi \, d\rho \, d\phi \, d\theta.$$

To see this, observe that the infinitesimal coordinate box bounded by the coordinate surfaces corresponding to values $\rho, \rho + d\rho, \phi, \phi + d\phi, \theta$, and $\theta + d\theta$ has dimensions

$d\rho$, $\rho\,d\phi$, and $\rho\sin\phi\,d\theta$. (See Figure 14.43.) Alternatively, the Jacobian of the transformation can be calculated:

$$\frac{\partial(x,y,z)}{\partial(\rho,\phi,\theta)} = \begin{vmatrix} \sin\phi\cos\theta & \rho\cos\phi\cos\theta & -\rho\sin\phi\sin\theta \\ \sin\phi\sin\theta & \rho\cos\phi\sin\theta & \rho\sin\phi\cos\theta \\ \cos\phi & -\rho\sin\phi & 0 \end{vmatrix}$$

$$= \cos\phi \begin{vmatrix} \rho\cos\phi\cos\theta & -\rho\sin\phi\sin\theta \\ \rho\cos\phi\sin\theta & \rho\sin\phi\cos\theta \end{vmatrix}$$

$$+ \rho\sin\phi \begin{vmatrix} \sin\phi\cos\theta & -\rho\sin\phi\sin\theta \\ \sin\phi\sin\theta & \rho\sin\phi\cos\theta \end{vmatrix}$$

$$= \cos\phi(\rho^2\sin\phi\cos\phi) + \rho\sin\phi(\rho\sin^2\phi)$$

$$= \rho^2\sin\phi.$$

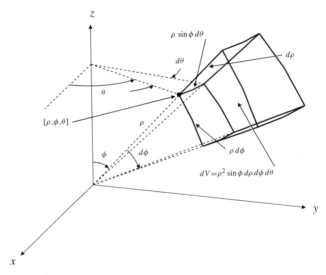

**Figure 14.43** The volume element in spherical coordinates

Spherical coordinates are suited to problems involving spherical symmetry, and in particular, to regions bounded by spheres centred at the origin, circular cones with axes along the $z$-axis, and vertical planes containing the $z$-axis.

**Example 6** A solid half-ball $H$ of radius $a$ has density depending on the distance $\rho$ from the centre of the base disk. The density is given by $k(2a-\rho)$, where $k$ is a constant. Find the mass of the half-ball.

**Solution** Choosing coordinates with origin at the centre of the base and, so that the half-ball lies above the $xy$-plane, we calculate the mass $m$ as follows:

$$m = \iiint_H k(2a-\rho)\,dV = \iiint_H k(2a-\rho)\,\rho^2\sin\phi\,d\rho\,d\phi\,d\theta$$

$$= k\int_0^{2\pi} d\theta \int_0^{\pi/2}\sin\phi\,d\phi \int_0^a (2a-\rho)\,\rho^2\,d\rho$$

$$= 2k\pi \times 1 \times \left(\frac{2a}{3}\rho^3 - \frac{1}{4}\rho^4\right)\bigg|_0^a = \frac{5}{6}\pi ka^4 \text{ units.}$$

**Remark**   In the above example both the integrand and the region of integration exhibited spherical symmetry, so that the choice of spherical coordinates to carry out the integration was most appropriate. The mass could have been evaluated in cylindrical coordinates. The iteration in that system is

$$m = \int_0^{2\pi} d\theta \int_0^a r\, dr \int_0^{\sqrt{a^2 - r^2}} k(2a - \sqrt{r^2 + z^2})\, dz$$

and is much harder to evaluate. It is even more difficult in Cartesian coordinates:

$$m = 4 \int_0^a dx \int_0^{\sqrt{a^2 - x^2}} dy \int_0^{\sqrt{a^2 - x^2 - y^2}} k(2a - \sqrt{x^2 + y^2 + z^2})\, dz.$$

The choice of coordinate system can greatly affect the difficulty of computation of a multiple integral.

Many problems will have elements of spherical and axial symmetry. In such cases it may not be clear whether it would be better to use spherical or cylindrical coordinates. In such doubtful cases the integrand is usually the best guide. Use cylindrical or spherical coordinates according to whether the integrand involves $x^2 + y^2$ or $x^2 + y^2 + z^2$.

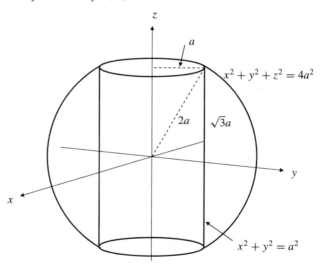

**Figure 14.44**    A solid ball with a cylindrical hole through it

**Example 7**    The moment of inertia about the $z$-axis of a solid of density $\delta$ occupying the region $R$ is given by the integral

$$I = \iiint_R (x^2 + y^2)\delta\, dV.$$

(See Section 14.7.) Calculate that moment of inertia for a solid of unit density occupying the region inside the sphere $x^2 + y^2 + z^2 = 4a^2$ and outside the cylinder $x^2 + y^2 = a^2$.

**Solution**   See Figure 14.44. In terms of spherical coordinates the required moment of inertia is

$$I = 2 \int_0^{2\pi} d\theta \int_{\pi/6}^{\pi/2} \sin\phi \, d\phi \int_{a/\sin\phi}^{2a} \rho^2 \sin^2\phi \, \rho^2 \, d\rho.$$

In terms of cylindrical coordinates it is

$$I = 2 \int_0^{2\pi} d\theta \int_a^{2a} r \, dr \int_0^{\sqrt{4a^2-r^2}} r^2 \, dz.$$

The latter formula looks somewhat easier to evaluate. We continue with it. Evaluating the $\theta$ and $z$ integrals, we get

$$I = 4\pi \int_a^{2a} r^3 \sqrt{4a^2 - r^2} \, dr.$$

Making the substitution $u = 4a^2 - r^2$, $du = -2r \, dr$, we obtain

$$I = 2\pi \int_0^{3a^2} (4a^2 - u)\sqrt{u} \, du = 2\pi \left( 4a^2 \frac{u^{3/2}}{3/2} - \frac{u^{5/2}}{5/2} \right) \Big|_0^{3a^2} = \frac{44}{5} \sqrt{3}\pi a^5.$$

# Exercises 14.6

1. Convert the spherical coordinates $[4, \pi/3, 2\pi/3]$ to Cartesian coordinates and to cylindrical coordinates.

2. Convert the Cartesian coordinates $(2, -2, 1)$ to cylindrical coordinates and to spherical coordinates.

3. Convert the cylindrical coordinates $[2, \pi/6, -2]$ to Cartesian coordinates and to spherical coordinates.

4. A point $P$ has spherical coordinates $[1, \phi, \theta]$ and cylindrical coordinates $[r, \pi/4, r]$. Find the Cartesian coordinates of the point.

Describe the sets of points in 3-space that satisfy the equations in Exercises 5–14. Here 4, $\theta$, $\rho$, and $\phi$ denote the appropriate cylindrical or spherical coordinates.

5. $\theta = \pi/2$

6. $\phi = 2\pi/3$

7. $\phi = \pi/2$

8. $\rho = 4$

9. $r = 4$

10. $\rho = z$

11. $\rho = r$

12. $\rho = 2x$

13. $\rho = 2\cos\phi$

14. $r = 2\cos\theta$

In Exercises 15–23, find the volumes of the indicated regions.

15. Inside the cone $z = \sqrt{x^2 + y^2}$ and inside the sphere $x^2 + y^2 + z^2 = a^2$

16. Above the surface $z = (x^2 + y^2)^{1/4}$ and inside the sphere $x^2 + y^2 + z^2 = 2$

17. Between the paraboloids $z = 10 - x^2 - y^2$ and $z = 2(x^2 + y^2 - 1)$

18. Inside the paraboloid $z = x^2 + y^2$ and inside the sphere $x^2 + y^2 + z^2 = 12$

19. Above the $xy$-plane, inside the cone $z = 2a - \sqrt{x^2 + y^2}$ and inside the cylinder $x^2 + y^2 = 2ay$

20. Above the $xy$-plane, under the paraboloid $z = 1 - x^2 - y^2$ and in the wedge $-x \leq y \leq \sqrt{3}x$

21. In the first octant, between the planes $y = 0$ and $y = x$ and inside the ellipsoid $\dfrac{x^2}{a^2} + \dfrac{y^2}{b^2} + \dfrac{z^2}{c^2} = 1$. *Hint:* use the change of variables suggested in Example 1.

* 22. Bounded by the hyperboloid $\dfrac{x^2}{a^2} + \dfrac{y^2}{b^2} - \dfrac{z^2}{c^2} = 1$ and the planes $z = -c$ and $z = c$

23. Above the $xy$-plane and below the paraboloid $z = 1 - \dfrac{x^2}{a^2} - \dfrac{y^2}{b^2}$

24. Evaluate $\displaystyle\iiint_R (x^2 + y^2 + z^2) \, dV$, where $R$ is the cylinder

$0 \le x^2 + y^2 \le a^2, 0 \le z \le h.$

**25.** Find $\displaystyle\iiint_B (x^2 + y^2)\, dV$, where $B$ is the ball given by

$x^2 + y^2 + z^2 \le a^2$.

**26.** Find $\displaystyle\iiint_B (x^2 + y^2 + z^2)\, dV$, where $B$ is the ball of Exercise 25.

**27.** Find $\displaystyle\iiint_R (x^2 + y^2 + z^2)\, dV$, where $R$ is the region which lies above the cone $z = c\sqrt{x^2 + y^2}$ and inside the sphere $x^2 + y^2 + z^2 = a^2$.

**28.** Evaluate $\displaystyle\iiint_R (x^2 + y^2)\, dV$ over the region $R$ of Exercise 27.

**29.** Find $\displaystyle\iiint_R z\, dV$ over the region $R$ satisfying

$x^2 + y^2 \le z \le \sqrt{2 - x^2 - y^2}$.

**30.** Find $\displaystyle\iiint_R x\, dV$ and $\displaystyle\iiint_R z\, dV$ over that part of the hemisphere $0 \le z \le \sqrt{a^2 - x^2 - y^2}$ that lies in the first octant.

**\* 31.** Find $\displaystyle\iiint_R x\, dV$ and $\displaystyle\iiint_R z\, dV$ over that part of the cone

$$0 \le z \le h\left(1 - \frac{\sqrt{x^2 + y^2}}{a}\right)$$

that lies in the first octant.

**\* 32.** Find the volume of the region inside the ellipsoid

$\dfrac{x^2}{a^2} + \dfrac{y^2}{b^2} + \dfrac{z^2}{c^2} = 1$ and above the plane $z = b - y$.

**\* 33.** Show that for cylindrical coordinates the Laplace equation

$$\frac{\partial^2 u}{\partial x^2} + \frac{\partial^2 u}{\partial y^2} + \frac{\partial^2 u}{\partial z^2} = 0$$

is given by

$$\frac{\partial^2 u}{\partial r^2} + \frac{1}{r}\frac{\partial u}{\partial r} + \frac{1}{r^2}\frac{\partial^2 u}{\partial \theta^2} + \frac{\partial^2 u}{\partial z^2} = 0.$$

**\* 34.** Show that in spherical coordinates the Laplace equation is given by

$$\frac{\partial^2 u}{\partial \rho^2} + \frac{2}{\rho}\frac{\partial u}{\partial \rho} + \frac{\cot\phi}{\rho^2}\frac{\partial u}{\partial \phi} + \frac{1}{\rho^2}\frac{\partial^2 u}{\partial \phi^2} + \frac{1}{\rho^2 \sin^2\phi}\frac{\partial^2 u}{\partial \theta^2} = 0.$$

**\* 35.** If $x$, $y$, and $z$ are functions of $u$, $v$, and $w$ with continuous first partial derivatives and nonvanishing Jacobian at $(u, v, w)$ show that they map an infinitesimal volume element in $uvw$-space bounded by the coordinate planes $u$, $u + du$, $v$, $v + dv$, $w$, and $w + dw$ into an infinitesimal "parallelepiped" in $xyz$-space having volume

$$dx\, dy\, dz = \left|\frac{\partial(x, y, z)}{\partial(u, v, w)}\right|\, du\, dv\, dw.$$

*Hint:* adapt the two-dimensional argument given in Section 14.4. What three vectors from the point $P = (x(u, v, w), y(u, v, w), z(u, v, w))$ span the parallelepiped?

# 14.7 Applications of Multiple Integrals

When we express the volume $V$ of a region $R$ in 3-space as an integral,

$$V = \iiint_R dV,$$

we are regarding $V$ as a "sum" of infinitely many *infinitesimal elements of volume,* that is, as the limit of the sum of volumes of smaller and smaller, nonoverlapping subregions into which we subdivide $R$. This idea of representing sums of infinitesimal elements of quantities by integrals has many applications.

For example, if a rigid body of constant density $\delta$ g/cm$^3$ occupies a volume $V$ cm$^3$ then its mass is $m = \delta V$ grams. If the density is not constant but varies continuously over the region $R$ of 3-space occupied by the rigid body, say $\delta = \delta(x, y, z)$, we can still regard the density as being constant on an infinitesimal element of $R$ having volume $dV$. The mass of this element is therefore $dm = \delta(x, y, z)\, dV$ and the mass of the whole body is calculated by integrating these mass elements over $R$:

$$m = \iiint_R \delta(x, y, z)\, dV.$$

Similar formulas apply when the rigid body is one- or two-dimensional and its density is given in units of mass per unit length or per unit area. In such cases single or double integrals are needed to sum the individual elements of mass. All this works because mass is "additive," that is, the mass of a composite object is the sum of the masses of the parts that comprise the object. The surface areas, gravitational forces, moments, and energies we consider in this section all have this additivity property.

## The Surface Area of a Graph

We can use a double integral over a domain $D$ in the $xy$-plane to add up surface area elements and thereby calculate the total area of the surface $z = f(x, y)$ defined for $(x, y)$ in $D$. We assume that $f$ has continuous first partial derivatives in $D$, so that its graph, the surface $S$ with equation $z = f(x, y)$, has a nonvertical tangent plane at $P = (x, y, f(x, y))$ if $(x, y)$ belongs to $D$. The vector

$$\mathbf{n} = -f_1(x, y)\mathbf{i} - f_2(x, y)\mathbf{j} + \mathbf{k}$$

is an upward normal to $S$ at $P$. An area element $dA$ at position $(x, y)$ in the $xy$-plane has a *vertical projection* onto $S$, in the form of a parallelogram whose area $dS$ is $\sec\gamma$ times the area $dA$, where $\gamma$ is the angle between $\mathbf{n}$ and $\mathbf{k}$. (See Figure 14.45.)

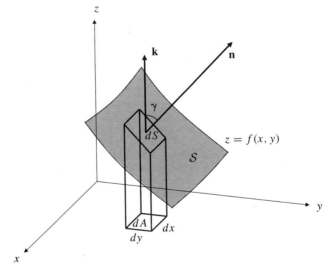

**Figure 14.45** The surface area element $dS$ on the surface $z = f(x, y)$ is $\sec\gamma$ times as large as its projection $dA$ onto the $xy$-plane

Since

$$\cos\gamma = \frac{\mathbf{n} \bullet \mathbf{k}}{|\mathbf{n}||\mathbf{k}|} = \frac{1}{\sqrt{1 + \big(f_1(x, y)\big)^2 + \big(f_2(x, y)\big)^2}},$$

we have

$$dS = \sqrt{1 + \left(\frac{\partial z}{\partial x}\right)^2 + \left(\frac{\partial z}{\partial y}\right)^2}\, dA.$$

Therefore the area of $S$ is

$$S = \iint_D \sqrt{1 + \left(\frac{\partial z}{\partial x}\right)^2 + \left(\frac{\partial z}{\partial y}\right)^2}\, dA.$$

**Example 1**    Find the area of that part of the hyperbolic paraboloid $z = x^2 - y^2$ that lies inside the cylinder $x^2 + y^2 = a^2$.

**Solution**    Since $\partial z/\partial x = 2x$ and $\partial z/\partial y = -2y$, the surface area element is

$$dS = \sqrt{1 + 4x^2 + 4y^2}\, dA = \sqrt{1 + 4r^2}\, r\, dr\, d\theta.$$

The required surface area is the integral of $dS$ over the disk $r \le a$:

$$S = \int_0^{2\pi} d\theta \int_0^a \sqrt{1 + 4r^2}\, r\, dr \qquad\qquad \text{(Let } u = 1 + 4r^2.\text{)}$$

$$= (2\pi)\frac{1}{8} \int_1^{1+4a^2} \sqrt{u}\, du$$

$$= \frac{\pi}{4}\left(\frac{2}{3}\right) u^{3/2}\Big|_1^{1+4a^2} = \frac{\pi}{6}\left((1 + 4a^2)^{3/2} - 1\right) \text{ square units.}$$

## The Gravitational Attraction of a Disk

Newton's universal law of gravitation asserts that two point masses $m_1$ and $m_2$, separated by a distance $s$, attract one another with a force

$$F = \frac{km_1 m_2}{s^2},$$

$k$ being a universal constant. The force on each mass is directed towards the other, along the line joining the two masses. Suppose that a flat disk $D$ of radius $a$, occupying the region $x^2 + y^2 = a^2$ of the $xy$-plane, has constant *areal density* $\sigma$ (units of mass per unit area). Let us calculate the total force of attraction that this disk exerts upon a mass $m$ located at the point $(0, 0, b)$ on the positive $z$-axis. The total force is a vector quantity. Although the various mass elements on the disk are in different directions from the mass $m$, symmetry indicates that the net force will be in the direction towards the centre of the disk, that is, towards the origin. Thus the total force will be $-F\mathbf{k}$, where $F$ is the magnitude of the force.

We will calculate $F$ by integrating the vertical component $dF$ of the force of attraction on $m$ due to the mass $\sigma\, dA$ in an area element $dA$ on the disk. If the area element is at the point with polar coordinates $(r, \theta)$, and if the line from this point to $(0, 0, b)$ makes angle $\Psi$ with the $z$-axis as shown in Figure 14.46, then the vertical component of the force of attraction of the mass element $\sigma\, dA$ on $m$ is

$$dF = \frac{km\sigma\, dA}{r^2 + b^2}\cos\Psi = km\sigma b\, \frac{dA}{(r^2 + b^2)^{3/2}}.$$

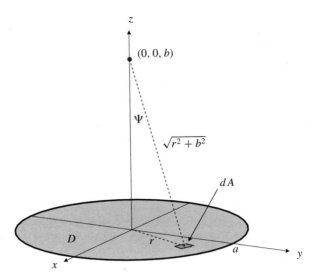

**Figure 14.46**    Each mass element $\sigma\,dA$ attracts $m$ along a different line

Accordingly, the total vertical force of attraction of the disk on $m$ is

$$
\begin{aligned}
F &= km\sigma b \iint_D \frac{dA}{(r^2 + b^2)^{3/2}} \\
&= km\sigma b \int_0^{2\pi} d\theta \int_0^a \frac{r\,dr}{(r^2 + b^2)^{3/2}} \qquad \text{(Let } u = r^2 + b^2.\text{)} \\
&= \pi km\sigma b \int_{b^2}^{a^2+b^2} u^{-3/2}\,du \\
&= 2\pi km\sigma \left( 1 - \frac{b}{\sqrt{a^2 + b^2}} \right).
\end{aligned}
$$

**Remark**    If we let $a$ approach infinity in the above formula we obtain the formula $F = 2\pi km\sigma$ for the force of attraction of a plane of areal density $\sigma$ on a mass $m$ located at distance $b$ from the plane. Observe that $F$ does not depend on $b$. Try to reason on physical grounds why this should be so.

**Remark**    The force of attraction on a point mass due to suitably symmetric solid objects (such as balls, cylinders, and cones) having constant density $\delta$ (units of mass per unit volume) can be found by integrating elements of force contributed by thin, disk-shaped slices of the solid. See Exercises 14–17 at the end of this section.

## Moments and Centres of Mass

In Sections 7.4 and 7.5 we considered certain problems concerning moments, centres of mass, and centroids that could be solved using single integrals. You might wish to review the introductions to those sections before continuing with the material below.

The centre of mass of a rigid body is that point (fixed in the body) at which the body can be supported so that in the presence of a constant gravitational field it will not experience any unbalanced torques that will cause it to rotate. The torques experienced by a mass element $dm$ in the body can be expressed in terms of the **moments** of $dm$ about the three coordinate planes. If the body occupies a region $R$ in 3-space and has continuous volume density $\delta(x, y, z)$, then the mass element $dm = \delta(x, y, z)\,dV$ that occupies the volume element $dV$ is said to have **moments** $(x - x_0)\,dm$, $(y - y_0)\,dm$, and $(z - z_0)\,dm$ about the planes $x = x_0$, $y = y_0$, and

$z = z_0$, respectively. Thus the total moments of the body about these three planes are

$$M_{x=x_0} = \iiint_R (x - x_0)\delta(x, y, z)\, dV = M_{x=0} - x_0 m$$

$$M_{y=y_0} = \iiint_R (y - y_0)\delta(x, y, z)\, dV = M_{y=0} - y_0 m$$

$$M_{z=z_0} = \iiint_R (z - z_0)\delta(x, y, z)\, dV = M_{z=0} - z_0 m$$

where $m = \iiint_R \delta\, dV$ is the mass of the body and $M_{x=0}$, $M_{y=0}$, and $M_{z=0}$ are the moments about the coordinate planes $x = 0$, $y = 0$, and $z = 0$, respectively. The **centre of mass** $\bar{P} = (\bar{x}, \bar{y}, \bar{z})$ of the body is that point for which $M_{x=\bar{x}}$, $M_{y=\bar{y}}$, and $M_{z=\bar{z}}$ are all equal to zero. Thus

---

**Centre of mass**

The centre of mass of a solid occupying region $R$ of 3-space and having continuous density $\delta(x, y, z)$ (units of mass per unit volume) is the point $(\bar{x}, \bar{y}, \bar{z})$ with coordinates given by

$$\bar{x} = \frac{M_{x=0}}{m} = \frac{\iiint_R x\delta\, dV}{\iiint_R \delta\, dV}, \qquad \bar{y} = \frac{M_{y=0}}{m} = \frac{\iiint_R y\delta\, dV}{\iiint_R \delta\, dV},$$

$$\bar{z} = \frac{M_{z=0}}{m} = \frac{\iiint_R z\delta\, dV}{\iiint_R \delta\, dV}.$$

---

These formulas can be combined into a single vector formula for the position vector $\bar{\mathbf{r}} = \bar{x}\mathbf{i} + \bar{y}\mathbf{j} + \bar{z}\mathbf{k}$ of the centre of mass in terms of the position vector $\mathbf{r} = x\mathbf{i} + y\mathbf{j} + z\mathbf{k}$ of an arbitrary point in $R$,

$$\bar{\mathbf{r}} = \frac{M_{x=0}\mathbf{i} + M_{y=0}\mathbf{j} + M_{z=0}\mathbf{k}}{m} = \frac{\iiint_R \delta\mathbf{r}\, dV}{\iiint_R \delta\, dV},$$

where the integral of the vector function $\delta\,\mathbf{r}$ is understood to mean the vector whose components are the integrals of the components of $\delta\,\mathbf{r}$.

**Remark** Similar expressions hold for distributions of mass over regions in the plane, or over intervals on a line. We use the appropriate areal or line densities and double or single definite integrals.

**Remark** If the density is constant it cancels out of the expressions for the centre of mass. In this case the centre of mass is a *geometric* property of the region $R$, and is called the **centroid** or **centre of gravity** of that region.

**Figure 14.47** Iteration diagram for triple integrals over a tetrahedron

**Example 2** Find the centroid of the tetrahedron $T$ bounded by the coordinate planes and the plane $\dfrac{x}{a} + \dfrac{y}{b} + \dfrac{z}{c} = 1$.

**Solution** The density is assumed to be constant, and so we may take it to be unity. The mass of $T$ is thus equal to its volume: $m = V = abc/6$. The moment of $T$ about the $yz$-plane is (see Figure 14.47):

$$M_{x=0} = \iiint_T x\, dV$$

$$= \int_0^a x\, dx \int_0^{b\left(1-\frac{x}{a}\right)} dy \int_0^{c\left(1-\frac{x}{a}-\frac{y}{b}\right)} dz$$

$$= c \int_0^a x\, dx \int_0^{b\left(1-\frac{x}{a}\right)} \left(1 - \frac{x}{a} - \frac{y}{b}\right) dy$$

$$= c \int_0^a x \left[\left(1 - \frac{x}{a}\right)y - \frac{y^2}{2b}\right]\Bigg|_0^{b\left(1-\frac{x}{a}\right)} dx$$

$$= \frac{bc}{2} \int_0^a x \left(1 - \frac{x}{a}\right)^2 dx$$

$$= \frac{bc}{2} \left[\frac{x^2}{2} - \frac{2}{3}\frac{x^3}{a} + \frac{x^4}{4a^2}\right]\Bigg|_0^a = \frac{a^2 bc}{24}.$$

Thus $\bar{x} = M_{x=0}/m = a/4$. By symmetry, the centroid of $T$ is $\left(\dfrac{a}{4}, \dfrac{b}{4}, \dfrac{c}{4}\right)$.

**Example 3** Find the centre of mass of a solid occupying the region $S$ that satisfies $x \geq 0$, $y \geq 0$, $z \geq 0$, and $x^2 + y^2 + z^2 \leq a^2$, if the density at distance $\rho$ from the origin is $k\rho$.

**Solution**    The mass of the solid is distributed symmetrically in the first octant part of the ball $\rho \leq a$ so that the centre of mass, $(\bar{x}, \bar{y}, \bar{z})$, must satisfy $\bar{x} = \bar{y} = \bar{z}$. The mass of the solid is

$$m = \iiint_S k\rho \, dV = \int_0^{\pi/2} d\theta \int_0^{\pi/2} \sin\phi \, d\phi \int_0^a (k\rho)\rho^2 \, d\rho = \frac{\pi k a^4}{8}.$$

The moment about the $xy$-plane is

$$M_{z=0} = \iiint_S zk\rho \, dV = \iiint_S (k\rho)\rho \cos\phi \, \rho^2 \sin\phi \, d\rho \, d\phi \, d\theta$$

$$= \frac{k}{2} \int_0^{\pi/2} d\theta \int_0^{\pi/2} \sin(2\phi) \, d\phi \int_0^a \rho^4 \, d\rho = \frac{k\pi a^5}{20}.$$

Hence $\bar{x} = \dfrac{k\pi a^5}{20} \Big/ \dfrac{k\pi a^4}{8} = \dfrac{2a}{5}$ and the centre of mass is $\left( \dfrac{2a}{5}, \dfrac{2a}{5}, \dfrac{2a}{5} \right)$.

## Moment of Inertia

The **kinetic energy** of a particle of mass $m$ moving with speed $v$ is

$$\text{KE} = \frac{1}{2} m v^2.$$

The mass of the particle measures its *inertia*, which is twice the energy it has when its speed is one unit.

If the particle is moving in a circle of radius $D$, its motion can be described in terms of its **angular speed**, $\Omega$, measured in radians per unit time. In one revolution the particle travels a distance $2\pi D$ in time $2\pi/\Omega$. Thus its (translational) speed $v$ is related to its angular speed by

$$v = \Omega D.$$

Suppose that a rigid body is rotating with angular speed $\Omega$ about an axis $L$. If (at some instant) the body occupies a region $R$ and has density $\delta = \delta(x, y, z)$ then each mass element $dm = \delta \, dV$ in the body has kinetic energy

$$d\text{KE} = \frac{1}{2} v^2 \, dm = \frac{1}{2} \delta \Omega^2 D^2 \, dV,$$

where $D = D(x, y, z)$ is the perpendicular distance from the volume element $dV$ to the axis of rotation $L$. The total kinetic energy of the rotating body is therefore

$$\text{KE} = \frac{1}{2} \Omega^2 \iiint_R D^2 \delta \, dV = \frac{1}{2} I \Omega^2,$$

where

$$I = \iiint_R D^2 \delta \, dV.$$

$I$ is called the **moment of inertia** of the rotating body about the axis $L$. The moment of inertia plays the same role in the expression for kinetic energy of rotation (in terms of angular speed) that the mass does in the expression for kinetic energy of translation (in terms of linear speed). The moment of inertia is twice the kinetic energy of the body when it is rotating with unit angular speed.

If the entire mass of the rotating body were concentrated at a distance $D_0$ from the axis of rotation, then its kinetic energy would be $\frac{1}{2}mD_0^2\Omega^2$. The **radius of gyration** $\bar{D}$ is the value of $D_0$ for which this energy is equal to the actual kinetic energy $\frac{1}{2}I\Omega^2$ of the rotating body. Thus $m\bar{D}^2 = I$ and the radius of gyration is

$$\bar{D} = \sqrt{I/m} = \left( \frac{\iiint_R D^2 \delta \, dV}{\iiint_R \delta \, dV} \right)^{1/2}.$$

### Example 4  (The acceleration of a rolling ball)

(a) Find the moment of inertia and radius of gyration of a solid ball of radius $a$ and constant density $\delta$ about a diameter of that ball.

(b) With what linear acceleration will the ball roll (without slipping) down a plane inclined at angle $\alpha$ to the horizontal?

**Solution**

(a) We take the $z$-axis as the diameter and integrate in cylindrical coordinates over the ball $B$ of radius $a$ centred at the origin. Since the density $\delta$ is constant we have

$$I = \delta \iiint_B r^2 \, dV$$

$$= \delta \int_0^{2\pi} d\theta \int_0^a r^3 \, dr \int_{-\sqrt{a^2-r^2}}^{\sqrt{a^2-r^2}} dz$$

$$= 4\pi\delta \int_0^a r^3 \sqrt{a^2 - r^2} \, dr \qquad (\text{Let } u = a^2 - r^2.)$$

$$= 2\pi\delta \int_0^{a^2} (a^2 - u)\sqrt{u} \, du$$

$$= 2\pi\delta \left( \frac{2}{3}a^2 u^{3/2} - \frac{2}{5}u^{5/2} \right)\Bigg|_0^{a^2} = \frac{8}{15}\pi\delta a^5.$$

Since the mass of the ball is $m = \frac{4}{3}\pi\delta a^3$, the radius of gyration is

$$\bar{D} = \sqrt{\frac{I}{m}} = \sqrt{\frac{2}{5}}\,a.$$

(b) We can determine the acceleration of the ball by using conservation of total (kinetic plus potential) energy. When the ball is rolling down the plane with speed $v$ its centre is moving with speed $v$ and losing height at a rate $v\sin\alpha$. (See Figure 14.48.) Since the ball is not slipping, it is rotating about a horizontal axis through its centre with angular speed $\Omega = v/a$. Hence its kinetic energy (due to translation and rotation) is

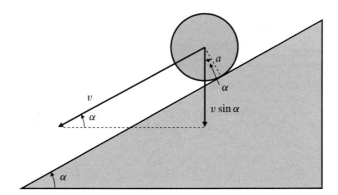

**Figure 14.48**   The actual velocity
and the vertical velocity of a ball rolling
down an incline

$$KE = \frac{1}{2}mv^2 + \frac{1}{2}I\Omega^2$$

$$= \frac{1}{2}mv^2 + \frac{1}{2}\frac{2}{5}ma^2\frac{v^2}{a^2} = \frac{7}{10}mv^2.$$

When the centre of the ball is at height $h$ (above some reference height) the ball has
(gravitational) potential energy

$$PE = mgh.$$

(This is the work that must be done against a constant gravitational force $F = mg$
to raise it to height $h$.)   Since total energy is conserved,

$$\frac{7}{10}mv^2 + mgh = \text{constant}.$$

Differentiating with respect to time $t$, we obtain

$$0 = \frac{7}{10}m\,2v\frac{dv}{dt} + mg\frac{dh}{dt} = \frac{7}{5}mv\frac{dv}{dt} - mgv\sin\alpha.$$

Thus the ball rolls down the incline with acceleration $\dfrac{dv}{dt} = \dfrac{5}{7}g\sin\alpha$.   ∎

## Exercises 14.7

**Surface area problems**

Use double integrals to calculate the areas of the surfaces in
Exercises 1–9.

1. The part of the plane $z = 2x + 2y$ inside the cylinder
   $x^2 + y^2 = 1$

2. The part of the plane $5z = 3x - 4y$ inside the elliptic
   cylinder $x^2 + 4y^2 = 4$

3. The hemisphere $z = \sqrt{a^2 - x^2 - y^2}$

4. The half-ellipsoidal surface $z = 2\sqrt{1 - x^2 - y^2}$

5. The conical surface $3z^2 = x^2 + y^2, 0 \le z \le 2$

6. The paraboloid $z = 1 - x^2 - y^2$ in the first octant

7. The part of the surface $z = y^2$ above the triangle with
   vertices $(0, 0)$, $(0, 1)$, and $(1, 1)$

8. The part of the surface $z = \sqrt{x}$ above the region $0 \le x \le 1$,

$0 \le y \le \sqrt{x}$

**9.** The part of the cylindrical surface $x^2 + z^2 = 4$ that lies above the region $0 \le x \le 2, 0 \le y \le x$

**10.** Show that the parts of the surfaces $z = 2xy$ and $z = x^2 + y^2$ that lie in the same vertical cylinder have the same area.

**11.** Show that the area $S$ of the part of the paraboloid $z = \frac{1}{2}(x^2 + y^2)$ lying above the square $-1 \le x \le 1$, $-1 \le y \le 1$ is given by

$$S = \frac{8}{3} \int_0^{\pi/4} (1 + \sec^2 \theta)^{3/2} \, d\theta - \frac{2\pi}{3},$$

and use numerical methods to evaluate the area to three decimal places.

**＊12.** The canopy shown in Figure 14.49 is the part of the hemisphere of radius $\sqrt{2}$ centred at the origin that lies above the square $-1 \le x \le 1, -1 \le y \le 1$. Find its area. *Hint:* it is possible to get an exact solution by first finding the area of the part of the sphere $x^2 + y^2 + z^2 = 2$ that lies above the plane $z = 1$. If you do the problem directly by integrating the surface area element over the square, you may encounter an integral that you can't evaluate exactly and will have to use numerical methods.

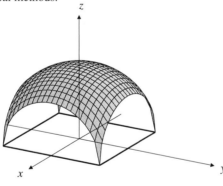

**Figure 14.49**

### Mass and gravitational attraction

**13.** Find the mass of a spherical planet of radius $a$ whose density at distance $R$ from the centre is $\delta = A/(B + R^2)$.

In Exercises 14–17, find the gravitational attraction that the given object exerts on a mass $m$ located at $(0, 0, b)$. Assume the object has constant density $\delta$. In each case you can obtain the answer by integrating the contributions made by disks of thickness $dz$, making use of the formula for the attraction exerted by the disk obtained in the text.

**14.** The ball $x^2 + y^2 + z^2 \le a^2$, where $a < b$.

**15.** The cylinder $x^2 + y^2 \le a^2, 0 \le z \le h$, where $h < b$.

**16.** The cone $0 \le z \le b - (\sqrt{x^2 + y^2})/a$.

**17.** The half-ball $0 \le z \le \sqrt{a^2 - x^2 - y^2}$, where $a < b$.

### Centres of mass and centroids

**18.** Find the centre of mass of an object occupying the cube $0 \le x, y, z \le a$ with density given by $\delta = x^2 + y^2 + z^2$.

Find the centroids of the regions in Exercises 19–22.

**19.** The prism $x \ge 0, y \ge 0, x + y \le 1, 0 \le z \le 1$

**20.** The unbounded region $0 \le z \le e^{-(x^2+y^2)}$

**21.** The first octant part of the ball $x^2 + y^2 + z^2 \le a^2$

**22.** The region inside the cube $0 \le x, y, z \le 1$ and under the plane $x + y + z = 2$

### Moments of inertia

**23.** Explain *in physical terms* why the acceleration of the ball rolling down the incline in Example 4 does not approach $g$ (the acceleration due to gravity) as the angle of incline, $\alpha$, approaches $90°$.

Find the moments of inertia and radii of gyration of the solid objects in Exercises 24–32. Assume constant density in all cases.

**24.** A circular cylinder of base radius $a$ and height $h$ about the axis of the cylinder

**25.** A circular cylinder of base radius $a$ and height $h$ about a diameter of the base of the cylinder

**26.** A right circular cone of base radius $a$ and height $h$ about the axis of the cone

**27.** A right circular cone of base radius $a$ and height $h$ about a diameter of the base of the cone

**28.** A cube of edge length $a$ about an edge of the cube

**29.** A cube of edge length $a$ about a diagonal of a face of the cube

**30.** A cube of edge length $a$ about a diagonal of the cube

**31.** The rectangular box $-a \le x \le a, -b \le y \le b, -c \le z \le c$ about the $z$-axis

**32.** The region between the two concentric cylinders $x^2 + y^2 = a^2$ and $x^2 + y^2 = b^2$ (where $0 < a < b$) and between $z = 0$ and $z = c$ about the $z$-axis

**33.** A ball of radius $a$ has constant density $\delta$. A cylindrical hole of radius $b < a$ is drilled through the centre of the ball. Find the mass of the remaining part of the ball, and its moment of inertia about the axis of the hole.

**34.** With what acceleration will a solid cylinder of base radius $a$, height $h$ and constant density $\delta$ roll (without slipping) down a plane inclined at angle $\alpha$ to the horizontal?

**35.** Repeat Exercise 34 for the ball with the cylindrical hole in Exercise 33. Assume that the axis of the hole remains horizontal while the ball rolls.

**＊36.** A rigid pendulum of mass $m$ swings about point $A$ on a horizontal axis. Its moment of inertia about that axis is $I$. The centre of mass $C$ of the pendulum is at distance $a$ from $A$. When the pendulum hangs at rest, $C$ is directly under $A$. (Why?) Suppose the pendulum is swinging. Let $\theta = \theta(t)$ measure the angular displacement of the line $AC$ from the vertical at time $t$. ($\theta = 0$ when the pendulum is in its rest position.) Use a conservation of energy argument similar to that in Example 4 to show that

$$\frac{1}{2} I \left(\frac{d\theta}{dt}\right)^2 - mga \cos\theta = \text{constant},$$

and hence, differentiating with respect to $t$, that

$$\frac{d^2\theta}{dt^2} + \frac{mga}{I} \sin\theta = 0.$$

This is a nonlinear differential equation, and it is not easily solved. However, for small oscillations ($|\theta|$ small) we can use the approximation $\sin\theta \approx \theta$. In this case the differential equation is that of simple harmonic motion. What is the period?

* **37.** Let $L_0$ be a straight line passing through the centre of mass of a rigid body $B$ of mass $m$. Let $L_k$ be a straight line parallel to and $k$ units distant from $L_0$. If $I_0$ and $I_k$ are the moments of inertia of $B$ about $L_0$ and $L_k$, respectively, show that $I_k = I_0 + k^2 m$. Hence a body always has smallest moment of inertia about an axis through its centre of mass. *Hint:* assume that the $z$-axis coincides with $L_0$, and that $L_k$ passes through the point $(k, 0, 0)$.

* **38.** Re-establish the expression for the total kinetic energy of the rolling ball in Example 4 by regarding the ball at any instant as rotating about a horizontal line through its point of contact with the inclined plane. Use the result of Exercise 37.

* **39.** (**Products of inertia**) A rigid body with density $\delta$ is placed with its centre of mass at the origin and occupies a region $R$ of 3-space. Suppose the six second moments $P_{xx}$, $P_{yy}$, $P_{zz}$, $P_{xy}$, $P_{xz}$, and $P_{yz}$ are all known, where

$$P_{xx} = \iiint_R x^2 \delta \, dV, \quad P_{xy} = \iiint_R xy\delta \, dV, \quad \cdots.$$

(There exist tables giving these six moments for bodies of many standard shapes. They are called products of inertia.) Show how to express the moment of inertia of the body about any axis through the origin in term of these six second moments. (If this result is combined with that of Exercise 37, the moment of inertia about *any* axis can be found.)

# Chapter Review

## Key Ideas

- **What do the following sentences or phrases mean?**
  ◇ a Riemann sum for $f(x, y)$ on $a \le x \le b, c \le y \le d$
  ◇ $f(x, y)$ is integrable on $a \le x \le b, c \le y \le d$
  ◇ the double integral of $f(x, y)$ over $a \le x \le b, c \le y \le d$
  ◇ iteration of a double integral
  ◇ the average value of $f(x, y)$ over region $R$
  ◇ the area element in polar coordinates
  ◇ a triple integral
  ◇ the volume element in cylindrical coordinates
  ◇ the volume element in spherical coordinates
  ◇ the surface area of the graph of $z = f(x, y)$
  ◇ the moment of inertia of a solid about an axis
- **Describe how to change variables in a double integral.**
- **How do you calculate the centroid of a solid region?**
- **How do you calculate the moment of inertia of a solid about an axis?**

## Review Exercises

**1.** Evaluate $\displaystyle\iint_R (x + y) \, dA$ over the first-quadrant region lying under $x = y^2$ and above $y = x^2$.

**2.** Evaluate $\displaystyle\iint_P (x^2 + y^2) \, dA$ where $P$ is the parallelogram with vertices $(0, 0)$, $(2, 0)$, $(3, 1)$, and $(1, 1)$.

**3.** Find $\displaystyle\iint_S (y/x) \, dA$ where $S$ is the part of the disk $x^2 + y^2 \le 4$ in the first quadrant and under the line $y = x$.

**4.** Consider the iterated integral

$$I = \int_0^{\sqrt{3}} dy \int_{y/\sqrt{3}}^{\sqrt{4-y^2}} e^{-x^2-y^2} \, dx.$$

(a) Write $I$ as a double integral $\displaystyle\iint_R e^{-x^2-y^2} \, dA$, and sketch the region $R$ over which the double integral is taken.

(b) Write $I$ as an iterated integral with the order of integrations reversed from that of the given iteration.

(c) Write $I$ as an iterated integral in polar coordinates.

(d) Evaluate $I$.

**5.** Find the constant $k > 0$ such that the volume of the region lying inside the sphere $x^2 + y^2 + z^2 = a^2$ and above the cone $z = k\sqrt{x^2 + y^2}$ is one-quarter of the volume contained by the whole sphere.

**6.** Reiterate the integral

$$I = \int_0^2 dy \int_0^y f(x, y) \, dx + \int_2^6 dy \int_0^{\sqrt{6-y}} f(x, y) \, dx$$

with the $y$ integral on the inside.

**7.** Let $J = \int_0^1 dz \int_0^z dy \int_0^y f(x, y, z)\,dx$. Express $J$ as an iterated integral where the integrations are to be performed in the order $z$ first, then $y$, and then $x$.

**8.** An object in the shape of a right-circular cone has height 10 m and base radius 5 m. Its density is proportional to the square of the distance from the base, and equals 3,000 kg/m³ at the vertex.

(a) Find the mass of the object.

(b) Express the moment of inertia of the object about its central axis as an iterated integral.

**9.** Find the average value of $f(t) = \int_t^a e^{-x^2}\,dx$ over the interval $0 \le t \le a$.

**10.** Find the average value of the function $f(x, y) = \lfloor x + y \rfloor$ over the quarter disk $x \ge 0$, $y \ge 0$, $x^2 + y^2 \le 4$. (Recall that $\lfloor x \rfloor$ denotes the greatest integer less than or equal to $x$.)

**11.** Let $D$ be the smaller of the two solid regions bounded by the surfaces

$$z = \frac{x^2 + y^2}{a} \quad \text{and} \quad x^2 + y^2 + z^2 = 6a^2,$$

where $a$ is a positive constant. Find $\iiint_D (x^2 + y^2)\,dV$.

**12.** Find the moment of inertia about the $z$-axis of a solid $V$ of density 1 if $V$ is specified by the inequalities $0 \le z \le \sqrt{x^2 + y^2}$ and $x^2 + y^2 \le 2ay$, where $a > 0$.

**13.** The rectangular solid $0 \le x \le 1$, $0 \le y \le 2$, $0 \le z \le 1$ is cut into two pieces by the plane $2x + y + z = 2$. Let $D$ be the piece that includes the origin. Find the volume of $D$ and $\bar{z}$, the $z$-coordinate of the centroid of $D$.

**14.** A solid $S$ consists of those points $(x, y, z)$ that lie in the first octant and satisfy $x + y + 2z \le 2$ and $y + z \le 1$. Find the volume of $S$ and the $x$-coordinate of its centroid.

**15.** Find $\iiint_S z\,dV$, where $S$ is the portion of the first octant that is above the plane $x + y - z = 1$ and below the plane $z = 1$.

**16.** Find the area of that part of the plane $z = 2x$ that lies inside the paraboloid $z = x^2 + y^2$.

**17.** Find the area of that part of the paraboloid $z = x^2 + y^2$ that lies below the plane $z = 2x$. Express the answer as a single integral and evaluate it to three decimal places.

**\* 18.** Find the volume of the smaller of the two regions into which the plane $x + y + z = 1$ divides the interior of the ellipsoid $x^2 + 4y^2 + 9z^2 = 36$. *Hint:* first change variables so that the ellipsoid becomes a ball. Then replace the plane by a plane with a simpler equation passing the same distance from the origin.

## Challenging Problems

**1.** The plane $(x/a) + (y/b) + (z/c) = 1$ (where $a > 0$, $b > 0$, and $c > 0$) divides the solid ellipsoid

$$\frac{x^2}{a^2} + \frac{y^2}{b^2} + \frac{z^2}{c^2} \le 1$$

into two unequal pieces. Find the volume of the smaller piece.

**2.** Find the area of the part of the plane $(x/a) + (y/b) + (z/c) = 1$ (where $a > 0$, $b > 0$, and $c > 0$) that lies inside the ellipsoid

$$\frac{x^2}{a^2} + \frac{y^2}{b^2} + \frac{z^2}{c^2} \le 1.$$

**3.** (a) Expand $1/(1 - xy)$ as a geometric series, and hence show that

$$\int_0^1 \int_0^1 \frac{1}{1 - xy}\,dx\,dy = \sum_{n=1}^{\infty} \frac{1}{n^2}.$$

(b) Similarly, express the following integrals as sums of series:

(i) $\displaystyle \int_0^1 \int_0^1 \frac{1}{1 + xy}\,dx\,dy,$

(ii) $\displaystyle \int_0^1 \int_0^1 \int_0^1 \frac{1}{1 - xyz}\,dx\,dy\,dz,$

(iii) $\displaystyle \int_0^1 \int_0^1 \int_0^1 \frac{1}{1 + xyz}\,dx\,dy\,dz.$

**4.** (a) Calculate $G'(y)$ if $G(y) = \int_0^{\infty} \frac{\tan^{-1}(xy)}{x}\,dx.$

(b) Evaluate $\int_0^{\infty} \frac{\tan^{-1}(\pi x) - \tan^{-1} x}{x}\,dx$. *Hint:* this integral is $G(\pi) - G(1)$.

**\* 5.** Let $P$ be the parallelepiped bounded by the three pairs of parallel planes $\mathbf{a} \cdot \mathbf{r} = 0$, $\mathbf{a} \cdot \mathbf{r} = d_1 > 0$, $\mathbf{b} \cdot \mathbf{r} = 0$, $\mathbf{b} \cdot \mathbf{r} = d_2 > 0$, $\mathbf{c} \cdot \mathbf{r} = 0$, and $\mathbf{c} \cdot \mathbf{r} = d_3 > 0$, where $\mathbf{a}$, $\mathbf{b}$, and $\mathbf{c}$ are constant vectors, and $\mathbf{r} = x\mathbf{i} + y\mathbf{j} + z\mathbf{k}$. Show that

$$\iiint_P (\mathbf{a} \cdot \mathbf{r})(\mathbf{b} \cdot \mathbf{r})(\mathbf{c} \cdot \mathbf{r})\,dx\,dy\,dz = \frac{(d_1 d_2 d_3)^2}{8|\mathbf{a} \cdot (\mathbf{b} \times \mathbf{c})|}.$$

*Hint:* Make the change of variables $u = \mathbf{a} \cdot \mathbf{r}$, $v = \mathbf{b} \cdot \mathbf{r}$, $w = \mathbf{c} \cdot \mathbf{r}$.

**6.** A hole whose cross-section is a square of side 2 is punched through the middle of a ball of radius 2. Find the volume of the remaining part of the ball.

**\* 7.** Find the volume bounded by the surface with equation $x^{2/3} + y^{2/3} + z^{2/3} = a^{2/3}$.

**\* 8.** Find the volume bounded by the surface $|x|^{1/3} + |y|^{1/3} + |z|^{1/3} = |a|^{1/3}$.

CHAPTER 15

# CHAPTER 15
# Vector Fields

***Introduction*** This chapter and the next are concerned mainly with vector-valued functions of a vector variable, typically functions whose domains and ranges lie in the plane or in 3-space. Such functions are frequently called *vector fields*. Applications of vector fields often involve integrals taken, not along axes or over regions in the plane or 3-space, but rather over curves and surfaces. We will introduce such line and surface integrals in this chapter. The next chapter will be devoted to developing analogues of the Fundamental Theorem of Calculus for integrals of vector fields.

## 15.1 Vector and Scalar Fields

A function whose domain and range are subsets of Euclidean 3-space, $\mathbb{R}^3$, is called a **vector field.** Thus a vector field, **F**, associates a vector $\mathbf{F}(x, y, z)$ with each point $(x, y, z)$ in its domain. The three components of **F** are scalar-valued (real-valued) functions $F_1(x, y, z)$, $F_2(x, y, z)$, and $F_3(x, y, z)$, and $\mathbf{F}(x, y, z)$ can be expressed in terms of the standard basis in $\mathbb{R}^3$ as

$$\mathbf{F}(x, y, z) = F_1(x, y, z)\mathbf{i} + F_2(x, y, z)\mathbf{j} + F_3(x, y, z)\mathbf{k}.$$

(Note that the subscripts here represent *components* of a vector, *not* partial derivatives.) If $F_3(x, y, z) = 0$ and $F_1$ and $F_2$ are independent of $z$ then **F** reduces to

$$\mathbf{F}(x, y) = F_1(x, y)\mathbf{i} + F_2(x, y)\mathbf{j}$$

and so is called a **plane vector field** or a vector field in the $xy$-plane. We will frequently make use of position vectors in the arguments of vector fields. The position vector of $(x, y, z)$ is $\mathbf{r} = x\mathbf{i} + y\mathbf{j} + z\mathbf{k}$, and we can write $\mathbf{F}(\mathbf{r})$ as a shorthand for $\mathbf{F}(x, y, z)$. In the context of discussion of vector fields, a scalar-valued function of a vector variable (that is, a function of several variables in the context of chapters 12–14) is frequently called a **scalar field.** Thus the components of a vector field are scalar fields.

Many of the results we prove about vector fields require that the field be smooth in some sense. We will call a vector field **smooth** wherever its component scalar fields have continuous partial derivatives of all orders. (For most purposes, however, second order would be sufficient.)

Vector fields arise in many situations in applied mathematics. Let us list some:

(a) The gravitational field $\mathbf{F}(x, y, z)$ due to some object is the force of attraction that the object exerts on a unit mass located at position $(x, y, z)$.

(b) The electrostatic force field $\mathbf{E}(x, y, z)$ due to an electrically charged object is the electrical force that the object exerts on a unit charge at position $(x, y, z)$. (The force may be either an attraction or a repulsion.)

(c) The velocity field $\mathbf{v}(x, y, z)$ in a moving fluid (or solid) is the velocity of motion of the particle at position $(x, y, z)$. If the motion is not "steady state" then the velocity field will also depend on time: $\mathbf{v} = \mathbf{v}(x, y, z, t)$.

(d) The gradient $\nabla f(x, y, z)$, of any scalar field $f$ gives the direction and magnitude of the greatest rate of increase of $f$ at $(x, y, z)$. In particular, a *temperature gradient*, $\nabla T(x, y, z)$, is a vector field giving the direction and magnitude of the greatest rate of increase of temperature $T$ at the point $(x, y, z)$ in a heat conducting medium. *Pressure gradients* provide similar information about the variation of pressure in a fluid such as an air mass or an ocean.

(e) The unit radial and unit transverse vectors $\hat{\mathbf{r}}$ and $\hat{\boldsymbol{\theta}}$ are examples of vector fields in the $xy$-plane. Both are defined at all points of the plane except the origin.

---

**Example 1** (**The gravitational field of a point mass**) The gravitational force field due to a point mass $m$ located at point $P_0$ having position $\mathbf{r}_0$ is

$$\mathbf{F}(x, y, z) = \mathbf{F}(\mathbf{r}) = \frac{-km}{|\mathbf{r} - \mathbf{r}_0|^3}(\mathbf{r} - \mathbf{r}_0)$$

$$= -km\frac{(x - x_0)\mathbf{i} + (y - y_0)\mathbf{j} + (z - z_0)\mathbf{k}}{\left((x - x_0)^2 + (y - y_0)^2 + (z - z_0)^2\right)^{3/2}}.$$

Observe that $\mathbf{F}$ points towards the point $\mathbf{r}_0$ and has magnitude

$$|\mathbf{F}| = km/|\mathbf{r} - \mathbf{r}_0|^2.$$

Some of the vectors in a plane section of the field are shown graphically in Figure 15.1. Each represents the value of the field at the position of its tail. Observe how the lengths of the vectors indicate that the strength of the force increases the closer you get to $P_0$. ∎

---

**Remark** The electrostatic field $\mathbf{F}$ due to a point charge $q$ at $P_0$ is given by the same formula as the gravitational field above, except with $-m$ replaced by $q$. The reason for the opposite sign is that like charges repel each other whereas masses attract each other.

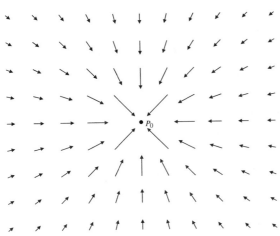

**Figure 15.1** The gravitational field of a point mass located at $P_0$

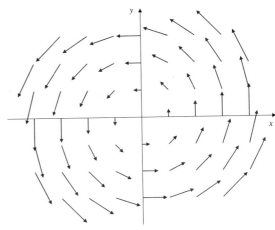

**Figure 15.2** The velocity field of a rigid body rotating about the $z$-axis

**Example 2**    The velocity field of a solid rotating about the $z$-axis with angular velocity $\boldsymbol{\Omega} = \Omega\mathbf{k}$ is

$$\mathbf{v}(x, y, z) = \mathbf{v}(\mathbf{r}) = \boldsymbol{\Omega} \times \mathbf{r} = -\Omega y\mathbf{i} + \Omega x\mathbf{j}.$$

Being the same in all planes normal to the $z$-axis, $\mathbf{v}$ can be regarded as a plane vector field. Some vectors of the field are shown in Figure 15.2.    ∎

## Field Lines (Integral Curves)

The graphical representation of vector fields such as those shown in Figures 15.1 and 15.2, and the wind velocity field over a hill shown in Figure 15.3 suggest a pattern of motion through space or in the plane. Whether or not the field is a velocity field, we can interpret it as such and ask what path will be followed by a particle, initially at some point, whose velocity is given by the field. The path will be a curve to which the field is tangent at every point. Such curves are called **field lines** or **integral curves** for the given vector field. In the specific case where the vector field gives the velocity in a fluid flow, the field lines are called **streamlines** of the flow; some of these are shown for the air flow in Figure 15.3. For a force field the field lines are called **lines of force.**

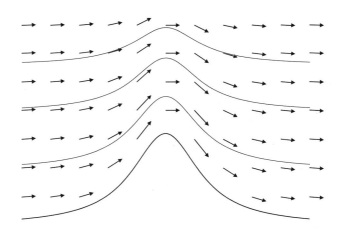

**Figure 15.3**    The velocity field of wind blowing over a hill

The field lines of $\mathbf{F}$ do not depend on the magnitude of $\mathbf{F}$ at any point, but only on the direction of the field. If the field line through some point has parametric equation $\mathbf{r} = \mathbf{r}(t)$ then its tangent vector $d\mathbf{r}/dt$ must be parallel to $\mathbf{F}(\mathbf{r}(t))$ for all $t$. Thus

$$\frac{d\mathbf{r}}{dt} = \lambda(t)\mathbf{F}\bigl(\mathbf{r}(t)\bigr).$$

For *some* vector fields this differential equation can be integrated to find the field lines. If we break the equation into components,

$$\frac{dx}{dt} = \lambda(t)F_1(x, y, z), \quad \frac{dy}{dt} = \lambda(t)F_2(x, y, z), \quad \frac{dz}{dt} = \lambda(t)F_3(x, y, z),$$

we can obtain equivalent differential expressions for $\lambda(t)\,dt$ and hence write the differential equation for the field lines in the form

$$\frac{dx}{F_1(x, y, z)} = \frac{dy}{F_2(x, y, z)} = \frac{dz}{F_3(x, y, z)}.$$

If multiplication of these differential equations by some function puts them in the form

$$P(x)\, dx = Q(y)\, dy = R(z)\, dz,$$

then we can integrate all three expressions to find the field lines.

**Example 3** Find the field lines of the gravitational force field of Example 1:

$$\mathbf{F}(x, y, z) = -km\, \frac{(x - x_0)\mathbf{i} + (y - y_0)\mathbf{j} + (z - z_0)\mathbf{k}}{\left((x - x_0)^2 + (y - y_0)^2 + (z - z_0)^2\right)^{3/2}}.$$

**Solution** The vector in the numerator of the fraction gives the direction of $\mathbf{F}$. Therefore, the field lines satisfy the system

$$\frac{dx}{x - x_0} = \frac{dy}{y - y_0} = \frac{dz}{z - z_0}.$$

Integrating all three expressions leads to

$$\ln|x - x_0| + \ln C_1 = \ln|y - y_0| + \ln C_2 = \ln|z - z_0| + \ln C_3,$$

or, on taking exponentials,

$$C_1(x - x_0) = C_2(y - y_0) = C_3(z - z_0).$$

This represents two families of planes all passing through $P_0 = (x_0, y_0, z_0)$. The field lines are the intersections of planes from each of the families and so are straight lines through the point $P_0$. (This is a *two-parameter* family of lines — any one of the constants $C_i$ that is nonzero can be divided out of the equations above.) The nature of the field lines should also be apparent from the plot of the vector field in Figure 15.1. ∎

**Example 4** Find the field lines of the velocity field $\mathbf{v} = \Omega(-y\mathbf{i} + x\mathbf{j})$ of Example 2.

**Solution** The field lines satisfy the differential equation

$$\frac{dx}{-y} = \frac{dy}{x}.$$

We can separate variables in this equation to get $x\, dx = -y\, dy$. Integration then gives $x^2/2 = -y^2/2 + C/2$, or $x^2 + y^2 = C$. Thus the field lines are circles centred at the origin in the $xy$-plane, as is also apparent from the vector field plot in Figure 15.2. If we regard $\mathbf{v}$ as a vector field in 3-space, we find that the field lines are horizontal circles centred on the $z$-axis:

$$x^2 + y^2 = C_1, \qquad z = C_2.$$

Our ability to find field lines depends on our ability to solve differential equations, and, in 3-space, systems of differential equations.

---

**Example 5** Find the field lines of $\mathbf{F} = xz\mathbf{i} + 2x^2z\mathbf{j} + x^2\mathbf{k}$.

**Solution** The field lines satisfy $\dfrac{dx}{xz} = \dfrac{dy}{2x^2z} = \dfrac{dz}{x^2}$, or, equivalently

$$dy = 2x\,dx \qquad \text{and} \qquad dy = 2z\,dz.$$

The level curves are the curves of intersection of the two families $y = x^2 + C_1$ and $y = z^2 + C_2$ of parabolic cylinders. ∎

---

## Vector Fields in Polar Coordinates

A vector field in the plane can be expressed in terms of polar coordinates in the form

$$\mathbf{F} = \mathbf{F}(r, \theta) = F_r(r, \theta)\hat{\mathbf{r}} + F_\theta(r, \theta)\hat{\boldsymbol{\theta}},$$

where $\hat{\mathbf{r}}$ and $\hat{\boldsymbol{\theta}}$, defined everywhere except at the origin by

$$\hat{\mathbf{r}} = \phantom{-}\cos\theta\,\mathbf{i} + \sin\theta\,\mathbf{j}$$
$$\hat{\boldsymbol{\theta}} = -\sin\theta\,\mathbf{i} + \cos\theta\,\mathbf{j},$$

are unit vectors in the direction of increasing $r$ and $\theta$ at $[r, \theta]$. Note that $\hat{\boldsymbol{\theta}}$ is just $\hat{\mathbf{r}}$ rotated 90° counterclockwise. Also note that we are using $F_r$ and $F_\theta$ to denote the components of $\mathbf{F}$ with respect to the basis $\{\hat{\mathbf{r}}, \hat{\boldsymbol{\theta}}\}$; the subscripts do not indicate partial derivatives. Here $F_r(r, \theta)$ is called the *radial* component of $\mathbf{F}$, and $F_\theta(r, \theta)$ is called the *transverse* component.

A curve with polar equation $r = r(\theta)$ can be expressed in vector parametric form

$$\mathbf{r} = r\hat{\mathbf{r}},$$

as we did in Section 11.6. This curve is a field line of $\mathbf{F}$ if its differential tangent vector

$$d\mathbf{r} = dr\,\hat{\mathbf{r}} + r\,\frac{d\hat{\mathbf{r}}}{d\theta}\,d\theta = dr\,\hat{\mathbf{r}} + r\,d\theta\,\hat{\boldsymbol{\theta}}$$

is parallel to the field vector $\mathbf{F}(r, \theta)$ at any point except the origin, that is, if $r = f(\theta)$ satisfies the differential equation

$$\frac{dr}{F_r(r, \theta)} = \frac{r\,d\theta}{F_\theta(r, \theta)}.$$

In specific cases we can find the field lines by solving this equation.

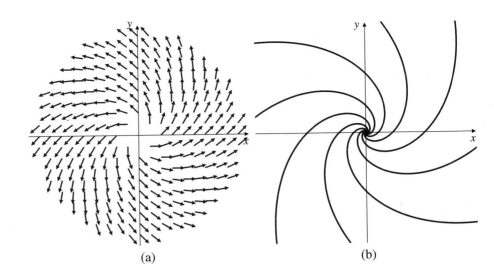

**Figure 15.4**

(a) The vector field $\mathbf{F} = \hat{\mathbf{r}} + \hat{\boldsymbol{\theta}}$

(b) Field lines of $\mathbf{F} = \hat{\mathbf{r}} + \hat{\boldsymbol{\theta}}$

(a)

(b)

---

**Example 6**  Sketch the vector field $\mathbf{F}(r, \theta) = \hat{\mathbf{r}} + \hat{\boldsymbol{\theta}}$, and find its field lines.

**Solution**  At each point $[r, \theta]$, the field vector bisects the angle between $\hat{\mathbf{r}}$ and $\hat{\boldsymbol{\theta}}$, and so makes a counterclockwise angle of $45°$ with $\hat{\mathbf{r}}$. All of the vectors in the field have the same length, $\sqrt{2}$. Some of the field vectors are shown in Figure 15.4(a). They suggest that the field lines will spiral outward from the origin. Since $F_r(r, \theta) = F_\theta(r, \theta) = 1$ for this field, the field lines satisfy $dr = r\,d\theta$, or, dividing by $d\theta$, $dr/d\theta = r$. This is the differential equation of exponential growth, and has solution $r = Ke^\theta$, or, equivalently, $r = e^{\theta + \alpha}$, where $\alpha = \ln K$ is a constant. Several such curves are shown in Figure 15.4(b). ∎

---

## Exercises 15.1

In Exercises 1–8, sketch the given plane vector field and determine its field lines.

**1.** $\mathbf{F}(x, y) = x\mathbf{i} + x\mathbf{j}$

**2.** $\mathbf{F}(x, y) = x\mathbf{i} + y\mathbf{j}$

**3.** $\mathbf{F}(x, y) = y\mathbf{i} + x\mathbf{j}$

**4.** $\mathbf{F}(x, y) = \mathbf{i} + \sin x\,\mathbf{j}$

**5.** $\mathbf{F}(x, y) = e^x\mathbf{i} + e^{-x}\mathbf{j}$

**6.** $\mathbf{F}(x, y) = \nabla(x^2 - y)$

**7.** $\mathbf{F}(x, y) = \nabla\ln(x^2 + y^2)$

**8.** $\mathbf{F}(x, y) = \cos y\,\mathbf{i} - \cos x\,\mathbf{j}$

In Exercises 9–16, describe the streamlines of the given velocity fields.

**9.** $\mathbf{v}(x, y, z) = y\mathbf{i} - y\mathbf{j} - y\mathbf{k}$

**10.** $\mathbf{v}(x, y, z) = x\mathbf{i} + y\mathbf{j} - x\mathbf{k}$

**11.** $\mathbf{v}(x, y, z) = y\mathbf{i} - x\mathbf{j} + \mathbf{k}$

**12.** $\mathbf{v}(x, y, z) = \dfrac{x\mathbf{i} + y\mathbf{j}}{(1 + z^2)(x^2 + y^2)}$

**13.** $\mathbf{v}(x, y, z) = xz\mathbf{i} + yz\mathbf{j} + x\mathbf{k}$

**14.** $\mathbf{v}(x, y, z) = e^{xyz}(x\mathbf{i} + y^2\mathbf{j} + z\mathbf{k})$

**15.** $\mathbf{v}(x, y) = x^2\mathbf{i} - y\mathbf{j}$

**\* 16.** $\mathbf{v}(x, y) = x\mathbf{i} + (x + y)\mathbf{j}$   *Hint:* let $y = xv(x)$.

In Exercises 17–20, determine the field lines of the given polar vector fields.

**17.** $\mathbf{F} = \hat{\mathbf{r}} + r\hat{\boldsymbol{\theta}}$

**18.** $\mathbf{F} = \hat{\mathbf{r}} + \theta\hat{\boldsymbol{\theta}}$

**19.** $\mathbf{F} = 2\hat{\mathbf{r}} + \theta\hat{\boldsymbol{\theta}}$

**20.** $\mathbf{F} = r\hat{\mathbf{r}} - \hat{\boldsymbol{\theta}}$

---

## 15.2 Conservative Fields

Since the gradient of a scalar field is a vector field it is natural to ask whether every vector field is the gradient of a scalar field. Given a vector field $\mathbf{F}(x, y, z)$, does there exist a scalar field $\phi(x, y, z)$ such that

$$\mathbf{F}(x, y, z) = \boldsymbol{\nabla}\phi(x, y, z) = \frac{\partial\phi}{\partial x}\mathbf{i} + \frac{\partial\phi}{\partial y}\mathbf{j} + \frac{\partial\phi}{\partial z}\mathbf{k}\,?$$

The answer in general is "no." Only special vector fields can be written in this way.

**DEFINITION 1**

> If $\mathbf{F}(x, y, z) = \boldsymbol{\nabla}\phi(x, y, z)$ in a domain $D$, then we say that $\mathbf{F}$ is a **conservative** vector field in $D$, and we call the function $\phi$ a **(scalar) potential** for $\mathbf{F}$ on $D$. Similar definitions hold in the plane or in $n$-space.

Like antiderivatives, potentials are not determined uniquely; arbitrary constants can be added to them. Note that $\mathbf{F}$ is **conservative in a domain** $D$ if and only if $\mathbf{F} = \boldsymbol{\nabla}\phi$ at *every* point of $D$; the potential $\phi$ cannot have any singular points in $D$.

The equation $F_1(x, y, z)\, dx + F_2(x, y, z)\, dy + F_3(x, y, z)\, dz = 0$ is called an **exact** differential equation if the left side is the differential of a scalar function $\phi(x, y, z)$:

$$d\phi = F_1(x, y, z)\, dx + F_2(x, y, z)\, dy + F_3(x, y, z)\, dz.$$

In this case the differential equation has solutions given by $\phi(x, y, z) = C$ (constant). Observe that the differential equation is exact if and only if the vector field $\mathbf{F} = F_1\mathbf{i} + F_2\mathbf{j} + F_3\mathbf{k}$ is conservative, and that $\phi$ is the potential of $\mathbf{F}$. Exact differential equations in two variables are discussed in Section 17.3.

Being scalar fields rather than vector fields, potentials for conservative vector fields are easier to manipulate algebraically than are the vector fields themselves. For instance, a sum of potential functions is the potential function for the sum of the corresponding vector fields. A vector field can always be computed from its potential function by taking the gradient.

**Example 1**    **(The gravitational field of a point mass is conservative)** Show that the gravitational field $\mathbf{F}(\mathbf{r}) = -km(\mathbf{r} - \mathbf{r}_0)/|\mathbf{r} - \mathbf{r}_0|^3$ of Example 1 in Section 15.1 is conservative wherever it is defined (that is, everywhere in $\mathbb{R}^3$ except at the origin), by showing that

$$\phi(x, y, z) = \frac{km}{|\mathbf{r} - \mathbf{r}_0|} = \frac{km}{\sqrt{(x - x_0)^2 + (y - y_0)^2 + (z - z_0)^2}}$$

is a potential function for $\mathbf{F}$.

**Solution**    Observe that

$$\frac{\partial\phi}{\partial x} = \frac{-km(x - x_0)}{\left((x - x_0)^2 + (y - y_0)^2 + (z - z_0)^2\right)^{3/2}} = \frac{-km(x - x_0)}{|\mathbf{r} - \mathbf{r}_0|^3} = F_1(\mathbf{r}),$$

and similar formulas hold for the other partial derivatives of $\phi$. It follows that $\boldsymbol{\nabla}\phi(x, y, z) = \mathbf{F}(x, y, z)$ for $(x, y, z) \neq (0, 0, 0)$, and $\mathbf{F}$ is conservative except at the origin.

**Remark**   It is not necessary to write the expression $km/|\mathbf{r} - \mathbf{r}_0|$ in terms of the components of $\mathbf{r} - \mathbf{r}_0$ as we did in Example 1 in order to calculate its partial derivatives. Here is a useful formula for the derivative of the length of a vector function $\mathbf{F}$ with respect to a variable $x$:

$$\frac{\partial}{\partial x}|\mathbf{F}| = \frac{\mathbf{F} \bullet \left(\frac{\partial}{\partial x}\mathbf{F}\right)}{|\mathbf{F}|}.$$

To see why this is true, express $|\mathbf{F}| = \sqrt{\mathbf{F} \bullet \mathbf{F}}$ and calculate its derivative using the Chain Rule and the Product Rule:

$$\frac{\partial}{\partial x}|\mathbf{F}| = \frac{\partial}{\partial x}\sqrt{\mathbf{F} \bullet \mathbf{F}} = \frac{1}{2\sqrt{\mathbf{F} \bullet \mathbf{F}}} 2\mathbf{F} \bullet \left(\frac{\partial}{\partial x}\mathbf{F}\right) = \frac{\mathbf{F} \bullet \left(\frac{\partial}{\partial x}\mathbf{F}\right)}{|\mathbf{F}|}.$$

Compare this with the derivative of an absolute value of a function of one variable:

$$\frac{d}{dx}|f(x)| = \operatorname{sgn}(f(x)) f'(x) = \frac{f(x)}{|f(x)|} f'(x).$$

In the context of Example 1 we have

$$\frac{\partial}{\partial x}\frac{km}{|\mathbf{r} - \mathbf{r}_0|} = \frac{-km}{|\mathbf{r} - \mathbf{r}_0|^2}\frac{\partial}{\partial x}|\mathbf{r} - \mathbf{r}_0| = \frac{-km(x - x_0)}{|\mathbf{r} - \mathbf{r}_0|^3},$$

with similar expressions for the other partials of $km/|\mathbf{r} - \mathbf{r}_0|$.

**Example 2**   Show that the velocity field $\mathbf{v} = -\Omega y\mathbf{i} + \Omega x\mathbf{j}$ of rigid body rotation about the $z$ axis (see Example 2 of Section 15.1) is not conservative.

**Solution**   There are two ways to show that no potential for $\mathbf{v}$ can exist. One way is to try to find a potential $\phi(x, y)$ for the vector field. We require

$$\frac{\partial \phi}{\partial x} = -\Omega y \quad \text{and} \quad \frac{\partial \phi}{\partial y} = \Omega x.$$

The first of these equations implies that $\phi(x, y) = -\Omega xy + C_1(y)$. (We have integrated with respect to $x$; the constant can still depend on $y$.) Similarly, the second equation implies that $\phi(x, y) = \Omega xy + C_2(x)$. Therefore, we must have $-\Omega xy + C_1(y) = \Omega xy + C_2(x)$, or $2\Omega xy = C_1(y) - C_2(x)$ for all $(x, y)$. This is not possible for any choice of the single-variable functions $C_1(y)$ and $C_2(x)$ unless $\Omega = 0$.

Alternatively, if $\mathbf{v}$ has a potential $\phi$, then we can form the mixed partial derivatives of $\phi$ from the two equations above and get

$$\frac{\partial^2 \phi}{\partial y \partial x} = -\Omega \quad \text{and} \quad \frac{\partial^2 \phi}{\partial x \partial y} = \Omega.$$

This is not possible if $\Omega \neq 0$ because the smoothness of $\mathbf{v}$ implies that its potential should be smooth, so the mixed partials should be equal. Thus no such $\phi$ can exist; $\mathbf{v}$ is not conservative. ∎

Example 2 suggests a condition that must be satisfied by any conservative plane vector field.

> **Necessary condition for a conservative plane vector field**
>
> If $\mathbf{F}(x, y) = F_1(x, y)\mathbf{i} + F_2(x, y)\mathbf{j}$ is a conservative vector field in a domain $D$ of the $xy$-plane, then the condition
>
> $$\frac{\partial}{\partial y} F_1(x, y) = \frac{\partial}{\partial x} F_2(x, y)$$
>
> must be satisfied at all points of $D$.

To see this, observe that

$$F_1\mathbf{i} + F_2\mathbf{j} = \mathbf{F} = \nabla\phi = \frac{\partial\phi}{\partial x}\mathbf{i} + \frac{\partial\phi}{\partial y}\mathbf{j}$$

implies the two scalar equations

$$F_1 = \frac{\partial\phi}{\partial x} \quad \text{and} \quad F_2 = \frac{\partial\phi}{\partial y},$$

and since the mixed partial derivatives of $\phi$ should be equal,

$$\frac{\partial F_1}{\partial y} = \frac{\partial^2\phi}{\partial y\partial x} = \frac{\partial^2\phi}{\partial x\partial y} = \frac{\partial F_2}{\partial x}.$$

A similar condition obtains for vector fields in 3-space.

> **Necessary conditions for a conservative vector field in 3-space**
>
> If $\mathbf{F}(x, y, z) = F_1(x, y, z)\mathbf{i} + F_2(x, y, z)\mathbf{j} + F_3(x, y, z)\mathbf{k}$ is a conservative vector field in a domain $D$ in 3-space, then we must have, everywhere in $D$,
>
> $$\frac{\partial}{\partial y} F_1 = \frac{\partial}{\partial x} F_2, \qquad \frac{\partial}{\partial z} F_1 = \frac{\partial}{\partial x} F_3, \qquad \frac{\partial}{\partial z} F_2 = \frac{\partial}{\partial y} F_3.$$

## Equipotential Surfaces and Curves

If $\phi(x, y, z)$ is a potential function for the conservative vector field $\mathbf{F}$ then the *level surfaces* $\phi(x, y, z) = C$ of $\phi$ are called **equipotential surfaces** of $\mathbf{F}$. Since $\mathbf{F} = \nabla\phi$ is normal to these surfaces (wherever it does not vanish), the field lines of $\mathbf{F}$ always intersect the equipotential surfaces at right angles. For instance, the equipotential surfaces of the gravitational force field of a point mass are spheres centred at the point; these spheres are normal to the field lines, which are straight lines passing through the point. Similarly, for a conservative plane vector field, the *level curves* of the potential function are called **equipotential curves** of the vector field. They are the *orthogonal trajectories* of the field lines; that is, they intersect the field lines at right angles.

**Example 3** Show that the vector field $\mathbf{F}(x, y) = x\mathbf{i} - y\mathbf{j}$ is conservative and find a potential function for it. Describe the field lines and the equipotential curves.

**Solution** Since $\partial F_1/\partial y = 0 = \partial F_2/\partial x$ everywhere in $\mathbb{R}^2$, we would expect $\mathbf{F}$ to be conservative. Any potential function $\phi$ must satisfy

$$\frac{\partial \phi}{\partial x} = F_1 = x \qquad \text{and} \qquad \frac{\partial \phi}{\partial y} = F_2 = -y.$$

The first of these equations gives

$$\phi(x, y) = \int x \, dx = \frac{1}{2}x^2 + C_1(y).$$

Observe that, since the integral is taken with respect to $x$, the "constant" of integration is allowed to depend on the other variable. Now we use the second equation to get

$$-y = \frac{\partial \phi}{\partial y} = C_1'(y) \quad \Rightarrow \quad C_1(y) = -\frac{1}{2}y^2 + C_2.$$

Thus $\mathbf{F}$ is conservative and, for any constant $C_2$,

$$\phi(x, y) = \frac{x^2 - y^2}{2} + C_2$$

is a potential function for $\mathbf{F}$. The field lines of $\mathbf{F}$ satisfy

$$\frac{dx}{x} = -\frac{dy}{y} \quad \Rightarrow \quad \ln|x| = -\ln|y| + \ln C_3 \quad \Rightarrow \quad xy = C_3.$$

The field lines of $\mathbf{F}$ are thus rectangular hyperbolas with the coordinate axes as asymptotes. The equipotential curves constitute another family of rectangular hyperbolas, $x^2 - y^2 = C_4$, with the lines $x = \pm y$ as asymptotes. Curves of the two families intersect at right angles. (See Figure 15.5.) Note, however, that $\mathbf{F}$ does not specify a direction at the origin and the orthogonality breaks down there; in fact neither family has a unique curve through that point. ∎

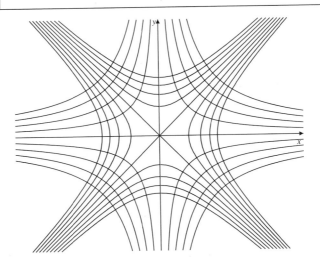

**Figure 15.5** The field lines (black) and equipotential curves (colour) for the field $\mathbf{F} = x\mathbf{i} - y\mathbf{j}$

**Remark** In the above example we constructed the potential $\phi$ by first integrating $\partial \phi/\partial x = F_1$. We could equally well have started by integrating $\partial \phi/\partial y = F_2$, in which case the constant of integration would have depended on $x$. In the end the same $\phi$ would have emerged.

**Example 4**    Decide whether the vector field

$$\mathbf{F} = \left(xy - \sin z\right)\mathbf{i} + \left(\frac{1}{2}x^2 - \frac{e^y}{z}\right)\mathbf{j} + \left(\frac{e^y}{z^2} - x\cos z\right)\mathbf{k}$$

is conservative in $D = \{(x, y, z) : z \neq 0\}$, and find a potential if it is.

**Solution**    Note that $\mathbf{F}$ is not defined when $z = 0$. However, since

$$\frac{\partial F_1}{\partial y} = x = \frac{\partial F_2}{\partial x}, \quad \frac{\partial F_1}{\partial z} = -\cos z = \frac{\partial F_3}{\partial x}, \quad \text{and} \quad \frac{\partial F_2}{\partial z} = \frac{e^y}{z^2} = \frac{\partial F_3}{\partial y},$$

$\mathbf{F}$ may still be conservative in domains not intersecting the $xy$-plane $z = 0$. If so, its potential $\phi$ should satisfy

$$\frac{\partial \phi}{\partial x} = xy - \sin z, \quad \frac{\partial \phi}{\partial y} = \frac{1}{2}x^2 - \frac{e^y}{z}, \quad \text{and} \quad \frac{\partial \phi}{\partial z} = \frac{e^y}{z^2} - x\cos z. \qquad (*)$$

From the first equation of $(*)$,

$$\phi(x, y, z) = \int (xy - \sin z)\, dx = \frac{1}{2}x^2 y - x \sin z + C_1(y, z).$$

(Again note that the constant of integration can be a function of any parameters of the integrand; it is constant only with respect to the variable of integration.) Using the second equation of $(*)$, we obtain

$$\frac{1}{2}x^2 - \frac{e^y}{z} = \frac{\partial \phi}{\partial y} = \frac{1}{2}x^2 + \frac{\partial C_1(y, z)}{\partial y}.$$

Thus

$$C_1(y, z) = -\int \frac{e^y}{z}\, dy = -\frac{e^y}{z} + C_2(z)$$

and

$$\phi(x, y, z) = \frac{1}{2}x^2 y - x \sin z - \frac{e^y}{z} + C_2(z).$$

Finally, using the third equation of $(*)$,

$$\frac{e^y}{z^2} - x\cos z = \frac{\partial \phi}{\partial z} = -x\cos z + \frac{e^y}{z^2} + C_2'(z).$$

Thus $C_2'(z) = 0$ and $C_2(z) = C$ (a constant). Indeed, $\mathbf{F}$ is conservative and, for any constant $C$,

$$\phi(x, y, z) = \frac{1}{2}x^2 y - x \sin z - \frac{e^y}{z} + C$$

is a potential function for $\mathbf{F}$ in the given domain $D$. $C$ may have different values in the two regions $z > 0$ and $z < 0$ whose union constitutes $D$. ∎

**Remark** If, in the above solution, the differential equation for $C_1(y, z)$ had involved $x$, or if that for $C_2(z)$ had involved either $x$ or $y$, we would not have been able to find $\phi$. This did not happen because of the three conditions on the partials of $F_1$, $F_2$, and $F_3$ verified at the outset.

**Remark** The existence of a potential for a vector field depends on the *topology* of the domain of the field (that is, whether the domain has *holes* in it, and what kind of holes) as well as on the structure of the components of the field itself. (Even if the necessary conditions given above are satisfied, a vector field may not be conservative in a domain that has *holes*.) We will be probing further into the nature of conservative vector fields in Section 15.4 and in the next chapter, and will eventually show that the above *necessary conditions* are also *sufficient* to guarantee that **F** is conservative if the domain of **F** satisfies certain conditions. At this point, however, we give an example in which a plane vector field fails to be conservative on a domain where the necessary condition is, nevertheless, satisfied.

**Example 5** For $(x, y) \neq (0, 0)$ define a vector field $\mathbf{F}(x, y)$ and a scalar field $\theta(x, y)$ as follows:

$$\mathbf{F}(x, y) = \left(\frac{-y}{x^2 + y^2}\right)\mathbf{i} + \left(\frac{x}{x^2 + y^2}\right)\mathbf{j}$$

$\theta(x, y) = $ the polar angle $\theta$ of $(x, y)$ such that $0 \leq \theta < 2\pi$.

Thus $x = r\cos\theta(x, y)$ and $y = r\sin\theta(x, y)$, where $r^2 = x^2 + y^2$. Verify the following:

(a) $\dfrac{\partial}{\partial y}F_1(x, y) = \dfrac{\partial}{\partial x}F_2(x, y)$ for $(x, y) \neq (0, 0)$.

(b) $\nabla\theta(x, y) = \mathbf{F}(x, y)$, for all $(x, y) \neq (0, 0)$ such that $0 < \theta < 2\pi$.

(c) **F** is not conservative on the whole $xy$-plane excluding the origin.

**Solution**

(a) We have $F_1 = \dfrac{-y}{x^2 + y^2}$ and $F_2 = \dfrac{x}{x^2 + y^2}$. Thus

$$\frac{\partial}{\partial y}F_1(x, y) = \frac{\partial}{\partial y}\left(-\frac{y}{x^2 + y^2}\right) = \frac{y^2 - x^2}{(x^2 + y^2)^2} = \frac{\partial}{\partial x}\left(\frac{x}{x^2 + y^2}\right)$$

$$= \frac{\partial}{\partial x}F_2(x, y)$$

for all $(x, y) \neq (0, 0)$.

(b) Differentiate the equations $x = r\cos\theta$ and $y = r\sin\theta$ implicitly with respect to $x$ to obtain

$$1 = \frac{\partial x}{\partial x} = \frac{\partial r}{\partial x}\cos\theta - r\sin\theta\frac{\partial\theta}{\partial x},$$

$$0 = \frac{\partial y}{\partial x} = \frac{\partial r}{\partial x}\sin\theta + r\cos\theta\frac{\partial\theta}{\partial x}.$$

Eliminating $\partial r/\partial x$ from this pair of equations and solving for $\partial\theta/\partial x$ leads to

$$\frac{\partial\theta}{\partial x} = -\frac{r\sin\theta}{r^2} = -\frac{y}{x^2 + y^2} = F_1.$$

Similarly, differentiation with respect to $y$ produces

$$\frac{\partial \theta}{\partial y} = \frac{x}{x^2 + y^2} = F_2.$$

These formulas hold only if $0 < \theta < 2\pi$; $\theta$ is not even continuous on the positive $x$-axis; if $x > 0$, then

$$\lim_{y \to 0+} \theta(x, y) = 0 \qquad \text{but} \qquad \lim_{y \to 0-} \theta(x, y) = 2\pi.$$

Thus $\nabla \theta = \mathbf{F}$ holds everywhere in the plane except at points $(x, 0)$ where $x \geq 0$.

(c) Suppose that $\mathbf{F}$ is conservative on the whole plane excluding the origin. Then $\mathbf{F} = \nabla \phi$ there, for some scalar function $\phi(x, y)$. Then $\nabla(\theta - \phi) = 0$ for $0 < \theta < 2\pi$, and $\theta - \phi = C$ (constant), or $\theta = \phi + C$. The left side of this equation is discontinuous along the positive $x$-axis but the right side is not. Therefore the two sides cannot be equal. This contradiction shows that $\mathbf{F}$ cannot be conservative on the whole plane excluding the origin. ∎

**Remark**    Observe that the origin $(0, 0)$ is a *hole* in the domain of $\mathbf{F}$ in the above example. While $\mathbf{F}$ satisfies the necessary condition for being conservative everywhere except at this hole, you must remove from the domain of $\mathbf{F}$ a half-line (ray), or, more generally, a curve from the origin to infinity in order to get a potential function for $\mathbf{F}$. $\mathbf{F}$ is *not* conservative on any domain containing a curve which surrounds the origin. Exercises 22–24 of Section 15.4 will shed further light on this situation.

## Sources, Sinks, and Dipoles

Imagine that 3-space is filled with an incompressible fluid emitted by a point source at the origin at a volume rate $dV/dt = 4\pi m$. (We say that the origin is a **source** of strength $m$.) By symmetry, the fluid flows outward on radial lines from the origin with equal speed at equal distances from the origin in all directions, and the fluid emitted at the origin at some instant $t = 0$ will at later time $t$ be spread over a spherical surface of radius $r = r(t)$. All the fluid inside that sphere was emitted in the time interval $[0, t]$, so we have

$$\frac{4}{3}\pi r^3 = 4\pi m t.$$

Differentiating this equation with respect to $t$ we obtain $r^2(dr/dt) = m$, and the outward speed of the fluid at distance $r$ from the origin is $v(r) = m/r^2$. The velocity field of the moving fluid is therefore

$$\mathbf{v}(\mathbf{r}) = v(r)\frac{\mathbf{r}}{|\mathbf{r}|} = \frac{m}{r^3}\mathbf{r}.$$

This velocity field is conservative (except at the origin) and has potential

$$\phi(\mathbf{r}) = -\frac{m}{r}.$$

A **sink** is a negative source. A sink of strength $m$ at the origin (which annihilates or sucks up fluid at a rate $dV/dt = 4\pi m$) has velocity field and potential given by

$$\mathbf{v}(\mathbf{r}) = -\frac{m}{r^3}\mathbf{r} \qquad \text{and} \qquad \phi(\mathbf{r}) = \frac{m}{r}.$$

The potentials or velocity fields of sources or sinks located at other points are obtained by translation of these formulas; for instance, the velocity field of a source of strength $m$ at the point with position vector $\mathbf{r}_0$ is

$$\mathbf{v}(\mathbf{r}) = -\nabla\left(\frac{m}{|\mathbf{r} - \mathbf{r}_0|}\right) = \frac{m}{|\mathbf{r} - \mathbf{r}_0|^3}(\mathbf{r} - \mathbf{r}_0).$$

This should be compared with the gravitational force field due to a mass $m$ at the origin. The two are the same except for sign and a constant related to units of measurement. For this reason we regard a point mass as a sink for its own gravitational field. Similarly, the electrostatic field due to a point charge $q$ at $\mathbf{r}_0$ is the field of a source (or sink if $q < 0$) of strength proportional to $q$; if units of measurement are suitably chosen we have

$$\mathbf{E}(\mathbf{r}) = -\nabla\left(\frac{q}{|\mathbf{r} - \mathbf{r}_0|}\right) = \frac{q}{|\mathbf{r} - \mathbf{r}_0|^3}(\mathbf{r} - \mathbf{r}_0).$$

In general, the field lines of a vector field converge at a source or sink of that field.

A **dipole** is a system consisting of a source and a sink of equal strength $m$ separated by a short distance $\ell$. The product $\mu = m\ell$ is called the **dipole moment** and the line containing the source and sink is called the **axis** of the dipole. Real physical dipoles, such as magnets, are frequently modelled by *ideal* dipoles that are the limits of such real dipoles as $m \to \infty$ and $\ell \to 0$ in such a way that the dipole moment $\mu$ remains constant.

**Example 6**   Calculate the velocity field, $\mathbf{v}(x, y, z)$, associated with a dipole of moment $\mu$ located at the origin and having axis along the $z$-axis.

**Solution**   We start with a source of strength $m$ at position $(0, 0, \ell/2)$ and a sink of strength $m$ at $(0, 0, -\ell/2)$. The potential of this system is

$$\phi(\mathbf{r}) = -m\left(\frac{1}{|\mathbf{r} - \frac{1}{2}\ell\mathbf{k}|} - \frac{1}{|\mathbf{r} + \frac{1}{2}\ell\mathbf{k}|}\right).$$

The potential of the ideal dipole is the limit of the potential of this system as $m \to \infty$ and $\ell \to 0$ in such a way that $m\ell = \mu$:

$$\phi(\mathbf{r}) = \lim_{\substack{\ell \to 0 \\ m\ell=\mu}} -m\left(\frac{|\mathbf{r} + \frac{1}{2}\ell\mathbf{k}| - |\mathbf{r} - \frac{1}{2}\ell\mathbf{k}|}{|\mathbf{r} + \frac{1}{2}\ell\mathbf{k}|\,|\mathbf{r} - \frac{1}{2}\ell\mathbf{k}|}\right)$$

$$= -\frac{\mu}{|\mathbf{r}|^2}\lim_{\ell \to 0}\frac{|\mathbf{r} + \frac{1}{2}\ell\mathbf{k}| - |\mathbf{r} - \frac{1}{2}\ell\mathbf{k}|}{\ell}$$

(now use l'Hôpital's Rule and the rule for differentiating lengths of vectors)

$$= -\frac{\mu}{|\mathbf{r}|^2}\lim_{\ell \to 0}\frac{\dfrac{(\mathbf{r} + \frac{1}{2}\ell\mathbf{k}) \bullet \frac{1}{2}\mathbf{k}}{|\mathbf{r} + \frac{1}{2}\ell\mathbf{k}|} + \dfrac{(\mathbf{r} - \frac{1}{2}\ell\mathbf{k}) \bullet (-\frac{1}{2}\mathbf{k})}{|\mathbf{r} - \frac{1}{2}\ell\mathbf{k}|}}{1}$$

$$= -\frac{\mu}{|\mathbf{r}|^2}\lim_{\ell \to 0}\left(\frac{\frac{1}{2}z + \frac{1}{4}\ell}{|\mathbf{r} + \frac{1}{2}\ell\mathbf{k}|} + \frac{\frac{1}{2}z - \frac{1}{4}\ell}{|\mathbf{r} - \frac{1}{2}\ell\mathbf{k}|}\right)$$

$$= -\frac{\mu z}{|\mathbf{r}|^3}.$$

The required velocity field is the gradient of this potential. We have

$$\frac{\partial \phi}{\partial x} = \frac{3\mu z}{|\mathbf{r}|^4} \frac{\mathbf{r} \bullet \mathbf{i}}{|\mathbf{r}|} = \frac{3\mu xz}{|\mathbf{r}|^5}$$

$$\frac{\partial \phi}{\partial y} = \frac{3\mu yz}{|\mathbf{r}|^5}, \qquad \text{similarly}$$

$$\frac{\partial \phi}{\partial z} = -\frac{\mu}{|\mathbf{r}|^3} + \frac{3\mu z^2}{|\mathbf{r}|^5} = \frac{\mu(2z^2 - x^2 - y^2)}{|\mathbf{r}|^5}$$

$$\mathbf{v}(\mathbf{r}) = \nabla \phi(\mathbf{r}) = \frac{\mu}{|\mathbf{r}|^5} \Big( 3xz\mathbf{i} + 3yz\mathbf{j} + (2z^2 - x^2 - y^2)\mathbf{k} \Big).$$

Some streamlines for a plane cross-section containing the $z$-axis are shown in Figure 15.6. ∎

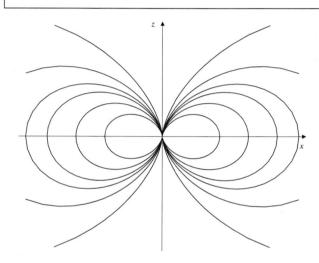

**Figure 15.6**   Streamlines of a dipole

## Exercises 15.2

In Exercises 1–6, determine whether the given vector field is conservative and find a potential if it is conservative.

1. $\mathbf{F}(x, y, z) = x\mathbf{i} - 2y\mathbf{j} + 3z\mathbf{k}$

2. $\mathbf{F}(x, y, z) = y\mathbf{i} + x\mathbf{j} + z^2\mathbf{k}$

3. $\mathbf{F}(x, y) = \dfrac{x\mathbf{i} - y\mathbf{j}}{x^2 + y^2}$   4. $\mathbf{F}(x, y) = \dfrac{x\mathbf{i} + y\mathbf{j}}{x^2 + y^2}$

5. $\mathbf{F}(x, y, z) = (2xy - z^2)\mathbf{i} + (2yz + x^2)\mathbf{j} - (2zx - y^2)\mathbf{k}$

6. $\mathbf{F}(x, y, z) = e^{x^2+y^2+z^2}(xz\mathbf{i} + yz\mathbf{j} + xy\mathbf{k})$

7. Find the three-dimensional vector field with potential
$\phi(\mathbf{r}) = \frac{1}{|\mathbf{r}-\mathbf{r}_0|^2}$.

8. Calculate $\nabla \ln |\mathbf{r}|$, where $\mathbf{r} = x\mathbf{i} + y\mathbf{j} + z\mathbf{k}$.

9. Show that the vector field

$$\mathbf{F}(x, y, z) = \frac{2x}{z}\mathbf{i} + \frac{2y}{z}\mathbf{j} - \frac{x^2 + y^2}{z^2}\mathbf{k}$$

is conservative, and find its potential. Describe the equipotential surfaces. Find the field lines of $\mathbf{F}$.

10. Repeat Exercise 9 for the field

$$\mathbf{F}(x, y, z) = \frac{2x}{z}\mathbf{i} + \frac{2y}{z}\mathbf{j} + \left(1 - \frac{x^2 + y^2}{z^2}\right)\mathbf{k}.$$

11. Find the velocity field due to two sources of strength $m$, one located at $(0, 0, \ell)$ and the other at $(0, 0, -\ell)$. Where is the velocity zero? Find the velocity at any point $(x, y, 0)$ in the $xy$-plane. Where in the $xy$-plane is the speed greatest?

* 12. Find the velocity field for a system consisting of a source of strength 2 at the origin and a sink of strength 1 at $(0, 0, 1)$. Show that the velocity is vertical at all points of a certain sphere. Sketch the streamlines of the flow.

Exercises 13–18 provide an analysis of two-dimensional sources and dipoles similar to that developed for three dimensions in the text.

13. In 3-space filled with an incompressible fluid we say that the $z$-axis is a **line source** of strength $m$ if every interval $\Delta z$

along that axis emits fluid at volume rate $dV/dt = 2\pi m \Delta z$. The fluid then spreads out symmetrically in all directions perpendicular to the $z$-axis. Show that the velocity field of the flow is

$$\mathbf{v} = \frac{m}{x^2 + y^2}(x\mathbf{i} + y\mathbf{j}).$$

14. The flow in Exercise 13 is two-dimensional because $\mathbf{v}$ depends only on $x$ and $y$ and has no component in the $z$ direction. Regarded as a *plane* vector field, it is the field of a two-dimensional point source of strength $m$ located at the origin. (That is, fluid is emitted at the origin at the *areal rate* $dA/dt = 2\pi m$.) Show that the vector field is conservative and find a potential function $\phi(x, y)$ for it.

*15. Find the potential, $\phi$, and the field, $\mathbf{F} = \nabla\phi$, for a two-dimensional dipole at the origin, with axis in the $y$ direction and dipole moment $\mu$. Such a dipole is the limit of a system consisting of a source of strength $m$ at $(0, \ell/2)$ and a sink of strength $m$ at $(0, -\ell/2)$, as $\ell \to 0$ and $m \to \infty$ such that $m\ell = \mu$.

*16. Show that the equipotential curves of the two-dimensional dipole in Exercise 15 are circles tangent to the $x$-axis at the origin.

*17. Show that the streamlines (field lines) of the two-dimensional dipole in Exercises 15 and 16 are circles tangent to the $y$-axis at the origin. *Hint:* it is possible to do this geometrically. If you choose to do it by setting up a differential equation you may find the change of dependent variable

$$y = vx, \qquad \frac{dy}{dx} = v + x\frac{dv}{dx}$$

useful for integrating the equation.

*18. Show that the velocity field of a line source of strength $2m$ can be found by integrating the (three-dimensional) velocity field of a point source of strength $m\,dz$ at $(0, 0, z)$ over the whole $z$-axis. Why does the integral correspond to a line source of strength $2m$ rather than strength $m$? Can the potential of the line source be obtained by integrating the potentials of the point sources?

19. Show that the gradient of a function expressed in terms of polar coordinates in the plane is

$$\nabla\phi(r, \theta) = \frac{\partial\phi}{\partial r}\hat{\mathbf{r}} + \frac{1}{r}\frac{\partial\phi}{\partial\theta}\hat{\boldsymbol{\theta}}.$$

(This is a repeat of Exercise 20 in Section 12.7.)

20. Use the result of Exercise 19 to show that a necessary condition for the vector field

$$\mathbf{F}(r, \theta) = F_r(r, \theta)\hat{\mathbf{r}} + F_\theta(r, \theta)\hat{\boldsymbol{\theta}}$$

(expressed in terms of polar coordinates) to be conservative is that

$$\frac{\partial F_r}{\partial\theta} - r\frac{\partial F_\theta}{\partial r} = F_\theta.$$

21. Show that $\mathbf{F} = r\sin 2\theta\hat{\mathbf{r}} + r\cos 2\theta\hat{\boldsymbol{\theta}}$ is conservative, and find a potential for it.

22. For what values of the constants $\alpha$ and $\beta$ is the vector field

$$\mathbf{F} = r^2\cos\theta\hat{\mathbf{r}} + \alpha r^\beta\sin\theta\hat{\boldsymbol{\theta}}$$

conservative? Find a potential for $\mathbf{F}$ if $\alpha$ and $\beta$ have these values.

# 15.3 Line Integrals

If a wire stretched out along the $x$-axis from $x = a$ to $x = b$ has constant *line density* $\delta$ (units of mass per unit length), then the total mass of the wire will be $m = \delta(b - a)$. If the density of the wire is not constant, but varies continuously from point to point (for example, because the thickness of the wire is not uniform), then we must find the total mass of the wire by "summing" (that is, integrating) differential elements of mass $dm = \delta(x)\,dx$:

$$m = \int_a^b \delta(x)\,dx.$$

In general, the definite integral, $\int_a^b f(x)\,dx$, represents the *total amount* of a quantity distributed along the $x$-axis between $a$ and $b$ in terms of the *line density*, $f(x)$, of that quantity at point $x$. The amount of the quantity in an *infinitesimal* interval of length $dx$ at $x$ is $f(x)\,dx$, and the integral adds up these infinitesimal contributions (or *elements*) to give the total amount of the quantity. Similarly, the integrals $\iint_D f(x, y)\,dA$ and $\iiint_R f(x, y, z)\,dV$ represent the total amounts of quantities distributed over regions $D$ in the plane and $R$ in 3-space in terms of the *areal* or *volume* densities of these quantities.

It may happen that a quantity is distributed with specified line density along a *curve* in the plane or in 3-space, or with specified areal density over a *surface* in 3-space. In such cases we require *line integrals* or *surface integrals* to add up the contributing elements and calculate the total quantity. We examine line integrals in these next two sections and surface integrals in Sections 15.5 and 15.6.

Let $\mathcal{C}$ be a bounded, continuous parametric curve in $\mathbb{R}^3$. Recall (from Section 11.1) that $\mathcal{C}$ is a *smooth curve* if it has a parametrization of the form

$$\mathbf{r} = \mathbf{r}(t) = x(t)\mathbf{i} + y(t)\mathbf{j} + z(t)\mathbf{k}, \qquad a \le t \le b,$$

with "velocity" vector $\mathbf{v} = d\mathbf{r}/dt$ continuous and nonzero. We will call $\mathcal{C}$ a **smooth arc** if it is a smooth curve with *finite* parameter interval $[a, b]$.

In Section 11.3 we saw how to calculate the length of $\mathcal{C}$ by subdividing it into short arcs using points corresponding to parameter values

$$a = t_0 < t_1 < t_2 < \cdots < t_{n-1} < t_n = b,$$

adding up the lengths $|\Delta\mathbf{r}_i| = |\mathbf{r}_i - \mathbf{r}_{i-1}|$ of line segments joining these points, and taking the limit as the maximum distance between adjacent points approached zero. The length was denoted

$$\int_{\mathcal{C}} ds$$

and is a special example of a line integral along $\mathcal{C}$ having integrand 1.

The line integral of a general function $f(x, y, z)$ can be defined similarly. We choose a point $(x_i^*, y_i^*, z_i^*)$ on the $i$th subarc and form the Riemann sum

$$S_n = \sum_{i=1}^{n} f(x_i^*, y_i^*, z_i^*)\,|\Delta\mathbf{r}_i|.$$

If this sum has a limit as $\max |\Delta\mathbf{r}_i| \to 0$, independent of the particular choices of the points $(x_i^*, y_i^*, z_i^*)$, then we call this limit the line integral of $f$ along $\mathcal{C}$, and denote it by

$$\int_{\mathcal{C}} f(x, y, z)\,ds.$$

If $\mathcal{C}$ is a smooth arc and if $f$ is continuous on $\mathcal{C}$, then the limit will certainly exist; its value is given by a definite integral of a continuous function, as shown in the next paragraph. It will also exist (for continuous $f$) if $\mathcal{C}$ is **piecewise smooth**, consisting of finitely many smooth arcs linked end to end; in this case the line integral of $f$ along $\mathcal{C}$ is the sum of the line integrals of $f$ along each of the smooth arcs. Improper line integrals can also be considered, where $f$ has discontinuities or where the length of an arc or curve is not finite.

## Evaluating Line Integrals

The length of $C$ was evaluated by expressing the arc length element $ds = |d\mathbf{r}/dt|\,dt$ in terms of a parametrization $\mathbf{r} = \mathbf{r}(t)$, $(a \le t \le b)$ of the curve, and integrating this from $t = a$ to $t = b$:

$$\text{length of } C = \int_C ds = \int_a^b \left|\frac{d\mathbf{r}}{dt}\right|\,dt.$$

More general line integrals are evaluated similarly:

$$\int_C f(x, y, z)\,ds = \int_a^b f(\mathbf{r}(t))\left|\frac{d\mathbf{r}}{dt}\right|\,dt.$$

Of course, all of the above discussion applies equally well to line integrals of functions $f(x, y)$ along curves $C$ in the $xy$-plane.

**Remark**   It should be noted that the value of the line integral of a function $f$ along a curve $C$ depends on $f$ and $C$, but not on the particular way $C$ is parametrized. If $\mathbf{r} = \mathbf{r}^*(u)$, $\alpha \le u \le \beta$, is another parametrization of the same smooth curve $C$, then any point $\mathbf{r}(t)$ on $C$ can be expressed in terms of the new parametrization as $\mathbf{r}^*(u)$, where $u$ depends on $t$: $u = u(t)$. If $\mathbf{r}^*(u)$ traces $C$ in the same direction as $\mathbf{r}(t)$ then $u(a) = \alpha$, $u(b) = \beta$, and $du/dt \ge 0$; if $\mathbf{r}^*(u)$ traces $C$ in the opposite direction then $u(a) = \beta$, $u(b) = \alpha$, and $du/dt \le 0$. In either event,

$$\int_a^b f(\mathbf{r}(t))\left|\frac{d\mathbf{r}}{dt}\right|\,dt = \int_a^b f(\mathbf{r}^*(u(t)))\left|\frac{d\mathbf{r}^*}{du}\frac{du}{dt}\right|\,dt = \int_\alpha^\beta f(\mathbf{r}^*(u))\left|\frac{d\mathbf{r}^*}{du}\right|\,du.$$

Thus the line integral is *independent of parametrization* of the curve $C$. The following example illustrates this fact.

**Example 1**   A circle of radius $a > 0$ has centre at the origin in the $xy$-plane. Let $C$ be the half of this circle lying in the half-plane $y \ge 0$. Use two different parametrizations of $C$ to find the moment of $C$ about $y = 0$.

**Solution**   We are asked to calculate $\displaystyle\int_C y\,ds$.

$C$ can be parametrized $\mathbf{r} = a\cos t\mathbf{i} + a\sin t\mathbf{j}$, $(0 \le t \le \pi)$. Therefore

$$\frac{d\mathbf{r}}{dt} = -a\sin t\mathbf{i} + a\cos t\mathbf{j} \qquad \text{and} \qquad \left|\frac{d\mathbf{r}}{dt}\right| = a,$$

and the moment of $C$ about $y = 0$ is

$$\int_C y\,ds = \int_0^\pi a\sin t\,a\,dt = -a^2\cos t\,\Big|_0^\pi = 2a^2.$$

$C$ can also be parametrized $\mathbf{r} = x\mathbf{i} + \sqrt{a^2 - x^2}\mathbf{j}$, $(-a \leq x \leq a)$, for which we have

$$\frac{d\mathbf{r}}{dx} = \mathbf{i} - \frac{x}{\sqrt{a^2 - x^2}}\mathbf{j},$$

$$\left|\frac{d\mathbf{r}}{dx}\right| = \sqrt{1 + \frac{x^2}{a^2 - x^2}} = \frac{a}{\sqrt{a^2 - x^2}}.$$

Thus the moment of $C$ about $y = 0$ is

$$\int_C y\, ds = \int_{-a}^{a} \sqrt{a^2 - x^2}\, \frac{a}{\sqrt{a^2 - x^2}}\, dx = a \int_{-a}^{a} dx = 2a^2.$$

It is comforting to get the same answer using different parametrizations. Unlike the line integrals of vector fields considered in the next section, the line integrals of scalar fields considered here do not depend on the direction (orientation) of $C$. The two parametrizations of the semicircle were in opposite directions but still gave the same result.

Line integrals frequently lead to definite integrals that are very difficult or impossible to evaluate without using numerical techniques. Only very simple curves and ones that have been contrived to lead to simple expressions for $ds$ are amenable to exact calculation of line integrals.

**Example 2**   Find the centroid of the circular helix $C$ given by

$$\mathbf{r} = a\cos t\, \mathbf{i} + a\sin t\, \mathbf{j} + bt\, \mathbf{k}, \qquad 0 \leq t \leq 2\pi.$$

**Solution**   As we observed in Example 5 of Section 11.3, $ds = \sqrt{a^2 + b^2}\, dt$ for this helix. On the helix we have $z = bt$, so its moment about $z = 0$ is

$$M_{z=0} = \int_C z\, ds = b\sqrt{a^2 + b^2} \int_0^{2\pi} t\, dt = 2\pi^2 b\sqrt{a^2 + b^2}.$$

Since the helix has length $L = 2\pi\sqrt{a^2 + b^2}$, the $z$-component of its centroid is $M_{z=0}/L = \pi b$. The moment of the helix about $x = 0$ is

$$M_{x=0} = \int_C x\, ds = a\sqrt{a^2 + b^2} \int_0^{2\pi} \cos t\, dt = 0,$$

$$M_{y=0} = \int_C y\, ds = a\sqrt{a^2 + b^2} \int_0^{2\pi} \sin t\, dt = 0.$$

Thus the centroid is $(0, 0, \pi b)$.

Sometimes a curve, along which a line integral is to be taken, is specified as the intersection of two surfaces with given equations. It is normally necessary to parametrize the curve in order to evaluate a line integral. Recall from Section 11.3 that if one of the surfaces is a cylinder parallel to one of the coordinate axes it is usually easiest to begin by parametrizing that cylinder. (Otherwise, combine the equations to eliminate one variable and thus obtain such a cylinder on which the curve lies.)

**Example 3**   Find the mass of a wire lying along the first octant part $C$ of the curve of intersection of the elliptic paraboloid $z = 2 - x^2 - 2y^2$ and the parabolic cylinder $z = x^2$ between $(0, 1, 0)$ and $(1, 0, 1)$ if the density of the wire at position $(x, y, z)$ is $\delta(x, y, z) = xy$.

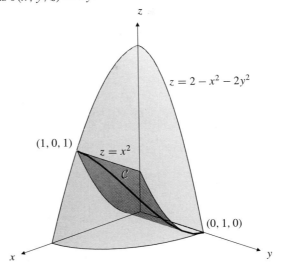

**Figure 15.7**   The curve of intersection of $z = x^2$ and $z = 2 - x^2 - 2y^2$

**Solution**   We need a convenient parametrization of $C$. Since the curve $C$ lies on the cylinder $z = x^2$ and $x$ goes from 0 to 1, we can let $x = t$ and $z = t^2$. Thus $2y^2 = 2 - x^2 - z = 2 - 2t^2$, so $y^2 = 1 - t^2$. Since $C$ lies in the first octant, it can be parametrized by

$$x = t, \qquad y = \sqrt{1 - t^2}, \qquad z = t^2, \qquad (0 \le t \le 1).$$

Then $dx/dt = 1, dy/dt = -t/\sqrt{1 - t^2}$, and $dz/dt = 2t$ so

$$ds = \sqrt{1 + \frac{t^2}{1 - t^2} + 4t^2}\, dt = \frac{\sqrt{1 + 4t^2 - 4t^4}}{\sqrt{1 - t^2}}\, dt.$$

Hence the mass of the wire is

$$
\begin{aligned}
m &= \int_C xy\, ds = \int_0^1 t\sqrt{1 - t^2}\,\frac{\sqrt{1 + 4t^2 - 4t^4}}{\sqrt{1 - t^2}}\, dt \\
&= \int_0^1 t\sqrt{1 + 4t^2 - 4t^4}\, dt && \text{Let } u = t^2. \\
&= \frac{1}{2}\int_0^1 \sqrt{1 + 4u - 4u^2}\, du \\
&= \frac{1}{2}\int_0^1 \sqrt{2 - (2u - 1)^2} && \text{Let } v = 2u - 1. \\
&= \frac{1}{4}\int_{-1}^1 \sqrt{2 - v^2}\, dv = \frac{1}{2}\int_0^1 \sqrt{2 - v^2}\, dv \\
&= \frac{1}{2}\left(\frac{\pi}{4} + \frac{1}{2}\right) = \frac{\pi + 2}{8}.
\end{aligned}
$$

(The final integral above was evaluated by interpreting it as the area of part of a circle. You are invited to supply the details. It can also be done by the substitution $v = \sqrt{2}\sin w$.) ∎

## Exercises 15.3

**1.** Show that the curve $C$ given by

$$\mathbf{r} = a\cos t\sin t\,\mathbf{i} + a\sin^2 t\,\mathbf{j} + a\cos t\,\mathbf{k}, \quad (0 \le t \le \tfrac{\pi}{2}),$$

lies on a sphere centred at the origin. Find $\displaystyle\int_C z\,ds$.

**2.** Let $C$ be the conical helix with parametric equations

$x = t\cos t$, $y = t\sin t$, $z = t$, $(0 \le t \le 2\pi)$. Find $\displaystyle\int_C z\,ds$.

**3.** Find the mass of a wire along the curve

$$\mathbf{r} = 3t\mathbf{i} + 3t^2\mathbf{j} + 2t^3\mathbf{k}, \quad (0 \le t \le 1),$$

if the density at $\mathbf{r}(t)$ is $1 + t$ gm/unit length.

**4.** Show that the curve $C$ in Example 3 also has parametrization $x = \cos t$, $y = \sin t$, $z = \cos^2 t$, $(0 \le t \le \pi/2)$, and recalculate the mass of the wire in that example using this parametrization.

**5.** Find the moment of inertia about the $z$-axis (that is, the value of $\displaystyle\int_C (x^2 + y^2)\,ds$), for a wire of constant density $\delta$ lying along the curve $C$: $\mathbf{r} = e^t\cos t\mathbf{i} + e^t\sin t\mathbf{j} + t\mathbf{k}$, from $t = 0$ to $t = 2\pi$.

**6.** Evaluate $\displaystyle\int_C e^z\,ds$, where $C$ is the curve in Exercise 5.

**7.** Find $\displaystyle\int_C x^2\,ds$ along the line of intersection of the two planes $x - y + z = 0$, and $x + y + 2z = 0$, from the origin to the point $(3, 1, -2)$.

**8.** Find $\displaystyle\int_C \sqrt{1 + 4x^2z^2}\,ds$, where $C$ is the curve of intersection of the surfaces $x^2 + z^2 = 1$ and $y = x^2$.

**9.** Find the mass and centre of mass of a wire bent in the shape of the circular helix $x = \cos t$, $y = \sin t$, $z = t$, $(0 \le t \le 2\pi)$ if the wire has line density given by $\delta(x, y, z) = z$.

**10.** Repeat Exercise 9 for the part of the wire corresponding to $0 \le t \le \pi$.

**11.** Find the moment of inertia about the $y$-axis, that is,

$$\int_C (x^2 + z^2)\,ds,$$

of the curve $x = e^t$, $y = \sqrt{2}\,t$, $z = e^{-t}$, $(0 \le t \le 1)$.

**12.** Find the centroid of the curve in Exercise 11.

**\* 13.** Find $\displaystyle\int_C x\,ds$ along the first octant part of the curve of intersection of the cylinder $x^2 + y^2 = a^2$ and the plane $z = x$.

**\* 14.** Find $\displaystyle\int_C z\,ds$ along the part of the curve $x^2 + y^2 + z^2 = 1$, $x + y = 1$, where $z \ge 0$.

**\* 15.** Find $\displaystyle\int_C \frac{ds}{(2y^2 + 1)^{3/2}}$, where $C$ is the parabola $z^2 = x^2 + y^2$, $x + z = 1$. *Hint:* use $y = t$ as parameter.

**16.** Express as a definite integral, but do not try to evaluate, the value of $\displaystyle\int_C xyz\,ds$, where $C$ is the curve $y = x^2$, $z = y^2$ from $(0, 0, 0)$ to $(2, 4, 16)$.

**\* 17.** The function

$$E(k, \phi) = \int_0^\phi \sqrt{1 - k^2\sin^2 t}\,dt$$

is called the **elliptic integral function of the second kind**. The **complete elliptic integral** of the second kind is the function $E(k) = E(k, \pi/2)$. In terms of these functions, express the length of one complete revolution of the elliptic helix

$$x = a\cos t, \qquad y = b\sin t, \qquad z = ct,$$

where $0 < a < b$. What is the length of that part of the helix lying between $t = 0$ and $t = T$ where $0 < T < \pi/2$?

**\* 18.** Evaluate $\displaystyle\int_\mathcal{L} \frac{ds}{x^2 + y^2}$, where $\mathcal{L}$ is the entire straight line with equation $Ax + By = C$, $(C \ne 0)$. *Hint:* use the symmetry of the integrand to replace the line with a line having a simpler equation but giving the same value to the integral.

# 15.4  Line Integrals of Vector Fields

In elementary physics the **work** done by a constant force of magnitude $F$ in moving an object a distance $d$ is defined to be the product of $F$ and $d$: $W = Fd$. There is, however, a catch to this; it is understood that the force is exerted in the direction of motion of the object. If the object moves in a direction different from that of the force (because of some other forces acting on it) then the work done by the particular force is the product of the distance moved and the component of the force in the direction of motion. For instance, the work done by gravity in causing a 10 kg crate to slide 5 m down a ramp inclined at $45°$ to the horizontal is $W = 50g/\sqrt{2}$ N·m, (where $g = 9.8$ m/s$^2$), since the scalar projection of the $10g$ N gravitational force on the crate in the direction of the ramp is $10g/\sqrt{2}$ N.

The work done by a *variable* force $\mathbf{F}(x, y, z) = \mathbf{F}(\mathbf{r})$, that depends continuously on position, in moving an object along a smooth curve $\mathcal{C}$ is the integral of *work elements* $dW$. The element $dW$ corresponding to arc length element $ds$ at position $\mathbf{r}$ on $\mathcal{C}$ is $ds$ times the tangential component of the force $\mathbf{F}(\mathbf{r})$ along $\mathcal{C}$ in the direction of motion:

$$dW = \mathbf{F}(\mathbf{r}) \bullet \hat{\mathbf{T}} \, ds.$$

Thus the total work done by $\mathbf{F}$ in moving the object along $\mathcal{C}$ is

$$W = \int_{\mathcal{C}} \mathbf{F} \bullet \hat{\mathbf{T}} \, ds = \int_{\mathcal{C}} \mathbf{F} \bullet d\mathbf{r},$$

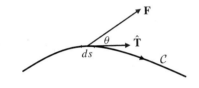

**Figure 15.8**    $dW = \mathbf{F} \cos\theta \, ds$
$\qquad\qquad = \mathbf{F} \bullet \hat{\mathbf{T}} \, ds$

where we use the vector differential $d\mathbf{r}$ as a convenient shorthand for $\hat{\mathbf{T}} \, ds$. Since

$$\hat{\mathbf{T}} = \frac{d\mathbf{r}}{ds} = \frac{dx}{ds}\mathbf{i} + \frac{dy}{ds}\mathbf{j} + \frac{dz}{ds}\mathbf{k}$$

we have

$$d\mathbf{r} = \hat{\mathbf{T}} \, ds = dx\,\mathbf{i} + dy\,\mathbf{j} + dz\,\mathbf{k},$$

and the work $W$ can be written in terms of the components of $\mathbf{F}$ as

$$W = \int_{\mathcal{C}} F_1 \, dx + F_2 \, dy + F_3 \, dz.$$

In general, if $\mathbf{F} = F_1\mathbf{i} + F_2\mathbf{j} + F_3\mathbf{k}$ is a continuous vector field, and $\mathcal{C}$ is a smooth curve, then the **line integral of the tangential component of F along an oriented curve $\mathcal{C}$** is

$$\int_{\mathcal{C}} \mathbf{F} \bullet d\mathbf{r} = \int_{\mathcal{C}} \mathbf{F} \bullet \hat{\mathbf{T}} \, ds$$
$$= \int_{\mathcal{C}} F_1(x, y, z) \, dx + F_2(x, y, z) \, dy + F_3(x, y, z) \, dz.$$

Such a line integral is sometimes called, somewhat improperly, the line integral of $\mathbf{F}$ along $\mathcal{C}$. (It is not the line integral of $\mathbf{F}$, which should have a vector value, but rather the line integral of the *tangential component* of $\mathbf{F}$, which has a scalar value.) Unlike the line integral considered in the previous section, this line integral depends on the direction of the orientation of $\mathcal{C}$; reversing the direction of $\mathcal{C}$ causes this line integral to change sign.

If $C$ is a closed curve, the line integral of the tangential component of $\mathbf{F}$ around $C$ is also called the *circulation* of $\mathbf{F}$ around $C$. The fact that the curve is closed is often indicated by a small circle drawn on the integral sign;

$$\oint_C \mathbf{F} \bullet d\mathbf{r}$$

denotes the circulation of $\mathbf{F}$ around the closed curve $C$.

Like the line integrals studied in the previous section, a line integral of a continuous vector field is converted into an ordinary definite integral by using a parametrization of the path of integration. For a smooth arc

$$\mathbf{r} = \mathbf{r}(t) = x(t)\mathbf{i} + y(t)\mathbf{j} + z(t)\mathbf{k}, \qquad (a \le t \le b),$$

we have

$$\int_C \mathbf{F} \bullet d\mathbf{r} = \int_a^b \mathbf{F} \bullet \frac{d\mathbf{r}}{dt}\, dt$$

$$= \int_a^b \left[ F_1\big(x(t), y(t), z(t)\big)\frac{dx}{dt} + F_2\big(x(t), y(t), z(t)\big)\frac{dy}{dt} \right.$$

$$\left. + F_3\big(x(t), y(t), z(t)\big)\frac{dz}{dt} \right] dt.$$

Although this type of line integral changes sign if the orientation of $C$ is reversed, it is otherwise independent of the particular parametrization used for $C$. Again, a line integral over a piecewise smooth path is the sum of the line integrals over the individual smooth arcs constituting that path.

**Example 1**    Let $\mathbf{F}(x, y) = y^2\mathbf{i} + 2xy\mathbf{j}$. Evaluate the line integral

$$\int_C \mathbf{F} \bullet d\mathbf{r}$$

from $(0, 0)$ to $(1, 1)$ along

(a) the straight line $y = x$,

(b) the curve $y = x^2$, and

(c) the piecewise smooth path consisting of the straight line segments from $(0, 0)$ to $(0, 1)$ and from $(0, 1)$ to $(1, 1)$.

**Solution**    The three paths are shown in Figure 15.9. The straight path (a) can be parametrized $\mathbf{r} = t\mathbf{i} + t\mathbf{j}, 0 \le t \le 1$. Thus $d\mathbf{r} = dt\mathbf{i} + dt\mathbf{j}$ and

$$\mathbf{F} \bullet d\mathbf{r} = (t^2\mathbf{i} + 2t^2\mathbf{j}) \bullet (\mathbf{i} + \mathbf{j})dt = 3t^2\, dt.$$

Therefore,

$$\int_C \mathbf{F} \bullet d\mathbf{r} = \int_0^1 3t^2\, dt = t^3 \Big|_0^1 = 1.$$

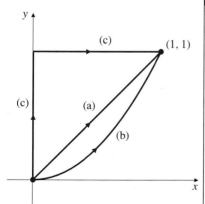

**Figure 15.9**    Three paths from $(0, 0)$ to $(1, 1)$

The parabolic path (b) can be parametrized $\mathbf{r} = t\mathbf{i} + t^2\mathbf{j}$, $0 \le t \le 1$, so that $d\mathbf{r} = dt\mathbf{i} + 2t\,dt\mathbf{j}$. Thus

$$\mathbf{F} \bullet d\mathbf{r} = (t^4\mathbf{i} + 2t^3\mathbf{j}) \bullet (\mathbf{i} + 2t\mathbf{j})\,dt = 5t^4\,dt,$$

and

$$\int_C \mathbf{F} \bullet d\mathbf{r} = \int_0^1 5t^4\,dt = t^5\Big|_0^1 = 1.$$

The third path (c) is made up of two segments, and we parametrize each separately. Let us use $y$ as parameter on the vertical segment (where $x = 0$ and $dx = 0$), and $x$ as parameter on the horizontal segment (where $y = 1$ and $dy = 0$):

$$\int_C \mathbf{F} \bullet d\mathbf{r} = \int_C y^2\,dx + 2xy\,dy$$
$$= \int_0^1 (1)\,dx + \int_0^1 (0)\,dy = 1.$$

In view of these results, we might ask whether $\int_C \mathbf{F} \bullet d\mathbf{r}$ is the same along *every* path from $(0, 0)$ to $(1, 1)$.

---

**Example 2**   Let $\mathbf{F} = y\mathbf{i} - x\mathbf{j}$. Find $\int_C \mathbf{F} \bullet d\mathbf{r}$ from $(1, 0)$ to $(0, -1)$ along
(a) the straight line segment joining these points, and
(b) three-quarters of the circle of unit radius centred at the origin and traversed counterclockwise.

**Solution**   Both paths are shown in Figure 15.10. The straight path (a) can be parametrized

$$\mathbf{r} = (1 - t)\mathbf{i} - t\mathbf{j}, \qquad 0 \le t \le 1.$$

Thus $d\mathbf{r} = -dt\mathbf{i} - dt\mathbf{j}$, and

$$\int_C \mathbf{F} \bullet d\mathbf{r} = \int_0^1 \left((-t)(-dt) - (1 - t)(-dt)\right) = \int_0^1 dt = 1.$$

The circular path (b) can be parametrized

$$\mathbf{r} = \cos t\,\mathbf{i} + \sin t\,\mathbf{j}, \qquad 0 \le t \le \frac{3\pi}{2},$$

so that $d\mathbf{r} = -\sin t\,dt\mathbf{i} + \cos t\,dt\mathbf{j}$. Therefore

$$\mathbf{F} \bullet d\mathbf{r} = -\sin^2 t\,dt - \cos^2 t\,dt = -dt,$$

and we have

$$\int_C \mathbf{F} \bullet d\mathbf{r} = -\int_0^{3\pi/2} dt = -\frac{3\pi}{2}.$$

In this case the line integral depends on the path to from $(1, 0)$ to $(0, -1)$ along which the integral is taken.

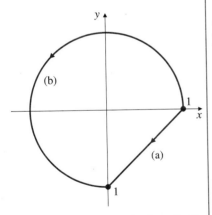

**Figure 15.10**   Two paths from $(1, 0)$ to $(0, -1)$

Some readers may have noticed that in Example 1 above the vector field **F** is conservative, while in Example 2 it is not. Theorem 1 below confirms the link between *independence of path* for a line integral of the tangential component of a vector field and the existence of a scalar potential function for that field. This and subsequent theorems require specific assumptions on the nature of the domain of the vector field **F**, so we need to formulate some definitions.

## Connected and Simply Connected Domains

Recall that a set $S$ in the plane (or in 3-space) is open if every point in $S$ is the centre of a disk (or a ball) having positive radius and contained in $S$. If $S$ is open and $B$ is a set (possibly empty) of boundary points of $S$, then the set $D = S \cup B$ is called a **domain**. A domain cannot contain isolated points. It may be closed, but it must have interior points near any of its boundary points. (See the early part of Section 10.1 for a discussion of open and closed sets, and interior and boundary points.)

**DEFINITION 2**

A domain $D$ in the plane (or in 3-space) is said to be **connected**, if for every pair of points $P$ and $Q$ in $D$, there exists a piecewise smooth curve in $D$ from $P$ to $Q$.

For instance, the set of points $(x, y)$ in the plane satisfying $x > 0$, $y > 0$, and $x^2 + y^2 \le 4$ is a connected domain, but the set of points satisfying $|x| > 1$ is not connected. (There is no path from $(-2, 0)$ to $(2, 0)$ lying entirely in $|x| > 1$.) The set of points $(x, y, z)$ in 3-space satisfying $0 < z < 1/(x^2 + y^2)$ is a connected domain but the set satisfying $z \ne 0$ is not.

**DEFINITION 3**

A connected domain $D$ is said to be **simply connected** if every *closed curve* in $D$ can be continuously shrunk to a point in $D$ without any part ever passing out of $D$.

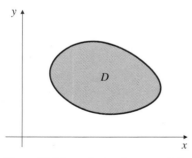

**Figure 15.11**    A simply connected domain

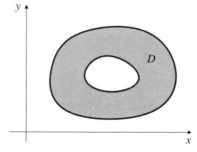

**Figure 15.12**    A connected domain that is not simply connected

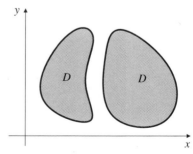

**Figure 15.13**    A domain that is not connected

Figure 15.11 shows a simply connected domain in the plane. Figure 15.12 shows a connected but not simply connected domain. (A closed curve surrounding the hole cannot be shrunk to a point without passing out of $D$.) The domain in Figure 15.13 is not even connected. It has two *components*, and if $P$ and $Q$ are points in different components they cannot be joined by a curve that lies in $D$.

In the plane, a simply connected domain $D$ can have no holes, not even a hole consisting of a single point. The interior of every non-self-intersecting closed curve in such a domain $D$ lies in $D$. For instance, the domain of the function $1/(x^2 + y^2)$

is not simply connected because the origin does not belong to it. (The origin is a "hole" in that domain.) In 3-space simple connectedness is a little more tricky to characterize. The set of points $(x, y, z)$ satisfying $1 < x^2 + y^2 + z^2 < 4$ (the region between two concentric spheres) is simply connected because any closed curve in that set can be shrunk to a point without having to leave that set. However, the interior of a doughnut (a torus) is not simply connected; a circle in the doughnut surrounding the doughnut hole cannot be shrunk to a point without part of it passing outside the doughnut. The set of all points in $\mathbb{R}^3$ different from the origin is simply connected, but the set of all points not lying on the $z$-axis is not. In general, each of the following conditions characterizes simply connected domains $D$:

(i)   Any closed curve in $D$ is the boundary of a "surface" lying wholly in $D$.

(ii)  If $\mathcal{C}_1$ and $\mathcal{C}_2$ are two curves in $D$ having the same endpoints, then $\mathcal{C}_1$ can be continuously deformed into $\mathcal{C}_2$, remaining in $D$ throughout the deformation process.

## Independence of Path

**THEOREM 1**

**Independence of path**

Let $D$ be an open, connected domain, and let $\mathbf{F}$ be a smooth vector field defined on $D$. Then the following three statements are *equivalent* in the sense that if any one of them is true, then so are the other two:

(a) $\mathbf{F}$ is conservative in $D$.

(b) $\displaystyle\oint_{\mathcal{C}} \mathbf{F} \bullet d\mathbf{r} = 0$ for every piecewise smooth, closed curve $\mathcal{C}$ in $D$.

(c) Given any two points $P_0$ and $P_1$ in $D$, $\displaystyle\int_{\mathcal{C}} \mathbf{F} \bullet d\mathbf{r}$ has the same value for all piecewise smooth curves in $D$ starting at $P_0$ and ending at $P_1$.

**PROOF**   We will show that (a) implies (b), that (b) implies (c), and that (c) implies (a). It then follows that any one implies the other two.

Suppose (a) is true. Then $\mathbf{F} = \nabla\phi$ for some scalar potential function $\phi$ defined in $D$. Therefore,

$$\mathbf{F} \bullet d\mathbf{r} = \left(\frac{\partial\phi}{\partial x}\mathbf{i} + \frac{\partial\phi}{\partial y}\mathbf{j} + \frac{\partial\phi}{\partial z}\mathbf{k}\right) \bullet \left(dx\,\mathbf{i} + dy\,\mathbf{j} + dz\,\mathbf{k}\right)$$

$$= \frac{\partial\phi}{\partial x}\,dx + \frac{\partial\phi}{\partial y}\,dy + \frac{\partial\phi}{\partial y}\,dz = d\phi.$$

If $\mathcal{C}$ is any piecewise smooth, closed curve, parametrized, say, by $\mathbf{r} = \mathbf{r}(t)$, $(a \le t \le b)$, then $\mathbf{r}(a) = \mathbf{r}(b)$, and

$$\int_{\mathcal{C}} \mathbf{F} \bullet d\mathbf{r} = \int_a^b \frac{d\phi(\mathbf{r}(t))}{dt}\,dt = \phi(\mathbf{r}(b)) - \phi(\mathbf{r}(a)) = 0.$$

Thus (a) implies (b).

**Figure 15.14**
$\mathcal{C}_1 - \mathcal{C}_2 = \mathcal{C}_1 + (-\mathcal{C}_2)$ is a closed curve

Now suppose (b) is true. Let $P_0$ and $P_1$ be two points in $D$, and let $\mathcal{C}_1$ and $\mathcal{C}_2$ be two piecewise smooth curves in $D$ from $P_0$ to $P_1$. Let $\mathcal{C} = \mathcal{C}_1 - \mathcal{C}_2$ denote the closed curve going from $P_0$ to $P_1$ along $\mathcal{C}_1$ and then back to $P_0$ along $\mathcal{C}_2$ in the opposite direction. (See Figure 15.14.) Since we are assuming that (b) is true, we have

$$0 = \oint_{\mathcal{C}} \mathbf{F} \bullet d\mathbf{r} = \int_{\mathcal{C}_1} \mathbf{F} \bullet d\mathbf{r} - \int_{\mathcal{C}_2} \mathbf{F} \bullet d\mathbf{r}.$$

Therefore,

$$\int_{C_1} \mathbf{F} \cdot d\mathbf{r} = \int_{C_2} \mathbf{F} \cdot d\mathbf{r},$$

and we have proved that (b) implies (c).

Finally, suppose that (c) is true. Let $P_0 = (x_0, y_0, z_0)$ be a fixed point in the domain $D$, and let $P = (x, y, z)$ be an arbitrary point in that domain. Define a function $\phi$ by

$$\phi(x, y, z) = \int_C \mathbf{F} \cdot d\mathbf{r},$$

where $C$ is some piecewise smooth curve in $D$ from $P_0$ to $P$. (Under the hypotheses of the theorem such a curve exists, and, since we are assuming (c), the integral has the same value for all such curves. Therefore, $\phi$ is well defined in $D$.) We will show that $\nabla\phi = \mathbf{F}$ and thus establish that $\mathbf{F}$ is conservative and has potential $\phi$.

It is sufficient to show that $\partial\phi/\partial x = F_1(x, y, z)$; the other two components are treated similarly. Since $D$ is open there is a ball of positive radius centred at $P$ and contained in $D$. Pick a point $(x_1, y, z)$ in this ball having $x_1 < x$. Note that the line from this point to $P$ is parallel to the $x$-axis. Since we are free to choose the curve $C$ in the integral defining $\phi$, let us choose it to consist of two segments, $C_1$ which is piecewise smooth and goes from $(x_0, y_0, z_0)$ to $(x_1, y, z)$, and $C_2$, a straight line segment from $(x_1, y, z)$ to $(x, y, z)$. (See Figure 15.15.) Then

$$\phi(x, y, z) = \int_{C_1} \mathbf{F} \cdot d\mathbf{r} + \int_{C_2} \mathbf{F} \cdot d\mathbf{r}.$$

The first integral does not depend on $x$, so its derivative with respect to $x$ is zero. The straight line path for the second integral is parametrized by $\mathbf{r} = t\mathbf{i} + y\mathbf{j} + z\mathbf{k}$, where $x_1 \le t \le x$ so $d\mathbf{r} = dt\mathbf{i}$ and

$$\frac{\partial\phi}{\partial x} = \frac{\partial}{\partial x} \int_{C_2} \mathbf{F} \cdot d\mathbf{r} = \frac{\partial}{\partial x} \int_{x_1}^{x} F_1(t, y, z)\, dt = F_1(x, y, z),$$

which is what we wanted. Thus $\mathbf{F} = \nabla\phi$ is conservative, and (c) implies (a).

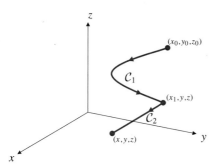

**Figure 15.15**    A special path from $P_0$ to $P_1$

**Remark**    It is very easy to evaluate the line integral of the tangential component of a *conservative* vector field along a curve $C$, when you know a potential for $\mathbf{F}$. If $\mathbf{F} = \nabla\phi$, and $C$ goes from $P_0$ to $P_1$, then

$$\int_C \mathbf{F} \cdot d\mathbf{r} = \int_C d\phi = \phi(P_1) - \phi(P_0).$$

As noted above, the value of the integral depends only on the endpoints of $C$.

**Remark**    In the next chapter we will add another item to the list of three conditions shown to be equivalent in Theorem 1, provided that the domain $D$ is *simply connected*. For such a domain each of the above three conditions in the theorem is equivalent to

$$\frac{\partial F_1}{\partial y} = \frac{\partial F_2}{\partial x}, \qquad \frac{\partial F_1}{\partial z} = \frac{\partial F_3}{\partial x} \qquad \text{and} \qquad \frac{\partial F_2}{\partial z} = \frac{\partial F_3}{\partial y}.$$

We already know that these equations are satisfied on a domain where $\mathbf{F}$ is conservative. Theorem 4 of Section 16.2 states that if these three equations hold on a simply connected domain, then $\mathbf{F}$ is conservative on that domain.

**Example 3** For what values of the constants $A$ and $B$ is the vector field

$$\mathbf{F} = Ax \sin(\pi y)\mathbf{i} + \left(x^2 \cos(\pi y) + Bye^{-z}\right)\mathbf{j} + y^2 e^{-z}\mathbf{k}$$

conservative? For this choice of $A$ and $B$, evaluate $\int_C \mathbf{F} \bullet d\mathbf{r}$, where $C$ is

(a) the curve $\mathbf{r} = \cos t\mathbf{i} + \sin 2t\mathbf{j} + \sin^2 t\mathbf{k}$, $(0 \leq t \leq 2\pi)$, and
(b) the curve of intersection of the paraboloid $z = x^2 + 4y^2$ and the plane $z = 3x - 2y$ from $(0, 0, 0)$ to $(1, 1/2, 2)$.

**Solution** $\mathbf{F}$ cannot be conservative unless

$$\frac{\partial F_1}{\partial y} = \frac{\partial F_2}{\partial x}, \qquad \frac{\partial F_1}{\partial z} = \frac{\partial F_3}{\partial x}, \qquad \text{and} \qquad \frac{\partial F_2}{\partial z} = \frac{\partial F_3}{\partial y},$$

that is, unless

$$A\pi x \cos(\pi y) = 2x \cos(\pi y), \qquad 0 = 0, \qquad \text{and} \quad -Bye^{-z} = 2ye^{-z}.$$

Thus we require that $A = 2/\pi$ and $B = -2$. In this case, it is easily checked that

$$\mathbf{F} = \nabla\phi, \qquad \text{where} \quad \phi = \left(\frac{x^2 \sin(\pi y)}{\pi} - y^2 e^{-z}\right).$$

For the curve (a) we have $\mathbf{r}(0) = \mathbf{i} = \mathbf{r}(2\pi)$, so this curve is a closed curve, and

$$\int_C \mathbf{F} \bullet d\mathbf{r} = \oint_C \nabla\phi \bullet d\mathbf{r} = 0.$$

Since the curve (b) starts at $(0, 0, 0)$ and ends at $(1, 1/2, 2)$, we have

$$\int_C \mathbf{F} \bullet d\mathbf{r} = \left(\frac{x^2 \sin(\pi y)}{\pi} - y^2 e^{-z}\right)\Bigg|_{(0,0,0)}^{(1,1/2,2)} = \frac{1}{\pi} - \frac{1}{4e^2}.$$

The following example shows how to exploit the fact that

$$\int_C \mathbf{F} \bullet d\mathbf{r}$$

is easily evaluated for conservative $\mathbf{F}$ even if the $\mathbf{F}$ we want to integrate isn't quite conservative.

**Example 4** Evaluate $I = \oint_C (e^x \sin y + 3y)dx + (e^x \cos y + 2x - 2y)dy$ counterclockwise around the ellipse $4x^2 + y^2 = 4$.

**Solution** $I = \oint_C \mathbf{F} \bullet d\mathbf{r}$, where $\mathbf{F}$ is the vector field

$$\mathbf{F} = \left(e^x \sin y + 3y\right)\mathbf{i} + \left(e^x \cos y + 2x - 2y\right)\mathbf{j}.$$

This vector field is not conservative, but it would be if the $3y$ term in $F_1$ were $2y$ instead; specifically, if

$$\phi(x, y) = e^x \sin y + 2xy - y^2,$$

then $\mathbf{F} = \nabla\phi + y\mathbf{i}$, the sum of a conservative part and a non-conservative part. Therefore, we have

$$I = \oint_C \nabla\phi \bullet d\mathbf{r} + \oint_C y \, dx.$$

The first integral is zero since $\nabla\phi$ is conservative and $C$ is closed. For the second integral we parametrize $C$ by $x = \cos t$, $y = 2 \sin t$, $(0 \le t \le 2\pi)$, and obtain

$$I = \oint_C y \, dx = -2 \int_0^{2\pi} \sin^2 t \, dt = -2 \int_0^{2\pi} \frac{1 - \cos(2t)}{2} \, dt = -2\pi.$$

## Exercises 15.4

In Exercises 1–6, evaluate the line integral of the tangential component of the given vector field along the given curve.

1. $\mathbf{F}(x, y) = xy\mathbf{i} - x^2\mathbf{j}$ along $y = x^2$ from $(0, 0)$ to $(1, 1)$

2. $\mathbf{F}(x, y) = \cos x \, \mathbf{i} - yj$ along $y = \sin x$ from $(0, 0)$ to $(\pi, 0)$

3. $\mathbf{F}(x, y, z) = y\mathbf{i} + zj - xk$ along the straight line from $(0, 0, 0)$ to $(1, 1, 1)$

4. $\mathbf{F}(x, y, z) = z\mathbf{i} - yj + 2xk$ along the curve $x = t$, $y = t^2$, $z = t^3$ from $(0, 0, 0)$ to $(1, 1, 1)$

5. $\mathbf{F}(x, y, z) = yz\mathbf{i} + xzj + xyk$ from $(-1, 0, 0)$ to $(1, 0, 0)$ along either direction of the curve of intersection of the cylinder $x^2 + y^2 = 1$ and the plane $z = y$

6. $\mathbf{F}(x, y, z) = (x - z)\mathbf{i} + (y - z)j - (x + y)k$ along the polygonal path from $(0, 0, 0)$ to $(1, 0, 0)$ to $(1, 1, 0)$ to $(1, 1, 1)$

7. Find the work done by the force field

$$\mathbf{F} = (x + y)\mathbf{i} + (x - z)j + (z - y)k$$

in moving an object from $(1, 0, -1)$ to $(0, -2, 3)$ along any smooth curve.

8. Evaluate $\oint_C x^2 y^2 \, dx + x^3 y \, dy$ counterclockwise around the square with vertices $(0, 0)$, $(1, 0)$, $(1, 1)$ and $(0, 1)$.

9. Evaluate

$$\int_C e^{x+y} \sin(y + z) \, dx + e^{x+y} \Big( \sin(y + z) + \cos(y + z) \Big) \, dy$$
$$+ e^{x+y} \cos(y + z) \, dz$$

along the straight line segment from $(0,0,0)$ to $(1, \frac{\pi}{4}, \frac{\pi}{4})$.

10. The field $\mathbf{F} = (axy + z)\mathbf{i} + x^2 j + (bx + 2z)k$ is conservative. Find $a$ and $b$, and find a potential for $\mathbf{F}$. Also, evaluate $\int_C \mathbf{F} \bullet d\mathbf{r}$, where $C$ is the curve from $(1, 1, 0)$ to $(0, 0, 3)$ that lies on the intersection of the surfaces $2x + y + z = 3$ and $9x^2 + 9y^2 + 2z^2 = 18$ in the octant $x \ge 0$, $y \ge 0$, $z \ge 0$.

11. Determine the values of $A$ and $B$ for which the vector field

$$\mathbf{F} = Ax \ln z \, \mathbf{i} + By^2 z \, j + \left( \frac{x^2}{z} + y^3 \right) k$$

is conservative. If $C$ is the straight line from $(1, 1, 1)$ to $(2, 1, 2)$, find

$$\int_C 2x \ln z \, dx + 2y^2 z \, dy + y^3 \, dz.$$

12. Find the work done by the force field

$$\mathbf{F} = (y^2 \cos x + z^3)\mathbf{i} + (2y \sin x - 4)j + (3xz^2 + 2)k$$

in moving a particle along the curve $x = \sin^{-1} t$, $y = 1 - 2t$, $z = 3t - 1$, $(0 \le t \le 1)$.

13. If $C$ is the intersection of $z = \ln(1 + x)$ and $y = x$ from $(0, 0, 0)$ to $(1, 1, \ln 2)$, evaluate

$$\int_C (2x \sin(\pi y) - e^z) \, dx + (\pi x^2 \cos(\pi y) - 3e^z) \, dy - xe^z \, dz.$$

14. Is each of the following sets a domain? a connected domain? a simply connected domain?

(a) the set of points $(x, y)$ in the plane such that $x > 0$ and $y \geq 0$

(b) the set of points $(x, y)$ in the plane such that $x = 0$ and $y \geq 0$

(c) the set of points $(x, y)$ in the plane such that $x \neq 0$ and $y > 0$

(d) the set of points $(x, y, z)$ in 3-space such that $x^2 > 1$

(e) the set of points $(x, y, z)$ in 3-space such that $x^2 + y^2 > 1$

(f) the set of points $(x, y, z)$ in 3-space such that $x^2 + y^2 + z^2 > 1$

In Exercises 15–19, evaluate the closed line integrals

(a) $\displaystyle\oint_C x\,dy$, (b) $\displaystyle\oint_C y\,dx$

around the given curves, all oriented counterclockwise.

**15.** The circle $x^2 + y^2 = a^2$

**16.** The ellipse $\dfrac{x^2}{a^2} + \dfrac{y^2}{b^2} = 1$

**17.** The boundary of the half disk $x^2 + y^2 \leq a^2$, $y \geq 0$

**18.** The boundary of the square with vertices $(0, 0)$, $(1, 0)$, $(1, 1)$, and $(0, 1)$

**19.** The triangle with vertices $(0, 0)$, $(a, 0)$ and $(0, b)$

**20.** On the basis of your results for Exercises 15–19, guess the values of the closed line integrals

(a) $\displaystyle\oint_C x\,dy$, (b) $\displaystyle\oint_C y\,dx$

for any non-self-intersecting closed curve in the $xy$-plane. Prove your guess in the case that $C$ bounds a region of the plane which is both $x$-simple and $y$-simple. (See Section 14.1.)

**21.** If $f$ and $g$ are scalar fields with continuous first partial derivatives in a connected domain $D$, show that

$$\int_C f\nabla g \bullet d\mathbf{r} + \int_C g\nabla f \bullet d\mathbf{r} = f(Q)g(Q) - f(P)g(P)$$

for any piecewise smooth curve in $D$ from $P$ to $Q$.

**22.** Evaluate

$$\frac{1}{2\pi} \oint_C \frac{-y\,dx + x\,dy}{x^2 + y^2}$$

(a) counterclockwise around the circle $x^2 + y^2 = a^2$,

(b) clockwise around the square with vertices $(-1, -1)$, $(-1, 1)$, $(1, 1)$, and $(1, -1)$,

(c) counterclockwise around the boundary of the region $1 \leq x^2 + y^2 \leq 2$, $y \geq 0$.

**23.** Review Example 5 in Section 15.2, in which it was shown that

$$\frac{\partial}{\partial y}\left(\frac{-y}{x^2 + y^2}\right) = \frac{\partial}{\partial x}\left(\frac{x}{x^2 + y^2}\right),$$

for all $(x, y) \neq (0, 0)$. Why does this result, together with that of Exercise 22, not contradict the final assertion in the remark following Theorem 1?

\* **24.** Let $C$ be a piecewise smooth curve in the $xy$-plane which does not pass through the origin. Let $\theta = \theta(x, y)$ be the polar angle coordinate of the point $P = (x, y)$ on $C$, not restricted to an interval of length $2\pi$, but varying continuously as $P$ moves from one end of $C$ to the other. As in Example 5 of Section 15.2, it happens that

$$\nabla\theta = -\frac{y}{x^2 + y^2}\mathbf{i} + \frac{x}{x^2 + y^2}\mathbf{j}.$$

If, in addition, $C$ is a closed curve, show that

$$w(C) = \frac{1}{2\pi} \oint_C \frac{x\,dy - y\,dx}{x^2 + y^2}$$

has an integer value. $w$ is called the **winding number** of $C$ about the origin.

# 15.5 Surfaces and Surface Integrals

This section and the next are devoted to integrals of functions defined over surfaces in 3-space. Before we can begin, it is necessary to make more precise just what is meant by the term "surface." Until now we have been treating surfaces in an intuitive way, either as the graphs of functions $f(x, y)$, or as the graphs of equations $f(x, y, z) = 0$.

A smooth curve is considered to be a *one-dimensional* object because points on it can be located by giving *one coordinate* (for instance, the distance from an endpoint). Therefore the curve can be defined as the range of a vector-valued function of a single real variable. Similarly, a surface is a *two-dimensional* object; points on it can be located by using *two coordinates*, and the surface can be defined as the range of a vector-valued function of two real variables. We will call certain such functions parametric surfaces.

## Parametric Surfaces

**DEFINITION 4**

A **parametric surface** in 3-space is a continuous function $\mathbf{r}$ defined on some rectangle $R$ given by $a \le u \le b$, $c \le v \le d$ in the $uv$-plane, and having values in 3-space:

$$\mathbf{r}(u, v) = x(u, v)\mathbf{i} + y(u, v)\mathbf{j} + z(u, v)\mathbf{k}, \qquad (u, v) \text{ in } R.$$

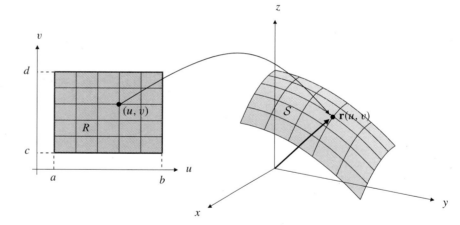

**Figure 15.16**    A parametric surface $S$ defined on parameter region $R$. The *contour curves* on $S$ correspond to the rulings of $R$

Actually, we think of the *range* of the function $\mathbf{r}(u, v)$ as being the parametric surface. It is a set $S$ of points $(x, y, z)$ in 3-space whose position vectors are the vectors $\mathbf{r}(u, v)$ for $(u, v)$ in $R$. (See Figure 15.16.) If $\mathbf{r}$ is one-to-one, then the surface does not intersect itself. In this case $\mathbf{r}$ maps the boundary of the rectangle $R$ (the four edges) onto a curve in 3-space, which we call the **boundary of the parametric surface**. The requirement that $R$ be a rectangle is made only to simplify the discussion. Any connected, closed, bounded set in the $uv$-plane, having well-defined area, and consisting of an open set together with its boundary points would do as well. Thus, we will from time to time consider parametric surfaces over closed disks, triangles, or other such domains in the $uv$-plane. Being the range of a continuous function defined on a closed, bounded set, a parametric surface is always bounded in 3-space.

**Example 1**    The graph of $z = f(x, y)$, where $f$ has the rectangle $R$ as its domain, can be represented as the parametric surface

$$\mathbf{r} = \mathbf{r}(u, v) = u\mathbf{i} + v\mathbf{j} + f(u, v)\mathbf{k}$$

for $(u, v)$ in $R$. Its scalar parametric equations are

$$x = u, \qquad y = v, \qquad z = f(u, v), \qquad (u, v) \text{ in } R.$$

For such graphs it is sometimes convenient to identify the $uv$-plane with the $xy$-plane, and write the equation of the surface in the form

$$\mathbf{r} = x\mathbf{i} + y\mathbf{j} + f(x, y)\mathbf{k}, \qquad (x, y) \text{ in } R.$$

**Example 2**    Describe the surface

$$\mathbf{r} = a \cos u \sin v\, \mathbf{i} + a \sin u \sin v\, \mathbf{j} + a \cos v\, \mathbf{k}, \qquad (0 \le u \le 2\pi, \quad 0 \le v \le \pi/2),$$

where $a > 0$. What is its boundary?

**Solution**    Observe that if $x = a \cos u \sin v$, $y = a \sin u \sin v$, and $z = a \cos v$, then $x^2 + y^2 + z^2 = a^2$. Thus the given parametric surface lies on the sphere of radius $a$ centred at the origin. (Observe that $u$ and $v$ are the spherical coordinates $\theta$ and $\phi$ on the sphere.) The restrictions on $u$ and $v$ allow $(x, y)$ to be any point in the disk $x^2 + y^2 \le a^2$, but force $z \ge 0$. Thus the surface is the *upper half* of the sphere. The given parametrization is one-to-one on the open rectangle $0 < u < 2\pi$, $0 < v < \pi/2$, but not on the closed rectangle, since the edges $u = 0$ and $u = 2\pi$ get mapped onto the same points, and the entire edge $v = 0$ collapses to a single point. The boundary of the surface is still the circle $x^2 + y^2 = a^2$, $z = 0$, and corresponds to the edge $v = \pi/2$ of the rectangle.

**Remark**    Surface parametrizations that are one-to-one only in the interior of the parameter domain $R$ are still reasonable representations of the surface. However, as in Example 2, the boundary of the surface may be obtained from only part of the boundary of $R$, or there may be no boundary at all, in which case the surface is called a **closed surface**. For example, if the domain of $\mathbf{r}$ in Example 2 is extended to allow $0 \le v \le \pi$, then the surface becomes the entire sphere of radius $a$ centred at the origin. The sphere is a closed surface, having no boundary curves.

**Remark**    Like parametrizations of curves, parametrizations of surfaces are not unique. The hemisphere in Example 2 can also be parametrized

$$\mathbf{r}(u, v) = u\mathbf{i} + v\mathbf{j} + \sqrt{a^2 - u^2 - v^2}\,\mathbf{k} \qquad \text{for} \quad u^2 + v^2 \le a^2.$$

Here the domain of $\mathbf{r}$ is a closed disk of radius $a$.

**Figure 15.17** A tube in the shape of a trefoil knot

**Example 3** (**A tube around a curve**) If $\mathbf{r} = \mathbf{F}(t)$, $a \le t \le b$, is a parametric curve $\mathcal{C}$ in 3-space having unit normal $\hat{\mathbf{N}}(t)$ and binormal $\hat{\mathbf{B}}(t)$, then the parametric surface

$$\mathbf{r} = \mathbf{F}(u) + s \cos v\, \hat{\mathbf{N}}(u) + s \sin v\, \hat{\mathbf{B}}(u), \quad a \le u \le b, \quad 0 \le v \le 2\pi,$$

is a tube-shaped surface of radius $s$ centred along the curve $\mathcal{C}$. (Why?) Figure 15.17 shows such a tube, having radius $s = 0.25$, around the curve

$$\mathbf{r} = (1 + 0.3\cos(3t))(\cos(2t)\mathbf{i} + \sin(2t)\mathbf{j}) + 0.35\sin(3t)\mathbf{k}, \quad 0 \le t \le 2\pi.$$

This closed curve is called a **trefoil knot**. ∎

## Composite Surfaces

If two parametric surfaces are joined together along part or all of their boundary curves, the result is called a **composite surface**, or, thinking geometrically, just a **surface**. For example, a sphere can be obtained by joining two hemispheres along their boundary circles. In general, composite surfaces can be obtained by joining a finite number of parametric surfaces pairwise along edges. The surface of a cube consists of the six square faces joined in pairs along the edges of the cube. This surface is closed since there are no unjoined edges to comprise the boundary. If the top square face is removed, the remaining five form the surface of a cubical box with no top. The top edges of the four side faces now constitute the boundary of this composite surface. (See Figure 15.18.)

**Figure 15.18** A composite surface obtained by joining five smooth parametric surfaces (squares) in pairs along edges. The four unpaired edges at the tops of the side faces make up the boundary of the composite surface

## Surface Integrals

In order to define integrals of functions defined on a surface as limits of Riemann sums, we need to refer to the *areas* of regions on the surface. It is more difficult to define the area of a curved surface than it is to define the length of a curve. However, you will likely have a good idea of what area means for a region lying in a plane, and we examined briefly the problem of finding the area of the graph of a function $f(x, y)$ in Section 14.7. We will avoid difficulties by assuming that all the surfaces we will encounter are "smooth enough" that they can be subdivided into small pieces each of which is approximately planar. We can then approximate the surface area of each piece by a plane area and add up the approximations to get a Riemann sum approximation to the area of the whole surface. We will make more precise definitions of "smooth surface" and "surface area" later in this section. For the moment, we assume the reader has an intuitive feel for what they mean.

Let $\mathcal{S}$ be a smooth surface of finite area in $\mathbb{R}^3$, and let $f(x, y, z)$ be a bounded function defined at all points of $\mathcal{S}$. If we subdivide $\mathcal{S}$ into small, nonoverlapping pieces, say $S_1, S_2, \ldots, S_n$, where $S_i$ has area $\Delta S_i$ (see Figure 15.19), we can form a **Riemann sum** $R_n$ for $f$ on $\mathcal{S}$ by choosing arbitrary points $(x_i, y_i, z_i)$ in $S_i$ and letting

$$R_n = \sum_{i=1}^{n} f(x_i, y_i, z_i)\, \Delta S_i.$$

If such Riemann sums have a unique limit as the diameters of all the pieces $S_i$ approach zero, independently of how the points $(x_i, y_i, z_i)$ are chosen, then we say that $f$ is **integrable** on $\mathcal{S}$, and denote the limit by

$$\iint_{\mathcal{S}} f(x, y, z) \, dS.$$

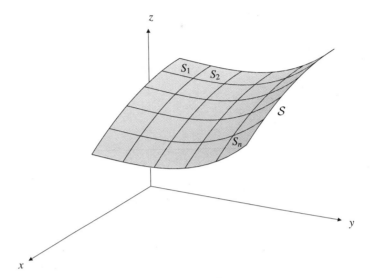

**Figure 15.19** A partition of a parametric surface into many nonoverlapping pieces

## Smooth Surfaces, Normals, and Area Elements

A surface is smooth if it has a unique tangent plane at any nonboundary point $P$. A nonzero vector $\mathbf{n}$ normal to that tangent plane at $P$ is said to be normal to the surface at $P$. The following somewhat technical definition makes this precise.

**DEFINITION 5**

A set $\mathcal{S}$ in 3-space is a **smooth surface** if any point $P$ in $\mathcal{S}$ has a neighbourhood $N$ (an open ball of positive radius centred at $P$) which is the domain of a smooth function $g(x, y, z)$ satisfying:

(i) $N \cap \mathcal{S} = \{Q \in N \,:\, g(Q) = 0\}$, and

(ii) $\nabla g(Q) \neq \mathbf{0}$, if $Q$ is in $N \cap \mathcal{S}$.

For example, the cone $x^2 + y^2 = z^2$, with the origin removed, is a smooth surface. Note that $\nabla(x^2 + y^2 - z^2) = \mathbf{0}$ at the origin, and the cone is not smooth there, since it does not have a unique tangent plane.

A parametric surface cannot satisfy the condition of the smoothness definition at its boundary points, but will be called **smooth** if that condition is satisfied at all non-boundary points.

We can find the normal to a smooth parametric surface defined on parameter domain $R$ as follows. If $(u_0, v_0)$ is a point in the interior of $R$, then $\mathbf{r} = \mathbf{r}(u, v_0)$ and $\mathbf{r} = \mathbf{r}(u_0, v)$ are two curves on $\mathcal{S}$, intersecting at $\mathbf{r}_0 = \mathbf{r}(u_0, v_0)$, and having, at that point, tangent vectors

$$\left.\frac{\partial \mathbf{r}}{\partial u}\right|_{(u_0, v_0)} \quad \text{and} \quad \left.\frac{\partial \mathbf{r}}{\partial v}\right|_{(u_0, v_0)}$$

respectively. Assuming these two tangent vectors are not parallel, their cross product $\mathbf{n}$, which is not zero, is *normal* to $S$ at $\mathbf{r}_0$. Furthermore, the *area element* on $S$ bounded by the four curves $\mathbf{r} = \mathbf{r}(u_0, v)$, $\mathbf{r} = \mathbf{r}(u_0 + du, v)$, $\mathbf{r} = \mathbf{r}(u, v_0)$, and $\mathbf{r} = \mathbf{r}(u, v_0 + dv)$, (see Figure 15.20), is an infinitesimal parallelogram spanned by the vectors $(\partial \mathbf{r}/\partial u)\, du$, and $(\partial \mathbf{r}/\partial v)\, dv$ (at $(u_0, v_0)$), and hence has area

$$dS = \left| \frac{\partial \mathbf{r}}{\partial u} \times \frac{\partial \mathbf{r}}{\partial v} \right| du\, dv.$$

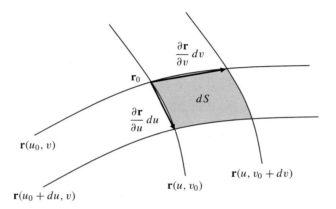

**Figure 15.20**    An area element $dS$ on a parametric surface

Let us express the normal vector $\mathbf{n}$ and the area element $dS$ in terms of the components of $\mathbf{r}$. Since

$$\frac{\partial \mathbf{r}}{\partial u} = \frac{\partial x}{\partial u}\mathbf{i} + \frac{\partial y}{\partial u}\mathbf{j} + \frac{\partial z}{\partial u}\mathbf{k} \qquad \text{and} \qquad \frac{\partial \mathbf{r}}{\partial v} = \frac{\partial x}{\partial v}\mathbf{i} + \frac{\partial y}{\partial v}\mathbf{j} + \frac{\partial z}{\partial v}\mathbf{k},$$

the **normal vector** to $S$ at $\mathbf{r}(u, v)$ is

$$\mathbf{n} = \frac{\partial \mathbf{r}}{\partial u} \times \frac{\partial \mathbf{r}}{\partial v} = \begin{vmatrix} \mathbf{i} & \mathbf{j} & \mathbf{k} \\ \dfrac{\partial x}{\partial u} & \dfrac{\partial y}{\partial u} & \dfrac{\partial z}{\partial u} \\ \dfrac{\partial x}{\partial v} & \dfrac{\partial y}{\partial v} & \dfrac{\partial z}{\partial v} \end{vmatrix}$$

$$= \frac{\partial(y, z)}{\partial(u, v)}\mathbf{i} + \frac{\partial(z, x)}{\partial(u, v)}\mathbf{j} + \frac{\partial(x, y)}{\partial(u, v)}\mathbf{k}.$$

Also, the **area element** at a point $\mathbf{r}(u, v)$ on the surface is given by

$$dS = \left| \frac{\partial \mathbf{r}}{\partial u} \times \frac{\partial \mathbf{r}}{\partial v} \right| du\, dv$$

$$= \sqrt{\left(\frac{\partial(y, z)}{\partial(u, v)}\right)^2 + \left(\frac{\partial(z, x)}{\partial(u, v)}\right)^2 + \left(\frac{\partial(x, y)}{\partial(u, v)}\right)^2}\, du\, dv.$$

The area of the surface itself is the "sum" of these area elements.

$$\text{Area of } S = \iint_S dS.$$

**Example 4**   The graph $z = g(x, y)$ of a function $g$ with continuous first partial derivatives in a domain $D$ of the $xy$-plane can be regarded as a parametric surface $S$ with parametrization

$$x = u, \qquad y = v, \qquad z = \dot{g}(u, v), \qquad (u, v) \text{ in } D.$$

In this case

$$\frac{\partial(y, z)}{\partial(u, v)} = -g_1(u, v), \qquad \frac{\partial(z, x)}{\partial(u, v)} = -g_2(u, v), \qquad \text{and} \qquad \frac{\partial(x, y)}{\partial(u, v)} = 1,$$

and, since the parameter region coincides with the domain $D$ of $g$, the surface integral of $f(x, y, z)$ over $S$ can be expressed as a double integral over $D$:

$$\iint_S f(x, y, z) \, dS$$
$$= \iint_D f\left(x, y, g(x, y)\right) \sqrt{1 + \left(g_1(x, y)\right)^2 + \left(g_2(x, y)\right)^2} \, dx \, dy.$$

As observed in Section 14.7, this formula can also be justified geometrically. The vector $\mathbf{n} = -g_1(x, y)\mathbf{i} - g_2(x, y)\mathbf{j} + \mathbf{k}$ is normal to $S$ and makes angle $\gamma$ with the positive $z$-axis, where

$$\cos \gamma = \frac{\mathbf{n} \bullet \mathbf{k}}{|\mathbf{n}|} = \frac{1}{\sqrt{1 + \left(g_1(x, y)\right)^2 + \left(g_2(x, y)\right)^2}}.$$

The surface area element $dS$ must have area $1/\cos \gamma$ times the area $dx \, dy$ of its perpendicular projection onto the $xy$-plane. (See Figure 15.21.) ∎

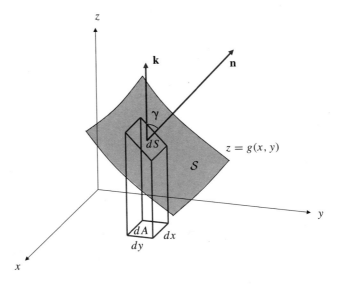

**Figure 15.21**    The surface area element $dS$ and its projection onto the $xy$-plane

## Evaluating Surface Integrals

We illustrate the use of the formulas given above for $dS$ in calculating surface integrals.

**Example 5**    Evaluate $\displaystyle\iint_{\mathcal{S}} z\, dS$ over the conical surface $z = \sqrt{x^2 + y^2}$ between $z = 0$ and $z = 1$.

**Solution**    Since $z^2 = x^2 + y^2$ on the surface $\mathcal{S}$, we have $\partial z/\partial x = x/z$ and $\partial z/\partial y = y/z$. Therefore

$$dS = \sqrt{1 + \frac{x^2}{z^2} + \frac{y^2}{z^2}}\, dx\, dy = \sqrt{\frac{z^2 + z^2}{z^2}}\, dx\, dy = \sqrt{2}\, dx\, dy.$$

(Note that we could have anticipated this result since the normal to the cone always makes an angle of $\gamma = 45°$ with the positive $z$-axis; see Figure 15.22. Therefore $dS = dx\, dy/\cos 45° = \sqrt{2}\, dx\, dy$.) Since $z = \sqrt{x^2 + y^2} = r$ on the conical surface, it is easiest to carry out the integration in polar coordinates:

$$\iint_{\mathcal{S}} z\, dS = \sqrt{2} \iint_{x^2 + y^2 \leq 1} z\, dx\, dy$$

$$= \sqrt{2} \int_0^{2\pi} d\theta \int_0^1 r^2\, dr = \frac{2\sqrt{2}\pi}{3}.$$

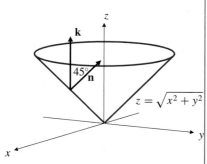

**Figure 15.22**    $dS = \sqrt{2}\, dx\, dy$ on this cone

**Example 6**    Find the moment of inertia about the $z$-axis of the parametric surface $x = 2uv$, $y = u^2 - v^2$, $z = u^2 + v^2$, where $u^2 + v^2 \leq 1$.

**Solution**    We are asked to find $\displaystyle\iint_{\mathcal{S}} (x^2 + y^2)\, dS$. We have

$$\frac{\partial(x, y)}{\partial(u, v)} = \begin{vmatrix} 2v & 2u \\ 2u & -2v \end{vmatrix} = -4(u^2 + v^2),$$

$$\frac{\partial(x, z)}{\partial(u, v)} = \begin{vmatrix} 2v & 2u \\ 2u & 2v \end{vmatrix} = 4(v^2 - u^2),$$

$$\frac{\partial(y, z)}{\partial(u, v)} = \begin{vmatrix} 2u & -2v \\ 2u & 2v \end{vmatrix} = 8uv.$$

Therefore, the surface area element on $\mathcal{S}$ is given by

$$dS = 4\sqrt{(u^2 + v^2)^2 + (v^2 - u^2)^2 + 4u^2v^2}\, du\, dv$$

$$= 4\sqrt{2(u^4 + v^4 + 2u^2v^2)}\, du\, dv = 4\sqrt{2}(u^2 + v^2)\, du\, dv.$$

Now $x^2 + y^2 = 4u^2v^2 + (u^2 - v^2)^2 = (u^2 + v^2)^2$. Thus

$$\iint_{\mathcal{S}} (x^2 + y^2)\, dS = \iint_{u^2 + v^2 \leq 1} (u^2 + v^2)^2\, 4\sqrt{2}(u^2 + v^2)\, du\, dv$$

$$= 4\sqrt{2} \int_0^{2\pi} d\theta \int_0^1 r^6 r\, dr \qquad \text{(using polar coordinates)}$$

$$= \sqrt{2}\pi.$$

This is the required moment of inertia.

Even though most surfaces we encounter can be easily parametrized, it is usually possible to obtain the surface area element $dS$ geometrically rather than relying on the parametric formula. As we have seen above, if a surface has a one-to-one projection onto a region in the $xy$-plane, then the area element $dS$ on the surface can be expressed as

$$dS = \left| \frac{1}{\cos \gamma} \right| dx\, dy = \frac{|\mathbf{n}|}{|\mathbf{n} \bullet \mathbf{k}|} dx\, dy,$$

where $\gamma$ is the angle between the normal vector $\mathbf{n}$ to $S$ and the positive $z$-axis. This formula is useful no matter how we obtain $\mathbf{n}$.

Consider a surface $S$ with equation of the form $F(x, y, z) = 0$. As we discovered in Section 12.7, if $F$ has continuous first partial derivatives that do not all vanish at a point $(x, y, z)$ on $S$, then the nonzero vector

$$\mathbf{n} = \nabla F(x, y, z)$$

is normal to $S$ at that point. Since $\mathbf{n} \bullet \mathbf{k} = F_3(x, y, z)$, if $S$ has a one-to-one projection onto the domain $D$ in the $xy$-plane, then

$$dS = \left| \frac{\nabla F(x, y, z)}{F_3(x, y, z)} \right| dx\, dy,$$

and the surface integral of $f(x, y, z)$ over $S$ can be expressed as a double integral over the domain $D$:

$$\iint_S f(x, y, z)\, dS = \iint_D f\big(x, y, g(x, y)\big) \left| \frac{\nabla F(x, y, z)}{F_3(x, y, z)} \right| dx\, dy.$$

Of course, there are analogous formulas for area elements of surfaces (and integrals over surfaces) with one-to-one projections onto the $xz$-plane or the $yz$-plane. ($F_3$ is replaced by $F_2$ and $F_1$, respectively.)

---

**Example 7**   Find the moment about $z = 0$, that is, $\iint_S z\, dS$, where $S$ is the hyperbolic bowl $z^2 = 1 + x^2 + y^2$ between the planes $z = 1$ and $z = \sqrt{5}$.

**Solution**   $S$ is given by $F(x, y, z) = 0$, where $F(x, y, z) = x^2 + y^2 - z^2 + 1$. It lies above the disk $x^2 + y^2 \le 4$ in the $xy$-plane. We have $\nabla F = 2x\mathbf{i} + 2y\mathbf{j} - 2z\mathbf{k}$, and $F_3 = -2z$. Hence, on $S$, we have

$$z\, dS = z \frac{\sqrt{4x^2 + 4y^2 + 4z^2}}{2z} dx\, dy = \sqrt{1 + 2(x^2 + y^2)}\, dx\, dy,$$

and the required moment is

$$\iint_S z\, dS = \iint_{x^2 + y^2 \le 4} \sqrt{1 + 2(x^2 + y^2)}\, dx\, dy$$

$$= \int_0^{2\pi} d\theta \int_0^2 \sqrt{1 + 2r^2}\, r\, dr = \frac{\pi}{3}(1 + 2r^2)^{3/2} \Big|_0^2 = \frac{26\pi}{3}.$$

---

The next example illustrates a technique that can often reduce the effort needed to integrate over a cylindrical surface.

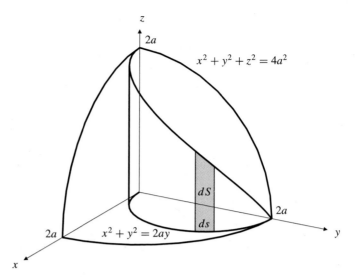

**Figure 15.23**   An area element on a cylinder. The $z$-coordinate has already been integrated

**Example 8**   Find the area of that part of the cylinder $x^2 + y^2 = 2ay$ that lies inside the sphere $x^2 + y^2 + z^2 = 4a^2$.

**Solution**   One quarter of the required area lies in the first octant. (See Figure 15.23.) Since the cylinder is generated by vertical lines, we can express an area element $dS$ on it in terms of the length element $ds$ along the curve $C$ in the $xy$-plane having equation $x^2 + y^2 = 2ay$:

$$dS = z\,ds = \sqrt{4a^2 - x^2 - y^2}\,ds.$$

In expressing $dS$ this way, we have already integrated $dz$, so only a single integral is needed to sum these area elements. Again it is convenient to use polar coordinates in the $xy$-plane. In terms of polar coordinates the curve $C$ has equation $r = 2a \sin\theta$. Thus $dr/d\theta = 2a \cos\theta$ and $ds = \sqrt{r^2 + (dr/d\theta)^2}\,d\theta = 2a\,d\theta$. Therefore the total surface area of that part of the cylinder that lies inside the sphere is given by

$$
\begin{aligned}
A &= 4 \int_0^{\pi/2} \sqrt{4a^2 - r^2}\, 2a\,d\theta \\
&= 8a \int_0^{\pi/2} \sqrt{4a^2 - 4a^2 \sin^2\theta}\,d\theta \\
&= 16a^2 \int_0^{\pi/2} \cos\theta\,d\theta = 16a^2 \text{ square units.}
\end{aligned}
$$

∎

**Remark**   The area calculated in Example 8 can also be calculated by projecting the cylindrical surface in Figure 15.23 into the $yz$-plane. (This is the only coordinate plane you can use. Why?) See Exercise 6 below.

In spherical coordinates, $\phi$ and $\theta$ can be used as parameters on the spherical surface $R = a$. The area element on that surface can therefore be expressed in terms of these coordinates.

Area element on the sphere $\rho = a$:  $dS = a^2 \sin\phi \, d\phi \, d\theta.$

(See Figure 14.43 in Section 14.6 and Exercise 2 at the end of this section.)

**Example 9**  Find $\displaystyle\iint_{\mathcal{S}} z^2 \, dS$ over the hemisphere $z = \sqrt{a^2 - x^2 - y^2}$.

**Solution**  Since $z = a \cos\phi$ and the hemisphere corresponds to $0 \le \theta \le 2\pi$, and $0 \le \phi \le \dfrac{\pi}{2}$, we have

$$\iint_{\mathcal{S}} z^2 \, dS = \int_0^{2\pi} d\theta \int_0^{\pi/2} a^2 \cos^2\phi \, a^2 \sin\phi \, d\phi$$

$$= 2\pi a^4 \left( -\frac{1}{3} \cos^3\phi \right) \Bigg|_0^{\pi/2} = \frac{2\pi a^4}{3}.$$

Finally, if a composite surface $\mathcal{S}$ is composed of *smooth parametric surfaces* joined pairwise along their edges, then we call $\mathcal{S}$ a **piecewise smooth surface**. The surface integral of a function $f$ over a piecewise smooth surface $\mathcal{S}$ is the sum of the surface integrals of $f$ over the individual smooth surfaces comprising $\mathcal{S}$. We will encounter an example of this in the next section.

## The Attraction of a Spherical Shell

In Section 14.7 we calculated the gravitational attraction of a disk in the $xy$-plane on a mass $m$ located at position $(0, 0, b)$ on the $z$-axis. Here we undertake a similar calculation of the attractive force exerted on $m$ by a spherical shell of radius $a$ and areal density $\sigma$ (units of mass per unit area) centred at the origin. This calculation would be more difficult if we try to do it by integrating the vertical component of the force on $m$ as we did in Section 14.7. It is greatly simplified if, instead, we use an integral to find the total *gravitational potential* $\Phi(0, 0, z)$ due to the sphere at position $(0, 0, z)$, and then calculate the force on $m$ as $\mathbf{F} = m \nabla\Phi(0, 0, b)$.

The distance from the point with spherical coordinates $[a, \phi, \theta]$ to the point $(0, 0, z)$ on the positive $z$-axis is (see Figure 15.24)

$$D = \sqrt{a^2 + z^2 - 2az \cos\phi}.$$

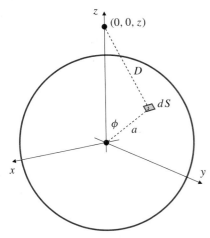

**Figure 15.24**  The attraction of a sphere

The area element $dS = a^2 \sin\phi \, d\phi \, d\theta$ at $[a, \phi, \theta]$ has mass $dm = \sigma \, dS$, and its gravitational potential at $(0, 0, z)$ is (see Example 1 in Section 15.2)

$$d\Phi(0, 0, z) = \frac{k \, dm}{D} = \frac{k\sigma a^2 \sin\phi \, d\phi \, d\theta}{\sqrt{a^2 + z^2 - 2az \cos\phi}}.$$

For the total potential at $(0, 0, z)$ due to the sphere, we integrate $d\Phi$ over the surface of the sphere. Making the change of variables $u = a^2 + z^2 - 2az \cos\phi$,

$du = 2az \sin\phi \, d\phi$, we obtain

$$\Phi(0, 0, z) = k\sigma a^2 \int_0^{2\pi} d\theta \int_0^\pi \frac{\sin\phi \, d\phi}{\sqrt{a^2 + z^2 - 2az\cos\phi}}$$

$$= 2\pi k\sigma a^2 \int_{(z-a)^2}^{(z+a)^2} \frac{1}{\sqrt{u}} \frac{du}{2az}$$

$$= \frac{2\pi k\sigma a}{z} \sqrt{u} \Big|_{(z-a)^2}^{(z+a)^2}$$

$$= \frac{2\pi k\sigma a}{z} \left( z + a - |z - a| \right) = \begin{cases} 4\pi k\sigma a^2/z & \text{if } z > a \\ 4\pi k\sigma a & \text{if } z < a. \end{cases}$$

The potential is constant inside the sphere, and decreases proportionally to $1/z$ outside. The force on a mass $m$ located at $(0, 0, b)$ is, therefore,

$$\mathbf{F} = m\nabla\Phi(0, 0, b) = \begin{cases} -(4\pi km\sigma a^2/b^2)\mathbf{k} & \text{if } b > a \\ \mathbf{0} & \text{if } b < a. \end{cases}$$

We are led to the somewhat surprising result that if the mass $m$ is anywhere inside the sphere, the net force of attraction of the sphere on it is zero. This is to be expected at the centre of the sphere, but away from the centre it appears that the larger forces due to parts of the sphere close to $m$ are exactly cancelled by smaller forces due to parts further away; these further parts have larger area and therefore larger total mass. If $m$ is outside the sphere, the sphere attracts it with a force of magnitude

$$F = \frac{kmM}{b^2},$$

where $M = 4\pi\sigma a^2$ is the total mass of the sphere. This is the same force that would be exerted by a point mass with the same mass as the sphere and located at the centre of the sphere.

**Remark**  A solid ball of constant density, or density depending only on the distance from the centre (for instance, a planet), can be regarded as being made up of mass elements that are concentric spheres of constant density. Therefore, the attraction of such a ball on a mass $m$ located outside the ball will also be the same as if the whole mass of the ball were concentrated at its centre. However, the attraction on a mass $m$ located somewhere inside the ball will be that produced by only the part of the ball that is closer to the centre than $m$ is. The maximum force of attraction will occur when $m$ is right at the surface of the ball. If the density is constant, the magnitude of the force increases linearly with the distance from the centre (why?) up to the surface and then decreases with the square of the distance as $m$ recedes from the ball. (See Figure 15.25.)

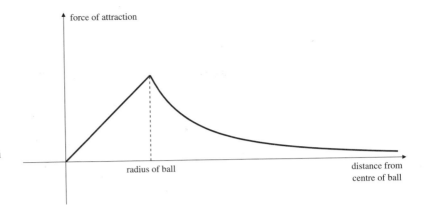

**Figure 15.25**    The force of attraction of a homogeneous solid ball on a particle located at varying distances from the centre of the ball

*Remark*    All of the above discussion also holds for the electrostatic attraction or repulsion of a point charge by a uniform charge density over a spherical shell, which is also governed by an inverse square law. In particular there is no net electrostatic force on a charge located inside the shell.

## Exercises 15.5

**1.** Verify that on the curve with polar equation $r = g(\theta)$ the arc length element is given by

$$ds = \sqrt{(g(\theta))^2 + (g'(\theta))^2}\, d\theta.$$

What is the area element on the vertical cylinder given in terms of cylindrical coordinates by $r = g(\theta)$?

**2.** Verify that on the spherical surface $x^2 + y^2 + z^2 = a^2$ the area element is given in terms of spherical coordinates by $dS = a^2 \sin\phi\, d\phi\, d\theta$.

**3.** Find the area of the part of the plane $Ax + By + Cz = D$ lying inside the elliptic cylinder

$$\frac{x^2}{a^2} + \frac{y^2}{b^2} = 1.$$

**4.** Find the area of the part of the sphere $x^2 + y^2 + z^2 = 4a^2$ that lies inside the cylinder $x^2 + y^2 = 2ay$.

**5.** State formulas for the surface area element $dS$ for the surface with equation $F(x, y, z) = 0$ valid for the case where the surface has a one-to-one projection on (a) the $xz$-plane, and (b) the $yz$-plane.

**6.** Repeat the area calculation of Example 8 by projecting the part of the surface shown in Figure 15.23 onto the $yz$-plane, and using the formula in Exercise 5(b).

**7.** Find $\displaystyle\iint_S x\, dS$ over the part of the parabolic cylinder $z = x^2/2$ that lies inside the first octant part of the cylinder $x^2 + y^2 = 1$.

**8.** Find the area of the part of the cone $z^2 = x^2 + y^2$ that lies inside the cylinder $x^2 + y^2 = 2ay$.

**9.** Find the area of the part of the cylinder $x^2 + y^2 = 2ay$ that lies outside the cone $z^2 = x^2 + y^2$.

**10.** Find the area of the part of the cylinder $x^2 + z^2 = a^2$ that lies inside the cylinder $y^2 + z^2 = a^2$.

**11.** A circular cylinder of radius $a$ is circumscribed about a sphere of radius $a$ so that the cylinder is tangent to the sphere along the equator. Two planes, each perpendicular to the axis of the cylinder, intersect the sphere and the cylinder in circles. Show that the area of that part of the sphere between the two planes is equal to the area of the part of the cylinder between the two planes. Thus the area of the part of a sphere between two parallel planes that intersect it depends only on the radius of the sphere and the distance between the planes, and not on the particular position of the planes.

**12.** Let $0 < a < b$. In terms of the elliptic integral functions defined in Exercise 17 of Section 15.3, find the area of that part of each of the cylinders $x^2 + z^2 = a^2$ and $y^2 + z^2 = b^2$ that lies inside the other cylinder.

**13.** Find $\displaystyle\iint_S y\, dS$, where $S$ is the part of the plane $z = 1 + y$ that lies inside the cone $z = \sqrt{2(x^2 + y^2)}$.

**14.** Find $\displaystyle\iint_S y\, dS$, where $S$ is the part of the cone $z = \sqrt{2(x^2 + y^2)}$ that lies below the plane $z = 1 + y$.

**15.** Find $\displaystyle\iint_S xz\, dS$, where $S$ is the part of the surface $z = x^2$ that lies in the first octant of 3-space and inside the paraboloid $z = 1 - 3x^2 - y^2$.

**16.** Find the mass of the part of the surface $z = \sqrt{2xy}$ that lies above the region $0 \le x \le 5, 0 \le y \le 2$, if the areal density of the surface is $\sigma(x, y, z) = kz$.

**17.** Find the total charge on the surface

$$\mathbf{r} = e^u \cos v \mathbf{i} + e^u \sin v \mathbf{j} + u\mathbf{k},$$

$0 \le u \le 1, 0 \le v \le \pi$, if the charge density on the surface is $\delta = \sqrt{1 + e^{2u}}$.

Exercises 18–19 concern **spheroids**, which are ellipsoids with two of their three semi-axes equal, say $a = b$:

$$\frac{x^2}{a^2} + \frac{y^2}{a^2} + \frac{z^2}{c^2} = 1.$$

**∗ 18.** Find the surface area of a **prolate spheroid** where $0 < a < c$. A prolate spheroid has its two shorter semi-axes equal, like a "pro football."

**∗ 19.** Find the surface area of an **oblate spheroid** where $0 < c < a$. An oblate spheroid has its two longer semi-axes equal, like the earth.

**20.** Describe the parametric surface

$$x = au \cos v, \qquad y = au \sin v, \qquad z = bv,$$

$(0 \le u \le 1, \ 0 \le v \le 2\pi)$, and find its area.

**∗ 21.** Evaluate $\displaystyle\iint_{\mathcal{P}} \frac{dS}{(x^2 + y^2 + z^2)^{3/2}}$, where $\mathcal{P}$ is the plane with equation $Ax + By + Cz = D, (D \neq 0)$.

**22.** A spherical shell of radius $a$ is centred at the origin. Find the centroid of that part of the sphere that lies in the first octant.

**23.** Find the centre of mass of a right-circular conical shell of base radius $a$, height $h$, and constant areal density $\sigma$.

**∗ 24.** Find the gravitational attraction of a hemispherical shell of radius $a$ and constant areal density $\sigma$ on a mass $m$ located at the centre of the base of the hemisphere.

**∗ 25.** Find the gravitational attraction of a circular cylindrical shell of radius $a$, height $h$, and constant areal density $\sigma$ on a mass $m$ located on the axis of the cylinder $b$ units above the base.

In Exercises 26–28, find the moment of inertia and radius of gyration of the given object about the given axis. Assume constant areal density $\sigma$ in each case.

**26.** A cylindrical shell of radius $a$ and height $h$ about the axis of the cylinder

**27.** A spherical shell of radius $a$ about a diameter

**28.** A right-circular conical shell of base radius $a$ and height $h$ about the axis of the cone

**29.** With what acceleration will the spherical shell of Exercise 27 roll down a plane inclined at angle $\alpha$ to the horizontal? (Compare your result with that of Example 4(b) of Section 14.7.)

## 15.6 Oriented Surfaces and Flux Integrals

Surface integrals of normal components of vector fields play a very important role in vector calculus, similar to the role played by line integrals of tangential components of vector fields. Before we consider such surface integrals we need to define the *orientation* of a surface.

### Oriented Surfaces

A smooth surface $\mathcal{S}$ in 3-space is said to be **orientable** if there exists a unit vector field $\hat{\mathbf{N}}(P)$ defined on $\mathcal{S}$, which varies continuously as $P$ ranges over $\mathcal{S}$ and which is everywhere normal to $\mathcal{S}$. Any such vector field $\hat{\mathbf{N}}(P)$ determines an **orientation** of $\mathcal{S}$. The surface must have two sides since $\hat{\mathbf{N}}(P)$ can have only one value at each point $P$. The side out of which $\hat{\mathbf{N}}$ points is called the **positive side**; the other side is the **negative side**. An **oriented surface** is a smooth surface together with a particular choice of orienting unit normal vector field $\hat{\mathbf{N}}(P)$.

For example, if we define $\hat{\mathbf{N}}$ on the smooth surface $z = f(x, y)$ by

$$\hat{\mathbf{N}} = \frac{-f_1(x, y)\mathbf{i} - f_2(x, y)\mathbf{j} + \mathbf{k}}{\sqrt{1 + (f_1(x, y))^2 + (f_2(x, y))^2}},$$

then the top of the surface is the positive side. (See Figure 15.26.)

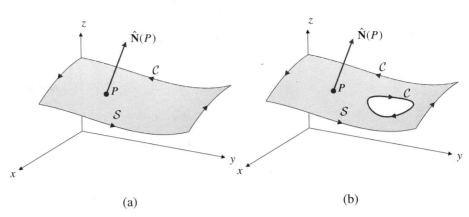

**Figure 15.26** The boundary curves of an oriented surface are themselves oriented with the surface on the left

(a)　　　　　　(b)

A smooth or piecewise smooth surface may be **closed** (that is, it may have no boundary), or it may have one or more boundary curves. (The unit normal vector field $\hat{\mathbf{N}}(P)$ need not be defined at points of the boundary curves.)

An oriented surface $S$ **induces an orientation** on any of its boundary curves $C$; if we stand on the positive side of the surface $S$ and walk around $C$ in the direction of its orientation, then $S$ will be on our left side. (See Figure 15.26(a) and (b).)

A *piecewise smooth* surface is **orientable** if, whenever two smooth component surfaces join along a common boundary curve $C$, they induce *opposite* orientations along $C$. This forces the normals $\hat{\mathbf{N}}$ to be on the same side of adjacent components. For instance, the surface of a cube is a piecewise smooth, closed surface, consisting of six smooth surfaces (the square faces) joined along edges. (See Figure 15.27.) If all of the faces are oriented so that their normals $\hat{\mathbf{N}}$ point out of the cube (or if they all point into the cube), then surface of the cube itself is oriented.

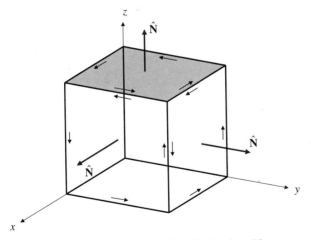

**Figure 15.27** The surface of the cube is orientable; adjacent faces induce opposite orientations on their common edge

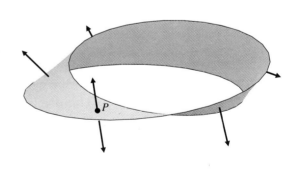

**Figure 15.28** The Möbius band is not orientable; it has only one "side"

Not every surface can be oriented, even if it appears smooth. An orientable surface must have two sides. For example, a Möbius band, consisting of a strip of paper with ends joined together to form a loop, but with one end given a half twist before the ends are joined, has only one side (make one and see), so it cannot be oriented. (See Figure 15.28.) If a nonzero vector is moved around the band, starting at point $P$, so that it is always normal to the surface, then it can return to its starting position pointing in the opposite direction.

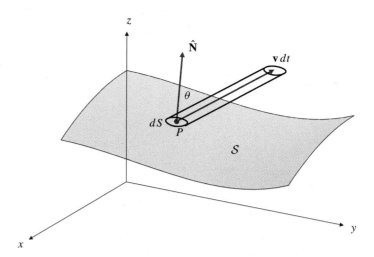

**Figure 15.29** The fluid crossing $dS$ in time $dt$ fills the tube

## The Flux of a Vector Field Across a Surface

Suppose 3-space is filled with an incompressible fluid that is flowing around with velocity field $\mathbf{v}$. Let $S$ be an imaginary, smooth, oriented surface in 3-space. (We called $S$ *imaginary* because it does not provide a barrier to the motion of the fluid. It is fixed in space, not moving with the fluid, and the fluid can freely move through it.) Let us calculate the rate at which fluid flows across $S$. Let $dS$ be a small area element at point $P$ on the surface. The fluid crossing that element between time $t$ and time $t + dt$ occupies a cylinder of base area $dS$ and height $|\mathbf{v}(P)|\, dt \, \cos\theta$, where $\theta$ is the angle between $\mathbf{v}(P)$ and the normal $\hat{\mathbf{N}}(P)$. (See Figure 15.29.) This cylinder has volume $\mathbf{v}(P) \bullet \hat{\mathbf{N}}(P)\, dS\, dt$. The rate at which fluid is crossing $dS$ is $\mathbf{v}(P) \bullet \hat{\mathbf{N}}(P)\, dS$, and the total rate at which it is crossing $S$ is given by the surface integral

$$\iint_S \mathbf{v} \bullet \hat{\mathbf{N}}\, dS \qquad \text{or} \qquad \iint_S \mathbf{v} \bullet d\mathbf{S},$$

where we use $d\mathbf{S}$ to represent the vector surface area element $\hat{\mathbf{N}}\, dS$.

**DEFINITION 6**

> **Flux of a vector field across an oriented surface**
>
> Given any continuous vector field $\mathbf{F}$, the integral of the normal component of $\mathbf{F}$ over the oriented surface $S$,
>
> $$\iint_S \mathbf{F} \bullet \hat{\mathbf{N}}\, dS \qquad \text{or} \qquad \iint_S \mathbf{F} \bullet d\mathbf{S}$$
>
> is called the **flux** of $\mathbf{F}$ across $S$.

When the surface is closed, the flux integral can be denoted by

$$\oiint_S \mathbf{F} \bullet \hat{\mathbf{N}}\, dS \qquad \text{or} \qquad \oiint_S \mathbf{F} \bullet d\mathbf{S}.$$

In this case we refer to the flux of $\mathbf{F}$ *out of* $S$ if $\hat{\mathbf{N}}$ is the unit *exterior* normal, and the flux *into* $S$ if $\hat{\mathbf{N}}$ is the unit *interior* normal.

**Example 1** Find the flux of the vector field $\mathbf{F} = m\mathbf{r}/|\mathbf{r}|^3$ out of a sphere $\mathcal{S}$ of radius $a$ centred at the origin. (Here $\mathbf{r} = x\mathbf{i} + y\mathbf{j} + z\mathbf{k}$.)

**Solution** Since $\mathbf{F}$ is the field associated with a source of strength $m$ at the origin (which produces $4\pi m$ units of fluid per unit time at the origin), the answer must be $4\pi m$. Let us calculate it anyway. We use spherical coordinates. At any point $\mathbf{r}$ on the sphere, with spherical coordinates $[a, \phi, \theta]$, the unit outward normal is $\hat{\mathbf{r}} = \mathbf{r}/|\mathbf{r}|$. Since the vector field is $\mathbf{F} = m\hat{\mathbf{r}}/a^2$ on the sphere, and since an area element is $dS = a^2 \sin\phi \, d\phi \, d\theta$, the flux of $\mathbf{F}$ out of the sphere is

$$\oiint_{\mathcal{S}} \left(\frac{m}{a^2}\hat{\mathbf{r}}\right) \bullet \hat{\mathbf{r}} \, a^2 \sin\phi \, d\phi \, d\theta = m \int_0^{2\pi} d\theta \int_0^{\pi} \sin\phi \, d\phi = 4\pi m.$$

$\blacksquare$

**Figure 15.30**   The three components of the surface of a solid cylinder with their outward normals

**Example 2**   Calculate the total flux of $\mathbf{F} = x\mathbf{i} + y\mathbf{j} + z\mathbf{k}$ outward through the surface of the solid cylinder $x^2 + y^2 \le a^2$, $-h \le z \le h$.

**Solution**   The cylinder is shown in Figure 15.30. Its surface consists of top and bottom disks and the cylindrical side wall. We calculate the flux of $\mathbf{F}$ out of each. Naturally, we use cylindrical coordinates.

On the top disk we have $z = h$, $\hat{\mathbf{N}} = \mathbf{k}$, and $dS = r \, dr \, d\theta$. Therefore, $\mathbf{F} \bullet \hat{\mathbf{N}} \, dS = hr \, dr \, d\theta$ and

$$\iint_{\text{top}} \mathbf{F} \bullet \hat{\mathbf{N}} \, dS = h \int_0^{2\pi} d\theta \int_0^a r \, dr = \pi a^2 h.$$

On the bottom disk we have $z = -h$, $\hat{\mathbf{N}} = -\mathbf{k}$, and $dS = r \, dr \, d\theta$. Therefore, $\mathbf{F} \bullet \hat{\mathbf{N}} \, dS = hr \, dr \, d\theta$ and

$$\iint_{\text{bottom}} \mathbf{F} \bullet \hat{\mathbf{N}} \, dS = \iint_{\text{top}} \mathbf{F} \bullet \hat{\mathbf{N}} \, dS = \pi a^2 h.$$

On the cylindrical wall $\mathbf{F} = a\cos\theta \, \mathbf{i} + a\sin\theta \, \mathbf{j} + z\mathbf{k}$, $\hat{\mathbf{N}} = \cos\theta \, \mathbf{i} + \sin\theta \, \mathbf{j}$, and $dS = a \, d\theta \, dz$. Thus $\mathbf{F} \bullet \hat{\mathbf{N}} \, dS = a^2 \, d\theta \, dz$ and

$$\iint_{\text{cylwall}} \mathbf{F} \bullet \hat{\mathbf{N}} \, dS = a^2 \int_0^{2\pi} d\theta \int_{-h}^h dz = 4\pi a^2 h.$$

The total flux of $\mathbf{F}$ out of the surface $\mathcal{S}$ of the cylinder is the sum of these three contributions:

$$\oiint_{\mathcal{S}} \mathbf{F} \bullet \hat{\mathbf{N}} \, dS = 6\pi a^2 h.$$

$\blacksquare$

Let $\mathcal{S}$ be a smooth, oriented surface with a one-to-one projection onto a domain $D$ in the $xy$-plane, and with equation of the form $G(x, y, z) = 0$. In Section 15.5 we showed that the surface area element on $\mathcal{S}$ could be written in the form

$$dS = \left|\frac{\nabla G}{G_3}\right| \, dx \, dy,$$

and hence surface integrals over $\mathcal{S}$ could be reduced to double integrals over the domain $D$. Flux integrals can be treated likewise. Depending on the orientation of $\mathcal{S}$, the unit normal $\hat{\mathbf{N}}$ can be written as

$$\hat{\mathbf{N}} = \pm \frac{\nabla G}{|\nabla G|}.$$

Thus the vector area element $d\mathbf{S}$ can be written

$$d\mathbf{S} = \hat{\mathbf{N}} \, dS = \pm \frac{\nabla G(x, y, z)}{G_3(x, y, z)} \, dx \, dy.$$

The sign must be chosen to give $\mathcal{S}$ the desired orientation. If $G_3 > 0$ and we want the positive side of $\mathcal{S}$ to face upward, we should use the "+" sign. Of course, similar formulas apply for surfaces with one-to-one projections onto the other coordinate planes.

**Example 3**  Find the flux of $z\mathbf{i} + x^2\mathbf{k}$ upward through that part of the surface $z = x^2 + y^2$ lying above the square $R$ defined by $-1 \le x \le 1$ and $-1 \le y \le 1$.

**Solution**  For $F(x, y, z) = z - x^2 - y^2$ we have $\nabla F = -2x\mathbf{i} - 2y\mathbf{j} + \mathbf{k}$ and $F_3 = 1$. Thus

$$d\mathbf{S} = (-2x\mathbf{i} - 2y\mathbf{j} + \mathbf{k}) \, dx \, dy,$$

and the required flux is

$$\iint_{\mathcal{S}} (z\mathbf{i} + x^2\mathbf{k}) \bullet d\mathbf{S} = \iint_R \left( -2x(x^2 + y^2) + x^2 \right) dx \, dy$$

$$= \int_{-1}^{1} dx \int_{-1}^{1} (x^2 - 2x^3 - 2xy^2) \, dx \, dy$$

$$= \int_{-1}^{1} 2x^2 \, dx = \frac{4}{3}.$$

(Two of the three terms in the double integral had zero integrals because of symmetry.)

For a surface $\mathcal{S}$ with equation $z = f(x, y)$ we have

$$\hat{\mathbf{N}} = \pm \frac{-\dfrac{\partial f}{\partial x}\mathbf{i} - \dfrac{\partial f}{\partial y}\mathbf{j} + \mathbf{k}}{\sqrt{1 + \left(\dfrac{\partial f}{\partial x}\right)^2 + \left(\dfrac{\partial f}{\partial y}\right)^2}}, \qquad \text{and}$$

$$dS = \sqrt{1 + \left(\frac{\partial f}{\partial x}\right)^2 + \left(\frac{\partial f}{\partial y}\right)^2} \, dx \, dy,$$

so that the vector area element on $\mathcal{S}$ is given by

$$d\mathbf{S} = \hat{\mathbf{N}} \, dS = \pm \left( -\frac{\partial f}{\partial x}\mathbf{i} - \frac{\partial f}{\partial y}\mathbf{j} + \mathbf{k} \right) dx \, dy.$$

Again, the $+$ sign corresponds to an upward normal.

For a general parametric surface $\mathbf{r} = \mathbf{r}(u, v)$, the unit normal $\hat{\mathbf{N}}$ and area element $dS$ were calculated in Section 15.5:

$$\hat{\mathbf{N}} = \pm \frac{\dfrac{\partial \mathbf{r}}{\partial u} \times \dfrac{\partial \mathbf{r}}{\partial v}}{\left| \dfrac{\partial \mathbf{r}}{\partial u} \times \dfrac{\partial \mathbf{r}}{\partial v} \right|}, \qquad dS = \left| \frac{\partial \mathbf{r}}{\partial u} \times \frac{\partial \mathbf{r}}{\partial v} \right| \, du \, dv.$$

Thus the vector area element is

$$d\mathbf{S} = \hat{\mathbf{N}} \, dS = \pm \left( \frac{\partial \mathbf{r}}{\partial u} \times \frac{\partial \mathbf{r}}{\partial v} \right) du \, dv.$$

**Example 4**  Find the flux of $\mathbf{F} = y\mathbf{i} - x\mathbf{j} + 4\mathbf{k}$ upward through $\mathcal{S}$, where $\mathcal{S}$ is the part of the surface $z = 1 - x^2 - y^2$ lying in the first octant of 3-space.

**Solution**  The vector area element corresponding to the upward normal on $\mathcal{S}$ is

$$d\mathbf{S} = \left( -\frac{\partial z}{\partial x}\mathbf{i} - \frac{\partial z}{\partial y}\mathbf{j} + \mathbf{k} \right) dx \, dy = (2x\mathbf{i} + 2y\mathbf{j} + \mathbf{k}) \, dx \, dy.$$

The projection of $\mathcal{S}$ onto the $xy$-plane is the quarter-circular disk $Q$ given by $x^2 + y^2 \le 1$, $x \ge 0$, and $y \ge 0$. Thus the flux of $\mathbf{F}$ upward through $\mathcal{S}$ is

$$\iint_{\mathcal{S}} \mathbf{F} \cdot d\mathbf{S} = \iint_{Q} (2xy - 2xy + 4) \, dx \, dy$$
$$= 4 \times (\text{area of } Q) = \pi.$$

**Example 5**  Find the flux of

$$\mathbf{F} = \frac{2x\mathbf{i} + 2y\mathbf{j}}{x^2 + y^2} + \mathbf{k}$$

downward through the surface $\mathcal{S}$ defined parametrically by

$$\mathbf{r} = u \cos v\mathbf{i} + u \sin v\mathbf{j} + u^2\mathbf{k}, \qquad (0 \le u \le 1, \ 0 \le v \le 2\pi).$$

**Solution**  First we calculate $d\mathbf{S}$.

$$\frac{\partial \mathbf{r}}{\partial u} = \cos v\mathbf{i} + \sin v\mathbf{j} + 2u\mathbf{k}$$

$$\frac{\partial \mathbf{r}}{\partial v} = -u \sin v\mathbf{i} + u \cos v\mathbf{j}$$

$$\frac{\partial \mathbf{r}}{\partial u} \times \frac{\partial \mathbf{r}}{\partial v} = -2u^2 \cos v\mathbf{i} - 2u^2 \sin v\mathbf{j} + u\mathbf{k}$$

Since $u \geq 0$ on $\mathcal{S}$, the latter expression is an upward normal. We want a downward normal, so we use

$$d\mathbf{S} = (2u^2 \cos v\,\mathbf{i} + 2u^2 \sin v\,\mathbf{j} - u\mathbf{k})\,du\,dv.$$

On $\mathcal{S}$ we have

$$\mathbf{F} = \frac{2x\mathbf{i} + 2y\mathbf{j}}{x^2 + y^2} + \mathbf{k} = \frac{2u \cos v\,\mathbf{i} + 2u \sin v\,\mathbf{j}}{u^2} + \mathbf{k},$$

so the downward flux of $\mathbf{F}$ through $\mathcal{S}$ is

$$\iint_{\mathcal{S}} \mathbf{F} \bullet d\mathbf{S} = \int_0^{2\pi} dv \int_0^1 (4u - u)\,du = 3\pi.$$

∎

# Exercises 15.6

**1.** Find the flux of $\mathbf{F} = x\mathbf{i} + z\mathbf{j}$ out of the tetrahedron bounded by the coordinate planes and the plane $x + 2y + 3z = 6$.

**2.** Find the flux of $\mathbf{F} = x\mathbf{i} + y\mathbf{j} + z\mathbf{k}$ outward across the sphere $x^2 + y^2 + z^2 = a^2$.

**3.** Find the flux of the vector field of Exercise 2 out of the surface of the rectangular box $0 \leq x \leq a, 0 \leq y \leq b$, $0 \leq z \leq c$.

**4.** Find the flux of the vector field $\mathbf{F} = y\mathbf{i} + z\mathbf{k}$ out across the boundary of the solid cone $0 \leq z \leq 1 - \sqrt{x^2 + y^2}$.

**5.** Find the flux of $\mathbf{F} = x\mathbf{i} + y\mathbf{j} + z\mathbf{k}$ upward through the part of the surface $z = a - x^2 - y^2$ lying above the plane $z = b$, where $b < a$.

**6.** Find the flux of $\mathbf{F} = x\mathbf{i} + x\mathbf{j} + \mathbf{k}$ upward through the part of the surface $z = x^2 - y^2$ lying inside the cylinder $x^2 + y^2 = a^2$.

**7.** Find the flux of $\mathbf{F} = y^3\mathbf{i} + z^2\mathbf{j} + x\mathbf{k}$ downward through the part of the surface $z = 4 - x^2 - y^2$ that lies above the plane $z = 2x + 1$.

**8.** Find the flux of $\mathbf{F} = z^2\mathbf{k}$ upward through the part of the sphere $x^2 + y^2 + z^2 = a^2$ in the first octant of 3-space.

**9.** Find the flux of $\mathbf{F} = x\mathbf{i} + y\mathbf{j}$ upward through the part of the surface $z = 2 - x^2 - 2y^2$ that lies above the $xy$-plane.

**10.** Find the flux of $\mathbf{F} = 2x\mathbf{i} + y\mathbf{j} + z\mathbf{k}$ upward through the surface $\mathbf{r} = u^2v\mathbf{i} + uv^2\mathbf{j} + v^3\mathbf{k}$, $(0 \leq u \leq 1, 0 \leq v \leq 1)$.

**11.** Find the flux of $\mathbf{F} = x\mathbf{i} + y\mathbf{j} + z^2\mathbf{k}$ upward through the surface $u \cos v\,\mathbf{i} + u \sin v\,\mathbf{j} + u\,\mathbf{k}$, $(0 \leq u \leq 2, 0 \leq v \leq \pi)$.

**12.** Find the flux of $\mathbf{F} = yz\mathbf{i} - xz\mathbf{j} + (x^2 + y^2)\mathbf{k}$ upward through the surface $\mathbf{r} = e^u \cos v\,\mathbf{i} + e^u \sin v\,\mathbf{j} + u\,\mathbf{k}$, where $0 \leq u \leq 1$ and $0 \leq v \leq \pi$.

**13.** Find the flux of $\mathbf{F} = m\mathbf{r}/|\mathbf{r}|^3$ out of the surface of the cube $-a \leq x, y, z \leq a$.

**\* 14.** Find the flux of the vector field of Exercise 13 out of the box $1 \leq x, y, z \leq 2$. *Note:* This problem can be solved very easily using the Divergence Theorem of Section 16.3; the required flux is, in fact, zero. However, the object here is to do it by direct calculation of the surface integrals involved, and as such it is quite difficult. By symmetry, it is sufficient to evaluate the net flux out of the cube through any one of the three pairs of opposite faces; that is, you must calculate the flux through only two faces, say $z = 1$ and $z = 2$. Be prepared to work very hard to evaluate these integrals! When they are done you may find the identities

$$2\arctan a = \arctan\left(\frac{2a}{1 - a^2}\right), \quad \text{and}$$

$$\arctan a + \arctan\left(\frac{1}{a}\right) = \frac{\pi}{2}$$

useful for showing that the net flux is zero.

**15.** Define the flux of a *plane* vector field across a piecewise smooth *curve*. Find the flux of $\mathbf{F} = x\mathbf{i} + y\mathbf{j}$ outward across

(a) the circle $x^2 + y^2 = a^2$.

(b) the boundary of the square $-1 \leq x, y \leq 1$.

**16.** Find the flux of

$$\mathbf{F} = -\frac{x\mathbf{i} + y\mathbf{j}}{x^2 + y^2}$$

*inward* across each of the two curves in the previous exercise.

**17.** If $\mathcal{S}$ is a smooth, oriented surface in 3-space and $\hat{\mathbf{N}}$ is the unit vector field determining the orientation of $\mathcal{S}$, show that the flux of $\hat{\mathbf{N}}$ across $\mathcal{S}$ is the area of $\mathcal{S}$.

*18. The Divergence Theorem presented in Section 16.3 implies that the flux of a constant vector field across any oriented, piecewise smooth, closed surface is zero. Prove this now for (a) a rectangular box and (b) a sphere.

# Chapter Review

## Key Ideas

- **What do the following phrases mean?**
  - ⋄ a vector field
  - ⋄ a scalar field
  - ⋄ a field line
  - ⋄ a conservative field
  - ⋄ a scalar potential
  - ⋄ an equipotential
  - ⋄ a source
  - ⋄ a dipole
  - ⋄ a connected domain
  - ⋄ a simply connected domain
  - ⋄ a parametric surface
  - ⋄ an orientable surface
  - ⋄ the line integral of $f$ along curve $C$
  - ⋄ the line integral of the tangential component of $\mathbf{F}$ along $C$
  - ⋄ the flux of a vector field through a surface
- **How are the field lines of a conservative field related to its equipotential curves or surfaces?**
- **How is a line integral of a scalar field calculated?**
- **How is a line integral of the tangential component of a vector field calculated?**
- **When is a line integral between two points independent of the path joining those points?**
- **How is a surface integral of a scalar field calculated?**
- **How is the flux of a vector field through a surface calculated?**

## Review Exercises

1. Find $\int_C \dfrac{1}{y}\,ds$, where $C$ is the curve

$$x = t, \quad y = 2e^t, \quad z = e^{2t}, \quad (-1 \le t \le 1).$$

2. Let $C$ be the part of the curve of intersection of the surfaces $z = x + y^2$ and $y = 2x$ from the origin to the point $(2, 4, 18)$. Evaluate $\int_C 2y\,dx + x\,dy + 2\,dz$.

3. Find $\iint_S x\,dS$ where $S$ is that part of the cone $z = \sqrt{x^2 + y^2}$ in the region $0 \le x \le 1 - y^2$.

4. Find $\iint_S xyz\,dS$ over the part of the plane $x + y + z = 1$ lying in the first octant.

5. Find the flux of $x^2 y\mathbf{i} - 10xy^2\mathbf{j}$ upward through the surface $z = xy$, $0 \le x \le 1$, $0 \le y \le 1$.

6. Find the flux of $x\mathbf{i} + y\mathbf{j} + z\mathbf{k}$ downward through the part of the plane $x + 2y + 3z = 6$ lying in the first octant.

7. A bead of mass $m$ slides down a wire in the shape of the curve

$$x = a\sin t, \quad y = a\cos t, \quad z = bt,$$

where $0 \le t \le 6\pi$.
   (a) What is the work done by the gravitational force $\mathbf{F} = -mg\mathbf{k}$ on the bead during its descent?
   (b) What is the work done against a resistance of constant magnitude $R$ which directly opposes the motion of the bead during its descent?

8. For what values of the constants $a$, $b$, and $c$ can you determine the value of the integral $I$ of the tangential component of

$$\mathbf{F} = (axy + 3yz)\mathbf{i} + (x^2 + 3xz + by^2 z)\mathbf{j} + (bxy + cy^3)\mathbf{k}$$

along a curve from $(0, 1, -1)$ to $(2, 1, 1)$ without knowing exactly which curve? What is the value of the integral?

9. Let $\mathbf{F} = (x^2/y)\mathbf{i} + y\mathbf{j} + \mathbf{k}$.
   (a) Find the field line of $\mathbf{F}$ that passes through $(1, 1, 0)$ and show that it also passes through $(e, e, 1)$.
   (b) Find $\int_C \mathbf{F} \bullet d\mathbf{r}$ where $C$ is the part of the field line in (a) from $(1, 1, 0)$ to $(e, e, 1)$.

10. Consider the vector fields

$$\mathbf{F} = (1 + x)e^{x+y}\mathbf{i} + (xe^{x+y} + 2y)\mathbf{j} - 2z\mathbf{k},$$
$$\mathbf{G} = (1 + x)e^{x+y}\mathbf{i} + (xe^{x+y} + 2z)\mathbf{j} - 2y\mathbf{k}.$$

   (a) Show that $\mathbf{F}$ is conservative by finding a potential for it.
   (b) Evaluate $\int_C \mathbf{G} \bullet d\mathbf{r}$, where $C$ is given by

$$\mathbf{r} = (1 - t)e^t\mathbf{i} + t\mathbf{j} + 2t\mathbf{k}, \quad (0 \le t \le 1),$$

   by taking advantage of the similarity between $\mathbf{F}$ and $\mathbf{G}$.

11. Find a plane vector field $\mathbf{F}(x, y)$ that satisfies all of the following conditions:
   (i) The field lines of $\mathbf{F}$ are the curves $xy = C$.
   (ii) $|\mathbf{F}(x, y)| = 1$ if $(x, y) \ne (0, 0)$.
   (iii) $\mathbf{F}(1, 1) = (\mathbf{i} - \mathbf{j})/\sqrt{2}$.
   (iv) $\mathbf{F}$ is continuous except at $(0, 0)$.

12. Let $S$ be the part of the surface of the cylinder $y^2 + z^2 = 16$ that lies in the first octant and between the planes $x = 0$ and $x = 5$. Find the flux of $3z^2 x\mathbf{i} - x\mathbf{j} - y\mathbf{k}$ away from the $x$-axis through $S$.

## Challenging Problems

* **1.** Find the centroid of the surface

$$\mathbf{r} = (2 + \cos v)(\cos u\,\mathbf{i} + \sin u\,\mathbf{j}) + \sin v\,\mathbf{k},$$

where $0 \le u \le 2\pi$ and $0 \le v \le \pi$. Describe this surface.

* **2.** A smooth surface $\mathcal{S}$ is given parametrically by

$$\mathbf{r} = (\cos 2u)(2 + v \cos u)\mathbf{i}$$
$$+ (\sin 2u)(2 + v \cos u)\mathbf{j} + v \sin u\,\mathbf{k},$$

where $0 \le u \le 2\pi$ and $-1 \le v \le 1$. Show that for *every* smooth vector field $\mathbf{F}$ on $\mathcal{S}$,

$$\iint_{\mathcal{S}} \mathbf{F} \bullet \hat{\mathbf{N}}\, dS = 0,$$

where $\hat{\mathbf{N}} = \hat{\mathbf{N}}(u, v)$ is a unit normal vector field on $\mathcal{S}$ that depends continuously on $(u, v)$. How do you explain this? *Hint:* try to describe what the surface $\mathcal{S}$ looks like.

* **3.** Recalculate the gravitational force exerted by a sphere of radius $a$ and areal density $\sigma$ centred at the origin on a point mass located at $(0, 0, b)$ by directly integrating the vertical component of the force due to an area element $dS$, rather than by integrating the potential as we did in the last part of Section 15.5. You will have to be quite creative in dealing with the resulting integral.

# CHAPTER 16

# Vector Calculus

*Introduction* In this chapter we develop two- and three-dimensional analogues of the one-dimensional Fundamental Theorem of Calculus. These analogues — Green's Theorem, Gauss's Divergence Theorem, and Stokes's Theorem — are of great importance both theoretically and in applications. They are phrased in terms of certain differential operators, divergence and curl, which are related to the gradient operator encountered in Section 12.7. The operators are introduced and their properties are derived in Sections 16.1 and 16.2. The rest of the chapter deals with the generalizations of the Fundamental Theorem of Calculus and their applications.

## 16.1 Gradient, Divergence, and Curl

First-order information about the rate of change of a 3-dimensional scalar field, $f(x, y, z)$, is contained in the three first partial derivatives $\partial f/\partial x$, $\partial f/\partial y$, and $\partial f/\partial z$. The gradient,

$$\mathbf{grad}\, f(x, y, z) = \nabla f(x, y, z) = \frac{\partial f}{\partial x}\mathbf{i} + \frac{\partial f}{\partial y}\mathbf{j} + \frac{\partial f}{\partial z}\mathbf{k},$$

collects this information into a single vector-valued "derivative" of $f$. We would like to develop similar ways of conveying information about the rate of change of vector fields.

First-order information about the rate of change of the vector field

$$\mathbf{F}(x, y, z) = F_1(x, y, z)\mathbf{i} + F_2(x, y, z)\mathbf{j} + F_3(x, y, z)\mathbf{k}$$

is contained in nine first partial derivatives, three for each of the three components of $\mathbf{F}$:

$$\frac{\partial F_1}{\partial x} \qquad \frac{\partial F_1}{\partial y} \qquad \frac{\partial F_1}{\partial z}$$

$$\frac{\partial F_2}{\partial x} \qquad \frac{\partial F_2}{\partial y} \qquad \frac{\partial F_2}{\partial z}$$

$$\frac{\partial F_3}{\partial x} \qquad \frac{\partial F_3}{\partial y} \qquad \frac{\partial F_3}{\partial z}.$$

(Again we stress that $F_1$, $F_2$, and $F_3$ denote the components of $\mathbf{F}$, not partial derivatives.) Two special combinations of these derivatives organize this information in particularly useful ways, as the gradient does for scalar fields. These are the **divergence** of $\mathbf{F}$ (**div F**), and the **curl** of $\mathbf{F}$ (**curl F**), defined as follows:

**Divergence and curl**

$$\textbf{div}\,\mathbf{F} = \nabla \bullet \mathbf{F} = \frac{\partial F_1}{\partial x} + \frac{\partial F_2}{\partial y} + \frac{\partial F_3}{\partial z},$$

$$\textbf{curl}\,\mathbf{F} = \nabla \times \mathbf{F}$$

$$= \left(\frac{\partial F_3}{\partial y} - \frac{\partial F_2}{\partial z}\right)\mathbf{i} + \left(\frac{\partial F_1}{\partial z} - \frac{\partial F_3}{\partial x}\right)\mathbf{j} + \left(\frac{\partial F_2}{\partial x} - \frac{\partial F_1}{\partial y}\right)\mathbf{k}$$

$$= \begin{vmatrix} \mathbf{i} & \mathbf{j} & \mathbf{k} \\ \frac{\partial}{\partial x} & \frac{\partial}{\partial y} & \frac{\partial}{\partial z} \\ F_1 & F_2 & F_3 \end{vmatrix}.$$

Note that the divergence of a vector field is a scalar field, while the curl is another vector field. Also observe the notation $\nabla \bullet \mathbf{F}$ and $\nabla \times \mathbf{F}$, which we will sometimes use instead of **div F** and **curl F**. This makes use of the *vector differential operator*,

$$\nabla = \mathbf{i}\,\frac{\partial}{\partial x} + \mathbf{j}\,\frac{\partial}{\partial y} + \mathbf{k}\,\frac{\partial}{\partial z}$$

frequently called *del* or *nabla*. Just as the gradient of the scalar field $f$ can be regarded as *formal scalar multiplication* of $f$ by $\nabla$, so also can the divergence and curl of $\mathbf{F}$ be regarded as *formal dot* and *cross* products of $\nabla$ with $\mathbf{F}$. When using $\nabla$ the order of "factors" is important; the quantities on which $\nabla$ acts must appear to the right of $\nabla$. For instance, $\nabla \bullet \mathbf{F}$ and $\mathbf{F} \bullet \nabla$ do not mean the same thing; the former is a scalar field, the latter is a scalar differential operator.

**Example 1** Find the divergence and curl of the vector field

$$\mathbf{F} = xy\mathbf{i} + (y^2 - z^2)\mathbf{j} + yz\mathbf{k}.$$

*Solution* We have

$$\textbf{div}\,\mathbf{F} = \nabla \bullet \mathbf{F} = \frac{\partial}{\partial x}(xy) + \frac{\partial}{\partial y}(y^2 - z^2) + \frac{\partial}{\partial z}(yz) = y + 2y + y = 4y,$$

$$\textbf{curl}\,\mathbf{F} = \nabla \times \mathbf{F} = \begin{vmatrix} \mathbf{i} & \mathbf{j} & \mathbf{k} \\ \frac{\partial}{\partial x} & \frac{\partial}{\partial y} & \frac{\partial}{\partial z} \\ xy & y^2 - z^2 & yz \end{vmatrix}$$

$$= \left[\frac{\partial}{\partial y}(yz) - \frac{\partial}{\partial z}(y^2 - z^2)\right]\mathbf{i} + \left[\frac{\partial}{\partial z}(xy) - \frac{\partial}{\partial x}(yz)\right]\mathbf{j}$$

$$+ \left[\frac{\partial}{\partial x}(y^2 - z^2) - \frac{\partial}{\partial y}(xy)\right]\mathbf{k} = 3z\mathbf{i} - x\mathbf{k}.$$

The divergence and curl of a two-dimensional vector field are also defined: if $\mathbf{F}(x, y) = F_1(x, y)\mathbf{i} + F_2(x, y)\mathbf{j}$, then

$$\text{div } \mathbf{F} = \frac{\partial F_1}{\partial x} + \frac{\partial F_2}{\partial y},$$

$$\text{curl } \mathbf{F} = \left( \frac{\partial F_2}{\partial x} - \frac{\partial F_1}{\partial y} \right) \mathbf{k}.$$

Note that the curl of a two-dimensional vector field is still a 3-vector and is perpendicular to the plane of the field. Although **div** and **grad** are defined in all dimensions, **curl** is defined only in three dimensions and in the plane (provided we allow values in three dimensions).

**Example 2**  Find the divergence and curl of $\mathbf{F} = xe^y\mathbf{i} - ye^x\mathbf{j}$.

*Solution*  We have

$$\text{div } \mathbf{F} = \nabla \bullet \mathbf{F} = \frac{\partial}{\partial x}(xe^y) + \frac{\partial}{\partial y}(-ye^x) = e^y - e^x,$$

$$\text{curl } \mathbf{F} = \nabla \times \mathbf{F} = \left( \frac{\partial}{\partial x}(-ye^x) - \frac{\partial}{\partial y}(xe^y) \right) \mathbf{k}$$
$$= -(ye^x + xe^y)\mathbf{k}.$$

## Interpretation of the Divergence

The value of the divergence of a vector field $\mathbf{F}$ at point $P$ is, loosely speaking, a measure of the rate at which the field "diverges" or "spreads away" from $P$. This spreading away can be measured by the flux out of a small closed surface surrounding $P$. For instance, $\text{div } \mathbf{F}(P)$ is the limit of the *flux per unit volume* out of smaller and smaller spheres centred at $P$.

**THEOREM 1**

**The divergence as flux density**

If $\hat{\mathbf{N}}$ is the unit outward normal on the sphere $\mathcal{S}_\epsilon$ of radius $\epsilon$ centred at point $P$, and if $\mathbf{F}$ is a smooth three-dimensional vector field, then

$$\text{div } \mathbf{F}(P) = \lim_{\epsilon \to 0^+} \frac{3}{4\pi\epsilon^3} \iint_{\mathcal{S}_\epsilon} \mathbf{F} \bullet \hat{\mathbf{N}} \, dS.$$

**PROOF**  Without loss of generality we assume that $P$ is at the origin. We want to expand $\mathbf{F}$ in a Taylor series about the origin (a Maclaurin series). As shown in Section 12.9 for a function of two variables, the Maclaurin series for a scalar-valued function of three variables takes the form

$$f(x, y, z) = f(0, 0, 0) + \frac{\partial f}{\partial x}\bigg|_{(0,0,0)} x + \frac{\partial f}{\partial y}\bigg|_{(0,0,0)} y + \frac{\partial f}{\partial z}\bigg|_{(0,0,0)} z + \cdots,$$

where "$\cdots$" represents terms of second and higher degree in $x$, $y$, and $z$. If we apply this formula to the components of $\mathbf{F}$, we obtain

$$\mathbf{F}(x, y, z) = \mathbf{F}_0 + \mathbf{F}_1 x + \mathbf{F}_2 y + \mathbf{F}_3 z + \cdots$$

where

$$\mathbf{F}_0 = \mathbf{F}(0, 0, 0)$$

$$\mathbf{F}_1 = \frac{\partial \mathbf{F}}{\partial x}\bigg|_{(0,0,0)} = \left(\frac{\partial F_1}{\partial x}\mathbf{i} + \frac{\partial F_2}{\partial x}\mathbf{j} + \frac{\partial F_3}{\partial x}\mathbf{k}\right)\bigg|_{(0,0,0)}$$

$$\mathbf{F}_2 = \frac{\partial \mathbf{F}}{\partial y}\bigg|_{(0,0,0)} = \left(\frac{\partial F_1}{\partial y}\mathbf{i} + \frac{\partial F_2}{\partial y}\mathbf{j} + \frac{\partial F_3}{\partial y}\mathbf{k}\right)\bigg|_{(0,0,0)}$$

$$\mathbf{F}_3 = \frac{\partial \mathbf{F}}{\partial z}\bigg|_{(0,0,0)} = \left(\frac{\partial F_1}{\partial z}\mathbf{i} + \frac{\partial F_2}{\partial z}\mathbf{j} + \frac{\partial F_3}{\partial z}\mathbf{k}\right)\bigg|_{(0,0,0)}$$

and again the "$\cdots$" represents the second- and higher-degree terms in $x$, $y$, and $z$. The unit normal on $\mathcal{S}_\epsilon$ is $\hat{\mathbf{N}} = (x\mathbf{i} + y\mathbf{j} + z\mathbf{k})/\epsilon$, so we have

$$\begin{aligned}
\mathbf{F} \bullet \hat{\mathbf{N}} = \frac{1}{\epsilon}\Big(&\mathbf{F}_0 \bullet \mathbf{i}\, x + \mathbf{F}_0 \bullet \mathbf{j}\, y + \mathbf{F}_0 \bullet \mathbf{k}\, z \\
&+ \mathbf{F}_1 \bullet \mathbf{i}\, x^2 + \mathbf{F}_1 \bullet \mathbf{j}\, xy + \mathbf{F}_1 \bullet \mathbf{k}\, xz \\
&+ \mathbf{F}_2 \bullet \mathbf{i}\, xy + \mathbf{F}_2 \bullet \mathbf{j}\, y^2 + \mathbf{F}_2 \bullet \mathbf{k}\, yz \\
&+ \mathbf{F}_3 \bullet \mathbf{i}\, xz + \mathbf{F}_3 \bullet \mathbf{j}\, yz + \mathbf{F}_3 \bullet \mathbf{k}\, z^2 + \cdots\Big).
\end{aligned}$$

We integrate each term within the parentheses over $\mathcal{S}_\epsilon$. By symmetry,

$$\oiint_{\mathcal{S}_\epsilon} x\, dS = \oiint_{\mathcal{S}_\epsilon} y\, dS = \oiint_{\mathcal{S}_\epsilon} z\, dS = 0,$$

$$\oiint_{\mathcal{S}_\epsilon} xy\, dS = \oiint_{\mathcal{S}_\epsilon} xz\, dS = \oiint_{\mathcal{S}_\epsilon} yz\, dS = 0.$$

Also, by symmetry,

$$\begin{aligned}
\oiint_{\mathcal{S}_\epsilon} x^2\, dS &= \oiint_{\mathcal{S}_\epsilon} y^2\, dS = \oiint_{\mathcal{S}_\epsilon} z^2\, dS \\
&= \frac{1}{3}\oiint_{\mathcal{S}_\epsilon} (x^2 + y^2 + z^2)\, dS = \frac{1}{3}(\epsilon^2)(4\pi\epsilon^2) = \frac{4}{3}\pi\epsilon^4,
\end{aligned}$$

and the higher-degree terms have surface integrals involving $\epsilon^5$ and higher powers. Thus

$$\frac{3}{4\pi\epsilon^3} \oiint_{\mathcal{S}_\epsilon} \mathbf{F} \bullet \hat{\mathbf{N}}\, dS = \mathbf{F}_1 \bullet \mathbf{i} + \mathbf{F}_2 \bullet \mathbf{j} + \mathbf{F}_3 \bullet \mathbf{k} + \epsilon\,(\cdots)$$

$$= \nabla \bullet \mathbf{F}(0, 0, 0) + \epsilon\,(\cdots)$$

$$\to \nabla \bullet \mathbf{F}(0, 0, 0)$$

as $\epsilon \to 0^+$. This is what we wanted to show.

**Remark**   The spheres $\mathcal{S}_\epsilon$ in the above theorem can be replaced by other contracting families of piecewise smooth surfaces. For instance, if $B$ is the surface of a rectangular box with dimensions $\Delta x$, $\Delta y$, and $\Delta z$ containing $P$ then

$$\mathbf{div}\, \mathbf{F}(P) = \lim_{\Delta x, \Delta y, \Delta z \to 0} \frac{1}{\Delta x \Delta y \Delta z} \oiint_B \mathbf{F} \bullet \hat{\mathbf{N}}\, dS.$$

See Exercise 12 at the end of this section.

**Remark** In two dimensions, the value **div F**$(P)$ represents the limiting *flux per unit area* outward across small, non-self-intersecting closed curves which enclose $P$. See Exercise 13 at the end of this section.

Let us return again to the interpretation of a vector field as a velocity field of a moving incompressible fluid. If the total flux of the velocity field outward across the boundary surface of a domain is positive (or negative), then the fluid must be produced (or annihilated) within that domain.

The vector field $\mathbf{F} = x\mathbf{i} + y\mathbf{j} + z\mathbf{k}$ of Example 2 in Section 15.6 has constant divergence, $\boldsymbol{\nabla} \bullet \mathbf{F} = 3$. In that example we showed that the flux of **F** out of a certain cylinder of base radius $a$ and height $2h$ is $6\pi a^2 h$, which is three times the volume of the cylinder. Exercises 2 and 3 of Section 15.6 confirm similar results for the flux of **F** out of other domains. This leads to another interpretation for the divergence; **div F**$(P)$ is the *source strength per unit volume* of **F** at $P$. With this interpretation, we would expect, even for a vector field **F** with nonconstant divergence, that the total flux of **F** out the surface $\mathcal{S}$ of a domain $D$ would be equal to the total source strength of **F** within $D$, that is,

$$\oiint_{\mathcal{S}} \mathbf{F} \bullet \hat{\mathbf{N}} \, dS = \iiint_{D} \boldsymbol{\nabla} \bullet \mathbf{F} \, dV.$$

This is the **Divergence Theorem**, which we will prove in Section 16.4.

---

**Example 3** Verify that the vector field $\mathbf{F} = m\mathbf{r}/|\mathbf{r}|^3$, due to a source of strength $m$ at $(0, 0, 0)$, has zero divergence at all points in $\mathbb{R}^3$ except the origin. What would you expect to be the total flux of **F** outward across the boundary surface of a domain $D$ if the origin lies outside $D$? if the origin is inside $D$?

**Solution** Since

$$\mathbf{F}(x, y, z) = \frac{m}{r^3}\left(x\mathbf{i} + y\mathbf{j} + z\mathbf{k}\right), \quad \text{where} \quad r^2 = x^2 + y^2 + z^2,$$

and since $\partial r/\partial x = x/r$, we have

$$\frac{\partial F_1}{\partial x} = m\frac{\partial}{\partial x}\left(\frac{x}{r^3}\right) = m\frac{r^3 - 3xr^2\left(\frac{x}{r}\right)}{r^6} = m\frac{r^2 - 3x^2}{r^5}.$$

Similarly,

$$\frac{\partial F_2}{\partial y} = m\frac{r^2 - 3y^2}{r^5} \quad \text{and} \quad \frac{\partial F_3}{\partial z} = m\frac{r^2 - 3z^2}{r^5}.$$

Adding these up, we get $\boldsymbol{\nabla} \bullet \mathbf{F}(x, y, z) = 0$ if $r > 0$.

If the origin lies outside the domain $D$ then the source density of **F** in $D$ is zero, so we would expect the total flux of **F** out of $D$ to be zero. If the origin lies inside $D$ then $D$ contains a source of strength $m$ (producing $4\pi m$ cubic units of fluid per unit time) and so we would expect the flux out of $D$ to be $4\pi m$. See Example 1 and Exercises 9 and 10 of Section 15.6 for specific examples.

## Distributions and Delta Functions

If $\ell(x)$ represents the line density (mass per unit length) of mass distributed on the $x$-axis, then the total mass so distributed is

$$m = \int_{-\infty}^{\infty} \ell(x)\,dx.$$

Now suppose that the only mass on the axis is a "point mass" $m = 1$ located at the origin. Then at all other points $x \neq 0$, the density is $\ell(x) = 0$, but we must still have

$$\int_{-\infty}^{\infty} \ell(x)\,dx = m = 1,$$

so $\ell(0)$ must be infinite. This is an ideal situation, a mathematical model. No real *function* $\ell(x)$ can have such properties; if a function is zero everywhere except at a single point then any integral of that function will be zero. (Why?) (Also, no real mass can occupy just a single point.) Nevertheless, it is very useful to model real, isolated masses as point masses, and to model their densities using **generalized functions** (also called **distributions**).

We can think of the density of a point mass 1 at $x = 0$ as the limit of large densities concentrated on small intervals. For instance, if

$$d_n(x) = \begin{cases} n/2 & \text{if } |x| \leq 1/n \\ 0 & \text{if } |x| > 1/n \end{cases}$$

(see Figure 16.1), then for any smooth function $f(x)$ defined on $\mathbb{R}$ we have

$$\int_{-\infty}^{\infty} d_n(x)\,f(x)\,dx = \frac{n}{2} \int_{-1/n}^{1/n} f(x)\,dx.$$

Replace $f(x)$ in the integral on the right with its Maclaurin series:

$$f(x) = f(0) + \frac{f'(0)}{1!}x + \frac{f''(0)}{2!}x^2 + \cdots.$$

Since

$$\int_{-1/n}^{1/n} x^k\,dx = \begin{cases} 2/((k+1)n^{k+1}) & \text{if } k \text{ is even} \\ 0 & \text{if } k \text{ is odd,} \end{cases}$$

we can take the limit as $n \to \infty$ and obtain

$$\lim_{n\to\infty} \int_{-\infty}^{\infty} d_n(x)\,f(x)\,dx = f(0).$$

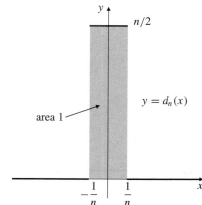

**Figure 16.1**   The functions $d_n(x)$ converge to $\delta(x)$ as $n \to \infty$

**DEFINITION 1**

The **Dirac distribution** $\delta(x)$ (also called the **Dirac delta function**, although it is really not a function) is the limit of the sequence $d_n(x)$ as $n \to \infty$. It is defined by the requirement that

$$\int_{-\infty}^{\infty} \delta(x)f(x)\,dx = f(0)$$

for every smooth function $f(x)$.

A formal change of variables shows that the delta function also satisfies

$$\int_{-\infty}^{\infty} \delta(x - t) f(t)\, dt = f(x).$$

**Example 4** In view of the fact that $\mathbf{F}(\mathbf{r}) = m\mathbf{r}/|\mathbf{r}|^3$ satisfies $\text{div } \mathbf{F}(x, y, z) = 0$ for $(x, y, z) \neq (0, 0, 0)$ but produces a flux of $4\pi m$ out of any sphere centred at the origin, we can regard $\text{div } \mathbf{F}(x, y, z)$ as a distribution

$$\text{div } \mathbf{F}(x, y, z) = 4\pi m \delta(x)\delta(y)\delta(z).$$

In particular, integrating this distribution against $f(x, y, z) = 1$ over $\mathbb{R}^3$, we have

$$\iiint_{\mathbb{R}^3} \text{div } \mathbf{F}(x, y, z)\, dV = 4\pi m \int_{-\infty}^{\infty} \delta(x)\, dx \int_{-\infty}^{\infty} \delta(y)\, dy \int_{-\infty}^{\infty} \delta(z)\, dz$$

$$= 4\pi m.$$

The integral can equally well be taken over *any domain* in $\mathbb{R}^3$ that contains the origin in its interior, and the result will be the same. If the origin is outside the domain the result will be zero. We will reexamine this situation after establishing the Divergence Theorem in Section 16.4. ∎

A formal study of distributions is beyond the scope of this book and is usually undertaken in more advanced textbooks on differential equations and engineering mathematics.

## Interpretation of the Curl

Roughly speaking, $\text{curl } \mathbf{F}(P)$ measures the extent to which the vector field $\mathbf{F}$ "swirls" around $P$.

**Example 5** Consider the velocity field,

$$\mathbf{v} = -\Omega y \mathbf{i} + \Omega x \mathbf{j},$$

of a solid rotating with angular speed $\Omega$ about the $z$-axis, that is, with angular velocity $\boldsymbol{\Omega} = \Omega \mathbf{k}$. Calculate the circulation of this field around a circle $C_\epsilon$ in the $xy$-plane centred at any point $(x_0, y_0)$, having radius $\epsilon$, and oriented counterclockwise. What is the relationship between this circulation and the curl of $\mathbf{v}$?

**Solution** The indicated circle has parametrization

$$\mathbf{r} = (x_0 + \epsilon \cos t)\mathbf{i} + (y_0 + \epsilon \sin t)\mathbf{j}, \qquad (0 \leq t \leq 2\pi),$$

and the circulation of $\mathbf{v}$ around it is given by

$$\oint_{\mathcal{C}_\epsilon} \mathbf{v} \bullet d\mathbf{r} = \int_0^{2\pi} \Big( -\Omega(y_0 + \epsilon \sin t)(-\epsilon \sin t) + \Omega(x_0 + \epsilon \cos t)(\epsilon \cos t) \Big) \, dt$$

$$= \int_0^{2\pi} \Big( \Omega\epsilon(y_0 \sin t + x_0 \cos t) + \Omega\epsilon^2 \Big) \, dt$$

$$= 2\Omega\pi\epsilon^2.$$

Since

$$\mathbf{curl}\,\mathbf{v} = \nabla\times\mathbf{v} = \Big( \frac{\partial}{\partial x}(\Omega x) - \frac{\partial}{\partial y}(-\Omega y) \Big)\mathbf{k} = 2\Omega\mathbf{k} = 2\boldsymbol{\Omega},$$

the circulation is the product of $(\mathbf{curl}\,\mathbf{v}) \bullet \mathbf{k}$ and the area bounded by $\mathcal{C}_\epsilon$. Note that this circulation is constant for circles of any fixed radius; it does not depend on the position of the centre.

The calculations in the example above suggest that the curl of a vector field is a measure of the *circulation per unit area* in planes normal to the curl. A more precise version of this conjecture is stated in Theorem 2 below. We will not prove this theorem now, because a proof at this stage would be quite complicated. (However, see Exercise 14 at the end of this section for a special case.) A simple proof can be based on Stokes's Theorem. (See Exercise 13 in Section 16.5.)

**THEOREM 2**

**The curl as circulation density**

If $\mathbf{F}$ is a smooth vector field and $\mathcal{C}_\epsilon$ is a circle of radius $\epsilon$ centred at point $P$ and bounding a disk $\mathcal{S}_\epsilon$ with unit normal $\hat{\mathbf{N}}$ (and orientation inherited from $\mathcal{C}_\epsilon$ — see Figure 16.2), then

$$\lim_{\epsilon\to0^+} \frac{1}{\pi\epsilon^2} \oint_{\mathcal{C}_\epsilon} \mathbf{F} \bullet d\mathbf{r} = \hat{\mathbf{N}} \bullet \mathbf{curl}\,\mathbf{F}(P).$$

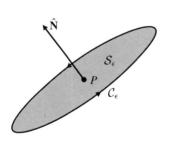

**Figure 16.2**

Example 5 also suggests the following definition for the *local* angular velocity of a moving fluid.

> The **local angular velocity** at point $P$ in a fluid moving with velocity field $\mathbf{v}(P)$ is given by
>
> $$\boldsymbol{\Omega}(P) = \tfrac{1}{2}\mathbf{curl}\,\mathbf{v}(P).$$

Theorem 2 states that the local angular velocity $\boldsymbol{\Omega}(P)$ is that vector whose component in the direction of any unit vector $\hat{\mathbf{N}}$ is one-half of the limiting circulation per unit area around the (oriented) boundary circles of small disks centred at $P$ and having normal $\hat{\mathbf{N}}$.

Not all vector fields with nonzero curl *appear* to circulate. The velocity field for the rigid body rotation considered in Example 5 appears to circulate around the axis of rotation, but the circulation around a circle in a plane perpendicular to that axis turned out to be independent of the position of the circle; it depended only on its area. The circle need not even surround the axis of rotation. The following example investigates a fluid velocity field whose streamlines are *straight lines*, but which still has nonzero, constant curl, and therefore constant local angular velocity.

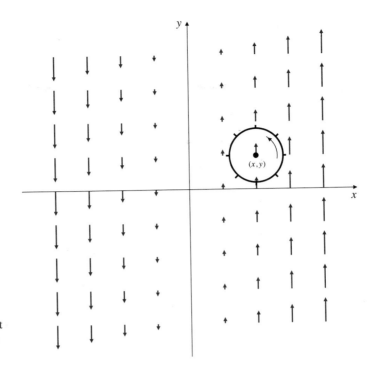

**Figure 16.3**    The paddle wheel is not only carried along but is set rotating by flow

**Example 6**    Consider the velocity field $\mathbf{v} = x\mathbf{j}$ of a fluid moving in the $xy$-plane. Evidently, particles of fluid are moving along lines parallel to the $y$-axis. However, **curl** $\mathbf{v}(x, y) = \mathbf{k}$, and $\Omega(x, y) = \frac{1}{2}\mathbf{k}$. A small "paddle wheel" of radius $\epsilon$ placed with its centre at position $(x, y)$ in the fluid (see Figure 16.3) will be carried along with the fluid at velocity $x\mathbf{i}$, but will also be set rotating with angular velocity $\Omega(x, y)$ which is independent of its position. This angular velocity is due to the fact that the velocity of the fluid along the right side of the wheel exceeds that along the left side.    ■

## Exercises 16.1

In Exercises 1–11, calculate **div F** and **curl F** for the given vector fields.

1. $\mathbf{F} = x\mathbf{i} + y\mathbf{j}$

2. $\mathbf{F} = y\mathbf{i} + x\mathbf{j}$

3. $\mathbf{F} = y\mathbf{i} + z\mathbf{j} + x\mathbf{k}$

4. $\mathbf{F} = yz\mathbf{i} + xz\mathbf{j} + xy\mathbf{k}$

5. $\mathbf{F} = x\mathbf{i} + x\mathbf{k}$

6. $\mathbf{F} = xy^2\mathbf{i} - yz^2\mathbf{j} + zx^2\mathbf{k}$

7. $\mathbf{F} = f(x)\mathbf{i} + g(y)\mathbf{j} + h(z)\mathbf{k}$    8. $\mathbf{F} = f(z)\mathbf{i} - f(z)\mathbf{j}$

9. $\mathbf{F}(r, \theta) = r\mathbf{i} + \sin\theta\mathbf{j}$, where $(r, \theta)$ are polar coordinates in the plane

10. $\mathbf{F} = \hat{\mathbf{r}} = \cos\theta\mathbf{i} + \sin\theta\mathbf{j}$

11. $\mathbf{F} = \hat{\boldsymbol{\theta}} = -\sin\theta\mathbf{i} + \cos\theta\mathbf{j}$.

* 12. Let $\mathbf{F}$ be a smooth, 3-dimensional vector field. If $B_{a,b,c}$ is the surface of the box $-a \leq x \leq a,\ -b \leq y \leq b,\ -c \leq z \leq c$, with outward normal $\hat{\mathbf{N}}$, show that

$$\lim_{a,b,c \to 0^+} \frac{1}{8abc} \oiint_{B_{a,b,c}} \mathbf{F} \bullet \hat{\mathbf{N}}\, dS = \nabla \bullet \mathbf{F}(0, 0, 0).$$

* 13. Let $\mathbf{F}$ be a smooth 2-dimensional vector field. If $\mathcal{C}_\epsilon$ is the circle of radius $\epsilon$ centred at the origin, and $\hat{\mathbf{N}}$ is the unit outward normal to $\mathcal{C}_\epsilon$, show that

$$\lim_{\epsilon \to 0^+} \frac{1}{\pi\epsilon^2} \oint_{\mathcal{C}_\epsilon} \mathbf{F} \bullet \hat{\mathbf{N}}\, ds = \mathbf{div\, F}(0, 0).$$

* 14. Prove Theorem 2 in the special case that $\mathcal{C}_\epsilon$ is the circle in the $xy$-plane with parametrization $x = \epsilon\cos\theta$, $y = \epsilon\sin\theta$,    $(0 \leq \theta \leq 2\pi)$. In this case $\hat{\mathbf{N}} = \mathbf{k}$. *Hint:* expand $\mathbf{F}(x, y, z)$ in a vector Taylor series about the origin as in the proof of Theorem 1, and calculate the circulation of individual terms around $\mathcal{C}_\epsilon$.

## 16.2 Some Identities Involving Grad, Div, and Curl

There are numerous identities involving the functions

$$\mathbf{grad}\, f(x, y, z) = \nabla f(x, y, z) = \frac{\partial f}{\partial x}\mathbf{i} + \frac{\partial f}{\partial y}\mathbf{j} + \frac{\partial f}{\partial z}\mathbf{k},$$

$$\mathbf{div}\, \mathbf{F}(x, y, z) = \nabla \bullet \mathbf{F}(x, y, z) = \frac{\partial F_1}{\partial x} + \frac{\partial F_2}{\partial y} + \frac{\partial F_3}{\partial z},$$

$$\mathbf{curl}\, \mathbf{F}(x, y, z) = \nabla \times \mathbf{F}(x, y, z) = \begin{vmatrix} \mathbf{i} & \mathbf{j} & \mathbf{k} \\ \frac{\partial}{\partial x} & \frac{\partial}{\partial y} & \frac{\partial}{\partial z} \\ F_1 & F_2 & F_3 \end{vmatrix},$$

and the **Laplacian operator**, $\nabla^2 = \nabla \bullet \nabla$, defined for a scalar field $\phi$ by

$$\nabla^2 \phi = \nabla \bullet \nabla \phi = \mathbf{div}\,\mathbf{grad}\,\phi = \frac{\partial^2 \phi}{\partial x^2} + \frac{\partial^2 \phi}{\partial y^2} + \frac{\partial^2 \phi}{\partial z^2},$$

and for a vector field $\mathbf{F} = F_1\mathbf{i} + F_2\mathbf{j} + F_3\mathbf{k}$ by

$$\nabla^2 \mathbf{F} = (\nabla^2 F_1)\mathbf{i} + (\nabla^2 F_2)\mathbf{j} + (\nabla^2 F_3)\mathbf{k}.$$

(The Laplacian operator, $\nabla^2 = (\partial^2/\partial x^2) + (\partial^2/\partial y^2) + (\partial^2/\partial z^2)$, is denoted by $\Delta$ in some books.) Recall that a function $\phi$ is called **harmonic** in a domain $D$ if $\nabla^2 \phi = 0$ throughout $D$. (See Section 12.4.)

We collect the most important identities together in the following theorem. Most of them are forms of the Product Rule. We will prove a few of the identities to illustrate the techniques involved (mostly brute-force calculation) and leave the rest as exercises. Note that two of the identities involve quantities like $(\mathbf{G} \bullet \nabla)\mathbf{F}$; this represents the vector obtained by applying the scalar differential operator $\mathbf{G} \bullet \nabla$ to the vector field $\mathbf{F}$:

$$(\mathbf{G} \bullet \nabla)\mathbf{F} = G_1\frac{\partial \mathbf{F}}{\partial x} + G_2\frac{\partial \mathbf{F}}{\partial y} + G_3\frac{\partial \mathbf{F}}{\partial z}.$$

**THEOREM 3**

**Vector differential identities**

Let $\phi$ and $\Psi$ be scalar fields and $\mathbf{F}$ and $\mathbf{G}$ be vector fields, all assumed to be sufficiently smooth that all the partial derivatives in the identities are continuous. Then the following identities hold:

(a)    $\nabla(\phi\Psi) = \phi\nabla\Psi + \Psi\nabla\phi$

(b)    $\nabla \bullet (\phi\mathbf{F}) = (\nabla\phi) \bullet \mathbf{F} + \phi(\nabla \bullet \mathbf{F})$

(c)    $\nabla \times (\phi\mathbf{F}) = (\nabla\phi) \times \mathbf{F} + \phi(\nabla \times \mathbf{F})$

(d)    $\nabla \bullet (\mathbf{F} \times \mathbf{G}) = (\nabla \times \mathbf{F}) \bullet \mathbf{G} - \mathbf{F} \bullet (\nabla \times \mathbf{G})$

(e)    $\nabla \times (\mathbf{F} \times \mathbf{G}) = (\nabla \bullet \mathbf{G})\mathbf{F} + (\mathbf{G} \bullet \nabla)\mathbf{F} - (\nabla \bullet \mathbf{F})\mathbf{G} - (\mathbf{F} \bullet \nabla)\mathbf{G}$

(f)  $\nabla(\mathbf{F} \bullet \mathbf{G}) = \mathbf{F} \times (\nabla \times \mathbf{G}) + \mathbf{G} \times (\nabla \times \mathbf{F}) + (\mathbf{F} \bullet \nabla)\mathbf{G} + (\mathbf{G} \bullet \nabla)\mathbf{F}$

(g)  $\nabla \bullet (\nabla \times \mathbf{F}) = 0$ $\qquad$ **(div curl = 0)**

(h)  $\nabla \times (\nabla \phi) = \mathbf{0}$ $\qquad$ **(curl grad = 0)**

(i)  $\nabla \times (\nabla \times \mathbf{F}) = \nabla(\nabla \bullet \mathbf{F}) - \nabla^2\mathbf{F}$

$\qquad\qquad\qquad$ **(curl curl = grad div** − Laplacian)

Identities (a)–(f) are versions of the Product Rule, and are first-order identities involving only one application of $\nabla$. Identities (g)–(i) are second-order identities. Identities (g) and (h) are equivalent to the equality of mixed partial derivatives and are especially important for understanding of **div** and **curl**.

**PROOF** We will prove only identities (c), (e), and (g). The remaining proofs are similar to these.

(c) The first component (**i** component) of $\nabla \times (\phi\mathbf{F})$ is

$$\frac{\partial}{\partial y}(\phi F_3) - \frac{\partial}{\partial z}(\phi F_2) = \frac{\partial \phi}{\partial y}F_3 - \frac{\partial \phi}{\partial z}F_2 + \phi\frac{\partial F_3}{\partial y} - \phi\frac{\partial F_2}{\partial z}.$$

The first two terms on the right constitute the first component of $(\nabla\phi) \times \mathbf{F}$ and the last two terms constitute the first component of $\phi(\nabla \times \mathbf{F})$. Therefore the first components of both sides of identity (c) are equal. The equality of the other components follows similarly.

(e) Again, it is sufficient to show that the first components of the vectors on both sides of the identity are equal. To calculate the first component of $\nabla \times (\mathbf{F} \times \mathbf{G})$ we need the second and third components of $\mathbf{F} \times \mathbf{G}$, which are

$$(\mathbf{F} \times \mathbf{G})_2 = F_3 G_1 - F_1 G_3 \quad \text{and} \quad (\mathbf{F} \times \mathbf{G})_3 = F_1 G_2 - F_2 G_1.$$

The first component of $\nabla \times (\mathbf{F} \times \mathbf{G})$ is therefore

$$\frac{\partial}{\partial y}(F_1 G_2 - F_2 G_1) - \frac{\partial}{\partial z}(F_3 G_1 - F_1 G_3)$$

$$= \frac{\partial F_1}{\partial y}G_2 + F_1\frac{\partial G_2}{\partial y} - \frac{\partial F_2}{\partial y}G_1 - F_2\frac{\partial G_1}{\partial y} - \frac{\partial F_3}{\partial z}G_1$$

$$- F_3\frac{\partial G_1}{\partial z} + \frac{\partial F_1}{\partial z}G_3 + F_1\frac{\partial G_3}{\partial z}.$$

The first components of the four terms on the right side of identity (e) are

$$((\nabla \bullet \mathbf{G})\mathbf{F})_1 = F_1\frac{\partial G_1}{\partial x} + F_1\frac{\partial G_2}{\partial y} + F_1\frac{\partial G_3}{\partial z}$$

$$((\mathbf{G} \bullet \nabla)\mathbf{F})_1 = \frac{\partial F_1}{\partial x}G_1 + \frac{\partial F_1}{\partial y}G_2 + \frac{\partial F_1}{\partial z}G_3$$

$$-((\nabla \bullet \mathbf{F})\mathbf{G})_1 = -\frac{\partial F_1}{\partial x}G_1 - \frac{\partial F_2}{\partial y}G_1 - \frac{\partial F_3}{\partial z}G_1$$

$$-((\mathbf{F} \bullet \nabla)\mathbf{G})_1 = -F_1\frac{\partial G_1}{\partial x} - F_2\frac{\partial G_1}{\partial y} - F_3\frac{\partial G_1}{\partial z}.$$

When we add up all the terms in these four expressions some cancel out and we are left with the same terms as in the first component of $\nabla \times (\mathbf{F} \times \mathbf{G})$.

(g) This is a straightforward calculation involving the equality of mixed partial derivatives:

$$\nabla \bullet (\nabla \times \mathbf{F}) = \frac{\partial}{\partial x}\left(\frac{\partial F_3}{\partial y} - \frac{\partial F_2}{\partial z}\right) + \frac{\partial}{\partial y}\left(\frac{\partial F_1}{\partial z} - \frac{\partial F_3}{\partial x}\right)$$
$$+ \frac{\partial}{\partial z}\left(\frac{\partial F_2}{\partial x} - \frac{\partial F_1}{\partial y}\right)$$
$$= \frac{\partial^2 F_3}{\partial x \partial y} - \frac{\partial^2 F_2}{\partial x \partial z} + \frac{\partial^2 F_1}{\partial y \partial z} - \frac{\partial^2 F_3}{\partial y \partial x} + \frac{\partial^2 F_2}{\partial z \partial x} - \frac{\partial^2 F_1}{\partial z \partial y}$$
$$= 0.$$

**Remark** Two *triple product* identities for vectors were previously presented in Exercises 18 and 23 of Section 10.3:

$$\mathbf{a} \bullet (\mathbf{b} \times \mathbf{c}) = \mathbf{b} \bullet (\mathbf{c} \times \mathbf{a}) = \mathbf{c} \bullet (\mathbf{a} \times \mathbf{b}),$$
$$\mathbf{a} \times (\mathbf{b} \times \mathbf{c}) = (\mathbf{a} \bullet \mathbf{c})\mathbf{b} - (\mathbf{a} \bullet \mathbf{b})\mathbf{c}.$$

While these are useful identities, they *cannot* be used to give simpler proofs of the identities in Theorem 3 by replacing one or other of the vectors with $\nabla$. (Why?)

## Vector Calculus Using Computer Algebra

Computer algebra programs are capable of doing vector algebra and vector calculus, though with some systems the commands that must be used to achieve a desired effect can only be discovered by patient reading of on-line and off-line documentation. For example, the following comands will cause Maple V to verify identity (i) of Theorem 3.

```
with(linalg);
f := vector([f1(x,y,z), f2(x,y,z), f3(x,y,z)]);
v := [x, y, z];
evalm(curl(curl(f,v),v) - grad(diverge(f,v),v)
    + map(laplacian,f,v));
```

Maple will respond with the result [0  0  0], indicating its agreement that

$$\nabla \times (\nabla \times \mathbf{F}) - \nabla(\nabla \bullet \mathbf{F}) + \nabla^2 \mathbf{F} = \mathbf{0}.$$

The `evalm` function causes the operators of addition and scalar multiplication of vectors to be applied to the components. Maple requires that the variables with respect to which the operators curl, grad, div (which Maple calls `diverge`), and laplacian are taken must be shown explicitly. These variables are stored in the vector `v`. Because Maple's `laplacian` function is defined for scalar rather than vector arguments, it must be `mapped` onto the components of `f`.

## Scalar and Vector Potentials

Two special terms are used to describe vector fields for which either the divergence or the curl is identically zero.

**Solenoidal and irrotational vector fields**

A vector field $\mathbf{F}$ is called **solenoidal** in a domain $D$ if $\mathbf{div}\,\mathbf{F} = 0$ in $D$.
A vector field $\mathbf{F}$ is called **irrotational** in a domain $D$ if $\mathbf{curl}\,\mathbf{F} = 0$ in $D$.

Part (h) of Theorem 3 says that $\mathbf{F} = \mathbf{grad}\,\phi \implies \mathbf{curl}\,\mathbf{F} = 0$. Thus

> **Every conservative vector field is irrotational.**

Part (g) of Theorem 3 says that $\mathbf{F} = \mathbf{curl}\,\mathbf{G} \implies \mathbf{div}\,\mathbf{F} = 0$. Thus

> **The curl of any vector field is solenoidal.**

The *converses* of these assertions hold if the domain of $\mathbf{F}$ satisfies certain conditions.

If $\mathbf{F}$ is a smooth, irrotational vector field on a simply connected domain $D$ then $\mathbf{F} = \nabla\phi$ for some scalar potential function defined on $D$, and so $\mathbf{F}$ is conservative.

If $\mathbf{F}$ is a smooth, solenoidal vector field on a domain $D$ with the property that every closed surface in $D$ bounds a domain contained in $D$, then $\mathbf{F} = \mathbf{curl}\,\mathbf{G}$ for some vector field $\mathbf{G}$ defined on $D$. Such a vector field $\mathbf{G}$ is called a **vector potential** of the vector field $\mathbf{F}$.

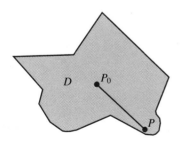

**Figure 16.4**    The line segment from $P_0$ to any point in $D$ lies in $D$

We cannot prove these result in their full generality at this point. However, both theorems have simple proofs in the special case where the domain $D$ is **star-like**. A star-like domain is one for which there exists a point $P_0$ such that the line segment from $P_0$ to any point $P$ in $D$ lies wholly in $D$. (See Figure 16.4.) Both proofs are *constructive* in that they tell you how to find a potential.

***Proof of Theorem 4 for star-like domains.*** Without loss of generality, we can assume that $P_0$ is the origin. If $P = (x, y, z)$ is any point in $D$, then the straight line segment

$$\mathbf{r}(t) = tx\mathbf{i} + ty\mathbf{j} + tz\mathbf{k}, \qquad 0 \le t \le 1$$

from $P_0$ to $P$ lies in $D$. Define the function $\phi$ on $D$ by

$$\phi(x, y, z) = \int_0^1 \mathbf{F}\big(\mathbf{r}(t)\big) \bullet \frac{d\mathbf{r}}{dt}\, dt$$

$$= \int_0^1 \Big(x F_1(\xi, \eta, \zeta) + y F_2(\xi, \eta, \zeta) + z F_3(\xi, \eta, \zeta)\Big)\, dt,$$

where $\xi = tx$, $\eta = ty$, and $\zeta = tz$. We calculate $\partial\phi/\partial x$, making use of the fact that $\mathbf{curl}\, \mathbf{F} = \mathbf{0}$ to replace $(\partial/\partial\xi)F_2(\xi, \eta, \zeta)$ with $(\partial/\partial\eta)F_1(\xi, \eta, \zeta)$ and $(\partial/\partial\xi)F_3(\xi, \eta, \zeta)$ with $(\partial/\partial\zeta)F_1(\xi, \eta, \zeta)$:

$$
\begin{aligned}
\frac{\partial\phi}{\partial x} &= \int_0^1 \left( F_1(\xi, \eta, \zeta) + tx\frac{\partial F_1}{\partial\xi} + ty\frac{\partial F_2}{\partial\xi} + tz\frac{\partial F_3}{\partial\xi} \right) dt \\
&= \int_0^1 \left( F_1(\xi, \eta, \zeta) + tx\frac{\partial F_1}{\partial\xi} + ty\frac{\partial F_1}{\partial\eta} + tz\frac{\partial F_1}{\partial\zeta} \right) dt \\
&= \int_0^1 \frac{d}{dt}\left( t\, F_1(\xi, \eta, \zeta) \right) dt \\
&= \left( t\, F_1(tx, ty, tz) \right)\Big|_0^1 = F_1(x, y, z).
\end{aligned}
$$

Similarly, $\partial\phi/\partial y = F_2$ and $\partial\phi/\partial z = F_3$. Thus $\nabla\phi = \mathbf{F}$.

The details of the proof of Theorem 5 for star-like domains are similar to those of Theorem 4, and we relegate the proof to Exercise 18 at the end of this section.

Note that vector potentials, when they exist, are *very* non-unique. Since **curl grad** $\phi$ is identically zero (Theorem 3(h)), an arbitrary conservative field can be added to **G** without changing the value of **curl G**. The following example illustrates just how much freedom you have in making simplifying assumptions when trying to find a vector potential.

**Example 1**   Show that the vector field $\mathbf{F} = (x^2 + yz)\mathbf{i} - 2y(x + z)\mathbf{j} + (xy + z^2)\mathbf{k}$ is solenoidal in $\mathbb{R}^3$ and find a vector potential for it.

**Solution**   Since $\mathbf{div}\, \mathbf{F} = 2x - 2(x + z) + 2z = 0$ in $\mathbb{R}^3$, **F** is solenoidal. A vector potential **G** for **F** must satisfy $\mathbf{curl}\, \mathbf{G} = \mathbf{F}$, that is,

$$
\begin{aligned}
\frac{\partial G_3}{\partial y} - \frac{\partial G_2}{\partial z} &= x^2 + yz, \\
\frac{\partial G_1}{\partial z} - \frac{\partial G_3}{\partial x} &= -2xy - 2yz, \\
\frac{\partial G_2}{\partial x} - \frac{\partial G_1}{\partial y} &= xy + z^2.
\end{aligned}
$$

The three components of **G** have nine independent first-partial derivatives, so there are nine "degrees of freedom" involved in their determination. The three equations above use up three of these nine degrees of freedom. That leaves six. Let us try to find a solution **G** with $G_2 = 0$ identically. This means that all three first partials of $G_2$ are zero so we have used up three degrees of freedom in making this assumption. We have three left. The first equation now implies that

$$
G_3 = \int (x^2 + yz)\, dy = x^2 y + \frac{1}{2} y^2 z + M(x, z).
$$

(Since we were integrating with respect to $y$, the constant of integration can still depend on $x$ and $z$.) We make a second simplifying assumption, that $M(x, z) = 0$. This uses up two more degrees of freedom, leaving one. From the second equation we have

$$\frac{\partial G_1}{\partial z} = \frac{\partial G_3}{\partial x} - 2xy - 2yz = 2xy - 2xy - 2yz = -2yz,$$

so

$$G_1 = -2 \int yz\, dz = -yz^2 + N(x, y).$$

We can not assume that $N(x, y) = 0$ identically, because that would require two degrees of freedom and we have only one. However, the third equation implies

$$xy + z^2 = -\frac{\partial G_1}{\partial y} = z^2 - \frac{\partial N}{\partial y}.$$

Thus, $(\partial/\partial y)N(x, y) = -xy$; observe that the terms involving $z$ have cancelled out. This happened because **div F** $= 0$. Had **F** not been solenoidal, we could not have determined $N$ as a function of $x$ and $y$ only from the above equation. As it is, however, we have

$$N(x, y) = -\int xy\, dy = -\frac{1}{2}xy^2 + P(x).$$

We can use our last degree of freedom to choose $P(x)$ to be identically zero, and hence obtain

$$\mathbf{G} = -\left(yz^2 + \frac{xy^2}{2}\right)\mathbf{i} + \left(x^2 y + \frac{y^2 z}{2}\right)\mathbf{k}$$

as the required vector potential for **F**. You can check that **curl G** $=$ **F**. Of course, other choices of simplifying assumptions would have led to very different functions **G**, which would have been equally correct.

# Exercises 16.2

1. Prove part (a) of Theorem 3.

2. Prove part (b) of Theorem 3.

3. Prove part (d) of Theorem 3.

4. Prove part (f) of Theorem 3.

5. Prove part (h) of Theorem 3.

6. Prove part (i) of Theorem 3.

* 7. Given that the field lines of the vector field $\mathbf{F}(x, y, z)$ are parallel straight lines, can you conclude anything about **div F**? about **curl F**?

8. Let $\mathbf{r} = x\mathbf{i} + y\mathbf{j} + z\mathbf{k}$ and let $\mathbf{c}$ be a constant vector. Show that $\nabla \bullet (\mathbf{c} \times \mathbf{r}) = 0$, $\nabla \times (\mathbf{c} \times \mathbf{r}) = 2\mathbf{c}$, and $\nabla(\mathbf{c} \bullet \mathbf{r}) = \mathbf{c}$.

9. Let $\mathbf{r} = x\mathbf{i} + y\mathbf{j} + z\mathbf{k}$ and let $r = |\mathbf{r}|$. If $f$ is a differentiable function of one variable show that

$$\nabla \bullet (f(r)\mathbf{r}) = rf'(r) + 3f(r).$$

Find $f(r)$ if $f(r)\mathbf{r}$ is solenoidal for $r \neq 0$.

10. If the smooth vector field **F** is both irrotational and solenoidal on $\mathbb{R}^3$, show that the three components of **F** and the scalar potential for **F** are all harmonic functions in $\mathbb{R}^3$.

11. If $\mathbf{r} = x\mathbf{i} + y\mathbf{j} + z\mathbf{k}$ and **F** is smooth, show that

$$\nabla \times (\mathbf{F} \times \mathbf{r}) = \mathbf{F} - (\nabla \bullet \mathbf{F})\mathbf{r} + \nabla(\mathbf{F} \bullet \mathbf{r}) - \mathbf{r} \times (\nabla \times \mathbf{F}).$$

In particular, if $\mathbf{\nabla} \bullet \mathbf{F} = 0$ and $\mathbf{\nabla} \times \mathbf{F} = \mathbf{0}$, then

$$\mathbf{\nabla} \times (\mathbf{F} \times \mathbf{r}) = \mathbf{F} + \mathbf{\nabla}(\mathbf{F} \bullet \mathbf{r}).$$

**12.** If $\phi$ and $\Psi$ are harmonic functions, show that $\phi \mathbf{\nabla} \Psi - \Psi \mathbf{\nabla} \phi$ is solenoidal.

**13.** If $\phi$ and $\Psi$ are smooth scalar fields show that

$$\mathbf{\nabla} \times (\phi \mathbf{\nabla} \Psi) = -\mathbf{\nabla} \times (\Psi \mathbf{\nabla} \phi) = \mathbf{\nabla} \phi \times \mathbf{\nabla} \Psi.$$

**14.** Verify the identity

$$\mathbf{\nabla} \bullet \Big( f(\mathbf{\nabla} g \times \mathbf{\nabla} h) \Big) = \mathbf{\nabla} f \bullet (\mathbf{\nabla} g \times \mathbf{\nabla} h),$$

for smooth scalar fields $f$, $g$, and $h$.

**15.** If the vector fields $\mathbf{F}$ and $\mathbf{G}$ are smooth and conservative, show that $\mathbf{F} \times \mathbf{G}$ is solenoidal. Find a vector potential for $\mathbf{F} \times \mathbf{G}$.

**16.** Find a vector potential for $\mathbf{F} = -y\mathbf{i} + x\mathbf{j}$.

**17.** Show that $\mathbf{F} = xe^{2z}\mathbf{i} + ye^{2z}\mathbf{j} - e^{2z}\mathbf{k}$ is a solenoidal vector field, and find a vector potential for it.

**\* 18.** Suppose $\operatorname{div} \mathbf{F} = 0$ in a domain $D$ any point $P$ of which can by joined to the origin by a straight line segment in $D$. Let $\mathbf{r} = tx\mathbf{i} + ty\mathbf{j} + tz\mathbf{k}$, $(0 \le t \le 1)$, be a parametrization of the line segment from the origin to $(x, y, z)$ in $D$. If

$$\mathbf{G}(x, y, z) = \int_0^1 t\mathbf{F}(\mathbf{r}(t)) \times \frac{d\mathbf{r}}{dt}\, dt,$$

show that $\operatorname{\mathbf{curl}} \mathbf{G} = \mathbf{F}$ throughout $D$. *Hint:* it is enough to check the first components of $\operatorname{\mathbf{curl}} \mathbf{G}$ and $\mathbf{F}$. Proceed in a manner similar to the proof of Theorem 4.

# 16.3 Green's Theorem in the Plane

The Fundamental Theorem of Calculus,

$$\int_a^b \frac{d}{dx} f(x)\, dx = f(b) - f(a),$$

expresses the integral, taken over the interval $[a, b]$, of the derivative of a single-variable function, $f$, as a *sum* of values of that function at the *oriented boundary* of the interval $[a, b]$, that is, at the two endpoints $a$ and $b$, the former providing a *negative* contribution and the latter a *positive* one. The line integral of a conservative vector field over a curve $\mathcal{C}$ from $A$ to $B$,

$$\int_{\mathcal{C}} \mathbf{\nabla} \phi \bullet d\mathbf{r} = \phi(B) - \phi(A),$$

has a similar interpretation; $\mathbf{\nabla} \phi$ is a derivative, and the curve $\mathcal{C}$, although lying in a two- or three-dimensional space, is intrinsically a one-dimensional object, and the points $A$ and $B$ constitute its boundary.

Green's Theorem is a two-dimensional version of the Fundamental Theorem of Calculus that expresses the *double integral* of a certain kind of derivative of a two-dimensional vector field $\mathbf{F}(x, y)$, namely the $\mathbf{k}$-component of $\operatorname{\mathbf{curl}} \mathbf{F}$, over a region $R$ in the $xy$-plane as a line integral (that is a "sum") of the tangential component of $\mathbf{F}$ around the curve $\mathcal{C}$ which is the oriented boundary of $R$:

$$\iint_R \operatorname{\mathbf{curl}} \mathbf{F} \bullet \mathbf{k}\, dA = \oint_{\mathcal{C}} \mathbf{F} \bullet d\mathbf{r},$$

or, more explicitly,

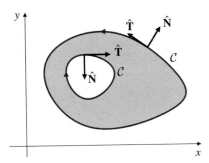

**Figure 16.5** A plane domain with positively oriented boundary

$$\iint_R \left( \frac{\partial F_2}{\partial x} - \frac{\partial F_1}{\partial y} \right) dx\, dy = \oint_C F_1(x, y)\, dx + F_2(x, y)\, dy.$$

For this formula to hold, $C$ must be the oriented boundary of $R$ considered as a surface with orientation provided by $\hat{\mathbf{N}} = \mathbf{k}$. Thus $C$ is oriented with $R$ on the left as we move around $C$ in the direction of its orientation. We will call such a curve positively oriented with respect to $R$. In particular, if $C$ is a simple closed curve bounding $R$, then $C$ is oriented counterclockwise. Of course, $R$ may have holes, and the boundaries of the holes will be oriented clockwise. In any case, the unit tangent $\hat{\mathbf{T}}$ and unit exterior (pointing out of $R$) normal $\hat{\mathbf{N}}$ on $C$ satisfy $\hat{\mathbf{N}} = \hat{\mathbf{T}} \times \mathbf{k}$. See Figure 16.5.

**THEOREM** **6**

### Green's Theorem

Let $R$ be a regular, closed region in the $xy$-plane whose boundary, $C$, consists of one or more piecewise smooth, non-self-intersecting, closed curves that are positively oriented with respect to $R$. If $\mathbf{F} = F_1(x, y)\mathbf{i} + F_2(x, y)\mathbf{j}$ is a smooth vector field on $R$, then

$$\oint_C F_1(x, y)\, dx + F_2(x, y)\, dy = \iint_R \left( \frac{\partial F_2}{\partial x} - \frac{\partial F_1}{\partial y} \right) dA.$$

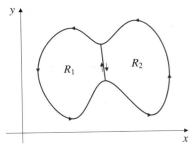

**Figure 16.6** Green's Theorem holds for the union of $R_1$ and $R_2$ if it holds for each of those regions

**PROOF** Recall that a regular region can be divided into non-overlapping subregions that are both $x$-simple and $y$-simple. (See Section 14.2.) When two such regions share a common boundary curve, they induce opposite orientations on that curve, so the sum of the line integrals over the boundaries of the subregions is just the line integral over the boundary of the whole region. (See Figure 16.6.) The double integrals over the subregions also add to give the double integral over the whole region. It therefore suffices to show the formula holds for a region $R$ that is both $x$-simple and $y$-simple.

Since $R$ is $y$-simple, it is specified by inequalities of the form $a \le x \le b$, $f(x) \le y \le g(x)$, with the bottom boundary $y = f(x)$ oriented left to right, and the upper boundary $y = g(x)$ oriented right to left. (See Figure 16.7.) Thus

$$-\iint_R \frac{\partial F_1}{\partial y}\, dx\, dy = -\int_a^b dx \int_{f(x)}^{g(x)} \frac{\partial F_1}{\partial y}\, dy$$

$$= \int_a^b \Big( -F_1\big(x, g(x)\big) + F_1\big(x, f(x)\big) \Big)\, dx.$$

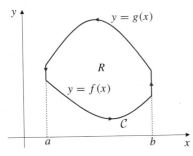

**Figure 16.7**
$\oint_C F_1\, dx = -\iint_R \dfrac{\partial F_1}{\partial y}\, dA$ for this $y$-simple region $R$

On the other hand, since $dx = 0$ on the vertical sides of $R$, and the top boundary is traversed from $b$ to $a$, we have

$$\oint_C F_1(x, y)\, dx = \int_a^b \Big( F_1\big(x, f(x)\big) - F_1\big(x, g(x)\big) \Big)\, dx = \iint_R -\frac{\partial F_1}{\partial y}\, dx\, dy.$$

Similarly, since $R$ is $x$-simple, $\displaystyle\oint_C F_2\, dy = \iint_R \frac{\partial F_2}{\partial x}\, dx\, dy$, and so

$$\oint_C F_1(x, y)\, dx + F_2(x, y)\, dy = \iint_R \left( \frac{\partial F_2}{\partial x} - \frac{\partial F_1}{\partial y} \right) dA.$$

**Example 1** (**Area bounded by a simple closed curve**) For any of the three vector fields

$$\mathbf{F} = x\mathbf{j}, \qquad \mathbf{F} = -y\mathbf{i}, \qquad \text{and} \qquad \mathbf{F} = \frac{1}{2}(-y\mathbf{i} + x\mathbf{j}),$$

we have $(\partial F_2/\partial x) - (\partial F_1/\partial y) = 1$. If $C$ is a positively oriented, piecewise smooth, simple closed curve bounding a region $R$ in the plane, then by Green's Theorem,

$$\oint_C x\,dy = -\oint_C y\,dx = \frac{1}{2}\oint_C x\,dy - y\,dx = \iint_R 1\,dA = \text{area of } R.$$

■

**Example 2** Evaluate $I = \oint_C (x - y^3)\,dx + (y^3 + x^3)\,dy$, where $C$ is the positively oriented boundary of the quarter-disk $Q$: $0 \le x^2 + y^2 \le a^2$, $x \ge 0$, $y \ge 0$.

**Solution** We use Green's Theorem to calculate $I$:

$$I = \iint_Q \left( \frac{\partial}{\partial x}(y^3 + x^3) - \frac{\partial}{\partial y}(x - y^3) \right) dA$$

$$= 3 \iint_Q (x^2 + y^2)\,dA = 3 \int_0^{\pi/2} d\theta \int_0^a r^3\,dr = \frac{3}{8}\pi a^4.$$

■

**Example 3** Let $C$ be a positively oriented, simple closed curve in the $xy$-plane, bounding a region $R$ and not passing through the origin. Show that

$$\oint_C \frac{-y\,dx + x\,dy}{x^2 + y^2} = \begin{cases} 0 & \text{if the origin is outside } R \\ 2\pi & \text{if the origin is inside } R. \end{cases}$$

**Solution** First, if $(x, y) \ne (0, 0)$, then, by direct calculation,

$$\frac{\partial}{\partial x}\left( \frac{x}{x^2 + y^2} \right) - \frac{\partial}{\partial y}\left( \frac{-y}{x^2 + y^2} \right) = 0.$$

If the origin is not in $R$, then Green's Theorem implies that

$$\oint_C \frac{-y\,dx + x\,dy}{x^2 + y^2} = \iint_R \left[ \frac{\partial}{\partial x}\left( \frac{x}{x^2 + y^2} \right) - \frac{\partial}{\partial y}\left( \frac{-y}{x^2 + y^2} \right) \right] dx\,dy = 0.$$

Now suppose the origin is in $R$. Since it is assumed that the origin is not on $C$, it must be an interior point of $R$. The interior of $R$ is open, so there exists $\epsilon > 0$ such that the circle $C_\epsilon$ of radius $\epsilon$ centred at the origin is in the interior of $R$. Let $C_\epsilon$ be oriented negatively (clockwise). By direct calculation (see Exercise 22(a) of Section 15.4) it is easily shown that

$$\oint_{C_\epsilon} \frac{-y\,dx + x\,dy}{x^2 + y^2} = -2\pi.$$

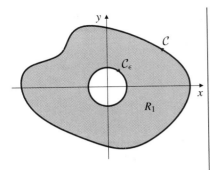

**Figure 16.8**

Together $C$ and $C_\epsilon$ form the positively oriented boundary of a region $R_1$ that excludes the origin. (See Figure 16.8.) So, by Green's Theorem,

$$\oint_C \frac{-y\,dx + x\,dy}{x^2 + y^2} + \oint_{C_\epsilon} \frac{-y\,dx + x\,dy}{x^2 + y^2} = 0.$$

The desired result now follows:

$$\oint_C \frac{-y\,dx + x\,dy}{x^2 + y^2} = -\oint_{C_\epsilon} \frac{-y\,dx + x\,dy}{x^2 + y^2} = -(-2\pi) = 2\pi. \quad\blacksquare$$

## The Two-Dimensional Divergence Theorem

The following theorem is an alternative formulation of the two-dimensional Fundamental Theorem of Calculus. In this case we express the double integral of **div F** (a derivative of **F**) over $R$ as a single integral of the outward normal component of **F** on the boundary $C$ of $R$.

**THEOREM** ⑦

### The Divergence Theorem in the Plane

Let $R$ be a regular, closed region in the $xy$-plane whose boundary, $C$, consists of one or more piecewise smooth, non-self-intersecting, closed curves. Let $\hat{\mathbf{N}}$ denote the unit outward (from $R$) normal field on $C$. If $\mathbf{F} = F_1(x, y)\mathbf{i} + F_2(x, y)\mathbf{j}$ is a smooth vector field on $R$, then

$$\iint_R \operatorname{div} \mathbf{F}\, dA = \oint_C \mathbf{F} \bullet \hat{\mathbf{N}}\, ds.$$

**PROOF** As observed in the second paragraph of this section, $\hat{\mathbf{N}} = \hat{\mathbf{T}} \times \mathbf{k}$, where $\hat{\mathbf{T}}$ is the unit tangent field in the positive direction on $C$. If $\hat{\mathbf{T}} = T_1\mathbf{i} + T_2\mathbf{j}$, then $\hat{\mathbf{N}} = T_2\mathbf{i} - T_1\mathbf{j}$. Now let **G** be the vector field with components $G_1 = -F_2$ and $G_2 = F_1$. Then $\mathbf{G} \bullet \hat{\mathbf{T}} = \mathbf{F} \bullet \hat{\mathbf{N}}$ and, by Green's Theorem

$$\iint_R \operatorname{div} \mathbf{F}\, dA = \iint_R \left( \frac{\partial F_1}{\partial x} + \frac{\partial F_2}{\partial y} \right) dA$$

$$= \iint_R \left( \frac{\partial G_2}{\partial x} - \frac{\partial G_1}{\partial y} \right) dA$$

$$= \oint_C \mathbf{G} \bullet d\mathbf{r} = \oint_C \mathbf{G} \bullet \hat{\mathbf{T}}\, ds = \oint_C \mathbf{F} \bullet \hat{\mathbf{N}}\, ds. \quad\blacksquare$$

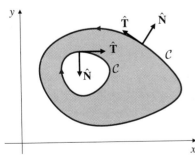

**Figure 16.9** $\quad \hat{\mathbf{N}} = \hat{\mathbf{T}} \times \mathbf{k}$

## Exercises 16.3

**1.** Evaluate $\oint_C (\sin x + 3y^2)\, dx + (2x - e^{-y^2})\, dy$, where $C$ is the boundary of the half-disk $x^2 + y^2 \le a^2$, $y \ge 0$, oriented counterclockwise.

**2.** Evaluate $\oint_C (x^2 - xy)\, dx + (xy - y^2)\, dy$ clockwise around the triangle with vertices $(0, 0)$, $(1, 1)$ and $(2, 0)$.

3. Evaluate $\oint_{\mathcal{C}} \left( x \sin(y^2) - y^2 \right) dx + \left( x^2 y \cos(y^2) + 3x \right) dy$, where $\mathcal{C}$ is the counterclockwise boundary of the trapezoid with vertices $(0, -2)$, $(1, -1)$, $(1, 1)$, and $(0, 2)$.

4. Evaluate $\oint_{\mathcal{C}} x^2 y \, dx - xy^2 \, dy$, where $\mathcal{C}$ is the clockwise boundary of the region $0 \le y \le \sqrt{9 - x^2}$.

5. Use a line integral to find the plane area enclosed by the curve $\mathbf{r} = a \cos^3 t \, \mathbf{i} + b \sin^3 t \, \mathbf{j}$, $0 \le t \le 2\pi$.

6. We deduced the two-dimensional Divergence Theorem from Green's Theorem. Reverse the argument and use the two-dimensional Divergence Theorem to prove Green's Theorem.

7. Sketch the plane curve $\mathcal{C}$: $\mathbf{r} = \sin t \, \mathbf{i} + \sin 2t \, \mathbf{j}$, $(0 \le t \le 2\pi)$.

Evaluate $\oint_{\mathcal{C}} \mathbf{F} \bullet d\mathbf{r}$, where $\mathbf{F} = y e^{x^2} \mathbf{i} + x^3 e^y \mathbf{j}$.

8. If $\mathcal{C}$ is the positively oriented boundary of a plane region R having area $A$ and centroid $(\bar{x}, \bar{y})$, interpret geometrically the line integral $\oint_{\mathcal{C}} \mathbf{F} \bullet d\mathbf{r}$, where:

(a) $\mathbf{F} = x^2 \mathbf{j}$,　(b) $\mathbf{F} = xy\mathbf{i}$, and　(c) $\mathbf{F} = y^2 \mathbf{i} + 3xy\mathbf{j}$.

* 9. **(Average values of harmonic functions)** If $u(x, y)$ is harmonic in a domain containing a disk of radius $r$ with boundary $C_r$ then the average value of $u$ around the circle is the value of $u$ at the centre. Prove this by showing that the derivative of the average value with respect to $r$ is zero (use divergence theorem and harmonicity of $u$), and the fact that the limit of the average value as $r \to 0$ is the value of $u$ at the centre.

# 16.4 The Divergence Theorem in 3-Space

**The Divergence Theorem** (also called **Gauss's Theorem** is one of two important versions of the Fundamental Theorem of Calculus in $\mathbb{R}^3$. (The other is Stokes's Theorem, presented in the next section.)

In the Divergence Theorem, the integral of the *derivative* **div F** $= \nabla \bullet \mathbf{F}$ over a domain in 3-space is expressed as the flux of **F** out of the surface of that domain. It herefore closely resembles the two-dimensional version Theorem 7 given in the previous section. The theorem holds, and we therefore state it, for a general class of domains in $\mathbb{R}^3$ that are bounded by piecewise smooth closed surfaces. However, we will restrict our proof to domains of a special type. Extending the concept of an $x$-simple plane domain defined in Section 14.2, we say the three-dimensional domain $D$ is $x$-**simple** if it is bounded by a piecewise smooth surface $S$ and if every straight line parallel to the $x$-axis and passing through an interior point of $D$ meets $S$ at exactly two points. Similar definitions hold for $y$-simple and $z$-simple, and we call the domain $D$ **regular** if it is a union of finitely many, nonoverlapping subdomains, each of which is $x$-simple, $y$-simple, and $z$-simple.

**THEOREM 8**

### The Divergence Theorem (Gauss's Theorem)

Let $D$ be a regular, 3-dimensional domain whose boundary $S$ is an oriented, closed surface with unit normal field $\hat{\mathbf{N}}$ pointing out of $D$. If $\mathbf{F}$ is a smooth vector field defined on $D$ then

$$\iiint_D \operatorname{div} \mathbf{F} \, dV = \oiint_S \mathbf{F} \bullet \hat{\mathbf{N}} \, dS.$$

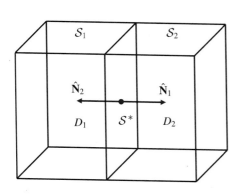

**Figure 16.10**    A union of abutting domains

**Figure 16.11**    A $z$-simple domain

**PROOF**   Since the domain $D$ is a union of finitely many non-overlapping domains that are $x$-simple, $y$-simple, and $z$-simple, it is sufficient to prove the theorem for a subdomain of $D$ with this property. To see this, suppose, for instance, that $D$ and $S$ are each divided into two parts, $D_1$ and $D_2$, and $S_1$ and $S_2$, by a surface $S^*$ slicing through $D$. (See Figure 16.10.) $S^*$ is part of the boundary of both $D_1$ and $D_2$, but the exterior normals, $\hat{\mathbf{N}}_1$ and $\hat{\mathbf{N}}_2$, of the two subdomains point in opposite directions on either side of $S^*$. If the formula in the theorem holds for both subdomains,

$$\iiint_{D_1} \operatorname{div} \mathbf{F} \, dV = \oiint_{S_1 \cup S^*} \mathbf{F} \bullet \hat{\mathbf{N}}_1 \, dS$$

$$\iiint_{D_2} \operatorname{div} \mathbf{F} \, dV = \oiint_{S_2 \cup S^*} \mathbf{F} \bullet \hat{\mathbf{N}}_2 \, dS,$$

then, adding these equations, we get

$$\iiint_{D} \operatorname{div} \mathbf{F} \, dV = \oiint_{S_1 \cup S_2} \mathbf{F} \bullet \hat{\mathbf{N}} \, dS = \oiint_{S} \mathbf{F} \bullet \hat{\mathbf{N}} \, dS;$$

the contributions from $S^*$ cancel out because on that surface $\hat{\mathbf{N}}_2 = -\hat{\mathbf{N}}_1$.

For the rest of this proof we assume, therefore, that $D$ is $x$-, $y$-, and $z$-simple. Since $D$ is $z$-simple, it lies between the graphs of two functions defined on a region $R$ in the $xy$-plane; if $(x, y, z)$ is in $D$ then $(x, y)$ is in $R$ and $f(x, y) \leq z \leq g(x, y)$. (See Figure 16.11.) We have

$$\iiint_{D} \frac{\partial F_3}{\partial z} \, dV = \iint_{R} dx \, dy \int_{f(x,y)}^{g(x,y)} \frac{\partial F_3}{\partial z} \, dz$$

$$= \iint_{R} \left( F_3\big(x, y, g(x, y)\big) - F_3\big(x, y, f(x, y)\big) \right) dx \, dy.$$

Now

$$\oiint_{S} \mathbf{F} \bullet \hat{\mathbf{N}} \, dS = \oiint_{S} \left( F_1 \mathbf{i} \bullet \hat{\mathbf{N}} + F_2 \mathbf{j} \bullet \hat{\mathbf{N}} + F_3 \mathbf{k} \bullet \hat{\mathbf{N}} \right) dS.$$

Only the last term involves $F_3$, and it can be split into three integrals, over the top surface $z = g(x, y)$, the bottom surface $z = f(x, y)$, and vertical side wall lying above the boundary of $R$:

$$\oiint_S F_3(x, y, z) \, \mathbf{k} \bullet \hat{\mathbf{N}} \, dS = \left( \iint_{\text{top}} + \iint_{\text{bottom}} + \iint_{\text{side}} \right) F_3(x, y, z) \, \mathbf{k} \bullet \hat{\mathbf{N}} \, dS.$$

On the side wall, $\mathbf{k} \bullet \hat{\mathbf{N}} = 0$, so that integral is zero. On the top surface, $z = g(x, y)$, the vector area element is

$$\hat{\mathbf{N}} \, dS = \left( -\frac{\partial g}{\partial x}\mathbf{i} - \frac{\partial g}{\partial y}\mathbf{j} + \mathbf{k} \right) dx \, dy.$$

Accordingly,

$$\iint_{\text{top}} F_3(x, y, z) \, \mathbf{k} \bullet \hat{\mathbf{N}} \, dS = \iint_R F_3\big(x, y, g(x, y)\big) \, dx \, dy.$$

Similarly, we have

$$\iint_{\text{bottom}} F_3(x, y, z) \, \mathbf{k} \bullet \hat{\mathbf{N}} \, dS = - \iint_R F_3\big(x, y, f(x, y)\big) \, dx \, dy;$$

the negative sign occurs because $\hat{\mathbf{N}}$ points down rather than up on the bottom. Thus we have shown that

$$\iiint_D \frac{\partial F_3}{\partial z} \, dV = \oiint_S F_3 \, \mathbf{k} \bullet \hat{\mathbf{N}} \, dS.$$

Similarly, because $D$ is also $x$-simple and $y$-simple,

$$\iiint_D \frac{\partial F_1}{\partial x} \, dV = \oiint_S F_1 \, \mathbf{i} \bullet \hat{\mathbf{N}} \, dS$$

$$\iiint_D \frac{\partial F_2}{\partial y} \, dV = \oiint_S F_2 \, \mathbf{j} \bullet \hat{\mathbf{N}} \, dS.$$

Adding these three results we get

$$\iiint_D \operatorname{div} \mathbf{F} \, dV = \oiint_S \mathbf{F} \bullet \hat{\mathbf{N}} \, dS.$$

The Divergence Theorem can be used in both directions to simplify explicit calculations of surface integrals or volumes. We give examples of each.

**Example 1**    Let $\mathbf{F} = bxy^2\mathbf{i} + bx^2y\mathbf{j} + (x^2 + y^2)z^2\mathbf{k}$, and let $S$ be the closed surface bounding the solid cylinder $R$ defined by $x^2 + y^2 \le a^2$ and $0 \le z \le b$. Find $\oiint_S \mathbf{F} \bullet d\mathbf{S}$.

**Solution**    By the Divergence Theorem,

$$\oiint_S \mathbf{F} \bullet d\mathbf{S} = \iiint_R \operatorname{div} \mathbf{F} \, dV = \iiint_R (x^2 + y^2)(b + 2z) \, dV$$

$$= \int_0^b (b + 2z) \, dz \int_0^{2\pi} d\theta \int_0^a r^2 \, r \, dr$$

$$= (b^2 + b^2) 2\pi (a^4/4) = \pi a^4 b^2.$$

**Example 2**  Evaluate $\displaystyle\oiint_{S} (x^2 + y^2)\, dS$ where $S$ is the sphere $x^2 + y^2 + z^2 = a^2$. Use the Divergence Theorem.

**Solution**  On $S$ we have

$$\hat{\mathbf{N}} = \frac{\mathbf{r}}{a} = \frac{x\mathbf{i} + y\mathbf{j} + z\mathbf{k}}{a}.$$

We would like to choose $\mathbf{F}$ so that $\mathbf{F} \bullet \hat{\mathbf{N}} = x^2 + y^2$. Observe that $\mathbf{F} = a(x\mathbf{i} + y\mathbf{j})$ will do. If $B$ is the ball bounded by $S$, then

$$\oiint_{S} (x^2 + y^2)\, dS = \oiint_{S} \mathbf{F} \bullet \hat{\mathbf{N}}\, dS = \iiint_{B} \operatorname{div} \mathbf{F}\, dV$$
$$= \iiint_{B} 2a\, dV = (2a)\frac{4}{3}\pi a^3 = \frac{8}{3}\pi a^4.$$

**Example 3**  By using the Divergence Theorem with $\mathbf{F} = x\mathbf{i} + y\mathbf{j} + z\mathbf{k}$, calculate the volume of a cone having base area $A$ and height $h$. The base can be any smoothly bounded plane region.

**Solution**  Let the vertex of the cone be at the origin and the base in the plane $z = h$ as shown in Figure 16.12. The solid cone $C$ has surface consisting of two parts, the conical wall $S$, and the base region $D$ that has area $A$. Since $\mathbf{F}(x, y, z)$ points directly away from the origin at any point $(x, y, z) \neq (0, 0, 0)$, we have $\mathbf{F} \bullet \hat{\mathbf{N}} = 0$ on $S$. On $D$, we have $\hat{\mathbf{N}} = \mathbf{k}$ and $z = h$, so $\mathbf{F} \bullet \hat{\mathbf{N}} = z = h$ on the base of the cone. Since $\operatorname{div} \mathbf{F}(x, y, z) = 1 + 1 + 1 = 3$, we have, by the Divergence Theorem,

$$3V = \iiint_{C} \operatorname{div} \mathbf{F}\, dV = \iint_{S} \mathbf{F} \bullet \hat{\mathbf{N}}\, dS + \iint_{D} \mathbf{F} \bullet \hat{\mathbf{N}}\, dS$$
$$= 0 + h \iint_{D} dS = Ah.$$

Thus $V = \frac{1}{3}Ah$, the well-known formula for the volume of a cone.

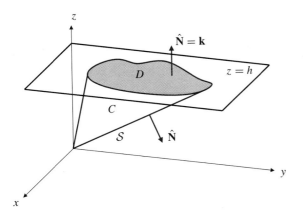

**Figure 16.12**    A cone with an arbitrarily shaped base

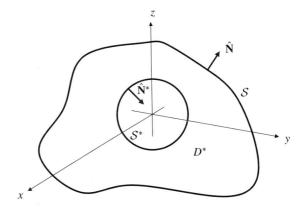

**Figure 16.13**    A solid domain with a spherical cavity

**Example 4** Let $S$ be the surface of an arbitrary regular domain $D$ in 3-space that contains the origin in its interior. Find

$$\oiint_S \mathbf{F} \bullet \hat{\mathbf{N}} \, dS,$$

where $\mathbf{F}(\mathbf{r}) = m\mathbf{r}/|\mathbf{r}|^3$ and $\hat{\mathbf{N}}$ is the unit outward normal on $S$.

**Solution** Since $\mathbf{F}$, and therefore $\mathbf{div}\,\mathbf{F}$, is undefined at the origin, we cannot apply the Divergence Theorem directly. To overcome this problem we use a little trick. Let $S^*$ be a small sphere centred at the origin bounding a ball contained wholly in $D$. (See Figure 16.13.) Let $\hat{\mathbf{N}}^*$ be the unit normal on $S^*$ pointing *into* the sphere, and let $D^*$ be that part of $D$ which lies outside $S^*$. As shown in Example 3 of Section 16.1, $\mathbf{div}\,\mathbf{F} = 0$ on $D^*$. Also,

$$\oiint_{S^*} \mathbf{F} \bullet \hat{\mathbf{N}}^* \, dS = -4\pi m,$$

is the flux of $\mathbf{F}$ *inward* through the sphere $S^*$. (See Example 1 of Section 15.6.) Therefore

$$0 = \iiint_{D^*} \mathbf{div}\,\mathbf{F} \, dV = \oiint_S \mathbf{F} \bullet \hat{\mathbf{N}} \, dS + \oiint_{S^*} \mathbf{F} \bullet \hat{\mathbf{N}}^* \, dS$$

$$= \oiint_S \mathbf{F} \bullet \hat{\mathbf{N}} \, dS - 4\pi m,$$

and so $\oiint_S \mathbf{F} \bullet \hat{\mathbf{N}} \, dS = 4\pi m$.

**Example 5** Find the flux of $\mathbf{F} = x\mathbf{i} + y^2\mathbf{j} + z\mathbf{k}$ upwards through the first-octant part $S$ of the cylindrical surface $x^2 + y^2 = a^2$, $0 \le y \le b$.

**Solution** $S$ is one of five surfaces that form the boundary of the solid region $D$ shown in Figure 16.14. The other four surfaces are planar: $S_1$ lies in the plane $z = 0$, $S_2$ lies in the plane $x = 0$, $S_3$ lies in the plane $y = 0$, and $S_4$ lies in the plane $y = b$. Orient all these surfaces with normal $\hat{\mathbf{N}}$ pointing out of $D$. On $S_1$ we have $\hat{\mathbf{N}} = -\mathbf{k}$, so $\mathbf{F} \bullet \hat{\mathbf{N}} = -z = 0$ on $S_1$. Similarly, $\mathbf{F} \bullet \hat{\mathbf{N}} = 0$ on $S_2$ and $S_3$. On $S_4$, $y = b$ and $\hat{\mathbf{N}} = \mathbf{j}$, so $\mathbf{F} \bullet \hat{\mathbf{N}} = y^2 = b^2$ there. If $S_{\text{tot}}$ denotes the whole boundary of $D$, then

$$\oiint_{S_{\text{tot}}} \mathbf{F} \bullet \hat{\mathbf{N}} \, dS = \iint_S \mathbf{F} \bullet \hat{\mathbf{N}} \, dS + 0 + 0 + 0 + \iint_{S_4} \mathbf{F} \bullet \hat{\mathbf{N}} \, dS$$

$$= \iint_S \mathbf{F} \bullet \hat{\mathbf{N}} \, dS + \frac{\pi a^2 b^2}{4}.$$

On the other hand, by the Divergence Theorem,

$$\oiint_{S_{\text{tot}}} \mathbf{F} \bullet \hat{\mathbf{N}} \, dS = \iiint_D \mathbf{div}\,\mathbf{F} \, dV = \iiint_D (2 + 2y) \, dV = 2V + 2V\bar{y},$$

where $V = \pi a^2 b/4$ is the volume of $D$, and $\bar{y} = b/2$ is the $y$-coordinate of the centroid of $D$. Combining these results, the flux of $\mathbf{F}$ upward through $S$ is

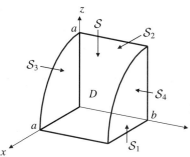

**Figure 16.14** The boundary of domain $D$ has five faces, one curved and four planar

$$\oiint_{\mathcal{S}} \mathbf{F} \bullet \hat{\mathbf{N}}\,dS = \frac{2\pi a^2 b}{4}\left(1 + \frac{b}{2}\right) - \frac{\pi a^2 b^2}{4} = \frac{\pi a^2 b}{2}.$$

Among the examples above, Example 4 is the most significant, and the one that best represents the way that the Divergence Theorem is used in practice. It is predominantly a theoretical tool, rather than a tool for calculation. We will look at some applications in the Section 16.5.

## Variants of the Divergence Theorem

Other versions of the Fundamental Theorem of Calculus can be derived from the Divergence Theorem. Two are given in the following theorem:

**THEOREM 9**

If $D$ satisfies the conditions of of the Divergence Theorem and has surface $\mathcal{S}$, and if $\mathbf{F}$ is a smooth vector field, and $\phi$ is a smooth scalar field, then:

(a)   $$\iiint_D \mathbf{curl\,F}\,dV = -\oiint_{\mathcal{S}} \mathbf{F}\times\hat{\mathbf{N}}\,dS,$$

(b)   $$\iiint_D \mathbf{grad}\,\phi\,dV = \oiint_{\mathcal{S}} \phi\hat{\mathbf{N}}\,dS.$$

**PROOF**   Observe that both of these formulas are equations of *vectors*. They are derived by applying the Divergence Theorem to $\mathbf{F}\times\mathbf{c}$ and $\phi\mathbf{c}$, respectively, where $\mathbf{c}$ is an arbitrary constant vector. We give the details for formula (a) and leave (b) as an exercise.

Using Theorem 3(d), we calculate

$$\nabla\bullet(\mathbf{F}\times\mathbf{c}) = (\nabla\times\mathbf{F})\bullet\mathbf{c} - \mathbf{F}\bullet(\nabla\times\mathbf{c}) = (\nabla\times\mathbf{F})\bullet\mathbf{c}.$$

Also, by the scalar triple product identity (see Exercise 18 of Section 10.3),

$$(\mathbf{F}\times\mathbf{c})\bullet\hat{\mathbf{N}} = (\hat{\mathbf{N}}\times\mathbf{F})\bullet\mathbf{c} = -(\mathbf{F}\times\hat{\mathbf{N}})\bullet\mathbf{c}.$$

Therefore

$$\left(\iiint_D \mathbf{curl\,F}\,dV + \oiint_{\mathcal{S}} \mathbf{F}\times\hat{\mathbf{N}}\,dS\right)\bullet\mathbf{c}$$
$$= \iiint_D (\nabla\times\mathbf{F})\bullet\mathbf{c}\,dV - \oiint_{\mathcal{S}} (\mathbf{F}\times\mathbf{c})\bullet\hat{\mathbf{N}}\,dS$$
$$= \iiint_D \mathbf{div}\,(\mathbf{F}\times\mathbf{c})\,dV - \oiint_{\mathcal{S}} (\mathbf{F}\times\mathbf{c})\bullet\hat{\mathbf{N}}\,dS = 0.$$

Since $\mathbf{c}$ is arbitrary, the vector in the large parentheses must be the zero vector. (If $\mathbf{c}\bullet\mathbf{a} = 0$ for every vector $\mathbf{c}$ then $\mathbf{a} = \mathbf{0}$.) This establishes formula (a).

## Exercises 16.4

In Exercises 1–4, use the Divergence Theorem to calculate the flux of the given vector field out of the sphere $\mathcal{S}$ with equation $x^2 + y^2 + z^2 = a^2$, where $a > 0$.

1. $\mathbf{F} = x\mathbf{i} - 2y\mathbf{j} + 4z\mathbf{k}$
2. $\mathbf{F} = ye^z\mathbf{i} + x^2e^z\mathbf{j} + xy\mathbf{k}$
3. $\mathbf{F} = (x^2 + y^2)\mathbf{i} + (y^2 - z^2)\mathbf{j} + z\mathbf{k}$

**4.** $\mathbf{F} = x^3\mathbf{i} + 3yz^2\mathbf{j} + (3y^2z + x^2)\mathbf{k}$

In Exercises 5–8, evaluate the flux of $\mathbf{F} = x^2\mathbf{i} + y^2\mathbf{j} + z^2\mathbf{k}$ outward across the boundary of the given solid region.

**5.** The ball $(x - 2)^2 + y^2 + (z - 3)^2 \le 9$

**6.** The solid ellipsoid $x^2 + y^2 + 4(z - 1)^2 \le 4$

**7.** The tetrahedron $x + y + z \le 3$, $x \ge 0$, $y \ge 0$, $z \ge 0$

**8.** The cylinder $x^2 + y^2 \le 2y$, $0 \le z \le 4$

**9.** Let $A$ be the area of a region $D$ forming part of the surface of a sphere of radius $R$ centred at the origin, and let $V$ be the volume of the solid cone $C$ consisting of all points on line segments joining the centre of the sphere to points in $D$. Show that

$$V = \frac{1}{3}AR$$

by applying the Divergence Theorem to $\mathbf{F} = x\mathbf{i} + y\mathbf{j} + z\mathbf{k}$.

**10.** Let $\phi(x, y, z) = xy + z^2$. Find the flux of $\nabla\phi$ upwards through the triangular planar surface $S$ with vertices at $(a, 0, 0)$, $(0, b, 0)$, and $(0, 0, c)$.

**11.** A conical domain with vertex $(0, 0, b)$ and axis along the $z$-axis has as base a disk of radius $a$ in the $xy$-plane. Find the flux of

$$\mathbf{F} = (x + y^2)\mathbf{i} + (3x^2y + y^3 - x^3)\mathbf{j} + (z + 1)\mathbf{k}$$

upward through the conical part of the surface of the domain.

**12.** Find the flux of $\mathbf{F} = (y + xz)\mathbf{i} + (y + yz)\mathbf{j} - (2x + z^2)\mathbf{k}$ upward through the first octant part of the sphere $x^2 + y^2 + z^2 = a^2$.

**13.** Let $D$ be the region $x^2 + y^2 + z^2 \le 4a^2$, $x^2 + y^2 \ge a^2$. The surface $S$ of $D$ consists of a cylindrical part, $S_1$, and a spherical part, $S_2$. Evaluate the flux of

$$\mathbf{F} = (x + yz)\mathbf{i} + (y - xz)\mathbf{j} + (z - e^x \sin y)\mathbf{k}$$

out of $D$ through (a) the whole surface $S$, (b) the surface $S_1$, and (c) the surface $S_2$.

**14.** Evaluate $\iint_S (3xz^2\mathbf{i} - x\mathbf{j} - y\mathbf{k}) \bullet \hat{\mathbf{N}}\,dS$, where $S$ is that part of the cylinder $y^2 + z^2 = 1$ which lies in the first octant and between the planes $x = 0$ and $x = 1$.

**15.** A solid region $R$ has volume $V$ and centroid at the point $(\bar{x}, \bar{y}, \bar{z})$. Find the flux of

$$\mathbf{F} = (x^2 - x - 2y)\mathbf{i} + (2y^2 + 3y - z)\mathbf{j} - (z^2 - 4z + xy)\mathbf{k}$$

out of $R$ through its surface.

**16.** The plane $x + y + z = 0$ divides the cube $-1 \le x \le 1$, $-1 \le y \le 1$, $-1 \le z \le 1$ into two parts. Let the lower part (with one vertex at $(-1, -1, -1)$) be $D$. Sketch $D$. Note that it has seven faces, one of which is hexagonal. Find the flux of $\mathbf{F} = x\mathbf{i} + y\mathbf{j} + z\mathbf{k}$ out of $D$ through each of its faces.

**17.** Let $\mathbf{F} = (x^2 + y + 2 + z^2)\mathbf{i} + (e^{x^2} + y^2)\mathbf{j} + (3 + x)\mathbf{k}$. Let $a > 0$ and let $S$ be the part of the spherical surface $x^2 + y^2 + z^2 = 2az + 3a^2$ that is above the $xy$-plane. Find the flux of $\mathbf{F}$ outward across $S$.

**18.** A pile of wet sand having total volume $5\pi$ covers the disk $x^2 + y^2 \le 1$, $z = 0$. The momentum of water vapour is given by $\mathbf{F} = \mathbf{grad}\,\phi + \mu\mathbf{curl}\,\mathbf{G}$, where $\phi = x^2 - y^2 + z^2$ is the water concentration, $\mathbf{G} = \frac{1}{3}(-y^3\mathbf{i} + x^3\mathbf{j} + z^3\mathbf{k})$, and $\mu$ is a constant. Find the flux of $\mathbf{F}$ upward through the top surface of the sand pile.

In Exercises 19–29, $D$ is a three-dimensional domain satisfying the conditions of the Divergence Theorem, and $S$ is its surface. $\hat{\mathbf{N}}$ is the unit outward (from $D$) normal field on $S$. The functions $\phi$ and $\Psi$ are smooth scalar fields on $D$. Also, $\partial\phi/\partial n$ denotes the first directional derivative of $\phi$ in the direction of $\hat{\mathbf{N}}$ at any point on $S$:

$$\frac{\partial\phi}{\partial n} = \nabla\phi \bullet \hat{\mathbf{N}}.$$

**19.** Show that $\oiint_S \mathbf{curl}\,\mathbf{F} \bullet \hat{\mathbf{N}}\,dS = 0$, where $\mathbf{F}$ is an arbitrary smooth vector field.

**20.** Show that the volume $V$ of $D$ is given by

$$V = \frac{1}{3}\oiint_S (x\mathbf{i} + y\mathbf{j} + z\mathbf{k}) \bullet \hat{\mathbf{N}}\,dS.$$

**21.** If $D$ has volume $V$ show that

$$\bar{\mathbf{r}} = \frac{1}{2V}\oiint_S (x^2 + y^2 + z^2)\hat{\mathbf{N}}\,dS$$

is the position vector of the centre of gravity of $D$.

**22.** Show that $\oiint_S \nabla\phi\times\hat{\mathbf{N}}\,dS = 0$.

**23.** If $\mathbf{F}$ is a smooth vector field on $D$, show that

$$\iiint_D \phi\,\mathbf{div}\,\mathbf{F}\,dV + \iiint_D \nabla\phi \bullet \mathbf{F}\,dV = \oiint_S \phi\mathbf{F} \bullet \hat{\mathbf{N}}\,dS.$$

*Hint:* use Theorem 3(b) from Section 16.2.

**Properties of the Laplacian operator**

**24.** If $\nabla^2\phi = 0$ in $D$ and $\phi(x, y, z) = 0$ on $S$, show that $\phi(x, y, z) = 0$ in $D$. *Hint:* Let $\mathbf{F} = \nabla\phi$ in Exercise 23.

**25.** **(Uniqueness for the Dirichlet problem)** The Dirichlet problem for the Laplacian operator is the boundary-value problem

$$\begin{cases} \nabla^2 u(x, y, z) = f(x, y, z) & \text{on } D \\ \quad u(x, y, z) = g(x, y, z) & \text{on } S, \end{cases}$$

where $f$ and $g$ are given functions defined on $D$ and $S$, respectively. Show that this problem can have at most one solution $u(x, y, z)$. *Hint:* suppose there are two solutions, $u$ and $v$, and apply Exercise 24 to their difference $\phi = u - v$.

**26. (The Neumann problem)** If $\nabla^2 \phi = 0$ in $D$ and $\partial\phi/\partial n = 0$ on $S$, show that $\nabla\phi(x, y, z) = 0$ on $D$. The Neumann problem for the Laplacian operator is the boundary-value problem

$$\begin{cases} \nabla^2 u(x, y, z) = f(x, y, z) & \text{on } D \\ \dfrac{\partial}{\partial n} u(x, y, z) = g(x, y, z) & \text{on } S \end{cases}$$

where $f$ and $g$ are given functions defined on $D$ and $S$ respectively. Show that if $D$ is connected, then any two solutions of the Neumann problem must differ by a constant on $D$.

**27.** Verify that $\displaystyle\iiint_D \nabla^2\phi \, dV = \iint_S \frac{\partial\phi}{\partial n} \, dS.$

**28.** Verify that

$$\iiint_D \left( \phi\nabla^2\Psi - \Psi\nabla^2\phi \right) dV$$

$$= \iint_S \left( \phi\frac{\partial\Psi}{\partial n} - \Psi\frac{\partial\phi}{\partial n} \right) dS.$$

**29.** By applying the Divergence Theorem to $\mathbf{F} = \phi\mathbf{c}$, where $\mathbf{c}$ is an arbitrary constant vector, show that

$$\iiint_D \nabla\phi \, dV = \iint_S \phi\hat{\mathbf{N}} \, dS.$$

**★ 30.** Let $P_0$ be a fixed point and for each $\epsilon > 0$ let $D_\epsilon$ be a domain with boundary $S_\epsilon$ satisfying the conditions of the Divergence Theorem. Suppose that the maximum distance from $P_0$ to points $P$ in $D_\epsilon$ approaches zero as $\epsilon \to 0^+$. If $D_\epsilon$ has volume $\text{vol}(D_\epsilon)$, show that

$$\lim_{\epsilon\to0^+} \frac{1}{\text{vol}(D_\epsilon)} \iint_{S_\epsilon} \mathbf{F}\bullet\hat{\mathbf{N}} \, dS = \text{div}\,\mathbf{F}(P_0).$$

This generalizes Theorem 1 of Section 16.1.

# 16.5 Stokes's Theorem

If we regard a region $R$ in the $xy$-plane as a surface in 3-space with normal field $\hat{\mathbf{N}} = \mathbf{k}$, the Green's Theorem formula can be written in the form

$$\oint_C \mathbf{F}\bullet d\mathbf{r} = \iint_R \text{curl}\,\mathbf{F}\bullet\hat{\mathbf{N}} \, dS.$$

Stokes's Theorem given below generalizes this to nonplanar surfaces.

**THEOREM 10**

### Stokes's Theorem

Let $S$ be a piecewise smooth, oriented surface in 3-space, having unit normal field $\hat{\mathbf{N}}$, and having boundary $C$ consisting of one or more piecewise smooth, closed curves with orientation inherited from $S$. If $\mathbf{F}$ is a smooth vector field defined on an open set containing $S$, then

$$\oint_C \mathbf{F}\bullet d\mathbf{r} = \iint_S \text{curl}\,\mathbf{F}\bullet\hat{\mathbf{N}} \, dS.$$

**PROOF** An argument similar to those given in the proofs of Green's Theorem and the Divergence Theorem shows that if $S$ is decomposed into finitely many nonoverlapping subsurfaces then it is sufficient to prove that the formula above holds for each of them. (If subsurfaces $S_1$ and $S_2$ meet along the curve $C^*$, then $C^*$ inherits opposite orientations as part of the boundaries of $S_1$ and $S_2$, so the line integrals along $C^*$ cancel out. See Figure 16.15(a).) We can subdivide $S$ into enough smooth subsurfaces that each one has a one-to-one normal projection onto a coordinate plane. We will establish the formula for one such subsurface, which we will now call $S$.

**Figure 16.15**

(a) Stokes's Theorem holds for a composite surface comprised of non-overlapping subsurfaces for which it is true

(b) A surface with a one-to-one projection on the $xy$-plane

(a)          (b)

Without loss of generality, assume that $S$ has a one-to-one normal projection onto the $xy$-plane, and that its normal field $\hat{\mathbf{N}}$ points upward. Therefore, on $S$, $z$ is a smooth function of $x$ and $y$ defined for $(x, y)$ in a region $R$ of the $xy$-plane. The boundaries $C$ of $S$, and $C^*$ of R, are both oriented counterclockwise as seen from a point high on the $z$-axis. (See Figure 16.15(b).) The normal field on $S$ is

$$\hat{\mathbf{N}} = \frac{-\dfrac{\partial z}{\partial x}\mathbf{i} - \dfrac{\partial z}{\partial y}\mathbf{j} + \mathbf{k}}{\sqrt{1 + \left(\dfrac{\partial z}{\partial x}\right)^2 + \left(\dfrac{\partial z}{\partial y}\right)^2}},$$

and the surface area element on $S$ is expressed in terms of the area element $dA = dx\,dy$ in the $xy$-plane as

$$dS = \sqrt{1 + \left(\frac{\partial z}{\partial x}\right)^2 + \left(\frac{\partial z}{\partial y}\right)^2}\, dA.$$

Therefore,

$$\iint_{S} \operatorname{curl} \mathbf{F} \bullet \hat{\mathbf{N}}\, dS = \iint_{R} \left[ \left(\frac{\partial F_3}{\partial y} - \frac{\partial F_2}{\partial z}\right)\left(-\frac{\partial z}{\partial x}\right) + \left(\frac{\partial F_1}{\partial z} - \frac{\partial F_3}{\partial x}\right)\left(-\frac{\partial z}{\partial y}\right) \right.$$
$$\left. + \left(\frac{\partial F_2}{\partial x} - \frac{\partial F_1}{\partial y}\right) \right] dA.$$

Since $z$ is a function of $x$ and $y$ on $C$, we have $dz = \dfrac{\partial z}{\partial x}\, dx + \dfrac{\partial z}{\partial y}\, dy$. Thus

$$\oint_{C} \mathbf{F} \bullet d\mathbf{r} = \oint_{C^*} \left[ F_1(x, y, z)\, dx + F_2(x, y, z)\, dy \right.$$
$$\left. + F_3(x, y, z)\left(\frac{\partial z}{\partial x}\, dx + \frac{\partial z}{\partial y}\, dy\right) \right]$$
$$= \oint_{C^*} \left( \left[ F_1(x, y, z) + F_3(x, y, z)\frac{\partial z}{\partial x} \right] dx \right.$$
$$\left. + \left[ F_2(x, y, z) + F_3(x, y, z)\frac{\partial z}{\partial y} \right] dy \right).$$

We now apply Green's Theorem in the $xy$-plane to obtain

$$\oint_C \mathbf{F} \bullet d\mathbf{r} = \iint_R \left( \frac{\partial}{\partial x} \left[ F_2(x, y, z) + F_3(x, y, z) \frac{\partial z}{\partial y} \right] \right.$$
$$\left. - \frac{\partial}{\partial y} \left[ F_1(x, y, z) + F_3(x, y, z) \frac{\partial z}{\partial x} \right] \right) dA$$
$$= \iint_R \left( \frac{\partial F_2}{\partial x} + \frac{\partial F_2}{\partial z} \frac{\partial z}{\partial x} + \frac{\partial F_3}{\partial x} \frac{\partial z}{\partial y} + \frac{\partial F_3}{\partial z} \frac{\partial z}{\partial x} \frac{\partial z}{\partial y} + F_3 \frac{\partial^2 z}{\partial x \partial y} \right.$$
$$\left. - \frac{\partial F_1}{\partial y} - \frac{\partial F_1}{\partial z} \frac{\partial z}{\partial y} - \frac{\partial F_3}{\partial y} \frac{\partial z}{\partial x} - \frac{\partial F_3}{\partial z} \frac{\partial z}{\partial y} \frac{\partial z}{\partial x} - F_3 \frac{\partial^2 z}{\partial y \partial x} \right) dA.$$

Observe that four terms in the final integrand cancel out, and the remaining terms are equal to the terms in the expression for $\iint_S \mathbf{curl\, F} \bullet \hat{\mathbf{N}} \, dS$ calculated above. Therefore the proof is complete.

**Remark**   If $\mathbf{curl\, F} = \mathbf{0}$ on a domain $D$ with the property that every piecewise smooth, non-self-intersecting, closed curve in $D$ is the boundary of a piecewise smooth surface in $D$, then Stokes's Theorem assures us that $\oint_C \mathbf{F} \bullet d\mathbf{r} = 0$ for *every* such curve $C$, and therefore $\mathbf{F}$ must be conservative. A simply connected domain $D$ does have the property specified above. We will not attempt a formal proof of this topological fact here, but it should seem plausible if you recall the definition of simple connectedness. A closed curve $C$ in a simply connected domain $D$ must be able to shrink to a point in $D$ without ever passing out of $D$. In so shrinking, it traces out a surface in $D$. This is why Theorem 4 of Section 16.2 is valid for simply connected domains.

**Example 1**   Evaluate $\oint_C \mathbf{F} \bullet d\mathbf{r}$, where $\mathbf{F} = -y^3 \mathbf{i} + x^3 \mathbf{j} - z^3 \mathbf{k}$, and $C$ is the curve of intersection of the cylinder $x^2 + y^2 = 1$ and the plane $2x + 2y + z = 3$ oriented so as to have a counterclockwise projection onto the $xy$-plane.

**Solution**   $C$ is the oriented boundary of an elliptic disk $S$ that lies in the plane $2x + 2y + z = 3$ and has the circular disk $R$: $x^2 + y^2 \leq 1$ as projection onto the $xy$-plane. (See Figure 16.16.) On $S$ we have

$$\hat{\mathbf{N}} \, dS = (2\mathbf{i} + 2\mathbf{j} + \mathbf{k}) \, dx \, dy.$$

Also,

$$\mathbf{curl\, F} = \begin{vmatrix} \mathbf{i} & \mathbf{j} & \mathbf{k} \\ \frac{\partial}{\partial x} & \frac{\partial}{\partial y} & \frac{\partial}{\partial z} \\ -y^3 & x^3 & -z^3 \end{vmatrix} = 3(x^2 + y^2)\mathbf{k}.$$

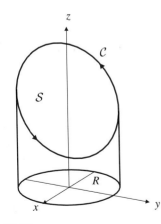

**Figure 16.16**

Thus, by Stokes's Theorem,

$$\oint_C \mathbf{F} \bullet d\mathbf{r} = \iint_S \mathbf{curl\, F} \bullet \hat{\mathbf{N}} \, dS$$
$$= \iint_R 3(x^2 + y^2) \, dx \, dy = 2\pi \int_0^1 3r^2 \, r \, dr = \frac{3\pi}{2}.$$

As with the Divergence Theorem, the principal importance of Stokes's Theorem is as a theoretical tool. However, it can also simplify the calculation of circulation integrals such as the one in the previous example. It is not difficult to imagine integrals whose evaluation would be impossibly difficult without the use of Stokes's Theorem or the Divergence Theorem. In the following example we use Stokes's Theorem twice, but the result could be obtained just as easily by using the Divergence Theorem.

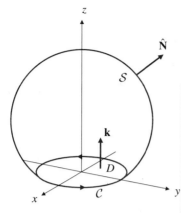

**Figure 16.17**    Part of a sphere, and a disk with the same boundary

**Example 2**    Find $I = \iint_S \text{curl}\,\mathbf{F} \bullet \hat{\mathbf{N}}\,dS$ where $S$ is that part of the sphere $x^2 + y^2 + (z-2)^2 = 8$ that lies above the $xy$-plane, $\hat{\mathbf{N}}$ is the unit outward normal field on $S$, and

$$\mathbf{F} = y^2 \cos xz\,\mathbf{i} + x^3 e^{yz}\mathbf{j} - e^{-xyz}\mathbf{k}.$$

**Solution**    The boundary, $C$, of $S$ is the circle $x^2 + y^2 = 4$ in the $xy$-plane, oriented counterclockwise as seen from the positive $z$-axis. (See Figure 16.17.) This curve is also the oriented boundary of the plane disk $D$: $x^2 + y^2 \le 4$, $z = 0$, with normal field $\hat{\mathbf{N}} = \mathbf{k}$. Thus, two applications of Stokes's Theorem give

$$I = \iint_S \text{curl}\,\mathbf{F} \bullet \hat{\mathbf{N}}\,dS = \oint_C \mathbf{F} \bullet d\mathbf{r} = \iint_D \text{curl}\,\mathbf{F} \bullet \mathbf{k}\,dA.$$

On $D$ we have

$$\text{curl}\,\mathbf{F} \bullet \mathbf{k} = \left( \frac{\partial}{\partial x}\left(x^3 e^{yz}\right) - \frac{\partial}{\partial y}\left(y^2 \cos xz\right) \right)\bigg|_{z=0}$$

$$= 3x^2 - 2y.$$

By symmetry, $\iint_D y\,dA = 0$, so

$$I = 3\iint_D x^2\,dA = 3\int_0^{2\pi} \cos^2\theta\,d\theta \int_0^2 r^3\,dr = 12\pi.$$

∎

**Remark**    A surface $S$ satisfying the conditions of Stokes's Theorem may no longer do so if a single point is removed from it. An isolated boundary point of a surface is not an orientable curve, and Stokes's Theorem may therefore break down for such a surface. Consider, for example, the vector field

$$\mathbf{F} = \frac{\hat{\theta}}{r} = -\frac{y}{x^2 + y^2}\mathbf{i} + \frac{x}{x^2 + y^2}\mathbf{j},$$

which is defined on the *punctured disk* $D$ satisfying $0 < x^2 + y^2 \le a^2$. (See Example 3 above.) If $D$ is oriented with upward normal $\mathbf{k}$, then its boundary consists of the oriented, smooth, closed curve, $C$, given by $x = a\cos\theta$, $y = a\sin\theta$, $(0 \le \theta \le 2\pi)$, and the isolated point $(0, 0)$. We have

$$\oint_C \mathbf{F} \bullet d\mathbf{r} = \int_0^{2\pi} \left( \frac{-\sin\theta}{a}\mathbf{i} + \frac{\cos\theta}{a}\mathbf{j} \right) \bullet (-a\sin\theta\mathbf{i} + a\cos\theta\mathbf{j})\,d\theta$$

$$= \int_0^{2\pi} (\sin^2\theta + \cos^2\theta)\,d\theta = 2\pi.$$

However,

$$\text{curl } \mathbf{F} = \frac{\partial}{\partial x}\left(\frac{x}{x^2 + y^2}\right) - \frac{\partial}{\partial y}\left(-\frac{y}{x^2 + y^2}\right) = 0$$

identically on $D$. Thus

$$\iint_D \text{curl }\mathbf{F} \bullet \hat{\mathbf{N}}\, dS = 0,$$

and the conclusion of Stokes's Theorem fails in this case.

## Exercises 16.5

1. Evaluate $\oint_C xy\, dx + yz\, dy + zx\, dz$ around the triangle with vertices $(1, 0, 0)$, $(0, 1, 0)$ and $(0, 0, 1)$, oriented clockwise as seen from the point $(1, 1, 1)$.

2. Evaluate $\oint_C y\, dx - x\, dy + z^2\, dz$ around the curve $C$ of intersection of the cylinders $z = y^2$ and $x^2 + y^2 = 4$, oriented counterclockwise as seen from a point high on the $z$-axis.

3. Evaluate $\iint_S \text{curl }\mathbf{F} \bullet \hat{\mathbf{N}}\, dS$, where $S$ is the hemisphere $x^2 + y^2 + z^2 = a^2$, $z \geq 0$ with outward normal, and $\mathbf{F} = 3y\mathbf{i} - 2xz\mathbf{j} + (x^2 - y^2)\mathbf{k}$.

4. Evaluate $\iint_S \text{curl }\mathbf{F} \bullet \hat{\mathbf{N}}\, dS$, where $S$ is the surface $x^2 + y^2 + 2(z - 1)^2 = 6$, $z \geq 0$, $\hat{\mathbf{N}}$ is the unit outward (away from the origin) normal on $S$, and

$$\mathbf{F} = (xz - y^3\cos z)\mathbf{i} + x^3 e^z\mathbf{j} + xyz\, e^{x^2+y^2+z^2}\mathbf{k}.$$

5. Use Stokes's Theorem to show that

$$\oint_C y\, dx + z\, dy + x\, dz = \sqrt{3}\,\pi a^2,$$

where $C$ is the suitably oriented intersection of the surfaces $x^2 + y^2 + z^2 = a^2$ and $x + y + z = 0$.

6. Evaluate $\oint_C \mathbf{F} \bullet d\mathbf{r}$ around the curve

$$\mathbf{r} = \cos t\,\mathbf{i} + \sin t\,\mathbf{j} + \sin 2t\,\mathbf{k}, \quad (0 \leq t \leq 2\pi),$$

where

$$\mathbf{F} = (e^x - y^3)\mathbf{i} + (e^y + x^3)\mathbf{j} + e^z\mathbf{k}.$$

*Hint:* show that $C$ lies on the surface $z = 2xy$.

7. Find the circulation of $\mathbf{F} = -y\mathbf{i} + x^2\mathbf{j} + z\mathbf{k}$ around the oriented boundary of the part of the paraboloid $z = 9 - x^2 - y^2$ lying above the $xy$-plane and having normal field pointing upward.

8. Evaluate $\oint_C \mathbf{F} \bullet d\mathbf{r}$, where

$$\mathbf{F} = ye^x\mathbf{i} + (x^2 + e^x)\mathbf{j} + z^2 e^z\mathbf{k},$$

and $C$ is the curve

$$\mathbf{r}(t) = (1 + \cos t)\mathbf{i} + (1 + \sin t)\mathbf{j} + (1 - \cos t - \sin t)\mathbf{k}$$

for $0 \leq t \leq 2\pi$. *Hint:* use Stokes's Theorem, observing that $C$ lies in a certain plane, and has a circle as its projection onto the $xy$-plane. The integral can also be evaluated by using the techniques of Section 15.4.

9. Let $C_1$ be the straight line joining $(-1, 0, 0)$ to $(1, 0, 0)$ and let $C_2$ be the semicircle $x^2 + y^2 = 1$, $z = 0$, $y \geq 0$. Let $S$ be a smooth surface joining $C_1$ to $C_2$ having upward normal, and let

$$\mathbf{F} = (\alpha x^2 - z)\mathbf{i} + (xy + y^3 + z)\mathbf{j} + \beta y^2(z + 1)\mathbf{k}.$$

Find the values of $\alpha$ and $\beta$ for which $I = \iint_S \mathbf{F} \bullet d\mathbf{S}$ is independent of the choice of $S$, and find the value of $I$ for these values of $\alpha$ and $\beta$.

10. Let $C$ be the curve $(x - 1)^2 + 4y^2 = 16$, $2x + y + z = 3$, oriented counterclockwise when viewed from high on the $z$-axis. Let

$$\mathbf{F} = (z^2 + y^2 + \sin x^2)\mathbf{i} + (2xy + z)\mathbf{j}) + (xz + 2yz)\mathbf{k}.$$

Evaluate $\oint_C \mathbf{F} \bullet d\mathbf{r}$.

**11.** If $C$ is the oriented boundary of surface $S$ and $\phi$ and $\Psi$ are arbitrary smooth scalar fields, show that

$$\oint_C \phi \nabla \Psi \bullet d\mathbf{r} = -\oint_C \Psi \nabla \phi \bullet d\mathbf{r}$$

$$= \iint_S (\nabla \phi \times \nabla \Psi) \bullet \hat{\mathbf{N}} \, dS.$$

Is $\nabla \phi \times \nabla \Psi$ solenoidal? Find a vector potential for it.

**12.** Let $C$ be a closed, non-self-intersecting, piecewise smooth plane curve in $\mathbb{R}^3$, which lies in a plane with unit normal $\hat{\mathbf{N}} = a\mathbf{i} + b\mathbf{j} + c\mathbf{k}$, and which has orientation inherited from that of the plane. Show that the plane area enclosed by $C$ is

$$\frac{1}{2} \oint_C (bz - cy) \, dx + (cx - az) \, dy + (ay - bx) \, dz.$$

**13.** Use Stokes's Theorem to prove Theorem 2 of Section 16.1.

# 16.6 Some Physical Applications of Vector Calculus

In this section we will show how the theory developed in this chapter can be used to model concrete applied mathematical problems. We will look at two areas of application, fluid dynamics and electromagnetism, and will develop a few of the fundamental vector equations underlying these disciplines. Our purpose is to illustrate the techniques of vector calculus in applied contexts, rather than to provide any complete or even coherent introductions to the disciplines themselves.

## Fluid Dynamics

Suppose that a region of 3-space is filled with a fluid (liquid or gas) in motion. Two approaches can be taken to describe the motion. We could attempt to determine the position, $\mathbf{r} = \mathbf{r}(a, b, c, t)$ at any time $t$, of a "particle" of fluid which was located at the point $(a, b, c)$ at time $t = 0$. This is the Lagrange approach. Alternatively, we could attempt to determine the velocity, $\mathbf{v}(x, y, z, t)$, the density, $\delta(x, y, z, t)$, and other physical variables such as the pressure, $p(x, y, z, t)$, at any time $t$ at any point $(x, y, z)$ in the region occupied by the fluid. This is the Euler approach.

We will examine the latter method, and describe how the Divergence Theorem can be used to translate some fundamental physical laws into equivalent mathematical equations. We assume throughout that the velocity, density, and pressure vary smoothly in all their variables, and that the fluid is an *ideal fluid*, that is, nonviscous (it doesn't stick to itself), homogeneous, and isotropic (it has the same properties at all points and in all directions). Such properties are not always shared by real fluids, and so we are dealing with a simplified mathematical model that does not always correspond exactly to the behaviour of real fluids.

Consider an imaginary closed surface $S$ in the fluid, bounding a domain $D$. We call $S$ "imaginary" because it is not a barrier that impedes the flow of the fluid in any way—it is just a means to concentrate our attention on a particular part of the fluid It is fixed in space and does not move with the fluid. Let us assume that the fluid is being neither created nor destroyed anywhere (in particular, there are no sources or sinks), and so the law of **conservation of mass** tells us that the rate of change of the mass of fluid in $D$ equals the rate at which fluid enters $D$ across $S$.

The mass of fluid in volume element $dV$ at position $(x, y, z)$ at time $t$ is $\delta(x, y, z, t) \, dV$, so the mass in $D$ at time $t$ is $\iiint_D \delta \, dV$. This mass changes at rate

$$\frac{\partial}{\partial t} \iiint_D \delta \, dV = \iiint_D \frac{\partial \delta}{\partial t} \, dV.$$

As we noted in Section 15.6, the volume of fluid passing *out* of $D$ through area element $dS$ at position $(x, y, z)$ in the interval from time $t$ to $t + dt$ is given by $\mathbf{v}(x, y, z, t) \bullet \hat{\mathbf{N}} \, dS \, dt$, where $\hat{\mathbf{N}}$ is the unit normal at $(x, y, z)$ on $S$ pointing out of $D$. Hence the mass crossing $dS$ outwards in that time interval is $\delta\mathbf{v} \bullet \hat{\mathbf{N}} \, dS \, dt$, and the *rate* at which mass is flowing out of $D$ across $S$ at time $t$ is

$$\oiint_{S} \delta\mathbf{v} \bullet \hat{\mathbf{N}} \, dS.$$

The rate at which mass is flowing *into* $D$ is the negative of the above rate. Since mass is conserved, we must have

$$\iiint_{D} \frac{\partial\delta}{\partial t} \, dV = -\oiint_{S} \delta\mathbf{v} \bullet \hat{\mathbf{N}} \, dS = -\iiint_{D} \mathbf{div}\,(\delta\mathbf{v}) \, dV,$$

where we have used the Divergence Theorem to replace the surface integral with a volume integral. Thus

$$\iiint_{D} \left( \frac{\partial\delta}{\partial t} + \mathbf{div}\,(\delta\mathbf{v}) \right) dV = 0.$$

This equation must hold for *any* domain $D$ in the fluid.

If a continuous function $f$ satisfies $\iiint_{D} f(P) \, dV = 0$ for every domain $D$, then $f(P) = 0$ at all points $P$, for if there were a point $P_0$ such that $f(P_0) \neq 0$, (say $f(P_0) > 0$), then, by continuity, $f$ would be positive at all points in some sufficiently small ball $B$ centred at $P_0$, and $\iiint_{B} f(P) \, dV$ would be greater than 0. Applying this principle, we must have

$$\frac{\partial\delta}{\partial t} + \mathbf{div}\,(\delta\mathbf{v}) = 0$$

throughout the fluid. This is called the **equation of continuity** for the fluid. It is equivalent to conservation of mass. Observe that if the fluid is **incompressible** then $\delta$ is a constant, independent of both time and spatial position. In this case $\partial\delta/\partial t = 0$ and $\mathbf{div}\,(\delta\mathbf{v}) = \delta \, \mathbf{div}\,\mathbf{v}$. Therefore the equation of continuity for an incompressible fluid is simply

$$\mathbf{div}\,\mathbf{v} = 0.$$

The motion of the fluid is governed by Newton's second law, which asserts that the rate of change of momentum of any part of the fluid is equal to the sum of the forces applied to that part. Again let us consider the part of the fluid in a domain $D$. At any time $t$ its momentum is $\iiint_{D} \delta\mathbf{v} \, dV$, and is changing at rate

$$\iiint_{D} \frac{\partial}{\partial t}(\delta\mathbf{v}) \, dV.$$

This change is due partly to momentum crossing $S$ into or out of $D$ (the momentum of the fluid crossing $S$), partly to the pressure exerted on the fluid in $D$ by the fluid outside, and partly to any external *body forces* (such as gravity or electromagnetic forces) acting on the fluid. Let us examine each of these causes in turn.

Momentum is transferred across $S$ into $D$ at rate

$$-\oiint_S \mathbf{v}(\delta \mathbf{v} \bullet \hat{\mathbf{N}}) \, dS.$$

The pressure on the fluid in $D$ is exerted across $S$ in the direction of the inward normal $-\hat{\mathbf{N}}$. Thus this part of the force on the fluid in $D$ is

$$-\oiint_S p\hat{\mathbf{N}} \, dS.$$

The body forces are best expressed in terms of the *force density* (force per unit mass), $\mathbf{F}$. The total body force on the fluid in $D$ is therefore

$$\iiint_D \delta \mathbf{F} \, dV.$$

Newton's second law now implies that

$$\iiint_D \frac{\partial}{\partial t}(\delta \mathbf{v}) \, dV = -\oiint_S \mathbf{v}(\delta \mathbf{v} \bullet \hat{\mathbf{N}}) \, dS - \oiint_S p\hat{\mathbf{N}} \, dS + \iiint_D \delta \mathbf{F} \, dV.$$

Again we would like to convert the surface integrals to triple integrals over $D$. If we use the results of Exercise 29 of Section 16.4 and Exercise 2 at the end of this section, we get

$$\oiint_S p\hat{\mathbf{N}} \, dS = \iiint_D \nabla p \, dV,$$

$$\oiint_S \mathbf{v}(\delta \mathbf{v} \bullet \hat{\mathbf{N}}) \, dS = \iiint_D \left(\delta(\mathbf{v} \bullet \nabla)\mathbf{v} + \mathbf{v}\,\mathbf{div}\,(\delta \mathbf{v})\right) \, dV.$$

Accordingly, we have

$$\iiint_D \left(\delta \frac{\partial \mathbf{v}}{\partial t} + \mathbf{v}\frac{\partial \delta}{\partial t} + \mathbf{v}\,\mathbf{div}\,(\delta \mathbf{v}) + \delta(\mathbf{v} \bullet \nabla)\mathbf{v} + \nabla p - \delta \mathbf{F}\right) \, dV = 0.$$

The second and third terms in the integrand cancel out by virtue of the continuity equation. Since $D$ is arbitrary, we must therefore have

$$\delta \frac{\partial \mathbf{v}}{\partial t} + \delta(\mathbf{v} \bullet \nabla)\mathbf{v} = -\nabla p + \delta \mathbf{F}.$$

This is the **equation of motion** of the fluid. Observe that it is not a *linear* partial differential equation; the second term on the left is not linear in $\mathbf{v}$.

## Electromagnetism

In 3-space there are defined two vector fields which determine the electric and magnetic forces that would be experienced by a unit charge at a particular point if it is moving with unit speed. (These vector fields are determined by electric charges and currents present in the space.) A charge $q_0$ at position $\mathbf{r} = x\mathbf{i} + y\mathbf{j} + z\mathbf{k}$ moving with velocity $\mathbf{v}_0$ experiences an electric force $q_0\mathbf{E}(\mathbf{r})$, where $\mathbf{E}$ is the **electric field**, and a magnetic force $\mu_0 q_0 \mathbf{v}_0 \times \mathbf{H}(\mathbf{r})$, where $\mathbf{H}$ is the **magnetic field** and $\mu_0 \approx 1.26 \times 10^{-6}$ N/ampere$^2$ is a physical constant called the **permeability of free space**. We will look briefly at each of these fields, but will initially restrict ourselves to considering *static* situations. Electric fields produced by static charge distributions, and magnetic fields produced by static electric currents do not depend on time. Later we will consider the interaction between the two fields when they are time dependent.

## Electrostatics

Experimental evidence shows that the value of the electric field at any point $\mathbf{r}$ is the vector sum of the fields caused by any elements of charge located in 3-space. A "point charge" $q$ at position $\mathbf{s} = \xi\mathbf{i} + \eta\mathbf{j} + \zeta\mathbf{k}$ generates the electric field

$$\mathbf{E}(\mathbf{r}) = \frac{q}{4\pi\epsilon_0} \frac{\mathbf{r} - \mathbf{s}}{|\mathbf{r} - \mathbf{s}|^3}, \qquad \text{(Coulomb's Law)},$$

where $\epsilon_0 \approx 8.85 \times 10^{-12}$ coulombs$^2$/N-m$^2$ is a physical constant called the **permittivity of free space**. This is just the field due to a point source of strength $q/4\pi\epsilon_0$ at $\mathbf{s}$. Except at $\mathbf{r} = \mathbf{s}$ the field is conservative, with potential

$$\phi(\mathbf{r}) = -\frac{q}{4\pi\epsilon_0} \frac{1}{|\mathbf{r} - \mathbf{s}|},$$

so for $\mathbf{r} \neq \mathbf{s}$ we have **curl E** $= \mathbf{0}$. Also **div E** $= 0$, except at $\mathbf{r} = \mathbf{s}$ where it is infinite; in terms of the Dirac distribution, **div E** $= kq\delta(x - \xi)\delta(y - \eta)\delta(z - \zeta)$. (See Section 16.1.) The flux of $\mathbf{E}$ outward across the surface $\mathcal{S}$ of any region $R$ containing $q$ is

$$\oiint_{\mathcal{S}} \mathbf{E} \bullet \hat{\mathbf{N}}\, dS = \frac{q}{\epsilon_0},$$

by analogy with Example 4 of Section 16.4.

Given a *charge distribution* of density $\rho(\xi, \eta, \zeta)$ in 3-space (so that the charge in volume element $dV = d\xi\, d\eta\, d\zeta$ at $\mathbf{s}$ is $dq = \rho\, dV$), the flux of $\mathbf{E}$ out of $\mathcal{S}$ due to the charge in $R$ is

$$\oiint_{\mathcal{S}} \mathbf{E} \bullet \hat{\mathbf{N}}\, dS = \frac{1}{\epsilon_0} \iiint_R dq = \frac{1}{\epsilon_0} \iiint_R \rho\, dV.$$

If we apply the Divergence Theorem to the surface integral, we obtain

$$\iiint_R \left( \mathbf{div\, E} - \frac{\rho}{\epsilon_0} \right) dV = 0,$$

and since R is an arbitrary region,

$$\mathbf{div\, E} = \frac{\rho}{\epsilon_0}.$$

This is the differential form of Gauss's Law. See Exercise 3 below.

The potential due to a charge distribution of density $\rho(\mathbf{s})$ in the region R is

$$\phi(\mathbf{r}) = -\frac{1}{4\pi\epsilon_0} \iiint_R \frac{dq}{|\mathbf{r} - \mathbf{s}|}\, dV$$

$$= -\frac{1}{4\pi\epsilon_0} \iiint_R \frac{\rho(\xi, \eta, \zeta)\, d\xi\, d\eta\, d\zeta}{\sqrt{(x - \xi)^2 + (y - \eta)^2 + (z - \zeta)^2}}.$$

If $\rho$ is continuous and vanishes outside a bounded region, the triple integral is convergent everywhere (see Exercise 4), so $\mathbf{E} = \nabla\phi$ is conservative throughout 3-space. Thus, at all points,

$$\operatorname{curl}\mathbf{E} = \mathbf{0}.$$

Since $\operatorname{div}\mathbf{E} = \operatorname{div}\nabla\phi = \nabla^2\phi$, the potential $\phi$ satisfies **Poisson's equation**

$$\nabla^2\phi = \frac{\rho}{\epsilon_0}.$$

In particular, $\phi$ is a harmonic function in regions of space where no charge is distributed.

## Magnetostatics

Magnetic fields are produced by moving charges, that is, by currents. Suppose that a constant electric current, $I$, is flowing in a filament along the curve $\mathcal{F}$. It has been determined experimentally that the magnetic field produced at position $\mathbf{r} = x\mathbf{i} + y\mathbf{j} + z\mathbf{k}$ by the elements of current $dI = I\,ds$ along the filament add vectorially, and that the element at position $\mathbf{s} = \xi\mathbf{i} + \eta\mathbf{j} + \zeta\mathbf{k}$ produces the field

$$d\mathbf{H}(\mathbf{r}) = \frac{I}{4\pi}\frac{d\mathbf{s}\times(\mathbf{r} - \mathbf{s})}{|\mathbf{r} - \mathbf{s}|^3}, \qquad \text{(the Biot–Savart Law),}$$

where $d\mathbf{s} = \hat{\mathbf{T}}\,ds$, $\hat{\mathbf{T}}$ being the unit tangent to $\mathcal{F}$ in the direction of the current. Under the reasonable assumption that charge is not created or destroyed anywhere, the filament $\mathcal{F}$ must form a closed circuit, and the total magnetic field at $\mathbf{r}$ due to the current flowing in the circuit is

$$\mathbf{H} = \frac{I}{4\pi}\oint_{\mathcal{F}}\frac{d\mathbf{s}\times(\mathbf{r} - \mathbf{s})}{|\mathbf{r} - \mathbf{s}|^3}.$$

Let $\mathbf{A}$ be a vector field be defined by

$$\mathbf{A}(\mathbf{r}) = \frac{I}{4\pi}\oint_{\mathcal{F}}\frac{d\mathbf{s}}{|\mathbf{r} - \mathbf{s}|},$$

for all $\mathbf{r}$ not on the filament $\mathcal{F}$. If we make use of the fact that

$$\nabla\left(\frac{1}{|\mathbf{r} - \mathbf{s}|}\right) = -\frac{\mathbf{r} - \mathbf{s}}{|\mathbf{r} - \mathbf{s}|^3},$$

and the vector identity $\nabla\times(\phi\mathbf{F}) = (\nabla\phi)\times\mathbf{F} + \phi(\nabla\times\mathbf{F})$, (with $\mathbf{F}$ the vector $d\mathbf{s}$, which does not depend on $\mathbf{r}$), we can calculate the curl of $\mathbf{A}$:

$$\nabla\times\mathbf{A} = \frac{I}{4\pi}\oint_{\mathcal{F}}\nabla\left(\frac{1}{|\mathbf{r} - \mathbf{s}|}\right)\times d\mathbf{s} = \frac{I}{4\pi}\oint_{\mathcal{F}}-\frac{\mathbf{r} - \mathbf{s}}{|\mathbf{r} - \mathbf{s}|^3}\times d\mathbf{s} = \mathbf{H}(\mathbf{r}).$$

Thus $\mathbf{A}$ is a vector potential for $\mathbf{H}$, and $\operatorname{div}\mathbf{H} = 0$ at points off the filament. We can also verify by calculation that $\operatorname{curl}\mathbf{H} = \mathbf{0}$ off the filament. (See Exercises 9–11 at the end of this section.)

Imagine a circuit consisting of a straight filament along the $z$-axis with return at infinite distance. The field $\mathbf{H}$ at a finite point will then just be due to the current along the $z$-axis, where the current $I$ is flowing in the direction of $\mathbf{k}$, say. The currents in all elements $d\mathbf{s}$ produce at $\mathbf{r}$ fields in the same direction, normal to the plane containing $\mathbf{r}$ and the $z$-axis. (See Figure 16.18.) Therefore the field strength $H = |\mathbf{H}|$ at a distance $a$ from the $z$-axis is obtained by integrating the elements

$$dH = \frac{I}{4\pi}\frac{\sin\theta\,d\zeta}{a^2 + (\zeta - z)^2} = \frac{I}{4\pi}\frac{a\,d\zeta}{\left(a^2 + (\zeta - z)^2\right)^{3/2}}.$$

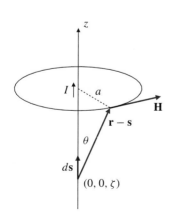

**Figure 16.18**

We have

$$H = \frac{Ia}{4\pi} \int_{-\infty}^{\infty} \frac{d\zeta}{\left(a^2 + (\zeta - z)^2\right)^{3/2}} \qquad \text{(Let } \zeta - z = a \tan\phi.)$$

$$= \frac{I}{4\pi a} \int_{-\pi/2}^{\pi/2} \cos\phi \, d\phi = \frac{I}{2\pi a}.$$

The field lines of **H** are evidently horizontal circles centred on the $z$-axis. If $C_a$ is such a circle, having radius $a$, then the circulation of **H** around $C_a$ is

$$\oint_{C_a} \mathbf{H} \bullet d\mathbf{r} = \frac{I}{2\pi a} 2\pi a = I.$$

Observe that the circulation calculated above is independent of $a$. In fact, if $C$ is any closed curve that encircles the $z$-axis once counterclockwise (as seen from above), then $C$ and $-C_a$ comprise the oriented boundary of a washer-like surface $S$ with a hole in it through which the filament passes. Since **curl H** $= \mathbf{0}$ on $S$, Stokes's Theorem guarantees that

$$\oint_{C} \mathbf{H} \bullet d\mathbf{r} = \oint_{C_a} \mathbf{H} \bullet d\mathbf{r} = I.$$

Furthermore, when $C$ is very small (and therefore very close to the filament) most of the contribution to the circulation of **H** around it comes from that part of the filament which is very close to $C$. It therefore does not matter whether the filament is straight, or infinitely long. For any closed-loop filament carrying a current, the circulation of the magnetic field around the oriented boundary of a surface through which the filament passes is equal to the current flowing in the loop. This is **Ampère's circuital law**. The surface is oriented with normal on the side out of which the current is flowing.

Now let us replace the filament with a more general current specified by a vector density, **J**. This means that at any point **s** the current is flowing in the direction **J(s)**, and that the current crossing an area element $dS$ with unit normal $\hat{\mathbf{N}}$ is $\mathbf{J} \bullet \hat{\mathbf{N}} \, dS$. The circulation of **H** around the boundary $C$ of surface $S$ is equal to the total current flowing across $S$, so

$$\oint_{C} \mathbf{H} \bullet d\mathbf{r} = \iint_{S} \mathbf{J} \bullet \hat{\mathbf{N}} \, dS.$$

By using Stokes's Theorem, we can replace the line integral with another surface integral, and so obtain

$$\iint_{S} (\mathbf{curl\,H} - \mathbf{J}) \bullet \hat{\mathbf{N}} \, dS = 0.$$

Since $S$ is arbitrary, we must have, at all points,

$$\mathbf{curl\,H} = \mathbf{J},$$

which is the pointwise version of Ampère's circuital law. It can be readily checked that if

$$\mathbf{A}(\mathbf{r}) = \frac{1}{4\pi} \iiint_{R} \frac{\mathbf{J}(\mathbf{s})}{|\mathbf{r} - \mathbf{s}|} \, dV,$$

then $\mathbf{H} = \mathbf{curl}\,\mathbf{A}$ (so that $\mathbf{A}$ is a vector potential for the magnetic field $\mathbf{H}$). Here $R$ is the region of 3-space where $\mathbf{J}$ is nonzero. If $\mathbf{J}$ is continuous and vanishes outside a bounded set, then the triple integral converges for all $\mathbf{r}$ (Exercise 4), and $\mathbf{H}$ is everywhere solenoidal:

$$\mathbf{div}\,\mathbf{H} = 0.$$

## Maxwell's Equations

The four equations obtained above for static electric and magnetic fields,

$$\mathbf{div}\,\mathbf{E} = \rho/\epsilon_0 \qquad \mathbf{div}\,\mathbf{H} = 0$$
$$\mathbf{curl}\,\mathbf{E} = \mathbf{0} \qquad \mathbf{curl}\,\mathbf{H} = \mathbf{J}$$

require some modification if the fields $\mathbf{E}$ and $\mathbf{H}$ depend on time. Gauss's Law $\mathbf{div}\,\mathbf{E} = \rho/\epsilon_0$ remains valid, as does $\mathbf{div}\,\mathbf{H} = 0$, which expresses the fact that there are no *known* magnetic *sources* or *sinks* (that is, magnetic *monopoles*). The field lines of $\mathbf{H}$ must be closed curves.

It was observed by Michael Faraday that the circulation of an electric field around a simple closed curve $\mathcal{C}$ corresponds to a change in the magnetic flux

$$\Phi = \iint_{\mathcal{S}} \mathbf{H} \bullet \hat{\mathbf{N}}\,dS$$

through any oriented surface $\mathcal{S}$ having boundary $\mathcal{C}$, according to the formula

$$\frac{d\Phi}{dt} = -\frac{1}{\mu_0} \oint_{\mathcal{C}} \mathbf{E} \bullet d\mathbf{r}.$$

Applying Stokes's Theorem to the line integral, we obtain

$$\iint_{\mathcal{S}} \mathbf{curl}\,\mathbf{E}\bullet\hat{\mathbf{N}}\,dS = \oint_{\mathcal{C}} \mathbf{E}\bullet d\mathbf{r} = -\mu_0 \frac{d}{dt}\iint_{\mathcal{S}} \mathbf{H}\bullet\hat{\mathbf{N}}\,dS = -\mu_0 \iint_{\mathcal{S}} \frac{\partial\mathbf{H}}{\partial t}\bullet\hat{\mathbf{N}}\,dS.$$

Since $\mathcal{S}$ is arbitrary, we obtain the differential form of Faraday's Law

$$\mathbf{curl}\,\mathbf{E} = -\mu_0 \frac{\partial\mathbf{H}}{\partial t}.$$

The electric field is irrotational only if the magnetic field is constant in time.

The differential form of Ampère's Law, $\mathbf{curl}\,\mathbf{H} = \mathbf{J}$, also requires modification. If the electric field depends on time, then so will the current density $\mathbf{J}$. Assuming conservation of charge (charges are not produced or destroyed) we can show by an identical argument to that used to obtain the continuity equation for fluid motion earlier in this section, that the rate of change of charge density satisfies

$$\frac{\partial\rho}{\partial t} = -\mathbf{div}\,\mathbf{J}.$$

(See Exercise 5 below.) This is inconsistent with Ampère's law because $\mathbf{div}\,\mathbf{curl}\,\mathbf{H} = 0$, while $\mathbf{div}\,\mathbf{J} \neq 0$ when $\rho$ depends on time. Note, however, that $\rho = \epsilon_0\mathbf{div}\,\mathbf{E}$ implies that

$$-\mathbf{div}\,\mathbf{J} = \frac{\partial\rho}{\partial t} = \epsilon_0\mathbf{div}\,\frac{\partial\mathbf{E}}{\partial t},$$

and so $\mathbf{div}\left(\mathbf{J} + \epsilon_0 \partial\mathbf{E}/\partial t\right) = 0$. This suggests that for the nonstatic case, Ampère's law becomes

$$\mathbf{curl\,H} = \mathbf{J} + \epsilon_0 \frac{\partial\mathbf{E}}{\partial t}$$

which indicates (as was discovered by Maxwell) that magnetic fields are not just produced by currents, but also by changing electric fields.

Together the four equations

$$\mathbf{div\,E} = \rho/\epsilon_0 \qquad\qquad \mathbf{div\,H} = 0$$

$$\mathbf{curl\,E} = -\mu_0\frac{\partial\mathbf{H}}{\partial t} \qquad \mathbf{curl\,H} = \mathbf{J} + \epsilon_0\frac{\partial\mathbf{E}}{\partial t}$$

are known as **Maxwell's equations**. They govern the way electric and magnetic fields are produced in 3-space by the presence of charges and currents.

# Exercises 16.6

1. **(Archimedes' principle)** A solid occupying region $R$ with surface $S$ is immersed in a liquid of constant density $\delta$. The pressure at depth $h$ in the liquid is $\delta gh$, so the pressure satisfies $\nabla p = \delta\mathbf{g}$, where $\mathbf{g}$ is the (vector) constant acceleration of gravity. Over each surface element $dS$ on $S$ the pressure of the fluid exerts a force $-p\hat{\mathbf{N}}\,dS$ on the solid.

   (a) Show that the resultant "buoyancy force" on the solid is

   $$\mathbf{B} = -\iiint_R \delta\mathbf{g}\,dV.$$

   Thus the buoyancy force has the same magnitude as, and opposite direction to, the weight of the liquid displaced by the solid. This is Archimedes' principle.

   (b) Extend the above result to the case where the solid is only partly submerged in the fluid.

2. By breaking the vector $\mathbf{F}(\mathbf{G}\bullet\hat{\mathbf{N}})$ into its separate components and applying the Divergence Theorem to each separately, show that

   $$\oiint_S \mathbf{F}(\mathbf{G}\bullet\hat{\mathbf{N}})\,dS = \iiint_D \left(\mathbf{F\,div\,G} + (\mathbf{G}\bullet\nabla)\mathbf{F}\right)dV,$$

   where $\hat{\mathbf{N}}$ is the unit outward normal on the surface $S$ of the domain $D$.

3. **(Gauss's Law)** Show that the flux of the electric field $\mathbf{E}$ outward through a closed surface $S$ in 3-space is $1/\epsilon_0$ times the total charge enclosed by $S$.

4. If $\mathbf{s} = \xi\mathbf{i} + \eta\mathbf{j} + \zeta\mathbf{k}$ and $f(\xi, \eta, \zeta)$ is continuous on $\mathbb{R}^3$ and vanishes outside a bounded region, show that for any fixed $\mathbf{r}$,

   $$\iiint_{\mathbb{R}^3} \frac{|f(\xi,\eta,\zeta)|}{|\mathbf{r}-\mathbf{s}|}\,d\xi\,d\eta\,d\zeta \le \text{constant}.$$

This shows that the potentials for the electric and magnetic fields corresponding to continuous charge and current densities that vanish outside bounded regions exist everywhere in $\mathbb{R}^3$. *Hint:* without loss of generality you can assume $\mathbf{r} = \mathbf{0}$ and use spherical coordinates.

5. The electric charge density, $\rho$, in 3-space depends on time as well as position if charge is moving around. The motion is described by the current density, $\mathbf{J}$. Derive the **continuity equation**

   $$\frac{\partial\rho}{\partial t} = -\mathbf{div\,J},$$

   from the fact that charge is conserved.

6. If $\mathbf{b}$ is a constant vector, show that

   $$\nabla\left(\frac{1}{|\mathbf{r}-\mathbf{b}|}\right) = -\frac{\mathbf{r}-\mathbf{b}}{|\mathbf{r}-\mathbf{b}|^3}.$$

7. If $\mathbf{a}$ and $\mathbf{b}$ are constant vectors, show that for $\mathbf{r}\ne\mathbf{b}$,

   $$\mathbf{div}\left(\mathbf{a}\times\frac{\mathbf{r}-\mathbf{b}}{|\mathbf{r}-\mathbf{b}|^3}\right) = 0.$$

   *Hint:* use identities (d) and (h) from Theorem 3 of Section 16.2.

8. Use the result of Exercise 7 to give an alternative proof that

   $$\mathbf{div}\oint_{\mathcal{F}}\frac{d\mathbf{s}\times(\mathbf{r}-\mathbf{s})}{|\mathbf{r}-\mathbf{s}|^3} = 0.$$

   Note that **div** refers to the $\mathbf{r}$ variable.

9. If **a** and **b** are constant vectors, show that for $\mathbf{r} \neq \mathbf{b}$,

$$\mathbf{curl}\left(\mathbf{a} \times \frac{\mathbf{r} - \mathbf{b}}{|\mathbf{r} - \mathbf{b}|^3}\right) = -(\mathbf{a} \bullet \nabla)\frac{\mathbf{r} - \mathbf{b}}{|\mathbf{r} - \mathbf{b}|^3}.$$

*Hint:* use identity (e) from Theorem 3 of 16.2.

10. If **F** is any smooth vector field, show that

$$\oint_{\mathcal{F}} (d\mathbf{s} \bullet \nabla)\mathbf{F}(\mathbf{s}) = \mathbf{0}$$

around any closed loop $\mathcal{F}$. *Hint:* the gradients of the components of **F** are conservative.

11. Verify that if **r** does not lie on $\mathcal{F}$, then

$$\mathbf{curl} \oint_{\mathcal{F}} \frac{d\mathbf{s} \times (\mathbf{r} - \mathbf{s})}{|\mathbf{r} - \mathbf{s}|^3} = \mathbf{0}.$$

Here **curl** is taken with respect to the **r** variable.

12. Verify the formula $\mathbf{curl\,A} = \mathbf{H}$, where **A** is the magnetic vector potential defined in terms of the steady state current density **J**.

13. If **A** is the vector potential for the magnetic field produced by a steady current in a closed-loop filament, show that $\mathbf{div\,A} = 0$ off the filament.

14. If **A** is the vector potential for the magnetic field produced by a steady, continuous current density, show that $\mathbf{div\,A} = 0$ everywhere. Hence show that **A** satisfies the vector Poisson equation: $\nabla^2\mathbf{A} = -\mathbf{J}$.

15. Show that in a region of space containing no charges ($\rho = 0$) and no currents ($\mathbf{J} = \mathbf{0}$), both $\mathbf{U} = \mathbf{E}$ and $\mathbf{U} = \mathbf{H}$ satisfy the wave equation

$$\frac{\partial^2 \mathbf{U}}{\partial t^2} = c^2 \nabla^2 \mathbf{U}$$

where $c = \sqrt{1/(\epsilon_0 \mu_0)} \approx 3 \times 10^8$ m/s.

* 16. **(Heat flow in 3-space)** The heat content of a volume element $dV$ within a homogeneous solid is $\delta c T\, dV$, where $\delta$ and $c$ are constants (the density and specific heat of the solid material) and $T = T(x, y, z, t)$ is the temperature at time $t$ at position $(x, y, z)$ in the solid. Heat always flows in the direction of the negative temperature gradient, and at a rate proportional to the size of that gradient. Thus the rate of flow of heat energy across a surface element $dS$ with normal $\hat{\mathbf{N}}$ is $-k\nabla T \bullet \hat{\mathbf{N}}\,dS$, where $k$ is also a constant depending on the material of the solid (the coefficient of thermal conductivity). Use "conservation of heat energy" to show that for any region $R$ with surface $S$ within the solid

$$\delta c \iiint_R \frac{\partial T}{\partial t}\,dV = k \oiint_S \nabla T \bullet \hat{\mathbf{N}}\,dS,$$

where $\hat{\mathbf{N}}$ is the unit outward normal on $S$. Hence show that heat flow within the solid is governed by the partial differential equation

$$\frac{\partial T}{\partial t} = \frac{k}{\delta c}\nabla^2 T = \frac{k}{\delta c}\left(\frac{\partial^2 T}{\partial x^2} + \frac{\partial^2 T}{\partial y^2} + \frac{\partial^2 T}{\partial z^2}\right).$$

## 16.7 Orthogonal Curvilinear Coordinates

In this optional section we will derive formulas for the gradient of a scalar field and the divergence and curl of a vector field in terms of coordinate systems more general than the Cartesian coordinate system used in the earlier sections of this chapter. In particular, we will express these quantities in terms of the cylindrical and spherical coordinate systems introduced in Section 14.6.

We denote by $xyz$-space, the usual system of Cartesian coordinates $(x, y, z)$ in $\mathbb{R}^3$. A different system of coordinates $[u, v, w]$ in $xyz$-space can be defined by a continuous transformation of the form

$$x = x(u, v, w), \qquad y = y(u, v, w), \qquad z = z(u, v, w).$$

If the transformation is one-to-one from a region $D$ in $uvw$-space onto a region $R$ in $xyz$-space, then a point in $R$ can be represented by a triple $[u, v, w]$, the (Cartiesian) coordinates of the unique point $Q$ in $uvw$-space which the transformation maps to $P$. In this case we say that the transformation defines a **curvilinear coordinate system** in $R$, and call $[u, v, w]$ the **curvilinear coordinates** of $P$ with respect to that system. Note that $[u, v, w]$ are Cartesian coordinates in their own space ($uvw$-space); they are curvilinear coordinates in $xyz$-space.

Typically, we relax the requirement that the transformation defining a curvilinear coordinate system be one-to-one, that is, that every point $P$ in $R$ should have a unique set of curvilinear coordinates. It is reasonable to require the transformation to be only *locally one-to-one*. Thus there may be more than one point $Q$ that gets mapped to a point $P$ by the transformation, but only one in any suitably small subregion of $D$. For example, in the plane polar coordinate system

$$x = r\cos\theta, \qquad y = r\sin\theta,$$

the transformation is locally one-to-one from $D$, the half of the $r\theta$-plane where $0 < r < \infty$, to the region $R$ consisting of all points in the $xy$-plane except the origin. Although, say, $[1, 0]$ and $[1, 2\pi]$ are polar coordinates of the same point in the $xy$-plane, they are not close together in $D$. Observe, however, that there is still a problem with the origin, which can be represented by $[0, \theta]$ for *any* $\theta$. Since the transformation is not even locally one-to-one at $r = 0$, we regard the origin of the $xy$-plane as a **singular point** for the polar coordinate system in the plane.

**Example 1**    The **cylindrical coordinate system** $[r, \theta, z]$ in $\mathbb{R}^3$ is defined by the transformation

$$x = r\cos\theta, \qquad y = r\sin\theta, \qquad z = z,$$

where $r \geq 0$. (See Section 14.6.) This transformation maps the half-space $D$ given by $r > 0$ onto all of $xyz$-space excluding the $z$-axis, and is locally one-to-one. We regard $[r, \theta, z]$ as cylindrical polar coordinates in all of $xyz$-space but regard points on the $z$-axis as singular points of the system since the points $[0, \theta, z]$ are identical for any $\theta$.

**Example 2**    The **spherical coordinate system** $[\rho, \phi, \theta]$ is defined by the transformation

$$x = \rho\sin\phi\cos\theta, \qquad y = \rho\sin\phi\sin\theta, \qquad z = \rho\cos\phi,$$

where $\rho \geq 0$ and $0 \leq \phi \leq \pi$. (See Section 14.6.) The transformation maps the region $D$ in $\rho\phi\theta$-space given by $\rho > 0$, $0 < \phi < \pi$ in a locally one-to-one way onto $xyz$-space excluding the $z$-axis. The point with Cartesian coordinates $(0, 0, z)$ can be represented by the spherical coordinates $[0, \phi, \theta]$ for arbitrary $\phi$ and $\theta$ if $z = 0$, by $[z, 0, \theta]$ for arbitrary $\theta$ if $z > 0$, and by $[|z|, \pi, \theta]$ for arbitrary $\theta$ if $z < 0$. Thus all points of the $z$-axis are singular for the spherical coordinate system.

## Coordinate Surfaces and Coordinate Curves

Let $[u, v, w]$ be a curvilinear coordinate system in $xyz$-space, and let $P_0$ be a nonsingular point for the system. Thus the transformation

$$x = x(u, v, w), \qquad y = y(u, v, w), \qquad z = z(u, v, w)$$

is locally one-to-one near $P_0$. Let $P_0$ have curvilinear coordinates $[u_0, v_0, w_0]$. The plane with equation $u = u_0$ in $uvw$-space gets mapped by the transformation to a surface in $xyz$-space passing through $P_0$. We call this surface a $u$-surface and still refer to it by the equation $u = u_0$; it has parametric equations

$$x = x(u_0, v, w), \qquad y = y(u_0, v, w), \qquad z = z(u_0, v, w)$$

with parameters $v$ and $w$. Similarly, the $v$-surface $v = v_0$ and the $w$-surface $w = w_0$ pass through $P_0$; they are the images of the planes $v = v_0$ and $w = w_0$ in $uvw$-space.

> **Orthogonal curvilinear coordinates**
>
> We say that $[u, v, w]$ is an **orthogonal curvilinear coordinate system** in $xyz$-space if, for every nonsingular point $P_0$ in $xyz$-space, the three **coordinate surfaces**, $u = u_0$, $v = v_0$, and $w = w_0$, intersect at $P_0$ at mutually right angles.

It is tacitly assumed that the coordinate surfaces are smooth at all nonsingular points, so we are really assuming that their normal vectors are mutually perpendicular. Figure 16.19 shows the coordinate surfaces through $P_0$ for a typical orthogonal curvilinear coordinate system.

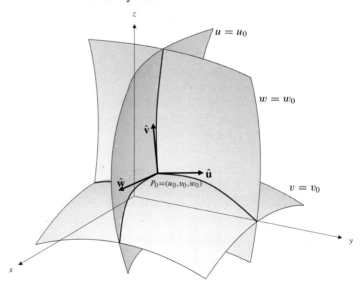

**Figure 16.19**   $u$-, $v$-, and $w$-coordinate surfaces

Pairs of coordinate surfaces through a point intersect along a **coordinate curve** through that point. For example, the coordinate surfaces $v = v_0$ and $w = w_0$ intersect along the $u$-**curve** with parametric equations

$$x = x(u, v_0, w_0), \quad y = y(u, v_0, w_0), \quad \text{and} \quad z = z(u, v_0, w_0),$$

where the parameter is $u$. A unit vector $\hat{\mathbf{u}}$ tangent to the $u$-curve through $P_0$ is normal to the coordinate surface $u = u_0$ there. Similar statements hold for unit vectors $\hat{\mathbf{v}}$ and $\hat{\mathbf{w}}$. For an orthogonal curvilinear coordinate system, the three vectors $\hat{\mathbf{u}}$, $\hat{\mathbf{v}}$, and $\hat{\mathbf{w}}$, form a basis mutually perpendicular unit vectors at any nonsingular point $P$. (See Figure 16.19.) We call this basis the **local basis** at $P$.

**Example 3** For the cylindrical coordinate system, (Figure 16.20), the coordinate surfaces are:

| | |
|---|---|
| circular cylinders with axis along the z-axis | ($r$-surfaces), |
| vertical half-planes radiating from the $z$-axis | ($\theta$-surfaces), |
| horizontal planes | ($z$-surfaces). |

The coordinate curves are:

| | |
|---|---|
| horizontal straight half-lines radiating from the $z$-axis | ($r$-curves), |
| horizontal circles with centres on the $z$-axis | ($\theta$-curves), |
| vertical straight lines | ($z$-curves). |

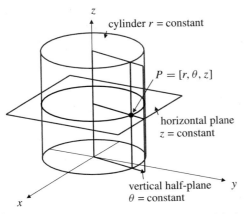

**Figure 16.20** The coordinate surfaces for cylindrical coordinates

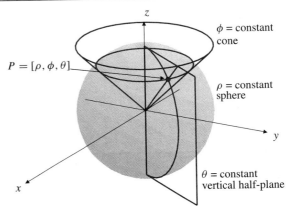

**Figure 16.21** The coordinate surfaces for spherical coordinates

**Example 4** For the spherical coordinate system, (see Figure 16.21), the coordinate surfaces are:

| | |
|---|---|
| spheres centred at the origin | ($\rho$-surfaces), |
| vertical circular cones with vertices at the origin | ($\phi$-surfaces), |
| vertical half-planes radiating from the $z$-axis | ($\theta$-surfaces). |

The coordinate curves are:

| | |
|---|---|
| half-lines radiating from the origin | ($\rho$-curves), |
| vertical semi-circles with centres at the origin | ($\phi$-curves), |
| horizontal circles with centres on the $z$-axis | ($\theta$-curves). |

## Scale Factors and Differential Elements

For the rest of this section we assume that $[u, v, w]$ are **orthogonal** curvilinear coordinates in $xyz$-space defined via the transformation

$$x = x(u, v, w), \qquad y = y(u, v, w), \qquad z = z(u, v, w).$$

We also assume that the the coordinate surfaces are smooth at any nonsingular point, and that the local basis vectors $\hat{\mathbf{u}}$, $\hat{\mathbf{v}}$, and $\hat{\mathbf{w}}$ at any such point form a right-handed triad. This is the case for both cylindrical and spherical coordinates. For spherical coordinates, this is the reason we chose the order of the coordinates as $[\rho, \phi, \theta]$ rather than $[\rho, \theta, \phi]$.

The **position vector** of a point $P$ in $xyz$-space can be expressed in terms of the curvilinear coordinates:

$$\mathbf{r} = x(u, v, w)\mathbf{i} + y(u, v, w)\mathbf{j} + z(u, v, w)\mathbf{k}.$$

If we hold $v = v_0$ and $w = w_0$ fixed and let $u$ vary, then $\mathbf{r} = \mathbf{r}(u, v_0, w_0)$ defines a $u$-curve in $xyz$-space. At any point $P$ on this curve, the vector

$$\frac{\partial \mathbf{r}}{\partial u} = \frac{\partial x}{\partial u}\mathbf{i} + \frac{\partial y}{\partial u}\mathbf{j} + \frac{\partial z}{\partial u}\mathbf{k}$$

is tangent to the $u$-curve at $P$. In general, the three vectors

$$\frac{\partial \mathbf{r}}{\partial u}, \qquad \frac{\partial \mathbf{r}}{\partial v}, \qquad \text{and} \qquad \frac{\partial \mathbf{r}}{\partial w}$$

are tangent, respectively, to the the $u$-curve, the $v$-curve, and the $w$-curve through $P$. They are also normal, respectively, to the $u$-surface, the $v$-surface, and the $w$-surface through $P$, so they are mutually perpendicular. (See Figure 16.19.) The lengths of these tangent vectors are called the **scale factors** of the coordinate system.

> The **scale factors** of the orthogonal curvilinear coordinate system $[u, v, w]$ are the three functions
>
> $$h_u = \left|\frac{\partial \mathbf{r}}{\partial u}\right|, \qquad h_v = \left|\frac{\partial \mathbf{r}}{\partial v}\right|, \qquad h_w = \left|\frac{\partial \mathbf{r}}{\partial w}\right|.$$

The scale factors are nonzero at a nonsingular point $P$ of the coordinate system, so the local basis at $P$ can be obtained by dividing the tangent vectors to the coordinate curves by their lengths. As noted previously, we denote the local basis vectors by $\hat{\mathbf{u}}, \hat{\mathbf{v}},$ and $\hat{\mathbf{w}}$. Thus

$$\frac{\partial \mathbf{r}}{\partial u} = h_u\hat{\mathbf{u}}, \qquad \frac{\partial \mathbf{r}}{\partial v} = h_v\hat{\mathbf{v}}, \qquad \text{and} \qquad \frac{\partial \mathbf{r}}{\partial w} = h_w\hat{\mathbf{w}}.$$

The basis vectors $\hat{\mathbf{u}}, \hat{\mathbf{v}},$ and $\hat{\mathbf{w}}$ will form a right-handed triad provided we have chosen a suitable order for the coordinates $u, v,$ and $w$.

**Example 5** For cylindrical coordinates we have $\mathbf{r} = r\cos\theta\,\mathbf{i} + r\sin\theta\,\mathbf{j} + z\mathbf{k}$, so

$$\frac{\partial \mathbf{r}}{\partial r} = \cos\theta\,\mathbf{i} + \sin\theta\,\mathbf{j}, \qquad \frac{\partial \mathbf{r}}{\partial\theta} = -r\sin\theta\,\mathbf{i} + r\cos\theta\,\mathbf{j}, \qquad \text{and} \qquad \frac{\partial \mathbf{r}}{\partial z} = \mathbf{k}.$$

Thus the scale factors for the cylindrical coordinate system are given by

$$h_r = \left|\frac{\partial \mathbf{r}}{\partial r}\right| = 1, \qquad h_\theta = \left|\frac{\partial \mathbf{r}}{\partial\theta}\right| = r, \qquad \text{and} \qquad h_z = \left|\frac{\partial \mathbf{r}}{\partial z}\right| = 1,$$

and the local basis consists of the vectors

$$\hat{\mathbf{r}} = \cos\theta\,\mathbf{i} + \sin\theta\,\mathbf{j}, \qquad \hat{\boldsymbol{\theta}} = -\sin\theta\,\mathbf{i} + \cos\theta\,\mathbf{j}, \qquad \hat{\mathbf{z}} = \mathbf{k}.$$

See Figure 16.22. The local basis is right-handed.

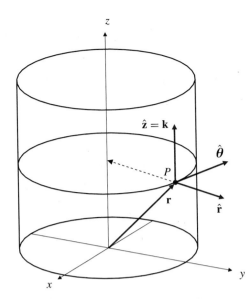

**Figure 16.22**   The local basis for cylindrical
coordinates

**Figure 16.23**   The local basis for spherical
coordinates

---

**Example 6**   For spherical coordinates we have

$$\mathbf{r} = \rho \sin \phi \cos \theta \mathbf{i} + \rho \sin \phi \sin \theta \mathbf{j} + \rho \cos \phi \mathbf{k}.$$

Thus the tangent vectors to the coordinate curves are

$$\frac{\partial \mathbf{r}}{\partial \rho} = \sin \phi \cos \theta \, \mathbf{i} + \sin \phi \sin \theta \, \mathbf{j} + \cos \phi \, \mathbf{k},$$

$$\frac{\partial \mathbf{r}}{\partial \phi} = \rho \cos \phi \cos \theta \, \mathbf{i} + \rho \cos \phi \sin \theta \, \mathbf{j} - \rho \sin \phi \, \mathbf{k},$$

$$\frac{\partial \mathbf{r}}{\partial \theta} = -\rho \sin \phi \sin \theta \, \mathbf{i} + \rho \sin \phi \cos \theta \, \mathbf{j},$$

and the scale factors are given by

$$h_\rho = \left| \frac{\partial \mathbf{r}}{\partial \rho} \right| = 1, \quad h_\phi = \left| \frac{\partial \mathbf{r}}{\partial \phi} \right| = \rho, \quad \text{and} \quad h_\theta = \left| \frac{\partial \mathbf{r}}{\partial \theta} \right| = \rho \sin \phi.$$

The local basis consists of the vectors

$$\hat{\rho} = \sin \phi \cos \theta \, \mathbf{i} + \sin \phi \sin \theta \, \mathbf{j} + \cos \phi \, \mathbf{k}$$

$$\hat{\phi} = \cos \phi \cos \theta \, \mathbf{i} + \cos \phi \sin \theta \, \mathbf{j} - \sin \phi \, \mathbf{k}$$

$$\hat{\theta} = -\sin \theta \, \mathbf{i} + \cos \theta \, \mathbf{j}.$$

See Figure 16.23. The local basis is right-handed.   ∎

---

The volume element in an orthogonal curvilinear coordinate system is the volume
of an infinitesimal *coordinate box* bounded by pairs of $u$-, $v$-, and $w$-surfaces
corresponding to values $u$ and $u+du$, $v$ and $v+dv$, and $w$ and $w+dw$, respectively.

See Figure 16.24. Since these coordinate surfaces are assumed smooth, and since they intersect at right angles, the coordinate box is rectangular, and is spanned by the vectors

$$\frac{\partial \mathbf{r}}{\partial u}\, du = h_u\, du\, \hat{\mathbf{u}}, \qquad \frac{\partial \mathbf{r}}{\partial v}\, dv = h_v\, dv\, \hat{\mathbf{v}}, \quad \text{and} \quad \frac{\partial \mathbf{r}}{\partial w}\, dw = h_w\, dw\, \hat{\mathbf{w}}.$$

Therefore the volume element is given by

$$dV = h_u h_v h_w\, du\, dv\, dw.$$

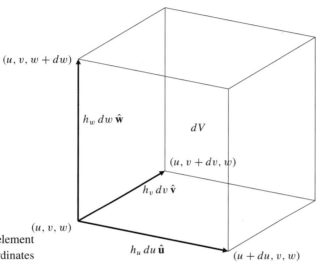

**Figure 16.24**   The volume element for orthogonal curvilinear coordinates

Furthermore, the surface area elements on the $u$-, $v$-, and $w$-surfaces are the areas of the appropriate faces of the coordinate box:

**Area elements on coordinate surfaces**

$$dS_u = h_v h_w\, dv\, dw, \qquad dS_v = h_u h_w\, du\, dw, \qquad dS_w = h_u h_v\, du\, dv.$$

The arc length elements along the $u$-, $v$-, and $w$-coordinate curves are the edges of the coordinate box:

**Arc length elements on coordinate curves**

$$ds_u = h_u\, du, \qquad ds_v = h_v\, dv, \qquad ds_w = h_w\, dw.$$

**Example 7**   For cylindrical coordinates, the volume element, as shown in Section 14.6, is

$$dV = h_r h_\theta h_z\, dr\, d\theta\, dz = r\, dr\, d\theta\, dz.$$

The surface are elements on the cylinder $r$ = constant, the half-plane $\theta$ = constant, and the plane $z$ = constant are, respectively,

$$dS_r = r\, d\theta\, dz, \qquad dS_\theta = dr\, dz, \qquad \text{and} \qquad dS_z = r\, dr\, d\theta.$$

**Example 8** For spherical coordinates, the volume element, as shown in Section 14.6, is

$$dV = h_\rho h_\phi h_\theta \, d\rho \, d\phi \, d\theta = \rho^2 \sin\phi \, d\rho \, d\phi \, d\theta.$$

The area element on the sphere $\rho$ = constant is

$$dS_\rho = h_\phi h_\theta \, d\phi \, d\theta = \rho^2 \sin\phi \, d\phi \, d\theta.$$

The area element on the cone $\phi$ = constant is

$$dS_\phi = h_\rho h_\theta \, d\rho \, d\theta = \rho \sin\phi \, d\rho \, d\theta.$$

The area element on the half-plane $\theta$ = constant is

$$dS_\theta = h_\rho h_\phi \, d\rho \, d\phi = \rho \, d\rho \, d\phi.$$

## Grad, Div, and Curl in Orthogonal Curvilinear Coordinates

The gradient $\nabla f$ of a scalar field $f$ can be expressed in terms of the local basis at any point $P$ with curvilinear coordinates $[u, v, w]$ in the form

$$\nabla f = F_u \hat{\mathbf{u}} + F_v \hat{\mathbf{v}} + F_w \hat{\mathbf{w}}.$$

In order to determine the coefficients $F_u$, $F_v$, and $F_w$ in this formula, we will compare two expressions for the directional derivative of $f$ along an arbitrary curve in $xyz$-space.

If the curve $\mathcal{C}$ has parametrization $\mathbf{r} = \mathbf{r}(s)$ in terms of arc length, then the directional derivative of $f$ along $\mathcal{C}$ is given by

$$\frac{df}{ds} = \frac{\partial f}{\partial u}\frac{du}{ds} + \frac{\partial f}{\partial v}\frac{dv}{ds} + \frac{\partial f}{\partial w}\frac{dw}{ds}.$$

On the other hand, this directional derivative is also given by $df/ds = \nabla f \bullet \hat{\mathbf{T}}$, where $\hat{\mathbf{T}}$ is the unit tangent vector to $\mathcal{C}$. We have

$$\hat{\mathbf{T}} = \frac{d\mathbf{r}}{ds} = \frac{\partial \mathbf{r}}{\partial u}\frac{du}{ds} + \frac{\partial \mathbf{r}}{\partial v}\frac{dv}{ds} + \frac{\partial \mathbf{r}}{\partial w}\frac{dw}{ds}$$

$$= h_u \frac{du}{ds}\hat{\mathbf{u}} + h_v \frac{dv}{ds}\hat{\mathbf{v}} + h_w \frac{dw}{ds}\hat{\mathbf{w}}.$$

Thus

$$\frac{df}{ds} = \nabla f \bullet \hat{\mathbf{T}} = F_u h_u \frac{du}{ds} + F_v h_v \frac{dv}{ds} + F_w h_w \frac{dw}{ds}.$$

Comparing these two expressions for $df/ds$ along $\mathcal{C}$, we see that

$$F_u h_u = \frac{\partial f}{\partial u}, \qquad F_v h_v = \frac{\partial f}{\partial v}, \qquad F_w h_w = \frac{\partial f}{\partial w}.$$

Therefore, we have shown that

**The gradient in orthogonal curvilinear coordinates**

$$\nabla f = \frac{1}{h_u}\frac{\partial f}{\partial u}\hat{\mathbf{u}} + \frac{1}{h_v}\frac{\partial f}{\partial v}\hat{\mathbf{v}} + \frac{1}{h_w}\frac{\partial f}{\partial w}\hat{\mathbf{w}}.$$

**Example 9**   In terms of cylindrical coordinates, the gradient of the scalar field $f(r, \theta, z)$ is

$$\nabla f(r, \theta, z) = \frac{\partial f}{\partial r}\,\hat{\mathbf{r}} + \frac{1}{r}\frac{\partial f}{\partial \theta}\,\hat{\boldsymbol{\theta}} + \frac{\partial f}{\partial z}\,\mathbf{k}.$$

∎

**Example 10**   In terms of spherical coordinates, the gradient of the scalar field $f(\rho, \phi, \theta)$ is

$$\nabla f(\rho, \phi, \theta) = \frac{\partial f}{\partial \rho}\,\hat{\boldsymbol{\rho}} + \frac{1}{\rho}\frac{\partial f}{\partial \phi}\,\hat{\boldsymbol{\phi}} + \frac{1}{\rho \sin \phi}\frac{\partial f}{\partial \theta}\,\hat{\boldsymbol{\theta}}.$$

∎

Now consider a vector field $\mathbf{F}$ expressed in terms of the curvilinear coordinates:

$$\mathbf{F}(u, v, w) = F_u(u, v, w)\hat{\mathbf{u}} + F_v(u, v, w)\hat{\mathbf{v}} + F_w(u, v, w)\hat{\mathbf{w}}.$$

The flux of $\mathbf{F}$ out of the infinitesimal coordinate box of Figure 16.24 is the sum of the fluxes of $\mathbf{F}$ out of the three pairs of opposite surfaces of the box. The flux out of the $u$-surfaces corresponding to $u$ and $u + du$ is

$$\mathbf{F}(u + du, v, w) \bullet \hat{\mathbf{u}}\,dS_u - \mathbf{F}(u, v, w) \bullet \hat{\mathbf{u}}\,dS_u$$
$$= \big(F_u(u + du, v, w)h_v(u + du, v, w)h_w(u + du, v, w)$$
$$\quad - F_u(u, v, w)h_v(u, v, w)h_w(u, v, w)\big)\,dv\,dw$$
$$= \frac{\partial}{\partial u}\big(h_v h_w F_u\big)\,du\,dv\,dw.$$

Similar expressions hold for the fluxes out of the other pairs of coordinate surfaces.

The divergence at $P$ of $\mathbf{F}$ is the flux *per unit volume* out of the infinitesimal coordinate box at $P$. Thus it is given by

**The divergence in orthogonal curvilinear coordinates**

$$\operatorname{div} \mathbf{F}(u, v, w) = \frac{1}{h_u h_v h_w}\bigg[\frac{\partial}{\partial u}\big(h_v h_w F_u(u, v, w)\big)$$
$$+ \frac{\partial}{\partial v}\big(h_u h_w F_v(u, v, w)\big) + \frac{\partial}{\partial w}\big(h_u h_v F_w(u, v, w)\big)\bigg].$$

**Example 11**   For cylindrical coordinates, $h_r = h_z = 1$, and $h_\theta = r$. Thus the divergence of $\mathbf{F} = F_r\hat{\mathbf{r}} + F_\theta\hat{\boldsymbol{\theta}} + F_z\mathbf{k}$ is

$$\operatorname{div} \mathbf{F} = \frac{1}{r}\bigg[\frac{\partial}{\partial r}\big(r F_r\big) + \frac{\partial}{\partial \theta}F_\theta + \frac{\partial}{\partial z}\big(r F_z\big)\bigg]$$
$$= \frac{\partial F_r}{\partial r} + \frac{1}{r}F_r + \frac{1}{r}\frac{\partial F_\theta}{\partial \theta} + \frac{\partial F_z}{\partial z}.$$

∎

**Example 12**    For spherical coordinates, $h_\rho = 1$, $h_\phi = \rho$, and $h_\theta = \rho \sin \phi$. The divergence of the vector field $\mathbf{F} = F_\rho \hat{\boldsymbol{\rho}} + F_\phi \hat{\boldsymbol{\phi}} + F_\theta \hat{\boldsymbol{\theta}}$ is

$$
\begin{aligned}
\mathbf{div}\,\mathbf{F} &= \frac{1}{\rho^2 \sin \phi} \left[ \frac{\partial}{\partial \rho}\left( \rho^2 \sin \phi\, F_\rho \right) + \frac{\partial}{\partial \phi}\left( \rho \sin \phi\, F_\phi \right) + \frac{\partial}{\partial \theta}\left( \rho\, F_\theta \right) \right] \\
&= \frac{1}{\rho^2}\frac{\partial}{\partial \rho}\left( \rho^2 F_\rho \right) + \frac{1}{\rho \sin \phi}\frac{\partial}{\partial \phi}\left( \sin \phi\, F_\phi \right) + \frac{1}{\rho \sin \phi}\frac{\partial F_\theta}{\partial \theta} \\
&= \frac{\partial F_\rho}{\partial \rho} + \frac{2}{\rho} F_\rho + \frac{1}{\rho}\frac{\partial F_\phi}{\partial \phi} + \frac{\cot \phi}{\rho} F_\phi + \frac{1}{\rho \sin \phi}\frac{\partial F_\theta}{\partial \theta}.
\end{aligned}
$$

∎

To calculate the curl of a function expressed in terms of orthogonal curvilinear coordinates we can make use of some previously obtained vector identities. First, observe that the gradient of the scalar field $f(u, v, w) = u$ is $\hat{\mathbf{u}}/h_u$, so that $\hat{\mathbf{u}} = h_u \nabla u$. Similarly, $\hat{\mathbf{v}} = h_v \nabla v$ and $\hat{\mathbf{w}} = h_w \nabla w$. Therefore, the vector field

$$
\mathbf{F} = F_u \hat{\mathbf{u}} + F_v \hat{\mathbf{v}} + F_w \hat{\mathbf{w}}
$$

can be written in the form

$$
\mathbf{F} = F_u h_u \nabla u + F_v h_v \nabla v + F_w h_w \nabla w.
$$

Using the identity $\mathbf{curl}\,(f\nabla g) = \nabla f \times \nabla g$ (see Exercise 13 of Section 16.2) we can calculate the curl of each term in the expression above. We have

$$
\begin{aligned}
\mathbf{curl}\,\left( F_u h_u \nabla u \right) &= \nabla (F_u h_u) \times \nabla u \\
&= \left[ \frac{1}{h_u}\frac{\partial}{\partial u}(F_u h_u)\hat{\mathbf{u}} + \frac{1}{h_v}\frac{\partial}{\partial v}(F_u h_u)\hat{\mathbf{v}} + \frac{1}{h_w}\frac{\partial}{\partial w}(F_u h_u)\hat{\mathbf{w}} \right] \times \frac{\hat{\mathbf{u}}}{h_u} \\
&= \frac{1}{h_u h_w}\frac{\partial}{\partial w}(F_u h_u)\hat{\mathbf{v}} - \frac{1}{h_u h_v}\frac{\partial}{\partial v}(F_u h_u)\hat{\mathbf{w}} \\
&= \frac{1}{h_u h_v h_w}\left[ \frac{\partial}{\partial w}(F_u h_u)(h_v \hat{\mathbf{v}}) - \frac{\partial}{\partial v}(F_u h_u)(h_w \hat{\mathbf{w}}) \right].
\end{aligned}
$$

We have used the facts that $\hat{\mathbf{u}} \times \hat{\mathbf{u}} = \mathbf{0}$, $\hat{\mathbf{v}} \times \hat{\mathbf{u}} = -\hat{\mathbf{w}}$, and $\hat{\mathbf{w}} \times \hat{\mathbf{u}} = \hat{\mathbf{v}}$ to obtain the result above. This is why we assumed that the curvilinear coordinate system was right-handed.

Corresponding expressions can be calculated for the other two terms in the formula for $\mathbf{curl}\,\mathbf{F}$. Combining the three terms, we conclude that the curl of

$$
\mathbf{F} = F_u \hat{\mathbf{u}} + F_v \hat{\mathbf{v}} + F_w \hat{\mathbf{w}}
$$

is given by

**The curl in orthogonal curvilinear coordinates**

$$
\mathbf{curl}\,\mathbf{F}(u, v, w) = \frac{1}{h_u h_v h_w}
\begin{vmatrix}
h_u \hat{\mathbf{u}} & h_v \hat{\mathbf{v}} & h_w \hat{\mathbf{w}} \\
\dfrac{\partial}{\partial u} & \dfrac{\partial}{\partial v} & \dfrac{\partial}{\partial w} \\
F_u h_u & F_v h_v & F_w h_w
\end{vmatrix}.
$$

**Example 13**    For cylindrical coordinates, the curl of $\mathbf{F} = F_r\hat{\mathbf{r}} + F_\theta\hat{\boldsymbol{\theta}} + F_z\mathbf{k}$ is given by

$$\mathbf{curl\,F} = \frac{1}{r}\begin{vmatrix} \hat{\mathbf{r}} & r\hat{\boldsymbol{\theta}} & \mathbf{k} \\ \dfrac{\partial}{\partial r} & \dfrac{\partial}{\partial \theta} & \dfrac{\partial}{\partial z} \\ F_r & rF_\theta & F_z \end{vmatrix}$$

$$= \left(\frac{1}{r}\frac{\partial F_z}{\partial \theta} - \frac{\partial F_\theta}{\partial z}\right)\hat{\mathbf{r}} + \left(\frac{\partial F_r}{\partial z} - \frac{\partial F_z}{\partial r}\right)\hat{\boldsymbol{\theta}} + \left(\frac{\partial F_\theta}{\partial r} + \frac{F_\theta}{r} - \frac{1}{r}\frac{\partial F_r}{\partial \theta}\right)\mathbf{k}.$$

∎

**Example 14**    For spherical coordinates, the curl of $\mathbf{F} = F_\rho\hat{\boldsymbol{\rho}} + F_\phi\hat{\boldsymbol{\phi}} + F_\theta\hat{\boldsymbol{\theta}}$ is given by

$$\mathbf{curl\,F} = \frac{1}{\rho^2\sin\phi}\begin{vmatrix} \hat{\boldsymbol{\rho}} & \rho\hat{\boldsymbol{\phi}} & \rho\sin\phi\,\hat{\boldsymbol{\theta}} \\ \dfrac{\partial}{\partial \rho} & \dfrac{\partial}{\partial \phi} & \dfrac{\partial}{\partial \theta} \\ F_\rho & \rho F_\phi & \rho\sin\phi\,F_\theta \end{vmatrix}$$

$$= \frac{1}{\rho\sin\phi}\left[\frac{\partial}{\partial \phi}(\sin\phi\,F_\theta) - \frac{\partial F_\phi}{\partial \theta}\right]\hat{\boldsymbol{\rho}}$$

$$+ \frac{1}{\rho\sin\phi}\left[\frac{\partial F_\rho}{\partial \theta} - \sin\phi\frac{\partial}{\partial \rho}(\rho F_\theta)\right]\hat{\boldsymbol{\phi}}$$

$$+ \frac{1}{\rho}\left[\frac{\partial}{\partial \rho}(\rho F_\phi) - \frac{\partial F_\rho}{\partial \phi}\right]\hat{\boldsymbol{\theta}}$$

$$= \frac{1}{\rho\sin\phi}\left[(\cos\phi)F_\theta + (\sin\phi)\frac{\partial F_\theta}{\partial \phi} - \frac{\partial F_\phi}{\partial \theta}\right]\hat{\boldsymbol{\rho}}$$

$$+ \frac{1}{\rho\sin\phi}\left[\frac{\partial F_\rho}{\partial \theta} - (\sin\phi)F_\theta - (\rho\sin\phi)\frac{\partial F_\theta}{\partial \rho}\right]\hat{\boldsymbol{\phi}}$$

$$+ \frac{1}{\rho}\left[F_\phi + \rho\frac{\partial F_\phi}{\partial \rho} - \frac{\partial F_\rho}{\partial \phi}\right]\hat{\boldsymbol{\theta}}.$$

∎

## Exercises 16.7

In Exercises 1–2, calculate the gradients of the given scalar fields expressed in terms of cylindrical or spherical coordinates.

**1.** $f(r, \theta, z) = r\theta z$      **2.** $f(\rho, \phi, \theta) = \rho\phi\theta$

In Exercises 3–8, calculate **div F** and **curl F** for the given vector fields expressed in terms of cylindrical coordinates or spherical coordinates.

**3.** $\mathbf{F}(r, \theta, z) = r\hat{\mathbf{r}}$      **4.** $\mathbf{F}(r, \theta, z) = r\hat{\boldsymbol{\theta}}$

**5.** $\mathbf{F}(\rho, \phi, \theta) = \sin\phi\,\hat{\boldsymbol{\rho}}$      **6.** $\mathbf{F}(\rho, \phi, \theta) = \rho\hat{\boldsymbol{\phi}}$

**7.** $\mathbf{F}(\rho, \phi, \theta) = \rho\hat{\boldsymbol{\theta}}$      **8.** $\mathbf{F}(\rho, \phi, \theta) = \rho^2\hat{\boldsymbol{\rho}}$

**9.** Let $x = x(u, v)$, $y = y(u, v)$ define orthogonal curvilinear coordinates $(u, v)$ in the $xy$-plane. Find the scale factors, local basis vectors, and area element for the system of coordinates $(u, v)$.

**10.** Continuing the previous exercise, express the gradient of a scalar field $f(u, v)$, and the divergence and curl of a vector field $\mathbf{F}(u, v)$ in terms of the curvilinear coordinates.

**11.** Express the gradient of the scalar field $f(r, \theta)$, and the divergence and curl of a vector field $\mathbf{F}(r, \theta)$ in terms of plane polar coordinates $(r, \theta)$.

12. The transformation

$$x = a \cosh u \cos v \qquad y = a \sinh u \sin v$$

defines **elliptical coordinates** in the $xy$-plane. This coordinate system has singular points at $x = \pm a$, $y = 0$.

(a) Show that the $v$-curves, $u$ = constant, are ellipses with foci at the singular points.

(b) Show that the $u$-curves, $v$ = constant, are hyperbolas with foci at the singular points.

(c) Show that the $u$-curve and the $v$-curve through a nonsingular point intersect at right angles.

(d) Find the scale factors $h_u$ and $h_v$, and the area element $dA$ for the elliptical coordinate system.

13. Describe the coordinate surfaces and coordinate curves of the system of elliptical cylindrical coordinates in $xyz$-space

defined by

$$x = a \cosh u \cos v, \qquad y = a \sinh u \sin v, \qquad z = z.$$

14. The Laplacian $\nabla^2 f$ of a scalar field $f$ can be calculated as **div** $\nabla f$. Use this method to calculate the Laplacian of the function $f(r, \theta, z)$ expressed in terms of cylindrical coordinates. (This repeats Exercise 22 of Section 14.5.)

15. Calculate the Laplacian $\nabla^2 f = $ **div** $\nabla f$ for the function $f(\rho, \phi, \theta)$, expressed in terms of spherical coordinates. (This repeats Exercise 34 of Section 14.6, but is now much easier.)

16. Calculate the Laplacian $\nabla^2 f = $ **div** $\nabla f$ for a function $f(u, v, w)$ expressed in terms of arbitrary orthogonal curvilinear coordinates $(u, v, w)$.

# Chapter Review

## Key Ideas

• **What do the following phrases and statements mean?**

◇ the divergence of a vector field **F**

◇ the curl of a vector field **F**

◇ **F** is solenoidal.          ◇ **F** is irrotational.

◇ a scalar potential          ◇ a vector potential

◇ orthogonal curvilinear coordinates

• **State the following theorems:**

◇ the Divergence Theorem          ◇ Green's Theorem

◇ Stokes's theorem

## Review Exercises

1. If $\mathbf{F} = x^2 z \mathbf{i} + (y^2 z + 3y)\mathbf{j} + x^2 \mathbf{k}$, find the flux of **F** across the part of the ellipsoid $x^2 + y^2 + 4z^2 = 16$, where $z \geq 0$, oriented with upward normal.

2. Let $S$ be the part of the cylinder $x^2 + y^2 = 2ax$ between the horizontal planes $z = 0$ and $z = b$, where $b > 0$. Find the flux of $\mathbf{F} = x\mathbf{i} + \cos(z^2)\mathbf{j} + e^z \mathbf{k}$ outward through $S$.

3. Find $\oint_C (3y^2 + 2xe^{y^2})dx + (2x^2 ye^{y^2})dy$ counterclockwise around the boundary of the parallelogram with vertices $(0, 0)$, $(2, 0)$, $(3, 1)$, and $(1, 1)$.

4. If $\mathbf{F} = -z\mathbf{i} + x\mathbf{j} + y\mathbf{k}$, what are the possible values of $\oint_C \mathbf{F} \bullet d\mathbf{r}$ around circles of radius $a$ in the plane $2x + y + 2z = 7$?

5. Let **F** be a smooth vector field in 3-space and suppose that for every $a > 0$, the flux of **F** out of the sphere of radius $a$ centred at the origin is $\pi(a^3 + 2a^4)$. Find the divergence of **F** at the origin.

6. Let $\mathbf{F} = -y\mathbf{i} + x\cos(1 - x^2 - y^2)\mathbf{j} + yz\mathbf{k}$. Find the flux of **curl F** upward through a surface whose boundary is the curve $x^2 + y^2 = 1$, $z = 2$.

7. Let $\mathbf{F}(\mathbf{r}) = r^\lambda \mathbf{r}$, where $\mathbf{r} = x\mathbf{i} + y\mathbf{j} + z\mathbf{k}$ and $r = |\mathbf{r}|$. For what value(s) of $\lambda$ is **F** solenoidal on an open subset of 3-space? Is **F** solenoidal on all of 3-space for any value of $\lambda$?

8. Given that **F** satisfies **curl F** $= \mu \mathbf{F}$ on 3-space, where $\mu$ is a nonzero constant, show that $\nabla^2 \mathbf{F} + \mu^2 \mathbf{F} = \mathbf{0}$.

9. Let $P$ be a polyhedron in 3-space having $n$ planar faces, $F_1$, $F_2$, ..., $F_n$. Let $\mathbf{N}_i$ be normal to $F_i$ in the direction outward from $P$ and let $\mathbf{N}_i$ have length equal to the area of face $F_i$. Show that

$$\sum_{i=1}^{n} \mathbf{N}_i = \mathbf{0}.$$

Also, state a version of this result for a plane polygon $P$.

10. Around what simple, closed curve $C$ in the $xy$-plane does the vector field

$$\mathbf{F} = (2y^3 - 3y + xy^2)\mathbf{i} + (x - x^3 + x^2 y)\mathbf{j}$$

have the greatest circulation?

11. Through what closed, oriented surface in $\mathbb{R}^3$ does the vector field

$$\mathbf{F} = (4x + 2x^3 z)\mathbf{i} - y(x^2 + z^2)\mathbf{j} - (3x^2 z^2 + 4y^2 z)\mathbf{k}$$

have the greatest flux?

**12.** Find the maximum value of

$$\oint_C \mathbf{F} \bullet d\mathbf{r}$$

where $\mathbf{F} = xy^2\mathbf{i} + (3z - xy^2)\mathbf{j} + (4y - x^2y)\mathbf{k}$, and $C$ is a simple closed curve in the plane $x + y + z = 1$ oriented counterclockwise as seen from high on the $z$-axis. What curve $C$ gives this maximum?

## Challenging Problems

**1. (The expanding universe)** Let $\mathbf{v}$ be the large-scale velocity field of matter in the universe. (*Large-scale* means on the scale of intergalactic distances — *small-scale* motion such as that of planetary systems about their suns, and even stars about galactic centres has been averaged out.) Assume that $\mathbf{v}$ is a smooth vector field. According to present astronomical theory, the distance between any two points is increasing, and the rate of increase is proportional to the distance between the points. The constant of proportionality, $C$, is called *Hubble's constant*. In terms of $\mathbf{v}$, if $\mathbf{r}_1$ and $\mathbf{r}_2$ are two points then

$$\big(\mathbf{v}(\mathbf{r}_2) - \mathbf{v}(\mathbf{r}_1)\big) \bullet (\mathbf{r}_2 - \mathbf{r}_1) = C|\mathbf{r}_2 - \mathbf{r}_1|^2.$$

Show that **div** $\mathbf{v}$ is constant, and find the value of the constant in terms of Hubble's constant. *Hint:* find the flux of $\mathbf{v}(\mathbf{r})$ out of a sphere of radius $\epsilon$ centred at $\mathbf{r}_1$ and take the limit as $\epsilon$ approaches zero.

**2. (Solid angle)** Two rays from a point $P$ determine an angle at $P$ whose measure in radians is equal to the length of the arc of the circle of radius 1 with centre at $P$ lying between the two rays. Similarly, an arbitrarily shaped half-cone $K$ with vertex at $P$ determines a **solid angle** at $P$ whose measure in **steradians** (*stereo* + *radians*) is the area of that part of the sphere of radius 1 with centre at $P$ lying within $K$. For example, the first octant of $\mathbb{R}^3$ is a half-cone with vertex at the origin. It determines a solid angle at the origin measuring

$$4\pi \times \frac{1}{8} = \frac{\pi}{2} \text{ steradians,}$$

since the area of the unit sphere is $4\pi$.

(a) Find the steradian measure of the solid angle at the vertex of a right-circular half-cone whose generators make angle $\alpha$ with its central axis.

(b) If a smooth, oriented surface intersects the general half-cone $K$, but not at its vertex $P$, let $S$ be the part of the surface lying within $K$. Orient $S$ with normal pointing away from $P$. Show that the steradian measure of the solid angle at $P$ determined by $K$ is the flux of $\mathbf{r}/|\mathbf{r}|^3$ through $S$, where $\mathbf{r}$ is the vector from $P$ to the point $(x, y, z)$.

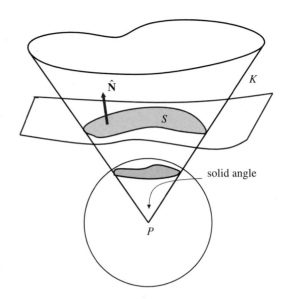

**Figure 16.25**

### Integrals over moving domains

By the Fundamental Theorem of Calculus, the derivative with respect to time $t$ of an integral of $f(x, t)$ over a "moving interval" $[a(t), b(t)]$ is given by

$$\frac{d}{dt} \int_{a(t)}^{b(t)} f(x, t)\, dx = \int_{a(t)}^{b(t)} \frac{\partial}{\partial t} f(x, t)\, dx$$
$$+ f(b(t), t)\frac{db}{dt} - f(a(t), t)\frac{da}{dt}.$$

The next three problems, suggested to the author by Luigi Quartapelle of the Politecnico di Milano, provide various extensions of this one-dimensional result to higher dimensions. As the calculations are somewhat lengthy, you may want to try to get some help from a computer algebra system. 💻

**3. (Rate of change of circulation along a moving curve)**

(a) Let $\mathbf{F}(\mathbf{r}, t)$ be a smooth vector field in $\mathbb{R}^3$ depending on a parameter $t$, and let

$$\mathbf{G}(s, t) = \mathbf{F}(\mathbf{r}(s, t), t) = \mathbf{F}(x(s, t), y(s, t), z(x, t), t),$$

where $\mathbf{r}(s, t) = x(s, t)\mathbf{i} + y(s, t)\mathbf{j} + z(s, t)\mathbf{k}$ has continuous partial derivatives of second order. Show that

$$\frac{\partial}{\partial t}\left(\mathbf{G} \bullet \frac{\partial \mathbf{r}}{\partial s}\right) - \frac{\partial}{\partial s}\left(\mathbf{G} \bullet \frac{\partial \mathbf{r}}{\partial t}\right)$$
$$= \frac{\partial \mathbf{F}}{\partial t} \bullet \frac{\partial \mathbf{r}}{\partial s} + \left((\nabla \times \mathbf{F}) \times \frac{\partial \mathbf{r}}{\partial t}\right) \bullet \frac{\partial \mathbf{r}}{\partial s}.$$

Here the curl $\nabla \times \mathbf{F}$ is taken with respect to the position vector $\mathbf{r}$.

(b) For fixed $t$ (which you can think of as time), $\mathbf{r} = \mathbf{r}(s, t)$, ($a \leq s \leq b$), represents parametrically a curve $C_t$ in $\mathbb{R}^3$. The curve moves as $t$ varies; the velocity of any point on $C_t$ is $\mathbf{v}_C(s, t) = \partial \mathbf{r}/\partial t$. Show that

$$\frac{d}{dt} \int_{C_t} \mathbf{F} \bullet d\mathbf{r} = \int_{C_t} \frac{\partial \mathbf{F}}{\partial t} \bullet d\mathbf{r} + \int_{C_t} \left( (\nabla \times \mathbf{F}) \times \mathbf{v}_C \right) \bullet d\mathbf{r}$$
$$+ \mathbf{F}(\mathbf{r}(b, t), t) \bullet \mathbf{v}_C(b, t) - \mathbf{F}(\mathbf{r}(a, t), t) \bullet \mathbf{v}_C(a, t).$$

*Hint:* write

$$\frac{d}{dt} \int_{C_t} \mathbf{F} \bullet d\mathbf{r} = \int_a^b \frac{\partial}{\partial t} \left( \mathbf{G} \bullet \frac{\partial \mathbf{r}}{\partial s} \right) ds$$
$$= \int_a^b \left[ \frac{\partial}{\partial s} \left( \mathbf{G} \bullet \frac{\partial \mathbf{r}}{\partial t} \right) \right.$$
$$\left. + \left( \frac{\partial}{\partial t} \left( \mathbf{G} \bullet \frac{\partial \mathbf{r}}{\partial s} \right) - \frac{\partial}{\partial s} \left( \mathbf{G} \bullet \frac{\partial \mathbf{r}}{\partial t} \right) \right) \right] ds.$$

Now use the result of (a).

* **4. (Rate of change of flux through a moving surface)** Let $S_t$ be a moving surface in $\mathbb{R}^3$ smoothly parametrized (for each $t$) by

$$\mathbf{r} = \mathbf{r}(u, v, t) = x(u, v, t)\mathbf{i} + y(u, v, t)\mathbf{j} + z(u, v, t)\mathbf{k},$$

where $(u, v)$ belongs to a parameter region $R$ in the $uv$-plane. Let $\mathbf{F}(\mathbf{r}, t) = F_1\mathbf{i} + F_2\mathbf{j} + F_3\mathbf{k}$ be a smooth 3-vector function and let $\mathbf{G}(u, v, t) = \mathbf{F}(\mathbf{r}(u, v, t), t)$.

(a) Show that

$$\frac{\partial}{\partial t} \left( \mathbf{G} \bullet \left[ \frac{\partial \mathbf{r}}{\partial u} \times \frac{\partial \mathbf{r}}{\partial v} \right] \right) - \frac{\partial}{\partial u} \left( \mathbf{G} \bullet \left[ \frac{\partial \mathbf{r}}{\partial t} \times \frac{\partial \mathbf{r}}{\partial v} \right] \right)$$
$$- \frac{\partial}{\partial v} \left( \mathbf{G} \bullet \left[ \frac{\partial \mathbf{r}}{\partial u} \times \frac{\partial \mathbf{r}}{\partial t} \right] \right)$$
$$= \frac{\partial \mathbf{F}}{\partial t} \bullet \left[ \frac{\partial \mathbf{r}}{\partial u} \times \frac{\partial \mathbf{r}}{\partial v} \right] + (\nabla \bullet \mathbf{F}) \frac{\partial \mathbf{r}}{\partial t} \bullet \left[ \frac{\partial \mathbf{r}}{\partial u} \times \frac{\partial \mathbf{r}}{\partial v} \right].$$

(b) If $C_t$ is the boundary of $S_t$ with orientation corresponding to that of $S_t$, use Green's Theorem to show that

$$\iint_R \left[ \frac{\partial}{\partial u} \left( \mathbf{G} \bullet \left[ \frac{\partial \mathbf{r}}{\partial t} \times \frac{\partial \mathbf{r}}{\partial v} \right] \right) \right.$$
$$\left. + \frac{\partial}{\partial v} \left( \mathbf{G} \bullet \left[ \frac{\partial \mathbf{r}}{\partial u} \times \frac{\partial \mathbf{r}}{\partial t} \right] \right) \right] du\, dv$$
$$= \oint_{C_t} \left( \mathbf{F} \times \frac{\partial \mathbf{r}}{\partial t} \right) \bullet d\mathbf{r}.$$

(c) Combine the results of (a) and (b) to show that

$$\frac{d}{dt} \iint_{S_t} \mathbf{F} \bullet \hat{\mathbf{N}}\, dS$$
$$= \iint_{S_t} \frac{\partial \mathbf{F}}{\partial t} \bullet \hat{\mathbf{N}}\, dS + \iint_{S_t} (\nabla \bullet \mathbf{F}) \mathbf{v}_S \bullet \hat{\mathbf{N}}\, dS$$
$$+ \oint_{C_t} (\mathbf{F} \times \mathbf{v}_C) \bullet d\mathbf{r},$$

where $\mathbf{v}_S = \partial \mathbf{r}/\partial t$ on $S_t$ is the velocity of $S_t$, $\mathbf{v}_C = \partial \mathbf{r}/\partial t$ on $C_t$ is the velocity of $C_t$, and $\hat{\mathbf{N}}$ is the unit normal field on $S_t$ corresponding to its orientation.

* **5. (Rate of change of integrals over moving volumes)** Let $S_t$ be the position at time $t$ of a smooth, closed surface in $\mathbb{R}^3$ that varies smoothly with $t$, and bounds at any time $t$ a region $D_t$. If $\hat{\mathbf{N}}(\mathbf{r}, t)$ denotes the unit outward (from $D_t$) normal field on $S_t$, and $\mathbf{v}_S(\mathbf{r}, t)$ is the velocity of the point $\mathbf{r}$ on $S_t$ at time $t$, show that

$$\frac{d}{dt} \iiint_{D_t} f\, dV = \iiint_{D_t} \frac{\partial f}{\partial t}\, dV + \oiint_{S_t} f \mathbf{v}_S \bullet \hat{\mathbf{N}}\, dS$$

holds for smooth functions $f(\mathbf{r}, t)$.

*Hint:* let $\Delta D_t$ consist of the points through which $S_t$ passes as $t$ increases to $t + \Delta t$. The volume element $dV$ in $\Delta D_t$ can be expressed in terms of the area element $dS$ on $S_t$ by

$$dV = \mathbf{v} \bullet \hat{\mathbf{N}}\, dS\, \Delta t.$$

Show that

$$\frac{1}{\Delta t} \left[ \iiint_{D_{t+\Delta t}} f(\mathbf{r}, t + \Delta t)\, dV - \iiint_{D_t} f(\mathbf{r}, t)\, dV \right]$$
$$= \iiint_{D_t} \frac{f(\mathbf{r}, t + \Delta t) - f(\mathbf{r}, t)}{\Delta t}\, dV$$
$$+ \frac{1}{\Delta t} \iiint_{\Delta D_t} f(\mathbf{r}, t)\, dV$$
$$+ \iiint_{\Delta D_t} \frac{f(\mathbf{r}, t + \Delta t) - f(\mathbf{r}, t)}{\Delta t}\, dV,$$

and show that the last integral $\to 0$ as $\Delta t \to 0$.

# CHAPTER 17

# Ordinary Differential Equations

*Introduction*  A **differential equation** (or **DE**) is an equation that involves one or more derivatives of an unknown function. Solving the differential equation means finding a function (or every such function) that satisfies the differential equation.

Many physical laws and relationships between quantities studied in various scientific disciplines are expressed mathematically as differential equations. For example, Newton's second law of motion ($F = ma$) states that the position $x(t)$ at time $t$ of an object of constant mass $m$ subjected to a force $F(t)$ must satisfy the differential equation (equation of motion):

$$m\frac{d^2x}{dt^2} = F(t).$$

Similarly, the biomass $m(t)$ at time $t$ of a bacterial culture growing in a uniformly supporting medium changes at a rate proportional to the biomass:

$$\frac{dm}{dt} = km(t),$$

which is the differential equation of exponential growth (or, if $k < 0$, exponential decay). Because differential equations arise so extensively in the abstract modelling of concrete phenomena, such equations and techniques for solving them are at the heart of applied mathematics. Indeed, most of the existing mathematical literature is either directly involved with differential equations, or is motivated by problems arising in the study of such equations. Because of this, we have introduced various differential equations, terms for their description, and techniques for their solution at several places in the development of calculus throughout this book. This final chapter provides a more unified framework for a brief introduction to the study of ordinary differential equations. Some material from earlier sections (especially Sections 7.9 and 3.7) is repeated here for the sake of unity. This chapter is, of necessity, relatively short. Students of mathematics and its applications usually take one or more full courses on differential equations, and even then hardly scratch the surface of the subject.

## 17.1  Classifying Differential Equations

Differential equations are classified in several ways. The most significant classification is based on the number of variables with respect to which derivatives appear in the equation. An **ordinary differential equation (ODE)** is one that involves derivatives with respect to only one variable. Both of the examples given above are ordinary differential equations. A **partial differential equation (PDE)** is one that involves partial derivatives of the unknown function with respect to more than one variable. For example, the **one-dimensional wave equation**

$$\frac{\partial^2 u}{\partial t^2} = c^2 \frac{\partial^2 u}{\partial x^2}$$

models the lateral displacement $u(x, t)$ at position $x$ at time $t$ of a stretched vibrating string. (See Section 12.4.) We will not discuss partial differential equations in this chapter.

Differential equations are also classified with respect to **order**. The order of a differential equation is the order of the highest order derivative present in the equation. The one-dimensional wave equation is a second-order PDE. The following example records the order of two ODEs.

| **Example 1** | $$\frac{d^2y}{dx^2} + x^3 y = \sin x \qquad \text{has order 2,}$$ |
| --- | --- |

$$\frac{d^3y}{dx^3} + 4x \left(\frac{dy}{dx}\right)^2 = y \frac{d^2y}{dx^2} + e^y \qquad \text{has order 3.}$$

Like any equation, a differential equation can be written in the form $F = 0$, where $F$ is a function. For an ODE, the function $F$ can depend on the independent variable (usually called $x$ or $t$), the unknown function (usually $y$), and any derivatives of the unknown function up to the order of the equation. For instance, an $n$th-order ODE can be written in the form

$$F(x, y, y', y'', \ldots, y^{(n)}) = 0.$$

An important special class of differential equations are those that are **linear**. An $n$th-order linear ODE has the form

$$a_n(x)y^{(n)}(x) + a_{n-1}(x)y^{(n-1)}(x) + \cdots$$
$$+ a_2(x)y''(x) + a_1(x)y'(x) + a_0(x)y(x) = f(x).$$

Each term in the expression on the left side is the product of a *coefficient* that is a function of $x$, and a second factor that is either $y$ or one of the derivatives of $y$. The term on the right does not depend on $y$; it is called the **nonhomogeneous term**. Observe that no term on the left side involves any power of $y$ or its derivatives other than the first power, and $y$ and its derivatives are never multiplied together.

A linear ODE is said to be **homogeneous** if all of its terms involve the unknown function $y$, that is, if $f(x) = 0$. If $f(x)$ is not identically zero, the equation is **nonhomogeneous**.

| **Example 2** | In Example 1 the first DE, |
| --- | --- |

$$\frac{d^2y}{dx^2} + x^3 y = \sin x,$$

is linear. Here the coefficients are $a_2(x) = 1$, $a_1(x) = 0$, $a_0(x) = x^3$, and the nonhomogeneous term is $f(x) = \sin x$. Although it can be written in the form

$$\frac{d^3y}{dx^3} + 4x \left(\frac{dy}{dx}\right)^2 - y \frac{d^2y}{dx^2} - e^y = 0,$$

the second equation is *not linear* (we say it is **nonlinear**) because the second term involves the square of a derivative of $y$, the third term involves the product of $y$ and one of its derivatives, and the fourth term is not $y$ times a function of $x$. The equation

$$(1 + x^2)\frac{d^3y}{dx^3} + \sin x \frac{d^2y}{dx^2} - 4\frac{dy}{dx} + y = 0$$

is a linear equation of order 3. The coefficients are $a_3(x) = 1 + x^2$, $a_2(x) = \sin x$, $a_1(x) = -4$, and $a_0(x) = 1$. Since $f(x) = 0$, this equation is *homogeneous*.

The following theorem states that any *linear combination* of solutions of a linear, homogeneous DE is also a solution. This is an extremely important fact about linear, homogeneous DEs.

**THEOREM 1**

If $y = y_1(x)$ and $y = y_2(x)$ are two solutions of the linear, homogeneous DE

$$a_n y^{(n)} + a_{n-1} y^{(n-1)} + \cdots + a_2 y'' + a_1 y' + a_0 y = 0$$

then so is the linear combination

$$y = A y_1(x) + B y_2(x)$$

for any values of the constants $A$ and $B$.

**PROOF** We are given that

$$a_n y_1^{(n)} + a_{n-1} y_1^{(n-1)} + \cdots + a_2 y_1'' + a_1 y_1' + a_0 y_1 = 0, \qquad \text{and}$$
$$a_n y_2^{(n)} + a_{n-1} y_2^{(n-1)} + \cdots + a_2 y_2'' + a_1 y_2' + a_0 y_2 = 0.$$

Multiplying the first equation by $A$ and the second by $B$ and adding the two gives

$$a_n (A y_1^{(n)} + B y_2^{(n)}) + a_{n-1} (A y_1^{(n-1)} + B y_2^{(n-1)})$$
$$+ \cdots + a_2 (A y_1'' + B y_2'') + a_1 (A y_1' + B y_2') + a_0 (A y_1 + B y_2) = 0.$$

Thus $y = A y_1(x) + B y_2(x)$ is also a solution of the equation.

The same kind of proof can be used to verify the following theorem.

**THEOREM 2**

If $y = y_1(x)$ is a solution of the linear, homogeneous equation

$$a_n y^{(n)} + a_{n-1} y^{(n-1)} + \cdots + a_2 y'' + a_1 y' + a_0 y = 0$$

and $y = y_2(x)$ is a solution of the linear, nonhomogeneous equation

$$a_n y^{(n)} + a_{n-1} y^{(n-1)} + \cdots + a_2 y'' + a_1 y' + a_0 y = f(x),$$

then $y = y_1(x) + y_2(x)$ is also a solution of the same linear, nonhomogeneous equation.

We will make extensive use of the two theorems above when we discuss second-order linear equations in Sections 17.6–17.8.

**Example 3** Verify that $y = \sin 2x$ and $y = \cos 2x$ satisfy the DE $y'' + 4y = 0$. Find a solution $y(x)$ of that DE that satisfies $y(0) = 2$ and $y'(0) = -4$.

**Solution** If $y = \sin 2x$, then $y'' = \dfrac{d}{dx}(2\cos 2x) = -4\sin 2x = -4y$. Thus $y'' + 4y = 0$. A similar calculation shows that $y = \cos 2x$ also satisfies the DE. Since the DE is linear and homogeneous, the function

$$y = A\sin 2x + B\cos 2x$$

is a solution for any values of the constants $A$ and $B$. We want $y(0) = 2$, so we need $2 = A\sin 0 + B\cos 0 = B$. Thus $B = 2$. Also,

$$y' = 2A\cos 2x - 2B\sin 2x.$$

We want $y'(0) = -4$, so $-4 = 2A\cos 0 - 2B\sin 0 = 2A$. Thus $A = -2$ and the required solution is $y = -2\sin 2x + 2\cos 2x$. ∎

**Remark** Let $P_n(r)$ be the $n$th-degree polynomial in the variable $r$ given by

$$P_n(r) = a_n(x)r^n + a_{n-1}(x)r^{n-1} + \cdots + a_2(x)r^2 + a_1(x)r + a_0(x),$$

with coefficients depending on the variable $x$. We can write the $n$th-order linear ODE with coefficients $a_k(x)$, $(0 \le k \le n)$, and nonhomogeneous term $f(x)$ in the form

$$P_n(D)y(x) = f(x)$$

where $D$ stands for the *differential operator* $d/dx$. The left side of the equation above denotes the application of the $n$th-order differential operator

$$P_n(D) = a_n(x)D^n + a_{n-1}(x)D^{n-1} + \cdots + a_2(x)D^2 + a_1(x)D + a_0(x)$$

to the function $y(x)$. For example,

$$a_k(x)D^k y(x) = a_k(x)\frac{d^k y}{dx^k}.$$

It is often useful to write linear DEs in terms of differential operators in this way.

**Remark** Unfortunately, the term *homogeneous* is used in more than one way in the study of differential equations. Certain ODEs that are not necessarily linear are called homogeneous for a different reason than the one applying for linear equations above. We will encounter equations of this type in Section 17.2.

# Exercises 17.1

In Exercises 1–10, state the order of the given DE, and whether it is linear or nonlinear. If it is linear, is it homogeneous or nonhomogeneous?

**1.** $\dfrac{dy}{dx} = 5y$

**2.** $\dfrac{d^2 y}{dx^2} + x = y$

**3.** $y\dfrac{dy}{dx} = x$

**4.** $y''' + xy' = x\sin x$

**5.** $y'' + x\sin x \, y' = y$

**6.** $y'' + 4y' - 3y = 2y^2$

**7.** $\dfrac{d^3y}{dt^3} + t\dfrac{dy}{dt} + t^2 y = t^3$

**8.** $\cos x\dfrac{dx}{dt} + x\sin t = 0$

**9.** $y^{(4)} + e^x y'' = x^3 y'$

**10.** $x^2 y'' + e^x y' = \dfrac{1}{y}$

**11.** Verify that $y = \cos x$ and $y = \sin x$ are solutions of the DE $y'' + y = 0$. Are any of the following functions solutions? (a) $\sin x - \cos x$, (b) $\sin(x + 3)$, (c) $\sin 2x$? Justify your answers.

**12.** Verify that $y = e^x$ and $y = e^{-x}$ are solutions of the DE $y'' - y = 0$. Are any of the following functions solutions? (a) $\cosh x = \frac{1}{2}(e^x + e^{-x})$, (b) $\cos x$, (c) $x^e$? Justify your answers.

**13.** $y_1 = \cos(kx)$ is a solution of $y'' + k^2 y = 0$. Guess and verify another solution $y_2$ that is not a multiple of $y_1$. Then find a solution that satisfies $y(\pi/k) = 3$ and $y'(\pi/k) = 3$.

**14.** $y_1 = e^{kx}$ is a solution of $y'' - k^2 y = 0$. Guess and verify another solution $y_2$ that is not a multiple of $y_1$. Then find a solution that satisfies $y(1) = 0$ and $y'(1) = 2$.

**15.** Find a solution of $y'' + y = 0$ that satisfies $y(\pi/2) = 2y(0)$ and $y(\pi/4) = 3$. *Hint:* see Exercise 11.

**16.** Find two values of $r$ such that $y = e^{rx}$ is a solution of $y'' - y' - 2y = 0$. Then find a solution of the equation that satisfies $y(0) = 1$, $y'(0) = 2$.

**17.** Verify that $y = x$ is a solution of $y'' + y = x$, and find a solution $y$ of this DE that satisfies $y(\pi) = 1$ and $y'(\pi) = 0$. *Hint:* use Exercise 11 and Theorem 2.

**18.** Verify that $y = -e$ is a solution of $y'' - y = e$, and find a solution $y$ of this DE that satisfies $y(1) = 0$ and $y'(1) = 1$. *Hint:* use Exercise 12 and Theorem 2.

---

## 17.2  First-Order Separable and Homogeneous Equations

In the next three sections we will develop techniques for solving first-order ODEs, mostly ones of the form

$$\frac{dy}{dx} = f(x, y).$$

Solving differential equations typically involves integration; indeed, the process of solving a DE is called *integrating* the DE. Nevertheless, solving DEs is usually more complicated than just writing down an integral and evaluating it. The only kind of DE that can be solved that way is the simplest kind of first-order, linear DE that can be written in the form

$$\frac{dy}{dx} = f(x).$$

The solution is then just the antiderivative of $f$:

$$y = \int f(x)\,dx.$$

### Separable Equations

The next most simple kind of equation to solve is a so-called **separable equation**. Such equations were examined in Section 7.9; part of that material is reproduced here. A separable equation is one of the form

$$\frac{dy}{dx} = f(x)g(y).$$

We can rewrite such an equation in the equivalent differential form

$$\frac{dy}{g(y)} = f(x)\,dx,$$

where the variables have been *separated* on opposite sides of the equation. This was possible because the derivative $dy/dx$ was a product of a function of $x$ alone times a function of $y$ alone, rather than a more general function of the two variables $x$ and $y$. We solve the separated equation by integrating both sides,

$$\int \frac{dy}{g(y)} = \int f(x)\,dx,$$

remembering to include a constant of integration on one side after the integrals are evaluated. (Why don't you need to include a constant of integration on both sides?) Assuming both of the integrals can be evaluated, we can obtain a **general solution** to the given DE in the form

$$G(y) = F(x) + C.$$

It may or may not be possible to solve this equation *explicitly* for $y$ as a function of $x$, but the equation will normally define $y(x)$ at least *implicitly*. The graphs of the curves $G(y) = F(x) + C$ for various values of $C$ are called **solution curves** of the given DE. A single **initial condition** of the form $y = b$ when $x = a$ (or, equivalently, $y(a) = b$) will serve to single out a unique solution curve from among those given by the general solution.

**Example 1**   Solve the equation $\dfrac{dy}{dx} = \dfrac{x}{y}$.

**Solution**   We rewrite the equation in the form

$$y\,dy = x\,dx,$$

and integrate both sides to get

$$\frac{1}{2}y^2 = \frac{1}{2}x^2 + C,$$

or $y^2 - x^2 = C_1$ where $C_1 = 2C$ is an arbitrary constant. The solution curves are rectangular hyperbolas with asymptotes $y = x$ and $y = -x$.

**Example 2**   Solve the initial-value problem

$$\begin{cases} \dfrac{dy}{dx} = x^2 y^3 \\ y(1) = 3. \end{cases}$$

**Solution**   The differential equation becomes $\dfrac{dy}{y^3} = x^2\,dx$, so $\displaystyle\int \frac{dy}{y^3} = \int x^2\,dx$ and

$$\frac{-1}{2y^2} = \frac{x^3}{3} + C.$$

Since $y = 3$ when $x = 1$, we have $-\frac{1}{18} = \frac{1}{3} + C$ and $C = -\frac{7}{18}$. Substituting this value into the above solution and solving for $y$, we obtain

$$y(x) = \frac{3}{\sqrt{7 - 6x^3}}.$$

This solution is valid for $x \le \left(\frac{7}{6}\right)^{1/3}$.

---

**Example 3**  Solve the logistic equation introduced in Section 3.4:

$$\frac{dy}{dt} = ky\left(1 - \frac{y}{L}\right).$$

Here $y(t)$ is the size of an animal population at time $t$, $k$ is a positive constant related to the fertility of the animals, and $L$ is the steady-state population size that can be sustained by the available food supply. Assume that $0 < y(t) < L$ for all $t$.

**Solution**  The equation can be separated to give

$$\frac{L\,dy}{y(L - y)} = k\,dt,$$

and solved by integrating both sides. Expanding the left side in partial fractions and integrating, we get

$$\int\left(\frac{1}{y} + \frac{1}{L - y}\right) dy = kt + C.$$

Assuming that $0 < y < L$, we therefore obtain

$$\ln y - \ln(L - y) = kt + C,$$

$$\ln\left(\frac{y}{L - y}\right) = kt + C.$$

We can solve this equation for $y$ by first taking exponentials of both sides:

$$\frac{y}{L - y} = e^{kt+C} = C_1 e^{kt}$$

$$y = (L - y)C_1 e^{kt}$$

$$y = \frac{C_1 L e^{kt}}{1 + C_1 e^{kt}},$$

where $C_1 = e^C$.

---

Many applied problems lead to separable equations. Three extended examples are given in Section 7.9. These concern solution concentrations, the rate of a chemical reaction, and determining a family of curves each of which intersects all the curves of a second family at right angles.

## First-Order Homogeneous Equations

A first-order DE of the form

$$\frac{dy}{dx} = f\left(\frac{y}{x}\right)$$

is said to be **homogeneous**. This is a *different* use of the term homogeneous from that in the previous section, which applied only to linear equations. Here homogeneous refers to the fact that $y/x$, and therefore $g(x, y) = f(y/x)$ is *homogeneous of degree 0* in the sense described after Example 7 in Section 12.5. Such a homogeneous equation can be transformed into a separable equation (and therefore solved) by means of a change of dependent variable. If we set

$$v = \frac{y}{x}, \qquad \text{or equivalently} \qquad y = xv(x),$$

then we have

$$\frac{dy}{dx} = v + x\frac{dv}{dx}$$

and the original differential equation transforms into

$$\frac{dv}{dx} = \frac{f(v) - v}{x},$$

which is separable.

**Example 4** Solve the equation

$$\frac{dy}{dx} = \frac{x^2 + xy}{xy + y^2}.$$

**Solution** The equation is homogeneous. (Divide the numerator and denominator of the right-hand side by $x^2$ to see this.) If $y = vx$ the equation becomes

$$v + x\frac{dv}{dx} = \frac{1 + v}{v + v^2} = \frac{1}{v},$$

or

$$x\frac{dv}{dx} = \frac{1 - v^2}{v}.$$

Separating variables and integrating, we calculate

$$\int \frac{v\,dv}{1 - v^2} = \int \frac{dx}{x} \qquad \text{let } u = 1 - v^2$$

$$-\frac{1}{2}\int \frac{du}{u} = \int \frac{dx}{x}$$

$$-\ln|u| = 2\ln|x| + C_1 = \ln C_2 x^2 \qquad (C_1 = \ln C_2)$$

$$\frac{1}{|u|} = C_2 x^2$$

$$|1 - v^2| = \frac{C_3}{x^2} \qquad (C_3 = 1/C_2)$$

$$\left|1 - \frac{y^2}{x^2}\right| = \frac{C_3}{x^2}.$$

The solution is best expressed in the form $x^2 - y^2 = C_4$. However, near points where $y \neq 0$, the equation can be solved for $y$ as a function of $x$.

## Exercises 17.2

Solve the differential equations in Exercises 1–20.

**1.** $\dfrac{dy}{dx} = \dfrac{y}{2x}$

**2.** $\dfrac{dy}{dx} = \dfrac{3y-1}{x}$

**3.** $\dfrac{dy}{dx} = \dfrac{x^2}{y^2}$

**4.** $\dfrac{dy}{dx} = x^2 y^2$

**5.** $\dfrac{dy}{dx} = \dfrac{x^2}{y^3}$

**6.** $\dfrac{dy}{dx} = x^2 y^3$

**7.** $\dfrac{dY}{dt} = tY$

**8.** $\dfrac{dx}{dt} = e^x \sin t$

**9.** $\dfrac{dy}{dx} = 1 - y^2$

**10.** $\dfrac{dy}{dx} = 1 + y^2$

**11.** $\dfrac{dy}{dt} = 2 + e^y$

**12.** $\dfrac{dy}{dx} = y^2(1-y)$

**13.** $\dfrac{dy}{dx} = \sin x \cos^2 y$

**14.** $x \dfrac{dy}{dx} = y \ln x$

**15.** $\dfrac{dy}{dx} = \dfrac{x+y}{x-y}$

**16.** $\dfrac{dy}{dx} = \dfrac{xy}{x^2 + 2y^2}$

**17.** $\dfrac{dy}{dx} = \dfrac{x^2 + xy + y^2}{x^2}$

**18.** $\dfrac{dy}{dx} = \dfrac{x^3 + 3xy^2}{3x^2 y + y^3}$

**19.** $x \dfrac{dy}{dx} = y + x \cos^2\left(\dfrac{y}{x}\right)$

**20.** $\dfrac{dy}{dx} = \dfrac{y}{x} - e^{-y/x}$

**21.** Find an equation of the curve in the $xy$-plane which passes through the point $(2,3)$, and has, at every point $(x, y)$ on it, slope equal to $2x/(1 + y^2)$.

**22.** Repeat the previous problem for the point $(1, 3)$ and slope $1 + (2y/x)$.

**23.** Show that the change of variables $\xi = x - x_0$, $\eta = y - y_0$ transforms the equation

$$\frac{dy}{dx} = \frac{ax + by + c}{ex + fy + g}$$

into the homogeneous equation

$$\frac{d\eta}{d\xi} = \frac{a\xi + b\eta}{e\xi + f\eta}$$

provided $(x_0, y_0)$ is the solution of the system

$$ax + by + c = 0$$
$$ex + fy + g = 0.$$

**24.** Use the technique of the previous exercise to solve the equation $\dfrac{dy}{dx} = \dfrac{x + 2y - 4}{2x - y - 3}$.

## 17.3 Exact Equations and Integrating Factors

A first-order differential equation expressed in differential form as

$$M(x, y)\, dx + N(x, y)\, dy = 0,$$

which is equivalent to $\dfrac{dy}{dx} = -\dfrac{M(x, y)}{N(x, y)}$, is said to be **exact** if the left-hand side is the differential of a function $\phi(x, y)$:

$$d\phi(x, y) = M(x, y)\, dx + N(x, y)\, dy.$$

The function $\phi$ is called an **integral function** of the differential equation. The level curves $\phi(x, y) = C$ of $\phi$ are the **solution curves** of the differential equation. For example, the differential equation

$$x\, dx + y\, dy = 0$$

has solution curves given by

$$x^2 + y^2 = C$$

since $d(x^2 + y^2) = 2(x\, dx + y\, dy) = 0.$

***Remark*** The condition that the differential equation $M\,dx + N\,dy = 0$ should be exact is just the condition that the vector field

$$\mathbf{F} = M(x, y)\,\mathbf{i} + N(x, y)\,\mathbf{j}$$

should be *conservative*; the integral function of the differential equation is then the potential function of the vector field. (See Section 15.2.)

A **necessary condition** for the exactness of the DE $M\,dx + N\,dy = 0$ is that

$$\frac{\partial M}{\partial y} = \frac{\partial N}{\partial x};$$

this just says that the mixed partial derivatives $\dfrac{\partial^2 \phi}{\partial x \partial y}$ and $\dfrac{\partial^2 \phi}{\partial y \partial x}$ of the integral function $\phi$ must be equal.

Once you know that an equation is exact, you can often guess the integral function. In any event, $\phi$ can always be found by the same method used to find the potential of a conservative vector field in Section 15.2.

**Example 1** Verify that the DE

$$(2x + \sin y - ye^{-x})\,dx + (x \cos y + \cos y + e^{-x})\,dx = 0$$

is exact and find its solution curves.

***Solution*** Here $M = 2x + \sin y - ye^{-x}$ and $N = x \cos y + \cos y + e^{-x}$. Since

$$\frac{\partial M}{\partial y} = \cos y - e^{-x} = \frac{\partial N}{\partial x},$$

the DE is exact. We want to find $\phi$ so that

$$\frac{\partial \phi}{\partial x} = M = 2x + \sin y - ye^{-x} \quad \text{and} \quad \frac{\partial \phi}{\partial y} = N = x \cos y + \cos y + e^{-x}.$$

Integrate the first equation with respect to $x$, being careful to allow the constant of integration to depend on $y$:

$$\phi(x, y) = \int (2x + \sin y - ye^{-x})\,dx = x^2 + x \sin y + ye^{-x} + C_1(y).$$

Now substitute this expression into the second equation:

$$x \cos y + \cos y + e^{-x} = \frac{\partial \phi}{\partial y} = x \cos y + e^{-x} + C_1'(y).$$

Thus $C_1'(y) = \cos y$, and $C_1(y) = \sin y + C_2$. (It is because the original DE was exact that the equation for $C_1'(y)$ turned out to be independent of $x$; this had to happen or we could not have found $C_1$ as a function of $y$ only.) Choosing $C_2 = 0$, we find that $\phi(x, y) = x^2 + x \sin y + ye^{-x} + \sin y$ is an integral function for the given DE. The solution curves for the DE are the level curves

$$x^2 + x \sin y + ye^{-x} + \sin y = C.$$

## Integrating Factors

Any ordinary differential equation of order 1 and degree 1 can be expressed in differential form: $M\,dx + N\,dy = 0$. However, this latter equation will usually not be exact. It *may* be possible to multiply the equation by an **integrating factor** $\mu(x, y)$ so that the resulting equation

$$\mu(x, y)\,M(x, y)\,dx + \mu(x, y)\,N(x, y)\,dy = 0$$

is exact. In general such integrating factors are difficult to find; they must satisfy the partial differential equation

$$M(x, y)\,\frac{\partial \mu}{\partial y} - N(x, y)\,\frac{\partial \mu}{\partial x} = \mu(x, y)\left(\frac{\partial N}{\partial x} - \frac{\partial M}{\partial y}\right),$$

which follows from the necessary condition for exactness stated above. We will not try to solve this equation here.

Sometimes it happens that a differential equation has an integrating factor depending on only one of the two variables. Suppose, for instance, that $\mu(x)$ is an integrating factor for $M\,dx + N\,dy = 0$. Then $\mu(x)$ must satisfy the ordinary differential equation

$$N(x, y)\,\frac{d\mu}{dx} = \mu(x)\left(\frac{\partial M}{\partial y} - \frac{\partial N}{\partial x}\right),$$

or,

$$\frac{1}{\mu(x)}\frac{d\mu}{dx} = \frac{\dfrac{\partial M}{\partial y} - \dfrac{\partial N}{\partial x}}{N(x, y)}.$$

This equation can be solved (by integration) for $\mu$ as a function of $x$ alone *provided that the right-hand side is independent of $y$*.

**Example 2** Show that $(x + y^2)\,dx + xy\,dy = 0$ has an integrating factor depending only on $x$, find it, and solve the equation.

**Solution** Here $M = x + y^2$ and $N = xy$. Since

$$\frac{\dfrac{\partial M}{\partial y} - \dfrac{\partial N}{\partial x}}{N(x, y)} = \frac{2y - y}{xy} = \frac{1}{x}$$

does not depend on $y$, the equation has an integrating factor depending only on $x$. This factor is given by $d\mu/\mu = dx/x$. Evidently $\mu = x$ is a suitable integrating factor; if we multiply the given differential equation by $x$, we obtain

$$0 = (x^2 + xy^2)\,dx + x^2 y\,dy = d\left(\frac{x^3}{3} + \frac{x^2 y^2}{2}\right).$$

The solution is therefore $2x^3 + 3x^2 y^2 = C$.

**Remark** Of course, it may be possible to find an integrating factor depending on $y$ instead of $x$. See Exercises 7–9 below. It is also possible to look for integrating factors that depend on specific combinations of $x$ and $y$, for instance, $xy$. See Exercise 10.

# Exercises 17.3

Show that the DEs in Exercises 1–4 are exact, and solve them.

**1.** $(xy^2 + y)\,dx + (x^2y + x)\,dy = 0$

**2.** $(e^x \sin y + 2x)\,dx + (e^x \cos y + 2y)\,dy = 0$

**3.** $e^{xy}(1 + xy)\,dx + x^2e^{xy}\,dy = 0$

**4.** $\left(2x + 1 - \dfrac{y^2}{x^2}\right)\,dx + \dfrac{2y}{x}\,dy = 0$

Show that the DEs in Exercises 5–6 admit integrating factors that are functions of $x$ alone. Then solve the equations.

**5.** $(x^2 + 2y)\,dx - x\,dy = 0$

**6.** $(xe^x + x \ln y + y)\,dx + \left(\dfrac{x^2}{y} + x \ln x + x \sin y\right)\,dy = 0$

**7.** What condition must the coefficients $M(x, y)$ and $N(x, y)$ satisfy if the equation $M\,dx + N\,dy = 0$ is to have an integrating factor of the form $\mu(y)$, and what DE must the integrating factor satisfy?

**8.** Find an integrating factor of the form $\mu(y)$ for the equation

$$2y^2(x + y^2)\,dx + xy(x + 6y^2)\,dy = 0,$$

and hence solve the equation. *Hint:* see Exercise 7.

**9.** Find an integrating factor of the form $\mu(y)$ for the equation $y\,dx - (2x + y^3e^y)\,dy = 0$, and hence solve the equation. *Hint:* see Exercise 7.

**10.** What condition must the coefficients $M(x, y)$ and $N(x, y)$ satisfy if the equation $M\,dx + N\,dy = 0$ is to have an integrating factor of the form $\mu(xy)$, and what DE must the integrating factor satisfy?

**11.** Find an integrating factor of the form $\mu(xy)$ for the equation

$$\left(x \cos x + \dfrac{y^2}{x}\right)\,dx - \left(\dfrac{x \sin x}{y} + y\right)\,dy = 0,$$

and hence solve the equation. *Hint:* see Exercise 10.

## 17.4 First-Order Linear Equations

A first-order **linear** differential equation is one of the type

$$\frac{dy}{dx} + p(x)y = q(x),$$

where $p(x)$ and $q(x)$ are given functions, which we assume to be continuous. The equation is *homogeneous* (in the sense described in Section 17.1) provided that $q(x) = 0$. In that case, the given linear equation is separable:

$$\frac{dy}{y} = -p(x)\,dx,$$

which can be solved by integrating both sides. Nonhomogeneous first-order linear equations can be solved by the following procedure involving the calculation of an integrating factor. This material repeats part of Section 7.9.

Let $\mu(x)$ be any antiderivative of $p(x)$:

$$\mu(x) = \int p(x)\,dx, \qquad \text{and} \qquad \frac{d\mu}{dx} = p(x).$$

If $y = y(x)$ satisfies the given equation, then we calculate, using the Product Rule,

$$\frac{d}{dx}\left(e^{\mu(x)}y(x)\right) = e^{\mu(x)}\frac{dy}{dx} + e^{\mu(x)}\frac{d\mu}{dx}y(x)$$

$$= e^{\mu(x)}\left(\frac{dy}{dx} + p(x)y\right) = e^{\mu(x)}q(x).$$

Therefore,

$$e^{\mu(x)} y(x) = \int e^{\mu(x)} q(x)\, dx$$

or

$$y(x) = e^{-\mu(x)} \int e^{\mu(x)} q(x)\, dx.$$

We reuse this method, rather than the final formula, in the following examples. We call $e^{\mu(x)}$ an **integrating factor** of the given differential equation because, if we multiply the equation by $e^{\mu(x)}$, the left side becomes the derivative of $e^{\mu(x)} y(x)$.

**Example 1**  Solve $\dfrac{dy}{dx} + \dfrac{y}{x} = 1$ for $x > 0$.

**Solution**  Here $p(x) = 1/x$ and $q(x) = 1$. We want $d\mu/dx = 1/x$, so that $\mu(x) = \ln x$ (for $x > 0$). Thus $e^{\mu(x)} = x$ and we calculate

$$\frac{d}{dx}(e^{\mu(x)} y) = \frac{d}{dx}(xy) = x\frac{dy}{dx} + y = x\left(\frac{dy}{dx} + \frac{y}{x}\right) = x.$$

Therefore

$$xy = \int x\, dx = \frac{1}{2}x^2 + C.$$

Finally,

$$y = \frac{1}{x}\left(\frac{1}{2}x^2 + C\right) = \frac{x}{2} + \frac{C}{x}.$$

This function is a solution of the given equation for any value of the constant $C$.

**Example 2**  Solve $\dfrac{dy}{dx} + xy = x^3$.

**Solution**  Here $p(x) = x$ so $\mu(x) = x^2/2$. We calculate

$$\frac{d}{dx}\left(e^{x^2/2} y\right) = e^{x^2/2}\frac{dy}{dx} + e^{x^2/2} xy = x^3 e^{x^2/2}.$$

Thus,

$$e^{x^2/2} y = \int x^3 e^{x^2/2}\, dx \qquad \text{Let} \quad U = x^2 \qquad dV = x\,e^{x^2/2}\, dx$$
$$\text{Then } dU = 2x\, dx \qquad V = e^{x^2/2}$$
$$= x^2 e^{x^2/2} - 2\int x\,e^{x^2/2}\, dx$$
$$= x^2 e^{x^2/2} - 2 e^{x^2/2} + C,$$

and, finally,

$$y = x^2 - 2 + Ce^{-x^2/2}.$$

Example 8 of Section 7.9 illustrates the application of the technique for solving a first-order linear DE to a typical *stream of payments* problem. We conclude this section with an applied example involving current flow in an electric circuit.

**Example 3**   **(An inductance-resistance circuit)** An electric circuit (Figure 17.1) contains a constant DC voltage source of $V$ volts, a switch, a resistor of size $R$ ohms, and an inductor of size $L$ henrys. The circuit has no capacitance. The switch, initially open so that no current is flowing, is closed at time $t = 0$ so that current begins to flow at that time. If the inductance $L$ were zero, the current would suddenly jump from 0 amperes when $t < 0$ to $I = V/R$ amperes when $t > 0$. However, if $L > 0$ the current cannot change instantaneously; it will depend on time $t$. Let the current $t$ seconds after the switch is closed be $I(t)$ amperes. It is known that $I(t)$ satisfies the initial-value problem

$$\begin{cases} L\dfrac{dI}{dt} + RI = V \\[2mm] \qquad\quad I(0) = 0. \end{cases}$$

**Figure 17.1**   An inductance-resistance circuit

Find $I(t)$. What is $\lim_{t \to \infty} I(t)$? How long does it take after the switch is closed for the current to rise to 90% of its limiting value?

**Solution**   The DE can be written in the form $\dfrac{dI}{dt} + \dfrac{R}{L}I = \dfrac{V}{L}$. It is linear and has integrating factor $e^{\mu(t)}$, where

$$\mu(t) = \int \frac{R}{L}\, dt = \frac{Rt}{L}.$$

Therefore

$$\frac{d}{dt}\left(e^{Rt/L}I\right) = e^{Rt/L}\left(\frac{dI}{dt} + \frac{R}{L}I\right) = e^{Rt/L}\frac{V}{L}$$

$$e^{Rt/L}I = \frac{V}{L}\int e^{Rt/L}\, dt = \frac{V}{R}e^{Rt/L} + C$$

$$I(t) = \frac{V}{R} + Ce^{-Rt/L}.$$

Since $I(0) = 0$, we have $0 = (V/R) + C$, so $C = -V/R$. Thus the current flowing at any time $t > 0$ is

$$I(t) = \frac{V}{R}\left(1 - e^{-Rt/L}\right).$$

It is clear from this solution that $\lim_{t \to \infty} I(t) = V/R$; the *steady state* current is the current that would flow if the inductance were zero.

$I(t)$ will be 90% of this limiting value when

$$\frac{V}{R}\left(1 - e^{-Rt/L}\right) = \frac{90}{100}\frac{V}{R}.$$

This equation implies that $e^{-Rt/L} = 1/10$, or $t = (L\ln 10)/R$. The current will grow to 90% of its limiting value in $(L\ln 10)/R$ seconds.

## Exercises 17.4

Solve the differential equations in Exercises 1–6.

**1.** $\dfrac{dy}{dx} - \dfrac{2y}{x} = x^2$

**2.** $\dfrac{dy}{dx} + \dfrac{2y}{x} = \dfrac{1}{x^2}$

**3.** $\dfrac{dy}{dx} + 2y = 3$

**4.** $\dfrac{dy}{dx} + y = e^x$

**5.** $\dfrac{dy}{dx} + y = x$

**6.** $\dfrac{dy}{dx} + 2e^x y = e^x$

Solve the initial-value problems in Exercises 7–10.

**7.** $\begin{cases} \dfrac{dy}{dt} + 10y = 1 \\ y(1/10) = 2/10 \end{cases}$

**8.** $\begin{cases} \dfrac{dy}{dx} + 3x^2 y = x^2 \\ y(0) = 1 \end{cases}$

**9.** $\begin{cases} y' + (\cos x)y = 2xe^{-\sin x} \\ y(\pi) = 0 \end{cases}$

**10.** $\begin{cases} x^2 y' + y = x^2 e^{1/x} \\ y(1) = 3e \end{cases}$

**11.** An object of mass $m$ falling near the surface of the earth is retarded by air resistance proportional to its velocity so that, according to Newton's second law of motion,

$$m\frac{dv}{dt} = mg - kv,$$

where $v = v(t)$ is the velocity of the object at time $t$, and $g$ is the acceleration of gravity near the surface of the earth. Assuming that the object falls from rest at time $t = 0$, that is, $v(0) = 0$, find the velocity $v(t)$ for any $t > 0$ (up until the object strikes the ground). Show $v(t)$ approaches a limit as $t \to \infty$. Do you need the explicit formula for $v(t)$ to determine this limiting velocity?

**12.** Repeat Exercise 11 assuming that the air resistance is proportional to the square of the velocity so that the equation of motion is

$$m\frac{dv}{dt} = mg - kv^2.$$

(Note that this equation is not linear, but it is separable.)

**13.** A savings account has an initial balance of $900 and earns interest at a nominal annual rate of 5% paid continuously into the account. The account is also being continuously depleted by service charges at a rate of $50/year. Write a DE satisfied by the balance $y(t)$ in the account after $t$ years. Find the balance in the account after five years.

## 17.5  Existence, Uniqueness, and Numerical Methods

A general first-order differential equation of the form

$$\frac{dy}{dx} = f(x, y)$$

specifies a slope $f(x, y)$ at every point $(x, y)$ in the domain of $f$, and therefore represents a **slope field**. Such a slope field can be represented graphically by drawing short line segments of the indicated slope at many points in the $xy$-plane. Slope fields resemble vector fields, but the segments are usually drawn having the same length and without arrowheads. Figure 17.2 portrays the slope field for the differential equation

$$\frac{dy}{dx} = x - y.$$

Solving a typical initial-value problem

$$\begin{cases} \dfrac{dy}{dx} = f(x, y) \\ y(x_0) = y_0 \end{cases}$$

involves finding a function $y = \phi(x)$ such that

$$\phi'(x) = f\big(x, \phi(x)\big) \qquad \text{and} \qquad \phi(x_0) = y_0.$$

The graph of the equation $y = \phi(x)$ is a curve passing through $(x_0, y_0)$ that is tangent to the slope-field at each point. Such curves are called **solution curves** of the differential equation. Figure 17.2 shows four solution curves for $y' = x - y$ corresponding to the initial conditions $y(0) = C$, where $C = -2, -1, 0$, and $1$.

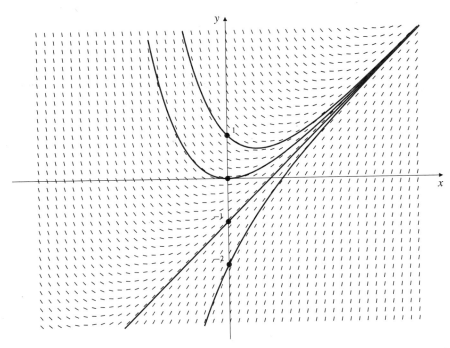

**Figure 17.2**    The slope field for the DE $y' = x - y$ and four solution curves for this DE

The DE $y' = x - y$ is linear, and can be solved explicitly by the method of Section 17.4. Indeed, the solution satisfying $y(0) = C$ is $y = x - 1 + (C + 1)e^{-x}$. Most differential equations of the form $y' = f(x, y)$ cannot be solved for $y$ as an explicit function of $x$, so we must use numerical approximation methods to find the value of a solution function $\phi(x)$ at particular points.

## Existence and Uniqueness of Solutions

Even if we cannot calculate an explicit solution of an initial-value problem, it is important to know when the problem has a solution, and whether that solution is unique.

**THEOREM  3**

**An existence and uniqueness theorem for first-order initial-value problems**

Suppose that $f(x, y)$ and $f_2(x, y) = (\partial/\partial y)f(x, y)$ are continuous on a rectangle $R$ of the form $a \le x \le b, c \le y \le d$, containing the point $(x_0, y_0)$ in its interior. Then there exists a number $\delta > 0$ and a *unique* function $\phi(x)$ defined and having a continuous derivative on the interval $(x_0 - \delta, x_0 + \delta)$ such that $\phi(x_0) = y_0$ and $\phi'(x) = f\big(x, \phi(x)\big)$ for $x_0 - \delta < x < x_0 + \delta$. In other words, the initial-value problem

$$\begin{cases} \dfrac{dy}{dx} = f(x, y) \\ y(x_0) = y_0 \end{cases} \qquad (*)$$

has a unique solution on $(x_0 - \delta, x_0 + \delta)$.

We give only an outline of the proof here. Any solution $y = \phi(x)$ of the initial-value problem $(*)$ must also satisfy the **integral equation**

$$\phi(x) = y_0 + \int_{x_0}^{x} f\big(t, \phi(t)\big)\, dt, \qquad (**)$$

and, conversely, any solution of the integral equation (**) must also satisfy the initial-value problem (*). A sequence of approximations $\phi_n(x)$ to a solution of (**) can be constructed as follows:

$$\phi_0(x) = y_0$$

$$\phi_{n+1}(x) = y_0 + \int_{x_0}^{x} f\big(t, \phi_n(t)\big)\, dt \qquad \text{for} \quad n = 0, 1, 2, \ldots$$

(These are called **Picard iterations.**) The proof of Theorem 3 involves showing that

$$\lim_{n \to \infty} \phi_n(x) = \phi(x)$$

exists on an interval $(x_0 - \delta, x_0 + \delta)$, and that the resulting limit $\phi(x)$ satisfies the integral equation (**). The details can be found in more advanced texts on differential equations and analysis.

**Remark** Some initial-value problems can have non-unique solutions. For example, the functions $y_1(x) = x^3$ and $y_2(x) = 0$ both satisfy the initial-value problem

$$\begin{cases} \dfrac{dy}{dx} = 3y^{2/3} \\ y(0) = 0. \end{cases}$$

In this case $f(x, y) = 3y^{2/3}$ is continuous on the whole $xy$-plane. However $\partial f / \partial y = 2y^{-1/3}$ is not continuous on the $x$-axis, and is therefore not continuous on any rectangle containing $(0, 0)$ in its interior. The conditions of Theorem 3 are not satisfied and the initial-value problem has a solution but not a unique one.

**Remark** The unique solution $y = \phi(x)$ to the initial-value problem (*) guaranteed by Theorem 3 may not be defined on the whole interval $[a, b]$, because it can "escape" from the rectangle $R$ through the top or bottom edges. Even if $f(x, y)$ and $(\partial / \partial y) f(x, y)$ are continuous on the whole $xy$-plane, the solution may not be defined on the whole real line. For example

$$y = \frac{1}{1 - x} \qquad \text{satisfies the initial-value problem} \qquad \begin{cases} \dfrac{dy}{dx} = y^2 \\ y(0) = 1 \end{cases}$$

but only for $x < 1$. Starting from $(0, 1)$, we can follow the solution curve as far as we want to the left of $x = 0$, but to the right of $x = 0$ the curve recedes to $\infty$ as $x \to 1-$. It makes no sense to regard the part of the curve to the right of $x = 1$ as part of the solution curve to the initial-value problem.

## Numerical Methods

Suppose that the conditions of Theorem 3 are satisfied, so we know that the initial-value problem

$$\begin{cases} \dfrac{dy}{dx} = f(x, y) \\ y(x_0) = y_0 \end{cases}$$

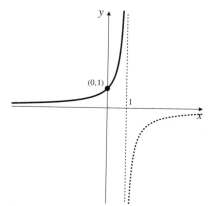

**Figure 17.3** The solution to $y' = y^2$, $y(0) = 1$ is the part of the curve $y = 1/(1 - x)$ to the left of the vertical asymptote at $x = 1$

has a unique solution $y = \phi(x)$ on some interval containing $x_0$. Even if we cannot solve the differential equation and find $\phi(x)$ explicitly, we can still try to find approximate values $y_n$ for $\phi(x_n)$ at a sequence of points

$$x_0, \quad x_1 = x_0 + h, \quad x_2 = x_0 + 2h, \quad x_3 = x_0 + 3h, \quad \cdots$$

starting at $x_0$. Here $h > 0$ (or $h < 0$) is called the **step size** of the approximation scheme. In the remainder of this section we will describe three methods for constructing the approximations $\{y_n\}$, namely

1. The Euler method,
2. The improved Euler method, and
3. The fourth-order Runge–Kutta method.

Each of these methods starts with the given value of $y_0$ and provides a formula for constructing $y_{n+1}$ when you know $y_n$. The three methods are listed above in increasing order of the complexity of their formulas, but the more complicated formulas produce much better approximations for any given step size $h$.

**The Euler method** involves approximating the solution curve $y = \phi(x)$ by a polygonal line (a sequence of straight line segments joined end to end), where each segment has horizontal length $h$, and has slope determined by the value of $f(x, y)$ at the end of the previous segment. Thus, if $x_n = x_0 + nh$, then

$$y_1 = y_0 + f(x_0, y_0)h$$
$$y_2 = y_1 + f(x_1, y_1)h$$
$$y_3 = y_2 + f(x_2, y_2)h$$

and, in general,

**Iteration formulas for Euler's method**

$$x_{n+1} = x_n + h, \qquad y_{n+1} = y_n + hf(x_n, y_n).$$

**Example 1**  Use Euler's method to find approximate values for the solution of the initial-value problem

$$\begin{cases} \dfrac{dy}{dx} = x - y \\ y(0) = 1 \end{cases}$$

on the interval $[0, 1]$ using

(a) five steps of size $h = 0.2$, and

(b) ten steps of size $h = 0.1$.

Calculate the error at each step, given that the problem (which involves a linear equation and so can be solved explicitly) has solution $y = \phi(x) = x - 1 + 2e^{-x}$.

**Solution**

(a) Here we have $f(x, y) = x - y$, $x_0 = 0$, $y_0 = 1$, and $h = 0.2$, so that

$$x_n = \frac{n}{5}, \qquad y_{n+1} = y_n + 0.2(x_n - y_n),$$

and the error is $e_n = \phi(x_n) - y_n$ for $n = 0, 1, 2, 3, 4$, and 5. The results of the calculation, which was done easily using a computer spreadsheet program, are presented in Table 1.

**Table 1.** Euler approximations with $h = 0.2$

| $n$ | $x_n$ | $y_n$ | $f(x_n, y_n)$ | $y_{n+1}$ | $e_n = \phi(x_n) - y_n$ |
|---|---|---|---|---|---|
| 0 | 0.0 | 1.000000 | −1.000000 | 0.800000 | 0.000000 |
| 1 | 0.2 | 0.800000 | −0.600000 | 0.680000 | 0.037462 |
| 2 | 0.4 | 0.680000 | −0.280000 | 0.624000 | 0.060640 |
| 3 | 0.6 | 0.624000 | −0.024000 | 0.619200 | 0.073623 |
| 4 | 0.8 | 0.619200 | 0.180800 | 0.655360 | 0.079458 |
| 5 | 1.0 | 0.655360 | 0.344640 | | 0.080399 |

The exact solution $y = \phi(x)$ and the polygonal line representing the Euler approximation are shown in Figure 17.4. The approximation lies below the solution curve, as is reflected in the positive values in the last column of Table 1, representing the error at each step.

(b) Here we have $h = 0.1$, so that

$$x_n = \frac{n}{10}, \qquad y_{n+1} = y_n + 0.1(x_n - y_n)$$

for $n = 0, 1, \ldots, 10$. Again we present the results in tabular form:

**Table 2.** Euler approximations with $h = 0.1$

| $n$ | $x_n$ | $y_n$ | $f(x_n, y_n)$ | $y_{n+1}$ | $e_n = \phi(x_n) - y_n$ |
|---|---|---|---|---|---|
| 0 | 0.0 | 1.000000 | −1.000000 | 0.900000 | 0.000000 |
| 1 | 0.1 | 0.900000 | −0.800000 | 0.820000 | 0.009675 |
| 2 | 0.2 | 0.820000 | −0.620000 | 0.758000 | 0.017462 |
| 3 | 0.3 | 0.758000 | −0.458000 | 0.712200 | 0.023636 |
| 4 | 0.4 | 0.712200 | −0.312200 | 0.680980 | 0.028440 |
| 5 | 0.5 | 0.680980 | −0.180980 | 0.662882 | 0.032081 |
| 6 | 0.6 | 0.662882 | −0.062882 | 0.656594 | 0.034741 |
| 7 | 0.7 | 0.656594 | 0.043406 | 0.660934 | 0.036577 |
| 8 | 0.8 | 0.660934 | 0.139066 | 0.674841 | 0.037724 |
| 9 | 0.9 | 0.674841 | 0.225159 | 0.697357 | 0.038298 |
| 10 | 1.0 | 0.697357 | 0.302643 | | 0.038402 |

Observe that the error at the end of the first step is about one-quarter of the error at the end of the first step in part (a), but the final error at $x = 1$ is only about half as large as in part (a). This behaviour is characteristic of Euler's Method.

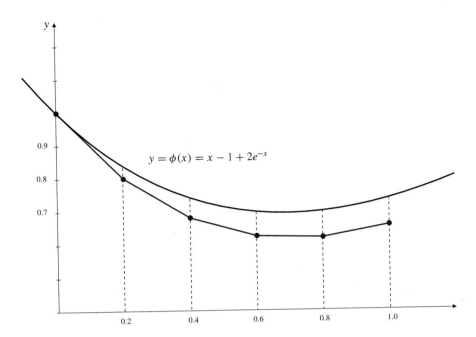

**Figure 17.4**   The solution $y = \phi(x)$ to $y' = x - y$, $y(0) = 1$, and an Euler approximation to it on $[0, 1]$ with step size $h = 0.2$

If we decrease the step size $h$ it takes more steps ($n = |x - x_0|/h$) to get from the starting point $x_0$ to a particular value $x$ where we want to know the value of the solution. For Euler's method it can be shown that the error at each step decreases on average proportionally to $h^2$, but the errors can accumulate from step to step, so the error at $x$ can be expected to decrease proportionally to $nh^2 = |x - x_0|h$. This is consistent with the results of Example 1. Decreasing $h$ and so increasing $n$ is costly in terms of computing resources, so we would like to find ways of reducing the error without increasing the step size. This is similar to developing better techniques than the Trapezoid Rule for evaluating definite integrals numerically.

**The improved Euler method** is a step in this direction. The accuracy of the Euler method is hampered by the fact that the slope of each segment in the approximating polygonal line is determined by the value of $f(x, y)$ at one endpoint of the segment. Since $f$ varies along the segment, we would expect to do better by using, say, the average value of $f(x, y)$ at the two ends of the segment, that is, by calculating $y_{n+1}$ from the formula

$$y_{n+1} = y_n + h\frac{f(x_n, y_n) + f(x_{n+1}, y_{n+1})}{2}.$$

Unfortunately, $y_{n+1}$ appears on both sides of this equation, and we can't usually solve the equation for $y_{n+1}$. We can get around this difficulty by replacing $y_{n+1}$ on the right side by its Euler approximation $y_n + hf(x_n, y_n)$. The resulting formula is the basis for the improved Euler method.

**Iteration formulas for the improved Euler method**

$$x_{n+1} = x_n + h$$
$$u_{n+1} = y_n + h\,f(x_n, y_n)$$
$$y_{n+1} = y_n + h\frac{f(x_n, y_n) + f(x_{n+1}, u_{n+1})}{2}.$$

**Example 2**   Use the improved Euler method with $h = 0.2$ to find approximate values for the solution to the initial-value problem of Example 1 on $[0, 1]$. Compare the errors with those obtained by the Euler method.

**Solution**   Table 3 summarizes the calculation of five steps of the improved Euler method for $f(x, y) = x - y$, $x_0 = 0$, and $y_0 = 1$.

**Table 3.**   Improved Euler approximations with $h = 0.2$

| $n$ | $x_n$ | $y_n$ | $u_{n+1}$ | $y_{n+1}$ | $e_n = \phi(x_n) - y_n$ |
|---|---|---|---|---|---|
| 0 | 0.0 | 1.000000 | 0.800000 | 0.840000 | 0.000000 |
| 1 | 0.2 | 0.840000 | 0.712000 | 0.744800 | $-0.002538$ |
| 2 | 0.4 | 0.744800 | 0.675840 | 0.702736 | $-0.004160$ |
| 3 | 0.6 | 0.702736 | 0.682189 | 0.704244 | $-0.005113$ |
| 4 | 0.8 | 0.704244 | 0.723395 | 0.741480 | $-0.005586$ |
| 5 | 1.0 | 0.741480 | 0.793184 | | $-0.005721$ |

Observe that the errors are considerably less than one-tenth those obtained in Example 1(a). Of course, more calculations are necessary at each step, but the number of evaluations of $f(x, y)$ required is only twice the number required for Example 1(a). As for numerical integration, if $f$ is complicated, it is these function evaluations that constitute most of the computational "cost" of computing numerical solutions.

**Remark**   It can be shown for well-behaved functions $f$ that the error at each step in the improved Euler method is bounded by a multiple of $h^3$, rather than $h^2$ as for the (unimproved) Euler method. Thus the cumulative error at $x$ can be bounded by a constant times $|x - x_0|h^2$. If Example 2 is repeated with ten steps of size $h = 0.1$, the error at $n = 10$ (that is, at $x = 1$) is $-0.001323$, which is about one-fourth the size of the error at $x = 1$ with $h = 0.2$.

   **The fourth-order Runge–Kutta method** further improves upon the improved Euler method, but at the expense of requiring more complicated calculations at each step. It requires four evaluations of $f(x, y)$ at each step, but the error at each step is less than a constant times $h^5$, so the cumulative error decreases like $h^4$ as $h$ decreases. Like the improved Euler method, this method involves calculating a certain kind of average slope for each segment in the polygonal approximation to the solution to the initial-value problem. We present the appropriate formulas below, but cannot derive them here.

**Iteration formulas for the Runge–Kutta method**

$$x_{n+1} = x_n + h$$
$$p_n = f(x_n, y_n)$$
$$q_n = f\left(x_n + \frac{h}{2}, y_n + \frac{h}{2}p_n\right)$$
$$r_n = f\left(x_n + \frac{h}{2}, y_n + \frac{h}{2}q_n\right)$$
$$s_n = f(x_n + h, y_n + hr_n)$$
$$y_{n+1} = y_n + h\frac{p_n + 2q_n + 2r_n + s_n}{6}.$$

**Example 3** Use the fourth-order Runge-Kutta method with $h = 0.2$ to find approximate values for the solution to the initial-value problem of Example 1 on $[0, 1]$. Compare the errors with those obtained by the Euler and improved Euler methods.

**Solution** Table 4 summarizes the calculation of five steps of the Runge–Kutta method for $f(x, y) = x - y$, $x_0 = 0$, and $y_0 = 1$ according to the formulas above. The table does not show the values of the intermediate quantities $p_n$, $q_n$, $r_n$, and $s_n$, but columns for these quantities were included in the spreadsheet in which the calculations were made.

**Table 4.** Fourth-order Runge–Kutta approximations with $h = 0.2$

| $n$ | $x_n$ | $y_n$ | $e_n = \phi(x_n) - y_n$ |
|---|---|---|---|
| 0 | 0.0 | 1.000000 | 0.0000000 |
| 1 | 0.2 | 0.837467 | −0.0000052 |
| 2 | 0.4 | 0.740649 | −0.0000085 |
| 3 | 0.6 | 0.697634 | −0.0000104 |
| 4 | 0.8 | 0.698669 | −0.0000113 |
| 5 | 1.0 | 0.735770 | −0.0000116 |

The errors here are about 1/500 of the size of the errors obtained with the improved Euler method, and about 1/7,000 of the size of the errors obtained with the Euler method. This great improvement was achieved at the expense of doubling the number of function evaluations required in the improved Euler method and quadrupling the number required in the Euler method. If we use 10 steps of size $h = 0.1$ in the Runge–Kutta method, the error at $x = 1$ is reduced to $-6.66482 \times 10^{-7}$, which is less than 1/16 of its value when $h = 0.2$.

Our final example shows what can happen with numerical approximations to a solution that is unbounded.

**Example 4** Obtain solutions at $x = 0.4$, $x = 0.8$, and $x = 1.0$ for solutions to the initial-value problem

$$\begin{cases} y' = y^2 \\ y(0) = 1 \end{cases}$$

using all three methods described above, and using step sizes $h = 0.2$, $h = 0.1$, and $h = 0.05$ for each method. What do the results suggest about the values of the solution at these points? Compare the results with the actual solution $y = 1/(1-x)$.

**Solution** The various approximations are calculated using the various formulas described above for $f(x, y) = y^2$, $x_0 = 0$, and $y_0 = 1$. The results are presented in Table 5.

**Table 5.**   Comparing methods and step sizes for $y' = y^2$, $y(0) = 1$

|  | $h = 0.2$ | $h = 0.1$ | $h = 0.05$ |
|---|---|---|---|
| **Euler** | | | |
| $x = 0.4$ | 1.488000 | 1.557797 | 1.605224 |
| $x = 0.8$ | 2.676449 | 3.239652 | 3.793197 |
| $x = 1.0$ | 4.109124 | 6.128898 | 9.552668 |
| **Improved Euler** | | | |
| $x = 0.4$ | 1.640092 | 1.658736 | 1.664515 |
| $x = 0.8$ | 4.190396 | 4.677726 | 4.897519 |
| $x = 1.0$ | 11.878846 | 22.290765 | 43.114668 |
| **Runge–Kutta** | | | |
| $x = 0.4$ | 1.666473 | 1.666653 | 1.666666 |
| $x = 0.8$ | 4.965008 | 4.996663 | 4.999751 |
| $x = 1.0$ | 41.016258 | 81.996399 | 163.983395 |

Little useful information can be read from the Euler results. The improved Euler results suggest that the solution exists at $x = 0.4$ and $x = 0.8$, but likely not at $x = 1$. The Runge–Kutta results confirm this, and suggest that $y(0.4) = 5/3$ and $y(0.8) = 5$, which are the correct values provided by the actual solution $y = 1/(1-x)$. They also suggest very strongly that the solution "blows up" at (or near) $x = 1$. ∎

## Exercises 17.5

A computer is almost essential for doing most of these exercises. The calculations are easily done with a spreadsheet program in which formulas for calculating the various quantities involved can be replicated down columns to automate the iteration process.

1. Use the Euler method with step sizes (a) $h = 0.2$, (b) $h = 0.1$, and (c) $h = 0.05$ to approximate $y(2)$ given that $y' = x + y$ and $y(1) = 0$.

2. Repeat Exercise 1 using the improved Euler method.

3. Repeat Exercise 1 using the Runge–Kutta method.

4. Use the Euler method with step sizes (a) $h = 0.2$, and (b) $h = 0.1$ to approximate $y(2)$ given that $y' = xe^{-y}$ and $y(0) = 0$.

5. Repeat Exercise 4 using the improved Euler method.

6. Repeat Exercise 4 using the Runge–Kutta method.

7. Use the Euler method with (a) $h = 0.2$, (b) $h = 0.1$, and (c) $h = 0.05$ to approximate $y(1)$ given that $y' = \cos y$ and $y(0) = 0$.

8. Repeat Exercise 7 using the improved Euler method.

9. Repeat Exercise 7 using the Runge–Kutta method.

10. Use the Euler method with (a) $h = 0.2$, (b) $h = 0.1$, and (c) $h = 0.05$ to approximate $y(1)$ given that $y' = \cos(x^2)$ and $y(0) = 0$.

11. Repeat Exercise 10 using the improved Euler method.

12. Repeat Exercise 10 using the Runge–Kutta method.

Solve the integral equations in Exercises 13–14 by rephrasing them as initial-value problems.

13. $y(x) = 2 + \displaystyle\int_1^x \big(y(t)\big)^2 \, dt$. *Hint:* find $\dfrac{dy}{dx}$ and $y(1)$.

14. $u(x) = 1 + 3 \displaystyle\int_2^x t^2 u(t) \, dt$. *Hint:* find $\dfrac{du}{dx}$ and $u(2)$.

15. The methods of this section can be used to approximate definite integrals numerically. For example,

$$I = \int_a^b f(x) \, dx$$

is given by $I = y(b)$, where

$$y' = f(x), \qquad \text{and} \qquad y(a) = 0.$$

Show that one step of the Runge–Kutta method with $h = b - a$ gives the same result for $I$ as does Simpson's Rule (Section 6.7) with two subintervals of length $h/2$.

16. If $\phi(0) = A \ge 0$ and $\phi'(x) \ge k\phi(x)$ on $[0, X]$, where $k > 0$ and $X > 0$ are constants, show that $\phi(x) \ge Ae^{kx}$ on $[0, X]$. *Hint:* calculate $(d/dx)(\phi(x)/e^{kx})$.

**✳ 17.** Consider the three initial-value problems

(A): $\quad u' = u^2 \qquad\qquad u(0) = 1$

(B): $\quad y' = x + y^2 \qquad\; y(0) = 1$

(C): $\quad v' = 1 + v^2 \qquad\; v(0) = 1$

(a) Show that the solution of (B) remains between the solutions of (A) and (C) on any interval $[0, X]$ where solutions of all three problems exist. *Hint:* we must

have $u(x) \geq 1$, $y(x) \geq 1$, and $v(x) \geq 1$ on $[0, X]$. (Why?) Apply the result of Exercise 16 to $\phi = y - u$ and to $\phi = v - y$.

(b) Find explicit solutions for problems (A) and (C). What can you conclude about the solution to problem (B).

(c) Use the Runge–Kutta method with $h = 0.05$, $h = 0.02$, and $h = 0.01$ to approximate the solution to (B) on $[0, 1]$. What can you conclude now?

# 17.6 Differential Equations of Second Order

The general second-order ordinary differential equation is of the form

$$F\left(\frac{d^2y}{dx^2}, \frac{dy}{dx}, y, x\right) = 0$$

for some function $F$ of four variables. When such an equation can be solved explicitly for $y$ as a function of $x$, the solution typically involves two integrations and therefore two arbitrary constants. A unique solution usually results from prescribing the values of the solution $y$ and its first derivative $y' = dy/dx$ at a particular point. Such a prescription constitutes an **initial-value problem** for the second-order equation.

## Equations Reducible to First Order

A second-order equation of the form

$$F\left(\frac{d^2y}{dx^2}, \frac{dy}{dx}, x\right) = 0$$

that does not involve the unknown function $y$ explicitly (except through its derivatives) can be reduced to a first order equation by a change of dependent variable; if $v = dy/dx$, then the equation can be written

$$F\left(\frac{dv}{dx}, v, x\right) = 0.$$

This first-order equation in $v$ may be amenable to the techniques described in earlier sections. If an explicit solution $v = v(x)$ can be found and integrated, then the function

$$y = \int v(x)\, dx$$

is an explicit solution of the given equation.

**Example 1**   Solve the initial-value problem

$$\frac{d^2y}{dx^2} = x\left(\frac{dy}{dx}\right)^2, \qquad y(0) = 1, \qquad y'(0) = -2.$$

**Solution**   If we let $v = dy/dx$, the given differential equation becomes

$$\frac{dv}{dx} = xv^2,$$

which is a separable first-order equation. Thus

$$\frac{dv}{v^2} = x\,dx$$

$$-\frac{1}{v} = \frac{x^2}{2} + \frac{C_1}{2}$$

$$v = -\frac{2}{x^2 + C_1}.$$

The initial condition $y'(0) = -2$ implies that $v(0) = -2$ and so $C_1 = 1$. Therefore

$$y = -2\int \frac{dx}{x^2 + 1} = -2\tan^{-1}x + C_2.$$

The initial condition $y(0) = 1$ implies that $C_2 = 1$ so the solution of the given initial-value problem is $y = 1 - 2\tan^{-1}x$.   ∎

A second-order equation of the form

$$F\left(\frac{d^2y}{dx^2}, \frac{dy}{dx}, y\right) = 0$$

that does not explicitly involve the independent variable $x$ can be reduced to a first-order equation by a change of both dependent and independent variables. Again let $v = dy/dx$, but regard $v$ as a function of $y$ rather than $x$; $v = v(y)$. Then

$$\frac{d^2y}{dx^2} = \frac{dv}{dx} = \frac{dv}{dy}\frac{dy}{dx} = v\frac{dv}{dy}$$

by the Chain Rule. Hence the given differential equation becomes

$$F\left(v\frac{dv}{dy}, v, y\right) = 0$$

which is a first-order equation for $v$ as a function of $y$. If this equation can be solved for $v = v(y)$ there still remains the problem of solving the separable equation $(dy/dx) = v(y)$ for $y$ as a function of $x$.

**Example 2** Solve the equation $y\dfrac{d^2y}{dx^2} = \left(\dfrac{dy}{dx}\right)^2$.

**Solution** The change of variable $dy/dx = v(y)$ leads to the equation

$$yv\frac{dv}{dy} = v^2,$$

which is separable, $dv/v = dy/y$, and has solution $v = C_1 y$. The equation

$$\frac{dy}{dx} = C_1 y$$

is again separable, and leads to

$$\frac{dy}{y} = C_1\, dx$$

$$\ln|y| = C_1 x + C_2$$

$$y = \pm e^{C_1 x + C_2} = C_3 e^{C_1 x}.$$

## Second-Order Linear Equations

The most frequently encountered ordinary differential equations arising in applications are second-order linear equations. The general second-order linear equation is of the form

$$a_2(x)\frac{d^2y}{dx^2} + a_1(x)\frac{dy}{dx} + a_0(x)y = f(x).$$

As remarked in Section 17.1, if $f(x) = 0$ identically, then we say that the equation is **homogeneous**. If the coefficients $a_2(x)$, $a_1(x)$, and $a_0(x)$ are continuous on an interval and $a_2(x) \neq 0$ there, then the homogeneous equation

$$a_2(x)\frac{d^2y}{dx^2} + a_1(x)\frac{dy}{dx} + a_0(x)y = 0$$

has a general solution of the form

$$y_h = C_1 y_1(x) + C_2 y_2(x)$$

where $y_1(x)$ and $y_2(x)$ are two **independent** solutions, that is, two solutions with the property that $C_1 y_1(x) + C_2 y_2(x) = 0$ for all $x$ in the interval only if $C_1 = C_2 = 0$. (We will not prove this here.)

Whenever one solution, $y_1(x)$, of a homogeneous linear second-order equation is known, another independent solution (and therefore the general solution) can be found by substituting $y = v(x)y_1(x)$ into the differential equation. This leads to a first-order, linear, separable equation for $v'$.

**Example 3**   Show that $y_1 = e^{-2x}$ is a solution of $y'' + 4y' + 4y = 0$, and find the general solution of this equation.

**Solution**   Since $y_1' = -2e^{-2x}$ and $y_1'' = 4e^{-2x}$, we have

$$y_1'' + 4y_1' + 4y_1 = e^{-2x}(4 - 8 + 4) = 0,$$

so $y_1$ is indeed a solution of the given differential equation. To find the general solution, try $y = y_1 v = e^{-2x} v(x)$. We have

$$y' = -2e^{-2x}v + e^{-2x}v'$$
$$y'' = 4e^{-2x}v - 4e^{-2x}v' + e^{-2x}v''.$$

Substituting these expressions into the given DE we obtain

$$0 = y'' + 4y' + 4y$$
$$= e^{-2x}(4v - 4v' + v'' - 8v + 4v' + 4v) = e^{-2x}v''.$$

Thus $y = y_1 v$ is a solution provided $v''(x) = 0$. This equation for $v$ has the general solution $v = C_1 + C_2 x$, so the given equation has the general solution

$$y = C_1 e^{-2x} + C_2 x e^{-2x} = C_1 y_1(x) + C_2 y_2(x),$$

where $y_2 = x e^{-2x}$ is a second solution of the DE, independent of $y_1$.

By Theorem 2 of Section 17.1, the general solution of the second-order, linear, nonhomogeneous equation (with $f(x) \neq 0$) is of the form

$$y = y_p(x) + y_h(x)$$

where $y_p(x)$ is any particular solution of the nonhomogeneous equation and $y_h(x)$ is the general solution (as described above) of the corresponding homogeneous equation. In Section 17.8 we will discuss the solution of nonhomogeneous linear equations. First, however, in Section 17.7 we concentrate on some special classes of homogeneous, linear equations.

# Exercises 17.6

**1.** Show that $y = e^x$ is a solution of $y'' - 3y' + 2y = 0$, and find the general solution of this DE.

**2.** Show that $y = e^{-2x}$ is a solution of $y'' - y' - 6y = 0$, and find the general solution of this DE.

**3.** Show that $y = x$ is a solution of $x^2 y'' + 2xy' - 2y = 0$ on the interval $(0, \infty)$, and find the general solution on this interval.

**4.** Show that $y = x^2$ is a solution of $x^2 y'' - 3xy' + 4y = 0$ on the interval $(0, \infty)$, and find the general solution on this interval.

**5.** Show that $y = x$ is a solution of the differential equation

$x^2 y'' - (2x + x^2)y' + (2 + x)y = 0$, and find the general solution of this equation.

**6.** Show that $y = x^{-1/2} \cos x$ is a solution of the Bessel equation with $v = 1/2$:

$$x^2 y'' + xy' + \left(x^2 - \frac{1}{4}\right) y = 0.$$

Find the general solution of this equation.

**First-order systems**

**7.** A system of $n$ first order, linear, differential equations in $n$ unknown functions $y_1$, $y_2$, $\cdots$, $y_n$ is written

$$y_1' = a_{11}(x)y_1 + a_{12}(x)y_2 + \cdots + a_{1n}(x)y_n + f_1(x)$$
$$y_2' = a_{21}(x)y_1 + a_{22}(x)y_2 + \cdots + a_{2n}(x)y_n + f_2(x)$$
$$\vdots$$
$$y_n' = a_{n1}(x)y_1 + a_{n2}(x)y_2 + \cdots + a_{nn}(x)y_n + f_n(x).$$

Such a system is called an $n \times n$ **first-order linear system** and can be rewritten in vector-matrix form as $\mathbf{y}' = \mathcal{A}(x)\mathbf{y} + \mathbf{f}(x)$, where

$$\mathbf{y}(x) = \begin{pmatrix} y_1(x) \\ \vdots \\ y_n(x) \end{pmatrix}, \quad \mathbf{f}(x) = \begin{pmatrix} f_1(x) \\ \vdots \\ f_n(x) \end{pmatrix},$$

$$\mathcal{A}(x) = \begin{pmatrix} a_{11}(x) & \cdots & a_{1n}(x) \\ \vdots & \ddots & \vdots \\ a_{n1}(x) & \cdots & a_{nn}(x) \end{pmatrix}.$$

Show that the second-order, linear equation $y'' + a_1(x)y' + a_0(x)y = f(x)$ can be transformed into a $2 \times 2$ first-order system with $y_1 = y$ and $y_2 = y'$ having

$$\mathcal{A}(x) = \begin{pmatrix} 0 & 1 \\ -a_0(x) & -a_1(x) \end{pmatrix}, \quad \mathbf{f}(x) = \begin{pmatrix} 0 \\ f(x) \end{pmatrix}.$$

**8.** Generalize the previous exercise to transform an $n$th-order linear equation

$$y^{(n)} + a_{n-1}(x)y^{(n-1)} + a_{n-2}(x)y^{(n-2)} + \cdots + a_0(x)y = f(x)$$

into an $n \times n$ first-order system.

**9.** If $\mathcal{A}$ is an $n \times n$ constant matrix, and if there exists a scalar $\lambda$ and a nonzero constant vector $\mathbf{v}$ for which $\mathcal{A}\mathbf{v} = \lambda\mathbf{v}$ show that $\mathbf{y} = C_1 e^{\lambda x}\mathbf{v}$ is a solution of the homogeneous system $\mathbf{y}' = \mathcal{A}\mathbf{y}$.

**10.** Show that the determinant $\begin{vmatrix} 2-\lambda & 1 \\ 2 & 3-\lambda \end{vmatrix}$ is zero for two distinct values of $\lambda$. For each of these values find a nonzero vector $\mathbf{v}$ which satisfies the condition $\begin{pmatrix} 2 & 1 \\ 2 & 3 \end{pmatrix}\mathbf{v} = \lambda\mathbf{v}$. Hence solve the system

$$y_1' = 2y_1 + y_2, \qquad y_2' = 2y_1 + 3y_2.$$

## 17.7 Linear Differential Equations with Constant Coefficients

In this section we will solve second-order, linear, homogeneous differential equations of the form

$$a\,y'' + b\,y' + cy = 0, \qquad (*)$$

where $a$, $b$, and $c$ are constants and $a \neq 0$, as well as higher-order equations of this type. Some of the material in this section repeats Section 3.7.

Equations of type $(*)$ arise in many applications of mathematics. The equation of simple harmonic motion,

$$y'' + \omega^2 y = 0,$$

is of type $(*)$ (with $a = 1$, $b = 0$ and $c = \omega^2$) and governs the vibration of a mass suspended by an elastic spring. If the motion of the mass is impeded by a resistance proportional to its velocity (supplied, say, by friction with the air) then the equation of motion would look like $(*)$ with a positive constant $b$. The mass will still vibrate if the resistance is not too large ($b^2 < 4ac$) but the amplitude of the vibration will decay exponentially with time, as we will see shortly.

Equation $(*)$ also arises in the study of current flowing in an electric circuit having resistors, capacitors, and inductors connected in series. In this case $a$ measures the inductance, $b$ the resistance, and $c$ the reciprocal of the capacitance of the circuit. In both the mechanical (vibrating mass) and electrical applications the constants $a$, $b$, and $c$ are positive. Also, the independent variable in all these applications is time, so we will regard the function $y$ in $(*)$ as a function of $t$ rather than $x$. Thus a *prime* $(')$ denotes $d/dt$ rather than $d/dx$.

Let us try to find a solution of equation $(*)$ having the form $y = e^{rt}$. Substituting this expression into the equation, we obtain

$$ar^2e^{rt} + bre^{rt} + ce^{rt} = 0.$$

Since $e^{rt}$ does not vanish, $y = e^{rt}$ will be a solution of the differential equation $(*)$ if and only if $r$ satisfies the quadratic **auxiliary equation**

$$ar^2 + br + c = 0, \qquad (**)$$

which has roots given by the **quadratic formula:**

$$r = \frac{-b \pm \sqrt{b^2 - 4ac}}{2a} = -\frac{b}{2a} \pm \frac{\sqrt{D}}{2a}.$$

Here $D = b^2 - 4ac$ is the **discriminant** of the auxiliary equation $(**)$.

There are three cases to consider, depending on whether $D$ is positive, zero, or negative.

**CASE I**   Suppose $D = b^2 - 4ac > 0$. Then the auxiliary equation has two different real roots, $r_1$ and $r_2$, given by

$$r_1 = \frac{-b - \sqrt{D}}{2a}, \qquad r_2 = \frac{-b + \sqrt{D}}{2a}.$$

(Often these roots can be found easily by factoring the left side of the auxiliary equation.) In this case $y = y_1(t) = e^{r_1 t}$ and $y = y_2(t) = e^{r_2 t}$ are independent solutions of the differential equation $(*)$, and so

$$y = C_1 e^{r_1 t} + C_2 e^{r_2 t},$$

is also a solution for any choice of the constants $C_1$ and $C_2$ by Theorem 1 of Section 17.1. Since the differential equation is of second order and this solution involves two arbitrary constants we suspect it is the **general solution**, that is, that every solution of the differential equation can be written in this form. Exercise 18 at the end of this section outlines a way to prove this fact.

---

**Example 1**   Find the general solution of $y'' + y' - 2y = 0$.

**Solution**   The auxiliary equation is $r^2 + r - 2 = 0$, or $(r + 2)(r - 1) = 0$. The auxiliary roots are $r_1 = -2$ and $r_2 = 1$. Hence the general solution of the differential equation is

$$y = C_1 e^{-2t} + C_2 e^t.$$

---

In order to facilitate discussion of the other two cases, $D = 0$, and $D < 0$, let us make a change of dependent variable in the differential equation $(*)$ to transform it into a similar equation with no first-order derivative. This can be achieved by letting

$$y(t) = e^{kt}u(t), \qquad \text{where} \quad k = -\frac{b}{2a}.$$

Differentiating twice, using the Product Rule, we obtain

$$y'(t) = e^{kt}[u'(t) + ku(t)]$$
$$y''(t) = e^{kt}[u''(t) + 2ku'(t) + k^2u(t)].$$

Substituting these expressions into the differential equation (∗), we obtain

$$e^{kt}\left(au'' + (2ak + b)u' + (ak^2 + bk + c)u\right) = 0.$$

Since $2ak = -b$, the coefficient of $u'$ in this equation is 0 and that of $u$ is

$$ak^2 + bk + c = \frac{ab^2}{4a^2} - \frac{b^2}{2a} + c = \frac{4ac - b^2}{4a} = \frac{-D}{4a}.$$

Dividing by $a$ and $e^{kt}$, we are left with the **reduced equation**

$$u'' - \left(\frac{D}{4a^2}\right)u = 0.$$

**CASE II**  Suppose $D = b^2 - 4ac = 0$. Then the auxiliary equation has two equal roots $r_1 = r_2 = -b/(2a) = k$. Moreover, the reduced equation is $u'' = 0$, which has general solution $u = A + Bt$, where $A$ and $B$ are any constants. Thus, the general solution to the differential equation (∗) for the case $D = 0$ is

$$y = A\,e^{kt} + Bt\,e^{kt}, \qquad \text{where } k = -\frac{b}{2a}.$$

---

**Example 2**  Find the general solution of $y'' + 6y' + 9y = 0$.

**Solution**  The auxiliary equation is $r^2 + 6r + 9 = 0$, or $(r + 3)^2 = 0$, which has identical roots $r = -3$. According to the discussion above, the general solution of the differential equation is

$$y = C_1 e^{-3t} + C_2 t\, e^{-3t}.$$

---

**CASE III**  Suppose $D = b^2 - 4ac < 0$. Then the auxiliary equation has complex roots, and the reduced equation for $u$ is

$$u'' + \omega^2 u = 0, \qquad \text{where} \quad \omega = \frac{\sqrt{-D}}{2a}.$$

We recognize this as the equation of simple harmonic motion. It has the general solution

$$u = C_1 \cos(\omega t) + C_2 \sin(\omega t),$$

where $C_1$ and $C_2$ are arbitrary constants. Therefore the given differential equation has the general solution

$$y = C_1 e^{kt} \cos(\omega t) + C_2 e^{kt} \sin(\omega t),$$
$$\text{where} \quad k = -\frac{b}{2a} \text{ and } \omega = \frac{\sqrt{4ac - b^2}}{2a}.$$

**Remark**   In Case III, where $b^2 - 4ac < 0$, the auxiliary equation (**) has complex conjugate roots given by

$$r = \frac{-b \pm \sqrt{b^2 - 4ac}}{2a} = k \pm i\omega,$$

where $i$ is the imaginary unit ($i^2 = -1$). Thus the values of $k$ and $\omega$ needed to write the general solution of (*) can be read immediately from the solution of the auxiliary equation.

**Example 3**   Find the general solution of $y'' + 4y' + 13y = 0$.

**Solution**   The auxiliary equation is $r^2 + 4r + 13 = 0$, which has solutions

$$r = \frac{-4 \pm \sqrt{16 - 52}}{2} = \frac{-4 \pm \sqrt{-36}}{2} = -2 \pm 3i.$$

Thus $k = -2$ and $\omega = 3$. The general solution of the given differential equation is

$$y = C_1 e^{-2t} \cos(3t) + C_2 e^{-2t} \sin(3t).$$

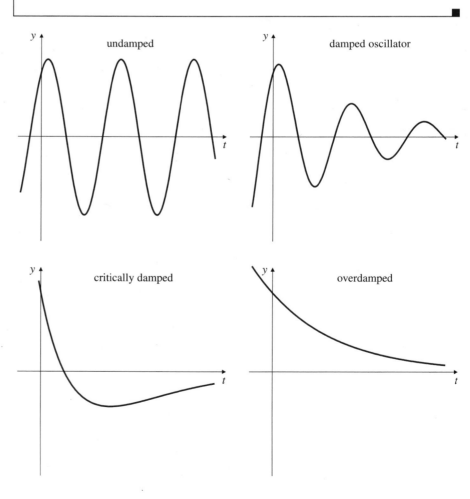

**Figure 17.5**
Undamped oscillator ($b = 0$)
Damped oscillator ($b > 0, b^2 < 4ac$)
Critically damped case ($b^2 = 4ac$)
Overdamped case ($b^2 > 4ac$)

**Remark** If $a$ and $c$ are positive, but $b = 0$, then equation (∗) is the differential equation of simple harmonic motion and has oscillatory solutions of fixed amplitude as shown in Section 3.7. If $b > 0$ but $b^2 < 4ac$ (Case III), the solutions still oscillate, but the amplitude diminishes exponentially as $t \to \infty$ because of the factor $e^{kt} = e^{-(b/2a)t}$. A system whose behaviour is modelled by such an equation is said to exhibit **damped harmonic motion**. If $b^2 = 4ac$ (Case II) the system is said to be **critically damped**, and if $b^2 > 4ac$ (Case I) it is **overdamped**. In these latter cases the behaviour is no longer oscillatory. (See Figure 17.5. Imagine a mass suspended by a spring in a jar of molasses.)

Here is a typical initial-value problem for a differential equation of type (∗).

**Example 4** Solve the initial-value problem $\begin{cases} y'' + 2y' + 2y = 0 \\ y(0) = 2 \\ y'(0) = -3. \end{cases}$

**Solution** The auxiliary equation for the DE $y'' + 2y' + 2y = 0$ is $r^2 + 2r + 2 = 0$, and has roots

$$r = \frac{-2 \pm \sqrt{4 - 8}}{2} = -1 \pm i.$$

Thus Case III applies: $k = -1$ and $\omega = 1$. The differential equation has the general solution

$$y = C_1 e^{-t} \cos t + C_2 e^{-t} \sin t.$$

Also

$$\begin{aligned} y' &= e^{-t}\left(-C_1 \cos t - C_2 \sin t - C_1 \sin t + C_2 \cos t\right) \\ &= (C_2 - C_1) e^{-t} \cos t - (C_1 + C_2) e^{-t} \sin t. \end{aligned}$$

Applying the initial conditions $y(0) = 2$ and $y'(0) = -3$, we obtain $C_1 = 2$ and $C_2 - C_1 = -3$. Hence $C_2 = -1$ and the initial-value problem has the solution

$$y = 2 e^{-t} \cos t - e^{-t} \sin t.$$

## Constant-Coefficient Equations of Higher Order

The techniques described above for solving second-order linear, homogeneous DEs with constant coefficients can also be applied to equations of higher order. We describe the procedure without offering any proofs. If

$$P_n(r) = a_n r^n + a_{n-1} r^{n-1} + \cdots + a_2 r^2 + a_1 r + a_0$$

is a polynomial of degree $n$ with constant coefficients $a_j$ $(0 \le j \le n)$ and $a_n \ne 0$, then the DE

$$P_n(D)y = 0, \qquad (∗)$$

where $D = d/dt$ (or $D = d/dx$) can be solved by substituting $y = e^{rx}$ and obtaining the *auxiliary equation* $P_n(r) = 0$. This polynomial equation has $n$ roots (see Appendix II) some of which may be equal, and some or all of which can be complex. If the coefficients of the polynomial $P_n(r)$ are all real then any complex roots must occur in complex conjugate pairs $a \pm ib$, where $a$ and $b$ are real.

The general solution of (∗) can be expressed as a *linear combination* of $n$ independent particular solutions

$$y = C_1 y_1(x) + C_2 y_2(x) + \cdots + C_n y_n(x),$$

where the $C_j$ are arbitrary constants. The independent solutions $y_1, y_2, \ldots, y_n$ are constructed as follows:

1. If $r_1$ is an $k$-fold root of the auxiliary equation (that is, if $(r - r_1)^k$ is a factor of $P_n(r)$), then

$$e^{r_1 t}, \quad te^{r_1 t}, \quad t^2 e^{r_1 t}, \quad \ldots, \quad t^{k-1} e^{r_1 t}$$

are $k$ independent solutions of (∗).

2. If $r = a + ib$ and $r = a - ib$ ($a$ and $b$ real) constitute a $k$-fold pair of complex conjugate roots of the auxiliary equation (that is, if $[(r - a)^2 + b^2]^k$ is a factor of $P_n(r)$), then

$$e^{at} \cos bt, \quad te^{at} \cos bt, \quad \ldots, \quad t^{k-1} e^{at} \cos bt,$$
$$e^{at} \sin bt, \quad te^{at} \sin bt, \quad \ldots, \quad t^{k-1} e^{at} \sin bt$$

are $2k$ independent solutions of (∗).

**Example 5**    Solve    (a) $y^{(4)} - 16y = 0$, and    (b) $y^{(5)} - 2y^{(4)} + y^{(3)} = 0$.

**Solution**    The auxiliary equation for (a) is $r^4 - 16 = 0$, which factors down to $(r - 2)(r + 2)(r^2 + 4) = 0$, and hence has roots $r = 2, -2, 2i,$ and $-2i$. Thus the DE (a) has general solution

$$y = C_1 e^{2t} + C_2 e^{-2t} + C_3 \cos(2t) + C_4 \sin(2t)$$

for arbitrary constants $C_1, C_2, C_3,$ and $C_4$.

The auxiliary equation for (b) is $r^5 - 2r^4 + r^3$, which factors to $r^3 (r - 1)^2 = 0$, and so has roots $r = 0, 0, 0, 1, 1$. The general solution of the $DE$ (b) is

$$y = C_1 + C_2 t + C_3 t^2 + C_4 e^t + C_5 te^t,$$

where $C_1, \ldots, C_5$ are arbitrary constants.

**Example 6** What is the order and the general solution of the constant-coefficient, linear, homogeneous DE whose auxiliary equation is

$$(r+4)^3(r^2+4r+13)^2 = 0?$$

**Solution** The auxiliary equation has degree 7 so the DE is of seventh order. Since $r^2+4r+13 = (r+2)^2+9$, which has roots $-2\pm3i$, the DE must have the general solution

$$y = C_1e^{-4t} + C_2te^{-4t} + C_3t^2e^{-4t}$$
$$+ C_4e^{-2t}\cos(3t) + C_5e^{-2t}\sin(3t) + C_6te^{-2t}\cos(3t) + C_7te^{-2t}\sin(3t).$$

## Euler (Equidimensional) Equations

A homogeneous, linear equation of the form

$$ax^2\frac{d^2y}{dx^2} + bx\frac{dy}{dx} + cy = 0$$

is called an **Euler equation** or an **equidimensional equation**, the latter term being appropriate since all the terms in the equation have the same dimension (that is, they are measured in the same units), provided that the constants $a$, $b$, and $c$ all have the same dimension. The coefficients of an Euler equation are *not constant*, but the technique for solving these equations is similar to that for solving equations with constant coefficients, so we include these equations in this section. As in the case of constant coefficient equations, we assume that the constants $a$, $b$, and $c$ are real numbers, and that $a \neq 0$. Even so, the leading coefficient, $ax^2$, does vanish at $x = 0$ (which is called a **singular point** of the equation), and this can cause solutions to fail to be defined at $x = 0$. We will solve the equation in the interval $x > 0$; the same solution will also hold for $x < 0$ provided we replace $x$ by $|x|$ in the solution.

Let us search for solutions in $x > 0$ given by powers of $x$; if

$$y = x^r, \qquad \frac{dy}{dx} = rx^{r-1}, \qquad \frac{d^2y}{dx^2} = r(r-1)x^{r-2},$$

then the Euler equation becomes

$$\big(ar(r-1) + br + c\big)x^r = 0.$$

This will be satisfied for all $x > 0$ provided that $r$ satisfies the **auxiliary equation**

$$ar^2 + (b-a)r + c = 0.$$

As for constant coefficient equations, there are three possibilities.

If $(b-a)^2 \geq 4ac$ then the auxiliary equation has two real roots

$$r_1 = \frac{a-b+\sqrt{(b-a)^2 - 4ac}}{2a},$$
$$r_2 = \frac{a-b-\sqrt{(b-a)^2 - 4ac}}{2a}.$$

In this case, the Euler equation has the general solution

$$y = C_1 x^{r_1} + C_2 x^{r_2}, \qquad (x > 0).$$

The general solution is usually quoted in the form

$$y = C_1 |x|^{r_1} + C_2 |x|^{r_2},$$

which is valid in any interval not containing $x = 0$, and may even be valid on intervals containing the origin if, for example, $r_1$ and $r_2$ are nonnegative integers.

**Example 7**    Solve the initial-value problem

$$2x^2 y'' - x y' - 2y = 0, \qquad y(1) = 5, \qquad y'(1) = 0.$$

**Solution**    The auxiliary equation is $2r(r-1) - r - 2 = 0$, that is, $2r^2 - 3r - 2 = 0$, or $(r - 2)(2r + 1) = 0$, and has roots $r = 2$ and $r = -(1/2)$. Thus, the general solution of the differential equation (valid for $x > 0$) is

$$y = C_1 x^2 + C_2 x^{-1/2}.$$

The initial conditions imply that

$$5 = y(1) = C_1 + C_2, \quad \text{and} \quad 0 = y'(1) = 2C_1 - \frac{1}{2} C_2.$$

Therefore $C_1 = 1$ and $C_2 = 4$, and the initial-value problem has solution

$$y = x^2 + \frac{4}{\sqrt{x}}, \qquad (x > 0).$$

If $(b - a)^2 = 4ac$ then the auxiliary equation has one double root, namely the root $r = (a - b)/2a$. It is left to the reader to verify that in this case the transformation $y = x^r v(x)$ leads to the general solution

$$y = C_1 x^r + C_2 x^r \ln x, \qquad (x > 0),$$

or, more generally,

$$y = C_1 |x|^r + C_2 |x|^r \ln |x|, \qquad (x \neq 0).$$

If $(b - a)^2 < 4ac$ then the auxiliary equation has complex conjugate roots

$$r = \alpha \pm i\beta, \qquad \text{where} \quad \alpha = \frac{a - b}{2a}, \quad \beta = \frac{\sqrt{4ac - (b - a)^2}}{2a}.$$

The corresponding powers $x^r$ can be expressed in real form in a manner similar to that used for constant coefficient equations; we have

$$x^{\alpha \pm i\beta} = e^{(\alpha \pm i\beta) \ln x} = e^{\alpha \ln x} \left[ \cos(\beta \ln x) \pm i \sin(\beta \ln x) \right]$$
$$= x^\alpha \cos(\beta \ln x) \pm i x^\alpha \sin(\beta \ln x).$$

Accordingly, the Euler equation has the general solution

$$y = C_1|x|^\alpha \cos(\beta \ln |x|) + C_2|x|^\alpha \sin(\beta \ln |x|).$$

**Example 8** Solve the DE $x^2 y'' - 3xy' + 13y = 0$.

**Solution** The DE has the auxiliary equation $r(r-1) - 3r + 13 = 0$, that is, $r^2 - 4r + 13 = 0$, which has roots $r = 2 \pm 3i$. The DE, therefore, has the general solution

$$y = C_1 x^2 \cos(3 \ln |x|) + C_2 x^2 \sin(3 \ln |x|).$$

## Exercises 17.7

In Exercises 1–12, find the general solutions for the given equations.

**1.** $y'' + 7y' + 10y = 0$

**2.** $y'' - 2y' - 3y = 0$

**3.** $y'' + 2y' = 0$

**4.** $4y'' - 4y' - 3y = 0$

**5.** $y'' + 8y' + 16y = 0$

**6.** $y'' - 2y' + y = 0$

**7.** $y'' - 6y' + 10y = 0$

**8.** $9y'' + 6y' + y = 0$

**9.** $y'' + 2y' + 5y = 0$

**10.** $y'' - 4y' + 5y = 0$

**11.** $y'' + 2y' + 3y = 0$

**12.** $y'' + y' + y = 0$

In Exercises 13–16, solve the given initial-value problems.

**13.** $\begin{cases} 2y'' + 5y' - 3y = 0 \\ y(0) = 1 \\ y'(0) = 0. \end{cases}$

**14.** $\begin{cases} y'' + 10y' + 25y = 0 \\ y(1) = 0 \\ y'(1) = 2. \end{cases}$

**15.** $\begin{cases} y'' + 4y' + 5y = 0 \\ y(0) = 2 \\ y'(0) = 2. \end{cases}$

**16.** $\begin{cases} y'' + y' + y = 0 \\ y(2\pi/\sqrt{3}) = 0 \\ y'(2\pi/\sqrt{3}) = 1. \end{cases}$

**17.** If $a > 0, b > 0$, and $c > 0$, prove that all solutions of the differential equation $ay'' + by' + cy = 0$ satisfy $\lim_{t \to \infty} y(t) = 0$.

**∗ 18.** Prove that the solution given in the discussion of Case I, namely $y = A e^{r_1 t} + B e^{r_2 t}$, is the general solution for that case as follows: first let $y = e^{r_1 t} u$ and show that $u$ satisfies the equation

$$u'' - (r_2 - r_1)u' = 0.$$

Then let $v = u'$, so that $v$ must satisfy $v' = (r_2 - r_1)v$. The general solution of this equation is $v = C e^{(r_2 - r_1)t}$. Hence find $u$ and $y$.

Find general solutions of the DEs in Exercises 19–22.

**19.** $y''' - 4y'' + 3y' = 0$

**20.** $y^{(4)} - 2y'' + y = 0$

**21.** $y^{(4)} + 2y'' + y = 0$

**22.** $y^{(4)} + 4y^{(3)} + 6y'' + 4y' + y = 0$

**23.** Show that $y = e^{2t}$ is a solution of

$$y''' - 2y' - 4y = 0$$

(where $'$ denotes $d/dt$) and find the general solution of this DE.

**24.** Write the general solution of the linear, constant-coefficient DE having auxiliary equation $(r^2 - r - 2)^2(r^2 - 4)^2 = 0$.

Find general solutions to the Euler equations in Exercises 25–30.

**25.** $x^2 y'' - xy' + y = 0$

**26.** $x^2 y'' - xy' - 3y = 0$

**27.** $x^2 y'' + xy' - y = 0$

**28.** $x^2 y'' - xy' + 5y = 0$

**29.** $x^2 y'' + xy' = 0$

**30.** $x^2 y'' + xy' + y = 0$

**∗ 31.** Solve the DE $x^3 y''' + xy' - y = 0$ in the interval $x > 0$.

# 17.8 Nonhomogeneous Linear Equations

We now consider the problem of solving the nonhomogeneous second-order differential equation

$$a_2(x)\frac{d^2y}{dx^2} + a_1(x)\frac{dy}{dx} + a_0(x)y = f(x). \tag{$*$}$$

We assume that two independent solutions, $y_1(x)$ and $y_2(x)$, of the corresponding homogeneous equation

$$a_2(x)\frac{d^2y}{dx^2} + a_1(x)\frac{dy}{dx} + a_0(x)y = 0$$

are known. The function $y_h(x) = C_1y_1(x) + C_2y_2(x)$, which is the general solution of the homogeneous equation, is called the **complementary function** for the nonhomogeneous equation. Theorem 2 of Section 17.1 suggests that the general solution of the nonhomogeneous equation is of the form

$$y = y_p(x) + y_h(x) = y_p(x) + C_1y_1(x) + C_2y_2(x),$$

where $y_p(x)$ is any **particular solution** of the nonhomogeneous equation. All we need to do is find *one solution* of the nonhomogeneous equation and we can write the general solution.

There are two common methods for finding a particular solution $y_p$ of the nonhomogeneous equation ($*$):

1. The method of undetermined coefficients, and

2. The method of variation of parameters.

The first of these hardly warrants being called a *method*; it just involves making an educated guess about the form of the solution as a sum of terms with unknown coefficients and substituting this guess into the equation to determine the coefficients. This method works well for simple DEs, especially ones with constant coefficients. The nature of the *guess* depends on the nonhomogeneous term $f(x)$, but can also be affected by the solution of the corresponding homogeneous equation. A few examples will illustrate the ideas involved.

**Example 1**   Find the general solution of $y'' + y' - 2y = 4x$.

**Solution**   Because the nonhomogeneous term $f(x) = 4x$ is a first-degree polynomial, we "guess" that a particular solution can be found which is also such a polynomial. Thus we try

$$y = Ax + B, \qquad y' = A, \qquad y'' = 0.$$

Substituting these expressions into the given DE we obtain

$$0 + A - 2(Ax + B) = 4x \qquad \text{or}$$
$$-(2A + 4)x + (A - 2B) = 0.$$

This latter equation will be satisfied for all $x$ provided $2A + 4 = 0$ and $A - 2B = 0$. Thus we require $A = -2$ and $B = -1$; a particular solution of the given DE is

$$y_p(x) = -2x - 1.$$

Since the corresponding homogeneous equation $y'' + y' - 2y = 0$ has auxiliary equation $r^2 + r - 2 = 0$ with roots $r = 1$ and $r = -2$, the given DE has the general solution

$$y = y_p(x) + C_1 e^x + C_2 e^{-2x} = -2x - 1 + C_1 e^x + C_2 e^{-2x}.$$ ∎

---

**Example 2** Find general solutions of the equations (where $'$ denotes $d/dt$)

(a) $y'' + 4y = \sin t$,

(b) $y'' + 4y = \sin(2t)$,

(c) $y'' + 4y = \sin t + \sin(2t)$.

**Solution**

(a) Let us look for a particular solution of the form

$$y = A \sin t + B \cos t \qquad \text{so that}$$
$$y' = A \cos t - B \sin t$$
$$y'' = -A \sin t - B \cos t.$$

Substituting these expressions into the DE $y'' + 4y = \sin t$, we get

$$-A \sin t - B \cos t + 4A \sin t + 4B \cos t = \sin t,$$

which is satisfied for all $x$ if $3A = 1$ and $3B = 0$. Thus $A = 1/3$ and $B = 0$. Since the homogeneous equation $y'' + 4y = 0$ has general solution $y = C_1 \cos(2t) + C_2 \sin(2t)$, the given nonhomogeneous equation has the general solution

$$y = \frac{1}{3} \sin t + C_1 \cos(2t) + C_2 \sin(2t).$$

(b) Motivated by our success in part (a), we might be tempted to try for a particular solution of the form $y = A \sin(2t) + B \cos(2t)$, but that won't work, because this function is a solution of the homogeneous equation, so we would get $y'' + 4y = 0$ for any choice of $A$ and $B$. In this case it is useful to try

$$y = At \sin(2t) + Bt \cos(2t).$$

We have

$$y' = A \sin(2t) + 2At \cos(2t) + B \cos(2t) - 2Bt \sin(2t)$$
$$= (A - 2Bt) \sin(2t) + (B + 2At) \cos(2t)$$
$$y'' = -2B \sin(2t) + 2(A - 2Bt) \cos(2t) + 2A \cos(2t)$$
$$\quad - 2(B + 2At) \sin(2t)$$
$$= -4(B + At) \sin(2t) + 4(A - Bt) \cos(2t).$$

Substituting into $y'' + 4y = \sin(2t)$ leads to

$$-4(B + At) \sin(2t) + 4(A - Bt) \cos(2t) + 4At \sin(2t) + 4Bt \cos(2t)$$
$$= \sin(2t).$$

Observe that the terms involving $t \sin(2t)$ and $t \cos(2t)$ cancel out and we are left with

$$-4B \sin(2t) + 4A \cos(2t) = \sin(2t),$$

which is satisfied for all $x$ if $A = 0$ and $B = -1/4$. Hence, the general solution for part (b) is

$$y = -\frac{1}{4}t \cos(2t) + C_1 \cos(2t) + C_2 \sin(2t).$$

(c) Since the homogeneous equation is the same for (a), (b), and (c), and the nonhomogeneous term in equation (c) is the sum of the nonhomogeneous terms in equations (a) and (b), the sum of particular solutions of (a) and (b) is a particular solution of (c). (This is because the equation is *linear.*) Thus the general solution of equation (c) is

$$y = \frac{1}{3} \sin t - \frac{1}{4}t \cos(2t) + C_1 \cos(2t) + C_2 \sin(2t).$$

We summarize the appropriate forms to try for particular solutions of constant-coefficient equations as follows:

**Trial solutions for constant-coefficient equations**

Let $A_n(x)$, $B_n(x)$, and $P_n(x)$ denote the $n$th-degree polynomials

$$A_n(x) = a_0 + a_1 x + a_2 x^2 + \cdots + a_n x^n$$
$$B_n(x) = b_0 + b_1 x + b_2 x^2 + \cdots + b_n x^n$$
$$P_n(x) = p_0 + p_1 x + p_2 x^2 + \cdots + p_n x^n$$

To find a particular solution $y_p(x)$ of the second-order linear, constant-coefficient, nonhomogeneous DE

$$a_2 \frac{d^2 y}{dx^2} + a_1 \frac{dy}{dx} + a_0 y = f(x)$$

use the following forms:

If $f(x) = P_n(x)$          try $y_p = x^m A_n(x)$.

If $f(x) = P_n(x)e^{rx}$       try $y_p = x^m A_n(x)e^{rx}$.

If $f(x) = P_n(x)e^{rx} \cos(kx)$ try $y_p = x^m e^{rx}[A_n(x) \cos(kx) + B_n(x) \sin(kx)]$.

If $f(x) = P_n(x)e^{rx} \sin(kx)$ try $y_p = x^m e^{rx}[A_n(x) \cos(kx) + B_n(x) \sin(kx)]$.

where $m$ is the smallest of the integers 0, 1, and 2, that ensures that no term of $y_p$ is a solution of the corresponding homogeneous equation

$$a_2 \frac{d^2 y}{dx^2} + a_1 \frac{dy}{dx} + a_0 y = 0.$$

## Resonance

For $\lambda > 0$, $\lambda \neq 1$, the solution $y_\lambda(t)$ of the initial-value problem

$$\begin{cases} y'' + y = \sin(\lambda t) \\ \quad y(0) = 0 \\ \quad y'(0) = 1. \end{cases}$$

can be determined by first looking for a particular solution of the DE having the form $y = A \sin(\lambda t)$, and then adding the complementary function $y = B \cos t + C \sin t$. The calculations give $A = 1/(1 - \lambda^2)$, $B = 0$, $C = (1 - \lambda - \lambda^2)/(1 - \lambda^2)$, so

$$y_\lambda(t) = \frac{\sin(\lambda t) + (1 - \lambda - \lambda^2) \sin t}{1 - \lambda^2}.$$

For $\lambda = 1$ the nonhomogeneous term in the DE is a solution of the homogeneous equation $y'' + y = 0$, so one must try for a particular solution of the form $y = At \cos t + Bt \sin t$. In this case, the solution of the initial-value problem is

$$y_1(t) = \frac{3 \sin t - t \cos t}{2}.$$

(This solution can also be found by calculating $\lim_{\lambda \to 1} y_\lambda(t)$ using l'Hôpital's rule.) Observe that this solution is unbounded; the amplitude of the oscillations become larger and larger as $t$ increases. In contrast, the solutions $y_\lambda(t)$ for $\lambda \neq 1$ are bounded for all $t$, though they can become quite large for some values of $t$ if $\lambda$ is close to 1. The graphs of the solutions $y_{0.9}(t)$, $y_{0.95}(t)$, and $y_1(t)$ on the interval $-10 \leq t \leq 100$ are shown in Figure 17.6.

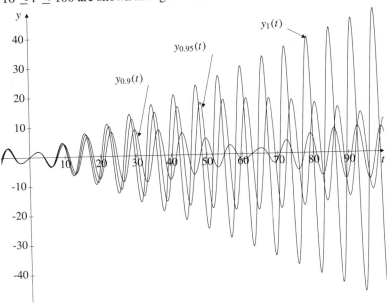

**Figure 17.6**    Resonance

The phenomenon illustrated here is called **resonance**. Vibrating mechanical systems have natural frequencies at which they will vibrate. If you try to force them to vibrate at a different frequency, the amplitude of the vibrations will themselves vary sinusoidally over time, producing an effect known as **beats**. The amplitudes of the beats can grow quite large, and the period of the beats lengthens as the forcing frequency approaches the natural frequency of the system. If the system has no resistive damping (the one illustrated here has no damping) then forcing vibrations at the natural frequency will cause the system to vibrate at ever increasing amplitudes.

As a concrete example, if you push a child on a swing, the swing will rise highest if your pushes are timed to have the same frequency as the natural frequency of the swing. Resonance is used in the design of tuning circuits of radios; the circuit is tuned (ususally by a variable capacitor) so that its natural frequency of oscillation is the frequency of the station being tuned in. The circuit then responds much more strongly to the signal received from that station than to others on different frequencies.

## Variation of Parameters

A more formal method for finding a particular solution $y_p(x)$ of the nonhomogeneous equation when we know two independent solutions, $y_1(x)$ and $y_2(x)$, of the homogeneous equation is to replace the constants in the complementary function by functions, that is, search for $y_p$ in the form

$$y_p = u_1(x)y_1(x) + u_2(x)y_2(x).$$

Requiring $y_p$ to satisfy the given nonhomogeneous DE provides one equation that must be satisfied by the two unknown functions $u_1$ and $u_2$. We are free to require them to satisfy a second equation also. To simplify the calculations below, we choose this second equation to be

$$u_1'(x)y_1(x) + u_2'(x)y_2(x) = 0.$$

Now we have

$$y_p' = u_1'y_1 + u_1y_1' + u_2'y_2 + u_2y_2' = u_1y_1' + u_2y_2'$$
$$y_p'' = u_1'y_1' + u_1y_1'' + u_2'y_2' + u_2y_2''.$$

Substituting these expressions into the given DE we obtain

$$a_2(u_1'y_1' + u_2'y_2') + u_1(a_2y_1'' + a_1y_1' + a_0y_1) + u_2(a_2y_2'' + a_1y_2' + a_0y_2)$$
$$= a_2(u_1'y_1' + u_2'y_2') = f(x),$$

because $y_1$ and $y_2$ satisfy the homogeneous equation. Therefore $u_1'$ and $u_2'$ satisfy the pair of equations

$$u_1'(x)y_1(x) + u_2'(x)y_2(x) = 0$$
$$u_1'(x)y_1'(x) + u_2'(x)y_2'(x) = \frac{f(x)}{a_2(x)}.$$

We can solve these two equations for the unknown functions $u_1'$ and $u_2'$ by Cramer's Rule (Theorem 5 of Section 10.6), or otherwise, and obtain

$$u_1' = -\frac{y_2(x)}{W(x)}\frac{f(x)}{a_2(x)}, \qquad u_2' = \frac{y_1(x)}{W(x)}\frac{f(x)}{a_2(x)},$$

where $W(x)$, called the **Wronskian** of $y_1$ and $y_2$, is the determinant

$$W(x) = \begin{vmatrix} y_1(x) & y_2(x) \\ y_1'(x) & y_2'(x) \end{vmatrix}.$$

Then $u_1$ and $u_2$ can be found by integration.

**Example 3**  Find the general solution of $y'' - 3y' + 2y = 4x$.

**Solution**  First we solve the homogeneous equation $y'' - 3y' + 2y = 0$, which has auxiliary equation $r^2 - 3r + 2 = 0$ with roots $r = 1$ and $r = 2$. Therefore two independent solutions of the homogeneous equation are $y_1 = e^x$ and $y_2 = e^{2x}$, and the complementary function is

$$y_h = C_1 e^x + C_2 e^{2x}.$$

A particular solution $y_p(x)$ of the nonhomogeneous equation can be found in the form

$$y_p = u_1(x)e^x + u_2(x)e^{2x}$$

where $u_1$ and $u_2$ satisfy

$$u_1' e^x + 2u_2' e^{2x} = 4x$$
$$u_1' e^x + u_2' e^{2x} = 0.$$

We solve these linear equations for $u_1'$ and $u_2'$ and then integrate to obtain

$$u_1' = -4xe^{-x} \qquad u_2' = 4xe^{-2x}$$
$$u_1 = 4(x+1)e^{-x} \qquad u_2 = -(2x+1)e^{-2x}.$$

Hence $y_p = 4x + 4 - (2x + 1) = 2x + 3$ is a particular solution of the nonhomogeneous equation, and the general solution is

$$y = 2x + 3 + C_1 e^x + C_2 e^{2x}.$$

**Remark**  This method for solving the nonhomogeneous equation is called the **method of variation of parameters**. It is completely general and extends to higher order equations in a reasonable way, but it is computationally somewhat difficult. We could have found $y_p$ more easily had we "guessed" that it would be of the form $y_p = Ax + B$ and substituted this into the differential equation to get

$$-3A + 2(Ax + B) = 4x$$
$$\text{or} \quad 2Ax + (2B - 3A) = 4x.$$

The only way this latter equation can be satisfied for all $x$ is to have $2A = 4$ and $2B - 3A = 0$, that is, $A = 2$ and $B = 3$.

## Exercises 17.8

Find general solutions for the nonhomogeneous equations in Exercises 1–12 by the method of undetermined coefficients.

**1.** $y'' + y' - 2y = 1$

**2.** $y'' + y' - 2y = x$

**3.** $y'' + y' - 2y = e^{-x}$

**4.** $y'' + y' - 2y = e^x$

**5.** $y'' + 2y' + 5y = x^2$

**6.** $y'' + 4y = x^2$

**7.** $y'' - y' - 6y = e^{-2x}$

**8.** $y'' + 4y' + 4y = e^{-2x}$

**9.** $y'' + 2y' + 2y = e^x \sin x$

**10.** $y'' + 2y' + 2y = e^{-x} \sin x$

**11.** $y'' + y' = 4 + 2x + e^{-x}$   **12.** $y'' + 2y' + y = xe^{-x}$

**13.** Repeat Exercise 3 using the method of variation of parameters.

**14.** Repeat Exercise 4 using the method of variation of parameters.

**15.** Find a particular solution of the form $y = Ax^2$ for the Euler equation $x^2y'' + xy' - y = x^2$, and hence obtain the general solution of this equation on the interval $(0, \infty)$.

**16.** For what values of $r$ can the Euler equation $x^2y'' + xy' - y = x^r$ be solved by the method of the previous exercise. Find a particular solution for each such $r$.

**17.** Try to guess the form of a particular solution for $x^2y'' + xy' - y = x$ and hence obtain the general solution for this equation on the interval $(0, \infty)$.

**18.** Use variation of parameters to solve $x^2y'' + xy' - y = x$.

**19.** Consider the nonhomogeneous, linear equation

$$x^2y'' - (2x + x^2)y' + (2 + x)y = x^3.$$

Use the fact that $y_1(x) = x$ and $y_2(x) = xe^x$ are independent solutions of the corresponding homogeneous equation (see Exercise 5 of Section 17.6) to find the general solution of this nonhomogeneous equation.

**20.** Consider the nonhomogeneous, Bessel equation

$$x^2y'' + xy' + \left(x^2 - \frac{1}{4}\right)y = x^{3/2}.$$

Use the fact that $y_1(x) = x^{-1/2}\cos x$ and $y_2(x) = x^{-1/2}\sin x$ are independent solutions of the corresponding homogeneous equation (see Exercise 6 of Section 17.6) to find the general solution of this nonhomogeneous equation.

## 17.9   Series Solutions

Many of the second-order, linear, differential equations that arise in applications do not have constant coefficients and are not Euler equations. If the coefficient functions of such an equation are sufficiently well behaved we can often find solutions in the form of power series (Taylor series). Such series solutions are frequently used to define new functions, whose properties are deduced partly from the fact that they solve particular differential equations. For example, Bessel functions of order $\nu$ are defined to be certain series solutions of Bessel's differential equation

$$x^2y'' + xy' + (x^2 - \nu^2)y = 0.$$

Series solutions for second-order homogeneous linear differential equations are most easily found near an **ordinary point** of the equation. This is a point $x = a$ such that the equation can be expressed in the form

$$y'' + p(x)y' + q(x)y = 0$$

where the functions $p(x)$ and $q(x)$ are **analytic** at $x = a$. (A function $f$ is analytic at $x = a$ if $f(x)$ can be expressed as the sum of its Taylor series in powers of $x - a$ in an interval of positive radius centred at $x = a$.) Thus we assume

$$p(x) = \sum_{n=0}^{\infty} p_n(x - a)^n, \qquad q(x) = \sum_{n=0}^{\infty} q_n(x - a)^n$$

with both series converging in some interval of the form $a - R < x < a + R$. Frequently $p(x)$ and $q(x)$ are polynomials and so are analytic everywhere. A change of independent variable $\xi = x - a$ will put the point $x = a$ at the origin $\xi = 0$, so we can assume that $a = 0$.

The following example illustrates the technique of series solution around an ordinary point.

**Example 1** Find two independent solutions in powers of $x$ for the Hermite equation

$$y'' - 2xy' + vy = 0.$$

For what values of $v$ does the equation have a polynomial solution?

**Solution** We try

$$y = \sum_{n=0}^{\infty} a_n x^n = a_0 + a_1 x + a_2 x^2 + a_3 x^3 + \cdots,$$

$$y' = \sum_{n=1}^{\infty} n a_n x^{n-1}$$

$$y'' = \sum_{n=2}^{\infty} n(n-1)a_n x^{n-2} = \sum_{n=0}^{\infty} (n+2)(n+1)a_{n+2} x^n.$$

(We have replaced $n$ by $n+2$ in order to get $x^n$ in the sum for $y''$.) We substitute these expressions into the differential equation to get

$$\sum_{n=0}^{\infty} (n+2)(n+1)a_{n+2} x^n - 2 \sum_{n=1}^{\infty} n a_n x^n + v \sum_{n=0}^{\infty} a_n x^n = 0$$

or $\quad 2a_2 + v a_0 + \sum_{n=1}^{\infty} \Big[ (n+2)(n+1)a_{n+2} - (2n-v)a_n \Big] x^n = 0.$

This identity holds for all $x$ provided that the coefficient of every power of $x$ vanishes; that is,

$$a_2 = -\frac{v a_0}{2}, \qquad a_{n+2} = \frac{(2n-v)a_n}{(n+2)(n+1)}, \qquad (n = 1, 2, \cdots).$$

The latter of these formulas is called a **recurrence relation**.

We can choose $a_0$ and $a_1$ to have any values; then the above conditions determine all the remaining coefficients $a_n$, $(n \ge 2)$. We can get one solution by choosing, for instance, $a_0 = 1$ and $a_1 = 0$. Then, by the recurrence relation,

$$a_3 = 0, \quad a_5 = 0, \quad a_7 = 0, \quad \cdots, \quad \text{and}$$

$$a_2 = -\frac{v}{2}$$

$$a_4 = \frac{(4-v)a_2}{4 \times 3} = -\frac{v(4-v)}{2 \times 3 \times 4} = -\frac{v(4-v)}{4!}$$

$$a_6 = \frac{(8-v)a_4}{6 \times 5} = -\frac{v(4-v)(8-v)}{6!}$$

$$\cdots$$

The pattern is obvious here:

$$a_{2n} = -\frac{v(4-v)(8-v)\cdots(4n-4-v)}{(2n)!}, \qquad (n = 1, 2, \cdots).$$

One solution to the Hermite equation is

$$y_1 = 1 + \sum_{n=1}^{\infty} -\frac{\nu(4-\nu)(8-\nu)\cdots(4n-4-\nu)}{(2n)!} x^{2n}.$$

We observe that if $\nu = 4n$ for some nonnegative integer $n$ then $y_1$ is an even polynomial of degree $2n$, for $a_{2n+2} = 0$ and all subsequent even coefficients therefore also vanish.

The second solution, $y_2$, can be found in the same way, by choosing $a_0 = 0$ and $a_1 = 1$. It is

$$y_2 = x + \sum_{n=1}^{\infty} \frac{(2-\nu)(6-\nu)\cdots(4n-2-\nu)}{(2n+1)!} x^{2n+1},$$

and is an odd polynomial of degree $2n + 1$ if $\nu = 4n + 2$.

Both of these series solutions converge for all $x$. The Ratio Test can be applied directly to the recurrence relation. Since consecutive nonzero terms of each series are of the form $a_n x^n$ and $a_{n+2} x^{n+2}$, we calculate

$$\rho = \lim_{n\to\infty} \left| \frac{a_{n+2} x^{n+2}}{x_n x^n} \right| = |x|^2 \lim_{n\to\infty} \left| \frac{a_{n+2}}{a_n} \right| = |x|^2 \lim_{n\to\infty} \left| \frac{2n - \nu}{(n+2)(n+1)} \right| = 0$$

for every $x$, so the series converges by the Ratio Test. ∎

If $x = a$ is not an ordinary point of the equation

$$y'' + p(x)y' + q(x)y = 0$$

then it is called a **singular point** of that equation. This means that at least one of the functions $p(x)$ and $q(x)$ is not analytic at $x = a$. If, however, $(x-a)p(x)$ and $(x-a)^2 q(x)$ are analytic at $x = a$ then the singular point is said to be a **regular singular point**. For example, the origin $x = 0$ is a regular singular point of any Euler equation, and also of Bessel's equation,

$$x^2 y'' + xy' + (x^2 - \nu^2)y = 0,$$

since $p(x) = 1/x$ and $q(x) = (x^2 - \nu^2)/x^2$ satisfy $xp(x) = 1$ and $x^2 q(x) = x^2 - \nu^2$, which are both polynomials and therefore analytic.

The solutions of differential equations are usually not analytic at singular points. However, it is still possible to find at least one series solution about such a point. The method involves searching for a series solution of the form $x^\mu$ times a power series, that is,

$$y = \sum_{n=0}^{\infty} a_n (x-a)^{n+\mu}, \qquad \text{where } a_0 \neq 0.$$

Substitution into the differential equation produces a quadratic **indicial equation**, which determines one or two values of $\mu$ for which such solutions can be found, and a **recurrence relation** enabling the coefficients $a_n$ to be calculated for $n \geq 1$. If the indicial roots are not equal, and do not differ by an integer, two independent solutions can be calculated. If the indicial roots are equal, or differ by an integer, one such solution can be calculated (corresponding to the larger indicial root), and a second solution can be found by using the technique described in Section 17.6. These calculations can be quite difficult. The reader is referred to standard texts on differential equations for more discussion and examples. We will content ourselves here with one final example.

**Example 2**   Find one solution, in powers of $x$, of Bessel's equation of order $v = 1$, namely

$$x^2 y'' + xy' + (x^2 - 1)y = 0.$$

**Solution**   We try

$$y = \sum_{n=0}^{\infty} a_n x^{\mu+n}$$

$$y' = \sum_{n=0}^{\infty} (\mu + n) a_n x^{\mu+n-1}$$

$$y'' = \sum_{n=0}^{\infty} (\mu + n)(\mu + n - 1) a_n x^{\mu+n-2}.$$

Substituting these expressions into the Bessel equation, we get

$$\sum_{n=0}^{\infty} \left[ \left( (\mu + n)(\mu + n - 1) + (\mu + n) - 1 \right) a_n x^n + a_n x^{n+2} \right] = 0$$

$$\sum_{n=0}^{\infty} \left[ (\mu + n)^2 - 1 \right] a_n x^n + \sum_{n=2}^{\infty} a_{n-2} x^n = 0$$

$$(\mu^2 - 1) a_0 + \left( (\mu + 1)^2 - 1 \right) a_1 x + \sum_{n=2}^{\infty} \left[ \left( (\mu + n)^2 - 1 \right) a_n + a_{n-2} \right] x^n = 0.$$

All of the terms must vanish. Since $a_0 \neq 0$ (we may take $a_0 = 1$) we obtain

$$\mu^2 - 1 = 0, \qquad\qquad \text{the indicial equation}$$

$$[(\mu + 1)^2 - 1] a_1 = 0,$$

$$a_n = -\frac{a_{n-2}}{(\mu + n)^2 - 1}, \quad (n \geq 2). \qquad \text{the recurrence relation}$$

Evidently $\mu = \pm 1$, and therefore $a_1 = 0$. If we take $\mu = 1$, then the recurrence relation is $a_n = -a_{n-2}/(n)(n + 2)$. Thus

$$a_3 = 0, \quad a_5 = 0, \quad a_7 = 0, \quad \cdots$$

$$a_2 = \frac{-1}{2 \times 4}, \quad a_4 = \frac{1}{2 \times 4 \times 4 \times 6}, \quad a_6 = \frac{-1}{2 \times 4 \times 4 \times 6 \times 6 \times 8}, \quad \cdots.$$

Again the pattern is obvious:

$$a_{2n} = \frac{(-1)^n}{2^{2n} n! (n + 1)!}$$

and one solution of the Bessel equation of order 1 is

$$y = \sum_{n=0}^{\infty} \frac{(-1)^n}{2^{2n} n! (n + 1)!} x^{2n+1}.$$

By the Ratio Test, this series converges for all $x$. ∎

**Remark**   Observe that if we tried to calculate a second solution using $\mu = -1$ we would get the recurrence relation

$$a_n = -\frac{a_{n-2}}{n(n - 2)},$$

and we would be unable to calculate $a_2$. This shows what can happen if the indicial roots differ by an integer.

# Exercises 17.9

**1.** Find the general solution of $y'' = (x-1)^2 y$ in the form of a power series $y = \sum_{n=0}^{\infty} a_n (x-1)^n$.

**2.** Find the general solution of $y'' = xy$ in the form of a power series $y = \sum_{n=0}^{\infty} a_n x^n$ with $a_0$ and $a_1$ arbitrary.

**3.** Find the first three nonzero terms in a power series solution in powers of $x$ for the initial-value problem $y'' + (\sin x)y = 0$, $y(0) = 1$, $y'(0) = 0$.

**4.** Find the solution, in powers of $x$, for the initial-value problem
$$(1 - x^2)y'' - xy' + 9y = 0, \quad y(0) = 0, \quad y'(0) = 1.$$

**5.** Find two power series solutions in powers of $x$ for $3xy'' + 2y' + y = 0$.

**6.** Find one power series solution for the Bessel equation of order $v = 0$, that is, the equation $xy'' + y' + xy = 0$.

# Chapter Review

## Key Ideas

- **What do the following phrases mean?**
  ◇ an ordinary DE       ◇ a partial DE
  ◇ the general solution of a DE
  ◇ a linear combination of solutions of a DE
  ◇ the order of a DE      ◇ a linear DE
  ◇ a separable DE        ◇ an exact DE
  ◇ an integrating factor    ◇ a constant coefficient DE
  ◇ an Euler equation       ◇ an auxiliary equation
- **Describe how to solve:**
  ◇ a separable DE        ◇ a first-order, linear DE
  ◇ a homogeneous, first-order DE
  ◇ a constant coefficient DE    ◇ an Euler equation
- **What conditions imply that an initial-value problem for a first-order DE has a unique solution near the initial point?**
- **Describe the following methods for solving first-order DEs numerically:**
  ◇ the Euler method      ◇ the improved Euler method
  ◇ the fourth-order Runge–Kutta method
- **Describe the following methods for solving a nonhomogeneous, linear DE:**
  ◇ undetermined coefficients   ◇ variation of parameters
- **What are an ordinary point and a regular singular point of a linear, second-order DE? Describe how series can be used to solve such an equation near such a point.**

## Review Exercises

Find general solutions of the differential equations in Exercises 1–16.

**1.** $\dfrac{dy}{dx} = 2xy$

**2.** $\dfrac{dy}{dx} = e^{-y} \sin x$

**3.** $\dfrac{dy}{dx} = x + 2y$

**4.** $\dfrac{dy}{dx} = \dfrac{x^2 + y^2}{2xy}$

**5.** $\dfrac{dy}{dx} = \dfrac{x+y}{y-x}$

**6.** $\dfrac{dy}{dx} = -\dfrac{y + e^x}{x + e^y}$

**7.** $\dfrac{d^2 y}{dt^2} = \left(\dfrac{dy}{dt}\right)^2$

**8.** $2\dfrac{d^2 y}{dt^2} + 5\dfrac{dy}{dt} + 2y = 0$

**9.** $4y'' - 4y' + 5y = 0$

**10.** $2x^2 y'' + y = 0$

**11.** $t^2 \dfrac{d^2 y}{dt^2} - t\dfrac{dy}{dt} + 5y = 0$

**12.** $\dfrac{d^3 y}{dt^3} + 8\dfrac{d^2 y}{dt^2} + 16\dfrac{dy}{dt} = 0$

**13.** $\dfrac{d^2 y}{dx^2} - 5\dfrac{dy}{dx} + 6y = e^x + e^{3x}$

**14.** $\dfrac{d^2 y}{dx^2} - 5\dfrac{dy}{dx} + 6y = xe^{2x}$

**15.** $\dfrac{d^2 y}{dx^2} + 2\dfrac{dy}{dx} + y = x^2$  **16.** $x\dfrac{d^2 y}{dx^2} - 2y = x^3$

Solve the initial-value problems in Exercises 17–26.

**17.** $\begin{cases} \dfrac{dy}{dx} = \dfrac{x^2}{y^2} \\ y(2) = 1 \end{cases}$  **18.** $\begin{cases} \dfrac{dy}{dx} = \dfrac{y^2}{x^2} \\ y(2) = 1 \end{cases}$

**19.** $\begin{cases} \dfrac{dy}{dx} = \dfrac{xy}{x^2 + y^2} \\ y(0) = 1 \end{cases}$  **20.** $\begin{cases} \dfrac{dy}{dx} + (\cos x)y = 2\cos x \\ y(\pi) = 1 \end{cases}$

**21.** $\begin{cases} y'' + 3y' + 2y = 0 \\ y(0) = 1 \\ y'(0) = 2 \end{cases}$

**22.** $\begin{cases} y'' + 2y' + (1 + \pi^2)y = 0 \\ y(1) = 0 \\ y'(1) = \pi \end{cases}$

**23.** $\begin{cases} y'' + 10y' + 25y = 0 \\ y(1) = e^{-5} \\ y'(1) = 0 \end{cases}$

**24.** $\begin{cases} x^2y'' - 3xy' + 4y = 0 \\ y(e) = e^2 \\ y'(e) = 0 \end{cases}$

**25.** $\begin{cases} \dfrac{d^2y}{dt^2} + 4y = 8e^{2t} \\ y(0) = 1 \\ y'(0) = -2 \end{cases}$

**26.** $\begin{cases} 2\dfrac{d^2y}{dx^2} + 5\dfrac{dy}{dx} - 3y = 6 + 7e^{x/2} \\ y(0) = 0 \\ y'(0) = 1 \end{cases}$

**27.** For what values of the constants $A$ and $B$ is the equation

$$[(x + A)e^x \sin y + \cos y]\,dx + x[e^x \cos y + B \sin y]\,dy = 0$$

exact? What is the general solution of the equation if $A$ and $B$ have these values?

**28.** Find a value of $n$ for which $x^n$ is an integrating factor for

$$(x^2 + 3y^2)\,dx + xy\,dy = 0$$

and solve the equation.

**29.** Show that $y = x$ is a solution of

$$x^2y'' - x(2 + x \cot x)y' + (2 + x \cot x)y = 0$$

and find the general solution of this equation.

**30.** Use the method of variation of parameters and the result of the previous exercise to find the general solution of the nonhomogeneous equation

$$x^2y'' - x(2 + x \cot x)y' + (2 + x \cot x)y = x^3 \sin x.$$

**31.** Suppose that $f(x, y)$ and $\dfrac{\partial}{\partial y}f(x, y)$ are continuous on the whole $xy$-plane, and that $f(x, y)$ is bounded there, say $|f(x, y)| \le K$. Show that no solution of $y' = f(x, y)$ can have a vertical asymptote. Describe the region in the plane in which the solution to the initial-value problem

$$\begin{cases} y' = f(x, y) \\ y(x_0) = y_0 \end{cases}$$

must remain.

# Appendix I
# Complex Numbers

Many of the problems to which mathematics is applied involve the solution of equations. Over the centuries the number system had to be expanded many times to provide solutions for more and more kinds of equations. The natural numbers

$$\mathbb{N} = \{1,\ 2,\ 3,\ 4,\ \ldots\}$$

are inadequate for the solutions of equations of the form

$$x + n = m, \qquad (m,\ n \in \mathbb{N}).$$

Zero and negative numbers can be added to create the integers

$$\mathbb{Z} = \{\ldots,\ -3,\ -2,\ -1,\ 0,\ 1,\ 2,\ 3,\ \ldots\}$$

in which that equation has the solution $x = m - n$ even if $m < n$. (Historically, this extension of the number system came much later than some of those mentioned below.) Some equations of the form

$$nx = m, \qquad (m,\ n \in \mathbb{Z}, \quad n \neq 0)$$

cannot be solved in the integers. Another extension is made to to include numbers of the form $m/n$, thus producing the set of rational numbers

$$\mathbb{Q} = \left\{ \frac{m}{n} \ :\ m,\ n \in \mathbb{Z}, \quad n \neq 0 \right\}.$$

Every linear equation

$$ax = b, \qquad (a,\ b \in \mathbb{Q}, \quad a \neq 0)$$

has a solution $x = b/a$ in $\mathbb{Q}$, but the quadratic equation

$$x^2 = 2$$

has no solution in $\mathbb{Q}$, as was shown in Section P.1. Another extension enriches the rational numbers to the real numbers $\mathbb{R}$ in which some equations like $x^2 = 2$ have solutions. However, other quadratic equations, for instance,

$$x^2 = -1$$

do not have solutions even in the real numbers, so the extension process is not complete. In order to be able to solve any quadratic equation, we need to extend the real number system to a larger set, which we call **the complex number system**. In this appendix we will define complex numbers and develop some of their basic properties. In Appendix II we will briefly consider functions whose domains and ranges consist of sets of complex numbers. We will also show that every polynomial equation can be solved in the complex number system.

The calculus of functions of a complex variable, called **complex analysis**, is very rich and leads to discoveries which have no counterparts for functions of real variables. We will hardly scratch the surface of the subject in these pages.

## Definition of Complex Numbers

We begin by defining the symbol $i$, called **the imaginary unit**[1], to have the property

$$i^2 = -1.$$

Thus we could also call $i$ the **square root of** $-1$ and denote it $\sqrt{-1}$. Of course, $i$ is not a real number; no real number has a negative square.

---

**DEFINITION** **1**

> A **complex number** is an expression of the form
>
> $$a + bi \qquad \text{or} \qquad a + ib$$
>
> where $a$ and $b$ are *real numbers*, and $i$ is the imaginary unit.

---

For example $3 + 2i$, $\frac{7}{2} - \frac{2}{3}i$, $i\pi = 0 + i\pi$, and $-3 = -3 + 0i$ are all complex numbers. The last of these examples shows that every real number can be regarded as a complex number. (We will normally use $a + bi$ unless $b$ is a complicated expression, in which case we will write $a + ib$ instead. Either form is acceptable.)

It is often convenient to represent a complex number by a single letter; $w$ and $z$ are frequently used for this purpose. If $a, b, x$, and $y$ are real numbers, and

$$w = a + bi, \qquad \text{and} \qquad z = x + yi,$$

then we can refer to the complex numbers $w$ and $z$. Note that $w = z$ if and only if $a = x$ and $b = y$. Of special importance are the complex numbers

$$0 = 0 + 0i, \qquad 1 = 1 + 0i, \qquad \text{and} \qquad i = 0 + 1i.$$

---

**DEFINITION** **2**

> If $z = x + yi$ is a complex number (where $x$ and $y$ are real), we call $x$ the **real part** of $z$ and denote it Re $(z)$. We call $y$ the **imaginary part** of $z$ and denote it Im $(z)$:
>
> $$\text{Re}\,(z) = \text{Re}\,(x + yi) = x, \qquad \text{Im}\,(z) = \text{Im}\,(x + yi) = y.$$

---

Note that both the real and imaginary parts of a complex number are real numbers.

$$\text{Re}\,(3 - 5i) = 3 \qquad\qquad \text{Im}\,(3 - 5i) = -5$$
$$\text{Re}\,(2i) = \text{Re}\,(0 + 2i) = 0 \qquad\qquad \text{Im}\,(2i) = \text{Im}\,(0 + 2i) = 2$$
$$\text{Re}\,(-7) = \text{Re}\,(-7 + 0i) = -7 \qquad\qquad \text{Im}\,(-7) = \text{Im}\,(-7 + 0i) = 0.$$

## Graphical Representation of Complex Numbers

Since complex numbers are constructed from pairs of real numbers (their real and imaginary parts) it is natural to represent complex numbers graphically as points in a Cartesian plane. We use the point with coordinates $(a, b)$ to represent the complex number $w = a + ib$. In particular, the origin $(0, 0)$ represents the complex number $0$, the point $(1, 0)$ represents the complex number $1 = 1 + 0i$, and the point $(0, 1)$ represents the point $i = 0 + 1i$. (See Figure I.1.)

---

[1]   In some fields, for example, electrical engineering, the imaginary unit is denoted $j$ instead of $i$. Like "negative", "surd", and "irrational", the term "imaginary" suggests the distrust which greeted the new kinds of numbers when they were first introduced.

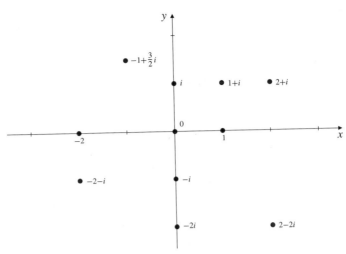

**Figure I.1**  An Argand diagram representing the complex plane

Such a representation of complex numbers as points in a plane is called an **Argand diagram**. Since each complex number is represented by a unique point in the plane, the set of all complex numbers is often referred to as **the complex plane**. The symbol $\mathbb{C}$ is used to represent the set of all complex numbers and, equivalently, the complex plane:

$$\mathbb{C} = \{x + yi \ : \ x, \ y, \ \in \mathbb{R}\}.$$

The points on the $x$-axis of the complex plane correspond to real numbers ($x = x + 0i$), so the $x$-axis is called the **real axis**. The points on the $y$-axis correspond to **pure imaginary** numbers ($yi = 0 + yi$), so the $y$-axis is called the **imaginary axis**.

It is often useful to consider the *polar coordinates* of a point in the complex plane.

**DEFINITION 3**

The distance from the origin to the point $(a, b)$ corresponding to the complex number $w = a + bi$ is called the **modulus** of $w$ and is denoted by $|w|$ or $|a + bi|$:

$$|w| = |a + bi| = \sqrt{a^2 + b^2}.$$

If the line from the origin to $(a, b)$ makes angle $\theta$ with the positive direction of the real axis (with positive angles measured counterclockwise), then we call $\theta$ an **argument** of the complex number $w = a + bi$, and denote it by $\arg(w)$ or $\arg(a + bi)$.

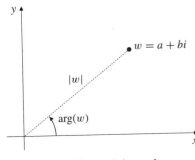

**Figure I.2**  The modulus and argument of a complex number

The modulus of a complex number is always real and nonnegative. It is positive unless the complex number is 0. Modulus plays a similar role for complex numbers that absolute value does for real numbers. Indeed, sometimes modulus is called absolute value.

Arguments of complex numbers are not unique. If $w = a + bi \neq 0$ then any two possible values for $\arg(w)$ differ by an integer multiple of $2\pi$. The symbol $\arg(w)$ actually represents not a single number, but a set of numbers. When we write $\arg(w) = \theta$ we are saying that the set $\arg(w)$ contains all numbers of the form $\theta + 2k\pi$, where $k$ is an integer.

If $w = a + bi$ where $a = \text{Re}\,(w) \neq 0$, then

$$\tan \arg(w) = \tan \arg(a + bi) = \frac{b}{a}.$$

This means that $\tan \theta = b/a$ for every $\theta$ in the set $\arg(w)$.

It is sometimes convenient to restrict $\theta = \arg(w)$ to an interval of length $2\pi$, say, the interval $0 \leq \theta < 2\pi$, or $-\pi < \theta \leq \pi$, so that nonzero complex numbers will have unique arguments. The value of $\arg(w)$ in the interval $0 \leq \theta < 2\pi$ is called the **principal argument** of $w$ and is denoted $\text{Arg}\,(w)$. Every complex number $w$ except 0 has a unique principal argument $\text{Arg}\,(w)$.

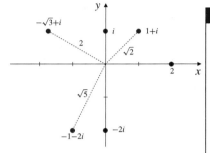

**Figure I.3**    Some complex numbers with their moduli

---

**Example 1**    (Some moduli and principal arguments)  See Figure I.3.

| | |
|---|---|
| $|2| = 2$ | $\text{Arg}\,(2) = 0$ |
| $|1 + i| = \sqrt{2}$ | $\text{Arg}\,(1 + i) = \pi/4$ |
| $|i| = 1$ | $\text{Arg}\,(i) = \pi/2$ |
| $|-2i| = 2$ | $\text{Arg}\,(-2i) = 3\pi/2$ |
| $|-\sqrt{3} + i| = 2$ | $\text{Arg}\,(-\sqrt{3} + i) = 5\pi/6$ |
| $|-1 - 2i| = \sqrt{5}$ | $\text{Arg}\,(-1 - 2i) = \pi + \tan^{-1}(2)$. |

---

**Remark**   The function atan2$(x, y)$ defined in many spreadsheet computer programs is a version of the argument of $x + yi$. This function usually takes values in the interval $(-\pi, \pi]$, so its relationship to $\text{Arg}\,(x + yi)$ is given by

$$\text{Arg}\,(x + yi) = \begin{cases} \text{atan2}(x, y) & \text{if } y \geq 0, \\ \text{atan2}(x, y) + 2\pi & \text{if } y < 0 \end{cases}$$

Given the modulus $r = |w|$ and any value of the argument $\theta = \arg(w)$ of a complex number $w = a + bi$, we have $a = r \cos\theta$ and $b = r \sin\theta$, so $w$ can be expressed in terms of its modulus and argument as

$$w = r \cos\theta + i\, r \sin\theta.$$

The expression on the right side is called the **polar representation** of $w$.

---

**DEFINITION 4**

The **conjugate** or **complex conjugate** of a complex number $w = a + bi$ is another complex number, denoted $\overline{w}$, given by

$$\overline{w} = a - bi.$$

---

**Example 2**    $\overline{2 - 3i} = 2 + 3i$,    $\overline{3} = 3$,    $\overline{2i} = -2i$.

Observe that

$$\text{Re}\,(\overline{w}) = \text{Re}\,(w) \qquad\qquad |\overline{w}| = |w|$$
$$\text{Im}\,(\overline{w}) = -\text{Im}\,(w) \qquad \arg\,(\overline{w}) = -\arg\,(w) \quad \text{or } 2\pi - \arg\,(w).$$

In an Argand diagram the point $\overline{w}$ is the reflection of the point $w$ in the real axis. (See Figure I.4.)

Note that $w$ is real ($\text{Im}\,(w) = 0$) if and only if $\overline{w} = w$. Also, $w$ is pure imaginary ($\text{Re}\,(w) = 0$) if and only if $\overline{w} = -w$. (Here $-w = -a - bi$ if $w = a + bi$.)

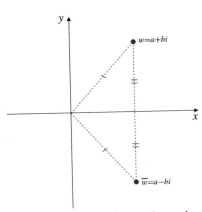

**Figure I.4** A complex number and its conjugate are mirror images of each other in the real axis

## Complex Arithmetic

Like real numbers, complex numbers can be added, subtracted, multiplied, and divided. Two complex numbers are added or subtracted as though they are two-dimensional vectors whose components are their real and imaginary parts.

---

**The sum and difference of complex numbers**

If $w = a + bi$ and $z = x + yi$ where $a, b, x,$ and $y$ are real numbers, then

$$w + z = (a + x) + (b + y)i$$
$$w - z = (a - x) + (b - y)i.$$

---

In an Argand diagram the points $w + z$ and $w - z$ are the points whose position vectors are, respectively, the sum and difference of the position vectors of the points $w$ and $z$. (See Figure I.5.) In particular, the complex number $a + bi$ is the sum of the real number $a = a + 0i$, and the pure imaginary number $bi = 0 + bi$.

Complex addition obeys the same rules as real addition: if $w_1, w_2,$ and $w_3$ are three complex numbers, the following are easily verified:

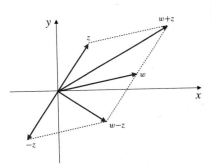

**Figure I.5** Complex numbers are added and subtracted vectorially. Observe the parallelograms

$$w_1 + w_2 = w_2 + w_1 \qquad\qquad \text{Addition is commutative.}$$
$$(w_1 + w_2) + w_3 = w_1 + (w_2 + w_3) \qquad \text{Addition is associative.}$$
$$|w_1 \pm w_2| \le |w_1| + |w_2| \qquad\qquad \text{the triangle inequality}$$

Note that $|w_1 - w_2|$ is the distance between the two points $w_1$ and $w_2$ in the complex plane. Thus the triangle inequality says that in the triangle with vertices $w_1, \mp w_2$ and 0, the length of one side is less than the sum of the other two.

It is also easily verified that the conjugate of a sum (or difference) is the sum (or difference) of the conjugates:

$$\overline{w + z} = \overline{w} + \overline{z}.$$

---

**Example 3**

(a) If $w = 2 + 3i$ and $z = 4 - 5i$, then

$$w + z = (2 + 4) + (3 - 5)i = 6 - 2i$$
$$w - z = (2 - 4) + (3 - (-5))i = -2 + 8i.$$

(b) $3i + (1 - 2i) - (2 + 3i) + 5 = 4 - 2i.$

Multiplication of the complex numbers $w = a + bi$ and $z = x + yi$ is carried out by formally multiplying the binomial expressions and replacing $i^2$ by $-1$:

$$wz = (a + bi)(x + yi) = ax + ayi + bxi + byi^2$$
$$= (ax - by) + (ay + bx)i.$$

**The product of complex numbers**

If $w = a + bi$ and $z = x + yi$ where $a, b, x$, and $y$ are real numbers, then

$$wz = (ax - by) + (ay + bx)i.$$

**Example 4**

(a) $(2 + 3i)(1 - 2i) = 2 - 4i + 3i - 6i^2 = 8 - i.$

(b) $i(5 - 4i) = 5i - 4i^2 = 4 + 5i.$

(c) $(a + bi)(a - bi) = a^2 - abi + abi - b^2i^2 = a^2 + b^2.$

Part (c) of the example above shows that the square of the modulus of a complex number is the product of that number with its complex conjugate:

$$w\,\overline{w} = |w|^2.$$

Complex multiplication has many properties in common with real multiplication. In particular, if $w_1$, $w_2$, and $w_3$ are complex numbers, then

$$w_1 w_2 = w_2 w_1 \qquad \text{Multiplication is commutative.}$$
$$(w_1 w_2)w_3 = w_1(w_2 w_3) \qquad \text{Multiplication is associative.}$$
$$w_1(w_2 + w_3) = w_1 w_2 + w_1 w_3 \qquad \text{Multiplication distributes over addition.}$$

The conjugate of a product is the product of the conjugates:

$$\overline{wz} = \overline{w}\,\overline{z}.$$

To see this, let $w = a + bi$ and $z = x + yi$. Then

$$\overline{wz} = \overline{(ax - by) + (ay + bx)i}$$
$$= (ax - by) - (ay + bx)i$$
$$= (a - bi)(x - yi) = \overline{w}\,\overline{z}.$$

It is particularly easy to determine the product of complex numbers expressed in polar form. If

$$w = r(\cos\theta + i\sin\theta), \qquad \text{and} \qquad z = s(\cos\phi + i\sin\phi),$$

where $r = |w|$, $\theta = \arg(w)$, $s = |z|$, and $\phi = \arg(z)$, then

$$wz = rs(\cos\theta + i\sin\theta)(\cos\phi + i\sin\phi)$$
$$= rs\big((\cos\theta\cos\phi - \sin\theta\sin\phi) + i(\sin\theta\cos\phi + \cos\theta\sin\phi)\big)$$
$$= rs\big(\cos(\theta + \phi) + i\sin(\theta + \phi)\big).$$

(See Figure I.6.) Since arguments are only determined up to integer multiples of $2\pi$, we have proved that

> **The modulus and argument of a product**
>
> $$|wz| = |w||z|, \qquad \text{and} \qquad \arg(wz) = \arg(w) + \arg(z).$$

The second of these equations says that the set $\arg(wz)$ consists of all numbers $\theta + \phi$ where $\theta$ belongs to the set $\arg(w)$ and $\phi$ to the set $\arg(z)$.

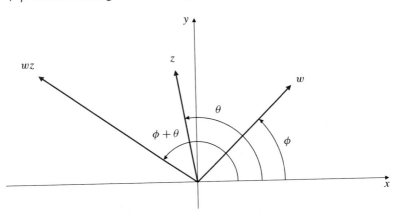

**Figure I.6** The argument of a product is the sum of the arguments of the factors

More generally, if $w_1, w_2, \ldots w_n$ are complex numbers then

$$|w_1 w_2 \cdots w_n| = |w_1||w_2| \cdots |w_n|$$
$$\arg(w_1 w_2 \cdots w_n) = \arg(w_1) + \arg(w_2) + \cdots + \arg(w_n).$$

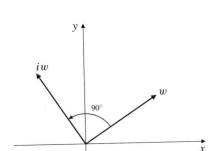

**Figure I.7** Multiplication by $i$ corresponds to counterclockwise rotation by $90°$

Multiplication of a complex number by $i$ has a particularly simple geometric interpretation in an Argand diagram. Since $|i| = 1$ and $\arg(i) = \pi/2$, multiplication of $w = a + bi$ by $i$ leaves the modulus of $w$ unchanged but increases its argument by $\pi/2$. Thus multiplication by $i$ rotates the position vector of $w$ counterclockwise by $90°$ about the origin.

Let $z = \cos\theta + i\sin\theta$. Then $|z| = 1$ and $\arg(z) = \theta$. Since the modulus of a product is the product of the moduli of the factors and the argument of a product is the sum of the arguments of the factors, we have $|z^n| = |z|^n = 1$ and $\arg(z^n) = n\arg(z) = n\theta$. Thus

$$z^n = \cos n\theta + i\sin n\theta,$$

and we have proved

**THEOREM 1**   **De Moivre's Theorem**

$$\big(\cos\theta + i\sin\theta\big)^n = \cos n\theta + i\sin n\theta.$$

**Example 5**    Express $(1 + i)^5$ in the form $a + bi$.

**Solution**    Since $|(1 + i)^5| = |1 + i|^5 = (\sqrt{2})^5 = 4\sqrt{2}$, and $\arg\left((1 + i)^5\right) = 5 \arg(1 + i) = \dfrac{5\pi}{4}$, we have

$$(1 + i)^5 = 4\sqrt{2}\left(\cos\frac{5\pi}{4} + i\sin\frac{5\pi}{4}\right) = 4\sqrt{2}\left(-\frac{1}{\sqrt{2}} - \frac{1}{\sqrt{2}}i\right) = -4 - 4i.$$

De Moivre's Theorem can be used to generate trigonometric identities for multiples of an angle. For example, for $n = 2$ we have

$$\cos 2\theta + i\sin 2\theta = \left(\cos\theta + i\sin\theta\right)^2 = \cos^2\theta - \sin^2\theta + 2i\cos\theta\sin\theta,$$

Thus $\cos 2\theta = \cos^2\theta - \sin^2\theta$, and $\sin 2\theta = 2\sin\theta\cos\theta$.

The **reciprocal** of the nonzero complex number $w = a + bi$ can be calculated by multiplying the numerator and denominator of the reciprocal expression by the conjugate of $w$:

$$w^{-1} = \frac{1}{w} = \frac{1}{a + bi} = \frac{a - bi}{(a + bi)(a - bi)} = \frac{a - bi}{a^2 + b^2} = \frac{\overline{w}}{|w|^2}.$$

Since $|\overline{w}| = |w|$, and $\arg(\overline{w}) = -\arg(w)$, we have

$$\left|\frac{1}{w}\right| = \frac{|\overline{w}|}{|w|^2} = \frac{1}{|w|}, \qquad \text{and} \qquad \arg\left(\frac{1}{w}\right) = -\arg(w).$$

The **quotient** $z/w$ of two complex numbers $z = x + yi$ and $w = a + bi$ is the product of $z$ and $1/w$, so

$$\frac{z}{w} = \frac{z\overline{w}}{|w|^2} = \frac{(x + yi)(a - bi)}{a^2 + b^2} = \frac{xa + yb + i(ya - xb)}{a^2 + b^2}.$$

We have:

**The modulus and argument of a quotient**

$$\left|\frac{z}{w}\right| = \frac{|z|}{|w|} \qquad \text{and} \qquad \arg\left(\frac{z}{w}\right) = \arg(z) - \arg(w).$$

The set $\arg(z/w)$ consists of all numbers $\theta - \phi$ where $\theta$ belongs to the set $\arg(z)$ and $\phi$ to the set $\arg(w)$.

**Example 6**    Simplify   (a) $\dfrac{2 + 3i}{4 - i}$.   (b) $\dfrac{i}{1 + i\sqrt{3}}$.

**Solution**

(a) $\dfrac{2 + 3i}{4 - i} = \dfrac{(2 + 3i)(4 + i)}{(4 - i)(4 + i)} = \dfrac{8 - 3 + (2 + 12)i}{4^2 + 1^2} = \dfrac{5}{17} + \dfrac{14}{17}i.$

(b) $\dfrac{i}{1+i\sqrt{3}} = \dfrac{i(1-i\sqrt{3})}{(1+i\sqrt{3})(1-i\sqrt{3})} = \dfrac{\sqrt{3}+i}{1^2+3} = \dfrac{\sqrt{3}}{4} + \dfrac{1}{4}i.$

Alternatively, since $|1+i\sqrt{3}| = 2$ and $\arg(1+i\sqrt{3}) = \tan^{-1}\sqrt{3} = \dfrac{\pi}{3}$, the

quotient in (b) has modulus $\dfrac{1}{2}$ and argument $\dfrac{\pi}{2} - \dfrac{\pi}{3} = \dfrac{\pi}{6}$. Thus

$$\frac{i}{1+i\sqrt{3}} = \frac{1}{2}\left(\cos\frac{\pi}{6} + i\sin\frac{\pi}{6}\right) = \frac{\sqrt{3}}{4} + \frac{1}{4}i.$$

## Roots of Complex Numbers

If $a$ is a positive real number, there are two distinct real numbers whose square is $a$. These are usually denoted

$\sqrt{a}$    (the positive square root of $a$), and

$-\sqrt{a}$    (the negative square root of $a$).

Every nonzero complex number $z = x + yi$ (where $x^2 + y^2 > 0$) also has two square roots; if $w_1$ is a complex number such that $w_1^2 = z$ then $w_2 = -w_1$ also satisfies $w_2^2 = z$. Again we would like to single out one of these roots and call it $\sqrt{z}$.

Let $r = |z|$, so that $r > 0$. Let $\theta = \text{Arg}(z)$. Thus $0 \le \theta < 2\pi$. Since

$$z = r(\cos\theta + i\sin\theta),$$

the complex number

$$w = \sqrt{r}\left(\cos\frac{\theta}{2} + i\,\sin\frac{\theta}{2}\right)$$

clearly satisfies $w^2 = z$. We call this $w$ the **principal square root** of $z$, and denote it $\sqrt{z}$. The two solutions of the equation $w^2 = z$ are, thus, $w = \sqrt{z}$ and $w = -\sqrt{z}$.

Observe that the imaginary part of $\sqrt{z}$ is always nonnegative since $\sin\dfrac{\theta}{2} \ge 0$

for $0 \le \theta < 2\pi$; in fact, since $\dfrac{\theta}{2} < \pi$, $\sqrt{z}$ is real only if $z$ is real ($\theta = 0$) in which case $\sqrt{z}$ is positive.

**Example 7**

(a) $\sqrt{4} = \sqrt{4(\cos 0 + i\sin 0)} = 2.$

(b) $\sqrt{i} = \sqrt{1\left(\cos\dfrac{\pi}{2} + i\sin\dfrac{\pi}{2}\right)} = \cos\dfrac{\pi}{4} + i\sin\dfrac{\pi}{4} = \dfrac{1}{\sqrt{2}} + \dfrac{1}{\sqrt{2}}i.$

(c) $\sqrt{-4i} = \sqrt{4\left(\cos\dfrac{3\pi}{2} + i\sin\dfrac{3\pi}{2}\right)} = 2\left(\cos\dfrac{3\pi}{4} + i\sin\dfrac{3\pi}{4}\right) = -\sqrt{2} + i\sqrt{2}.$

(d) $\sqrt{-\dfrac{1}{2} + i\dfrac{\sqrt{3}}{2}} = \sqrt{\cos\dfrac{2\pi}{3} + i\sin\dfrac{2\pi}{3}} = \cos\dfrac{\pi}{3} + i\sin\dfrac{\pi}{3} = \dfrac{1}{2} + \dfrac{\sqrt{3}}{2}i.$

Given a nonzero complex number $z$ we can find $n$ distinct complex numbers $w$ that satisfy $w^n = z$. These $n$ numbers are called $n$th roots of $z$. For example, if $z = 1 = \cos 0 + i \sin 0$, then each of the numbers

$$w_1 = 1$$
$$w_2 = \cos \frac{2\pi}{n} + i \sin \frac{2\pi}{n}$$
$$w_3 = \cos \frac{4\pi}{n} + i \sin \frac{4\pi}{n}$$
$$w_4 = \cos \frac{6\pi}{n} + i \sin \frac{6\pi}{n}$$
$$\vdots$$
$$w_n = \cos \frac{2(n-1)\pi}{n} + i \sin \frac{2(n-1)\pi}{n}$$

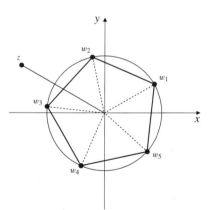

satisfies $w^n = 1$ so is an $n$th root of 1. (These numbers are usually called the **$n$th roots of unity**.) Figure I.8 shows the three cube roots of 1. Observe that they are at the three vertices of an equilateral triangle with centre at the origin and one vertex at 1. In general, the $n$ $n$th roots of unity lie on a circle of radius 1 centred at the origin, at the vertices of a regular $n$-sided polygon with one vertex at 1.

If $z$ is any nonzero complex number, and $\theta$ is the principal argument of $z$ ($0 \le \theta < 2\pi$) then the number

$$w_1 = |z|^{1/n} \left( \cos \frac{\theta}{n} + i \sin \frac{\theta}{n} \right)$$

is called the **principal $n$th root** of $z$. All the $n$th roots of $z$ are on the circle of radius $|z|^{1/n}$ centred at the origin, and are at the vertices of a regular $n$-sided polygon with one vertex at $w_1$. (See Figure I.9.) The other $n$th roots are

$$w_2 = |z|^{1/n} \left( \cos \frac{\theta + 2\pi}{n} + i \sin \frac{\theta + 2\pi}{n} \right)$$
$$w_3 = |z|^{1/n} \left( \cos \frac{\theta + 4\pi}{n} + i \sin \frac{\theta + 4\pi}{n} \right)$$
$$\vdots$$
$$w_n = |z|^{1/n} \left( \cos \frac{\theta + 2(n-1)\pi}{n} + i \sin \frac{\theta + 2(n-1)\pi}{n} \right).$$

We can obtain all $n$ of the $n$th roots of $z$ by multiplying the principal $n$th root by the $n$th roots of unity.

**Figure I.8**    The cube roots of unity

**Figure I.9**    The five 5th roots of $z$

---

**Example 8**    Find the 4th roots of $-4$. Sketch them in an Argand diagram.

**Solution**    Since $|-4|^{1/4} = \sqrt{2}$ and $\arg(-4) = \pi$, the principal 4th root of $-4$ is

$$w_1 = \sqrt{2} \left( \cos \frac{\pi}{4} + i \sin \frac{\pi}{4} \right) = 1 + i.$$

The other three 4th roots are at the vertices of a square with centre at the origin and one vertex at $1 + i$. (See Figure I.10.) Thus the other roots are

$$w_2 = -1 + i, \qquad w_3 = -1 - i, \qquad w_4 = 1 - i.$$

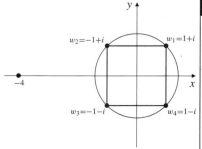

**Figure I.10**    The four 4th roots of $-4$

# Exercises

In Exercises 1–4, find the real and imaginary parts (Re $(z)$ and Im $(z)$) of the given complex numbers $z$, and sketch the position of each number in the complex plane (i.e. in an Argand diagram).

**1.** $z = -5 + 2i$

**2.** $z = 4 - i$

**3.** $z = -\pi i$

**4.** $z = -6$

In Exercises 5–15, find the modulus $r = |z|$ and the principal argument $\theta = \mathrm{Arg}\,(z)$ of each given complex number $z$, and express $z$ in terms of $r$ and $\theta$.

**5.** $z = -1 + i$

**6.** $z = -2$

**7.** $z = 3i$

**8.** $z = -5i$

**9.** $z = 1 + 2i$

**10.** $z = -2 + i$

**11.** $z = -3 - 4i$

**12.** $z = 3 - 4i$

**13.** $z = \sqrt{3} - i$

**14.** $z = -\sqrt{3} - 3i$

**15.** $z = 3\cos\dfrac{4\pi}{5} + 3i\sin\dfrac{4\pi}{5}$

**16.** If $\mathrm{Arg}\,(z) = 7\pi/4$ and $\mathrm{Arg}\,(w) = \pi/2$, find $\mathrm{Arg}\,(zw)$.

**17.** If $\mathrm{Arg}\,(z) = \pi/6$ and $\mathrm{Arg}\,(w) = \pi/4$, find $\mathrm{Arg}\,(z/w)$.

In Exercises 18–23, express in the form $z = x + yi$ the complex number $z$ whose modulus and argument is given.

**18.** $|z| = 2$,    $\arg(z) = \pi$

**19.** $|z| = 5$,    $\arg(z) = \tan^{-1}\dfrac{3}{4}$

**20.** $|z| = 1$,    $\arg(z) = \dfrac{3\pi}{4}$

**21.** $|z| = \pi$,    $\arg(z) = \dfrac{\pi}{6}$

**22.** $|z| = 0$,    $\arg(z) = 1$

**23.** $|z| = \dfrac{1}{2}$,    $\arg(z) = -\dfrac{\pi}{3}$

In Exercises 24–27, find the complex conjugates of the given complex numbers.

**24.** $5 + 3i$

**25.** $-3 - 5i$

**26.** $4i$

**27.** $2 - i$

Describe, geometrically, (or make a sketch of) the set of points $z$ in the complex plane satisfying the given equations or inequalities in Exercises 28–33.

**28.** $|z| = 2$

**29.** $|z| \le 2$

**30.** $|z - 2i| \le 3$

**31.** $|z - 3 + 4i| \le 5$

**32.** $\arg z = \dfrac{\pi}{3}$

**33.** $\pi \le \arg(z) \le \dfrac{7\pi}{4}$

Simplify the expressions in Exercises 34–43.

**34.** $(2 + 5i) + (3 - i)$

**35.** $i - (3 - 2i) + (7 - 3i)$

**36.** $(4 + i)(4 - i)$

**37.** $(1 + i)(2 - 3i)$

**38.** $(a + bi)(\overline{2a - bi})$

**39.** $(2 + i)^3$

**40.** $\dfrac{2 - i}{2 + i}$

**41.** $\dfrac{1 + 3i}{2 - i}$

**42.** $\dfrac{1 + i}{i(2 + 3i)}$

**43.** $\dfrac{(1 + 2i)(2 - 3i)}{(2 - i)(3 + 2i)}$

**44.** Prove that $\overline{z + w} = \overline{z} + \overline{w}$.

**45.** Prove that $\overline{\left(\dfrac{z}{w}\right)} = \dfrac{\overline{z}}{\overline{w}}$.

**46.** Express each of the complex numbers $z = 3 + i\sqrt{3}$ and $w = -1 + i\sqrt{3}$ in polar form (that is, in terms of its modulus and argument). Use these expressions to calculate $zw$ and $z/w$.

**47.** Repeat the previous exercise for $z = -1 + i$ and $w = 3i$.

**48.** Use de Moivre's Theorem to find a trigonometric identity for $\cos 3\theta$ in terms of $\cos\theta$, and one for $\sin 3\theta$ in terms of $\sin\theta$.

**49.** Describe the solutions, if any, of the equations (a) $\overline{z} = 2/z$, and (b) $\overline{z} = -2/z$.

**50.** For positive real numbers $a$ and $b$ it is always true that $\sqrt{ab} = \sqrt{a}\sqrt{b}$. Does a similar identity hold for $\sqrt{zw}$, where $z$ and $w$ are complex numbers? *Hint:* consider $z = w = -1$.

**51.** Find the three cube roots of $-1$.

**52.** Find the three cube roots of $-8i$.

**53.** Find the three cube roots of $-1 + i$.

**54.** Find all the fourth roots of $4$.

**55.** Find all complex solutions of the equation $z^4 + 1 - i\sqrt{3} = 0$.

**56.** Find all solutions of $z^5 + a^5 = 0$, where $a$ is a positive real number.

∗ **57.** Show that the sum of the $n$ $n$th roots of unity is zero. *Hint:* show that these roots are all powers of the principal root.

# Appendix II

# Complex Functions

Most of this book is concerned with developing the properties of **real functions**, that is, functions of one or more real variables, having values that are themselves real numbers or vectors with real components. The definition of *function* given in Section P.4 can be paraphrased to allow for complex-valued functions of a complex variable.

**DEFINITION 1**

> A **complex function** $f$ is a rule that assigns a unique complex number $f(z)$ to each number $z$ in some set of complex numbers (called the **domain** of the function).

Typically, we will use $z = x + yi$ to denote a general point in the domain of a complex function, and $w = u + vi$ to denote the value of the function at $z$; if $w = f(z)$ then the real and imaginary parts of $w$ ($u = \text{Re}(w)$ and $v = \text{Im}(w)$) are real-valued functions of $z$, and hence real-valued functions of the two real variables $x$ and $y$:

$$u = u(x, y), \qquad v = v(x, y).$$

For example, the complex function $f(z) = z^2$, whose domain is the whole complex plane $\mathbb{C}$, assigns the value $z^2$ to the complex number $z$. If $w = z^2$ (where $w = u+vi$ and $z = x + yi$), then

$$u + vi = (x + yi)^2 = x^2 - y^2 + 2xyi,$$

so that

$$u = \text{Re}(z^2) = x^2 - y^2, \qquad \text{and} \qquad v = \text{Im}(z^2) = 2xy.$$

It is not convenient to draw the *graph* of a complex function. The graph of $w = f(z)$ would have to be drawn in a four-dimensional (real) space since two dimensions (a $z$-plane) are required for the independent variable and two more dimensions (a $w$-plane) are required for the dependent variable. Instead, we can graphically represent the behaviour of a complex function $w = f(z)$ by drawing the $z$-plane and the $w$-plane separately, and showing the image in the $w$-plane of certain, appropriately chosen sets of points in the $z$-plane. For example, Figure II.1 illustrates the fact that for the function $w = z^2$ the image of the quarter-disk $|z| \leq a, 0 \leq \arg(z) \leq \frac{\pi}{2}$ is the half-disk $|w| \leq a^2, 0 \leq \arg(w) \leq \pi$. To see why this is so, observe that if $z = r(\cos\theta + i\sin\theta)$, then $w = r^2(\cos 2\theta + i\sin 2\theta)$. Thus the function maps the circle $|z| = r$ onto the circle $|w| = r^2$, and the radial line $\arg(z) = \theta$ onto the radial line $\arg(w) = 2\theta$.

## Limits and Continuity

The concepts of limit and continuity carry over from real functions to complex functions in an obvious way providing we use $|z_1 - z_2|$ as the distance between the complex numbers $z_1$ and $z_2$. We say that

$$\lim_{z \to z_0} f(z) = \lambda$$

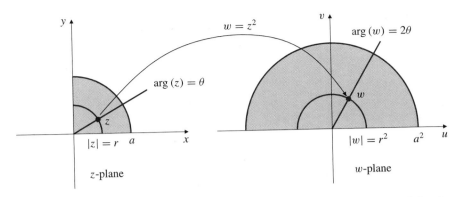

**Figure II.1** The function $w = z^2$ maps a quarter-disk of radius $a$ to a half-disk of radius $a^2$ by squaring the modulus and doubling the argument of each point $z$

provided we can ensure that $|f(z) - \lambda|$ is as small as we wish by taking $z$ sufficiently close to $z_0$. Formally,

**DEFINITION 2**

We say that $f(z)$ tends to the **limit** $\lambda$ as $z$ approaches $z_0$, and we write

$$\lim_{z \to z_0} f(z) = \lambda,$$

if for every positive real number $\epsilon$ there exists a positive real number $\delta$, (depending on $\epsilon$), such that

$$0 < |z - z_0| < \delta \quad \Longrightarrow \quad |f(z) - \lambda| < \epsilon.$$

**DEFINITION 3**

The complex function $f(z)$ is **continuous** at $z = z_0$ if $\lim_{z \to z_0} f(z)$ exists and equals $f(z_0)$.

All the laws of limits and continuity apply as for real functions. Polynomials, that is, functions of the form

$$P(z) = a_0 + a_1 z + a_2 z^2 + \cdots + a_n z^n,$$

are continuous at every point of the complex plane. Rational functions, that is, functions of the form

$$R(z) = \frac{P(z)}{Q(z)},$$

where $P(z)$ and $Q(z)$ are polynomials, are continuous everywhere except at points where $Q(z) = 0$. Integer powers $z^n$ are continuous except at the origin if $n < 0$. The situation for fractional powers is more complicated. For example, $\sqrt{z}$ (the principal square root) is continuous except at points $z = x > 0$. The function $f(z) = \overline{z}$ is continuous everywhere, because

$$|\overline{z} - \overline{z_0}| = |\overline{z - z_0}| = |z - z_0|.$$

## The Complex Derivative

The definition of derivative is the same as for real functions:

**DEFINITION 4**

The complex function $f$ is **differentiable** at $z$, and has **derivative** $f'(z)$ there, provided

$$\lim_{h \to 0} \frac{f(z+h) - f(z)}{h} = f'(z)$$

exists.

Note, however, that in this definition $h$ is a complex number. The limit must exist no matter how $h$ approaches 0 in the complex plane. This fact has profound implications. The existence of a derivative in this sense forces the function $f$ to be much better behaved than is necessary for a differentiable real function. For example, it can be shown that if $f'(z)$ exists for all $z$ in an open region $D$ in $\mathbb{C}$, then $f$ has derivatives of *all* orders throughout $D$. Moreover, such a function is the sum of its Taylor series

$$f(z) = f(z_0) + f'(z_0)(z - z_0) + \frac{f''(z_0)}{2!}(z - z_0)^2 + \cdots$$

about any point $z_0$ in $D$; the series has positive radius of convergence $R$, and converges in the disk $|z - z_0| < R$. For this reason, complex functions that are differentiable on open sets in $\mathbb{C}$ are usually called **analytic functions**. It is beyond the scope of this introductory appendix to prove these assertions. They are proved in courses and texts on complex analysis.

The usual differentiation rules apply:

$$\frac{d}{dz}\big(Af(z) + Bg(z)\big) = Af'(z) + Bg'(z)$$

$$\frac{d}{dz}\big(f(z)g(z)\big) = f'(z)g(z) + f(z)g'(z)$$

$$\frac{d}{dz}\left(\frac{f(z)}{g(z)}\right) = \frac{g(z)f'(z) - f(z)g'(z)}{\big(g(z)\big)^2}$$

$$\frac{d}{dz}f\big(g(z)\big) = f'\big(g(z)\big)g'(z).$$

As one would expect, the derivative of $f(z) = z^n$ is $f'(z) = nz^{n-1}$.

**Example 1**    Show that the function $f(z) = \overline{z}$ is not differentiable at any point.

**Solution**    We have

$$f'(z) = \lim_{h \to 0} \frac{\overline{z+h} - \overline{z}}{h}$$

$$= \lim_{h \to 0} \frac{\overline{z} + \overline{h} - \overline{z}}{h} = \lim_{h \to 0} \frac{\overline{h}}{h}.$$

But $\overline{h}/h = 1$ if $h$ is real, and $\overline{h}/h = -1$ if $h$ is pure imaginary. Since there are real and pure imaginary numbers arbitrarily close to 0, the limit above does not exist, and so $f'(z)$ does not exist.

The following theorem links the existence of the derivative of a complex function $f(z)$ with certain properties of its real and imaginary parts $u(x, y)$ and $v(x, y)$.

**THEOREM 1**

### The Cauchy–Riemann equations

If $f(z) = u(x, y) + iv(x, y)$ is differentiable at $z = x + yi$ then $u$ and $v$ satisfy the Cauchy-Riemann equations

$$\frac{\partial u}{\partial x} = \frac{\partial v}{\partial y}, \qquad \frac{\partial v}{\partial x} = -\frac{\partial u}{\partial y}.$$

Conversely, if $u$ and $v$ are sufficiently smooth, (say, if they have continuous second partial derivatives near $(x, y)$), and if $u$ and $v$ satisfy the Cauchy–Riemann equations at $(x, y)$, then $f$ is differentiable at $z = x + yi$, and

$$f'(z) = \frac{\partial u}{\partial x} + i\frac{\partial v}{\partial x}.$$

**PROOF** First assume that $f$ is differentiable at $z$. Letting $h = s + ti$, we have

$$f'(z) = \lim_{h \to 0} \frac{f(z + h) - f(z)}{h}$$

$$= \lim_{(s,t) \to (0,0)} \left[ \frac{u(x + s, y + t) - u(x, y)}{s + it} + i\frac{v(x + s, y + t) - v(x, y)}{s + it} \right].$$

The limit must be independent of the path along which $h$ approaches 0. Letting $t = 0$, so that $h = s$ approaches 0 along the real axis, we obtain

$$f'(z) = \lim_{s \to 0} \left[ \frac{u(x + s, y) - u(x, y)}{s} + i\frac{v(x + s, y) - v(x, y)}{s} \right] = \frac{\partial u}{\partial x} + i\frac{\partial v}{\partial x}.$$

Similarly, letting $s = 0$, so that $h = ti$ approaches 0 along the imaginary axis, we obtain

$$f'(z) = \lim_{t \to 0} \left[ \frac{u(x, y + t) - u(x, y)}{it} + i\frac{v(x, y + t) - v(x, y)}{it} \right]$$

$$= \lim_{t \to 0} \left[ \frac{v(x, y + t) - v(x, y)}{t} - i\frac{u(x, y + t) - u(x, y)}{t} \right]$$

$$= \frac{\partial v}{\partial y} - i\frac{\partial u}{\partial y}.$$

Equating these two expressions for $f'(z)$ we see that

$$\frac{\partial u}{\partial x} = \frac{\partial v}{\partial y}, \qquad \frac{\partial v}{\partial x} = -\frac{\partial u}{\partial y}.$$

To prove the converse, we use the result of Exercise 18 of Section 12.6. Since $u$ and $v$ are assumed to have continuous second partial derivatives, we must have

$$u(x + s, y + t) - u(x, y) = s\frac{\partial u}{\partial x} + t\frac{\partial u}{\partial y} + O(s^2 + t^2)$$

$$v(x + s, y + t) - v(x, y) = s\frac{\partial v}{\partial x} + t\frac{\partial v}{\partial y} + O(s^2 + t^2),$$

where we have used the "Big-O" notation (see Definition 5 of Section 4.8); the expression $O(\lambda)$ denotes a term satisfying $|O(\lambda)| \leq K|\lambda|$ for some constant $K$. Thus, if $u$ and $v$ satisfy the Cauchy–Riemann equations, then

$$
\frac{f(z+h) - f(z)}{h} = \frac{s\dfrac{\partial u}{\partial x} + t\dfrac{\partial u}{\partial y} + i\left(s\dfrac{\partial v}{\partial x} + t\dfrac{\partial v}{\partial y}\right) + O(s^2 + t^2)}{s + it}
$$

$$
= \frac{(s+it)\dfrac{\partial u}{\partial x} + i(s+it)\dfrac{\partial v}{\partial x}}{s+it} + O(\sqrt{s^2 + t^2})
$$

$$
= \frac{\partial u}{\partial x} + i\frac{\partial v}{\partial x} + O(\sqrt{s^2 + t^2}).
$$

Thus we may let $h = s + ti$ approach 0 and obtain

$$
f'(z) = \frac{\partial u}{\partial x} + i\frac{\partial v}{\partial x}.
$$

It follows immediately from the Cauchy–Riemann equations that the real and imaginary parts of a differentiable complex function are real harmonic functions:

$$
\frac{\partial^2 u}{\partial x^2} + \frac{\partial^2 u}{\partial y^2} = 0, \qquad \frac{\partial^2 v}{\partial x^2} + \frac{\partial^2 v}{\partial y^2} = 0.
$$

(See Exercise 15 of Section 12.4.)

## The Exponential Function

Consider the function

$$
f(z) = e^x \cos y + ie^x \sin y,
$$

where $z = x + yi$. The real and imaginary parts of $f(z)$,

$$
u(x, y) = \text{Re}\,(f(z)) = e^x \cos y, \qquad \text{and} \qquad v(x, y) = \text{Im}\,(f(z)) = e^x \sin y
$$

satisfy the Cauchy–Riemann equations

$$
\frac{\partial u}{\partial x} = e^x \cos y = \frac{\partial v}{\partial y}, \qquad \text{and} \qquad \frac{\partial v}{\partial x} = e^x \sin y = -\frac{\partial u}{\partial y}
$$

everywhere in the $z$-plane. Therefore $f(z)$ is differentiable (analytic) everywhere, and satisfies

$$
f'(z) = \frac{\partial u}{\partial x} + i\frac{\partial v}{\partial x} = e^x \cos y + ie^x \sin y = f(z).
$$

Evidently $f(0) = 1$, and $f(z) = e^x$ if $z = x$ is a real number. It is therefore natural to denote the function $f(z)$ as the exponential function $e^z$.

**The complex exponential function**

$$
e^z = e^x(\cos y + i \sin y) \qquad \text{for} \quad z = x + yi.
$$

In particular, if $z = yi$ is pure imaginary, then

$$e^{yi} = \cos y + i \sin y,$$

a fact that can also be obtained by separating the real and imaginary parts of the Maclaurin series for $e^{yi}$:

$$e^{yi} = 1 + (yi) + \frac{(yi)^2}{2!} + \frac{(yi)^3}{3!} + \frac{(yi)^4}{4!} + \frac{(yi)^5}{5!} + \cdots$$

$$= \left(1 - \frac{y^2}{2!} + \frac{y^4}{4!} - \cdots\right) + i\left(y - \frac{y^3}{3!} + \frac{y^5}{5!} - \cdots\right)$$

$$= \cos y + i \sin y.$$

Observe that

$$|e^z| = \sqrt{e^{2x}\left(\cos^2 y + \sin^2 y\right)} = e^x,$$

$$\arg(e^z) = \arg(e^{yi}) = \arg(\cos y + i \sin y) = y,$$

$$\overline{e^z} = e^x \cos y - i e^x \sin y = e^x \cos(-y) + i e^x \sin(-y) = e^{\bar{z}}.$$

In summary:

> **Properties of the exponential function**
>
> If $z = x + yi$ then $\overline{e^z} = e^{\bar{z}}$. Also,
>
> $$\operatorname{Re}(e^z) = e^x \cos y, \qquad\qquad |e^z| = e^x,$$
> $$\operatorname{Im}(e^z) = e^x \sin y, \qquad\qquad \arg(e^z) = y.$$

**Example 2**   Sketch the image in the $w$-plane of the rectangle

$$R : a \le x \le b, \quad c \le y \le d$$

in the $z$-plane under the transformation $w = e^z$.

**Solution**   The vertical lines $x = a$ and $x = b$ get mapped to the concentric circles $|w| = e^a$ and $|w| = e^b$. The horizontal lines $y = c$ and $y = d$ get mapped to the radial lines $\arg(w) = c$ and $\arg(w) = d$. Thus the rectangle $R$ gets mapped to the polar region $P$ shown in Figure II.2.

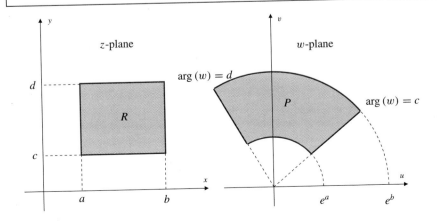

**Figure II.2**   Under the exponential function $w = e^z$ vertical lines get mapped to circles centred at the origin and horizontal lines get mapped to half-lines radiating from the origin

Note that if $d - c \geq 2\pi$, then the image of $R$ will be the entire annular region $e^a \leq |w| \leq e^b$, which may be covered more than once. The exponential function $e^z$ is periodic with period $2\pi i$:

$$e^{z+2\pi i} = e^z \qquad \text{for all} \quad z,$$

and is therefore not one-to-one on the whole complex plane. However, $w = e^z$ is one-to-one from any horizontal strip of the form

$$-\infty < x < \infty, \qquad c < y \leq c + 2\pi$$

onto the whole $w$-plane.

## The Fundamental Theorem of Algebra

As observed at the beginning of this appendix, extending the number system to include complex numbers allows larger classes of equations to have solutions. We conclude this appendix by verifying that polynomial equations always have solutions in the complex numbers.

A **complex polynomial** of degree $n$ is a function of the form

$$P_n(z) = a_n z^n + a_{n-1} z^{n-1} + \cdots + a_2 z^2 + a_1 z + a_0,$$

where $a_0, a_1, \ldots, a_n$ are complex numbers and $a_n \neq 0$. The numbers $a_i$ ($0 \leq i \leq n$) are called the **coefficients** of the polynomial. If they are all real numbers then $P_n(x)$ is called a real polynomial.

A complex number $z_0$ that satisfies the equation $P(z_0) = 0$ is called a **zero** or **root** of the polynomial. Every polynomial of degree 1 has a zero: if $a_1 \neq 0$, then $a_1 z + a_0$ has zero $z = -a_0/a_1$. This zero is real if $a_1$ and $a_0$ are both real.

Similarly, every complex polynomial of degree 2 has two zeros. If the polynomial is

$$P_2(z) = a_2 z^2 + a_1 z + a_0$$

(where $a_2 \neq 0$), then the zeros are given by the *quadratic formula*

$$z = z_1 = \frac{-a_1 - \sqrt{a_1^2 - 4a_2 a_0}}{2a_2}, \quad \text{and} \quad z = z_2 = \frac{-a_1 + \sqrt{a_1^2 - 4a_2 a_0}}{2a_2}.$$

In this case $P_2(z)$ has two linear factors:

$$P_2(z) = a_2(z - z_1)(z - z_2).$$

Even if $a_1^2 - 4a_2 a_0 = 0$, so that $z_1 = z_2$, we still regard the polynomial as having two (equal) zeros, one corresponding to each factor. If the coefficients $a_0$, $a_1$, and $a_2$ are all real numbers the zeros will be real provided $a_1^2 \geq 4a_2 a_0$. When real coefficients satisfy $a_1^2 < 4a_2 a_0$ then the zeros are complex, in fact complex conjugates: $z_2 = \overline{z_1}$.

**Example 3** Solve the equation $z^2 + 2iz - (1 + i) = 0$.

**Solution** The zeros of this equation are

$$z = \frac{-2i \pm \sqrt{-4 + 4(1 + i)}}{2}$$

$$= -i \pm \sqrt{i}$$

$$= -i \pm \frac{1 + i}{\sqrt{2}} = \frac{1}{\sqrt{2}}\left(1 + (1 - \sqrt{2})i\right) \quad \text{or} \quad -\frac{1}{\sqrt{2}}\left(1 + (1 + \sqrt{2})i\right).$$

∎

The Fundamental Theorem of Algebra asserts that every complex polynomial of positive degree has a complex zero.

**THEOREM 2**

**The Fundamental Theorem of Algebra**

If $P(z) = a_n z^n + a_{n-1} z^{n-1} + \cdots + a_1 z + a_0$ is a complex polynomial of degree $n \geq 1$, then there exists a complex number $z_1$ such that $P(z_1) = 0$.

**PROOF** (We will only give an informal sketch of the proof.) We can assume that the coefficient of $z^n$ in $P(z)$ is $a_n = 1$ since we can divide the equation $P(z) = 0$ by $a_n$ without changing its solutions. We can also assume that $a_0 \neq 0$; if $a_0 = 0$, then $z = 0$ is certainly a zero of $P(z)$. Thus we deal with the polynomial

$$P(z) = z^n + Q(z)$$

where $Q(z)$ is a polynomial of degree less than $n$ having a nonzero constant term. If $R$ is sufficiently large, then $|Q(z)|$ will be less than $R^n$ for all numbers $z$ satisfying $|z| = R$. As $z$ moves around the circle $|z| = R$ in the $z$-plane, $w = z^n$ moves around the circle $|w| = R^n$ in the $w$-plane ($n$ times). Since the distance from $z^n$ to $P(z)$ is equal to $|P(z) - z^n| = |Q(z)| < R^n$, it follows that the image of the circle $|z| = R$ under the transformation $w = P(z)$ is a curve that winds around the origin $n$ times. (If you walk around a circle of radius $r$ $n$ times, with your dog on a leash of length less than $r$, and your dog returns to his starting point, then he must also go around the centre of the circle $n$ times.) This situation is illustrated for the particular case

$$P(z) = z^3 + z^2 - iz + 1, \qquad |z| = 2$$

in Figure II.3. The image of $|z| = 2$ is the large curve in the $w$-plane that winds around the origin three times. As $R$ decreases, the curve traced out by $w = P(z)$ for $|z| = R$ changes continuously. For $R$ close to 0, it is a small curve staying close to the constant term $a_0$ of $P(z)$. For small enough $R$ the curve will not enclose the origin. (In Figure II.3 the image of $|z| = 0.3$ is the small curve staying close to the point 1 in the $w$-plane.) Thus for some value of $R$, say $R = R_1$, the curve must pass through the origin. That is, there must be a complex number $z_1$, with $|z_1| = R_1$, such that $P(z_1) = 0$.

**Remark** The above proof suggests that there should be $n$ such solutions of the equation $P(z) = 0$; the curve has to go from winding around the origin $n$ times to winding around the origin 0 times as $R$ decreases towards 0. We can establish this as follows. $P(z_1) = 0$ implies that $z - z_1$ is a factor of $P(z)$:

$$P(z) = (z - z_1)P_{n-1}(z),$$

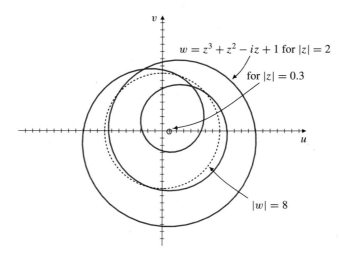

**Figure II.3**    The image of the circle $|z| = 2$ winds around the origin in the $w$-plane three times, but the image of $|z| = 0.3$ does not wind around the origin at all

where $P_{n-1}$ is a polynomial of degree $n - 1$. If $n > 1$, then $P_{n-1}$ must also have a zero, $z_2$, by the Fundamental Theorem. We can continue this argument inductively to obtain $n$ zeros and factor $P(z)$ into a product of the constant $a_n$ and $n$ linear factors:

$$P(z) = a_n(z - z_1)(z - z_2) \cdots (z - z_n).$$

Of course, some of the zeros can be equal.

***Remark***   If $P$ is a real polynomial, that is, one whose coefficients are all real numbers, then $\overline{P(z)} = P(\overline{z})$. Therefore, if $z_1$ is a non-real zero of $P(z)$, then so is $z_2 = \overline{z_1}$:

$$P(z_2) = P(\overline{z_1}) = \overline{P(z_1)} = \overline{0} = 0.$$

Real polynomials can have complex zeros, but they must always occur in complex conjugate pairs. Every real polynomial of odd degree must have at least one real zero.

**Example 4**    Show that $z_1 = -i$ is a zero of the polynomial

$$P(z) = z^4 + 5z^3 + 7z^2 + 5z + 6,$$

and find all the other zeros of this polynomial.

***Solution***   First observe that $P(z_1) = P(-i) = 1 + 5i - 7 - 5i + 6 = 0$, so $z_1 = -i$ is indeed a zero. Since the coefficients of $P(z)$ are real, $z_2 = i$ must also be a zero. Thus $z + i$ and $z - i$ are factors of $P(z)$, and so is

$$(z + i)(z - i) = z^2 + 1.$$

Dividing $P(z)$ by $z^2 + 1$ we obtain

$$\frac{P(z)}{z^2 + 1} = z^2 + 5z + 6 = (z + 2)(z + 3).$$

Thus the four zeros of $P(z)$ are $z_1 = -i$, $z_2 = i$, $z_3 = -2$, and $z_4 = -3$.

# Exercises

In Exercises 1–12, the $z$-plane region $D$ consists of the complex numbers $z = x + yi$ that satisfy the given conditions. Describe (or sketch) the image $R$ of $D$ in the $w$-plane under the given function $w = f(z)$.

**1.** $0 \le x \le 1$, $0 \le y \le 2$;    $w = \overline{z}$.

**2.** $x + y = 1$;    $w = \overline{z}$.

**3.** $1 \le |z| \le 2$, $\dfrac{\pi}{2} \le \arg z \le \dfrac{3\pi}{4}$;    $w = z^2$.

**4.** $0 \le |z| \le 2$, $0 \le \arg(z) \le \dfrac{\pi}{2}$;    $w = z^3$.

**5.** $0 < |z| \le 2$, $0 \le \arg(z) \le \dfrac{\pi}{2}$;    $w = \dfrac{1}{z}$.

**6.** $\dfrac{\pi}{4} \le \arg(z) \le \dfrac{\pi}{3}$;    $w = -iz$.

**7.** $\arg(z) = -\dfrac{\pi}{3}$;    $w = \sqrt{z}$.

**8.** $x = 1$;    $w = z^2$.    **9.** $y = 1$;    $w = z^2$.

**10.** $x = 1$;    $w = \dfrac{1}{z}$.

**11.** $-\infty < x < \infty$, $\dfrac{\pi}{4} \le y \le \dfrac{\pi}{2}$;    $w = e^z$.

**12.** $0 < x < \dfrac{\pi}{2}$, $0 < y < \infty$;    $w = e^{iz}$.

In Exercises 13–16, verify that the real and imaginary parts of each function $f(z)$ satisfy the Cauchy–Riemann equations, and thus find $f'(z)$.

**13.** $f(z) = z^2$.    **14.** $f(z) = z^3$.

**15.** $f(z) = \dfrac{1}{z}$.    **16.** $f(z) = e^{z^2}$.

**17.** Use the fact that $e^{yi} = \cos y + i \sin y$ (for real $y$) to show that

$$\cos y = \frac{e^{yi} + e^{-yi}}{2} \quad \text{and} \quad \sin y = \frac{e^{yi} - e^{-yi}}{2i}.$$

The previous exercise suggests that we define complex functions

$$\cos z = \frac{e^{zi} + e^{-zi}}{2} \quad \text{and} \quad \sin z = \frac{e^{zi} - e^{-zi}}{2i},$$

as well as extend the definitions of the hyperbolic functions to

$$\cosh z = \frac{e^z + e^{-z}}{2} \quad \text{and} \quad \sinh z = \frac{e^z - e^{-z}}{2}.$$

Exercises 18–26 develop properties of these functions and relationships among them.

**18.** Show that $\cos z$ and $\sin z$ are periodic with period $2\pi$, and that $\cosh z$ and $\sinh z$ are periodic with period $2\pi i$.

**19.** Show that $(d/dz) \sin z = \cos z$, and $(d/dz) \cos z = -\sin z$. What are the derivatives of $\sinh z$ and $\cosh z$?

**20.** Verify the identities $\cos z = \cosh(iz)$, and $\sin z = -i \sinh(iz)$. What are the corresponding identities for $\cosh z$ and $\sinh(z)$ in terms of $\cos$ and $\sin$?

**21.** Find all complex zeros of $\cos z$ (that is, all solutions of $\cos z = 0$).

**22.** Find all complex zeros of $\sin z$.

**23.** Find all complex zeros of $\cosh z$ and $\sinh z$.

**24.** Show that $\mathrm{Re}\,(\cosh z) = \cosh x \cos y$ and $\mathrm{Im}\,(\cosh z) = \sinh x \sin y$.

**25.** Find the real and imaginary parts of $\sinh z$.

**26.** Find the real and imaginary parts of $\cos z$ and $\sin z$.

Find the zeros of the polynomials in Exercises 27–32.

**27.** $P(z) = z^2 + 2iz$    **28.** $P(z) = z^2 - 2z + i$

**29.** $P(z) = z^2 + 2z + 5$    **30.** $P(z) = z^2 - 2iz - 1$

**31.** $P(z) = z^3 - 3iz^2 - 2z$    **32.** $P(z) = z^4 - 2z^2 + 4$

**33.** The polynomial $P(z) = z^4 + 1$ has two pairs of complex conjugate zeros. Find them, and hence express $P(z)$ as a product of two quadratic factors with real coefficients.

In Exercises 34–36, check that the given number $z_1$ is a zero of the given polynomial, and find all the zeros of the polynomial.

**34.** $P(z) = z^4 - 4z^3 + 12z^2 - 16z + 16$;    $z_1 = 1 - \sqrt{3}\,i$.

**35.** $P(z) = z^5 + 3z^4 + 4z^3 + 4z^2 + 3z + 1$;    $z_1 = i$.

**36.** $P(z) = z^5 - 2z^4 - 8z^3 + 8z^2 + 31z - 30$;    $z_1 = -2 + i$.

**37.** Show that the image of the circle $|z| = 2$ under the mapping $w = z^4 + z^3 - 2iz - 3$ winds around the origin in the $w$-plane four times.

# Appendix III
# Continuous Functions

The development of calculus depends in an essential way on the concept of limit of a function and thereby on properties of the real number system. In Chapter 1 we presented these notions in an intuitive way and did not attempt to prove them except in Section 1.5 where the *formal* definition of limit was given and used to verify some elementary limits and prove some simple properties of limits.

Many of the results on limits and continuity of functions stated in Chapter 1 may seem quite obvious; most students and users of calculus are not bothered by applying them without proof. Nevertheless, mathematics is a highly logical and rigorous discipline, and any statement, however obvious, that cannot be proved by strictly logical arguments from acceptable assumptions must be considered suspect. In this appendix we build upon the formal definition of limit given in Section 1.5, and combine it with the notion of *completeness* of the real number system first encountered in Section P.1 to give formal proofs of the very important results about continuous functions stated in Theorems 8 and 9 of Section 1.4, the Max-Min Theorem and the Intermediate-Value Theorem. Most of our development of calculus in this book depends essentially on these two theorems.

The branch of mathematics that deals with proofs such as these is called mathematical analysis. This subject is usually not pursued by students in introductory calculus courses, but is postponed to higher years and studied by students in majors or honours programs in mathematics. It is hoped that some of this material will be of value to honours-level calculus courses and individual students with a deeper interest in understanding calculus.

## Limits of Functions

At the heart of mathematical analysis is the formal definition of limit, Definition 9 in Section 1.5, which we restate as follows:

---

**The formal definition of limit**

We say that $\lim_{x \to a} f(x) = L$ if for every positive number $\epsilon$ there exists a positive number $\delta$, depending on $\epsilon$, (that is, $\delta = \delta(\epsilon)$), such that

$$0 < |x - a| < \delta \quad \Longrightarrow \quad |f(x) - L| < \epsilon.$$

---

Section 1.5 was marked "optional" because understanding the material presented there was not essential for learning calculus. However, that material is an *essential* prerequisite for this appendix. It is highly recommended that you go back to Section 1.5 and read it carefully, paying special attention to Examples 2 and 4, and attempt at least Exercises 31–36. These exercises provide proofs for the standard laws of limits stated in Section 1.2.

## Continuous Functions

Consider the following definitions of continuity, which are equivalent to those given in Section 1.4.

**DEFINITION 1**

**Continuity of a function at a point**

A function $f$, defined on an open interval containing the point $a$, is said to be continuous at the point $a$ if

$$\lim_{x \to a} f(x) = f(a),$$

that is, if for every $\epsilon > 0$ there exists $\delta > 0$ such that if $|x - a| < \delta$, then $|f(x) - f(a)| < \epsilon$.

**DEFINITION 2**

**Continuity of a function on an interval**

A function $f$ is continuous on an interval if it is continuous at every point of that interval. In the case of an endpoint of a closed interval, $f$ need only be continuous on one side. Thus, $f$ is continuous on the interval $[a, b]$ if

$$\lim_{t \to x} f(t) = f(x)$$

for each $x$ satisfying $a < x < b$, and

$$\lim_{t \to a+} f(t) = f(a) \qquad \text{and} \qquad \lim_{t \to b-} f(t) = f(b).$$

These concepts are illustrated in Figure III.1.

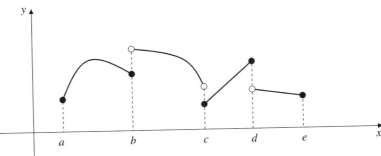

**Figure III.1**    $f$ is continuous on the intervals $[a, b]$, $(b, c)$, $[c, d]$, and $(d, e)$

Some important results about continuous functions are collected in Theorems 6 and 7 of Section 1.4, which we restate here:

**THEOREM 1**

**Combining continuous functions**

(a) If $f$ and $g$ are continuous at the point $a$, then so are $f + g$, $f - g$, $fg$, and, if $g(a) \neq 0$, $f/g$.

(b) If $f$ is continuous at the point $L$ and if $\lim_{x \to a} g(x) = L$, then we have

$$\lim_{x \to a} f\big(g(x)\big) = f(L) = f\big(\lim_{x \to a} g(x)\big).$$

In particular, if $g$ is continuous at the point $a$ (so that $L = g(a)$), then $\lim_{x \to a} f(g(x)) = f(g(a))$, that is, $f \circ g(x) = f(g(x))$ is continuous at $x = a$.

(c) The function $f(x) = C$ (constant) and $g(x) = x$ are continuous on the whole real line.

(d) For any rational number $r$ the function $f(x) = x^r$ is continuous at every real number where it is defined.

**PROOF**    Part (a) is just a restatement of various rules for combining limits; for example,

$$\lim_{x \to a} f(x)g(x) = (\lim_{x \to a} f(x))(\lim_{x \to a} g(x)) = f(a)g(a).$$

Part (b) can be proved as follows. Let $\epsilon > 0$ be given. Since $f$ is continuous at $L$, there exists $k > 0$ such that $|f(g(x)) - f(L)| < \epsilon$ whenever $|g(x) - L| < k$. Since $\lim_{x \to a} g(x) = L$, there exists $\delta > 0$ such that if $0 < |x - a| < \delta$, then $|g(x) - L| < k$. Hence if $0 < |x - a| < \delta$, then $|f(g(x)) - f(L)| < \epsilon$, and $\lim_{x \to a} f(g(x)) = f(L)$.

The proofs of (c) and (d) are left to the student in Exercises 3–9 at the end of this appendix.

## Completeness and Sequential Limits

**DEFINITION 3**

A real number $u$ is said to be an **upper bound** for a nonempty set $S$ of real numbers if $x \leq u$ for every $x$ in $S$.
The number $u^*$ is called the **least upper bound** of $S$ if $u^*$ is an upper bound for $S$ and $u^* \leq u$ for every upper bound $u$ of $S$.
Similarly, $\ell$ is a **lower bound** for $S$ if $\ell \leq x$ for every $x$ in $S$. The number $\ell^*$ is the **greatest lower bound** of $S$ if $\ell^*$ is a lower bound and $\ell \leq \ell^*$ for every lower bound $\ell$ of $S$.

**Example 1**    Set $S_1 = [2, 3]$ and $S_2 = (2, \infty)$. Any number $u \geq 3$ is an upper bound for $S_1$. $S_2$ has no upper bound; we say that it is not bounded above. The least upper bound of $S_1$ is 3. Any real number $\ell \leq 2$ is a lower bound for both $S_1$ and $S_2$. $\ell^* = 2$ is the greatest lower bound for each set. Note that the least upper bound and greatest lower bound of a set may or may not belong to that set.

We now recall the completeness axiom for the real number system, which we discussed briefly in Section P.1.

**The completeness axiom for the real numbers**

A nonempty set of real numbers that has an upper bound must have a least upper bound.
Equivalently, a nonempty set of real numbers having a lower bound must have a greatest lower bound.

We stress that this is an *axiom* to be assumed without proof. It cannot be deduced from the more elementary algebraic and order properties of the real numbers. These other properties are shared by the rational numbers, a set that is not complete. The completeness axiom is essential for the proof of the most important results about continuous functions, in particular for the Max-Min Theorem and the Intermediate-Value Theorem. Before attempting these proofs, however, we must develop a little more machinery.

In Section 9.1 we stated a version of the completeness axiom that pertains to *sequences* of real numbers; specifically, that an increasing sequence that is bounded above converges to a limit. We begin by verifying that this follows from the version stated above. (Both statements are, in fact, equivalent.) As noted in Section 9.1, the sequence

$$\{x_n\} = \{x_1, \ x_2, \ x_3, \ \ldots\}$$

is a function on the positive integers, that is, $x_n = x(n)$. We say that the sequence converges to the limit $L$, and we write $\lim x_n = L$, if the corresponding function $x(t)$ satisfies $\lim_{t \to \infty} x(t) = L$ as defined above. More formally,

**DEFINITION 4**

> **Limit of a sequence**
> We say that $\lim x_n = L$ if for every positive number $\epsilon$ there exists a positive number $N = N(\epsilon)$ such that $|x_n - L| < \epsilon$ holds whenever $n \geq N$.

**THEOREM 2**

If $\{x_n\}$ is an increasing sequence that is bounded above, that is,

$$x_{n+1} \geq x_n \qquad \text{and} \qquad x_n \leq K \qquad \text{for } n = 1, 2, 3, \ldots,$$

then $\lim x_n = L$ exists. (Equivalently, if $\{x_n\}$ is decreasing and bounded below, then $\lim x_n$ exists.)

**PROOF** Let $\{x_n\}$ be increasing and bounded above. The set $S$ of real numbers $x_n$ has an upper bound, $K$, and so has a least upper bound, say $L$. Thus $x_n \leq L$ for every $n$, and if $\epsilon > 0$ then there exists a positive integer $N$ such that $x_N > L - \epsilon$. (Otherwise $L - \epsilon$ would be an upper bound for $S$ that is lower than the least upper bound.) If $n \geq N$, then we have $L - \epsilon < x_N \leq x_n \leq L$, so $|x_n - L| < \epsilon$. Thus $\lim x_n = L$. The proof for a decreasing sequence that is bounded below is similar.

**THEOREM 3**

If $a \leq x_n \leq b$ for each $n$, and if $\lim x_n = L$, then $a \leq L \leq b$.

**PROOF** Suppose that $L > b$. Let $\epsilon = L - b$. Since $\lim x_n = L$, there exists $n$ such that $|x_n - L| < \epsilon$. Thus $x_n > L - \epsilon = L - (L - b) = b$, which is a contradiction since we are given that $x_n \leq b$. Thus $L \leq b$. A similar argument shows that $L \geq a$.

**THEOREM 4**

If $f$ is continuous on $[a, b]$, if $a \leq x_n \leq b$ for each $n$, and if $\lim x_n = L$, then $\lim f(x_n) = f(L)$.

The proof is similar to that of Theorem 1(b), and is left to the student as Exercise 15 at the end of this appendix.

## Continuous Functions on a Closed, Finite Interval

We are now in a position to prove the main results about continuous functions on closed, finite intervals.

**THEOREM 5**

### The Boundedness Theorem

If $f$ is continuous on $[a, b]$, then $f$ is bounded there; that is, there exists a constant $K$ such that $|f(x)| \leq K$ if $a \leq x \leq b$.

**PROOF**   We show that $f$ is bounded above; a similar proof shows that $f$ is bounded below. For each positive integer $n$ let $S_n$ be the set of points $x$ in $[a, b]$ such that $f(x) > n$:

$$S_n = \{x : a \leq x \leq b \quad \text{and} \quad f(x) > n\}.$$

We would like to show that $S_n$ is empty for some $n$. It would then follow that $f(x) \leq n$ for all $x$ in $[a, b]$; that is, $n$ would be an upper bound for $f$ on $[a, b]$.

Suppose, to the contrary, that $S_n$ is nonempty for every $n$. We will show that this leads to a contradiction. Since $S_n$ is bounded below ($a$ is a lower bound), by completeness $S_n$ has a greatest lower bound; call it $x_n$. (See Figure III.2.) Evidently $a \leq x_n$. Since $f(x) > n$ at some point of $[a, b]$ and $f$ is continuous at that point, $f(x) > n$ on some interval contained in $[a, b]$. Hence $x_n < b$. It follows that $f(x_n) \geq n$. (If $f(x_n) < n$ then by continuity $f(x) < n$ for some distance to the right of $x_n$, and $x_n$ could not be the greatest lower bound of $S_n$.)

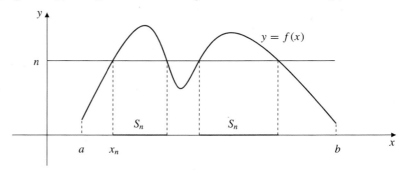

**Figure III.2**    The set $S_n$

For each $n$ we have $S_{n+1} \subset S_n$. Therefore $x_{n+1} \geq x_n$ and $\{x_n\}$ is an increasing sequence. Being bounded above ($b$ is an upper bound) this sequence converges, by Theorem 2. Let $\lim x_n = L$. By Theorem 3, $a \leq L \leq b$. Since $f$ is continuous at $L$, $\lim f(x_n) = f(L)$ exists by Theorem 4. But since $f(x_n) \geq n$, $\lim f(x_n)$ cannot exist. This contradiction completes the proof.

**THEOREM 6**

### The Max-Min Theorem

If $f$ is continuous on $[a, b]$, then there are points $v$ and $u$ in $[a, b]$ such that for any $x$ in $[a, b]$ we have

$$f(v) \leq f(x) \leq f(u);$$

that is, $f$ assumes maximum and minimum values on $[a, b]$.

**PROOF**   By Theorem 5 we know that the set $S = \{f(x) : a \le x \le b\}$ has an upper bound and therefore, by the completeness axiom, a least upper bound. Call this least upper bound $M$. Suppose that there exists no point $u$ in $[a, b]$ such that $f(u) = M$. Then by Theorem 1(a), $1/(M - f(x))$ is continuous on $[a, b]$. By Theorem 5, there exists a constant $K$ such that $1/(M - f(x)) \le K$ for all $x$ in $[a, b]$. Thus $f(x) \le M - 1/K$, which contradicts the fact that $M$ is the *least* upper bound for the values of $f$. Hence there must exist some point $u$ in $[a, b]$ such that $f(u) = M$. Since $M$ is an upper bound for the values of $f$ on $[a, b]$, we have $f(x) \le f(u) = M$ for all $x$ in $[a, b]$.

The proof that there must exist a point $v$ in $[a, b]$ such that $f(x) \ge f(v)$ for all $x$ in $[a, b]$ is similar.

---

**THEOREM  7**

### The Intermediate-Value Theorem

If $f$ is continuous on $[a, b]$ and $s$ is a real number lying between the numbers $f(a)$ and $f(b)$, then there exists a point $c$ in $[a, b]$ such that $f(c) = s$.

**PROOF**   To be specific, we assume that $f(a) < s < f(b)$. (The proof for the case $f(a) > s > f(b)$ is similar.) Let $S = \{x : a \le x \le b \text{ and } f(x) \le s\}$. $S$ is nonempty ($a$ belongs to $S$) and bounded above ($b$ is an upper bound) so by completeness $S$ has a least upper bound; call it $c$.

Suppose that $f(c) > s$. Then $c \ne a$ and, by continuity, $f(x) > s$ on some interval $(c - \delta, c]$ where $\delta > 0$. But this says $c - \delta$ is an upper bound for $S$ lower than the least upper bound, which is impossible. Thus $f(c) \le s$.

Suppose $f(c) < s$. Then $c \ne b$ and, by continuity, $f(x) < s$ on some interval of the form $[c, c + \delta)$ for some $\delta > 0$. But this says that $[c, c + \delta) \subset S$, which contradicts the fact that $c$ is an upper bound for $S$. Hence we cannot have $f(c) < s$. Therefore, $f(c) = s$.

---

For more discussion of these theorems, and some applications, see Section 1.4.

---

# Exercises

**1.** Let $a < b < c$ and suppose that $f(x) \le g(x)$ for $a \le x \le c$. If $\lim_{x \to b} f(x) = L$ and $\lim_{x \to b} g(x) = M$, prove that $L \le M$. *Hint:* assume that $L > M$ and deduce that $f(x) > g(x)$ for all $x$ sufficiently near $b$. This contradicts the condition that $f(x) \le g(x)$ for $a \le x \le b$.

**2.** If $f(x) \le K$ on the intervals $[a, b)$ and $(b, c]$, and if $\lim_{x \to b} f(x) = L$, prove that $L \le K$.

**3.** Use the formal definition of limit to prove that $\lim_{x \to 0+} x^r = 0$ for any positive, rational number $r$.

Prove the assertions in Exercises 4–9.

**4.** $f(x) = C$ (constant) and $g(x) = x$ are both continuous on the whole real line.

**5.** Every polynomial is continuous on the whole real line.

**6.** A rational function (quotient of polynomials) is continuous everywhere except where the denominator is 0.

**7.** If $n$ is a positive integer and $a > 0$, then $f(x) = x^{1/n}$ is continuous at $x = a$.

**8.** If $r = m/n$ is a rational number, then $g(x) = x^r$ is continuous at every point $a > 0$.

**9.** If $r = m/n$ where $m$ and $n$ are integers and $n$ is odd, show that $g(x) = x^r$ in continuous at every point $a < 0$. If $r \ge 0$ show that $g$ is continuous at 0 also.

**10.** Prove that $f(x) = |x|$ is continuous on the real line.

Use the definitions from Chapter 3 for the functions in Exercises 11–14 to show that these functions are continuous on their respective domains.

**11.** $\sin x$

**12.** $\cos x$

**13.** $\ln x$

**14.** $e^x$

**15.** Prove Theorem 4.

**16.** Suppose that every function that is continuous and bounded on $[a, b]$ must assume a maximum value and a minimum value on that interval. Without using Theorem 5, prove that every function $f$ that is continuous on $[a, b]$ must be bounded on that interval. *Hint:* show that $g(t) = t/(1 + |t|)$ is continuous and increasing on the real line. Then consider $g\big(f(x)\big)$.

# Appendix IV
# The Riemann Integral

In Section 5.3 we defined the definite integral $\int_a^b f(x)\,dx$ of a function $f$ that is continuous on the finite, closed interval $[a, b]$. The integral was defined as a kind of "limit" of Riemann sums formed by partitioning the interval $[a, b]$ into small subintervals. In this appendix we will reformulate the definition of the integral so that it can be used for functions that are not necessarily continuous; in the following discussion we assume only that $f$ is **bounded** on $[a, b]$. Later we will prove Theorem 2 of Section 5.3 which asserts that any continuous function is integrable.

Recall that a **partition** $P$ of $[a, b]$ is a finite, ordered set of points $P = \{x_0, x_1, x_2, \ldots, x_n\}$ where $a = x_0 < x_1 < x_2 < \cdots < x_{n-1} < x_n = b$. Such a partition subdivides $[a, b]$ into $n$ subintervals $[x_0, x_1]$, $[x_1, x_2]$, $\ldots$, $[x_{n-1}, x_n]$ where $n = n(P)$ depends on the partition. The length of the $j$th subinterval $[x_{j-1}, x_j]$ is $\Delta x_j = x_j - x_{j-1}$.

Suppose that the function $f$ is bounded on $[a, b]$. Given any partition $P$, the $n$ sets $S_j = \{f(x) : x_{j-1} \le x \le x_j\}$ have least upper bounds $M_j$ and greatest lower bounds $m_j$, $(1 \le j \le n)$, so that

$$m_j \le f(x) \le M_j \qquad \text{on } [x_{j-1}, x_j].$$

We define upper and lower Riemann sums for $f$ corresponding to the partition $P$ to be

$$U(f, P) = \sum_{j=1}^{n(P)} M_j \Delta x_j, \qquad \text{and}$$

$$L(f, P) = \sum_{j=1}^{n(P)} m_j \Delta x_j.$$

(See Figure IV.1.) Note that if $f$ is continuous on $[a, b]$, then $m_j$ and $M_j$ are in fact the minimum and maximum values of $f$ over $[x_{j-1}, x_j]$ (by Theorem 6 of Appendix III); that is, $m_j = f(l_j)$ and $M_j = f(u_j)$, where $f(l_j) \le f(x) \le f(u_j)$ for $x_{j-1} \le x \le x_j$.

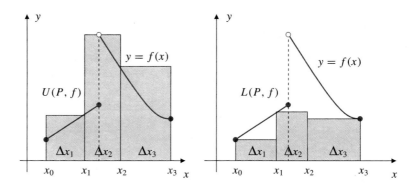

**Figure IV.1**  Upper and lower sums corresponding to the partition $P = \{x_0, x_1, x_2, x_3\}$

If $P$ is any partition of $[a, b]$ and we create a new partition $P^*$ by adding subdivision points to those of $P$, thus subdividing the subintervals of $P$ into sma ones, then we call $P^*$ a **refinement** of $P$.

**THEOREM 1**

If $P^*$ is a refinement of $P$, then $L(f, P^*) \geq L(f, P)$ and $U(f, P^*) \leq U(f, P)$

**PROOF**  If $S$ and $T$ are sets of real numbers, and $S \subset T$, then any lower bound upper bound) of $T$ is also a lower bound (or upper bound) of $S$. Hence the grea lower bound of $S$ is at least as large as that of $T$, and the least upper bound of no greater than that of $T$.

Let $P$ be a given partition of $[a, b]$ and form a new partition $P'$ by add one subdivision point to those of $P$, say the point $k$ dividing the $j$th subinter $[x_{j-1}, x_j]$ of $P$ into two subintervals $[x_{j-1}, k]$ and $[k, x_j]$. (See Figure IV.2.) Let $m'_j$, and $m''_j$ be the greatest lower bounds of the sets of values of $f(x)$ on the inter $[x_{j-1}, x_j]$, $[x_{j-1}, k]$, and $[k, x_j]$, respectively. Then $m_j \leq m'_j$ and $m_j \leq m''_j$. T $m_j(x_j - x_{j-1}) \leq m'_j(k - x_{j-1}) + m''_j(x_j - k)$, and so $L(f, P) \leq L(f, P')$.

If $P^*$ is a refinement of $P$, it can be obtained by adding one point at a tim those of $P$ and thus $L(f, P) \leq L(f, P^*)$. We can prove that $U(f, P) \geq U(f,$ in a similar manner.

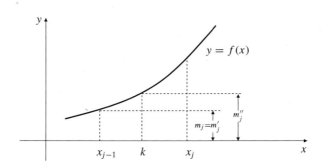

**Figure IV.2**    Adding one point to a partition

**THEOREM 2**

If $P$ and $P'$ are any two partitions of $[a, b]$, then $L(f, P) \leq U(f, P')$.

**PROOF**  Combine the subdivision points of $P$ and $P'$ to form a new partition which is a refinement of both $P$ and $P'$. Then by Theorem 1,

$$L(f, P) \leq L(f, P^*) \leq U(f, P^*) \leq U(f, P').$$

No lower sum can exceed any upper sum.

Theorem 2 shows that the set of values of $L(f, P)$ for fixed $f$ and various partiti $P$ of $[a, b]$ is a bounded set; any upper sum is an upper bound for this set. completeness, the set has a least upper bound, which we shall denote $I_*$. T $L(f, P) \leq I_*$ for any partition $P$. Similarly, there exists a greatest lower bound for the set of values of $U(f, P)$ corresponding to different partitions $P$. It follo that $I_* \leq I^*$. (See Exercise 4 at the end of this appendix.)

**DEFINITION**  **1**

**The Riemann integral**

If $f$ is bounded on $[a, b]$ and $I_* = I^*$, then we say that $f$ is **Riemann integrable**, or simply **integrable** on $[a, b]$ and denote by

$$\int_a^b f(x)\, dx = I_* = I^*$$

the **(Riemann) integral** of $f$ on $[a, b]$.

The following theorem provides a convenient test for determining whether a given bounded function is integrable.

**THEOREM** **3**

The bounded function $f$ is integrable on $[a, b]$ if and only if for every positive number $\epsilon$ there exists a partition $P$ of $[a, b]$ such that $U(f, P) - L(f, P) < \epsilon$.

**PROOF** Suppose that for every $\epsilon > 0$ there exists a partition $P$ of $[a, b]$ such that $U(f, P) - L(f, P) < \epsilon$, then

$$I^* \le U(f, P) < L(f, P) + \epsilon \le I_* + \epsilon.$$

Since $I^* < I_* + \epsilon$ must hold for every $\epsilon > 0$, it follows that $I^* \le I_*$. Since we already know that $I^* \ge I_*$, we have $I^* = I_*$ and $f$ is integrable on $[a, b]$.

Conversely, if $I^* = I_*$ and $\epsilon > 0$ are given, we can find a partition $P'$ such that $L(f, P') > I_* - \epsilon/2$, and another partition $P''$, such that $U(f, P'') < I^* + \epsilon/2$. If $P$ is a common refinement of $P'$ and $P''$, then by Theorem 1, we have that $U(f, P) - L(f, P) \le U(f, P'') - L(f, P') < (\epsilon/2) + (\epsilon/2) = \epsilon$, as required.

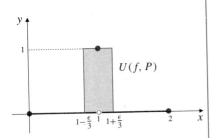

**Figure IV.3**

**Example 1** Let $f(x) = \begin{cases} 0 & \text{if } 0 \le x < 1 \text{ or } 1 < x \le 2 \\ 1 & \text{if } x = 1. \end{cases}$

Show that $f$ is integrable on $[0, 2]$ and find $\int_0^2 f(x)\, dx$.

**Solution** Let $\epsilon > 0$ be given. Let $P = \{0, 1 - \epsilon/3, 1 + \epsilon/3, 2\}$. Then $L(f, P) = 0$ since $f(x) = 0$ at points of each of these subintervals into which $P$ subdivides $[0, 2]$. (See Figure IV.3.) Since $f(1) = 1$, we have

$$U(f, P) = 0\left(1 - \frac{\epsilon}{3}\right) + 1\left(\frac{2\epsilon}{3}\right) + 0\left(2 - \left(1 - \frac{\epsilon}{3}\right)\right) = \frac{2\epsilon}{3}.$$

Hence $U(f, P) - L(f, P) < \epsilon$ and $f$ is integrable on $[0, 2]$. Since $L(f, P) = 0$ for every partition, $\int_0^2 f(x)\, dx = I_* = 0$.

**Example 2**    Let $f(x)$ be defined on $[0, 1]$ by

$$f(x) = \begin{cases} 1 & \text{if } x \text{ is rational} \\ 0 & \text{if } x \text{ is irrational.} \end{cases}$$

Show that $f$ is not integrable on $[0, 1]$.

**PROOF**    Every subinterval of $[0, 1]$ having positive length contains both ratic and irrational numbers. Hence, for any partition $P$ of $[0, 1]$ we have $L(f, P) =$ and $U(f, P) = 1$. Thus $I_* = 0$ and $I^* = 1$, and so $f$ is not integrable on $[0, 1]$

## Uniform Continuity

When we assert that a function $f$ is continuous on the interval $[a, b]$, we in that for every $x$ in that interval, and every $\epsilon > 0$, we can find a positive numb (depending on *both* $x$ and $\epsilon$) such that $|f(y) - f(x)| < \epsilon$ whenever $|y - x|$ and $y$ lies in $[a, b]$. In fact, however, it is possible to find a number $\delta$ *depend* *only on* $\epsilon$ such that $|f(y) - f(x)| < \epsilon$ holds whenever $x$ and $y$ belong to $[a, b]$ satisfy $|y - x| < \delta$. We describe this phenomenon by saying that $f$ is **unifor** **continuous** on the interval $[a, b]$.

**THEOREM 4**

If $f$ is continuous on the closed, finite interval $[a, b]$, then $f$ is uniformly continu on that interval.

**PROOF**    Let $\epsilon > 0$ be given. Define numbers $x_n$ in $[a, b]$ and subsets $S_n$ of $[a$ as follows:

$$x_1 = a$$
$$S_1 = \left\{ x : x_1 < x \leq b \text{ and } |f(x) - f(x_1)| \geq \frac{\epsilon}{3} \right\}.$$

If $S_1$ is empty, stop; otherwise let

$$x_2 = \text{the greatest lower bound of } S_1$$
$$S_2 = \left\{ x : x_2 < x \leq b \text{ and } |f(x) - f(x_2)| \geq \frac{\epsilon}{3} \right\}.$$

If $S_2$ is empty, stop; otherwise proceed to define $x_3$ and $S_3$ analogously. We proc in this way as long as we can; if $x_n$ and $S_n$ have been defined and $S_n$ is not em we define

$$x_{n+1} = \text{the greatest lower bound of } S_n$$
$$S_{n+1} = \left\{ x : x_{n+1} < x \leq b \text{ and } |f(x) - f(x_{n+1})| \geq \frac{\epsilon}{3} \right\}.$$

At any stage where $S_n$ is not empty, the continuity of $f$ at $x_n$ assures us $x_{n+1} > x_n$ and $|f(x_{n+1}) - f(x_n)| = \epsilon/3$.

We must consider two possibilities for the above procedure: either $S_n$ is em for some $n$, or $S_n$ is nonempty for every $n$.

Suppose $S_n$ is nonempty for every $n$. Then we have constructed an infir increasing sequence $\{x_n\}$ in $[a, b]$ that, being bounded above (by $b$), must h a limit by completeness (Theorem 2 of Appendix III). Let $\lim x_n = x^*$. have $a \leq x^* \leq b$. Since $f$ is continuous at $x^*$ there exists $\delta > 0$ such

$|f(x) - f(x^*)| < \epsilon/8$ whenever $|x - x^*| < \delta$ and $x$ lies in $[a, b]$. Since $\lim x_n = x^*$, there exists a positive integer $N$ such that $|x_n - x^*| < \delta$ whenever $n \geq N$. For such $n$ we have

$$\frac{\epsilon}{3} = |f(x_{n+1}) - f(x_n)| = |f(x_{n+1}) - f(x^*) + f(x^*) - f(x_n)|$$
$$\leq |f(x_{n+1}) - f(x^*)| + |f(x_n) - f(x^*)|$$
$$< \frac{\epsilon}{8} + \frac{\epsilon}{8} = \frac{\epsilon}{4},$$

which is clearly impossible. Thus $S_n$ must, in fact, be empty for some $n$.

Suppose that $S_N$ is empty. Thus $S_n$ is nonempty for $n < N$, and the procedure for defining $x_n$ stops with $x_N$. Since $S_{N-1}$ is not empty, $x_N < b$. In this case define $x_{N+1} = b$ and let

$$\delta = \min\{x_2 - x_1, \, x_3 - x_2, \, \ldots, \, x_{N+1} - x_N\}.$$

The minimum of a finite set of positive numbers is a positive number, and so $\delta > 0$. If $x$ lies in $[a, b]$, then $x$ lies in one of the intervals $[x_1, x_2], [x_2, x_3], \ldots, [x_N, x_{N+1}]$. Suppose $x$ lies in $[x_k, x_{k+1}]$. If $y$ is in $[a, b]$ and $|y - x| < \delta$, then $y$ lies in either the same subinterval as $x$ or in an adjacent one; that is, $y$ lies in $[x_j, x_{j+1}]$ where $j = k - 1, k,$ or $k + 1$. Thus

$$|f(y) - f(x)| = |f(y) - f(x_j) + f(x_j) - f(x_k) + f(x_k) - f(x)|$$
$$\leq |f(y) - f(x_j)| + |f(x_j) - f(x_k)| + |f(x_k) - f(x)|$$
$$< \frac{\epsilon}{3} + \frac{\epsilon}{3} + \frac{\epsilon}{3} = \epsilon,$$

which was to be proved.

We are now in a position to prove that a continuous function is integrable.

**THEOREM  5**   If $f$ is continuous on $[a, b]$, then $f$ is integrable on $[a, b]$.

**PROOF**   By Theorem 4, $f$ is uniformly continuous on $[a, b]$. Let $\epsilon > 0$ be given. Let $\delta > 0$ be such that $|f(x) - f(y)| < \epsilon/(b - a + 1)$ whenever $|x - y| < \delta$ and $x$ and $y$ belong to $[a, b]$. Choose a partition $P = \{x_0, x_1, \ldots, x_n\}$ of $[a, b]$ for which each subinterval $[x_{j-1}, x_j]$ has length $\Delta x_j < \delta$. Then the greatest lower bound, $m_j$, and the least upper bound, $M_j$, of the set of values of $f(x)$ on $[x_{j-1}, x_j]$ satisfy $M_j - m_j \leq \epsilon/(b - a + 1) < \epsilon/(b - a)$. Accordingly,

$$U(f, P) - L(f, P) < \frac{\epsilon}{b - a} \sum_{j=1}^{n(P)} \Delta x_j = \frac{\epsilon}{b - a}(b - a) = \epsilon.$$

Thus $f$ is integrable on $[a, b]$, as asserted.

# Exercises

**1.** Let $f(x) = \begin{cases} 1 & \text{if } 0 \le x \le 1 \\ 0 & \text{if } 1 < x \le 2 \end{cases}$. Prove that $f$ is integrable on $[0, 2]$ and find the value of $\int_0^2 f(x)\, dx$.

**2.** Let $f(x) = \begin{cases} 1 & \text{if } x = 1/n, \qquad n = 1, 2, 3, \ldots \\ 0 & \text{for all other values of } x \end{cases}$.
Show that $f$ is integrable over $[0, 1]$ and find the value of the integral $\int_0^1 f(x)\, dx$.

**∗ 3.** Let $f(x) = 1/n$ if $x = m/n$ where $m, n$ are integers having no common factors, and let $f(x) = 0$ if $x$ is an irrational number. Thus $f(1/2) = 1/2$, $f(1/3) = f(2/3) = 1/3$, $f(1/4) = f(3/4) = 1/4$, etc. Show that $f$ is integrable on $[0, 1]$ and find $\int_0^1 f(x)\, dx$. *Hint:* show that for any $\epsilon > 0$, only finitely many points of the graph of $f$ over $[0, 1]$ lie above the line $y = \epsilon$.

**4.** Prove that $I_*$ and $I^*$ defined in the paragraph following Theorem 2 satisfy $I_* \le I^*$ as claimed there.

**5.** Prove parts (c), (d), (e), (f), (g), and (h) of Theorem 3 in Section 5.4 for the Riemann Integral.

**6.** Use the definition of uniform continuity given in the paragraph preceding Theorem 4 to prove that $f(x) = \sqrt{x}$ uniformly continuous on $[0, 1]$. Do not use Theorem Theorem 4 itself.

**7.** Show directly from the definition of uniform continuity (without using Theorem 5 of Appendix III) that a functi uniformly continuous on a closed, finite interval is necessarily bounded there.

**8.** If $f$ is bounded and integrable on $[a, b]$, prove that $F(x) = \int_a^x f(t)\, dt$ is uniformly continuous on $[a, b]$. ( were continuous we would have a stronger result; $F$ wo be differentiable on $(a, b)$ and $F'(x) = f(x)$, (which is Fundamental Theorem of Calculus.))

# Answers to Odd-Numbered Exercises

**Chapter P**

**Preliminaries**

**1.** $0.\overline{2}$        **3.** $4/33$

**5.** $1/7 = 0.\overline{142857}$, $2/7 = 0.\overline{285714}$,
$3/7 = 0.\overline{428571}$, $4/7 = 0.\overline{571428}$,
$5/7 = 0.\overline{714285}$, $6/7 = 0.\overline{857142}$

**7.** $[0, 5]$      **9.** $(-\infty, -6) \cup (-5, \infty)$

**11.** $(-2, \infty)$      **13.** $(-\infty, -2)$

**15.** $(-\infty, 5/4]$      **17.** $(0, \infty)$

**19.** $(-\infty, 5/3) \cup (2, \infty)$      **21.** $[0, 2]$

**23.** $(-2, 0) \cup (2, \infty)$      **25.** $[-2, 0) \cup [4, \infty)$

**27.** $x = -3, 3$      **29.** $t = -1/2, -9/2$

**31.** $s = -1/3, 17/3$      **33.** $(-2, 2)$

**35.** $[-1, 3]$      **37.** $\left(\dfrac{5}{3}, 3\right)$

**39.** $[0, 4]$      **41.** $x > 1$

**43.** true if $a \geq 0$, false if $a < 0$

**1.** $\Delta x = 4$, $\Delta y = -3$, dist $= 5$

**3.** $\Delta x = -4$, $\Delta y = -4$, dist $= 4\sqrt{2}$

**5.** $(2, -4)$

**7.** circle, centre $(0, 0)$, radius 1

**9.** points inside and on circle, centre $(0, 0)$, radius 1

**11.** points on and above the parabola $y = x^2$

**13.** (a) $x = -2$, (b) $y = 5/3$

**15.** $y = x + 2$      **17.** $y = 2x + b$

**19.** above      **21.** $y = 3x/2$

**23.** $y = (7 - x)/3$      **25.** $y = \sqrt{2} - 2x$

**27.** 4, 3,

**29.** $\sqrt{2}, -2/\sqrt{3}$

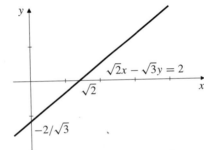

**31.** (a) $y = x - 1$, (b) $y = -x + 3$

**33.** $(2, -3)$      **37.** 5

**39.** $\$23,000$      **43.** $(-2, -2)$

**45.** $\left(\frac{1}{3}(x_1 + 2x_2), \frac{1}{3}(y_1 + 2y_2)\right)$

**47.** circle, centre $(2, 0)$, radius 4

**49.** perp. if $k = -8$, parallel if $k = 1/2$

**1.** $x^2 + y^2 = 16$      **3.** $x^2 + y^2 + 4x = 5$

**5.** $(1, 0), 2$      **7.** $(1, -2), 3$

**9.** exterior of circle, centre $(0, 0)$, radius 1

**11.** closed disk, centre $(-1, 0)$, radius 2

**13.** washer shaped region between the circles of radius 1 and 2 centred at $(0, 0)$

**15.** first octant region lying inside the two circles of radius 1 having centres at $(1, 0)$ and $(0, 1)$

**17.** $x^2 + y^2 + 2x - 4y < 1$

**19.** $x^2 + y^2 < 2$, $x \geq 1$      **21.** $x^2 = 16y$

**23.** $y^2 = 8x$

**25.** $(0, 1/2), y = -1/2$

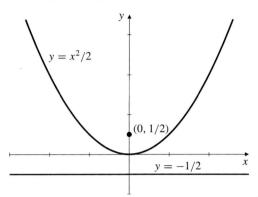

**27.** $(-1, 0), x = 1$

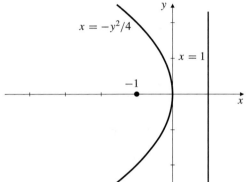

**29.** (a) $y = x^2 - 3$, (b) $y = (x - 4)^2$, (c) $y = (x - 3)^2 + 3$, (d) $y = (x - 4)^2 - 2$

**31.** $y = \sqrt{(x/3) + 1}$    **33.** $y = \sqrt{(3x/2) + 1}$

**35.** $y = -(x + 1)^2$    **37.** $y = (x - 2)^2 - 2$

**39.** $(2, 7)$,    $(1, 4)$    **41.** $(4, -3)$,    $(-4, 3)$

**43.** ellipse, centre $(0, 0)$, semiaxes 2, 1

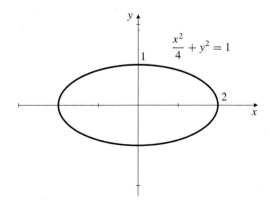

**45.** ellipse, centre $(3, -2)$, semiaxes 3, 2

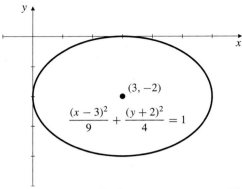

**47.** hyperbola, centre $(0, 0)$, asymptotes $x = \pm 2y$, vertices $(\pm 2, 0)$

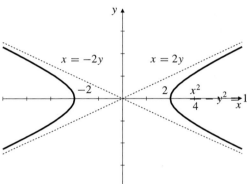

**49.** rectangular hyperbola, asymptotes $x = 0$ and $y = 0$, vertices $(2, -2)$ and $(-2, 2)$

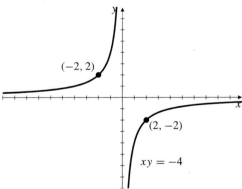

**51.** (a) reflecting the graph in the $y$-axis, (b) reflecting the graph in the $x$-axis.

**53.**

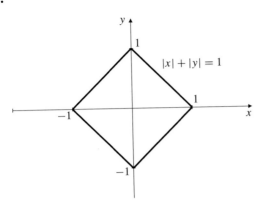

$|x| + |y| = 1$

**35.**

**37.**

$y = \dfrac{2}{x+2}$

$x = -2$

$y = \dfrac{x}{x+1}$

### Section P.4   (page 34)

**1.** $D(f) = \mathbb{R},\ R(f) = [1, \infty)$

**3.** $D(G) = (-\infty, 4],\ R(g) = [0, \infty)$

**5.** $D(h) = (-\infty, 2),\ R(h) = (-\infty, \infty)$

**7.** Only (ii) is the graph of a function. Vertical lines can meet the others more than once.

**11.** even, sym. about $y$-axis

**13.** odd, sym. about $(0, 0)$     **15.** sym. about $(2, 0)$

**17.** sym. about $x = 3$

**19.** even, sym. about $y$-axis

**21.** no symmetry

**23.**

$y = -x^2$

**25.**

$y = (x - 1)^2$

**27.**

$y = 1 - x^3$

**29.**

$y = \sqrt{x} + 1$

**31.**

$y = -|x|$

**33.**

$y = |x - 2|$

$y = f(x) + 2$     $(2, 2)$

$y = f(x + 2)$

**39.** $D = [0, 2],\ R = [2, 3]$

**41.** $D = [-2, 0],\ R = [0, 1]$

**43.** $D = [0, 2],\ R = [-1, 0]$

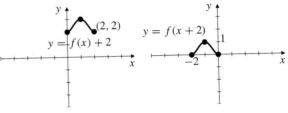

$y = -f(x)$

$y = f(4 - x)$     $(3, 1)$

**45.** $D = [2, 4],\ R = [0, 1]$     **47.** $[-0.18, 0.68]$

**49.** $y = 3/2$

**51.** $(2, 1),\ y = x - 1,\ y = 3 - x$

**53.** $f(x) = 0$

### Section P.5   (page 40)

**1.** The domains of $f + g$, $f - g$, $fg$, and $g/f$ are $[1, \infty)$. The domain of $f/g$ is $(1, \infty)$.

$(f + g)(x) = x + \sqrt{x - 1}$

$(f - g)(x) = x - \sqrt{x - 1}$

$(fg)(x) = x\sqrt{x - 1}$

$(f/g)(x) = x/\sqrt{x - 1}$

$(g/f)(x) = \sqrt{x - 1}/x$

**3.**

**5.**

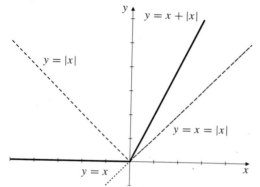

**7.** (a) 2, (b) 22, (c) $x^2 + 2$, (d) $x^2 + 10x + 22$, (e) 5, (f) $-2$, (g) $x + 10$, (h) $x^4 - 6x^2 + 6$

**9.** (a) $(x - 1)/x, x \neq 0, 1$,
  (b) $1/(1 - \sqrt{x - 1})$ on $[1, 2) \cup (2, \infty)$,
  (c) $\sqrt{x/(1 - x)}$, on $[0, 1)$
  (d) $\sqrt{\sqrt{x - 1} - 1}$, on $[2, \infty)$

**11.** $(x + 1)^2$      **13.** $x^2$

**15.** $1/(x - 1)$      **19.** $D = [0, 2], R = [0, 2]$

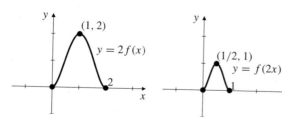

**21.** $D = [0, 1], R = [0, 1]$

**23.** $D = [-4, 0], R = [1, 2]$

**25.**

**27.** (a) $A = 0$, $B$ arbitrary, or $A = 1$, $B = 0$
  (b) $A = -1$, $B$ arbitrary, or $A = 1$, $B = 0$

**29.** all integers

**31.**

**33.** $f^2,\ g^2,\ f \circ f,\ f \circ g,\ g \circ f$ are even
  $fg,\ f/g,\ g/f, g \circ g$ are odd
  $f + g$ is neither, unless either $f(x) = 0$ or $g(x) = 0$.

### Section P.6    *(page 53)*

**1.** $-1/\sqrt{2}$      **3.** $\sqrt{3}/2$

**5.** $(\sqrt{3} - 1)/(2\sqrt{2})$      **7.** $-\cos x$

**9.** $-\cos x$      **11.** $1/(\sin x \cos x)$

**17.** $3 \sin x - 4 \sin^3 x$

**19.** period $\pi$

**21.** period 2

**23.**

**25.** $\cos \theta = -4/5,\ \tan \theta = -3/4$

**27.** $\sin \theta = -2\sqrt{2}/3,\ \tan \theta = -2\sqrt{2}$

**29.** $\cos \theta = -\sqrt{3}/2,\ \tan \theta = 1/\sqrt{3}$

**31.** $a = 1$, $b = \sqrt{3}$
**33.** $b = 5/\sqrt{3}$, $c = 10/\sqrt{3}$
**35.** $a = b \tan A$     **37.** $a = b \cot B$
**39.** $c = b \sec A$     **41.** $\sin A = \sqrt{c^2 - b^2}/c$
**43.** $\sin B = 3/(4\sqrt{2})$     **45.** $\sin B = \sqrt{135}/16$
**47.** $6/(1 + \sqrt{3})$
**49.** $b = 4 \sin 40°/ \sin 70° \approx 2.736$
**51.** approx. 16.98 m

# Chapter 1
# Limits and Continuity

## Section 1.1 *(page 59)*

**1.** $((t + h)^2 - t^2)/h$ ft/s     **3.** 4 ft/s

**5.** $-3$ m/s, 3 m/s, 0 m/s
**7.** to the left, stopped, to the right

**9.** height 2, moving down
**11.** $-1$ ft/s, weight moving downward

**13.** day 43–47

## Section 1.2 *(page 68)*

**1.** (a) 1, (b) 0, (c) 1     **3.** 1
**5.** 0     **7.** 1
**9.** 2/3     **11.** 0
**13.** 0     **15.** does not exist
**17.** 1/6     **19.** 0
**21.** $-1$     **23.** does not exist
**25.** 2     **27.** 3/8
**29.** $-1/2$     **31.** 8/3
**33.** 1/4     **35.** $1/\sqrt{2}$
**37.** $2x$     **39.** $-1/x^2$
**41.** $1/(2\sqrt{x})$     **43.** 1
**45.** 1/2     **47.** 1
**49.** 0     **51.** 2
**53.** does not exist     **55.** does not exist
**57.** $-1/(2a)$     **59.** 0
**61.** $-2$     **63.** $\pi^2$
**65.** (a) 0, (b) 8, (c) 9, (d) $-3$
**67.** 5     **69.** 1
**71.** 0.7071     **73.** $\lim_{x \to 0} f(x) = 0$
**75.** 2

**77.** $x^{1/3} < x^3$ on $(-1, 0)$ and $(1, \infty)$,
$x^{1/3} > x^3$ on $(-\infty, -1)$ and $(0, 1)$,
$\lim_{x \to a} h(x) = a$ for $a = -1, 0$, and 1

## Section 1.3 *(page 74)*

**1.** 1/2     **3.** $-3/5$
**5.** 0     **7.** $-3$
**9.** $-2/\sqrt{3}$     **11.** does not exist
**13.** $+\infty$     **15.** 0
**17.** $-\infty$     **19.** $-\infty$
**21.** $-\infty$     **23.** $\infty$
**25.** $-\infty$     **27.** $\infty$
**29.** $-\sqrt{2}/4$     **31.** $-2$
**33.** $-1$
**35.** horiz: $y = 0$, $y = -1$, vert: $x = 0$
**37.** 1     **39.** 1
**41.** $-\infty$     **43.** 2
**45.** $-1$     **47.** 1
**49.** 3     **51.** does not exist
**53.** 1
**55.** $C(t)$ has a limit at every real $t$ except at the integers.
$\lim_{t \to t_0 -} C(t) = C(t_0)$ everywhere, but
$$\lim_{t \to t_0 +} C(t) = \begin{cases} C(t_0) & \text{if } t_0 \text{ not integral} \\ C(t_0) + 1.5 & \text{if } t_0 \text{ an integer} \end{cases}$$

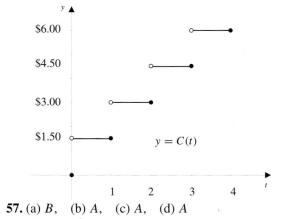

**57.** (a) $B$,   (b) $A$,   (c) $A$,   (d) $A$

## Section 1.4 *(page 85)*

**1.** at $-2$, right cont. and cont., at $-1$ disc., at 0 disc. but left cont., at 1 disc. and right cont., at 2 disc.
**3.** no abs. max, abs. min 0
**5.** no     **7.** cont. everywhere
**9.** cont. everywhere except at $x = 0$, disc. at $x = 0$

**11.** cont. everywhere except at the integers, discontinuous but left-continuous at the integers

**13.** 4, $x + 2$     **15.** $1/5$, $(t - 2)/(t + 2)$

**17.** $k = 8$     **19.** no max, min = 0

**21.** 16     **23.** 5

**25.** $f$ positive on $(-1, 0)$ and $(1, \infty)$; $f$ negative on $(-\infty, -1)$ and $(0, 1)$

**27.** $f$ positive on $(-\infty, -2)$, $(-1, 1)$ and $(2, \infty)$; $f$ negative on $(-2, -1)$ and $(1, 2)$

**35.** max 1.593 at $-0.831$, min $-0.756$ at 0.629

**37.** max $31/3 \approx 10.333$ at $x = 3$, min 4.762 at $x = 1.260$

**39.** 0.682     **41.** $-0.657$,    1.410

### Section 1.5   *(page 91)*

**1.** between 12°C and 20°C

**3.** (1.99, 2.01)     **5.** (0.81, 1.21)

**7.** $\delta = 0.01$     **9.** $\delta \approx 0.0165$

### Review Exercises   *(page 93)*

**1.** 13     **3.** 12

**5.** 4     **7.** does not exist

**9.** does not exist     **11.** $-\infty$

**13.** $12\sqrt{3}$     **15.** 0

**17.** does not exist     **19.** $-1/3$

**21.** $-\infty$     **23.** $\infty$

**25.** does not exist     **27.** 0

**29.** 2     **31.** no disc.

**33.** disc. and left cont. at 2

**35.** disc. and right cont. at $x = 1$

**37.** no disc.

### Challenging Problems   *(page 93)*

**1.** to the right     **3.** $-1/4$

**5.** 3     **7.** $T, F, T, F, F$

# Chapter 2
# Differentiation

### Section 2.1   *(page 100)*

**1.** $y = 3x - 1$     **3.** $y = 8x - 13$

**5.** $y = 12x + 24$     **7.** $2x + 27y = 9$

**9.** $x - 4y = -5$     **11.** $x - 4y = -2$

**13.** $y = 2x_0 x - x_0^2$     **15.** no

**17.** yes, $x = -2$     **19.** yes, $x = 0$

**21.** (a) $3a^2$;   (b) $y = 3x - 2$ and $y = 3x + 2$

**23.** (1, 1),  (−1, 1)     **25.** $k = 3/4$

**27.** horiz. tangent at (0, 0), (3, 108), (5, 0)

**29.** horiz. tangent at $(-0.5, 1.25)$, no tangents at $(-1, 1)$ and $(1, -1)$

**31.** horiz. tangent at $(0, -1)$

**33.** no, consider $y = x^{2/3}$ at (0, 0)

### Section 2.2   *(page 108)*

**1.**     **3.**

**5.** on $[-2, 2]$ except at $x = -1$ and $x = 1$

**7.** slope positive for $x < 1.5$, negative for $x > 1.5$; horizontal tangent at $x = 1.5$

**9.** singular points at $x = -1, 0, 1$, horizontal tangents at about $x = \pm 0.57$

**11.** $2x - 3$     **13.** $3x^2$

**15.** $-\dfrac{4}{(2 + x)^2}$     **17.** $\dfrac{1}{\sqrt{2t + 1}}$

**19.** $1 - \dfrac{1}{x^2}$     **21.** $-\dfrac{x}{(1 + x^2)^{3/2}}$

**23.** $-\dfrac{1}{2(1 + x)^{3/2}}$

**25.** Define $f(0) = 0$, $f$ is not differentiable at 0

**27.** at $x = -1$ and $x = -2$

**29.**

| $x$ | $\dfrac{f(x) - f(2)}{x - 2}$ | | $x$ | $\dfrac{f(x) - f(2)}{x - 2}$ |
|---|---|---|---|---|
| 1.9 | −0.26316 | | 2.1 | −0.23810 |
| 1.99 | −0.25126 | | 2.01 | −0.24876 |
| 1.999 | −0.25013 | | 2.001 | −0.24988 |
| 1.9999 | −0.25001 | | 2.0001 | −0.24999 |

$$\frac{d}{dx}\left(\frac{1}{x}\right)\Bigg|_{x=2} = -\frac{1}{4}$$

**31.** $x - 6y = -15$

**33.** $y = \dfrac{2}{a^2 + a} - \dfrac{2(2a + 1)}{(a^2 + a)^2}(t - a)$

**35.** $22t^{21}$, all $t$     **37.** $-(1/3)x^{-4/3}$, $x \neq 0$

**39.** $(119/4)s^{115/4}$, $s \geq 0$     **41.** $-16$

**43.** $1/(8\sqrt{2})$     **45.** $y = a^2 x - a^3 + \dfrac{1}{a}$

**47.** $y = 6x - 9$ and $y = -2x - 1$

**49.** $\dfrac{1}{2\sqrt{2}}$       **53.** $f'(x) = \frac{1}{3}x^{-2/3}$

### Section 2.3   (page 117)

**1.** $6x - 5$       **3.** $x^2 - x + 1$

**5.** $2Ax + B$       **7.** $\frac{1}{3}s^4 - \frac{1}{5}s^2$

**9.** $\frac{1}{3}t^{-2/3} + \frac{1}{2}t^{-3/4} + \frac{3}{5}t^{-4/5}$

**11.** $x^{2/3} + x^{-8/5}$       **13.** $\dfrac{5}{2\sqrt{x}} - \frac{3}{2}\sqrt{x} - \frac{5}{6}x^{3/2}$

**15.** $-\dfrac{2x + 5}{(x^2 + 5x)^2}$       **17.** $\dfrac{\pi^2}{(2 - \pi t)^2}$

**19.** $(4x^2 - 3)/x^4$

**21.** $-t^{-3/2} + (1/2)t^{-1/2} + (3/2)\sqrt{t}$

**23.** $-\dfrac{3(2\sqrt{x} + 1)}{2\sqrt{x}(x + \sqrt{x})^2}$       **25.** $-\dfrac{24}{(3 + 4x)^2}$

**27.** $\dfrac{1}{\sqrt{t}(1 - \sqrt{t})^2}$       **29.** $\dfrac{ad - bc}{(cx + d)^2}$

**31.** $10 + 70x + 150x^2 + 96x^3$

**33.** $2x(\sqrt{x} + 1)(5x^{2/3} - 2) + \dfrac{1}{2\sqrt{x}}(x^2 + 4)(5x^{2/3} - 2)$

      $+ \dfrac{10}{3}x^{-1/3}(x^2 + 4)(\sqrt{x} + 1)$

**35.** $\dfrac{6x + 1}{(6x^2 + 2x + 1)^2}$       **37.** $-1$

**39.** $20$       **41.** $-\dfrac{1}{2}$

**43.** $-\dfrac{1}{18\sqrt{2}}$       **45.** $y = 4x - 6$

**47.** $(1, 2)$ and $(-1, -2)$       **49.** $\left(-\frac{1}{2}, \frac{4}{3}\right)$

**51.** $y = b - \dfrac{b^2 x}{4}$

**53.** $y = 12x - 16$, $y = 3x + 2$

**55.** $x/\sqrt{x^2 + 1}$

### Section 2.4   (page 123)

**1.** $12(2x + 3)^5$       **3.** $-20x(4 - x^2)^9$

**5.** $\dfrac{30}{t^2}\left(2 + \dfrac{3}{t}\right)^{-11}$       **7.** $\dfrac{12}{(5 - 4x)^2}$

**9.** $-2x\,\mathrm{sgn}\,(1 - x^2)$       **11.** $\begin{cases} 8 & \text{if } x > 1/4 \\ 0 & \text{if } x < 1/4 \end{cases}$

**13.** $\dfrac{-3}{2\sqrt{3x + 4}(2 + \sqrt{3x + 4})^2}$

**15.** $-\dfrac{5}{3}\left(1 - \dfrac{1}{(u - 1)^2}\right)\left(u + \dfrac{1}{u - 1}\right)^{-8/3}$

**17.**

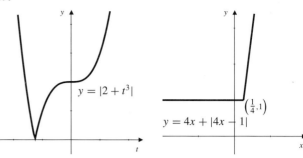

$y = |2 + t^3|$

$y = 4x + |4x - 1|$    $\left(\frac{1}{4}, 1\right)$

**23.** $(5 - 2x)f'(5x - x^2)$       **25.** $\dfrac{f'(x)}{\sqrt{3 + 2f(x)}}$

**27.** $\dfrac{1}{\sqrt{x}}f'(3 + 2\sqrt{x})$

**29.** $15f'(4 - 5t)f'(2 - 3f(4 - 5t))$

**31.** $\dfrac{3}{2\sqrt{2}}$       **33.** $102$

**35.** $-6\left(1 - \frac{15}{2}(3x)^4\left((3x)^5 - 2\right)^{-3/2}\right)$

      $\times \left(x + \left((3x)^5 - 2\right)^{-1/2}\right)^{-7}$

**37.** $y = 2^{3/2} - \sqrt{2}(x + 1)$       **39.** $y = \frac{1}{27} + \frac{5}{162}(x + 2)$

**41.** $\dfrac{x(x^4 + 2x^2 - 2)}{(x^2 + 1)^{5/2}}$       **43.** $857{,}592$

**45.** no; yes; both functions are equal to $x^2$.

### Section 2.5   (page 129)

**3.** $-3\sin 3x$       **5.** $\pi \sec^2 \pi x$

**7.** $-\dfrac{2}{\pi}\sin\dfrac{2x}{\pi}$       **9.** $3\csc^2(4 - 3x)$

**11.** $r\sin(s - rx)$       **13.** $2\pi x \cos(\pi x^2)$

**15.** $\dfrac{-\sin x}{2\sqrt{1 + \cos x}}$

**17.** $-(1 + \cos x)\sin(x + \sin x)$

**19.** $(3\pi/2)\sin^2(\pi x/2)\cos(\pi x/2)$

**21.** $a\cos 2at$       **23.** $2\cos(2x) + 2\sin(2x)$

**25.** $\sec^2 x - \csc^2 x$       **27.** $\tan^2 x$

**29.** $-t\sin t$       **31.** $1/(1 + \cos x)$

**33.** $2x\cos(3x) - 3x^2\sin(3x)$

**35.** $2x[\sec(x^2)\tan^2(x^2) + \sec^3(x^2)]$

**37.** $-\sec^2 t \sin(\tan t)\cos(\cos(\tan t))$

**41.** $y = \pi - x$, $y = x - \pi$

**43.** $y = 1 - (x - \pi)/4$, $y = 1 + 4(x - \pi)$

**45.** $y = \dfrac{1}{\sqrt{2}} + \dfrac{\pi}{180\sqrt{2}}(x - 45)$

**47.** $\pm(\pi/4, 1)$       **51.** yes, $(\pi, \pi)$

**53.** yes, $(2\pi/3, (2\pi/3) + \sqrt{3})$, $(4\pi/3, (4\pi/3) - \sqrt{3})$

**55.** 2        **57.** 1

**59.** 1/2

**61.** infinitely many, 0.336508, 0.161228

## Section 2.6 (page 137)

**1.** $c = \dfrac{a+b}{2}$       **3.** $c = \pm\dfrac{2}{\sqrt{3}}$

**9.** inc. on $\left(-\infty, -\dfrac{2}{\sqrt{3}}\right)$ and $\left(\dfrac{2}{\sqrt{3}}, \infty\right)$, dec. on $\left(-\dfrac{2}{\sqrt{3}}, \dfrac{2}{\sqrt{3}}\right)$

**11.** inc. on $(-2, 0)$ and $(2, \infty)$; dec. on $(-\infty, -2)$ and $(0, 2)$

**13.** inc. on $(-\infty, 3)$ and $(5, \infty)$; dec. on $(3, 5)$

**15.** inc. on $(-\infty, \infty)$

## Section 2.7 (page 144)

**1.** 4%           **3.** $-4\%$

**5.** 1%           **7.** 6%

**9.** $8 \text{ ft}^2/\text{ft}$

**11.** $1/\sqrt{\pi A}$ units/square unit

**13.** $16\pi \text{ m}^3/\text{m}$

**15.** $\dfrac{dC}{dA} = \sqrt{\dfrac{\pi}{A}}$ length units/area unit

**17.** CP. $x = 0$, incr. $x > 0$, decr. $x < 0$

**19.** CP. $x = 0$, $x = -4$, incr. on $(-\infty, -4)$ and $(0, \infty)$, decr. on $(-4, 0)$

**23.** 0.535898       **25.** 0, -0.518784

**27.** (a) 10,500 l/min, 3,500 l/min, (b) 7,000 l/min

**29.** decreases at 1/8 pound/mi

**31.** (a) \$300, (b) $C(101) - C(100) = \$299.50$

**33.** (a) $-\$2.00$, (b) \$9.11

## Section 2.8 (page 149)

**1.** $\begin{cases} y' = -14(3 - 2x)^6, \\ y'' = 168(3 - 2x)^5, \\ y''' = -1680(3 - 2x)^4 \end{cases}$

**3.** $\begin{cases} y' = -12(x - 1)^{-3}, \\ y'' = 36(x - 1)^{-4}, \\ y''' = -144(x - 1)^{-5} \end{cases}$

**5.** $\begin{cases} y' = \frac{1}{3}x^{-2/3} + \frac{1}{3}x^{-4/3}, \\ y'' = -\frac{2}{9}x^{-5/3} - \frac{4}{9}x^{-7/3} \\ y''' = \frac{10}{27}x^{-8/3} + \frac{28}{27}x^{-10/3} \end{cases}$

**7.** $\begin{cases} y' = \frac{5}{2}x^{3/2} + \frac{3}{2}x^{-1/2} \\ y'' = \frac{15}{4}x^{1/2} - \frac{3}{4}x^{-3/2} \\ y''' = \frac{15}{8}x^{-1/2} + \frac{9}{8}x^{-5/2} \end{cases}$

**9.** $y' = \sec^2 x$, $y'' = 2\sec^2 x \tan x$, $y''' = 4\sec^2 x \tan^2 x + 2\sec^4 x$

**11.** $y' = -2x\sin(x^2)$, $y'' = -2\sin(x^2) - 4x^2\cos(x^2)$, $y''' = -12x\cos(x^2) + 8x^3\sin(x^2)$

**13.** $(-1)^n n! x^{-(n+1)}$     **15.** $n!(2 - x)^{-(n+1)}$

**17.** $(-1)^n n! b^n (a + bx)^{-(n+1)}$

**19.** $f^{(n)} = \begin{cases} (-1)^k a^n \cos(ax) & \text{if } n = 2k \\ (-1)^{k+1} a^n \sin(ax) & \text{if } n = 2k + 1 \end{cases}$ where $k = 0, 1, 2, \ldots$

**21.** $f^{(n)} = (-1)^k [a^n x \sin(ax) - na^{n-1}\cos(ax)]$ if $n = 2k$, or $(-1)^k [a^n x \cos(ax) + na^{n-1}\sin(ax)]$ if $n = 2k + 1$, where $k = 0, 1, 2, \ldots$

**23.** $-\dfrac{1 \times 3 \times 5 \times \cdots \times (2n-3)}{2^n} 3^n (1 - 3x)^{-(2n-1)/2}$, $(n = 2, 3, \ldots)$

## Section 2.9 (page 154)

**1.** $\dfrac{1-y}{2+x}$       **3.** $\dfrac{2x+y}{3y^2 - x}$

**5.** $\dfrac{2 - 2xy^3}{3x^2 y^2 + 1}$       **7.** $-\dfrac{3x^2 + 2xy}{x^2 + 4y}$

**9.** $2x + 3y = 5$       **11.** $y = x$

**13.** $y = 1 - \dfrac{4}{4 - \pi}\left(x - \dfrac{\pi}{4}\right)$

**15.** $y = 2 - x$       **17.** $\dfrac{2(y-1)}{(1-x)^2}$

**19.** $\dfrac{(2 - 6y)(1 - 3x^2)^2}{(3y^2 - 2y)^3} - \dfrac{6x}{3y^2 - 2y}$

**21.** $-a^2/y^3$

## Section 2.10 (page 160)

**1.** $5x + C$       **3.** $\frac{2}{3}x^{3/2} + C$

**5.** $\frac{1}{4}x^4 + C$       **7.** $-\cos x + C$

**9.** $a^2 x - \frac{1}{3}x^3 + C$       **11.** $\frac{4}{3}x^{3/2} + \frac{9}{4}x^{4/3} + C$

**13.** $\frac{1}{12}x^4 - \frac{1}{6}x^3 + \frac{1}{2}x^2 - x + C$

**15.** $\frac{1}{2}\sin(2x) + C$       **17.** $\dfrac{-1}{1+x} + C$

**19.** $\frac{1}{3}(2x + 3)^{3/2} + C$       **21.** $-\cos(x^2) + C$

**23.** $\tan x - x + C$

**25.** $(x + \sin x \cos x)/2 + C$

**27.** $y = \frac{1}{2}x^2 - 2x + 3$, all $x$

**29.** $y = 2x^{3/2} - 15$, $(x > 0)$

**31.** $y = \dfrac{A}{3}(x^3 - 1) + \dfrac{B}{2}(x^2 - 1) + C(x - 1) + 1$, (all $x$)

**33.** $y = \sin x + (3/2)$, (all $x$)

**35.** $y = 1 + \tan x$, $-\pi/2 < x < \pi/2$

**37.** $y = x^2 + 5x - 3$, (all $x$)

**39.** $y = \dfrac{x^5}{20} - \dfrac{x^2}{2} + 8$, (all $x$)

**41.** $y = 1 + x - \cos x$, (all $x$)

**43.** $y = 3x - \dfrac{1}{x}$, $(x > 0)$

**45.** $y = -\dfrac{7\sqrt{x}}{2} + \dfrac{18}{\sqrt{x}}$, $(x > 0)$

### Section 2.11   (page 167)

**1.** (a) $t > 2$,   (b) $t < 2$,   (c) all $t$,   (d) no $t$,
    (e) $t > 2$,   (f) $t < 2$,   (g) 2,   (h) 0

**3.** (a) $t < -2/\sqrt{3}$ or $t > 2/\sqrt{3}$,
    (b) $-2/\sqrt{3} < t < 2/\sqrt{3}$,   (c) $t > 0$,   (d) $t < 0$,
    (e) $t > 2/\sqrt{3}$ or $-2/\sqrt{3} < t < 0$,
    (f) $t < -2/\sqrt{3}$ or $0 < t < 2/\sqrt{3}$,
    (g) $\pm 12/\sqrt{3}$ at $t = \pm 2/\sqrt{3}$,   (h) 12

**5.** acc $= 9.8$ m/s$^2$ downward at all times;
    max height $= 4.9$ m; ball strikes ground at $9.8$ m/s

**7.** time 27.8 s; distance 771.6 m

**9.** $4h$ m, $\sqrt{2}v_0$ m/s          **11.** 400 ft

**13.** 0.833 km

**15.** $v = \begin{cases} 2t & \text{if } 0 < t \le 2 \\ 4 & \text{if } 2 < t < 8 \\ 20 - 2t & \text{if } 8 \le t < 10 \end{cases}$
    $v$ is continuous for $0 < t < 10$.
    $a = \begin{cases} 2 & \text{if } 0 < t < 2 \\ 0 & \text{if } 2 < t < 8 \\ -2 & \text{if } 8 < t < 10 \end{cases}$
    $a$ is continuous except at $t = 2$ and $t = 8$.
    Maximum velocity 4 is attained for $2 \le t \le 8$.

**17.** 7 s          **19.** 448 ft

### Review Exercises   (page 168)

**1.** $18x + 6$          **3.** $-1$

**5.** $6\pi x + 12y = 6\sqrt{3} + \pi$

**7.** $\dfrac{\cos x - 1}{(x - \sin x)^2}$          **9.** $x^{-3/5}(4 - x^{2/5})^{-7/2}$

**11.** $-2\theta \sec^2 \theta \tan \theta$          **13.** $20x^{19}$

**15.** $-\sqrt{3}$          **17.** $-2xf'(3 - x^2)$

**19.** $2f'(2x)\sqrt{g(x/2)} + \dfrac{f(2x)\,g'(x/2)}{4\sqrt{g(x/2)}}$

**21.** $f'(x + (g(x))^2)(1 + 2g(x)g'(x))$

**23.** $\cos x\, f'(\sin x)\, g(\cos x) - \sin x\, f(\sin x)\, g'(\cos x)$

**25.** $7x + 10y = 24$          **27.** $\dfrac{x^3}{3} - \dfrac{1}{x} + C$

**29.** $2\tan x + 3\sec x + C$          **31.** $4x^3 + 3x^4 - 7$

**33.** $I_1 = x\sin x + \cos x + C$, $I_2 = \sin x - x\cos x + C$

**35.** $y = 3x$

**37.** points $k\pi$ and $k\pi/(n+1)$ where $k$ is any integer

**39.** $(0, 0)$, $(\pm 1/\sqrt{2}, 1/2)$, dist. $= \sqrt{3}/2$ units

**41.** (a) $k = g/R$          **43.** 15.3 m

**45.** 80 ft/s or about 55 mph

### Challenging Problems   (page 170)

**3.** (a) 0, (b) 3/8, (c) 12, (d) $-48$, (e) 3/7, (f) 21

**13.** $f(m) = C - (m - B)^2/(4A)$

**17.** (a) $3b^2 > 8ac$

# Chapter 3
# Transcendental Functions

### Section 3.1   (page 178)

**1.** $f^{-1}(x) = x + 1$
    $\mathcal{D}(f^{-1}) = \mathcal{R}(f) = \mathcal{R}(f^{-1}) = \mathcal{D}(f) = \mathbb{R}$

**3.** $f^{-1}(x) = x^2 + 1$, $\mathcal{D}(f^{-1}) = \mathcal{R}(f) = [0, \infty)$,
    $\mathcal{R}(f^{-1}) = \mathcal{D}(f) = [1, \infty)$

**5.** $f^{-1}(x) = x^{1/3}$
    $\mathcal{D}(f^{-1}) = \mathcal{R}(f) = \mathcal{R}(f^{-1}) = \mathcal{D}(f) = \mathbb{R}$

**7.** $f^{-1}(x) = -\sqrt{x}$,   $\mathcal{D}(f^{-1}) = \mathcal{R}(f) = [0, \infty)$,
    $\mathcal{R}(f^{-1}) = \mathcal{D}(f) = (-\infty, 0]$

**9.** $f^{-1}(x) = \dfrac{1}{x} - 1$,   $\mathcal{D}(f^{-1}) = \mathcal{R}(f) = \{x : x \ne 0\}$,
    $\mathcal{R}(f^{-1}) = \mathcal{D}(f) = \{x : x \ne -1\}$

**11.** $f^{-1}(x) = \dfrac{1 - x}{2 + x}$,
    $\mathcal{D}(f^{-1}) = \mathcal{R}(f) = \{x : x \ne -2\}$,
    $\mathcal{R}(f^{-1}) = \mathcal{D}(f) = \{x : x \ne -1\}$

**13.** $g^{-1}(x) = f^{-1}(x + 2)$    **15.** $k^{-1}(x) = f^{-1}\left(-\dfrac{x}{3}\right)$

**17.** $p^{-1}(x) = f^{-1}\left(\dfrac{1}{x} - 1\right)$

**19.** $r^{-1}(x) = \dfrac{1}{4}\left(3 - f^{-1}\left(\dfrac{1 - x}{2}\right)\right)$

**21.** $f^{-1}(x) = \begin{cases} \sqrt{x - 1} & \text{if } x \ge 1 \\ x - 1 & \text{if } x < 1 \end{cases}$

**25.** $1/[6(f^{-1}(x))^2]$          **27.** 2.23362

**29.** $\mathbb{R}$,  1

**31.** $c = 1$, $a$, $b$ arbitrary, or $a = b = 0$, $c = -1$.

### Section 3.2   (page 182)

**1.** $\sqrt{3}$          **3.** $x^6$

**5.** 3

**7.** $-2x$

**9.** $x$

**11.** 1

**13.** 1

**15.** 2

**17.** $\log_a(x^4 + 4x^2 + 3)$

**19.** $4.728804\ldots$

**21.** $x = (\log_{10} 5)/(\log_{10}(4/5)) \approx -7.212567$

**23.** $x = 3^{1/5} = 10^{(\log_{10} 3)/5} \approx 1.24573$

**29.** $1/2, -2$

### Section 3.3  (page 192)

**1.** $\sqrt{e}$

**3.** $x^5$

**5.** $-3x$

**7.** $\ln \dfrac{64}{81}$

**9.** $\ln(x^2(x-2)^5)$

**11.** $x = \dfrac{\ln 2}{\ln(3/2)}$

**13.** $x = \dfrac{\ln 5 - 9\ln 2}{2\ln 2}$

**15.** $0 < x < 2$

**17.** $3 < x < 7/2$

**19.** $5e^{5x}$

**21.** $(1 - 2x)e^{-2x}$

**23.** $\dfrac{3}{3x-2}$

**25.** $\dfrac{e^x}{1+e^x}$

**27.** $\dfrac{e^x - e^{-x}}{2}$

**29.** $e^{x+e^x}$

**31.** $\dfrac{e^x}{(1+e^x)^2}$

**33.** $e^x(\sin x - \cos x)$

**35.** $\dfrac{1}{x\ln x}$

**37.** $2x\ln x$

**39.** $(2\ln 5)5^{2x+1}$

**41.** $t^x x^t \ln t + t^{x+1}x^{t-1}$

**43.** $\dfrac{b}{(bs+c)\ln a}$

**45.** $x^{\sqrt{x}}\left(\dfrac{1}{\sqrt{x}}\left(\tfrac{1}{2}\ln x + 1\right)\right)$

**47.** $\sec x$

**49.** $-\dfrac{1}{\sqrt{x^2+a^2}}$

**51.** $f^{(n)}(x) = e^{ax}(na^{n-1} + a^n x), \quad n = 1, 2, 3, \ldots$

**53.** $y' = 2xe^{x^2}, \quad y'' = 2(1+2x^2)e^{x^2},$
$y''' = 4(3x+2x^3)e^{x^2}, \quad y^{(4)} = 4(3+12x^2+4x^4)e^{x^2}$

**55.** $f'(x) = x^{x^2+1}(2\ln x + 1),$
$g'(x) = x^{x^x} x^x \left(\ln x + (\ln x)^2 + \dfrac{1}{x}\right);$
$g$ grows more rapidly than does $f$.

**57.** $f'(x) = f(x)\left(\dfrac{1}{x-1} + \dfrac{1}{x-2} + \dfrac{1}{x-3} + \dfrac{1}{x-4}\right)$

**59.** $f'(2) = \dfrac{556}{3675}, \quad f'(1) = \dfrac{1}{6}$

**61.** $f$ inc. for $x < 1$, dec. for $x > 1$

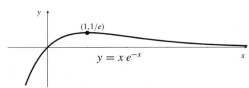

**63.** $y = ex$

**65.** $y = 2e\ln 2(x - 1)$

**67.** $-1/e^2$

**69.** $f'(x) = (A + B)\cos \ln x + (B - A)\sin \ln x,$
$\int \cos \ln x \, dx = \dfrac{x}{2}(\cos \ln x + \sin \ln x),$
$\int \sin \ln x \, dx = \dfrac{x}{2}(\sin \ln x - \cos \ln x)$

**71.** (a) $F_{2B,-2A}(x)$;  (b) $-2e^x(\cos x + \sin x)$

### Section 3.4  (page 200)

**1.** 0

**3.** 2

**5.** 0

**7.** 0

**9.** 566

**11.** 29.15 years

**13.** 160.85 years

**15.** 4,139 gm

**17.** \$7,557.84

**19.** about 14.7 years

**21.** (a) $f(x) = Ce^{bx} - (a/b)$,
(b) $y = (y_0 + (a/b))e^{bx} - (a/b)$

**23.** 22.35°C

**25.** 6.84 min

**29.** $(0, -(1/k)\ln(y_0/(y_0 - L)))$, solution $\to -\infty$

**31.** about 7,671 cases, growing at about 3,028 cases/week

### Section 3.5  (page 209)

**1.** $\pi/3$

**3.** $-\pi/4$

**5.** 0.7

**7.** $-\pi/3$

**9.** $\dfrac{\pi}{2} + 0.2$

**11.** $2/\sqrt{5}$

**13.** $\sqrt{1 - x^2}$

**15.** $\dfrac{1}{\sqrt{1+x^2}}$

**17.** $\dfrac{\sqrt{1-x^2}}{x}$

**19.** $\dfrac{1}{\sqrt{2+x-x^2}}$

**21.** $\dfrac{-\operatorname{sgn} a}{\sqrt{a^2 - (x-b)^2}}$

**23.** $\tan^{-1} t + \dfrac{t}{1+t^2}$

**25.** $2x\tan^{-1} x + 1$

**27.** $\dfrac{\sqrt{1-4x^2}\sin^{-1} 2x - 2\sqrt{1-x^2}\sin^{-1} x}{\sqrt{1-x^2}\sqrt{1-4x^2}\left(\sin^{-1} 2x\right)^2}$

**29.** $\dfrac{x}{\sqrt{(1-x^4)}\sin^{-1} x^2}$

**31.** $\sqrt{\dfrac{a-x}{a+x}}$

**33.** $\dfrac{\pi - 2}{\pi - 1}$

**37.** $\dfrac{d}{dx} \csc^{-1} x = -\dfrac{1}{|x|\sqrt{x^2 - 1}}$

**39.** $\tan^{-1} x + \cot^{-1} x = -\dfrac{\pi}{2}$ for $x < 0$

**41.** cont. everywhere, differentiable except at $n\pi$ for integers $n$

**43.** continuous and differentiable everywhere except at odd multiples of $\pi/2$

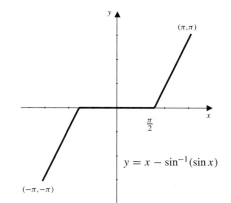

**49.** $\tan^{-1}\left(\dfrac{x-1}{x+1}\right) - \tan^{-1} x = \dfrac{3\pi}{4}$ on $(-\infty, -1)$

**51.** $f'(x) = 1 - \operatorname{sgn}(\cos x)$

**53.** $y = \dfrac{1}{3} \tan^{-1} \dfrac{x}{3} + 2 - \dfrac{\pi}{12}$

**55.** $y = 4 \sin^{-1} \dfrac{x}{5}$

## Section 3.6 (page 215)

**3.** $\tanh(x + y) = \dfrac{\tanh x + \tanh y}{1 + \tanh x \tanh y}$

$\tanh(x - y) = \dfrac{\tanh x - \tanh y}{1 - \tanh x \tanh y}$

**5.** $\dfrac{d}{dx} \sinh^{-1}(x) = \dfrac{1}{\sqrt{x^2 + 1}}$,

$\dfrac{d}{dx} \cosh^{-1}(x) = \dfrac{1}{\sqrt{x^2 - 1}}$,

$\dfrac{d}{dx} \tanh^{-1}(x) = \dfrac{1}{1 - x^2}$,

$\displaystyle\int \dfrac{dx}{\sqrt{x^2 + 1}} = \sinh^{-1}(x) + C$,

$\displaystyle\int \dfrac{dx}{\sqrt{x^2 - 1}} = \cosh^{-1}(x) + C \quad (x > 1)$,

$\displaystyle\int \dfrac{dx}{1 - x^2} = \tanh^{-1}(x) + C \quad (-1 < x < 1)$

**7.** (a) $\dfrac{x^2 - 1}{2x}$; (b) $\dfrac{x^2 + 1}{2x}$; (c) $\dfrac{x^2 - 1}{x^2 + 1}$; (d) $x^2$

**9.** $\coth^{-1} x = \tanh^{-1} \dfrac{1}{x} = \dfrac{1}{2} \ln\left(\dfrac{x+1}{x-1}\right)$, domain: all $x$ such that $|x| > 1$, range: all $y \neq 0$, derivative: $-1/(x^2 - 1)$

**11.** $f_{A,B} = g_{A+B, A-B}$; $\quad g_{C,D} = f_{(C+D)/2, (C-D)/2}$

**13.** $y = y_0 \cosh k(x - a) + \dfrac{v_0}{k} \sinh k(x - a)$

## Section 3.7 (page 223)

**1.** $y = Ae^{-5t} + Be^{-2t}$   **3.** $y = A + Be^{-2t}$

**5.** $y = (A + Bt)e^{-4t}$

**7.** $y = (A \cos t + B \sin t)e^{3t}$

**9.** $y = (A \cos 2t + B \sin 2t)e^{-t}$

**11.** $y = (A \cos \sqrt{2}t + B \sin \sqrt{2}t)e^{-t}$

**13.** $y = \dfrac{6}{7}e^{t/2} + \dfrac{1}{7}e^{-3t}$

**15.** $y = e^{-2t}(2 \cos t + 6 \sin t)$

**23.** $y = \dfrac{3}{10} \sin(10t)$, circ freq 10, freq $\dfrac{10}{2\pi}$, per $\dfrac{2\pi}{10}$, amp $\dfrac{3}{10}$

**25.** $\begin{aligned} A &= \alpha \cos(\omega c) - \beta \sin(\omega c) \\ B &= \alpha \sin(\omega c) + \beta \cos(\omega c) \end{aligned}$

**27.** $y = A \cos(\omega(t - a)) + \dfrac{B}{\omega} \sin(\omega(t - a))$

**29.** 16 Hz for 900gm; 48 Hz for 100 gm

## Review Exercises (page 224)

**1.** $1/3$    **3.** both limits are 0

**5.** max $1/\sqrt{2e}$, min $-1/\sqrt{2e}$

**7.** $f(x) = 3e^{(x^2/2) - 2}$

**9.** (a) about 13.863%, (b) about 68 days

**11.** $e^{2x}$    **13.** y=x

**15.** 13.8165% approx.

**17.** $\cos^{-1}x = \frac{\pi}{2}-\sin^{-1}x, \cot^{-1}x = \operatorname{sgn} x \sin^{-1}(1/\sqrt{x^2+1})$,
$\csc^{-1}x = \sin^{-1}(1/x)$

**19.** 15°C

# Chapter 4
# Some Applications of Derivatives

## Section 4.1   *(page 231)*

**1.** 32 cm²/min

**3.** increasing at $160\pi$ cm²/s

**5.** (a) $1/(6\pi r)$ km/hr, (b) $1/(6\sqrt{\pi A})$ km/hr

**7.** 2 cm²/s    **9.** increasing at 2 cm³/s

**11.** increasing at rate 12

**13.** increasing at rate $2/\sqrt{5}$

**15.** $45\sqrt{3}$ km/h    **17.** 1/3 m/s, 5/6 m/s

**19.** 100 tons/day    **21.** $16\frac{4}{11}$ min after 3:00

**23.** $1/(18\pi)$ m/min

**25.** $9/(6250\pi)$ m/min, 4.64 m

**27.** 8 m/min    **29.** dec. at 126.9 km/h

**31.** 1/8 units/s    **33.** $\sqrt{3}/16$ m/min

**35.** (a) down at 24/125 m/s, (b) right at 7/125 m/s

**37.** dec. at 0.0197 rad/s    **39.** 0.047 rad/s

## Section 4.2   *(page 240)*

**1.** abs min 1 at $x = -1$; abs max 3 at $x = 1$

**3.** abs min 1 at $x = -1$; no max

**5.** abs min $-1$ at $x = 0$; abs max 8 at $x = 3$; loc max 3 at $x = -2$

**7.** abs min $a^3 + a - 4$ at $x = a$; abs max $b^3 + b - 4$ at $x = b$

**9.** abs max $b^5 + b^3 + 2b$ at $x = b$; no min value

**11.** no max or min values

**13.** max 3 at $x = -2$, min 0 at $x = 1$

**15.** abs max 1 at $x = 0$; no min value

**17.** no max or min value

**19.** loc max at $x = -1$; loc min at $x = 1$

**21.** loc max at $x = \frac{1}{3}$; loc min at $x = 1$

**23.** loc max at $x = \frac{3}{5}$; loc min at $x = 1$; critical point $x = 0$ is neither max nor min

**25.** loc max at $x = -1$ and $x = 1/\sqrt{5}$; loc min at $x = 1$ and $x = -1/\sqrt{5}$

**27.** abs min at $x = 0$

**29.** loc min at CP $x = -1$ and endpoint SP $x = \sqrt{2}$; loc max at CP $x = 1$ and endpoint SP $x = -\sqrt{2}$

**31.** loc max at $x = 2n\pi - \dfrac{\pi}{3}$; loc min at $x = 2n\pi + \dfrac{\pi}{3}$
$(n = 0, \pm1, \pm2, \ldots)$

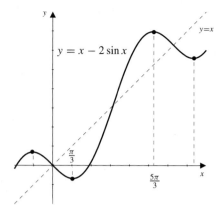

**33.** loc max at CP $x = \sqrt{3}/2$ and endpoint SP $x = -1$;
loc min at CP $x = -\sqrt{3}/2$ and endpoint SP $x = 1$

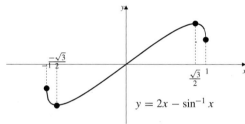

**35.** abs max at $x = 1/\ln 2$

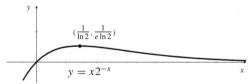

**37.** abs max at $x = e$

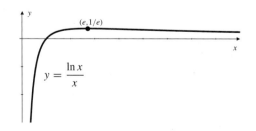

**39.** loc max at CP $x = 0$; abs min at SPs
$x = \pm1$

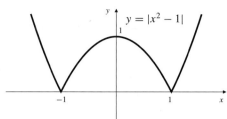

**41.** abs max at CPs $x = (2n+1)\pi/2$; abs min at SPs
$x = n\pi$ $(n = 0, \pm1, \pm2, \ldots)$

**43.** no max or min; SPs at $x = \pm1$

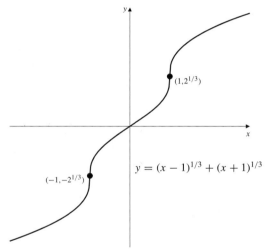

**45.** no max or min          **47.** max 2, min $-2$

**49.** has min, no max          **51.** yes, no

## Section 4.3   *(page 245)*

**1.** conc down on $(0, \infty)$          **3.** conc up on $\mathbb{R}$

**5.** conc down on $(-1, 0)$ and $(1, \infty)$; conc up on $(-\infty, -1)$ and $(0, 1)$; infl $x = -1, 0, 1$

**7.** conc down on $(-1, 1)$; conc up on $(-\infty, -1)$ and $(1, \infty)$; infl $x = \pm1$

**9.** conc down on $(-2, -2/\sqrt{5})$ and $(2/\sqrt{5}, 2)$; conc up on $(-\infty, -2)$, $(-2/\sqrt{5}, 2/\sqrt{5})$ and $(2, \infty)$; infl $x = \pm2, \pm2/\sqrt{5}$

**11.** conc down on $(2n\pi, (2n+1)\pi)$; conc up on $((2n-1)\pi, 2n\pi)$, $(n = 0, \pm1, \pm2, \ldots)$; infl $x = n\pi$

**13.** conc down on $\left(n\pi, (n+\frac{1}{2})\pi\right)$;
conc up on $\left((n-\frac{1}{2})\pi, n\pi\right)$; infl $x = n\pi/2$,
$(n = 0, \pm 1, \pm 2, \ldots)$

**15.** conc down on $(0, \infty)$, up on $(-\infty, 0)$; infl $x = 0$

**17.** conc down on $(-1/\sqrt{2}, 1/\sqrt{2})$, up on $(-\infty, -1/\sqrt{2})$
and $(1/\sqrt{2}, \infty)$; infl $x = \pm 1/\sqrt{2}$

**19.** conc down on $(-\infty, -1)$ and $(1, \infty)$; conc up on
$(-1, 1)$; infl $x = \pm 1$

**21.** conc down on $(-\infty, 4)$, up on $(4, \infty)$; infl $x = 4$

**23.** no concavity, no inflections

**25.** loc min at $x = 2$; loc max at $x = \frac{2}{3}$

**27.** loc min at $x = 1/\sqrt[4]{3}$; loc max at $-1/\sqrt[4]{3}$

**29.** loc max at $x = 1$; loc min at $x = -1$ (both abs)

**31.** loc (and abs) min at $x = 1/e$

**33.** loc min at $x = 0$; inflections at $x = \pm 2$ (not discernible by Second Derivative Test)

**35.** abs min at $x = 0$; abs max at $x = \pm 1/\sqrt{2}$

**39.** If $n$ is even, $f_n$ has a min and $g_n$ has a max at $x = 0$.
If $n$ is odd both have inflections at $x = 0$.

**Section 4.4   (page 255)**

**1.** (a) $g$, (b) $f''$, (c) $f$, (d) $f'$

**3.** (a) $k(x)$, (b) $g(x)$, (c) $f(x)$, (d) $h(x)$

**5.**

**7.**

**9.**

**11.**

**13.**

**15.**

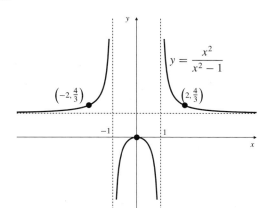

$$y = \frac{x^2}{x^2 - 1}$$

$\left(-2, \frac{4}{3}\right)$    $\left(2, \frac{4}{3}\right)$

**17.**

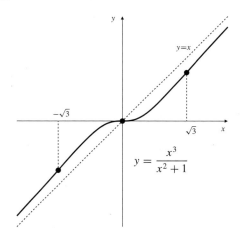

$y=x$

$-\sqrt{3}$    $\sqrt{3}$

$$y = \frac{x^3}{x^2 + 1}$$

**19.**

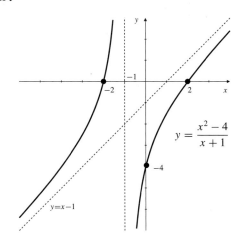

$$y = \frac{x^2 - 4}{x + 1}$$

$y=x-1$

**21.**

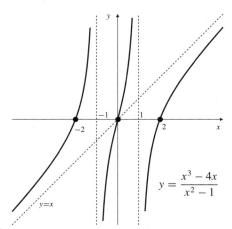

$y=x$

$$y = \frac{x^3 - 4x}{x^2 - 1}$$

**23.**

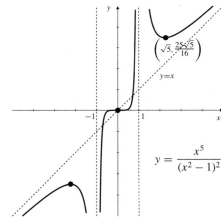

$\left(\sqrt{5}, \frac{25\sqrt{5}}{16}\right)$

$y=x$

$$y = \frac{x^5}{(x^2 - 1)^2}$$

**25.**

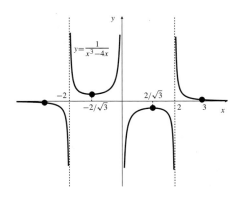

$y = \dfrac{1}{x^3 - 4x}$

$-2/\sqrt{3}$    $2/\sqrt{3}$

**27.**

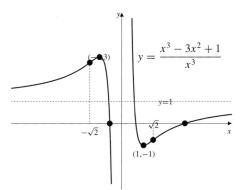

$$y = \frac{x^3 - 3x^2 + 1}{x^3}$$

**29.**

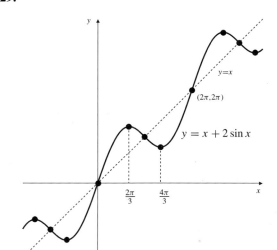

$$y = x + 2\sin x$$

**31.**

$$y = x\,e^x$$

**33.**

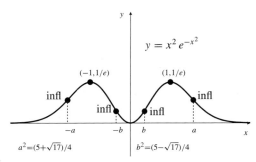

$$y = x^2\,e^{-x^2}$$

$a^2 = (5 + \sqrt{17})/4$     $b^2 = (5 - \sqrt{17})/4$

**35.**

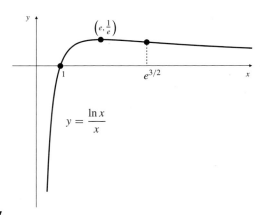

$$y = \frac{\ln x}{x}$$

**37.**

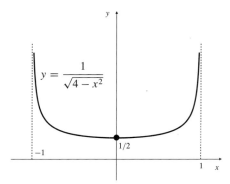

$$y = \frac{1}{\sqrt{4 - x^2}}$$

**39.**

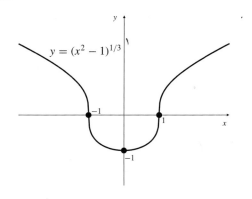

$$y = (x^2 - 1)^{1/3}$$

**41.**

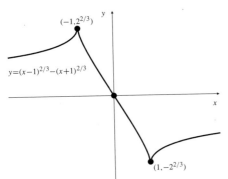

$(-1, 2^{2/3})$

$y$

$y = (x-1)^{2/3} - (x+1)^{2/3}$

$x$

$(1, -2^{2/3})$

**43.** $y = 0$. curve crosses asymptote at $x = n\pi$ for every integer $n$.

## Section 4.5 (page 263)

**1.** $49/4$

**3.** 20 and 40

**5.** 71.45

**11.** $R^2$ sq. units

**13.** $2ab$ un$^2$

**15.** width $8 + 10\sqrt{2}$ m, height $4 + 5\sqrt{2}$ m

**17.** rebate \$250

**19.** point 5 km east of $A$

**21.** (a) 0 m, (b) $\pi/(4 + \pi)$ m

**23.** $8\sqrt{3}$ units

**25.** $\left[(a^{2/3} + b^{2/3})^3 + c^2\right]^{1/2}$ units

**27.** $3^{1/2}/2^{1/3}$ units

**29.** height $\dfrac{2R}{\sqrt{3}}$, radius $\sqrt{\dfrac{2}{3}} R$ units

**31.** base 2m×2m, height 1m

**33.** width $\dfrac{20}{4 + \pi}$ m, height $\dfrac{10}{4 + \pi}$ m

**37.** width $R$, depth $\sqrt{3}R$

**39.** $Q = 3L/8$

**41.** $2\sqrt{6}$ ft

**43.** $\dfrac{2\pi}{9\sqrt{3}} R^3$ cubic units

## Section 4.6 (page 272)

**1.** 1.41421356237

**3.** 0.453397651516

**5.** 1.64809536561, 2.352392647658

**7.** 0.510973429389

**9.** infinitely many, 4.49340945791

**13.** max 1, min $-0.11063967219\ldots$

**15.** $x_1 = -a$, $x_2 = a = x_0$. Look for a root half way between $x_0$ and $x_1$

**17.** $x_n = (-1/2)^n \to 0$ (root) as $n \to \infty$.

**19.** 0.95025

**21.** 0.45340

**23.** $N(x_n)$ is the Newton's method approximation $x_{n+1}$

## Section 4.7 (page 279)

**1.** $6x - 9$

**3.** $2 - (x/4)$

**5.** $(7 - 2x)/27$

**7.** $\pi - x$

**9.** $(1/4) + (\sqrt{3}/2)(x - (\pi/6))$

**11.** about 8 cm$^2$

**13.** about 62.8 mi

**15.** $\sqrt{50} \approx \frac{99}{14} \approx 7.071429$, error $< 0$, $|\text{error}| < \frac{1}{2744} \approx 0.0003644$, (7.07106, 7.071429)

**17.** $\sqrt[4]{85} \approx \frac{82}{27}$, error $< 0$, $|\text{error}| < \frac{1}{2 \times 3^6}$, (3.03635, 3.03704)

**19.** $\cos 46° \approx \dfrac{1}{\sqrt{2}}\left(1 - \dfrac{\pi}{180}\right) \approx 0.694765$, error $< 0$, $|\text{error}| < \dfrac{1}{2\sqrt{2}}\left(\dfrac{\pi}{180}\right)^2$, (0.694658, 0.694765)

**21.** $\sin(3.14) \approx \pi - 3.14$, error $< 0$, $|\text{error}| < (\pi - 3.14)^3/2 < 2.02 \times 10^{-9}$, $(\pi - 3.14 - (\pi - 3.14)^3/2, \pi - 3.14)$

**23.** (7.07106, 7.07108), $\sqrt{50} \approx 7.07107$

**25.** (3.03635, 3.03641), $\sqrt[4]{85} \approx 3.03638$

**27.** $3 \le f(3) \le 13/4$

**29.** $g(1.8) \approx 0.6$, $|\text{error}| < 0.0208$

**31.** about 1005 cm$^3$

## Section 4.8 (page 287)

**1.** $1 - x + \frac{1}{2}x^2 - \frac{1}{6}x^3 + \frac{1}{24}x^4$

**3.** $1 + \dfrac{x - e}{e} - \dfrac{(x - e)^2}{2e^2} + \dfrac{(x - e)^3}{3e^3} - \dfrac{(x - e)^4}{4e^4}$

**5.** $2 + \dfrac{x - 4}{4} - \dfrac{(x - 4)^2}{64} + \dfrac{3(x - 4)^3}{1536}$

**7.** $x^{1/3} \approx 2 + \frac{1}{12}(x - 8) - \frac{1}{288}(x - 8)^2$, $9^{1/3} \approx 2.07986$, $0 < \text{error} \le 5/(81 \times 256)$, $2.07986 < 9^{1/3} < 2.08010$

**9.** $\dfrac{1}{x} \approx 1 - (x - 1) + (x - 1)^2$, $\dfrac{1}{1.02} \approx 0.9804$, $-(0.02)^3 \le \text{error} < 0$, $0.980392 \le \dfrac{1}{1.02} < 0.9804$

**11.** $e^x \approx 1 + x + \frac{1}{2}x^2$, $e^{-0.5} \approx 0.625$, $-\frac{1}{6}(0.5)^3 \le \text{error} < 0$, $0.604 \le e^{-0.5} < 0.625$

**13.** $\sin x = x - \dfrac{x^3}{3!} + \dfrac{x^5}{5!} - \dfrac{x^7}{7!} + R_7$; $R_7 = \dfrac{\sin X}{8!} x^8$ for some $X$ between 0 and $x$

**15.** $\sin x = \dfrac{1}{\sqrt{2}}\left[1 + \left(x - \dfrac{\pi}{4}\right) - \dfrac{1}{2!}\left(x - \dfrac{\pi}{4}\right)^2 - \dfrac{1}{3!}\left(x - \dfrac{\pi}{4}\right)^3 + \dfrac{1}{4!}\left(x - \dfrac{\pi}{4}\right)^4\right] + R_4$; where $R_4 = \dfrac{\cos X}{5!}\left(x - \dfrac{\pi}{4}\right)^5$ for some $X$ between $x$ and $\pi/4$

17. $\ln x = (x - 1) - \dfrac{(x-1)^2}{2} + \dfrac{(x-1)^3}{3} - \dfrac{(x-11)^4}{4}$

$+ \dfrac{(x-1)^5}{5} - \dfrac{(x-1)^6}{6} + R_6;$

where $R_6 = \dfrac{(x-1)^7}{7X^7}$ for some $X$ between 1 and $x$

19. $\dfrac{1}{e^3} + \dfrac{3}{e^3}(x+1) + \dfrac{9}{2e^3}(x+1)^2 + \dfrac{9}{2e^3}(x+1)^3$

21. $x^2 - \frac{1}{3}x^4$     23. $1 - 2x^2 + 4x^4 - 8x^6$

25. $x + \dfrac{x^3}{3!} + \dfrac{x^5}{5!} + \cdots + \dfrac{x^{2n+1}}{(2n+1)!}$

27. $e^{-x} = 1 - x + \dfrac{x^2}{2!} - \dfrac{x^3}{3!} + \cdots + (-1)^n \dfrac{x^n}{n!} + R_n;$

where $R_n = (-1)^{n+1} \dfrac{e^{-X}x^{n+1}}{(n+1)!}$ for some $X$ between

0 and $x$;

$\dfrac{1}{e} \approx \dfrac{1}{2!} - \dfrac{1}{3!} + \cdots + \dfrac{1}{8!} \approx 0.36788$

29. $1 - 2x + x^2$ ($f$ is its own best quadratic approximation; (error = 0). $g(x) \approx 4 + 3x + 2x^2$; error = $x^3$; since $g'''(x) = 6 = 3!$, therefore error = $\dfrac{g'''(X)}{3!}x^3$; no improvement possible.

## Section 4.9    (page 294)

1. 3/4                3. $a/b$
5. 1                  7. 1
9. 0                  11. $-3/2$
13. 1                 15. $-1/2$
17. $\infty$          19. $2/\pi$
21. $-2$             23. $a$
25. 1                 27. $-1/2$
29. $e^{-2}$          31. 0
33. $f''(x)$

## Review Exercises    (page 295)

1. 15.3 m
3. 80 ft/s or about 55 mph
5. (a) $-1,600$ ohms/min, (b) $-1,350$ ohms/min
7. 2, 000                9. $32\pi R^3/81$ un$^3$
11. 9000 cm$^3$          13. approx 0.057 rad/s
15. about 9.69465 cm      17. 2.06%
19. $\frac{\pi}{4} + 0.0475 \approx 0.83290$, |error| $< 0.00011$
21. 0, 1.4055636328
23. approx. $(-1.1462, 0.3178)$

## Challenging Problems    (page 297)

1. (a) $\dfrac{dx}{dt} = \dfrac{k}{3}(x_0^3 - x^3)$, (b) $V_0/2$
3. (b) 11
5. (c) $y_0(1 - (t/T))^2$, (d) $(1 - (1/\sqrt{2}))T$
7. $P^2(3 - 2\sqrt{2})/4$
9. (a) $\cos^{-1}(r_2/r_1)^2$, (b) $\cos^{-1}(r_2/r_1)^4$.
11. approx 921 cm$^3$

# Chapter 5
# Integration

## Section 5.1    (page 303)

1. $1^3 + 2^3 + 3^3 + 4^3$     3. $3 + 3^2 + 3^3 + \cdots + 3^n$
5. $\dfrac{(-2)^3}{1^2} + \dfrac{(-2)^4}{2^2} + \dfrac{(-2)^5}{3^2} + \cdots + \dfrac{(-2)^n}{(n-2)^2}$
7. $\sin\dfrac{\pi}{3k} + \sin\dfrac{2\pi}{3k} + \sin\dfrac{3\pi}{3k} + \cdots + + \sin\dfrac{k\pi}{3k}$
9. $\sum_{i=5}^{9} i$          11. $\sum_{i=1}^{200} 2$
13. $\sum_{i=2}^{99} (-1)^i i^2$     15. $\sum_{i=0}^{n} x^i$
17. $\sum_{i=1}^{n} (-1)^{i-1}/i^2$     19. $\sum_{i=1}^{100} \sin(i - 1)$
21. 15                23. $n(n+1)(2n+7)/6$
25. $n^2 - (m-1)^2$     27. $\dfrac{\pi(\pi^n - 1)}{\pi - 1} - 3n$
29. $\ln(n!)$          31. 400
33. $(x^{2n+1} + 1)/(x+1)$     35. $-4,949$
39. $2^m - 1$          41. $n/(n+1)$

## Section 5.2    (page 310)

1. 3/2 sq. un.         3. 6 sq. un.
5. 26/3 sq. un.        7. 15 sq. un.
9. 4 sq. un.           11. 32/3 sq. un.
13. $3/(2\ln 2)$ sq. un.
15. $\ln(b/a)$, follows from definition of ln
17. 0                  19. $\pi/4$

## Section 5.3    (page 316)

1. $L(f, P_8) = 7/4$,    $U(f, P_8) = 9/4$
3. $L(f, P_4) = \dfrac{e^4 - 1}{e^2(e-1)} \approx 4.22,$

$U(f, P_4) = \dfrac{e^4 - 1}{e(e-1)} \approx 11.48$

**5.** $L(f, P_6) = \dfrac{\pi}{6}(1 + \sqrt{3}) \approx 1.43$,

$U(f, P_6) = \dfrac{\pi}{6}(3 + \sqrt{3}) \approx 2.48$

**7.** $L(f, P_n) = \dfrac{n-1}{2n}$, $U(f, P_n) = \dfrac{n+1}{2n}$,

$\int_0^1 x\, dx = \dfrac{1}{2}$

**9.** $L(f, P_n) = \dfrac{(n-1)^2}{4n^2}$, $U(f, P_n) = \dfrac{(n+1)^2}{4n^2}$,

$\int_0^1 x^3\, dx = \dfrac{1}{4}$

**11.** $\int_0^1 \sqrt{x}\, dx$     **13.** $\int_0^\pi \sin x\, dx$

**15.** $\int_0^1 \tan^{-1} x\, dx$

### Section 5.4 (page 323)

**1.** 0           **3.** 8

**5.** $(b^2 - a^2)/2$     **7.** $\pi$

**9.** 0           **11.** $2\pi$

**13.** 0         **15.** $(2\pi + 3\sqrt{3})/6$

**17.** 16        **19.** 12

**21.** 32/3     **23.** $(4 + 3\pi)/12$

**25.** $\ln 2$      **27.** $\ln 3$

**29.** 4          **31.** 1

**33.** $\pi/2$       **35.** 1

**37.** 11/6     **39.** $\dfrac{\pi}{3} - \sqrt{3}$

**41.** 41/2     **43.** 3/4

**45.** $k = \bar{f}$

### Section 5.5 (page 329)

**1.** 4           **3.** 1

**5.** $\dfrac{5}{12}$         **7.** 9

**9.** $80\frac{4}{5}$      **11.** $\dfrac{2 - \sqrt{2}}{2\sqrt{2}}$

**13.** $(1/\sqrt{2}) - (1/2)$    **15.** $e^\pi - e^{-\pi}$

**17.** $(a^e - 1)/\ln a$     **19.** $\pi/2$

**21.** $\dfrac{\pi}{3}$         **23.** $\frac{1}{5}$ sq. un.

**25.** $\frac{32}{3}$ sq. un.     **27.** $\frac{1}{6}$ sq. un.

**29.** $\frac{1}{3}$ sq. un.     **31.** $\frac{1}{12}$ sq. un.

**33.** $2\pi$ sq. un.      **35.** 3`

**37.** $\frac{16}{3}$           **39.** $e - 1$

**41.** $\dfrac{\sin x}{x}$        **43.** $-2\,\dfrac{\sin x^2}{x}$

**45.** $\dfrac{\cos t}{1 + t^2}$      **47.** $(\cos x)/(2\sqrt{x})$

**49.** $f(x) = \pi e^{\pi(x-1)}$

**51.** $1/x^2$ is not continuous (or even defined) at $x = 0$ so the Fundamental Theorem cannot be applied over $[-1, 1]$; Since $1/x^2 > 0$ on its domain, we would expect the integral to be positive if it exists at all. (It doesn't.)

**53.** $F(x)$ has a maximum value at $x = 1$ but no minimum value.

**55.** 2

### Section 5.6 (page 337)

**1.** $-\frac{1}{2}e^{5-2x} + C$     **3.** $\frac{2}{9}(3x + 4)^{3/2} + C$

**5.** $-\frac{2}{9}(x^3 + 2)^{-3/2} + C$    **7.** $-\frac{1}{32}(4x^2 + 1)^{-4} + C$

**9.** $\frac{1}{2}e^{x^2} + C$     **11.** $\frac{1}{2}\tan^{-1}\left(\frac{1}{2}\sin x\right) + C$

**13.** $2\ln\left|e^{x/2} - e^{-x/2}\right| + C = \ln\left|e^x - 2 + e^{-x}\right| + C$

**15.** $-\frac{2}{5}\sqrt{4 - 5s} + C$    **17.** $\frac{1}{2}\sin^{-1}\left(\dfrac{t^2}{2}\right) + C$

**19.** $-\ln\left(1 + e^{-x}\right) + C$    **21.** $-\frac{1}{2}(\ln\cos x)^2 + C$

**23.** $\tan^{-1}\dfrac{t}{3} + \ln(t^2 + 9) + C$

**25.** $\frac{1}{2}\tan^{-1}\dfrac{x+3}{2} + C$

**27.** $\frac{1}{8}\cos^8 x - \frac{1}{6}\cos^6 x + C$

**29.** $-\dfrac{1}{3a}\cos^3 ax + C$

**31.** $\frac{5}{16}x - \frac{1}{4}\sin 2x + \frac{3}{64}\sin 4x + \frac{1}{48}\sin^3 2x + C$

**33.** $\frac{1}{5}\sec^5 x + C$

**35.** $\frac{2}{3}(\tan x)^{3/2} + \frac{2}{7}(\tan x)^{7/2} + C$

**37.** $\frac{3}{8}\sin x - \frac{1}{4}\sin(2\sin x) + \frac{1}{32}\sin(4\sin x) + C$

**39.** $\frac{1}{3}\tan^3 x + C$

**41.** $-\frac{1}{9}\csc^9 x + \frac{2}{7}\csc^7 x - \frac{1}{5}\csc^5 x + C$

**43.** $\frac{14}{3}\sqrt{17} + \frac{2}{3}$     **45.** $3\pi/16$

**47.** $\ln 2$        **49.** $2$, $2(\sqrt{2} - 1)$

**51.** $\pi/32$ sq. un.

### Section 5.7 (page 342)

**1.** $\dfrac{1}{6}$ sq. units     **3.** $\dfrac{64}{3}$ sq. units

**5.** $\dfrac{125}{12}$ sq. units     **7.** $\dfrac{1}{2}$ sq. units

**9.** $\dfrac{5}{12}$ sq. units     **11.** $\dfrac{15}{8} - 2\ln 2$ sq. units

**13.** $\dfrac{\pi}{2} - \dfrac{1}{3}$ sq. units

**15.** $\dfrac{4}{3}$ sq. units

**17.** $2\sqrt{2}$ sq. units

**19.** $1 - \pi/4$ sq. units

**21.** $(\pi/8) - \ln\sqrt{2}$ sq. units

**23.** $(4\pi/3) - 2\ln(2 + \sqrt{3})$ sq. units

**25.** $(4/\pi) - 1$ sq. units

**27.** $0.60906561$ sq. units

**29.** $\dfrac{4}{3}$ sq. units

**31.** $\dfrac{e}{2} - 1$ sq. units

### Review Exercises    (page 343)

**1.** sum is $n(n + 2)/(n + 1)^2$

**3.** $20/3$

**5.** $4\pi$

**7.** $0$

**9.** $2$

**11.** $\sin(t^2)$

**13.** $-4e^{\sin(4s)}$

**15.** $f(x) = -\frac{1}{2}e^{(3/2)(1-x)}$

**17.** $9/2$ sq. units

**19.** $3/10$ sq. units

**21.** $(3\sqrt{3}/4) - 1$ sq. units

**23.** $(\frac{1}{6}\sin(2x^3 + 1) + C$

**25.** $98/3$

**27.** $(\pi/8) - (1/2)\tan^{-1}(1/2)$

**29.** $-\cos\sqrt{2s + 1} + C$

**31.** min $-\pi/4$, no max

# Chapter 6
# Techniques of Integration

### Section 6.1    (page 351)

**1.** $x\sin x + \cos x + C$

**3.** $\dfrac{1}{\pi}x^2\sin\pi x + \dfrac{2}{\pi^2}x\cos\pi x - \dfrac{2}{\pi^3}\sin\pi x + C$

**5.** $\frac{1}{4}x^4\ln x - \frac{1}{16}x^4 + C$

**7.** $x\tan^{-1}x - \frac{1}{2}\ln(1 + x^2) + C$

**9.** $\left(\frac{1}{2}x^2 - \frac{1}{4}\right)\sin^{-1}x + \frac{1}{4}x\sqrt{1 - x^2} + C$

**11.** $\frac{7}{8}\sqrt{2} + \frac{3}{8}\ln(1 + \sqrt{2})$

**13.** $\frac{1}{13}e^{2x}(2\sin 3x - 3\cos 3x) + C$

**15.** $\ln(2 + \sqrt{3}) - \dfrac{\pi}{6}$

**17.** $x\tan x - \ln|\sec x| + C$

**19.** $\dfrac{x}{2}\left[\cos(\ln x) + \sin(\ln x)\right] + C$

**21.** $\ln x\left(\ln(\ln x) - 1\right) + C$

**23.** $x\cos^{-1}x - \sqrt{1 - x^2} + C$

**25.** $\dfrac{2\pi}{3} - \ln(2 + \sqrt{3})$

**27.** $\frac{1}{2}(x^2 + 1)\left(\tan^{-1}x\right)^2 - x\tan^{-1}x + \frac{1}{2}\ln(1 + x^2) + C$

**29.** $\dfrac{1 + e^{-\pi}}{2}$ square units

**31.** $I_n = x(\ln x)^n - nI_{n-1}$,
$I_4 = x\left[(\ln x)^4 - 4(\ln x)^3 + 12(\ln x)^2 - 24(\ln x) + \right.$

**33.** $I_n = -\dfrac{1}{n}\sin^{n-1}x\cos x + \dfrac{n - 1}{n}I_{n-2}$,
$I_6 = \dfrac{5x}{16} - \cos x\left[\frac{1}{6}\sin^5 x + \frac{5}{24}\sin^3 x + \frac{5}{16}\sin x\right] + C$,
$I_7 = -\cos x\left[\frac{1}{7}\sin^6 x + \frac{6}{35}\sin^4 x + \frac{8}{35}\sin^2 x + \frac{16}{35}\right] -$

**35.** $I_n = \dfrac{x}{2a^2(n - 1)(x^2 + a^2)^{n-1}} + \dfrac{2n - 3}{2a^2(n - 1)}I_{n-1}$,
$I_3 = \dfrac{x}{4a^2(x^2 + a^2)^2} + \dfrac{3x}{8a^4(x^2 + a^2)} + \dfrac{3}{8a^5}\tan^{-1}\dfrac{x}{a}$

**37.** Any conditions which guarantee that
$f(b)g'(b) - f'(b)g(b) = f(a)g'(a) - f'(a)g(a)$
will suffice.

### Section 6.2    (page 359)

**1.** $\frac{1}{2}\sin^{-1}(2x) + C$

**3.** $\frac{9}{2}\sin^{-1}\dfrac{x}{3} - \frac{1}{2}x\sqrt{9 - x^2} + C$

**5.** $-\dfrac{\sqrt{9 - x^2}}{9x} + C$

**7.** $-\sqrt{9 - x^2} + \sin^{-1}\dfrac{x}{3} + C$

**9.** $\frac{1}{3}(9 + x^2)^{3/2} - 9\sqrt{9 + x^2} + C$

**11.** $\dfrac{1}{a^2}\dfrac{x}{\sqrt{a^2 - x^2}} + C$

**13.** $\dfrac{x}{\sqrt{a^2 - x^2}} - \sin^{-1}\dfrac{x}{a} + C$

**15.** $\dfrac{1}{a}\sec^{-1}\dfrac{x}{a} + C$

**17.** $\frac{1}{3}\tan^{-1}\dfrac{x + 1}{3} + C$

**19.** $\frac{1}{32}\tan^{-1}\dfrac{2x + 1}{2} + \dfrac{1}{16}\dfrac{2x + 1}{4x^2 + 4x + 5} + C$

**21.** $a\sin^{-1}\dfrac{x - a}{a} - \sqrt{2ax - x^2} + C$

**23.** $\dfrac{3 - x}{4\sqrt{3 - 2x - x^2}} + C$

**25.** $\frac{3}{8}\tan^{-1}x + \dfrac{3x^3 + 5x}{8(1 + x^2)^2} + C$

**27.** $\frac{1}{2}\ln\left(1 + \sqrt{1 - x^2}\right) - \frac{1}{2}\ln|x| - \dfrac{\sqrt{1 - x^2}}{2x^2} + C$

**29.** $2\sqrt{x} - 4\ln(2 + \sqrt{x}) + C$

**31.** $\frac{6}{7}x^{7/6} - \frac{6}{5}x^{5/6} + \frac{3}{2}x^{2/3} + 2x^{1/2} - 3x^{1/3} - 6x^{1/6} + 3\ln(1 + x^{1/3}) + 6\tan^{-1}x^{1/6} + C$

**33.** $\dfrac{\pi}{6} - \dfrac{\sqrt{3}}{8}$

**35.** $\pi/3$

**37.** $\dfrac{2}{\sqrt{3}}\tan^{-1}\left(\dfrac{2\tan(\theta/2) + 1}{\sqrt{3}}\right) + C$

**39.** $\dfrac{2}{\sqrt{5}}\tan^{-1}\left(\dfrac{\tan(\theta/2)}{\sqrt{5}}\right) + C$

**41.** $\dfrac{9}{2\sqrt{2}}\tan^{-1}\dfrac{1}{\sqrt{2}}-\dfrac{1}{2}$ square units

**43.** $a^2\cos^{-1}\left(\dfrac{b}{a}\right)-b\sqrt{a^2-b^2}$ square units

**45.** $\dfrac{25}{2}\left(\sin^{-1}\dfrac{4}{5}-\sin^{-1}\dfrac{3}{5}\right)-12\ln\dfrac{4}{3}$ square units

**47.** $\dfrac{\ln(Y+\sqrt{1+Y^2})}{2}$ sq. units

### Section 6.3  (page 368)

**1.** $\ln|2x-3|+C$

**3.** $\dfrac{x}{\pi}-\dfrac{2}{\pi^2}\ln|\pi x+2|+C$

**5.** $\dfrac{1}{6}\ln\left|\dfrac{x-3}{x+3}\right|+C$

**7.** $\dfrac{1}{2a}\ln\left|\dfrac{a+x}{a-x}\right|+C$

**9.** $x-\frac{4}{3}\ln|x+2|+\frac{1}{3}\ln|x-1|+C$

**11.** $3\ln|x+1|-2\ln|x|+C$

**13.** $\dfrac{1}{3(1-3x)}+C$

**15.** $-\frac{1}{9}x-\dfrac{13}{54}\ln|2-3x|+\frac{1}{6}\ln|x|+C$

**17.** $\dfrac{1}{2a^2}\ln\dfrac{|x^2-a^2|}{x^2}+C$

**19.** $x+\dfrac{a}{3}\ln|x-a|-\dfrac{a}{6}\ln(x^2+ax+a^2)$
$\qquad-\dfrac{a}{\sqrt{3}}\tan^{-1}\dfrac{2x+a}{\sqrt{3}a}+C$

**21.** $\frac{1}{3}\ln|x|-\frac{1}{2}\ln|x-1|+\frac{1}{6}\ln|x-3|+C$

**23.** $\dfrac{1}{4}\ln\left|\dfrac{x+1}{x-1}\right|-\dfrac{x}{2(x^2-1)}+C$

**25.** $\dfrac{1}{27}\ln\left|\dfrac{x-3}{x}\right|+\dfrac{1}{9x}+\dfrac{1}{6x^2}+C$

**27.** $\dfrac{t-1}{4(t^2+1)}-\frac{1}{4}\ln|t+1|+\frac{1}{8}\ln(t^2+1)+C$

**29.** $\dfrac{1}{3}\ln\left|\dfrac{1-\sqrt{1-x^2}}{x}\right|+\dfrac{1}{12}\ln\left(\dfrac{(2+\sqrt{1-x^2})^2}{3+x^2}\right)+C$

**31.** $\dfrac{1}{\sqrt{1+x^2}}+\dfrac{1}{2}\ln\left|\dfrac{1-\sqrt{1+x^2}}{1+\sqrt{1+x^2}}\right|+C$

**33.** $\dfrac{1-2x^2}{x\sqrt{x^2-1}}+C$

### Section 6.4·  (page 373)

**5.** $\dfrac{x\sqrt{x^2-2}}{2}+\ln|x+\sqrt{x^2-2}|+C$

**7.** $-\sqrt{3t^2+5}/(5t)+C$

**9.** $(x^5/3125)(625(\ln x)^4-500(\ln x)^3+300(\ln x)^2$
$\qquad-120\ln x+24)+C$

**11.** $(1/6)(2x^2-x-3)\sqrt{2x-x^2}-(1/2)\sin^{-1}(1-x)+C$

**13.** $(x-2)/(4\sqrt{4x-x^2})+C$

### Section 6.5  (page 381)

**1.** $1/2$  **3.** $1/2$

**5.** $3\times2^{1/3}$  **7.** $3/2$

**9.** $3$  **11.** $\pi$

**13.** $1/2$  **15.** diverges to $\infty$

**17.** $2$  **19.** diverges

**21.** $0$  **23.** $1$ sq. units

**25.** $2\ln 2$ square units  **29.** $2$

**31.** diverges to $\infty$  **33.** converges

**35.** diverges to $\infty$  **37.** diverges to $\infty$

**39.** diverges  **41.** diverges to $\infty$

### Section 6.6  (page 389)

**1.** $T_4=4.75$,
$\quad M_4=4.625$,
$\quad T_8=4.6875$,
$\quad M_8=4.65625$,
$\quad T_{16}=4.671875$,
$\quad$ Actual errors:
$\quad I-T_4\approx-0.0833333$,
$\quad I-M_4\approx\ \ 0.0416667$,
$\quad I-T_8\approx-0.0208333$,
$\quad I-M_8\approx\ \ 0.0104167$,
$\quad I-T_{16}\approx-0.0052083$
$\quad$ Error estimates:
$\quad |I-T_4|\le0.0833334$,
$\quad |I-M_4|\le0.0416667$,
$\quad |I-T_8|\le0.0208334$,
$\quad |I-M_8|\le0.0104167$,
$\quad |I-T_{16}|\le0.0052084$

**3.** $T_4 = 0.9871158$,
$M_4 = 1.0064545$,
$T_8 = 0.9967852$,
$M_8 = 1.0016082$,
$T_{16} = 0.9991967$,
Actual errors:
$I - T_4 \approx \ \ 0.0128842$,
$I - M_4 \approx -0.0064545$,
$I - T_8 \approx \ \ 0.0032148$,
$I - M_8 \approx -0.0016082$,
$I - T_{16} \approx \ \ 0.0008033$
Error estimates:
$|I - T_4| \le 0.020186$,
$|I - M_4| \le 0.010093$,
$|I - T_8| \le 0.005047$,
$|I - M_8| \le 0.002523$,
$|I - T_{16}| \le 0.001262$

**5.** $T_4 = 46$, $T_8 = 46.7$
**7.** $T_4 = 3,000$ km$^2$, $T_8 = 3,400$ km$^2$
**9.** $T_4 \approx 2.02622$, $\quad M_4 \approx 2.03236$,
$T_8 \approx 2.02929$, $\quad M_8 \approx 2.02982$,
$T_{16} \approx 2.029555$
**11.** $M_8 \approx 1.3714136$, $\quad T_{16} \approx 1.3704366$, $I \approx 1.371$

### Section 6.7  (page 394)

**1.** $S_4 = S_8 = I$, $\quad$ Errors $= 0$
**3.** $S_4 \approx 1.0001346$, $\quad S_8 \approx 1.0000083$,
$I - S_4 \approx -0.0001346$, $\quad I - S_8 \approx -0.0000083$
**5.** $46.93$
**7.** For $f(x) = e^{-x}$:
$|I - S_4| \le 0.000022$, $|I - S_8| \le 0.0000014$;
for $f(x) = \sin x$,
$|I - S_4| \le 0.00021$,
$|I - S_8| \le 0.000013$
**9.** $S_4 \approx 2.0343333$, $\quad S_8 \approx 2.0303133$,
$S_{16} \approx 2.0296433$

### Section 6.8  (page 401)

**1.** $3 \displaystyle\int_0^1 \dfrac{u \, du}{1 + u^3}$
**3.** $\displaystyle\int_{-\pi/2}^{\pi/2} e^{\sin\theta} \, d\theta$, $\quad$ or $\quad 2 \displaystyle\int_0^1 \dfrac{e^{1-u^2} + e^{u^2-1}}{\sqrt{2 - u^2}} \, du$

**5.** $4 \displaystyle\int_0^1 \dfrac{dv}{\sqrt{(2 - v^2)(2 - 2v^2 + v^4)}}$
**7.** $T_2 \approx 0.603553$ $\quad T_4 \approx 0.643283$,
$T_8 \approx 0.658130$, $\quad T_{16} \approx 0.663581$;
Errors: $I - T_2 \approx 0.0631$, $\quad I - T_4 \approx 0.0234$,
$I - T_8 \approx 0.0085$, $\quad I - T_{16} \approx 0.0031$;
Errors do not decrease like $1/n^2$ because the second
derivative of $f(x) = \sqrt{x}$ is not bounded on $[0, 1]$.

**9.** $I \approx 0.74684$ with error less than $10^{-4}$; seven terms
of the series are needed.
**11.** $A = 1, u = 1/\sqrt{3}$
**13.** $A = 5/9, B = 8/9, u = \sqrt{3/5}$
**15.** $R_1 \approx 0.7471805$, $\quad R_2 \approx 0.7468337$,
$R_3 \approx 0.7468241$, $\quad I \approx 0.746824$
**17.** $R_2 = \dfrac{2h}{45} \left(7y_0 + 32y_1 + 12y_2 + 32y_3 + 7y_4\right)$

### Review Exercises on Techniques of Integration  (pag

**1.** $\frac{2}{3} \ln|x + 2| - \frac{1}{6} \ln|2x + 1| + C$

**3.** $\frac{1}{4} \sin^4 x - \frac{1}{6} \sin^6 x + C$ $\quad$ **5.** $\dfrac{3}{4} \ln\left|\dfrac{2x-1}{2x+1}\right| + C$

**7.** $-\dfrac{1}{3}\left(\dfrac{\sqrt{1-x^2}}{x}\right)^3 + C$ $\quad$ **9.** $\frac{1}{5}\left(5x^3 - 2\right)^{1/3} + C$

**11.** $\frac{1}{16} \tan^{-1} \dfrac{x}{2} + \dfrac{x}{8(4 + x^2)} + C$

**13.** $\dfrac{1}{2\ln 2}\left(2^x \sqrt{1 + 4^x} + \ln(2^x + \sqrt{1 + 4^x})\right) + C$
**15.** $\frac{1}{4} \tan^4 x + \frac{1}{6} \tan^6 x + C$
**17.** $-e^{-x}\left(\frac{2}{5} \cos 2x + \frac{1}{5} \sin 2x\right) + C$
**19.** $\dfrac{x}{10}\left(\cos(3 \ln x) + 3 \sin(3 \ln x)\right) + C$

**21.** $\frac{1}{4}\left(\ln(1 + x^2)\right)^2 + C$
**23.** $\sin^{-1} \dfrac{x}{\sqrt{2}} - \dfrac{x\sqrt{2 - x^2}}{2} + C$

**25.** $\dfrac{1}{64}\left(-\dfrac{1}{7(4x+1)^7} + \dfrac{1}{4(4x+1)^8} - \dfrac{1}{9(4x+1)^9}\right) +$
$C$
**27.** $-\frac{1}{4} \cos 4x + \frac{1}{6} \cos^3 4x - \frac{1}{20} \cos^5 4x + C$
**29.** $-\frac{1}{2} \ln(2e^{-x} + 1) + C$
**31.** $-\frac{1}{2} \sin^2 x - 2 \sin x - 4 \ln(2 - \sin x) + C$

**33.** $-\dfrac{\sqrt{1 - x^2}}{x} + C$

**35.** $\frac{1}{48}(1 - 4x^2)^{3/2} - \frac{1}{16}\sqrt{1 - 4x^2} + C$
**37.** $\sqrt{x^2 + 1} + \ln(x + \sqrt{x^2 + 1}) + C$
**39.** $x + \frac{1}{3} \ln|x| + \frac{4}{3} \ln|x - 3| - \frac{5}{3} \ln|x + 3| + C$
**41.** $-\frac{1}{10} \cos^{10} x + \frac{1}{6} \cos^{12} x - \frac{1}{14} \cos^{14} x + C$
**43.** $\dfrac{1}{2} \ln|x^2 + 2x - 1| - \dfrac{1}{2\sqrt{2}} \ln\left|\dfrac{x + 1 - \sqrt{2}}{x + 1 + \sqrt{2}}\right| + C$

**45.** $\frac{1}{3}x^3 \sin^{-1} 2x + \frac{1}{24}\sqrt{1 - 4x^2} - \frac{1}{72}(1 - 4x^2)^{3/2} + C$
**47.** $\frac{1}{128}\left(3x - \sin(4x) + \frac{1}{8} \sin(8x)\right)$

**49.** $\tan^{-1} \dfrac{\sqrt{x}}{2} + C$

**51.** $\dfrac{x^2}{2} - 2x + \dfrac{1}{4}\ln|x| + \dfrac{1}{2x} + \dfrac{15}{4}\ln|x+2| + C$

**53.** $-\dfrac{1}{2}\cos(2\ln x) + C$    **55.** $\dfrac{1}{2}\exp(2\tan^{-1} x) + C$

**57.** $\dfrac{1}{4}\left(\ln(3 + x^2)\right)^2 + C$    **59.** $\dfrac{1}{2}\left(\sin^{-1}(x/2)\right)^2 + C$

**61.** $\sqrt{x^2 + 6x + 10} - 2\ln(x + 3 + \sqrt{x^2 + 6x + 10}) + C$

**63.** $\dfrac{2}{5(2 + x^2)^{5/2}} - \dfrac{1}{3(2 + x^2)^{3/2}} + C$

**65.** $\dfrac{6}{7}x^{7/6} - \dfrac{6}{5}x^{5/6} + 2\sqrt{x} - 6x^{1/6} + 6\tan^{-1} x^{1/6} + C$

**67.** $\dfrac{2}{3}x^{3/2} - x + 4\sqrt{x} - 4\ln(1 + \sqrt{x}) + C$

**69.** $\dfrac{1}{2(4 - x^2)} + C$

**71.** $\dfrac{1}{3}x^3 \tan^{-1} x - \dfrac{1}{6}x^2 + \dfrac{1}{6}\ln(1 + x^2) + C$

**73.** $\dfrac{1}{5}\ln\left|\dfrac{3\tan(x/2) - 1}{\tan(x/2) + 3}\right| + C$

**75.** $\dfrac{1}{2}\ln|\tan(x/2)| - \dfrac{1}{4}\left(\tan^{-1}(x/2)\right)^2 + C$
$= \dfrac{1}{4}\left(\ln\left|\dfrac{1 - \cos x}{1 + \cos x}\right| - \dfrac{1 - \cos x}{1 + \cos x}\right) + C$

**77.** $2\sqrt{x} - 2\tan^{-1}\sqrt{x} + C$

**79.** $\dfrac{1}{2}x^2 + \dfrac{4}{3}\ln|x - 2| - \dfrac{2}{3}\ln(x^2 + 2x + 4)$
$+ \dfrac{4}{\sqrt{3}}\tan^{-1}\dfrac{x + 1}{\sqrt{3}} + C$

### Review Exercises (Other) (page 404)

**1.** $I = \dfrac{1}{2}\left(xe^x \cos x + (x - 1)e^x \sin x\right),$
$J = \dfrac{1}{2}\left((1 - x)e^x \cos x + xe^x \sin x\right)$

**3.** diverges to $\infty$    **5.** $-4/9$

**9.** 367,000 m$^3$

**11.** $T_8 = 1.61800,\ S_8 = 1.62092,\ I \approx 1.62$

**13.** (a) $T_4 = 5.526,\ S_4 = 5.504$; (b) $S_8 = 5.504$; (c) yes, because $S_4 = S_8$, and Simpson's Rule is exact for cubics.

### Challenging Problems (page 404)

**1.** (c) $I = \dfrac{1}{630}, \dfrac{22}{7} - \dfrac{1}{630} < \pi < \dfrac{22}{7} - \dfrac{1}{1260}.$

**3.** (a) $\dfrac{1}{\sqrt{3}}\tan^{-1}\left(\dfrac{2x + 1}{\sqrt{3}}\right) + \dfrac{1}{\sqrt{3}}\tan^{-1}\left(\dfrac{2x - 1}{\sqrt{3}}\right),$

(b) $\dfrac{1}{\sqrt{2}}\tan^{-1}(\sqrt{2}x + 1) + \dfrac{1}{\sqrt{2}}\tan^{-1}(\sqrt{2}x - 1)$

**7.** (a) $a = 7/90,\ b = 16/45,\ c = 2/15.$
(b) one interval: approx 0.6321208750, two intervals: approx 0.6321205638, true val: 0.6321205588

## Chapter 7
## Applications of Integration

### Section 7.1 (page 415)

**1.** $\dfrac{\pi}{5}$ cu. units    **3.** $\dfrac{3\pi}{10}$ cu. units

**5.** (a) $\dfrac{16\pi}{15}$ cu. units,  (b) $\dfrac{8\pi}{3}$ cu. units

**7.** (a) $\dfrac{27\pi}{2}$ cu. units,  (b) $\dfrac{108\pi}{5}$ cu. units

**9.** (a) $\dfrac{15\pi}{4} - \dfrac{\pi^2}{8}$ cu. units,  (b) $\pi(2 - \ln 2)$ cu. units

**11.** $\dfrac{10\pi}{3}$ cu. units    **13.** about 35%

**15.** $\dfrac{\pi h}{3}\left(b^2 - 3a^2 + \dfrac{2a^3}{b}\right)$ cu. units

**17.** $\dfrac{\pi}{3}(a - b)^2(2a + b)$ cu. units

**19.** $\dfrac{4\pi ab^2}{3}$ cu. units

**21.** (a) $\pi/2$ cu. units,  (b) $2\pi$ cu. units

**23.** $k > 2$    **25.** about 1,524 cu. units

**27.** $8192\pi/105$ cu. units    **29.** $R = \dfrac{h\sin\alpha}{\sin\alpha + \cos 2\alpha}$

### Section 7.2 (page 419)

**1.** 6 m$^3$    **3.** 132 ft$^3$

**5.** $\pi a^2 h/2$ cm$^3$    **7.** $3z^2$ sq. units

**9.** $\dfrac{16r^3}{3}$ cu. units    **11.** $72\pi$ cm$^3$

**13.** $\pi r^2(a + b)/2$ cu. units    **15.** $\dfrac{16,000}{3}$ cu. units

**17.** $12\pi\sqrt{2}$ in$^3$    **19.** approx 97.28 cm$^3$

### Section 7.3 (page 427)

**1.** $2\sqrt{5}$ units    **3.** $52/3$ units

**5.** $(2/27)(13^{3/2} - 8)$ units

**7.** 6 units    **9.** $(e^2 + 1)/4$ units

**11.** $2\ln 3 - 1$ units
**13.** $\ln(1 + \sqrt{2}) - \ln\sqrt{3}$ units

**15.** $\ln(e^2 + e^{-2})$ units    **17.** $6a$ units

**19.** 1.0338 units    **21.** 1.0581

**23.** $(10^{3/2} - 1)\pi/27$ sq. units

**25.** $\dfrac{64\pi}{81}\left[\dfrac{(13/4)^{5/2} - 1}{5} - \dfrac{(13/4)^{3/2} - 1}{3}\right]$ sq. units

**27.** $2\pi \left( \sqrt{2} + \ln(1 + \sqrt{2}) \right)$ sq. units

**29.** $2\pi \left( \dfrac{255}{16} + \ln 4 \right)$ sq. units

**31.** $4\pi^2 ab$ sq. units

**33.** $8\pi \left( 1 + \dfrac{\ln(2 + \sqrt{3})}{2\sqrt{3}} \right)$ sq. units

**35.** $s = \dfrac{5}{\pi}\sqrt{4 + \pi^2}\, E \left( \dfrac{\pi}{\sqrt{4 + \pi^2}} \right)$

**37.** $k > -1$

**39.** (a) $\pi$ cu. units;  (c) "Covering" a surface with paint requires putting on a layer of constant thickness. Far enough to the right, the horn is thinner than any prescribed constant, so it can contain less paint than would be necessary to paint its surface.

### Section 7.4   (page 435)

**1.** mass $\dfrac{2L}{\pi}$; centre of mass at $\bar{s} = \dfrac{L}{2}$

**3.** $m = \frac{1}{4}\pi\delta_0 a^2$;   $\bar{x} = \bar{y} = \dfrac{4a}{3\pi}$

**5.** $m = \dfrac{256k}{15}$;   $\bar{x} = 0$,   $\bar{y} = \dfrac{16}{7}$

**7.** $m = \dfrac{ka^3}{2}$;   $\bar{x} = \dfrac{2a}{3}$,   $\bar{y} = \dfrac{a}{2}$

**9.** $m = \int_a^b \delta(x)\big(g(x) - f(x)\big)\,dx$;
$M_{x=0} = \int_a^b x\delta(x)\big(g(x) - f(x)\big)\,dx$,   $\bar{x} = M_{x=0}/m$,
$M_{y=0} = \frac{1}{2}\int_a^b \delta(x)\big((g(x))^2 - (f(x))^2\big)\,dx$,
$\bar{y} = M_{y=0}/m$

**11.** Mass is $\frac{8}{3}\pi R^4$ kg. The centre of mass is along the line through the centre of the ball perpendicular to the plane, at a distance $R/10$ m from the centre of the ball on the side opposite the plane.

**13.** $m = \frac{1}{8}\pi\delta_0 a^4$;   $\bar{x} = 16a/(15\pi)$,   $\bar{y} = 0$,
$\bar{z} = 8a/15$

**15.** $m = \frac{1}{3}k\pi a^3$;   $\bar{x} = 0$,   $\bar{y} = \dfrac{3a}{2\pi}$

**17.** about $5.57C/k^{3/2}$

### Section 7.5   (page 441)

**1.** $\left( \dfrac{4r}{3\pi}, \dfrac{4r}{3\pi} \right)$

**3.** $\left( \dfrac{\sqrt{2} - 1}{\ln(1 + \sqrt{2})}, \dfrac{\pi}{8\ln(1 + \sqrt{2})} \right)$

**5.** $\left( 0, \dfrac{9\sqrt{3} - 4\pi}{4\pi - 3\sqrt{3}} \right)$

**7.** $\left( \dfrac{19}{9}, -\dfrac{1}{3} \right)$

**9.** The centroid is on the axis of symmetry of the hemisphere half way between the base plane and the vertex.

**11.** The centroid is on the axis of the cone, one quarter of the cone's height above the base plane.

**13.** $(8/9, 11/9)$

**15.** $(0, 2/(3(\pi + 2)))$

**17.** $\left( \dfrac{\pi}{2}, \dfrac{\pi}{8} \right)$

**19.** $\left( \dfrac{2r}{\pi}, \dfrac{2r}{\pi} \right)$

**21.** $(1, -2)$

**23.** $\dfrac{5\pi}{3}$ cu. units

**25.** $(0.71377, 0.26053)$

**27.** $\left( 1, \frac{1}{5} \right)$

**29.** $\bar{x} = \dfrac{M_{x=0}}{A}, \bar{y} = \dfrac{M_{y=0}}{A}$,
where $A = \displaystyle\int_c^d \big(g(y) - f(y)\big)\,dy$,
$M_{x=0} = \frac{1}{2}\displaystyle\int_c^d \big((g(y))^2 - (f(y))^2\big)\,dy$,
$M_{y=0} = \displaystyle\int_c^d y\big(g(y) - f(y)\big)\,dy$

**31.** diamond orientation, edge upward

### Section 7.6   (page 449)

**1.** (a) 235,200 N,   (b) 352,800 N

**3.** $6.12 \times 10^8$ N

**5.** $8.92 \times 10^6$ N

**7.** $7.056 \times 10^5$ N-m

**9.** $2450\pi a^3 \left( a + \dfrac{8h}{3} \right)$ N-m

### Section 7.7   (page 453)

**1.** \$11,000

**3.** $\$8(\sqrt{x} - \ln(1 + \sqrt{x}))$

**5.** \$9,063.46

**7.** \$5,865.64

**9.** \$50,000

**11.** \$11,477.55

**13.** \$64,872.10

**15.** $\int_0^T e^{-\lambda(t)} P(t)\,dt$

**17.** about 23,300, \$11,890

### Section 7.8   (page 464)

**1.** (a) $\dfrac{2}{9}$,   (b) $\mu = 2$, $\sigma^2 = \dfrac{1}{2}$, $\sigma = \dfrac{1}{\sqrt{2}}$,
(c) $\dfrac{8}{9\sqrt{2}} \approx 0.63$

**3.** (a) 3,   (b) $\mu = \dfrac{3}{4}$, $\sigma^2 = \dfrac{3}{80}$, $\sigma = \sqrt{\dfrac{3}{80}}$,
(c) $\dfrac{69}{20}\sqrt{\dfrac{3}{80}} \approx 0.668$

**5.** (a) 6,   (b) $\mu = \dfrac{1}{2}$, $\sigma^2 = \dfrac{1}{20}$, $\sigma = \sqrt{\dfrac{1}{20}}$,
(c) $\dfrac{7}{5\sqrt{5}} \approx 0.626$

**7.** (a) $\dfrac{2}{\sqrt{\pi}}$, (b) $\mu = \dfrac{1}{\sqrt{\pi}} \approx 0.0.564$, $\sigma^2 = \dfrac{\pi - 2}{2\pi}$,

$\sigma = \sqrt{\dfrac{\pi - 2}{2\pi}} \approx 0.426$, (c) $\text{Pr} \approx 0.68$

**11.** (a) 0, (b) $e^{-3} \approx 0.05$, (c) $\approx 0.046$

**13.** approximately 0.006

### Section 7.9 (page 472)

**1.** $y^2 = Cx$

**3.** $x^3 - y^3 = C$

**5.** $Y = Ce^{t^2/2}$

**7.** $y = \dfrac{Ce^{2x} - 1}{Ce^{2x} + 1}$

**9.** $y = -\ln\left(Ce^{-2t} - \frac{1}{2}\right)$

**11.** $y = x^3 + Cx^2$

**13.** $y = \frac{3}{2} + Ce^{-2x}$

**15.** $y = x - 1 + Ce^{-x}$

**17.** $y = \sqrt{4 + x^2}$

**19.** $y = \dfrac{2x}{1 + x}$, $(x > 0)$

**21.** If $a = b$ the given solution is indeterminate $0/0$; in this case the solution is $x = a^2kt/(1 + akt)$.

**23.** $v = \sqrt{\dfrac{mg}{k}} \dfrac{e^{2\sqrt{kg/mt}} - 1}{e^{2\sqrt{kg/mt}} + 1}$, $v \to \sqrt{\dfrac{mg}{k}}$

**25.** the hyperbolas $x^2 - y^2 = C$

### Review Exercises (page 473)

**1.** about 833

**3.** $a \approx 1.1904$, $b \approx 0.0476$

**5.** $a = 2.1773$

**7.** $\left(\frac{8}{3\pi}, \frac{4}{3\pi}\right)$

**9.** about 27,726 N-cm

**11.** $y = 4(x - 1)^3$

**13.** \$8, 798.85

### Challenging Problems (page 473)

**1.** (b) $\ln 2/(2\pi)$, (c) $\pi/(4k(k^2 + 1))$

**3.** $y = (r/h^3)x^3 - 3(r/h^2)x^2 + 3(r/h)x$

**7.** $y = L \ln \dfrac{x}{L - \sqrt{L^2 - x^2}} - \sqrt{L^2 - x^2}$

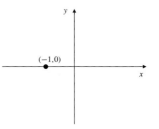

**9.** ellipse, centre $(0,2)$

**11.** parabola, vertex $(-1, -4)$

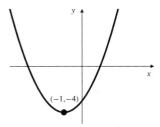

**13.** hyperbola, centre $\left(-\frac{3}{2}, 1\right)$
asymptotes
$2x + 3 = \pm 2^{3/2}(y - 1)$

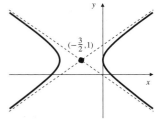

**15.** ellipse, centre $(1, -1)$

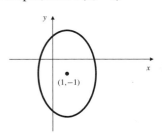

**17.** $y^2 - 8y = 16x$ or $y^2 - 8y = -4x$

# Chapter 8

# Conics, Parametric Curves, and Polar Curves

### Section 8.1 (page 487)

**1.** $(x^2/5) + (y^2/9) = 1$

**3.** $(x - 2)^2 = 16 - 4y$

**5.** $3y^2 - x^2 = 3$

**7.** single point $(-1, 0)$

**19.** rectangular hyperbola, centre $(1, -1)$,
semiaxes $a = b = \sqrt{2}$,
eccentricity $\sqrt{2}$,
foci $(\sqrt{2} + 1, \sqrt{2} - 1)$,
$(-\sqrt{2} + 1, -\sqrt{2} - 1)$,
asymptotes $x = 1$, $y = -1$

**21.** hyperbola, centre $(0,0)$,
semiaxes
$a = 1/\sqrt{2 + \sqrt{2}}$,
$b = 1/\sqrt{2 - \sqrt{2}}$,
foci
$\pm\big(\cos(\pi/8), \sin(\pi/8)\big)$,
asymptotes $x = 0$, $y = -x$

**23.** ellipse, centre $(0,0)$,
semi-axes $a = 2$, $b = 1$,
foci $\pm\left(2\sqrt{\frac{3}{5}}, -\sqrt{\frac{3}{5}}\right)$

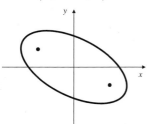

**25.** $(1 - \varepsilon^2)x^2 + y^2 - 2p\varepsilon^2 x = \varepsilon^2 p^2$

---

*Section 8.2   (page 494)*

**1.** $y = (x - 1)^2/4$

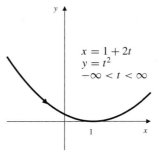

**3.** $y = (1/x) - 1$    **5.** $x^2 + y^2 = 9$

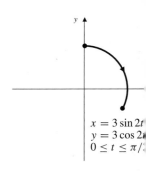

**7.** $\dfrac{x^2}{9} + \dfrac{y^2}{16} = 1$

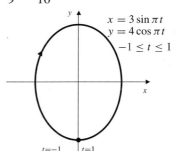

**9.** $x^{2/3} + y^{2/3} = 1$

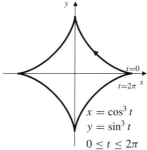

**11.** the right half of the hyperbola $x^2 - y^2 = 1$

**13.** the curve starts at the origin and spirals twice counterclockwise around the origin to end at $(4\pi, 0)$

**15.** $x = m/2$,    $y = m^2/4$,    $(-\infty < m < \infty)$

**17.** $x = a \sec t, \; y = a \sin t$;
$y^2 = a^2(x^2 - a^2)/x^2$

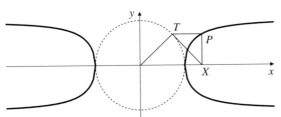

**19.** $x^3 + y^3 = 3xy$

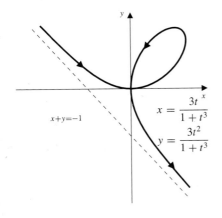

### Section 8.3 (page 499)

**1.** vertical at $(1, -4)$

**3.** horizontal at $(0, -16)$ and $(8, 16)$; vertical at $(-1, -11)$

**5.** horizontal at $(0, 1)$, vertical at $(\pm 1/\sqrt{e}, 1/e)$

**7.** horiz. at $(0, \pm 1)$, vert. at $(\pm 1, 1/\sqrt{2})$ and $(\pm 1, -1/\sqrt{2})$

**9.** $-3/4$      **11.** $-1/2$

**13.** $x = t - 2, \; y = 4t - 2$    **15.** slopes $\pm 1$

**17.** not smooth at $t = 0$

**19.** not smooth at $t = 0$

**21.**            **23.**

**25.**

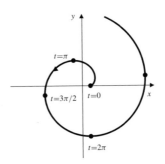

### Section 8.4 (page 504)

**1.** $4\sqrt{2} - 2$ units      **3.** $6a$ units

**5.** $\frac{8}{3}\left((1 + \pi^2)^{3/2} - 1\right)$ units

**7.** 4 units      **9.** $8a$ units

**11.** $2\sqrt{2}\pi(1 + 2e^\pi)/5$ sq. units

**13.** $72\pi(1 + \sqrt{2})/15$ sq. units

**15.** $256/15$ sq. units      **17.** $1/6$ sq. units

**19.** $9\pi/2$ sq. units

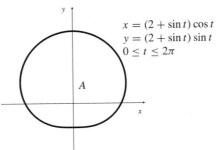

**23.** $32\pi a^3/105$ cu. units

### Section 8.5 (page 511)

**1.** $x = 3$, vertical straight line

**3.** $3y - 4x = 5$, straight line

**5.** $2xy = 1$, rectangular hyperbola

**7.** $y = x^2 - x$, a parabola

**9.** $y^2 = 1 + 2x$, a parabola

**11.** $x^2 - 3y^2 - 8y = 4$, a hyperbola

**13.**

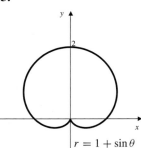

$r = 1 + \sin\theta$

**15.**

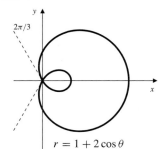

$r = 1 + 2\cos\theta$

**Section 8.6** *(page 515)*

**1.** $\pi^2$ sq. units

**3.** $a^2$ sq. units

$r = \sqrt{\theta}$

$A$

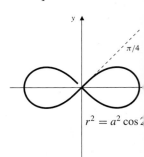

$\pi/4$

$r^2 = a^2\cos 2\theta$

**5.** $\pi/2$ sq. units

**7.** $2 + (\pi/4)$ sq. units

**17.**

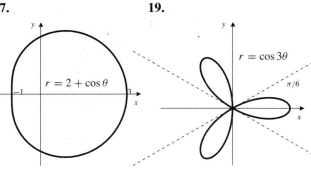

$r = 2 + \cos\theta$

$-1$   $3$

**19.**

$r = \cos 3\theta$

$\pi/6$

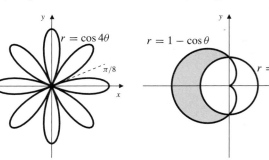

$r = \cos 4\theta$

$\pi/8$

$r = 1 - \cos\theta$

$r =$

**9.** $\pi/4$ sq. units

**11.** $\pi - \frac{3}{2}\sqrt{3}$ sq. units

**21.**

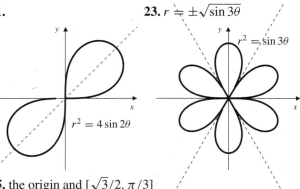

$r^2 = 4\sin 2\theta$

**23.** $r = \pm\sqrt{\sin 3\theta}$

$r^2 = \sin 3\theta$

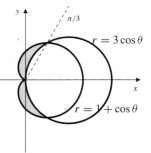

$\pi/3$

$r = 3\cos\theta$

$r = 1 + \cos\theta$

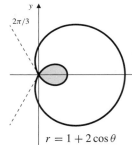

$2\pi/3$

$r = 1 + 2\cos\theta$

**25.** the origin and $[\sqrt{3}/2, \pi/3]$

**27.** the origin and $[3/2, \pm\pi/3]$

**29.** asymptote $y = 1$,
$r = 1/(\theta - \alpha)$ has
asymptote $(\cos\alpha)y - (\sin\alpha)x = 1$

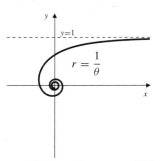

$y=1$

$r = \dfrac{1}{\theta}$

**13.** $\dfrac{\sqrt{1+a^2}}{a}\left(e^{a\pi} - e^{-a\pi}\right)$ units

**17.** $67.5°, -22.5°$

**19.** $90°$ at $(0,0)$,
$\pm 45°$ at $\left(1 - \dfrac{1}{\sqrt{2}}, \dfrac{\pi}{4}\right)$,
$\pm 135°$ at $\left(1 + \dfrac{1}{\sqrt{2}}, \dfrac{5\pi}{4}\right)$

**21.** horizontal at $\left(\pm\frac{\pi}{4}, \sqrt{2}\right)$, vertical at $(2, 0)$ and the origin

**23.** horizontal at $(0,0)$, $\left(\frac{2}{3}\sqrt{2}, \pm\tan^{-1}\sqrt{2}\right)$,
$\left(\frac{2}{3}\sqrt{2}, \pi \pm \tan^{-1}\sqrt{2}\right)$,
vertical at $\left(0, \frac{\pi}{2}\right)$, $\left(\frac{2}{3}\sqrt{2}, \pm\tan^{-1}(1/\sqrt{2})\right)$,
$\left(\frac{2}{3}\sqrt{2}, \pi \pm \tan^{-1}(1/\sqrt{2})\right)$

**31.** $x = f(\theta)\cos\theta, \quad y = f(\theta)\sin\theta$

**39.** $\ln\theta_1 = 1/\theta_1$, point $(-0.108461, 0.556676)$; $\ln\theta_2 = -1/(\theta_2 + \pi)$, point $(-0.182488, -0.178606)$

**25.** horizontal at $\left(4, -\frac{\pi}{2}\right)$, $\left(1, \frac{\pi}{6}\right)$, $\left(1, \frac{5\pi}{6}\right)$,
vertical at $\left(3, -\frac{\pi}{6}\right)$, $\left(3, -\frac{5\pi}{6}\right)$, no tangent at $\left(0, \frac{\pi}{2}\right)$

### Review Exercises   (page 516)

**1.** ellipse, foci $(\pm1, 0)$, semi-major axis $\sqrt{2}$, semi-minor axis 1
**3.** parabola, vertex $(4, 1)$, focus $(15/4, 1)$
**5.** straight line from $(0, 2)$ to $(2, 0)$
**7.** the parabola $y = x^2 - 1$ left to right
**9.** first quadrant part of ellipse $16x^2 + y^2 = 16$ from $(1, 0)$ to $(0, 4)$
**11.** horizontal tangents at $(2, \pm2)$ (i.e. $t = \pm1$)
vertical tangent at $(4, 0)$ (i.e. $t = 0$)

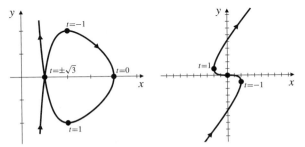

**13.** horizontal tangent at $(0, 0)$ (i.e. $t = 0$)
vertical tangents at $(2, -1)$ and $(-2, 1)$
(i.e. $t = \pm1$)
**15.** $1/2$ sq. units
**17.** $1 + e^2$ units

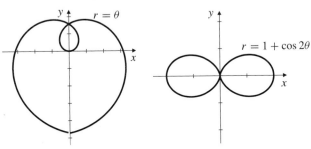

**19.** $r = \theta$
**21.** $4 = 1 + \cos 2\theta$
**23.** $r = 1 + 2\cos 2\theta$

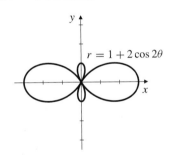

**25.** $\pi + (3\sqrt{3}/4)$ sq. units    **27.** $(\pi - 3)/2$ sq. units

### Challenging Problems   (page 516)

**1.** $16\pi \sec\theta$ cm$^2$    **5.** $40\pi/3$ ft$^3$
**7.** about 84.65 minutes
**9.** $r^2 = \cos(2\theta)$ is the inner curve; area between curves is $1/3$ sq. units

# Chapter 9

# Sequences, Series, and Power Series

### Section 9.1   (page 526)

**1.** bounded, positive, increasing, convergent to 2
**3.** bounded, positive, convergent to 4
**5.** bounded below, positive, increasing, divergent to infinity
**7.** bounded below, positive, increasing, divergent to infinity
**9.** bounded, positive, decreasing, convergent to 0
**11.** divergent          **13.** divergent
**15.** $\infty$          **17.** 0
**19.** 1          **21.** $e^{-3}$
**23.** 0          **25.** $1/2$
**27.** 0          **29.** 0
**31.** $\lim_{n\to\infty} a_n = 5$
**33.** If $\{a_n\}$ is (ultimately) decreasing, then either it is bounded below, and therefore convergent, or else it is unbounded below and therefore divergent to negative infinity.

### Section 9.2   (page 533)

**1.** $\dfrac{1}{2}$
**3.** $\dfrac{1}{(2 + \pi)^8\big((2 + \pi)^2 - 1\big)}$
**5.** $\dfrac{25}{4,416}$          **7.** $\dfrac{8e^4}{e - 2}$
**9.** diverges to $\infty$          **11.** $\dfrac{3}{4}$
**13.** $\dfrac{1}{3}$          **15.** div. to $\infty$
**17.** div. to $\infty$          **19.** diverges
**21.** 14 m

**25.** If $\{a_n\}$ is ultimately negative, then the series $\sum a_n$ must either converge (if its partial sums are bounded below), or diverge to $-\infty$ (if its partial sums are not bounded below).

**27.** false, e.g. $\sum \dfrac{(-1)^n}{2^n}$          **29.** true

**31.** true

## Section 9.3   (page 545)

**1.** converges                    **3.** diverges to $\infty$

**5.** converges                    **7.** diverges to $\infty$

**9.** converges                    **11.** diverges to $\infty$

**13.** diverges to $\infty$              **15.** converges

**17.** converges                   **19.** diverges to $\infty$

**21.** converges                   **23.** converges

**25.** converges

**27.** $s_n + \dfrac{1}{3(n+1)^3} \le s \le s_n + \dfrac{1}{3n^3};\quad n = 6$

**29.** $s_n + \dfrac{2}{\sqrt{n+1}} \le s \le s_n + \dfrac{2}{\sqrt{n}};\quad n = 63$

**31.** $0 < s - s_n \le \dfrac{n+2}{2^n(n+1)!(2n+3)};\quad n = 4$

**33.** $0 < s - s_n \le \dfrac{2^n(4n^2 + 6n + 2)}{(2n)!(4n^2 + 6n)};\quad n = 4$

**39.** converges, $a_n^{1/n} \to (1/e) < 1$

**41.** no info from ratio test, but series diverges to infinity since all terms exceed 1.

**43.** (b) $s \le \dfrac{2}{k(1-k)},\ k = \tfrac{1}{2}$,

(c) $0 < s - s_n < \dfrac{(1+k)^{n+1}}{2^n k(1-k)},\ k = \dfrac{n+2 - \sqrt{n^2 + 8}}{2(n-1)}$

for $n \ge 2$

## Section 9.4   (page 552)

**1.** conv. conditionally          **3.** conv. conditionally

**5.** diverges                     **7.** conv. absolutely

**9.** conv. conditionally          **11.** diverges

**13.** 999                         **15.** 13

**17.** converges absolutely if $-1 < x < 1$, conditionally if $x = -1$, diverges elsewhere

**19.** converges absolutely if $0 < x < 2$, conditionally if $x = 2$, diverges elsewhere

**21.** converges absolutely if $-2 < x < 2$, conditionally if $x = -2$, diverges elsewhere

**23.** converges absolutely if $-\tfrac{7}{2} < x < \tfrac{1}{2}$, conditionally if $x = -\tfrac{7}{2}$, diverges elsewhere

**25.** AST does not apply directly, but does if we remove all the 0 terms; series converges conditionally

**27.** (a) false, e.g. $a_n = \dfrac{(-1)^n}{n}$,

(b) false, e.g. $a_n = \dfrac{\sin(n\pi/2)}{n}$, (see Exercise 25),

(c) true

**29.** converges absolutely for $-1 < x < 1$, conditionally if $x = -1$, diverges elsewhere

## Section 9.5   (page 564)

**1.** centre 0, radius 1, interval $(-1, 1)$

**3.** centre $-2$, radius 2, interval $[-4, 0)$

**5.** centre $\tfrac{3}{2}$, radius $\tfrac{1}{2}$, interval $(1, 2)$

**7.** centre 0, radius $\infty$, interval $(-\infty, \infty)$

**9.** $\dfrac{1}{(1-x)^3} = \sum\limits_{n=0}^{\infty} \dfrac{(n+1)(n+2)}{2} x^n,\ (-1 < x < 1)$

**11.** $\dfrac{1}{(1-x)^2} = \sum\limits_{n=0}^{\infty}(n+1)x^n,\quad (-1 < x < 1)$

**13.** $\dfrac{1}{(2-x)^2} = \sum\limits_{n=0}^{\infty} \dfrac{n+1}{2^{n+2}} x^n,\quad (-2 < x < 2)$

**15.** $\ln(2-x) = \ln 2 - \sum_{n=1}^{\infty} \dfrac{x^n}{2^n n},\quad (-2 \le x < 2)$

**17.** $\dfrac{1}{x^2} = \sum\limits_{n=0}^{\infty} \dfrac{n+1}{2^{n+2}}(x+2)^n,\quad (-4 < x < 0)$

**19.** $\dfrac{x^3}{1 - 2x^2} = \sum\limits_{n=0}^{\infty} 2^n x^{2n+3},\quad \left(-\dfrac{1}{\sqrt{2}} < x < \dfrac{1}{\sqrt{2}}\right)$

**21.** $\left(-\tfrac{1}{4}, \tfrac{1}{4}\right);\quad \dfrac{1}{1+4x}$

**23.** $[-1, 1);\quad \tfrac{1}{3}$ if $x = 0$,

$-\dfrac{1}{x^3}\ln(1-x) - \dfrac{1}{x^2} - \dfrac{1}{2x}$ otherwise

**25.** $(-1, 1);\quad \dfrac{2}{(1-x^2)^2}$          **27.** $3/4$

**29.** $\pi^2(\pi + 1)/(\pi - 1)^3$          **31.** $\ln(3/2)$

## Section 9.6   (page 573)

**1.** $e^{3x+1} = \sum_{n=0}^{\infty} \dfrac{3^n e}{n!} x^n,\ (\text{all } x)$

**3.** $\sin\left(x - \dfrac{\pi}{4}\right)$

$= \dfrac{1}{\sqrt{2}} \sum\limits_{n=0}^{\infty}(-1)^n\left[-\dfrac{x^{2n}}{(2n)!} + \dfrac{x^{2n+1}}{(2n+1)!}\right],\ (\text{all } x)$

**5.** $x^2 \sin\left(\dfrac{x}{3}\right) = \sum_{n=0}^{\infty} \dfrac{(-1)^n}{3^{2n+1}(2n+1)!} x^{2n+3},\ (\text{all } x)$

**7.** $\sin x \cos x = \sum_{n=0}^{\infty} \dfrac{(-1)^n 2^{2n}}{(2n+1)!} x^{2n+1},\ (\text{all } x)$

**9.** $\dfrac{1+x^3}{1+x^2} = 1 - x^2 + \sum\limits_{n=2}^{\infty}(-1)^n\left(x^{2n-1}+x^{2n}\right)$,
$(-1 < x < 1)$

**11.** $\ln\dfrac{1-x}{1+x} = -2\sum\limits_{n=1}^{\infty}\dfrac{x^{2n-1}}{2n-1}$, $\quad(-1 < x < 1)$

**13.** $\cosh x - \cos x = 2\sum\limits_{n=0}^{\infty}\dfrac{x^{4n+2}}{(4n+2)!}$, (all $x$)

**15.** $e^{-2x} = e^2\sum\limits_{n=0}^{\infty}\dfrac{(-1)^n 2^n}{n!}(x+1)^n$, (all $x$)

**17.** $\cos x = \sum\limits_{n=0}^{\infty}\dfrac{(-1)^{n+1}}{(2n)!}(x-\pi)^{2n}$, (all $x$)

**19.** $\ln 4 + \sum\limits_{n=1}^{\infty}\dfrac{(-1)^{n-1}}{4^n n}(x-2)^n$, $\quad(-2 < x \le 6)$

**21.** $\sin x - \cos x =$
$\sqrt{2}\sum\limits_{n=0}^{\infty}\dfrac{(-1)^n}{(2n+1)!}\left(x-\dfrac{\pi}{4}\right)^{2n+1}$, (all $x$)

**23.** $\dfrac{1}{x^2} = \dfrac{1}{4}\sum\limits_{n=0}^{\infty}\dfrac{n+1}{2^n}(x+2)^n$, $\quad(-4 < x < 0)$

**25.** $(x-1)+\sum\limits_{n=2}^{\infty}\dfrac{(-1)^n}{n(n-1)}(x-1)^n$, $\quad(0 \le x \le 2)$

**27.** $1+\dfrac{x^2}{2}+\dfrac{5x^4}{24}$ **29.** $x+\dfrac{x^2}{2}-\dfrac{x^3}{6}$

**31.** $1+\dfrac{x}{2}-\dfrac{x^2}{8}$ **33.** $e^{x^2}$ (all $x$)

**35.** $\dfrac{e^x - e^{-x}}{2x} = \dfrac{\sinh x}{x}$ if $x \ne 0$, 1 if $x = 0$

**37.** (a) $1+x+x^2$, (b) $3+3(x-1)+(x-1)^2$

### Section 9.7 (page 577)

**1.** 1.22140 **3.** 3.32011

**5.** 0.99619 **7.** $-0.10533$

**9.** 0.42262 **11.** 1.54306

**13.** $I(x) = \sum\limits_{n=0}^{\infty}\dfrac{(-1)^n}{(2n+1)(2n+1)!}x^{2n+1}$, (all $x$)

**15.** $K(x) = \sum\limits_{n=0}^{\infty}\dfrac{(-1)^n}{(n+1)^2}x^{n+1}$, $\quad(-1 \le x \le 1)$

**17.** $M(x) = \sum\limits_{n=0}^{\infty}\dfrac{(-1)^n}{(2n+1)(4n+1)}x^{4n+1}$,
$(-1 \le x \le 1)$

**19.** 0.946 **21.** 2

**23.** $-3/25$ **25.** 0

**27.** $y = \sum\limits_{n=0}^{\infty}(-1)^n\left[\dfrac{2^n n!}{(2n)!}x^{2n}+\dfrac{1}{2^{n-1}n!}x^{2n+1}\right]$

### Section 9.8 (page 581)

**1.** $\dfrac{1}{720}(0.2)^7$ **3.** $\dfrac{1}{120}(0.5)^5$

**5.** $\dfrac{4\sec^2(0.1)\tan^2(0.1)+2\sec^4(0.1)}{4!\,10^4}$

**7.** $\dfrac{24}{120(1.95)^5(20)^5}$

**9.** $2^x = \sum\limits_{n=0}^{\infty}\dfrac{(x\ln 2)^n}{n!}$, all $x$

**11.** $\sin x = \sum\limits_{n=0}^{\infty}(-1)^n\dfrac{x^{2n+1}}{(2n+1)!}$, all $x$

**13.** $\dfrac{1}{1-x} = \sum\limits_{n=0}^{\infty}x^n$, $-1 < x < 1$

**15.** $\dfrac{x}{2+3x} = \sum\limits_{n=1}^{\infty}(-1)^{n-1}3^{n-1}\left(\dfrac{x}{2}\right)^n$, $-\dfrac{2}{3} < x < \dfrac{2}{3}$

**17.** $\sin x = \dfrac{1}{2}\sum\limits_{n=0}^{\infty}\dfrac{c_n}{n!}\left(x-\dfrac{\pi}{6}\right)^n$, (for all $x$), where
$c_n = (-1)^{n/2}$ if $n$ is even, and $c_n = (-1)^{(n-1)/2}\sqrt{3}$ if
$n$ is odd

**19.** $\ln x = \sum\limits_{n=1}^{\infty}(-1)^{n-1}\dfrac{(x-1)^n}{n}$, $0 < x \le 2$

**21.** $\dfrac{1}{x} = -\dfrac{1}{2}\sum\limits_{n=0}^{\infty}\left(\dfrac{x+2}{2}\right)^n$, $-4 < x < 0$

### Section 9.9 (page 585)

**1.** $\sqrt{1+x}$
$=$
$1+\sum\limits_{n=1}^{\infty}\dfrac{(-1)^{n-1}1\times 3\times 5\times\cdots\times(2n-3)}{2^n n!}x^n$
$|x| < 1$

**3.** $\sqrt{4+x}$
$= 2+\dfrac{x}{4}+2\sum\limits_{n=2}^{\infty}(-1)^{n-1}\dfrac{1\times 3\times 5\times\cdots\times(2n-3)}{2^{3n}n!}x^n$,
$(-4 < x \le 4)$

**5.** $\sum\limits_{n=0}^{\infty}(n+1)x^n$, $\quad|x| < 1$

### Section 9.10 (page 592)

**1.** $2\pi/3$ **3.** $\pi$

**5.** $2\sum\limits_{n=1}^{\infty}(-1)^{n-1}(\sin(nt))/n$

**7.** $\dfrac{1}{4}-\sum\limits_{n=1}^{\infty}\left(\dfrac{2\cos((2n-1)\pi t)}{(2n-1)^2\pi^2}+\dfrac{(-1)^n\sin(n\pi t)}{n\pi}\right)$

**9.** 1

**11.** $2\sum\limits_{n=1}^{\infty}\dfrac{(-1)^n}{n\pi}\sin(n\pi t)$

**13.** $\pi^2/8$

### Review Exercises (page 593)

**1.** conv. to 0 **3.** div. to $\infty$

**5.** $\lim_{n\to\infty}a_n = \sqrt{2}$ **7.** $4\sqrt{2}/(\sqrt{2}-1)$

**9.** 2 **11.** converges

**13.** converges

**15.** converges

**17.** conv. abs.

**19.** conv. cond.

**21.** conv. abs.  for $x$ in $(-1, 5)$, cond. for $x = -1$, div. elsewhere

**23.** 1.202

**25.** $\sum_{n=0}^{\infty} x^n/3^{n+1}$, $|x| < 3$

**27.** $1 + \sum_{n=1}^{\infty}(-1)^{n-1}x^{2n}/(ne^n)$, $-\sqrt{e} < x \le \sqrt{e}$

**29.** $x + \sum_{n=1}^{\infty}(-1)^n 2^{2n-1}x^{2n+1}/(2n)!$, all $x$

**31.** $(1/2) + \sum_{n=1}^{\infty} \dfrac{(-1)^n\, 1 \times 4 \times 7 \times \cdots \times (3n-2)x^n}{2 \times 24^n\, n!}$,

$-8 < x \le 8$

**33.** $\sum_{n=0}^{\infty}(-1)^n(x - \pi)^n/\pi^{n+1}$, $0 < x < 2\pi$

**35.** $1 + 2x + 3x^2 + \frac{10}{3}x^3$   **37.** $1 - \frac{1}{2}x^2 + \frac{5}{24}x^4$

**39.** $\begin{cases} \cos\sqrt{x} & \text{if } x \ge 0 \\ \cosh\sqrt{|x|} & \text{if } x < 0 \end{cases}$   **41.** $\pi^2/(\pi - 1)^2$

**43.** $\ln(e/(e - 1))$   **45.** $1/14$

**47.** 3, 0.49386   **49.** $\sum_{n=1}^{\infty} \dfrac{2}{n}\sin(nt)$

**Challenging Problems   (page 594)**

**5.** (c) 1.645

**7.** (a) $\infty$, (c) $e^{-x^2}$, (d) $f(x) = e^{x^2}\int_0^x e^{-t^2}\, dt$

# Chapter 10
# Vectors and Coordinate Geometry in 3-Space

**Section 10.1   (page 603)**

**1.** 3 units   **3.** $\sqrt{6}$ units

**5.** $|z|$ units;   $\sqrt{y^2 + z^2}$ units

**7.** $\cos^{-1}(-4/9) \approx 116.39°$

**9.** $\sqrt{3}/2$ sq. units   **11.** $\sqrt{n - 1}$ units

**13.** the half-space containing the origin and bounded by the plane passing through $(0, -1, 0)$ perpendicular to the $y$-axis.

**15.** the vertical plane (parallel to the $z$-axis) passing through $(1, 0, 0)$ and $(0, 1, 0)$.

**17.** the sphere of radius 2 centred at $(1, -2, 3)$.

**19.** the solid circular cylinder of radius 2 with axis along the $x$-axis.

**21.** the parabolic cylinder generated by translating the parabola $z = y^2$ in the $yz$-plane in the direction of the $x$-axis .

**23.** the plane through the points $(6, 0, 0)$, $(0, 3, 0)$ and $(0, 0, 2)$.

**25.** the straight line through $(1, 0, 0)$ and $(1, 1, 1)$.

**27.** the circle in which the sphere of radius 2 centred at the origin intersects the sphere of radius 2 with centre $(2, 0, 0)$.

**29.** the ellipse in which the plane $z = x$ intersects the circular cylinder of radius 1 and axis along the $z$-axis.

**31.** the part of the solid circular cylinder of radius 1 and axis along the $z$-axis lying above or on the plane $z = y$.

**33.** bdry $(0,0)$ and $x^2 + y^2 = 1$; interior$= S$; $S$ open

**35.** bdry of $S$ is $S$; interior empty; $S$ is closed

**37.** bdry — the spheres $x^2 + y^2 + z^2 = 1$ and $x^2 + y^2 + z^2 = 4$; interior — points between these spheres; $S$ is closed

**39.** bdry of $S$ is $S$, namely the line $x = y = z$; interior is empty; $S$ closed

**Section 10.2   (page 613)**

**1.** (a) $3\mathbf{i} - 2\mathbf{j}$, (b) $-3\mathbf{i} + 2\mathbf{j}$,
(c) $2\mathbf{i} - 5\mathbf{j}$, (d) $-2\mathbf{i} + 4\mathbf{j}$,(e) $-\mathbf{i} - 2\mathbf{j}$, (f) $4\mathbf{i} + \mathbf{j}$, (g) $-7\mathbf{i} + 20\mathbf{j}$, (h) $2\mathbf{i} - (5/3)\mathbf{j}$

**3.** a) $6\mathbf{i} - 10\mathbf{k}$,   $8\mathbf{j}$,   $-3\mathbf{i} + 20\mathbf{j} + 5\mathbf{k}$
b) $5\sqrt{2}$,   $5\sqrt{2}$
c) $\frac{3}{5\sqrt{2}}\mathbf{i} \pm \frac{4}{5\sqrt{2}}\mathbf{j} - \frac{1}{\sqrt{2}}\mathbf{k}$   d) 18
e) $\cos^{-1}(9/25) \approx 68.9°$   f) $18/5\sqrt{2}$
g) $(27/25)\mathbf{i} + (36/25)\mathbf{j} - (9/5)\mathbf{k}$

**9.** from southwest at $50\sqrt{2}$ km/h.

**11.** head at angle $\theta$ to the east of $AC$, where
$$\theta = \sin^{-1}\frac{3}{2\sqrt{1 + 4k^2}}.$$
The trip not possible if $k < \frac{1}{4}\sqrt{5}$. If $k > \frac{1}{4}\sqrt{5}$ there is a second possible heading, $\pi - \theta$, but the trip will take longer.

**13.** $t = 2$

**15.** $\cos^{-1}(2/\sqrt{6}) \approx 35.26°$, $90°$

**17.** $(\mathbf{i} + \mathbf{j} + \mathbf{k})/\sqrt{3}$

**19.** $\lambda = 1/2$, midpoint, $\lambda = 2/3$, 2/3 of way from $P_1$ to $P_2$, $\lambda = -1$, $P_1$ is midway between this point and $P_2$.

**21.** plane through point with position vector $(b/|\mathbf{a}|^2)\mathbf{a}$ perpendicular to $\mathbf{a}$.

**23.** $\mathbf{x} = 2\mathbf{i} - 3\mathbf{j} - 4\mathbf{k}$

**25.** $(|\mathbf{u}|\mathbf{v} + |\mathbf{v}|\mathbf{u})/\||\mathbf{u}|\mathbf{v} + |\mathbf{v}|\mathbf{u}\|$

**31.** $\mathbf{u} = (\mathbf{w} \bullet \mathbf{a}/|\mathbf{a}|^2)\mathbf{a}$, $\mathbf{v} = \mathbf{w} - \mathbf{u}$

**33.** $\mathbf{x} = (\mathbf{a} + K\hat{\mathbf{u}})/(2r)$, $\mathbf{y} = (\mathbf{a} - K\hat{\mathbf{u}})/(2s)$, where $K = \sqrt{|\mathbf{a}|^2 - 4rst}$ and $\hat{\mathbf{u}}$ is any unit vector

**35.** about 12.373 m   **37.** about 19 m

**Section 10.3   (page 622)**

**1.** $5\mathbf{i} + 13\mathbf{j} + 7\mathbf{k}$   **3.** $\sqrt{6}$ sq. units

**5.** $\pm\frac{1}{3}(2\mathbf{i} - 2\mathbf{j} + \mathbf{k})$   **15.** 4/3 cubic units

**17.** $k = -6$

**19.** $\lambda = \dfrac{\mathbf{x}\bullet(\mathbf{v}\times\mathbf{w})}{\mathbf{u}\bullet(\mathbf{v}\times\mathbf{w})}$, $\mu = \dfrac{\mathbf{x}\bullet(\mathbf{w}\times\mathbf{u})}{\mathbf{u}\bullet(\mathbf{v}\times\mathbf{w})}$, $\nu = \dfrac{\mathbf{x}\bullet(\mathbf{u}\times\mathbf{v})}{\mathbf{u}\bullet(\mathbf{v}\times\mathbf{w})}$

**21.** $\mathbf{u}\times(\mathbf{v}\times\mathbf{w}) = -2\mathbf{i}+7\mathbf{j}-4\mathbf{k}$, $(\mathbf{u}\times\mathbf{v})\times\mathbf{w} = \mathbf{i}+9\mathbf{j}+9\mathbf{k}$; the first is in the plane of $\mathbf{v}$ and $\mathbf{w}$, the second is in the plane of $\mathbf{u}$ and $\mathbf{v}$.

## Section 10.4   (page 631)

**1.** a) $x^2+y^2+z^2=z^2$;  b) $x+y+z=x+y+z$;
  c) $x^2+y^2+z^2=-1$

**3.** $x-y+2z=0$    **5.** $7x+5y-z=12$

**7.** $x-5y-3z=-7$    **9.** $x+6y-5z=17$

**11.** $(\mathbf{r}_1-\mathbf{r}_2)\bullet[(\mathbf{r}_1-\mathbf{r}_3)\times(\mathbf{r}_1-\mathbf{r}_4)]=0$

**13.** planes passing through the line $x=0$, $y+z=1$ (except the plane $y+z=1$ itself)

**15.** $\mathbf{r}=(1+2t)\mathbf{i}+(2-3t)\mathbf{j}+(3-4t)\mathbf{k}$,
  $(-\infty<t<\infty)$
  $x=1+2t$, $y=2-3t$, $z=3-4t$,   $(-\infty<t<\infty)$
  $\dfrac{x-1}{2}=\dfrac{y-2}{-3}=\dfrac{z-3}{-4}$

**17.** $\mathbf{r}=t(7\mathbf{i}-6\mathbf{j}-5\mathbf{k})$;   $x=7t$,   $y=-6t$,
  $z=-5t$;   $x/7=-y/6=-z/5$

**19.** $\mathbf{r}=\mathbf{i}+2\mathbf{j}-\mathbf{k}+t(\mathbf{i}+\mathbf{j}+\mathbf{k})$;
  $x=1+t$,   $y=2+t$,   $z=-1+t$;
  $x-1=y-2=z+1$

**21.** $\dfrac{x-4}{-5}=\dfrac{y}{3}$,   $z=7$

**23.** $\mathbf{r}_i\neq\mathbf{r}_j$, $(i,j=1,\cdots,4,\ i\neq j)$,
  $\mathbf{v}=(\mathbf{r}_1-\mathbf{r}_2)\times(\mathbf{r}_3-\mathbf{r}_4)\neq 0$, $(\mathbf{r}_1-\mathbf{r}_3)\bullet\mathbf{v}=0$.

**25.** $7\sqrt{2}/10$ units    **27.** $18/\sqrt{69}$ units

**29.** all lines parallel to the $xy$-plane and passing through $(x_0,y_0,z_0)$.

**31.** $(x,y,z)$ satisfies the quadratic if either
  $A_1x+B_1y+C_1z=D_1$ or $A_2x+B_2y+C_2z=D_2$.

## Section 10.5  (page 635)

**1.** ellipsoid centred at the origin with semiaxes 6, 3 and 2 along the $x$-, $y$- and $z$-axes respectively.

**3.** circle with centre $(1,-2,3)$ and radius $1/\sqrt{2}$.

**5.** elliptic paraboloid with vertex at the origin, axis along the $z$-axis, and cross-section $x^2+2y^2=1$ in the plane $z=1$.

**7.** hyperboloid of two sheets with vertices $(\pm 2,0,0)$ and circular cross-sections in planes $x=c$, $(c^2>4)$.

**9.** hyperbolic paraboloid — same as $z=x^2-y^2$ but rotated $45°$ about the $z$-axis (counterclockwise as seen from above).

**11.** hyperbolic cylinder parallel to the $y$-axis, intersecting the $xz$-plane in the hyperbola $(x^2/4)-z^2=1$.

**13.** parabolic cylinder parallel to the $y$-axis.

**15.** circular cone with vertex $(2,3,1)$, vertical axis, and semi-vertical angle $45°$.

**17.** circle in the plane $x+y+z=1$ having centre $(1/3,1/3,1/3)$ and radius $\sqrt{11/3}$.

**19.** a parabola in the plane $z=1+x$ having vertex at $(-1/2,0,1/2)$ and axis along the line $z=1+x$, $y=0$.

**21.** $\dfrac{y}{b}-\dfrac{z}{c}=\lambda\left(1-\dfrac{x}{a}\right)$,   $\dfrac{y}{b}+\dfrac{z}{c}=\dfrac{1}{\lambda}\left(1+\dfrac{x}{a}\right)$;
  $\dfrac{y}{b}-\dfrac{z}{c}=\mu\left(1+\dfrac{x}{a}\right)$,   $\dfrac{y}{b}+\dfrac{z}{c}=\dfrac{1}{\mu}\left(1-\dfrac{x}{a}\right)$

**23.** $\mathbf{a}=\mathbf{i}\pm\mathbf{k}$ (or any multiple)

## Section 10.6  (page 643)

**1.** $\begin{pmatrix} 6 & 7 \\ 5 & -3 \\ 1 & 1 \end{pmatrix}$

**3.** $\begin{pmatrix} aw+by & ax+bz \\ cw+dy & cx+dz \end{pmatrix}$

**5.** $\mathcal{A}\mathcal{A}^T=\begin{pmatrix} 4 & 3 & 2 & 1 \\ 3 & 3 & 2 & 1 \\ 2 & 2 & 2 & 1 \\ 1 & 1 & 1 & 1 \end{pmatrix}$   $\mathcal{A}^2=\begin{pmatrix} 1 & 2 & 3 & 4 \\ 0 & 1 & 2 & 3 \\ 0 & 0 & 1 & 2 \\ 0 & 0 & 0 & 1 \end{pmatrix}$

**7.** 36    **15.** $\begin{pmatrix} 1 & -1 & 0 \\ 0 & 1 & -1 \\ 0 & 0 & 1 \end{pmatrix}$

**17.** $x=1$, $y=2$, $z=3$

**19.** $x_1=1$, $x_2=2$, $x_3=-1$, $x_4=-2$

## Review Exercises   (page 644)

**1.** plane parallel to $y$-axis through $(3,0,0)$ and $(0,0,1)$

**3.** all points on or above the plane through the origin with normal $\mathbf{i}+\mathbf{j}+\mathbf{k}$

**5.** circular paraboloid with vertex at $(0,1,0)$ and axis along the $y$-axis, opening in the direction of increasing $y$

**7.** hyperbolic paraboloid

**9.** points inside the ellipsoid with vertices at $(\pm 2,0,0)$, $(0,\pm 2,0$, and $(0,0,\pm 1)$

**11.** cone with axis along the $x$-axis, vertex at the origin, and elliptical cross-sections perpendicular to its axis

**13.** oblique circular cone (elliptic cone). Cross-sections in horizontal planes $z=k$ are circles of radius 1 with centres at $(k,0,k)$

**15.** horizontal line through $(0,0,3)$ and $(2,-1,3)$

**17.** circle of radius 1 centred at $(1,1,1)$ in plane normal to $\mathbf{i}+\mathbf{j}+\mathbf{k}$

**19.** $2x-y+3z=0$    **21.** $2x+5y+3z=2$

**23.** $7x+4y-8z=6$

**25.** $\mathbf{r}=(2+3t)\mathbf{i}+(1+t)\mathbf{j}-(1+2t)\mathbf{k}$

**27.** $x=3t$, $y=-2t$, $z=4t$

**29.** $(\mathbf{r}_2-\mathbf{r}_1)\times(\mathbf{r}_3-\mathbf{r}_1)=\mathbf{0}$

**31.** $(3/2)\sqrt{34}$ sq. units

**33.** $A^{-1} = \begin{pmatrix} 1 & 0 & 0 & 0 \\ -2 & 1 & 0 & 0 \\ 1 & -2 & 1 & 0 \\ 0 & 1 & -2 & 1 \end{pmatrix}$

### Challenging Problems (page 645)

**5.** condition: $\mathbf{a} \cdot \mathbf{b} = 0$,

$$\mathbf{x} = \frac{\mathbf{b} \times \mathbf{a}}{|\mathbf{a}|^2} + t\mathbf{a} \text{ (for any scalar } t)$$

# Chapter 11
# Vector Functions and Curves

### Section 11.1 (page 652)

**1.** $\mathbf{v} = \mathbf{j}$, $v = 1$, $\mathbf{a} = \mathbf{0}$, path is the line $x = 1$, $z = 0$

**3.** $\mathbf{v} = 2t\mathbf{j} + \mathbf{k}$, $v = \sqrt{4t^2 + 1}$, $\mathbf{a} = 2\mathbf{j}$, path is the parabola $y = z^2$, in the $yz$-plane

**5.** $\mathbf{v} = 2t\mathbf{i} - 2t\mathbf{j}$, $v = 2\sqrt{2}t$, $\mathbf{a} = 2\mathbf{i} - 2\mathbf{j}$, path is the straight half-line $x + y = 0$, $z = 1$, $(x \geq 0)$

**7.** $\mathbf{v} = -a\sin t\mathbf{i} + a\cos t\mathbf{j} + c\mathbf{k}$, $v = \sqrt{a^2 + c^2}$, $\mathbf{a} = -a\cos t\mathbf{i} - a\sin t\mathbf{j}$, path is a circular helix

**9.** $\mathbf{v} = -3\sin t\mathbf{i} - 4\sin t\mathbf{j} + 5\cos t\mathbf{k}$, $v = 5$, $\mathbf{a} = -\mathbf{r}$, path is the circle of intersection of the plane $4x = 3y$ with the sphere $x^2 + y^2 + z^2 = 25$

**11.** $\mathbf{a} = \mathbf{v} = \mathbf{r}$, $v = \sqrt{a^2 + b^2 + c^2}\, e^t$, path is the straight line $\dfrac{x}{a} = \dfrac{y}{b} = \dfrac{z}{c}$

**13.** $\mathbf{v} = -(e^{-t}\cos e^t + \sin e^t)\mathbf{i}$
$\quad +(-e^{-t}\sin e^t + \cos e^t)\mathbf{j} - e^t\mathbf{k}$
$v = \sqrt{1 + e^{-2t} + e^{2t}}$
$\mathbf{a} = [(e^{-t} - e^t)\cos e^t + \sin e^t]\mathbf{i}$
$\quad +[(e^{-t} - e^t)\sin e^t - \cos e^t]\mathbf{j} - e^t\mathbf{k}$
The path is a spiral lying on the surface $z = -1/\sqrt{x^2 + y^2}$

**15.** $\mathbf{a} = -3\pi^2\mathbf{i} - 4\pi^2\mathbf{j}$　　　**17.** $\sqrt{3/2}(-\mathbf{i} + \mathbf{j} - 2\mathbf{k})$

**19.** $\mathbf{v} = 2\mathbf{i} + 4\mathbf{j} + 4\mathbf{k}$, $\mathbf{a} = \frac{-8}{9}(2\mathbf{i} + \mathbf{j} - 2\mathbf{k})$

**29.** $\dfrac{d}{dt}(\mathbf{u} \times (\mathbf{v} \times \mathbf{w})) = \dfrac{d\mathbf{u}}{dt} \times (\mathbf{v} \times \mathbf{w})$
$\quad +\mathbf{u} \times \left(\dfrac{d\mathbf{v}}{dt} \times \mathbf{w}\right) + \mathbf{u} \times \left(\mathbf{v} \times \dfrac{d\mathbf{w}}{dt}\right)$

**31.** $\mathbf{u}''' \cdot (\mathbf{u} \times \mathbf{u}')$

**33.** $\mathbf{r} = \mathbf{r}_0 e^{2t}$, $\mathbf{a} = 4\mathbf{r}_0 e^{2t}$; the path is a straight line through the origin in the direction of $\mathbf{r}_0$

**35.** $\mathbf{r} = \mathbf{r}_0 + \dfrac{1 - e^{-ct}}{c}\mathbf{v}_0 - \dfrac{g}{c^2}(ct + e^{-ct} - 1)\mathbf{k}$

### Section 11.2 (page 659)

**1.** $\dfrac{e - 1}{e}$, $\dfrac{e^2 - 1}{e^2}$

**3.** $\mathbf{r} = \cos t\mathbf{i} + \sin t\mathbf{j} + \mathbf{k}$; the curve is a circle of radius 1 in the plane $z = 1$

**5.** $\dfrac{\pi^2 R}{72}$ towards the ground, where $R$ is the radius of the earth; $4.76°$ west of south

### Section 11.3 (page 666)

**1.** $x = \sqrt{a^2 - t^2}$, $y = t$, $0 \leq t \leq a$

**3.** $x = a\sin\theta$, $y = -a\cos\theta$, $\frac{\pi}{2} \leq \theta \leq \pi$

**5.** $\mathbf{r} = -2t\mathbf{i} + t\mathbf{j} + 4t^2\mathbf{k}$

**7.** $\mathbf{r} = 3\cos t\mathbf{i} + 3\sin t\mathbf{j} + 3(\cos t + \sin t)\mathbf{k}$

**9.** $\mathbf{r} = (1 + 2\cos t)\mathbf{i} - 2(1 - \sin t)\mathbf{j}$
$\quad + (9 + 4\cos t - 8\sin t)\mathbf{k}$

**11.** Choice (b) leads to $\mathbf{r} = \dfrac{t^2 - 1}{2}\mathbf{i} + t\mathbf{j} + \dfrac{t^2 + 1}{2}\mathbf{k}$, which represents the whole parabola. Choices (a) and (c) lead to separate parametrizations for the halves $y \geq 0$ and $y \leq 0$ of the parabola. For (a) these are $\mathbf{r} = t\mathbf{i} \pm \sqrt{1 + 2t}\mathbf{j} + (1 + t)\mathbf{k}$, $(t \geq -1/2)$

**13.** $(17\sqrt{17} - 16\sqrt{2})/27$ units

**15.** $\displaystyle\int_1^T \dfrac{\sqrt{4a^2t^4 + b^2t^2 + c^2}}{t}\, dt$ units;
$a(T^2 - 1) + c\ln T$ units

**17.** $\pi\sqrt{2 + 4\pi^2} + \ln(\sqrt{2}\pi + \sqrt{1 + 2\pi^2})$ units

**19.** $\sqrt{2e^{4\pi} + 1} - \sqrt{3} + \frac{1}{2}\ln\dfrac{e^{4\pi} + 1 - \sqrt{2e^{4\pi} + 1}}{e^{4\pi}}$
$\quad -\frac{1}{2}\ln(2 - \sqrt{3})$ units

**21.**

**23.** $\mathbf{r} = \dfrac{1}{\sqrt{A^2 + B^2 + C^2}}(As\mathbf{i} + Bs\mathbf{j} + Cs\mathbf{k})$

**25.** $\mathbf{r} = a\left(1 - \dfrac{s}{K}\right)^{3/2}\mathbf{i} + a\left(\dfrac{s}{K}\right)^{3/2}\mathbf{j} + b\left(1 - \dfrac{2s}{K}\right)\mathbf{k}$,
$0 \leq s \leq K$, $K = (\sqrt{9a^2 + 16b^2})/2$

### Section 11.4 (page 675)

**1.** $\hat{\mathbf{T}} = \dfrac{1}{\sqrt{1 + 16t^2 + 81t^4}}(\mathbf{i} - 4t\mathbf{j} + 9t^2\mathbf{k})$

**3.** $\hat{\mathbf{T}} = \dfrac{1}{\sqrt{1 + \sin^2 t}}(\cos 2t\mathbf{i} + \sin 2t\mathbf{j} - \sin t\mathbf{k})$

### Section 11.5 (page 681)

**1.** $1/2$, $27/2$　　　　　　　**3.** $27/(4\sqrt{2})$

**5.** $\hat{\mathbf{T}} = (\mathbf{i} + 2\mathbf{j})/\sqrt{5}$, $\hat{\mathbf{N}} = (-2\mathbf{i} + \mathbf{j})/\sqrt{5}$, $\hat{\mathbf{B}} = \mathbf{k}$

**7.** $\hat{\mathbf{T}} = \dfrac{1}{\sqrt{1 + t^2 + t^4}}(\mathbf{i} + t\mathbf{j} + t^2\mathbf{k})$,

$\hat{\mathbf{B}} = \dfrac{1}{\sqrt{t^4 + 4t^2 + 1}}(t^2\mathbf{i} - 2t\mathbf{j} + \mathbf{k})$,

$\hat{\mathbf{N}} = \dfrac{-(t + 2t^3)\mathbf{i} + (1 - t^4)\mathbf{j} + (t^3 + 2t)\mathbf{k}}{\sqrt{t^4 + 4t^2 + 1}\sqrt{1 + t^2 + t^4}}$,

$\kappa = \dfrac{\sqrt{t^4 + 4t^2 + 1}}{(t^4 + t^2 + 1)^{3/2}}$, $\tau = \dfrac{2}{t^4 + 4t^2 + 1}$

**9.** $\kappa(t) = 1/\sqrt{2}$, $\tau(t) = 0$, curve is a circle in the plane $y + z = 4$, having centre $(2, 1, 3)$ and radius $\sqrt{2}$

**11.** i) $\hat{\mathbf{T}} = \mathbf{i}$, $\quad \hat{\mathbf{N}} = \dfrac{2\mathbf{j} - \mathbf{k}}{\sqrt{5}}$,

$\hat{\mathbf{B}} = \dfrac{\mathbf{j} + 2\mathbf{k}}{\sqrt{5}}$, $\quad \kappa = \sqrt{5}$, $\quad \tau = 0$

ii) $\hat{\mathbf{T}} = \sqrt{\frac{2}{3}}(\mathbf{j} - \frac{1}{\sqrt{2}}\mathbf{k})$, $\quad \hat{\mathbf{B}} = \frac{1}{\sqrt{13}}(-\mathbf{i} + 2\mathbf{j} + 2\sqrt{2}\mathbf{k})$,

$\hat{\mathbf{N}} = -\frac{1}{\sqrt{39}}(6\mathbf{i} + \mathbf{j} + \sqrt{2}\mathbf{k})$, $\quad \kappa = \frac{2\sqrt{39}}{9}$, $\quad \tau = -\frac{6\sqrt{2}}{13}$

**13.** max $a/b^2$, min $b/a^2$

**15.** $\kappa = \dfrac{e^x}{(1 + e^{2x})^{3/2}}$,

$\mathbf{r} = (x - 1 - e^{2x})\mathbf{i} + (2e^x + e^{-x})\mathbf{j}$

**17.** $\frac{3}{2\sqrt{2ar}}$

**21.** $\mathbf{r} = -4x^3\mathbf{i} + (3x^2 + \frac{1}{2})\mathbf{j}$

**23.** $f(x) = \frac{1}{8}(15x - 10x^3 + 3x^5)$

## Section 11.6 (page 691)

**3.** velocity: $1/\sqrt{2}$, $1/\sqrt{2}$;
   acceleration: $-e^{-\theta}/2$, $e^{-\theta}/2$.

**5.** $|\text{force}| = mh^2/r^3$ where $r^2\dot{\theta} = h$ is constant.

**7.** 42,777 km, the equatorial plane

**9.** $\frac{T}{4\sqrt{2}}$     **13.** $3/4$

**15.** $(1/2) - (\epsilon/\pi)$

**19.** $r = A \sec \omega(\theta - \theta_0)$, $\quad \omega^2 = 1 - (k/h^2)$ if $k < h^2$,
   $r = 1/(A + B\theta)$ if $k = h^2$,
   $r = A e^{\omega\theta} + B e^{-\omega\theta}$, $\quad \omega^2 = (k/h^2) - 1$, $\quad$ if $k > h^2$;
   there are no bounded orbits that do not approach the origin except in the case $k = h^2$ if $B = 0$ when there are circular orbits. (Now aren't you glad gravitation is an inverse square rather than an inverse cube attraction?)

**21.** centre $\left(\dfrac{\ell\epsilon}{\epsilon^2 - 1}, 0\right)$;

asymptotes in directions $\theta = \pm \cos^{-1}\left(-\dfrac{1}{\epsilon}\right)$;

semi-transverse axis $a = \dfrac{\ell}{\epsilon^2 - 1}$;

semi-conjugate axis $b = \dfrac{\ell}{\sqrt{\epsilon^2 - 1}}$;

semi-focal separation $c = \dfrac{\ell\epsilon}{\epsilon^2 - 1}$.

## Review Exercises (page 692)

**3.** $\mathbf{v} = 2(\mathbf{i} + 2\mathbf{j} + 2\mathbf{k})$, $\mathbf{a} = (8/3)(-2\mathbf{i} - \mathbf{j} + 2\mathbf{k})$

**5.** $\kappa = \tau = \sqrt{2}/(e^t + e^{-t})^2$

**9.** $4a(1 - \cos(T/2))$ units

**11.** $\mathbf{r}_C(t) = a(t - \sin t)\mathbf{i} + a(1 - \cos t)\mathbf{j}$

**13.** $\hat{\rho} = \sin\phi \cos\theta\mathbf{i} + \sin\phi \sin\theta\mathbf{j} + \cos\phi\mathbf{k}$

$\hat{\phi} = \cos\phi \cos\theta\mathbf{i} + \cos\phi \sin\theta\mathbf{j} - \sin\phi\mathbf{k}$

$\hat{\theta} = -\sin\theta\mathbf{i} + \cos\theta\mathbf{j}$
right-handed

## Challenging Problems (page 693)

**1.** (a) $\Omega = \Omega\dfrac{\mathbf{j} + \mathbf{k}}{\sqrt{2}}$, $\Omega \approx 7.272 \times 10^{-5}$.
   (b) $\mathbf{a}_C = -\sqrt{2}\Omega v\mathbf{i}$.
   (c) about 15.5 cm west of $P$.

**3.**
   (c) $\mathbf{v}(t) = (\mathbf{v}_0 - (\mathbf{v}_0 \bullet \mathbf{k})\mathbf{k}) \cos(\omega t) + (\mathbf{v}_0 \times \mathbf{k}) \sin(\omega t) + (\mathbf{v}_0 \bullet \mathbf{k})\mathbf{k}$.
   (d) Straight line if $\mathbf{v}_0$ is parallel to $\mathbf{k}$, circle if $\mathbf{v}_0$ is perpendicular to $\mathbf{k}$.

**5.** (a) $y = (48 + 24x^2 - x^4)/64$

**7.** (a) Yes, time $\pi a/(v\sqrt{2})$, (b) $\phi = \dfrac{\pi}{2} - \dfrac{vt}{a\sqrt{2}}$,

$\theta = \ln\left[\sec\left(\dfrac{vt}{a\sqrt{2}}\right) + \tan\left(\dfrac{vt}{a\sqrt{2}}\right)\right]$.
   (c) infinitely often

# Chapter 12
# Partial Differentiation

## Section 12.1 (page 701)

**1.** all $(x, y)$ with $x \neq y$    **3.** all $(x, y)$ except $(0, 0)$

**5.** all $(x, y)$ satisfying $4x^2 + 9y^2 \geq 36$

**7.** all $(x, y)$ with $xy > -1$

**9.** all $(x, y, z)$ except $(0, 0, 0)$

**11.** $z = f(x, y) = x$    **13.** $z = f(x, y) = y^2$

**15.** $f(x, y) = \sqrt{x^2 + y^2}$   **17.** $f(x, y) = |x| + |y|$

**19.** $f(x, y) = x - y = C$   **21.** $f(x, y) = xy = C$

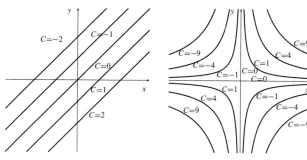

**23.** $f(x, y) = \dfrac{x - y}{x + y} = C$   **25.** $f(x, y) = xe^{-y} = C$

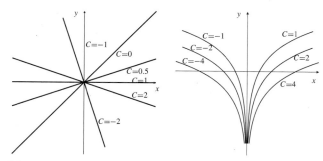

**27.** At $B$, because the contours are closer together there

**29.** a plane containing the $y$-axis, sloping uphill in the $x$ direction

**31.** a right-circular cone with base in the $xy$-plane and vertex at height 5 on the $z$-axis

**33.** No, different curves of the family must not intersect in the region.

**35.** (a) $\sqrt{x^2 + y^2}$,   (b) $(x^2 + y^2)^{1/4}$,
(c) $x^2 + y^2$,   (d) $e^{\sqrt{x^2 + y^2}}$

**37.** spheres centred at the origin

**39.** circular cylinders with axis along the $z$-axis

**41.** regular octahedra with vertices on the coordinate axes

## Section 12.2 (page 706)

**1.** 2                            **3.** does not exist

**5.** $-1$                         **7.** 0

**9.** does not exist               **11.** 0

**13.** $f(0, 0) = 1$

**15.** all $(x, y)$ such that $x \neq \pm y$; no; yes; yes $f(x, x) = \frac{1}{2x}$ makes $f$ continuous at $(x, x)$ for $x \neq 0$; no, $f$ has no continuous extension to the line $x + y = 0$.

**17.** no, yes                     **19.** $a = c = 0$,   $b \neq 0$

**23.** a surface having no tears in it, meeting vertical lines through points of the region exactly once

## Section 12.3 (page 714)

**1.** $f_1(x, y) = f_1(3, 2) = 1$,  $f_2(x, y) = f_2(3, 2) = -1$

**3.** $f_1 = 3x^2 y^4 z^5$,   $f_2 = 4x^3 y^3 z^5$,   $f_3 = 5x^3 y^4 z^4$
All three vanish at $(0, -1, -1)$.

**5.** $\dfrac{\partial z}{\partial x} = \dfrac{-y}{x^2 + y^2}$,   $\dfrac{\partial z}{\partial y} = \dfrac{x}{x^2 + y^2}$,
At $(-1, 1)$: $\dfrac{\partial z}{\partial x} = -\dfrac{1}{2}$,  $\dfrac{\partial z}{\partial y} = -\dfrac{1}{2}$

**7.** $f_1 = \sqrt{y} \cos(x\sqrt{y})$,   $f_2 = \dfrac{x \cos(x\sqrt{y})}{2\sqrt{y}}$,
At $(\pi/3, 4)$: $f_1 = -1$,   $f_2 = -\pi/24$

**9.** $\dfrac{\partial w}{\partial x} = y \ln z \, x^{(y \ln z - 1)}$,   $\dfrac{\partial w}{\partial y} = \ln x \, \ln z \, x^{y \ln z}$,
$\dfrac{\partial w}{\partial z} = \dfrac{y \ln x}{z} x^{y \ln z}$
At $(e, 2, e)$: $\dfrac{\partial w}{\partial x} = \dfrac{\partial w}{\partial z} = 2e$,   $\dfrac{\partial w}{\partial y} = e^2$.

**11.** $f_1(0, 0 = 2$,  $f_2(0, 0) = -1/3$

**13.** $z = -4x - 2y - 3$;   $\dfrac{x+2}{-4} = \dfrac{y-1}{-2} = \dfrac{z-3}{-1}$

**15.** $z = \dfrac{1}{\sqrt{2}}\left(1 - \dfrac{x - \pi}{4} + \dfrac{\pi}{16}(y - 4)\right)$;
$\dfrac{x - \pi}{-1/4\sqrt{2}} = \dfrac{y - 4}{\pi/16\sqrt{2}} = \dfrac{z - 1/\sqrt{2}}{-1}$

**17.** $z = \frac{2}{5} + \frac{3x}{25} - \frac{4y}{25}$;   $\dfrac{x-1}{3} = \dfrac{y-2}{-4} = \dfrac{z-1/5}{-25}$

**19.** $z = \ln 5 + \frac{2}{5}(x - 1) - \frac{4}{5}(y + 2)$;
$\dfrac{x - 1}{2/5} = \dfrac{y + 2}{-4/5} = \dfrac{z - \ln 5}{-1}$

**21.** $z = \dfrac{x + y}{2} - \dfrac{\pi}{4}$;   $2(x - 1) = 2(y + 1) = -z - \dfrac{\pi}{4}$

**23.** $(0, 0)$,   $(1, 1)$,   $(-1, -1)$

**33.** $w = f(a, b, c) + f_1(a, b, c)(x - a) + f_2(a, b, c)(y - b) + f_3(a, b, c)(z - c)$

**35.** $\sqrt{7}/4$ units

**37.** $f_1(0, 0) = 1$, $f_2(0, 0)$ does not exist.

**39.** $f$ is continuous at $(0, 0)$; $f_1$ and $f_2$ are not.

## Section 12.4 (page 719)

**1.** $\dfrac{\partial^2 z}{\partial x^2} = 2(1 + y^2)$,   $\dfrac{\partial^2 z}{\partial x \partial y} = 4xy$,   $\dfrac{\partial^2 z}{\partial y^2} = 2x^2$

**3.** $\dfrac{\partial^2 w}{\partial x^2} = 6xy^3 z^3$,   $\dfrac{\partial^2 w}{\partial y^2} = 6x^3 yz^3$,
$\dfrac{\partial^2 w}{\partial z^2} = 6x^3 y^3 z$,   $\dfrac{\partial^2 w}{\partial x \partial y} = 9x^2 y^2 z^3$,
$\dfrac{\partial^2 w}{\partial x \partial z} = 9x^2 y^3 z^2$,   $\dfrac{\partial^2 w}{\partial y \partial z} = 9x^3 y^2 z^2$

**5.** $\dfrac{\partial^2 z}{\partial x^2} = -y e^x$,   $\dfrac{\partial^2 z}{\partial x \partial y} = e^y - e^x$,   $\dfrac{\partial^2 z}{\partial y^2} = x e^y$

**7.** 27, 10, $x^2 e^{xy}(xz \sin xz - (3 + xy)\cos xz)$

**19.** $u(x, y, z, t) = t^{-3/2}e^{-(x^2+y^2+z^2)/4t}$

### Section 12.5   (page 730)

**1.** $\dfrac{\partial w}{\partial t} = f_1 g_2 + f_2 h_2 + f_3 k_2$

**3.** $\dfrac{\partial z}{\partial u} = g_1 h_1 + g_2 f' h_1$

**5.** $\dfrac{dw}{dz} = f_1 g_1 h' + f_1 g_2 + f_2 h' + f_3,$

$\dfrac{\partial w}{\partial z}\Big|_x = f_2 h' + f_3,$

$\dfrac{\partial w}{\partial z}\Big|_{x,y} = f_3$

**7.** $\dfrac{\partial z}{\partial x} = \dfrac{-5y}{13x^2 - 2xy + 2y^2}$

**9.** $2 f_1(2x, 3y)$          **11.** $2x\, f_2(y^2, x^2)$

**13.** $dT/dt = e^{-t}\big(f'(t) - f(t)\big); dT/dt = 0$ if $f(t) = e^t$: in this case the decrease in $T$ with time (at fixed depth) is exactly balanced by the increase in $T$ with depth.

**15.** $4f_{11} + 12f_{12} + 9f_{22},\quad 6f_{11} + 5f_{12} - 6f_{22},$
$9f_{11} - 12f_{12} + 4f_{22}$

**17.** $f_1 \cos s - f_2 \sin s + f_{11}\, t\, \cos s \, \sin s$
$+ f_{12}\, t(\cos^2 s - \sin^2 s) - f_{22}\, t\, \sin s\, \cos s$

**19.** $f_2 + 2y^2 f_{12} + xy f_{22} - 4xy f_{31} - 2x^2 f_{32};$
all derivatives at $(y^2, xy, -x^2)$

**27.** $\sum_{i,j=1}^{n} x_i x_j\, f_{ij}(x_1, \cdots, x_n) = k(k-1)\, f(x_1, \cdots, x_n)$

**31.** $u(x, y) = f(x + ct)$

### Section 12.6   (page 738)

**1.** 6.9                      **3.** 0.0814

**5.** 2.967                   **7.** (a) 3%, (b) 2%, (c) 1%

**9.** 8.88 ft$^2$

**11.** 169 m, 24 m, most sensitive to angle at $B$

**13.** $\begin{pmatrix} \cos\theta & -r\sin\theta \\ \sin\theta & r\cos\theta \end{pmatrix}$

**15.** $\begin{pmatrix} 2x & z & y \\ -\ln z & 2y & -x/z \end{pmatrix},$     (5.99, 3.98)

### Section 12.7   (page 749)

**1.** $4\mathbf{i} + 2\mathbf{j};\quad z = 4x + 2y - 3;\quad 2x + y = 3$

**3.** $(3\mathbf{i} - 4\mathbf{j})/25;\quad 3x - 4y - 25z + 10 = 0;$
$3x - 4y + 5 = 0$

**5.** $(2\mathbf{i} - 4\mathbf{j})/5;\ 2x - 4y - 5z = 10 - 5\ln 5;\ x - 2y = 5$

**7.** $x + y - 3z = -3$          **9.** $\sqrt{3}y + z = \sqrt{3} + \pi/3$

**11.** $\dfrac{4}{\sqrt{5}}$                      **13.** $1 - 2\sqrt{3}$

**17.** in directions making angles $-30°$ or $-150°$ with positive $x$-axis; no; $-\mathbf{j}$.

**19.** $7\mathbf{i} - \mathbf{j}$

**21.** a)

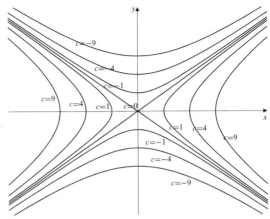

b) in direction $-\mathbf{i} - \mathbf{j}$
c) $4\sqrt{2}k$ deg/unit time
d) $12k/\sqrt{5}$ deg/unit time
e) $x^2 y = -4$

**23.** $3x^2 - 2y^2 = 10$          **25.** $-4/3$

**27.** $\mathbf{i} - 2\mathbf{j} + \mathbf{k}$

**33.** $D_{\mathbf{v}}(D_{\mathbf{v}} f) = v_1^2 f_{11} + v_2^2 f_{22} + v_3^2 f_{33} + 2v_1 v_2 f_{12} + 2v_1 v_3 f_{13}$
$+ 2v_2 v_3 f_{23}.$
This is the second derivative of $f$ as measured by an observer moving with velocity $\mathbf{v}$.

**35.** $\dfrac{\partial^2 T}{\partial t^2} + 2D_{\mathbf{v}(t)}\left(\dfrac{\partial T}{\partial t}\right) + D_{\mathbf{a}(t)}T + D_{\mathbf{v}(t)}(D_{\mathbf{v}(t)}T)$

### Section 12.8   (page 760)

**1.** $-\dfrac{x^4 + 3xy^2}{y^3 + 4x^3 y},\ y \neq 0,\ y^2 \neq -4x^3$

**3.** $\dfrac{3xy^4 + xz}{xy - 2y^2 z},\ y \neq 0,\ x \neq 2yz$

**5.** $\dfrac{x - 2t^2 w}{2xy^2 - w},\ w \neq 2xy^2$     **7.** $-\dfrac{\partial G/\partial x}{\partial G/\partial u},\ \dfrac{\partial G}{\partial u} \neq 0$

**9.** $-\dfrac{v^2 H_2 + w H_3}{u^2 H_1 + t H_3},\quad u^2 H_1 + t H_3 \neq 0,$
all derivatives at $(u^2 w, v^2 t, wt)$

**11.** $\dfrac{2w - 4y}{4x - w},\ 4x \neq w$     **13.** $\dfrac{1}{6},\ \dfrac{1}{2},\ \dfrac{1}{6},\ -\dfrac{1}{2},\ -\dfrac{1}{6}$

**15.** $r$; all points except the origin

**17.** $-3/2$

**19.** $-\dfrac{\partial(F, G, H)}{\partial(y, z, w)} \Big/ \dfrac{\partial(F, G, H)}{\partial(x, z, w)}$

**21.** 15;     $-\dfrac{\partial(F, G, H)}{\partial(x_2, x_3, x_5)} \Big/ \dfrac{\partial(F, G, H)}{\partial(x_1, x_3, x_5)}$

**23.** $2(u + v)$, $\quad -2$, $\quad 0$

**3.**

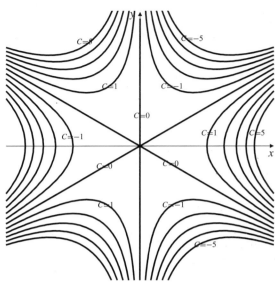

### Section 12.9 (page 766)

**1.** $\displaystyle\sum_{n=0}^{\infty}(-1)^n \frac{x^n y^{2n}}{2^{n+1}}$

**3.** $\displaystyle\sum_{n=0}^{\infty}(-1)^n \frac{x^{2n+1}(y+1)^{2n+1}}{2n+1}$

**5.** $\displaystyle\sum_{n=0}^{\infty}\sum_{k=0}^{n} \frac{1}{k!(n-k)!} x^{2k} y^{2n-2k}$

**7.** $\frac{1}{2} - \frac{1}{4}(x-2) + \frac{1}{2}(y-1) + \frac{1}{8}(x-2)^2$
$\quad - \frac{1}{2}(x-2)(y-1) + \frac{1}{2}(y-1)^2 - \frac{1}{16}(x-2)^3$
$\quad + \frac{3}{8}(x-2)^2(y-1) - \frac{3}{4}(x-2)(y-1)^2 + \frac{1}{2}(y-1)^3$

**9.** $x + y^2 - \frac{x^3}{3}$

**11.** $1 - (y-1) + (y-1)^2 - \frac{1}{2}(x - \frac{\pi}{2})^2$

**13.** $-x - x^2 - (5/6)x^3$

**15.** $-\dfrac{x}{3} - \dfrac{2y}{3} - \dfrac{2x^2}{27} - \dfrac{8xy}{27} - \dfrac{8y^2}{27}$

**17.** $\dfrac{[(2n)!]^3}{(n!)^2}$

### Review Exercises (page 767)

**1.**

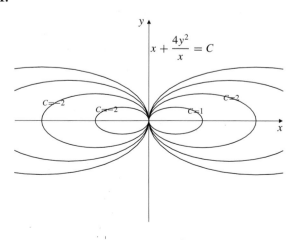

$$x + \frac{4y^2}{x} = C$$

**5.** cont. except on lines $x = \pm y$; can be extended to $x = y$ except at the origin; if $f(0,0) = 0$ then $f_1(0,0) = f_2(0,0) = 1$

**7.** (a) $ax + by + 4cz = 16$,
(b) the circle $z = 1$, $x^2 + y^2 = 12$, (c) $\pm(2, 2, \sqrt{2})$

**9.** 7,500 m$^2$, 7.213%

**11.** (a) $-1/\sqrt{2}$, (b) dir. of $\pm(\mathbf{i} + 3\mathbf{j} - 4\mathbf{k})$, (c) dir. of $-7\mathbf{i} + 5\mathbf{j} + 2\mathbf{k}$

**15.** (a) $\partial u/\partial x = -5$, $\partial u/\partial y = 1$, (b) $-1.13$

## Chapter 13
## Applications of Partial Derivatives

### Section 13.1 (page 778)

**1.** $(2, -1)$, loc. (abs) min.

**3.** $(0, 0)$, saddle pt; $(1, 1)$, loc. min.

**5.** $(-4, 2)$, loc. max.

**7.** $(0, n\pi)$, $n = 0, \pm1, \pm2, \cdots$, all saddle points

**9.** $(0, a)$, $(a > 0)$, loc min; $(0, a)$, $(a < 0)$, loc max; $(0, 0)$ saddle point; $(\pm1, 1/\sqrt{2})$, loc. (abs) max; $(\pm1, -1/\sqrt{2})$, loc. (abs) min.

**11.** $(3^{-1/3}, 0)$, saddle pt.

**13.** $(-1, -1)$, $(1, -1)$, $(-1, 1)$, saddle pts; $(-3, -3)$, loc. min.

**15.** $(1, 1, \frac{1}{2})$, saddle pt.

**17.** $(0, 0)$, saddle pt; $(\frac{1}{\sqrt{2}}, \frac{1}{\sqrt{2}})$, $(-\frac{1}{\sqrt{2}}, -\frac{1}{\sqrt{2}})$, loc. (abs) max; $(\frac{1}{\sqrt{2}}, -\frac{1}{\sqrt{2}})$, $(-\frac{1}{\sqrt{2}}, \frac{1}{\sqrt{2}})$, loc. (abs) min.

**19.** max $e^{-3/2}/2\sqrt{2}$, min $-e^{-3/2}/2\sqrt{2}$; $f$ is continuous everywhere, and $f(x, y, z) \to 0$ as $x^2 + y^2 + z^2 \to \infty$.

**21.** length = width = $(2V)^{1/3}$, height = $(V/4)^{1/3}$

**23.** $8abc/(3\sqrt{3})$ cu. units

**25.** CPs are $(\sqrt{\ln 3}, -\sqrt{\ln 3})$ and $(-\sqrt{\ln 3}, \sqrt{\ln 3})$.

**27.** $f$ does not have a local minimum at $(0, 0)$; the second derivative test is inconclusive $(B^2 = AC)$.

**29.** $A > 0$, $\begin{vmatrix} A & E \\ E & B \end{vmatrix} > 0$, $\begin{vmatrix} A & E & F \\ E & B & H \\ F & H & C \end{vmatrix} > 0$,

$\begin{vmatrix} A & E & F & G \\ E & B & H & I \\ F & H & C & J \\ G & I & J & D \end{vmatrix} > 0$

### Section 13.2 (page 785)

**1.** max $5/4$, min $-2$

**3.** max $(\sqrt{2} - 1)/2$, min $-(\sqrt{2} + 1)/2$.

**5.** max $2/3\sqrt{3}$, min $0$     **7.** max $1$, min $-1$

**9.** max $1/\sqrt{e}$, min $-1/\sqrt{e}$

**11.** max $4/9$, min $-4/9$

**13.** no limit; yes, max $f = e^{-1}$ (at all points of the curve $xy = 1$)

**15.** \$625,000,     \$733,333

**17.** max $37/2$ at $(7/4, 5)$

**19.** 6667 kg deluxe, 6667 kg standard

### Section 13.3 (page 793)

**1.** 84, 375          **3.** 1 unit

**5.** max 4 units, min 2 units

**7.** $a = \pm\sqrt{3}$, $b = \pm2\sqrt{3}$, $c = \pm\sqrt{3}$

**9.** max 8, min $-8$     **11.** max 2, min $-2$

**13.** max 7, min $-1$     **15.** $\dfrac{2\sqrt{6}}{3}$ units

**17.** $\pm\sqrt{\dfrac{3n(n+1)}{2(2n+1)}}$     **19.** $\frac{1}{6} \times \frac{1}{3} \times \frac{2}{3}$

**21.** width = $\left(\dfrac{2V}{15}\right)^{1/3}$, depth = $3 \times$ width,

height = $\dfrac{5}{2} \times$ width

**23.** max 1, min $-\frac{1}{2}$

**27.** method will not fail if $\nabla f = \mathbf{0}$ at extreme point; but we will have $\lambda = 0$.

### Section 13.4 (page 800)

**1.** at $(\bar{x}, \bar{y})$ where $\bar{x} = \left(\sum_{i=1}^{n} x_i\right)/n$, $\bar{y} = \left(\sum_{i=1}^{n} y_i\right)/n$

**3.** $a = \left(\sum_{i=1}^{n} y_i e^{x_i}\right) \Big/ \left(\sum_{i=1}^{n} e^{2x_i}\right)$

**5.** If $A = \sum x_i^2$, $B = \sum x_i y_i$, $C = \sum x_i$, $D = \sum y_i^2$, $E = \sum y_i$, $F = \sum x_i z_i$, $G = \sum y_i z_i$, and $H = \sum z_i$, then

$\Delta = \begin{vmatrix} A & B & C \\ B & D & E \\ C & E & n \end{vmatrix}$,     $a = \dfrac{1}{\Delta}\begin{vmatrix} F & B & C \\ G & D & E \\ H & E & n \end{vmatrix}$,

$b = \dfrac{1}{\Delta}\begin{vmatrix} A & F & C \\ B & G & E \\ C & H & n \end{vmatrix}$,     $c = \dfrac{1}{\Delta}\begin{vmatrix} A & B & F \\ B & D & G \\ C & E & H \end{vmatrix}$

**7.** Use linear regression to fit $\eta = a + bx$ to the data $(x_i, \ln y_i)$. Then $p = e^a$, $q = b$. These are not the same values as would be obtained by minimizing the expression $\sum(y_i - pe^{qx_i})^2$.

**9.** Use linear regression to fit $\eta = a + b\xi$ to the data $\left(x_i, \dfrac{y_i}{x_i}\right)$. Then $p = a$, $q = b$. Not the same as minimizing $\sum(y_i - px_i - qx_i^2)^2$.

**11.** Use linear regression to fit $\eta = a + b\xi$ to the data $\left(e^{-2x_i}, \dfrac{y_i}{e^{x_i}}\right)$. Then $p = a$, $q = b$. Not the same as minimizing $\sum(y_i - pe^{x_i} - qe^{-x_i})^2$. Other answers possible.

**13.** If $A = \sum x_i^4$, $B = \sum x_i^3$, $C = \sum x_i^2$, $D = \sum x_i$, $H = \sum x_i^2 y_i$, $I = \sum x_i y_i$, and $J = \sum y_i$, then

$\Delta = \begin{vmatrix} A & B & C \\ B & C & D \\ C & D & n \end{vmatrix}$,     $a = \dfrac{1}{\Delta}\begin{vmatrix} H & B & C \\ I & C & D \\ J & D & n \end{vmatrix}$,

$b = \dfrac{1}{\Delta}\begin{vmatrix} A & H & C \\ B & I & D \\ C & J & n \end{vmatrix}$,     $c = \dfrac{1}{\Delta}\begin{vmatrix} A & B & H \\ B & C & I \\ C & D & J \end{vmatrix}$

**15.** $a = 5/6$, $I = 1/252$

**17.** $a = 15/16$, $b = -1/16$, $I = 1/448$

**19.** $a = \dfrac{20}{\pi^5}(\pi^2 - 16)$,     $b = \dfrac{12}{\pi^4}(20 - \pi^2)$

**21.** $a_k = \dfrac{2}{\pi}\int_0^\pi f(x)\cos kx\, dx$,    $(k = 0, 1, 2, \cdots)$

**23.** $\pi - \dfrac{4}{\pi}\sum_{k=0}^{\infty}\dfrac{\cos((2k+1)x)}{(2k+1)^2}$;    $-x$

### Section 13.5 (page 810)

**1.** $\dfrac{(-1)^n n!}{(x+1)^{n+1}}$       **3.** $2\sqrt{\pi}(\sqrt{y} - \sqrt{x})$

**5.** $\dfrac{2x}{(1+x^2)^2}$;    $\dfrac{(6x^2-2)}{(1+x^2)^3}$

**7.** $\dfrac{\pi}{2x}$, assume $x > 0$;   $\dfrac{\pi}{4x^3}$;   $\dfrac{3\pi}{16x^5}$

**9.** $n!$       **11.** $f(x) = \int_0^x e^{-t^2/2}\, dt$

**13.** $y = x^2$       **15.** $x^2 + y^2 = 1$

**17.** $y = x - \frac{1}{4}$       **19.** no

**21.** no; a line of singular points

**23.** $x^2 + y^2 + z^2 = 1$

**25.** $y = x - \epsilon\sin(\pi x) + \dfrac{\pi\epsilon^2}{2}\sin(2\pi x) + \cdots$

**27.** $y = \frac{1}{2} - \frac{2}{5}\epsilon x - \dfrac{16}{125}\epsilon^2 x^2 + \cdots$

**29.** $x \approx 1 - \frac{1}{100e} - \frac{1}{30000e^2}$, $y \approx 1 - \frac{1}{30000e^2}$

### Section 13.6  (page 815)

**1.** (0.797105, 2.219107)

**3.** ($\pm 0.2500305$, $\pm 3.9995115$), ($\pm 1.9920783$, $\pm 0.5019883$)

**5.** (0.3727730, 0.3641994), ($-1.4141606$, $-0.9877577$)

**7.** $x = x_0 - \frac{\Delta_1}{\Delta}$, $y = y_0 - \frac{\Delta_2}{\Delta}$, $z = z_0 - \frac{\Delta_3}{\Delta}$,

where $\Delta = \dfrac{\partial(f, g, h)}{\partial(x, y, z)}\Big|_{(x_0, y_0, z_0)}$

and $\Delta_i$ is $\Delta$ with the $i$th column replaced with $\begin{matrix} f \\ g \\ h \end{matrix}$

**9.** 18 iterations near $(0, 0)$, 4 iterations near $(1, 1)$; the two curves are tangent at $(0,0)$, but not at $(1,1)$.

### Review Exercises  (page 816)

**1.** $(0, 0)$ saddle pt., $(1, -1)$ loc. min.

**3.** $(2/3, 4/3)$ loc. min; $(2, -4)$ and $(-1, 2)$ saddle points

**5.** yes, 2, on the sphere $x^2 + y^2 + z^2 = 1$

**7.** max $1/(4e)$, min $-1/(4e)$

**9.** (a) $L^2/48$ cm$^2$, (b) $L^2/16$ cm$^2$

**11.** $4\pi$ sq. units

**13.** $16\pi$ cu. units

**15.** 1,688 widgets, $2.00 each

**17.** $y \approx -2x - \epsilon x e^{-2x} + \epsilon^2 x^2 e^{-4x}$

### Challenging Problems  (page 817)

**3.** $\frac{1}{2}\ln(1 + x^2)\tan^{-1}x$

# Chapter 14
# Multiple Integration

### Section 14.1  (page 823)

**1.** 15

**3.** 21

**5.** 15

**7.** 96

**9.** 80

**11.** 36.6258

**13.** 20

**15.** 0

**17.** $5\pi$

**19.** $\frac{\pi a^3}{3}$

**21.** $\frac{1}{6}$

### Section 14.2  (page 830)

**1.** 5/24

**3.** 4

**5.** $\frac{ab(a^2 + b^2)}{3}$

**7.** $\pi$

**9.** $\frac{3}{56}$

**11.** $\frac{33}{8}\ln 2 - \frac{45}{16}$

**13.** $\frac{e-2}{2}$

**15.** $\frac{1}{2}\left(1 - \frac{1}{e}\right)$; region is a triangle with vertices $(0,0)$, $(1,0)$ and $(1,1)$

**17.** $\frac{\pi}{4\lambda}$; region is a triangle with vertices $(0,0)$, $(0,1)$ and $(1,1)$

**19.** 1/4 cu. units

**21.** 1/3 cu. units

**23.** $\ln 2$ cu. units

**25.** $\frac{\pi}{2\sqrt{2}}$ cu. units

**27.** $\frac{16a^3}{3}$ cu. units

### Section 14.3  (page 836)

**1.** converges to 1

**3.** converges to $\pi/2$

**5.** diverges to $\infty$

**7.** converges to 4

**9.** converges to $1 - \frac{1}{e}$

**11.** diverges to $\infty$

**13.** converges to $2\ln 2$

**15.** $k > a - 1$

**17.** $k < -1 - a$

**19.** $k > -\frac{1+a}{1+b}$ (provided $b > -1$)

**21.** $\frac{1}{2}$, $-\frac{1}{2}$ (different answers are possible because the *double integral* does not exist.)

**23.** $\frac{a^2}{3}$

**25.** $\frac{4\sqrt{2}a}{3\pi}$

**27.** yes, $1/(2\pi)$

### Section 14.4  (page 846)

**1.** $\pi a^4/2$

**3.** $2\pi a$

**5.** $\pi a^4/4$

**7.** $a^3/3$

**9.** $\pi(e^{a^2} - 1)/4$

**11.** $\frac{(\sqrt{3}+1)a^3}{6}$

**13.** $\frac{1}{3}$

**15.** $\frac{2a}{3}$

**17.** $k < 1$; $\frac{\pi}{1-k}$

**19.** $\frac{a^4}{16}$

**21.** $\frac{2\pi}{3}$ cu. units

**23.** $\frac{4\pi(2\sqrt{2}-1)a^3}{3}$ cu. units

**25.** $16[1 - (1/\sqrt{2})]a^3$ cu. units

**27.** $1 - \frac{4\sqrt{2}}{3\pi}$ units

**29.** $\frac{4}{3}\pi abc$ cu. units

**31.** $2a \sinh a$

**33.** $\frac{3\ln 2}{2}$ sq. units

**35.** $\frac{1}{4}(e - e^{-1})$

### Section 14.5  (page 854)

**1.** $8abc$

**3.** $16\pi$

**5.** 2/3

**7.** 1/15

**9.** $2/(3\pi)$

**11.** $\frac{3}{16} \ln 2$

**13.** $\pi\sqrt{\frac{\pi}{6}}$

**15.** $1/8$

**17.** $\int_0^1 dx \int_0^1 dy \int_0^{1-y} f(x, y, z)\, dz$

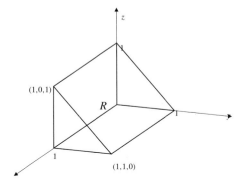

**19.** $\int_0^1 dx \int_0^x dy \int_0^{x-y} f(x, y, z)\, dz$

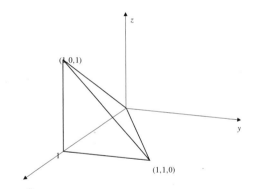

**27.** $(e-1)/3$

**29.** $\overline{f} = \dfrac{1}{\text{vol}(R)} \iiint_R f\, dV;\ 1$

## Section 14.6  (page 863)

**1.** Cartesian: $(-\sqrt{3}, 3, 2)$; cylindrical: $[2\sqrt{3}, 2\pi/3, 2]$

**3.** Cartesian: $(\sqrt{3}, 1, -2)$; spherical: $[2\sqrt{2}, 3\pi/4, \pi/6]$

**5.** the half-plane $x = 0$, $y > 0$

**7.** the $xy$-plane

**9.** the circular cylinder of radius 4 with axis along the $z$-axis

**11.** the $xy$-plane

**13.** sphere of radius 1 with centre $(0, 0, 1)$

**15.** $\frac{2}{3}\pi a^3 \left(1 - \frac{1}{\sqrt{2}}\right)$ cu. units

**17.** $24\pi$ cu. units

**19.** $(2\pi - \dfrac{32}{9})a^3$ cu. units.

**21.** $\dfrac{abc}{3} \tan^{-1} \dfrac{a}{b}$ cu. units  **23.** $\frac{\pi ab}{2}$ cu. units

**25.** $\dfrac{8\pi a^5}{15}$

**27.** $\dfrac{2\pi a^5}{5}\left(1 - \dfrac{c}{\sqrt{c^2+1}}\right)$

**29.** $\dfrac{7\pi}{12}$

**31.** $\dfrac{ha^3}{12};\quad \dfrac{\pi a^2 h^2}{48}$

## Section 14.7  (page 872)

**1.** $3\pi$ sq. units  **3.** $2\pi a^2$ sq. units

**5.** $24\pi/\sqrt{3}$ sq. units

**7.** $(5\sqrt{5} - 1)/12$ sq. units

**9.** 4 sq. units  **11.** $5.123$

**13.** $4\pi A\left[a - \sqrt{B}\tan^{-1}\left(\frac{a}{\sqrt{B}}\right)\right]$ units

**15.** $2\pi km\delta(h + \sqrt{a^2 + (b-h)^2} - \sqrt{a^2 + b^2})$

**17.** $2\pi km\delta(h + \sqrt{a^2 + (b-h)^2} - \sqrt{a^2 + b^2})$

**19.** $\left(\frac{1}{3}, \frac{1}{3}, \frac{1}{2}\right)$  **21.** $\left(\frac{3a}{8}, \frac{3a}{8}, \frac{3a}{8}\right)$

**23.** The model still involves angular acceleration to spin the ball—it doesn't just fall. Part of the gravitational energy goes to producing this spin even in the limiting case.

**25.** $I = \pi\delta a^2 h \left(\frac{h^2}{3} + \frac{a^2}{4}\right), \quad \bar{D} = \left(\frac{h^2}{3} + \frac{a^2}{4}\right)^{1/2}$

**27.** $I = \frac{\pi\delta a^2 h}{3}\left(\frac{2h^2 + 3a^2}{20}\right), \quad \bar{D} = \left(\frac{2h^2 + 3a^2}{20}\right)^{1/2}$

**29.** $I = \frac{5a^5\delta}{12}, \quad \bar{D} = \sqrt{\frac{5}{12}}\,a$

**31.** $I = \frac{8}{3}\delta abc(a^2 + b^2), \quad \bar{D} = \sqrt{\frac{a^2 + b^2}{3}}$

**33.** $m = \frac{4\pi}{3}\delta(a^2 - b^2)^{3/2}, \quad I = \frac{1}{5}m(2a^2 + 3b^2)$

**35.** $\frac{5a^2 g \sin\alpha}{7a^2 + 3b^2}$

**39.** The moment of inertia about the line
$\mathbf{r}(t) = At\mathbf{i} + Bt\mathbf{j} + Ct\mathbf{k}$ is

$$\frac{1}{A^2 + B^2 + C^2}\big((B^2 + C^2)P_{xx} + (A^2 + C^2)P_{yy}$$
$$+ (A^2 + B^2)P_{zz} - 2ABP_{xy} - 2ACP_{xz} - 2BCP_{yz}\big).$$

## Review Exercises  (page 874)

**1.** $3/10$  **3.** $\ln 2$

**5.** $k = 1/\sqrt{3}$

**7.** $\displaystyle\int_0^1 dx \int_x^1 dy \int_y^1 f(x, y, z)\, dz$

**9.** $(1 - e^{-a^2})/(2a)$  **11.** $\dfrac{8\pi}{15}(18\sqrt{6} - 41)a^5$

**13.** vol $= 7/12,\ \bar{z} = 11/28$  **15.** $17/24$

**17.** $\dfrac{1}{6}\displaystyle\int_0^{\pi/2}\left[(1 + 16\cos^2\theta)^{3/2} - 1\right] d\theta \approx 7.904$ sq. units

**Challenging Problems** **(page 875)**

**1.** $\pi abc \left( \dfrac{2}{3} - \dfrac{8}{9\sqrt{3}} \right)$ cu. units

**3.** (b) (i) $\sum_{n=1}^{\infty} (-1)^{n-1} 1/n^2$, (ii) $\sum_{n=1}^{\infty} 1/n^3$, (iii) $\sum_{n=1}^{\infty} (-1)^{n-1} 1/n^3$

**7.** $4\pi |a|^3/35$ cu. units

# Chapter 15
# Vector Fields

## Section 15.1 (page 881)

**1.** field lines: $y = x + C$

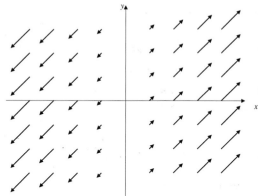

**3.** field lines: $y^2 = x^2 + C$

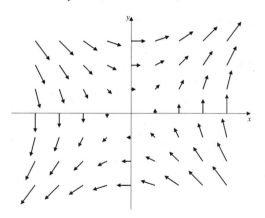

**5.** field lines: $y = -\frac{1}{2} e^{-2x} + C$

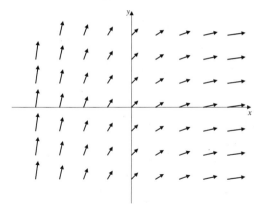

**7.** field lines: $y = Cx$

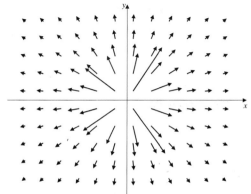

**9.** streamlines are lines parallel to $\mathbf{i} - \mathbf{j} - \mathbf{k}$

**11.** streamlines: $x^2 + y^2 = a^2$, $x = a \sin(z - b)$ (spirals)

**13.** $y = C_1 x$, $2x = z^2 + C_2$

**15.** $y = Ce^{1/x}$          **17.** $r = \theta + C$

**19.** $r = C\theta^2$

## Section 15.2 (page 890)

**1.** conservative; $\dfrac{x^2}{2} - y^2 + \dfrac{3z^2}{2}$

**3.** not conservative

**5.** conservative; $x^2 y + y^2 z - z^2 x$

**7.** $-2 \dfrac{\mathbf{r} - \mathbf{r}_0}{|\mathbf{r} - \mathbf{r}_0|^4}$

**9.** $(x^2 + y^2)/z$; equipotential surfaces are paraboloids $z = C(x^2 + y^2)$; field lines are ellipses $x^2 + y^2 + 2z^2 = $ ⸁ $y = Bx$ in vertical planes through the origin

**11.** $\mathbf{v} = \dfrac{m(x\mathbf{i} + y\mathbf{j} + (z - \ell)\mathbf{k})}{[x^2 + y^2 + (z - \ell)^2]^{3/2}} + \dfrac{m(x\mathbf{i} + y\mathbf{j} + (z + \ell)\mathbf{k}}{[x^2 + y^2 + (z + \ell)^2]^3}$

$\mathbf{v} = \mathbf{0}$ only at the origin; $\mathbf{v}(x, y, 0) = \dfrac{2m(x\mathbf{i} + y\mathbf{j})}{(x^2 + y^2 + \ell^2)^3}$

speed maximum on the circle $x^2 + y^2 = \ell^2/2$, $z = 0$

**15.** $\phi = -\dfrac{\mu y}{r^2}$,

$\mathbf{F} = \dfrac{\mu(2xy\mathbf{i} + (y^2 - x^2)\mathbf{j})}{r^4}$,   $(r^2 = x^2 + y^2)$

**21.** $\phi = \frac{1}{2}r^2 \sin 2\theta$

**ection 15.3   (page 896)**

**1.** $\frac{a^2}{2}(\sqrt{2} + \ln(1 + \sqrt{2}))$   **3.** 8 gm

**5.** $\frac{8}{6}((2e^{4\pi} + 1)^{3/2} - 3^{3/2})$

**7.** $3\sqrt{14}$

**9.** $m = 2\sqrt{2}\pi^2$, $(0, -1/\pi, 4\pi/3)$

**11.** $(e^6 + 3e^4 - 3e^2 - 1)/(3e^3)$

**13.** $(\sqrt{2} + \ln(\sqrt{2} + 1))a^2/2$

**15.** $\pi/\sqrt{2}$

**17.** $4\sqrt{b^2 + c^2}\, E\left(\sqrt{\dfrac{b^2 - a^2}{b^2 + c^2}}\right)$;

$\sqrt{b^2 + c^2}\, E\left(\sqrt{\dfrac{b^2 - a^2}{b^2 + c^2}}, T\right)$

**Section 15.4   (page 904)**

**1.** $-1/4$                   **3.** $1/2$

**5.** $0$                       **7.** $19/2$

**9.** $e^{1+(\pi/4)}$

**11.** $A = 2$, $B = 3$; $4\ln s - \frac{1}{2}$

**13.** $-13/2$               **15.** a) $\pi a^2$, b) $-\pi a^2$

**17.** a) $\dfrac{\pi a^2}{2}$, b) $-\dfrac{\pi a^2}{2}$   **19.** a) $ab/2$, b) $-ab/2$

**23.** The plane with origin removed is not simply connected.

**Section 15.5   (page 917)**

**1.** $dS = ds\, dz = \sqrt{(g(\theta))^2 + (g'(\theta))^2}\, d\theta\, dz$

**3.** $\frac{\pi ab\sqrt{A^2+B^2+C^2}}{|C|}$ sq. units $(C \neq 0)$

**5.** (a) $dS = |\nabla F/F_1|\, dx\, dz$,   (b) $dS = |\nabla F/F_2|\, dy\, dz$

**7.** $\frac{\pi}{8}$                       **9.** $16a^2$ sq. units

**13.** $2\pi$                     **15.** $1/96$

**17.** $\pi(3e + e^3 - 4)/3$

**19.** $2\pi a^2 + \dfrac{2\pi ac^2}{\sqrt{a^2 - c^2}} \ln\left(\dfrac{a + \sqrt{a^2 - c^2}}{c}\right)$ sq. units

**21.** $2\pi\sqrt{A^2 + B^2 + C^2}/|D|$

**23.** one-third of the way from the base to the vertex on
the axis

**25.** $2\pi k\sigma ma\left(\dfrac{1}{\sqrt{a^2+(b-h)^2}} - \dfrac{1}{\sqrt{a^2+b^2}}\right)$

**27.** $I = \frac{8}{3}\pi\sigma a^4$; $\bar{D} = \sqrt{\frac{2}{3}}a$

**29.** $\frac{3}{5} g \sin\alpha$

**Section 15.6   (page 924)**

**1.** 6                           **3.** $3abc$

**5.** $\pi(3a^2 - 4ab + b^2)/2$   **7.** $4\pi$

**9.** $2\sqrt{2}\pi$                 **11.** $4\pi/3$

**13.** $4\pi m$                   **15.** a) $2\pi a^2$,   b) 8

**Review Exercises   (page 925)**

**1.** $(3e/2) - (3/(2e))$   **3.** $8\sqrt{2}/15$

**5.** 1

**7.** (a) $6\pi mgb$, (b) $6\pi R\sqrt{a^2 + b^2}$

**9.** (b) $e^2$                   **11.** $(x\mathbf{i} - y\mathbf{j})/\sqrt{x^2 + y^2}$

**Challenging Problems   (page 926)**

**1.** centroid $(0, 0, 2/\pi)$; upper half of the surface of the
torus obtained by rotating the circle $(x-2)^2 + z^2 = 1$,
$y = 0$, about the $z$-axis

# Chapter 16
# Vector Calculus

**Section 16.1   (page 935)**

**1.** $\operatorname{div}\mathbf{F} = 2$, $\operatorname{curl}\mathbf{F} = 0$

**3.** $\operatorname{div}\mathbf{F} = 0$, $\operatorname{curl}\mathbf{F} = -\mathbf{i} - \mathbf{j} - \mathbf{k}$

**5.** $\operatorname{div}\mathbf{F} = 1$, $\operatorname{curl}\mathbf{F} = -\mathbf{j}$

**7.** $\operatorname{div}\mathbf{F} = f'(x) + g'(y) + h'(z)$,   $\operatorname{curl}\mathbf{F} = 0$

**9.** $\operatorname{div}\mathbf{F} = \cos\theta\left(1 + \dfrac{1}{r}\cos\theta\right)$;

$\operatorname{curl}\mathbf{F} = -\sin\theta\left(1 + \dfrac{1}{r}\cos\theta\right)\mathbf{k}$

**11.** $\operatorname{div}\mathbf{F} = 0$;   $\operatorname{curl}\mathbf{F} = (1/r)\mathbf{k}$

**Section 16.2   (page 941)**

**7.** $\operatorname{div}\mathbf{F}$ can have any value,
$\operatorname{curl}\mathbf{F}$ must be normal to $\mathbf{F}$

**9.** $f(r) = Cr^{-3}$

**15.** If $\mathbf{F} = \nabla\phi$ and $\mathbf{G} = \nabla\Psi$ then $\nabla\times(\phi\nabla\Psi) = \mathbf{F}\times\mathbf{G}$.

**17.** $\mathbf{G} = ye^{2z}\mathbf{i} + xye^{2z}\mathbf{k}$ is one possible vector potential.

**Section 16.3   (page 945)**

**1.** $\pi a^2 - 4a^3$             **3.** 9

**5.** $\dfrac{3\pi ab}{8}$ sq. units

**7.** 0

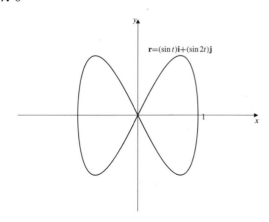

$r=(\sin t)\mathbf{i}+(\sin 2t)\mathbf{j}$

### Section 16.4 (page 951)

**1.** $4\pi a^3$      **3.** $(4/3)\pi a^3$

**5.** $360\pi$      **7.** $81/4$

**11.** $\frac{2}{3}\pi a^2 b + \frac{3}{10}\pi a^4 b + \pi a^2$

**13.** (a) $12\sqrt{3}\pi a^3$,   (b) $-4\sqrt{3}\pi a^3$,   (c) $16\sqrt{3}\pi a^3$

**15.** $(6 + 2\bar{x} + 4\bar{y} - 2\bar{z})V$    **17.** $9\pi a^2$

### Section 16.5 (page 957)

**1.** $1/2$      **3.** $-3\pi a^2$

**7.** $9\pi$

**9.** $\alpha = -\frac{1}{2}$, $\beta = -3$, $I = -\frac{3}{8}\pi$

**11.** yes, $\phi\nabla\Psi$

### Section 16.7 (page 976)

**1.** $\nabla f = \theta z\hat{\mathbf{r}} + z\hat{\boldsymbol{\theta}} + r\theta\mathbf{k}$

**3.** div $\mathbf{F} = 2$,   curl $\mathbf{F} = 0$

**5.** div $\mathbf{F} = \dfrac{2\sin\phi}{\rho}$,   curl $\mathbf{F} = -\dfrac{\cos\phi}{\rho}\hat{\boldsymbol{\theta}}$

**7.** div $\mathbf{F} = 0$,   curl $\mathbf{F} = \cot\phi\,\hat{\boldsymbol{\rho}} - 2\hat{\boldsymbol{\phi}}$

**9.** scale factors: $h_u = \left|\dfrac{\partial\mathbf{r}}{\partial u}\right|$, $h_v = \left|\dfrac{\partial\mathbf{r}}{\partial v}\right|$

     local basis: $\hat{\mathbf{u}} = \dfrac{1}{h_u}\dfrac{\partial\mathbf{r}}{\partial u}$, $\hat{\mathbf{v}} = \dfrac{1}{h_v}\dfrac{\partial\mathbf{r}}{\partial v}$

     area element: $dA = h_u h_v\,du\,dv$

**11.**   $\nabla f(r,\theta) = \dfrac{\partial f}{\partial r}\hat{\mathbf{r}} + \dfrac{1}{r}\dfrac{\partial r}{\partial\theta}\hat{\boldsymbol{\theta}}$

     $\nabla\bullet\mathbf{F}(r,\theta) = \dfrac{\partial F_r}{\partial r} + \dfrac{1}{r}F_r + \dfrac{1}{r}\dfrac{\partial F_\theta}{\partial\theta}$

     $\nabla\times\mathbf{F}(r,\theta) = \left(\dfrac{\partial F_\theta}{\partial r} + \dfrac{1}{r}F_\theta - \dfrac{1}{r}\dfrac{\partial F_r}{\partial\theta}\right)\mathbf{k}$

**13.** $u$-surfaces: vertical elliptic cylinders with focal ax[es] at $x = \pm a$, $y = 0$
     $v$-surfaces: vertical hyperbolic cylinders with foc[al] axes at $x = \pm a$, $y = 0$
     $z$-surfaces: horizontal planes
     $u$-curves: horizontal hyperbolas with foci $x = \pm a$ $y = 0$
     $v$-curves: horizontal ellipses with foci $x = \pm a$, $y = 0$
     $z$-curves: vertical straight lines

**15.** $\nabla f = \dfrac{\partial^2 f}{\partial\rho^2} + \dfrac{2}{\rho}\dfrac{\partial f}{\partial\rho} + \dfrac{1}{\rho^2}\dfrac{\partial^2 f}{\partial\phi^2}$

     $+ \dfrac{\cot\phi}{\rho^2}\dfrac{\partial f}{\partial\phi} + \dfrac{1}{\rho^2\sin^2\phi}\dfrac{\partial^2 f}{\partial\theta^2}$

### Review Exercises (page 977)

**1.** $128\pi$      **3.** $-6$

**5.** $3/4$      **7.** $\lambda = -3$, no

**11.** the ellipsoid $x^2 + 4y^2 + z^2 = 4$ with outward normal

### Challenging Problems (page 978)

**1.** div $\mathbf{v} = 3C$

# Chapter 17
# Ordinary Differential Equations

### Section 17.1 (page 983)

**1.** 1, linear, homogeneous    **3.** 1, nonlinear

**5.** 2, linear, homogeneous
**7.** 3, linear, nonhomogeneous

**9.** 4, linear, homogeneous
**11.** (a) and (b) are solutions, (c) is not
**13.** $y_2 = \sin(kx)$, $y = -3(\cos(kx) + (3/k)\sin(kx))$
**15.** $y = \sqrt{2}(\cos x + 2\sin x)$
**17.** $y = x + \sin x + (\pi - 1)\cos x$

### Section 17.2 (page 988)

**1.** $y^2 = Cx$      **3.** $x^3 - y^3 = C$

**5.** $\dfrac{y^4}{4} = \dfrac{x^3}{3} + C$      **7.** $Y = Ce^{t^2/2}$

**9.** $y = \dfrac{Ce^{2x} - 1}{Ce^{2x} + 1}$    **11.** $y = -\ln\left(Ce^{-2t} - \frac{1}{2}\right)$

**13.** $y = \tan^{-1}(C - \cos x) + n\pi$
**15.** $2\tan^{-1}(y/x) = \ln(x^2 + y^2) + C$
**17.** $\tan^{-1}(y/x) = \ln|x| + C$

**19.** $y = x\tan^{-1}(\ln|Cx|)$    **21.** $y^3 + 3y - 3x^2 = 24$

### Section 17.3  (page 991)

**1.** $2xy + x^2 y^2 = C$        **3.** $xe^{xy} = C$

**5.** $\ln|x| - \frac{y}{x^2} = C$

**7.** $\dfrac{1}{M}\left(\dfrac{\partial N}{\partial x} - \dfrac{\partial M}{\partial y}\right)$ must depend only on $y$

**9.** $\dfrac{1}{M}\left(\dfrac{\partial N}{\partial x} - \dfrac{\partial M}{\partial y}\right)$ must depend only on $y$.

$x - y^2 e^y = Cy^2$

**11.** $\dfrac{1}{\mu}\dfrac{d\mu}{dx} = \dfrac{\dfrac{\partial N}{\partial x} - \dfrac{\partial M}{\partial y}}{xM - yN}$ must depend only on $xy$;

$\dfrac{\sin x}{y} - \dfrac{y}{x} = C$

### Section 17.4  (page 994)

**1.** $y = x^3 + Cx^2$         **3.** $y = \frac{3}{2} + Ce^{-2x}$

**5.** $y = x - 1 + Ce^{-x}$      **7.** $y = (1 + e^{1-10t})/10$

**9.** $y = (x^2 - \pi^2)e^{-\sin x}$

**13.** $\dfrac{dy}{dt} = 0.05y - 50$,   \$871.60

### Section 17.5  (page 1002)

**1.** (a) 1.97664, (b) 2.187485, (c) 2.306595

**3.** (a) 2.436502, (b) 2.436559, (c) 2.436563

**5.** (a) 1.097897, (b) 1.098401

**7.** (a) 0.89441, (b) 0.87996, (c) 0.872831

**9.** (a) 0.865766, (b) 0.865769, (c) 0.865769

**11.** (a) 0.898914, (b) 0.903122, (c) 0.904174

**13.** $y = 2/(3 - 2x)$

**17.** (b) $u = 1/(1-x)$, $v = \tan(x + \frac{\pi}{4})$. $y(x)$ is defined at least on $[0, \pi/4)$ and satisfies $1/(1-x) \le y(x) \le \tan(x + \frac{\pi}{4})$ there.

### Section 17.6  (page 1006)

**1.** $y = C_1 e^x + C_2 e^{2x}$      **3.** $y = C_1 x + \dfrac{C_2}{x^2}$

**5.** $y = C_1 x + C_2 x e^x$

### Section 17.7  (page 1015)

**1.** $y = C_1 e^{-5t} + C_2 e^{-2t}$    **3.** $y = C_1 + C_2 e^{-2t}$

**5.** $y = (C_1 + C_2 t)e^{-4t}$

**7.** $y = (C_1 \cos t + C_2 \sin t)e^{3t}$

**9.** $y = (C_1 \cos 2t + C_2 \sin 2t)e^{-t}$

**11.** $y = (C_1 \cos \sqrt{2}t + C_2 \sin \sqrt{2}t)e^{-t}$

**13.** $y = \frac{6}{7}e^{t/2} + \frac{1}{7}e^{-3t}$

**15.** $y = e^{-2t}(2\cos t + 6\sin t)$

**19.** $y = C_1 + C_2 e^t + C_3 e^{3t}$

**21.** $y = C_1 \cos t + C_2 \sin t + C_3 t \cos t + C_4 t \sin t$

**23.** $y = C_1 e^{2t} + C_2 e^{-t}\cos t + C_3 e^{-t}\sin t$

**25.** $y = Ax + Bx\ln x$      **27.** $y = Ax + \dfrac{B}{x}$

**29.** $y = A + B\ln x$

**31.** $y = C_1 x + C_2 x\ln x + C_3 x(\ln x)^2$

### Section 17.8  (page 1021)

**1.** $y = -\dfrac{1}{2} + C_1 e^x + C_2 e^{-2x}$

**3.** $y = -\dfrac{1}{2}e^{-x} + C_1 e^x + C_2 e^{-2x}$

**5.** $y = -\frac{2}{125} - \frac{4x}{25} + \frac{x^2}{5} + C_1 e^{-x}\cos(2x) + C_2 e^{-x}\sin(2x)$

**7.** $y = -\dfrac{1}{5}xe^{-2x} + C_1 e^{-2x} + C_2 e^{3x}$

**9.** $y = \dfrac{1}{8}e^x(\sin x - \cos x) + e^{-x}(C_1 \cos x + C_2 \sin x)$

**11.** $y = 2x + x^2 - xe^{-x} + C_1 + C_2 e^{-x}$

**15.** $y_p = \dfrac{x^2}{3}$,   $y = \dfrac{x^2}{3} + C_1 x + \dfrac{C_2}{x}$

**17.** $y = \dfrac{1}{2}x\ln x + C_1 x + \dfrac{C_2}{x}$

**19.** $y = -x^2 + C_1 x + C_2 x e^x$

### Section 17.9  (page 1026)

**1.** $y = a_0\left(1 + \displaystyle\sum_{k=1}^{\infty} \dfrac{(x-1)^{4k}}{4(k!)(3)(7)\cdots(4k-1)}\right)$

$+ a_1\left(x - 1 + \displaystyle\sum_{k=1}^{\infty}\dfrac{(x-1)^{4k+1}}{4(k!)(5)(9)\cdots(4k+1)}\right)$

**3.** $y = 1 - \frac{1}{6}x^3 + \frac{1}{120}x^5 + \cdots$

**5.** $y_1 = 1 + \displaystyle\sum_{k=1}^{\infty}\dfrac{(-1)^k x^k}{(k!)(2)(5)(8)\cdots(3k-1)}$,

$y_2 = x^{1/3}\left(1 + \displaystyle\sum_{k=0}^{\infty}\dfrac{(-1)^k x^k}{(k!)(4)(7)\cdots(3k+1)}\right)$

### Review Exercises  (page 1026)

**1.** $y = Ce^{x^2}$          **3.** $y = Ce^{2x} - \dfrac{x}{2} - \dfrac{1}{4}$

**5.** $x^2 + 2xy - y^2 = C$    **7.** $y = C_1 - \ln|t + C_2|$

**9.** $y = e^{x/2}(C_2 \cos x + C_2 \sin x)$

**11.** $y = C_1 t\cos(2\ln|t|) + C_2 t\sin(2\ln|t|)$

**13.** $y = \frac{1}{2}e^x + xe^{3x} + C_1 e^{2x} + C_2 e^{3x}$

**15.** $y = x^2 - 4x + 6 + C_1 e^{-x} + C_2 x e^{-x}$

**17.** $y = (x^3 - 7)^{1/3}$      **19.** $y = e^{x^2/2y^2}$

**21.** $y = 4e^{-t} - 3e^{-2t}$      **23.** $y = (5t - 4)e^{-5t}$

**25.** $y = e^{2t} - 2\sin(2t)$

**27.** $A = 1,\; B = -1,\quad x(e^x \sin y + \cos y) = C$

**29.** $y = C_1 x + C_2 x \cos x$

## Appendix I   Complex Numbers
### (page A-11)

**1.** $\Re(z) = -5,\; \Im(z) = 2$      **3.** $\Re(z) = 0,\; \Im(z) = -\pi$

**5.** $|z| = \sqrt{2},\; \theta = 3\pi/4$      **7.** $|z| = 3,\; \theta = \pi/2$

**9.** $|z| = \sqrt{5},\; \theta = \tan^{-1}2$

**11.** $|z| = 5,\; \theta = \pi + \tan^{-1}(4/3)$

**13.** $|z| = 2,\; \theta = 11\pi/6$      **15.** $|z| = 3,\; \theta = 4\pi/5$

**17.** $23\pi/12$      **19.** $4 + 3i$

**21.** $\frac{\pi\sqrt{3}}{2} + \frac{\pi}{2}i$      **23.** $\frac{1}{4} - \frac{\sqrt{3}}{4}i$

**25.** $-3 + 5i$      **27.** $2 + i$

**29.** closed disk, radius 2, centre 0

**31.** closed disk, radius 5, centre $3 - 4i$

**33.** closed plane sector lying under $y = 0$ and to the left of $y = -x$

**35.** $4$      **37.** $5 - i$

**39.** $2 + 11i$      **41.** $-\frac{1}{5} + \frac{7}{5}i$

**43.** $1$

**47.** $zw = -3 - 3i,\; \dfrac{z}{w} = \dfrac{1+i}{3}$

**49.** (a) circle $|z| = \sqrt{2}$, (b) no solutions

**51.** $-1,\; \frac{1}{2} \pm \frac{\sqrt{3}}{2}\mathbf{i}$

**53.** $2^{1/6}(\cos\theta + i\sin\theta)$ where $\theta = \pi/4, 11\pi/12, 19\pi/$

**55.** $\pm 2^{1/4}\left(\frac{\sqrt{3}}{2} + \frac{1}{2}i\right),\; \pm 2^{1/4}\left(\frac{1}{2} - \frac{\sqrt{3}}{2}i\right)$

## Appendix II   Complex Functions
### (page A-21)

**1.** $0 \le \Re(w) \le 1,\; -2 \le \Im(w) \le 0$

**3.** $1 \le |w| \le 4,\; \pi \le \arg w \le \dfrac{3\pi}{2}$

**5.** $\dfrac{1}{2} \le |w| < \infty,\; -\dfrac{\pi}{2} \le \arg w \le 0$

**7.** $\arg(w) = 5\pi/6$      **9.** parabola $v^2 = 4u + 4$

**11.** $u \ge 0,\; v \ge u$      **13.** $f'(z) = 2z$

**15.** $f'(z) = -1/z^2$

**19.** $\dfrac{d}{dz}\sinh z = \cosh z,\; \dfrac{d}{dz}\cosh z = \sinh z$

**21.** $z = \dfrac{\pi}{2} + k\pi,\; (k \in \mathbb{Z})$

**23.** zeros of $\cosh z$: $z = i\left(\dfrac{\pi}{2} + k\pi\right)\,(k \in \mathbb{Z})$

  zeros of $\sinh z$: $z = k\pi i\,(k \in \mathbb{Z})$

**25.** $\Re(\sinh z) = \sinh x \cos y,\; \Im(\sinh z) = \cosh x \sin y$

**27.** $z = 0,\; -2i$      **29.** $z = -1 \pm 2i$

**31.** $z = 0,\; i,\; 2i$

**33.** $z = \dfrac{1 \pm i}{\sqrt{2}},\; z = \dfrac{-1 \pm i}{\sqrt{2}}$

  $z^4 + 1 = (z^2 + \sqrt{2}z + 1)(z^2 - \sqrt{2}z + 1)$

**35.** $z = -1,\; -1,\; -1,\; i,\; -i$

# Index

## INTEGRATION RULES

$$\int (Af(x) + Bg(x))\, dx = A \int f(x)\, dx + B \int g(x)\, dx$$

$$\int f'(g(x))\, g'(x)\, dx = f(g(x)) + C$$

$$\int U(x)\, dV(x) = U(x)\, V(x) - \int V(x)\, dU(x)$$

$$\int_a^b f'(x)\, dx = f(b) - f(a)$$

$$\frac{d}{dx} \int_a^x f(t)\, dt = f(x)$$

## ELEMENTARY INTEGRALS

$$\int x^r\, dx = \frac{1}{r+1} x^{r+1} + C \text{ if } r \neq -1$$

$$\int \frac{dx}{x} = \ln|x| + C$$

$$\int e^x\, dx = e^x + C$$

$$\int a^x\, dx = \frac{a^x}{\ln a} + C$$

$$\int \sin x\, dx = -\cos x + C$$

$$\int \cos x\, dx = \sin x + C$$

$$\int \sec^2 x\, dx = \tan x + C$$

$$\int \csc^2 x\, dx = -\cot x + C$$

$$\int \sec x \tan x\, dx = \sec x + C$$

$$\int \csc x \cot x\, dx = -\csc x + C$$

$$\int \tan x\, dx = \ln|\sec x| + C$$

$$\int \cot x\, dx = \ln|\sin x| + C$$

$$\int \sec x\, dx = \ln|\sec x + \tan x| + C$$

$$\int \csc x\, dx = \ln|\csc x - \cot x| + C$$

$$\int \frac{dx}{\sqrt{a^2 - x^2}} = \sin^{-1}\frac{x}{a} + C$$

$$\int \frac{dx}{a^2 + x^2} = \frac{1}{a} \tan^{-1}\frac{x}{a} + C$$

$$\int \frac{dx}{a^2 - x^2} = \frac{1}{2a} \ln\left|\frac{x+a}{x-a}\right| + C$$

$$\int \frac{dx}{x\sqrt{x^2 - a^2}} = \frac{1}{a} \sec^{-1}\left|\frac{x}{a}\right| + C$$

## TRIGONOMETRIC INTEGRALS

$$\int \sin^2 x\, dx = \frac{x}{2} - \frac{1}{4} \sin 2x + C$$

$$\int \cos^2 x\, dx = \frac{x}{2} + \frac{1}{4} \sin 2x + C$$

$$\int \tan^2 x\, dx = \tan x - x + C$$

$$\int \cot^2 x\, dx = -\cot x - x + C$$

$$\int \sec^3 x\, dx = \frac{1}{2} \sec x \tan x + \frac{1}{2} \ln|\sec x + \tan x| + C$$

$$\int \csc^3 x\, dx = -\frac{1}{2} \csc x \cot x + \frac{1}{2} \ln|\csc x - \cot x| + C$$

$$\int \sin ax \sin bx\, dx = \frac{\sin(a-b)x}{2(a-b)} - \frac{\sin(a+b)x}{2(a+b)} + C \text{ if } a^2 \neq b^2$$

$$\int \cos ax \cos bx\, dx = \frac{\sin(a-b)x}{2(a-b)} + \frac{\sin(a+b)x}{2(a+b)} + C \text{ if } a^2 \neq b^2$$

$$\int \sin ax \cos bx\, dx = -\frac{\cos(a-b)x}{2(a-b)} - \frac{\cos(a+b)x}{2(a+b)} + C \text{ if } a^2 \neq b^2$$

$$\int \sin^n x\, dx = -\frac{1}{n} \sin^{n-1} x \cos x + \frac{n-1}{n} \int \sin^{n-2} x\, dx$$

$$\int \cos^n x\, dx = \frac{1}{n} \cos^{n-1} x \sin x + \frac{n-1}{n} \int \cos^{n-2} x\, dx$$

$$\int \tan^n x\, dx = \frac{1}{n-1} \tan^{n-1} x - \int \tan^{n-2} x\, dx \text{ if } n \neq 1$$

$$\int \cot^n x\, dx = \frac{-1}{n-1} \cot^{n-1} x - \int \cot^{n-2} x\, dx \text{ if } n \neq 1$$

$$\int \sec^n x\, dx = \frac{1}{n-1} \sec^{n-2} x \tan x + \frac{n-2}{n-1} \int \sec^{n-2} x\, dx \text{ if } n \neq 1$$

$$\int \csc^n x\, dx = \frac{-1}{n-1} \csc^{n-2} x \cot x + \frac{n-2}{n-1} \int \csc^{n-2} x\, dx \text{ if } n \neq 1$$

$$\int \sin^n x \cos^m x\, dx = -\frac{\sin^{n-1} x \cos^{m+1} x}{n+m} + \frac{n-1}{n+m} \int \sin^{n-2} x \cos^m x\, dx \text{ if } n \neq -m$$

$$\int \sin^n x \cos^m x\, dx = \frac{\sin^{n+1} x \cos^{m-1} x}{n+m} + \frac{m-1}{n+m} \int \sin^n x \cos^{m-2} x\, dx \text{ if } m \neq -n$$

$$\int x \sin x\, dx = \sin x - x \cos x + C$$

$$\int x \cos x\, dx = \cos x + x \sin x + C$$

$$\int x^n \sin x\, dx = -x^n \cos x + n \int x^{n-1} \cos x\, dx$$

$$\int x^n \cos x\, dx = x^n \sin x - n \int x^{n-1} \sin x\, dx$$

## INTEGRALS INVOLVING $\sqrt{x^2 \pm a^2}$

$$\int \sqrt{x^2 \pm a^2}\, dx = \frac{x}{2}\sqrt{x^2 \pm a^2} \pm \frac{a^2}{2}\ln|x + \sqrt{x^2 \pm a^2}| + C$$

$$\int \frac{dx}{\sqrt{x^2 \pm a^2}} = \ln|x + \sqrt{x^2 \pm a^2}| + C$$

$$\int \frac{\sqrt{x^2 + a^2}}{x}\, dx = \sqrt{x^2 + a^2} - a\ln\left(\frac{a + \sqrt{x^2 + a^2}}{x}\right) + C$$

$$\int \frac{\sqrt{x^2 - a^2}}{x}\, dx = \sqrt{x^2 - a^2} - a\sec^{-1}\frac{x}{a} + C$$

$$\int x^2\sqrt{x^2 \pm a^2}\, dx = \frac{x}{8}(2x^2 \pm a^2)\sqrt{x^2 \pm a^2} - \frac{a^4}{8}\ln|x + \sqrt{x^2 \pm a^2}| + C$$

$$\int \frac{x^2}{\sqrt{x^2 \pm a^2}}\, dx = \frac{x}{2}\sqrt{x^2 \pm a^2} \mp \frac{a^2}{2}\ln|x + \sqrt{x^2 \pm a^2}| + C$$

$$\int \frac{\sqrt{x^2 \pm a^2}}{x^2}\, dx = -\frac{\sqrt{x^2 \pm a^2}}{x} + \ln|x + \sqrt{x^2 \pm a^2}| + C$$

$$\int \frac{dx}{x^2\sqrt{x^2 \pm a^2}} = \mp\frac{\sqrt{x^2 \pm a^2}}{a^2 x} + C$$

$$\int \frac{dx}{(x^2 \pm a^2)^{3/2}} = \frac{\pm x}{a^2\sqrt{x^2 \pm a^2}} + C$$

$$\int (x^2 \pm a^2)^{3/2}\, dx = \frac{x}{8}(2x^2 \pm 5a^2)\sqrt{x^2 \pm a^2} + \frac{3a^4}{8}\ln|x + \sqrt{x^2 \pm a^2}| + C$$

## INTEGRALS INVOLVING $\sqrt{a^2 - x^2}$

$$\int \sqrt{a^2 - x^2}\, dx = \frac{x}{2}\sqrt{a^2 - x^2} + \frac{a^2}{2}\sin^{-1}\frac{x}{a} + C$$

$$\int \frac{\sqrt{a^2 - x^2}}{x}\, dx = \sqrt{a^2 - x^2} - a\ln\left|\frac{a + \sqrt{a^2 - x^2}}{x}\right| + C$$

$$\int \frac{x^2}{\sqrt{a^2 - x^2}}\, dx = -\frac{x}{2}\sqrt{a^2 - x^2} + \frac{a^2}{2}\sin^{-1}\frac{x}{a} + C$$

$$\int x^2\sqrt{a^2 - x^2}\, dx = \frac{x}{8}(2x^2 - a^2)\sqrt{a^2 - x^2} + \frac{a^4}{8}\sin^{-1}\frac{x}{a} + C$$

$$\int \frac{dx}{x^2\sqrt{a^2 - x^2}} = -\frac{\sqrt{a^2 - x^2}}{a^2 x} + C$$

$$\int \frac{\sqrt{a^2 - x^2}}{x^2}\, dx = -\frac{\sqrt{a^2 - x^2}}{x} - \sin^{-1}\frac{x}{a} + C$$

$$\int \frac{dx}{x\sqrt{a^2 - x^2}} = -\frac{1}{a}\ln\left|\frac{a + \sqrt{a^2 - x^2}}{x}\right| + C$$

$$\int \frac{dx}{(a^2 - x^2)^{3/2}} = \frac{x}{a^2\sqrt{a^2 - x^2}} + C$$

$$\int (a^2 - x^2)^{3/2}\, dx = \frac{x}{8}(5a^2 - 2x^2)\sqrt{a^2 - x^2} + \frac{3a^4}{8}\sin^{-1}\frac{x}{a} + C$$

## INTEGRALS OF INVERSE TRIGONOMETRIC FUNCTIONS

$$\int \sin^{-1}x\, dx = x\sin^{-1}x + \sqrt{1 - x^2} + C$$

$$\int \tan^{-1}x\, dx = x\tan^{-1}x - \frac{1}{2}\ln(1 + x^2) + C$$

$$\int \sec^{-1}x\, dx = x\sec^{-1}x - \ln|x + \sqrt{x^2 - 1}| + C$$

$$\int x\sin^{-1}x\, dx = \frac{1}{4}(2x^2 - 1)\sin^{-1}x + \frac{x}{4}\sqrt{1 - x^2} + C$$

$$\int x\tan^{-1}x\, dx = \frac{1}{2}(x^2 + 1)\tan^{-1}x - \frac{x}{2} + C$$

$$\int x\sec^{-1}x\, dx = \frac{x^2}{2}\sec^{-1}x - \frac{1}{2}\sqrt{x^2 - 1} + C$$

$$\int x^n\sin^{-1}x\, dx = \frac{x^{n+1}}{n+1}\sin^{-1}x - \frac{1}{n+1}\int \frac{x^{n+1}}{\sqrt{1 - x^2}}\, dx + C \text{ if } n \neq -1$$

$$\int x^n\tan^{-1}x\, dx = \frac{x^{n+1}}{n+1}\tan^{-1}x - \frac{1}{n+1}\int \frac{x^{n+1}}{1 + x^2}\, dx + C \text{ if } n \neq -1$$

$$\int x^n\sec^{-1}x\, dx = \frac{x^{n+1}}{n+1}\sec^{-1}x - \frac{1}{n+1}\int \frac{x^n}{\sqrt{x^2 - 1}}\, dx + C \text{ if } n \neq -1$$

## EXPONENTIAL AND LOGARITHMIC INTEGRALS

$$\int xe^x\, dx = (x - 1)e^x + C$$

$$\int x^n e^x\, dx = x^n e^x - n\int x^{n-1}e^x\, dx$$

$$\int \ln x\, dx = x\ln x - x + C$$

$$\int x^n \ln x\, dx = \frac{x^{n+1}}{n+1}\ln x - \frac{x^{n+1}}{(n+1)^2} + C$$

$$\int x^n(\ln x)^m\, dx = \frac{x^{n+1}}{n+1}(\ln x)^m - \frac{m}{n+1}\int x^n(\ln x)^{m-1}\, dx$$

$$\int e^{ax}\sin bx\, dx = \frac{e^{ax}}{a^2 + b^2}(a\sin bx - b\cos bx) + C$$

$$\int e^{ax}\cos bx\, dx = \frac{e^{ax}}{a^2 + b^2}(a\cos bx + b\sin bx) + C$$

## INTEGRALS OF HYPERBOLIC FUNCTIONS

$$\int \sinh x\, dx = \cosh x + C$$

$$\int \cosh x\, dx = \sinh x + C$$

$$\int \tanh x\, dx = \ln(\cosh x) + C$$

$$\int \coth x\, dx = \ln|\sinh x| + C$$

$$\int \operatorname{sech} x\, dx = \tan^{-1}|\sinh x| + C$$

$$\int \operatorname{csch} x\, dx = \ln\left|\tanh\frac{x}{2}\right| + C$$

$$\int \sinh^2 x\, dx = \frac{1}{4}\sinh 2x - \frac{x}{2} + C$$

$$\int \cosh^2 x\, dx = \frac{1}{4}\sinh 2x + \frac{x}{2} + C$$

$$\int \tanh^2 x\, dx = x - \tanh x + C$$

$$\int \coth^2 x\, dx = x - \coth x + C$$

$$\int \operatorname{sech}^2 x\, dx = \tanh x + C$$

$$\int \operatorname{csch}^2 x\, dx = -\coth x + C$$

$$\int \operatorname{sech} x \tanh x\, dx = -\operatorname{sech} x + C$$

$$\int \operatorname{csch} x \coth x\, dx = -\operatorname{csch} x + C$$